Irrigation Efficiency 467
Irrigation Impact on River Flows 473
Irrigation Management in Humid Regions 478
Irrigation Management for the Tropics 483
Irrigation Mechanical Systems: Sprinkler 490
Irrigation Metering 495
Irrigation, Preplant 498
Irrigation Return Flow, Quality and 502
Irrigation Sagacity 507
Irrigation with Saline Water 510
Irrigation Scheduling with Plant Indicators:
    Field Applications 512
Irrigation Scheduling with Plant Indicators:
    Measurement 519
Irrigation Scheduling by Remote Sensing
    Technologies 523
Irrigation Scheduling by Soil Water Status 528
Irrigation Scheduling by Water Budgeting 532
Irrigation, Sewage Effluent Use for 535
Irrigation, Supplemental 537
Irrigation, Surface 540
Irrigation Systems, Drip 546
Irrigation Systems, History of Farm 549
Irrigation Systems, Subsurface Drip 560
Isotopes 565
Journals 569
La Niña 574
Land Drainage, Subsurface 576
Leaf Water Potential 579
Livestock and Poultry Production, Water
    Consumption for 588
Livestock, Water Harvesting Methods for 593
Livestock Water Quality Standards 596
Manure Management, Beef Cattle Industry
    Requirements 601
Manure Management, Dairy 603
Manure Management, Poultry 607
Manure Management, Swine 610
Marketing 612
Matric Potential 615
Microbial Sampling 618
Neuse River 622
Nitrogen Measurement 625
Nutrient Best Management Practices 630
Observation Wells 633
Open-Channel Spillways 636
Oxygen Measurement: Biological–Chemical
    Oxygen Demand 642
Pathogens in Water 645
Pesticide Contamination, Groundwater 650
Pesticide Contamination, Surface Water 654
Pfiesteria Piscicida 658
pH 663
Phosphorus Measurement 666
Plant Available Soil Water 669
Plant Exposure to Water Stress During
    Specific Growth Stages 673
Plant Water Stress: Optional Parameters for
    Stress Relief 676
Plant Water Use, Stomatal Control 680
Plant Yield and Water Use 686
Plants, Critical Growth Periods 689
Plants, Osmotic Adjustment 692
Plants, Osmotic Potential 696

Plants, Salt Tolerance 701
Pollution, Nonpoint Source 704
Pollution, Point Source 707
Precipitation Distribution 713
Precipitation, Forms 719
Precipitation Measurement 721
Precipitation Measurement
    Sensors 724
Precipitation Modification 727
Precipitation Simulation Models 729
Precipitation, Stochastic Properties 734
Precipitation Storms 737
Precision Agriculture, and Water Use 740
Professional Societies 744
Psychrometry for Measuring Plant and Soil Water Status:
    Accuracy, Interpretation, and Sampling 748
Psychrometry for Measuring Plant and Soil Water Status:
    Theory, Types, and Uses 751
Pump Powering with Internal Combustion Engines 756
Pumps, Displacement 759
Quality Influenced by Livestock Production on Range
    and Pasture Land 764
Quality Modeling 768
Quality Sampling of Runoff from
    Agricultural Fields 771
Rainfall Shelters 777
Rainfed Farming 780
Rangeland Management for Enhanced
    Water Utilization 783
Rangeland Water Yield, Influence of
    Brush Clearing on 788
Rangelands, Water Balance on 791
Research Centers for Dryland and
    Semiarid Regions 795
Research Organizations 800
Reverse Osmosis 803
Richards' Equation 809
Ring and Tension Infiltrometers 812
Rural Water Supply, Water Harvesting for 816
Sacramento–San Joaquin Delta 823
Saline Seeps 826
Saline Water 829
Salinity and Solute Measurement by Time
    Domain Reflectometry 832
Salton Sea 836
Selenium 840
Soil Macropores, Water and Solute Movement in 843
Soil Moisture Measurement by Feel
    and Appearance 847
Soil Salinity Measurement 852
Soil Water, Antecedent 858
Soil Water, Capillary Rise of 861
Soil Water Diffusion 865
Soil Water Energy Concepts 868
Soil Water Flow Under Saturated Conditions 871
Soil Water Flow Under Unsaturated Conditions ... 875
Soil Water, Gravimetric Measurement 879
Soil Water Hysteresis 882
Soil Water Measurement by Capacitance 885
Soil Water Measurement by Neutron
    Thermalization 889
Soil Water Measurement by Time Domain
    Reflectometry 894

Encyclopedia of
Water Science

Encyclopedia of Water Science

edited by

B. A. Stewart
West Texas A&M University
Canyon, Texas, U.S.A.

Terry A. Howell
United States Department of Agriculture
Agricultural Research Service (USDA-ARS)
Bushland, Texas, U.S.A.

MARCEL DEKKER, INC.　　　　NEW YORK • BASEL

Library of Congress Cataloging-in-Publication Data
A catalog record for this book is available from the Library of Congress.

ISBN
Print: **0-8247-0948-9**
Print/Online: **0-8247-4241-9**
Online: **0-8247-0947-0**

Cover photograph courtesy Dr. Katharina Helming

This book is printed on acid-free paper.

Headquarters
Marcel Dekker, Inc.
270 Madison Avenue, New York, NY 10016
tel: 212-696-9000; fax: 212-685-4540

Distribution and Customer Service
Marcel Dekker, Inc.
Cimarron Road, Monticello, New York 12701, U.S.A.
tel: 800-228-1160; fax: 845-796-1772

Eastern Hemisphere Distribution
Marcel Dekker AG
Hutgasse 4, Postfach 812, CH-4001 Basel, Switzerland
tel: 41-61-260-6300; fax: 41-61-260-6333

World Wide Web
http://www.dekker.com

The publisher offers discounts on this book when ordered in bulk quantities. For more information, write to Special Sales/Professional Marketing at the headquarters address above.

PRINTED IN THE UNITED STATES OF AMERICA

James S. Schepers   *United States Department of Agriculture, Agricultural Research Service (USDA-ARS), University of Nebraska, Lincoln, Nebraska, U.S.A.*

David Seckler   *Department of Agriculture and Resource Economics, Colorado State University, Fort Collins, Colorado, U.S.A.*

Thomas R. Sinclair   *United States Department of Agriculture, Agricultural Research Service (USDA-ARS), Crop Genetics and Environmental Research Unit, Gainesville, Florida*

Wayne R. Skaggs   *Department of Biological and Agricultural Engineering, North Carolina State University, Raleigh, North Carolina, U.S.A.*

Martin Smith   *Land and Water Development Division, Food and Agriculture Organization of the United Nations (FAO), Rome, Italy*

Jean L. Steiner   *United States Department of Agriculture, Agricultural Research Service (USDA-ARS), Grazinglands Research Laboratory, El Reno, Oklahoma, U.S.A.*

Donald L. Suarez   *United States Department of Agriculture, Agricultural Research Service (USDA-ARS), Salinity Laboratory, Riverside, California, U.S.A.*

Ian White   *Centre for Resource and Environmental Studies, Australian National University Canberra, Australian Capital Territory, Australia*

Topical Editors

Gary Bubenzer / *College of Agriculture and Life Sciences, University of Wisconsin, Madison, Wisconsin, U.S.A.*

William C. G. Burns / *American Society of International Law–Wildlife Interest Group El Cerrito, California, U.S.A.*

R. Nolan Clark / *United States Department of Agriculture, Agricultural Research Service (USDA-ARS), Conservation and Production Research Laboratory (CPRL), Bushland, Texas, U.S.A.*

Elias Fereres / *Instituto de Agricultura Sostenible, Consejo Superior de Investigaciones Científicas (CSIC), Cordoba, Spain*

Marcel Fuchs / *Agricultural Research Organization, Institute of Soil, Water, and Environmental Sciences, The Volcani Center, Bet Dagan, Israel*

M. B. Kirkham / *Department of Agronomy, Kansas State University, Manhattan, Kansas, U.S.A.*

Bruce J. Lesikar / *Department of Agricultural Engineering, Texas A&M University, College Station, Texas, U.S.A.*

Miguel A. Mariño / *Department of Land, Air, and Water Resources, University of California, Davis, California, U.S.A.*

Derrel L. Martin / *Department of Biological Systems Engineering, University of Nebraska, Lincoln, Nebraska, U.S.A.*

James D. Oster / *Department of Environmental Sciences, University of California, Riverside, California, U.S.A.*

Clarence W. Richardson / *United States Department of Agriculture, Agricultural Research Service (USDA-ARS), Grassland, Soil, and Water Research Laboratory, Blackland, Texas, U.S.A.*

William J. Rogers / *Department of Life, Earth, and Environmental Science, West Texas A&M University, Canyon, Texas, U.S.A.*

James S. Schepers / *United States Department of Agriculture, Agricultural Research Service (USDA-ARS), University of Nebraska, Lincoln, Nebraska, U.S.A.*

Wayne R. Skaggs / *Department of Biological and Agricultural Engineering, North Carolina State University, Raleigh, North Carolina, U.S.A.*

Jean L. Steiner / *United States Department of Agriculture, Agricultural Research Service (USDA-ARS), Grazinglands Research Laboratory, El Reno, Oklahoma, U.S.A.*

Raymond J. Supalla / *Department of Agricultural Economics, University of Nebraska, Lincoln, Nebraska, U.S.A.*

John M. Sweeten / *Agricultural Research and Extension Center, Texas A&M University, Amarillo, Texas, U.S.A.*

Darrel Temple / *United States Department of Agriculture, Agricultural Research Service (USDA-ARS), Plant Science and Water Conservation, Stillwater, Oklahoma, U.S.A.*

Daniel L. Thomas / *Department of Biological and Agricultural Engineering, University of Georgia, Tifton, Georgia, U.S.A.*

Paul W. Unger / *United States Department of Agriculture, Agricultural Research Service (USDA-ARS), retired, Conservation and Production Research Laboratory (CPRL), Bushland, Texas, U.S.A.*

List of Contributors

S. G. K. Adiku / *University of Ghana, Accra, Ghana*

L. R. Ahuja / *United States Department of Agriculture (USDA), Fort Collins, Colorado, U.S.A.*

J. David Aiken / *University of Nebraska, Lincoln, Nebraska, U.S.A.*

Richard G. Allen / *University of Idaho Research and Extension Center, Kimberly, Idaho, U.S.A.*

L. H. Allen, Jr. / *United States Department of Agriculture (USDA), Gainesville, Florida, U.S.A.*

Lal K. Almas / *West Texas A&M University, Canyon, Texas, U.S.A.*

Faye Anderson / *University of Maryland, College Park, Maryland, U.S.A.*

Michael A. Anderson / *University of California, Riverside, California, U.S.A.*

Randy L. Anderson / *United States Department of Agriculture (USDA), Brookings, South Dakota, U.S.A.*

A. N. Angelakis / *National Foundation for Agricultural Research, Heracleion, Greece*

Ramon Aragüés / *Agronomic Research Service, Diputación General de Aragón, Zaragoza, Spain*

Jeff Arnold / *United States Department of Agriculture (USDA), Temple, Texas, U.S.A.*

Francisco Arriaga / *University of Wisconsin, Madison, Wisconsin, U.S.A.*

B. W. Auvermann / *Texas A&M University, Amarillo, Texas, U.S.A.*

James E. Ayars / *United States Department of Agriculture (USDA), Parlier, California, U.S.A.*

James L. Baker / *Iowa State University, Ames, Iowa, U.S.A.*

John Baker / *United States Department of Agriculture (USDA), St. Paul, Minnesota, U.S.A.*

Khaled M. Bali / *University of California, Holtville, California, U.S.A.*

J. Barragan / *University of Lleida, Lleida, Spain*

Dorothea Bartels / *University of Bonn, Bonn, Germany*

Philip J. Bauer / *United States Department of Agriculture (USDA), Florence, South Carolina, U.S.A.*

R. Louis Baumhardt / *United States Department of Agriculture (USDA), Bushland, Texas, U.S.A.*

Steven Bednarz / *United States Department of Agriculture (USDA), Temple, Texas, U.S.A.*

M. Hossein Behboudian / *Massey University, Palmerston North, New Zealand*

George W. Bomar / *Texas Department of Licensing and Regulation, Austin, Texas, U.S.A.*

John Borrelli / *Texas Tech University, Lubbock, Texas, U.S.A.*

David D. Bosch / *United States Department of Agriculture (USDA), Tifton, Georgia, U.S.A.*

Thierry Boulard / *Institut National de la Recherche Agronomique (INRA), Avignon, France*

John Van Brahana / *University of Arkansas, Fayetteville, Arkansas, U.S.A.*

V. Bralts / *Purdue University, West Lafayette, Indiana, U.S.A*

James Brandle / *University of Nebraska, Lincoln, Nebraska, U.S.A.*

D. R. Bray / *University of Florida, Gainesville, Florida, U.S.A.*

David D. Breshears / *Los Alamos National Laboratory, Los Alamos, New Mexico, U.S.A.*

Sherry L. Britton / *United States Department of Agriculture (USDA), Stillwater, Oklahoma, U.S.A.*

Michael S. Brown / *West Texas A&M University, Canyon, Texas, U.S.A.*

Gary Bubenzer / *University of Wisconsin, Madison, Wisconsin, U.S.A.*

Graeme D. Buchan / *Lincoln University, Canterbury, New Zealand*

Robert D. Burman / *University of Wyoming, Laramie, Wyoming, U.S.A.*

Charles M. Burt / *California Polytechnic State University, San Luis Obispo, California, U.S.A.*

Amnon Bustan / *Ben-Gurion University of the Negev, Beer-Sheva, Israel*

Keith C. Cameron / *Lincoln University, Canterbury, New Zealand*

Carl R. Camp, Jr. / *United States Department of Agriculture (USDA), Florence, South Carolina, U.S.A.*

Kenneth L. Campbell / *University of Florida, Gainesville, Florida, U.S.A.*

Lucila Candela / *Technical University of Catalonia-UPC, Barcelona, Spain*

Peter S. Carberry / *Commonwealth Scientific and Industrial Research Organisation (CSIRO), Toowoomba, Queensland, Australia*

Jesús Carrera / *Technical University of Catalonia (UPC), Barcelona, Spain*

Dipankar Chakraborti / *Jadavpur University, Calcutta, India*

J. W. Chandler / *University of Cologne, Cologne, Germany*

Andrew C. Chang / *University of California, Riverside, California, U.S.A.*

André Chanzy / *Institut National de la Recherche Agronomique (INRA), Avignon, France*

Philip Charlesworth / *Commonwealth Scientific and Industrial Research Organisation (CSIRO), Townsville, Queensland, Australia*

Alexander H.-D. Cheng / *University of Delaware, Newark, Delaware, U.S.A.*

George M. Chescheir / *North Carolina State University, Raleigh, North Carolina, U.S.A.*

Uttam Kumar Chowdhury / *Jadavpur University, Calcutta, India*

Gary A. Clark / *Kansas State University, Manhattan, Kansas, U.S.A.*

Sharon A. Clay / *South Dakota State University, Brookings, South Dakota, U.S.A.*

Albert J. Clemmens / *United States Department of Agriculture (USDA), Phoenix, Arizona, U.S.A.*

Wayne Clyma / *Water Management Consultant, Fort Collins, Colorado, U.S.A.*

Bonnie G. Colby / *University of Arizona, Tucson, Arizona, U.S.A.*

W. Arden Colette / *West Texas A&M University, Canyon, Texas, U.S.A.*

Ray Correll / *Commonwealth Scientific and Industrial Research Organisation (CSIRO), Adelaide, South Australia, Australia*

Dennis L. Corwin / *United States Department of Agriculture (USDA), Riverside, California, U.S.A.*

Richard Cruse / *Iowa State University, Ames, Iowa, U.S.A.*

Seth M. Dabney / *United States Department of Agriculture (USDA), Oxford, Mississippi, U.S.A.*

Jacob H. Dane / *Auburn University, Auburn, Alabama, U.S.A.*

Philippe Debaeke / *Institut National de la Recherche Agronomique (INRA), Castanet-Tolosan, France*

Sherri L. DeFauw / *University of Arkansas, Fayetteville, Arkansas, U.S.A.*

G. R. Diak / *University of Wisconsin, Madison, Wisconsin, U.S.A.*

George Di Giovanni / *Texas A&M University Agricultural Research and Extension Center, El Paso, Texas, U.S.A.*

Peter Dillon / *Commonwealth Scientific and Industrial Research Organisation (CSIRO), Adelaide, South Australia, Australia*

William A. Dugas / *Texas Agricultural Experiment Station, Temple, Texas, U.S.A.*

Tim Dybala / *United States Department of Agriculture (USDA), Temple, Texas, U.S.A.*

Gabriel Eckstein / *International Water Law Project, Washington, District of Columbia, U.S.A.*

Bahman Eghball / *United States Department of Agriculture (USDA), Lincoln, Nebraska, U.S.A.*

F. A. El-Awar / *United Nations Office of the Humanitarian Coordinator for Iraq (UNOHCI)—Baghdad, New York, New York, U.S.A.*

Krisztina Eleki / *Iowa State University, Ames, Iowa, U.S.A.*

Robert O. Evans / *North Carolina State University, Raleigh, North Carolina, U.S.A.*

Steven R. Evett / *United States Department of Agriculture (USDA), Bushland, Texas, U.S.A.*

Floron C. Faries, Jr. / *West Texas A&M University, Canyon, Texas, U.S.A.*

Norman R. Fausey / *United States Department of Agriculture (USDA), Columbus, Ohio, U.S.A.*

C. W. Fetter, Jr. / *C. W. Fetter, Jr., Associates, Oshkosh, Wisconsin, U.S.A.*

Guy Fipps / *Texas A&M University, College Station, Texas, U.S.A.*

Graham E. Fogg / *University of California, Davis, California, U.S.A.*

James L. Fouss / *United States Department of Agriculture (USDA), Baton Rouge, Louisiana, U.S.A.*

Thomas G. Franti / *University of Nebraska, Lincoln, Nebraska, U.S.A.*

Gary W. Frasier / *United States Department of Agriculture (USDA), Fort Collins, Colorado, U.S.A.*

J. R. Frederick / *Clemson University, Florence, South Carolina, U.S.A.*

Marcel Fuchs / *Agricultural Research Organization, Bet Dagan, Israel*

Stuart R. Gagnon / *University of Maryland, College Park, Maryland, U.S.A.*

David K. Gattie / *University of Georgia, Athens, Georgia, U.S.A.*

Glendon W. Gee / *Pacific Northwest National Laboratory, Richland, Washington, U.S.A.*

Timothy R. Ginn / *University of California, Davis, California, U.S.A.*

David A. Goldhamer / *University of California, Parlier, California, U.S.A.*

Noel Gollehon / *United States Department of Agriculture (USDA), Washington, District of Columbia, U.S.A.*

Stephen R. Grattan / *University of California, Davis, California, U.S.A.*

Patricia K. Haan / *Texas A&M University, College Station, Texas, U.S.A.*

Ardell D. Halvorson / *United States Department of Agriculture (USDA), Fort Collins, Colorado, U.S.A.*

Dorota Z. Haman / *University of Florida, Gainesville, Florida, U.S.A.*

Joel R. Hamilton / *University of Idaho, Moscow, Idaho, U.S.A.*

Clayton L. Hanson (Retired) / *United States Department of Agriculture (USDA), Boise, Idaho, U.S.A.*

Gregory J. Hanson / *United States Department of Agriculture (USDA), Stillwater, Oklahoma, U.S.A.*

Mohamed Hantush / *United States Environmental Protection Agency (US EPA), Cincinnati, Ohio, U.S.A.*

Michael J. Hayes / *National Drought Mitigation Center and University of Nebraska, Lincoln, Nebraska, U.S.A.*

Phillip D. Hays / *University of Arkansas, Fayetteville, Arkansas, U.S.A.*

Richard W. Healy / *United States Geological Survey (USGS), Lakewood, Colorado, U.S.A.*

Robert W. Hill / *Utah State University, Logan, Utah, U.S.A.*

Michael C. Hirschi / *University of Illinois, Urbana, Illinois, U.S.A.*

Kyle D. Hoagland / *University of Nebraska, Lincoln, Nebraska, U.S.A.*

Laurie Hodges / *University of Nebraska, Lincoln, Nebraska, U.S.A.*

Glenn J. Hoffman / *University of Nebraska, Lincoln, Nebraska, U.S.A.*

James E. Hook / *University of Georgia, Tifton, Georgia, U.S.A.*

Jan W. Hopmans / *University of California, Davis, California, U.S.A.*

Terry A. Howell / *United States Department of Agriculture (USDA), Bushland, Texas, U.S.A.*

Joel M. Hubbell / *Idaho National Engineering and Environmental Laboratory, Idaho Falls, Idaho, U.S.A.*

Paul F. Hudak / *University of North Texas, Denton, Texas, U.S.A.*

Ray G. Huffaker / *Washington State University, Pullman, Washington, U.S.A.*

Rodney L. Huffman / *North Carolina State University, Raleigh, North Carolina, U.S.A.*

Frank Humenik / *North Carolina State University, Raleigh, North Carolina, U.S.A.*

John Hutson / *Flinders University, Adelaide, South Australia, Australia*

Keith E. Idso / *Center for the Study of Carbon Dioxide and Global Change, Tempe, Arizona, U.S.A.*

Sherwood B. Idso (Retired) / *United States Department of Agriculture (USDA), Phoenix, Arizona, U.S.A.*

W. Jabre / *American University of Beirut, Beirut, Lebanon*

C. Rhett Jackson / *University of Georgia, Athens, Georgia, U.S.A.*

Bruce R. James / *University of Maryland, College Park, Maryland, U.S.A.*

Dan Jaynes / *United States Department of Agriculture (USDA), Ames, Iowa, U.S.A.*

Kevin Jeanes / *ASK Laboratories, Inc., Amarillo, Texas, U.S.A.*

Gregory L. Johnson / *United States Department of Agriculture (USDA), Portland, Oregon, U.S.A.*

Ordie R. Jones (Retired) / *United States Department of Agriculture (USDA), Bushland, Texas, U.S.A.*

A. J. Karamanos / *Agricultural University of Athens, Athens, Greece*

Timothy O. Keefer / *United States Department of Agriculture (USDA), Tucson, Arizona, U.S.A.*

Keith A. Kelling / *University of Wisconsin, Madison, Wisconsin, U.S.A.*

Rami Keren / *Institute of Soil, Water and Environmental Sciences, Bet Dagan, Israel*

Dennis Kincaid / *United States Department of Agriculture (USDA), Kimberly, Idaho, U.S.A.*

M. B. Kirkham / *Kansas State University, Manhattan, Kansas, U.S.A.*

Peter Kleinman / *United States Department of Agriculture (USDA), University Park, Pennsylvania, U.S.A.*

Rai Kookana / *Commonwealth Scientific and Industrial Research Organisation (CSIRO), Adelaide, South Australia, Australia*

John Kost / *Iowa State University, Ames, Iowa, U.S.A.*

D. Koutsoyiannis / *National Technical University of Athens, Zographou, Greece*

Jacek A. Koziel / *Texas A&M University, Amarillo, Texas, U.S.A.*

Timothy P. Krantz / *University of Redlands, Redlands, California, U.S.A.*

William L. Kranz / *University of Nebraska, Norfolk, Nebraska, U.S.A.*

W. P. Kustas / *United States Department of Agriculture (USDA), Beltsville, Maryland, U.S.A.*

John M. Laflen / *Purdue University, Buffalo Center, Iowa, U.S.A.*

Rattan Lal / *The Ohio State University, Columbus, Ohio, U.S.A.*

Freddie R. Lamm / *Kansas State University, Colby, Kansas, U.S.A.*

Matthias Langensiepen / *Humboldt University of Berlin, Berlin, Germany*

Robert J. Lascano / *Texas A&M University, Lubbock, Texas, U.S.A.*

Brian Leib / *Washington State University, Pullman, Washington, U.S.A.*

Jay A. Leitch / *North Dakota State University, Fargo, North Dakota, U.S.A.*

Rick P. Leopold / *United States Department of Agriculture (USDA), Bryan, Texas, U.S.A.*

Daniel G. Levitt / *Science & Engineering Associates, Inc., Santa Fe, New Mexico, U.S.A.*

Hong Li / *University of Florida, Lake Alfred, Florida, U.S.A.*

Hugo A. Loáiciga / *University of California, Santa Barbara, California, U.S.A.*

Sally D. Logsdon / *United States Department of Agriculture (USDA), Ames, Iowa, U.S.A.*

Guy H. Loneragan / *West Texas A&M University, Canyon, Texas, U.S.A.*

Birl Lowery / *University of Wisconsin, Madison, Wisconsin, U.S.A.*

Richard Lowrance / *United States Department of Agriculture (USDA), Tifton, Georgia, U.S.A.*

Stephan J. Maas / *Texas Tech University, Lubbock, Texas, U.S.A.*

Babs Makinde-Odusola / *Riverside Public Utilities, Riverside, California, U.S.A.*

Joseph R. Makuch / *National Agricultural Library, Beltsville, Maryland, U.S.A.*

Kyle R. Mankin / *Kansas State University, Manhattan, Kansas, U.S.A.*

Thomas Marek / *Texas A&M University, Amarillo, Texas, U.S.A.*

Miguel A. Mariño / *University of California, Davis, California, U.S.A.*

Dean A. Martens / *United States Department of Agriculture (USDA), Tucson, Arizona, U.S.A.*

Luciano Mateos / *Instituto de Agricultura Sostenible, Consejo Superior de Investigaciones Científicas, Córdoba, Spain*

Donald K. McCool / *United States Department of Agriculture (USDA), Pullman, Washington, U.S.A.*

Kevin B. McCormack / *United States Environmental Protection Agency (US-EPA), Washington, District of Columbia, U.S.A.*

Richard McDowell / *Ag Research Ltd., Invermay Agricultural Centre, Mosglel, New Zealand*

Marshall J. McFarland / *Texas A&M University, College Station, Texas, U.S.A.*

W. Allan McGinty / *Texas A&M University Research and Extension Center, San Angelo, Texas, U.S.A.*

Gregory McIsaac / *University of Illinois, Urbana, Illinois, U.S.A.*

Mallavarapu Megharaj / *Commonwealth Scientific and Industrial Research Organisation (CSIRO), Adelaide, South Australia, Australia*

Tessa Marie Mills / *The Horticulture and Food Research Institute of New Zealand (HortResearch), Palmerston North, New Zealand*

J. Kent Mitchell / *University of Illinois, Urbana, Illinois, U.S.A.*

R. H. Mohtar / *Purdue University, West Lafayette, Indiana, U.S.A.*

Philip A. Moore, Jr. / *United States Department of Agriculture (USDA), Fayetteville, Arkansas, U.S.A.*

M. S. Moran / *United States Department of Agriculture (USDA), Tucson, Arizona, U.S.A.*

James M. Morgan / *Tamworth Centre for Crop Improvement, Tamworth, New South Wales, Australia*

James I. L. Morison / *University of Essex, Colchester, United Kingdom*

John R. Morse / *United States Department of Agriculture (USDA), Pullman, Washington, U.S.A.*

Miranda Y. Mortlock / *Office of Economic and Statistical Research (OESR), Brisbane, Queensland, Australia*

Saqib Mukhtar / *Texas A&M University, College Station, Texas, U.S.A.*

Ranjan S. Muttiah / *Texas Agricultural Experiment Station, Temple, Texas, U.S.A.*

M. H. Nachabe / *University of South Florida, Tampa, Florida, U.S.A.*

Ravendra Naidu / *Commonwealth Scientific and Industrial Research Organisation (CSIRO), Adelaide, South Australia, Australia*

Jerry Neppel / *Iowa State University, Ames, Iowa, U.S.A.*

John R. Nimmo / *United States Geological Survey (USGS), Menlo Park, California, U.S.A.*

Janice G. Norris / *Texas A&M University, College Station, Texas, U.S.A.*

Iwao Ohtsu / *Nihon University College of Science and Technology, Tokyo, Japan*

Derrick M. Oosterhuis / *University of Arkansas, Fayetteville, Arkansas, U.S.A.*

William J. Orts / *United States Department of Agriculture (USDA), Albany, California, U.S.A.*

James D. Oster / *University of California, Riverside, California, U.S.A.*

Lloyd B. Owens / *United States Department of Agriculture (USDA), Coshocton, Ohio, U.S.A.*

Ioan C. Paltineanu / *Paltin International, Inc., Laurel, Maryland, U.S.A.*

Charalambos Papelis / *Desert Research Institute, Las Vegas, Nevada, U.S.A.*

David B. Parker / *West Texas A&M University, Canyon, Texas, U.S.A.*

Henry S. Parker / *United States Department of Agriculture (USDA), Wyndmoor, Pennsylvania, U.S.A.*

Dov Pasternak / *International Crops Research Institute for the Semi-Arid Tropics, Niamey, Niger*

Kunal Paul / *Jadavpur University, Calcutta, India*

William A. Payne / *Texas A&M University Agricultural Research and Extension Center, Amarillo, Texas, U.S.A.*

Kurt D. Pennell / *Georgia Institute of Technology, Atlanta, Georgia, U.S.A.*

Alain Perrier / *Institut National Agronomique de Paris-Grignon (INAPG), Paris, France*

Steven P. Phillips / *United States Geological Survey (USGS), Sacramento, California, U.S.A.*

Joseph L. Pikul, Jr. / *United States Department of Agriculture (USDA), Brookings, South Dakota, U.S.A.*

Suresh D. Pillai / *Texas A&M University, College Station, Texas, U.S.A.*

Zvi Plaut / *Agricultural Research Organization, Bet Dagan, Israel*

Daniel H. Pote / *United States Department of Agriculture (USDA), Booneville, Arkansas, U.S.A.*

Quazi Quamruzzaman / *Dhaka Community Hospital, Dhaka, Bangladesh*

Nageswararao C. Rachaputi / *Queensland Department of Primary Industries, Kingaroy, Queensland, Australia*

David E. Radcliffe / *University of Georgia, Athens, Georgia, U.S.A.*

Mohammad Mahmudur Rahman / *Jadavpur University, Calcutta, India*

John C. Reagor / *Texas A&M University, College Station, Texas, U.S.A.*

John A. Replogle / *United States Department of Agriculture (USDA), Phoenix, Arizona, U.S.A.*

Marty B. Rhoades / *West Texas A&M University, Canyon, Texas, U.S.A.*

Mark J. Roberson / *Keller–Bleisner Engineering, LLC, Sacramento, California, U.S.A.*

Paul D. Robillard / *The Pennsylvania State University, University Park, Pennsylvania, U.S.A.*

Clay A. Robinson / *West Texas A&M University, Canyon, Texas, U.S.A.*

Mark Robinson / *Centre for Ecology and Hydrology, Oxfordshire, United Kingdom*

Judy A. Rogers / *University of Arkansas, Fayetteville, Arkansas, U.S.A.*

William J. Rogers / *West Texas A&M University, Canyon, Texas, U.S.A.*

Renee Rokicki / *University of South Florida, Tampa, Florida, U.S.A.*

Wes Rosenthal / *Texas Agricultural Experiment Station, Temple, Texas, U.S.A.*

James F. Ruff / *Colorado State University, Fort Collins, Colorado, U.S.A.*

Loret M. Ruppe / *University of California, Davis, California, U.S.A.*

John Ryan / *International Center for Agricultural Research in the Dry Areas (ICARDA), Aleppo, Syria*

David W. Rycroft / *Southampton University, Southampton, United Kingdom*

Edward John Sadler / *United States Department of Agriculture (USDA), Florence, South Carolina, U.S.A.*

R. Sakthivadivel / *International Water Management Institute (IWMI), Colombo, Sri Lanka*

Hilmy Sally / *International Water Management Institute (IWMI), Pretoria, South Africa*

Gary Sands / *University of Minnesota, St. Paul, Minnesota, U.S.A.*

Arland D. Schneider / *United States Department of Agriculture (USDA), Bushland, Texas, U.S.A.*

Lawrence J. Schwankl / *University of California, Davis, California, U.S.A.*

Mark S. Seyfried / *United States Department of Agriculture (USDA), Boise, Idaho, U.S.A.*

M. S. Shafique (Retired) / *United Nations Office for Project Services, Lahore, Pakistan*

Michael C. Shannon / *United States Department of Agriculture (USDA), Riverside, California, U.S.A.*

Andrew Sharpley / *United States Department of Agriculture (USDA), University Park, Pennsylvania, U.S.A.*

Brenton S. Sharratt / *United States Department of Agriculture (USDA), Pullman, Washington, U.S.A.*

Roger Shaw / *University of California, Davis, California, U.S.A.*

J. Sheehy / *International Rice Research Institute, Metro Manila, Philippines*

Adel Shirmohammadi / *University of Maryland, College Park, Maryland, U.S.A.*

Clinton C. Shock / *Oregon State University, Ontario, Oregon, U.S.A.*

Steven Shultz / *North Dakota State University, Fargo, North Dakota, U.S.A.*

Thomas R. Sinclair / *United States Department of Agriculture (USDA), Gainesville, Florida, U.S.A.*

James "Buck" Sisson / *Idaho National Engineering and Environmental Laboratory, Idaho Falls, Idaho, U.S.A.*

Gaylord V. Skogerboe / *Utah State University, Logan, Utah, U.S.A.*

Jeffrey G. Skousen / *West Virginia University, Morgantown, West Virginia, U.S.A.*

Douglas R. Smith / *United States Department of Agriculture (USDA), West Lafayette, Indiana, U.S.A.*

James E. Smith / *McMaster University, Hamilton, Ontario, Canada*

Matt C. Smith / *University of Georgia, Tifton, Georgia, U.S.A.*

Richard L. Snyder / *University of California, Davis, California, U.S.A.*

Robert E. Sojka / *United States Department of Agriculture (USDA), Kimberly, Idaho, U.S.A.*

Kenneth H. Solomon / *California Polytechnic State University, San Luis Obispo, California, U.S.A.*

Marios Sophocleous / *University of Kansas, Lawrence, Kansas, U.S.A.*

Roy F. Spalding / *University of Nebraska, Lincoln, Nebraska, U.S.A.*

James L. Starr / *United States Department of Agriculture (USDA), Beltsville, Maryland, U.S.A.*

Jean L. Steiner / *United States Department of Agriculture (USDA), El Reno, Oklahoma, U.S.A.*

B. A. Stewart / *West Texas A&M University, Canyon, Texas, U.S.A.*

Richard J. Stirzaker / *Commonwealth Scientific and Industrial Research Organisation (CSIRO), Canberra, Australian Capital Territory, Australia*

Claudio O. Stockle / *Washington State University, Pullman, Washington, U.S.A.*

David A. Stonestrom / *United States Geological Survey (USGS), Menlo Park, California, U.S.A.*

David E. Stooksbury / *University of Georgia, Athens, Georgia, U.S.A.*

Rosemary Streatfeild / *Washington State University, Pullman, Washington, U.S.A.*

Scott J. Sturgul / *University of Wisconsin, Madison, Wisconsin, U.S.A.*

John M. Sweeten / *Texas A&M University, Amarillo, Texas, U.S.A.*

Donald L. Tanaka / *United States Department of Agriculture (USDA), Mandan, North Dakota, U.S.A.*

Kenneth K. Tanji / *University of California, Davis, California, U.S.A.*

Daniel L. Thomas / *University of Georgia, Tifton, Georgia, U.S.A.*

Thomas Lee Thurow / *University of Wyoming, Laramie, Wyoming, U.S.A.*

Jeanette Thurston-Enriquez / *United States Department of Agriculture (USDA), Lincoln, Nebraska, U.S.A.*

Judy A. Tolk / *United States Department of Agriculture (USDA), Bushland, Texas, U.S.A.*

Neil C. Turner / *Commonwealth Scientific and Industrial Research Organisation (CSIRO), Wembley (Perth), Western Australia, Australia*

Ronald W. Tuttle (Retired) / *United States Department of Agriculture (USDA), Washington, District of Columbia, U.S.A.*

Andree Tuzet / *Institut National de la Recherche Agronomique (INRA), Thiverval Grignon, France*

Melvin T. Tyree / *United States Department of Agriculture (USDA), Burlington, Vermont, U.S.A.*

Darrell N. Ueckert / *Texas A&M University Research and Extension Center, San Angelo, Texas, U.S.A.*

Paul W. Unger (Retired) / *United States Department of Agriculture (USDA), Bushland, Texas, U.S.A.*

H. H. Van Horn / *University of Florida, Gainesville, Florida, U.S.A.*

Jan van Schilfgaarde (Retired) / *United States Department of Agriculture (USDA), Fort Collins, Colorado, U.S.A.*

George F. Vance / *University of Wyoming, Laramie, Wyoming, U.S.A.*

Curtis J. Varnell / *University of Arkansas, Fayetteville, Arkansas, U.S.A.*

J. C. V. Vu / *United States Department of Agriculture (USDA), Gainesville, Florida, U.S.A.*

Wynn R. Walker / *Utah State University, Logan, Utah, U.S.A.*

Ivan A. Walter / *Ivan's Engineering, Inc., Denver, Colorado, U.S.A.*

Dong Wang / *University of Minnesota, St. Paul, Minnesota, U.S.A.*

Pao K. Wang / *University of Wisconsin, Madison, Wisconsin, U.S.A.*

Anderson L. Ward / *Pacific Northwest National Laboratory, Richland, Washington, U.S.A.*

Walter Wenzel / *University of Agricultural Sciences, Vienna, Austria*

Hal Werner / *South Dakota State University, Brookings, South Dakota, U.S.A.*

Mark E. Westgate / *Iowa State University, Ames, Iowa, U.S.A.*

French Wetmore / *French & Associates, Ltd., Park Forest, Illinois, U.S.A.*

Norman K. Whittlesey / *Washington State University, Pullman, Washington, U.S.A.*

Keith Wiebe / *United States Department of Agriculture (USDA), Washington, District of Columbia, U.S.A.*

Bradford P. Wilcox / *Texas A&M University, College Station, Texas, U.S.A.*

Mark Wilf / *Hydranautics, Oceanside, California, U.S.A.*

Donald A. Wilhite / *National Drought Mitigation Center and University of Nebraska, Lincoln, Nebraska, U.S.A.*

Lyman S. Willardson / *Utah State University, Logan, Utah, U.S.A.*

Dennis E. Williams / *Geoscience Support Services, Inc., Claremont, California, U.S.A.*

David A. Woolhiser / *Colorado State University, Fort Collins, Colorado, U.S.A.*

Stephen R. Workman / *University of Kentucky, Lexington, Kentucky, U.S.A.*

Jon M. Wraith / *Montana State University, Bozeman, Montana, U.S.A.*

Graeme C. Wright / *Queensland Department of Primary Industries, Kingaroy, Queensland, Australia*

I. P. Wu / *University of Hawaii, Honolulu, Hawaii, U.S.A.*

Laosheng Wu / *University of California, Riverside, California, U.S.A.*

Richard T. Wynne / *National Oceanic and Atmospheric Administration (NOAA), Amarillo, Texas, U.S.A.*

Youichi Yasuda / *Nihon University College of Science and Technology, Tokyo, Japan*

C. Dean Yonts / *University of Nebraska, Scottsbluff, Nebraska, U.S.A.*

Michael H. Young / *Desert Research Institute, Las Vegas, Nevada, U.S.A.*

Robert A. Young / *Colorado State University, Fort Collins, Colorado, U.S.A.*

Bofu Yu / *Griffith University, Nathan, Queensland, Australia*

Xinhua Zhou / *University of Nebraska, Lincoln, Nebraska, U.S.A.*

Contents

Academic Disciplines / *Joseph R. Makuch and Paul D. Robillard* 1

Acid Rain and Precipitation Chemistry / *Pao K. Wang* 4

Acidification for Prevention of Clogging of Drip Lines and Emitters / *Andrew C. Chang* 7

African Market Garden / *Dov Pasternak and Amnon Bustan* 9

Agricultural Runoff Characteristics / *Matt C. Smith, Daniel L. Thomas,*
and David K. Gattie 15

Agroforestry for Enhancing Water Use Efficiency / *James Brandle,*
Xinhua Zhou, and Laurie Hodges 19

Aquifer Recharge / *John R. Nimmo, David A. Stonestrom, and Richard W. Healy* 22

Aquifer Transmissivity / *Mohamed Hantush* 26

Aquifers / *Miguel A. Mariño* 30

Aquifers, Artificial Recharge of / *Steven P. Phillips* 33

Aquifers, Karst / *John Van Brahana* 37

Aquifers, Karst: Water Quality and Water-Resource Problems / *John Van Brahana* 41

Aquifers, Ogallala / *B. A. Stewart* 43

Aral Sea Disaster / *Guy Fipps* 45

Boron / *Rami Keren* 49

Canal Automation / *Albert J. Clemmens* 53

Carbon Dioxide, Plants and Transpiration / *L. H. Allen, Jr., J. C. V. Vu,*
and J. Sheehy 57

Chemical Measurement / *William J. Rogers and Kevin Jeanes* 62

Chemigation / *William L. Kranz* 67

Chlorination for Disinfection and Prevention of Clogging of Drip Lines
and Emitters / *Andrew C. Chang* 72

Chromium / *Bruce R. James* 75

Conservation Tillage and No-Tillage / *Paul W. Unger* 80

Consumptive Water Use / *Freddie R. Lamm* 83

Crop Coefficients / *Richard G. Allen* 87

Crop Development Models / *Peter. S. Carberry* 91

Crop Plants: Critical Developmental Stages of Water Stress / *Zvi Plaut* 95

Crop Residues, Snow Capture by / *Donald K. McCool, Brenton S. Sharratt,*
and John R. Morse 101

Crops, Water Harvesting for / *Gary W. Frasier* 105

Darcy's Law / *Graeme D. Buchan and Keith C. Cameron* 109

Databases / *Rosemary Streatfeild* 113

Drainage Coefficient / *Gary Sands* 118

Drainage, Controlled / *Robert O. Evans* 121

Drainage, Hydrologic Impacts of / *Mark Robinson and David W. Rycroft* ... 128

Drainage: Inadequacy and Crop Response / *Norman R. Fausey* ... 132

Drainage of Irrigated Land / *James E. Ayars* ... 135

Drainage, Land Shaping for / *Rodney L. Huffman* ... 138

Drainage Materials / *James L. Fouss* ... 142

Drainage Modeling / *George M. Chescheir* ... 147

Drainage for Soil Salinity Management / *Glenn J. Hoffman* ... 152

Drainage and Water Quality / *James L. Baker* ... 156

Drought / *Donald A. Wilhite* ... 160

Drought Avoidance and Drought Adaptation / *J. W. Chandler and Dorothea Bartels* ... 163

Drought Hardening and Presowing Seed Hardening / *Neil C. Turner* ... 166

Drought Management / *Michael J. Hayes and Donald A. Wilhite* ... 170

Drought Resistance / *M. B. Kirkham* ... 173

Dryland Cropping Systems / *William A. Payne* ... 178

Dryland Farming / *Clay A. Robinson* ... 183

Dust Bowl Era / *R. Louis Baumhardt* ... 187

El Niño / *David E. Stooksbury* ... 192

Energy Dissipation Structures / *Youichi Yasuda and Iwao Ohtsu* ... 195

Erosion, Accelerated / *J. Kent Mitchell and Michael C. Hirschi* ... 199

Erosion Control, Tillage/Residue Methods / *Richard Cruse, Jerry Neppel, John Kost, and Krisztina Eleki* ... 205

Erosion Control, Vegetative / *Seth M. Dabney* ... 209

Erosion and Precipitation / *Bofu Yu* ... 214

Erosion Process Modeling / *John M. Laflen* ... 218

Erosion and Productivity / *Francisco Arriaga and Birl Lowery* ... 222

Erosion Research, History of / *Rattan Lal* ... 225

Erosion Research, Instrumentation for / *Gary Bubenzer* ... 229

Eutrophication / *Thomas G. Franti and Kyle D. Hoagland* ... 232

Evaporation and Eddy Correlation / *Roger Shaw and Richard L. Snyder* ... 235

Evaporation and Energy Balance / *Matthias Langensiepen* ... 238

Evaporation from Lakes and Large Bodies of Water / *John Borrelli* ... 242

Evaporation as a Process / *Alain Perrier and Andree Tuzet* ... 245

Evaporation from Soils / *André Chanzy* ... 249

Evapotranspiration Formulas / *Robert D. Burman* ... 253

Evapotranspiration in Greenhouses / *Thierry Boulard* ... 258

Evapotranspiration, Reference and Potential / *Marcel Fuchs* ... 264

Evapotranspiration, Remote Sensing of / *W. P. Kustas, G. R. Diak, and M. S. Moran* ... 267

Everglades / *Kenneth L. Campbell* ... 275

Farm Ponds / *Ronald W. Tuttle* ... 278

Fertilizer/Pesticide Leaching, Irrigation Management and / *Luciano Mateos* ... 282

Field Water Supply and Balance / *Jean L. Steiner* ... 285

Filtration and Particulate Removal / *Lawrence J. Schwankl* ... 289

Floodplain Management / *French Wetmore* ... 294

Floods and Flooding / *Jay A. Leitch and Steven Shultz* ... 300

Flow Measurement, History of / *James F. Ruff* ... 306

Fluoride / *Judy A. Rogers* ... 312

Frozen Soil, Water Movement in / *John Baker* 314

Furrow Dikes / *Ordie R. Jones and R. Louis Baumhardt* 317

Global Temperature Change and Terrestrial Ecology / *Sherwood B. Idso*
    and Keith E. Idso 321

Groundwater Arsenic Contamination / *Dipankar Chakraborti, Mohammad Mahmudur*
    Rahman, Kunal Paul, Uttam Kumar Chowdhury, and Quazi Quamruzzaman 324

Groundwater Contamination / *C. W. Fetter, Jr.* 330

Groundwater Law of the Western United States / *J. David Aiken* 333

Groundwater Levels, Mapping / *Marios Sophocleous* 336

Groundwater Levels, Measuring / *Paul F. Hudak* 342

Groundwater Mining / *Hugo A. Loáiciga* 345

Groundwater Modeling / *Jesús Carrera* 350

Groundwater Modeling: How Codes Work / *Jesús Carrera* 358

Groundwater Pollution from Mining / *George F. Vance and Jeffrey G. Skousen* 363

Groundwater Pollution by Nitrogen Fertilizers / *Lloyd B. Owens* 369

Groundwater Pollution by Phosphorus Fertilizers / *Bahman Eghball* 374

Groundwater Pumping Methods / *Dennis E. Williams* 376

Groundwater Quality / *Loret M. Ruppe and Timothy R. Ginn* 389

Groundwater Quality, Irrigated Agriculture and / *Stephen R. Grattan* 392

Groundwater, Regulation of / *Kevin B. McCormack* 398

Groundwater, Saltwater Intrusion in / *Alexander H.-D. Cheng* 404

Groundwater, World Resources / *Lucila Candela* 407

Hydrologic Cycle / *John Van Brahana* 412

Hydrologic and Hydraulic Science and Technology in Ancient Greece / *D. Koutsoyiannis*
    and A. N. Angelakis 415

Hydrologic Process Modeling / *Daniel L. Thomas and Matt C. Smith* 418

Hydrology Research Centers / *Daniel L. Thomas* 421

Internet / *Janice G. Norris* 424

Irrigated Agriculture, Economic Impacts of Investments in / *Robert A. Young* 428

Irrigated Agriculture and Endangered Species Policy / *Ray G. Huffaker,*
    Norman K. Whittlesey, and Joel R. Hamilton 431

Irrigated Agriculture, Historical View of / *Lyman S. Willardson* 434

Irrigated Agriculture: Managing Toward Sustainability / *Wayne Clyma,*
    M. S. Shafique, and Jan van Schilfgaarde 437

Irrigated Agriculture, Social Impacts of / *Gaylord V. Skogerboe* 443

Irrigated Water, Market Role in Reallocating / *Bonnie G. Colby* 446

Irrigated Water, Polymer Application in / *William J. Orts and Robert E. Sojka* 449

Irrigation Design Steps and Elements / *Gary A. Clark* 454

Irrigation Economics, Global / *Keith Wiebe and Noel Gollehon* 459

Irrigation Economics, United States / *Noel Gollehon* 463

Irrigation Efficiency / *Terry A. Howell* 467

Irrigation Impact on River Flows / *Robert W. Hill and Ivan A. Walter* 473

Irrigation Management in Humid Regions / *Edward John Sadler,*
    Carl R. Camp, Jr., and James E. Hook. 478

Irrigation Management for the Tropics / *R. Sakthivadivel and Hilmy Sally* 483

Irrigation Mechanical Systems: Sprinkler / *Dennis Kincaid* 490

Irrigation Metering / *Albert J. Clemmens and John A. Replogle* 495

Irrigation, Preplant / *James E. Ayars* 498

Irrigation Return Flow, Quality and / *Ramon Aragüés and Kenneth K. Tanji* 502

Irrigation Sagacity / *Kenneth H. Solomon and Charles M. Burt* 507

Irrigation with Saline Water / *B. A. Stewart* 510

Irrigation Scheduling with Plant Indicators: Field Applications /
David A. Goldhamer 512

Irrigation Scheduling with Plant Indicators: Measurement / *David A. Goldhamer* 519

Irrigation Scheduling by Remote Sensing Technologies / *Stephan J. Maas* 523

Irrigation Scheduling by Soil Water Status / *Philip Charlesworth*
and Richard J. Stirzaker 528

Irrigation Scheduling by Water Budgeting / *Claudio O. Stockle and Brian Leib* 532

Irrigation, Sewage Effluent Use for / *B. A. Stewart* 535

Irrigation, Supplemental / *Philippe Debaeke* 537

Irrigation, Surface / *Wynn R. Walker* 540

Irrigation Systems, Drip / *I. P. Wu, J. Barragan, and V. Bralts* 546

Irrigation Systems, History of Farm / *Wayne Clyma* 549

Irrigation Systems, Subsurface Drip / *Carl R. Camp, Jr., and Freddie R. Lamm* 560

Isotopes / *Michael A. Anderson* 565

Journals / *Joseph R. Makuch and Stuart R. Gagnon* 569

La Niña / *David E. Stooksbury* 574

Land Drainage, Subsurface / *Stephen R. Workman* 576

Leaf Water Potential / *A. J. Karamanos* 579

Livestock and Poultry Production, Water Consumption for / *David B. Parker*
and Michael S. Brown 588

Livestock, Water Harvesting Methods for / *Gary W. Frasier* 593

Livestock Water Quality Standards / *John M. Sweeten, Floron C. Faries, Jr.,*
Guy H. Loneragan, and John C. Reagor 596

Manure Management, Beef Cattle Industry Requirements / *John M. Sweeten*
and B. W. Auvermann 601

Manure Management, Dairy / *H. H. Van Horn and D. R. Bray* 603

Manure Management, Poultry / *Saqib Mukhtar and Patricia K. Haan* 607

Manure Management, Swine / *Frank Humenik* 610

Marketing / *Lal K. Almas and W. Arden Colette* 612

Matric Potential / *Melvin T. Tyree* 615

Microbial Sampling / *Suresh D. Pillai and George Di Giovanni* 618

Neuse River / *Curtis J. Varnell and John Van Brahana* 622

Nitrogen Measurement / *Jacek A. Koziel* 625

Nutrient Best Management Practices / *Scott J. Sturgul and Keith A. Kelling* 630

Observation Wells / *Phillip D. Hays and John Van Brahana* 633

Open-Channel Spillways / *Gregory J. Hanson and Sherry L. Britton* 636

Oxygen Measurement: Biological–Chemical Oxygen Demand / *David B. Parker*
and Marty B. Rhoades 642

Pathogens in Water / *Jeanette Thurston-Enriquez* 645

Pesticide Contamination, Groundwater / *Roy F. Spalding* 650

Pesticide Contamination, Surface Water / *Sharon A. Clay* 654

Pfiesteria Piscicida / *Henry S. Parker* 658

pH / *William J. Rogers* 663

Phosphorus Measurement / *Bahman Eghball and Daniel H. Pote* 666

Plant Available Soil Water / *Judy A. Tolk* 669

Plant Exposure to Water Stress During Specific Growth Stages / *Zvi Plaut* 673

Plant Water Stress: Optional Parameters for Stress Relief / *Zvi Plaut* 676

Plant Water Use, Stomatal Control / *James I. L. Morison* 680

Plant Yield and Water Use / *M. Hossein Behboudian and Tessa Marie Mills* 686

Plants, Critical Growth Periods / *Tessa Marie Mills and M. Hossein Behboudian* 689

Plants, Osmotic Adjustment / *James M. Morgan* 692

Plants, Osmotic Potential / *Mark E. Westgate* 696

Plants, Salt Tolerance of / *Michael C. Shannon* 701

Pollution, Nonpoint Source / *Ravendra Naidu, Mallavarapu Megharaj, Peter Dillon,
Rai Kookana, Ray Correll, and Walter Wenzel* 704

Pollution, Point Source / *Ravendra Naidu, Mallavarapu Megharaj, Peter Dillon,
Rai Kookana, Ray Correll, and Walter Wenzel* 707

Precipitation Distribution Patterns / *Gregory L. Johnson* 713

Precipitation, Forms of / *Richard T. Wynne* 719

Precipitation Measurement / *Marshall J. McFarland* 721

Precipitation Measurements with Remote Sensors / *Marshall J. McFarland* 724

Precipitation Modification / *George W. Bomar* 727

Precipitation Simulation Models / *Timothy O. Keefer* 729

Precipitation, Stochastic Properties / *David A. Woolhiser* 734

Precipitation Storms / *Clayton L. Hanson* 737

Precision Agriculture and Water Use / *Robert J. Lascano and Hong Li* 740

Professional Societies / *Faye Anderson* 744

**Psychrometry for Measuring Plant and Soil Water Status: Accuracy,
Interpretation, and Sampling** / *Derrick M. Oosterhuis* 748

Psychrometry for Measuring Plant and Soil Water Status: Theory, Types, and Uses /
Derrick M. Oosterhuis 751

Pump Powering with Internal Combustion Engines / *Hal Werner* 756

Pumps, Displacement / *Dorota Z. Haman* 759

Quality Influenced by Livestock Production on Range and Pasture Land /
Thomas Lee Thurow 764

Quality Modeling / *Richard Lowrance* 768

Quality Sampling of Runoff from Agricultural Fields / *John M. Laflen* 771

Rainfall Shelters / *Arland D. Schneider* 777

Rainfed Farming / *Philip J. Bauer, Edward John Sadler, and J. R. Frederick* 780

Rangeland Management for Enhanced Water Utilization / *Darrell N. Ueckert
and W. Allan McGinty* 783

Rangeland Water Yield, Influence of Brush Clearing on / *William A. Dugas, Steven Bednarz,
Tim Dybala, Ranjan S. Muttiah, Wes Rosenthal, and Jeff Arnold* 788

Rangelands, Water Balance on / *Bradford P. Wilcox, David D. Breshears,
and Mark S. Seyfried* 791

Research Centers for Dryland and Semiarid Regions / *John Ryan* 795

Research Organizations / *Gabriel Eckstein* 800

Reverse Osmosis / *Mark Wilf* 803

Richards' Equation / *Graeme D. Buchan* 809

Ring and Tension Infiltrometers / *Dong Wang and Laosheng Wu* 812

Rural Water Supply, Water Harvesting for / *R. H. Mohtar, F. A. El-Awar, and W. Jabre* 816

Sacramento–San Joaquin Delta / *Mark J. Roberson* 823

Saline Seeps / *Ardell D. Halvorson* 826

Saline Water / *Khaled M. Bali* 829

Salinity and Solute Measurement by Time Domain Reflectometry / *Jon M. Wraith* 832

Salton Sea / *Timothy P. Krantz* 836

Selenium / *Dean A. Martens* 840

Soil Macropores, Water and Solute Movement in / *David E. Radcliffe* 843

Soil Moisture Measurement by Feel and Appearance / *Rick P. Leopold* 847

Soil Salinity Measurement / *Dennis L. Corwin* 852

Soil Water, Antecedent / *Sally D. Logsdon* 858

Soil Water, Capillary Rise of / *James E. Smith* 861

Soil Water Diffusion / *Laosheng Wu* 865

Soil Water Energy Concepts / *Sally D. Logsdon* 868

Soil Water Flow Under Saturated Conditions / *Jan W. Hopmans and Graham E. Fogg* 871

Soil Water Flow Under Unsaturated Conditions / *Jan W. Hopmans and Jacob H. Dane* 875

Soil Water, Gravimetric Measurement / *Joseph L. Pikul, Jr.* 879

Soil Water Hysteresis / *Dan Jaynes* 882

Soil Water Measurement by Capacitance / *James L. Starr and Ioan C. Paltineanu* 885

Soil Water Measurement by Neutron Thermalization / *Steven R. Evett* 889

Soil Water Measurement by Time Domain Reflectometry / *Steven R. Evett* 894

Soil Water Potential Measurement by Granular Matrix Sensors / *Clinton C. Shock* 899

Soil Water Potential Measurement by Tensiometers / *Joel M. Hubbell and James "Buck" Sisson* 904

Soil Water Storage Measurement by Soil Probes / *Clay A. Robinson* 908

Soil, Waterborne Chemicals Leaching Through / *John Hutson* 911

Soils, Field Capacity of Water in / *M. H. Nachabe, L. R. Ahuja, and Renee Rokicki* 915

Soils, Hydraulic Conductivity Rates in / *David D. Bosch and Adel Shirmohammadi* 919

Soils, Hygroscopic Water Content in / *Daniel G. Levitt and Michael H. Young* 923

Soils, Permanent Wilting Points / *Judy A. Tolk* 927

Soils, Water Infiltration and / *Sally D. Logsdon* 930

Soils, Water Percolation Through / *Kurt D. Pennell* 934

Storativity and Specific Yield / *Hugo A. Loáiciga and Paul F. Hudak* 937

Summer Fallow / *Donald L. Tanaka and Randy L. Anderson* 942

Surface Water Law in the Western United States / *J. David Aiken* 946

Surface Water Pollution by Nitrogen Fertilizers / *Gregory McIsaac* 950

Surface Water Pollution by Surface Mines / *Jeffrey G. Skousen and George F. Vance* 956

Surface Water Quality and Phosphorus Applications / *Richard McDowell, Andrew Sharpley, and Peter Kleinman* 961

Surface Water Quality Protection for Concentrated Animal Feeding Operations / *Douglas R. Smith and Philip A. Moore, Jr.* 965

Tailwater Recovery and Reuse / *C. Dean Yonts* 969

Timber Harvesting—Influence on Water Yield and Water Quality / *C. Rhett Jackson* 973

Transpiration / *Thomas R. Sinclair* 977

Transpiration Efficiency / *Graeme C. Wright and Nageswararao C. Rachaputi* 982

Transpiration and Water Use Efficiency / *Miranda Y. Mortlock* 989

Uptake by Plant Roots / *S. G. K. Adiku* 992

Uptake by Plant Roots, Modeling Water Extraction / *S. G. K. Adiku* 995

Urban Water Engineering and Management in Ancient Greece /
    A. N. Angelakis and D. Koutsoyiannis 999

Vadose Zone and Groundwater Protection / *Michael H. Young*
    and Charalambos Papelis 1008

Vapor Transport in Dry Soils / *Glendon W. Gee and Anderson L. Ward* 1012

Water Properties / *James D. Oster* 1017

Well Drilling / *Thomas Marek* 1020

Wellhead Protection / *Babs Makinde-Odusola* 1024

Wells, Hydraulics of / *Mohamed Hantush* 1029

Wetland Ecosystems / *Sherri L. DeFauw* 1034

Wetlands as Treatment Systems / *Kyle R. Mankin* 1038

Preface

All living things require water. More specifically, they require it daily and often in a nearly pure state. As world population increased from 2.5 billion people in 1950 to more than 6 billion in 2000, water demand escalated. One in five people on this planet does not have access to safe and affordable drinking water, and half do not have access to sanitation. With global population predicted to reach almost 8 billion by 2025, water management and treatment will become critical.

Most of the world's water is salt water, unsuitable for most uses. Fresh water makes up only 2.5% of the water supply and two-thirds of this is in the form of glaciers and permafrost, leaving less than 1% of the world's water available for use. Of this remaining 1%, agriculture is the biggest user of water withdrawals from groundwater and surface water supplies: Agriculture comprises 69% of water use compared with industrial and domestic users, who consume 21% and 10%, respectively. But as population growth continues and industrialization expands, there will be greater competition among all users and an increasing need for more efficient water use.

Food production during the past 50 years has kept pace with population growth. The increase in per capita grain production from 247 kg in 1950 to more than 300 kg today is due largely to increased irrigation requiring large amounts of water:

- Approximately 500 kg of water is required to produce 1 kg of potato, while wheat, maize, and rice require approximately 900, 1400, and 2000 kg, respectively, for each kilogram of grain.
- Around 95% of all agricultural land and 83% of the cropland depend entirely on precipitation to meet plant needs.
- Although the 17% of cropland that is irrigated uses an enormous amount of water, it produces almost 40% of the world's food and fiber needs.

Future food requirements will require even additional irrigated lands. However, while irrigation greatly increases food production, it can simultaneously degrade soil and water resources, leading to serious environmental problems.

In recent years, there has been a dramatic change in the way meat and dairy animals are handled. Large, concentrated animal feeding operations are becoming commonplace and require large amounts of water for both livestock consumption and manure and waste handling. These facilities can also present a potential pollution hazard for surface and groundwater resources.

Industrial and domestic water users often degrade water to the point that it must be treated before it can be used for other purposes or even returned to the environment. Wastewater treatment usually requires extensive facilities and expenditures to ensure environmental protection.

Efficient water use and water resource protection can be accomplished only by informed producers and policymakers with access to state-of-the-art information. The *Encyclopedia of Water Science* is designed and compiled to meet this need. An international team of hundreds of dedicated scientists, policymakers, educators, and others involved with water use have prepared over 250 entries addressing important topics ranging from water composition to irrigation water application to agricultural fields. An advisory board was important in planning the scope, and topic editors reviewed entries and offered advice to the editors and authors. We thank all these individuals for their efforts. The initial edition, to be updated quarterly online, addresses critical issues of water use. Perhaps more importantly, the authors have identified additional sources of information for readers who need further, in-depth resources. Published in both online and print formats, the encyclopedia's features will appeal to a wide range of users.

Thanks are also due to the staff of Marcel Dekker, Inc., for their efforts in handling the thousands of communications required to invite authors, review drafts, and manage other matters necessary to produce a publication of this magnitude. It was a great pleasure to work with Ellen Lichtenstein and Sapna Maloor. Their professionalism, commitment to excellence, and dedication are much appreciated and were the key to accomplishing this task. The information assembled will be a useful tool in helping humanity address and meet water use challenges of the 21st century.

B. A. Stewart
Terry A. Howell

Encyclopedia of
Water Science

Academic Disciplines

Joseph R. Makuch
National Agricultural Library, Beltsville, Maryland, U.S.A.

Paul D. Robillard
The Pennsylvania State University, University Park, Pennsylvania, U.S.A.

INTRODUCTION

Several academic disciplines from the agricultural, environmental, and social sciences contribute to the field of agricultural water management. Successful agricultural water management enhances the prospects for farm profitability by efficiently meeting the water needs of crops and livestock while protecting the natural environment. Achieving these results requires a multidisciplinary approach: knowledge from many disciplines must be integrated into agricultural water management decisions.

OVERVIEW OF AGRICULTURAL WATER MANAGEMENT

Agricultural water management encompasses a wide range of activities that relate to managing water efficiently to grow food and fiber, while protecting water, and other elements of the natural environment, from degradation. The focus is on managing the quantity and quality of water resources/supplies. Agricultural water management encompasses:

- Crop irrigation, drainage, erosion control, nutrient and pest management.
- Animal production, especially manure management.
- Provision of safe drinking water for humans and animals.
- Groundwater and surface water quality protection.

AGRICULTURAL WATER MANAGEMENT DISCIPLINES

Academic disciplines are fields of study characterized by academic departments, scholarly journals, and professional societies. Those disciplines that address agricultural water management are as varied as agricultural water management issues. Academic disciplines usually associated with managing agricultural water quality and quantity are described below.

Descriptions of Specific Disciplines

Agricultural economics

Agricultural economics contributes to agricultural water management by addressing both farm-level business decisions and broader policy decisions related to water. Agricultural economists examine issues such as farm profitability under different irrigation systems; costs and benefits of government conservation programs; and the influence of cost-share rates on the adoption of best management practices to reduce nonpoint-source pollution.

Agricultural engineering

Agricultural engineering is concerned with the design and development of systems to identify, analyze, treat, and remediate water resources, particularly as they relate to agricultural activities. Agricultural engineers are also involved in the operation of environmental quality protection and control systems. Engineering principles are applied to monitoring, design, and operation of systems to evaluate the environmental impact of agricultural practices on surface water and groundwater quantity and quality. Emphasis areas include engineering design and operation of irrigation and drainage systems; erosion and sedimentation control structures; modeling of contaminant transport processes; and design of watershed monitoring and control systems utilizing simulation models, geographic information systems (GIS), and remote-sensing data. Specialized fields in agricultural engineering include land application of wastewater and controlled environments (such as water and wastewater treatment and operation systems for greenhouses; confined livestock facilities; and aquaculture). In all cases, the impact of these operations on water quality is of concern.

Encyclopedia of Water Science
DOI: 10.1081/E-EWS 120010295

Agricultural law

Agricultural law covers the legal aspects of agricultural water use, including environmental impacts. The issues may relate to individual farm enterprises or to larger public policy questions. At the farm level, legal issues might include rights to water for irrigation, or liability for degradation of a stream. Broader public policy issues could include developing, implementing, and enforcing regulations specifying manure management practices that are protective of water quality.

Agricultural meteorology and climatology

Agricultural meteorology relates to agricultural water management in a number of ways. The type, timing, amount, intensity, and duration of precipitation are vitally important to agriculture. For example, long-term rainfall records are used in water quality models that examine agricultural contaminants in runoff or those that leach into groundwater. Agricultural meteorology also provides a knowledge base for irrigation scheduling.

The related field of climatology helps identify possible changes in precipitation patterns attributable to global climate change that could have an effect on agricultural water management. The study of climate variability also helps broaden our understanding of droughts and floods. Understanding of this variability can help agricultural water managers plan and prepare for extreme weather events.

Agronomy and soil science

Agronomy, and the associated disciplines of crop science, plant science, and horticulture, are all concerned with growing plants. Plant types range from alfalfa to tomatoes to turf grass, but all require water in the proper amount and at the proper time. These disciplines study plant water requirements and evaluate irrigation systems for their efficiency and effectiveness. The disciplines are also concerned with long-term effects of irrigation, particularly the buildup of salinity in soils and contaminated return flows. Land application of municipal wastewater and biosolids is also under the purview of these disciplines. Understanding the quality of irrigation water and its effects on plants and the environment is especially important in such waste management/crop production systems.

Nutrients in fertilizers and manures—applied to promote plant growth—and pesticides—applied to combat diseases and kill weeds and damaging insects—can harm water quality if not managed effectively. Agronomists study ways to optimize the amount, timing, and application methods of fertilizers and manures to ensure efficient plant growth while protecting groundwater and surface water quality. Similarly, agronomists help design integrated pest management programs that protect plant health while avoiding unnecessary applications of synthetic pesticides that may result in water contamination. To reduce sedimentation of water bodies by soil erosion, agronomists study cover crops and plant-related aspects of conservation tillage systems.

A closely related discipline, soil science, focuses on soil characteristics, responses to management, and effects of use. With regard to agricultural water management, soil scientists study the water-holding capacities and drainage characteristics of soils, the transport of nutrients and pesticides through the soil profile, and the effectiveness of conservation tillage systems in controlling erosion.

Animal science

Animal science deals with raising livestock and poultry. Animal scientists study the water quality and quantity requirements of various farm animals for optimum production. Since nutrients in manures can cause water quality problems, animal scientists address this problem by finding ways for animals to use nutrients in their feed more efficiently, thereby excreting less in the manure.

Aquatic biology

Aquatic biology is the study of living resources—and the factors affecting them—in freshwater systems. Agricultural activities may produce unintended, negative effects on aquatic organisms. For example, expanding cropland by clearing trees from a riparian area removes shading from the stream. Without this protection, stream temperatures may rise to a level unsuitable for cold-water fish. Aquatic biologists can assess the effect agricultural practices may have on aquatic life, both on the farm and downstream.

Environmental engineering

Environmental engineering is concerned with the physical, chemical, and biological control of water and wastewater treatment systems. All aspects of water and wastewater treatment systems are related to environmental engineering practices including collection, storage, stabilization, advanced chemical and biological treatment, and distribution systems. Environmental engineers are typically responsible for the design and operation of water and wastewater treatment plants. Specialized areas include engineering design and evaluation of conduit (pressurized) and open channel flow systems as well as hazardous waste treatment and disposal systems. Environmental engineers also model and evaluate the impact of municipal and industrial wastewater discharges on surface water and groundwater quality.

Forestry and natural resources conservation

Forestry and natural resources conservation also have relevance to agricultural water management. Specialists in these fields may study the establishment, growth, maintenance, and species composition of woody and herbaceous plants in riparian buffers; the effects of stream bank plantings to provide shade and cover and reduce stream bank erosion and other management methods to reduce sediment inputs to streams.

Hydrology and hydrogeology

Hydrology and hydrogeology are focused on water availability and movement as they relate to both surface and groundwater flow regimes. Hydrology is the scientific building block for scientific and engineering water management disciplines. Monitoring of precipitation, stream flows and groundwater flows, in space and time, are the basis for both short-term and long-term quantity and quality records. These records are used to evaluate agricultural and other land-use impacts on water resources, as well as to develop tools that provide predictive capabilities for impact analysis.

Hydrogeology—the integrated sciences of geology and hydrology—focuses on the various geologic formations and their hydraulic characteristics, providing opportunities to develop tools for the estimation of aquifer recharge and yield under different climates, land use, and water withdrawal scenarios. Hydrogeologists are also concerned with the vulnerability of aquifers to contamination from land uses in recharge areas. Studies in this area involve monitoring and evaluation of land-use practices and how they impact the natural aquifer quality parameters; water quality changes derived from land-use impacts; and the effect these quality changes may have on water supply withdrawals and return flows to streams.

Range science

Range science is concerned with rangeland ecology and the use of rangelands to meet human needs, including raising livestock for food. When animals graze in rangeland riparian areas, they may increase erosion and sedimentation, deposit manure and urine in or very near streams, and reduce stream shading by damaging adjacent vegetation. Stream water quality can suffer, and aquatic habitats can be negatively altered. Range scientists investigate ways to utilize rangelands for livestock grazing in ways that protect water quality and associated natural resources.

CONCLUSION

The broad spectrum of agricultural water management activities provides opportunities for many disciplines to contribute to the field. Some disciplines focus on biological, chemical, and physical aspects, while others address social and economic issues. Each discipline has tools and technologies that can be applied to improving agricultural water management, while addressing related environmental, economic, and social problems.

FURTHER READING

Ag Career Database; National FFA Organization, 2000; http://209.206.185.114/careersearch.cfm (accessed May 2001)

Brown, J.A. Disciplines in Agriculture. Agrologist **1983**, *12*(4), 12–13

Careers in Agricultural Science; Florida Department of Agriculture and Consumer Services; http://www.fl-ag.com/PlanetAg/careers.htm (accessed May 2001)

Exploring Careers in Agronomy, Crops, Soils, and Environmental Sciences; American Society of Agronomy; Crop Science Society of America; Soil Science Society of America: Madison, WI, 1996; 27 pp.

Kunkel, H.O. The Issue of Academic Disciplines Within Agricultural Research. In *Agricultural Research Policy: Selected Issues*, Papers Presented to the 150th National Meeting of the American Association for the Advancement of Science, New York, NY, May 24–29, 1984; Batie, S.S.; Marshall, J.P., Eds.; Virginia Polytechnic Institute and State University: Blacksburg, VA, 1986; 19–32

Acid Rain and Precipitation Chemistry

Pao K. Wang
University of Wisconsin, Madison, Wisconsin, U.S.A.

INTRODUCTION

Acid rain is a phenomenon of serious environmental concern. By definition, acid rain refers to rainwater that is acidic. But in reality, it is more accurate to use the term *acid deposition* since not only rain but also snow, sleet, hail, and even fog can become acidic. In addition to the process where acids become associated with precipitation (called *wet deposition*), acid gases and particles can also be deposited on the earth surface directly (called *dry deposition*). However, since the name "acid rain" has become a household term and its formation is better understood than other types of acid deposition, the following discussions will focus on acid rain.

Whereas an aqueous solution is acidic if its pH value is less than 7.0, acid rain refers to rainwater with pH less than 5.6. This is because, even without the presence of man-made pollutants, natural rainwater is already acidic as CO_2 in the atmosphere reacts with water to produce carbonic acid:

$$CO_2 + H_2O(l) \Leftrightarrow H_2CO_3(l) \tag{1}$$

The pH value of this solution is around 5.6. Even though the carbonic acid in rain is fairly dilute, it is sufficient to dissolve minerals in the Earth's crust, making them available to plant and animal life, yet not acidic enough to cause damage. Other atmospheric substances from volcanic eruptions, forest fires, and similar natural phenomena also contribute to the natural sources of acidity in rain. Still, even with the enormous amounts of acids created annually by nature, normal rainfall is able to assimilate them to the point where they cause little, if any, known damage.

However, large-scale human industrial activities have the potential of throwing off this acid balance, and converting natural and mildly acidic rain into precipitation with stronger acidity and far-reaching environmental effects. This is the root of the acid rain problem, which is not only of national but also international concern. This problem may have existed for more than 300 yr starting at the time when the industrial revolution demanded a large scale burning of coal in which sulfur was a natural contaminant. Several English scholars, such as Robert Boyle in the 17th century and Robert A. Smith of the 19th century, wrote about the acids in air and rain; though, there was a lack of appreciation of the magnitude of the problem at that time. Individual studies of the acid rain phenomenon in North America started in the 1920s, but the true appreciation of the problem came only in the 1970s.

To address this problem, the U.S. Congress established the National Acid Precipitation Assessment Program (NAPAP) to study the causes and impacts of the acid deposition. This research established that the acid rain does cause broad environmental and health effects, the pollution causing acid deposition can travel hundreds of miles, and the electric power generation is mainly responsible for SO_2 ($\sim 65\%$) and NO_x emissions ($\sim 30\%$). Subsequently, Congress created the Acid Rain Program under Title IV (Acid Deposition Control) of the 1990 Clean Air Act Amendments. Electric utilities are required to reduce their emissions of SO_2 and NO_x significantly. By 2010, they need to lower their emissions by 8.5 million tons compared to their 1980 levels. They also need to reduce their NO_x emissions by 2 million tons each year compared to the levels before the Clean Air Act Amendments.

However, it may not be adequate to solve the acid emission merely at the national level. With increasing industrialization of the Third World countries in the coming century, one can expect great increase of the atmospheric loading of SO_2 and NO_x because many of these countries will burn fossil fuels to satisfy their energy needs. Clearly, some form of international agreements need to be forged to prevent serious environmental degradation due to acid rain.

THE CHEMISTRY OF ACID RAIN

Sulfuric acid (H_2SO_4) and nitric acid (HNO_3) are the two main acid species in the rain. The partitioning of acids in rain may be different in different places. In the United States, the partitioning is H_2SO_4 ($\sim 65\%$), HNO_3 ($\sim 30\%$), and others ($\sim 5\%$). While there are many possible chemicals that may serve as the precursors of acid rain, the two main substances are SO_2 and NO_x (and NO_x consists of NO and NO_2), and both are released to the atmosphere via the industrial combustion process. While power generation is the predominant source of these

Encyclopedia of Water Science
DOI: 10.1081/E-EWS 120010318

precursors, industrial boilers and automobiles also contribute substantially. When these precursors enter the cloud and precipitation systems, acid rain occurs. Fig. 1 shows a schematic of the acid rain formation process.

Once airborne, these chemicals can be involved in milliards of chemicals reactions. The main paths that lead to acid rain formation are described as follows.

Sulfuric Acid

SO_2 is believed to be the main precursor for the formation of sulfuric acid drops. Its main source in the atmosphere is the combustion of fossil fuels. This is because sulfur is a natural contaminant in coal (especially the low grade ones) and oil. The following reactions are thought to occur when SO_2 is absorbed by a water drop (see e.g., Refs. [1,2]):

$$SO_2(g) + H_2O(l) \Leftrightarrow SO_2 \cdot H_2O(l) \qquad (2)$$

$$SO_2 \cdot H_2O \Leftrightarrow H^+ + HSO_3^- \qquad (3)$$

$$HSO_3^- \Leftrightarrow H^+ + SO_3^{-2} \qquad (4)$$

$$SO_3^{-2} \overset{\text{oxidation}}{\Rightarrow} SO_4^{-2} \qquad (5)$$

The oxidant of the last step can be H_2O_2, O_3, OH, and others. There are still controversies about the identity of the oxidants.

Note that the equilibrium of the above reaction system is controlled by the pH values of the drop, and the presence of ammonia is often considered together with these reactions since it affects the pH of the drop. A detailed discussion of these reactions and their rates is given in Chapter 17 of Ref. [1].

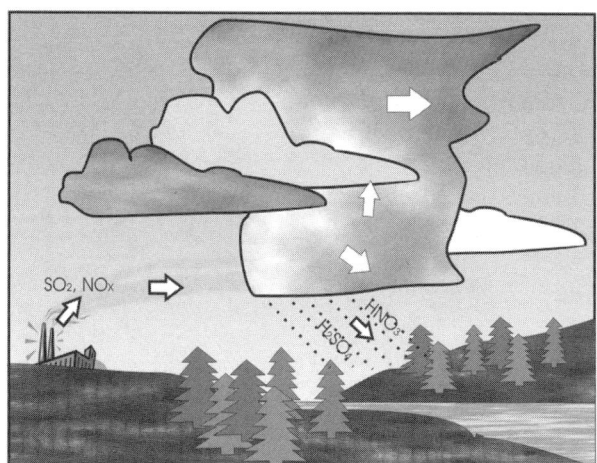

Fig. 1 A schematic of the acid rain formation process.

Nitric Acid

The main ingredients for the formation of nitric acid are NO and NO_2 (and are often combined into one category, NO_x). It is commonly thought that the main path of nitric acid found in clouds and raindrops is the formation of gas phase, HNO_3, followed by its uptake by liquid water. Although there are reactions of NO_x with liquid water that can lead to nitric acid, they are thought to be unimportant due to their slow reaction rates.

The main reaction for HNO_3 formation is

$$NO_2 + OH + M \rightarrow HNO_3 + M \qquad (6)$$

where M can be any neutral molecule. NO can be converted to NO_2 by the following reaction:

$$2NO + O_2 \rightarrow 2NO_2 \qquad (7)$$

DROP-SCALE TRANSPORT PROCESSES OF ACID RAIN

The chemical reactions described earlier must be considered together with the transport processes to obtain a quantitative picture of the acid rain formation. This is especially true for SO_2 because absorption and reactions occur simultaneously. The convective transport influences the concentrations of different species and hence the reaction rates. Fig. 2 illustrates these processes schematically. These include the following.

External Transport

This refers to the transport of SO_2 gas toward the surface of the drop. It is a convective diffusion process (both convective transport and diffusional transport occur) and is influenced by the flow fields created by the falling drop and atmospheric conditions (pressure and temperature).

Interfacial Transport

Once SO_2 is adsorbed on the surface of the drop, it must be transferred into the interior for further reactions to occur. The time for establishing phase equilibrium is controlled by Henry's law constant and mass accommodation coefficient of SO_2.

Internal Transport

In the interior of the drop, reactions 2–5 occur. At the same time, these species are transported by both diffusion and internal circulation. The latter is caused by the motion of the liquid drop in a viscous medium and can influence the production rates of these species (see Ref. [1]).

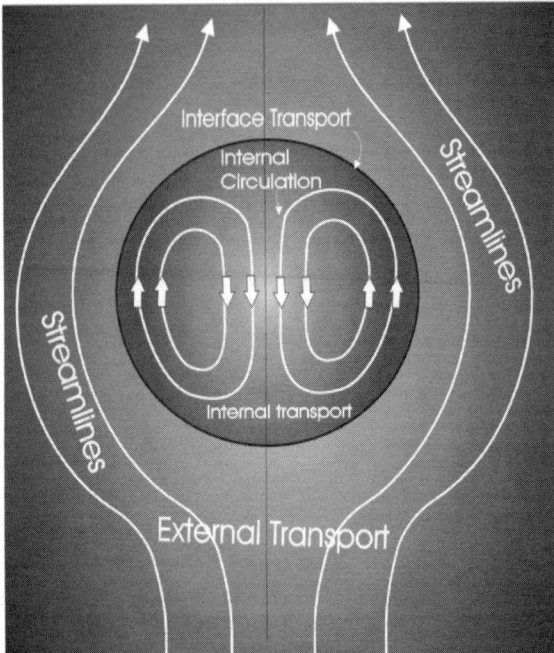

Fig. 2 A schematic of the drop-scale transport process of sulfur species involved in the acid rain.

ENVIRONMENTAL FACTORS INFLUENCING THE ACID RAIN FORMATION AND IMPACTS

Like many environmental hazards, the acid rain process is not driven by a few well-controlled physical and chemical processes, but involves complicated interactions between the chemicals and the environments they exist in. While the main ingredients of acid rain come from industrial activities, many other factors may influence the formation of acid rain and its impacts. The following are some of the most important.

Meteorological Factors

Acid rain occurs in the atmosphere and hence is greatly influenced by meteorological factors such as wind direction and speed, amount and frequency of precipitation, pressure patterns, and temperature. For example, in drier climates, such as the western United States, wind blown alkaline dust is abundant and tends to neutralize the acidity in the rain. This is the *buffering effect* of the dust. In humid climates, like the Eastern Seaboard, less dust is in the air, and precipitation tends to be more acidic.

Seasonality may also influence acid precipitation. For example, while it is true that rain may be more acidic in summer (because of higher demands for energy and hence more fossil fuel used), the snow in winter can also pick up substantial amount of acids. These snow-borne acids can accumulate throughout winter (if the weather is cold enough) and then are released in large doses during the spring thaw. These large doses of acid may have more significant effects during fish spawning or seed germination than the same doses at some less critical time.

Topography and Geology

The topography and geology of an area have marked influence on acid rain effects. Research from the U.S. EPA pointed out that areas most sensitive to acid precipitation are those with hard, crystalline bedrock and very thin surface soils. Here, in the absence of buffering properties of soil, acid rains will have direct access to surface waters and their delicate ecosystem. Areas with steep topography, such as mountainous areas, generally have thin surface soils and hence are very vulnerable to acid rain. In contrast, a thick soil mantle or one with high buffering capacity, such as most flatlands, helps keep acid rain damage down.

The location of water bodies is also important. Headwater lakes and streams are especially vulnerable to acidification. Lake depth, the ratio of water-shed area to lake area, and the residence time in lakes all play a part in determining the consequent threat posed by acids. The transport mode of the acid (rains or runoff) also influences the effects.

Biota

Acid rain may fall on trees causing damages. The kinds of trees and plants in an area, their heights, and whether they are deciduous or evergreen may all play a part in the potential effects of acid rain. Without a dense leaf canopy, more acid may reach the earth to impact on soil and water chemistries. Stresses on the plants will also affect the balance of local ecosystem. Additionally, the rate at which different types of plants carry on their normal life processes influences an area's ratio of precipitation to evaporation. In locales with high evaporation rates, acids will concentrate on leaf surfaces. Another factor is that leaf litter decomposition may add to the acidity of the soil due to normal biological actions.

REFERENCES

1. Pruppacher, H.R.; Klett, J.D. *Microphysics of Clouds and Precipitation*; Kluwer Academic Publishers, 1997; 954.
2. Seinfeld, J.H.; Pandis, S.N. *Atmospheric Chemistry and Physics*; John Wiley and Sons, 1998.

Acidification for Prevention of Clogging of Drip Lines and Emitters

Andrew C. Chang
University of California, Riverside, California, U.S.A.

INTRODUCTION

Water from both surface and underground sources pick up particulate matter during conveyance—sands, silts, plant fragments, algae, diatoms, larvae, snails, fishes, etc. While the majority of the suspended solids may be removed in preirrigation treatments such as sedimentation and filtration, some of the fine silts and colloidal clay particles inadvertently remain and settle inside the lateral lines or emitters impeding the water flow. As the flow slows down and/or the chemical background of the water changes, chemical precipitates and/or microbial flocs and slimes begin to form and grow, thus microirrigation emitter clogging occurs. This section delineates the occurrences of chemical precipitates and the chemistry of acidification that is employed to mitigate clogging caused by chemical precipitates. Clogging resulting from formation of microbial flocs and slimes is controllable by acidification as well as chlorination.

OVERVIEW

Within the extensive network of a drip irrigation system, it is difficult to predict where or when clogging will take place. The hydraulic characteristics such as flow velocity, path length, orifice diameter, and pressure compensation all affect the flow rate and thus the clogging. The lower end of an operating irrigation system (laterals and emitters) should be visually inspected for build-up of deposits, and the flow rates and pressures of the systems should be regularly tested. Routine examinations will identify segments of the network that are potentially problematic and isolate them for corrective measures. The clogging, once formed in the distribution system, is difficult to mitigate. Prevention is by far the preferred measure.

Clogging caused by the deposition of inorganic suspended substances may be overcome by regular flushing of the system and by employment of self-cleaning emitters. High-dosage, short-duration shock treatment with acidification and chlorination may be necessary to dissolve the chemical precipitates and to inactivate the microorganisms. Many publications outlined the practical and operational aspects of acidification and chlorination processes for drip irrigation.[1-4]

CHEMICAL PRECIPITATION

Calcium (Ca^{2+}) and bicarbonate (HCO_3^-) ions in water have a tendency to form calcium carbonate precipitates when the temperature and pH rise and CO_2 partial pressure changes. As the pH rises, the HCO_3^- ion in the bicarbonate–carbonate equilibrium shifts toward the CO_3^{2-} ion. The HCO_3^- ion in water is also in equilibrium with CO_2 in the atmosphere. When the temperature of water rises, the dissolved CO_2 escapes and again the equilibrium shifts toward the CO_3^{2-} ion. The reactions result in the precipitation of calcium carbonate:

$$Ca(HCO_3)_{2(aq)} \rightarrow CaCO_{3(s)} + H_2O + CO_{2(g)}$$

Water, high in hardness, is especially susceptible to the precipitation reaction.

After an irrigation event, water left behind in the emitter and the laterals will evaporate. The evaporation leaves behind mineral deposits that are carbonate as well as chloride and sulfate salts of calcium, magnesium, sodium, and potassium near an emitter outlet, or orifice. The chloride and sulfate salts may be dissolved in subsequent irrigation. Because of their low solubility, minerals such as calcite (calcium carbonate), gypsum (calcium sulfate), and magnesium hydroxide are likely to accumulate over time on and around the emitter openings. For saline water, the deposits will build up rapidly. Ground water may also contain reduced forms of iron and manganese. Upon exposure to oxygen in the atmosphere, they are oxidized and the oxidized iron and manganese ions form precipitates with hydroxide, carbonate, and phosphate in water. The deposits accumulate in and around the microirrigation line, and emitters invariably are mixtures of precipitates of different chemical nature.

ACIDIFICATION

Calcium carbonate is by far the most common chemical precipitate causing clogging in drip irrigation systems.[5] Acids are frequently added to irrigation water to prevent formation of precipitates or to dissolve precipitates when they form in the drip irrigation lines and emitters.

In water, the solubility of calcium carbonate is a function of the Ca^{2+} concentration, alkalinity, and pH of the water.[6] The pH at which the calcium carbonate solubility in the water reaches saturation is designated as the saturation pH, pH_s, and it may be calculated as:[7]

$$pH_s = pK_2 + p[Ca^{2+}] - pK_{sp} - \log(2[\text{Alkalinity}])$$
$$- \log \gamma_m$$

where p denotes $-\log$ operator, [ ] denotes molar concentration of the chemical species specified inside the brackets, K_2 is the dissociation constant of HCO_3^- to CO_3^{2-}, K_{sp} is the solubility product of calcium carbonate, and γ_m is the activity coefficient of monovalent ion. Alkalinity refers to the ability of the water to resist the change of pH when acid or base is added. It is measured as moles of H^+ required for reducing the pH of 1 L of water to 4.5. If the pH of the water is maintained at less than the calculated pH_s, calcium carbonate precipitation will not take place in the water.

To dissolve precipitates in and weaken the attachments on drip lines and emitters, acids are added to reduce the pH of water to approximately 2, and they should remain in the affected sections for at least 24 hr. Any strong acid such as sulfuric, hydrochloric, or nitric acid will serve the purpose. Afterwards, the treated section of the drip lines is flushed to remove the dissolved and loosened deposits.

When acid is added into water, the pH does not change at a constant rate with the addition. The volume of acid required to lower the pH to a given level is dependent on the alkalinity of the water. It may be necessary to perform a titration trial on a water sample to determine the acid addition required for achieving the desired pH level.

REFERENCES

1. Gilbert, R.G.; Ford, H.W. Operation Principles. In *Trickle Irrigation for Crop Production, Design, Operation, and Management*, Development in Agricultural Engineering 9; Nakayama, F.S., Bucks, D.A., Eds.; Elsevier: Amsterdam, 1986; 142–163.
2. Schwankl, L.; Prichard, T. Chlorination. In *Drip Irrigation for Row Crops*; Hanson, B., Schwankl, L., Grattan, S.R., Prichard, T., Eds.; University of California Irrigation Program, University of California: Davis, 1994; 129–139.
3. English, S.D. Filtration and Water Treatment for Micro-irrigation. *Drip/Trick Irrigation in Action*, Proceedings of the 3rd International Drip/Trickle Irrigation Congress, ASAE Publication 10-85; American Society of Agricultural Engineers: St. Joseph, MI, 1985; Vol. I, 129–139.
4. Meyer, J.L. Cleaning Drip Irrigation Systems. *Drip/Trick Irrigation in Action*, Proceedings of the 3rd International Drip/Trickle Irrigation Congress, ASAE Publication 10-85; American Society of Agricultural Engineers: St. Joseph, MI, 1985; Vol. I, 41–44.
5. Bucks, D.A.; Nakayama, F.S.; Warrick, A.W. Principles, Practices, and Potentialities of Trickle (Drip) Irrigation. Adv. Irrig. **1982**, *1*, 219–297.
6. Langelier, W.E. The Analytical Control of Anti-corrosion Water Treatment. J. Am. Water Works Assoc. **1936**, *28*, 1500–1521.
7. Loewenthal, R.E.; Marais, G.V.R. *Carbonate Chemistry of Aquatic Systems: Theory and Application*; Ann Arbor Science Publishers: Ann Arbor, MI, 1976.

African Market Garden

Dov Pasternak
*International Crops Research Institute for the Semi-Arid Tropics, Niamey,
Niger*

Amnon Bustan
Ben-Gurion University of the Negev, Beer-Sheva, Israel

INTRODUCTION

Agricultural production in sub-Saharan Africa relies mainly on rain-fed systems.[1] In the semiarid and dry subhumid regions of Africa, these systems are neither sustainable nor profitable. Areas of monocultures of grains and legumes exhibit severe land degradation, mostly as a result of water and wind erosion.[2] Crop yields are very small, the commercial value of the common grains (millet and sorghum) is very low, and revenue is thus meager. On average, crop failures occur in two out of five years as a result of droughts. The final outcome of these processes is severe poverty.

Irrigated agriculture can help alleviate poverty and reduce the stress on natural resources, especially since many countries in dry Africa are rich in nonutilized water resources. For example, the combined annual discharge of the Niger and the Senegal rivers is about 40 billion $m^3\,yr^{-1}$, a value comparable with the 75 billion $m^3\,yr^{-1}$ of the Nile river. In addition, rich shallow aquifers underlie much of Sahelian Africa,[3] and there are freatic shallow aquifers in the proximity of seasonal rivers. The large-scale utilization of such water reserves for irrigation is hampered by the relatively high costs of large water projects and of inputs and by low crop yields.[4] Furthermore, there is little motivation to initiate large-scale schemes, since international prices for irrigated commodities are relatively low.

In many areas of dry Africa, market gardens are the only form of irrigated agriculture.[4] These gardens are sustainable and profitable, because they supply perishable products such as fruit and vegetables that, for obvious reasons, cannot be imported from elsewhere. The importance of market gardens is growing steadily due to the rapid increase of urban population that can afford to buy fruits and vegetables in city markets. However, for the market gardens to be successful, many problems associated with climatic and social conditions and the lack of appropriate technologies have to be addressed first. One of the most important problems is that of irrigation. Most market gardens are irrigated by hand with watering cans. This activity is obviously extremely labor demand-

ing and inefficient. In some places, surface irrigation (mainly basin irrigation) is also practiced. The drawbacks of this latter system are the relatively high energy input required for motorized water pumps and the low water use efficiency, particularly in sandy soils, from which a large proportion of the water is lost through seepage. Yields and quality of fruits and vegetables are low because of the inefficient supply of water and nutrients, the low quality of seeds, and ineffective pest and disease control methods.

BACKGROUND INFORMATION

In many countries (particularly those towards the north of the continent), market gardens operate only five months of the year (November–March). During the rainy season (July–October) there is shortage of labor for irrigation, since, at that time, all labor is directed to rain-fed fields. Furthermore, in the rainy season, it is difficult to control diseases and pests. The April–July period is very hot, and most vegetable species do not give good yields under conditions of severe heat. Furthermore, in the hot season, it is difficult to carry watering cans or to do any sort of work under the scorching sun.

Drip irrigation can provide the solution to the above-described problems. Drip systems can easily deliver water to the field in daily quantities required on the basis of transpiration demands. With this system, all plants receive the same quantity of water and fertilizers, and very low soil water tension is maintained throughout the day. These advantages lead to significant increases in yield and in improved product quality as compared with other systems.[5–9] An additional advantage is that drip irrigation is particularly suitable for saline water irrigation.[10]

Conventional drip-irrigation systems were designed for large surfaces and are too costly and difficult to maintain for small fields. In response to the need for systems suitable for small areas, Israeli drip-irrigation companies have recently developed low-pressure drip-irrigation (LPDI) systems for the small traditional greenhouses of China. Although the LPDI concept was first tested in Israel

Encyclopedia of Water Science
DOI: 10.1081/E-EWS 120010255

as long ago as the mid-1980s,[11] the system was not commercialized at that time, mainly because of problems of drip clogging. The recently developed LPDIs have large drip orifices that minimize clogging by impurities in the irrigation water.

The LPDI system has been adapted by us to become the basis of an integrated production system, designated the African Market Garden (AMG) [or in French, Jardin Potager Africain (JPA)] described below.

THE AFRICAN MARKET GARDEN

Technical Description and Operation of the LDPI System

The hydraulic performance of three different LPDI systems manufactured in Israel (Ein Tal, Netafim, and Plastro-Gvat) has been tested in the field.[12] In all the tested systems, the variation in water discharge from the first to the last dripper in a 12.5-m long drip lateral at a water pressure of 1.3 m did not fall below 90%, and at 1.8 m of pressure there were no differences in water discharge among individual drippers along a 12.5-m drip lateral.

A schematic presentation of the LPDI manufactured by Netafim is given in Fig. 1.

The system consists of the following elements: a water reservoir positioned at least 1 m above the field level, a water valve, a filter, distribution lines, and drip laterals. Additional details of the system are given in Fig. 1. The operation of the system is very simple. It involves filling of the reservoir to a particular level, cleaning the filter, and opening the tap. Irrigation is completed when the reservoir is empty. Once a week, the ends of all laterals are opened; the system is flushed for about 5 min to clean it from possible accumulated impurities; and the reservoir is drained.

The relationship between the reservoir size and the irrigated area is constant and depends on the daily evapotranspiration (ET) of the particular region. As an example, the average daily potential ET in Niamey for 12 mon of the year is given in Fig. 2. Two distinct seasons can be observed. A season of high ET of about 8 mm day^{-1} (February–May) and a season of lower ET of about 6 mm day^{-1} (June–February). The volume of the reservoir should be planned to accommodate the maximum daily quantity of irrigation water required in the particular region. In places with two distinct seasons, such as Niamey, two lines are drawn on the reservoir. In Niamey, the upper line indicates the volume that is needed to give the field a daily irrigation rate of 8 mm. The lower line indicates the volume that will supply 6 mm day^{-1}.

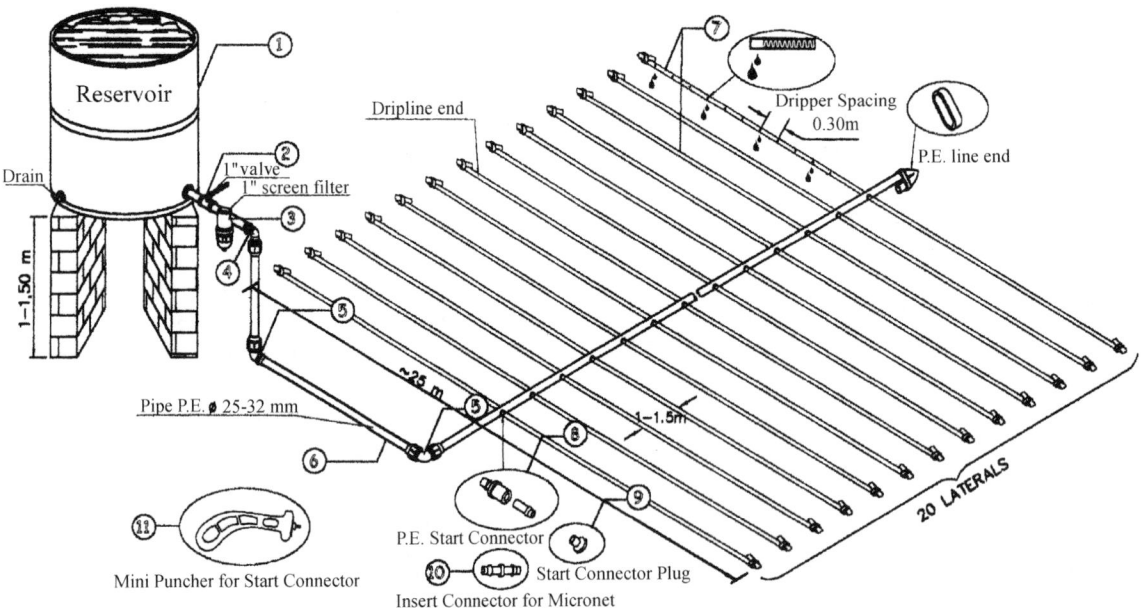

Fig. 1 Schematic presentation of the LDPI system. 1. Water reservoir; 2. plastic ball valve 1″ female thread; 3. plastic filter 1″ male thread (120 mesh); 4. P.E. quick-coupling elbow—25 mm × 1″, female thread; 5. P.E. quick-coupling elbow—25 mm; 6. main line—LDPE pipe 25 mm class 2.5; 7. dripline—LDPE integrated FDS dripline; 8. start connector—barbed type, for FDS dripline; 9. start connector plug; 10. insert connector for 8-mm dripline; 11. mini puncher for start connector. (Note: The illustration of the Netafim design does not infer any preference of the authors for the design of that company.)

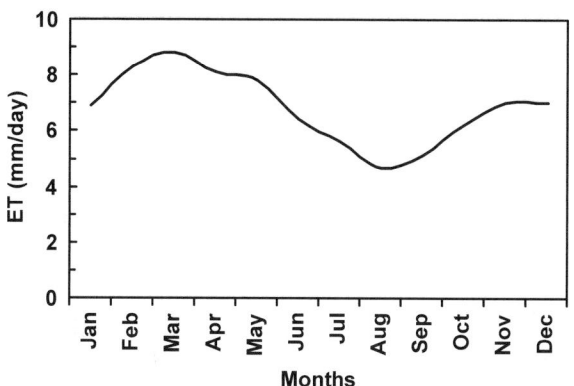

Fig. 2 Average monthly values of potential ET in Niamey, Niger. (From ICRISAT Sahelian Center-Niger.)

The "basic unit" concept was introduced to describe a system with a fixed ratio between the reservoir volume and the field size. For example, in Niamey, an area of 500 m² requires a reservoir with a volume of 4000 L (4 m³) to supply a peak daily water requirement of 8 mm. Two systems were developed: 1) a "thrifty" system, which caters for farmers with limited resources (in many instances these are women who operate small backyard gardens or small plots in communal gardens); and 2) a "commercial" system, which caters to larger-scale producers, particularly periurban farmers. The basic unit for the thrifty system consists of an old 200-L oil drum (these are available everywhere in Africa at low prices), supplying an area of 40 m² and giving a daily irrigation rate of 5 mm (compromises in the quantity of water have to be made for the sake of simplification). The system is very versatile. For example, the same reservoir and distribution lines can serve an area of 80 m² or an area of 120 m² by filling the reservoir two or three times a day, as required. Likewise, two or three (or more) barrels can be interconnected to provide a basic unit of 80 m² or 120 m². The same principles are applicable to the commercial version of the AMG.

Water, Salt Buildup, and Nutrition Management

The problem of salt buildup in the soil usually accompanies irrigation systems in various degrees of severity. Among the parameters involved in salt buildup are the ratio between the evaporative demands and precipitations (including irrigation), water quality, soil texture, and the ability of irrigation water to leach salts.

In arid and semiarid regions, evaporative demands are high (e.g., Fig. 2). To prevent rapid salt buildup it is recommended to keep the daily amount of irrigation water above (10%–20%) plant water requirements. The excess amount of water is needed to leach salts away from the rhizosphere. In cases of saline irrigation water, leaching requirements increase. In general, salt accumulation tends to be more intensive and fast in fine- than in coarse-textured soils due to the much larger specific surface area (SSA) of the former, that provides tight interactions with soil water solution.

Water, as the vector of salt in soils, determines salt distribution in the soil profile. Water moves in the soil in three directions: 1) upward, due to evaporation and capillarity; 2) horizontally, by capillary action; and 3) vertically (downward), by gravitational forces. In conventional drip irrigation, with water discharge rates above 2 L hr⁻¹, salts are leached thus creating a gradient of salt concentration that increases from beneath the emitter to the margins of the wetted zone. It has been suspected that with the lower water discharge rates (0.3 L hr⁻¹–0.7 L hr⁻¹) of LPDI the vertical water vector is too weak to provide sufficient salt leaching. To clarify this point, we conducted a series of experiments, in which salt distribution in the soil profile was compared between water discharge rates of 0.3 L hr⁻¹ (1.3 mm hr⁻¹) and 2 L hr⁻¹ (8.0 mm hr⁻¹) in fine- and coarse-textured soils, respectively (Fig. 3).

The crop was sweet corn. The field was irrigated daily based on 80% evaporation from a class A USWB evaporation pan adjusted to the leaf area index (LAI) of the crop. The electrical conductivity (EC) of irrigation water containing soluble fertilizers was 1.5 dS m⁻¹. Soil samples were taken 75 days after sowing. There was no rain during the experimental period. Large differences occurred, as expected, between fine- and coarse-textured soils. In fine-textured soil, salt accumulation at the upper layer of soil profile (0 cm–10 cm) was remarkably fast, with no influence of water discharge rates (Fig. 3(A) and (B)). Below that layer, the pattern of salt distribution was a reflection of water distribution. Indeed, salt leaching to the margins was less pronounced under 0.3 as compared with 2-L hr⁻¹ emitters. Similar patterns of salt distribution, but to a much lesser extent, were observed in the coarse-textured soil (Fig. 3(C) and (D)).

Two conclusions can be derived from these observations. First, to ensure sustainability, the use of the LPDI system in fine-textured soils should be restricted to regions with a considerable rainy season (above 400 mm yr⁻¹), during which accumulated salts are discarded either by water runoff or leaching to deep soil layers.

The second conclusion relates to fertilizing management. The relatively weak salt leaching under LPDI emitters implies that it is not necessary to add expensive soluble fertilizers to every irrigation event, as practiced with conventional drip irrigation. The prevailing horizontal movement of water under the low-discharge dripper can become an advantage when a heavy dose of manure

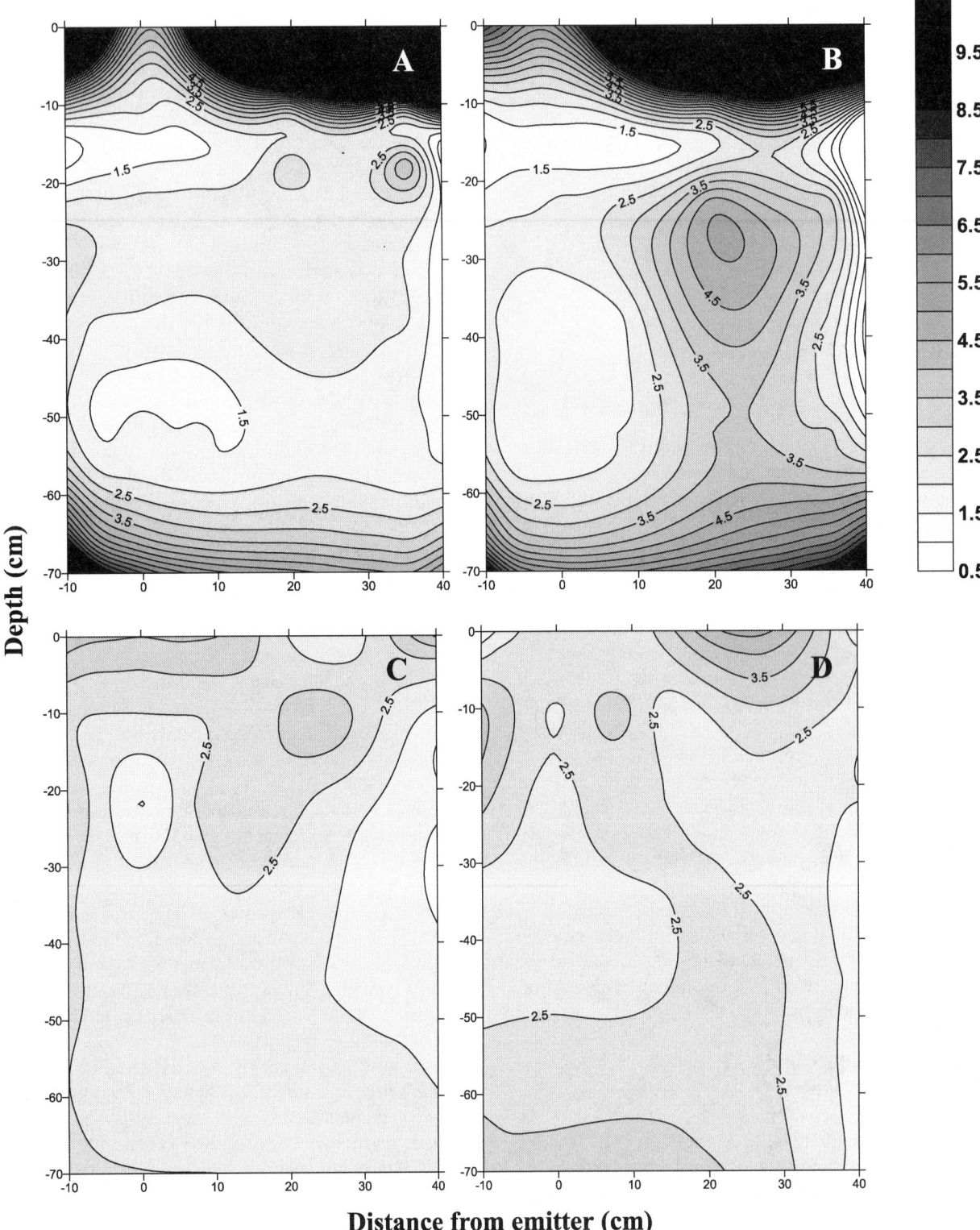

Fig. 3   Salt distribution (EC [dS m^{-1}] of saturated soil extracts) in the soil profile. Soil was sampled from 30 places at various depths and distances from the emitter. Drawings (produced using Surfer 7.0, Golden Software Ltd) represent soil profiles under 0.3-L hr^{-1} (A, C) and 2-L hr^{-1} (B, D) water discharge rates in fine- (A, B) and coarse-textured (C, D) soils, 75 day after sowing sweet corn. Amounts of irrigation water (EC$_I$ ~ 1.5 dS m^{-1}) were equal. The fine-textured soil was a silty clay loam containing 30% silt, 45% clay, and 25% sand. The coarse-textured soil contained 10% silt, 10% clay, and 80% sand.

("side dressing") is applied between the drip lines. The roots will move towards the buried manure and draw most of the required nutrients from this source. Nevertheless, to guarantee a minimal supply of nitrogen to the crops, it is recommended that nitrogen, in the form of urea, be applied once a week at a rate of $0.5 \, g \, m^{-2}$. The minimization of fertilizer application through water will prevent the formation of bacterial slime in the laterals, an important cause of drip clogging.

In a preliminary trial on lettuce in which soluble fertilizer was replaced by an application of organic manure, there was no difference in yield between the manured plots and the plots that received daily application of fertilizer through the irrigation system.

Effect of Rate of Emitter Discharge on Yield

At an average discharge of $1.5 \, mm \, hr^{-1}$ and a daily ET of 6 mm, an irrigation cycle in the AMG lasts for 4 hr. In the light of reports that continuous irrigation may be beneficial to crop yields,[13-15] the effect of two different irrigation rates on sweet corn yield was tested (Table 1). It is evident from Table 1 that in both soil types prolonged irrigation has a slight positive effect on dry matter yield but does not affect ear yield. Thus, for corn (which under desert conditions suffers from water stress[15]), continuous application of water does not offer a significant advantage.

Crop Management

The typical market garden of Africa is characterized by the production of a mix of vegetables and fruit trees in a relatively small plot. The AMG thus also incorporates a mixture of crops. In hot dry areas, date palms are added to produce a three-layered production system (date palms, fruit trees, and annual crops). In a typical $500 \, m^2$ plot, 9 date palms are planted (1 male and 8 female) in a $9 \times 11 \, m$ configuration. Date palms, through their high transpiration

rates and shading effects, produce a microclimate that facilitates reasonable growth of fruit trees and vegetables during the hot dry season. This date-palm-based production system is known as "oasis agriculture." Date palms also improve the profitability of the system. At present-day prices, the income from 8 female plants is about $800 \, yr^{-1}$. To prevent competition for water between the date palms and the other crops in the AMG, the drip laterals are looped around the stems of the palms to triple the amount of water given to the palms.

CONCLUSION

The AMG—a new production system conceived in 1998—incorporates all the advantages of the conventional drip-irrigation system at a fraction of its cost. It is simple to operate and to maintain, and it provides significant increases in yield as well as considerable savings of energy and labor (this aspect is particularly important for women who operate small gardens). This system is applicable to all developing countries that require small-scale irrigation schemes.

The International Program for Arid Land Crops (IPALAC, which is managed by Ben-Gurion University of the Negev, Beer-Sheva, Israel) and Desert Margins Program (DMP, managed by the International Crops Research Institute for the Semi-Arid Tropics) have joined hands to disseminate the AMG system in semiarid Africa, starting in Ethiopia and in the Sahel. Recently, such systems were also installed in Rajasthan, India.

The AMG can therefore serve as a platform for the improvement of the small-scale irrigated agriculture in Africa. Its introduction will facilitate year-round production of irrigated fruit and vegetables, the incorporation of quality vegetable and fruit tree varieties, and the application of modern cost-effective methods for pest and disease management. An adoption of the AMG should significantly contribute to the alleviation of poverty—the most serious problem plaguing sub-Saharan Africa at the beginning of the 21st Century.

Table 1 Effect of two rates of irrigation on stover and ear yield of sweet corn fine- and coarse-textured soils

Irrigation intensity ($mm \, hr^{-1}$)	Stover DM yield[a] ($kg \, m^{-2}$)	Ear yield ($kg \, m^{-2}$)	Ears (m^{-2})
Fine-textured soil			
1.3	1.67a	2.45	6.17
8.0	1.25b	2.36	6.21
Coarse-textured soil			
1.3	1.25	2.06	6.32
8.0	1.18	1.89	5.97

[a] No significant difference between values denoted with same letter at $P \leq 0.05$.

REFERENCES

1. Leisinger, K.M.; Shcmitt, K. *Survival in the Sahel. An Ecological and Development Challenge*, International Service for National Agriculture Research (ISNAR), ISBN 9.2-9118-020-3, 1995.
2. Goorse, J.E.; Steeds, D.R. *Desertification in the Sahelian and the Sudanian Zones of West Africa*, World Bank Technical Paper No 61; Washington, DC, 1988.
3. Issar, S.A.; Nativ, R. Water Beneath Deserts; Key to the Past, Resource for the Present. Episodes **1988**, *2*, 256–269.

4. Postel, S. *Pillar of Sand*; W. W. Norton & Company: New York, 1999; 313.

5. Bernstein, L.; Francois, L.E. Comparison of Drip, Furrow and Sprinkler Irrigation. Soil Sci. **1973**, *115*, 73–86.

6. Kadam, J.R. Effects of Irrigation Methods on Root Shoot Biomass and Yield of Tomatoes. J. Maharashtra Agric. Univ. **1993**, *18*, 493–494.

7. Bresler, E. Trickle-Drip Irrigation Principles and Applications to Soil Water Management. Adv. Agron. **1975**, *29*, 343–393.

8. Shmueli, M.; Goldberg, D. Emergence, Early Growth and Salinity of Five Vegetable Crops Germinated by Sprinkle and Trickle Irrigation in an Arid Zone. Hortic. Sci. **1971**, *6*, 563–565.

9. Sanders, D.C.; Howell, T.A.; Hile, M.M.S.; Hodges, L.; Meek, D.; Phene, C.J. Yield and Quality of Processing Tomatoes in Response to Irrigation Rate and Schedule. J. Am. Soc. Hortic. Sci. **1989**, *114*, 904–908.

10. DeMalach, Y.; Pasternak, D. Drip Irrigation a Better Solution for Crop Production with Brackish Water in Deserts. In *Conference on Alternative Strategies for Desert Development and Management*, Sacramento, California, 1/5-10/6, 1973.

11. Or, U.; Rimon, D. Advanced Technologies in Traditional Agriculture. J. Sustainable Agric. **1991**, *2*, 1–10.

12. Pasternak, D.; Bustan, A.; Ventura, M.; Klotz, H.; Eshetu, F.; Mpuisang, T. Use of Low-Pressure Drip Irrigation to Produce Dates in the Market Gardens of Semi Arid Africa. In *Proceedings of the Date Palm International Symposium*, Windhoek, Namibia, 2000; 322–327.

13. Pasternak, D.; DeMalach, Y. Irrigation with Brackish Water Under Desert Conditions (X). Irrigation Management of Tomatoes (*Lycopersicon esculentum*) on Desert Sand Dunes. Agric. Water Manag. **1995**, *26*, 121–132.

14. Levinson, B.; Adato, I. Influence of Reduced Rate of Water and Fertilizer Application using Daily Intermittent Drip Irrigation on the Water Requirement, Root Development and Responses of Avocado Trees. J. Hortic. Sci. **1991**, *66*, 449–463.

15. Pasternak, D.; De Malach, Y.; Borovic, I. Irrigation with Brackish Water Under Desert Conditions II. The Physiological and Yield Response of Corn (*Zea mays*) to Continuous Irrigation with Brackish Water and to Alternating Brackish–Fresh–Brackish Water Irrigation. Agric. Water Manag. **1985**, *10*, 47–60.

Agricultural Runoff Characteristics

Matt C. Smith
Daniel L. Thomas
University of Georgia, Tifton, Georgia, U.S.A.

David K. Gattie
University of Georgia, Athens, Georgia, U.S.A.

INTRODUCTION

Agricultural runoff is surface water leaving farm fields as a result of receiving water in excess of the infiltration rate of the soil. Excess water is primarily due to precipitation, but it can also be due to irrigation and snowmelt on frozen soils. In the early 20th century, there was considerable concern about erosion of farm fields due to rainfall. The concern was primarily related to the loss of valuable topsoil from the fields and the resulting loss in productivity (see Erosion and Productivity). With the passage of the Federal Water Pollution Control Act Amendments of 1972, the potential for pollution of surface water features such as rivers and lakes due to agricultural runoff was officially recognized and an assessment of the nature and extent of such pollution was mandated.[1,2]

Agricultural runoff is grouped into the category of nonpoint source pollution because the potential pollutants originate over large, diffuse areas and the exact point of entry into water bodies cannot be precisely identified (see Pollution, Point and Nonpoint Source). Nonpoint sources of pollution are particularly problematic in that it is difficult to capture and treat the polluted water before it enters a stream. Point sources of pollution such as municipal sewer systems usually enter the water body via pipes and it is comparatively easy to collect that water and run it through a treatment system prior to releasing it into the environment. Because of the nonpoint source nature of agricultural runoff, efforts to minimize or eliminate pollutants are, by necessity, focused on practices to be applied on or near farm fields themselves. In other words, we usually seek to prevent the pollution rather than treating the polluted water.

Due to the great successes made in treating polluted water from point sources such as municipal and industrial wastewater treatment plants, the relative significance of pollution from agricultural runoff has increased. Agricultural runoff is now considered to be the primary source of pollutants to the streams and lakes in the United States. It is also the third leading source of pollution in U.S. estuaries.[3] The water pollutants that occur in agricultural runoff include eroded soil particles (sediments), nutrients, pesticides, salts, viruses, bacteria, and organic matter.

AGRICULTURAL RUNOFF QUANTITY

Agricultural runoff occurs when the precipitation rate exceeds the infiltration rate of the soil. Small soil particles that have been dislodged by the impact of raindrops can fill and block soil pores with a resulting decrease in infiltration rate throughout the duration of the storm. As the excess precipitation builds up on the soil surface it flows in thin layers from higher areas of the field towards lower areas. This diffuse surface runoff quickly starts to concentrate in small channels called rills. The concentrated flow will generally have a higher velocity than the flow in thin films over the surface. The concentrated flow velocity may become rapid enough to cause scouring of the soil that makes up the channel sides and bottom. The dislodged soil particles can then be carried by the flowing water to distant locations in the same field or be carried all the way to a receiving water body. If the quantity of flow and the velocity of flow are large enough, the rills can grow so large that they cannot be easily repaired by typical earth moving machinery. When this happens, the rill has become a gulley.

The quantity of runoff from agricultural fields is not usually listed explicitly as a concern separate from the quality of the runoff. However, it should be considered because it transports the pollutants and can cause erosion of receiving streams due to excessive flows. If less runoff is allowed to leave a field, there is less flow available to transport pollutants to the stream. Also, if more water is retained on the field, there is likely to be a corresponding reduction in the amount of supplemental water that will need to be added through irrigation. Runoff quantity varies significantly due to factors such as soil type, presence of vegetation and plant residue, physical soil structures such as contoured rows and terraces, field topography, and the timing and intensity of the rainfall event.

Some agricultural practices increase the infiltration capacity of the soil while other practices can result in decreases. The presence of vegetation and plant residues on a field reduce runoff due to improving and maintaining soil infiltration capacity. Actively growing plants also reduce the amount of water in the soil due to evapotranspiration, thus making more room for infiltrating water to be stored in the soil profile. Bare soils increase runoff because there is nothing except the soil surface to absorb the energy of the falling raindrops. The rain, therefore, dislodges soil particles that will tend to seal the surface and reduce infiltration.

SOIL EROSION AND ASSOCIATED POLLUTANTS

One of the primary pollutants in agricultural runoff is eroded soil. In 1975, 223 million acres of cropland produced 3700 million tons of eroded sediments or an average of 17 tons of soil lost per acre of cropland per year (see various *Erosion* articles). It is estimated that cropland, pasture, and rangeland contributed over 50% of the sediments discharged to surface waters in 1977.[4] As noted above, the energy of raindrops can dislodge and transport soil particles. In the aquatic environment the eroded soil is called sediment. There are several concerns related to excessive sediments in aquatic systems. These include loss of field productivity, habitat destruction, reduced capacity in reservoirs, and increased dredging requirements in shipping channels.

Eroded sediments represent a loss of fertile topsoil from the field, which can reduce the productivity of the field itself. Soil formation is an extremely slow process occurring over periods ranging from decades to centuries.[5] Possible results to a grower from excessive erosion of their fields include increasing fertilizer and water requirements, planting more tolerant crops, and possibly abandoning the field for agricultural production (see the article *Erosion and Productivity*).

A second concern is that many of these sediments are heavy and will settle out in slow moving portions of streams or in reservoirs. The settled sediments can dramatically alter the ecology of the streambed. Aquatic plants, insects, and fish all have specific requirements related to composition of the streambed for them to live and reproduce.[6] Sediments in reservoirs reduce the volume of the reservoir available to store water. This may result in reduced production of hydroelectric power, reduced water availability for municipal supply, interference with navigation and recreation, and increased dredging requirements to maintain harbor navigability.

Another concern with eroded sediments is that they can transport other pollutants into receiving waters. The plant nutrient phosphorus, for example, is most often transported from the fields where it was applied as fertilizer by

chemically bonding to clay minerals. Many agricultural pesticides also bond to eroded clays and organic matter. Once these chemicals have entered the aquatic ecosystem, many processes occur that can result in the release of the pollutants from their sediment carriers. Phosphorus, when released, can contribute to the eutrophication of lakes and reservoirs (see the articles *Eutrophication* and *Surface Water Pollution by Phosphorus Fertilizers*). Pesticides and their degradation products can be toxic to aquatic life and must be removed from municipal water supplies (see the article *Pesticide Contamination, Surface Water*).

Erosion from animal agriculture such as feedlots and pastures can also result in the transport of sediments composed of animal manures (see the various *Manure Management* articles). These sediments can transport significant quantities of potential pathogens (viruses and bacteria). The animal manures are primarily organic in nature and can serve as a food source for natural bacteria in the receiving water. When these naturally occurring bacteria begin to utilize the organic matter in this way they may lower or deplete the water of dissolved oxygen as they respire and multiply. This use of oxygen by aquatic bacteria is known as biochemical oxygen demand (BOD). High levels of BOD can reduce stream oxygen level to the point that fish and other organisms that require dissolved oxygen suffer, die, or relocate, when possible, to more suitable habitats.[6]

DISSOLVED POLLUTANTS

Agricultural runoff can carry with it many pollutants that are dissolved in the runoff water itself. These may include plant nutrients, pesticides, and salts. Since these pollutants are dissolved in the runoff, control measures are most often aimed at reducing the volume of runoff leaving an agricultural field, or making the pollutants less available to be dissolved into the runoff water.

One of the major pollutants of concern in agricultural runoff is the plant nutrient nitrogen. Nitrogen is a relatively cheap component of most fertilizers and is necessary for plant growth. Unfortunately, nitrogen in the form of nitrate is highly soluble in water. Thus nitrate can be easily dissolved in runoff water. Just as it does in an agricultural field, nitrogen can promote growth of aquatic vegetation. Excess nitrogen and phosphorus in runoff can lead to the eutrophication of lakes, reservoirs, and estuaries (see the articles *Eutrophication* and *Surface Water Pollution by Nitrogen Fertilizers*). Nitrogen in the form of ammonia can be dissolved into runoff from pastures and feedlots. Ammonia is toxic to many aquatic organisms, thus it is important to minimize ammonia in runoff.[7]

Many agriculturally applied pesticides are also soluble in water. They can be dissolved in runoff and transported into aquatic ecosystems where there is a potential for toxic effects. These pesticides must also be removed from drinking water supplies and, if concentrations are high or persistent, such treatment can be difficult and expensive. Stable, persistent pesticides can bioaccumulate in the food chain with the result that consumers of fish from contaminated waters might be exposed to higher concentrations than exist in the water itself.[8]

Runoff from agricultural fields can contain significant concentrations of dissolved salts. These salts originate in precipitation, irrigation water, fertilizers and other agricultural chemicals, and from the soil minerals. Plants generally exclude ions of chemicals that they do not need. In this way, dissolved salts in irrigation water, for example, can be concentrated in the root zone of the growing crop. Runoff can redissolve these salts and transport them into aquatic ecosystems where some, naturally occurring selenium for example, can be toxic to fish and other wildlife.[9]

Transport of fertilizers and pesticides from their point of application can result in significant environmental costs. This transport, or loss from the field, can also have significant negative economic impacts on the grower. Fertilizers lost from the field are not available to promote crop growth. Agricultural chemicals lost from the field, likewise, are not available to protect the plants from pests and diseases. In both cases the grower is paying for expensive inputs and paying to apply them. It is always in the growers' and the environment's best interests, therefore, to keep agricultural chemicals in the field where they are needed and where they were applied.

CONTROL OF AGRICULTURAL RUNOFF

One of the most direct methods of controlling pollution by agricultural runoff is to minimize the potential for runoff to occur. Other methods can be employed to reduce the amounts of sediments and dissolved chemicals in runoff. As a whole, management practices designed to minimize the potential for environmental damage from agricultural runoff are called best management practices (BMPs), (see the article *Nutrient Best Management Practices*). Many times, practices aimed at controlling one aspect of agricultural runoff are also effective at reducing other components. This is due to the interrelationships between runoff volume, erosion, transport, dissolution, and delivery.

Maintaining good soil tilth and healthy vegetation can minimize runoff. This will promote increased infiltration and a resultant decrease in runoff. Other management practices such as terracing, contour plowing, and using vegetated waterways to convey runoff can result in decreased quantities of runoff by slowing the water leaving the field and allowing more time for infiltration to

occur. Construction of farm ponds to receive runoff can result in less total runoff from the farm, lowered peak rates of runoff, and storage of runoff for use in irrigation or livestock watering.[2]

Control of water pollution by the mineral and organic sediments and associated chemicals in agricultural runoff is most effectively achieved by reducing erosion from the field. The primary method of reducing erosion is by maintaining a vegetative or plant residue cover on the field at all times or minimizing areas of the field that are bare. Techniques utilized to accomplish these tasks include conservation tillage, strip tillage, and the use of cover crops (see the article *Erosion Control, Tillage/Residue Methods*). Additional measures that can be employed at the edge of the field, or off-site include vegetative filter strips and farm ponds (see the article *Farm Ponds*).

Methods to control the loss of nitrogen and other plant nutrients from cropland include applying nitrogen in the quantity required by the crop and at the time the crop needs it (see the article *Nutrient Best Management Practices*). This requires multiple applications and can be difficult for tall crops. For this reason, most, or all, of the nitrogen required by the crop is often applied at planting. Nitrogen fertilizers have often been applied based on general recommendations for the type of crop to be grown. Since nitrogen fertilizers are relatively inexpensive, growers have tended to over apply rather than under apply. Soil tests can tell a grower how much nitrogen is already in the soil and how much needs to be applied for a specific crop. Efforts have been made to make the nitrogen less soluble by changing the form of nitrogen applied to the field so that it becomes available to the plants (and, thus available for loss in runoff) more slowly.[10]

One method of controlling the loss of agricultural chemicals is to minimize their solubility in water. Another is to minimize their use through programs such as integrated pest management (IPM) where some crop damage is allowed until it reaches a point that it becomes economically justified to apply pesticides.[11] And a third approach is to make the chemicals more easily degraded so that they do their job and then degrade into other, less harmful, chemicals so that they do not stay around long enough to be influenced by runoff-producing rainfall events.

CONCLUSION

Agricultural runoff is one of the leading causes of water quality impairment in streams, lakes, and estuaries in the United States. It can transport large quantities of sediments, plant nutrients, agricultural chemicals, and natural occurring minerals from farm fields into receiving water bodies. In many cases the loss of these substances from the field represent an economic loss to the grower as well as a potential environmental contaminants. There are many methods by which the quantity of agricultural runoff can be reduced. Many of these methods are referred to generically as BMPs. Adoption of BMPs can also improve the quality (reduce contaminant concentrations) of the runoff that does leave the farm. By reducing the quantity and improving the quality of agricultural runoff, it will be possible to improve the water quality in our streams, river, lakes, and estuaries.

REFERENCES

1. U.S. Environmental Protection Agency. *EPA Releases Guidelines for New Water Quality Standards*; 2002; http://www.epa.gov/history/topics/fwpca/02.htm (accessed July 2002).
2. Stewart, B.A.; Woolhiser, D.A.; Wischmeier, W.H.; Caro, J.H.; Frere, M.H. *Control of Water Pollution from Cropland, Volume II—An Overview*, EPA-600/2-75-026b; U.S. Environmental Protection Agency: Washington, DC, 1976.
3. U.S. Environmental Protection Agency. *Nonpoint Source Pollution: The Nation's Largest Water Quality Problem*; 2002; http://www.epa.gov/OWOW/NPS/facts/point1.htm (accessed July 2002).
4. Leeden, Van der *The Water Encyclopedia*; Lewis Publishers: Chelsea, MI, 1990.
5. Foth, H.D. *Fundamentals of Soil Science*, 8th Ed.; John Wiley & Sons, Inc.: New York, NY, 1990.
6. Gordon, N.D.; McMahon, T.A.; Finlayson, B.L. *Stream Ecology: An Introduction for Ecologists*; John Wiley & Sons Inc.: New York, NY, 1992.
7. Abel, P.D. *Water Pollution Biology*, 2nd Ed.; Taylor & Francis, Inc.: Bristol, 1996.
8. U.S. Environmental Protection Agency. *The Persistent Bioaccumulators Project*; 2002; http://www.epa.gov/chemrtk/persbioa.htm (accessed 15 July 2002).
9. U.S. Geological Survey. *Public Health and Safety: Element Maps of Soils*; http://minerals.cr.usgs.gov/gips/na/0elemap.htm#elemap (accessed 15 July 2002).
10. Owens, L.B. Impacts of Soil N Management on the Quality of Surface and Subsurface Water. In *Soil Process and Water Quality*; Lal, R., Stewart, B.A., Eds.; Lewis Publishers, Inc.: Boca Raton, FL, 1994.
11. U.S. Department of Agriculture. *National Integrated Pest Management Network*; 2002; http://www.reeusda.gov/agsys/nipmn/ (accessed 15 July 2002).

Agroforestry for Enhancing Water Use Efficiency

James Brandle
Xinhua Zhou
Laurie Hodges
University of Nebraska, Lincoln, Nebraska, U.S.A.

A

INTRODUCTION

Agroforestry is the intentional integration of trees and shrubs into agricultural systems. Windbreaks, riparian forest buffers, alley-cropping, silvopastoral grazing systems, and forest farming are the primary agroforestry practices found in temperate regions of North America.[1] Placing trees and shrubs on the landscape changes the surface energy balance, influences the surrounding microclimate, and has the potential to alter water use and productivity of adjacent crops.[2,3]

In agricultural systems, water is often the major factor limiting growth. When water availability is limited as a result of limited supply or high cost, its efficient use becomes critical to successful production systems. For example, proper irrigation at the appropriate stage of crop development minimizes pumping costs and increases yield; reducing soil tillage conserves soil water and may enhance yield, and reducing surface runoff or trapping snow improves soil water storage for future crop use. These water conservation efforts contribute to the efficient use of available water and are determined primarily by management practices. In contrast, Tanner and Sinclair[4] distinguish between the efficient use of water and water use efficiency (WUE). WUE is primarily a function of physiological responses of plants to environmental conditions. This review focuses on WUE defined as the amount of biomass (or grain) produced per unit of land area for each unit of water consumed.[4]

Soil water may be consumed by evaporation from the soil surface or by the transport of water through the plant and subsequent evaporation from the leaf surface. The rate of water consumption is determined by the microclimate of the crop. Because agroforestry practices alter the microclimate of adjacent fields, they affect WUE of plants growing in those fields.

DISCUSSION

Windbreaks, riparian forest buffers or alley-cropping systems are the practices most likely to be integrated into crop production systems. In all three practices, trees and shrubs tend to be arranged in narrow barriers adjacent to the crop field. Microclimate responses downwind of any of these types of barriers are similar and the following discussion applies to all three types of barriers. As wind approaches these barriers, it is diverted up and over the barrier creating two zones of protection, a larger zone to the lee of the barrier (the side away from the wind) and a smaller zone on the windward side of the barrier. In these zones, wind speed is reduced and turbulence and eddy structure in the vicinity of the barrier are altered. As a result of these changes, the transfer coefficients for heat and mass between the crop and the atmosphere are altered; the gradients of temperature, humidity, and carbon dioxide concentration above the soil and canopy are changed;[5] and the plant processes of transpiration and photosynthesis are altered.[6]

McNaughton[5] defined two regions within the leeward zone of protection: the *quiet zone*, extending from the top of the barrier down to a point in the field located approximately $8H$ leeward (H is the height of the barrier) and a *wake zone*, lying beyond the quiet zone and extending from approximately $8H$ to a distance of $20H$ to $25H$ from the barrier. Within the quiet zone where turbulence is reduced, we expect conditions to be such that the canopy is "uncoupled" from the atmospheric conditions above the sheltered zone, while in the wake zone where turbulence is increased, we expect the canopy to become more strongly "coupled" to the atmosphere above. In both locations we would expect the rates of photosynthesis and transpiration to be altered and WUE to change.

The magnitude of change in wind speed, as well as the extent of microclimate modifications within the quiet and wake zones, are largely determined by the structure of the windbreak or barrier and the underlying meteorological conditions. Structure refers to the amounts of solid material and open space and their arrangement within the barrier. Dense barriers, for example, multiple rows of conifers, generally result in greater wind speed reduction but more turbulence. More porous barriers, for example, single rows of deciduous species, result in less wind speed reduction but also less turbulence. The downwind extent of the protected area is generally greater for more porous

Encyclopedia of Water Science
DOI: 10.1081/E-EWS 120010098

barriers. As a result, narrow, less dense barriers (40%–60% density) are typically used to protect crop fields.

The overall influence of wind protection on plant water relations is complex and linked to temperature, humidity, wind speed, and other meteorological conditions found in the protected zone, the amount of available soil water, crop size, and stage of development.[2,3,7] Until recently, the major effect of wind protection and its influence on crop growth and yield were assumed to be due primarily to soil water conservation and reduced water stress of sheltered plants.[8,9] There is little question that the evaporation rate from bare soil is reduced in the protected zone.[3] However, the effect of reduced wind speed on transpiration, evaporation from the plant canopy, and overall plant water status is less clear.[2,3,7]

According to Grace,[9] transpiration rates may increase, decrease, or remain unaffected by wind protection depending on wind speed, atmospheric resistance, and saturation vapor pressure deficit. Cleugh[3] suggests that as stomatal resistance increases, evaporation from the canopy may actually be increased with a reduction in wind speed. When stomatal resistance is high and water is limited, stomatal resistance controls the rate of evaporation from the leaf surface, not the amount of turbulence. Under these conditions a decrease in wind speed and turbulent mixing may increase the potential for evaporation from the leaf surface.[3]

Evaporation from the leaf surface consists of two phases, an energy driven phase and a diffusion driven phase. Movement of water through the plant and out the stomata is driven by the water potential gradient within the plant. This gradient is influenced by the plant's energy balance. On the lee side of the buffer, reduced wind speed and turbulent mixing lead to increases in leaf temperature and transpiration to meet the increased energy load on the plant. If adequate water is available, it is moved through the plant to the leaf surface and the potential for evaporation from the leaf surface is increased. If water is limited, the stomata partially or completely close, transpiration is reduced, and evaporation from the leaf surface declines.

In contrast, movement of water vapor across the leaf boundary layer is controlled by the vapor pressure gradient and the thickness of the leaf boundary layer. As wind speed decreases, the thickness of this boundary layer increases, the vapor pressure gradient decreases, and the rate of evaporation from the leaf surface decreases. The relative magnitude of the two processes determines whether or not transpiration and subsequent evaporation from the canopy are increased, decreased, or remain unchanged.[7,9,10]

While these theoretical considerations are important in understanding the process, several studies[11–13] have demonstrated a good correlation between wind protection,

conservation of soil water, and enhanced crop yield. Even so, the effect of wind protection on WUE is neither constant throughout the growing period[7] nor is it consistent over varying meteorological conditions.

Agroforestry practices impact the water relations of the crop by affecting the loss of water through damaged leaves. On soils subject to wind erosion, windbreaks or other agroforestry buffers provide significant reductions in the amount of wind blown soil and subsequent abrasion of plant parts and cuticular damage.[9,14] Loss of cuticular integrity or direct tearing of the leaves[15] reduces the ability of the plant to control water loss.

Agroforestry buffers have a direct effect on the distribution of precipitation, both rain and snow. In the case of snow, a porous barrier will result in a more uniform distribution of snow across the field, providing additional soil water for the crop.[16] In the case of rain, the barrier has minimal influence on the distribution of precipitation across the field; however, in the area immediately adjacent to the barrier a rain shadow may occur on the leeward side. On the windward side, the barrier may lead to slightly higher levels of measured precipitation at or near the base of the trees due to increased stem flow or dripping from the canopy.

Trees and shrubs used in agroforestry practices also consume a portion of the available water. In the area immediately adjacent to the barrier, competition for water between the crop and the barrier has a negative impact on yield. These same areas are also subject to some degree of shading depending on the orientation of the barrier. These changes in radiation load influence the energy balance and thus the growth and development of the crop and the utilization of water.[2]

SUMMARY

In summary, agroforestry practices such as windbreaks, riparian forest buffers and alley-cropping systems generally improve both the efficient use of water by the agricultural system and the WUE of the individual crop. In the case of efficient water use, the evidence is clear. In the case of crop WUE, the evidence leaves some unanswered questions. How do we account for the varied crop yield responses reported in the literature? In many cases yields are increased but no clear relationship to crop water budget is shown. In other cases crop yield response is minimal. Under what meteorological conditions are the effects of agroforestry practices most valuable to water balance questions? Final crop yield is a integration of the environmental conditions over the entire growing season. Many different combinations of environmental conditions may result in similar plant responses. How do we address the numerous combinations of plant stress and plant growth to determine "a response" to wind protection? To answer many of these questions it will be necessary to intensify the numerical modeling methods developed by Wilson[17] and Wang and Takle.[18] With a better model to describe the turbulence fields and the transport of water, heat, and carbon dioxide as influenced by agroforestry practices, it should be possible to assess the numerous combinations of environmental factors influencing crop growth in these systems.

REFERENCES

1. Lassoie, J.P.; Buck, L.E. Development of Agroforestry as an Integrated Land Use Management Strategy. In *North American Agroforestry: An Integrated Science and Practice*; Garrett, H.E., Rietveld, W.J., Fisher, R.F., Eds.; American Society of Agronomy, Inc.: Madison, WI, 2000; 1–29.
2. Brandle, J.R.; Hodges, L.; Wight, B. Windbreak Practices. In *North American Agroforestry: An Integrated Science and Practice*; Garrett, H.E., Rietveld, W.J., Fisher, R.F., Eds.; American Society of Agronomy, Inc.: Madison, WI, 2000; 79–118.
3. Cleugh, H.A. Effects of Windbreaks on Airflow, Microclimates and Crop Yields. Agrofor. Syst. **1998**, *41*, 55–84.
4. Tanner, C.B.; Sinclair, T.R. Efficient Water Use in Crop Production: Research or Re-search? In *Limitations to Efficient Water Use in Crop Production*; Taylor, H.M., Jordan, W.R., Sinclair, T.R., Eds.; American Society of Agronomy, Inc.: Madison, WI, 1983; 1–27.
5. McNaughton, K.G. Effects of Windbreaks on Turbulent Transport and Microclimate. Agric. Ecosyst. Environ. **1988**, *22/23*, 17–39.
6. Grace, J. Some Effects of Wind on Plants. In *Plants and Their Atmospheric Environment*; Grace, J., Ford, E.D., Jarvis, P.G., Eds.; Blackwell Scientific Publications: Oxford, 1981; 31–56.
7. Nuberg, I.K. Effect of Shelter on Temperate Crops: A Review to Define Research for Australian Conditions. Agrofor. Syst. **1998**, *41*, 3–34.
8. Caborn, J.M. *Shelterbelts and Microclimate*, Forestry Commission Bulletin No. 29; Her Majesty's Stationery Office: Edinburgh, 1957; 135.
9. Grace, J. Plant Response to Wind. Agric. Ecosyst. Environ. **1988**, *22/23*, 71–88.
10. Thornley, J.H.M.; Johnson, I.R. *Plant and Crop Modeling: A Mathematical Approach to Plant and Crop Physiology*; Clarendon Press: New York, 1990; 669.
11. Song, Z.M.; Wei, L. The Correlation between Windbreak Influenced Climate and Crop Yield. In *Agroforestry Systems in China*; Zhu, Z.H., Cai, M.T., Wang, S.J., Jiang, Y.X., Eds.; International Development Research Centre (IDRC, Canada), Regional Office for Southeast and East Asia, published jointly with the Chinese Academy of Forestry: Singapore, 1991; 21–115.
12. Wu, Y.Y.; Dalmacio, R.V. Energy Balance, Water Use and Wheat Yield in a Paulownia-Wheat Intercropped Field. In *Agroforestry Systems in China*; Zhu, Z.H., Cai, M.T., Wang, S.J., Jiang, Y.X., Eds.; International Development Research Centre (IDRC, Canada), Regional Office for Southeast and East Asia, published jointly with the Chinese Academy of Forestry: Singapore, 1991; 54–65.
13. Huxley, P.A.; Pinney, A.; Akunda, E.; Muraya, P. A Tree/Crop Interface Orientation Experiment with a *Grevillea robusta* Hedgerow and Maize. Agrofor. Syst. **1994**, *26*, 23–45.
14. Kort, J. Benefits of Windbreaks to Field and Forage Crops. Agric. Ecosyst. Environ. **1988**, *22/23*, 165–190.
15. Miller, J.M.; Böhm, M.; Cleugh, H.A. *Direct Mechanical Effects of Wind on Selected Crops: A Review*, Technical Report Number 67; CSIRO Center for Environmental Mechanics: Canberra, Australia, 1995; 68.
16. Scholten, H. Snow Distribution on Crop Fields. Agric. Ecosyst. Environ. **1988**, *22/23*, 363–380.
17. Wilson, J.D. Numerical Studies of Flow Through a Windbreak. J. Wind Eng. Ind. Aerodyn. **1985**, *21*, 119–154.
18. Wang, H.; Takle, E.S. A Numerical Simulation of Boundary-Layer Flows Near Shelterbelts. Boundary-Layer Meteorol. **1995**, *75*, 141–173.

Aquifer Recharge

John R. Nimmo
David A. Stonestrom
United States Geological Survey (USGS), Menlo Park, California, U.S.A.

Richard W. Healy
United States Geological Survey (USGS), Lakewood, Colorado, U.S.A.

INTRODUCTION

Aquifer recharge was defined by Meinzer[1] and Heath[2] as water that moves from the land surface or the unsaturated zone into the saturated zone. This definition excludes saturated flow between aquifers, which avoids double-accounting in large-scale studies, so it might be more precisely called "aquifer-system" or "saturated-zone" recharge. *Recharge rate* designates either a flux [L^3/T] into a specified portion of aquifer, or a flux density [L/T] into an aquifer at a point. Sources of water for recharge include precipitation that infiltrates, permanent or ephemeral surface water, irrigation, and artificial recharge ponds. Recharge may reach the aquifer directly from portions of rivers, canals, or lakes,[3] though usually it first travels by various means through the unsaturated zone.

Recharge varies considerably with time and location. Temporal variation occurs, for example, with seasonal or short-term variations in precipitation and evapotranspiration (ET). This variability is especially evident in thin unsaturated zones, where recharge may occur within a short time of infiltration. In deep unsaturated zones, recharge may be homogenized over several years so that it may occur with essentially constant flux even though fluxes at shallow depths are erratic. Spatial variation occurs with climate, topography, soils, geology, and vegetation. For example, a decrease of slope or increase of soil permeability may lead to greater infiltration and greater recharge. Many applications use a concept of recharge that is time-averaged or areally averaged.

Both the amount of infiltration and the fraction of it that becomes recharge tend to be greater with more abundant water, so the recharge process is most efficient if infiltration is concentrated in space and time. Because ET may extract most or all of the water that infiltrates, water is more likely to become recharge if it moves rapidly below the root zone. Temporal concentration occurs during storms, floods, and snowmelt, when ongoing processes such as ET are overwhelmed. Spatial concentration typically occurs in depressions and channels, where higher water contents promote rapid movement by

increasing the hydraulic conductivity (K), the amount of preferential flow, and the downward driving force at a wetting front. Quantitative estimation of recharge rate contributes to the understanding of large-scale hydrologic processes. It is important for evaluating the sustainability of ground water supplies, though it does not equate with a sustainable rate of extraction.[4] Because it represents a first approximation to the rate of solute transport to the aquifer, the recharge rate is also important to estimate contaminant fluxes and travel times from sources near the land surface. Methods for obtaining a quantitative estimate of recharge mostly require a combination of various types of data which themselves may be hard to estimate, so in general it is wise to apply multiple methods and compare their results.

WATER BUDGET METHODS

The water balance for a basin can be stated as

$$P + Q_{on}^{sw} + Q_{on}^{gw} = ET^{sw} + ET^{uz} + ET^{gw} + Q_{off}^{sw}$$
$$+ Q_{off}^{gw} + Q_{bf} + \Delta S^{snow} + \Delta S^{sw}$$
$$+ \Delta S^{uz} + \Delta S^{gw} \qquad (1)$$

where P is precipitation and irrigation; Q_{on} and Q_{off} are water flow on and off of the site, respectively; Q_{off}^{sw} is runoff; Q_{bf} is baseflow (ground water discharge to streams or springs); and ΔS is change in water storage. Superscripts refer to surface water, ground water, unsaturated zone, or snow, and all parameters are in units of L/T (or volume per unit surface area per unit time). For the saturated zone only, a water balance can be written for a defined area as

$$R = \Delta Q^{gw} + Q_{bf} + ET^{gw} + \Delta S^{gw} \qquad (2)$$

where R is recharge and ΔQ^{gw} is the difference between ground water flow off of and onto the basin. This equation implies that water arriving at the water table: 1) flows out of the basin as ground water flow; 2) discharges to the

Encyclopedia of Water Science
DOI: 10.1081/E-EWS 120010040

surface; 3) is evapotranspired; or 4) goes into storage. Substitution in Eq. 1 produces a simpler form of the water balance:

$$R = P + Q_{on}^{sw} - Q_{off}^{sw} - ET^{sw} - ET^{uz} - \Delta S^{snow}$$
$$- \Delta S^{sw} - \Delta S^{uz} \qquad (3)$$

Water budget methods include all techniques based, in one form or another, on one of these water balance equations.

The most common water budget method is the "residual" approach: all other components in the water budget are measured or estimated and R is set equal to the residual. Water budget methods can be applied over a wide range of space and time scales. The major limitation of the residual approach is that the accuracy of the recharge estimate depends on the accuracy with which other components can be measured. This limitation can become significant when the magnitude of R is small relative to other variables. The time scale for applying water budget methods is important, with more frequent tabulations likely to improve accuracy. If the water budget is calculated daily, P can greatly exceed ET on a single day, even in arid settings. Averaging over longer time periods tends to dampen out extreme precipitation events and hence underestimate recharge. Annual recharge estimated with water budgets range from 23 mm in a region of India[5] to 400 mm at a site in the eastern United States.[6]

Watershed, surface water flow, and ground water flow models constitute an important class of water budget methods that have been used to estimate of recharge (e.g., Ref. [7]). An attractive feature of models is their predictive capability. They can be used to gauge the effects of future climate or land-use changes on recharge rates.

METHODS BASED ON SURFACE WATER OR GROUND WATER DATA

Fluctuations in ground water levels can be used to estimate recharge to unconfined aquifers according to

$$R = S_y \, dh/dt = S_y \Delta h/\Delta t \qquad (4)$$

where S_y is specific yield, h is water table height, and t is time. The method is best applied over short time periods in regions with shallow water tables that display sharp water-level rises and declines. Analysis of water-level fluctuations can also, however, be useful for determining the magnitude of long-term change in recharge rates caused by climate or land-use change. The method is only appropriate for estimating recharge for transient events; recharge occurring under steady flow conditions cannot be estimated. Difficulties lie in

determining S_y and ensuring that fluctuations are due to recharge, not to changes in pumping rates or atmospheric pressure or other phenomena. Recharge rates estimated by this technique range from 11 mm over a 26-month period in Saudi Arabia[8] to 541 mm yr^{-1} over 1 yr for a small basin in the United States.[9]

Ground water levels can also be used to estimate flow, Q, through a cross-section of an aquifer that is aligned with an equipotential line. Multiplying K by the hydraulic gradient normal to the section times the area of the section calculates Q. Recharge is determined by dividing Q by the surface area of the aquifer upgradient from the section.

Methods of estimating recharge based on surface-water data include the Channel Water Balance Method (CWBM) and determination of baseflow by hydrograph separation. The CWBM involves measuring discharge at two gauges on a stream; the difference in discharge between the upstream and downstream gages is the transmission loss. This loss may become recharge, ET, or bank storage. Hydrograph separation involves identifying what portion of gauged stream flow is derived from ground water discharge. Rutledge and Daniel[10] developed an automated technique for this purpose and applied the method to estimate recharge at 15 sites. Drainage areas for the sites ranged from less than 52 km^2 to more than 5200 km^2; estimated annual recharge was between about 13 cm and 64 cm.

DARCIAN METHODS

Applied in the unsaturated zone, Darcy's law gives a flux density equal to K times the driving force, which equals the recharge rate if certain conditions apply. Matric-pressure gradients must be measured or demonstrated to be negligible. Some types of preferential flow are inherently nondarcian and if important would need to be determined separately. Accurate measurements are necessary to know K adequately under field conditions at the point of interest. For purposes requiring areal rather than point estimates, additional interpretation and calculation are necessary.

In the simplest cases, in a region of constant downward flow in a deep unsaturated zone, gravity alone drives the flow. With a core sample from this zone, laboratory K measurements at the original field water content directly indicate the long-term average recharge rate.[11]

In the general case, transient water contents and matric pressures must be measured in addition to K.[12] Transient recharge computed with Darcy's law can relate to storms or other short-term events, or provide data for integration into temporal averages.

TRACER METHODS

Increasing availability and precision of physical and chemical analytical techniques have led to a proliferation in applications of tracer methods for recharge estimation. Isotopic and chemical tracers include tritium, deuterium, oxygen-18, bromide, chloride, chlorine-36, carbon-14, agricultural chemicals, dyes, chlorofluorocarbons, and noble gases. In practice, concentrations measured in pore water are related to recharge by applying chemical mass-balance equations, by matching patterns inherited from infiltrating water, or by determining the age of the water. Tracer methods provide point and areal estimates of recharge. Multiple tracers used together can test assumptions and constrain estimates.

The most common tracer for estimating recharge is chloride. Chloride continually arrives at the land surface in precipitation and dust. Chloride is conservative in many environments and is nonvolatile. Under suitable conditions,

$$R = P[Cl_p]/[Cl_r] \tag{5}$$

where P is precipitation and $[Cl_p]$ and $[Cl_r]$ are chloride concentrations in precipitation and pore water, respectively. Chloride mass-balance methods can be applied to unsaturated profiles[13] and entire basins.[14]

Isotopic composition of water provides a useful tracer of the hydrologic cycle. The isotopic makeup of precipitation varies with altitude, season, storm track, and other factors. Recharge estimates using isotopic varieties of water usually employ temporal or geographic trends in infiltrating water.

Nonconservative tracers can indicate the length of time that water is isolated from the atmosphere, that is, its "age." Recharge rates can be inferred from water ages if mixing is small. If ages are known along a flow line,

$$R = \theta L/(A_2 - A_1) \tag{6}$$

where θ is volumetric water content, A_1 and A_2 are ages at two points, and L is separation length. One point is often located at the water table. Preindustrial water can be dated by decay of predominately cosmogenic radioisotopes, including carbon-14 and chlorine-36. The abundance of tritium and other radioisotopes increased greatly during atmospheric weapons testing, labeling recent precipitation. Additional compounds for dating modern recharge include chlorofluorocarbons, krypton-85, and agricultural chemicals.[15,16]

Heat is yet another tracer of ground water recharge. Daily, seasonal, and other temperature fluctuations at the land surface produce thermal signals that can be traced through shallow profiles.[17,18] Water moving through deeper profiles alters geothermal gradients, which can be used in inverse modeling to obtain recharge rates.[19]

OTHER METHODS

Additional geophysical techniques provide recharge estimates based on the water-content dependence of gravitational, seismic, and electromagnetic properties of earth materials. Repeated high-precision gravity surveys can indicate changes in the quantity of subsurface water from recharge events.[20] Similarly, repeated surveys using seismic or ground-penetrating-radar equipment can resolve significant changes in water-table elevation associated with transient recharge.[21] In addition to surface-based techniques, cross-bore tomographic imaging can provide detailed three-dimensional reconstructions of water distribution and movement during periods of recharge.[22]

REFERENCES

1. Meinzer, O.E. The Occurrence of Ground Water in the United States, with a Discussion of Principles. U.S. Geological Survey Water-Supply Paper 489, 1923; 321.
2. Heath, R.C. Basic Ground-Water Hydrology. U.S. Geological Survey Water-Supply Paper 2220; 1983; 84.
3. Winter, T.C.; Harvey, J.W.; Franke, L.O.; Alley, W.M. Ground Water and Surface Water a Single Resource. U.S. Geological Survey Circular 1139, 1998; 79.
4. Bredehoeft, J.D.; Papadopoulos, S.S.; Cooper, H.H., Jr. Groundwater: The Water-Budget Myth, Scientific Basis of Water-Resource Management: Studies in Geophysics; National Academy Press: Washington, DC, 1982; 51–57.
5. Narayanpethkar, A.B.; Rao, V.V.S.G.; Mallick, K. Estimation of Groundwater Recharge in a Basaltic Aquifer. Hydrol. Proc. 1994, 8, 211–220.
6. Steenhuis, T.S.; Jackson, C.D.; Kung, S.K.; Brutsaert, W. Measurement of Groundwater Recharge in Eastern Long Island, New York, U.S.A. J. Hydrol. 1985, 79, 145–169.
7. Bauer, H.H.; Mastin, M.C. Recharge from Precipitation in Three Small Glacial-Till-Mantled Catchments in the Puget Sound Lowland, Washington. U.S. Geological Survey Water-Resources Inv. Rep. 96-4219, 1997; 119.
8. Abdulrazzak, M.J.; Sorman, A.U.; Alhames, A.S. Water Balance Approach Under Extreme Arid Conditions—A Case Study of Tabalah Basin, Saudi Arabia. Hydrol. Proc. 1989, 3, 107–122.
9. Rasmussen, W.C.; Andreasen, G.E. Hydrologic Budget of the Beaverdam Creek Basin, Maryland. U.S. Geol. Survey. Water-Supply Paper 1472, 1959; 106.
10. Rutledge, A.T.; Daniel, C.C. Testing an Automated Method to Estimate Ground-Water Recharge from Streamflow Records. Ground Water 1994, 32 (No. 2), 180–189.

11. Nimmo, J.R.; Stonestrom, D.A.; Akstin, K.C. The Feasibility of Recharge Rate Determinations Using the Steady-State Centrifuge Method. Soil Sci. Soc. Am. J. **1994**, *58*, 49–56.

12. Freeze, A.R.; Banner, J. The Mechanism of Natural Ground-Water Recharge and Discharge 2. Laboratory Column Experiments and Field Measurements. Water Resour. Res. **1970**, *6* (No. 1), 138–155.

13. Bromley, J.; Edmunds, W.M.; Fellman, E.; Brouwer, J.; Gaze, S.R.; Sudlow, J.; Taupin, J.-D. Estimation of Rainfall Inputs and Direct Recharge to the Deep Unsaturated Zone of Southern Niger Using the Chloride Profile Method. J. Hydrol. **1997**, *188–189* (No. 1–4), 139–154.

14. Anderholm, S.K. Mountain-Front Recharge Along the Eastern Side of the Middle Rio-Grande Basin, Central New Mexico. U.S. Geological Survey Water-Resources Investigations Report 00-4010, 2000; 36.

15. Ekwurzel, B.S.P.; Smethie, W.M., Jr.; Plummer, L.N.; Busenberg, E.; Michel, R.L.; Weppernig, R.; Stute, M. Dating of Shallow Groundwater: Comparison of the Transient Tracers 3H/3He, Chlorofluorocarbons, and 85Kr. Water Resour. Res. **1994**, *30* (No. 6), 1693–1708.

16. Davisson, M.L.; Criss, R.E. Stable Isotope and Groundwater Flow Dynamics of Agricultural Irrigation Recharge into Groundwater Resources of the Central Valley, California. In *Proc. Isotopes in Water Resources Management*, Vienna (Austria), Mar 20–24, 1995; International Atomic Energy Agency: Vienna, 1996; 405–418.

17. Lapham, W.W. Use of Temperature Profiles Beneath Streams to Determine Rates of Vertical Ground-Water Flow and Vertical Hydraulic Conductivity. U.S. Geological Survey Water-Supply Paper 2337, 1989; 34.

18. Constantz, J.; Thomas, C.L.; Zellweger, G. Influence of Diurnal Variations in Stream Temperature on Streamflow Loss and Groundwater Recharge. Water Resour. Res. **1994**, *30*, 3253–3264.

19. Rousseau, J.P., Kwicklis, E.M., Gillies, D.C., Eds.; Hydrogeology of the Unsaturated Zone, North Ramp Area of the Exploratory Studies Facility, Yucca Mountain, Nevada. U.S. Geological Survey Water Resources Investigations Report 98-4050, 1999.

20. Pool, D.R.; Schmidt, W. Measurement of Ground-Water Storage Change and Specific Yield Using the Temporal-Gravity Method Near Rillito Creek, Tucson, Arizona. U.S. Geological Survey Water-Resources Investigations Report 97-4125, 1997; 30.

21. Haeni, F.P. Application of Seismic-Refraction Techniques to Hydrologic Studies. U.S. Geological Survey Open File Report 84-746, 1986; 144.

22. Daily, W.; Ramirez, A.; LaBrecque, D.; Nitao, J. Electrical Resistivity Tomography of Vadose Water Movement. Water Resour. Res. **1992**, *28* (No. 5), 1429–1442.

Aquifer Transmissivity

Mohamed Hantush
United States Environmental Protection Agency (US EPA), Cincinnati, Ohio, U.S.A.

INTRODUCTION

Evaluation of groundwater resources requires the knowledge of the capacity of aquifers to store and transmit ground water. This requires estimates of key hydraulic parameters, such as the transmissivity, among others. The transmissivity T (m^2/sec) is a hydraulic property, which measures the ability of the aquifer to transmit ground water throughout its entire saturated thickness. It is defined as the product of the hydraulic conductivity K (m/sec) and the saturated thickness B (m), in the direction normal to the base of the aquifer:

$$T = KB \qquad (1)$$

CONCEPTS

Figure 1 illustrates a confined unit, or permeable unit sandwiched between impervious or semipervious layers. The hydraulic (or piezometric) head gradient in the two piezometers tapping the aquifer and separated by a unit distance is unity, since they measure a drop in the hydraulic head of magnitude one. The flow through the shaded window of height B and unit width normal to the flow direction is the aquifer transmissivity T. This follows from Darcy's law, which requires that groundwater flow rate per unit area normal to the flow direction is equal to the hydraulic conductivity, if the hydraulic head gradient is unity.

In unconfined aquifers, however, the transmissivity is not as well defined as in confined units. The saturated thickness h (m) extends from the water table vertically down to the aquifer bed in an unconfined aquifer. The transmissivity varies in time in unconfined aquifers, since the water table often fluctuates in response to recharge from the overlying vadose zone or dewatering of the aquifer by pumping. It decreases during pumping and increases during recharge.

In stratified formations the hydraulic conductivity distribution is also stratified and can actually vary from one location to another by orders of magnitude. With this variability and that of the saturated thickness, $B(x,y)$, the transmissivity is given by the integral of the hydraulic conductivity over the saturated thickness[1]

$$T(x, y) = \int_0^{B(x,y)} K(\mathbf{x})\, dz \qquad (2)$$

where $\mathbf{x}$ denotes the Cartesian coordinates (x,y,z). As an illustration of the use of Eq. 2, for a layered confined aquifer composed of N distinct layers, each with thickness b_i and constant hydraulic conductivity K_i, the transmissivity at any given point in the horizontal plane is given by

$$T = BK_A, \quad B = \sum_{i=1}^{N} b_i, \quad K_A = (1/B)\sum_{i=1}^{N} b_i K_i \qquad (3)$$

where K_A is the arithmetic mean of the hydraulic conductivity.

TRANSMISSIVITY OF AQUIFERS

Table 1 shows range of values of T which may be encountered in common aquifers of thicknesses in the range of 5 m – 100 m. In general, transmissivities greater than 0.015 m^2/sec represent good aquifers for water-well exploitation.[6] Karstic limestones, in which sizable proportions of the original rock has been dissolved and removed, are highly transmissive aquifers. Alluvial valleys, which are predominantly unconsolidated sand and gravel, are among the most productive aquifers in the United States.[3] Permeable basalts and fractured igneous and metamorphic rocks have a relatively large transmissivity and serve as good aquifers. Nonkarstic limestones, silt, glacial till, and solid igneous and metamorphic rocks are the least transmissive and make poor aquifers. Sandstone aquifers have a low transmissivity, but they are significant sources of potable water.

RELATIONSHIP TO GROUNDWATER FLOW

The concept of aquifer transmissivity is widely used in the analysis of hydraulics of water wells. It is introduced when the groundwater flow in aquifers is essentially horizontal. This is commonly the case in aquifers whose lateral extensions are much greater than their thicknesses and where the equipotential lines are nearly vertical. The groundwater flow rate $\mathbf{Q}$ (m^2/sec) integrated over the saturated thickness of the aquifer, per unit aquifer width, is related to the transmissivity

Encyclopedia of Water Science
DOI: 10.1081/E-EWS 120010158

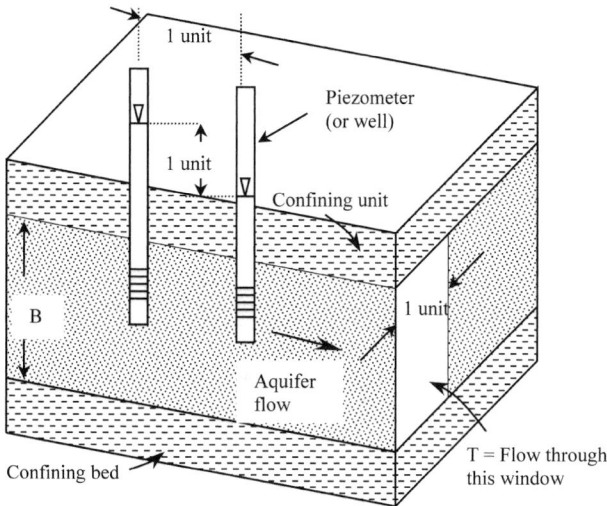

Fig. 1 Illustration of transmissivity in a confined aquifer.

through the Darcy relationship:

$$\mathbf{Q} = -T\,\nabla\varphi \qquad (4)$$

where $\mathbf{Q} = Q_x\mathbf{i} + Q_y\mathbf{j}$ and $\nabla\varphi = (\partial\varphi/\partial x)\mathbf{i} + (\partial\varphi/\partial y)\mathbf{j}$. Q_x and Q_y are the flow rates per unit width in the x and y directions, respectively, and $\mathbf{i}$ and $\mathbf{j}$ are the unit length orthogonal vectors in the x and y directions, respectively. Eq. 4 in combination with the groundwater flow balance equation describe essentially horizontal groundwater flow in aquifers.

ANISOTROPY AND PRINCIPAL DIRECTIONS

Implicit in Eq. 4 is that the transmissivity is invariant to the orientation in the (x,y) plane. If the transmissivity is

dependent on the direction of flow in an aquifer, the aquifer is said to be anisotropic. This anisotropy stems from the anisotropy of the hydraulic conductivity, and when the latter is the same in all directions, the aquifer is said to be isotropic. In natural aquifers the transmissivity is anisotropic, and groundwater flow rate, in this case, is given by the following general form:

$$\mathbf{Q} = -\mathbf{T}\cdot\nabla\varphi, \qquad \mathbf{T} = \begin{bmatrix} T_{xx} & T_{xy} \\ T_{yx} & T_{yy} \end{bmatrix} \qquad (5)$$

in which $\mathbf{T}$ is the second order symmetric tensor of transmissivity of an anisotropic aquifer.[1] It is equal to the product of the hydraulic conductivity tensor and the aquifer saturated thickness. As an example, the component T_{xy} gives the contribution of a unit hydraulic gradient in the y-direction to the flow rate in the x-direction Q_x, and the component T_{xx} gives the contribution of a unit hydraulic gradient in the x-direction to Q_x. The four elements appearing in Eq. 5 depend on the chosen coordinate system. The principal directions of anisotropy are defined as the orientation θ from the original x–y coordinates to the new ξ–η coordinate system (Fig. 2), such that the off diagonal elements in the transformed system are zero,

$$\mathbf{T} = \begin{bmatrix} T_{\xi\xi} & 0 \\ 0 & T_{\eta\eta} \end{bmatrix} \qquad (6)$$

where $T_{\xi\xi}$ and $T_{\eta\eta}$ are the principal transmissivities. Both are related to the transmissivities in the original x–y coordinates (T_{xx}, T_{yy}, and T_{xy}) and the orientation θ by simple algebraic relationships.[1] The transmissivity in the major direction of anisotropy $T_{\xi\xi}$ is greater than in the minor direction $T_{\eta\eta}$. Fig. 2 illustrates the transmissivity

Table 1 Representative values of transmissivity for aquifers of thicknesses 5 m–100 m

Material	Transmissivity (m²/sec)[a]
Unconsolidated	
Gravel	5×10^{-3}–100
Sand	5×10^{-7}–1
Silt	5×10^{-9}–2×10^{-3}
Glacial till	5×10^{-12}–2×10^{-4}
Rocks	
Karst limestone	5×10^{-6}–2
Permeable basalt	1×10^{-6}–2
Fractured igneous and metamorphic rocks	4×10^{-8}–3×10^{-2}
Limestone, dolomite	5×10^{-9}–2×10^{-4}
Sandstone	5×10^{-10}–6×10^{-4}

[a] Values are estimated from representative values of hydraulic conductivity.
Adapted from Refs. [3,6].

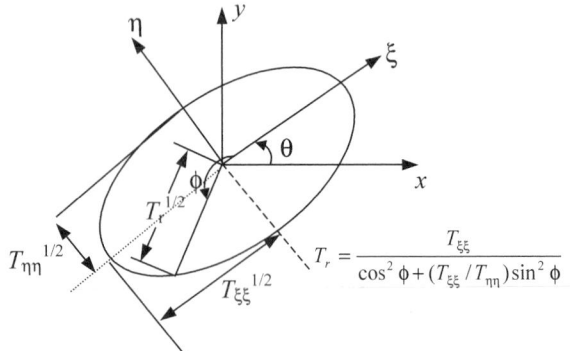

Fig. 2 Ellipse of directional transmissivity.

ellipse in the principal direction coordinates and the relationship between the directional transmissivity T_r and the principal components $T_{\xi\xi}$ and $T_{\eta\eta}$. Anisotropy with respect to transmissivity can be estimated using aquifer test analysis.[5]

METHODS OF ESTIMATION

Estimation of aquifer transmissivity through analysis of aquifer test data is a standard practice in the evaluation of groundwater resources. Aquifer test is an in-situ method for estimating field scale transmissivity in which the water bearing materials are tested under natural conditions. In this test, the well is pumped at a given (usually constant) rate and drawdown (i.e., drop in elevation of the water level in the well from its initial static position) is observed and recorded in time in the pumping well itself and possibly in at least one observation well in the vicinity. In this method, a logarithmic plot of the applicable well-flow equation, called a "type curve," is superimposed on a logarithmic drawdown-time plot, called a "data curve," and the transmissivity is then estimated using a graphical matching technique.[4,8,10] This technique assumes the aquifer is homogeneous and requires experience and judgment of qualified persons conducting these tests, because observed drawdown variations with time and distance from the pumped well may be interpreted in several ways and therefore subject to uncertainty. The performance of graphical matching techniques depends largely on the selection of the well-flow equation that most accurately resembles the flow system under consideration.

Evaluation of groundwater resources in regional aquifers requires estimates of the transmissivity at locations where it is not available. This can be achieved by solving the "inverse problem" for the unknown parameters. In this method, the flow domain is overlain by a discrete mesh of nodal points, and the groundwater flow equation is approximated by a set of algebraic equations, one for each nodal point. Unknown T values at the nodal points are identified by trial and error or automatically, by a gradient-based search technique.[11] The optimal set of transmissivities is the one that produces the best match, such as minimizing the sum of weighted least square errors between the observed hydraulic heads and those obtained from the solution of the algebraic equations at the measurement points. Many of the restrictive assumptions often made in aquifer test analysis are relaxed in the numerical methods for solving the inverse problem. However, these techniques may suffer from nonuniqueness and instability of the solutions, and can result in unrealistic estimates for T, e.g., negative values or solutions fluctuating between imposed lower and upper bounds of the transmissivity.[11] Current trends in hydrology account for local-scale spatial variation of the transmissivity using statistical methods in which this property can be idealized as a space random function. Field evidence support that transmissivity is lognormally distributed.[2] The geostatistical approach[7] combines process understanding of groundwater flow in aquifers with statistical estimation methods to provide an effective tool for mapping transmissivities over regional aquifers, by making use of their in-situ estimates inferred from aquifer tests and hydraulic head measurements. The literature on aquifer properties estimation is rich in innovative approaches for estimating both the transmissivity spatial structure and its values at locations where measurements are not available.[7,9,11]

Notice: The U.S. Environmental Protection Agency through its Office of Research and Development funded and managed the research described here through in-house effort. It has been subjected to Agency review and approved for publication.

REFERENCES

1. Bear, J. The Equation of Motion of a Homogeneous Fluid. In *Dynamics of Fluids in Porous Media*; American Elsevier: New York, 1972; 764.
2. Delhomme, J.P. Spatial Variability and Uncertainty in Groundwater Flow Parameters: A Geostatistical Approach. Water Resour. Res. **1979**, *15*, 269–280.
3. Domenico, P.A.; Scwartz, F.W. Ground Water Movement. In *Physical and Chemical Hydrogeology*; John Wiley & Sons: New York, 1990; 824.
4. Hantush, M.S. Hydraulics of Wells. In *Advances in Hydroscience*; Ven Te Chow, Ed.; Academic Press: New York, 1964; 281–442.
5. Hantush, M.S. Analysis of Data from Pumping Tests in Anisotropic Aquifers. J. Geophys. Res. **1966**, *71*, 421–426.
6. Freeze, R.A.; Cherry, J.A. Physical Properties and Principles. In *Groundwater*; Prentice Hall: Englewood Cliffs, NJ, 1979; 15–79.

7. Kitanidis, P.K. *Introduction to Geostatistics*; Cambridge University Press: New York, 1997; 249.

8. Neuman, S.P. Theory of Flow in Unconfined Aquifers Considering Delayed Response of the Water Table. Water Resour. Res. **1972**, *9*, 1031–1045.

9. Sun, N.-Z. *Inverse Problems in Groundwater Modeling*; Kluwer: Norwell, MA, 1994; 352.

10. Theis, C.V. The Relation Between Lowering of the Piezometric Surface and the Rate and Duration of Discharge of a Well Using Ground-Water Storage. Am. Geophys. Union Trans. **1935**, *16*, 519–524.

11. Willis, R.; Yeh, W.W.-G. The Inverse Problem in Groundwater Systems. In *Groundwater Systems Planning & Management*; Prentice-Hall: Englewood Cliffs, NJ, 1987; 347–412.

Aquifers

Miguel A. Mariño
University of California, Davis, California, U.S.A.

INTRODUCTION

An aquifer is a geologic formation that can transmit, store, and yield significant quantities of water. An aquifer can be confined (one bounded above and below by impervious formations), unconfined (one with a water table serving as its upper boundary), or leaky (one that can gain or lose water through adjacent semipervious formations). In addition to the aforementioned porous-media aquifers, there are karst aquifers in which water flow is concentrated along fractures, fissures, conduits, and other interconnected openings. The hydraulic properties of aquifers (e.g., transmissivity and storativity) are best determined by testing the aquifers in place. Water contained in an aquifer is called ground water and is commonly extracted by means of a well. The quality of ground water determines, to a large extent, its suitability for a particular use, such as irrigation, public water supply, etc. Human activity poses a threat to ground-water quality and already has resulted in incidents of ground-water pollution or contamination. Because ground water tends to move slowly, it may take many years after the start of pollution before contaminated water shows up in a well. The best way to protect the quality of ground water is to prevent its contamination.

AQUIFERS

An aquifer is a geologic formation that yields significant quantities of water (e.g., coarse sand and gravel formation). In contrast, an aquiclude is a formation that may contain water but cannot transmit it in significant quantities. A clay stratum is an example. For all practical purposes, an aquiclude can be considered an impervious formation. An aquitard is a semipervious formation, transmitting water very slowly compared with an aquifer. It can, however, permit the passage of large quantities of water over a large (horizontal) area. An aquitard is often called a semipervious layer or a leaky formation. An aquifuge is an impervious formation that neither contains nor transmits water. Solid granite is an example.

An aquifer can be regarded as an underground-storage reservoir. Water enters the aquifer naturally through precipitation or influent streams—and artificially through wells or other recharge methods. Water leaves the aquifer naturally through springs or effluent streams—and artificially through pumping wells. Fig. 1 is a schematic representation of several aquifers and observation wells.[1]

A confined aquifer, also called artesian aquifer or pressure aquifer, is bounded above and below by impervious formations. Water in a well penetrating such an aquifer will rise above the base of the upper confining formation; it may or may not reach the ground surface. A well that penetrates a confined aquifer is called an artesian well—it is called a flowing well if the water level in the well reaches, or exceeds the elevation of, the ground surface. The water levels in a number of wells penetrating a confined aquifer are the hydrostatic-pressure levels of the water in the aquifer at the well sites. The water levels define an imaginary surface called the piezometric or potentiometric surface.

An unconfined aquifer, also called phreatic aquifer or water-table aquifer, is one with a water table (phreatic surface or surface of atmospheric pressure) serving as its upper boundary. Actually, above the water table is a capillary fringe often neglected in ground-water studies. A special case of an unconfined aquifer is the perched aquifer. It occurs wherever an impervious (or relatively impervious) stratum of limited horizontal area supports a ground-water body that is above the main water table. A well that taps an unconfined aquifer is called a water-table (or gravity) well. The water level in such a well corresponds approximately to the position of the water table at that location.

Ground-water levels can be measured to estimate the piezometric-surface or water-table distribution in an aquifer or to determine fluctuations in hydraulic head over time. Maps of ground-water levels are used to estimate ground-water flow direction and velocity, to assess ground-water vulnerability, to locate landfills and wastewater disposal sites, etc.

Aquifers, whether confined or unconfined, that can gain or lose water through adjacent aquitards or semipervious formations are called leaky aquifers. A confined aquifer that has at least one semipervious confining bed is called a leaky-confined aquifer. An unconfined aquifer that rests on a semipervious stratum is called a leaky-unconfined aquifer. Leakage across semipervious formations can be significant.

Encyclopedia of Water Science
DOI: 10.1081/E-EWS 120010038

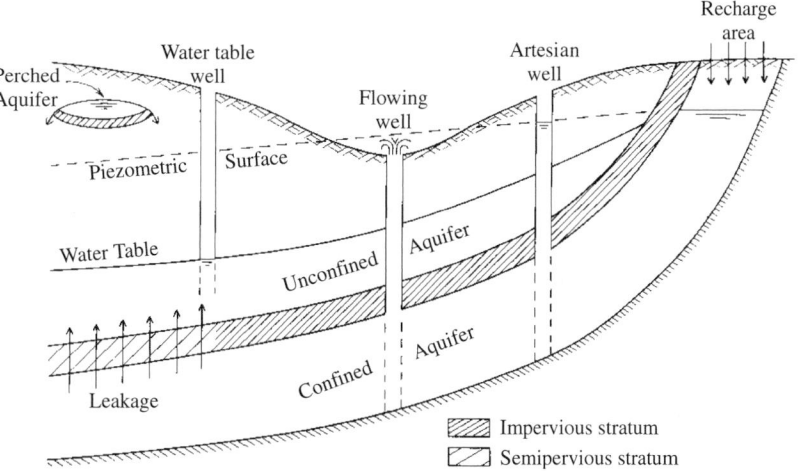

Fig. 1 Schematic of several aquifers and observation wells.[1]

Fig. 1 shows an unconfined aquifer underlain by a confined one. In the recharge area, the confined aquifer becomes unconfined. A portion of the confined and unconfined aquifers is leaky, with the amount and direction of leakage governed by the difference in piezometric head across the semipervious stratum.

In addition to the aforementioned types of aquifers, there are karst aquifers made up of soluble rock strata at or near the earth's surface in which water flow is concentrated along fractures, fissures, conduits, and other interconnected openings.

Aquifer elasticity is the main mechanism responsible for volumes of water released from or added to storage in aquifers. In confined aquifers, water is derived from storage primarily by the elastic properties of both the aquifer matrix and the water. In unconfined aquifers, water released from storage is due mainly to gravity drainage (drainage of the pore space above the lowered water table) and partly to elastic storage (as in confined aquifers). In leaky aquifers, water is derived from storage in the main (confined) aquifer, gravity drainage if the aquifer is unconfined, and elastic storage in aquitards, and induced vertical leakage across these units.

The general properties of aquifers to transmit, store, and yield water (e.g., transmissivity, storativity, and leakage factor) are usually referred to as hydraulic properties of aquifers, or simply aquifer parameters. Because of the many factors on which these parameters depend, numerical values must depend on experimental determination. Although various laboratory techniques are available,[2,3] more reliable results are obtained from field tests[1–6] of the aquifers in place.

The transmissivity of the aquifer determines the ability of the aquifer to transmit water through its entire thickness. In confined aquifers, the transmissivity is represented as the product of hydraulic conductivity and aquifer thickness, in the direction normal to the base of the aquifer (the hydraulic conductivity expresses the ease with which a fluid is transported through a porous medium). Because in unconfined aquifers the saturated thickness extends from the water table to the base of the aquifer, the transmissivity varies in time as the water table often fluctuates in response to recharge and pumping.

The storage capacity of an aquifer is quantified by its storativity, also called the storage coefficient. The storativity indicates the relationship between changes in the volume of water stored in an aquifer and corresponding changes in the elevations of the piezometric surface, or the water table. It can be defined as the volume of water that a column of the aquifer, of unit cross-section, releases from or adds to storage per unit decline or rise of piezometric surface (confined aquifers) or water table (unconfined aquifers). In a confined aquifer, the storativity is caused by the compressibility of the water and the elastic properties of the aquifer. In an unconfined aquifer, the storativity is due mostly to dewatering or refilling the zone through which the water table moves (e.g., water removed by gravity drainage) and partly to water and aquifer compressibility in the saturated zone. A certain amount of water, however, is held in place against gravity in the pores between grains under molecular and surface-tension forces. Thus, the storativity of an unconfined aquifer is less than the porosity by a factor called specific retention (the ratio between the volume of water that a soil will retain against gravity and the total volume of the soil). Reflecting this phenomenon, the storativity of an unconfined aquifer is often called specific yield (the ratio between the volume of water that a soil will yield by gravity and the total volume of the soil). Also often used in this context is the term, effective porosity.

A parameter characterizing a leaky aquifer is the leakance, or coefficient of leakage, of the semipervious formation. It is a measure of the ability of this formation to transmit vertical leakage and is defined by the ratio of the hydraulic conductivity of this formation to its thickness. The reciprocal of the leakance can be thought of as the resistance of the semipervious formation. Another parameter, the leakage factor, is the root of the ratio of the transmissivity of the aquifer to the leakance of the semipervious formation. It determines the areal distribution of the leakage.

The quality of water contained in an aquifer determines, to a large extent, the suitability of the water for a particular use, such as irrigation, public water supply, etc. The quality of ground water is a consequence of all processes and reactions that have acted on the water from the moment it condensed in the atmosphere to the time it is discharged by a well.[2] Human activity poses a threat to ground-water quality and already has resulted in incidents of ground-water pollution or contamination.[2,6] The latter refers to the presence of a chemical or biological agent in the ground water in such a concentration that it renders the water unfit for a certain use. Because ground water tends to move slowly, it may take many years after the start of pollution before contaminated water shows up in a well. The best way to protect the quality of ground water is to prevent its contamination.

REFERENCES

1. Mariño, M.A.; Luthin, J.N. *Seepage and Groundwater*; Elsevier Scientific Publishing Company: New York, 1982.
2. Bouwer, H. *Groundwater Hydrology*; McGraw-Hill: New York, 1978.
3. Todd, D.K. *Groundwater Hydrology*, 2nd Ed.; John Wiley: New York, 1980.
4. Freeze, R.A.; Cherry, J.A. *Groundwater*; Prentice Hall: Englewood Cliffs, NJ, 1979.
5. Domenico, P.A.; Schwartz, F.W. *Physical and Chemical Hydrogeology*, 2nd Ed.; John Wiley: New York, 1998.
6. Fetter, C.W. *Applied Hydrogeology*, 4th Ed.; Prentice Hall: Upper Saddle River, NJ, 2000.

Aquifers, Artificial Recharge of

Steven P. Phillips
United States Geological Survey (USGS), Sacramento, California, U.S.A.

INTRODUCTION

Limited freshwater resources in many parts of the world have led to the development of artificial recharge techniques for conveying surface water and reclaimed wastewater to groundwater reservoirs for later use and for other applications. Other applications include using artificial recharge to create a barrier to saltwater intrusion, reduce land subsidence, raise water levels, and improve water quality by using the natural filtering capabilities of aquifer systems. Subsurface storage of water has many advantages over surface storage, and often is the more physically and economically viable alternative. The worldwide use of artificial recharge likely will increase in the future with continued growth in population and associated competition for finite freshwater resources.

DEFINITION

Artificial recharge has been defined in many ways, varying with points of view and the evolution of applications and methods. It is generally defined by Todd[1] as "the practice of increasing by artificial means the amount of water that enters a groundwater reservoir." These artificial means include various forms of surface infiltration and direct well injection. For this discussion, forms of enhanced, induced, and incidental recharge, as defined by Bouwer,[2] will be excluded. Enhanced recharge is the increased infiltration of precipitation; it is controlled primarily through vegetation management. Induced recharge is the increased flow of surface water into the aquifer system caused by the placement of wells or other collectors near surface-water bodies. Incidental recharge is caused by leakage of water and sewer pipes, excess irrigation, and other human activities not designed to cause groundwater recharge. Also excluded from this discussion is a recharge from the injection of saltwater used to enhance petroleum recovery, and from deep disposal of wastes.

APPLICATIONS

Artificial recharge programs began in the late 19th century in the United States, and well before that in Europe. It was recognized early on that storing water in the groundwater system held certain advantages over traditional surface storage, including proximity to water sources and points of use, limited engineering and construction costs, and little or no evaporative losses. Prior to the mid-20th century, the primary application of these early programs was enhancement of groundwater resources for drinking and agricultural purposes. Surface water generally was captured when it was available and stored in the groundwater system for later use during high-demand periods. An annotated bibliography of early artificial recharge work is provided by Todd.[1]

Research during the early period of artificial recharge spawned a number of modern applications.[3] These applications include purification of wastewater or poor-quality surface water, creation of barriers against the intrusion of saltwater and other contaminants, abatement of land subsidence, and other environmentally or economically driven applications.

METHODS

Surface Infiltration

The most common form of artificial recharge is surface infiltration, whereby engineered systems allow an increased infiltration of water through subsurface materials to the water table. Although surface infiltration systems are subject to losses from evaporation and may unintentionally attract insects and waterfowl, they often are an efficient and economical means of artificial recharge. Common surface infiltration systems are shown in Fig. 1.

Surface infiltration systems can be divided into two categories: in-channel and off-channel.[2] In-channel systems use temporary or permanent dams or levees designed to impede flow and raise the water surface in existing surface-water channels. The raised water surface increases in-channel storage and the area of the streambed through which infiltration occurs, thereby increasing recharge. Often, in-channel systems are self-cleaning, as fine particles that impede infiltration are removed during high flows, but associated dams and levees require maintenance.

Encyclopedia of Water Science
DOI: 10.1081/E-EWS 120010072

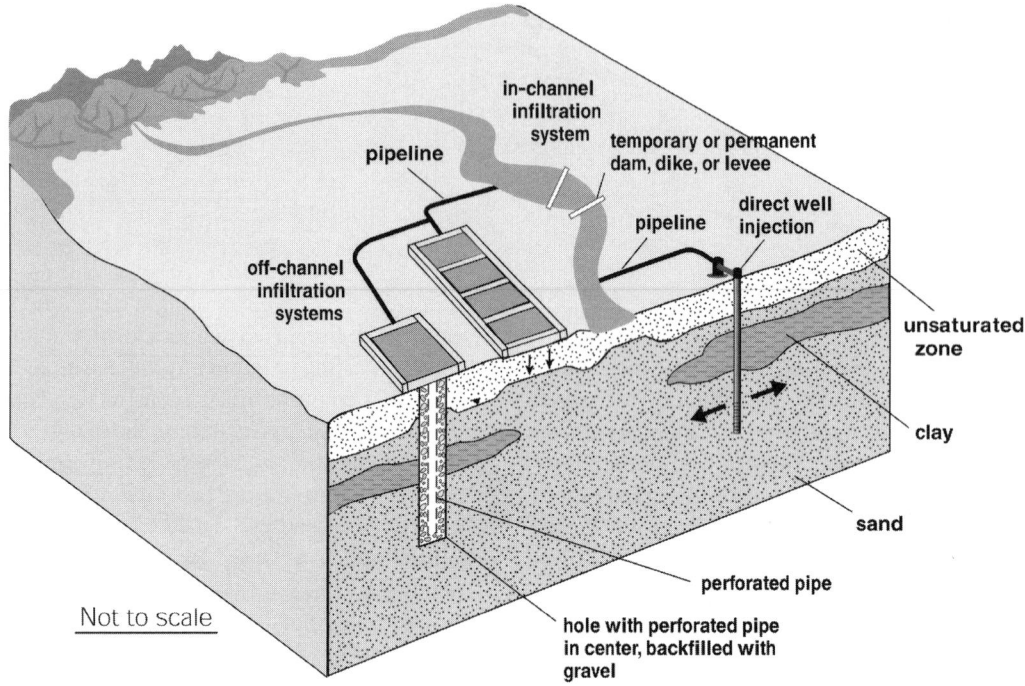

Fig. 1 Diagram showing examples of artificial recharge systems.

Off-channel systems involve the use of water from any source to fill existing or constructed basins, pits, ponds, and other structures for surface infiltration. Favorable site conditions for off-channel systems include available and affordable land, an unconfined aquifer that has sufficient transmissivity to keep the water table below the infiltration surface, permeable surface soils for sufficient infiltration rates, a lack of poorly permeable subsurface units that impede downward flow to the water table, and a source of water that has low suspended solids and compatible chemical and biological characteristics.

The site conditions for off-channel systems, if unfavorable, can be altered in some circumstances. Poorly permeable surface soils and (or) deeper fine-grained units above the water table can be penetrated with trenches or holes backfilled with sand or gravel to enhance their capacity to transmit water vertically. These trenches or holes can be used alone, with water delivered through perforated pipes, or in combination with basins or other surface impoundments.

If sediment concentrations or other suspended solids in the source water are too high, rapid clogging of infiltration basins can occur. Pretreatment of the source water in a reservoir or dedicated basin allows solids to settle, sometimes with the help of coagulants. Some chemical and biological processes can also cause clogging, which often is addressed by filtering and disinfecting during

pretreatment. Post-treatment clogging generally is controlled by maintenance of infiltration basins, which involves periodic cleaning of the basin bottom.[2]

Additional considerations for surface infiltration systems include the presence of soluble or dissolved potential contaminants in the subsurface, and desirable soil types for various applications. Knowledge of the presence and distribution of natural and anthropogenic contaminants in the vicinity of a proposed artificial recharge site is required to avoid introducing contaminants to the groundwater system and (or) transporting them to undesired locations. Specific soil types are preferable for some applications. For example, a coarse-grained homogeneous soil is preferable for achieving maximum recharge. A finer-grained soil that has higher capacity for sorption may be preferable for sites that use poor-quality source water.

Direct Well Injection

Direct well injection is becoming a common method of artificial recharge. For this method, water is pumped or gravity-fed through wells into confined and unconfined aquifers, and sometimes into the unsaturated zone.[2,4] The same wells used for injection often are used to recover the water. Direct well injection generally is more expensive than surface infiltration because of costs of well construction and water treatment requirements. However, injection allows water to bypass poorly permeable soils and subsurface units to allow rapid recharge of deep aquifers through thick unsaturated zones, avoids transport of near-surface contaminants to the saturated zone, and requires little land, enabling strategic well placement. These attributes make direct well injection particularly useful in urban settings and in areas where the unconfined aquifer is too contaminated to use surface infiltration methods.

Clogging is a key design and operational consideration for direct well injection because recharge water must pass through a very small area, the well screen and borehole wall, to enter the aquifer system. Clogging is one of the reasons that injection rates typically are about one-third to one-half of the extraction rate, though this ratio varies widely from site to site.[5] Suspended solids are a common cause of clogged injection wells. Most suspended solids can be removed in reservoirs or basins as described previously and (or) through specially designed piping systems and filters. However, the small amount that typically remains in the recharge water often is the primary clogging agent. The management strategy most often employed is periodic extraction from the well, which removes much of the caked solids.

Clogging can also be caused by biological growth, mineral precipitates, and air or gas bubbles. Growth of existing or introduced microorganisms can rapidly clog an injection well, and generally is managed by using continual low-level disinfection and periodic shock treatments with chlorine.[4] Mixing of water types during injection can cause precipitation of minerals on the well screen and within the gravel pack and aquifer materials. Commonly, geochemical modeling is done during the design phase to determine the potential for adverse geochemical reactions; adjustments of pH or other properties of the recharge water are sometimes made during operations to avoid mineral precipitation. Air introduced into the well through free-falling water or cavitation in pipes can enter the aquifer system and lodge in pore spaces, effectively reducing the hydraulic conductivity of the aquifer materials and thus the rate of injection. Dissolved gases coming out of solution have a similar effect. These reductions in hydraulic conductivity can be avoided through proper system design and pretreatment of source water.

ISSUES

Public Health

Public health issues associated with artificial recharge are most often raised in connection with the use of treated wastewater. There is much interest in expanding the use of

wastewater for artificial recharge, because it is a continuous and increasing source of water and more stringent regulations have increased disposal costs.[5,6] Artificial recharge can improve the quality of treated wastewater through microbial degradation and sorption of some organic constituents; however, there are well-understood and emerging pathogens and other potential toxicants for which diligent monitoring and active research are required to protect public health.[7] However, defined, data from ongoing studies and active projects, which include potable and nonpotable uses, suggest that wastewater is a viable source of water for artificial recharge.[5,6]

Environmental

Artificial recharge can have environmental effects that may be important to predict prior to implementation.[5] Flow in streams and other water bodies can increase with a rise in the water table, or decrease with diversions to recharge facilities. The quality of surface water and groundwater can be improved, degraded, or changed in some way that affects the end use of the water. Land subsidence can be reduced by slowing or reversing water-level declines. Development of a shallow water table can cause waterlogging and increased salinity. Reduced pumping lifts saves energy, reducing the environmental effects of energy production. A good understanding of these and other potential environmental effects of artificial recharge projects is a key to their long-term viability.

THE ROLE OF SCIENCE

Artificial recharge involves complex hydraulic, chemical, and biological responses and interactions in a groundwater system.[7] The role of science is to generate an understanding of these complexities through monitoring and analysis, and to develop tools to aid in the planning and management of artificial recharge projects.

Monitoring and Analysis

A set of methods has evolved over the years for monitoring and analyzing the effects of artificial recharge projects.[2,4,8] However, shortcomings in these methods and a number of existing and emerging issues continue to drive research efforts. Microgravity surveying is a geophysical technique that has been used recently to better define water-table changes associated with artificial recharge.[9] New tracer methods using existing and introduced chemicals, isotopes, and heat are improving our ability to track recharged water.[7] Recent research shows that the introduction of organic matter through artificial recharge and the composition of the organic matter may affect the evolution of groundwater chemistry and the formation of disinfection byproducts. A large body of research is focused on addressing the fate of compounds introduced through artificial recharge of treated wastewater, which is needed to safeguard public health. Other research efforts include development of more sophisticated modeling techniques for predicting complex biochemical processes and microbial transport; improving methods for detecting microbial pathogens; and determining the potential for mobilization of arsenic and other trace elements.[7]

Planning and Management Tools

Successful planning and management of an artificial recharge project often requires consideration of many water-management objectives, water-routing capabilities, economics, and hydraulic effects. Simultaneous consideration of these diverse factors can be accomplished by using optimization techniques designed to identify an efficient way to meet an objective, given a set of constraints. The linkage of a predictive groundwater flow model with optimization techniques (a simulation/optimization model) allows for simultaneous consideration of the flow system and physical and (or) economic constraints determined by water-resource managers. Simulation/optimization models have been applied to groundwater problems for decades[10,11] and have been used in the planning and management of artificial recharge projects (e.g., Refs. 12 and 13).

REFERENCES

1. Todd, D.K. *Artificial Recharge of Ground Water Through 1954*; Water-Supply Paper 1477; U.S. Geological Survey: Washington, DC, 1954; 115 p.

2. Bouwer, H. Artificial Recharge of Groundwater: Hydrogeology and Engineering. Hydrogeol. J. **2002**, *10* (1), 121–142.

3. Signor, D.C.; Growitz, D.J.; Kam, W. *Annotated Bibliography on Artificial Recharge of Ground Water, (1955–67)*; Water-Supply Paper 1990; U.S. Geological Survey: Washington, DC, 1970; 141 p.

4. Pyne, R.D.G. *Groundwater Recharge and Wells*; Lewis Publishers: Boca Raton, FL, 1995; 376 p.

5. National Research Council, *Ground Water Recharge Using Waters of Impaired Quality*; National Academy Press: Washington, DC, 1994; 283 p.

6. Asano, T. *Artificial Recharge of Groundwater*; Butterworth Publishers: Stoneham, MA, 1985; 767 p.

7. Aiken, G.R.; Kuniensky, E.L., Eds. *U.S. Geological Survey Artificial Recharge Workshop Proceedings, Sacramento, CA, April 2–4, 2002*; Open-File Report 02-89; US Geological Survey: Washington, DC, 2002; 85 p.

8. Fowler, L.C., Ed. *Operation and Maintenance of Ground Water Facilities*; ASCE Manuals and Reports on Engineering Practice No. 86; American Society of Civil Engineers: New York, 1996.

9. Howle, J.F.; Phillips, S.P.; Ikehara, M.E. Estimating Water-Table Change Using Microgravity Surveys During an ASR Program in Lancaster, Antelope Valley, California. In *Management of Aquifer Recharge for Sustainability*. Proceedings of the Fourth International Symposium on Artificial Recharge of Groundwater, Adelaide, Australia, Sept 22-26, 2002, Dillon, P.J., Ed.; Balkema, 2002; 567.

10. Gorelick, S.M. A Review of Distributed Parameter Groundwater Management Modeling Methods. Water Resour. Res. **1983**, *19* (2), 305–319.

11. Ahlfeld, D.P.; Mulligan, A.E. *Optimal Management of Flow in Groundwater Systems*; Academic Press: San Diego, 2000; 185 p.

12. Phillips, S.P.; Carlson, C.S.; Metzger, L.F.; Sneed M.; Galloway, D.L.; Ikehara, M.E.; Hudnut, K.W. Optimal Management of an ASR Program to Control Land Subsidence in Lancaster, Antelope Valley, California. In *Management of Aquifer Recharge for Sustainability*. Proceedings of the Fourth International Symposium on Artificial Recharge of Groundwater, Adelaide, Australia, Sept 22-26, 2002, Dillon, P.J., Ed.; Balkema, 2002; 567.

13. Reichard, E.G. Groundwater–Surface Water Management with Stochastic Surface Water Supplies: A Simulation-Optimization Approach. Water Resour. Res. **1995**, *31* (11), 2845–2865.

Aquifers, Karst

John Van Brahana
University of Arkansas, Fayetteville, Arkansas, U.S.A.

A

INTRODUCTION

Karst aquifers are water-bearing, soluble rock layers at or near the earth's surface in which groundwater flow is concentrated along secondarily enlarged fractures, fissures, conduits, and other interconnected openings. They are formed by the chemical dissolving action of slightly acidic water on highly soluble rocks, most notably limestone and dolomite, and to a lesser degree, gypsum, anhydrite, and halite. For the processes of karst to be active, water must dynamically circulate through these soluble rocks— exposing the rock to interaction with water and enabling transport of solutes, and the water must be undersaturated with respect to the chemical constituents of the rock— enabling dissolution to occur. This interplay of flow (hydrology) and dissolution (geochemistry) removes rock, creating increasingly larger voids along the pathways the water follows through time. Karst aquifer development commonly results in distinctive landforms, but visible surface features are not an essential attribute of karst, because in many instances, the surface features may be covered by soil or regolith. Although karst and karst aquifers most commonly are recognized as having distinctive landforms and topography (e.g., closed depres sions, sinkholes, sinking streams, dry valleys, caves, dissolutionally enlarged joints or bedding planes, grikes, karren, and springs), these are indicators of karst rather than definitive elements. Whereas most karst areas express part or all of these features, the key essential element of the karst definition must include "distinctive subsurface hydrol-ogy,"[1] characterized by secondarily enlarged flow path-ways. Other distinctive hydrology components include a high degree of interconnectivity between surface and groundwater, relatively rapid groundwater flow, great areal and temporal variability of aquifer properties, numerous springs, great susceptibility for contamination from human and natural activities at the land surface, and lack of filtration and attenuation of contamination. Water-bearing rocks with these attributes are likely karst aquifers if they are highly soluble, even if no karst landforms are present.[1]

Karst aquifers are widespread and intensively utilized. Their worldwide occurrence ranges from 12% to 25% of the earth's surface.[1–4] About 25% of the world's population is estimated to rely on freshwater supplies from these aquifers.[3] Fig. 1 shows the dominant regional karst aquifers in the United States in terms of water use, well yields, and spring discharge.[4] Globally, karst aquifers comprise important water resources in, e.g., southern China, southeast Asia, western Europe and the Mediterranean basin, and the Caribbean islands.

MAJOR SOURCES OF KARST INFORMATION

Karst-forming processes can be some of the most dynamic, erosive forces that counterbalance the uplifting forces of tectonics.[5] Karst can be responsible for the most active surface water/groundwater interactions of all aquifer types.[6] Coupled with more than 60 controlling influences (e.g., lithologic, structural, hydrologic, geochemical, and geomorphic), these processes result in the most variable hydrogeology within a single aquifer type of all the earth's rocks.[7] These facts notwithstanding, the hydrology of these aquifers is not unpredictable. Refs. 2, 3, and 8–20 synthesize the current state of understanding of the hydrology of karst aquifers as well as numerical simulation as a hydrogeologic tool. Websites include the National Speleological Society,[22] the Karst Waters Institute,[23] and the Karst Commission of the International Association of Hydrogeologists.[21,24]

UNIQUE ATTRIBUTES OF KARST AQUIFERS

Notable differences between karst and porous media aquifers are: groundwater flow is commonly turbulent in karst, and Darcy's law is not appropriate for quantification in a karst aquifer, except at large scales exceeding several kilometers; groundwater in karst aquifers contains multiple components of flow types, a mix of fastflow and slowflow conditions from the epikarst, the vadose zone, and the phreatic zone; fastflow in karst aquifers dominates advec-tive transport; fastflow paths in karst aquifers are difficult to predict without tracing studies; the interaction between surface water and groundwater in karst lands is more pronounced than in nonkarst terrane; allogenic recharge to karst aquifers by subsurface capture of streamflow derived outside of the groundwater basin is common in karst; spring recharge boundaries in karst commonly do not coincide with surface water boundaries; karst aquifers generally exhibit a much wider range of variability of hydraulic char-acteristics than porous media aquifers (Table 1); hydraulic characteristics can vary areally, in orientation, and tempo-rally, increasing as dissolution progresses, and decreasing as deposition of insoluble sediments occur or chemical precipitation occludes zones of enhanced permeability;

Encyclopedia of Water Science
DOI: 10.1081/E-EWS 120010039

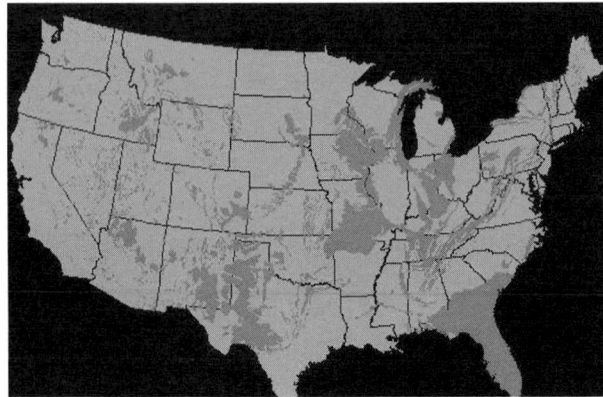

Fig. 1 Distribution of karst aquifers in the U.S. (From Ref. [4].)

karst aquifers commonly have greater hydraulic conductivity but lower storativity than comparably sized nonkarst aquifers; water quality in karst aquifers varies temporally, a result of mixing from the different flow components; karst aquifers typically have more permeable, open flow systems, and less chance for filtration and solute retardation; karst aquifers are generally more prone to contamination; karst aquifers tend to concentrate and accumulate flow, whereas porous-media aquifers tend to retain diffusive conditions; springs integrate flow from all parts of a karst aquifer; springs and cave streams are excellent locations to sample characteristic water quality from a karst aquifer; and wells are not a representative way to sample karst aquifers, unless it has been documented that the wells intersect the zones of fastflow.

FACTORS CONTROLLING THE FLOW OF GROUNDWATER IN KARST AQUIFERS

The laws of physics that govern the flow of groundwater in other aquifers apply equally to karst (water flows from areas of high energy to areas of low energy). Groundwater

obeys the laws of thermodynamics, always following the path of least resistance.

While all factors responsible for the creation of a karst aquifer are important (Table 2), there is a general hierarchy of relative importance of geologic factors as they affect karst-aquifer formation. From most important to less important, one version of this ranking follows.[7] Lithology of the hydrogeologic framework of aquifers and confining beds, including bulk chemical purity, grain size, original porosity and permeability inherited from early diagenesis, layer thickness, sequence thickness, caprock integrity and strength, and vertical variability in permeability, plays an essential role in karst aquifer formation. Simply stated, if no soluble rocks exist, there is no karst. As well as defining the aquifer potential of the rock matrix to be dissolved, lithologic factors influence the anisotropy and heterogeneity of the rock mass, and define zones potentially more permeable and therefore more favorable for flow. For given groundwater conditions, there is a preferential pathway of flow, one that successfully captures increasingly large volumes of flow in the karst system at the expense of other competing pathways. Structural and tectonic factors, including brittle fracture and folding processes, uplift, tilting, and metamorphism, can be very important in karst aquifer development.[2,3,7,9] Structural processes not only influence the orientation of preexisting permeability zones through folding, tilting, and uplift, they also create new pathways in the form of secondary rock fractures (such as joints and faults), both in confining beds and aquifers. Under the extreme conditions of metamorphism, structural processes can obliterate previous permeability by recrystallizing carbonate aquifers into marble. The role of structure in creating vertical short circuits along subsurface flowpaths is almost as important as lithology, but the dimensions are typically smaller by several orders of magnitude, and structure is ranked lower only for this reason. Other factors (Table 2), including hydrologic, geochemical, diagenetic, weathering-geomorphic, and historical

Table 1 Comparison of typical flow and hydraulic characteristics of karst aquifers

Rock type	Primary porosity and permeability	Secondary porosity and permeability	Flow regime	Discharge from large springs
Cavernous limestone	Low porosity low average but highly variable permeability	Low porosity anisotropic; huge permeability in conduits, very low permeability in rock mass	Laminar to turbulent, if deeply buried, may be stagnant	Regional aquifers $> 10^3$ L sec^{-1} at major springs
Chalk	High porosity, low permeability, intergranular	Usually not significant, usually undeveloped regionally, jointing can be important	Usually stagnant, very sluggish, commonly only local if at all	Commonly confining layers— $> 10^3$ L sec^{-1} or less; 0 L sec^{-1} is common, may be local aquifer if fractured
Dolomite	Low porosity, low to moderate permeability	Variable from diagenetic pin-point porosity to conduit flow. Joints and faults can be important	Variable usually laminar, can be turbulent to sluggish	Vary from confining layers to regional aquifers: 10^1 to 10^2 L sec^{-1}
Marble	Very low, essentially none	Significant if jointed or fractured, solution conduits can develop if flow system complete	Usually laminar where secondary permeability developed—can be variable, turbulent to sluggish	Typically local aquifers yield 10^2 to 10^1 L sec^{-1}, 0 is common

(Adapted from Ref. 7.)

Table 2 Geologic factors, processes and controls that affect porosity and permeability of karst aquifers

Factor	Processes	Controls	General influence
Diagenetic	Compaction Cementation Pressure solution Solution (includes recrystallization, inversion, micritization)	Original porosity and permeability; original mineralogy; grain size/surface area; proximity to sea level (uplift or burial); volume and rate of water movement; fluid chemistry: pH, pCO_2 salts in solution; temperature, pressure	Influences initial distribution of porosity and permeability of indurated rock mass. Many of these are geochemical in nature; they occur very early in the history of the rock
Geochemical	Solution (dissolution) Dolomitization Dedolomitization Precipitation Sulfate reduction Redox	Groundwater flux; original porosity and permeability; mineralogy; fluid chemistry: pH, pCO_2, salts in solution, temperature, pressure, mineral-water saturation	Influences later development of porosity and permeability; influences water chemistry
Lithologic-stratigraphic		Layer thickness; sequence thickness; variability in texture (vertical); variability in permeability (vertical); original porosity and permeability inherited from diagensis; bulk chemical purity; grain size	Influences anisotropy of rock mass, thereby resulting in zones potentially more permeable if other geologic factors are favorable
Structural-tectonic	Uplift Tilting Folding Jointing Faulting Metamorphism	Fracture density; openness of fractures; layer (permeability) orientation	Influences orientation of permeability zones. Influences integrity of confining layers. In extreme instances (metamorphism), influences existence of permeability zones
Hydrologic	Dynamic groundwater	Climatic—temperature; climatic—precipitation; depth of circulation; location of boundaries; existence of complete flow systems; flux; initial anisotropy–vertical variation; springs; surface water/groundwater relation; recharge; hydraulic gradient; size of groundwater basin	Influences existence of flow systems. Influences rate of flow system evolution
Weathering geomorphic	Infilling (fluvial and glacial) Unloading	Topography; relief; soil development by sedimentation; cap rock; degree of karstification; base level; surface slope	Influences development of flow systems. Influences destruction of permeability. Influences shallow porosity–permeability development
Historical geologic–chronologic		Sequence of events; duration of events	Influences stage of development of specific permeability zones

(From Ref. 7.)

geologic–chronologic influences,[7] can be important at a local scale, but owing to the extensive variability of karst aquifers, their role and influence is not ubiquitous regionally.

SINKHOLE FLOODING

Sinkhole flooding is a common problem around karst aquifers, especially in areas where there is rapid development by humans. Construction in a watershed locally modifies the hydrologic budget, inhibiting infiltration in some areas, diverting infiltration to overland flow which is in turn pirated underground at focused recharge points elsewhere. When storms generate more discharge than the karst aquifer can transmit, water levels in the aquifer rise rapidly. Unlike surface streams that have wide flood plains, karst aquifers have relatively narrow, confined boundaries within the rock. The water level of these aquifers reflects the least-constrained dimension, and when more flow is introduced than can pass, water levels rise precipitously (> 100 ft), and flood low-lying areas of the karst flow system. Some flood-prone sinkholes are miles from the nearest surface stream or flood plain, and property owners may not realize they are at risk until a flood occurs.[20]

CAVITY COLLAPSE, SUBSIDENCE, AND OTHER ENGINEERING AND CONSTRUCTION PROBLEMS

Collapse of cave passages, underground voids, and more commonly, collapse of the soils and sediments overlying karst bedrock is a natural process that occurs in response to the continual evolution of the enlarging karst ground water-flow system. This process is exacerbated by human activity, including rapid lowering or raising of the water level, changes in runoff chemistry (pH), increased vibrations, and loading the land above the aquifer with buildings and other massive structures. Highways, bridges, dams, buildings, cars, and homes have been lost as karst rocks equilibrate to the loads they support. Millions of dollars in construction have been spent repairing cracks, filling openings, grouting leaks, and otherwise restoring the structures. The most catastrophic sinkhole event in recorded history occurred in December 1962 in West Driefontein, South Africa. Twenty-nine lives were lost by the sudden disappearance of a building into a huge sinkhole that measured more than 180 ft across.[4] Loss of life is not common, but this incident reflects the dynamic, rapid changes that are associated with water resources and hydrogeologic framework in areas underlain by soluble rocks.

CONCLUSION

Karst aquifers occur beneath a significant portion of the earth's surface (from 12% to 25%, depending on one's definition of karst). Karst aquifers are more highly soluble than other rock types, and most commonly include limestones, dolomites, marbles, and evaporites. In fact, given enough time, karst-like dissolution features have been observed in most lithologies. Karst aquifers are unique compared to porous media aquifers in that they are highly variable in water transmission properties, and if flow is concentrated along focused pathways, and the water in the aquifer is aggressive, these aquifers evolve, flowpaths become enlarged, and rapid movement of groundwater becomes the norm. Karst aquifers are more easily contaminated than porous media aquifers; attenuation of contamination in these aquifers typically is minimal, owing to significantly reduced surface area of solid aquifer/water contact Collapse of the cave passages, underground voids, and more commonly, collapse of the soils and sediments overlying the soluble bedrock is a natural process, and the resulting topography that overlies these aquifers typically is characterized by internally drained depressions called sinkholes (dolines).

REFERENCES

1. Quinlan, J.F.; Smart, P.L.; Schindel, G.M.; Alexander, E.C., Jr.; Edwards, A.J.; Smith, A.R. Recommended Administrative/Regulatory Definition of Karst Aquifer, Principles for Classification of Carbonate Aquifers, Practical Evaluation of Vulnerability of Karst Aquifers, and Determination of Optimum Sampling Frequency at Springs. Proceedings Volume of the Third Conference on Hydrogeology, Ecology, Monitoring and Management of Ground Water in Karst Terranes; Nashville, Tennessee Association of Ground Water Scientists and Engineers, 1991; 573–635.
2. Ford, D.C.; Williams, P.C. *Karst Geomorphology and Hydrology*; Unwin Hyman: London, 1989; 601 pp.
3. White, W.B. *Geomorphology and Hydrology of Karst Terrains*; Oxford University Press: New York, 1988; 464 pp.
4. Veni, G.; DuChene, H.; Crawford, N.C.; Groves, C.G.; Huppert, G.N.; Kastning, E.H.; Olson, R.; Wheeler, B.J. *Living with Karst—A Fragile Foundation*; American Geological Institute Environmental Awareness Series 4: Alexandria, VA, 2001; 64 pp.
5. White, W.B.; Culver, D.C.; Herman, J.S.; Kane, T.C.; Mylroie, J.E. Karst Lands. Am. Sci. **1995**, *83*, 450–459.
6. Winter, T.C.; Harvey, J.W.; Franke, O.L.; Alley, W.M. *Ground Water and Surface Water—A Single Resource*; U.S. Geological Survey Circular 1139: Denver, CO, 1999; 1–79.
7. Brahana, J.V.; Thrailkill, J.; Freeman, T.; Ward, W.C. Carbonate Rocks. In *Hydrogeology*; Back, W., Rosenshein, J.S., Seaber, P.R., Eds.; Geological Society of America: Boulder, CO, 1988; Vol. O-2, 333–352.
8. Dreybrodt, W. *Processes in Karst Systems—Physics, Chemistry, and Geology*; Springer-Verlag: Berlin, 1988; 288 pp.
9. Palmer, A.N., Palmer, M.V., Sasowsky, I.D., Eds. *Speleogenesis—Evolution of Karst Aquifers*; National Speleological Society, Inc.: Huntsville, AL, 2000; 527 pp.
10. Kresic, N. *Karst Water Resources*; Lewis Publishers: Boca Raton, FL, 1998; 384 pp.
11. Worthington, S.R.H. Karst Hydrogeology of the Canadian Rocky Mountains. Unpublished Ph.D. dissertation, McMaster University, Hamilton, Ontario, Canada, 1994; 227 pp.
12. Palmer, A.N. Groundwater Processes in Karst Teranes. In *Groundwater Geomorphology*; Higgins, C.G., Cates, D.R., Eds.; Special Paper 252; Geological Society of America: Boulder, CO, 1990; 177–209.
13. Ford, D.C.; Palmer, A.N.; White, W.B. Landform Development—Karst. In *Hydrogeology*; Back, W., Rosenshein, J.S., Seaber, P.R., Eds.; Geological Society of America: Boulder, CO, 1988; Vol. O-2, 401–412.
14. Palmer, A.N., Palmer, M.V., Sasowsky, I.D., Eds. *Groundwater Flow and Contaminant Transport in Carbonate Aquifers*; A. A. Balkema: Rotterdam, Netherlands, 2000; 288 pp.
15. Palmer, A.N., Palmer, M.V., Sasowsky, I.D., Eds. *Karst Modeling*; Special Publication 5; Karst Waters Institute: Charles Town, West Virginia, 1999; 1–294.
16. Quinlan, J.F.; Davies, G.J.; Jones, S.W.; Huntoon, P.W. The Applicability of Numerical Models to Adequately Characterize Ground-water Flow in Karstic and Other Triple-Porosity Aquifers. In *Subsurface Fluid-Flow Modeling*; Ritchy, J.D., Rumbaugh, J.O., Eds.; Special Technical Publication 1288; American Society for Testing and Materials: West Conshohocken, PA, 1996; 114–133.
17. Labat, A.; Ababou, R.; Mangin, A. Introduction of Wavelet Analyses to Rainfall/Runoffs Relationship for a Karstic Basin: The Case of Licq-atherey Karstic System (France). Ground Water **2001**, *39* (4), 605–615.
18. Dreiss, S. Regional Scale Transport in a Karst Aquifer. 1. Component Separation of Spring Flow Hydrographs. Water Resour. Res. **1989**, *25* (1), 117–125.
19. Dreiss, S. Regional Scale Transport in a Karst Aquifer. 2. Linear Systems and Time Moment Analysis. Water Resour. Res. **1989**, *25* (1), 126–134.
20. Pinault, J.-L.; Plagnes, V.; Aquilina, L.; Bakalowicz, M. Inverse Modeling of the Hydrological and Hydrochemical Behavior of Hydrosystems: Characterization of Karst System Functioning. Water Resour. Res. **2001**, *37* (8), 2191–2204.
21. Drew, D.; Hotzl, H. *Karst Hydrogeology and Human Activities: Impacts, Consequences, and Implications*; International Association of Hydrogeologists, A. A. Balkema: Rotterdam, Holland, 1999; 322 pp.
22. http://www.caves.org (accessed February 2002).
23. http://www.karstwaters.org (accessed February 2002).
24. http://www.iah.org (accessed February 2002).

Aquifers, Karst: Water Quality and Water-Resource Problems

John Van Brahana
University of Arkansas, Fayetteville, Arkansas, U.S.A.

INTRODUCTION

Owing to the rapid flow, the open nature of the flow paths, and the general lack of aquifer surface area for water/rock interaction, karst aquifers are more at risk to contamination than porous-media aquifers. Problems typically include lack of attenuation, little or no response time, and unexpected flow directions and plume transport. In addition to water quality problems, catastrophic subsidence and flooding are common problems where soluble rocks occur near land surface.

FACTORS CONTROLLING GROUNDWATER QUALITY IN KARST AQUIFERS

Dissolution is the dominant chemical process in karst aquifers, and water type is controlled by the inorganic reactions of the recharge water and the aquifer matrix.[1-5] In carbonate aquifers, bicarbonate (HCO_3^-) is the dominant anion. Where limestone is the dominant lithology, calcium (Ca^{+2}) is the prevailing cation. Where dolomite aquifers predominate, magnesium (Mg^{+2}) and Ca^{+2} are the dominant cations, typically in a ratio of about 1:3, respectively. Carbonate aquifers yield hard water, hardness being a measure of the soap-consuming properties of water, a function of the concentration of Mg^{+2} and Ca^{+2} ions in solution. The chemistry of meteoric recharge water to the aquifer is predominantly a weak (pH 5–7) carbonic acid (H_2CO_3), formed from H_2O mixing with carbon dioxide (CO_2) in the atmosphere and from soil gas picked up in solution during recharge through organic-rich soil and regolith. At some locations, however, such as the area of the Pecos River in west Texas and eastern New Mexico (see Fig. 1) in the vicinity of Carlsbad Caverns, meteoric H_2O has mixed with hydrogen sulfide gas (H_2S) generated by organic matter degradation and oil formation from basins to the east. This reaction formed sulfuric acid (H_2SO_4), resulting in groundwater of a mixed sulfate/bicarbonate type.

Karst aquifers formed in evaporite rocks are much less common than karst aquifers in carbonate rocks, primarily because their solubility is much greater than carbonates, and in humid climates the chemical rates of erosion are extremely rapid. Evaporite karst is preserved in arid and semiarid climates of the western United States (see Fig. 1) and other areas of the world, but these aquifers typically yield highly-mineralized, nonpotable water and are seldom used for water supply. Gypsum and anhydrite yield calcium sulfate type waters (Ca^{+2}, SO_4^{-2}), and halite produces brines of sodium chloride (Na^+, Cl^-) character.[1]

The overall water quality of karst aquifers is a reflection of the geochemistry of all the sources of recharge, weighted by the proportion of the percentage contribution of that source to the total hydrologic budget, coupled with water/rock reactions within soils and rocks upgradient from the aquifer.[1-5] Mixing of karst waters is a major chemistry-controlling process. If land use in the recharge area of the aquifer allows undesirable constituents, water quality in the aquifer will be degraded, as elaborated under the pollution section that follows.[6-11]

Sediment is an important part of water quality of karst aquifers. Turbulent flow, coupled with open flow systems allows clay-sized particles of soil to be transported through these groundwater systems.[12] The importance of sediment to water quality is that these particles have huge surface areas, and they typically have electrical charges that attract and hold potentially hazardous substances, such as trace metals and organic chemicals. Sediments also provide a substrate for the preservation and transport of bacteria, virus, and other pathogens. Therefore, contaminants that otherwise would not be present within the aquifer may be mobilized and attached to sediment particles. Contaminants here are sorbed to particles in suspension, and are not in solution; they are in transit through the aquifer just the same, and are available for reaction. If these are ingested in any form by organisms (such as mercury attached to clay may be ingested by bottom-feeding fish) the contaminant may become biologically active (methylated) and bioaccumulated.

WATER-RESOURCE PROBLEMS IN KARST AQUIFERS

The most recurring problems humans experience when dealing with karst aquifers are those related to pollution,[6-11] flooding,[3,4,6] and subsidence.[3,4,6] This is because human activity creates disequilibrium. Changes

Encyclopedia of Water Science
DOI: 10.1081/E-EWS 120012955

Fig. 1 Distribution of Karst aquifers in the U.S. (Modified from Ref. [6].)

within the aquifer occur rapidly in response to changes at the surface, flow velocities can typically transport sediment and other potentially large conduit-plugging debris into the aquifer, and there is less chance for filtration or sorption of unwanted solutes and constituents in karst. Thus, land use in the area contributing recharge to a karst aquifer needs to be closely monitored to minimize negative impacts.

Pollution

Pollution is common in karst aquifers near areas with high human or animal population densities. Unwanted constituents can move into an aquifer and mix with the native ground water, polluting the aquifer, rendering it unsuitable for further use. Many pollutants move through fast-flow karst aquifers rapidly, and the duration of the impact is short, a matter of days, weeks, or months. Some contaminants, however, such as dense, nonaqueous phase liquids (DNAPLs), sink to the bottom of flow zones in karst aquifers and remain until decomposed by natural processes; the duration of pollution of this sort may be hundreds of years or more. Manufacturing spills, illegal dumping in sinkholes, spills from transporting hazardous wastes, leakage from underground storage tanks, leachate plumes from landfills, excess nutrients from animal manure and inorganic fertilizers, pharmaceuticals and endocrine disruptors from water-treatment facilities, and microbial pathogens from leaky sewers, improperly functioning septic systems, and confined animal feeding operations (CAFOs) are but a few of the well-documented pollution problems found in karst aquifers.[4,6-11] Wherever concentrations of constituents exist at the surface in karst terrane, the chance for movement into the subsurface

is great.[13] These risks, combined with the susceptibility of this type of aquifer to contamination, call for vigilance in conducting most human activities in all karst areas, and long-term planning and zoning for those areas that are particularly susceptible. Websites include the the Karst Waters Institute,[14] and the Karst Commisssion of the International Association of Hydrogeologists.[15]

REFERENCES

1. Hanshaw, B.B.; Back, W. Major Geochemical Processes in the Evolution of Carbonate Aquifer Systems. J. Hydrol. **1979**, *43*, 287–312.
2. Thrailkill, J.; Robl, T.R. Carbonate Geochemistry of Vadose Water Recharging Limestone Aquifers. J. Hydrol. **1981**, *54*, 195–208.
3. Ford, D.C.; Williams, P.C. *Karst Geomorphology and Hydrology*; Unwin Hyman: London, 1989; 601.
4. White, W.B. *Geomorphology and Hydrology of Karst Terrains*; Oxford University Press: New York, 1988; 464.
5. Worthington, S.R.H. Karst Hydrogeology of the Canadian Rocky Mountains. unpublished Ph.D. dissertation, McMaster University, Hamilton, Ontario, Canada, 1994; 227.
6. Veni, G.; DuChene, H.; Crawford, N.C.; Groves, C.G.; Huppert, G.N.; Kastning, E.H.; Olson, R.; Wheeler, B.J. *Living with Karst—A Fragile Foundation*; American Geological Institute Environmental Awareness Series 4: Alexandria, Virginia, 2001; 64.
7. White, W.B.; Culver, D.C.; Herman, J.S.; Kane, T.C.; Mylroie, J.E. Karst Lands. Am. Sci. **1995**, *83*, 450–459.
8. Winter, T.C.; Harvey, J.W.; Franke, O.L.; Alley, W.M. Ground Water and Surface Water—A Single Resource. U.S. Geological Survey Circular **1999**, *1139*, 1–79.
9. Kresic, N. *Karst Water Resources*; Lewis Publishers, 1998; 384.
10. Sasowsky, I.D. Wicks, C.M., Eds. *Groundwater Flow and Contaminant Transport in Carbonate Aquifers*; A.A. Balkema: Rotterdam, Netherlands, 2000; 288.
11. Drew, D.; Hotzl, H. *Karst Hydrogeology and Human Activities: Impacts, Consequences, and Implications*; International Association of Hydrogeologists: A.A. Balkema: Rotterdam, Holland, 1999; 322.
12. Mahler, B.J.; Bennett, P.C.; Zimmerman, M. Lanthanide-Labeled Clay: A New Method for Tracing Sediment Transport in Karst. Ground Water **1997**, *35*, 835–843.
13. Katz, B.G.; Coplen, T.B.; Bullen, T.D.; Davis, J.H. Use of Chemical and Isotopic Tracers to Characterize the Interactions Between Ground Water and Surface Water in Mantled Karst. Ground Water **1997**, *35*, 1014–1028.
14. http://www.karstwaters.org (accessed February 2002).
15. http://www.iah.org (accessed February 2002).

Aquifers, Ogallala

B. A. Stewart
West Texas A&M University, Canyon, Texas, U.S.A.

INTRODUCTION

The Ogallala aquifer underlies about $174,000\,\text{mi}^2$ of land in the U.S. Great Plains. States with the largest land areas above the aquifer are Nebraska, Texas, Kansas and Colorado with smaller amounts in New Mexico, Oklahoma, South Dakota, and Wyoming (Fig. 1). The aquifer extends approximately 800 mi north to south and 400 mi east to west. The aquifer consists of sand and gravel beds that generally lie from 50 to 300 ft beneath the surface and are 150–300 ft thick. In some areas, however, the saturated thickness of the aquifer exceeds 1000 ft.[1]

The water contained in the Ogallala aquifer is essentially fossil water taken 10,000 yr–25,000 yr ago from the glacier-laden Rocky Mountains before it was geologically cut off by the Pecos River and the Rio Grande.[2] More than 3 billion acre-feet (an acre-foot is a foot of water on 1 acre, or 325,851 gal) are stored in the aquifer.

The amount of water that can be extracted from an aquifer is primarily a function of two characteristics—hydraulic conductivity and specific yield. Hydraulic conductivity is the rate of water flow in m^3/sec through a cross-sectional area of $1\,\text{m}^2$ under a hydraulic gradient of 1 m/m at a temperature of about 15°C. Specific yield is the ratio of the volume of water that the saturated aquifer will yield by gravity drainage to the total volume of the aquifer. The hydraulic conductivity governs how fast water can be pumped from an aquifer and the specific yield indicates how much water can be pumped. Gutentag et al.[1] reported that 76% of the aquifer had specific yields of 10–20%, although the values ranged from less than 5 to 30% and averaged 15.1%. Thus, as an average, pumping enough to drop the water level of the aquifer 100 cm will yield 15.1 cm of water at the surface.

USE OF WATER FROM THE AQUIFER

The area underlain by the Ogallala aquifer is semiarid with average annual precipitation ranging from about 375 in. to 625 in. There is also wide variability both within and between years. The annual precipitation ranges for any given year from about 50% of average to 200% of average. Because of the low and varied precipitation, much of the area has no year around water supply from surface supplies. The Ogallala aquifer, therefore, became the lifeblood for development of the Great Plains. Opie[2] tells of Gropp homesteading 20 mi northwest of Garden City, Kansas in 1887. For 2 yr, Gropp rolled a large barrel of water three-fourths of a mile from a neighbor's well until he hand-dug his own 70-m-deep well. Windmills became widely used through the Great Plains to supply water for livestock and homesteads.

The Great Plains was largely developed by livestock producers and dryland farmers. Crop production, however, was highly variable because of the erratic precipitation. As technologies for pumping water improved, irrigation became common and now uses more than 90% of the water pumped from the Ogallala aquifer.

Although some irrigation from the Ogallala aquifer using windmills occurred in the late 1800s, it was limited and sporadic. The drought of the 1930s spurred irrigation development because the precipitation was so low that crops could not be produced otherwise. The major expansion occurred during the 1950s when another drought occurred at the same time that there were technological advances in well drilling and pumping plants, readily available nitrogen fertilizer, inexpensive energy, profitable crop prices, development of hybrid grain sorghum, and available financing. Irrigation from the Ogallala aquifer increased from about 800,000 ha in 1949 to 5.2 million ha in 1978. During this period, approximately 60 cm of water was pumped annually from the aquifer for every irrigated hectare.[1]

Irrigation developed most rapidly in the beginning in the southern part of the Ogallala aquifer area because this area was more drought prone, the depth to the water table was less, and the land was generally flat making it relatively easy to apply water and run it down the rows. Many of the soils in the northern plains were too rolling and too sandy to irrigate by running water down furrows so irrigation development was slowed until reliable center-pivot sprinkler systems became available.

The major problem facing the Ogallala aquifer region today is not knowing how long the water supply will last. Also, the increase in energy prices has also had major impacts on irrigation at different times. In some cases and in some years, the cost of pumping the water is greater than the benefits derived.

Encyclopedia of Water Science
DOI: 10.1081/E-EWS 120010217

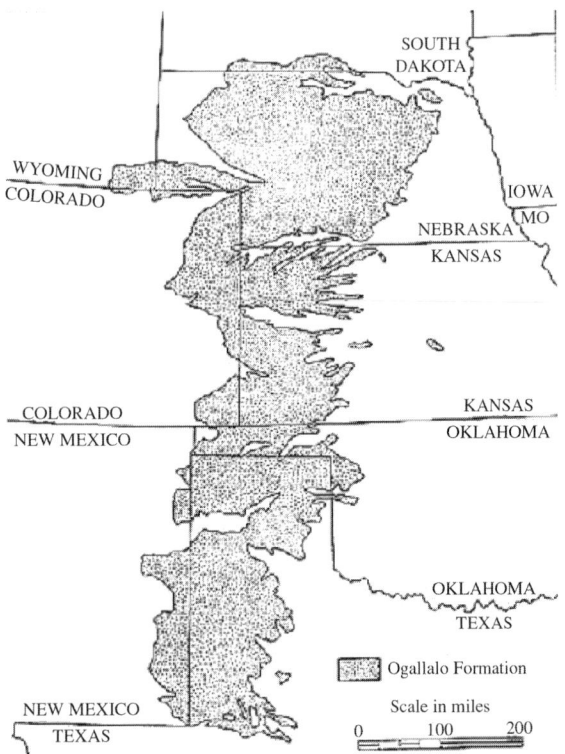

Fig. 1   The area underlain by the Ogallala Aquifer.

The distribution of the water in the Ogallala is very different among the states. Nebraska has 36% of the land (but most of it unsuitable for cultivation) above the aquifer and in 1980 had 66% of the drainable water in storage. Texas had 20% of the land but only 12% of the drainable water comparing closely with Kansas that had 17% of the land and 10% of the drainable water.

The water rights that govern the pumping of water from the aquifer vary for different states. In Nebraska, for example, the water is public property. However, the owner of land is entitled to appropriate the ground waters found under the land, but the owner cannot extract and appropriate them in excess of a reasonable and beneficial use upon the land owned, especially if such is injurious to others who have substantial rights to the waters, and if the natural underground supply is insufficient for all users, each is entitled to a reasonable proportion of the whole. In contrast, Texas law gives the overlying landowner the right to capture and use ground water beneath the land, regardless of the impact on adjoining or more distant users of the supply.[3]

The future of irrigation from the Ogallala aquifer is not clear. While only a relatively small percentage of the total water in the aquifer has been removed, essentially all of the recoverable water has been removed in some portions of the aquifer. The ratio of the volume of drainable water remaining to the volume of water depleted by 1980 was about 19 for the entire aquifer area. Even in Texas, the State with the greatest depletion, 3.4 times as much water remains as has been removed. In some areas, however, the volume of water remaining in the aquifer was less than the volume that had been removed.[1] Depletion since 1980 has continued, although at a somewhat slower rate in most areas because of declining well capacities, higher energy prices, and lower grain prices. It is generally believed that irrigation will decline in much of the Ogallala aquifer region both in areas and in the amount of water applied per hectare. Irrigation in the Texas High Plains has already declined from more than 2.4 million ha to about 1.8 million ha. The efficiency of water use, however, is increasing as a result of improved irrigation systems and crop management practices. This will enable irrigation to continue longer than otherwise would be feasible and irrigation will continue for many portions of the Ogallala aquifer area even in Texas nearly indefinitely.

Many cities, towns, and rural homes depend on water from the Ogallala aquifer. For the most part, there will be sufficient water for domestic uses even when there is no longer sufficient water for irrigation. Irrigation requires huge amounts of water and pumping for irrigation usually stops well before all the water that can be pumped is removed. Thus, there is usually adequate water remaining for domestic use even in areas where irrigation is no longer feasible.

The future of the Ogallala aquifer is unknown. Although some of the early settlers in the area thought there were underground rivers flowing beneath the land, it is clearly understood today that the water stored in the aquifer accumulated over eons of time. The rate of recharge is almost negligible in relation to the rate that water has been pumped since the 1950s. The way that the aquifer is used in the future will be shaped by many factors, but most importantly by the ability and willingness of the people to manage the water in the Ogallala as a nonrenewable resource.

REFERENCES

1. Gutentag, E.D.; Heimes, F.J.; Krothe, N.C.; Luckey, R.R.; Weeks, J.B. *Geohydrology of the High Plains Aquifer in Parts of Colorado, Kansas, Nebraska, New Mexico, Oklahoma, South Dakota, Texas and Wyoming*, U.S. Geological Survey Professional Paper 1400-B; U.S. Government Printing Office: Washington, DC, 1984.

2. Opie, J.O. *Water for a Dry Land*; University of Nebraska Press: Lincoln, NE, 1993.

3. Templer, O.W. The Legal Context for Ground Water Use. In *Groundwater Exploitation in the High Plains*; Kromm, D.E., White, S.E., Eds.; University Press of Kansas: Lawrence, 1992; 64–87.

Aral Sea Disaster

Guy Fipps
Texas A&M University, College Station, Texas, U.S.A.

INTRODUCTION

The Aral Sea is one of the worst ecological disasters on our planet. What was once the world's fourth largest inlet sea, the Aral Sea has lost over 60% of its surface area, two-third of its volume, declined 40 m in depth, and has fallen to the eighth largest inland body of water in the world.

The cause is attributed to a vast expansion of irrigation in the Central Asian Republics beginning in the 1950s, which greatly reduced inflows to the Sea. The diversion of water for massive irrigation development was done deliberately by Soviet Union officials, unconcerned about the consequences of their actions.

The environmental, social, and economic damage has been immense. Winds pick up dust from the dry seabed and deposit it over a large populated area. The dust likely contains pesticide and chemical residues that are blamed for the serious rise in mortality and health problems in the region. The Sea, and the now exposed dry seabed, may also be contaminated by runoff from a former Soviet military base and a biological weapons lab. The ecosystem of the Aral Sea has collapsed, and climate changes in the Aral Sea Basin have been documented. Hundreds of agreements have been signed since 1980s on programs designed to address the "Aral Sea Problem" which, to date, have not been effective at preventing the continuing shrinking of the sea.

THE ARAL SEA BASIN

The Aral Sea is located in Central Asia and lies between Uzbekistan and Kazakhstan in a vast geological depression, the Turan lowlands, in the Kyzylkum and Karakum Deserts. In the 1950s, the sea covered 66,000 km^2, contained about 1090 km^3 of water, and had a maximum depth of about 70 m. The Aral Sea supported vast fisheries and shipping industries. At that time the sea was fed by two rivers, the Amu Darya (2540 km) and the Syr Darya (2200 km), which originate in the mountain ranges of central Asia and flow through the five republics of Uzbekistan, Kazakhstan, Kyrgyzstan, Tajikistan, and Turkmenistan.

The two rivers provide most of the fresh water used in Central Asia. In the last 50 years, about 20 dams and reservoirs and 60 major irrigation schemes have been constructed. About 82% of river diversions are for agricultural use and 14% is for municipal and industrial use (Table 1).

Water demand due to population growth and industrial expansion continues to increase (Table 2). Since 1960, the population of the Central Asian republics has increased 140% and totals over 50 million. Likewise, industrial production using large amounts of water has also increased. Examples include steel production which rose 200%, cement production by 170%, and electricity generation by a factor of 12.

The total inflows to the Aral Sea began decreasing rapidly in the 1960s, and by 1990 the storage volume of the sea has decreased by 600 km^3 (Table 3). As the water level fell, salinity levels have tripled, rising from about 1000 ppm to just under 3000 ppm today. By the 1980s, as the Aral Sea problem became well known in the Soviet Union, government officials proposed ambitious projects to divert water from other rivers, including ones in South Russia and Siberia, to be transported to the Aral Sea in massive canals. However, these plans died with the breakup of the Soviet Union.

The decrease in sea level has now split the Aral Sea into two separate water bodies: the Small and Large Aral Seas (Maloe More and Bol'shiye More) each separately fed by the Syr Darya and the Amu Darya, respectively. The once vast Amu Darya delta which once covered 550,000 ha has now shrunk to less than 20,000 ha.

IRRIGATION AND COTTON

For thousands of years, Central Asian farmers diverted water from the Amu Darya and Syr Darya Rivers, transforming desert into green oases and supporting great civilizations. Historically, irrigation water use was conducted at a sustainable level. The creation of the Soviet Union and the collectivization of farmlands resulted in the end for traditional agricultural practices. Beginning as early as 1918, Soviet leaders began expanding irrigated land in Central Asia for export and hard currency. Cotton was known as "white gold." The USSR became a net exporter of cotton by the 1930s, and

Table 1 Average water supply and demand in the Aral Sea Basin

Total water available	km^3	%
Amu Darya Basin	84.3	64
Syr Darya Basin	47.8	36
Total	132.1	100
Water Demand		
Agriculture		
Amu Darya Basin	44.8	81.6
Syr Darya Basin	34.6	
Municipal Water		
Amu Darya	3	6.5
Syr Darya	3.3	
Industry		
Amu Darya	3	8.2
Syr Darya	5	
Livestock		
Amu Darya	0.2	0.2
Syr Darya	0	
Fishery		
Amu Darya	2.6	3.5
Syr Darya	0.8	
Total	97.3	100

Table 3 Decline of the Aral Sea during the 1980s and total estimated inflows from the Amu Darya and Syr Darya rivers

Year	Inflows (km^3)	Aral Sea Volume (km^3)	Aral Sea Surface area (km^2)
1911–1960	56.0	1064	66,100
1981	6.0	618	50,500
1982	0.04	583	49,300
1983	2.3	539	47,700
1984	7.9	501	46,100
1986	0.0	424	41,100
1987	9.0		
1988	23.0		41,000
1989		300	30,000

by the 1980s, was ranked fourth in the world in cotton production.

The policy of emphasizing cotton production was accelerated in the 1950s as Central Asia's irrigated agriculture was expanded and mechanized. In 1956, the Kara Kum Canal was opened, diverting one-third of the flow in the Amu Darya to new cultivated areas in the deserts of Uzbekistan and Turkmenistan. The year 1960 represents the critical junction when the Aral Sea began to drop. Irrigated cotton production and water diversions continued to be expanded until the break-up of the Soviet Union (Table 4).

Estimates are that upwards of 80% of the workforce is employed in agriculture. The main agricultural crops in the basin are cotton (6.4 million ha), forage (1.7 million ha), rice (0.4 million ha), and tree crops (0.4 million ha).

Some Central Asian irrigation experts estimate that only 20%–25% of the water diverted from the rivers is actually used by the crops, the rest being lost in the canals that transport the water to the fields and due to inefficient irrigation practices used on-farm. It is believed that over the past decade, adequate maintenance, repair, and renovation of the irrigation infrastructure were not performed at a meaningful level, and water losses from deteriorating canals, gates, and other facilities have increased.

Most land is under furrow irrigation, with drip irrigation accounting for about 5% of the irrigated cropland (used primarily on orchard crops), and sprinkler irrigation accounts for about 3%. Even though the water saving benefits of gated pipe are well known in the region, less than one-sixth of the farms use this technology. Reasons may include costs and product availability. Most farms follow the centuries' old practice of cutting earthen canals with shovels in order to divert water into the field. The volume of water available at these farm ditches is not sufficient to provide an even distribution of water over the field. As a result, water logging and soil salinity now affects about 40% of all the cultivated land in the region.

Table 2 General statistics of the Aral Sea Basin countries in 1995

	Kazakhstan	Uzbekistan	Turkmenistan	Kyrgyzstan	Tajikistan
Area, km^2	2,717,300	447,400	488,100	198,500	143,100
Irrigated land, km^2	23,080	41,500	12,450	10,320	6.940
Population	17,376,615	23,089,261	4,075,316	4,769,877	6,155,474
Population growth rate, %	0.62	2.08	2.5	1.5	2.6

Table 4 Cultivated land along the Amu Darya and Syr Darya rivers

	Before 1917	1960	1980	1992
Millions of hectares	5.2	10	15	18.3

MUYNAK AND ARALSK

Of all the villages affected by the drying of the Aral Sea, Muynak is the best known. Historically, Muynak was located on an island of the vast Aral Sea delta at the convergence of the Amu Darya River in Karakalpakstan (a semi-autonomist republic in Uzbekistan). In 1962, the island became a peninsula. By 1970 the former seaport was 10 km from the sea. The retreat of the sea accelerated and the town was 40 km from the sea by 1980, 70 km in 1995, and close to 100 km today.

Over 3000 fishermen once worked the abundant waters around Muynak which supported 22 different commercial species of fish. In 1957, Muynak fishermen harvested 26,000 tons of fish, about half of the total catch that year taken from the Aral Sea. Muynak also produced 1.1 million farmed muskrat skins which were used to produce coats and hats.

The Kazakhstan city of Aralsk, was once located on the northern edge of the Aral Sea, and like Muynak, had major fisheries and commerce industries. A major shipping and transport industry existed between these two cities. As the Aral Sea skunk, Aralsk found itself farther and farther from the shore which had retreated nearly 129 km by the 1980s. In the early 1990s, a dam was built just to the south of the mouth of the Syr Darya, to protect the northern part of the Aral Sea, letting the southern portion of the Aral Sea evaporate. Although only 10% of the water in the Syr Darya River reaches the northern part of the Aral Sea, the Little Aral has risen 3 m since the construction of the dam, and the shoreline has crept to within 16 km of the town.

ENVIRONMENTAL PROBLEMS

The Aral Sea is an unfortunate example of an old Uzbek proverb: "at the beginning you drink water, at the end you drink poison." As the rivers flow through cultivated areas, they pick up fertilizers, pesticides, and salts from runoff, drainage water and groundwater flow. In the 1960s, it was common for about 550 kg ha^{-1} of chemicals to be applied to cotton fields in Central Asia, compared to an average of 25 kg used for other crops in the Soviet Union. Residues of these chemicals are now found on the dry seabed. Estimates are that millions of tons of dust are picked up from the seabed and distributed over the Aral Sea region.

The Sea may have been contaminated from runoff from by two former USSR military installations in the area. A chemical weapons testing facility was located on the Ust-Jurt Plateau (north shore), and was closed in the mid-1980s. Renaissance Island (Vorzrozhdeniya Island), located in the central Aral Sea, was the site of the former USSR Government's Microbiological Warfare Group which produced the deadly Anthrax virus. Some scientists believe that some containers holding the virus were not properly stored or destroyed. As the Aral Sea continues to dry and water levels recede, the ever-expanding island will soon connect to the surrounding land. Scientists fear that reptiles, including snakes that have been exposed to the various viruses, will move onto the surrounding land and

Fig. 1 This NASA photograph (STS085-503-119) was taken in August 1997 and looks toward the southeast. The Amu Darya River is visible to the right and the Syr Darya on the left. The Aral Sea is now separated into the Small Aral to the north and the Large Aral to the south. Shown are the approximate extent of the Aral Sea in 1957 before a massive expansion of irrigation diversions from the rivers.

possibly infect the humans living around the shores of the Aral Sea.

The Area Sea once supported a complex ecosystem, an oasis in the vast desert. Over 20 species of fish are now extinct. Karakalpakstan scientists believe that a total of about 100 species of fish and animals that once flourished in the region are now extinct, as are many unique plants.

Residents believe that there is a direct correlation between the drying of the sea and changes in climate of the Aral Sea Basin. The moderating effect of the sea has diminished and temperatures are now about 2.5°C higher in the summer and lower in the winter. Rainfall in the already arid basin has decreased by about 20 mm.

THE HUMAN TRAGEDY

Over the last 50 years, there has been a large increase in mortality, illnesses, and poor health in the region. Some estimate that 70%–90% of the population of Karakalpakstan suffer some an environmentally induced malady. Tuberculosis is rampant. Hardest hit are women and children. Common health problems include kidney diseases, thyroid dysfunctions, anemia, bronchitis, and cancers.

CONCLUSION

Some accounts are that since 1984, hundreds of international agreements have been signed to address Aral Sea problems. The early agreements had the goals of first stabilizing the Sea, then slowly increasing flows to restore its ecosystem. In 1992, the Interstate Commission for Water Coordination was formed by the five central Asian republics, which also accepted, in principle, to adhere to the limits on water diversions as set during the Soviet era in 1984 and 1987. To date, however, no progress has been made on stabilizing or reversing the declining inflows. With no water reaching the Aral Sea from the Amu Darya, scientists predict that this portion of the sea (the Large Aral Sea) will disappear by 2020 (Fig. 1).

FURTHER READING

The Aral Sea Homepage: http://www.dfd.dlr.de/app/land/aralsee/

Requiem for a dying sea: http://www.oneworld.org/patp/pap_aral.html/

Disappearance of the Aral Sea: http://www.grida.no/aral/maps/aral.htm

Earth from Space: http://earth.jsc.nasa.gov/categories.html

Boron

Rami Keren

Institute of Soil, Water and Environmental Sciences,
Bet Dagan, Israel

INTRODUCTION

Boric acid is moderately soluble in water. Its solubility increases markedly with temperature due to the large negative heat of dissolution. Boron is considered as a typical metalloid having properties intermediate between the metals and the electronegative nonmetals. Boron has a tendency to form anionic rather than cationic complexes. Boron chemistry is of covalent B compounds and not of B^{3+} ions because of its very high ionization potentials. Boron has five electrons, two in the inner spherical shell ($1s^2$), two in the outer spherical shell ($2s^2$), and one in the dumbbell shaped shell ($2p_x^1$).[1] In the hybrid orbital state, the three electrons in the 2s and 2p orbitals form a hybrid orbital state ($2s^1 2p_x^1 2p_y^1$), where each electron is alone in an orbit whose shape has both spherical and dumbbell characteristics. Each of these three orbits can hold one electron from another element to form a covalent bond between the element and B (BX_3). This leaves one 2p electron orbit that can hold two electrons, which if filled would completely fill the eight electron positions (octet) associated with the second electron shell around B. BX_3 compounds behave as acceptor Lewis acids toward many Lewis bases such as amines and phosphines. The acceptance of two electrons from a Lewis base completes the octet of electrons around B. Boron also completes its octet by forming both anionic and cationic complexes.[1] Therefore, tri-coordinate B compounds have strong electron-acceptor properties and may form tetra-coordinate B structures. The charge in tetra-coordinate derivatives may range from negative to neutral and positive, depending upon the nature of the ligands.

For the unshared oxygen atoms bound to B, they are, probably, always OH groups. Thus, in accordance with the electron configuration of B, boric acid acts as a weak Lewis acid:

$$B(OH)_3 + 2H_2O = B(OH)_4^- + H_3O^+ \qquad (1)$$

The formation of borate ion is spontaneous. The first hydrolysis constant of $B(OH)_3$, K_{h1}, is 5.8×10^{-10} at 20°C,[2] and the other K_{h2} and K_{h3} values are 5.0×10^{-13} and 5.0×10^{-14}, respectively.[3] A dissociation beyond $B(OH)_4^-$ is not necessary to explain the experimental data, at least below pH 13.[4,5] Boron species other than

$B(OH)_3$ and $B(OH)_4^-$, however, can be ignored in soils for most practical purposes. The first hydrolysis constant of $B(OH)_3$ varies with temperature from 3.646×10^{-10} at 178 K to 7.865×10^{-10} at 318 K.[6]

Both $B(OH)_3$ and $B(OH)_4^-$ ion species are essentially monomeric in aqueous media at low B concentration ($\leq 0.025 \, mol \, L^{-1}$). However, at high B concentration, polyborate ions exist in appreciable amount.[7] The equilibria between boric acid, monoborate ions, and polyborate ions in aqueous solution are rapidly reversible. In aqueous solution, most of the polyanions are unstable relative to their monomeric forms $B(OH)_3$ and $B(OH)_4^-$.[8] Results of nuclear magnetic resonance[9] and Raman spectrometry[10] lead to the conclusion that $B(OH)_3$ has a trigonal-planar structure, whereas the $B(OH)_4^-$ ion in aqueous solution has a tetrahedral structure. This difference in structure can lead to differences in the affinity of clay for these two B species.

BORON–SOIL INTERACTION

The elemental form of boron (B) is unstable in nature and found combined with oxygen in a wide variety of hydrated alkali and alkaline earth-borate salts and borosilicates as tourmaline. The total B content in soils, however, has little bearing on the status of available B to plants.

Boron can be specifically adsorbed by different clay minerals, hydroxy oxides of Al, Fe, and Mg, and organic matter.[11] Boron is adsorbed mainly on the particle edges of the clay minerals rather than the planar surfaces. The most reactive surface functional group on the edge surface is the hydroxyl exposed on the outer periphery of the clay mineral. This functional group is associated with two types of sites that are available for adsorption: Al(III) and Si(IV), which are located on the octahedral and tetrahedral sheets, respectively. The hydroxyl group associated with this site can form an inner sphere surface complex with a proton at low pH values or with a hydroxyl at high pH values. The B adsorption process can be explained by the surface complexation approach, in which the surface is considered as a ligand.[12] Such specific adsorption, which occurs irrespective of the sign of the net surface charge, can occur theoretically for any species capable of coordination with the surface metal ions. However, because oxygen is the ligand commonly coordinated to the metal ions in clay minerals, the B species $B(OH)_3$ and $B(OH)_4^-$ are particularly involved in such reactions. Possible surface complex configurations for B—broken edges of clay minerals—were suggested by Keren et al.[12]

Keren and Bingham[11] reviewed the factors that affect the adsorption and desorption of B by soil constituents and the mechanisms of adsorption. Soil pH is one of the most important factors affecting B adsorption. Increasing pH enhances B adsorption on clay minerals, hydroxy-Al and soils, showing a maximum in the alkaline pH range (Fig. 1).

The response of B adsorption on clays to variations in pH can be explained as follows. Below pH 7, $B(OH)_3$ predominates and since the affinity of the clay for this species is relatively low, the amount of adsorption is small. Both $B(OH)_4^-$ and OH^- concentrations are low at this pH; thus, their contribution to total B adsorption is small despite their relatively strong affinity for the clay. As the pH is increased to about 9, the $B(OH)_4^-$ concentration increases rapidly. Since the OH^- concentration is still low relative to the B concentration, the amount of adsorbed B increases rapidly. Further increases in pH result in an enhanced OH^- concentration relative to $B(OH)_4^-$, and B adsorption decreases rapidly due to the competition by OH^- at the adsorption sites. Adsorption models for soils, clays, aluminum oxide, and iron oxide minerals have been derived by various workers.[13–17]

In assessing B concentration in irrigation water, however, the physicochemical characteristics of the soil must be taken into consideration because of the interaction between B and soil. Boron sorption and desorption from soil adsorption sites regulate the B concentration in soil solution depending on the changes in solution B concentration and the affinity of soil for B. Thus, adsorbed B may buffer fluctuations in solution B concentration, and B concentration in soil solution may change insignificantly by changing the soil-water content (Fig. 2). When irrigation with water high in B is planned, special attention should be paid to this interaction because of the narrow difference between levels causing deficiency and toxicity symptoms in plants.

BORON–PLANT INTERACTION

Boron is an essential micronutrient element required for growth and development of plants.

Many of the experimental data suggest that B uptake in plants is probably a passive process. There are clear evidences, however, that B uptake differs among species.[18] Several mechanisms have been postulated to explain this apparent paradox.[18–20] Boron deficiency in plants initially affects meristematic tissues, reducing or terminating growth of root and shoot apices, sugar

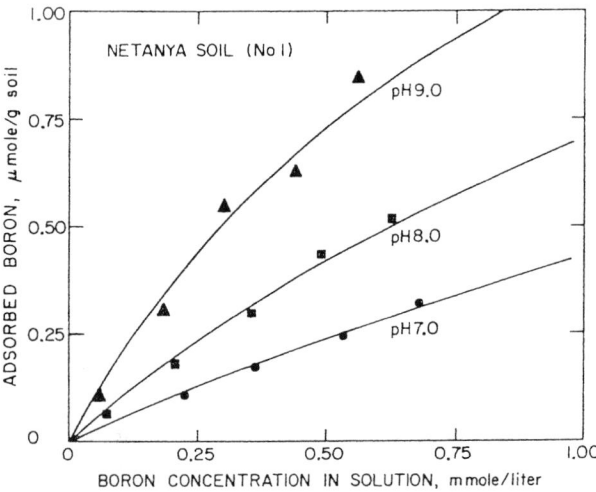

Fig. 1 Boron adsorption isotherms for a soil as a function of solution B concentration and pH. Bold lines—calculated values. (From Ref. [28].)

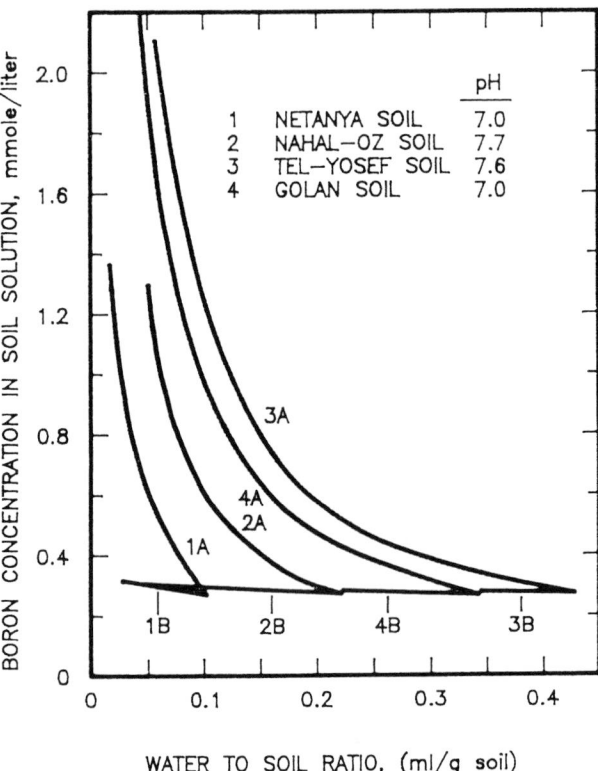

Fig. 2 Boron concentration in soil solution as a function of solution-to-soil ratio for a given total amount of B. (From Ref. [28].) (A) No interaction between B and soil, (B) Boron adsorption account for.

transport, cell-wall synthesis and structure, carbohydrate metabolism and many biochemical reactions.[21,22] Tissue B concentrations associated with the appearance of vegetative deficiency symptoms have been identified in many crop species. It is essential to remember that for B, as for phosphorus and several other plant nutrient elements, deficiency may be present long before visual deficiency symptoms occur.

Excess and toxicity of boron in soils of semi-arid and arid areas are more of a problem than deficiency. Boron toxicity occurs in these areas either due to high levels of B in soils or due to additions of B in irrigation water. A summary of B tolerance data based upon plant response to soluble B is given by Maas.[23] Bingham et al.[24] showed that yield decrease of some crops (wheat, barley, and sorghum) due to B toxicity could be estimated by using a model for salinity response, suggested by Maas and Hoffman.[25]

There is a relatively small difference between the B concentration in soil solution causing deficiency and that resulting in toxicity symptoms in plants.[11] A consequence of this narrow difference is the difficulty posed in management of appropriate B levels in soil solution.

The suitability of irrigation water has been evaluated on the basis of criteria that determine the potential of the water to cause plant injury and yield reduction. In assessing the B in irrigation water, however, the physicochemical characteristics of the soil must be taken into consideration because the uptake by plants is dependent only on B activity in soil solution.[26,27] Boron uptake by plants grown in a soil of low-clay content is significantly greater than that of plants grown in a soil of high-clay content at the same given level of added B (Fig. 3). This knowledge may improve the efficacy of using water of different qualities, whereby water with relatively high B levels could be used to irrigate B-sensitive crops in soils that show a high affinity to B. Such water can be used for irrigation as long as the equilibrium B concentration in soil solution is below the toxic concentration threshold (the maximum permissible concentration for a given crop species that does not reduce yield or lead to injury symptoms) for the irrigated crop. The existing criteria for irrigation water, however, make no reference to differences in soil type.

REFERENCES

1. Cotton, F.A.; Wilkinson, G. *Advanced Inorganic Chemistry*, 5th Ed; Wiley & Sons: New York, NY, 1988.
2. Owen, B.B. The Dissociation Constant of Boric Acid from 10 to 50°. J. Am. Chem. Soc. **1934**, *56*, 1695–1697.
3. Konopik, N.; Leberl, O. Colorimetric Determination of PH in the Range of 10 to 15. Monatsh **1949**, *80*, 420–429.
4. Ingri, N. Equilibrium Studies of the Polyanions Containing B^{III}, Si^{IV}, Ge^{IV} and V^{V}. Svensk. Kem. Tidskr. **1963**, *75*, 199–230.
5. Mesmer, R.E.; Baes, C.F., Jr.; Sweeton, F.H. Acidity Measurements at Elevated Temperature. VI. Boric Acid Equilibria. Inorg. Chem. **1972**, *11*, 537–543.
6. Owen, B.B.; King, E.J. The Effect of Sodium Chloride Upon the Ionozation of Boric Acid at Various Temperatures. J. Am. Chem. Soc. **1943**, *65*, 1612–1620.
7. Adams, R.M. *Boron, Metallo-Boron Compounds and Boranes*; John Wiley & Sons: New York, 1964.
8. Onak, T.P.; Landesman, H.; Williams, R.E.; Shapiro, I. The B^{II} Nuclear Magnetic Resonance Chemical Shifts and Spin Coupling Values for Various Compounds. J. Phys. Chem. **1959**, *63*, 1533–1535.
9. Good, C.D.; Ritter, D.M. Alkenylboranes: II. Improved Preparative Methods and New Observations on Methylvinylboranes. J. Am. Chem. Soc. **1962**, *84*, 1162–1166.
10. Servoss, R.R.; Clark, H.M. Vibrational Spectra of Normal and Isotopically Labeled Boric Acid. J. Chem. Phys. **1957**, *26*, 1175–1178.
11. Keren, R.; Bingham, F.T. Boron in Water, Soil and Plants. Adv. Soil Sci. **1985**, *1*, 229–276.
12. Keren, R.; Grossl, P.R.; Sparks, D.L. Equilibrium and Kinetics of Borate Adsorption–Desorption on Pyrophyllite in Aqueous Suspensions. Soil Sci. Soc. Am. J. **1994**, *58*, 1116–1122.
13. Keren, R.; Gast, R.G.; Bar-Yosef, B. pH-Dependent Boron Adsorption by Na-Montmorillonite. Soil Sci. Soc. Am. J. **1981**, *45*, 45–48.

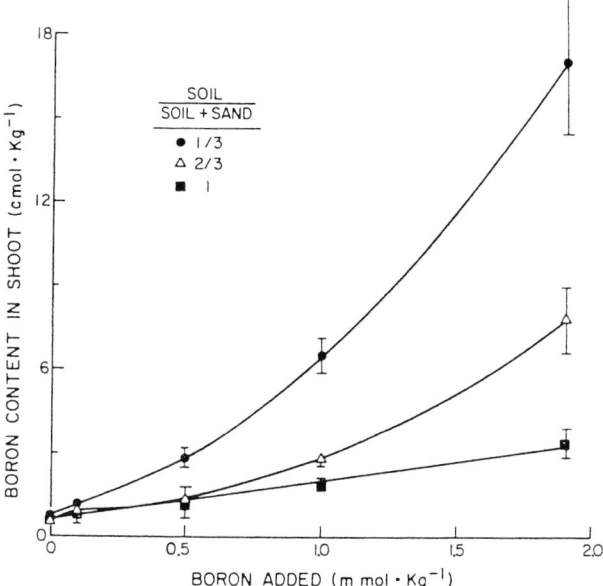

Fig. 3 Relationship between B content in wheat shoot and the amount of B added to soil, for three ratios of soil–sand mixtures. (From Ref. [26].)

14. Keren, R.; Gast, R.G. pH Dependent Boron Adsorption by Montmorillonite Hydroxy-Aluminum Complexes. Soil Sci. Soc. Am. J. **1983**, *47*, 1116–1121.

15. Goldberg, S.; Glaubig, R.A. Boron Adsorption on Aluminum and Iron Oxide Minerals. Soil Sci. Soc. Am. J. **1985**, *49*, 1374–1379.

16. Goldberg, S.; Glaubig, R.A. Boron Adsorption on California Soils. Soil Sci. Soc. Am. J. **1986**, *50*, 1173–1176.

17. Goldberg, S.; Forster, H.S.; Heick, E.L. Boron Adsorption Mechanisms on Oxides, Clay Minerals and Soils Inferred from Ionic Strength Effects. Soil Sci. Soc. Am. J. **1993**, *57*, 704–708.

18. Nable, R.O. Effects of B Toxicity Amongst Several Barley Wheat Cultivars: A Preliminary Examination of the Resistance Mechanism. Plant Soil **1988**, *112*, 45–52.

19. Nable, R.O.; Lance, R.C.M.; Cartwright, B. Uptake of Boron and Silicon by Barley Genotypes with Differing Susceptibilities to Boron Toxicity. Ann. Bot. **1990**, *66*, 83–90.

20. Brown, P.H.; Hu, H. Boron Uptake by Sunflower, Squash and Cultured Tobacco Cells. Physiol. Plant **1994**, *91*, 435–441.

21. Loomis, W.D.; Durst, R.W. Chemistry and Biology of Boron. BioFactors **1992**, *3*, 229–239.

22. Marschner, H. *Mineral Nutrition of Higher Plants*, 2nd Ed.; Academic Press: London, 1995.

23. Maas, E.V. Salt Tolerance of Plants. In *Handbook of Plant Science in Agriculture*; Christie, B.R., Ed.; CRC Press, Inc.: Cleveland, Ohio, 1984.

24. Bingham, F.T.; Strong, J.E.; Rhoades, J.D.; Keren, R. An Application of the Maas–Hoffman Salinity Response Model for Boron Toxicity. Soil Sci. Soc. Am. J. **1985**, *49*, 672–674.

25. Maas, E.V.; Hoffman, G.J. Crop Salt Tolerance—Current Assessment. ASCE J. Irrig. Drainage Div. **1977**, *103*, 115–134.

26. Keren, R.; Bingham, F.T.; Rhoades, J.D. Effect of Clay Content on Soil Boron Uptake and Yield of Wheat. Soil Sci. Soc. Am. J. **1985**, *49*, 1466–1470.

27. Keren, R.; Bingham, F.T.; Rhoades, J.D. Plant Uptake of Boron as Affected by Boron Distribution between Liquid and Solid Phases in Soil. Soil Sci. Soc. Am. J. **1985**, *49*, 297–302.

28. Mezuman, U.; Keren, R. Boron Adsorption by Soils Using a Phenomenological Adsorption Equation. Soil Sci. Soc. Am. J. **1981**, *45*, 722–726.

Canal Automation

Albert J. Clemmens
United States Department of Agriculture (USDA), Phoenix, Arizona, U.S.A.

INTRODUCTION

Canal automation refers to a wide variety of hydraulic structures, mechanical and electronic hardware, communications, and software used to improve the operation of canals that transmit water and deliver it to users. Early canal automation consisted of hydro/mechanical devices used to adjust a single canal gate, with the intent of controlling the adjacent water level or flow rate. These local devices evolved over time to include mechanical/electric controllers and finally to electronic control, although many hydro/mechanical gates are still used successfully. A major shift in canal automation resulted from the use of radio or hardwire communication to control all canal structures from a single location. Today, commercially available supervisory control and data acquisition (SCADA) systems are used for remote, manual supervisory control of canals. The use of a centralized control station (particularly SCADA systems) has also led to the use of computers for the automatic remote control of entire canals. A variety of devices, methods, and control algorithms have been developed for canal automation. These are summarized in Refs. [1,2].

BACKGROUND

The objective of canal operations is to deliver a certain rate or volume of water to a particular location, for example to a reservoir, to a farm canal, etc. Canals differ radically in operation from pressurized pipelines, where users simply open an outlet to receive water. If an outlet to a canal is opened, the flow generally causes the water level (pressure or head) to drop, but only in the vicinity of the outlet. That pressure drop moves upstream only gradually and may never reach the upstream source of the canal. The increase in demand can literally empty the canal. A canal can operate as a demand system only if there is sufficient storage within the canal to handle immediate changes in demand. Even so, if an increase in demand is not matched by an increase in the canal inflow, canal volume and water levels will drop. More often, demand changes are prearranged or scheduled.

Check structures are used in canals to provide a higher head on outlet structures (e.g., users' delivery gates), and a head that is independent of canal flow rate. These check structures usually consist of a series of gates and/or weirs (Fig. 1). Methods to control the flow rates through a canal outlet usually consist of either an automatic flow-rate-controlled outlet (discussed later) or a manually controlled outlet with (manual or automatic) control of the water level upstream from the outlet structure, which if held constant usually provides constant flow (Fig. 2). The later method is the most common approach.

The canal section between two check structures is often called a canal pool. Automatic control methods differ in where within the canal pool the water level is to be controlled, at the upstream end, at the downstream end, or some average value (and associated pool volume). Even when outlet devices have some automatic controls, they generally only maintain constant (or near constant) flow within a range of upstream water levels.

FLOW-RATE CONTROL

Flow-rate control is most frequently applied to the head of a canal. (The outlet of one canal is the head of another.) Automatic control of flow rate at a canal outlet (or headgate) has been accomplished primarily by two methods; hydro/mechanical flow-rate control devices and mechanical/electric or electronic feedback control from a flow-measurement device, where the gate itself can serve as the measuring device. If the measurement device is a weir or flume, then a constant water-level device (discussed in the next section) can be used to adjust the outlet to maintain constant flow. Flow-rate control at canal check structures is also used with some volume and downstream-water-level controllers. However, this assumes that mismatches between inflow and outflow will be adjusted by other control actions upstream. Otherwise, flow-rate control is not sustainable.

UPSTREAM WATER-LEVEL CONTROL

The most common method of manual canal control is upstream water-level control, where check gates are adjusted to maintain a constant water level on the upstream side of each check structure. Canal inflow is set to match

Fig. 1 Check structure with motorized check gates and a manual outlet gate shown as part of SCADA automatic control screen.

the demand, usually manually. Upstream flow changes are automatically passed through each check structure as it maintains its upstream level. However, upstream water-level control will cause all errors in canal inflow and

outflow to move to the downstream end of the canal, causing either shortages or surpluses there, regardless of the type of automatic control. In addition, a series of automatic upstream water-level controllers can, if not properly set or adjusted, cause the flow rate at the downstream end of the canal to oscillate.

A variety of automatic methods have also been developed for upstream control. The simplest is a duckbill weir, which is a very long fixed (no moving parts) weir, where the change in water level for a large change in flow is very small (a decrement). Neyrpic gates use a float on the upstream side of the gate to adjust gate position to maintain a constant upstream level, again with a decrement (Fig. 3). Several other hydro/mechanical gates have also been used, e.g., controlled leak gates. More details and references can be found in Ref. [2]. Electrical/mechanical devices have also been used to maintain constant upstream water levels.[3] In general, these have not proven to be reliable and have been

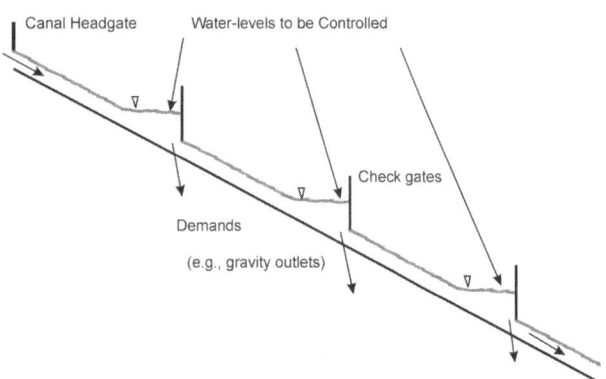

Fig. 2 Profile drawing of canal with check gates and outlets.

essentially replaced with electronic devices. Programmable logic controllers (PLCs) and remote terminal units (RTUs) have been used to maintain constant upstream levels locally at the structure with PI or PID type logic (proportional, integral, derivative). Such control can usually be conveniently programmed into SCADA software for remote operation.

DOWNSTREAM WATER-LEVEL CONTROL

Under downstream control, the controller adjusts the check gate to maintain the water-level downstream from the check structure. It is far easier to control the level immediately below the check structure; however, in general outlets are located at the downstream end of each canal pool, making the level to be controlled far from the check structure. This complicates the control and makes downstream control difficult to apply without a thorough control-engineering approach. Early attempts at downstream water-level control adjusted each check structure based on one downstream water level[3]—as a series of local controllers (Fig. 4). This has proven not to be very effective, and a more centralized approach can dramatically improve control.[4]

CONSTANT VOLUME CONTROL

Canal water levels and volumes are related. If precise control of water levels is not critical, it is sometimes easier to control the canal pool volumes. Volume control methods measure one or more water levels and convert this to a pool volume. When the pool volume deviates from a target, a volume error is determined, and an adjustment is made to the pool inflow rate. This change in pool inflow can be computed pool by pool with simple logic (such as Bival[1]) or can be determined from a centralized perspective (such as Dynamic Regulation[1]). The rate of change of pool volume can also be used to determine the difference in pool inflow and outflow, and pool inflow volume can then be adjusted to bring inflow into balance with outflow. Target pool volumes can also be varied to provide more balanced control of the canal, e.g., all pools reduced in volume in the response to canal inflow limitations. Volume control is very effective for the control of pools with pumped outlets that are not very sensitive to level. This control method is not as effective for gravity outlets, unless the outlets themselves have automatic controls.

ROUTING DEMAND CHANGES (GATE STROKING)

As discussed above, demand changes in a canal cannot be handled strictly by feedback because downstream water-level response may never reach the upstream end of the canal. Thus, most canal flow changes are prescheduled. Flow changes made at the head of a canal arrive at downstream outlets at some later time. Knowledge of this time delay allows operators to schedule a change in flow at the canal headgate so that the flow change will arrive at the outlet gate at the desired time. Unfortunately, the sudden flow change made upstream arrives only gradually at the downstream checks, making the exact timing difficult to predict. A variety of schemes have been developed to compute these flow change schedules automatically. Some use numerical methods to solve the governing equations of flow. These have proven to be unreliable and difficult to implement. More simple, volume-based procedures have proven more effective.

Fig. 3 Amil gate for constant upstream level control.

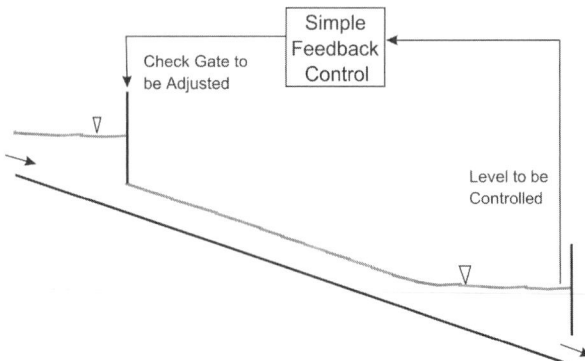

Fig. 4 Schematic drawing of downstream water-level control for a single canal pool.

CENTRALIZED CONTROL

Remote monitoring of a canal, or a canal network, from a central location allows an operator to provide more timely control at check gates than if that control were done by the same individual locally, traveling from check to check. Supervisory control and data acquisition systems provide automatic data collection (including communication with remote sites), display, and archiving. In addition, all of the other automatic control feature discussed earlier can be implemented from a central control station. This has some advantages in terms of reliability, accountability, and safety. Some functions can be performed locally, but in general control is improved if they are based on operations from a centralized perspective.

REFERENCES

1. Malaterre, P.-O.; Rogers, D.C.; Schuurmans, J. Classification of Canal Control Algorithms. J. Irrig. Drain. Eng. **1998**, *124* (1), 3–10.
2. Rogers, D.C.; Goussard, J. Canal Control Algorithms Currently in Use. J. Irrig. Drain. Eng. **1998**, *124* (1), 11–15.
3. Rogers, D.C.; Ehler, D.G.; Falvey, H.T.; Serfozo, E.A.; Voorheis, P.; Johansen, R.P.; Arrington, R.M.; Rossi, L.J. *Canal Systems Automation Manual*; U.S. Bureau of Reclamation: Denver, CO, 1995; Vol. 2.
4. Clemmens, A.J.; Anderson, S.S., (Eds.) *Modernization of Irrigation Water Delivery Systems*, Proc. USCID Workshop, Scottsdale, AZ, October 17–21, 1999; USCID: Denver, CO, 1999; 714.

Carbon Dioxide, Plants and Transpiration

L. H. Allen, Jr.
J. C. V. Vu
United States Department of Agriculture (USDA),
Gainesville, Florida, U.S.A.

J. Sheehy
International Rice Research Institute, Metro Manila, Philippines

INTRODUCTION

Atmospheric carbon dioxide (CO_2) concentration has varied throughout the life history of the earth. During the last two million years of cyclic formation and partial melting of ice caps, the CO_2 concentration has ranged from as low as about 180 ppm (mole fraction basis) during the coldest periods to as high as about 300 ppm during interglacial warm periods.[1] The atmospheric CO_2 concentration increased from about 280 ppm in pre-industrial times to 315 ppm in 1958 when the first careful continuous measurements were made at Mauna Loa, Hawaii.[2] Since then, CO_2 has continued to increase and the concentration is about 370 ppm currently. This increase has been due primarily to burning of fossil fuels and secondarily to deforestation and land-use changes. This increase of CO_2 (and other greenhouse-effect gases) is expected to cause global warming and other climate changes, but rising CO_2 will also affect plants directly.

Carbon dioxide is the first molecular link in the food chain of most life on Earth. Through the process of photosynthesis in green plants, carbon of CO_2 is incorporated into simple sugars which enter eventually into other biochemical reactions in the creation of living matter. There are three primary types of photosynthetic metabolism pathways used by green plants; C_3, C_4, and crassulation acid metabolism (CAM). The most abundant in terms of number of plants is the C_3 photosynthetic biochemical pathway, in which CO_2 binds onto the enzyme ribulose 1,5-bisphosphate carboxylase/oxygenase (Rubisco) in the chloroplasts, and enters into a biochemical cycle in which a 3-carbon sugar, phospho-glyceric acid, is the first product. Wheat (*Triticum aestivum* L.), rice (*Oryza sativa* L.), pulses (various bean and pea species), and most vegetable crops, fruit crops, and trees have the C_3 pathway of photosynthesis. In the C_4 photosynthetic pathway plants, CO_2 first binds with phosphoenolpyruvate carboxylase (PEPcase) in mesophyll cells of leaves, thereby forming a 4-carbon sugar (e.g., malic acid) that is translocated to bundle sheath cells

surrounding vascular tissue. Photosynthesis is then completed via the C_3 pathway. The most familiar C_4 plants are maize (*Zea mays* L.), sugarcane (*Saccharum officinarium* L.), sorghum (*Sorghum bicolor* [L.] Moench), and many tropical and subtropical grass species. The CAM plants typically open their stomata at night and take up CO_2 by incorporating it into phosphoenolpyruvate via PEPcase and sequestering CO_2 as 4-carbon compounds such as malic acid.[3] As a group, CAM plants have less economic importance than C_3 and C_4 plants, but they are prevalent in arid and semiarid tropical zones. Pineapple (*Ananas comosus* [L.] Merr.) is the best known CAM plant, but *Agave* species are used for leaf fibre and prickly pear cactus (*Opuntia ficus-indica* [L.] P. Mill) and other *Opuntia* species are important as feed material for cattle and other uses.

Photosynthetic plants are found in habitats ranging from aquatic to xerophytic environments. Responses to increasing atmospheric CO_2 can vary among these habitats, so the focus of this discussion will be on mesophytic terrestrial green plants where rising atmospheric CO_2 and predicted concomitant climate changes are likely to have important impacts.

PLANT GAS EXCHANGE

Photosynthesis

The quantitative effects of CO_2, light, and temperature on leaf photosynthetic carbon assimilation were established about 100 years ago.[4] Plant scientists developed renewed interest in these effects, especially CO_2 effects, when it became apparent that carbon dioxide concentration of the atmosphere was increasing. Carbon dioxide has two direct effects on plant leaves. Increased CO_2 both increases rates of photosynthesis and decreases stomatal conductance of CO_2 and water vapor. Studies of photosynthetic rates of most C_3 plants show that leaves increase their uptake of CO_2 when concentrations are increased to at least

Encyclopedia of Water Science
DOI: 10.1081/E-EWS 120010285

1000 ppm and beyond. Increases in CO_2 uptake of C_4 plants rise rapidly and flatten out above 400 ppm because of their CO_2-concentrating mechanism. Using a Michaelis–Menten type rectangular hyperbola model fitted to a set of C_3 plant, soybean (*Glycine max* [L.] Merr.) responses to a wide range of CO_2 concentrations, Allen et al.[5] predicted that canopy net photosynthetic rates would increase about 53% with a doubling of CO_2 concentration. Increases in leaf photosynthetic rates as external CO_2 concentrations increase result from higher influx rates of CO_2 through the stomata of leaves and on to the sites of carboxylation in the leaf chloroplasts.

Transpiration

Leaves are porous inside, occupied by cells with air gaps and passageways to the leaf stomata. Transpiration is the process whereby liquid water evaporates from the surface of cells within the leaf into the leaf air-space and diffuses to the outside through stomata. Stomata are pores on the surface of the leaf that are formed by two guard cells. These guard cells distort to cause the stomata to open in the presence of light and decreased intercellular leaf concentration of CO_2.

Morison[6] reported that leaf stomatal conductance of many plants decreased about 40% with a doubling of atmospheric CO_2 concentration. However, most experiments have shown that whole plants or plant communities decrease transpiration rate (TR) only to about 10% with a doubling of CO_2. The explanation of this difference is related to the energy balance in a real-world environment. As the stomatal conductance decreases with increasing CO_2 concentration, the TR decreases. However, as TR decreases, leaf temperature will begin to rise because of less evaporational cooling. As leaf temperature rises, the vapor pressure of water inside the leaves increases, and thus increases the driving force for evaporation of water. The resultant effect of these processes is that, although stomatal conductance decreases as CO_2 concentration is increased, the energy balance feedback effects cause the vapor pressure to increase, and thereby largely counter-balances the expected reduction in transpirational water use by plants, as described below.

Energy Balance of a Leaf

The coupled energy and mass exchange of a leaf can be expressed in an energy balance equation:

$$R_n = H + \ell E + \lambda \ell$$

where R_n = net radiation flux density; H = sensible heat flux density; ℓE = latent heat flux density, where ℓ is the latent heat of evaporation and E is the water vapor flux density in mass units; $\lambda \ell$ = photochemical heat flux density, with λ being the heat of CO_2 fixation and ℓ being the CO_2 flux density in mass units. Note that $\lambda \ell$ can be ignored to simplify the energy balance equation.

Sensible heat flux density can be expressed in a resistance and temperature difference form, and latent heat flux density can be expressed in a resistance and vapor pressure difference form, as:

$$R_n = \rho C_p (T_L - T_A)/r_{a,h}$$

$$+ (\rho C_p/\Gamma)(e*[T_L] - e_a)/(r_{a,v} + r_{s,v})$$

where ρ = air density; C_p = air heat capacity; T_L = leaf temperature; T_A = ambient air temperature; $r_{a,h}$ = aerodynamic boundary layer resistance for heat transfer; Γ = psychometric constant for converting vapor flux density to energy flux density; $e*[T_L]$ = saturation vapor pressure at leaf temperature; e_a = ambient air vapor pressure; $r_{a,v}$ = aerodynamic boundary layer resistance for vapor exchange and $r_{s,v}$ = stomatal diffusion resistance for water vapor.

Decreasing stomatal conductance caused by elevated CO_2 would decrease latent heat flux density and create an unbalance in the energy balance equation. (Note: $r_{s,v}$ is the reciprocal of stomatal conductance.) Evaporative cooling would decrease, but this would cause T_L and $e*[T_L]$ to increase and thus increase the driving force for transpiration. The resultant effect is that a 40% decrease in leaf conductance will result in about 10% decrease in whole-crop transpiration.

Water-Use Efficiency (WUE)

WUE is defined here as the ratio of net photosynthetic CO_2 exchange rate (CER) to TR.

$$\text{WUE} = \text{CER}/\text{TR}$$

The WUE is impacted by increasing atmospheric CO_2 in two primary ways. First, decrease of stomatal conductance caused by the effects of increasing CO_2 will decrease TR to a limited extent. Secondly, and more importantly, elevated CO_2 increases plant photosynthetic CER. The ratio of WUE at elevated CO_2 (WUE_e) to ambient CO_2 (WUE_a) is given by:

$$\text{WUE}_e/\text{WUE}_a = (\text{CER}_e/\text{CER}_a)(\text{TR}_a/\text{TR}_e)$$

$$= (1 + \Delta\text{CER}/\text{CER}_a)(1 + \Delta\text{TR}/\text{TR}_e)$$

where ΔCER and ΔTR are the differences, $\text{CER}_e - \text{CER}_a$ and $\text{TR}_a - \text{TR}_e$, respectively. Furthermore, the percentage contribution of CER changes to WUE ratios is:

$$(\Delta CER/CER_a)/(\Delta CER/CER_a + \Delta TR/TR_e) \times 100$$

and the percentage contribution of TR changes to WUE ratio is:

$$(\Delta TR/TR_e)/(\Delta CER/CER_a + TR/TR_e) \times 100$$

Allen[7] reported that the contribution of CER to the increase in WUE ratio ranged from 60% to 90% for several C_3 plants, but was only about 25% for a C_4 plant (maize).

Respiration

There have been reports of direct suppression of plant respiration by elevated CO_2. However, this effect is not always found. Regardless of the putative direct effect of CO_2 on respiration, plants grown in elevated CO_2 are frequently found to have greater respiration rates because they are larger plants. Probably the best statement is that respiration is directly proportional to the amount of plant nitrogen or the amount of proteinaceous metabolic components of the plant.

PLANT GROWTH AND DEVELOPMENT

Vegetative Growth Responses

Most plants, both C_3 and C_4 species, show an increase in vegetative growth when exposed to elevated CO_2.[8,9] The

largest responses seem to be from legume species that can fix their own nitrogen symbiotically. These increases are generally 30% or greater for C_3 species (Table 1) and about 10% for C_4 species. The exact nature of the biomass growth responses will depend on how the study is implemented and the state in the life cycle when the measurements are made. The greatest relative responses occur in isolated plants in early stages of growth.[9,10] In cropping systems, the space available for individual plant expansion becomes limited when complete ground cover is achieved, and responses to elevated CO_2 become somewhat limited by available light per unit land area.

Plants generally accumulate photoassimilates (non-structural carbohydrates) in leaves and other vegetative structures under elevated CO_2 treatments. In addition, some plants acclimate to elevated CO_2 by down-regulating the synthesis of Rubisco, which can lead to less response to elevated CO_2. The combination of these two effects tends to increase the carbon-to-nitrogen ratio in the aboveground biomass. The acclimation response to elevated CO_2 has been clearly shown for rice, whereas soybean has little down-regulation response.[11]

Reproductive Growth and Seed Yield

Reproductive growth and seed yield increase with increasing CO_2, but generally not as much as biomass accumulation. Most crop plants do not appear to be adapted to take full advantage of the increased photosynthetic rates under elevated CO_2. For example, Allen and Boote[12] reported that soybean seed yields

Table 1 Total biomass yield and seed yield response of soybean plants (21 plants per square meter at final harvest) to temperature and CO_2 with day/night, maximum/minimum cycles of 28/18, 32/22, 36/26, 40/30, 44/34, and 48/38 °C. The CO_2 concentrations were maintained at 700 ppm for each temperature treatment and at 350 ppm for 28/18 and 40/30 °C

CO_2 concentration (ppm)	Temperature, day maximum/night minimum (°C)					
	28/18	32/22	36/26	40/30	44/34	48/38
Biomass yield (grams per plant)						
700	23.0	24.1	27.0	25.5	26.1	1.7
350	15.1			16.6		
Ratio	1.52			1.54		
Seed yield (grams per plant)						
700	10.0	11.4	12.3	8.7	0.5	0.0
350	7.6			6.7		
Ratio	1.32			1.30		
Harvest index						
700	0.43	0.47	0.45	0.34	0.02	
350	0.50			0.40		
Ratio	0.86			0.85		

Source: Allen and Boote.[12]

increased by about 30% with doubled CO_2 concentration although total biomass increased by about 50% (Table 1). Thus, the harvest index (the ratio of seed yield to total above-ground biomass accumulation) was decreased from 0.50 at 350 ppm to 0.43 at 700 ppm for a harvest index ratio of 0.86 when plants were grown at day/night maximum/minimum temperatures of 28/18 °C. When grown at 40/30 °C, harvest index values were lower with 0.43 at 350 ppm and 0.34 at 700 ppm, with a ratio of 0.85. In general, the evidence for $CO_2 \times$ temperature interaction on either vegetative biomass accumulation or seed yield has been weak, although there are some reports of a strong interaction.[13,14]

TEMPERATURE AND DROUGHT EFFECTS

Along with the direct effects of increasing atmospheric CO_2, there may be indirect effects on plants because of associated global warming and other climatic changes. First, increasing temperatures would increase the TR of plants. Both experimental data and crop models indicate that TR would increase from about 4% to 8% per 1 °C rise in temperature (the specific increase depends on other environmental and plant conditions). The Third Assessment Report of the Intergovernmental Panel on Climate Change (IPCC) predicts that global warming could be 1.4EC–5.8 °C between now and 2100.[15] Using a midrange value of 6% increase in TR per 1 °C, these data indicate that TR could increase from 8% to 35%, depending on the global warming scenario. Thus, global warming might override any savings in plant water use that would arise from a doubling of CO_2 concentration. On a local scale, drought is harder to predict, but decreases in rainfall could make climate change stresses on plants even more severe.

Increases in temperature would generally not cause serious reductions in photosynthesis or vegetative biomass growth. However, increases in temperature can cause serious problems with reproductive development and seed yield (Table 1). Generally, pollen development and pollination (fertilization) are decreased seriously by temperature increases above the optimum for seed production.[16] In general, seed productivity appears to decrease about 10% per °C from the optimum temperature to essentially zero seed production at about 10EC above the optimum. Again, using the IPCC estimates of 1.4 °C–5.8 °C warming, seed productivity might be decreased from 14% to 58% depending on the severity of global warming.

Recently, the International Rice Research Institute, Philippines[17] found that rice cultivars carry on pollination processes over a 9–12-hr period during the daytime, and that "early pollinators" are more successful than "late pollinators" in rice seed production under elevated daytime temperatures. Plant selection and incorporation of early morning pollination is thus a strategy that might be employed in adapting to global warming that might be induced by increasing atmospheric CO_2 concentration.

CONCLUSIONS

Increasing atmospheric CO_2 concentration will likely increase plant photosynthesis, growth, and seed productivity. A doubling of CO_2 is expected to increase seed yields about 30% and decrease transpirational water use about 10%. However, global warming and climatic changes could alter this scenario (Table 1). Predicted levels of global warming might increase plant transpiration by 8–35% and override the small water savings due to decreased stomatal conductance. Furthermore, predicted levels of global warming could decrease seed yield by 14–58%, unless progress can be made in plant selections to avoid the detrimental effects of high temperature on pollination, seed development, and growth processes.

REFERENCES

1. Carbon Dioxide Information Analysis Center, CDIAC, http://cdiac.esd.ornl.gov/home.html (accessed October 2001).
2. Pales, J.C.; Keeling, C.D. The Concentrations of Atmospheric Carbon Dioxide in Hawaii. J. Geophys. Res. 1965, 70 (24), 6053–6076.
3. Nobel, P.S. Crop Ecosystem Responses to Climatic Change: Crassulacean Acid Metabolism Crops. In Climate Change and Global Crop Productivity; Reddy, K.R., Hodges, H.F., Eds.; CABI Publishing: Wallingford, Oxon, United Kingdom, 2000; 315–331.
4. Blackman, F.F.; Matthaei, G.L.C. Experimental Researches in Vegetable Assimilation and Respiration. IV. A Quantitative Study of Carbon-Dioxide Assimilation. Proc. R. Soc. Lond. Ser. B 1905, 198, 402–460.
5. Allen, L.H., Jr. Boote, K.J.; Jones, J.W.; Jones, P.H.; Valle, R.R.; Acock, B.; Rogers, H.H.; Dahlman, R.C. Response of Vegetation to Rising Carbon Dioxide: Photosynthesis, Biomass, and Seed Yield of Soybean. Global Biogeochem. Cycles 1987, 1 (1), 1–14.
6. Morison, J.I.L. Sensitivity of Stomata and Water Use Efficiency to High CO_2. Plant Cell Environ. 1985, 8 (6), 467–474.
7. Allen, L.H., Jr. Carbon Dioxide Increase: Direct Impacts on Crops and Indirect Effects Mediated Through Anticipated Climatic Changes. In Physiology and Determination of Crop Yield; Boote, K.J., Bennett, J.M., Sinclair, T.R., Paulsen, G.M., Eds.; ASA-CSSA-SSSA: Madison, Wisconsin, 1994; 425–459.

8. Kimball, B.A. Carbon Dioxide and Agricultural Yield: An Assemblage and Analysis of 430 Prior Observations. Agron. J. **1983**, *75* (5), 779–788.

9. Poorter, H. Interspecific Variation in the Growth Responses of Plants to an Elevated Ambient CO_2 Concentration. Vegetatio **1993**, *104/105*, 77–99.

10. Poorter, H. Interspecific Variation in the Growth Responses of Plants to an Elevated Ambient CO_2 Concentration. In *CO_2 and Biosphere*; Rozema, J., Lambers, H., van de Geijn, S.C., Cambridge, M.L., Eds.; Kluwer Academic Publishers: Dordrecht, 1993; 77–99.

11. Vu, J.C.V.; Allen, L.H., Jr. Boote, K.J.; Bowes, G. Effects of Elevated CO_2 And Temperature on Photosynthesis and Rubisco in Rice and Soybean. Plant Cell Environ. **1997**, *20* (1), 68–76.

12. Allen, L.H., Jr.; Booke, K.J. Crop Ecosystem Responses to Climatic Change: Soybean. In *Climate Change and Global Crop Productivity*; Reddy, K.R., Hodges, H.F., Eds.; CABI Publishing: Wallingford Oxon, United Kingdom, 2000; 133–160.

13. Kimball, B.A.; Mauney, J.R.; Nakayama, F.S.; Idso, S.B. Effects of Increasing Atmospheric CO_2 on Vegetation. Vegetatio **1993**, *104/105*, 65–75.

14. Kimball, B.A.; Mauney, J.R.; Nakayama, F.S.; Idso, S.B. Effects of Increasing Atmospheric CO_2 on Vegetation. In *CO_2 and Biosphere*; Rozema, J., Lambers, H., van de Geijn, S.C., Cambridge, M.L., Eds.; Kluwer Academic Publishers: Dordrecht, 1993; 65–75.

15. Schneider, S.H. What Is "Dangerous" Climate Change? Nature **2001**, *411* (6833), 17–19.

16. Prasad, P.V.V.; Craufurd, P.Q.; Kankani, V.G.; Wheeler, T.R.; Boote, K.J. Influence of High Temperature During Pre- and Post-anthesis Stages of Floral Development on Fruit-Set and Pollen Germination in Peanut. Aust. J. Plant Physiol. **2001**, *28* (3), 233–240.

17. Sheehy, J.; Elmido, A.; Mitchell, P. Are There Time-of-Day Clock Genes for Flowering? In *Annual Meeting Abstracts 2001*. Annual Meetings of the American Society of Agronomy, Charlotte, NC, Oct. 21–25, 2001; American Society of Agronomy: Madison, Wisconsin, 2001. (J.Sheehy@cigar.org).

FURTHER READING

Advances in Carbon Dioxide Effects Research; Allen, L.H., Jr., Kirkham, M.B., Olszyk, D.M., Whitman, C.E., Eds.; American Society of Agronomy Special Pub. No. 61; ASA-CSSA-SSSA: Madison, Wisconsin, 1997.

Global Climate Change and Agricultural Production; Bazzaz, F., Sombroek, W., Eds.; John Wiley and Sons: Chichester, England, 1996.

Physiology and Determination of Crop Yield; Boote, K.J., Bennett, J.M., Sinclair, T.R., Paulsen, G.M., Eds.; ASA-CSSA-SSSA: Madison, Wisconsin, 1994.

Agricultural Dimensions of Global Climate Change; Kaiser, H.M., Drennen, T.E., Eds.; St. Lucie Press: Delray Beach, Florida, 1993.

Water Use in Crop Production; Kirkham, M.B., Ed.; The Haworth Press, Inc.: New York, 1999.

Carbon Dioxide and Terrestrial Ecosystems; Koch, G.W., Mooney, H.A., Eds.; Academic Press, Inc.: San Diego, 1996.

National Research Council. *Managing Water Resources in the West Under Conditions of Climate Uncertainty*; National Academy Press: Washington, DC, 1991.

Climate Change and Rice; Peng, S., Ingram, K.T., Neue, H.-U., Ziska, L.H., Eds.; Springer-Verlag: Berlin, 1995.

Climate Change and Global Crop Productivity; Reddy, K.R., Hodges, H.F., Eds.; CABI Publishing: Wallingford, Oxon, United Kingdom, 2000.

CO_2 and Biosphere; Rozema, J., Lambers, H., van de Geijn, S.C., Cambridge, M.L., Eds.; Kluwer Academic Publishers: Dordrecht, 1993.

Climate Change and the Global Harvest; Rosenzweig, C., Hillel, D., Eds.; Oxford University Press: New York, 1998.

Climate Change and U.S. Water Resources; Waggoner, P.E., Ed.; John Wiley and Sons: New York, 1990.

Food, Climate, and Carbon Dioxide; Wittwer, S.H., Ed.; Lewis Publishers, CRC Press, Inc.: Boca Raton, Florida, 1995.

Biotic Feedbacks in the Global Climatic System; Woodwell, G.M., Mackenzie, F.T., Eds.; Oxford University Press: New York, 1995.

Chemical Measurement

William J. Rogers
West Texas A&M University, Canyon, Texas, U.S.A.

Kevin Jeanes
ASK Laboratories, Inc., Amarillo, Texas, U.S.A.

INTRODUCTION

The analysis of water has become a common and important task for industry, municipalities, and agriculture in response to today's increased public awareness and participation in conserving, protecting, and improving the quality of water resources. Analytical chemistry methods are used for the identification of one or more analytes or constituents and properties in a sample and the determination of the concentration of these components. The identification of components in water is called a qualitative analysis while the determination of relative amounts is a quantitative analysis.

Quantitative analyses provide data describing the quantity of the analyte in a measured amount of sample. The results of environmental analysis for water are commonly expressed in relative terms as parts of analytes per unit of sample, such as percent (%), parts per million (ppm), milligrams per liter ($mg\,L^{-1}$) or parts per billion (ppb), and micrograms per liter ($\mu g\,L^{-1}$). With recent improved technology, parts per trillion detections can be obtained for some constituents. Special methods allow detections of individual molecules in some cases.

DATA QUALITY

To produce quality data the sample must accurately reflect the matrix or media from where it was collected. Sampling is more efficient in liquid than solid matrices, such as soil, due to the chemical properties of water and the homogeneity of solutes. However, the increased risk to biological systems and public health associated with contaminated water complicates the decision of which methods will provide the data necessary to make informed decisions about the quality of the water, the protection of human health, and the protection of the environment. Data Quality Objectives (DQO) must be formulated that outline considerations such as regulatory compliance and guidance, accuracy, target analytes, required analytical method performance, availability and reliability of field measurement, number of samples needed to be analyzed,

and cost of analysis. In summary, DQOs can be defined as those sampling and analytical objectives that provide the number of samples and the quality of results needed to satisfy the decision making process. Because the success or failure of an analysis is often critically dependent on the proper method selection, the decision of which analytical method to request is difficult and is only made easier with experience. A good working relationship with a qualified laboratory can prove an invaluable asset to the data collector.

The analysis of water can be divided into common groupings, organic and inorganic analytes, which require varied analytical methods. A clear establishment of DQOs to support the decision making process requires an understanding of those methods and the methods which meet the regulatory requirements. The U.S. Environmental Protection Agency has developed a series of documents that outline the DQO process which can be accessed on the internet.[1–3] Gilbert[4] has published a very useful handbook, "*Statistical Methods for Environmental Pollution*" which addresses statistical implications of sampling and data quality analysis.

Before developing a sampling plan, due to the large number of methods and variability of techniques, care must be taken to determine the accuracy and sensitivity requirements for the application and then determine holding times, preservatives, sample size, and containers required by the method chosen to meet the DQO. A listing of approved methods for National Pollutant Discharge Elimination System (NPDES) permits including sample containers, preservatives, and holding times is available in 40 CFR part 136 and can be easily downloaded from the web by following the links from the site www.access.gpo.gov/nara/cfr. The regulation, which governs the activity that is being analyzed or monitored, will provide guidance and identify the method required to meet regulatory compliance. It is important to carefully research the project requirements, because several analytes have no holding time (such as pH; refer to pH entry) and must be analyzed in the field. Field methods offer flexibility and reduced cost; however, accuracy is still dependent on the equipment, personnel, and the quality assurance plan that

Encyclopedia of Water Science
DOI: 10.1081/E-EWS 120010183

outlines training, calibration, maintenance, and procedure. Due to the large number of methods and complexity of their quality assurance requirements, a qualified laboratory should be consulted to determine which method will yield data of sufficient quantity and quality to meet the DQOs.

ANALYTICAL METHODS AND PROTOCOLS

It is important to understand that the various regulations require specific analytical methods. Within these regulations, different methods maybe required based on the media type (soil, water, sewage, etc.) Specific methods [5–8] have been established by the EPA, professional organizations, and even the state agencies for these various matrices (i.e., soil, water, and sewage). EPAs SW846 methods[5] for the evaluation of solid waste are used for the evaluation of solid waste which includes water and are used in this discussion because they are commonly used in water analysis and the methods are easily downloaded from the web at (http://www.epa.gov/epaoswer/hazwaste/test/main.htm). EPA also published EPA-600[6] series for water and EPA-500[7] series for drinking water, but the method's advantages and disadvantages remain the same for each series of methods. It is important to note that subtle differences do exist for method procedures and quality control and the method series prescribed in the regulations should be utilized.

ANALYTICAL METHODS FOR INORGANIC CONSTITUENTS

The analytical methods for inorganic constituents in water for environmental monitoring are commonly subdivided into wet chemistry and metals. These methods are used for analyzing nutrients and elemental analytes, and are associated with water quality parameters and metal contamination. These methods are usually requested for waters that are accepting treated municipal and industrial effluents or that have suspected impacts from sewage, agriculture, and industry.

WET CHEMISTRY

Wet chemistry methods are the classical bench methods and include common water quality parameters [i.e., pH, biochemical demand (BOD), hardness, and alkalinity] utilizing colorimetric, potentiometric, gravimetric, titration, and chromatographic determinations. These methods have varying degrees of sensitivity and accuracy.

METALS

Techniques for the analysis of trace-metal concentrations include direct-aspiration or flame atomic absorption spectrometry (FLAA), graphite-furnace atomic absorption spectrometry (GFAA), inductively coupled argon plasma atomic emission spectrometry (ICP-AES), inductively coupled argon mass spectrometry (ICP-MS), and cold-vapor atomic absorption spectrometry (CVAA). Each of these methods has advantages and disadvantages that should be addressed before selection of an analytical procedure.

FLAA

FLAA is the most common method utilized by small commercial laboratories, municipalities, and industrial laboratories because of the affordability and ease of operation. FLAA commonly uses an acetylene/air flame as an energy source for dissociating the aspirated sample into the elemental state enabling the analyte to absorb light from a specific wavelength. Since each element has to be analyzed separately using that metal's specific wavelength, there is reduced risk for matrix interference. The sensitivity is usually acceptable for most applications, but currently technological improvements are lowering allowable analyte limits that will strain the FLAA capabilities. SW846-7000[5] outlines the general method and associated digestions, interferences, and sensitivity.

GFAA

GFAA replaces the flame with a heated graphite furnace that allows the experienced analyst to remove matrix interferences and concentrate the analyte of concern by using temperature profiles and matrix modifiers. This method requires more exacting analyst intervention and interpretation making it more difficult than FLAA. The method does increase sensitivity and when coupled with ZEEMAN has a very low background interference. GFAA greatly enhances the ability to analyze Selenium and Arsenic that can prove difficult to analyze with other methods. The method has the potential of positive interferences from memory effect, smoke producing matrices, organic materials, and carbide formation and is extremely dependent on the skill of the analyst. GFAA requires a stringent QC program to ensure that matrix interferences have no adverse effect on the analyte of concern. SW846-7000[5] outlines the general method and associated digestions, interferences, and sensitivity.

ICP

ICP allows rapid simultaneous or sequential analysis of many metals making it the major method utilized by large commercial laboratories. ICP instruments are expensive and complicated analytical instruments requiring skilled analysts and exacting quality control procedures. ICP-AES methods are susceptible to high single element interferences in matrices high in salts or other elements making trace analysis of other elements in these matrices problematic. Arsenic and selenium lack sensitivity due to physical properties but can be enhanced using a hydride aspiration system. Lead, antimony, and thallium also have sensitivity problems on the ICP-AES but can be analyzed at low levels using ICP-MS or GFAA. ICP-MS greatly enhances sensitivity for metals making it the preferred method when the analyses of very low concentrations are required. The main interference is selenium that has mass interference with the argon dimer. SW846-6010[5] outlines the general method and associated digestions, interferences, and sensitivity.

CVAA

CVAA is the technique used for the analysis of mercury by using a selective digestion method. Although this method is extremely sensitive, it is subject to interferences from sulfide, chlorine, and organic compounds. There are several models of instruments for mercury but all require a thorough quality assurance plan to ensure the accuracy of the results. SW846-7470[5] outlines the general method and associated digestions, interferences, and sensitivity.

ANALYTICAL METHODS FOR ORGANIC CONSTITUENTS

The analysis of organics in water offers the data collector a variety of challenges due to the sheer number of methods and analytes. The decision of which method to utilize is determined by regulations, desired data quality, and cost. It should be noted that there are usually multiple methods that can quantify a compound requiring an experienced data collector to determine which method meets the requirements in the DQOs.

Organic chemical methods can be divided into those that determine total organic matter present and individual organic compounds or groups of compounds.

Total Organic Matter Present

Total organic methods measure such parameters as BOD,[6] chemical oxygen demand (COD),[6] total organic carbon (TOC),[8] oil and grease,[5,6] total recoverable petroleum hydrocarbons (TRPH),[5,6] oil and grease in sludge,[5] total phenols,[5,6] and surfactants.[6] A detailed discussion is included on BOD and COD in the article titled *Oxygen Measurement: Biological—Chemical Oxygen Demand*.

Individual Organic Constituents or Groups

A very detailed discussion would be needed to address the methods used to analyze the vast numbers of naturally occurring and man-made organic compounds found in water. EPA SW-846 provides a good listing of methods. Internet services, such as Toxnet (http://toxnet.nlm.nih.gov/), can also be useful tools. Toxnet is a cluster of databases on toxicology, hazardous chemicals, and related areas. One database, the Hazardous Substances Data Base, provides detailed descriptions of specific compounds as well as analytical methods. Numerous methods are typically presented in the HSDB (http://toxnet.nlm.nih.gov/cgi-bin/sis/htmlgen?HSDB). If the data are to be used to satisfy a regulatory requirement, the investigator must select the method that is approved by the appropriate regulatory agency.

Organic chemicals are typically analyzed using one or more of the following technologies: gas chromatographic technique (GC), halogen sensitive detector (HALL), photoionization detector (PID), flame ionization detector (FID), electron capture detector (ECD), nitrogen phosphorous detector (NPD), flame photometric detector (FPD), high performance liquid chromatography (HPLC), and various extraction methods such as purge and trap (P and T), separatory funnel extraction, and continuous liquid–liquid extraction.

The environmental field separates organic contaminants by their physical properties into volatile and semivolatile components. If the compound can be purged from an aqueous sample using an inert gas, the compound is considered volatile, and if the compound requires extraction using a solvent, it is considered a semivolatile compound. The separation of compounds into these groups greatly affects the sampling techniques utilized. Volatile components require sampling with zero headspace in 40 mL vials[2] with a Teflon septum. It is important that there is no air in these samples because it can significantly alter the results. Semivolatiles can be collected in liter glass jars with Teflon liners and are not as sensitive to air as volatiles. A qualified laboratory should be consulted to obtain proper sample containers, volumes, and preservatives and can also assist in the determination of which method best matches the project requirements.

CHROMATOGRAPHY

Although there are numerous detectors utilized in the quantitation of organic compounds, the methods introduced in this article will all use GC to separate and isolate individual components. In GC the vaporized components of a sample are separated as a result of partition between the mobile and stationary phases in the column. An analyst controls the separation of individual components by choice of column, method of injection, method of extraction, volume of injection, temperature program, and carrier gas flow. This separation allows the analyst to assign a retention time that a particular compound will elute off the column and identify that compound on the chromatogram. This proves to be an effective qualitative procedure if the matrix contains a small number of analytes. The identification of compounds from a highly contaminated or complex matrix tests the ability of the chromatography and requires intervention of the analyst to interpret the chromatogram or to manipulate the sample to produce more highly resolved chromatography. The effects of analyst intervention will usually reduce the limit of quantification by introducing dilution factors to achieve good baseline separations of the individual for identification. The actions of the analyst must be strictly controlled by experience and the laboratory's quality assurance plan to maintain the integrity of the data. Chromatography is the method of separation for identification, but the eluted sample is then passed onto a detector to quantify the amount of an analyte present in the original sample. The ability to meet the required analytical limits for organic compounds is extremely dependent on the effects of the matrix and the experience of the laboratory performing the analysis.

TWO-DIMENSIONAL DETECTORS

The choice of detectors is varied and is determined by the regulatory requirements and the cost of the analysis. Two-dimensional detectors utilize the retention time and response of a component to identify and quantify the compound and are extremely sensitive if operated by an experienced analyst. There are multiple types of detectors to consider when making a decision, but the regulatory requirements usually identify the detector required for compliance. These detectors all have unique qualities that aid in isolating the types of compounds that are being analyzed and the proper detector can increase sensitivity and decrease interferences. Some commonly used detectors include FID which measures all hydrocarbons, PID which measures aromatic hydrocarbons, and ECDs which are sensitive to halogens, peroxides, and nitro groups. These types of detectors provide sensitivity for

quantitation but rely on the chromatography for identification increasing the risk of miss identifying compounds which coelute or miss quantifying compounds in highly contaminated matrices.

THREE-DIMENSIONAL DETECTORS

The mass selective detector (MS) is a 3-D detector that utilizes retention time, response, and mass spectrum to identify and quantify the analyte of concern. Like the 2-D detectors, MS detectors rely on the chromatography to separate the compounds and the response to quantify the compounds but then utilized the mass spectral data to confirm identification. This technique is not as sensitive as the 2-D detectors but is much more accurate removing the possibility of miss identification. Utilizing the MS is preferred when analyzing a complex matrix, but usually increases the cost of analysis. As in all analytical methods, the quality of the data is heavily dependent on the quality of the analyst and the laboratory. SW846-8000 series outlines the general methods and associated extractions, interferences, and sensitivity of organic analysis.

When analyzing an aqueous sample the decision of which analytical method to utilize is difficult and critical to the success of the project. The data collector must meet the requirements established in DQOs to ensure that the data collected is of sufficient quality and quantity to make supportable decisions. The quality of data is determined by sampling, analytical methods, and the quality plans of all the parties involved in collection of the data. The manager must make informed decisions when selecting an analytical method that will fulfill the requirements of the project while balancing cost, time, and risk.

CONCLUSION

With increased industrialization and agriculture, the quality of our water resources will become a critical issue. The need to monitor water resources such as agricultural effluents, groundwater resources, and surface water resources will increase dramatically. The existing and emerging technology will allow analysis of chemicals in water at levels approaching molecular levels. These developments, however, can be very costly. The challenge in developing a plan for water analysis sampling is to develop a plan that satisfies the data quality needed to satisfy the decision making process. This includes balancing both the quality and quantity of samples to characterize the water to address protection of human health and the environment and regulatory standards.

REFERENCES

1. USEPA. *Guidance for the Data Quality Objectives Process (G-4)*; EPA/600/R-96/055; Office of Environmental Information: Washington, D.C., August 2000 (http://www.epa.gov/quality/qs-docs/g4-final.pdf).

2. USEPA. *EPA Guidance for Quality Assurance Project Plans (G-5)*; EPA/600/R-98/018; Office of Research and Development: Washington, D.C., February 1998 (http://www.epa.gov/quality/qs-docs/g5-final.pdf).

3. USEPA. *EPA Guidance for Data Quality Assessment (G-9)*; EPA/600/R-96/084; Office of Environmental Information: Washington, D.C., July 2000 (http://erb.nfesc.navy.mil/erb_a/restoration/methodologies/g9-final.pdf).

4. Gilbert, R.O. *Statistical Methods for Environmental Pollution*; John Wiley and Sons, Inc.: New York, 1987.

5. USEPA. *SW-846 Test Methods for Evaluating Solid Waste, Physical/Chemical Methods*; Office of Solid Waste: Washington, D.C., 2002 (http://www.epa.gov/epaoswer/hazwaste/test/sw846.htm).

6. USEPA. *Methods for Chemical Analysis of Water and Wastes*; EPA 600/4-79-020; Office of Environmental Information: Washington, D.C., 2002 (http://www.epa.gov/OGWDW/methods/epachem.html).

7. USEPA. *Methods for the Determination of Organic compounds in Drinking Water*; EPA 600/4-88-039; Office of Environmental Information: Washington, D.C., 2002 (http://www.epa.gov/OGWDW/methods/epachem.html).

8. American Public Health Association, American Water Works Association, Water Environmental Federation. *Standard Methods for Examination of Water and Wastewater*, 20th Ed.; Washington, D.C., 1999.

Chemigation

William L. Kranz
University of Nebraska, Norfolk, Nebraska, U.S.A.

INTRODUCTION

Chemigation is the practice of distributing approved agricultural chemicals such as fertilizers, herbicides, insecticides, fungicides, nematicides, and growth regulators by injecting them into water flowing through a properly designed and managed irrigation system. The term chemigation was originally coined to describe the concept of applying commercial fertilizers that were needed for crop production. Field research, and advances in sprinkler and chemical injection technology have stimulated the use of chemigation as a major crop production tool. Today chemigation is one of the more efficient, economical, and environmentally safe methods of applying chemicals needed for successful crop, orchard, turf, greenhouse, and landscape operations.

Chemigation began with the application of commercial fertilizers through irrigation systems in the late 1950s.[1] Later tests were initiated on sprinkler application of herbicides to selectively control weeds in field crops, fruit and nut orchards, rice, and potatoes.[2,3] These research efforts led the way for what has become a major research topic to identify management and equipment required for chemical application in agricultural and nonagricultural production settings.

The primary use of chemigation is to apply chemical directly to the soil using a range of irrigation water distribution systems. For example, drip/trickle, sprinklers, and some surface irrigation systems are commonly used to apply commercial fertilizers. However, federal regulations limit application of restricted use pesticides to systems that can safely and uniformly apply a chemical to a specific site at a rate specified on a chemical label. Though estimates vary greatly, chemigation is used to apply fertilizers on nearly four million hectares in the United States.[4] Specialists in Florida, Texas, and Wyoming report that more than 50% of their irrigated land received at least one chemigation application.[5]

ADVANTAGES OF CHEMIGATION

Chemigation offers producers of food and fiber many advantages that result from using existing equipment and timeliness of chemical applications. Advantages of chemigation include the following:[6,7]

- Uniformity of chemical application is equal to or greater than other means of application.
- Timeliness and flexibility of application are greater.
- Improved efficacy of some chemicals.
- Potential for reduced environmental risks.
- Lower application costs in some cases.
- Less mechanical damage to plants.
- Less soil compaction.
- Potential reduction in chemical applications.
- Reduced operator hazards.
- Application cost savings for multiple applications.

DISADVANTAGES OF CHEMIGATION

Chemigation also requires additional equipment and management to obtain successful results. Some of the disadvantages of chemigation include:[6,7]

- Chemical application accuracy depends on water application uniformity.
- Longer time of application than other methods.
- Some pesticide labels prohibit chemigation as a means of application.
- Potential for source water contamination.
- Additional capital costs for equipment.
- Potential for increased legal requirements in some states.
- Increased management requirements by the operator.

CHEMIGATION EQUIPMENT

Safe and efficient chemigation requires that the irrigation equipment, injection device, and safety equipment be properly installed and maintained. Fig. 1 provides an overview of equipment necessary for chemigation systems using groundwater. State and federal regulations specify the type of irrigation water distribution system that can be used and the required safety equipment. It is up to the

Encyclopedia of Water Science
DOI: 10.1081/E-EWS 120010031

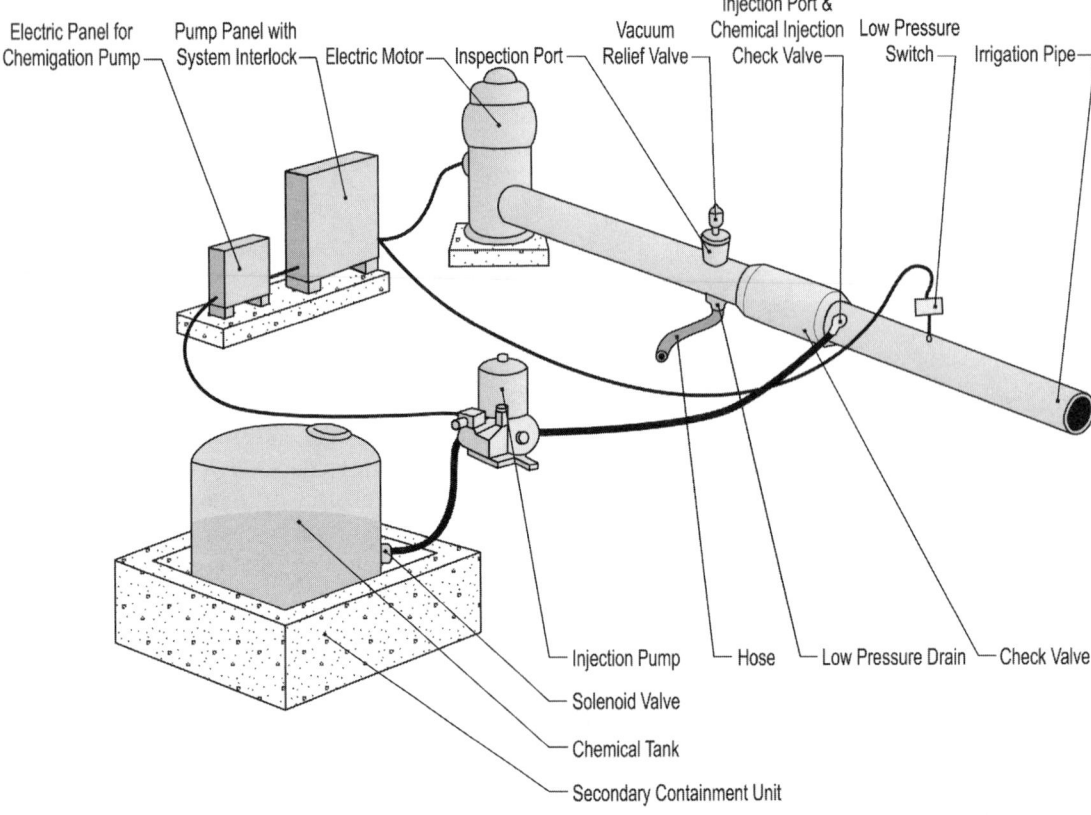

Electric Panel for Chemigation Pump · Pump Panel with System Interlock · Electric Motor · Inspection Port · Vacuum Relief Valve · Injection Port & Chemical Injection Check Valve · Low Pressure Switch · Irrigation Pipe

Injection Pump · Hose · Low Pressure Drain · Check Valve

Solenoid Valve

Chemical Tank

Secondary Containment Unit

Fig. 1 Chemigation injection and safety equipment commonly required when pumping groundwater. (Drawing courtesy of Midwest Plan Service, Ames, IA.)

irrigator to ensure the use of appropriate equipment and procedures.

Irrigation Equipment

Chemigation requires equipment capable of applying chemicals uniformly and with differing amounts of water, accurate and dependable injection equipment, and safety equipment for source water and worker protection. Appropriate sprinkler design and high application volumes can solve problems associated with canopy penetration and deposition that impact some aerial applications. Uniform water application can precisely place and incorporate chemicals in the soil and limit leaching of soluble chemicals from the zone of application.

Several different types of irrigation equipment can and are being used to distribute chemicals via chemigation. Most chemigation is conducted using either sprinkler or drip/trickle irrigation systems. Center pivot and linear-move systems are most commonly used for chemigation since prescription applications can be made with a high degree of uniformity. Drip/trickle systems are commonly used to place precise amounts of plant nutrients near the

zone of plant uptake thus increasing chemical use efficiency.

In general, surface irrigation systems have limited potential for chemigation. Water distribution in furrow systems is typically nonuniform along the row and among rows. Thus, in-field variation in water infiltration results in chemical application uniformity that is below levels desired for chemigation. Development of surge-flow systems can improve distribution uniformity, however, the question remains whether consistent results are possible and whether producers have sufficient experience to make equipment adjustments when necessary. Level basin irrigation systems offer improved uniformity of water application, but water quality concerns have limited the use of chemigation.

Injection Equipment

Chemical injection can occur using either active or passive devices. Active devices use an external energy supply to create pressures at the injector outlet that exceed the irrigation pipeline pressure. Injection pumps are often powered by constant speed or variable speed electric motors. Typical examples include piston, diaphragm, rotary, and gear pumps. However, most new installations use either piston or diaphragm pumps (Fig. 2). These injection devices are relatively expensive. Component selection allows the injection of commercial fertilizers, acids, or pesticides. Intermittent end guns and corner systems can lead to variable chemical application by constant rate injectors due to changes in the irrigation rate per hour.[8] Application errors of approximately 20% are possible when corner systems are used with a constant rate injection device. When the irrigation rate will change during a chemigation event, it is preferable to use a variable rate injection device.

Passive devices take advantage of pressure differentials that result from using a throttling valve or pitot tube unit to add chemical to water flowing through a pipeline. Chemicals are metered into the system using a venturi meter or orifice plate. These systems have low capital cost requirements. However, pumping cost may be greater since irrigation pump outlet pressure must be equal to the water distribution system pressure plus the friction loss associated with the throttling valve. In addition, changes in pumping pressure directly impact chemical injection rates which can lead to nonuniform chemical applications.

Selection criteria for injection devices include potential injection rates, available power supply, and the type of chemical to be injected. A single injection device is typically not capable of covering the range of injection

rates and chemical types that could conceivably be applied via chemigation. Hence, if plant nutrients and pesticides are to be applied, two injection devices are desirable. Diaphragm injection devices offer greater chemical compatibility, ease of calibration, and precise injection rates which make them good choices for pesticide injection. Commercial fertilizers are less caustic and require relatively high injection rates which make high capacity piston and diaphragm injection devices good options. Research has noted that injection equipment calibration was necessary for each injection device and operating pressure.[9] Manufacturing tolerances and pipeline pressure impacted the rate of chemical injection. Further, performance tests conducted on new and used diaphragm pumps found that proper maintenance is required to ensure long-term accuracy of chemical injection rates.[10]

Safety Equipment

State and federal regulations differ regarding safety equipment that is required for chemigation. For example, the Nebraska Chemigation Act requires the safety equipment also found in many state regulations.[11] Most requirements are met through installation of a backflow protection device. Requirements typically include (Fig. 3):

1. A mainline check valve to prevent concentrated chemical and/or dilute chemical solution from flowing back into the water source.
2. A chemical injection line check valve to prevent flow of chemical from the chemical supply tank into the irrigation pipeline and to prevent flow of water through the injection system into the chemical supply tank.
3. Vacuum relief valve to prevent back siphoning of concentrated chemical and/or dilute chemical solution into the water source.
4. Low pressure drain to prevent back flow of chemical and/or dilute chemical solution into the water source should the mainline check valve fail.
5. An inspection port to ensure that the mainline check valve and low pressure drain are functioning properly.
6. An interlock between the injection system and the irrigation pumping plant to prevent injection of concentrated chemical into the irrigation pipeline should there be an unexpected shutdown of the irrigation pump.

American Society of Agricultural Engineers have published EP409.1 Safety Devices for Chemigation[12] which recommends the addition of a two-way interlock between the injection system and irrigation pumping plant and a normally-closed solenoid valve on the outlet of

Fig. 2 Typical portable injection equipment for center pivot installations. (Photo courtesy of Agri-Inject, Inc., Yuma, CO.)

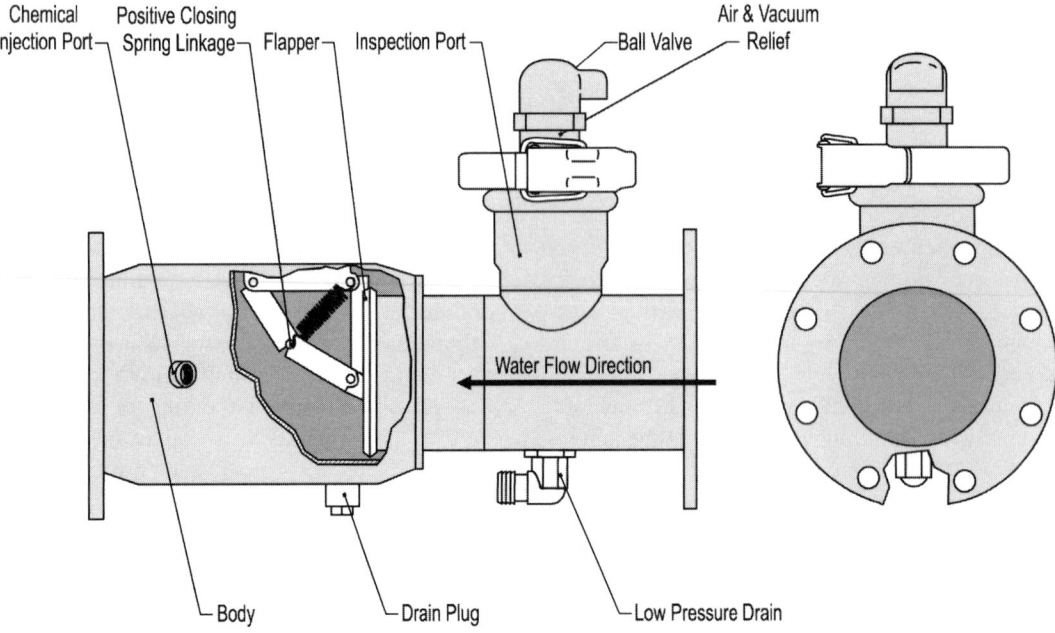

Fig. 3 Mainline check valve used to prevent flow of chemicals into a water source. (Drawing courtesy of Midwest Plan Service, Ames, IA.)

the chemical supply tank to prevent chemical spills attributed to chemical injection line or injection device failures. The engineering practice also encourages the positioning of a fresh water source near the chemical supply tank for washing chemicals that may contact skin, the use of a strainer on the chemical tank outlet to prevent fouling of injection equipment, the grading of the soil surface to direct flow away from the water supply, the location of mixing tanks and injection equipment safely away from potential sources of electrical sparks to prevent explosions, and the use of components that are well suited to a range of chemical formulations.

MANAGEMENT PRACTICES

Management flexibility based on chemical placement, application rate and mobility in the soil, water quality, application cost, and weather factors make chemigation a unique and effective production tool. Chemigation provides the opportunity to synchronize fertilizer applications to match plant needs and incorporate and, if needed, activate pesticides to increase efficacy. Equally important, chemigation provides the opportunity to reduce chemical applications by eliminating the need for insurance-type applications. Fields can be scouted for disease or pests and chemical applied only if damage or pest numbers exceed economic thresholds. Soil and plants can be monitored to determine fertilizer needs, making

near real-time adjustments in the time of application and chemical formulation possible. Individual nozzle controls make site-specific applications well within reach.[13] However, a considerable amount of work remains to ascertain if site-specific applications are economical and to incorporate management tools into system controls.

CONCLUSION

Chemigation has gradually become one of the most effective means of chemical application available for crop production and landscape systems. Advantages of highly uniform prescription applications outweigh the potential disadvantages in most cases. Effective chemigation hinges on the selection of appropriate irrigation systems, chemical injection devices, and safety equipment. Through proper management, chemigation is poised to be a production practice that can help increase the quality and quantity of food produced worldwide.

REFERENCES

1. Bryan, B.B.; Thomas, E.L., Jr. *Distribution of Fertilizer Materials Applied with Sprinkler Irrigation Systems,* Agricultural Experiment Station Research Bulletin 598; University of Arkansas: Fayetteville, AK, 1958; 12 pp.

2. Ogg, A.G., Jr. Applying Herbicides in Irrigation Water—A Review. Crop Prod. **1986**, *5* (1), 53–65.

3. Cary, P.J. Applying Herbicides and Other Chemicals Through Sprinkler Systems. Proceedings of the 21st Annual Conference of the Washington State Weed Association; Yakima, WA, 1971.

4. USDA, Application of Chemicals in Irrigation and Times Irrigated by Selected Crop: 1998 and 1994. *1998 Farm and Ranch Irrigation Survey: Table 24*; U.S. National Agricultural Statistics Service, Washington, DC, 1999; 102–124.

5. Adams Business Media; 2000 Annual Irrigation Survey Continues Steady Growth. Irrig. J. **2001**, *51* (1), 12–40.

6. Scherer, T.F.; Kranz, W.L.; Pfost, D.; Werner, H.D.; Wright, J.A.; Yonts, C.D. Chemigation. *MWPS-30 Sprinkler Irrigation Systems*; Midwest Plan Service: Ames, IA, 1999; 145–166.

7. Threadgill, E.D. Introduction to Chemigation: History, Development, and Current Status. In *Proceedings of the Chemigation Safety Conference, Lincoln, NE, April 17–18, 1985*; Vitzthum, E.F., Hay, D.R., Eds.; University of Nebraska Cooperative Extension: Lincoln, NE, 1985.

8. Eisenhauer, D.E. Irrigation System Characteristics Affecting Chemigation. In *Proceedings of the Chemigation Safety Conference, Lincoln, NE, April 17–18, 1985*; Vitzthum, E.F., Hay, D.R., Eds.; University of Nebraska Cooperative Extension: Lincoln, NE, 1985.

9. Kranz, W.L.; Eisenhauer, D.E.; Parkhurst, A.M. Calibration Accuracy of Chemical Injection Devices. Appl. Engr. Agric. **1996**, *12* (2), 189–196.

10. Cochran, D.L.; Threadgill, E.D. Injection Devices for Chemigation: Characteristics and Comparisons. National Meeting of the American Society of Agricultural Engineers, Paper No. 86-2587; ASAE: St. Joseph, MI, 49805, 1986.

11. Vitzthum, E.F. *Using Chemigation Safely and Effectively*; University of Nebraska Cooperative Extension Division: Lincoln, NE, 2000; 1–59.

12. ASAE, *EP409.1: Safety Devices for Chemigation*; American Society of Agricultural Engineers: St. Joseph, MI, 2001.

13. Evans, R.G.; Buchleiter, G.W.; Sadler, E.J.; King, B.A.; Harting, G.H. Controls for Precision Irrigation with Self-propelled Systems. Proceedings of the 4th Decennial National Irrigation Symposium, Phoenix, AZ, 2000; 322–331.

Chlorination for Disinfection and Prevention of Clogging of Drip Lines and Emitters

Andrew C. Chang
University of California, Riverside, California, U.S.A.

INTRODUCTION

Water from both surface and underground sources picks up particulate matter during conveyance—sands, silts, plant fragments, algae, diatoms, larvae, snails, fishes, etc. While the majority of the suspended solids may be removed in preirrigation treatments such as sedimentation and filtration, some of the fine silts and colloidal clay particles inadvertently remain and settle inside the lateral lines or emitters impeding the water flow. As the flow slows down and/or the chemical background of the water changes, chemical precipitates and/or microbial flocs and slimes begin to form and grow, thus microirrigation emitter clogging occurs. This section delineates the occurrences of chemical precipitates and the chemistry of acidification that is employed to mitigate clogging caused by chemical precipitates. Clogging resulting from formation of microbial flocs and slimes is controllable by acidification as well as chlorination.

OVERVIEW

Within the extensive network of a drip irrigation system, it is difficult to predict where or when clogging will take place. The hydraulic characteristics such as flow velocity, path length, orifice diameter, and pressure compensation all affect the flow rate and thus the clogging. The lower end of an operating irrigation system (laterals and emitters) should be visually inspected for build-up of deposits, and the flow rates and pressures of the systems should be regularly tested. Routine examinations will identify segments of the network that are potentially problematic and isolate them for corrective measures. The clogging, once formed in the distribution system is difficult to mitigate. Prevention is by far the preferred measure.

Clogging caused by the deposition of inorganic suspended substances may be overcome by regular flushing of the system and by employment of self-cleaning emitters. High-dosage, short-duration shock treatment with acidification and chlorination may be necessary to dissolve the chemical precipitates and to inactivate the microorganisms. Many publications have outlined the practical and operational aspects of acidification and chlorination processes for drip irrigation.[1–4]

BIOLOGICAL GROWTH AND BIOLOGICAL GROWTH INDUCED CHEMICAL PRECIPITATION

Many microorganisms may grow within the water delivery network in the absence of light producing slime and causing iron and sulfur to precipitate in the water. Organic substrates and nutrients will enhance bacterial growth. The aggregates resulting from microbial slimes adhering to the suspended solids in the water are the primary causes of clogging.

When the iron present in water is in the ferrous form (Fe^{2+}), it may be oxidized to the ferric iron (Fe^{3+}) that in turn form precipitates. The oxidation reactions are often mediated by filamentous (*Gallionella, Leptothrix, Toxothrix, Crenothrix,* and *Sphaerotilus* spp.) and nonfilamentous (*Psedomonas* and *Enterobactor* spp.) bacteria.[1] In the presence of dissolved oxygen, Fe^{2+} is oxidized to (Fe^{3+}) according to the reaction

$$Fe^{2+} + \frac{1}{4}O_2 + H^+ = Fe^{3+} + \frac{1}{2}H_2O$$

At pH 7, the solubility of Fe^{3+} is approximately 6 orders of magnitude lower than that of the ferrous iron Fe^{2+}. Solution Fe concentrations as low as $0.1\,mg\,L^{-1}$ may result in significant deposition of Fe precipitates in the distribution systems.

When hydrogen sulfite is present in the water, *Thiothrix* spp. bacteria oxidize the S^{2-} in H_2S to insoluble elemental sulfur, S^0, in the presence of dissolved oxygen:

$$H_2S + \frac{1}{2}O_2 = H_2O + S^0$$

The potential for clogging is related to the quality of irrigation water. Based on field experience, concentrations that exceed the levels labeled as low given in Table 1 can cause clogging; the more the parameters exceed the limits, the higher the clogging potential.

CHLORINATION

Chlorination is by far the most common method used to disinfect water. It involves the addition of chlorine or

Encyclopedia of Water Science
DOI: 10.1081/E-EWS 120010376

Table 1 Water quality criteria for drip irrigation

Parameter	Clogging potential		
	Low	Moderate	High
Suspended solids (mg L^{-1})	< 50	50–100	> 100
pH	< 7	7–8	> 8
Dissolved solids (mg L^{-1})	< 500	500–2000	> 2000
Manganese (mg L^{-1})	< 0.1	0.1–1.5	> 1.5
Iron (mg L^{-1})	< 0.1	0.1–1.5	> 1.5
Calcium and magnesium (mg L^{-1})	< 20	20–50	> 50
Hydrogen sulfite (mg L^{-1})	< 0.5	0.5–2	> 2
Bacterial population (count mL^{-1})	< 10^5	10^5–5 × 10^5	> 5 × 10^5

chlorine compounds to produce chemical species that have ability to inactivate microorganisms present in water. The method was first introduced over 100 yr ago and has remained as an effective, cost-effective, and easy to operate process.

When chlorine is dissolved in water, it hydrolyzes to hypochlorous acid (HOCl) that subsequently ionizes to hypochlorite (OCl$^-$).

$$Cl_{2(g)} + H_2O \rightarrow HOCl_{(aq)} + H^+ + Cl^-$$

$$HOCl_{(aq)} \rightarrow H^+ + OCl^-$$

For other chlorine chemicals (such as the active ingredients found in household bleach), the reactions are similar:

$$Ca(OCl)_2 + 2H_2O \rightarrow 2HOCl + Ca(OH)_2$$

$$NaOCl + H_2O \rightarrow HOCl + NaOH$$

$$CaClOCl + 2H_2O \rightarrow HOCl + H^+ + Cl^- + Ca(OH)_2$$

HOCl and OCl$^-$ are the chlorine species active in the disinfection actions. The ratio of HOCl and OCl$^-$ species in water is dependent on pH (Fig. 1). This relationship is significant, as HOCl$_{(aq)}$ is a far more effective chemical species for disinfection than OCl$^-$.

HOCl and OCl$^-$ are strong oxidants and may be dissipated before significant disinfection occurs because of their reactions with various impurities in water. Chemical species such as H$_2$S, SO$_3$$^{2-}$, NO$_2$$^-$, Fe^{2+}, and Mn^{2+} react rapidly with HOCl according to the following reactions:

$$H_2S + 4HOCl \rightarrow H_2SO_4 + 4HCl$$

$$2Fe^{2+} + Cl_{2(g)} + 6H_2O \rightarrow 2Fe(OH)_3 + 2Cl^- + 6H^+$$

$$Mn^{2+} + Cl_{2(g)} + 2H_2O \rightarrow MnO_2 + 2Cl^- + 4H^+$$

In these reactions, the disinfecting power of the added chlorine (i.e., Cl$_2$, HOCl, and OCl$^-$) is spent in oxidizing the reduced forms of sulfur (S^{2-}), iron (Fe^{2+}), and manganese (Mn^{2+}) present in the water. If Fe^{2+} and Mn^{2+} are present in water, as in some ground waters, they will be oxidized through the chlorination process and the oxidized Fe and Mn species are considerably less soluble in the water. It is preferable that the reduced Fe and Mn species are oxidized at the pre-irrigation treatment to prevent the formation of precipitates in the distribution system.

If ammonia is present in the water it will also react with HOCl$_{(aq)}$ and result in the formation of chloroamines:

$$NH_3 + HOCl \rightarrow NH_2Cl + H_2O$$

$$NH_2Cl + HOCl \rightarrow NHCl_2 + H_2O$$

$$NHCl_2 + HOCl \rightarrow NCl_3 + H_2O$$

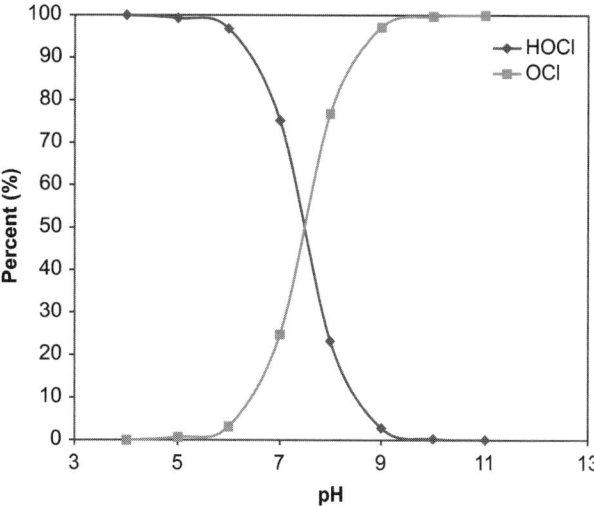

Fig. 1 Distribution of HOCl and OCl^{-1} in water.

These combined chlorine species are considerably less efficient in inactivating microorganisms than $HOCl_{(aq)}$. If the addition of chlorine continues after the conversions, combined chlorines will be further oxidized to form gaseous nitrogen species and chloride:

$$NH_2Cl + NHCl_2 + HOCl \rightarrow N_2O + 4HCl$$

$$HN_2Cl + NCl_3 + HOCl \rightarrow N_2 + 4HCl$$

In addition, organic reducing agents such as phenols and unsaturated organic compounds also react with free chlorine. Notably, $HOCl_{(aq)}$ reacts with dissolved organic matter in water to form trihalomethanes. Trihalomethanes are chemical species that have the structure of a methane molecule in which three of the hydrogens are substituted by permutations of halogens (I, Cl, and Br). Because of their similarity in chemical structure, trihalomethanes are categorized along with chloroform ($CHCl_3$), the most common trihalomethane species among them, as potential carcinogens. Some irrigation waters, such as those obtained from the Sacramento Delta in California, can contain significant amounts of humic and fulvic acids, which are precursors of trihalomethanes. Unlike in drinking water through which consumers may be exposed to potentially harmful chemicals, the chlorination-induced trihalomethanes are not expected to be absorbed by plants. However, additional chlorine must be spent to satisfy the reactions before the sufficient concentrations of effective chlorine species are present in the treated water to satisfy the disinfection needs.

The disinfecting potential of the water may be represented by the chlorine residue that sums up the free and combined chlorine species present in the water (Fig. 2). In this diagram, the added chlorine at first does not result in any chlorine residue as it reacts with the reducible substances in the water. Subsequently, the chlorine residue rises as the added chlorine reacts with ammonia in water to form the combined chlorine species and then falls as these compounds decompose to chloride and gaseous nitrogen. The most effective disinfecting chlorine species will not be present in the water until all reactions are completed and the dosage reaches beyond the break point marked in Fig. 2.

The amount of chlorine required to reach the break point is dependent on the amounts of reducible substances and ammonia present in the water. This dosage is empirically determined for each water.

In chlorination, the effectiveness of microbial kill is in proportion to the concentration of disinfectant and time of contact. Low-chlorine dosages may be compensated by longer time of contact and vice versa. In practical

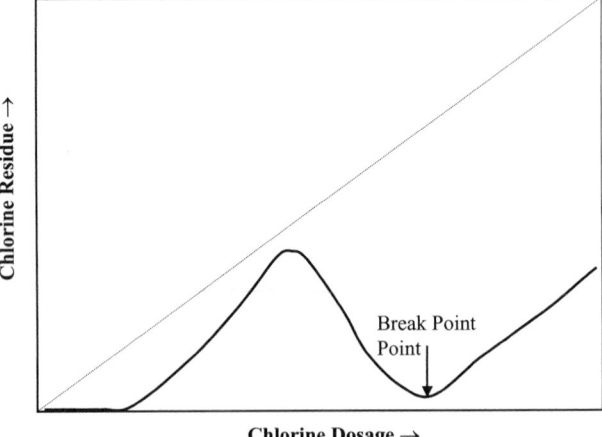

Fig. 2 A schematic depiction on the chemical reactions of chlorination and the formation of chlorine residues for disinfections.

applications, the microbial growth in drip-irrigation lines may be controlled by continuous chlorination at rates that result a chlorine residue concentration of $1\,mg\,L^{-1}$–$2\,mg\,L^{-1}$, or at intermittent basis with a chlorine residue concentration of $10\,mg\,L^{-1}$–$20\,mg\,L^{-1}$ for 30 min–60 min once each day of irrigation. When severe blockages caused by microbial growth occur, super-chlorination at chlorine residue concentrations of $500\,mg\,L^{-1}$–$1000\,mg\,L^{-1}$ may be necessary until the blockage is removed.

REFERENCES

1. Gilbert, R.G.; Ford, H.W. Operation Principles. In *Trickle Irrigation for Crop Production, Design, Operation, and Management*, Development in Agricultural Engineering 9; Nakayama, F.S., Bucks, D.A., Eds.; Elsevier: Amsterdam, 1986; 142–163.

2. Schwankl, L.; Prichard, T. Chlorination. In *Drip Irrigation for Row Crops*; Hanson, B., Schwankl, L., Grattan, S.R., Prichard, T., Eds.; University of California Irrigation Program, University of California: Davis, 1994; 129–139.

3. English, S.D. Filtration and Water Treatment for Micro-irrigation. *Trickle Irrigation in Action*, Proceedings of the 3rd International Drip/Trickle Irrigation Congress; American Society of Agricultural Engineers: St. Joseph, MI, 1985; ASAE Publication 10-85, Vol. I, 129–139.

4. Meyer, J.L. Cleaning Drip Irrigation Systems. *Drip/Trick Irrigation in Action*, Proceedings of the 3rd International Drip/Trickle Irrigation Congress; American Society of Agricultural Engineers: St. Joseph, MI, 1985; ASAE Publication 10-85, Vol. I, 41–44.

Chromium

Bruce R. James
University of Maryland, College Park, Maryland, U.S.A.

INTRODUCTION

Chromium is a heavy metal that is essential for human health in its trivalent form [Cr(III)], but may cause cancer if inhaled in the hexavalent state [Cr(VI)]. Trivalent Cr is only sparingly-soluble in neutral to alkaline natural waters, but it can be oxidized to Cr(VI) by manganese (III,IV) (hydr)oxides, hydrogen peroxide, ozone, chlorine gas, hypochlorite, and other electron acceptors. Hexavalent Cr can be reduced to Cr(III) by elemental iron and iron(II), sulfides, easily-oxidized organic compounds, and other electron donors. Both oxidation and reduction reactions of chromium are governed by redox potential (Eh) and acidity of natural waters (pH).

OCCURRENCE OF CHROMIUM IN NATURAL WATERS AND WATER SUPPLIES

Concerns surrounding the presence of chromium (Cr) in natural waters and drinking water supplies must address a paradox of this heavy metal related to the contrasting solubilities and toxicities of its common oxidation states in natural environments: Cr(III) and Cr(VI). Chromium(III) is essential for human health in trace amounts as an activator of insulin,[1] but it exists predominantly in nature in cationic forms that are typically only sparingly-soluble in near-neutral pH soils, plants, cells, and natural waters.[2] In contrast, Cr(VI) is anionic and much more soluble than Cr(III) over the pH range of natural systems. It is toxic to many cells, is classified by USEPA as a Class A carcinogen by inhalation, and is a regulated contaminant of drinking water supplies.[3] When soluble Cr is detected in natural waters, especially at high concentrations, it is usually Cr(VI) derived from industrial wastes containing Cr(VI) or possibly resulting from the oxidation of certain forms of Cr(III) in soils or sediments.[4,5]

The balance of the different forms and the solubilities of Cr(III) and Cr(VI) in natural waters is governed by pH, aeration status (Eh or oxidation–reduction potential), and other environmental conditions (Table 1). Understanding and predicting the oxidation state, solubility, mobility, and bioavailability of Cr in water are further complicated by the fact that Cr(III) can be oxidized (lose three electrons) to form Cr(VI); whereas Cr(VI) can gain three electrons and be reduced to Cr(III).[6,7] Natural variation and human-induced changes in pH and the oxidation–reduction status of soil and water can control the solubility of Cr. As a result, purification of drinking water supplies and treatment of waste waters contaminated with Cr are possible through chemical and microbiological processes that modify the acidity and the relative abundance of oxidizing and reducing agents for Cr.[8,9]

Chromium is the seventh most abundant metal on earth with an average content of 100 mg/kg in the earth's crust and 3700 mg/kg for the earth as a whole,[10] principally as Cr(III) in unreactive, insoluble minerals, such as chromite ($FeO \cdot Cr_2O_3$). Roasting chromite ore under alkaline, high temperature conditions oxidizes Cr_2O_3 to soluble Cr(VI), a widely-used starting material for production of stainless steel, pressure-treated lumber, chrome-tanned leather, pigments, chrome-plated metals, and other common products used in modern societies.[11] As a result, Cr(VI) remaining in chromite ore processing residue, chrome plating bath waste, paint aerosols, and other industrial wastes may enrich soils and contaminate surface waters and groundwater that are supplies for domestic uses, irrigation, and industrial processes.

In contrast to these concentrated, anthropogenic sources of Cr(VI); naturally-occurring sources of Cr are predominantly Cr(III) and occur at low concentrations. Ultramafic and basaltic rocks (and soils developed from these parent materials), however, may contain up to 2400 mg Cr/kg, and can release small fractions of the Cr contained in them as Cr(VI), either through dissolution of Cr(VI) minerals or possibly via oxidation of Cr(III). As a result, Cr(VI) has been detected in groundwater (< 0.05 mg/L–0.5 mg/L) in arid regions dominated by these alkaline, Cr-rich rocks and soils. A concentration of Cr(VI) of 7.5 mg/L in pH 12.5 groundwater from Jordan is the highest known level that is not due to human influence. Naturally-occurring Cr in alkaline, aerobic ocean water exists principally as Cr(VI) at concentrations in the range of 3 nM–7.3 nM (0.16–0.38 µg/L).[12]

Based on the known carcinogenicity of Cr(VI) to humans by inhalation, and due to uncertainty about its

Encyclopedia of Water Science
DOI: 10.1081/E-EWS 120010209

Table 1 Oxidation states and forms of chromium in natural waters

Oxidation state	Form	Name	Chemical conditions of water under which it is found and pertinent reactions in natural waters
Chromium (III) (trivalent chromium)	$Cr(H_2O)_6^{3+}$	Hexaquochromium(III)	pH < 3.5; strong affinity for negatively-charged ions (e.g., phosphate) and colloid surfaces (e.g., living cells and phyllosilicate clays or fulvic and humic acids); green color
	$Cr(H_2O)_5OH^{2+}$	Monohydroxychromium(III)	First hydrolysis product formed at pH > 3.5 upon dilution of or addition of base to solutions of Cr(III); green
	$Cr(H_2O)_4(OH)_2^+$	Dihydroxychromium(III)	Second hydrolysis product of Cr(III); may dimerize and polymerize to form large molecular weight cations in planes of octahedra; green
	$Cr(H_2O)_3(OH)_3^0$	Chromium hydroxide	Metastable, uncharged hydrolysis product that precipitates as the sparingly-soluble $Cr(OH)_3$
	$Cr(H_2O)_2(OH)_4^-$	Hydroxochromate	Fourth hydrolysis product of Cr(III) that may form at pH > 11; may oxidize to Cr(VI) by O_2
	Cr(III)–organic acid complexes and chelates	For example: chromium citrate, chromium picolinate, chromium fulvate	Soluble complexes and chelates in which water molecules of hydration surrounding $Cr(H_2O)_6^{3+}$ are displaced by carboxylic acid and N-containing ligands; formation is pH- and concentration-dependent; blue–green–purple colors, depending on ligand binding Cr(III)
Chromium (VI) (hexavalent chromium)	H_2CrO_4	Chromic acid	Fully-protonated form of Cr(VI) formed at pH < 1; see Fig. 2 for key Eh values for redox
	$HCrO_4^-$	Bichromate	Form of Cr(VI) that predominates at 1 < pH < 6.4; yellow; see Fig. 2 for key Eh values for redox
	CrO_4^{2-}	Chromate	Form of Cr(VI) that predominates at pH > 6.4; yellow; see Fig. 2 for key Eh values for redox
	$Cr_2O_7^{2-}$	Dichromate	Form of Cr(VI) that predominates at pH < 3 and in concentrated solutions (> 1.0 mM); rapidly reverts to $HCrO_4^-$ or CrO_4^{2-} upon dilution or pH change; orange

long-term effects on human health via ingestion in drinking water, the USEPA has set a maximum contaminant level for total Cr [Cr(III)-plus-Cr(VI)] in drinking water in the United States of 100 μg/L.[3] This valence-independent standard is based on research results that showed no observed adverse effects of Cr(III) or Cr(VI) at 25,000 μg/L in drinking water given to rats, and after factoring in "uncertainty" and "safety" factors. The standard is based on total, soluble Cr (rather than Cr(VI) alone) because USEPA assumed that (a) Cr(III) is in dynamic equilibrium with Cr(VI) and could be oxidized, (b) the reduction of Cr(VI) to Cr(III) in the stomach and digestive tract is incomplete, and (c) despite the low toxicity of Cr(III), it may react with DNA in cells. The State of California has proposed the first valence-specific drinking water standard (public health goal) for Cr(VI) at 2.5 μg/L, an action based on a desire to be highly-protective of human health and drinking water quality.[13]

SOLUBILITY CONTROLS OF CHROMIUM CONCENTRATIONS IN WATER

Most inorganic compounds of Cr(III) are less soluble in water than are those of Cr(VI) because Cr(III) cations have high ionic potentials (charge-to-size ratio) and hydrolyze to form covalent bonds with OH^- ions (Table 1). When three OH^- anions surround the Cr^{3+} cation, it is particularly stable in water as the sparingly-soluble compound, $Cr(OH)_3$ (Table 2). Upon aging and dehydration, $Cr(OH)_3$ slowly converts to the more crystalline, less soluble Cr_2O_3.[12] Incorporation of Fe(III) or Fe(II) into solid phases and precipitates containing Cr(III) renders the Cr(III) less soluble, often by a factor of 1000 in the solubility product (Ksp).[14,15] In the pH range of 5.5–8, Cr(III) reaches minimum solubility in water due to this hydrolysis and precipitation reaction, an important process that controls the movement of Cr(III) in soils enriched with industrial waste waters and solid materials. Under strongly acidic conditions (pH < 4), unhydrolyzed

Table 2 Solubility in water at pH 7 of selected chromium compounds

Oxidation state of Cr	Compound name	Formula	Approximate solubility (moles Cr/L)
Chromium (III)	Chromium(III)hydroxide	$Cr(OH)_{3\ (am)}$	10^{-12}
	Chromium(III) oxide	$Cr_2O_{3\ (cr)}$	10^{-17}
	Chromite	$FeO \cdot Cr_2O_{3\ (cr)}$	10^{-20}
	Chromium chloride	$CrCl_3$	Highly soluble
	Chromium sulfate	$Cr_2(SO_4)_3$	Highly soluble
	Chromium phosphate	$CrPO_4$	10^{-10}
	Chromium fluoride	CrF_3	1.2×10^{-3}
	Chromium arsenate	$CrAsO_4$	10^{-10}
Chromium (VI)	Potassium chromate	K_2CrO_4	3.2
	Sodium chromate	Na_2CrO_4	5.4
	Calcium chromate	$CaCrO_4$	0.14
	Barium chromate	$BaCrO_4$	1.7×10^{-3}
	"Zinc yellow" pigment	$3ZnCrO_4 \cdot K_2CrO_4 \cdot Zn(OH)_2 \cdot 2H_2O$	8.2×10^{-3}
	Strontium chromate	$SrCrO_4$	5.9×10^{-3}
	Lead chromate	$PbCrO_4$	1.8×10^{-6}
	Chromium jarosite	$KFe_3(CrO_4)_2(OH)_{6\ (cr)}$	10^{-30}

$Cr(H_2O)_6^{3+}$ cations exist in solution; while $Cr(OH)_4^-$ forms under strongly alkaline conditions (pH > 11), particularly in response to adding base to solutions of soluble salts of Cr(III), e.g., $CrCl_3$, $Cr(NO_3)_3$, or $Cr_2(SO_4)_3$.

Other anions besides OH^- coordinate with $Cr(H_2O)_6^{3+}$ and displace water molecules of hydration to form sparingly-soluble compounds and soluble chelates (Table 2). In water treatment facilities and in natural waters; phosphate ($H_2PO_4^-$ HPO_4^{2-}, PO_4^{3-}), arsenate ($H_2AsO_4^-$, $HAsO_4^{2-}$, AsO_4^{3-}) and fluoride (F^-) may form low solubility compounds with Cr(III). Organic complexes of Cr(III) with carboxylic acids (RCOOH, e.g., citric, oxalic, tartaric, fulvic) remain soluble at pH values above which $Cr(OH)_3$ forms. By increasing the solubility of Cr(III) in neutral and alkaline waters, such organic complexes enhance the potential for absorption of Cr(III) by cells. Stable, insoluble complexes of Cr(III) also form with humic acids and other high molecular aggregate weight organic moieties in soils, sediments, wastes, and natural waters.[16]

With the exception of chromium jarosite (Table 2), Cr(VI) compounds are more soluble over the pH range of natural waters than are those of Cr(III); thereby leading to the greater concern about the potential mobility and bioavailability of Cr(VI) than Cr(III) in natural waters. The alkali salts of Cr(VI) are highly soluble, $CaCrO_4$ is moderately soluble, and $PbCrO_4$ and $BaCrO_4$ are only sparingly-soluble. In colloidal environments containing aluminosilicate clays and (hydr)oxides of Al(III), Fe(II,III), and Mn(III,IV) (e.g., in soils and sediments), Cr(VI) anions may be adsorbed similarly to SO_4^{2-}. Low pH and high ionic strength promote retention of $HCrO_4^-$ and CrO_4^{2-} on

positively-charged sites, especially those associated with colloidal surfaces dominated by pH-dependent charge. Such electrostatic adsorption may be reversible, or the sorbed Cr(VI) species may gradually become incorporated into the structure of the mineral surface (chemisorption). Recently-precipitated $Cr(OH)_3$ can adsorb Cr(VI) or incorporate Cr(VI) within its structure as it forms, thereby forming a Cr(III)–Cr(VI) compound.[17]

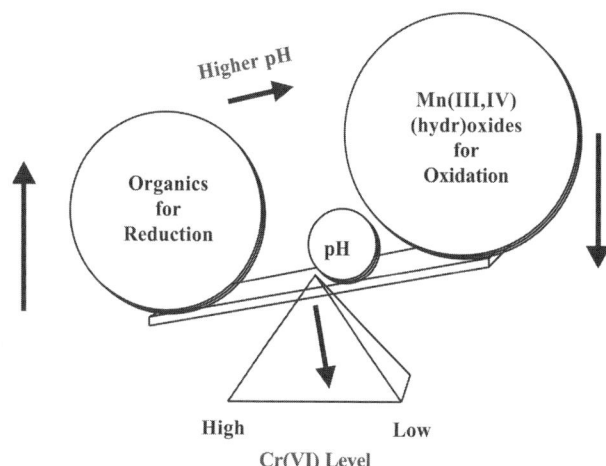

Fig. 1 Seesaw model depicting a balance of the oxidation of Cr(III) by Mn(III,IV)(hydr)oxides and the reduction of Cr(VI) by organic compounds, with the pH acting as a sliding control (master variable) on the seesaw to set the redox balance for given quantities and reactivities of oxidants and reductants. The equilibrium quantity of Cr(VI) in the water is indicated by the pointing arrow from the fulcrum.

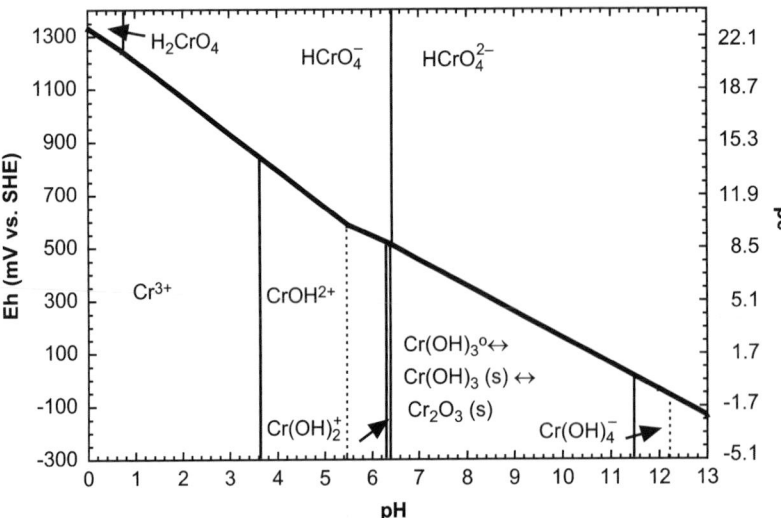

Fig. 2 Eh–pH diagram illustrating the stability fields defined by Eh (redox potential relative to the standard hydrogen electrode, SHE) and pH for Cr(VI) and Cr(III) at 10^{-4} M total Cr. The vertical dashed lines indicate semi-quantitatively the pH range in which $Cr(OH)_3$ is expected to control Cr(III) cation activities in the absence of other ligands besides OH^-.

OXIDATION–REDUCTION CHEMISTRY OF CHROMIUM IN NATURAL WATERS

The paradox of the contrasting solubilities and toxicities of Cr(III) and Cr(VI) in natural waters and living systems is complicated by two electron transfer reactions: Cr(III) can oxidize to Cr(VI) in soils and natural waters; and Cr(VI) can

reduce to Cr(III) in the same systems, and at the same time. Understanding the key electron transfer processes (redox) and predicting environmental conditions governing them are central to treatment of drinking water, waste waters, contaminated soils, and to predicting the hazard of Cr in natural systems.[18] The metaphor of a seesaw (Fig. 1) is useful in picturing the undulating nature of the changes in Cr

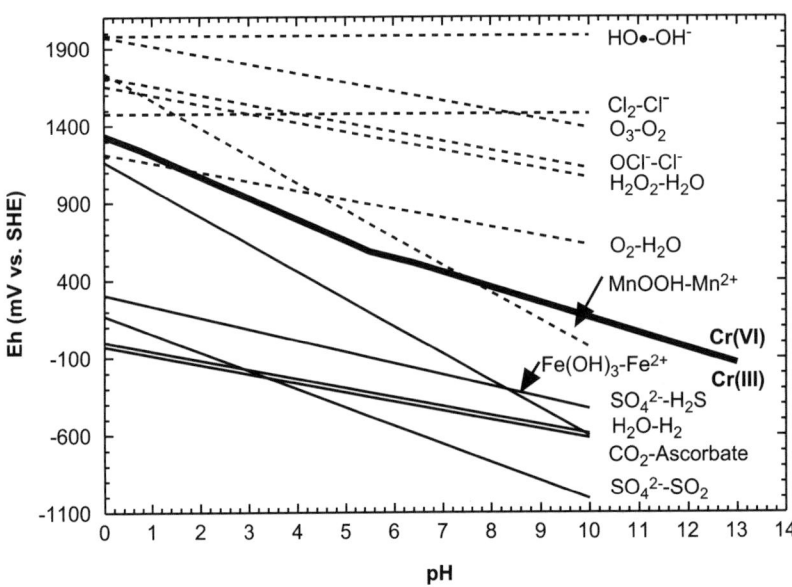

Fig. 3 Eh–pH diagram showing potential oxidants for Cr(III) in natural waters as dashed lines above the bold Cr(VI)–Cr(III) line; and potential reductants for Cr(VI) below the line. Each line for an oxidant (first species of the pair) and reductant (second species) combination represents the reduction potential (in mV) at a given pH established by that oxidant–reductant pair (e.g., O_3–O_2). The oxidant member of a pair for a higher line is expected to oxidize the reductant member of the lower line, thereby establishing the area and species between the lines as thermodynamically favored to exist at chemical equilibrium.

speciation in water due to oxidation of Cr(III) and reduction of Cr(VI). A balance for the two redox reactions is achieved in accordance with the quantities and reactivities of reductants and oxidants in the system (e.g., organic matter and Mn(III,IV) (hydr)oxides), as modulated by pH as a master variable.[8]

The thermodynamics (energetics predicting the relative stability of reactants and products of a chemical reaction) of the interconversions of Cr(III) and Cr(VI) compared to other redox couples can be used to predict the predominance of Cr(III) or Cr(VI) in water supplies (Fig. 2). The Eh variable defines the predicted voltage (electron pressure) that must be applied at a given pH to reduce Cr(VI) to Cr(III), and this pressure increases (lower Eh values) as pH increases, as shown in an Eh–pH diagram (Fig. 2). Certain electron-poor species may act as oxidants (electron acceptors) for Cr(III), especially soluble forms of Cr(III), in the treatment of water supplies or in soils enriched with Cr(III) (Ref. 19, Fig. 3). Examples are those above the bold line for Cr(VI)–Cr(III) on the Eh–pH diagram: Cl_2, OCl^-, H_2O_2, O_3, and MnOOH. In contrast, electron-rich species may donate electrons to electron-poor Cr(VI) and reduce it to Cr(III): Fe^{2+} [or Fe(0)], H_2S, H_2, ascorbic acid (and organic compounds, generally), and SO_2. Sunlight may affect the kinetics of both oxidation and reduction reactions for Cr, a relevant fact for natural processes in lakes and streams and for treatment technologies for drinking water purification. Depending on pH, temperature, and the concentrations of oxidants and reductants, Cr(VI)-to-Cr(III) ratios in natural waters may be predicted.

Predictions of the likelihood of Cr(III) oxidation and Cr(VI) reduction occurring are important for water treatment and for establishing health-based regulations and allowable limits for Cr(VI) and Cr(III) in water supplies. In agricultural soil–plant–water systems, Cr(VI) added in irrigation water or formed via oxidation of Cr(III) will reduce to Cr(VI) if electron donors (e.g., Fe^{2+}, H_2S, and organic matter) and Eh–pH conditions are sufficiently reducing (Refs. [4,17] Fig. 2). If not reduced, Cr(VI) may leach from surface soils to subsoils and groundwater. Therefore, predictions of Cr bioavailability and mobility in natural waters must consider redox reactions of this heavy metal.

REFERENCES

1. Anderson, R.A. Essentiality of Chromium in Humans. Sci. Tot. Environ. **1989**, *86*, 75–81.
2. Kimbrough, D.E.; Cohen, Y.; Winer, A.M.; Creelman, L.; Mabuni, C. A Critical Assessment of Chromium in the Environment. Crit. Rev. Environ. Sci. Technol. **1999**, *29* (1), 1–46.
3. Goldhaber, S.; Vogt, C. Development of the Revised Drinking Water Standard for Chromium. Sci. Tot. Environ. **1989**, *86*, 43–51.
4. Bartlett, R.; James, B. Behavior of Chromium in Soils: III. Oxidation. J. Environ. Qual. **1979**, *8*, 31–35.
5. James, B.R.; Petura, J.C.; Vitale, R.J.; Mussoline, G.R. Oxidation–Reduction Chemistry of Chromium: Relevance to the Regulation and Remediation of Chromate-Contaminated Soils. J. Soil Contam. **1997**, *6* (6), 569–580.
6. Bartlett, R.J. Chromium Redox Mechanisms in Soils: Should We Worry About Cr(VI)? *Chromium Environmental Issues*; FrancoAngeli: Milan, 1997; 1–20.
7. James, B.R. Remediation-by-Reduction Strategies for Chromate-Contaminated Soils. Environ. Geochem. Health **2001**, *23* (3), 175–179.
8. James, B.R. The Challenge of Remediating Chromium-Contaminated Soil. Environ. Sci. Technol. **1996**, *30*, 248A–251A.
9. Fendorf, S.; Wielenga, B.W.; Hansel, C.M. Chromium Transformations in Natural Environments: The Role of Biological and Abiological Processes in Chromium(VI) Reduction. Intern. Geol. Rev. **2000**, *42*, 691–701.
10. Nriagu, J.O. Production and Uses of Chromium. In *Chromium in the Natural and Human Environments*; Nriagu, J.O., Nieboer, E., Eds.; Wiley-Interscience: New York, 1988; 81–103.
11. Barnhart, J. Chromium Chemistry and Implications for Environmental Fate and Toxicity. J. Soil Contam. **1997**, *6* (6), 561–568.
12. Ball, J.W.; Nordstrom, D.K. Critical Evaluation and Selection of Standard State Thermodynamic Properties for Chromium Metal and Its Aqueous Ions, Hydrolysis Species, Oxides, and Hydroxides. J. Chem. Eng. Data **1998**, *43*, 895–918.
13. Morry, D. Public Health Goal for Chromium in Drinking Water. Office of Environmental Health Hazard Assessment, California Environmental Protection Agency, 1999, 1–20.
14. Rai, D.; Sass, B.M.; Moore, D.A. Chromium(III) Hydrolysis Constants and Solubility of Chromium(III) Hydroxide. Inorg. Chem. **1987**, *26*, 345–349.
15. Sass, B.M.; Rai, D. Solubility of Amorphous Chromium(III)–Iron(III) Hydroxide Solid Solutions. Inorg. Chem. **1987**, *26*, 2228–2232.
16. Nieboer, E.; Jusys, A.A. Biologic Chemistry of Chromium. In *Chromium in the Natural and Human Environments*; Nriagu, J.O., Nieboer, E., Eds.; Wiley-Interscience: New York, 1988; 21–80.
17. James, B.R.; Bartlett, R.J. Behavior of Chromium in Soils: VII. Adsorption and Reduction of Hexavalent Forms. J. Environ. Qual. **1983**, *12*, 177–181.
18. James, B.R. Redox Phenomena. In *Encyclopedia of Soil Science*, 1st Ed.; Lal, R., Ed.; Marcel Dekker, Inc.: New York, 2002; 1098–1100.
19. James, B.R.; Bartlett, R.J. Redox Phenomena. In *Handbook of Soil Science*; Sumner, M.E., Ed.; CRC Press: Boca Raton, 2000; B169–B194.

Conservation Tillage and No-Tillage

Paul W. Unger (Retired)
United States Department of Agriculture (USDA),
Bushland, Texas, U.S.A.

INTRODUCTION

Conservation tillage is any tillage or tillage and planting system that results in at least a 30% cover of crop residues on the soil surface after planting the next crop.[1] It is used mainly to control soil erosion, but it also helps conserve water. In comparison, conventional tillage refers to tillage operations normally used for crop production that bury most residues and result in $< 30\%$ cover after planting. Tillage that incorporates all residues into soil is clean tillage.

Tillage methods such as sweep, chisel, paraplow, subsoiling, slit, and strip rotary can usually qualify as conservation tillage. Even disk tillage may qualify, provided adequate residues are retained on the surface. The ultimate conservation tillage method is no-tillage (or zero tillage) for which the next crop is planted without any soil disturbance since harvesting the previous crop. A special planter usually is needed to prepare a narrow, shallow seedbed for the seed being planted.[1] Sometimes, no-tillage is used in combination with a subsoiling operation that facilitates crop seeding and early plant root growth, but which leaves the surface residues virtually undisturbed, except for the slot caused by the subsoiling implement.[1]

Adequate residues are not always produced to provide 30% cover [e.g., dryland (nonirrigated) crops]. Also, a crop such as cotton (*Gossypium hirsutum* L.) may not produce enough residue under some conditions to satisfy the required ground cover for conservation tillage. Under such conditions, some conventional or even clean tillage methods can provide for soil and water conservation. Any tillage method that results in a rough or ridged surface helps reduce soil erosion by wind. Listing (ridge-forming tillage) commonly is used to help control wind erosion in the cotton-producing area of West Texas where residue amounts usually are low (personal observation). Even plowing that brings erosion resistant clods to the surface helps control wind erosion on some sandy soils.[2] Any tillage method that impedes or prevents water flow across the surface helps reduce soil erosion by water and usually helps conserve water. Listing on the contour retains water on the surface, thus reducing erosion and conserving water. Furrow diking in conjunction with listing improves water retention where contour tillage is not used.[3] Graded-furrow tillage allows excess water to flow slowly from land, thus reducing the potential for erosion; it also provides water conservation benefits.[4]

ADVANTAGES AND DISADVANTAGES OF USING CONSERVATION TILLAGE AND NO-TILLAGE

Advantages

Compared with clean tillage, advantages of different conservation tillage types, including no-tillage, include improved erosion control, a cleaner environment, greater water conservation, equal or greater crop yields, less equipment and maintenance cost, lower energy and labor requirements, and greater net returns. Erosion control benefits with conservation tillage result from retaining more residues on the soil surface. For controlling erosion by wind, residues shield the surface and reduce wind speed at the surface to below the threshold required for erosion to occur. Erosion by water is reduced because residues reduce the rate and amount of water flow across the surface. Residues also result in less soil particle detachment and transport due to raindrop splash and flowing water. The value of surface cover provided by crop residues for controlling erosion by wind and water is illustrated in Fig. 1.[5]

Greater water conservation with conservation tillage results from residues retarding the rate of water flow across the surface, thus providing more time for infiltration. Residues also shield the surface against raindrop impact, thus dissipating the energy of raindrops, reducing surface sealing, and maintaining favorable infiltration rates. Residues reduce soil water evaporation by shading the soil and slowing the wind at the soil surface. Of course, the soil must have adequate storage capacity for the water to be retained for later use by crops.

Use of conservation tillage reduces erosion, thus resulting in a cleaner environment. Erosion by wind damages crops, causes health and visibility problems,

Encyclopedia of Water Science
DOI: 10.1081/E-EWS 120010096
Published 2003 by Marcel Dekker, Inc.

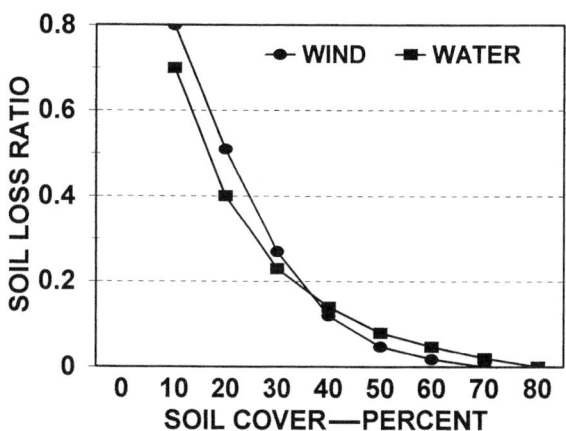

Fig. 1 Relationship between soil loss ratio (soil loss with cover divided by soil loss from bare soil). (Redrawn from Fig. 10 in Ref. [5].)

clogs roads and waterways, damages machinery and homes, and pollutes the air. Erosion by water damages crops, roads, machinery, and homes. It also pollutes water with soil particles, chemicals adhering to the particles, and chemicals dissolved in water.

Crop yields are affected by numerous factors. Yields with conservation tillage systems often are greater than with clean tillage, provided no major problems are encountered. Yield increases, especially with no-tillage, usually are attributable to greater soil water conservation, especially in subhumid and semiarid regions without irrigation. More favorable soil temperatures may be involved also. In warm or hot regions, high soil temperatures may injure plants, and surface residues with no-tillage result in temperature decreases of up to 10°C,[6,7] which result in better crop performance. In cool regions, low temperatures with no-tillage usually are detrimental to crop yields because planting is delayed beyond the optimum date.[8]

Advantages of lower equipment inventories, equipment maintenance, and energy and labor requirements with conservation tillage are interrelated. With most conservation tillage methods, and especially no-tillage, tillage frequency and intensity are lower than with clean tillage. As a result, less equipment may be needed, smaller tractors may be satisfactory (for no-tillage), and the tractors and equipment are used less frequently. This results in less equipment maintenance and in lower fuel and labor requirements. Some fuel energy savings, however, may be partially offset by the energy required to produce herbicides and fertilizer, especially where no-tillage is used. The no-tillage system is based on using herbicides for weed control, and more nitrogen fertilizer is used under some conditions, especially when first converting to the system.

As for yields, many factors affect net returns for a crop production system. However, if production costs are not

greater and yields are equal to or exceed those with clean tillage, then net returns should be equal or greater with conservation tillage, especially with no-tillage, because equipment inventories and maintenance and labor and energy requirements are lower.[9,10]

Disadvantages

Problems with conservation tillage, especially no-tillage, occur under some conditions.[8,11–13] A greater use of herbicides results in concern regarding the potential for polluting soil and water resources. Lower soil temperatures in cool regions delay crop planting, thereby potentially reducing crop yields. On poorly drained soils, additional water retained by using no-tillage aggravates the excess soil water problem, thus generally reducing crop yields. Some weeds are difficult to control with herbicides, which, along with the high cost of some herbicides, may increase production costs. The possible need for new equipment may also increase production costs, especially when a change to a no-tillage system is first made. Because crop residues are retained on the surface when a no-tillage system is used, there is the potential for increased pest problems (insects, diseases, rodents). Problems are greater with some insects and less with others, indicating that insect populations must be closely evaluated regardless of tillage system used. Organisms of some plant diseases are carried over to the next crop when residues are retained. Surface residues also provide shelter for rodents, which may be detrimental for the production of some crops. Other possible disadvantages include limited residue availability, greater soil compaction, and a need for greater managerial ability. Certainly, conservation tillage and no-tillage are not suitable for all conditions. However, with good management, most problems (real or potential) can be minimized or avoided.

RESULTS ACHIEVED BY USING CONSERVATION OR NO-TILLAGE

The value of conservation and no-tillage farming methods for controlling erosion, conserving water, and increasing crop yields has been shown in numerous studies. Because of space limitations, however, only few examples will be given. Probably the most dramatic example regarding the value of no-tillage for controlling erosion occurred during a rainstorm on watersheds planted to corn (*Zea mays* L.) in Ohio.[14] Treatments were clean tillage with sloping rows (land slope 6.6%), clean tillage with contour rows (land slope 5.8%), and no-tillage with contour rows (land slope 20.7%). On the respective treatment areas, rainfall was 140, 140, and 129 mm; runoff was 112, 58, and 64 mm; and sediment loss was 50.7, 7.2, and 0.07 Mg ha^{-1}. Even though the slope was much greater, soil loss was negligible from the no-tillage area. Runoff also was low, which provided an opportunity to store more soil water, but soil water information was not given.

After harvesting irrigated winter wheat (*Triticum aestivum* L.), moldboard-, rotary-, disk-, sweep-, and no-tillage treatments were imposed to manage the residues during the fallow period until planting dryland grain sorghum [*Sorghum bicolor* L. (Moench)] 10–11 mo later at Bushland, Texas. Weed control was similar with all treatments. Plant available soil water contents averaged 149, 143, 158, 179, and 207 mm at sorghum planting and sorghum grain yields averaged 2.56, 2.19, 2.37, 2.77, and 3.34 Mg ha^{-1} with the respective treatments. Greater water contents and yields with conservation tillage (sweep and especially no-tillage) resulted from more residues retained on the surface than with other treatments. The residues resulted in greater infiltration and lower evaporation, but the effect of the different processes could not be determined.[15]

A field study at Akron, Colorado, clearly showed the value of surface residues with conservation tillage (minimum- and no-tillage) for reducing evaporation. Soil water contents 1 day after a 13.5-mm rain were similar to the 15-cm depth where conventional-, minimum-, and no-tillage treatments were imposed after harvesting winter wheat. The treatments resulted in 1.2, 2.2, and 2.7 Mg ha^{-1} of surface residues, respectively. After 34 rainless days, the soil had dried to a $<0.1\,\mathrm{m}^3\,\mathrm{m}^{-3}$ water content to 12-, 9-, and 5-cm depths, respectively.[16] The value of surface residues for reducing evaporation also was shown under laboratory conditions.[17,18]

CONCLUSION

Conservation tillage and no-tillage farming systems are based on retaining sufficient crop residues on the soil surface, mainly to control erosion. Other benefits include water conservation; environmental protection; equipment, energy, and labor savings; and often greater net returns to the producer. Some disadvantages occur under some conditions and the systems, especially no-tillage, may not be suitable for all conditions. Most disadvantages, however, can be overcome or minimized by careful management.

REFERENCES

1. SSSA (Soil Science Society of America), *Glossary of Soil Science Terms, 1996*; Soil Science Society of America: Madison, WI, 1997.
2. Fryrear, D.W. Wind Erosion: Mechanics, Prediction, and Control. In *Dryland Agriculture, Strategies for Sustainability*, Advances in Soil Science; Singh, R.P., Parr, J.F.,

Stewart, B.A., Eds.; Springer-Verlag: New York, 1990; Vol. 13, 187–199.
3. Clark, R.N.; Jones, O.R. Furrow Dams for Conserving Rainwater in a Semiarid Climate. *Proceeding of Conference on Crop Production with Conservation in the 80s, Chicago, IL, December 1980*; Am. Soc. Agric. Eng.: St. Joseph, MI, 1981; 198–206.
4. Richardson, C.W. Runoff, Erosion, and Tillage Efficiency on Graded-Furrow and Terraced Watersheds. J. Soil Water Conserv. **1973**, *28* (4), 162–164.
5. Papendick, R.I.; Parr, J.F.; Meyer, R.E. Managing Crop Residues to Optimize Crop/Livestock Production Systems for Dryland Agriculture. In *Dryland Agriculture, Strategies for Sustainability*, Advances in Soil Science; Singh, R.P., Parr, J.F., Stewart, B.A., Eds.; Springer-Verlag: New York, 1990; Vol. 13, 253–272.
6. Allen, R.R.; Musick, J.T.; Wiese, A.F. *No-Till Management of Furrow Irrigated Continuous Grain Sorghum. Prog. Rpt. PR-3332 C*; Texas Agric. Exp. Stn.: College Station, 1975.
7. Rockwood, W.G.; Lal, R. Mulch Tillage: A Technique for Soil and Water Conservation in the Tropics. Span **1974**, *17*, 77–79.
8. Radke, J.K. Managing Early Season Soil Temperatures in the Northern Corn Belt Using Configured Soil Surfaces and Mulches. Soil Sci. Soc. Am. J. **1982**, *46* (5), 1067–1071.
9. Crosson, P.; Hanthorn, M.; Duffy, M. The Economics of Conservation Tillage. In *No-Tillage and Surface Tillage Agriculture*; Sprague, M.A., Triplett, G.B., Eds.; John Wiley and Sons: New York, 1986; 409–436.
10. Harman, W.L.; Martin, J.R. Economics of Conservation Tillage in Texas. In *Conservation Tillage: Today and Tomorrow*, Misc. Publ. MP-1634; Gerik, T.J., Harris, B.L., Eds.; Texas Agric. Exp. Stn.: College Station, 1987; 24–37.
11. Phillips, R.E.; Phillips, S.H. *No-Tillage Agriculture, Principles and Practices*; Van Norstrand-Reinhold: New York, 1984.
12. Sprague, M.A.; Triplett, G.B. *No-Tillage and Surface-Tillage Agriculture*; John Wiley and Sons: New York, 1986.
13. Unger, P.W. Reduced Tillage. In *Semiarid Lands and Deserts, Soil Resource and Reclamation*; Skujinš, J., Ed.; Marcel Dekker, Inc.: New York, 1991; 387–422.
14. Harrold, L.L.; Edwards, W.M. A Severe Rainstorm Test of No-Till Corn. J. Soil Water Conserv. **1972**, *27* (1), 30.
15. Unger, P.W. Tillage and Residue Effects on Wheat, Sorghum, and Sunflower Grown in Rotation. Soil Sci. Soc. Am. J. **1984**, *48* (4), 885–891.
16. Smika D.E. Seed Zone Soil Water Conditions with Reduced Tillage in the Semi-arid Central Great Plains. Proceedings 7th Conference of the International Soil Tillage Research Organization, Sweden; 1976.
17. Unger, P.W.; Parker, J.J. Evaporation Reduction from Soil with Wheat, Sorghum, and Cotton Residues. Soil Sci. Soc. Am. J. **1976**, *40* (6), 938–942.
18. Ji, Shangning; Unger, P.W. Soil Water Accumulation Under Different Precipitation, Potential Evaporation, and Straw Mulch Conditions. Soil Sci. Soc. Am. J. **2002**, in press.

Consumptive Water Use

Freddie R. Lamm
Kansas State University, Colby, Kansas, U.S.A.

C

INTRODUCTION

Consumptive water use is defined as the total quantity of water used in a given period of time as transpiration from the crop shoots and leaves, the water evaporated from the wetted soil or crop surfaces, and the small amount of water used in the building of plant tissue. In general, less than 1% of the consumptive water use is incorporated into plant tissue (i.e., split in the light reaction of photosynthesis and then incorporated), so consumptive water use is often used synonymously with the term evapotranspiration (sum of evaporation and transpiration). The principal factors affecting the magnitude of consumptive water use are the amount and orientation of actively transpiring plant tissues, atmospheric conditions, soil-water reserves, and soil texture.

ORIGIN

The term consumptive water use apparently originated in the United States during the early part of the twentieth century[1,2] to describe water and/or irrigation requirements of crops. One of the earliest recorded documentations of the term was by the American Society of Civil Engineers in 1930.[3] Although worldwide, evapotranspiration is probably a more highly utilized term, consumptive water use is still used in the United States, particularly in federal and state management agencies and legal institutions. In European countries, evaporation is sometimes used instead of evapotranspiration in a context that covers evaporative losses from water surfaces, soil, or plants.[4]

UTILIZATION

Information about consumptive water use is utilized in the planning, development, and management of almost all water resources and supply projects, not just irrigation projects. For example, water-resource planners must have estimates of consumptive water use of forests and rangelands when determining long-term yield (runoff) from such lands in planning for reservoirs. Consumptive water use estimates are utilized in planning of wastewater-reuse systems, so that a given parcel of land is not overloaded hydraulically with water, resulting in excessive runoff or deep percolation. Government agencies often rely on estimated consumptive water use values to develop interstate river and stream compacts and to mediate disputes arising from these compacts. Legal institutions may carefully differentiate consumptive water use from the total water diverted from a resource, to determine what water is truly lost from a surface and/or ground water basin. Conversely, a legal institution might be more keenly interested in promoting crops that maximize the consumptive water use, if evapotranspiration is being utilized to clean up or reduce a contaminated water source.

The time scale for which consumptive water use is determined depends on the needs of the end-user. A modern irrigator may schedule irrigation based on hourly, daily, or weekly estimates of consumptive water use. The same irrigator, in planning for a new irrigation system, might need to extend these estimates to include monthly and seasonal estimates. In planning, irrigation system application amounts for a single irrigation event, it is good design practice to match the peak consumptive water use for the critical crop growth periods. In planning the overall irrigation system size, it is necessary to consider the consumptive water use over the entire season to ensure that sufficient seasonal water is available for the planned irrigated area. Similarly, a wastewater-reuse system operator may need to know the consumptive water use of a crop during distinct short periods of time to prevent hydraulically overloading the soil. That same operator might use annual-consumptive water use to size the wastewater-storage reservoirs and land area used for application. Hydrologists and other water-resource planners may use time scales ranging from hourly to as much as a decade, depending on their accuracy needs and the intended use of the information.

PARTITIONING OF CONSUMPTIVE WATER USE

It is difficult to make generalizations about the partitioning of consumptive water use into the major components of evaporation and transpiration. The amount of evaporation from the wetted soil and the wetted cropped surfaces

Encyclopedia of Water Science
DOI: 10.1081/E-EWS 120010164

depends heavily on how often those surfaces are wetted by precipitation or irrigation and the ratio of soil to crop surface. Although there is no single value that can adequately describe the evaporation fraction of consumptive water use, a seasonal value of 20% may be of sufficient accuracy in many cases. Transpiration depends more heavily on the amount of actively growing leaves and shoots, their exposure to atmospheric conditions, and the ability of the plant roots to extract water from the soil layers. Nevertheless, the partitioning of these two major components is of great importance in managing water resources.

Evaporative Component

Evaporation from soil surfaces is generally described as occurring in two or three phases, the energy-limiting stage, the rapidly-falling stage, and the slowly-falling rate stage (Fig. 1). Early work described the process as three phases.[5] It was later recognized that the latter two stages could be adequately described as one soil-limiting stage by expressing evaporation as decreasing with the square root of time.[6,7] The first stage of soil evaporation when the soil surface is wet occurs at a rate that is only limited by the energy available (atmospheric demand and latent heat in soil storage) to evaporate water. The soil-limiting stage begins when water does not diffuse to the soil surface in sufficient quantity to meet the evaporative demand of the available energy.

Direct evaporation from plant tissues encompasses water that is temporarily trapped (canopy interception storage) on the plant leaves and shoots following rainfall or sprinkler irrigation. This water may be in the form of droplets on leaves or larger amounts trapped in leaf whorls

or joints between the leaf and the shoot. This evaporation generally occurs at a rate that is only limited by the amount of energy available for evaporation. However, this evaporative loss will temporarily suppress plant transpiration during the evaporation period.[8] Canopy interception storage and the resultant evaporative loss will vary with plant type and structure and with the ratio of evaporative demand to precipitation rate. On an annual basis, these evaporative losses can be a significant factor in forest hydrology, ranging from 20% to 40% for conifer forests and 10% to 20% for hardwood forests.[9] Direct evaporative losses from interception storage following a single precipitation or sprinkler irrigation event for a fully developed corn canopy is approximately 1.5 mm–2.5 mm.[10–13]

Transpiration

The other major component of consumptive water use, transpiration, is usually larger than the evaporation component because the plant has multiple transpiring surfaces exposed to the atmospheric demand, and the plant roots and stem can also transport water from deeper soil layers to the transpiring crop surfaces. Attempts to directly measure transpiration also have their limitations, similar to difficulties with measuring the evaporative component. These attempts include measurements of plant water use from large pots,[14,15] alternate lysimeter comparisons,[16,8] using portable translucent field chambers in the field,[17] measuring water flowrate in plant stems,[18] modeling of evaporation and transpiration processes,[19,20] and algebraic manipulations of the E and T components using a combination of measurement methods that partially define a given component.

MODELING OF CONSUMPTIVE WATER USE

Although the various evaporation and transpiration processes have been studied for centuries,[21] Penman[22] is often credited with the pioneering research in establishing a modern physical basis for modeling evaporation and transpiration. Even recent efforts to encourage adoption of a more standardized method of calculating evapotranspiration[23,24] use the basic framework outlined by Penman. Because crop type, size, and leaf orientation all can affect evapotranspiration, the term reference evapotranspiration is often used to express evapotranspiration based on atmospheric demand for a given reference crop under specific growth conditions. A modern equation to calculate reference evapotranspiration gaining credibility and acceptance is referred to as the FAO-56 Penman–Monteith equation.[23] Major atmospheric variables in this equation are net radiation, air

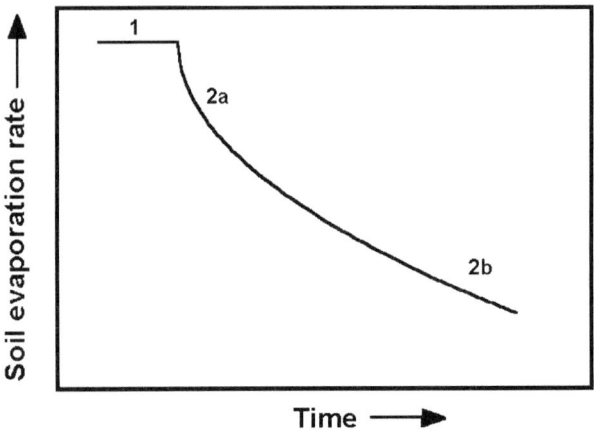

Fig. 1 Typical soil evaporation as related to time since wetting. Stage 1 is only limited by available energy. The soil limiting stages 2a (rapidly falling) and 2b (slowly falling) are a function of the square root of time since the end of stage 1.

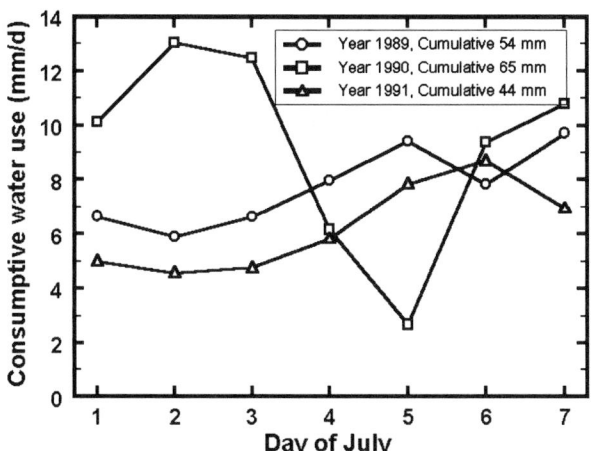

Fig. 2 Variation in consumptive water use of corn (*Zea mays* L.) over a 7-day period in July for the years 1989–1991 at Colby, Kansas. Note that cumulative consumptive water use amounts for the three years varied from 44 mm to 65 mm for the 7-day period.

In some regions of the world, consumptive water use may not vary much on a daily or annual basis. In other areas, such as the U.S. Great Plains, moving atmospheric fronts may drastically change values from one day to the next (Fig. 2) and general climatic conditions may result in large cumulative differences between years (Fig. 3). These temporal variations emphasize that using historical averages for consumptive water use may have limited value in some regions for accurate short-term forecasting or irrigation scheduling.

temperature, wind speed, and the saturation vapor pressure deficit. Soil heat flux is another variable in the equation, but can sometimes be neglected, depending on the time step of the calculation. Crop and soil coefficients are then used to modify the reference evapotranspiration to determine the consumptive water use for the period of interest. Crop coefficients generally vary with crop and stage of growth and often are empirically derived or calibrated for a given locale. Soil coefficients are used to decrease the calculated consumptive water use when soil water redistribution begins limiting water transport to the transpiring surfaces.

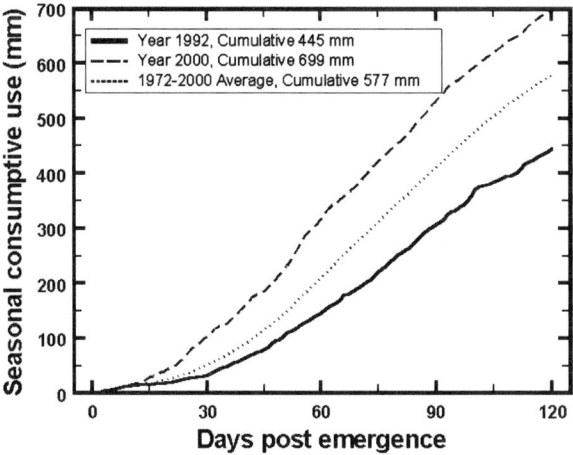

Fig. 3 Variation in seasonal consumptive water use of corn (*Zea mays* L.) at Colby, Kansas between an extremely cool, damp year (1992) and an extremely hot, dry year (2000) as compared to the long term-average value.

REFERENCES

1. Jensen, M.E.; Burman, R.D.; Allen, R.G., (Eds.) Introduction. *Evapotranspiration and Irrigation Water Requirements*, ASCE Manuals and Rep. on Engr. Practice No. 70; Amer. Soc. Civil Engr.: New York, NY, 1990; Chap. 1. 1–8.
2. Howell, T.A.; Cuenca, R.H.; Solomon, K.H. Crop Yield Response. In *Management of Farm Irrigation Systems*; Hoffman, G.J., Howell, T.A., Solomon, K.H., Eds.; Amer. Soc. Agric. Engr.: St. Joseph, MI, 1990; Chap. 5, 93–122.
3. ASCE; Consumptive Use of Water in Irrigation. Progess Report. Trans. Am. Soc. Civil Eng. **1930**, *94*, 1349–1399.
4. Monteith, J.L. Evaporation from Land Surfaces: Progress in Analysis and Prediction Since 1948. In *Proc. Nat'l Conf. on Advances in Evapotranspiration, Chicago, IL, Dec 16–17, 1985*; Amer. Soc. Agric. Engr.: St. Joseph, MI, 1985; 4–12.
5. Kolasew, F.E. Ways of Suppressing Evaporation of Soil Moisture. Sb. Rab. Agr. Fix. **1941**, *3*, 67.
6. Gardner, W.R. Solutions of the Flow Equation for the Drying of Soils and Other Porous Media. Soil Sci. Soc. Am. Proc. **1959**, *23*, 183–187.
7. Ritchie, J.T. Model for Predicting Evaporation from a Row Crop with Incomplete Cover. Water Resour. Res. **1972**, *8* (5), 1024–1213.
8. Tolk, J.A.; Howell, T.A.; Steiner, J.L.; Krieg, D.R.; Schneider, A.D. Role of Transpiration Suppression by Evaporation of Intercepted Water in Improving Irrigation Efficiency. Irrig. Sci. **1995**, *16*, 89–95.
9. Zinke, P.J. Forest Interception Studies in the United States. In *Forest Hydrology*; Sopper, W.E., Lull, H.W., Eds.; Pergamon: Oxford, England, 1967; 137–161.
10. Lamm, F.R.; Manges, H.L. Partitioning of Sprinkler Irrigation Water by a Corn Canopy. Trans. Am. Soc. Agric. Eng. (ASAE) **2000**, *43* (4), 909–918.
11. Steiner, J.L.; Kanemasu, E.T.; Clark, R.N. Spray Losses and Partitioning of Water Under a Center Pivot Sprinkler System. Trans. Am. Soc. Agric. Eng. (ASAE) **1983**, *26* (4), 1128–1134.
12. Seginer, I. Net Losses in Sprinkler Irrigation. Agric. Meteorol. **1967**, *4*, 281–291.
13. Smajstrla, A.G.; Hanson, R.S. Evaporation Effects on Sprinkler Irrigation Efficiencies. Proc. Soil Crop Sci. Soc., Florida **1980**, *39*, 28–33.

14. Briggs, L.J.; Shantz, H.L. The Water Requirements of Plants: I. Investigations in the Great Plains in 1910 and 1911. USDA Bur. Plant Indr. Bull. **1913**, *284*, 49.

15. Kiesselbach, T.A. Transpiration as a Factor in Crop Production. Res. Bull. No. 6; Nebraska Ag. Expt. Station, 1916; 214.

16. Klocke, N.L.; Heermann, D.F.; Watts, D.G. Daily Trends in Soil Evaporation and Plant Transpiration. In *Proc. Specialty Conf. Advances in Irrigation and Drainage. Surviving External Pressures, Jackson, WY, Jul 20–22, 1983; Amer. Soc. Civil Engr.: New York, NY, 1983*; 55–62.

17. Reicosky, D.C. Advances in Evapotranspiration Measured Using Portable Field Chambers. In *Proc. Nat'l Conf. on Advances in Evapotranspiration, Chicago, IL, Dec 16–17, 1985*; Amer. Soc. Agric. Engr.: St. Joseph, MI, 1985; 79–86.

18. Peressotti, A.; Ham, J.M. A Dual-Heater Gage for Measuring Sap Flow with an Improved Heat-Balance Method. Agron. J. **1996**, *88* (2), 149–155.

19. Thompson, A.L.; Martin, D.L.; Norman, J.M.; Howell, T.A. Scheduling Effects on Evapotranspiration with Overhead and Below Canopy Application. In *Proc. Int'l Conf. Evapotranspiration and Irrigation Scheduling, San Antonio, TX, Nov 3–6, 1996*; Camp, C.R., Sadler, E.J., Yoder, R.E., Eds.; Amer. Soc. Agric. Engr.: St. Joseph, MI, 1996; 182–188.

20. Norman, J.M.; Campbell, G.S. Application of a Plant-Environment Model to Problems in Irrigation. In *Advances in Irrigation*; Hillel, D., Ed.; Academic Press: New York, NY, 1983; Vol. 2, 155–188.

21. Brutsaert, W.H. *Evaporation into the Atmosphere*; D. Reidel Publishing Co.: Dordrecht, Holland, 1982; 299.

22. Penman, H.L. Natural Evaporation from Open Water, Bare Soil and Grass. Proc. R. Soc. Lond., Ser. A **1948**, *193*, 120–146.

23. Allen, R.G.; Pereira, L.S.; Raes, D.; Smith, M. Crop Evapotranspiration: Guidelines for Computing Crop Water Requirements. United Nations FAO, Irrig. and Drain. Paper No. 56; Rome, Italy, 1998; 300.

24. Allen, R.G.; Elliot, R.; Mecham, B.; Jensen, M.E.; Itensifu, D.; Howell, T.A.; Snyder, R.; Brown, P.; Echings, S.; Spofford, T.; Hattendorf, M.; Cuenca, R.H.; Wright, J.L.; Martin, D. Issues, Requirements, and Challenges in Selecting and Specifying a Standardized ET Equation. In *Proc. 4th Decennial National Irrig. Symp., Phoenix, AZ, Nov 14–16, 2000*; Amer. Soc. Agric. Engrs.: St. Joseph, MI, 2000; 201–208.

Crop Coefficients

Richard G. Allen
University of Idaho Research and Extension Center, Kimberly, Idaho, U.S.A.

C

INTRODUCTION

Crops and vegetation on the earth's surface vary in height, amount of leaf area, amount of soil shaded, color, amount of stomatal control to evaporation, and amount of soil wetness beneath the canopy. All of these factors affect, to some degree, the amount of evapotranspiration (ET) from the crop or vegetation. Rather than assigning parameters for all of these terms during the process of predicting ET from a specific type of vegetation using an ET equation, as covered in the entry on Evapotranspiration Formulas, the impacts of these variables are often lumped into a single parameter, termed the crop coefficient, K_c. This approach is done to reduce the complexity and time requirement for predicting ET for each type of crop or vegetation, and relies upon a common "reference ET" for a defined type of reference vegetation to represent the change in ET caused by variation in weather parameters. K_c is defined as the ratio of ET from a crop or soil surface to ET from the reference surface. Reference ET is the ET from a fully vegetated surface covering the soil, and normally represents ET from clipped grass (termed ET_o) or alfalfa (termed ET_r).

OVERVIEW

In general, four primary characteristics distinguish crop ET from reference ET: 1) crop cover density and total leaf area; 2) resistance of foliage epidermis and soil surface to the flow of water vapor; 3) aerodynamic roughness of the crop canopy; and 4) reflectance of the crop and soil surface to short wave radiation.

When the K_c is known, crop ET (ET_c) is calculated for a specific time period as:

$$ET_c = K_{co}ET_o \text{ and } ET_c = K_{cr}ET_r \qquad (1)$$

where K_{co} is the K_c for the grass ET_o basis and K_{cr} is the K_c for the alfalfa ET_r basis. Because reference ET represents nearly all effects of weather, K_c varies predominately with specific crop characteristics and only a small amount with climate. This enables the transfer of standard values and curves for K_c between locations and climates. This transfer has led to the widespread acceptance and usefulness of the K_c approach. K_c has been primarily developed and applied

to agricultural situations. However, K_c is generally valid for natural vegetation and conditions including open water, although it can have large spatial variability. In situations where K_c has not been derived by ET measurement, it can be estimated from fraction of ground cover or leaf area index (LAI), using procedures in Refs. 1 and 2.

K_c varies during the growing season as: the plants develop, the fraction of ground covered by vegetation changes, and the plants age and mature (Fig. 1). K_c varies according to the wetness of the soil surface, especially when there is little vegetation cover. Under bare soil conditions, K_c has a high value when soil is wet and its value steadily decreases as the soil dries (Fig. 2).

CROP COEFFICIENT CURVES

Two different approaches are used to calculate K_c. The simpler approach uses a single K_c curve that represents time-averaged effects of evaporation from the soil surface. The result is a relatively smooth, consistently increasing or decreasing K_c curve (Fig. 2). The second K_c approach separates the K_c into two coefficients, with one coefficient, the basal crop coefficient, termed K_{cb}, representing K_c for a dry soil surface (with or without vegetation) having little evaporation but full transpiration. The second coefficient, the evaporation coefficient, K_e, represents the evaporation component from the soil surface (Fig. 2). The value for K_e changes daily as the soil surface wets or dries, whereas the value for K_{cb} is more consistent day-to-day:

$$K_c = K_s K_{cb} + K_e \qquad (2)$$

where K_s [0–1] represents the reduction in K_c due to environmental stresses, primarily from soil water shortage or soil salinity. All four terms are dimensionless. In application of the dual $K_{cb} + K_e$ procedure, a daily calculation must be made to estimate water content and associated evaporation rate from the soil surface, so that the approach is relatively computationally intensive. However, estimates can be up to 50% more accurate for any particular day, as compared with the single K_c approach, especially for the first few days following soil wetting during initial and development periods. The dual

Encyclopedia of Water Science
DOI: 10.1081/E-EWS 120010037

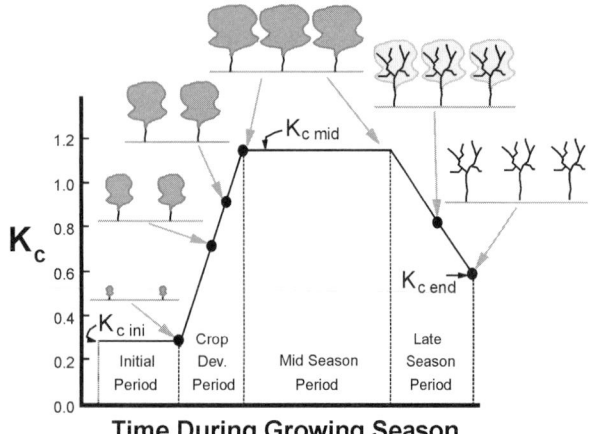

Fig. 1 General K_c curve showing relationship between stage of growth and K_c. (After Ref. 1.)

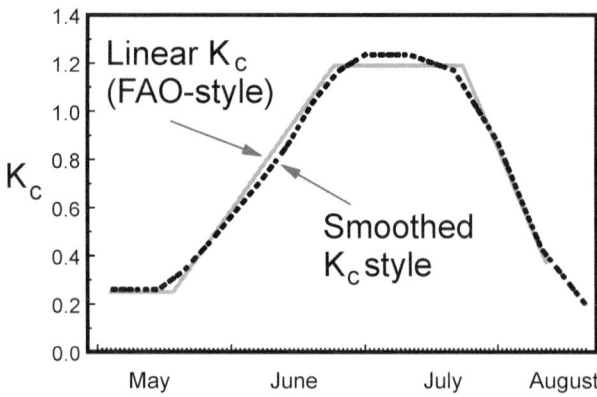

Fig. 3 Typical styles of crop coefficient curves.

procedure is applied on a daily timestep and is readily adapted to spreadsheet programs.

Time-averaged (single) K_c is used for planning studies and irrigation or water resources systems design where averaged effects of soil wetting are appropriate. The dual K_c approach is better for irrigation scheduling, soil water balance computations, and research where specific effects of day-to-day variation in soil wetness are important.

K_c (or K_{cb}) changes during a growing season, reflecting changes in the vegetation and ground cover. Initially, K_c is small, generally between 0.1 and 0.4, and K_{cb} is between 0.0 and 0.2. K_c increases during the period of rapid plant growth until it reaches a maximum value at the time of near maximum ground cover. Towards the end of the growing cycle, K_c decreases as plants age, ripen, or die due to natural or cultural practices.

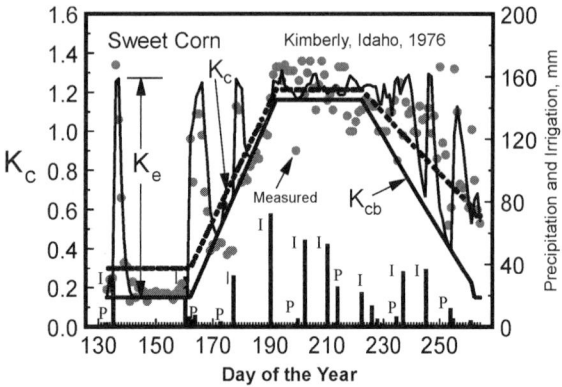

Fig. 2 Basal K_{cb}, soil evaporation coefficient K_e, and time-averaged (single) K_c (dotted line) curves for a crop of sweet corn grown near Kimberly, Idaho during 1976. Also shown are actual measurements of K_c (dots) determined from weighing lysimeters. (Data from Dr. J.L. Wright, USDA-ARS, Kimberly.)

Styles of Crop Coefficient Curves

Fig. 3 illustrates two common shapes used to represent K_c curves for growing seasons. Smooth curves as in Ref. 3 exhibit a smoothed change in K_c with time, whereas linearly shaped K_c curves as in Ref. 1 are constructed using four line segments. Both shapes are useful and valid for predicting K_c.

Definition of Growing Periods Within the Growing Season

A growing season can be divided into four basic periods as shown in Fig. 1. The initial period represents the period following planting of annuals until about 10% ground cover or following initiation of leaves for perennials. The development period extends from the end of the initial period until the crop reaches "effective full cover." Mid-season extends from effective full cover to when plant vigor or greenness begin to decrease. The late-season period extends from end of mid-season until harvest or crop death. Information on relative lengths of growing periods of crops is found in Ref. 1.

Effective full cover for row crops occurs when leaves between rows of plants begin to intermingle, or when plants reach nearly full size, if no intermingling occurs. For crops taller than 0.5 m, effective full cover is reached when the average fraction of ground surface shaded by vegetation at solar noon is about 0.7–0.8. Effective full cover for many crops begins at flowering. Plants may continue to grow in both height and leaf area after the attainment of effective full cover. Effective full cover can be predicted when the crop reaches an LAI of 3, where LAI is defined as the total area of leaves (one side only) per unit area of ground. The beginning of the late season is generally signaled by the beginning of yellowing or senescence of leaves for annual crops, leaf drop, or browning of fruit.

Construction of a Linear K_c Curve

Only three defined values for K_c are required to construct the linear K_c curve: K_c during the initial period ($K_{c\,ini}$), K_c during the mid-season period ($K_{c\,mid}$), and K_c at the time of harvest or crop death ($K_{c\,end}$). In addition, lengths of the four growing season periods, in days, are needed.

Grass-based K_cs. General values for $K_{c\,ini}$, $K_{c\,mid}$, and $K_{c\,end}$ and basal $K_{cb\,ini}$, $K_{cb\,mid}$, and $K_{cb\,end}$ for primary types of crops and conditions are listed in Table 1 from Ref. 1. These values are K_{co} based on grass reference ET_o as defined by the FAO-56 Penman–Monteith equation. Details on calculating ET_o are given under the entry on Evapotranspiration Formulas and in Ref. 1. The Penman–Monteith method was selected by FAO-56 as the best method for standardized calculation of reference ET from a clipped cool-season grass. Cool-season grass is a standard for ET_o worldwide because it can be grown over a wide range of climates and is relatively easy to maintain. Generally, ET_o is computed by ET equation rather than measured. K_c and K_{cb} are listed for specific crops in Refs. 1–4.

There is close similarity in K_c among crops having similar characteristics, e.g. among crops in the vegetable groups, since plant height, leaf area, ground coverage, and water management are similar. $K_{c\,ini}$ values in Table 1 are approximate. Graphs and equations in Ref. 1 provide better estimates for $K_{c\,ini}$ that account for frequency of wetting and soil type.

Alfalfa-based K_cs. Wright[3,4] established crop coefficients for crops common to central and northern latitudes of the Western United States. These coefficients are based on the alfalfa reference ET_r represented by the 1982 Kimberly Penman Equation.[3] Alfalfa is sometimes preferred as the reference crop rather than clipped grass because it is taller than grass and has ET that is more similar to maximum ET from many agricultural crops.[3] Therefore, K_{cr}s based on ET_r generally peak at values of 1.0. Values for K_{cr} cannot be interchanged with values for K_{co} and vice versa. Values for K_{co} average about 15%–30% higher than K_{cr}.

Crop Coefficients Applied to Hourly Time Periods

For many crops the ratio of ET_c to ET_o or ET_r is relatively constant during the day. Therefore, K_c is relatively constant during the day, also, as shown in Fig. 4 for a sugar

Table 1 Time-averaged single crop coefficients, and basal crop coefficients for well-managed crops in subhumid climates, for use with ET_o

Crop	Single K_c			Basal K_{cb}		
	$K_{c\,ini}$	$K_{c\,mid}$	$K_{c\,end}$	$K_{cb\,ini}$	$K_{cb\,mid}$	$K_{cb\,end}$
Small vegetables	0.7	1.05	0.95	0.15	0.95	0.85
Vegetables—roots	0.5	1.10	0.95	0.15	1.00	0.85
Vegetables—legumes	0.4	1.15	0.55	0.15	1.10	0.50
Vegetables—solanum family	0.4	1.15	0.80	0.15	1.10	0.70
Vegetables—cucumber family	0.4	1.00	0.80	0.15	0.95	0.70
Fiber crops	0.35	1.15	0.70	0.15	1.10	0.60
Oil crops	0.35	1.15	0.30	0.15	1.10	0.25
Cereals	0.3	1.15	0.4	0.15	1.10	0.25
Forages	0.60	1.15	1.10	0.60	1.10	1.05
Sugar cane	0.40	1.25	0.75	0.15	1.20	0.70
Grapes and berries	0.30	1.00	0.50	0.20	0.95	0.45
Fruit trees	0.60	0.95	0.75	0.50	0.90	0.70
Bare soil						
Wet	1.00	1.20	1.20	—	—	—
Dry	0.15	0.15	0.15	0.00	0.00	0.00
Wetlands	0.60	1.20	0.60	0.50	1.15	0.50
Open water						
< 2 m depth or in subhumid clim. or tropics	—	1.05	1.05	—	—	—
> 5 m depth, clear	—	0.75	1.25	—	—	—

(After Ref. 1.)

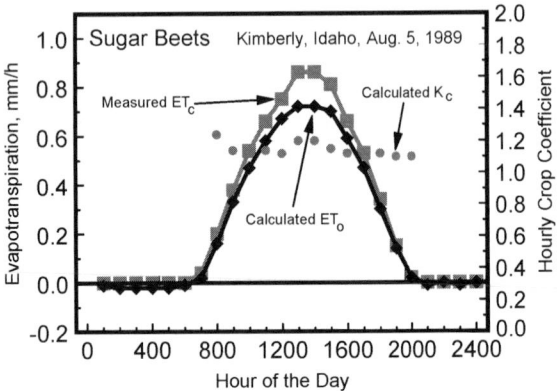

Fig. 4 Measured ET_c (by precision lysimeter) and calculated ET_o and K_c for a sugar beet crop near Kimberly, Idaho for hourly periods during August 5, 1989. (Data from Dr. J.L. Wright, USDA-ARS, Kimberly.)

beet crop near Kimberly, Idaho. ET_o was calculated using the FAO Penman–Monteith ET_o method.

ADJUSTMENT OF K_{CO} TO ACCOUNT FOR EFFECTS OF CLIMATE

K_cs based on grass ET_o (K_{co}) are somewhat impacted by general climate. Under humid conditions, K_{co} does not exceed about 1.05–1.10 because the vapor pressure deficit (VPD) driving ET is small and K_{co} becomes less dependent on the differences between the aerodynamic characteristics of crop and reference. Under arid conditions, the effect of differences in aerodynamic characteristics between crop and grass reference become more pronounced because the VPD of the air is relatively large. Hence, K_{co} for tall crops under arid conditions can be as high as 1.2 or more. Because alfalfa ET_r is more aerodynamically rough, values for K_{cr} generally do not vary with climate.

K_C DURING NONGROWING PERIODS

The value for K_c for periods following crop harvest or death will depend on the average water content of the soil surface and amount of vegetation or mulched cover remaining. When the soil surface is mostly bare, K_c can be set equal to $K_{c\,ini}$, and figures and equations for $K_{c\,ini}$ from Ref. 1 can be applied. When dead and dry vegetation or mulch covers the soil surface, K_c will be less than $K_{c\,ini}$. K_c following harvest can be estimated using guidelines in Chapters 9 and 11 of Ref. 1.

COEFFICIENTS FOR LIMITED WATER

The value for K_c is reduced when soil water content of the plant root zone is too low to sustain transpiration at the level predicted by Eq. 1. The reduction is accomplished by multiplying K_{cb} (in Eq. 2) or the single K_c (in Eq. 1) by the water stress coefficient, K_s, predicted for effects of limited water as

$$K_s = \frac{\theta - \theta_{WP}}{\theta_t - \theta_{WP}} \tag{3}$$

where θ is mean volumetric soil water content in the root zone ($m^3\,m^{-3}$), θ_t is the threshold θ for the root zone, below which transpiration is decreased ($m^3\,m^{-3}$), and θ_{WP} is the soil water content at the wilting point ($m^3\,m^{-3}$). Equation 3 is applied when $\theta \geq \theta_t$, and $K_s = 1.0$ for $\theta \exists \theta_t$.

REFERENCES

1. Allen, R.G.; Pereira, L.S.; Raes, D.; Smith, M. *Evapotranspiration and Crop Water Requirements*; Irrigation and Drainage Paper No. 56; FAO: Rome, 1998.
2. Allen, R.G.; Pruitt, W.O.; Businger, J.A.; Fritschen, L.J.; Jensen, M.E.; Quinn, F.H. Evaporation and Transpiration. *Hydrology Handbook*, 2nd Ed.; ASCE: New York, 1996; 125–252.
3. Wright, J.L. New Evapotranspiration Crop Coefficients. J. Irrig. Drain. Div., ASCE **1982**, *108*, 57–74.
4. Wright, J.L. Crop coefficients for estimates of daily crop evapotranspiration. In *Irrigation Scheduling for Water and Energy Conservation in the 80's*, Proceedings of the National Irrigation and Drainage Conference of ASCE, Albuquerque, NM, July 14–16, 1981; 18–26.

Crop Development Models

Peter S. Carberry
*Commonwealth Scientific and Industrial Research Organisation (CSIRO),
Toowoomba, Queensland, Australia*

INTRODUCTION

With crop development models there exists a hierarchy of approaches, which operate at varying levels of complexity, both in terms how component processes are addressed and in the way these processes are represented mathematically and within software products. The complexity of a model is partly determined by the nature of the issue that motivated model development, and partly by the data available to run the model. Nevertheless, crop simulation models generally simulate the changing state of a crop–soil system given initial system conditions, management interventions to the system, and values for the environmental variables that drive the system. Timesteps for data input and output are generally daily, although shorter durations are sometimes used for component processes. While crop yield is a primary output of such models, changes in other state variables, such as soil water or fertility status, are also often of interest. There are a number of reviews of crop simulation models and their make-up (e.g., Ref. [1]).

The majority of crop models can be described as stand-alone software, where growth of a single crop is simulated in response to climatic and soil conditions and to information on crop management. Less common are cropping systems models, which simulate multiple crop species growing in sequence or in combination. The soil component mostly consists of a soil water balance but, in some cases, it may also include a soil nutrient balance. Daily maximum and minimum temperatures, solar radiation, and rainfall are the most common climatic inputs, although pan evaporation, wind speed, and relative humidity are also sometimes used. In response to these inputs, most models simulate key physiological processes, including phenological development, leaf canopy development, radiation interception, conversion of absorbed energy into photosynthates, and partitioning of assimilates between plant components, including yield (Fig. 1).

The simulation of crop transpiration, soil water extraction by roots, water evaporation from the soil surface, and reduced growth under conditions of water deficit result in almost all crop models being responsive to variable soil water contents. This basic framework, or close derivatives of it, have formed the basis of much of the quantitative analysis of crop growth and resulted in the integration of this knowledge into many of the current crop simulation models.

PHYSIOLOGICAL DETERMINANTS OF CROP GROWTH

Thermal Time

Crop duration is often highly correlated with temperature such that crops will take different times from sowing to maturity under different temperature regimes. The concept of thermal time is the mechanism used to represent a crop's evolved requirement to accumulate a minimum time for development through each essential growth stage. Thermal time is also referred to as heat units, day-degrees, or growing degree days and has units of °C day.

Thermal time each day (δTT, °C day) is calculated from a broken linear function of temperature (T), using the following three equations:

$$\delta TT = 0 \quad T < T_b \text{ or } T > T_m$$

$$\delta TT = T - T_b \quad T_b < T < T_o$$

$$\delta TT = (T_o - T_b)[1 - (T - T_o)/(T_m - T_o)]$$

$$T_o < T < T_m$$

where T_b is a base temperature, T_o an optimum temperature, and T_m a maximum temperature beyond which development ceases. Values for T_b, T_o, and T_m differ for different crops, although as a general rule summer-growing crops have values in the order of 10°C, 30°C, and 40°C, respectively, while the values for winter-growing crops are closer to 0°C, 20°C, and 35°C.

Crop Phenology

The phenology of most crops can be described using distinct developmental phases—e.g., 1) sowing to germination; 2) germination to emergence; 3) a period of vegetative growth after emergence during which the plant

Encyclopedia of Water Science
DOI: 10.1081/E-EWS 120010282

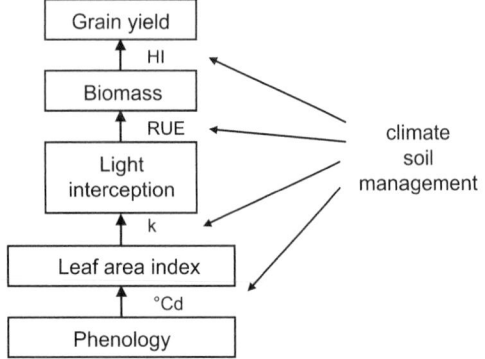

Fig. 1 A basic framework describing the physiological determinants of crop growth, development and yield—terms are described in the text.

is unresponsive to photoperiod; 4) a photoperiod-induced phase (PIP), which ends at floral initiation; 5) a flower development phase, which ends at 50% flowering; 6) a lag phase prior to commencement of grain filling; 7) a linear phase of grain filling; and 8) a period between the end of grain filling and physiological maturity. These phases are generally modeled as functions of temperature 1)–8) and photoperiod 4). The sequence of these phases provides the developmental time course against which growth processes such as carbon accumulation and partitioning can be mapped.

Daily thermal time (δTT) is accumulated during each phase of development until accumulated thermal time thresholds (θ) are satisfied and then development progresses to the next phase. A set, cultivar-specific, thermal time is often required to complete most developmental phases. As most crops are photoperiod-sensitive, the duration of the PIP changes as photoperiod changes. Photoperiod (hr) is equal to daylength plus civil twilight. The thermal time requirement for PIP (θ_{PIP}) is recalculated each day during PIP as a function of daily photoperiod (p), photoperiod sensitivity of the cultivar (p_s, °C day hr^{-1}) and its maximum optimal photoperiod (p_b). For plants where flowering is hastened under short daylength, i.e., short-day plants (generally summer growing crops),

$$\theta_{PIP} = p_s(p - p_b)$$

where, for photoperiods less than or equal to p_b, θ_{PIP} equals zero. For plants where flowering is hastened under long daylengths, i.e., long-day plants (generally winter-growing crops),

$$\theta_{PIP} = p_s(p_b - p)$$

and θ_{PIP} equals zero for photoperiods greater than or equal to p_b. Progress (r) through PIP can be calculated as

$$r = \Sigma(\delta TT / \theta_{PIP})$$

and PIP ends when r is greater than or equal to 1.

Leaf Area Development and Light Interception

A crop's canopy can be defined in terms of its photosynthetically active or green leaf area, made up of the total lamina area of emerged leaves less the area already senesced. The daily change in plant leaf area (ΔA, mm^2 plant^{-1}) can be described using functions in the form

$$\Delta A = \lambda \delta TT - \Delta S$$

where δTT is daily thermal time (°C day), λ is increase in plant leaf area per unit of thermal time (mm^2 plant^{-1} °C day^{-1}) and ΔS is change in leaf area senescence (mm^2 plant^{-1}).

Leaf area index (L, mm^2 mm^{-2}) is the ratio of green leaf area of the crop per unit of ground area and is the parameter required to describe the light relations of crop canopies. Thus the amount of light intercepted (I) has been adequately described using Beer's law,[2] such that

$$I = I_o(1 - e^{-kL})$$

where I_o is incoming daily solar radiation (MJ m^{-2}) and k is the light extinction coefficient of the canopy. The value of k increases for situations where efficacy of light interception increases, for example with narrow row spacing or for genotypes with horizontal leaf inclination.

Assimilate Accumulation

Biomass accumulation under optimal growth conditions can be linearly related to cumulative light interception for a number of crops.[3] The slope of this relationship, the amount of dry matter produced per unit of solar radiation intercepted, is termed the crop radiation use efficiency (RUE, g MJ^{-1}). Radiation use efficiency is used as a species-specific parameter to approximate the net result from the processes of photosynthesis and respiration, which many earlier-developed crop models simulated explicitly.[4]

The daily increase in crop biomass (ΔW, g m^{-2}) can be estimated by

$$\Delta W = \varepsilon I$$

where ε is RUE (g MJ^{-1}) and I is the amount of intercepted solar radiation (MJ m^{-2}). This equation assumes that photosynthetic gains and respiratory losses are in balance. Values of RUE are higher for C$_4$ species

$(1.2\,\mathrm{g\,MJ^{-1}}\text{–}1.6\,\mathrm{g\,MJ^{-1}})$ than C_3 species $(0.8\,\mathrm{g\,MJ^{-1}}\text{–}1.2\,\mathrm{g\,MJ^{-1}})$.

Assimilate Partitioning and Crop Yield

Crop biomass results from the daily accumulation of the increase in above-ground biomass (ΔW) over the duration of the crop. Biomass is partitioned into plant components (leaf, stem, flower, grain, root) using partitioning coefficients (η_L, η_S, η_F, η_G, η_R) the values of which are dependent on developmental stage. The daily increase in dry weight of grain $(\Delta W_G, \mathrm{g\,m^{-2}})$ can be estimated as

$$\Delta W_G = \eta_G \Delta W + \tau_G W$$

where W and ΔW are total plant biomass and the amount of its daily increase, and η_G and τ_G are, respectively, the proportions of new assimilates partitioned and existing assimilates remobilized to the grain component. Most increase in grain yield depends on $\eta_G \Delta W$ during the grain-filling period, but assimilate reserves $(\tau_G W)$ can also contribute to grain yield, especially in maintaining grain growth when assimilate supply is limited.

Some models predict grain sink demand by predicting values for grain number (grains plant^{-1}) and grain growth rate (mg grain^{-1}) (e.g., Ref. [5]). Alternatively, other models employ an input parameter that sets the potential daily increase in harvest index (HI) to predict demand.[6] Harvest index is the ratio of grain yield to above-ground biological yield (W_G/W).

Actual grain weight is predicted from the balance between assimilate supply and grain sink demand. For instance, when ΔW during grain filling is greater than sink demand, $\eta_G < 1$ and $\tau_G = 0$, whereas, if ΔW is less than sink demand, $\eta_G = 1$ and $\tau_G > 0$.

Plant Water Relations

Crop water uptake, soil evaporation, rainfall infiltration and runoff, and soil water redistribution and drainage are simulated within a crop model's soil water balance. Daily crop water uptake from the soil is a consequence of the balance between crop water demand and soil water supply.

Crop water use is strongly correlated with biomass production[7]—as the leaf stomata open in order to take up CO_2 for photosynthesis, water is also lost in transpiration. This relationship between the potential amount of daily biomass produced relative to the amount of water transpired represents an apparent transpiration efficiency (TE, $\mathrm{g\,m^{-2}\,mm^{-1}}$). However, any direct measure of TE will change from day to day depending upon the humidity of the atmosphere, quantified as a vapor pressure deficit (VPD, kPa). On low humidity (high VPD) days, more water is required to be transpired to produce the same amount of biomass as on high humidity (low VPD) days. In many crop models, a T coefficient (TE$_C$, kPa) is set for each crop species and it represents the inherent efficiency of biomass production per water use—its units are $\mathrm{kPa/g\,m^{-2}/mm^{-1}}$, which collapses to simply kPa if one considers that $1\,\mathrm{kg} = 1\,\mathrm{m^3}$ water. Values of TE$_C$ are generally higher in C_4 crops (~ 0.009 kPa) than in C_3 crops (~ 0.005 kPa).

Daily crop transpiration demand $(\Delta E_p, \mathrm{mm})$ can thus be estimated by

$$\Delta E_p = \Delta W \times \nu/(\tau \times 1000)$$

where τ is TE$_C$ (kPa), ν is VPD (kPa), ΔW ($\mathrm{g\,m^{-2}}$) is the potential daily increase in crop biomass and 1000 is the factor for converting weight:volume of water.

An alternative approach to estimating crop transpiration demand is to assume that atmospheric demand, set by daily potential evapotranspiration (E_O), drives crop transpiration such that

$$\Delta E_p = E_O I/I_o$$

where I/I_o is proportional daily light interception as calculated from Beer's Law. Potential evapotranspiration can be calculated from climatic parameters by either the Penman–Monteith[8] or the Priestley–Taylor[9] or simply derived from measurements of pan evaporation.

Soil water supply to a crop is defined as the maximum amount of soil water that can be extracted on a daily basis from the root zone. Crop water supply is therefore a function of rooting depth, the amount of plant available water in each soil layer, and the ability of the crop roots to extract soil water. Depth of rooting is generally assumed to increase from soon after emergence at a constant rate ($\sim 10\text{–}30\,\mathrm{mm\,day^{-1}}$) until either the maximum depth of the soil profile is reached or until root extension ceases at a nominated phenological stage (usually around flowering). The calculation of available water for each soil layer is the difference between the soil water content (SW, $\mathrm{mm\,mm^{-1}}$) on a day and the crop lower limit (CLL, $\mathrm{mm\,mm^{-1}}$) of soil water content.

The potential root water uptake (ω, mm) by plants from each soil layer can be calculated as a function of available soil water, root length density, and the diffusivity of water per unit of root length ($\mathrm{mm^{-3}\,mm^{-1}}$).[5] However, neither root length density nor water uptake per unit of root length is easily determined experimentally. Alternatively, soil water supply from a layer (S_i, mm) can be simulated as

$$S_i = \phi_i k l_i$$

where ϕ_i (mm) is the available soil water and kl_i is the rate constant for layer i. The kl constant for each layer is empirically derived from experimental data on crop water extraction and it amalgamates the effects of both root length density and soil water diffusivity, which limit the rate of water uptake.[10] Values of kl typically vary between 0.01 for deep layers with low root length densities to 0.10 for surface layers with high root length densities.

Daily assimilate accumulation, transpiration, and leaf development are decreased below potential values when water deficits occur by using the ratio of potential root water uptake to actual plant evaporative demand. Similar methods of simulating water deficit have been used to predict delays in plant phenology and seedling mortality.

APPLICATION OF CROP DEVELOPMENT MODELS

Modeling is not new to research on cropping systems. In fact, modeling goes back to at least the 1950–60s.[11,12] Since then, investment in simulation modeling has grown in line with the rapid advances in computers themselves. Well-known examples of significant and sustained modeling efforts would include the models developed at the University of Wageningen,[13] the CERES,[5] and CROPGRO[14] suite of crop models contained within the DSSAT software[15] and the APSIM systems simulation model.[16]

Crop development models can be used to simulate the effects of agronomic management on crop growth and development—the comparison of alternative production scenarios using simulated crop performance is a key component of agricultural operations research. The effects of site selection, crop genotype, sowing time, sowing depth, plant population, irrigation regime, nitrogen fertilizer rate, previous cropping history, and fallowing may all be dealt with by many crop models. However, not all determinants of system performance (e.g., the incidence of pest and disease) may be addressed in any one crop model. Nevertheless, there are numerous examples of the use of simulation models in the assessment of agricultural production strategies (e.g., Ref. [17]).

REFERENCES

1. Whisler, F.D.; Acock, B.; Baker, D.N.; Fye, R.E.; Hodges, H.F.; Lambert, J.R.; Lemmon, H.E.; McKinion, J.M.; Reddy, V.R. Crop Simulation Models in Agronomic Systems. Adv. Agron. 1986, 40, 141–208.

2. Monsi, M.; Saeki, T. Über den Lichtfaktor in den Pflanzengesellschaften und Sein Bedeutung für die Stoffproduction. Jpn. J. Bot. 1953, 14, 22–52.

3. Sinclair, T.R.; Muchow, R.C. Radiation Use Efficiency. Adv. Agron. 1999, 65, 215–265.

4. Duncan, W.G.; Loomis, R.S.; Williams, W.A.; Hanau, R. A Model for Simulating Photosynthesis in Plant Communities. Hilgardia 1967, 38, 181–205.

5. Jones, C.A.; Kiniry, J.R., Eds.; CERES-Maize: A Simulation Model of Maize Growth and Development; Texas A&M University Press: College Station, Texas, 1986.

6. Sinclair, T.R. Water and Nitrogen Limitations in Soybeans Crop Production. I. Model Development. Field Crops Res. 1986, 15, 125–141.

7. Tanner, C.B.; Sinclair, T.R. Efficient Water Use in Crop Production: Research or Re-search? In Limitations to Efficient Water Use in Crop Production; Taylor, H.M., Jordon, W.R., Sinclair, T.R., Eds.; American Society of Agronomy: Madison, WI, 1983; 1–27.

8. Monteith, J.L. Evaporation and Environment. Symp. Soc. Exp. Biol. 1965, 19, 205–234.

9. Priestley, C.H.B.; Taylor, R.J. On the Assessment of Surface Heat Flux and Evaporation Using Large-Scale Parameters. Mon. Weather Rev. 1972, 100, 81–92.

10. Monteith, J.L. How Do Crops Manipulate Supply and Demand? Phil. Trans. R. Soc. Lond. A 1986, 316, 245–259.

11. van Bavel, C.H.M. A Drought Criterion and Its Application in Evaluating Drought Incidence and Hazard. Agron. J. 1953, 45, 167–172.

12. De Wit, C.T. Dynamic Concepts in Biology; IBP/PP Technical Meeting, Productivity of Photosynthetic Systems Models and Methods, Trebon: Czechoslovakia, 1969.

13. Penning de Vries, F.W.T.; van Laar, H.H., Eds.; Simulation of Plant Growth and Crop Production. Purdoc, Wageningen, 1982.

14. Boote, K.J.; Jones, J.W.; Hoogenboom, G. Simulation of Crop Growth: CROPGRO Model. In Agricultural Systems Modeling and Simulation; Peart, R.M., Curry, R.B., Eds.; Marcel Dekker: New York, 1998; 651–692.

15. Uehara, G. Technology Transfer in the Tropics. Outlook Agric. 1989, 18, 38–42.

16. McCown, R.L.; Hammer, G.L.; Hargreaves, J.N.G.; Holzworth, D.P.; Freebairn, D.M. APSIM: A Novel Software System for Model Development, Model Testing, and Simulation in Agricultural Systems Research. Agric. Syst. 1996, 50, 255–271.

17. Muchow, R.C.; Bellamy, J.A., (Eds.) Climatic Risk in Crop Production: Models and Management in the Semiarid Tropics and Subtropics; CAB International: Wallingford, 1991.

Crop Plants: Critical Developmental Stages of Water Stress

Zvi Plaut
Agricultural Research Organization, Bet Dagan, Israel

INTRODUCTION

Developmental stages at which crop plants are more sensitive to water deficit as compared to others are known as critical stages. Restricting water supply during these stages may affect productivity more severely than during other periods. Early irrigation timing studies[1] demonstrated that stress sensitivity was greatest from floral development through pollination. The possibility to increase water-use efficiency with minimal damage to crops by determining sensitive growth stages will be outlined based on recent studies. Most studies are concerned with two major objectives: 1) to determine sensitive growth stages in order to avoid any stress during this period; and 2) to determine insensitive growth stages in order to save water and cause minimal damage to the crop. Differences in sensitivity at different developmental stages of a given crop may depend on growing conditions, environmental factors, and crop cultivars and may thus result in disagreement among investigators.

FIELD CROPS

Wheat

Wheat was shown to be most sensitive to water stress during booting through early grain filling.[2] Water application prior to boot stage and during advanced grain filling was found to have limited effect on grain yield. The decrease in grain yield was most marked when the optimal water availability, which was 70% of total available soil water within the root zone, was reduced by 33% during the sensitive growth stage.[3] Similar results were obtained with various cultivars as well as at very different locations.[4–7]

The effect of water stress at the vegetative stage was not only relatively low, but plants could easily recover from this stress.[6] However, the stage of tillering, which is prior to or at the beginning of the vegetative stage, was found to be quite sensitive to water stress.[8,9] Water stress at this stage may reduce the number of tillers, which can result in severe yield losses.

Corn

Full irrigation of corn throughout its entire growing season was claimed to be more profitable than any irrigation regime applying deficit irrigation.[10] It was, thus, advised to reduce the area of grown corn rather than the application of water, if water is rate limiting. Other investigators claimed that corn is able to tolerate short periods of water deficit during its vegetative stage, and is more sensitive between late vegetative growth and grain filling. Optimal irrigation scheduling, at this critical stage, decreased the consumption of water with minimal yield losses.[11] Insufficient water application during this period resulted mainly in a decrease in kernel number.

It seems that the discrepancies shown for corn response to irrigation timing may be explained on the basis of different plant biomass available to support grain yield. It was found that grain yield of corn was closely linked to the accumulated biomass.[12] Fig. 1 presents the range of grain yield and crop biomass of two different field experiments, which differ markedly in initial biomass production, but show similar relationship between biomass production and grain yield. In order to produce maximal biomass, sufficient water supply is needed throughout the season as shown earlier. When the produced biomass is limiting, optimal water supply is especially needed at the critical time of flowering and grain filling, when the utilization of constituents stored in the vegetative organs takes place.

Sorghum

Sorghum is sensitive to water stress at equivalent growth stages as corn, but its sensitivity is lower. The sensitive stage is from heading through grain filling.[13] A stress sensitivity index was introduced by Meyer et al.,[14] which can be calculated from the following equation:

$$Y/Y_{\mathrm{p}} = \Pi\left(\sum \mathrm{ET}_i \Big/ \sum \mathrm{ET}_{\mathrm{p}i}\right)^{\lambda_i} \qquad (1)$$

in which Y and Y_{p} are actual and potential yields, when moisture is not limiting. ET_i and $\mathrm{ET}_{\mathrm{p}i}$ are actual and potential evapotranspiration rates for the i growth stages. The symbol Π is a multiplication factor and λ_i is the sensitivity index of the crop to water stress at the i stage. The calculated values of λ_i for sorghum at different growth stages were 0.04, 0.20, and 0.18 for vegetative, from panicle initiation through anthesis, and grain filling, respectively.[15] These values were much lower than those calculated for corn: 0.06–0.18, 1.54, and 0.03 for

Encyclopedia of Water Science
DOI: 10.1081/E-EWS 120010364

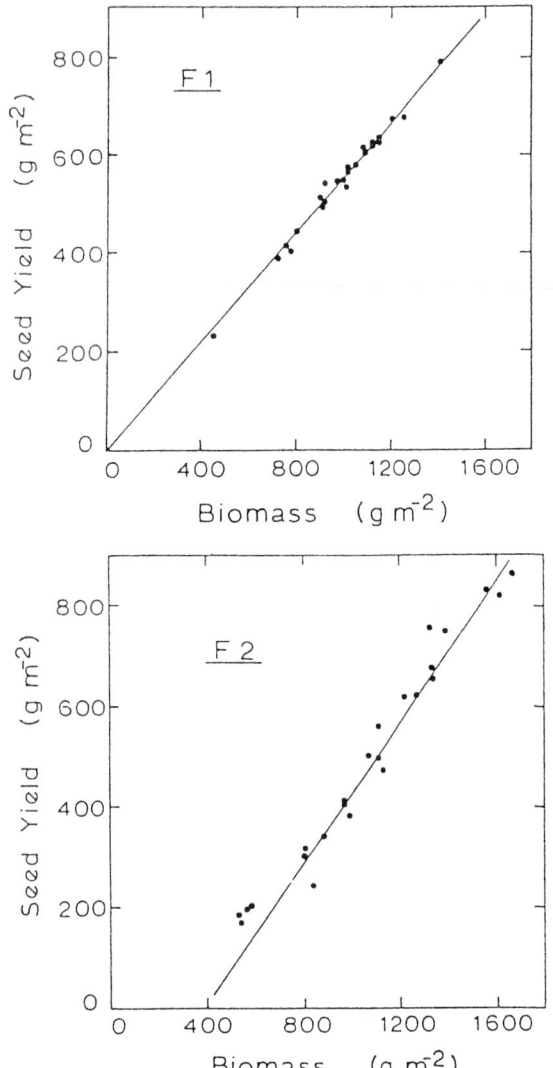

Fig. 1 Maize grain yield vs. crop biomass for individual plots in Florida Exp. Solid line was obtained from linear regression analysis.

vegetative, reproductive, and ripening growth stages, respectively.[14]

In sweet sorghum, the critical stage of water stress was found to be at a much earlier stage, namely during leaf growth.[16] Water stress at this stage reduced biomass by 30% when compared with unstressed plants.

Rice

Water stress occurring during the vegetative stage had a relatively small effect on grain yield in rice.[17] The effect of water stress on yield was most severe when it occurred during panicle development. Stress at this stage delayed anthesis and reduced the number of spikiests per panicle to 60% of the fully irrigated control while drought during grain filling decreased yield by 40% of the control only.

Soybean

No distinct sensitivity of specific developmental stages was outlined for soybean. Internode length and plant height were affected when plants were stressed during both vegetative and flowering stages.[18] This may repress the number of flowers and pods, but could also be a result for enhanced flower and pod abortion.[19] Stress at the stage of seed filling reduced the size of the seeds.

Beans

The analysis of crop sensitivity to water stress at different growth stages is more appropriate if stress intensity is comparable and the quantities of water, which are applied, are similar. Typical results of such an experiment are presented in Table 1 for Black beans.[20] Seed yield was mostly decreased in both years, when water was eliminated during the reproductive stage and all yield components (pods per plant, seeds per pod, and seed weight) were affected. Moreover, plants did not make use of later applied water.

Application of water to field beans (*Vicia faba*), during or shortly after flowering, increased bean yield even under high rates of rainfall.[21] There are other indications,[22] showing that a period of mild water stress during flowering of field beans followed by large quantities of water after flowering was optimal for achieving high grain yield. Conditions of extremely low water stress throughout the growing period may result in decreased dry-matter partitioning to reproductive organs.

The sensitivity of mung bean to water stress at different growth stages was not well defined. It was outlined that irrigation at vegetative and pod development stages was needed in order to assure and improve yield.[23] However, when irrigation is applied during one developing stage only, flowering seems to be the most critical stage.[24] Nonetheless, it was recommended to avoid water stress during two out of three growth stages—vegetative, flowering, and pod filling and maturation.

Peas

Field peas responded positively to irrigation at the vegetative growth stage.[21] This was found, in particular, for more drought-sensitive cultivars. Irrigation of field peas throughout the growing season gave, however, the highest yield but a lower water-use efficiency.

Table 1 Seed yield, yield components, water use, and water use efficiency for black bean for 1995 and 1996 at Akron, Colorado

Treatment	Water withheld during[a]	Population (plants ha^{-1})	Pods plant^{-1}	Seeds pod^{-1}	Seed wt (mg)	Seed yield[b] (kg ha^{-1})	Water use (cm)	Water use efficiency (kg ha^{-1} cm^{-1})
1997								
1	—	176,700	20.3	3.1	182.6	1,975	45.5	43.4
2	GF	160,300	15.4	3.1	156.1	1,280	45.0	28.4
3	R	158,500	12.5	2.4	219.2	1,035	44.7	23.2
4	V	154,800	19.4	3.4	188.3	2,511	41.7	60.2
P[c]		0.5757	0.2190	0.0030	0.000	0.009	0.011	0.005
LSD(0.05)		39,500	8.9	0.4	13.9	719	2.00	16.2
1996								
1	—	196,700	14.6	4.2	203.9	2,758	31.7	87.2
2	GF	202,200	14.5	4.4	186.7	2,672	35.2	76.5
3	R	202,200	12.1	3.8	180.2	1,881	31.9	59.2
4	V	176,700	15.1	3.5	215.2	2,197	26.5	83.3
P		0.0494	0.0154	0.0763	0.0030	0.0012	0.053	0.0101
LSD(0.05)		19,200	1.6	0.7	13.9	303	5.74	13.7

[a] V = vegetative state, R = reproductive stage, GF = grain-filling stage.
[b] Yield reported at moisture content of 0.14 kg H_2O/kg dry matter.
[c] P is the probability level of significant differences due to water stress treatments.

Groundnuts

Groundnuts seem to be most sensitive to water stress during pod development rather than during flowering or even pegging.[25] Water stress during this stage reduced pod yield by 56% as compared to 27% and 45% at the other two stages, respectively.

Moreover, moderate water stress during the preflowering phase was outlined by other investigators to enhance subsequent pod growth.[26] Greater synchrony of pod set in moderately stressed plants resulted in a greater proportion of mature pods at the final harvest.

Cotton

Cotton is a crop of different water requirements throughout its growing season. The requirement is quite low during vegetative growth but increases during the reproductive period, which occurs in most cotton growing regions when transpiration demand is also maximal. The peak in water requirement remains throughout flowering until maturation (opening) of the first boll, when the requirement decreases again (Fig. 2). In order to avoid stress during flowering, it was recommended to apply 3–4 irrigations according to ET.[27] Exposure of the crop to water stress during vegetative growth will avoid excessive vegetative growth, and minimize the consumption of stored assimilates that may compete with reproductive growth. The timing of the first irrigation, which terminates the early stress period, can be set according to several parameters, out of which plant water potential is mostly recommended.[28] The optimal leaf water potential for the initiation of the irrigation season was about −1.8 MPa.[28]

Sunflowers

Sunflowers are known as being fairly tolerant to water stress and may produce an acceptable yield under dryland farming.[29] A relief of water stress by irrigation will, however, enhance production of achenes and of achene oil considerably.[30] The most sensitive growth stage was found to be between stem elongation and the end of flowering. All yield components were found to be affected by irrigation.

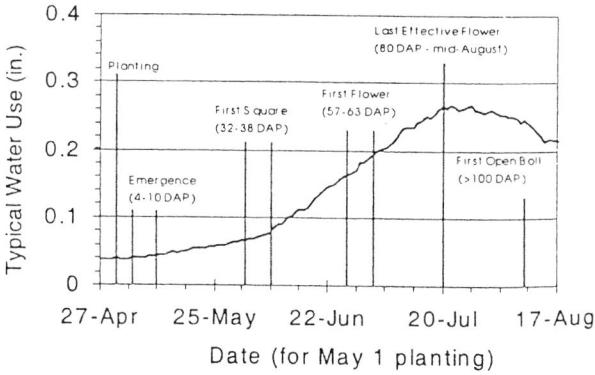

Fig. 2 Estimated typical water use for cotton based on 30-yr-mean maximum temperatures recorded at the University of Arkansas Northeast Research and Extension Center.

VEGETABLE CROPS AND POTATOES

Flowering and fruit set are known to be the most sensitive stage for all crops in which the fruit or grains are the edible organs like garden peas, fresh corn, tomatoes, oilseed rape, watermelon, and others. Water stress at this stage reduced yield by 43%–60%, but only by 32%–50% at the second most sensitive stage, which was fruit enlargement or grain filling. The vegetative and ripening stages were much less sensitive.[31,32] In vegetable crops, which are grown for their vegetative organs (as cabbage and onion), no distinct sensitivity was found at any particular developmental stage. Katerji et al.,[33] who maintained similar stress intensities (according to predawn Ψ_p) at different growth stages, came to a similar conclusion for pepper.

In potatoes, which are essentially parts of the group in which vegetative organs are consumed, withholding water during tuberization caused a sharp decrease in yield and hindered physiological processes.[34] Continuous drought between tuber initiation and final tuber growth reduced yield as well. As far as quality is concerned, even slight water stress during tuber bulking was found to have deleterious effects on specific tuber gravity, dry matter and starch content, and chip yield. This implies that frequent irrigations are essential at this stage.[35]

Cassava, which is mainly grown for animal feed and alcohol production from its starchy tubers, is also part of this group. Water stress during bulking of the underground tubers was found to be very detrimental.[36] This was in fact related to the leaf area and assimilates production rather than to tuber bulking.

FRUIT TREE CROPS

Apples

Reducing the amount of applied water during the growing season was shown to have only slight affects on fruit yield and quality.[37,38] Since restriction of shoot growth is desirable, mainly due to practical reasons, this may even be considered as a positive effect of stress. Water stress may also improve quality of apple fruits depending on timing. It was shown that withholding irrigation during the last 90 day prior to harvest resulted in advanced fruit maturity, more yellow skin color, higher total soluble solids (TSS), and increased flesh firmness during storage, and had hardly any effect on yield. Application of a similar stress during the initial 100 day after full bloom hardly improved the fruit quality. The response of irrigation withdrawal throughout the season gave similar results to late withdrawal, while the control (fully irrigated) trees responded similarly to the early withdrawal of irrigation.[39]

Prunes

It was shown many years ago that prunes are tolerant to water stress, as it took 4 yr of no irrigation to decrease trunk growth and 5 yr to decrease fruit yield.[40] These results were obtained with widely spaced trees on a deep soil. It was found much later that prunes responded by decreased fruit load and fruit hydration, smaller fruits, and increased flowering when subjected to water stress for 75 day of the lag phase during fruit growth. This effect on fruit load was even more critical under conditions of shallow soil, as water storage was limiting.[41]

Apricots

Two critical periods of water stress were outlined: the second exponential fruit growth period and immediately after harvest.[42] Water stress during the first period resulted in decreased yield and quality, as the harvested fruits were smaller. The response to the second period of stress was a reduction in yield, in the following year, due to increased drop of young fruits.

Citrus

Fruit yield of clementine citrus was especially sensitive to water stress during flowering and fruit set. This was in addition to the effect of water stress on fruit yield throughout the year.[43] Fruit quality and vegetative growth were affected by water stress during the ripening period. Clementines seem to be more sensitive to water stress as compared to other citrus species, which were studied earlier.[44] Fruit size seems to be the most sensitive parameter responding to stress. Valencia oranges were also found to be stress sensitive and their critical sensitivity periods were flowering and fruit cell enlargement. In grapefruit and naval oranges, flowering and fruit cell division were most sensitive.[45] Juice content and the ratio of TSS to acid (TSS/acid) of the juice were also reduced if conditions of water stress prevailed during cell enlargement.

REFERENCES

1. Salter, P.J.; Goode, J.E. *Crop Responses to Water at Different Stages of Growth*; Ommonw. Agric. Bur.: Farnham Royal, Buck, England, 1967.
2. Dusek, D.A.; Musick, J.T. *Deficit Irrigation of Winter Wheat: Southern Plains*; ASAE Paper No. 92-2608, ASAE: St. Joseph, MI, 1992.
3. Schneider, A.D.; Howell, T.A. Methods, Amounts and Timing of Sprinkling Irrigation for Winter Wheat. Trans. ASAE **1997**, *40* (1), 137–142.

4. Hamdy, A.; Lacirignola, C.; Akl, A. *Competing Drought Conditions Through Supplementary Irrigation with Saline Water.* Proceedings of 6th Drainage Workshop on Drainage and the Environment, Ljubljana, Slovenia, Apr 21–29, 1996; 718–726.

5. Ravichandran, V.; Mungse, H.B. Response of Wheat to Moisture Stress At Critical Growth Stages. Ann. Plant Physiol. **1997**, *11* (2), 208–211.

6. Abayomi, Y.A.; Wright, D. Effects of Water Stress on Growth and Yield of Spring Wheat (*Tricticum Aestivum* L.) Cultivars. Trop. Agric. **1999**, *76* (2), 120–125.

7. Zhang, H.; Wang, X.; You, M.; Liu, C. Water-Yield Relations and Water-Use Efficiency of Winter Wheat in North China. Irrig. Sci. **1999**, *19* (1), 37–45.

8. Christen, O.; Sieling, K.; Richterharder, H.; Hanus, H. Effects of Temporary Water-Stress Before Anthesis on Growth, Development and Grain-Yield of Spring Wheat. Eur. J. Agron. **1995**, *4* (1), 27–36.

9. Pal, S.K.; Verma, U.N.; Thakur, R.; Singh, M.K.; Upasani, R.R. Dry-Matter Partitioning of Late Sown Wheat Under Different Irrigation Schedules. Indian J. Agric. Sci. **2000**, *70* (12), 831–834.

10. Lamm, F.R.; Nelson, M.E.; Rogers, D.H. Resource Allocation in Corn Production with Water Resource Constraints. Trans. ASAE **1993**, *9* (4), 379–385.

11. D'Andria, R.; Chiaranda, F.Q.; Lavini, A.; Mori, M. Grain Yield and Water Consumption of Ethephon-Treated Corn Under Different Irrigation Regimes. Agron. J. **1997**, *89*, 104–112.

12. Sinclair, T.R.; Bennett, J.M.; Muchow, R.C. Relative Sensitivity of Grain Yield and Biomass Accumulation to Drought in Field-Grown Maize. Crop Sci. **1990**, *30*, 690–693.

13. Krieg, D.R.; Lascano, R.J. Sorghum. In *Irrigation of Agricultural Crops*; Agr. Monogr. 30, Stewart, E.A., Nielsen, D.R., Eds.; ASA, CSSA, SSSA: Madison, WI, 1990.

14. Meyer, S.J.; Hubbard, K.G.; Wilhite, D.A. A Crop Special Drought Index for Corn. I. Model Development and Validation. Agron. J. **1993**, *85*, 388–395.

15. Paes de Camargo, M.B.; Hubbard, K.G. Drought Sensitivity Indices for a Sorghum Crop. J. Prod. Agric. **1999**, *12* (2), 312–316.

16. Mastrorilli, M.; Katerji, N.; Rana, G. Productivity and Water Use Efficiency of Sweet Sorghum as Affected by Soil Water Deficits Occurring At Different Vegetative Growth Stages. Eur. J. Agron. **1999**, *11* (3/4), 207–215.

17. Boonjung, H.; Fukai, S. Effects of Soil Water Deficit At Different Growth Stages on Rice Growth and Yield Under Upland Conditions. 2. Phenology, Biomass Production and Yield. Field Crops Res. **1996**, *48* (1), 47–55.

18. Desclaux, D.; Huynh, T.T.; Roumet, P. Identification of Soybean Characteristics that Indicate the Timing of Drought Stress. Crop Sci. **2000**, *40* (3), 716–722.

19. Saitoh, K.; Mahmood, T.; Kuroda, T. Effect of Stress At Different Growth Stages on Flowering and Pod Set in Determinate and Indeterminate Soybean Cultivars. Jpn. J. Crop Sci. **1999**, *68* (4), 537–544.

20. Nielsen, D.C.; Nelson, N.A. Black Bean Sensitivity to Water Stress At Various Growth Stages. Crop Sci. **1998**, *38*, 422–427.

21. Knott, C.M. Irrigation of Spring Field Beans (*Vicia Faba*): To Timing At Different Crop Growth Stages. J. Agric. Sci. **1999**, *132*, 407–415.

22. Grashoff, C. Effect of Pattern of Water Supply On *Vicia Faba* L. 2. Pod Retention and Filling, and Dry Matter Partitioning, Production and Water Use. Neth. J. Agric. Sci. **1990**, *38*, 131–143.

23. Haqqani, A.M.; Pandey, R.K. Response of Mung Bean to Water Stress and Irrigation At Various Growth Stages and Plant Densities: Ii. Yield and Yield Components. Trop. Agric. **1994**, *71* (4), 289–294.

24. De Costa, W.A.J.M.; Shanmugathsan, K.N. Effects of Irrigation At Different Growth Stages on Vegetative Growth of Mung Bean, *Vigna Radiata* (L.) *Wilczek*, in Dry and Intermediate Zones of Sri Lanka. J. Agron. Crop Sci. **1999**, *183*, 137–143.

25. Golakiya, B.A. Drought Response of Groundnut: VII. Identification of Critical Growth Stages Most Acceptable to Water Stress. Adv. Plant Sci. **1993**, *6* (1), 20–27.

26. Nageswara Rao, R.C.; Williams, J.H.; Sivakumar, M.V.K.; Wadia, K.D.R. Effect of Water Deficit At Different Growth Phases of Peanut. II. Response to Drought during Preflowering Phase. Agron. J. **1988**, *80*, 431–438.

27. Vories, E.D.; Glover, R.E. Effect of Irrigation Timing on Cotton Yield and Earliness. Proc. Beltwide Cotton Conf. **2000**, 1439–1440.

28. Wrona, A.F.; Kerbt, T.; Shouse, P. Effect of Irrigation Timing on Yield and Earliness of Five Cotton Varieties. Proc. Beltwide Cotton Conf. **1995**, 1108–1109.

29. Cox, W.J.; Jolliff, G.D. Growth and Yield of Sunflower and Soybean Under Soil Water Deficit. Agron. J. **1986**, *78*, 226–230.

30. Hedge, M.R.; Havanagi, G.V. Effect of Moisture Stress At Different Growth Phases on Seed Setting and Yield of Sunflower. Karnataka Jour. Agric. Sci. **1989**, *2* (3), 147–150.

31. Morse, R. Efficient Water Use—The Role of Irrigation At Critical Growth Stages. Vegetable Growers News **1983**, *38* (3), 1–3.

32. Champolivier, L.; Merrien, A. Effects of Water Stress Applied At Different Growth Stages To *Brassica Napus* L. Var *Oleifera* On Yield Components and Seed Quality. Eur. J. Agron. **1996**, *5* (3–4), 153–160.

33. Katerji, N.; Mastrorolli, M.; Hamdy, A. Effect of Water Stress At Different Growth Stages on Pepper Yield. Acta Hort. **1993**, *335*, 165–171.

34. DallaCosta, L.; DelleVedove, G.; Gianquinto, G.; Giovanardi, R.; Peressoti, A. Yield, Water Use Efficiency and Nitrogen Uptake in Potato: Influence of Drought Stress. Potato Res. **1997**, *40* (1), 19–34.

35. Gunel, E.; Karadogan, T. Effect of Irrigation Applied At Different Growth Stages and Length of Irrigation Period on Quality Characters of Potato Tubers. Potato Res. **1998**, *41*, 9–19.

36. Baker, G.R.; Fukai, A.C.; Wilson, G.L. The Response of Cassava to Water Deficits At Various Stages of Growth in the Subtropics. Aust. J. Agric. Res. **1989**, *40*, 517–528.

37. Ebel, R.C.; Proebsting, E.L.; Patterson, M.E. Regulated Deficit Irrigation May Alter Apple Maturity, Quality and Storage Life. Hortscience **1993**, *28*, 141–143, E. L.

38. Ebel, R.C.; Proebsting, E.L.; Evans, R.G. Deficit Irrigation to Control Vegetative Growth in Apples and Monitoring Fruit Growth to Schedule Irrigation. Hortscience **1995**, *30*, 1229–1232.

39. Kilili, A.W.; Behboudian, M.H.; Mills, T.M. Postharvest Performance of "Braeburn" Apples in Relation to Withholding of Irrigation At Different Stages of the Growing Season. J. Hort. Sci. **1996**, *71*, 693–701.

40. Hendrickson, A.H.; Veihmeyer, F.J. Irrigation Experiments with Prunes. Univ. Calif. Agric. Exp. Sta. Bull. **1934**, *573*, 1–44.

41. Lampinen, B.D.; Shackel, K.A.; Southwick, S.M.; Olson, B.; Goldhamer, D. Sensitivity of Yield and Fruit Quality of French Prune to Water Deprivation At Different Fruit Growth Stages. J. Am. Soc. Hort. Sci. **1995**, *120* (2), 139–147.

42. Torrecillas, A.; Domingo, R.; Galego, R.; Ruiz-Sanchez, M.C. Apricot Tree Response to Withholding Irrigation At Different Phenological Periods. Sci. Hort. **2000**, *85* (3), 201–215.

43. Ginestar, C.; Castel, J.R. Responses of Young Clementine Citrus Trees to Water Stress during Different Phenological Periods. J. Hort. Sci. **1996**, *71*, 551–559.

44. Shalhevet, J.; Levy, Y. Citrus Trees. In *Irrigation of Agricultural Crops*; Agr. Monogr. 30, Stewart, B.A., Nielsen, D.R., Eds.; ASA, CSSA, SSSA: Madison, WI, 1990.

45. Mostert, P.G.; Zyl, J.L.; Proft, M.P.; Verhoyn, M.N.J. Gains in Citrus Fruit Quality Through Regulated Irrigation. Acta Hort. **2000**, *516*, 123–130.

Crop Residues, Snow Capture by

Donald K. McCool
Brenton S. Sharratt
John R. Morse
United States Department of Agriculture (USDA), Pullman,
Washington, U.S.A.

INTRODUCTION

The primary purpose of retaining or capturing snow by residue management in semiarid regions is to increase the supply of soil moisture for subsequent crops. Snow retention and capture are of major benefit in regions, such as the North American Great Plains, where blowing snow can result in considerable loss of moisture from the landscape. This moisture is often needed to supplement seasonal rainfall and to bolster spring crop production. In the Great Plains, rainfall during the growing season is typically low and is quickly lost from the surface or shallow depth in the soil due to high evaporative demand.[1]

Managing snow on the landscape is also beneficial for moderating soil temperatures and protecting dormant plants from freezing and desiccating. This is particularly important in cold regions where lethal soil temperatures can occur in the absence of snow cover during winter. Snow cover aids in reducing the depth of soil freezing which, in combination with duration of soil freezing, can affect hydrological processes such as infiltration and runoff. Snow retention or trapping also reduces the loss of water associated with sublimation of blowing snow. Sublimation can result in a substantial loss of precipitation in cold regions. Indeed, sublimation of blowing snow has been found to comprise 15%–40% of the annual snowfall in the Canadian Prairies and up to 45% of annual snowfall in the arctic region of North America.[2]

POTENTIAL FOR SOIL WATER INCREASE

A relatively large portion of the annual precipitation occurs in the form of snow in the cold, semiarid regions of the United States and Canada. In the Northern Great Plains of the United States, about 20% of the annual precipitation of 250 mm to over 500 mm occurs in the form of snow.[3,4] Steppuhn[1] indicated that in the Canadian Prairies, 20%–30% of the annual precipitation of 300 mm yr^{-1} to over 500 mm yr^{-1} occurs as snowfall. Staple and

Lehane,[5] Zentner et al.,[6] and Pomeroy and Gray[7] report somewhat different values for precipitation falling as snow on the Canadian Prairies. These studies suggest that from 30% to nearly 40% of annual precipitation occurs in the form of snow. The shortage of precipitation to sustain maximum crop production in cold, semiarid regions requires the use of management techniques that conserve winter precipitation. Conservation of winter precipitation is vital in the Great Plains where economical production of spring wheat requires the annual withdrawal of about 250 mm–400 mm of water from the soil.[8] Indeed, Greb[9] suggested that sustainable crop production can only be attained in the Great Plains as a result of soil water recharge occurring from snowmelt.

Crop residues are only effective in retaining or trapping blowing snow to a depth equivalent to the height of the residue.[10] Surface winds are moderated by exposed residue elements, thus deposition will occur within the residue canopy as long as the elements can effectively retard wind velocity. Once the residue canopy is filled with snow, there is little or no obstruction of the horizontal wind by the residue elements; thus, no further deposition will occur. Implicit in the capture of snow by crop residues is the recharge of moisture within the soil profile. Staple, Lehane, and Wenhardt[11] found in Saskatchewan that recharge of the soil profile from snowmelt was greater in fields covered with stubble than in low-residue fallow. Stubble height also influences snow depth and therefore soil water recharge during winter. Soil water recharge is generally accentuated by taller stubble as a result of greater retention or trapping of snow in taller stubble during winter.[12,13] The extent of recharge, however, is dependent on the rapidity of snowmelt as well as on other soil physical properties such as soil water content and frost depth. Soil frost may prevent infiltration of snowmelt and enhance runoff. Thus, snow retention does not always impact soil water. In fact, Sharratt[10] found, in the northern U.S. Corn Belt, that stubble height had little effect on over-winter changes in soil water content despite large differences in snow depth. Soil water recharge was greater for stubble cut near the soil surface than at either a 30-cm or 60-cm height despite a thicker snow pack in

Encyclopedia of Water Science
DOI: 10.1081/E-EWS 120010102

the 30-cm and 60-cm stubble. Although Greb, Smika, and Black[14] found that loose stubble, as a result of autumn tillage, was less efficient at retaining blowing snow than undisturbed, well-anchored stubble at several locations across the Great Plains, soil water storage in early spring was the same for the loose and well-anchored stubble.

EFFECT OF ADDITIONAL WATER ON CROP YIELD

In the semiarid region of Saskatchewan, yield of wheat is dependent on precipitation received during the growing season as well as on soil water reserves at the time of sowing. de Jong and Rennie[15] found that spring wheat yield was influenced to a greater extent by precipitation than by soil water reserves. They reported a yield increase of $70 \, kg \, ha^{-1}$–$200 \, kg \, ha^{-1}$ for every 25-mm increase in growing season precipitation, but did not find a positive yield response to an increase in soil water storage at the time of sowing. In contrast, a 10-yr study at Swift Current, Saskatchewan indicated an average spring wheat yield increase of $80 \, kg \, ha^{-1}$ for a 13-mm increase in soil water storage due to enhanced snow cover from a trap strip maintained in a wheat stubble field. However, the yield response to soil water storage doubled during years with a dry growing season.[6] Staple and Lehane[16] summarized 12 yr of data on water use by spring wheat at seven Experimental Substations in southern Saskatchewan and found an average yield increase of $235 \, kg \, ha^{-1}$ for each additional 25 mm of water used by the crop.

In the Northern Great Plains of the United States, wheat yield is also dependent on precipitation and soil water storage. Cole[17] summarized results from studies at a number of sites of the relationship of spring wheat yield to annual precipitation and found yield increases of $145 \, kg \, ha^{-1}$–$215 \, kg \, ha^{-1}$ from each 25-mm increase in precipitation above a base of 205 mm–255 mm. A positive yield response of wheat to stored soil moisture was found by Johnson[18] who reported an increase in yield of $50 \, kg \, ha^{-1}$–$290 \, kg \, ha^{-1}$ for every 25 mm of water stored over winter.

Crop residues may not always bolster wheat yield as a result of enhancing soil water recharge over winter. Indeed, Cutworth and McConkey[19] found that yield was greater when wheat was grown in taller stubble due to a more favorable microclimate in taller rather than shorter stubble during the growing season. In the absence of differences in soil water content in the early growing season, greater yield in taller stubble was attributed to lower evaporative demands in taller than in shorter stubble.

EFFECT ON SOIL TEMPERATURE, FROST, AND RUNOFF

A series of studies on the effect of corn stubble height and residue cover on soil temperature, frost depth, and spring thaw have been conducted by the Agricultural Research Service in West-Central Minnesota. In a 3-yr study by Benoit et al.,[20] tillage and residue practices, which retained more residue on the soil surface, resulted in greater snow cover and therefore reduced frost depth and hastened thaw and warming of the soil in early spring. They found practices that retained or trapped an additional 0.1 m of snow on the surface reduced seasonal frost penetration by 0.21 m. In a subsequent study, Sharratt, Benoit, and Vorhees[21] found that standing corn stubble trapped more snow during winter and hastened warming and thawing (by as much as 20 day) of the soil as compared with prostrate corn residue or a bare soil. They later found that soil under 60-cm height corn stubble thawed as much as 15 day earlier than under 30-cm height stubble, and at least 25 day earlier than under 0-cm height stubble. However, net radiation and maximum soil temperature in the early spring were generally higher for soils without residue cover vs. soils with residue cover.[10]

Taller stubble not only has a larger capacity to trap more snow during winter, but also prolongs the period of snow cover during winter.[10] The potential for runoff, therefore, is likely greater for taller stubble. Indeed, Willis, Haas, and Carlson[3] found that taller stubble not only hastened snowmelt runoff, but also partitioned more of the snowmelt to runoff than did shorter stubble.

TECHNIQUES TO ENHANCE SNOW CAPTURE

Uniform Height Stubble

Leaving a uniform cover of standing stubble is the simplest technique of retaining snow on the soil surface in windy regions. If sufficient snow is available and the residue is dense, snow will fill to the top of the stubble.[10] The water density of freshly fallen snow is typically about 10% of its depth (e.g., 300 mm of snow will contain 30 mm of liquid water), but when blown and settled into stubble or drifts, the water density of snow may range from 18% to 35%.[7] Thus, stubble of 40-cm height might capture as much as 140 mm of water. The quality of standing stubble will also influence snow capture. Stubble weakened by subsurface tillage in autumn will likely bend as a result of the weight of the snow load or shear stresses exerted by the wind and thus reduce the efficiency of snow capture compared to undisturbed, well-anchored stubble.[14]

Alternate Height Stubble

Stubble can be cut at alternate heights, high (30 cm–60 cm) and low (15 cm–30 cm), in subsequent passes of a swather. Snow might fill all of the short stubble and part of the tall stubble. The taller stubble might not fill completely. Pomeroy and Gray[7] indicated that using the technique added 31 mm of water to the soil via snowmelt as compared to a uniform canopy of short stubble.

Leave Strips

Narrow (0.30 m) barriers of a standing crop can be left without harvesting. Under proper conditions, the loss of income from the grain would be offset by the cost of additional water gained from snow trapped between barriers. The barriers should be oriented perpendicular to the prevailing wind direction and spacing would be about 20 times the height, based on results generated from several locations.[1]

Trap Strips

Trap strips are similar to leave strips in that narrow (0.40 m–0.60 m) strips of tall stubble are left in the field with stubble rows oriented perpendicular to the prevailing wind direction. Trap strips are formed by the use of a deflector attachment on the swather or combine, which bends the stems sideways, and only the heads and a small part of the stem are removed at harvest. The strips are 250 mm–350 mm taller than adjacent stubble. The strips are more effective at trapping snow when strips and spacing are relatively narrow (width 0.75 m and spacing 5 m) than when strips are wider and more widely spaced (width 1.5 m and spacing 10 m).[7] Results of a 10-yr study at Swift Current, Saskatchewan showed trap strips conserved 13 mm more soil water than short standing stubble (45 mm vs. 32 mm). Increase ranged from 0 mm in years with minimal snowfall to 48 mm in years when snow accumulations were favorable.[6]

Permanent Vegetation Barriers

Permanent vegetation barriers consist of rows of trees, shrubs, or plants. When used for snow management, they are often referred to as living snow fences. The amount of snow trapped by this type of barrier is influenced by the porosity and height of the vegetation. Porosity varies with species and spacing between individual trees, shrubs, or plants. For uniform spreading of snow across a field, vegetation density should be no more than 40%.[22] Snow capture is optimized when the porosity of the vegetation approaches 50%; at this porosity, the snowdrift can be expected to extend as much as 25 times the barrier

height.[22] Greater barrier densities result in shorter, deeper deposits and less benefited area. Greater nonuniformity of depth of snow results in greater differences in soil drying and problems in seeding; part of the area may be too dry while other areas may be too wet to be seeded in a timely manner.[22]

Snow Ridges

Snow ridges formed mechanically and perpendicular to the prevailing wind can act as wind barriers and collect blowing snow. The practice has potential value but it has not been widely used. The snow trapping effect is much the same as from a solid fence.[7] If the ridge does not consolidate after plowing it can be removed by high winds,[1] and success will be diminished if melting occurs before the snow fall period is concluded.

CONCLUSION

Snow capture by standing stubble has potential to increase soil moisture to enhance spring cropping in cold semiarid regions of the United States and Canada. Snow capture techniques such as permanent vegetation barriers or snow ridges are useful but require maintenance or construction each winter and are not widely used.

REFERENCES

1. Steppuhn, H. *Snow Management for Crop Production on the Canadian Prairies*; Proceedings of the 48th Annual Meeting, Western Snow Conference, Laramie, WY, Apr 15–17, 1980; 50–61.
2. Pomeroy, J.W.; Essery, R.L.H. Turbulent Fluxes during Blowing Snow: Field Tests of Model Sublimation Predictions. Hydrol. Proc. **1999**, *13*, 2963–2975.
3. Willis, W.O.; Haas, H.J.; Carlson, C.W. Snowpack Runoff as Affected by Stubble Height. Soil Sci. **1969**, *107* (4), 256–259.
4. Chow, V.T. *Handbook of Applied Hydrology*; McGraw-Hill Book Company: New York, NY, 1964; 1–1418.
5. Staple, W.J.; Lehane, J.J. The Conservation of Soil Moisture in Southern Saskatchewan. Sci. Agric. **1952**, *32*, 36–47.
6. Zentner, B.; Campbell, C.; Seles, F.; McConkey, B. Benefits of Enhanced Snow Trapping, Agriculture Canada Research Newsletter No. 8; Swift Current, SK, Oct 15, 1993.
7. Pomeroy, J.W.; Gray, D.M. *Snowcover Accumulation, Relocation and Management*, National Hydrology Research Institute Science Report No. 7; 1995.
8. Gray, D.M.; Male, D.H. *Handbook of Snow: Principles, Processes, Management and Use*; Pergamon Press: Willowdale, Ontario, Canada, 1981; 1–776.

9. Greb, B.W. Snowfall Characteristics and Snowmelt Storage At Akron, Colorado. Proc. Symp. Snow Manag. Great Plains, Univ. Nebraska Agric. Exp. Stn Publ. **1975**, *73*, 45–64.

10. Sharratt, B.S. Corn Stubble Height and Residue Placement in the Northern US Corn Belt. Part I. Soil Physical Environment during Winter. Soil Tillage Res. **2002**, *64* (3–4), 243–252.

11. Staple, W.J.; Lehane, J.J.; Wenhardt, A. Conservation of Soil Moisture from Fall and Winter Precipitation. Can. J. Soil Sci. **1960**, *40*, 80–88.

12. Aase, J.K.; Siddoway, F.H. Stubble Height Effects on Seasonal Microclimate, Water Balance, and Plant Development of No-Till Winter Wheat. Agric. Meteorol. **1980**, *21*, 1–20.

13. Smika, D.E.; Whitfield, C.J. Effect of Standing Wheat Stubble on Storage of Winter Precipitation. J. Soil Water Conserv. **1966**, *21*, 138–141.

14. Greb, B.W.; Smika, D.E.; Black, A.L. Effect of Straw Mulch Rates on Soil Water Storage during Summer Fallow in the Great Plains. Proc. Soil Sci. Soc. Am. **1967**, *31*, 556–559.

15. de Jong, E.; Rennie, D.A. Effect of Soil Profile Type and Fertilizer on Moisture Use by Wheat Grown on Fallow or Stubble. Can. J. Soil Sci. **1969**, *49*, 189–197.

16. Staple, W.J.; Lehane, J.J. Wheat Yield and Use of Moisture on Substations in Southern Saskatchewan. Can. J. Agric. Sci. **1954**, *34*, 460–468.

17. Cole, J.S. *Correlations Between Annual Precipitation and Yield of Spring Wheat in the Great Plains*; USDA Tech. Bull. No. 636, 1938.

18. Johnson, W.C. Some Observations on the Contribution of an Inch of Seeding-Time Soil Moisture to Wheat Yield in the Great Plains. Agron. J. **1964**, *56*, 29–35.

19. Cutworth, H.W.; McConkey, B.G. Stubble Height Effects on Microclimate, Yield and Water Use Efficiency of Spring Wheat Grown in a Semiarid Climate on the Canadian Prairies. Can. J. Plant Sci. **1997**, *77*, 359–366.

20. Benoit, G.R.; Mostaghimi, S.; Young, R.A.; Lindstrom, M.J. Tillage-Residue Effects on Snow Cover, Soil Water, Temperature and Frost. Trans. ASAE **1986**, *29* (2), 473–479.

21. Sharratt, B.S.; Benoit, G.R.; Vorhees, V.B. Winter Soil Microclimate Altered by Corn Residue Management in the Northern Corn Belt of the USA. Soil Tillage Res. **1998**, *49* (3), 243–248.

22. Brandle, J.R.; Nickerson, H.D. *Windbreaks for Snow Management*; EC96-1770-X, University of Nebraska: Lincoln, 1996; 1–6.

Crops, Water Harvesting for

Gary W. Frasier
United States Department of Agriculture (USDA), Fort Collins, Colorado, U.S.A.

INTRODUCTION

A significant portion of the world's land surface is too dry for intensive agriculture without supplemental water, usually in some form of irrigation using surface water diversion or pumped groundwater. There are many parts of these same arid and semiarid lands where irrigation water is inadequate, unavailable, or unsuitable. Yet, many of these lands have in the past, or currently, supported some form of cultivated agriculture, even in areas that receive less than 200 mm of rainfall per year. How can there be intensive agriculture in areas where rainfall quantities are less than 200 mm/yr? The answer is; the crops are grown using a technique of water supply called water harvesting. Even most arid lands have relatively large quantities of water available in the form of precipitation that is potentially available for some beneficial use if it can be collected or concentrated and stored.

What is water harvesting? Water harvesting is a technique of water supply that can be used where conventional surface or groundwater sources are unavailable or unsuitable. The basic premise of water harvesting is the collection of precipitation from a specific area for some beneficial use. Precipitation runoff is collected from a relatively large area and stored or concentrated onto a smaller area. This provides a multiplication factor for maximizing the benefits of the limited precipitation. The water collection area may be a natural undisturbed hillslope or some type of prepared impermeable surface. The collected water can be used for growing crops, drinking water for human and animals, or other domestic uses. The collected water may be used immediately by placement in the soil (infiltration) or stored in an appropriate container for later use.

BACKGROUND

There is evidence of water harvesting structures being used over 9000 yr ago in the Edom Mountains of Southern Jordan.[1] The people of Ur practiced water harvesting as early as 4500 BC.[2] Studies have shown that extensive agricultural systems using water harvesting techniques existed in some areas 3000 yr–4000 yr ago. One such area

is the vast arid zone adjacent to the "Fertile Crescent" of the Middle East. The Fertile Crescent stretches from Israel through Lebanon and Syria ending in Mesopotamia along the Tigris–Euphrates Valley. In historical times this was a major agricultural area utilizing water from various streams and rivers in the area for irrigating crops. With increasing population pressures there was an exodus of the population from within the "Crescent" into the adjacent arid deserts outside the Crescent.[3] These desert areas were considerably less desirable for agricultural production than the areas the people had left. There were no perennial surface water supplies, streams, or groundwater.[4] Even so, extensive agricultural communities were developed in the desert lands that, in some areas, still flourish today.[3] There is evidence that similar techniques were used over 400 yr ago in Southwestern United States where Mesa Verde National Park is located.[5]

A common concept is that water harvesting has only been used in, or is most suitable for, arid lands. In reality water harvesting can be used almost anywhere where other water sources are inadequate, unavailable, or unsuitable. In recent times water harvesting has been used for growing crops in many places in the world such as Israel, Egypt, Jordan, Mexico, Australia, and the United States. It has been used to supply drinking water for domestic use and animals in places such as Hawaii (United States), Thailand, Mexico, Australia, and Egypt. It is most effective where there is a predictable quantity and timing of precipitation during the period when the water is needed.

TYPES OF WATER HARVESTING FOR CROPS

Crop production using water harvesting techniques is commonly referred to as runoff farming. In runoff farming the collected water can be applied directly to fields from the catchment area during the precipitation event or stored for later application using some system of irrigation.

One method of runoff farming is called floodwater farming. The precipitation runoff flowing down a channel during a storm event is directed or diverted onto a field or cropping area. A second method is called microcatchment farming. With microcatchments, each plant or small group of plants has a small runoff contributing area directly upslope of the growing area. Typically the runoff area is 5–20 times larger than the cropping area. This technique has been used very extensively for various tree crops such as pistachio, olives, and almonds (Fig. 1).

A third method of runoff farming encompasses a combination of both direct application of the runoff water and later irrigation with excess runoff water from a stored source. The land is formed into a series of large ridges and furrows. Crops such as fruit trees or grapes are planted in the bottom of the furrows. Water from the side slopes of the ridges drains onto the crop area in the bottom of the furrows. Runoff water that is not directly infiltrated into the planted area continues down the center of the furrow into some storage pond or container. At some later date the water is pumped back onto the crop area as needed using some form of an irrigation system (Fig. 2).

POTENTIAL OF WATER HARVESTING

If all the water that falls as precipitation on a given piece of land can be collected and put to beneficial use, there is usually adequate water to sustain life and support some form of agriculture. This can be illustrated using an example from the Negev desert of Israel. While historical precipitation records are not available from prehistoric days, various experts believe there have been no major changes in the past 2000 yr.[6] Current yearly precipitation records for a typical area in the Negev desert show that precipitation ranges from 28 mm/yr to 168 mm/yr, with an average of about 86 mm/yr. Most of the precipitation occurs during the winter months, November–March, with about 16 rainy days/yr, 12 days with precipitation greater than 1 mm, 3 days with precipitation greater than 10 mm, with only a single storm event greater than 25 mm/day every 2 yr. Average hourly intensities are relatively low, less than 5 mm/hr, but for short periods of 5 min to 10 min, precipitation intensities up to 20 mm/hr–50 mm/hr have been recorded.[7] These precipitation characteristics are similar to other arid lands in the world.

Even with a low annual precipitation occurring as infrequent storms, considerable water can be collected and utilized. One millimeter of precipitation per square meter is equal to 1 L of water. Using the Negev desert data; if all the annual precipitation (85 mm) occurring on 10 m^2 of land can be collected and used to irrigate 1 m^2, it is equivalent to 850 mm of water of precipitation. Collecting the precipitation runoff from large areas for use on smaller areas can provide adequate quantities of water for growing crops, even if it occurs only on a few days each year. Table 1 gives some estimated quantities of water from other areas of the world, which can potentially be collected from precipitation. All that is necessary is to have some means for storing the collected water until it is needed.

Even if only a portion of the total precipitation is collected and effectively used, crops can be grown in areas that would normally be considered unsuitable. Researchers have been able to re-construct some of the ancient farms in the Negev desert and grow various fruits and nuts, like olives, pomegranates, figs, almonds, pistachio, apricots, peaches, plums, and vegetables-like onions, peas, artichokes, and asparagus. Fields of wheat and barley have produced adequate food grain crops.[8]

Fig. 1 Microcatchment water harvesting for growing jojoba near Phoenix, Arizona, USA in a 230 mm annual precipitation zone.

Fig. 2 Ridge and furrow water harvesting system for growing pistachios near Saltillo, Coahuila, Mexico. Excess precipitation runoff is collected in a storage pond at the lower edge of the field for later application to the trees by a drip irrigation system.

Table 1 Water quantities potentially available from precipitation for selected locations in the world

Location	Latitude	Longitude	Elevation (m)	Record length (yr)	Precipitation (L/m²)[a]
Kabul, Afghanistan	34° 30'N	69° 13'E	1,815	45	320
Cipolletti, Argentina	38° 57'S	67° 59'W	270	24	160
Alice Springs, Australia	23° 48'S	133° 53'E	545	30	250
Arica, Chile	18° 28'S	70° 20'W	30	25	0
Alexandria, Egypt	31° 12'N	29° 53'E	32	61	180
New Delhi, India	28° 35'N	77° 12'E	210	75	640
Baghdad, Iraq	33° 20'N	44° 24'E	35	15	140
Tehran, Iran	35° 41'N	51° 19'E	1,200	33	250
Jerusalem, Israel	31° 47'N	35° 13'E	810	50	500
Amman, Jordan	31° 58'N	35° 59'E	775	25	280
Kuwait, Kuwait	29° 21'N	48° 00'E	5	10	130
Chihuahua, Mexico	28° 42'N	105° 57'W	1,350	22	390
Marrakech, Morocco	31° 36'N	08° 01'W	460	31	240
Karachi, Pakistan	24° 48'N	66° 59'E	4	59	200
Riyadh, Saudi Arabia	24° 39'N	46° 42'E	590	3	80
Khartoum, Sudan	15° 37'N	32° 33'E	390	46	160
Aleppo, Syria	36° 14'N	37° 08'E	390	10	390
Tunis, Tunisia	36° 47'N	10° 12'E	65	50	420
United States					
Albuquerque, NM	35° 03'N	106° 37'W	1,620	30	210
El Paso, TX	31° 48'N	106° 34'W	1,190	30	200
Las Vegas, NV	36° 05'N	115° 10'W	660	30	100
Phoenix, AZ	33° 26'N	112° 01'W	340	30	180
Reno, NV	39° 30'N	119° 47'W	1,340	30	180

[a] 1 mm of precipitation equals 1 L/m².
From "Climates of the World" Historical Climatology Series 6-4, U.S. Dept. Commerce, NOAA, Asheville, North Carolina, January 1969. www.ncdc.noaa.gov (accessed November 2000).

To maximize the benefits of water harvesting for growing crops under limited precipitation conditions, several other factors are desirable. These include:

Soil type—in the cropping area, it is desirable to have deep soils with a high water-holding capacity that will retain the water within the plant rooting depth. The water collecting area (catchment) should have impermeable soils or a surface that prevents the water from infiltrating and maximizes the runoff. Catchment areas should have sufficient slope and a topography that rapidly carries the runoff water from the area.

Precipitation—maximum benefits of water harvesting are achieved if the precipitation occurs during cooler weather when evapotranspiration rates are the lowest. Precipitation intensities must be greater than the infiltration rate of the catchment area. When growing crops there is an added benefit if the precipitation occurs during the cropping season. This reduces the period of time necessary to store the collected water.

Crop type—crop species must be drought tolerant or capable of surviving extended dry periods. Cropping practices must include plants that are capable of utilizing the available water efficiently yet can withstand prolonged time intervals when water may be limited or nonexistent. Good cropping practices must also recognize that water requirements for plant establishment are frequently different than the water requirements for mature established plants. During the plant establishment phase, rooting systems are usually shallow, which necessitates the water be available in the upper layers of the soil profile. Under these conditions there is also the potential for significant losses of the soil water by evaporation from the unprotected (nonshaded) soil surface.

ADVANTAGES AND DISADVANTAGES OF WATER HARVESTING

If there is some precipitation, water harvesting can be a method of water supply. In most instances the collected water is of a very good quality (pure).

While water harvesting can supply water in most areas, it should not be considered an inexpensive means of water supply. There are appreciable costs of preparing catchment areas and water storage facilities. Maximum runoff efficiency is obtained by sealing or covering the soil surface. These techniques are relatively expensive and may not be cost effective in many locations. At sites where land area and labor are relatively inexpensive and readily available, smoothing of the soil surface may be the most effective means of collecting the required quantities of water. Runoff per unit area from a smoothed soil catchment surface may be relatively low. Using a larger catchment area can offset the lower runoff efficiency.

In many locations, the cost of the water storage can represent the major expense of a water harvesting facility. In these instances it may be desirable to design the storage supply to meet the needs even if there is excess water during part of the year.

For maximum long-term effectiveness, water harvesting systems must have scheduled and timely maintenance and repair. Many systems have been adequately designed and constructed, and yet have failed to supply the anticipated quantities of water within a relatively short time interval because of inadequate maintenance. Usually the required maintenance or repair can be accomplished in a relatively short period of time without a lot of expense.

REFERENCES

1. Bruins, H.M.; Evenari, M.; Nessler, U. Rainwater-Harvesting Agriculture for Food Production in Arid Zones: The Challenges of the African Famine. Appl. Geogr. **1986**, *6*, 13–33.

2. Frasier, G.W. Forward. In *Proc. Water Harvesting Symposium*; Phoenix, Arizona 26–28 March 1974, USDA-ARS W-22, Frasier, G.W., Ed.; Agricultural Research Service U.S. Department of Agriculture, Washington, D.C., 1975; 1.

3. Evenari, M.; Shanan, L.; Tadmor, N.H. "Runoff Farming" In the Desert. I. Experimental Layout. Agron. J. **1968**, *60*, 29–32.

4. Kedar, E.Y. Water Utilization by Roman-Period Cities of the Negev, Israel. *Proc. Symposium on Urbanization in the Arid Land*; mimo. American Association for the Advancement of Science: Chicago, IL, 1970; 38.

5. Myers, L.E. Precipitation Runoff Inducement. In *Proc. Symposium on Water Supplies for Arid Regions*; Contribution No. 10 Committee on Desert and Arid Zones Research, May 1–2 1967, Tucson, Arizona, Gardner, J.L., Myers, L.E., Eds.; American Association for the Advancement of Science and the University of Arizona, University of Arizona Press, Tucson, AZ, 1967; 22–30.

6. Evenari, M.; Shanan, L.; Tadmor, N.H.; Aharoni, Y. Ancient Agriculture in the Negev. Science **1961**, *133* (3457), 979–996.

7. Anon. *Ancient and Modern Water-Harvesting in the Negev Desert*; mimo. Dept. of Botany, Hebrew University: Jerusalem, Israel, 1967; 17.

8. Agarwal, A. Coaxing the Barren Desert Back to Life. New Sci. **1977**, 674–678.

Darcy's Law

Graeme D. Buchan
Keith C. Cameron
Lincoln University, Canterbury, New Zealand

D

INTRODUCTION

In simple flow systems in nature, the fundamental law of flow is linear, i.e., the flow rate increases in direct proportion to the "driving force" for the flow, where the driving "force" is the gradient (rate of change with distance) of some so-called "potential." For example, the flow of heat obeys Fourier's law, with the heat flux proportional to the gradient of temperature, T. (T is sometimes called the *thermodynamic potential*). For the flow of liquid water in permeable materials, it fell to the French hydraulics engineer, Henry Philibert Gaspard Darcy (1800–1858) to show that the flow obeys a similar law ("Darcy's law"), with the flow rate proportional to the gradient of the *hydraulic potential*. More generally, we now know that Darcy's law applies to flow of most simple liquids in any porous (and permeable) medium.

Historical

In the early 19th century, the city of Dijon in France had a water supply among the worst in Europe.[1] Darcy was a civil engineer, born in Dijon, and set himself the task of improving the city's water supply. He decided to investigate the filtration of water by sands and gravels, reported in 1856 in his report *Les Fontaines Publiques de la Ville de Dijon*.[2] Using simple but ingenious equipment (see Fig. 1), he arrived at his universal law for the mass flow of liquids in permeable materials.[3] Thus, Darcy joined the "famous four" who revealed the simplicity (yet paradoxically the complexity) of Nature's most basic laws of flow: Fourier's law for heat flow; Ohm's law for electric current; Fick's law for gas diffusion; and Darcy's law for liquid flow in materials.

THEORY OF LIQUID FLOW IN PERMEABLE MEDIA

Saturated Flow: Darcy's Law

Darcy[2] first established his flow equation for water flow in saturated sand (Fig. 1). He found that the flow rate per unit cross-sectional area, q, through the pipe in Fig. 1 is proportional to the difference in head h between the ends, and inversely proportional to length L:

$$q = K(h_B - h_C)/L \qquad (1)$$

where K is a proportionality constant.

In more general form, for flow along the x-direction:

$$q = -K \, dh/dx \qquad (2)$$

Here q (in units of $m \, sec^{-1}$) is the "Darcy velocity." This is the average apparent velocity of the water, as if it were flowing across the entire area, solids as well as pores. $K = K_{sat}$ (also in $m \, sec^{-1}$) is the saturated *hydraulic conductivity*, and dh/dx is the driving force for flow, i.e., the gradient of the hydraulic head h (meters head of water) in the direction of flow, x (m).

K_{sat} is strongly controlled by the pore space of the permeable medium, especially the pore sizes.[4] For soils, K_{sat} varies enormously with texture. See Table 1. K_{sat} is approximately a measure of the maximum drainage rate of a soil. For example, a sandy soil may have $K_{sat} \approx 5 \times 10^{-5} \, m \, sec^{-1} = 180 \, mm \, hr^{-1}$ while a clay soil may have $K_{sat} \approx 10^{-8} \, m \, sec^{-1} = 0.036 \, mm \, hr^{-1}$. Thus, a rainfall of only $1 \, mm \, hr^{-1}$ would drain freely into the sand, but would cause surface ponding on the clay. Soil structure also controls pore sizes and hence K_{sat}, e.g., a soil with a tightly packed platy structure will have lower K_{sat} than one with open, porous "granular" structure.

Unsaturated Flow: Buckingham–Darcy Equation

Soils are mostly unsaturated, and the generalization of Darcy's law to unsaturated flow was developed by Buckingham.[5] He reasoned that hydraulic conductivity $K(\theta)$ is a function of the soil volume occupied by the conducting water, i.e., the volumetric water content θ. Also, the pressure potential h becomes negative, since water is now under suction. Thus we can again assume Darcy's law, Eq. 2, but now $K = K(\theta)$ decreases very rapidly as soil loses water. See Fig. 2.

The actual water velocity between the grains is greater than the Darcy velocity, with an average value $v = q/\theta$. Note that v is an average value, and masks a microscopically complex flow pattern in the tortuous, multisized pore space, with a range of speeds and directions. An extreme example of

Encyclopedia of Water Science
DOI: 10.1081/E-EWS 120010271

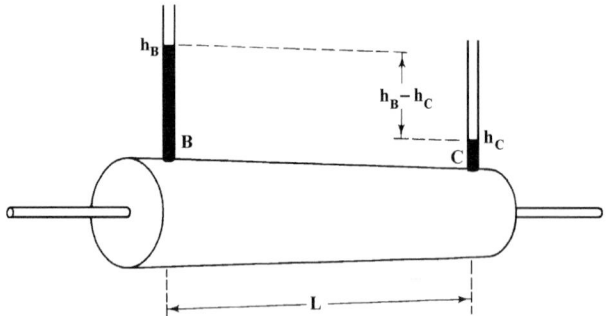

Fig. 1 Horizontal pipe filled with sand to demonstrate the experiment of Darcy (1856).His original equipment was actually vertically oriented, which would add an extra term to the head difference in Eq. 1, equal to the height difference $z_B - z_C$. (From Ref. [6].)

the microscopic variability of flow velocity occurs in water flow through soils with strong structure development, e.g., with large cracks or wormholes. Here, water can be fast-tracked along the large macropores by so-called "bypass flow." (This phenomenon can fast-track contaminants to groundwater, and also contributes to the *dispersion* of solutes during their transport.)

Note that the hydraulic conductivity depends not only on the architecture of the soil pore space, but also on the properties of water itself, especially its viscosity. For example, water's viscosity almost doubles from 30°C to 0°C. So soil water flows more slowly in winter than in summer. In order to extend Darcy's law to other liquids, it is desirable to transform the hydraulic conductivity (which is specific to water only) to a more absolute measure of the "conductivity" of the permeable material, independent of the fluid. This leads to the "intrinsic permeability":

$$\kappa = K(\mu/\rho g) \tag{3}$$

Here μ is the viscosity of water, ρ is the density of water, and g is the acceleration due to gravity. In Eq. 3, the dependence of K on the properties specific to water has

been factored out. (ρg represents the "heaviness" of water, or the amount of pressure produced by a water column of unit height). κ is now controlled only by the properties of the permeable medium, and can be used for any other liquid (e.g., oil) that might flow in the medium. Incidentally, like many famous scientists, Darcy has a unit named after him. The "darcy" is a unit of the intrinsic permeability κ, but is in such antiquated (pre-SI) units, that it is now little used, except by petroleum engineers.[6]

APPLICATIONS

Soil Water Flow

Darcy's law is used extensively in soil science, in drainage theory, and in most models of water and solute transport in soil. Fig. 2 shows how the hydraulic conductivity function $K(\theta)$ is strongly influenced by soil texture. Paradoxically, coarse-texture soils (e.g., sands), while more conductive than fine-texture soils (e.g., silt loams) at saturation (Table 1), lose their conductive capacity at low suctions, due to rapid desaturation of their large pores. This has interesting practical applications, e.g., in sports turf soils. A gravel layer beneath a sandy root zone provides excellent drainage at saturation, but then desaturates, becomes almost nonconductive, arrests drainage from the root zone, and hence enhances root zone moisture retention.

Groundwater Flow

Groundwater is a major source of irrigation and municipal water in many parts of the world, and is commonly pumped either from shallow groundwater layers, or from deeper aquifers, which are either permeable gravels or rock layers. Darcy's law again applies to the saturated flow. However, groundwater hydrologists are interested in the ability of an aquifer, with thickness d, to deliver water to a well which cuts across the aquifer. See Fig. 3. The aquifer's supply capacity is thus controlled by the product

Table 1 Typical values of saturated hydraulic conductivity for soils, ranging from sand to clay

Drainage class of soil	Saturated hydraulic conductivity, K_{sat} (mm hr^{-1})	Approximate soil texture class
Class 1: very slow	< 1	Clay
Class 2: slow	1–5	Clay loam
Class 3: moderately slow	5–20	Silty clay loam
Class 4: moderate	20–60	Silt loam
Class 5: moderately rapid	60–125	Loam
Class 6: rapid	125–250	Sandy loam
Class 7: very rapid	> 250	Sand

(From Ref. 10.)

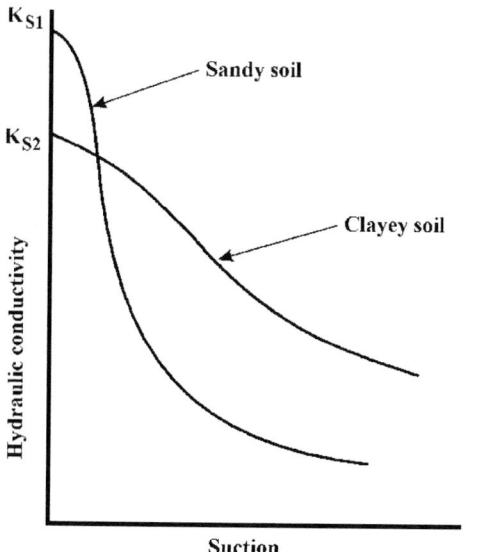

Fig. 2 Hydraulic conductivity $K(\theta)$ of two soils of contrasting texture. K_{s1} and K_{s2} are the conductivities at saturation. K decreases dramatically as water content θ decreases. (From Hillel, D. *Environmental Soil Physics*; Academic Press: San Diego, CA, 1998; 208.)

$K \times d$ of aquifer conductivity K and its thickness d, a quantity called "transmissivity."[7]

Two-Phase Flow: Flow in Oil Reservoirs

Above, we described Darcy's law for a single fluid. Multiphase flow occurs where several (usually two) nonmiscible liquids share the pore space. It occurs in oil

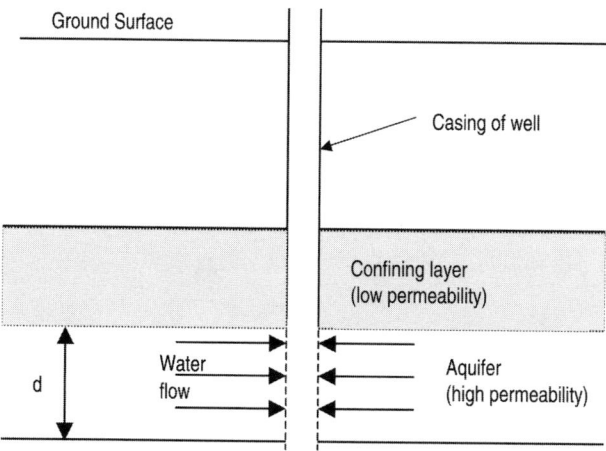

Fig. 3 Schematic of an aquifer layer supplying water to a groundwater well. The aquifer's supply capacity depends on its "transmissivity," i.e., its saturated hydraulic conductivity K_{sat} multiplied by its thickness d. The aquifer is "confined" by an impermeable layer above.

reservoirs if they contain both oil and water. Darcy's law was generalized in 1949 by Muskat (see Ref. [8]) to describe the more complex flow of two phases together. The generalized form of Darcy's law is used by oil exploration scientists. This form also applies in groundwater that has been contaminated by so-called NAPLs: "nonaqueous phase liquids," or organic liquids immiscible in water. Thus, Darcy's law can be applied to oil extraction, and to the remediation of polluted groundwater.

Further Complexities

Anisotropic materials

Some properties of materials may be anisotropic, i.e., for directional phenomena (such as water flow) the controlling property depends on the direction. For example, in sedimentary deposits K perpendicular to the stratification is usually less than K parallel to it. In clay subsoils, the clay particles (typically plate-shaped) may be preferentially orientated in the horizontal plane, and vertical conductivity will then be much less than horizontal conductivity, improving the layer's ability to act as a barrier layer to vertical contaminant transport. A result of anisotropy is that in 3-D flow, the velocity flow lines are not parallel to the head gradient.

Vapor flow

Fluids can be transported in the *vapor* as well as the liquid phase. However, Darcy's law applies strictly only to the mass flow of liquids, not to the flow of vapor, which moves by gas diffusion. Vapor diffusion also obeys a simple proportional law (Fick's Law), but is driven most strongly by gradients of temperature T. This leads to "coupled flows." For example, in a moist soil near the ground surface, the vertical (z-) gradient dT/dz can be very strong. Then heat and moisture flows occur simultaneously and in linkage: water is distilled from warmer to cooler regions, and carries latent heat energy, as well as the water itself.[9]

REFERENCES

1. Brown, G. Darcy's Law: Basics and More. 2000. http:// biosystems.okstat.edu/darcy/LaLoi/index.htm (accessed May 2001).
2. Darcy, H. *Les Fontaines Publiques de la Ville de Dijon*; V. Dalmont: Paris, 1856.
3. Lage, J.L. The Fundamental Theory of Flow Through Permeable Media, from Darcy to Turbulence. In *Transport Phenomena in Porous Media*, 1st Ed.; Ingham, D.B., Pop, I., Eds.; Pergamon: Oxford, UK, 1998; 1–4.

4. Cameron, K.C.; Buchan, G.D. Porosity and Pore-Size Distribution. In *Encyclopedia of Soil Science*; Lal, R., Ed.; Marcel Dekker: NY, 2001.

5. Buckingham, E. *Studies on the Movement of Soil Moisture*; USDA Bur. Soils Bull., USDA, 1907; 38.

6. Fetter, C.W. *Applied Hydrogeology*, 4th Ed.; Prentice Hall: NJ, 2001.

7. De Wiest, R.J.M. Fundamental Principles of Groundwater Flow. In *Flow Through Porous Media*; De Weist, R.J.M., Ed.; Academic Press: NY, 1969; Chap. 1.

8. Morel-Seytoux, H.J. Flow of Immiscible Liquids in Porous Media. In *Flow Through Porous Media*; De Wiest, R.J.M., Ed.; Academic Press: NY, 1969; Chap. 11.

9. Buchan, G.D. Soil Temperature Regime. In *Soil and Environmental Analysis: Physical Methods*; Smith, K.A., Mullins, C., Eds.; Marcel Dekker, Inc.: New York, 2001; 539–594.

10. McLaren, R.G.; Cameron, K.C. *Soil Science. Sustainable Production and Environmental Protection*; Oxford University Press: New Zealand, 1996; 122.

Databases

Rosemary Streatfeild
Washington State University, Pullman, Washington, U.S.A.

D

INTRODUCTION

A database is an electronic bibliographic or full-text product providing access to publications in either a general or subject specific area. Information on water science can be accessed through a number of different databases, some with a focus on water resources, while others provide relational coverage. Most researchers, looking for as much information on their topics as available, will use more than one product to locate materials.

Databases are available through vendors on different platforms and through variable subscription options. They differ in coverage, content, search and retrieval methods, source materials, and access modes. Most provide more than one search level, varying in the number of searchable fields and level of expertise, and employ search features such as Boolean operators, truncation and proximity methods, phrasing and limiting that differ by vendor. Help pages, indexes, and controlled-language thesauri accompany most packages to assist with searching.

Other features commonly engaged by database vendors include display options, selection of citations ("marking") for printing, downloading and emailing, saving and retrieving searches, combining searches, linking to online catalogs and full-text electronic journals, and alerting services. The addition of new materials ("updating") occurs regularly, but each vendor follows its own schedule.

DATABASES

Agnic[1] offers agricultural information provided by the National Agricultural Library (NAL), Land-Grant Universities, and other institutions, each providing information on a discrete area of emphasis. Subject coverage includes Water Quality; Government, Law and Regulations; Plant Sciences; Aquaculture and Fisheries; Earth and Environmental Sciences, and Forestry. Descriptions of each record are included, along with accessibility links. A thesaurus and search tips are provided. Update schedules and years of coverage are established by the respective institutions that provide the information.

Agricola,[2] a database provided free by the NAL, includes citations from materials relating to agriculture, forestry, life sciences, and other related disciplines dating from the 15th century. It is available also on CSA (2), Community of Science (4), Dialog (5), Ebsco (6), OCLC (9), Ovid (10), Silverplatter (11), and STN (12) platforms. The NAL version is divided into the Online Public Access Catalog (Books, etc.) and Journal Article Citation Index (Articles, etc.). It is updated daily from over 5000 sources.

AGRIS,[3] sponsored by the Food and Agricultural Organization of the United Nations and available in English, Spanish, and French, offers an international perspective on agriculture and related subjects. Available on CD-ROM (updated quarterly) and on the Internet (updated monthly) from 1975, there are over 144,000 records. Sources include journal articles, books, technical reports, and gray literature from all countries that participate in the UN program. A thesaurus (AGROVOC) is provided. Coverage is in many subject areas including water resources and management, aquatic sciences and fisheries, agriculture, natural resources, and pollution. The database is bibliographic only, but information is provided about document availability. There is also a personalized search profiling service provided. Dialog (5) and Silverplatter (11) supply *AGRIS* on their platforms also.

AgroBase, produced by the Government Research Center[4] at the National Technical Information Service, combines *AGRIS* and *AGRICOLA* into one database, with a total of over 5.5 million records, many including abstracts, with coverage from 1970 to present.[5] Topics include water quality, aquatic sciences and fisheries, hydrology, and hydroponics. It is available on the NISC (6) *BiblioLine* platform and is updated monthly.

Applied Science and Technology Abstracts, from H. W. Wilson Co. (13), provides over 1,000,000 records from more than 600 sources in the fields of chemistry, engineering, physics, and others. The dates covered are from 1983, with abstracts provided from 1994. The database is updated weekly (*WilsonWeb*) or monthly (*WilsonDisc*). There is also a full-text version dating from 1997, which is updated four times weekly. Other platforms include Dialog (5), Ebsco (6), OCLC (9), Ovid (10), and Silverplatter (11).

Aquatic Biology, Aquaculture and Fisheries Resources is an anthology of thirteen files available through NISC (6) on their *BiblioLine* platform, and on CD-ROM. The database dates from 1971 and is updated quarterly. It

Encyclopedia of Water Science
DOI: 10.1081/E-EWS 120010125

contains over 888,600 records with abstracts on the science and management of aquatic organisms and environments. Subject coverage includes the ecology, biology, nutritional and environmental aspects of fish and aquatic environments. There are 13 file sources, which include *ASFA: Aquatic Sciences and Fisheries Abstracts Part*1: *Biological Sciences and Living Resources*, and several abstracting services relating to aquaculture.

ASFA: Aquatic Sciences and Fisheries Abstracts, produced by Cambridge Scientific Abstracts (2), is available through the Internet and on CD-ROM. It provides international information in the science, technology, and management of marine, freshwater, and brackish water environments and organisms. The database dates from 1978 and is updated monthly. Internet platforms include Dialog (5), Ovid (10), SilverPlatter (11), and STN (12). There are over 717,000 bibliographic records, many with abstracts, in aquaculture, biology, ecology, the environment, marine sciences, oceanography, pollution, and water. A thesaurus is available.

BIOSIS Previews, from the BIOSIS organization,[6] combines Biological Abstracts with Biological Abstracts/-Reports, Reviews, Meetings (RRM) to provide references from journals, books, meetings, reviews, and other publications. Platforms include Dialog (5), Ebsco (*Biological Abstracts* and *BasicBIOSIS* only) (6), ISI (7), Ovid (10), Silverplatter (11), and STN (12). It has approximately 13 million records dating from 1969 to the present from over 5500 international sources, and is updated weekly. Subject coverage is life science topics including sources relating to water. The records are bibliographic with abstracts.

Biological and Agricultural Index, from H. W. Wilson Co. (13), covers the literature in the fields of biology and agriculture since 1983, including subject areas relating to water. It is available on CD-ROM (updated monthly) and through the Internet (updated weekly) on OCLC (9), Ovid (10), Silverplatter (11), and *WilsonWeb* and *WilsonDisc* platforms and is bibliographic only. *Biological and Agricultural Index Plus* is also available, providing full-text and abstracts of 45 journals from 1994, and updated weekly.

CAB ABSTRACTS, available on CD-ROM and the Internet, is a bibliographic database compiled by CAB International (1) of its print abstracts. It covers the international literature in areas such as agriculture, forestry, horticulture, and the management and conservation of natural resources dating back at least 10 yr. Sources include scientific journals, monographs, books, technical reports, theses, reviews, conference proceedings, patents, annual reports, bibliographies and guides, and translated journals. There are over 3.5 million records in the database, which is updated weekly, and a thesaurus is

provided. Other platforms include Dialog (5), Ovid (10), and Silverplatter (11).

Chemical Abstracts is produced by Chemical Abstracts Service (CAS) (2) and made available electronically on their *SciFinder Scholar* platform. It is also available through Dialog (5). The database contains over 16 million citations and abstracts from journal articles, patents, reviews, technical reports, monographs, conference and symposium proceedings, dissertations, and books dating back to 1967 from over 8000 sources worldwide. Subject coverage is of all facets of pure and applied chemistry, including water and aquatic sciences. The database is updated weekly.

Current Contents is available in CD-ROM and Internet formats from ISI (7) in seven discipline areas including *Agriculture, Biology and Environmental Sciences, Life Sciences,* and *Physical, Chemical and Earth Sciences*. It provides access to complete bibliographic information for a 2-yr backfile, which is updated daily, from articles, editorials, meeting abstracts, commentaries, and other significant items of over 8000 scholarly journals and more than 2000 books. Dialog (5), Ovid (10), and Silverplatter (11) make *Current Contents* available on their platforms.

EiCompendex, available on the Internet through Engineering Village, is produced by Engineering Information, Inc.[7] and provides comprehensive interdisciplinary bibliographic information relating to engineering and technology dating back to 1970. It is updated monthly. There are over three million summaries from over 5000 sources including journal articles, technical reports, conference papers and proceedings, and Web sites. This database is also available on Dialog (5), Ebsco (6), Ovid (10), and Silverplatter (11) platforms.

Environmental Sciences and Pollution Management, produced by CSA (2) and available on OCLC (9) and Ovid (10) platforms, contains over one million records about aquatic pollution, water resource issues, and other subject coverage from over 4000 sources since 1981. It is updated monthly. Abstracts and bibliographic citations are provided.

General Science Abstracts, from the H.W. Wilson Co. (13), provides indexing since 1984, and abstracts since 1993 of 191 periodicals. The database focuses on student and nonspecialist coverage of several fields, many relating to water resources. There are 615,000 records which are updated weekly (*WilsonWeb*) or monthly (*WilsonDisc*). A full-text version is also available for 57 periodicals dating from 1996, which is updated four times weekly. *General Science Abstracts* is available as well on the OCLC (9), Ovid (10), and Silverplatter (11) platforms with abstracts only.

GeoArchive, produced by Geosystems (United Kingdom) and available on circular CD-Rom and the Internet from Oxmill Publishing,[8] contains international information covering geological, hydrological, and environmental sciences dating from 1974. It is also available through Dialog (5) with over 628,000 bibliographic records and is updated monthly. It is also included in NISC's *Marine, Oceanographic and Freshwater Resources* database. Sources include journals, magazines, conference proceedings, doctoral dissertations, technical reports, maps, and books. Indexing is by the thesaurus, *Geosaurus.*

GeoRef, provided by the American Geological Institute,[9] covers the geology of North America since 1785 and international geology since 1933. It is available on CD-ROM (with monthly updates) and through the Internet (updated twice monthly) from many vendors including CSA (2), Community of Science, Inc. (4), Dialog (5), Ebsco (6), OCLC (9), SilverPlatter (11), and STN (12). There are over 2.2 million bibliographic records with abstracts from journals, books, maps, conference papers, reports, and theses, and a thesaurus is provided. The print equivalent is *Bibliography and Index of Geology.*

Groundwater and Soil Contamination Database, provided by the American Geological Institute,[9] provides complete bibliographic information for over 60,000 references from 2500 serial titles published since 1975. It is updated quarterly with worldwide coverage of the literature in geology, hydrology, and the environment, with emphasis on reports of the U.S. Geological Survey and other US government departments.

Hydrology InfoBase, produced by Geosystems (United Kingdom) and available on CD-Rom and the Internet from Oxmill Publishing,[8] is a database subset of *GeoArchive* (above). It covers information from international sources in all fields relating to hydrology, including geomorphology, soil science, water–rock interactions, water resources, energy, pollution, agriculture, forestry, engineering, and environment dating from 1970. It includes bibliographical references with abstracts.

Marine, Oceanographic and Freshwater Resources contains materials dating back to 1964 from 14 different sources, and is available from NISC (8) on *BiblioLine* or CD-ROM. It is updated quarterly, with over 1,000,000 records, and provides coverage on international marine and oceanic information, and estuarine, brackish water, and freshwater environments. Subject areas include environmental quality; limnology and freshwater environments; physical oceanography; pollution, acid rain, and global warming; sea-level fluctuations; biological oceanography and ecology. Sources include *Part 1: Aquatic Sciences and Fisheries Abstracts* (above), *Part 2: Ocean Technology, Policy and Non-Living Resources* and *Part 3: Aquatic Pollution and Environmental Quality, Oceanic Abstracts, National Oceanic and Atmospheric*

Administration, and others available from United States and international institutions.

National Oceanic Atmospheric Administration's Library and Information Network[10] is a 23-institution consortium providing nine collections dating back to 1820. Subject areas include hydrographic surveying, oceanography, meteorology, hydrology, living marine resources, and meteorological satellite applications. The database contains more than 127,000 bibliographic records from over 9000 serial titles, 1500 active journal subscriptions, 35,000 reports, and meteorological data publications from 100 countries, as well as 1000 rare books. Materials can be requested for loan.

NTIS (National Technical Information Service) database[11] is available from the Government Research Center (URL: http://grc.ntis.gov/) and is a central source for government information. It provides access to over 2,000,000 titles produced by government agencies since 1964. Subjects include agriculture, energy, the environment, and other science and technology areas. Descriptive summaries and bibliographic records are provided. Ebsco (6), NISC (8), Ovid (10), Silverplatter (11), and USGovSearch[12] also offer access to this database through their platforms.

Science Citation Index Expanded, from ISI (7) and available on CD-ROM and through the Internet as part of the *Web of Science,* is a citation index dating back to 1945, updated weekly. It provides bibliographic information, author abstracts, and cited references found in 3500 of the world's leading scholarly science and technical journals covering more than 150 disciplines. The *Web of Science* covers more than 8000 international journals in the sciences, social sciences, arts and humanities, and offers access to electronic full-text journal articles.

Selected Water Resources Abstracts SWRA)[13] provides over 10,000 abstracts dating from 1977 (and earlier) to 1997 from worldwide technical literature covering a wide variety of topics relating to water resources. Sources include journals, monographs, conference proceedings, reports, and U.S. Government documents. Subject coverage includes groundwater, water quality, water planning, and water law and rights. The complete SWRA print records dating back to 1967 are available through *Water Resources Abstracts.* See also the USGS WRSIC Research Abstracts database[14] and the Universities Water Information Network.[15]

Waternet, a bibliographic database provided by the American Water Works Association (AWWA),[16] provides over 50,000 citations and abstracts from journals, books, proceedings, government reports, and technical papers from publishers around the world. It is available in CD-ROM format, and updated twice a year. Document delivery services are offered.

Water Resources Abstracts is a database maintained by CSA (2). It was formerly produced by the Water Resources Scientific Information Center[14] of the U.S. Geological Survey.[17] It is available in CD-ROM and Internet formats on Dialog (5), NISC (8), and Silverplatter (11) platforms. With over 360,000 records, the database provides citations and abstracts for water resources from 1967 and is updated monthly. Print sources include *Water Resources Abstracts* (1994–present), *Selected Water Resources Abstracts* (1967–1994), *Water Quality Instructional Resources Information System* (1979–1989), and the *WRSIC Thesaurus*.

Water Resources Worldwide, available from NISC (8) on its *BiblioLine* platform and CD-ROM, provides coverage of industrial and environmental aspects of water, wastewater, and sanitation from international sources including South Africa's *WATERLIT*, Canada's *AQUAREF*, CAB Abstract's *Aquatic Subset* and the Netherlands' *DELFT HYDRO* databases. Emphases include water in arid lands, aquatic information relevant to agricultural practice, and engineering and related technological disciplines. Updated quarterly, there are over 531,300 citations and abstracts dating back to 1970.

PLATFORMS

(1) CAB International:[18] *CAB Abstracts*.
(2) CAS: Chemical Abstracts Service:[19] *SciFinder Scholar*.
(3) CSA: Cambridge Scientific Abstracts:[20] *Agricola; ASFA: Aquatic Sciences and Fisheries Abstracts; Biotechnology and Bioengineering Abstracts; Environmental Sciences and Pollution Management; GeoRef; Water Resources Abstracts*.
(4) Community of Science, Inc.:[21] *Agricola; GeoRef*.
(5) Dialog:[22] *Agricola; Agris; ASFA: Aquatic Sciences and Fisheries Abstracts; Applied Science and Technology Abstracts; BIOSIS Previews; CAB Abstracts; Chemical Abstracts; Current Contents; EiCompendex; GeoArchive; GeoRef; Water Resources Abstracts*.
(6) Ebsco Information Services[23] *Ebscohost: Agricola; Applied Science and Technology Abstracts; BasicBIOSIS and Biological Abstracts; GeoRef; NTIS*.
(7) ISI: Institute of Science Information:[24] *BIOSIS; Current Contents; Science Citation Index Expanded (Web of Science)*.
(8) NISC: National Information Services Corporation[25] *BiblioLine: AgroBase; Aquatic Biology, Aquaculture and Fisheries Resources; Marine, Oceanography and Freshwater Resources; NTIS;*

Water Resources Abstracts; Water Resources Worldwide.
(9) OCLC[26] *FirstSearch: Agricola; Applied Science and Technology Abstracts; Biological and Agricultural Index; Environmental Sciences and Pollution Management; General Science Abstracts; GeoRef*.
(10) Ovid Technologies:[27] *Agricola; ASFA: Aquatic Sciences and Fisheries Abstracts; Applied Science and Technology Abstracts; Biological and Agricultural Index; BIOSIS Previews; CAB Abstracts; Current Contents; EiCompendex; Environmental Sciences and Pollution Management; General Science Abstracts; NTIS*.
(11) Silverplatter Information, Inc.[28] *SPIRS: Agricola; Agris; ASFA: Aquatic Sciences and Fisheries Abstracts; Applied Science and Technology Abstracts; Biological and Agricultural Index; BIOSIS Previews; CAB Abstracts; Current Contents; EiCompendex; GeoRef; NTIS; Water Resources Abstracts*.
(12) STN:[29] *Agricola; ASFA: Aquatic Sciences and Fisheries Abstracts; BIOSIS Previews; GeoRef*.
(13) H. W. Wilson Co.[30] *WilsonDisc and WilsonWeb: Applied Science and Technology Abstracts; Biological and Agricultural Index; General Science Abstracts*.

REFERENCES

1. URL: http://www.agnic.org/ (accessed 3/02/2001).
2. URL: http://www.nal.usda.gov/ag98/ (accessed 9/17/2001). A list of journals indexed by *Agricola* is available at URL: http://www.nal.usda.gov/indexing/ljiarch.htm.
3. URL: http://www.fao.org/agris/ (accessed 9/17/2001).
4. URL: http://grc.ntis.gov/ (accessed 9/04/2001).
5. Email confirmation of coverage dates 9/05/2001 from GRC Help Desk: grchelp@ntis.gov
6. URL: http://www.biosis.org/ (accessed 9/07/2001).
7. URL: http://www.ei.org/ (accessed 9/10/2001).
8. URL: http://www.oxmill.com/ (accessed 9/10/2001).
9. URL: http://www.agiweb.org/ (accessed 9/17/2001).
10. URL: http://www.lib.noaa.gov/ (accessed 3/03/2001).
11. URL: http://grc.ntis.gov/ntisdb.htm (accessed 9/10/2001).
12. URL: http://govsearch.northernlight.com/ (accessed 9/10/2001).
13. URL: http://water.usgs.gov/swra/ (accessed 3/03/2001).
14. URL: http://www.uwin.siu.edu/databases/wrsic/
15. URL: http://www.uwin.siu.edu/dir_database/index.html
16. URL: http://www.awwa.org/waternet/index.html (accessed 9/26/2001).
17. URL: http://www.usgs.gov/
18. URL: http://www.cabi.org/ (accessed 9/17/2001).
19. URL: http://www.cas.org/ (accessed 9/04/2001).

20. URL: http://www.csa.com/ (accessed 9/17/2001).
21. URL: http://www.cos.com/ (accessed 9/17/2001).
22. URL: http://www.dialog.com (accessed 2/24/2001).
23. URL: http://www.ebsco.com/ (accessed 9/17/2001).
24. URL: http://www.isinet.com/ (accessed 9/17/2001).
25. URL: http://www.nisc.com/Frame/NISC_products-f.htm (accessed 3/02/2001 and 9/04/2001).
26. URL: http://www.oclc.org/firstsearch/ (accessed 3/02/2001).
27. URL: http://www.ovid.com (accessed 9/07/2001).
28. URL: http://www.silverplatter.com/ (accessed 9/07/2001).
29. URL: http://www.cas.org/stn.html (accessed 9/18/2001).
30. URL: http://www.hwwilson.com/ (accessed 9/12/2001).

ADDITIONAL RESOURCES

The National Agricultural Library maintains a list with short descriptions of additional water resources databases at its Water Quality Information Center web site (URL: http://www.nal.usda.gov/wqic/dbases.html).

Ahn, Myeonghee Lee; Walker, R.D. Retrieval Effectiveness of Water Resources Literature. *Proceedings of the 5th International Conference on Geoscience Information*, Prague, Czech Republic, 1994; Geoscience Information Soc., 1996; 1–23.

Clark, Katharine, E. Water Resources Abstracts: A Review. CD-ROM Professional *3* (May 1990), 102–107.

Haas, Stephanie; Mae Clark. Research Journals and Databases Covering the Field of Agrochemicals and Water Pollution. Sci. Technol Libr. *13* (Winter 1992), 57–63.

Keeley, Kurt, M. AWWA's (American Water Works Association) Private Files. *Private File Creation/Database Construction: A Proceeding with Five Case Studies*; Hlava, M.M.K., Ed.; Special Libraries Association: New York, NY, 1984; 59–69.

Pretorius, Martha. Water Research, Electronic Journals and Databases: The South African Way. *Electronic Information and Publications: Looking to the Electronic Future, Let's Not Forget the Archival Past*, Proceedings of the 24th Annual Conference of the International Association of Aquatic and Marine Science Libraries and Information Centers (IAMSLIC) and the 17th Polar Libraries Colloquy (PLC), Reykjavik, Iceland, Fort Pierce, Sept 20–25, 1998; Markham, J.W., Duda, A.L., Martha Andrews, Eds.; IAMSLIC: FL, 1999; 171–174.

Drainage Coefficient

Gary Sands
University of Minnesota, St. Paul, Minnesota, U.S.A.

INTRODUCTION

Artificial (surface and subsurface) drainage systems are designed for the timely removal of excess water from land to reduce the risk of water damage to crops, soils, or structures. The term drainage coefficient represents the quantity or rate at which water is removed by the drainage system to either lower a water table (saturated portion of the soil profile) or accommodate surface runoff. For subsurface drainage systems, drainage coefficients are usually expressed as a depth of water removed per 24 hr over the drained area (mm/day), and for surface drainage systems, as a rate of flow per unit area drained. Drainage practitioners—farmers, contractors, engineers—routinely use the term drainage coefficient as an important criterion in design, operation, and management of drainage systems.

ESTIMATION AND SELECTION OF DRAINAGE COEFFICIENTS

The estimation and selection of drainage coefficients for a drainage project require an understanding of soil properties, surface and subsurface hydrology, and involve economic and risk decision-making. Estimation of soil drainage rates involves the process of determining the nature and extent of excess water, and how soil and drainage design factors influence the water removal rate for the soil of interest. Knowledge of the hydrology of the area to be drained (field, farm, building site, watershed) such as, the amount of excess water that occurs in response to various rainfall events and the drainage characteristics of the soil, is very important. Because drainage coefficients depend on both climatic and soil/watershed characteristics, they should be regarded as site/region specific values.

The rate at which water can be drained from the soil depends on soil hydraulic properties (e.g., hydraulic conductivity or the ability to transmit water), various drainage design parameters, and water table depth. Table 1 shows the effect that some of these parameters have on drainage rate.

The rate at which water can be removed from the field depends not only on the previous factors, but also the size and slope of the drains. Some practitioners consider the term drainage coefficient to indicate the maximum capacity of the drainage system, in mm/day. The capacity of the drainage pipe network may in fact, exceed the rate at which water can move through the soil to the drains. Hence, the actual drainage coefficient is typically less than the capacity of the drainage pipe network.

It is through the judicious selection of the design parameters that the drainage practitioner influences the rate at which water is removed from the drainage area. For drainage in irrigated regions, additional water for irrigation leaching requirements (application of excess water to "flush" the rooting zone) must also be factored into drainage coefficients and system capacities.

The *selection* of the appropriate design drainage coefficient from a set of possible values comprises elements of both economics and risk. The appropriate coefficient depends not only on the estimated cost of drainage measures or techniques, but also on the relative value of that which is to be protected from water damage. In the case of agricultural crops and other plants, the relationship of plant growth and performance to excess water stress is paramount to selecting appropriate drainage coefficients. Plants that are more sensitive to excess water stress or are of higher value may justify the selection of higher drainage coefficients. The effects of inadequate

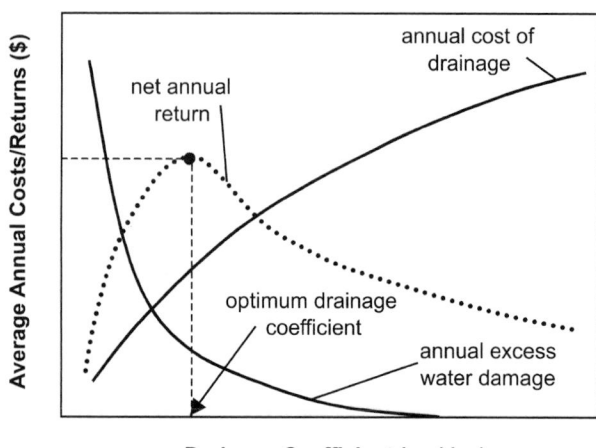

Fig. 1 Economic consequences of drainage coefficient (system capacity) selection.

Encyclopedia of Water Science
DOI: 10.1081/E-EWS 120010051

Table 1 Effect of various drainage parameters on drainage rate

Parameter	Change in parameter	Effect on drainage rate (all other factors unchanged)
Drain spacing (for systems with parallel drains)	Increase	Decrease
Drain depth	Increase	Increase
Soil texture	Lighter	Increase
Soil hydraulic conductivity	Increase	Increase
Water table depth (over the drains)	Increase	Increase

Table 2 Criteria for water table depth and drainage coefficients for various countries

Country	Water table depth (m)	Drainage coefficient (mm/day)
Humid regions		
South China (Jiangsu province)	Concentrated root system layer plus the height of the capillary moisture saturation of the soil wheat 0.5–1.2 cotton 0.5–1.5	
Germany		7–18
Hungary	0.5–1.2	3.5–5.2
Ireland	0.4–0.6	10–15
Netherlands	0.3–0.5	7–10 (in greenhouses, 20–30)
Poland	Planting season 0.4–0.6 Growing season 0.45–0.8	5–8
Portugal		9–18
France	Drawdown from 0.20 to 0.45–0.5 within one day for intensive arable land, 3 days–5 days for grassland or less intensive arable land	10–20 (in mountainous areas up to 50)
Japan	Paddy monoculture 0.3–0.4 after 2 days–3 days of rainfall and 0.4–0.5 after 7 days of rainfall Permanent crops 0.5–0.6 after 2 days–3 days, and 0.6–1.0 after 7 days of rainfall	10–50
U.S.A.		10–38
Semiarid and arid regions		
India	1.2	2.5
Iraq	1.2	2.5–3.0
Pakistan	Cultivated land 1.0 Fallow land 1.5	2.5–3.5
Romania	Sandy soils 0.6–0.8 Intermediate soils 1.0–1.2 Heavy soils 1.3–1.2	3.5
USSR		0.8–3.5
Australia	For moderately saline groundwater (2.0 ds/m) One week after irrigation 0.8–1.1 Horticulture 0.45–0.75	0.8–3.5 horticultural crops 2.5–5.0

Source: From Ref. [7].

drainage on crop growth and yield were recently summarized by Evans and Fausey.[1]

As a design criterion, the drainage coefficient plays the primary role in the size and extent (capacity), and ultimately, the cost of a drainage system. For a given area of interest, use of a larger drainage coefficient typically results in a more extensive drainage system, greater system capacity, and cost. These costs must be considered together with the expected benefits of the system to produce a design as close to the economic optimum (maximum net return) as possible. Fig. 1 shows the concept of balancing increased drainage system cost with increased levels of excess water control. The selection of drainage coefficients must also incorporate risk because, as a precipitation-driven process, drainage needs from year-to-year are uncertain. Thus, risk and economics are inherent elements in the selection of drainage coefficients, implying that the design of a drainage system that reduces the risk of damage to zero can rarely, if ever, be justified.

A number of different approaches to estimating and selecting drainage coefficients have been taken over the years. These approaches can be generalized into the following categories: 1) mathematical models; 2) field experimentation/measurement; and 3) computer modeling. Mathematical models and field experiments focus on estimating drainage rates based on soil, rainfall, and drainage design parameters. Computer models such as DRAINMOD[2] have been used to both estimate drainage coefficients and select optimal design parameters for drainage systems based on economics and risk. Detailed examples of these approaches can be found in Skaggs and Van Schilfgaarde:[3] mathematical models (Chapters 4–8); computer modeling (Chapters 13–15) and Skaggs and Tabrizi;[4] and field measurement.[5,6]

DRAINAGE COEFFICIENTS ADOPTED IN DIFFERENT COUNTRIES

Framji et al.[7] surveyed drainage practices world-wide and produced the following summary table (Table 2) of water table depth requirements and drainage coefficients used in various countries for subsurface drainage systems.

REFERENCES

1. Evans, R.O.; Fausey, N.R. Effects of Inadequate Drainage on Crop Growth and Yield. In *Drainage of Agricultural Land*; Skaggs, R.W., Van Schilfgaarde, J., Eds.; American Society of Agronomy: Madison, WI, 1999; 13–54.
2. Skaggs, R.W. A Water Management Model for Shallow Water Table Soils. University of North Carolina Water Resources Research Institute Tech. Rep. 134; 1978.
3. Framji, K.K.; Garg, B.C.; Kaushish, S.P., (Eds.) *Drainage of Agricultural Lands*; American Society of Agronomy: Madison, WI, 1999; 1328.
4. Skaggs, R.W.; Tabrizi, A. Design Drainage Rates for Estimating Drain Spacings in North Carolina. Trans. ASAE **1986**, 29 (6), 1631–1640.
5. Irwin, R.W.; Bryant, G.J.; Toombs, M.R.; Stone, J.A. Evaluation of a Drainage Coefficient for Brookston Clay Soil. Trans. ASAE **1987**, 30 (5), 1343–1346.
6. Ahmadi, M.Z. A Field Approach to Estimation of Humid Area Drainage Coefficients. Agric. Water Manag. **1995**, 29, 101–105.
7. Framji, K.K.; Garg, B.C.; Kaushish, S.P., (Eds.) *Design Practices for Covered Drains in an Agricultural Land Drainage System—A World-Wide Survey*; International Commision on Irrigation and Drainage: Chanakyapuri, New Delhi, India, 1987; 438.

Drainage, Controlled

Robert O. Evans
North Carolina State University, Raleigh, North Carolina, U.S.A.

INTRODUCTION

Excessive soil water is a major concern on soils with seasonally shallow water tables. Drainage is the practice of removing excess water from land in order to facilitate seedbed preparation and planting and to provide adequate aeration following excessive rainfall. Several techniques are available to improve drainage and reduce excess water-related crop stress. These include both surface practices[1] and subsurface practices.[2–4] While wetness is the major concern, soil moisture under rainfed conditions varies such that crops periodically suffer from drought stresses even on traditionally shallow water table soils. Intensive drainage systems that are often necessary to remove excess water during extreme wet periods, tend to remove more water than necessary during drier periods, a condition referred to as temporary overdrainage.[5] To reduce the occurrence of overdrainage and improve crop utilization of rainfall, a water control structure may be installed in the drainage outlet to regulate or "control" the rate and amount of drainage, Fig. 1. The decline in the drainage volume often results in a reduction in the nutrient load being discharged with the drainage water.[7,8] While recent growth in the use of controlled drainage has been to conserve water and enhance drainage water quality, controlled drainage has been used historically to reduce subsidence in drained organic soil.[9] This application continues in places such as the Everglades agricultural area in Florida, the Wester Johor area in Malaysia, and several other locations around the world.[10]

HOW CONTROLLED DRAINAGE WORKS

Controlled drainage involves the use of some type of adjustable, flow-retarding structure placed in the drainage outlet that allows the water level in the outlet to be artificially set. Many types of structures can be used depending on the layout of the drainage system. Controlled drainage may be practiced with either surface or subsurface drainage systems, although the benefits of drainage control are closely correlated to subsurface drainage intensity. In other words, controlled drainage effectiveness increases as the subsurface drainage intensity increases. Where drain tubing or field ditches

outlet directly to an open channel such as a canal or stream, the system is referred to as an open system. Water control structures for open systems may range from simple, stop-log, weir type structures often referred to as flashboard risers,[11] Fig. 2, to automated inflatable dam type structures.[5] Where drain tubes outlet to main drains rather than open channels, the system is referred to as a closed system.[12] Several tubing manufacturers have designed and marketed barrel type structures for use in closed systems that function as a weir in the main drain line and allow the water level to be controlled.

When operated in the controlled drainage mode, drainage occurs as long as the water table in the field is at a higher elevation than the weir elevation at the control structure. As the water table in the field recedes, the rate of drainage decreases. Once the water table drops below the weir setting, drainage stops; however, the water table will continue to recede as the crop removes water by evapotranspiration. Once the field water table drops below the water level in the outlet, the process may reverse and water stored in the outlet ditch flows back through the drains into the soil profile. The amount of water stored in the outlet depends on the dimensions of the outlet. Large canals may supply the equivalent of 5 mm–10 mm while tubing outlets store very little water. In either case, water stored in the soil profile that would otherwise drain is typically of greater magnitude than the amount of water stored in the outlet. In the controlled drainage mode, the water level in the outlet typically fluctuates several times during the growing season between the weir setting and the bottom of the outlet, Fig. 3, in response to daily fluctuation in rainfall and evapotranspiration.

The control structure is normally sized to convey the full capacity of the ditch or waterway during high flow periods. For a flashboard riser type structure, the flashboards function as a rectangular weir, Fig. 4, and the flow over the weir is computed by the equation:

$$Q = CH^{3/2}(L - 0.2H) \qquad (1)$$

where Q is the discharge in cubic meters per second neglecting velocity of approach, L the length of weir in meters, H the head on the weir in meters measured at a point no less than $4H$ upstream from the weir, and C is 3.33 for rectangular weir.

Encyclopedia of Water Science
DOI: 10.1081/E-EWS 120010231

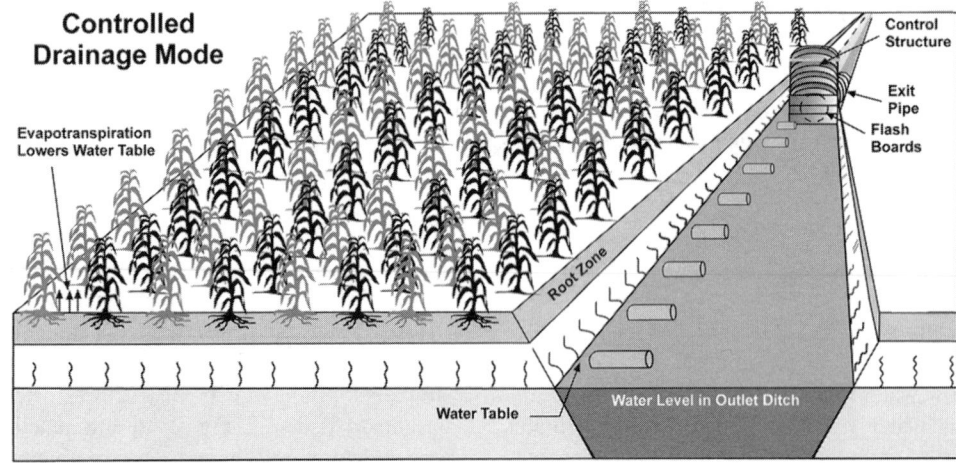

Fig. 1 Schematic of the controlled drainage operational mode. Drainage stops when the water table drops to the same level as the top of the control structure (weir). The water table may continue to drop due to evapotranspiration. (From Ref. [6].)

The weir design is normally based on fully contracted flow, which means that the weir crest and sides are far enough removed from the bottom and sides of the weir box or channel that "fully contracted" flow is developed. The discharge pipe for a flashboard riser structure is sized as a culvert (i.e., boards are out and ditch is flowing full), although the weir is normally sized as though boards are in place. This usually eliminates the need for the farmer to rush out to the structure and remove boards each time a high flow event occurs (flash flood type event). The design head on the weir is typically assumed to be between 150 mm and 300 mm. These design constraints result in a weir length that is about 1.5 times the diameter of the culvert or outlet pipe. Similar design guidelines are used for barrel type structures.[15] The backfill over the outlet pipe must be of suitable texture and compaction to

function as a dam. The outlet pipe often serves as a road crossing so the pipe length typically varies from 6 m to 12 m depending on depth of the ditch and whether or not head walls are constructed. Water pressure acting against the upstream side of the flashboard riser results in uplift, which tends to cause the structure to "float up." For small structures, typically structures with weir lengths less than 0.5 m, the weight of the soil over the outlet pipe is adequate to counteract the buoyancy of the water being held by the structure. For structures larger than 0.5 m, concrete should be poured around the base of the structure to offset the buoyancy of the upstream water.

PRODUCTION BENEFITS OF CONTROLLED DRAINAGE

In shallow water table soils, crop yield is roughly related to water table depth as shown in Fig. 5. Under highly controlled environmental conditions with a static water table, there is an optimum water table depth, typically 0.6 m–1 m deep, where yield will be maximized. This optimum depth is associated primarily with the type of crop and the soil physical properties affecting soil-water and aeration. Under field conditions, the water table position is constantly fluctuating such that an absolute optimum rarely exists. When the water table is close to the soil surface, conditions are typically too wet for optimum crop growth and yields are often suppressed due to wet stress. Holding the water table too high can result in root pruning and nitrogen deficiency as high water levels promote rapid loss of nitrogen through denitrification. Similarly, when the water table drops "too low" below the root zone, capillary rise (as was shown in Fig. 3) is not

Fig. 2 Flashboard riser type water control structure used to manage the outlet water level in an open ditch system.

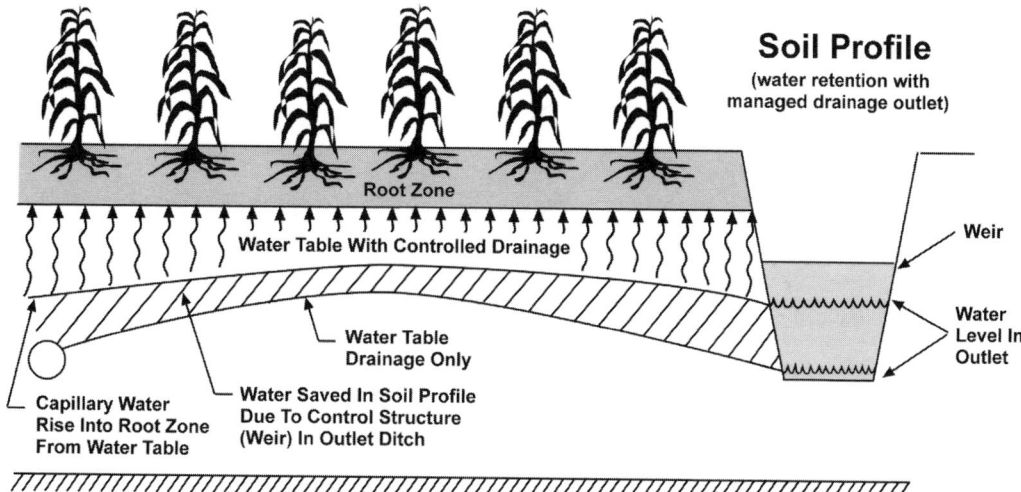

Fig. 3 Water level fluctuation with a controlled drainage system. The cross hatched area represents the amount of water saved during one cycle. Once the water table drops below the weir, it does not rise again until the next rainfall event large enough to cause percolation below the root zone. (From Ref. [13].)

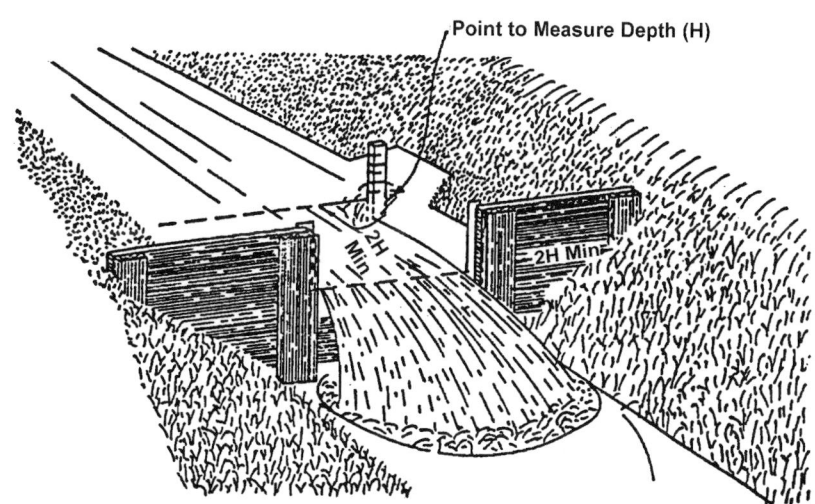

Fig. 4 Schematic of a rectangular contracted weir representing a flashboard riser type water control structure. (From Ref. [14].)

adequate to supply evapotranspiration requirements leading to crop yield reduction due to drought stress. The objective and challenge with controlled drainage is to manage the water table within these two extremes.

Controlled drainage has the greatest production benefit where drought conditions are intermittent and of short duration. For a single event, controlled drainage may retain up to 25 mm of water in a sandy soil profile that would otherwise drain from the system. The water saved could delay drought stress for a period of 3–7 days depending on evapotranspiration. Over the course of a growing season, drainage control may conserve upwards

of 75 mm that would otherwise drain from the soil.[13] Actual storage depends on the drainage intensity, drainage system layout, and soil drainable porosity. The benefit of the water saved depends on the rainfall amount and distribution during the growing season coupled with the water requirements of the crop.

Crop yield response to water table depth and subirrigation has been studied extensively.[17] Although controlled drainage has been practiced with a variety of crops, there are only a few field studies documenting yield response. Most studies have involved corn, soybean, or wheat. In a watershed scale study in North Carolina,

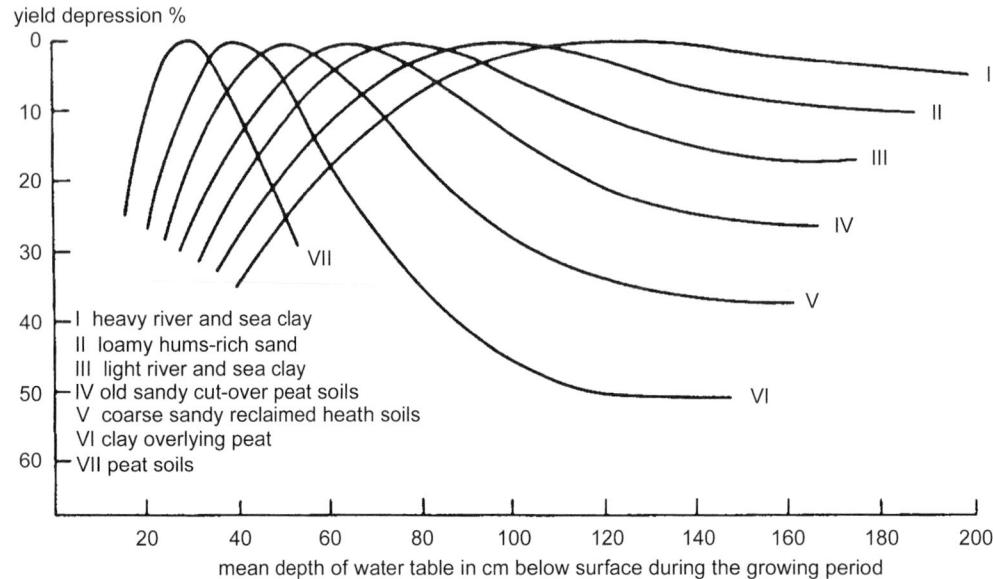

Fig. 5 Yield depression as a function of the mean depth of the water table during the growing season for various soil types (From Ref. [16].)

Parsons and Evans[18] reported a 15%–25% yield increase with water level control on corn, Table 1. In a 10-yr study, Cozier et al. (unpublished data, N.C. State University, Department of Soil Science) observed yield responses ranging from − 16 to 13% for corn and 5–21% for soybean with controlled drainage compared to conventional drainage. They observed considerable year to year variation that was closely correlated to rainfall with controlled drainage being most beneficial in dry years. Winter wheat yield was generally suppressed by controlled drainage, a result they concluded was due to periods of excess moisture occurring during the winter and early spring. While yield increases have been observed, results demonstrate that controlled drainage must be closely managed to obtain consistent yield benefits.

Table 1 Summary of corn yields at the Mitchel Creek Stream Control Project, 1981–1985

	Corn yield, kg/ha			
	No stream control		Stream water control	
Year	Nonirrigated	Irrigated	Nonirrigated	Irrigated
1981	6,460	10,662	—	—
1982	6,899	8,279	8,279	10,286
1983	3,199	7,840	5,394	9,784
1984	7,401	9,533	7,338	10,411
1985	6,899	8,844	9,847	11,038
Mean	6,147	9,032	7,715	10,412

(From Ref. [18].)

WATER QUALITY BENEFITS OF CONTROLLED DRAINAGE

Fertilized cropland is a potential source of nitrogen and phosphorus, which can contribute to the nutrient enrichment of surface water ecosystems. Many artificially drained soils are adjacent to environmentally sensitive and ecologically important surface water resources. Often natural streams and surface water bodies provide the outlet for artificial drainage systems. Research has shown that agricultural drainage water may contain fertilizer nutrients. In many of the surface water bodies, nutrient levels, particularly nitrogen and phosphorus, have become high enough that a very delicate balance exists between undesirable species such as blue-green algae and other desirable flora.[19] Controlled drainage has been recognized in some states as a best management practice (BMP) to reduce the transport and delivery of nitrogen and phosphorus to surface waters.[7]

The first suggested use of controlled drainage for the purpose of reducing nitrate-nitrogen losses in drainage water came from experiments on drainage from irrigated land.[20,21] Both groups of researchers were successful but the practice apparently was not adopted in either location.[22] Research on the water quality benefits of controlled drainage was begun in North Carolina in 1974 and have continued since that time. Evans et al.[23] summarized drainage water quality studies representing approximately 125 site years of drainage and controlled drainage water quality data collected at 14 locations in North Carolina. Skaggs et al.[24] presented a comprehensive review of research on hydrology and water quality effects of agricultural drainage, citing studies from several countries. Gilliam et al.[22] explained the processes by which nutrients are transported in drainage waters and how drainage control could be utilized to reduce drainage losses. Collectively, these reviews represent more than 200 published articles on the hydrology and water quality of drainage and controlled drainage practices. General conclusions derived from these reviews are summarized later. The reader is encouraged to refer to the earlier reviews for details and citations from the original work.

The original idea of using controlled drainage to reduce nitrate-nitrogen transport was that holding the water table closer to the soil surface would encourage more rapid and complete denitrification. Several studies have documented modest decreases (typically less than 15%) in nitrate-nitrogen concentration resulting from controlled drainage. However, the most dominant factor affecting the reduction in nitrate effluxes appears to be associated with the reduction in drainage volume sometimes on the order of 30% per year. The combined effect of concentration and outflow reduction resulted in a net decrease in nitrogen efflux of 45% in the North Carolina studies, Fig. 6. The reduction in outflow also resulted in a reduction in phosphorus efflux, Fig. 7, although controlled drainage did not cause a change in P concentration.

APPLICATION AND MANAGEMENT CONSIDERATIONS

The successful management of controlled drainage systems rests on two important objectives. The first is achieving optimum production efficiency and maximum nutrient utilization by the crop. The second is attaining maximum water quality benefits. A major challenge for

controlled drainage is determination of the optimum water control level and then maintenance of the water table within that range. Typically, the costs of additional structures needed to maintain a suitable water level becomes prohibitive when the land slope exceeds 0.5%. Thus, controlled drainage is most practical on relatively flat fields. As noted earlier, potential production benefits are greatest in coarse textured drained soils sometimes prone to overdrainage and drought. Several studies have documented that the nitrogen reduction benefits increase at higher control levels up to about 300 mm from the soil surface. Ideal yields result when water levels are in the range 600 mm–1000 mm. Under some conditions, productivity, water quality, or both goals may need to be mutually compromised for the benefit of the other. At other times, productivity and water quality goals may be compatible at least seasonally. Gilliam et al.[26] present general management recommendations that attempt to achieve a balance between production and water quality goals. They suggest that for most mineral soils, the water table should be maintained between 300 mm and 1000 mm, depending on the crop and its stage of development, the need to access fields with equipment, and prevailing weather conditions. As a guide, crop production goals can be satisfied during the growing season with only modest compromise to water quality. They suggest that some water quality benefit will be realized, although not necessarily optimized, whenever the water level is maintained within 1 m of the soil surface. Water levels in the range 500 mm–750 mm will satisfy crop requirements for most crops grown on most mineral soils during nonextreme wet periods. Water control levels should be lowered to 1000 mm to accommodate field operations involving heavy equipment. By holding the water table high (within 300 mm of the surface) during noncropping periods, water quality goals can be optimized

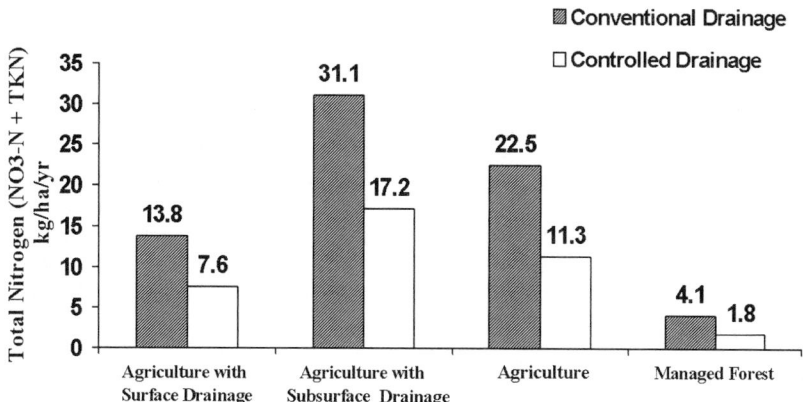

Fig. 6 Average annual total nitrogen transport in drainage outflow as measured at the field edge of 14 sites in eastern North Carolina. Controlled drainage resulted in a net 45% reduction compared to conventional drainage. (From Ref. [25].)

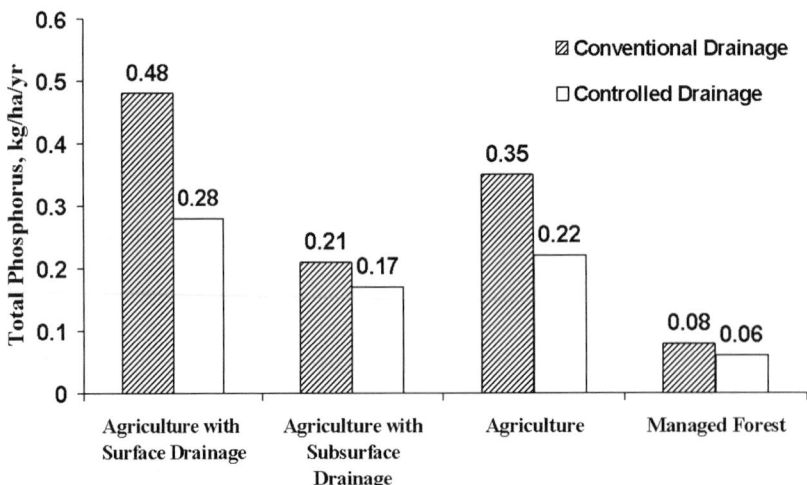

Fig. 7 Average annual total phosphorus transport in drainage outflow as measured at the field edge of 12 sites in eastern North Carolina. Controlled drainage resulted in a net 35% reduction compared to conventional drainage. (From Ref. [25].)

with no adverse production impacts. It should be noted that many of the management indicators are hidden from view and the response to adjustments is not always immediate. Thus, intensive management with long-term monitoring is necessary to develop a site-specific understanding of the system.

SUMMARY

The technical feasibility of controlled drainage is well documented. Controlled drainage can increase crop yields, reduce overdrainage, reduce the transport of fertilizer nutrients and other potential pollutants, and improve water use efficiency. The magnitude of the benefits vary among fields and watersheds as well as from year to year. The success of controlled drainage at any scale is influenced by soils, crops, topography, seasonal rainfall, hydraulic properties within the controlled area, and overall management of the system. While research over the past 30 yr has lead to significant improvements in design and operational methods, there still remains a need to improve and fine tune management strategies to optimize the net benefits of controlled drainage.

REFERENCES

1. Carter, C.E. Surface Drainage. In *Agricultural Drainage*, Agron. Monogr. 38; Skaggs, R.W., vanSchilfgaarde, J., Eds.; ASA, CSSA, and SSSA: Madison, WI, 1999; 503–507.

2. Schwab, G.O.; Fouss, J.L. Drainage Materials. In *Agricultural Drainage*, Agron. Monogr. 38; Skaggs, R.W., vanSchilfgaarde, J., Eds.; ASA, CSSA and SSSA: Madison, WI, 1999; 911–926.

3. Broughton, R.S.; Fouss, J.L. Subsurface Drainage Installation Machinery and Methods. In *Agricultural Drainage*, Agron. Monogr. 38; Skaggs, R.W., vanSchilfgaarde, J., Eds.; ASA, CSSA, and SSSA: Madison, WI, 1999; 967–1004.

4. Spoor, G.; Leeds-Harrison, P. Nature of Heavy Soils and Potential Drainage Problems. In *Agricultural Drainage*, Agron. Monogr. 38; Skaggs, R.W., vanSchilfgaarde, J., Eds.; ASA, CSSA and SSSA: Madison, WI, 1999; 1051–1182.

5. Doty, C.W.; Parsons, J.E.; Tabrizi, A.N.; Skaggs, R.W.; Badr, A.W. Deep Ditch Overdrainage Affects Water Table Depth and Crop Yield. In Proceedings of the Specialty Conference, Environmentally Sound Water and Soil Management, ASCE, 1982; 113–121.

6. Evans, R.O.; Skaggs, R.W. Agricultural Water Management for Coastal Plain Soils. North Carolina Agricultural Extension Service Bulletin AG-355, 1985; 12.

7. Evans, R.O.; Skaggs, R.W.; Gilliam, J.W. Controlled Versus Conventional Drainage Effects on Water Quality. J. Irr. Drain Eng. ASCE **1995**, *121* (4), 271–276.

8. Zucker, L.A.; Brown, L.C., Eds. Agricultural Drainage: Water Quality Impacts and Subsurface Drainage Studies in the Midwest. Ohio State University Extension Bulletin 871; The Ohio State University, 1998; 40.

9. Stevens, J.C. Drainage of Peat and Muck Lands. *Water: The 1955 Yearbook of Agriculture*; U.S. Government Printing Office: Washington, DC, 1955; 539–557.

10. Skaggs, R.W. Water Table Management: Subirrigation and Controlled Drainage. In *Agricultural Drainage*, Agron. Monogr. 38; Skaggs, R.W., vanSchilfgaarde, J., Eds.; ASA, CSSA and SSSA: Madison, WI, 1999; 695–718.

11. Evans, R.O.; Parsons, J.E.; Stone, K.; Wells, W.B. Water Table Management on a Watershed Scale. J. Soil Water Conserv. **1992**, *47* (1), 58–64.

12. Belcher, H.W.; Merva, G.E.; Shayya, W.H. SI-DESIGN—a Simulation Model to Assist with the Design of Subirrigation Systems. In *Water Management in the Next Century*, Trans Workshop Subsurface Drainage Models; 15th Int. Congress ICID. The Hague, Lorre, E., Ed.; Cemagref-Dicova: Antony Cedex, France, 1993; 295–308.

13. Evans, R.O.; Skaggs, R.W. Operating Controlled Drainage and Subirrigation Systems. North Carolina Agricultural Extension Service Bulletin AG-356, 1985; 10.

14. Extension Agricultural Engineering. N. C. State College, Engineering Drawing No. 2503; 1942.

15. Hancor *Subsurface Irrigation and Drainage Systems Contractor Design Manual*; Hancor, Inc.: Findlay, OH, 1985.

16. Visser, W.C. De Landbouwwaterhuishouding in Nederland. Comm. Onderz. landb. Waterhuish. Ned. TNO, Rapport nr. 1. 231. (as cited by Wesseling, J. Crop Growth and Wet Soils. In *Drainage for Agriculture*; van Schilfgaarde, J., (Ed.); ASA Monograph No. 17; Madison, WI, Chap. 2, 1974), 1958.

17. Evans, R.O.; Fausey, N.R. Effects of Inadequate Drainage on Crop Growth and Yield. In *Agricultural Drainage*, Agron. Monogr. 38; Skaggs, R.W., vanSchilfgaarde, J., Eds.; ASA, CSSA and SSSA: Madison, WI, 1999; 13–54.

18. Parsons, J.E.; Evans, R.O. Stream Water Level Control for Irrigation Supplies. In *Managment of Farm Irrigation Systems*; Hoffman, G.J., Howell, T.A., Solomon, K.H., Eds.; ASAE: St. Joseph, MI, 1990; 971–981.

19. Paerl, H.W. Dynamics of Blue-Green Algal (*Microcystis aeruginosa*) Blooms in the Lower Neuse River, North Carolina: Causative Factors and Potential Controls, Report No. 229; North Carolina Water Resources Research Institute: Raleigh, 1987; 164.

20. Meek, B.D.; Grass, L.B.; Willardson, L.S.; MacKenzie, A.J. Applied Nitrogen Losses in Relation to Oxygen Status of Soils. Soil Sci. Soc. Am. Proc. **1970**, *33*, 575–578.

21. Willardson, L.S.; Meek, B.D.; Grass, L.B.; Dickey, G.L.; Bailey, J.W. Nitrate Reduction with Submerged Drains. Trans. ASAE **1972**, *15*, 84–85, 90.

22. Gilliam, J.W.; Baker, J.L.; Reddy, K.R. Water Quality Effects of Drainage in Humid Regions. In *Agricultural Drainage*, Agron. Monogr. 38; Skaggs, R.W., vanSchilfgaarde, J., Eds.; ASA, CSSA and SSSA: Madison, WI, 1999; 801–830.

23. Evans, R.O.; Gilliam, J.W.; Skaggs R.W. Controlled Drainage Management Guidelines for Improving Drainage Water Quality. N.C. Cooperative Extension Service Bulletin AG-443, 1991; 15.

24. Skaggs, R.W.; Breve, M.A.; Gilliam, J.W. Hydrologic and Water Quality Impacts of Agricultural Drainage. Environ. Sci. Technol. **1994**, *24* (1), 1–32.

25. Evans, R.O.; Skaggs, R.W.; Gilliam, J.W. Rural Land Use, Water Movement and Coastal Water Quality. N.C. Cooperative Extension Service Bulletin AG-605, 2000; 12.

26. Gilliam, J.W.; Osmond, D.L.; Evans, R.O. *Selected Agricultural Best Management Practices to Control Nitrogen in the Neuse River Basin*, North Carolina Agricultural Research Service Technical Bulletin 311; North Carolina State University: Raleigh, 1997; 53.

Drainage, Hydrologic Impacts of

Mark Robinson
Centre for Ecology and Hydrology, Oxfordshire, United Kingdom

David W. Rycroft
Southampton University, Southampton, United Kingdom

INTRODUCTION

Land drainage is the practice of removing excess water from the land, and it is one of the most important land management tools for improving crop production in many parts of the world. Drainage systems may be broadly divided into surface drainage (comprising land grading and open ditches), shallow drainage (such as subsoiling to mechanically loosen the upper layer of soil), subsurface or groundwater drainage (buried perforated pipes or deep ditches), and the main drainage systems (commonly open channels) used to convey the drain water away.[1] Drainage will inevitably affect the pattern of water flows from the land and into the receiving watercourses. It is these downstream impacts of farmland drainage on the timing and magnitude of peak flows which are considered here, using the results of experimental studies and computer simulations, to present a coherent picture, and to answer most of the apparent anomalies and conflicts.

HYDROLOGIC IMPACTS

Concern about the possible downstream effects of drainage is shown by many published papers worldwide, in North America,[2,3] Great Britain,[4,5] and continental Europe, including France,[6] Netherlands,[7] Ireland,[8,9] Finland,[10] and Germany.[11] The role of drainage has been highlighted by recent flood events—for example in the Midwest of the United States in 1993, and across Europe in 1997—which reawakened concerns that drainage could aggravate flooding downstream.

There has been a debate about the effects of drainage on streamflow for well over a century, but until recently due to the lack of appropriate data, the debate has been largely speculation. Too often, the absence of evidence has erroneously been taken as evidence of an absence of effect. The earliest published account[12] was a report of a 4-day meeting held at the Institution of Civil Engineers in London in 1861. Many of the arguments and opinions expressed have resonance today, but due to the absence of objective measurements the participants were unable to reach any conclusions and the meeting was inconclusive.

These conflicting opinions resulted from differences in the emphasis given to the two processes of water storage and routing. Considering the former, it may be argued that because drainage lowers the water table, the available storage capacity in the soil is enlarged and able to absorb more storm rainfall, thereby reducing peak flow rates. In contrast, according to the routing argument, the purpose of drainage is to "remove water from the land more quickly" than under natural conditions, so peak outflows must necessarily increase.

Probably more work has been carried out in Britain upon the effects of agricultural drainage upon streamflow than in any other country. Britain was the originator of modern field drainage[13] and so became the first country where concern arose about its downstream effects, it is also one of the most extensively drained countries in the world.

It is only in the last few years that it has been possible to obtain a coherent picture based on observations of field processes, and supported and extended by computer modeling. This has shown that general statements that drainage "causes" or "reduces" flood risk downstream are oversimplifications of the complex processes involved, and that any consideration of the impact of drainage on streamflow must identify the point of interest, whether at the outfall from the field, along the main channel, or a combination of both at the catchment scale.

Experimental studies indicate that the provision of surface drainage will result in higher peak flows downstream. This was shown by a long-term experiment at Sandusky in Northern Ohio,[14] and is a result of the reduction/elimination of surface storage capacity, as well as the provision of more efficient faster flow routes. This has been demonstrated conclusively both by experimental studies and by computer simulations.

In contrast, there seems to be general agreement from experimental studies that subsurface drainage of waterlogged, poorly permeable clay soils reduces peak outflows.[15–17] Since this is one of the most common situations where artificial drainage is used, it might be

Encyclopedia of Water Science
DOI: 10.1081/E-EWS 120010232

considered to represent the most general result of field drainage.

There are, however, instances where even on heavy soils this result may not apply. Due to their low hydraulic conductivity, most water movement in clay soils is confined to flow through macropores, such as cracks. As a result of clay shrinkage and cracking in warm, dry summers, rapid macropore flow can result in larger peak flows from the drained land than from the undrained land. The role of macropores on the seasonality of peak flows from drained land was demonstrated in detail.[18]

More permeable, drier soils may also be drained where there is an economic justification—for example, drainage of land producing high value crops. In contrast to clay soils, relatively few scientific field studies have investigated the impact of draining lighter, more permeable soils. This may be partly due to the emphasis on draining clay soils, but also, no doubt, results from the greater practical difficulty encountered in plot definition where the soils are more permeable. Nevertheless, data are available from several drainage experiments on permeable soils. At Withernwick[19] flow peaks were increased in the first year after drainage and there was then a reduction in the following years due to the progressive deterioration of the secondary system of subsoiling designed to improve the soil structure. Supporting evidence of increased peak flows following the drainage of more permeable soils also comes from studies at Cockle Park in northern Britain[20] and Ellingen in central Germany.[21]

To identify factors influencing drainage response, the results of field drainage experiments under temperate northern European climates were analyzed in terms of their site characteristics.[22,23] This included topography, precipitation, drainage depth and spacing, natural (i.e., predrainage) soil water regime, and the soil properties. The only characteristics distinguishing sites, where drainage increased peak flows from those where they were reduced, were those relating to the soil water regime before drainage. The experimental sites all had similar land practices on the drained and the undrained land.

Drainage reduced peak flows on sites, which had wetter soils, with poor natural drainage, and significant amounts of storm runoff were generated as overland flow and near-surface flow in the thin upper layers of the soil. These sites had higher topsoil clay contents, and shallower depths to a poorly permeable subsoil horizon. When artificially drained, the surface saturation was largely eliminated, greatly increasing the soil water storage capacity.

In contrast, at sites with more permeable, loamy soils which were not routinely saturated before drainage, natural stormflow occurred predominantly by slower subsurface flow, the artificial drainage pipes provided more rapid flow routes leading to increases in peak outflows.

The findings are summarized in Fig. 1. This shows the topsoil texture, together with the effect of drainage on peak flows, and provides the engineer or conservationist with an initial guide to predict the effect on flows of the drainage of a site, based on a knowledge of the predrainage site characteristics.

Further insights into the factors controlling the impact of drainage may be obtained by the application of modeling techniques to investigate the important interaction between soil properties and climate in determining soil water regimes. DRAINMOD[24] was applied to two of the field sites with similar climates: a heavy clay soil at Grendon and a more permeable loam at Withernwick. The model was applied to each site using actual field values of drain and soil parameters, and the simulated peak flows from drained and undrained land were compared for similar rainfall inputs. The results showed a 70% lower median peak flow after drainage of clay soil and an increase of 40% in the median peak flow from the more permeable land.[23]

The modeled fluxes and water stores confirmed that the reduction in peaks from the clay soil after drainage was achieved by a change in storm runoff generation from overland flow (caused by soil saturation) to subsurface drainflow. For the loamy soil, the model indicates that the increase in peak subsurface flow rates was due to the steeper hydraulic gradients created by the closer spaced artificial drains.

The model also demonstrated the effect of different climatic conditions. If the loam soil site at Withernwick had double the normal rainfall (1200 mm yr^{-1} instead of 600 mm yr^{-1}) the resulting increase in ground wetness would be sufficient to generate substantial amounts of

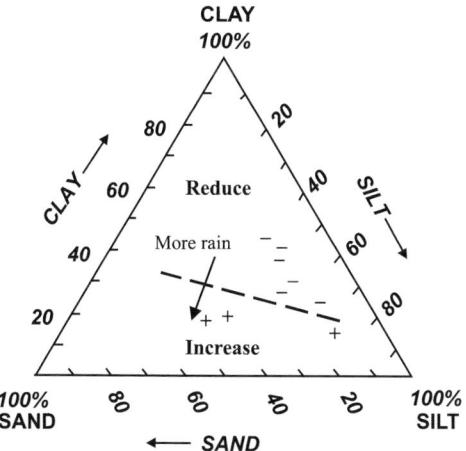

Fig. 1 Observed impact of pipe drainage on downstream peak flows (increase/reduce), showing the importance of soil texture. Model simulations of climate changes indicate that higher rainfall and wetter ground conditions will shift the balance towards drainage schemes reducing peak flows. See text for details.

overland flow on the undrained land. Artificial drainage in this case would then reduce peak flows—exactly as happens for a clay soil (where in contrast the ground wetness is caused by the low soil permeability). Using the model in this way enables these effects of site characteristics to be explored in an objective manner. The overall dominant criterion—the amount and frequency of surface runoff from undrained land—can be assessed in terms of both soil properties and climatic characteristics.

CONCLUSIONS

The effect of subsurface drainage on peak flows depends upon site wetness. If the water table is close to the surface (due to high rainfall or poor permeability), natural flows occur either over the surface or through the upper, more permeable layers of the soil. Drainage will increase soil water storage capacity and hence the amount of water that can infiltrate, thereby reducing surface runoff and peak storm flows. If the water table is deeper, due to a dry climate or due to more permeable soils, natural flows will occur through the body of the soil. In this case, artificial drainage will increase peak flows as a result of the shorter flow paths and steeper hydraulic gradients.

It must be noted that these conclusions depend upon the scale of the drainage considered. At the river catchment scale, main channel improvements will undoubtedly increase the speed of flow routing, and the timing of arrival of flows from different subcatchments will influence the peak discharge at the point of interest. The relative importance of field drainage and main drainage channels will vary with storm size: field drainage being dominant for small and medium storms, but main channel improvements becoming dominant for large events. In extreme situations where the rainfall intensity exceeds the infiltration capacity of the soil, the effects of the subsurface drains will be minimal but the associated improved watercourses will rapidly carry away the surface runoff.

Overall, it seems likely that in large catchments, drainage schemes with substantial associated surface drainage and main channel improvements will lead to higher flow peaks downstream, even though locally the effect of drainage may be to lower the peak flows.

REFERENCES

1. Skaggs, R.W.; van Schilfgaarde, J., (Eds.) *Agricultural Drainage*, Agron Monograph 38; ASA, CSSA, and SSSA: Madison, WI, 1999.

2. Whiteley, H.R. Hydrologic Implications of Land Drainage. Can. Water Res. J. **1979**, *4*, 12–19.

3. Serrano, S.E.; Whiteley, H.R.; Irwin, R.W. Effects of Agricultural Drainage on Streamflow in the Middle Thames River, Ontario, 1949–1980. Can. J. Civil Eng. **1985**, *12*, 875–885.

4. Bailey, A.D.; Bree, T. Effect of Improved Land Drainage on River Flows. *Flood Studies Report—5 Years on*; Thomas Telford: London, 1981; 131–142.

5. Rycroft, D.W. The Hydrological Impact of Land Drainage. 4th International Drainage Workshop, Cairo; ICID-CHD, CEMAGREF, 1990; 189–197.

6. Oberlin, G. Influence du Drainage et de l'Assainissement rural sur l'Hydrologie. CEMAGREF Bull. **1981**, *285*, 45–56.

7. Warmerdam, P.M.M. The Effect of Drainage Improvement on the Hydrological Regime of a Small Representative Catchment Area in the Netherlands. *Application of Results from Representative and Experimental Basins*; UNESCO Press: Paris, 1982; 318–338.

8. Burke, W. Aspects of the Hydrology of Blanket Peat in Ireland. Int. Assoc. Hydrol. Sci. **1975**, *105*, 171–181.

9. Wilcock, D.N. The Hydrology of a Peatland Catchment in N Ireland Following Channel Clearance and Land Drainage. In *Man's Impact on the Hydrological Cycle in the UK*; Hollis, G.E., Ed.; Geo Abstracts: Norwich, 1979; 93–107.

10. Seuna, P.; Kauppi, L. *Influence of Subdrainage on Water Quantity and Quality in a Cultivated Area in Finland*, Water Research Institute Publ. No. 43; Nat. Board of Waters: Helsinki, Finland, 1981.

11. Harms, R.W. The Effects of Artificial Subsurface Drainage on Flood Discharge. In *Hydraulic Design in Water Resources Engineering: Land Drainage*; Smith, K.V.H., Rycroft, D.W., Eds.; Computational Mechanics Publication: Southampton, 1986; 189–198.

12. Bailey Denton, J. On the Discharge from Underdrainage and Its Effects on the Arterial Channels and Outfalls of the Country. Proc. Inst. Civil Eng. **1862**, *21*, 48–130.

13. Van Der Beken The Development of the Theory and Practice of Land Drainage in the Nineteenth Century. In *Water for the Future*; Wunderlich, W.O., Prins, J.E., Eds.; A.A. Balkema: Rotterdam, 1987; 91–99.

14. Schwab, G.O.; Thiel, T.J.; Taylor, G.S.; Fouss, J.L. Tile and Surface Drainage of Clay Soils 1. Hydrologic Performance with Grass Cover USDA. Agric. Res. Serv. Bull. **1963**, 935.

15. Robinson, M.; Beven, K.J. The Effect of Mole Drainage on the Hydrological Response of a Swelling Clay Soil. J. Hydrol. **1983**, *63*, 205–223.

16. Harris, G.L.; Goss, M.J.; Dowdell, R.J.; Howse, K.P.; Morgan, P. A Study of Mole Drainage with Simplified Cultivation for Autumn Sown Crops on a Clay Soil. II. Soil Water Regimes, Water Balances and Nutrient Loss in Drain Water, 1978–80. J. Agric. Sci. **1984**, *102*, 561–581.

17. Armstrong, A.C.; Garwood, E.A. Hydrological Consequences of Artificial Drainage of Grassland. Hydrol. Process. **1991**, *5*, 157–174.

18. Robinson, M.; Mulqueen, J.; Burke, W. On Flows from a Clay Soil—Seasonal Changes and the Effect of Mole Drainage. J. Hydrol. **1987**, *91*, 339–350.

19. Robinson, M.; Ryder, E.L.; Ward, R.C. Influence on Streamflow of Field Drainage in a Small Agricultural Catchment. J. Agric. Water Manag. **1985**, *10*, 145–148.

20. Armstrong, A.C. *The Hydrology and Water Quality of a Drained Clay Catchment—Cockle Park, Northumberland,* Report RD/FE/10; MAFF: London, UK, 1983.

21. Schuch, M. Regulation of Water Regime of Heavy Soils by Drainage, Subsoiling and Liming and Water Movement in This Soil. Paper 1.14. In *Proc. International Drainage Workshop,* May, 1978; Wesseling, J., Ed.; International Institute for Land Reclamation and Improvement: Wageningen, 1978; 253–267

22. Robinson, M. *Impact of Improved Land Drainage on River Flows,* Institute of Hydrology Report 113; Wallingford, UK, 1990. ISBN 0-948540-24-9.

23. Robinson, M.; Rycroft, D.W. The Impact of Drainage on Streamflow. In *Agricultural Drainage*; Agron Monograph 38, Chap. 23. Skaggs, R.W., van Schilfgaarde, J., Eds.; ASA, CSSA, and SSSA: Madison, WI, 1999; 767–800.

24. Skaggs, R.W. Drainage Simulation Models. In *Agricultural Drainage,* Agron Monograph 38; Skaggs, R.W., van Schilfgaarde, J., Eds.; ASA, CSSA, and SSSA: Madison, WI, 1999; Chap. 13, 469–500.

Drainage: Inadequacy and Crop Response

Norman R. Fausey
United States Department of Agriculture (USDA), Columbus, Ohio, U.S.A.

INTRODUCTION

Drainage is an agricultural water management practice that has been used for many centuries.[1] In early times, development of cities and commerce was dependent upon stable and bountiful agriculture requiring fertile soils and adequate rainfall or irrigation water. There is evidence of failure of early irrigation-based agriculture due to salt accumulation in the soils because of not understanding how to use drainage to leach the salt from the soil.[2] Areas with adequate rainfall to support permanent agriculture frequently also need drainage to manage excess water in the soil. Soils in low-lying areas were recognized by early farmers as the more fertile and productive soils, but these soils were also subject to periodic flooding and crop loss or damage. The goals for early drainage works seem to be centered on removing standing water from crops. Archeological evidence from the Mayan culture in Central America indicates that ridges or raised beds were constructed and used as planting zones to avoid inundation of crops, a very early form of surface drainage. Ancient Greek and Roman writings included instructions for construction of both surface and subsurface drains.[1] Agriculture has evolved to a highly mechanized industry, and this has intensified the demands on drainage. Modern goals for drainage include a trafficable soil surface for timely planting and harvesting of crops using large machines; an aerated root zone that promotes good crop nutrition and minimizes disease organisms; sustained high crop yields; and an ability to maintain the salt balance within the soil profile.

INADEQUATE DRAINAGE

Inadequate drainage results when excess water (or salt) in or on the soil causes economic impairment to the present or intended use of the soil. For agriculture, this definition allows for excess water (or salt) in the soil during times when crop yield is not reduced or reduced by an amount less than the cost of improving the drainage. Some factors that affect the adequacy of drainage at any given time are the type of crop and its stage of growth, the type of soil, the current weather pattern, and the time required to complete field activities (including salt leaching). The adequacy of drainage involves a complex interdependence among soil, climate, crop, and economic factors.

TRAFFICABILITY

Excess soil water causes loss of soil strength leading to an inability to support and to provide traction for the equipment used to plant, tend, and harvest the crop. Poor trafficability may cause delays at critical times for planting, applying fertilizers and pesticides, and harvesting. Timeliness is important to both the quantity and quality of crops.

Delays in planting shorten the growing season, alter the plants' responses to rainfall and temperature patterns and day-length changes, and affect the plants' ability to compete with weeds and resist attack by insects and disease. For most spring-seeded crops, there is a critical or threshold date after which yield is reduced by delay of planting. Evans and Fausey[3] have given a very good recent review.

Delays in applying fertilizers and pesticides can cause serious economic effects. Lack of nutritional requirements in readily available form and sufficient quantity can severely reduce biomass accumulation and the harvestable yield. Disease, insects, and weeds can totally overwhelm a crop if not managed or controlled in a timely manner. Delays in harvesting can lead to loss of quality and value for most products and missed windows of economic opportunity for niche market crops.

Crop response to trafficability and timely fieldwork, made possible with drainage, could be very significant in terms of the quantity and quality of yield and also economically important.

ROOT ZONE AERATION

While water by itself is not harmful to plants, excess water interferes with soil aeration, especially the adequate supply of oxygen for root growth and respiration and for beneficial soil microbial and biological activity. Gaseous byproducts of root respiration and organic matter decomposition by microorganisms accumulate, sometimes to toxic levels, when excess water is present. The excess

Encyclopedia of Water Science
DOI: 10.1081/E-EWS 120010234

water fills the pores spaces in the soil and blocks the pathways for the exchange or equilibration of gases between the soil and the atmosphere. When these pathways are blocked, diffusion of gases between the soil and the above-ground atmosphere declines or ceases completely and oxygen in the soil can be depleted rapidly. The rate of decline in soil oxygen content is dependent upon the metabolic activity of the microorganisms and plant roots and the soil temperature.

Poor soil aeration can suppress or prevent seed germination; slow or terminate root growth; and, depending on the duration, cause wilting, poor growth, early maturation, or even death of the above-ground plant parts. The impact on the above-ground plant parts is a direct result of the effects on the roots.

Seed germination requires both water and oxygen. Water imbibition through the seed coat initiates germination, after which both water and oxygen are necessary to sustain the process. An excess of water in the soil surrounding the seed can cause an insufficient supply of oxygen reaching the rapidly dividing and growing cells. Cell division and growth rate are reduced when the supply of oxygen is inadequate, even for a few hours. If no oxygen can reach the seed, germination cannot continue, and, once terminated, will not resume.

Root elongation is slowed or terminated by an inadequate supply of oxygen. Low oxygen concentrations reduce the rate of root elongation, but do not result in root death. Total lack of oxygen for as little as a few hours can kill roots. Root elongation is vital to bring roots to the vicinity of nutrients and water that are needed to sustain plant growth and development. Under low oxygen conditions, increased resistance at the root impedes water movement into roots. McDaniel[4] reported the recovery of corn roots to normal growth rates if the duration of excess water was less than three days; otherwise the total root mass, maximum root depth, and seasonal consumptive water use were significantly reduced.

These observations lead to establishing drainage system design and performance criteria that are intended to avoid prolonged periods of excess water in the vicinity of germinating seeds and within the root zone of plants. Generally, under rain-fed agriculture, it is recommended that the drainage system has the capacity to lower the water table from the soil surface to a depth of 30 cm within 24 hr in order to adequately aerate the root zone.

SALT LEACHING

Drainage is required in irrigated agriculture to provide a means to manage the salt balance in the soil. Irrigation waters, whether from surface of subsurface sources,

contain salts such as sodium, chlorine, and bromine. These salts originate from rock during the ongoing process of weathering, and are transported by water to streams and groundwater. Irrigation water, after being applied to the soil, is taken up by the plants largely to transport nutrients into the plant and to cool the plants during transpiration, or is evaporated directly from the soil into the atmosphere. In either case, the salts are left behind in the soil and accumulate over time as more irrigation water is added to the soil.

In order to manage the salt balance in the soil, subsurface drainage is necessary and additional water is required to dissolve the salt and transport it out of the root zone. This additional water is known as the leaching requirement. In some cases, natural drainage rates are sufficient; in others, subsurface drainage must be installed to provide the drainage requirements. Hoffman and Durnford[5] discuss the design of drainage systems for salinity control in detail.

CROP RESPONSE

The response of plants to excess water stress resulting from inadequate drainage varies greatly with the stage of plant development and growth. Plants are very fragile during the germination stage. Once water has been imbibed through the seed coat and the germination process has been initiated, even 2 hr to 3 hr of flooding are enough to interrupt the process and kill the developing embryos.[6] As plants grow, specialized tissues and structures develop that help the plants cope with their environment. Once the shoots emerge from the soil and photosynthesis begins, the plants are no longer dependent solely on stored energy and have a direct connection with the above-ground atmosphere. At this stage, the plant is a much more complex system that is capable of tolerating extended periods of root zone flooding without death. Vegetative growth and yield are affected by the duration of flooding, the stage of growth at the time of flooding, and the prevailing temperature during the flooding. Depending upon the plant species, physiological adaptations may occur that allow the plant to survive prolonged flooding; however, significant reductions in growth and yield typically accrue. Plants tolerate flooding stress better under cool and cloudy conditions than under hot and sunny conditions. Tolerance to flooding tends to increase with plant age.

Flooding is a result of inadequate drainage and can cause a decrease in photosynthesis,[7] in biomass accumulation,[8] and in seed yield.[9] Damaged and dead roots in flooded plants[10] have been attributed to the lack of oxygen to support root respiration.[11,12] Flooding causes premature senescence, which results in leaf chlorosis, necrosis, defoliation, cessation of growth, and reduced yield.[13] While the lack of oxygen has been proposed as the main problem associated with flooding,[14] growth reduction and yield loss during and after flooding could also arise from root rot diseases,[15] nitrogen deficiency,[16] or nutrient imbalance.[17,18]

The common plant response to excess salt is a general stunting of growth. As salt concentrations increase above a threshold level, both the growth rate and ultimate size of the plants progressively decrease.[19] The threshold and rate of growth reduction vary widely among crop species. Some begin to exhibit injury symptoms and growth reductions at salt concentrations only twice that are present in nonsaline soil. Others actually grow better in moderately saline environments.

CONCLUSION

Yield reductions may occur as a result of excess water (or salt) on undrained or inadequately drained soils. These yield reductions may be due to factors related to trafficability or root zone aeration in the case of excess water, or inadequate leaching in the case of salt. Drainage is an effective management tool for minimizing these reductions.

REFERENCES

1. Beaucamp, K.H. A History of Drainage and Drainage Methods. In *Farm Drainage in the United States: History, Status and Prospects*; Pavelis, G.A., Ed.; USDA Econ. Res. Serv. Misc. Publ. 1455, Econ. Res. Serv.: Washington, DC, 1987; 13–29.

2. Donnan, W.W. An Overview of Drainage Worldwide. Proceedings of the 3rd National Drainage Symposium, Chicago, IL, Dec 13–14, 1976; Hoffman, G.J., Ed.; American Society of Agricultural Engineers: St. Joseph, MI, 1976; 6–9.

3. Evans, R.O.; Fausey, N.R. Effects of Inadequate Drainage on Crop Growth and Yield. In *Agricultural Drainage*; Skaggs, R.W., van Schilfgaarde, J., Eds.; American Society of Agronomy: Madison, WI, 1999; 13–54.

4. McDaniel, V. Effects of Shallow Water Tables on Corn Roots, Crop Yield and Hydrology, Ph.D. Dissertation. North Carolina State Univ.: Raleigh, 1995.

5. Hoffman, G.J.; Durnford, D.S. Drainage Design for Salinity Control. In *Agricultural Drainage*; Skaggs, R.W., van Schilfgaarde, J., Eds.; American Society of Agronomy: Madison WI, 1999; 579–614.

6. VanToai, T.T.; Bolles, C.S. Postanoxic Injury in Soybean (*Glycine max*) Seedlings. Plant Physiol. **1991**, *97*, 588–592.

7. Musgrave, M.E. Waterlogging Effect on Yield and Photosynthesis in Eight Winter Wheat Cultivars. Crop Sci. **1994**, *34*, 1314–1318.

8. Oosterhuis, D.M.; Scott, H.D.; Hampton, R.E.; Wullscheger, S.D. Physiological Response of Two Soybean (*Glycine max* (L) Merr) Cultivars to Short-Term Flooding. Environ. Exp. Bot. **1990**, *30*, 85–92.

9. VanToai, T.T.; Beuerline, J.E.; Schmitthenner, A.F.; St. Martin, S.K. Genetic Variability for Flooding Tolerance in Soybeans. Crop Sci. **1994**, *34*, 1112–1115.

10. Huck, M.G. Variation in Taproot Elongation Rate as Influenced by Composition of the Soil Air. Agron. J. **1970**, *62*, 815–818.

11. Crawford, R.M.M. Oxygen Availability as an Ecological Limit to Plant Distribution. Adv. Ecol. Res. **1992**, *23*, 93–185.

12. Kozlowski, T.T. *Flooding and Plant Growth*; Academic Press: New York, 1984.

13. Zhang, J.; VanToai, T.; Huynh, L.; Preiszner, J. Development of Flooding-Tolerant *Arabidosis thaliana* by Auto-regulated Cytokinin Production. Mol. Breed. **2000**, *6*, 135–144.

14. Trought, M.C.T.; Drew, M.C. Effects of Waterlogging on Young Wheat Plants (*Triticum aestivum* L.) and on Soil Solutes at Different Soil Temperatures. Plant Soil **1982**, *69*, 311–326.

15. Schmitthenner, A.F. Problems and Progress in Control of Phytophthora Root Rot of Soybean. Plant Dis. **1985**, *69*, 362–368.

16. Fausey, N.R.; VanToai, T.T.; McDonald, M.B., Jr. Responses of Ten Com Cultivars to Flooding. Trans. Am. Soc. Agric. Eng. **1985**, *28*, 1794–1797.

17. Hendry, G.A.F.; Broklebank, K.J. Iron-Induced Oxygen Radical Metabolism in Waterlogged Plants. New Phytol. **1985**, *101*, 199–206.

18. Thomson, C.J.; Atwell, B.J.; Greenway, H. Response of Wheat Seedlings to Low O_2 Concentration in Nutrient Solution: II. J. Exp. Bot. **1989**, *40*, 993–999.

19. Mass, E.V.; Grattan, S.R. Crop Yields as Affected by Salinity. In *Agricultural Drainage*; Skaggs, R.W., van Schilfgaarde, J., Eds.; American Society of Agronomy: Madison, WI, 1999; 55–108.

Drainage of Irrigated Land

James E. Ayars
United States Department of Agriculture (USDA), Parlier, California, U.S.A.

INTRODUCTION

Food production statistics indicate that approximately 30% of the world's food supply is produced by irrigated agriculture and that this percentage will increase in future. As such, irrigated agriculture has an important role to play in meeting the world's future food demand. Irrigated agriculture is practiced in humid areas to supplement rainfall, particularly during droughts, and in arid and semi-arid areas of the world as the sole water supply during crop production. Recent statistics compiled by the International Commission on Irrigation and Drainage (ICID) for 97 member countries of the Commission show there are approximately 271.1 Mha of irrigated land.[1]

However, there are no statistics to indicate what percentage of this irrigated area requires drainage. The most recent statistics on irrigation methods show that about 6% (15 Mha) of this irrigated area is irrigated by either sprinkler or micro-irrigation implying that the remainder of the area is irrigated with some other method, most probably using surface irrigation techniques. This is significant, since surface irrigation methods used on 94% of the world's irrigated area are generally considered to be less efficient than sprinkler or micro-irrigation. Areas where inefficient irrigation takes place are more likely to require artificial drainage to sustain crop production.

The classic example of the need for drainage is the decline of Mesopotamia in the area between the Tigris and Euphrates Rivers. This was a rich agricultural area that relied on irrigation to sustain itself. However, the area had no drainage other than the existing natural drainage capacity of the soil. As a result of poor irrigation practice, the water table rose and the soil gradually salinized resulting in poorer yields and ultimately no production and considerable desertification. Several things were attempted to stave off the inevitable but nothing was successful and villages were abandoned and agriculture ended because of the lack of drainage.[2]

Irrigation is the application of water to meet the crop water requirement. The systems used include sprinklers, micro-irrigation systems, and surface methods such as furrows, level basins, flood, and combinations of these. No irrigation system is 100% efficient in the application of water, so there are losses resulting from soil variations and man's inability to meet crop water requirements and maintain salt balance in the crop root zone. These losses have been termed deep percolation and have been defined as the water that moves past the root zone into the groundwater. The magnitude of the loss will be determined by the selected irrigation system, its design and management, and the soil and crop being irrigated. Irrigation efficiencies are in the range of 70%–85% for surface systems, 80%–90% for sprinkler systems, and in excess of 90% for micro-irrigation systems for reasonably well-managed systems. The consequence of poor efficiency is that more water has been applied to meet the crop water requirement than has been determined as being needed. This excess water then becomes deep percolation and has to be removed or the soil will become water logged and aeration will be a problem.

NEED FOR DRAINAGE

Soil drainage is needed to provide adequate aeration and salinity control for agricultural production in areas where crops are grown under conditions of natural precipitation or artificial irrigation.

Aeration

Growing crops need a well-aerated root zone to survive and meet yield potential. If the natural drainage capacity of the soil is inadequate to remove the excess water, then the soil will eventually become saturated, either from precipitation and/or irrigation, and artificial subsurface drainage will be required to provide a well-aerated soil. Soils that have low saturated hydraulic conductivities or impeding layers that are either compacted or contain soil with low hydraulic conductivity will have limited natural drainage capacity. Investigations that are needed as part of the design process for irrigation systems are generally required to determine the need for artificial drainage.

Salinity Control

Maintaining an aerated root zone is a problem that is common to both arid and semi-arid areas and to humid areas, while salinity control is generally a problem only in arid and semi-arid areas. Salinity is found in both the soil

Encyclopedia of Water Science
DOI: 10.1081/E-EWS 120010177

and irrigation water in arid and semi-arid areas and has to be controlled to prevent salination of the soil and the eventual loss of production. Salt accumulates in the soil as crops use pure water leaving behind salts that are in the water. Also, when crops use water from shallow groundwater, the salt is left behind in the crop root zone.

Another chapter discusses the use of drainage in the management of soil salinity. It is important to note that the design of both irrigation and drainage systems includes consideration of the leaching fraction for salinity control. The leaching fraction is a component of the deep percolation loss from irrigation inefficiency. A separate leaching fraction may or may not be required, depending on the quality of the irrigation water and the efficiency of the irrigation practices.

DRAINAGE SYSTEM DESIGN

Drainage systems can be characterized as either horizontal or vertical. The horizontal systems are made up of clay or concrete tile or plastic pipes that are installed parallel to the soil surface to collect water and let it flow by gravity to an outlet. A vertical system is a pumped well that is used for drainage. Vertical drainage is discussed in another section. In arid areas, deep open ditches are often used as drains to collect subsurface drainage water as well as surface water losses and then discharge this water to a surface water body. Economics is often the consideration involved in which method is selected as best to use for the conditions involved.

The design objective for a good drainage system is to remove water from the soil; that is to either lower the water table to specific depth in a given period of time or to prevent the water from rising in the soil above a specified depth. A well-designed horizontal drainage system results in a specification of the drain lateral size, depth, and spacing to provide adequate aeration for the crop and to control the salinity in the crop root zone. The two basic design methods that are currently applied in irrigated areas are labeled transient and steady state.

Transient Design

The transient method was developed by the U.S. Bureau of Reclamation[3] and accounts for the soil type, crop, and uses intermittent application of irrigation water and rainfall. The design is an iterative process, where a drain lateral, depth, and spacing are specified, then the deep percolation calculated from the irrigation and rainfall sequence is applied to the existing water table. Each application of water results in the water table rising closer to the soil surface. A drain out period following the application removes water from the soil and lowers the

water table. The yearly water table response is then calculated as a succession of drain out periods following the addition of the deep percolation. The drain spacing is adjusted for a given depth until the depth to the water table at the mid-point between the drains meets the design criteria specified to occur at the end of the irrigation season. For a crop rotation, the deep percolation used in the analysis is based on the crop with the largest water requirement and deep percolation losses.

Steady State

The steady state method has been adapted from procedures used in humid areas. The deep percolation losses are calculated based on the crop water requirement and rainfall and are distributed uniformly throughout the year with the lateral spacing being based on this average rate. The criteria are set to remove a specified volume of water and to lower the water table to a given depth in a specified number of days. This is called the drainage coefficient and is discussed in more detail in another article. A mid-point water table depth is specified in the design and assumed to remain relatively constant at this depth throughout the year. This is significantly different from the transient design where the depth to water table varies over a wide range during the year.

In recent years, environmental concerns over the disposal of drainage water from agricultural land have had impacts on the design criteria for drainage systems. The designs have changed to account for crop water use from shallow groundwater and to consider water quality.[4,5] The new design recommendations result in the installation of drain laterals at shallower depths than used in the past with either the transient or the steady state design criteria. The shallower placement of the drain lines allows the water table to become closer to the soil surface, makes the shallow groundwater available for plant use, and reduces the depth of the flow lines to the laterals. A reduction in the depth of the flow lines reduces the salt concentration of the drainage water in arid areas where there is increasing soil salinity with depth in the profile.

DRAINAGE SYSTEM MANAGEMENT

Active management of subsurface drainage systems is a relatively new concept for drains installed in irrigated agricultural areas. Managed drains are contrasted with free flowing drains. Controlled drainage has been used extensively in humid areas, but concerns over salinity management have limited the application of this concept in arid areas.

Free Flowing

In the past, the management of horizontal subsurface drains and open drains has assumed that the drains would be free flowing and that all water removed from the soil would be discharged and disposed of to either a stream or river or an evaporation basin. However, environmental concerns related to water quality issues have resulted in significant changes in the way drainage water is managed. In many areas, surface and subsurface drainage are not mixed and surface water that runs off the field is mixed back into the irrigation supply. Subsurface drainage water is either used as a supplemental source of irrigation water or discharged into an evaporation basin. Reuse of drainage water on progressively more salt tolerant crops or other vegetation is used to increase the salt concentration and reduce the drainage volume prior to discharging the drainage water into an evaporation basin for disposal.

Controlled

This is a relatively new option for managing drainage water in irrigated agriculture and is only suitable for application when the drain laterals are installed perpendicular to the grade of the soil surface. This permits adequate control of the groundwater depth over a significant portion of the field similar to the conditions found in humid areas. The depth to water table is controlled by installing a control structure at the outlet of the drainage system or strategically in the field. The height of the control structure can be varied to regulate the water table depth in the field. Adoption of controlled drainage will increase crop water uptake from shallow groundwater in cases where the crop salt tolerance and groundwater salinity are compatible. It will also alter flow patterns and reduce water discharge and salt load from the system.

WATER QUALITY IMPACTS

Irrigation and drainage in arid and semi-arid areas has a dual impact on water quality in surface water. The diversion of irrigation water from a stream or river reduces the total flow of that watercourse thus decreasing the dilution potential of the stream. When drainage water is returned to the stream it may have been degraded by salt, fertilizers, pesticides, herbicides, and other elements that are in solution. Fertilizers, particularly nitrate fertilizer, which is very soluble and mobile in water, and phosphorus, contribute to the growth of aquatic vegetation. Nitrate is a problem when considering drinking water standards. Depending on the parent material of the soil and the level of leaching trace elements such as selenium, boron, arsenic, and molybdenum are problems in addition to the sodium, calcium, bicarbonate, and sulfate routinely found in drainage water.

Because of the potential for transport of fertilizers, salts, trace elements, pesticides, and herbicides, it is important that the drainage system is designed and managed with the irrigation system to improve the total water management and reduce drainage flow. This is also a change in the way drainage design and management has been approached in irrigated agriculture.

REFERENCES

1. ICID. Database—Important Data of ICID Member Countries. URL: www.ICID.org/index_e.html 2002.
2. Gelbrud, D.E. Managing Salinity, Lessons from the Past. J. Soil Water Conserv. **1985**, *40*, 329–331.
3. U.S. Department of Interior, *Drainage Manual*; U.S. Department of Interior: Denver, CO, 1993.
4. Ayars, J.E.; McWhorter, D.B. *Incorporating Crop Water Use in Drainage Design in Arid Areas*; Proceedings, Specialty Conference, Development and Management Aspects of Irrigation and Drainage Systems, Irrigation and Drainage Division, ASCE, Keyes, C.G., Ward, T.J., Eds.; American Society of Civil Engineers: New York, NY, 1985; 380–389.
5. Ayars, J.E.; Hutmacher, R.B. Crop Coefficients for Irrigation Cotton in the Presence of Groundwater. Irrig. Sci. **1994**, *15* (1), 45–52.

Drainage, Land Shaping for

Rodney L. Huffman
North Carolina State University, Raleigh, North Carolina, U.S.A.

INTRODUCTION

The natural drainage network does not always provide adequate outlets for runoff. Land shaping can alter the surface configuration to permit surface water to flow easily by gravity to outlets, usually ditches or natural streams.[1] Sumps and pumping stations may be required to lift drainage water into a ditch or stream if the drained area is either lower than the outlet or so flat that the natural gradient is inadequate to achieve the required discharge.[2] Land shaping speeds removal of surface water, thereby improving access for field operations and promoting healthier crops and higher yields.

Poor surface drainage is common in landscapes such as glaciated areas, coastal plains, floodplains, deltas, and old lake beds. Problem areas are typically flat to gently rolling and may contain numerous small depressions. Slowly permeable soils and relatively large distances to discharge areas reduce internal drainage rates, exacerbating the problem of excess water. With clayey soils, surface drainage usually provides a better cost–benefit ratio than subsurface drainage[3] and may eliminate the need for subsurface drains in some cases.

Land shaping for drainage entails modification of the surface of the land to facilitate the flow of water. In some cases, only a small percentage of the land surface must be modified. In others, the entire land surface must be reshaped. Factors that must be considered include: existing topography, intended use of the land, characteristics of the soil profile, local climate, and the intended outlet for the drainage water.

METHODS

Any but the most minor land shaping operation may remove all of the topsoil from a cut (borrow) area. To maintain productivity, it may be necessary to remove and stockpile topsoil for redistribution over the project area after the primary shaping work is completed.

Grading and Smoothing

Land grading is the shaping of the land surface to predetermined grades. (*Land leveling* is a special case where the final grade is a level surface.) *Land smoothing* is the removal of irregular, uneven, broken, mounded, and jagged surfaces without the use of survey information.[4] Very shallow and/or small depressions may be filled by minor scraping and smoothing if there is no need to have specific final grades.

Entire fields or portions of fields can be graded to facilitate water movement. Utilization of any of the following practices must consider soil erodibility, slope steepness, slope length, adjacent land surfaces, outlet location and capacity, and volumes of earthwork required.

Uniform Slopes

A field can be graded to a planar surface with a uniform slope (Fig. 1). This may include major and minor slopes, i.e., along the crop rows and across the crop rows. The slope may be zero to permit uniform flooding, e.g., where rice (*Oryza sativa* L.) is grown. In other cases, a slope of about 0.1–0.5% is desirable.[5,6] Maximum recommended slopes depend on the soils, slope lengths, and location.

Nonuniform Slopes

Where planar surfaces are desired but uniform slopes would require excessive earthwork, nonuniform slopes are employed. Nonuniform slopes are composed of two or more piece-wise uniform sections (Fig. 2). The upslope sections are generally steeper than the downslope sections. Where long slopes would permit excessive soil loss, erosion control measures such as terraces should be considered.

Warped Surfaces

Warped surfaces are nonplanar surfaces with smoothly varying slopes (Fig. 3). They range from relatively simple to very complex. Warped surfaces may be designed where planar surfaces are unnecessary and would require excessive earthwork. Warped surfaces take advantage of the existing topography and tend to follow existing grades fairly closely to minimize cut and fill volumes.[7]

Encyclopedia of Water Science
DOI: 10.1081/E-EWS 120010228

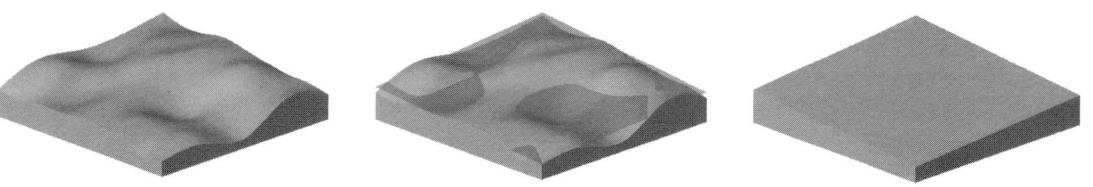

Fig. 1 Land grading to uniform slopes. Left: existing grade. Center: existing grade with final grade superimposed. Right: final grade.

Fig. 2 Land grading to nonuniform slopes. Left: existing grade. Center: existing grade with final grade superimposed. Right: final grade.

Fig. 3 Land grading to warped surfaces. Left: existing grade. Center: existing grade with final grade superimposed. Right: final grade.

Crowned Surfaces

Bedding and *crowning* (Fig. 4) are very similar in concept, differing mainly in scale and sophistication. Bedding is the practice of using the deadfurrows between lands (resulting from moldboard or turn plowing) as small field drains. With bedding, field operations are typically parallel to the deadfurrows, which must be oriented somewhat up-down slope to facilitate drainage. The only equipment needed for construction and maintenance is a plow. The elevation difference between the top of the bed and the bottom of the deadfurrow is typically 15 cm–45 cm. In the Corn Belt region of the United States, the width of beds ranges from 7 m for very slow internal drainage to 28 m for fair internal drainage.[8]

Crowning is the practice of grading land between parallel drains to an approximately parabolic convex shape. If the drains have side slopes of 8:1 or flatter, planting, cultivating, and harvesting operations may run perpendicular to the drains, which allows runoff to flow easily toward the drains between the rows. Plowing should be done parallel to the drains. The spacing between drains

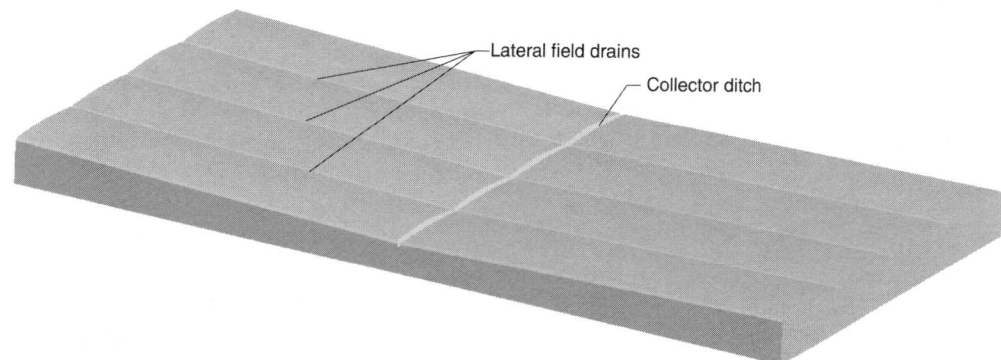

Fig. 4 Bedding and crowning. Land surfaces between parallel field drains or deadfurrows are sloped slightly toward the drains.

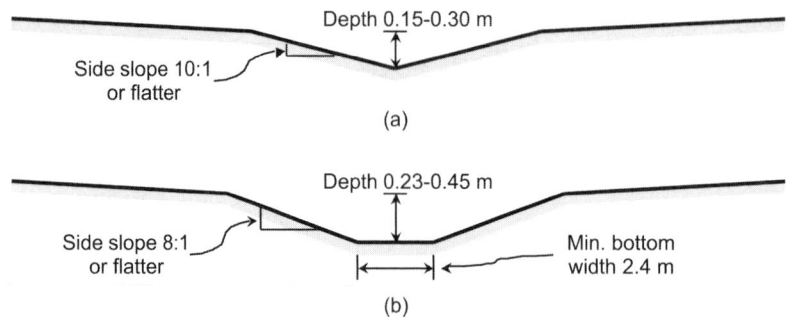

Fig. 5 Cross-sections of field drains: (a) triangular or "vee" cross-section; (b) trapezoidal cross-section.

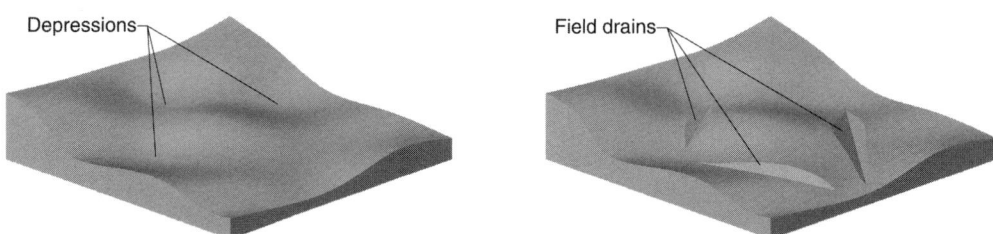

Fig. 6 Random field drains. Left: existing land with undrained depressions. Right: final configuration with field drains providing outlets for occasional depressions.

may be as much as 360 m if rows drain in both directions and the soils are erosion resistant. On highly erodible soils, it is recommended that slopes not exceed 90 m in length.[8] The slope of the land should generally not exceed 0.3% but may be up to 0.5% on erosion-resistant soils.[7]

Surface Drains

Surface drainage can be improved by creating shallow channels to collect and/or convey water across the surface toward an outlet. Field drains are typically constructed with either triangular (vee) or trapezoidal cross-sections (Fig. 5). The side slopes of the drains should be 10:1 or flatter for triangular and 8:1 or flatter for trapezoidal cross-sections. The minimum recommended bottom width for trapezoidal field drains is 2.4 m. Typical depths are 0.15 m–0.3 m for triangular and 0.23 m–0.45 m for trapezoidal drains.[6]

Field drains may be constructed individually as needed to drain occasional depressions or in parallel systems to drain entire fields.

Random Field Drains

Where depressional areas are too large or deep to simply fill, random field drains can be installed to provide outlets as needed (Fig. 6). Side slopes of 10:1 or flatter are recommended to permit normal field traffic. Discharge

capacity is not considered in design of random field drains unless the drained area exceeds 2 ha. Grades should not be less than 0.05%[6] and should not exceed 0.2% for sandy soils or 0.5% for clay soils.[8]

Random field drains are often constructed such that a depression drains through one or more other depressions along the way toward an outlet.

Parallel Field Drains

On flat to very gently sloping terrains, a system of parallel field drains may be installed. The land between drains is often crowned (see "Crowned Surfaces" above) to aid water movement toward the drains (Fig. 4).

CONCLUSION

Land shaping is a cost-effective way to improve surface drainage. Properly designed grading, smoothing, and/or field drains can enhance productivity while requiring minimal maintenance.

REFERENCES

1. American Society of Agricultural Engineers (ASAE), *Agricultural Drainage Outlets—Open Channels*, ASAE Engr. Practice 407.1 DEC98; ASAE: St. Joseph, Michigan, 1998.

2. American Society of Agricultural Engineers (ASAE), *Design of Agricultural Drainage Pumping Plants*, ASAE Engr. Practice 369.1 DEC 94; ASAE: St. Joseph, Michigan, 1994.

3. Schwab, G.O.; Fausey, N.R.; Desmond, E.D.; Holman, J.R. *Tile and Surface Drainage of Clay Soils*, Research Bulletin 1166; Ohio Agricultural Research and Development Center: Wooster, Ohio, 1985.

4. American Society of Agricultural Engineers (ASAE), *Soil and Water Terminology*, ASAE Standard 526.2 JAN 01; St. Joseph, Michigan: 2001, 2001.

5. U.S. Dept. of Agriculture, Soil Conservation Service. *National Engineering Handbook*; USDA-SCS, 1971; Sect. 16, Chap. 3, p. 15.

6. American Society of Agricultural Engineers (ASAE), *Design and Construction of Surface Drainage Systems on Agricultural Lands in Humid Areas*, ANSI/ASAE Engr. Practice 302.4 JUN 00; ASAE: St. Joseph, Michigan, 2000.

7. Carter, C.E. Surface Drainage. In *Agricultural Drainage*, Agronomy Monograph No. 38; Skaggs, R.W., van Schilfgaarde, J., Eds.; American Society of Agronomy, Crop Science Society of America, Soil Science Society of America: Madison, Wisconsin, 1999; 1023–1048.

8. Schwab, G.O.; Fangmeier, D.D.; Elliot, W.J.; Frevert, R.K. *Soil and Water Conservation Engineering*, 4th Ed.; John Wiley & Sons, Inc.: New York, 1993.

Drainage Materials

James L. Fouss
United States Department of Agriculture (USDA), Baton Rouge, Louisiana, U.S.A.

INTRODUCTION

Subsurface drainage technology changed and modernized more during the 1965–1980 period than in the previous 100 yr. The inefficient and slow installation of heavy rigid drainage conduit materials (clay and concrete draintile) gave way by the early 1970s to light-weight flexible corrugated plastic drain tubing installed with laser-beam-controlled high-speed trenchers and plow-type equipment. In fact, the developments of the modern drainage plow equipment and the laser-beam automatic grade control were a direct result of the technological developments for corrugated plastic drainage tubing, and the need for a rapid and accurate method to install the new drainage material.[a]

BACKGROUND

The development of a rapid and low-cost technique for subsurface drainage had challenged engineers and inventors for centuries. Many ideas emerged over time, but very few found widespread use or application. With the development of the power trenching machine in 1875, the goal of mechanized drain installation seemed to have been reached—and it lasted around 100 yr. However, the extraordinarily large amount of drainage work that was needed around the world required even less labor, more speed, and lower costs. Efforts to modify the mole drainage concept and installation methods were particularly important. The goal was to use the inherent high speed of installation of mole drainage and its elimination of relatively slow ditching and backfilling operations associated with conventional drainage methods. Because the mole drain collapsed after a short time in many soils, most of the research focused on stabilizing the mole channel with structural support, using a tube or mole-liner; this approach, although showing some promise, was not satisfactory for adoption or use.[1] This investigative work

with the mole plow did lead, however, to the eventual development of the drain-tube plow equipment for installing subsurface plastic drains that is in common use today throughout the world.

Corrugated-wall polyethylene plastic tubing, originally developed and used in the United States in the mid-1960s for underground electrical and telephone line conduit applications, was modified and perforated to serve as a subsurface drainage tube in early experiments.[1] By the latter half of the 1960s almost all the research and development on drainage materials and methods of materials handling and installation for agricultural drainage applications had begun to focus on corrugated-wall plastic tubing, primarily because of the advantages of low material requirement vs. high-strength ratio and flexibility for ease of coiling and handling. Continuous extrusion and molding machinery for manufacturing the new plastic tubing, with primarily polyethylene and polyvinyl chloride resins, had been perfected earlier in Germany for small diameter drain tubing. Underground drainage with the new conduit (about 50 mm in diameter) caught on rapidly in Germany and soon spread to other regions of Europe.

Research in the United States on developing polyethylene corrugated-wall plastic tubing, of 100 mm diameter, for agricultural subsurface drainage began in 1965.[1,2] The corrugated-wall tube structure developed for polyethylene plastic (Fig. 1) provided high strength to resist deflection by radial type loads from over-burden soil, but with a considerably reduced requirement for wall thickness as compared with smooth-wall tubing. Both tubing unit weight and unit cost are reduced significantly by pipe-wall corrugations.

By 1967, corrugated plastic drainage tubing was being manufactured commercially in the United States for the agricultural market, and the new industry grew rapidly.[3,4] By the mid-1970s, corrugated plastic drainage tubing had wide acceptance for agricultural drainage, highway berm drainage, septic tank leach field, and construction site applications. By 1983, 95% of all agricultural subsurface drains installed annually in the United States, and more than 80% of Canada, were corrugated plastic tubing.[5,6]

---

[a]Detailed reports on the innovations in drainage technology are given by: Fouss,[3] Fouss and Reeve,[10] and Schwab and Fouss.[6]

Encyclopedia of Water Science
DOI: 10.1081/E-EWS 120010052

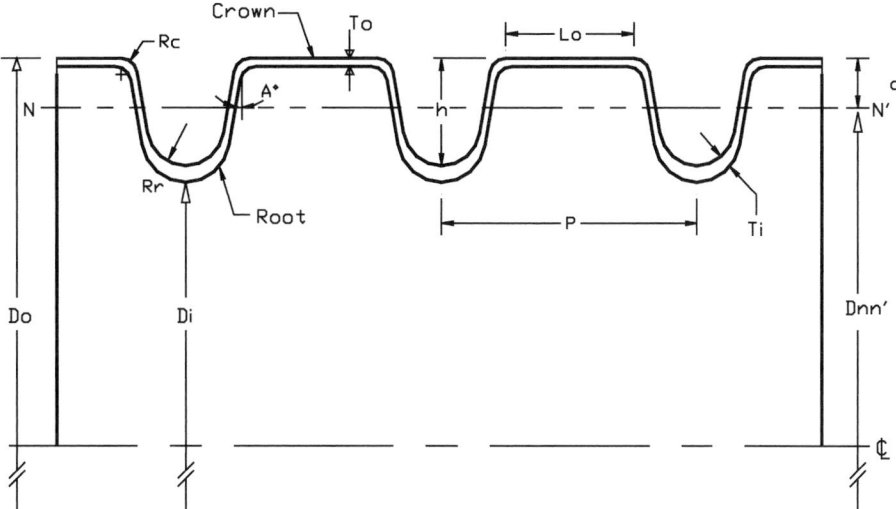

Fig. 1 Cross-section of corrugated-wall polyethylene plastic drain tubing.

CORRUGATED PLASTIC TUBING

The structural strength of a plastic pipe can be expressed as a function of its deflection resistance when loaded between parallel plates (see Fig. 2). The parallel-plate test method is required in ASTM Standard Specifications F-405 and F-667 for corrugated-polyethylene tubing. The strength-deflection characteristic determined for a conduit tested by this method and defined as the "pipe stiffness," is expressed in units of applied load per unit length of pipe sample per unit of vertical deflection (flattening) of the pipe (i.e., F/L/L or F/L^2). The parallel-plate pipe stiffness is expressed mathematically in terms of the geometrical, physical, and pipe-wall material properties of the conduit structure, given as:

$$Pipe\ Stiffness = (W/\Delta Y) = 53.6EI/(D_{NN})^3 \qquad (1)$$

where, W = parallel-plate load on a sample length of pipe (F/L); Y = vertical pipe deflection under parallel-plate

load (L); E = modulus of elasticity for pipe-wall material (F/L^2); I = moment of inertia of pipe-wall cross-section (L^4/L; i.e., per unit of pipe length); D_{NN} = diameter of pipe to the neutral axis (NN) of pipe-wall cross-section (L); and 53.6 = dimensionless constant related to angular position of parallel-plate loads on pipe circumference and to convert from pipe radius to pipe diameter.

Eq. 1 applies to the linear range of deflection between parallel-plates for plastic corrugated-wall pipe, which typically occurs from 0 to between 5% and 10% deflection of the inside pipe diameter.[7] At a specified pipe stiffness (W/Y) for a regular corrugated-wall or smooth-core corrugated plastic pipe of given inside diameter (D_i) and assumed neutral axis diameter (D_{NN}), and for a given plastic resin material of known modulus of elasticity (E), the only term unknown in Eq. 1 is I, which represents the moment of inertia of the pipe-wall cross-section. Corrugation shape and smooth interior wall features govern the magnitude of I, the major structural parameter of the plastic pipe determined or controlled through product design and fabrication.

The cost of corrugated tubing is almost directly proportional to tubing weight. The longitudinal flexibility of the corrugated-wall tubing makes it coilable for ease of handling, but the coilability characteristic also makes it stretchable. Thus, a compromise in design of corrugation shape has been necessary, and special materials handling procedures and equipment have been developed to prevent stretch during handling and installation.[b]

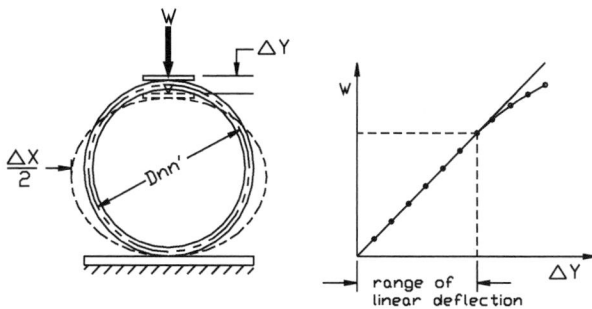

Fig. 2 Parallel-plate load/deflection method of measuring drainpipe stiffness.

---

[b]The reader is referred to Fouss,[3,7] and Schwab and Fouss,[6] for detailed discussions on optimal design procedures for corrugated-wall plastic drainage tubing for resistance to both deflection and stretch.

TUBING STANDARDS

Specifications and performance standards were developed during the early 1970s for these new corrugated plastic drainage products under the auspices of ASTM, which involves voluntary and cooperative efforts among industry, government, and public groups. This resulted in an ASTM Standard Designation F405 entitled "Standardization Specification for Corrugated Polyethylene Tubing."[8] A major step in the development of this standard was the recognition by the cooperating groups that corrugated plastic tubing is a flexible-type conduit with properties substantially different from the classical rigid draintile such as clay, shale, or concrete.

Under field conditions, a flexible conduit gains most of its vertical soil load-carrying capacity from the support provided by the soil compressed at the sides of the conduit. The density of this sidefill material is the key element in load-carrying capability of the pipe–soil composite structure. The sidefill material provides lateral support to the conduit to give it more rigidity and acts in combination with the conduit to form a vertical load-carrying arch (Ref. 6).

The parallel-plate method for measuring the deflection resistance of the corrugated plastic tube was adopted as an integral part of the ASTM F405 Standard Specification. This standard was developed to provide minimum values for physical and chemical properties as related to product performance, including handling and installation. Minimum deflection resistance is specified for 5%–10% deflection of tubing diameter. The standard also included a requirement on elongation (stretch) resistance, which limited elongation to 5% when a specified tensile load is applied. Table 1 gives the ASTM recommended minimum values for pipe stiffness and elongation for various drain diameters. In 1978, a revision of the standard specified a falling "Tup" impact test, conducted at a cold temperature to detect brittle or poor-quality plastic resin.

FABRICATION AND MARKETING

Water entry openings are made in the corrugated-wall drain tube wall during the manufacturing operation by punching or drilling holes, sawing short narrow slots, or other means of perforation. Typically, the openings are formed in the corrugation roots (valleys) rather than on the crowns (outside diameter), and are positioned in three or more rows along the length of the tubing. The cross-sectional area of openings for water entry to the drain varies among manufacturers, but ranges from 21 to more than 148 square centimeters per linear meter of drain. ASTM Standard F405 requires a minimum of 21 square centimeters per linear meter of pipe. Because the drainwall openings are controlled in the manufacturing operation, the quality of installation improved significantly with corrugated tubing compared with ceramic tile. The crack spacing between ceramic draintile sections had to be controlled during installation, thus giving rise to great variability in drain quality among contractors.

Most of the early corrugated plastic drainage tubing was black, but by the mid-1970s, tubing was produced in lighter colors such as white, yellow, gray, and red. Ultraviolet stabilizers and antioxidants were incorporated in the plastic resin to increase its resistance to weathering when tubing was stored outside and exposed to sunlight. The lighter color tubing was developed partially for marketing purposes, but improved performance during handling and installation was also realized because strength and stretch resistance were maintained, even when exposed to the hot sun. The darker tubing was more prone to absorbing the sunlight, which elevated the tube-wall temperature, thus reducing the tubing's stretch resistance during handling and installation.

Corrugated plastic tubing larger than 300 mm in diameter is generally more expensive than the same-size clay or concrete tile, but the market demand and use for the lighter and easier to handle corrugated plastic is increasing significantly. These large-size corrugated conduits (300 mm–600 mm) are also used extensively for culvert

Table 1 Physical test requirements for corrugated plastic tubing

Physical requirement	Standard quality[a] MPa (psi)	Heavy duty[b] MPa (psi)
Pipe stiffness at 5% deflection, minimum	0.17 (24)	0.21 (30)
Pipe stiffness at 10% deflection, minimum	0.13 (19)	0.175 (25)
Elongation, maximum %	10	5

[a] ASTM F405-97 (75 mm–200 mm diameter). Ref. [8].
[b] ASTM F667-97 (250 mm, 300 mm, and 380 mm diameters). Ref. [12].

applications (Watkins and Colleagues[13]) which was an area formerly thought to be reserved for concrete and steel pipe. The noncorrosive nature of the product and the advances in the structural performance of plastics for this use are milestones in the drainage industry.

MATERIALS HANDLING

The use of corrugated tubing greatly reduced labor and energy requirements in drainage materials handling. Initially, the typical 100-mm diameter tubing used for laterals was supplied in 76-m coiled lengths and weighed about 36 kg. This compared with a weight of about 900 kg for clay or concrete tile of the same diameter and total length.

As the demand of, and use for, corrugated plastic drainage tubing grew in the United States, contractors desired larger and larger coils to make the materials handling operation even more efficient. In 1984, typical coil sizes available for 100-mm diameter tubing were 915-m "maxi-coils" and 1525-m "jumbo coils." The 76-m coil is still commonly used for many small agricultural jobs, for industrial installations, and around housing projects. Several types of self-loading trailers and wagons became available to string tubing in the field. Special reels were developed for mounting directly onboard the drainage equipment to uncoil the tubing as it was installed.[c]

Use of the maxi-coils and special reels for stringing tubing reduced tubing stretch problems during installation, even for black tubing on hot, sunny days. The 2% carbon-black used as the ultraviolet light inhibitor in black tubing is superior in performance and lower in cost than the light pigments, permitting outdoor storage of the product. The power tubing feeder designed to eliminate the natural stretch-producing drag at the top of the tubing chute was one of the most significant developments in minimizing the adverse effects of stretch.[9]

Diameters of corrugated plastic pipe increased from the original 100 mm in the mid-1960s to 600 mm by 1982. Sizes up through 254 mm are commonly coiled for shipment and handling. Drain sizes larger than 300 mm are typically manufactured and shipped in 20-ft lengths. There is a noteworthy market for 76-mm corrugated tubing, which is typically shipped in 105-m standard coils, or 1525-m coils. The 100-mm tubing is considered the minimum tube size for lateral drains in most areas of the United States and Canada. A 127-mm diameter is specified as the minimum-size lateral drain in Iowa, and in Minnesota, a 150-mm drain is the preferred minimum diameter.

SYNTHETIC DRAIN ENVELOPE MATERIALS

Although graded sand and gravel envelopes have distinct performance advantages, the cost is generally prohibitive in areas where natural sands and gravels are not readily available. For this reason and because thin-membrane fabrics are easily handled and installed, especially in conjunction with corrugated plastic pipe, synthetic envelopes have become widely used throughout the major drainage areas of the United States and Canada. With the rapid adoption and widespread use of corrugated plastic drainage tubing, the development of synthetic fabrics as envelopes to protect these drains against sedimentation advanced rapidly. Because subsurface drain envelopes are used primarily to protect the drain from the inflow of sediment and still maintain free open flow of gravity water from the soil profile into the drain, the development of envelopes has been mostly centered around the performance of thin membranes with fine sand and coarse silt-size particles (0.005 mm–0.125 mm). Understanding of the basics and development of improved practices in the use of drain-synthetic envelopes have both advanced significantly in the past two decades.

Fabrics that were developed by major chemical and oil companies for other engineering applications were readily available from the 1960s to the 1980s and thus were quickly adopted for use as materials for subsurface drain envelopes. Many of these materials have been tested for use as drain envelopes, including polyester, nylon, and polypropylene, which were commercially available in North America. While woven, knitted, and spun-bonded productions of the above materials have been used, the most commonly used products from among these are knitted polyester (sock), spin-bonded nylon (Cerex™, Drainguard™), and spun-bonded polypropylene (Typar™, Remay™).[d]

By the early 1980s, as much as 8% of the corrugated plastic drainage tubing installed had a synthetic fabric envelope. These synthetic envelopes are light in weight and compact for ease of handling during transportation and installation. They are also relatively low cost compared with sand or gravel envelopes. The synthetic fabrics may be placed directly onto the tubing during manufacturing, or the envelope is placed on the tubing during installation.

Standards and specifications for synthetic fabric envelopes or drainpipe filter materials were still not developed by the early 2000s, even though various commercial products had been available and in use for nearly 30 yr. Developing performance standards for these products was complicated by the many variables involved

---

[c]The reader is referred to Broughton and Fouss[9] for detailed discussions on modernized materials handling and installation equipment.

[d]Trade and company names are included in this article for the benefit of the reader and do not imply endorsement or preferential treatment of the product listed by USDA.

in installation and hydraulic variables encountered in the field. Research had been conducted to determine why fabric materials plug up in some soil types, particularly in clays and silty clay loams, but results were not definitive. In other cases where the fabric mesh size was too large and the sediments were extremely fine, such as in very fine sand and/or silt loams, envelopes failed by allowing excess sediment to pass through the fabric and into the drain tubing. Fortunately, technical information about the past research efforts and their applications are available in the literature in a suitable form to permit the proper selection or design of envelopes for subsurface drains installed in various type of soils.[e]

The performance of a thin-membrane envelope depends primarily on the conditions of the soil at the time of installation, the imposed hydraulics on the system, and the method of installation. Failures are more common when the drain is installed under extremely wet conditions, where the soils are unstable and subject to "quick" conditions, and where the initial hydraulic head imposed on the drain during water table draw down is much higher at or soon after installation than that likely to occur once the soil surrounding the drainpipe has settled and stabilized. Drains installed with envelopes, even in very fine sandy or silty soils, have performed satisfactorily when installed where the water table had been low, the surface soil had been dry for better machine operation, and excessive hydraulic heads were not imposed on the system during installation. After the drain is installed and functioning, the soil near the drain stabilizes and the hydraulic head at the drain then becomes a function of head conditions as modified by the head loss of resistance to flow in the soil.

Experience and research have shown that favorable installation conditions and extreme care on the part of the contractor are both very important to obtaining trouble-free performance of subsurface drains with thin-membrane envelopes.

REFERENCES

1. Fouss, J.L. Plastic Drains and Their Installation. Proceedings of ASAE Conference on Drainage for Efficient Crop Production, Chicago, December, 1965; 55–57.

2. Fouss, J.L. Corrugated Plastic Drains Plowed-In Automatically. Trans. ASAE **1968**, *11*, 804–808.

3. Fouss, J.L. Drain Tube Materials and Installation. In *Drainage for Agriculture*; Monograph No. 17, van Schilfgaarde, J., Ed.; 197–200 AM. Soc. Agronomy: Madison, WI, 1974; 147–177.

4. Reeve, R.C.; Slicker, R.E.; Land, T.J. Corrugated Plastic Tubing. Proceedings, Am. Soc. Civil Engineering International Conference on Underground Plastic Pipe, 1981; 277–242.

5. Schwab, G.O.; Fouss, J.L. Plastic Drainage Tubing: Successor to Shale Tile. Agric. Eng. **1985**, *66* (7), 23–26.

6. Schwab, G.O.; Fouss, J.L. Drainage Material. In *Agricultural Drainage*; Agronomy Monograph 38, Skaggs, R.W., van Schilfgaarde, J., Eds.; ASA, CSSA, and SSSA: Madison, WI, 1999; 911–926.

7. Fouss, J.L. *Structural Design Procedure for Corrugated Plastic Drainage Tubing*, Technical Bulletin No. 1466; U.S. Dept. of Agriculture, Agricultural Research Service, 1973.

8. American Society for Testing and Materials (ASTM), *Standard Specification for Corrugated Polyethylene (PE) Tubing and Fittings (F405-97). ASTM Subcommittee F-17.65 on Land Drainage Systems*; ASTM: West Conshohocken, PA, 2001.

9. Broughton, R.S.; Fouss, J.L. Subsurface Drainage Installation Machinery and Methods. In *Agricultural Drainage*; Agronomy Monograph 38, Skaggs, R.W., van Schilfgaarde, J., Eds.; ASA, CSSA, and SSSA: Madison, WI, 1999; 967–1004.

10. Fouss, J.L.; Reeve, R.C. Advances in Drainage Technology: 1955–1985, USDA Misc. Publ. 1455; Farm drainage in the United States: History, Status, and Prospects, 1987; Chap. 3, 30–47.

11. Vlotman, W.F.; Willardson, L.S.; Dierickx, W. *Envelope Design for Subsurface Drains*, International Institute for Land Reclamation and Improvement (ILRI) Publication 56; Wageningen, The Netherlands, 2000; ISBN 90-70754-53-3, pp. 358.

12. American Society for Testing and Materials (ASTM), *Standard Specification for Large Diameter Corrugated Polyethylene Pipe and Fittings (F667-97). ASTM Subcommittee F-17.65 on Land Drainage Systems*; ASTM: West Conshohocken, PA, 2001.

13. Watkins, R.K.; Reeve, R.C.; Goddard, J.B. *Effect of Heavy Loads on Buried Corrugated Polyethylene Pipe*; Transportation Research Board, National Academy of Sciences: Washington, DC, 1983; 99–108.

---

[e]The reader is referred to Vlotman et al.[11] for detailed discussions on past research and the accepted procedures and methods for selection or design of subsurface drain envelopes.

Drainage Modeling

George M. Chescheir
North Carolina State University, Raleigh, North Carolina, U.S.A.

D

INTRODUCTION

Methods have been used over the centuries to guide the design and installation of drainage systems. These methods range from simple guidelines that relate drainage requirements to soil types, to computer programs that simulate long-term day-to-day performance of drainage systems in response to weather conditions and management practices. Our discussion here will be limited to groundwater equations that are applied to drainage and to the computer models that use these equations to simulate drainage systems. More detailed reviews of drainage models and their applications are available to the reader.[1–3]

The primary objective of drainage is to provide a favorable environment for crop production; therefore, development of drainage models has depended on formulations of equations to describe movement of shallow groundwater. Models have progressed with advances in math, science, and computer technology toward more accurate solutions of increasingly complex equations and boundary conditions. Technological advances have also made possible more rigorous treatment of other important processes such as crop growth, evapotranspiration, and rainfall that vary with time. The resulting models can simulate the performance of various drainage system designs over long periods of time and evaluate system performance in terms of specified objective functions, such as crop yield and profit.

Drainage models have also evolved in response to changing needs and concerns of the communities affected by drainage. The primary objective of drainage during early model development was to enable land development and increase crop production. Since the 1970s, communities in the United States and Europe have been concerned about the impact of agriculture and drainage on the quality of water draining to sensitive environments. Recent developments in drainage models have therefore focused on the fate and transport of nutrients and pesticides in drainage systems. The resulting computer programs integrate routines for groundwater flow, solute transport, crop response, and climatological processes into comprehensive simulation models.

EQUATIONS FOR DRAINAGE MODELING

Simple Analytic Equations

The simplest models are the analytic equations that relate steady state flow to drain depth, hydraulic conductivity, and drain spacing. The ellipse equation is commonly used for the case of parallel drainage ditches (Fig. 1).

$$R = \frac{4K}{L^2}(b^2 - D^2)$$

where R is the steady recharge rate (often defined as the drainage coefficient), K, the effective lateral hydraulic conductivity, b, the water table height above the impermeable layer, D, the water level in the ditches above the impermeable layer, and L, the spacing between the ditches.

The ellipse equation was derived assuming that all flow lines are horizontal (Dupuit–Forschheimer assumptions), which is reasonable for most cases of flow to ditches. These assumptions, however, do not apply for the flow lines as they converge to a drain tile. Methods have been developed that account for the convergence of flow near the drain by calculating an effective depth (d_e) from the drain tile to the impermeable layer and replacing D with d_e in the ellipse equation. Discussion of steady state drainage equations and their derivations can be found in Ritzema,[4] and van der Ploeg Harton and Kirkham.[5]

Nonsteady drainage equations have been developed to determine the time required for the water table drawdown from an initial elevation to a lower elevation. Development and applications of drawdown equations are discussed by Ritzema,[4] and Youngs.[6] These equations, however, have not been as widely used as the steady state equations.

Boussinesq Equation

The simple analytic equations are limited to specific cases for parallel drains at normal spacings and where site and boundary conditions are relatively uniform. There are, however, many situations in which soils, crops, and topography vary in the horizontal direction or in which quantifying the horizontal variation of the water table is important. An example of this type of situation is shown in

Encyclopedia of Water Science
DOI: 10.1081/E-EWS 120010113

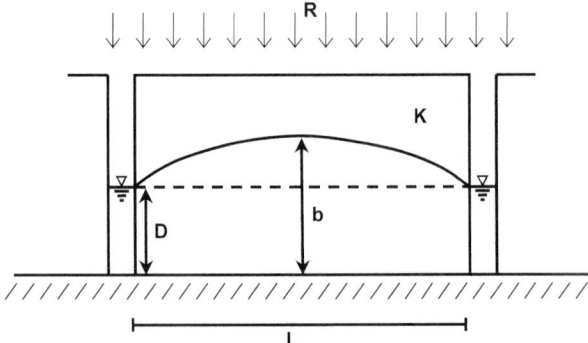

Fig. 1 Schematic of steady state drainage to parallel ditches. The ellipse equation would be used to describe saturated flow in these conditions.

Fig. 2. For this example, one would need to determine the depth to the water table at any horizontal point along the soil profile, which varies in thickness and hydraulic conductivity in addition to having nonuniform boundary conditions.

The simplest and most common approach is the use of the Boussinesq equation to characterize flow in the saturated zone only (see Ref. 6). The Boussinesq equation is based on the DF assumptions and the principle of continuity. Referring to Fig. 2, the Boussinesq equation may be written as,

$$f(h)\frac{\partial h}{\partial t} = \frac{\partial}{\partial x}\left[K(h)\frac{\partial h}{\partial x}\right] + R(x,t)$$

where h is the water table height above the impermeable layer, $f(h)$ is the drainable porosity and $K(h)$ is the effective lateral hydraulic conductivity, both written as a function of h, $R(x,t)$ is the vertical recharge rate, x is the horizontal position, and t is the time.

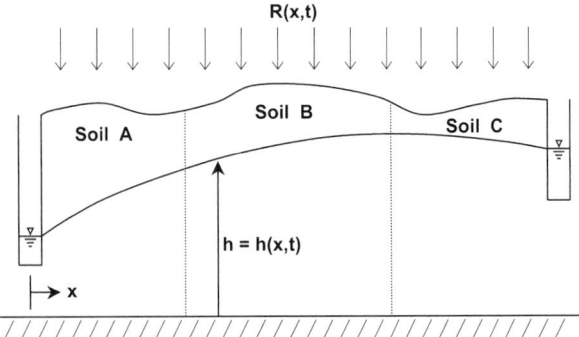

Fig. 2 Schematic of a drainage case where there is nonuniform variations in the horizontal direction. The Boussinesq equation would be used to describe saturated flow in these conditions.

Many of the simple analytical equations were derived from the Boussinesq equation for uniform boundary conditions. For nonuniform conditions, the Boussinesq equation is solved numerically using finite element or finite difference methods.

Richards Equation

The Boussinesq equation describes groundwater flow only in horizontal directions and only in the saturated zone, with the vertical recharge rate term, $R(x,t)$, being a lumped term representing the net effect of vertical movement in the unsaturated zone. The Richards' equation is a more exact description of water movement and storage in the unsaturated zone. This adds levels of complexity since both hydraulic conductivity and soil water content are related to soil water pressure head. The Richards equation is presented in a separate entry in this Encyclopedia (see *Richards' Equation*). Most drainage conditions of interest can be described by solving the Richards' equation subject to appropriate boundary and initial conditions. Solutions provide soil water contents and pressure heads as functions of time and space as well as the position of the water table and flux rates due to drainage, subirrigation, infiltration, and evapotranspiration.

SIMULATION MODELS

Solutions to the equations discussed thus far assume some idealized conditions driving the system such as steady rainfall, evapotranspiration, or an initial condition for transient drainage. In reality, drainage systems are subject to many perturbations that occur randomly though time. The most notable random perturbation is precipitation. Although processes involving plant root growth, evaporation, and transpiration also have very important random impacts on drainage systems. With increased speed and power of computers, the performance of drainage system designs for longer and more representative conditions can be simulated. The resulting simulation models are therefore multiple solutions to groundwater equations in response to variable temporal and boundary conditions. Simulation models can be combined with methods to predict crop yield and solute transport to evaluate system designs in terms of multiple objective functions such as crop yield, profit, or drainage water quality.

Two-Dimensional Richards Equation

General simulation models for solving the 2-D Richards equation subject to changing boundary conditions are available. The integrated program, HYDRUS-2D[7,8] was developed at the U.S. Salinity Laboratory to simulate water flow, heat transfer, and solute movement in variably saturated soils. HYDRUS-2D is a combination of the SWMS_2D and CHAIN_2D models that use finite element methods to solve the Richards' equation and the convective–dispersion equations. The flow equation incorporates a sink value to account for water uptake by plant roots. The models can handle a wide range of boundary conditions, including ditches and drain tubes, as well as boundaries controlled by atmospheric conditions. HYDRUS-2D also includes programs for generating finite element grids, for organizing input, and displaying output. The U.S. Geological Survey has also developed an integrated program, VS2DI[9,10] to simulate water flow, heat transfer, and solute movement in variably saturated soils. VS2DI is an integration of the VS2DT and VS2DH models based on finite difference solutions to the Richards equation with pre and post processing programs.

One-Dimensional Richards Equation

Drainage simulation models have been developed that use the 1-D Richards' equation to describe vertical water movement in a soil column subject to variable surface boundary conditions such as atmospheric and crop uptake conditions. Lateral flow to the drains is usually calculated with simple analytic equations for saturated flow or with tabular flux–groundwater relationships. The widely used model, SWATRE,[11] has been combined with other models and routines for describing plant growth, solute transport, and soil heat flux to create the comprehensive model, SWAP.[12,13] Vertical water flow calculations can consider the effect of hysteresis and preferential flow due to soil cracking or water repellent soil. Solute transport calculations consider convection, diffusion and dispersion, nonlinear adsorption, first-order decomposition, and root uptake. A soil heat flow equation is solved analytically assuming uniform thermal conductivity and soil heat capacity, or solved numerically from soil composition and moisture content. Plant growth is simulated based on the calculated radiation energy absorbed by the plant canopy.

Several other drainage simulation models (see Ref. 1) have been developed based on the 1-D Richards equation. The Root Zone Water Quality Model (RZWQM) uses a mass-conservation technique[14] to solve the Richards equation. Like the SWAP model, the RZWQM[15] has become an integrated model that simulates major physical, chemical, and biological processes in an agricultural crop production system. The Root Zone Water Quality Model considers water and solute movement through the soil profile including macropores, soil heat flux, crop growth, nutrient and pesticide transformations, and agricultural management practices.

Water Balance Models

The water balance models discussed in this section perform water balances at one or two points in the soil profile using analytically or numerically calculated values for saturated flow, ET, infiltration, seepage, and other inflows or outflows. The widely used water balance model, DRAINMOD[16,17] was developed for the design and evaluation of multicomponent drainage and related water management systems. The model conducts a water balance on an hour-by-hour, day-by-day basis and calculates infiltration, ET, drainage, surface runoff, subirrigation, deep seepage, water table depth, and soil water status at each time step. Lateral saturated flow to and from the drains is calculated by simple analytic equations. Soil water is distributed vertically assuming a drained-to-equilibrium profile above the water table. Water content can be as low as the wilting point in a separate dry zone that can form in the crop root zone. As with other currently available drainage models, DRAINMOD is now an integrated model that considers the major processes occurring in a drained crop production system. Routines have been added to the model to predict crop yield and to calculate heat flux. Additional routines have been added to consider the effects of drainage and water management on losses of nitrogen and on soil salinity.

Several models that were originally developed to predict losses of sediment, nutrients, and pesticides from sloping upland soils have been modified for use on more poorly drained flatland soils. These models include EPIC-WT,[18] WEPP,[19] ADAPT,[20] and GLEAMS-WT.[21] The modifications to these models usually involved addition of algorithms similar to those used in DRAINMOD to predict drainage rates, infiltration, and water table response. In other cases, the output calculated by DRAINMOD were used as input to other models such as CREAMS.[22]

Boussinesq Equation

Drainage simulation models based on the 1-D Richards equation and water balance methods examine the soil column at the midpoint between parallel drains or ditches. This is due to the use of analytic equations for calculating saturated flow to the drains. For evaluations of most drainage designs, these models are very practical; however, there are situations where boundary conditions or flow domains are complex and models based on parallel drainage equations will not suffice. Possible scenarios may be similar to the case shown in Fig. 2, or may be best represented in 3-D such as when ditches are perpendicular or serpentine.

Parsons, Skaggs, and Doty.[23] developed the simulation model, WATRCOM, using finite element solutions to the Boussinesq equation. Solutions to the 1-D form facilitated a quasi 2-D model while solutions to the 2-D form facilitated a quasi 3-D model. A water balance similar to the one in DRAINMOD was conducted at each node and coupled to the finite element solutions. Other routines were added to route surface water, to determine ditch water levels for controlled drainage situations and to calculate crop yield. A similar approach was used by De Laat et al.[24] to develop GELGAM, which was used for regional water resource planning.

CONCLUSION

Drainage models have been developed in response to advances in math, science and technology, and to the changing needs and concerns of society. Many drainage models have integrated routines for describing plant growth, solute transport, and soil heat flux to create comprehensive models able to predict crop yield and the quantity and quality of drainage water for a wide range of field and climatological conditions. Consequently, a wide variety of drainage models are now available to design and evaluate drainage and water management systems for agriculture and other purposes. The potential user is faced with the challenge of selecting which model to use for their particular situation. Obviously, the most complex model could be used for almost any situation; however, many expenses come with the most complex model. Most notably are the expenses required to gather and process large amounts of detailed input data and the expenses required for training the model user or hiring a qualified expert user. In many cases, a simpler and less expensive model can be used to obtain satisfactory designs. The wise project manager and model user will clearly define the objectives of their system, assess the capabilities and limitations of the available models, and select a drainage model that is suitable for their needs.

REFERENCES

1. Skaggs, R.W. Drainage Simulation Models. In *Agricultural Drainage*; Agronomy Monograph 38, Skaggs, R.W., van Schilfgaarde, J., Eds.; ASA: Madison, WI, 1999; 469–500.
2. Skaggs, R.W.; Chescheir, G.M. Application of Drainage Simulation Models. In *Agricultural Drainage*; Agronomy Monograph 38, Skaggs, R.W., van Schilfgaarde, J., Eds.; ASA: Madison, WI, 1999; 537–564.
3. Feddes, R.A.; Kabat, P.; van Bavel, P.J.T.; Bronswijk, J.J.B.; Halbertsma, J. Modeling Soil Water Dynamics in the Unsaturated Zone—State of the Art. J. Hydrol. 1988, 100, 69–111.
4. Ritzema, H.P. Subsurface Flow to Drains. In *Drainage Principles and Applications*; Ritzema, H.P., Ed.; Wageningen: Netherlands, 1994; Vol. 16, 263–304.
5. van der Ploeg, R.R.; Horton, R.; Kirkham, D. Steady Flow to Drains and Wells. In *Agricultural Drainage*; Agronomy Monograph 38, Skaggs, R.W., van Schilfgaarde, J., Eds.; ASA: Madison, WI, 1999; 213–264.
6. Youngs, E.G. Non-steady Flow to Drains. In *Agricultural Drainage*; Agronomy Monograph 38, Skaggs, R.W., van Schilfgaarde, J., Eds.; ASA: Madison, WI, 1999; 265–296.
7. Simùnek, J.; Sejna, M.; van Genuchten, M.Th. The HYDRUS-2D software package for simulating two-dimensional movement of water, heat, and multiple solutes in variably saturated media. Version 2.0, IGWMC-TPS-53, International Ground Water Modeling Center, Colorado School of Mines, Golden, CO, 1999; 251pp.
8. HYDRUS_2D (http://www.ussl.ars.usda.gov/MODELS/HYDRUS2D.HTM) (accessed January 2003).
9. Hsieh, P.A.; Wingle, W.; Healy, R.W. *VS2DI—a Graphical Software Package for Simulating Fluid Flow and Solute or Energy Transport in Variably Saturated Porous Media*; U.S. Geol. Surv. Water Resource Invest. Rep. 99-4130; U.S. Geol. Surv.: Denver, CO, 1999; 16.
10. VS2DI (http://wwwbrr.cr.usgs.gov/projects/GW_Unsat/vs2di/index.html) (accessed January 2003).
11. Belmans, C.; Wesseling, J.G.; Feddes, R.A. Simulation of the Water Balance of a Cropped Soil: SWATRE. J. Hydrol. 1983, 63, 271–286.
12. Kroes, J.G.; van Dam, J.C.; Huygen, J.; Vervoort, R.W. *User's Guide of SWAP version 2.0; Simulation of Water Flow, Solute Transport and Plant Growth in the Soil–Water–Atmosphere–Plant Environment*, Technical Document 53; DLO Winand Staring Centre, 1999; 128.
13. SWAP (http://www.alterra.nl/models/swap/index.htm) (accessed January 2003).
14. Johnson, K.E.; Liu, H.H.; Dane, J.H.; Ahuja, L.R.; Workman, S.R. Simulating Fluctuating Water Tables and Tile Drainage with a Modified Root Zone Water Quality Model and a New Model WAFLOWM. Trans. ASAE 1995, 38, 75–83.
15. RZWQM (http://gpsr.ars.usda.gov/products/rzwqm.htm) (accessed January 2003).
16. Skaggs, R.W. *A Water Management Model for Shallow Water Table Soils*, Water Resour. Res. Inst. Tech Rep. 134.; Univ. of North Carolina, 1976.
17. DRAINMOD (http://www.bae.ncsu.edu/soil_water/drainmod/) (accessed January 2003).
18. EPIC (http://www.brc.tamus.edu/epic/) (accessed January 2003).
19. WEPP (http://topsoil.nserl.purdue.edu/nserlweb/weppmain/wepp.html) (accessed January 2003).
20. ADAPT (http://www.ag.ohio-state.edu/%7Emsea/BESTAQUA.html) (accessed January 2003).
21. Reyes, M.R.; Bengston, J.L.; Fouss, J.L.; Rogers, J.S. GLEAMS Hydrology Submodel Modified for Shallow Water Table Conditions. Trans. ASAE 1993, 36, 1771–1778.

22. Parsons, J.E.; Skaggs, R.W.; Gilliam, J.W. Pesticide Fate with DRAINMOD/CREAMS. Proceedings of CREAMS/GLEAMS Symposium, Tifton, GA, September 27–29; Univ. of Georgia: Tifton, GA, 1991; 123–125.

23. Parsons, J.E.; Skaggs, R.W.; Doty, C.W. Development and Testing of a Water Management Simulation Model (WATRCOM): Development. Trans. ASAE **1991**, *34*, 120–128.

24. De Laat, P.J.M.; Atwater, R.H.C.M.; van Bakel, P.J.T. GELGAM—a Model for Regional Water Management. In *Proceeding of Technical Meeting 37*, Versl. Meded. Comm. Hydrol. Onderz. TN027, The Hague, The Netherlands, 1981.

Drainage for Soil Salinity Management

Glenn J. Hoffman
University of Nebraska, Lincoln, Nebraska, U.S.A.

INTRODUCTION

Soil water must drain through the crop root zone when salinity is a hazard to prevent salts from increasing to levels detrimental to crop production. Drainage occurs whenever irrigation and rainfall provide soil water in excess of the soil's storage capacity. In humid regions, rainfall normally satisfies crop water requirements and precipitation infiltrating into the soil in excess of this requirement leaches (drains) salts present below the crop root zone. In subhumid areas, rainfall is often inadequate in amount or temporal distribution to satisfy crop needs and irrigation is implemented. For arid regions, rainfall is never abundant and the preponderance of the crop water requirement must be provided by irrigation. Regardless of the climate, if soluble salts are present, water in excess of that needed to satisfy the crop water requirement must be provided to leach excess salts. Leaching may be accomplished continuously or at intervals, depending on the degree of salinity control required. It may take decades or as little as one season, depending on the hydrogeology of the area, but without drainage, agricultural productivity cannot be sustained where salinity is a threat. For a more complete discussion on drainage design for salinity control, the reader is referred to Hoffman and Durnford.[1]

DRAINAGE CONDITIONS

All soils have an inherent ability to transmit soil water provided a hydraulic gradient exists. If the hydraulic gradient is positive downward, drainage occurs. Soils with compacted layers, fine texture, or layers of low hydraulic conductivity may be so restrictive to downward water movement that drainage is insufficient to remove excess salts. In some areas, the hydrogeology may be such that the hydraulic gradients are predominantly upward. This leads to water logging and salination.

Before designing a man-made drainage system, the natural drainage rate should be determined. If the natural hydraulic gradient causes soil water to drain out of the crop root zone, the capacity of the artificial system can be reduced, thereby decreasing the cost for drainage. In some situations, upward flow into the crop root zone from a shallow aquifer can significantly increase the drainage requirement. The upward movement of groundwater leads to salination as the water evaporates at the soil surface, leaving salts behind. If upward flow is ignored, the drainage system may be inadequate. Regardless of the source, an artificial drainage system will not function unless it is below the surface of the water table.

DRAINAGE REQUIREMENT

Saline Soils

The amount of drainage required to maintain a viable irrigated agriculture depends on the salt content of the irrigation water, soil, and groundwater; crop salt tolerance; climate; soil properties; and management. At present, the only economical means of controlling soil salinity is to ensure an adequate net downward flow of water through the crop root zone to a suitable disposal site. If drainage is inadequate, harmful amounts of salt can accumulate.

In irrigated agriculture, water is supplied to the crop from irrigation, rainfall, snow melt, and upward flow from groundwater. Water is lost through evaporation, transpiration, and drainage. The difference between water inflows and outflows is the change in soil water storage. A water balance, expressed in terms of equivalent depths (D) of water, can be written as

$$D_s = D_i + D_r + D_g - D_e - D_t - D_d \qquad (1)$$

where the subscripts s, i, r, g, e, t, and d designate storage, irrigation, rainfall and snow melt, groundwater, evaporation, transpiration, and drainage, respectively. The corresponding salt balance, where S is the amount of salt and C is salt concentration, can be expressed as

$$S_s = D_i C_i + D_r C_r + D_g C_g + S_m + S_f - D_d C_d - S_p - S_c \qquad (2)$$

with S_s being salt storage, S_m is the salt dissolved from minerals in the soil, S_f indicates salt added as fertilizer or amendment, S_p is precipitated salts, and S_c is the salt removed in the harvested crop.

Rarely do conditions prevail long enough for steady state to exist in the crop root zone. However, it is instructive to assume steady state to understand

Encyclopedia of Water Science
DOI: 10.1081/E-EWS 120010112

the relationship between drainage and salinity. If upward movement of salt, the term $(S_m + S_f - S_p - S_c)$, and the change in salt storage are all essentially zero, then the salt balance Eq. 2 can be reduced to

$$D_d C_d = D_i C_i + D_r C_r \qquad (3)$$

The leaching fraction, L, is the ratio of the amount of water draining below the crop root zone, D_d, and the amount applied, $D_i + D_r$. The ratio of the salt concentration entering and leaving the root zone can also be used to estimate L. Since C_r is essentially zero.

$$L = C_i/C_d = D_d/D_i + D_r \qquad (4)$$

The concept in Eq. 4 is important because it illustrates the relationship between leaching fraction and salinity.

The minimum leaching fraction that a crop can endure without yield reduction is termed the leaching requirement, L_r. The leaching requirement is the minimum amount of drainage required to prevent excess accumulations of salt that result in loss of crop yield. Several models have been proposed to estimate the drainage (leaching) requirement. Of the four models tested,[2] the one presented in Fig. 1 agrees well with measured values of the drainage requirement through the range of agricultural interest. The drainage requirement given in Fig. 1 is the fraction of the volume of applied water that

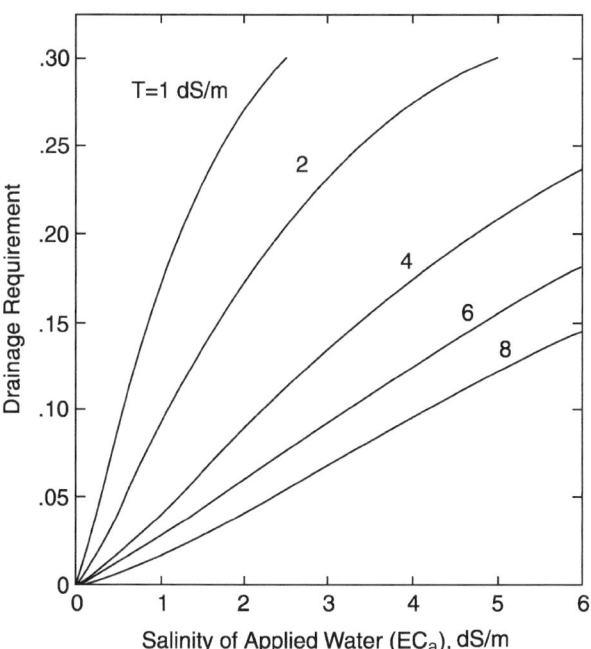

Fig. 1 Drainage requirement as a function of the salinity of the applied water (reported as the volume weighted electrical conductivity) and the salt tolerance threshold value for the crop (T). (Adapted from Ref. [15].)

must pass through the crop root zone as a function of the salinity of the applied water and the salt tolerance of the crop.

Sodic Soils

A soil is said to be sodic if an excessive concentration of sodium causes a deterioration of soil structure. The impact of excess sodium is a reduction in hydraulic conductivity and crust formation. Sodic conditions decrease the rate of drainage. Before a sodic soil can be restored to full productivity the excess sodium in the soil must be replaced with calcium or magnesium. This process frequently requires copious amounts of leaching to reclaim the soil. The design of an artificial drainage system that may be required, however, is based upon the long-term requirement for drainage as estimated in Fig. 1 rather than the anticipated high drainage requirements for reclaiming a sodic soil.

DRAINAGE SYSTEM DESIGN

There are three types of subsurface systems used to control soil salinity: relief drains, shallow wells, and interceptor drains. Relief drains, usually consisting of perforated corrugated plastic tubes buried in a regularly spaced pattern, is the most common subsurface system. Laterals for relief drains are typically placed 2.0 m–3.5 m deep and are spaced horizontally ten to hundreds of meters apart where salinity is a hazard. Shallow wells, called tube wells in some regions, can also be used to lower the water table by allowing pumping from shallow, unconfined aquifers. Tube wells are spaced at distances of a few hundred meters to several kilometers and may be a few meters to a hundred meters deep. Interceptor drains are used to remove excess soil water from saline seeps. Frequently, one subsurface drain, properly located at the upslope side of the seep, is sufficient. Regardless of the type of drainage system, the depth of the water table must be maintained low enough that (1), salts in the soil profile move to the water table (2), the rate of water movement by capillary flow to the soil surface because of evaporation is minimal, and (3), upflow of saline groundwater into the root zone is prevented.

Relief Drains

A relief drainage system consists of a main drain, collector drains, and field drains (laterals). The main drain is frequently a surface stream or an open drainage canal. Collectors and laterals are usually buried in a regular parallel pattern. Either open ditches or perforated pipes can serve as collectors and laterals. Open ditches are not normally installed now because they occupy land, are difficult to maintain, and are only capable of shallow drainage. Laterals are up to 300 m long and terminate in a collector drain. Both single- and double-sided entries by laterals into a collector are common.

Drain Depth

Subsurface drains are installed much deeper for salinity control in arid regions than drains for water table control in humid regions. The goal for salinity control is to place the drains deep to limit salination of the root zone by capillary upflow. Drains are placed at depths of 2.0 m–3.5 m in arid regions.[3] The appropriate drain depth depends upon the depth capacity of the installation machinery, the location of a shallow soil layer that impedes water movement, and anticipated benefits compared to additional costs of deeper installation.

Drain Spacing

The spacing between laterals is often estimated using simple drainage design equations. Drain spacing determinations can be based on criteria of steady-state, falling-water-table, or fluctuating-water-table conditions.[4] For large drainage projects or where more accurate values are desired, computerized drainage design models are available. An early computer model developed by Skaggs[5] has been altered by several for irrigated conditions.[6,7] Other models present drainage designs for irrigated areas based on optimization,[8] decision support systems,[9] or reuse of drainage water.[10]

Drainage Wells

Shallow or tube wells offer a viable alternative to relief drains when the aquifer has sufficient transmissivity to provide a significant yield of drain water and the vertical permeability between the crop root zone and the aquifer is adequate. Under these conditions, tube wells have the advantages of being able to lower the water table to greater depths than relief drains and also provide supplemental water for irrigation if the quality is appropriate.

Because drainage wells can be installed at convenient locations within the area to be drained and can be operated either continuously or intermittently, the management of a system of drainage wells is more versatile than relief drains. Relief drains are typically a passive drainage system relying on gravity and designed to operate continuously.

Economic comparisons between the costs of drainage wells and relief drains vary. It is generally found that relief drains have lower construction and operation costs.[11] However, Mohtadullah[12] showed tube wells were a better economic choice than relief drains for the Indus Basin.

Saline Seeps

The occurrence of saline water at the soil surface downslope from a recharge area is referred to as a saline seep. Saline seeps can occur because of the reduction of evapotranspiration that occurs when grasses or forests are converted to cropland in the upland (recharge) areas of a watershed. Dryland farming practices that include fallow periods tend to aggravate the seepage problem. Salination occurs as water infiltrating in the upper elevations of the watershed moves through salt-laden substrate on its path to a discharge site at a lower elevation. In the discharge area of the seep, crop growth is reduced or the plants killed by an intolerable level of salinity. Saline seeps can be distinguished from other saline soil conditions by their recent origin, relatively local extent, saturated soil profile, and sensitivity to precipitation and cropping systems.[13] Saline seeps occur throughout the Great Plains of North American and in Australia, India, Iran, Turkey, and Latin America.[14]

Planting crops in the recharge area that consume soil water before it percolates below the crop root zone will prevent saline seeps. Failing this, improved drainage may provide a solution. Installing an interceptor subsurface drain immediately upslope from the saline seep is frequently a successful solution. Interceptor drains to control seepage should be installed as deep as practical. If the layer restricting soil water flow is not too deep, placing the interceptor drain just above this layer is the most effective location.

REFERENCES

1. Hoffman, G.J.; Durnford, D.S. Drainage Design for Salinity Control. *Agricultural Drainage*, Agronomy Monograph No. 38; American Society of Agronomy: Madison, WI, 1999; Chap. 17, 579–614.
2. Hoffman, G.J. Drainage Required to Manage Salinity. J. Irrig. Drain. Div., American Society of Civil Engineers, New York **1985**, *111*, 199–206.
3. Ochs, W.J. *Project Drainage Issues*, Proceedings of the 3rd International Workshop on Land Drainage, Columbus, OH, December 7–11, 1987; E-83 to E-88.
4. Bouwer, H. Developing Drainage Design Criteria. *Drainage for Agriculture*, Agronomy Monograph No. 17; American Society of Agronomy: Madison, WI, 1974; 67–79.
5. Skaggs, R.W. *A Water Management Model for Artificially Drained Soils*; Report 267, Water Resources Research Institute, North Carolina State University: Raleigh, NC, 1980.
6. Chang, A.C.; Hermsmeir, L.F.; Johnston, W.R. *Application of DRAINMOD on Irrigated Cropland*, Paper no. 81-2543;

American Society of Agricultural Engineers: St. Joseph, MI, 1981.

7. Skaggs, R.W.; Chescheir, G.M. Application of Drainage Simulation Models. *Agricultural Drainage*, Agronomy Monograph No. 38; American Society of Agronomy: Madison, WI, 1999; Chap. 15, 529–556.

8. Knapp, K.C.; Wichlens, D. Dynamic Optimization Models for Salinity and Drainage Management. In *Agricultural Salinity Assessment and Management*; American Society of Civil Engineers: New York, 1990; Chap. 25, 530–548.

9. Gates, T.K.; Wets, R.J.-B; Grisiner, M.E. Stochastic Approximation Applied to Optimal Irrigation and Drainage Planning. J. Irrig. Drain. Div., Am. Soc. Civil Eng. **1989**, *115*, 488–510.

10. El-Din El-Quosy, D.; Rijtema, P.E.; Boels, D.; Abdel-Khalik, M.; Roest, C.W.J.; Adbel-Gawad, S. Prediction of the Quantity and Quality of Drainage Water by the Use of Mathematical Modeling. In *Land Drainage in Egypt*; Drainage Research Institute: Cairo, Egypt, 1989; 207–241.

11. Zhang, W. Drainage Inputs and Analysis in Water Master Plans, Proceedings of the 4th International Drainage Workshop, Cairo, Egypt, February 1990; 181–187.

12. Mohtadullah, K. Interdisciplinary Planning, Data Needs and Evaluation for Drainage Projects. Proceedings of the 4th International Drainage Workshop, Cairo, Egypt, February 1990; 127–140.

13. Brown, P.L.; Halvorson, A.D.; Siddoway, F.H.; Mayland, H.F.; Miller, M.R. *Saline-Seep Diagnosis, Control, and Reclamation*, USDA Conservation Research Report No. 30; 1983.

14. Halvorson, A.D. Management of Dryland Saline Seeps. *Agricultural Salinity Assessment and Management*; American Society of Civil Engineers: New York, 1990; Chap. 15, 372–392.

15. Hoffman, G.J.; Van Genuchten, M.Th. Soil Properties and Efficient Water Use: Water Management for Salinity Control. *Limitations to Efficient Water Use in Crop Production*; American Society of Agronomy: Madison, WI, 1983; Chap. 2C, 73–85.

Drainage and Water Quality

James L. Baker
Iowa State University, Ames, Iowa, U.S.A.

INTRODUCTION

Artificial subsurface drainage is required on many agricultural lands to remove excess precipitation and/or irrigation water in order to provide a suitable soil environment for plant growth and a soil surface capable of physically supporting necessary traffic, e.g., for tillage, planting, and harvesting. Although this drainage makes otherwise wet soils very productive, subsurface drainage alters the time and route by which excess water reaches surface waters and can carry nutrients, pesticides, bacteria, and suspended solids to surface waters and cause nonpoint source pollution problems. The net water quality impact of subsurface drainage considered here is determined by comparison to the same cropping system not having subsurface drainage (not on the fact that the existence of adequate drainage will affect land use). The degree of pollutant transport with surface runoff and subsurface drainage is determined by the product of the volumes of water and the pollutant concentrations in the water. These are both influenced by environmental conditions, pollutant properties, and management factors, and their interactions with subsurface drainage are discussed.

ENVIRONMENTAL CONDITIONS

Soils/Hydrology

Whether water added to the soil surface as precipitation or irrigation infiltrates or becomes surface runoff is critical to water quality. Topography/slope, soil moisture content, texture, and soil structure, including the existence of preferential flow paths or "macropores," can affect both the rate and route of water infiltration (as shown in Fig. 1). While different drain spacings are used to provide a desired "drainage coefficient" (e.g., 0.5 in./day) based on the internal drainage characteristics of the subsoils, the conditions of the surface soil will determine what percentage of applied water (rain or irrigation) will infiltrate. In general, the existence of subsurface drainage increases the volume of infiltration and thus decreases the volume of surface runoff (and pollutant loss with that

runoff), increases shallow percolation, and lowers water tables. This effect will be more pronounced for "lighter" soils that have higher infiltration rates and lower water holding capacities. These changes affect the final quality of cropland drainage because of differences in time and type of soil-water–chemical interactions. Surface runoff allows water to come in contact only with the surface soil and materials such as crop residue and surface-applied fertilizers, manures, and pesticides present on it for short periods of time. On the other hand, water that infiltrates and percolates through the soil comes in intimate contact with all the soils in the profile to at least the depth of the tile drain, and generally, for much longer periods of time. The soil residence time is shorter for that portion of infiltrated water intercepted by the tile drains than for water that must flow through the underground strata to appear as base flow.

Climate/Precipitation

The timing, amount, and intensity of water inputs, including precipitation and irrigation, relative to evapotranspiration (ET) determines the timing and amount of excess water that will leave agricultural land as surface runoff and subsurface drainage. Inputs at low intensities generally will totally infiltrate, but as intensities increase past the rate of infiltration, surface runoff begins. As just discussed, because decreased antecedent soil moisture contents, which result from the existence of artificial subsurface drainage, increase infiltration rates, the volume of surface runoff generally is decreased when subsurface drainage exists. This effect will be more pronounced for areas where input intensities often exceed the rates of infiltration.

For a given crop and climatic region, the amount of ET is relatively constant, so the volume of subsurface drainage is very much dependent on the amount of input water above that value. As an example, in a 4-yr Iowa tile drainage experiment,[1] during a low rainfall year, only a trace of subsurface flow occurred; in a wet year (116 cm of precipitation vs the average of 80 cm), there was 29 cm of flow; and the overall average flow for the four yr was 15 cm.

Encyclopedia of Water Science
DOI: 10.1081/E-EWS 120010233

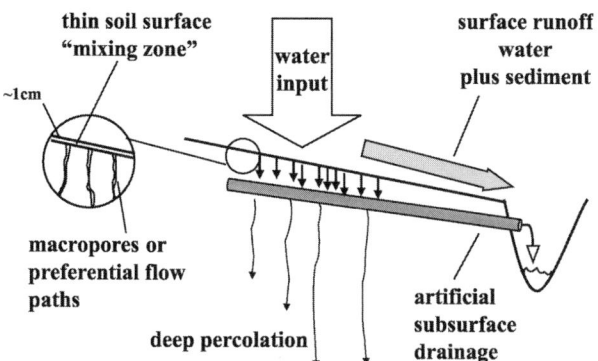

Fig. 1 Schematic of transport processes.

POLLUTANT PROPERTIES

Persistence

Pollutants that have a limited existence in agricultural lands because of plant uptake/chemical transformation (i.e., nutrients), degradation (i.e., pesticides), and die-off (i.e., micro–organisms) have different potentials for off-site transport with water based on their persistence. Because surface runoff is much more of an immediate process than subsurface drainage, concentrations of nonpersistent pollutants in subsurface drainage are usually lower than in surface runoff. This effect will be more pronounced, the lesser the persistence is. However, the "route" of infiltration, where in some cases the drainage water moves quickly through the soil profile because of macropores (see Fig. 1), plays a role and can somewhat negate the expected dissipation effect.

Adsorption/Filtration

Based on their chemical properties and/or their physical size, some pollutants transported down into and through the soil with subsurface drainage are removed from the flow stream by the soil. In particular, pollutants that are positively charged (e.g., ammonium-nitrogen (NH_4-N)) and larger, less soluble, organic compounds (e.g., some pesticides) can be attenuated by adsorption to soil clay and organic matter. Inorganic P ions can be removed by complexation or precipitation with soil cations. Micro-organisms and sediment can be filtered out by small soil pores. Thus, in general, with the exception of soluble salts of nitrate-nitrogen (NO_3-N), sulfate (SO_4), and chloride (Cl) anions, pollutant concentrations in subsurface drainage are lower than in surface runoff. This effect will be more pronounced, the greater the interaction between the pollutant and soil is.[2]

For sediment itself, because soil erosion is dependent on the erosive ability and transport capacity of surface

runoff, reducing the runoff flow rate and volume with subsurface drainage reduces sediment loss. For example, in a 6-yr study in the lower Mississippi Valley, surface runoff volumes from plots on a clay loam soil with subsurface drainage were 34% less than for plots without subsurface drainage, and the corresponding soil loss of 3500 kg/ha/yr represented a decrease of 30%. For the plots with subsurface drainage, both sediment concentrations and losses in subsurface drainage were about one-tenth those in surface runoff.[3]

For nitrogen (N), loss from poorly drained soils is usually much less than that from soils with improved drainage systems.[4] While significant N can be transported with sediment (often sediment has at least 1000 ppm N), land needing subsurface drainage is usually not highly susceptible to erosion, and N loss is dominated by soluble inorganic-N loss. In a 3-yr tile-drained watershed study in northeast Iowa,[5] NO_3-N losses in solution represented over 85% of the total N losses, including NH_4-N, organic-N in solution, and N associated with sediment. In a 5-yr study in east-central Iowa,[6,7] where nutrient concentrations in both surface runoff and subsurface drainage from cropland were monitored, NO_3-N concentrations in subsurface drainage averaged about 12 mg/L and 2–3 times those in surface runoff. While NH_4-N concentrations were usually 2–10 times higher in surface runoff, on an absolute scale, the concentrations were overall much lower than those for NO_3-N and constituted only a fraction of the total N loss.

For phosphorus (P), transport is primarily in surface runoff with sediment (often sediment has at least 500 ppm P) and dissolved in surface runoff. For conventionally tilled cropland, about 75%–90% of P transported in surface runoff is with sediment. In areas where soil erosion is minimal, soluble P in surface runoff water can dominate transport. The soluble P in surface runoff (and subsurface drainage) is regulated by adsorption/desorption characteristics of soil. Therefore, with P in surface soils generally much higher than in subsoils, subsurface drainage usually has much lower soluble P concentrations than in surface runoff.

For pesticides, because of their adsorption characteristics, like with P, concentrations in surface runoff water are usually much greater than in subsurface drainage. However, this effect is even more dramatic for pesticides because unlike P, pesticides have a limited persistence, which decreases their potential for movement with subsurface flows with long travel times. In a series of studies on herbicides in surface and subsurface drainage,[8–10] atrazine concentrations in May (the period of application) were 75 μg/L in surface runoff for plots with surface drainage only, and were 51 μg/L and 1 μg/L in surface runoff and subsurface drainage, respectively, for plots that also had subsurface drainage. Not only were there much lower atrazine concentrations in subsurface drainage water, but the presence of the subsurface drains delayed and reduced surface runoff such that concentrations in surface runoff from the plots with subsurface drainage were reduced about one-third.

For bacteria, a review by Crane et al.[11] showed that fecal coliform counts in surface runoff from manured lands often were greater than 10,000/100 mL. In a comparison of surface runoff and subsurface drainage, Culley and Phillips[12] found similar counts (>10,000/100 mL) for fecal coliform in surface runoff from both manured and fertilized plots, but with much, much lower counts (<5/100 mL) in subsurface drainage.

MANAGEMENT PRACTICES

Tillage Systems

The hydrologic interactions between tillage systems and subsurface drainage that affects water quality involves how tillage affects the timing, route, and volume of infiltration (and hence the relative volumes of subsurface drainage and surface runoff). In a review of the hydrologic effects of conservation tillage, Baker[13] noted that changing the relative volumes of subsurface drainage and surface runoff also can affect chemical concentrations in those carriers. In general, increasing infiltration increases the time for beginning of surface runoff, which in turn reduces the concentrations of chemicals at the soil surface (shown as a thin mixing zone in Fig. 1) and therefore in surface runoff. However, the effect of conservation tillage on infiltration is time-dependent. For the first storm after any tillage there is usually less runoff from the tilled soil, although on an annual basis, conservation (or less) tillage often results in lower total surface runoff volumes.

Cropping

As with tillage, the hydrologic interactions between cropping systems and subsurface drainage that affects water quality involves how cropping affects the timing, route, and volume of infiltration (and hence the relative volumes of subsurface drainage and surface runoff). A major difference between perennial crops such as forages, and row-crops such as corn and soybeans, is the volume and timing of ET demands. In general, with higher, more consistent ET, perennial crops would have lower total drainage volumes. A bigger effect of cropping would be the effect of needed chemical applications and their potential losses. For example, the large amounts of N needed (added, recycled, and/or fixed) in a continuous

corn or corn–soybean rotation means there is usually high NO_3-N concentrations in the soil profile, and hence in subsurface drainage when it occurs. However, for grasses and alfalfa, NO_3-N concentrations in subsurface drainage are much lower.[14,15]

Controlled Drainage

In areas where subsurface drainage exists, controlling the timing of outflows has been suggested as one method to reduce chemical losses. This controlled drainage could reduce losses by reducing both subsurface volumes and chemical concentrations. The potential for reduced concentrations is probably the greatest for NO_3-N, where the process of denitrification would reduce NO_3 to N gases in the soil profile where high water tables and the presence of organic matter drives the system anaerobic. The results summarized from 125 site-years of data from North Carolina[16] showed that controlled drainage reduced subsurface drainage volumes an average of 30% compared to uncontrolled drainage systems. Reductions in N and P lost with subsurface drainage were 45% and 35%, respectively. While almost all the reduction of P loss was due to decrease in drainage volume; for N, reductions in NO_3-N concentrations also contributed to the reduction.

CONCLUSION

The total effect of surface drainage on surface water resources receiving drainage from agricultural lands involves the relative volumes of surface runoff and subsurface drainage, and the relative concentrations of sediment, nutrients, pesticides, and bacteria. In general, the existence of subsurface drainage increases infiltration rates which delays and reduces the volume of surface runoff. For pollutants lost mostly with surface runoff, which include sediment, NH_4-N, P, pesticides, and bacteria, not only is the volume of the carrier reduced, but also the concentrations. This is because delayed runoff and more water moving through the surface-mixing zone reduce the amounts of contaminants at the soil surface available to interact with added water and overland flow. Thus the only real water quality negative to subsurface drainage is the increased volume of water moving thorough the soil profile carrying the soluble unadsorbed NO_3-N anion. The use of improved in-field N management in the way of rate, method, and timing of N applications has some potential to reduce this problem.[17] However, other practices such as controlled drainage or construction/reconstruction of wetlands may be needed to provide some NO_3-N reduction treatment. Alternatively, reducing the amount of row-crops grown on subsurface-drained lands could have a large impact although the current economics of doing that would be quite negative.

REFERENCES

1. Baker, J.L.; Campbell, K.L.; Johnson, H.P.; Hanway, J.J. Nitrate, Phosphorus and Sulfate in Subsurface Drainage Water. J. Environ. Qual. **1975**, *4*, 406–412.

2. Gilliam, J.W.; Baker, J.L.; Reddy, K.R. Water Quality Effects of Drainage in Humid Regions. *Drainage Monograph*; Am. Soc. Agron.: Madison, WI, 1999, Chap. 24.

3. Bengston, R.L.; Carter, C.E.; Morris, H.F.; Bartkiewicz, S.A. The Influence of Subsurface Drainage Practices on Nitrogen and Phosphorus Losses in a Warm, Humid Climate. Trans. ASAE **1988**, *31*, 729–733.

4. Gambrell, R.P.; Gilliam, J.W.; Weed, S.B. Nitrogen Losses from Soils of the North Carolina Coastal Plain. J. Environ. Qual. **1975**, *4*, 317–323.

5. Baker, J.L.; Melvin, S.W.; Agua, M.M.; Rodecap, J. *Collection of Water Quality Data for Modeling the Upper Maquoketa River Watershed*, Paper No. 99-2222; ASAE: St. Joseph, MI, 1999.

6. Johnson, H.P.; Baker, J.L. *Field-to-Stream Transport of Agricultural Chemicals and Sediment in an Iowa Watershed: Part I. Data Base for Model Testing (1976–1978)*, Rep. No. EPA-600/S3-82-032; USEPA Environ. Res. Lab.: Athens, GA, 1982.

7. Johnson, H.P.; Baker, J.L. *Field-to-Stream Transport of Agricultural Chemicals and Sediment in an Iowa Watershed: Part II. Data Base for Model Testing (1979–1980)*, Rep. No. EPA-600/S3-84-055; USEPA Environ. Res. Lab.: Athens, GA, 1984.

8. Southwick, L.M.; Willis, G.H.; Bengston, R.L.; Lormand, T.J. Effect of Subsurface Drainage on Runoff Losses of Atrazine and Metolachlor in Southern Louisiana. Bull. Environ. Contam. Toxicol. **1990**, *45*, 113–119.

9. Southwick, L.M.; Willis, G.A.; Bengston, R.L.; Lormand, T.J. Atrazine and Metolacholor in Subsurface Drain Water in Louisiana. J. Irrig. Drain. Eng. **1990**, *116*, 16–23.

10. Bengston, R.L.; Southwick, J.M.; Willis, G.H.; Carter, C.E. The Influence of Subsurface Drainage Practices on Herbicide Losses. Trans. ASAE **1990**, *33*, 415–418.

11. Crane, S.R.; Moore, J.A.; Grismer, M.E.; Miner, J.R. Bacterial Pollution from Agricultural Source: A Review. Trans. ASAE **1983**, *26*, 858–866, 872.

12. Culley, J.L.B.; Phillips, P.A. Bacteriological Quality of Surface and Subsurface Runoff from Manured Sandy Clay Loam Soil. J. Environ. Qual. **1982**, *11*, 155–158.

13. Baker, J.L. Hydrologic Effects of Conservation Tillage and Their Importance Relative to Water Quality. In *Effects of Conservation Tillage and Groundwater Quality*; Logan, T.J., Davidson, J.M., Baker, J.L., Overcash, M.R., Eds.; Lewis Publishers, Inc.: Chelsea, MI, 1987, Chap. 6.

14. Baker, J.L.; Melvin, S.W. Chemical Management, Status, and Findings. *Agricultural Drainage Well Research and Demonstration Project—Annual Rep. and Project Summary*; Iowa Dep. of Agric. and Land Stewardship and Iowa State Univ.: Ames, 1994; 27–60.

15. Weed, D.A.J.; Kanwar, R.S. Nitrate and Water Present in and Flowing from Root-Zone Soil. J. Environ. Qual. **1996**, *25*, 709–719.

16. Evans, R.O.; Skaggs, R.W.; Gilliam, J.W. Management Practice Effects on Water Quality. Proc. Irrig. Drain. 1990 Natl. Conf. ASCE Irrig. Div., Durango, CO, 1990; 182–191.

17. Baker, J.L. Limitations of Improved Nitrogen Management to Reduce Nitrate Leaching and Increase Use Efficiency. Optimizing Nitrogen Management in Food and Energy Production and Environmental Protection: Proceedings of the 2nd International Nitrogen Conference on Science and Policy. Sci. World **2001**, *1* (S2), 10–16.

Drought

Donald A. Wilhite
*National Drought Mitigation Center and University
of Nebraska, Lincoln, Nebraska, U.S.A.*

INTRODUCTION

Drought differs from other natural hazards in several ways.[1] First, drought is a slow-onset, creeping natural hazard. Its effects often accumulate slowly over a considerable period of time and may linger for years after the termination of the event. Therefore, the onset and end of drought is difficult to determine. Second, the absence of a precise and universally accepted definition of drought adds to the confusion about whether or not a drought exists and, if it does, its degree of severity. Realistically, definitions of drought must be region and application (or impact) specific. This is one explanation for the scores of definitions that have been developed. Third, drought impacts are nonstructural and spread over a larger geographical area than are damages that result from other natural hazards. Because drought can affect such large areas, it is far more difficult to quantify impacts and respond effectively.

Although many people consider drought a natural or physical event, it has both a natural and social component. The risk associated with drought for any region is a product of both the region's exposure to the event (i.e., probability of occurrence at various severity levels) and the vulnerability of that area or region to the event. The natural event (i.e., meteorological drought) is a result of the occurrence of persistent large-scale disruptions in the global circulation pattern of the atmosphere. Exposure to drought varies spatially, and we can do little to alter drought occurrence. Vulnerability, on the other hand, is determined by social factors such as population growth, population shifts (regional and rural to urban), demographic characteristics, technology, policy, environmental awareness, and social behavior. These factors change over time and thus vulnerability will increase or decrease in response to these changes.

DROUGHT DEFINITION AND TYPES

Drought is an insidious natural hazard that results from a departure of precipitation from expected or "normal" that, when extended over a season or longer period of time, is insufficient to meet the demands of human activities. Drought is normally grouped by type as follows: meteorological, hydrological, agricultural, and socioeconomic.[1] Fig. 1 explains the relationship between these various types of drought and the duration of the event. Meteorological drought is commonly defined on the basis of the degree of precipitation deficiency, compared to "normal" or average, and the duration of the dry period. Thus, intensity and duration are the key characteristics of these definitions. Agriculture is usually the first economic sector affected by drought because soil moisture supplies are often quickly depleted. Agricultural drought links various characteristics of meteorological drought to agricultural impacts, focusing on precipitation shortages, differences between actual and potential evapotranspiration, and soil water deficits. Agricultural drought would develop more quickly on sandy soils because of lower soil water-holding capacity. A plant's demand for water depends on prevailing weather conditions, biological characteristics of the specific plant, its stage of growth, and the physical and biological properties of the soil. A definition of agricultural drought should account for the variable susceptibility of crops at different stages of crop development.

Hydrological droughts are associated with the effects of periods of precipitation deficits on surface or subsurface water supply (i.e., streamflow, reservoir and lake levels, groundwater). Extended drought periods may result in serious depletion of these components of the hydrological system. Hydrological droughts are usually out of phase or lag the occurrence of meteorological and agricultural droughts. More time elapses before precipitation deficiencies are detected in surface and subsurface water supplies. As a result, impacts are out of phase with those in other economic sectors. Also, water in hydrological storage systems is often used for multiple and competing purposes (e.g., power generation, flood control, irrigation, recreation), further complicating the sequence and quantification of impacts. Competition for water in these storage systems escalates during drought, and conflicts between water users increase significantly. Hydrological drought is also likely to continue long after the end of meteorological drought because of the time necessary to recharge surface and subsurface water supplies.

Finally, socioeconomic drought is associated directly with the supply of some commodity or economic good (e.g., hay, hydroelectric power) and that supply is related

Encyclopedia of Water Science
DOI: 10.1081/E-EWS 120010109

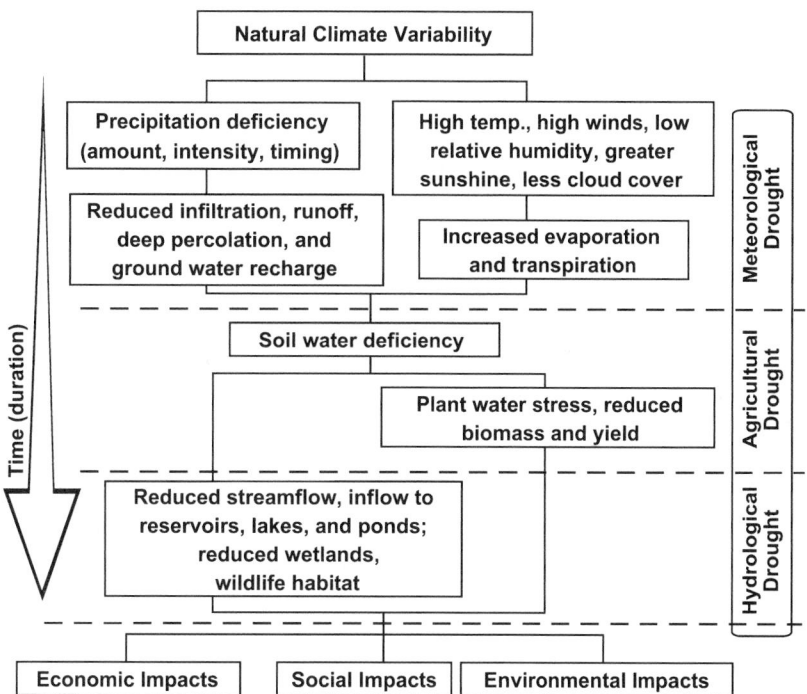

Fig. 1 Relationship between various types of drought and duration of drought events. (From Ref. [1].)

directly to precipitation levels. Increases in population can substantially alter the demand for these economic goods over time. Thus, the incidence of drought could increase because of a change in the frequency of meteorological drought, a change in societal vulnerability to water shortages, or both. For example, poor land-use practices such as overgrazing can decrease animal carrying capacity and increase soil erosion, which exacerbates the impacts of and vulnerability to future droughts.

DROUGHT CHARACTERISTICS

Droughts differ from one another in three essential characteristics: intensity, duration, and spatial coverage. Intensity refers to the degree of the precipitation shortfall and/or the severity of impacts associated with the shortfall. It is generally measured by the departure of some climatic index from normal and is closely linked to duration in the determination of impact. Many indices exist and are used to detect the onset and severity of drought conditions. One of the principal difficulties with any index is the determination of the threshold between nondrought and drought conditions or levels of severity (i.e., moderate, severe, extreme). These thresholds are important because they are used to determine when emergency response or mitigation actions are triggered.

Another distinguishing feature of drought is its duration. Droughts usually require a minimum of two to three months to become established but then can continue for months or years. Drought impacts are magnified as dry conditions extend through multiple seasons or years.

Droughts also differ in terms of their spatial characteristics. The areas affected by severe drought evolve gradually, and regions of maximum intensity shift from season to season. As drought emerges and intensifies, its core area or epicenter shifts and its spatial extent expands and contracts throughout the duration of the event.

THE IMPACTS OF DROUGHT

The impacts of drought are diverse and often ripple through the economy. Impacts are often referred to as direct or indirect. Because of the number of affected groups and sectors associated with drought, its spatial extent, and the difficulties connected with quantifying environmental damages and personal hardships, the precise determination of the financial costs of drought is an arduous task.

The impacts of drought can be classified into three principal areas: economic, environmental, and social.[2] Economic impacts range from direct losses in the broad agricultural and agriculturally related sectors, including forestry and fishing, to losses in recreation, transportation, banking, and energy. Other economic impacts would include added unemployment and loss of revenue to local, state, and federal government. Environmental losses are the result of damages to plant and animal species, wildlife habitat, and air and water quality; forest and range fires; degradation of landscape quality; and soil erosion. Although these losses are difficult to quantify, growing public awareness and concern for environmental quality has forced public officials to focus greater attention on these effects. Social impacts mainly involve public safety, health, conflicts between water users, and inequities in the distribution of impacts and disaster relief programs. As with all natural hazards, the economic impacts of drought are highly variable within and between economic sectors and geographic regions, producing a complex assortment of winners and losers with the occurrence of each disaster.

DROUGHT PLANNING AND MITIGATION

Drought planning is defined as actions taken by individual citizens, industry, government, and others in advance of drought for the purpose of mitigating some of the impacts and conflicts associated with its occurrence.[3] Because drought is a normal part of climate variability for virtually all regions, it is important to develop plans to deal with these extended periods of water shortage in a timely, systematic manner. This planning process needs to occur at various levels of government and be integrated between levels of government.

The purpose of a drought plan is to reduce the impacts of drought by identifying the principal sectors, groups, or regions most at risk and developing mitigation actions and programs that can reduce these risks in advance.[3] Plans will also improve coordination within and between levels of government. Generally, drought plans have three basic components: monitoring and early warning; risk and impact assessment; and response and mitigation.[4] Substantial progress in state-level drought planning has been made in the United States in recent years. States with drought plans have increased from 3 in 1982 to 31 in 2001. Drought plans are at the foundation of improved drought management, but only if they emphasize risk assessment and mitigation programs and actions.

CONCLUSION

Drought is an insidious natural hazard that is a normal part of the climate of virtually all regions. It should not be viewed as merely a physical phenomenon. Rather, drought is the result of an interplay between a natural event and the demand placed on water supply by human-use systems.

Many definitions of drought exist; it is unrealistic to expect a universal definition to be derived. The three characteristics that differentiate one drought from another are intensity, duration, and spatial extent. The impacts of drought are diverse and generally classified as economic, social, and environmental. Impacts ripple through the economy and may linger for years after the termination of the drought episode. It appears that societal vulnerability to drought is escalating in both developing and developed countries, and at a significant rate. It is imperative that increased emphasis be placed on mitigation, preparedness, and prediction and early warning if society is to reduce the economic and environmental damages associated with drought and its personal hardships. This will require improved coordination within and between levels of government and the active participation of stakeholders.

REFERENCES

1. Wilhite, D.A. Drought as a Natural Hazard: Concepts and Definitions. In *Drought: A Global Assessment*, Hazards and Disasters: A Series of Definitive Major Works; Wilhite, D.A., Ed.; Routledge: London, 2000; Vol. 1, 3–18.
2. Wilhite, D.A.; Vanyarkho, O. Drought: Pervasive Impacts of a Creeping Phenomenon. In *Drought: A Global Assessment*, Hazards and Disasters: A Series of Definitive Major Works; Wilhite, D.A., Ed.; Routledge: London, 2000; Vol. 1, 245–255.
3. Wilhite, D.A.; Hayes, M.J.; Knutson, C.; Smith, K.H. Planning for Drought: Moving from Crisis to Risk Management. J. Am. Water Res. Assoc. **2000**, *36* (4), 697–710.
4. Wilhite, D.A.; Sivakumar, M.K.V.; Wood, D.A. (Eds.). Early Warning Systems for Drought Preparedness and Management. Proceedings of an Expert Group Meeting, Lisbon, Portugal, Sept 5–7, 2000; World Meteorological Organization: Geneva, Switzerland, 2000.

Drought Avoidance and Drought Adaptation

J. W. Chandler
University of Cologne, Cologne, Germany

Dorothea Bartels
University of Bonn, Bonn, Germany

D

INTRODUCTION

Surviving periods without water is one of the greatest challenges faced by many plants. To cope with this challenge, plants have developed several strategies: either adaptation mechanisms which allow them to survive the adverse drought conditions, or the possession of particular growth habits to circumvent or avoid drought. Both mechanisms must have arisen under the same evolutionary constraints to enable plants to cope with low water availability. Plants avoid drought by completing their life cycle during the wet season when sufficient water is available. This strategy has been adopted by many flowering annuals. Another drought-avoidance mechanism is the formation of deep roots which allows better access to groundwater resources. The development of deep roots is an example where it is difficult to distinguish between avoidance and adaptation mechanisms. Most adaptation mechanisms are constitutive and are also present during nonstressful conditions. The objective of drought-adaptation mechanisms is to decrease transpiration and to improve water up-take. The development of succulence in leaves and roots, sunken stomata, reduction of transpiring surfaces even by the shedding of leaves or the presence of specialized photosynthetic pathways (C_4 and CAM plants) are examples of drought-avoidance mechanisms. In summary, adaptation includes modifications of a plant on the morphological, anatomical and/or biochemical level to cope better with water-deficit.

Acclimation is a third mechanism, different from avoidance and adaptation. Acclimation is a response of plants to changing water conditions, and it is generally associated with the synthesis of a specific set of transcripts. During acclimation, plants acquire resistance to stress conditions which may otherwise be lethal. This review will focus on avoidance and adaptation mechanisms.

MECHANISMS OF DROUGHT AVOIDANCE AND ADAPTATION

Anatomical Mechanisms

Most anatomical adaptations to drought conditions contribute to the maintenance of a positive water balance, either by maximizing water up-take, or minimizing water loss. To maximize water absorption from the soil, desert plants often have well-developed xylem tissue, which helps rapid water conduction in times of water supply. For example, in the case of *Lygeum spartam*, the roots are extremely hygroscopic to maximize water absorption. At the level of minimizing water loss, roots of plants from arid environments often develop thick bark, show sclerification of the cortical cells or ensure that the vascular cylinder is protected by periderm formation or necrosis of the cortical parenchyma. In addition, a feature mainly specific to desert grasses is the production of sheath roots which exude a mucilage to cement particles around the root and protect against dehydration. In aerial parts of the plant, transpiration is the major process through which water is lost. Structural characteristics specific to desert plants, which contribute to reducing water loss, logically include a reduction in total leaf surface area and a concomitantly low surface area/volume ratio. This tendency towards succulence often reflects a reduction in leaf cell size.[1] Many diverse mechanisms have evolved to reduce water loss in plants where the vapor-pressure gradient between vegetative organs and the air is extreme. One common mechanism is the development of either a thick waxy epidermis or a multiple epidermis in leaves. Equally important is the chemical composition of the epidermis, which will regulate how effective water loss is minimized. Many desert plants have a thick covering of hairs or trichomes on the leaves, which in the same way as cuticular wax or resin, serves to reduce solar radiation to the leaves and increase solar reflectance.[2] In addition to

Encyclopedia of Water Science
DOI: 10.1081/E-EWS 120010161

creating a boundary layer of still air close to the leaf surface, this traps moisture and reduces evapotranspiration. In the same way, sunken stomata also minimize transpiration through the creation of a pocket of moist still air above the stomatal pore. A more specialized characteristic of some xerophytes, e.g., *Achillea fragrantissima* is to produce annual interxylary cork rings. This helps to reduce water loss and limits water flow to a narrow zone of xylem.[2]

Morphological Characteristics

The major ways plants avoid drought stress is to alter their whole morphology to minimize transpirational water loss. Plants growing in arid environments have a low shoot/root ratio, which means that each unit of aerial transpiring surface is provided with water by many more roots than is the case for mesophytes, thereby increasing the potential for a positive water balance. Individual plants may show a remarkable ability to modulate leaf area, depending on environmental conditions. For example, the desert shrub *Lycium shawii* can produce broad thin leaves in wet conditions, or small leaves in progressively drier habitats, or even no leaves at all. Many desert plants avoid desiccation by shedding their leaves during the dry season, such as broom-like xerophytes. Under extreme conditions, leaf shedding may also be accompanied by branch shedding in some species. Other plants shed large winter leaves at the start of the dry season and form increasingly smaller leaves throughout the dry season, such as *Artemisia herba-alba*. *Zygophyllum dumosum* reduces its transpiring surface up to 96% by shedding its leaf blades to leave only the petioles. The ability of a plant to orientate its aerial parts to minimize the effect of sunlight on transpiration include epinastic or hyponastic growth responses, or nyctinasty, the endogenously-controlled circadian rhythm of leaf movement. Another mechanism to reduce water loss from exposed surfaces is that of leaf rolling, shown by many desert grasses, such as *Sporobolus arabicus*. This ensures that the adaxial side of the leaf, containing the stomata faces inwards and is protected from the direct impact of the climate and reduces transpiration loss. Although plants in arid environments have a high root/shoot ratio, it is hard to generalize about the nature of the root systems of plants adapted to drought. Root depth is often significantly increased in response to drought[3] and phraeatophytes have extremely deep tap root systems. However, some plants (e.g., cacti) have superficial root systems. It is at least accepted that the capacity of the root system to develop early and rapidly in the life cycle is an important factor in drought resistance.[4] Some xerophytes and geophytes produce ephemeral roots or fine rootlets in response to rain very rapidly, even within a few hours for some cacti,[5] just below the soil surface, which absorb dew as well as ground water.

Physiological Mechanisms

Xerophytes usually have higher osmotic pressures in their roots and shoots than mesophytes, which increases the efficiency of water absorption. Maintaining osmotic pressure in drought conditions may be achieved by adjusting the cytoplasmic content of either organic acids (malate), inorganic cations (K^+), carbohydrates (glucose, fructose, sugar alcohols) or amino acids (proline). This osmotic adjustment, together with cell wall elasticity, can maintain osmotic pressure and appropriate cell volume as cellular water is lost. The ability to osmoregulate not only serves to prevent further water loss in dry environments, but allows for the continued uptake of water against large negative water potentials.

C_4 and CAM pathways of photosynthesis are usually found in plants growing in environments where high temperatures predominate, and are both considered as physiological mechanisms to avoid water loss. This is a result for CAM plants of the bulk of carbon being fixed during the night, when leaf temperatures and water vapor-pressure differences are low, so that stomatal opening minimizes transpirational water loss and maintains a high water use efficiency. In C_4 species, the biochemical steps of CO_2 assimilation are spatially separated: CO_2 is first fixed into oxaloacetic acid in the mesophyll cells and is then transferred to the bundle sheath cells for entry into the Calvin cycle. The morphologically distinct bundle sheath and mesophyll cells represent a Krantz anatomy and C_4 photosynthesis is associated with a higher photosynthetic efficiency and a higher water use efficiency than for C_3 plants.

Growth Habits

Many angiosperms have evolved a drought-avoidance strategy by altering their life cycles to evade potential dry seasons. These winter annuals and ephemerals are therefore abundant in arid environments with seasonal droughts, since they have very short life cycles, as rapid as 2–3 weeks (e.g., *Linaria haelava*) and can successfully reproduce and die before the onset of drought and leave a reserve of viable dormant seeds in the soil. Perennials, as already mentioned, evade desiccation by partly or completely (as in geophytes having corms or bulbs) losing the vegetative structures at the onset of drought. Ephemerals and winter annuals have been shown to have much higher values of reproductive resource allocation than other plants,[6] since they need not retain reserves for perennation. In general, most plants avoid drought by accelerating the transition from vegetative to reproductive growth or at least reducing reproductive growth in general, so that resource allocation does not outstrip resource availability.

Many aspects of growth form contribute to the avoidance of desiccation; an increase in bushiness to minimize water loss and increase shading of leaves. The tree *Ocotea foetens*, which grows in areas of low rainfall, shows a spectacular adaptation to drought by forming a very dense canopy which aids the condensation of water from fog, which then runs to the tree base to irrigate itself and other species.[7] Otherwise, direct water absorption from the air is limited to lichen and algae, which can obtain water from air with a relative humidity of more than 70%.[8] Foliar uptake of precipitation, dew or water vapor by vascular plants is an extremely debatable phenomenon according to strict defining criteria. Many poikilohydric plants, including some ferns and "resurrection" plants[9] may equilibrate their water content with the relative humidity of the air during drought periods, and fully rehydrate in plentiful water supply.

FUTURE PERSPECTIVES

Further study of the molecular biology underlying many of the adaptation and avoidance responses will contribute to our understanding of the genetic bases of these processes. If traits contributing to drought tolerance are determined by a single gene, the possibility is raised that biotechnology and gene transfer techniques may be able to engineer plants better adjusted or better able to respond to water-deficit conditions. Some preliminary successes towards obtaining plants genetically enhanced for coping with drought stress are emerging. For example, manipulating the expression of the transcription factor cbf from *Arabidopsis* results in the alteration of a drought-stress pathway and results in *Arabidopsis* plants with an improved drought tolerance.[10] The identification of quantitative trait loci involved in heightened drought tolerance may also facilitate the introduction of loci from one species into another and improve more complex traits.

REFERENCES

1. Cutler, J.M.; Rains, R.S.; Loomis, R.S. The Importance of Cell Size in the Water Relations of Plants. Physiol. Plant. **1977**, *40*, 255–260.
2. Fahn, A. *Plant Anatomy*, 3rd Ed.; Pergamon Press: Oxford, 1989; 544.
3. Reader, R.J.; Jalili, A.; Grime, J.P.; Spencer, R.E.; Matthews, N. A Comparative Study of Plasticity in Seedling Rooting Depth in Drying Soil. J. Ecol. **1992**, *81*, 543–550.
4. Abdel Rahman, A.A.; Batouny, K.H. Root Development and Establishment Under Desert Conditions. Bull. Inst. Désert d'Egypte **1959**, *9*, 21–40.
5. Kausch, W. Relation between Root Growth, Transpiration and CO_2-Exchange in Several Cacti. Planta **1965**, *66*, 229–238.
6. Bell, K.L.; Hiatt, H.D.; Niles, W.E. Seasonal Changes in Biomass Allocation in Eight Winter Annuals of the Mohave Desert. J. Ecol. **1979**, *67*, 781–787.
7. Gioda, A.; Acosta Baladon, A.; Fontanel, P.; Hernandez Martinez, Z.; Santos, A. L'Arbre Fontaine. La Recherche **1993**, *249*, 1400–1406.
8. Orshan, G. Plant Form as Describing Vegetation and Expressing Adaptation to Environment. Ann. Bot. **1986**, *44*, 7–38.
9. Gaff, D.F.; Latz The Occurrence of Resurrection Plants in the Australian Flora. Aust. J. Bot. **1978**, *26*, 485–492.
10. Kasuga, M.; Liu, Q.; Miura, S.; Yamaguchi-Shinozaki, K.; Shinozaki, K. Improving Plant Drought, Salt and Freezing Tolerance by Gene Transfer of a Single Stress-Inducible Transcription Factor. Nature Biotechnol. **1999**, *17*, 287–291.

Drought Hardening and Presowing Seed Hardening

Neil C. Turner
Commonwealth Scientific and Industrial Research Organisation (CSIRO),
Wembley (Perth), Western Australia, Australia

INTRODUCTION

Drought hardening is the process whereby a plant is subjected to partial drying so that when it is exposed to a subsequent drought event, the plant is able to withstand a greater severity of drought. During drought hardening, morphological, physiological, and chemical changes are induced within the plant as a result of phytohormone activity that enable the cells to withstand greater dehydration. Drought hardening also confers greater cold tolerance to the plant. Drought hardening is widely used in seedlings to increase their survival rate when transplanted. The wetting and drying of seed, known as presowing seed hardening, is also used to increase the germination, emergence, and drought tolerance of seedlings.

The acclimation of plants is the *nonheritable* modification by the plant in response to exposure to new climatic conditions such as drought.[1] It relies on the occurrence of temporary phenotypic modifications induced by the change in environment. Acclimation differs from adaptation in that the latter refers to the *heritable* modifications in structure or function that increase the probability of a plant surviving and reproducing in a particular environment.[1] Hardening is the equivalent of acclimation,[1] that is, it depends on phenotypic modifications. Seedlings are "hardened" prior to transplanting by exposure to full sunlight and transient water stress to ensure greater survival.

EFFECTS OF DROUGHT HARDENING

The changes that take place during drought hardening are several. First, drought hardening induces morphological and anatomical changes in the plant. In wheat, exposure to water stress during the growth of the flag leaf resulted in a reduction in leaf area and thickness, smaller cells with thicker walls, and an increased stomatal frequency.[2] Similarly, in cotton a reduced frequency of irrigation resulted in smaller leaves with smaller cells and thicker cell walls.[3] These morphological changes resulting from drought preconditioning make the plant more xeromorphic and less sensitive to subsequent drought.[4,5] Drought

hardening also influences the physiological responses of plants. In cotton, drought hardening made leaf expansion and stomatal conductance less sensitive to a subsequent water deficit.[4] Also, the proportion of root dry weight to shoot dry weight shifts in favor of the root during drought hardening,[4] thereby enabling increased water uptake during subsequent water shortage.

Additionally, drought hardening induces a lowering of the osmotic potential and turgor maintenance.[6] This is illustrated, for cotton, in Fig. 1. The turgor pressure was higher and osmotic pressure was lower at a particular leaf relative water content in drought-hardened cotton plants exposed to three cycles of water stress, compared to continuously well-watered cotton plants.[5] The lower osmotic potential in the hardened plants may arise from solute accumulation by osmotic adjustment or by changes in tissue elasticity.[5,6] In cotton, while solutes increased during a subsequent stress, solute accumulation was similar in drought hardened and nonhardened plants suggesting that it was the change in cell size and tissue elasticity that resulted in the changes observed in Fig. 1.[5] Certainly, the accumulation of solutes by osmotic adjustment[6,7] during drought hardening can play a role in maintenance of turgor and physiological activity in a subsequent cycle of drought.[8] One consequence of osmotic adjustment is the continued growth of roots and deeper extraction of water from the soil,[9] thereby maintaining the water status of the plant high during a subsequent drying cycle and delaying the onset of plant dehydration, a mechanism recognized as drought avoidance. The solutes that accumulate during drought hardening are soluble sugars and amino acids, especially proline.[7,10] Table 1 gives the relative concentrations of amino acids and soluble sugars in the phyllodes (leaves) of *Acacia cyanophylla* seedlings during drought hardening for periods up to 13 mo. Reducing the water available to one-sixth in the well-watered tree seedlings induced, after the first month, a gradual increase in the concentration of aspartic acid, glutamine acid, proline, and soluble sugars.

When soils dry, it is now recognized that the leaves and roots synthesize the phytohormone, abscisic acid (ABA). Root-synthesized ABA is quickly transferred to the leaves in the xylem sap.[11] ABA closes stomata, reduces leaf

Encyclopedia of Water Science
DOI: 10.1081/E-EWS 120010163

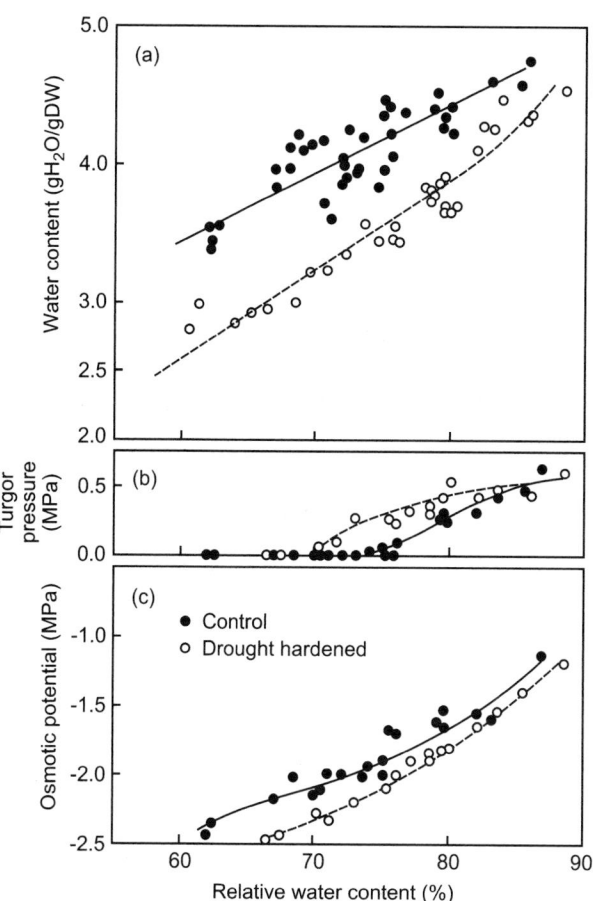

Fig. 1 The water content (a), turgor pressure (b), and osmotic potential (c), of cotton leaves subjected to drought hardening or kept well watered. Drought hardening was induced by three cycles of withdrawing water until the stomata closed during the day. (Adapted from Ref. [5].)

Table 1 Influence of drought hardening for various periods of time on the composition of the phyllodes (leaves) of *Acacia cyanophylla* Lindl. The results are presented as drought hardened as a fraction of the unhardened plant. Drought hardening was induced by providing one-sixth the water of that given to the unhardened well-watered plants

Phyllode composition	Duration of hardening (mo)			
	1	3	5	13
Free amino acids	0.63	1.60	2.53	5.73
Aspartic acid	0.44	1.55	5.53	5.00
Glutamic acid	0.46	1.49	1.29	2.69
Proline	0.53	2.00	3.31	9.00
Soluble sugars	1.00	1.13	1.39	1.94

Source: (Adapted from Ref. [10].)

growth, increases leaf senescence and abscission, and promotes root growth.[11,12] Thus, ABA acting alone or possibly in concert with other phytohormones such as cytokinins, sap pH, and sap mineral composition,[13] induces changes in the plant's water loss and water harvesting capabilities that enable it to withstand subsequent drought better. This suggests that ABA plays an important role in the induction of drought hardening.

ABA is also known to induce production of specific proteins, dehydrins, which are part of the late embryogenesis abundant (LEA) class of proteins that are also stimulated by dehydration in a wide range of plants.[14–16] However, the role of these proteins in drought hardening is still unclear since the overexpression and downregulation of the genes had no influence on a plants' abilities to withstand subsequent water deficits.

Water deficits also induce accumulation of compatible solutes such as glycine betaine, proline betaine, and other quaternary ammonium compounds that may play a role in drought hardening by acting as osmoprotectants and to store energy for use when stress is relieved.[17] Genes for their overproduction have been identified and genetic modification to increase the content of compatible solutes has been shown to increase salinity tolerance.[18]

The changes induced by drought hardening also reduce the chilling injury of plants. The morphological, physiological, and chemical changes induced during drought hardening such as smaller cells, thicker cell walls, lower stomatal conductances and transpiration, high soluble sugar, and amino acid levels appear to confer greater chilling resistance to the plant. However, chilling resistance and drought hardening appear to be induced by separate mechanisms. Chilling of the roots does induce water deficits in the leaves because of the decreased hydraulic conductance of roots at low temperature, provided shoot transpiration is high.[19] ABA increased when plants were chilled[20] at high relative humidities that induced a water deficit in the chilled plants, but not in plants chilled at high relative humidities designed to minimize transpiration.[21] Indeed, low temperature appeared to minimize ABA production even in water-stressed plants[21] suggesting that ABA does not induce chilling resistance like it induces drought hardening.

PRESOWING HARDENING

Henckel and his coworkers concluded that presowing hardening of seeds conferred greater drought resistance on a plant after germination.[22] Presowing hardening is achieved by soaking the seed in water for a period of about 2 days so that imbibition of water occurs and then slowly air drying the seed until it reaches the initial water content.[22] Henckel[22] suggested that the presowing hardening induced a number of physicochemical changes to the cytoplasm, including greater hydration of colloids, higher viscosity and elasticity, increased bound water, increased hydrophilic and decreased lipophilic colloids, and an increased temperature for protein coagulation. Other studies showed that presowing hardening of carrot increased the embryo size by about 50%[23] and increased the speed of germination and emergence of seedlings[23,24] such that in a drying soil a greater proportion of seeds germinated and emerged.[25] The benefits that the presowing hardening had on the drought resistance of the plant have been disputed.[26] Both benefits[23] and lack of benefit[25] have been reported. Presowing hardening of rice seed was shown to increase the percentage emergence of the seed in three of eight cultivars of rice, but not in another five.[27] The increased emergence was associated with longer coleoptiles and greater root length, suggesting that in the cultivars in which presowing hardening had a beneficial effect it was from the faster initial growth, as in previous studies.[23,25]

CONCLUSION

Drought hardening is a recognized method that is widely used when transplanting seedlings into the field and also provides widespread benefit to crops grown under dryland conditions or with limited irrigation. Presowing hardening of seed, on the other hand, has had mixed results and has not been widely adopted, particularly as soaking and drying the seed is a costly and exacting practice.

REFERENCES

1. Kramer, P.J. Drought, Stress and the Origin of Adaptations. In *Adaptation of Plants to Water and High Temperature Stress*; Turner, N.C., Kramer, P.J., Eds.; Wiley: New York, 1980; 7–20.
2. Zagdanska, B.; Kozdoj, J. Water Stress-Induced Changes in Morphology and Anatomy of Flag Leaf of Spring Wheat. Acta Societatis Botanicorum Poloniae **1994**, *63*, 61–66.
3. Cutler, J.M.; Rains, D.W.; Loomis, R.S. The Importance of Cell Size in the Water Relations of Plants. Physiologia Plantarum **1977**, *40*, 255–260.
4. Cutler, J.M.; Rains, D.W. Effects of Irrigation History on Responses of Cotton to Subsequent Water Stress. Crop Science **1977**, *17*, 329–335.
5. Cutler, J.M.; Rains, D.W. Effects of Water Stress and Hardening on the Internal Water Relations and Osmotic Constituents of Cotton Leaves. Physiologia Plantarum **1978**, *42*, 261–268.
6. Turner, N.C.; Jones, M. Turgor Maintenance by Osmotic Adjustment: A Review and Evaluation. In *Adaptation of Plants to Water and High Temperature Stress*; Turner, N.C., Kramer, P.J., Eds.; Wiley: New York, 1980; 87–103.
7. Jones, M.M.; Osmond, C.; Turner, N.C. Accumulation of Solutes in Leaves of Sorghum and Sunflower in Response to Water Deficits. Australian Journal of Plant Physiology **1980**, *7*, 193–205.
8. Brown, K.W.; Jordan, W.R.; Thomas, J.C. Water Stress Induced Alterations of the Stomatal Response to Decreases in Leaf Water Potential. Physiologia Plantarum **1976**, *37*, 1–5.
9. Morgan, J.M.; Condon, A.G. Water Use, Grain Yield and Osmoregulation in Wheat. Australian Journal of Plant Physiology **1986**, *13*, 523–532.
10. Albouchi, A.; Ghrir, R.; El Aouni, M.H. Drought Hardening, Soluble Carbohydrates and Free Amino Acid Accumulation in *Acacia cyanophylla* Lindl. Phyllodes. Annales des Sciences Forestieres **1997**, *54*, 155–168.
11. Tardieu, F. Drought Perception by Plants. Do Cells of Droughted Plants Experience Water Stress? Plant Growth Regulation **1996**, *20*, 93–104.
12. Spollen, W.G.; Sharp, R.E.; Saab, I.N.; Wu, Y. Regulation of Cell Expansion in Roots and Shoots at Low Water Potentials. In *Water Deficits: Plant Responses from Cell to Community*; Smith, J.A.C., Griffiths, H., Eds.; Bios: Oxford, U.K., 1993; 37–52.
13. Schurr, U.; Gollan, T.; Schulze, E.-D. Stomatal Response to Drying Soil in Relation to Changes in Sap Composition of *Helianthus annuus*. II. Stomatal Sensitivity to Abscisic Acid Imported from the Xylem Sap. Plant, Cell and Environment **1992**, *15*, 561–567.
14. Dure, L. Structural Motifs in Lea Proteins. In *Plant Responses to Cellular Dehydration During Environmental Stress*; Close, T.J., Bray, E.A., Eds.; American Society of Plant Physiologists: Rockville, MD, 1993; 91–103.
15. Bartels, D.; Alexander, R.; Schneider, K.; Elster, R.; Velasco, R.; Alamillo, J.; Bianchi, G.; Nelson, D.; Salamini, F. Desiccation-Related Gene Products Analyzed in a Resurrection Plant and in Barley Embryos. In *Plant Responses to Cellular Dehydration During Environmental Stress*; Close, T.J., Bray, E.A., Eds.; American Society of Plant Physiologists: Rockville, MD, 1993; 119–127.
16. Bohnert, H.J.; Sheveleva, E. Plant Stress Adaptions—Making Metabolism Move. Current Opinion in Plant Biology **1998**, *1*, 267–274.
17. Hanson, A.D. Accumulation of Quaternary Ammonium and Tertiary Sulfunium Compounds. In *Plant Responses to Cellular Dehydration During Environmental Stress*; Close, T.J., Bray, E.A., Eds.; American Society of Plant Physiologists: Rockville, MD, 1993; 30–36.
18. Kavi Kishor, P.B.; Hong, Z.; Miao, G.H.; Hu, C.A.A.; Verma, D.P.S. Over Expression of Δ^1-Pyrroline-5-carboxylate Synthetase Increases Proline Production and Confers Osmotolerance in Transgenic Plants. Plant Physiology **1995**, *108*, 1387–1394.
19. Turner, N.C.; Jarvis, P.J. Photosynthesis in Sitka Spruce (*Picea sitchensis* (Bong.) Carr). IV. Response to Soil Temperature. Journal of Applied Ecology **1975**, *12*, 561–576.

20. Capell, B.; Dörffling, K. Genotype-Specific Differences in Chilling Tolerance of Maize in Relation to Chilling-Induced Changes in Water States and Abscisic Acid Accumulation. Physiologia Plantarum **1993**, *88*, 638–646.

21. Vernieri, P.; Pardossi, A.; Tognoni, F. Influence of Chilling and Drought on Water Relations and Abscisic Acid Accumulation in Bean. Australian Journal of Plant Physiology **1991**, *18*, 25–35.

22. Henckel, P.A. Physiology of Plants Under Drought. Annual Review of Plant Physiology **1964**, *15*, 363–386.

23. Austin, R.B.; Longden, P.C.; Hutchinson, J. Some Effects of "Hardening" Carrot Seed. Annals of Botany **1969**, *33*, 883–895.

24. Lush, W.M.; Groves, R.H.; Kaye, P.E. Presowing Hydration–Dehydration Treatments in Relation to Seed Germination and Early Seedling Growth of Wheat and Ryegrass. Australian Journal of Plant Physiology **1981**, *8*, 409–425.

25. Lush, W.M.; Groves, R.M. Germination, Emergence and Surface Establishment of Wheat and Ryegrass in Response to Natural and Artificial Hydration–Dehydration Cycles. Australian Journal of Agricultural Research **1981**, *32*, 731–739.

26. Bewley, J. D Physiological Aspects of Desiccation Toleration. Annual Review of Plant Physiology **1979**, *30*, 195–238.

27. Andoh, H.; Kobata, T. Does Wetting and Redrying the Seed Before Sowing Improve Rice Germination and Emergence Under Low Soil Moisture Conditions? Plant Production Science **2000**, *3*, 161–163.

Drought Management

Michael J. Hayes
Donald A. Wilhite
National Drought Mitigation Center and University of Nebraska, Lincoln, Nebraska, U.S.A.

INTRODUCTION

Some would argue that drought cannot be "managed." Yes, it is true droughts are a normal part of climate for virtually all areas of the world (e.g., Fig. 1), and that droughts affect more people worldwide than any other natural hazard.[1] It is also true that officials from both developing and developed nations struggle to deal with the wide range of economic, environmental, and social impacts related to droughts. However, these officials are not powerless to reduce the impacts of drought. Rather, there are important management actions that officials at local, regional, and national levels can take to reduce the impacts from droughts. The approach taken to address drought impacts and reduce their effects is called drought management, or perhaps more appropriately, drought risk management. The long-term goal is to reduce the impacts of drought through the adoption of drought preparedness plans.

SHIFTING THE EMPHASIS FROM CRISIS TO DROUGHT RISK MANAGEMENT

Traditionally, droughts have been viewed as unusual occurrences that creep up on officials who are typically unprepared to deal with the impacts droughts create. This is why drought has been called the "creeping phenomenon."[1] In reality, drought is a normal feature for virtually all climates. Officials often react to the occurrence of drought through "crisis management." After a drought is over, officials turn their attention to the next crisis, and any lessons learned about responding to the drought are most likely lost and forgotten. This crisis management approach is illustrated in the "Hydro-Illogical Cycle" (Fig. 2). Crisis management approaches to dealing with droughts are reactive, poorly coordinated and targeted, untimely, and generally too late. As a result, they are largely ineffective.

In order to break the Hydro-Illogical Cycle, officials around the world at local, regional, and national scales need to adopt a drought risk management approach. Drought risk management involves taking actions before droughts occur in order to reduce the drought impacts. It has three main

components: 1) a comprehensive drought monitoring and early warning system; 2) planning and building the institutional capacity to respond to droughts; and 3) identification and implementation of mitigation actions and policies that can be taken before the next drought. These components will be discussed in greater detail.

A comprehensive drought monitoring and early warning system is a critical component of drought risk management because effective, timely decisions related to droughts can only be made if officials have an accurate assessment of the potential or developing drought event. This early warning system must incorporate all of the critical components of the hydrologic system (e.g., precipitation, streamflow, groundwater, snowpack, soil moisture, and reservoir and lake levels) because drought severity cannot be defined by precipitation deficiencies alone. A comprehensive system will assist officials by providing appropriate "triggers" for actions that the officials need to take, or by identifying when particular impacts are going to occur. An effective drought monitoring and early warning system requires synthesis and analysis of timely data and an efficient dissemination system to communicate this information (e.g., the media, extension services, or the World Wide Web).

Drought planning is a very important component of drought risk management because it establishes and preserves the institutional capacity with which officials can respond to droughts and reduce drought impacts. There are many benefits of a drought plan. A drought plan serves as the organizational framework for dealing with droughts and improving the coordination between and within levels of government. In addition, drought plans enable proactive mitigation and response to droughts; enhance early warning through integrated monitoring efforts; involve stakeholders, which are necessary for successful programs; identify areas, groups, and sectors particularly at risk; improve information dissemination by outlining the information delivery systems and strategies; and build public awareness of the need for improved drought and water management.

Several methodologies exist for assisting officials with the development of drought plans. One of these methodologies, described by Wilhite et al.,[2] is a 10-step

Encyclopedia of Water Science
DOI: 10.1081/E-EWS 120010110

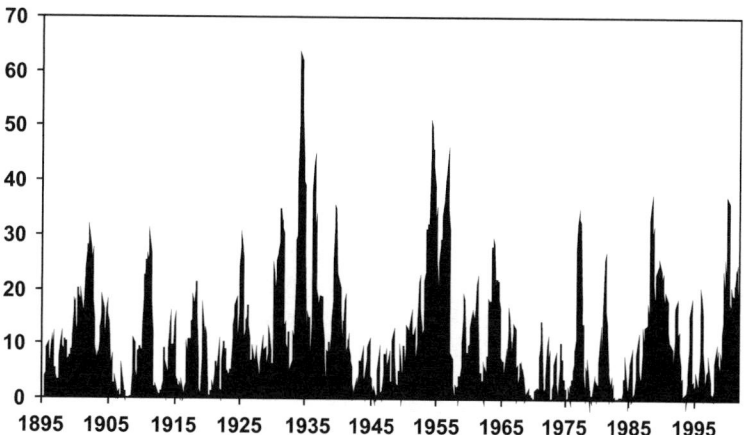

Fig. 1 The percent area of the United States in severe to extreme drought by month from 1895 through March 2002. Similar periodic patterns appear on graphs depicting regional hydrological basins in the United States, and would likely appear for most regions in the world. (From: National Climatic Data Center, Asheville, North Carolina, U.S.A.)

drought planning process that targets drought planners in the United States and elsewhere (http://drought.unl.edu/center/pdfpubs/10step.pdf) (accessed April 2002). The process was designed to be generic and adaptable because it is important for planners to develop a plan appropriate for their regional and governmental structures. These plans must be dynamic, reflecting changing government policies, technologies, personnel, and natural resources management practices.

The third important component of drought risk management is mitigation. Mitigation is defined as

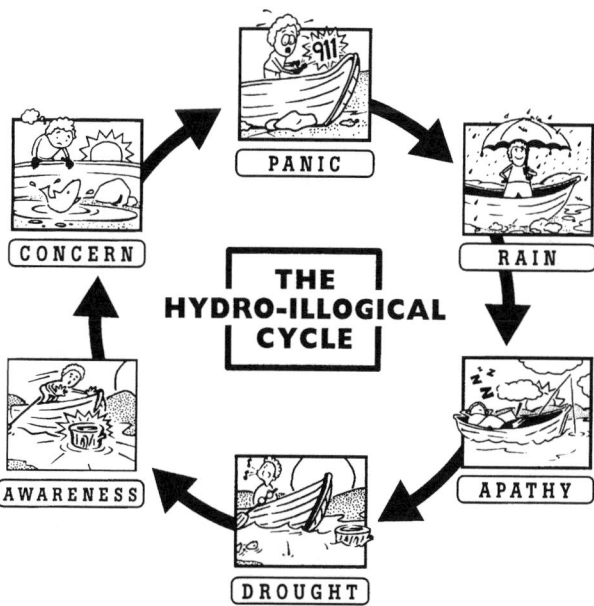

Fig. 2 The Hydro-Illogical Cycle. (Source and copyright: National Drought Mitigation Center, University of Nebraska, Lincoln, Nebraska, U.S.A.)

the policies and actions taken before a drought that will reduce drought impacts. Sometimes, if officials are alert enough and can see the development of drought in its early stages, mitigation can take place during the drought's early stages and may be very effective in reducing impacts as the drought becomes more severe. Otherwise, actions taken during a drought are generally responses directly related to the drought's severity and impacts. These responses are important, of course, and need to be well documented ahead of time within a drought plan. But it is important to keep in mind that mitigation is most effective if it takes place during times when drought is not occurring and officials are not responding to drought during a crisis. Mitigation actions should address vulnerabilities associated with drought with the goal of reducing impacts in future events.

What are some examples of drought mitigation? Certainly the development of a comprehensive drought monitoring and early warning system and the development of a drought plan, as described above, are two examples of mitigation. Both of these actions should be taken before a region is experiencing drought. Other broad categories for potential drought mitigation actions include revising or developing legislation or public policies related to drought and water supplies; water supply augmentation and the development of new supplies; demand reduction and the development of water conservation programs; public education and awareness programs; specific priorities for water allocations; and water use conflict resolution.[3]

As with drought planning, there are several methodologies for identifying the appropriate mitigation actions to take in a region. In 1998, as part of the activities of the Western Drought Coordination Council (WDCC), a methodology was developed to look at drought risk. An important part of this methodology was the identification of mitigation actions and how these actions would be implemented.[4] The methodology also involves identifying and understanding the people and sectors that are vulnerable to droughts and why, allowing officials to target their mitigation efforts more effectively.

DROUGHT RISK CHALLENGES AND OPPORTUNITIES

Serious challenges still remain in drought risk management. One of these challenges is the acceptance of drought as a natural hazard, and a hazard that needs to be prepared for. Fragmented resource management and numerous federal programs present challenges, as do the declining financial and human resources. Confusion over the difference between mitigation and response is a challenge, and many times officials have a difficult time identifying innovative mitigation actions and implementing new policies and programs. Stakeholder involvement and acceptance still needs improvement. Perhaps one of the biggest challenges is to maintain the momentum for risk management in a changing political climate.

Some progress toward drought risk management is being made around the world. Australia and New Zealand, for example, have had national drought policies and strategies to reduce drought impacts.[5,6] Other nations are looking at establishing national drought policies. A global drought preparedness network is in the development stages; this network would assist nations by promoting drought risk management and sharing lessons learned about drought monitoring, planning, and mitigation. The network, based at the National Drought Mitigation Center/International Drought Information Center at the University of Nebraska, would be made up of regional networks coordinated by institutions around the world. Collectively this network of regional networks may enhance the drought management capability of many nations.

In the United States, three states had drought plans in 1982. As of 2002, 33 states have drought plans, and 6 of those states incorporate mitigation actions into their plans. In 1998, New Mexico became the first state to develop drought plan that emphasizes mitigation. Five states are currently in the process of developing drought plans, and it is hoped that mitigation will be a major component of each of these new plans. A number of Native American nations in the southwestern United States have developed drought mitigation plans recently as well. In addition, improved coordination has occurred within federal agencies and between federal and state governments.

New drought monitoring efforts and products have been developed in recent years. One of the best examples of progress in this area is the Drought Monitor product, developed to assess current drought conditions in the United States. The first Drought Monitor map was issued in August 1999, and a weekly update is posted every Thursday morning (http://drought.unl.edu/dm/) (accessed April 2002). The unique feature of this product is that four agencies rotate creating the map: the National Drought Mitigation Center, the United States Department of Agriculture, the Climate Prediction Center, and National Climatic Data Center of the National Oceanic and Atmospheric Administration. In addition, a feedback network of more than 160 local experts provides input about the map's portrayal of drought conditions before the map is released each week.

CONCLUSION

Clearly, there is reason for optimism about drought risk management and reducing drought impacts in the future. But it is also clear that officials around the world need to take proactive steps to develop comprehensive and integrated drought monitoring and early warning systems, determine who and what is at risk to droughts and why, and create drought mitigation plans with specific actions that address these risks with the goal of reducing the impacts of future drought events. There is a growing recognition that drought risk management is a critical ingredient of sustainable development planning and must be addressed systematically through risk-based policies and plans.

REFERENCES

1. Wilhite, D.A. Drought as a Natural Hazard: Concepts and Definitions. In *Drought: A Global Assessment*; Wilhite, D.A., Ed.; Routledge: New York, 2000; Vol. 1, 1–18.
2. Wilhite, D.A.; Hayes, M.J.; Knutson, C.; Smith, K.H. Planning for Drought: Moving from Crisis to Risk Management. J. Am. Water Res. Assoc. **2000**, *36* (4), 697–710.
3. Wilhite, D.A. State Actions to Mitigate Drought: Lessons Learned. In *Drought: A Global Assessment*; Wilhite, D.A., Ed.; Routledge: New York, 2000; Vol. 2, 149–157.
4. Knutson, C.; Hayes, M.J.; Phillips, T. How to Reduce Drought Risk, Preparedness and Mitigation Working Group of the Western Drought Coordination Council: Lincoln, NE; 1998; 1–43, http://drought.unl.edu/handbook/risk.pdf (accessed April 2002).
5. O'Meagher, B.; Smith, M.S.; White, D.H. Approaches to Integrated Drought Risk Management: Australia's National Drought Policy. In *Drought: A Global Assessment*; Wilhite, D.A., Ed.; Routledge: New York, 2000; Vol. 2, 115–128.
6. Haylock, H.J.K.; Ericksen, N.J. From State Dependency to Self-Reliance: Agricultural Drought Policies and Practices in New Zealand. In *Drought: A Global Assessment*; Wilhite, D.A., Ed.; Routledge: New York, 2000; Vol. 2, 105–114.

Drought Resistance

M. B. Kirkham
Kansas State University, Manhattan, Kansas, U.S.A.

D

INTRODUCTION

Agricultural drought is defined as "a climatic excursion involving a shortage of precipitation sufficient to adversely affect crop production or range productivity."[1, p. 2] For centuries, plants have been classified based on their response to drought. Theophrastus, the Greek philosopher and botanist (371/370–288/287 BC), divided plants into groups according to their need for water. He said, "For there are some plants which cannot live except in wet; and again these are distinguished from one another by their fondness for different kinds of wetness … Others … seek out dry places."[2, p. 31–33] In the early 1900s, researchers tried to define "drought resistance" exactly. Maximov[3] followed the definition of Kearney and Shantz[4] and Shantz,[5] and defined drought-resisting plants as those that "resist drought by storing up a supply of water in their fleshy bodies, to be used when none can be obtained from the soil … To this type belong succulents, such as cacti and *Agave*, and many epiphytes. Plants of a nonsucculent type, but with large water reservoirs in their stems or in their underground organs, e.g., many trees of the African grasslands, which spring into bloom before the rains, are also included by Shantz in this group."[3, p. 309] Levitt's[6] definition of drought-resisting plants is widely taught today. He divided them into two groups: drought-avoiding and drought-tolerating plants.[6, p. 355] Drought avoidance can be achieved through restriction of water loss or by expansion of the root system to reach a greater supply of water.[7, p. 3] Tolerance is the ability of an organism to perform well, or survive, despite the existence of a stressed condition within its tissues.[7, p. 35] The distinction between drought-avoiding and drought-tolerant plants is not always clear, and Levitt[6, p. 418] added groups such as "tolerant avoiders" to cover "more complicated" situations.

NEED FOR QUANTITATIVE DEFINITION

These definitions are not quantitative. Since the work of Philip,[8] who pioneered the concept of the soil, plant, and

atmosphere as a thermodynamic continuum for water transfer, we know that water moves in soil and plants along a potential energy gradient. The potential energy can be measured and compared between drought-resistant and drought-sensitive plants. Such measurements, because they are quantitative, can be replicated by others, negating the need for vague terminology.

DEFINITION OF WATER POTENTIAL AND ITS COMPONENTS

Under equilibrium conditions, the state of water at a particular point in a plant or in soil can be written in terms of the various components of the potential energy, as follows:[9]

$$\psi = \psi_s + \psi_p + \psi_m + \psi_g, \tag{1}$$

where ψ is the water potential, ψ_s the osmotic (solute)-potential component, ψ_p the pressure (turgor)-potential component, ψ_m the matric component due to capillary or adsorption forces such as those in the cell wall, and ψ_g is the component due to gravity. For plants, the matric potential and the gravitational potential usually are neglected, and Eq. 1 reduces to

$$\psi = \psi_s + \psi_p. \tag{2}$$

We now have a relationship that can define the state of water at any point in a plant and that can be compared and provide a value that can be compared with values for other plants. Descriptive terminology, like drought tolerance and drought avoidance, is no longer necessary because we can quantify drought resistance by measuring water potential and its components.

MEASUREMENT OF WATER POTENTIAL AND ITS COMPONENTS

The most accurate way to measure water potential is by using thermocouple psychrometers, because they measure relative humidity in soil or plants, from which water potential energy can be related by using the Kelvin

Contribution No. 01-235-B from the Kansas Agricultural Experiment Station.

Encyclopedia of Water Science
DOI: 10.1081/E-EWS 120010162

equation:[10]

$$\psi = (RT/V_w^o)\ln(e/e_o), \qquad (3)$$

where ψ is water potential, R the ideal gas constant, T the absolute temperature, V_w^o the molar volume of pure water, e the partial pressure of water vapor in air, e_o the saturated vapor pressure, and e/e_o is the relative humidity. Thermocouple psychrometers were not used routinely to measure plant water potential until 1960s, because microvoltmeters were not on the market until then. They enabled measurement of the small voltages necessary in the technique. Osmotic pressure can be measured using thermocouple psychrometers after breaking cell membranes, for example, by freezing. Turgor potential is obtained by subtracting osmotic potential from water potential. However, pressure chambers are used more frequently to measure water potential, because they are easier to use and do not require careful temperature

control. Boyer[11] showed that the amount of pressure necessary to force water out of the leaf cells into the xylem tissue is a function of the water potential of the leaf cells.

DROUGHT RESISTANCE OF C_3 AND C_4 PLANTS

Crops vary in drought resistance. In particular, plants with C_4 type of photosynthesis have a lower transpiration ratio (250–350 g H_2O/g dry weight) than those with C_3 type of photosynthesis (450–950 g H_2O/g dry weight).[12, p. 165] That is, C_4 plants use less water to produce a certain amount of dry matter or grain than C_3 plants. The difference in water requirements of plants was noted almost 100 yr ago by Briggs and Shantz,[13] even though the photosynthetic pathways had not been discovered then.

In a study of six row crops grown in Kansas (corn, *Zea mays* L.; millet, *Pennisetum americanum* L.; sorghum, *Sorghum bicolor* (L.) Moench; pinto bean, *Phaseolus vulgaris* L.; soybean, *Glycine max* (L.) Merr.; and sunflower, *Helianthus annuus* L.), sunflower had the highest evapotranspiration, and sorghum generally had the lowest water use.[14] Despite sunflower's high use of water, it is considered a drought-resistant crop because it has a deep root system that can use water at depths that are unavailable to crops like sorghum.[15] However, in comparison to sorghum, sunflower has: 1) a higher transpiration rate; 2) a lower stomatal resistance; and 3) a lower hydraulic resistance, all of which are nonwater-conserving characteristics.

Even though elevated levels of carbon dioxide close stomata (it is an excellent antitranspirant), studies show that high concentrations of carbon dioxide in the air, which occur now and are predicted to get even higher, do not save water used by plants. Leaves become larger with higher amounts of carbon dioxide (because of a higher rate of photosynthesis for growth), so they have a greater number of stomata through which water is lost. But water-use efficiency increases with elevated carbon dioxide.[16] That is, less water is needed to produce a certain amount of grain with elevated carbon dioxide than with an ambient level of carbon dioxide. For example, elevated carbon dioxide (about two times the ambient concentration) reduced the water requirement (reciprocal of water-use efficiency) of C_4 plants (big bluestem; *Andropogon gerardii* Vitman) by about 35%[17] and reduced the water requirement of C_3 plants (winter wheat; *Triticum aestivum* L.) by about 30%[16] under both well-watered and dry conditions. The increased water-use efficiency is of great importance in a semi-arid region. Research also has shown that augmented levels of carbon dioxide compensate for reductions in growth by drought both in a C_3 species (winter wheat)[16] and in a C_4 species (big bluestem).[17]

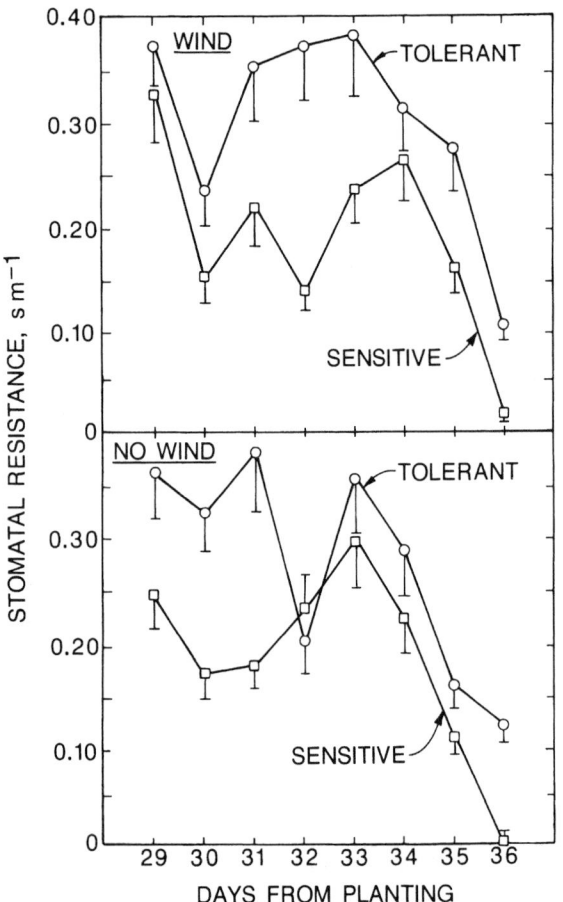

Fig. 1 Stomatal resistance of a drought-resistant (tolerant, 'KanKing') and drought-sensitive ('Ponca') cultivar of winter wheat grown in wind or still air. Vertical lines show standard deviations. Only half the bar has been shown to avoid cluttering the figure. (Adapted from Ref. [18], with permission.)

D

DROUGHT RESISTANCE OF DIFFERENT CULTIVARS

Not only do different crops vary in drought resistance, but also cultivars (cultivated varieties) of the same crop vary. Intensive work comparing two cultivars of winter wheat—one known to be drought resistant ('KanKing') and the other known to be drought sensitive ('Ponca')—has shown that the drought-resistant cultivar has a higher stomatal resistance (Fig. 1),[18] usually has a lower water potential and turgor potential (Fig. 2),[18] is more efficient in utilization of mineral elements in the soil, more salt tolerant, has a different hormonal regulation (e.g., is insensitive to abscisic acid), has a higher hydraulic resistance, and is better able to extract water from drying soil than the drought-sensitive cultivar. Studies with drought-resistant cultivars of corn and sorghum have substantiated those results with winter wheat. A drought-resistant genotype of sorghum ('IA 28') produced more ethylene, a gaseous hormone, than a drought-sensitive genotype ('Redlan') (Fig. 3).[19]

GROWTH OF DROUGHT-RESISTANT PLANTS

Drought-resistant varieties usually do not grow as well and yield as much as drought-sensitive varieties under well-watered conditions (Fig. 4), because the stomata of the drought-resistant varieties are more closed (Fig. 1). If less carbon dioxide is taken up, then growth is reduced. Stomatal conductance is related directly to growth.[20] To achieve a high yield under optimal conditions, a variety with a high stomatal conductance (a drought-sensitive cultivar) should be planted.

Planting a drought-resistant and a drought-sensitive variety of a crop together might be advantageous in a sustainable farming system to assure some yield under

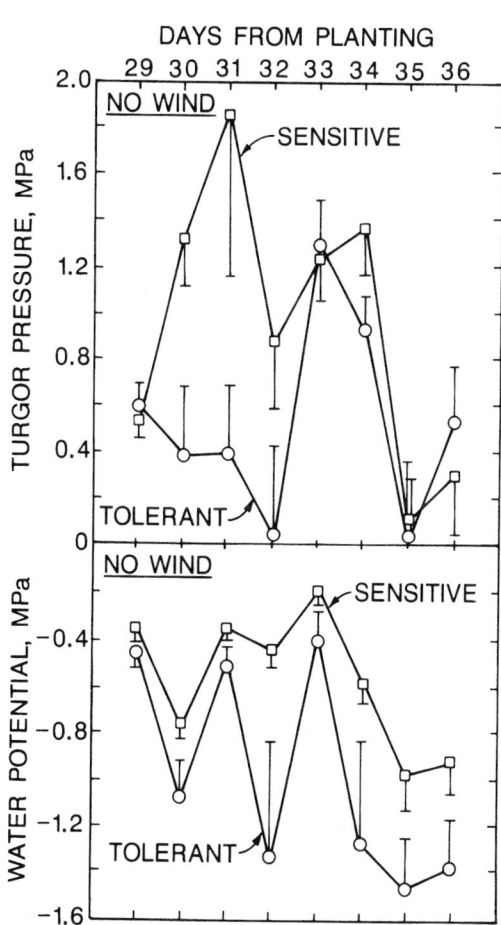

Fig. 2 Water potential and turgor potential (pressure) of a drought-resistant (tolerant, 'KanKing') and drought-sensitive ('Ponca') cultivar of winter wheat grown in still air. For vertical lines, see legend to Fig. 1. (Adapted from Ref. [18], with permission.)

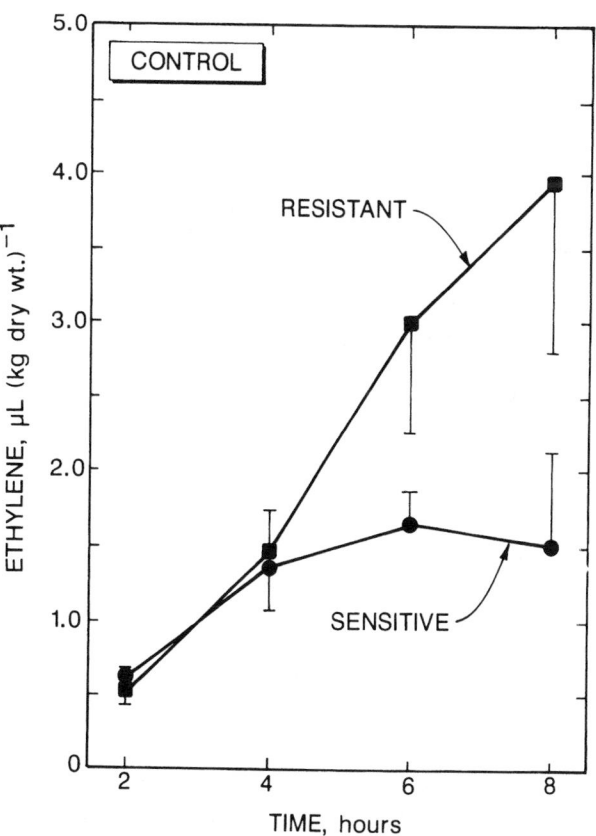

Fig. 3 Ethylene production by drought-resistant ('IA 28') and drought-sensitive ('Redlan') genotypes of sorghum. Plants were grown in sand culture and watered with nutrient solution. For vertical lines, see legend of Fig. 1. (Adapted from Ref. [19], with permission.)

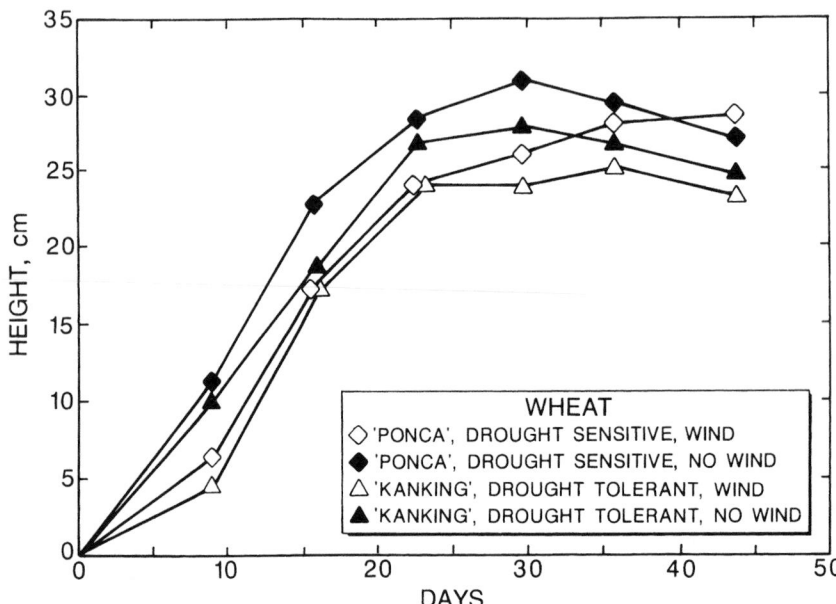

Fig. 4 Height of a drought-resistant ('KanKing') and a drought-sensitive ('Ponca') cultivar of winter wheat grown in wind or still air. Same study as shown in Figs. 1 and 2. (From author's files; never published.)

drought. In dry years, the drought-resistant variety should survive and produce some grain. In wet years, the drought-sensitive variety should yield well. However, when a drought-sensitive (Ponca) and a drought-resistant (KanKing) cultivar of winter wheat were grown together, the drought-sensitive cultivar used up water, causing the drought-resistant cultivar to die. When the drought-resistant cultivar was grown alone, it was able to survive drought.[21]

REFERENCES

1. Rosenberg, N.J. Introduction. In *Drought in the Great Plains. Research on Impacts and Strategies*; Rosenberg, N.J., Ed.; Water Resources Pub.: Littleton, CO, 1980; 1–17.

2. Theophrastus. *Enquiry into Plants and Minor Works on Odours and Weather Signs*, with an English Translation by Sir Arthur Hort; Harvard University Press: Cambridge, MA, 1948; Vol. I, 1–475.

3. Maximov, N.A. *The Plant in Relation to Water. A Study of the Physiological Basis of Drought Resistance*; George Allen and Unwin, Ltd.: London, 1929; 1–451.

4. Kearney, T.H.; Shantz, H.L. The Water Economy of Dry-Land Crops. *Year Book U.S. Dep. Agr.* U.S. Dep. Agr.: Washington, DC, 1911; 351–362.

5. Shantz, H. Drought Resistance and Soil Moisture. Ecology **1927**, *8*, 145–157.

6. Levitt, J. *Responses of Plants to Environmental Stresses*; Academic Press: New York, 1972; 1–697 (see esp. 353–424).

7. Barnes, R.F.; Beard, J.B., (Eds.) *Glossary of Crop Science Terms*; Crop Science Society of America: Madison, WI, 1992; 1–88.

8. Philip, J.R. Plant Water Relations: Some Physical Aspects. Annu. Rev. Plant Physiol. **1966**, *17*, 245–268.

9. Kirkham, M.B. Soil–Water Relationships. In *Encyclopedia of Agricultural Science*; Arntzen, C.J., Ritter, E.M., Eds.; Academic Press: San Diego, 1994; Vol. 4, 151–168.

10. Rawlins, S.L. Theory of Thermocouple Psychrometers for Measuring Plant and Soil Water Potential. In *Psychrometry in Water Relations Research*; Brown, R.W., van Haveren, B.P., Eds.; Utah Agr. Exp. Sta.: Logan, UT, 1972; 43–50.

11. Boyer, J.S. Leaf Water Potentials Measured with a Pressure Chamber. Plant Physiol. **1967**, *42*, 133–137.

12. Salisbury, F.B.; Ross, C.W. *Plant Physiology*, 2nd Ed.; Wadsworth Pub. Co., Inc.: Belmont, CA, 1978; 1–436.

13. Briggs, L.J.; Shantz, H.L. The Water Requirement of Plants. I. Investigations in the Great Plains in 1910 and 1911. U.S. Dep. Agr., Bur. Plant Ind. Bull. No. 284. 1913, 1–49.

14. Kirkham, M.B.; Redelfs, M.S.; Stone, L.R.; Kanemasu, E.T. Comparison of Water Status and Evapotranspiration of Six Row Crops. Field Crops Res. **1985**, *10*, 257–268.

15. Rachidi, F.; Kirkham, M.B.; Stone, L.R.; Kanemasu, E.T. Soil Water Depletion by Sunflower and Sorghum Under Rainfed Conditions. Agr. Water Manag. **1993**, *24*, 49–62.

16. Chaudhuri, U.N.; Kirkham, M.B.; Kanemasu, E.T. Carbon Dioxide and Water Level Effects on Yield and Water Use of Winter Wheat. Agron. J. **1990**, *82*, 637–641.

17. Kirkham, M.B.; He, H.; Bolger, T.P.; Lawlor, D.J.; Kanemasu, E.T. Leaf Photosynthesis and Water Use of Big Bluestem Under Elevated Carbon Dioxide. Crop Sci. **1991**, *31*, 1589–1594.

18. Kirkham, M.B. Water Potential and Turgor Pressure as a Selection Basis for Wind-Grown Winter Wheat. Agr. Water Manage. **1978**, *1*, 343–349.

19. Zhang, J.; Kirkham, M.B. Ethylene Production by Two Genotypes of Sorghum Varying in Drought Resistance. Cereal Res. Commun. **1991**, *19*, 357–360.

20. Gardner, W.R. Internal Water Status and Plant Response in Relation to the External Water Régime. In *Plant Response to Climatic Factors*; Slatyer, R.O., Ed.; U.N. Educ. Sci. Cult. Organ.: Paris, 1973; 221–225.

21. Kirkham, M.B. Growth and Water Relations of Two Wheat Cultivars Grown Separately and Together. Biol. Agr. Hort. **1989**, *6*, 35–46.

Dryland Cropping Systems

William A. Payne
Texas A&M University Agricultural Research and Extension Center, Amarillo, Texas, U.S.A.

INTRODUCTION

Agriculture involves the management of land, water, energy, labor, and other resources by humans for the purpose of producing food and fiber. An *agricultural system* refers to the regional classification of a particular type of *farming system,* which is a combination of crops, animals, and management practices. The fundamental social, ecological, and economic unit of a farming system is the *farm*, which is of course managed by farmers.

Farmers manage their pastures and fields according to a *cropping system*, which is characterized by specific management practices, e.g., for soil tillage, rotation of field crops within a field from one season to the next, or maintenance of soil fertility. In developed countries, farmers tend to manage cropping systems with a view towards generating profit and, to a certain extent, maintaining a lifestyle. In developing countries, the majority of farmers tend to be small landholders who manage cropping systems more to avoid risk of crop failure and hunger. There are of course exceptions to these general tendencies. Relationships among agricultural systems, farming systems, cropping systems, and farms are explored in greater detail by Loomis and Connor[1]

In *dryland cropping systems*, crops depend upon rainfall for water supply rather than upon irrigation. Dryland cropping systems may therefore be viewed as a subset of rainfed systems. They are commonly found in semi-arid environments, where precipitation tends to be low and erratic, and other environmental stresses, such as high temperature, are common. In dryland cropping systems, water supply is usually the factor that limits crop production most. For this reason, whether for profit or subsistence, farmers must use water supply efficiently for crop production. Some illustrative dryland cropping systems of the world are summarized in Tables 1–3. Pearson et al.[2] give a more detailed discussion of dryland cropping systems of the world.

FUTURE PERSPECTIVE

In the coming decades, dryland cropping systems will play an increasingly important role in maintaining global food security, because of dwindling land and especially water resources.[3] It follows that the challenge of meeting growing food demand, while protecting environmentally sensitive lands from agricultural expansion, will fall increasingly upon dryland cropping systems. Efficient water use in dryland cropping systems should therefore be of interest to society as a whole, and not merely to the farmer.

EFFICIENT WATER USE AND CROP PRODUCTION

The amount of water required to produce a crop has always interested farmers, and has occupied scientists for much of the last century. However, as Tanner and Sinclair[4] pointed out, the description of this relationship with such terms as "efficient water use" and "water-use efficiency," (WUE) can be ambiguous. There exist different perspectives on what constitutes yield (e.g., marketable yield, total biomass, above-ground biomass, or photosynthate) and "used" water. Because of these different perspectives, there are different definitions of efficiency. Indeed, it has been held that, strictly speaking, the ratio of growth to water use is not an efficiency at all, since it lacks a theoretical maximum value.

To the farmer managing an irrigated cropping system, any measure that reduces the amount of water for which he has to pay has achieved increased WUE. However, an irrigation engineer might define WUE as the ratio of yield increase from irrigation to the amount of water applied. An agronomist might define it as the ratio of crop yield to the sum of water transpired by the plant plus that which evaporates from the soil surface (termed "evapotranspiration" or "total evaporation"). To a plant scientist, it may refer to the ratio of yield to plant transpiration over part or all of the growth cycle, whereas a more basic physiologist might define it in terms of diffusion of CO_2 and H_2O molecules during a period of a few seconds.

Here, we adopt the convention used by Tanner and Sinclair.[4] Any management practice that conserves soil water to increase crop production, such as runoff capture or storage of rainfall for eventual crop use, constitutes an

Encyclopedia of Water Science
DOI: 10.1081/E-EWS 120010047

Table 1 Major crop components of dryland cropping systems of Africa, Southern Asia and Australia, and eastern Oregon and Washington (U.S.)

Location	Crops
West Africa	Millet/sorghum, maize, groundnuts, cowpea, sesame, cassava, yams, and tree legumes
East Africa	Maize/barley, sorghum, millet, and teff (Ethiopia)
Southern Asia	Sorghum/millet (India), maize/rice (other); cassava; kenaf, wheat, groundnut, soybean, and chickpea
S. Australia, East WA, and OR (U.S.)	Wheat/barley; lupins, peas, mustard, and improved pastures

(Adapted from Ref. [2].)

efficient use of water. On the other hand, WUE is an index of crop or individual plant performance. The ratio of yield (Y) or biomass (DM) production to evapotranspiration (ET), or WUE_{ET}, may be viewed as an index of field performance of the crop with regard to water use, whereas the ratio of Y or DM to transpiration (T), or WUE_T, can be viewed as an index of plant performance. However, this is a simplified discussion. Readers are referred to Tanner and Sinclair[4] or a recent group of papers edited by Payne[5] for a more advanced presentation.

It is important to keep in mind that efficient water use and WUE are concepts mostly used by scientists. Most farmers have more practical goals, such as yield stability and sustainability, because these facilitate economic planning. However, there are many perspectives on what constitutes stability and sustainability.[6]

WATER USE AND CROP YIELD

Plants grow by using solar energy intercepted by leaves to fix atmospheric CO_2 as part of the process of photosynthesis. When CO_2 diffuses into the leaf through open stomata to be fixed by specialized enzymes, there is a simultaneous export of water vapor (transpiration) in response to a concentration gradient of H_2O, caused by high humidity within the leaf and lower humidity in the atmosphere. Because of this simultaneous import of CO_2 and export of H_2O, crop growth, biomass production, and yield are roughly proportional to transpiration.

A plant's environment affects WUE_T for a number of reasons. Atmospheric humidity, e.g., determines the size of the concentration gradient that drives H_2O export from the leaf. If atmospheric humidity decreases, then the

Table 2 Distribution of major dryland cropping systems in Mediterranean countries

Country	Main crop rotations
Italy	Cereal–hay crops–cereal
	Fallow–cereal–cereal–fallow (has grazing value)
	Cereal–tobacco, sugar beet, grain legumes
Greece	70% cereal–cereal
	10% cereal–hay crops
	2% cereal–grain legumes
	18% with other alternatives (cotton, sugar beet, tobacco)
Algeria	80% fallow–cereal (fallow has grazing value)
	15% cereal–hay crops
	5% cereal–grain legumes
Morocco:	
low rainfall zones	25% fallow (has grazing value)
	75% continuous cereal
High rainfall zones	15% continuous cereal
	70% grain legumes
	15% fallow (grazed)
Lebanon	Little remaining fallow. Where it exists it has grazing value except in low rainfall zone
	Rotations are wheat–barley–wheat–alfalfa and wheat–lentils

(Adapted from Ref. [2].)

Table 3 Change of dryland cropping systems with rainfall amount, growing season length, soil type, and local preferences in India

Environment	Intercrop system	Sequential system
Jodhpur 380 mm rainfall, 11 week growing season, Cambisol soil	Green gram or cluster bean grown with pearl millet	Pearl millet followed by fallow
Hisar 400 mm rainfall, 13 week growing season, and Cambisol soil	Pearl millet/mung bean or Pearl millet/cowpea (for animal fodder)	Pearl millet followed by chickpea or Mung bean followed by mustard
Hyderabad 770 mm rainfall, 25 week growing season, and deep vertisol soil	Sorghum/pigeonpea	Sorghum followed by safflower, sorghum followed by chickpea, or maize followed by chickpea
Bangalore 890 mm rainfall, 32 weeks growing season, and deep luvisol soil	Finger millet/soybean, groundnut/pigeonpea, or finger millet/maize	Cowpea followed by finger millet

(Adapted from Ref. [2].)

gradient increases, causing the plant to expend more water to fix the same amount of carbon. It thereby also affects the ratio between biomass production and transpiration. Other environmental factors, such as nutrient and water availability, can also affect WUE_T.

Importantly, WUE_T is also under genetic influence. To a large extent, this is governed by the photosynthetic pathway of the particular crop species; "C_3" species (e.g., wheat, beans, or rice) generally have lower WUE_T than "C_4" species (e.g., maize or sorghum), particularly in warm climates. However, within species, there is also considerable genetic variability for WUE_T.[7] There is, therefore, scope for modest increases in WUE_T through modern plant breeding methods.

Because of the same proportionality between biomass accumulation, or growth, and T, farmers and agriculturalists increase yield by increasing T. Under irrigated conditions, T is increased because of greater total amount of water available to the cropping system, which translates into greater yield. Under dryland conditions, however, the total amount of water available to the cropping system is limited by precipitation. Increasing T, therefore, requires that as much precipitation as possible becomes available for use by the crop. An understanding of how this is done requires a basic understanding of the soil water balance.

THE SOIL WATER BALANCE

When precipitation falls upon the soil surface, it remains there ponded, evaporates, runs off, or infiltrates into the soil. Because soils are porous, water can be stored and transmitted within a given volume of soil, much as it can within a sponge. Water is extracted from soil pores by plant roots, and then conducted through the stems to leaves through specialized conducting tissue. As described in the previous section, water is then transmitted via transpira-

tion into the atmosphere through leaf stomata. Water moves through the soil, plant, and atmosphere continuum in response to energy gradients, much as it runs down hill in response to potential energy gradients imposed by gravity. This thermodynamic process is described in more detail by Nobel.[8]

To understand how precipitation can be managed such that transpiration is maximal, we use the soil water balance, which is a restatement of the fundamental principle of conservation of mass. Any change in the amount of water stored within a specified soil volume (usually the crop root zone) must be equal to the difference between any inputs and outputs. That is,

$$\Delta S = \text{Inputs} - \text{Outputs}, \tag{1}$$

where ΔS represents the change in the amount of water storage (S) in the root zone of the plant. Hillel[9] gives a much more thorough description of the soil water balance and related subjects.

Water inputs and outputs of a dryland agricultural field are generally restricted to precipitation (P) and run-on (R_{on}), while outputs include drainage from the root zone (D), evaporation from the soil surface (E), run-off (R_{off}) and plant transpiration (T). We can substitute these into Eq. 1 to get

$$\Delta S = (P + R_{on}) - (E + T + D + R_{off}) \tag{2}$$

Recalling that yield is proportional to T, we rearrange Eq. 2 to view the variables representing processes that determine T, and therefore yield:

$$T = (P + R_{on}) - \Delta S - (E + D + R_{off}) \tag{3}$$

The degree to which the terms on the right side of Eq. 3 can be managed varies with cropping system features, and depends upon such factors as soil physical and chemical properties, slope, weather patterns, water table depth, landscape position, crops grown, and availability of machinery and other inputs.

Four common means of managing the soil water balance terms to maximize T include 1) use of appropriate crops and crop sequence; 2) addition of soil amendments or other inputs; 3) soil surface management; and 4) water harvesting. Volumes have been written on each of these four subjects. A general review of the third topic can be found in Chan's review article, and of the fourth topic in the chapter by Frasier and Tanaka in this encyclopaedia. Here, therefore, an overview of the first two is given. More detailed discussions of all four topics are available from Hillel,[9] Unger and Stewart,[10] and Loomis and Connor.[1]

APPROPRIATE CROP AND CROP SEQUENCE

The daily water demand of a crop varies with its size and growth stage. As plants grow, so do the size and number of their leaves, which constitute assimilatory (for CO_2) and evaporative (for H_2O) surfaces. Generally speaking, the larger the total leaf area of a crop, the greater its water demand, until an approximately constant ratio of leaf area to water use (T or ET) is reached. For many crops, this constant ratio is reached at leaf area indices (LAI, i.e., leaf area divided by land area) greater than 3, but this is crop- and site-specific. Many dryland crops never reach an LAI value of 3, particularly where soil nutrient status is poor. For most agricultural plants, the largest LAI occurs near flowering which, for most crops, is also the growth stage during which yield is most sensitive to drought or high temperature.

Crop daily water demand also depends on the evaporative demand of the atmosphere, which is determined by temperature, humidity, windspeed, and solar radiation. The magnitude and annual patterns of evaporative demand change from region to region. Furthermore, the evaporative demand to which the crop canopy is subjected can be modified by canopy and soil properties, because these affect how radiation is intercepted and windspeed momentum is transferred, which in turn affect temperature and humidity profiles.

A fundamental strategy of efficient water use in dryland cropping systems is to match the pattern of crop water demand to that of soil water storage, or S of Eq. 6. In many environments, this simply means that plant growth and water demand should approximately match rainfall patterns. Short-duration crops, e.g., should be grown where rainy seasons are short, medium-duration crops should be grown where rainy seasons are of medium length, and so on. This principle is illustrated for the various cropping systems of India in Table 3. The use of crops with growth cycles that are too long in relation to

rainfall patterns or seasonal patterns of S usually results in yield loss because of the onset of drought and unmet water demands of the crop toward the end of the growth cycle. On the other hand, the use of crops with growth cycles that are too short usually results in reduced yield loss because T is less than it potentially could be.

Unfortunately, in many semi-arid regions, rainfall is erratic as well as low. Indeed, rainfall variability often limits yield more than amount per se. Farmers use a number of strategies suited to their particular setting to cope with rainfall variability. In general, under variable rainfall environments, drought should be the least probable when crop demand and vulnerability are greatest.

In many tropical countries, multicropping systems, in which two or more crops with different flowering and maturity dates are grown together in the same season, are used as a method of reducing risk of total crop failure. Multicropping systems include "intercropping" systems, in which rows of one crop are alternated with those of another, "relay cropping" systems, in which an early-seeded crop is later inter-sown with a second, later-maturing crop, and "alley-cropping" or "agroforestry" systems, in which crop species are grown between woody or tree species. In addition to reducing risk, these cropping systems also improve use of sunlight, water, nutrients, and labor in low-input farming systems. Examples of risk averse, multicropping systems are given in Tables 1–3. Francis[11] explores multicropping more thoroughly.

Adjusting plant population, or spacing between plants, is another strategy by which farmers maximize crop T and, thereby, yield. By increasing plant population, E is reduced because more sunlight is intercepted by leaves rather than by the soil surface. Crops that have the ability to tiller profusely, such as wheat, tend to attain the same leaf area and yield over a range of plant population. Crops that do not tiller tend to have much lower plasticity, and therefore yield and WUE_{ET} are much more sensitive to plant population. In semi-arid environments in which the probability of rainfall is very low during the growing season, risk-averse farmers decrease plant population to reduce LAI, and therefore the rate of decrease in S. Optimal plant spacing therefore varies from region to region due to weather pattern, soil type, and farmers' perception of and tolerance to risk.

ADDITION OF SOIL AMENDMENTS AND OTHER INPUTS

Farmers apply a number of amendments to their fields in order to affect chemical, biotic, or physical soil properties, which in turn affect crop growth. Perhaps, the most important soil amendment is organic and mineral fertilizer, which is added to increase or maintain soil fertility. Fertilizer is added in many different ways, ranging from manure deposition by grazing animals, to rotation with leguminous crops that fix atmospheric nitrogen, to sophisticated precision mineral fertilizer applicators.

The importance of proper soil fertility to efficient water use, WUE_{ET} and, in some cases, WUE_T cannot be overemphasized. Among other things, it increases rooting depth and density, and the soil volume to which roots have access. Maintenance of soil fertility therefore can increase the amount of water to which plants have access (S), and decrease losses of water to drainage from the roots zone (D). Additionally, it increases WUE_{ET} by increasing plant growth and in particular crop leaf area, which shades the soil surface and thereby decreases E. Under highly infertile soil conditions, such as those found in many parts of Africa, addition of relatively small amounts of fertilizer can increase WUE_T as well.

Inputs other than fertilizer include insecticides, fungicides, and herbicides. Insects and disease must be controlled to efficiently use water in dryland cropping systems because they directly attack grain or reproductive organs of the plant, which obviously reduces yield, and therefore WUE_{ET}. Weeds compete for the same resources that crops use, including water, sunlight, and nutrients. Disease, insects, and some parasitic plants may also affect efficient use of water by damaging conductive tissue of the roots, stems, and leaves, thereby decreasing T and growth.

CONCLUSION

Farmers manage their fields and pastures according to a particular cropping system, which contributes to an overall farming and agricultural system. Dryland cropping systems tend to predominate in semi-arid systems with undependable rainfall. The sustained trends of continued global population growth, diminished land availability, and growing competition for fresh water will increasingly place the challenge of meeting food demands and protecting environmentally sensitive land upon dryland cropping systems. Since crop yield is proportional to T, this requires managing the soil water balance such that as much precipitation as possible is ultimately used as transpiration. Four basic methods of achieving this are the use of appropriate crops and crop sequence, soil surface management, addition of soil amendments, and water harvesting. The best method will depend upon specific characteristics of the particular cropping system. Finally, there is some potential for genetically increasing WUE_T.

REFERENCES

1. Loomis, R.S.; Connor, D.J. *Crop Ecology: Productivity and Management in Agricultural Systems*; Cambridge University Press: Cambridge, UK, 1992.
2. Pearson, C.J.; Norman, D.W.; Dixon, J., (Eds.) *Sustainable Dryland Cropping in Relation to Soil Productivity—FAO Soils Bulletin 72*; FAO: Rome, 1995.
3. Evans, L.T. *Feeding the 10 billion Plants and Population Growth*; Cambridge University Press: Cambridge, UK, 1998.
4. Tanner, C.B.; Sinclair, T.R. Efficient Water Use in Crop Production: Research or Re-search. In *Limitations to Efficient Water Use in Production*; Taylor, H.M., Jordan, W.R., Sinclair, T.R., Eds.; ASA, CSSA, and SSSA: Madison, WI, 1983; 1–27.
5. Payne, W.A. Water Relations of Sparse Canopied Crops. Agron. J. **2000**, *92*, 807.
6. Payne, W.A.; Keeney, D.R.; Rao, S.C. Sustainability of Agricultural Systems in Transition. *Proceedings of an International Symposium*; American Society of Agronomy: Madison, WI, 2002.
7. Condon, A.G.; Richards, R.A. Broad Sense Heritability and Genotype × Environment Interaction for Carbon Isotope Discrimination in Field-Grown Wheat. Aust. J. Agric. Res. **1992**, *43*, 921–934.
8. Nobel, P.S. *Biophysical Plant Physiology and Ecology*; W.H. Freeman and Co: New York, 1983.
9. Hillel, D. *Environmental Soil Physics*; Academic Press: San Diego, CA, 1998.
10. Unger, P.W.; Stewart, B.A. Soil Management for Efficient Water Use: An Overview. In *Limitations to Efficient Water Use in Production*; Taylor, H.M., Jordan, W.R., Sinclair, T.R., Eds.; ASA, CSSA, and SSSA: Madison, WI, 1983; 419–460.
11. Francis, C.A., Ed.; *Multiple Cropping Systems*, Macmillan Publishing Company: New York, 1986.

Dryland Farming

Clay A. Robinson
West Texas A&M University, Canyon, Texas, U.S.A.

D

INTRODUCTION

Dryland farming is the use of land for crop production in regions where growing season precipitation alone is usually inadequate to produce a summer grain crop. Droughts of varying intensity and duration are common in these regions. Dryland farming systems are dependent on natural precipitation, so the primary management concern in dryland farming systems is the capture and efficient use of water.[1]

Sometimes the term, dryland, is used in humid regions to mean "not irrigated." *Rainfed farming* or agriculture is the preferred term in regions where growing season precipitation alone is usually adequate to produce annual summer crops, and other management issues (fertility, pests, etc.) are more important than water conservation.

DRYLAND FARMING CHARACTERISTICS

Irrigation is practiced in many dryland regions when surface or groundwater is available to provide water for growing crops. Supplemental irrigation is also practiced in some humid regions to enhance production and limit losses due to drought. Irrigated land area will continue to decrease due to declining water levels, water quality and salinity problems, rising energy prices, and increased water demand for industrial, municipal, development, and other uses. As irrigated land area decreases, principles of dryland farming become more important.

Dryland farming is practiced worldwide with a diversity of mechanization and specific technologies. All dryland farming systems utilize some common principles:

- Using fallow (allowing land to lie idle during a growing season), tillage systems, residues, mulch, and/or structures to increase soil-water storage.
- Using tillage systems and mulch to limit evaporative water loss.
- Selecting shorter season, drought-resistant, and/or drought-tolerant crop genotypes.
- Selecting crops and rotations based on precipitation patterns and growing seasons.

- Manipulating plant density and geometry to optimize the evaporation (water lost from soil) to transpiration (water used by crops) ratio.
- Water harvesting.

Every dryland farming system does not incorporate all these principles, but all dryland farming systems use some combination of these principles. In addition to conserving water, many of these principles limit erosion by wind and water, and some enhance soil organic matter levels. These benefits are important since most dryland farming regions exist in fragile ecosystems. Long-term productivity in dryland regions depends on maintaining or enhancing the soil resource. Any management system that does not control erosion and limit soil degradation is not sustainable.

Every major continent has regions suitable for dryland crop production. Most dryland crop production occurs in areas classified as arid, semiarid, and subhumid. Though dryland cropping systems are diverse, they share one characteristic: Evapotranspiration (ET, combined water loss from crops and soil) exceeds precipitation during the growing season. In much of the North American Great Plains, monthly precipitation never exceeds one-half the ET.[2] Other dryland regions have some period during the year when soil-water storage is possible because monthly precipitation exceeds monthly ET.[2]

Fallow is used to store water from precipitation in the soil. Even in the Great Plains where monthly ET exceeds monthly precipitation, there are several days each year when precipitation exceeds ET and water can be stored in the soil. The efficiency of water storage during fallow depends on tillage choices and climate (Table 1). Tillage choices determine the intensity and depth of soil disturbance, and the quantity of residue remaining on the surface. Greater tillage intensity or tillage depth increases soil drying and decreases soil-water storage efficiency. Soil-water storage is directly related to residue quantities remaining on the surface.[3] Regional climate determines atmospheric demand for water and thus PET. In the central Great Plains, seasonal ET decreases with increasing latitude.

Using fallow to increase stored soil water decreases cropping intensity (number of crops per year). Increased soil-water storage has no economic benefit unless that

Encyclopedia of Water Science
DOI: 10.1081/E-EWS 120010042

Table 1 Tillage and water storage efficiency during fallow at Akron, Colorado and Bushland, Texas

	Precipitation stored as soil water (%)	
Tillage method	Akron, Colorado[a]	Bushland, Texas[b]
Disk, conventional	19	15
Sweep, stubble-mulch	33	23
No-till	48	35

[a] Adapted from Ref. [5].
[b] Adapted from Ref. [6].

Fig. 1 Blade (sweep) plows used in the United States (a) and China (b).

water can be used to produce a crop. Table 2 identifies some of the common cropping systems in the Great Plains. Research in the central Great Plains shows that less intensive tillage systems store more soil water and allow cropping intensity to increase. In the southern Great Plains, adoption of less intensive tillage systems has not altered cropping intensity, but yields have increased and crop failures due to drought are less common.

Many developing countries practice dryland cropping without large equipment. The principles of water conservation still work: disturb less soil (limit tillage), expose less soil surface (use mulch), and catch more water. Fig. 1 shows blade (sweep) plows used in the United States (Fig. 1a) and China (Fig. 1b). The same principle is at work, both limit soil disturbance and leave residues on the surface. Figure 2 shows wheat residue as a mulch in Texas (Fig. 2a) and a stone mulch in Gansu, China (Fig. 2b). Both mulches decrease evaporation, slow water movement across the surface, increase infiltration, and protect the soil from erosion. Fig. 3 shows the impact of furrow dykes (also called tied ridges) on precipitation capture and storage. The soil probe in Fig. 3a was inserted 30 cm into a furrow without dykes, but in the adjacent row with dykes (Fig. 3b), the soil probe was inserted to 120 cm, indicating an increase in plant-available water of about 12 cm.

Furrow dykes capture precipitation, limit runoff, and increase infiltration into the soil.

A dust mulch may limit evaporation under certain conditions. Shallow tillage is practiced immediately following a rain, leaving the surface loose and unconsolidated. The loose soil limits upward capillary movement of water as the soil surface dries, thus limiting evaporation. Dust mulching probably works in developing countries where farmers use light equipment and draft

Table 2 Crop production intensity and precipitation use efficiency of some common Great Plains cropping systems with a stubble-mulch (sweep) tillage system

Crop-fallow sequence[a]	Cropping intensity	Land use intensity	Precipitation used in crop production (%)
WW-F	1 crop in 2 yr	0.50	39
WW-F-S-F (WSF)	2 crops in 3 yr	0.67	45
Annual cropping[b]	1 crop in 1 yr	< 1.0[c]	60

Source: Adapted from Ref. [7].
[a] WW—winter wheat, F—fallow, S—sorghum.
[b] Other summer crops used are corn, cotton, millet, sorghum, soybean, and sunflower.
[c] Includes crop failures in drought years.

Fig. 2 Wheat residue mulch in Bushland, Texas (a) and stone mulch in Gansu, China (b).

Fig. 3 Soil probe in furrows without (a) and with (b) furrow dykes.

animals, and tillage begins as soon as the rain stops. In mechanized systems, the field must be dry enough to support a tractor. Shallow tillage in these fields probably increases water loss because the evaporation that dust mulch can prevent has already occurred, and subsequent tillage further dries the soil.

Crop calendars are another important dryland-management tool. Winter wheat is common in the central and northern Great Plains, giving way to spring wheat in the Prairie Provinces of Canada. The winter wheat-growing season matches the precipitation and evaporative demand of the climate. There is usually fall precipitation to establish the crop. Wheat is dormant much of the winter, allowing some water storage from snow and precipitation events. Precipitation increases in the spring as wheat breaks dormancy. In the central Great Plains, wheat is harvested before the highest summer temperatures. Dryland corn is becoming more common in the central Great Plains as reduced tillage practices increase the soil-water stored. The single precipitation peak matches the corn-growing season. Dryland corn in the southern Great Plains is not a viable option because the precipitation distribution is bimodal, and the valley occurs when corn

reaches pollination and grain fill. Cotton has been historically limited to the southern Great Plains because the growing season is too short further north although shorter season varieties are being developed.

Another method to limit evaporation from soil is to achieve a closed plant canopy sooner. Recent research into planting geometries recommends using narrower row spacings, higher plant populations, and shorter season hybrids. This combination allows a more rapid canopy development, which decreases weed competition and evaporation. The result is that more water can be used by the plants, producing greater yields.[4] The higher plant populations induce more rapid development and maturity. Short-season hybrids are used so the crop does not deplete the soil water during vegetative growth.

Most crop varieties used in dryland production have a drought tolerance mechanism, enabling them to endure short droughts. Some crops slow metabolic activity and essentially go dormant to avoid the drought. Other crops reduce metabolic activity and water use during the drought. Both mechanisms allow the crop to resume normal growth when the water stress is alleviated.

Fig. 4   Water harvesting project in Gansu, China.

The benefits of water harvesting are easily seen beside every road. The plants in the ditch are greener, taller, and lusher than those in nearby pastures. Many cultures have long used water-harvesting techniques to improve crop yields. Fig. 4 shows a water-harvesting system in Gansu, China, in which one-sixth of the land is covered with plastic and used as a watershed. The water collected from the plastic-covered watershed is stored in cisterns and used to irrigate the cropland on the remaining five-sixths of the land. Conservation bench terraces use a 2 or 3 to 1 watershed to bench ratio. This supplies enough water to the bench area to allow annual cropping. A wheat–sorghum-fallow system is used on the watershed, increasing the cropping intensity from 0.67 to 0.78.

CONCLUSION

Dryland farming systems are diverse, but all emphasize the capture and efficient use of precipitation through fallow, tillage, and residue management systems, crop selection, row spacing, plant populations, and/or water harvesting. Specific farming technologies are not universally applicable, but the basic principles of water conservation can be applied across all levels of technology.

REFERENCES

1. Clark, N.P. Welcome and Overview. *Challenges in Dryland Agriculture: A Global Perspective*, Proceedings of the International Conference on Dryland Farming, Amarillo/Bushland, TX, Aug 15–19, 1988; Unger, P.W., Jordan, W.R., Sneed, T.V., Jensen, R.W., Eds.; Texas Agricultural Experiment Station: College Station, TX, 1988; 2–5.
2. Stewart, B.A. Dryland Farming: The North American Experience. *Challenges in Dryland Agriculture: A Global Perspective*, Proceedings of the International Conference on Dryland Farming, Amarillo/Bushland, TX, Aug 15–19, 1988; Unger, P.W., Jordan, W.R., Sneed, T.V., Jensen, R.W., Eds.; Texas Agricultural Experiment Station: College Station, TX, 1988; 54–59.
3. Unger, P.W. Straw-Mulch Rate Effect on Soil Water Storage and Sorghum Yield. Soil Sci. Soc. Am. J. **1978**, *42* (3), 486–491.
4. Stewart, B.A.; Steiner, J.L. Water Use Efficiency. *Dryland Agriculture: Strategies for Sustainability*; Advances in Soil Science, Springer-Verlag: New York, 1990; Vol. 13, 151–173.
5. Greb, B.W.; Smika, D.E.; Welsh, J.R. Technology and Wheat Yields in the Central Great Plains: Experiment Station Advances. J. Soil Water Conserv. **1979**, *34* (6), 264–268.
6. Unger, P.W.; Wiese, A.F. Managing Irrigated Winter Wheat Residues for Water Storage and Subsequent Dryland Grain Sorghum Production: Effects of Tillage, Residue Levels, and Soil Water Content. Soil Sci. Soc. Am. J. **1979**, *43* (3), 582–588.
7. Johnson, W.C.; Davis, R.G. *Research on Stubble-Mulch Farming of Winter Wheat*, Conservation Research Report 16; U.S. Dept. of Agriculture—Agriculture Research Service: Washington, DC, 1972; 1–29.

FURTHER READING

Singh, R.P.; Parr, J.F.; Stewart, B.A. *Dryland Agriculture: Strategies for Sustainability*; Advances in Soil Science; Springer-Verlag: New York, 1990; Vol. 13.

Stewart, B.A.; Robinson, C.A. Are Agroecosystems Sustainable in Semiarid Regions? Adv. Agron. **1990**, *60*, 191–228.

Dust Bowl Era

R. Louis Baumhardt
United States Department of Agriculture (USDA), Bushland, Texas, U.S.A.

INTRODUCTION

The Dust Bowl era was the period of drought from 1931 to 1939 that was coupled with severe wind-driven soil erosion of overgrazed rangeland and soil exposed by the use of farming practices not adapted to the semiarid U.S. Great Plains. The eroding soil from once productive range and crop lands filled the air with billowing clouds of dust that subsequently buried farm equipment, buildings, and even barbed-wire fences (Fig. 1); thus, making the living conditions of many Great Plains inhabitants unbearable. On the Great Plains wind is common and drought recurrent; therefore, farm implements and management methods were developed for producing crops under these conditions. Likewise, farmers have evolved into innovative practitioners of soil and water conservation techniques that rely on residue management practices and crop rotations with fallow periods to store precipitation in the soil for later crop use.

HISTORY

During a sustained drought beginning in 1931 and continuing until 1939, wind erosion of range and farmlands filled the air with clouds of dust for days at a time. The Dust Bowl shifted annually over the Great Plains to affect different areas and grew with the expanding drought to damage an annual peak of about 20 million hectares.[1] However, the overall affected area (Fig. 2) encompassed almost 40 million hectares that extended from south of Lubbock, Texas (33° 34′ N, 101° 52′ W) to north of Colby, Kansas (39° 23′ N, 101° 3′ W) into Nebraska and from Great Bend, Kansas (38° 22′ N, 98° 50′ W) west to near Pueblo, Colorado (38° 16′ N, 104° 37′ W). The most severely affected farmland was located within a 160-km radius of Liberal, Kansas (37° 2′ N, 100° 55′ W), the center of the Dust Bowl.

The Dust Bowl land was native range for the North American bison and home to Native Americans prior to Euro-American settlement. It had been labeled the "Great American Desert" by explorer Stephen Long following his expedition to the area about 1820.[2] The challenges of this region, whether invoked by the perceptions of "Desert"-life or by Native Americans protecting their homes and

hunting interests, limited cultivation. For example, in 1879 or about five years after the Red River Indian wars, only 264 ha were cultivated in all of the 26 counties that make up the Texas Panhandle,[2] but cultivation expanded with favorable rains during 1882–1887 and 1895–1906.[3] Native rangeland was typically cultivated by tillage methods adapted from the more humid U.S. regions, which buried most of the plant residues, e.g., a Lacrosse disc breaking plow that relied on as many as 12 horses and mules.[4] Draft animal requirements for forage crops and native range limited some soil disturbance and provided, incidental, residues that protected the land. These farming practices that indirectly conserved soil were replaced by agricultural mechanization, which expanded tillage and allowed a single farmer to manage increasingly more land.

Agricultural mechanization and increased demand for wheat by Europe during World War I nearly doubled the amount of land cultivated from 1910 to 1920.[1] However, mean annual rain during the period 1918–1929 averaged about 100 mm above the 515 mm norm[3] and promoted continued farm expansion to about 16 million hectares that were largely placed into a wheat monoculture. The booming wheat market, beneficial rains, and increasing agricultural mechanization placed in motion rapidly expanding cultivation that exposed millions of hectares of land with potentially erodible soil. It was the climatic conditions of drought from 1930 to 1940 (Fig. 3) that ultimately triggered wind erosion of excessively tilled land and the Dust Bowl.[3]

DUST BOWL LESSONS

In a 1936 report to President Roosevelt from the drought area committee, Morris Cooke and others outlined the nature, causes, and recommended lines of action to ameliorate factors resulting in the Dust Bowl.[5] They noted that Great Plains agriculture had developed a dependency on over grazing and excessive plowing, which exposed loose soil to the wind. These farming practices did not conform to natural conditions of the Great Plains and resulted in an unstable agriculture and unsafe economy. The basic problem causing the Dust Bowl was identified as the attempt to impose farming practices suitable for humid regions on the semiarid Great Plains. The committee

Encyclopedia of Water Science
DOI: 10.1081/E-EWS 120010100
Published 2003 by Marcel Dekker, Inc.

Fig. 1 The devastation imparted by dust storms to Great Plains farmsteads from Texas shown at the bottom (1938 USDA Photo by: B. C. McLean, Image # 01D11486) north to South Dakota (1936 USDA Photo by: Sloan, Image # 00D10971).

further recognized, as unrealistic, the expectations of climate changes toward improved temperature, precipitation, and wind conditions. Therefore, in a region of limited annual precipitation, farming practices to reduce run-off and increase water storage in the soil were critical to agricultural success.

The drought area committee further stated that the 1862 federal homesteading policy exacerbated land degradation by offering unrealistically small farm allotments for the semiarid Great Plains west of the 100th meridian.[5] That is, the government policy actually encouraged over utilization of pasture and cultivated land. Subsequent

efforts to correct the homesteading policy by increasing land allotments as late as 1916[6] were heralded by the often-cited 1909 Bureau of Soils Bulletin 55 claim of an "indestructible and immutable soil resource."[7] The hazard of over cultivation and grazing was the exposure of loose soil to wind and erosion. This damage was aggravated further by volatile wheat markets that encouraged speculative production by absentee landowners relying on tenant farmers. In some cases, the tenants were transient farmers that only custom planted and harvested crops without remaining on the land. The proportion of land farmed by tenants increased from about

The Dust Bowl

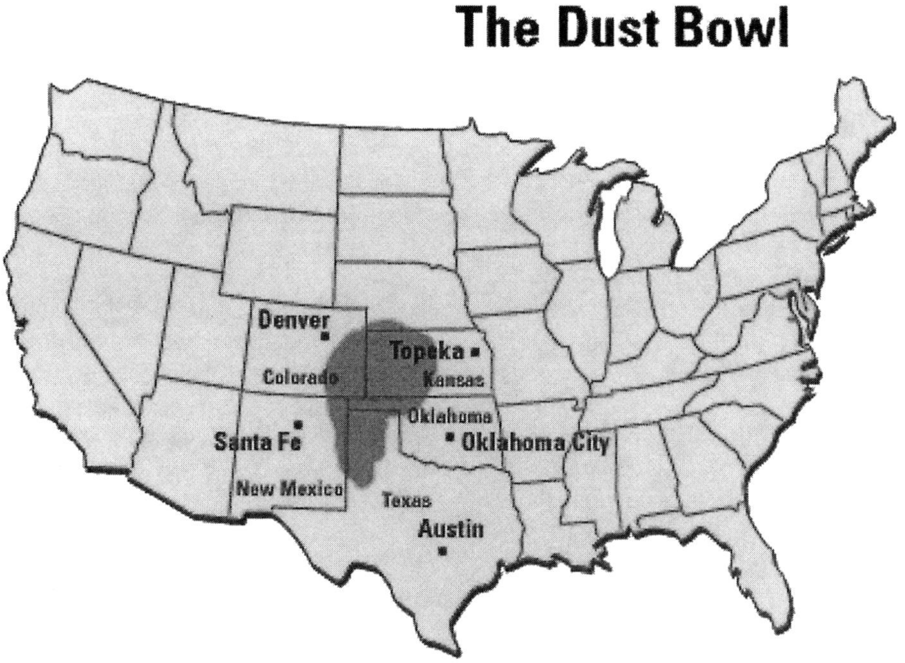

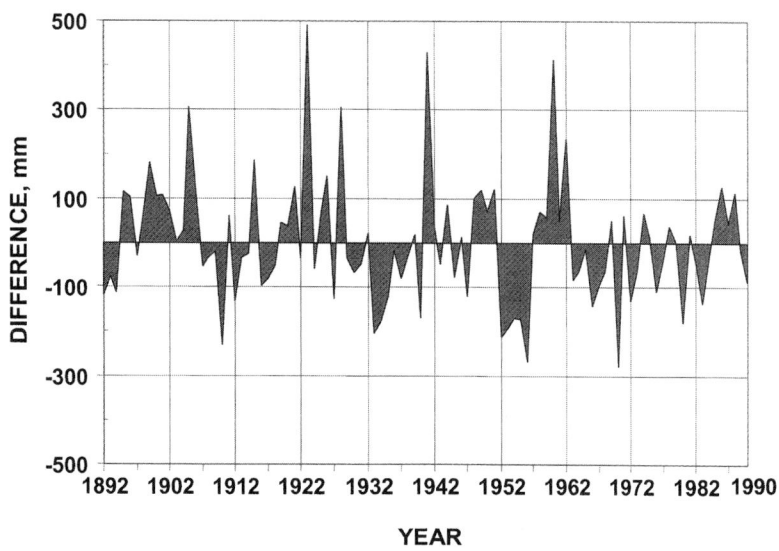

Fig. 2 The United States and the overall affected "Dust Bowl" area, from the "American Experience."[13]

Fig. 3 Deviation from the mean annual precipitation (515 mm) at Amarillo plotted for the period 1892–1990.

16% in 1880 to over 40% in 1935,[5] but the transient tenant farmers abandoned the land when commodity markets collapsed.

Agriculture capable of withstanding recurrent drought periods replaced the excessive tillage practices that incorporated crop residues and degraded the structure or natural cohesiveness of soil. Alternative tillage practices were developed to control weeds and the use of precipitation stored as soil water. These tillage practices also undercut rather than inverted the soil, thus reducing soil disturbance and increasing crop residues retained at the surface to conserve soil and water.[8] Revised land policies promoted conservation practices by rewarding farmers for using contour plowing, listing, and strip cropping methods.[9] The Dust Bowl wheat monoculture required timely fall and winter precipitation for crop establishment and growth; however, in much of the southern Great Plains mean monthly precipitation is limited during this critical period (see example for Amarillo area, Fig. 4). In lieu of wheat monocultures, practical wheat and summer crop rotations with an intervening fallow (i.e., two crops in three years) were developed to take advantage of summer rain (Fig. 4) and to provide sufficient opportunity for storing precipitation as soil water during fallow and improve crop establishment.

The damaging effect of excessive tillage contributed significantly to soil erosion throughout the Dust Bowl, but it may have been overstated as in Rexford Tugwell's film *The Plow that Broke the Plains.*[1] Soil erosion was also triggered by overgrazing and drought conditions, which were reduced through improved cattle management and the use of irrigation. Depressed commodity prices, however, virtually eliminated irrigation of crops, e.g., the Texas Panhandle had some 170 irrigation wells in 1930 or 60 to 80 fewer wells than a decade earlier in 1920.[10] Irrigation expanded slowly until drought conditions of the 1950s promoted rapid growth from Texas to Kansas.[11,12] Irrigation as a solution to drought in the Dust Bowl region almost exclusively depends on the Ogallala aquifer,[12,7] which has now dramatically declined. If irrigation was the dominant factor preventing soil erosion during the 1950s by offsetting drought conditions, it would follow that the Dust Bowl miseries may eventually return when irrigation from the southern Ogallala becomes impractical.[7]

AGRICULTURE—DUST BOWL VICTIM OR VILLAIN

In 1933, the director of new Soil Erosion Service, Hugh H. Bennett, indicted Americans as great destroyers of land as substantiated by the Dust Bowl conditions and called for awakening to improved farming practices.[13] Farmers and

their children likewise recognized the fragility of the land and the inappropriate nature of their farming practices in laments that "All the good soil will blow off this land if these sand storms continue"[14] and "It would be better if the sod had never been broke ..."[15] Many farmers expanded production to offset lower prices and passively relied on luck to "hit big" with a crop that would change their fortune even as the commodity market collapsed in the 1920s.[12] The resulting economy was unstable and led to a general depopulation trend and agricultural collapse that was squarely in line with the creation of a "Buffalo Commons"[16] whereupon the government would step in to buy abandoned Great Plains farmland and restore it to an undisturbed range condition.

In response to the disastrous effects of the Dust Bowl, government programs were redesigned to encourage diversified agricultural crop production using tested practices and improved tools. That is, agriculture was empowered with new noninverting tillage implements capable of penetrating the hard dry soils like the Graham-Hoeme plow for controlling weeds while retaining crop residue at the soil surface.[8] Innovative wheat-sorghum cropping sequences optimized soil water storage opportunities and increased the probability of capturing rain for crop use. A growing number of managers now farming the Great Plains minimize soil disturbance and protect their crop residues as vital resources to optimize the storage of precipitation as soil water.[17] The efficiency of precipitation storage in the soil has improved from about 20% during the Dust Bowl to more than 40% by using innovative crop sequences with fallow periods and no or reduced tillage.[18] Farmers now utilize preplanned alternative rotation sequences to optimize crop water use during periods of beneficial rain and include other production inputs like fertilizers in response to specific

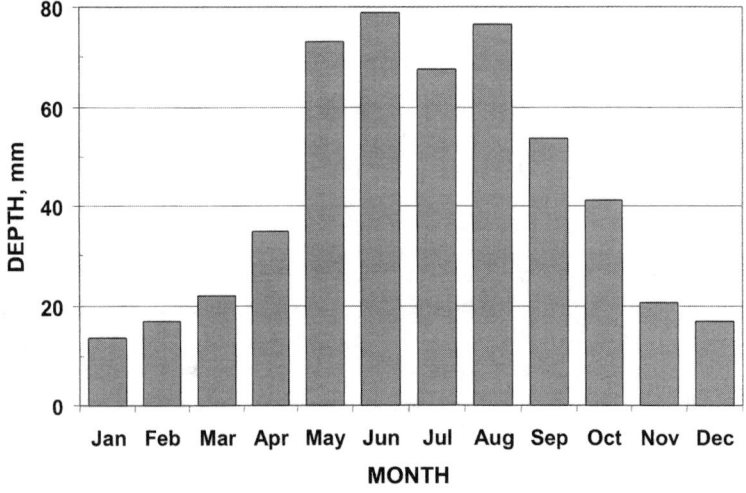

Fig. 4 Mean, 1892–1990, monthly precipitation at Amarillo.

needs.[17] These innovations, in contrast to Dust Bowl soil management using inversion tillage and wheat monocultures, have resulted in substantially more stable economies and slowed the depopulation trend.

In contrast to the farmers of the Dust Bowl hoping to "hit-big" on a crop, many of today's Great Plains farmers are more proactive managers that respond to adverse growing conditions with alternative technology.[12] For example, when drought conditions reappeared during the early 1950s, Kansas farmers widely adopted irrigation to stabilize production. Since that time fluctuating irrigation costs and the competition for and depletion of water resources have driven innovation in irrigation. These innovations include irrigation scheduling methods to meet plant demand and improved application technologies such as low-pressure center pivot systems. While these innovations will prolong the use of irrigation to offset recurrent drought conditions, the finite nature of Ogallala water supply[7,10–12] focuses concern on the potential of a recurrent Dust Bowl. The development and application of new soil and crop management practices not available during the 1930s will determine if the Dust Bowl is as recurrent as drought.

REFERENCES

1. Hurt, R.D. *The Dust Bowl, An Agricultural and Social History*; Nelson-Hall: Chicago, IL, 1981.
2. Price, B.B.; Rathjen, F.W. *The Golden Spread: An Illustrated History of Amarillo and the Texas Panhandle*; Windsor Publications: Northridge, CA, 1986.
3. Johnson, W.C.; Davis, R.G. *Research on Stubble-Mulch Farming of Winter Wheat*, USDA Conservation Research Report No. 16; U.S. Government Printing Office: Washington, DC, 1972.
4. Steiert, J. The Ulrich Steiert Family. Hart Beat **2000**, *38* (9), C8–C9.
5. Cooke, M.L. (Chair); Bennett, H.H.; Fowler, F.H.; Harrington, F.C.; Moore, R.C.; Page, J.C.; Wallace, H.A.; Tugwell, R.G. *Report of the Great Plains Drought Area Committee*, Box 13, Hopkins Papers, Franklin D. Roosevelt Lib., 1936, http://newdeal.feri.org/texts/450.htm (accessed Dec 2000).
6. Gray, L.C.; Bennett, J.B.; Kraemer, E.; Sparhawk, W.N. The Causes: Traditional Attitudes and Institutions. *Soils and Men USDA Yearbook of Agriculture*; U.S. Government Printing Office: Washington D.C., 1938; 111–136.
7. Sachs, A. Dust to Dust: Forgotten Lessons of America's Great Agricultural Catastrophe. World Watch **1994**, *7* (1), 32–35.
8. Allen, R.R.; Fenster, C.R. Stubble-Mulch Equipment for Soil and Water Conservation in the Great Plains. J. Soil Water Cons. **1986**, *41* (1), 11–16.
9. USDA-AAA, *Better-Balanced Farming for the Oklahoma and Texas Wheat and Grain-Sorghum Area: Including Special Provisions for Wind Erosion Area*, SR Leaflet No. 102; Government Printing Office: Washington D.C., 1937; 11.
10. Green, D.E. *Land of the Underground Rain: Irrigation on the Texas High Plains, 1910–1970*; University of Texas Press: Austin, TX, 1973.
11. Musick, J.T.; Pringle, F.B.; Harman, W.L.; Stewart, B.A. Long-Term Irrigation Trends—Texas High Plains. Appl. Eng. Agric. **1990**, *6* (6), 717–724.
12. Williams, D.D.; Bloomquist, L.E. *From Dust Bowl to Green Circles: A Case Study of Haskell County, Kansas*, SB 662; Agric. Exp. Stn., Kansas State Univ.: Manhattan, KS, 1996.
13. Gazit, C. *Surviving the Dust Bowl*, The American Experience, WGBH Educational Foundation, 1998. http://www.pbs.org/wgbh/amex/dustbowl/index.html (accessed Dec 2000).
14. Neugebauer, J.M. *Plains Farmer: The Diary of William G. DeLoach, 1914–1964*; Texas A&M University Press: College Station, TX, 1991.
15. FSA-OWI Collection. *LC-USF34-018264-C DLC*, Library of Congress, Prints and Photographs Division. 1938.
16. Popper, D.E.; Popper, F.J. The Great Plains: From Dust to Dust. Planning **1987**, *53*, 12–18.
17. Hackett, T. Dietrick Drops "Cast-Iron Farming" Before It's Too Late. Natl Conserv. Tillage Dig. **1999**, *6* (2), 4–7.
18. Greb, B.W.; Smika, D.E.; Welsh, J.R. Technology and Wheat Yields in the Central Great Plains: Experiment Station Advances. J. Soil Water Cons. **1979**, *34* (6), 264–268.

El Niño

David E. Stooksbury
University of Georgia, Athens, Georgia, U.S.A.

INTRODUCTION

The term El Niño refers to a number of related oceanic and atmospheric phenomena. A general definition of El Niño is a change in weather patterns associated with warmer than normal sea surface water temperatures in the central and eastern equatorial Pacific Ocean (see Fig. 1). El Niño is the best known example of interannual variation in the earth's weather and climate patterns.

Originally, El Niño referred to warmer than normal surface water temperatures off the coast of Peru. The name El Niño comes from the appearance of warm surface water temperatures around Christmas. This relationship between surface water warming and Christmas led the locals to call the phenomena El Niño, Spanish for boy, the Christ child. The appearance of an El Niño event inhibits the upwelling of cold and nutrient-rich water along the Ecuadorian and Peruvian coasts. Without the upwelling of cold, nutrient-rich waters, the fish migrate to more favorable locations and thus there are few fish for local communities.

El Niño changes the position of the subtropical jet stream, the steering current for weather systems. Changes in the subtropical jet stream cause changes in weather patterns. These weather pattern changes cause some regions of the earth to experience above-normal rainfall while other areas experience below-normal rainfall. El Niño weather patterns also cause some regions of the earth to be warmer than normal while other regions are cooler than normal. While the change in weather patterns associated with El Niño can be dramatic, most regions of the earth experience minimal to no direct impact from El Niño. El Niño events occur every three to seven years lasting from a few months to a year or more.

IMPACTS

In the United States, El Niño weather patterns usually mean a warm and wet fall in the central and northern plains while the Pacific Northwest and Middle Atlantic states experience drier than normal conditions. During winter, much-above normal rainfall usually occurs from southern California to the Gulf of Mexico and South Atlantic states. Across the northern two-thirds of the lower 48 states, El Niño winters are usually much warmer than normal.

Across the Pacific Northwest states, winter precipitation is much below normal during El Niño winters making the region vulnerable to droughts. During an El Niño spring, the region east of the Mississippi River usually experiences below normal to much-below normal rainfall. The Pacific Northwest remains dry during the spring. Springtime temperatures are usually below to much below normal across the south while above normal to much-above normal temperatures are expected in the Pacific Northwest, the northern Rockies, and the northern plains making these regions vulnerable to drought. The southwestern United States usually experiences above-normal precipitation in an El Niño spring.

The impacts of El Niño weather patterns vary from one event to another. The impacts depend on the warmth of the surface water, the exact location of the warm surface water, the areal extent of the warm surface water, and other regional and global weather patterns. An ocean–atmosphere linkage that mitigates the impacts of El Niño is the Pacific Decadal Oscillation (PDO). The PDO has similar impacts as El Niño and can increase or decrease the impacts of El Niño. Unlike El Niño, the PDO cycle is decades long and not a few years.

During the summer monsoon season (late summer into early fall), the intermountain region of the western United States normally has above-normal rainfall during an El Niño weather pattern. The region has an increased probability of experiencing flash floods. The impacts of El Niño on the summer monsoon can be mitigated by the PDO.

From southern Mexico to northern South America, El Niño weather patterns normally increase rainfall and can lead to major flooding, especially in mountainous regions.

Not all regions impacted by El Niño have increased precipitation. The El Niño weather pattern usually brings drier than normal conditions to northern Australia, Indonesia, and the Philippines, often causing drought conditions.

Mechanism

It is now known that there is a linkage between the appearance of warm surface water temperatures and atmospheric phenomena. For El Niño events, this means linking eastern and central equatorial Pacific Ocean

Encyclopedia of Water Science
DOI: 10.1081/E-EWS 120010221

December - February La Niña Conditions

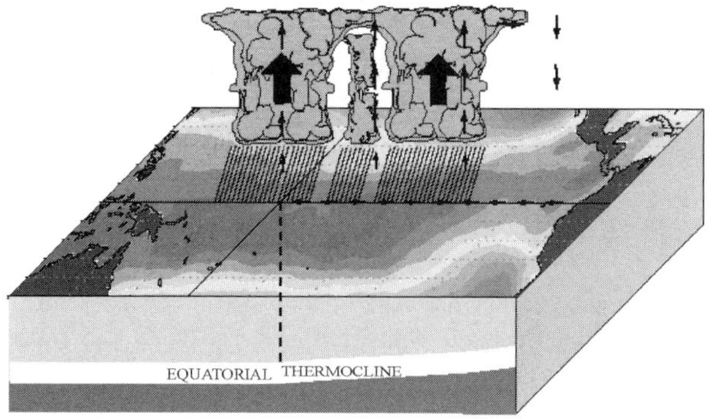

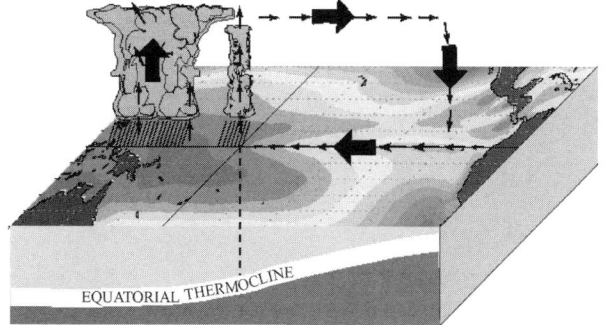

Fig. 1 Atmospheric and oceanic patterns during an El Niño www.cpc.ncep.noaa.gov/products/analysis_monitoring/ensocycle/enso_schem.html (From the National Weather Service Climate Prediction Center, Camp Springs, MD.)

surface temperatures with atmospheric pressure patterns across the Pacific Ocean (see Fig. 2). The atmospheric pressure patterns linked with El Niño is called the Southern Oscillation, SO. The combination of El Niño and SO is called ENSO (El Niño-Southern Oscillation). The term ENSO is often used interchangeably with the term El Niño.

The strength of SO is calculated by the surface atmospheric pressure anomaly differences between Tahiti and Darwin, Australia (Tahiti anomaly minus Darwin anomaly). This measure of SO strength is called the Southern Oscillation Index, SOI. A surface atmospheric pressure anomaly is calculated by subtracting the mean atmospheric surface pressure from the observed atmospheric surface pressure. Thus, if the observed atmospheric surface pressure is less than the mean, the anomaly has a negative value. When the SOI has a negative value, it means that the surface atmospheric pressure is less than

December - February Normal Conditions

Fig. 2 Atmospheric and oceanic patterns during "neutral" or normal conditions www.cpc.ncep.noaa.gov/products/analysis_-monitoring/ensocycle/meanrain.html (From the National Weather Service Climate Prediction Center, Camp Springs, MD.)

normal at Tahiti and above normal at Darwin (negative anomaly at Tahiti minus a positive anomaly at Darwin). A negative SOI is correlated with warming of the surface water in the eastern and central equatorial Pacific Ocean.

The linkage between SO and El Niño is complex. At the most basic level, sea surface temperature patterns influence atmospheric pressure patterns, and atmospheric pressure patterns influence wind speed and direction and thus the sea surface temperature patterns. Warm surface temperatures over the western Pacific Ocean lead to increased convection and lower surface pressure across the western Pacific Ocean. The normal cold surface water of the eastern Pacific Ocean is associated with relatively high surface atmospheric pressure. Air moves (wind) from areas of high atmospheric pressure to areas of low atmospheric pressure. The greater the pressure gradient (pressure difference between two locations divided by the distance between the two locations), the greater the wind speed. The moving air in contact with the ocean surface causes ocean surface currents, which redistribute the ocean surface temperature pattern. The stronger the wind, the more the occurrence of redistribution of surface water temperatures.

With ENSO, the linkage between the ocean and the atmosphere results in decreasing or increasing easterly trade-wind (wind from the east to the west) speeds over the equatorial Pacific Ocean. When the SOI is negative, the pressure gradient across the eastern and western Pacific Ocean is decreased. With a decreased pressure gradient, the speed of the easterly trade-winds decreases, and warm surface water from the western Pacific Ocean is able to "slosh back" over the colder surface water in the eastern Pacific Ocean. The decreased easterly winds also leads to a decreased upwelling of cold, nutrient-rich water along the coast of Ecuador and Peru. When the SOI is positive, the easterly trade-wind speed increases. The increased wind speed "piles-up" warm surface water in the western Pacific Ocean leading to below normal surface temperatures in the eastern and central equatorial Pacific Ocean due to strong upwelling.

Below normal surface temperatures in the eastern and central equatorial Pacific Ocean is the opposite of an El Niño event and is called either a La Niña (Spanish for girl) event or El Viejo (Spanish for old man) event.

Changes in the equatorial Pacific surface temperature patterns impact the weather patterns in other regions of the earth. During an El Niño pattern, sea surface temperature patterns change. These sea surface temperature pattern changes impact the locations of evaporative heat movement from the ocean surface to the atmosphere. With different evaporative heat patterns, there are changes in the locations of wintertime jet streams and thus the storm tracks. With changes in jet stream patterns and storm tracks, weather patterns across numerous regions can change. An example is the change in storm tracks that brings Pacific Ocean storms into southern California and across the southern-tier of states instead of the Pacific Northwest.

Since the 1990s scientists have used Pacific Ocean surface temperature data and computer models to predict the occurrence of an El Niño event months in advance. While these predictions are not perfect, they allow for planning to mitigate or take advantage of a shift in weather patterns. Thus regions that normally experience flooding during an El Niño event can plan to mitigate the impacts. For regions like Indonesia or the Pacific Northwest of the United States, drought mitigation plans can be activated months in advance.

For more detailed information about El Niño, see Ref. [1].

REFERENCE

1. National Oceanic and Atmospheric Administration Climate Prediction Center. El Niño/La Niña website. http://www.cpc. ncep.noaa.gov/products/analysis_monitoring/lanina/ (accessed Aug 2002).

Energy Dissipation Structures

Youichi Yasuda
Iwao Ohtsu
Nihon University College of Science and Technology, Tokyo, Japan

E

INTRODUCTION

Hydraulic structures such as dams, weirs, and drop structures have energy dissipators as a means of dissipating the excess energy of high-velocity flows, in order to protect the riverbed and banks downstream.

In energy-dissipation structures, there are stilling basins with horizontal or sloping aprons, stilling basins with baffles or sills, bucket-type stilling basins, and baffled or stepped chutes. The type of energy-dissipation structure to be selected depends on the kind of hydraulic structure, the discharge, the magnitude of the energy head upstream of the hydraulic structure, the tailwater conditions, and the topographical and geological characteristics of the river or channel.[1−3]

In order to properly complete the hydraulic design of an energy-dissipation structure, it is important to know the downstream flow conditions. In addition, information regarding flow conditions around the hydraulic structures might help in improving the landscape and other features of the river environment and in preserving the ecosystem for aquatic animals in the river.

OVERVIEW

The flow conditions that are used as energy dissipators introduced here are: hydraulic jumps on horizontal aprons, hydraulic jumps forced by a vertical sill, hydraulic jumps on sloping aprons, hydraulic jumps below abrupt expansions, transition flows over drop structures, and stepped-channel flows.

ENERGY DISSIPATORS IN STILLING BASINS

Hydraulic Jumps in Prismatic Horizontal Channels

In a stilling basin, the formation of a hydraulic jump is the most effective method of dissipating the kinetic energy of a high-velocity flow. A hydraulic jump is a transitional phenomenon from high velocity supercritical flow to lower velocity subcritical flow. The flow conditions of the hydraulic jump in a horizontal channel changes according to the inflow conditions and the shape of the channel. The formation of a symmetric jump with a surface roller is an effective energy dissipator.

A hydraulic jump in a horizontal smooth rectangular channel is referred to as a classical jump. Classical jumps have been classified into undular jumps, weak jumps, oscillating jumps, steady jumps, and strong jumps.[1,4] Steady and strong jumps can be utilized in a stilling basin. However, the position of a classical jump is very sensitive to changes in the downstream depth.[5] Some kind of elements are needed in order to stabilize the jump location (see "Forced Hydraulic Jumps by a Vertical Sill").

In a trapezoidal horizontal channel, a submerged jump is recommended as an energy dissipator because a free jump becomes asymmetric for a mild side slope, and a submerged jump keeps a symmetric flow for any side slope (Fig. 1).

The sequent-depth ratio (ratio of downstream to upstream depth) and the energy loss of a free or submerged jump can be predicted theoretically. The jump length has been discussed by many researchers[1,6−8] and may be predicted for a free or submerged jump in a rectangular or trapezoidal channel.[9,10]

Forced Hydraulic Jumps by a Vertical Sill

When the downstream flow depth is less than the sequent depth required for a classical hydraulic jump, sills and blocks have been utilized in order to stabilize the jump location in a stilling basin. Standard designs for stilling basins employing sills and blocks have been published by the U.S.B.R.[1,8]

The flow conditions of a forced jump change according to the inflow Froude number, the sill height, the position of the sill, the boundary-layer development at the toe of jump, and the upstream and downstream depths. As illustrated in Fig. 2, the flow configuration upstream of the sill depends on the downstream depth in some cases (Type-I jump). In other cases, the flow configuration upstream of the sill is independent of the downstream

Encyclopedia of Water Science
DOI: 10.1081/E-EWS 120010344

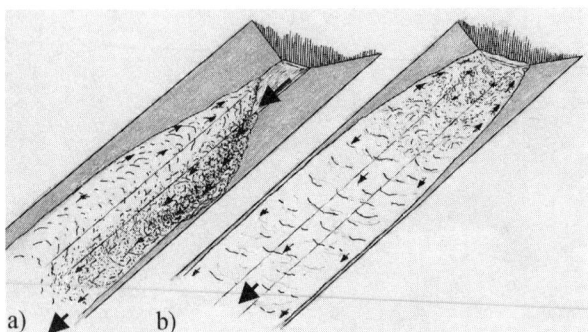

Fig. 1 Flow conditions of hydraulic jump in a trapezoidal channel with a mild side slope: (a) asymmetric flow; (b) symmetric flow.

depth (Type-II jump). If a forced jump is not formed, the supercritical flow splashes over the sill (this is referred to as a splashing flow) (Fig. 2).

The hydraulic conditions required to form each type of flows have been documented.[11,12] When the discharge and the upstream and downstream depths are given, the height and position of the sill required to form a forced jump can be predicted. The drag force acting on the vertical sill in a forced jump has also been investigated.[12,13] An experimental equation for the length of a stilling basin required for the formation of a forced jump has been developed.[8,11]

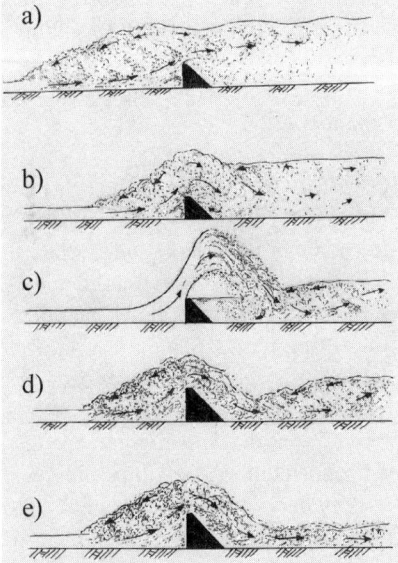

Fig. 2 Flow conditions of flow over a sill: (a) and (b) Type I forced hydraulic jumps; (c) splashing flow; and (d) and (e) Type II forced hydraulic jumps.

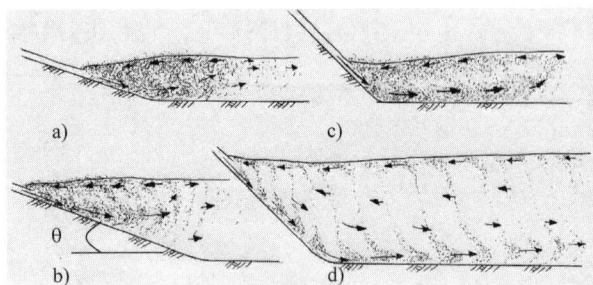

Fig. 3 Flow conditions in sloping channels with a horizontal channel portion: (a) and (b): degree of channel slope θ is smaller than 19° (a) B-type hydraulic jump; (b) D-type hydraulic jump, (c) and (d): degree of channel slope θ is larger than 40° (c) B-type hydraulic jump; (d) plunging flow.

Hydraulic Jumps on Sloping Aprons

If the downstream depth is greater than the sequent depth of a classical hydraulic jump, control of the hydraulic jump by a sloping apron is effective as an energy dissipator.[1]

The flow conditions of the hydraulic jump change according to the inflow Froude number, the channel slope, and the upstream and downstream depths (Fig. 3). When the degree of channel slope is small, the jump occurs on the sloping channel apron, and the high velocity decays in a short distance (Fig. 3b). This flow condition is favorable as an energy dissipator.[1] If the degree of channel slope and the downstream depth become large, the flow becomes a plunging flow. For plunging flow, the high-velocity flow along the channel bed continues far downstream, and the effect of the surface eddy on velocity decay is negligibly small (Fig. 3d). This condition is less effective for energy dissipation.

The hydraulic conditions for the formation of various types of jumps and the length of the jumps have been documented and are predictable for a wide range of inflow Froude numbers, channel slopes, and downstream depths.[1,8,14]

Hydraulic Jumps Below Abrupt Expansions

Both symmetric and asymmetric flows can exist when an outlet conduit is connected to a wide open-channel (Fig. 4). An asymmetric flow is also observed in an open-channel having an abrupt expansion.[8] With asymmetric flow conditions, high-velocity flow may exist along one sidewall for a significant distance downstream. Maintenance of conditions suitable for the formation of a symmetrical jump is recommended for energy dissipation below an abrupt expansion.[3,15]

The minimum downstream depth required to form symmetric flow at an abrupt-symmetrical expansion has been documented for a wide range of expansion ratios,

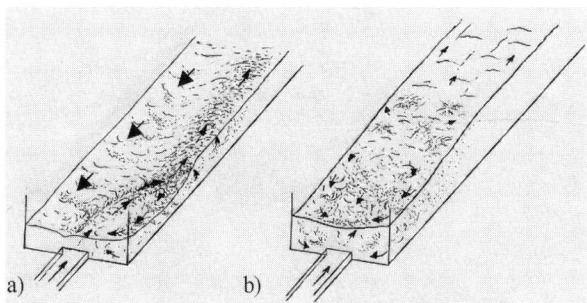

Fig. 4 Flow conditions of submerged hydraulic jump below an abrupt expansion: (a) asymmetric flow; (b) symmetric flow.

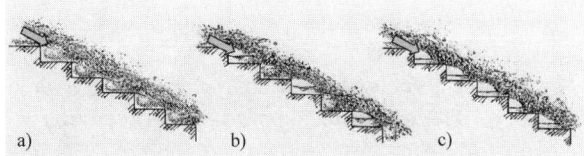

Fig. 6 Flow conditions on a stepped channel: (a) skimming flow; (b) transition flow; (c) nappe flow.

aspect ratios, and inflow Froude numbers.[15] In addition, empirical equations for predicting the jump length for symmetric flow conditions have been developed. A flow chart for designing a stilling basin with an abrupt-symmetrical expansion is available.[15]

Transitional Flows at Abrupt Drops

An abrupt drop in a channel may be used to stabilize the jump position effectively for a change of the flow depth.

A plunging flow with a surface roller is not always formed at the downstream region of drop structures. When the flow passing over a drop structure transits from supercritical flow to subcritical flow, various types of flow conditions are formed according to the inflow Froude number, the drop height, and the upstream and down-

stream depths.[8,16,17] For example, if the downstream depth is increased, the flow condition might change from a plunging flow to a wave train where the undular surface with a main flow propagates far downstream as illustrated in Fig. 5.

The hydraulic conditions required to form each type flow condition have been presented for a wide range of inflow Froude numbers, drop heights, and downstream depths.[17] Also, low-drop and high-drop structures have been defined according to the differences of flow patterns and design criteria for each type of drop developed.

ENERGY DISSIPATORS ON SPILLWAYS

A baffle chute that dissipates energy along the entire length of the channel is useful as an energy dissipator of high-velocity spillway flows.[1] Recently, stepped spillways have been utilized in connection with the roller-compacted concrete dam-construction method, and stepped-channel flows have been investigated by many researchers.[18,19]

The flow conditions on a stepped surface change according to the discharge, the step height, the slope angle of the stepped channel, and the total drop. Flow conditions have been classified as skimming flow (the main flow skims above a stepped channel, and a corner eddy is formed without an air-pocket in each step), nappe flow (an air-pocket is always formed in an aerated-flow region below the nappe), and transition flow (a transition between

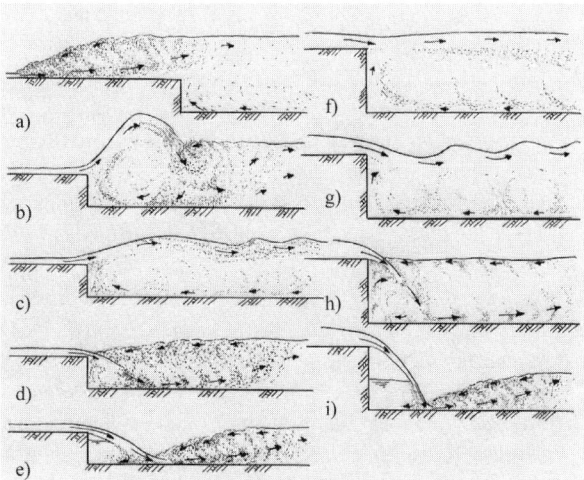

Fig. 5 Flow conditions at abrupt drops: (a)–(e): approaching flow on the step is supercritical (a) A-type hydraulic jump; (b) wave-type flow; (c) wave train; (d) B-type hydraulic jump; and (e) minimum B-type hydraulic jump (f)–(i): critical flow exists on the step (f) surface-jet flow; (g) wave train; (h) plunging condition; (i) limited jump.

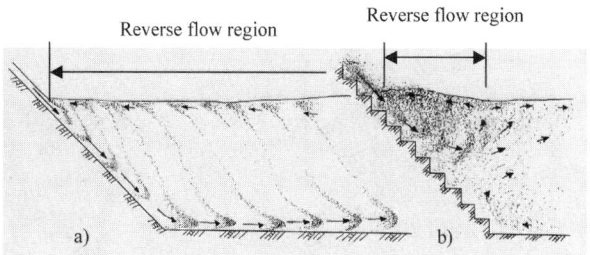

Fig. 7 Comparison of reverse flow region of plunging flows: (a) plunging flow on a smooth sloping channel; (b) plunging flow on a stepped channel.

a skimming flow and a nappe flow with an air-pocket partly formed) (Fig. 6).

Experimental investigations have revealed the hydraulic conditions for the formation of each flow condition and the energy loss due to stepped flows.[20,21]

The utilization of stepped surface in approach channel is effective for the energy dissipation of plunging flow region. Especially, when a stepped channel is used for the steep section of a spillway, there is less tendency to develop plunging flow with a reverse-flow region on transition to the downstream channel[22] (Fig. 7). This increases the effectiveness of the energy dissipation in the transition region.

CONCLUSION

For design of an energy-dissipation structure, it is important to know the downstream flow conditions. A variety of flow conditions may be used for energy dissipation according to the type of the hydraulic structure, the discharge, and the downstream flow depth.

A stabilized jump with a surface roller is an effective energy dissipator in jump-type stilling basins. Baffle chutes and stepped spillways are effective in dissipating energy along the length of steep channels such as spillways.

REFERENCES

1. Peterka, A.J. Hydraulic Design of Stilling Basins and Energy Dissipators. United States Department of the Interior Bureau of Reclamation, Engineering Monograph, a Water Resources Technical Publication, 1978; 25.
2. The US Army Corps of Engineers. Hydraulic Design of Spillways. Technical Engineering and Design Guides as Adapted from the US Army Corps of Engineers, ASCE, 1995; 12.
3. Visher, D.L.; Hager, W.H. Energy Dissipators. Hydraulic Structures Design Manual; IAHR, A.A. Balkema: Rotterdam, The Netherlands, 1995; 9.
4. Chow, V.T. *Open Channel Hydraulics*; McGraw-Hill International: New York, 1959.
5. Ohtsu, I.; Yasuda, Y. Characteristics of Supercritical Flow Below Sluice Gate. J. Hydraul. Eng., ASCE **1994**, *120* (3), 332–346.
6. Rajaratnam, N. Hydraulic Jumps. In *Advances in Hydroscience*; Chow, V.T., Ed.; Academic Press: New York, 1967; 4.
7. Ohtsu, I.; Yasuda, Y.; Awazu, S. *Free and Submerged Hydraulic Jumps in Rectangular Channels*; Report of Research Institute of Science and Technology, Nihon University: Tokyo, Japan, 1990; 35, 1–50.
8. Hager, W.H. *Energy Dissipators and Hydraulic Jump*; Water Science and Technology Library, Kluwer: Dordrecht, The Netherlands, 1992; 8.
9. Ohtsu, I. Free Hydraulic Jump and Submerged Hydraulic Jump in Trapezoidal and Rectangular Channels. Trans. Jpn Soc. Civil Eng. **1976**, *8*, 122–125.
10. Ohtsu, I.; Yasuda, Y. Discussion of Hydraulic Jump in Triangular Channel. J. Hydraul. Res., IAHR **1989**, *127* (1), 178–188.
11. Ohtsu, I. Forced Hydraulic Jump by a Vertical Sill. Trans. Jpn Soc. Civil Eng. **1981**, *13*, 165–168.
12. Ohtsu, I.; Yasuda, Y.; Yamanaka, Y. Drag on Vertical Sill of Forced Jump. J. Hydraul. Res., IAHR **1991**, *19* (1), 29–47.
13. Ohtsu, I.; Yasuda, Y.; Hashiba, H. Drag on Vertical Sill of Forced Hydraulic Jump. Proc. 25th IAHR Congr. **1993**, *Theme A*,329–336.
14. Ohtsu, I.; Yasuda, Y. Hydraulic Jump in Sloping Channels. J. Hydraul. Eng., ASCE **1991**, *117* (7), 905–921.
15. Ohtsu, I.; Yasuda, Y.; Ishikawa, M. Submerged Hydraulic Jump Below Abrupt Expansions. J. Hydraul. Eng., ASCE **1999**, *125* (5), 492–499.
16. Rand, W. Flow Geometry at Straight Drop Spillways. J. Hydraul. Div., Proc. ASCE, **1955**, *81* (HY5), 1–13; **1956**, *82* (HY1), 57–62; and **1956**, *82* (HY3), 7–9.
17. Ohtsu, I.; Yasuda, Y. Transition from Supercritical to Subcritical Flow at an Abrupt Drop. J. Hydraul. Res., IAHR **1991**, *29* (3), 309–328.
18. Chanson, H. *The Hydraulics of Stepped Chutes and Spillways*; Swets & Zeitlinger, Lisse, A.A. Balkema: Lisse, The Netherlands, 2001.
19. Minor, H.E.; Hager, W.H. *Hydraulics of Stepped Spillways*; Proceedings of the International Workshop on Hydraulics of Stepped Spillways; A.A. Balkema: Rotterdam, The Netherlands, 2000.
20. Ohtsu, I.; Yasuda, Y. Characteristics of Flow Conditions on Stepped Channels. Proc. 26th IAHR Congr. **1997**, *Theme D*,583–588.
21. Yasuda, Y.; Takahashi, M.; Ohtsu, I. Energy Dissipation of Skimming Flows on Stepped Channels. XXIX IAHR Congr. Proc. **2001**, *Theme D*,531–536.
22. Yasuda, Y.; Ohtsu, I. *Characteristics of Plunging Flows in Stepped Channel Chutes*; Proc. of the International Workshop on Hydraulics of Stepped Spillways; A.A. Balkema: Rotterdam, The Netherlands, 2000; 95–102.

Erosion, Accelerated

J. Kent Mitchell
Michael C. Hirschi
University of Illinois, Urbana, Illinois, U.S.A.

INTRODUCTION

In 1997, the United States lost almost 970 million metric tons of soil through erosion by water.[1] Soil erosion has always taken place and always will. It is a natural process. The surface of the earth is continually undergoing what might be called a "face lift in slow motion." Slowly, the coastline is receding, the hills and mountaintops are being carried down to the valleys, and the river deltas are being enlarged. The form of erosion that occurs naturally, without man's influence, is called geologic erosion. Some of the best examples of geologic erosion are the Grand Canyon, the Badlands of South Dakota, the canyons of Utah, and the great river valleys. Without human interference, geologic erosion would occur at a low rate on level land; and on gentle slopes, erosion would only be a minor problem. In humid areas there is an ideal environment for plant growth, and thus, there would be protective cover for the soil, thereby further slowing the rate of erosion.

Agriculture and urban development has replaced protective cover with plants that are of more value to man. However, such plants often do not cover the soil as effectively as the natural growth. Some farmers leave the soil totally bare through much of the year. The result is that accelerated erosion may increase to destructive proportions on some soils, carrying away topsoil and nutrients, washing pollutants into streams, filling waterways with sediment, and reducing the natural productivity of the land.

Compared with the magnitude of the problem, the basic cause (raindrops and resulting runoff) may seem insignificant. Yet, falling raindrops strike the ground with surprising force and the cumulative effect is immense. With no vegetative cover or mulch to absorb the impact, rain is especially erosive on cropland left bare between plantings. To understand the problem, let us look at some of the mechanisms of erosion.

RAINFALL

Soil erosion is the detachment of particles from the soil mass and their transport downstream. When it rains, drops up to 6 mm in diameter bombard the soil surface at impact velocities of up to 9 m/sec.[2] In general, the more intense the rainfall, the larger the drop size will be.[3]

The constantly pounding raindrops dislodge soil particles and aggregates and splash them up to 1 m away.[4] When rain hits vertically on a horizontal surface, the splash is equal in all directions. On a slope, more of the splash goes downhill than uphill[5] (Fig. 1). In wind-driven rainfall, splash movement depends on slope and wind direction.

To observe the effects of splash erosion, look at a white fence or building next to bare soil just after a rain. Most likely, rain will have splashed soil as high as 1 m on the fence or building. Other visual reminders of erosion's impact are soil pedestals, which can be created during a heavy rain when particles underneath a stone or piece of residue remain protected (Fig. 2). Meanwhile erosion batters and washes away unprotected soil from around the object, sculpting a soil pedestal that corresponds to the shape of the stone or residue.

A raindrop falling on a thin sheet of water detaches soil particles more readily than one falling on dry soil. Splash erosion increases with surface water depth, but only up to a depth about equal to the raindrop diameter. Once the water becomes deeper, the splash effect is reduced.[6]

Actually, if water did not accumulate on and run off the soil surface, the splashing of soil particles would not be a major concern. In most cases, splashed particles are not moved far enough to greatly disturb the soil surface. But water does accumulate on the soil surface. If rain falls hard enough and long enough, the soil eventually will become saturated and the surface will seal. The ground will have trouble absorbing more water; and in low spots, water will collect in small ponds. If rain continues, these ponds ultimately will overflow and water will move downhill.

TRANSPORT

The concern about splashed soil becomes clearer because of transport processes. Raindrops dislodge particles from the soil mass and runoff water transports the eroded soil (Fig. 3). Unless there is runoff water, raindrops cannot do much damage. But by the same token, runoff water is

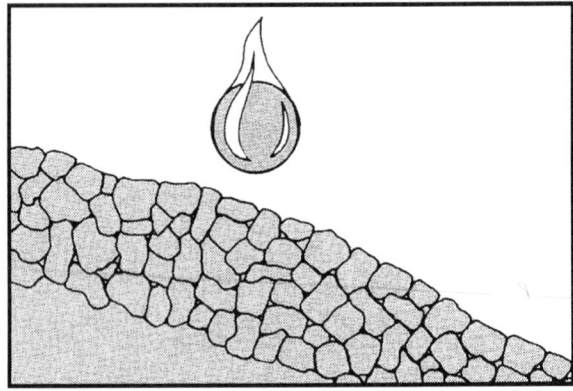

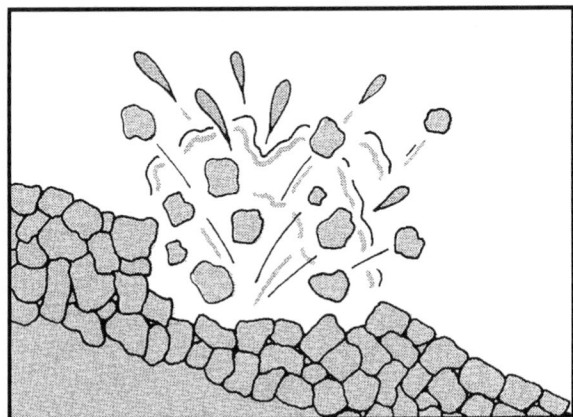

Fig. 1 Soil particles and aggregates are detached by raindrops.

Fig. 2 Raindrop erosion has removed soil particles on all sides of this piece of crop residue creating a soil pedestal.

dependent upon soil dislodged by raindrops for material to transport. An exception to this occurs when runoff moves as concentrated flow with sufficient energy to both dislodge and transport soil particles.

Water flowing off the soil surface (Fig. 3) provides the mechanism for transporting particles loosened by rainfall (Fig. 1). Although described as sheet flow, this type of flow seldom occurs in an uninterrupted sheet. Usually the water detours around clods, spills out of small depressions, and in general moves with sluggish irregularity. Even so, the water is able to carry soil particles. This type of erosion and transport is more properly called inter-rill erosion. The transport ability is influenced by the energy level of the flow, which in turn is dependent on the depth of flow and slope of the land. Flat areas have little or no runoff and low runoff velocities; consequently, little or no transport occurs. Runoff from steeper areas flows at greater velocities and may have considerable transport capability.

Sheet or inter-rill erosion is difficult to see, but its damage can be extensive. The destruction is more obvious when a plow turns up light colored subsoil on sloping land; which indicates that much of the topsoil has been eroded. A typical soil is made up of clay, silt, and sand particles. Erosion has a greater tendency to remove the finest

material, the clay, than it does the coarsest material, the sand particles; however, most plant nutrients are attached to the fine, clay particles. So erosion, a selective process, steals the most valuable part of the soil, as well as important organic matter.

RILLS

When the thin layer of water moves downhill, it tends to concentrate in tiny channels called "rills." Rills look like miniature streams; bending and cutting through the soil. Raindrops continue to break apart the soil, but runoff also has built up enough momentum to break loose particles. More importantly, rills have an excellent ability to transport soil particles. This type of flow usually occurs on only a small percentage of a field, but because the flow is concentrated, it can cause erosion. The rills thus created leave small channels that can be obliterated by normal tillage operations (Fig. 4). Energy levels of water flowing

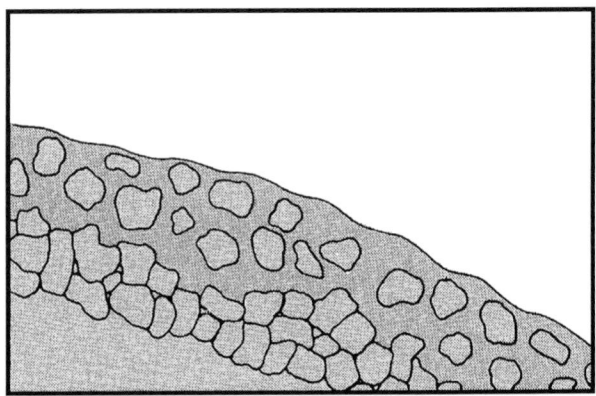

Fig. 3 Soil particles and aggregates are transported downslope with runoff water.

in rills vary somewhat, depending on the depth of flow and slope of the channel. Long, steep slopes allow rivulets with considerable erosive power to develop.

In many situations, rill flow detaches less material than does splash erosion. However, while a rill is forming, raindrops continue to detach soil within shallow rills and from the surrounding soil surface. Eroded material is transported to the rills by sheet or inter-rill flow. Rill flow has an exceptional capacity to transport the detached particles. Because the flow is concentrated, material can be transported within these small channels. A few soils are very susceptible to rill erosion; thus any rill flow that develops can easily detach soil particles or aggregates.

GULLIES

Eventually, the rills in a field will merge to form larger channels. These may form even larger channels and can become deep enough to be labeled "gullies." Channels are defined as gullies when they cannot be obliterated with normal tillage operations. They are large, noticeable scars on the land. In many areas of the Midwest, gully erosion has divided fields into small parcels that are inefficient to farm (Fig. 5). Deep gullies with vertical side walls are a phenomenon found in the deep loess soils along the bluffs of the Illinois and Mississippi Rivers. Some can reach depths of 9 m (30 ft or more). Rills and gullies often progress upstream at a head-cut or overfall (small waterfall). As the pool below the overfall enlarges, the turbulent water undercuts the overfall; eventually the soil sloughs off and is transported downstream.

Gully erosion can be deceiving. Although it is the most obvious form of erosion, it usually does not remove as much soil as the other, less visible forms of erosion. Gully erosion has been shown to provide from 0% to 89% of the sediment yield to streams.[7] Thus, gully erosion may be a significant problem depending upon the soil and topography. Data from Illinois taken in 2000 indicates that 22% of the fields are affected by various degrees of gully erosion.[8]

A very wide rill that has eroded soil through the tilled layer has been termed an "ephemeral gully." Although an ephemeral gully can be obliterated with modern tillage equipment, such a rill has been named a gully because of the great mass of material removed (Fig. 6). Ephemeral gullies are somewhat transitory rather than permanent like classical gullies.

Streambank erosion is a process similar to rill and gully erosion that occurs along the edge of perennial and ephemeral streams. Undercutting and sloughing are the primary agents of detachment, with sediment falling directly into the flowing water that transports the sediment downstream.

DEPOSITION

Sedimentation from soil or other materials carried by moving water may occur with sheet, rill, gully, and stream flow. Natural or artificial dams are a prime place for runoff to collect. Large particles settle in quiet pools formed at these sites. When the water is slowly released, much of the material is deposited as sediment (Figs. 7 and 8).

Ponding is apt to occur in small depressions or above contour furrows in inter-rill areas. It may also occur above small debris dams formed from residue in rills and gullies, terrace channels, or reservoirs in large streams. In addition, dense vegetation can reduce the flow velocity, thereby allowing soil material to be deposited. Effects of this process are sometimes seen in grassed waterways where the center gradually fills with sediment.

Fig. 4 An example of rill erosion.

Fig. 5 Gully erosion on sloped land.

Fig. 6 An ephemeral gully with sheet or interrill erosion deposition at the edges.

EROSION PROCESSES

All three processes of detachment, transport, and deposition occur during an erosive rainfall event. The extent that these processes occur are determined by the amount and intensity of rainfall, topography of the land surface, vegetative cover, and character of soil.

Each type of soil has its own inherent susceptibility to the forces of erosion; in large part because of chemical composition and organic matter content. Large-grained materials are easily detached by raindrop splash or flowing water, however, they are not easily transported. On the other hand, fine soils such as clays and mixtures of clays and silts that bond together tightly are not easily detached, but once free, they are transported with little difficulty. For this reason, fine materials can be carried considerable distances, whereas larger particles are deposited somewhere along the flow path.

Mulch and vegetative covers play an important role in hindering the erosion process. Without protective ground cover, raindrops may splash soil particles up to 1 m (Fig. 9). However, when mulch lays directly on the ground and completely covers the soil surface, the force from falling raindrops is absorbed and, thus, eliminates or reduces splash erosion (Fig. 10).

Canopy cover will also reduce drop erosion to a great extent. Close growing crops such as corn and soybeans catch raindrops and keep them from hitting the soil directly. Much of the water runs down the plant stem, although some runs off the leaves. Falling on bare soil, these drops cause a small amount of detachment, but since they have fallen from a lesser height, detachment is less than with no canopy cover (Fig. 11). Trees provide less protection for bare soil because of the greater height from which the drops fall. However, forests usually contain protective ground cover in the form of leaf or needle mulch.

Not only do ground covers intercept raindrops and keep them from detaching soil particles, but these covers also prevent soil compaction which restricts infiltration of

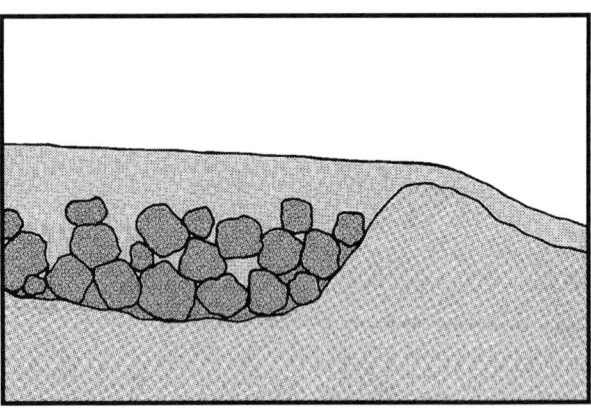

Fig. 7 Small irregularities in the soil surface, acting as small dams, cause ponding of runoff; then, some of the soil aggregates and particles are deposited as sediment and remain after flow ends.

water into the soil. With greater infiltration, there is less runoff. However, some runoff with transport capacity will occur.

Even when no particles are detached by raindrop splash, the flow itself, forming larger and larger rivulets, can eventually loosen particles. By slowing down the velocity of flowing water, vegetation is helpful in reducing flow erosion. In a highly susceptible soil, some rill erosion may occur beneath the mulch cover, but the flow is impeded and the degree of erosion reduced.

EROSION MODELING

Many factors, among them rainfall, soil, topography, and vegetative cover, affect the erosion process. Although many of these processes are recognized and understood, scientists do not yet have enough detail for developing complete physically based mathematical models.

Fig. 8 An example of sediment deposition near the edge of a field.

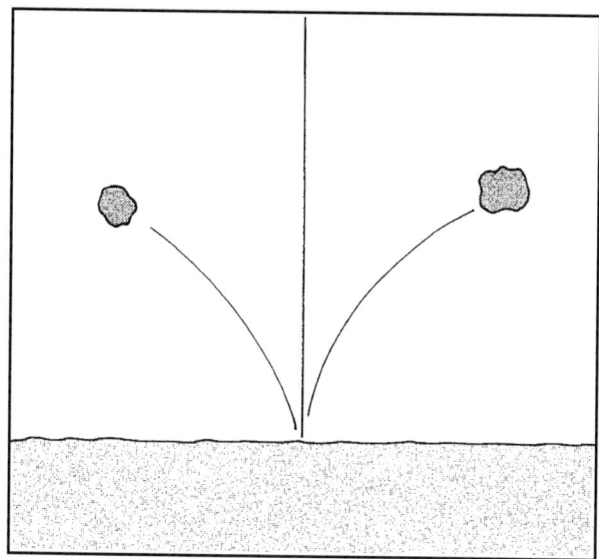

Fig. 9 With no protective ground cover, raindrops splash soil particles up to 1 m.

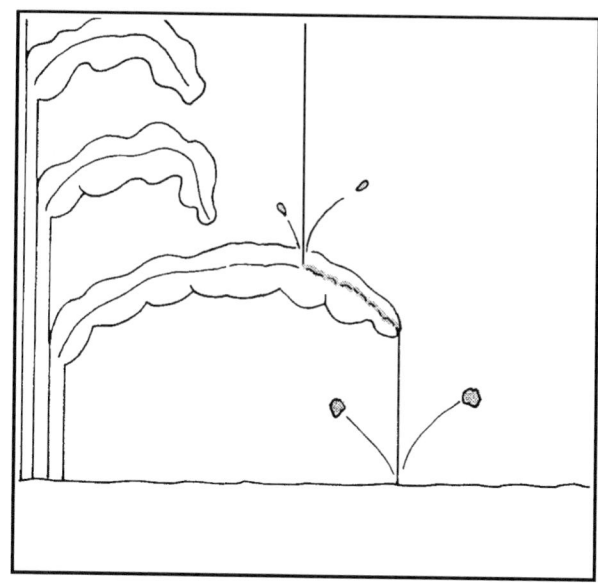

Fig. 11 The leaves of close-growing crops absorb the force of falling raindrops, thus minimizing the splash.

Some investigations have advanced more rapidly than others. For example, U.S. Department of Agriculture scientists at Purdue University and at Oxford, Mississippi, have made considerable progress in defining splash detachment and inter-rill transport mechanisms. Work in defining rill flow detachment and transport mechanisms continues, and progress is being made.

At present, the Universal Soil Loss Equation (USLE)[9] and the Revised Universal Soil Loss Equation (RUSLE)[10] are the most widely accepted methods of

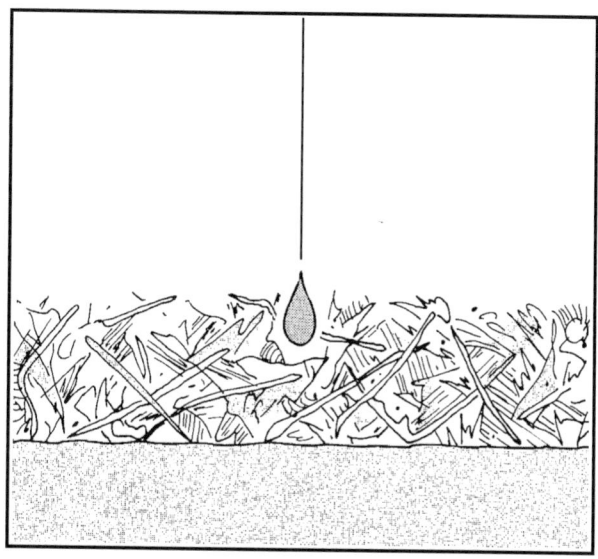

Fig. 10 Mulch cover dissipates the raindrop energy and hinders or eliminates splash erosion.

estimating soil loss from land surfaces. The equations include the effects of rainfall erosivity, soil erodibility, slope gradient and length, ground cover management, and erosion control practices. Although empirical, the equations provide the best estimates available for these complex phenomena. Using either equation, conservationists can estimate soil loss from a field and recommend alternative cultural practices for bringing excessive erosion to within tolerable limits. Many researchers are working on ways to improve estimates of the various parameters in this equation. A more advanced model, the Water Erosion Prediction Project (WEPP) equation,[11] has been developed. It is currently being validated by erosion scientists and used by researchers. It will become available for use by practitioners in the field.

We still need models that describe the erosion process in precise physical terms, but this is a long-range project. We also need to evaluate parameters for methods now used for estimating erosion. Both types of studies must continue concurrently. The new knowledge gained will help researchers develop exact descriptions for clearly defining erosion and sediment transport. The evaluation of parameters for current prediction methods enable conservationists to define causes and suggest cures for erosion problems.

REFERENCES

1. National resources inventory, USDA, NRCS: Washington, DC, 1997; http://www.nhq.nrcs.usda.gov/land/meta/m5112.html (accessed July 2001).

2. Laws, J.O. Measurements of the Fall-Velocity of Water-Drops and Raindrops. Trans. AGU **1941**, *22*, 709–721.

3. Laws, J.O.; Parson, D.A. The Relation of Rain Drop Size to Intensity. Trans. AGU **1943**, *24*, 452–459.

4. Ellison, W.D. Soil Erosion Studies—Part II. Agric. Eng. **1947**, *28*, 197–201.

5. Ellison, W.D. *Soil Erosion by Rainstorms*, Navdocks P-42; Bureau of Yards and Docks, Department of the Navy: Washington, DC, 1950; 1–5.

6. Palmer, R.S. Waterdrop Impact Forces. Trans. ASAE **1965**, *8* (1), 69–70.

7. Glymph, L.M. Importance of Sheet Erosion as a Source of Sediment. Trans. AGU **1957**, *38*, 903–907.

8. Illinois Soil Conservation Transect Survey Summary. Illinois Department of Agriculture: Springfield, IL, 2000; http://www.agr.state.il.us/Transect%20survey2000.htm (accessed July 2001).

9. Wischmeier, W.H.; Smith, D.D. *Predicting Rainfall Erosion Losses: A Guide to Conservation Planning*, Agric. Handb. No. 537; USDA, ARS: Washington, DC, 1978; 1–58.

10. Renard, K.G.; Foster, G.R.; Weesies, G.A.; McCool, D.K.; Yoder, D.C. *Predicting Soil Erosion by Water: A Guide to Conservation Planning with the Revised Universal Soil Loss Equation (RUSLE)*, Agric. Handb. No. 703; USDA, ARS: Washington, DC, 1997; 1–384.

11. Flanagan, D.C.; Nearing, M.A., (Eds.) *USDA—Water Erosion Prediction Project Hillslope Profile and Watershed Model Documentation*, NSERL Report No. 10; NSERL, USDA, ARS, MWA: West Lafayette, IN, 1995; 1.1–14.28.

Erosion Control, Tillage/Residue Methods

Richard Cruse
Jerry Neppel
John Kost
Krisztina Eleki
Iowa State University, Ames, Iowa, U.S.A.

E

INTRODUCTION

Historically, tillage has been used for seedbed preparation, weed control, residue burial, and fertilizer/manure incorporation. Modern technology, i.e., improved planters and pesticides, has reduced tillage requirements for crop production with equipment adapted to higher crop residue conditions. Tillage to improve the seedbed for planter performance and crop production also increases the soil susceptibility to erosion from both wind and water.

TILLAGE AND SOIL EROSION PROCESSES

The processes of soil detachment, transport, and deposition occur during erosion. Soil surface conditions created by tillage greatly influence these processes and therefore soil erosion losses. Tillage practices that increase (or at least do not decrease) soil structural stability, leave plant residues on the soil surface, slow surface-water flow velocity, and/or promote high infiltration rates favor soil conservation.

Surface residue cover greatly influences soil erosion. When 30% of the surface is covered, soil erosion losses are reduced by approximately 50% compared with a bare, tilled soil (Fig. 1). Conservation tillage is considered to be any tillage system that has at least 30% of the soil surface covered by plant residues after planting[1] (Fig. 1).

TILLAGE EFFECT ON SOIL PROPERTIES CRITICAL TO EROSION

Residues intercept raindrops and minimize soil detachment. This reduces soil available for transport and also limits surface seal development. This improves infiltration, which in turn reduces the amount of water runoff and transport potential. Residues also slow surface flow velocity, by acting like little dams on the surface. This causes soil deposition to occur on the upslope side of the residue pieces where water flow slows.[2]

Surface roughness and structural stability play dual roles. A rough surface stores water in the surface depressions between the clods or aggregates during heavy rainfall. This slows runoff and limits transport. The large pores of a rough surface also require more soil detachment to create a surface seal than a smooth surface, minimizing transport. Similarly, contour tillage, tillage occurring across the slope, reduces runoff by increasing surface water storage and slowing water runoff velocity.

Open pores from the subsurface to the soil surface are critical for high infiltration rates and therefore low transport potential. Tillage practices that promote stable structure and result in surface residues to intercept raindrop impact promote stable open pores. Most tillage practices, however, weaken structure and therefore promote soil detachment from raindrop impact. Tillage also disrupts earthworm activity. Earthworms can play a major role in producing large open pores on the soil surface and very high infiltration rates with selected management systems. Four basic tillage/management systems will be discussed. Many variations of each system exist. Also, other systems using the principles described in this paper have been developed and can be located in other literatures, for example see Ref. [3].

NO-TILL

No-till results in minor soil disturbance only during planting, leaving the greatest possible amount of surface residue after planting (Fig. 2).

Compared to cleanly tilled systems, no-till can reduce erosion by as much as 95%.[4] The accumulation of residue from season to season reduces erosion by protecting the soil surface from impacting raindrops, as well as improving structural stability, pore size, and pore stability. Earthworm activity is promoted by no-till. Where large earthworm populations exist and the population is active on or near the surface, runoff can be very low or nonexistent, even for large rainfall events.

Encyclopedia of Water Science
DOI: 10.1081/E-EWS 120010094

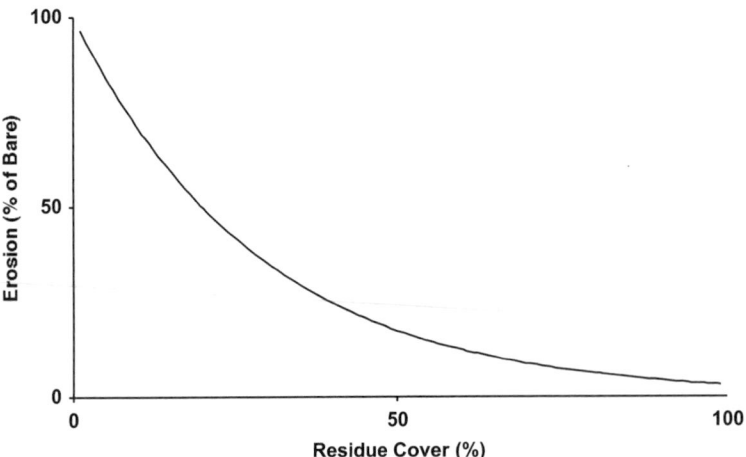

Fig. 1 Effect of surface residue cover on soil erosion by water, expressed as the percent of erosion observed for bare soil. (Adapted from Ref. [2].)

No-till is best adapted for semiarid regions, sloping soils, and/or soils with good internal drainage. Low soil temperatures can be a problem in cooler climates, especially in early spring if soil is wet. One variation of no-till creates a residue-free band over the planted row with the planter. This helps the row zone warm and dry faster than if the soil is residue covered. In hotter climates, the reduced soil temperature caused by residue can be advantageous. In the absence of tillage, weed control is typically done through herbicide application. However, a combination of herbicide and cultivation for weed control can also be practiced.

Residue-covered surfaces also reduce soil water evaporation. No-till practices may conserve sufficient water under semiarid conditions (or on droughty soils) to significantly increase crop yield relative to that for other tillage methods (Fig. 3).

RIDGE TILLAGE

With ridge tillage the soil surface "is left undisturbed from harvest to planting" except for strips up to one-third of the row-width. Planting is completed on the ridge and usually involves removal of the ridge top. Planting is completed with sweeps, disk openers, coulters, or row cleaners. Residue is left on the surface between ridges. Weed control is accomplished with crop protection products

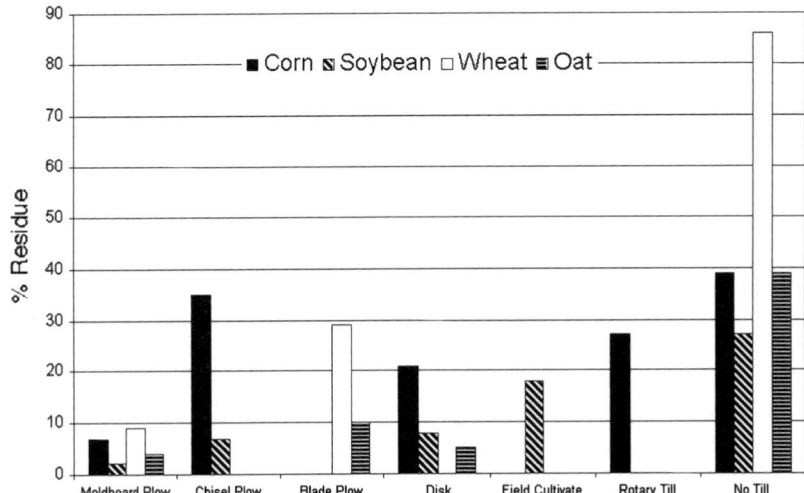

Fig. 2 Percent residue by tillage method. (Adapted from *Conservation Tillage Systems and Management*, Midwest Plan Service, Iowa State University, 1992.)

Fig. 3 No-till corn following wheat.

(frequently banded) and/or cultivation. Ridges are rebuilt during row cultivation.[1]

Ridge tillage historically is practiced on wetter, poorly drained soils in northern climatic row cropping regions.[6] However, it is also a viable option in semiarid, rain-fed row cropping regions where soil moisture conservation is a necessity.

Selected aspects of ridge tillage systems protect the soil from soil erosion. Crop residues remain on the soil surface from harvest until planting of the succeeding crop. Only a portion of the soil surface is disturbed at planting. This leaves a high percentage of crop residues on the soil surface and maintains the large pores for rapid water infiltration. Crop residues from the row (ridge) are placed in the interrow (valley) area. This "extra" residue decreases detachment and slows water runoff, enhancing infiltration in the interrow area. Ridge tillage promotes controlled traffic, the practice of maintaining a fixed traffic pattern in the field, such that only certain interrow areas experience wheel-caused compaction. Controlled wheel

traffic limits soil compaction to preselected interrows. The nontraffic areas maintain large pores and stable structure, enhancing infiltration. Ridge tillage is much more effective at conserving soil if ridging and planting are done on the contour than if up and down hill management is used.

Ridge till also offers opportunities to reduce weed control costs through banding of herbicides and row cultivation for weed management (Fig. 4). Nutrient losses may be reduced by injection and/or subsurface application of fertilizers.[7] Also, ridged soil warms quicker in the spring than no-till soil, permitting earlier planting in many situations.[8]

MULCH TILLAGE

Mulch tillage is a full width conservation tillage system involving one or more soil loosening operations prior to planting. Mulch tillage maintains a substantial amount of plant residue cover before and after crop establishment. Tillage tools such as chisel plows, field cultivators, disks, or blades are typically used for primary tillage. Secondary tillage is minimized to conserve surface residue.[9]

Mulch can be from any crop material. It is normally retained on the surface during harvest of the previous crop. The amount of mulch left on the surface depends on the sequence of the tillage, tool(s) used, and the mulch material of the previous crop. In general, the higher the crop yield, the more surface residue will exist.[10]

Surface mulch reduces the evaporation of soil water, increasing soil water content, relative to that occurring with a bare surface. Consequently soil warming can be slower in the spring, which can slow plant emergence and early development. The higher soil water content can also favorably affect crop yield under dry conditions. Mulch tillage increases soil organic matter content compared to

Fig. 4 Harvested corn on ridge-till soil.[5]

Fig. 5 Mulch tillage procedure.

more intensive tillage systems (Fig. 5). Weed control can be done mechanically, with herbicides, or with a combination of the two.

STRIP TILLAGE

Strip tillage involves tilling only in the crop row zone. The interrow area is untilled with surface residue left undisturbed. Tillage may be done in fall or spring. However, fall tillage is more commonly done. Typically a row cleaner, coulter, shank, and covering disks till each row area. This combination of components is equally spaced to match planter row-spacing so that the succeeding crop is planted in the tilled zones.[11] This system offers the combined advantages of no-till in the interrow zone and conventional tillage in the planted zone. The tilled zone is normally warmer and drier than if no tillage were performed.

Weed control can be through herbicide application, cultivation, or a combination of these methods. Fertilizer can be applied during the tillage operation at the base of the tilled depth and/or during the cropping season. Strip tillage is used only for row crop production. Strip tillage on the contour is much more effective at conserving soil than planting up and down hill.

SUMMARY

Tillage systems that leave residues on the surface, promote stable soil structure, and/or result in open pores to the soil surface favor water infiltration and soil conservation. Surface residue management is closely related to stable soil structure development and open surface pores. No-till, ridge tillage, mulch tillage, and strip tillage exemplify

management systems that use these principles for favorable crop production and reduced soil erosion rates.

REFERENCES

1. http://www.ctic.purdue.edu/Core4/CT/Definitions.html (accessed Aug 29, 2001).
2. Renard, K.G.; Foster, G.R.; Weesies, G.A.; McCool, D.K.; Yoder, D.C. *Predicting Soil Erosion by Water: A Guide to Conservation Planning with the Revised Universal Soil Loss Equation (RUSLE)*, Agricultural Handbook No. 703; U.S. Department of Agriculture: Washington, DC, 1997; 384.
3. http://www.inform.umd.edu/EdRes/Topic/AgrEnv/ndd/agronomy/DEFINITIONS_OF_TILLAGE_SYSTEMS_FOR_CORN.html (accessed Aug 27, 2001).
4. Midwest Plan Service, Water Erosion. *Conservation Tillage Systems and Management*, 1st Ed.; Iowa State University: Ames, 1992; 9.
5. http://www.ctic.purdue.edu/Core4/CT/PhotosGraphics.html (accessed Aug 29, 2001).
6. Uri, N.D. Introduction. *Conservation Tillage in U.S. Agriculture: Environmental, Economic, and Policy Issues*; Food Products Press: Binghamton, NY, 1999; 1–4.
7. Kanwar, R.S.; Colvin, T.S.; Karlen, D.L. Ridge, Moldboard, Chisel, and No-Till Effects on Tile Water Quality Beneath Two Cropping Systems. J. Prod. Agric. **1997**, *10* (2), 227–234.
8. Radke, J.K. Managing Early Season Soil Temperatures in the Northern Corn Belt Using Configured Soil Surfaces and Mulches. Soil Sci. Soc. Am. J. **1982**, *46*, 1067–1071.
9. http://www.agric.gov.ab.ca/crops/wheat/wtmgt06.html (accessed Aug 27, 2001).
10. http://muextension.missouri.edu/xplor/agguides/agengin/g01650.htm (accessed Aug 27, 2001).
11. http://www.ag.ohio-state.edu/~ohioline/aex-fact/0507.html (accessed Aug 29, 2001).

Erosion Control, Vegetative

Seth M. Dabney
United States Department of Agriculture (USDA), Oxford, Mississippi, U.S.A.

E

INTRODUCTION

Vegetation controls erosion by dissipating the erosive forces of rainfall and runoff (erosivity) and by reducing the susceptibility of soil to erosion (erodibility). Vegetation alters the partitioning of rainfall between infiltration, surface storage, and surface runoff. Erosivity is reduced because rainfall kinetic energy is absorbed, runoff volume is reduced due to increased infiltration, and runoff velocity is slowed through increased surface detention and reduced development of areas of concentrated flow. Vegetation reduces soil erodibility by increasing soil aggregation, binding aggregates together with roots, and lowering soil matric potential. Vegetation may cover the entire soil surface, as with crops, cover crops, or forests; or it may be limited to specific critical areas, as with various types of conservation buffers. This chapter reviews the mechanisms and processes by which vegetation reduces soil erosion by water, with emphasis on vegetative buffers. Crop residue effects are considered in another entry.

GENERAL MECHANISMS

Slower Runoff

Theoretically, if runoff occurs uniformly over a plane, its depth increases in a predictable manner as slope length increases. In practice, the development of concentrated flow areas of high velocity limits the depth of sheet flows. By slowing runoff, vegetation can reduce or delay the development of rills and associated concentrated-flow erosion. Vegetation may increase runoff depth 10-fold compared to an equivalent discharge over a smooth surface or fivefold deeper than rainfall-impacted flow over a natural bare soil surface.[1] By increasing water depth fivefold, average velocity, V, is reduced fivefold. Since erosivity of runoff is proportional to V^2 and its sediment transport capacity is proportional to V^5, (see Ref. [2]) vegetation reduces concentrated-flow erosion.

The retardation of surface runoff is a critical aspect of the functioning of conservation buffers. Fig. 1 shows the situation where sediment-laden runoff encounters a vegetated buffer. Because of the additional hydraulic resistance of stems and leaves, flow depth within the buffer, D_2, is greater than upslope of the buffer's influence, D_0. The depth at the upslope edge of the buffer, D_1, however, is greater even than that within the buffer (D_2) because of: 1) enhanced vegetation growth at the buffer margin; 2) compression of stems into a denser barrier; and 3) loading of the buffer edge with trapped residues and thatch. In many studies, more than half of the sediment trapped by vegetated buffers is deposited in the ponded area upslope of the buffer. Where the ponded area is deep and slow-flowing, transport capacity is negligible and the water surface approaches horizontal. In these circumstances, the fraction of particles with fall velocity V_{si} that will be trapped (T_i) is given by Ref. [3]:

$$T_i = 1 - \exp[-V_{si}L/q] \tag{1}$$

where q is the specific discharge and L is the length of the pond (Fig. 1). When the ponded area retains significant transport capacity, trapping efficiency is reduced and a transport capacity or sediment re-entrainment term must be added.[4]

Increased Infiltration of Water into Soil

Vegetation increases infiltration by: 1) reducing the development of surface seals that limit infiltration rates; 2) increasing soil water storage capacity through evapotranspiration; and 3) developing soil macroporosity through root growth and enhanced activities mesofauna such as earthworms and ants. By covering the soil and absorbing the kinetic energy of raindrops, vegetation can prevent the detachment and rearrangement of soil particles that result in the creation of soil seals[5] and thus increases infiltration. Although water use varies with species and climate, vegetation transpires approximately $0.3\,\text{m}^3$ of water for each kg of above-ground dry matter produced.[6] This transpiration leaves more capacity in the soil for infiltration of subsequent rains and thus reduces runoff and erosion.[7] Vegetation increases soil macroporosity directly through root growth[8] and indirectly by improving the habitat and activity of mesofauna.[9] By slowing runoff, vegetation increases the depth of ponded water and the area of soil that that is submerged, thus increasing opportunities for macropore flow.

Encyclopedia of Water Science
DOI: 10.1081/E-EWS 120010092

209

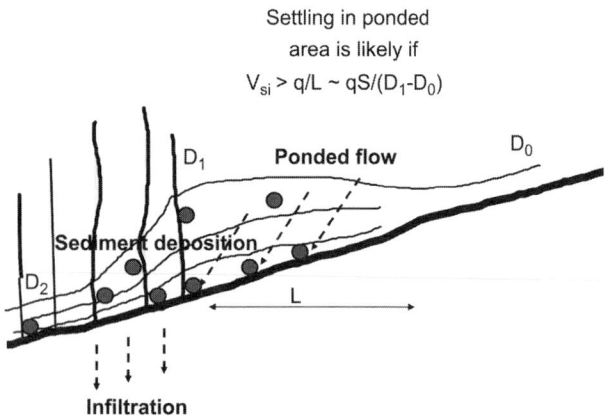

Settling in ponded
area is likely if

$V_{si} > q/L \sim qS/(D_1-D_0)$

Fig. 1 Schematic illustration of how vegetative buffers slow runoff, increasing flow depth and trapping sediment.

Reduced Soil Erodibility

Soil erodibility refers to the ease with which soil particles (primary or aggregates) can be detached and transported by the shear forces associated with raindrop splash or flowing water. Soil with increased organic matter content has greater aggregate stability,[10] and hence greater resistance to detachment and transport. The effects of vegetation on reducing erodibility include consolidation of soil with time after tillage and binding together of soil particles by roots and by microorganisms that use plant biomass and exudates as a food source.[11]

VEGETATIVE BUFFERS

Buffer Types

Conservation buffers designed to reduce soil erosion and/or sediment delivery are usually areas of perennial vegetation placed at critical points in a landscape. These buffers may be located along stream banks, along the edges of fields, or may be placed within fields. To distinguish among these buffer types, the nomenclature of the U.S. Department of Agriculture—Natural Resources Conservation Service (NRCS) is adopted.

The seven conservation buffers types that reduce sediment delivery in runoff are summarized in Table 1. Practices normally located at the edges of fields are listed first, and those usually placed within fields are listed last. In addition to controlling erosion and/or reducing sediment delivery, many of these buffers can also serve additional purposes such as improving water quality and providing wildlife habitat. Current national standards for these practices are given in the NRCS National Handbook of Conservation Practices, which is available on the internet:

http://www.ftw.nrcs.usda.gov/nhcp_2.html. Descriptive information about each practice can be found in the CORE4 training materials: http://www.nhq.nrcs.usda.gov/technical/ECS/agronomy/core4.pdf. Local specifications criteria can be found in the local NRCS Field Office Technical Guide.

The edge-of-field buffers are: Riparian forest buffer (RFB), filter strip (FS), and field border (FB). An RFB is a forested area adjacent to a water body and is frequently combined with grass buffers. A field boarder is a grassed field margin. Because it may be used for parking and turning equipment, a FB is also usually wider than the minimum indicated in Table 1. In contrast to an FB, traffic is usually excluded from an FS and vegetation and slope requirements are far more stringent (Table 1). Generally, edge-of-field buffers are designed primarily to trap sediment and infiltrate water, not to control in-field erosion. The RFB is an exception in that it can control concentrated flow erosion caused by out-of-bank flood flows. The FB controls local scour on sloping head lands where concentrated water flows enter or exit a field. To properly function, these edge-of-field buffers require that runoff pass through them as diffuse, sheet flow.

The other four buffer types in Table 1 function within fields and are designed to control in-field erosion. Three of these buffers, alley cropping (AC), contour buffer strip (CBS), and vegetative barrier (VB) control sheet-and-rill erosion by interrupting hillslopes with strips of permanent vegetation aligned close to the contour (Fig. 2). The widths of these buffers are often varied so that the edges of each cropped zone stay parallel and within strip gradient specifications (Table 1). Alley cropping involves growing crops and forages between strips of trees. Vegetative barriers are usually narrow strips of large stiff-stemmed grasses (Fig. 2). Contour buffer strips are somewhat wider strips with less stringent vegetation and contour alignment requirements (Table 1).

Only two buffer practices, grassed waterway (GW) and VB, may be specifically designed to control in-field concentrated-flow erosion. Grassed waterways are oriented up-and-down the slope and are planted with vegetation that is intended to be submerged while functioning. In contrast, VB designed to controlling concentrated-flow erosion are planted perpendicular to the flow direction and are intended to remain unsubmerged while retarding runoff.

Buffer Hydraulic Resistance

The hydraulic resistance of vegetation frequently is parameterized with Manning's equation:

$$V = \frac{1}{n}R^{2/3}S^{1/2} \tag{2}$$

Table 1 Comparison of water erosion control purposes and selected criteria of buffers types in the USDA-NRCS National Handbook of Conservation Practices that can be used to reduce sediment

Buffer type	NRCS code	Erosion control purposes		Criteria					
		Sheet-and-rill erosion	Concentrated flow erosion	Field slope (%)	Maximum strip gradient	Minimum strip width (SW) (m)	Strip spacing	Maximum field length	Minimum stem density
Riparian forest buffer	391		+		Along stream corridor	11			
Field border	386		+		Along field edge	6			
Filter strip	393			1–10	< 0.5%	6		50 × SW	1500 m^{-2}
Grassed waterway	412		+		Along flow gradient		In concentrated flow areas		n-VR curve and permissible velocity
Alley cropping	311	+			Contour	6	Species light requirements		
Contour buffer strip	332	+		2–8	< 2%	5 (Grass)	1/2 of RUSLE critical slope length (CSL)	RUSLE	540 m^{-2} (Grass)
						9 (Legume)		CSL	320 m^{-2} (Legume)
Vegetative barrier	601	+	+		< 1%	1	1.3–2.0 m		Depends on stem diameter (Table)

Source: http://www.ftw.nrcs.usda.gov/nhcp_2.html.

where V is the average flow velocity, R is the hydraulic radius (flow-area divided by wetted perimeter), S is the land slope gradient, and n is a hydraulic resistance parameter. Fig. 3 shows how Manning's n varies with the product V and R for three kinds of buffer vegetation. At low flows with unsubmerged vegetation, the hydraulic radius reduces to the flow depth, H, and VR equals the specific discharge. When the dominant component of hydraulic resistance is drag on emergent stems that are uniform with height, such as with the simulated FSs made of brush bristles (Fig. 3) in a flume with a smooth floor,[13] average velocity remains constant with increasing flow and n increases in proportion to the 2/3 power of discharge.[14]

Fig. 2 Vegetative barriers of vetiver grass (*Vetiveria zizanioides*) planted in rows on contour lines to hold the soil in St. Vincent, British West Indies, during the 1950s.[12]

At high flows, all of the vegetation is submerged and the main factor determining hydraulic resistance is the length of the stems that are dragging in the flow.[15] As discharge increases, more and more of the flow occurs in the zone above the submerged vegetation until eventually the hydraulic resistance of the vegetation becomes a constant. The vegetal retardance curve labeled "A" in Fig. 3 represents 0.9–1.0 m tall vegetation while "E" reflects vegetation that had been burned or mowed at about 4 cm height. In designing a GW, the erodibility of the underlying soil and the growth characteristics of the vegetal cover determine a maximum permissible velocity or the allowable hydraulic stress on the soil, and the channel is designed with dimensions great enough that, with expected vegetation, the permissible velocity or stress will not be exceeded at the design discharge.

Vegetative barriers have application at specific discharges that span the range between those of FS and GW (Fig. 3) and can thus be used to complement other buffer types by spreading out concentrated runoff. At low flows, the hydraulic resistance of VB increases more rapidly than the 2/3 power of discharge because stems and leaves become less clumped together, increasing projected area with increasing height in the lower canopy. At greater discharges, flow-depth increases to the point where stems begin to thin out or bend. Then average velocity increases, the flow resistance, expressed as Manning's n, ceases to increase and begins to decline, even while flow depth may continue to increase with increasing discharge.[16] The stiff grasses used to form VB remain erect and emergent at greater flows than other vegetation types in Fig. 3 because the large-diameter stems are stiffer and are on the order of 2 m tall. The enhanced growth and residue loading noted to occur at the edge of all buffers are also important factors

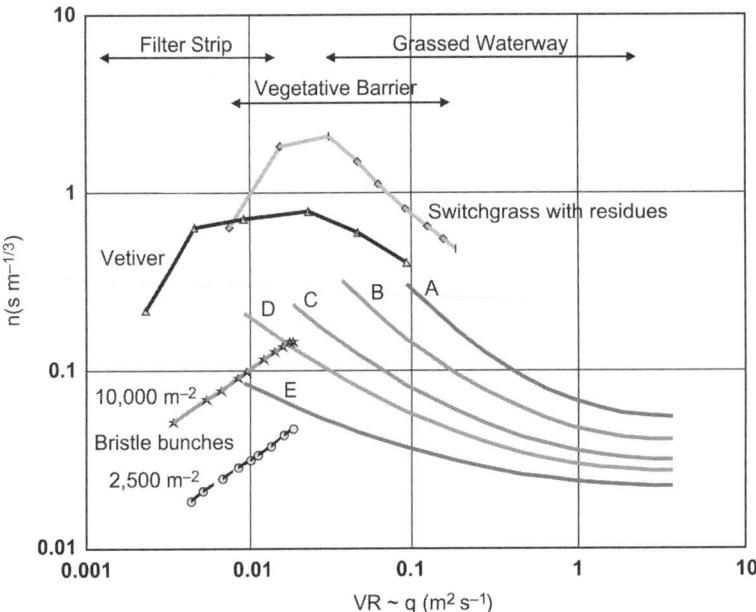

Fig. 3 Hydraulic roughness of vegetated areas first increases with increasing flow as more vegetation interacts with the flow, then decreases with increasing flow as flow approaches the height of the vegetation and submerges it. Data for A–E from Ref. [15]; brush bristle data from Ref. [13]; switchgrass (*Panicum virgatum*) data from Ref. [16]; vetiver from Dabney (unpublished).

that give VB greater hydraulic resistance than retardance class A vegetation. Riparian forest buffer vegetation, of course, remains erect at even greater flows than does VB vegetation, but usually offers less hydraulic resistance at low flows.

REFERENCES

1. Parsons, D.A. *Depths of Overland Flow*, SCS-TP-82; U.S. Dept. of Agriculture-Soil Conservation Service: Washington, DC, 1949; 1–33.
2. Meyer, L.D.; Wischmeier, W.H. Mathematical Simulation of the Process of Soil Erosion by Water. Trans. ASAE **1969**, *12* (754–758), 762.
3. Dabney, S.M.; Meyer, L.D.; Harmon, W.C.; Alonso, C.V.; Foster, G.R. Depositional Patterns of Sediment Trapped by Grass Hedges. Trans. ASAE **1995**, *38* (6), 1719–1729.
4. Beuselinck, L.; Govers, G.; Steegen, A.; Harisine, P.B.; Poesen, J. Evaluation of the Simple Settling Theory for Predicting Sediment Deposition by Overland Flow. Earth Surf. Process. Landforms **1999**, *24*, 993–1007.
5. Römkens, M.J.M.; Prasad, S.N.; Whisler, F.D. Surface Sealing and Infiltration. In *Process Studies in Hillslope Hydrology*; Anderson, M.G., Burt, T.P., Eds.; John Wiley & Sons, Ltd.: New York, 1990; 127–172.
6. Tanner, C.B.; Sinclair, T.R. Efficient Water Use in Crop Production: Research or Re-search. In *Limitations to Efficient Water Use in Crop Production*; Taylor, H.M., Ed.; Am. Soc. Agron.: Madison, WI, 1983; 1–27.
7. Zhu, J.C.; Gantzer, C.J.; Anderson, S.H.; Alberts, E.E.; Beuselinck, P.R. Runoff, Soil and Dissolved Nutrient Losses from No-Till Soybean with Winter Cover Crops. Soil Sci. Soc. Am. J. **1989**, *53*, 1210–1214.
8. Tomlin, A.D.; Shipitalo, M.J.; Edwards, W.M.; Protz, R. Earthworms and Their Influence on Soil Structure and Infiltration. *Earthworm Ecology and Biogeography in North America*; Lewis Pub.: Boca Raton, FL, 1995; 159–183.
9. Elkins, C.B.; Harland, R.L.; Homeland, C.S. Grass Roots as a Tool for Penetrating Soil Hardpans and Increasing Crop Yields. Proc. South. Pasture Forage Crop Imp. Conf. **1977**, *34*, 21–26.
10. Angers, D.A. Changes in Soil Aggregation and Organic Carbon Under Corn and Alfalfa. Soil Sci. Soc. Am. J. **1992**, *56*, 1244–1249.
11. Renard, K.G.; Foster, G.R.; Weesies, G.A.; McCool, D.K.; Yoder, D.C.; (coordinators) *Predicting Soil Erosion by Water: A Guide to Conservation Planning with the Revised Universal Soil Loss Equation (RUSLE)*, Agric. Handbook 703; U.S. Department of Agriculture-Agricultural Research Service: Washington, DC, 1997; 1–384.
12. Vélez, I. Soil Conservation Practices in the Caribbean Archipelago. Sci. Monthly **1952**, *74* (3), 183–185.
13. Jin, C.X.; Römkens, M.J.M.; Griffioen, F. Estimating Manning's Roughness Coefficient for Shallow Overland Flow in Non-submerged Vegetative Filter Strips. Trans. ASAE **2000**, *43* (6), 1459–1466.
14. Petryk, S.; Bosmajian, G. Analysis of Flow Through Vegetation. J. Hydr. Div. ASCE **1975**, *101* (HY7), 871–884.

15. Temple, D.M.; Robinson, K.M.; Ahring, R.M.; Davis, A.G. *Stability Design of Grass-Lined Open Channels*, Agric. Handbook 667; U.S. Department of Agriculture-Agricultural Research Service: Washington, DC, 1987.

16. Temple, D.; Dabney, S. *Hydraulic Performance Testing of Stiff Grass Hedges*, Proceedings of the Seventh Federal Interagency Sedimentation Conference, Reno, Nevada, 25–29 March, 2001; Vol. 2 (XI), 118–124.

Erosion and Precipitation

Bofu Yu
Griffith University, Nathan, Queensland, Australia

INTRODUCTION

Erosion is a natural process to detach soil and rock fragments for subsequent removal, or transportation, of these materials to areas of lower elevation on the surface of the earth. In the context of agriculture, the primary agents for erosion are water and wind. Climate, precipitation in particular, plays a critical role in determining where and when erosion occurs and the magnitude of erosion rate. Rainfall erosivity, i.e., the ability of rain to cause erosion, is largely a function of rain amount and peak intensity. Rainfall erosivity and its seasonal variation in relation to the Universal Soil Loss Equation (USLE) and the revised USLE (RUSLE) can be estimated from mean annual rainfall and daily rain amount. Low rainfall, dry soil surface, and poor ground cover are the necessary conditions for wind erosion to prevail.

RAINFALL EROSIVITY

Rainfall erosivity is a measure of the climatic influence on water erosion. When other variables such as topography and vegetation cover are held constant, the rate of erosion is directly related to the level of rainfall erosivity. A number of rainfall erosivity indices have been proposed so that the amount of soil eroded is linearly proportional to the rainfall erosivity index *ceteris paribus*. The most commonly used rainfall erosivity index is EI_{30}, where E is the total kinetic energy per unit area for a storm ($MJ\,ha^{-1}$) and I_{30} is its peak 30-min intensity ($mm\,h^{-1}$). Wischmeier and Smith[1] found that the combination of kinetic energy and peak intensity is most closely related to the observed amount of soil loss. The *R*-factor in the USLE and RUSLE is the mean annual sum of these EI_{30} values.[2,3] Other measures of rainfall erosivity worthy of note include the modified Fournier Index,[4,5] $KE > 1$ index,[6] and the so-called Universal Index of Onchev.[7] Numerous other attempts have been made to search for a rainfall-based estimator of the observed amount of erosion that is superior to EI_{30}. Most of these studies have relied on restricted databases that have limited their applicability. Most of these other indices or estimators are highly correlated with each other and with EI_{30}.

Although the definition of EI_{30} is straightforward, its calculation requires long-term rainfall data at short time intervals ($< 30\,min$) that are not widely available for most parts of the world. To develop a better understanding of what is exactly involved in EI_{30}, it is helpful to examine how this index is calculated. I_{30} is the maximum intensity for any 30-min interval in a storm, while the storm energy depends on how rainfall intensity varies during the event:

$$E = \int_T e(I)I\,\mathrm{d}t \tag{1}$$

where I is the rainfall intensity, T the rain duration, and $e(I)$ a function of rain intensity called the unit energy equation. The consensus is that the unit energy as a function of rain intensity assumes the following functional form[3,8]

$$e(I) = e_{\max}(1 - \alpha e^{-I/I_o}) \tag{2}$$

For RUSLE, the following was recommended: $e_{\max} = 0.29\,MJ\,ha^{-1}\,mm^{-1}$; $\alpha = 0.72$; $I_o = 20\,mm\,h^{-1}$.[3] It can be shown from Eqs. 1 and 2 that the storm energy is bounded:

$$0.28e_{\max}P < E < e_{\max}P \tag{3}$$

where P is the total rain (mm). The theoretical upper and lower bounds are related to zero and infinite intensity, respectively. Analyzing 6-min rain data for a number of sites around Australia shows that the ratio of storm energy to $e_{\max}P$ ranges mostly from 0.5 to 0.8, and the ratio is slightly higher in tropical/subtropical than in temperate regions (Table 1). From Table 1, it is also clear that the storm energy is always highly correlated with rain amount. Given that storm energy is primarily a function of rain total, it follows that rainfall erosivity, as defined in relation to USLE/RUSLE, depends mainly on rain total and peak intensity, and to a much lesser extent on rain duration.

For areas where long-term high-resolution rain data are unavailable, the simpler method to estimate rainfall erosivity in the context of USLE/RUSLE is to use the fairly consistent relationship between the mean annual rainfall and the *R*-factor:[9,10]

$$R\text{-factor} = 0.05(MAR)^{1.6} \quad R^2 = 0.82 \tag{4}$$

Encyclopedia of Water Science
DOI: 10.1081/E-EWS 120010323

Table 1 Linear relationship between rain amount (P) and storm energy (E) as in $E = \alpha e_{max} P$ for selected sites in Australia (n = number of storms analyzed; R^2—coefficient of determination, representing the fraction of the total variation in the observed E values that can be explained by rain amount)

Location	Climate	α	n	R^2
Perth	Temperate, winter rain	0.521	2354	0.96
Melbourne	Temperate, uniform rain	0.530	1800	0.93
Brisbane	Subtropical, summer rain	0.626	4088	0.96
Darwin	Tropical, summer rain	0.742	3701	0.98

where MAR is the mean annual rainfall (mm). The regression (Eq. 4) is based on a combined database for 161 sites (132 sites in the United States and 29 sites in Australia).[9,10] MAR ranges from 67 mm to 2060 mm for these sites. The nonlinear relationship suggests that a 10% change to MAR would lead to 16% change to rainfall erosivity. This highly sensitive nature of rainfall erosivity to rainfall would have important implications for the impacts of climate change on soil erosion. Reasonably good relationships between the Modified Fournier Index and the R-factor have also been noted.[5,9] The difference between the two estimates, however, is small, and little is gained by using the Modified Fournier Index.[10] If we need to estimate the seasonal distribution of rainfall erosivity, daily rain data can be used, especially, in areas with a marked wet season in winter. Monthly and annual rain total are no longer adequate because summer rain with high peak intensity can lead to higher rainfall erosivity in the relatively drier months. Rainfall erosivity can be related to rain amount using a power function in the form:

$$EI_{30} = aP^{\beta} \qquad (5)$$

The calibrated values of β for a number of sites around the world are summarized in Table 2. The β value mostly varies in the range from 1.5 to 1.8 with higher values found largely at higher latitudes. Such relationships for daily

erosivity are sufficient for determining the seasonal variation of rainfall erosivity for USLE/RUSLE.

Rain total and peak rainfall intensity are also key precipitation variables for a physical description of water erosion processes.[22–24] Mass balance dictates that in an area of net erosion, the amount of soil loss, SL, is given by

$$SL = Qc \qquad (6)$$

where Q is the runoff amount and c is the sediment concentration. In this context, the effects of rain on erosion manifest themselves in terms of the amount of surface runoff generated and the level of sediment concentration in the runoff water. With nonclimatic variables held constant, the amount of runoff is largely determined by rainfall amount and to a lesser extent by the rainfall intensity. Sediment concentration is related to both rainfall intensity and runoff rate. Rainfall detachment is linearly related to rainfall intensity. Shear stress or stream power commonly used to quantify flow detachment is intrinsically related to the runoff rate. Thus, in this physical framework for soil erosion, rainfall intensity plays a direct role in rain detachment. Rain amount and intensity also play an indirect role in flow detachment and transport of eroded sediments by determining the magnitude of runoff amount and runoff rate.

Precipitation is important to water erosion because soil particles and aggregates are detached by raindrops and surface runoff. A lack of precipitation, on the other hand, leads to low moisture levels near the soil surface, and thus renders the soil particularly susceptible to wind erosion. Wind speed, precipitation, and potential evaporation were used to develop indices of wind erosion.[25–27] For given wind speed and potential evaporation, wind erosivity is inversely related to precipitation. Fig. 1 shows schematic relationships between precipitation and vegetation cover, rainfall and wind erosivity, and predominant erosion processes. In high rainfall areas, the rate of actual erosion is not necessarily high in spite of high rainfall erosivity unless the usually good vegetation cover is removed and

Table 2 The average exponent and its one standard deviation in the power function relating daily rain (P) to rainfall erosivity (EI_{30}) as in $EI_{30} = aP^{\beta}$

Country	Latitude range	Number of sites	β ± 1 s.d.	References
Finland	60°N–66°N	8	1.77 ± 0.06	[11]
Canada	49°N–53°N	12	1.75 ± 0.13	[12]
The United States	31°N–43°N	11	1.81 ± 0.16	[13]
Italy	36°N–42°N	35	1.53 ± 0.19	[14]
Equatorial (Malaysia, Indonesia, Brazil)	4°N–10°S	4	1.64 ± 0.18	[15–17]
Australia (tropical region)	10°S–25°S	41	1.49 ± 0.28	[18]
South Africa	31°S–33°S	4	1.47 ± 0.17	[19]
Australia (temperate region)	28°S–35°S	33	1.49 ± 0.25	[20,21]

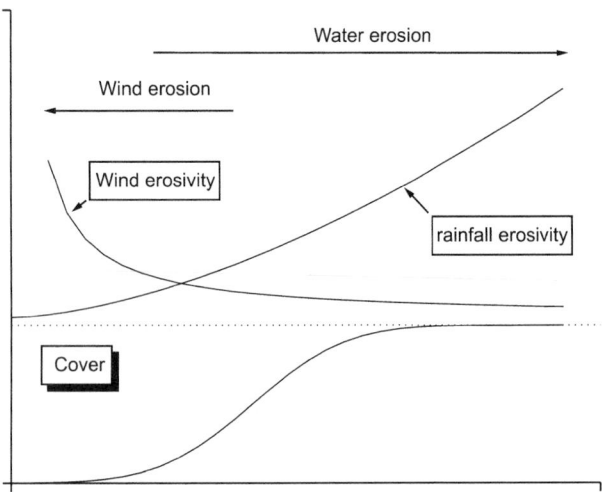

Fig. 1 Schematic relationships between precipitation, vegetation cover, rainfall, and wind erosivity.

the soil surface exposed. In arid and semiarid areas with low rainfall, the combined effects of poor ground cover and dry soil surface make the land particularly vulnerable to wind erosion.

CONCLUSION

Precipitation is a key climatic variable that determines the type and magnitude of erosion. In the context of water erosion, rain amount and peak intensity are the most important variables in determining the erosion rate. For areas without high-resolution rainfall intensity data, the R-factor and its seasonal variation for USLE/RUSLE can be estimated from mean annual rainfall and daily rain amount. Absence of rain, concomitant dry soil surface, and poor ground cover are the necessary conditions for wind to become the dominant erosion agent.

REFERENCES

1. Wischmeier, W.H.; Smith, D.D. Rainfall Energy and Its Relationship to Soil Loss. Trans. Am. Geophys. Union **1958**, *39*, 285–291.
2. Wischmeier, W.H.; Smith, D.D. *Predicting Rainfall Erosion Losses—A Guide to Conservation Planning*, Agriculture Handbook No. 537; U.S. Department of Agriculture, Washington, D.C., 1978.
3. Renard, K.G.; Foster, G.R.; Weesies, G.A.; McCool, D.K.; Yoder, D.C. Coordinators. *Predicting Soil Erosion by Water: A Guide to Conservation Planning with the Revised Universal Soil Loss Equation (RUSLE)*, Agriculture Handbook No. 703; U.S. Department of Agriculture, Washington, D.C., 1997.
4. Fournier, F. *Climat et Erosion: La Relation Entre l'Erosion du Sol par l'Eau et les Precipitations Atmospheriques*; Presses Universitaires de France: Paris, 1960.
5. Arnoldus, J.M.J. Methodology Used to Determine the Maximum Potential Average Annual Soil Loss Due to Sheet and Rill Erosion in Morocco. FAO Soils Bull. **1977**, *34*, 39–51.
6. Hudson, N.W. *Soil Conservation*; B. T. Batsford: London, 1976.
7. Onchev, N.G. Universal Index for Calculating Rainfall Erosivity. In *Soil Erosion and Conservation*; El-Swaify, S.A., Moldenhauer, W.C., Lo, A., Eds.; Soil and Water Conservation Society: Ankeny, Iowa, 1985; 423–431.
8. van Dijk, A.I.J.M.; Bruijnzeel, L.A.; Rosewell, C.J. Rainfall Intensity–Kinetic Energy Relationships: A Critical Literature Appraisal. J. Hydrol. **2002**, *26* (1), 1–23.
9. Renard, K.G.; Freimund, J.R. Using Monthly Precipitation Data to Estimate The R-Factor in the Revised USLE. J. Hydrol. **1994**, *157*, 287–306.
10. Yu, B.; Rosewell, C.J. A Robust Estimator of The R-Factor for the Universal Soil Loss Equation. Trans. ASAE **1996**, *39*, 559–561.
11. Posch, M.; Rekolainen, S. Erosivity Factor in the Universal Soil Loss Equation Estimated from Finnish Rainfall Data. Agric. Sci. Finl. **1993**, *2*, 271–279.
12. Bullock, P.R.; de Jong, E.; Kiss, J.J. An Assessment of Rainfall Erosion Potential in Southern Saskatchewan from Daily Rainfall Records. Can. Agric. Eng. **1989**, *32*, 17–24.
13. Richardson, C.W.; Foster, G.R.; Wright, D.A. Estimation of Erosion Index from Daily Rainfall Amount. Trans. ASAE **1983**, *26*, 153–157, 160.
14. Bagarello, V.; D'Asaro, F. Estimating Single Storm Erosion Index. Trans. ASAE **1994**, *37*, 785–791.
15. Yu, B.; Hashim, G.M.; Eusof, Z. Estimating The R-Factor using Limited Rainfall Data: A Case Study from Peninsular Malaysia. J. Soil Water Conserv. **2001**, *56* (2), 101–105.
16. van der Linden, P. Soil Erosion in Central-Java (Indonesia): A Comparative Study of Erosion Rates Obtained by Erosion Plots and Catchment Discharges. In *Rainfall Simulation, Runoff and Soil Erosion*; de Ploey, J., Ed.; 1983; Catena Supplement 4, Elsevier, Amsterdam, 141–160.
17. Elsenbeer, H.; Cassel, D.K.; Tinner, W. A Daily Rainfall Erosivity Model for Western Amazonia. J. Soil Water Conserv. **1993**, *48*, 439–444.
18. Yu, B. Rainfall Erosivity and Its Estimation for Australia's Tropics. Aust. J. Soil Res. **1998**, *36*, 143–165.
19. van Breda Weaver, A. Rainfall Erosivity in Ciskei: Its Estimation and Relationship with Observed Soil Erosion. SA Geographer **1989-1990**, *17* (1/2), 13–23.
20. Yu, B.; Rosewell, C.J. An Assessment of a Daily Rainfall Erosivity Model for New South Wales. Aust. J. Soil Res. **1996**, *34*, 139–152.

21. Yu, B.; Rosewell, C.J. Rainfall Erosivity Estimation Using Daily Rainfall Amounts for South Australia. Aust. J. Soil Res. **1996**, *34*, 721–733.

22. Foster, G.R.; Flanagan, D.C.; Nearing, M.A.; Lane, L.J.; Risse, L.M.; Finkner, S.C. Hillslope Erosion Component. In *USDA-Water Erosion Prediction Project: Hillslope Profile and Watershed Model Documentation*, NSERL Report No. 10; Flanagan, D.C., Nearing, M.A., Eds.; USDA-ARS Nat. Soil Erosion Research Lab.: West Lafayette, Ind., 1995, Chap. 11.

23. Rose, C.W.; Coughlan, K.J.; Ciesiolka, C.A.A.; Fentie, B. Program GUEST (Griffith University Erosion System Template). In *A New Soil Conservation Methodology and Application to Cropping Systems in Tropical Steeplands*, Technical Report, No. 40; Coughlan, K.J., Rose, C.W., Eds.; Australian Centre for International Agricultural Research: Canberra, 1997; 34–58.

24. Morgan, R.P.C.; Quenton, J.N.; Smith, R.E.; Govers, G.; Poesen, J.W.A.; Auerswald, K.; Chisci, G.; Torri, D.; Styczen, M.E. The European Soil Erosion Model (EUR-OSEM): A Dynamic Approach for Predicting Sediment Transport from Fields and Small Catchments. Earth Surf. Process. Landforms **1998**, *23*, 527–544.

25. Chepil, W.S.; Siddoway, F.H.; Armbrush, D.V. Climatic Factor for Estimating Wind Erodibility of Farm Fields. J. Soil Water Conserv. **1962**, *9*, 257–265.

26. FAO (Food and Agriculture Organization, United Nations). Soil Erosion by Wind and Measuring for Its Control on Agricultural Lands. Development Paper No. 71. Rome, Italy, 1979.

27. Skidmore, E.L. Wind-Erosion Climatic Erosivity. Clim. Change **1986**, *9*, 195–208.

Erosion Process Modeling

John M. Laflen
Purdue University, Buffalo Center, Iowa, U.S.A.

INTRODUCTION

Erosion is the removal (detachment) of a mass of soil from one part of the earth and its relocation (transport and deposition) to other parts of the earth. Water erosion is that portion of erosion caused by water.

Modeling of water erosion includes modeling the state of the soil and biomass system on and below the land surface in addition to modeling the detachment, transport, and deposition of eroded material. The state of the system when rainfall/snowmelt/irrigation occurs determines the reaction to the forces applied by rainfall; the magnitude of and reaction to the forces applied by surface runoff; and to a great degree, the total detachment, transport, and deposition of soil during rainfall events.

Modeling the state of the system involves modeling of many processes. These would include hydrologic processes—including those related to water movement, use, and storage above and below ground; those related to the accumulation, decomposition, and use of biomass-plant growth, root growth, biomass decomposition, grazing, and addition or removal of biomass. While such modeling is critical to accurately model soil loss from most lands, it is a comprehensive subject beyond the scope of this chapter.

A model is defined[1] as "a system of postulates, data, and inferences presented as a mathematical description of an entity or state of affairs." The objective here is to describe these processes and how they might be mathematically described and modeled. The processes described are those that occur on source areas on relatively small tracts of land.

MODELING THE EROSION PROCESS

The erosion process is usually visualized as detachment and transport by rainfall and detachment, transport, and deposition by flowing water. The detachment and transport by rainfall is usually termed interrill erosion. The detachment, transport, and deposition by flowing water, depending on scale, is referred to as rill erosion, channel erosion, ephemeral gully erosion, or gully erosion—in this chapter they are lumped appropriately as channel erosion.

Interrill Erosion

Interrill erosion is the detachment and transport of soil by raindrops and very shallow flow. It is constant down a slope as long as soil and surface properties remain constant.[2] Interrill processes generally occur within a meter or so of the point of impact of a water drop, and deliver the detached material to nearby channels called rills. If there is no flow in a channel, the detached interrill material stays close to the point of detachment. Interrill erosion is usually most apparent on row sideslopes.

The forces and energies in interrill processes are derived from waterdrops (rainfall and irrigation) and the shallow flows near where these drops impact the soil surface. Interrill erosion is not positionally sensitive, being relatively constant over an entire surface where cover, microtopography, soil, and waterdrops remain constant.

Interrill erosion has been modeled a number of ways. The detachment of soil due to individual raindrops has been mathematically modeled.[3,4] Numerous equations, empirical in nature, have been developed that express the detachment and transport of raindrops as a function of slope steepness, and raindrop and runoff characteristics. In the development of the Universal Soil Loss Equation, soil erosion for a rainfall event was expressed as a function of rainfall energy and a maximum 30-min rainfall intensity.[5] In a classic modeling of the erosion process[6] interrill detachment (D_i) was expressed as a function of the intensity (I) squared:

$$D_i = \alpha I^2 \tag{1}$$

Recently,[7] interrill erosion detachment rate was related to interrill slope and the intensity squared as:

$$D_i = K_i I^2 (1.05 - 0.85\,e^{-4\sin\Phi}) \tag{2}$$

where K_i is the interrill soil erodibility and Φ is the slope angle.

In the Water Erosion Prediction Project (WEPP) model[8] interrill erosion is modeled as the product of intensity and flow rate. Interrill detachment is written as

$$D_i = K_i I q \mathrm{adj} \tag{3}$$

where q is the flow rate and adj is a series of adjustment factors including adjustments for slope (as given in Eq. 2),

Encyclopedia of Water Science
DOI: 10.1081/E-EWS 120010224

sealing and crusting, residue cover, canopy cover, and canopy height. The major difference between Eqs. 2 and 3, in addition to the expansion of adjustment factors beyond the slope, is the use of an Iq term rather than the I^2 term in Eq. 2. This change was based on Australian research[9,10] and on the observation that interrill detachment was low for soils that had low runoff rates.

The energy of raindrops does the greatest damage in interrill areas, causing crusts on the soil surface that greatly increase surface runoff on interrill areas.[11] This runoff then drives the erosion and sediment transport process in channels. Thus, soil erosion control must begin in interrill areas in the control of rates and volumes of surface runoff.

Interrill erosion occurs at the soil surface—the region of the soil that is most biologically and chemically active. Interrill soil erosion removes a disproportionate amount of the soil's fertility, chemicals for the control of weeds, insects and diseases, and organic matter. These losses can eventually have serious consequences for the soil, and for receiving waters. The loss of fertility was the basis for establishing soil tolerance values in the United States,[12] and interrill erosion rates under clean tillage are often near the allowable soil loss.

Channel Erosion

Channel erosion is distinctly and visibly different than interrill erosion. Because they are distinctly different processes, they are modeled separately from interrill erosion. Channels are the visible erosion process that points to the existence of a threat to the sustainability of a land resource. Interrill erosion scarcely leaves a visible

mark on the land, channel erosion causes ditches, gullies, and serious impediments to farming. Channel processes are positionally sensitive. Until the hydraulic forces that detach channel material exceed a limiting value, channel erosion does not occur, an important element in stable channel design.

Channel erosion takes many forms. First, it may take the form of rill erosion—the channel that interrill material is usually delivered to. Rill erosion is viewed as a channel that receives only interrill material. In rills, a common expression for representing the detachment due to flowing water is in an excess hydraulic shear model (Fig. 1) with adjustment for sediment in transport as shown in Eq. 4

$$D_r = K_r(\tau - \tau_c)(1 - g/T_c) \qquad (4)$$

where D_r is the rill detachment, K_r, the rill erodibility, τ, the hydraulic shear, τ_c, the critical hydraulic shear, g, the sediment load, and T_c, the sediment transport capacity. There has been much discussion in the literature about whether or not hydraulic shear is the detaching mechanism, and whether or not the sediment in transport affects sediment detachment. Rill erosion and rill initiation is apparently greatly influenced by seepage forces.[13]

Larger channels are formed where more than one rill intersect, or where flow concentrates. Some of these channels may be called ephemeral gullies, such gullies can be obliterated by tillage implements, hence the term "ephemeral." The process in these channels and gullies are mathematically described in CREAMS,[14] and are represented in a model for estimating erosion in ephemeral gullies.[15,16] They are modeled similarly in the WEPP model.[8] In these models, erosion occurred over some width (computed or input) with material removed until a soil layer was reached that would not erode (usually the bottom of the latest tillage depth), then the channel would widen until it had reached an ultimate width where the flow depth was so low that the hydraulic shear was less than the critical shear. In studies of ephemeral gullies in the United States, the ratio of erosion from these small channels to sheet and rill erosion ranged from 0.24 to 1.47.

Gully erosion is found on many lands. Gullies may destroy the land, making it unusable for intense agricultural production. In the 1930s, it was reported that 20 million ha of former U.S. cropland was useless for further production because it had been stripped of topsoil or riddled with gullies and that most of this land had been abandoned.[17] Much of the Southern Piedmont has been stripped of its topsoil, and dissected and gullied so badly that the land is unsuitable for agriculture, with the entire area having lost an average of 0.17 m of topsoil[18] in the 270 yr of settlement—about 800 t/km²/yr. The erosion was attributed to the use of clean-cultivated cash crops, and the

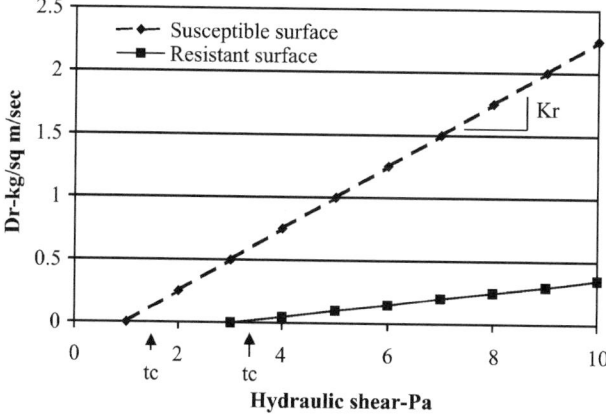

Fig. 1 Visual representation of an excess hydraulic shear model. The slope of the line is the rill erodibility, and the intercept with the x axis is the critical hydraulic shear. Soil differences and management differences may drastically change both rill erodibility and critical hydraulic shear.

exploitative nature of the land clearing and farming methods, quite similar in effect to those of Africa.[19]

The failure sequence for gully formation begins with a channel deepening when the force of flowing water exceeds the channels resisting force.[20] Then banks fail, and material is deposited in the gully and is carried away by subsequent flows. When a headcut occurs, material is deposited in the gully, and is carried away by subsequent flows. The location of gullies and the rate of gully formation is difficult to predict and model.

The failure sequence of gullies in the western Iowa loessial soil area begins with a weakening of the soil material at the base of the gully wall[21] attributable to wetting of the soil at the base of the gully wall. Once the base failed, overhanging material sloughed and then eroded material was transported downstream. The depth to water table in relation to the geometry of the gully bank played an important role in gully head and gully bank failure. It was observed that soil strength decreases with increasing moisture content, that seepage forces might be important, and that the increased unit weight of the soil mass with greater water content exerts more force.

A good gully erosion model has not been developed and accepted by the modeling community.

Deposition

Only a small portion of material detached and transported by interrill and channel processes reach major water bodies. Much of the detached and transported material deposits within a short distance from where it was detached. Major deposition sites are where slopes flatten, where flow velocities are reduced, or where temporary pondage occurs. Footslopes, culverts, fence lines, and small impounds are major deposition sites near sources of eroded material.

Deposition is a very selective process, with larger, denser particles depositing more readily than smaller, less dense particles. Usually Stokes law is used to estimate fall velocities of sediments based on eroded sediment sizes and densities.[21] This approach has been used in modeling impoundments.[22,23]

Deposition in channels is generally modeled when the sediment transport capacity is less than the sediment load in suspension. It has been modeled similarly to settling tanks in water treatment plants, with sediment fall velocity being modeled based on sediment size and density, and deposition based on these fall velocities and the channel flow velocity

$$\text{Deposition} = \beta(V_f/q)(T_c - g) \qquad (5)$$

where β is a rain induced turbulence coefficient (0.5 when runoff due to rainfall or sprinkler irrigation, otherwise 1.0

for snowmelt and furrow irrigation), V_f, the fall velocity of sediment (m/sec), q, the flow rate per unit width (m²/sec), T_c, the transport capacity (kg/sec/m), and g, the sediment load (kg/sec/m).[24] According to Eq. 5, when transport and sediment load are equal, there is no deposition computed. When the fall velocity is small, and rain is occurring on the channel, deposition rates would be expected to be small. Deposition rates are calculated for individual sediment sizes and densities, then deposition rates can be integrated across the sediment sizes to estimate total deposition amounts within a channel reach.

CONCLUSION

Erosion processes are usually visualized as interrill and channel processes, although channel processes include an extremely broad range of channels—including small rills, larger channels, ephemeral gullies, and classical gullies. The processes described herein have been represented mathematically in a number of different ways. In this paper, we have limited the discussion to those that occur on smaller tracts of land, choosing to avoid continuously flowing streams on large areas.

The erosion processes are driven by both rainfall and runoff. Interrill processes are due almost entirely to rainfall and shallow runoff while channel processes are driven by runoff. Hence, control practices must be tailored to meet the conditions for controlling detachment in both areas, and to induce deposition in selected areas.

REFERENCES

1. *Merriam Webster's Collegiate Dictionary*, 10th Ed.; Merriam-Webster, Inc.: Springfield, MA, 1996.
2. Young, R.A.; Wiersma, J.L. The Role of Raindrop Impact in Soil Detachment and Transport. Water Resour. Res. **1973**, *9* (6), 1629–1636.
3. Huang, C.; Bradford, J.M.; Cushman, J.H. A Numerical Study of Raindrop Impact Phenomena: The Rigid Case. Soil Sci. Soc. Am. Proc. **1982**, *46*, 14–19.
4. Huang, C.; Bradford, J.M.; Cushman, J.H. A Numerical Study of Raindrop Impact Phenomena: The Elastic Deformation Case. Soil Sci. Soc. Am. Proc. **1983**, *47*, 855–861.
5. Wischmeier, W.H.; Smith, D.D. Rainfall Energy and Its Relationship to Soil Loss. T. Am. Geophys. U. **1958**, *39*, 285–291.
6. Meyer, L.D.; Wischmeier, W.H. Mathematical Simulation of the Process of Soil Erosion by Water. T. Am. Soc. Agric. Eng. **1969**, *12* (6), 754–762.
7. Liebenow, A.; Elliot, W.J.; Laflen, J.M.; Kohl, K.D. Interrill Erodibility: Collection and Analysis of Data from

E

Cropland Soils. T. Am. Soc. Agric. Eng. **1990**, *33(6)*, 1882–1888.

8. Laflen, J.M.; Elliot, W.J.; Flanagan, D.C.; Meyer, C.R.; Nearing, M.A. WEPP-Predicting Water Erosion using a Process-Based Model. J. Soil Water Conserv. **1997**, *52* (2), 96–102.

9. Kinnell, P.I.A. Interrill Erodibilities Based on the Rainfall Intensity–Flow Discharge Erosivity Factor. Aust. J. Soil Res. **1993a**, *31* (3), 319–332.

10. Kinnell, P.I.A. Runoff as a Factor Influencing Experimentally Determined Interrill Erodibilities. Aust J. Soil Res. **1993b**, *31* (3), 333–342.

11. Duley, F.L. Surface Factors Affecting Rate of Intake of Water. Soil Sci. Soc. Am. Proc. **1939**, *4* (1), 60–64.

12. Smith, D.D. Interpretation of Soil Conservation Data for Field Use. Agric. Eng. **1941**, *22* (5), 173–175.

13. Huang, C.; Laflen, J.M. Seepage and Soil Erosion for a Clay Loam Soil. Soil Sci. Soc. Am. J. **1996**, *60*, 408–416.

14. Foster, G.R.; Lane, L.J.; Nowlin, J.D.; Laflen, J.M.; Young, R.A. A Model to Estimate Sediment from Field-Sized Areas. In *CREAMS: A field-Scale Model for Chemicals, Runoff, and Erosion from Agricultural Management Systems*; Knisel, W.A., Ed.; USDA Conservation Research Report-26, Washington, D.C., 1980; 36–64.

15. Laflen, J.M.; Watson, D.A.; Franti, T.G. Ephemeral Gully Erosion. *Proc. 4th Federal Interagency Sedimentation Conf., March 24–27, 1986, Las Vegas NV*; Section 3, 1986; Vol. 1, 299–307.

16. Watson, D.A.; Laflen, J.M.; Franti, T.G. Estimating Ephemeral Gully Erosion. Am. Soc. Agric. Eng. **1986**, Paper No. 86-2020 American Society of Agricultural Engineering, St. Joseph, MI.

17. Bennett, H.H. *Soil Conservation*; McGraw-Hill Book Company, Inc.: New York, 1939.

18. Trimble, S.W. *Man-Induced Soil Erosion on the Southern Piedmont*; Soil and Water Conservation Society: Ankeny, IA, 1974; 1700–1970.

19. Lal, R. Low-Resource Agriculture Alternatives in Sub-Saharan Africa. J. Soil Water Conserv. **1990**, (4), 437–445.

20. Bradford, J.M.; Farrell, D.A.; Larson, W.E. Mathematical Evaluation of Factors Affecting Gully Stability. Soil Sci. Soc. Am. Proc. **1973**, *37* (1), 103–107.

21. Foster, G.R.; Young, R.A.; Neibling, W.H. Sediment Composition for Nonpoint Source Pollution Analyses. T. Am. Soc. Agric. Eng. **1985**, *28*, 133–139.

22. Lindley, M.R.; Barfield, B.J.; Wilson, B.N. Surface Impoundment Element Model Description. In *USDA-Water Erosion Prediction Project (WEPP) Hillslope Profile and Watershed Model Documentation, NSERL Report No. 10*; Flanagan, D.C., Nearing, M.A., Eds.; National Soil Erosion Research Laboratory, USDA-Agricultural Service: West Lafayette, IN, 1995.

23. Laflen, J. M. Simulation of Sedimentation in Tile-Outlet Terraces PhD Dissertation, Iowa State University, Ames, IA.

24. Foster, G.R.; Flanagan, D.C.; Nearing, M.A.; Lane, L.J.; Risse, L.M.; Finkner, S.C. Hillslope Erosion Component. In *USDA-Water Erosion Prediction Project (WEPP) Hillslope Profile and Watershed Model Documentation, NSERL Report No. 10*; Flanagan, D.C., Nearing, M.A., Eds.; National Soil Erosion Research Laboratory, USDA-Agricultural Service: West Lafayette, IN, 1995.

Erosion and Productivity

Francisco Arriaga
Birl Lowery
University of Wisconsin, Madison, Wisconsin, U.S.A.

INTRODUCTION

Land productivity is influenced by many factors including sunlight and precipitation, but the most productive land can be altered by a simple process like erosion. Although soil erosion is a natural process, it creates serious problems, both environmental and economical, worldwide. Soil erosion and deposition of eroded material have a detrimental effect on soil and crop production and on surface water quality. Erosion causes soil degradation by removing topsoil, which is often rich in organic matter, and by reducing the total depth of the soil profile. Additionally, erosion causes off-site water contamination by transporting agricultural chemicals, such as pesticides, fertilizers, and naturally occurring minerals or biologically derived nutrients, to rivers and lakes. Therefore, it is essential that we reduce soil erosion and understand what effects it may have, so that to the extent possible we can minimize the harm caused by erosion.

The erosional process alters important soil physical, chemical, and biological properties necessary for optimal crop production.[1] It is often agreed that the main impact of erosion on soil productivity is caused by changes in soil chemical properties (i.e., fertility); however, soil physical (i.e., water holding capacity) properties undergo significant changes that are often overlooked. Fertilizers and manures have been used with varying levels of success to restore the fertility of eroded land, and manures might restore some physical properties such as water holding capacity and structure. However, the total soil depth is irreplaceable. It is universally accepted that the long-term productivity potential of an eroded soil is lower than that of an uneroded one. Simulation models of soil erosion and changes in long-term crop productivity for various regions of western Europe estimate that productivity could drop as much as 30% for soils with a shallow profile, less than 75 cm, by 2100.[2] However, these estimates are somewhat conservative since they only take into account soil depth, and not changes in soil organic matter and nutrient losses.[3]

EROSION AND CROP PRODUCTION: SOIL PHYSICAL PROPERTIES

Erosion is defined as the detachment and movement of soil by water, wind, or ice. Many factors affect the erosional process; however, the type of soil, ground cover, and landscape are considered the most important ones. One of the most noticeable effects of soil erosion is the reduction in organic matter of the surface soil layers.[4–8] Since organic matter plays a crucial role in soil structure and in the formation of soil aggregates,[9–12] it is not surprising that researchers have found a decrease in aggregation and aggregate stability in eroded soils.[5–7] A reduction in aggregate stability can result in decreased water infiltration rates, and thus reduced water recharge of the soil profile for plant use and groundwater recharge. Additionally, a decrease in aggregation can hamper crop-seedling emergence, root growth and development, and tillage operations through the formation of soil surface crusts and increases in soil bulk density.[13–17]

Scientists have found a correlation between reduced crop yields and decline in organic matter contents in eroded soils. 20 yr after soil desurfacing, Lindstrom et al.[4] found a decrease in organic matter levels in the Ap horizon (surface soil) with increasing depth of topsoil removal. This decrease in organic matter was accompanied by a decrease in corn grain and stover yields, as well as an increase in soil bulk density of surface and subsurface horizons. Similarly, Schumacher et al.[14] found a reduction in organic carbon, in the Ap horizon, of about 10% from moderate to severe erosion areas in a study conducted to examine properties of 11 soils in the North Central Region of the United States. However, scientists have also reported an increase in organic carbon from moderate to severe erosion in 2 of the 11 soils studied. Increases in organic carbon with increasing erosion level are infrequent, but can be attributed to increased clay contents in the surface of eroded soil (from the exposure of subsoil rich in clayey materials), and consequently to an increased interaction between soil particles and organic carbon, making organic carbon more stable in the soil.[18–20] Nevertheless, reductions in corn yields on eroded areas were observed for the 11 soils in the Schumacher et al.[14] study. Lowery et al.[16] found a significant increase in bulk density of the Ap horizon, as well as an increase in clay content, decreases in plant available water, and decrease in hydraulic conductivity of saturated soil for the same 11 soils investigated by Schumacher et al.[14] Corn grain yield decreased by 30% following removal of the surface 20 cm of a silty clay loam soil to simulate erosion.[7] Since

Encyclopedia of Water Science
DOI: 10.1081/E-EWS 120010093

fertilizer was applied at twice the rate in the desurfaced areas, reduction in grain production was attributed to decreased soil organic carbon, crack formation, drought stress, and corn disease.

Crop yield is generally related to the amount of water that is available to a crop from the soil. Greater capacity to hold water because of greater clay content can result in greater crop yields on eroded land in years when rainfall is less than normal.[21,22] Since the amount and time of precipitation have great effects on crop yield, the effects of erosion are more pronounced in some years than others.[23]

Because of the impact of soil water on yield, position in the landscape has an influence on productivity.[21–25] In general, linear slopes are more eroded than foot and head slopes. This relationship between landscape position and erosion adds to the difficulty of assessing the effects of erosion on crop productivity. On sloping terrain, landscape variations contribute to the many factors determining where water infiltrates and where it flows after a rainfall event. In general, water tends to run off steep sloping areas and infiltrate in lower landscape positions. Thus, lower landscape positions tend to be more productive than steeper slopes.[8,24]

In addition to landscape position, poor plant production can be attributed to changes in soil-water holding characteristics which can be altered by erosion.[26] Water is held in the soil under greater negative pressure, making it less available for crop use, with increasing level of erosion because of increases in clay content in the exposed lower horizons. Damage to soil physical properties caused by erosion has a significant negative impact on crop production.[6,13,27]

EROSION AND CROP PRODUCTION: SOIL CHEMICAL PROPERTIES

Organic matter not only plays an important role in shaping the soil physical characteristics, but also affects soil chemical properties. It serves as a source of plant nutrients and aids in the soil pH buffering capacity. Humus, or stable soil organic matter, is one of the most chemically active components in soil and serves as a major reservoir for charged molecules, reducing the loss of nutrients and pesticides by leaching.[18–20] When organic matter is reduced by erosion, there is a greater potential for leaching of nutrients which leads to a decline in soil productivity.

Lack of phosphorus (P) has been linked to reduced crop yields in eroded soils. Delays in emergence, plant development, and yield have been recorded in eroded areas.[4,28] This has been attributed to reduced P uptake by plants grown on eroded land.[29]

Nutrient loss from erosion has been described as one of the major causes of soil fertility depletion in Kenya[30] and in the Phillippines.[8] Soil-water erosion is associated with plant nutrient removal, especially P. Sediment collected from eroded areas is usually richer in P than the original soil. Changes in soil pH, organic carbon, and total nitrogen can also be correlated to soil loss by erosion. Thus, soil erosion removes necessary plant nutrients. However, when nutrients are lost by erosion, the loss can be compensated for by fertilizer application, but loss of soil organic matter is not easily replaceable and affects soil chemical and physical properties. As previously noted, organic matter improves soil-water holding capacity and aggregate stability.

CONCLUSIONS

Since important soil properties for plant production are degraded by soil erosional processes, crop productivity is often reduced in eroded soils. Even though intensive farming practices can mask some of the effects of erosion on crop production, erosion effects are still real and detrimental to long-term soil quality and production. Soil erosion mainly impacts and changes soil chemical and physical properties. Most of these changes are caused by the removal of surface soil layers and the subsequent exposure of lower soil horizons. Major changes in soil properties include soil particle size distribution and organic matter content. Changes in soil particle size distribution depend on the existing soil conditions, but in most cases, clay content increases with increasing erosion. Since surface soil rich in organic matter is removed during the erosional process, organic matter content is reduced in eroded soils. Changes in these two soil characteristics usually create changes in other important soil properties, such as bulk density, aggregation, water retention, hydraulic conductivity, CEC, pH, and nutrient availability, among others. Changes in soil particle size distribution are difficult, if not impossible, to reverse, and can be considered more or less permanent. However, organic matter contents can potentially be increased by applying organic matter sources. One such source is animal manure. Increases in organic matter can help to ameliorate the effects of erosion on soil properties, especially soil physical properties. Therefore, cattle manure has been proposed for use on eroded soil as an amendment to ameliorate the effects of erosion. Furthermore, as already discussed, aggregate formation and stability are aided by soil organic matter. Thus, organic matter can potentially increase a soil's resistance to erosion.

REFERENCES

1. Lal, R. Effects of Soil Erosion on Crop Productivity. CRC Crit. Rev. Plant Sci. **1987**, *5*, 303–367.

2. de la Rosa, D.; Moreno, J.A.; Mayol, F.; Bonson, T. Assessmentof Soil Erosion Vulnerability in Western Europe and Potential Impact on Crop Productivity Due to Loss of Soil Depth Using ImpelERO Model. Agric. Ecosyst. Environ. **2000**, *81*, 179–190.

3. Hoag, D.L. TheIntertemporal Impact of Soil Erosion on Non-uniform Soil Profiles: A New Direction in Analyzing Erosion Impacts. Agric. Syst. **1998**, *56*, 415–429.

4. Lindstrom, M.J.; Schumacher, T.E.; Lemme, G.D.; Gollany, H.M. SoilCharacteristics of a Mollisol and Corn (*Zea mays* L.) Growth 20 Years After Topsoil Removal. Soil Tillage Res. **1986**, *7*, 51–62.

5. Dormaar, J.F.; Lindwall, C.W.; Kozub, G.C. Effectivenessof Manure and Commercial Fertilizer in Restoring Productivity of an Artificially Eroded Dark Brown Chernozemic Soil Under Dryland Conditions. Can. J. Soil Sci. **1988**, *68*, 669–679.

6. Olson, K.R.; Nizeyimana, E. Effects of Soil Erosion on Corn Yields of Seven Illinois Soils. J. Prod. Agric. **1988**, *1*, 13–19.

7. Chengere, A.; Lal, R. SoilDegradation by Erosion of a Typic Hapludalf in Central Ohio and Its Rehabilitation. Land Degrad. Rehab. **1995**, *6*, 223–238.

8. Poudel, D.D.; Midmore, D.J.; West, L.T. Erosionand Productivity of Vegetable Systems on Sloping Volcanic Ash-Derived Philippine Soils. Soil Sci. Soc. Am. J. **1999**, *63*, 1366–1376.

9. Chaney, K.; Swift, R.S. Studieson Aggregate Stability. II. The Effect of Humic Substances on the Stability of Re-formed Soil Aggregates. J. Soil Sci. **1986**, *37*, 337–343.

10. Drury, C.F.; Stone, J.A.; Findlay, W.I. MicrobialBiomass and Soil Structure Associated with Corn, Grasses, and Legumes. Soil Sci. Soc. Am. J. **1991**, *55*, 805–811.

11. Jordahl, J.L.; Karlen, D.L. Comparisonof Alternative Farming Systems. III. Soil Aggregate Stability. Am. J. Altern. Agric. **1993**, *8*, 27–33.

12. Juma, N.G. A Conceptual Framework to Link Carbon and Nitrogen Cycling to Soil Structure Formation. Agric. Ecosyst. Environ. **1994**, *51*, 257–267.

13. Frye, W.W.; Ebelhar, S.A.; Murdock, L.W.; Blevins, R.L. SoilErosion Effects on Properties and Productivity of Two Kentucky Soils. Soil Sci. Soc. Am. J. **1982**, *46*, 1051–1055.

14. Schumacher, T.E.; Lindstrom, M.J.; Mokma, D.L.; Nelson, W.W. CornYield: Erosion Relationships of Representative Loess and Till Soils in the North Central United States. J. Soil Water Conserv. **1994**, *49*, 77–81.

15. Fahnestock, P.; Lal, R.; Hall, G.F. LandUse and Erosional Effects on Two Ohio Alfisols: I. Soil Properties. J. Sustain. Agric. **1995a**, *7*, 63–84.

16. Lowery, B.; Swan, J.; Schumacher, T.; Jones, A. PhysicalProperties of Selected Soils by Erosion Class. J. Soil Water Conserv. **1995**, *50*, 306–311.

17. Shaffer, M.J.; Schumacher, T.E.; Ego, C.L. Simulatingthe Effects of Erosion on Corn Productivity. Soil Sci. Soc. Am. J. **1995**, *59*, 672–676.

18. Mortland, M.M. Clay–Organic Complexes and Inter-actions. Adv. Agron. **1970**, *22*, 75–117.

19. Bohn, H.L.; McNeal, B.L.; O'Connor, G.A. *Soil Chemistry*; 2nd Ed. John Wiley & Sons Inc.: New York, 1985.

20. Sparks, D.L. *Environmental Soil Chemistry*; Academic Press: New York, 1995.

21. Stone, J.R.; Gilliam, J.W.; Cassel, D.K.; Daniels, R.B.; Nelson, L.A.; Kleiss, H.J. Effectsof Erosion and Landscape Position on the Productivity of Piedmont Soils. Soil Sci. Soc. Am. J. **1985**, *49*, 987–991.

22. Ebeid, M.M.; Lal, R.; Hall, G.F.; Miller, E. ErosionEffects on Soil Properties and Soybean Yield of a Miamian Soil Western Ohio in a Season with Below Normal Rainfall. Soil Technol. **1995**, *8*, 97–108.

23. Swan, J.B.; Shaffer, M.J.; Paulson, W.H.; Peterson, A.E. Simulatingthe Effects of Soil Depth and Climatic Factors on Corn Yield. Soil Sci. Soc. Am. J. **1987**, *51*, 1025–1032.

24. Pierce, F.J.; Dowdy, R.H.; Larson, W.E.; Graham, W.A.P. SoilProductivity in the Corn Belt: An Assessment of Erosion's Long-Term Effects. J. Soil Water Conserv. **1984**, *39*, 131–136.

25. Daniels, R.B.; Gilliam, J.W.; Cassel, D.K.; Nelson, L.A. SoilErosion Class and Landscape Position in the North Carolina Piedmont. Soil Sci. Soc. Am. J. **1985**, *49*, 991–995.

26. Andraski, B.J.; Lowery, B. ErosionEffects on Soil Water Storage, Plant Water Uptake, and Corn Growth. Soil Sci. Soc. Am. J. **1992**, *56*, 1911–1919.

27. Fahnestock, P.; Lal, R.; Hall, G.F. LandUse and Erosional Effects on Two Ohio Alfisols: II. Crop Yields. J. Sustain. Agric. **1995b**, *7*, 85–100.

28. Larney, F.J.; Olson, B.M.; Janzen, H.H.; Lindwall, C.W. EarlyImpact of Topsoil Removal and Soil Amendments on Crop Productivity. Agron. J. **2000**, *92*, 948–956.

29. Tanaka, D.L. SpringWheat Straw Production and Compo-sition as Influenced by Topsoil Removal. Soil Sci. Soc. Am. J. **1995**, *59*, 649–654.

30. Gachene, C.K.K.; Jarvis, N.J.; Linner, H.; Mbuvi, J.P. SoilErosion Effects on Soil Properties in a Highland Area of Central Kenya. Soil Sci. Soc. Am. J. **1997**, *61*, 559–564.

Erosion Research, History of

Rattan Lal
The Ohio State University, Columbus, Ohio, U.S.A.

INTRODUCTION

Soil erosion implies detachment, transport, and deposition of soil by energy from water, wind, or gravity. Specific sources of energy to detach and transport soil are called "agents" of erosion. Soil erosion is a natural process and is responsible for formation of the most fertile soils, such as alluvial soils of the river valleys (e.g., Indus, Ganges, Euphrates, Yangtze, Nile) and loess soils of the savannas (e.g., Loess Plateau in China, the Palouse region of northwestern United States). The natural rate of erosion may be less than $0.5\,\text{mm}\,\text{yr}^{-1}$, and often as low as $0.1\,\text{mm}\,\text{yr}^{-1}$. The natural processes, however, can be accelerated by anthropogenic activities drastically exacerbating the rate of soil detachment, transport, and deposition. In contrast to the natural process, the accelerated soil erosion is an extremely destructive process leading to severe adverse effects on long-term productivity on-site, and pollution of natural waters and sedimentation of waterways and reservoirs off-site. Anthropogenic activities that accelerate the soil erosion process include deforestation, biomass burning, conversion of natural to agricultural ecosystems, and plowing especially up and down the slope for monoculture of open-canopy crops (e.g., corn) without protective ground cover of crop residue or a cover crop. The accelerated rate of erosion may be $0.5\,\text{mm}\,\text{yr}^{-1}$ to $10\,\text{mm}\,\text{yr}^{-1}$. For loess-derived soils, such as those in the Yangtze basin in China, the accelerated rate may be several $\text{cm}\,\text{yr}^{-1}$ causing severe problems of sedimentation off-site.

On-site effects of accelerated erosion on reduction in long-term soil productivity are attributed to decline in effective rooting depth, reduction in plant available water capacity, depletion of soil organic matter content and the attendant adverse effects on soil structure, and loss of plant nutrients. Whereas the loss of plant nutrients (e.g., N, P, K) can be replenished by addition of fertilizers, that of the available water capacity is difficult to compensate. Thus, the problem of accelerated soil erosion is closely linked to the issue of sustainability. Some land uses and farming/cropping systems are not sustainable because of the severe problem of accelerated soil erosion.[1]

SOIL EROSION AND HUMAN CIVILIZATION

Settled agriculture originated some 10 to 13 millennia ago in major river valleys by the so-called hydric civilizations. Simple tools were developed between 5000 and 4000 BC to place and cover seed in the soil, to eradicate weeds, and bury the crop residue. A written record of plow (or ard) is found in Mesopotamia about 3000 BC.[2] Archaeological evidence shows the use of animal-driven plows dating back to 2500 BC in the Indus Valley.[3] Since their humble beginnings from 5000 to 4000 BC, the tools used to turn over, mix, and pulverize the soil, have been drastically transformed to suit the soil-specific needs for mechanized farm operations. Soil can now be plowed deeper, pulverized more, and disturbed more than ever before. In fact, plowing renders the soil in a state of an unstable equilibrium that exacerbates risks of soil erosion by water and wind.

Where the natural soil erosion created the most fertile soils in river valleys that were the cradle of modern civilization, the on-set of accelerated erosion by plowing toppled many of the same civilizations by washing/blowing away the mere foundation on which they developed. Accelerated erosion caused some of the thriving civilizations to vanish.[4] Indeed, it was the accelerated soil erosion in the Mediterranean Basin that destroyed the Roman Empire and toppled the Phoenicians.[5] Siltation of the irrigation systems in ancient Mesopotamia ruined the once thriving agriculture established since 10,000 BC.[6] The ancient kingdoms of Lydia and Sardis were ruined by severe soil erosion.[6] The demise of Harappan–Kalibangan culture in the Indus Valley[7] and that of Incas in Central America[4] has been attributed to soil erosion and the attendant degradation. This "quiet crisis," analogous to "cancer" of the land, has "plagued" the earth and challenged farmers ever since the time they began to use the land for settled and intensive agriculture.[8]

SOIL EROSION RESEARCH

Managing and controlling soil erosion has been a challenge since the dawn of settled agriculture. An attempt to controll erosion on sloping lands led to introduction of an innovative "terraced" agriculture.

Encyclopedia of Water Science
DOI: 10.1081/E-EWS 120010087

"Terracing" has been a cultural tradition in many ancient civilizations around the world including the Middle East (The Phoenicians), East and Southeast Asia, West Asia (Yemen), and Central and South America. The Incas designed elaborate systems of stonewalled terraces in Peru.[9,10]

Modern research on soil erosion process and technologies to control it began in the United States during the 1930s. Since that time, both basic and applied aspects of soil erosion research have been conducted throughout the world.[11]

(1) *Soil Erosion Research in the CGIAR System*: Some applied issues of soil erosion research are addressed at several international agricultural research centers (IARCs) managed by the Consultative Group on International Agricultural Research (CGIAR). Relevant among these are four natural resources management centers including International Institute of Tropical Agriculture (IITA) in Ibadan, Nigeria, established in 1967; Centro Internacional de Agricultura Tropical (CIAT) in Cali, Colombia, also established in 1967; International Crops Research Institute for the Semi-Arid Tropics (ICRISAT) near Hyderabad, India, and the International Board for Soil Research and Management (IBSRAM) in Bangkok, Thailand, established in 1984. Extensive research on plot and watershed scales was done during the 1970s and 1980s at IITA.[12,13] Similar watershed management research was conducted at ICRISAT during the 1970s and 1980s.[14] Plot-scale experiments on erosional impacts on soil quality and productivity were conducted during the 1980s and 1990s at CIAT[15] and IBSRAM.[16] Soil erosion research at IBSRAM was sponsored by the Australian Center for International Agricultural Research (ACIAR). An important aspect of the erosion research at IARCs involves development of networks to establish cooperative programs with national agricultural research institutes (NARIs) in ecoregions of their mandate. An international conference on "Soil Conservation and Management in the Humid Tropics" held at IITA in 1975[17] brought together a group of scientists that eventually created a network that periodically organizes conferences around the world under the auspices of "International Soil Conservation Organization" (ISCO).[18] Closely related with ISCO is the World Association of Soil and Water Conservation (WAS-WAC).[19]

(2) *United Nations and Related Organizations*: Other international organizations that have soil erosion research at international scales include the Food and Agricultural Organization (FAO) of the United Nations in Rome, Italy, and the United Nations Environment Program (UNEP) in Nairobi, Kenya. The FAO attempted to develop a methodology for assessment of soil degradation by erosion and other processes,[20] and organized a network to assess erosional effects on productivity.[21] Global research on desertification and its control has been organized by UNEP.[22]

In addition to ISCO and WASWAC, there are other international professional societies whose members are involved in research on soil erosion. Two important organizations among these are the International Association of Hydrological Sciences (IAHS), and the International Soil Tillage Research Organization (ISTRO). Activities of IAHS have been sponsored by the United Nations Educational, Scientific and Cultural Organization (UNESCO).[23,24] Members of ISTRO are primarily involved in soil tillage research[25] but also address the problem of soil erosion. The International Union of Soil Sciences (IUSS) has established a special commission dealing with soil erosion and conservation.[26]

(3) *National Research Organization*: Soil erosion and its control is among priority research issues with most national research organizations in soil, agronomy, hydrology, and agricultural engineering. Many countries have special departments dealing with the issue of soil conservation such as the Soil Conservation Service (SCS) now Natural Resource Conservation Service (NRCS) in the United States. The SCS was established during the "dust bowl" era in the 1930s by H.H. Bennett. Accordingly, national societies have been established in several countries to address the issue. Some examples of such societies are Soil and Water Conservation Society (SWCS) of the United States and Canada; Australian Society of Soil and Water Conservation; Indian Society of soil and Water Conservation, etc. In addition to annual conferences, some of these societies also publish journals and books devoted to the relevant theme of soil erosion and its impact on productivity, water quality, and the greenhouse effect.

There are also national laboratories and institutions involved in both basic and applied research. Two examples of such institutions are the National Soil Erosion Research Laboratory, West Lafayette, Indiana, U.S.A.; and Central soil and Water Conservation Research and Training Institute, Dehra Dun, U.P., India.[27] Similar institutions exist in China and elsewhere. There are also regional organizations involved in soil and water conservation research. The East African Agricultural and Forestry Research Organization (EAFRO) established long-term experiments on watershed management in east Africa.[28] Similar, long-term experiments were established in Francophone Africa by ORSTOM and IRA.[29,30] Long-term experiments were also established in southern Africa by Hudson.[31] A regional project in the United States entitled "Soil Erosion and Productivity" (NC-174) was established in early 1980s to assess the impact of erosion on crop yields.[32]

FUTURE RESEARCH NEEDS

Soil erosion research is now at the crossroads. The focus during the 20th century has been on measurement and prediction of the rate of erosion and on-site effects of erosion on loss of agronomic productivity. Erosion effects on water quality, as affected by dissolved and suspended loads through nonpoint source pollution, have also been addressed. The emphasis has been on the study of erosional processes at the plot scale or landscape level. Soil erosion will continue to be an important and challenging process in relation to sustainable management of soil and water resources during the 21st century. However, there is a strong need for a paradigm shift. In addition to understanding basic processes at aggregate and soilscape level, it is also important to study processes at watershed and river basin scales.[33] There is a strong need to link erosional processes with water and energy balance, and cycling of elements with particular reference to C, N, and P. Linking erosional processes with C balance at the watershed scale is a high priority in order to assess the fate of C and N redistributed over the landscape and transported to the aquatic ecosystems. It is important to establish the cause–effect relationship between emissions of greenhouse gases (CO_2, CH_4, N_2O, NO_x) and erosional processes. The projected global warming has raised new issues related to soil erosion and emission of greenhouse gases. Do erosion and deposition cycles exacerbate emission of greenhouse gases from soil?[34] What is the fate of carbon in erosion-displaced sediments?[35] Does sedimentation and burial of C in depressional sites and aquatic ecosystems take C out of circulation and a long-term sequestration?[36] What may be the effects of predicted global warming, estimated to be $1-4°C$ by the end of the 21st century, on soil erodibility and soil's susceptibility to erosion at soilscape, landscape, and the watershed scales? These environmental concerns are over and above the on-site effects of erosion on decline in long-term productivity. A study of such processes necessitates establishment of long-term coordinated research at regional and international scales.

CONCLUSION

Soil erosion will remain a serious issue during the 21st century. Traditionally soil erosion research has been conducted by the CGIAR, FAO, UNEP, and national organizations. The empirical research conducted during the second half of he 20th century focused on measurement of erosion risks, and on the on-site loss in productivity. There is a need for a paradigm shift in conducting research at watershed scale and linking erosional processes to water and energy balance and cycling of C, N, P, and other elements. It is important to establish links between soil erosion and the emission of greenhouse gases into the atmosphere. The impact of projected global warming, and that of increase in atmospheric concentration of CO_2 and other greenhouse gases, on erosional processes need to be assessed. The much needed paradigm shift will necessitate establishment of regional and international networks to address issues of global importance including erosional effects on soil quality, water quality, and emission of greenhouse gases.

REFERENCES

1. Lal, R. Soil Erosion Impact on Agronomic Productivity and Environment Quality. Crit. Rev. Plant Sci. **1998**, *17*, 319–364.
2. Hillel, D. *Environmental Soil Physics*; Academic Press: San Diego, 1998; 771.
3. Archeology book in India British Museum.
4. Olson, G.W. Archaeology: Lessons on Future Soil Use. J. Soil Water Conserv. **1981**, *36*, 261–264.
5. Eckholm, E.P. Losing Ground Norton, New York, 1976.
6. Lowdermilk, W.C. Conquest of Land Through 7000 years. USDA-ARS Bulletin 99, Washington, DC, 1939.
7. Singh, H.P. Management of Desertic Soils. *Review of Soil Research in India*; ICAR: New Delhi, India, 1982; 676–699.
8. Brown, L.R.; Wolf, E. *Soil Erosion: Quiet Crisis in the World Economy*; Worldwatch Paper 60, Worldwatch Institute: Washington, DC, 1984.
9. Denevan, W. Terrace Abandonment in the Peruvian Andes: Extent, Causes and Progress for Restoration. Proc. Consejo Nacional de Ciencia y Tecnologia, Seminario-Taller, Lima, Peru, 1985.
10. Williams, L.S. *Inca Terraces and Controlled Erosion*; Proc. Conf. Latin American Geographers: Merida, Mexico, 6–10 January, 1987.
11. Lal, R. Soil Degradation by Erosion. Land Degrad. Dev. **2001**, *12*, 1–21.
12. Lal, R. Soil Erosion Problems on Alfisols in Western Nigeria and Their Control. IITA, Monograph 1, 1976; 208.
13. Lal, R. Deforestation of Tropical Rainforest and Hydrological Problems. In *Tropical Agricultural Hydrology*; Lal, R., Russell, E.W., Eds.; John Wiley & Sons: Chichester, U.K., 1981; 131–140.
14. Kampen, J.; Krishna, H.; Pathak, P. Rainy Season Cropping on Deep Vertisols in the Semi-arid Tropics—Effects on Hydrology and Soil Erosion. In *Tropical Agricultural Hydrology*; Lal, R., Russell, E.W., Eds.; John Wiley & Sons: Chichester, U.K., 1981; 257–271.
15. Ruppenthal, M. *Soil Conservation in Andean Cropping Systems: Soil Erosion and Crop Productivity in Traditional and Forage-Legume-Based Cassava Cropping Systems in the South Colombian Andes*; CIAT: Cali, Colombia, 1995; 110.

16. Penning de Vries, F.W.T., Agus, F., Kerr, J., Eds. *Soil Erosion at Multiple Scales: Principles and Methods for Assessing Causes and Impacts*; IBSRAM, CABI Publishing: Wallingford, Oxon, U.K., 1998; 390.

17. Greenland, D.J., Lal, R., Eds. *Soil Conservation and Management in the Humid Tropics*; John Wiley & Sons: Chichester, U.K., 1977; 283.

18. El-Swaify, S.A; Moldenhauer, W.C.; Lo, A. *Soil Erosion and Conservation*; ISCO/SWCS: Ankeny, IA, 1985.

19. Moldenhauer, W.C., Hudson, N.W., Eds. *Conservation Farming on Steeplands*; World Association of Soil and Water Conservation, SWCS: Ankeny, IA, 1988; 296.

20. FAO/UNEP/UNESCO. *A Provisional Methodology for Soil Degradation Assessment*; FAO: Rome, Italy, 1979.

21. Stocking, M.A.; Sanders, D.W. The Impact of Erosion on Soil Productivity. Contour (Jakarta) **1993**, *5*, 12–16.

22. UNEP. *Status of Desertification and the Implementation of United Nations Plan of Action to Combat Desertification*; UNEP: Nairobi, Kenya, 1991.

23. Hadley, R.F.; Lal, R.; Onstad, C.A.; Walling, D.-E.; Yair, A. *Recent Developments in Erosion and Sediment Yield Studies*; IAHS/UNESCO: Paris, France, 1985; 127.

24. IAHS. *Erosion and Sediment Transport Measurement*; IAHS Publication No. 133; IUGG: Paris, France, 1981; 527.

25. Lal, R. Thematic Evolution of ISTRO: Transition in Scientific Issues and Research Focus from 1955 to 2000. Soil Tillage Res. **2001**, *61*, 3–12.

26. Lal, R., Ed. *Soil Erosion Research Methods*, 2nd Ed.; IUSS/SWCS: Ankeny, IA, 1994; 340.

27. Singh, G.; Babu, R.; Narain, P.; Bhushan, L.S.; Abrol, I.P. Soil Erosion Rates in India. J. Soil Water Conserv. **1992**, *47*, 93–95.

28. Edwards, K.A.; Blackie, J.R. Results of East African Catchment Experiments. In *Tropical Agricultural Hydrology*; Lal, R., Russell, E.W., Eds.; John Wiley & Sons: Chichester, U.K., 1981; 153–161.

29. Roose, E.J. Application of the USLE of Wischmeier and Smith in West Africa. In *Soil Conservation and Management in the Humid Tropics*; Greenland, D.J., Lal, R., Eds.; John Wiley & Sons: Chichester, U.K., 1977; 177–187.

30. Forest, F.; Poulain, J.F. A Study of Runoff and Its Effects on Water Balance in Rainfed Agriculture in Upper Volta. In *Soil Physical Properties and Crop Production in the Tropics*; Lal, R., Greenland, D.J., Eds.; John Wiley & Sons: Chichester, U.K., 1979; 521–527.

31. Hudson, N.W., Jackson, D.C. Results Achieved in the Measurement of Erosion and Runoff in Southern Rhodesia. Third Inter-African Soils Conf., Daloba, Central African Republic, 1959.

32. Lal, R., Ed. *Soil Quality and Soil Erosion*; SWCS/CRC Press: Boca Raton, FL, 1999; 329.

33. Lal, R., Ed. *Integrated Watershed Management in the Global Ecosystems*; Soil Water Cons. Soc./CRC Press: Boca Raton, FL, 2000; 395.

34. Lal, R. World Cropland Soils as Source or Sink for Atmospheric Carbon. Adv. Agron. **2001**, *71*, 145–191.

35. Lal, R. Influence of Soil Erosion on Carbon Dynamics in the World. Symp. Proc. "Land Use, Erosion and Carbon Dynamics," IRD, Montpellier, France, Sept 23–28, 2002.

36. Smith, S.V.; Renwick, W.H.; Buddemeier, R.W.; Crossland, C.J. Budgets of Soil Erosion and Deposition for Sediments and Sedimentary Organic Carbon Across the Conterminous United States. Global Biogeochem. Cycles **2001**, *15*, 697–707.

Erosion Research, Instrumentation for

Gary Bubenzer
University of Wisconsin, Madison, Wisconsin, U.S.A.

E

INTRODUCTION

Soil erosion and infiltration research may be enhanced by the use of simulated rainfall. Simulators make it possible to control rainfall characteristics such as intensity, duration, and energy levels. Storms may be duplicated at any time or location. Raindrops may be formed on the tips of drop formers or by nozzles. Except for specialized applications, simulators using nozzles to form drops are preferred because of the ability to produce high impact energy from relatively short fall heights.

OVERVIEW

Sediment laden runoff is often sampled by depth integrating or depth integrating samplers. Depth integrating samplers are preferred whenever there is a nonuniform distribution of sediment in the runoff; while flow with uniformly distributed sediment loads may be sampled form a single point. Samples may be collected in specified time increments or in constant runoff volume increments.

Rainfall Simulation

Early soil erosion researchers were dependent upon natural rainfall for their work. As a result, they had no control over the nature or timing of the rainfall event. Without this control it was impossible to predict accurately, soil conditions or crop stage at the time of the storm. It was impossible to duplicate storm events at various times and locations. Valuable data was often forgone simply because it did not rain at the desired time and location.

Approximately 70 yr ago, researchers began to use simulated rainfall on erosion plots. Earlier attempts consisted of covering the plots with water from sprinkling cans and other crude water application sprinklers. While these systems allowed researchers to apply given amounts of water to plots under controlled soil–plant conditions, the water applied had little similarity with naturally occurring rainfall.

Over time criteria were developed for improved rainfall simulators.[1] Among the criteria were:

- Drop size distribution near that of natural rainstorms of the geographical area.
- Drop impact velocities near those of natural raindrops.
- Intensities in the range of storm of interest.
- Plot area of sufficient size to satisfactorily represent the treatment and conditions being evaluated.
- Rainfall characteristics fairly uniform over the plot area.
- Rainfall application nearly continuous.
- Angle of impact nearly vertical.
- Capability of reproducing storm durations at selected intensities.
- Portability of movement from site to site.
- Satisfactory operation under a wide range of climatic conditions.

Two types of rainfall simulators were developed. The first were simulators that produced rainfall by forming drops on the tips of yarn, hollow glass tubes, hypodermic needles, or plastic tubing. Size and rate of drop formation (intensity) were controlled by the size and length of the dripper and the water head. Drops broke from the end of the tube when the weight of the drop was sufficient to overcome the surface tension. The primary advantage of the drip type simulators is the researchers' ability to accurately control drop sizes. Drop size was primarily a function of the tube material and diameter with little change in diameter over a relatively wide range of intensities. Drip simulators, however, had two serious limitations. Since the drops form on the tip of a dripper, they have no initial fall velocity. This means, they must fall from a considerable height in order to reach terminal velocity at the soil to simulate natural rainfall's droplet velocity. Such fall height are difficult to obtain, especially for field research and difficult to control under adverse climatic conditions. Secondly, a very large number of drippers per unit area are needed to provide uniform application. Therefore, drip simulators are very difficult to use in plots larger than a few square feet or under outdoor conditions.

The second general type of rainfall simulators uses spray nozzles to produce raindrops. Early nozzles, such as those used for watering gardens tended to produce large drops, high intensities, and low impact energies.

Encyclopedia of Water Science
DOI: 10.1081/E-EWS 120010090

Over time, these nozzles have been replaced by ones that produce drop size distributions, more similar to those of natural storms, at pressures that produce near terminal impact velocities. However, most of these nozzles produce intensities well above those found in natural storms of interest. To overcome these high intensities, intermittent rainfall is produced by intercepting a portion of the water before it strikes the soil surface. Mechanical systems for rainfall interception have been replaced by computer-controlled systems that make it possible to produce storms with varying intensity patterns.

Modern rainfall simulators can reproduce storms with desired rainfall characteristics on command. This means that soil conditions and cover conditions at the time of rainfall can be predetermined. Identical storms can be produced at different locations and at different times. Such simulators have provided researchers with a tool for greatly accelerating the knowledge of erosion processes.

Sampling Eroded Sediments

Sediment sampling is necessary to determine the concentration of sediment in the runoff water and the sediment load of a storm or series of storms. In erosion research the primary parameters of interest are the sediment load (total sediment lost per storm or series of storms) and the sedigraph (sediment eroded as a function of time). In order to determine these two parameters, it is necessary to measure both the flow and the sediment. Flow measurements are usually made through flumes, weirs, or other rated cross sections. Only instrumentation for the collection of sediment will be discussed in this article; for information on flow measurement the reader is referred to Dendy et al.[2]

Total flow samplers consist of a collection container large enough to collect all of the runoff and sediment from the design storm. Because of limitations on container size, use is limited to small plots. Such plots are usually not applicable to erosion studies because they are not large enough to allow the erosion process to develop. Therefore, most erosion studies use partial flow samplers where only a fraction of the total runoff is collected.

One of the simplest partial flow samplers is the Multislot divider.[2] A plate with multiple slots is inserted into the runoff stream. Flow through a portion of the slots, depending upon the fraction of the flow to be sampled, is diverted into the sampling container. The remainder of the flow is bypassed. The devices are simple to construct and install and require no power source. They may be used to determine sediment load, but do not provide data for the development of a sedigraph.

The Coshocton runoff sampler consists of a small flume and a slotted rotating disk.[2] As water exits the flume, the flow across the face of the disk causes the slotted disk to rotate. Only that portion of the water and sediment striking the slot is diverted into the sampling container. The remainder of the water and sediment is bypassed. The Coshocton samplers do not require an external power source. Like the slotted samplers, they are used primarily for determining total load.

Pump samplers are commonly used to collect sediment and runoff samples for modern erosion research. Major components of a pump sampler include:

- The intake that collects water from one or more points in the flow system.
- The pumping system to move the water from the intake to a series of collection containers.
- A flushing system to ensure that samples and successive samples are not contaminated by water remaining in the system from previous samplings.
- A collection system usually consisting of a series of jars for holding the samples.
- A control system for determining the sampling interval.
- A battery or a.c. current power supply.

Intake units may be either depth integrating or time integrating units. Depth integrating units consist of multiple inlets placed at various depths throughout the flow profile or of a single intake unit that is moved vertically during sampling. Such units are used to ensure accurate sampling when the sediment in the flow is stratified. Point integrating samplers consist of a single intake unit place in the flow stream. They provide satisfactory results when the sediment is well mixed in the runoff stream.

The control system is used to determine the sampling pattern. Two approaches are used. Samples may be collected at time intervals or at volumetric intervals. Samples may be collected at predetermined time intervals. Usually each sample is placed in a separate container. By combining the sediment concentration found in each sample with the hydrograph it is possible to determine not only the total load, but also the sedigraph for the storm event. Composing of sample for determination of sediment load or pollutant transport is not possible since each sample represents a different portion of the total flow.

Volumetric interval sampling is based upon flow volume rather than time. Samples are collected at predetermined flow volume intervals. Sediment load and the sedigraph are easily determined since each sample represents the same volume of flow. Volumetric sampling enables researchers to compose samples only when the total sediment load is needed.

REFERENCES

1. Meyer, L.D. Methods for Attaining Desired Rainfall Characteristics. *Proc. of the Rainfall Simulator Workshop*; ARM-W-10, U.S. Department of Agriculture, 1979; 35–44.

2. Dendy, F.E.; Allen, P.B.; Piest, R.F. Sedimentation. In *Field Manual for Research in Agricultural Hydrology*; Agricultural Handbook 224, Brakensiek, D.L., Osborn, H.B., Rawls, W.J., Eds.; U.S. Department of Agriculture, 1979; Chap. 4.

Eutrophication

Thomas G. Franti
Kyle D. Hoagland
University of Nebraska, Lincoln, Nebraska, U.S.A.

INTRODUCTION

Eutrophication is the nutrient enrichment of surface water and the subsequent impacts on water quality and the aquatic ecosystem. The over abundance of plant nutrients, usually nitrogen and phosphorus—but sometimes silicon, potassium, calcium, iron, or manganese—creates the conditions for excessive plant growth.[1] Algal blooms are an example of excessive growth caused by over supply of nutrients. A body of water is classified by its trophic state based on the amount of nutrients supplied to it. An oligotrophic state is low in nutrients, a mesotrophic state is intermediate, and a eutrophic or hypereutrophic state is high in nutrients.[2] Eutrophication is a natural process. However, as a natural process it takes generations, or even thousands of years, for eutrophication to cause significant changes. It is the acceleration of the process, known as cultural eutrophication, that is of the greatest concern to water quality and the health of aquatic systems.

CULTURAL EUTROPHICATION

Cultural eutrophication is the result of excess nutrients—primarily nitrogen and phosphorus—delivered to rivers, lakes, and estuaries by the activities of humans. Nutrient loading can come from both point sources and diffuse sources. Point sources of nutrients are generally urban sewage treatment or industrial water treatment. Diffuse sources (or nonpoint sources) of nutrients include agriculture, deforestation, and urban lawn runoff. The use of commercial fertilizer in crop production can lead to losses of nutrients to surface water. Concentrated animal feeding operations are also a potential source of excess nutrients.[2,3] Finally, deforestation and subsequent soil erosion, can deliver excess nutrients, sediment, and organic matter to surface water, creating the conditions for eutrophication.

The natural aging process of lakes in some climates can lead from an oligotrophic state, with clean open water supporting usually cold water fish species, to a eutrophic state of warmer, shallow water supporting different species of plants and fish, and ultimately, to a filled lake closed over by a bog or fen. This process usually takes thousands of years. In contrast, cultural eutrophication can occur in a matter of one or two generations.

The last 50 years of the Twentieth Century saw the greatest impact of eutrophication on the world's waters. Cultural eutrophication has accelerated the lake aging process, reduced water quality, and impacted aquatic plant and animal populations, at economic costs. These impacts are being felt worldwide.[1]

Modern society's use of phosphorus additives in detergents has lead to excessive phosphorus loading to surface waters.[1] Modern agriculture's use of commercial fertilizer and fertilizer's relative abundance, low cost, and over use, has lead to excessive nitrogen loading to surface water. Agriculture also contributes excessive phosphorus from fertilizer and from intensive livestock operations and land application of manure.[3,4] Urban sources, such as sewage treatment plants, storm water runoff, industrial water treatment facilities, and food processing contribute excess nutrients. Sewage treatment loading to waters, including treated gray water, is generally considered the greater source of phosphorus. The greater contributor of nitrogen is agricultural crop production. Other significant agricultural sources are animal confinement operations, forestry, and atmospheric inputs.

RATE OF EUTROPHICATION

The rate of eutrophication is controlled by the rate at which nitrogen and phosphorus are delivered to a body of water. It is generally accepted that phosphorous is the limiting nutrient for algal growth in lakes.[1,5] Another limiting factor for algal growth is light. As excess phosphorus enters a lake it can trigger high levels of algal growth. Excessive growth, or blooms, can reduce water clarity. Aquatic macrophytes may also thrive under these conditions. This process can increase the productivity of a lake to an extent. When it becomes excessive, the process of cultural eutrophication has begun.

The nutrients supplied to an aquatic system is the most important factor that determines the species and amount of plant material, which in turn controls available oxygen and the animal species that thrive.[1] As plants and macrophytes grow and die, organic matter accumulates

Encyclopedia of Water Science
DOI: 10.1081/E-EWS 120010332

on the lake bottom. Decomposition of excessive amounts of organic matter can consume available oxygen and create anoxic conditions. An anoxic condition drives away plant and animal species dependent on oxygen. Slowly a lake's original community changes.

This anoxic state creates another problem if a lake becomes stratified. Temperature differences in the water at the lake's surface and at depth can create thermal stratification. In this condition the cold, low oxygen water in the hypolimnion, or lower layer, does not mix with the epilimnion, the warmer surface water. In this condition the anoxic conditions in the hypolimnion creates a condition where the sediment releases phosphorus, which is transported upward to the epilimnion and contributes to increased algal growth. This internal cycling of phosphorus continues the eutrophication process even if no additional phosphorus enters the lake.[5,6]

EFFECTS OF EUTROPHICATION

There are various biological effects of eutrophication. Besides the problems of algal blooms and anoxic conditions previously described, there are changes in water temperature, reduction in water clarity, increased macrophyte production, population shifts in both plants and animals, and accelerated aging of lakes. Fish kills and the coastal "red tides" and "brown tides" are also potential environmental impacts from eutrophication.[5,7]

There are many impacts on human use of water from eutrophication. Potable water—water used for human consumption—can be significantly degraded by eutrophication. Besides the excess nutrients themselves, there is excess algae and other plant growth which can contribute to unwanted odors and tastes. Thus, additional water treatment is needed to make the water drinkable.[1,4]

The decline of commercial and recreational fisheries is also an indirect result of eutrophication. Fish populations may move away, or over time shift to less desirable fish species as water temperature, clarity, and quality change.[1] There are also negative impacts on recreational use of surface waters such as boating and swimming as a result of floating plant growth, smell, and the overall loss of aesthetic quality.[4] Finally, the general decline of an aquatic ecosystem and the loss of biodiversity are impacts of eutrophication.[1,3,7]

Some of these impacts can cause significant economic losses as well. Water treatment costs rise as water quality decreases. Algal blooms can contribute to shutting down water treatment plants. Advanced water treatment of municipal water to remove nutrients (such as alum addition)—hence removing a contributing factor to eutrophication—adds another expense to water treatment. Loss of commercial fisheries is an economic impact to a region. Finally, loss of recreation income from tourists, boaters, and recreational fisherman can have significant impacts on a local economy.[4,7]

REDUCTION AND MANAGEMENT

Two approaches can be taken to the reduction and management of eutrophication: 1) reduction of nutrient loads; and 2) managing the existing high-nutrient state.[1] Reduction of loading is clearly the more robust approach. Reduced loads are necessary if long-term improvement is expected. However, because of the ecosystem changes brought on by eutrophication and the potential problem of phosphorus cycling previously described, water quality and ecosystem improvements may not respond quickly to reduced loading.[3] A combination of reduced loads and in-lake controls may be needed.

Reducing nutrient loads requires reduction of both point and diffuse sources. To reduce point sources, sewage and industrial water treatment plants need advanced water treatment to remove nutrients.[1,7] Advanced water treatment is expensive, but is the only way to reduce these point source nutrients. Most diffuse sources contributing nutrients to surface water come from agricultural operations. Loading is dependent upon the type of crops grown, soil type, climate, cultural practices, fertilizer use, and whether animal waste management practices are sufficient to reduce loading. Nutrient reduction from these sources will require improved management of fertilizer and animal waste, and continued reduction of soil erosion throughout the watershed.[1]

There are several methods used to manage nutrient levels in lakes. These fall into two categories: those that 1) remove nutrients; or 2) manage nutrient levels without nutrient removal. Methods to remove nutrients include: 1) lake flushing; 2) hypolimnetic water withdrawal; 3) sediment removal (e.g., dredging); and 4) nutrient inactivation by precipitation (e.g., alum treatment). Methods to manage nutrients without removing them include: 1) artificial mixing and/or aeration; 2) dilution by the addition of water lower in nutrients; 3) bottom sealing to prevent internal nutrient cycling; 4) manipulation of lake biological communities (e.g., selective fish harvesting or the introduction of fish predators such as largemouth bass, walleye, or brown trout); 5) introduction of biological controls for unwanted macrophyte growth (e.g., weevils or grass carp); and 6) herbicide or other chemical treatments (e.g., copper sulfate) of excessive algal or macrophyte growth.[1,6]

Clearly the best way to reduce or reverse eutrophication is to reduce nutrient loading, that is, targeting the source of the problem.[1] This long-term solution involves participation and management by people throughout the entire watershed. In-lake nutrient management can be done, but may require annual inputs and regular management.

REFERENCES

1. Harper, D. *Eutrophication of Freshwaters: Principles, Problems and Restoration*; Chapman & Hall: London, 1992.
2. Smith, V.H.; Tilman, G.D.; Nekola, J.C. Eutrophication: Impacts of Excess Nutrient Inputs on Freshwater, Marine, and Terrestrial Ecosystems. Environ. Pollut. **1999**, *100*, 179–196.
3. Carpenter, S.R.; Caraco, N.F.; Correll, D.L.; Howarth, R.W.; Sharpley, A.N.; Smith, V.H. Nonpoint Pollution of Surface Waters with Phosphorus and Nitrogen. Ecological Applications **1998**, *8* (3), 559–568.
4. Daniel, T.C.; Sharpley, A.N.; Lemunyon, J.L. Agricultural Phosphorus and Eutrophication: A Symposium Overview. J. Environ. Qual. **1998**, *27* (4), 251–257.
5. Correll, D.L. The Role of Phosphorus in the Eutrophication of Receiving Waters: A Review. J. Environ. Qual. **1998**, *27* (2), 261–266.
6. Cooke, G.D.; Welch, E.B.; Peterson, S.A.; Newroth, P.R. *Lake and Reservoir Restoration*; Butterworths Publishers: Stoneham, MA, 1986.
7. Klapper, H. *Control of Eutrophication in Inland Waters*; Ellis Horwood Limited: Chichester, West Sussex, England, 1991.

Evaporation and Eddy Correlation

Roger Shaw
Richard L. Snyder
University of California, Davis, California, U.S.A.

E

INTRODUCTION

Except within the first few millimeters of the surface, turbulence in the atmospheric boundary layer greatly dominates molecular diffusion with respect to the mixing of the variety of materials that are exchanged with the underlying surface. To a large extent, and using appropriate instrumentation, these turbulent "eddy flux" motions are measurable with a high level of precision and with a high degree of spatial and temporal resolution. If the property being transported is also measured with equivalent precision and spatial and temporal resolution, it is possible to monitor the flux density of the property continuously across any plane of interest.

EDDY FLUX

Consider the measurement of this eddy flux across a plane parallel to and a few meters above a flat horizontal surface. It is only the vertical component of the fluctuating wind velocity that is responsible for the flux across the plane. The fact that there is a net transport of some specific entity across that plane implies a correlation between the vertical wind component and that entity. For example, if water vapor is released into the atmosphere from the surface beneath, updrafts will generally contain higher vapor content than will downdrafts, and vertical velocity (positive upwards) will be positively correlated with vapor content. Fig. 1 illustrates this with data collected above a deciduous forest. The origin of the term *eddy correlation* is thus quite apparent, although *eddy covariance* is becoming more popular because it is the covariance of the velocity and scalar that is actually used. Since short-period fluxes are not a concern, a suitable averaging period between, say, 10 min and 30 min is typically selected. The averaging period is constrained to substantially exceed the duration of the largest eddy involved in the transport process, and yet be short enough to be unaffected by any lack of stationarity in the environmental conditions.

It is normal to separate the perturbation and the time-averaged components of the quantity of interest. For example, $\rho_v = \bar{\rho}_v + \rho_v{}'$ is an expression for absolute humidity (water vapor density, $kg\,m^{-3}$), where the overbar signifies a time average over a specified interval of time and the prime indicates a departure from the mean. The vertical velocity component w ($m\,s^{-1}$) can be treated similarly, such that $w = \bar{w} + w'$. This separation into mean and perturbation parts is referred to as Reynolds notation. By definition, means of the fluctuating parts are equal to zero (e.g., $\bar{w}' = 0$, $\bar{\rho}'_v = 0$). If the mean flow is horizontal, $\bar{w} = 0$.

Using Reynolds decomposition, the flux density for water vapor E ($kg\,m^{-2}s^{-1}$) is written as

$$E = \bar{w}\,\bar{\rho}_v + \overline{w'\rho'_v}$$

However, if there is no convergence or divergence of air due to sloping surface, the mean vertical velocity ($\bar{w}$) and hence the first term on the right equals zero. This simplifies the equation to $E = \overline{w'\rho'_v}$. The term on the right hand side contains the covariance of vertical velocity and absolute humidity fluctuations, and is an unambiguous expression for the flux of water vapor that does not depend on any assumptions about the mixing properties of atmospheric turbulence.

The eddy covariance technique is direct to the extent that it requires no assumptions about the mixing properties of the air. It is assumed that the measurement made at a small distance above the surface (one to several meters) is representative of the underlying surface. Because the airflow is mainly horizontal, with imposed 3-D perturbations, the signals from which the covariance is derived are representative of an area upwind of the measurement point. This is best defined in terms of the *footprint* of the source distribution,[1] which describes in a statistical manner the source probability distribution. The footprint depends on instrument height, becoming more distant as the instrument height increases. It is also dependent on surface roughness and atmospheric stability. Under nocturnal or otherwise stable conditions, the footprint might be far removed from the measurement point, whereas nearby footprints are expected under unstable conditions. Another common descriptor of the source region uses the term *fetch* to describe the upwind distance of uniform features required to ensure that the measurement is representative of the underlying surface and not

Encyclopedia of Water Science
DOI: 10.1081/E-EWS 120010306

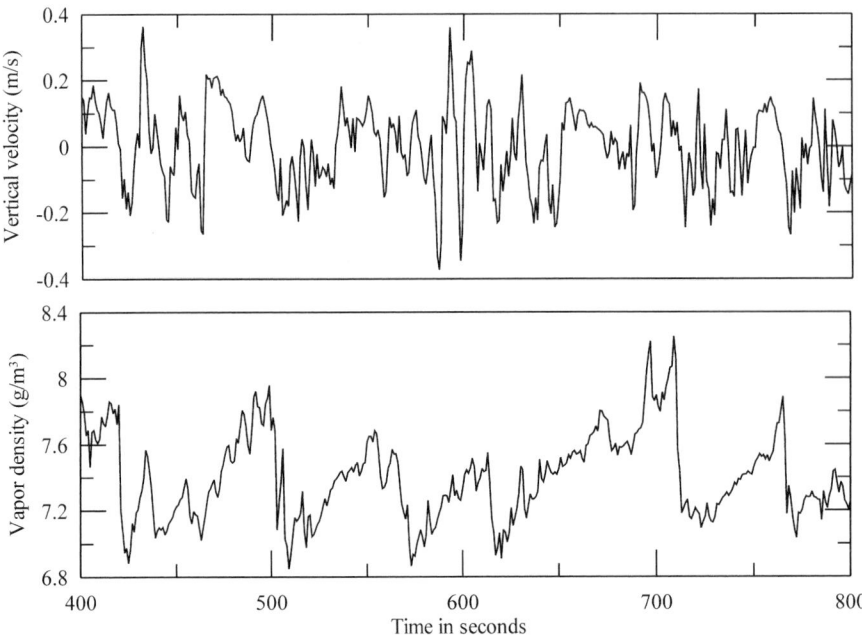

Fig. 1 Time traces of vertical velocity and humidity fluctuations above a deciduous forest showing positive correlation during a period when the foliage was actively transpiring.

contaminated by the flux from a distant surface. A general rule-of-thumb is that a uniform fetch is required that equals at least 100 times the height of the instrument above the effective surface. Such a fetch may be more than sufficient under unstable conditions but is likely to be inadequate when the atmosphere is very stable.

CORRECTIONS TO FLUX MEASUREMENT

Corrections are needed to the eddy covariance flux of a minor constituent in the presence of a flux of sensible heat and/or of a more major gaseous flux.[2] This arises because of density perturbations in the minor constituent imposed by the presence of the major flux. (No correction is needed if the mixing ratio of the constituent is measured instead of its density.) In a relative sense, the flux of a trace gas may require a large correction, and it may also be necessary to adjust a calculation of the flux density of water vapor in the presence of a considerable sensible heat flux. Webb, Pearman, and Leuning[2] estimated that corrections to the vapor flux vary from a few percent to more than 10% on occasion. They proposed the expression $E = 1.010(1 + 0.051\beta_r)E_r$, where β_r is the uncorrected Bowen ratio ($\beta_r = H/\lambda E_r$) and E_r is the uncorrected vapor flux density. H is the sensible heat flux density ($W\,m^{-2}$) and λ is the latent heat of evaporation ($J\,kg^{-1}$).

Often, eddy covariance measurements are made over surfaces that are not horizontal or over tall forests where variations in tree height create local departures from horizontal mean flow. Additional problems arise if the sensor is misaligned or if the tower or mast that supports the sensors create aerodynamic interference, or, indeed, if the sensors themselves distort the flow. Common practice is to perform coordinate rotations in a two-pass operation to force the mean lateral and vertical component velocities to zero ($\overline{v} = \overline{w} = 0$). A sensor misalignment of one degree can cause errors on the order of 3%–4% for water vapor flux. Rotating the coordinates of the wind velocity vectors so that the vertical axis is orthogonal to the mean wind streamline will minimize tilt errors but procedures such as this are not without their problems and the reader is referred to the text by Kaimal and Finnigan[3] for further discussion.

EDDY COVARIANCE SENSORS

Sensors must measure vertical velocity and water vapor concentration with sufficient frequency response to record the most rapid fluctuations important to the diffusion process. Typically, a frequency response of the order of 10 Hz–20 Hz is sufficient, but the response-time requirement depends on wind speed, atmospheric stability, and on the height of the instrumentation. The outputs are sampled digitally at a sufficient rate to obtain a statistically stable value for the covariance; typically, this rate is several samples per second.

Because collocation of sensors is usually not practical, instruments are placed apart but as close to each other as

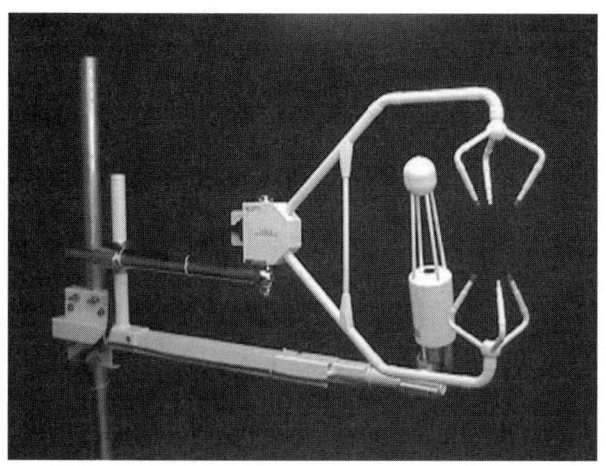

Fig. 2 Photograph of a triaxial sonic anemometer and an open-path IR hygrometer. (Photo courtesy of Campbell Scientific, Inc.)

possible without unnecessary interference. Physical separation can lead to an underestimate of the flux. For example, Lee and Black[4] calculated an underestimate in the flux density of 3% when the ratio of the sensor separation distance to the difference between the measurement height and the zero plane displacement (effective level of momentum sink inside the canopy) was about 5%.

High-frequency wind vector data are usually obtained with tri-axial sonic anemometer (Fig. 2) in which ultrasound pulses ($\geq 40\,\text{kHz}$) are transmitted between an array of transducer pairs. The axial wind velocity (V_d) over the transducer separation distance (d) is given by $V_d = d/2(1/t_1 - 1/t_2)$, where t_1 and t_2 are pulse transit times in each of the two directions. The instrument performs an internal coordinate rotation to provide signals of three orthogonal velocities from a nonorthogonal transducer path array. Since the pulse transit time is usually only a fraction of a millisecond, the procedure of sending pulses back and forth is typically repeated up to 200 times per second and an output presented 10 to 20 times per second.

A range of humidity sensors has been employed for eddy covariance measurements of evaporation, including thermocouple psychrometry in some of the very earliest devices. In modern applications, high-frequency measurements of water vapor density are most commonly made with optical absorption devices operating in either ultraviolet (UV) or infrared (IR) wavelengths. The former utilize water vapor absorption in the spectral region of about $0.12\,\mu\text{m}$ and open path commercial units are available as Lyman-alpha and krypton hygrometers. Lyman-alpha hygrometers use an excited hydrogen

source, magnesium fluoride windows, and a nitric oxide detector. Strong absorption by water vapor allows for short paths ($\sim 1\,\text{cm}$) but the source ages, and the surfaces of the windows are subject to etching by water, and degrade with time. Such degradation is reversible, however, with appropriate cleaning. The krypton hygrometer uses a krypton glow tube as source. It operates much the same as the Lyman-alpha hygrometer and has the advantage of a more stable source but suffers to some degree from greater sensitivity to the gases: oxygen and ozone.

IR hygrometers generally operate in a differential mode at two nearby wavelengths: one with strong water vapor absorption and the other where absorption is weak. Longer optical paths are needed than in the case of UV-wavelength sensors and folding of the path is common. IR hygrometers are either closed or open path. In the case of the former, air is sampled by a tube at the site of the velocity measurement and drawn at high speed to a chamber of the hygrometer. A mechanical chopper switches the optics between the sample and the reference cells to allow amplification of the signal.

CONCLUSION

Eddy covariance is commonly used to determine sensible and latent heat fluxes from crop canopies, from rangeland, and from forests. Measurements of evapotranspiration are used to estimate crop coefficients, and are used in irrigation management and planning. In addition, eddy covariance is used to calibrate other less costly and more robust methods such as the surface renewal method for estimating energy and scalar fluxes.

REFERENCES

1. Schuepp, P.H.; Leclerc, M.Y.; MacPherson, J.I.; Desjardins, R.L. Footprint Prediction of Scalar Fluxes from Analytical Solutions of the Diffusion Equation. Boundary-Layer Meteorol. **1990**, *50*, 355–373.
2. Webb, E.K.; Pearman, G.I.; Leuning, R. Correction of Flux Measurements for Density Effects Due to Heat and Water Vapor Transfer, Q. J. R. Meteorol. Soc. **1980**, *106*, 85–100.
3. Kaimal, J.C.; Finnigan, J.J. *Atmospheric Boundary Layer Flows*; Oxford University Press: Oxford, 1994, 289 pp.
4. Lee, X.; Black, T.A. Relating Eddy Correlation Sensible Heat Flux to Horizontal Sensor Separation in the Unstable Atmospheric Surface Layer, J. Geophys. Res. **1994**, 18545–18553.

Evaporation and Energy Balance

Matthias Langensiepen
Humboldt University of Berlin, Berlin, Germany

INTRODUCTION

Various forms of energy drive water transport through the hydrological cycle. Radiant energy, originating from the sun, provides the input energy for the cycle. Once matter absorbs this energy, it is converted into sensible heat that elevates the temperature of the air and the ground, and latent heat that causes evaporation, driving thereby the cycle against the pull of gravity. Further transport is generated by kinetic energy and pressure energy of the moving air masses. Translocation of vapor is accompanied by continuous interchanges among radiant, thermal, kinetic, and pressure energy. Large amounts of latent heat are released when water condenses in the clouds and falls as precipitation on the earth surface. It carries kinetic energy while flowing through watersheds. Vertical movement and percolation through the earth's crust finally causes changes in potential and pressure energies.

The first law of thermodynamics states that energy is neither created nor destroyed, only converted from one form into another. This effectively means that the input and output energies of a completely defined system must balance. Storage effects may temporarily disturb this equilibrium condition. The energy balance must thus be expressed in its most general form as:

Energy Input = Energy Output + Energy Storage

The water balance of the earth–atmosphere system can be treated analogically as the mass of water is conserved at all times. Evaporation is the connecting link between the system's water and energy balances. It is a surface process, which takes place at the lower boundary of the atmosphere and is an important component of the surface energy balance (see Fig. 1):

$$R_n = LE + H + G - A + S - L_pF_p$$

where R_n is the flux of net allwave-radiation, L the latent heat, E the evaporation rate, H the flux of sensible heat, G the heat flux at the lower boundary of the surface, and A the energy advected to the surface when the ground properties have horizontal discontinuities. The energy balance is sometimes parameterized for a volume of surface material (for example water body, soil, or canopy volume). As the solar energy input undergoes diurnal and annual fluctuations, heat storage S may become an important component of the balance when it is applied at time intervals shorter than the fluctuation period. When the layer includes vegetation, biochemical energy storage due to photosynthesis can also be considered. L_p is then the thermal conversion factor of carbon dioxide, and F_p is the flux of CO_2.

Shortwave radiation from the sun is the sole energy input of the earth–atmosphere system. Its net amount available for heat conversion is related to geographical location, time, atmospheric transparency, atmospheric path length, geometrical distribution of the surface elements, and their optical properties. Complementary longwave radiation exchange is governed by surface to air temperature differences and cloudiness. Net shortwave and longwave radiation form the net allwave radiation R_n.

The partitioning of R_n into the remaining terms of the surface energy balance determines the rate of surface evaporation and depends on the availability of surface water.

ENERGY BALANCE AND WATER AVAILABILITY

Unlimited Water Availability

When water availability is unlimited on a large scale, such as in oceans, vertical temperature gradients within the atmosphere tend to be very close to the adiabatic value, and most of the available energy (R_n) is diverted to latent heat (L_vE) from moisture flux at the surface. Wind gradients near the surface are typically very steep under such conditions and quickly approach values that remain nearly constant throughout the convective boundary layer. Vertical motion is damped out by strong subsidence inversion at the upper boundary of this well mixed layer. Heat Storage (S) has a dominant effect on the diurnal course of the ocean's energy balance leaving only little energy for transport ($L_vE + H$) into the air until late in the afternoon. This situation is reversed during the night, where heat released from the ocean surface becomes the major source of energy. Large-scale advection (A) partly disturbs the thermal inertia of oceans, which has considerable effects on the global weather systems

Encyclopedia of Water Science
DOI: 10.1081/E-EWS 120010307

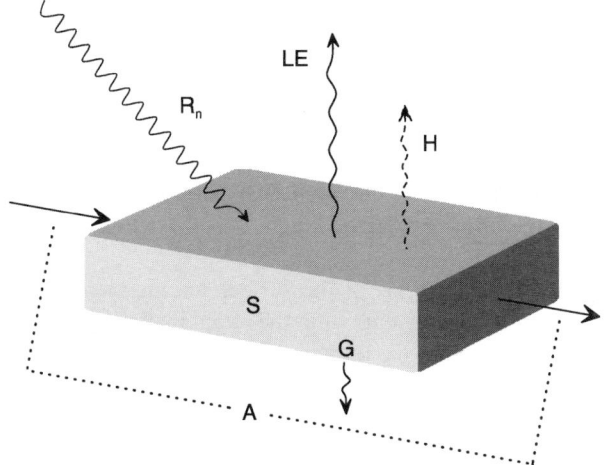

Fig. 1 Schematic illustration of the surface energy balance. (R_n = net radiation, LE = latent heat flux, H = sensible heat flux, G = ground heat flux, S = heat storage, A = advection. All symbols are expressed in $W\,m^{-2}$.)

(Gulfstream, Humboldt-Stream, El-Nino). The smaller the water volume becomes the more it is likely that its thermal inertia is disturbed by local advection due to horizontal discontinuities of thermal surface properties. Since radiation can penetrate into water bodies it is also possible that the underlying floor becomes a source of heat.

Snow Covered and Frozen Surface Layers

When water is bound in snow covered or frozen surface layers, energy partitioning is affected by the penetration of shortwave radiation, phase changes, and internal distribution of water. Net radiation gain (R_n) is commonly restricted by a high surface albedo. The optical depth of snow and ice also affect radiation absorption and penetration. The available energy is mainly partitioned between storage energy (S) and energy required to allow water to change between frozen and liquid states (L_f). Only little energy is consumed by atmospheric transport ($LE + H$). Phase changes of water within the layer (freezing, melting, condensation, evaporation, sublimation) are accompanied by the continuous consumption or liberation of energy. The internal partitioning of available energy is thus influenced by the physical states of water: Water has a high specific heat of $4216\,J\,kg^{-1}\,K^{-1}$ at $0°C$ due to a strong intermolecular bonding force. Fifteen percent of the hydrogen bonds break when water changes from a solid to liquid state. The energy required to effect this change is $0.334\,MJ\,kg^{-1}$ and is called the latent heat of fusion L_f. Nearly 7.5 times as much energy is required at this temperature level to cause water to further change from a liquid to a gaseous state. The corresponding energy is called latent heat of vaporization L_v and is temperature

dependant ($2.5\,MJ\,kg^{-1}$ at $0°C$, $2.45\,MJ\,kg^{-1}$ at $20°C$, and $2.41\,MJ\,kg^{-1}$ at $40°C$). In the event that water changes directly from a solid to a gaseous state (sublimation) the required latent heat of sublimation L_s is the algebraic sum of L_f and L_v. Freezing or condensation liberates energy, the amount depending on the corresponding phase shift. When the surface layer is below the freezing point and the sky is clear, net radiation can become negative under conditions of decreased radiation availability (high latitudes). It becomes positive, however, when the sky window is obstructed by clouds or surface emission is exceeded by incoming radiation. When the surface melts, both, radiation and convection act as energy sources, sometimes accompanied by additional heat input from rainfall. Surface temperatures change only little during this process, because most energy is stored as latent heat of fusion.

Water Scarcity

When water is scarce, as is the case in deserts, most of the available energy (R_n) is consumed by surface heating, which can be sensed as a rise in surface temperature. Sensible heat (H) dissipation from dry surfaces lowers the density of air increasing its instability and tendency to rise. The instable air parcels form plumes (thermals) that progressively cool down as they mix with the surrounding air and are finally capped off by the inversion layer. Additional air is entrained from the top of the capping inversion layer and dragged to the ground by sinking motion of the cooling air masses. The height of the inversion layer is dependant on the amount of energy available for surface heating. At night and during early morning, winds in deserts are light, turbulence is low, air is stable or neutral, and the inversion layer is close to the ground. Net radiation (R_n) is partitioned into surface heat-flux (G) and heat storage (S) under such conditions. However, low thermal admittance of the barren dry soil diverts a major portion of the available energy to sensible heat (H), increasing air instability and turbulence. They promote the build up of miniature whirlwinds known as dust devils. The situation is reversed during afternoons, where sinking radiation energy input stabilizes air masses. High differences between day and night temperatures are a consequence of lacking water, the magnitude depending on the diurnal evolution of net radiation (R_n). Sloping terrain and thermal surface heterogeneities induce horizontal heat transport, known as advection (A). It causes the buildup of wind gusts and turbulence, which act as kinetic energy sources in soil erosion. Advection also plays a significant role in the energy balance of wet surface islands in dry areas (Oasis effect). The evaporative demand of the atmosphere is generally high under conditions of elevated air temperature and limited water availability.

Vegetation Control of Water Availability

When vegetation cover controls water availability, energy partitioning is affected by the physiological state of the plants. The sites of regulation are *stomata* (from Greek "mouth"), tiny pores serving as pathways between the plant interior and the atmosphere. Each pore is surrounded by a pair of specialized cells (guard cells), which control its aperture and respond to plant internal and external signals. Light, vapor pressure deficit, and water potential are the principal controlling signals. Carbon dioxide, hormones (abscisic acid and cytokinins), and photosynthetic assimilation capacity have so far been detected as additional regulating factors. Signals and plant responses are acting in an integrated manner and form the canopy resistance against water loss. Development and growth determine the evolution of plant stand architecture and hence the spatial distribution of exchange surfaces. The more the surfaces are vertically exposed against airflow, the higher is their capacity to absorb momentum. In neutral transport conditions, the logarithmic portion of the wind-profile above a canopy extrapolates downward to a height where wind speed becomes zero. This level is called *zero plane displacement* and is defined as the average height of mass and heat exchange within a canopy volume. This height changes in accordance with foliage density distribution, form drag, and wind speed. The type of surface vegetation cover thus influences the magnitude of heat and mass exchange. When determining the energy balance of a plant stand, two sources of water have to be considered, canopy and soil. If the vegetation cover is sparse or is at an early development stage, significant portions of the available energy (R_n) can reach the soil level. In this case, the availability of water depends on biological factors and the soil hydraulic properties (water retention, hydraulic conductivity, and soil water diffusivity). The partitioning of available energy (R_n) into the heat terms of the energy balance is largely determined by the water status of the soil–plant system. Latent heat ($L_v E$) from transpiration is the major energy sink when soil water is abundantly available. In case radiation reaches the canopy floor, latent heat ($L_v E$) from soil evaporation as well as soil heat flux (G) are additional sinks of energy. Advection (A) may become an additional source of energy in hot climates. Energy storage due to photosynthesis ($L_p F_p$) is very small in comparison with the other components of the energy balance and is therefore often neglected. Heat storage (S) becomes important in massive canopies like forests. When water becomes limited, surface regulation restricts latent heat loss ($L_v E$) and sensible heat (H) becomes the principal sink of energy

causing rises in surface temperature. Plants have flexible capabilities to optimize production in response to such conditions.

DETERMINATION OF THE SURFACE ENERGY BALANCE

Model determinations of the surface energy balance are commonly carried out with the combination equation, which emphasizes the mutual relation between latent and sensible heat fluxes. Practical methods assume equality, either between scalars and momentum (aerodynamic method) or between the eddy diffusivities for heat and vapor. The ratio of sensible to latent heat is then proportional to the ratio of air temperature over vapor concentration (Bowen ratio $\beta = H/L_v E$). Instrumentation can be categorized in accordance to their application. Surface parameters are commonly measured with net-radiometers (R_n), heat flow sensors (G), and lysimeters ($L_v E$). Gradient measurements above exchange surfaces involve determinations of wind speed (anemometers), air temperature (thermometers), air humidity (hygrometers, psychrometers), and CO_2 (infrared gas analyzers). Sonic anemometers, quartz thermometers, Lyman-alpha, and Krypton hygrometers are applied with the eddy correlation method. Remote sensors can be used to deduce turbulence parameters, heat and momentum fluxes from backscattered or forward-propagated signals (sodars, radars, and lidars).

CONCLUSION

The first law of thermodynamics states that the input and output energies of any given system must balance. Solar radiation is the sole energy input of the earth–atmosphere system. Its partitioning into surface fluxes of latent and sensible heat is determined by the physical properties and availability of surface water, the "evaporative demand" of the atmosphere, and the nature of the surface. Evaporation is the connecting link between the system's energy and water balances. The quantification of a system's energy balance requires a definition of its boundary conditions. They consist of the spatial and temporal dimensions of the system and its exchange surfaces, their physical transport properties, the energy states across the system boundaries, and possible modes of energy transfer.

FURTHER READING

Brutsaert, W. *Evaporation into the Atmosphere*, 1st Ed.; Kluwer: Dordrecht, 1982; 299 pp.

Campbell, G.S.; Norman, J. *Introduction to Environmental Biophysics*, 2nd Ed.; Springer: Berlin, 1998; 286 pp.

Hillel, D. *Environmental Soil Physics*, 1st Ed.; Academic Press: San Diego, 1998; 771 pp.

Jones, H.G. *Plants and Microclimate*, 2nd Ed.; Cambridge University Press: Cambridge, 1992; 428 pp.

Kaimal, J.C.; Finnigan, J.J. *Atmospheric Boundary Layer Flows: Their Structure and Measurement*, 1st Ed.; Oxford University Press: Oxford, 1994; 289 pp.

Oke, T.R. *Boundary Layer Climates*, 2nd Ed.; Routledge: London, 1987; 435 pp.

Ross, J. *The Radiation Regime and Architecture of Plant Stands*, 1st Ed.; Junk: Den Haag, 1982; 391 pp.

Zeiger, E.; Farquhar, G.D.; Cowan, I.R. *Stomatal Function*, 1st Ed.; Stanford University Press: Stanford, 1987.

E

Evaporation from Lakes and Large Bodies of Water

John Borrelli
Texas Tech University, Lubbock, Texas, U.S.A.

INTRODUCTION

The conversion of water from liquid state to vapor state is called evaporation. Evaporation requires energy—approximately $540\,\text{cal/cm}^3$ of water ($\approx 2.45\,\text{MJ/kg}$). Research has shown that the rate of evaporation is primarily a function of temperature, solar energy, wind velocity, vapor pressure deficit, and advected energy. The energy for evaporation comes predominately from solar radiation and wind. Evaporation is a major component of the hydrologic cycle, second only to precipitation. As such, precise documentation of evaporation from lakes and other water bodies is required for wise management of our water resources.

Annual lake evaporation across the United States has been estimated to range from 60 cm/yr to over 200 cm/yr.[1] The annual evaporation rates for several typical lakes vary from 51 cm/yr for Hungary Horse Reservoir (cool northern climate) to 223 cm/yr for Lake Mead (desert southwest) (Table 1). Consequently, there is a necessity to accurately measure evaporation rates and provide numerical models for estimating evaporation from numerous lakes and reservoirs where direct measurement is too costly to undertake.

TECHNIQUES FOR MEASURING LAKE EVAPORATION

There are three widely accepted methods for measuring the evaporation rates of lakes: a water budget, an energy budget, and the eddy correlation method. The water budget and the energy budget require a considerable amount of investment in personnel, instruments, and time. As a result, these methods are applied sparingly to calibrate numerical models.[2] With today's dependable computer technology, the eddy correlation method has become widely used in recent years. Most studies will employ two or all of the methods.

Water Budget

If all components of the water budget could be measured accurately, it is the only method that directly measures evaporation. The water budget for a lake is as follows:

$$Evap = [(SW_{\text{in}} - SW_{\text{out}} + GW_{\text{in}} - GW_{\text{out}} + S_{\text{b}}$$
$$- S_{\text{e}})/\text{Area} + \text{PPT}]/\text{Time}$$

where $Evap$ $[\text{LT}^{-1}]$ is evaporation, SW_{in} $[\text{L}^3]$ is surface water inflow, SW_{out} $[\text{L}^3]$ is surface water outflow, GW_{in} $[\text{L}^3]$ is groundwater inflow, GW_{out} $[\text{L}^3]$ is groundwater outflow, S_{b} $[\text{L}^3]$ is lake storage at the beginning of the time period, S_{e} $[\text{L}^3]$ is the lake storage at the end of the time period, Area $[\text{L}^2]$ is the surface area of the lake, PPT $[\text{L}]$ is precipitation, and Time $[\text{T}]$ is the time period over which the measurements are made.[3] Evaporation is the residual of several measured terms and contains the errors included in the measurement of all those terms. Precipitation, for example, can have a bias error of up to 20% due to wind currents around the orifice of a rain gauge.[4] To use the water budget to measure evaporation, the inflow and outflow from the lake must be relatively small compared to the storage; otherwise, the errors in measurement will dominate the determination of evaporation. Overall, the error of measurement is $\pm 5\%\text{--}10\%$.

Energy Budget

The energy budget uses the conservation of energy principle to determine net transfer of energy into and out of a lake. Like the water budget, the evaporation rate is computed as the residual of all other terms; thus, it will contain residual measurement errors. Sturrock et al.[5] used the following energy budget equation in the study of Williams Lake:

$$Q_{\text{x}} = Q_{\text{s}} - Q_{\text{r}} + Q_{\text{a}} - Q_{\text{ar}} - Q_{\text{bs}} + Q_{\text{v}} - Q_{\text{e}} - Q_{\text{h}} - Q_{\text{w}}$$
$$+ Q_{\text{b}}$$

where Q_{x} is the change in energy content of the body of water, Q_{s} is incoming short-wave radiation, Q_{r} is reflected short-wave radiation, Q_{a} is incoming long-wave radiation, Q_{ar} is reflected long-wave radiation, Q_{bs} is long-wave radiation emitted from the body of water, Q_{v} is net energy advected to the body of water, Q_{e} is energy used for evaporation, Q_{h} is energy conducted from the water as sensible heat, Q_{w} is energy advected from the body of

Encyclopedia of Water Science
DOI: 10.1081/E-EWS 120010365

Table 1 Annual evaporation from lakes

Lake	Annual evaporation (cm)	Longitude	Latitude	Area (ha)	Average depth (m)
Pyramid Lake[2]	128	119°40'	40°00'	46,640	61
Salton Sea[2]	179	116°10'	33°05'	88,100	8
Lake Ontario[2]	73	77°00'	44°00'	1,940,000	86
Hyco Lake[2]	94	79°05'	36°15'	1,760	6
Hungary Horse Reservoir[2]	51	113°55'	46°00'	9,700	15
Lake Kerr[2]	118	81°50'	29°20'	1,040	5
Lake Mead[2]	223	114°30'	36°05'	51,400	54
Lake Okeechobee[9]	147	80°55'	27°00'	182,130	3
Amistad Reservoir[2]	203	101°20'	29°20'	27,900	16
Great Salt Lake[2]	101	112°30'	41°00'	388,900	10

water by the evaporated water, and Q_b is heat transfer to the water from the bottom sediments. All terms are expressed in W/m^2.

Eddy Correlation

At the surface of the water, the water vapor in the air is nearly saturated. As air moves across the surface, small eddies transport the water vapor vertically at a net air movement of zero in the vertical direction. With current instrumentation, it is now possible to measure the vertical flux of water vapor or evaporation above the surface of a lake. The eddy correlation method directly measures the evaporative flux as presented by Shuttleworth[6] in the following formula:

$$E = 86.4 \overline{\rho_a w' q'}$$

where E is the evaporation rate (mm/day), ρ_a is the air density (g/m^3), w' is the vertical wind velocity (m/sec^1), and q' is the specific humidity (g of water/g of air). The overbar denotes a mean value over a specific interval and the prime denotes an instantaneous deviation from the mean. Kizer and Elliot[7] provide a complete procedure to measure and calculate all terms needed to use the eddy correlation method. The accuracy for the eddy correlation measurements is 5–10%. [6] This compares favorably with the energy and water budget methods, which have the same range of accuracy. Measurements are taken at a point but are used to represent a large area of a lake. This causes some error because there are different microclimates over a large lake.

ESTIMATION OF EVAPORATION

Evaporation cannot be measured at all lakes and reservoirs by using the methods described above. Thus, researchers have developed several equations that use climatological data for estimating evaporation. The most widely used equation is the modified Penman equation that was originally developed for evaporation as well as to estimate evapotranspiration from vegetation.[8] The modified Penman equation requires data on wind, net solar radiation, humidity, and temperature. There are many equations called modified Penman. The following is a good example of a modified Penman equation:[6]

$$E_p = \frac{\Delta}{\Delta + \gamma}(R_n + A_h) + \frac{\gamma}{\Delta + \gamma} \frac{6.43(1 + 0.536U_2)D}{\lambda}$$

where E_p is estimated potential evaporation (mm/d), R_n is net radiation exchange for the free water surface (mm/d), A_h is significant energy advected to the water body (mm/d), U_2 is wind speed at 2 m (m/sec), D is vapor pressure deficit (kPa), λ is latent heat of vaporization (MJ/kg), Δ the gradient of the saturation vapor–temperature curve (kPa/°C), and γ is the psychrometric constant (kPa/°C). Please refer to Shuttleworth[6] for details on the calculation of different variables.

Investigators have found that the modified Penman equation (not necessarily the same modifications as above) has estimated evaporation within the accuracy of measured evaporation rates.[9–11] The modified Penman equation does not take into account the heat stored in a lake, which can be significant. The Penman equation will overpredict evaporation during warmer months and underpredict evaporation during the colder months.[11] On an annual basis, the modified Penman has proven reliable over a wide range of locations and climatic conditions.

Pan evaporation rates have been widely used to estimate lake evaporation. Kohler et al.[12] reported on an extensive study at Lake Hefner in Oklahoma comparing lake evaporation and pan evaporation. They reported that the annual ratio for a U.S. Weather Bureau Class A pan evaporation to lake evaporation was 0.7. This proportional constant is called the pan coefficient. The USGS[13] reported that monthly pan coefficients varied from 0.13 in February to 1.32 in November. Annual pan coefficients have been reported as low as 0.51 at Lake Mead[14] to 0.75 at Lake Okeechobee.[9] Evaporation pans provide reliable results if several stations are used. However, pan evaporation records are often erratic and often trend downward with time because of environmental changes of surroundings and poor maintenance of the pan.

There are many other equations that have been developed to estimate evaporation. They include mass-transfer equations,[5] radiation equation,[9] temperature equations,[10] etc. The applicability of these equations is generally limited to their use in environments similar to those in which the equations were calibrated.

REFERENCES

1. Farnsworth, R.K.; Thompson, E.S.; Peck, E.L. *Evaporation Atlas for the Contiguous 48 United State*, NOAA Technical Report NWS-33; National Weather Service Reports: Washington, DC, 1982; 1–28.

2. Anderson, M.E.; Jobson, H.E. Comparison of Techniques for Estimating Annual Lake Evaporation Using Climatological Data. Water Resour. Res. **1982**, *18* (3), 630–636.

3. Rose, W.J.; Robertson, D.M. *Hydrology, Water Quality, and Phosphorus Loading of Kirby Lake, Barron County, Wisconsin*, Fact Sheet FS-066-98; U.S. Geological Survey, U.S. Department of the Interior: Washington, DC, 1998; 1–4.

4. Wanielista, M.; Kersten, R.; Eaglin, R. *Hydrology—Water Quantity and Quality Control*; John Wiley & Sons, Inc.: New York, 1997; 68–117.

5. Sturrock, A.M.; Winter, T.C.; Rosenberry, D.O. Energy Budget Evaporation from Williams Lake: A Closed Lake in North Central Minnesota. Water Resour. Res. **1992**, *28* (6), 1605–1617.

6. Shuttleworth, W.J. Evaporation. In *Handbook of Hydrology*; David, R.M., Ed.; McGraw-Hill, Inc.: New York, 1993; 4.1–4.53.

7. Kizer, M.A.; Elliott, R.L. Eddy Correlation Measurement in Crop Water Use. The 1987 International Winter Meeting of the American Society of Agricultural Engineers, Chicago, IL, Dec 15–18, 1987; American Society of Agricultural Engineers: St. Joseph, MI; Paper No. 87-2504, 1–14.

8. Penman, H.L. Natural Evaporation from Open Water, Bare Soil, and Grass. Proc. R. Soc., Ser. A **1948**, *193*, 120–145.

9. Abtew, W. Evaporation Estimation for Lake Okeechobee in South Florida. J. Irrig. Drain. Eng., Am. Soc. Civil Eng. **2001**, *127* (3), 140–147.

10. Hill, R.W. Consumptive Use of Irrigated Crops in Utah. Research Report 145; Utah Agricultural Experiment Station: Utah State University: Logan, Utah, 1994; 1–361.

11. Vardavas, I.M.; Fountoulakis, A. Estimation of Lake Evaporation from Standard Meteorological Measurements: Application to Four Australian Lakes in Different Climatic Regions. Ecol. Model. **1996**, *84*, 139–150.

12. Kohler, M.A.; Nordenson, T.J.; Fox, W.W. Evaporation from Pans and Lakes, Research Paper No. 38; Weather Bureau, U.S. Department of Commerce: Washington, DC, 1955; 1–20.

13. U.S. Geological Survery (USGS). Water-Loss Investigations: Lake Hefner Studies, USGS Professional Paper 269; U.S. Geological Survey, Department of the Interior: Washington, DC, 1954; 1–158.

14. Hughes, G.H. Analysis of Techniques Used to Measure Evaporation from Salton Sea, California, Geological Survey Professional Paper 272-H; U.S. Geological Survey: Washington, DC, 1967; 151–176.

Evaporation as a Process

Alain Perrier
*Institut National Agronomique de Paris-Grignon (INAPG),
Paris, France*

Andree Tuzet
*Institut National de la Recherche Agronomique (INRA),
Thiverval Grignon, France*

INTRODUCTION

Evaporation results from complex energy and mass exchanges and can occur on any humid surfaces in contact with air. The change of liquid water to vapor consumes energy (latent heat of vaporization $2.46 \times 10^6 \, \text{J} \, \text{kg}^{-1}$). The water vapor diffuses in the air and is taken away by air convection. This process cools the surface heated by radiation (net radiation) or eventually by convection mostly during the day. Then evaporation increases with surface availability of water and energy. The reverse of this process is called condensation (water and energy gain). Unit used for evaporation flux density (or condensation) is mass of water by unit of surface and unit of time (mass flux: $\text{kg}^{-1} \, \text{m}^{-2} \, \text{sec}^{-1}$ or $\text{mm} \, \text{day}^{-1}$; or energy flux $\text{W} \, \text{m}^{-2}$).

HISTORICAL APPROACH

Since the sixth century BC, Greek antiquity has recognized evaporation as a main basic process of all meteorological knowledge: "rains are generated from evaporation that is sent up from the earth toward under the sun" according to Anaximander of Miletos.[1] Among the first direct measurements, Perrault (1670) and Sedileau (1730) analyze water balance between evaporation ($825 \, \text{mm} \, \text{yr}^{-1}$) and rain ($515 \, \text{mm} \, \text{yr}^{-1}$) to supply Versailles's ornamental lakes and fountains. This observation raises for the first time the question: "how with such water deficit, most of the rivers continue to flow in summer and plant canopies maintain transpiration and growth?" The given explanation arrived later and was that evaporation is a process under control of regional water balance. This water balance must include deep water flows, soil water content changes, and plant evapotranspiration widely reduced compared to free water evaporation (development of hydrology with Darcy's law 1880 and later of soil physics, then soil–plant–atmosphere continuum).

HYDROLOGIC CYCLE

At earth's global scale and with interannual mean, the water cycle dominates climates and influences meteorology. The radiative energy budget of earth (incoming solar radiation and outgoing infrared radiation with all their complex radiative interactions between earth surface and atmosphere such as the greenhouse effect, etc.) must balance to zero. The resulting radiative energy supply at the earth surface (from long-wave and short-wave radiation balance) amounts to 30% of the mean extraterrestrial solar irradiance (the mean extraterrestrial irradiance is equal to one-fourth of solar constant $\approx 1380/4 \, \text{W} \, \text{m}^{-2}$). Furthermore, the mean energy radiative budget of the atmosphere leads to a same energy loss (30% of the mean extraterrestrial irradiance). Then, convective fluxes (sensible heat flux, 6%, latent heat flux or water evaporation, 24%) restored the equilibrium between the heating earth surface and the cooling atmosphere. The energy consumed at the surface by latent heat flux can be released in the atmosphere through the reverse process of vapor condensation. Then, the processes of evaporation and condensation of the water cycle are the main energy exchanges in earth surface energy budget.[2]

The volume of water exchange between the earth and the atmosphere is so huge ($420 \times 10^{12} \, \text{m}^3 \, \text{yr}^{-1}$) compared with the atmosphere reservoir ($13 \times 10^{12} \, \text{m}^3 \, \text{yr}^{-1}$) that the time period in the atmosphere for water vapor is no more than 12 days. As a consequence, rains appear more on oceans where there is a constant total water availability than on continents where most of the time there is only more or less bound water ($P_c = 0.6 P_o$ and $E_c = 0.4 E_o$ by unit of surface). Furthermore, the water balance of oceans is negative and that of continents is positive. The reverse occurs in the atmosphere (above oceans and continents) that leads to an atmospheric water advection from oceans to continents (one-third of continental rains originates from oceanic advection and two-third from continental evaporation).

Encyclopedia of Water Science
DOI: 10.1081/E-EWS 120010036

As continental evaporation supplies a significant proportion of atmospheric water vapor, anthropogenic activity that reduces evaporation (deforestation for example) tends to diminish rain. This activity initiates a positive feedback loop that lowers evaporation, and further desiccation leading to aridification and/or desertification. Other meteorological and pedological processes, like increase of drying surface albedo, lowered surface roughness, elevated surface temperature, soil crusting and erosion, accelerate the degradation.

EVAPORATION UNDER SURFACE ENERGY BALANCE

For any component of a physical or biological system, the balance of all energy fluxes is achieved by adjustments of temperature. The equation describing this energy balance is based on the principle of energy conservation , meaning that in-and-out flux of all energy fluxes are equal with no sink or source of energy at the surface. For deriving a simplified equation for the energy balance near the surface, we assume the surface to be a finite-depth interfacial layer, which must have finite mass and heat capacity. Depending on the nature of the surface, this layer may consist of soil, canopy, or some other substrate like water or snow.

Energy Balance

Then, a 1-D energy balance equation for this layer can be expressed as:

$$R_{N} = H + LE + G \qquad (1)$$

where R_{N} is the net radiation flux, H and LE are the sensible and latent heat fluxes to or from the air, and G is the ground heat flux to or from the subsurface medium (all fluxes in $W\,m^{-2}$). Here we used the sign convention that all the radiative fluxes directed towards the surface are positive, while other energy fluxes (convective or conductive) directed away from the surface are positive and vice versa.

1. The net radiation flux R_{N} is a result of radiation balance between short-wave and long-wave radiation received at or emitted by the surface which can be written as:

$$R_{N} = (1 - a)R_{g} + \varepsilon(R_{a} - \sigma T_{S}^{4}) \qquad (2)$$

   where R_{g}, global solar radiation, and R_{a}, long-wave atmospheric radiation, are the two terms of incident radiation ($W\,m^{-2}$); a is albedo (proportion of solar radiation reflected by surface) and ε is emissivity

defining the radiative properties of the subsurface (proportion of long-wave radiation emitted compared to a black body emission); so that εR_{a} is the absorbed downward long-wave radiation ($W\,m^{-2}$) and $\varepsilon \sigma T_{S}^{4}$ is the emitted long-wave radiation ($W\,m^{-2}$) with T_{S} the subsurface temperature (K).

2. The conductive ground heat flux G to or from the subsurface medium depends on physical properties of the soil and other factors including surface temperature (hence time of day) and soil moisture content, which, in turn, depend on whether it is a bare or vegetated surface.

3. The balance of energy fluxes at the surface places a constraint upon the sum of the convective fluxes, $(H + LE)$, thus emphasizing the importance of partitioning the available energy $(R_{N} - G)$ between the sensible and latent heat fluxes. These convective fluxes depend on surface characteristics, wind speed, and temperature or vapor pressure gradients.

$$H = \rho c_{p} \frac{T_{S} - T_{a}}{r_{a}}$$

$$LE = \frac{\rho c_{p}}{\gamma} \frac{P(T_{dS}) - P(T_{d})}{r_{a}} \qquad (3)$$

where ρ is the volumetric mass of air ($kg\,m^{-3}$); c_{p}, the heat capacity of air ($J\,kg^{-3}\,K^{-1}$); r_{a}, aerodynamic resistance to diffusion between the surface z_{s}, and the reference height, z_{r}, ($s\,m^{-1}$); T_{a} and T_{d}, the air temperature and dew point temperature at level z_{r} (K); T_{dS}, the dew point temperature at the surface (K); γ, psychometric constant ($P\,K^{-1}$); $P(T_{d})$, saturation vapor pressure at T_{d} (P).[3]

Evaporation

Equations 1 and 3 can be combined to yield the combination equation:

$$E, LE = \frac{\Delta}{\Delta + \gamma} \left[(R_{N} - G) + \rho c_{p} \frac{\Theta_{r} - \Theta_{S}}{r_{a}} \right] \qquad (4)$$

where Δ is the slope of the saturation vapor pressure ($P\,K^{-1}$), and Θ_{z}, the air hygrometry temperature deficit (K) at level z [$\Theta_{z} = T(z) - T_{d}(z)$].

The combination equation neatly displays the two essential physical controls on evaporation: the supply of energy and the diffusion of water vapor from the surface. Depending on the value of the relevant parameters five different cases may occur:[4]

1. $\Theta_S = 0$ defines potential evaporation (EP): evaporation from any large uniform moist or wet (after rain) area so that the surface vapor pressure is saturated. This potential value mostly under climatic forcing is called climatic demand. The resulting surface temperature is always the lowest (near air temperature or even lower) for given air temperature, humidity, wind speed, and incoming radiation.

2. $\Theta_S = \Theta_r$ is a situation corresponding to a long exchange over an extended area; that is known as "equilibrium evaporation, E_o." In this case, wind speed and consequently convection have no effects and limit evaporation to a proportion of the radiant (R_N) and conductive (G) energy supply to the surface ($R_N - G$). This "equilibrium evaporation" is also an asymptotic regional value when air characteristics (the air hygrometry temperature deficit, Θ_z) tend towards surface characteristics (Θ_S) and may be considered as the climatic evaporation.

3. $\Theta_{S\max} = \Theta_r + (R_N - G)r_a/\rho c_p$ occurs on a dry surface where the evaporation is equal to 0. In these conditions, the surface temperature is maximum ($T_{S\max} = T_a + (R_N - G)r_a/\rho c_p$).

4. $0 < \Theta_S < \Theta_r$ is referred to as "oasis effect"; the air is drier at the reference level z_r than it is at the surface. In this situation, the strong availability of water at the surface allows relatively high evaporation ($E_o < E < EP$). Sometimes, in this case, energy consumed by evaporation exceeds energy supplied by radiation; that implies a surface temperature cooler than the surrounding air and the atmosphere supplies sensible heat to the surface. The actual value of evaporation, E, is the real offer.

5. $\Theta_r < \Theta_S < \Theta_{S\max}$ is referred to as "island effect"; the air is wetter at the reference level z_r than it is at the surface, and evaporation is low ($0 < E < E_o$), decreasing the real offer. In this case, surface temperature increases to the maximum value ($\Theta_{S\max}$) as evaporation decreases to zero.

REGIONAL EVAPORATION

At the regional scale following several days of stable conditions, air boundary layer conditions characteristics control surface convective exchanges. These fluxes modify energy and mass budget of the planetary boundary layer. Most often, sensible heat flux releases energy to the boundary layer increasing air temperature and simultaneously evaporation adds water vapor. So, under wind direction according to distance or on a same point according to time, mean air and dew point temperature of boundary layer are changing. These time and space modifications induce evaporation changes by feedback. As a result, at regional scale under given net radiation (R_N) and soil water storage available for water flux, ΔQ, this mean air and dew point temperature difference ($\bar{\Theta}_Z$) of the boundary layer moves in few days toward a limit; this limit is the equilibrium value, Θ_S.[5] The analytical solution for this limit shows a value directly proportional to net radiation and to the soil water storage deficit [difference between maximum possible storage and actual storage, ($\Delta Q_{\max} - \Delta Q$)]. With high irradiance and low rainfall, the unavailability of soil moisture limits evaporation but induces dry air conditions in the boundary layer (high level of the limit, $\bar{\Theta}_Z$) that enhances potential demand (EP). In a first approximation, this approach describes how regional evaporation decreases and how climatic demand simultaneously increases (giving the relation EP + E = $2E_o$) and how local vegetation faces higher temperature and greater water stress (high demand EP and low offer ET). These conditions of aridification reduce plant cover and consequently evaporation enhancing aridification by positive feedback toward desert conditions.

CANOPY AND SOIL EVAPORATION

When local conditions allow the existence of a full vegetation cover, equilibrium evaporation (with $\Theta_S \approx \Theta_r$) provides an acceptable estimate of the vapor phase of the hydrological cycle. It depends on net radiation (mostly solar radiation), ground heat flux, and slightly on air temperature through the slope of saturation vapor function (Δ), as quoted in many scientific publications $\{E_o = [\Delta/(\Delta + \gamma)](R_N - G)\}$.[6] It is always convenient to analyze or to calculate evaporation for a given surface under given climatic conditions as a proportion of this equilibrium value $\{E = (1 + \beta)E_o\}$ introducing a discrepancy term β. As shown in Eq. 4, this coefficient is widely dependent on the difference between the air water hygrometry deficit, Θ_r, and Θ_s, that of the surface. With abundant water supply, discrepancy coefficient β may reach values around $0.3-0.4$ due to effective water uptake by the plant roots.[7] For vegetation submitted to water shortage this coefficient may decrease to -0.4. When the surface is completely dry, usually bare soil, the coefficient drops more till -1. In fact with bare soil, in response to strong climatic demand, only water diffusion from deeper soil (slow process of diffusion) can supply water for evaporation and the soil surface dries quickly building a growing dry layer called "mulch," which reduces strongly evaporation.[8]

According to seasons, the water balance is positive or negative. Generally, vegetation grows when the balance is positive. Its increasing leaf area index allows evaporation to pass climatic equilibrium and to go beyond (β varying from -0.1 to 0.3). Later in the season, lack of available water in soil appears, accelerating vegetation senescence and hamper evaporation (β dropping from 0.3 to -0.2 or less as -0.6, then reaches -1 as stored soil moisture falls to zero).

ANIMAL TRANSPIRATION

Animals need to be fed with water in order to supply their excretion and evaporation (transpiration). Most of them have developed very impermeable skin to fight water losses (Θ_s, near $\Theta_{s\,max}$), but respiration may remain a main loss of water (internal evaporation). Even if an acceleration of blood circulation carrying energy to the surface of bodies occurs, homeotherms may have difficulties to regulate their internal temperature without substantial evaporation when ambient temperature exceeds the survival limit (between 37 and 43°C). In this case, because evaporation is an effective mean to consume energy, sweat from glands wets the skin surface, which returns to a small Θ_s inducing strong evaporation. Animals can also accelerate their respiration rhythm and evaporation from lungs, or animals without sweat like pachyderm can wet their skin with water or fresh mud.

CONCLUSION

With radiative balance, evaporation is the main term of any system energy budget in the biosphere and the main cooling process for ecosystem. Furthermore evaporation (or condensation) is also the fundamental phenomenon into water cycle. So, plant plays a particular part as component of water cycle and benefits from these efficient processes. Although plants evaporate less efficiently than free water, they are more efficient than bare soil through their ability to extract water from the deep layers of soil; some trees can reach down to several meters, even decameters. Animals have to protect themselves against excess loss, and respiration as well as sweat and blood circulation tends to cool instead of surface skin evaporation.

REFERENCES

1. Brutsaert, W.H. *Evaporation into the Atmosphere*; R Deidel Publishing Company: Dordrecht, Holland, 1982.
2. Perrier, A.; Tuzet, A. Généralitiés: Léau dans la biosphére. In *Traité d'irrigation*; Tiercelin, J.R., Ed.; Lavoisier, Paris, 1998; 7–43.
3. Monteith, J.L. Evaporation and Environment. Symp. Soc.; Exp. Biol. **1965**, XIX, 205–234.
4. Perrier, A.; Tuzet, A. Land Surface Processes: Description, Theoretical Approaches, and Physical Laws Underlying Their Measurements. In *Land Surface Evaporation: Measurement and Parameterization*; Schmugge, T.J., Andre, J.C., Eds.; Springer-Verlag: Berlin, 1992; 145–155.
5. Perrier, A. Land Surface Processes: Vegetation. In *Land Surface Processes in Atmospheric General Models*; Eagleson, P.S., Ed.; Cambridge Univ. Press: Cambridge, 1982; 395–448.
6. Priestley, C.H.B.; Taylor, R.J. On the Assessment of Surface Heat Flux and Evaporation Using Large Scale Parameters. Month. Weath. Rev. **1972**, *100* (2), 81–92.
7. Allen R.G.; Pereira L.S.; Raes D.; Smith M. Crop Evaporation. FAO Irrigation and Drainage Paper N°56, FAO ed.; 1998; 300pp.
8. Tuzet, A.; Perrier, A. Bases Conceptuelles de l'analyse des Besoins. In *Traité d'irrigation*; Technique et Documentation, Thiercelin, J.R., Ed.; Lavoisier, 1998; 147–162.

Evaporation from Soils

André Chanzy
Institut National de la Recherche Agronomique (INRA), Avignon, France

INTRODUCTION

Evaporation is defined as the water-vapor flux from a surface towards the atmosphere. Evaporation from soil is an important component in soil water and energy balances. The rate of the soil evaporation flux varies commonly from $0 \, \text{kg/m}^2 \, \text{day}$ to $15 \, \text{kg/m}^2 \, \text{day}$ (also expressed as the equivalent depth of a water layer covering the surface from 0 mm/day to 15 mm/day).

Soil evaporation presents a strong variability depending on the climate conditions, the surface, and soil hydraulic properties. Fig. 1 represents evaporation sequences of two different soils under the same climate. The cases wet and dry differ by the strategy to maintain or not the soil water. When the soil is well watered, variations of the evaporation follow roughly those of the climatic demand characterized by reference evapotranspiration corresponding to a well-watered short grass surface. The differences between two wet soils or between the wet soils and the reference evapotranspiration are caused by surface properties. As soil dries evaporation decreases at rates depending on soil hydraulic characteristics.

Fig. 1 also shows the three evaporation phases. During phase I, the surface is wet enough to maintain an evaporation similar to that of a permanently watered soil (in Fig. 1 see the first day for the silty clay loam and the four first days for the loam). Phase II corresponds to the period of decreasing evaporation which does not depend on the climatic demand. Phase III occurs at the end of an evaporation period and is characterized by low and almost constant evaporation (in Fig. 1 see the silty clay loam after day 15).

WHY WATER EVAPORATES FROM SOILS

Evaporation occurs when the vapor concentration in equilibrium with the soil surface (C_s, kg m^{-3}) is higher than that of the air (C_a) above the soil (see A in Fig. 2). The vapor-flux intensity that results from this difference depends on the vapor-transport processes in the lower part of the atmosphere. The transport mechanisms are vapor diffusion and turbulence generated by the airflow over a rough surface (here the soil) and/or the air temperature

differences between the soil and the air. In most cases turbulence is the dominant transport mechanism. Thus, when the soil surface is wet, evaporation increases with the wind velocity, the temperature difference between the soil surface and the air, and the surface roughness. At the surface, when vapor moves towards the atmosphere, water vaporization occurred to maintain a water-vapor concentration that respect the thermodynamic equilibrium of the water between the liquid and vapor phases. As soil looses water vapor, its surface cools to supply the heat required for the liquid to vapor phase change.

HOW SOIL CONTROLS EVAPORATION

Soil controls the vapor concentration (C_s) at the surface level (see B in Fig. 2).

The water thermodynamic equilibrium at a liquid–vapor interface is described by the Gibbs relationship that relates the water chemical potential to the temperature and C_s. From this relationship one can demonstrate that:

$$\psi = (RT/M) \cdot \rho_w \cdot \text{Log}(C_s/C_{sat}(T)) \qquad (1)$$

where ψ is the soil-water surface potential (Pa), T the surface temperature (K), R the ideal gas constant, M (kg) the water molar mass, ρ_w (kg m^{-3}) the volumetric mass of liquid water, and C_{sat} the saturated vapor concentration which depends on the temperature. The soil potential ψ is linked to the soil moisture by a soil dependant relationship (commonly named retention curve). In wet condition ($\psi > -1 \, \text{MPa}$), $C_s/C_{sat}(T) > 0.99$ and thus C_s is controlled by the surface temperature ($C_s \cong C_{sat}(T)$). For dry soils ($\psi < -1 \, \text{MPa}$) C_s is controlled by both surface moisture and temperature.

Soil controls the water supply of the evaporative surface (see C in Fig. 2).

As a consequence of the water vaporization, the soil surface dries. So, the water-potential gradient increases near the surface and an upward water flux tends to homogenize the water potential between the surface and the upper soil layers. Such an upward water flux partly balances the water loss and thus contributes to maintain a wetness at the soil surface. The flux intensity depends on soil characteristics such as the retention curve and the

Encyclopedia of Water Science
DOI: 10.1081/E-EWS 120010071

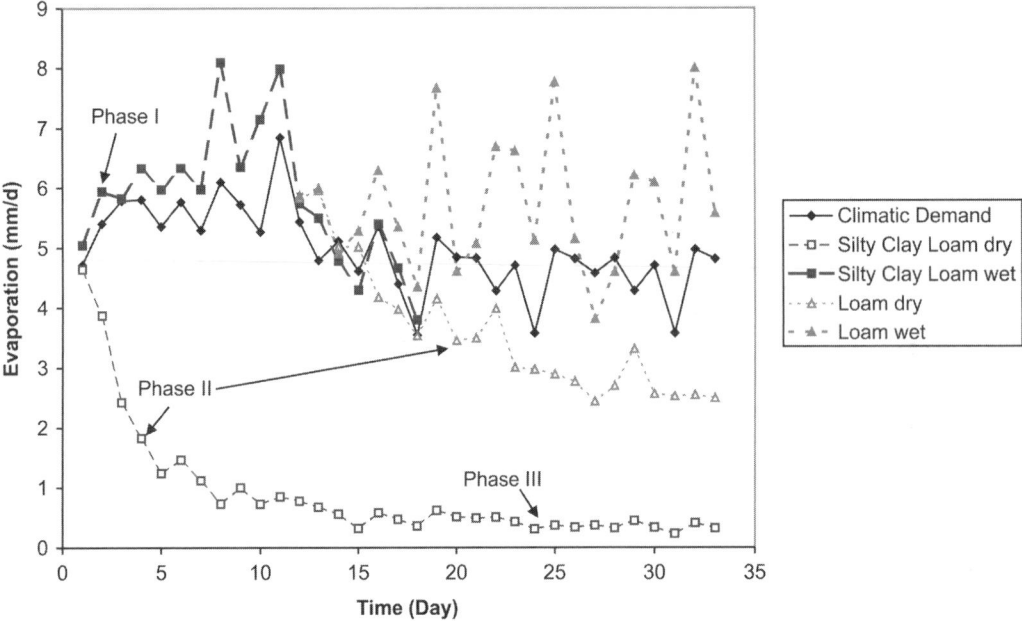

Fig. 1 Daily evaporation sequences for two different soils. For the dry condition no water supply was provided during the sequence whereas the wet conditions correspond to an irrigated surface where the wetness was maintained at saturation.

hydraulic conductivity. When hydraulic conductivity is high (i.e., a wet and/or conductive soil) the upward flux is sufficient to maintain wet conditions (i.e., $\psi > -1\,\mathrm{MPa}$) at the surface. This situation corresponds typically to the evaporation phase I. When the soil conductivity is low, the upward flux does not balance the water loss and then surface layers dries. Such a drying explains the

evaporation decrease observed during the evaporation phase II.

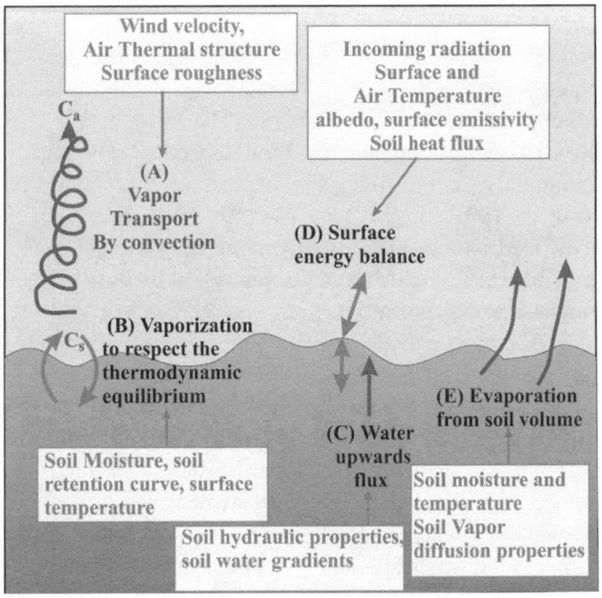

Vaporization occurs within soil (see E in Fig. 2).

There are soil/air interfaces within the soil when it is not saturated with liquid water. Vapor-concentration gradients can produce vapor fluxes by diffusion and convection. It is then possible that the soil volume contributes to the evaporation. In very dry condition corresponding to phase III, evaporation from the soil volume is the dominant contribution.[1] The soil heating by external sources and the soil vapor diffusive characteristics are the main factors affecting the vapor flux whereas atmospheric convection has little influence.[1] The soil thickness that contributes to evaporation is variable. It can reach several meters in case of desert areas where the water table is the main source of evaporation.[2]

ENERGY PROCESSES DURING EVAPORATION

As the soil provides the energy required for converting liquid water into vapor, evaporation lowers soil temperature. As a consequence, C_s decreases since $C_{sat}(T)$ monotonously increases with temperature. Without an external source of energy, C_s decreases until equilibrating with C_a, at which point evaporation stops. Therefore, the energy supply is a key factor for the evaporation (see D in

Fig. 2 Main physical processes involved in soil evaporation.

Fig. 2). The energy fluxes at the soil surface are linked by the surface energy conservation law:

$$R_n + H + LE + G = 0 \qquad (2)$$

where R_n ($W\,m^{-2}$) is the net radiation, H ($W\,m^{-2}$) the sensible heat flux, E ($kg\,m^{-2}\,sec^{-1}$) the evaporation flux, L ($J\,kg^{-1}$) the water vaporization latent heat, and G ($W\,m^{-2}$) the soil energy flux including both conductive and latent heat fluxes. The net radiation term is quantitatively the most important term. It can be written by the following equation:

$$R_n = (1 - a)R_s + \varepsilon R_a - \varepsilon\sigma T_s^4 \qquad (3)$$

where R_s is the incoming solar radiation ($W\,m^{-2}$), R_a the atmospheric radiation ($W\,m^{-2}$), ε the soil emissivity, a the surface albedo, and T_s the surface temperature. The two last terms of the Eq. 3 right side generally balances each other and thus, solar radiation is the main source of energy. Soil albedo, a, defined as the fraction of reflected solar radiation has a determinant effect on the surface energy balance. It varies from 0.1 to 0.4 according to the soil (soil chemical composition and roughness) and decreases when soil moisture increases.[3]

MEASUREMENTS OF SOIL EVAPORATION

Evaporation can be measured either by a soil-water balance or by micrometeorological observations. The soil-water balance approach consists in monitoring the water storage from the surface to a given depth and the water flux at that depth. This can be implemented by in situ soil moisture measurements or by using weighing lysi-meters.[4] These methods are appropriate to assess the evaporation at a local scale.

With the micrometeorological approach the evaporation turbulent flux above the surface is inferred directly or as a residual term of the surface energy balance equation (Eq. 1 by measuring the three other terms).[5] Measurements of turbulent fluxes (H, LE) have to be achieved over homogeneous plots and at a distance of about 50 m to 100 m from the plot boundary. Micrometeorological methods are then suitable to assess the fluxes at a field scale with a time resolution of approximately 10 min to 30 min.

SOIL EVAPORATION MODELING

Evaporation can be physically represented in mechanistic models that couple the soil heat and water flows with atmospheric fluxes.[2] Simpler approaches are available. Evaporation during phase I (also called

potential evaporation PE) can be assessed using the Penman Equation:[6]

$$LPE = \frac{\gamma}{\gamma + \Delta}f(U)(C_{sat}(T_a) - C_a) + \frac{\Delta}{\gamma + \Delta}(R_n + G) \qquad (4)$$

where γ is psychrometric constant ($\cong 67\,Pa\,K^{-1}$), Δ is the slope of the "saturation vapor pressure–air temperature (T_a)" relation and f is the turbulent vapor exchange coefficient which depends on the wind velocity (U). At a daily time step empirical relationships are given for the $f(U)$ and $(R_n + G)$ terms[7] allowing an estimation of LPE from standard climatic measurements (T_a, U, C_a, and incoming radiation).

For evaporation phase II and III numerous models are available. However, all of these models link the actual evaporation to the PE with a parameterization that involves the soil surface moisture. This quantity is either explicitly introduced in the evaporation models (Eq. 1) or estimated by a cumulative time or PE[8,9] from the beginning of the phase II period.

HOW CAN WE ACT ON SOIL EVAPORATION?

By modifying the soil properties and surface properties, it is possible to act on the rate of evaporation. Covering the soil surface with a plastic film or crop residues (mulch) suppresses or limits the vapor flux from the surface to atmosphere. This is a very efficient way to limit soil-water loss by evaporation. Soil tillage practices modify the surface roughness, the albedo, and the hydraulic conductivity. These modifications act on evaporation in different ways but it is difficult to foresee the resulting impact. Tillage is often used to break the porosity continuity. The unsaturated conductivity is then reduced accelerating drying of the soil surface layers that act like a mulch reducing further evaporation.

REFERENCES

1. Chanzy, A.; Bruckler, L. Significance of Soil Surface Moisture with Respect to Daily Bare Soil Evaporation. Water Resour. Res. **1993**, *29* (4), 1113–1125.
2. Menenti, M. Physical Aspects and Determination of Evaporation in Deserts Applying Remote Sensing Techniques. Report 10 (Special Issue), Institute of Land and Water Management Research (ICW): Wageningen, The Netherlands, 1984; 202.
3. Idso, S.B.; Jackson, R.D.; Reginato, R.J.; Kimball, B.A.; Nakayama, F.S. The Dependence of Bare Soil Albedo on Soil Water Content. J. Appl. Meteorol. **1975**, *14*, 109–113.

4. Boast, C.W. Evaporation from Bare Soil Measured with High Spatial Resolution. In *Method of Soil Analysis. Part 1 Physical and Mineralogical Methods*, 2nd Ed.; Klute, A., Ed.; Soil Science Society of America, Inc.: Madison, Wisconsin USA, 1986; 889–900.

5. Arya, S.P. *Introduction to Micrometeorology*; Academic Press Inc.: San Diego, California, 1988; 307.

6. Brutsaert, W. *Evaporation into Atmosphere, Theory, History and Applications*; Kluwer Academic Publishers: The Netherlands, 1982; 299.

7. Seguin, B. Etude Comparée des Méthodes d'Estimation d'ETP en Climat Méditerranéen du Sud de la France (Région d'Avignon). Ann. Agron. **1975**, *26* (6), 671–691.

8. Ritchie, J.T. Model for Predicting Evaporation from a Row Crop with Incomplete Cover. Water Resour. Res. **1972**, *8* (5), 1204–1213.

9. Brisson, N.; Perrier, A. A Semiempirical Model of Bare Soil Evaporation for Crop Simulation Model. Water Resour. Res. **1991**, *27* (5), 719–727.

Evapotranspiration Formulas

Robert D. Burman
University of Wyoming, Laramie, Wyoming, U.S.A.

INTRODUCTION

An agricultural scientist or hydrologist often needs to be able to calculate numerical values of evapotranspiration, ET, in equivalent depth per time units. ET is commonly defined as the transfer of water vapor to the atmosphere through evaporation from the earth's surface and transpiration from plants. This article is intended to be useful to both the engineering disciplines and the general area described by the words agricultural sciences.

The measurement of ET is both complex and expensive. First there is the requirement that much climatic data must be collected so that measurements can be associated with climatic data useful in predicting ET. Then usually measurements are taken using sensitive lysimeters. It is very possible to use simpler approaches such as nonweighing lysimeters or neutron meters but these require careful management. Study sites must have proper fetch. Water table conditions are technically difficult to properly include. The necessary calculations can usually be made using spreadsheets though often computations require computer programming using advanced programming languages. Originally measurements of ET were made on simple monocultures involving commercial crops. The need for ET measurements on more complex plant communities such as native vegetation or greenhouse plants has added complexity to ET measurements. To summarize the ET measurements are expensive, difficult, and proper data analysis is difficult.

The calculation of ET involves the process of evaporation and is obviously related to climatic variables such as solar radiation, humidity, wind movement, and temperature. ET also involves plants, which have growth cycles and may involve a single plant in the case of monocultures such as a commercial crop such as corn or maize but may also involve many kinds of plants in the case of pastures.

BACKGROUND INFORMATION

Many formulas have been developed that can be used to compute ET using climatic data. The formulas range from computation using supple correlations usually for monthly calculations to much more complex formulas for calculating ET for daily time periods or even shorter periods as short as hourly calculations. Climatic data come from existing data in databases or climatic data can be collected using automated electronic systems that are very flexible and may be programmed to collect very specific data. The automated systems can be programmed to do a considerable amount of summaries and data processing. Often the climatic data needs to be extrapolated a considerable distance. When using historic climatic data it must be analyzed for equipment operation, placement, and other things involving data suitability.

The determination of ET is a very complex process. Early methods of estimating ET involved empirical but intuitively logical correlations such as those using day length. Later the combination approach, which has an easy to understand theoretical basis was developed. Still many parts of calculating ET involve a great many empirical correlations.

The professional who needs to determine evapotranspiration can easily find a large number of formulas that are available. It is assumed here that virtually all professionals have computer skills and that a reasonable computer is available. The choice of an ET formula is difficult involving several factors. The first is consideration of good professional practice. The question of acceptance of a method is very important. Then available or collectable climatic data is also very important.

The question of consistency may be impartment. When possible it is desirable to estimate ET, net radiation, crop coefficients, and corrections for limited soil water using the same methods used by the primary reference. The final ET estimates usually involve ET calculations, various radiation components, and existing plant factors. Then to properly apply a method it is necessary to answer several questions. First, is the resulting ET calculation for the direct calculation, potential, or reference definitions? Next the professional must know the time period for the resulting ET estimates. Are estimates suitable for monthly, daily, or short-term estimates such as hourly time periods? It is necessary to properly classify and identify the purposes of calculations.

Formulas for only a few selected methods of estimating ET are shown in detail because of space limitations. For historical reasons, the Blaney Criddle, BC, method is

Encyclopedia of Water Science
DOI: 10.1081/E-EWS 120010310

shown in detail.[1] Three versions of the Penman method are shown. One is an early version of the Penman method[2] and the later addition to the Penman method identified as the Penman–Monteith, PM, method.[3,4] The version of the PM method as described in FAO 56[5] will be discussed in more detail.

SI units are used exclusively except for the very limited use of English units used for historic reasons. When each method is discussed the recommended use of the method will be shown including limitations.

DIRECT, POTENTIAL OR REFERENCE DEFINITIONS IN ET

Early formulas for estimating ET were intended for specific crops at a given time, thus the definition ET_d is used to represent this quantity. A crop coefficient, K_c is not used in the estimation. An example of this is the BC method, which involved simple empirical terms.

The Penman method then followed using a more fundamental application of physics through the radiation and energy balance concepts of the evaporative process. The concept of potential evapotranspiration, ET_p, was developed from this concept. The definition of ET_p has been changeable over time. For example a water surface was used by Penman.[2] Then ET_p was defined as ET from various crops whose growth was not limited by reduced soil water amounts in the root zone. ET_p has largely been replaced by the concept of reference ET, ET_r, following the concepts now used in FAO 56.[5]

The defined quantity ET_r is now widely used. The vegetative surfaces that define ET_r are often hypothetically based on physical characteristics of grass or alfalfa.[3] The current reference definition is a combination of the definitions for ET_p and ET_r plus calculation details often describing specific methods of calculating various parameters. Reference ET is currently based on either a short, smooth crop like grass or a more aerodynamically rougher crop like alfalfa.[3] ET_o is used in FAO 56 for ET_r which form a short (0.12 m tall), cool-season grass.

N FUNDAMENTAL EQUATION

The following equation illustrates the overall methods of calculating ET. The form of the equation is intended to illustrate direct estimation or those using crop coefficients. Detailed methods of determining "crop coefficients" appear in a separate article in this encyclopedia (See the article *crop coefficients*).

$$\mathrm{ET} = (\mathrm{ET_r})_{\text{climatic data}}^{\text{uses meas or est}} K_c K_{sw}$$

$$= (\mathrm{ET_d})_{\text{climatic data}}^{\text{uses meas or est}} K_{sw} \tag{1}$$

where K_{sw} is a correction for dry soil water amounts, K_c if used is a crop coefficient, and ET_d is ET calculated without the use of crop coefficients using a method like the earlier versions of the BC method or using the PM method as used by the extensive British MORECS system.[1]

SPECIFIC FORMULAS

Detailed discussions of specific methods of estimating ET follow. Often calculating methods are complicated and good backgrounds in thermodynamics and meteorology are helpful to follow their developments. Some classifications of ET formulas follow even though any classification scheme is by nature somewhat arbitrary.

Temperature Methods

Air temperature is intuitively related to the evaporation process. Most of us assume that evaporation is greater when air temperature is greater than when the air temperature is lower. Many ET formulas use air temperature as a major input data.

Blaney–Criddle method

The BC method[2] became widely accepted in the 1950s and marked the start of widespread evapotranspiration calculations. Due to its simplicity and easily understood concepts, it was often adopted in the western United States for legal water rights determinations. The following is intended to estimate ET by direct calculations only. The suitable time period is for monthly calculations.

$$U = \sum k_{BC} f \tag{2}$$

$$T_F = 1.8 T_C + 32 \tag{3}$$

$$f = T_F p / 100^{[3]} \tag{4}$$

where U is defined as the consumptive use of water for the growing season in inches, T_F is mean monthly air temperature in Fahrenheit, p is the monthly percent of daylight hours in the year, and k_{BC} is the monthly BC consumptive use coefficient (not the same as a crop coefficient as now used).

Hargreaves method

The Hargreaves method is described in various publications involving Hargreaves and is described in detail in Ref. [3]. The method is said to be suitable for computing

ET$_r$ for 10-day periods for a grass reference crop.

$$ET_o = 0.0023 R_A TD^{1/2}(T + 17.8) \qquad (5)$$

where TD is the mean monthly maximum air temperature − the mean monthly minimum air temperature in °C and R_A is extraterrestrial radiation MJ m^{-2} day^{-1}.

Turc method

The Turc Method is from France and is thoroughly discussed in Ref. [2]. The method was originated in the humid parts of Europe and earlier versions have a correction for dry conditions where the relative humidity is less than 50%.

$$\lambda ET_p = 0.013 \frac{T}{T + 15}(R_s + 50) \qquad (6)$$

where T is the average daily air temperature in °C and R_s is solar radiation in cal cm^{-2} day^{-1}. Calculations are suitable for 10-day periods.

Combination Methods

The use of the word combination arises from the use of an energy balance and an evaporation function to derive the basic Penman ET formula. Three variations of the Penman method follow.

Penman method

The Penman method, which also is known as a combination method, was first introduced in 1948[2] and later simplified by Penman in 1963. The original version used sunshine duration to estimate radiation. A detailed discussion of the Penman method and many of its variations is found in Ref. [3]. The origin and development of the combination equation represented a major step forward in the science of predicting ET. Many derivations exist, and it is easy to see the assumptions made in the derivations. The method has been widely used for monthly or daily calculations. Determinations have been for direct, potential, or reference crops. Most of the calculations using the Penman method have utilized monthly or daily time periods.

Many empirical wind functions have been used, and the Penman method has been used with both grass and alfalfa reference crops. The reader is urged to look for locally calibrated versions that may be applicable for the area in question. The version explained in detail here is credited to

Jensen and Wright, 1972.[3] The equation follows:

$$DET = \frac{\Delta}{\Delta + \delta}(R_n - G)$$
$$+ \frac{\delta}{\Delta + \delta} 6.43(e_s - e_a)(0.75 + 0.00115u_2) \qquad (7)$$

where u_2 is wind movement in km day^{-1} at a height of 2 m.

Priestley–Taylor method

The Priestley–Taylor Method was developed in 1972 and is a truncated version of the Penman combination ET equation. The wind term was dropped and the radiation term multiplied by a constant α, which is greater than 1. The value of the constant determines the type of ET calculated. The Priestley–Taylor Method has often been used to calculate potential ET.

$$\lambda E_p \quad \text{or} \quad \lambda ET = \alpha \frac{\Delta}{\Delta + \gamma}(R_n - G) \qquad (8)$$

where α is an empirical constant ($\alpha = 1.26$ is common and represents wet or humid conditions) and the remainder of the variables are defined elsewhere. The value of the constant α determines the kind of output from the equation.

Penman–Monteith method

The PM method[4] is a major addition to the Penman method, which was not originally developed for reference crop ET calculations. The use of this refinement of the PM method is discussed in many places including Refs. [3,5,6].

Historically determinations using the PM equation have been for direct ET estimates and for reference crop ET estimates. The following equation has been adapted and used for grass referable crops and is described by Allen et al.[6] Suitable time periods are monthly, daily, or even hourly calculations.

$$\lambda ET = \frac{\Delta(R_n - G) + \rho_a C_p \frac{(e_s - e_a)}{r_a}}{\Delta + \gamma\left(1 + \frac{r_s}{r_a}\right)} \qquad (9)$$

The following equations are used in Ref. [6] and in FAO 56.[5] Many different equations can be used to compute r_a, the canopy aerodynamic surface resistance. Different formulas and approaches are used in MORECS.[7]

$$r_a = \frac{\ln\left[\frac{z_m - d}{z_{om}}\right]\ln\left[\frac{z_h - d}{z_{oh}}\right]}{k^2 u_z} \qquad (10)$$

$$d = 2/3h \qquad (11)$$

$$z_{om} = 0.123h \qquad (12)$$

$$z_{oh} = 0.1 z_{om} \qquad (13)$$

where h is the height of vegetation in m, k is von Karman's constant (commonly taken as 0.41), u_z is wind velocity in m sec^{-1} at a height of z meters. The bulk surface and canopy surface resistance are calculated by the following

$$r_s = \frac{r_1}{0.5\,\text{LIA}} \qquad (14)$$

from FAO 56[5] defines a grass reference as a hypothetical crop with a height $h = 0.12$ m, a constant leaf surface resistance, r_1, of 70 sec m^{-1} and with an albedo of 0.23. At this point different assumptions will result in an alfalfa reference crop or direct calculations of ET. The definition of grass reference ET[6] results in the following equation.

$$\text{ET} = \frac{0.408\Delta(R_n - G) - \gamma\frac{C_{int}}{T+273}u_2(e_s - e_a)}{\Delta + \gamma(1 + 0.34u_2)} \qquad (15)$$

For daily calculations

$$C_{int(daily)} = 900 \qquad (16)$$

and for hourly calculations

$$C_{int(hourly)} = 37 \qquad (17)$$

Parameters, Combination Methods

The following parameters apply to various versions of combination methods. For example the version of the PM method used in MORECS is very different.

General

λ is latent heat of vaporization in MJ kg^{-1}, Δ is the slope of the vapor pressure temperature relationship in kPa °C^{-1}, R_n is net radiation in MJ m^{-2} day^{-1}, and G is soil heat flux in MJ m^{-2} day^{-1}, ρ_a is air density in kg m^{-3}, C_p is the specific heat of dry air (1.013 MJ kg^{-1}°C^{-1}), e_s is saturation vapor pressure in kPa, e_a is actual vapor pressure of the air in kPa, r_a is aerodynamic resistance in sec m^{-1}, r_s is bulk surface resistance in sec m^{-1}, and γ is the psychomotor constant kPa °C^{-1}.

R_n and G should be estimated by the best available methods. For detailed descriptions and examples see Refs. [3,5].

Vapor pressure

Standard values of saturation vapor pressure appear in thermodynamic steam tables. Many empirical equations have been developed to predict saturation vapor pressure. The following equation has been adopted as a standard equation for ET estimation.[5,6]

$$e^o = 0.6018\exp\left[\frac{17.27T}{T + 237.3}\right] \qquad (18)$$

where e^o is the saturation vapor pressure of the air in kPa and T is temperature in centigrade units.

Vapor pressure deficit (VPD)

$$\text{VPD} = e^s - e^a \qquad (19)$$

where VPD is defined as the vapor pressure gradient, e^s is the saturation vapor pressure, and e^a is the actual vapor pressure of the air. All vapor pressures are in kPa.

The calculation of VPD appears to be quite simple because of its relatively simple definition. However, actual calculations involve many assumptions depending upon the data available. For example available data may include average, maximum, or minimum relative humidifies. For these data limitations the estimator of ET should carefully follow the recommendations of the principal reference used.

CORRECTIONS DUE TO LIMITED SOIL WATER

A very intuitive notion is the idea that actual crop ET, ET_a, is reduced by limited soil water. Corrections of many kinds

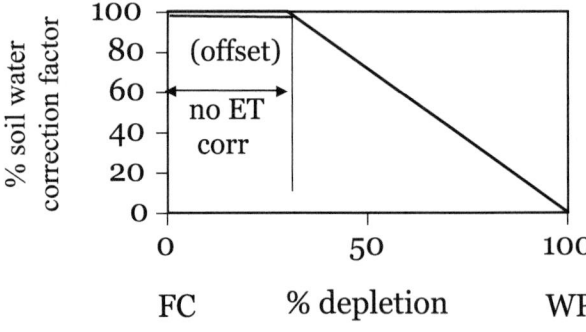

Fig. 1   Soil water correction vs. depletion.

have been used. Where crop coefficients are used the corrections are usually incorporated into the K_c values. Functions leading to a great many types of functions or relationships have been used. At times corrections have been incorporated into ET calculations yielding relationships that are often difficult to predict. Burman and Pochop[8] discuss the many limited soil water corrections that are available and widely used. Only relationships using segments of straight lines are discussed here (Fig. 1).

REFERENCES

1. Blaney, H.F.; Criddle, W.D. *Determining Consumptive Use and Irrigation Water Requirements*; Utah State Engineer: Salt Lake City, UT, 1962; 59.

2. Penman, H.L. Natural Evaporation from Open Water, Bare Soil and Grass. Proc. R. Soc. London, Ser. A **1948**, *193*, 120–146.

3. Jensen, M.E., Burman, R.D., Allen, R.G., Eds. *Evapotranspiration and Irrigation Water Requirements*; American Society of Civil Engineers: New York, NY, 1990; 332.

4. Monteith, J.L. *Evaporation and Environment*; Cambridge University Press: Cambridge, 1965; 205–223.

5. Allen, R.G.; et al. *Crop Evapotranspiration, Guidelines for Computimg Crop Water Requirements, FAO 56*; Food and Agriculture Organization of the United Nations: Rome, Italy, 1998; 300.

6. Allen, R.G.; et al. Operational Estimates of Reference Evapotranspiration. Agron. J. **1989**, *81* (4), 650–662.

7. Hough, M.; et al. *Rainfall and Evaporation Calculation System MORECS Version 2*; British Meterological Society: Bracknell, Berkshire, England, 1997; 80.

8. Burman, R.; Pochop, L.O. *Evaporation Evapotranspiration and Climatic Data*, Developments in Atmospheric Science; Elseiver: Amsterdam, 1994; Vol. 22, 278.

Evapotranspiration in Greenhouses

Thierry Boulard
Institut National de la Recherche Agronomique (INRA), Avignon, France

INTRODUCTION

Crop transpiration is the most important energy dissipation mechanism determining the thermal environment of the protected crops. Through the transpiration mechanism, the crop builds its own climate that in turn influences the transpiration.[1] As already noted by different authors,[2–4] protected crop transpiration analysis is coupled to the energy balance of the whole system and depends strongly on the greenhouse characteristics (cladding material) and on the climate control equipment (shading screen, fog system, heating, and ventilation). Therefore, reliable estimations for plant requirements must take these factors into account and conversely we must consider the mechanisms of coupling between crop transpiration and the greenhouse climate.

THEORY

Crop Transpiration Estimation from Inside Climate

Water vapor conductance (or resistance) between the leaves and the bulk of inside air, regulated by physical and physiological processes, governs greenhouse crop transpiration. With a leaf-air saturation vapor pressure deficit D_l (Pa), the transpiration Φ (W m^{-2}) of a crop characterized by a leaf area index LAI and a total resistance r_t (m sec^{-1}) to water vapor transfer is given by:

$$\Phi = \frac{\rho C_p}{\gamma} LAI \frac{D_l}{r_t} \tag{1}$$

In Eq. 1: ρ (kg m^{-3}) is the density of air, C_p (J kg^{-1} °C^{-1}) its specific heat, and γ (Pa K^{-1}) is the psychrometric constant. This simple formulation requires the leaf temperature measurement (T_l) for the determination of the leaf air saturation vapor pressure deficit D_l ($D_l = w^*(T_l) - w_i$), where w_i is inside air humidity and $w^*(T_l)$ the saturation pressure at leaf temperature. Difficulties with surface temperature measurements make Eq. 1 inconvenient for practical use. The Penman–Monteith

equation or big leaf equation[5] eliminates crop surface temperature:

$$\Phi = \frac{\delta(R_n - S_h) + \rho C_p(D_i/r_a)}{\delta + \gamma(r_c/r_a)} \tag{2}$$

Here δ is the slope of the saturated vapor pressure curve at the mean air temperature, R_n is the net radiation, S_h is the soil heat flux, r_a is the aerodynamic resistance, r_c is the total canopy resistance ($r_c = r_t/$LAI), and D_i is the inside air water vapor deficit ($w_i^* - w_i$).

As net radiation and soil heat flux are seldom measured in greenhouses, ($R_n - S_h$) can be replaced by G_a, the radiation absorbed by the crop, which can be estimated from the incident global radiation G_i and the crop leaf area index LAI.[6,7] If r'_a is the aerodynamic resistance of only one face of a leaf ($r'_a = 2r_a$), Eq. 2 can be rearranged as follows:

$$\Phi = \frac{\delta(r'_a/2)}{\delta(r'_a/2) + \gamma r_t} G_a + \frac{\rho C_p \text{LAI}}{\delta(r'_a/2) + \gamma r_t} D_i \tag{3}$$

In this equation, the transpiration rate (Φ) is the sum of a radiative component proportional to the radiation (G_a) and an advective component, proportional to the inside air vapor pressure deficit (D_i). This model was first applied to compute greenhouse tomato crop transpiration,[7] but pertains also to other greenhouse crops.

Water Vapor Transfers Between Leaf Surface and Greenhouse Air

The resistance to water vapor flow transfers between the leaf stomatal chambers and the air is a critical parameter of the model. The total canopy resistance r_t is the sum of the aerodynamic resistance between leaf surface and bulk greenhouse air r'_a, plus the leaf resistance r_s, which is the parallel connection of stomatal and cuticular resistances. Water vapor transfer through the stomata occurs mainly under the leaf surface but also partly at the upper leaf surface for amphistomatic leaves (tomato leaves for example). In this case (Fig. 1), the ratio A of the upper to under leaf surface stomatal resistance ($A = r_{ss}/r_{si}$) allows

Encyclopedia of Water Science
DOI: 10.1081/E-EWS 120010311

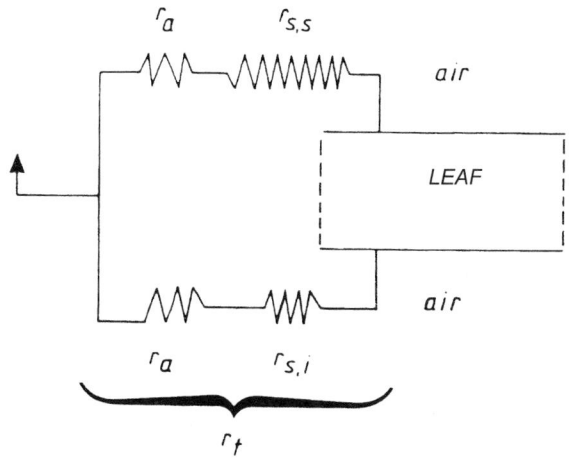

Fig. 1 Scheme of the resistances to water vapor transfers between leaf and air: r'_a: aerodynamic resistance of one face of the leaf, r_{ss}: "stomatal" resistance of the upper leaf surface, r_{si}: "stomatal" resistance of the lower leaf surface, r_t: "total" air leaf resistance.

the determination of the total leaf resistance:

$$r_t = \frac{r'_a + Ar_{si}^2 + (1+A)r'_a r_{si}}{2r'_a + (1+A)r_{si}} \tag{4}$$

As the stomatal density is higher on the lower side of the leaves, A value can vary with the stomata opening, depending on light intensity. For a tomato crop in greenhouse conditions, Boulard et al.[8] found the following relation between A and the inside global

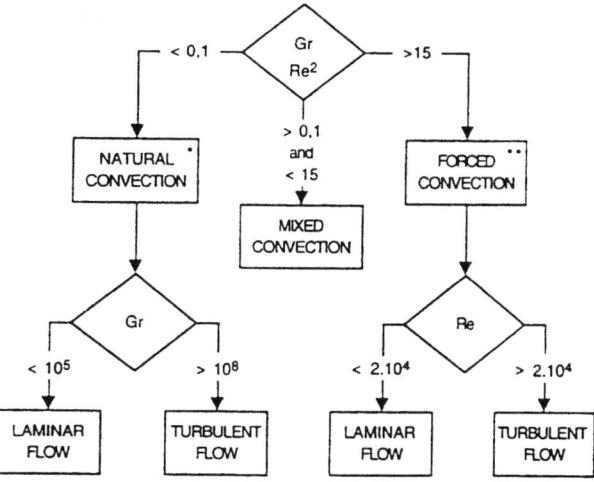

Fig. 2 Convective transfer regimes in greenhouse with respect to the Reynolds and Grashoff numbers. (After Pieters et al.[18])

radiation G_i (W m^{-2}), over the crop cover:

$$A = \log(2.7 + 0.057G_i) \tag{5}$$

Climate dependence of the stomatal resistance

As the crop water demand can generally be satisfied in greenhouse conditions, leaf stomatal resistance mainly depends on climate conditions, including solar radiation, G_i,[9] leaf air saturation deficit, D_i,[10] and temperature, T_i.[11] Following Jarvis,[12] many authors[7,8] have expressed the stomatal resistance of greenhouse crops as a function of the greenhouse air climate parameters following a general form of multiplicative models:

$$r_s = r_{smin}f_1(G_i)f_2(D_i)f_3(T_i) \tag{6}$$

where r_{smin} is the minimum stomatal resistance of the leaf (for tomato leaves, $r_{smin} \approx 100\,\mathrm{s\,m^{-1}}$), and f_{1-3}, the response functions.

For describing the response functions to the different environmental variables, three main types of relations have already been used:

- Exponential relation,[13] as for the dependence of r_s on global radiation:[8]

$$r_s = r_{smin}\left(1 + \frac{1}{\exp(0.05(G_i - 50))}\right) \tag{7}$$

- Polynomial models,[7]
- Homographic functions.[14]

With the exception of a few plants (lettuce for example), the Penman–Monteith formulation applies to most greenhouse crops with specific parameters and functions for the climate dependence of the stomatal resistance: cucumbers,[15] ornamental species,[16] or roses.[17]

Determination of the aerodynamic resistance

The aerodynamic resistance of the leaf, r'_a, depends on the aerodynamic regime prevailing in the greenhouse. Pieters et al.[18] summarized these different regimes according to the Reynolds (ul/ν) and Grasshof ($g\beta\Delta Tl^3/\nu^2$) numbers (Fig. 2), where u and l are the characteristic air speed and leaf length, respectively, β the coefficient of thermal expansion, g the gravity constant, ΔT the leaf air temperature gap, and ν the kinematic viscosity of air.

Air speed $< 0.2\,\mathrm{m\,sec^{-1}}$[19] in Venlo type greenhouses and $< 0.3\,\mathrm{m\,sec^{-1}}$ in multispan plastic houses with roof openings[20] justifies the laminar flow assumptions of many authors.[7,21] For wind force only, r'_a relates to average interior air speed following the classical relation:

$$r'_a = \rho C_p l/(0664\lambda\,Pr^{1/3}\,Re^{1/2}) = 305(l/u)^{1/2} \quad (8)$$

where λ is the air thermal conductivity and Pr the Prandl number of air.

Following Wang et al.,[20] u the characteristic interior air speed (m sec^{-1}), is proportional to the ventilation flux ϕ_v (m^3 sec^{-1}) divided by A_c (m^2), the vertical cross section area perpendicular to the average direction of the inside air flux:

$$u = \frac{\phi_v}{A_c} \quad (9)$$

With thermal stratification, transport must combine forced and free convection as discussed in details by numerous authors, various formula based on the combination of forced and free convection being proposed by Seginer,[22] Stanghellini,[7] Yang et al.,[23] and Zhang and Lemeur.[21]

Simplified Penman–Monteith Formulation

As several parameters are needed for the application of the complete model (relations 1–9), one can consider that the leaf stomatal and aerodynamics resistances can be considered as roughly constants in greenhouse conditions. Consequently Eq. 3 can be expressed in a much simpler form:

$$\Phi = AG_a + BD_i \quad (10)$$

where A and B are constant values for a given greenhouse crop stage (Table 1).

Crop Transpiration Estimation from Outside Climate Parameters

For ventilated greenhouses used in Mediterranean regions, crop transpiration can derive directly from outside climate[30] and greenhouse ventilation characteristics by solving the energy balance:

$$\lambda\Phi + H = \Pi G + Q_h - S_h - K_S\Delta T \quad (11)$$

where Φ is the latent heat of canopy transpiration (W m^{-2}), H the sensible heat exchange by ventilation (W m^{-2}), G(W m^{-2}) the outside global solar radiation, Π the solar absorption by the greenhouse-crop system, Q_h (W m^{-2}) the heating flux density provided by the heating system, S_h (W m^{-2}) the heat storage or retrieval rate of the greenhouse-soil system, K_S the overall heat transfer coefficient through the cover between inside and outside and ΔT (K) the air temperature difference between inside and outside.

At the equilibrium, if we neglect the other evapocondensative phenomena, the latent heat exchange due to canopy transpiration is proportional to the difference of air humidity between indoors and outdoors:

$$\Phi = K_v\Delta e \quad (12)$$

where Δe is the water vapor pressure gap between the interior and exterior air (Pa). K_v (W m^{-2} Pa^{-1}) is the latent heat transfer coefficient proportional to the ventilation flux V_f (m^3 sec^{-1}):

$$K_v = \lambda\xi\rho V_f/A_g \quad (13)$$

λ (J kg^{-1}) is the latent heat of water vaporization; ξ (6.25 × 10^{-6} kg$_w$kg$_a^{-1}$ Pa^{-1}) the conversion factor between the air water vapor content and the air water vapor pressure, and A_g (m^2) the greenhouse area.

Table 1 Identified values of the PM coefficients (Eq. 10) from different sources for different crops and stages

Source	Crop	A	B (W kg$_a$/kg$_w$m^2)
Jolliet and Bailey[2]	Tomato	0.34	45
Doorenbos and Pruitt[24]	Tomato	0.54	20
Jemaa[25]	Tomato ($L = 1.33$)	0.32	11
—	Tomato ($L = 3.5$)	0.36	8
—	Tomato ($L = 3.8$)	0.37	15
Pollet[26]	Lettuce	0.28	35
Stanghellini[7]	Tomato ($LAI = 1$)	0.30	72
Jolliet[27]	Tomato ($LAI = 1$)	0.28	40
Kittas et al.[28]	Rose	0.24	29
Lorenzo et al.[29]	Cucumber ($LAI = 1$)	0.23	90

The sensible heat exchange by ventilation can also be expressed with respect to the difference of air temperature between indoors and outdoors ΔT:

$$H = K_H \Delta T \qquad (14)$$

K_H (W m^{-2} K^{-1}) the sensible heat transfer coefficient is proportional to the ventilation flux V_f (m^3 sec^{-1}):

$$K_H = \rho C_p V_f / A_g \qquad (15)$$

The water vapor pressure deficit of the interior air (D_i) is a linear function of the water vapor pressure deficit and temperature of exterior air, D_o and T_o:

$$D_i = \delta(T_o)(\Delta T) - \Delta e + D_o \qquad (16)$$

where $\delta(T_o)$ is the slope of the water vapor saturation curve at T_o (Pa K^{-1}).

The system composed of Eq. 2 and Eqs. 11–16 constitutes a linear system of five equations with five unknowns (Φ, H, D_i, ΔT, and Δe), which can be solved analytically and Φ can be deduced from both outside climate and greenhouse-crop:

$$\Phi = \frac{\Pi G + Q_h - S_h + \frac{(K_S + K_H)K_2}{K_1 K_H + \delta K_2} D_o}{1 + \frac{(K_S + K_H)(1 - K_1 + K_2/K_v)}{K_1 K_H + \delta K_2}} \qquad (17)$$

with K_S the overall energy loss coefficient can be considered as dependent on external wind speed V (m sec^{-1}) following the simple relation:[31]

$$K_S = C + DV \qquad (18)$$

where C and D depend on the greenhouse design (ratio of the soil surface- to the greenhouse cover: S_s/S_c), on the type of the cover material (glass, polyethylene, PVC) and on the presence of a single or double cover.

K_1 and K_2 are combinations of crop characteristic parameters:

$$K_1 = \frac{\delta}{\delta + \gamma(r_t/r_a)} \qquad (19)$$

$$K_2 = \frac{LAI\rho C_p}{\delta r_a + \gamma r_t} \qquad (20)$$

K_v and K_h given, respectively, by Eqs. 13 and 15, are the most crucial parameters of this model because they describe the coupling of the crop with the atmosphere through the ventilation flux V_f.

THE COUPLING BETWEEN THE CROP AND THE ATMOSPHERE

As indicated by relations 11–15, greenhouse air temperature and humidity depend on solar absorption and on the balance between crop transpiration (main source of water vapor) and the losses of sensible heat and water vapor by ventilation. Eq. 17 suggests a strong coupling between the crop and the outside atmosphere when the ventilation flux is important as confirmed by Boulard et al.[32] who studied the dependence of a mature greenhouse tomato crop transpiration on outside climate when using natural ventilation, evaporative cooling, and a shading screen (Fig. 3).

This coupling between greenhouse crop transpiration and the environmental control was analyzed both for Northern Europe climate conditions[2,3] and Mediterranean conditions.[32,33] Environmental control models and strategies were derived from these studies, pertaining both for irrigation and for climate control in hot and dry conditions[34] and winter conditions.[35] For a mature

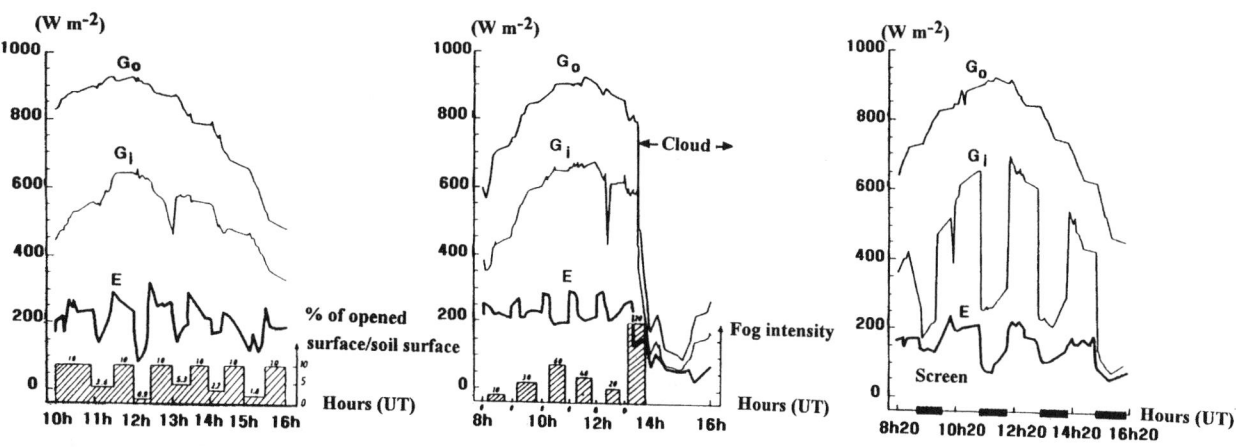

Fig. 3 Effect of ventilation, fog system, and shading screens on the transpiration rate of a tomato greenhouse crop in summer. (After Boulard et al.[32])

greenhouse crop and buoyancy driven ventilation in hot and arid conditions, Arbel et al.[36] have performed a similar study and proposed a numerical treatment of the interactions between cooling, ventilation and crop transpiration.

When modeling the case of sparsely planted seedlings in greenhouses, Seginer[37] shows that sparse plants transpire more per unit surface, due to micro-advection of energy "surplus" from the surrounding dry soil. He determined that, if water supply to the stomata is not limiting, the canopy temperature of a sparse crop is normally similar to that of a dense crop. However, high foliar potential transpiration may lead to water stress to be corrected by artificial evaporative cooling and increased ground albedo.

CONCLUSIONS

The Penman-Monteith model provides accurate estimates of crop transpiration in greenhouses. Proper evaluation of radiation balance, heat, and water vapor transfer allows to link the rate of transpiration with climate control operations to improve crop-growing conditions. As water availability becomes an increasing constraint on horticultural production, the refined tuning of irrigation based on development of the transpiration models opens perspectives for the improved efficiency of water use.

REFERENCES

1. Stanhill, G.; Scholte-Albers, J. Solar Radiation and Water Loss from Glasshouse Roses. J. Am. Soc. Hortic. Sci. **1974**, *99* (2), 107–110.
2. Jolliet, O.; Bailey, B.J. The Effect of Climate on Tomato Transpiration in Greenhouses: Measurements and Models Comparison. Agric. For. Meteorol. **1992**, *58*, 43–62.
3. Stanghellini, C.; van Meurs, W.T. Environmental Control of Greenhouse Crop Transpiration. J. Agric. Eng. Res. **1992**, *51*, 297–311.
4. Fuchs, M. Transpiration and Foliage Temperature in a Greenhouse. International Workshop on Cooling Systems for Greenhouses, May 2–5, 1993; Fuchs, M, Segal, I, Nevo, A, Teitel, M, Eds.; Agritech: Tel-Aviv, 55–68.
5. Monteith, J.L. Principles of Environmental Physics. Arnold, 1973.
6. Goudriaan, J. *Crop Micrometeorology: A Simulation Study. Simulation Monographs*; PUDOC: Wageningen, 1977; 250.
7. Stanghellini, C. Transpiration of Greenhouse Crops. An Aid to Climate Management. Ph.D. Dissertation, Wageningen Agricultural University, The Netherlands, 1977.
8. Boulard, T.; Baille, A.; Mermier, M.; Vilette, F. Mesures et Modélisation de la Résistance Stomatique Foliaire et de la Transpiration d'un Couvert de Tomate de Serre. Agronomie **1991**, *11*, 259–274.
9. Meidner, H.; Mansfield, T.A. *Physiology of Stomata*; McGraw-Hill: London, 1968; 179.
10. Lange, O.L.; Losche, R.; Schultze, E.D.; Kappen, L. Response of Stomata to Changes in Humidity. Planta **1971**, *100*, 76–86.
11. Nielson, R.E.; Jarvis, P.G. Photosynthesis in Stika Spruce. 6: Response of Stomata to Temperature. J. Appl. Ecol. **1975**, *12*, 536–610.
12. Jarvis, P.G. The Interpretation of the Variations in Leaf Water Potential and Stomatal Conductance Found in Canopies in the Field. Philos. Trans. R. Soc. **1976**, *273*, 593–610.
13. Avissar, R.; Avissar, P.; Mahrer, Y.; Bravdo, B.A. A Model to Simulate Response of Plant Stomata to Environmental Conditions. Agric. For. Meteorol. **1985**, *34*, 21–29.
14. Farquhar, G.D. Feedforward Response of Stomata to Humidity. Aust. J. Plant Physiol. **1978**, *5*, 787–800.
15. Bakker, J.C. Leaf Conductance of Four Glasshouse Vegetable Crops as Affected by Air Humidity. Agric. For. Meteorol. **1991**, *55*, 23–36.
16. Baille, M.; Baille, A.; Laury, J.C. A Simplified Model for Predicting Evapotranspiration Rate of Nine Ornamental Species vs. Climate Factors and Leaf Area. Sci. Hortic. **1994**, *59*, 217–232.
17. Baille, M.; Baille, A.; Delmon, D. Micro Climate and Transpiration of Greenhouse Rose Crops. Agric. For. Meteorol. **1994**, *72*, 83–97.
18. Pieters, J.G.; Deltour, J.M.; Debruyckere, M.J. Condensation and Static Heat Transfer Through Greenhouse Conver during Night. Trans. Am. Soc. Agric. Eng. **1994**, *37* (6), 1965–1972.
19. Wang, S.; Deltour, J. Lee Side Ventilation Induced Air Movements in a Large Scale Multi Span Greenhouse. J. Agric. Eng. Res. **1999**, *74*, 103–110.
20. Wang, S.; Boulard, T.; Haxaire, R. Air Speed Profiles in a Naturally Ventilated Greenhouse. Agric. For. Meteorol. **1999**, *96*, 181–188.
21. Zhang, L.; Lemeur, R. Effect of Aerodynamic Resistance on Energy Balance and PM Estimates of Evaporation in Greenhouse Conditions. Agric. For. Meteorol. **1992**, *58*, 209–228.
22. Seginer, I. On the Night Transpiration of Greenhouse Roses Under Glass or Plastic Cover. Agric. For. Meteorol. **1984**, *30*, 257–268.
23. Yang, X.; Short, T.H.; Robert, D.F.; Bauerle, W.L. Transpiration, Leaf Temperature and Stomatal Resistance of a Cucumber Crop. Agric. For. Meteorol. **1990**, *5*, 197–209.
24. Doorenbos, J.; Pruitt, W.O. Crop Water Requirements. FAO Irrigation and Drainage Paper; 1975; 24, 15–50.
25. Jemaa, R. Mise au Point et Validation de Modèles de Transpiration de Cultures de Tomate Hors Sol sous Serre. Application à la Conduite de la Fert-Irrigation. Thèse de l'ENSA Rennes, 1995.
26. Pollet, S. Application of the Penman-Monteith Model to Calculate the Evapotranspiration of Head Lettuce *Lactuca*

E

sativa L. var Capitata in Glasshouse Conditions. Acta Hortic. **1999**, *519*, 151–161.

27. Jolliet, O. HORTITRANS, a Model for Predicting and Optimizing Humidity and Transpiration in Greenhouses. J. Agric. Eng. Res. **1994**, *57*, 23–37.

28. Kittas, C.; Katsoulas, N.; Baille, A. Transpiration and Canopy Resistance of Greenhouse Soilless Roses: Measurements and Modeling. Acta Hortic. **1999**, *507*, 61–68.

29. Lorenzo, P.; Medrano, E.; Sanchez-Guerrero, M.C. Greenhouse Crop Transpiration: An Implement to Soilless Irrigation Management. Acta Hortic. **1998**, *458*, 113–119.

30. Boulard, T.; Wang, S. Greenhouse Crop Transpiration Simulation from External Climate Conditions. Agric. For. Meteorol. **2000**, *100*, 25–34.

31. Bailey, B.; Cotton, B. Glasshouse Thermal Screen: Influence of Single and Double Screens on Heat Loss and Crop Environment. NJAE Department Note DN/G/821; 1980; 15.

32. Boulard, T.; Baille, A.; Le Gall, F. Etude des Effets de Différents Équipements de Climatisation sur la Transpiration d'un Couvert de Tomates de Serre. Agronomie **1991**, *11*, 543–553.

33. Papadakis, G.; Frangoudakis, A.; Kiritsis, S. Experimental Investigation and Modelling of Heat and Mass Transfer between a Tomato Crop and the Greenhouse Environment. J. Agric. Eng. Res. **1994**, *57*, 217–227.

34. Boulard, T.; Baille, A. A Simple Greenhouse Climate Control Model Incorporating Effects of Aeration and Evaporative Cooling. Agric. For. Meteorol. **1993**, *65*, 145–157.

35. Stanghellini, C.; de Jong, T. A Model of Humidity and Its Application in Greenhouse. Agric. For. Meteorol. **1995**, *76*, 129–148.

36. Arbel, A.; Shlykar, A.; Barak, M. Buoyancy Driven Ventilation in a Greenhouse Cooled by a Fogging System. Acta Hortic. **1999**, *534*, 327–334.

37. Seginer, I. Transpirational Cooling of a Greenhouse Crop with Partial Ground Cover. Agric. For. Meteorol. **1994**, *71*, 265–281.

Evapotranspiration, Reference and Potential

Marcel Fuchs
Agricultural Research Organization, Bet Dagan, Israel

INTRODUCTION

Water vapor loss from land surfaces depends on meteorological factors that provide the energy required to transform liquid into vapor and disperse the free water molecules into the atmosphere. The process can occur only if continuity in the gaseous phase between the liquid water and the atmosphere is established. More specifically, it depends also on the availability of liquid water in the vegetation and in the soil. The purpose of the *potential* and *reference evapotranspiration* concepts is to standardize liquid water availability.

DEFINITIONS

Potential evapotranspiration is the rate of water vapor loss from vegetation-covered ground when its entire surface in contact with the atmosphere is wet. Maintaining a continuous and persistent presence of liquid water the interface between the plants and the atmosphere is not feasible in real situations. Therefore, potential evapotranspiration is a theoretical concept specifying an asymptotic upper limit of actual evapotranspiration.

Reference evapotranspiration is the rate of water vapor loss from ground fully covered with actively growing short grass of uniform height whose root system is unrestrictedly supplied with water. This quantity can be determined by measuring the water vapor loss from vegetation growing on a soil that is frequently and uniformly wetted by natural rainfall or irrigation. The frequency of wetting should be such that the soil moisture content remains in the range for which the flow of water towards the roots is unimpeded, a condition characterized by the occurrence of gravitational drainage. This evapotranspiration rate can be measured with weighing lysimeters.

OVERVIEW

The evaporation of liquid water from plant communities involves both supply and demand controlled processes: supply as the flow of water to the plant organs and soil pores where the liquid to vapor transformation occurs and demand as the weather driven delivery of heat that converts liquid into vapor. This link to the heat balance of the surface sets the process in the realm of meteorology. As open water surfaces have infinite supply, meteorologists initially considered that evaporation from free water could provide a measure of the demand. However, evaporation from free water surfaces does not provide universal relationships with soil moisture withdrawal by vegetation because the energy and mass exchanges of water surfaces and vegetation or soil respond differently to radiation, wind, air humidity, and temperature.[1]

In search of a more appropriate measure of atmospheric evaporative demand, Thornthwaite[2] introduced the concept of *potential evapotranspiration* to classify climates according to their effect on the water balance of vegetation. He used the terms *evapotranspiration* to include water vapor sources in the plants and in the soil and *potential* to indicate unlimited liquid flow toward the surface. He formulated a temperature-based empirical estimate of the water vapor loss, but results did not relate correctly with measured evapotranspiration. Consequently, the conceptual impact of Thornthwaite's method surpassed its practical significance on hydrology, climatology, and irrigation science. Penman[3] developed a similar concept, based on an approximate linear solution of the energy balance for a short green lawn fully shading the ground and never lacking water. Setting the water vapor pressure of the vegetation to its saturated value at the vegetation surface temperature fulfilled the condition of unlimited water supply. The resulting formula was a linear combination of a radiation term and a wind function with empirical coefficients fitting estimated evaporation of open water to the evapotranspiration of well-watered lawn. Penman's approach elegantly eliminated explicit reference to surface temperature and surface water vapor pressure from the energy balance solution, enabling potential evapotranspiration to be calculated from standard data measured in meteorological stations.

The method became the standard for determining potential evapotranspiration and found numerous applications in irrigated agriculture. However, measured evapotranspiration occasionally exceeded the calculated potential value. Reasons for these apparent anomalies were easily identified. As liquid to vapor conversion occurs below the epidermis of transpiring organs, fitted

Encyclopedia of Water Science
DOI: 10.1081/E-EWS 120010312

parameters included implicitly the diffusive resistance of stomatal pores in the epidermis. Some plants have lower stomatal resistance than the lawn for which the parameters of Penman's formula were derived. Many vegetation canopies have higher radiation absorption and stronger aerodynamic exchanges, and therefore, are capable of higher evapotranspiration rates than grass.

The need to invoke an additional resistance in the pathway between the source of the vapor and the free atmosphere prevented the definition of the unambiguous upper bound implied in the original potential evapotranspiration concept. Still, the calculated values provided a *reference* for comparing water use by plants growing under widely diverse climatic conditions. Irrigation oriented scientists introduced the term *reference evapotranspiration* to reflect the conceptual change in the meaning of potential evapotranspiration.[4,5] Initially, alfalfa served as the reference surface, but irregular growth following repeated mowing and the limited range of climates where it could be grown favored the use of ubiquitous grass kept at a height between 0.08 m and 0.15 m. Lysimeter measurements of grass evapotranspiration served to recalibrate Penman's formula to determine reference evapotranspiration.[5] Thus, reference evapotranspiration became the water vapor loss of a well-watered grass surface, as in Penman's original operational definition of potential evapotranspiration.

The convergence of definitions led to the interchangeable use potential and reference evapotranspiration, confusing novices and generating futile controversy among experts. The term potential evapotranspiration should be kept for the theoretical upper limit of water vapor loss from a given vegetation-type when resistance of the vapor pathway in the plant tissues approaches zero. Reference evapotranspiration should designate the water vapor loss of an extended, actively growing, well-watered grass, fully covering the ground and mown to remain between 0.08 m and 0.15 m high. The resistance of vapor pathway inside the plants assumes the minimum value experimentally determined for what is believed to be unrestricted water supply to the roots. This experimental minimum resistance used to define reference evapotranspiration replaces the theoretical condition of unlimited liquid flow to the surface in the definition of potential evapotranspiration. Therefore, reference evapotranspiration represents the closest experimental realization of potential evapotranspiration.

ADDITIONAL INSIGHT

Potential and reference conditions imply evapotranspiration rates higher than those from vegetation without free water on its surface and undergoing periodic water shortage between watering events. The difference increases with the aridity of the climate and the moisture deficit in the soil. The larger latent heat dissipation increases the water vapor content of the air and reduces the energy available for air and soil heating, leading to a wetter and cooler microclimate. Thus, creating the conditions for realizing potential or reference evapotranspiration decreases its value. This paradox is reminiscent of Schrödinger's cat in quantum mechanics. It led Bouchet[6] to formulate the concept of complementary evapotranspiration based on the hypothesis that the sum of potential evapotranspiration and actual evapotranspiration is a constant. The idea was adapted to derive climatological estimates of regional evapotranspiration,[7,8] without requiring values for soil moisture availability and stomatal resistance of plants. As regional evapotranspiration is extremely difficult to measure, applications deriving from the complementary evapotranspiration concept did not gain acceptance. Furthermore, the approach could not evaluate water use of agricultural fields at the spatial and temporal scale required to control soil moisture by irrigation. For these applications, the climatic factors determining evapotranspiration had to consider the specific radiometric, aerodynamic, and stomatal resistance properties of the crop surface.

The radiation balance and aerodynamic transport terms in Penman's original derivation of potential evapotranspiration assumes empirical functions adapted to the standard data recorded in meteorological stations: daily hours of sunshine, average air temperature and vapor pressure deficit, and wind run. The increased availability of pyranometers giving a direct measurement of incident solar energy improved the accuracy of the method. Net pyrradiometers measuring the total radiant energy absorbed by the vegetation surface constituted an important additional step to the accurate calculation of potential evapotranspiration for real vegetation surfaces. The calculation of potential evapotranspiration became even more specific when Businger[9] introduced turbulent transport characteristics of the air surface layer to parameterize explicitly the aerodynamic properties of the vegetation, using the roughness length to quantify the drag exerted by the vegetation. Adjustments accounting for buoyancy[10] and separation between the sink-source lengths dimensions for momentum and water vapor[11] further fine-tuned the aerodynamic function.

Monteith[12] realized the full potential of Penman's contribution by relating potential evapotranspiration to actual evapotranspiration in terms of a surface resistance that lumped the stomatal resistance of transpiring organs and the resistance of soil to water vapor diffusion. Surface resistance quantified field scale soil moisture availability and established a link with laboratory studies of physiological indicators of plant water stress. Setting the surface resistance to its minimum value allowed the calculation of reference evapotranspiration. However, selecting the value of minimum resistance proved to be difficult, because stomatal resistance varies with radiation, carbon dioxide concentration, air humidity, and temperature. Some of its variability is related to the physiology and biochemistry of plants. As mechanisms regulating stomatal resistance are still only partially understood its value remains unpredictable. This biological uncertainty affects reference evapotranspiration and weakens its reliability as an objective indicator of atmospheric evaporative demand. The widely accepted definition of plant surface characteristics for realizing reference evapotranspiration[13] has retained the fuzziness of a 1956 published statement about "extended surface of short green crop, actively growing, completely shading the ground, of uniform height and not short of water."[14] By contrast, potential evapotranspiration sets unequivocally the minimum resistance to zero. With proper modeling of radiation and momentum absorption, it can parameterize specifically any vegetation geometry. Therefore, despite the lack of experimental validation, it provides the most consistent climatic measure of evaporative conditions.

REFERENCES

1. Tanner, C.B. Measurement of Evapotranspiration. In *Irrigation of Agricultural Lands*; Hagan, R.M., Haise, H.R., Edminster, T.W., Eds.; American Society of Agronomy: Madison, WI, 1967; Vol. 11, 534–574.

2. Thornthwaite, C.W. An Approach Toward a Rational Classification of Climate. Geogr. Rev. **1948**, *38*, 55–94.

3. Penman, H.L. Natural Evaporation from Open Water, Bare Soil, and Grass. Roy. Soc. Lond., Proc. Ser. A **1948**, *193*, 120–146.

4. Jensen, M.E.; Robb, D.C.N.; Franzoy, C.E. Scheduling Irrigation Using Climate–Crop–Soil Data. J. Irrig. Drain. Eng.—ASCE **1970**, *96*, 25–38.

5. Doorenbos, J.; Pruitt, W.O. *Guidelines for Predicting Crop Water Requirements*; FAO: Rome, 1977; 24, 1–144.

6. Bouchet, R.J. Signification et Portée Agronomique de l'Évapotranspiration Potentielle. Ann. Agron. **1961**, *12*, 51–63.

7. Morton, F.I. Climatological Estimates of Evapotranspiration. J. Hydraul. Div. Proc. ASCE **1976**, *102*, 275–291.

8. Brutsaert, W.; Stricker, H. An Advection-Aridity Approach to Estimate Actual Regional Evapotranspiration. Water Resour. Res. **1979**, *15*, 443–450.

9. Businger, J.A. Some Remarks on Penman's Equations for the Evapotranspiration. Neth. J. Agric. Sci. **1956**, *4*, 77–80.

10. Fuchs, M.; Tanner, C.B. Evaporation from a Drying Soil. J. Appl. Meteorol. **1967**, *6*, 852–857.

11. Garratt, J.R.; Hicks, B.B. Momentum, Heat and Water Vapour Transfer to and from Natural and Surfaces. Quart. J. R. Meteorol. Soc. **1973**, *99*, 680–687.

12. Monteith, J.L. Evaporation and Environment. In *The State and Movement of Water in Living Organisms*, Symp. Soc. Exp. Biol.; Fogg, G.E., Ed.; Cambridge University Press: Cambridge, 1965; Vol. 19, 205–234.

13. Allen, R.G.; Pereira, L.S.; Raes, D.; Smith, M. *Crop Evapotranspiration. Guidelines for Computing Crop Water Requirements, FAO Irrigation and Drainage paper*; FAO: Rome, 1998; Vol. 56, 300.

14. Anonymous; Conclusions Reached After Discussions Concerning Evaporation. Neth. J. Agric. Sci. **1956**, *4*, 95–97.

Evapotranspiration, Remote Sensing of

W. P. Kustas
United States Department of Agriculture (USDA), Beltsville, Maryland, U.S.A.

G. R. Diak
University of Wisconsin, Madison, Wisconsin, U.S.A.

M. S. Moran
United States Department of Agriculture (USDA), Tucson, Arizona, U.S.A.

INTRODUCTION

For over two decades, approaches to sense evapotranspiration (ET) remotely have made use of radiometric surface temperatures [$T_R(\theta)$, where θ is the radiometer viewing angle] as a key surface boundary condition in the land–surface energy balance. Such methods include simple flux–profile (single-level) models of surface exchange, statistical/analytical schemes, and other techniques that are based on more complex physical models of the land surface, including the so-called soil–vegetation–atmosphere–transfer (SVAT) schemes.[1]

Typically, these methods estimate fluxes through the evaluation of a surface–air temperature gradient at a single time. The aerodynamic resistance to heat transfer is largely defined by the aerodynamic roughness length, and the land surface is treated as a single effective surface in contact with the atmosphere. Any factor that introduces errors into the evaluation of this gradient, as well as the simplifications of the model, may introduce significant errors in the resulting flux estimates.

This article gives a brief overview of some of the modeling schemes that have utilized remotely sensed surface temperature data. Some recent modeling efforts will be described that address the limitations described below. These include 1) uncertainty in T_R (θ); 2) observations of T_A at regional scales; and 3) non uniqueness of the radiometric–aerodynamic temperature relationship. The resulting modeling framework leads to a more reliable scheme for quantifying ET at regional scales using satellite remote sensing.

SOURCES OF ERROR IN ET ESTIMATION

Even after performing the corrections for atmospheric attenuation and surface emissivity required to obtain a radiometric surface temperature from a satellite-measured brightness temperature, there remains 1–3° uncertainty in $T_R(\theta)$. Compounding this is the fact that vegetation density, architecture, and angle of view of the radiometer also have significant effects on brightness temperature observations (the angle-of-view "effect" being most pronounced for surfaces with partial canopy cover). As a result of these error sources, estimates of the surface–air temperature gradient and resulting fluxes are likely to have large uncertainties.[2]

An additional complication is the significant differences that exist between the radiative and the so-called "aerodynamic" (single level, "effective") surface temperature.[3] Unfortunately, this aerodynamic temperature is a construct that cannot be measured and many of the factors affecting the radiometric temperature are not well correlated to the aerodynamic roughness, making radiometric–aerodynamic temperature relationships somewhat ambiguous to begin with.

For applications over regional scales, deriving the required meteorological upper boundary conditions [i.e., shelter-or anemometer-level (2 m–10 m) air temperature and wind speed] for each satellite pixel may also lead to significant errors in flux evaluations. Typically, these meteorological quantities come from an analysis of hourly weather observations (observations typically spaced on the order of 100 km apart), and may not be representative of actual conditions at a given location.

OVERVIEW OF REMOTE SENSING METHODS

The most common way to estimate ET is to solve for the latent heat flux, LE, as a residual in the energy balance equation for the land surface:

$$LE = R_N - G - H \tag{1}$$

where R_N is the net radiation, G, the soil heat flux, and H,

Encyclopedia of Water Science
DOI: 10.1081/E-EWS 120010313

the sensible heat flux all usually given in $W\,m^{-2}$. The quantity $R_N - G$ is commonly called the "available energy"; remote sensing methods for estimating these components are described in Kustas and Norman.[1] Typically with reliable estimates of remotely sensed solar radiation (e.g., Ref. 4), differences between remote sensing estimates and observed $R_N - G$ are within 10%.

The largest uncertainty in estimating LE comes from computing H. A simple form to express and examine the relationship between H and the surface–air temperature difference is via a resistance relationship (e.g., Ref. 5),

$$H = \rho C_P \frac{T_R(\theta) - T_A}{R_A + R_{EX}} \qquad (2)$$

In this equation, T_A is the near-surface air temperature, ρ, the air density, C_P, the specific heat of air, R_A, the aerodynamic resistance and R_{EX}, the so-called "excess resistance," which addresses the fact that momentum and heat transport from the roughness elements differ.[6] The method offers the possibility of mapping surface heat fluxes on a regional scale by using radiometric temperature observations, $T_R(\theta)$ (converted from satellite brightness temperatures) if R_A and R_{EX} can be estimated appropriately. R_{EX} has been related to the ratio of roughness lengths for momentum, z_{OM}, and heat, z_{OH}, and the friction velocity $u*$ having the form[5,6]

$$R_{EX} = k^{-1}\ln\left(\frac{z_{OM}}{z_{OH}}\right)u*^{-1} \qquad (3)$$

where $k = 0.4$ is von Karman's constant. While addressing the well-known differences in efficiency between momentum and heat transport from natural surfaces, this model is just one of several that have been developed (e.g., Refs. 5 and 7). There have been numerous efforts in recent years to apply Eq. 2 and hence determine the behavior of R_{EX} or z_{OH} for different surfaces, but no universal relation exists for land surfaces with large spatial and temporal variations in the magnitude of z_{OH} having been documented.[1] These results are due, in part, to the fact that this formulation lumps view angle dependency of $T_R(\theta)$ into the excess resistance, which makes the relation useless for any conditions except those similar to the training data.[8] Nevertheless, the method for estimating ET using the approach summarized in Eqs. 1–3 is still widely applied.

Satellite observations are essentially "instantaneous" or merely "snap shots" of the surface conditions. For many practical applications, LE estimates over longer time scales (daily values or longer) are needed. This was the impetuous for an empirical scheme for estimating daily LE, LE_D, suggested by Jackson, Reginato, and Idso[9] using observations of $T_R(\theta)$ and T_A near mid-day or

maximum heating:

$$LE_D = R_{N,D} - B(T_{R,i}(\theta) - T_{A,i})^n \qquad (4)$$

where the subscript i and D represent "instantaneous" and daily values, respectively. The coefficients B and n have been related to physical properties of the land surface and atmosphere, such as z_{OM} and stability, respectively.[10] Both theoretical and experimental studies have evaluated Eq. 4 lending further support for its utility as a simple technique for estimating LE_D.[11–13] In fact, studies have applied Eq. 4 to meteorological satellites for longer term regional ET monitoring.[14]

A major drawback with these approaches summarized above, however, is that there is no distinction made between soil and vegetation canopy contributions to land-surface fluxes or to satellite-measured brightness temperatures used to diagnose the fluxes. Hence, vegetation water use or stress cannot be evaluated. Furthermore, as evidence from many previous studies both the resistances in Eq. 2 and consequently the B parameter in Eq. 4 are not uniquely defined by surface roughness parameters. In addition to experimental evidence (e.g., Refs. 15 and 16), Kustas et al.[8] using SVAT simulations, have shown the lack of a unique relationship between $T_R(\theta)$ and the aerodynamic surface temperature, T_O, (satisfying the flux relationship in Eq. 2 when used with traditional expressions for the resistances; see Ref. 2).

An alternative approach proposed recently considers the soil and vegetation contribution to the total or composite heat fluxes and soil and vegetation temperatures to the radiometric temperature measurements in the so-called "Two-Source" Modeling (TSM) scheme.[17] This allows for Eq. 2 to be recast into the following expression:

$$H = \rho C_P \frac{T_R(\theta) - T_A}{R_R} \qquad (5)$$

where R_R is the radiometric–convective resistance given by[17]

$$R_R = \frac{T_R(\theta) - T_A}{\dfrac{T_C - T_A}{R_A} + \dfrac{T_S - T_A}{R_A + R_S}} \qquad (6)$$

where T_C is the canopy temperature, T_S, the soil temperature, and R_S, the soil resistance to heat transfer. An estimate of leaf area index or fractional vegetation cover, f_C, is used to estimate T_C and T_S from $T_R(\theta)$:

$$T_R(\theta) \approx \left(f_C(\theta)T_C^4 + (1 - f_C(\theta))T_S^4\right)^{1/4} \qquad (7)$$

where $f_C(\theta)$ is the fractional vegetative cover at radiometer viewing angle θ, and R_S is computed from a relatively simple formulation predicting wind speed near the soil surface.[17] With some additional formulations for estimating canopy transpiration, and the dual requirement of energy, and radiative balance of the soil and vegetation components, closure in the set of equations is achieved. Through model validation studies, revisions to the original two-source formulations have been made improving its utility under a wider range of the environmental conditions.[8,18]

Several relatively early studies recognized the need to assess the impact of vegetation cover on remote methods for deriving ET. For example, Price[19] used information provided in the Vegetation Index–radiometric temperature, VI–$T_R(\theta)$, space. This work involved the use of an energy balance model for computing spatially distributed fluxes from the variability within the Normalized Difference Vegetation Index, NDVI–$T_R(\theta)$ space from a single satellite scene. NDVI was used to estimate the fraction of a pixel covered by vegetation and showed how one could derive bare soil and vegetation temperatures and, with enough spatial variation in surface moisture, estimate daily ET for the limits of full cover vegetation, dry and wet bare soils.

Following Price,[19] Carlson, Gillies, and Perry[20] combined an Atmospheric Boundary Layer (ABL) model with a SVAT for mapping surface soil moisture, vegetation cover, and surface fluxes. Model simulations are run for two conditions: 100% vegetative cover with the maximum NDVI being known a priori, and with bare soil conditions knowing the minimum NDVI. Using ancillary data, including a morning atmospheric sounding, vegetation and soil type information, root-zone and surface

soil moisture are varied, respectively, until the modeled and measured $T_R(\theta)$ are closely matched for both cases so that fractional vegetated cover and surface soil moisture are derived. Comparisons between modeled–derived fluxes and observations have been made recently by Gillies et al.[21] indicating approximately 90% of the variance in the fluxes was captured by the model.

In a related approach, Moran et al.[22] defined theoretical boundaries in VI–($T_R(\theta)-T_A$) space using the Penman–Monteith equation. The boundaries define a trapezoid, which has at the upper two corners unstressed and stressed 100% vegetated cover and at the lower two corners, wet and dry bare soil conditions (Fig. 1). In order to calculate the vertices of the trapezoid, measurements of R_N, vapor pressure, T_A, and wind speed are required as well as vegetation specific parameters; these include maximum and minimum VI for the full-cover and bare soil case, maximum leaf area index, and maximum and minimum stomatal resistance. Moran et al.[22] analyze and discuss several of the assumptions underlying the model, especially those concerning the linearity between variations in canopy–air temperature and soil–air temperatures and transpiration and evaporation. Information about ET rates are derived from the location of the VI–[$T_R(\theta)-T_A$] measurements within the date and time-specific trapezoid. This approach permits the technique to be used for both heterogeneous and uniform areas and thus does not require having a range of NDVI and surface temperature in the scene of interest as required by Carlson, Gillies, and Perry[20] and Price.[19] Moran[23] compared the method for estimating relative rates of ET with

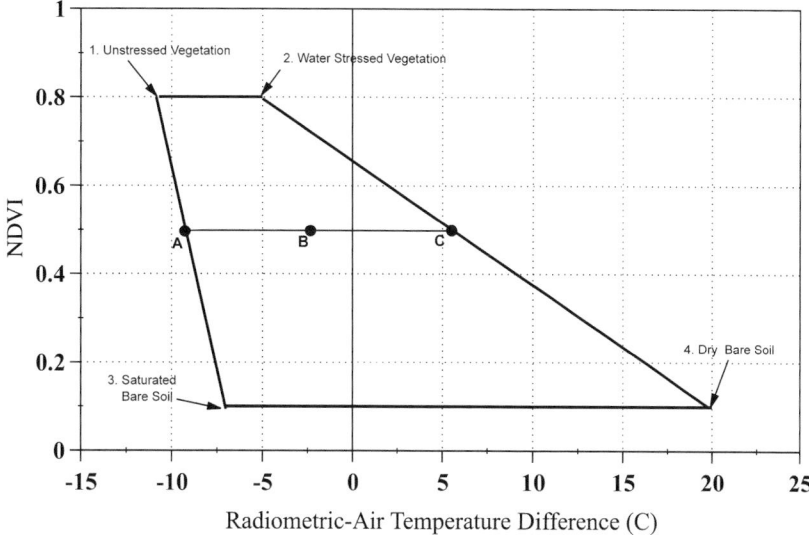

Fig. 1 The trapezoidal shape that results from the theoretical relation between radiative temperature minus air temperature [$T_R(\theta) - T_A$] and the NDVI from Moran et al.[22] With a measurement of ($T_R(\theta) - T_A$) at point C, it would be possible to equate the ratio of actual to potential LE with the ratio of distances CB and AB.

observations over agricultural fields and showed it could be used for irrigation scheduling purposes.

These modeling schemes, however, are vulnerable to errors in the radiometric temperature observations and most require screen level meteorological inputs (primarily wind speed, u, and air temperature, T_A, observations) which at regional scales suffer from errors of representativeness (observation not taken at the same location where flux estimates are performed). Approaches using remotely sensed data for estimating the variation of these quantities are being developed and tested.[24,25] How reliable the algorithms are for different climatic regimes needs to be evaluated.

A robust modeling framework to address some of these limitations was proposed early on in the application of satellite observations by Wetzel Atlas, and Woodward[26] Strictly speaking, the Wetzel et al. study was aimed at the estimation of soil moisture from remotely sensed data, but an evaluation of surface fluxes is implicit in the scheme. The study recognized that using a time rate of change in $T_R(\theta)$ from a geostationary satellite such as from the Geosynchronous Operational Environmental Satellite (GOES) coupled to an ABL model could mitigate some of the inherent problems arising from the use of single-time-level data, such as atmospheric corrections, emissivity, and instrument calibration. By using time rate of change of $T_R(\theta)$, one reduces the need for absolute accuracy in satellite calibration, and atmospheric and emissivity corrections, all significant challenges (see Refs. 1 and 8). Diak and Whipple[27] implemented this approach with a method for partitioning the available energy into LE and H by using the rate of rise of $T_R(\theta)$ from GOES and ABL growth and included a procedure to account for effects of horizontal and vertical temperature advection and vertical motions above the ABL.

Further refinements to these time-rate-of-change schemes have been recently developed[28,29] that use an energy closure scheme based on energy conservation within the ABL. The so-called Atmospheric–LandEXchange-Inverse (ALEXI) model uses a simple slab model of the time-development of the ABL in response to heat input to the lower atmosphere. A profile of atmospheric temperature at the initial time (usually from an analysis of synoptic data) serves as the upper boundary condition in atmospheric temperature. Through surface–ABL energy balance considerations and implementation of the TSM scheme for the land surface component of the model,[17] ALEXI couples ABL development to the temporal changes in surface radiometric temperature from GOES and fraction vegetation cover from Advanced Very High Resolution Radiometer, AVHRR–NDVI. The advantages of using temporal changes in brightness temperature measurements have been noted. With an energy balance method utilizing the temporal change of ABL structure, errors that arise in

schemes utilizing shelter-level ($\sim 2\,m$ above ground level) measurements of air temperature (to estimate the surface–air temperature gradient) for estimating the heat fluxes are also mitigated. Approaches that utilize this surface–air temperature gradient, typically evaluated within 10 m of the surface, are very sensitive to errors in the evaluation of the gradient arising from errors both in the representativeness of the air temperature measurements, and errors in evaluating radiometric temperatures.

Another much simpler scheme, which also uses the TSM framework, employs the time rate of change in radiometric temperature and air temperature observations from a nearby weather station in a simple formulation for computing regional heat fluxes, called the Dual-Temperature-Difference (DTD) approach.[30] Although this technique requires air temperature observations, by using a time difference in air temperature, errors caused by using local shelter level observations for representing a region are still reduced. Moreover, the scheme is simple, thus it is computationally efficient and does not require atmospheric sounding data for initialization.

APPLICATION OF ALEXI AND DTD METHODS

An example of the utility of the DTD approach is presented at the field scale using ground-based $T_R(\theta)$ observations and regional weather station data from sites in subhumid and semiarid climatic regions (i.e., Oklahoma and Arizona). In addition, a comparison of regional scale heat fluxes between the more rigorous ALEXI model and the simple DTD method using satellite data over the U.S. Great Plains is presented.

With the field scale $T_R(\theta)$ observations, the comparisons in Fig. 2 are LE estimates using the original TSM approach and the DTD scheme with regional weather station data (T_A and u) collected 50 km–100 km away from the site compared to on-site flux tower observations.[30] There is considerably more scatter using the TSM vs. the DTD approach with nonlocal meteorological inputs resulting in a Root Mean Square Error (RMSE) on the order of $100\,W\,m^{-2}$. Using the DTD scheme, there is a significant reduction in scatter with the flux observations yielding almost a 40% reduction in error with a RMSE $\sim 65\,W\,m^{-2}$.

To illustrate a regional application of the DTD and ALEXI approaches, GOES brightness temperature data and NOAA–AVHRR satellite observations were used with surface synoptic data for July 2, 1997 over the U.S. Great Plains, same case study used by Mecikalski et al.[29]. The domain investigated was divided into 10 km × 10 km grid cells, with 223 cells east-to-west and 201 in the meridional direction, a total of 44,823 cells. NOAA–AVHRR–NDVI product for the region was utilized to estimate fractional

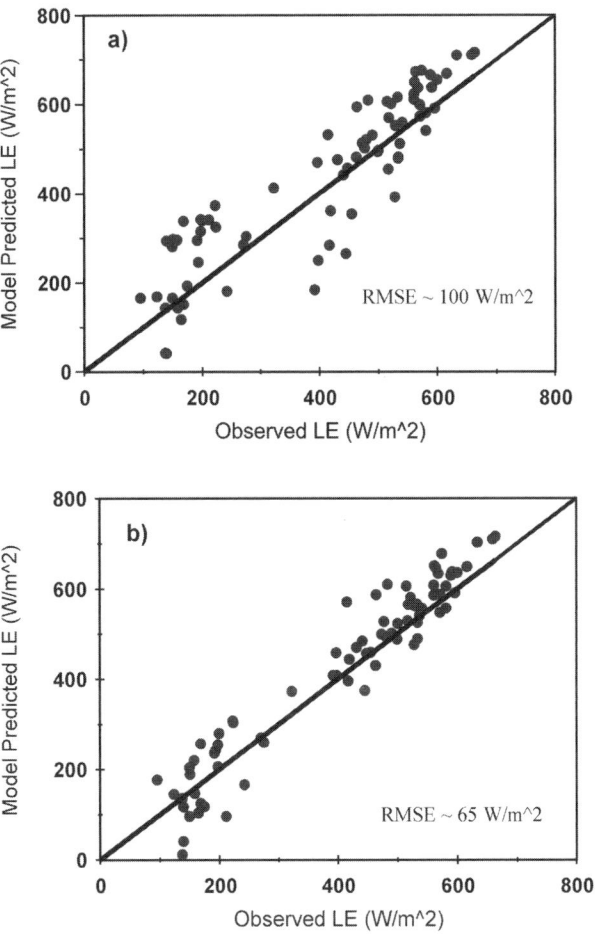

Fig. 2 Comparison between observed and modeled mid-day latent heat flux, LE, using (a) original TSM scheme and (b) DTD approach. Regional T_A and u observations are from weather stations $\sim 50\,km$ to $\sim 100\,km$ away from study site. Line represents perfect agreement with observations.

vegetation cover. Hourly GOES brightness temperature measurements for the region were cloud screened and subsequently linearly time-interpolated to 1.5 hr and 5.5 hr after local sunrise. These top-of-atmosphere brightness temperatures were then atmospherically corrected to estimate surface radiometric surface temperatures and corrected for emissivity using land surface classification data (for details, see Ref. 29).

The estimates of LE for 5.5 hr after local sunrise for the domain are shown in Fig. 3 from the DTD and ALEXI schemes. Areas that are white in this figure were either those identified as cloudy by screening procedures, and thus were not evaluated in either method, or did not achieve model convergence (primarily ALEXI). The DTD method displays very similar spatial features as the ALEXI output, although, as shown, there is a systematic difference between the two, with the DTD method showing overall higher values of LE.

Unlike ALEXI, in which air temperature is dynamically determined within the scheme, in the DTD method, air temperature is a measured (from surface synoptic data) and invariant upper boundary condition for the model. The horizontal spacing of hourly synoptic air temperature measurements is roughly 100 km, while the satellite data and the DTD grid on which the $T_R(\theta)$ and NDVI data are applied have a significantly higher resolution. With fixed boundary conditions measured on the scale of 100 km, DTD cannot account for the sub-synoptic-scale interactions between surface radiometric temperatures and air temperature, as does ALEXI. Nevertheless, results from the DTD procedure are encouraging in their ability to duplicate the spatial patterns from ALEXI, a much more complicated and data-intensive parameterization. Computer processing time for the domain shown in Fig. 3 for the ALEXI model was

DTD Output

Latent Heat Flux (W m^{-2}) 2 July 1997

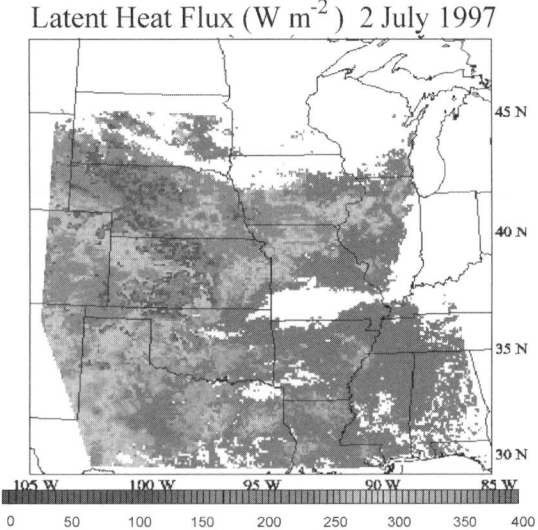

ALEXI Output

Latent Heat Flux (W m^{-2}) 2 July 1997

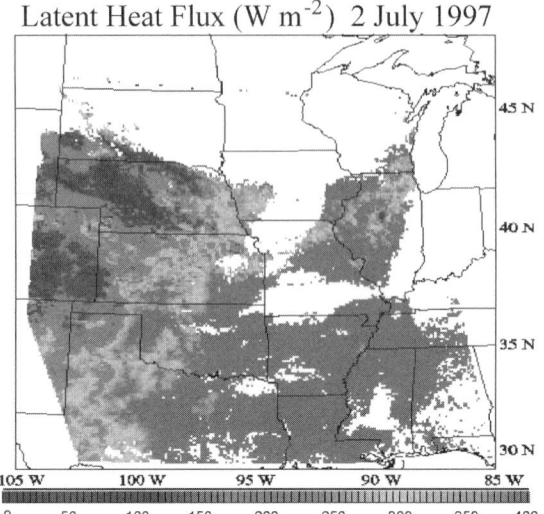

Fig. 3   Regional scale latent heat flux maps from DTD and ALEXI ~5.5 hr after local sun rise for the U.S. Great Plains region on July 2, 1997.

about 35 min, while the DTD scheme required less than 1 min of processing time on the same UNIX workstation.

CONCLUSION

Current efforts incorporating remote sensing data into SVAT modeling schemes that accommodate the fundamental differences between aerodynamic and radiometric temperatures and that are not sensitive to measurement errors should greatly enhance the prospect of quantifying ET at regional scales with remote sensing. The measurement errors with the largest impact on ET estimation are atmospheric and emissivity effects in converting satellite brightness temperatures to radiometric surface temperatures and assigning meteorological variables, primarily air temperature, for each satellite pixel from regional weather station observations.[28] Due to limited spatial observations of atmospheric properties, the uncertainty in the surface–air temperature difference is likely to be several degrees resulting in unreliable ET estimation, which have significantly hampered many past modeling approaches.

Although the current approaches described here, ALEXI and DTD, address most of these limitations, there is a drawback to these schemes in that the source of radiometric temperatures (GOES), and the atmospheric boundary layer closure and weather station network

dictate an output resolution of 5 km–10 km. For many applications, particularly evaluating ET for individual fields, these 5 km–10 km estimates are at a much coarser spatial scale. Unfortunately, temporal changes (1/2-hourly) of satellite brightness temperatures are only available from GOES at a minimum resolution of ~5 km. Other satellites have much finer spatial resolution, such as the Land Remote-Sensing Satellite (Landsat) and the Advanced Spaceborne Thermal Emission Reflectance Radiometer (ASTER), but have much coarser temporal coverage (~16 days).

Kustas and Norman[31] found subpixel variability in surface properties can result in large errors in pixel-average heat flux estimation, using pixel-average inputs when there is a significant discontinuity in surface conditions, particularly under low winds. A solution to the problem of spatial resolution was introduced by Norman et al.,[32] who developed a scheme for "disaggregating" ALEXI 5 km flux estimates (called DisALEXI) to the 30 m scale using high-resolution NDVI and $T_{\mathrm{R}}(\theta)$ data, and the local 50 m air temperature estimate provided by ALEXI as the important atmospheric boundary condition in temperature. Although, this scheme makes use of energy conservation principles applied to ABL dynamics to deduce air temperature via ALEXI, it still does not consider local variability in mean air properties. However, the preliminary results are encouraging, suggesting disaggregation of coarse spatial resolution ET output may be feasible periodically with high resolution data from Landsat or ASTER.

REFERENCES

1. Kustas, W.P.; Norman, J.M. Use of Remote Sensing for Evapotranspiration Monitoring Over Land Surfaces. Hydrol. Sci. J. des Sci. Hydrologiques **1996**, *41*, 495–516.

2. Hall, F.G.; Huemmrich, K.F.; Geotz, S.J.; Sellers, P.J.; Nickerson, J.E. Satellite Remote Sensing of Surface Energy Balance: Success, Failures and Unresolved Issues in FIFE. J. Geophys. Res. **1992**, *97* (D17), 19061–19090.

3. Norman, J.M.; Becker, F. Terminology in Thermal Infrared Remote Sensing of Natural Surfaces. Remote Sens. Rev. **1995**, *12*, 159–173.

4. Diak, G.R.; Bland, W.L.; Mecikalski, J.R. A Note on First Estimates of Surface Insolation from GOES8 Visible Satellite Data. Agric. Forest Meteorol. **1996**, *82*, 219–226.

5. Garratt, J.R.; Hicks, B.B. Momentum, Heat and Water Vapour Transfer to and from Natural and Surfaces. Quart. J. R. Meteorol. Soc. **1973**, *99*, 680–687.

6. Brutsaert, W. *Evaporation into the Atmosphere*; D. Reidel: Dordrecht, 1982; 299 pp.

7. McNaughton, K.G.; Van den Hurk, B.J.J.M. A 'Lagrangian' Revision of the Resistors in the Two-Layer Model for Calculating the Energy Budget of a Plant Canopy. Boundary-Layer Meteorol. **1995**, *74*, 262–288.

8. Kustas, W.P.; Norman, J.M.; Schmugge, T.J.; Anderson, M.C. Mapping Surface Energy Fluxes with Radiometric Temperature. In *Thermal Remote Sensing in Land Surface Processes*; Quattrochi, D.A., Luvall, J.C., Eds.; Taylor and Francis: London, UK., 2002.

9. Jackson, R.D.; Reginato, R.J.; Idso, S.B. Wheat Canopy Temperature: A Practical Tool for Evaluating Water Requirements. Water Resour. Res. **1977**, *13*, 651–656.

10. Seguin, B.; Itier, B. Using Midday Surface Temperature to Estimate Daily Evaporation from Satellite Thermal IR Data. Int. J. Remote Sens. **1983**, *4*, 371–383.

11. Carlson, T.N.; Buffum, M.J. On Estimating Total Daily Evapotranspiration from Remote Surface Measurements. Remote Sens. Environ. **1989**, *29*, 197–207.

12. Lagouarde, J.-P. Use of NOAA AVHRR Data Combined with an Agrometeorological Model for Evaporation Mapping. Int. J. Remote Sens. **1991**, *12*, 1853–1864.

13. Carlson, T.N.; Capehart, W.J.; Gillies, R.R. A New Look at the Simplified Method for Remote Sensing of Daily Evapotranspiration. Remote Sens. Environ. **1995**, *54*, 161–167.

14. Seguin, B.; Lagouarde, J. P.; Saranc, M. The Assessment of Regional Crop Water Conditions from Meteorological Satellite Thermal Infrared Data. Remote Sens. Environ. **1991**, *35*, 141–148.

15. Vining, R.C.; Blad, B.L. Estimation of Sensible Heat Flux from Remotely Sensed Canopy Temperatures. J. Geophys. Res. **1992**, *97* (D17), 18951–18954.

16. Verhoef, A.; De Bruin, H.A.R.; Van den Hurk, B.J.J.M. Some Practical Notes on the Parameter kB^{-1} for Sparse Vegetation. J. Appl. Meteorol. **1997**, *36*, 560–572.

17. Norman, J.M.; Kustas, W.P.; Humes, K.S. A Two-Source Approach for Estimating Soil and Vegetation Energy Fluxes from Observations of Directional Radiometric Surface Temperature. Agric. Forest Meteorol. **1995**, *77*, 263–293.

18. Kustas, W.P.; Norman, J.M. Evaluation of Soil and Vegetation Heat Flux Predictions Using a Simple Two-Source Model with Radiometric Temperatures for Partial Canopy Cover. Agric. Forest Meteorol. **1999**, *94*, 13–29.

19. Price, J.C. Using Spatial Context in Satellite Data to Infer Regional Scale Evapotranspiration. IEEE Trans. Geosci. Remote Sens. **1990**, GE-*28*, 940–948.

20. Carlson, T.N.; Gillies, R.R.; Perry, E.M. A Method to Make Use of Thermal Infrared Temperature and NDVI Measurements to Infer Soil Water Content and Fractional Vegetation Cover. Remote Sens. Rev. **1994**, *52*, 45–59.

21. Gillies, R.R.; Carlson, T.N.; Cui, J.; Kustas, W.P.; Humes, K.S. Verification of the 'Triangle' Method for Obtaining Surface Soil Water Content and Energy Fluxes from Remote Measurements of Normalized Difference Vegetation Index (NDVI) and Surface Radiant Temperature. Int. J. Remote Sens. **1997**, *18*, 3145–3166.

22. Moran, M.S.; Clarke, T.R.; Inoue, Y.; Vidal, A. Estimating Crop Water Deficit Using the Relation Between Surface–Air Temperature and Spectral Vegetation Index. Remote Sens. Environ. **1994**, *49*, 246–263.

23. Moran, M.S. Irrigation Management in Arizona Using Satellites and Airplanes. Irrig. Sci. **1994**, *15*, 35–44.

24. Bastiaanssen, W.G.M.; Feddes, R.A.; Holtslag, A.A.M. A Remote Sensing Surface Energy Balance Algorithm for Land (SEBAL) Part 1: Formulation. J. Hydrol. **1998**, *212–213*, 200–213.

25. Gao, W.; Coultier, R.L.; Lesht, B.M.; Qui, J.; Wesely, M.L. Estimating Clear-Sky Regional Surface Fluxes in the Southern Great Plains Atmospheric Radiation Measurement Site with Ground Measurements and Satellite Observations. J. Appl. Meteorol. **1998**, *37*, 5–22.

26. Wetzel, P.J.; Atlas, D.; Woodward, R. Determining Soil Moisture from Geosynchronous Satellite Infrared Data: A Feasibility Study. J. Climatol. Appl. Meteorol. **1984**, *23*, 375–391.

27. Diak, G.R.; Whipple, M.A. Improvements to Models and Methods for Evaluating the Land-Surface Energy Balance and "Effective" Roughness Using Radiosonde Reports and Satellite-Measured "Skin" Temperatures. Agric. Forest Meteorol. **1993**, *63*, 189–218.

28. Anderson, M.C.; Norman, J.M.; Diak, G.R.; Kustas, W.P.; Mecikalski, J.R. A Two-Source Time-Integrated Model for Estimating Surface Fluxes from Thermal Infrared Satellite Observations. Rem. Sens. Environ. **1997**, *60*, 195–216.

29. Mecikalski, J.R.; Diak, G.R.; Anderson, M.C.; Norman, J.M. Estimating Fluxes on Continental Scales Using

Remotelysensed Data in an Atmosphericland Exchange Model. J. Appl. Meteorol. **1999**, *38*, 1352–1369.

30. Norman, J.M.; Kustas, W.P.; Prueger, J.H.; Diak, G.R. Surface Flux Estimation Using Radiometric Temperature: A Dual Temperature Difference Method to Minimize Measurement Error. Water Resour. Res. **2000a**, *36*, 2263–2274.

31. Kustas, W.P.; Norman, J.M. Evaluating the Effects of Sub-pixel Heterogeneity on Pixel Average Fluxes. Remote Sens. Environ. **2000**, *74*, 327–342.

32. Norman, J.M.; Daniel, L.C.; Diak, G.R.; Twine, T.E.; Kustas, W.P.; French, A.N.; Schmugge, T.J. Satellite Estimates of Evapotranspiration on the 100-m Pixel Scale. IEEE IGARRS 2000 Proc. **2000b**, IV,1483–1485.

Everglades

Kenneth L. Campbell
University of Florida, Gainesville, Florida, U.S.A.

E

INTRODUCTION

The Everglades of south Florida was originally a broad, shallow "River of Grass"[1] that extended from the south shore of Lake Okeechobee to Florida Bay at the southern tip of the state, east to the Coastal Ridge, and west to the Immokalee Ridge. Historically, the area was a vast sawgrass marsh, dotted with tree islands and interspersed with wet prairies and sloughs covering an area about 40 mi wide by 100 mi long. One of the unique regions of the world, it has steadily decreased in size and declined in health during the past century. Half its wetland area has been lost to agriculture and urban development and the remaining segments are impacted by lack of a clean, dependable water supply. Natural water flows have been diverted for irrigation, drinking water, and flood protection. The conveyance system of canals, levees, structures, and pumps developed for flood control has altered natural patterns of water flow and storage, adversely affecting food webs that supported a diverse ecosystem. Nutrient runoff from urban and agricultural sources is transported by the conveyance system to the remaining natural wetland areas, causing undesirable changes in flora and fauna. Hydroperiod changes have altered natural fire patterns and stimulated invasion of exotic species. A multi-agency state and federal task force has developed a Comprehensive Everglades Restoration Plan (CERP)[2] to address and reverse these major changes to this unique wet-land ecosystem. The major hydrologic modifications to be addressed in the Everglades restoration include: 1) regain lost storage capacity; 2) restore more natural hydropatterns; 3) improve timing and quantities of fresh water deliveries to estuaries; and 4) restore water quality conditions. The Comprehensive Plan, considered the world's largest such project, includes more than 60 components proposed for implementation over a period of four decades with an estimated investment approaching $8 billion. State and federal legislation provides for a 50/50 cost share between the federal and state governments to implement the plan.

EVERGLADES WATER MANAGEMENT—PAST, PRESENT, AND FUTURE

History

Primitive canals were dug in portions of the Everglades as early as the late 1800s in attempts to reclaim fertile swampland for agriculture.[3] Early promoters and developers led people to believe that a productive subtropical agriculture was possible in the entire Everglades region. These early attempts at land reclamation were largely unsuccessful until the 1920s when a period of less than normal rainfall helped dry the region around Lake Okeechobee for farming. Following severe hurricane damage in the region in the late 1920s and again in 1947, the focus was shifted from land reclamation to flood protection and the Central and Southern Florida Flood Control Project was authorized and implemented beginning in 1948. Over the next 15 yr, this project resulted in a perimeter dike around Lake Okeechobee and the extensive conveyance system of canals, levees, structures, and pumps currently in place. It also allowed development of the Everglades Agricultural Area (EAA), a highly productive, 700,000-acre region of organic soils in the northern Everglades used primarily for sugar cane and winter vegetable production.[4]

Environmental Issues

By the mid-1960s, concerns were already growing about conservation issues and adverse environmental impacts. Additional areas along the eastern border of the Everglades have since experienced urban encroachment. A total of about 1 million acres, roughly 50% of the Everglades wetlands, have been transformed for human uses during the past half-century. The 1700 mi of canals and levees in the region have interrupted connections between the central Everglades and the adjacent wetlands, resulting in over-drainage in some areas and excessive

Encyclopedia of Water Science
DOI: 10.1081/E-EWS 120010374

flooding in others. This system provides water supply, flood protection, water management, and other benefits to south Florida, but it must be modified to reduce the negative impacts on the environment. The current canal system works very effectively, discharging an average of 1.7 billion gal of water per day to the ocean and gulf. This discharge must be reduced if future urban, agricultural, and environmental demands for water are to be met.

Today's remaining Everglades have been significantly affected by the current water management system. Wading birds and other wildlife populations are greatly decreased. Tree islands, with their unique combination of wetland and terrestrial vegetation and wildlife, are considered to be an excellent indicator of the overall health of the Everglades. Many of these tree islands have disappeared from the northern Everglades over the past 50 yr, and many others have been taken over by exotic vegetation. These effects are mainly due to changes in the quantity, quality, timing, and distribution of water that have occurred over the years as a result of changed water management. Water depth, duration, and timing are important to both wildlife and vegetation. The sawgrass wetlands of the Everglades developed under very low nutrient conditions with rainfall as the main source of phosphorus. Nutrient inflows, especially phosphorus, as a result of development and modified water management have influenced changes in vegetation type.[5] Where phosphorus concentrations have increased, sawgrass and spike rush have been replaced by cattail causing undesirable changes in the ecosystem. Native vegetation remains healthy where phosphorus concentrations are low.

Restoration

Restoration of the remaining Everglades depends upon a knowledge and understanding of the original conditions. Efforts are focusing on improving upstream water quality and the distribution, timing, depth, and flow of surface water into and through the Everglades. Early historical information sources, combined with further interpretation and analysis, are being used to estimate original drainage patterns and soil, topographic and vegetation conditions before canal drainage began in the late 1800s. Results of these studies indicate that the predrainage landscape of the Everglades probably was configured in subtle ridges and sloughs with two major flow pathways: a flow path southeastward to the Atlantic Ocean, and a southwestward flow path along Shark Slough to the Gulf of Mexico.[6] These flow patterns may have influenced the ridge and slough landscape configuration that is important to the health of the ecosystem. Redevelopment of these flow patterns and landscape configuration will be important to

the restoration process. About 70% less water flows through the Everglades today compared to the historic Everglades system.

The main goal of Everglades restoration is to deliver the correct amount of water, with the correct quality, to the correct locations, and at the correct time.[7] Most of the water currently lost to the ocean or gulf will be stored in surface and subsurface storage areas until needed, when 80% of it will be allocated to the environment and 20% to increase urban and agricultural water supplies. Water to be stored for future use will be routed through surface storage reservoirs and wetland-based stormwater treatment areas to improve its quality. Additional water quality improvements can be expected from comprehensive integrated water quality planning efforts currently in progress. To restore water flow paths, more than 240 mi of canals and levees will be removed in the Everglades. This will allow more natural overland water flow in the remaining natural areas of the Everglades. Water held and released will be managed to match natural discharge patterns more closely. Operational plans will be developed in some areas to simulate natural rainfall patterns with water releases to improve the timing of water flowing through the Everglades ecosystem. These strategies are all being designed to enhance not only ecosystem restoration, but also urban and agricultural water supply and flood protection as part of the process of moving toward a more sustainable south Florida.

CONCLUSION

The Everglades landscape is a unique combination of subtropical wetlands and uplands, including sawgrass marshes, sloughs, wet prairies, tree islands, tropical hardwood hammocks, pinelands, and mangroves. It provides important habitat for many threatened and endangered species. Water management for flood control and water supply purposes has caused some areas to become drier and others to become wetter than normal. More than half of the original wetland area has been lost to agricultural and urban development. The introduction of increased nutrients resulting from this development has caused undesirable shifts in vegetation communities. Hydrologic changes have altered the extent of naturally occurring fires and promoted the growth of exotic species. While the current water management system performs well for flood protection it must be modified to reduce adverse environmental impacts and conserve more fresh water to meet a variety of needs. A Comprehensive Everglades Restoration Plan received initial authorization in 2000 to begin the restoration of the south Florida ecosystem and provide for water-related needs of the region. This plan addresses the quantity, quality, distribution, and timing of water to the Everglades. A large amount of additional information regarding the Everglades is available on the web at http://www.sfwmd.gov/koe_section/2_everglades.html and http://www.evergladesplan.org/.

The following quote from the Comprehensive Everglades Restoration Plan web site[7] conveys the importance of the Everglades and the current restoration program.

> The significance of the remaining Everglades to the nation and the world has been affirmed time and again. Congress established Everglades National Park. The Everglades have also been designated an International Biosphere Reserve, a World Heritage Site, and a Wetland of International Significance. Identified as one of the world's major ecosystem types, the Everglades are home to 68 threatened or endangered plant and animal species. The benefits and functions of these plants and animals may never be known if we do not restore and protect their habitat. Saving the Everglades requires us to save the entire south Florida ecosystem. The ecological and cultural significance of the Everglades is equal to the Grand Canyon, the Rocky Mountains, or the Mississippi River. As responsible stewards of our natural and cultural resources, we cannot sit idly by and watch any of these disappear. The Everglades deserves the same recognition and support.

REFERENCES

1. Douglas, M.S. *The Everglades: River of Grass*; Hurricane House, Coconut Grove, FL, 1947.
2. U.S. Army Corps of Engineers, Jacksonville District. *Central and Southern Florida Project Comprehensive Review Study Final Integrated Feasibility Report and Programmatic Environmental Impact Statement*, April 1999; The Corps: Jacksonville, FL, South Florida Water Management District: West Palm Beach, FL, 1999; 10 Vols.; http://www.evergladesplan.org/pub/restudy_eis.shtml (accessed March 2002).
3. Blake, N.M. *Land into Water—Water into Land: A History of Water Management in Florida*; University Presses of Florida: Tallahassee, FL, 1980.
4. Bottcher, A.B.; Izuno, F.T. *Everglades Agricultural Area (EAA): Water, Soil, Crop, and Environmental Management*; University Press of Florida: Gainesville, FL, 1994.
5. Reddy K.R., O'Connor G.A., Schelske C.L., Eds.; *Phosphorus Biogeochemistry in Subtropical Ecosystems*; Lewis Publishers: Boca Raton, FL, 1999.
6. South Florida Water Management District. *Watershed Management: Everglades/Florida Bay*; http://www.sfwmd.gov/org/wrp/wrp_evg/ (accessed March 2002).
7. U.S. Army Corps of Engineers, Jacksonville District, South Florida Water Management District. *Rescuing an Endangered Ecosystem—the Journey to Restore America's Everglades*; http://www.evergladesplan.org/ (accessed March 2002).

Farm Ponds

Ronald W. Tuttle (Retired)
United States Department of Agriculture (USDA),
Washington, District of Columbia, U.S.A.

INTRODUCTION

The rural American landscape is a rich tapestry of interdependent ecosystems. It is also a working landscape of over 900 million privately owned acres devoted to cropland, pastureland, or rangeland.[1] Scattered across this diverse matrix are countless farm ponds, reflecting the light of day and night (Fig. 1[2]). Although there is no accurate count of the total number of ponds in the United States, a conservative estimate is well over 2 million. The United States Department of Agriculture (USDA), Natural Resources Conservation Service (NRCS) reported more than 2.1 million ponds had been built on privately owned lands by 1980.[3] The Soil Conservation Service [SCS (now NRCS)] assisted in the planning and construction of approximately 2 million farm ponds during the 30-yr period between 1945 and 1975. According to SCS historical records, in 1974 the South region of the United States led in the cumulative number of ponds built with 1,108,959. The Midwest region was next with 450,847 ponds. The West and Northeast regions followed with 278,360 and 134,327 ponds respectively. Texas, Oklahoma, Mississippi, and Kentucky were the leading states in number of ponds built by SCS at that time.[4]

Farm ponds continue to be much in demand with many constructed each year in the United States. Iowa, for example, reports 87,000 farm ponds with an additional 1000 being added yearly.[5] A conservative estimate suggests over 50,000 ponds ranging in size from less than 1 acre to over 30 acre in Virginia.[6] The list goes on with Mississippi reporting more than 280,000 farm ponds ranging in size from 1/2 acre to 40 acre.[7]

POND CHARACTERISTICS

Farm ponds are commonly described as water impoundments used for agricultural or domestic farm uses and enjoyment. NRCS defines them as a water impoundments made by constructing a dam or by excavating a pit or dugout.[8] There are two general types of farm ponds, largely determined by topography. Embankment ponds are formed by impounding water behind a dam built across a watercourse. Good sites occur in gently sloping valleys with steep side slopes to provide adequate pond depth and discourage the establishment of aquatic vegetation. NRCS recommends dams that are less than 35-ft high and located where their failure will not result in loss of life; damage to buildings, highways, and other infrastructure elements; or in interrupted use of public utilities.[3] Excavated ponds, as the name implies, are constructed by removing soil to create a pond basin at an elevation below the surrounding ground level. Unlike embankment ponds, they are typically constructed on relatively level areas where a source of water may be more limited.

Water adds variety to a landscape, thereby enhancing its aesthetic quality. Many terrestrial species as well as fish, amphibians, and waterfowl are dependent upon the habitat offered by farm ponds. They attract songbirds, small and large mammals, osprey, heron, and other species in a linked web of life. In addition to their intrinsic value for wildlife and aesthetic quality, farm ponds are important for livestock, recreation, energy conservation, fish production, and water supply for irrigation or farmstead fire protection. Properly managed ponds also reduce storm runoff, aid in erosion control, and improve water quality (Fig. 2).

PLANNING AND DESIGN CONSIDERATIONS

Farm ponds should be properly planned, designed, and constructed if they are to function as intended. A basic problem-solving process involving inventory and analysis of prevailing natural resource conditions, identification of related problems and opportunities, and evaluation of alternatives will lead to appropriate decisions when planning and designing a pond.

There are many useful references on the subject of planning, designing, and constructing ponds. The NRCS publication *Ponds—Planning, Design, and Construction* (see Ref. [3]) is available at NRCS offices and describes basic requirements for building a pond. The many details covered in this and other sources of information are beyond the scope of this publication; however, major considerations include location, water supply, and soil type.

Encyclopedia of Water Science
DOI: 10.1081/E-EWS 120010055

Fig. 1   A typical farm pond in rural America.

Location

Site and watershed investigation will determine if an area is suitable for the type of pond desired. The relationship of alternative pond sites to prevailing ecological structure and functions within the larger landscape or watershed is critical to achieving a properly functioning farm pond. Dams, for example, are proven to have significant detrimental impacts upon the equilibrium of stream corridors.[9] Therefore, it is generally not advisable to dam streams for the purpose of constructing an embankment pond. Locating them nearby will protect the stream and prevent damaging floodwaters and silt from entering the pond.

Farm ponds should be planned and designed to fulfill their intended use as an integral part of the surrounding landscape. This is achieved with minimum disturbance to existing landform, vegetation, water, and structures. The desired principal use(s) of a pond will determine its best location. A pond intended for aesthetic quality and fire protection, for example, should be sited near farmstead structures and easily visible from important viewpoints near the home or elsewhere.[10] This proximity to major viewpoints will also help to prevent misuse of the farm pond and ensure greater safety.

Water Supply

High quality water from either a surface or groundwater source is important to properly functioning farm ponds. Watersheds with good vegetative cover and conservation systems installed to protect the land are best suited for an appropriate supply of water.

Fig. 2   A farm pond managed to reduce storm runoff, aid in erosion control, and improve water quality.

The amount, intensity, and duration of surface runoff should be evaluated to determine if the watershed above the pond site is large enough to provide an adequate water supply. In place of local runoff information, NRCS offers a general guide for estimating the approximate size of drainage area needed based upon the desired capacity of an embankment or excavated pond (Fig. 3). The acreage needed in a drainage area for each acre-foot of pond storage can be determined by using this guide; however, some adjustments may be necessary to account for extremes in local runoff conditions. One acre-foot is equal to the amount of water to be stored to a depth of one foot over one acre (a total of 325,851 gal). Details for estimating storm runoff and recommended minimum depths of ponds to account for normal seepage and evaporation are also available.[3]

Soil Type

Many ponds fail because they are built in the wrong type of soil. A properly functioning pond must, by definition, hold water. The soil must be suitable for the pond bottom as well as for the dam in the case of embankment ponds. Deep soil that has slowly permeable subsoil containing lots of clay or silty clay is the best. Sites containing coarse-textured sand, gravel, sand–gravel mixtures are generally unsuitable unless an adequate clay content is present. Areas containing limestone or gypsum are especially hazardous due to crevices, sinkholes, or channels that can drain water from a pond very rapidly. A clue to suitability of the site is the degree of success of nearby ponds.[3]

POND MANAGEMENT

Management and maintenance are just as important to a properly functioning farm pond as good planning, design, and construction. Appropriate management prescriptions, which can be diverse in their nature, are often dictated by the principal use(s) of a pond. Fencing to exclude livestock from ponds and installing a gravity-fed watering trough nearby, for example, will often increase the value of ponds for multiple use and enjoyment.

A diverse plant community will also increase a pond's value for multiple uses. Trees, shrubs, herbs, and grasses established and maintained as a buffer will provide wildlife habitat, improve aesthetic quality, and increase the life expectancy of ponds by reducing erosion. However, deep-rooting trees or shrubs should be prevented

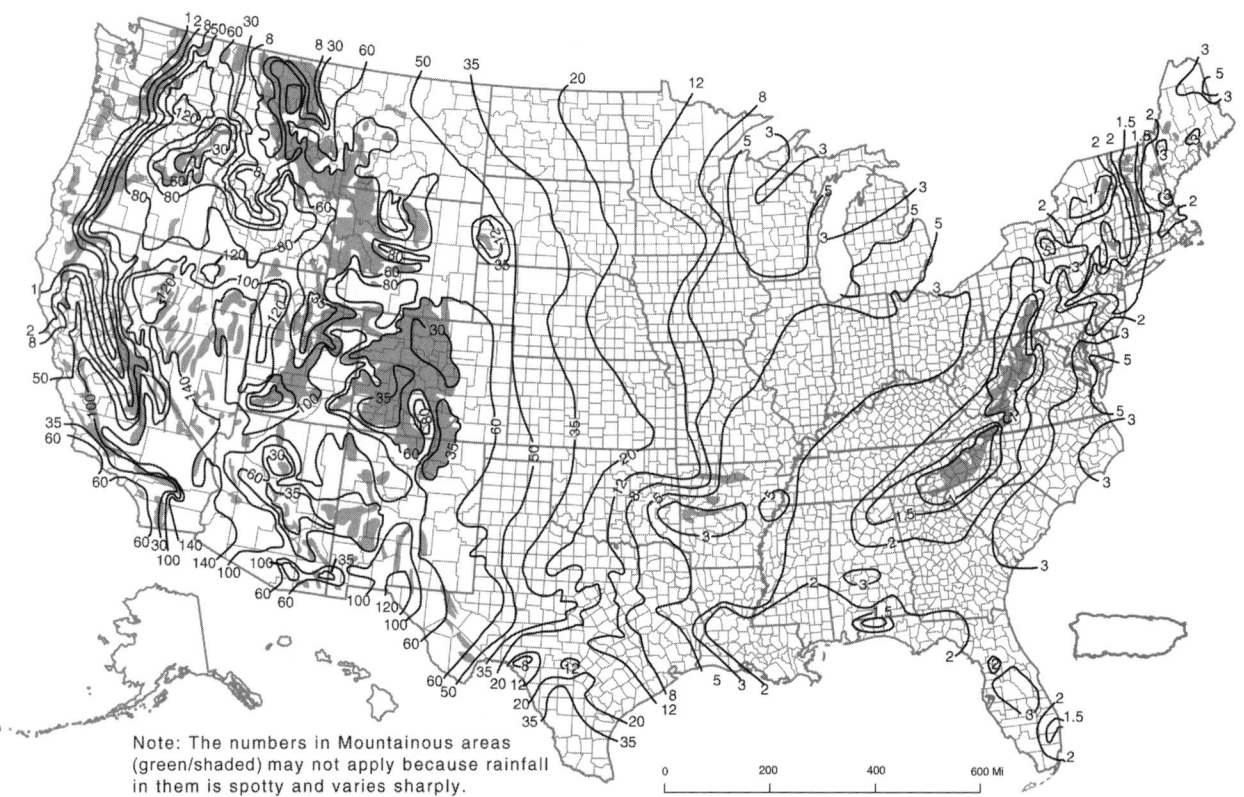

Note: The numbers in Mountainous areas (green/shaded) may not apply because rainfall in them is spotty and varies sharply.

Fig. 3   A guide for estimating acres of drainage area required for an acre-foot of pond storage.

from growing on dam embankments, where their roots can endanger the integrity of the structure.

Damage from erosion, burrowing animals, livestock, silting, aquatic vegetation, overflow, undercutting, and other sources can occur rapidly and should be corrected promptly. Occasionally, a pond will begin to leak water at an excessive rate and require corrective measures. Clay blankets, bentonite, chemical additives to reduce soil permeability, and waterproof linings are common alternatives for sealing ponds.[3] Excessive aquatic plant growth is another maintenance problem that will hasten eutrophication and seriously degrade conditions for use and enjoyment of the pond. Adequate pond depth will discourage undesired plants from becoming established; otherwise, mechanical removal or chemical methods may become necessary.

Owners have an obligation to ensure their farm ponds are as safe as possible. Signs warning of dangers should be installed and hazards to swimmers removed. Rules regulating recreational use of the pond and lifesaving devices such as ring buoys and ropes or poles stationed at ponds will provide additional protection.

SOURCES OF ASSISTANCE

Most states and other governing entities have regulations pertaining to pond construction. Those planning and designing ponds must comply with these requirements by contacting the local planning board or other appropriate governing body before building a pond. Landowners are responsible for obtaining permits, performing necessary maintenance, and ensuring pond safety.

The local county Soil and Water Conservation District (SWCD) should be the first stop for those interested in planning and constructing a farm pond. The SWCD works closely with other local, state, and federal government entities, including State Departments of Natural Resources, Cooperative Extension Service, and the USDA, NRCS. Collectively, they can provide helpful technical and financial assistance.

Cost-share funds also may be available for the construction of ponds. The Environmental Quality Incentives Program administered by the NRCS is one of the several similar programs available to farmers and ranchers. It offers financial, educational, and technical assistance to install or implement conservation practices, including farm ponds.[11] Program managers indicate that over 8000 ponds were contracted from 1997 to 2000 under this program.

REFERENCES

1. United States Department of Agriculture, Natural Resources Conservation Service (USDA-NRCS), *Americas Private Land—A Geography of Hope*, Program Aid 1548; United States Department of Agriculture: Washington, DC, 1996; 7.

2. Lean, V. Collectanea, 1902. In *A Dictionary of Environmental Quotations*; Rodes, B., Odell, R., Eds.; The Johns Hopkins University Press: Baltimore, MD, 1998; 134.

3. United States Department of Agriculture, Natural Resources Conservation Service (USDA, NRCS), *Ponds—Planning, Design, Construction*, Agricultural Handbook 590, 2nd Ed.; Vol. 1; United States Department of Agriculture: Washington, DC, 1997; 1, 9–23; 24–26; 62–65.

4. United States Department of Agriculture, Soil Conservation Service (USDA-SCS), *Practice on the Land*; United States Department of Agriculture: Washington, DC, 1974.

5. Iowa State University, Department of Animal Ecology. *Iowa's Natural Resources*; Iowa State University: Ames, IA; http://www.ag.iastate.edu/departments/aecl/resource.htm (accessed Sept 2001).

6. Helfrich, L.A.; Pardue, G.B. *Pond Construction—Some Practical Considerations*; Publication # 420-011, http://www.ext.vt.edu/pubs/fisheries/420-011/420-011.html (accessed Nov 2001) Department of Fisheries and Wildlife Sciences, Virginia Tech: Blacksburg, VA, 1996.

7. Brunson, M.W.; Riecke, D.; Hubbard, W., (Eds.) *Managing Mississippi Farm Ponds and Small Lakes*, Publication 1428, 3rd Ed.; http://www.msucares.com/pubs/pub1428.htm (accessed Sept 2001) Mississippi State University Extension Service: Mississippi State, MS, 2001.

8. United States Department of Agriculture, Natural Resources Conservation Service (USDA-NRCS). *National Handbook of Conservation Practices*; United States Department of Agriculture: Washington, DC; http://www.ftw.usda.gov/practice_stds.html (accessed Sept 2001).

9. The Federal Interagency Stream Restoration Working Group, *Stream Corridor Restoration—Principles, Processes, and Practices*; http://www.usda.gov/stream_restoration (accessed Sept 2001) United States Department of Agriculture: Washington, DC, 1998; Chap. 3, 7–8.

10. Wells, G. *Landscape Design—Ponds*, Landscape Architecture Note 2; United States Department of Agriculture, Soil Conservation Service: Washington, DC, 1988; 5–7.

11. United States Department of Agriculture, Natural Resources Conservation Service (USDA-NRCS), *USDA Farm Bill Conservation Provisions*; Environmental Quality Incentives Program Fact Sheet, http://www.nhq.nrcs.usda.gov/CCS/FB96OPA/eqipfact.html (accessed Nov, 2001); United States Department of Agriculture: Washington, DC, 1996, Chap. 3.

Fertilizer/Pesticide Leaching, Irrigation Management and

Luciano Mateos
Instituto de Agricultura Sostenible, Consejo Superior de Investigaciones Científicas, Córdoba, Spain

INTRODUCTION

Good irrigation management begins by selecting the appropriate irrigation method and strategy according to the water availability, the characteristics of the climate, soil and crop, and to the economic and social circumstances. Good irrigation management continues with the actual application of the scheduled water, its even distribution over the field, and the storage in the root zone of as much of the applied water as possible. During the cropping season, the irrigation schedule must be adjusted to the weather variations and to other cropping practices such as fertilization and pesticides application.

Depending on the irrigation method, the water is distributed through pipes or overland channels. The water that is not stored in the root zone percolates or runs off the field. The percolated water can leach solutes; the water that runs off can carry away solutes or chemical components adsorbed in suspended soil aggregates.

Therefore, since irrigation water can be one mean by which fertilizers and pesticides are transported out of the root zone, good irrigation management must be integrated with the application and subsequent dynamic of the fertilizers and pesticides.

FACTORS AFFECTING LEACHING

The amount of percolation is determined by the average infiltrated depth in relation to the previous soil water content, by the spatial distribution of the infiltrated depth, the hydrologic characteristics of the soil, and the concentration, location, and chemical characteristics of the solutes. The potential for leaching is especially large on coarse-textured soils because of their low cation-exchange and water-holding capacity and their high water permeability.

Most of the water below field capacity is held in the soil, i.e., a uniform depth that refills a homogeneous root zone up to field capacity can be potentially consumed by the crop. On the contrary, if the average infiltrated depth is deeper than that require to take the root zone water content up to field capacity, the excess of water will percolate. Preferential flow through the soil macropores may cause

drainage before the water content is risen up to field capacity. A nonuniform water infiltration and distribution within the soil profile can cause percolation at certain locations in the field while other locations may suffer water deficit.

The dynamics of water flow in soils influences the transport behavior of reactive and nonreactive solutes. The solute velocity for low-frequency intermittent flow may be larger than for continuous flow because the infiltration rate increases after the drying cycles. When the soil is initially wet, the leaching of a solute applied in a solution just before the irrigation is lower than when the soil is initially relatively dry. The reason for this behavior is that the large pressure gradients under dry conditions pull the solution into the small pores.[1] Preferential transport is more likely to occur under ponded conditions, where flow occurs under saturation, than under application intensities limiting the infiltration rate.[2] In addition, under ponded conditions, the natural spatial variation of soil infiltration characteristics should increase the field scale dispersion of a leaching chemical in comparison to transport under flux-controlled boundary conditions.[3]

IRRIGATION SCHEDULING AND LEACHING

Irrigation scheduling is the determination of the next irrigation date and the depth of water to apply. Proper irrigation scheduling controls drainage and thus leaching of fertilizers and pesticides. One accepted irrigation practice to reduce leaching of fertilizers and pesticides is to apply the water necessary to bring the soil to field capacity. Even with the right amount of water, significant leaching occurs if rainfall events come soon after irrigation. An option under these conditions is to allow the soil to become drier between irrigation events; thus the probability of rainfall on a water full soil decreases; but the chances of the crop running into water stress are higher. Alternatively, irrigation depths smaller than that required to fill the soil to field capacity leave soil storing capacity for unforeseen rains.

A reduction in fertilizer or pesticide input generally results in a leaching reduction of that agrochemical. However, a reduction in irrigation amount does not

Encyclopedia of Water Science
DOI: 10.1081/E-EWS 120010229

necessarily imply a reduction in leaching. It may be found that crop production is optimized and N losses to the environment are minimized when the crop is irrigated for full evapotranspiration replacement. Under conditions of deficit irrigation, water stress restricts crop growth; thus nutrients uptake and fertilizer recovery are lower. Nitrate leaching between the harvesting of one crop and the planting of the next may be more important than nitrate leaching induced directly by the irrigation water.

IRRIGATION UNIFORMITY AND LEACHING

The inherent and management-induced nonuniformity of the irrigation systems implies that some water deficit and/or drainage must occur after the irrigations. There is a trade-off between uniformity, water deficit, and percolation. To avoid water deficit at any point in the field, excess of water must be applied. The infiltrated water that is not used to refill the root zone will percolate. This amount must be larger as more nonuniform are the water application and infiltration and lesser the allowable crop water deficit.

The occurrence of drainage due to nonuniformity of the water application implies leaching. Experimentally quantifying the effect on leaching of irrigation scheduling and uniformity in relation to fertilization is very complex. Pang et al.[4] used a crop model to simulate the combined effects of these factors on crop yield and nitrogen leaching. These authors found that high corn yield under low nitrate leaching constraints is possible only with irrigation systems that have a Christiansen uniformity coefficient of 90 or greater. Vickner et al.[5] added an economic analysis to the yield and nitrate leaching responses to irrigation uniformity. They used a dynamic model to appraise policy options for regulating groundwater quality in the western region of United States of America, finding that a limit on leaching due to corn production is economically feasible by increasing the uniformity of center-pivot irrigation systems.

IRRIGATION METHOD AND LEACHING

The irrigation methods are classified under three major groups: surface, drip/micro, and sprinkler.

Surface Irrigation

The efficiency and uniformity of surface irrigation depends on the control of the relationship inflow-soil infiltration rate-application time, soil heterogeneity, and on land grading and field microtopography. The distinctive feature of surface irrigation is that the soil surface is the transportation medium. Therefore, field water distribution occurs simultaneously to and it is controlled by infiltration. Furthermore, the infiltration rate varies spatially and temporally.

Infiltration in a surface-irrigated field is usually higher at its upstream end, where water flows for longer time (i.e., has a greater opportunity time for infiltration). The risk for leaching is, thus, higher at the upper part of the field. For a target depth at a given location in the field, percolation can be reduced by increasing the inflow rate and decreasing the application time. Advance will be faster, thus opportunity time variability along the field will be less and so the infiltrated water at the field head.

Surge flow (the application of water in intermittent pulses) and compacted furrows—e.g., wheel furrows—also increase uniformity and decrease head percolation by reducing the infiltration rate. But if the field is open at the downstream end, the higher stream size can result in excessive run-off. In this case, inflow cutback after completion of the advance phase or tailwater recovery are options for restricting run-off.

Alternate-furrow irrigation combined with fertilizer placement in the nonirrigated furrow has the potential to reduce fertilizer leaching. However, adequate root development in the nonirrigated furrow is required to allow nutrients uptake, and avoid residual fertilizer that can be potentially leached.[6]

Microtopography also affects opportunity time variability in basin irrigation. Laser leveling reduces the soil surface microrelief, thus infiltration is more uniform and percolation and leaching can be better controlled.

In addition to the opportunity time nonuniformity, the natural heterogeneity of the soil infiltration characteristics enhances the infiltration variability and the risk of percolation below the root zone. Some authors claim that the variability of the soil infiltration characteristics is damped under surge flow.

The solute displacement under the continuous flow typical of paddy rice can be expected to be lower than under drying-irrigation cycles because of the larger infiltration rate in the later situation. Fertilization timing in relation to irrigation timing may also affect leaching. For instance, as it was pointed out earlier, a slug of liquid nitrogen fertilizer applied to a relatively dry topsoil will be less prone to subsequent leaching than if applied to wet soil. If the slug is applied to wet soil, delaying irrigation for several days will reduce leaching.[1] For similar reasons, preferential solute movement is more likely to occur under flood irrigation, where water ponds on the soil surface, than under sprinkler or drip/micro irrigation, where the application rate is usually lower than the soil infiltration rate.

Drip/Micro Irrigation

Under drip/micro irrigation, water is applied directly to small areas adjacent to the plants through emitters placed along a water delivery line. High application frequency and partial soil wetting distinguish drip/micro irrigation. The use of drip irrigation leaving dry part of the soil should be beneficial in reducing N leaching in regions where rain occurs during the crop growing season.

The hydraulic features of the drip/micro irrigation systems allow very uniform emitter flow and water application control. Relevant nonuniformity can only stem from poor system design and/or maintenance. Irrigation scheduling can be easily implemented if the crop water requirements are properly estimated. Therefore, percolation and leaching out of the root zone of drip/micro irrigated crops can be minimized.

Sprinkler Irrigation

In sprinkler irrigation, water is distributed using a pressurized system with nozzles or jets that apply the water through the air. The water distribution patterns of the sprinklers in a system can be slightly different due to pressure differences. The individual patterns are not uniform, neither compositions of arranged stationary or moving patterns. Moreover, the distribution patterns can be distorted by the effect of the wind. Therefore, certain nonuniformity is inherent to sprinkler irrigation and the risk of percolation exists under most of the irrigation management scenarios. The nonuniformity and percolation risk can be at different spatial scales. The redistribution of water within the soil and the extent of the crop roots can contribute to damp the small-scale variability and thus the reduction of the percolation risk. However, neither the soil nor the crop will be able to damp the part of the nonuniformity due to differences among laterals and along individual laterals.

CHEMIGATION

Chemigation can be an efficient way of applying pesticides and fertilizers. Best management practices endorse split applications of fertilizer to match nutrient supply and demand and to reduce the potential of leaching. Applying fertilizers with irrigation water (fertigation) expands opportunities for timed applications.

Chemigation in drip/micro irrigation is a well-developed practice.[7] Also sprinkler fertigation (chemigation) has technical basis. Recent modeling attempts have tried to develop scientific criteria for surface fertigation. Boldt et al.[8] used a surface irrigation model to simulate the distribution of N during surge irrigation—a promising means of furrow fertigation.

Despite the potential advantages of chemigation, experimental results indicate that flood chemigation or chemigation with high-rate sprinkler irrigation may actually increase rather than decrease deep leaching of agricultural chemicals. Jaynes et al.[2] observed that a tracer applied with the irrigation water moves deeper into the soil than a tracer sprayed on the soil surface immediately before irrigation because the former is better able to use preferential pathways and move deeply into the soil. Therefore, caution on timing of the chemical application and flow regime are important aspects to consider when fertigation is practiced.

REFERENCES

1. Tillman, R.W.; Scotter, D.R.; Clothier, B.E.; White, R.E. Solute Movement During Intermittent Water Flow in a Field Soil and Some Implications for Irrigation and Fertilizer Application. Agric. Water Manag. 1991, 20 (2), 119–133.
2. Jaynes, D.B.; Rice, R.C.; Hunsaker, D.J. Solute Transport During Chemigation of a Level Basin. Trans. ASAE 1992, 35 (6), 1809–1815.
3. Jaynes, D.B.; Rice, R.C. Transport of Solutes as Affected by Irrigation Method. Soil Sci. Soc. Am. J. 1993, 57 (5), 1348–1353.
4. Pang, X.P.; Letey, J.; Wu, L. Irrigation Quantity and Uniformity and Nitrogen Application Effects on Crop Yield and Nitrogen Leaching. Soil Sci. Soc. Am. J. 1997, 61 (1), 257–261.
5. Vickner, S.S.; Hoag, D.L.; Frasier, W.M.; Ascough, J.C. A Dynamic Economic-Analysis of Nitrate Leaching in Corn Production Under Nonuniform Irrigation Conditions. Am. J. Agric. Econ. 1998, 80 (2), 397–408.
6. Benjamin, J.G.; Porter, L.K.; Duke, H.R.; Ahuja, L.R. Corn Growth and Nitrogen Uptake with Furrow Irrigation and Fertilizer Bands. Agron. J. 1997, 89 (4), 609–612.
7. Bar-Yosef, B. Advances in Fertigation. Adv. Agron. 1999, 65, 1–77.
8. Bolt, A.L.; Watts, D.G.; Eisenhauer, D.E.; Schepers, J.S. Simulation of Water Applied Nitrogen Distribution Under Surge Irrigation. Trans. ASAE 1994, 37 (4), 1157–1165.

Field Water Supply and Balance

Jean L. Steiner
United States Department of Agriculture (USDA), El Reno, Oklahoma, U.S.A.

INTRODUCTION

Field water supply has been a major focus of agricultural research and management. The soil water balance is a widely used method of tracking soil water supply in a field. This approach provided some of the earliest information available about the amount of water required to produce a crop, the relationship between water use and plant production, and water stress impacts on plant water use, and remains an important approach to research and management today.

The soil water balance (Fig. 1) can be given as:

$$SW_t = SW_i + P + I - R - E - T - D \tag{1}$$

where SW is soil water content within a defined root zone, t and i subscripts represent the end and beginning of a time period, respectively, P is precipitation, I is irrigation, R is runoff, D is drainage below the root zone, E is evaporation from the soil surface, and T is transpiration, with all terms in the same units and over the time period defined by the t and i subscripts. The R term might be modified to include any horizontal movement of surface water or shallow water table flow, which can be either imported to or exported from the defined soil volume. In most circumstances, the D term is downward flux below the root zone, but can be defined to include vertical flux across the bottom of the root zone that could include upward movement from a shallow water table to deep rooted plants. The E term can be considered to include water evaporated from any wetted surface (e.g., ponded water, wetted plants, evaporation or sublimation of accumulated snow), as well as evaporation of water from the soil profile. Frequently, the soil water balance is used to determine terms of Eq. 1 (e.g., $E + T$, $SW_t - SW_i$,) by measuring or estimating the remaining terms. Infiltration is often estimated by measuring P and R, if the other terms can be considered negligible during the precipitation event. Gardner[1] provides additional detail about the soil water balance.

WATER BALANCE COMPONENTS

Soil Water Content

Many different methods have been used to measure soil water content. A direct method that has been used since the early days of soil and agricultural research and remains common today is the gravimetric method. For the gravimetric method, soil samples, often cores, are collected and the water content is determined by weighing the sample before and after oven drying to determine the quantity of water lost by evaporation. Gravimetric sampling offers the benefit of providing a direct measurement of soil water content using simple equipment. However, it is time consuming and cannot be used to provide repeated measurement at the same location because it is destructive sampling. Given the high degree of spatial variability in most field soils, this limits the ability to determine temporal changes in soil moisture precisely, limiting the application of the soil water balance to relatively longer time periods. Methods that provide repeated measures of soil water content at the same location, such as using neutron probe or other technologies, reduce the problem of spatial variability and allow the soil water balance to be applied over shorter time periods.

Precipitation and Irrigation

Water is added to the system through precipitation or irrigation. Since precipitation is often highly variable, the rain gauge should be as near the site of the investigation as possible and should use standard weather gauges, properly sited away from tall buildings or vegetation that can distort rainfall catch. At field scales, irrigation applications are not totally uniform, so the gross irrigation amount may have to be adjusted by efficiency and uniformity factors to determine the net input to the soil water balance. However, well-managed, modern irrigation techniques such as low pressure applicators on center pivots and drip or subsurface irrigation methods, can provide very uniform distributions of water in a field with high efficiency.

Horizontal Movement of Water

The horizontal surface movement of water is largely runoff. In some situations, there may be a net gain through run-on of water from another portion of the landscape. However, there is normally a net loss to the water balance of a particular field when horizontal movement occurs. The proportion of precipitation that runs off depends on

Encyclopedia of Water Science
DOI: 10.1081/E-EWS 120010155

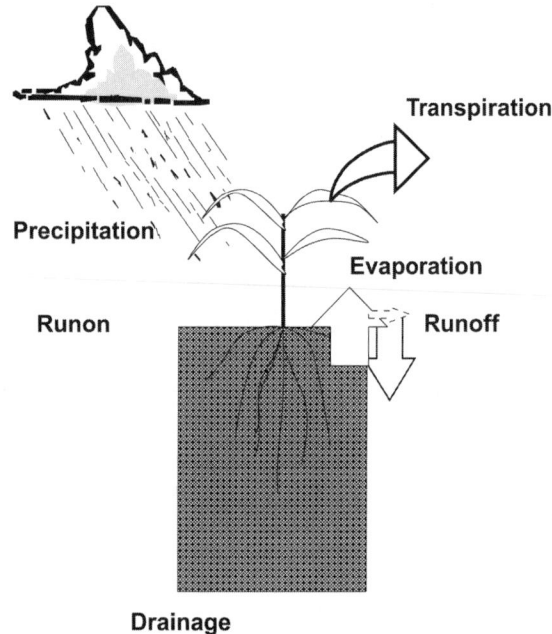

Transpiration

Precipitation

Evaporation

Runon

Runoff

Drainage

Fig. 1 Soil water balance for an agricultural plant.

soil properties, vegetation, and topography, as well as the intensity and duration of rainfall. In some soil water balance calculations, runoff is assumed to be negligible because of level topography, or in some cases berms are constructed around small experimental areas to restrict horizontal movement. When a large amount of precipitation falls as snow, there can be considerable movement caused by the wind and drifting. In some field situations, there is a substantial horizontal movement of water below the surface, making it difficult to use the water balance method.

Evaporation and Transpiration

Because it is difficult in field studies to separate loss of water by evaporation from soil and transpiration from plants, these terms are often linked into a single term, usually called evapotranspiration, E_t. From a management perspective, options for influencing soil evaporation and transpiration are different, so the terms are presented separately in Eq. 1.

Energy Balance

The energy balance approach, often used to estimate E_t, was developed for a plant–soil system under conditions where transpiration dominated the total loss. This approach has been successful because a large portion of energy that enters the earth's atmosphere is required to

transform water from liquid to vapor (the evaporation process). The combination evaporation equation derived by Penman[2] and Monteith[3] can be expressed as

$$E_t = \{\Delta(R_n + G) + [\rho C_p(e_s - e_a)/r_{av}]\}/L[\Delta$$

$$+ \gamma(1 + r_s/r_{av})] \tag{2}$$

where Δ is the slope of the saturation vapor pressure curve, R_n is net radiation, G is soil heat flux, ρ is air density, C_p is the specific heat of dry air, e_s and e_a are the vapor pressures at the evaporating surface (assumed to be the saturation vapor pressure at the surface temperature) and of the atmosphere, respectively; r_{av} is the aerodynamic resistance, L is the latent heat of vaporization, γ is the psychrometric constant, and r_s is surface resistance. Because it is difficult to measure some of these terms, particularly the surface resistance term, methods have been developed to evaluate the equation for potential evaporation conditions using standard weather data, applying assumptions about a well-watered, vegetated surface; and then relating the potential evaporation to actual evaporation for a particular vegetative surface (such as early or late in the season with low vegetative cover or under water-stressed conditions) using crop coefficients and other adjustments. These methods are discussed thoroughly in Allen et al.[4] and Allen.[5]

Radiation Balance and Soil Heat Flux

Net radiation is the balance resulting from the incoming and outgoing fluxes of short and long-wave radiation and is affected by several aspects of the soil and plant cover at the surface.[6] Short wave radiation is strongly affected by surface roughness, color, soil water content, and solar angle. For soils and plants, emitted long-wave radiation is largely determined by surface temperature, so wetter soils (which are generally cooler) have less outgoing long-wave radiation than drier, warmer soils. Incident long-wave radiation is influenced by sky conditions (cloudiness, cloud type, etc.), and it will generally decline with increased sky cover. As vegetative cover increases, the effect of soil conditions on net radiation decreases to negligible levels. For calculation of short-term evaporation, soil heat flux is sometimes taken as a fraction of net radiation ($\sim 10\%$). For bare soils, such as early or late in a growing season, the fraction can be much higher. For longer time periods, changes in air temperature can be used to estimate the flux of heat into or out of the soil over the period.[4] As vegetative cover or crop residue cover increases, the flow of heat in the soil is reduced because temperature gradients in the soil decrease as the soil is shaded.

Drainage of Water Across the Lower Boundary

In some cropping systems, the flux across the lower boundary of the root zone can be considered to be negligible. This is true primarily in semiarid or arid climatic regimes and in soils with a high water holding capacity and/or a low hydraulic conductivity, such as clay loams, silty clay loams, or in some cases silt loams and loams. It may also be true where there is a restrictive layer in the soil profile that prevents or slows the flow of water below the root zone. However, when considering year round water balances, soil water is often lost below the root zone during at least part of the annual cycle in periods of high precipitation or low evapotranspiration. Water losses below the root zone can be measured using lysimeters, or can be calculated using measurements of soil water tension or content over time along with knowledge of soil hydraulic properties. In some situations, water can move upward from a shallow water table into the root zone of deep rooted plants.

APPLICATIONS

Irrigation Scheduling

Irrigated agriculture is one of the most intensive forms of agriculture. Having control of the water supply to ensure adequate water for crop growth allows a producer to invest more in other inputs that ensure a high yield and quality of the crop. However, irrigation applications can be expensive and excessive water application can result in loss of nutrients and other production inputs as well as causing environmental problems. Therefore, it is important to apply enough, but not too much, irrigation water. One way to do this is to monitor soil water content during the growing season and apply knowledge of the soil water balance to guide timing and amount of irrigation applications. Some irrigation scheduling models use weather data to simulate the evapotranspiration and maintain a soil water budget to predict changes in the soil water content. Depending on the rate of water use by the crop and the amount of soil water storage capacity, the producer can project the upcoming needs for irrigation applications.

Rainfed Cropping

Rainfed cropping is subject to great risks because of the high variability of rainfall in most agricultural regions. In many regions, water stored in the soil at planting time is an important component of the seasonal water supply. If there is a large amount of water stored at planting time, that stored water provides a buffer against dry periods during the growing season, and the producer might plan for an average or good yield level and invest in inputs to support those yield levels. If the water storage at planting is low, then the risk of crop losses due to growing season drought is high and the producer may decide to reduce or delay investment in some inputs until later in the season when more is known about growing season precipitation and forecasts. Analysis of long-term climatological records using a soil water balance approach can be used to evaluate alternative crops or rotations for a region. In some cases, high soil water levels at the end of the growing season will result in a low faction of off-season precipitation being stored for the next crop with greater losses to percolation (with some possible nutrient leaching) or runoff (with possible greater erosion).

Plant Growth and Natural Resource Modeling

Soil water balance calculations are an integral part of plant growth and hydrologic models. Many of these models have been developed to operate at a daily time step and have been applied to a wide range of analyses. Some examples of such models that are available for downloading from the internet include crop growth, erosion, and hydrology models developed at the Grassland Soil and Water Research Laboratory at Temple, Texas, http://arsserv0.tamu.edu/intro.htm; the soil organic matter model, CENTURY, developed at the Natural Resource Ecology Laboratory at Colorado State University, http://www.nrel.colostate.edu/projects/century5/; water balance, irrigation management, and soils models developed by Dr. J. T. Ritchie and colleagues, http://nowlin.css.msu.edu/; the Decision Support System for Agrotechnology Transfer, DSSAT, which is a series of crop models and associated weather, crop, and soil data bases, http://icasanet.org/dssat/ and a suite of models, ranging from nutrient management and water quality models to an operational tool for whole farm/ranch strategic planning developed by the Great Plains System Research Unit at Fort Collins, Colorado, http://gpsr.ars.usda.gov/products/. These models, and many more, include a soil water balance as an integral part of the system.

CONCLUSION

The soil water balance approach has played an important role in improving our understanding and management of plant, water, and soil resources. The field soil water balance involves accounting for inputs of water to the system, such as precipitation and irrigation, as well as water leaving the system via evapotranspiration, runoff, and drainage below the root zone. The way a field is managed can have a large impact on the magnitude of the components of the water balance, such as runoff and drainage, as well as patterns of evapotranspiration and partitioning of the water loss into soil evaporation and transpiration. In agriculture, increasing the amount of transpiration increases productivity of the system. Soil water balance approaches can be applied to irrigated and rainfed agriculture and are an integral part of all plant growth and natural resource model, so understanding the basic concepts and principles is important for sound water management.

REFERENCES

1. Gardner, W.R. Soil Properties and Efficient Water Use: An Overview. In *Limitations to Efficient Water Use in Crop Production*; Taylor, H.M., Jordon, W.R., Sinclair, T.R., Eds.; Am. Soc. Agron., Crop Sci. Soc. Am., Soil Sci. Soc. Am.: Madison, WI, 1983; 45–71.

2. Penman, H.L. Natural Evaporation from Open Water, Bare Soil, and Grass. Proc. R. Soc. Lond. **1948**, 120–145.

3. Monteith, J.L. *Evaporation and Environment*, Symp. Soc. Exp. Biol.; Academic Press: New York, 1965; Vol. 19, 205–234.

4. Allen, R.G.; Pereira, L.S.; Raes, R.; Smith, M. *Crop Evapotranspiration—Guidelines for Computing Crop Water Requirements—FAO Irrigation and Drainage Paper 56*; Food and Agriculture Organization of the United Nations: Rome, 1998; 300.

5. Allen, R.G. Using the FAO-56 Dual Crop Coefficient Method Over an Irrigated Region as Part of an Evapotranspiration Intercomparison Study. J. Hydrol. **2000**, *229*, 27–41.

6. Steiner, J.L. Crop Residue Effects on Water Conservation. In *Managing Agricultural Residues*; Unger, P.W., Ed.; Lewis Publishers: Boca Raton, Florida, 1994; 41–76.

Filtration and Particulate Removal

Lawrence J. Schwankl
University of California, Davis, California, U.S.A.

INTRODUCTION

Filtration of water to remove particulate matter and biological contaminants is critical to the efficient operation of many pressurized irrigation systems. Filtration for sprinkler irrigation systems, where there are large-size water contaminants that can clog the sprinkler nozzles, is usually done with screen or disk filters. Microirrigation systems, with small flow passageways in the drip emitters and microsprinklers, may use screen, disk, or sand media filters to remove small particulates and organic contaminants to prevent clogging. The choice of which filter to use is often based on the water quality.

FILTRATION REQUIREMENTS

The degree of filtration for sprinkler and microirrigation systems is significantly different. Sprinkler irrigation, with its larger nozzle openings can pass all but the larger particulates. Thus, filtration treatment to remove the trash and larger sand particles is adequate. Unless heavy organic contaminant loadings occur, organic materials are passed through the sprinkler system with little clogging hazard. Microirrigation systems, with their small passageways, require more extensive filtration systems. The degree of filtration recommended for specific drip emitters and microsprinklers is available from the manufacturer and should be followed.

The degree of filtration of screen and disk filters is designated by their mesh size. The mesh size is the number of openings per inch of screen. The degree of filtration of sand media filters is determined by the size of the sand media particles with the sand media sizes referenced to equivalent mesh size (Table 1).

Mineral particulates in irrigation water range in size from sands to silts to clays. The equivalent mesh sizes for these mineral particles are given in Table 2. Few microirrigation systems require greater than 200-mesh filtration. Note that small sand particles, silts, and clays will pass through a 200-mesh screen. These very small particles can pass though drip emitters or microsprinklers, or they may settle out in the pipelines or lateral lines requiring flushing to be removed (discussed later).

TYPES OF FILTERS

Suction Screen Filters

Suction screens (Fig. 1)[3,4] are used on centrifugal pump intakes, where there is a significant problem with large particulates and trash in the water as can be the case from surface water sources such as rivers and streams. Used by themselves, they may provide adequate filtration for sprinkler irrigation systems, but not for microirrigation systems. Rather, they may be the first filtration step for microirrigation systems, removing the large particulates which would quickly overwhelm the screen, disk, or media filters also being used.

To be effective, suction screen filters should filter out the contaminants and keep themselves clean. Some suction screen filters continually rotate and use water jets to clean the contaminants off the screen. The water flowing by the intake screen carries the contaminant's downstream.

Centrifugal Sand Separators

Centrifugal sand separators (Fig. 2)[3,5] are well suited to removing larger sand particles which may be present in both surface water sources and in groundwater. They are designed to "swirl" the water passing through them, using centrifugal forces to remove the sand particles. While sand particles may not clog sprinkler systems, they may cause wear to the sprinkler nozzles and should be removed. In sprinkler irrigation systems, centrifugal sand separation may be the only filtration required, particularly when groundwater is used.

Larger sand particles must be removed from microirrigation systems since they will cause clogging. While screen, disk, or sand media filters can all remove sand particles, large volumes of sand may clog these filters quickly. In microirrigation systems, centrifugal sand separators are often used as the first stage filtration method, followed by screen, disk, or sand media filters.

Screen Filters

Two types of screen filters are common—pressurized screen filters (Fig. 3)[6–10] and gravity flow screen filters (Fig. 4).[3,8,10] In a gravity flow screen filter, water is

Encyclopedia of Water Science
DOI: 10.1081/E-EWS 120010202

Table 1 Sand media size and screen mesh designation

Sand no.	Effective sand size (in.)	Screen mesh designation
8	0.059	70
11	0.031	140
16	0.026	170
20	0.018	230
30	0.011	400

(Adapted from Ref. [1].)

Table 2 Particle size classifications by mesh size

Particle size	Mesh equivalent
Very coarse sand	10–18 mesh
Coarse sand	18–35 mesh
Medium sand	35–60 mesh
Fine sand	60–160 mesh
Very fine sand	160–270 mesh
Silt	270–400 mesh
Clay	Smaller than 400 mesh

(Adapted from Ref. [2].)

allowed to run over the screen filter, open to the atmosphere, with the filtered water falling through the screen and being collected. The contaminants caught on the screen are either washed off the screen by the water flowing across the steeply inclined screen, or, in another design a slightly inclined screen is continually washed clean by a rotating jet which moves the contaminants into a collection trough. The use of a gravity screen filter requires the irrigation water to be pressurized following filtration.

A pressurized screen filter is plumbed into the irrigation system, and filtration is accomplished as the pressurized water passes through it. Pressurized screen filters are used in sprinkler irrigation to remove larger particles, which may clog the sprinkler nozzle or cause excessive wear. Screen filters are widely used in microirrigation systems, particularly where groundwater is used. Pressurized screen filters may not be appropriate for use with water high in organic matter. The organic contaminants may quickly clog the screen and be difficult to remove. Once the screen is clogged, there may be a significant pressure loss across the screen and the flow rate through the screen may be substantially reduced. Installation of pressurized screen

filters with upstream and downstream pressure gauges is recommended so that the manager can easily note when the screen needs cleaning.

Some pressurized screen filters require the screen element to be manually removed for cleaning. Others have a backwash system so that the screens can be cleaned without disassembling the filter. Some of these backwash systems are operated manually while others allow the backwash to be done automatically, either on a set time interval and/or on a pressure loss across the screen, sensing system.

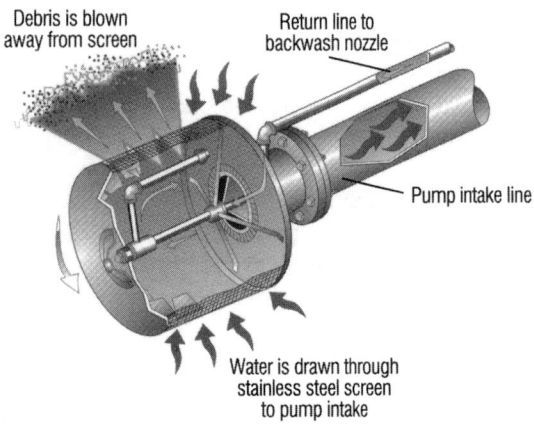

Fig. 1 Suction screen filter on the intake to a pump. (Courtesy of the Claude Laval Corporation.)

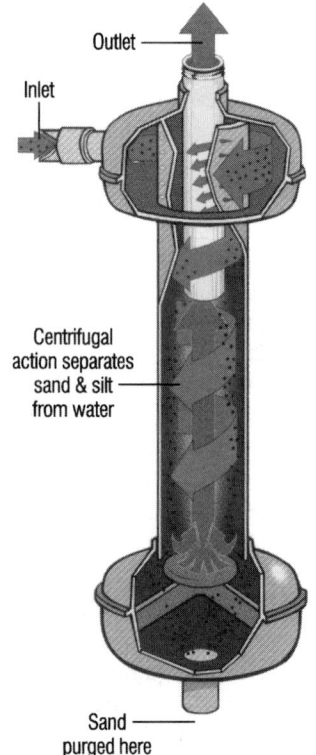

Fig. 2 Centrifugal sand separator. (Courtesy of the Claude Laval Corporation.)

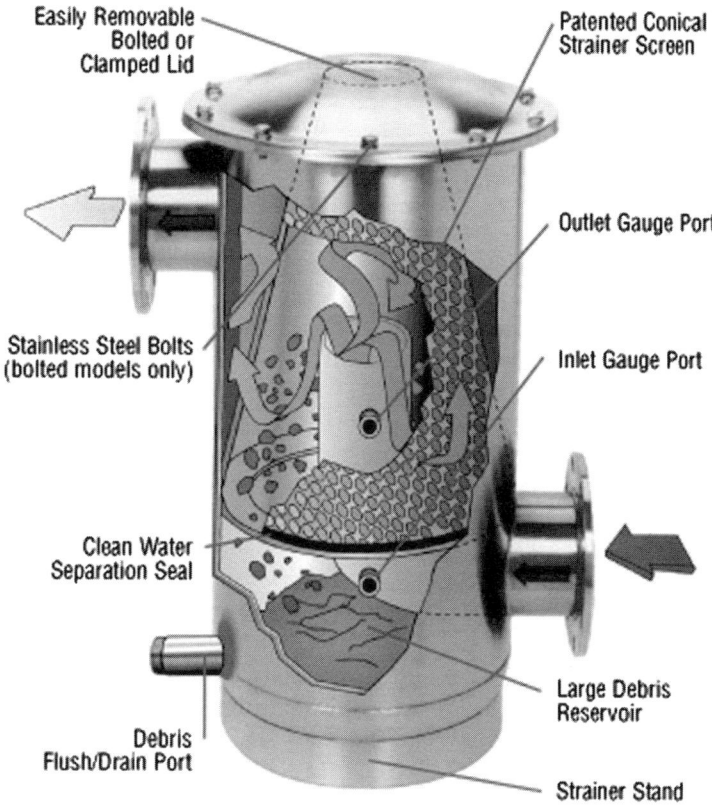

Easily Removable Bolted or Clamped Lid

Patented Conical Strainer Screen

Outlet Gauge Port

Inlet Gauge Port

Stainless Steel Bolts (bolted models only)

Clean Water Separation Seal

Debris Flush/Drain Port

Large Debris Reservoir

Strainer Stand

Fig. 3 Pressurized screen filter. (Courtesy of Miller-Leaman, Inc.)

The recommended, maximum flow rate through the screen filter will be specified by its manufacturer. Waters high in contaminants will clog the filter more quickly. Automatic backwash filters may be advantageous under these conditions, or an alternative would be a larger filter element (or more filters plumbed in parallel) to increase the interval between manual cleanings.

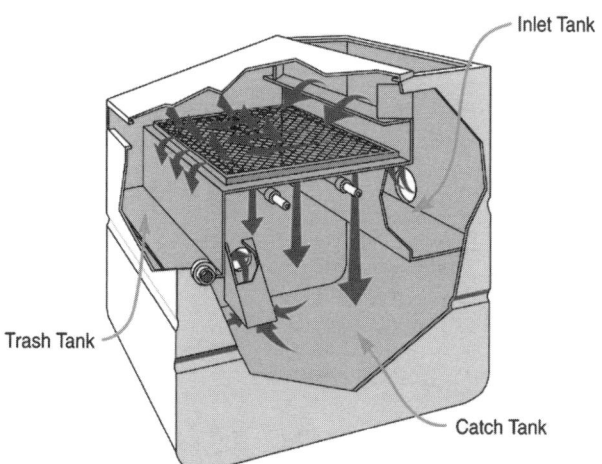

Inlet Tank

Trash Tank

Catch Tank

Fig. 4 Gravity flow screen filter. (Courtesy of Fresno Valves and Castings, Inc.)

Disk Filters

Disk filters (Fig. 5)[9–11] consist of a stack of thin disks, tightly held together, each having a series of very small grooves along their sides. Water is filtered as it flows through the grooves. The degree of filtration is measured as mesh size. Disk filters effectively filter particulate matter, and they will remove organic contaminants from the water but the organic contaminants tend to clog the disk filter quickly, necessitating frequent cleaning. Most disk filters must be disassembled and cleaned manually, but there are automatic backwash disk filters available. Where the water is high in organic matter, a disk filter with an automatic backwash system may be advantageous. The water required for backwashing disk filters is less than that for sand media filters.

Sand Media Filters

Sand media filters (Fig. 6)[5,6,9,10,12] are tanks made of epoxy-coated metal or stainless steel. They are filled with a filtering media, often silica sand. The particle size of the

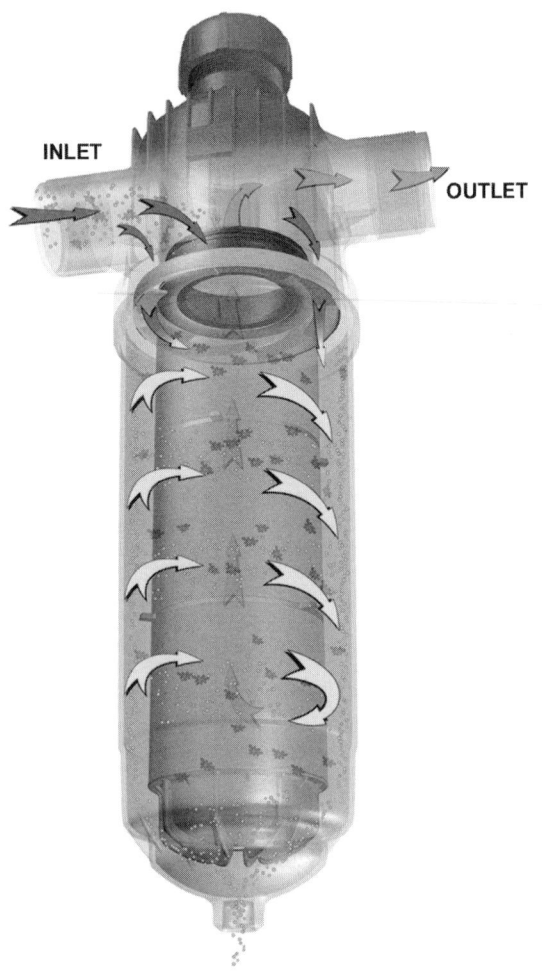

Fig. 5 Disk filter. (Courtesy of Miller-Leaman, Inc.)

media is selected according to the desired degree of filtration (Table 1). Water contaminants are filtered from the water as the water flows down through the media. An under-drain, made from either an epoxy cake or perforated pipe at the bottom of the tank, collects the filtered water and retains the filtering media during filtration.

Sand media filters have a greater filtering capacity than screen or disk filters and can be used to remove both organic contaminants and particulate matter,[13,14] making them well suited for filtering surface waters. At least two media filter tanks, plumbed in parallel, are required at a site so that as one filter is being backwashed, the other filter(s) can continue to provide water for the backwashing and for irrigation. Additional sand media filter tanks can be added if increased filtration capacity is needed. Frequently, a backup screen filter is placed downstream of the sand media filters to catch any sand escaping the media filters, either from routine operation or from failure of the media filter's under-drain system.

The recommended flow rate for sand media filters is $35 \, m^3 \, hr^{-1} m^{-2}$–$60 \, m^3 \, hr^{-1} m^{-2}$ ($15 \, gal \, min^{-1} \, ft^{-2}$–$25 \, gal \, min^{-1} \, ft^{-2}$) of filter surface area. The higher flow rates can be used where the water contains less than 10 ppm of suspended material. If the water has 100 ppm or more of suspended material, the lower filter flow rates should be used to avoid the need for frequent backwashing. Manufacturers of sand media filters provide recommended filter flow rates both for filtration and for backwashing of filters. These recommendations should be followed.

Backwashing of sand media filters can either be done manually or automatically. When backwashing, a three-way valve at the top of the filter changes position, and clean water passes upward from the under-drain system.

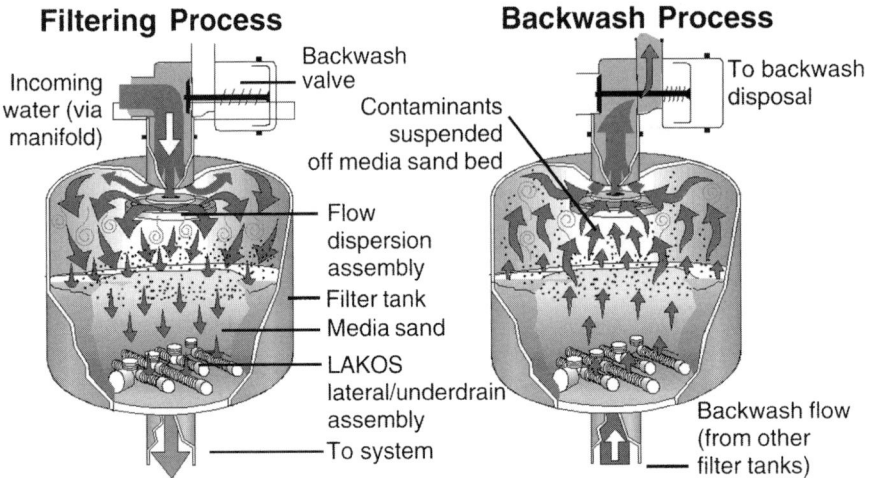

Fig. 6 Sand media filters. (Courtesy of the Claude Laval Corporation.)

This suspends and agitates the filter media with contaminants being flushed out of the filter with the backwash water. Pressure gauges should be installed upstream and downstream of the filters and backwashing should be done when the pressure drop across the filters (approximately 70 kPa) indicates that they are dirty. Automatic backwashing systems allow the media filters to be cleaned on a desired time interval or when the pressure drop across the filter exceeds a selected value.

Disposal of backwash water can be a problem when using sand media filters. The backwash flow rate is nearly $45 \, m^3 \, hr^{-1}$ ($200 \, gal \, min^{-1}$) for a typical 48-in. (1.2 m) sand media filter, so a substantial volume of backwash water is generated. Some microirrigation system managers are even constrained to disposing of backwash water by using reservoirs or tile drain systems.

FLUSHING

Small sand, silt, and clay particles pass through the filters used in microirrigation systems. These fine particles frequently settle in the pipelines and polyethylene lateral lines of microirrigation systems and, unless they are flushed out, can lead to clogging of drip emitters or microsprinklers.

Appropriately sized flush-out valves should be located at the end of pipelines. These valves can be opened and the particles that have settled in the pipelines flushed out. Following flushing of the pipelines, the ends of the lateral lines should be opened, a few at a time, and allowed to flush clear. In drip irrigation systems designed for row crops, the lateral lines may be manifolded together to allow more convenient flushing. An alternative to manual flushing of lateral lines is to use self-flushing end caps on the lateral lines. These end caps allow a short flush at the beginning and end of the irrigation event.

REFERENCES

1. Schwankl, L.; Hanson, B.; Prichard, T. Filtration Equipment. *Microirrigation of Trees and Vines*. University of California, DANR Publication No. 3378; 1996; 142 pp.
2. Hanson, B.; Schwankl, L.; Grattan, S.R.; Prichard, T. Sand-Media Filtration. *Drip Irrigation for Row Crops*. University of California, DANR Publication No. 3376; 1997; 238 pp.
3. Website for the Claude Laval Corporation, html: www.lakos.com (accessed January, 2002).
4. Website for Sure-Flo fittings, html: www.sure-flo.com (accessed January, 2002).
5. Website for Yardney Water Filtration Systems, html: www.yardneywater.thomasregister.com (accessed January, 2002).
6. Website for Miller-Leaman Incorporated, html: www.millerleaman.com (accessed January, 2002).
7. Website for Filtomat Self Cleaning Filters, html: www.filtomat.com (accessed January, 2002).
8. Website for Everfilt, html: www.everfilt.com (accessed January, 2002).
9. Website for Amiad North America, html: www.amiadusa.com (accessed January, 2002).
10. Website for Fresno Valves and Castings, Inc., html: www.fresnovalves.com (accessed January, 2002).
11. Website for Arkal Filtration Systems, html: www.arkal-filters.com (accessed January, 2002).
12. Website for Netafim Irrigation, www.netafim-usa.com/ag/products/filtration.asp (accessed January, 2002).
13. Burt, C.M. Media Tanks for Filtration, Parts I and II. Irrig. J. **1994**, July/Augustand September/October.
14. Burt, C.M.; Styles, S. *Drip and Micro Irrigation for Trees, Vines, and Row Crops*; Cal Poly San Luis Obispo, ITRC Publication, 1999.

Floodplain Management

French Wetmore
French & Associates, Ltd., Park Forest, Illinois, U.S.A.

INTRODUCTION

Throughout time, floods have altered the landscape. Flooding is a natural process and floodplains are created and altered by that process. Floodplains have also been altered by human development, with consequences to those who live in them.

During the early settlement of the United States, locations near water provided required access to transportation, a water supply, and water power. These areas had fertile soils, making them prime agricultural lands. In recent decades, development along waterways and shorelines has been spurred by the recreational value of these sites.

The result has been an increasing level of damage and destruction wrought by the natural forces of flooding on human development. Flooding has become the nation's number one natural hazard. It affects more property each year and has accounted for over 70% of the Presidential disaster declarations since 1970.

HISTORICAL APPROACHES

During the 1920s, the insurance industry concluded that flood insurance could not be a profitable venture because the only people who would want flood coverage would be those who lived in floodplains. As they were sure to be flooded, the rates would be too high to attract customers. Unlike other hazards, such as wind and hail, where the risk can be spread, private industry opted out of playing a role in flood protection.

With the great Mississippi River flood of 1927, the federal government became a major player in flooding. As defined by several Flood Control Acts, the role of government agencies was to build massive flood control structures to control the great rivers, protect coastal areas, and prevent flash flooding.

Until the 1960s, such structural flood control projects were seen as the primary way to reduce flood losses. In some areas, they still are. However, starting in the 1960s, people questioned the effectiveness of this single solution. Disaster relief expenses were going up, making all taxpayers pay more to provide relief to those with property in floodplains. Studies during the 1960s concluded that flood losses were increasing, in spite of the number of flood control structures that had been built.

One of the main reasons structural flood control projects failed to reduce flood losses was that people continued to build in floodplains. In response, federal, state, and local agencies began to develop policies and programs with a "non-structural" emphasis, ones that did not prescribe projects to control or redirect the path of floods.

A milestone in this effort was the creation of the National Flood Insurance Program (NFIP) in 1968. The NFIP is based on a mutual agreement between the Federal government [represented by the Federal Emergency Management Agency (FEMA)] and local governments. Federally guaranteed flood insurance is made available in those communities that agree to regulate development in their mapped floodplains.

If the communities do their part in making sure future floodplain development meets certain criteria, FEMA will provide flood insurance for properties in the community. The Federal government is willing to support insurance because, over time, local practices will reduce the exposure to flood damage.

Also during the 1960s and 1970s, interest increased in protecting and restoring the environment, including the natural resources and functions of floodplains. Coordinating flood loss reduction programs with environmental protection and watershed management programs has since become a major goal of federal, state, and local programs. This evolution is shown graphically in Fig. 1. Now, we no longer depend solely on structural projects to control floodwater. Instead of "flood control," we now speak of "floodplain management."

FLOODPLAIN MANAGEMENT

Floodplain management is officially defined by the Federal Government's *Unified National Program for Floodplain Management* as "a decision-making process that aims to achieve the wise use of the nation's floodplains." (see Ref. [1], p. 8) "Wise use" means both reduced flood losses and protection of the natural resources and functions of floodplains. This is accomplished through different tools, including, but not limited to:

- Floodplain mapping
- Land use regulations.

Encyclopedia of Water Science
DOI: 10.1081/E-EWS 120010108

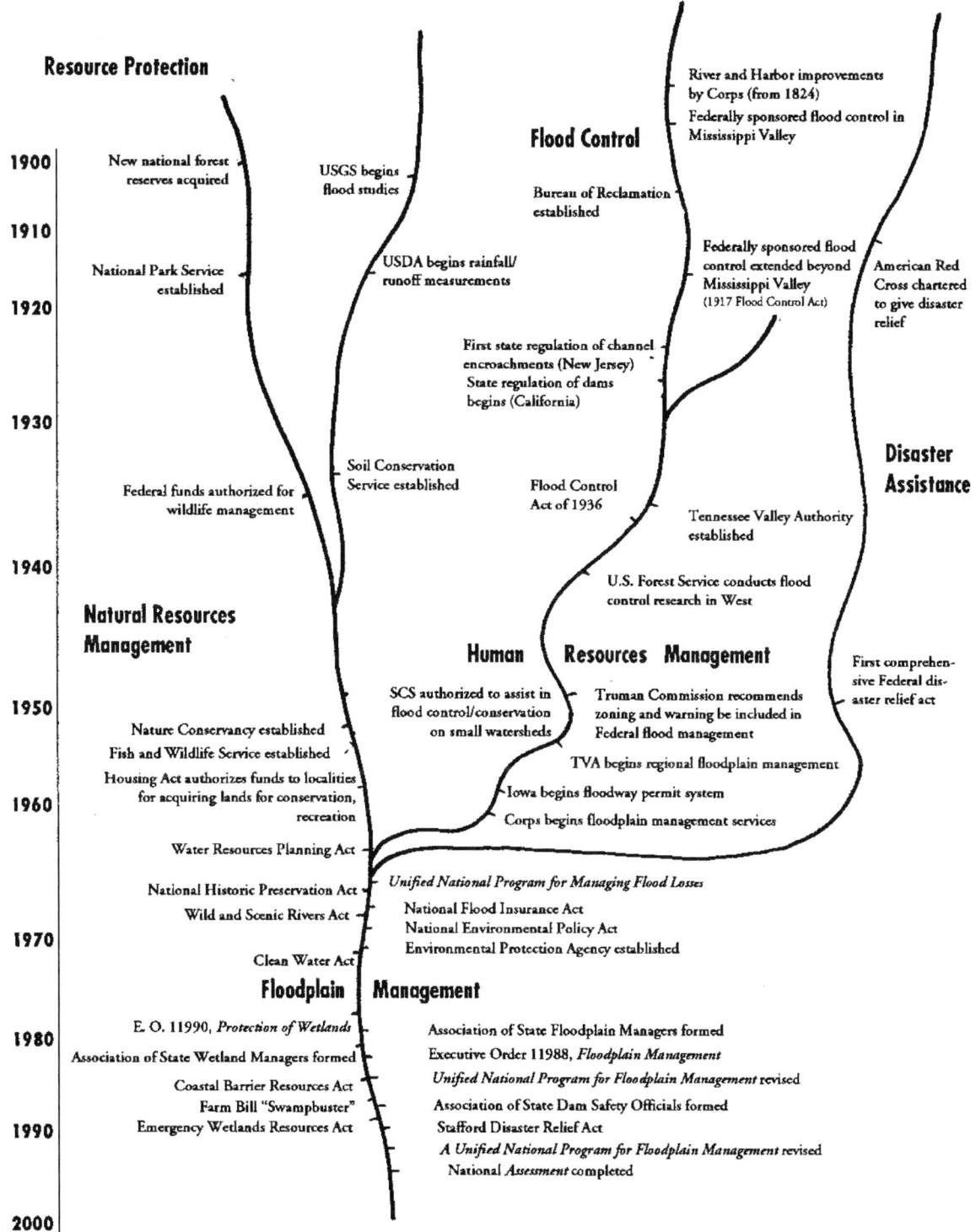

Fig. 1 Evolution of floodplain management in the United States.

- Preservation of floodprone open space.
- Flood control (levees, reservoirs, channel modifications, etc.)
- Acquiring and clearing damaged or damage-prone areas.
- Floodproofing buildings to reduce their susceptibility to damage by floodwaters.
- Flood insurance.
- Water quality best management practices.
- Flood warning and response.
- Wetland protection programs.
- Public information.

There are a variety of Federal, state, and local programs that administer these tools. Private organizations and property owners also have roles.

THE NATIONAL FLOOD INSURANCE PROGRAM

The nation's focal floodplain management program is the NFIP. It has prepared floodplain maps for 22,000 communities. FEMA sets the minimum land use development standards that participating communities must administer within the floodplains designated on their Flood Insurance Rate Maps. These standards are summarized in Fig. 2.

While participation is voluntary, communities that decide not to join or not to enforce those regulations do not receive Federal financial assistance for insurable buildings in their floodplains. Rather than face the loss of Federal aid (including VA home loans, HUD housing help, and disaster assistance), just about every community with a significant flood problem has joined. By 2002, 19,700 cities and counties were participating.

Within participating communities, Federal law requires the purchase of a flood insurance policy as a condition of receiving Federal aid, including mortgages and home improvement loans from Federally regulated or insured lenders. This requirement, coupled with personal experiences with flooding, has convinced over four million property owners to buy flood insurance. Unfortunately, it is estimated that only half of the properties in the FEMA mapped floodplains are insured.

OTHER FEDERAL PROGRAMS

FEMA administers other floodplain management programs, including:

- Disaster assistance programs that help flooded communities and property owners recover after a flood.
- Mitigation assistance programs that fund local projects to acquire and clear floodprone properties.
- Research and technical assistance activities in the fields of mapping, planning, mitigation, and floodproofing.
- The National Dam Safety Program which assists state programs that regulate dams (dam failures were a factor in three of the four largest killer floods since 1970).

The U.S. Army Corps of Engineers is the second largest participant in Federal floodplain management programs. While it is best known as the builder of structural flood control projects, it has its own authority to regulate new development in navigable waterways and wetlands. It is also the leader in the technical aspects of floodproofing and river basin planning.

The U.S. Department of Agriculture's Natural Resources Conservation Service has a role in planning and building flood control projects, similar to the Corps', but limited to smaller watersheds. Through local soil and water conservation districts, NRCS staff can be valuable advisors to local officials reviewing floodplain or watershed development proposals.

Just as rivers traverse many lands, floodplain management pervades many government programs. Other agencies with floodplain management responsibilities include:

- Tennessee Valley Authority (where floodplain management got its start)
- Bureau of Reclamation (water control projects in the west)
- U.S. Geological Survey (river data and mapping)
- Environmental Protection Agency (water quality programs)
- Small Business Administration (disaster assistance for private property owners)
- National Oceanic and Atmospheric Administration (coastal zone policies)
- National Weather Service (the lead in flood warning programs)

OTHER PROGRAMS

State and local agencies are also into a variety of floodplain management activities. Their regulatory programs often exceed the NFIP requirements. Many states set additional minimum standards for mapping, floodplain

The National Flood Insurance Program (NFIP) is administered by the Federal Emergency Management Agency (FEMA). As a condition of making flood insurance available for their residents, communities that participate in the NFIP agree to regulate new construction in the area subject to inundation by the 100-year (base) flood.

There are four major floodplain regulatory requirements. Additional floodplain regulatory requirements may be set by state and local law.

1. All development in the 100-year floodplain must have a permit from the community. The NFIP regulations define "development" as any manmade change to improved or unimproved real estate, including but not limited to buildings or other structures, mining, dredging, filling, grading, paving, excavation or drilling operations or storage of equipment or materials.

2. Development should not be allowed in the floodway. The NFIP regulations define the floodway as the channel of a river or other watercourse and the adjacent land areas that must be reserved in order to discharge the base flood without cumulatively increasing the water surface elevation more than one foot. The floodway is usually the most hazardous area of a riverine floodplain and the most sensitive to development. At a minimum, no development in the floodway may cause an obstruction to flood flows. Generally an engineering study must be performed to determine whether an obstruction will be created.

3. New buildings may be built in the floodplain, but they must be protected from damage by the base flood. In riverine floodplains, the lowest floor of residential buildings must be elevated to or above the base flood elevation (BFE). Nonresidential buildings must be either elevated or floodproofed.

4. Under the NFIP, a "substantially improved" building is treated as a new building. The NFIP regulations define "substantial improvement" as any reconstruction, rehabilitation, addition, or other improvement of a structure, the cost of which equals or exceeds 50 percent of the market value of the structure before the start of construction of the improvement. This requirement also applies to buildings that are substantially damaged.

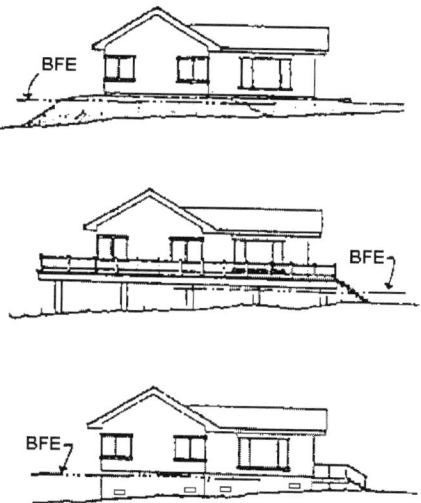

Communities are encouraged to adopt local ordinances that are more comprehensive or provide more protection than the Federal criteria. This is especially important in areas with older Flood Insurance Rate Maps that may not reflect the current hazard. Such ordinances could include prohibiting certain types of highly damage-prone uses from the floodway or requiring that structures be elevated 1 or more feet above the BFE. The NFIP's Community Rating System provides insurance premium credits to recognize the additional flood protection benefit of higher regulatory standards.

Fig. 2   Minimum National Flood Insurance Program regulatory requirements.

and wetland regulations and water quality. Some state agencies require their own permits, in addition to local permits, for new construction on waterways, lakes, shorelines, and floodplains.

In addition to being the lead regulators, most flood control projects are built and operated by local governments: cities, towns, counties, and special districts. The trend at the local level is toward special purpose authorities at the county or multicounty level to tackle problems holistically at the watershed level.

Private organizations have become more directly involved, too. Groups like the Nature Conservancy and land trusts work to preserve floodprone areas that have natural benefits. Others, like the National Wildlife Federation and American Rivers, are active on the political scene, reminding government agencies of their responsibilities and working to strengthen or expand their programs.

Over time, the distinction between what is done by what level of government has blurred. There are more and more cooperative and coordinated approaches, especially with increased nonfederal cost sharing requirements and regional and river basin organizations. A recent example of this is FEMA's Cooperating Technical Partners program where a state or local government can contribute to the cost of floodplain mapping and have a say on the techniques and standards used to prepare their Flood Insurance Rate Maps.

Another reason for the blurring of the distinction is the increased professionalization of the field. Most people active in floodplain management are members of the Association of State Floodplain Managers. Private practitioners and staff from all levels of government work together on solving common problems, rather than debating authorities or funding. There is also a new program that certifies floodplain managers. In less than 3 years, over 1000 professionals have earned the right to put "CFM" after their names.

PROGRESS

The impact of these efforts can be measured in three ways: threat to life, property damage, and the environment. Statistics have shown that the loss of life due to floods decreased during the last century, primarily due to better warning and public information programs.

Progress in the other two fields has not been as encouraging. Property damage is still increasing, although at a slower rate than if there were no NFIP and other floodplain management efforts (Fig. 3). It is harder to see improvements in water quality and habitat protection, but it is generally concluded that while things are better than if there were no programs, we have a long way to go.

AGRICULTURAL CONCERNS

Farmers, ranchers, and other agricultural interests are likely to be involved in floodplain management in several different ways. First, as landowners, their freedom to develop the floodplain portions of their properties may be limited by floodplain management or wetland regulations.

Federal, state, and/or local regulations require permits for the following:

- Regrading in the floodway.
- Construction of a levee.
- Modifications to a channel.
- Filling in a wetland.
- Construction of a new building in the floodplain.

This is the controversial part of floodplain management: activities on one's own property are subject to government restrictions in order to prevent diverting flood flows to other properties or adversely affecting wetlands or habitat or to reduce government disaster response and assistance expenses. While many state laws exempt some agricultural activities from local zoning or building codes, FEMA has ensured that in every state, agricultural buildings will be regulated as a condition for a city or county to participate in the NFIP.

A loan or Federal financial assistance to purchase, improve or repair a building in the floodplain will likely be accompanied by a requirement to purchase a flood insurance policy on that building. However, by taking certain protection measures, such as elevating the building above flood levels, insurance premiums can be reduced.

Federal and state programs are not all about restrictive regulations. Federal disaster assistance, flood insurance and crop insurance can come to one's aid after a flood. After the Great Flood of 1993 in the Mississippi River basin, many farmers accepted Federal funds to set aside wetlands and marginal farmland as a start to allowing Mother Nature to reclaim the natural floodplains.

Hopefully, farmers, ranchers, and other agricultural interests will become involved in floodplain management activities voluntarily and in a broader extent. They can reduce their own exposure to flood losses, help their communities and neighbors protect themselves, and improve their environment. Good places to learn more are the following websites:

- FEMA—www.FEMA.gov
- Association of State Floodplain Managers— www.floods.org

Both have links to other agencies and organizations. The latter has links to state floodplain management associations.

REFERENCE

1. *A Unified National Program for Floodplain Management*; FEMA 248, Federal Interagency Floodplain Management Task Force: Washington, DC, 1994.

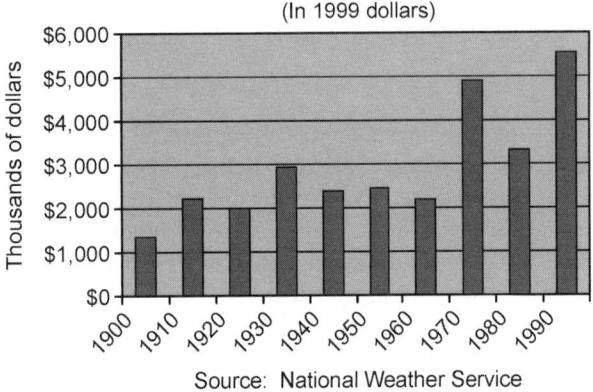

Fig. 3 Dollar damage caused by flooding.

(In 1999 dollars)

Source: National Weather Service

FURTHER READING

Addressing Your Community's Flood Problems: A Guide for Elected Officials; Association of State Floodplain Managers, Inc.: Madison, WI, 1996.

Answers to Questions About the National Flood Insurance Program; FEMA-387, http://www. fema.gov/nfip/qanda.htm (accessed February 2002) Federal Emergency Management Agency: Washington, DC, 2001.

National Flood Programs in Review; http://www.floods.org/PDF%20files/2000-fpm.pdf (accessed February 2002) Association of State Floodplain Managers: Madison, WI, 2000.

Floodplain Management in the United States: An Assessment Report; FIA-18, Federal Interagency Floodplain Management Task Force: Washington, DC, 1992.

Using Multi-Objective Management to Reduce Flood Losses in Your Watershed; Association of State Floodplain Managers, Inc.: Madison, WI, 1996.

Floods and Flooding

Jay A. Leitch
Steven Shultz
North Dakota State University, Fargo, North Dakota, U.S.A.

INTRODUCTION

Floods and flooding—the temporary condition of too much water—continue to plague many parts of the earth. The Yangtze River in China, for example, has flooded more than 1000 times in the past 2000 yr.[1] Floods occur on a large portion of the earth's land mass, and affect a significant percentage of its human inhabitants, since flood plains attract human activity. For example, three floods (1987, 1931, 1938) along the Hwang Ho (Yellow) River in China resulted in nearly 7 million deaths[1] (Table 1). Just what constitutes a flood depends somewhat on the perspective. From a human perspective, a flood is the act of getting one's person or property inundated (wet) (Fig. 1). From a purely physical perspective, a flood is the naturally occurring, temporary inundation of normally dry land. From either perspective, floods are fairly easy to describe.

PHYSICAL ASPECTS OF FLOODS

Physically, a flood occurs when the land surface is temporarily covered with water. Types of flooding include *flash floods* that result from rapid accumulation of water usually due to an intense rain storm (infrequently due to dam failure); *channel flooding* that results when water flows exceed the capacity of a waterway; and *overland or sheet flooding* that can occur when snow melt, storm water, or tidal surges (e.g., a tsunami) inundate large areas of relatively flat land that is normally dry. Simply put, flooding results from intense or prolonged precipitation, from rapid snow melt, from coastal surges, or, rarely, from dam failure.[2]

Floods are commonly characterized by their frequency, or expected frequency, which is based on the record of past events and hydrologic modeling. For example, a flood that has occurred only once in 100 yr of record is a 100-yr flood and has an expected return frequency of every 100 yr. Unfortunately, current and future conditions do not always match past records. In fact, 100-yr floods could occur two or even three years in a row. The expected frequency of floods is also referred to as the recurrence interval.

Likewise, a 10-yr flood is relatively common, expected to happen once every ten years, or have a 1 in 10 chance of happening in any one year. At the other extreme, 500-yr floods are rare, low frequency, high volume events. These frequency extremes are in large part related to the physical dimensions of floods, which can be explained by examining watershed maps, river cross-sections, and flood hydrographs.

Watersheds

A watershed (also known as a basin or catchment) is that portion of the earth's surface where runoff terminates or accumulates in a common hydrologic feature, such as a lake or river (Fig. 2). At a localized level, the watershed of a pond or stream includes all the land area that contributes runoff to the pond or stream. On a regional level, the watershed of a river includes all the land area that contributes runoff to the river or its tributaries. Examples of regional watersheds would be the Colorado River basin or the Ohio River basin. Finally, at the largest scale, the continental divide separates large, continental watersheds whose runoff ultimately flows to the oceans surrounding continents. Examples of some of the largest watersheds that drain significant portions of continents include the Amazon, Mississippi, and Nile River basins.

Watershed shape, drainage patterns, and runoff routing help to determine stream flow and flooding within the watershed. Shapes range from circular to elongated; drainage patterns range from dendritic (i.e., tree-like) to ditch;[3] and runoff routing ranges from natural flow to artificial flow and detention basins. A long, narrow watershed would be less likely to experience high peak flooding than a more circular watershed, all other things being equal, largely because water does not enter the main stem of the river at the same time.

Overall topographic relief within a watershed also affects the type of flooding. Relatively flat (i.e., low relief) watersheds may experience more sheetwater flooding, while steeper watersheds might experience more flash flooding. Proximity to oceans or seas is necessary for flooding to occur as a result of tsunamis and tropical storms.

Encyclopedia of Water Science
DOI: 10.1081/E-EWS 120010107

Table 1 Examples of Severe Flooding Worldwide

Date	Location	Impacts
1861	Sacramento River (California)	7,000 deaths, 300 villages destroyed, 2 million homeless
1887	Huang Ho (Yellow) River (China)	900,000 deaths
1889	Conemaugh River (Pennsylvania)	2,000 deaths, $10 million property damage
1900	Galveston, Texas	6,000 deaths, 3,000 buildings destroyed
1931	Huang Ho (Yellow) River (China)	3,700,000 deaths
1936–37	Mississippi River	800,000 injured, 500 deaths, $200 million property damage
1955	Atlantic Coast (hurricane Hasel)	$1.6 billion property damage
1960	Bangladesh	6,000 deaths
7/71–6/72	77 flood events in the United States	519 deaths, 141, 151 dwellings destroyed or damaged
1979	Zambezi River (Mozambique)	45 deaths, 250,000 homeless
1979	Morvi (India)	As many as 15,000 deaths
1981	Northern India	1,500 deaths, extensive crop losses
1982	El Salvador, Guatemala	More than 1,300 deaths
1985	Northern Italy	361 deaths
1993	Mississippi River	40 deaths, $10 billion property damage, 42,000 homes destroyed, 20 million acres of farm land disrupted
1997	Red River of the North (U.S.A.)	45,000 people evacuated, downtown Grand Forks burns

Sources: Frits van der Leeden, Fred L. Troise, and David Keith Todd. 1991. *The Water Encyclopedia*, Second Edition, Lewis Publishers: Chelsa, Michigan, and Owen Oliver. S, Daniel D. Chiras, and John P. Reganold. 1998. *Natural Resource Conservation, Seventh Edition*. Prentice Hall: Upper Saddle River, New Jersey.

Fig. 1 Urban flooding as a result of runoff.

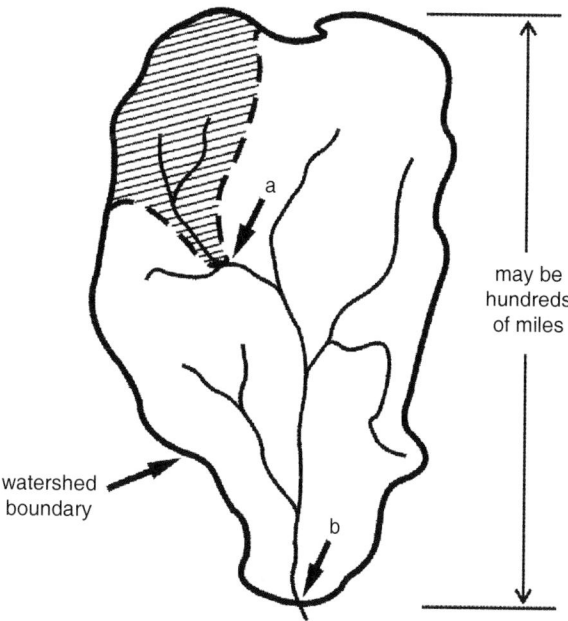

Fig. 2 Generalized watershed with subwatershed (shaded). Points a and b are watershed outlet locations for typical stream flow hydrographs (Fig. 3) and cross-sections (Fig. 4).

A number of other factors affect flooding within watersheds including climate and geographic orientation, soil types and land use/cover, and man-made alterations.

Climate and Geographic Orientation

Watersheds at higher latitudes that slope generally toward the Equator flood less frequently during spring snow melt than watersheds that slope toward the earth's poles. The snow pack in high latitude watersheds that slope toward the poles usually melts first in the headwaters, causing ice jams and flooding as it flows downstream and coincides with the timing of local snow melt and runoff. This is the case with the Red River of the North in central North America and many other north-flowing rivers in North America and asia.

Soil Types and Land Use/Cover

Watersheds consisting of more permeable soils are generally less prone to precipitation-based flooding than those with impervious soils. Watersheds with land uses or land cover that promote infiltration, evapo-transpiration, or that simply impede runoff, are less prone to flooding. Conversely, watersheds with impervious soils and/or land uses and cover that accelerate runoff may be more prone to flooding. The role of soils in flooding is far more complicated than this, since soils play a major role in topographic relief, the development of drainage patterns, and water-borne deposition.

Man-Made Alterations

Both the frequency and the severity of floods are affected by man-made changes in land use, such as converting forested land to cultivated crop land. Clearly, urbanization, the process of converting areas that may have good potential for infiltration to impervious surfaces (such as roads, parking lots, or buildings), can increase the likelihood of flooding. Human alterations of drainage, such as channelization and retention basins, impact runoff and stream flows. Finally, human attempts to control floods (e.g., dams and dikes) may change flooding regimes; but this effect is far more pronounced on low volume, high frequency floods than it is on high volume, low frequency floods. In spite of all that humans have done to control flooding, it is widely accepted that there is no way to completely eradicate flooding or flood damages.[4]

Flood Hydrographs

A flood hydrograph is a two-dimensional graph depicting how much water flows by a given point during a certain time period (Fig. 3).[3] A hydrograph could be constructed for any point in a watershed with adequate data.

The vertical axis of a hydrograph depicts the volume of stream flow expressed as cubic feet per second, liters per second, cubic meters per day, or acre feet per day. The vertical axis may also depict the river stage at a certain point, i.e., feet/meters above "flood stage" or another benchmark. Flood stage is when stream flow is sufficient to exceed the normal channel and spill over to the flood way (Fig. 3).

Floods can occur in smaller watersheds with flows of only a few $100\,m^3$/sec, while the maximum flood flow

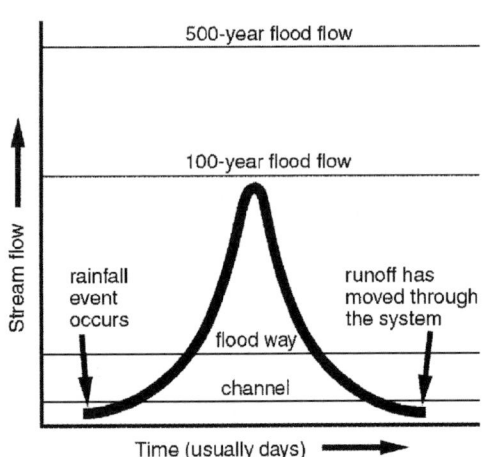

Flood hydrograph showing typical flow thresholds.

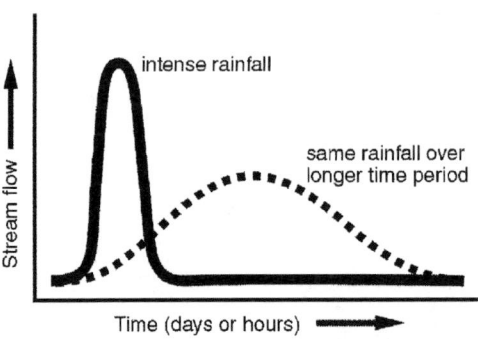

Subwatershed hydrograph showing the difference between intense, rapid runoff and prolonged runoff.

Fig. 3 Generalized stream flow hydrographs.

estimated in the Amazon Basin, one of the world's largest watersheds, is 370,000 m³/sec[4]

The horizontal axis of a hydrograph depicts time, usually in 24-hr increments or less, since floods generally occur over a period of several days. However, a spring snow melt flood at high latitudes may occur over weeks, while heavy rainfall in a mountainous region may result in a flash flood within hours.

River Cross-Section

A river cross-section is a profile of where the river flows at various river stages. As a river floods, more area is covered, and areas outside the main channel become part of the river. A cross-section depicts the "normal" channel, the flood way, and flood plains for several floods of various recurrence intervals (Fig. 4). The channel is where we would expect the river to be most of the time.

River cross-sections and their flood plain characteristics may be changed by structural measures such as dikes, levees, and dams. For example, there are 29 locks and dams, hundreds of runoff canals, and many miles of levees along the 2400 miles of the Mississippi River; each of which has an impact on downstream cross sections of the river and the associated hydrographs.[2] Prior to these control measures the Mississippi River typically flooded large areas every year.

Flood Way

The flood way is land immediately adjacent to rivers and streams that regularly (often annually) becomes inundated by channel overflow.

Flood Plain

The year-specific flood plain is the extent of land that is inundated with each frequency of flood. For example, the 10-yr flood plain, which we expect to be inundated about every ten years, is narrower than the 100-yr flood plain.

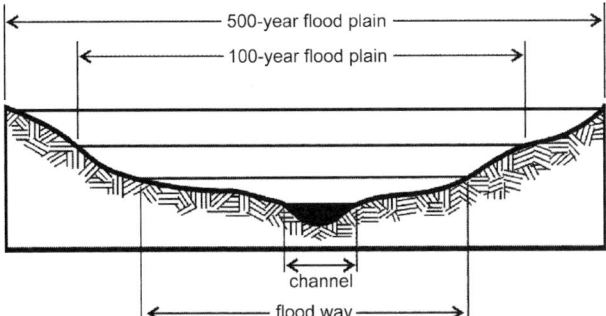

Fig. 4 River valley cross-section showing flood plains.

The 500-yr flood plain, which represents extreme, highly unlikely events, stretches well beyond the 100-yr flood plain.

HUMAN ASPECTS OF FLOODS

Throughout history, floods have caused disruptions in human activity, from inconvenience to property damage to loss of life. In addition to the obvious direct impacts of high water, floods affect human activity by depositing sediments, changing stream channels, uprooting trees and moving boulders, and altering fish and wildlife habitats. Some of these impacts can be positive, as in the case of the Nile before it was dammed, where annual flooding was referred to as the "Gift of the Nile," because of the fertility it added to the soil.[2]

Some of the most severe floods rank high among the world's greatest natural disasters in terms of their impact on humans. For example, the Mississippi River flood of 1993 is considered by many to be the greatest natural disaster in the history of the United States with estimated loses of $10 billion.[4] Flood flows at Hannibal, Missouri, were measured at 2 ft higher than the 500-yr flood mark. Virtually no region of the United States is immune from potential flooding[5] or, for that matter, in the world.

The greatest loss of human lives due to floods has occurred in China (Table 1). In 1931, 3.7 million people lost their lives when the Huang Ho (Yellow) River in China flooded, where only 44 yr earlier, nearly 1 million people perished in a flood. Flood prevention, flood warning, and flood fighting have greatly reduced the numbers of deaths due to flooding. However, over 500 people lost their lives in 77 separate flood events in the United States in 1971–72, the highest annual number of flood-related casualties in the past four decades.[6] Nearly half of those deaths occurred in one event near Rapid City, South Dakota, on June 9, 1972, when campers were caught in a flash flood. Overall, nearly three-fourths of the flood-related deaths in the United States involve automobiles.

While flood-related fatalities are decreasing on a per capita basis worldwide, flood-related property damage is increasing. The decrease in fatalities is due to better forecasting and warning, while the increase in property damage is largely due to increased density of urban development, much of which is subject to some degree of flooding risk. More than 2000 cities in the United States are located at least partially in a floodplain.[4] Flood damages in the United States have grown from an average of about $2 billion/year in the first part of the 20th century to about $4 billion/yr in the last quarter of the 20th century, with nearly $20 billion in damages estimated for all U.S. floods in 1993.[3]

Over time, people have adapted to flooding with varying degrees of success. The first, and still the simplest, way for humans to avoid flood problems is to minimize their activity in flood plains. However, since some of the world's best land resources, busiest transportation corridors, and most populated built-up areas are adjacent to waterways, abandonment of flood plains is neither likely nor feasible. Other mechanisms for dealing with floods can be categorized as either structural or nonstructural.

Structural Measures to Control Flooding

Structural measures for mitigating the adverse impacts of flooding on humans include dams, dikes, levees and flood walls, channel modifications, diversions, flood proofing, and pumping systems. These measures are intended to reduce the severity, frequency, duration, or geographic extent of flooding by physically altering the flow of water in space and time. Dams retain water for later release when it will not contribute to flood flows. Many of the world's largest dams have been built at least in part to control flooding. Dikes, levees, and flood walls protect property by blocking water from reaching structures (e.g., ring dikes) or raising the river bank to keep higher flows within the flood way. Channel modifications are used to straighten, shorten, or deepen channels to accelerate the flow of water. Flood water diversions or bypass channels may be used to route water around urban or built-up areas where it is not feasible to enlarge the existing channel,[3] for example the Red River flood water bypass around the city of Winnipeg, Manitoba. Flood proofing may involve waterproofing, de-watering, or elevating structures within a flood plain. Pumping systems may be used to remove excess water from low lying areas or from the "wrong side" of dikes and levees when water overtops them or excess runoff occurs.

The success, in physical or economic terms, of structural measures to control flooding has been mixed. Economic effectiveness of structural measures to control floods is most commonly assessed using benefit-cost analysis.[7] Criticisms of economic efficiency analysis include a failure to include all the costs or all the benefits, insufficient time-series data for predicting flood frequencies and severities, and not adequately accounting for the human pain, suffering, and anxiety involved with all types

Fig. 5 Flooding.

of floods. Nonetheless, benefit-cost analysis is a helpful tool to identify and quantify effects and to systematically evaluate a project's feasibility.

Nonstructural Measures to Control Flooding

Nonstructural flood control measures include mechanisms to modify the severity of flooding through runoff retarding land stewardship practices, enhanced flood prediction and warning systems, disaster preparedness, and flood plain awareness and zoning. The human impact of flooding can be mitigated through flood insurance, tax adjustments, flood emergency measures, and post flood recovery assistance.[4]

Flood Fighting

Once a flood is imminent or occurring, various measures are taken to minimize the negative impacts. Flood fighting in larger events is usually led by government or domestic and international NGO relief agencies. Evacuation, rescue, and last minute measures to protect life and property are carried out under emergency conditions. The U.S. Federal Emergency Management Agency (FEMA) and the U.S. Army Corps of Engineers are the principal players in organized flood fighting efforts in the United States, working closely with state and substate government units and NGOs.

Flood Recovery

Disaster relief agencies-public, private, local or international-routinely provide assistance following major flood events by helping to get individuals, businesses, and infrastructure back to normal. However, the time immediately after a major flood is the best time to begin to prepare for the next major flood by providing incentives to discourage rebuilding or relocating in flood prone areas. In recent years, aggressive government buyout of flood prone structures usually followed major flood events in the United States.

CONCLUSIONS

The temporary inundation of normally dry land-flooding-is a natural phenomena that occurs worldwide in spite of ongoing efforts to control it. Floods of all sizes and types can be described in physical terms using flood hydrographs and other fairly basic tools. Flood characteristics are a function of watershed shape, weather and climate, land use and land cover, and man-made alterations.

The human dimensions of flooding—primarily preparation for floods, flood fighting, and flood recovery—have more qualitative aspects than the physical dimensions, which are more quantitative. Floods will never be completely controlled, especially the largest ones. However, the more that is known about floods, climate, and human behavior related to flooding (Fig. 5), the better prepared humans will be to minimize the damages caused by flooding.

ACKNOWLEDGMENTS

The authors appreciate helpful review comments received from Terry Howell, B. A. Stewart, and Daniel Thomas. Thanks to Deborah Tanner for her assistance with the manuscript, including preparation of graphics.

REFERENCES

1. NOVA Online: Dealing with the Deluge, www.pbs.org/wgbh/nova/flood/deluge. (accessed November, 2002)
2. Ward, R. *Floods: A Geographical Perspective*; The Macmillian Press Ltd: New York, 1978.
3. Wurbs, R.A.; James, W.P. *Water Resources Engineering*; Prentice Hall: Upper Saddle River, NJ, 2002.
4. Owen, O.S.; Chiras, D.D.; Reganold, J.P. *Natural Resource Conservation*, 7th Ed.; Prentice Hall: Upper Saddle River, NJ, 1998.
5. USGS Fact Sheet 024-00, March 2000 (ks.water.usgo.gov/kansas/pubs/fact-sheets/fs.024-00.html). (accessed November, 2002)
6. van der Leeden, F.; Troise, F.L.; Todd, D.K. *The Water Encyclopedia*, 2nd Ed.; Lewis Publishers: Chelsa, MI, 1991.
7. Abelson, P. *Project Appraisal and Valuation of the Environment: General Principles and Six Case-Studies in Developing Countries*; St. Martin's Press, Inc.: New York, 1996.

Flow Measurement, History of

James F. Ruff
Colorado State University, Fort Collins, Colorado, U.S.A.

INTRODUCTION

Hydraulic structures existed before recorded history. Archeologists have found irrigation systems in Mesopotamia and check and diversion dams on the Arabian Peninsula dating to about 5800 BC. The first water level records on the Nile River appeared about 3050 BC. The Romans, even though they did not fully comprehend hydraulic principles relating to discharge, devised a method based on pipe areas in order to charge for water supplied to baths and private residences. Hero, a Greek of the first century AD, was the first to express the basis for flow measurement as we know it today. This important finding went unnoticed, however, for about 1500 yr until Leonardo da Vinci extended the relationship to the continuity equation, but even da Vinci's work went unknown until his manuscripts were found in 1690. The German engineer, Reinhard Woltman, developed the spoke-vane current meter in 1790, a breakthrough for measuring velocities in rivers and canals. During the 18th and 19th centuries development and installation of weirs and flumes made flow measurements possible on irrigation canals, and gaging stations were constructed on many rivers to provide records of flows. New technology has provided various water measurement techniques, and stream flow data now can be accessed at over 4200 gaging stations in the United States.

ANCIENT HYDRAULIC STRUCTURES

Hydraulic structures such as diversion dams, irrigation canals, and ditches were conceived and built when humans began farming on arid lands and needed supplemental water to nourish their plants. They used crude implements or sticks to dig ditches, the intake being just a cut in a stream bank. As a stream level dropped and rose over the yearly cycle, stones probably were placed in the stream as a dike or dam to raise the water level. Construction method was trial and error. Early humans developed some intuitive understanding of construction techniques, and of water quantity and application rates, which was passed on generation to generation. This is evident from archaeological studies of Mesopotamian irrigation systems dating back to about 5800 BC.[1,2] On the southern tip of

the Arabian peninsula[3] at about this same time, check dams were constructed in Wadi Shumlya to divert some of the river flow into canals, and these structures represent some of the oldest known water management structures.

Early water management also developed in Egypt. Water for the Nile River depends on runoff from the highlands of east central Africa. The flood reaches Egypt starting about July, peaks about mid September, and recedes until January, providing sufficient water in normal years to produce an ample harvest. Basin irrigation[4], which evolved as a result of this cycle, is a process of building dikes around agricultural fields starting in January to allow rising floodwaters to flow into diked fields. When a flood reaches its peak and begins to recede, openings in the dike are closed and water remains on the fields from six to eight weeks. Since famines could result from improper water levels, timing of inundation of the fields had to be matched with the water level in the river. Basin irrigation was possible for small groups of farmers near the river, but about 3200 BC a strong unified government headed by King Menes expanded the cultivated area by making numerous larger basins between the river and the desert, thus expanding the scope of water management.

No records exist of any attempt to measure water levels, volumes, or flow in all of these early systems until about 3050 BC.[5,6] By then in Egypt, however, water levels were measured on gages (nilometers) at several sites along the Nile River between Nubia and the Nile delta. Rising water levels were observed at the nilometers, and runners carried the information north from station to station. Nilometers had a two-fold purpose: to predict the area of inundated fields, and, thus, a year's harvest; and to establish water level as a tax basis.

In Egypt, by 2600 BC, dams and embankments were constructed for river training, river diversion for land reclamation, flood protection, and irrigation. Evidence of Egyptian skill in rock-filled dam construction may be seen in the remains of the right abutment of the Al Kufra Dam[4] discovered on Wadi Algarawi near Helwan about 30 km south of Cairo and constructed between 2700 and 2600 BC.

In 641 AD the Arabs conquered Egypt and ruled until 1250 AD. During this reign they reestablished nilometers between Aswan and Cairo, some on earlier nilometer sites. One nilometer on the southern tip of the Isle of Roda at

Encyclopedia of Water Science
DOI: 10.1081/E-EWS 120010345

Cairo is the best known of the Arab nilometers and was built in 715 AD. The Roda nilometer consists of a tower constructed with the foundation below the river and three openings in the walls to convey water from the Nile. A measurement pillar was placed in the center of the tower and is shown in Fig. 1. Maximum and minimum Nile water levels were recorded at the Roda nilometer for over 1000 yr until 1890.[7] No hydrologic records on other rivers are comparable. Many present day gaging station stilling wells, consisting of cased well or sump on the riverbank attached to the river with pipes and containing a permanently fixed staff gage to read the water level, are very similar to the Roda nilometer.

HYDRAULIC CONCEPTS RELATED TO FLOW MEASUREMENTS

The Romans devised an early method of flow measurement to be able to charge baths and private homes at a flat rate for a regulated water discharge. The flow from a standardized distributing pipe made of lead originally was taken as the discharge.[8,9] The Romans measured a cross-sectional area of such a pipe and referred to it as a *quinaria*. The quinaria was not a measure of volume but was the capacity of a lead pipe five-fourth digits in diameter flowing constantly under pressure. They believed that the sum of all pipe areas supplied from an aqueduct should equal the cross sectional area of the supply canal. The Romans compared streams of water merely by their cross sectional areas and did not comprehend that velocity of a stream had any part in the quantity of water supplied. It appears the Romans did not fully comprehend hydraulic principles relating discharge, area, velocity, and time, even though they made great advances in distributing water.

Fig. 1   Roda Nilometer on the Nile River at Cairo. (From a lantern slide of T. H. McAllister, Manufacturing Optician, New York, c1900.)

Hero of Alexandria,[7,8,10,11] a Greek, lived sometime in the first or second century AD and was the first to express correctly the relationship for flow by using the time element along with cross-sectional area, velocity, and volume. His description of how to determine the quantity of water a spring can deliver, taken from his book *Dioptra*, became the basis for flow measurement as we know it today.

This important finding was ignored or went unnoticed for about 1500 yr until Leonardo da Vinci (1452–1519) and Castelli (c1577–1644) rediscovered the relationship and extended it to the continuity equation. Leonardo's treatment of hydraulics was almost completely unknown until after 1690, however, when his manuscript on hydraulics was found in a trunk in Rome.

Rouse[10] in *History of Hydraulics* describes many engineers and scientists in the hydraulics field who contributed to an understanding of flow and the need for measurement of discharge and velocity. Robert Hooke (1635–1703) presented a paper on feathering of windmill blades to the Royal Society of Britain and suggested that a similar machine could be used in water. The Italian Giovanni Poleni in 1717 analyzed flow through a rectangular opening extending to a free surface as a series of horizontal strips with the velocity of each assumed proportional to the square root of the distance of the strip from the original free surface. Later the same approach was used to derive the head-discharge relationship for sharp-crested weirs. The basic weir equation given below is often named after Poleni.

$$Q = (2/3)Cb(2g)^{1/2}h^{3/2}$$

In the weir equation, Q is the discharge; C, discharge coefficient; b, width of weir; g, acceleration of gravity; and h, head measured from the crest of the weir to the upstream water surface.

About 1768 the French engineer Antoine Chezy (1718–1798) developed the resistance formula for velocity in a stream, when he was required to determine the cross section and the discharge for a canal to be constructed from the Yvette River to Paris. This formula, widely used in Europe today, relates discharge, Q, to the hydraulic radius, R, the slope of the energy grade line, S, and a dimensional coefficient, C, that depends on the bed roughness or resistance. The equation is:

$$Q = C(RS)^{1/2}$$

A German engineer, Reinhard Woltman (1757–1837), provided a significant contribution to flow measurement when he published a treatise in 1790 describing the application of the spoke-vane type of current meter with a revolution counter to the measurement of river flow. This was a major breakthrough in quantifying flow in rivers and canals and led to the use of stream velocity measurements to establish the relationship between head and discharge.

Robert Manning (1816–1897), an Irish engineer, presented a paper in 1889 to the Institution of Civil Engineers of Ireland on the resistance equation in the form given below that was in better agreement with available data than any relationship then in general use.

$$V = KR^{2/3}S^{1/2}$$

Although Manning did not realize it, this form of the equation was one of two equations Philippe Gaspard

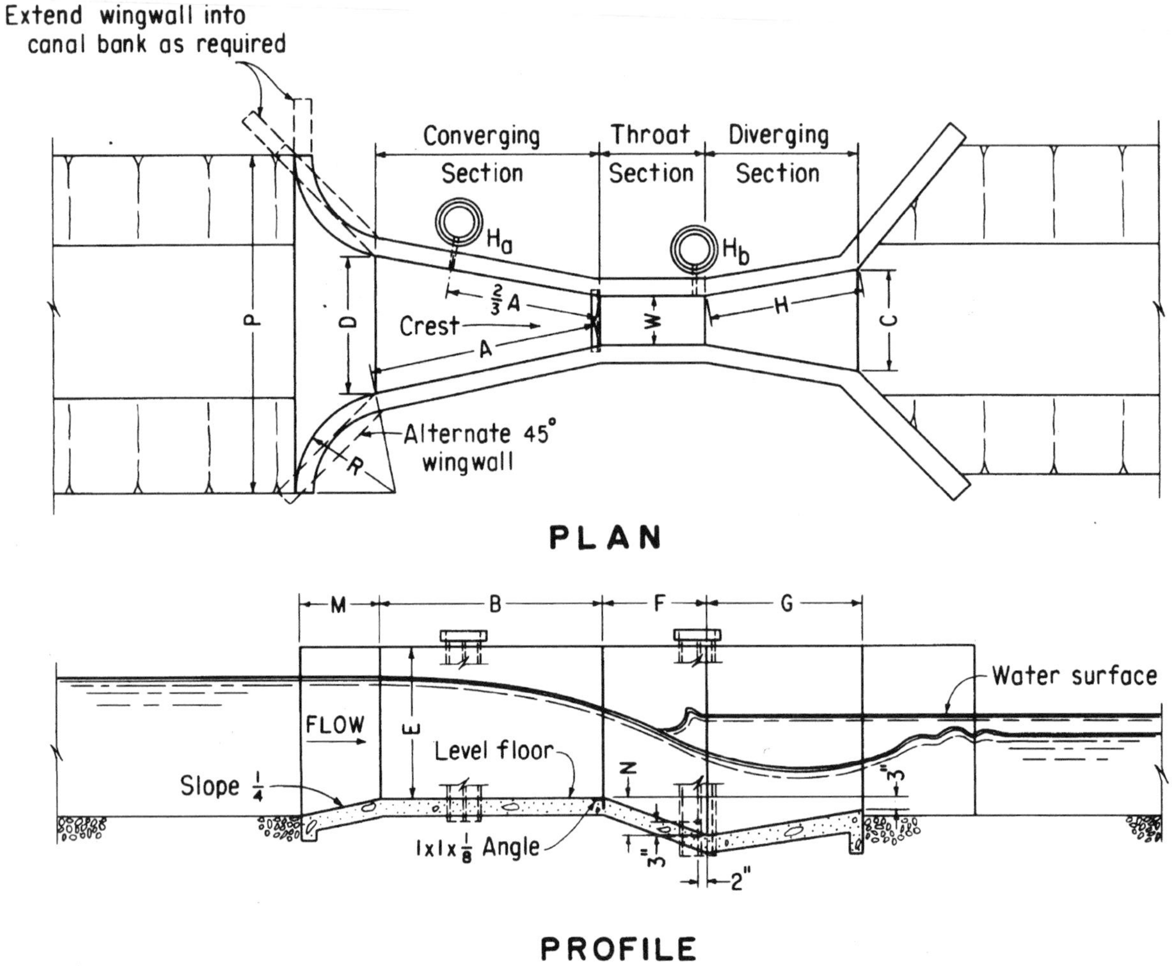

W		A		⅔A		B		C		D		E		F		G		M		N		P		R		FREE-FLOW CAPACITY MINIMUM	FREE-FLOW CAPACITY MAXIMUM
FT.	IN.	FT.	IN.	FT.	IN.	FT.	IN.	FT.	IN.	FT.	IN.	FT.	IN.	FT.	IN.	FT.	IN.	FT.	IN.	FT.	IN.	FT.	IN.	FT.	IN.	CFS	CFS
0	6	2	7/16	1	4 5/16	2	0	1	3 1/2	1	3 5/8	2	0	1	0	2	0	1	0	0	4 1/2	2	11 1/2	1	4	.05	3.9
	9	2	10 5/8	1	11 1/8	2	10	1	3	1	10 5/8	2	6	1	0	1	6	1	0		4 1/2	3	6 1/2	1	4	.09	8.9
1	0	4	6	3	0	4	4 7/8	2	0	2	9 1/4	3	0	2	0	3	0	1	3		9	4	10 3/4	1	8	.11	16.1
1	6	4	9	3	2	4	7 7/8	2	6	3	4 3/8	3	0	2	0	3	0	1	3		9	5	6	1	8	.15	24.6
2	0	5	0	3	4	4	10 7/8	3	0	3	11 1/2	3	0	2	0	3	0	1	3		9	6	1	1	8	.42	33.1
3	0	5	6	3	8	5	4 3/4	4	0	5	1 7/8	3	0	2	0	3	0	1	3		9	7	3 1/2	1	8	.61	50.4
4	0	6	0	4	0	5	10 5/8	5	0	6	4 1/4	3	0	2	0	3	0	1	6		9	8	10 3/4	2	0	1.3	67.9
5	0	6	6	4	4	6	4 1/2	6	0	7	6 5/8	3	0	2	0	3	0	1	6		9	10	1 1/4	2	0	1.6	85.6
6	0	7	0	4	8	6	10 3/8	7	0	8	9	3	0	2	0	3	0	1	6		9	11	3 1/2	2	0	2.6	103.5
7	0	7	6	5	0	7	4 1/4	8	0	9	11 3/8	3	0	2	0	3	0	1	6		9	12	6	2	0	3.0	121.4
8	0	8	0	5	4	7	10 1/8	9	0	11	1 3/4	3	0	2	0	3	0	1	6		9	13	8 1/4	2	0	3.5	139.5

Fig. 2 Standard Parshall flume plan, profile, and dimensions.

Gauckler (1826–1905), a French engineer, had proposed in 1868. In the above equation, K must have the dimension of length to the one-third power over time to be dimensionally homogeneous. Manning himself proposed use of another dimensionally correct equation in place of the previous equation, but it was not widely accepted. When K is replaced by the term $1/n$, the equation is now commonly referred to as the Manning equation and n is a roughness coefficient found empirically. The coefficient, n, is generally given only as a number without dimensional units.

FLOW MEASUREMENT DEVICES FOR DAMS AND HYDRAULIC STRUCTURES

By the end of the 19th century, many hydraulic engineers continued to work on ways to measure large flows more accurately. At this same time, the Western United States was being settled and developed with a great need for irrigation water. In the arid West water is scarce, and natural river flows vary greatly over the course of a year. Prior to 1900, there were only twenty-four large dams in the United States. By 1998, the United States had 75,000 dams over 2 m high with 6375 categorized as large dams (over 15 m high). The need existed for flow measurement structures ranging from those for small irrigation ditches to those for large rivers, and a variety of measurement devices have been developed. Orifices, venturi meters, and magnetic or sonic flow meters can be installed in smaller outlet pipes to measure small flows. These pressure conduit devices are part of the flow from a dam and will not be discussed here, but the flow measurement theory and operation can be found in many books.[12–14] Weirs, flumes, or gaging stations are frequently found downstream from dams where a measurement of all discharges passing over or through the dam can be made. Weirs and

Fig. 3 Rating party stream-gaging on the Cache la Poudre River near Fort Collins, Colorado, 1902.

flumes have been installed for flow measurement in open channel situations, because they are relatively easy to construct and to maintain and provide a satisfactory degree of accuracy. To make discharge measurements using weirs and flumes requires a water depth or head relative to the crest at a point in the flow a short distance upstream from the crest.

Weirs and flumes have significant developmental histories. Weirs have been studied since the 18th century and many different shapes of broad and sharp-crested weirs were studied to determine discharge coefficients for many flow conditions. The entrance to most spillways on dams forms a weir. Most flumes evolved from broad-crested weirs. Flumes with different wall and floor constrictions also have received considerable attention as flow measuring structures. Many books describe weirs and flumes and provide flow equations, discharge coefficients, and/or rating tables relating upstream head to discharge.[14–16] Discharge coefficients[17] for uncontrolled and gate-controlled ogee crests and weirs at the entrance to spillways also have been determined so discharge passing through the spillway can be estimated.

A venturi flume[18,19] shown in Fig. 2, developed by Ralph Leroy Parshall (1881–1959) and named for him, became the standard of irrigation measurement in the 1930s, because it had been extensively tested and laboratory-calibrated for sizes of Parshall flumes with throat widths from 3 in. to 8 ft. No other flumes had been tested so extensively. Parshall flumes have been constructed with throat widths from 10 ft to 50 ft, and information on discharge characteristics can be found in the Water Measurement Manual published by the Bureau of Reclamation.[16] Parshall flumes were used extensively by the Bureau of Reclamation in canals supplied by diversion dams.

The long-throated flume or ramp flume[20,22] developed by John A. Replogle of the U.S. Department of Agriculture often provides economy of construction and more flexible capabilities for open-channel flow situations than other flume types. The simplest long-throated flume consists of a ramp constructed from the channel bed up to a horizontal broad-crested weir. Construction of concrete long-throated flumes requires less forming than for other types of flumes. When installed, these flumes can be computer calibrated within ± 2%. The Bureau of Reclamation and Agricultural Research Service have developed a new computer program,[20,21] WinFlume, to assist in retrofitting existing installations and designing new installations. The ramp flume is beginning to replace the Parshall flume for open channel flow measurements.

Gaging stations, as well, have a developmental history. A gaging station is a point in a canal, river, or stream where numerous current meter traverses have been made to develop a relationship between the measured head and the discharge. Fig. 3 shows a rating party in 1902 stream gaging the Cache la Poudre River. For more detailed information on measurement of stage, area, and velocity, and on equipment and gaging stations, refer to Chapter 1 in the National Handbook of Water Data Acquisitions.[14] The gaging station consists of a stilling well on the bank of the river. The water level in the river is measured in the well that generally is connected to the water in the river by pipes. The station can be a recording or a nonrecording station, and the water level is referenced to a specific datum. At a recording gaging station, a water-stage recorder produces a graphic, punched or printed record of the rise and fall of the water surface in an open channel with respect to time. By comparing the head with the rating curve, a discharge can be established. With the advent of the digital recorders and telemetry systems, stream flow data now can be accessed in real-time at over 4200 stations in the U.S. Geological Survey network.[22]

At nonrecording gaging stations, the water level in a stilling well most often is measured by directly reading a staff gage, a rod or rigid board, precisely graduated and accurately located for scalar measurement. This is not too different than when the flood level of the Nile River was measured by the Egyptians five thousand years ago at the nilometer. Some progress has been made, because now we also can relate the water level to a discharge passing that point in the river.

As the history of water transport, diversion, storage, and measurement evolved, one of the most prominent conclusions to be made is that this evolution has occurred largely in places where water quantity is insufficient. From the aridity of Mesopotamia and Egypt, to the aridity of parts of the Roman Empire and the North American West, when water quantity is insufficient, inventors and engineers of their respective cultures undertake to manage water with great care. Need produces motivation which produces inspiration, a process continuing to this day.

REFERENCES

1. Khuzistan Province in Southwestern Iran Is One of the Sites Thought to Be Where Agriculture Developed, http://www.anthro.mnsu.edu/archaeology/sites/middle_east/khuzistan.html (accessed Sept 2001).
2. Ancient Irrigation in Mesopotamia and Egypt, http://www-geology.ucdavis.edu/~GEL115/115CH17oldirrigation.html (accessed Sept 2001).
3. McCoriston, J. The Place of Ancient Agricultural Practices and Techniques in Yemen Today: Problems and Perspectives, http://www.aiys.org/webdate/peljoy.html (accessed Sept 2001).
4. El-Kady, M.; Abdel-Ghani Seoudi, M.; El-Bahei Esawy, M.; Abdel-Aliem, T. The Nile and History of Irrigation in Egypt; Egyptian National Committee of Irrigation and

Drainage, Ministry of Public Works and Water Resources: Cairo, Egypt, 1995; 51–146.
5. Bell, B. The Oldest Records of the Nile Floods. Geogr. J. 1970, 136 (4), 569–573.
6. Breasted, J.H. Ancient Records of Egypt, First Published 1906; Russell and Russell, Inc.: New York, 1962; Vol. 1, 51–72.
7. Garbrecht, G. Hydrologic and Hydraulic Concepts in Antiquity. In Hydraulics and Hydraulics Research: A Historical Perspective; Garbrecht, G., Ed.; A. A. Balkema: Rotterdam/Boston, 1987; 1–22.
8. Fahlbusch, H. Vitruvius and Frontinus—Hydraulics in the Roman Period. In Hydraulics and Hydraulics Research: A Historical Perspective; Garbrecht, G., Ed.; A. A. Balkema: Rotterdam/Boston, 1987; 23–32.
9. Frontinus, S.J. The Stratagems and the Aqueducts of Rome. With an English Translation by Bennett, C. E., (the Translation of the Aqueducts Being a Revision of That of Hershel, C.), McElwain, M.B., Ed.; Harvard University Press: Cambridge, Massachusetts, 1925; William Heinemann Ltd.: London, 1950; 363–421.
10. Rouse, H.; Ince, S. History of Hydraulics; Dover Publications Inc.: New York, 1963; 8–138.
11. Cohen, M.R.; Drabkin, I.E. A Source Book in Greek Science; McGraw-Hill Book Company Inc.: New York, 1948; 240–241.
12. Bean, H.S., (Ed.) Fluid Meters, 6th Ed.; The American Society of Mechanical Engineers: New York, 1971; 19–236.
13. Howe, J.W. Flow Measurement. In Engineering Hydraulics; Rouse, H., Ed., 6th Printing; John Wiley and Sons, Inc.: New York, 1967; 200–206.
14. U.S. Geological Survey, Office of Water Data Coordination, National Handbook of Recommended Methods of Water Data Acquisition; Prepared Cooperatively by Agencies of the U.S. Government Government Printing Office: Washington, D. C., 1980; Chap. 1.
15. Bos, M.G., (Ed.) Discharge Measurement Structures; International Institute for Land Reclamation and Improvement: Wageningen, The Netherlands, 1976; 15–289.
16. United States Department of the Interior, Bureau of Reclamation, Water Measurement Manual; A Water Resources Technical Publication, 3rd Ed., http://www.usbr.gov/wrrl/fmt/wmm/indexframe.html (accessed Sept 2001); U.S. Government Printing Office: Washington, 1997; 7-1–11-16.
17. United States Department of the Interior, Bureau of Reclamation, Design of Small Dams; A Water Resources Technical Publication, 2nd Ed. (Revised); U.S. Government Printing Office: Washington, 1977; 345–390.
18. Parshall, R.L. The Parshall Measuring Flume; Prepared Under the Direction of W. W. McLaughlin, Chief, Division of Irrigation, Bureau of Agricultural Engineering, United States Department of Agriculture Colorado Experiment Station: Fort Collins, Colorado, 1936; 1–57.
19. Aisenbrey, A.J., Jr.; Hayes, R.B.; Warren, H.J.; Winsett, D.L.; Youong, R.B. United States Department of the Interior. Bureau of Reclamation. Design of Small Canal

Structures, A Water Resources Technical Publication; U.S. Government Printing Office: Denver, Colorado, 1974; 243–297.

20. Wahl, T.L.; Clemmems, A.J.; Replogle, J.A.; Bos, M.G. WINFLUME—Windows-Based Software for the Design of Long-Throated Measuring Flumes, http://www.usbr.gov/wrrl/twahl/winflume-asae.pdf (accessed Sept 2001).

21. Wahl, T.L.; Clemmems, A.J.; Bos, M.G.; Replogle, J.A. The WinFlume Home Page, http://www.usbr.gov/wrrl/winflume/ (accessed Sept 2001).

22. United States Geological Survey; Streamflow Information for the Nation, http://water.usgs.gov/pubs/FS/FS-006-97/ (accessed Sept 2001).

Fluoride

Judy A. Rogers
University of Arkansas, Fayetteville, Arkansas, U.S.A.

INTRODUCTION

Fluoride is an ion of the element, fluorine and is found dissolved in natural waters, commonly in concentrations less than $1.0\,\mathrm{mg\,L^{-1}}$, and seldom outside the range from about $0.01\,\mathrm{mg\,L^{-1}}$ to $10.0\,\mathrm{mg\,L^{-1}}$. Fluoride is incorporated by humans into bone and tooth structure; public health attempts to add low concentrations (less than $1\,\mathrm{mg\,L^{-1}}$) of dissolved fluoride into drinking water to strengthen teeth and minimize cavities (dental caries) has been characterized by rancor and controversy. At fluoride concentrations of $2\,\mathrm{mg\,L^{-1}}$–$4\,\mathrm{mg\,L^{-1}}$, mottling and otherwise aesthetically unappealing tooth discoloration may occur; at greater concentrations, fluoride poisoning, or fluorosis, can cause structural damage to teeth and bone.

SOURCES OF FLUORIDE IN WATER

The major source of fluoride in water is dissolution of minerals, including amphiboles, fluorite, apatite, and mica. Rocks rich in alkali metals, obsidian, volcanic condensates, and volcanic ash are generally higher in fluoride content than most other igneous rocks. Sources ascribe concentrations from $2\,\mathrm{mg\,L^{-1}}$ to $3\,\mathrm{mg\,L^{-1}}$ fluoride in ground water from coastal plain sediments in South Carolina to the dissolution of fluorapatite in fossil sharks' teeth in the aquifer material.

Geochemically, fluoride ions have the same charge and nearly the same radius as hydroxide ions, thereby facilitating the replacement of each other in mineral structures.

The form of fluoride that is most commonly added in water-treatment applications is hydrofluorosilicic acid (HSD), also referred to as fluorosilicic acid. In this aqueous form, the compound is a transparent, water-white to straw-yellow solution. At 60°F, a 25% solution of HSD typically possesses a specific gravity of 1.224 and weighs $10.2\,\mathrm{lb\,gal^{-1}}$.

HISTORY

Dr. H. Trendley Dean identified the beneficial dental health effects of adjusting the level of fluoride in drinking water in 1931. While researching the cause of tooth enamel mottling, Dean "discovered" that in those individuals exhibiting signs of mottling there tended to be a higher than normal background level of fluoride in their drinking water. Consequently, Dean termed this condition "fluorosis." Comparing the prevalence of fluorosis and the incidence of dental caries (cavities), Dean discovered a strong inverse relationship. The greater the level of fluoride in a community's water supply, the lower the incidence of dental caries in the children living there. Realizing the health benefits of fluoride, many public water agencies in the United States have included it as part of their water treatment through a process known as fluoridation.

However, fluoridation has not been accomplished without its fair share of controversy over the years. Since its adoption as a public health measure, fluoridation of U.S. water supplies has met opposition from various groups. As is the case in many controversies, fact and fiction sometimes becomes blurred. For example, since the "Red Scare" associated with the fear of communism in the 1950s and 1960s, there have been groups who have suggested that fluoridation of public water systems is a means of "mass medication." Conspiracy theorists have cited the fact that fluoride was an essential element found in Zyklon B, the infamous "gassing" agent used by the Nazis as a part of their horrific "Final Solution" measure in the Death Camps. While it is true the Zyklon B did contain a derivative of fluoride, there is no scientific evidence to substantiate the claim that HSD can affect any level of "mind control."

Today's opponents of public water fluoridation have also cited the case of unsolicited mass medication. However, mind control is rarely mentioned, especially when highly trained and educated scientists present the arguments. In June 2000, the National Treasury Employees Union, Chapter 280, voiced their opposition to the fluoridation of U.S. public drinking water supplies. This group asserted that the long-term effects of fluoride exposure needed to be further investigated in order to determine whether or not the possibility of toxicity is of issue.

Relying on data gathered over the last half of the 20th Century, the Center for Disease Control and Prevention (CDC) in Atlanta, Georgia has documented the overall

Encyclopedia of Water Science
DOI: 10.1081/E-EWS 120010210

decline in decayed, missing, and filled teeth in children who live in communities that fluoridate their drinking water supplies. During the first half of the 20th Century, American children were plagued by tooth loss due to dental caries. As Dental practices at that time were not as sophisticated as today, many people were beset with extensive tooth loss. Indeed, many young men were rejected for military service during both World Wars because they failed to meet the minimum standard of having six opposing teeth, a problem that is largely unheard of today. Leading dental organizations have attributed the decline in tooth loss due to caries to the fluoridation of public drinking water supplies. According to the latest figures released by the CDC, more than 144 million citizens in greater than 10,000 communities in the United States have access to fluoridated drinking water supplies.

As is the case in most chemicals used in the production of public drinking water, the Environmental Protection Agency (EPA) has assessed an optimal range for fluoride. This range has been set between 0.7 and 1.2 million parts per million (ppm) or milligrams per liter ($mg L^{-1}$). A level of 4 ppm of fluoride in public drinking water sources has been set as the Maximum Contaminant Level by the EPA. In an effort to minimize the possibility of any undesirable effects, the EPA has also established a Secondary Maximum Contaminant Level (SMCL) of 2 ppm. It should be noted that while this SMCL is a nonenforceable limit, suppliers of public drinking water are encouraged to notify the public should that level be exceeded. This is due to the fact that children are more susceptible to the negative effects of any chemical, including fluoride.

In those areas where higher levels of fluoride occur naturally, community water systems are not required to attempt to reduce those levels to what would be considered therapeutic. This is of particular importance in water consumption by infants, toddlers, and small children. Due to their smaller body mass, these individuals are more susceptible to any possible negative effects of fluoridation. The most common negative effect of higher levels of fluoride in drinking water is that of tooth mottling. Only in extremely rare cases have any cases of skeletal fluorosis been seen or reported. Therefore, in those communities where there is a naturally higher level of fluoride in the water, parents are encouraged to substitute bottled drinking water for tap water for infants and children less than ten years of age.

FURTHER READING

Centers for Disease Control and Prevention (CDC). Engineering and Administrative Recommendations for Water Fluoridation. MMWR **1995**, *44*, 1–40.

Centers for Disease Control and Prevention (CDC). Achievements in Public Health 1900–1999: Fluoridation of Drinking Water to Prevent Dental Caries. MMWR **1999**, *48*, 933–940.

Hem, J. Study and Interpretation of the Chemical Characteristics of Natural Water. U.S. Geological Survey Water-Supply Paper 2254, 3rd Ed.; 1992.

Klotter, J. Union of EPA Professionals Opposes Fluoridation. The Townsend Letter for Doctors and Patients, 2001.

Sweeney, R. CDC Releases New Guidelines on Fluoride Use to Prevent Tooth Decay. American Family Physician, Sept 15, 2001.

Zack, A. Geochemistry of Fluoride in the Black Creek Aquifer System of Horry and Georgetown Counties, South Carolina and Its Physiological Implications. U.S. Geological Survey Water-Supply Paper 2067, 1980.

Frozen Soil, Water Movement in

John Baker
United States Department of Agriculture (USDA), St. Paul, Minnesota, U.S.A.

INTRODUCTION

Water movement in freezing and thawing soils can have important physical and physiological consequences. Water movement toward a freezing front can, under the proper circumstances, cause vertical displacement of the soil. This condition, known as frost heave, causes millions of dollars of damage to roads and structures. Water movement in frozen soil is also important during the thawing process; the impaired hydraulic conductivity of frozen soil sometimes causes catastrophic flooding when snow melts rapidly while the soil beneath is still frozen. An understanding of the principles of water movement in frozen soil is helpful in preventing or minimizing the damage associated with these hazards.

THEORY

The hydrodynamics of frozen soils differ from those of unfrozen soils primarily because the hydrostatic relationships are different. The most fundamental difference is that water and heat flow are much more strongly linked in frozen soils. Soil water does not freeze at a single temperature, but rather freezes incrementally over a temperature range. Within this range, water and ice coexist in thermodynamic equilibrium, the proportions of each dependent on temperature, solute content of the water, and retention properties of the medium. As the temperature decreases and more ice is formed, the water potential of the remaining liquid decreases as well. Once ice nucleation has occurred in a freezing soil, the pressure in the liquid phase and the temperature are related through the Clapeyron equation:

$$\frac{p_w - \pi}{\rho_w} - \frac{p_i}{\rho_i} = L_f \frac{T - T_0}{T_0}$$

where p_i and p_w are the gauge pressures within the ice and water phases, ρ_i and ρ_w are the densities of the respective phases, and π is the osmotic pressure of the soil solution. L_f is the latent heat of fusion ($334 \, kJ \, kg^{-1}$), T is the temperature, and T_0 is the temperature at which bulk water freezes, both in K.

Evaluation of this equation reveals that the quantity on the left side has a temperature dependence of approximately $1.2 \, kJ \, kg^{-1} \, K^{-1}$. The osmotic pressure depends on the solute concentration of the soil water, but its temperature dependence is quite small, on the order of π/T. In many cases, and particularly in unsaturated soil, the gauge pressure within the ice phase, p_i, should be negligible. Thus, the change in p_w (more commonly known as matric potential) with respect to T in a freezing soil will be about $1.2 \, MPa \, K^{-1}$. The relationship between the temperature of a frozen soil and its liquid water content is graphically expressed in a freezing characteristic curve, analogous to the moisture characteristic curve that describes water retention in unfrozen soil.[1,2] It has been shown that the moisture characteristic and the freezing characteristic are superimposable for porous media that are completely colloidal, i.e., clay suspensions, where surface tension effects are negligible. For such materials, the liquid water content corresponding to a specific gauge pressure should be the same whether its cause is drying or freezing (ignoring the issue of hysteresis). For media that are devoid of colloids, i.e., pure sands and silts, the rules for similarity are also clear, but different. Here, the ratio of the surface tensions of an air–water interface (σ_{aw}) and an ice–water interface (σ_{iw}) must be taken into account. For materials of this sort, it has been demonstrated that for similar water contents during drying and freezing the pore water pressure will be more negative in the drying soil by a factor of 2.2, the ratio of σ_{aw} to σ_{iw}. This means that at a specific pore water pressure, there will be less liquid water in the frozen soil than in the drying soil. Unfortunately, most soils contain both colloidal and noncolloidal particles, so direct scaling of a freezing characteristic curve from known moisture characteristic data is not possible, and the freezing characteristic must be determined empirically.

REDISTRIBUTION OF WATER DURING FREEZING

The decline in water potential during freezing creates a gradient favoring water flow toward the freezing front. The extent of freezing-induced water movement depends on the balance between heat flow and water flow. If the delivery of latent heat (the product of the water flow rate and the latent heat of fusion) to the freezing front matches

Encyclopedia of Water Science
DOI: 10.1081/E-EWS 120010330

the (sensible) heat flow rate away from it, the downward movement of the freezing front will stall as ice accumulates, filling available pore space. Under the proper circumstances ice can continue to form even after all the pore spaces are filled, resulting in the formation of lenses of pure ice and displacement of the soil above, a process known as frost heave.[3] Frost heave can cause tremendous structural damage to buildings and roadways, and can also harm plants and trees.

Since thermodynamic similarity exists between freezing and drying, i.e., both are functions of pore size, it is often assumed that for similar *liquid* water contents, the hydraulic conductivity of a frozen soil and an unfrozen soil will also be similar.[4] Models that employ this assumption sometimes overestimate water movement during freezing, leading some to posit additional, unspecified impedance to unsaturated flow in frozen soil.[5] Conclusive data remain elusive, due primarily to experimental difficulties, but some generalizations are possible. Coarse-textured, sandy soils, when unfrozen, generally have high saturated hydraulic conductivities, but since their pores drain at gauge pressures close to zero their conductivities decrease dramatically with desaturation. Finer textured soils generally have lower saturated hydraulic conductivities, but since the decrease in water content with declining pore water pressure is more gradual, their conductivities decrease more slowly, so that they can often sustain more water movement in the frozen state than sandy soils. For this reason, they are more prone to redistribution and frost heave during freezing.

Despite the substantial decrease in water potential during freezing, there often is minimal movement of water as the freezing front penetrates. Unless there is a ready supply of water close to the plane of freezing, the soil beneath will soon become desiccated, causing a sharp decrease in hydraulic conductivity, to the point that the delivery of latent heat cannot match the rate of sensible heat loss, so the freezing front moves downward. Thus, initially dry soils may freeze with little or no redistribution of moisture. Consistent with the Clapeyron equation, the largest water-filled pores freeze first, at temperatures closest to 0°C, and as the temperature decreases the water in progressively smaller pores freezes. Even in relatively moist soils, the hydraulic conductivity is often insufficient to support anything more than local redistribution of moisture. This is manifested in ice crystal formation in large pores and cracks, without significant change in water distribution profile at a scale detectable by traditional methods of soil moisture measurement.

INFILTRATION

Infiltration of water into frozen soil is a critical issue, due to the sometimes catastrophic flooding that can occur following snowmelt or rainfall on frozen soil. It is widely accepted that freezing dramatically lowers the infiltration capacity of a soil. This is generally, but not always, true, for reasons alluded to earlier. In wet soils, and soils with water tables near the surface, water movement during freezing fills large pores, and in extreme cases creates lenses of pure ice. Just as the largest water-filled pores are the first to freeze, they are also the last to melt, at temperatures closest to 0°C. These are the pores that are the most important in infiltration, so that infiltration rates are much lower if they are ice-filled. Even in well-drained, unsaturated soils, local redistribution during freezing is often sufficient to fill large pores at the soil surface with ice, retarding subsequent infiltration. However, in drier soils and in coarse textured soils, the large pores can remain air-filled and infiltration rates may approach those measured under unfrozen conditions.[6] Some evidence suggests that snowmelt infiltration in agricultural soils can be improved by the creation of large pores through tillage, either before or during freezing.[7,8]

CONCLUSION

Water movement in frozen soils remains rather less understood than the hydrology of unfrozen soil. A primary problem is the inability to separately measure water and ice contents at the spatial scales necessary to resolve water flow processes without inadvertently affecting them. This situation is exacerbated by the fact that water and heat flow are much more strongly coupled in frozen soil than in unfrozen soil. A clearer picture of the subject depends upon the development and application of innovative experimental methods.

REFERENCES

1. Koopmans, R.W.R.; Miller, R.D. Soil Freezing and Soil Water Characteristic Curves. Soil Sci. Soc. Am. Proc. **1966**, *30*, 680–685.
2. Spaans, E.J.A.; Baker, J.M. The Soil Freezing Characteristic: Its Measurement and Similarity to the Soil Moisture Characteristic. Soil Sci. Soc. Am. J. **1996**, *60*, 13–19.
3. Miller, R.D. Freezing Phenomena in Soils. In *Applications of Soil Physics*; Hillel, D., Ed.; Academic Press: New York, NY, 1980; 254–299.
4. Flerchinger, G.L.; Saxton, K.E. Simultaneous Heat and Water Model of a Freezing Snow-residue-soil System I. Theory and Development. Trans. ASAE **1989**, *32*, 565–571.
5. Lundin, L.-C. Hydraulic Properties in an Operational Model of Frozen Soil. J. Hydrol. **1990**, *118*, 289–310.
6. Granger, R.J.; Gray, D.M.; Dyck, G.E. Snowmelt Infiltration to Frozen Prairie Soils. Can. J. Earth Sci. **1984**, *21*, 669–677.

7. Pikul, J.L.; Aase, J.K. Fall Contour Ripping Increases Water Infiltration into Frozen Soil. Soil Sci. Soc. Am. J. **1998**, *62*, 1017–1024.

8. Van Es, H.M.; Schinddelbeck, R.R. Frost Tillage for Soil Management in the Northeastern USA. J. Minn. Acad. Sci. **1995**, *59*, 37–39.

Furrow Dikes

Ordie R. Jones (Retired)
R. Louis Baumhardt
United States Department of Agriculture (USDA),
Bushland, Texas, U.S.A.

INTRODUCTION

Furrow dikes are small earthen dams formed periodically between the ridges of a ridge-furrow tillage system or, alternatively, small basins created in the loosened soil behind a ripper shank or chisel. The furrow diking practice is known by many names, including tied ridges, furrow damming, basin tillage, basin listing, and microbasin tillage.[1] The dikes or basins store potential runoff on the soil surface, allowing the water to infiltrate (Fig. 1) thus, decreasing storm or irrigation runoff and increasing storage and plant available water in the soil. Furrow diking is a soil and water conservation practice that is adaptable to both dryland and irrigated crop production. It is most often used on gently sloping terrain in arid and semiarid areas where crops are grown under water deficit conditions. This practice has become widely adopted due to new herbicide technologies to control weeds, herbicide tolerant crops, and improved mechanical equipment for constructing the dikes.

HISTORY

Furrow diking was first used on the Great Plains, U.S.A., in 1931 by C.T. Peacock, a wheat farmer at Arriba, Colorado.[1] By the late 1930s, commercial diking equipment was available and furrow diking was practiced extensively in the central Great Plains.[2] Research on the effectiveness of furrow diking for conserving soil and water and increasing crop yields was conducted at several central Great Plains sites, including Colby Kansas,[3] Hayes, Kansas,[4] Woodward, Oklahoma,[5] and at other locations. Most research involved the wheat–fallow rotation, and no consistent increases in yield due to diking were shown. Yield responses were more consistent for systems involving summer row crops.

Concurrent with development of furrow diking in the U.S. Great Plains, the practice was adapted for use in the arid and semiarid tropics, mostly in Africa. Farmers in the cotton (*Gossypium hirsutum* L.) growing regions of Tanzania used hand-tied basins in the 1940s to retain runoff. Research on tied ridges was conducted in Tanzania and Nigeria.[6–8] The U.K. National Institute of Agricultural engineering (NIAE) pioneered the development of mechanized methods of constructing tied ridges in the tropics.[9]

By 1950, the practice of furrow diking on the Great Plains had been abandoned because of the slow operating speed of basin forming equipment, poor weed control, erratic yield responses, and difficulty with seedbed preparation and subsequent tillage.[1] Another factor in the demise of furrow diking was the rapid adoption of stubble-mulch tillage for wheat production in the 1940s and 1950s. Stubble-mulch tillage also leaves the surface flat with crop residues remaining to protect the soil against wind erosion, a prevalent problem in the Great Plains.[2]

A resurgence in furrow diking began in the 1970s and 1980s when diking equipment improved,[10] and herbicides achieved more effective weed control. Favorable responses to furrow diking were obtained with cotton grain sorghum [*Sorghum bicolor* L. (Moench)], and sunflower (*Helianthus annuus* L.).[1,11,12] The furrow diking practice was rapidly adopted by farmers of the Great Plains, and by 1984, an estimated 800,000 ha were being furrow diked, mostly on land cropped to cotton. The practice continues to be widely used with dryland cotton and sorghum, and is used extensively with center pivot irrigation systems to reduce irrigation runoff and to improve the efficiency of irrigation application.

EQUIPMENT

Equipment for constructing dikes or basins ranges from hand hoes and shovels to complex hydraulic motor-tripped mechanical units. Commercially available diking equipment includes the raising shovel, tripping shovel, basin implantation, and "chain" diker types.[13] Currently, the most commonly used equipment is the tripping shovel type, which has one, two, or three paddles that trip when filled with soil, thus depositing the soil and forming a small basin and dike between rows (Fig. 2). Most units trip independently due to the pressure of soil accumulating in

Encyclopedia of Water Science
DOI: 10.1081/E-EWS 120010226

Fig. 1 Runoff of rain is retained by furrow dikes for continued infiltration (right), but this water is lost from undiked (left) fields.

Fig. 2 The most common type of furrow diker is the tripping paddle type, which is often used concurrently with cultivation of ridge till fields after planting.

front of the paddle and work well in loose, mellow, sandy, or loamy soils. Spacing between dikes within the row depends on soil conditions and tractor speed, but a 1 m– 2 m spacing is common.

Furrow diking with the commonly used tripping shovel units is usually performed in conjunction with another tillage operation such as listing, planting, or cultivation in row crop production. Thus, a separate tillage operation is not required, and furrow diking can be performed very economically.[14] Some operators do not construct dikes in traffic furrows, thus facilitating cultivation, spraying, and other cultural operations.

Another type of basin forming equipment, applicable to row cropping with flat tillage, is the Dammer-diker,[a] which uses blades (shovels) mounted on spikes in a wheel-type arrangement to "implant" small reservoirs or basins in loose soil as they rotate behind a ripper or chisel shank. The action of the blades would be similar to inserting a hand shovel into the ground and pivoting the handle forward, thus forming a depression in the soil. This rather intense tillage operation increases infiltration, reduces runoff, and is particularly applicable to crop production on sloping land under sprinkler irrigation.[15]

Another type of basin tillage equipment, applicable to flat tillage for small grain production and to range seeding or renovation, in the "chain" diker has been developed in Australia.[16] This device, called the "Conservation King,"[a] forms basins by using special shaped metal

paddles welded onto links of ship anchor chain, lengths of which rotate between bearings spaced about 5 m apart. In field tests of a 5-m wide unit, the authors found that the equipment performed well on a flat, sweep-plowed field, creating numerous small basins with an estimated surface depression storage capacity of 25 mm. On a no-till fallow field, with consolidated surface soil (clay loam), indentations formed with the chain diker were small and ineffective for water storage.

DRYLAND APPLICATIONS

Crop yield responses to furrow diking are highly variable under dryland crop conditions. When rain was not timely for crop use or was insufficient to produce runoff, the benefits of diking were masked.[17] Negative responses usually result from poor weed control or from poor aeration due to ponding of excess water. The need to reduce runoff must be balanced with the need for surface drainage during wet periods, especially on soils that have low intake or water holding capacity.[18] A possible solution to this problem is to dike alternate furrows. This method proved highly successful in increasing the yield of cotton in Africa.[7]

Cotton responds well to the additional water provided by furrow diking since it is a deep-rooted crop usually grown under water deficit conditions on dryland. In Texas, Gerard et al.[12] reported a 82 mm decrease in storm runoff and a cotton lint increase of 116 kg ha^{-1} (32%) due to furrow diking. Clark[19] reported a 36% increase in cotton lint yield, also in Texas. Increased cotton yield in response

---

[a]The mention of trade or manufacturer names is made for information only and does not imply an endorsement, recommendation, or exclusion by USDA—Agricultural Research Service. Mention of a pesticide neither constitutes a recommendation for use nor it implies registration under FIFRA as amended.

to furrow diking was also demonstrated in Tanzania and Nigeria.[7,20]

Grain sorghum also responds well to runoff conservation with furrow diking. In tests at Bushland, Texas, furrow diking and land leveling were equally effective in preventing runoff and increasing sorghum yield with an annual cropping system. The maximum yield increase due to furrow diking in this six-year study was 2460 kg ha^{-1} and averaged 760 kg ha^{-1}. The environmental and crop management factors that resulted in large sorghum yield responses to furrow diking were: 1) continual (annual) cropping that did not allow the soil water content of the root zone to be replenished during the noncrop period; 2) large rainfall/runoff events that occurred immediately before or early in the sorghum growing season with dikes in place to capture runoff; and 3) limited growing season precipitation that increased reliance on stored soil water.[1]

IRRIGATED APPLICATIONS

Furrow diking can be used with graded furrow and sprinkler irrigation systems. Operators often dike alternate furrows and irrigate the nondiked furrow, thus 50% of the land area can capture and store storm runoff. Stewart et al.[21] developed a limited irrigated-dryland (LID) farming system for the conjunctive use of rainfall and irrigation on graded furrows. The LID system uses a limited water supply to irrigate the upper-half of the field fully, which is fully fertilized and seeded for maximum production. The next quarter of the field has reduced inputs and is managed as a tailwater runoff section, with the lower quarter of the field used as a "sink" to capture and utilize both rainfall and irrigation runoff from the wetter sections of the field. Furrow diking was used to capture precipitation on alternate (nonirrigated) furrows in the fully irrigated and tailwater runoff sections, and to capture and prevent rainfall and irrigation runoff from all furrows in the dryland section. The LID system was not widely adopted by farmers because of the different seeding rates and management requirements of the system, but it used both precipitation and a limited amount of irrigation water very effectively for increased sorghum yield.

The primary use of furrow dikes in irrigated agriculture is to improve water application efficiencies of sprinkler and low energy precision application (LEPA) irrigation systems by reducing or eliminating surface runoff. These irrigation systems are linear or center pivots that use drop tubes with low-pressure orifice-controlled emitters. Water is delivered on to the soil surface over a small area as the system moves through the field in a circular fashion. Required furrow dikes prevent LEPA applied irrigation water from moving down the furrow, thus increasing infiltration and distribution uniformity across the field.

Irrigation water application efficiencies can exceed 95% with the LEPA system.[22] With center-pivot irrigation, an LEPA system requires the furrow diked rows to run in a circular pattern for all growing crops.

CONCLUSION

Furrow diking is a soil and water conservation practice that is versatile and can be adapted to dryland or irrigated crop production. Reasonably priced equipment is available so that furrow diking can be used on most soils and with many crops. Cotton, sorghum, sunflower, and corn have responded well to furrow diking in field tests. Conditions conducive to positive crop responses to furrow diking on dryland are: 1) annual or intensive cropping, 2) large rainfall/runoff events occurring before or early in the growing season; and 3) limited growing season precipitation. Negative crop responses to furrow diking are usually due to poor weed control or to retention of excessive water on the soil surface, which may cause aeration problems or restrict timely planting and tillage.

REFERENCES

1. Jones, O.R.; Clark, R.N. Effects of Furrow Dikes on Water Conservation and Dryland Crop Yields. Soil Sci. Soc. Am. J. **1987**, *51* (5), 1307–1314.
2. Musick, J.T. Precipitation Management Techniques—New and Old. In 8th Annual Groundwater Management District Conference, Lubbock, TX, December 1981; Smith, Ed.; High Plains Underground Water Conservation District No. 1; Lubbock, TX, 1981; 13–19.
3. Kruska, J.B.; Mathews, O.R. *Dryland Crop-Rotation and Tillage Experiments at the Colby (Kansas) Branch Experiment Station*; USDA Circular 979; U.S. Printing Office: Washington, DC, 1956; 51–52.
4. Luebs, R.E. *Investigations of Cropping Systems, Tillage Methods, and Cultural Practices for Dryland Farming at the Fort Hays (Kansas Branch Experiment Station)*; Bull. 449; Kansas Agric. Exp. Stn.: Manhattan, 1962; 32–33.
5. Locke, L.F.; Mathews, O.R. *Relation of Cultural Practices to Winter Wheat Production, Southern Great Plains Field Station, Woodward, Oklahoma*; USDA Circular 917; U.S. Government Printing Office: Washington, DC, 1953; 15–17.
6. Faulkner, O.T. Experiments on Ridged Cultivation in Tanganyika and Nigeria. Trop. Agric. **1944**, *21* (9), 177–178.
7. Lawes, D.A. A New Cultivation Technique in Tropical Africa. Nature (London) **1963**, *198*, 1328.
8. Lawes, D.A. Rainfall Conservation and Yield of Sorghum and Groundnut in Northern Nigeria. Exp. Agric. **1966**, *2*, 139–146.

9. Dagg, M.; McCartney, J.C. The Agronomic Efficiency of the N.I.A.E. Mechanized Tied Ridge System of Cultivation. Exp. Agric. **1968**, *4*, 279–294.

10. Lyle, W.M.; Dixon, D.R. Basin Tillage for Rainfall Retention. Trans. Am. Soc. Agric. Eng. **1977**, *20* (6), 1013–1017.

11. Gerard, C.J.; Sexton, P.D.; Conover, D.M. Effect of Furrow Diking, Subsoiling and Slope Position on Crop Yield. Agron. J. **1984**, *76* (6), 945–950.

12. Gerard, C.J.; Sexton, P.; Clark, L.E.; Gilmore, E.C., Jr. *Sorghum for Grain: Production Strategies in the Rolling Plains*; Bull. 1428; Texas Agric. Exp. Stn.: College Station, 1983.

13. Jones, O.R.; Stewart, B.A. Basin Tillage. Soil. Tillage Res. **1990**, *18* (2–3), 249–265.

14. Harris, B.L.; Krishna, J.H. Furrow Diking to Conserve Moisture. J. Soil Water Conserv. **1989**, *44* (4), 271–273.

15. Longley, T.S. Reservoir Tillage for Center Pivot Irrigation. PNR 84-209. *Proc. Pacific Northwest Region Annual Meeting, Kennewick, WA, September 1984*; Am. Soc. Agric. Eng.: St. Joseph, MI, 1984.

16. Wiedemann, H.T.; Smallacombe, B.A. Chain Diker—A New Tool to Reduce Runoff. Agric. Eng. **1989**, *70* (5), 12–15.

17. Baumhardt, R.L.; Wendt, C.W.; Keeling, J.W. Tillage and Furrow Diking Effects on Water Balance and Yields of Sorghum and Cotton. Soil Sci. Soc. Am. J. **1993**, *57* (4), 1077–1083.

18. El-Swaify, S.A.; Pathak, P.; Rego, T.J.; Singh, S. Soil Management of Optimized Productivity Under Rainfed Conditions in the Semiarid Tropics. In *Advances in Soil Science*; Sterwart, B.A., Ed.; Springer-Verlag: New York, Berlin, Heidelberg, Tokyo, 1985; Vol. 1, 1–64.

19. Clark, L.E. *Response of Cotton to Cultural Practices*; Prog. Rpt. PR4175; Texas Agric. Exp. Stn.: College Station, 1983.

20. Peat, J.E.; Prentice, A.N. The Maintenance of Soil Productivity in Sukumaland and Adjacent Areas, Tanganyika. East Afr. Agric. J. **1949**, *15*, 48.

21. Stewart, B.A.; Musick, J.T.; Dusek, D.A. Yield and Water Use Efficiency of Grain Sorghum in a Limited Irrigation-Dryland Farming System. Agron. J. **1983**, *75* (4), 629–634.

22. Lyle, W.M.; Bordovsky, J.P. Low Energy Precision Application (LEPA) Irrigation System. Trans. ASAE **1981**, *24* (5), 1241–1245.

Global Temperature Change and Terrestrial Ecology

Sherwood B. Idso (Retired)
United States Department of Agriculture (USDA),
Phoenix, Arizona, U.S.A.

Keith E. Idso
Center for the Study of Carbon Dioxide and Global Change,
Tempe, Arizona, U.S.A.

INTRODUCTION

Over the past two centuries, the earth has experienced significant increases in surface air temperature and atmospheric CO_2 concentration, as the planet has recovered from the global chill of the Little Ice Age, and the engines of the Industrial Revolution have burned ever greater quantities of coal, gas, and oil. Many people have imputed a number of negative biological consequences to these environmental changes. However, surveys of the shifting ranges of butterfly and bird species tell a vastly different story, while studies of the net effect of concomitant changes in the air's temperature and CO_2 concentration on plant physiological processes reveal positive consequences as well. In light of these observations, earth's terrestrial ecosystems appear destined to experience increases in stability and biodiversity in areas where they are not adversely affected by the local activities of man.

EARTH IN TRANSITION

Perceived Problems of Global Warming

It has been claimed that earth's temperature throughout the 1990s was higher than it had been at any other time in the past millennium, due largely to an enhancement of the atmosphere's greenhouse effect that is believed by many to have resulted from the historical increase in the air's carbon dioxide (CO_2) concentration.[1] Furthermore, it is repeatedly charged that this change in climate is causing many species of plants to migrate to higher latitudes and altitudes in search of cooler weather. It has also been claimed that the globe is warming at such a rapid rate that it will soon be impossible for much of the world's vegetation to migrate fast enough to avoid extinction; and it is warned that this phenomenon will raise havoc with the planet's ecology and lead to the destruction of much of its biodiversity.[2]

Climatic Complexities

On the surface, these contentions sound plausible. Digging a bit deeper, however, they are found to be highly debatable. With respect to the global warming aspect of the issue, most of the temperature increase the earth has experienced during what we could call the Age of Fossil Fuels did not occur over the past half-century or so, when atmospheric CO_2 concentrations rose most dramatically. Rather, it occurred in the latter part of the nineteenth century and the few decades that followed. Over this time period—which preceded humanity's most prodigious mining and burning of coal, gas, and oil—the earth, on its own, gradually recovered from the global chill of the Little Ice Age, which had not been produced by a decline in atmospheric CO_2 and, therefore, did not require an increase in atmospheric CO_2 to be ameliorated; and these facts suggest that the burning of fossil fuels may not have been the cause of any warming that is evident in the historical record, as has finally been acknowledged by the scientist who set in motion all the concern about the subject several years ago.[3]

There is also a considerable controversy about the precise nature of climate change over the past millennium. In contradiction of the claim that the last decade of the twentieth century was the warmest period of the last thousand years, numerous studies suggest that the Medieval Warm Period of the first part of the millennium—when there was much less CO_2 in the air than there is now—was the warmest,[4] while others contend that the alleged warming of the last two decades of the twentieth century was more virtual than real.[5] Hence, there is by no means any scientific consensus about the climatic significance of the ongoing rise in the air's CO_2 content.

Biological Complexities

Questions about the biological aspects of the issue are even more complex, though not as contentious, as direct experimentation can be employed to investigate most of

Encyclopedia of Water Science
DOI: 10.1081/E-EWS 120010284

the concerns that have been raised. One thing we have learned, e.g., is that it is not just the potential increase in air temperature that could influence the future ecology of the planet; there is also the ongoing rise in the air's CO_2 content, which exerts a number of important influences on the world's vegetation, not the least of which is the documented tendency for elevated levels of atmospheric CO_2 to change the many ways in which plants respond to rising temperatures.

THE MITIGATING ROLE OF CO_2

CO_2–Temperature Interactions

The story begins with the well-established fact that CO_2 is a powerful aerial fertilizer, which when added to the air can substantially increase the vegetative productivity of nearly all plants.[6,7] It continues with the fact that numerous studies have demonstrated that the percent increase in growth produced by an increase in the air's CO_2 content typically rises with an increase in air temperature.[8] In addition, at the species-specific upper-limiting air temperature at which plants typically die from thermal stress under current atmospheric CO_2 concentrations, higher CO_2 concentrations have been shown to protect plants and help them stave off thermal death.[9]

Another effect of atmospheric CO_2 enrichment that influences the biosphere's response to global warming is its ability to increase the species-specific temperature at which plants grow best.[10] Indeed, it has been experimentally demonstrated that the typical CO_2-induced increase in plant optimum temperature is as great as, if not greater than, the CO_2-induced global warming typically predicted by the state-of-the-art climate models.[10,11] Hence, an increase in the air's CO_2 concentration—even if it did have a tendency to warm the earth (which is hotly debated)—would not produce an impetus for plants to migrate to places of cooler air temperature, for they would grow equally well, if not better, in a warmer and CO_2-enriched environment. In seven different studies where this phenomenon was experimentally investigated, in fact, it was found that a 300 part-per-million increase in the air's CO_2 concentration resulted in the rate of net photosynthesis at the greater CO_2-induced optimum plant temperature, which was 5.9°C higher, being nearly twice as great as the rate that prevailed at the reduced CO_2 concentration and lower optimum plant temperature.[11]

Effects on Ecosystem Biodiversity

As a consequence of these observations, we would expect that if the air's temperature and CO_2 concentration rose in unison—as happened globally during the demise of the Little Ice Age and as is happening currently in specific regions of the world—there would be no major changes in the locations of the high-temperature boundaries of the geographical ranges of various plants. The locations of their low-temperature boundaries, however, would clearly be able to move towards higher latitudes and altitudes, which would expand the sizes of their ranges. Hence, with the greater overlapping of ranges that would result, ecosystem plant biodiversity would be expected to increase everywhere. Also, if the herbivores that feed on the plants—and the predators that feed on them—moved with the plants, we would expect to see an increase in the local biodiversity of animals as well, which is, in fact, exactly what is happening in various parts of the world.[12]

In a study of more than half a hundred European butterfly species, for example, Parmesan et al.[13] found that most of them moved northward in response to a regional warming of 0.8°C over the past century. However, in almost all of these northward "migrations," only the northern boundaries of the ranges moved. Furthermore, the northward range expansions did not displace other butterfly species residing in the newly acquired territories, for essentially none of the southern boundaries of any species shifted. Hence, because of the consequent increased overlapping of ranges, butterfly biodiversity must have increased in many areas of Europe over the past century in response to the warming and atmospheric CO_2 increase experienced there.

Moving another step up the trophic ladder of the food chain, Thomas and Lennon,[14] in a study of an equally large number of British bird species, found that from 1970 to 1990 the northern boundaries of species residing in the southern part of Britain shifted northward by an average of 19 km, while the southern boundaries of species residing in the northern part of the country shifted not at all. Consequently, there has been a measurable increase in the overlapping of British bird ranges over the latter part of the twentieth century, along with a concomitant increase in ecosystem biodiversity. Also, in a study of all the passerine (perching) bird species of North and South America, Manne et al.[15] determined that the fraction of endangered species, i.e., those threatened with extinction, drops off significantly as range size increases, which appears to be the result of simultaneous increases in air temperature and atmospheric CO_2 concentration.

CONCLUSION

In view of these real-world observations, there is a strong likelihood that if the air's CO_2 concentration continues to rise as it has in the past, and if air temperature also rises, both ecosystem biodiversity and stability will increase, in contradiction of many simplistic predictions. Perhaps that is why Cowling[16] has stated "we should be less concerned about rising CO_2 and rising temperatures and more worried about the possibility that future atmospheric CO_2 will suddenly stop increasing, while global temperatures continue rising." Clearly, these are areas of deep societal concern, where more research is needed to help clarify the issues for policymakers who are agonizing over what to do (or not do!) about the ongoing rise in the air's CO_2 content.

REFERENCES

1. Mann, M.E.; Bradley, R.S.; Hughes, M.K. Northern Hemisphere Temperatures During the Past Millennium: Inferences, Uncertainties and Limitations. Geophys. Res. Lett. **1999**, *26*, 759–762.
2. Root, T.L.; Schneider, S.H. Can Large-Scale Climatic Models Be Linked with Multiscale Ecological Studies? Conserv. Biol. **1993**, *7*, 256–270.
3. Hansen, J.; Sato, M.; Ruedy, R.; Lacis, A.; Oinas, V. Global Warming in the Twenty-First Century: An Alternative Scenario. Proc. Natl. Acad. Sci. U.S.A. **2000**, *97*, 9875–9880.
4. http://www.co2science.org/subject/medievalwarmperiod.htm (accessed February 2001).
5. http://www.co2science.org/edit/v3 edit/v3n13edit.htm (accessed February 2001).
6. Kimball, B.A. Carbon Dioxide and Agricultural Yield: An Assemblage and Analysis of 430 Prior Observations. Agron. J. **1983**, *75*, 779–788.
7. Poorter, H. Interspecific Variation in the Growth Response of Plants to an Elevated Ambient CO_2 Concentration. Vegetatio **1993**, *104/105*, 77–97.
8. Idso, S.B. *CO_2 and the Biosphere: The Incredible Legacy of the Industrial Revolution*; Department of Soil, Water & Climate, University of Minnesota: St. Paul, MN, 1995; 1–60.
9. Idso, S.B.; Allen, S.G.; Anderson, M.G.; Kimball, B.A. Atmospheric CO_2 Enrichment Enhances Survival of *Azolla* At High Temperatures. Environ. Exp. Bot. **1989**, *29*, 337–341.
10. Long, S.P. Modification of the Response of Photosynthetic Productivity to Rising Temperature by Atmospheric CO_2 Concentrations: Has Its Importance Been Underestimated? Plant Cell Environ. **1991**, *14*, 729–739.
11. Idso, K.E.; Idso, S.B. Plant Responses to Atmospheric CO_2 Enrichment in the Face of Environmental Constraints: A Review of the Past 10 Years' Research. Agric. For. Meteorol. **1994**, *69*, 153–203.
12. Idso, K.E.; Idso, S.B.; Idso, C.D. Atmospheric CO_2 Enrichment: Implications for Ecosystem Biodiversity. Technology **2000**, *7S*, 57–69.
13. Parmesan, C.; Ryrholm, N.; Stefanescu, C.; Hill, J.K.; Thomas, C.D.; Descimon, H.; Huntley, B.; Kaila, L.; Kullberg, J.; Tammaru, T.; Tennent, W.J.; Thomas, J.A.; Warren, M. Poleward Shifts in Geographical Ranges of Butterfly Species Associated with Regional Warming. Nature **1999**, *399*, 579–583.
14. Thomas, C.D.; Lennon, J.J. Birds Extend Their Ranges Northwards. Nature **1999**, *399*, 213.
15. Manne, L.L.; Brooks, T.M.; Pimm, S.L. Relative Risk of Extinction of Passerine Birds on Continents and Islands. Nature **1999**, *399*, 258–261.
16. Cowling, S.A. Plants and Temperature–CO_2 Uncoupling. Science **1999**, *285*, 1500–1501.

Groundwater Arsenic Contamination

Dipankar Chakraborti
Mohammad Mahmudur Rahman
Kunal Paul
Uttam Kumar Chowdhury
Jadavpur University, Calcutta, India

Quazi Quamruzzaman
Dhaka Community Hospital, Dhaka, Bangladesh

INTRODUCTION

Arsenic is an element with atomic number 33 and atomic weight 74.92. It exists throughout the earth's crust and is the 20th abundant element in nature. For centuries, arsenic has been used as a drug and as a poison. Arsenic is thought to exert its toxicity by combining with certain enzymes and thereby interfering with cellular metabolism.

Groundwater arsenic contamination and sufferings of people have been reported in 20 countries in different parts of the world (Fig. 1). The magnitude is considered highest in four Asian countries, and the severity order is Bangladesh > West Bengal—India > P.R. China > Taiwan.

GROUNDWATER ARSENIC CONTAMINATION IN WEST BENGAL—INDIA

West Bengal—India's groundwater arsenic contamination in villages was first reported in 1982, and arsenical skin lesions were first detected in 1983. Twenty-two patients with arsenical skin lesions were known from five villages in four districts.

About 50% of the districts in West Bengal—India reported groundwater arsenic concentration above $50\,\mu g/L$. Six million people are drinking arsenic-contaminated water above $50\,\mu g/L$ from 74 police stations/blocks in 9 arsenic affected districts including a part of Calcutta city in West Bengal (Fig. 2). In 2600 villages/wards, arsenic in groundwater has been found above $50\,\mu g/L$. In a preliminary study from 255 villages, 86,000 people were examined and 8500 people have been registered with arsenical skin lesions. Fig. 3 shows an arsenic patient with severe keratosis. In affected villages, the following skin manifestations and other symptoms of arsenic toxicity were detected—diffuse melanosis;

mucous membrane pigmentation on tongue, gum, and lips; spotted melanosis; leuco-melanosis; spotted and diffuse keratosis; and dorsal and limb keratosis. The following nondermatological complications were also observed in victims suffering from arsenic toxicity—weakness and anemia, muscle pain, nonpetting oedema, conjunctival congestion, laryngitis, myopathy, neurological problem, chronic bronchitis, asthmatic bronchitis, hepatomegaly, splenomegaly, ascitis, and various types of external and internal cancer.

Arsenical skin lesions from nine affected districts of West Bengal affect an estimated 300,000 people.[1] From arsenic-affected areas of West Bengal, over 99,000 water samples from hand tubewells have been analyzed by flow injection hydride generation atomic absorption spectrometry. Fifty-five percent had arsenic concentrations above $10\,\mu g/L$ and 25% above $50\,\mu g/L$. The highest concentration of arsenic found in a hand tubewell was $3880\,\mu g/L$. About 25,000 biological samples (hair, nail, urine, skin scale) have been analyzed from villagers living in arsenic-affected villages (about 40% samples of total 25,000 are from arsenic patients) and on average 80% of the biological samples had arsenic above normal arsenic level in human body. This indicates many more are subclinically affected.

GROUNDWATER ARSENIC CONTAMINATION IN BANGLADESH

Groundwater arsenic contamination and sufferings of people in Bangladesh surfaced in 1995. At that time, there was information about three affected villages in two police stations of two districts (Narayanganj and Faridpur). During the last 7 yr, a tremendous amount of survey work was done to determine the magnitude of the arsenic calamity in Bangladesh. Present survey

Encyclopedia of Water Science
DOI: 10.1081/E-EWS 120010367

NAME OF COUNTRIES

1. POLAND
2. ONTARIO, CANADA
3. NEW ZEALAND
4. SPAIN
5. HUNGARY
6. LANE COUNTY, WESTERN
   OREGON, USA
7. MONTE QUEMODO, CORDOBA,
   NORTH REGION LAGUNERA,
   NORTH MEXICO
8. REGION LAGUNERA,
   NORTH MEXICO
9. TAIWAN
10. ANTOFAGASTA, CHILE
11. LASSEN COUNTY,
    CALIFORNIA, USA
12. SRILANKA
13. NOVA SCOTIA, CANADA
14. FAIRBANKS, ALASKA, USA
15. MILLARD COUNTY, UTAH, USA
16. FALLON, NEVADA
17. INNER MONGOLIAN A.R. CHINA
18. XINJIANG UIGHUR A.R. CHINA
19. BANGLADESH
20. WEST BENGAL, INDIA

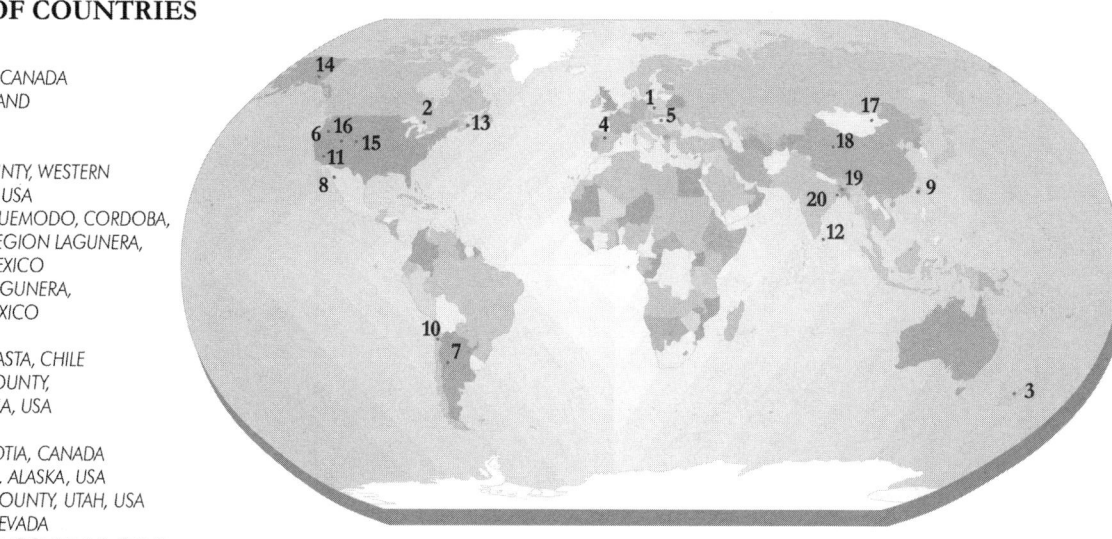

Fig. 1   Shows groundwater arsenic incidents round the world.

reports indicate that 2000 villages in 178 police stations of 50 districts out of total 64 districts in Bangladesh, groundwater contains arsenic above 50 μg/L. Bangladesh comprises four existing geo-morphological regions: 1) Deltaic region (including coastal region); 2) Flood Plain; 3) Tableland; and 4) Hill Tract. Of these four regions, Hill Tract is free of arsenic contamination. Most of the Tableland region is also contamination-free (except Flood Plain deposition on the eroded surface of Tableland). The highly arsenic-contaminated areas of Bangladesh are Deltaic region followed by Flood Plain (Fig. 4). Huge arsenic-free groundwater aquifers remain in selected areas of Bangladesh.[2] Arsenic-contaminated areas of Bangladesh belong to arsenic-bearing holocene sediments. Bangladesh's arsenic calamity is considered the worst in the world. The World Bank and World Health Organization (WHO) described the magnitude of arsenic contamination in Bangladesh.[3] The World Bank's local chief stated that tens of millions of people are at risk from health effects, and that 43,000 of the 68,000 villages are presently at risk or could be at risk in future. According to the prediction of WHO, within a few years, death across much of southern Bangladesh (1 in 10 adults) could be from cancers triggered by arsenic.[3] The area and population of Bangladesh are 148,393 km^2 and 120 million, respectively. Thirty-four thousand hand tubewell water samples from 64 districts in Bangladesh have been analyzed and 56% contained

arsenic above 10 μg/L, that the WHO recommended level of arsenic in drinking water with 37% contained more than 50 μg/L, the WHO maximum permissible limit. Maximum concentration of arsenic found in groundwater of Bangladesh was 4730 μg/L. Overall result shows only 25% and 37% of hand tubewells contain arsenic above 50 μg/L in arsenic-affected areas of West Bengal and Bangladesh, respectively, but there are many villages in West Bengal and Bangladesh where 80%–90% of hand tubewells contain arsenic above 50 μg/L. It has been estimated that at the present time, more than 25 million people in Bangladesh are drinking arsenic-contaminated water above 50 μg/L, and 51 million people are drinking water above 10 μg/L. Analyses of more than 9900 biological samples from arsenic-affected villages of Bangladesh indicate that 95% of samples contain arsenic above normal level. So far in a preliminary survey, over 10,000 people have been identified with arsenical skin lesions from 222 out of 253 villages surveyed for patients. Fig. 5 shows an arsenic patient with squamous cell carcinoma.

SOCIAL PROBLEM AND IGNORANCE

Arsenic poisoning in villages of West Bengal and Bangladesh are causing social problems that are the biggest curse.

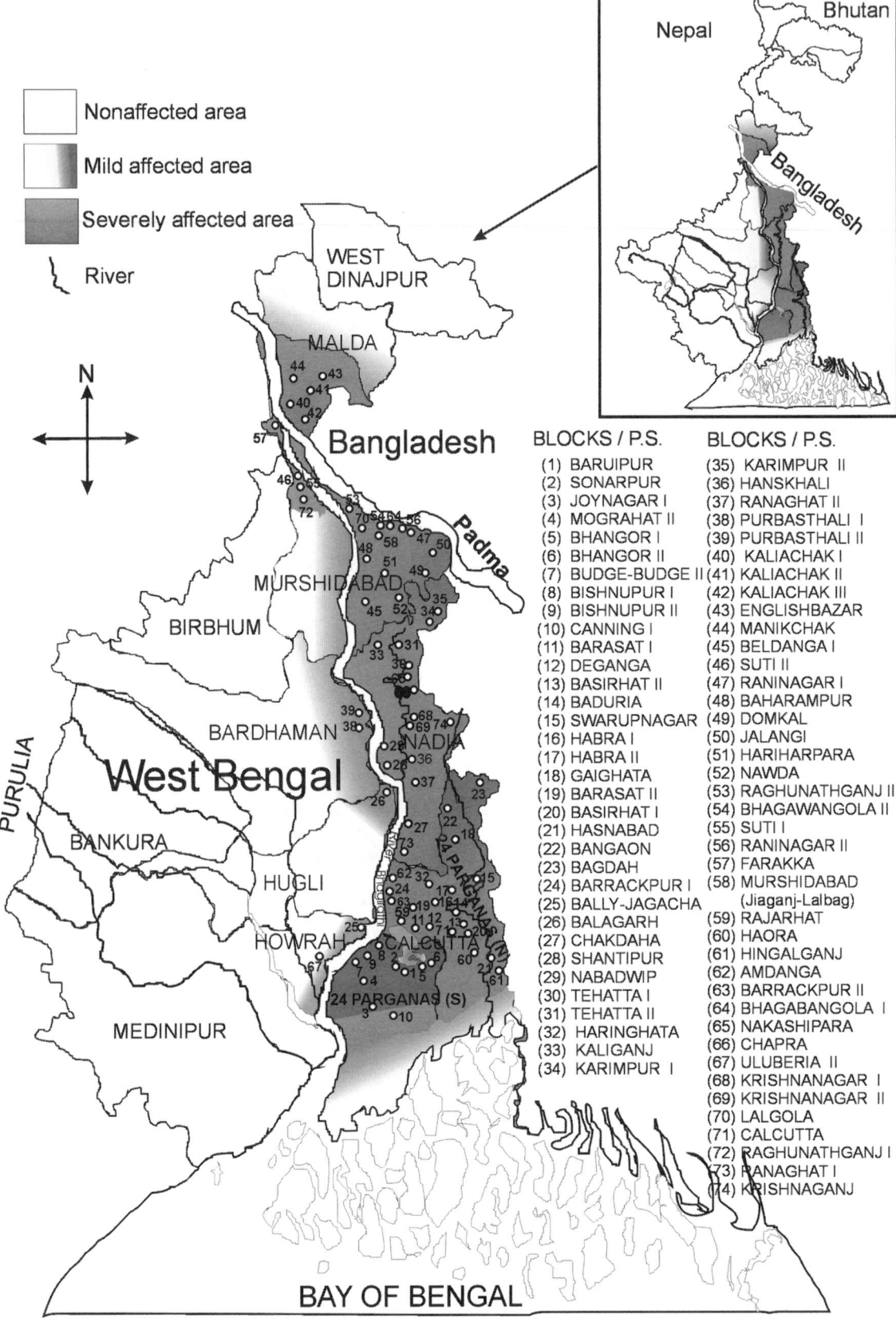

Fig. 2   Map shows the present arsenic affected areas and blocks of West Bengal—India.

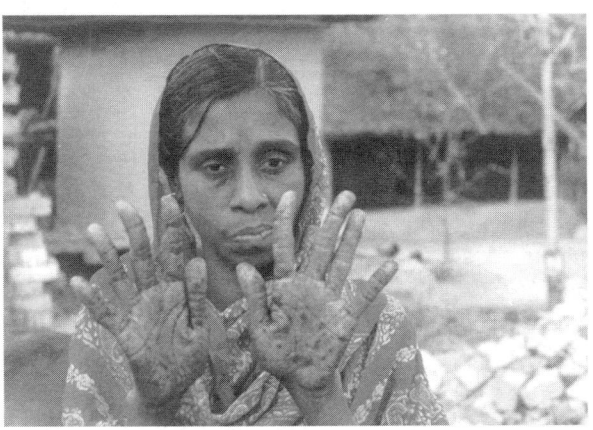

Fig. 3 Shows an arsenic patient with severe keratosis.

The prevailing social problems in the villages are as follows.

1. Due to ignorance, the villagers assume the arsenical skin lesions are a case of leprosy and force arsenic patients to maintain an isolated life or avoid them socially. It is a social curse and human tragedy.
2. Affected wives are sent back to their parents and often with their children.
3. Marriages in the affected villages have become a serious problem because of skin lesions.
4. Jobs/services have been denied/ignored to the arsenic-affected people.
5. When a husband or a wife has been singled out as an arsenic patient, the social problem has increased and destroyed the social fabric.

Most of the people in the affected villages are not aware of the serious consequence of arsenic toxicity. People think arsenical skin lesions are just a single skin disease and will be cured with ointment. Some of them also think the skin lesions are the "Wrath of God or Curse of God." A group also think the skin manifestation is due to the sin committed in their last birth.

SOURCE OF ARSENIC

A single Rural Water Supply Scheme (RWSS) from Malda, one of the arsenic-affected districts of West Bengal—India, is withdrawing 147 kg of arsenic with groundwater in a year and 6.4 t of arsenic is being withdrawn in a year from 3000 shallow, large-diameter tubewells in use for agricultural irrigation in Deganga police station of North 24-Parganas district, West Bengal.[4] It indicates that the source of arsenic is not antropogenic and is geologic. Although the source of

arsenic is believed to be aquifer sediments, the chemistry and mineralogy of the sediments of Ganges–Brahmaputra–Meghna (GBM) delta and arsenic leaching from the aquifer are not well understood. Reports[5,6] show existence of arsenic-rich pyrite in sediments of the delta region of Gangetic West Bengal. A probable explanation of arsenic contamination to the aquifer was predicted due to breakdown of arsenic-rich pyrite that occurred due to heavy groundwater withdrawal (i.e., underground aquifer is aerated and oxygen causes degradation of pyrite, the arsenic rich source). The cause of groundwater arsenic contamination in West Bengal and Bangladesh was also predicted due to reduction of arsenic-rich iron oxyhydroxide in anoxic groundwater.[7,8,9]

Whatever may be the mechanism of arsenic leaching to the aquifer, in West Bengal—India 38,865 km^2 and in Bangladesh 118,849 km^2 are arsenic affected areas, and population in West Bengal—India living in arsenic affected areas is 42.7 million and 104.9 million in Bangladesh. This does not mean that the total population (147.6 million) is drinking arsenic-contaminated water in West Bengal and Bangladesh and will suffer from arsenic toxicity, but it does indicate the risk levels. Our knowledge about long-term effects on those who have stopped drinking arsenic-contaminated water, those drinking contaminated water, and those suffering from arsenical skin lesions is not complete. A limited follow-up study for the last 10 yr indicates that a percentage of those suffering from severe skin lesions are getting internal/external cancers. A future danger to those living in West Bengal—India and Bangladesh is that arsenic is entering the food chain. Of great concern is the huge amount of arsenic applied to agricultural land from contaminated water from hand tube-wells used for irrigation.

HOW TO COMBAT THE PRESENT ARSENIC CRISIS

The mistakes made in the past and that are persisting even today are due to the exploitation of groundwater for irrigation without even trying to adopt effective watershed management to harness the huge surface water resources and rain water. In West Bengal—India and Bangladesh, huge amounts of arsenic-free surface water is in ponds, canals, rivers, wetlands, flooded river basins, and ox-bow lakes. Per capita available surface water in Bangladesh is about 11,000 m^3. West Bengal—India and Bangladesh are known as the land of rivers and have approximately 2000 mm annual rainfall. Instead of using those resources, groundwater is being pumped without proper management. Proper watershed

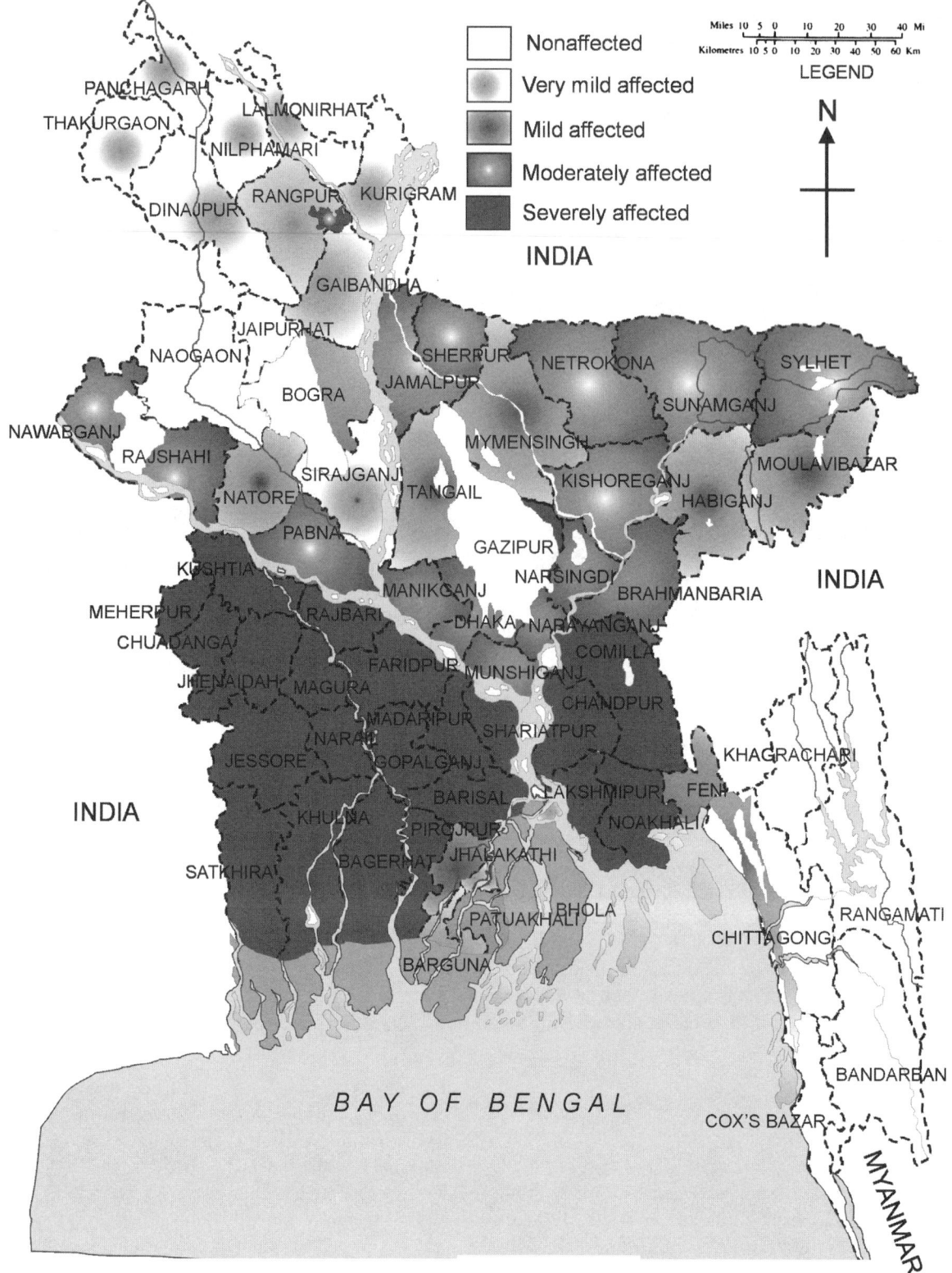

Fig. 4   Map shows the status of arsenic in groundwater in all 64 districts of Bangladesh and in four geo-morphological regions.

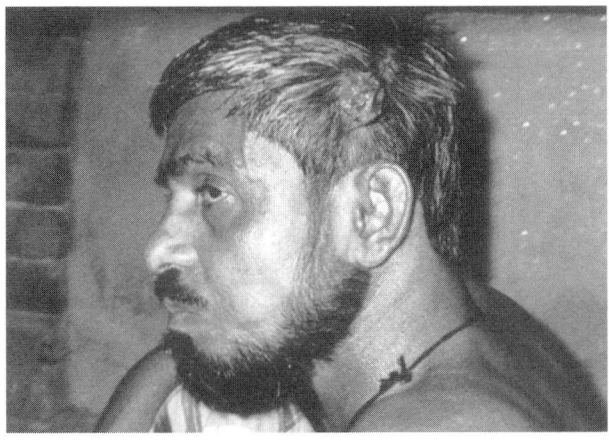

Fig. 5 Shows an arsenic patient with Squamous cell carcinoma on head.

management and villager participation are needed to combat the present arsenic crisis.

REFERENCES

1. Chowdhury, U.K.; Biswas, B.K.; Roy Chowdhury, T.; Samanta, G.; Mandal, B.K.; Basu, G.K.; Chanda, C.R.; Lodh, D.; Saha, K.C.; Mukherjee, S.C.; Roy, S.; Kabir, S.; Quamruzzaman, Q.; Chakraborti, D. Groundwater Arsenic Contamination in Bangladesh and West Bengal, India. Environ. Health Perspect. **2000**, *108* (5), 393–397.
2. Chakraborti, D.; Biswas, B.K.; Basu, G.K.; Chowdhury, U.K.; Roy Chowdhury, T.; Lodh, D.; Chanda, C.R.; Mandal, B.K.; Samanta, G.; Chakraborti, A.K.; Rahman, M.M.; Paul, K.; Roy, S.; Kabir, S.; Ahmed, B.; Das, R.; Salim, M.; Quamruzzaman, Q. Possible Arsenic Contamination Free Groundwater Source in Bangladesh. J. Surf. Sci. Technol. **1999**, *15* (3–4), 180–188.
3. Pearce, F. Arsenic in the Water. The Guardian (UK) 1999, 19/25 Feb, 2–3.
4. Mandal, B.K.; Roy Chowdury, T.; Samanta, G.; Basu, G.K.; Chowdhury, P.P.; Chanda, C.R.; Lodh, D.; Karan, N.K.; Tamili, D.K.; Das, D.; Saha, K.C.; Chakraborti, D. Arsenic in Groundwater in Seven Districts of West Bengal, India— The Biggest Arsenic Calamity in the World. Curr. Sci. **1996**, *70* (11), 976–986.
5. Das, D.; Samanta, G.; Mandal, B.K.; Roy Chowdhury, T.; Chanda, C.R.; Chowdhury, P.P.; Basu, G.K.; Chakraborti, D. Arsenic in Groundwater in Six Districts of West Bengal, India. Environ. Geochem. Health **1996**, *18* (1), 5–15.
6. Roy Chowdhury, T.; Basu, G.K.; Mandal, B.K.; Biswas, B.K.; Samanta, G.; Chowdhury, U.K.; Chanda, C.R.; Lodh, D.; Roy, S.L.; Saha, K.C.; Roy, S.; Kabir, S.; Quamruzzaman, Q.; Chakraborti, D. Arsenic Poisoning in the Ganges Delta. Nature **1999**, *401*, 545–546.
7. Nickson, R.; McArthur, J.M.; Burgess, W.G.; Ahmed, K.M.; Ravenscroft, P.; Rahman, M. Arsenic Poisoning in Bangladesh Groundwater. Nature (Lond.) **1998**, *395*, 338.
8. Bhattacharya, P.; Larson, M.; Leiss, A.; Jacks, G.; Sracek, A.; Chatterjee, D. Genesis of Arseniferous Groundwater in the Alluvial Aquifers of Bengal Delta Plains and Strategies for Low Cost Remediation. International Conference on Arsenic Pollution of Groundwater in Bangladesh: Cause, Effects and Remedies, Dhaka, Bangladesh, February 8–12, 1998; 120.
9. Nickson, R.T.; McArthur, J.M.; Ravenscroft, P.; Burgess, W.G.; Ahmed, K.M. Mechanism of Arsenic Release to Groundwater, Bangladesh and West Bengal. Appl. Geochem. **2000**, *15*, 403–413.

Groundwater Contamination

C. W. Fetter, Jr.

C. W. Fetter, Jr., Associates, Oshkosh, Wisconsin, U.S.A.

INTRODUCTION

Contamination can be defined as the presence of a biological or chemical agent in groundwater in such a concentration that it renders water unfit for a particular use.[1] Agricultural uses of water include domestic drinking water, stock watering, and irrigation. Water that is contaminated for purposes of drinking might be perfectly suitable for use in irrigation.

Contaminants can be from both anthropogenic and natural sources, for example, arsenic. Arsenic found in groundwater in northeastern Wisconsin comes from a naturally occurring mineral, arsenopyrite, present in aquifer. Arsenic has also become a contaminant in groundwater due to use its use in agriculture as a pesticide as well as industrial sites where arsenic was used as a wood preservative.[1] The drinking water standard for arsenic in the United States for many years was 50 μg/L (micrograms per liter). However, as of May 2000 the U.S. Environmental Protection Agency was reviewing the standard and it will most likely be lowered, possibly to as low as 10 μg/L.

TYPES OF CONTAMINANTS

Groundwater contaminants fall into two broad categories, biological and chemical. Biological contaminants include bacteria, viruses, and protozoa. Chemical contaminants can be classified as organic or inorganic. Organic chemicals are based on a framework of carbon and hydrogen atoms. Inorganic compounds include all other chemicals, although some will have carbon present in an inorganic form, such as carbonate (CO_3^{2-}) and bicarbonate (HCO_3^-).

Organic chemicals include fuels and most pesticides. Fuels such as gasoline and diesel are composed of hundreds of different organic chemicals in varying proportions depending upon the source, and their composition will vary depending upon the season. Fuels do not mix with groundwater, rather if present in the ground they will float on the water table. They are sometimes referred to as Light Non-Aqueous Phase Liquid (LNAPL) as they are less dense than water. However, some of the chemicals that comprise gasoline and diesel

will separate from fuel into a dissolved form in the groundwater. The most soluble of these chemicals are benzene, toluene, ethylbenzene, and xylenes. They are referred to by acronym BTEX.[2] Some organic pesticides may be soluble in water as they may be mixed with water prior to application to a field.

Inorganic chemicals found in groundwater are salts that dissociate into cations and anions when in contact with water. The cations include heavy metals such as iron, lead, manganese, cadmium, chromium, zinc, and mercury. The anions include nitrate (NO_3^-), nitrite (NO_2^-), sulfate (SO_4^{2-}), fluoride (F^-), chloride (Cl^-), arsenate (AsO_4^{3-}) and arsenite (AsO_3^{3-}).

SOURCES OF CONTAMINATION

Sources of contamination can be divided into point sources and nonpoint sources. As the name implies, point sources can be traced to a very specific location. An example of a point source might be a septic tank, a landfill, or a pesticide mixing area. Nonpoint sources are dispersed across the landscape. Fertilizer and pesticides applied to fields are examples of nonpoint sources.

Human- and animal wastes are sources of potential groundwater contamination due to the presence of bacteria and viruses as well as nitrogen compounds. One chemical compound frequently found in groundwater in rural areas is nitrate. This can come from cesspools and septic tanks, barnyards, manure spread as fertilizer, and chemical fertilizers. Nitrate and nitrite in drinking water in excess of 10 mg/L (milligrams per liter) as nitrogen have been implicated in infant methamoglobanemia or "blue baby syndrome." Another salt found in animal waste is chloride. This will impart a salty taste in drinking water if present in amounts in excess of 250 mg/L.

Pesticides are also a potential source of groundwater contamination. They can be found concentrated in areas where pesticides are mixed or equipment is washed. Likewise pesticides can be a nonpoint source of contamination when they are spread on a field. For example, atrazine has been found in groundwater in Wisconsin as a result of use on corn crops. Not only can pesticides occur in the environment, but breakdown products called metabolites can also occur in groundwater.

Encyclopedia of Water Science
DOI: 10.1081/E-EWS 120010305

When water is used for irrigation, some will evaporate. This will concentrate the soluble salts in the remaining water, which will drain down to the water table. As a result, toxic salts may build up in the soil and groundwater. This situation has developed in some areas of California with selenium.

Fuels used on the farm can leak from underground storage tanks resulting in the formation of a pool of LNAPL on the water table below the tank and dissolved BTEX chemicals in the groundwater. The federal drinking water standard for benzene in the United States is 5 μg/L. In some states the groundwater standard is even lower, 1 μg/L.

Chemicals used for degreasing equipment can also contaminate groundwater if improperly disposed.[3] Many degreasers contain chlorinated organic compounds such as trichloroethylene (TCE) and 1,1,1-trichlorethane (TCA). These liquids are denser than water and mix poorly with water. They are referred to by acronym DNAPL. If disposed into the environment, for example by spilling on soil, they can migrate vertically to the water table and then sink below the water table into the underlying aquifers. These compounds are sparingly soluble in water, but even small amounts are dangerous. The federal drinking water standard for trichloroethylene in the United States is 5 μg/L. In some states the groundwater standard is even lower, 1 μg/L.

Chemicals used in wood preservatives are also potential groundwater contaminants. These include creosote and CCA (copper, chromium, arsenic). Treated wood itself would most likely not contaminate groundwater, but spilled or improperly disposed wood-treating chemicals could contaminate the groundwater.

EFFECT OF CLIMATE

In humid climates, the water table may be close to the surface and frequent rains can leach contaminants from the soil and transport them down to the water table. If the climate is more arid, contaminants in the soil zone are less likely to be transported to the water table, which itself is likely to be deeper than in a corresponding area that is more humid. However, evaporation of irrigation water in arid climates may result in a build up of soluble salts in the soil and the excess irrigation water that may eventually reach the water table.

TRANSPORT OF CONTAMINANTS

Dissolved contaminants are carried by flowing groundwater through a process called *advection*. If the contaminant is *conservative*, it will move at the same rate as the groundwater in which it is dissolved. An example of a dissolved salt that is conservative is chloride. Water flowing through an aquifer will not all be moving at the same rate. Groundwater moves through pores and cracks in the ground. Some of these openings in the ground are larger than others, and water in the larger openings will be moving faster than water in the smaller openings. As a result, the faster moving water will spread out in front of the rest of the mass of the water. If a contaminant is present in a low concentration, the closer the contaminant gets to the moving front the faster moving water mixes with uncontaminated water. This process is called *longitudinal dispersion*. Through dispersion and diffusion a *plume* of groundwater contamination is formed. This plume is nothing more than a contiguous zone where the contaminant is present in the groundwater. If there is an ongoing source of contamination at the start of the plume, the greatest concentration of the contaminant will be found there and the concentration will decrease in the direction of the groundwater flow. The contaminant plume will extend along the direction of groundwater flow, but also spread sideways through a process called *lateral dispersion*. This is due to the flowing groundwater taking branching pathways.[4]

Nonaqueous phase liquids also have the potential to move through the soil and underlying aquifers. Their movement is dependent upon the ability of the nonaqueous phase liquid to overcome capillary forces and displace air in the pores above the water table and water in the pores of the earth below the water table.

FATE OF CONTAMINANTS

Biological agents are particles of protoplasm. As such they can travel through large pores and cracks in the earth, but not small ones. Fine-grained soils can remove bacteria and viruses by *filtration*, usually within a few hundred meters or less of the source. Some aquifers such as coarse gravel, fractured rock, and carbonate rock have larger openings. Bacteria and viruses can travel for significant distances in such aquifers.

Ionic substances can be removed from groundwater by *ion exchange*. In this phenomenon, ions such as sodium and calcium, which are loosely bound to clay particles can be exchanged for other cations, such as lead, mercury, cadmium, and manganese. The heavy metal contaminants will thus be removed from the groundwater. The ability of a soil to remove contaminants by ion exchange is measured as the ion-exchange capacity of the soil.

Dissolved organic compounds can be removed from groundwater by *adsorption* onto organic matter contained in the soil or rock. The rate of adsorption is inversely proportional to the water solubility of the organic compound. Those that have a low solubility are more tightly bound to the soil organic matter than those that are more soluble. This propensity to be absorbed is measured by a property known as the octanol–water partition coefficient. The other important factor is the percentage of organic matter in the soil. Obviously, the greater the percentage of organic matter, the more of dissolved organic compounds it can absorb.

Finally, many of the dissolved organic compounds can potentially be broken down into simpler compounds by the action of microbes in the soil and aquifer. This process is known as *biodegradation*. The components of petroleum based fuels can be degraded by soil bacteria. The end result is either carbon dioxide or methane, depending upon the presence or absence of dissolved oxygen in the aquifer. BTEX compounds are most readily degraded under aerobic conditions, i.e., with dissolved oxygen present. However, under certain geochemical conditions in the aquifer they can also be degraded in the absence of oxygen, but at a slower rate. Many other organic chemicals

dissolved in groundwater, such as the chlorinated solvents, can be degraded either biologically or abiotically under the right geochemical conditions.[5]

REFERENCES

1. Fetter, C.W., Jr. *Contaminant Hydrogeology*, 2nd Ed.; Prentice-Hall, Inc.: Upper Saddle River, NJ, 1999; 500 pp.
2. Bedient, P.B.; Rifai, H.S.; Newell, C.J. *Ground Water Contamination, Transport and Remediation*, 2nd Ed.; Prentice-Hall PTR: Upper Saddle River, NJ, 1999; 604 pp.
3. Pankow, J.F.; Cherry, J.A. *Dense Chlorinated Solvents*; Waterloo Press: Portland, Oregon, 1996; 522 pp.
4. Charbeneau, R.J. *Groundwater Hydraulics and Pollutant Transport*; Prentice-Hall, Inc.: Upper Saddle River, NJ, 2000; 593 pp.
5. Chapelle, F.H. *Ground-Water Microbiology and Geochemistry*, 2nd Ed.; Wiley and Sons, Inc.: New York, 2000; 475 pp.

Groundwater Law of the Western United States

J. David Aiken
University of Nebraska, Lincoln, Nebraska, U.S.A.

INTRODUCTION

Groundwater law in the United States is a bewildering mix of state court decisions and state statutes. While some generalization is possible, each state's groundwater law is unique.

COMMON LAW STATES

The common law doctrines of absolute ownership, reasonable use, correlative rights, and eastern correlative rights, are based on state court decisions and are implemented through litigation or private negotiation. While prior appropriation was initially adopted in a few western states by court decision, it will be discussed separately as a statutory rather than a judicial doctrine.

Absolute Ownership

The earliest judicial theory of groundwater rights is the doctrine of absolute ownership, also referred to as the English rule. Under the absolute ownership doctrine the landowner is, by virtue of land ownership, considered owner of the groundwater in place, similar to mineral ownership. Thus in absolute ownership jurisdictions, a landowner may pump as much groundwater as he is able to, without regard to the effect of his pumping on neighboring landowners.

The English rule of absolute ownership reflected 19th century judicial observations that the movement of groundwater was unknowable and thus it was unfair to hold a landowner liable for interfering with a neighbor's well when it was not knowable whether the defendant's pumping actually affected plaintiff's well or not. The English rule was once quite popular in the United States, but now only Texas, among the western states still is an absolute ownership jurisdiction.

Reasonable Use

The reasonable use rule, or American rule, was developed in the 19th century. Under the American rule, a landowner is entitled to use groundwater on his own land without waste. If his use exceeds this "reasonable use," he is liable for damages. The American rule is followed in a few eastern states where it is being judicially replaced by the eastern correlative rights doctrine. The reasonable use doctrine is part of the groundwater jurisprudence of Nebraska, Arizona, and California.

Correlative Rights

The California doctrine of correlative rights also initially developed in the 19th century but has continued to develop to this day. Under the correlative rights doctrine, if the groundwater supply is inadequate to meet the needs of all users, each user could be judicially required to proportionally reduce his use until the overdraft ends. The policy significance of correlative rights is that each well owner is treated as having an equal right to groundwater regardless of when first use was initiated.

The correlative rights doctrine is part of the groundwater jurisprudence of California and Nebraska, although its sharing feature has been incorporated into the groundwater depletion statutes of a few other western states as well.

APPROPRIATION STATES

Most western states (except Texas, Nebraska, Arizona, and California) apply the doctrine of prior appropriation to groundwater. This means that the right itself is dependent upon obtaining a state permit rather than simply owning land overlying the groundwater supply. Between groundwater users, priority of appropriation gives the better right. This means that first in time is first in right.

GROUNDWATER RIGHTS

In the common law states, groundwater rights are based upon owning land overlying the groundwater supply and are defined by court decision. In appropriation states, groundwater rights are based upon obtaining a state permit and complying with its terms. In appropriation states, state statutes generally define the extent of groundwater rights.

Encyclopedia of Water Science
DOI: 10.1081/E-EWS 120010121

WELL INTERFERENCE CONFLICTS

Well interference is where the cone of depression of one well intersects with the cone of depression of another well, reducing the yield of both wells. In an artesian aquifer, well interference may occur when the pumping from one well drops the water level below the pumps of another well. Well interference may occur even when there is sufficient water available to supply all users—it may be the result of inadequate wells rather than an inadequate supply. Most groundwater disputes have tended to be well interference disputes.

Common Law States

In absolute ownership states, a landowner is not liable for interfering with a neighbor's well. Thus the neighbor's only recourse is to drill a new well deeper than the neighbor's well. This has been described as "the race to the pumphouse." In reasonable use states, a landowner complaining of well interference is entitled to relief only if the complained-of use is wasteful or not on overlying land. Thus, plaintiffs complaining of well interference have little legal remedy in the absence of gross waste or nonoverlying uses. The courts' definition of what constitutes a wasteful use is rather generous. Arizona courts have defined overlying land to include only the tract of land where the well is located. Nebraska, a reasonable use state, minimizes well interference conflicts between high-capacity wells through statutory well-spacing restrictions. In correlative rights states, competing pumpers have equal rights during shortages.

Appropriation States

In appropriation states, well interference conflicts may be reduced through permit conditions, such as well-spacing restrictions and pumping restrictions. Prior appropriation is primarily a surface water doctrine that has been applied rather uncritically to groundwater. As groundwater problems developed, the principles of prior appropriation were modified to better apply to the groundwater context. Two modifications that were made in response to well interference conflicts are, establishment of reasonable pumping depths and problem area regulations.

Reasonable pumping depths

Sometimes the senior or oldest wells may not be fully penetrating. To allow senior appropriators to insist upon original pumping depths being maintained could seriously constrain groundwater development. Thus several appropriation states do not strictly maintain priority during well interference disputes, but only protect "reasonable pumping depths" through well permit restrictions on pumping. If a senior's well cannot pump at that depth, typically the senior appropriator is responsible for replacing the well at his own expense.

Problem area regulations

In some appropriation states, groundwater development and use has resulted in chronic well interference problems. In some appropriation states, special pumping and development restrictions may be imposed by the state engineer in designated problem areas. Regulations include a ban on new high-capacity wells and pumping restrictions to maintain reasonable pumping depths and reduce interference conflicts.

GROUNDWATER DEPLETION

Safe Yield

Groundwater depletion may be defined as the situation where average annual withdrawals from the aquifer exceed average annual recharge. This is sometimes referred to as groundwater overdraft. Overdraft is significant in the Ogallala aquifer region, including Nebraska, Kansas, Colorado, Texas, and New Mexico, as well as in California and Arizona. The amount of water that may be safely withdrawn without leading to long-term aquifer depletion is sometimes referred to as the "safe-yield" amount.

Common Law Doctrines

Of the overlying rights doctrines, only correlative rights doctrine addresses depletion. Pumpers can completely ignore depletion in absolute ownership states, and need be concerned about depletion only to the extent their uses are wasteful or nonoverlying in reasonable use states. In eastern correlative rights states, courts can apportion water between competing users. However, Florida (a permit state) is the primary eastern state with significant groundwater depletion concerns.

In theory courts in correlative rights states can limit withdrawals to the aquifer's safe yield, thus preventing depletion. In practice, in California safe-yield adjudications are used primarily to define baseline pumping rights so that groundwater recharge agencies can charge pumpers a pumping fee for using more than their safe-yield allocation.

Problem Area Regulations

In western states, the most common way to deal with depletion is to establish special problem area regulations. Once the problem area has been administratively defined, typically no new high-capacity wells may be drilled within the problem area. Less frequently are the uses of existing appropriators limited, a significant policy failing. Initial groundwater appropriation allocations are typically generous, not requiring a high degree of water use efficiency. Where problem area allocations have been established, they typically are high enough to allow current irrigation practices to be maintained with little or no change. Any changes in irrigation management typically come only as well yields decline.

CONJUNCTIVE USE

In California, the courts have recognized the rights of entities storing water underground to control the use of that water. As a result, when groundwater pumpers have received their safe-yield allocation through a court adjudication, they typically are required to pay a fee to the recharge entity for pumping water stored underground, i.e. for pumping groundwater in excess of their safe-yield allocation. Where both surface water and groundwater are available to groundwater pumpers, the recharge entity can raise or lower groundwater pumping fees to encourage surface water use during periods of ample surface supplies, or to discourage surface water use during periods of surface water shortage.

SURFACE–GROUNDWATER INTERFERENCE

Where ground and surface water supplies are hydrologically connected, courts typically have followed the "underground stream" doctrine to interrelate surface and groundwater rights of use. This means that wells will be treated as surface diversions and governed by surface water law. In the West, priority would govern surface–groundwater disputes (except in Nebraska). Under the "Templeton" doctrine, the New Mexico State Engineer has required a junior groundwater appropriator to purchase and retire sufficient surface appropriations to compensate for the expected stream depletion effect of his proposed well. Colorado has an elaborate system for integrating surface appropriations and appropriations of subflow and tributary groundwater. Generally junior groundwater appropriators are expected through plans of augmentation to compensate the stream for their expected stream depletion effects of well pumping.

FURTHER READING

Ashley, S.A.; Smith, Z.A. *Groundwater Management in the West*; University of Nebraska Press: Lincoln NE, 1999.
Tarlock, A. D. *Law of Water Rights and Resources*; West Group: St Paul MN, 2002; Chap. 4, 6.

Groundwater Levels, Mapping

Marios Sophocleous
University of Kansas, Lawrence, Kansas, U.S.A.

INTRODUCTION

Maps of groundwater levels are used to estimate groundwater flow direction and velocity, to assess groundwater vulnerability, to locate landfills and wastewater disposal sites, and as input to hydrologic and pollutant transport models. Because groundwater is hidden from view beneath the land surface, groundwater can only be directly observed through monitoring wells. However, because these observations are limited to specific points, mapping groundwater levels requires hydrogeologically appropriate techniques to generalize the point measurements. Rules or models for spatially and temporally generalizing monitoring (sample) data across the groundwater system are inherent and essential to hydrogeologic science. Our understanding of ground water is the product of a long history of hypothesis and model development, testing, and refinement.[1]

The position of the water table is the product of a wide range of static and dynamic environmental conditions and processes affecting the rate at which water enters and leaves the saturated zone of the aquifer. The water table rises if the rate of water added (recharge) exceeds the rate of water leaving (discharge); conversely, the water table falls if discharge exceeds recharge. The water-table surface is therefore not static, nor flat (as the name implies), but responsive to climatic, vegetative, geomorphic, and geologic conditions.

As Matson and Fels[1] also pointed out, traditional water-table mapping uses graphical methods to interpolate between water-table measurements and hydrogeologic boundaries, with professional judgment and experience filling the gaps in sampling. Computer assisted approaches may incorporate surface mapping methods such as trend surface interpolation and *kriging*;[2] many of these tools are currently provided in Geographic Information Systems (GIS) software. Other methods employ mathematical modeling to predict water-table elevation from hydrogeologic conditions and processes.

DESIGNING A MONITORING SYSTEM

Setting up a monitoring system requires careful consideration of both the hydrogeologic setting and the data needed. It is premature and wasteful to locate monitoring wells without first synthesizing what is known about the setting—in other words, without formulating a sound conceptual model of the system under study. For example, water-supply wells drilled without understanding area hydrogeology may be placed where 1) the aquifer is thin or missing altogether, 2) the aquifer is present but not very productive, or 3) the aquifer contains water of poor quality.[3]

Areas in which the geology is highly variable require more extensive (and costly) water-level monitoring systems than comparatively more homogeneous areas. The degree of geologic complexity is often not known or appreciated during the early phases of a testing program, and it may require several stages of drilling, well installation, water-level measurement, and analysis of hydrogeologic data before the required understanding is achieved. Due to space limitations, the design for an optimal spacing of groundwater-level monitoring wells cannot be covered in this article; however, the reader is referred to Refs. [4,5] for examples of such an observation well network design.

NATURAL PROCESSES CAUSING GROUNDWATER-LEVEL FLUCTUATIONS

To interpret the monitored water levels, one needs to understand the various processes causing fluctuations in groundwater level. These are the effects of hydrologic processes active in the atmosphere, land surface, and subsurface, the groundwater movement in hydrodynamic flow systems, groundwater recharge and discharge processes, atmospheric pressure changes, plant transpiration, aquifer compression and dilation, and others.

In addition to natural processes, human activities also cause groundwater-level fluctuations. Major among them are: 1) groundwater withdrawals from wells; 2) artificial recharge; 3) irrigation; 4) land clearing; 5) pumping of hydrocarbons and brine from reservoirs; 6) construction of water reservoirs; 7) mining; and 8) loading and unloading by heavy equipment, such as freight trains.

Encyclopedia of Water Science
DOI: 10.1081/E-EWS 120010085

ANALYSIS, INTERPRETATION, AND PRESENTATION OF WATER-LEVEL DATA

Primary uses of groundwater-level data are to understand and predict water-level changes and to assess the direction of flow beneath an area. The usual procedure is to plot the location of wells on a base map, convert the depth-to-water measurements to elevations, plot the water-level elevations on the base map, and then construct a groundwater elevation contour map. Constructing a water-level change map, as will be explained later on (see section on "Examples of Groundwater-level Data Interpretation"), will indicate the extent and severity of water-level declines resulting from a variety of factors, including human development and droughts. The direction of ground-water flow is estimated by drawing ground-water flow lines perpendicular to the ground-water elevation contours (Figs. 1 and 2) if the aquifer can be considered *homogeneous* and *isotropic*.

The relatively simple approach to estimating ground-water flow directions described above is suitable where wells are screened in the same zone and the flow of groundwater is predominantly horizontal. However, as attention has focused on detecting the subsurface position of contaminant plumes or predicting possible contaminant migration pathways, this simple approach has been shown to be not always valid.[6] Increasingly, flow lines shown on vertical sections are required to complement the planar maps showing horizontal flow directions to illustrate how groundwater is flowing either upward or downward beneath a site.

Groundwater flows in three dimensions, and as such can have both horizontal and vertical (either upward or downward) flow components. The magnitude of either the horizontal or the vertical flow component and the direction of groundwater flow are dependent on several factors: recharge and discharge conditions, aquifer heterogeneity, and aquifer anisotropy. Dalton et al.[6] summarized these factors, and the following draws on their summary.

In recharge areas, groundwater flows downward (or away from the water table), whereas in discharge areas groundwater flows upward (or toward the water table). Groundwater migrates nearly horizontally in areas where neither recharge nor discharge conditions prevail. For example, in Fig. 1 well cluster A is located in a recharge area, well cluster B is located in an area where flow is predominantly lateral, and well cluster C is located in a discharge area.[7] Note in Fig. 1 that wells located adjacent to one another, but finished at different depths, may display different water-level elevations.

In a heterogeneous aquifer, hydrogeologic properties are dependent on position within a geologic formation,[8] and thus the geology needs to be considered in evaluating water-level data. While recharge or discharge may cause vertical gradients to be present within a discrete geologic zone, vertical gradients may also be caused by the contrast in hydraulic conductivity between aquifer zones. This is especially evident where a deposit of low hydraulic conductivity overlies a deposit of relatively higher hydraulic conductivity.

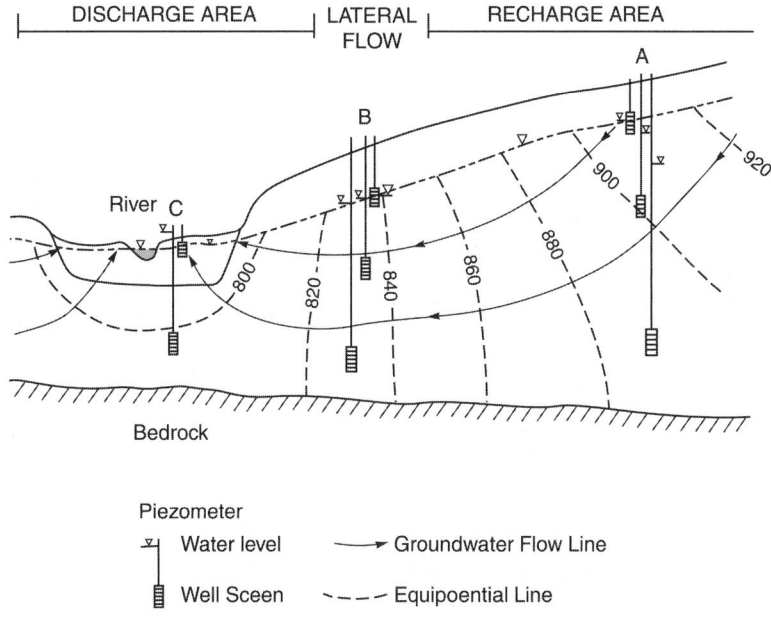

Fig. 1 Ideal flow system showing recharge and discharge relationships (adapted from Saines, 1981[7]).

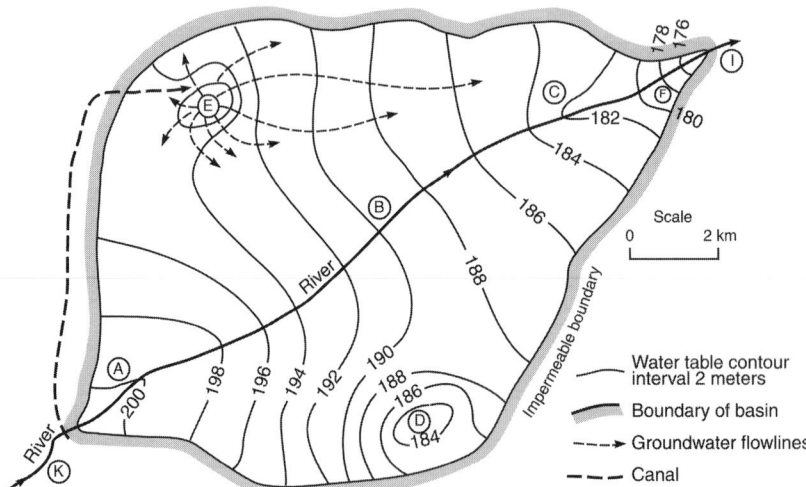

Fig. 2 Contour map of the water table in a small hypothetical groundwater basin. If the aquifer is homogeneous and isotropic and if the slope of the water table is not large, the map can be used to construct a flow net. A small number of flowlines (shown as dash lines) have been drawn on the map. Excessive convergence of the flow lines suggests a changing transmissivity of the aquifer (from Ref. [9]).

Aquifer anisotropy refers to an aquifer condition in which aquifer properties vary with direction at a point within a geologic formation.[8] For example, many aquifer zones were deposited in more or less horizontal layers, causing the horizontal hydraulic conductivity to be greater than the vertical hydraulic conductivity. In anisotropic zones, where the horizontal component of hydraulic conductivity is higher than the vertical one, flow will be restricted to higher elevations compared to an equivalent flow system in isotropic zones showing the same water-level conditions.

The practical significance of the three factors discussed earlier is that groundwater levels can be a function of either well-screen depth or of well position along a groundwater flow line or, more commonly, a combination of the two.[6] For these reasons, considerable care needs to be taken in evaluating water-level data.

Interpreting Water-Level Data

Dalton et al.[6] also summarized the various steps in groundwater-level data interpretation. The first step in interpreting groundwater-level data is to make a thorough assessment of the site geology. The vertical and horizontal extent and relative positions of aquifer zones and the hydrologic properties of each zone should be determined to the fullest extent possible. It is extremely important to have as detailed an understanding of the site geology as possible. Detailed surficial geologic maps and geologic sections should be constructed to provide the framework to interpret data on groundwater levels.

The next step in interpreting these data is to review monitoring wells with respect to screen elevations and the various zones in which the screens are situated. The objective of this review is to identify whether vertical hydraulic gradients are present beneath the site and to determine the probable cause of the gradients.

Once the presence and magnitude of vertical gradients and the distribution of data with respect to each zone are established, the direction of groundwater flow can be assessed. If the geologic system is relatively simple and substantial vertical gradients are not present, a planar groundwater elevation contour map can be prepared which shows the direction of groundwater flow. However, if multiple zones of differing hydraulic conductivity are present beneath the site, several planar maps may be required to show the horizontal component of flow within each zone (typically the zones of relatively higher hydraulic conductivity) and vertical sections are required to illustrate how groundwater flows between each zone.[6] The presence of vertical gradients can be anticipated in areas where sites are underlain by a layered (heterogeneous) geologic sequence, especially where deposits of lower hydraulic conductivity overlie deposits of substantially higher hydraulic conductivity; or are located within recharge or discharge areas.

Site activities can modify local conditions to such an extent that groundwater flows in directions contrary to what would be expected for "natural" conditions. For example, drainage ditches can modify flow within near-surface deposits, and facility-induced recharge can create local downward gradients in regional discharge areas.[6]

As mentioned previously, groundwater flow directions and water levels are not static and can change in response to a variety of factors, such as seasonal precipitation, irrigation, well pumping, changing river stage, and fluctuations caused by tides. Fluctuations caused by these factors can modify, or even reverse, horizontal and vertical flow gradients and thus alter groundwater flow directions.

Contouring of Water-Level Elevation Data

Typically, as Dalton et al.[6] also outlined, groundwater flow directions are assessed by preparing groundwater elevation contour maps. Water-level elevations are plotted on base maps and linear interpolations of data between measuring points are made to construct contours of equal elevation (Fig. 2). These maps should be prepared using data from wells screened in the same zone, where the horizontal component of the groundwater flow gradient is greater than the vertical gradient. The greatest amount of interpretation is typically required at the periphery of the data set. A reliable interpretation requires that at least a conceptual analysis of the hydrogeologic system be made. The probable effects of aquifer boundaries, such as valley walls or drainage features, need to be considered.

Computer contouring and statistical analysis (such as kriging) of water-level elevation data are becoming more popular. These tools offer several advantages, especially for large data sets. However, the approach and assumptions that underlie these methods should be thoroughly understood before they are applied, and the computer output should be critically reviewed. The most desirable approach would be to interpret the water-level data using both manual and computer techniques.[6] If different interpretations result, then the discrepancy between the interpretations should be resolved by further analysis of the geologic and water-level data.

Examples of Groundwater-Level Data Interpretation

Several common errors in interpreting and contouring groundwater-level data are summarized by Davis and DeWiest.[9] Fig. 2 presents a number of water-table configurations related to common geologic or hydrologic causes. Area A is an area of recharge within an alluvial fan where the surface is 24 m above the water table. Here the stream continually loses water to the permeable substrata. Streams with this relationship to the water table are called *influent* or *losing* streams. In such cases, ground-water contours form a V, pointing downstream when they cross a losing stream. At point B, the water in the stream is at the same elevation as the water table. The water-table contour is normal to the stream at this point because there is no flow from the stream and groundwater flowlines are therefore tangent to the direction of the stream. At C the surface of the stream is below the water table, and the stream receives groundwater discharge. At C the stream is called an *effluent* or *gaining* stream. Groundwater contours

bend upstream when they cross a gaining stream. At F the stream is still an effluent stream, but most of the groundwater has already been discharged into the stream so the contours no longer bend sharply upstream.[9] Point D is an area of heavy pumping in which the water has been lowered to 6 m below the stream level at B. After a short period, the pumping at D should make the contours shift so the river will be influent at B. Area E is an area of recharge in which surplus irrigation water has produced a ground-water mound 3 m above the stream surface at B. The stream at K and I is flowing in an impervious channel. The difference between the discharges at K and I is equal to the water lost or gained within the ground-water basin.

Common mistakes in mapping groundwater levels are a failure to distinguish between the water levels of different aquifers and to identify wells that have contact with more than one aquifer (Fig. 3). If the area is one of complex stratigraphy or structure, the data should be interpreted with maximum use of geologic information. Similar problems occur if observation wells completed at different depths in recharge and/or discharge areas are all combined to produce a groundwater elevation contour map. In such areas, vertical flow components are significant (Fig. 1) and water levels in wells completed at different depths will be at different elevations. In such cases, only shallow wells screened at or near the water table should be used for constructing water-table maps.

Surface-water features such as springs, ponds, lakes, streams, and rivers can interact with the water table. In addition, the water table is often a subdued reflection of the surface topography. All this must be taken into account when preparing a water-table map.[10] A base map showing the surface topography and the locations of surface-water features should be prepared. The elevations of lakes and ponds can be helpful information. The locations of the wells are then plotted on the base map, and the water-level elevations are noted. The datum for the water level in wells should be the same as the datum for the surface topography. Interpolation of contours between data points is strongly influenced by the surface topography and surface-water features. For example, groundwater contours cannot be higher than the surface topography. The depth to groundwater will typically be greater beneath hills than beneath valleys. If a lake is present, the lake surface is flat and the water table beneath it is also flat.[10] Hence, groundwater contours must go around it (Fig. 4a). The only exception to this rule is when the lake is perched on low-permeability sediments and has a surface elevation above the main water table.[10] Mistakes in constructing water-table maps are often associated with purely mechanical extrapolation of contours between measured water levels. The water table thus can be placed mistakenly above the land surface

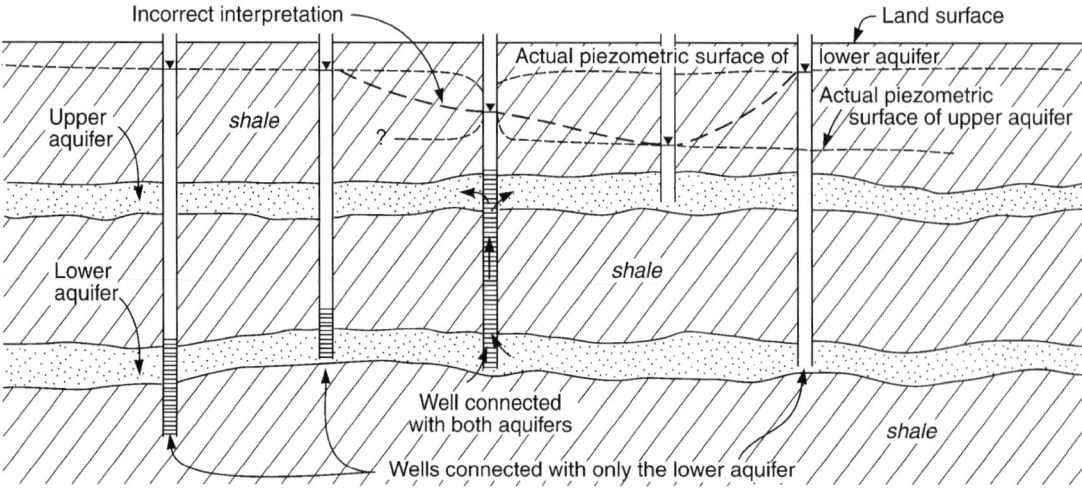

Fig. 3 Observation wells in a region having two confined aquifers under separate pressures. Correct interpretation of water levels is almost impossible unless details of well construction are known (from Ref. [9]).

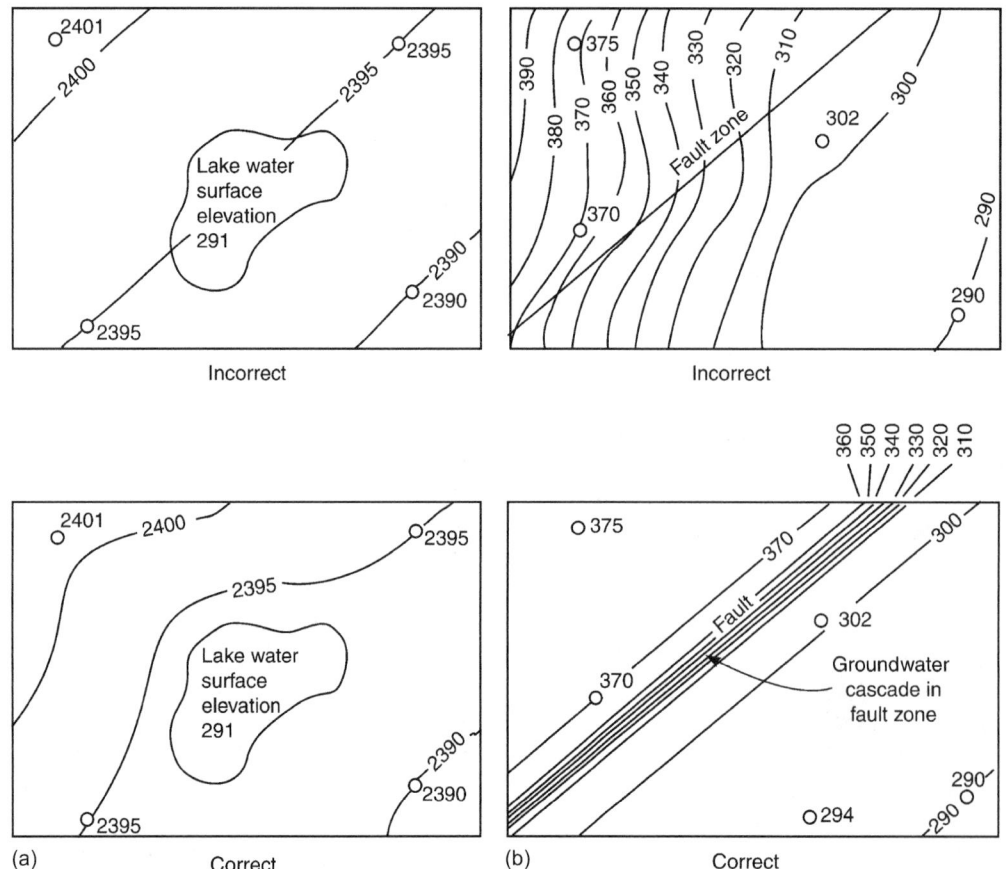

Fig. 4 Common errors encountered in contouring water-table maps in areas of (a) topographic depressions occupied by lakes, and (b) fault zones (from Ref. [9]).

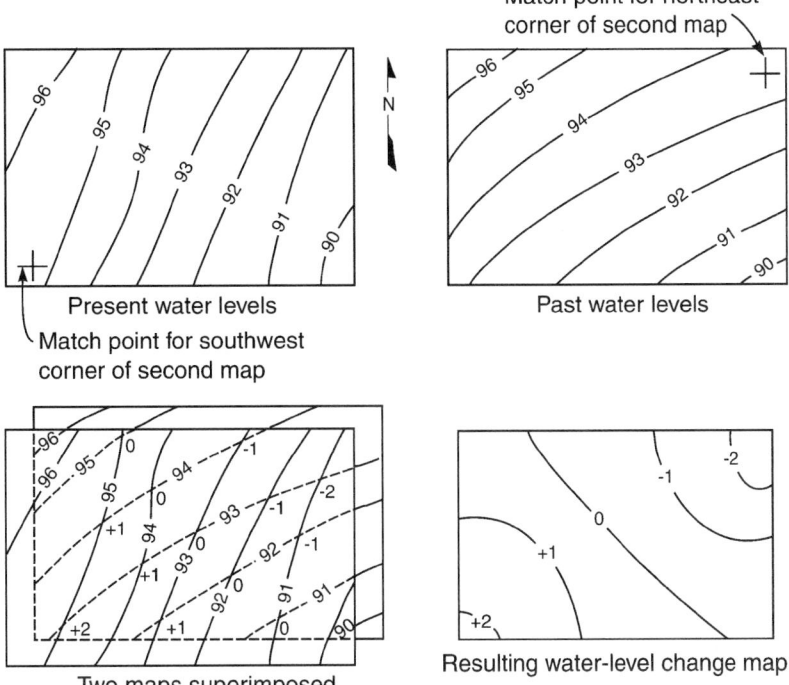

Fig. 5   Construction of a water-level change map by superimposing water-level contour maps (from Ref. [9]).

(Fig. 4a), or obvious geologic structures are ignored (Fig. 4b; Ref. [9]).

In areas where the groundwater levels exhibit a gentle gradient, the groundwater contours will be spaced well apart. If the gradient is steep, the groundwater contours will be closer together. Groundwater will flow in the general direction that the water-level surface is sloping.

Water-level change maps are constructed by plotting the change of water levels in wells during a given span of time. If the study is of a short span of time, data from the same wells can be used. If, however, the time span is long (of the order of 50 yr or more), it is impossible in some areas to measure the same wells, owing to their rather rapid destruction or failure. The best procedure in this case is to draw two water-table maps of the years of interest.[9] The maps are then superimposed and the water-level changes at contour intersections are recorded. The values can then be transferred to a separate map and lines of equal water-level change can be drawn (Fig. 5). Modern technology, especially the use of GIS, has made such procedures much easier and faster.

REFERENCES

1. Matson, K.C.; Fels, J.E.; 1996. Approaches to Automated Water Table Mapping. In: Proceedings, Third International Conference/Workshop on Integrating GIS and Environmental Modeling. Santa Fe, New Mexico, January 21-26, 1996. Santa Barbara, CA: National Center for Geographic Information and Analysis, http://www.ncgia. ucsb.edu/conf/SANTA_FE_CD-ROM/sf_papers/matson_kris/santa-fe.2.html (accessed May 11, 2001).

2. Olea, R.A. *Geostatistics for Engineers and Earth Scientists*; Kluwer Academic Publishers: Boston, MA, 1999.

3. Stone, W.J. *Hydrogeology in Practice: A Guide to Characterizing Groundwater Systems*; Prentice Hall: Upper Saddle River, NJ, 1999.

4. Olea, R.A. Sampling Design Optimization for Spatial Functions. Math. Geol. **1984**, *16* (4), 369–392.

5. Sophocleous, M.A. Groundwater Observation Network Design for the Kansas Groundwater Management Districts, USA. J. Hydrol. **1983**, *61* (4), 371–389.

6. Dalton, M.G.; Huntsman, B.E.; Bradbury, K. Acquisition and Interpretation of Water-Level Data. In *Practical Handbook of Ground-Water Monitoring*; Nielsen, D.M., Ed.; Lewis Publishers: Chelsea, MI, 1991; 367–395.

7. Saines, M. Errors in Interpretation of Ground-Water Level Data. Ground Water Monit. Rev. **1981**, *1* (1), 56–61.

8. Freeze, R.A.; Cherry, J.A. *Groundwater*; Prentice Hall: Englewood Cliffs, NJ, 1979.

9. Davis, S.N.; DeWiest, R.J.M. *Hydrogeology*; John Wiley and Sons, Inc.: New York, 1966.

10. Fetter, C.W. *Applied Hydrogeology*, 4th Ed.; Prentice-Hall, Inc.: Upper Saddle River, NJ, 2001.

Groundwater Levels, Measuring

Paul F. Hudak
University of North Texas, Denton, Texas, U.S.A.

INTRODUCTION

A "groundwater level" is the elevation of water in a well tapping an aquifer. Well construction in addition to hydraulic conditions in the aquifer influence measured groundwater levels. Hydraulic head, the mechanical energy per unit weight of water,[1] is equal to the elevation to which water rises in a cased well open to a "point" in an aquifer. The hydraulic head measurement pertains only to that point and is normally expressed in units of length above mean sea level. Groundwater levels from several cased wells, at several points in time, illustrate spatial and temporal patterns in hydraulic head within an aquifer.

WELL CASING AND SCREENED INTERVAL

For the purpose of monitoring hydraulic head, wells should have as short a screened interval (intake) as possible, generally less than 3 m long.[2] Short intakes are especially important if there are strong vertical flow components. Under these conditions, piezometers (wells with intakes less than 0.3 m long) provide more accurate hydraulic head data.[3]

A hydraulic head measurement can be obtained by subtracting the depth to water in a well from the elevation of a reference point at the top of the well casing. The well casing should be permanently marked at the reference point—depth to water measurements should always be made from that point. The reference point must be accurately surveyed, to within 0.01 ft (3 mm). It should be resurveyed every 5 years to account for settling. Unstable terrain, such as expansive clay soils or bogs, requires more frequent surveying. The initial survey should also establish $x–y$ coordinates of each well. Each well at a field site should be permanently marked with a unique identifier (ID).

The water table represents the surface of an unconfined aquifer. Wells used to measure water table elevations should be screened across or just beneath the water table. In a well tapping a confined aquifer, groundwater levels will rise higher than the top of the aquifer (where it contacts an overlying confining layer). A flowing artesian condition exists if the water level rises above the land surface.

Measuring hydraulic head at flowing wells requires an extension pipe or pressure gage.[3] An extension pipe, tightly fitted to the top of a well casing, must be tall enough to contain the rising water. Alternatively, a pressure gage can be attached to the top of the well casing. The gage measures pressure head (height of water level above gage) or water pressure (pressure head times specific weight of water). The pressure head measurement should be added to the height of the gage above the reference point.

MEASURING DEVICES

Groundwater levels can be measured with several devices, including measuring tapes and poppers, chalk-coated tapes, acoustic probes, electrical sensors, pressure transducers, air lines, time domain reflectometry (TDR), floats and pulleys, and vibrating wire (VW) piezometers.

Poppers (Fig. 1) make an audible sound when dropped onto a water column.[4] The tape should be read at the reference point when the popper just reaches the water column. Length of the popper should be accounted for in the water depth measurement.

Weighted, chalk-coated tapes are similarly lowered down a well, but should penetrate the water column. The tape should be marked where it touches the reference point and then withdrawn from the well. Depth to water equals the distance from the marked point to the top of the wetted portion of the tape. Chalk-coated tapes are one of the most common and accurate methods for measuring groundwater levels.[5]

Acoustic probes transmit sound waves from the top of a well casing to the water level in the well. They measure sound-wave travel times and convert them to distance (depth to water). Electrical sensors (Fig. 1) transmit sound or light signals when a probe enters the water column. The measurement should be made as the probe enters the water column. Submerging and raising the probe, and taking a measurement as the signal stops, is less accurate because dripping water may prolong the signal. False signals from water condensed on the sides of a well should also be considered when using electrical sensors.

Pressure transducers (Fig. 1), air lines, and TDR measure the height of a water column above a submerged probe or tube. They can be connected to data loggers that store water level measurements and corresponding times. Pressure transducers are often used to obtain frequent water level measurements in observation wells during pumping tests.

Encyclopedia of Water Science
DOI: 10.1081/E-EWS 120010081

Fig. 1 Pressure transducer (left), electrical sensor (middle), and vinyl tape and popper (right).

Air lines are less accurate and used mainly in wells being pumped.[5] The TDR devices transmit pulses down a coaxial cable and analyze the reflected voltage signature.[6] A strong voltage drop at the air–water interface is produced by the difference in dielectric constant between air and water. Time domain reflectometry cables can be used in riser pipes as small as 12 mm in diameter.

Floats sit on the water column in a well. One end of a cable is attached to the float, and the other to a counter weight. The cable is draped over a pulley at ground level, and the pulley rotates as water levels in the well rise or fall. Water levels can be recorded with a pen-and-chart or digital system.

Vibrating wire piezometers can be lowered down wells, buried in boreholes, or pushed into unconsolidated sediment. One end of a stretched magnetic wire is anchored and the other attached to a diaphragm, which deflects in proportion to pore-water pressure. Any deflection of the diaphragm changes the tension in the wire, thus affecting the resonant frequency of the VW. Measured pore-water pressures can be converted to pressure head by dividing by the specific weight of water. Adding the pressure head measurement to the elevation of the sensor gives the hydraulic head at the sensor.

FIELD CONSIDERATIONS

Prior to measuring groundwater levels, they should be allowed to recover a minimum of 24 hr following any well construction, development, purging and sampling, or aquifer testing.[5] Recovery may take longer in aquifers with a low hydraulic conductivity.

Measuring devices should be inert and regularly calibrated, taking into account stretch of tapes, wires, or cables. Water level measurements should be made to the nearest 0.01 ft (3 mm) and repeated for accuracy. Well depths should also be measured during each field visit. These do not require as much accuracy as water level measurements and can be accomplished with a weighted tape measure.

Water levels should be measured before collecting water samples or performing aquifer tests, which disturb static water elevations. Ideally, the same device should be used to measure all wells (except in pumping tests requiring frequent or simultaneous measurements at different wells). In a contaminated aquifer, the first water level measurement should be made at the cleanest well, and subsequent measurements should be made at progressively more contaminated wells. The measuring device should be thoroughly cleaned between wells.

At wells with floating immiscible contaminants, both depth to the immiscible layer and depth to water should be measured. This can be done with interface probes or tapes coated with reactive paste, which transmit different signals or colors when contacting different fluids. An immiscible layer depresses the water column in a well—measured depth to water should be corrected by subtracting the product of immiscible layer thickness and specific gravity.

Unless a data logger is being used, each water level measurement should be recorded in a field book, along with the well ID, time of measurement, and device used.

Weather conditions and the name of the person making the measurements should also be recorded in the field book.

MAPS AND GRADIENTS

When using water level measurements to construct a contour map of the water table (unconfined aquifer) or potentiometric surface (confined aquifer), the wells should be measured during the same time interval (typically less than 24 hr) and open to the same hydrostratigraphic interval.[2] As many wells as possible should be measured, without sacrificing the above considerations. Moreover, the wells should be spread throughout the study area to avoid inaccurate hydraulic head extrapolations.

Hydraulic gradient, change in hydraulic head with distance, should be calculated along a flow line in a water table or potentiometric surface map. Flow lines should be constructed perpendicular to hydraulic head contours (equipotential lines), unless the aquifer is anisotropic. A minimum of three wells defines a sloping plane and local flow direction. However, three wells allow for only a local, linear approximation of the groundwater flow direction.

Vertical gradients in groundwater can be computed from hydraulic head measurements at adjacent wells open at different depths. A vertical gradient can also indicate the gaining or losing status of a surface water body such as a lake or stream. This can be accomplished by driving a narrow steel pipe with a slotted conical tip about 0.5 m into the bottom of the water body. The vertical gradient is the difference between water levels in the piezometer and water body, divided by the distance between the bottom of the water body and bottom of the piezometer. A higher water level in the piezometer indicates an upward gradient and gaining condition, whereas a lower level in the piezometer indicates a downward gradient and that surface water is seeping into the ground.

REFERENCES

1. Fetter, C.W. *Applied Hydrogeology*, 4th Ed.; Prentice Hall: Upper Saddle River, NJ, 2000.
2. EPA (U.S. Environmental Protection Agency), *RCRA Ground Water Monitoring: Draft Technical Guidance*; Government Institutes: Rockville, MD, 1994.
3. Sanders, L.L. *A Manual of Field Hydrogeology*; Prentice Hall: Upper Saddle River, NJ, 1998.
4. Hudak, P.F. *Principles of Hydrogeology*, 2nd Ed.; Lewis Publishers: Boca Raton, FL, 2000.
5. Driscoll, F.G. *Groundwater and Wells*; Johnson Division: St. Paul, MN, 1986.
6. Dowding, C.H.; Nicholson, G.A.; Taylor, P.A.; Agoston, A.; Pierce, C.E. Recent Advancements in TDR Monitoring of Ground Water Levels and Piezometric Pressures. In *Proceedings of the 2nd North American Rock Mechanics Symposium, Montreal, Quebec, June 19–21, 1996*; Aubertin, M., Hassani, F., Mitri, H., Eds; Balkema: Rotterdam, The Netherlands, 1996.

Groundwater Mining

Hugo A. Loáiciga
University of California, Santa Barbara, California, U.S.A.

INTRODUCTION

Groundwater mining is defined as the extraction of ground water from aquifers by humans. This definition is analogous to that concerning the mining of mineral resources. There is, however, a fundamental difference between groundwater mining and the mining of minerals. Groundwater is, in most cases, a renewable resource. On the other hand, mineral resources, such as silver and gold ores, are nonrenewable. Groundwater is a renewable resource because it is replenished naturally by fluxes that arise in the hydrologic cycle. The sum of the fluxes that replenish ground water is called recharge. During periods of plentiful precipitation, and in the absence of human intervention, aquifers are replenished by recharge. During droughts, groundwater storage and groundwater levels decline due to low levels of recharge. There are also groundwater deposits of "fossil" ground waters that have become isolated from the hydrologic cycle. These deposits resemble in many respects oil reservoirs. One important shared characteristic is that extraction of the resource, be it ground water or oil, produces an irreversible reduction in its stock. Continued mining of such fossil deposits leads to their eventual depletion.

This article is devoted to an analysis of the effects of groundwater mining on renewable ground water. The latter constitutes most of the ground water used by humans. Principles of sustainable groundwater mining are presented and illustrated with data from one of the most productive aquifers in the world.

GROUNDWATER MINING AND THE WATER BALANCE

Let us consider an aquifer that is subject to groundwater mining. Assume that the amount of groundwater storage is denoted by S, and that recharge (R), groundwater pumping (W), and outflow (G) affect the status of storage as shown in Fig. 1. Groundwater pumping is the means by which ground water is mined. The recharge is the net water flux into groundwater storage from surface water sources. It includes percolation, seepage (from rivers and lakes), and artificial recharge (by wells and spreading basins). Groundwater uptake by plants, baseflow, and spring flow

abstractions from groundwater are also included in the calculation of aquifer recharge. The groundwater outflow (G) term is the net of subsurface fluxes in and out of groundwater storage across the (subsurface) aquifer boundaries. From water-balance considerations for a period of duration T, it is evident that the change in groundwater storage is given by the following equation:

$$S(T) - S(0) = \int_{t=0}^{t=T} [R(t) - W(t) - G(t)] \quad T \geq 0 \qquad (1)$$

in which $S(0)$ and $S(T)$ are the storages at time zero (initial storage) and time T, respectively.

Pumping may be measured accurately with well meters. It commonly exhibits a strong seasonal pattern, rising during periods of low precipitation (i.e., during dry seasons) and subsiding during wet seasons. This is true for urban and agricultural groundwater uses. In addition, as a result of population growth, urban groundwater mining typically exhibits an increasing trend over time.[1] The recharge flux in Eq. 1 is, in general, difficult to estimate. Recharge depends strongly on the amount of precipitation, and, thus, it tends to replicate the seasonality and inter-annual variability observed in the climate specific to the region where the aquifer is found.[2] The groundwater outflow term (G) is not amenable to direct measurement. Instead, it must be estimated by indirect methods.[3] When the aquifer boundaries coincide with groundwater divides, the outflow term (G) is negligible.

Fig. 2 shows the evolution of annual groundwater recharge, pumping, and spring flow from 1934 to 1995 in the Edwards Aquifer of Texas, one of the most productive groundwater systems in the world.[1] The subsurface outflow term (G) in the Edwards aquifer is negligible.[4] In this instance, it is advantageous to treat separately the flux of water into the aquifer (i.e., recharge) from the discharge of ground water at several large springs (i.e., spring flow).

Recharge takes place primarily by means of stream seepage along aquifer outcrops. It is seen in Fig. 2 that recharge shows large inter-annual fluctuations, and that those fluctuations appear to become larger and larger over time. Groundwater pumping displays a long-term increasing trend until year 1985, even during the drought of 1936–1959. The intermittent lows in groundwater pumping after 1985 were caused by Court orders imposed

Encyclopedia of Water Science
DOI: 10.1081/E-EWS 120010157

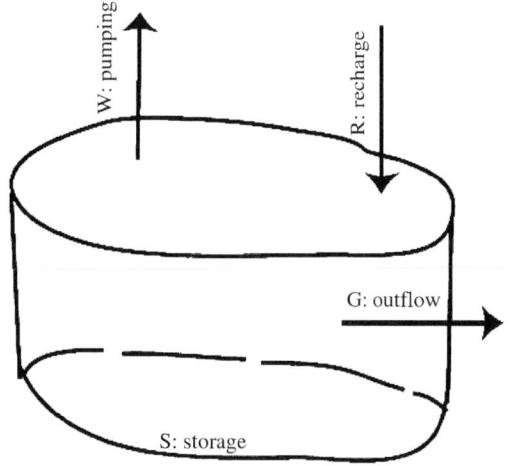

Fig. 1 Schematic of a mined aquifer.

on the mining of the Edwards Aquifer to protect aquatic habitats in the discharge zone near springs.[1]

Spring flow is concentrated along large fault springs that define the discharge zone of the Edwards aquifer.[1]. As shown in Fig. 2, it is a smoothed-out and dampened replica of annual recharge. It lags recharge by a short period of time, typically less than 2 yr. The time series of spring flow shown in Fig. 2 does not represent natural groundwater discharge because of the effect that groundwater pumping had on spring flow. If the Edwards aquifer had not been mined in the period 1934–1995, the amount of spring flow would have been roughly equal to the amount of recharge. The latter is demonstrated in Fig. 3, where the cumulative recharge and the cumulative pumping plus spring flow time series are plotted. By adding pumping to spring flow, the latter is reconstructed

to what would have been its natural value during the period of analysis. The differences between the two time series plotted in Fig. 3 arise from unequal beginning and ending aquifer storages.

Fig. 4 shows the change in storage, $S(T) - S(0)$, calculated from Eq. 1 for the Edwards Aquifer data shown in Fig. 2. It is seen there that during the drought period between 1936 (point 1) and 1956 (point 2) aquifer storage dropped by $3500 \times 10^6 \, m^3$ as a result of groundwater mining. Between 1956 and 1992 (point 3) ground water continued to be mined, yet, there was a recovery of aquifer storage equal to $5100 \times 10^6 \, m^3$. Since the Edwards Aquifer was severely de-watered in 1956—demonstrated by the drying of major springs—and in 1992 water levels rose to historically high levels after heavy El Niño rainfall, it can be concluded that the Edwards Aquifer extractable storage must be on the order of $5100 \times 10^6 \, m^3$. The evolution of storage $S(T)$ captures one important aspect of groundwater mining in any aquifer. For a full grasp of groundwater mining, however, one must broaden the scope of its analysis.

GROUNDWATER MINING AND SUSTAINABLE AQUIFER USE

The key question regarding groundwater mining is how to pump ground water from an aquifer without compromising the availability of ground water in storage, while maintaining its natural water quality and protecting water bodies that may depend on the status of groundwater storage (e.g., influent streams and lakes, springs, and wetlands). Groundwater mining concerns must go beyond the amount of ground water extracted or left in aquifer storage. Other environmental considerations must be taken

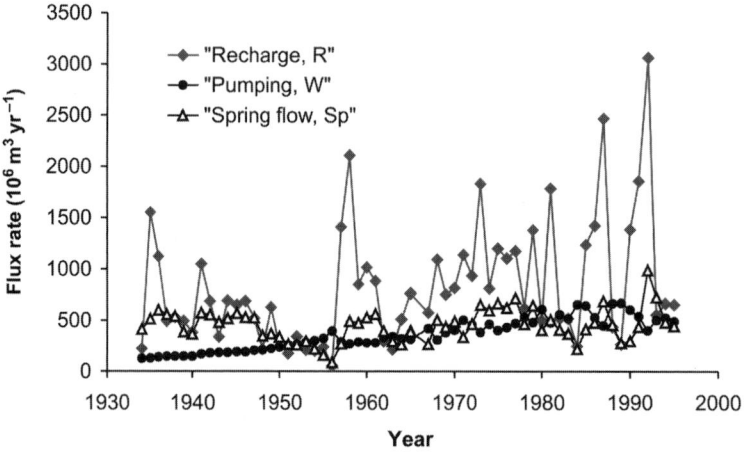

Fig. 2 Groundwater pumping (W), recharge (R), and spring flow (Sp) in the Edwards Aquifer.

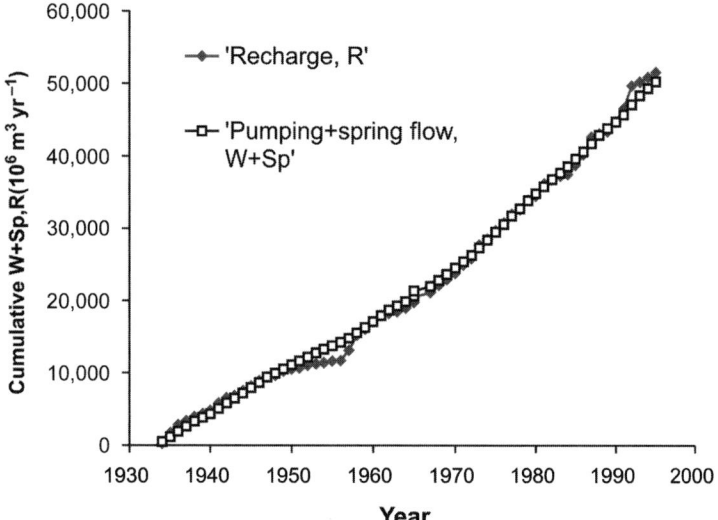

Fig. 3 Cumulative recharge and cumulative pumping plus spring flow from 1934 to 1995.

into account in determining the best way to mine an aquifer. Groundwater mining that ensures a long-term supply of good-quality water while protecting the environment is what we call sustainable groundwater mining. Simplistic rules such as "groundwater pumping shall not exceed the long-term average recharge" are inadequate to cope with the spectrum of impacts associated with groundwater mining. This is so because even if pumping does not exceed the long-term recharge (which, by the way, may be difficult to estimate accurately), groundwater storage may still reach levels that are detrimental from the perspective of water-quality protection and environmental conservation. The cyclic and variable nature of recharge, and the instinctive drive to intensify groundwater mining during periods of low

precipitation (to irrigate crops for example, or to water lawns and gardens[5]) pose serious challenges to sustainable groundwater mining during periods of low precipitation, be they seasonal or associated with protracted drought.[6]

Fig. 5 shows a graph of the cumulative recharge in the Edwards Aquifer. The cumulative or mass recharge is expressed by:

$$\text{Mass recharge} = \sum_{t=0}^{T} \text{recharge}(t) \quad T \geq 0 \qquad (2)$$

The curve shown in Fig. 5 is called a "mass curve" for the Edwards Aquifer. Mass curves are widely used in the analysis of stream flow time series for the purpose of

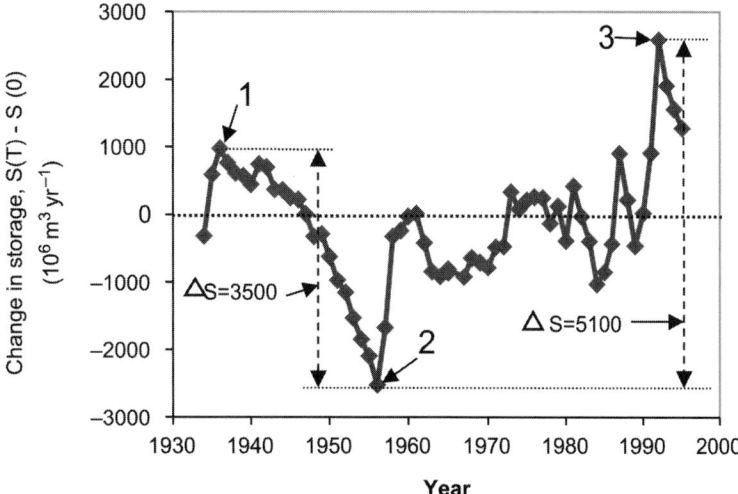

Fig. 4 Changes in aquifer storage as a result of groundwater mining and climate.

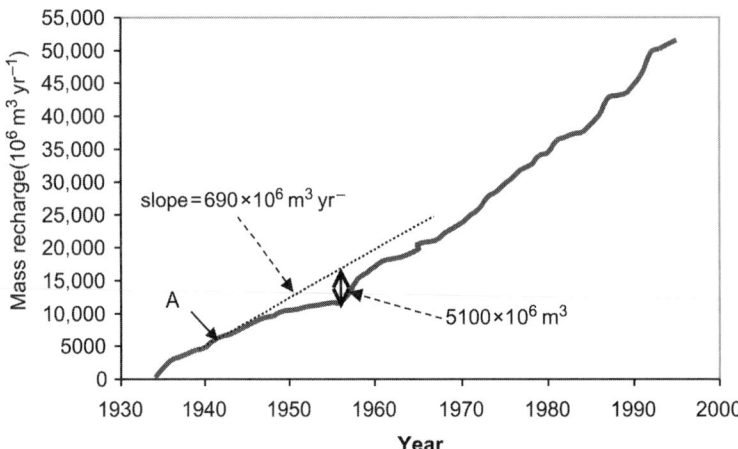

Fig. 5 Mass curve and the estimation of an average groundwater mining rate.

sizing surface reservoirs or determining reservoir releases.[7] The mass curve is used herein to provide a first estimate of long-term groundwater pumping. Assume an extractable groundwater storage of $5100 \times 10^6 \, m^3$, which was estimated from Fig. 4. One finds that the minimum-slope tangent to the mass curve (in this case drawn through point A in Fig. 5) that encompasses the estimated groundwater storage of $5100 \times 10^6 \, m^3$ has a slope of $690 \times 10^6 \, m^3 \, yr^{-1}$ (see Fig. 5). Ignoring spring flow and related impacts associated with groundwater mining, the magnitude of that slope equals the average long-term groundwater pumping that would be consistent with an usable aquifer storage of $5100 \times 10^6 \, m^3$. It turns out, however, that a pumping rate of $690 \times 10^6 \times m^3 \times yr^{-1}$ would cause exceedingly low spring flow values during low-recharge years and adverse and irreversible impacts on aquatic ecosystems supported by the Edwards Aquifer springs.[1] More detailed simulations by the author, which were carried out with a specially calibrated numerical groundwater model for the Edwards Aquifer,[1] indicated that during low-recharge periods (1947–1956, for example) the aquifer may not be mined at a rate greater than $123 \times 10^6 \, m^3 \, yr^{-1}$ in order to protect minimum spring flow levels and aquatic habitats. For comparison, during the period 1934–1995, the average pumping in the Edwards Aquifer was on the order of $360 \times 10^6 \, m^3 \, yr^{-1}$, while during the high-growth period 1970–1995 pumping averaged $514 \times 10^6 \, m^3 \, yr^{-1}$, a mining strategy that has left a legacy of adverse ecological impacts.

CONCLUSIONS

The former example illustrates important factors that must be considered in the planning of sustainable groundwater

mining. The first is the long-term behavior of aquifer recharge and aquifer discharge (besides artificial pumping). Secondly, one must have an in-depth understanding of the hydraulic and ecological linkages of aquifer storage and discharge to dependent ecosystems. The rate of pumping must be adjusted to the natural fluctuations of recharge. This requires detailed numerical simulations of aquifer response to pumping under specific recharge conditions. Although water-quality deterioration effected by groundwater mining was not specifically addressed in this work, it is another consideration that must be taken into account in planning sustainable groundwater mining strategies. The excessive lowering of aquifer storage may induce the upwelling of poor-quality groundwater and/or the intrusion of saltwater.

ACKNOWLEDGMENTS

This article was supported in part by U.S. Geological Survey grant HQ-96-GR-02657 and by California Water Resources Center grant WR-952.

REFERENCES

1. Loáiciga, H.A.; Maidment, D.; Valdes, J.B. Climate-Change Impacts in a Regional Karst Aquifer. J. Hydrol. **2000**, *227*, 173–194.
2. Loáiciga, H.A. Climate-Change Impacts in Regional-Scale Aquifer: Principles and Field Application. In *Ground Water Updates*; Sato, K., Iwasa, Y., Eds.; Springer-Verlag: Tokyo, 2000; 247–252.
3. Zektser, I.S.; Loáiciga, H.A. Ground-Water Fluxes in the Global Hydrologic Cycle: Past, Present, and Future. J. Hydrol. **1993**, *144*, 405–427.

4. Edwards Underground Water District, *Edwards/Glen Rose Hydrologic Communication, San Antonio Region, Texas*, Report 95-03; Edwards Underground Water District: San Antonio, Texas, USA, 1995.

5. Loáiciga, H.A.; Renehan, S. Municipal Water Use and Water Rates Driven by Severe Drought: A Case Study. J. Am. Water Res. Assoc. **1997**, *33*, 1313–1326.

6. Loáiciga, H.A.; Michaelsen, J.; Garver, S.; Haston, L.; Leipnik, R.B. Droughts in River Basins of the Western United States. Geophys. Res. Lett. **1992**, *19* (20), 2051–2054.

7. Linsley, R.K.; Franzini, J.B. *Water Resources Engineering*, 3rd Ed.; McGraw-Hill: New York, 1979.

Groundwater Modeling

Jesús Carrera
Technical University of Catalonia (UPC), Barcelona, Spain

INTRODUCTION

A model is an entity built to reproduce some aspect of the behavior of a natural system. In the context of groundwater, aspects to be reproduced may include: groundwater flow (heads, water velocities, etc.); solute transport (concentrations, solute fluxes, etc.); reactive transport (concentrations of chemical species reacting among themselves and with the solid matrix, minerals dissolving or precipitating, etc.); multiphase flow (fractions of water, air, nonaqueous phase liquids, etc.); energy (soil temperature, surface radiation, etc.); and so forth.

Depending on the type of description of reality that one is seeking (qualitative or quantitative), models can be classified as conceptual or mathematical. A conceptual model is a qualitative description of "some aspect of the behavior of a natural system." This description is usually verbal, but may also be accompanied by figures and graphs. In the groundwater flow context, a conceptual model involves defining the origin of water (areas and processes of recharge) and the way it flows through and exits the aquifer. In contrast, a mathematical model is an abstract description (abstract in the sense that it is based on variables, equations, and the like) of "some aspect of the behavior of a natural system." However, the motivation of mathematical models is not abstraction, but rather quantification. For example, a groundwater flow mathematical model should yield the time evolution of heads and fluxes (water movements) at every point in the aquifer.

Both conceptual and mathematical models seek understanding. Some would argue that understanding is not possible without quantification. Reversely, one cannot even think of writing equations without some sort of qualitative understanding. The methods of conceptual modeling are those of conventional hydrogeology (study geology, measure heads and hydraulic parameters, hydrochemistry, etc). On the other hand, the methods of mathematical modeling (discretization, calibration, etc.) are more specific. Yet, it should be clear from the outset that conceptualization is the first step in modeling and that mathematical modeling helps in building firm conceptual models.

Depending on the manner in which equations are solved, models can be classified as: analog, analytical, and numerical. Analog models are based on a physical simulation of a phenomenon governed by the same equation(s) as that of our natural system. For example, because of the equivalence between electrostatics and steady state flow, one may use conductive paper subject to an electrical current to solve the flow equation (a parallelism can be established between electric potential and hydraulic head). This kind of application, however, is restricted mainly to teaching. Boxes of resistances and condensators were used in the 1950s and 1960s as analog aquifer models, but they have become inefficient compared to computers. As a result, analog models are no longer used in practice.

Analytical models are based on closed-form solutions to the groundwater flow and transport equations. They are convenient in the sense that they are easy to evaluate and intuitive (visual inspection of the equation may yield an idea of the phenomenon). As a result, they are used very frequently. Examples include solutions of problems in well hydraulics, tracer movement, etc.

Numerical models are based on discretizing the partial differential equations governing flow and transport. This leads to linear systems of equations that can only be solved with the aid of computers. The advantage of numerical models lies in their generality. Analytical models are constrained to homogeneous domains and very simple geometry and boundary conditions. Numerical models, on the other hand, can handle spatially and temporally variable properties, arbitrary geometry and boundary conditions, and complex processes. The price to pay is methodological singularity. Analytical models are easy to use. Numerical models can be complex and, often, difficult.

Because of the methodological singularity mentioned above, this chapter concentrates on mathematical numerical models. Analytical solutions are not discussed. In addition, conceptual modeling will be discussed as the first step in modeling, but not by itself.

Encyclopedia of Water Science
DOI: 10.1081/E-EWS 120010080

WHAT CAN BE MODELED AND WHAT FOR?

Modeled Phenomena

The most basic phenomenon is groundwater flow (Fig. 1) because of its intrinsic importance and because it is needed for subsequent processes. In essence, the flow equation expresses two things. First, groundwater moves according to Darcy's law. Second, a mass balance must be satisfied in the whole aquifer and in each of its parts. Therefore, the main output from flow models is a mass balance: classified inflows, outflows, and storage variations. The output also includes where water flows through the aquifer (water fluxes) and heads (water levels in the aquifer). In essence, input data are a thorough description of hydraulic conductivity (and/or transmissivity), storativity, recharge/discharge throughout the model domain, as well as conditions at the model boundaries. Obviously, these data are never available, and the modeler has to use a good deal of ingenuity to generate them. This is where the conceptual model becomes important.

Specific cases of flow phenomena are unsaturated and multiphase flow. In the first case, one models water flow in the vadose zone or, in general, in areas where water does not fill all the pores.[1] Therefore, besides heads and fluxes, one must work with water contents (volume of water per unit volume of aquifer), capillary pressures and suctions

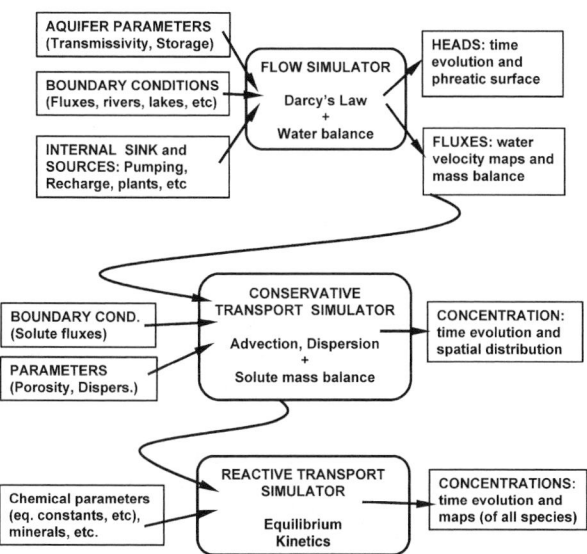

Fig. 1 A groundwater flow model involves using a flow simulator to take aquifer parameters, boundary conditions, and internal sink and sources as inputs and obtain heads and water fluxes as output. Water fluxes are used in conservative transport models, together with porosity, diffusivity, and solute mass inflows, to yield time evolution and spatial distribution of inert tracers.

(difference between water and air pressure). From the input viewpoint, the main singularity of unsaturated flow is the need to specify the retention curve (water content vs. suction) and relative permeability (permeability vs. water content). The multiphase flow case is similar, but includes several fluids (phases). It is used to represent the flow of air or mixtures of liquids, singularly nonaqueous phase liquids (NAPLs), which have been the subject of much research in recent years.[2]

Conservative transport refers to the movement of inert substances dissolved in water. Solutes are affected by advection (displacement of the solute as linked to flowing water) and dispersion (dilution of contaminated water with clean water, which causes the size of the contaminated area to grow while reducing peak concentrations). The main input to a solute transport model is the output of a flow model (water fluxes). Additionally, porosity and dispersivity need to be specified (Fig. 1). The output is the time evolution and spatial distribution of concentrations. While the amount of data needed for solute transport modeling is relatively small, it must be stressed that solute transport is extremely sensitive to variability and errors in water fluxes. A flow model may be good enough for flow results (heads and water balances) but insufficiently detailed to yield water fluxes good enough for solute transport. Therefore, modeling solute transport ends up being rather difficult.

Reactive transport refers to the movement of solutes that react among themselves and with the soil phase. Reactions can be of many kinds, ranging from sorption of a contaminant onto a solid surface to redox phenomena controlling the degradation of an organic pollutant. Input for reactive transport modeling includes not only the output of flow and conservative transport models but also the equilibrium constants of the reactions (usually available from chemistry databases) and the parameters controlling reaction kinetics. However, the most difficult input is the proper identification of relevant chemical processes. Model output includes the concentrations of all chemical species, the reaction rates, etc.

Coupled models refer to models in which different phenomena are affected reciprocally. Density dependent flow is a typical example. Variations in density affect groundwater flow (e.g., dense sea water sinks under light fresh water), which in turn affects solute transport and, hence, density distribution. Other coupled phenomena are the nonisothermal flow of water (coupling flow and energy transport) and the mechanically driven flow of water (coupling flow and mechanical deformation equations).

What Are Models Built For?

While discussing the usage of models, it is convenient to distinguish between site-specific models and generic models. The former are aimed at describing a specific aquifer while the latter emphasize processes, regardless of where they take place.

Groundwater management is the ideal use of site-specific models. Management involves deciding where to extract and/or inject water to satisfy water needs while ensuring water quality and other constraints. In this context, it is important to point out that a model is essentially a system for accounting water fluxes and stores (Fig. 2) in the same way that the accounting system of a company keeps track of money fluxes and reserves. No one would imagine a well-managed company without a proper accounting system. Aquifers will not be managed accurately until they have a model running on real time. Unfortunately, at present, this is still a dream. Because of the difficulties in building and maintaining models and because of legal and practical difficulties to manage aquifers in real time, models are rarely, if ever, used in this fashion.

Instead, models are often used as decision support tools. Building an accurate model is very difficult and time consuming. As a result, one can rarely expect models to yield exact predictions. However, approximate models are much easier to build. These do not result in precise forecasts but normally allow reasonable assessments of the outcome of different management alternatives, i.e., the relative advantages and disadvantages of each alternative can be evaluated and the options ranked. This is usually all one needs for decision making.

This type of use is very frequent in aquifer rehabilitation, where one has to choose among several alternatives, including the option of doing nothing.[3]

Models are also used for supporting aquifer exploration policies, i.e., for answering questions such as "how much water can be extracted?," "where should one pump to minimize environmental impact?," etc. In fact, a large body of literature is devoted to this kind of questions in an optimal fashion.[4]

Site-specific models are most frequently used, however, as a tool to support aquifer characterization efforts. This is somewhat ironic because a model is an essentially quantitative tool while site characterization is rather qualitative. Yet, experience dictates that modeling is the only way to consistently integrate the kind of data available in site characterization. These data are very diverse and range from geologic maps to isotope concentrations. One can use vastly different models to verbally explain all observations. Quantitative consistency is not so easy to check and requires the use of a model. Because of the difficulties in fully describing all data, this kind of model use is rarely described in the scientific literature.

Models can also be used in generic fashion as teaching or research tools to gain understanding on physico-chemical phenomena. In these cases, they do not aim at representing a specific aquifer, but at evaluating the role of some processes under idealized conditions. A classical example of this type of use is the analysis of flow on regional basins.[5] Models are used in this fashion to explain geological processes.[6,7] Much emphasis has been placed in recent years on the evaluation of the effects of spatial variability. This involves issues such as upscaling, i.e., finding the relationship between large-scale effective parameters and small-scale measurements;[8] or analysis of hydraulic tests.[9]

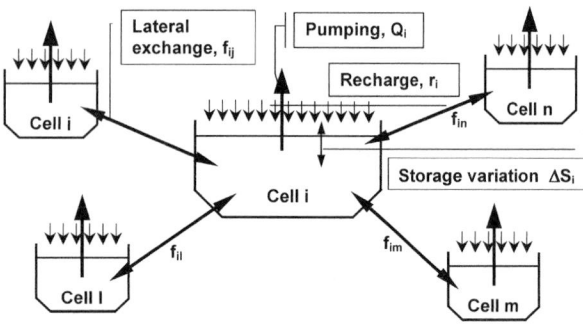

Fig. 2 A groundwater model is the accounting system of an aquifer. It keeps track of the balance of each section (cells or compartments in the groundwater language) by evaluating exchanges with the outside (pumping Q_i, recharge, r_i, etc.) and with the adjacent sections (f_{ij}). The difference between inflow and outflow is equal to the variation in reserves (storage variation, ΔS_i). A well-managed company needs an accounting system, and so does an aquifer.

HOW ARE MODELS BUILT: THE MODELING PROCESS

The procedure to build a model is outlined in Fig. 3. First, one defines a conceptual model (i.e., zones of recharge, boundaries of aquifers, etc). Second, one discretizes the model domain into a finite element or finite difference grid. This can be entered as input data for a simulation code. Unfortunately, output data will rarely fit the observed aquifer heads and concentrations. This is what motivates calibration, i.e., the modification of model parameters to ensure that model output is indeed similar to what has been observed in reality. The model thus calibrated can be considered a "representation of the natural system" and can be used for management or simulation purposes.

The above procedure is formally described in Fig. 4. This section is devoted to discussing in detail the modeling

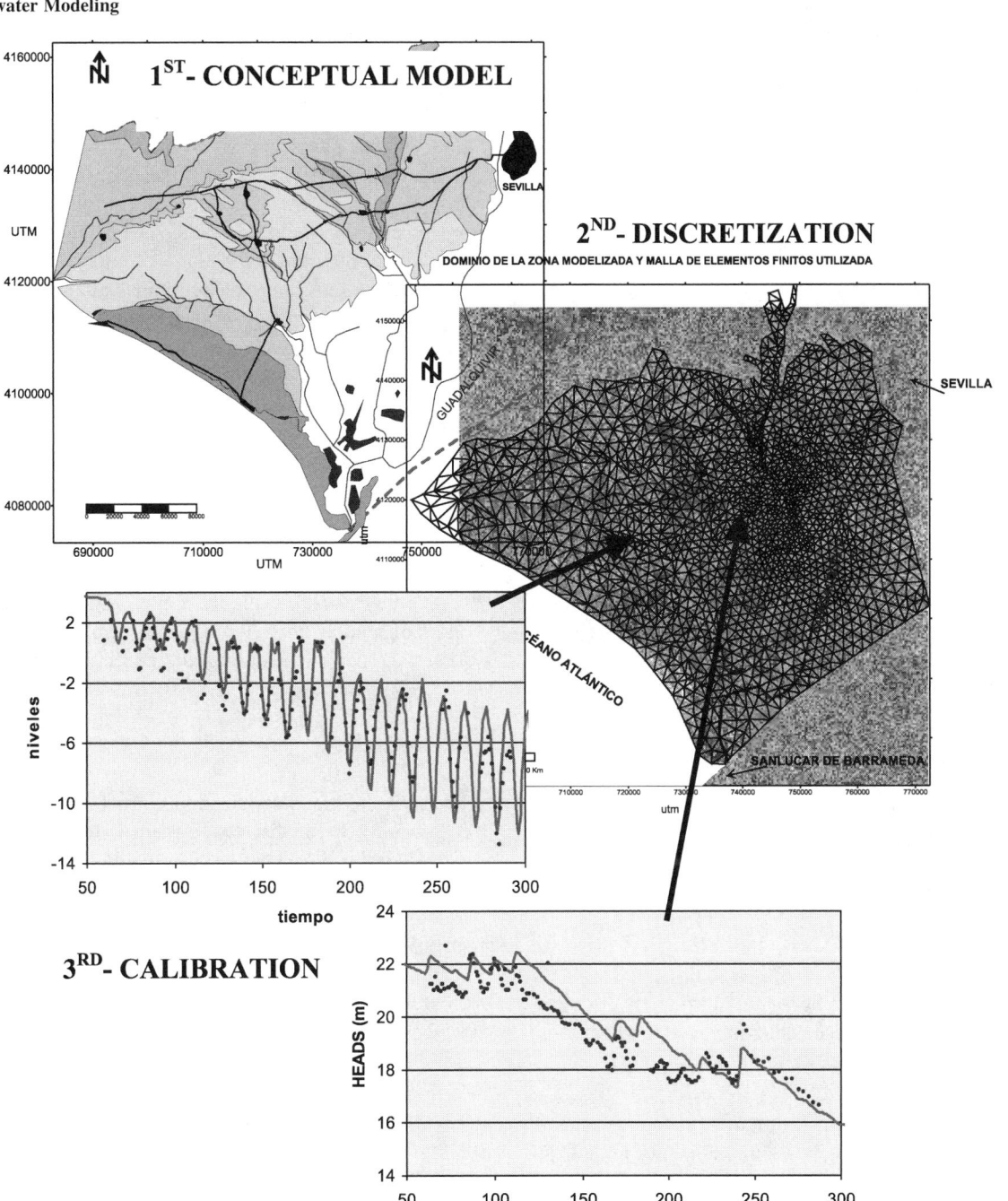

Fig. 3 Building a model involves three basic steps: conceptualization, discretization, and calibration. Example from the Almonte-Marismas aquifer.

steps as previously described.[10–12] In practice, the effort behind each of these tasks may be very sensitive to the objectives of the studies and model. For the time being, we will assume that one is building a model aimed at describing reality in detail for the purpose making predictions.

Conceptualization

Modeling starts by defining which processes are important and how they are represented in the model. Definition of the relevant processes is termed "process identification" and it is needed for several reasons. First, the number of

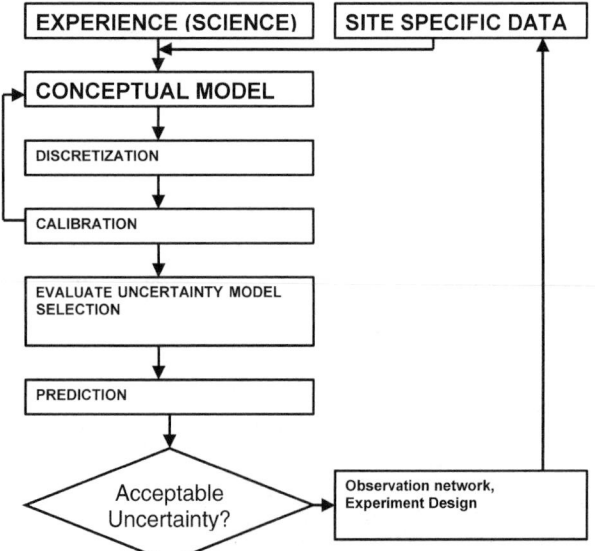

Fig. 4 A formal description of the modeling process. Modeling starts with an understanding of the natural system (conceptual model), which is based on experience about such kind of systems (science) and on data from the site. Writing the conceptual model in a manner adequate for computer solution requires discretization. The resulting model is still dependent on many parameters that are uncertain. During calibration, these parameters are adjusted so that model outputs are close to measurements (recall Fig. 3). Model predictions may be uncertain because so are the fitted parameters or because different models are consistent with observations. If uncertainty is unacceptably high, one should perform additional measurements or experiments and redo the whole process.

processes that may affect flow and transport is very large. For practical reasons, the modeler is forced to select those that affect the phenomenon under study, most significantly. Second, not all processes are well understood and they have to be treated in a simplified manner. In short, process identification involves simplifications, both in the choice of the processes and in the way they are implemented in the model.

Model structure identification refers to the definition of parameter variability, boundary conditions, etc. In a somewhat narrower but more systematic sense, model structure identification implies expressing the model in terms of a finite number of unknowns called model parameters. Parameters controlling the above processes are variable in space. In some cases, they also vary in time or depend on heads and/or concentrations. As discussed earlier, data are scarce so that such variability cannot be expressed accurately. Therefore, the modeler is also forced to make numerous simplifications to express the patterns of parameter variations, boundary conditions, etc. These

assumptions are reflected on what is denoted as model structure.

The conceptualization step of any modeling effort is somewhat subjective and dependent on the modeler's ingenuity, experience, scientific background, and way of looking at the data. Selection of the physico-chemical processes to be included in the model is only rarely the most difficult issue. The most important processes affecting the movement of water and solutes underground (advection, dispersion, sorption, etc.) are relatively well known. Ignoring a relevant process will only be caused by misjudgments and should be pointed out by reviewers, which illustrates why reviewing by others is important. Difficulties arise when trying to characterize those processes and, more specifically, the spatial variability of controlling parameters.

In spite of the large amount of data usually available, their qualitative nature prevents a detailed definition of the conceptual model. Thus, more than one description of the system may result from the conceptualization step. Selecting one conceptual model among several alternatives is sometimes performed during calibration, as discussed later.

Discretization

Strictly speaking, discretization consists of substituting a continuum by a discrete system. However, we are extending this term here to describe the whole process of going from mathematical equations, derived from the conceptual model, to numerical expressions that can be solved by a computer. Closely related is the issue of verification, which refers to ensuring that a code accurately solves the equations that it is claimed to solve. As such, verification is a code-dependent concept. However, using a verified code is not sufficient for mathematical correctness. One should also make sure that time and space discretization is adequate for the problem being addressed. Moreover, numerical implementation of a conceptual model is not always straightforward international code comparison projects; INTRANCOIN and HYDROCOIN have shown the need for sound conceptual models and independent checks of calculation results. Even well-posed mathematical problems lead to widely different solutions when solved by different people, because of slight variations in the solution methodology or misinterpretations in the formulation.[13] The reasons behind these differences and ways to solve them only become apparent after discussions among them.

The main concern during discretization is accuracy. In this sense, it is not conceptually difficult, although it can be complex. Accuracy is not only restricted to numerical errors (differences between numerical and exact solutions of the involved equations) but also refers to the precision with which the structure of spatial variability reproduces the natural system.

Calibration and Error Analysis

The choice of numerical values for model parameters is made during calibration, which consists of finding those values that grant a good reproduction of head and concentration data (Fig. 3) and are consistent with prior independent information.

Calibration is rarely straightforward. Data come from various sources, with varying degrees of accuracy and levels of representativeness. Some parameters can be measured directly in the field, but such measurements are usually scarce and prone to error. Furthermore, since measurements are most often performed on scales and under conditions different from those required for modeling purposes, they tend to be both numerically and conceptually different from model parameters. The most dramatic example of this is dispersivity, whose representative value increases with the scale of measurement so that dispersivities derived from tracer tests cannot be used directly in a large-scale model. As a result, model parameters are calibrated by ensuring that simulated heads and concentrations are close to the corresponding field measurements.

Calibration can be tedious and time consuming because many combinations of parameters have to be evaluated, which also makes it prone to be incomplete. This, coupled to difficulties in taking into account the reliability of different pieces of information, makes it very hard to evaluate the quality of results. Therefore, it is not surprising that significant efforts have been devoted to the development of automatic calibration methods.[14–16]

Model Selection

The first step in any modeling effort involves constructing a conceptual model, describing it by means of appropriate governing equations, and translating the latter into a computer code. Model selection involves the process of choosing between alternative model forms. Methods for model selection can be classified into three broad categories. The first category is based on a comparative analysis of residuals (differences between measured and computed system responses) using objective as well as subjective criteria. The second category is denoted parameter assessment and involves evaluating whether or not computed parameters can be considered as "reasonable." The third category relies on theoretical measures of model validity known as "identification criteria." In practice, all three categories will be needed: residual analysis and parameter assessment suggest ways to modify an existing model and the resulting improvement in model performance is evaluated on the basis of

identification criteria. If the modified model is judged an improvement over the previous model, the former is accepted and the latter discarded.

The most widely used tool of model identification is residual analysis. In the groundwater context, the spatial and frequency distributions of head and concentration residuals are very useful in pointing towards aspects of the model that need to be modified. For example, a long tail in the breakthrough curve not properly simulated by a single porosity model may point to a need for incorporating matrix diffusion or a similar mechanism. These modifications should, whenever possible, be guided by independent information. Qualitative data such as lithology, geological structure, geomorphology, and hydrochemistry are often useful for this purpose. A particular behavioral pattern of the residuals may be the result of varied causes that are often difficult to isolate. Spatial and/or temporal correlation among residuals may be a consequence of not only improper conceptualization, but also measurement or numerical errors. Simplifications in simulating the stresses exerted over the system are always made and they lead to correlation among residuals. Distinguishing between correlations caused by improper conceptualization and measurement errors is not an easy matter. This makes analysis of residuals a limited tool for model selection.

An expedite way of evaluating a model concept is based on assessing whether or not the parameters representing physico-chemical properties can be considered "reasonable"; i.e., whether or not their values make sense and/or are consistent with those obtained elsewhere. Meaningless parameters can be a consequence of either poor conceptualization or instability. If a relevant process is ignored during conceptualization, the effect of such process may be reproduced by some other parameter. For example, the effect of sorption is to keep part of the solute attached to the solid phase, hence retarding the movement of the solute mass; in linear instantaneous sorption, this effect cannot be distinguished from standard storage in the pores. Therefore, if one needs an absent porosity (e.g., larger than one) to fit observation, one should consider the possibility of including sorption in the model. However, despite this example, parameter assessment tends to be more useful for ruling out some model concepts than for giving a hint on how to modify an inadequate model. Residual analysis is usually more helpful for this purpose.

Instability may also lead to unreasonable parameter estimates during automatic calibration, despite the validity of the conceptual model. When the number of data or their information content is low, small perturbations in the measurement or deviations in the model may lead to drastically different parameter estimates. When this happens, the model may obtain equally good fits with widely different parameter sets. Thus, one may converge to a senseless parameter set while missing other perfectly meaningful sets. This type of behavior can be easily identified by means of a thorough error analysis and corrected by fixing the values of one or several parameters.[14]

Predictions and Uncertainty

Formulation of predictions involves a conceptualization of its own. Quite often, the stresses, whose response is to be predicted, lead to significant changes in the natural system, so that the structure used for calibration is no longer valid for prediction. Changes in the hydrochemical conditions or in the flow geometry may have to be incorporated into the model. While numerical models can be used for network design or as investigation tools, most models are built in order to study the response of the medium to various scenario alternatives. Therefore, uncertainties on future natural and man-induced stresses also cause model predictions to be uncertain. Finally, even if future conditions and conceptual model are exactly known, errors in model parameters will still cause errors in the predictions. In summary, three types of prediction uncertainties can be identified: conceptual model uncertainties; stresses uncertainties; and parameters uncertainties.

The first group includes two types of problems. One is related to model selection during calibration. That is, more than one conceptual model may have been properly calibrated and data may not suffice to distinguish which one is the closest to reality. It is clear that such indetermination should be carried into the prediction stage because both models may lead to widely different results under future conditions. The second type of problems arise from improper extension of calibration to prediction conditions, i.e., from not taking into account changes in the natural system or in the scale of the problem. The only way we think about dealing with this problem consists of evaluating carefully whether or not the assumptions in which the calibration was based are still valid under future conditions. Indeed, model uncertainties can be very large.

We do not think that, strictly speaking, the second type of uncertainties, those associated with future stresses, falls in the realm of modeling. While future stresses may affect the validity of the model, they are external to it. In any case, this type of uncertainty is evaluated by carrying out simulations under a number of alternative scenarios, whose definition is an important subject in itself.

The last set of prediction uncertainties is the one associated with parameter uncertainties, which can be quantified quite well.

CONCLUSION

Groundwater modeling involves so many subjective decisions that it can be considered as an art. This is somewhat contrary to the widely accepted perception of models as something objective. The fact is that numerous assumptions need to be made both about the selection of relevant processes and about the manner of representing them in the computer. All these assumptions are specified in the conceptual model.

The result relies so heavily on conceptualization that models ought to be viewed as theories about the behavior of natural systems. Model predictions should rarely be viewed as firm statements about the future evolution of aquifers. Rather, they should be considered references against which actual data has to be compared. Codes do exist for modeling most processes affecting groundwater (flow, transport, reactions, thermomechanics, etc). It is lack of understanding and lack of data what limits the actual application of those codes.

Having specified a conceptual model, the remaining steps (discretization, calibration, uncertainty analysis, prediction) are relatively objective, in the sense that systematic procedures can be followed. This explains why conceptualization is so important. It also explains why modeling is the best way of integrating widely different data. Uncertain as it is, it may represent unambiguously the overall knowledge of the aquifer.

Models represent the water balance (or solute balance, or energy balance) at the overall aquifer and at each of its parts. Therefore, they can also be viewed as accounting systems. It is argued that well managed aquifers need real time models to help decision making, the same way that well managed companies need financial accounting systems. This is the challenge modelers must meet in the near future.

REFERENCES

1. Neuman, S.P. Saturated–Unsaturated Seepage by Finite Elements. ASCE, J. Hydraulics Div. **1973**, *99* (12), 2233–2250.
2. Abriola, L.M.; Pinder, G.F. A Multiphase Approach to the Modeling of Porous Media Contamination by Organic Compounds. Water Resour. Res. **1985**, *21* (1), 11–26.
3. Konikow, L.F.; Bredehoeft, J.D. Groundwater Models Cannot Be Validated. Adv. Water Resour. **1992**, *15* (1), 75–83.
4. Gorelick, S. A Review of Distributed Parameter Groundwater Management Modeling Methods. Water Resour. Res. **1983**, *19*, 305–319.
5. Freeze, R.A.; Witherspoon, P.A. Theoretical Analysis of Regional Groundwater Flow. II, Effect of Water Table

Configuration and Subsurface Permeability Variations. Water Resour. Res. **1966**, *3*, 623–634.

6. Garven, G.; Freeze, R.A. Theoretical Analysis of the Role of Groundwater Flow in the Genesis of Stratabound Ore Deposits. II, Quantitative Results. Am. J. Sci. **1984**, *284*, 1125–1174.

7. Ayora, C.; Tabener, C.; Saaltink, M.W.; Carrera, J. The Genesis of Dedolomites: A Discussion Based on Reactive Transport Modeling. J. Hydrol. **1998**, *209*, 346–365.

8. Sánchez-Vila, X.; Girardi, J.; Carrera, J. A Synthesis of Approaches to Upscaling of Hydraulic Conductivities. Water Resour. Res. **1995**, *31* (4), 867–882.

9. Meier, P.; Carrera, J.; Sánchez-Vila, X. An Evaluation of Jacob's Method Work for the Interpretation of Pumping Tests in Heterogeneous Formations. Water Resour. Res. **1998**, *34* (5), 1011–1025.

10. Carrera, J.; Mousavi, S.F.; Usunoff, E.; Sánchez-Vila, X.; Galarza, G. A Discussion on Validation of Hydrogeological Models. Reliab. Eng. Syst. Saf. **1993**, *42*, 201–216.

11. Anderson, M.P.; Woessner, W.W. *Applied Groundwater Modeling*; Academic Press: San Diego, 1992.

12. Konikow, L.F.; Bredehoeft, J.D. *Computer Model of Two-Dimensional Solute Transport and Dispersion in Groundwater: Reston, VA*; U.S. Geo. Surv. TWRI, Book 7, 1, 1978; Chap. C2, 40.

13. NEA-SKI, *The International HYDROCOIN Project, Level 2: Model Validation*; OECD: Paris, France, 1990; 194.

14. Carrera, J.; Neuman, S.P. Estimation of Aquifer Parameters Under Steady State and Transient Conditions: I Through III. Water Resour. Res. **1986**, *22* (2), 199–242.

15. Loaiciga, H.A.; Marino, M.A. The Inverse Problem for Confined Aquifer Flow: Identification and Estimation with Extensions. Water Resour. Res. **1973**, *23* (1), 92–104.

16. Hill, M.C. A Computer Program (MODFLOWP) for Estimating Parameters of a Transient, Three Dimensional, Groundwater Flow Model Using Nonlinear Regression, USGS Open-File Report, 1992; 91–484.

Groundwater Modeling: How Codes Work

Jesús Carrera
Technical University of Catalonia (UPC), Barcelona, Spain

INTRODUCTION

Numerical methods are tools used by people who develop codes for solving equations governing groundwater problems. All problems that we are interested in are governed by partial differential equations. The computer cannot directly solve these and one needs numerical methods to transform them into a solvable form. In essence, all numerical methods are based on, first, discretizing (i.e., substituting the continuum by a discrete medium) and, second, approximating the differential equation by a system of equations. Numerical methods differ in the way discretization and approximations are performed. To illustrate these two steps, we will first develop them in detail for a generic numerical method.

We will, then, introduce the classical numerical methods (finite element, finite differences, etc.). This section ends with a discussion on specific methods for solute transport.

A GENERIC NUMERICAL METHOD FOR SOLVING GROUNDWATER FLOW

As mentioned earlier, all methods require, first, discretizing and, second, approximating the physical phenomenon. For the generic method we are going to present here, discretization will be performed as shown in Fig. 1. That is, the continuum aquifer domain will be substituted by a discrete number of cells. Furthermore, the continuum aquifer heads, $h(x,y)$, are substituted by a discrete number of model heads, h_i.

The second step, approximation, can be made in different manners. For the purpose of this section, it is sufficient to bear in mind that the flow equation is nothing but a mass balance. Therefore, we will express the mass balance in cell i as change in storage equals inflows minus outflows.

$$\Delta S_i = f_{ij} + f_{il} + f_{im} + f_{in} + g_i \qquad (1)$$

where ΔS_i is the rate of change in storage during one time step (say, between time t^k and time t^{k+1}); f_{ij} is the inflow into cell i from cell j (and the same for f_{il}, f_{im}, and f_{in}); and g_i are external inflows into cell i (for example, recharge, minus pumping, minus evaporation, minus river outflow, etc.). Each of the terms in Eq. 1 is relatively easy to approximate.

Storage variation can be derived from the definition of storage coefficient, (S is the change in volume of water stored per unit surface area of aquifer and per unit change in head):

$$\Delta S_i = SA_i \frac{(h_i^{k+1} - h_i^k)}{\Delta t} \qquad (2)$$

where A_i is the surface area of cell i, h_i^k is the head in node i at time k and $\Delta t = t^{k+1} - t^k$ is the time step. Darcy's law gives lateral inflows

$$f_{ij} = -Tw_{ij} \frac{h_i - h_j}{L_{ji}} = a_{ij}(h_i - h_j) \qquad (3)$$

where T is transmissivity; w_{ij} is the width of the connection between nodes i and j; L_{ji} is the length of such connection; and a_{ij} is implicitly defined as $-Tw_{ij}/L_{ji}$. The remaining inflow terms, f_{il}, f_{im}, and f_{in} are defined likewise. Changing these terms to the left-hand side of Eq. 1 and rearranging terms yields:

$$SA_i \frac{(h_i^{k+1} - h_i^k)}{\Delta t} + a_{ii}h_i + a_{ij}h_j + a_{il}h_l$$
$$+ a_{im}h_m + a_{in}h_n = g_i \qquad (4)$$

where $a_{ii} = -a_{ij} - a_{il} - a_{im} - a_{in}$. If an equation like Eq. 4 is written for all cells from $i = 1$ through N, N being the number of cells (nodes), the resulting system of equations can be rewritten in matrix form as:

$$\mathbf{D} \frac{(\mathbf{h}^{k+1} - \mathbf{h}^k)}{\Delta t} + \mathbf{Ah} = \mathbf{g} \qquad (5)$$

where $\mathbf{D}$ is a diagonal matrix whose i-th diagonal term is precisely SA_i. This matrix is often called storage matrix. $\mathbf{A}$ is the conductance matrix, a square symmetric matrix whose components are a_{ij}. Finally, $\mathbf{g}$ is the source vector.

All the numerical methods to be outlined in subsequent sections lead to equations analogous to Eq. 5. Moreover, the meaning of the terms in such equations is always similar to that in Eq. 5. Namely, the system represents the mass balance at each of the N nodes (cells); specifically the i-th equation represents the mass balance at the i-th node. The first term, $\mathbf{D}(\mathbf{h}^{k+1} - \mathbf{h}^k)/\Delta t$, always represents storage variations. The second term, $\mathbf{Ah}$, represents outflows from minus inflows into the i-th cell from

Encyclopedia of Water Science
DOI: 10.1081/E-EWS 120018705

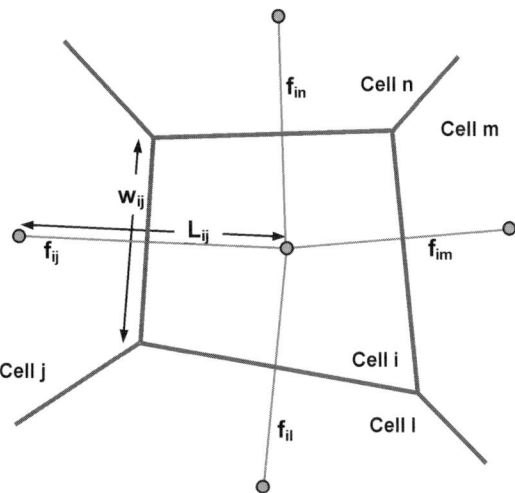

Fig. 1 The system of equations for a generic groundwater model is obtained by establishing a water balance (inputs minus outputs equal storage variations) at each cell. Inputs and outputs include water exchanges with the outside (pumping Q_i, recharge, r_i, etc.) and with adjacent cells (f_{ij}). The latter are expressed, using Darcy's law, as $f_{ij} = T_{ij}w_{ij}(h_j - h_i)/L_{ij}$.

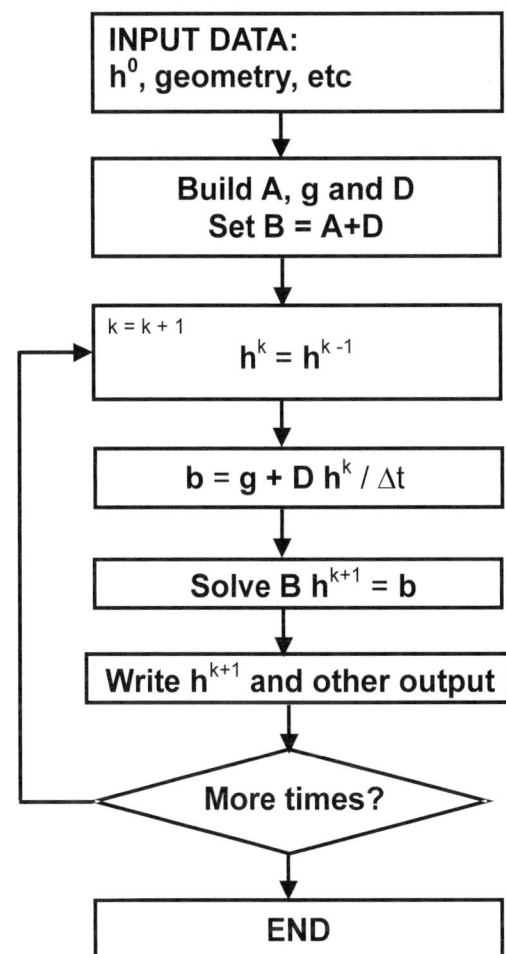

Fig. 2 Basic steps involved in simulating groundwater flow. Heads $\mathbf{h}^{k+1}$ are computed by solving equation $\mathbf{Bh}^{k+1} = \mathbf{b}$. They may be written for later drawings. They are used as initial head for the next time increment. These steps (time loop) are repeated sequentially until the last time is reached.

the adjacent cells. Finally, term **g** represents external inflows minus outflows (recharge, pumping, etc.) at all i.

Eq. 5 needs to be integrated in time. For this purpose, let us assume that **Ah** is evaluated at time $k + 1$ ($\mathbf{Ah}^{k+1}$). Then, Eq. 5 can be rewritten as:

$$\left(\mathbf{A} + \frac{\mathbf{D}}{\Delta t}\right)\mathbf{h}^{k+1} = g + \frac{\mathbf{D}}{\Delta t}\mathbf{h}^k \tag{6}$$

This is simply a linear system, which can be solved using conventional methods.

$$\mathbf{Bh}^{k+1} = \mathbf{b} \tag{7}$$

where $\mathbf{B} = \mathbf{A} + \mathbf{D}/\Delta t$ and $\mathbf{b} = \mathbf{g} + \mathbf{Dh}^k/\Delta t$. This system is solved sequentially in time.

That is, most codes solve Eq. 7 using the following steps (Fig. 2):

1. Input all data. Set $k = 0$
2. Compute **g**, **A** (Eq. 3); **D** (Eq. 5) and **B** (Eq. 7)
3. Set $k = k + 1$
4. Build **b** (Eq. 7)
5. Solve $\mathbf{Bh}^{k+1} = \mathbf{b}$
6. If $k = k_{\max}$ (maximum number of time steps), end. Otherwise, return to step 3.

Most codes follow a structure such as this, although each method displays specific features. Some of these are outlined below.

FINITE DIFFERENCES (FD)

As mentioned at the beginning, numerical methods differ in the way in which the domain is discretized and in the way in which the partial differential equation is transformed into a linear system of equations. In finite differences, the problem domain is discretized in a regular grid (Fig. 3a), usually rectangular (equilateral triangles or hexagons are possible, but very rare). The grid may be centered at the corners (nodes are located at the vertices of the squares) or at the cells (nodes are located at the center of the squares, such as in Fig. 3a).

Regarding the approximation of the partial differential equations, several alternatives are possible. The most intuitive consists of substituting all derivates by an incremental ratio. That is, the derivative between adjacent

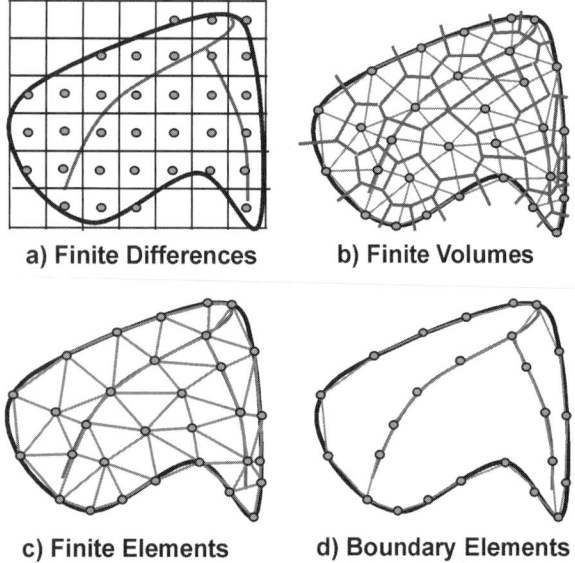

a) Finite Differences   **b) Finite Volumes**

c) Finite Elements   **d) Boundary Elements**

Fig. 3 The most widely spread methods of discretization are Finite Differences, which consists of subdividing the model domain into regular rectangles, and Finite Elements, which is based on dividing the aquifer region into elements of arbitrary shape (often triangles). Finite Volumes, also called Integrated Finite Differences, divides the region into polygons. The Boundary Element Method is very convenient, when applicable, because it only requires discretizing boundaries (both internal and external).

nodes i and j is approximated as:

$$\frac{\partial h}{\partial x} = \frac{h_i - h_j}{\Delta x} \tag{8}$$

where h_i and h_j are heads at nodes i and j, respectively, and Δx is the distance between them. Approximating all derivatives by means of equations analogous to Eq. 8 leads to a system identical to Eq. 5. In fact, the finite differences method is often introduced using a mass balance approach such as the one in the section "A Generic Numerical Method for Solving Groundwater Flow," only using a regular instead of a generic grid. This is the method used in MODFLOW,[1] HST3D,[2] and their children.

INTEGRATED FINITE DIFFERENCES (IFD)

The basic philosophy of this method is very similar to that of the generic method introduced in the generic numerical section. Basically, the domain is discretized in a number of cells centered around arbitrarily located nodes. Frequently, the cells are the Thiessen polygons of the set of nodes. This allows adapting the node density to the problem (e.g., increasing nodes density where accuracy is needed most).

Model equations can be derived using a mass balance approach, such as in the generic method section. Integrating the flow equation over each cell and applying Green's identity to transform volume integrals in boundary fluxes can also yield model equations. This type of approach is the basis of the finite volume method, which is widely used nowadays.

FINITE ELEMENT METHOD

Finite element method (FEM) discretization consists of elements and nodes. Elements are generalized polygons (normally triangles or curvilinear quadrilaterals). Nodes are points located at the vertices and, sometimes, at the sides or the middle of the element. Unlike FD or IFD, cells around the nodes are not defined. Still, in many cases, one may write the equations in such a way that the mass balance formulation of the generic method section is still valid. However, the most singular feature of the FEM is the way the solution is interpolated, so that it becomes defined at every point. That is, head (or concentrations) is approximated as:

$$h(x) \cong \hat{h}(x) = \sum_i h_i N_i(x) \tag{9}$$

where h_i are nodal heads and N_i are interpolation functions. Since $\hat{h}$ is not the exact solution, it would yield a residual if substituted in the flow equation. Minimizing this residual, which requires somewhat sophisticated maths, leads to a system similar to Eq. 5.

BOUNDARY ELEMENT METHOD

The idea behind the boundary element method (BEM) is similar to that of the FEM. The main difference stems from the choice of interpolation functions, which are taken as the fundamental solutions of the flow equation (or whatever equation is to be solved). As a result, when the corresponding $\hat{h}$ is substituted in the flow equation, the residuals are zero. Since the equation is satisfied exactly in the model domain, one is only left with boundary conditions. In fact, as shown in Fig. 3d, discretization is only required at the boundaries, where boundary heads are defined so as to satisfy approximately the boundary conditions. This method is extremely accurate, but its applicability is limited by the need of finding the fundamental solutions. This constrains the BEM to flow problems in relatively homogeneous domains.

SIMULATING SOLUTE TRANSPORT

All the methods discussed above can be used for simulating solute transport. They are called Eulerian methods because they are based on a fixed (as opposed to moving) grid and all derivatives are based on a fixed coordinate system. They work fine when dispersion is dominant. Otherwise, they may lead to numerical problems (Fig. 4). Two dimensionless numbers are used to anticipate numerical difficulties. Specifically, discretization must satisfy the following conditions.

Peclet number

$$Pe = \frac{v\Delta x}{D} \cong \frac{\Delta x}{\alpha} < \frac{1}{2} \qquad (10)$$

Courant number

$$Co = \frac{v\Delta t}{\Delta x} < 1 \qquad (11)$$

where v is the solute velocity, α is dispersivity, Δx is the distance between nodes and we have assumed that $D \cong \alpha v$. The condition on the Courant number implies that the solute at one node will not move beyond the following node downstream during one time step. This condition is easy to meet because usually groundwater moves slowly and also reducing Δt to satisfy Eq. 11 is not difficult. The condition on the Peclet number, Eq. 10, implies that Δx is smaller than $\alpha/2$, which may require very small elements, leading to a huge computational burden. Because of this, conventional Eulerian methods are not applicable to many groundwater problems, which have motivated the search of alternative methods.

Alternative methods can be Eulerian or Lagrangian. Among the former, the most popular is upstream weighting, introduced by Heinrich et al.[3] in the FEM, but with a huge number of papers thereafter. It consists of slightly modifying Eulerian equations so as to ensure stability. The problem is that, in doing so, it introduces numerical dispersion. As a result, the wiggles at the solute front in Fig. 4 are substituted by an artificially smeared front. However, the vast majority of alternative formulations for solute transport are Lagrangian in the sense that time variations are written in terms of the material derivative, which expresses the rate of change in concentration of a particle that moves with the water. In this way, the advective term, which is the cause of problems in eulerian formulations, disappears.

The number of Lagrangian methods is very large and many researchers have devoted much effort to find one, which is universal. The fact that so many methods have survived to date suggests that the effort has not been fully successful. Still, in practical problems, one can usually find a suitable method. Following is an outline of some of the most popular methods, with a discussion of their advantages and disadvantages and early references. The interested reader should seek further.

The most natural Lagrangian method is to write the equations on a moving grid, that is on a grid whose nodes move with water. This method is highly accurate, but expensive because the grid has to be updated every time step. Moreover, the grid can become highly deformed over time.

To avoid problems with moving grids it is frequent to work with particles. Displacing the particles with the moving water represents advection, while dispersion can be represented with a variety of methods. One such possibility is to add a random component to each particle basis displacement. This is statistically equivalent to each particle basis dispersion and is the basis of the "random walk" method.[4] The method requires careful implementation, but its main drawback is the fact that the solution is given in terms of number of particles per cell. If one is interested in spatial distributions of the solute, a huge number of particles may be needed.

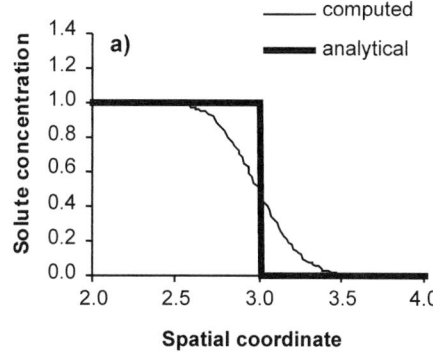
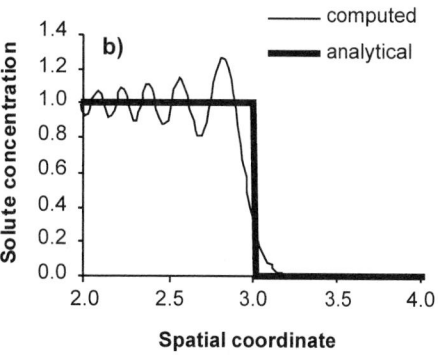

Fig. 4 Difficulties typically associated to numerical simulation of solute transport: a) front smearing: the concentration front is more dispersed than it should; b) instability oscillations: too high and/or too low (even negative) concentrations.

The method of characteristics (MOC) overcomes the above difficulty by assigning concentrations to particles and interpolating them onto a fixed grid, where dispersion and, possibly, other transport processes are modeled. Concentrations are then interpolated back onto the particles. The method has become very popular in groundwater because of the USGS MOC Code.[5] The method is very practical, but the interpolation back-and-forth between particles and grid may introduce numerical dispersion and mass balance errors.

The modified method of characteristics[6] tries to overcome the problems associated to interpolating particle concentrations by redefining them in each time interval so that at the end of the time step they coincide with a node location. The method is very accurate, although some interpolation errors still occur when the front is abrupt. Some of these problems are overcome by the Eulerian–Lagrangian Localized Adjoint Method,[7] which looks as the most promising method.

CONCLUSION

Computer codes are available for simulating all phenomena affecting groundwater. In essence, they represent the balance of water (or salt, contaminants, or energy) in a manner that can be solved by the computer. This is achieved by, first, discretizing the problem and, second, rewriting it as a system of equations. This type of approach has been successful for flow problems. Solute transport, on the other hand, remains ellusive. No single method is universally successful. Instead, one must seek the appropriate code in each case.

REFERENCES

1. McDonald, M.G.; Harbaugh, A.W. A Modular Three-Dimensional Finite Difference Groundwater Flow Model. *Techniques of Water Resources Inv.*, of the USGS, Book 6, Ch. A1, 1988.
2. Kipp, K.L. HST3D, A Computer Code for Simulation of Heat and Solute Transport in Three Dimensional Groundwater Flow Systems. USGS Water Recourses Inv. Rep. 86-4095; 1987.
3. Heinrich, J.C.; Huyakorn, P.S.; Mitchell, A.R.; Zienkiewicz, O.C. An Upwind Finite Element Scheme for Two-Dimensional Transport Equation. Int. J. Numer. Methods Eng. **1977**, *11*, 131–143.
4. Prickett, T.A.; Naymik, T.G.; Lonnquist, C.G. A "Random Walk" Solute Transport Model for Selected Groundwater Quality Evaluations. Illinois State Water Survey Bull. **1981**, *55*, 62.
5. Konikow, L.F.; Bredehoeft, J.D. *Computer Model of Two-Dimensional Solute Transport and Dispersion in Ground Water*; U.S. Geo. Surv. TWRI: Reston, Va., 1978; Book 7, 1 Chap. C2, 40p.
6. Neuman, S.P. A Eulerian–Lagrangian Numerical Scheme for the Dispersion–Convection Equation Using Conjugate Space–Time Grids. J. Comput. Phys. **1981**, *41*, 270–294.
7. Celia, M.A.; Russell, T.F.; Herrera, I.; Ewing, R.E. An Eulerian–Lagrangian Localized Adjoin Method for the Advection–Diffusion Equation. Adv. Water Resour. **1990**, *13*, 187–206.

Groundwater Pollution from Mining

George F. Vance
University of Wyoming, Laramie, Wyoming, U.S.A.

Jeffrey G. Skousen
West Virginia University, Morgantown, West Virginia, U.S.A.

G

INTRODUCTION

Surface and underground mining activities can have direct and indirect impacts on the quantity, quality, and usability of groundwater supplies. The nature of the mining activity, geological substrata, and re-distribution of surface and subsurface materials will determine to a large degree how groundwater supplies will be impacted. As waters interact and alter the disturbed geologic materials, constituents such as salts, metals, trace elements, and/or organic compounds become mobilized.[1,2] Once mobilized, the dissolved substances can leach into deep aquifers, resulting in groundwater quality impacts. In addition to concerns due to naturally occurring contaminants, mining activities may also contribute to groundwater pollution from leaking underground storage tanks, improper disposal of lubricants and solvents, contaminant spills as well as others.

In the United States, the *Clean Water Act* (CWA), which was enacted in 1948 as the *Water Pollution Control Act* (WPCA), and the CWA amendments in 1977, establishes the authority for all water pollution control actions at the federal level.[3] The *Safe Drinking Water Act* (SDWA), which was enacted in 1974 and amended in 1996, was promulgated to protect drinking water supplies by legislating maximum contaminant levels (MCLs) above which waters are considered unsafe for human consumption, and defined enforcement standards that states are required to use for determining minimum treatments needed to improve water quality.[4] Examples of some MCLs that may be associated with water quality issues relating to mining activities are listed in Table 1.

Because mining activities can result in poor quality groundwaters, enforcement of regulations is needed to minimize and/or eliminate potential problems. The Surface Mining Control and Reclamation Act (SMCRA) of 1977 specifies policies and practices for mining and reclamation to minimize water quality impacts.[6] The SMCRA requires that specific actions be taken to protect the quantity and quality of both on- and off-site groundwaters. All mines are required to meet either state or federal groundwater guidelines, which are generally related to priority pollutant standards described in the CWA.

GROUNDWATER RESOURCES

Our groundwater resources are the world's third largest source of water and represent 0.6% of the earth's water content. Approximately 53% of the U.S. population uses groundwater as a drinking water source, but this percentage increases to almost 97% for rural households. In areas of low rainfall, weathering and translocation of dissolved constituents is relatively slow compared to high rainfall areas. In addition, physical disruption of rocks into small particles can enhance mineral weathering that results in mineral dissolution and migration of dissolved substances. Transport of contaminants from surface and subsurface environments to groundwaters is generally accelerated as the amount of percolating water increases.

Infiltrating water moves through the vadose zone (unsaturated region) into groundwater zones (saturated region). The upper boundary of the groundwater system (e.g., water table) fluctuates depending on the amount of water received by, or depleted from, the groundwater zone. Groundwater movement is a function of hydraulic gradients and hydraulic conductivities, which represent the ease with which water moves as a function of gravitational forces and the permeability of substrata materials. Groundwater moves faster in coarse textured substrata and as the slope of the water table increases. Aquifers are groundwater systems that have sufficient porosity and permeability to supply enough water for a specific purpose. In order for an aquifer to be useful, it must be able to store, transmit, and yield sufficient amounts of good quality water. Important hydrogeological characteristics of a site that determine groundwater quantity and quality are listed in Table 2.

Encyclopedia of Water Science
DOI: 10.1081/E-EWS 120010062

Table 1 Select contaminants in drinking waters that may be influenced by mining activities

Contaminant	MCL[a] (mg/L)	MCLG[b] (mg/L)
Inorganics		
Arsenic	0.006	0.006
Cadmium	0.005	0.005
Chromium	0.1	0.1
Copper	LV[c]	1.3
Cyanide	0.2	0.2
Fluoride	4	4
Lead	0.015	0
Mercury	0.002	0.002
Nickel	0.1	0.1
Nitrate (NO_3-N)	10	10
Selenium	0.05	0.05
Sulfate	500	500
Thallium	0.002	0.0005
Radionuclides		
Radon	300 q/L	0
Uranium	0.02 q/L	0
Organics		
Benzene	0.005	0
Carbon tetrachloride	0.005	0
Pentachlorophenol	0.001	0
Toluene	1	1
Xylenes	10	10
Microbiological		
Total coliforms	LV	0
Viruses	LV	0

[a] MCL = maximum contaminant levels permissible for a contaminant in water that is delivered to any user of a public water system.
[b] MCLG = maximum contaminant level goals of a drinking water contaminant that is protective of adverse human health effects and which allows for an adequate margin of safety.
[c] LV = lowest value that can be achieved using best available technology.
Source: (From Ref. [5].)

GROUNDWATER CONTAMINANTS

There are several types of substances that can affect groundwater quality.[1,7] Water contaminants include inorganic, organic, and biological materials, of which some have a direct impact on water quality, whereas others indirectly cause physical, chemical, or biological changes. Substances that can impact groundwaters include nutrients, salts, heavy metals, trace elements, and organic chemicals, as well as contaminants such as radionuclides, carcinogens, pathogens, and petroleum wastes (Table 3). Some groundwaters are derived from mining activities contain natural (e.g., methane gas) and synthetic organic chemicals. Organic contamination may result from leaking gas tanks, oil spills, or run-off from equipment-servicing areas. In these cases, the source of the contamination must be identified and removed. Gasoline, diesel, or oil-soaked

Table 2 Important hydrogeological characteristics of a site that determine groundwater quantity and quality

Geological
    Type of water-bearing unit or aquifer (overburden, bedrock)
    Thickness, areal extent of water-bearing units and aquifers
    Type of porosity (primary, such as intergranular pore space, or secondary, such as bedrock discontinuities, e.g., fracture or solution cavities)
    Presence or absence of impermeable units or confining layers.
    Depths to water tables; thickness of vadose zone.

Hydraulic
    Hydraulic properties of water-bearing unit or aquifer (hydraulic conductivity, transmissivity, storability, porosity, dispersivity)
    Pressure conditions (confined, unconfined, leaky confined)
    Groundwater flow directions (hydraulic gradients, both horizontal and vertical), volumes (specific discharge), rate (average linear velocity)
    Recharge and discharge areas
    Groundwater or surface water interactions; areas of ground water discharge to surface water
    Seasonal variations of groundwater conditions

Groundwater use
    Existing or potential underground sources of drinking water
    Existing or near-site use of groundwater

areas should be immediately excavated and disposed of by approved methods.

The chemistry of groundwaters and potential levels of naturally occurring contaminants are related to: 1) groundwater hydrologic conditions; 2) mineralogy of the mined and locally impacted geological material; 3) mining operation (e.g., extent of disturbed materials and its exposure to atmospheric conditions); and 4) time. Movement of metal contaminants in groundwater varies depending on the chemical of concern, and include considerations such as with cobalt (Co), copper (Cu), nickel (Ni), and zinc (Zn) mobility being greater than silver (Ag) and lead (Pb), which tend to be more mobile than gold (Au) and tin (Sn).[1] As conditions such as pH, redox, and ionic strength change over time, dissolved constituents in groundwaters may decrease due to adsorption, precipitation, and chemical speciation reactions and transformations.

Acid mine drainage (AMD) is most prevalent at inactive and abandoned surfaces and underground mine sites. If geological substrata containing reduced S minerals [e.g., pyrite (FeS_2)] is exposed to oxygen (O_2), such as when pyritic overburden materials are brought to the earth surface during mining activities and then re-buried, high concentrations of sulfuric acid (H_2SO_4) can develop and form acid waters with pH levels below 2. Neutralization of some of the acidity produced during the oxidation of

Table 3 Different classes of groundwater contaminants and their origins

Water Contaminant Class	Contributions
Inorganic chemicals	Toxic metals and acidic substances from mining operations and various industrial wastes
Organic chemicals	Petroleum products, pesticides, and materials from organic wastes industrial operations
Infectious agents	Bacteria and viruses from sewage and other organic wastes
Radioactive substances	Waste materials from mining and processing of radioactive substances or from improper disposal of radioactive isotopes

Source: (From Ref. [8].)

reduced S-compounds occurs when silicate minerals dissolve; however, during this process, high levels of potentially toxic metals such as aluminum (Al), copper (Cu), cadmium (Cd), iron (Fe), manganese (Mn), Ni, Pb, Zn, may be released. For example, mining of coal in the Toms Run area of northwestern Pennsylvania resulted in groundwater contamination by AMD containing high concentrations of Fe and sulfate (SO_4) that leached into the underlying aquifer through joints, fractures, and abandoned oil and gas wells.

The Gwennap Mining District in the United Kingdom contained numerous mines that operated over several centuries to extract various mineral resources. One of these mines, the Wheal Jane metalliferous mine in Cornwall, extracted ores that included cassiterite (Sn-containing mineral), chalcopyrite (Cu), pyrite (Fe), wolframite [tungsten (W)], arsenopyrite [arsenic (As)], in addition to smaller deposits of Ag, galena (Pb), and other minerals.[9] After closure in the early 1990s, extensive voids remaining in the Wheal Jane mine that contained oxidized and weathered minerals were flooded. Initial groundwater quality was poor with a pH of 2.8 and a total metal concentrations close to 5000 mg/L, which contained high levels of Fe, Zn, Cu, and Cd. Water quality worsened with depth, and at 180 m the groundwater had a pH of 2.5 and metal concentrations of 2200, 1500, 44, and 5 mg/L for Fe, Zn, Cu, and Cd, respectively. Current treatment of discharge waters originating from the mine involves an expensive process and will continue to be long-term if environment quality in the region is to be preserved. A similar situation occurred when a Zn mine in southwestern France was closed. However, after flooding, discharge mine waters contained high concentrations of Zn, Cd, Mn, Fe, and SO_4 even though the solution pH was near neutral.

Within the Coeur D'Alene District of Idaho, location of the Bunker Hill Superfund site, groundwater samples have been found to contain high concentrations of Zn, Pb, and Cd.[10] The contamination was believed to originate from

the leaching of old mine tailings that were deposited on a sand and gravel aquifer. When settling ponds were developed nearby the old tailings, re-charge of the local groundwaters resulted in a rise in the water table that saturated the tailings causing considerable metal leaching to occur.

Gold mining operations have used cyanide as a leaching agent to solubilize Au from ores, which often contain arsenopyrite [As, Fe, sulfur (S)], and in some cases pyrite.[1] During the leaching process, Ag is also recovered as a by-product if present. Unfortunately, cyanide is a powerful nonselective solvent that will solubilize numerous substances that can be environmental contaminants. These ore waste materials are often stored in tailing ponds, and depending on the local geology and climate, cyanide present in the tailings can exist as free cyanide (CN^-, HCN), inorganic compounds (NaCN, $HgCN_2$), metal–cyanide complexes with Cu, Fe, Ni, and Zn, and/or the compound CNS. Because cyanide species are mobile and persistent under certain conditions, there is the potential for trace element and cyanide migration into groundwaters. For example, a tailings dam failure resulted in cyanide contamination of groundwater at a gold mining operation in British Columbia, Canada.[1]

Arsenic and uranium (U) contamination has resulted from extensive mining and smelting of ores containing various metals (Ag, Au, Co, Ni, Pb, and Zn) and/or nonmetals (As, phosphorus (P), and U). Contaminated As groundwaters have been a source of surface re-charge and drinking water supplies; As in a contaminated river of Canada were 7 and 13 times greater than the recommended national and local drinking water standards, respectively.[1] Arsenic is known as a carcinogen and has been the contributing cause of death to humans in several parts of the world that rely on As-contaminated drinking waters.[7] Waters from dewatering a U mine in New Mexico had elevated levels of U and radium (Ra) activities as well as high concentrations of dissolved molybdenum (Mo) and selenium (Se), which were detected in stream waters 140 km downstream from the mine.

GROUNDWATER ANALYSIS

Both remediation and prevention of groundwater contamination by nutrients, salts, heavy metals, trace elements, organic chemicals (natural and synthetic), pathogens, and other contaminants requires the evaluation of the composition and concentration of these constituents either in-situ or in groundwater samples.[2] Monitoring may require the analysis of physical properties, inorganic and organic chemical compositions, and/or microorganisms according to well-established protocols for sampling, storage, and analysis.[11] For example, if groundwaters will be used for human or animal consumption, the most appropriate tests would be nitrate-nitrogen (NO_3-N), trace metals, pathogens, and organic chemicals. Several common constituents measured in groundwaters are listed in Table 4; however, other tests can be conducted on waters including tests for hardness, chlorine, radioactivity, or water toxicity.[12]

Recommendations based on interpretation of the groundwater test results should be related to the ultimate use of the water.[2] The interpretation and recommendation processes may be as simple as determining that a drinking water well exceeds the established MCLs for NO_3-N and recommending the well should not be used as a drinking water source or that a purification system be installed. However, interpretations of most groundwater analyses can be quite complicated and require additional information for proper interpretation. If a contaminant exceeds an acceptable concentration, all potential sources contributing to the pollution and pathways by which the contaminant moves must be determined. In many cases, multiple groundwater contaminants are present at different concentrations. Because the interpretation of water analyses is a complex process, recommendations should be based on a complete evaluation of the water's physical, chemical, and biological properties. Integrating water analyses into predictive models that can assess the effects

of mining activities on water quality is needed in the long term to determine the most effective means to preserve and restore water quality.

STRATEGIES FOR REMEDIATING CONTAMINATED GROUNDWATERS

Mine sites that have been contaminated generally contain mixtures of inorganic and/or organic constituents, so it is important to understand these multicomponent systems in order to develop remediation strategies. Therefore, a proper remediation program must consider *identification*, *assessment*, and *correction* of the problem.[13] Identification of a potential problem site requires that either the past history of the area and activities that took place are known, or when a water analysis indicates a site has been contaminated. Assessment addresses questions such as is there a problem, where is the problem, and what is the extent of the problem? Afterwards a *remediation action* plan must be developed that will address the specific problems identified. A remediation action program may require that substrata materials (e.g., backfill) and groundwater be treated.

If remedial action is considered necessary, three general options are available—containment, in-situ treatment, or pump-and-treat (Fig. 1). The method(s) used for the containment of contaminants are beneficial for restricting contaminant movement. Of the remediation techniques, in-situ treatment measures are the most appealing because they generally do less surface damage, require a minimal amount of facilities, reduce the potential for human exposure to contaminants, and when effective, reduce or remove the contaminant. In-situ remediation can be achieved by physical, chemical, and/or biological techniques. Biological in-situ techniques used for groundwater bioremediation can either rely on the indigenous (native) microorganisms to degrade organic contaminants or on amending the groundwater environment with microorganisms (bioaugmentation). The pump-and-treat method, however, is one of the more commonly used processes for remediating contaminated

Table 4 Groundwater quality parameters and constituents measured in some testing programs

Physical parameters	Metals and trace elements	Nonmetallic constituents	Organic chemicals	Microbiological parameters
Conductivity, salinity, sodicity, dissolved solids, temperature, odors	Al, Ag, As, Ca, Cd, Cr, Cu, Fe, Mg, Mn, Na, Ni, Pb, Se, Sr, Zn	pH, acidity, alkalinity, dissolved O_2, B, CO_2, HCO_3, Cl, CN, F, I, NH_4, NO_2, NO_3, P, Si, SO_4	Methane, oil and grease, organic acids, volatile acids, organic C, pesticides, phenols, surfactants	Coliforms, bacteria, viruses

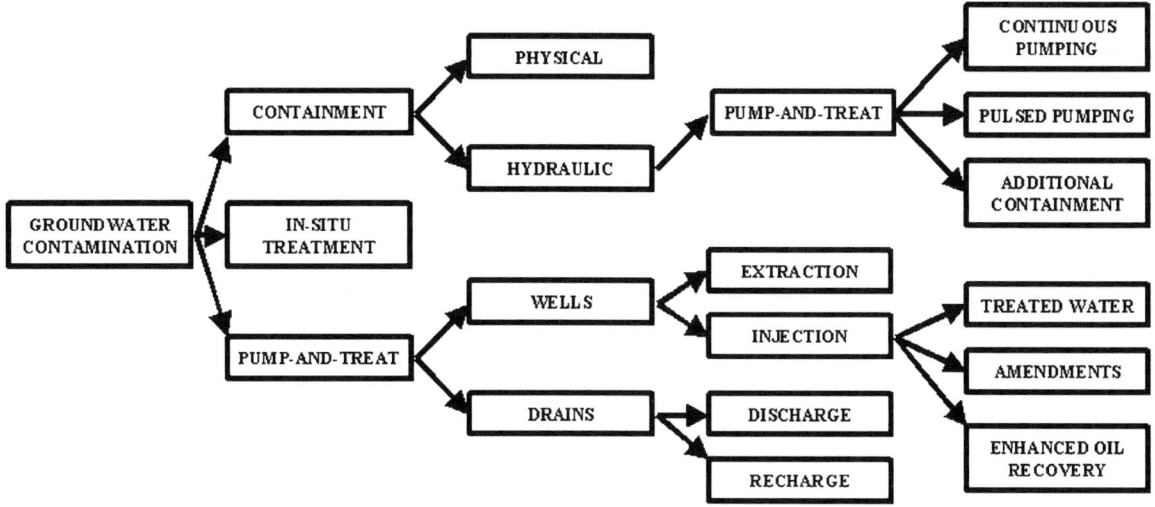

Fig. 1 Remediation options to consider if cleanup of contaminated groundwater is required.

groundwaters. With the pump-and-treat method, the contaminated waters are pumped to the surface where one of many treatment processes can be utilized. A major consideration in the pump-and-treat technology is the placement of wells, which is dependent on site characteristics (see Table 2). Extraction wells are used to pump the contaminated water to the surface where it can be treated and re-injected or discharged. Injection wells can be used to re-inject the treated water, water containing nutrients and other substances that increase the chances for chemical alteration or microbial degradation of the contaminants, or materials for enhanced oil recovery.

Treatment techniques can be grouped into three categories including physical, chemical, and biological methods.[2,13]

Physical methods include several techniques. *Adsorption* methods physically sorb or trap contaminants on various types of resins. *Separation* treatments include physically separating contaminants by forcing water through semipermeable membranes (e.g., reverse-osmosis). *Flotation*, or density separation, is commonly used to separate low-density organic chemicals from groundwaters. *Air and steam stripping* can remove volatile organic chemicals. *Isolation* utilizes barriers placed above, below, or around sites to restrict movement of the contaminant; containment systems should have permeabilities of 10^{-7} cm/sec (approximately 0.1 ft/yr) or less.

Chemical methods are also numerous. *Chemical treatment* involves addition of chemical agent(s) in an injection system to neutralize, immobilize and/or chemically modify contaminants. *Extraction (leaching)* of contaminants uses one of several different aqueous extracting agents such as an acid, base, detergent, or organic solvent miscible in water. *Oxidation and reduction*

of groundwater contaminants is commonly done using air, oxygen, ozone, chlorine, hypochlorite, and hydrogen peroxide. *Ionic and nonionic* exchange resins can adsorb contaminants, reducing their leaching potential.

Biological methods for contaminant remediation are less extensive than physical and chemical techniques. *Land treatment* is an effective method for treating groundwaters by applying the contaminated waters to lands using surface, overland flow, or subsurface irrigation. *Activated sludge and aerated surface impoundments* are used to precipitate or degrade contaminants present in water and include both aerobic and anaerobic processes. *Biodegradation* is one of several biological-mediated processes that transform contaminants and utilizes vegetation and microorganism.

REFERENCES

1. Ripley, E.A.; Redmann, R.E.; Crowder, A.A. *Environmental Effects of Mining*; St. Lucie Press: Delray Beach, FL, 1996; 356.
2. Pierzynski, G.M.; Sims, J.T.; Vance, G.F. *Soils and Environmental Quality*, 2nd Ed.; CRC Press, Inc.: Boca Raton, FL, 2000; 459.
3. U.S. Congress. *Clean Water Act*; Public Law 95-217; 1977.
4. U.S. Congress. *Safe Drinking Water Act* and Amendments; Public Law 104-182; 1996.
5. U.S. Environmental Protection Agency Office of Water. *Drinking Water Regulations and Health Advisories*, USEPA Report 822-B-96; USEPA: Washington, DC; 1996.
6. U.S. Congress. *Surface Mining Control and Reclamation Act*; Public Law 95-87; 1977.
7. Manahan, S.E. *Environmental Chemistry*, 5th Ed.; Lewis Publishers: Chelsea, MI, 1991.

8. Nielsen, D.M. *A Practical Handbook of Groundwater Monitoring*; Lewis Publishers: Boca Raton, FL, 1991.

9. Bowen, G.G.; Dussek, C.; Hamilton, R.M. Pollution Resulting from the Abandonment and Subsequent Flooding of Wheal Jane Mine in Cornwall, UK. In *Groundwater Contaminants and Their Migration*, Special Publication 128; Mather, J., Banks, D., Dumpleton, S., Fermor, M., Eds.; 1998; 93–99.

10. Todd, D.K.; McNulty, D.E.O. *Polluted Groundwater*; Water Information Center, Inc.: Huntington, NY, 1975; 32–33.

11. Clesceri, L.S.; Greenberg, A.E.; Eaton, A.D., (Eds.) *Standard Methods for the Examination of Water and Wastewater*, 20th Ed.; American Public Health Association: Washington, DC, 1998; 1220.

12. U.S. Environmental Protection Agency. Quality Criteria for Water, Document 440/5-86-001; EPA: Washington, DC, 1986.

13. Hyman, M.H. Groundwater and Soil Remediation. In *Encyclopedia of Environmental Pollution and Cleanup*; Meyers, R.A., Editor-in-Chief; John Wiley & Sons, Inc.: New York, NY, 1999; Vol. 1, 684–712.

Groundwater Pollution by Nitrogen Fertilizers

Lloyd B. Owens
*United States Department of Agriculture (USDA),
Coshocton, Ohio, U.S.A.*

INTRODUCTION

Why is nitrogen in groundwater a problem? High nitrate levels in water consumed by humans can cause adverse health problems, and groundwater is a major source of water for human consumption. A part of groundwater resurfaces to feed surface water from streams to oceans. High levels of nitrogen can cause excess plant and bacterial growth, which upon death and decay can deplete much of the oxygen in water. This causes fish kills and "dead zones," such as the area of hypoxia in the Gulf of Mexico. This encyclopedia entry discusses agricultural practices, e.g., row crops, grasslands/turf, container horticultural crops, that contribute to nitrogen in groundwater. Some agricultural practices to reduce the contributing factors are also presented.

WHY NITROGEN IN GROUNDWATER IS A PROBLEM

Human Health Impacts

Groundwater is a major source of water for human consumption. Nitrogen present in groundwater is usually in the form of nitrate (NO_3), and at high levels can pose major human health concerns, especially for infants. The link between high NO_3 in polluted water and serious blood changes in infants was first reported in 1945. From 1947 to 1950, 139 cases of methemoglobinemia were reported, including 14 deaths in Minnesota alone. Thus, a standard has been set that NO_3 in excess of $45 \, mg \, L^{-1}$ ($10 \, mg \, L^{-1}$ NO_3-N) is considered hazardous to human health.[1]

Environmental Impacts

A part of groundwater, especially shallow groundwater, resurfaces to feed streams, rivers, and reservoirs and eventually estuaries and oceans. Nutrients, pollutants, in the groundwater are carried via these routes as well and can cause excess plant and bacterial growth in aquatic systems. The decay of this organic matter can deplete much of the oxygen in the water causing fish kills and "dead zones" to occur. Phosphorus receives much of the attention in regards to eutrophication in fresh waters

because it often is the limiting nutrient. But as water systems become more brackish, there is a shift to N limitation.[2] A major example of this situation is the area of hypoxia in the Gulf of Mexico.[3] Hypoxia occurs when the concentration of dissolved oxygen is less than $2 \, mg \, L^{-1}$.

Nitrogen contributions to the Gulf of Mexico, and other large bodies of water, come via surface runoff and resurfacing of groundwater. There are also several agricultural sources of nitrogen, e.g., nitrogen fertilizer, surface application of manure, manure from grazing systems, and mineralization of organic matter. However, the focus of this chapter will be only on groundwater, and how it is impacted by the leaching of nitrogen fertilizers. Although several aspects of nitrate leaching will be addressed and accompanied by supporting references, space does not permit this chapter to be a comprehensive literature review.

AGRICULTURAL PRACTICES CONTRIBUTING TO THE NITROGEN IN GROUNDWATER PROBLEM

Row Crops

High levels of NO_3-N in subsurface drainage from row crops, especially corn (*Zea mays* L.), are well documented. Nitrate-N concentrations in tile lines draining silt loam soils in Iowa with fertilized, continuous corn, or corn in rotation already exceeded $10 \, mg \, L^{-1}$ two decades ago.[4,5] Other high NO_3-N levels have been reported in tile lines with clay loams in Minnesota;[6–8] with silty clay loams in Illinois;[9] with silt loams in Indiana;[10] with silt loams/silty clay loams in Ohio;[11] and with clay over silty clay loam and fine sand over clay in Ontario, Canada.[12] Analyses of subsurface water collected with monolith lysimeters[13,14] and ceramic porous-cup samplers[15,16] are in agreement with these findings.

Nitrate-N concentrations have been studied in tile drains frequently because of their wide spread use and the relative ease of collecting a sample. The majority of NO_3-N moves in the subsurface water during the winter recharge period.[9,10,13] There are several factors that impact the amount of N export from tiles, including timing

Encyclopedia of Water Science
DOI: 10.1081/E-EWS 120010199

and area of N fertilization.[9] Increasing the drain spacing decreases the NO$_3$-N losses in tiles[6,10] although the NO$_3$-N concentration in the tiles may change very little.[10] Even though the increased drain spacing should reduce the NO$_3$-N losses in the tile, it probably increases the NO$_3$-N losses in seepage below the drains. Model simulation studies show that reducing N fertilization rates will have much greater impact for reducing NO$_3$-N losses than changing tile drain spacing or depth.[6]

Too often, inexpensive N fertilizer has been applied in excess to crops to ensure that inadequate N will not limit crop yields. The difficulty in synchronizing N applications with crop needs contributes to such practices. It has been shown that there is a direct relationship between NO$_3$-N loss by leaching and application rates of N that exceed crop needs.[15,17] Excess N in soil can result from overapplication of N fertilizers or manure or from residual N from the previous year (as well as from mineralization of organic N). This can be a particular problem following a dry year because reduced crop growth will not utilize as much N fertilizer as during a year when a "normal" amount of water was available.[8] Therefore, there is an increased amount of residual N to begin the next cropping season. Even at economic optimum N (EON) levels, considering all sources of N, concentrations of NO$_3$-N in subsurface water have been found to exceed the $10 \, mg \, L^{-1}$ maximum contaminant level (MCL).[15,17] The conclusion can be drawn that optimum corn production will likely produce elevated NO$_3$-N concentrations in groundwater.[17]

In irrigated agriculture, a similar impairment to groundwater quality exists from N fertilizer management. High concentrations of NO$_3$-N were found in subsurface water under a sprinkler irrigated crop rotation in Spain,[18] a sprinkler irrigated corn–soybean rotation in Nebraska,[19] and flood irrigated wheat in Arizona.[20] Even with irrigation BMPs, NO$_3$-N concentrations in groundwater above the MCL can be expected.

Grasslands/Turf

Because of the animal component, NO$_3$-N leaching in grazed grassland is quite complex. Leaching of NO$_3$-N from grasslands is greatly increased with the presence of grazing livestock.[21,22] Even on highly fertilized pastures, much of the leached NO$_3$-N has been attributed to excreta.[23,24] Studies in England,[25,26] the Netherlands,[27] and the eastern USA[28,29] have shown that NO$_3$-N concentrations in subsurface water are often greater than $10 \, mg \, L^{-1}$ when $> 100 \, kg \, N \, ha^{-1}$ is applied annually to grazed grasslands. Other processes, such as the accumulation of fertilizer-N during drought or the release of N from decaying plant material, e.g., resulting from tilling or killing the sod in preparation for reseeding, may

influence N leaching from the pasture as a whole, rather than acting specifically on areas affected by urine.[23] In some nongrazed systems, NO$_3$ leaching from highly fertilized systems is low, e.g., $29 \, kg \, N \, ha^{-1}$ lost from ryegrass (*Lolium perenne* L.) receiving $420 \, kg \, N \, ha^{-1}$.[30]

Fertilized turf, whether it be home lawns or golf courses, raises environmental issues. Annual applications up to $244 \, kg \, N \, ha^{-1}$ to turfgrass on sandy loam soils in Rhode Island do not appear to pose a threat to drinking water aquifers,[31] although overwatering can cause increased N loadings to bays and estuaries in coastal areas. The excess N movement would be more prevalent with late summer N applications. Nitrate-N concentrations in subsurface water were the highest on an Ohio silt loam in the late summer and early autumn but did not exceed the MCL when $220 \, kg \, N \, ha^{-1}$ per year was applied to turfgrass.[32] The exception was the occurrence of high NO$_3$-N concentrations with the soil disturbance during establishment of the turf. Grass sod has the capacity to use large amounts of N; 85%–90% of fertilizer N can be retained in the turf–soil ecosystem.[33] Roots and thatch can represent a large N pool because it becomes available for mineralization and subsequent leaching if disturbed.[34] Reseeding and sod establishment within 2 mo of "turf death" can stabilize this N pool.[33] High rates of NO$_3$ leaching can occur at very high N fertilizer rates, e.g., $450 \, kg \, N \, ha^{-1}$ per year. Even though most of the NO$_3$ leaching occurred in the autumn and winter, it was an accumulation of all N fertilizer application and not just the autumn application.[35] Excess NO$_3$ in the fall is the driving force that causes NO$_3$ leaching, regardless of the N source or time of application. Therefore, high rates of N application to turf should be avoided in the fall, because it can result in high NO$_3$ leaching rates. A survey of several golf courses across the USA indicated that NO$_3$-N concentrations above the MCL occurred in only 4% of the samples;[36] most of these were apparently due to prior agricultural land use. Pollution of groundwater by NO$_3$ leaching from N fertilized turf should be minimal with good management, which includes consideration of soil texture, N source, rate and timing, and irrigation/rainfall.[37]

Container Horticultural Crops

Although the acreage for container horticultural crops is small compared to row crops or grasslands, the production intensity is great and "hot spots" of potential NO$_3$ leaching could develop. Assuming $80,000 \, pots \, ha^{-1}$ for a typical foliage plant nursery and using a soluble granular fertilizer, over $650 \, kg \, N \, ha^{-1}$ could be lost through leaching annually.[38] During a 10-week greenhouse study of potted flowers, average NO$_3$-N in the leachate ranged from $250\,mg\,N\,L^{-1}$ to $450\,mg\,N\,L^{-1}$.[39] As long as the amount of water applied to the plants did not exceed plant usage (and the greenhouse canopy remained intact to prevent precipitation inputs), there would be little NO_3 movement from the soil beneath the pots, unless there was a high water table. Nevertheless, this area of N accumulation would eventually need to be addressed. The use of controlled release fertilizers is one practice that can significantly reduce leaching losses.[38] Also, vegetable crops that have high N demand but low apparent N recovery, e.g., sweet peppers, can leave large amounts of N in the soil and residues at harvest.[40]

AGRICULTURAL PRACTICES TO MITIGATE THE NITROGEN IN GROUNDWATER PROBLEM

Use of Winter Cover Crops

Winter cover crops have been shown to be an effective strategy in reducing NO_3 leaching during the winter period.[41–44] A variety of crops, e.g., annual grasses, cereals, legumes, have been used with varying degrees of success depending on soils, climate, cropping sequences, etc. Care needs to be exercised with long-term cover crops, because if they are disturbed, some of the accumulated N may become mineralized and actually increase NO_3 leaching.[45] Sometimes cover crops cannot be counted on as a best management practice (BMP) to reduce NO_3 leaching.[46] On the Delmarva Peninsula in the Mid-Atlantic U.S.A, a rye winter cover crop following corn did not reduce NO_3 leaching. One factor was that the existing crop did not permit a sufficiently early seeding of the cover crop.

Use of Soil Nitrate Tests

Preplant N tests (PPNT) or presidedress N tests (PSNT) can assess the N stored in soil from cover crops and help to give adequate N credits for legume N carry-over in a crop rotation, such as a soybean N credit in a corn–soybean rotation. Nitrification inhibitors used with N fertilizer in the ammoniacal form can slow the rate of oxidation of reduced forms of N to NO_3-N, and subsequently decrease the amount of NO_3-N leaching,[47] especially with fall N applications.

Even with these improved practices, it may be necessary to reduce the N fertilization rate below the EON level to achieve NO_3-N concentrations in groundwater below the MCL.

Use of Alternate Grassland/Turf Management

Several management options to reduce nitrate leaching from grasslands include the use of grass–legume mixtures instead of highly fertilized grass;[24,48] coordinating the timing and N fertilizer application rate with other N sources, e.g., manure applications, to avoid excessive N application;[49] use of irrigation, especially during dry periods, to encourage N uptake;[24] and an integration of cutting forage and grazing, especially cutting in late summer areas that have been intensively grazed earlier in the year. In areas where NO_3 contamination from turf is a concern, late summer N fertilizer applications should be reduced and watering should be limited.[31]

REFERENCES

1. U.S. Public Health Service, *Public Health Drinking Water Standards*, U.S. Public Health Publ. 956; U.S. Government Printing Office: Washington, DC, 1996.
2. Correll, D.L. The Role of Phosphorus in the Eutrophication of Receiving Waters: A Review. J. Environ. Qual. **1998**, *27* (2), 261–266.
3. Burkart, M.R.; James, D.E. Agricultural-Nitrogen Contributions to Hypoxia in the Gulf of Mexico. J. Environ. Qual. **1999**, *28* (3), 850–859.
4. Baker, J.L.; Campbell, H.P.; Johnson, H.P.; Hanway, J.J. Nitrate, Phosphorus, and Sulfate in Subsurface Drainage Water. J. Environ. Qual. **1975**, *4* (3), 406–412.
5. Baker, J.L.; Johnson, H.P. Nitrate-Nitrogen in Tile Drainage as Affected by Fertilization. J. Environ. Qual. **1981**, *10* (4), 519–522.
6. Davis, D.M.; Gowda, P.H.; Mulla, D.J.; Randall, G.W. Modeling Nitrate Nitrogen Leaching in Response to Nitrogen Fertilizer Rate and Tile Drain Depth or Spacing for Southern Minnesota, USA. J. Environ. Qual. **2000**, *29* (5), 1568–1581.
7. Randall, G.W.; Huggins, D.R.; Russelle, M.P.; Fuchs, D.J.; Nelson, W.W.; Anderson, J.L. Nitrate Losses Through Subsurface Tile Drainage in Conservation Reserve Program, Alfalfa, and Row Crop Systems. J. Environ. Qual. **1997**, *26* (5), 1240–1247.
8. Randall, G.W.; Iragavarapu, T.K. Impact of Long-Term Tillage Systems for Continuous Corn on Nitrate Leaching to Tile Drainage. J. Environ. Qual. **1995**, *24* (2), 360–366.
9. Gentry, L.E.; David, M.B.; Smith, K.M.; Kovacic, D.A. Nitrogen Cycling and Tile Drainage Nitrate Loss in Corn/Soybean Watershed. Agric. Ecosyst. Environ. **1998**, *68*, 85–97.
10. Kladivko, E.J.; Grochulska, J.; Turco, R.F.; Van Scoyoc, G.E.; Eigel, J.D. Pesticide and Nitrate Transport into Subsurface Tile Drains of Different Spacings. J. Environ. Qual. **1999**, *28* (3), 997–1004.
11. Logan, T.J.; Schwab, G.O. Nutrient and Sediment Characteristics of Tile Effluent in Ohio. J. Soil Water Conserv. **1976**, *31* (1), 24–27.

12. Miller, M.H. Contribution of Nitrogen and Phosphorus to Subsurface Drainage Water from Intensively Cropped Mineral and Organic Soils in Ontario. J. Environ. Qual. **1979**, *8* (1), 42–48.

13. Cookson, W.R.; Rowarth, J.S.; Cameron, K.C. The Effect of Autumn Applied ^{15}N-Labelled Fertilizer on Nitrate Leaching in a Cultivated Soil During Winter. Nutr. Cycl. Agroecosyst. **2000**, *56*, 99–107.

14. Owens, L.B.; Malone, R.W.; Shipitalo, M.J.; Edwards, W.M.; Bonta, J.V. Lysimeter Study on Nitrate Leaching from a Corn–Soybean Rotation. J. Environ. Qual. **2000**, *29* (2), 467–474.

15. Andraski, T.W.; Bundy, L.G.; Brye, K.R. Crop Management and Corn Nitrogen Rate Effects on Nitrate Leaching. J. Environ. Qual. **2000**, *29* (4), 1095–1103.

16. Steinheimer, T.R.; Scoggin, K.D.; Kramer, L.A. Agricultural Chemical Movement Through a Field-Size Watershed in Iowa: Subsurface Hydrology and Distribution of Nitrate in Groundwater. Environ. Sci. Technol. **1998**, *32*, 1039–1047.

17. Jemison, J.M., Jr. Fox, R.H. Nitrate Leaching from Nitrogen-Fertilized and Manured Corn Measured with Zero-Tension Pan Lysimeters. J. Environ. Qual. **1994**, *23* (2), 337–343.

18. Diez, J.A.; Caballero, R.; Roman, R.; Tarquis, A.; Cartegena, M.C.; Vallejo, A. Integrated Fertilizer and Irrigation Management to Reduce Nitrate Leaching in Central Spain. J. Environ. Qual. **2000**, *29* (5), 1539–1547.

19. Klocke, N.L.; Watts, D.G.; Schneekloth, J.P.; Davison, D.R.; Todd, R.W.; Parkhurst, A.M. Nitrate Leaching in Irrigated Corn and Soybean in a Semi-arid Climate. Trans. ASAE **1999**, *42* (2), 1621–1630.

20. Ottman, M.J.; Tickes, B.R.; Husman, S.H. Nitrogen-15 and Bromide Tracers of Nitrogen Fertilizer Movement in Irrigated Wheat Production. J. Environ. Qual. **2000**, *29* (5), 1500–1508.

21. Ball, P.R.; Ryden, J.C. Nitrogen Relationships in Intensively Managed Temperate Grasslands. Plant Soil **1984**, *76*, 23–33.

22. Ryden, J.C.; Ball, P.R.; Garwood, E.A. Nitrate Leaching from Grassland. Nature **1984**, *311*, 50–53.

23. Cuttle, S.P.; Scurlock, R.V.; Davies, B.M.S. A 6-Year Comparison of Nitrate Leaching from Grass/Clover and N-Fertilized Grass Pastures Grazed by Sheep. J. Agric. Sci., Cambridge **1998**, *131*, 39–50.

24. Whitehead, D.C. Leaching of Nitrogen from Soils. In *Grassland Nitrogen*; CAB International: Wallingford, UK, 1995; 129–151.

25. Haigh, R.A.; White, R.E. Nitrate Leaching from a Small, Underdrained, Grassland, Clay Catchment. Soil Use Manag. **1986**, *2*, 65–70.

26. Roberts, G. Nitrogen Inputs and Outputs in a Small Agricultural Catchment in the Eastern Part of the United Kingdom. Soil Use Manag. **1987**, *3*, 148–154.

27. Steenvoorden, J.H.A.M.; Fonck, H.; Oosterom, H.P. Losses of Nitrogen from Intensive Grassland Systems by Leaching and Surface Runoff. In *Nitrogen Fluxes in Intensive Grassland Systems*; van der Meer, H.G., Ryden, J.C., Ennik, G.C., Eds.; Martinus Nijhoff Publ.: Dordrecht, The Netherlands, 1986; 85–97.

28. Owens, L.B.; Van Keuren, R.W.; Edwards, W.M. Nitrogen Loss from a High-Fertility, Rotational Pasture Program. J. Environ. Qual. **1983**, *12* (3), 346–350.

29. Owens, L.B.; Edwards, W.M.; Van Keuren, R.W. Nitrate Levels in Shallow Groundwater Under Pastures Receiving Ammonium Nitrate or Slow-Release Nitrogen Fertilizer. J. Environ. Qual. **1992**, *21* (4), 607–613.

30. Garwood, E.A.; Ryden, J.C. Nitrate Loss Through Leaching and Surface Runoff from Grassland: Effects of Water Supply, Soil Type and Management. In *Nitrogen Fluxes in Intensive Grassland Systems*; van der Meer, H.G., Ryden, J.C., Ennik, G.C., Eds.; Martinus Nijhoff Publ.: Dordrecht, The Netherlands, 1986; 99–113.

31. Morton, T.G.; Gold, A.J.; Sullivan, W.M. Influence of Overwatering and Fertilization on Nitrogen Losses from Home Lawns. J. Environ. Qual. **1988**, *17* (1), 124–130.

32. Geron, C.A.; Danneberger, T.K.; Traina, S.J.; Logan, T.J.; Street, J.R. The Effects of Establishment Methods and Fertilization Practices on Nitrate Leaching from Turfgrass. J. Environ. Qual. **1993**, *22* (1), 119–125.

33. Bushoven, J.T.; Jiang, Z.; Ford, H.J.; Sawyer, C.D.; Hull, R.J.; Amador, J.A. Stabilization of Soil Nitrate by Reseeding with Ryegrass Following Sudden Turf Death. J. Environ. Qual. **2000**, *29* (5), 1657–1661.

34. Jiang, Z.; Bushoven, J.T.; Ford, H.J.; Sawyer, C.D.; Amador, J.A.; Hull, R.J. Mobility of Soil Nitrogen and Microbial Responses Following the Sudden Death of Established Turf. J. Environ. Qual. **2000**, *29* (5), 1625–1631.

35. Roy, J.W.; Parkin, G.W.; Wagner-Riddle, C. Timing of Nitrate Leaching from Turfgrass After Multiple Fertilizer Applications. Water Qual. Res. J. Canada **2000**, *35* (4), 735–752.

36. Cohen, S.; Svrjcek, A.; Durborow, T.; Barnes, N.L. Water Quality Impacts by Golf Courses. J. Environ. Qual. **1999**, *28* (3), 798–809.

37. Petrovic, A.M. The Fate of Nitrogenous Fertilizers Applied to Turfgrass. J. Environ. Qual. **1990**, *19* (1), 1–14.

38. Broschat, T.K. Nitrate, Phosphorus, and Potassium Leaching from Container-Grown Plants Fertilized by Several Methods. Hort. Sci. **1995**, *30* (1), 74–77.

39. McAvoy, R.J. Nitrate Nitrogen Movement Through the Soil Profile Beneath a Containerized Greenhouse Crop Irrigated with Two Leaching Fractions and Two Wetting Agent Levels. J. Am. Soc. Hort. Sci. **1994**, *119* (3), 446–451.

40. Tei, F.; Benincasa, P.; Guiducci, M. Nitrogen Fertilisation of Lettuce, Processing Tomato and Sweet Pepper: Yield, Nitrogen Uptake and the Risk of Nitrate Leaching. Acta Hort. **1999**, *506*, 61–67.

41. McCracken, D.V.; Smith, M.S.; Grove, J.H.; MacKown, C.T.; Blevins, R.L. Nitrate Leaching as Influenced by Cover Cropping and Nitrogen Source. Soil Sci. Soc. Am. J. **1994**, *58* (5), 1476–1483.

42. Rasse, D.P.; Ritchie, J.T.; Peterson, W.R.; Wei, J.; Smucker, A.J. Rye Cover Crop and Nitrogen Fertilization

Effects on Nitrate Leaching in Inbred Maize Fields. J. Environ. Qual. **2000**, *29* (1), 298–304.

43. Zhou, X.; MacKenzie, A.F.; Madramootoo, C.A.; Kaluli, J.W.; Smith, D.L. Management Practices to Conserve Soil Nitrate in Maize Production Systems. J. Environ. Qual. **1997**, *26* (5), 1369–1374.

44. Ball-Coelho, B.R.; Roy, R.C. Overseeding Rye into Corn Reduces NO_3 Leaching and Increases Yield. Can. J. Soil Sci. **1997**, *77*, 443–451.

45. Hansen, E.M.; Djurhuus, J.; Kristensen, K. Nitrate Leaching as Affected by Introduction or Discontinuation of Cover Crop Use. J. Environ. Qual. **2000**, *29* (4), 1110–1116.

46. Ritter, W.F.; Scarborough, R.W.; Christie, A.E.M. Winter Cover Crops as a Best Management Practice for Reducing Nitrogen Leaching. J. Contam. Hydrol. **1998**, *34* (1), 1–15.

47. Owens, L.B. Nitrate Leaching Losses from Monolith Lysimeters as Influenced by Nitrapyrin. J. Environ. Qual. **1987**, *16* (1), 34–38.

48. Owens, L.B.; Edwards, W.M.; Van Keuren, R.W. Groundwater Nitrate Levels Under Fertilized Grass and Grass–Legume Pastures. J. Environ. Qual. **1994**, *23* (4), 752–758.

49. Jarvis, S.C. Progress in Studies of Nitrate Leaching from Grassland Soils. Soil Use Manag. **2000**, *16*, 152–156.

G

Groundwater Pollution by Phosphorus Fertilizers

Bahman Eghball
United States Department of Agriculture (USDA), Lincoln, Nebraska, U.S.A.

INTRODUCTION

Phosphorus (P) is a primary nutrient necessary for plant growth. When the P level in soil is below what is essential for plant needs, P is supplied to the soil by the addition of P fertilizer or organic residuals (i.e., manure). Because of the P fertilizer use in the past few decades or application of manure or other organic residuals, a greater portion of the soils in each state in the United States have soil test P levels that exceed the critical level for plant growth. The excess P in soil is then subjected to leaching loss or transport in surface runoff either in soluble or in particulate (sediment-bound) forms. Phosphorus that is moving downward in the soil profile can eventually reach the ground water, especially in areas with shallow or perched groundwater. Phosphorus moving downward in the soil may also be intercepted by artificial drainage systems (i.e., tile drains) that are located within $1\,m - 2\,m$ from the soil surface.

PHOSPHOROUS LEACHING AND FIXATION

Phosphorus leaching can occur as a slow process in the soil or rapidly with preferential flow, which is the movement of water and solute through cracks and earthworm holes in soil. The point at which P might come in contact with groundwater depends on soil properties and the proximity of the groundwater to the soil surface. The U.S. Environmental Protection Agency has no safe drinking water concentration limit for P. The major concern about P enrichment of groundwater is that groundwater frequently emerges as surface water and if it contains sufficient P, can cause eutrophication (nutrient enrichment that causes algae bloom and oxygen depletion in water).

Phosphorus movement in soil is primarily through the diffusion process with the rate influenced by the amount of P applied, soil water content, bulk density (i.e., porosity), and chemical reaction of P with the soil constituents. The average rate of diffusion in three Nebraska medium and fine-textured soils was $0.00011\,cm^2\,hr^{-1}$ ($0.000017\,in.^2\,hr^{-1}$) following the application of P fertilizer ($15\,kg\,P\,ha^{-1}$) to replace what is needed by a corn crop with the expected yield of $5600\,kg\,ha^{-1}$ ($90\,bu\,acre^{-1}$).[1] The P in soil is not usually an environmental concern until the soil test P is in the very high (excessive) category for plant needs (Fig. 1). High levels of P in soil can be a source of groundwater pollution when P is leached in soil. This is especially of concern when the groundwater is near the soil surface; groundwater has an upward movement toward the soil surface in certain times of the year, and in coarse-textured soils. In areas where groundwater is deep, the pollution of groundwater with P is of little concern even if the P level in soil is excessive. For example, P fertilizer applied at a rate of $100\,kg\,ha^{-1}\,yr^{-1}$ to a sandy loam soil with a P adsorption capacity of $150\,mg\,kg^{-1}$ and a bulk density of $1.4\,kg\,m^{-3}$ would take 21 yr to reach a 1 m (3.3 ft) soil depth, assuming no preferential flow. If the water table was located several meters deep, it would take many years for the applied P to reach the groundwater. Factors that can influence P leaching in soil are given in Table 1.

Applied fertilizer P interacts with various constituents in soil and can be readily immobilized. Phosphorus in all chemical fertilizers (except rock phosphate) is about 100% plant-available and therefore, their reaction in soil should be similar. The mechanism involved in reducing P movement in acid soils includes the reaction of ortho-phosphate ions ($H_2PO_4^-$ and HPO_4^{2-}) with iron and aluminum to form insoluble compounds. In alkaline soils, the P retention mechanisms include precipitation of calcium phosphate compounds, surface precipitation of P on solid phase calcium carbonate, and retention of P by clay particles that are saturated with calcium. Therefore, P leaching in soil is very limited since P interacts with the soil constituents. Eghball et al.[1] found that maximum fertilizer P movement from a band applied at $60\,kg\,P\,ha^{-1}$ was about 4 cm in 3 mo in three different soils. The size of the band was not expected to expand much after 3 mo. The most P movement occurred in the first few weeks after application. However, P can leach deep into the soil with preferential flow (a small number of pores is used to move water) where P moves with water through cracks and earthworm holes. Preferential flow is the primary mechanism for deep movement of P in fine-textured soils (clayey types) either in particulate form or in soluble form.

There are several factors that influence fixation of P in the soil. These include amount and type of clay, time of reaction, soil pH, temperature, and organic matter. The greater the clay content of a soil, the more P is adsorbed.

Encyclopedia of Water Science
DOI: 10.1081/E-EWS 120010200

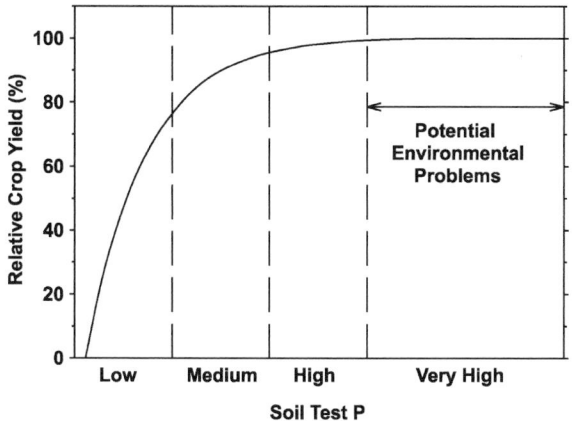

Fig. 1 Relative crop yield as influenced by soil test P level. (From Ref. [4].)

More P is retained by the 1:1 clays (i.e., kaolonitic type) that are found in the humid and tropic areas (high rainfall high temperature) than other clay types. Phosphorus fixation in soil increases with time after P addition indicating that P gradually becomes more insoluble with time unless plants remove P from soil. Phosphorus fixation also increases with increasing temperature. Increasing soil organic matter usually results in increased P solubility and thus reduces P fixation in soil. Soils that are primarily made up of muck and peat are subject to increased P leaching from added P fertilizer.

PHOSPHOROUS IN TILE DRAINS

Phosphorus can leach into the drainage water in fields with tile drains. Tiles are placed 1 m – 2 m (3.3 ft – 6.6 ft) deep in poorly drained soils, so there is a potential for the applied fertilizer P to leach into the soil and reach the tile drain.

Table 1 Factors that can influence phosphorus leaching in the soil

Factor	P leaching risk
Excessive soil P level	High
Shallow or perched water table, waterlogged conditions	High
Fine-textured soils	Low
Fine-textured soils with preferential flow	High
Coarse-textured soils (sandy)	High
Tile drain	High
Soils with high organic matter	Low
Organic soils (soil with organic matter > 20%)	High
High soil Ca, Al, or Fe contents	Low

The tile water usually empties into the surface water and if it contains sufficient P, can cause eutrophication. Since most of the tile drains are located in humid and semi-humid regions with large rainfall potential, leaching of P through thin soil layers above the tile-drains can especially be high. In a study conducted in Canada,[2] tile-drain water samples were collected from 27 fields that mainly received P fertilizer. Drain water from 14 out of 27 fields exceeded the Canadian standard of 0.03 mg total PL^{-1} in the water. More than 80% of the P in tile drain was particulate (sediment-bound) and dissolved organic P. Of those exceeding the standard, 10 out of 14 were clayey soils with medium to high soil P levels indicating loss of P through preferential flow.

PHOSPHOROUS SATURATION

Phosphorus leaching can also occur in areas with a naturally high P level in the soil. In these areas, P can leach deep into the soil, especially coarse-textured soils, and reach the groundwater. The closer the ground water is to the soil surface, the greater the potential of P reaching the water body. The capacity of a soil to retain P is limited. Repeated and/or heavy application of P fertilizer can saturate the upper soil layers with P. Usually the fraction of P saturated soil decreases with depth. However, when the upper layers of soil have been saturated, P movement to the subsoil can occur. Long-term (> 50 yr) application of beef cattle feedlot manure and chemical fertilizer resulted in P leaching to a maximum of 1.8 m (6 ft) depth in a sandy loam soil.[3] Phosphorus from manure source moved deeper in the soil than P from chemical fertilizer indicating that some of the manure P components were not subject to P fixation by the calcium carbonate layer located about 0.75 m (2.5 ft) deep in this soil.

REFERENCES

1. Eghball, B.; Sander, D.H.; Skopp, J. Diffusion, Adsorption and Predicted Longevity of Banded Phosphorus Fertilizer in Three Soils. Soil Sci. Soc. Am. J. **1990**, *54*, 1161–1165.
2. Beauchemin, S.; Simard, R.R.; Cluis, D. Forms and Concentration of Phosphorus in Drainage Water of Twenty-seven Tile-drained Soils. J. Environ. Qual. **1998**, *27*, 721–728.
3. Eghball, B.; Binford, G.D.; Baltensperger, D.D. Phosphorus Movement and Adsorption in a Soil Receiving Long-term Manure and Fertilizer Application. J. Environ. Qual. **1996**, *25*, 1339–1343.
4. Sharpsburg, A.N.; Daniel, T.C.; Edwards, D.R. Phosphorus Movement in the Landscape. J. Prod. Agric. **1999**, *6*, 492–500.

Groundwater Pumping Methods

Dennis E. Williams
Geoscience Support Services, Inc., Claremont, California, U.S.A.

INTRODUCTION

Groundwater has been used for municipal, industrial, irrigation, and other purposes since prehistoric times. In today's world, groundwater is becoming increasingly more important as a reserve against drought, especially in the arid and semiarid lands. Extraction of groundwater from aquifers beneath the earth is the subject of this section. Various methods of pumping groundwater will be discussed ranging from simple hand-powered systems to high-capacity deep-well turbine pumps.

REVIEW OF BASIC GEOHYDROLOGIC PRINCIPLES

Prior to any discussion of groundwater pumping methods, the reader should be familiar with some fundamental principles of groundwater—how it occurs, how it moves, and what governs its movement from areas of recharge to areas of discharge—irrespective of whether the discharge is natural or withdrawn through man-made devices (e.g., wells and pumps).

Aquifers are geologic formations or groups of formations capable of yielding water in usable quantities. Groundwater is the subsurface runoff component of the hydrologic cycle, which moves through and is stored in the interstitial spaces found between the solid particles of geologic formations. In unconsolidated materials such as sand and gravel, groundwater moves through the pore space, which occurs, between individual grains of solid material. This pore space is called primary porosity. In consolidated rocks (e.g., granite, volcanics, and limestone), groundwater moves through secondary porosity created as the result of fracturing, fissuring, or weathering. Groundwater flows from areas of recharge to areas of discharge with the rate and direction of flow governed by both the magnitude of the decreasing hydraulic head and the nature of the aquifer materials. Under the same hydraulic gradient, groundwater moves faster through more permeable materials (e.g., coarse sand and gravel) and slower through less permeable materials such as silty sands.

Aquifers may be grouped into three main types: confined, unconfined, and semiconfined, depending upon their subsurface layering and permeability. Confined aquifers, also known as artesian aquifers, are saturated formations found between low permeability materials. The low permeability materials prevent movement of water into or out of the saturated zone. Unconfined or water table aquifers have no upper confining layers. Semiconfined or "leaky" aquifers have semipervious layering either above or below and as a result, may allow water to flow vertically into or out of the aquifer depending on the difference in vertical hydraulic gradients.

As groundwater moves, it may discharge naturally to the earth's surface resulting in a spring, or contributing to the inflow of a lake or stream. Groundwater may also be artificially extracted from the subsurface through pumping or flowing wells. Flowing wells occur in confined (artesian) aquifers where the hydraulic head rises above the top of the well casing. Groundwater discharge to springs occurs when the groundwater surface intersects the land surface—usually in the sides of steep canyons (Fig. 1). Similarly, groundwater may flow into a subsurface drain or trench when groundwater levels in the aquifer are higher than that in the drain or trench. The Ghanats of Iran are an example of groundwater flowing into man-made subterranean tunnels. A Ghanat consists of a series of vertical (hand-dug) shafts typically spaced approximately 100 m ($\sim$300 ft) apart roughly paralleling the slope of alluvial fans located near the base of mountain ranges. Starting in the lowermost vertical shaft, a horizontal tunnel is dug which laterally connects the vertical shafts, working upslope until the water table is encountered. At this point, groundwater flows by gravity into the tunnel and is conveyed downslope for irrigation or domestic use. Ghanats are still used extensively throughout Iran as a method for tapping deep groundwater without the use of any type of pumping equipment. In ancient times, the Romans also used the technique in conjunction with aqueducts to serve urban water supply systems.[1]

Shallow, hand-dug wells have been used for centuries for irrigation and domestic use where surface water is not a reliable source. In modern times, deep vertical wells are used extensively in arid and semiarid environments to supply water for all purposes including domestic, industrial, agricultural, and municipal applications.

Encyclopedia of Water Science
DOI: 10.1081/E-EWS 120010075

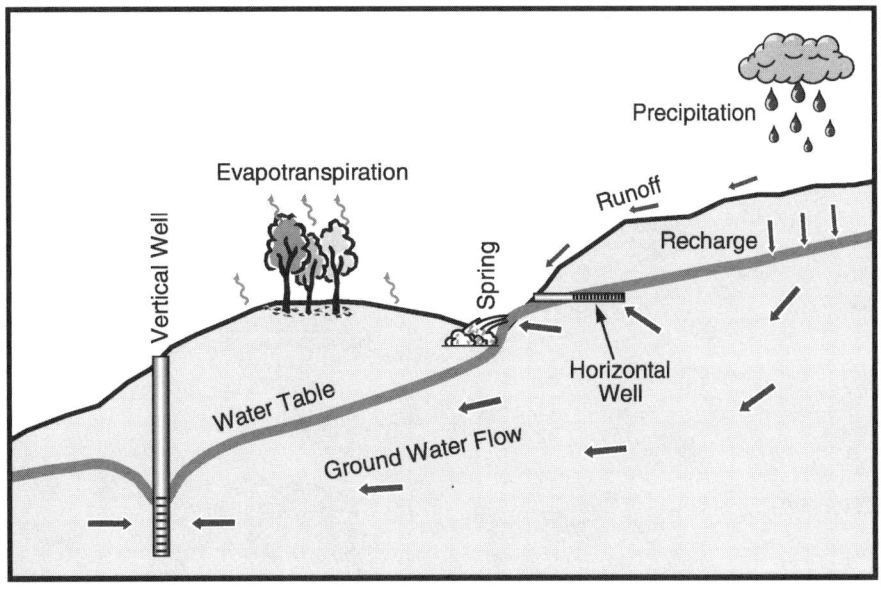

Fig. 1 Hydrologic cycle.

WELLS AND GROUND WATER PUMPING SYSTEMS

Water wells may be constructed in a variety of different aquifer materials in order to supply water for different uses. Most wells are vertical (Fig. 2); however, in specialized cases, wells may be horizontal (to enhance the flow from springs or seeps) or may include a central caisson with lateral "spokes" to induce greater infiltration from the aquifer (i.e., Ranney collector well—Fig. 3(a) and (b)).

To withdraw water from nonflowing water wells, a variety of pumping methods may be employed. The simplest of these do not require electrical energy or fuel-powered motors. These methods include positive displacement-type pumps such as windmills or hand pumps, or more simply a bucket attached to a rope used to raise water to the surface. Most pumps, however, require some sort of mechanical energy—supplied by a drive motor or engine—to lift water to the land surface.

When a well pump is turned on, groundwater first flows into the pump intake from the volume stored within the well casing and borehole area itself. As this volume is typically small compared to the capability of the pump to produce water, a hydraulic gradient forms between the pumping level inside the well and the groundwater level in the near-well zone. A "cone of depression" thus develops around the well, which assumes a general logarithmic shape (Figs. 4 and 5). As pumping continues, the cone of depression expands outward from the well until the

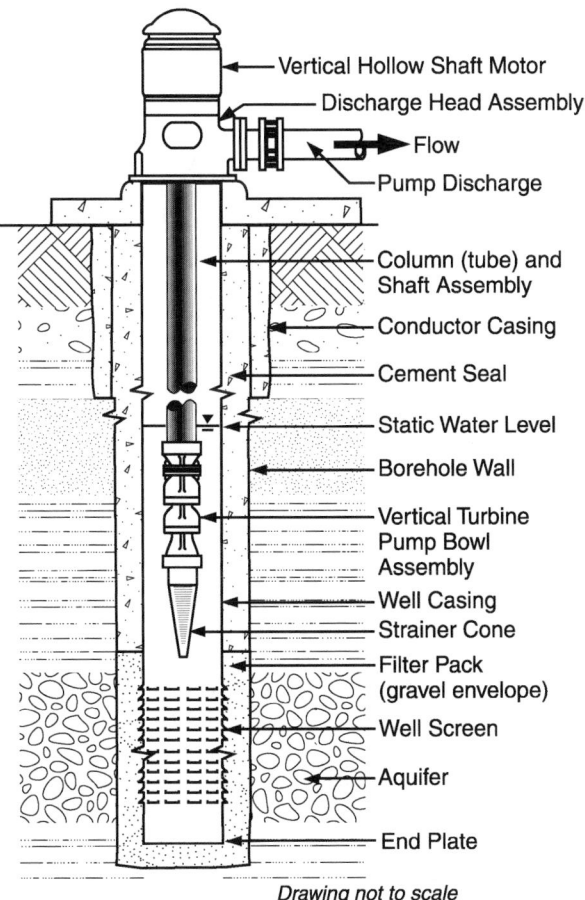

Fig. 2 Example of municipal water well showing deep-well turbine pump (Roscoe Moss Company).

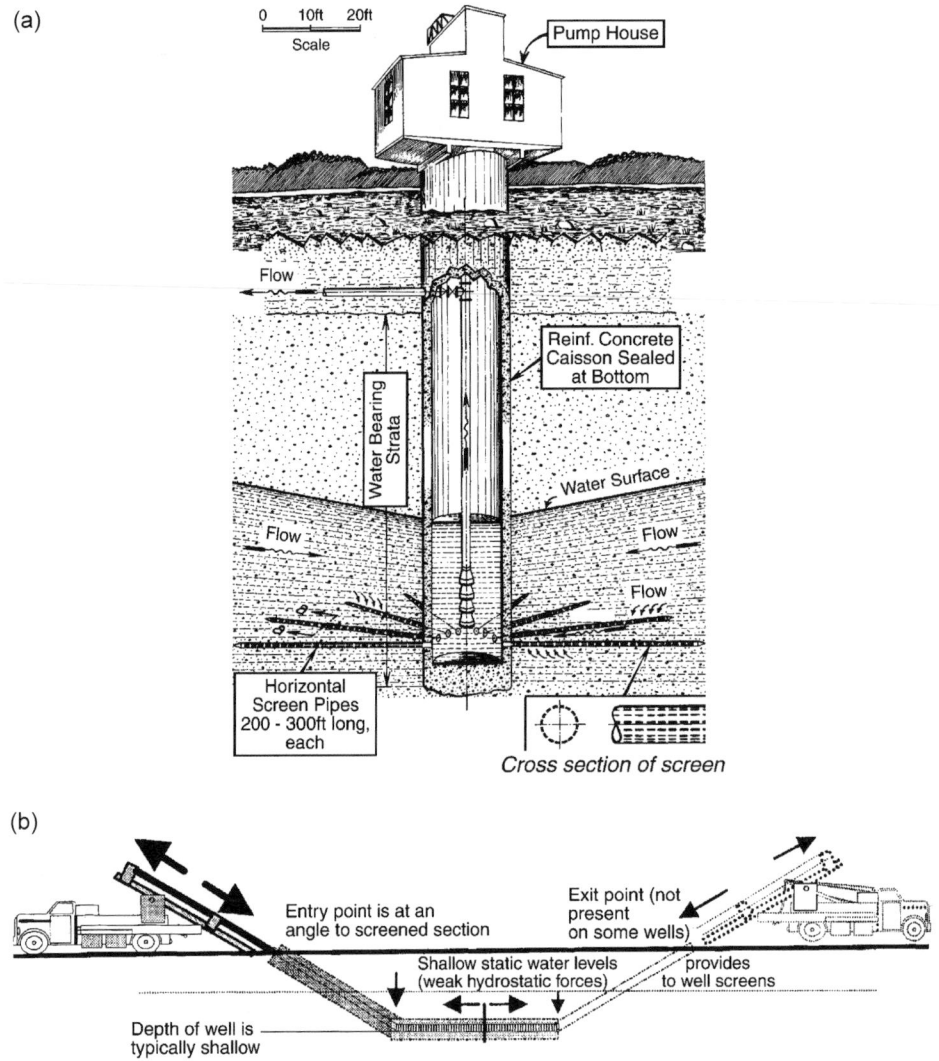

Fig. 3 (a) Typical radial collector well (Ranney Water Systems, Inc.). (b) Horizontal well (Ground Water Publishing Co.).

recharge captured by the cone of depression equals the discharge requirements of the well. When the cone of depression reaches a steady or nonchanging condition the well discharge rate is said to be in equilibrium with the recharge rate to the well. The type of aquifer materials and amount of water being pumped from the well determine the size and shape of the cone of depression. For example, domestic wells generally pump for short periods of time [measured as gallons per minute (gpm)]at rates of 1 gpm– 15 gpm. This results in small, poorly defined cones of depression. On the other hand, deep, large diameter, municipal water supply wells completed in coarse-grained alluvial aquifers can easily produce 2000 gpm–4000 gpm for long periods of time with cones of depressions extending several thousand feet. Fig. 5 defines pumping well terminology as related to the cone of depression.

COMMON GROUND WATER PUMPING METHODS

Types of Pumps

Pumps may be classified in accordance to use (e.g., shallow or deep wells), design (positive or variable displacement), and method of operation (rotary, recipro-cating, centrifugal, jet, or airlift).[2] Shallow-well pumps (suction-lift pumps) are generally installed above ground. Deep-well pumps are always installed in the well casing with the pump intake submerged below the pumping level. Intake areas to deep-well pumps are always under a positive head and do not require suction to pump the water. Fig. 6 shows examples of centrifugal, jet, and rotary pump types.

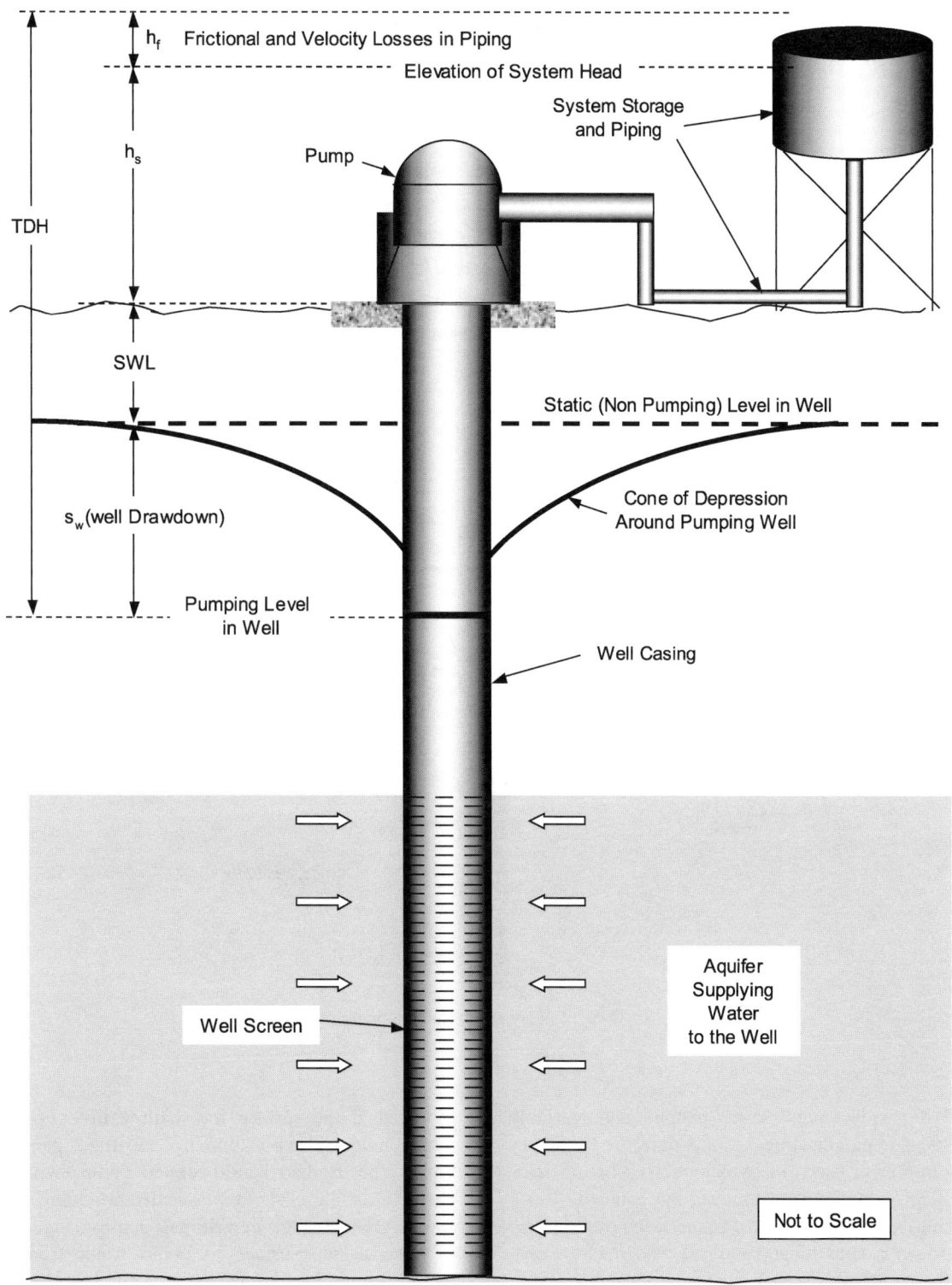

Fig. 4 Schematic of vertical deep water supply well, pump, and storage system.

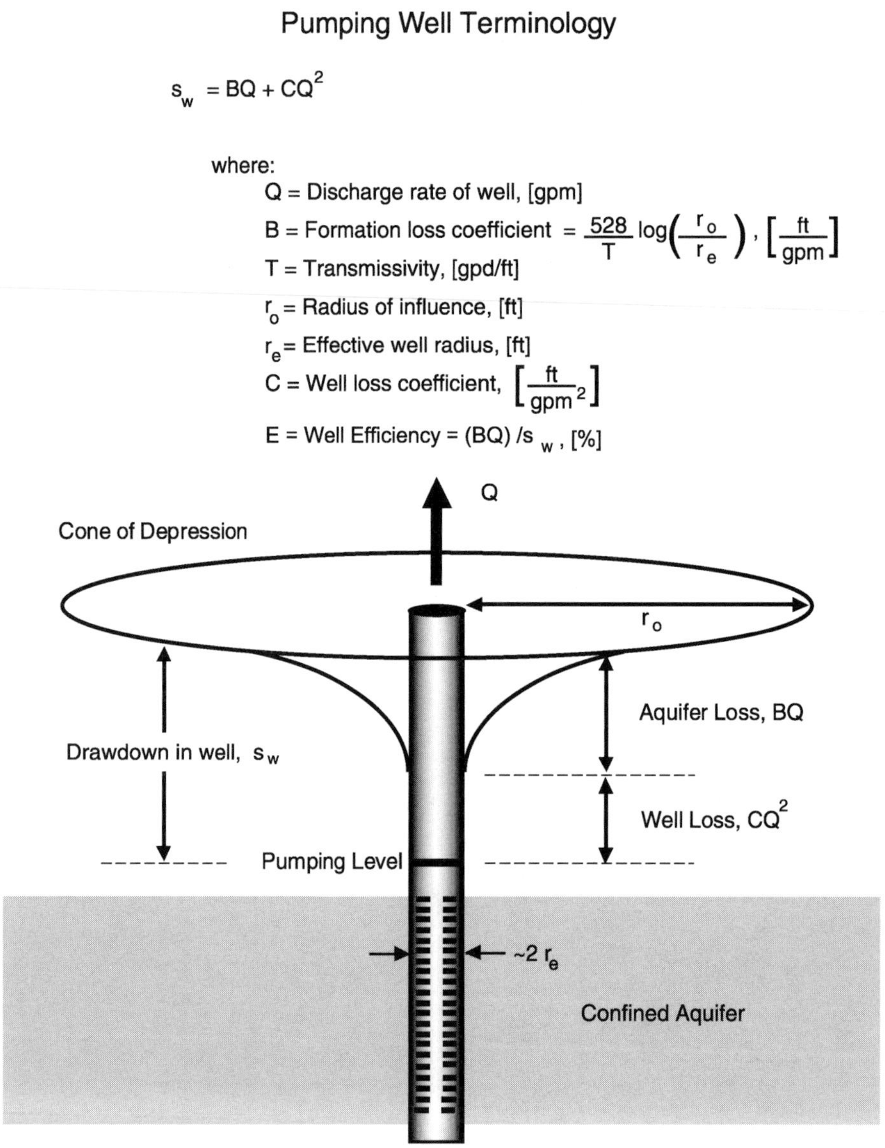
Pumping Well Terminology
$s_w = BQ + CQ^2$
where:
Q = Discharge rate of well, [gpm]
B = Formation loss coefficient = $\frac{528}{T}\log\left(\frac{r_o}{r_e}\right)$, $\left[\frac{ft}{gpm}\right]$
T = Transmissivity, [gpd/ft]
r_o = Radius of influence, [ft]
r_e = Effective well radius, [ft]
C = Well loss coefficient, $\left[\frac{ft}{gpm^2}\right]$
E = Well Efficiency = $(BQ)/s_w$, [%]
Q
Cone of Depression
r_o
Aquifer Loss, BQ
Drawdown in well, s_w
Well Loss, CQ^2
Pumping Level
~2 r_e
Confined Aquifer

Fig. 5 Cone of depression around a pumping well.

Positive displacement (e.g., piston) and variable displacement (e.g., centrifugal pumps) are the two types most commonly used in water wells. In positive displacement-pumps, water is moved mechanically (for a given pump) and directly related to the speed of the pump (e.g., hand strokes per minute) and independent of the total lift (i.e., head). Pump discharge rate is changed by varying pump speed and decreases only slightly with increasing head.[3] In variable displacement-pumps on the other hand (e.g., airlift or centrifugal), the discharge rate depends largely on the total dynamic head (TDH) and decreases as the head increases.

Positive displacement-pumps are used for hand-pumped wells or windmill type of power with cylinders mounted at the surface for shallow lifts (or down the well for deeper applications). Centrifugal pumps run at higher speeds than hand-operated pumps with electric, gasoline, or diesel motors typically providing the power source. Single-stage centrifugal pumps (Fig. 6(a)) can be used at the surface to pump water from shallow wells, but multiple stages are needed in wells where depths to pumping levels are deep.

Another type of pump is the jet pump. Jet pumps (Fig. 6(b)) may also be used at the surface for pumping shallow groundwater. A jet pump forces water down one pipe (through a high-pressure nozzle), and returns the water to the surface through a second pipe, where the discharge is used. Jet pumps are typically used for

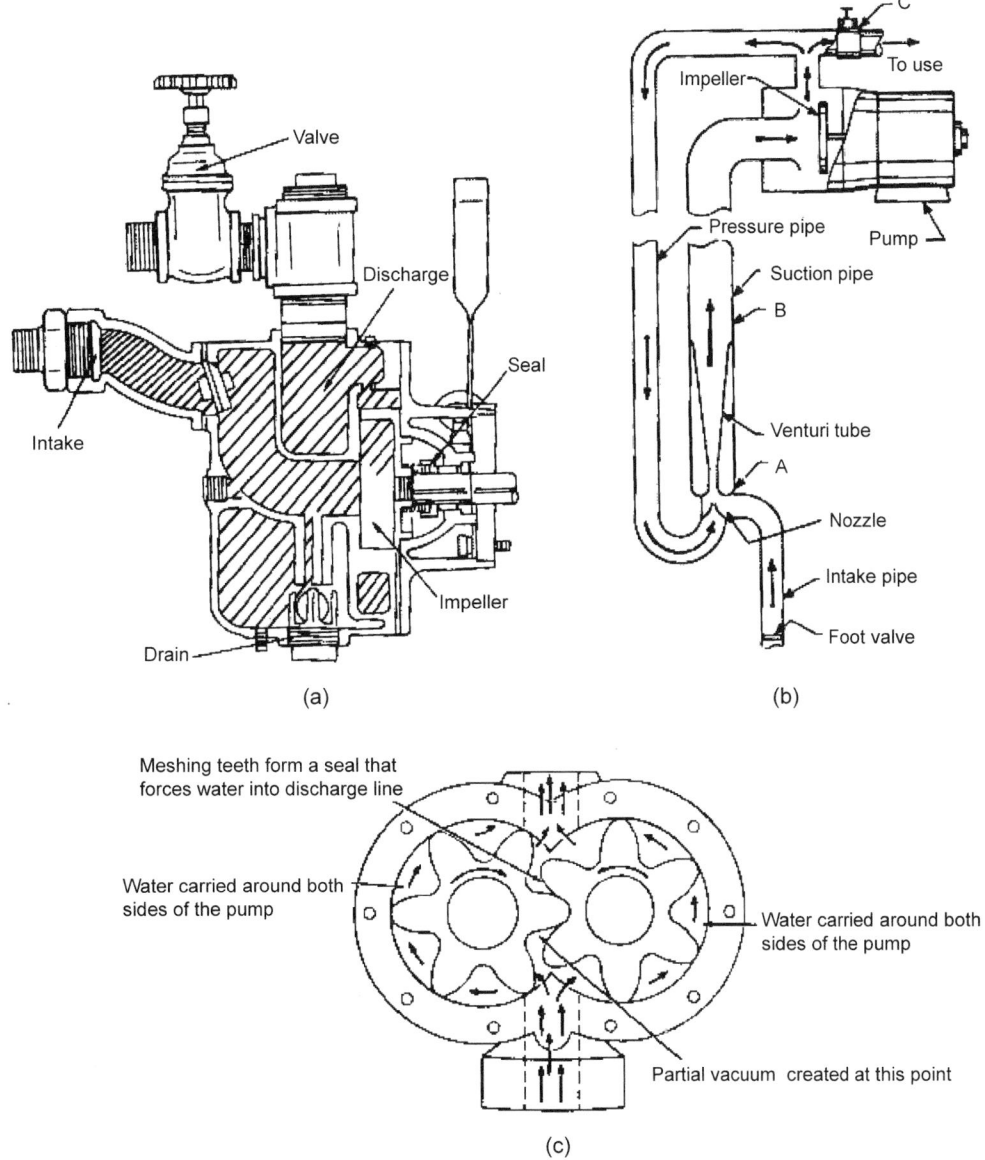

Fig. 6 (a) Single-stage centrifugal pump (U.S. Government Printing Office). (b) Jet pump (U.S. Government Printing Office). (c) Rotary pump (U.S. Government Printing Office).

applications where the depth to water is less than approximately 22 ft–25 ft.

Power is required to lift groundwater to the surface, either indirectly by suction pumps or directly from hand or mechanically driven pumps. Windmills may be a good choice for lifting water from shallow or deep-water wells in rural communities, or where conventional power supplies or fuel costs are either unavailable or very expensive. Modern technology has also produced solar cells that convert sunlight directly into electricity. One of the most important applications for solar cells in rural areas all over the world is for pumping water.[4]

Hand Pumps

A variety of inexpensive positive displacement-pumps are available to pump water from wells. One such type is called a "pitcher" pump (Fig. 7) and is commonly used when the water table is less than approximately 22 ft–25 ft below the surface.[2,5] (The theoretical maximum suction lift equal to atmospheric pressure ~ 34 ft or 14.7 pounds per square inch (psi), cannot be achieved with these pumps). Pitcher pumps are surface-mounted, reciprocating or single-acting piston pumps utilizing a hand-operated plunger inside a cylinder set on top of the well casing. The

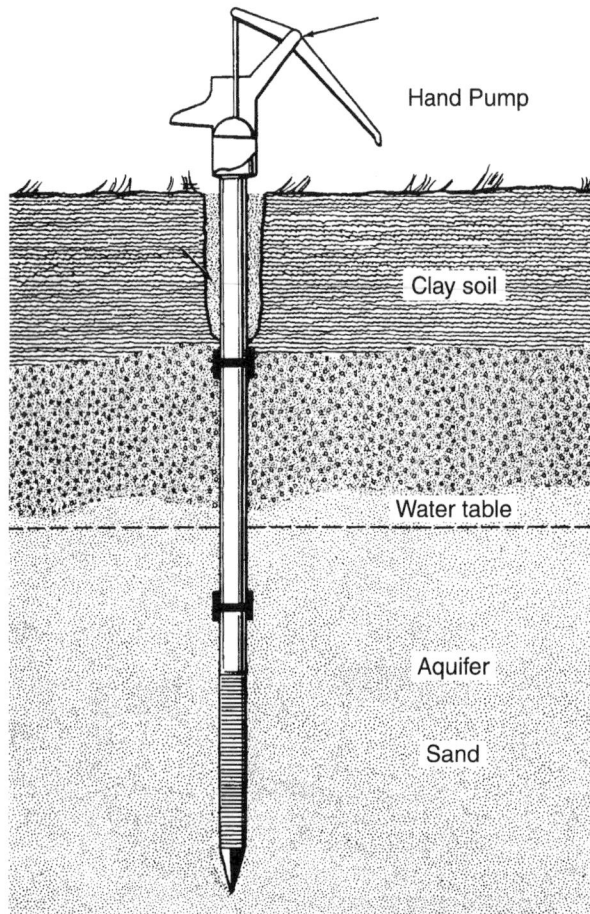

Fig. 7 Example of suction pump (U.S. Government Printing Office).

pump suction pipe is attached to the bottom of the cylinder. The plunger has a simple ball valve that opens on the down stroke and closes on the upstroke. A check valve at the lower end of the cylinder opens on the upstroke of the pump and closes on the down stroke. Through continuous upstroke and down stroke actions water flows out of the discharge pipe. For deeper depths (up to 250 ft), a surface pump stand and separate lift cylinder can be installed down the well, which effectively "pushes" the water to the surface with the help of interconnecting rods. The latter method is often employed in a windmill system. Fig. 7 illustrates a typical suction pump.

Another manually operated water pump is the Treadle Pump. The Treadle Pump was developed to provide low cost, sustainable, environment friendly technology, which can be sold at a fair market price, in rural areas.[6]

The Treadle Pump relies upon the basic suction lifting principle of the hand pump consisting of two barrels, plungers, and treadles. One person through use of foot pedals can operate it. The discharge rate of a Treadle Pump can achieve approximately 10 gpm–15 gpm depending on the size of the pump/suction depth etc.

Rotary Pumps

These pumps use a system of rotating gears to create a suction at the inlet and force a water stream out of the discharge. The gears' teeth move away from each other at the inlet port. This action causes a partial vacuum and the water in the suction pipe rises. In the pump, the water is carried between the gear teeth and around both sides of the pump case. At the outlet, the teeth moving together and meshing causes a positive pressure that forces the water into the discharge line. In a rotary gear pump, water flows continuously and steadily with very small pulsations. The pump size and shaft rotation speed determine how much water is pumped per hour. Gear pumps are generally intended for low-speed operation. The flowing water lubricates all internal parts. Therefore, the pumps should be used for pumping water that is free of sand or grit. If sand or grit does flow through the gears, the close-fitting gear teeth will wear, thus reducing pump efficiency or lifting capacity.

Airlift Pumps

Water can also be pumped from a well using an airlift pump (Fig. 8(a) and (b)). The airlift pump assembly consists of a vertical discharge pipe (eductor pipe) and a smaller air pipe which are both submerged below the well's pumping level for approximately two-third of their length.[7] Compressed air is forced through the air pipe to within a few feet of the bottom of the eductor pipe. The mixture of air bubbles and water formed inside the eductor pipe results in the air/water fluid being lighter than the water outside the eductor pipe (i.e., inside the well casing). This results in the air/water fluid flowing upward and out the top of the eductor pipe. Airlift pumps produce the best results when the submergence ratio of the air and eductor pipe is approximately 60%, however, reasonable results can be obtained with submergence as low as 30%.[8] An example of a 60% submergence is when the length of the air pipe is 200 ft (B on Fig. 8(a)) and the pumping water level depth is 80 ft (C on Fig. 8(a)). This results in a submergence of 120/200 (60%).

Centrifugal Pumps

Centrifugal pumps are variable displacement-pumps with the discharge rate being inversely related to the head supplied. That is, when the TDH increases, the discharge rate decreases.[9] A centrifugal pump contains a rotating impeller within a housing. The centrifugal forces generated by the spinning impeller impart kinetic energy to the water. This

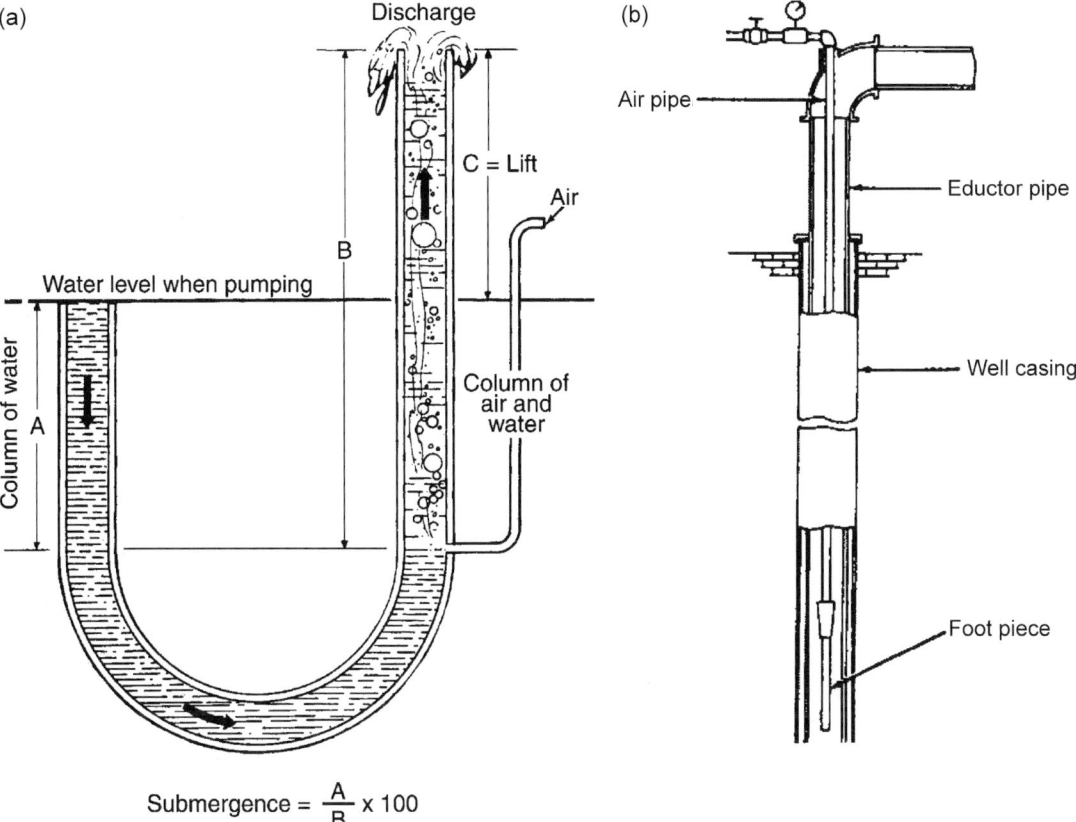

Fig. 8 (a) Principle of airlift pump (U.S. Government Printing Office). (b) Example of airlift pump (U.S. Government Printing Office).

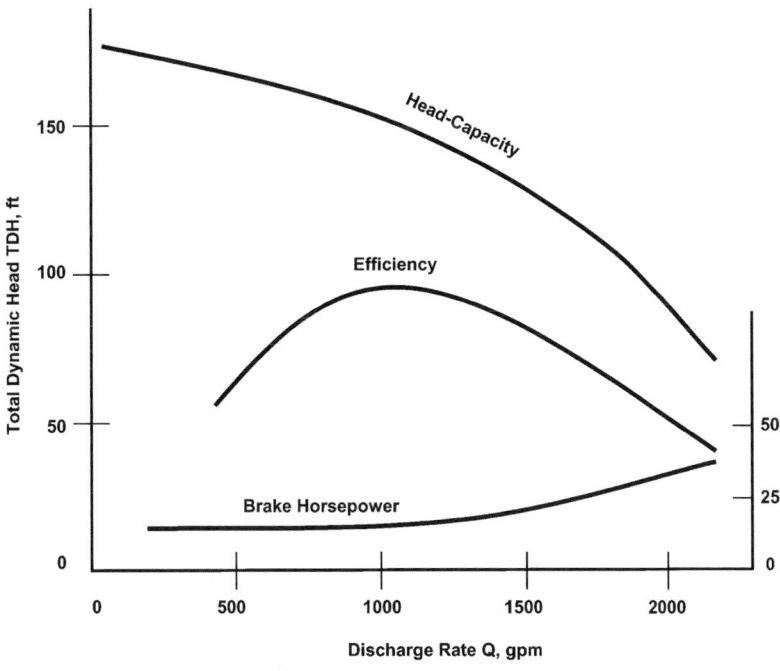

Fig. 9 Typical pump performance curves for centrifugal pumps.

kinetic energy is converted into pressure at the discharge side of the pump. The general characteristic pump performance curves for centrifugal pumps are shown in Fig. 9 and relate TDH, pump efficiency, and horsepower.

Centrifugal pumps may be used to pump water from shallow wells with high water tables and low drawdowns. Centrifugal pumps may also be used in deep wells (deep-well vertical centrifugal pumps or commonly known as deep-well turbine pumps), and are installed inside the well casing below the water level. These latter pumps consist of a number of pump bowls with impellers, each set above another, which are added so as to "build" the head required. The impellers may be driven by either a motor at the surface (typically, electric, gasoline, or diesel) and connected to the pump by a long shaft and tube assembly (surface drive). The impellers may also be powered by a submerged electric motor directly coupled to the pump (submersible drive).

When a centrifugal suction pump is used to produce water from shallow wells, all pump components and suction lines must be completely filled with water or "primed" in order to operate. Hand-operated or motor-powered vacuum pumps are typically used for priming.

Deep-Well Turbine Pumps

Deep-well turbine pumps are the most common type of pump used in cased water wells where the groundwater surface is below the practical limits of centrifugal suction pumps.

The turbine pump has three main parts: 1) the head assembly; 2) the column (tube) and shaft assembly; and 3) the pump bowl assembly. The discharge head is typically cast iron or fabricated steel, and is designed to be installed on a foundation. The discharge head supports the column (tube), shaft, and bowl assemblies and directs the discharge of water. Additionally, it also provides a base to support an electric motor, a right angle gear drive or a belt drive (Fig. 2).

The column and shaft assembly connects the head and pump bowls. The line shaft transfers the power from the motor to the pump impeller(s). The impellers lift the water and the column conveys the lifted water to the surface. The line shaft on a deep-well turbine pump may be either water- or oil-lubricated. The oil-lubricated pump has an enclosed shaft (oil tube) into which oil drips at the surface, lubricating the bearings by gravity. The water-lubricated pump has an open shaft, where the pumped water itself lubricates the bearings. If a high content of sand in the discharge is anticipated, an oil-lubricated pump should be selected in order to keep the bearings clean. The pump bowl encloses the impeller. In most deep-well turbine installations, several bowls (stages) are stacked in series. A four-stage bowl assembly contains four impellers attached by a common shaft, and will operate at four times the discharge head of a single-stage pump.[10]

Impellers used in turbine pumps may be either semiopen or enclosed. The vanes on semiopen impellers are open on the bottom and rotate with a very close tolerance to the bottom of the pump bowl (enclosure).

The operating characteristics of deep-well turbine pumps are determined by laboratory testing and depend largely on bowl design, impeller type, and speed. Vertical turbine pumps are generally designed for specific speeds [measured as revolutions per minute (rpm)]; generally, either 1800 rpm or 3500 rpm for deep-well turbine pump applications. Other speeds are used for specialized applications.[11] Pump performance curves at these speeds can be obtained from the manufacturer of each pump.

Submersible Pumps

A submersible pump is a turbine pump close-coupled to a submersible electric motor (Fig. 10). Both pump and motor are suspended below the water surface, eliminating the long drive shaft and bearing retainers required for a deep-well turbine pump. The pump bowl assembly is located above the motor. Water enters the pump through a screen located between the pump and motor. The pump curve for a submersible pump is very similar to a deep-well turbine pump.[10]

Submersible motors are smaller in diameter and much longer than ordinary motors. Because of their smaller diameter, they are lower in efficiency than those used for centrifugal or deep-well turbine pumps.

Most submersible pumps used for domestic purposes use either single or two-phase power, while larger pumps used for agricultural, industrial, or municipal purposes require three-phase power. Electrical wiring connecting the pump motor to the surface power supply must be watertight with all connections sealed. Submersible pumps can be selected to provide a wide range of flow rate and TDH combinations.

Pump Head and Power Requirements

Before selecting a pump, a careful and complete inventory of the conditions under which the pump will operate must take place. The discharge rate and TDH will be determined by the specific use and distribution system (Figs. 4 and 11). The TDH of a pump is the sum of the elevation and pressure heads plus head losses due to friction and velocity[3] (Fig. 4). Friction head is the sum of the energy loss due to the flow of water through a pipe, and is a function of the velocity and pipeline diameter including losses through fittings and valves as well as changes in flow direction and pipeline diameter. Values for these losses can be calculated or obtained from friction loss

G

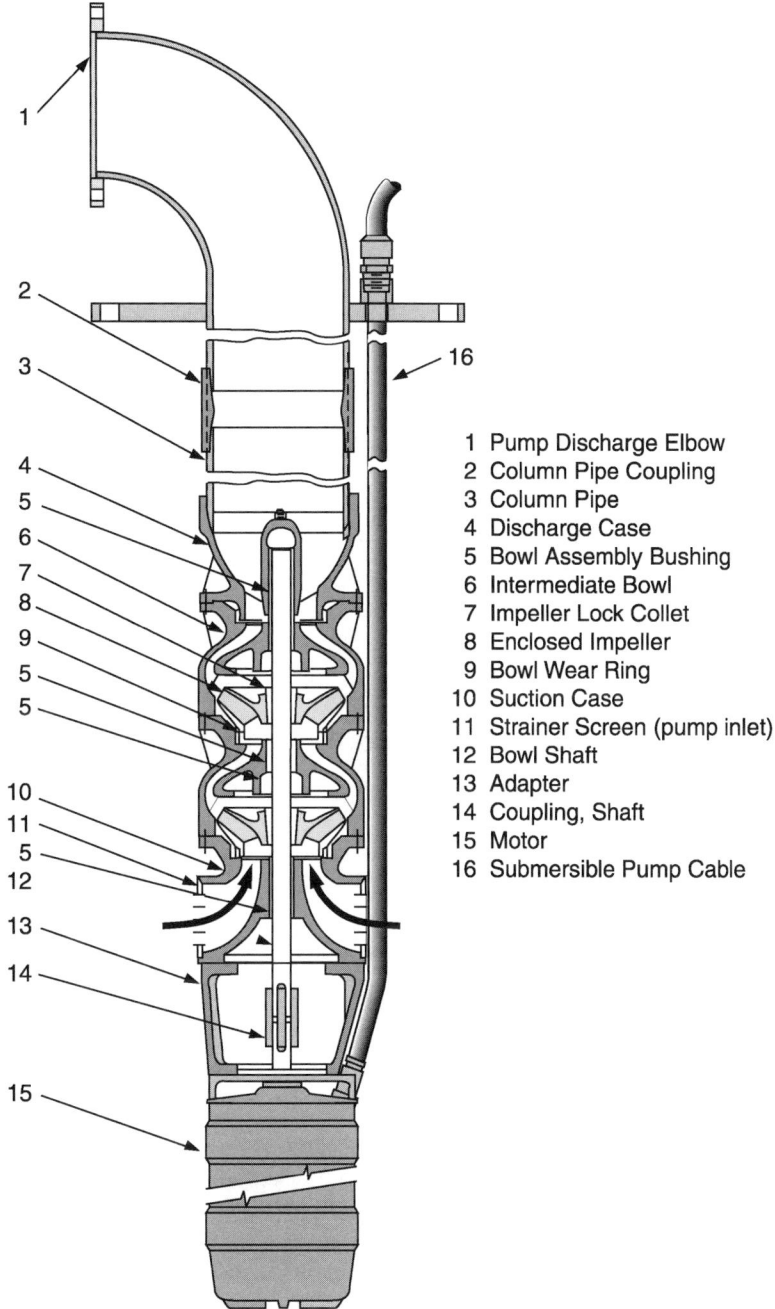

1 Pump Discharge Elbow
2 Column Pipe Coupling
3 Column Pipe
4 Discharge Case
5 Bowl Assembly Bushing
6 Intermediate Bowl
7 Impeller Lock Collet
8 Enclosed Impeller
9 Bowl Wear Ring
10 Suction Case
11 Strainer Screen (pump inlet)
12 Bowl Shaft
13 Adapter
14 Coupling, Shaft
15 Motor
16 Submersible Pump Cable

Fig. 10 Vertical turbine multi-stage submersible pump (Roscoe Moss Company).

tables. The discharge rate, which will be produced, is a function of both the system and pump characteristics. The intersection of the system head curve and the pump head curve determine the discharge rate at which the pump will operate (Fig. 11).

Cavitation (i.e., implosion of air bubbles and water vapor on the impeller, causing pitting) occurs when the hydraulic head at the pump intake is too low. The head must be high enough so that as velocity increases (and pressure decreases), within the pump, the pressure cannot drop below the vapor pressure of the water. The minimum head needed at the pump intake is termed the net positive suction head (NPSH) and is specific to the operation and pump design.

The power required to move water through a pump may be calculated using the following formula:

$$WHP = (Q \times \mathrm{TDH})/(3960)$$

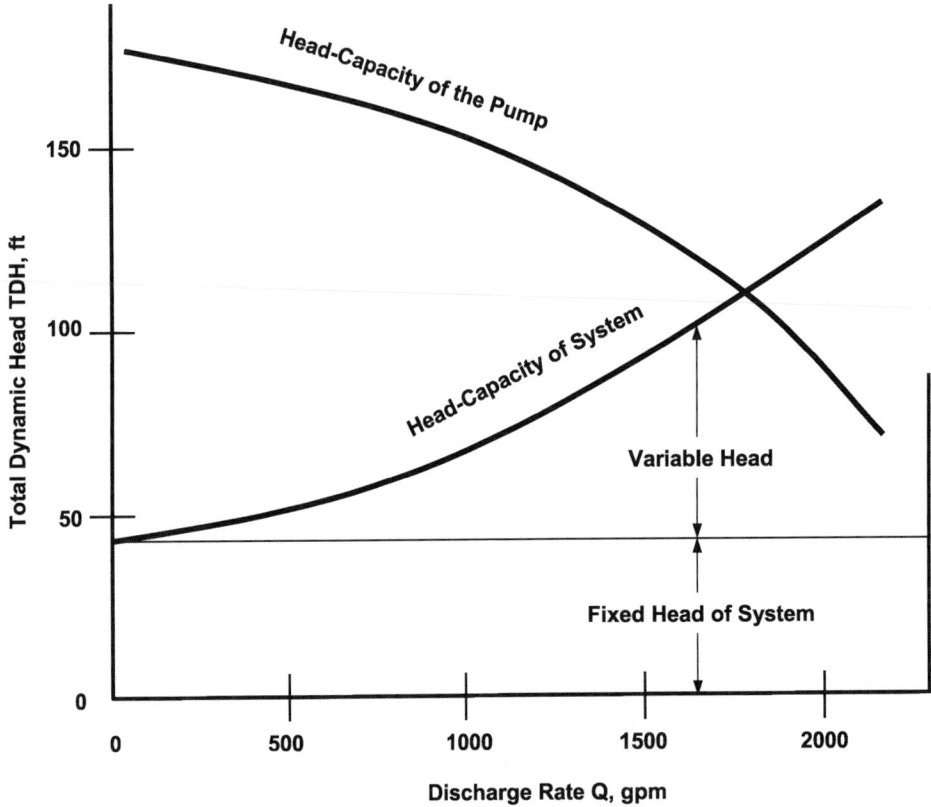

Fig. 11   Typical system head and pump curves.

where, WHP = water horsepower, Q = discharge rate of the well pump, [gpm], TDH = total dynamic head, [ft] = s + SWL + h_s + h_f, s = drawdown in the pumping well, [ft], SWL = depth to static water level below reference point, [ft], h_s = elevation of system head above reference point, [ft], h_f = friction (and velocity) head losses in piping system from pump to system storage, [ft]

However, the actual horsepower required to run a pump will be greater than the water horsepower as pumps and drivers are not 100% efficient. The horsepower required to pump a specified flow rate against a specified TDH therefore is the brake horsepower (BHP), and is calculated using the following formula:[10]

$$BHP = (Q \times \text{TDH})/(3960 \times e)$$

where, BHP = water horsepower (horsepower rating of the power unit), e = pump efficiency $\times$ drive efficiency (expressed as a decimal)

The pump efficiency (percentage) may be read directly from the pump curve provided by the manufacturer. The drive efficiency is the efficiency value (percentage) given by the driver unit (source of power to the pump) by the manufacturer.

Pump Selection Criteria

Proper selection of a pump must consider both anticipated pumping conditions and well type, which include the following main design parameters:[10]

- Diameter of the well
- Required discharge rate
- TDH
- Depth to static ground water
- Friction losses
- Power requirements
- Power source
- Water quality and/or potential for sand production

Centrifugal suction pumps are generally used for shallow groundwater levels (e.g., less than 22 ft–25 ft) while most installations utilize deep-well turbine or submersible pumps. The discharge rate of the well is often overlooked when selecting a pump for small wells. If too large a pump is installed in a small capacity well, the result will be to either

Table 1 Pump selection criteria

Type of pump	Practical suction lift[a]	Usual well-pumping depths	Usual pressure heads	Advantages	Disadvantages	Remarks
Reciprocating 1. Shallow Well 2. Deep Well	22 ft–26 ft 22 ft–25 ft	22 ft–26 ft Up to 600 ft	100 ft–200 ft	1. Positive action 2. Discharge against variable heads 3. Pumps water containing sand and silt 4. Especially adapted to low capacity and high lifts	1. Pulsating discharge 2. Subject to vibration and noise 3. Maintenance cost may be high 4. May cause destructive pressure if operated against closed valve	1. Best suited for capacities of 5 gpm–25 gpm against moderate to high heads 2. Adaptable to hand operation 3. Can be installed in very small diameter walls (2-in. casing) 4. Pump must be set directly over well (deep well only)
Centrifugal 1. Shallow well				1. Smooth, even flow 2. Pumps water containing sand and silt 3. Pressure on system is even and free from shock 4. Low-starting torque 5. Usually reliable and good service life		
Straight centrifugal (single stage)	20 ft maximum	10 ft–20 ft	100 ft–150 ft		1. Loses prime easily 2. Efficiency depends on operating under design heads and speed	Very efficient pump for capacities above 50 gpm and heads up to about 150 ft
Regenerative vane turbine type (single impeller)	28 ft maximum	28 ft	100 ft–200 ft		Same as straight centrifugal except maintains priming easily	Reduction in pressure with increased capacity not as severe as straight centrifugal
2. Deep well Vertical line shaft turbine (multi-stage)	Impellers submerged	50 ft–300 ft	100 ft–800 ft	Same as shallow-well turbine	1. Efficiency depends on operating under design head and speed 2. Requires straight well large enough for turbine bowl and housing 3. Lubrication and alignment of shaft critical 4. Abrasion from sand	
Submersible turbine (multi-stage)	Pump and motor submerged	50 ft–400 ft	80 ft–900 ft	1. Same as shallow-well turbine 2. Easy to frost proof installation 3. Short pump shaft to motor	1. Repair to motor or pump requires pulling from well 2. Sealing of electrical equipment from water vapor critical 3. Abrasion from sand	Difficulty with sealing has caused uncertainty as to service life in data
Jet 1. Shallow well	15 ft–20 ft below ejector	Up to 15 ft–20 ft below ejector	80 ft–150 ft	1. High capacity at low heads 2. Simple in operation 3. Does not have to be installed over the well 4. No moving parts in the well	1. Capacity reduces as lift increases 2. Air in suction or return line will stop pumping	
2. Deep well	15 ft–20 ft below ejector	25 ft–120 ft, 200 ft maximum	80 ft–150 ft	Same as shallow-well jet	Same as shallow-well jet	The amount of water returned to ejector increases with increased lift—50% of total water pumped at 50 ft lift and 75% at 100 ft lift
Rotary 1. Shallow well (gear type)	22 ft	22 ft	50 ft–250 ft	1. Positive action 2. Discharge constant under variable heads 3. Efficient operation	1. Subject to rapid wear if water contains sand or silt 2. Wear of gears reduces efficiency	
2. Deep well (Helical-rotary type)	Usually submerged	50 ft–500 ft	100 ft–500 ft	1. Same as shallow-well rotary 2. Only one moving pump device in well	Same as shallow well rotary except no gear wear	A rubber stator increases life of pump; flexible drive coupling has been weak point in pump; best adapted for low capacity and high heads

[a]Practical suction lift at sea level. Reduce lift 1 ft for each 1000 ft above sea level.

temporarily drain the well (i.e., "break suction"), or exceed the maximum possible suction lift. It is very important, therefore, to match pumping requirements and well characteristics when selecting the optimum pump for each installation. Table 1 provides general guidelines for pump selection.[2]

REFERENCES

1. Glick, T.F. Movement of Ideas and Techniques. *Islamic and Christian Spain in the Early Middle Ages*; Princeton University Press, 1979.

2. Multiservice Procedures for Well-Drilling Operations. Army Manual No. 5-484; Navy Facilities Engineering Command Pamphlet No. 1065; Air Force Manual No. 32-1072; Washington DC, March 8, 1994.

3. Mariño, M.A.; Luthin, J.N. *Seepage and Groundwater*; Elsevier Scientific Publishing Company: New York, 1982.

4. Dankoff Solar Products. www.dankoffsolar.com (accessed Jan 2001).

5. Ashe, William. *Understanding Water Wells*, Technical Paper #68; Volunteers in Technical Assistance: Arlington, Virginia, 1990; 1–14.

6. Lifewater Canada. http://www.lifewater.ca/ndexpump.htm (accessed Jan 2001).

7. *Wells*, Technical Manual No. 5-297; Air Force Manual No. 85-23; Departments of the Army and the Air Force: Washington DC, Aug 1, 1957.

8. Driscoll, F.G. *Ground Water and Wells*; Johnson Division, Universal Products Co.: St. Paul, MN, 1972.

9. Helweg, O.J.; Scalmanini, J.C.; Scott, V.H. *Improving Well and Pump Efficiency*; American Water Works Association: U.S.A., 1983.

10. Roscoe Moss Company, *Handbook of Ground Water Development*; John Wiley & Sons: New York, NY, 1990.

11. Driscoll, F.G. *Groundwater and Wells*, 2nd Ed.; Johnson Division: St. Paul, MN, 1986.

FURTHER READING

Campbell, M.D.; Lehr, J.H. *Water Well Technology*, McGraw-Hill Book Company: New York, NY, 1973.

Florida Cooperative Extension Service, Institute of Food and Agricultural Sciences, University of Florida. www.edis.ifas.u-fl.edu/BODY_w1001 (accessed Jan 2001).

Hix, G.L. More on the Development of Horizontal Wells. Water Well J. **1996**, *L* (2), 47–49.

Honors Program, University of Nevada, Reno, Tony Zuliani. www.honors.unr.edu/~tzuliani/wtpaper.html#pump (accessed Jan 2001).

Microsoft Encarta Online Encyclopedia 2000. http://encarta.msn. com (accessed Jan 2001).

Neptune Internet Services, Inc., John Yonge. www.neptune.on. ca/~jyonge/windmill.htm (accessed Jan 2001).

North Dakota State University. www.ext.nodak.edu/extpubs/a-geng/irrigate/ae1057w.htm#centrifugal (accessed Jan 2001).

Groundwater Quality

Loret M. Ruppe
Timothy R. Ginn
University of California, Davis, California, U.S.A.

G

INTRODUCTION

Groundwater quality refers to the type and concentration of constituents in a given source of groundwater. Constituents in groundwater may originate from the natural environment with which the groundwater comes in contact, or may be introduced as pollutants from external sources. Constituents can be dissolved solids and gases, suspended solids, hydrogen ions, and microorganisms. There is wide variation in the chemical and biological constituents in groundwater due to the differing qualities of water that recharge groundwater, and due to the different environments through which groundwater passes. The unique polar nature of the water molecule makes it a ready solvent, with the capacity to dissolve many solid-phase minerals into solution, and many of the elements comprising the subsurface environment dissolve into ground water.

Groundwater is never free from all impurities. A principal and ubiquitous constituent class is that of ions in solution, that have dissolved into the groundwater from the earth materials in which it flows. Total ionic concentration includes various dissolved salts and associated mineral species as well as hydrogen, and is quantified as total dissolved solids (TDS). The concentration of TDS in units of mass of ions per volume of water is used to classify water, with fresh water (0 mg/L – 1000 mg/L TDS) differing from brackish (1000 mg/L – 10,000 mg/L TDS), saline (10,000 mg/L – 100,000 mg/L TDS), and brine (greater than 100,000 mg/L TDS) waters in dissolved solids concentration. Potable water typically has less than 500 mg/L TDS, while the concentration in seawater is approximately 35,000 mg/L.[1]

TYPICAL CONSTITUENTS IN GROUNDWATER

Inorganic solids comprising the geologic material of the subsurface constitute the greatest concentrations of constituents in groundwater, with bicarbonate, calcium, chloride, magnesium, sodium, and sulfate typically 90% of the TDS in groundwater.[1] As groundwater ages in an aquifer, dominant ions tend to shift from calcium (Ca^{2+}) and bicarbonate (HCO_3^-) to sodium (Na^+) and chloride (Cl^-). Constituents generally found in uncontaminated groundwater are classified in Table 1 according to relative abundance.[2,3]

MECHANISMS INFLUENCING GROUND-WATER QUALITY

The processes by which chemicals are dissolved into the groundwater are primarily: mineral dissolution and precipitation, microbially mediated oxidation and reduction reactions, ion exchange and adsorption, and hydrolysis.[4] Each of these processes is described later.

Mineral Dissolution and Precipitation

Rainwater is slightly acidic (pH < 7), and more so in regions with acidic air pollutants that dissolve into water droplets. Once the rainwater reaches the ground and percolates through the root zone, the degree of acidity can increase further as the oxygen in the water is consumed by the decay of organic matter and by the respiration of plant roots. Oxygen removal creates an oxygen sink that is filled by the further dissociation of aqueous compounds containing oxygen such as bicarbonate, that also puts more hydrogen ions into solution, thus reducing pH. Water from rainfall, lakes, streams, and other sources travels through the unsaturated vadose zone, and accumulates in the saturated zone of an aquifer. As the water passes through the vadose zone and through the aquifer, the acidity of the water causes the dissolution of the geologic features into which it comes in contact. An example is the dissolution of calcite, $CaCO_3$, the basic constituent of limestone, marble, and chalk that is commonly found in sedimentary rock. In the presence of acidity in the groundwater, in the form of carbonic acid, H_2CO_3, calcium and bicarbonate become dissolved ions in the groundwater solution.

$$CaCO_3 + H_2CO_3 \rightarrow Ca^{2+} + 2HCO_3^-$$

Given an adequate supply of calcite and carbonic acid, calcium will continue to dissolve into solution until an equilibrium state is reached.

Encyclopedia of Water Science
DOI: 10.1081/E-EWS 120010304

Table 1 The dissolved constituents in potable groundwater classified according to relative abundance

Major constituents (greater than 5 mg/L)	
Bicarbonate	Silicon
Calcium	Sodium
Chloride	Sulfate
Magnesium	Carbonic acid
Nitrogen	
Minor constituents (0.01 mg/L–10.0 mg/L)	
Boron	Nitrate
Carbonate	Potassium
Fluoride	Strontium
Iron	Bromide
Oxygen	Carbon dioxide
Trace constituents (less than 0.1 mg/L)	
Aluminum	Nickel
Antimony	Niobium
Arsenic	Phosphate
Barium	Platinum
Beryllium	Radium
Bismuth	Rubidium
Cadmium	Ruthenium
Cerium	Scandium
Cesium	Selenium
Chromium	Silver
Cobalt	Thallium
Copper	Thorium
Gallium	Tin
Germanium	Titanium
Gold	Tungsten
Indium	Uranium
Iodide	Vanadium
Lanthanum	Ytterbium
Lead	Yttrium
Lithium	Zinc
Manganese	Zirconium
Molybdenum	
Organic compounds (shallow)	
Humic acid	Tannins
Fulvic acid	Lignins
Carbohydrates	Hydrocarbons
Amino acids	
Organic compounds (deep)	
Acetate	
Propionate	

Source: Domenico and Schwartz,[2] modified from Davis and DeWiest.[3]

Oxidation and Reduction Reactions

Microorganisms, primarily bacteria, are ubiquitous throughout the subsurface environment, with population diversity and density typically decreasing logarithmically with depth. The metabolic activities of the microorganisms catalyze reactions within the groundwater that involve transferring electrons to form different compounds.

Organic material can be converted to inorganic compounds such as carbon dioxide and water through this process. Such reactions are the basis for in situ bioremediation of organic contaminants.

Ion Exchange and Adsorption

Essentially all surfaces of aquifer materials are electrically charged. For instance sands are quartzitic materials generally carrying negative charges, but are often partly coated with mineral oxide (iron, manganese, aluminum) compounds that have a net positive charge. Ions in the groundwater attach (adsorb) to the charged surfaces, thereby altering the chemical make-up of the groundwater. Ion exchange refers to the preferential sorption due to electrostatic forces of multivalent ions over monovalent ions, that is reversible if the aqueous concentration of monovalent ions is high enough. Clay minerals have dense surface charges and are typically involved in ion exchange and adsorption. Ion exchange in clay minerals involves intra-particle sites, and exchange is associated with swelling or shrinking of the clay medium on the macroscale. For instance when a single bivalent ion is replaced by two monovalent ions, the clay particle swells.

Hydrolysis

As noted earlier, the polar structure of water facilitates reaction with chemical compounds to form new compounds. The replacement of ions in a compound with H^+ or OH^- ions of water is termed hydrolysis. The chemical make-up of groundwater will influence the degree to which hydrolysis will occur.

Some of these geochemical and biochemical reactions occur simultaneously, while others occur sequentially. The rates at which these reactions occur vary considerably, ranging from nearly instantaneously to slowly enough that the equilibrium is never reached. Knowledge about the processes is critical in predicting the way in which groundwater quality evolves and responds to treatment, and the rate of the change; efforts continue by scientists and engineers to accurately model the reactions occurring in the complex groundwater environment.

GROUND WATER QUALITY ISSUES

Groundwater contamination has a direct effect on the quality of drinking water for many people. More than fifty percent of the drinking water in the United States is groundwater.[5] Contaminants to groundwater can originate from a "point source" (PS; single specific discharge point) or from a "nonpoint source" (NPS; diffuse source that contributes a contaminant, or contaminants, to the environment). PS of groundwater contamination include industrial waste discharges, leaking petroleum storage facilities, and municipal wastewater treatment plant discharges. NPS discharges are often associated with rainfall runoff and snowmelt events from agricultural operations, roadways and vehicle emissions, construction sites, mining operations, landfills, and logging activities. Other sources of NPS pollution include soil erosion (sediment transfer), failing onsite wastewater treatment systems, animal wastes in feedlot runoff, or animal waste holding pond overflows.

Agricultural Wastes

Agricultural operations contribute many constituents that contaminate the groundwater. The chemical pesticides, herbicides, and fertilizers applied to crops can reach the groundwater through land application and in rainfall runoff. Animal manure wastewater, which harbors human pathogens such as *Cryptosporidium parvum* oocysts, can similarly migrate to the groundwater. NPS contamination can result from irrigation using animal manure wastewater, as well as land application of waste solids or liquids for nonirrigation purposes.

Industrial Wastes

The disposal of the chemical byproducts of industrial processes is regulated to varying degrees around the world. The wastes may not be adequately treated before being discharged to the environment, where they frequently migrate to groundwater. Additional sources of contamination are chemical spills and leaking storage tanks. Such sources include those from military and energy facilities, which often involve heavy metals and/or radionuclides in solution, as well as dissolved explosives and solid fuels for propellants.

Municipal Wastewater

The increase in synthetic chemical usage around the home results in the discharge of portions of these chemicals into the wastewater sewerage system. Treatment plants are designed to purify human wastewater, but are not designed to remove these additional chemicals, with the result that increasing concentrations of chemicals are bypassing the treatment process and are released into the environment, eventually being detected in groundwater.

Synthetic Chemicals

Increasingly, groundwater contamination results from the growing usage of synthetic chemicals. With over 65,000 synthetic chemicals in common use in the United States today, these chemicals are being detected in groundwater supplies with increasing frequency. Products that contain organic chemicals include solvents, pesticides, paints, inks, dyes, varnishes, and gasoline. The U.S. Environmental Protection Agency performed groundwater surveys in the 1990, that have confirmed the widespread presence of organic contaminants.[5]

Groundwater Salinization

In coastal regions, the pumping of groundwater from an aquifer can result in the intrusion of saline waters, thereby severely altering the quality of water. As groundwater mining increases, this is becoming an increasing problem in many regions as salinity can degrade the quality to the extent that it is no longer potable.

Groundwater is a resource that is increasing in value for the support of human activities and life itself. Increasing efforts are required to slow the degradation of groundwater quality in much of the United States.

REFERENCES

1. Freeze, R.A.; Cherry, J.A. *Groundwater*; Prentice-Hall Inc.: New Jersey, 1979; 1–604.
2. Domenico, P.A.; Schwartz, F.W. *Physical and Chemical Hydrogeology*, 2nd Ed.; John Wiley & Sons, Inc.: New York, 1998; 1–506.
3. Davis, S.N.; DeWiest, R.J.M. *Hydrogeology*; John Wiley & Sons, Inc.: New York, 1966; 1–463.
4. Watson, I.; Burnett, A.D. *Hydrology: An Environmental Approach*; CRC Press, Inc.: Boca Raton, FL, 1995; 1–702.
5. http://www.epa.gov/seahome/groundwater/src/overview. htm (accessed August, 2001).

Groundwater Quality, Irrigated Agriculture and

Stephen R. Grattan
University of California, Davis, California, U.S.A.

INTRODUCTION

Irrigation water quality can have a profound impact on crop production inasmuch as irrigated agriculture can affect groundwater quality. Irrigated agriculture not only involves the application of water, which contains dissolved mineral elements, but is often coupled with other inputs such as fertilizers and pesticides. Many of these constituents can leach past the crop rootzone and pollute the underlying aquifer. Other chapters in the Encyclopedia will address impacts of irrigated agriculture on groundwater quality. The emphasis of this chapter is on groundwater quality, and its potential impacts on irrigated agriculture.

GROUNDWATER QUALITY

All groundwater sources used for irrigation contain dissolved mineral salts, but the concentration and composition of the dissolved salts vary from one aquifer to another. Dissolved mineral salts form ions; either positively charged cations or negatively charged anions. The most common cations are calcium (Ca^{2+}), magnesium (Mg^{2+}), and sodium (Na^+) whereas the most abundant anions are chloride (Cl^-), sulfate (SO_4^{2-}), and bicarbonate (HCO_3^-). Potassium (K^+), carbonate (CO_3^{2-}), nitrate (NO_3^-), and trace elements also exist in groundwater supplies but most often concentrations of these constituents are comparatively low. On the other hand, some groundwater sources contain boron (B) at comparatively low concentrations but at levels that may be detrimental to certain crops.

An understanding of the quality of water used for irrigation and its potential negative impacts on the crop, soil, and irrigation system is essential to avoid problems and optimize production. The salinity of the water is important because too much salt can reduce crop production while too little salt or certain compositions of salt (i.e., sodic waters) can reduce water infiltration, which indirectly affects the crop. Certain elements or combination of elements in the groundwater can be toxic to sensitive crops or pose a management or maintenance problem. More detailed information on water quality and impacts on agriculture can be found in Ref. [1]. For more information on the nature and extent of agricultural salinity, see Ref. [2] or visit http://water.usgs.gov/nwis/gw for actual groundwater quality data in the United States.

Characterizing Salinity

There are two water quality parameters that characterize the salinity of the irrigation water: electrical conductivity (ECw) and total salt concentration or total dissolved solids (TDS). The units of TDS are usually in milligrams of salt per liter of water ($mg\,L^{-1}$). This term is used by many commercial analytical laboratories and represents the total mg of salt that would remain after a liter of water is evaporated to dryness. Often, TDS is reported as parts per million (ppm), which is numerically equivalent to $mg\,L^{-1}$. The higher the TDS, the higher is the salinity of water.

Electrical conductivity is a much more useful term because the measurement can be made instantaneous in the field. Salts that are dissolved in water conduct electricity and therefore the salt content in the water is directly related to the ECw. Units of EC reported by labs are usually in decisiemens per meter ($dS\,m^{-1}$) or millimhos or micromhos per centimeter ($mmhos\,cm^{-1}$ or $\mu mmhos\,cm^{-1}$). One $mmho\,cm^{-1} = 1000\,\mu mmhos\,cm^{-1} = 1\,dS\,m^{-1}$.

Often, a conversion between ECw and TDS is made based on guidelines from Ref. [3] (i.e., ECw × 640 = TDS) but caution is advised because this conversion is dependent on both salinity and composition of the water. The USDA-ARS Salinity Laboratory has a web site that has educational material, models, databases, and lists of publications on various chemical, physical, and phyto-biological aspects of salinity including a pdf version of Handbook 60 http://www.ussl.ars.usda.gov/.

Characterizing Sodicity

The sodicity or alkalinity of the groundwater is characterized on the basis of its Na^+ relative to Ca^{2+} and Mg^{2+} concentration. Sodicity[4] refers to either the exchangeable Na percentage (ESP), or the sodium adsorption ratio (SAR) of the soil solution. The SAR = $Na^+/(Ca^{2+} + Mg^{2+})^{0.5}$ where ion concentrations are millimolar and the ESP is the percentage of the soil's cation-exchange-capacity (CEC) occupied by Na^+.

Encyclopedia of Water Science
DOI: 10.1081/E-EWS 120010073

RELATIONS BETWEEN IRRIGATION WATER, SOIL SALINITY, AND LEACHING

Salts can accumulate in the root zone from the irrigation water due to insufficient leaching. To prevent salt accumulation in the root zone from the irrigation water, the soil must be adequately leached. Leaching is the process of applying more water to the field than can be held by the soil in the crop root zone such that the excess water drains below the root system carrying salts with it. The more water that is applied in excess of the crop water requirement, the less the salinity in the root zone will be despite the fact that more salt has been added to the field. The term "leaching fraction" (LF) is used to relate the fraction or percent of water infiltrated to the field that actually drains below the root zone.

Below are some useful relationships between the salinity in the irrigation water (ECw) and the average root zone salinity (ECe). The ECe is the electrical conductivity of the saturate soil paste (i.e., soil samples are saturated with distilled water, the soil water is then extracted, and the EC is measured on the extracted water). These relationships predict what would happen over the long-term if the LFs indicated are achieved, assuming steady-state conditions and a 40-30-20-10 root water extraction pattern where the top and bottom quarters of the root zone extracts 40% and 10% of the crops consumptive water use.

LF 10%   ECw × 2.1 = ECe

LF 15%–20%   ECw × 1.5 = ECe

LF 30%   ECw = ECe

The leaching requirement is an attractive concept but has limitations. First, the ET of the crop is assumed to be independent of the average root zone salinity.[5] Thus, calculated crop water requirements will be high where the average root zone salinity exceeds the threshold salinity of the crop, which corresponds to a yield potential less than 100%. Second, the leaching requirement is based on steady-state conditions and does not account for the initial salinity status in the root zone. Finally, applying irrigation water to a field to achieve a given LF is very difficult, if not impossible, particularly with fine textured soils in climates with high evaporative demand. Nevertheless, in order to control salinity, leaching must occur whether it is achieved before the season, midway through the season, or at the end of the season.[1]

In fields where salinity has increased in the root zone to damaging levels, "reclamation leaching" is recommended. Ref. [6] provides additional information on reclamation of soils.

For more information on relations between irrigation water salinity, leaching, and root zone salinity—see Refs. [7,8].

IMPACT ON CROPS

Salinity, caused by either too much salts in the groundwater supply and/or insufficient leaching, can directly affect the crop in two ways, by osmotic effects and by specific ion effects. The osmotic effects are responsible for growth reduction, the most common whole-plant response to salinity. Within limits, isosmotic concentrations of different combinations of salts cause nearly equal reductions in growth. On the other hand, specific ions such as Na^+, Cl^-, and B may be particularly injurious to certain crops or under specific management practices. A detailed discussion of mechanisms of salt tolerance and injury can be found in Refs. [9,10] and references cited therein.

Estimating Yield Potential

Crops vary widely in their response to salinity. Some crops such as bean and onion are very sensitive to salinity while others such as cotton and asparagus are tolerant. The salt tolerance of a crop is best described by plotting its relative yield as a function of the average root zone salinity (ECe). This response curve is represented by two line segments; one, a tolerance plateau with zero slope and the second, a concentration-dependent line whose slope indicates the yield reduction per unit increase in ECe[11] (Fig. 1).

Maas and Hoffman[11] assembled a table with salinity coefficients. The point where the first line segment meets the second line segment is referred to as the yield threshold coefficient (a). This represents the maximum soil salinity a crop can tolerate before its yield declines. The slope of the second line is the second salinity coefficient (b), which represents the percent decrease in yield per unit increase in ECe. Thus, the relative yield (%) = $100 - b(ECe - a)$. Additional salinity coefficients can be found in Ref. [12].

These salinity coefficients are particularly useful in predicting yield potentials based on either the average root zone salinity or based on the irrigation water itself by using the relations between ECw, ECe, and LF described earlier.

It is important to emphasize that these are only guidelines and assume that all other factors such as fertility, irrigation scheduling, and pest control are managed to maximize crop performance. It is also important to note that most of the experiments that were used to generate these guidelines were conducted in the interior of California where the climate is hot and dry during the summer. Crops grown in the coastal regions or where the climate is milder will likely tolerate greater

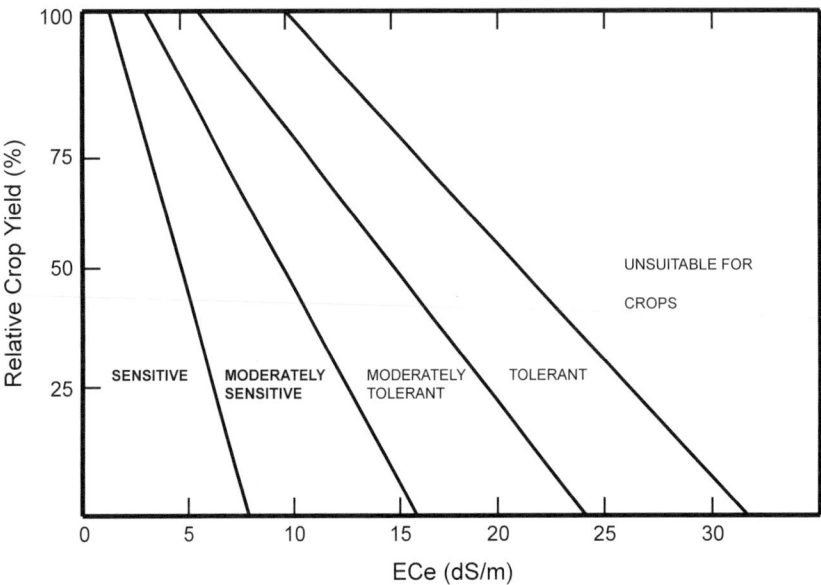

Fig. 1   Divisions for classifying crop tolerance to salinity. (From Ref. [12].)

salinities than indicated in these publications. For more detailed information on relations between crop yields and salinity, see Refs. [1,12].

Crop Toxicity to Specific Elements

In addition to salinity's general osmotic effect, some crops, particularly tree and vines, are injured by certain elements, specifically Na^+, Cl^-, and boron (B). These elements are absorbed by the root and move with the transpirational stream to the leaves where they concentrate. Thus, older leaves or older portions of leaves such as margins and tips transpire more than younger tissue and develop injury first. Injury usually begins as chlorosis and advances into necrosis as injury becomes more severe. Fig. 2 shows both Cl and B injury to tomato leaves. Although the injury is similar, Cl injury in most crops has more chlorosis (leaf yellowing) contiguous to the necrotic tissue whereas necrotic portions from B injury are often darker with a reddish-brown coloration.

This additional injury complicates salt-tolerance in that the combined osmotic and specific-ion effects may affect the yield potentials of the crop more than the salt-tolerance guidelines would indicate. Tables are provided that list the maximum concentration of Cl or B in the soil water that a crop can tolerate before it develops symptoms of ion toxicity.[12]

The irrigation method can affect crop sensitivity to water quality. With drip and furrow irrigation, chloride

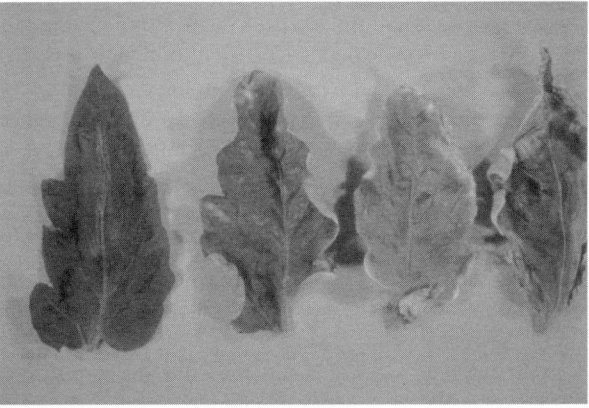

Fig. 2   Progression of injury to tomato leaves due to boron toxicity (upper) and chloride toxicity (lower).

and sodium injury does not generally occur in most vegetable and row crops unless salinity is severe. Under sprinkler irrigation, injury may develop on wetted leaves of susceptible plants such as peppers, potatoes, and tomatoes if the ECw exceeds $1.5\,dS\,m^{-1}$ (see Ref. [12]). Injury occurs due to direct foliar absorption of salts. Susceptibility to leaf injury is related to leaf wettability, leaf morphology, and the rate of foliar salt absorption and not tolerance to soil salinity. Increased frequency of sprinkler irrigation is usually more damaging that increased duration.

Some vegetable and row crops are sensitive to boron. Generally, leaf injury must be severe to cause reduced yields and crop quality. Long-term use of irrigation water containing more than $0.5\,mg\,L^{-1}$–$0.7\,mg\,L^{-1}$ boron can reduce the yields of bean, onion, garlic, strawberry, broccoli, carrot, potato, and lettuce and greater than $2\,mg\,L^{-1}$ can reduce yields of cabbage and cauliflower.

Unlike most annual crops, tree and vine crops are generally sensitive to boron, chloride, and sodium toxicity. Tolerances vary among varieties and rootstocks. Tolerant varieties and rootstocks resist the uptake and accumulation of toxic ions in the stem and leaf tissue. Continued use of irrigation water with boron concentrations in excess of $0.75\,mg\,L^{-1}$ can reduce the yields of grapes and many deciduous tree and fruit crops. This represents a threshold concentration and does not imply that irrigation water with boron at or slightly above this level cannot be used successfully.

Chloride moves readily with the soil water and is taken up by the roots. It is then transported to the stems and leaves. Sensitive berries and avocado rootstocks can tolerate only up to $120\,mg\,L^{-1}$ Cl while grapes can tolerate up to $700\,mg\,L^{-1}$ or more.

The ability of the tree to tolerate sodium varies considerably. Sodium injury on avocado, citrus, and stone-fruit trees has been reported at concentrations as low as $115\,mg\,L^{-1}$. Initially sodium is retained in the roots and lower trunk but after 3 or 4 yr the conversion of sapwood to heartwood apparently releases the accumulated sodium, which then moves to the leaves causing leaf burn. It is unclear how extensive sodium toxicity occurs because when injury is evident, levels of chloride are often high as well. Ref. [12] contains information on crops as they are affected by specific ion toxicity.

Climate and soil factors affect crops response to specific-ion injury. Under cool, moist climatic conditions, higher concentrations of B, Cl, or Na can be tolerated. Hot dry weather on the other hand could cause more severe injury at a given tissue-ion concentration. In addition, soil conditions influence the time it takes for injury to occur. The finer the soil texture, the longer it will take for injury to occur. Furthermore, there is an evidence that salinity may reduce boron's injurious effect so that plants can

tolerate a higher concentration of B than the guidelines indicate. For more information on boron, see Refs. [12,13].

Indirect Na Effects on Plants

In addition to osmotic and specific-ion toxicity, sodic or saline–sodic groundwater may also induce an indirect effect such as Na-induced Ca deficiency. Ca deficiency in the crop maybe obvious such as whip-like appearances in young emerging leaves, blackheart in celery, blossom end rot in tomato and pepper, but are more likely to be subtle where visual symptoms are absent.[14] Such an interaction has been described in Refs. [10,14].

IMPACT ON SOILS

Soil physical properties can be affected by irrigation with sodic or saline–sodic groundwater particularly when good quality water or rains follow.[15] Potential consequences include reduced infiltration and redistribution rates within the soil, poor soil tilth, and inadequate aeration resulting in anoxic conditions for roots. These negative impacts are enhanced with decreasing soil salinity and with increasing exchangeable Na (i.e., ESP).

At the soil surface, infiltration rates and soil tilth are particularly sensitive to salt and exchangeable Na levels. The mechanical impact and stirring action of the irrigation water, or rain, combined with the freedom for soil particle movement at the soil surface, can result in low infiltration rates when the soil is wet, and hard, dense soil crusts when the soil is dry. Crusts can block the emergence of seedlings thereby reducing stand establishment (Fig. 3). Tillage of crusted soils can result in hard soil clods that are difficult to reduce in size when the clod is dry. Extensive tillage can

Fig. 3 Reduced stand establishment in cotton in a field previously irrigated with saline–sodic water.

be required to prepare a seed bed with sufficient tilth to assure adequate soil/seed contact for seed germination.

Infiltration of Irrigation Water

There are two water quality parameters that are currently used to assess irrigation water quality for potential water infiltration problems. These are the ECw and the SAR. Both a low salt content (low ECw) and high SAR can cause permeability or water infiltration problems even on sandy soils.

A low ECw and/or high SAR can act separately or collectively to disperse soil aggregates, which in turn reduces the number of large pores in the soil. These large pores are responsible for adequate aeration and drainage. A negative effect from the breakdown of soil aggregates is soil sealing and crust formation. Table 1 provides guidelines that can be used to assess the potential likelihood of water infiltration problems based on ECw and SAR.

Table 1 indicates that water infiltration problems are likely if the ECw is less than $0.3 \, dS \, m^{-1}$ regardless of the SAR. For example, if the ECw falls below $0.4 \, dS \, m^{-1}$, infiltration rates can drop to less than $0.1 \, in. \, hr^{-1}$ ($2.5 \, mm \, hr^{-1}$). An infiltration rate of $2.5 \, mm \, hr^{-1}$ would require 30 hr for a full irrigation of 75 mm to infiltrate the soil. Thus, very high quality water can cause infiltration problems even when applied on soils with a high sand content. Soils may also be prone to water infiltration problems in late Fall and Winter months after high quality rainwater falls on fields previously irrigated with sodic or saline–sodic groundwater. For more information on soil response to saline and sodic conditions, see Refs. [15,16].

Fortunately, infiltration problems due to a low salt content or high SAR can easily be improved by the addition of amendments to either the irrigation water or soil that directly (e.g., gypsum) or indirectly (e.g., acidifying agents) supply free calcium (Ca^{2+}) to the soil water. When the irrigation water contacts gypsum, it dissolves into Ca^{2+} and SO_4^{2-} ions, which slightly increases the salinity of the water while simultaneously

reducing the SAR. The Ca^{2+} cations are then free to displace Na^+ cations adsorbed onto the negatively charged clay particles enhancing flocculation, improving soil structure, and increasing the water infiltration rate. Information on the management and reclamation of sodic soils is provided in Ref. [15].

IMPACTS ON IRRIGATION SYSTEMS

Irrigation water supplies, particularly those from wells, can contain other constituents that may affect water quality and its potential use for irrigated agriculture. Of particular concern are carbonates (HCO_3^- and CO_3^{2-}), nitrate (NO_3^-), and reduced iron (Fe^{2+}) and manganese (Mn^{2+}).

High pH and excessive amounts of bicarbonate can be problematic. In fields that are irrigated with low-pressure systems such as drip or minisprinklers, calcite or scale can build up near the orifice of the sprinkler or emitter, which can reduce the water discharge. This type of problem can be corrected by injecting acid-forming materials in the irrigation water. Unsightly white residues (calcium carbonates) can be left behind on leaves and fruits that have been sprinkler irrigated, potentially affecting the aesthetic quality. In addition, bicarbonate could increase the SAR of the soil water by precipitating calcium. This problem can usually be corrected by frequent gypsum applications. Bicarbonate has also been found to be toxic to some plants under certain conditions.[1]

Nitrates are often found in groundwater supplies particularly in areas where intensive irrigated agriculture has occurred over the years. From a public health perspective, there are concerns when excessive levels are found in domestic wells. From an irrigation perspective, NO_3^- in the groundwater can be viewed as a resource. For example, 27 lb of N can be applied to a field with each acre ft of water if the water supply contains $10 \, mg \, L^{-1}$ (or ppm) NO_3-N ($45 \, mg \, L^{-1}$ when expressed as NO_3^-). It is important that the grower with water of such a quality reduces the N application rates in the field accordingly to

Table 1 Likelihood of potential water infiltration problems based on ECw and SAR

SAR of irrigation or soil water	Potential water infiltration problem	
	Unlikely if ECw is (dS m^{-1})	Likely if ECw is (dS m^{-1})
0–3	> 0.6	< 0.3
3–6	> 1.0	< 0.4
6–12	> 2.0	< 0.5
12–20	> 3.0	< 1.0
20–40	> 5.0	< 2.0

(From Refs. [1,15].)

accommodate this extra input of nitrogen. Should this be ignored, excessive vegetative growth and re-contamination of the aquifer can occur. Certain shallow groundwaters such as drainage waters sources may contain enough nitrates to affect crop quality. Examples are delayed maturity or extensive vegetative growth in grape, citrus, and tomatoes or reduced sugar contents in sugar beet and grape.

Iron, manganese, and sulfur are often present in groundwater in the soluble yet chemically reduced forms (Fe^{2+}, Mn^{2+}, and sulfides). Certain bacteria in the water can oxidize these soluble reduced forms to insoluble oxidized forms. Bacterial colonies are associated with these oxidized constituents and form a gel or slime responsible for clogging filters and drip emitters. Reduced iron and manganese can create emitter-clogging problems at concentrations as low as $0.1 \, mg \, L^{-1}$–$0.2 \, mg \, L^{-1}$. High concentrations of iron and manganese can be reduced by chemical precipitation, which is enhanced by aerating the water and allowing the residue to settle out before it is used for irrigation. Low, yet still problematic, concentrations of iron and manganese may be maintained in a soluble form by reducing the pH of the irrigation water by injecting acid in the system.

REFERENCES

1. Ayers, R.S.; Westcot, D.W. *Water Quality for Agriculture*, FAO Irrigation and Drainage Paper 29, Rev. 1; Food and Agriculture Organization of the United Nations: Rome, 1985; 174.
2. Tanji, K.K. Nature and Extent of Agricultural Salinity. In *Agricultural Salinity Assessment and Management*, ASCE Manuals and Reports on Engineering Practice No. 71; Tanji, K.K., Ed.; ASCE: New York, 1990; 1–17.
3. United States Salinity Laboratory Staff, USDA Handbook 60. *Diagnosis and Improvement of Saline and Alkali Soils*; U.S. Printing Office, Washington, DC, 1954; 160.
4. Sumner, M.E.; Naidu, R., (Eds.) *Sodic Soils: Distribution, Properties, Management and Environmental Consequences*; Oxford University Press: New York, 1998; 207.
5. Letey, J.; Knapp, K.; Solomon, K. Crop-Water Production Functions Under Saline Conditions. In *Agricultural Salinity Assessment and Management*, ASCE Manuals and Reports on Engineering Practice No. 71; Tanji, K.K., Ed.; ASCE: New York, 1990; 305–326.
6. Keren, R.; Miyamoto, S. Reclamation of Saline, Sodic, and Boron-Affected Soils. In *Agricultural Salinity Assessment and Management*, ASCE Manuals and Reports on Engineering Practice No. 71; Tanji, K.K., Ed.; ASCE: New York, 1990; 410–431.
7. Hoffman, G.J. Leaching Fraction and Root Zone Salinity Control. In *Agricultural Salinity Assessment and Management*, ASCE Manuals and Reports on Engineering Practice No. 71; Tanji, K.K., Ed.; ASCE: New York, 1990; 237–261.
8. Pratt, P.F.; Suarez, D.L. Irrigation Water Quality Assessments. In *Agricultural Salinity Assessment and Management*, ASCE Manuals and Reports on Engineering Practice No. 71; Tanji, K.K., Ed.; ASCE: New York, 1990; 220–236.
9. Jacoby, B. Mechanisms Involved in Salt Tolerance of Plants. In *HandBook of Plant and Crop Stress*; Pessarakli, M., Ed.; Marcel Dekker, Inc.: New York, 1999; 97–123.
10. Lauchli, A.; Epstein, E. Plant Responses to Saline and Sodic Conditions. In *Agricultural Salinity Assessment and Management*, ASCE Manuals and Reports on Engineering Practice No. 71; Tanji, K.K., Ed.; ASCE: New York, 1990; 113–137.
11. Maas, E.V.; Hoffman, G.J. Crop Salt Tolerance—Current Assessment. J. Irrig. Drain. Div., ASCE **1977**, *103* (IR2), 115–134.
12. Maas, E.V.; Grattan, S.R. Crop Yields as Affected by Salinity. In *Agricultural Drainage*, Agron. Monograph 38; Skaggs, R.W., van Schilfgaarde, J., Eds.; ASA, CSSA, SSSA: Madison, WI, 1999; 55–108.
13. Grattan, S.R.; Oster, J.D. Use and Reuse of Saline–Sodic Waters for Irrigation of Crops. In *Crop Production in Saline Environments*; Goyal, S.S., Sharma, S.K., Rains, D.W., Eds.; Haworth Press: New York, 2001 (in press).
14. Grattan, S.R.; Grieve, C.M. Salinity-Mineral Nutrient Relations in Horticultural Crops. Sci. Hort. **1999**, *78*, 127–157.
15. Oster, J.D.; Jayawardane, N.S. Agricultural Management of Sodic Soils. In *Sodic Soils: Distribution, Processes, Management and Environmental Consequences*; Summer, M.E., Naidu, R., Eds.; Oxford Univ. Press: New York, 1998; Chap. 8, 125–147.
16. Shainberg, I.; Singer, M.J. Soil Response to Saline and Sodic Conditions. In *Agricultural Salinity Assessment and Management*, ASCE Manuals and Reports on Engineering Practice No. 71; Tanji, K.K., Ed.; ASCE: New York, 1990; 91–112.

Groundwater, Regulation of

Kevin B. McCormack
United States Environmental Protection Agency (US-EPA), Washington, District of Columbia, U.S.A.

INTRODUCTION

Much of the momentum for groundwater protection and remediation began in the late 1970s and continued to grow through the 1980s. Many environmental statutes and regulations that directly and indirectly concern groundwater protection were enacted at the federal, state, and local levels during this period.[1] At the time, groundwater protection remained a relatively new undertaking for many states and localities. Within the past 15 yrs, numerous reports have documented the need for more effective coordination of groundwater protection programs at the federal, state, and local levels.[28] National and local studies increasingly indicate that many activities adversely impact groundwater quality.[4] Contamination incidents and impairment from overpumping, such as permanent loss of aquifer storage capacity and land subsidence, remain a local problem because of the relatively slow rate at which groundwater travels. "What These Threats Mean to the Nation" describes a variety of agricultural, industrial/commercial, and waste disposal practices that are known to contaminate groundwater.

Based on the data that have been collected to date, groundwater quality appears to be generally good nationwide (that is, groundwater contaminant levels are usually below applicable drinking water standards). Locally, however, groundwater quality is being threatened by a variety of land uses.[21] Although groundwater appears to be of higher quality than surface water throughout the United States, contamination incidents and overpumping remain a problem for numerous localities. A variety of agricultural, industrial, commercial, and waste disposal practices are known to contaminate groundwater. The occurrence of nitrates, pesticides, organic chemicals, and other contaminants reveal the impact of certain land uses on groundwater quality. Overpumping can limit water availability to nearby wells; reduce groundwater flow to streams, lakes, and wetlands; permanently damage aquifer storage capacity; and induce salt-water intrusion to freshwater aquifers.[6,10] Because no one federal, state, or local authority can manage all these threats, a coordinated approach for groundwater management is needed.

BARRIERS TO SUCCESSFUL PREVENTION AND PROTECTION PROGRAMS

There are probably as many groundwater protection programs as there are states. States differ in the goals they set for groundwater, the standards they apply to it, and the mechanisms through which it is protected, and their approaches to drinking water protection of supplies drawn from groundwater sources. Groundwater quality is typically protected at the state level through programs, which control the potential sources of contamination and address remediation of contamination. States identify their maximum contaminant limit goals for groundwater, which function as ambient standards. Classification of groundwater, and of land uses which might affect it, are common tools. Discharge permits or other regulatory controls can be used to prevent groundwater contamination, through the imposition of performance or effluent-type limits on dischargers. States vary as to both the sources of groundwater contamination which they regulate, and the standards to which these sources are subject.

States have identified three primary barriers for achieving a more comprehensive approach:[31]

1. Fragmentation of groundwater programs among and within agencies impedes effective management. At the state level, authorities to manage the resource are often held among different state agencies with conflicting priorities and goals. Communicating and coordinating among departments with groundwater responsibilities can be difficult. In turn, these barriers can create

Disclaimer: This work consists of excerpts of material previously published by the U.S. Environmental Protection Agency, Washington, DC, in the form of Rules, Regulations, Guidance Documents, Position Papers, Scientific Studies, and national meetings and workshops sponsored by EPA and by private concerns, and as such is considered public domain. Any mistakes or inaccuracies are therefore in these excerpted parts and not my original writing. Copyright authorization is thereby waived. Further, the views expressed by the author are his own, and should in no way be construed as representing official EPA policy.

Encyclopedia of Water Science
DOI: 10.1081/E-EWS 120010219

an impediment for accessing funds for comprehensive planning efforts.

2. There is a lack of understanding of groundwater resources locally and regionally (e.g., the extent and condition of the resource, the physical nature of the aquifer, the behavior of contaminants within and their movement through aquifers, the influence of surface water to groundwater and vice versa). Better information to assess the effectiveness of groundwater protection efforts and to determine the impact of certain land uses on groundwater is needed to set priorities for groundwater protection efforts.

3. Lack of funding targeted directly to groundwater is the reason most often cited by states for limited efforts in undertaking a more comprehensive resource-based approach. Groundwater protection is often not a high priority for funding; mandated programs usually prevail for funding. Most states indicate that the mandates under other federal programs often preclude the state from exercising flexibility to use funds for nonmandated groundwater protection priorities.

THREATS TO GROUND WATER

Although groundwater quality in this country is generally good, many local activities threaten the resource by point and nonpoint contaminant sources as well as by overpumping. Sources, most frequently cited as being of greatest concern, include underground storage tanks (USTs), landfills, septic systems, hazardous waste sites, surface impoundments, above-ground storage tanks, industrial facilities, spills, fertilizer and pesticide applications, pipelines and sewer lines, agricultural chemical facilities, shallow injection wells, salt water intrusion, animal feedlots, land application, mining, urban runoff, salt storage and road salting, and hazardous waste generators.[14,15,27,29–32]

Various federal, state, and academic information relates agricultural, industrial, waste disposal, and other land uses with groundwater degradation. Certain land uses are known to impair groundwater quality, but the ability to predict the level of impairment from specific activities is difficult, especially over long periods of time.[5] The US Geological Survey (USGS) National Water Quality Assessment (NAWQA) Program is the principal source of information on groundwater quality available in the United States today. Under the NAWQA program, USGS collects new water quality data in 60 special study regions of the country, conducts retrospective analyses of existing data (such as state data), and prepares national-scale syntheses of the results.[22,27,29,31]

EPA is also developing a National Contaminant Occurrence Database (NCOD) to track contaminants in groundwater and surface water sources of drinking water supply.

WHAT THESE THREATS MEAN TO THE NATION

Public Health Impacts

Both short-term illness and chronic health impacts are associated with the consumption of contaminated drinking water. For example, the presence of pathogenic microorganisms can cause acute gastrointestinal illness, Hepatitis A, and other diseases. Carcinogenic chemicals can increase the incidence of cancer. Other chemicals can adversely impact the growth and development of children. For instance, high levels of nitrate in drinking water consumed by newborns can lead to a fatal condition known as "blue baby syndrome." Once groundwater is contaminated with certain compounds, certain treatment processes, such as disinfection with chlorination used by public water systems, can transform these compounds into chemicals that may also pose concern (such as trihalomethanes, a group of carcinogenic disinfection-byproducts), thereby exposing the population to other health risks. In addition, some contaminants, such as nitrates, are expensive to treat and may be very costly to remove through home treatment. Groundwater contamination in rural areas is a particular public health concern.[9]

Economic Impacts

Groundwater contamination can also impair the economic well-being of a nation. In 1995, EPA examined costs associated with 6 communities that had experienced actual or imminent contamination of the groundwater supplied through their public water systems. The costs associated with alternative water supplies, water treatment, and contaminant source removal or remediation ranged from over $0.5 million to about $2.4 million. A 1992 analysis by EPA indicated that for 51 selected communities with contaminated or threatened drinking water systems, the cost of remediation averaged $5.9 million per community water system, with most costing between $1 million and $10 million.[7,13,20,23,24]

Ecological Impacts

Groundwater is also critical to the ecological health of the country. Groundwater provides many ecological benefits through its linkage with surface water. The interrelationships of groundwater with wetlands, lakes, ponds, and streams are complex. In areas where groundwater has been contaminated (by domestic wastewater or industrial discharges), ecological impacts can be detected in the form of eutrophication and loss of native fish and plants.[12]

REGULATION OF GROUNDWATER

Over the past 25 years, federal laws, regulations, and programs have come to reflect the growing importance that the nation places on using groundwater wisely and protecting the resource. Beginning with the 1972 amendments to the federal Water Pollution Control Act, and followed by the Safe Drinking Water Act (SDWA) in 1974, the federal government's role in groundwater protection has increased. With the passage of the Resource Conservation and Recovery Act (RCRA) in 1976 and the Comprehensive Environmental Response, Compensation, and Liability Act (CERCLA) in 1980, the federal government's current focus on groundwater remediation was established.

The cleanup approach to groundwater protection at the federal level has been very costly, and has left the management of many contaminant threats to state and local government authorities, including Indian tribes.[11] In the absence of a federal regulatory framework, the degree to which states and local governments address groundwater concerns varies considerably.[8] Some states have well-coordinated, effective groundwater protection programs, while others have all they can do to maintain programs that are minimally protective of the public health.

Protection and Prevention Programs

Below is a chronological list of EPA's protection and prevention-related rules, regulations, and activities specifically targeted towards groundwater-based drinking water supplies:

1972 Federal Water Pollution Control Act Amendments

1972 Federal Insecticide, Fungicide, and Rodenticide Act (FIFRA)

1974 Safe Drinking Water Act (SDWA)

1976 Resource Conservation and Recovery Act (RCRA)

1980 Underground Injection Control Program established

1980 Comprehensive Environmental Response and Compensation and Liability Act (Superfund)

1984 Hazardous and Solid Waste Amendments to RCRA

1984 US EPA Ground Water Strategy and Office of Ground Water Protection established

1986 Superfund Amendments and Reauthorization Act: Underground Storage Tank Program

1986 SDWA Amendments: Wellhead Protection and Sole Source Aquifer Programs

1987 Clean Water Act

1991 EPA Ground Water Strategy Revised

1992 Comprehensive State Ground Water Protection Program Guidance

1992 Interagency Task Force on Monitoring Water Quality (through 1996)

1993 Pesticide State Management Plans under FIFRA

1996 SDWA Amendments: Source Water Assessment and Protection Program

1996 FIFRA Amendments under the Food Quality Control Act of 1996

1997 National Water Quality Monitoring Council formed

1998 Clean Water Action Plan

1998 Underground Storage Tank Closure/Upgrade Requirements

1999 Class V Underground Injection Control Final Rule

2000 Proposed Ground Water Rule

The most salient of these programs and activities are described briefly as follows.

Wellhead protection (WHP) programs

WHP is essentially a pollution prevention program oriented towards reducing threats to groundwater quality in sources destined for use as public drinking water supply. The basic elements of the WHP program are: 1) statement of purpose; 2) defining roles and duties or participating agencies; 3) delineation of WHP areas; 4) identification of potential contaminant sources within the delineated area; 5) development of differential management techniques to deal with these sources; 6) development of long- and short-range contingency planning for water supply replacement in the event of contamination or physical disruption; and 7) Development of a decision-making process for siting new wells.[17–19]

Comprehensive state groundwater protection programs (CSGWPPs)

About a dozen states have developed an EPA-approved CSGWPP that promotes a more strategic, resource-based approach to groundwater protection, and more than half the states are undertaking efforts that are essential to a comprehensive approach to groundwater protection. However, only a few states have been able to complete, or have begun to develop, a comprehensive list of groundwater protection priorities. Even fewer states have indicated that they have identified available program funding sources to address their comprehensive groundwater protection priorities in a systematic, consistent way.[3,25]

Source water assessment and prevention programs

Section 1453 of the SDWA as amended in 1996 established the source water assessment progam (SWAP), which requires all states to complete assessments of their public drinking water supplies. By 2003, each state and participating Indian tribe will delineate the boundaries of areas in the state (or on tribal lands) that supply water for each public drinking water system (PWS), identify significant potential sources of contamination, and determine how susceptible each system is to sources of contamination.[26]

Federal, State, and Local Regulations

Federal regulations

Clean Water Act (CWA). Groundwater protection is addressed in Section 102 of the CWA, providing for the development of federal, state, and local comprehensive programs for reducing, eliminating, and preventing groundwater contamination.

SDWA. Under the SDWA, EPA is authorized to ensure that water is safe for human consumption. To support this effort, SDWA gives EPA the authority to promulgate maximum contaminant levels (MCLs) that define safe levels for some contaminants in public drinking water supplies. One of the most fundamental ways to ensure consistently safe drinking water is to protect the source of that water (i.e., groundwater). Source water protection is achieved through four programs: the WHP Program, the Sole Source Aquifer (SSA) Program,[16] the Underground Injection Control (UIC) Program,[21] and, under the 1996 Amendments, the Source Water Assessment Program (SWAP).[26]

RCRA. The intent of RCRA is to protect human health and the environment by establishing a comprehensive regulatory framework for investigating and addressing past, present, and future environmental contamination or groundwater and other environmental media. In addition, management of USTs is also addressed under RCRA.

CERCLA. CERCLA provides a federal "Superfund" to clean-up soil and groundwater contaminated by uncontrolled or abandoned hazardous waste sites as well as by accidents, spill, and other emergency releases of pollutants and contaminants into the environment. Through the Act, EPA was given power to seek out those parties responsible for any release and assure their cooperation in the clean-up. The program is designed to recover costs, when possible, from financially viable individuals and companies when the clean-up is complete.[2]

FIFRA. FIFRA protects human health and the environment from the risks of pesticide use by requiring the testing and registration of all chemicals used as active ingredients of pesticides and pesticide products. Under the Pesticide Management Program, states and tribes wishing to continue use of chemicals of concern are required to prepare a prevention plan that targets specific areas vulnerable to groundwater contamination. Mandates may not address the most pressing groundwater protection concerns of a particular community or area.

State regulations

Although most states have begun implementing components of a comprehensive program, many states report that much work remains to be completed. Funding, lack of agency coordination, and an absence of priority-setting mechanisms are obstacles most frequently identified by the states to explain the lack of comprehensive planning and coordination. The 1999 GroundWater Protection Council report examined the state's level of achievement in implementing the components of a comprehensive groundwater protection program.[3]

OUTLOOK FOR THE FUTURE

Over the past 20 yr, thousands of local groundwater contamination incidents have been identified and the nation has devoted many billions of public and private dollars to clean-up these problems. Although these efforts have protected many people from exposure to groundwater contaminants released from sources, such as hazardous waste sites and leaking USTs, some incidences of groundwater contamination have not yet been fully cleaned up. In some instances, groundwater remediation can take a decade or more to be completed. Furthermore, in many parts of the country, we are using groundwater at a faster rate than it can be replenished through natural recharge, and, in some cases, we are permanently losing future storage capacity. Although many of these programs emphasize surface waters and need to integrate groundwater management for a truly comprehensive approach to water resource management, they provide models for better coordination and integration. Some examples follow:

The Clean Water Action Plan (CWAP)

At the federal level, CWAP emphasizes the importance of a comprehensive approach for restoring and protecting waters among nine federal agencies (EPA, Department of

Interior, Department of Defense, Department of Energy, Department of Agriculture, Department of Transportation, Department of Commerce, Department of Justice, and Tennessee Valley Authority). CWAP is both a vision statement and a blueprint for the future. It focuses on: 1) promoting water quality protection and restoration on a watershed basis and 2) strengthening core clean water programs to protect human health, increase natural resources stewardship, reduce polluted runoff, and provide citizens and officials with crucial information.

Intergovernmental Task Force on Monitoring Water Quality (ITFM)

The ITFM was established in 1992 and given the charge of reviewing water quality monitoring nationwide and developing an integrated national monitoring strategy. In 1995, ITFM produced The Strategy to Improve Water-Quality Monitoring in the United States. In 1997, the National Water Quality Monitoring Council (NWQMC) was formed as a successor to ITFM. During overall strategy development, a Ground Water Focus Group (GWFG) concentrated on issues related to groundwater and aquifer systems. The GWFG recommended that water quality monitoring must consider differences in spatial, temporal, and other characteristics between ground and surface water resources.

State Watershed Protection Frameworks

State Watershed Protection Frameworks are designed to coordinate existing resource management programs and build new partnerships that result in more effective and efficient management of land and water resources. These frameworks provide not only a mechanism for coordinating the point and nonpoint source management activities that have been the historic focus of state water quality programs, but also a forum for meeting the objectives of groundwater, wellhead, and drinking source water protection programs. Many State Watershed Protection Frameworks incorporate a priority-setting and targeting mechanism to focus resources on watersheds requiring the highest degree of management to remediate existing problems or address emerging threats.

SWAPs

SWAPs established under the 1996 Amendments to the SDWA provide an additional coordination mechanism for state programs. The states are required in their SWAPs to assess the degree to which all PWS in the state are susceptible to contamination. These assessments will be accomplished by: 1) delineating the sources of water supply to the PWS; 2) inventorying the contaminants and

contaminant sources within that delineated area; and 3) assessing how susceptible the PWS are to those sources of contamination. In many states, these assessments will be accomplished through cooperative efforts, involving several state agencies, local governments, and private water suppliers.

REFERENCES

1. Fort, D.D.; Gabin, V.L.; Pinnes, E. *Managing Groundwater Quality and Quantity in the Western States: A Report to the Environmental Protection Agency*; Natural Heritage Institute: San Francisco, CA, March 1993.
2. Ground Water Protection Council. Superfund Setaside: Prevention of Contamination (White Paper Report); Oklahoma City, OK, 1996.
3. Ground Water Protection Council. Survey Results on State's Comprehensive Approach to Ground Water Protection; Oklahoma City, OK, 1999.
4. Liner, E.B.; Morley, E.; Stanger, J. *Assessing the Experience of Local Groundwater Protection Programs*; The Urban Institute: Washington, DC, February 1994.
5. Liner, E.B.; Morley, E.; Hatry, H.P.; Dusenbury, P.P. *Protecting the Nation's Groundwater: Guidelines for State and Local Action*; The Urban Institute: Washington, DC, November 1991.
6. Liner, E.B.; Morley, E.; Hatry, H.P.; Dusenbury, P.P.; Hoch, S. *State Management of Groundwater: Assessment of Practices and Progress*; The Urban Institute: Washington, DC, December 1989.
7. National Association of Counties. Counties and the Takings Issue: How Far Can Government Go in Regulating Private Property? Washington, DC, 1998
8. National Environmental Education and Training Foundation. The National Report Card on Safe Drinking Water Knowledge, Attitudes and Behaviors; Washington, DC, July 1999.
9. National Environmental Education and Training Foundation. 1999 Environmental Education Leadership Roundtables on Drinking Water Source Protection: A Report; Washington, DC, 1999.
10. National Small Flows Clearinghouse, *Septic Stats: An Overview*; Environmental Services and Training Division: Morgantown, WV, 1999.
11. Schmidt, C. *Protecting Drinking Water: A Workbook for Tribes*; Water Education Foundation: Sacramento, CA, 1999.
12. Stone, A.W.; Stone, A.J.L. *Wetlands and Ground Water in the United States*; The American Ground Water Trust and The Audubon Society of New Hampshire: Dublin, OH, 1994.
13. The Freshwater Foundation. Economic Implications of Groundwater Contamination to Companies and Cities; Navarre, MN, 1998.
14. U.S. Bureau of Census. Estimates of Housing Units, Households, Households by Age of Householder, and

Persons per Household, Population Estimates Program; Washington, DC, July 1996.

15. U.S. Department of Agriculture and U.S. Environmental Protection Agency. Draft Unified Strategy for Animal Feeding Operations; Washington, DC, 1998.

16. USEPA; Office of Ground Water Protection. Sole Source Aquifer Designation Petitioner Guidance, EPA 440/6-87-003; Washington, DC, February 1987.

17. USEPA; Office of Water. Wellhead Protection Programs/Tools for Local Governments, EPA 440/6-89-002; Washington, DC, April 1989.

18. USEPA; Office of Water. Wellhead Protection, A Decision-Maker's Guide, EPA 440-6-87-009; May 1992.

19. USEPA; Office of Water. Why Do Wellhead Protection? Issues and Answers in Protecting Public Drinking Water Supply Systems, EPA 813-K-95-001; Washington, DC, May 1995.

20. USEPA; Office of Water and Office of Policy, Planning and Evaluation. A Framework for Measuring the Economic Benefits of Ground Water, EPA 230-B-95-003; Washington, DC, October 1995.

21. USEPA; Office of Water. Safe Drinking Water Act, Section 1429 Ground Water Report to Congress, EPA 816-R-99-016; Washington, DC, October 1999.

22. USEPA; Office of Water. National Water Quality Inventory: 1996 Report to Congress, Ground Water Chapters, EPA 816-R-98-011; Washington, DC, 1996.

23. USEPA; Office of Water. Benefits and Cost of Prevention: Case Studies of Community Wellhead Protection, EPA 813-B-95-005; Washington, DC, 1995.

24. USEPA; Office of Water. Clean Water and the American Economy, An Overview: Perspectives on Ground Water, Pre-Conference Papers; Arlington, VA, October 1992.

25. USEPA; Office of The Administrator. Final Comprehensive State Ground Water Protection Guidance; Washington, DC, 1992.

26. USEPA; Office of Water. State Source Water Assessment and Protection Programs Guidance, EPA 816-R-97-009; Washington, DC, August 1997.

27. USEPA; Office of Ground Water Protection. A Review of Sources for Ground Water Contamination from Light Industry—Technical Assistance Document, EPA 440/6-90-005; Washington, DC, 1990.

28. U.S. Geological Survey. Strategic Directions for the U.S. Geological Survey Ground Water Resources Program: A Report to Congress. In U.S. Geological Survey Report to Congress; Washington, DC, 1998.

29. U.S. Geological Survey. Volatile Organic Chemicals (VOCs) in Ground Water of the United States: Preliminary Results of the National Water-Quality Assessment (NAWQA) Program, VOC National Synthesis Project; U.S. Geological Survey: Washington, DC, January 1998.

30. U.S. Geological Survey. Pesticides in Surface and Ground Water of the United States: Summary of Results of the National Water-Quality Assessment (NAWQA) Program, Pesticides National Synthesis Project; U.S. Geological Survey: Washington, DC, July 1998.

31. U.S. Geological Survey. Occurrence of Pesticides in Shallow Ground Water of the United States: Initial Results of the National Water-Quality Assessment (NAWQA) Program, Pesticides National Synthesis Project; U.S. Geological Survey: Washington, DC, January 1998.

32. U.S. Geological Survey. MTBE in the Nation's Ground Water, National Water-Quality Assessment (NAWQA) Program Results; U.S. Geological Survey: Washington, DC, April 1999.

Groundwater, Saltwater Intrusion in

Alexander H.-D. Cheng
University of Delaware, Newark, Delaware, U.S.A.

INTRODUCTION

The origin of saltwater intrusion into freshwater aquifers can come from natural sources such as seawater, and deep formation brines, or from anthropogenic sources such as de-icing salt, agricultural return flow, and leachate from landfills. The most frequent occurrences are found in coastal regions where overexploitation of groundwater has caused the encroachment of seawater into freshwater aquifers. Once an aquifer is invaded, a part of the salt will adsorb onto the solid surface making it difficult to reverse the process and restore the aquifer. The slow movement of groundwater also makes the remediation time long. Salinity in water poses health hazard for human and livestock, damages crops, and corrode pipes and boilers in industrial uses. Hence, the invasion of saltwater into a freshwater aquifer means the loss of that aquifer for water sources.

MECHANISMS OF SALTWATER INTRUSION

Fig. 1 gives a schematic view of seawater intrusion into an *unconfined aquifer*. We observe that saltwater is heavier, hence tends to move underneath the freshwater layer. The freshwater, however, has a *hydraulic gradient* downward towards the coast, hence will flow to the sea. This outflow momentum force can counter balance the density-driven seawater. Without it, seawater will continue to move inland until the entire aquifer below sea level is occupied by it. Since such a hydraulic gradient always exists due to the precipitation recharge inland, an equilibrium position will establish, shown as the *interface* in Fig. 1. The *toe* then marks the maximum extent of intrusion.

A simple theory that allows a rule-of-thumb estimate of the salt–fresh water interface location is given by the Ghyben–Herzberg relation:[1]

$$\xi = \frac{\rho_f}{\rho_s - \rho_f} h_f \approx 40 h_f$$

where ξ is the interface location below sea level, h_f the freshwater head above sea level (referring to Fig. 1), ρ_f the freshwater density (1 g/cm^3), and ρ_s the saltwater density (approximately 1.025 g/cm^3). What the above relation says

is, for every meter of freshwater head above sea level, the interface is pushed down 40 m. When the interface touches the bottom of aquifer, the toe is located. This 40:1 ratio may sound like a good news for repelling saltwater; however, if pumping activity is increased inland, as quite often is the case due to increased population in coastal zones, the reduced freshwater head level close to the coast will allow saltwater to move a large distance landward. If a pumping well is situated above the interface in the freshwater zone, any small *drawdown* will cause the interface to rise up sharply to meet the well, known as *upconing*. This means that it is nearly impossible to sustain extraction of freshwater above the invaded saltwater wedge. This portion of freshwater, including the natural recharge, is considered lost.

GEOPHYSICAL AND GEOCHEMICAL INVESTIGATIONS

The presence of salinity in aquifers, its source, and the underlying physical, chemical, and geological processes leading to the intrusion can be detected or interpreted by a combination of geophysical and geochemical investigations. Geophysical methods measure the spatial distribution of physical properties of the earth, such as bulk electrical conductivity and seismic velocity. For investigation of saltwater intrusion in shallower depths, the *DC resistivity method*, which introduces electrical current into the ground through electrodes driven into soil, is most effective because the presence of salt increases the bulk conductivity of the soil. The *electromagnetic method* sends out a time varying magnetic field that generates electrical currents in the ground whose strength is dependent on the conductivity of the earth. The varying electrical field in turn generates a secondary electromagnetic wave that can be detected above ground. There are several variations of the electromagnetic method, including frequency-domain, airborne, loop–loop, time-domain, very low frequency, and ground penetration radar method.[2] These surface geophysical methods have the advantage of being able to map the salinity variation over a large horizontal area. Its resolution in the vertical direction, however, decreases with depth. The *borehole method* allows the introduction of tools into the formation

Encyclopedia of Water Science
DOI: 10.1081/E-EWS 120010083

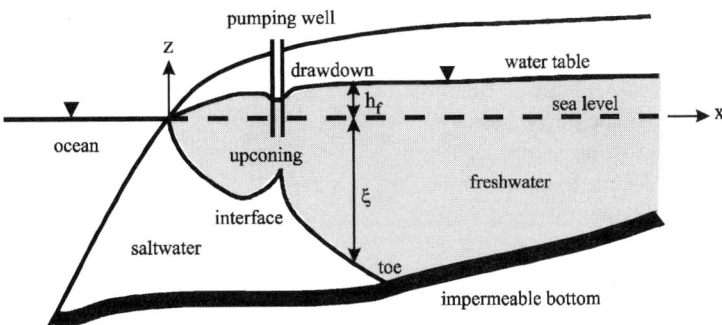

Fig. 1 Seawater intrusion into an unconfined aquifer.

at larger depths to produce higher resolution electrical resistivity, electromagnetic, and radiometric logs.

The geochemical method investigates the chemical composition of groundwater for not only the presence of chloride and sodium, but also other ions such as K, Mg, Ca, Br, SO_4, and HCO_3. Ratio of these ions, such as Cl/Br, Na/Cl, Ca/Mg, Ca/(HCO_3+SO_4), can often provide a chemical signature to the origin of salt contamination—whether it comes from seawater, fossil water, or anthropogenic sources.[3] Isotope studies can indicate the age of the water, hence can further help in identifying the source.

MATHEMATICAL MODELING

The use of field surveys, such as geophysical and geochemical studies, can reveal the present state of saltwater intrusion, and perhaps some insight into its history. It, however, cannot make prediction into the future, and particularly cannot be used for scenario building and impact assessment based on different levels of anthropogenic activities. Mathematical models are needed for these purposes.

The Ghyben–Herzberg relation is a highly simplified model. More rigorously, the dynamic movement of groundwater flow and the solute transport of salt needs to be considered. Generally speaking, there does not exist a sharp division between saltwater and freshwater zones, as implied in Fig. 1. The salt concentration continuously changes from that of seawater to that of freshwater. A solute transport model including advection and dispersion is needed for the modeling. In addition, the salt at higher concentration is an active solute, because it can affect the density of water and can drive the flow. Hence a *density-dependent solute transport model* should be used. There are occasions, however, when the predominant change of concentration from saltwater to freshwater takes place within a narrow region called the *transition zone*. In that case, a simplification using the *sharp interface model* can

be attempted. Furthermore, if the aquifer modeled is of regional scale, then the flow is often integrated in the depth direction to reduce the three-dimensional problems to two-dimensional ones. The governing equations, boundary conditions, and justification of using the various models can be found in Ref. [4].

COMPUTER MODELS

With the exception of some simple geometries of saltwater intrusion for which analytical solutions are available,[5] numerical solutions are needed for practical applications. Two of the most widely used computer codes are: SHARP[6]—for sharp interface model—and SUTRA[7]—for density-dependent solute transport model—both developed by the U.S. Geological Survey. However, like many complex engineering problems, there is no single code that can be most versatile, efficient, accurate, and stable at the same time, thus dominating the rest of the codes. Depending on the availability and reliability of input data, and the limited resources dedicated to modeling, different computer codes have been developed to offer a wide range of choices. A comprehensive survey of the computer codes can be found in Ref. [8].

COMBATING SALTWATER INTRUSION

One of the most effective ways of combating saltwater intrusion is to regulate pumping activities. Generally speaking, the amount of groundwater extraction should not exceed that of natural replenishment. Optimization of pumping patterns to maximize the yield and minimize the extent of intrusion is a high-priority management issue. Recharge of natural surface water or reclaimed wastewater into aquifers can increase the freshwater outflow rate to push back the saltwater wedge. A recharge near the coast can build a local freshwater mound that forms a barrier to protect the *water table* depression inland. Extraction of saltwater in an invaded saltwater wedge can also protect the freshwater behind, if a proper way can be found to dispose of the extracted saltwater. A similar method involving pumping simultaneously in the upper freshwater zone and the lower saltwater zone to prevent upconing, known as *double pumping*, has been attempted. Using *collector wells* (horizontal wells) to skim the thin layer of freshwater floating on top of the saltwater wedge has been effectively used in water-poor countries such as Israel. Land reclamation has the added effect of pushing saltwater to the sea. Finally, in places where large freshwater springs flowing to the sea can be identified, physical barriers, such as solid walls or slurry curtains, can be used to intercept freshwater.

REFERENCES

1. Bear, J. *Hydraulics of Groundwater*; McGraw-Hill: New York, 1979; 567 pp.
2. Stewart, M.T. Geophysical Investigations. In *Seawater Intrusion in Coastal Aquifers—Concepts, Methods and Practices*; Bear, J.; Cheng, A.H.-D.; Sorek, S.; Ouazar, D.; Herrera, I., Eds.; Chap. 2, Kluwer: Dordrecht, 1999; 9–50.
3. Jones, B.F.; Vengosh, A.; Rosenthal, E.; Yechieli, Y. Geochemical Investigations. In *Seawater Intrusion in Coastal Aquifers—Concepts, Methods and Practices*; Bear, J.; Cheng, A.H.-D.; Sorek, S.; Ouazar, D.; Herrera, I., Eds.; Kluwer: Dordrecht, 1999; Chap. 3, 51–71.
4. Bear, J. Conceptual and Mathematical Modeling. In *Seawater Intrusion in Coastal Aquifers—Concepts, Methods and Practices*; Bear, J.; Cheng, A.H.-D.; Sorek, S.; Ouazar, D.; Herrera, I., Eds.; Kluwer: Dordrecht, 1999; Chap. 5, 127–161.
5. Cheng, A.H.-D.; Ouazar, D. Analytical Solutions. In *Seawater Intrusion in Coastal Aquifers—Concepts, Methods and Practices*; Bear, J.; Cheng, A.H.-D.; Sorek, S.; Ouazar, D.; Herrera, I., Eds.; Kluwer: Dordrecht, 1999; Chap. 6, 163–191.
6. Essaid, H.I. USGS SHARP Model. In *Seawater Intrusion in Coastal Aquifers—Concepts, Methods and Practices*; Bear, J.; Cheng, A.H.-D.; Sorek, S.; Ouazar, D.; Herrera, I., Eds.; Kluwer: Dordrecht, 1999; Chap. 8, 213–247.
7. Voss, C.I. USGS SUTRA Code—History, Practical Use, and Application in Hawaii. In *Seawater Intrusion in Coastal Aquifers—Concepts, Methods and Practices*; Bear, J.; Cheng, A.H.-D.; Sorek, S.; Ouazar, D.; Herrera, I., Eds.; Kluwer: Dordrecht, 1999; Chap. 9, 249–313.
8. Sorek, S.; Pinder, G.F. Survey of Computer Codes and Case Histories. In *Seawater Intrusion in Coastal Aquifers—Concepts, Methods and Practices*; Bear, J.; Cheng, A.H.-D.; Sorek, S.; Ouazar, D.; Herrera, I., Eds.; Kluwer: Dordrecht, 1999; Chap. 12, 399–461.
9. Bear, J.; Cheng, A.H.-D.; Sorek, S.; Ouazar, D.; Herrera, I., (Eds.) *Seawater Intrusion in Coastal Aquifers—Concepts, Methods and Practices*; Kluwer: Dordrecht, 1999; 625 pp.

Groundwater, World Resources

Lucila Candela
Technical University of Catalonia-UPC, Barcelona, Spain

INTRODUCTION

Groundwater constitutes an important source of water supply for domestic, industrial, and irrigation uses in different countries due to its availability, which is not subject to multiannual and seasonal fluctuations. At present it is the main source of domestic water supply in most European countries, the United States, Australia, and some countries of Asia and Africa both in large and small towns and in rural areas.[1,2] Groundwater is used for irrigation in about one-third of all irrigated lands.[3]

Natural groundwater resources are understood to be the total amount of recharge (replenishment) of groundwater under natural conditions as a result of infiltration of precipitation, seepage from rivers and lakes, leakage from overlying and underlying aquifers, and inflow from adjacent areas. In some cases, the average annual recharge of aquifers, evaluated from average annual precipitation, equals groundwater runoff. Natural groundwater resources may be equated to groundwater discharge (runoff) when the evaporation from the water table may be ignored or estimated separately.[4] Under this assumption, groundwater runoff data are widely used to characterize regional groundwater resources and are an important component of the hydrologic cycle and environment.[5–7]

The role of groundwater in the water balance and water resources of regions is quantitatively characterized by the groundwater runoff/precipitation ratio or groundwater recharge. The runoff/precipitation ratio is extremely variable depending on meteorological factors, composition of rocks, etc. Distribution of groundwater recharge to river/total river runoff ratios shows the effect of geographical and altitudinal zonality. The quantity of recharge ranges widely. Analysis of conditions of generation of groundwater resources within continents shows that this global process depends on a complex combination of various natural factors. Principal amongst these are precipitation, vegetation, soil type and geology, and the hydrogeological features of the area.

In regions where aquifers are mainly composed of sands, specific groundwater discharge values are twice as large as in regions where the percentage of sands in aquifers is small. In this regard distribution of specific values of groundwater discharge on a global scale is subject to latitudinal zonality. Values generally increase from subartic regions to medium-latitude zones, in humid tropics and tropics, and decrease in semiarid and arid regions. Large groundwater discharge values may be found in karst limestones (up to $20\,L\,sec^{-1}\,km^{-2}$), sand quaternary deposits (up to $18\,L\,sec^{-1}\,km^{-2}$) or highly fractured rocks (up to $10\,L\,sec^{-1}\,km^{-2}$), although values are normally dependent on topographic elevations and annual precipitation. Marine sandy and clayey sediments show minimal discharge values ($0.1\,L\,sec^{-1}\,km^{-2}$ and smaller).[4]

The main task of areal hydrogeological subdivision when compiling groundwater runoff and resources maps is to distinguish territories which are sufficiently uniform in terms of groundwater distribution and particularities of groundwater generation.[8]

AQUIFER TYPES

The principal aquifers are found in six types of permeable geologic materials:[8] unconsolidated deposits of sand and gravel; semiconsolidated sand; sandstone; carbonate rocks interbedded sandstone and carbonate rocks; and basalt and other types of volcanic rocks. Large areas of the world are underlain by crystalline rocks permeable only where they are fractured or weathered, and generally yield only small amounts of water to wells. In many places, they are the only source of water supply. However, because these rocks extend over large areas, important volumes of groundwater are withdrawn from them.

Unconsolidated Sand and Gravel Aquifers

Unconsolidated sand and gravel aquifers are characterized by intergranular porosity and all contain water primarily under unconfined or water-table conditions, but locally confined conditions may exist where aquifers contain beds of low permeability. Different categories can be distinguished, which occupy different geologic settings. The sediments are mostly alluvial deposits, but locally may include windblown sand, coarse-grained glacial outwash, and fluvial sediments deposited by streams recharge. Large areas of the world are covered with sediments deposited during several advances and retreats of continental glaciers. The glacial sand and gravel deposits form numerous local but productive aquifers.

Encyclopedia of Water Science
DOI: 10.1081/E-EWS 120010218

Aquifers commonly receive direct recharge from precipitation and streamflow infiltration. Regional movement is down the valley in the direction of stream flow, lake or playa (located in the center of the basin). Basins in arid regions might contain deposits of salt, anhydrite, gypsum or borate produced by evaporation or mineralized water in their central parts. Also, much of the infiltrating water is lost by transpiration by riparian vegetation.

Consolidated/Fractured Sedimentary Aquifers

Aquifers in sandstones are more widespread than those in all other kinds of consolidated rocks. Sandstone retain some primary porosity unless cementation has filled all the pores, but most of the porosity in these consolidated rocks consists of secondary openings such as joints, fractures, and bedding planes. The water is not highly mineralized in areas were the aquifer outcrops or are buried to shallow depths, but mineralization generally increases as the water moves downgrading toward the structural basin.

Carbonate Rock Aquifers

The water-yielding properties of carbonate rocks are highly variable; some yield almost no water and are considered to be confining units, whereas others are among the most productive aquifers known. The original texture and porosity of carbonate deposits can range from 1% to more than 50%. Recharge water enters the aquifer through sinkholes, swallow holes, and sinking streams, some of which terminate at large depressions called blind valleys.

Basaltic and Other Volcanic-Rock Aquifers

Volcanic rocks have a wide range of chemical, mineralogic structural, and hydraulic properties due largely to rock type. Unaltered pyroclastic deposits have porosity and permeability characteristics like those of poorly sorted sediments; rhyolites have low permeability except where they are fractured.

AVAILABILITY AND USE

Except for widely scattered places, existing data are not uniformly distributed in space and time because hydrologic investigations have been mostly conducted in areas where water supply or water quality problems existed, or where large quantities of groundwater were withdrawn. Long-term hydrologic records are rare and usually collected only during the course of a study or perhaps for a few years after the study has ended. No systematic investigation on groundwater resources and exploitation in many regions of the world have been conducted.[3,9]

The annual groundwater use for the world as a whole can be placed at $750-800 \times 10^9 \, m^3$, a modest value when compared to overall water availability (Tables 1 and 2). But an overwhelming majority of the world's cities and towns depend on groundwater for municipal water supplies. Over 35 countries of the world use more than $1 \times 10^9 \, m^3$ of groundwater annually.[10] Because of spatial imbalances in the occurrence of groundwater and the pattern of demand, massive problems of groundwater overexploitation are found in areas where high population exist or under intensive agriculture development.

Table 1 Annual groundwater recharge and withdrawals in the world

World region	Average annual recharge ($km^3 \, yr^{-1}$)[a]	Annual groundwater withdrawals ($km^3 \, yr^{-1}$)[b]
Asia	2505	352
Europe	1368	78
Middle East and N. Africa	137	75
Sub-Saharan Africa	1548	9
North America	1884	110
C. America and Caribbean	344	29
South America	3693	14
Oceania	270	2

[a] Amount of water that is estimated to annually infiltrate into aquifers. It would represent the amount of water that could be annually withdrawn.
[b] Abstractions from aquifers. These data are scarce and not currently available for all countries in each region.
Source: WRI. Environmental Data Tables. World Resources 2000–2001 **2000**, (8).

Table 2 Groundwater use in selected areas of the world

Country	Annual recharge ($km^3\,yr^{-1}$)	Groundwater use (%)
Russian Federation	900	< 1
China	800	10
India	450	30

Source: (From Ref. [3].)

GROUNDWATER RESOURCES DISTRIBUTION

Europe

All types of aquifers are currently exploited: large well fields in artesian basins of platform type, such as Paris and London; river valleys (France, Volga region); cones and intermontane depressions (Italy, Switzerland).[11,12] In many cases their exploitation is accompanied by the generation of large and deep cones of depression.

Groundwater runoff in Europe is quite irregular, depending on the geostructural, climatic, and orographic conditions and the flow media generation: karst, porous fractured, and porous. Specific discharge values distribution is governed by the geological structure.

According to the available data (Table1), groundwater use estimation in Europe is $78\,km^3\,yr^{-1}$, which constitutes 21% of the total water consumption. Urban and rural population constitute the most important water consumers, accounting for 56% of the total water consumption. Groundwater is the main source for public water supply (more than 70% of total resources), especially on islands and some European countries like Denmark. More than 90% of big cities and towns are exclusively supplied by groundwater (among them Berlin, Rome). Although groundwater is mainly used for irrigation in Southern countries like Spain with values ranging between $0.7\,km^3\,yr^{-1}$ and $5\,km^3\,yr^{-1}$, other European countries, like the Netherlands, may also use it during dry years.

Africa

Africa is one of the regions of the world facing serious water shortages because of greater disparities in water availability and use, and because water resources are unevenly distributed. Groundwater, first considered as a main resource for water in urban, rural areas, and mining, especially in coastal areas and arid regions, is now tending to be extended to the most isolated desert and tropical regions. In Libya, groundwater accounts for 95% of country's freshwater withdrawals, while in some areas of North Africa it is a significant source for irrigated agriculture. In many parts of the continent, groundwater resources have not yet been fully explored and tapped. According to the geographic and climatic homogeneity, which has a direct influence on water resources, Africa can be divided into several regions: Northern, Sudano-Sahelian, Gulf of Guinea, Central, Eastern, Indian Ocean Islands and Southern. This vast territory can be subdivided into a number of large aquifer systems subject to very varied climatic conditions.[13,14]

Basement rocks cover most of the central territory and aquifers are not very productive except in few cases. Sedimentary formations of sandstones overlaying the basement areas may constitute good aquifers, such the Karoo basin. The coastal sedimentary basins are the most productive aquifers, being intensely exploited along the shoreline. Alluvials are among the most important and also serve large populations, especially in Northern Africa. Karstified limestones of North-West Africa and Madagascar can yield flow rates up to $100\,m^3\,hr^{-1}$. Also large fossil aquifers are present in the Saharan and Nubian deserts made of sedimentary basins and being largely exploited.

South America, Central America, and the Caribbean

This area extending from the Central America Isthmus to South America has the most abundant river flow. Groundwater is unevenly distributed in quantity, but quality is usually good for domestic and industrial supply, presently the highest priority. Total water withdrawal from the aquifers is difficult to estimate because most comes from uncontrolled private and public wells. Based on UN estimates, 50%–60% of total population domestic and industrial supply is from groundwater. Water withdrawn can be estimated at between $12\,km^3\,yr^{-1}$ and $14\,km^3\,yr^{-1}$, very low in comparison with the estimated renewable resources. Groundwater reserves estimation is $238,000\,km^3$, discharge to rivers being $3898\,km^3\,yr^{-1}$. Discharge values are high in the humid equatorial zone and minimum in the Atlantic Andean Cordillera and northeast Brazil.

According to geologic and tectonic features, four major water-bearing domains can be distinguished:[15,16] superficial deposits; deep aquifers in sedimentary basins; folded mountain chains; and precambrian basement bedrocks. Vast areas of South America are composed of Precambrian crystalline rocks which are not highly productive unless weathered or intensively fractured. The hydrogeologic map of South America[16] shows 16 hydrogeological provinces with similar characteristics including the previously-mentioned water-bearing domains. Some of the formations' resources are considered as the most important water-bearing formations, such as the Amazon Sedimentary Basin ($32,500\,km^3$), Parnaíba-Maranhao

($17,000 \, \mathrm{km}^3$), and the Paraná Sedimentary basin, where the Guarani aquifer extending over $1,500,000 \, \mathrm{km}^2$ has $50,000 \, \mathrm{km}^3$ of storage.

North America

Groundwater is an important source of water in the United States and Mexico, but it represents less than 5% of Canada's total water use. About 22% of the total water use in the United States ($290 \times 10^6 \, \mathrm{m}^3 \, \mathrm{day}^{-1}$) is supplied by groundwater; about 50% of the US population depends on groundwater for domestic uses and also major cities and metropolitan areas and irrigation has made the High Plains one of the most important agricultural areas. Half of the U.S. population draws its domestic water supply from groundwater.[1] In Mexico, where desert and semiarid conditions prevail over two-thirds of the country, groundwater is widely used. Urban areas of Mexico use groundwater as their sole or principal source.

Unconsolidated sand and gravel are the most widespread aquifers, with intergranular porosity, and water primarily under water-table conditions.[17,18] Some unconsolidated aquifers have supplied large amounts of water for irrigation, like the High Plains aquifer ($56 \times 10^6 \, \mathrm{m}^3 \, \mathrm{day}^{-1}$ withdrawn from the aquifer for irrigation in 1990); in the United States, about 20% of the groundwater withdrawn is derived from the High Plains aquifer.

Carbonate rock aquifers are most extensive in eastern United States and in the Bahamas, western Canada and Yucatan (Mexico), and some of them are considered among the most productive aquifers known. Most of them consist of limestone but dolomite and marble locally yield water. More than $13 \times 10^6 \, \mathrm{m}^3 \, \mathrm{day}^{-1}$ (1990 data) were withdrawn from the Floridan aquifer system, the sole source of water supply for the city of Miami.

Oceania

While groundwater resources in New Zealand, Pacific Islands, and New Guinea are difficult to quantify due to the limited information, the aridity of much of the Australian continent is a significant factor in the occurrence and assessment of groundwater resources.[19] A large part of western and central Australia is arid, with a mean annual rainfall below 250 mm. Total amount of groundwater used in Australia is estimated at $2460 \times 10^6 \, \mathrm{m}^3$ in 1983 from more than 500,000 tube-wells, 14% of the total amount of water used.[20] The greatest concentrations are near Perth, Adelaide, South Australia, western Victoria, and on the central Queensland coast. Surficial aquifers are the most important sources for irrigation, urban, and industrial supply. Fractured rock aquifers of igneous and meta-

morphic rocks are of relative importance, although they may locally provide high groundwater yields.

Most highly productive aquifers are the surficial sedimentary aquifers associated with inland or eastern coastal rivers, up to 100 m thick. Also sand dunes, coastal, and deltaic alluvium sediments form important aquifers along the east coast and in central Queensland. Australia's main arid-zone irrigation scheme is based on groundwater extracted from sands and gravel of Central Australia. Several large deep sedimentary basin aquifers (Amadeus, Canning, Great Artesian, Murray, Otway, Perth, Eucla, Officer), extending over more than $24,000 \, \mathrm{km}^2$, constitute a reliable source of old, good quality groundwater. The Great Artesian basin covers $1.7 \times 10^6 \, \mathrm{km}^2$, is up to 300 m thick and is one of the largest basins in the world.[21] More than 20,000 nonflowing and more than 4000 flowing artesian wells have been drilled. Individual well flows exceeding $100 \, \mathrm{L} \, \mathrm{sec}^{-1}$ have been recorded. Diffuse natural discharge from the Great Artesian Basin has been calculated to be about $1.4 \times 10^6 \, \mathrm{m}^3 \, \mathrm{day}^{-1}$.

Asia and Middle East

Continental Asia is an immense and complex geographical area of great extremes. Some parts of China and India are among the most populated in the world while the deserts of central Asia and the interior high plateau are extremely thinly populated. Most of the land of the Arabian Peninsula and in central and eastern Iran is a desert, reflecting different patterns of groundwater use.

In Asia and Middle East groundwater has been developed since ancient times, especially in arid regions where no other source of water supply is permanently available.[22,23] Large-scale developments are found in northern and coastal areas of China, where artesian aquifers and the loess and karst areas of the central south are tapped for urban and industrial supply. Groundwater irrigation distribution by subregions is, according to AQUASTAT,[3] in the Arabian peninsula 96.6%; Middle East 18.2%; and Central Asia 34% (although for Bangladesh it represents 69%).

Groundwater exists in the area as semiconfined, unconfined shallow, and deep aquifers. Recharge is faster in the Middle East countries although the aquifers of the Arabian Peninsula contain much larger reserves. The interior arid regions of the Middle East countries may include geological formations which can be considered as aquicludes or aquitards. Among the groundwater-bearing zones, the most important are alluvial of large rivers, vast, complex sedimentary formations holding artesian water and sedimentary basins in coastal areas (Israel). Some carbonate basins of importance are also present in the Mediterranean area (Lebanon, Syria) and Pakistan.

Weathered crystalline rocks and lava flows constitute important aquifers in peninsular India and Northern Syria.

Although groundwater quality is suitable for irrigation and domestic uses, salinization of groundwater has occurred in several areas of the Indus Plain and Pakistan.

REFERENCES

1. http://www.cgiar.org/iwmi/pubs/WWVisn/GrWater.htm
2. http://www.grida.no/geo2000/english
3. http://www.fao.org/ag/agl/aglw/Aquastatweb/Main/html/aquastat.htm
4. Zekster, I.S.; Dzhamalov, R.G. *Role of Ground Water in the Hydrological Cycle and in Continental Water Balance*; IHP-UNESCO: Paris, 1988; 133.
5. Van der Leeden, F.; Troise, F.L.; Todd, K.D. *Water Encyclopedia*; Lewis Publishers, Chelsea, Michigan, 1990; 808.
6. Gleick, P. *Water in Crisis: A Guide to the World's Fresh Water Resources*; Oxford University Press, New York, 1993; 1–493.
7. UN. *Comprehensive Assessment of the Freshwater Resources of the World: Report of the Secretary General*, United Nations, Commission on Sustainable Development, April 7–15, 1997; E/CN/17/1997/9, 1997; 1–35.
8. Dzhamalov, R.G.; Zektser, I. *Digital World Map of Hydrogeological Conditions and Groundwater Flow*; CDromVersion 2.0, HydroScience Press, 1999.
9. http://www.wri.org/water/index.html
10. Llamas, R.; Back, W.; Margat, J. Groundwater Use: Equilibrium between Social Benefits and Potential Environment Costs. Appl. Hydrol. **1992**, (1), 3–14.
11. UN, *Groundwater in the Western Hemisphere*; Natural Resources/Water Series No. 4., United Nations, Department for Technical Co-operation for Development: New York, 1976; 337.
12. UN, *Groundwater in Western and Central Europe*, Natural Resources/Water Series No. 27; United Nations, Department for Technical Co-operation for Development: New York, 1991; 363.
13. UN, *Les Eaux souterraines de l'Afrique Orientale, Centrale et Australe*, Ressources naturelles/Serie eau No. 19; United Nations, Department for Technical Co-operation for Development: New York, 1988; 341.
14. UN, *Groundwater in North and West Africa*, Natural Resources/Water Series No. 18; United Nations, Department for Technical Co-operation for Development: New York, 1988; 405.
15. Rebouças, A.C. Desarrollo y Tendencias de la Hidrogeología en America Latina. In *Hidrogeología, Estado Actual y Perspectivas*; Anguita, Aparicio, Candela, Zurbano, Eds.; CIMNE: Barcelona, 1991; 429–453.
16. UNESCO-IHP/DNPM/CPRM. *Hydrogeological Map of South America 1: 5,000,000. Exploratory tex*, Rio de Janeiro, 1996; 212.
17. http://www.water.usgs.gov/ogw
18. http://capp.water.usgs.gov/gwa/ch_h/H-text1.html
19. UN, *Groundwater in the Pacific Region*, Natural Resources/Water Series No. 12; United Nations, Department for Technical Co-operation for Development: New York, 1983; 289.
20. DPIE, *Review of Australia's Water Resources and Water Use. Vol. 1. Water Resources Data Set*; Dep. of Primary Industries and Energy. Australian Government Publishing Service: Canberra, 1987; 158.
21. Habermehl, M.A. The Great Artesian Basin Australia. BRM J. Aust. Geol. Geophys. **1980**, (5), 9–38.
22. UN, *Groundwater in the Mediterranean and Western Asia*, Natural Resources/Water Series No. 9; United Nations, Department for Technical Co-operation for Development: New York, 1982; 230.
23. UN, *Groundwater in Continental Asia (Central, Eastern, Southern, South-Eastern Asia)*, Natural Resources/Water Series No. 15; United Nations, Department for Technical Co-operation for Development: New York, 1986; 391.

Hydrologic Cycle

John Van Brahana
University of Arkansas, Fayetteville, Arkansas, U.S.A.

INTRODUCTION

The hydrologic cycle describes the dynamic, water-circulation system of the Earth. Water we see today is the same water that was originally derived from degassing of volcanoes as the Earth cooled from the molten mass that was our primordial planet several billion years ago. This water has been continuously recycled by natural processes, changing from liquid to solid or vapor and then back again, moving and flowing endlessly in response to the physical and chemical conditions of the environment of our planet.

PROCESSES AND PATHWAYS

The dominant processes of the hydrologic cycle, and the pathways along which we can trace water movement, include the following (Fig. 1): evaporation from oceans and open bodies of water on the Earth's surface into the atmosphere; evapotranspiration by plants of soil water into atmospheric vapor; condensation of water vapor into liquid or solid particles (clouds); precipitation as the condensed water or ice falls from the atmosphere back to the surface; infiltration of the water into subsurface (soil and groundwater) reservoirs; baseflow contribution to streams from groundwater; streamflow recharging of groundwater; streamflow runoff if the precipitation rate exceeds the infiltration rate; surface and subsurface flow back to oceans or intermediate reservoirs; and storage.[1–10]

In one sense, the hydrologic cycle is one of the most basic concepts of water science, yet in detail the concept is complex because it involves all forms of water of the hydrosphere, and it is affected by many influencing factors that are not always obvious. Because this circulation of water is intimately tied to energy transfer, it is helpful to start with the basic physics of the forces that drive this seemingly endless flow of water on our planet.

ENERGY SOURCES

The underlying source of energy that drives the movement of water throughout the hydrosphere of the Earth is solar radiation—thermal energy from our sun. Solar energy heats water, causes it to evaporate and change state from liquid to a gas, and in so doing, facilitates its movement through the atmosphere in response to wind and pressure changes. Every gram of liquid water at its point of vaporization requires an input of 540 cal of thermal energy to convert it to a gas. Worldwide, water vapor represents a huge source of energy storage and transport in a hydrologic link between atmosphere, oceans, and continents, which we call "climate." The residence time of water vapor in the atmosphere is short, usually no more than several days or weeks, until it condenses and falls as precipitation. In undergoing condensation to a liquid, the energy stored in the vapor is released.

As a liquid, water is controlled by gravity. If it can move, it will, always move downhill to a lower potential energy state, and always along the path of least resistance, down the steepest gradient. As water moves, it expends energy. Fast-flowing runoff, particularly in streams, is the single most dominant agent of erosion of the surface of our planet. Glaciers likewise are effective at sculpting the land surface, but owing to their limited occurrence on only 10% of the continents, their impact is not nearly as widespread as that of flowing streams. Thus, erosion and the Earth's landforms are intimately tied to the hydrologic cycle. Ultimately, water reaches the lowest accessible level possible, which for most places on the Earth is sea level. Internally drained basins that are isolated from the oceans by mountain ranges and other high divides may exist below sea level (i.e., Death Valley in California; Caspian Sea in Kazakhstan; Dead Sea in Jordan), but these represent local base-level conditions rather than regional or global conditions. These areas of internal drainage are typically formed by tensional tectonics, where blocks of rock are downfaulted (grabens) due to forces that tend to pull the continents apart. Water drains into these depressions under the force of gravity from the surrounding highlands, and escapes only by evaporation.

HYDROLOGIC RESERVOIRS AND IMPLICATIONS FOR HUMANS

Oceans form the largest of our hydrologic reservoirs (Table 1), covering about 70% of the Earth's surface, and including about 96% of all of its water.[8,11,12] Ocean water unfortunately is saline and nonpotable (undrinkable),

Encyclopedia of Water Science
DOI: 10.1081/E-EWS 120010048

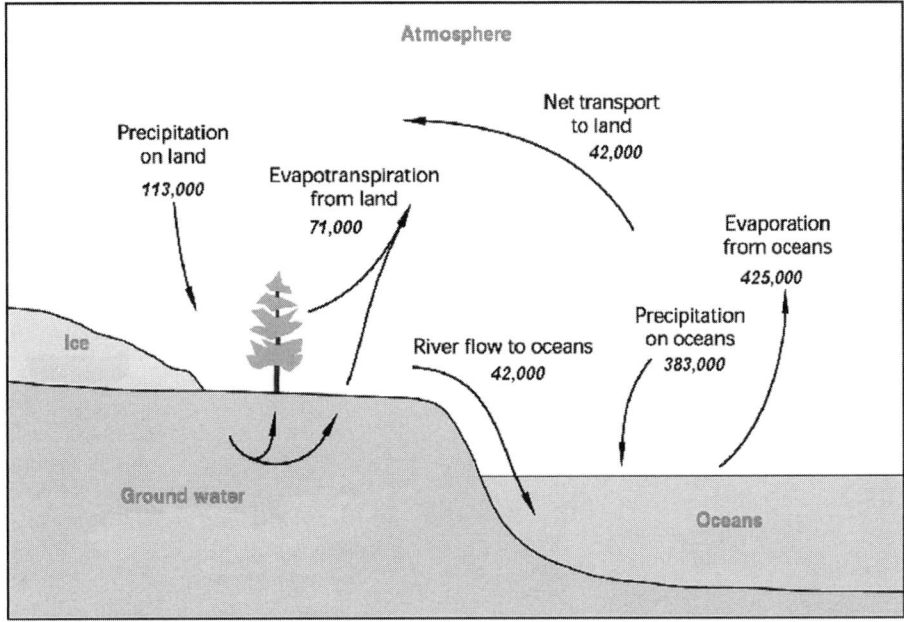

Fig. 1 Quantitative representation of processes in the hydrologic cycle showing transfer rates between reservoirs in units of cubic kilometers per year. (From Ref. 5.)

containing about 35,000 mg L^{-1} of dissolved solids,[13] much too salty for human consumption. Of the remaining 3%–4% of the Earth's water found in reservoirs on the continents, approximately 1% of this (principally saline lakes or deep, saline groundwater) also is nonpotable.[6,8,11,12,13] Thus, about 97% of all the water on this Earth is too mineralized for humans to drink without expensive desalinization. Of the approximately 3% of the total water that is fresh, the largest percentage, estimated as 1.7%–2.97% by different experts, is stored in icecaps and glaciers, far removed from most of the Earth's population and its water needs. Thus, the freshwater needs

of the world are served by a fraction of 1% of the total hydrologic budget, primarily water in storage and transit as shallow groundwater, freshwater lakes, soil moisture, water in manmade reservoirs, and rivers. These data are synthesized from several comprehensive studies, and although the values do not match exactly, they generally do not vary by more than 1% or 2%.[8,12]

As an integrated earth system, the hydrologic cycle has no discrete beginning or ending point, but from a consumptive human point of view, the oceans are the major source of water, the atmosphere is the deliverer, and the land is the user. In this system, no water is lost or gained,

Table 1 Comparisons of quantities and percentages of Earth's water in the major storage reservoirs of the hydrologic cycle, based on estimates from UNESCO and NRC

Storage reservoirs of the hydrologic cycle	UNESCO		NRC	
	Volume, in 10^6 km^3	%	Volume, in 10^6 km^3	%
Oceans	1,338	96.5	1,400	95.96
Icecaps and glaciers	24.3	1.73	43.4	2.97
Groundwater	23.4	1.69	15.3	1.05
Lakes	0.176	0.013	0.125	0.009
Atmospheric water	0.0129	0.001	0.0155	0.001
Soil moisture	0.0165	0.0012	0.065	0.004
Rivers	0.0021	0.0002	0.0017	0.00012
Biologic water	0.001	0.0001	0.002	0.00013
Total	1,386		1,459	

(From Refs. 8 and 12.)

but the amount of water available to the user may fluctuate mostly because of variations in the delivering agent. In the geologic past, large alterations in the cyclic roles of the atmosphere and the oceans produced deserts and glaciation across entire continents. During the last ice age, the colder climate resulted in a greater percentage of water being stored as snow and ice, with a decreased percentage of water being stored in the oceans. Scientists see evidence that major sea-level declines corresponded with maximum glacial development, and in fact, point to "drowned" valleys (e.g., Chesapeake Bay in the United States, and the fiords in the Scandanavian countries) that were eroded and formed when sea level was much lower, and have since been inundated and flooded by rising sea level as the glacial ice melted.

Historically, freshwater in the hydrologic cycle has been enough to serve human needs, but exponential population growth and water usage in regions with little freshwater are posing ever-increasing political and planning problems. For these areas, freshwater has to be imported from great distances, at great expense, and to the detriment of other regions that need the water for their own use. Many of the most pressing problems of the 21st century will be related to obtaining freshwater for the world's expanding population.

REFERENCES

1. http://observe.arc.nasa.gov/nasa/earth/hydrocycle/hydro2. html (accessed July 2002).

2. http://www.uwsp.edu/geo/faculty/ritter/geog101/modules/ hydrosphere/ hydrosphere_title_page.html (accessed July 2002).

3. http://ww2010.atmos.uiuc.edu/(Gh)/guides/mtr/hyd/home. rxml (accessed July 2002).

4. http://ga.water.usgs.gov/edu/mearth.html (accessed July 2002).

5. Winter, T.C.; Harvey, J.W.; Franke, O.L.; Alley, W.M. *Ground Water and Surface Water—A Single Resource*, U.S. Geological Survey Circular 1139; Denver, Colorado, 1999; 1–79.

6. Maidment, D.R., Ed. *Handbook of Hydrology*; McGraw-Hill, Inc., 1993; 1-1–1-15.

7. Driscoll, F.G., Ed. *Groundwater and Wells*, 2nd Ed.; Johnson Division: St. Paul, MN, 1986, Chapter 1.

8. U.S.S.R. Committee for the International Hydrological Decade, World Water Balance and Water Resources of the Earth, English translation. *Studies and Reports in Hydrology*; UNESCO: Paris, 1978; Vol. 25.

9. World Meteorological Organization–UNESCO. *International Glossary of Hydrology*, 2nd Ed.; WMO Report No. 385; Geneva, Switzerland, 1992.

10. Wilson, W.E., Moore, J.E., Eds. *Glossary of Hydrology*; American Geological Institute, 1998.

11. Bras, R.L. *Hydrology*; Addison-Wesley: Reading, MA, 1990.

12. National Research Council (NRC). *Global Change in the Geosphere–Biosphere*; National Academy Press: Washington, DC, 1986.

13. Hem, J.D. *Study and Interpretation of the Chemical Characteristics of Natural Water*, U.S. Geological Survey Water Supply Paper 2254, Washington, DC, 1992.

Hydrologic and Hydraulic Science and Technology in Ancient Greece

D. Koutsoyiannis
National Technical University of Athens, Zographou, Greece

A. N. Angelakis
National Foundation for Agricultural Research, Heracleion, Greece

INTRODUCTION

The approach typically followed in problem solving today is represented by the sequence in the order: Understanding—data—application. However, historical evolution in the development of water science and technology (and other scientific and technological fields) followed the reverse order: application preceded understanding.[1] Thus, technological application in water resources started in Greece as early as ca. 2000 B.C.. Specifically, in the Minoan civilization and later in the Mycenaean civilization several remarkably advanced technologies have been applied for groundwater exploitation, water transportation, water supply, stormwater and wastewater sewerage systems, flood protection, drainage, and irrigation of agricultural lands. Much later, around 600 B.C., Greek philosophers developed the scientific views of natural phenomena for the first time ever. In these, hydrologic and meteorological phenomena had a major role, given that water was considered by the Ionic school of philosophy (founded by Thales of Miletus; ca. 624–545 B.C.) as the primary substance from which all things were derived. Even later, during the Hellenistic period, significant developments were done in hydraulics, which along with progress in mathematics allowed the invention of advanced instruments and devices, like Archimedes' water screw pump.

SCIENTIFIC VIEWS OF HYDROLOGIC PHENOMENA AND HYDRAULICS

It has been believed by many contemporary water scientists that ancient Greeks did not have understanding of water related phenomena, and had a wrong conception of the hydrologic cycle. This belief is mainly based on views of Plato (ca. 429–347 B.C.), who in his dialogue Phaedo (14.112) expresses an erroneous theory (based on Homer's poetical view) of hydrologic cycle; notably, his wrong theory was adopted by many thinkers and scientists

from Seneca (ca. 4 B.C.–65 A.D.) to Descartes (1596–1650).

However, long before Plato, as well as much later, several Greek philosophers had developed correct explanations of hydrologic cycle, revealing good understanding of the related phenomena. In fact, as Koutsoyiannis and Xanthopoulos[2] note, the first civilization in which these phenomena were approached in an organized theoretical manner, through reasoning combined with observation, and without involving divine and other hyperphysical interventions, was the Greek civilization. The same authors catalog a number of ancient Greek contributions revealing correct understanding of water related phenomena. Thus, the Ionic philosopher, Anaximenes (585–525 B.C.) studied the meteorological phenomena and presented reasonable explanations for the formation of clouds, hail and snow, and the cause of winds and rainbow. The Pythagorean philosopher Hippon (5th century B.C.) recognizes that all waters originate from sea. Anaxagoras, who lived in Athens (500–428 B.C.) to Empedocles (ca. 493–433 B.C.) and is recognized equally as the founder of experimental research, clarified the concept of hydrologic cycle: the sun raises water from the sea into the atmosphere, from where it falls as rain; then it is collected underground and feeds the flow of rivers. He also studied several meteorological phenomena, generally supporting and complementing Anaximenes' theories; his theory about thunders, which was against the belief that they are thrown by Zeus, probably cost him imprisonment (ca. 430 B.C.). In particular, he correctly assumed that winds are caused by differences in the air density: the air, heated by the sun, moves towards the North pole leaving gaps that cause air currents. He also studied Nile's floods and attributed them to snowmelt in Ethiopia. The "enigma" of Nile's floods (which, contrary to the regime of Mediterranean rivers, occur in summer) was also thoroughly studied by Herodotus (480–430 B.C.), who seemed to have clear knowledge of hydrologic cycle and its mechanisms.

Aristotle (384–323 B.C.), in his treatise Meteorologica clearly states the principles of hydrologic cycle, clarifying that water evaporates by the action of sun and forms vapor, whose condensation forms clouds; he also recognizes indirectly the principle of mass conservation through hydrologic cycle. Theophrastus (372–287 B.C.) adopts and completes the theories of Anaximenes and Aristotle for formation of precipitation from vapor condensation and freezing; his contribution to the understanding of the relationship between wind and evaporation was significant. Epicurus (341–270 B.C.) contributed to physical explanations of meteorological phenomena, contravening the superstitions of his era.

Archimedes (287–212 B.C.), the famous Syracusan scientist and engineer considered by many as the greatest mathematician of antiquity or even of the entire history, was also the founder hydrostatics. He introduced the principle, named after him, that a body immersed in a fluid is subject to an upward force (buoyancy) equal in magnitude to the weight of fluid it displaces. Hero (Heron) of Alexandria, who lived after 150 B.C., in his treatise *Pneumatica* studied the air pressure, in connection to water pressure, recognizing that air is not void but a substance with mass consisting of small particles. He is recognized[3] as the first person who formulated the discharge concept in a water flow and made flow measurements.

Unfortunately, many of these correct explanations and theories were ignored or forgotten for many centuries, only to be re-invented during Renaissance or later. This was not restricted to water related phenomena. For example, the heliocentric model of the solar system was first formulated by the astronomer Aristarchus of Samos (310–230 B.C.), 1800 yr before Copernicus (who admits this in a note). Aristarchus also figured out how to measure the distances to the Sun and the Moon and their sizes. In addition, not only did ancient Greeks know that Earth is spherical, but also Eratosthenes (276–194 B.C.) calculated, 1700 yr before Columbus, the circumference of the earth, with an error of only 3%, by measuring the angle of the sun's rays at different places at the same time; in addition, the geographer Strabo (67 B.C. – 23 A.D.) had defined the five zones or belts of Earth's surface (torrid, two temperate, and two frigid) that we use even today.

HYDRAULIC MECHANISMS AND DEVICES

The foundation of hydraulics after Archimedes led to the invention of several hydraulic mechanisms and devices with significant contribution to diverse applications from lifting of water to musical instruments. Although in past several devices were in use to lift water to a higher elevation, the first device that had the characteristics of a

pump with the modern meaning is Archimede's helix or water screw. The invention of the water screw is based on the study of the spiral, for which Archimedes wrote a treatise entitled *On Spirals*, in 225 B.C.. This invention of Archimedes was first mentioned by Diodorus Siculus (first century B.C.; Bibliotheke, I 34.2, V 37.3) and Athenaeus of Naucratis (ca. 200 B.C.; Deipnosophistae, V) who transferred an earlier text (of the late 3rd century B.C.) by Moschion, describing a giant ship named Syracusia.

This pump is an ingenious device which functions in a simple and elegant manner by rotating an inclined cylinder bearing helical blades around its axis whose bottom is immersed in the water to be pumped. As the screw turns, water is trapped between the helical blades and the walls, and thus rises up to the length of the screw and drains out at the top (Fig. 1).

As mentioned by Athenaeus of Naucratis, the first use of the water screw must have been by Archimedes himself to remove the large amount of bilge water that would accumulate on the large ship *Syracusia*. There is historical and archaeological evidence that in past the use of the water screw was propagated to all Mediterranean countries as well as to the east up to India. It was rotated by a man or a draft animal. Its uses range from irrigation (e.g., in Egypt) to draining of water in mines (e.g., in Spain). In its original form, the screw of Archimedes is used even today in some parts of the world. For example, farmers in Egypt and other countries in Africa use it to raise irrigation water from the banks of rivers.

A modern version of the screw that is in industrial use today has two main differences from its original one: it is powered by a motor and the screw rotates inside the cylinder rather than the entire cylinder being rotated;

Fig. 1 Archimedes' water screw in its original form as depicted in an Italian stamp (not quite correctly from a technical point of view) along with a bust probably representing Archimedes (from http://www.mcs.drexel.edu/~crorres/Archimedes/Stamps/).

Fig. 2 Archimedes' water screw in its modern form, as implemented in the wastewater treatment plant of Athens (one of nine screws that pump 1 million m^3 per day).

the latter modification allows the top-half of the cylinder to be removed, which facilitates cleaning and maintenance. The modern screw is the best choice for pumping installations when water contains large sediments or debris, and when the discharge is large and the height small. Thus, the screw is used today mainly for pumping wastewater and stormwater runoff (Fig. 2). It has been also used in other types of applications such as pumping of oil and supporting blood circulation during surgical procedures.

Another pumping mechanism, the force pump was invented by engineer (initially barber) Ctesibius of Alexandria (ca. 285–222 B.C.) who was also the inventor of other instruments such as the hydraulic clock and hydraulis—a hydraulic musical instrument. The force pump has been described by Philon Byzantius (Pneumatica), Hero of Alexandria (Pneumatica, I 28), and Vitruvius (X 7, 1–3). This pump is composed of two cylinders with pistons that were moved by means of connecting rods attached to opposite ends of a single lever. The force pump was used in many applications, such as in wells for pumping water, boats for bilge-water pump, basement pump, mining apparatus, fire extinguisher, and water jets. Yet another pumping device, the chain pump was invented in Alexandria by an engineer Philon Byzantius (260–180 B.C.). This comprised a set of pots attached to a chain or belt that was moved by a rotating wheel. Several pneumatic devices and mechanisms including a steam boiler, a reactive motor, the organ (harmonium), and several jet springs have been invented by Hero of Alexandria.[4–6] Most of them were based on the siphon principle, or more generally, the combined action of air and water pressure. Ctesibius, Philon Byzantius, and Hero were the three most famous engineers of Hellenistic Alexandria, whose studies mark a significant progress in hydraulics. This progress allowed installation of advanced water supply systems like that of the citadel at Pergamon, in which pressure pipes (probably made of metal) were implemented. It also led to the great advances in the art of aqueducts during the Roman period.

REFERENCES

1. Dooge, J.C.I. Hydrology, Past and Present. J. Hydraulic Res. **1988**, 26 (1), 5–26.
2. Koutsoyiannis, D.; Xanthopoulos, T. Τεχνική Υδρολοία (Engineering Hydrology), 3rd Ed.; National Technical University of Athens: Athens, 1999; 1–418 (in Greek).
3. U.S. Committee on Opportunities in the Hydrological Sciences, Opportunities in the Hydrologic Sciences; National Academy Press: Washington, DC, 1991.
4. Lazos, C.D. Μηχανική και Τεχνολογία στην Αρχαία Ελλάδα (Mechanics and Technology in Ancient Greece); Aeolos: Athens, 1993 (in Greek).
5. Lazos, C.D. Υδραυλικά Όργανα και Μηχανισμοί στην Αίγυπτο των Πτολεμαίων (Hydraulic Instruments and Mechanisms in Egypt of Ptolemies); Aeolos: Athens, 1999 (in Greek).
6. Tokaty, G.A. A History and Philosophy of Fluid Mechanics; Henley on Thames: Foulis, 1971 (reprinted by Dover: New York, 1994).

Hydrologic Process Modeling

Daniel L. Thomas
Matt C. Smith
University of Georgia, Tifton, Georgia, U.S.A.

INTRODUCTION

Hydrologic process modeling has evolved in two basic forms: physical and numerical. Physical models of hydrologic processes provide a "scaled" representation of a particular watershed, field, ground water flow path, and atmospheric condition. These type models are typically used for river hydraulics and hydraulic structure design verification. For example, the U.S. Army Corp. of Engineers maintained a scale model of the Mississippi River on Mud Island near Memphis, Tennessee (U.S.A). The physical model provided a reasonable representation of the impacts of physical stream changes and other conditions (precipitation, snowmelt, flooding, etc.) on the hydrologic response of the river. The implementation of diversion structures and management alternatives were originally based on the physical model.

The advent of fast computers and numerical approaches to represent processes created greater flexibility to test different and new hydrologic conditions. The following discussion emphasizes computer-based process modeling. Many of the examples will be for surface water hydrologic processes, but the same ideas and approaches apply to other components within the hydrologic cycle.

CLASSIFICATION OF PROCESS MODELS

Models may be classified in a number of different ways. Singh[1] and Haan et al.[2] provide excellent descriptions of the different approaches to classifying models. Computer-based models are in two basic forms: lumped or distributed (Fig. 1). Lumped models do not take into account the spatial variability that is normally present in a hydrologic situation (watershed, field, soil, aquifer, atmosphere, etc.). The parameters, relationships, and results are "lumped" by averaging or using dominant characteristics in the area of interest. Models may also be lumped on a temporal scale (hour, day, month, year) to allow simpler modeling of dynamic and complicated processes. Distributed models do take into account the variability conditions that occur in the area of interest. For example, a watershed (the area where all runoff water flows to a common outlet) may include farmland, grass, forest, urban areas, streams, and

lakes. As precipitation falls on each of these areas, unique processes occur (interception, infiltration, runoff). As a water droplet passes through different areas in the watershed, other processes and conditions will apply (stream and groundwater flow, etc.).

Practically all models that are described as "distributed," have lumped characteristics (input parameters, spatial and boundary conditions, etc.). Rainfall and runoff processes may be distributed, but are likely not "fully distributed." For example, the ANSWERS-2000 model[3] is considered a distributed—parameter hydrologic and water quality model for watershed scale runoff and water quality evaluations. It breaks the watershed into "elements" that contain lumped parameters based on dominant characteristics. The size of the elements can be selected based on variability conditions and overall size of the watershed. Water is passed from one element to another in a distributed network.

Models may also be classified by the process approach used in the model. The three basic descriptions are deterministic, stochastic, and mixed (Fig. 1). If all of the variables in the model are considered to be free from random variation, then the model is deterministic.[2] For example, a surface hydrology model may take input from precipitation, determine how much of that water will be associated with storage (surface and subsurface), how much water is associated with flow (surface and subsurface), and how much water is lost (evaporation, transpiration, and interception). Each process component can be represented by a different "box" (or submodel), with water coming in, water being stored, and water going out. Bringing all the different "boxes" together creates a hydrologic model system.

If any of the variables in a mathematical relationship in a model may be regarded as having random values, then the model system is considered to be stochastic. For example, precipitation over a period of time represents a random sequence. We cannot predict "actual" precipitation into the future with any degree of certainty. However, precipitation can be described in a distribution with an expected probability of occurrence (based on past history).

Most model systems contain a combination of deterministic and stochastic processes. The direct results from simulation models may not reflect the stochastic nature of the inputs. Many "continuous" models are

Encyclopedia of Water Science
DOI: 10.1081/E-EWS 120010104

designed to run over a long period of time, so that a range in conditions (such as precipitation) can be simulated to produce a distribution of potential responses.

There are several other classifications of models. A physically based model has relationships that represent actual physical processes. An empirically based model has relationships that are derived from measured data, but have no direct relationship to actual physical processes (Fig. 1).

"Simple" or "complicated" is a classification that applies to the potential user. Originally, simple models tended to require few parameters and therefore yielded simple results. Examples of simple, empirical models are the Rational Method and the SCS Curve Number approach for calculating runoff from rainfall.[2,4] Complicated models are more likely to require extensive and detailed input data and are also more likely to yield more "detailed" results. Complicated models may be capable of responding to small changes in parameter values. The desire of most model developers is to make a complicated model easier to use (to increase the potential user-base). The use of interactive interfaces, geographic information systems, and standard data sets has helped "parameterize" a complicated model. However, if the user needs only general results, a simple model may suffice.

HISTORICAL DEVELOPMENT OF HYDROLOGIC PROCESS MODELS

One of the earliest computer-based hydrologic process models was the Stanford Watershed Model.[5] This model provided the hydrologic foundation for many later models. The Stanford Watershed Model used a lumped parameter approach, with daily or hourly rainfall and could simulate several years of runoff. The model did require calibration with some existing data to be effective.

The SCS curve number for calculating runoff from daily rainfall is probably the most widely used, empirically based, hydrologic process model within other model systems. Currently available water quality models such as GLEAMS,[6] EPIC,[7] SWRRB,[8] and PRZM[9] all use the SCS curve number to calculate runoff. The SCS curve number calculates runoff based on a "number" that reflects surface cover, soil characteristics, and antecedent soil moisture conditions.

CURRENT HYDROLOGIC PROCESS MODELS AND APPLICATIONS

A large number of hydrologic process models are currently available. Most of these models have specific applications (flood determination, river and ground water flow

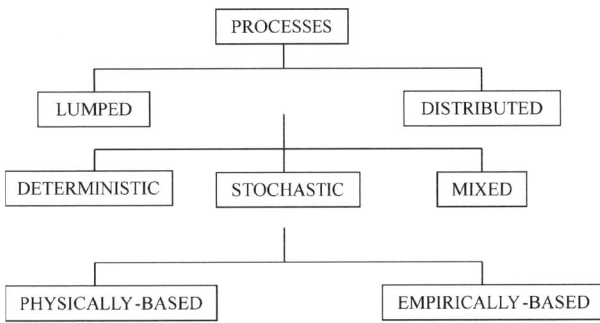

Fig. 1 Basic classification components for hydrologic process models.

characterization, atmospheric processing, and water quality evaluation). Many of the models that are currently in use have graphical interfaces for improved input of parameters and interpretation of results. One example of a model system that has evolved over time is HEC-1. HEC-1 (a product of the Hydrologic Engineering Center of the Corp. of Engineers; 1) was originally designed to create a river flow hydrograph for a watershed. The model has been modified and enhanced into commercial products with graphical displays. Some versions can evaluate flood hydrographs, breached dam conditions, and even estimates of flood damage. Statistical analysis packages are included in many commercial products to increase the benefits of the model system.

The continual increase in computer speed has created opportunities for more-complicated models to be used by a wider range of clientele. Significant efforts have been expended to convert research-oriented, process models into systems that can be used by essentially anyone. This availability increases the potential for abusing models and their results. If models are used for unintended purposes, or unrealistic parameters are selected, model results will be suspect. It is important to select the model that is most useful to the intended application, to be sure how to use the model, and to not attempt to "stretch" the model beyond its' intended range of conditions. Haan et al.,[2] Singh,[1] and Parsons et al.[10] provide information on many of the currently available models, including where they can be obtained, their intended use, and what can be expected as a result.

CONCLUSION

Hydrologic process modeling provides an economical approach to representing a hydrologic condition for analyzing the status of water. Computer-based models are available for many of the different processes within the hydrologic cycle and for many different applications.

REFERENCES

1. Sharpley, A.N.; Williams, J.R., (Eds.) *Computer Models of Watershed Hydrology*; Water Resources Publications: P.O. Box 260026, Highlands Ranch, CO 80126-0026 U.S.A., 1995.

2. Sharpley, A.N.; Williams, J.R., (Eds.) *Hydrologic Modeling of Small Watersheds*, ASAE Monograph Number 5; ASAE: St. Joseph, MI 49085 U.S.A., 1982.

3. Dillaha, T. ANSWERS-2000. 2002. http://dillaha.bse.vt.edu/answers/AboutAnswers.html (accessed Feb., 2002).

4. Sharpley, A.N.; Williams, J.R., (Eds.) *Soil and Water Conservation Engineering*, 2nd Ed.; John Wiley and Sons, Inc.: New York, New York. U.S.A., 1966.

5. Crawford, N.H.; Linsley, R.K., Jr. *Digital Simulation in Hydrology: Stanford Watershed Model IV*, Technical Report No. 39; Department of Civil Engineering, Stanford University: Stanford, CA, 1966.

6. Knisel, W.G.; Leonard, R.A.; Davis, F.M.; Nicks, A.D. *GLEAMS Version 2.10. Part III. User Manual*; USDA, ARS, Conservation Research Report Series, Coastal Plain Experiment Station, Southwest Watershed Research Laboratory, Tifton, GA 31793 U.S.A., 1993.

7. Sharpley, A.N.; Williams, J.R., (Eds.) *EPIC, Erosion Productivity Impact Calculator: 1. Model Documentation*, Technical Bulletin No. 1768; U.S. Department of Agriculture, Blackland Research Center, Texas A&M University, 808 Blackland Rd., Tempe, TX 76502 U.S.A., 1990.

8. Arnold, J.G.; Williams, J.R.; Nicks, A.D.; Sammons, N.B. *SWRRB: A Basin Scale Simulation Model for Soil and Water Resources Management*; Texas A&M University Press: College Station, 1990.

9. PRZM, *Users Manual for the Pesticide Root Zone Model. Release 1*; (EPA 600/3-84-109) NTIS. Accession Number: PB85 158913/AS, U.S. EPA, Environmental Research Laboratory, 960 College Station Road, Athens, GA 30605, 1984.

10. *Agricultural Non-point Source Water Quality Models: Their Use and Application*; Parsons, J.E., Thomas, D.L., Huffman, R., Eds.; Southern Cooperative Series Bulletin No. 398. ISBN: 1-58161-398-9; 2001; http://www3.bae.ncsu.edu/s273/ModelProj/index.html (accessed Feb., 2002).

Hydrology Research Centers

Daniel L. Thomas
University of Georgia, Tifton, Georgia, U.S.A.

H

INTRODUCTION

The study of water has been one of the most critical and beneficial needs since man understood the relationship between water and life. From the beginning of recorded history, water-related events (rainfall, snowfall, and runoff), patterns (stream flow, watershed contributions, and groundwater characteristics), and extremes (floods and droughts) have influenced hydrology research. As man impacted the environment and nature continued its process of weathering, approaches have been sought to better understand available resources, their properties, and how to minimize man's impacts on those resources.

Hydrology research encompasses fairly broad spatial and temporal components of water movement. Man has sought knowledge about the movement of water through shallow rivulets in a field to flows and conditions in major rivers and oceans. Man has been concerned about how water moves from melting snow to clouds and water vapor in atmospheric relationships. Research efforts have included attempts to understand water movement around soil particles to larger, more cavernous flows within major underground aquifers. Many of these research endeavors have included the development of physical and computer-based model systems as a way to better understand the phenomenon.

As man continues to influence watersheds and flow paths, research into hydrologic function will continue to be required. Significant research efforts are occurring within existing centers, laboratories, and institutes. This section is designed to identify many of the resource locations for hydrology research. This section will not include extensive listings of different centers for dryland and semi-arid research or centers for water quality investigation.

This document is neither designed to attest to the quality nor breadth of research programs within individual centers. The reader is encouraged to make contacts and determine those characteristics individually. One web-based resource for accessing different centers around the globe that are associated with hydrology is www.spatial-hydrology.com/researchcenter.html (accessed October 2001). This resource includes a dynamic resource listing of many different groups since new organizations will continue to emerge while others disappear.

Academic-Oriented Research Programs

Much of the existing hydrology research occurs within academic institutions without a specific designation as a center, laboratory, or institute. These resources should not be excluded from potential access. Obviously, some institutions have extensive hydrology research programs, while others do not. The breadth of the research programs at particular universities is a function of the individual faculty and their interests. The Universities Council on Water Resources[2] provides an extensive listing of many hydrology-oriented programs and is a resource to the member institutions. Also included in their programs is the Universities Water Information Network that allows searching for particular topics and programs of interest.

Research has been coordinated through a variety of organizations that could be classified as "clearing houses" or centers for access to hydrology-oriented information. Some groups are professional organizations while others provide an opportunity to work toward specific goals through regional approaches. The idea is that most organizations do not have all the expertise to address many critical hydrology issues; thus, organizational structures are required to bring experts together. Several organizations that encourage and support hydrology research include the American Institute of Hydrology, the International Association of Hydraulic Engineering and Research (IAHR), the American Geophysical Union (AGU), the American Society of Civil Engineers (ASCE, including the Environmental and Water Resources Institute, EWRI), the Society for engineering in agricultural, food, and biological systems (ASAE), and the Soil Science Society of America (SSSA). The Cooperative State Research, Education and Extension Service[1] also provides coordination of regional projects. Some of these projects encompass hydrology-related research.

Federally Coordinated Research Programs

The U.S. Department of Agriculture (USDA), Agricultural Research Service (ARS),[3] and the United States Geologic Survey (USGS)[4] are two agencies of the United States that have primary missions that include hydrology research. The USDA, ARS have several watershed laboratories, and have been involved in the development

Encyclopedia of Water Science
DOI: 10.1081/E-EWS 120010176

of field- and basin-scale models of hydrologic and water quality phenomenon. Most of their work has been limited to surface hydrology and shallow subsurface hydrologic investigation. However, some extensive groundwater work has occurred in the western United States. The designated watershed research laboratories (Tucson, Arizona; Tifton, Georgia; Boise, Idaho; Coshocton, Ohio; Watkinsville, Georgia) are all involved in some aspects of watershed hydrology for soil, topography, and climatic characteristics of the particular region. Additional USDA, ARS laboratories are involved in stream flow hydraulics and structural impacts on flow. In almost every USDA, ARS research location, some aspect of water (quantity and/or quality) is being investigated. For more complete and up-to-date access to current projects, they provide a searchable web site under "research" at their primary web site.

The USGS is involved in monitoring and modeling stream flow, groundwater, and their interactions. The USGS also coordinates programs on acid rain and national water use and quality issues. In many states, they are the designated agency providing statewide and regional statistics on water use. The USGS monitoring network provides real-time access to stream flow and groundwater conditions.[4]

Other federal organizations that support research into hydrology characteristics include the Army Corps. Of Engineers and the Bureau of Reclamation (river basin water management), the National Oceanic and Atmospheric Association (NOAA), and the USDA, Natural Resources Conservation Service (NRCS). NOAA has programs associated with atmospheric sciences and global climate conditions, while the NRCS works with small watershed hydrology, dams, and field scale impacts on hydrology and water quality.

Hydrology Research Centers in the United States

The number of centers and laboratories in the United States that have programs that involve hydrology is quite extensive. The listing below is not designed to catalog every one. The primary emphasis is toward indicating the diversity that exists by several examples. The National Institutes for Water Resources (http://wrri.nmsu.edu/niwr/ accessed November 2001) include the different state-level water resources research centers/institutes. These centers may be funded from federal and/or state resources with the primary goal (usually) of initiating water resources projects within that particular state. For example, the Water Resources Research Center at Arizona State University (http://ag.arizona.edu/azwater/ accessed November 2001) has funded research programs investigating riparian systems, flooding, and evaporation effects.

The Stanford Center for Reservoir Forecasting (http://ekofisk.standford.edu/SCRFweb/index.html accessed November 2001) is focused directly on programs that relate to reservoir characterization, performance, and modeling of processes within reservoirs. The Belle W. Baruch Institute for Marine Biology and Coastal Research in Columbia, South Carolina (http://inlet.geol.sc.edu/ accessed November 2001) has programs investigating coastal hydrology issues. The Center for Water Research and Policy in Columbia, South Carolina (http://watercenter.environ.sc.edu/ accessed November 2001) is involved in groundwater systems, contaminant transport processes, and water balance research. The Snow Hydrology Group at the University of California, Santa Barbara (www.icess.ucsb.edu/hydro/hydro.html/ accessed November 2001) is investigating watershed hydrology and modeling under snow conditions. The Florida Center for Environmental Studies in Palm Beach Garden, Florida (www.ces.fau.edu accessed November 2001) represents 10 state universities and four major private universities that are involved in research on river restoration, everglades hydrology, wetland functions, modeling and monitoring, and international programs to address hydrology issues in Central and South America.

International Hydrology Research Centers

The Center for Ecology and Hydrology in Wallingford, U.K. (www.nwl.ac.uk/ih/ accessed October, 2001) is the former Institute of Hydrology. They have very broad areas of interest including flooding, droughts, climate habitats, rivers, plants, and soils. The Center for Science and Environment in New Delhi, India (www.oneworld.org/cse accessed October 2001) has directed their current research emphasis toward the analysis of floods. The Watershed Science Center at Trent University in Ontario, Canada (www.trentu.ca/wsc/welcome.shtml accessed November 2001) has programs oriented toward forest hydrology, structural impacts on fish habitat, and watershed ecosystem management. The Commonwealth Scientific & Industrial Research Organisation (CSIRO, http://www.csiro.au/ accessed November 2001) has programs on natural and managed ecosystems, and weather and climate as they relate to social, economic, and ecological factors.

Centers Studying Drought, Weather, and Climate

Several organizations are involved in the study of drought and weather. The National Drought Mitigation Center in Lincoln, Nebraska (http://enso.unl.edu/ndmc accessed November 2001) is involved in the development of techniques to improve risk management during droughts and improve the forecasting and understanding of drought.

The Global Hydrology and Climate Center in Huntsville, Alabama (www.ghcc.msfc.nasa.gov accessed October 2001) has programs oriented toward the study of the global water cycle. The National Weather Service (www.nws.noaa.gov accessed November 2001) is involved in research associated with forecasts of river levels, floods, and water supply needs. The weather service has regional offices throughout the United States to allow more specific investigation into regional and local weather phenomenon. The Center for International Earth Science Information Network at Columbia University in New York (www.ciesin.org accessed November 2001) is an additional resource for research associated with global climate change.

Centers Studying Pollutant Transport

Pollutant transport is directly associated with hydrology of surface and groundwater resources. The ability of a particular water medium to move pollutants from one place to another and the integrated processes associated with those pollutants are directly tied to the hydrologic characteristics. Besides the many organizations mentioned earlier, Oak Ridge National Laboratory (www.esd.ornl.gov accessed October 2001), the Savannah River Ecology Laboratory (www.uga.edu/~srel/ accessed October 2001), the Center for Environmental Research and Training (CERT, associated with the National Environmental Health Association, www.nehacert.org accessed November 2001), and the Center for Water Research and Policy in Columbia, South Carolina (http://watercenter.environ.sc.edu/ accessed November 2001) all have programs that are directly associated with pollutant transport.

CONCLUSION

Hydrology Research Centers provide an excellent resource to current and future investigation into the characteristics of hydrology. Until man fully understands all hydrologic phenomena and ceases to impact the movement of water, there will need to be research into the future. Centers and institutes provide an appropriate method to focus on particular problems while reducing potential overlaps with other programs.

REFERENCES

1. CSREES. 2001. The Cooperative State Research, Education and Extension Service. www.reeusda.gov (accessed November 2001).
2. UCWR. 2001. Universities Council on Water Resources. www.uwin.siu.edu (accessed October 2001).
3. USDA, ARS. 2001. United States Department of Agriculture, Agricultural Research Service. www.ars.usda.gov (accessed October 2001).
4. USGS. 2001. United States Geologic Survey. www.usgs.gov (accessed October 2001). Go to "water."

Internet

Janice G. Norris
Texas A&M University, College Station, Texas, U.S.A.

INTRODUCTION

The Internet holds a wealth of water science information sites with free public access. This information can be predominantly found on the World Wide Web (WWW) and although a web search engine can locate millions of web pages related to water, finding specific information can be difficult due to the lack of central organization. The web is further complicated by its dynamism. Web pages change, relocate, and disappear frequently. Fortunately, government, organization and educational web sites have become more stable over the past decade and virtual libraries and subject-specific search engines have been created to assist the searcher. Full text documents, primary data, real-time data, interactive maps, and archived data are available through databases on the WWW, but the databases can be hard to locate. This article outlines methods for locating information in web sites that have moved or changed, describes some effective search engines, and discusses some of the virtual libraries databases that may be of particular value to the researcher who seeks water science information on the WWW.

INTERNET DEVELOPMENT HIGHLIGHTS

The technology developed in 1969 as a U.S. Defense Department project named the Advanced Research Projects Agency Network (ARPANET), which allowed computers to share information simultaneously, gave rise to the Internet. The development of Hypertext Markup Language (HTML) and web browser technology in 1989 gave easy access to the Internet, created the WWW, and with the introduction of the first search engine, Yahoo, in 1994, opened the Internet to the world.[1] The number of available web pages increased phenomenally with the advent of search engines. At the close of the 20th century there were approximately 2.1 billion static web pages[2] with an incredible number of these pages devoted to water science topics.

Water information originating around the world can be discovered and accessed in seconds within the vast library that is the WWW. However, the search for information can be overwhelming and frustrating due to the huge number of web sites and the absence of a central organizing or cataloging mechanism. The nebulous ever-changing nature of web sites is also problematic for Internet users. A 1996 study determined that the half-life of a web site is around 2.9 yr.[3] Fortunately virtual libraries, government sites, and many organizational sites have become relatively stable in recent years.

THE SEARCH FOR WATER INFORMATION ON THE WWW

The WWW, a subset of the Internet, is easily searched and accessed and is a primary location of information available for public access. Public access means that all users have the capability to view the information without having to paying for access. Of course the WWW also contains substantial amounts of information that requires the payment of a fee for access. Fortunately, a multitude of water information is freely available. Important sources of this information include government agencies, educational institutions, and water organizations. However, navigating through the millions of web pages devoted to water topics to find specific information is not easy or fast. Individual web sites can be massive and site navigation can be convoluted. Many sites contain links that lead one out of the site without warning. Web pages continually evolve by changing location within the web site, dividing into a number of pages, or they cease to exist. There are general methods that are useful in determining the contents of a web site or to find the new location of a specific page:

- View the entire page. Many times it is necessary to scroll down the page to locate the site's search engine or index.
- Use the site's help screens.
- Site maps provide a fast way to determine what information a web site contains.
- When a URL does not appear to be correct, reduce the indexing by one level (the next to last slash) until a page is accessed, then look for a hyperlink on that page to the information.
- Enter only the first part of the URL to the domain (.gov, .edu, .com, .org), then use a tool provided by the site to locate the desired information.

Encyclopedia of Water Science
DOI: 10.1081/E-EWS 120010126

Table 1 Selected subject specific search engines (all accessed February 2001)

Name	URL
AgriBiz Search Engines for Agriculture	http://www.agribiz.com/agInfo/seaAgri.html
Aqueous.com	http://www.aqueous.com/index.asp
ASAE Technical Information Library	http://asae.frymulti.com/
DataWeb	http://dataweb.usbr.gov/html/search.html
FirstGov	http://www.firstgov.gov/
Galaxy	http://www.galaxy.com/galaxy/Info/about.html
INFOMINE	http://infomine.ucr.edu/
Search4science	http://www.search4science.com/
Search Adobe PDF Online	http://searchpdf.adobe.com/
StudyWeb	http://www.studyweb.com/
Web-Agri: The First Agricultural Search Engine	http://www.web-agri.com/
Wetlands	http://www.sws.org/wetlands/

The WWW consists of sites that can be considered libraries or databases, though there are no uniform protocols that govern how sites are named. Common titles of web sites that organize web pages by subject are Digital Library, Virtual Library, Web Links, Database, Web Directory, Resources, etc. Search engines are actually indexes of web pages but could also be considered libraries.

SEARCH ENGINES

Search engines, metacrawlers, and directories are popular tools (all commonly denominated "search engines") used for locating information on the WWW. However, all three tools may lead to an overwhelming list of web pages if one uses a broad search term, such as "water." For example, over 8 million web pages were retrieved using the term "water" in the popular search engines Alta Vista, Northern Light, and Excite, supporting the proposition that there are vast resources on the topic to be found on the WWW, but also emphasizing the need for more precise queries. Even directories that provide subdivisions to sites with a more narrow focus, such as Yahoo, Infoseek, and Lycos retrieve hundreds to thousands of links.

Water science information retrieval on the Internet is further complicated by the cross-disciplinary nature of the subject. Water science can encompass the categories of hydraulics, hydrogeology, economics, chemistry, climate, weather, environment, ecology, agriculture, pollution, engineering, etc. Therefore, efficient information retrieval using search engines requires very specific search terms and the use of more than one search engine is required for thoroughness.

Fortunately, subject-specific search engines have been developed to assist the search for water-related information on the WWW. Selected subject-specific search engines are listed in Table 1. Considering the amount of

Table 2 Selected Internet libraries containing water information (all accessed February 2001)

Name	URL
Academic Info	http://www.academicinfo.net
Amazing Environmental Organization Web Directory	http://www.webdirectory.com/
BUBL Information Service	http://bubl.ac.uk/
CyberStacks	http://www.public.iastate.edu/~CYBERSTACKS/homepage.html
EEVL Edinburgh Engineering Virtual Library	http://www.eevl.ac.uk/welcome.html
Mel: The Michigan Electronic Library	http://mel.lib.mi.us/
Online Electronic Science Library	http://www.sc.edu/library/science/elibind.html
PubScience	http://pubsci.osti.gov/
The Water Librarians' Home Page	http://www.wco.com/~rteeter/waterlib.html
Web Links of the International Association of Hydrology	http://www.iah.org/weblinks.htm
World Wide Web Virtual Library: Earth Sciences	http://www-vl-es.geo.ucalgary.ca/VL-EarthSciences.html
WWW Virtual Library	http://vlib.org/Overview.html
Yahoo! Reference Library	http://dir.yahoo.com/Reference/Libraries/

Table 3 Selected databases and public access data (all accessed February 2001)

Name	URL
AES NWT Water Bibliography	http://www.aina.ucalgary.ca/aes/
Agricultural Research Data Directory	http://agros.usda.gov/
Aquastat	http://www.fao.org/ag/AGL/AGLW/aquastat/aquastat.htm
Association of American State Geologists "Links to State Geological Survey Pages"	http://www.kgs.ukans.edu/AASG/
Civil Engineering Database	http://www.pubs.asce.org/cedbsrch.html
Database of Online Documents Covering Water and Agriculture	http://www.nal.usda.gov/wqic/wqdb/esearch.html
Earth Observing System Data Gateway	http://edcimswww.cr.usgs.gov/pub/imswelcome/
Envirofacts: Queries, Maps, and Reports	http://www.epa.gov/enviro/html/qmr.html
Environmental Atlas	http://www.epa.gov/ceisweb1/ceishome/atlas/
Research Imagery and Data at the GHCC	http://wwwghcc.msfc.nasa.gov/ghcc_data.html
Global Hydrologic Archive and Analysis System	http://www.watsys.sr.unh.edu/
Global Change Master Directory	http://gcmd.nasa.gov/
GRID-Arendal's Online GIS and Map and Graphics Database	http://www.grida.no/db/
Hydro-Climatic Data Network (1874–1988)	http://wwwrvares.er.usgs.gov/hcdn_cdrom/1st_page.html
IWRN Directories of Water Resources Agencies/Organizations/ Institutions In The Americas	http://www.uwin.siu.edu/IWRN/orgs/
IRRISOFT: Database on Irrigation & Hydrology Software	http://www.wiz.uni-kassel.de/kww/irrisoft/
National Atlas of the United States of America	http://www.nationalatlas.gov/
NOAA Environmental Services Data Directory	http://www.esdim.noaa.gov/NOAA-Catalog/index.html
The Quality of Our Nation's Water Introduction State Fact Sheets	http://www.epa.gov/OW/resources/st_intro.html
Real-Time Water Data	http://water.usgs.gov/realtime.html
StreamNet: On-line Data	http://www.streamnet.org/online_data.html
Wateright: Reference Data and Glossary	http://www.wateright.org/site2/reference/index.asp
World Water & Climate Atlas	http://www.cgiar.org/iwmi/Watlas/atlas.htm
Selected Water-Resources Abstracts	http://water.usgs.gov/swra/
Texaset	http://texaset.tamu.edu/
Types of water-use data available from USGS	http://water.usgs.gov/watuse/wudata.html
Universities Water Information Network Databases	http://www.uwin.siu.edu/dir_database/index.html
Universities Water Information Network Table of Content	http://www.uwin.siu.edu/tocnoframes.html
Universities Water Information Network Water Experts Directory Search	http://www.uwin.siu.edu/dir_directory/expert/search.html
Water Supply Information within Reclamation	http://www.usbr.gov/main/watersupply.html
Water Resources Data	http://water.usgs.gov/data.html
WIN: Find Environmental Data and Maps	http://www.epa.gov/win/datamap.html

information generated by the U.S. government, one of the most useful search engines is FirstGov, which provides access to all online U.S. Federal Government resources. This site also provides links for state and local governments as well as interesting topic links for science/technology and agriculture/food and laws. Some of the search engines locate sites relevant to water and/or agriculture such as AgriBiz, Aqueous, Web-Agri, and Wetlands, while others look for full text Adobe Acrobat (PDF) files related to the search. All of the other search engines in the list provide a broader coverage of science or engineering but can be extremely useful in locating water information.

VIRTUAL LIBRARIES AND WATER SCIENCE

There is little distinction between web sites titled "virtual library" and sites that maintain lists of WWW links organized by subject or databases sites. All organize valuable WWW resources. If we consider web libraries to be those sites that organize information and provide access to the information, then subject-specific libraries exist under a variety of names such as gateways, web directories, web links, or personal/organization home pages. Numerous libraries are included within web sites of organizations and government agencies and can be denominated "library," "resources," "reference," etc.

Table 2 provides a selection of WWW libraries containing water information along with their URLs.

A standard model for virtual libraries does not exist because they are as varied as the individuals or organizations that have created them. Virtual library web sites include search engines and/or list water topics in sections labeled science, geology, earth science, agriculture, or environment. Some, like the Online Electronic Science Library created by the University of South Carolina, which contains approximately 2500 web resources, are set up and cataloged like traditional academic libraries complete with reference, book, journal, and tool categories. Other traditional library-based models include EEVL, the Edinburgh Engineering Virtual Library; Mel, the Michigan Electronic Library; and BUBL. Virtual libraries, such as the International Association of Hydrogeology site, categorize water information by geographic location, while the Water Librarians' Home Page includes a variety of categories that the author considers important in his or her work as a librarian in a water agency. Each library is designed to provide access to information important to users.

WATER SCIENCE DATABASES

The sharing of information is the primary function of the Internet, and the sharing of data in the world of water science is extremely valuable due to the interdisciplinary nature of water and its global importance. The WWW includes many water-related databases containing real-time data, interactive maps, or historical primary data and presents the scientist, student, and water professional with access to information that was previously neglected due to lack of accessibility or whose existence was unknown. Unfortunately, water data are not conveniently collected in one neat category and may be difficult to locate on the WWW. There is also a question of accuracy and reliability when discussing data. Table 3 presents a selection of fairly stable databases produced by government agencies, water organizations, or educational institutions that should provide accurate, reliable data. The web sites were chosen to provide a broad range of

data types and coverage or as major indexes to data sets and are only intended as a starting point for locating data on the web.

THE FUTURE OF WATER INFORMATION ON THE WEB

During the past decade we have witnessed an explosion of information on the Internet. Much of the information is of questionable value or commercial in nature. The virtual libraries have started to organize the web by evaluating and grouping relevant web pages into coherent catalogues, which eliminates the searcher's major frustration of sifting through thousands of irrelevant sites to find information. Searching for resources, then making them useful by cataloging/organizing and continuous maintenance is an extremely labor-intensive process. Librarian Assistants, a software package, currently being developed by one electronic library, may simplify the process[4] and make web organization more feasible. As developing countries' online presence increases over the next few years, new sources of water information will appear on the web. Web-masters will be challenged not only by the sheer volume of information to organize, but also with the daunting task of meeting the needs of searchers in areas with diverse technological capabilities and where change is the only certainty.

REFERENCES

1. Raphel, M. Untangling the Web: First of a Three Part Series. Art Business News **2000**, *27* (12), 100.
2. http://www.sims.berkeley.edu/how-much-info/ (accessed Feb 2001).
3. Koehler, W. An Analysis of Web Page and Web Site Constancy and Permanence. J. Am. Soc. Inf. Sci. **1999**, *50* (2), 162–180.
4. Coutinho, F.C.; Eastman, C.M.; Hare, C.B.; Skinder, R.F. Integrating Digital Resources into a Traditional University Research Library. Issues in Science and Technology Librarianship, 23 (summer); 1999, computer file accessible at: http://www.library.ucsb.edu/istl/99-summer/article3.html.

Irrigated Agriculture, Economic Impacts of Investments in

Robert A. Young
Colorado State University, Fort Collins, Colorado, U.S.A.

INTRODUCTION

Because crop irrigation represents the largest single consumptive user of water in the United States and in the world, and many governments in arid areas have encouraged irrigated agriculture, measures of the economic impacts of changes in irrigated agriculture are of interest for evaluation of proposed irrigation-related public policies. Such policies include potential investments in new irrigation water supplies, transfers of irrigation water to emerging urban, industrial, and environmental demands, and plans for long-term groundwater management policies. Space limitations restrict this discussion to investment issues, although the concepts and evidence are relevant to related topics. This contribution identifies several types of economic impacts of irrigation development, sketches the conventional economic framework for evaluating public policies relating to irrigation, presents evidence on the magnitude of impacts, and concludes with a skeptical assessment of the social returns on public investments in irrigated agriculture and the methods used by public agencies to evaluate such investments.

CONCEPTS FOR EVALUATING NET ECONOMIC IMPACTS OF IRRIGATION-RELATED POLICIES

Standard economic evaluation consists of making estimates in money terms of the beneficial or desired impacts (*benefits*) and the adverse or undesired impacts (*costs*) and balancing the one against the another to determine the net economic impact.[1] Evaluations are necessarily site-specific, because of varying local physical, biological, economic, and policy conditions.

An important initial concept in water policy evaluation is the *accounting stance*, which refers to the point of view or perspective from which the analysis takes place.[2] It can reflect either the *private* individual or the *public* or *social* viewpoint. The social viewpoint is normally from the national perspective, but a regional approach can also be identified. Under the private accounting stance, the private investor is assumed to take the policy environment and hence, prices of productive inputs and outputs as given. From the national social perspective, academic economic doctrine advocates that input and output prices should be adjusted for any distortions from ideal market conditions, such as for public subsidies or unpriced third party impacts.

Another important distinction is between *direct* (or *primary*) and *external* economic impacts.[1] Direct economic impacts accrue to the basic producing unit, the farm. Direct benefits are the *net* monetary value of the output of the water supply initiative, and are measured by the producers' *willingness to pay* for those outputs. Direct costs are the foregone benefits of using those resources in the best alternative use, and reflect the value of resources or inputs used to accomplish the project or initiative. External impacts (also called spillover impacts) arise in addition to the direct project impacts and are those unpriced effects registered on third parties, and can be either positive or negative and either *real* or *pecuniary*. Real indirect effects are due to physical linkages (usually through the hydrologic system) between the activities of two or more affected parties and reflect actual output changes. Pecuniary or *secondary* external impacts reflect income changes occurring via the price system linkages between and among the farms, firms, and households that make up the economy.

Consider now a simple framework (Model 1) that shows the conditions for economic feasibility of a potential investment in irrigated agriculture from the point of view of the private investor. (All benefit and cost elements in the models presented below are assumed to be expressed in annual equivalent terms, employing a consistent interest rate and planning period and reflecting the same general price level.)

Model 1: $DB_p > DC_p$

where the symbols represent the following concepts: The subscript p denotes the private perspective. DB_p is the direct private user benefit (willingness to pay for the initiative) and DC_p is the direct private cost. Direct benefit reflects the economic value of the physical increment in production due to the increment in water supply. Direct benefit is often called the *net return* to or the *value* of water, and is conventionally calculated as the estimated increment in gross revenues from crop sales minus the increment in nonwater costs of producing the crops.[2] Direct costs are the costs of bringing the irrigation water

Encyclopedia of Water Science
DOI: 10.1081/E-EWS 120010369

supply to the farm, which for example might be the annualized cost of installing and operating an irrigation well and pump or the annual assessment associated with accessing water from a community or public water storage and supply project. Model 1 asserts simply that the contemplated investment is economically feasible if, from the private irrigator's perspective, direct benefits exceed direct costs.

Turning to evaluation of the impacts of an irrigation investment from the public or social accounting stance, three types of adjustments and additions should be made to Model 1. First, benefits and costs are adjusted for subsidies or other government-induced market distortions. For example, crops produced with the aid of government support programs—such as cotton or rice in the southwestern United States—would be valued at lower price levels, derived from estimated free market prices (which task is a challenge itself). Costs would similarly be adjusted for public subsidies (such as low-cost credit, energy, or irrigation water) or penalties (e.g., minimum wage regulations). On balance, these adjustments usually make the social net benefit of added irrigation water less than the private net benefit.

The other adjustments needed for a shift to the public accounting stance are to incorporate monetary estimates of any external effects, both real and pecuniary. These steps are represented in Model 2, in which direct impacts are expressed in social prices (adjusted for market price distortions, denoted by introducing a subscript s) and external impacts (both real and pecuniary) are incorporated in the formula:

Model 2: $DB_s + IB + SB > DC_s + IC + SC$

The terms new in Model 2 are IB, representing indirect (real external) benefits, SB denoting secondary (pecuniary external) benefits, IC standing for real external costs, and SC denoting secondary external costs.

Secondary benefits, the multiplier effects arising from increased purchases of production inputs and consumption goods when a project comes into operation, are typically concentrated in the project region. They are normally measured by specialized economic models (such as regional interindustry models), which simulate the effects of an increment of resources on the economy. Secondary costs (SC) are the pecuniary benefits foregone when a public investment draws funds (via taxes) from the economy at large. Secondary costs typically spread throughout the national economy and are very difficult to measure. The conventional economic wisdom (embedded in public planning manuals) is that from the national accounting stance, secondary or pecuniary costs are at least as large or larger than secondary benefits. Hence, the two effects offset each other and except in special cases,

secondary economic impacts can be ignored for national irrigation investment planning purposes.[1]

Indirect costs and benefits, the other class of external effects, should also be incorporated into evaluations adopting a public accounting stance. Indirect benefits are seldom economically important, but indirect costs are typically very significant. Examples of indirect costs of irrigation water diversions include reduced downstream water supplies or adverse effects on water quality downstream for offstream (irrigators, industries, and households) and instream (hydroelectric power plants, recreational water users, and fish and wildlife habitat) water users.

EMPIRICAL EVIDENCE ON ECONOMIC IMPACTS OF IRRIGATED AGRICULTURE

A number of sources suggest that the direct economic benefits of irrigation, even from the private accounting stance, are not as large as assumed by nonspecialists or the lay public. One bit of evidence is that farmers are seldom able or willing to pay for public project costs, even if repayment requirements are but a small fraction of actual costs.[3] Econometric studies of land and water rights markets infer that direct benefits of irrigation investments are modest relative to costs.[4] River basin simulation models that adopt a public accounting stance by incorporating indirect costs show that indirect costs to instream water users (such as hydropower producers) may exceed the economic benefits of upstream crop irrigation.[5] Elsewhere, similar evidence is accumulating that when social costs are accounted for, net social benefits of public irrigation developments have been quite unimpressive. The large loss to Aral Sea fisheries and to regional environmental quality from diverting the inflow source for cotton production is one well-known example. Econometric studies in India and China report low rates of economic return to investments in irrigated agriculture (implying negative net social benefits when discounted at conventional interest rates) particularly as compared to the return on expenditures on agricultural research, on education, or on rural road construction.[6]

If the public feasibility studies were correct in concluding that substantial net economic benefits would flow from water resource developments, regional economic studies of ex post impacts would be expected to show corresponding positive impacts on economic growth indicators. Several statistical studies of the role of water investments in regional economic growth in the United States conducted over two decades ago were unable to find statistically significant positive effects of water development on regional incomes.[7] More recently, a regional economic model of the Sacramento Valley (a California agricultural region comprising nearly 2 million irrigated acres) simulated the effect of hypothetical drought scenarios which would reduce water availability by up to 25%. Even the most drastic scenario, and measuring both direct and secondary effects, was predicted to reduce employment by only 300 jobs and reduce total regional income by less than 1%.[8]

The large regional secondary (multiplier) effects from irrigation development sometimes assumed by non-economists are not substantiated on careful study. And, because labor-saving technologies have reduced the labor requirements in agriculture and related industries (dramatically so in the developed world), direct and secondary employment impacts of irrigated agriculture are found to be modest. These ex post studies have of necessity used private prices. If the data had permitted adjustments for subsidized product and input prices, and acknowledged the downstream indirect costs, the conclusions would be even more pessimistic from the social accounting stance.

CONCLUSION

Many early public investments in irrigated agriculture likely yielded an adequate social return. However, conceptual and empirical reasons combine to make it difficult to avoid the inference that the net economic benefits of investments in irrigated agriculture over, say, the last half century, have not been large in the United States and even elsewhere. This conclusion is particularly firm when a social accounting stance is adopted, so that negative indirect effects are accounted for and impact measures are adjusted for input and output subsides. By

making overoptimistic assumptions on crop productivity and prices, by ignoring the opportunity costs of certain inputs, or not properly accounting for public subsidies and external costs, public irrigation planning agencies have tended to systematically overstate net economic benefits of public investments in irrigated agriculture.

REFERENCES

1. Gittinger, J.P. *Economic Analysis of Agricultural Projects*; Johns Hopkins University Press (for the World Bank): Baltimore, 1982.
2. Young, R.A. *Measuring Economic Benefits of Water Investments and Policies*, Technical Report 338; World Bank: Washington, DC, 1996.
3. Wilson, P.N. Economic Discovery in Federally Supported Irrigation Districts. J. Agric. Res. Econ. **1997**, *22* (1), 61–77.
4. Faux, J.; Perry, G.M. Estimating Irrigation Water Value using Hedonic Price Analysis: A Case Study of Malheur County Oregon. Land Econ. **1999**, *75* (3), 440–452.
5. Booker, J.F.; Young, R.A. Modeling Intrastate and Interstate Markets for Colorado River Water Resources. J. Environ. Econ. Manag. **1994**, *26* (1), 66–87.
6. Fan, S.; Hazell, P. Returns to Public Investments in the Less-Favored Areas of India and China. Am. J. Agric. Econ. **2001**, *83* (5), 1217–1222.
7. Howe, C.W. Effects of Water Resource Development on Economic Growth: The Conditions for Success. Nat. Resour. J. **1976**, *6*, 939–956.
8. Lee, H.; Sumner, D.; Howitt, R. Potential Economic Impacts of Irrigation-Water Reductions Estimated for the Sacramento Valley. Calif. Agric. **2001**, *55* (2), 33–40.

Irrigated Agriculture and Endangered Species Policy

Ray G. Huffaker
Norman K. Whittlesey
Washington State University, Pullman, Washington, U.S.A.

Joel R. Hamilton
University of Idaho, Moscow, Idaho, U.S.A.

INTRODUCTION

Irrigated agriculture in the western United States (the West) holds the most senior appropriative water rights allocated pursuant to state statutes, and accounts for about 90% of the consumptive water use in the West.[1] Appropriation of the dependable flow of regional rivers into irrigation has altered natural flow regimes to the detriment of aquatic h
abitats, and consequently has contributed to the listing of several species relying on aquatic habitats as endangered or threatened pursuant to the federal Endangered Species Act (ESA).[2] Notable examples are the listings of several anadromous salmon species in the Columbia River Basin,[3] and of waterfowl and fish species in the Platte River Basin.[4]

The curtailment of state appropriative water rights pursuant to ESA-sanctioned species recovery plans has placed federal and state law on a collision course whose resolution will establish the legal parameters governing policy tradeoffs in allocating water between irrigated agricultural and endangered species protection. Our objective is to illuminate what these legal parameters might be. We begin with brief reviews of the ESA and of the prior appropriation doctrine that provides the foundation of state water law.

THE ENDANGERED SPECIES ACT

The Endangered Species Act (ESA) elevates the conservation of endangered species to the highest level of federal policy objectives, and sets forth a legal procedure affording them extensive protection. Section 1533 authorizes two federal agencies to "list" imperiled species as "endangered" (defined as species in danger of becoming extinct through all or a significant portion of their range) or "threatened" (defined as species likely to become endangered in the foreseeable future) based on solely biological criteria. The Secretary of Commerce—acting through the National Marine Fisheries Service (NMFS)—lists marine species, and the Secretary of the Interior—acting through the Fish and Wildlife Service (FWS)—lists all other species. The listing agencies are required to designate the species' critical habitat, and to prepare "recovery plans" detailing strategies to revive populations to healthy levels. Species recovery receives top priority in the formulation of recovery plans if conflicts arise with construction, development, or other economic activities.

Section 1536 directs federal agencies to consult with listing agencies to ensure that proposed federal actions do not jeopardize the continued existence of a listed (or proposed to be listed) species or adversely modify its critical habitat. If the proposed action is deemed to have an "incidental" impact, the listing agency can require that the consulting agency take "reasonable and prudent measures" to minimize the impact. A consulting agency is banned from making any "irreversible or irretrievable" action that would foreclose the implementation of such measures.

Section 1538 bans the "taking" (e.g., harassment, killing, or capturing) of listed species, and applies to all persons within the jurisdiction of the United States, with exceptions for "incidental takings" (defined as takings that are "incidental to, and not the purpose of, the carrying out of an otherwise lawful activity").

THE PRIOR APPROPRIATION DOCTRINE

Water allocation generally is governed by state law in the West. Variations of the prior appropriation doctrine provide the foundation of most western water law. Briefly, a person acquires the right to use some quantity of publicly owned water by diverting it to a beneficial use on a fixed tract of land (the "water duty"). The priority of the right is established by the date of first diversion. During water shortages, the longest-term (senior) appropriators receive their full water duties until no water remains at the source. The water rights of shorter-term (junior) appropriators are

Encyclopedia of Water Science
DOI: 10.1081/E-EWS 120010118

curtailed completely. Water that is not beneficially used is forfeited and available for re-appropriation by another person ("use it or lose it").

The prior appropriation doctrine ideally protects water-right holders from encroachment by other water-right holders taking water out-of-priority or enlarging their rights in a manner not prescribed by statute. This protection depends on the security of complicated interrelationships or "use-dependencies" created among water users due to the fugitive nature of water resources. Actual consumption of water ("consumptive use") in irrigation is often less than the full amount diverted from the stream ("diversion"). Unconsumed water may return to the stream ("return flow"). Return flows, along with natural stream flows, supply water available for appropriation by other irrigators, and thus constitute a portion of their water rights. Appropriators who modify water use from that prevailing when their water rights were granted may shift the timing, location, quantity, or quality of return/escape flows, and consequently may impair other use-dependent water rights.

CONFLICT

The extent to which federal environmental programs such as the ESA authorize federal regulators to disrupt state-created appropriative water rights is controversial. At one extreme, some observers contend that such programs establish "federal regulatory rights" that empower the federal government to "cancel the historic *de facto* assignment of property rights in commons to exploiters and reassign them to the government as agent for the public generally." (See Ref. [5, p. 3].) At the other extreme, some federal courts have held that the federal government must defer to state-created water rights in the absence of explicit congressional intent to pre-empt them.[6]

So far, the federal government has not used the ESA as authority to establish a new brand of "federal regulatory rights." However, it has curtailed state-created water rights under two sections of the law. Federal agencies supplying or distributing water to private irrigators have curtailed state-granted water rights for varying lengths of time in compliance with a Section 1536 consultation with the listing ESA agency. For example, the U.S. Forest Service shut down irrigation ditches operating on agency land in the Methow Valley in Washington State for much of the 1999 irrigation season.[7] In another example, the U.S. Bureau of Reclamation cut-off water to 90% of the 220,000 acres in the Klamath Project in Oregon for much of the 2001 irrigation season.[8] The potential for further curtailment of state-granted water rights under Section 1536 consultations is great because the Bureau of Reclamation is the largest supplier and manager of water in the West.[1]

The federal government also has relied on Section 1538 (banning the taking of listed species by private parties) to threaten the curtailment of state-created water rights of irrigators using nonfederally developed or delivered water. For example, NMFS officials warned the Methow Valley Irrigation District that water would be cut-off during the 2002 irrigation season for about 250 irrigators unless the district switched to a more efficient, fish-friendly means of distributing water.[9]

The Methow Valley and Klamath Project water curtailments, and the ensuing losses to the local agricultural economy, understandably have generated substantial ill-will toward the federal government among irrigators, irrigation districts, rural communities, and state governments. For example, the Klamath Project curtailment is estimated to have cost farmers approximately $200 million in lost crops.[10] Business in local communities also has suffered.[10] The prospect of these losses drove about 100 irrigators to risk arrest when they ran an irrigation line to divert water around a canal head gate that the federal government had closed to protect listed fish in Upper Klamath Lake.[11] Could the federal government avoid such confrontation by evolving toward the extreme of deferring to state prior appropriation statutes to satisfy ESA mandates? Unfortunately, the prior appropriation doctrine is not designed to protect aquatic habitat, and thus would be an ineffective replacement for the legal protection that endangered species receive under the ESA.[12] Nondiversionary water uses were not recognized as beneficial uses when traditional appropriative rights were being locked into irrigated agriculture in the late 19th and early 20th centuries. Consequently, irrigated agriculture currently has priority regardless of how little water remains for nondiversionary uses. For example, Wilkinson noted that, under the prior appropriation doctrine, the most senior water-rights holders need not share the water with emerging new water needs, but can "with impunity, flood deep canyons and literally dry up streams, as has happened with some regularity." (See Ref. [13, p. 21].)

Are there policies available for protecting endangered aquatic habitats while mitigating adverse impacts on state-created water rights? Economists have long recommended water marketing as a means of shifting water from prior appropriative uses to competing private and public uses. Unfortunately, while most state water statutes permit public interest groups to purchase water rights for the purpose of augmenting instream flows, state protection of such rights from appropriation by other water-rights holders generally is difficult. Moreover, states impose moderate to severe limits on water transfers to protect third-party water-rights holders from impairment due to changes in return flows. Perhaps the best policy for accommodating endangered species protection and state-created water rights is the use of specialized water transfers designed to limit the extent and duration of impairment to use-dependent rights. Examples are "trial transfers" (transfers that can be modified or revoked if actual impairment results), "one-time temporary transfers" (transfers whose short-term nature makes injuries short-lived), and "contingent transfers" (transfers that occur intermittently and are triggered only by some predetermined contingency such as instream flow below some critical level).[12,14]

REFERENCES

1. Michelsen, A.M.; Taylor, R.G.; Huffaker, R.G.; McGuckin, J.T. Emerging Agricultural Water Conservation Price Incentives. J. Agric. Resour. Econ. **1999**, *24* (1), 222–238.
2. Endangered Species Act, 16 U.S.C. §§1531–1543.
3. Federal Register, 1991, 56(224), 58619–58624.
4. Ring, R. Saving the Platte. High Country News **1999**, *31* (2), 1.
5. Tarlock, D. The Endangered Species Act and Western Water Rights. Land Water Law Rev. **1985**, *20* (1), 1–30.
6. California v. U.S., 438 U.S. 645, 1978.
7. Hicks, L. Ditch Problems Have No Easy Solutions. Methow Valley News **1999**, September *2*.
8. Bernard, J. Dozens Trespass in Irrigation Protest. Spokane Spokesman-Rev. **2001**, August *30*.
9. Hanson, D. Feds, Methow Facing off in Water Dispute. Spokane Spokesman-Rev. **2001**, September *3*.
10. Hold Your Breath, Suckers: A Fish That Has Divided the West May Help Define Gale Norton's Greenery. The Economist, February 9, 2002; 28–29.
11. Associated Press. Farmers Bypass Canal Head Gate to Water Crops. The Spokane Spokesman-Review, 2001, July 16
12. Huffaker, R.G.; Whittlesey, N.; Hamilton, J.R. The Role of Prior Appropriation in Allocating Water Resources into the 21st Century. Water Resour. Dev. **2000**, *16* (2), 265–273.
13. Wilkinson, C. *Crossing the Next Meridian*; Island Press: Washington, DC, 1992.
14. Hamilton, J.R.; Whittlesey, N.; Halverson, P. Interruptible Water Markets in the Pacific Northwest. Am. J. Agric. Econ. **1989**, *71* (1), 265–273.

Irrigated Agriculture, Historical View of

Lyman S. Willardson
Utah State University, Logan, Utah, U.S.A.

INTRODUCTION

There is no known record of the beginning of irrigated agriculture. It was most likely started on a very small scale by someone trying to keep a wilted plant alive by pouring water on the soil around its base. Then, ways were found to keep more plants supplied with water when they were remote from the water supply. However, carrying water from a spring or a stream to supply many plants is heavy work. By scraping small furrows from a stream to the plants, irrigation could be practiced with a greatly reduced labor input. The practitioners soon realized that they could produce more food by keeping an adequate supply of water available to their plants at all times. The availability of more food from irrigated plants meant that more people could live in a smaller area and communal living could be practiced. Communities could grow into cities and cities could grow into nations. When governments were organized, public resources were available to construct the necessary infrastructure to supply water to all suitable lands. It is certain that irrigation became a necessity as population increased in arid or semiarid areas. In many areas, the season of limited rainfall corresponds to the season of maturity of food crops. If the crops are short of water at that time, yields are severely depressed. It therefore became important to develop irrigation systems that could supply water to crops in seasons of rainfall shortage. With irrigation, it is possible to grow crops in areas of very low rainfall or areas where nearly all the precipitation falls in the nongrowing season. Irrigation is a means of taking advantage of the productive capacity of suitably fertile soils which lack only adequate water for crop plants in their normal growing season.

IRRIGATION IN ANCIENT TIMES

One of the earliest written records of irrigation practice was found in the Code of Hammurabi.[1] Various translations of the Code exist that include references to laws related to irrigation. Irrigation in Babylonian times was very important. One article in the law says "If the irrigator neglects to repair his dyke, or leaves his runnel open and causes a flood, he has to make good the damage done to his neighbor's crops, or be sold with his family to pay. The theft of a watering-machine, water-bucket, or other agricultural implement was heavily fined." This law was in effect in approximately 1750 B.C.

Another famous historical irrigation development occurred between the Tigris and Euphrates rivers in what is modern Iraq. A very large civilization developed in that area and then disappeared. Originally, it was thought that the developing population denuded the watersheds and the canals filled with silt from erosion of the watersheds, making continued irrigation impossible. It is more likely that a rising water table in the area caused salinization of the soils being irrigated. The loss of food production on the salt-affected soils caused the civilization to disappear. Another possible explanation is that the area was overrun by conquerors who had no appreciation for the need to maintain the irrigation systems. The irrigation works were allowed to deteriorate until they could no longer feed the population.

In a book, copyrighted in 1898 by King,[2] the author reported extensive irrigation developments in many areas of the world. He references a paper presented by Mr. Frederick S. Gipps before the Royal Society of New South Wales in 1887 claiming that the first authentic lake or reservoir was Lake Maeris. It was constructed by King Maeris or King Amenemhet III of the 12th Dynasty in 2084 B.C. Water was stored at the time of flooding on the Nile to relieve some flooding and was later released back into the river. Sesostris in 1491 B.C. built many canals in Lower Egypt for irrigation and transportation. Egypt claims the world's oldest dam[3] built in about 3000 B.C. The Phoenicians, about 1100 B.C.[2] were irrigating the "African shore" and had gardens and large plantations "abounding in canals." The Bible mentions many ditches in Second Kings 3:16–17.

In more modern times, the Romans built extensive aqueducts to supply water to cities. They therefore had the technology necessary to develop and transport water for irrigation and food production. The valley of the Po River in Italy is currently irrigated with many old, if not ancient, canals.

China[2] also has some ancient irrigation works on a grand scale. The Great Imperial Canal has a length of more than 1000 km (650 mi) and connects two rivers. The canal even crossed some lakes on elevated dikes.

Encyclopedia of Water Science
DOI: 10.1081/E-EWS 120010063

One of the most significant democratic institutional arrangements in the history of irrigation is the "Tribunal of Waters," which still exists, after more than 1000 yr, in the irrigation districts of eastern Spain.[4] Each Tribunal, which consists of locally elected canal presidents, meets at a fixed time every week to hear farmers' complaints about water use offenders and applies appropriate sanctions.

When the Spaniards began their conquest of Mexico and Peru, they found irrigation being practiced on a relatively large scale. In the ruins of older cities, they found evidence of irrigation canals that had long been abandoned. In Peru, in a district named Condesuyos, they found a comprehensive canal that passed through a number of basins and had a length of approximately 600 km.

In recent historical times, as engineering and science progressed, irrigation was practiced extensively in many of the countries of Europe, not considered to be arid or even semiarid. Much of the irrigation water was applied to pastures. Water wheels were used to lift water out of streams. By 1800, the total irrigated area of the world was about 8 million ha.[5]

The British, during their commonwealth period in the late 1800s, developed large irrigation systems in the areas that are now India and Pakistan. Water was diverted from rivers and carried in large canals for long distances. Many wells were also developed to irrigate localized areas. More than 2,000,000 ha were brought under irrigation. In Ceylon (Sri Lanka) at the same time, irrigation systems consisted of a system of tanks (reservoirs) that served as small an area as one farm. Runoff water from rainstorms was captured, stored, and used in times of water shortage, primarily for irrigation of paddy rice. Irrigation was just beginning in southeast Australia, which was under British rule during this same period. By 1900, the total irrigated land in the world was about 48 million ha.[5]

Extensive irrigation in the United States of America began in the mid-1800s in the West. At that time, there was limited irrigation of small gardens by native Mexicans living in the southern area of California. When the Mormons migrated from the humid east and central part of the United States and the rainfed areas of Europe to the arid valleys of the Rocky Mountains, the settlers found themselves in a low rainfall environment that would not support agricultural crops without irrigation. With limited information about irrigation gained from explorers of the West, they began diverting water from perennial streams onto the soil to make it possible to plow, cultivate, and plant food crops. Diversion of water from natural streams became a common practice and new settlements were established wherever there was a dependable water supply. The arid west was settled rapidly and extensively and a modern civilization was quickly established, based on irrigated agriculture. Ancient civilizations grew as a result

of irrigation. Western American civilization developed based on irrigation. Without irrigation, the arid western United States could only support a limited population of hunters and gatherers.

As the number of settlers in the West increased, the competition for water increased and it became necessary to determine how the water should be divided equitably among the potential users. Some crops needed more water than others and some soils needed more frequent irrigation to keep plants growing properly. New laws had to be developed specifically for management of the limited water supply. The system of law defining Riparian Water Rights, common to humid areas, gave legal use rights to the landowner touching the streams. This law was not appropriate for water-short areas where the water had to be taken away from the stream to nonriparian land. A new system of water law, called Prior Rights, was developed. The doctrine of prior rights states that first in time of use is first in right and that beneficial use of the water determines the right. The first person to use the water has a legally superior right to use the water and can maintain that right as long as he uses the water beneficially. Since the land was essentially worthless without a water right, water rights normally were sold with the land.

The establishment of the Land Grant University system by President Abraham Lincoln in 1862, gave emphasis to the development of agriculture. In the arid and semiarid areas of the country, agriculture depended on irrigation water management. Universities in those areas gave special emphasis to research on soil–plant–water relations. This information was very important to the effective and efficient use of water. In ancient times, water was applied to the soil until the irrigator felt that the soil was wet enough. He also learned that too much water was damaging to the plants. His only means of applying water was to carry water to the individual plants or bring the water from the source in a small ditch or furrow. He could also build a small dike around an area containing plants and flood irrigation water into the resulting basin. Rice can be irrigated by continually flooding it in a basin area. From ancient to modern times, the majority of the irrigation taking place in the world is by surface irrigation methods. Irrigation that uses the soil surface to transport as well as absorb the water is called surface irrigation. As a result of research, mostly at land grand universities, irrigation water can now be applied using mechanical systems. The water is transported in pipes and is applied to the soil using sprinklers or drip-irrigation systems. Surface irrigation, to be successful, requires a lot of hand labor and the good judgment of the irrigator. Mechanized irrigation can be accomplished by complete automation, including the decisions of when to irrigate and how much water to apply.

Irrigation on a large scale developed in the American West soon after the passage of the Reclamation Act in

1902. The U.S. Bureau of Reclamation of the U.S. Department of the Interior, backed by the U.S. Government financing, was able to build large dams that provided water for reclaiming many western desert lands. In recent years, controversy[6] has arisen over the environmental changes caused by using rivers to generate electric power, to supply water to remote cities in other drainage basins, and to make desert lands agriculturally productive, rather than leaving the rivers in their natural ecological state. Similar objections have not been raised in India, Pakistan, and China where there are even larger expanses of irrigated land.

CONCLUSION

There is an International Commission on Irrigation and Drainage headquartered in New Delhi, India. This organization maintains statistics about irrigation and encourages more efficient and effective use of irrigation water. There are 25 countries in Africa, 16 in the America, 29 in Asia and Oceana, and 28 in Europe that have significant areas under irrigation. In the year 1998, there were more than 271 million ha of land being irrigated in the world. At least 22 persons depend, in some degree, on the food produced on each hectare of land irrigated in the world. Irrigation plays a more important role in food production in the modern world than it did in the ancient world. During the 21st Century, most of the world's increased food supply will have to come from the higher agricultural productivity of existing irrigated lands.

REFERENCES

1. Babylonian Law—The Code of Hammurabi. *Encyclopedia Britannica 1910–1911*, 11th Ed.; Yale Law School Avalon Project.
2. King, F.H. *Irrigation and Drainage, Principles and Practice of Their Cultural Phases. 1912*; The Macmillan Company: New York, 1913; 66–72.
3. Israelsen, O.W.; Hansen, V.E. *Irrigation Principles and Practices*; John Wiley and Sons: New York, 1962; 1–3.
4. Glick, T.F. *Irrigation and Society in Medieval Valencia*; The Belknap Press of Harvard University: Cambridge, MA, 1970; 386 pp.
5. Fukuda, H. *Irrigation in the World—Comparative Developments*; University of Tokyo Press: Tokyo, Japan, 1976; 329 pp.
6. Reisner, M. *Cadillac Desert*; Penguin Books: New York, 1986; 582 pp.

Irrigated Agriculture: Managing Toward Sustainability

Wayne Clyma
Water Management Consultant, Fort Collins, Colorado, U.S.A.

M. S. Shafique (Retired)
United Nations Office for Project Services, Lahore, Pakistan

Jan van Schilfgaarde (Retired)
United States Department of Agriculture (USDA), Fort Collins, Colorado, U.S.A.

INTRODUCTION

Water management improvements in an irrigated valley improve productivity, reduce the environmental impacts of irrigation, enhance the environment of rivers, and usually provide additional water that can be used for other purposes. Long-standing traditional beliefs are that water cannot be saved by improving water management in an irrigated valley. These beliefs are shown to be based upon erroneous assumptions. Improving water management increases yields, area affected by waterlogging and salinity is reduced, return flows and salinity are lowered, and water saved in the reservoir improves the environment of the river when released. Organizations and individual actions need to be changed for successful water management improvements.

Irrigated agriculture plays an important role in food production. Commonly, irrigated agricultural production averages nearly twice the production of rainfed agriculture per unit area. Actually, effective irrigated agriculture easily produces 3–5 times rainfed production. The reduced level of average production under irrigation is a measure of the inadequacy of performance of irrigation.

The Food and Agricultural Organization (FAO)[1] has shown that water shortages are currently an issue in many countries. By 2030, these shortages will cause serious food shortages in many countries. Therefore, a key strategy for irrigated agriculture is to develop water conservation programs to conserve and enhance water supplies and substantially increase productivity worldwide.

IRRIGATION WATER MANAGEMENT STATUS

Irrigation water management practices need significant improvement worldwide.[2,3] Fig. 1a shows an irrigated valley with major waterlogging and salinity problems caused by poor water management practices. Fig. 1b shows evaporation from standing water and excess evapotranspiration from crops supported from a high water table. Poor or nonfunctioning drainage systems also cause excess evapotranspiration as shown in Fig. 1c. Water use for irrigation is often 2–4 times as much as good management achieves. Encroaching salinity and waterlogging remove millions of hectares from production each year. Worldwide, the irrigated area severely or moderately affected by waterlogging and salinity equals 30% of the irrigated area, and it increases by 1%–2% per year.[1] Poor water management practices throughout an irrigation project cause the waterlogging and salinity. With high water tables, from 15% to 85% of the rainfall plus irrigation supplied can evapotranspire. Evaporation and evapotranspiration losses from waterlogged areas represent a major source of water that can be conserved. Thus, water management improvements to increase productivity, reduce the impacts of waterlogging and salinity, and conserve water supplies are urgently needed practices.

Farm Water Management Assessment

Crop yields under irrigated agriculture can be assessed using many different strategies. Comparing average yields to record yields for a country or an irrigated valley is one strategy for assessing the potential yields. Often, average yields are one-third to one-fifth and even less of the record yields.

Causes of these lower yields are varied. Basic to yield improvement under surface irrigation is a precision-leveled field especially when level basins are the field system. Level basins should be within 15 mm of the average elevation for good water management and to achieve potential yields. Because fields are usually designed and leveled by farmers, most are not adequately leveled. Over application of water to fields or parts of fields is often the major factor causing waterlogging in irrigation projects. Farmers in Pakistan were thought to use

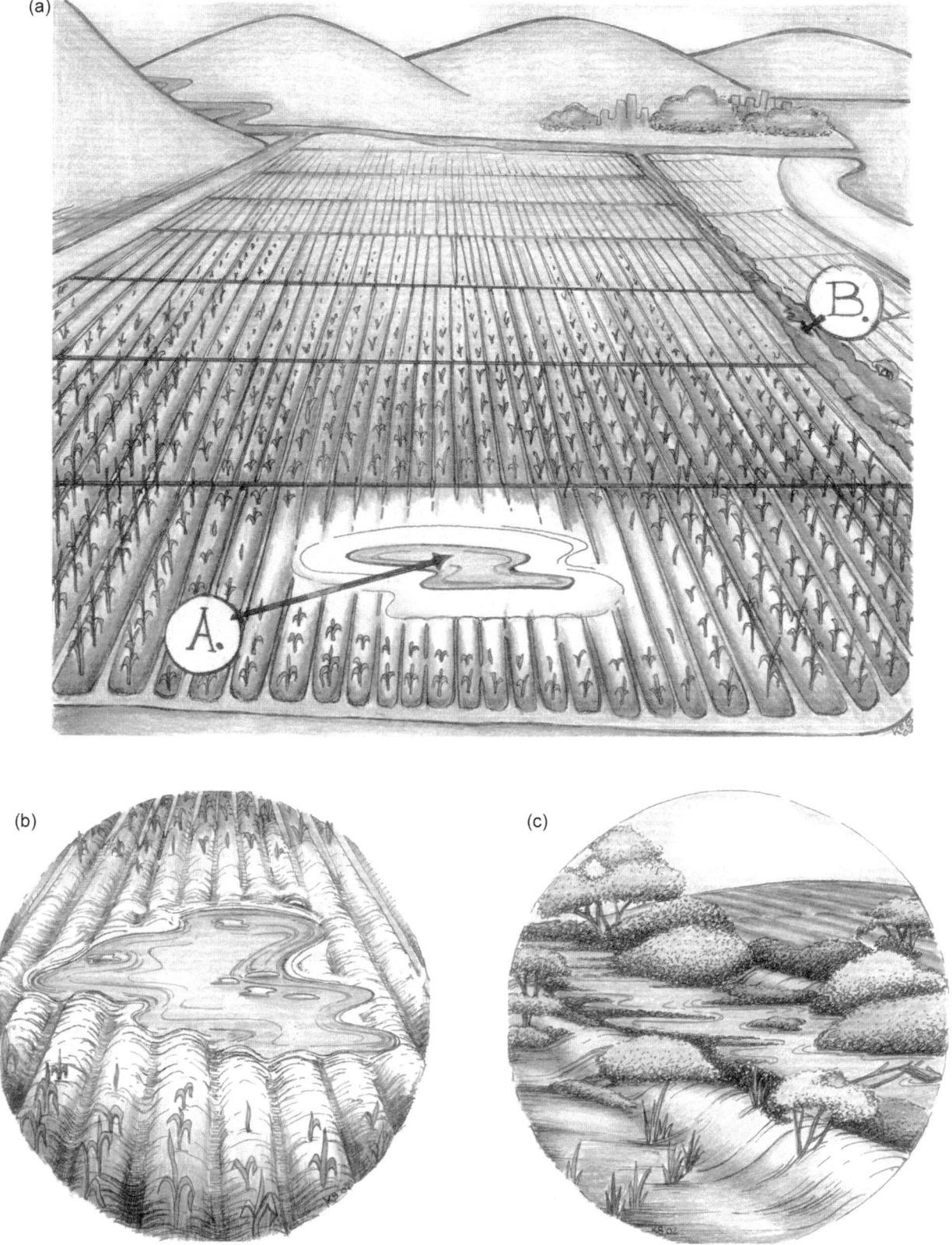

Fig. 1   (a) Irrigated valley with waterlogged area and poorly defined surface drainage system.[12] (b) Waterlogging and salinity is an integral part of irrigated agriculture in most projects and the water evaporated or evapotranspired is a nonbeneficial use.[12] (c) Drainage systems for many irrigation projects cause nonbeneficial use and reduce return flows.[12]

lower than recommended levels of fertilizer. More careful evaluation showed that the level of fertilizer used was near the optimum for the fields' potential yields.

Studies in several countries have shown that precision leveling (precision conventional leveling or laser leveling) with appropriate inputs contributed to many-fold yield increases while water requirements were reduced substantially, often by half or even more. Thus, water management improvements at the field level increase productivity while reducing the water required for irrigating the field. Water management and input improvements generally increase yields sufficiently to more than pay for the services when quality service is provided at an effective cost with credit available when needed. Productivity improvements and increases in the effective use of water supplies are often the highest priorities for a country.

Delivery of Water Supplies

Field studies in irrigation projects around the world have shown that adequacy, dependability, and equity are important, but often unattained, goals of water delivery. Canals and watercourses usually do not provide a target discharge of necessary duration for farmers to irrigate adequately. Undependable water supplies can cause farmers not to plant any crop, or to use traditional seeds with little or no fertilizer instead of high yielding varieties with adequate fertilizer. They also may not follow appropriate cropping practices, and limit the use of adequate weed and pest controls.[4] Therefore, undependable water supplies often create greater constraints on productivity and potential productivity than just those caused by inadequate water. Assessments of productivity in canal commands in many countries have shown that yields under undependable, inadequate water supplies are fractions of adjacent fields with dependable, adequate water.[4]

Farmers irrigate too frequently and apply too much water when supplies are undependable but available. They use the water when available because experience has taught them that it may not be available when they next need to irrigate. Frequent irrigations cause excess evapotranspiration and increase waterlogging. Inadequate water supplies reduce actual yields, and teach farmers that investments in inputs for higher yields are not profitable. Studies in India showed that the intensity of cash value crops was directly correlated with distance from the canal outlet with the type of crop grown determined by water supply.[5] In Nepal and Sri Lanka, when farmers were not adequately advised about expected water supplies, they planned for inadequate water supplies as described earlier.

Every extensive water distribution system studied had head to tail inequities in the adequacy and dependability of water supplies. Inequities in water delivery have been measured in India, Egypt, Sri Lanka, Nepal, Pakistan, Thailand, Somalia, and the United States of America to name a few countries. Upper reaches of most delivery systems or branches with influential farmers often have three or more times the target water supply while those at the end of the canals, laterals, and watercourses receive half or zero of the target.

Water Quality Management

Irrigation water management in recent decades has focused on salinity management—now urgently needed. Irrigating a field supplies water to be used by the crop through evapotranspiration. Since the water changes to vapor as it leaves the soil or plant surface, the salts it carries remain in the soil. All irrigation leaves salts in the soil that must be controlled by adding additional water. The additional water travels through the rootzone and removes salts by a process called leaching. The excess water becomes deep percolation that goes to groundwater or to a substratum. Some excess water, carrying the leached salt, often travels to the river as return flow. Additional salts can be in the return flow because the groundwater was more saline to start with or because the substratum had excess salts that were added to the water. Sometimes, natural return flows to the river are not sufficient and rapid enough, and drainage must be provided to remove the excess water and salts. Erosion of the fields during irrigation may cause sediment pickup. Erosion damages the irrigated fields, and the sediment may damage lands and waterways where the sediment is deposited. Sediment that remains in suspension in the water limits the value of the water for reuse.

Return flows from irrigation may also contain nitrogen, phosphorus, and other agrochemicals, used to control weeds or pests, from agricultural operations that contaminate the water for other uses. Small amounts of selenium, boron, or other elements toxic to plants (and humans) in low concentrations may also limit the value of the return flows for subsequent irrigation or other uses. Excess chloride, which causes foliage damage when the water is used in sprinkle irrigation, can be a serious problem especially on vegetable crops. Excess sodium in the irrigation water may cause difficulties with infiltration of the water into soils and thus limit the usefulness of the water. Reductions in water quality lower the value of water or increase the costs of using the water.

Traditional Views of Water Management

Water conservation is a widely misunderstood concept. Placed in an irrigated valley context, even greater misunderstandings exist. First, many believe that improving low irrigation efficiencies automatically make large amounts of water available. They believe the conserved water is available for the improved farm, the irrigated valley, or other water uses as an additional water supply. This understanding is not valid when the excess water returns to the river. The value assigned to the water saved must be reduced by the value of the return flows for reuse.

Second, another common misunderstanding, shared by the public and many professionals, is that water conservation has almost no place in irrigated valleys. They believe water can be conserved only when direct flows to salt sinks, such as saline water bodies and the ocean, are prevented. This view is supported by often erroneous assumptions that 100% of the excess irrigation water is available and 100% of this available water returns to the river as return flow.[6] The reduced value of the return flow caused by reductions in water quality is often not considered. This concept seriously hampers efforts to reduce the impacts of waterlogging and salinity and control the environmental effects of irrigation. A more balanced concept for achieving water conservation in an irrigated valley is the focus of the next section.

SUSTAINABLE WATER MANAGEMENT FOR IRRIGATED AGRICULTURE

Sustainable water management must achieve effective water management improvements and manage to limit salinity impacts. Institutional and attitudinal changes are essential to achieve such improvements.

Sustainable Water Management

A farmer irrigating a field is the focus of water management improvements. Reducing the volume of water required for irrigating the field is the objective.

Table 1 Depletion of total water supply from nonbeneficial evapotranspiration from a waterlogged area in an irrigation project

Waterlogged area (%)	Total water supply (irrigation + rainfall) ($m^3 m^{-2}$)	Depletion (%)
15	0.63	42
30	0.63	86
15	1.69	16
30	1.69	34

Project area is 500,000 ha, annual potential ET is $1.88 \, m^3 m^{-2}$, irrigation water supplies are $0.16 \, m^3 m^{-2}$ and $1.22 \, m^3 m^{-2}$, and rainfall is $0.47 \, m^3 m^{-2}$.

Water not supplied for irrigating the field is water saved. Water not released from a reservoir or diverted from the river is water conserved. When the reduced supply for irrigation results in reduced return flows to the river, then an appropriate volume of the saved water may be made available to replace the reduction in return flow. Water released to replace return flows was conserved because the water has a greater value since it was not used initially for irrigation. The remaining water is available for reallocation, whether in reservoir or groundwater storage, or is continuing flow in the river. Water available for reallocation can be allocated for other uses such as industrial, municipal, environmental, or even irrigation.

Water available for reallocation comes from reduction of nonbeneficial evapotranspiration within the irrigation project. The water conserved also can result from water supplies stored in groundwater systems or from returns to the river with major increases in salinity that materially reduces the value of the water. Water seriously contaminated when used for irrigation by one of the previously defined contaminants also can be conserved. Excess irrigation water that returns to salt sinks, such as the ocean or saline water bodies, also can be conserved.

Water conservation accomplished by improving water management that reduces nonbeneficial evapotranspiration, as an example, is now illustrated for a waterlogged irrigation project. Table 1 summarizes approximate data for an irrigation project in the central part of the Punjab in Pakistan.[7] Within the irrigation project, beneficial use comes from crop evapotranspiration. Nonbeneficial use comes from areas waterlogged, areas with a high water table that increases evaporation at the soil surface, poorly leveled fields with low areas where standing water evaporates, and poor drainage systems. Water is available for evaporation at the potential rate when water stands on the land surface as in a waterlogged area.

Total water supply is an important variable and rainfall is important particularly in monsoon climates. With the lower water supply (Table 1), waterlogging developed from irrigation in Pakistan over more than a 100 yr period because water losses persisted although the total water supply was less than potential evapotranspiration. Eighty-six percent of the lower water supply is lost to nonbeneficial evapotranspiration (Table 1) when waterlogging developed. When water supplies were increased because dams were completed, waterlogging continued because beneficial and nonbeneficial uses did not exceed the water supply. With the high water supply, as much as 34% of the total water supply is lost (Table 1). The key point is that large volumes of the water supply are lost through nonbeneficial evapotranspiration. These data do not assess the impacts of unlevel fields, and ineffective drainage systems. Improving water management can achieve major results through water conservation by eliminating nonbeneficial uses.

Water Quality Management

Irrigated agriculture is not sustainable unless salinity is managed. A strategy for improving management of salinity should include reducing leaching fractions to a minimum.[9] This strategy is consistent with the water management improvement focus because when irrigation efficiencies are 40% or less, then attempting to target a 5% or less leaching fraction is not appropriate. Lower leaching fractions reduce the total salts that return to the river in return flows, and may sometimes precipitate some salts before they enter the river.[10] More careful control of leaching can precipitate salts below the rootzone and minimize the pick up of salts from saline strata.

A useful strategy is to use good quality irrigation water to grow salt sensitive crops such as lettuce. Then, as salinity increases down an irrigated valley, to grow more salt tolerant crops such as wheat and cotton, and further down the valley even more tolerant crops such as barley.[9] Continuing to use water from the river for irrigation that increases in salinity reduces the return flows to the river and lowers the volume of saline water that must be managed for disposal. Evaporation ponds and pipelines to salt sinks such as the ocean are appropriate disposal alternatives.[9] These strategies for managing salinity are a key part of the water management improvements for increased water conservation.

A critical need is a system for costing and valuing changes in water quality. Traditionally, reductions in water quality are neither assigned a cost nor are improvements in water quality assigned a value. Keller et al.[8] suggested that return flow volumes be reduced by the volume of additional water required for leaching because of the salinity increases in the return flow. While this is an important step, the result is an inadequate approach for costing and valuing increases in salinity, and does not evaluate other important quality changes. A strategy for costing and valuing water quality changes is further described.

The amount of water used to replace the reduced return flows from water conservation should be based upon an effective volume or value assessment. Clyma and Shafique[6] suggest that the effective value of the return flow be determined. Then, the reduction in return flow is replaced by a volume from storage that equals the reduction in effective value caused by water conservation, if any. Keller and Keller's[8] approach does not consider reductions in yield or crop changes that result from increased salinity, or other reductions in value such as lifting costs for pumping. Clyma and Shafique[6] allow for increased value of the replacement water and decreases in

value from salinity, energy costs, impacts of other contaminants, and other factors. An important consideration is the trade-offs between water stored in a reservoir, water in groundwater storage, and return flows that vary in amount and the time when available in the river.

The valley water management strategy starts at the farm with improvements in water control that cause water applied to approach crop needs and minimum leaching requirements. Using minimum leaching requirements does reduce salt loads. Minimum return flows also reduce the total salt load in the river but may increase the concentration of salts in the return flows. Salinity increases down the river are reduced because of the lower total salts. Water releases from the upstream reservoir or remaining in the river provide additional flow volume for the river further reducing salinity in the river. Additional good quality water is made available for subsequent canal commands. When the return flow salinity reaches a critical level approaching zero value for the return flows, then disposal of the water from evaporation ponds or with a pipeline to the ocean or another salt sink can be considered. Such a management strategy approaches achieving a permanent irrigated agriculture.

Institutional and Attitudinal Changes

Farmers change their water management when they understand and experience the value of change. Supporting organizations and their professional personnel must appreciate the changes needed to enable them to support farmers in accomplishing such changes effectively. Institutions must modify their policies and programs to define and then support farmer needs for change. Then, they must support personnel that effectively provides the needed support to farmers.[11] Changing individuals and organizations—farmer, private, and local, county, state, and federal units—is difficult but can be successfully accomplished.[11] Changes to water laws will often be needed if water is to be conserved. Farmers must benefit financially when water conservation increases the effective water supply. Water banks, begun recently in some Western States in the United States, offer some of the changes needed. Water rights in many countries will need major redefinition.

SUMMARY AND CONCLUSIONS

Water management improvements provide major opportunities for increasing productivity, reducing the environmental impacts of irrigation, and increasing water supplies by accomplishing water conservation. Poorly managed water supplies and field irrigation limit productivity and create major increases in waterlogging and salinity. Water quality for irrigation decreases down a valley further reducing productivity.

Water management improvements reduce the amount of water required to irrigate a field. Water not supplied to the field is water saved. When return flows are reduced because of improvements, water saved can be released to replace the return flows. The remaining water saved is water available for reallocation. It is available for a variety of uses such as industrial, municipal, environmental, or even expanded irrigation.

Improvements in water management reduce water-logging and salinity. Because leaching volumes are reduced, total salt returning to the river is often reduced. Water released to replace return flows further improves the quality of water in the river and provides more water for fish and wild life at an improved quality. Careful management may achieve a permanent irrigated agriculture.

REFERENCES

1. FAO, *Food Production: The Critical Role of Water*; Food and Agricultural Organization, Rome, 1996.
2. Clyma, W. Changing Irrigated Agriculture for the New Millennium. *National Irrigation Symposium*; Proceedings of the Fourth Decennial Symposium, November 14–16, 2000, Phoenix, AZ, ASAE: St. Joseph, MI, 2000; 182–186.
3. Clyma, W.; Clemmens, A.J. Farmer Management Strategies for Level Basins using Advance Distance Criteria. In *National Irrigation Symposium*; Proceedings of the Fourth Decennial Symposium, November 14–16, 2000, Phoenix, AZ, Evans, R.G., Benham, B.L., Trooien, P.P., Eds.; ASAE: St. Joseph, MI, 2000; 573–578.
4. Clyma, W.; Katariya, S.R.; Nelson, L.J.; Tomar, S.P.; Reddy, J.M.; Bakliwal, S.K.; Haider, M.I.; Mehta, U.R.; Lowdermilk, M.K. *Diagnostic Analysis of Farm Irrigation Systems on the Gambhiri Irrigation Project, Rajasthan, India*, Water Management Synthesis Report No. 17; Colorado State University: Fort Collins, CO, 1983; Vol. I–V.
5. Jayaraman, T.K.; Lowdermilk, M.K.; Nelson, L.J.; Clyma, W.; Reddy, J.M.; Haider, M.I. *Diagnostic Analysis of Farm Irrigation Systems in the Mahi-Kadana Irrigation Project, Gujarat, India*, Water Management Synthesis Report No. 18; Colorado State University: Fort Collins, CO, 1983.
6. Clyma, W.; Shafique, M.S. *Basin-Wide Water Management Concepts for the New Millennium*, ASAE Paper No. 012051; ASAE: St. Joseph, MI, 2001; 1–16.
7. Clyma, W. Evapotranspiration and Irrigation Water Requirements for Pakistan. Bull. Irrig. Drain. Flood Control Res. Council **1973**, *3* (2), 1–8.
8. Keller, A.; Keller, J.; Seckler, D. *Integrated Water Resource Systems: Theory and Policy Implications*, Research Report 3; International Irrigation Management Institute: Colombo, Sri Lanka, 1996; 1–15.
9. van Schilfgaarde, J.; Rhoades, J.D. Coping with Salinity. In *Water Scarcity, Impacts in Western Agriculture*; Engelbert, E.A., Ed.; University of California Press: Berkeley, CA, 1984; Vol. 6, 157–179.
10. Suarez, D.L.; Rhoades, J.D. Effect of Leaching Fraction on River Salinity. J. Irrig. Drng. Div., ASCE **1977**, *103* (IR2), 245–257.
11. Dedrick, A.R.; Bautista, E.; Clyma, W.; Levine, D.B.; Rish, S.A. The Management Improvement Program: A Process for Improving the Performance of Irrigated Agriculture. Irrig. Drng. Syst. **2000**, *14* (1–2), 5–39.
12. Clyma, W.; Shafique, M.S. *Water Saved in Irrigated Valleys by Improving Classical Efficiency*; Colorado Institute for Irrigation Management, The Water Center: Fort Collins, CO, 2002.

Irrigated Agriculture, Social Impacts of

Gaylord V. Skogerboe
Utah State University, Logan, Utah, U.S.A.

INTRODUCTION

Ancient hydraulic civilizations had absolute power, including strong organizational coordination and complete control of resources, elaborate postal communication and intelligence networks for social control, along with the dominant religion being under the authority of the state.

In sharp contrast, small indigenous farmer-managed hydraulic societies have existed for centuries in many countries. These irrigation systems are operated by using democratic principles wherein all farmers participate in managing their system.

During the past century, the world's irrigated land has increased fivefold, but there will be limited expansion in the future. These systems are commonly managed by government agencies. But, water scarcity will require that some of the irrigation water supplies be transferred to meet increasing municipal and industrial water needs. In order to feed growing populations, much higher levels of irrigation water management will be required. This will necessitate transferring these government-managed irrigated systems to farmers so that they become more self-reliant and innovative. To significantly increase crop yields with less water requires: 1) a clearly defined water rights system; and 2) sustainable farmers organizations.

NATURE OF SOCIAL IMPACTS

The degree of productivity in irrigated agriculture is highly dependent upon the degree of cooperation among farmers and with the individuals responsible for managing the irrigation system. In addition, each farmer is impacted by the actions occurring upstream, such as unreliable water deliveries and water theft. Farmers who are more self-reliant are more likely to maintain their irrigation facilities. Cooperation and independence also foster innovations that lead to increased agricultural productivity, possibly more employment and better health. The dominant factor impacting these traits is the type of social organization employed for managing an irrigation system.

Worldwide irrigated agriculture has grown from 8 Mha in 1800 to 48 Mha a century later,[1] with the United Nations[2] estimating 255 Mha in 1995, which is most likely an overestimate. More importantly, the amount of irrigated land is expected to increase very little in the future. About 17% of the world's agricultural land is presently irrigated, which accounts for about 40% of the world's food production.[2]

For the future, planners place heavy emphasis upon 75% or more of increased food production coming from irrigated agriculture. A good case has been made[3] for doubling the productivity of water in order to feed 8 billion people within the next three decades, while protecting the world's ecosystems. With increasing water scarcity in many global locations, some of the present irrigation water supplies must be transferred to meet future urban and industrial water demands. This will necessitate much higher levels of water management, with a major emphasis on significantly improved social organization of irrigation systems in many parts of the world.

CENTRALLY-ADMINISTERED HYDRAULIC CIVILIZATIONS

Wittfogel[4] reports on the administrative management of numerous irrigation systems around the globe, but especially Asia, over many thousand of years. Small-scale irrigation is called a "hydraulic society," while a large-scale and government-managed irrigation network is a "hydraulic civilization." There are three paramount characteristics of a hydraulic civilization: 1) involves a division of labor; 2) intensifies cultivation; and 3) necessitates cooperation on a large scale.

Hydraulic civilizations used corvee forced labor, which was conscripted on a temporary, but recurring, basis. In Imperial China, every commoner family was expected on demand to provide labor for hydraulic and other public services. The writings of India, as well as the Incas and Aztecs, indicate a similar claim on corvee labor.

In terms of social control and natural resources development, the master builders of hydraulic civilizations had no equal in the nonhydraulic world because of control over the entire country's labor and materials. The dispersed castles of Medieval Europe are clear evidence of feudal society, just as huge administrative cities and colossal palaces, temples and tombs of Asia, Egypt, and ancient America express the organizational coordination and resources mobilization of the hydraulic civilizations.[4]

Encyclopedia of Water Science
DOI: 10.1081/E-EWS 120010242

Administrators and officers were placed in all major settlements, which virtually everywhere assumed the character of government-controlled administrative and garrison towns. In addition, almost all hydraulic civilizations enhanced their power by elaborate systems of postal communication and intelligence, which became a formidable weapon of social control. The masters of the empire in China combined state roads and man-made waterways in establishing a postal and intelligence system that lasted for more than 2000 yr, but with some disruptions.[4]

The government of a hydraulic civilization was an integral part of the irrigation management bureaucracy, with the dominant religion being closely attached to the state. Nowhere in hydraulic civilizations did the dominant religion place itself outside the authority of the state as a national or international autonomous church.[4] This formidable concentration of vital functions gave the government its genuinely absolutist power, where its rule was not effectively checked by nongovernmental forces.

Egypt was an important deviation. The central government imposed a tax on the farmers of 10%–20% of their harvest, but the administration of the irrigation system remained local.[3] An observation[5] is that "Egypt probably survived for so long because production did not depend on a centralized state; the collapse of government or the turnover of dynasties did little to undermine irrigation and agricultural production at the local level."

INDIGENOUS FARMER-MANAGED HYDRAULIC SOCIETIES

For many centuries, farmer-managed irrigation systems (FMISs) have existed at various locations around the world. In Asia, the systems in Nepal, Thailand, and the Philippines have been partially investigated. The social organizational arrangements are a sharp contrast with the despotic hydraulic civilizations described previously.

Two-thirds of the irrigated agriculture in Nepal is farmer-managed; mostly, these thousands of systems are autonomous, self-governing entities ranging in cultivated area from 10 ha to 15,000 ha. A comparative study of 21 FMISs has been reported,[6] along with institutional arrangements consisting of social organization and property rights in water.[7] These irrigation organizations perform tasks of water acquisition, water allocation and distribution, resource mobilization (people and tools), system maintenance, decision-making, communication, and conflict resolution.

Decisions regarding irrigation water management are made by the irrigators as a whole at their annual meeting, where the farmers review the performance of the previous year, audit and settle accounts, decide on the plan and program for each major task, and elect officeholders.

An irrigation management committee is elected to carry out the decisions of the general body of irrigators. Remuneration to committee members often consists of cash or kind, but sometimes nothing.

Resource mobilization (such as channel cleaning and replacement of low-cost structures that failed due to floods), may be based on the size of landholding, water shares, water outlet size, village units, or the number of households in the command area. Water allocation may be based on the size of landholding, labor contributions for maintenance, original investment, water shares, or the type of land. Water distribution needs intensive supervision, particularly when the water supply is barely sufficient to meet the crop needs.

When traveling throughout Thailand, it is readily apparent that the best agriculture occurs on roughly 2000 small FMISs located in the North (two-thirds of cultivated area), where irrigation has been practiced for at least 700 yr. A 10-yr multidisciplinary study of five FMISs[8] shows there is a high degree of acceptance among the farmers of the water rules and regulations, which in earlier times were considered sacred because they provided rice for everyone; thus, water theft was considered a severe crime against society.

In Thailand, the farmer leaders for a FMIS receive much respect because they are trusted by the farmers, which results from the leaders diligently doing what they promised, along with being very fair in their dealings, and placing a strong emphasis on bettering the community. These traits result[8] in equitable water distribution.

GOVERNMENT AGENCY-MANAGED IRRIGATION SYSTEMS

From the mid-19th century through the 20th century, about three-fourths of the developed irrigated lands are administered by government irrigation agencies. Majority of these irrigation systems in developing countries are not properly maintained, so they are unable to increase agricultural productivity for feeding a growing population. Rehabilitation is often considered a remedy, but usually results in another costly cycle of improvement and decay with no long-term benefits.[9] There is a growing perception that these public irrigation agencies lack the incentives and responsiveness to improve management performance and that a management system which is more accountable to farmers will be more equitable and responsive.

The argument can easily be made that farmers under these agency-managed irrigation systems (AMISs) are oppressed. Certainly, they have limited control of their destiny. They are not organized for administering their irrigation system, and they do not have meaningful water rights. Increasing agricultural productivity over time is highly dependent on farmers being empowered so that they become more self-reliant and innovative, thereby benefiting socially and economically from their ingenuity.

SELF-RELIANCE AND INNOVATION

The degree of independence and innovation demonstrated by the farmers in an irrigation system is a good indicator of social impacts. A healthy agricultural environment relies on farmers being innovative, which in turn is dependent upon farmers being able to benefit socially and economically from their inventive behavior. A major goal of agricultural development should be to establish an institutional environment that strongly supports innovations by farmers.

In order for farmers to become more confident, a highly participatory approach is required from all types of agricultural support services. Farmers must not only be treated as equals, but they must be recognized as the local experts. Thus, the attitudes and behavior by those individuals providing support services is crucial to successful agricultural development. The most under-utilized resource for improving irrigated agriculture, the farmers, can only be effectively strengthened by using participatory approaches that strongly emphasize farmers first.[10]

The most significant determinant of self-reliance will be the degree that farmers manage their own irrigation system. Farmers recognize the significance of controlling the entire canal network, including the canal headworks. This can be readily envisioned for relatively small irrigation systems, but farmer management is even more important for the much larger canal systems such as encountered in China, India, and Pakistan.

INSTITUTIONAL EMPOWERMENT

The major lesson from past hydraulic civilizations, numerous FMISs over many centuries, as well as Australia, Canada, and the American West in the last 150 yr, is that locally managed irrigation enterprises are to be preferred for long-term sustainability. The critical ingredients to highly productive agriculture for such systems are having the power to assess the beneficiaries for making improvements, along with water rights to encourage long-term investments. The success of irrigated agriculture requires fitting many pieces of the puzzle together, including social cohesion, but certainly institutional measures should lead, not follow technology, in this continual struggle for progress.

CONCLUSION

Entering the 21st century, there were about 25 countries experimenting with the transfer of irrigation system management from government agencies to farmers, which is encouraging. This is a time-consuming task requiring decades. The American West generally required three decades, or more, to develop effective irrigation institutions.

The necessity for doubling the water productivity of irrigated agriculture over the next three decades is strongly dependent upon enhancing the social impacts by having a clearly defined water rights system in each irrigated region, as well as sustainable farmers organizations as measured by: 1) equitable water distribution throughout the irrigation system; and 2) farmers feeling free to report offenders (such as stealing water) and their organization is capable of applying sanctions.[11]

REFERENCES

1. Fukuda, H. *Irrigation in the World—Comparative Developments*; University of Tokyo Press: Tokyo, 1976.
2. United Nations Food and Agriculture Organization, *1996 Production Yearbook*; FAO: Rome, 1997; Vol. 50.
3. Postel, S. *Pillar of Sand—Can the Irrigation Miracle Last?* W. W. Norton and Company: New York, 1999.
4. Wittfogel, K.A. *Oriental Despotism: A Comparative Study of Total Power*; Yale University Press: New Haven, CT, 1957.
5. Hassan, F.A. The Dynamics of a Riverine Civilization: A Geoarchaeological Perspective on the Nile River Valley, Egypt. World Archaeol. **1997**, *29* (1).
6. Pradhan, P. *Patterns of Irrigation Organization in Nepal: A Comparative Study of 21 Farmer-Managed Irrigation Systems*; Nepal No. 1 Country Paper, International Irrigation Management Institute: Colombo, Sri Lanka, 1989.
7. Martin, E.D.; Yoder, R. *Institutions for Irrigation Management in Farmer-Managed Systems: Examples from the Hills of Nepal*; Research Paper No. 5, International Irrigation Management Institute: Colombo, Sri Lanka, 1987.
8. Surarerks, V. *Historical Development and Management of Irrigation Systems in Northern Thailand*; Chareon Printing Ltd.: Bangkok, Thailand, 1986.
9. Skogerboe, G.V.; Merkley, G.P. *Irrigation Maintenance and Operations Learning Process*; Water Resources Publications LLC: Highlands Ranch, CO, 1996.
10. Chambers, R.; Pacey, A.; Thrupp, L.A., Eds. *Farmers First: Farmer Innovation and Agricultural Research*; Intermediate Technology Publications: London, 1989.
11. Skogerboe, G.V.; Bandaragoda, D.J. *Towards Environmentally Sustainable Agriculture in the Indus Basin Irrigation System*; Pakistan Report No. R-77, International Irrigation Management Institute: Lahore, 1998.

Irrigated Water, Market Role in Reallocating

Bonnie G. Colby
University of Arizona, Tucson, Arizona, U.S.A.

INTRODUCTION

Market transactions are an important strategy for responding to water scarcity and the conflicts among water users, communities, and governments that can be stimulated by water scarcity. This article outlines the policy and economic issues raised by water markets, drawing on several decades of experience with marketing water in the western United States.

WHY HAVE MARKETS DEVELOPED?

Market acquisitions of irrigation water rights are increasingly common worldwide in regions where existing water supplies are fully appropriated and development of new supplies is costly. Those needing additional water must bid supplies away from current water users, primarily irrigators. In the American West, urban growth, environmental disputes, and Native American water claims, all create incentives for acquisition of agricultural water supplies and a few active regional markets have developed. While market acquisitions of additional water supplies often are essential to economic development, they also are the subjects of controversy and complex regulatory systems exist to govern market transactions in water.

A water market consists of the interaction of individuals and organizations which buy, sell and lease water rights, use of water supplies, and access to water-related infrastructure (canals, pumps, reservoirs). The degree of market activity varies among and within the western United States in terms of numbers of buyers and sellers, frequency of transactions and prices. Only a few areas have well-developed water markets with many transactions occurring every year. In other areas, sales of water rights largely involve water exchanges among neighboring farmers, transactions occur sporadically and price information is difficult to obtain. The Southwest generally is perceived to be the most active region of the United States, with respect to market activity, although drought in the Pacific Northwest is stimulating transactions there.[1]

In the western United States throughout the 1900s, water development projects diverted vast amounts of water from streams in order to irrigate crops. Water quality, recreation, and wildlife benefits associated with water left instream were largely unacknowledged, as were Native American claims to water. Irrigation districts, farmers, ranchers, and towns were accorded property rights in the resource.

Water supplies are renewed by nature in a stochastic and seasonal manner so that policymakers and water users cannot predict river flows far in advance. This uncertainty has prompted investment in infrastructure to store and convey surface water and to recharge groundwater so that supplies are available in a more predictable manner. Public and private expenditures to reduce variability in water supplies have been immense. Federal subsidies for irrigation projects in the western United States have covered approximately eighty percent of the capital costs of providing irrigation water to farm lands receiving water from federal projects.[2,3]

The West's economic transition from ranching, irrigated farming, and mining to urban growth, services, tourism, and industry has brought strong pressure to transfer water out of agriculture. Agriculture still accounts for 85%–95% of water use in most western states, and the cost of reducing irrigated acreage so that water can be available for other uses generally is far less than the cost of developing new water supplies. Western U.S. cities pioneered water marketing by purchasing irrigated land, sometimes entire irrigation districts, to acquire water rights for urban development.[4] While urban growth still is the driving force behind water markets, water transfers to support wildlife, fisheries, and recreation have become more common.[5] Transfers have become more complex and innovative in order to respond to drought, and to environmental and community concerns.

PUBLIC GOODS AND EXTERNALITIES

The term "public good" refers to resources characterized by nonexcludability, meaning it is difficult or impossible to exclude those who do not pay from enjoying the benefits of the resource. Water for recreation and wildlife habitat provides public benefits for which beneficiaries cannot readily be charged a user fee. Streamflows also provide public good benefits through dilution and water quality enhancement. Many individuals who benefit from streams and wetlands may be "free riders," enjoying these resources but making no payments—because payments

Encyclopedia of Water Science
DOI: 10.1081/E-EWS 120010119

are not required. Due to nonexcludability and free rider tendencies, market transactions alone are unlikely to ensure that adequate flows remain in streams to preserve habitat, water quality, and recreational opportunities. Therefore, public agencies sometimes assume the task of protecting streamflows.

Water transfers can generate externalities, including reduced water supplies for other water right holders, diminished economic activity in areas from which water is taken, lower river flows and degradation of water quality, fish and wildlife habitat, and recreation. While water transfers create positive externalities in the area to which water is being moved, it is the negative impacts that create controversy and pressure to carefully regulate transactions.

Western U.S. state laws specifically exclude some parties who may experience significant externalities from formally objecting to water transfer approval. In general, only water right holders can force their concerns to be accounted for. Recreationists and environmental advocates typically have little bargaining power in the regulatory process. Broader access to property rights in water and to the transfer approval process can allow a wider array of externalities to be considered.

MARKETS AND LITIGATION: COMPLEMENTARY FORCES

Voluntary transfers of water are not the only mechanism used to move water out of agriculture. Complex legal proceedings, termed adjudications, are taking place in many areas to quantify and prioritize the competing claims of Native Americans, wilderness areas, cities, and farms. Litigation based on the Endangered Species Act, the Clean Water Act, federal reserved rights and the public-trust doctrine has successfully forced reallocation of water to enhance streamflows for recreation, fish, wildlife, and water quality. Voluntary and involuntary pressures for reallocation often work in a complementary manner. There is no incentive quite so effective in stimulating voluntary transfers as the looming threat of a protracted and costly court battle. The threat of judicial and administrative reallocations has provided impetus for numerous voluntary reallocations among parties embroiled in conflicts over water.[6]

DEVELOPING COST EFFECTIVE POLICIES

The key challenge in developing policies to govern water markets is to utilize the flexibility that markets offer, while protecting third parties and public interests that can be impaired by water transfers. The complex nature of water rights and the changing social values associated with water make instantaneous, faceless and standardized transactions in water improbable and undesirable. Nevertheless, market

incentives should play a significant role in water allocation; to move water to uses where it generates higher economic returns and to give water users incentives for efficient water use. A "command and control" bureaucratic allocation system is undesirable due to its inflexibility as new demands arise and water values change. While government policies must play a primary role in evaluating proposed water transfers to prevent uncompensated third party impacts, bureaucracies should not dictate how much water must be used by whom and for what purpose.

Every western U.S. state imposes conditions on water transfers and there is no "free market." Market transactions sometimes resemble complex diplomatic negotiations rather than commodity exchanges. Regulatory policies generate uncertainties and costs for transferors and these costs sometimes are perceived as unnecessary impositions on the market. However, public policies should not necessarily seek to minimize the cost of reallocating water because appropriately structured transaction costs may facilitate efficient reallocation, by giving transacting parties an incentive to account for social costs of transfers.[7]

Transaction costs are the costs of making a market system work. In western U.S. water markets, parties incur transaction costs in searching for water supplies, contacting willing buyers and sellers, ascertaining the characteristics of water rights, negotiating price, and obtaining legal approval for the proposed change in water use. This latter category of transactions costs can include attorneys' fees, engineering and hydrologic studies, court costs, and fees paid to state agencies.

Transaction costs incurred to comply with regulatory policies reflect the substantial and multiple economic benefits associated with water in various uses, benefits which can be impaired by a transfer. Transactions costs are an important issue in western water reallocation. If the costs of implementing a water transfer become too high, many beneficial transfers will not take place and water supplies will remain locked into suboptimal use patterns. On the other hand, the ability to impose transactions costs on those proposing to transfer water represents bargaining power in the water allocation process. Some transaction costs are necessary, justified by the need to better account for externalities and public goods. Transaction costs also reflect the absence of "free" information and the need for hydrologic, legal, and economic data to address externalities in an efficient manner (see Ref. [8], for a detailed discussion of balancing transactions costs and consideration of third-party impacts.)

CONCLUSION

In summary, market transactions are an essential response to water scarcity. Without the flexibility provided through voluntary transfers, water supplies would remain locked into outdated patterns of use. Markets allow water to move permanently out of agriculture and to be leased to alleviate temporary scarcities (as during drought). Water scarcity creates tensions worldwide and voluntary transactions are one important strategy for addressing such conflicts. However, to provide flexibility and increased economic returns from regional water supplies, markets must be governed by policies that carefully weigh the advantages of a proposed transfer against externalities, impairment of public goods, and the concerns of affected communities and governments.

REFERENCES

1. Smith, R., (Ed.) *Water Strategist*; StratEcon: Claremont, California, 2001.
2. Wahl, R. *Markets for Federal Water*; Resources for the Future: Washington, D.C., 1989.
3. Anderson, T.L.; Snyder, P. *Water Markets: Priming the Invisible Pump*; Cato Institute: Washington, D.C., 1997.
4. Saliba, B.C.; Bush, D.B. *Water Marketing in Theory and Practice: Market Transfers Water Values and Public Policy*; Westview Press: Boulder, 1987.
5. Colby, B.G. Benefits, Costs and Water Acquisition Strategies: Economic Considerations in Instream Flow Protection. In *Instream Flow Law and Policy*; MacDonnell, L., Ed.; University of Colorado Press: Boulder, 1993, Chap. 6.
6. Colby, B.G.; McGinnis, M.; Rait, K. Mitigating Environmental Externalities Through Voluntary and Involuntary Water Allocation: Nevada's Truckee-Carson River Basin. Nat. Res. J. **1991**, *31* (4), 757–784.
7. Colby, B.G. Transaction Costs and Efficiency in Western Water Allocation. Am. J. Agric. Econ. **1990**, *72*, 1184–1192.
8. National Research Council, *Water Transfers in the West: Efficiency, Equity and the Environment*; National Academy Press: Washington, D.C., 1992.

Irrigated Water, Polymer Application in

William J. Orts
United States Department of Agriculture (USDA), Albany, California, U.S.A.

Robert E. Sojka
United States Department of Agriculture (USDA), Kimberly, Idaho, U.S.A.

INTRODUCTION

In the past decade, water-soluble polyacrylamide (PAM) was identified as an environmentally safe and highly effective erosion preventing and infiltration enhancing polymer when applied in furrow irrigation water at $1\,mg\,L^{-1}$–$10\,mg\,L^{-1}$, i.e., 1 ppm–10 ppm.[1−9] Various polymers and biopolymers have long been recognized as viable soil conditioners because they stabilize soil surface structure and pore continuity. The new strategy of adding the conditioner, high molecular weight anionic PAM, *to irrigation water* in the first several hours of irrigation implies a significant costs savings over traditional application methods, in which hundreds of kilograms per hectare of soil additives are tilled into the entire (15 cm deep) soil surface layer. By adding PAM to the irrigation water, soil structure is improved in the important 1–5 mm thick layer at the soil/water interface of the 25%–30% of field surface contacted by flowing water.[7]

In 1995, the U.S. Natural Resource Conservation Service (NRCS) published a PAM-use conservation practice standard for PAM-use in irrigation water.[10] A 3-year study[2] applying these standards showed that PAM at dosage rates of $1\,kg\,ha^{-1}$–$2\,kg\,ha^{-1}$ per irrigation eliminated 94% (80%–99% range) of sediment loss in furrow irrigation runoff, while increasing infiltration 15%–50%. Seasonal application rates using the NRCS standard typically total $3\,kg\,ha^{-1}$–$5\,kg\,ha^{-1}$.

As PAM-use is one of the most effective and economical technologies for reducing soil-runoff, it has branched into stabilization of construction sites and road cuts, with formal statewide application standards set in Wisconsin and several southern states. Recent studies with biopolymers such as charged polysaccharides,[11−14] whey,[15] and industrial cellulose derivatives[11,14] introduce potential biopolymer alternatives to PAM.

POLYACRYLAMIDE

The term polyacrylamide and acronym "PAM" are chemistry jargon for a broad class of acrylamide-based polymers varying in chain length, charge type, charge concentration, and the number and types of side-group substitutions.[16−20] Typically, PAM for erosion control is a charged copolymer with one in five acrylamide chain segments replaced by an acrylic acid entity (Fig. 1), which generally exhibits a negative charge in water. Molecular weights of PAM used for irrigated agriculture range from $12\,million\,g\,mol^{-1}$ to $15\,million\,g\,mol^{-1}$ (over 150,000 monomer units per chain). As a result of its structure, PAM attracts soil particles via coulombic and Van der Waals forces.[11,17,21,22] Ionic bridging creates large stable aggregates of PAM and soil, in which charged entities on both the polymer and multiple soil particles are thought to interact with the aid of calcium counterions.[11,22−24] Chain bridging further stabilizes aggregates, whereby the long polymer chain spans between separate soil particles. Despite their large size, PAM copolymers used for erosion control are formulated to dissolve in water, although this sometimes requires vigorous agitation.

PAM Erosion Control

Lentz and Sojka[2] reported a 94% reduction in runoff sediment loss over 3 yr using the NRCS application standard.[10] The 1995 NRCS standard calls for dissolving 10 ppm (or $10\,g\,m^{-3}$) PAM in furrow inflow water as it first crosses a field—typically the first 10%–25% of an irrigation duration—then halting PAM dosing when runoff begins. Under many circumstances, applying PAM continuously at 1 ppm–2 ppm for the full irrigation cycle can be equally effective, although continuous application at 0.25 ppm PAM was a third less effective.[25−27]

PAM and Infiltration

The infiltration rate of PAM-treated furrows on medium to fine textured soil is usually higher than untreated furrows—typically 15% higher than for untreated water on silt loam soils and up to 50% higher on clays.[28] Bjorneberg[29] reported that in tube diameters > 10 mm, the PAM–water viscosity did not rise sharply until the PAM concentration in the water was > 400 ppm.

Encyclopedia of Water Science
DOI: 10.1081/E-EWS 120010141

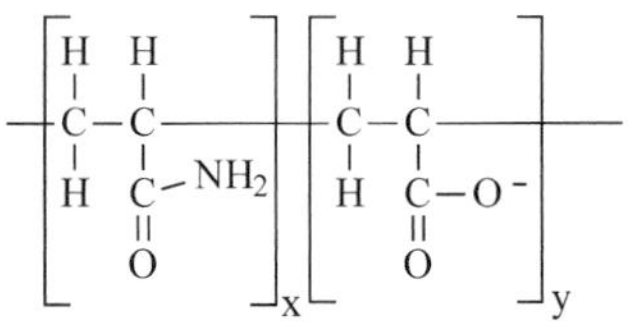

Fig. 1 PAM: Poly(acrylamide-co-acrylic acid).

However, in small soil pores, "apparent viscosity" increases significantly, even at the low PAM concentrations used for erosion control.[30] Most likely, PAM infiltration effects are a balance between prevention of surface sealing and apparent viscosity increases in soil pores.[30–34] In medium to fine textured soils, maintenance of pore continuity via aggregate stabilization is more important. In coarse textured soils, where PAM achieves little pore continuity enhancement, infiltration effects are nil or even slightly negative, particularly above 20 ppm.[28]

Because PAM prevents erosion of furrow bottoms and sealing of the wetted perimeter, water moves about 25% further laterally in silt loams compared to nontreated furrows.[1,2] This can be a significant water conserving effect for early irrigations. Farmers should take advantage of PAMs erosion prevention to improve field infiltration uniformity by increasing inflow rates two to threefold (compared to normal). This reduces infiltration opportunity time differences between inflow and outflow ends of furrows.[28,35]

Sprinkler Application of PAM

Farmers and agronomists are showing interest in PAM for sprinkler irrigation.[5,6,36–40] PAM may prevent runoff/runon problems and ponding effects on stand establishment and irrigation uniformity. Polyacrylamide sprinkler application rates of $2\,kg\,ha^{-1}$–$4\,kg\,ha^{-1}$ reduced runoff 70% and soil loss 75% compared with controls.[36] However, the effectiveness of sprinkler-applied PAM is more variable than for furrow irrigation because of application strategies and system variables that affect water drop energy, the rate of water and PAM delivery, and possible application timing scenarios. Multiple groups[6,36–40] report improved aggregate stability from sprinkler-applied PAM, leading to decreased runoff and erosion. Flanagan et al.[5,6] increased sprinkler infiltration with 10 ppm PAM, which they attributed to reduced surface sealing. Polyacrylamide effects under sprinkler irrigation have been more transitory, less predictable and have usually needed higher seasonal field application totals for efficacy. However, farmers with sprinkler infiltration uniformity problems (runoff or runon), e.g., with center pivots on steep or variable slopes, have begun to use PAM. Testimonials claim that PAM-use improves

stands because of reduced ponding, crusting and damping off (a plant seedling disease complex).

ENVIRONMENTAL IMPACT OF PAM

The overriding environmental impact of PAM is reduced erosion-induced sediment runoff,[1,2] with corresponding reductions of entrained chemical residue reaching riparian waterways.[41–43] For example, PAM prevents yearly topsoil runoff of up to $6.4\,tn\,acre^{-1}$[2] and at least three times that as on-field erosion.[34] Since toxic pesticides and herbicides are transported via soil sediment to open water and then eventually into the air there is an increasing need to prevent soil-runoff. Recently, PAM was shown to sequester biological and chemical contaminants of runoff, providing significant potential for reduced spread of phytopathogens, animal coliforms, and other organisms of public health concern.[44,45]

The main environmental concerns in PAM-use revolve around polymer purity,[46,47] and issues related to biodegradation/accumulation;[48–53] i.e., since PAM degrades slowly, the long-term, unknown effects on organisms must be considered. Biological degradation of PAM incorporated into soil is about 10% per year.[50] However, low application rates and shallow surface application is thought to accelerate degradation via various pathways, including deamination, shear-induced chain scission, and UV photosensitive chain scission.[50–53] Even at 10% annual degradation, PAM accumulation is insignificant at these application rates. Sojka and Lentz[26] showed that only 1%–3% of applied PAM leaves fields in runoff and that this is quickly adsorbed by entrained sediment or ditch surfaces. Barvenik[16,50] noted that anionic PAM is safe for aquatic organisms at surprisingly high concentrations, with $LC_{50} > 50$ times the inflow dosage rates. Water impurities further buffer environmental effects by quickly deactivating dissolved PAM.

Care must be taken by PAM supplies to ensure polymer purity, since the acrylamide monomer (AMD) used to synthesize PAM is a neurotoxin. The EPA recently reviewed the use of PAM with USDA and PAM industry scientists, and concluded that the AMD concentrations of $< 0.05\%$ found in products for use during furrow irrigation are acceptable, with minimal amounts of monomer released into the environment.[26,53] The first step in the biodegradation of PAM is early removal of the amine group from the polymer backbone,[46,47,54–56] with reversion to AMD thermodynamically unfavorable.[53] Although these environmental issues about PAM are raised, PAM is widely recognized as a safe, environmentally friendly, hygenically safe, and cost-effective flocculating agent. It has been used industrially for decades as a soil conditioner, in food processing, and in various water treatment processes.

BIOPOLYMER ALTERNATIVES TO PAM

PAMs successful use in irrigation water to reduce erosion and improve infiltration has raised questions of whether it is the "best" polymer for the application. There is increasing anecdotal and scientific evidence[57,58] that PAM efficacy varies with different soils and waters. Variations include sodicity, texture, bulk density, and surface charge-related properties. It would be beneficial to have a wide array of polymers with potentially different soil-stabilizing mechanisms, applicable to different soil types.

Of course, any reduction in price would also benefit farmers. The market price of PAM, i.e., several dollars per kilogram, is high relative to many commodity polymers, such as polyethylene, polypropylene, and polystyrene. Treatment for 1 year can cost up to $25 per hectare, which is still cost competetive with conventional erosion abating technologies such as straw bales, settling ponds, and underground or drip irrigation systems.

The increasing market pull of organic farming techniques is a strong reason to explore alternatives to PAM. Polyacrylamide cannot be used during organic farmering because it is a synthetic polymer derived from nonrenewable resources. Natural polymers, which often degrade via relatively benign routes, may be more suitable. Biopolymer alternatives to PAM would likely have marketing advantages due to public *perception* of being safer.

Cellulose and starch xanthates were among the first industrial biopolymers shown to stabilize soil.[11,14] Menefee and Hautala[14] reduced sediment runoff by nearly 98% by surface treating 20° sloped plots with cellulose xanthate solution (0.4%). Orts et al.[11] added cellulose xanthate to the irrigation water of lab-scale mini-furrows, and reduced erosion 80% when xanthate was applied at concentrations of 80 ppm or greater, which is well above the standard PAM application rate of 10 ppm and even 5 ppm.

Chitosan, the biopolymer derived from crab and shrimp shells, was shown to reduce erosion losses as effectively as PAM in lab-scale mini-furrow at concentrations of 20 ppm.[22] With such favorable lab test results, chitosan was further tested in a series of field tests at the USDA Northwest Irrigation and Soil Research Lab, Kimberly, Idaho.[22] In the field tests, chitosan reduced erosion-induced soil losses by, at best, half of the control, but far less effectively than PAM. Such poor comparative results, however, do not mean that chitosan had no effect on the irrigation. Observations of the furrows treated with chitosan revealed remarkable results in the first ~ 20 m of the furrow. In fact, chitosan acted as such an effective flocculating agent that it removed fine sediments, and even algae from the irrigation water. Perhaps chitosan binds so readily with sediment that it flocculates out of solution near the top of the furrow. The major drawback of chitosan is its market cost of over $3 kg^{-1}, roughly twice the price of PAM.

CONCLUSION

U.S. agricultural PAM-use for erosion control and infiltration improvement reached 400,000 ha in 1999,[59] with U.S. and worldwide markets expected to grow as farmers recognize PAMs efficacy, and as government-mandated water quality legislation is realized. The success of PAM in agriculture opens the possibility to explore other Ag-related uses for PAM,[45] as well as the potential to find alternatives to PAM. For example, modified polysaccharides[11-14] and cheese whey, the protein concentrate from cheese processing, are particularly interesting natural soil stabilizers, and could be used to treat irrigation water.

REFERENCES

1. Lentz, R.D.; Shainberg, I.; Sojka, R.E.; Carter, D.L. Preventing Irrigation Furrow Erosion with Small Applications of Polymers. Soil Sci. Soc. Am. J. **1992**, *56*, 1926–1932.
2. Lentz, R.D.; Sojka, R.E. Field Results Using Polyacrylamide to Manage Furrow Erosion and Infiltration. Soil Sci. **1994**, *158*, 274–282.
3. Paganyas, K.P. Results of the Use of Series "K" Compounds for the Control of Irrigational Soil Erosion. Sov. Soil Sci. **1975**, *5*, 591–598.
4. Ben-Hur, M.; Keren, R. Polymer Effects on Water Infiltration and Soil Aggregation. Soil Sci. Soc. Am. J. **1997**, *61*, 565–570.
5. Flanagan, D.C.; Norton, L.D.; Shainberg, I. Effects of Water Chemistry and Soil Amendments on a Silt Loam Soil—Part 1: Infiltration and Runoff. Trans. ASAE **1997**, *40*, 1549–1554.
6. Flanagan, D.C.; Norton, L.D.; Shainberg, I. Effects of Water Chemistry and Soil Amendments on a Silt Loam Soil—Part 1: Infiltration and Runoff. Trans. ASAE **1997**, *40*, 1555–1561.
7. Sojka, R.E.; Lentz, R.D. Time for Yet Another Look at Soil Conditioners. Soil Sci. **1994**, *158*, 233–234.
8. McElhiney, M.; Osterli, P. An Integrated Approach for Water Quality: The PAM Connection. In *Proceedings: Managing Irrigation-Induced Erosion and Infiltration with Polyacrylamide May 6, 7, and 8, 1996*; University of Idaho

Misc. Pub., 101-96, Sojka, R.E., Lentz, R.D., Eds.; College of Southern Idaho: Twin Falls, ID, 1996; 27–30.

9. Wallace, A. Use of Water-Soluble Polyacrylamide for Control of Furrow Irrigation-Induced Soil Erosion. In *Handbook of Soil Conditioners, Substances That Enhance the Physical Properties of Soil*; Wallace, A., Terry, R.E., Eds.; Marcel Dekker, Inc.: New York, 1997; 42–54.

10. Anonymous *Interim Conservation Practice Standard—Irrigation Erosion Control (Polyacrylamide)*,WNTC I-201; USDA-NRCS West Nat'l Tech. Center: Portland, OR, 1995.

11. Orts, W.J.; Sojka, R.E.; Glenn, G.M. Biopolymer Additives to Reduce Erosion-Induced Soil Losses During Irrigation. Ind. Crops Prod. **2000**, *11*, 19–29.

12. Singh, R.P.; Karmakar, G.P.; Rath, S.K.; Karmakar, N.C.; Pandey, S.R.; Tripathy, T.; Panda, J.; Kannan, K.; Jain, S.K.; Lan, N.T. Biodegradable Drag Reducing Agents and Flocculants Based on Polysaccharides: Materials and Applications. Polym. Eng. Sci. **2000**, *40*, 46–60.

13. Singh, R.P.; Tripathy, T.; Karmakar, G.P.; Rath, S.K.; Karmakar, N.C.; Pandey, S.R.; Kannan, K.; Jain, S.K.; Lan, N.T. Novel Biodegradable Flocculants Based on Polysaccharides. Curr. Sci. **2000**, *78*, 798–803.

14. Menefee, E.; Hautala, E. Application of Xanthate Solutions to Stabilize Soil Structure. Nature **1975**, *275*, 530–532.

15. Robbins, C.W.; Lehrsch, G.A. Cheese Whey as a Soil Conditioner. In *Handbook of Soil Conditioners, Substances That Enhance the Physical Properties of Soil*; Wallace, A., Terry, R.E., Eds.; Marcel Dekker, Inc.: New York, 1997.

16. Barvenik, F.W. Polyacrylamide Characteristics Related to Soil Applications. Soil Sci. **1994**, *158*, 235–243.

17. Bicerano, J. Predicting Key Polymer Properties to Reduce Erosion in Irrigated Soil. Soil Sci. **1994**, *158*, 255–266.

18. Lentz, R.D.; Sojka, R.E.; Carter, D.L. Influence of Polymer Charge Type and Density on Polyacrylamide Ameliorated Irrigated Furrow Erosion. Proceedings of the 24th Annual International Erosion Control Association Conference, Indianapolis, IN, February 23–26, 1993; 159–168.

19. Lentz, R.D.; Sojka, R.E. PAM Conformation Effects on Furrow Erosion Mitigation Efficacy. In *Proceedings: Managing Irrigation-Induced Erosion and Infiltration with Polyacrylamide May 6, 7, and 8, 1996*; University of Idaho Misc. Pub., 101-96, Sojka, R.E., Lentz, R.D., Eds.; College of Southern Idaho: Twin Falls, ID, 1996; 71–77.

20. Lentz, R.D.; Sojka, R.E.; Ross, C.W. Polyacrylamide Molecular Weight and Charge Effects on Sediment Loss and Infiltration in Treated Irrigation Furrows. Int. J. Sediment Res. **2000**, *15*, 17–30.

21. Levy, G.J.; Miller, W.P. Polyacrylamide Adsorption and Aggregate Stability. Soil Tillage Res. **1999**, *51*, 121–128.

22. Orts, W.J.; Sojka, R.E.; Glenn, G.M.; Gross, R.A. Preventing Soil Erosion with Polymer Additives. Polym. News **1999**, *24*, 406–413.

23. Wallace, A.; Wallace, G.A. Need for Solution or Exchangeable Calcium And/or Critical EC Level for Flocculation of Clay by Polyacrylamides. In *Proceedings: Managing Irrigation-Induced Erosion and Infiltration with Polyacrylamide May 6, 7, and 8, 1996*; University of Idaho

Misc. Pub., 101-96, Sojka, R.E., Lentz, R.D., Eds.; College of Southern Idaho: Twin Falls, ID, 1996; 59–63.

24. Wallace, B.H.; Reichert, J.M.; Eltz, L.F.; Norton, L.D. Conserving Topsoil in Southern Brazil with Polyacrylamide and Gypsum. Proceedings of the International Symposium: Soil Erosion Research for the 21st Century, Honolulu, Hawaii, January 3–5, 2001; ASAE publication 701P0007, 183–187.

25. Lentz, R.D. Irrigation (Agriculture): Using Polyacrylamide to Control Furrow Irrigation-Induced Erosion. In *Yearbook of Science and Technology*; Parker, S.P., Ed.; McGraw Hill, Inc.: New York, 1996; 162–165.

26. Sojka, R.E.; Lentz, R.D. Polyacrylamide for Furrow-Irrigation Erosion Control. Irrig. J. **1996**, *64*, 8–11.

27. Lentz, R.D.; Sojka, R.E. *Applying Polymers to Irrigation Water: Evaluating Strategies for Furrow Erosion Control*, ASAE Paper No. 992014; ASAE: St. Joseph, MI, 1999.

28. Sojka, R.E.; Lentz, R.D.; Trout, T.J.; Ross, C.W.; Bjorneberg, D.L.; Aase, J.K. Polyacrylamide Effects on Infiltration in Irrigated Agriculture. J. Soil Water Conserv. **1998**, *53*, 325–331.

29. Bjorneberg, D.L. Temperature, Concentration, and Pumping Effects on PAM Viscosity. Trans. ASAE **1998**, *41*, 1651–1655.

30. Malik, M.; Letey, J. Pore-Size-Dependent Apparent Viscosity for Organic Solutes in Saturated Porous Media. Soil Sci. Soc. Am. J. **1992**, *56*, 1032–1035.

31. Green, V.S.; Stott, D.E.; Norton, L.D.; Graveel, J.G. Polyacrylamide Molecular Weight and Charge Effects on Infiltration Under Simulated Rainfall. Soil Sci. Soc. Am. J. **2000**, *64*, 1786–1791.

32. Nadler, A.; Perfect, E.; Kay, B.D. Effect of Polyacrylamide Application on the Stability of Dry and Wet Aggregates. Soil Sci. Soc. Am. J. **1996**, *60*, 555–561.

33. Shainberg, I.; Levy, G.J. Organic Polymers and Soil Sealing in Cultivated Soils. Soil Sci. **1994**, *158*, 267–273.

34. Trout, T.J.; Sojka, R.E.; Lentz, R.D. Polyacrylamide Effect on Furrow Erosion and Infiltration. Trans. ASAE **1995**, *38* (3), 761–765.

35. Sojka, R.E.; Lentz, R.D. Reducing Furrow Irrigation Erosion with Polyacrylamide (PAM). J. Prod. Agric. **1997**, *10*, 1–2, 47–52.

36. Aase, J.K.; Bjorneberg, D.L.; Sojka, R.E. Sprinkler Irrigation Runoff and Erosion Control with Polyacrylamide—Laboratory Tests. Soil Sci. Soc. Am. J. **1998**, *62*, 1681–1687.

37. Ben-Hur, M. Runoff, Erosion, and Polymer Application in Moving-Sprinkler Irrigation. Soil Sci. **1994**, *158*, 283–290.

38. Bjorneberg, D.L.; Aase, J.K.; Sojka, R.E. Runoff and Erosion Control with Polyacrylamide Applied Through Sprinkler Irrigation. ASAE Meeting Paper No. 982103, 1998.

39. Bjorneberg, D.L.; Aase, J.K.; Sojka, R.E. Sprinkler Irrigation Runoff and Erosion Control with Polyacrylamide. Proceedings of the 4th Decennial Symposium: National Irrigation Symposium, Phoenix, Arizona, November 14–16, 2000; ASAE publication 701P0004, 513–522.

40. Sanderson, A.; Hewitt, J.; Huddleston, E.W.; Ross, J.B. Polymer and Invert Emulsifying Oil Effects upon Droplet Size Spectra of Sprays. J. Environ. Sci. Health B-Pestic. **1994**, *29* (4), 815–829.

41. Agassi, M.; Letey, J.; Farmer, W.J.; Clark, P. Soil Erosion Contribution to Pesticide Transport by Furrow Irrigation. J. Environ. Qual. **1995**, *24*, 892–895.

42. Bahr, G.L.; Steiber, T.D. Reduction of Nutrient and Pesticide Losses Through the Application of Polyacrylamide in Surface Irrigated Crops. In *Proceedings: Managing Irrigation-Induced Erosion and Infiltration with Polyacrylamide May 6, 7, and 8, 1996*; University of Idaho Misc. Pub., 101-96, Sojka, R.E., Lentz, R.D., Eds.; College of Southern Idaho: Twin Falls, ID, 1996; 41–48.

43. Singh, G.; Letey, J.; Hanson, P.; Osterli, P.; Spencer, W.F. Soil Erosion and Pesticide Transport from an Irrigated Field. J. Environ. Sci. Health **1996**, B*31* (1), 25–41.

44. Sojka, R.E.; Entry, J.A. Influence of Polyacrylamide Application to Soil on Movement of Microorganisms in Runoff Water. Environ. Pollut. **1999**, *108*, 405–412.

45. Entry, J.A.; Sojka, R.E. Polyacrylamide Compounds Remove Coliform Bacteria from Animal Wastewater. Proceedings of the 2nd International Symposium on Preferential Flow: Water Movement and Chemical Transport in the Environment, Honolulu, Hawaii, January 3–5, 2001, ASAE publication 701P0006; 277–280.

46. Shanker, R.; Ramakrishna, C.; Seth, P.K. Microbial Degradation of Acrylamide Monomer. Arch. Microbiol. **1990**, *154*, 192–198.

47. Shanker, R.; Seth, P.K. Toxic Effects of Acrylamide in a Freshwater Fish, *Heteropneustes fossilis*. Bull. Environ. Contam. Toxicol. **1986**, *37*, 274–280.

48. Smith, E.A.; Prues, S.L.; Oehme, F.W. Environmental Degradation of Polyacrylamides 1. Effects of Artificial Environmental Conditions: Temperature, Light, and pH. Ecotoxicol. Environ. Saf. **1996**, *35*, 121–135.

49. Smith, E.A.; Prues, S.L.; Oehme, F.W. Environmental Degradation of Polyacrylamides. 2. Effects of Environmental (Outdoor) Exposure. Ecotoxicol. Environ. Saf. **1997**, *37*, 76–91.

50. Barvenik, F.W.; Sojka, R.E.; Lentz, R.D.; Andrawes, F.F.; Messner, L.S. Fate of Acrylamide Monomer Following Application of Polyacrylamide to Cropland. In *Proceedings: Managing Irrigation-Induced Erosion and Infiltration with Polyacrylamide May 6, 7, and 8, 1996*; University of Idaho Misc. Pub., 101-96, Sojka, R.E., Lentz, R.D., Eds.; College of Southern Idaho: Twin Falls, ID, 1996; 103–110.

51. Hamilton, J.K.; Reinert, D.H.; Freeman, M.B. Aquatic Risk Assessment of Polymers. Environ. Sci. Technol. **1994**, *28* (4), 187A–192A.

52. Lande, S.S.; Bosch, S.J.; Howard, P.H. Degradation and Leaching of Acrylamide in Soil. J. Environ. Qual. **1979**, *8*, 133–137.

53. Bologna, L.S.; Andrawes, F.F.; Barvenik, F.W.; Lentz, R.D.; Sojka, R.E. Analysis of Residual Acrylamide in Field Crops. J. Chromatogr. Sci. **1999**, *37*, 240–244.

54. Kay-Shoemake, J.L.; Watwood, M.E.; Lentz, R.D.; Sojka, R.E. Polyacrylamide as an Organic Nitrogen Source for Soil Microorganisms with Potential Impact on Inorganic Soil Nitrogen in Agricultural Soil. Soil Biol. Biochem. **1998**, *30*, 1045–1052.

55. Kay-Shoemake, J.L.; Watwood, M.E.; Sojka, R.E.; Lentz, R.D. Polyacrylamide as a Substrate for Microbial Amidase. Soil Biol. Biochem. **1998**, *30* (13), 1647–1654.

56. Kay-Shoemake, J.L.; Watwood, M.E.; Sojka, R.E.; Lentz, R.D. Soil Amidase Activity in Polyacrylamide-Treated Soils and Potential Activity Toward Common Amidase-Containing Pesticides. Biol. Fertil. Soils **1998**, *31*, 183–186.

57. Ben-Hur, M.; Malik, M.; Letey, J.; Mingelgrin, U. Adsorption of Polymers on Clays as Affected by Clay Charge and Structure, Polymer Properties, and Water Quality. Soil Sci. **1992**, *153*, 349–356.

58. Teo, J.; Ray, C.; El-Swaify, S.A. Polymer Effect on Soil Erosion Reduction and Water Quality Improvement for Selected Tropical Soils. Proc. International Symp.: Soil Erosion Research for the 21st Century, Honolulu, Hawaii, January 3–5, 2001; ASAE Publication 701P0007, 42–45.

59. Sojka, R.E.; Lentz, R.D.; Shainberg, I.; Trout, T.J.; Ross, C.W.; Robbins, C.W.; Entry, J.A.; Aase, J.K.; Bjorneberg, D.L.; Orts, W.J.; Westermann, D.T.; Morishita, D.T.; Watwood, M.E.; Spofford, T.L.; Barvenik, F.W. Irrigating with Polyacrylamide (PAM)—Nine Years and a Million Acres of Experience. Proceedings Irrigation 2000 Symposium, Phoenix, AZ, November, ASAE, 2000.

Irrigation Design Steps and Elements

Gary A. Clark
Kansas State University, Manhattan, Kansas, U.S.A.

INTRODUCTION

In general, the primary objective of most irrigation systems is to provide water to a crop to meet the evapotranspiration demands in the absence of rainfall. While irrigation systems can be used for other purposes such as chemigation, cold protection, or heat stress relief, this discussion will focus on the design steps and elements associated with the objective of meeting crop water requirements.

An irrigation system typically includes a pump, various pipes, valves, and water emission or discharging devices such as sprinklers or drip emitters. Fig. 1 shows an example layout that could apply to a sprinkler irrigation or microirrigation system with a control head that shows many of the components that may or may not be used. The inclusion of filters or strainers and chemical injection systems will be dependent on the type of irrigation emission devices used, and characteristics of the water quality. In the design process, these components along with the other pipes and valves are sized, arranged, and connected together using a variety of fittings to create a working system that will transport water from a supply source to the water storage system (soil, potting media, etc.) for a crop in an efficient, timely, and cost-effective manner. Because many design scenarios are possible, the designer must be knowledgeable about the land or field characteristics, the cropping system, the water supply, the pipeline hydraulics, and the operational characteristics of different irrigation systems and associated components. While design texts and other references are available[1–8] and provide much greater detail and background on the design process, this discussion provides an overview of the elements of the irrigation system design process and recommended steps to follow. As an additional reference source, standards[9–13] have been developed to assist the designer with recommended practices and procedures for system design and evaluation.

INITIAL ASSESSMENT

The initial stages of the design process require some basic knowledge of the various elements that will influence the design and operation of the irrigation system. Thus, an initial assessment should be conducted to answer the following general questions:

1. What is the intended use and desired goal for the irrigation system? While many irrigation systems are used to meet the full, supplemental irrigation requirements of the crop, other systems are used to make reasonable use of limited water supplies. Another aspect to assess is whether the system is to be dedicated to a single field within a production season, or if the system needs to be portable.
2. Where is the water source located; what is the availability of the water source; and what is the quality of the water source?
3. What are the characteristics of the land area that is to be irrigated as well as climatic conditions of the geographic region?
4. What are the production system characteristics of the crop (or crops) that is (are) to be grown and irrigated, and what is the crop value?
5. Are irrigation supplies and services locally available?

With answers to the above questions, the designers use their engineering and general knowledge to synthesize the information into rough drafts of one or more design scenarios that can be presented to the client. During this initial phase, the designer and the client meet and discuss the proposed design scenarios identifying additional desired outcomes or system constraints that may be subsequently incorporated into a new design scenario. After an acceptable design scenario is identified, the more formal steps of the design process are conducted. The following sections discuss the elements of design and selection of components as influenced by the water source, field and cropping system characteristics, and the water supply system.

WATER SOURCES—QUANTITY AND QUALITY

The design process must include an assessment of how much water is available from the source for use (quantity and capacity or rate), and then to quantify how much water is needed or must be used for the desired goal. The total

Encyclopedia of Water Science
DOI: 10.1081/E-EWS 120010249

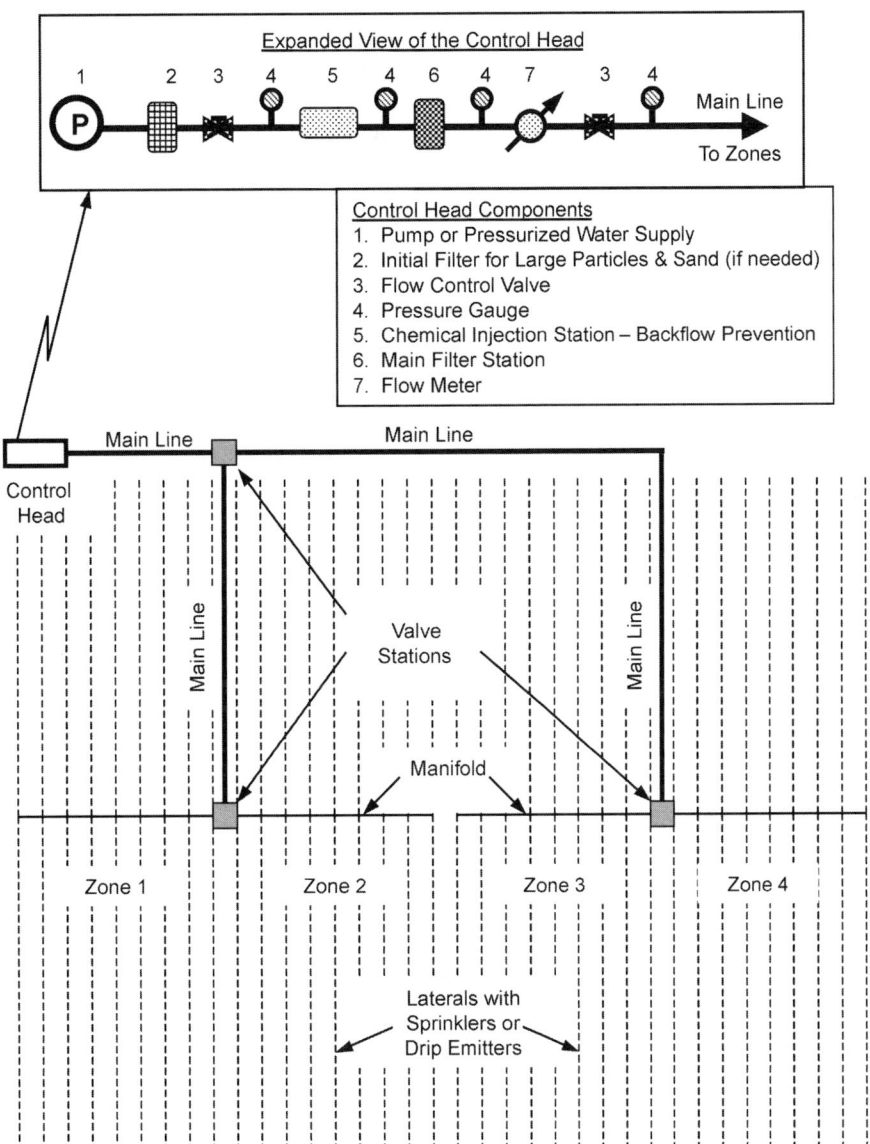

Fig. 1 General layout of a solid-set sprinkler irrigation system or microirrigation system showing the control head and associated components, main supply lines, manifold supply lines, and lateral lines that would contain sprinklers, microsprinklers, or drip emitters.

quantity of available water will depend on applicable laws or allocation procedures and the size and physical characteristics of the source. Water availability limitations, use restrictions, ownership, and uncertainty in water supply amounts will influence the total irrigated area, the irrigation scheduling decision process, and perhaps the choice of components and "permanency" of the system. For example, a limited water supply may be used to adequately irrigate a limited irrigated area or may be used to "deficit" irrigate a larger crop production area. While most irrigation systems are designed for long-term use (>10 yr) on a single site, some systems may have some "portability" included into the design in order to

accommodate uncertainty in a local water source or land-lease agreements. These situations are not necessarily the norm, but can provide some unique and challenging design scenarios.

The following equation represents a basic mass balance approach and is used to determine any one of the components given the values of the other three.

$$(Q_{\mathrm{sys}})(T_{\mathrm{c}}) = C(A)(I_{\mathrm{gr}}) \tag{1}$$

where Q_{sys} is either the flow rate of the irrigation system or the water system supply rate from the source (L/sec); T_{c}, the operational time (hr) of the system per cycle or period of time needed or desired to apply I_{gr}; A, the size of the area to be irrigated (ha); I_{gr}, the gross depth of irrigation water that is to be applied (mm) as a daily or seasonal amount and must correspond to T_c; and C, a constant of proportionality to adjust and properly cancel the units in the other four variables ($C = 2.78$ for these SI units).

The gross irrigation depth is related to the net irrigation depth, I_{net}, with the irrigation system efficiency as:

$$I_{gr} = \frac{I_{net}}{E_{sys}} \tag{2}$$

Irrigation system efficiency, E_{sys}, can range from as low as 20% to over 90% and characterizes water that is "lost" in the conveyance system (canals or pipes), the distribution system (sprinklers, emitters, orifices, etc.), and water that is lost in the field due to runoff and/or deep percolation below the root zone of the crop. The net irrigation depth, I_{net}, represents the amount of water that is needed for and directly useable by the crop. This may be expressed as the amount of water to refill the soil profile from a certain deficit level to the field capacity level, it may represent a daily peak or design evapotranspiration depth, or it may represent a seasonal (or specific time period) depth of water that is to be applied. For example, the last condition may represent the amount of water that needs to be applied from a lagoon to a cropped land area in a certain window of time, T_c. It may also represent the result of a seasonal water balance:

$$I_{net} = ET_c - P_e - ASW \tag{3}$$

where I_{net} is the seasonal net irrigation water requirement; ET_c, the seasonal crop evapotranspiration; P_e, the seasonal effective precipitation; and ASW, the available water in the soil profile at the beginning of the irrigation period that can be used by the crop during the irrigation period.

The quality of the water source must be assessed as to how physical, biological, and/or chemical constituents in the water may affect or interact with components of the delivery system (pump, pipes, valves, and emitters), the soil, and crop. Water treatment and amendment practices may need to be incorporated into the design to avoid clogging of certain irrigation components. The required level of treatment will depend on the quality of the water and the sensitivity of components to the various constituents in the water.

Water quality from both groundwater and surface water sources can range from excellent to very poor, and typical quality concerns include suspended solids, dissolved solids, and biological organisms. Poor well screening and/or well development problems can result in suspended sand, silt, or clay particles. Surface water sources may have suspended particles of silt and/or clay, aquatic plants, small fish, algae, larvae, or other organic debris. While these physical constituents can generally be controlled with proper filtration, chemical treatment may be necessary to neutralize related organic growths.

Dissolved solids such as calcium, iron, or other elements can precipitate under certain conditions and subsequently clog some microirrigation emitters. Biological growths include slimes (associated with iron and/or hydrogen sulfide), fungi, and algae. Such organic growths can grow within and clog pipelines, valves, and irrigation emission devices (sprinklers, drip emitters, etc.). Chemical treatment of the water is often necessary under these conditions in addition to filtration to prevent or "clean-up" these organic growths. Severe instances of several of the above water quality problems may require expensive remediation components and/or management practices to ensure proper and continual operation of the irrigation system. Because such conditions may result in a financially impractical design, or poor system performance or failure, a thorough assessment of the quality of the water source must be performed prior to completion of the final design.

Recycled water sources (municipal wastewater, live-stock wastewater lagoons, industry wastewater sources) should be thoroughly assessed for their physical, chemical, and biological constituents. Key concerns will include pH, salts, nitrogen, and phosphorus. Some "contaminated" water sources may contain heavy metals or organic compounds that may be of concern when applied to agricultural crops and fields. Water application and/or loading rates may need to be assessed with respect to the concentrations of certain key elements or compounds in the water. Allowable water application amounts may not be sufficient to meet peak or design crop water demands and supplemental "clean" water sources may be needed to augment the water supply.

FIELD AND CROPPING SYSTEM CHARACTERISTICS

The site for the planned irrigation systems needs to be assessed for dimensions, topography, physical features, soil characteristics, and climatic characteristics. The size and shape of the field must be measured and include lengths of boundaries, interior angles of adjoining sides, and any on-site, physical obstructions (trees, power poles, buildings, etc.). Obstructions should be identified as to whether they can be removed. It is also beneficial to identify on-site or nearby electrical power sources. Because land slope and surface conditions influence the type of irrigation systems that can be used and the design of the selected system, those elements should be characterized. A contour map can be very helpful for these purposes.

Soil characteristics should include physical and chemical assessments. Soil texture in the surface and subsoil components of the profile influence water holding capacity and can also influence rooting characteristics of the crop. Surface conditions should be assessed for infiltration rates and subsurface conditions should be assessed for high water table conditions or restrictive soil horizons. A chemical analysis of the soil should be conducted to evaluate the pH, salinity, and nutritional characteristics.

Climatic conditions have several impacts on system selection and design. The utilization, selection, spacing, and placement of sprinklers and microsprinklers will be influenced by local wind characteristics of speed, duration, and direction. While most systems are designed to meet the evapotranspiration demands on the crop, some systems may have heat stress relief or cold protection incorporated into the design. For example, if a citrus irrigation system needs to be used for general crop evapotranspiration-based irrigation requirements and for freeze protection, then the entire field must be irrigated simultaneously rather than sequentially in zones. Thus the resulting pump system and pipe network must be substantially larger.

Irrigation is used for a variety of crops and cropping systems that include traditional field crops (corn, cotton, etc.), tree crops (citrus, apples, cherries, etc.), vine crops (grapes and other berries), vegetable crops (tomatoes, melons, etc.), ornamental plants (flowers, shrubs, trees, etc.), and turf (landscapes, golf courses, and commercial production). These crops have various heights, plant densities, row spacing, plant spacing, sensitivity to water stress, bedding or soil tillage practices, artificial or natural mulches, and other cultural practices. These characteristics need to be considered by the designer and incorporated into the design. In addition, crop value and irrigated yield potential will also influence the type and complexity of the irrigation system.

WATER SUPPLY SYSTEM

The water supply system includes a pump and a network of pipes, valves, and fittings (Fig. 1) to deliver water to the infield distribution system, which may be a center pivot system, solid-set sprinklers, microirrigation laterals, or other distribution devices. The pump has two primary purposes: 1) it moves water at a desired flow rate; and 2) it provides energy to the water. Eq. 1 was discussed and presented as a method to determine the required pump capacity (Q_{sys}). The energy requirements of the pump are often referred to as "pump head" and expressed in units of a height (m) of a column of water.

Required pump head has three components that include elevation head, friction head, and pressure head. Elevation head refers to the vertical elevation difference between the pumping water level and the level of the highest irrigation

system outlet (sprinklers, etc.). As water flows through the pipes and other fittings, friction reduces the energy level of the water and is characterized as friction head. Finally, the irrigation system will have a specified water pressure for proper operation of the discharge devices and this is characterized using pressure head. Because most water-discharging devices provide different flow rates of water with respect to operational pressure, one of the design goals is to minimize pressure variations within the pipe network and to maintain variations within allowable design limits. Therefore, while the designer is sizing and configuring the main lines, header pipes, laterals, associated fittings and components, material costs, pipeline flow velocities, and pressure head variations due to elevation changes and friction head losses are also being computed and analyzed to achieve an economical, hydraulically balanced and uniform irrigation system.

CONCLUSION

Irrigation system design involves an assessment of the intended use and desired goal for the system; the location, availability, and quality of the water source; physical and climatic characteristics of the site; and production system characteristics of the crop. The designer sizes and configures the pump system, pipelines, water discharge devices, and the associated fittings and components into a system that will uniformly distribute water to meet the desired goals within the economic, cultural, and physical constraints associated with the site, the crops, and the production system. The final design and installed system should be evaluated to ascertain proper performance and operation.

ACKNOWLEDGMENTS

The author would like to acknowledge the following individuals for their mentorship, collaboration, and professional feedback, which have all contributed to his professional growth and development in the area of irrigation system design and management: Dr. Allen G. Smajstrla (deceased), Professor, Agricultural and Biological Engineering Department, University of Florida; Dr. Freddie Lamm, Professor, NW Research and Extension Center, Kansas State University; Dr. Dorota Haman, Professor, Agricultural and Biological Engineering Department, University of Florida; Mr. James Prochaska, P.E., President, JNM Technologies, Inc.; Dr. Danny Rogers, Professor, Department of Biological and Agricultural Engineering, Kansas State University; Dr. Muluneh Yitayew, Associate Professor, Department of Agricultural and Biosystems Engineering, University of

Arizona; and the many students, in the United States and abroad, who have participated in my irrigation system design courses, workshops, and seminars.

REFERENCES

1. Karmelli, D.; Keller, J. *Trickle Irrigation Design*, 1st Ed.; Rain Bird Sprinkler Mfg. Corp.: Glendora, CA, 1975; 133.

2. Pair, C.H.; Hinz, W.H.; Frost, K.R.; Sneed, R.E.; Schiltz, T.J. *Irrigation*, 5th Ed.; The Irrigation Association: Arlington, VA, 1983; 686.

3. Heermann, D.F.; Kohl, R.A. Fluid Dynamics of Sprinkler Systems. In *Design and Operation of Farm Irrigation Systems*, ASAE Monograph No. 3; Jensen, M.E., Ed.; ASAE: St. Joseph, MI, 1983; 583–620.

4. Addink, J.W.; Keller, J.; Pair, C.H.; Sneed, R.E.; Wolfe, J.W. Design and Operation of Sprinkler Systems. In *Design and Operation of Farm Irrigation Systems*, ASAE Monograph No. 3; Jensen, M.E., Ed.; ASAE: St. Joseph, MI, 1983; 621–662.

5. Howell, T.A.; Stevenson, D.S.; Aljibury, F.K.; Gitlin, H.M.; Wu, I.P.; Warick, A.W.; Raats, P.A.C. Design and Operation of Trickle (Drip) Systems. In *Design and Operation of Farm Irrigation Systems*, ASAE Monograph No. 3; Jensen, M.E., Ed.; ASAE: St. Joseph, MI, 1983; 663–720.

6. Keller, J.; Bliesner, R.D. *Sprinkle and Trickle Irrigation*; Van Nostrand Reinhold: New York, 1990; 652.

7. Boswell, M.J. *Micro-irrigation Design Manual*; James Hardie Irrigation: El Cajon, CA, 1990.

8. Scherer, T.; Kranz, W.; Pfost, D.; Werner, H.; Wright, J.; Yonts, C.D. *Sprinkler Irrigation Systems*, Midwest Plan Service Publication, MWPS-30; Iowa State University: Ames, IA, 1999; 250 p.

9. ASAE, Design and Installation of Microirrigation Systems. *ASAE Standards*, ASAE EP405.1 DEC99; ASAE Standards: St. Joseph, MI, 2000.

10. ASAE, Polyethylene Pipe Used for Microirrigation Laterals. *ASAE Standards*, ASAE S435 DEC99; ASAE Standards: St. Joseph, MI, 2000.

11. ASAE, Field Evaluation of Microirrigation Systems. *ASAE Standards*, ASAE EP458 DEC99; ASAE Standards: St. Joseph, MI, 2000.

12. ASAE, Design, Installation and Performance of Underground Thermoplastic Irrigation Pipelines. *ASAE Standards*, ASAE S376.2 JAN98; ASAE Standards: St. Joseph, MI, 2000.

13. ASAE, Graphic Symbols for Pressurized Irrigation System Design. *ASAE Standards*, ASAE S491 DEC99; ASAE Standards: St. Joseph, MI, 2000.

Irrigation Economics, Global

Keith Wiebe
Noel Gollehon
United States Department of Agriculture (USDA),
Washington, District of Columbia, U.S.A.

INTRODUCTION

Irrigation involves complex interactions among ecological, social, and economic processes at a variety of scales, with important implications for agricultural production, income generation, poverty reduction, and environmental quality. No simple measure can fully capture the global economic importance of irrigation. Nevertheless, a review of historic trends in agricultural demand and resource use indicates that irrigation has contributed to dramatic increases in global crop yields and production over the past four decades. Given projected trends in demand for agricultural commodities and in the availability and condition of land and other natural resources, irrigation will continue to play a critical role in the future. Improved management will be necessary, however, to balance public and private economic and environmental objectives.

TRENDS IN DEMAND FOR AGRICULTURAL COMMODITIES

Global demand for agricultural commodities has increased rapidly since the mid-20th century as a result of growth in population, income, and other factors. Based on continued growth in these factors, the Food and Agriculture Organization (FAO)[1] and the International Food Policy Research Institute[2] project that global demand for cereals will increase by 1.2%–1.3% per year over the next several decades, while demand for meat will increase slightly faster. Most of the increased demand is projected to come from developing countries, especially in Asia (most of which are already highly dependent on irrigation). Although demand growth rates are slowing and remain within the range of crop production growth rates achieved over the past several decades, demands on natural resources—including water—will increase.

TRENDS IN USE OF NATURAL RESOURCES

Land

The Food and Agriculture Organization reports that the total area devoted to annual and permanent crops worldwide has increased by about 0.3% per year since 1961, to 1.5 billion ha in 1998. Growth has slowed markedly in the past decade, to about 0.1% per year, as a result of weak grain prices, deliberate policy reforms (in North America and Europe), and institutional changes (e.g., those in the former Soviet Union). The Food and Agriculture Organization estimates that an additional 2.7 billion ha currently in other uses are suitable for crop production, but this land is unevenly distributed and includes land with relatively low yield potential and/or significant environmental value. Therefore, cropland area is expected to expand only slightly over the next several decades.

Genetic Resources

About half of all gains in crop yields over the past century are attributable to genetic improvements through scientific plant breeding.[3] By the 1990s, most developing countries' (and all the developed countries') cropland in wheat, rice, and maize was planted to scientifically bred varieties. Gains from genetic improvements will continue in future, but likely at slower rates and increasing research costs.[3]

Climate

The Intergovernmental Panel on Climate Change,[4] representing a broad scientific consensus, projects that the earth's climate will change significantly during the 21st century because of increasing concentrations of carbon dioxide (CO_2) and other "greenhouse" gases in the atmosphere. Given the adjustments that farmers would likely make in response to these climatic changes, aggregate global crop production may not be dramatically affected, but regional impacts may be significant: agricultural production would tend to increase in temperate latitudes and decrease in the tropics due to projected changes in precipitation and temperature (and thus in the spatial and temporal distribution of water).

Water

Fresh water is abundant globally, but only a small portion—about $10,000 \, km^3 \, yr^{-1}$—is renewable and available for human use. Furthermore, this portion is distributed unevenly between countries, within countries, and across seasons and years. Of this portion, about a third is currently withdrawn for human use.[5] Agriculture accounts for about 70% of water withdrawals worldwide, and over 90% of withdrawals in low-income, developing countries.[6]

TRENDS IN IRRIGATION

The extent of irrigated cropland worldwide has grown at an average annual rate of 1.8% since 1961 (six times the rate of total cropland expansion), from 139 million ha in 1961 (10% of total cropland) to 274 million ha in 1999 (18% of total cropland).[7] Growth in irrigated area has been especially rapid in India, West Asia, North Africa, Latin America and the Caribbean, and about two-thirds of the world's irrigated area is in Asia (Table 1).

Irrigation expansion has slowed significantly in recent decades, from 2.2% per year during 1967–1982 to 1.5% per year during 1982–1995, due to declining cereal prices, the lower quality of land available for new irrigation, and the increasing economic, social, and environmental costs of large-scale irrigation systems.[2] The Food and Agriculture Organization[1] projects that irrigation expansion will slow further to an average increase of 0.6% per year through 2030.

Water withdrawn for agriculture averaged about 1 m in depth over all irrigated area in 1990 in 118 countries studied by Seckler et al.[5] Wood, Sebastian, and Scherr[8] note that global estimates of irrigation efficiency (i.e., the proportion of water withdrawn for irrigation that is

actually consumed by crops) average about 43%, with most of the remainder being returned to the river or to the groundwater aquifer.

TRENDS IN AGRICULTURAL PRODUCTION

Growth in publicly-funded surface irrigation and in largely privately-funded tubewell irrigation contributed significantly to the food production increases and real food price declines of the Green Revolution.[9] Food and Agriculture Organization data indicate that cereal yields have increased in developing countries by an average of 2.3% per year since the early 1960s. Some of this increase is due to increased use of irrigation water (along with fertilizer and scientifically bred crop varieties); in developing countries, cereal yields are more than twice as high in irrigated areas ($3.8 \, \text{Mg ha}^{-1}$) as they are in rainfed areas ($1.7 \, \text{Mg ha}^{-1}$). Irrigated cropland now produces 30%–40% of the world's crop output, including nearly two-thirds of all rice and wheat; at international agricultural prices for 1989–1991, the irrigated share corresponds to a total value of roughly $400–530 billion per year.[8]

Over time, however, subsidized water delivery (whether via public infrastructure or subsidized fuel for private tubewell operation) and inadequate property rights in water have led to excessive and inefficient exploitation of water resources in some countries.[10] Barker and van Koppen[9] argue that these trends will adversely affect food production in key grain-producing areas (including India and China) in the coming decades. Waterlogging and salinization of irrigated land threaten crop yields in some

Table 1 Irrigation indicators, 1990

Region	Total irrigated area (million ha)	Area growth rate ($\% \, \text{yr}^{-1}$)	Irrigation depth (m)	Irrigation efficiency[a] (%)
World	243.0	1.6	1.0	43
Asia	154.4	1.9	1.0	39
China	48.0	1.3	1.0	39
India	45.1	2.7	1.1	40
Other Asia	61.3	1.7	0.9	32
West Asia and North Africa	22.6	2.5	1.2	60
North America	21.6	0.9	0.9	53
Europe	16.7	0.6	0.9	56
Latin America and Caribbean	16.2	2.4	1.2	45
Sub-Saharan Africa	4.8	1.2	1.6	50
South Africa	1.3	0.8	1.2	45
Oceania	2.1	3.6	0.3	66

[a] Irrigation efficiency is a complex concept, but is generally defined as the ratio of water actually used by crops (i.e., returned to the atmosphere via transpiration) to the gross amount of water extracted for irrigation use.[8] For further discussion, see Ref. 13. (From Ref. 8.)

areas, and are likely to become an increasing problem in the absence of appropriate management.

Meanwhile, population growth and the increasing cost of developing new sources of water will place increasing pressure on world water supplies in the coming decades. Even as demand for irrigation water increases, farmers face growing competition for water from urban and industrial users, and from demands to protect in-stream ecological functions by imposing minimum in-stream flows. In light of these conditions, Rosegrant et al.[2] argue that water will likely become a major constraint on increased food production and improved food security in many developing countries, especially in Central and Western Asia and in Africa. Seckler et al.[5] assert that in a growing number of countries, water has become the single most important constraint to increased food production.

OPTIONS FOR INCREASED SUPPLY AND IMPROVED MANAGEMENT OF WATER

Water storage is a key component of strategies to overcome spatial and temporal variability in precipitation and river flows. National governments, multilateral agencies, and local communities have invested heavily in dams over the past century, but such investments are becoming increasingly expensive in financial, environmental, and political terms.[6] Groundwater has been withdrawn at rates in excess of recharge and degraded through contamination in many areas. Interbasin transfers may be appealing in some areas, but are characterized by the same costs that limit new investment in dams. Water recycling and desalination remain too costly for extensive use.

Given limitations on increased supply, a variety of options for improved water management become important. Serageldin[11] notes that because of water's unique characteristics, governments have generally assumed central responsibility for its management. In seeking to assure access by all, however, governments generally price water as though it were an abundant resource rather than a scarce one, thereby encouraging excessive use in many countries.[12] As the costs of excessive use are recognized, e.g., in terms of groundwater depletion and salinization, increasing attention is being paid to policies that address market and government failures and provide incentives for more efficient water use. Key among these are efforts to price water at levels that better reflect costs, and to establish tradable water rights. Policy and technology also play a role in changing management practices to improve water infiltration and moisture-holding capacity on agricultural lands.

Seckler et al.[5] estimate that about half of the projected increase in global demand for water by the year 2025 can be met by reducing losses to evaporation and sinks, controlling salinity and pollution, reallocating water from lower-valued to higher-valued crops, and investing in genetic improvements that increase crop yields per unit of water. It is important to note, however, that reducing runoff or deep percolation to groundwater may affect the water supply for other water users or for environmental purposes, resulting in unintended and undesirable impacts.[13]

CONCLUSION

In considering strategies to improve irrigation supplies and management, it is essential that the full costs and benefits to all affected parties are recognized. Ultimately, the extent to which crop production keeps pace with future increases in demand at acceptable economic and environmental cost will depend on the institutions, market incentives, policy measures, and investments in research and infrastructure that influence how water and other resources are used.

REFERENCES

1. FAO. *Agriculture: Towards 2015/30 (Technical Interim Report)*; Food and Agriculture Organization of the United Nations: Rome, 2000.
2. Rosegrant, M.W.; Paisner, M.S.; Meijer, S.; Witcover, J. *Global Food Projections to 2020: Emerging Trends and Alternative Futures*; International Food Policy Research Institute: Washington, DC, 2001.
3. Byerlee, D.; Heisey, P.; Pingali, P. Realizing Yield Gains for Food Staples in Developing Countries in the Early Twenty-First Century: Prospects and Challenges. In *Food Needs of the Developing World in the Early Twenty-First Century*, Pontificiae Academiae Scientiarum Scripta Varia 97, Proceedings of the Study-Week of the Pontifical Academy of Sciences, January 27–30, 1999; 2000.
4. Intergovernmental Panel on Climate Change website http://www.ipcc.ch/index.html (accessed 28 February, 2002).
5. Seckler, D.; Amarasinghe, U.; Molden, D.; de Silva, R.; Barker, R. *World Water Demand and Supply, 1990 to 2025*; Scenarios and Issues, International Water Management Institute: Colombo, Sri Lanka, 1998.
6. Meinzen-Dick, R.S.; Rosegrant, M.W. Overview. In *Overcoming Water Scarcity and Quality Constraints*; Meinzen-Dick, R.S., Rosegrant, M.W., Eds.; 2020 Vision Focus 9; International Food Policy Research Institute: Washington, DC, 2001.
7. FAO. FAOSTAT online database http://apps.fao.org, Food and Agriculture Organization of the United Nations, (accessed 15 February, 2002).
8. Wood, S.; Sebastian, K.; Scherr, S.J. *Pilot Analysis of Global Ecosystems: Agroecosystems*; International Food

Policy Research Institute and World Resources Institute: Washington, DC, 2000.

9. Barker, R.; van Koppen, B. *Water Scarcity and Poverty*; IWMI Water Brief 3; International Water Management Institute: Colombo, Sri Lanka, 1999.

10. Rosegrant, M.W. Dealing with Water Scarcity in the 21st Century. In *The Unfinished Agenda: Perspectives on Overcoming Hunger, Poverty, and Environmental Degradation*; Pinstrup-Andersen, P., Pandya-Lorch, R., Eds.; International Food Policy Research Institute: Washington, DC, 2001.

11. Serageldin, I. *Toward Sustainable Management of Water Resources*; World Bank: Washington, DC, 1995.

12. Coyle, W.; Gilmour, B. Water Supply in the APEC Region: Scarcity or Abundance? *Agricultural Outlook*; USDA Economic Research Service (November): Washington, DC, 2001.

13. Aillery, M.; Gollehon, N. Irrigation Water Management. *Agricultural Resources and Environmental Indicators, 2000*; USDA Economic Research Service: Washington, DC, 2000, http://www.ers.usda.gov/Emphases/Harmony/issues/arei2000/.

Irrigation Economics, United States

Noel Gollehon
United States Department of Agriculture (USDA), Washington, District of Columbia, U.S.A.

I

INTRODUCTION

Irrigation is the defining characteristic of crop production in the American West and an increasingly important feature of crop production in the Eastern United States. The irrigated cotton fields of the Southwest, corn farms of the Plains, and citrus groves of Florida, all attest to the magnitude, extent, and importance of irrigation. This article provides an overview of the contribution of irrigation to crop production in the United States. It also focuses on the sales value of crops produced, but provides a context for those values by first considering irrigated area and water use in irrigation. Readers will gain insight into irrigation's importance both from an economic and a resource use perspective.

IRRIGATED AGRICULTURAL AREA

The National Agricultural Statistics Service (NASS), USDA conducts the Census of Agriculture on a 5-yr interval.[1] The 1997 Census data are the latest available; with the next Census of Agriculture due to be collected in 2003 for the 2002 calendar year. The long history, consistent methodology, and statistical reliability of the Census data series makes these data especially useful for capturing irrigation trends.

According to the 1997 Census of Agriculture,[1] 55.0 million acres of agricultural land were irrigated in the United States. This represents a new census-year high, with an additional 5.6 million acres (over 11%) over levels reported in the 1992 Census of Agriculture. The distribution of irrigated lands (both cropland and pastureland) shows that 78% (43 million acres) were located in the 19 Western states with the remaining 22% (12 million acres) in 31 Eastern states (Table 1).

Some cropland is irrigated in all 50 states. In 1997, irrigated land area ranged from about 2500 acres in Vermont, New Hampshire, and Alaska to about 8.7 million acres in California. Irrigated areas have historically been concentrated in the West (89% of U.S. irrigated area in 1969) because arid conditions required irrigation to supplement inadequate growing season rainfall. The West still retains the bulk of the irrigated land, but irrigated area is expanding in the more humid East. Since 1969, irrigated land in the East has increased by almost the same number of acres as in the West, with a much faster rate of growth (187%–23%). More recently (1987–1997), irrigated land in the West increased by about 5.3 million acres (14%) compared with 3.3 million acres (38%) in the East.

Of the 55 million irrigated acres reported in the 1997 Census, there were 50 million acres of "Cropland Harvested," and 5 million acres of "Pastureland and other land." Nationally, irrigated cropland represented about 16% of all harvested cropland. In the West, irrigated cropland harvested comprises a greater share of total cropland acres (about 27%), representing about 76% of the nation's total harvested irrigated cropland. While irrigation in the East accounts for the remaining 24% of the nation's total, only a small share (7%) of the harvested cropland in the East was irrigated (Table 1).

WATER USED FOR IRRIGATION

The U.S. Geological Survey, U.S. Department of the Interior, estimates both water withdrawals and consumptive use every 5 yr.[2] Estimates are made at a local level based on locally available information, including theoretical estimates of crop water use, crop area, delivery records of off-farm water suppliers, and details on conveyance losses, water application rates, and return flows.

Three measures can be used to characterize water use for agricultural irrigation: withdrawals, applications, and consumptive use. Withdrawals represent total water diverted from surface water sources and extracted from groundwater aquifers.[2] Applications measure that portion of the water withdrawn that is delivered to the field, excluding off-field conveyance system losses and gains.[3] Water applications represent the portion of withdrawals that are directly under producers' control and are thus impacted by on-farm irrigation management and

Views expressed are the author's and do not necessarily represent those of Economic Research Service or USDA.

DOI: 10.1081/E-EWS 120010123
Published 2003 by Marcel Dekker, Inc. All rights reserved.

Table 1 Irrigated area in the United States, by region, 1997

Region	Harvested cropland irrigated (million acres)	Pastureland irrigated (million acres)	Total irrigated area (million acres)
United States	50.0	5.0	55.0
Western states[a]	38.2	4.8	43.0
Eastern states	11.8	0.2	12.0

[a] Western states includes HI, AK, WA, OR, CA, ID, NV, MT, WY, UT, CO, AZ, NM, ND, SD, NE, KS, OK, and TX.
Source: Ref. [1].

technology choice decisions. Consumptive use refers to that portion of water withdrawn and applied that is actually consumed for plant needs.[2,4] Consumptive use is usually estimated based on plant water requirement models, and does not include excess water lost to percolation, runoff, or evaporation, other than that required for plant growth.

Measures of irrigation water use may be used to describe the impact of irrigation on hydrologic conditions. Withdrawals are the best indication of the water quantity impacts of irrigation water diversions. While withdrawn water that is not consumptively used may be available for future use, the location, quality, and timing of availability are often affected. Consumptive use is an indicator of the water quantity lost to the immediate hydrologic cycle. Irrigation withdrawals, as well as withdrawals for other out-of-stream uses, may be quantified and compared. None of these measures consider in-stream water uses, such as hydroelectric power generation, navigation, recreation flows, or flows to maintain ecosystems. In-stream uses may be more significant than off-stream uses in many locations, but specific quantities are difficult to measure.

Irrigated agriculture withdraws and consumes the most freshwater of any economic sector in the United States Irrigation accounts for withdrawals of 150 million acre feet (maf) nationally, almost 40% of total freshwater withdrawals (Table 2). When measured by consumptive use, irrigation uses about 91 maf of water, or more than 80% of the total consumptive use. Comparing across sectors,

irrigation consumes about 60% of irrigation water withdrawn—a much greater share relative to the average consumption rate of 9% in other sectors.[2]

The water use picture varies substantially between the 19-state western and 31-state eastern regions. In the West, irrigated agriculture accounts for 133 maf or 75% of total freshwater withdrawals in the region. The average water quantity withdrawn per acre of irrigated crop and pastureland is 2.7 acre ft. When measured by consumptive use, irrigation uses about 79 maf of water, or almost 90% of total water consumed in the West. Roughly two-thirds of the irrigation water in the West is supplied from surface water sources with groundwater accounting for the remaining supply. California has more than double the irrigation withdrawals of any other State (32 maf), while Idaho, Colorado, and Texas—all have withdrawals greater than 10 maf. Of these four states, only Texas withdraws more groundwater than surface water.[2]

By comparison, irrigation withdrawals in the 31 Eastern states account for 17 maf, or about 8% of total regional withdrawals. Withdrawal rates were substantially lower than in the West, at 1.3 acre ft per acre of irrigated crop and pastureland reflecting the greater natural precipitation in agricultural areas. Irrigation consumptive use (12 maf) accounts for 52% of total water consumed in the East. (Thermoelectric power generation, which withdraws a large quantity of water but consumes little, accounts for most Eastern withdrawals.) Groundwater is the primary source of irrigation water in the East. Arkansas withdraws

Table 2 Irrigation and water withdrawals and consumption in the United States, 1995

Region	Sector	Water withdrawals (maf)	Consumptive water use (maf)
United States	All	382	112
	Irrigation	150	91
Western states[a]	All	179	88
	Irrigation	133	79
Eastern states	All	203	24
	Irrigation	17	12

[a] Western states include HI, AK, WA, OR, CA, ID, NV, MT, WY, UT, CO, AZ, NM, ND, SD, NE, KS, OK, and TX.
Source: Ref. [2].

the largest quantity of irrigation water (6.6 maf) among eastern states, primarily from groundwater sources. Florida (almost 4 maf) and Mississippi (almost 2 maf) are also major irrigation withdrawal states in the region.[2]

Several factors will influence the extent of future increases or continuation of current withdrawal levels of water for irrigation use. Increasing demands from other sectors for both out-of-stream and in-stream use have recently limited irrigation water withdrawals in some areas, particularly under drought conditions. Increasing capital costs for new projects and recognition of environmental impacts have combined to limit large-scale new water developments to augment irrigation water supplies. It is unlikely that these trends will reverse, at least in the near future, implying that expansions in irrigated area will likely occur primarily through more efficient use of the water already dedicated to agricultural production.

VALUE OF IRRIGATION

Crop sales reports from the Census of Agriculture provide the basis for current estimates of irrigated crop values. Individual Census of Agriculture farms were classified into one of three irrigation groups for each commodity group: only irrigated, only nonirrigated, and combined irrigated and nonirrigated. Irrigated commodity sales was calculated as the sum of the only irrigated farms plus an apportioned share of the sales on combined farms in the 1997 Census of Agriculture.[1]

Crop sales, which measure the value of commodities leaving the farm gate, also serve as an indication of economic activity associated with farming and related income flows through rural areas. Preferred measures of irrigation's contribution to crop production would be profitability of irrigated agriculture or the direct value of total crop production, but those estimates are not available. The Census of Agriculture reports the amount of crop sales at the farm market gate, but does not capture the value of crops that are produced and consumed on-the-farm without entering a market channel, most prevalent with irrigated forage and feed crops used on the farm. Although this value adjustment is not known for 1997, the underestimation was about 15% of sales value in 1987.[5]

Based on calculations from the 1997 Census of Agriculture information, there were 309 million acre of harvested cropland that produced crop sales of $98 billion. Irrigated crops occupied 16% of that area, but accounted for 49% of the total value of sales from U.S. farms and ranches (Fig. 1, top row). Average sales per harvested acre were $950 for irrigated cropland compared with $200 for nonirrigated cropland. Irrigated crop sales were highest for orchards, vegetables, and nursery crops while irrigated

cropland area was dominated by grain and forage crops, primarily corn for grain and alfalfa hay.

In the West, the 1997 Census reported 142 million acre of harvested cropland that produced total crop sales of $45 billion. Irrigated crops in the West accounted for 27% of the area, but produced 72% of the total value of sales in the region (Fig. 1, middle row). The sales of Western irrigated crops totaled about $32 billion in 1997, or roughly one-third of all U.S. crop sales. Average sales per harvested acre in the West were $850 for irrigated and $122 for nonirrigated cropland. As was the case when examining the national values, irrigated crop sales were led by orchards, vegetables, and nursery crops while irrigated cropland area was dominated by grain and forage crops.

In the East, the 1997 Census reported 167 million acres of harvested cropland with total crop sales of almost $53 billion. Irrigated crops in the east occupied only 7% of the harvested cropland area, but produced $15 billion or 29% of the region's total value of sales (Fig. 1, bottom row). Average sales per harvested irrigated acre were greater than the national average at over $1200, while nonirrigated cropland averaged sales of $200 per acre. The greatest contribution to sales totals were made by irrigated nursery crops, orchard crops, and vegetables. Rice, soybeans, and corn for grain dominated irrigated cropland area in the region.

The wide differences in crop sales values, coupled with the fact that most of the crop sales comes from high-valued crops, provides significant flexibility for irrigated agriculture to adjust to changes in water availability. Farmers can adjust to physical water shortages by improving irrigation technology and/or adjusting cropping choices to maintain production of the higher-valued crops. This ability to substitute crops is an important response to water shortfalls. In addition, innovative water markets have increased the ability of farmers and water suppliers to transfer water, enabling maintenance of higher-valued crops during droughts.[6]

CONCLUSION

This article examined three measures of irrigation: irrigated agricultural area, water used in irrigation, and the sales value of crops produced. By all three measures, irrigation is an important contributor to the value of crop agriculture. In 1997, irrigated lands produced 49% of crop sales on only 16% of the harvested crop area, providing an important input to most of the nation's higher-valued crop production. Irrigation is also an important component in the nation's hydrologic picture by virtue of the spatial extent and volume of water involved. In 1995, irrigation accounted for 40% of the nation's water freshwater withdrawals from lakes, rivers, and aquifers. Irrigation

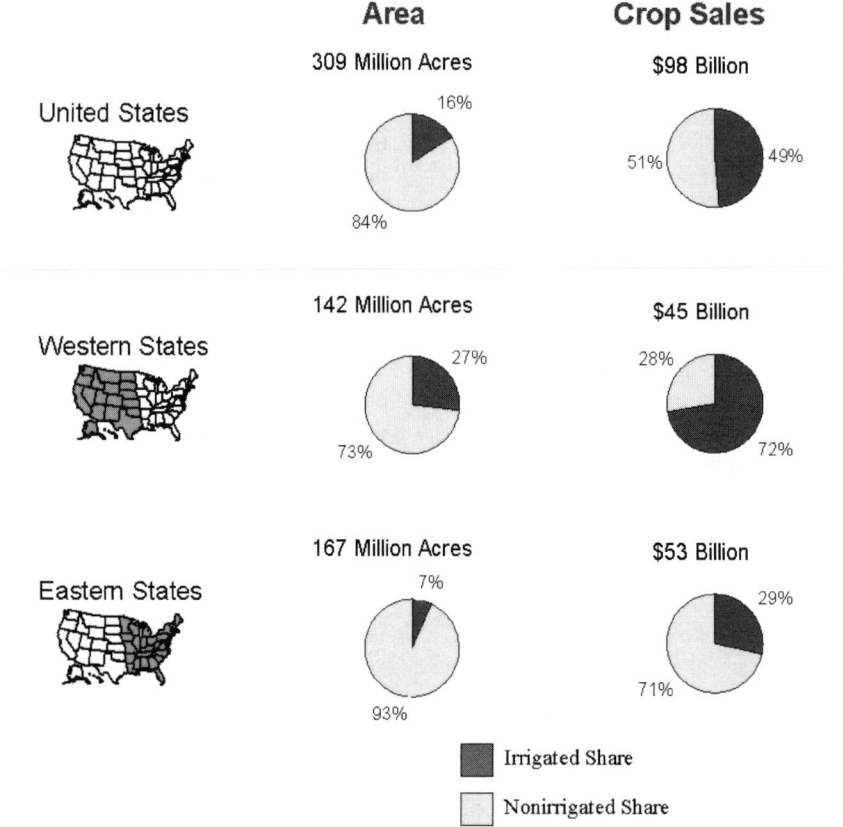

Fig. 1   Irrigated crop area and sales as a share of total, 1997. (from Ref. 1).

accounted for over 80% of the total consumptive use by all sectors of the economy. In future increased competition for water will affect irrigated agriculture's ability to withdraw and consume water at current levels.

REFERENCES

1. National Agricultural Statistics Service, *1997 Census of Agriculture*, AC97-A-51; U.S. Department of Agriculture, 1999.
2. Solley, W.B.; Pierce, R.R.; Perlman, H.A. *Estimated Use of Water in the United States in 1995*, U.S. Geological Survey Circular 1200; U.S. Geological Survey, U.S. Department of the Interior, 1998.
3. National Agricultural Statistics Service, *1998 Farm and Ranch Irrigation Survey*, AC97-SP-1 Vol. 3, Special Studies Part 1; U.S. Department of Agriculture, 1999.
4. Aillery, M.; Gollehon, N. Irrigation Water Management. *Agricultural Resources and Environmental Indicators, 1996–97*, Agricultural Handbook No. 712; Economic Research Service, U.S. Department of Agriculture, 1997; 225–240.
5. Bureau of the Census, *1987 Census of Agriculture*, AC87-A-51; U.S. Department of Commerce, 1989.
6. Gollehon, N. Water Markets: Implications for Rural Areas of the West. *Rural Development Perspectives*; Economic Research Service, U.S. Department of Agriculture, 1999; Vol. 14 (2, August), 57–63.

Irrigation Efficiency

Terry A. Howell
United States Department of Agriculture (USDA), Bushland, Texas, U.S.A.

INTRODUCTION

Irrigation efficiency is a critical measure of irrigation performance in terms of the water required to irrigate a field, farm, basin, irrigation district, or an entire watershed. The value of irrigation efficiency and its definition are important to the societal views of irrigated agriculture and its benefit in supplying the high quality, abundant food supply required to meet our growing world's population. "Irrigation efficiency" is a basic engineering term used in irrigation science to characterize irrigation performance, evaluate irrigation water use, and to promote better or improved use of water resources, particularly those used in agriculture and turf/landscape management.[1–4] Irrigation efficiency is defined in terms of: 1) the irrigation system performance, 2) the uniformity of the water application, and 3) the response of the crop to irrigation. Each of these irrigation efficiency measures is interrelated and will vary with scale and time. Fig. 1 illustrates several of the water transport components involved in defining various irrigation performance measures. The spatial scale can vary from a single irrigation application device (a siphon tube, a gated pipe gate, a sprinkler, a microirrigation emitter) to an irrigation set (basin plot, a furrow set, a single sprinkler lateral, or a microirrigation lateral) to broader land scales (field, farm, an irrigation canal lateral, a whole irrigation district, a basin or watershed, a river system, or an aquifer). The timescale can vary from a single application (or irrigation set), a part of the crop season (preplanting, emergence to bloom or pollination, or reproduction to maturity), the irrigation season, to a crop season, or a year, partial year (premonsoon season, summer, etc.), or a water year (typically from the beginning of spring snow melt through the end of irrigation diversion, or a rainy or monsoon season), or a

period of years (a drought or a "wet" cycle). Irrigation efficiency affects the economics of irrigation, the amount of water needed to irrigate a specific land area, the spatial uniformity of the crop and its yield, the amount of water that might percolate beneath the crop root zone, the amount of water that can return to surface sources for downstream uses or to groundwater aquifers that might supply other water uses, and the amount of water lost to unrecoverable sources (salt sink, saline aquifer, ocean, or unsaturated vadose zone).

The volumes of the water for the various irrigation components are typically given in units of depth (volume per unit area) or simply the volume for the area being evaluated. Irrigation water application volume is difficult to measure, so it is usually computed as the product of water flow rate and time. This places emphasis on accurately measuring the flow rate. It remains difficult to accurately measure water percolation volumes groundwater flow volumes, and water uptake from shallow groundwater.

IRRIGATION SYSTEM PERFORMANCE EFFICIENCY

Irrigation water can be diverted from a storage reservoir and transported to the field or farm through a system of canals or pipelines; it can be pumped from a reservoir on the farm and transported through a system of farm canals or pipelines; or it might be pumped from a single well or a series of wells through farm canals or pipelines. Irrigation districts often include small to moderate size reservoirs to regulate flow and to provide short-term storage to manage the diverted water with the on-farm demand. Some on-farm systems include reservoirs for storage or regulation of flows from multiple wells.

Water Conveyance Efficiency

The conveyance efficiency is typically defined as the ratio between the water that reaches a farm or field and that diverted from the irrigation water source.[1,3,4] It is defined as

$$E_c = 100 \frac{V_f}{V_t} \tag{1}$$

Encyclopedia of Water Science
DOI: 10.1081/E-EWS 120010252
Published 2003 by Marcel Dekker, Inc. All rights reserved.

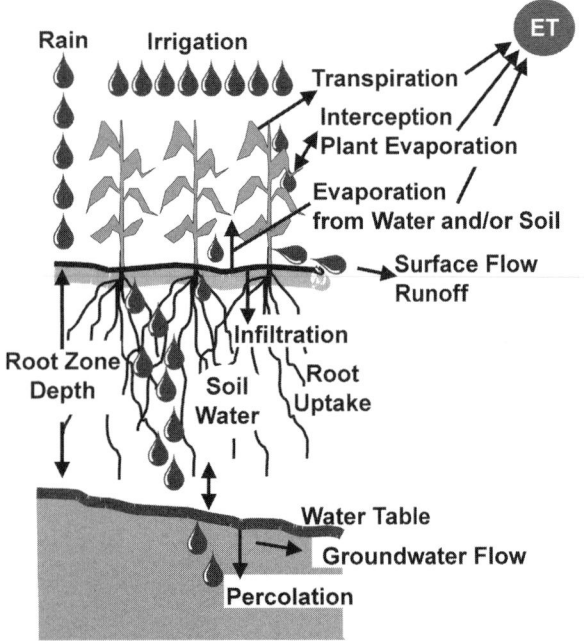

Fig. 1 Illustration of the various water transport components needed to characterize irrigation efficiency.

where E_c is the conveyance efficiency (%), V_f is the volume of water that reaches the farm or field (m³), and V_t is the volume of water diverted (m³) from the source. E_c also applies to segments of canals or pipelines, where the water losses include canal seepage or leaks in pipelines. The global E_c can be computed as the product of the individual component efficiencies, E_{ci}, where i represents the segment number. Conveyance losses include any canal spills (operational or accidental) and reservoir seepage and evaporation that might result from management as well as losses resulting from the physical configuration or condition of the irrigation system. Typically, conveyance losses are much lower for closed conduits or pipelines[4] compared with unlined or lined canals. Even the conveyance efficiency of lined canals may decline over time due to material deterioration or poor maintenance.

Application Efficiency

Application efficiency relates to the actual storage of water in the root zone to meet the crop water needs in relation to the water applied to the field. It might be defined for individual irrigation or parts of irrigations (irrigation sets).

Table 1 Example of farm and field irrigation application efficiency and attainable efficiencies

Irrigation method	Field efficiency (%)			Farm efficiency (%)		
	Attainable	Range	Average	Attainable	Range	Average
Surface						
Graded furrow	75	50–80	65	70	40–70	65
w/tailwater reuse	85	60–90	75	85	—	—
Level furrow	85	65–95	80	85	—	—
Graded border	80	50–80	65	75	—	—
Level basins	90	80–95	85	80	—	—
Sprinkler						
Periodic move	80	60–85	75	80	60–90	80
Side roll	80	60–85	75	80	60–85	80
Moving big gun	75	55–75	65	80	60–80	70
Center pivot						
Impact heads w/end gun	85	75–90	80	85	75–90	80
Spray heads wo/end gun	95	75–95	90	85	75–95	90
LEPAª wo/end gun	98	80–98	95	95	80–98	92
Lateral move						
Spray heads w/hose feed	95	75–95	90	85	80–98	90
Spray heads w/canal feed	90	70–95	85	90	75–95	85
Microirrigation						
Trickle	95	70–95	85	95	75–95	85
Subsurface drip	95	75–95	90	95	75–95	90
Microspray	95	70–95	85	95	70–95	85
Water table control						
Surface ditch	80	50–80	65	80	50–80	60
Subsurface drain lines	85	60–80	75	85	65–85	70

ª LEPA is low energy precision application.
(From Refs. 6,7,11.)

Application efficiency includes any application losses to evaporation or seepage from surface water channels or furrows, any leaks from sprinkler or drip pipelines, percolation beneath the root zone, drift from sprinklers, evaporation of droplets in the air, or runoff from the field. Application efficiency is defined as

$$E_a = 100 \frac{V_s}{V_f} \tag{2}$$

where E_a is the application efficiency (%), V_s is the irrigation needed by the crop (m^3), and V_f is the water delivered to the field or farm (m^3). The root zone may not need to be fully refilled, particularly if some root zone water-holding capacity is needed to store possible or likely rainfall. Often, V_s is characterized as the volume of water stored in the root zone from the irrigation application. Some irrigations may be applied for reasons other than meeting the crop water requirement (germination, frost control, crop cooling, chemigation, fertigation, or weed germination). The crop need is often based on the "beneficial water needs."[5] In some surface irrigation systems, the runoff water that is necessary to achieve good uniformity across the field can be recovered in a "tailwater pit" and recirculated with the current irrigation or used for later irrigations, and V_f should be adjusted to account for the "net" recovered tailwater. Efficiency values are typically site specific. Table 1 provides a range of typical farm and field irrigation application efficiencies[6–8] and potential or attainable efficiencies for different irrigation methods that assumes irrigations are applied to meet the crop need.

Storage Efficiency

Since the crop root zone may not need to be refilled with each irrigation, the storage efficiency has been defined.[4] The storage efficiency is given as

$$E_s = 100 \frac{V_s}{V_{rz}} \tag{3}$$

where E_s is the storage efficiency (%) and V_{rz} is the root zone storage capacity (m^3). The root zone depth and the water-holding capacity of the root zone determine V_{rz}. The storage efficiency has little utility for sprinkler or microirrigation because these irrigation methods seldom refill the root zone, while it is more often applied to surface irrigation methods.[4]

Seasonal Irrigation Efficiency

The seasonal irrigation efficiency is defined as

$$E_i = 100 \frac{V_b}{V_f} \tag{4}$$

where E_i is the seasonal irrigation efficiency (%) and V_b is the water volume beneficially used by the crop (m^3). V_b is somewhat subjective,[4,5] but it basically includes the required crop evapotranspiration (ET$_c$) plus any required leaching water (V_l) for salinity management of the crop root zone.

Leaching requirement (or the leaching fraction)

The leaching requirement,[9] also called the leaching fraction, is defined as

$$L_r = \frac{V_d}{V_f} = \frac{EC_i}{EC_d} \tag{5}$$

where L_r is the leaching requirement, V_d is the volume of drainage water (m^3), V_f is the volume of irrigation (m^3) applied to the farm or field, EC$_i$ is the electrical conductivity of the irrigation water (dS m^{-1}), and EC$_d$ is the electrical conductivity of the drainage water (dS m^{-1}). The L_r is related to the irrigation application efficiency, particularly when drainage is the primary irrigation loss component. The L_r would be required "beneficial" irrigation use ($V_l \equiv L_r V_i$), so only V_d greater than the minimum required leaching should reduce irrigation efficiency. Then, the irrigation efficiency can be determined by combining Eqs. (4) and (5)

$$E_i = 100 \left(\frac{V_b}{V_f} + L_r \right) \tag{6}$$

Burt et al.[5] defined the "beneficial" water use to include possible off-site needs to benefit society (riparian needs or wildlife or fishery needs). They also indicated that V_f should not include the change in the field or farm storage of water, principally soil water but it could include field (tailwater pits) or farm water storage (a reservoir) that wasn't used within the time frame that was used to define E_i.

IRRIGATION UNIFORMITY

The fraction of water used efficiently and beneficially is important for improved irrigation practice. The uniformity of the applied water significantly affects irrigation efficiency. The uniformity is a statistical property of the applied water's distribution. This distribution depends on many factors that are related to the method of irrigation, soil topography, soil hydraulic or infiltration characteristics, and hydraulic characteristics (pressure, flow rate, etc.) of the irrigation system. Irrigation application distributions are usually based on depths of water (volume per unit area); however, for microirrigation systems they are usually based on emitter flow volumes because the entire land area is not typically wetted.

Christiansen's Uniformity Coefficient

Christiansen[10] proposed a coefficient intended mainly for sprinkler system based on the catch volumes given as

$$C_U = 100 \left[\frac{1 - (\sum |X - \bar{x}|)}{\sum X} \right] \tag{7}$$

where C_U is the Christiansen's uniformity coefficient in percent, X is the depth (or volume) of water in each of the equally spaced catch containers in mm or ml, and $\bar{x}$ is the mean depth (volume) of the catch (mm or ml). For C_U values $> 70\%$, Hart[11] and Keller and Bliesner[8] presented

$$C_U = 100 \left[1 - \left(\frac{\sigma}{x} \right) \left(\frac{2}{\pi} \right)^{0.5} \right] \tag{8}$$

where σ is the standard deviation of the catch depth (mm) or volume (ml). Eq. 8 approximates the normal distribution for the catch amounts.

The C_U should be weighted by the area represented by the container[12] when the sprinkler catch containers intentionally represent unequal land areas, as is the case for catch containers beneath a center pivot. Heermann and Hein[12] revised the C_U formula (Eq. 8) to reflect the weighted area, particularly intended for a center pivot sprinkler, as follows:

$$C_{U(H\&H)} = 100 \left\{ 1 - \left[\frac{\left(\sum S_i \left| V_i - \left(\frac{\sum V_i S_i}{\sum S_i} \right) \right| \right)}{\sum (V_i S_i)} \right] \right\} \tag{9}$$

where S_i is the distance (m) from the pivot to the ith equally spaced catch container and V_i is the volume of the catch in the ith container (mm or ml).

Low-Quarter Distribution Uniformity

The distribution uniformity represents the spatial evenness of the applied water across a field or a farm as well as within a field or farm. The general form of the distribution

uniformity can be given as

$$D_{U_p} = 100 \left(\frac{\bar{V}_p}{\bar{V}_f} \right) \tag{10}$$

where D_{U_p} is the distribution uniformity (%) for the lowest p fraction of the field or farm (lowest one-half $p = 1/2$, lowest one-quarter $p = 1/4$), $\bar{V}_p$ is the mean application volume (m³), and $\bar{V}_f$ is the mean application volume (m³) for the whole field or farm. When $p = 1/2$ and $C_U > 70\%$, then the D_U and C_U are essentially equal.[13] The USDA-NRCS (formerly, the Soil Conservation Service) has widely used D_{Ulq} ($p = 1/4$) for surface irrigation to access the uniformity applied to a field, i.e., by the irrigation volume (amount) received by the lowest one-quarter of the field from applications for the whole field. Typically, D_{U_p} is based on the postirrigation measurement[5] of water volume that infiltrates the soil because it can more easily be measured and better represents the water available to the crop. However, the postirrigation infiltrated water ignores any water intercepted by the crop and evaporated and any soil water evaporation that occurs before the measurement. Any water that percolates beneath the root zone or the sampling depth will also be ignored.

The D_U and C_U coefficients are mathematically interrelated through the statistical variation (coefficient of variation, $\sigma/\bar{x}$, C_v) and the type of distribution. Warrick[13] presented relationships between D_U and C_U for normal, log-normal, uniform, specialized power, beta- and gamma-distributions of applied irrigations.

Emission Uniformity

For microirrigation systems, both the C_U and D_U concepts are impractical because the entire soil surface is not wetted. Keller and Karmeli[14] developed an equation for microirrigation design as follows

$$E_U = 100[1 - 1.27(C_{vm})n^{-1/2}] \left(\frac{q_m}{\bar{q}} \right) \tag{11}$$

where E_U is the design emission uniformity (%), C_{vm} is the manufacturer's coefficient of variability in emission device flow rate (1/h), n is the number of emitters per plant, q_m is the minimum emission device flow rate (1/h) at the minimum system pressure, and $\bar{q}$ is the mean emission device flow rate (1/h). This equation is based on the D_{Ulq} concept,[4] and includes the influence of multiple emitters per plant that each may have a flow rate from a population of random flow rates based on the emission device manufacturing variation. Nakayama, Bucks, and Clemmens[15] developed a design coefficient based more closely on the C_U concept for emission device flow rates from a normal distribution given as

$$C_{Ud} = 100(1 - 0.798(C_{vm})n^{-1/2}) \quad (12)$$

where C_{Ud} is the coefficient of design uniformity in percent and the numerical value, 0.798, is

$$\left(\frac{2}{\pi}\right)^{0.5}$$

from Eq. 8.

Many additional factors affect microirrigation uniformity including hydraulic factors, topographic factors, and emitter plugging or clogging.

WATER USE EFFICIENCY

The previous sections discussed the engineering aspects of irrigation efficiency. Irrigation efficiency is clearly influenced by the amount of water used in relation to the irrigation water applied to the crop and the uniformity of the applied water. These efficiency factors impact irrigation costs, irrigation design, and more important, in some cases, the crop productivity. Water use efficiency (WUE) has been the most widely used parameter to describe irrigation effectiveness in terms of crop yield. Viets[16] defined WUE as

$$WUE = \frac{Y_g}{ET} \quad (13)$$

where WUE is water use efficiency ($kg\,m^{-3}$), Y_g is the economic yield ($g\,m^{-2}$), and ET is the crop water use (mm). Water use efficiency is usually expressed by the economic yield, but it has been historically expressed as well in terms of the crop dry matter yield (either total biomass or aboveground dry matter). These two WUE bases (economic yield or dry matter yield) have led to some inconsistencies in the use of the WUE concept. The transpiration ratio (transpiration per unit dry matter) is a more consistent value that depends primarily on crop species and the environmental evaporative demand,[17] and it is simply the inverse of WUE expressed on a dry matter basis.

Irrigation Water Use Efficiency

The previous discussion of WUE does not explicitly explain the crop yield response to irrigation. Water use efficiency is influenced by the crop water use (ET). Bos[3] defined a term for WUE to characterize the influence of

irrigation on WUE as

$$WUE = \frac{(Y_{gi} - Y_{gd})}{(ET_i - ET_d)} \quad (14)$$

where WUE is irrigation water use efficiency ($kg\,m^{-3}$), Y_{gi} is the economic yield ($g\,m^{-2}$) for irrigation level i, Y_{gd} is the dryland yield ($g\,m^{-2}$; actually, the crop yield without irrigation), ET_i is the evapotranspiration (mm) for irrigation level i, and ET_d is the evapotranspiration of the dryland crops (or of the ET without irrigation). Although Eq. 14 seems easy to use, both Y_{gd} and ET_d are difficult to evaluate. If the purpose is to compare irrigation and dryland production systems, then dryland rather than nonirrigated conditions should be used. If the purpose is to compare irrigated regimes with an unirrigated regime, then appropriate values for Y_{gd} and ET_d should be used. Often, in most semiarid to arid locations, Y_{gd} may be zero. Bos[3] defined irrigation WUE as

$$IWUE = \frac{(Y_{gi} - Y_{gd})}{IRR_i} \quad (15)$$

where IWUE is the irrigation efficiency ($kg\,m^{-3}$) and IRR_i is the irrigation water applied (mm) for irrigation level i. In Eq. 15, Y_{gd} may be often zero in many arid situations.

CONCLUSION

Irrigation efficiency is an important engineering term that involves understanding soil and agronomic sciences to achieve the greatest benefit from irrigation. The enhanced understanding of irrigation efficiency can improve the beneficial use of limited and declining water resources needed to enhance crop and food production from irrigated lands.

REFERENCES

1. Israelsen, O.R.; Hansen, V.E. *Irrigation Principles and Practices*, 3rd Ed.; Wiley: New York, 1962; 447.
2. ASCE; Describing Irrigation Efficiency and Uniformity. J. Irrig. Drain. Div., ASCE **1978**, *104* (IRI), 35–41.
3. Bos, M.G. Standards for Irrigation Efficiencies of ICID. J. Irrig. Drain. Div., ASCE **1979**, *105* (IRI), 37–43.
4. Heermann, D.F.; Wallender, W.W.; Bos, M.G. Irrigation Efficiency and Uniformity. In *Management of Farm Irrigation Systems*; Hoffman, G.J., Howell, T.A., Solomon, K.H., Eds.; Am. Soc. Agric. Engrs.: St. Joseph, MI, 1990; 125–149.
5. Burt, C.M.; Clemmens, A.J.; Strelkoff, T.S.; Solomon, K.H.; Bliesner, R.D.; Hardy, L.A.; Howell, T.A.; Eisenhauer, D.E. Irrigation Performance Measures: Effi-

ciency and Uniformity. J. Irrig. Drain. Eng. **1997**, *123* (3), 423–442.

6. Howell, T.A. Irrigation Efficiencies. In *Handbook of Engineering in Agriculture*; Brown, R.H., Ed.; CRC Press: Boca Raton, FL, 1988; Vol. I, 173–184.

7. Merriam, J.L.; Keller, J. *Farm Irrigation System Evaluation: A Guide for Management*; Utah State Univ.: Logan, UT, 1978; 271.

8. Keller, J.; Bliesner, R.D. *Sprinkle and Trickle Irrigation*; The Blackburn Press: Caldwell, NJ, 2000; 652.

9. U.S. Salinity Laboratory Staff. *Diagnosis and Improvement of Saline and Alkali Soils*; Handbook 60; U.S. Govt. Printing Office: Washington, DC, 1954; 160.

10. Christiansen, J.E. *Irrigation by Sprinkling*; California Agric. Exp. Bull. No. 570; Univ. of Calif.: Berkeley, CA, 1942; 94.

11. Hart, W.E. Overhead Irrigation by Sprinkling. Agric. Eng. **1961**, *42* (7), 354–355.

12. Heermann, D.F.; Hein, P.R. Performance Characteristics of Self-Propelled Center-Pivot Sprinkler Machines. Trans. ASAE **1968**, *11* (1), 11–15.

13. Warrick, A.W. Interrelationships of Irrigation Uniformity Terms. J. Irrig. Drain. Eng., ASCE **1983**, *109* (3), 317–332.

14. Keller, J.; Karmeli, D. *Trickle Irrigation Design*; Rainbird Sprinkler Manufacturing: Glendora, CA, 1975; 133.

15. Nakayama, F.S.; Bucks, D.A.; Clemmens, A.J. Assessing Trickle Emitter Application Uniformity. Trans. ASAE **1979**, *22* (4), 816–821.

16. Viets, F.G. Fertilizers and the Efficient Use of Water. Adv. Agron. **1962**, *14*, 223–264.

17. Tanner, C.B.; Sinclair, T.R. Efficient Use of Water in Crop Production: Research or Re-Search? In *Limitations to Efficient Water Use in Crop Production*; Taylor, H.M., Jordan, W.R., Sinclair, T.R., Eds.; Am. Soc. Agron., Crop Sci. Soc. Am., Soil Sci. Soc. Am.: Madison, WI, 1983; 1–27.

Irrigation Impact on River Flows

Robert W. Hill
Utah State University, Logan, Utah, U.S.A.

Ivan A. Walter
Ivan's Engineering, Inc., Denver, Colorado, U.S.A.

INTRODUCTION

The practice of irrigation necessitates developing a water source, conveying the water to the field, application of the water to the soil and collection and reuse or disposal of tailwater and subsurface drainage. These processes alter river basin hydrology and water quality in space and time. To sum up the effect of irrigation on a watershed in a word, it would be: *DEPLETION*.

In hydrologic studies it is common engineering practice to quantify the impact upon the stream(s) from which the irrigation water is diverted. The impact upon the stream is actually of two kinds: 1) diversions that decrease the streamflow and 2) return flows that increase the stream-flow. The engineering term used to describe the overall impact is "streamflow depletion" which means the net reduction in streamflow resulting from diversion to irrigation uses. Actual stream depletions are a function of many factors including the amount and timing of diversions, the type of diversion structure (well vs. ditch), crops grown, soil type, depth to groundwater, irrigation method, irrigation efficiency, properties of the alluvial aquifer, area irrigated, and evapotranspiration of precipitation, groundwater, and irrigation water.

Depletion

Depletion, in this context, is the consumptive abstraction of water from the hydrologic system as a result of irrigation. It is in addition to consumptive water use that would have occurred in the unmodified natural situation. As an example, waters of the Bear River Basin of Southern Idaho, Northern Utah, and Western Wyoming, because it is an interstate system, are administered by a federally established commission under the authority of the Bear River Compact.[1] Depletion is the basis, in the compact, for allocating Bear River water use among the three states. It is defined by a "Commission Approved Procedure" which includes consideration of land use and incorporates an equation for estimating depletion based on evapotranspiration. In a study for the commission, Hill[5] defined crop

depletion as:

$$Dpl = Et - SMco - Pef \tag{1}$$

where Dpl is estimated depletion for a given site or sub-basin; Et is calculated crop water use; SMco is moisture which is "carried over" from the previous nongrowing season (October 1–April 30) as stored soil water in the root zone available for crop water use subsequent to May 1; and Pef is an estimate of that portion of precipitation measured at an NWS station during May–September, which could be used by crops.

The carry-over soil moisture (SMco) was estimated by assuming that 67% of adjusted precipitation from October through April could be stored in the root zone. If this exceeded 75% of the available soil water-holding capacity of the average root zone in the sub-basin, the excess was considered as lost to drainage or runoff and not available for crop use. Growing season precipitation was considered to be 80% effective in contributing to crop water use. The effectiveness factor of 80% allowed for precipitation depths throughout a sub-basin that might differ from NWS rain–gage amounts. It also included a reduction for mismatches in timing between rainfall events and irrigation scheduling.

HYDROGRAPH MODIFICATION

Diversion of significant amounts of water from rivers and streams for irrigation and subsequent return flows alters the shape and timing of downstream hydrographs. In watersheds where mountain snowmelt provides the irrigation supply, such as in the Western United States, diversion during the spring runoff attenuates the peak flow rate while later return flows extend the flow duration into late summer and early fall.

Reservoir Storage

Storage of water in reservoirs can significantly modify the natural stream hydrograph depending on the timing and

Encyclopedia of Water Science
DOI: 10.1081/E-EWS 120010370

quantity of the storage right. Irrigators with junior rights may only be able to store during time periods with low irrigation demand, such as during the winter, or during peak flow periods. Reductions of stream flow during the winter time may have considerable impact on downstream in-stream flows. Whereas, storage during periods of peak runoff may not affect minimum in-stream flow needs, but could deposit considerable amounts of sediment in the reservoir.

Irrigation Return Flows

Irrigation return flows are comprised of surface runoff and/or subsurface drainage that becomes available for subsequent rediversion from either a surface stream or a groundwater aquifer downstream (hydrologically) of the initial use. Reusable return flow can be estimated as irrigation diversion minus crop related depletions minus additional abstractions. Additional abstractions include incidental consumptive use from water surfaces as in open drains, along with noncrop vegetation. The timing of return flow varies from nearly instantaneous (recaptured tailwater) to delays of weeks and months or perhaps longer with deep percolation subsurface drainage. In a hydrologic model study of the Bear River Basin[4] delay times between diversion and subsequent appearance of the return flow at the next downstream river gage varied from 1.5 months to as long as 6 months. The delay appeared to be related to sub-basin shape and size.

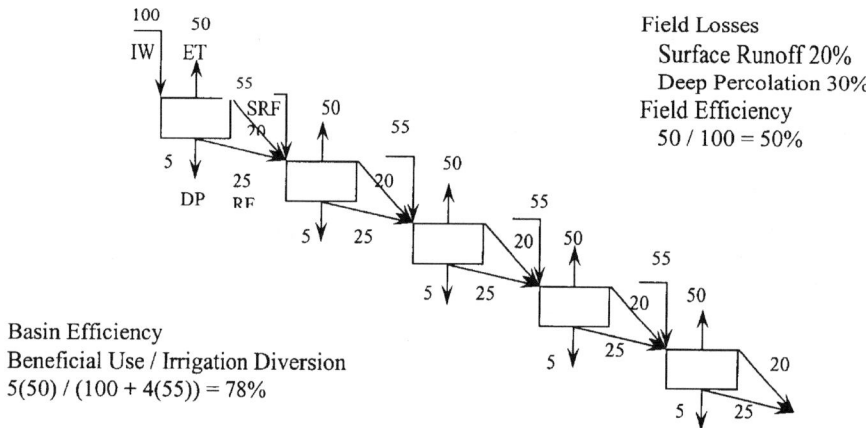

IW – Irrigation Water Supply
ET – Evapotranspiration (beneficial use) from irrigation water
DP – Deep Percolation below root zone to subsurface water
SRF – Surface Return Flow
RF – Return Flow of subsurface water
EV – Evaporation from droplets in air and wind drift losses with sprinklers

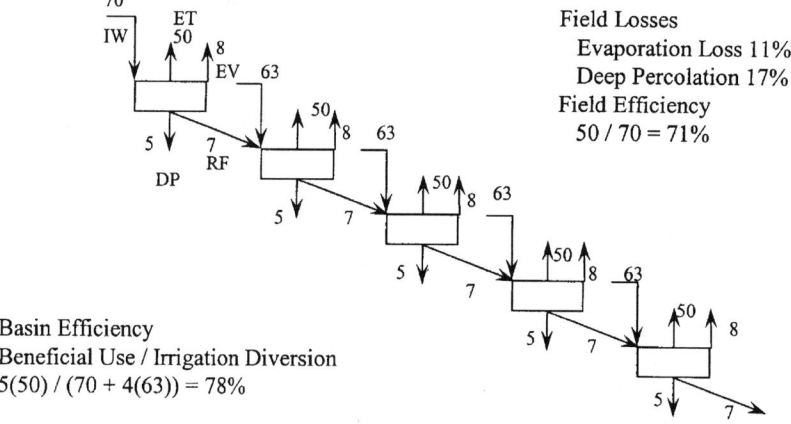

Fig. 1 Comparison of basin efficiencies between surface and sprinkler irrigation methods with four return flow reuse cycles.

Irrigation Methods

Four general irrigation methods are used: surface, subsurface, sprinkler, and trickle (also known as low flow or drip). Surface methods include wild or controlled flooding, furrow, border-strip, and ponded water (basin, paddy, or low-head bubbler). Hand move, wheel move, and center pivot are examples of sprinkler irrigation. Trickle irrigation includes point source emitters, microspray, bubbler, and linesource drip tape (above or below ground). Whereas the efficiency of surface irrigation is dependent upon the skills and experience of the irrigator, the performance of trickle and sprinkler systems is more dependent on the design. Generally, the more control that the system design (hardware) has on the irrigation system performance, the higher the application efficiency (E_a) can be. Thus, typical wheel move sprinklers have higher E_a values than surface irrigation, but lower values than for center pivots or trickle, assuming better than average management practices for each method.

The impact on river flows can be quite different among the various irrigation methods. The nature of furrow and border surface irrigation generally produces tail water runoff, which can be immediately recaptured and reused, as well as deep percolation, which may not be available for reuse until after a period of time. Tailwater is essentially eliminated and deep percolation reduced with sprinklers (Fig. 1) compared to conventional surface irrigation. Whereas, with drip methods, deep percolation can be further reduced. The reduction of deep percolation implies increased salt concentration in the root zone leachate, but, perhaps significant reduction in salt pick-up potential from geologic conditions.

Irrigation Efficiencies

Although a full discussion of the several variations of irrigation efficiency is beyond the scope herein, two terms will be defined and discussed. More complete discussions relating to irrigation efficiencies and water requirements are given elsewhere.[6–9,13] Keller and Bliesner[9] give a particularly thorough presentation of distribution uniformity and efficiencies.

Application efficiency (E_a) :

$$E_a = 100 \times \frac{\text{Volume of water stored in the root zone}(V_s)}{\text{Volume of water delivered to farm or field}(V_f)}$$

Distribution uniformity :

The distribution uniformity is a measure of how evenly the on-farm irrigation system distributes the water across the field. The definition of DU is:

$$DU = 100 \times \frac{\text{Average of the lowest 25\% of infiltrated water depth}}{\text{Average of all infiltrated water depths across the field}}$$

On-farm or field application efficiencies can be affected by the distribution uniformity and vary widely for both surface and sprinkle irrigation methods. This is largely due to difference in management practices, appropriateness of design in matching the site conditions (slope, soils, and wind), and the degree of maintenance. In addition, for a given system uniformity, the higher the proportion of the field that is adequately irrigated (i.e., infiltrated water refills the soil water deficit) the lower will be the application efficiency. This is due to greater deep percolation losses in the overirrigated portions of the distribution pattern. Some values determined in recent Utah field evaluations are:

Method	Observed		
	High (%)	Low (%)	Typical (%)
Surface irrigation			
E_a	72	24	50
Tailwater	55	5	20
Deep percolation	65	20	30
Sprinkler irrigation			
E_a	84	52	70
Evaporation	45	8	12
Deep percolation	37	8	18

The E_a for a particular field may vary greatly during the season. Cultivation practices, microconsolidation of the soil surface and vegetation will alter surface irrigation efficiency both up and down from the seasonal average. Seasonal and diurnal variations in wind, humidity, and temperature will also affect sprinkle application efficiencies.

BASIN IRRIGATION EFFICIENCY

The actual irrigation efficiency realized for several successive downstream fields where capture and reuse of return flows is experienced is higher than the E_a of an individual field. This notion of "Basin Irrigation Efficiency"[12,13] is illustrated in Fig. 1. This simple example comparison of surface and sprinkle methods assumes four reuse cycles. In each of the five "fields" Et is assumed to be 50 units. The surface runoff is captured for reuse on the next field. All of the irrigation-related evaporation is assumed "lost" as well as 5 units of deep percolation. After the fifth field, all surface and subsurface flows are lost. The basin efficiency for surface is 78%, which is the same as for sprinkle. The surface irrigation basin efficiency increase is dependent upon the surface return flow reuse, which is 20 units in this example. However, the depletion is greater for sprinkler due to the extra evaporation. In a Colorado field study, Walter and Altenhofen[11] found a progressive increase in irrigation efficiencies from field (average E_a of 45%), to farm, to efficiency of ditch or sectors (average of 83%). This was due to the reuse of tailwater (10%–20% of delivery) and deep percolation (46% of delivery).

ENVIRONMENTAL CONCERNS

The process of evapotranspiration, or crop water use, extracts pure water from the soil water reservoir, which leaves behind the dissolved solids (salts) contained in the applied irrigation water. The "evapoconcentration" of salts is an inevitable result of irrigation for crop production. As stated by Bishop and Peterson[2]:

> "…Other uses add something to the water, but irrigation basically takes some of the water away, concentrating the residual salts. Irrigation may also add substances by leaching natural salts or other materials from the soil or washing them from the surface. Irrigation return flow is a process by which the concentrated salts and other substances are conveyed from agricultural lands to the common stream or the underground water supply…"

Water Quality Implications for Agriculture

Irrigated agriculture is dependent upon adequate, reasonably good quality water supplies. As the level of salt increases in an irrigation source, the quality of water for plant growth decreases. Since all irrigation waters contain a mixture of natural salt, irrigated soils will contain a similar mix to that in the applied water, but generally at a higher concentration. This necessitates applying extra irrigation water, or taking advantage of nongrowing season precipitation, to leach the salts below the root zone.

Salt Loading Pick-Up

Water percolating below the root zone or leaking from canals and ditches may "pick-up" additional salts from mineral weathering or from salt–bearing geologic formations (such as the Mancos shale of Western Colorado and Eastern Utah). This salt pick-up will increase the salt load of return flows and consequently increase the salinity of receiving waters.

In the Colorado River Basin in the United States and Mexico salinity is a concern because of its adverse effects on agricultural, municipal, and industrial users.[10] The Salinity Control Act of 1974 (Public Law 93-320) created the Colorado River Basin Salinity Control Program to develop projects to reduce salt loading to the Colorado River. Salinity control projects include lining open canals and laterals (or replacing with pipe) and installing sprinklers in place of surface irrigation for the purpose of decreasing salt loading caused by canal leakage and irrigated crop deep percolation. Recently selenium in irrigation return flow has become a concern[3] and may also be reduced by salinity reduction projects.

In-Stream Flow Requirements

Diversions in some reaches in some Western United States streams are "dried up" immediately downstream of diversion structures during times of peak irrigation demand. This condition eliminates any use of the reach for fisheries and other uses which depend on in-stream flow. In some instances, negotiated agreements with senior water rights users have allowed for bypass of minimal amounts of water to sustain the fishery or habitat, and for control of tailwater runoff to reduce agricultural related chemicals in the receiving water.

REFERENCES

1. Bear River Commission. *Amended Bear River Compact.* S.B. 255, 1979 Utah Legislature Session. 1979.
2. Bishop, A.A.; Peterson, H.B. (team leaders). Characteristics and Pollution Problems of Irrigation Return Flow. Final Report Project 14-12-408, Fed. Water Pollution Control Adm. U.S. Dept. of Interior (Ada, Oklahoma). Utah State University Foundation, Logan, Utah. May 1969.
3. Butler, D.L. *Effects of Piping Irrigation Laterals on Selenium and Salt Loads, Montrose Arroyo Basin, Western Colorado*; U.S. Geological Survey Water-Resources Investigations Report 01-4204 (in cooperation with the U.S. Bureau of Reclamation): Denver, Colorado, 2001.
4. Hill, R.W.; Israelsen, E.K.; Huber, A.L.; Riley, J.P. *A Hydrologic Model of the Bear River Basin*; PRWG72-1, Utah Water Resource Laboratory, Utah State University: Logan, Utah, 1970.
5. Hill, R.W.; Brockway, C.E.; Burman, R.D.; Allen, L.N.; Robison, C.W. *Duty of Water Under the Bear River Compact: Field Verification of Empirical Methods for Estimating Depletion. Final Report*; Utah Agriculture Experiment Station Research Report No. 125, Utah State University: Logan, Utah, January 1989.
6. Hoffman, G.J.; Howell, T.; Solomon, K. *Management of Farm Irrigation Systems*; The American Society of Agricultural Engineers: St. Joseph, MI, 1990.

7. Jensen, M.E. *Design and Operation of Farm Irrigation Systems*; ASAE Monograph, American Society of Agricultural Engineers: St. Joseph, MI, 1983.

8. Jensen, M.E., Burman, R.D., Allen, R.G., Eds. *Evapotranspiration and Irrigation Water Requirements*; ASCE Manual No. 70; American Society of Civil Engineers: New York, NY, 1990.

9. Keller, J.; Bliesner, R.D. *Sprinkle and Trickle Irrigation*; Van Nostrand Reinhold: New York, NY, 1990.

10. U.S. Department of the Interior, *Quality of Water—Colorado River Basin: Bureau of Reclamation, Upper Colorado Region*; Progress Report no 19; Salt Lake City, Utah, 1999.

11. Walter, I.A.; Altenhofen, J. Irrigation Efficiency Studies—Northern Colorado. *Proceedings USCID Water Management Seminar. Sacramento, CA October 5–7*; USCID: Denver, CO, 1995.

12. Willardson, L.S.; Wagenet, R.J. *Basin-Wide Impacts of Irrigation Efficiency*; Proceedings of an ASCE Specialty Conference on Advances in Irrigation and Drainage, Jackson, WY, 1983.

13. Willardson, L.S.; Allen, R.G.; Frederiksen, H.D. *Elimination of Irrigation Efficiencies*; Proceedings 13th Technical Conference, USCID: Denver, CO, 1994.

Irrigation Management in Humid Regions

Edward John Sadler
Carl R. Camp, Jr.
*United States Department of Agriculture (USDA), Florence,
South Carolina, U.S.A.*

James E. Hook
University of Georgia, Tifton, Georgia, U.S.A.

INTRODUCTION

Irrigation management includes deciding how much irrigation water to apply, and when to start and stop the irrigation. For any management decision, the choice of operation depends on what one wants to do. The simple answer may appear to be "put on some water," but the choice is often more complex. For instance, one can attempt to maximize net return, minimize operating costs (especially labor), maximize yield, optimize limited water supply, minimize environmental risk, or optimize production under a limited irrigation system capacity. All of these may be constrained by regulations. In general, water supply and irrigation costs control the economics, so the best result is obtained by maximizing yield on all irrigated land, usually called the land-limiting case. In simple terms, one irrigates to avoid crop water stress.

Important irrigation system parameters for consideration include the irrigation application rate per unit area, the total system supply rate, and for moving systems, the velocity. At a given pumping rate, moving machines cover their entire irrigated area in a given return time. If more application depth is required, the system can be set for a slower velocity, which trades increasing depth for longer operating times. For solid-set systems, including sprinklers and drip, increasing application depth is achieved by operating the system longer.

All these considerations apply for both arid and humid areas, and are covered briefly here so that the reader may interpret the article without referring elsewhere. For the following, the discussion concentrates on the particular case of humid areas, contrasting with the conventional, more-arid case.

CONTEXT OF HUMID AREAS

Humid area climate differs from arid area climate in several ways. First is the defining characteristic, humidity. In humid areas, the dew point often equals the early morning air temperature, unlike most arid areas, and the difference in vapor pressure deficit affects crop temperature. Along with higher humidity comes, generally, more clouds. These reduce the total daily solar radiation, which reduces the evaporative demand. Finally, humid areas generally receive more rain than arid areas. The possibility of rain occurring just after an irrigation can complicate an irrigation manager's decision in two important ways beyond applying unnecessary water. One is the risk of waterlogging a crop and causing damage by lack of aeration. The other is risk of leaching nutrients or other chemicals.

On the other hand, during periods without rain, the weather in humid areas can be similar to that in arid areas. Rain-free days are generally less cloudy, which then means more nearly clear-sky solar radiation and higher evaporative demand. Also, air temperature may be higher and humidity lower than averages, which include the cooler wet days. Similarity between humid and arid regions during periods of drought, which can occur in as little as two rainless weeks, creates additional challenges for irrigation managers in the humid region. Strategies optimized assuming the next rainfall is imminent can fail if the next rain occurs four or six weeks later.

Another consequence of higher rainfall amounts in humid areas is a radically different water supply system than in arid regions. There are few large water projects with objectives to provide irrigation water, and no extensive water districts to manage the allocation. Therefore, most water supplies must be farmer-developed. Historically, these were farm ponds retaining runoff or streams from either of which farmers pumped directly. Where groundwater was available, wells were added to provide backup to ponds, and where extensive aquifers existed, irrigation expanded using high-capacity wells. Since they were developed individually, farmer water supplies were not regulated, or if so, minimal information was required for permits. As a direct result of this history of irrigation development, little knowledge exists regarding the water withdrawals, irrigation capabilities, or area

Encyclopedia of Water Science
DOI: 10.1081/E-EWS 120010033

irrigated in most humid areas. A useful case study of the difficulties caused by this lack of information can be found in southwestern Georgia.[1]

IRRIGATION SCHEDULING EMPLOYED IN HUMID AREAS

As mentioned earlier, in the usual (land-limiting) case, one achieves the optimum economic return by maximizing yield on all irrigated land, which is achieved by irrigating to avoid crop stress. This can be done in several ways, all of which have been used in humid areas of the eastern United States. One can sense the water status of the crop, measure soil moisture, or compute the soil water balance. There are many variations on these three approaches. Even the fixed time-clock control of lawn and other turf irrigation systems (if adjusted properly) is an attempt to maintain soil moisture in a range suitable for plant health.

Plant Stress Methods

The most well-known indicators of plant water stress are visual: rolled or drooping leaves and color change. However, by the time these conditions occur in crops, yield has already been reduced. Therefore, scientists looked for earlier indicators of water stress. One early stress measurement is the infrared thermometer, which is a noncontact device and is sensitive to longwave ($\sim 10\,\mu m$) radiation, that measures the average temperature in the field of view. For theoretical reasons, the difference between the canopy and the air temperature is the important measure, but the vapor pressure deficit is an important factor in the interpretation of the temperature difference. In humid areas, the canopy temperature may be somewhat higher relative to the air temperature than in arid areas. Both air temperature and vapor pressure deficit are taken into account with the crop water stress index, or CWSI, but research in the humid southeastern United States indicates that additional work is needed before this method can be widely applied.

Other plant stress monitors have been proposed and are included here for reference. Near-infrared (NIR, $\sim 1\,\mu m$) photography and remote sensing have been used in research environments, but other stresses, such as disease or poor nutrition, can also cause NIR responses. Some research has monitored leaf water potential using the pressure chamber or the leaf press. Sap flow devices have been used to measure water movement into plants, particularly for perennial plants. Since water flow is in response to plant water needs and soil water supply, the device can detect periods when these are limited. A recent report from Israel suggested that minute changes in leaf thickness could be sensed as an indication of plant water

status. All of these devices have been useful in research for assessing plant water stress, but expense and/or complexity have limited their application in production.

Successful use of any plant water status measure depends on being able to identify some trigger point at which to initiate irrigation. For the infrared-thermometer-based CWSI method, some have determined responses to initiation at one value or another, say 0.5 in the range 0 (no stress) to 1.0 (complete stress). However, as mentioned earlier, using any absolute value of the CWSI in humid areas requires additional research or possibly local calibration. In addition, scheduling irrigation based on observing stress is subject to an inherent limitation in that it cannot successfully predict what time in the future one needs to irrigate. In practice, experience can overcome this limitation.

These local calibrations may be avoided using an innovative approach to initiating irrigation using an indicator of the variation in soil water that exists across a field. In this approach, an infrared thermometer is read as it is moved across a field, as from the window of a moving vehicle. Irrigation is triggered when the variation in temperature in the series of measurements exceeds a certain amount. Basically, this approach uses the driest area of the field as an indicator; when it gets dry, the rest of the field would not be far behind, so irrigate soon. Because the air temperature, vapor pressure, and other factors are all reasonably constant during the scan, this method can use the actual crop temperature.

Soil Moisture Methods

Research has shown that plants can extract water from soil when it is held somewhere between the field capacity, which happens after free drainage following rain and is between $-0.01\,MPa$ and $-0.03\,MPa$ soil water potential, and the permanent wilting point, usually assumed to be at $-1.5\,MPa$ soil water potential. These concepts have been debated, but use has shown them to be useful approximations. Most of the water contained in the soil is held between the field capacity and $-0.1\,MPa$. For this reason, tensiometers, which can measure water between $0\,MPa$ and $-0.08\,MPa$, can monitor soil water over a range important to irrigation.

Researchers in the southeastern United States have employed tensiometers, with irrigation being triggered when the potential at 0.3-m depth gets drier from $-0.02\,MPa$ to $-0.05\,MPa$. Important considerations include the depth, the position of the sensor relative to the plant row and roots, and also the crop species. Tensiometers must be serviced periodically to remove air and ensure that they have not gotten out of range, which causes unpredictable readings. They must also be monitored frequently in sandy soils because the water removed from the soil in a day might cause the readings to go out of range.

Electrical resistance devices have been embedded in the root zone to measure soil water potential indirectly, using the known relationship between water content and electrical resistance of gypsum. Advantages of these devices are low maintenance and adaptability for reading with simple meters or data loggers. With experience, managers can use these simple devices to indicate that the soil is becoming too dry for continued plant growth.

The other measure of soil moisture is water content. As mentioned earlier, the water content and potential are related through the water-holding capacity function, which may differ for each soil and soil layer. If this relationship is known, soil water content (SWC) sensors can be used to sense soil moisture for irrigation purposes, by determining the water content corresponding to the field capacity and wilting point. Water content can be expressed per unit volume or per unit weight. The only practical method that produces a value per unit weight is the gravimetric technique, in which a sample is weighed, dried, and weighed again. If this technique is used, the mass of soil per unit volume in the original state, or bulk density, must be known to convert to a volume basis. Knowing the water content on a volume basis adds to the irrigation manager's tools because the difference between the water content and the wilting point is an indication of how much water remains, and the difference between field capacity and the water content is how much can be applied at the time of the measurement. Clearly, soil moisture measurements must represent the water content of the effective root zone for this technique to be useful, and the root zone thickness changes with type and usually increases with age of the crop.

Water Balance Techniques

The checkbook-type water balance method has been known for nearly 50 yr. This direct analog to a bank checkbook uses rain and irrigation as credits and evapotranspiration (ET) as a debit to maintain a water content between the field capacity and wilting point for the root zone. One can adjust the rainfall for runoff and drainage below the root zone. Availability of ET data has been the main problem using this technique in the southeastern United States. Evaporation pans have been used, with research testing whether screened or open pans are most reliable.

A physical model of the checkbook method has been implemented using an evaporation pan directly. For this, a calibrated scale is placed on the pan (usually a screened pan) with indications for a full and an empty rooting zone. An inexpensive, recent implementation uses a large washtub specially fitted with a float attached to a flag visible from some distance. When the water level has dropped to a point equivalent to the soil water refill amount, the flag passes a preset mark and irrigation is indicated. An overflow hole is set at a level representing full SWC. Since the device is placed in the field of sprinkler irrigated crops, it receives both irrigation and rainfall, filling to the overflow mark with excessive rain or

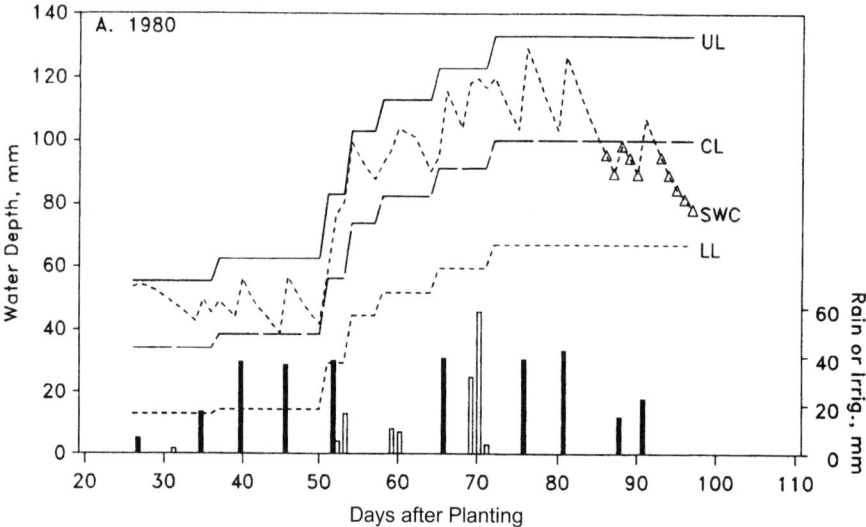

Fig. 1 Water balance technique illustrated. Source was computer-based method in Camp and Campbell, 1988. The lines labeled UL and LL are the upper and lower limits of available water within the root zone. The line CL was the irrigation control point, here at 50% management-allowed depletion. The line SWC is the computed SWC, with triangles flagging the need to irrigate. Solid bars indicate irrigation; open bars indicate rain.

simply adding to the pan as rain and irrigation partially refill the soil and pan.

A computer model of the water balance is simply an automated version of the manual and physical methods mentioned earlier. Usually, ET is calculated from weather station data for temperature and solar radiation, but if ET data were available in published reports, it could be entered. Some computer-based methods use the best available information, starting with measurements, then calculations from weather station data, then calculations from forecast weather, and for predictions beyond the forecast period, historical data. Fig. 1 illustrates the water balance technique. Note the increased water-holding capacity of the profile as the root zone expanded during the season. Up to ~ 85 days after planting, the irrigation in this case was scheduled to successfully control the SWC above the CL line. After that time, the SWC fell below the control limit, and the computer program flagged the line with triangles to indicate the need for irrigation.

Common Considerations

In all of the previous soil-based methods, some decision must be made about the allowable range of soil moisture. Seldom does an irrigator want to allow the soil moisture to drop very close to the wilting point; most trigger points are approximately 30%–50% depletion of available water-holding capacity, for the sake of insurance. Conversely, if irrigation fills the soil rooting zone to field capacity, and rain falls soon after, then the root zone can be subject to aeration problems, and this free water (rain) is lost through runoff or drainage. This choice of management-allowed depletion depends on the probability of rainfall and the likely amount that could be tolerated or used. In this regard, humid-area management is different than arid-area management. The range over which the manager can control soil moisture may need to be restricted to allow for the higher chance of rain.[2]

These requirements, a narrow range for deficit irrigation and inherent limits of the soil water storage, support using frequent, small irrigations rather than less-frequent, large ones. This can lead to a higher fraction of water evaporated directly from the soil surface, which is wet more frequently in the case of sprinkler irrigation, and to increased disease incidence when susceptible crops are frequently wetted. The combination of these considerations brings more interest in buried drip irrigation, which can, if so designed, irrigate frequently, yet keep the soil surface mostly dry.

All measurements represent the area where the measurements are made, but spatial variation will cause any measurement to be unrepresentative of the entire area in most fields. Assuming the field is irrigated the same throughout, how many measurements are needed to represent the entire field? There is no simple answer to this question, nor is there one for the trade offs when distinctly dissimilar soils exist all in one irrigation management unit.

Comparisons

A multi-state study of irrigation scheduling methods concluded that, if properly employed, tensiometers, evaporation pans, and computerized water balance methods, all could be used in the southeastern United States. Tensiometers had one advantage in that they were fairly universal, requiring little calibration. On the other hand, they required significant labor for reading and maintenance. Evaporation pans were also labor-intensive. Computerized water balance models were data-intensive and occasionally needed adjustment of the SWC to eliminate accumulated errors, but were much more amenable to forecasting future irrigation needs.

TECHNICAL ABSTRACT

During the long history of irrigation, management of irrigation systems, which includes deciding how much irrigation to apply and when to do it, has been the subject of much study. Most of this work has been done in primarily arid areas, where the development of irrigation started earlier. However, current trends include increasing irrigated areas in humid regions, for which the contrasting climatic conditions require correspondingly different management techniques. In addition to having higher humidity, humid areas are generally more cloudy (lower solar radiation and thus lower evaporative demand), receive more rainfall, and tend to be cooler on average. However, during even short droughts, the conditions may be quite similar to those in arid regions. Dynamic weather complicates the management of irrigation systems in humid regions, forcing managers to trade off the possibility of rain against the need to leave storage space for potential rain by controlling a relatively narrow range of management-allowed depletion. Doing so can be achieved more easily using frequent, light irrigations instead of less-frequent, heavy ones commonly used in arid regions. Case studies of irrigation management in the southeastern United States serve to illustrate the common management methods, which include tensiometers, evaporation pans, and computer-based water balances. Continuing trends of increasing irrigated area and increasing interest in precision agriculture may combine to focus on spatially variable irrigation management in humid regions.

INTERPRETIVE SUMMARY

Irrigation management includes deciding when to apply irrigation, and also how much to apply. Making these choices in humid regions is somewhat more complicated than in arid ones, primarily because of the possibility of receiving rain shortly after an irrigation. Besides being wasteful, this possibility also carries a risk of drainage and runoff carrying nutrients to groundwater or streams. Managing irrigation to save some room in the soil for possible rain requires a careful balance between crop needs and soil capacity, which can be limited by sandy soils or shallow rooting depths. Management methods leave storage space for potential rain by controlling a relatively narrow range of management-allowed depletion. Doing so in humid regions can be achieved more easily using frequent, light irrigations instead of less-frequent, heavy ones commonly used in arid regions. Case studies of irrigation management in the southeastern United States showed that common methods, which include tensiometers, evaporation pans, and computer-based water balances, can all work. Increases in irrigated area and interest in precision agriculture may combine to focus on spatially variable irrigation management in humid regions.

CONCLUSION

At the current time, two trends in irrigation are apparent. While they may also exist elsewhere, they are somewhat recent in the southeastern United States. The first is the simultaneous increase in irrigated area and increased competition with nonfarm users for water resources. This leads to both a less-than-optimal water supply and higher valuation of the water resource. Therefore, additional questions arise. If one cannot irrigate all the land, where should the water be used to greatest advantage? Should it be used only on high value crops? Should it be applied suboptimally to all the land? Or should the second trend, interest in precision (site-specific) agriculture, be extended to irrigation, so that each individual soil in a field could be irrigated optimally, or less-productive soils be left rainfed while productive soils are irrigated optimally? These questions are currently of increasing interest to researchers. Should the southeastern drought of 1998–2002 continue, they will likely be of increasing interest as well to producers.

REFERENCES

1. Hook, J.E. *Water Crisis in the Humid Southeast: Implications for Farm Irrigation*; Univ. Georgia NESPAL: Tifton, GA, 1999.
2. Camp, C.R.; Sadler, E.J.; Sneed, R.E.; Hook, J.E.; Ligetvari, F. Irrigation for Humid Areas. In *Management of Farm Irrigation Systems*; Hoffman, G.J., Howell, T.A., Solomon, K.H., Eds.; Am. Soc. Agric. Eng.: St. Joseph, MI, 1990; Chap. 15, 549–578.

Irrigation Management for the Tropics

R. Sakthivadivel
International Water Management Institute (IWMI), Colombo, Sri Lanka

Hilmy Sally
International Water Management Institute (IWMI), Pretoria, South Africa

INTRODUCTION

The tropics refer to "that part of the world located between 23.5 degrees north and south of the equator." The tropics make up 38% of the earth's land surface (approximately 5 billion ha) and 45% of the world's population, estimated at 6.1 billion in 2000,[1] live there. About 75 countries, most of them "developing," lie wholly or mostly in the tropics.* The data and statistics that will be presented in this article are based on entire countries. Hence, even though some parts of southern China are within the latitudinal definition of the tropics, China has not been included. On the other hand, India, Bangladesh, Mexico, and Brazil have been taken into account although parts of these countries lie outside the boundaries. With this provison, the major irrigated countries such as China, Iran, Pakistan, the Russian Federation, Turkey, and the United States of America fall outside the tropics.

This article will first present the climate and other key features of the tropics, with special emphasis on water, irrigation, and food production. Trends in irrigated agriculture and changes in land–water–people balances are then shown. Finally, strategies and conditions to promote effective and sustainable management of irrigation and water resources are discussed.

The need to adopt a holistic approach, integrating the technical, social, and institutional aspects of irrigation and water resources management, is highlighted.

---

*The main countries include: Angola, Bangladesh, Benin, Bolivia, Brazil, Burkina Faso, Burundi, Cambodia, Cameroon, Central African Republic, Chad, Colombia, Congo (Democratic Republic), Congo (Republic of), Costa Rica, Côte d'Ivoire, Cuba, Djibouti, Dominican Republic, Ecuador, El Salvador, Eritrea, Ethiopia, Gabon, Gambia, Ghana, Guatemala, Guinea, Guinea Bissau, Haiti, Honduras, India, Indonesia, Kenya, Korea (North), Korea (South), Liberia, Madagascar, Malawi, Malaysia, Mali, Mauritania, Mauritius, Mexico, Mozambique, Myanmar, Namibia, Nicaragua, Niger, Nigeria, Oman, Panama, Papua New Guinea, Paraguay, Peru, Philippines, Réunion, Rwanda, Sao Tome & Principe, Senegal, Sierra Leone, Somalia, Sri Lanka, Sudan, Swaziland, Tanzania, Thailand, Togo, Uganda, Uruguay, Venezuela, Vietnam, Yemen, Zambia, Zimbabwe.

CLIMATE AND CROPS

Tropical temperatures remain fairly constant throughout the year with a mean monthly temperature variation of 5°C or less between the average of the three warmest and the three coldest months.[2] Annual rainfall varies from 0 mm to 10,000 mm, decreasing with increasing latitude, with a high year-to-year and monthly variability. Indeed, the climate of the region is largely determined by the distribution of rainfall rather than the total amount. Three main zones can be delimited on this basis: 1) permanent humid zone, experiencing 9.5 mo–12 mo of rain, located very close to the equator, and occupying roughly one-fourth of the tropics; 2) seasonally humid zone (4.5 mo–9.5 mo of rain), covering one-half of the tropics and where most crops (including rice in the monsoon region of Asia) are grown; 3) semiarid zone, receiving 2 mo–4.5 mo of rainfall and providing good growing conditions for crops such as maize and cotton.

Although agriculture is the main economic activity in the tropical regions, the proportion of cultivated lands (about 10%) is virtually the same as in the temperate region. But tropical soils tend to be more fragile and failure to replace soil nutrients can ultimately undermine their productive capacity. Rice, cassava, corn, wheat, sorghum, and millet are among the main crops grown in the tropics. Ninety percent of rice production comes from tropical Asia.

IRRIGATION SYSTEMS AND RESOURCE ENDOWMENTS

Irrigation systems in the tropics are subject to both shortage and surplus of water at different times, posing special problems for managing irrigation water efficiently and productively. Differences in rainfall regimes can have a significant impact on irrigation management and irrigation performance. For example, in a situation where rainfed cropping is possible, farmers may cultivate extents of land beyond what the irrigation system had been designed to support. This would complicate management in the event of

Encyclopedia of Water Science
DOI: 10.1081/E-EWS 120010034

a drought because all the crops could need water irrespective of whether they were authorized for irrigation or not; while in more favorable rainfall conditions, little or no irrigation water may be needed even for authorized crops.

In contrast, farmers in arid areas would plan their activities in expectation of irrigation rather than depending on rainfall. As irrigation is more predictable and the demands relatively stable, management is simplified. Producers in such low-rainfall areas know what to expect by way of water supply in the rainy season and would accordingly limit the extent of high-value, water sensitive crops with the balance area under low-value crops or kept fallow. If there is ample rainfall, it will benefit all the planted crops. But if the rains fail, low-value crops would probably be lost, whereas high-input crops would survive due to irrigation. Access to alternate sources of water such as a well could allow high value crops over the whole farm. Given the costs associated with pumping, producers would also tend to make effective use of the surface water supply (subject to any legal constraints) and rainfall. Indeed, how well irrigation systems make use of rainfall will have decisive implications for water management and for efficiency of water use.

Land and labor endowments can also affect irrigation development and management, and the relative prosperity of a region. The eastern and western Gangetic plains in India, lying in two different climatic zones, provide a good example.

Population density in the eastern Gangetic plain has always been higher than in the west, with most suitable land farmed. Higher population pressure has resulted in extension of cultivation to less favorable areas with an attendant decline in farm size. But the persistence of yield-limiting constraints means that even in a good year there is little surplus. In contrast, in the western Gangetic plain, irrigated agriculture has greatly expanded, cropping patterns changed, and higher yields are obtained. This illustrates the fact that favorable shifts in the relative endowments of land and labor in association with technologies that improve the reliability and predictability of agricultural conditions can greatly enhance regional and national productivity.

TRENDS IN IRRIGATED AGRICULTURE

Population and its growth are crucial factors that drive water development strategies whether for food production, or for domestic or industrial purposes. In this section we will examine the trends in irrigation, food production and food consumption, and assess the impacts on water availability and water use in the tropics.

Fig. 1 shows the past and projected trends in population growth and per capita internally renewable water resources (IRWR) in the tropics for the 60-year period 1965–2025. Between 1965 and 1995 the population has nearly doubled and the per capita IRWR halved. If population growth in the tropics follows the United Nations (UN) medium projection path, this will result in a further 20% decrease in IRWR. The per capita IRWR

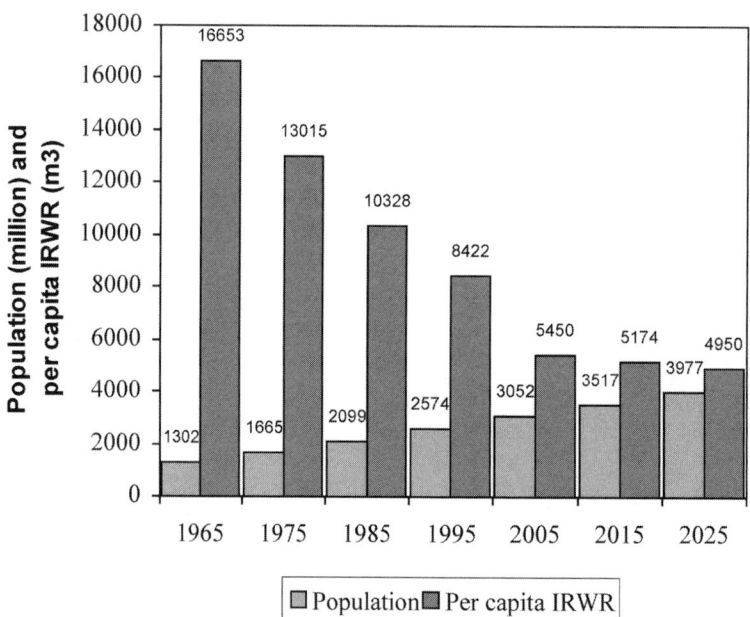

Fig. 1 Trends in population and internally renewable water resources in the tropics. (Adapted from Ref. [3].)

available in 2025 is projected to be 5000 m^3 (kiloliters), which is still considered enough to meet the water needs of each person in the tropics. However, this aggregate figure masks substantial spatial and temporal variations between countries and regions on one hand, and between different times in the year on the other.

Table 1 shows that in 1995, the average per capita calorie supply in the tropical countries was 2456 kcal. Assuming that a calorie supply of 2700 kcal–3000 kcal per person per day is needed to meet most of the nutritional requirements of people in developing countries, this indicates that a substantial number of poor people in this region suffer from nutritional deficiencies. Table 2 shows the production and productivity statistics for the same 30-year period between 1965 and 1995.

In 1995, the total cereal consumption was 529 million metric tons (Table 1), against a total cereal production of only 470 million metric tons (Table 2), indicating a production deficit of 59 million metric tons. According to IWMI,[3] this shortfall would increase to 98 million metric tons to meet a targeted calorific requirement of 2747 kcal in 2025, even though this is still lower than the global average. The above analysis brings to light the urgent need to increase cereal production in the tropics in the next few decades to meet the food and nutritional requirement of its people.

Table 1 also shows that although the total amount of cereal consumed increased by 147% during the 30-year period 1965–1995 (at an average annual growth rate of 3.1%), there has not been a commensurate increase in per capita calorie supply, which has recorded an overall increase of only 19% in the same period.

From Table 2 it can be observed that cereal production in the tropics has increased by 137% in the 30 yr since 1965 (at nearly 2.9% per year), largely as a result of the near doubling of both irrigated area and yield. While the cereal irrigated area in the tropics is only 30% of the cultivated area, it provides nearly 50% of the cereal production.

In terms of the water resources situation, the total water diverted in the tropics is 1057 km^3, of which the total primary water diverted is only 685 km^3. The rest is return flow. This 685 km^3 represents only 8% of potentially utilizable water resources (PUWR) of the tropics. But 89% of this water is used for irrigation. Hence, improving the productivity performance of existing irrigation systems must be given high priority in efforts to overcome the food and nutritional deficits confronting the tropics. Possible irrigation management strategies to help achieve this goal are discussed in the next section.

STRATEGIES FOR IMPROVING AND SUSTAINING IRRIGATED AGRICULTURE

While irrigated agriculture has made significant contributions to the food security of growing populations, concerns have also been expressed about its performance and some of its less desirable consequences: the poor returns to irrigation investments, the environmental impacts (such as waterlogging and salinization) brought about by poor design and operation of irrigation schemes, and the lack of attention paid to the needs of the poorer sections of society. So, the challenge is to find ways of improving and sustaining food security and livelihoods that make optimum use of available resources and do not degrade the productive capacity of land and water.

While increasing cropped area, either by expansion of irrigation facilities or increasing crop intensity remains a possibility, there must also be emphasis on improving land and water productivity through the adoption of appropriate agricultural and water management practices. Measures will include: developing high yielding crop varieties, promoting innovative low-cost techniques for water harvesting and water application, conjunctive use of surface and groundwater, and implementation of institutional and policy reforms to enable integrated water resources management at the basin level, recognizing the multiple uses and users of water.

Table 1 Total population, calorie supply, and cereal consumption in the tropics.

Year	Population		Per capita calorie supply		Cereal consumption	
	Population (million)	Annual growth (%)	Total (kcal)	Annual growth (%)	Total (M Mt)	Annual growth (%)
1965	1302	—	2061	—	214	—
1975	1665	2.5	2114	0.3	289	3.1
1985	2099	2.3	2277	0.7	399	3.3
1995	2574	2.1	2456	0.8	529	2.9
Growth 1965–1995	97.6%	2.3%	19%	0.6%	147%	3.1%

Source: (Adapted from Ref. [3].)

Table 2 Cereal production, cereal harvested area, cereal yield, and net irrigated area.

Year	Cereal Production		Area		Yield		Irrigated area	
	Total (M Mt)	Annual growth (%)	Total (Mha)	Annual growth (%)	Average (ton/ha)	Annual growth (%)	Net area (Mha)	Annual growth (%)
1965	198	—	220	—	0.90	—	47	—
1975	269	3.1	240	0.9	1.12	2.2	60	2.5
1985	370	3.2	258	0.7	1.43	2.5	76	2.4
1995	470	2.4	279	0.8	1.68	1.7	96	2.3
Growth 1965–1995	137%	2.9%	27%	0.8%	87%	2.1%	104%	2.4%

Source: (Adapted from Ref. [3].)

The Basin Perspective and Water Savings

It is generally recognized that the river basin is the appropriate unit of analysis to assess water availability, water use, and thereby, the scope for water savings. Essentially, water saving means diverting water from nonbeneficial or less beneficial uses and making it available for other more productive uses. For example, flows to saline sinks or unrecoverable water bodies can be minimized through interventions that reduce irrecoverable deep percolation and surface runoff. Similarly, the pollution caused by the movement of salts into recoverable irrigation return flows can be reduced by minimizing the passage of these flows through saline soils or saline groundwater.

Decreasing nonbeneficial depletion of water can also be achieved by: 1) reducing evaporation from water applied to irrigated fields by adopting appropriate precision irrigation technologies such as drip irrigation, or agronomic practices such as mulching, or by changing the crop planting dates to match the period of lower evaporative demand; 2) reducing the evaporation from fallow land; 3) decreasing the area of free water surfaces; 4) decreasing the amount of non- or less-beneficial vegetation, and controlling weeds; and 5) by diverting saline or otherwise polluted water directly to sinks without having to dilute it with fresh water.

Projects are quite often justified on the basis of water savings. But this is misleading because, the commonly used term "irrigation efficiency" ignores water recycling and reuse, phenomena that are prevalent in many irrigation systems. Therefore, one has to be careful in assuming that apparent water savings at field level will automatically result in real water savings. Only proper basin-level analysis will reveal if there are uncommitted outflows available and whether water savings are really possible.[4]

Role of Storage

Different annual rainfall regimes result in different levels of availability of fresh water resources. In the permanent humid climatic zone there is water surplus but because the rainfall distribution is regular, the run-off is quite stable and there are few problems with floods. In many parts of the seasonally humid and semiarid climates, river run-off is irregularly distributed within the year. Heavy rain alternates with dry spells resulting in alternating flood and drought periods. This is particularly serious in monsoon Asia and in the semiarid areas of Africa and India. In such areas, suitable measures to capture and store excess water for irrigation, industry, and domestic use must be adopted. In Sri Lanka and southern India, this has been done over thousands of years through the construction of small storage reservoirs (called "tanks") by building a bund at a strategic location in a catchment area.

In fact, much of the growth in irrigation in the last three decades has been made possible by water development projects ranging from multipurpose storage reservoirs to extensive groundwater extraction from underground aquifers. Storage, whether in reservoirs, small tanks, farm ponds, or groundwater aquifers helps to match water demand and supply especially in drier periods, in the face of spatial and temporal variations in natural water supply. A recent development in India has been the concept of "watershed-based systems for resource conservation, management and use," which involves the optimum use of precipitation through improved water, soil, and crop management. This is accomplished through improving infiltration of rainfall into the soil, run-off collection, and by recovery from wells after deep percolation resulting in the improvement and stabilization of agriculture in the watershed.

As stated in the preceding section, planning for storage is best done on the basis of water resources analysis in a basin perspective. One of the first steps is to ascertain whether the basin in question is open, closed, or semiclosed.[5] Current water use and productivity in the basin must be assessed to determine the extent to which increased demands for irrigated agricultural production can be met by increasing water productivity, and the degree to which increased demands will require increased consumption of water. Then plans to capture and use any uncommitted discharge from open or semiclosed basins can be drawn up. Combinations of small and large surface water storage and groundwater recharge are generally the best systems where they are feasible. In monsoon Asia, research and development are needed on how to manage water under monsoonal conditions, particularly on how to develop effective irrigation management responses to rainfall.

Increasing the Productivity of Water

Not only is per capita water availability decreasing in many developing countries but agriculture's share of water is also declining while water demands are increasing from the industrial and urban sectors. When managing water for agriculture, especially in areas where water rather than land is the limiting resource, it is useful to shift the focus from increasing the productivity of land to increasing the productivity of water. That is, to identify and adopt agricultural and water management practices that achieve more output per unit of water consumed. On one hand, these will include selecting crops or crop varieties that are less water-consuming, or which yield higher physical or economic productivity per unit of water, and improved land preparation and fertilization practices. On the other hand, techniques such as deficit, supplemental, or precision irrigation that allow better control, timing, and reliability of water supplies will enable farmers to apply limited amounts of water to their crops in the time and amount that help realize optimum crop response to water. Any water thereby freed up can, in turn, be reallocated to other uses with potentially dramatic increases in overall economic productivity of water.

Such techniques do not always imply high-tech options but will include simple bucket and drum kits for drip irrigation, pitcher irrigation, small sprinkler systems, level basins, as well as conventional drip and sprinkler systems. Innovative and affordable water management systems such as these are especially important for small farmers in situations where rainfall is limited and uncertain, and where one or two irrigation applications can have a big impact on crop yields and household food security and income.

Sustainable Management of Groundwater

In many tropical countries with high levels of rural poverty, groundwater development offers major opportunities for promoting food and improving livelihoods. Affordable innovations in manual irrigation technologies such as the treadle pump (costing US\$ 12–25 per unit and which can be operated even by children) have dramatically improved poor people's access to groundwater in Bangladesh, Eastern India, and Nepal.[6] The capital requirements to develop groundwater irrigation are generally low and its productivity higher compared to surface irrigation. It offers farmers

irrigation water "on-demand" and responds slower to drought. Farmers also tend to exercise more care in using it because of the costs involved in lifting water, thus maximizing application efficiencies.

The undoubted benefits of groundwater development have to be balanced with the risks of overexploitation and contamination. One of the most serious effects of groundwater depletion is seawater intrusion in coastal aquifers as in the Tamilnadu coast near Madras in India and in the Saurashtra coast of the Western Indian State of Gujarat. While there are only rough estimates of the amount of unsustainable groundwater use, the $1\,m-3\,m$ per year decline in water tables occurring in pump intensive areas of India and China clearly highlights the gravity and magnitude of the problem.

While reducing pumping for irrigation is an obvious response, this will have adverse effects on the outputs from this highly productive form of agriculture and ultimately affect the food security of the concerned countries. More desirable solutions are groundwater recharge and increasing water productivity to achieve the same production with less water.

Regulating groundwater overdraft is a far more complex and difficult issue compared with stimulating groundwater use where it is abundant. Enforcing groundwater laws in developing regions present formidable difficulties given the sheer numbers involved (for e.g., the total number of private tubewells in South Asia is thought to exceed 20 million, largely unregistered and unlicensed, and growing at a rate of 1 million per year). The challenge therefore is to identify appropriate institutional and legal frameworks compatible with the local environment through a careful learning process approach to combat the problem of groundwater overdraft.

Effective Irrigation Management Institutions and Policies

Sound institutions and policies are vital for effective irrigation management and increasing food production while sustaining land and water quality, especially in light of growing demands and competition for water from other sectors. Irrigation management must therefore be viewed within the overall context of an integrated and sustainable approach to water resources management that is sensitive to the requirements of all uses and users of water. This not only entails the formulation of adequate laws, institutions, and policies, but also the development of the requisite organizational capacity and skills for enforcement and regulation.

Key institutional attributes include the demarcation of the roles, rights and responsibilities of the various actors in the water sector, the promotion of new forms of public and private partnerships for investment, operation and maintenance, and the emergence of financially self-reliant service delivery organizations that are responsive and accountable to water users. In fact, the overarching concern must be to ensure meaningful participation of all stakeholders in the whole gamut of planning, operation, maintenance, and management of irrigation schemes.

Improving Irrigation Services

The level of irrigation services provided to farmers is influenced by the physical design of the system, its operation and maintenance, as well as the underlying institutional environment. When farmers are provided with reliable irrigation services, they would be more likely to invest in improved technologies and practices, generally resulting in increased production, higher incomes, and improved irrigation performance. But it is useful to remember that, in general, more flexible and sophisticated technology will require levels of management capability that are not always readily available in rural environments. Thus, in seeking to provide a stable and predictable water supply to farmers, it is quite important to find ways and means of doing so that are commensurate with the available skills and resources. The dilemma of rigidity vs. flexibility in irrigation design, and the interactions between design and system management have been discussed in detail by Horst.[7]

In discussing the large public irrigation schemes of monsoon Asia, Burns[8] has pointed out that structuring such systems to formalize the allocation of sporadic wet season scarcity, while protecting the civil works from producer damage, is a necessary task. He further states that unless these schemes are run as regulated and monitored public utilities, actual system performance will bear little resemblance to design intentions due to the rent-seeking activities of system designers, operators, and individual farmers.

Ensuring reliable irrigation services also implies the establishment of: 1) clear rules and agreements between providers and users giving details of the nature of the service and the compensation arrangements for providing and receiving such services; 2) mechanisms for monitoring and control of obligations; 3) modalities for conflict-resolution; and 4) procedures for modifying and updating agreements. These aspects take on added significance in light of the progressive disengagement of public agencies from irrigation management with attendant transfer of responsibilities to the beneficiaries.

CONCLUSION

Feeding the world's population presents a formidable challenge: there are more mouths to feed, less arable land available, increasing competition for water resources, and major concerns about deteriorating ecosystems. Irrigated agriculture makes undeniable contributions to the food security of people in the tropics, providing nearly 50% of its cereal production. But there is a gap between supply and demand and it is urgent to find sustainable ways to improve the performance of the irrigation sector and to satisfy the nutritional requirements of a growing population. The management of natural resources, especially water, takes on added significance as population increases and changes in resource utilization place greater pressures on land and water.

With fewer opportunities to expand irrigated areas by the development of new systems, and growing demands and competition for water from other sectors, the emphasis must shift to practicing more effective methods of management with particular focus on improving the productivity of water. A vital prerequisite is for irrigation management to be viewed within an overall framework of an integrated, holistic approach to water resources management. This is best done on the basin scale, taking into account the needs of all uses and users but also considering the conjunctive use of surface water, groundwater, and rainfall.

Given the multiple facets of irrigation development and management, a piecemeal and uni-dimensional focus on individual components is bound to produce outputs that fall short of expectations. For instance, concentrating only on the technological aspects of an innovation while neglecting other aspects such as maintenance, institutional support, training, and skills development, will yield disappointing overall results, however, well that individual component has been addressed. Hence, the need to take a holistic view and move towards service-oriented irrigation management with meaningful participation of all interested parties.

REFERENCES

1. http://www.census.gov/ipc/www/popwnote.html (accessed February 2002).

2. Sanchez, P.A. *Properties and Management of Soils in the Tropics*; John Wiley & Sons: New York, 1976.

3. IWMI, *Water for Rural Development: Background Paper on Water for Rural Development*; International Water Management Institute: Colombo, Sri Lanka, 2001.

4. Molden, D.J.; Sakthivadivel, R. Water Accounting to Assess Use and Productivity of Water. Int. J. Water Resour. Dev. **1999**, *15* (1/2), 55–71.

5. Keller, A.; Sakthivadivel, R.; Seckler, D. *Water Scarcity and the Role of Storage in Development*, Research Report 39; International Water Management Institute: Colombo, Sri Lanka, 2000.

6. Shah, T.; Alam, M.; Dinesh Kumar, A.; Nagar, R.K.; Mahendra Singh *Pedaling Out of Poverty; Social Impact of a*

Manual Irrigation Technology in South Asia, Research Report 45; International Water Management Institute: Colombo, Sri Lanka, 2000.

7. Horst, L. *The Dilemmas of Water Division: Considerations and Criteria for Irrigation System Design*; International Water Management Institute: Colombo, Sri Lanka, 1998.

8. Burns, E.R. Irrigated Rice Culture in Monsoon Asia: The Search for an Effective Water Control Technology. World Dev. **1993**, *212* (5), 771–789.

Irrigation Mechanical Systems: Sprinkler

Dennis Kincaid
United States Department of Agriculture (USDA), Kimberly, Idaho, U.S.A.

INTRODUCTION

Sprinkler irrigation can be defined as the controlled distribution of water as discrete droplets through air. Sprinkler devices were first invented during the late 19th and early 20th centuries, primarily for lawns and gardens.[1] Their widespread use for agricultural crops did not come about until the availability of lightweight aluminum pipe and low cost electricity following World War II. Sprinkler irrigation is particularly well suited to rolling topography, shallow soils, and sandy soils, which are difficult to irrigate efficiently with gravity flow surface irrigation systems. Sprinkler systems are now used on one-half of the 50 million irrigated acres in the United States.[2]

Early sprinkler systems used manually moved pipe and fixed or portable pumps and mainlines. Hand-move lines are relatively low cost in terms of equipment investment, but are labor intensive. Labor costs provided the primary incentive to develop mechanically moved sprinkler systems. Another factor has been the increase in farm size and the desire to create automated irrigation systems so that one person can irrigate more land. This article describes the mechanics of the various systems. Special sprinklers, mounting devices, and pressure regulators have been developed for these systems. Information on the sprinklers, system design and management, etc. can be found in the listed references.[3–6]

STATIONARY OR PERIODIC-MOVE LATERALS

A sprinkler lateral is a continuous length of pipe upon which sprinklers are mounted, usually equally spaced. Mechanically moved or hand-moved stationary laterals remain in a fixed position while irrigating, which are then drained and moved to a new predetermined position. They normally irrigate rectangular fields and require several sets to completely irrigate the field.

Sideroll Wheeline

The sideroll wheeline lateral consists of an aluminum pipe that serves as an axle for a series of rigidly attached wheels, the whole of which is rolled sideways to move the entire lateral simultaneously when drained (Fig. 1). The pipe is a special high-strength alloy tubing, 100 mm–125 mm in diameter, capable of withstanding considerable torque. The wheel spacing is usually about 12 m, with sprinklers located midway between the wheels. These laterals are typically 400 m in length, but may be as long as 800 m. The wheel radius must be at least as large as the height of the crops to be grown, typically about 0.75 m–1.0 m. Sideroll laterals are not used to irrigate tall crops such as corn. A powered mover unit located near the center of the lateral provides torque to roll the lateral. Some longer laterals use two movers located approximately one-fourth of the distance from each end and connected by a small rotating shaft to coordinate the movement. The movers are powered by a small gas engine, electric motors, or hydraulic motors. The lateral is moved 2–4 complete revolutions between sets (12 m–15 m). For convenience, some sideroll laterals can be moved by an operator standing at the inlet end of the lateral. A short flexible hose is used to connect the lateral to a water supply outlet. Water is supplied from a fixed mainline with outlets spaced at some unit multiple of the wheel circumference.

Trail-Line Lateral

The trail-line system, also called a movable solid-set, consists of a lateral mounted on two-wheel support towers (see "Continuous-Move Laterals"), which serves as a movable mainline for a set of trailing sublaterals. The trail-lines are lightweight aluminum sprinkler sublaterals typically spaced about 12 m–16 m apart and up to 120 m in length. The powered lateral drags the trail-lines between sets. When one irrigation event is complete, the trail-lines are disconnected, and the lateral is moved to the opposite end of the trail-lines, which are then reconnected to the lateral. The whole system is then moved back across the field dry to the first position, or can irrigate each set in turn as it is moved back to the initial position.

CONTINUOUS-MOVE LATERALS

Continuous-move laterals are those which travel while irrigating, either smoothly or intermittently in small

Encyclopedia of Water Science
DOI: 10.1081/E-EWS 120010064

Fig. 1 Sideroll lateral with mover unit in foreground. Note the small driveshaft parallel to the lateral pipe on the right, which transfers power to the mover mechanism.

increments. There are two main types: those called linears that travel in a nearly straight line path, irrigating rectangular areas; and pivoting laterals that rotate about one fixed end, and thus irrigate circular areas. Both types use the same hardware and differ mainly in the way they are controlled and supplied with water. Continuous-move laterals have been built in many variations and styles over the years, but the most common type in use today is shown in Figs. 2 and 3. The lateral pipe is steel or aluminum (100 mm–250 mm diameter) and is usually supported 3 m–4 m above ground to provide sufficient clearance for most crops. The lateral is made up of several 30 m–55 m long spans, where each span pipe is integrated into a rigid truss. The pipe joints between spans are flexible, and one end of each span is supported by a two-wheeled tower, both wheels being powered. The wheels are typically powered by electric motors, but fluid motors (oil or water) are also used. The movement of the wheels on intermediate towers is controlled by switches or valves, which automatically function to keep the entire lateral in a nearly straight line. Travel speed determines the water application depth.

Linear-Move Laterals

Linear-move laterals are usually supplied with water through a flexible drag hose (Fig. 3), or alternatively through a suction pipe moving in an open canal. In addition, automatic coupling systems have been built. The hose-drag system can travel twice the length of hose in one set. An on-board diesel powered generator or drag cable provides electricity. The wheels normally follow the same track each pass to minimize crop damage. The outermost towers determine travel speed and travel direction. Special

guidance systems must be used to keep the wheels traveling in the same path each pass. One method uses radio antennas to follow a buried cable, while another type follows an aboveground cable or small guide trench. Recently, Global Positioning System receivers have been employed to guide traveling laterals and swingspan pivots (Fig. 2). Traveling laterals up to 800 m in length have been built.

Center-Pivot Laterals

The pivoting lateral or center-pivot system is supplied with water and electrical power through simple swivel couplings at the fixed end. The lateral can rotate continuously about the pivot, so that when the first irrigation is completed the lateral is in position to begin the next irrigation. They require no coupling or uncoupling of hoses or pipes and a pivot without a swingspan requires no guidance system. The low labor requirement of this system has made it very popular, and they are used on about 50% of the sprinkler irrigated land.

The major disadvantage of the center-pivot is the circular irrigated area, which leaves the corners of a square field unirrigated. In large developments, the circles can be nested to minimize the unirrigated area. Where the economics of the situation dictate that the corners must be irrigated, several options exist. A large sprinkler mounted on the outer end of the lateral and controlled to turn only in the corner areas can irrigate a portion of the corners.

Another option is the swingspan corner system (Fig. 2) consisting of an additional span up to 70 m in length that pivots about the outer end of the lateral. As the lateral moves into a corner area, the swingspan pivots outward,

Fig. 2   Center-pivot lateral with swingspan corner system partly extended. Sprinklers are mounted on drop tubes below the lateral.

effectively extending the length of the lateral. The swingspan wheels are steerable, and a buried cable or GPS guidance system similar to the linear-move controls its movement. Sprinklers on the swingspan are automatically sequenced on or off as the swingspan moves outward or retracts, thus maintaining a nearly equal water application per unit area. A center-pivot equipped with a swingspan can irrigate up to 97% of a square field.

Recently, manufacturers have developed traveling laterals that can operate both as linears or pivoting laterals, making it possible to irrigate odd-shaped fields and cover more area with a given length of lateral (Fig. 3). Pivot laterals can also be made towable so that they can irrigate more than one field.

TRAVELING SPRINKLERS

Large traveling sprinklers, called big guns, can throw water up to 70 m, and can irrigate large areas. They require

relatively high-pressure (up to 1000 kPa) water supplies, so operating costs can be quite high.

Hard-Hose Travelers

The most popular traveler, called a hard-hose traveler, consists of a big-gun sprinkler mounted on a cart and supplied by a semirigid hose, which retains its round shape when wound upon a reel (Fig. 4). Water is supplied through the center of the reel from a mainline outlet. Initially, the hose is pulled out along the travel path by a tractor. The reel remains in a fixed position while irrigating and slowly rotates, dragging the hose and sprinkler across the field as the hose is reeled in. A water turbine, reciprocating cylinder, or small gas engine provides power to turn the reel. Reel speed is automatically adjusted to account for the change in diameter due to hose wrap. The hose can be up to 400 m in length and 115 mm diameter. The hose reel is mounted on a trailer for transport by a small tractor.

Fig. 3 Linear-traveling lateral, hose-drag type, with on-board engine. This lateral can also pivot about the inlet structure. Sprinklers are mounted on drop tubes.

Soft-Hose Travelers

A similar traveling sprinkler uses a soft, collapsible hose (Fig. 5). The sprinkler cart is pulled by a cable and winch, usually mounted on the cart itself, and powered by a water turbine or reciprocating cylinder. Initially, the hose is laid out alongside the travel path and the cable is reeled out and attached to a fixed anchor. The sprinkler can travel twice the hose length in one set. The hose must be drained and reeled up between sets.

Boom Travelers

Boom travelers are similar to big-gun travelers except that the gun sprinkler is replaced by a horizontal boom structure extending perpendicular to the travel path, and

Fig. 4 Hard-hose traveler completing a pass. Note inlet supply hose lower right.

Fig. 5 Soft-hose traveler beginning a pass. The sprinkler cart is towed by a cable (not visible).

high enough to clear the crop. This in effect creates a traveling lateral upon which sprinklers or spray heads are mounted. The advantage of this system is that it requires much less water pressure than the large gun sprinklers, and water application uniformity is improved.

CONCLUSION

Mechanically moved sprinkler systems will increase in sophistication as computerized controls are developed, particularly for precision variable water and chemical application. Center-pivot systems will likely predominate because of their inherent advantages, including ease of automation and continuous rotation capability.

REFERENCES

1. Morgan, R.M. *Water and the Land—A History of American Irrigation*; The Irrigation Association: Fairfax, VA, 1993.
2. Pair, C.H., (Ed.) Farm and Ranch Irrigation Survey. *1997 Census of Agriculture*; http://www.nass.usda.gov/ USDA: Washington, DC, 1998.
3. Pair, C.H., (Ed.) *National Irrigation Symposium*; Proceedings of the 4th Decennial Symposium, Phoenix, AZ, Nov 14–16, 2000, American Society of Agricultural Engineers: St. Joseph, MI, 2000.
4. Pair, C.H., (Ed.) *Design and Operation of Farm Irrigation Systems*, ASAE Monograph No. 3; American Society of Agricultural Engineers: St. Joseph, MI, 1980.
5. Keller, J.; Bliesner, R.D. *Sprinkle and Trickle Irrigation*; Chapman and Hall: New York, 1990.
6. Pair, C.H., (Ed.) *Irrigation*, 5th Ed.; The Irrigation Association: Fairfax, VA, 1983.

Irrigation Metering

Albert J. Clemmens
John A. Replogle
United States Department of Agriculture (USDA), Phoenix, Arizona, U.S.A.

INTRODUCTION

The measurement of applied irrigation water is one of the major links in efforts to achieve effective water management worldwide. Measurement of flow for irrigation differs from most municipal and industrial water metering requirements because the water is spread over very large areas. This usually results in the need to measure both very large flows in canals near water supply sources and small flows spread over very large areas, perhaps in small trickle irrigation lines near the points of use. Irrigation was often done with waters that were not needed by other uses, although competition by these other uses is increasing, causing much controversy in water-use planning, policy, and development; therefore enhancing the need for accurate flow metering. Traditionally, flow meters have been classified according to the physical principle or property exploited, such as those related to sound; magnetism; electricity; chemical reactions; mixing; and volume, mass, and energy relations.[1] The device that exploits these properties to interact with the water is called the *primary element*, and produces an indication that can be detected with a *secondary element* for the user to observe, or otherwise use. This classification of meters according to exploited properties can be broadly grouped into *flow-rate meters* or *quantity meters*, according to the effect that is first observable. For example, a weir is a flow-rate meter, and a bucket is a simple quantity meter. Not all meters are currently practical for use in irrigated agriculture. Major restrictions to irrigation applications are often the lack of electric power at the metering site, capital cost, and poor maintenance support. Thus, practical irrigation metering emphasizes low cost, reasonable accuracy, and simplicity and the ability to meter waters with high sediment and/or trash loads. Meters that meet these criteria for the irrigation setting are discussed later. A wide variety of meters are discussed in more detail in Refs. [2,3].

OPEN-CHANNEL FLOW METERING

The most common measurement methods for open channels are: current metering, weirs, and flumes. A few additional metering methods will be discussed, but they represent a small portion of irrigation meters.

Current Metering

Current metering is a common method for the measurement of flow in rivers, streams, and large irrigation canals. In this method, a series of velocity measurements are made at many selected points across the channel, usually with a small propeller meter or a cup-type meter similar in concept to the cup-anemometer used for wind velocity. Other types of velocity meters are entering the market that are based on electro-magnetic and ultrasonic concepts. The large number of velocity measurements, and their statistical averaging, helps to compensate for odd-shaped channels and the various flow velocities that may exist across the channel.

Details of current metering methods are given in Ref. [4]. Current meter measurements are labor intensive and represent only a sample of flow at a particular time. Thus, to be useful for continuous monitoring of flow, a relationship between water depth and flow is used to estimate flow during times between current meter measurements. This adds additional error to the measurement of flow volume over time.[4] For these large flows, measurements with flumes and weirs, discussed later, can be difficult due to the large size of the channel and the need for a drop in water surface as the water passes over the weir or flume.

More recently, ultrasonic meters have been used for stream gauging. Transit-time ultrasonic stream gauging is based on detecting stream flow velocity by using the difference in time for sound transmissions sent obliquely across the stream in opposite directions. This difference is translated into average velocity in the sound path that was sampled. Setting these meters so that the average velocity sampled is related to the average flow velocity is a limitation. Improved Doppler ultrasonic meters depend on reflected sound waves from flowing particles in the flow, rather similar to the action of the familiar Police-Radar units, and they are able to sample a series of locations within the flow profile. These meters are becoming more accurate and less expensive. Both types of ultrasonic meters are also applicable to pipelines.

Encyclopedia of Water Science
DOI: 10.1081/E-EWS 120010143

Weirs and Flumes

Flumes and weirs work on the principal of critical flow, which means that the flow rate is a maximum for a given energy (combination of depth and velocity functions). However, in order to cause critical flow, there must be a drop in water level through the flume or weir. A wide variety of flumes and weirs have been developed since the late 1800s. Most of these are gradually being replaced by a family of flumes and weirs that can be calibrated using computer techniques rather than relying on laboratory calibration. They are called long-throated flumes and broad-crested weirs. For these flumes and weirs, the channel size is reduced by contracting the sides or raising the floor to form what is called a throat section that is long relative to the flow depth, producing nearly parallel flow that can be treated mathematically; hence, these flumes are also called the "computable flumes." Additionally, the amount of water surface drop across the flume or weir needed to provide a measurement is not large, typically about 15% of the depth in the throat. Because they can be calibrated by computer, they can be made of essentially any prismatic shape and can be calibrated to as-built dimensions (Fig. 1). A number of portable and adjustable flumes have been developed for flow survey work, as opposed to permanent installation, and are available commercially (Fig. 2). Further details on these flumes and weirs can be found in Ref. [5]. Software for design and calibration is also available free-of-charge on the web: http://www.usbr.gov/wrrl/winflume/

Long-throated flumes and broad-crested weirs are often the simplest and most cost-effective method for measuring flow rates in open channels. They are used worldwide. However, in order to obtain flow volume, the instantaneous flow-rate measurements must be totaled over time. This requires either periodic observation or continuous recording. Devices for accomplishing this often cost more than the flume or weir itself. Also, there are locations where the opportunity for sufficient water surface drop is not available to produce flow measurements over the full range of desired discharges.

PIPELINE FLOW METERING

While advances in canal flow measurements have significantly aided irrigation water management worldwide, developments in pipe flow measurements have also impacted irrigation flow measurements. A common method for irrigation flow measurement in pipelines, and often the least expensive, has been the propeller meter. Similar to the current meter, the velocity of the turning propeller is related to the average velocity of flow in the pipe. Early meters used mechanical gears to turn a cumulative volume meter, while magnetic indicators of propeller rotation and digital electronics are now common. The biggest problem with these meters continues to be bearing wear from sand that is often in pumped wells, as well as errors in flow caused by poor approaching flow conditions. Pipe bends, elbows, valves, etc., can cause most pipeline flow meters to be inaccurate if they are closer than about 10 pipe diameters upstream and 2 pipe diameters downstream.[1,6]

Modern electronics have greatly improved the ability of secondary devices to monitor primary devices based on well-known primary elements, for example, the differential pressure across a Venturi meter or orifice meter, or the speed of sonic waves across a pipe. These electronic

Fig. 1 Flume in large canal. Note stilling well and the region of wavy water surface. Sill crest is located about midway between these two features. Canal flow depth is about 2.4 m (8 ft) and sill crest is about 1.4 m (4.5 ft) high.

Fig. 2 Adjustable flume being installed while channel continues to flow. Capacity 56 L/sec (2 cfs).

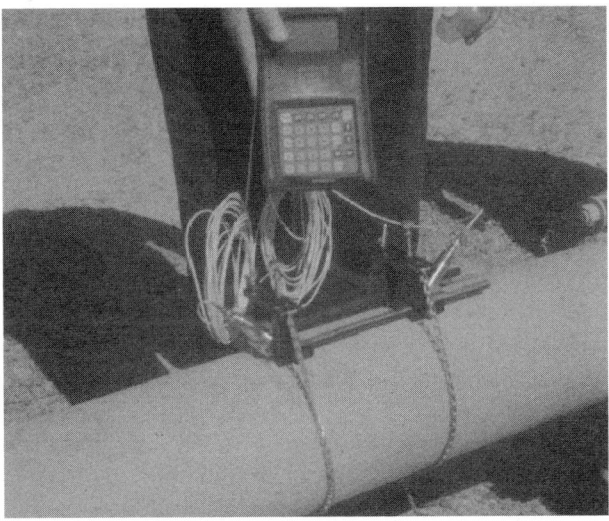

Fig. 3 Portable ultrasonic flow meter on outlet pipe of an irrigation well. Sensors are usually mounted on the side of pipe to avoid air bubbles that may be in the pipe.

advances have resulted in lower-cost metering systems, often with improved accuracy. Many metering techniques depend heavily on these advances in electronics, such as vortex-shedding meters, ultrasonic Doppler flow meters, and the ultrasonic transit-time flow meters. Many of the older, popular meters are described in a number of references.[6] Vortex-shedding meters are becoming more common because of their relatively low cost. However,

like propeller meters, they obstruct the flow and are not suitable where debris can enter the pipeline. Multipath ultrasonic meters are being used for large irrigation flows, and single path ultrasonic meters are used for smaller flows in a few locations. These meters are relatively immune to debris in the pipeline and so can be used in culverts or short pipe sections that are supplied by trash-filled open canals. However, they tend to have high cost and/or high maintenance. Doppler ultrasonic meters that measure point velocities are advancing and may prove more accurate and cost effective in the near future. Irrigation applications often require one-time flow surveys rather than permanent installations, and portable Doppler ultrasonic meters, portable transonic meters, and pitot-tube systems are useful for such surveys[7] (Fig. 3).

Sometimes the suggested 10 pipe diameters upstream from a meter are not available to assure proper meter functioning. It is sometimes more practical to attempt to modify the flow profile approaching the meter than to increase pipe length. Flow-conditioning devices, such as vanes to keep the flow from swirling and wall obstruction in the form of large-opening orifice to break up wall jetting have proven effective.

REFERENCES

1. Bean, H.S., (Ed.) Fluid Meters, Their Theory and Application. *Report of ASME Research Committee on Fluid Meters*, 6th Ed.; The American Society of Mechanical Engineers: New York, 1971; 273.
2. Bos, M.G. *Discharge Measurement Structures*, 3rd Revised Ed.; International Institute for Land Reclamation and Improvement/ILRI: Wageningen, The Netherlands; 1989.
3. *Water Measurement Manual*, 3rd Ed.; U.S. Department of Interior, Bureau of Reclamation. Superintendent of Documents, U.S. Government Printing Office: Washington, DC. 1997.
4. Herschy, R.W. *Streamflow Measurement*. Elsevier Applied Science Publishers: New York, 1985; 453.
5. Clemmens, A.J.; Wahl, T.L.; Bos, M.G.; Replogle, J.A. *Water Measurement with Flumes and Weirs*, Publication 58; International Institute for Land Reclamation and Improvement/ILRI: Wageningen, The Netherlands, 2001; 382.
6. Miller, R.W. *Flow Measurement Engineering Handbook*, 3rd Ed.; McGraw-Hill Books: New York, 1996; 553.
7. Replogle, J.A. Some Observations on Irrigation Flow Measurements at the End of the Millennium. Trans. ASAE **2002**, *18* (1), 6–14.

Irrigation, Preplant

James E. Ayars
United States Department of Agriculture (USDA), Parlier, California, U.S.A.

INTRODUCTION

Irrigation is the process of supplying supplemental water necessary for plant growth and development by several application techniques, e.g., microirrigation, surface irrigation, sprinkler irrigation, subirrigation (by raising a shallow water table), and subsurface irrigation. It is used not only in arid and semiarid areas of the world, but also in humid areas to supplement rainfall during periods of drought. In this context, water is applied after the crop has been planted and growth has begun. The depth and timing of the application are based on the crop water requirement, the irrigation water quality, the crop salt tolerance, and the available soil water storage capacity. These topics are discussed fully in other parts of the encyclopedia.

Preplant irrigation is the application of water to a field during a fallow period between crops to accomplish a goal to replenish soil water that is anticipated for future requirements for plant growth and development. Identified uses of preplant irrigation include germination, salinity management, soil water management, fumigation, weed control, and fertilizer placement. Preplant irrigation is used in arid, semiarid, and humid areas throughout the world with the largest applications being in arid and semiarid areas. The depth and timing of the application will depend on the following: the irrigation purpose, the irrigation water quality, the existing stored soil water, the soil salinity, the crop rotation, the crop salt tolerance, and the depth to shallow groundwater.

USES

Germination

Plant establishment and development are critical to achieving yield and production goals in annual cropping systems. Germination of the seed is the first step in this process followed by plant and root development and extension. An adequate supply of soil water is required in the zone of seed placement to assure that these processes occur and are sustained. In humid areas, rainfall is generally adequate for these purposes, but in arid and semiarid areas rainfall is insufficient to supply the necessary water either due lack of water or poor timing.

The San Joaquin Valley of California is typical of this situation. Rainfall occurs primarily during the winter months with the total varying from 150 mm to over 600 mm during this period. The effectiveness of the rainfall as a future water supply is a function of the depth and timing of the occurrences. Most of the rainfall can be lost to evaporation when only small 2 mm–3 mm amounts occur. If the rainfall is in excess of about 12 mm, some of this water will be stored and be available later to meet plant water requirements. If a crop is being planted in the fall after either a period of summer fallow or the harvest of a summer crop, there will be little or no soil water stored and available for germination. As a result, it is necessary to irrigate the field to provide the water necessary for germination and early plant growth.

When a crop is planted in the fall immediately following a summer crop, irrigation will probably be withheld until after the crop has been planted. If the crop is planted in the fall following a summer fallow, the field will be irrigated prior to planting, and the crop planted into soil water allowing the producer to schedule operations in a timely fashion.

Spring planted crops generally follow a winter fallow period that is used to prepare fields for planting by creating seed beds and preplant irrigating to restore soil water. After irrigation, the seedbed is tilled to provide a mulched surface and prevent further evaporation. At planting, the seed is placed in the soil and allowed to germinate. After germination has been completed, the soil over the seed is removed mechanically, and the seed can sprout and continue to grow. This process is called planting to soil water and is used extensively in cotton production in California. Organic producers often use this method to germinate and sprout direct seeded crops. Usually, care is required in the dry soil layer mechanical removal process and must be applied soon after germination before seedling elongation begins.

Weed Control

Preplant irrigation plays an important role in weed management in both conventional and organic farming practices. In sugar beet production, preplant irrigation is used to germinate weed seed and carryover crop seeds (barley, wheat, and oats) prior to planting the sugar beet

Encyclopedia of Water Science
DOI: 10.1081/E-EWS 120010145

seeds. Sugar beet is not a very competitive crop and significant production can be lost to excessive growth of other plants. Once the weed or volunteer seed has germinated, it can be destroyed by using either chemicals (paraquat or glyphosate) or cultivation. When pre-emergence herbicides are used, preplant irrigation with sprinklers has improved the selectivity and activity of the chemicals because less water is used and less chemical is lost to leaching.

Organic producers use preplant irrigation to germinate weed seeds and then eliminate them either through cultivation or by burning. Another technique for weed control uses large amounts of compost applied prior to bed preparation. The compost is preplant irrigated to speed the composting process, which generates large amounts of heat that kills the existing weed seed. This process has to be completed prior to planting to prevent damage to the crop.

Salinity Management

Managing soil salinity is critical to successful agriculture in arid and semiarid areas of the world. Salts naturally occur in the soil, in the water used for irrigation, and in the fertilizers used for production. Crop growth and production will be reduced and eventually eliminated unless salinity in the root zone is controlled. Plants have a wide range of salt tolerance,[1] ranging from sensitive to salt tolerant. In addition to the basic tolerance, plant tolerance varies depending on growth stage with germination being the most sensitive time and maturity being the least sensitive.

Salt accumulates in the root zone as crops extract essentially pure water from the stored soil water; thus, leaving salt in the soil. In addition to salt applied by the irrigation water, salt can be transported up to the root zone from a shallow saline water table as the crop uses this water for plant growth. Evaporation from the soil surface also moves salt up into the soil profile and often to the soil surface. This accumulation of salt can have a significant negative impact on germination unless it is removed prior to planting. Leaching is the term used for removal of salt from the soil profile, and it is accomplished by irrigating in excess of the total water needed simply to meet the water requirements of the crop. The leaching requirement is the term used to describe the excess water needed to control the accumulation of salt in the soil profile. This requirement can be met incrementally with each irrigation or once a season.

Preplant irrigation is an effective method of managing salt in the profile. Generally, the large application made to replenish the stored soil water in the profile is adequate to transport the salt beneath the upper part of the soil profile, critical for germination and early crop development. Data

in Fig. 1 show the salt mass in sprinkler irrigated and furrow irrigated plots to a depth of 120 cm for December 1997 and June 1998. In both cases, the salinity was higher at each depth increment in December than in June. The reductions in salinity through the profile was a result of deep percolation and salinity leaching from the preplant irrigation and rainfall.

In arid areas, there is generally inadequate rainfall to accomplish the necessary leaching, and preplant irrigation is required.

Soil Water Management

Crop yield and biomass are directly related to total crop water use. Soil is the reservoir that stores water for plant use, and rainfall and irrigation are the sources of supply of the stored soil water. Soil water holding capacity is a function of soil type with sandy soils being able to store small amounts of water and loams, silty clay loams, and clays storing larger amounts.

During the growing season, plants remove water from the stored soil water and irrigation replenishes the depleted water supply. Irrigation scheduling is used to determine when to irrigate and how much to irrigate. It is not a precise science because of the variability in climate, soils, and crop stand and development. The stored soil water acts as a buffer and reduces the impacts on plant growth due to water stress from errors in the scheduling process.

The total water available increases over the growing season as the root system extends and explores deeper into

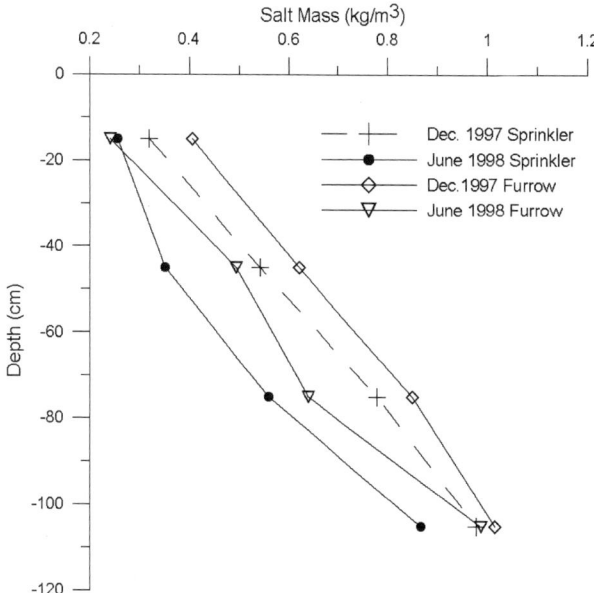

Fig. 1 Salt mass as a function of depth in sprinkler and furrow irrigated field.

the soil profile. If water extracted by deep-rooted crops is not replaced by irrigation or rainfall prior to planting the next crop, the following crop might suffer water stress late in the growing season. Preplant irrigation is an effective means to refill the soil profile and to store water for late season plant use in soils with large storage capacity.

Fertilizer Placement

Microirrigation systems are sometimes used to apply fertilizers in the seedbed prior to planting. This is accomplished by injecting fertilizers into the water and only applying a small amount of water. This is possible with microirrigation and not with other systems. Applying fertilizers with surface irrigation can result in excessive losses due to the operational characteristics of these systems.

Fumigation

When fumigants are applied with preplant irrigation it has to be done well in advance of planting to ensure no damage is done to the crop. Fumigants are applied by sprinkler and microirrigation systems during preplant irrigation to minimize the losses due to deep percolation. Subsurface drip irrigation is being used to apply fumigants prior to planting strawberries.

MANAGEMENT

Depth of Application

The depth of application will be determined by the intended use. If weed control was the intent, the irrigation has to be small enough to provide water to germinate the weed seed and still permit timely cultivation or create a situation where the weeds subsequently die from lack of water. This might require only 5 mm–10 mm of water. The same would be true with fumigation and fertilization applications. Only enough water would be applied to transport the chemical into the soil and position it in the correct portion of the crop root zone but not enough to transport it out of the root zone. The amount will be determined based on the existing soil water status, the irrigation system, and the size of the plot being irrigated.

Leaching and soil water management are often accomplished with a single application of water based on the existing soil water content. In arid areas with deep-rooted crops, the root zone often extends to a depth of 1 m–2 m below the soil surface. The soil water depletion can be in excess of 200 mm in loams to clay soils, and replenishing the water is adequate to transport the salt from the soil surface well into the profile. This occurs in

part because of the inefficiency of the irrigation systems. Surface irrigation systems and sprinkler systems have efficiencies of application in the range of 75%–85%. This means that for the field to receive at least 200 mm of water, an additional 15%–25% more water has to be applied. This inefficiency is generally adequate to provide the necessary leaching fraction when the water quality being used is considered. The depth of application can be determined by measuring the soil water content by soil sampling and a gravimetric determination, by neutron attenuation, by time domain reflectrometry, and by capacitance methods. The method selected will be a function of the crop, soils, and manager preference and experience. The closer the application is made to the time of planting, the more opportunity there is for rainfall to provide part of the water needed to replenish the root zone.

Timing of Application

Timing for the preplant irrigation will be determined by the intended use. For weed control, water has to be applied such that the weed will be dead prior to emergence of the crop and any irrigation associated with emergence. Fumigation has to occur early enough that the residual effect of the fumigants has dissipated prior to planting. This will be a function of the fumigant, the climate at the time of fumigation, the soil type, and the crop being grown following fumigation, whether it will be direct seeded or transplanted. Some crops, i.e., cotton, require a minimum soil temperature prior to planting to ensure good germination and stand establishment. In this instance, preplant irrigation needs to be done early to allow time for soil heating prior to the optimum planting date for the crop.

Managing soil water and leaching has to be done early enough that the soil has time to drain and return to a soil water content that is acceptable for cultivation. In large irrigated areas, this application of preplant irrigation occurs for several months prior to planting. This is possible because of the some of the planting techniques described in a previous section.

Water Quality Impacts

Improper management of preplant irrigation can have a significant negative impact on shallow groundwater quality and ultimately drainage water quality. This occurs primarily when preplant irrigation is used for soil water management, germination, and salinity control. Large amounts of water are applied in a single application during these operations and if the amount applied is significantly greater than what is required, deep percolation occurs. Deep percolation is the movement of water below the root zone and it is lost to crop production. This water carries along salt, fertilizers, and fumigants that are in the soil profile and mixes with the existing ground water. Depending on the chemical, this can create problems many years into the future.

Method of Application

Preplant irrigation can be done with any available irrigation system. The system normally used for irrigation during the season is the one that is most often used for preplant. Surface irrigation methods (furrow, flood, and basin) and sprinkler irrigation are the most common application methods when soil water management and salinity control are the goals. While surface systems are effective, they often have poor irrigation efficiency because the high infiltration rate as result of tillage

following the crop. This inefficiency is manifested by excessive deep percolation losses and poor distribution uniformity. Surface systems and sprinklers cover the entire surface area and are thus very effective on large areas. Microirrigation systems are used when fumigation and fertilization are important because of the ability to apply small depths of water and to precisely place the water and chemical.

REFERENCE

1. Maas, E.V. Crop Salt Tolerance. In *Agricultural Salinity Assessment and Management*, ASCE Manuals and Reports on Engineering Practice No. 71; Tanji, K.K., Ed.; American Society of Civil Engineers: New York, NY, 1990; 262–304.

Irrigation Return Flow, Quality and

Ramon Aragüés
*Agronomic Research Service, Diputación General de Aragón, Zaragoza,
Spain*

Kenneth K. Tanji
University of California, Davis, California, U.S.A.

INTRODUCTION

The return flows from irrigated agriculture (i.e., Irrigation
Return Flows, IRF) are considered the major diffuse or
"nonpoint" contributor to the pollution of surface and
groundwater bodies.[1] This off-the-farm discharge ("off-
site" contamination) is inevitable since irrigated agriculture
cannot survive if salts and other constituents accumulate in
excessive amounts in the crop's root zone ("on-site"
contamination), and so they must be reached and exported
with the drainage waters.[2] Thus, the major task concerning
the viability and the long-term sustainability of irrigated
agriculture is the attainment of a proper balance for
optimizing crop production while minimizing both the "on-
site" and the "off-site" environmental damages or impacts
and, ultimately, finding an acceptable disposal of the IRF.[3,4]

As a consequence of this increasing "off-site"
environmental problem, water pollution standards and
emerging policies regulating the discharge of the IRF are
being implemented in developed countries. The key
policies for mitigating the negative environmental impacts
of irrigation are incorporated in the Water Pollution
Control Act in United States,[5] and in the Nitrates,
Habitats and Environmental Impact Assessment, and
Water Framework directives in European Union.[6]

The degree of the "off-site" irrigation-induced pollution depends on the hydrogeological characteristics of
the irrigated land and substrata, the agricultural production
technologies used, and the water supply and drainage
conveyance systems.[1] This entry reviews these issues in
IRF, describes the main components and chemical
constituents of IRF, and summarizes recommended
management practices aimed at reducing the off-site
water quality impact from irrigated agriculture.

COMPONENTS OF IRF

Fig. 1 gives a schematic diagram of a typical irrigation-
crop-soil-drainage system, composed of the water
delivery, the farm, and the water removal subsystems.[2]

The water removal subsystem (i.e., the IRF) may be
divided into the surface drainage, consisting of the
overflow or bypass water and surface runoff or tailwater,
and the collected subsurface drainage components. Since
IRF are mixtures of these components, their proportions
determine the final quality of IRF. Table 1 summarizes the
expected water quality changes of the three IRF
components (overflow, tailwater, and subsurface drainage)
relative to the quality of the applied irrigation water.

Overflow is the result of operational spill waters from
distribution conveyances that are directly discharged into
the drainage system and its quality is generally similar to
that of the irrigation water (Table 1).

Tailwater is the portion of the applied irrigation water
that runs off over the soil and discharges from the lower end
of the field directly into the drain system. Because of its
limited contact and exposure to the soil surface, its quality
degradation is generally minor. Even so, these waters may
increase slightly in salinity and may pick up considerable
amounts of sediments and associated nutrients (phosphorus
in particular) as well as water-applied agricultural
chemicals such as pesticides and nitrogen fertilizers
(anhydrous ammonia in particular) (Table 1).

Subsurface drainage is the portion of the infiltrating
water that flows through the soil and is collected by the
under drainage system. Because of its more intimate
contact with the soil and the dynamic soil–plant–water
interactions, its quality degradation is generally substantial. These subsurface drain waters carry any anthropogenic chemicals present in a soluble form in the soil water
as well as any salts and other soluble elements present in
the soil and parent geologic material and intercepted
shallow groundwaters. The salinity and agrochemicals in
subsurface drainage are the primary source of pollution
associated with irrigated agriculture (Table 1).

WATER QUALITY CONSTITUENTS IN IRF

Irrigation return flows provide the vehicle for conveying
the pollutants to a receiving stream or groundwater

Encyclopedia of Water Science
DOI: 10.1081/E-EWS 120010262

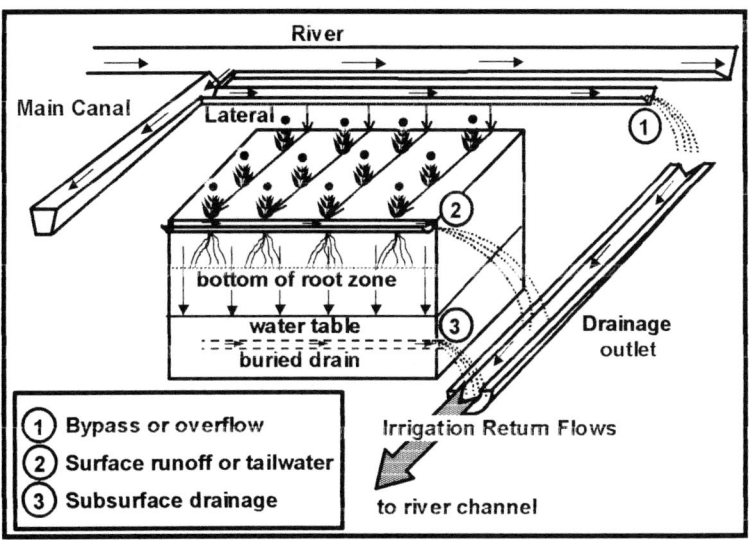

Fig. 1 Idealized sketch showing the diversion of irrigation water through a main canal, its distribution through a lateral, and its application to croplands. The three main components of the irrigation return flows (IRF) to the river channel are shown. The deeper groundwater zone, a second receiving water system, is not shown.

reservoir. It is therefore necessary to characterize their most important water quality constituents (namely, inorganic salts, agrochemicals and trace elements) and to develop management strategies aimed at alleviating their detrimental effects on the receiving water bodies.

Salts

Salts are a major quality factor since they can restrict the municipal, industrial and agricultural uses of water and can dramatically decrease the productivity and sustainability of irrigated agriculture in arid zones.

The primary source of dissolved mineral salts (also referred to as salinity) is the chemical weathering of rocks, minerals, and soils. Salinity is reported in terms of total dissolved solids (TDS in $mg\,L^{-1}$) or Electrical Conductivity (EC in $dS\,m^{-1}$ at 25°C). The main solutes contributing to salinity are the cations calcium (Ca), sodium (Na), and magnesium (Mg), and the anions chloride (Cl), sulfate (SO_4), and bicarbonate (HCO_3). These solutes are reactive in waters and soil solutions

Table 1 Quality parameters of the three irrigation return flow (IRF) components and their expected quality changes as related to the quality of the irrigation water

	Components of IRF		
Quality parameters	**Overflow**	**Tailwater**	**Subsurface drainage**
General quality degradation	0	+	+ +
Salinity	0	0, +	+ +
Nitrogen	0	0, +, + +	+ +, +
Phosphorus	0, +	+ +	0, −, +
Oxygen demanding organics	0	+, 0	0, −, − −
Sediments	0, +, −	+ +	− −
Pesticide residues	0	+ +	0, −, +
Trace elements	0	0, +	0, −, +
Pathogenic organisms	0	0, +	−, − −

0: Negligible quality changes expected.
+, −: Expected to be slightly higher (i.e., pick up), lower (deposition).
+ +: Expected to be significantly higher due to concentrating effects, application of agricultural chemicals, erosional losses, pick up of natural geochemical sources, etc.
− −: Expected to be significantly lower due to filtration, fixation, microbial degradation, etc.

participating, among others, in cation exchange and mineral solubility. The excessive accumulation of Na (i.e., sodicity, generally expressed by the Sodium Adsorption Ratio or SAR) in the soil solution and exchange complex may impair poor soil physical properties and is a critical factor in the sustainability of irrigated soils.[7]

Growing plants extract water through evapotranspiration and leave behind most of the dissolved salts, increasing its concentration in the soil water ("evapoconcentration effect"). Irrigation also adds to the salt load in IRF by leaching natural salts arising from weathered minerals occurring in the soil profile, or deposited below ("weathering effect").[2] As a consequence of both effects, it follows that the salinity and chemical composition of IRF depend basically on the characteristics of the irrigation water, the soil and subsoil, and the hydrogeology, as well as on the management of the irrigation water or Leaching Fraction (LF) defined as the fraction of infiltrated water that percolates out of the root zone. Thus, high LFs promote the weathering effect and the salt load carried out with the IRF (i.e., increased "off-site" pollution) whereas low LFs promote the evapoconcentration effect and the concentration of salts in the crop's root zone (i.e., increased "on-site" pollution).

In conclusion, the mass of salts or salt loading in IRF depends mainly on the salinity of the irrigation water, the minerals present in the soil and subsoil, and the water management (LF). The salt loading values may vary widely, from values similar to those of the irrigation water to values one order of magnitude higher. Thus, typical salt loading values in IRF from arid-land irrigated agriculture vary between $2\,Mg\,ha^{-1}\,yr^{-1}$ and $20\,Mg\,ha^{-1}\,yr^{-1}$.[1,2,7] The quantification of salt loading is critical to ascertain the "off-site" contamination of irrigated agriculture, since the prediction of the resultant salt concentration in a body of water after mixing with the IRF requires knowledge of the mass of salts (i.e., concentration and flow) in each contributing body.

Nitrogen

Nitrogen can be in either the organic or the inorganic (ammonium, nitrate and nitrite) form. Organic N is predominant in surface drainage (although it is not usually an issue in arid areas), whereas inorganic N is predominant in subsurface drainage water. Although nitrite is considered more hazardous than nitrate, it is in general a transient form of N present in small quantities. Nitrate is thus the dominant form of N in IRF and should be the focus of the water quality evaluation.[3]

High nitrate (NO_3) concentrations in IRF are a major concern since they may cause eutrophication (excessive algal growth) and hypoxia (decline in dissolved oxygen from decay of algae) problems. When nitrate is ingested in substantial amounts by humans and animals, it may cause methemoglobinemia (blue-bay like symptoms from oxygen starvation exhibited by infants and elderly) and certain cancers.[7] Thus, USEPA has set the maximum allowable concentration of nitrate in public water supplies at $45\,mg\,L^{-1}$,[5] whereas the European Union has limited it to $50\,mg\,L^{-1}$.[6]

The three major sources of nitrate found in IRF are leaching from croplands, land disposal of urban sewage, and concentrated animal feeding (beef feedlots, dairies, swine, chicken houses) wastes. The potential for nitrate leaching is a function of soil type, weather conditions and crop management system. In general, the higher the N application rate, the greater the amount of N available to be lost, since fertilizer N recovery by harvested crops averages about 50% and tends to be even lower when high N application rates are used. In addition, mineralization of organic N, followed by nitrification of NH_4 may also increase the N losses.[4]

Drainage has a large influence upon losses of nitrogen. The N loss from poorly drained soils is generally much less than from soils with improved drainage systems. As previously indicated, much of the N transported in surface runoff is organic N associated with the sediment, although the amount lost is usually small and poses little threat to the environment except in pristine waters. On the other hand, nitrate concentrations in subsurface drainage water are much higher and variable, depending on the N fertilization rates and time of applications, and on water and soil management.[4]

Phosphorus

Phosphorus (P), present in both organic and inorganic forms, is a relevant water quality constituent in IRF because of its contribution to eutrophication of surface waters. Most of the P in surface drainage is in particulate (i.e., sediment and organic matter-bound) form whereas most of the P in subsurface drainage water is in soluble phosphate form.

The release of P depends on such biogeochemical processes as adsorption/desorption of phosphate, precipitation/dissolution of inorganic P forms, and mineralization of organic P forms.[7] Phosphorus in subsurface drainage waters is typically low in concentration because of its strong adsorption in arid zone soils. Thus, although P discharge from agricultural fields vary considerably, it is usually in the range of $0.2\,kg\,P\,ha^{-1}\,yr^{-1}$ to less than $3\,kg\,P\,ha^{-1}\,yr^{-1}$. Even though P loading in IRF is minor, the P concentrations measured in many agricultural IRF may be orders of magnitude above the soluble ($10\,\mu g\,P\,L^{-1}$) and total ($20\,\mu g\,P\,L^{-1}$) critical levels assumed to accelerate the eutrophication of freshwater aquatic ecosystems.[5]

Table 2 Summary of recommended management practices at the water delivery, farm, and water removal subsystems to reduce off-site water quality impacts from irrigated agriculture

Water delivery subsystem
Designed to meet the farm water requirements while reducing undesirable water losses
Canal lining and/or closed conduits and reservoir lining: prevent seepage losses, phreatophyte ET losses, soil waterlogging, and groundwater recharge; improve irrigation water quality (i.e., suspended solids).
Installation of flow measuring devices: water control; appropriate water charges and penalties; reduce bypass losses; attain high water-conveyance efficiencies.
Construction of regulation reservoirs at the irrigation district level to increase flexibility in water delivery.
Implement an efficient institutional framework, service-oriented besides its regulatory character; scheduled maintenance programs.

Farm subsystem
Designed to maintain or increase crop productivity while improving source control
Improve cultural practices: rate and timing of fertilizers; slow-release fertilizers; fertigation; pest control; seeding and tillage practices.
Adopt less environmentally damaging agricultural practices: integrated management systems; mixed cropping practices; organic farming.
Increase irrigation application efficiency and uniformity: proper design of the farm irrigation layout; choice of irrigation system; optimum irrigation scheduling; reduce evaporation through mulching and reduced tillage.
Minimize the Leaching Fraction according to the leaching requirement of crops: reduce drainage volume; maximize mineral precipitation; minimize pick up of salts.
Provide training and technical services to farmers; eliminate institutional constraints.

Water removal subsystem
Designed to improve sink control and minimize loading in IRF
Constraints in disposal of IRF to meet quality objectives in the receiving water body.
Reuse for irrigation drainage waters, municipal wastewaters and sewage effluents; integrated on-farm drainage management (i.e., on-farm cycling of drainage waters through biological materials-agroforestry systems).
Ocean and inland (i.e., evaporation ponds; solar evaporators; deep well injection) disposal of drainage waters.
Design and management of drainage systems: include water quality as a design parameter; depth and distance of placement of drains; integrated drain flow and irrigation management; crop water use from shallow watertables (i.e., subirrigation); controlled drainage (i.e., management of the water level in the drainage outlet); reduce nitrate effluxes by maintaining a high water table to increase denitrification losses.
Pumping and disposal of groundwater to reduce intercepted groundwater by the drainage network.
Flowing of surface drainage water through vegetated filters and riparian vegetation (removal of sediments and sediments-associated contaminants), flowing of subsurface drainage water through riparian zones (removal of nitrate due to plant uptake and denitrification); flowing of drainage water through constructed wetlands (sink for sediment, nutrients, trace elements, and pesticides).
Physical, chemical, and biological treatment of drainage waters: particle removal; adsorption, air stripping; desalination (membrane processes and distillation); coagulation and flocculation; chemical precipitation; ion exchange; advanced oxidation processes; biofiltration (irrigation of specific crops that accumulate large quantities of undesirable constituents such as Se, Mo, B, NO_3, etc.); algal–bacterial treatment facilities (removal of NO_3 and Se).

Pesticide Residues

Pesticide contamination in IRF is of concern in some agricultural areas, although it is in general less significant than the salinity or nitrogen pollution problem.[3]

Pesticides used in irrigated agriculture include herbicides, insecticides, fungicides, and nematicides. These various types make it difficult to assess their potential impacts on water quality. Pesticide concentrations in surface drainage are usually much greater than those in subsurface drainage due to the filtering action of the soil. Thus, the total loss of pesticides via subsurface drainage is usually 0.15% or less of the amount applied, whereas losses via surface drainage can be up to 5% or more.[4]

The environmental fate of pesticides is quite complex. Chemical-specific properties influence the reactivity of pesticides. Pesticides can be degraded by microbes, chemical and photochemical reactions, adsorbed on to soil organic matter and clay minerals, lost to the atmosphere through volatilization, and lost through surface runoff and leaching.[4] Once a pesticide enters into the soil, its fate is largely dependent on sorption (evaluated by use of a sorption coefficient based on the organic carbon content of soils) and persistence (evaluated in terms of the half-life or the time it takes for 50% of the chemical to be degraded or transformed). Pesticides with low sorption coefficient (such as atrazine, DBCP, and aldicarb) are likely to leach readily, whereas pesticides with long half-lives (such as DDT, lindane, and endosulfan) are so persistent that many of them banned various decades ago are still found in stream sediments or are now being detected in the groundwaters.[7]

Trace Elements

High concentrations of trace elements in soils and waters pose a threat to agriculture, wildlife, drinking water, and human health. The trace elements of most importance, documented as pollutants associated with irrigated agriculture, are barium (Ba) and lithium (Li) (alkali and alkali earth metals), chromium (Cr), molybdenum (Mo), and vanadium (V) (transition metals), arsenic (As), boron (B), and selenium (Se) (nonmetals), and cadmium (Cd), copper (Cu), lead (Pb), mercury (Hg), nickel (Ni), and zinc (Zn) (heavy metals).[3] Those trace elements such as As, Cd, Hg, Pb, B, Cr, and Se are especially harmful to aquatic species because of biological magnification.[5] Due to the generally narrow window between deficiency and toxicity of trace elements, it is essential to have an adequate information on their concentrations in soils and waters.

The sources of trace element contamination may be divided into natural (i.e., geologic materials) and agricultural-induced (i.e., fertilizers, irrigation waters, soil and water amendments, animal manures, sewage effluent and sludge, and pesticides). Increases in trace element concentrations in surface runoff are generally not expected, whereas the presence of trace elements in groundwaters is influenced by the nature of the sources, the speciation and reactivity of the trace elements, and the mobility and transport processes. Thus, high concentrations of trace elements in subsurface drainage water appear to be strongly associated with the geologic setting of the irrigated area and may be affected by the same processes that affect the soil and groundwater salinity.[3,7]

An illustrative example of trace element contamination is the selenium toxicosis of waterfowl at Kesterson reservoir (California, U.S.A.), a terminal evaporation pond for drainage waters high in Se (300 ppb average) originating from the Moreno shale, a geologic formation of the Coast Range Mountains in the west side of the San Joaquin Valley.[7]

MANAGEMENT OPTIONS TO REDUCE OFF-SITE WATER QUALITY IMPACTS FROM IRRIGATED AGRICULTURE

The basic idea behind the control of irrigation-induced environmental problems is the change in focus from a "water resource development" to a "water resource management" approach. This new "thinking" involves both policy changes, such as reducing the applied water through economic and regulatory policies (i.e., water metering, water pricing, licenses and time-limited abstraction permits), and developing farmer's incentives for promoting best management practices (i.e., compensation and agri-environment payments for irrigated crops), and a variety of technical measures.[5,6,8]

Since a detailed description of the technical measures is too lengthy for this entry, Table 2 summarizes some of the recommended strategies aimed at reducing the off-site water quality impacts from irrigated agriculture. However, it should be cautioned that these measures should be applied in a "case-by-case" basis, since some of them could aggravate the "on-site" pollution problems. Typical examples will be (i) the "minimum leaching fraction concept," that could promote soil sodification and structural stability problems due to the precipitation of calcium minerals such as calcite and gypsum, (ii) the reuse of drainage water for irrigation, which is only sustainable if it is of sufficient good quality, and (iii) the disposal of drainage water in evaporation ponds, which may eventually lead to other environmental problems.

The reader is referred to the references given at the end of the entry for further information on the myriad of technical management options developed in the last decades and on details of their advantages and limitations.

REFERENCES

1. Law, J.P., Skogerboe, G.W., Eds.; *Irrigation Return Flow Quality Management*, Proceedings of National Conference, Fort Collins, Co, May 16–19, 1977; 451.
2. Yaron, D., (Ed.) *Salinity in Irrigation and Water Resources*, Civil Engineering No. 4; Marcel Dekker; New York, 1981; 432.
3. Food and Agriculture Organization of the United Nations, In *Management of Agricultural Drainage Water Quality*, FAO Water Report No. 13; Madramootoo, C.A., Johnston, W.R., Willardson, L.S., Eds.; FAO: Rome, 1997; 94.
4. Skaggs, R.W., van Schilfgaarde, J., Eds.; *Agricultural Drainage*, Agronomy No. 38; ASA, CSSA and SSSA, Inc.: Madison, WI, 1999; 1328.
5. National Research Council, *Soil and Water Quality, an Agenda for Agriculture*; Batie, S.A. Chair; Committee on Long-Range Soil and Water Conservation, National Academy Press: Washington, DC, 1993; 516.
6. Institute for European Environmental Policy, *The Environmental Impacts of Irrigation in the European Union*; Report to the Environment Directorate of the European Commission: London, 2000; 138.
7. Tanji, K.K., (Ed.) *Agricultural Salinity Assessment and Management*, ASCE Manuals and Reports on Engineering Practices No. 71; American Society of Civil Engineers: New York, 1990; 619.
8. Food and Agriculture Organization of the United Nations, In *Control of Water Pollution from Agriculture*, FAO Irrigation and Drainage Paper No. 55; Ongley, E.D., Ed.; FAO: Rome, 1996; 85.

Irrigation Sagacity

Kenneth H. Solomon
Charles M. Burt
California Polytechnic State University, San Luis Obispo, California, U.S.A.

INTRODUCTION

Throughout the Western United States, water rights are granted for *reasonable* and *beneficial* purposes. Engineers require an irrigation performance parameter that embodies the reasonable and beneficial standard to assess irrigation systems, practices, and competing uses, whether against each other, or against benchmark targets. Irrigation sagacity (IS), initially proposed by Solomon,[1] is such a parameter. The term *sagacity* comes from *sagacious*, meaning wise or prudent.

IS is fundamentally different from irrigation efficiency (IE), long used to quantify beneficial use of irrigation water.[2] Water is used beneficially if it contributes directly to the agronomic production of the crop. However, due to physical, economic, or managerial constraints, and various environmental requirements, some degree of nonbeneficial use is generally reasonable. IS goes beyond IE to incorporate quantification of reasonable uses: those uses that may not contribute to agronomic production, but are nonetheless justified under the particular circumstances at hand.

BENEFICIAL USES

Both IE and IS credit those portions of the irrigation water that are judged to be beneficially used. Examples of beneficial uses include: crop evapotranspiration (ET), water harvested with the crop, water used for salt control (leaching), climate control, seedbed preparation, softening the soil crust for seedling emergence, and ET from beneficial plants (windbreak, cover-crop, habitat for beneficial insects). Evaporation during regular and reclamation leaching, and evaporation during necessary irrigations are beneficial, because an agronomic objective is achieved during those events.

Examples of nonbeneficial uses at the farm level include: overirrigation due to nonuniformity, uncollected tailwater, deep percolation beyond that needed for salt removal, unnecessary evaporation from wet soil outside cropped area, spray drift beyond field boundaries, and evaporation associated with excessively frequent irrigations. At the irrigation district level, nonbeneficial uses include: spills, seepage, evaporation from canals or reservoirs, and ET from nonbeneficial plants such as weeds and phreatophytes.

REASONABLE USES

IS quantifies that portion of irrigation water going to sagacious (either beneficial or reasonable) uses.[1,2] Reasonable uses are those that, while not directly benefiting agronomic production, are nonetheless reasonable under prevailing economic and physical conditions. Examples of reasonable, though not beneficial, water uses include the following.

Losses that cannot be economically avoided are considered reasonable. For example, canal seepage may be reasonable if canal seepage rates are low and it is not economical to line the canal to avoid that seepage. No irrigation system can be designed to apply water with perfect uniformity, so some deep percolation due to nonuniformity is reasonable.

Losses tied to technical requirements may be reasonable. Reservoirs in the distribution system add flexibility and reduce canal spills. Evaporation from such reservoirs constitutes a reasonable use. Microirrigation systems generally require filtration, and filters need to be flushed periodically. Filter flush water may be a reasonable use. If sprinkler irrigation is the appropriate technology, spray evaporation and wind drift losses are an inevitable consequence of using that technology to irrigate, and hence are a reasonable use.

Losses due to the uncertainties associated with many aspects of water management may be reasonable. Exactly how much water is held in the soil? Exactly how much crop ET since the last irrigation? Exactly how much water is necessary for maintenance leaching? In the face of such uncertainties, it is reasonable for the farmers to err on the side of overapplication, so some deep percolation due to uncertainty may be reasonable.

Losses that contribute toward environmental goals may be reasonable. If canal seepage feeds a wetland or wildlife habitat area in a timely manner, that seepage may be deemed a reasonable use. (Even though feeding a wetlands/habitat area meets environmental goals, it is

Encyclopedia of Water Science
DOI: 10.1081/E-EWS 120010254

not considered a beneficial use because it does not directly aid the production of the crop being irrigated.) If tailwater blends with drainage water to meet water quality standards in receiving waters, then that tailwater may be a reasonable use.

Nonsagacious uses (neither beneficial nor reasonable) are those uses without economic, practical, or other justification. An example of a nonsagacious use is wet soil and spray evaporation associated with excessively frequent irrigations. No agronomic objective is served by irrigating more frequently than needed, and it is difficult to imagine an economic justification for doing so. Hence, these losses are without justification. They are unreasonable and nonsagacious.

CONSERVATION IMPLICATIONS

It is a common misunderstanding that $(1 - IE)$ represents the fraction of the applied irrigation water that is wasted, and therefore, the fraction that may be conserved or reallocated. However, as noted above, some degree of nonbeneficial use is generally reasonable, so the potential for conservation and reallocation consists only of water uses that are both nonbeneficial and unreasonable.

DETERMINATION OF SAGACITY

As with other irrigation performance parameters, application of the IS concept requires that boundaries be specified, flows into and out of the bounded area be quantified, fractions of the irrigation water flowing to various destinations be estimated, and judgments be made about whether those fractions are beneficial or reasonable. Whereas the determination of beneficial use involves only an agronomic criterion—direct contribution to the agronomic production of the crop—the determination of reasonable use involves more varied criteria.

Feasibility

Identifying a particular nonbeneficial water use as unreasonable requires that an alternate practice be identified that uses less water, and is practically, technically, economically, and environmentally feasible.

Practical feasibility considers physical constraints such as limitations due to climate, soil, terrain, water delivery schedules, or water travel time. Required resources, which can include labor (sufficient quantity and with suitable experience), infrastructure (maintenance of specialized equipment, extension advice on proposed crops, etc.), and information (precise knowledge, facts, data available when needed), are available. Even after identifying the

benefits of a new practice, there will be a lag time before implementation is possible. Decision-makers need to be convinced, approvals obtained, plans drawn, and financing arranged. Thus, a realistic time schedule for implementation is also required.

Technical feasibility has not only hardware but software and operational aspects. Equipment must be available, affordable, and perform reliably in an agricultural environment. It must satisfy requirements for accuracy and precision of flow, time or other quantities to be measured or controlled. Local, farm scale demonstration projects may be necessary to prove that equipment and plans are reliable, and operations feasible. A phased transition into any new practice should be planned.

Economic feasibility is an obvious but complex test. It is not enough to compare the costs of operating one way to the costs of operating another way. The proposition facing a farmer is to change from one practice to another. The costs involved in abandoning an old practice and adopting a new one, or converting from the old to the new, may well be greater than the cost if the new practice were started from scratch in a new operation. Further, even if the annualized cost of an alternate practice is favorable, it may not be possible for farmers to implement it unless additional resources such as financing and credit are available. Economic feasibility must also consider risk. It is not reasonable to ask farmers to undertake a large risk to actualize the potential of a small benefit.

Economic feasibility must include plans and mechanisms for properly allocating the costs of alternate practices to those who will ultimately reap the benefit. This may be particularly difficult in the case of alternate practices whose ultimate beneficiary is the environment, because it is often not clear who "ought" to pay on behalf of an environmental common good.

Environmental feasibility requires that the alternative practice must be environmentally benign or beneficial, or that the costs of any required environmental mitigation are considered.

If no alternate practice using less water meets all four feasibility tests, the current practice is reasonable. The current practice is unreasonable if a feasible alternate using less water exists. The amount of unreasonable (nonsagacious) use due to the current practice is the difference between the current use and the (reduced) use of the preferred alternate.

CONCLUSION

The results of a sagacity determination may vary with location, geographic scale, or time. Because sagacity includes economics, which can change as markets and prices do, sagacity can change with place and time. Technology and the availability of resources are also factors influencing sagacity that can change with place and time.

Results of the various feasibility checks can depend on scale. To a farmer, the district's water delivery policies and schedule are given. At the district level, these things may be considered adjustable. Districts can and should consider options that individual farmers cannot consider. Economics and the ability to absorb risk change with scale as well. What may not be economical to an individual farmer could be economical to a district, region, or to another competing water user, if there is a way for them to share in the costs as well as the benefits. While individual farmers are less able to bear risks due to uncertainty or reduced water use, the shift to a district or societal level offers the potential to "average" individual outcomes and pool risks.

So sagacity is very much a site-, scale-, and time-specific quantity. Therefore, a necessary preliminary to the determination of sagacity described above is to specify the boundaries and geographic extent of study area, the time frame for economic and technological determinations, and the perspective for feasibility checks (individual farmer, district, region, or society).

For a more complete discussion of IS and its application, the reader is referred to Ref. 3.

REFERENCES

1. Solomon, K.H. *Technical Memorandum*; Center for Irrigation Technology, California State University: Fresno, CA, 1993; 1–2.
2. Burt, C.M.; Clemmens, A.J.; Strelkoff, T.S.; Solomon, K.H.; Bliesner, R.D.; Hardy, L.A.; Howell, T.A.; Eisenhauer, D.E. Irrigation Performance Measures—Efficiency and Uniformity. J. Irrig. Drain. Eng. **1997**, *123* (6), 423–442.
3. Solomon, K.H.; Burt, C.M. Irrigation Sagacity: A Measure of Prudent Water Use. Irrig. Sci. **1999**, *18*, 135–140.

Irrigation with Saline Water

B. A. Stewart
West Texas A&M University, Canyon, Texas, U.S.A.

INTRODUCTION

As water becomes more limited, there is increasing use of saline waters for irrigation that were previously considered unsuitable. Rhoades et al.[1] classified saline waters as shown in Table 1. Electrical conductivity is a convenient and practical method for classifying saline waters because there is a direct relationship between the salt content of the water and the conductance of an electrical current through water containing salts. Electrical conductivity values are expressed in siemens (S) at a standard temperature of 25°C.

Most waters used for irrigation have electrical conductivities less than $2\,dS\,m^{-1}$.[1] When water higher than this level is used, there can be serious negative effects on both plants and soils. As salinity in the root zone increases, the osmotic potential of the soil solution decreases and therefore reduces the availability of water to plants. At some point, the concentration of salts in the root zone can become so great that water will actually move from the plant cells to the root zone because of the osmotic effect. Salts containing ions such as boron, chloride, and sodium can also be toxic to plants when accumulated in large quantities in the leaves. The extent that plant growth is affected by saline water is dependent on the crop species. Some plants, such as barley and cotton, are much more resistant to salt than crops like beans. Rhoades et al.[1] list the tolerance levels of a wide range of fiber, grain, and special crops; grasses and forage crops; vegetable and fruit crops; woody crops; and ornamental shrubs, trees, and ground cover. Soils are also negatively impacted by salt, particularly sodium salts. Sodium ions tend to disperse clay particles and this has deleterious effects on infiltration rate, structure, and other soil physical properties.

IRRIGATING WITH SALINE WATERS

Water limitations and the need to increase food and fiber production in many parts of the world have resulted in the use of water for irrigation containing increasing levels of salts. The United States, Israel, Tunisia, India, and Egypt have been particularly active in irrigating with saline waters.[1] Rhoades et al.[1] published an extensive paper on the use of saline waters for crop production and it is a valuable guide for anyone interested in the subject. They reported that many drainage waters, including shallow ground waters underlying irrigated lands, fall in the range of $2\,dS\,m^{-1}$ to $10\,dS\,m^{-1}$ in electrical conductivity. Such waters are in ample supply in many developed irrigated lands and have good potential even though they are often discharged to better quality surface waters or to waste outlets. These waters can be successfully used in many cases with proper management. Reuse of second-generation drainage waters with electrical conductivity values of $10\,dS\,m^{-1}$ to $25\,dS\,m^{-1}$ is also sometimes possible but to a much lesser degree because the crops that can be grown with these waters are atypical and much less experience exists upon which to base management recommendations.

Miller and Gardiner[2] suggest that successful irrigation with saline water requires three principles. First, the soil should be maintained near field capacity to keep the salt concentration as low as possible. Second, application techniques should avoid any wetting of the foliage. Third, salts accumulating in the soil should be periodically leached. To accomplish these objectives, Miller and Gardiner[2] recommend the following general rules:

- Apply water at or below soil surface. Sprinklers should be used only if they avoid wilting the foliage (such as sprinkling before plant emergence or below-canopy to avoid salt-burn damage).
- Keep water additions almost continuous, but at or below field capacity so that most flow is unsaturated. This maintains adequate aeration.
- Enough water should be added to keep salts moving downward, thus avoiding salt buildup in the root zone.

Miller and Gardiner[2] stress that these rules are difficult to meet and are best satisfied by some form of drip irrigation. They also state that due to the need for high water levels and because of high sodium ratios that sandy soils are more adaptable to the use of saline waters than soils containing high percentages of silt and clay particles.

Rhoades et al.[1] also list specific management practices for producing crops with salty waters. Their list includes the following guidelines:

Encyclopedia of Water Science
DOI: 10.1081/E-EWS 120014079

Table 1 Classification of saline waters

Water class	Electrical conductivity (dS m^{-1})	Salt concentration (mg L^{-1})	Type of water
Nonsaline	< 0.7	< 500	Drinking and irrigation
Slightly saline	0.7–2	500–1500	Irrigation
Moderately saline	2–10	1500–7000	Primary drainage and groundwater
Highly saline	10–25	7000–15,000	Secondary drainage and groundwater
Very highly saline	25–45	15,000–35,000	Very saline groundwater
Brine	> 45	> 45,000	Seawater

(From Ref. 1.)

- Selection of crops or crop varieties that will produce satisfactory yields under the existing or predicted conditions of salinity or sodicity.
- Special planting procedures that minimize or compensate for salt accumulation in the vicinity of the seed.
- Irrigation to maintain a relatively high level of soil moisture and to achieve periodic leaching of the soil.
- Use of land preparation to increase the uniformity of water distribution and infiltration, leaching and removal of salinity.
- Special treatments (such as tillage and additions of chemical amendments, organic matter and growing green manure crops) to maintain soil permeability and tilth. The crop grown, the quality of water used for irrigation, the rainfall pattern and climate, and the soil properties determine to a large degree the kind and extent of management practices needed.

BLENDING LOW-SALT AND SALTY WATERS

Miller and Gardiner[2] reported that countries such as Israel have developed extensive canal and reservoir systems where both low-salt and salty waters are mixed to obtain usable water. Rhoades et al.,[1] however, state that blending or diluting excessively saline waters with good quality water supplies should only be undertaken after consideration is given to how this affects the volumes of consumable water in the combined and separate supplies. They suggest that blending or diluting drainage waters with good quality waters in order to increase water supplies or to meet discharge standards may be inappropriate under certain situations. More crop production can usually be achieved from the total water supply by keeping the water components separated. Serious consideration should be given for keeping saline drainage waters separate from the good quality water, especially when the good quality waters are used for irrigation of salt-sensitive crops. The saline waters can be used more effectively by substituting them for good quality water to irrigate certain crops grown in the rotation after seeding establishment.

CONCLUSION

There is ample evidence that saline waters once considered unacceptable for irrigation can be used successfully provided that they are properly managed. There is also ample evidence, however, to show that these waters can be highly damaging to the environment and to the soil resource base when improperly managed. Therefore, saline waters should be only used for irrigation after careful study and considering as many factors as possible. Then, when the waters are used for irrigation, a careful monitoring program should be implemented of both the crops produced and of the resulting soil and environmental changes.

REFERENCES

1. Rhoades, J.D.; Kandiah, A.; Marshali, A.M. The Use of Saline Water for Crop Production. *FAO Irrigation and Drainage Paper 48*; Food and Agriculture Organization of the United Nations: Rome, 1992.
2. Miller, R.W.; Gardiner, D.T. *Soils in Our Environment*, 8th Ed.; Prentice Hall: Upper Saddle River, NJ, 1998.

Irrigation Scheduling with Plant Indicators: Field Applications

David A. Goldhamer
University of California, Parlier, California, U.S.A.

INTRODUCTION

While there has been extensive research in measuring plant water status and its impact on plant processes and aspects of crop production, relatively little work has been published on giving specific protocols for using plant-based measurements in irrigation scheduling. This is likely the result of plant water status being very dynamic since the plant is coupled to both its soil water and atmospheric surroundings. Thus, no single value can be used to indicate the onset of water stress. Plant water status changes diurnally and over the season, and its dynamic nature makes it difficult to identify threshold values for practical use. Plant water measurements, by themselves, mean little if they are not considered relative to the equivalent measurements representing fully irrigated plants in the same environment. This has been accomplished by developing "reference" or "baseline" values, representing the behavior of plants under nonlimiting soil water supply.

APPLICATIONS OF PLANT INDICATORS FOR THE MANAGEMENT OF WATER STRESS IN IRRIGATION

It is well documented that virtually all plant processes are affected by a reduction in plant water status.[1] Plant water deficits decrease leaf growth and thus, leaf area. With field and row crops, where the goal is to achieve a full canopy as soon as possible, early season water deficits translate into lower field photosynthesis and ultimately lower total biomass production. Thus, with crops harvested for biomass, such as alfalfa and silage corn, any water stress will reduce yield. The linear nature of the classical production function relating yield to crop water use illustrates this fact clearly.[2]

The effect of water deficits on production of crops where only the reproductive organ is harvested is much less straightforward. One example is cotton; an indeterminate row crop. Severe water stress can reduce leaf area, photosynthesis, the number of fruiting sites, and thus reduce lint yield.[3] On the other hand, mild water stress, enough to significantly reduce vegetative growth, did not reduce yield.[4] There are few documented cases of water stress increasing crop yields. Fereres[5] reported that mild

water deficits increased yields of cotton and sorghum over those under full irrigation. This was attributed to the increase in harvest index due to a greater partitioning of assimilate from vegetative to reproductive sinks. Chalmers et al.[6] reported that regulated deficit irrigation (RDI) significantly reduced unwanted vegetative growth in peach and consequently, an increase in harvest fruit size, again due to greater assimilate partitioning. It should be noted that others have tried to reproduce these results, specifically, increased fruit size in response to water stress, and failed.[7–9] However, these experiments occurred under different soil and atmospheric conditions and with different cultivars. There are far more instances reported of water stress improving some aspect of fruit quality than fruit size. For example, Goldhamer et al.[9] found that RDI can significantly reduce peel creasing in navel oranges without negatively affecting other yield components, thus increasing grower profit.

The interactions between irrigation and pest and disease management offer opportunities for the use of plant indicators for beneficial purposes. While water stress is usually associated with higher insect pressures, there are cases where water stress had beneficial effects. Leigh et al.[10] found that lygus bug levels in cotton were reduced by 50% by the imposition of water stress compared to fully irrigated plants. Goldhamer et al.[11] found that epicarp lesion on pistachio nuts, believed to be the result of feeding by leaf-footed plant bugs, was significantly reduced in severely stressed trees. Reduced insect pressures in response to water stress are usually attributed to a less favorable feeding environment due to higher canopy temperatures. However, high temperatures are also related with greater pest pressures due to higher insect development rates.[12] Teviotdale et al.[13] noted that preharvest water deficits in almond trees significantly reduced the fungal disease of hull rot in almonds. Further work resulted in a recommendation that water stress be imposed for a two-week period about 1 mo prior to harvest not to be less than a predawn leaf (Ψ) water potential of -1.5 MPa.[14] Although this approach provided a great amount of disease control, the authors advised that water stress could reduce kernel size (3%–5%) and thus, the grower must decide on whether the impact of the disease would be worse than that of water stress.

Encyclopedia of Water Science
DOI: 10.1081/E-EWS 120013504

Table 1 Values of midday stem Ψ in MPa expected for fully irrigated prune trees under different air temperature and relative humidity conditions

Temperature (°C)	Air relative humidity (RH) (%)						
	10	20	30	40	50	60	70
6.9	− 0.68	− 0.65	− 0.62	− 0.59	− 0.56	− 0.53	− 0.50
9.7	− 0.73	− 0.70	− 0.66	− 0.62	− 0.59	− 0.55	− 0.52
12.4	− 0.79	− 0.75	− 0.70	− 0.66	− 0.62	− 0.58	− 0.54
15.2	− 0.85	− 0.81	− 0.76	− 0.71	− 0.66	− 0.61	− 0.56
18.0	− 0.93	− 0.87	− 0.82	− 0.76	− 0.70	− 0.64	− 0.58
20.8	− 1.02	− 0.95	− 0.88	− 0.82	− 0.75	− 0.68	− 0.61
23.6	− 1.12	− 1.04	− 0.96	− 0.88	− 0.80	− 0.72	− 0.65
26.3	− 1.23	− 1.14	− 1.05	− 0.96	− 0.87	− 0.78	− 0.68
29.1	− 1.36	− 1.26	− 1.15	− 1.04	− 0.94	− 0.83	− 0.73
31.9	− 1.51	− 1.39	− 1.26	− 1.14	− 1.02	− 0.90	− 0.78

Source: Adapted from Ref. [15].

INTERPRETATION OF MEASUREMENTS FOR IRRIGATION DECISION MAKING

The fact that plant species and plant processes differ in their sensitivity to water stress complicates the issue of applying indicators of plant water status to irrigation management. For example, predawn water Ψ of leaves in fully irrigated pistachio tress in the central valley of California is from − 0.8 MPa to − 1.0 MPa compared with − 0.15 MPa to − 0.20 MPa for walnut trees in the same environment. Mild to moderate water stress from mid-May to early July has a little impact on any pistachio yield component while it reduces the size of harvested walnuts. In order to use plant Ψ in production agriculture, it is of paramount importance that both accurate measurements of normalized water stress and how they impact yield be known for successful application of any plant-based scheduling program. This two-phase knowledge is rare. Nevertheless, some workers have developed plant-based protocols that have achieved varying degrees of success in on-farm water management. Examples of the most promising of these approaches for tree crops are highlighted in the following discussion.

Shackel et al.[15] proposed using stem Ψ for irrigation scheduling of prunes. This crop is an ideal candidate for plant-based scheduling that accurately identifies stress in that it is a dried product and thus, lower fruit hydration at harvest resulting from water deficits during the season is actually beneficial. The influence of evaporative demand was addressed by providing a table showing stem Ψ for a range of relative humidity and air temperature conditions (Table 1). Additional work had identified periods of the season, which were most stress tolerant[16] resulting in specific, recommended protocols of desired stem Ψ over the season (Table 2). These protocols have been adopted by numerous prune growers in California.

One short coming of plant-based scheduling approaches is that they do not provide quantitative information on how much water should be applied at each irrigation as is possible with soil water monitoring and atmospheric-based methods. Thus, protocols for using plant Ψ should somehow address this issue. Here, both measurement and irrigation frequency are important; the more frequent the monitoring, the greater is the opportunity to adjust the irrigation amount. It is difficult to conduct frequent manual stem Ψ monitoring, and this measurement cannot be automated currently. There is little

Table 2 Suggested target levels of midday stem Ψ in MPa during the growing season in prunes

Period	Month						
	March	April	May	June	July	August	September
Early-	− 0.6	− 0.8	− 0.9	− 1.1	− 1.2	− 1.3	− 1.5
Mid-	− 0.6	− 0.8	− 0.9	− 1.0	− 1.2	− 1.3	− 1.4
Late-	− 0.7	− 0.9	− 1.0	− 1.1	− 1.2	− 1.4	− 1.5

Source: Adapted from Ref. [16].

chance of coupling these manual measurements to irrigation electronic controllers.

Goldhamer and Fereres[17] presented protocols based on experimental data for using trunk diameter measure-ment to schedule irrigations in almond trees. They suggested maximum daily trunk shrinkage (MDS) as the stress indicator parameter. These data were normalized for vapor pressure deficit (VPD) using reference values

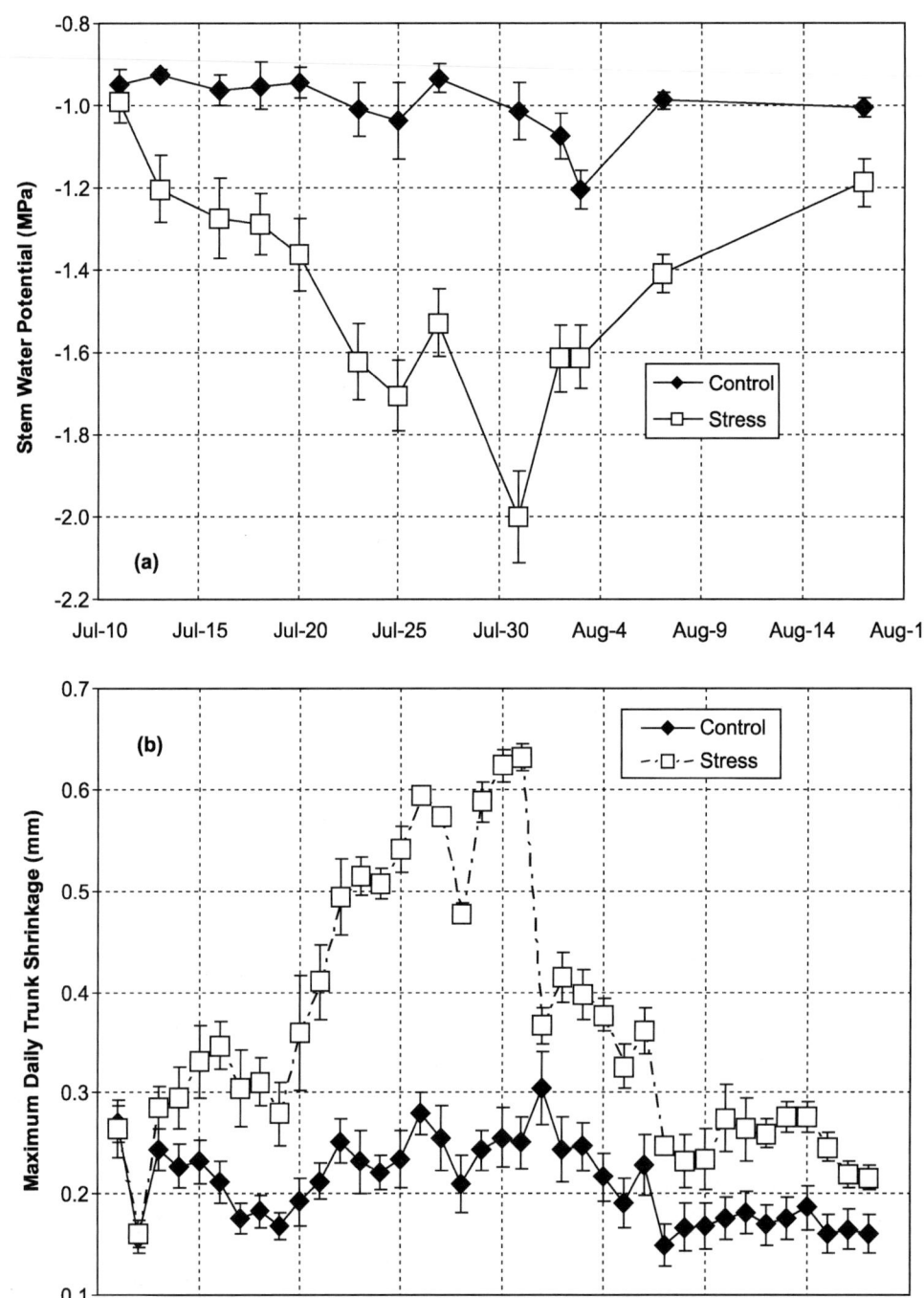

Fig. 1 For mature peach trees under both full (Control) and deficit irrigation (Stress) through July-31 followed by a return to full irrigation: a) stem water potential (SWP); b) maximum daily trunk shrinkage (MDS); c) MDS and SWP signal strength (Stress/Control); d) MDS coefficient of variation (noise); e) SWP coefficient of variation (noise), and f) MDS and SWP "signal/noise" ratio. Vertical bars are two standard errors. (From Ref. [18].)

(continued)

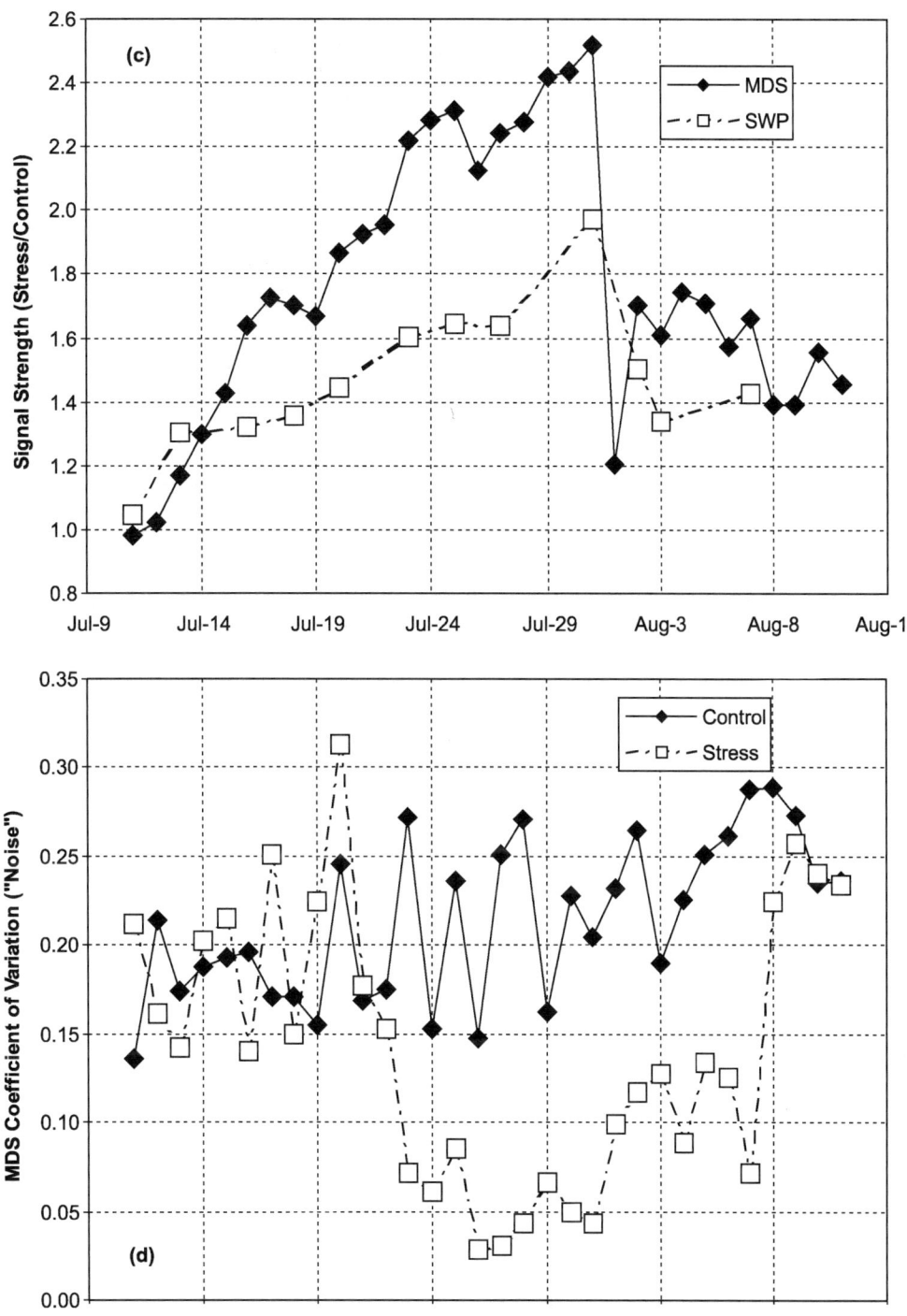

Fig. 1 Continued.

obtaining either from fully irrigated trees or relationships previously developed between VPD and fully irrigated trees. They used the term "signal" (ratio of measured to reference value) to represent the stress magnitude and suggested target signals for irrigation scheduling based on how much stress the grower desired during the season. If the measured signal consistently exceeds the target threshold, the irrigation rate is increased by 10% for the next cycle. If the measured signal is below the threshold, the irrigation rate is lowered by 10%. Their goal was to have the actual signal oscillate as close as possible around the selected threshold level. This approach was recently

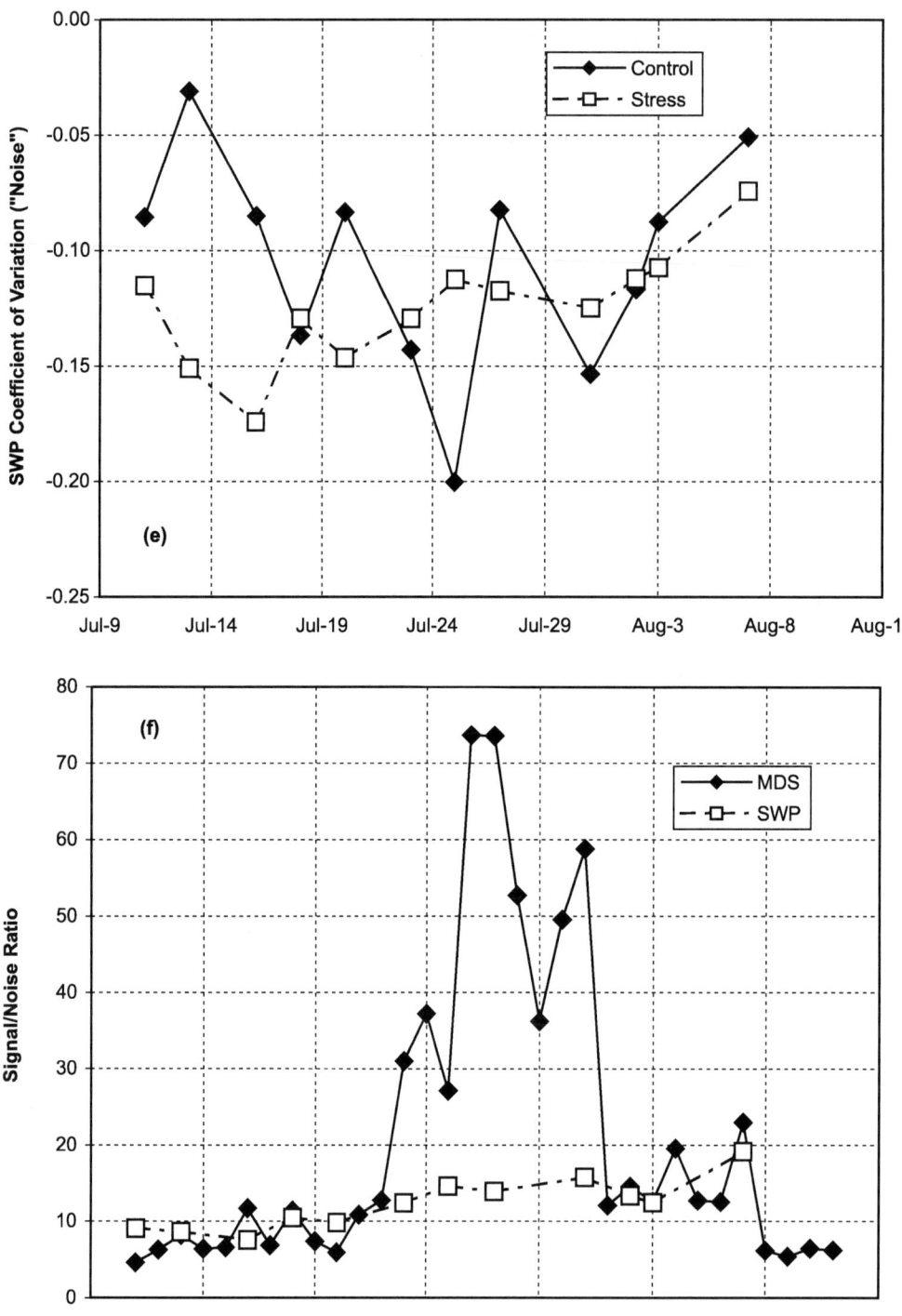

Fig. 1 Continued.

validated in a field study comparing different target threshold values.

Goldhamer et al.[18] conducted an analysis of the sensitivity of these scheduling approaches to progressively greater water deficits on mature peach trees under high-frequency irrigation. Reference values were determined using control trees under nonlimiting soil moisture conditions (irrigation based on a weighing lysimeter). Stem Ψ and MDS measurements during both the deficit irrigation period and when full irrigation was reintroduced to the stressed trees tracked each other well (Fig. 1a, b). Both stem Ψ and MDS were equal in identifying the onset of stress but after the first few days, the MDS signal was clearly higher than the stem Ψ signal for the remainder of the stress period (Fig. 1c). Following the return to full irrigation, both indicator signals remained well-above, 1.0, reflecting the fact that the soil water reservoir was not refilled.

In addition to signal strength, the variability of the indicator values will determine the usefulness of the signal in irrigation scheduling. There have been reports that MDS is more variable than SWP in tree crops.[17,19,20] High measurement variability ("noise") would require more tree measurements to decrease the uncertainty, increasing the monitoring costs. Goldhamer et al.[18] found that MDS variability was higher than that of SWP under mild to moderate stress but lower than MDS under severe water stress (Fig. 1d, e). During the entire stress range, SWP variability remained unchanged. The signal/noise ratio, which integrates both the indicator strength and variability, was equal for both stem Ψ and MDS under mild to moderate stress (Fig. 1f). In other words, even though the MDS signal was higher than stem Ψ in this stress range, the increased variability reduced the usefulness as a scheduling indicator. However, under severe stress, the MDS signal/noise ratio was about five times higher than the peach stem Ψ value. This example clearly illustrates that a multitude of factors must be taken into account in evaluating which plant-based monitoring technique performs best under a given set of conditions.

OVERCOMING BARRIERS FOR GROWER ADOPTION

While scientists may disagree over which plant-based monitoring approach provides the most sensitive indicator of water stress and how stress impact yield, on-farm personnel responsible for irrigation scheduling have much simpler question that requires an affirmative answer for method adoption to occur: "Will it make their jobs easier?" A secondary issue is whether it is cost effective and the time profitably used. It is believed that the current state-of-the-art in plant-based monitoring is stem Ψ for the following reasons: 1) it is grounded in proven scientific principles; and 2) it has been adopted by some growers. Its primary shortcoming is that the measurement is manually taken, requiring the technician to make trips to the field. Moreover, the measurements must be made within a 2 hr or so period around midday. Additional negatives are that a relatively bulky pressure chamber must be hauled to the field (although a lighter "pump-up" version is now commercially available), and data processing is manual. Continuously recorded trunk diameter measurements overcome the logistical problems and labor requirements of the stem Ψ measurements. Since they are electronically

recorded, graphical representation, which is a powerful incentive for grower adoption, is relatively easy.

Electronically recorded measurements also facilitate data analysis. One can envision algorithms in a computer/controller that not only derive indicator parameters, such as MDS, but also generate and execute system-operating times. Additionally, the rapid availability and ease of data transfer on from local to global networks offer the prospect of irrigation control and consultation from remote locations. To be sure, there are trade-offs associated with using trunk diameter measurements in irrigation scheduling. The data are most variable ("noisy") than corresponding stem Ψ values.[18] The special LVDT mounts currently cost $US 40–80 and LVDTs are $US 150–250. Repairing electronic devices is beyond the ability of most on-farm personnel. Additionally, dataloggers located in the field require periodic trips by a technician to download data unless wireless communication has been established.

The decreasing availability and increasing cost of water for agriculture will be incentives to improve irrigation management that growers will be unable to ignore in the future. Using plant-based methods to quantify stress and adjusting irrigation schedules accordingly should become more prevalent, especially in high water cost areas and with high value crops. They can be used as stand-alone methods or more likely, to augment and fine tune the more widely available types of atmospheric-based approaches. Indeed, a combination of scheduling techniques may prove optimal, especially for cropping situations where early season stress is desirable. For example, the quality of wine grapes has been shown to improve with deficit irrigation prior to veraison (berry color change). Williams (personal communication) recommend that leaf water Ψ be maintained at -1.0 MPa (mild stress) early in the season and then irrigated at 60% ET_c from veraison to harvest. Imposing early season stress based on irrigating at certain percentages of ET_c is difficult since it is necessary to accurately characterize and account for the amount of water available in the soil moisture reservoir. Early season plant-based monitoring overcomes this problem. As growers become more sophisticated in their irrigation programs, especially those that utilize water stress beneficially (RDI), plant-based monitoring, whether they taken manually or recorded electronically, will become a more valuable tool for use in production agriculture.

REFERENCES

1. Hsiao, T.C. Plant Responses to Water Stress. Annu. Rev. Plant Physiol. **1973**, *24*, 519–570.
2. Vaux, H.J.; Pruitt, W.O. Crop-Water Production Functions. In *Advances in Irrigation*; Hillel, D., Ed.; Academic Press: New York, 1983; Vol. 2, 61–93.

3. Turner, N.C. Adaptation to Water Deficits: A Changing Perspective. Aust. J. Plant Physiol. **1986**, *13*, 175–190.

4. Grimes, D.W.; Yamada, H.; Dickens, W.L. Functions for Cotton (*Gossypium Hirsutum* L.) Production from Irrigation and Nitrogen Fertilization Variables: I. Yield and Evapotranspiration. Agron. J. **1969**, *61*, 769–773.

5. Fereres, E. Variability in Adaptive Mechanisms to Water Deficits in Annual and Perennial Crop Plants. Bull. Soc. Bot. Fr. Actual Bot. **1984**, *131*, 17–32.

6. Chalmers, D.J.; Mitchell, P.D.; van Heek, L.A.G. Control of Peach Tree Growth and Productivity by Regulated Water Supply, Tree Density, and Summer Pruning. J. Am. Soc. Hort. Sci. **1981**, *106*, 307–312.

7. Girona, J.; Mata, M.; Goldhamer, D.A.; Johnson, R.S.; DeJong, T.M. Patterns of Soil and Tree Water Status and Leaf Functioning During Regulated Deficit Irrigation Scheduling in Peach. J. Am. Soc. Hort. Sci. **1993**, *118* (5), 580–586.

8. Marsal, J.; Girona, J. Relationship Between Leaf Water Potential and Gas Exchange Activity at Different Phenological Stages and Fruit Loads in Peach Trees. J. Am. Soc. Hort. Sci. **1997**, *122* (3), 415–421.

9. Goldhamer, D.A.; Salinas, M.; Crisosto, C.; Day, K.R.; Soler, M.; Moriana, A. Effects of Regulated Deficit Irrigation and Partial Root Zone Drying on Late Harvest Peach Tree Performance. Acta Hort. **2002**, in press.

10. Leigh, T.F.; Grimes, D.W.; Dickens, W.L.; Jackson, C.E. Planting Pattern, Plant Population, Irrigation, and Insect Interactions in Cotton. Environ. Entomol. **1974**, *3*, 492–496.

11. Goldhamer, D.A.; Phene, B.C.; Beede, R.; Scherlin, L.; Brazil, J.; Kjelgren, R.K.; Rose, D. Water Management Studies of Pistachio: I. Tree Performance After Two Years of Sustained Deficit Irrigation. California Pistachio Industry Annual Report 1985–86; 1986; 104–112.

12. Van de Urie, M.; McMutry, J.A.; Huffaker, C.B. Ecology of Tetranychid Mites and Their Natural Enemies: A Review.

III. Biology, Ecology, Pest Status, and Host–Plant Relations of Tetranychids. Hilgardia **1972**, *41*, 343–432.

13. Teviotdale, B.L.; Michailides, T.J.; Goldhamer, D.A.; Viveros, M. Reduction of Almond Hull Rot Disease Caused By *Rhizopus Stolonifer* By Early Termination of Preharvest Irrigation. Plant Dis. **1995**, *79* (4), 402–405.

14. Teviotdale, B.L.; Goldhamer, D.A.; Viveros, M. Effects of Deficit Irrigation on Hull Rot Disease of Almond Trees Caused by *Monilinia fructicola* and *Rhizopus stolonifer*. Plant Dis. **2001**, *85* (4), 399–403.

15. Shackel, K.A.; Ahmadi, H.; Biasi, W.; Buchner, R.; Goldhamer, D.; Gurusinghe, S.; Hasey, J.; Kester, D.; Krueger, B.; Lampinen, B.; McGourty, G.; Micke, W.; Mitcham, E.; Olson, B.; Pelletrau, K.; Philips, H.; Ramos, D.; Schwankl, L.; Sibbett, S.; Snyder, R.; Southwick, S.; Stevenson, M.; Thorpe, M.; Weinbaum, S.; Yeager, J. Plant Water Status as an Index of Irrigation Need in Deciduous Fruit Trees. HortTechnology **1997**, *7* (1), 23–29.

16. Shackel, K.; Lampinen, B.; Southwick, S.; Goldhamer, D.; Olson, W.; Sibbett, S.; Krueger, W.; Yeager, J. Deficit Irrigation in Prunes: Maintaining Productivity with Less Water. HortScience **2000**, *35*, 30–33.

17. Goldhamer, D.A.; Fereres, E. Irrigation Scheduling Protocols using Continuously Recorded Trunk Diameter Measurements. Irrig. Sci. **2001**, *20*, 115–125.

18. Goldhamer, D.A.; Fereres, E.; Cohen, M.; Girona, J.; Mata, M.; Soler, M.; Salinas, M. Comparison of Continuous and Discrete Plant-Based Monitoring for Detecting Tree Water Deficits and Barriers to Grower Adoption for Irrigation Management. Acta Hort. **2000**, *537*, 431–445.

19. Ginestar, C.; Castel, J.R. Utilitzacion de Dendrometro Como Indicadores de Estres Hidrico en Mandarinos Jovenes Regados por Poteo. Riegos Drenajes XXI **1996**, *89*, 40–46.

20. Naor, A. Midday Stem Water Potential as a Plant Water Stress Indicator for Irrigation Scheduling in Fruit Trees. Acta Hort. **2000**, *537*, 447–454.

Irrigation Scheduling with Plant Indicators: Measurement

David A. Goldhamer
University of California, Parlier, California, U.S.A.

INTRODUCTION

Even though optimizing plant biomass or fruit production is the goal of irrigation, the plant is rarely the primary focus in irrigation scheduling techniques. Atmospheric techniques, where the plant is considered only tangentially when evapotranspiration (ET_c) is estimated, and soil moisture monitoring with a wide variety of instruments are by far the dominant scheduling approaches in use today. The plant sits midway in the soil, plant, and atmospheric continuum and is the integrator of both the water status of the soil and the atmosphere. Moreover, virtually all plant processes that ultimately affect productivity are directly or indirectly linked to plant water status.[1] Although there are a variety of methods to measure or infer plant water status, few references in the literature propose the use of directly measured or inferred plant water status measurements in irrigation scheduling.[2–5] The primary reason for this may involve the difficulties in interpreting plant water status measurements due to their interactions with evaporative demand and crop specific physiological factors. On-farm use of plant-based scheduling is exceedingly low. However, the increasing importance of agricultural water productivity and recent advancements in equipment and sensors for plant-based monitoring focus renewed interest in this scheduling approach.

MEASURING PLANT WATER STATUS

The most common parameter used for characterizing the water status of plants is the water potential (Ψ). There are a variety of instruments that have been used to measure plant water potential that are covered in detail by Hsiao.[6] The instrument that best fit the requirements for use in irrigation programming is the pressure chamber. In addition, there are other techniques best suited for laboratory conditions such as thermocouple psychrometry[7] and the Shardakov dye method.[6]

The Pressure Chamber

The pressure chamber requires that a leaf be excised, placed, and sealed in a chamber with the cut petiole end sticking out, and then the chamber is pressurized with nitrogen gas until xylem sap just appears at the end of the petiole.[8] The "balancing pressure" created by the compressed gas in the chamber is, under reasonable assumptions, a measure of the leaf Ψ.

There are a variety of techniques used and precautions recommended when using the pressure chamber to determine leaf Ψ. Of particular importance is to minimize water loss from the leaf between excision and placement in the chamber. This can be accomplished by covering the leaf with damp cheesecloth or a small plastic bag prior to excision and placing the leaf/cloth/bag combination in the chamber. Some pressure chamber operators blow into the plastic bag just before placing it over the leaf in order to create high enough humidity to minimize transpiration. The rate that the chamber is pressurized also can influence the reading.[9] Hsiao[6] recommends pressurizing at a rate of less than $0.1\,\text{MPa sec}^{-1}$ at the beginning of the measurement and $0.02\,\text{MPa sec}^{-1}$ as the balancing pressure is approached. Turner[10] suggested a pressurization rate of $0.025\,\text{MPa sec}^{-1}$. Fast pressurization rates can also cause adiabatic heating in the chamber. This also can be controlled by using the aforementioned rates of pressurization and by covering the leaf with a plastic bag.

A variant in measuring leaf Ψ is stem Ψ. Here, an interior, shaded leaf is covered by a small plastic bag overlaid with aluminum foil for a period of time prior to excision. Shackel et al.[4] suggested that this period be a minimum of 2 hr while Fulton et al.[11] indicate that transpiration ceases with 15 min of bagging. The elimination of transpiration results in an equilibrium in Ψ between the leaf and the adjacent stem and presumably in a tree, the trunk. Thus, stem Ψ should be higher than leaf Ψ, less coupled to the aerial environment, more representative of the whole tree water status, and less variable than leaf Ψ measurements, that are influenced by stomatal behavior and leaf shade history. Naor[12] found that stem Ψ was a better plant water stress indicator than predawn or midday leaf Ψ.

Leaf and stem Ψ change with time over the day. Highest values occur at predawn, become more negative during the morning, sometimes reaching a "plateau" for 2 hr–3 hr just after solar noon, and then increasing in the late afternoon and evening. Thus, readings are usually taken during the plateau period for day-to-day comparisons.

Encyclopedia of Water Science
DOI: 10.1081/E-EWS 120010260

Indirect evaluation of plant water status may be accomplished by numerous plant-based parameters that are related to plant Ψ. Many of these parameters have logistical and operational advantages over direct measurement techniques. Following is a description of those considered most relevant for irrigation programming.

Stomatal Opening

Current models of steady-state porometers can accurately measure stomatal conductance, which is directly related to photosynthesis, and linked to plant water status. However, water deficits need to be moderate to severe to cause stomatal closure; thus, this parameter is not considered as a sensitive indicator for use in irrigation management.

Plant Organ Size Variations

Both herbaceous and woody plant stems undergo diurnal oscillations in size due to both hydration and growth.[13] In the short term, size variations due to changing levels of hydration within various plant tissues greatly exceed those resulting from growth.[14] It is generally agreed that the living cells of the phloem, cambium, and parenchyma that surround the xylem provide most of the stored water contributing to size variations.[15] While Irvine and Grace[16] believe that contractions occur within the xylem in response to varying xylem Ψp, the consensus is that xylem tissues are almost totally rigid and contribute very little (10%–30%) to daily stem size variations.[15,17] It is generally agreed that stem diameter fluctuations are directly related to changes in plant water status.[18,19] While most of the attention in using plant organs as indicators of water status has been on stems, leaf, and fruit size daily fluctuations also occur. However, there is no known commercial use of these techniques. In the case of leaves, this may be due to having to calibrate thickness against an independent measure of Ψ for each leaf.[20] Recent advancements in the size-measuring sensors and mounting hardware for large, well attached fruits bode well for future research in fruit size oscillations' relations to Ψ as well as simply using fruit growth as a scheduling indicator.

Linear variable displacement transducers (LVDTs) and strain gauges are the most common types of organ size measuring devices. Since daily stem diameter oscillations are very small (generally less than 300 μm in mature fruit tree trunks), care must be taken to prevent temperature effects on the sensors and mounting hardware. Measures include shading the instruments and using mounting materials with low thermal expansion properties. Stem diameter oscillations are identified by continuously recording organ size using dataloggers. Data are down-

loaded manually in the field or transmitted by cellular phone or radio signals to computers.

Canopy Temperature

This technique relies on the fact that water-stressed plants undergo stomatal closure and consequently higher canopy temperatures due to lower transpirational cooling. The development of the infrared thermometer (IRT) made the canopy temperature measurement possible without physically contacting the plant and signaled the start of a high level of research on using the measurement in irrigation management.[21] Theoretical analysis[22] and experimental work[23] evolved the concept of the crop water stress index (CWSI). The CWSI is based on the temperature difference between the canopy (T_c) and surrounding air (T_a), normalized for the vapor pressure deficit of the air. It is calculated based on the ratio of where the measured $T_c - T_a$ value falls between equivalent values for a fully irrigated canopy (lower baseline experimentally determined) and severely stressed canopy (upper baseline empirically determined) under equivalent evaporative demand conditions. Protocols for use involve irrigating as to not exceed CWSI threshold values during specific periods of the season.

The CWSI method has been tested primarily on field and row crops. The technique requires that canopy temperature be accurately determined with the IRT and thus, soil within the view of the instrument must be avoided, although a method for correcting for soil effects has been developed.[24] This method is not sensitive to detect water deficits so mild that do not reduce crop transpiration.

Expansive Growth

The growth rate of leaves and stems is one of the most sensitive of all plant processes to water stress.[1] With indeterminate crops such as cotton, growth rate evaluation is likely the most popular, and certainly the oldest, plant method used to time irrigations. This is facilitated by the fact that any reduction in the distance between the growing terminal of the main stem and the reddish color associated with mature main stem tissue is visually very apparent. In addition to visual observations of growth, sensors may be placed on vegetative or reproductive organs to quantify growth. Goldhamer and Fereres[5] reported that this is a viable technique in young peach trees. Regardless of whether the approach is visual or mechanical, physiological and environmental factors other than water status can affect expansive growth. For example, growth with determinate crops ceases shortly after the onset of anthesis.

Sap Flow

Sensors using both heat pulse[25] and heat balance[26] techniques have been developed to estimate transpiration in individual plants. Both techniques involve a heat source placed on the stem or trunk of the plant and then thermocouples either inserted into the conducting tissue or placed along the surface of the stem or trunk. One major difference in the techniques is that the heat pulse measures flow velocity and an estimate of the cross sectional area of flow is required to calculate transpiration. Quantifying flow cross sectional area is usually done by visually evaluating conducting tissue, which in the case of trees, requires a core taken with a trunk boring tool. This is difficult and can result in significant errors.[27] Nevertheless, sap flow measurements can be qualitatively useful in identifying differences in transpiration between plants of similar size and the same species. The primary drawback with using sap flow measurement for irrigation scheduling is the same as with stomatal conductance and canopy temperature—transpiration is not affected under mild stress levels likely to affect expansive growth. Many techniques have been developed in plant physiological research to characterize the water status of plants, but only a few, such as those discussed above, have promise in the development of relevant applications for irrigation scheduling.

REFERENCES

1. Hsiao, T.C. Plant Responses to Water Stress. Annu. Rev. Plant Physiol. **1973**, *24*, 519–570.
2. Grimes, D.W.; Yamada, H. Relation of Cotton Growth and Yield to Minimum Leaf Water Potential. Crop Sci. **1982**, *22*, 134–139.
3. Peretz, J.; Evans, R.G.; Proebsting, E.L. Leaf Water Potentials for Management of High Frequency Irrigation on Apples. Trans. ASAE **1984**, *84*, 437–442.
4. Shackel, K.A.; Ahmadi, H.; Biasi, W.; Buchner, R.; Goldhamer, D.; Gurusinghe, S.; Hasey, J.; Kester, D.; Krueger, B.; Lampinen, B.; McGourty, G.; Micke, W.; Mitcham, E.; Olson, B.; Pelletrau, K.; Philips, H.; Ramos, D.; Schwankl, L.; Sibbett, S.; Snyder, R.; Southwick, S.; Stevenson, M.; Thorpe, M.; Weinbaum, S.; Yeager, J. Plant Water Status as an Index of Irrigation Need in Deciduous Fruit Trees. HortTechnology **1997**, *7* (1), 23–29.
5. Goldhamer, D.A.; Fereres, E. Irrigation Scheduling Protocols Using Continuously Recorded Trunk Diameter Measurements. Irrig. Sci. **2001**, *20*, 115–125.
6. Hsiao, T.C. Measurements of Plant Water Status. In *Irrigation of Agricultural Crops*; Stewart, B.A., Nielsen, D.R., Eds.; American Society of Agronomy: Madison, WI, 1990; 30, 243–279.
7. Boyer, J.S. Thermocouple Psychrometry in Plant Research. *Proceedings of the International Conference on Measurement of Soil and Plant Water Status*; Utah State University: Logan, UT, July 1987; Vol. 3, 26–30.
8. Scholander, P.F.; Hammel, H.T.; Hemmingsen, E.A.; Bradstreet, E.D. Hydrostatic Pressure and Osmotic Potential in Leaves of Mangroves and Some Other Plants. Proc. Natl Acad. Sci. U.S.A. **1964**, *52*, 119–125.
9. Naor, A.; Peres, M. Pressure-Induced Rate Affects the Accuracy of Stem Water Potential Measurements in Deciduous Fruit Trees Using the Pressure-Chamber Technique. J. Hortic. Biotechnol. **2001**, *76* (6), 661–663.
10. Turner, N.C. Techniques and Experimental Approaches for Measurement of Plant Water Status. Plant Soil **1981**, *58*, 339–366.
11. Fulton, A.; Buchner, R.; Olson, W.; Shackel, K. Rapid Equilibration of Leaf and Stem Water Potential Under Field Conditions in Almonds, Walnuts, and Prunes. HortTechnology **2001**, *11* (4), 609–615.
12. Naor, A. Midday Stem Water Potential as a Plant Water Stress Indicator for Irrigation Scheduling in Fruit Trees. Acta Hortic. **2000**, *537*, 447–454.
13. Kozlowski, T.T. Diurnal Variations in Stem Diameter of Small Trees. Bot. Gaz. **1967**, *123*, 60–68.
14. Kozlowski, T.T. Shrinking and Swelling of Plant Tissues. In *Water Deficits and Plant Growth*; Kozlowski, T.T., Ed.; Academic Press: New York, 1972; Vol. 3, 1–64.
15. Brough, D.W.; Jones, H.G.; Grace, J. Diurnal Changes in Water Content of the Stems of Apple Trees, as Influenced by Irrigation. Plant Cell Environ. **1986**, *9*, 1–7.
16. Irvine, J.; Grace, J. Continuous Measurements of Water Tensions in the Xylem of Trees Based on the Elastic Properties of Wood. Planta **1997**, *202*, 455–461.
17. Ueda, M.; Shibata, E. Diurnal Changes in Branch Diameter as Indicator of Water Status of Hinoki Cypress, *Chamaecyparis obtusa*. Trees **2001**, *15*, 315–318.
18. Klepper, B.; Browning, V.D.; Taylor, H.M. Stem Diameter in Relation to Plant Water Status. Plant Physiol. **1971**, *48*, 683–685.
19. Panterne, P.; Burger, J.; Cruiziat, P. A Model of the Variation of Water Potential and Diameter Within a Woody Axis Cross-Section Under Transpiration Conditions. Trees **1998**, *12*, 293–301.
20. McBurney, T. The Relationship Between Leaf Thickness and Plant Water Potential. J. Exp. Bot. **1992**, *43* (248), 327–335.
21. Ehrler, W.L.; Isdo, S.B.; Jackson, R.D.; Reginato, R.J. Wheat Canopy Temperature: Relation to Plant Water Potential. Agron. J. **1978**, *70* (2), 251–256.
22. Jackson, R.D.; Idso, S.B.; Reginato, R.J.; Pinter, P.J. Canopy Temperature as a Crop Water Indicator. Water Resour. Res. **1981**, *17*, 1133–1138.
23. Idso, S.B. Non-water-Stressed Baselines: A Key to Measuring and Interpreting Plant Water Stress. Agric. Meteorol. **1981**, *29*, 213–217.

24. Moran, M.S.; Clarke, T.R.; Inoue, Y.; Vidal, A. Estimating Crop Water Deficit Using the Relation Between Surface-Air Temperature and Spectral Vegetation Index. Remote Sens. Environ. **1994**, *49*, 246–263.

25. Cohen, Y.; Fuchs, M.; Green, G.C. Improvement of the Heat Pulse Method for Measuring Sap Flow in the Stem of Trees and Herbaceous Plants. Agronomie **1981**, *9*, 321–325.

26. Sakuratani, T. A Heat Balance Method for Measuring Water Flow in the Stem of Intact Plant. J. Agric. Meteorol. **1981**, *37*, 9–17.

27. Cohen, M.; Goldhamer, D.A.; Fereres, E.; Girona, J.; Mata, M. Assessment of Peach Tree Responses to Irrigation Water Deficits by Continuous Monitoring of Trunk Diameter Changes. J. Hortic. Sci. Biotechnol. **2001**, *76* (1), 55–60.

Irrigation Scheduling by Remote Sensing Technologies

Stephan J. Maas
Texas Tech University, Lubbock, Texas, U.S.A.

INTRODUCTION

Remote sensing is a method of quantifying physical characteristics of an object through measurements that do not require physical contact with the object. The physical characteristics of the object are determined from measurements of electromagnetic radiation, often in specific wavelengths, reflected or emitted by the object's surface. Although we usually associate remote sensing with satellites or aircraft, some types of remote sensing involve measurements made with ground-based sensors.

Researchers have suggested that there are two main characteristics that favor the use of remote sensing as part of a procedure for scheduling the irrigation of agricultural crops. First, remote sensing provides a quantification of the degree of crop water stress derived from measurements made directly on the crop canopy. This is in contrast to inferring crop water stress from measurements of properties of the soil, such as soil water content or soil water potential, in which the crop is grown. Second, remote sensing imagery can show the detailed variability of crop water stress within a field. When field conditions are variable, basing an irrigation strategy on a limited number of measurements may lead to over- or underwatering some parts of the field. Remote sensing imagery can provide a depiction of the variability in crop water stress across a field with a spatial resolution much greater than what can typically be achieved through conventional field measurements, such as water content or leaf water potential.

APPROACHES

Attempts to use remote sensing in irrigation scheduling have concentrated on the observation of two physical characteristics of the plant canopy: leaf temperature and leaf water content. Each of these characteristics has given rise to distinct approaches in measuring and utilizing remote sensing data in this application.

Leaf Temperature

The temperature of the leaf surfaces in a plant canopy is the result of the balance between the energy gained from the surrounding environment and lost to the surrounding environment.[1] During the daytime, when most irrigation-related remote sensing observations are made, the energy balance of a plant canopy can be expressed,

$$R_{ab} + C = E + R_{em}$$

In this expression, R_{ab} is the longwave and shortwave radiant energy absorbed by the leaf canopy from the sun, the sky, and surrounding soil and plant surfaces. C is the sensible heat gained or lost to the surrounding air through convection. If the surrounding air is warmer than the plant canopy, the canopy will gain heat energy from the air. The opposite is true if the surrounding air is cooler than the plant canopy. E is the latent heat energy lost by the plant canopy through transpiration, i.e., the evaporation of water from the leaves. R_{em} is the longwave radiant energy emitted by the plant canopy, and is a function of the temperature of the leaf surfaces,

$$R_{em} = \varepsilon \sigma T^4$$

In this expression, ε is the emissivity of the canopy (approximately 1 for leaves), σ is the Stefan–Boltzmann constant, and T is the absolute temperature of the leaf canopy (in K). It is through this expression that remote sensing systems can measure leaf temperature, by measuring the longwave radiant energy emitted by the plant canopy.

In the energy balance for the plant canopy, the magnitude of E is determined by the ambient micrometeorological conditions (air temperature, humidity, and wind speed) and the degree to which the leaf stomata are open. As crop plants deplete the soil water below the level necessary for optimum growth, the stomata close to restrict the further loss of water from the plants through evaporation. As the magnitude of E is reduced through stomatal closure, the magnitude of R_{em} must correspondingly increase to maintain the energy balance. This increase is observed through remote sensing as an increase in leaf temperature.

Several different approaches have been described for using remote sensing observations of leaf temperature in irrigation scheduling. The most well-documented of these are described in the following three sections.

Encyclopedia of Water Science
DOI: 10.1081/E-EWS 120010261

Thermal Kinetic Window

As in most organisms, the biochemical reactions leading to the growth of crop plants are controlled by enzymes. The temperature dependence of enzyme function helps establish the growth rate of a crop plant in a given environment. If the environment is too warm or too cool, enzyme activity will be inhibited and the potential rate of growth will be reduced. A range of temperature has been identified for each of several crops within which the activities of many important enzymes are at optimum levels.[2] This range is called the Thermal Kinetic Window (TKW). In arid and semiarid regions, leaf temperatures may exceed the range specified by the TKW during part of the day, particularly if soil water is limited.[3] Under these circumstances, irrigating the crop can cause leaf temperatures to fall within the TKW by increasing the loss of heat energy from the canopy through evaporation (Fig. 1). Maintaining leaf temperatures within the TKW by this strategy insures that biochemical reactions proceed at optimum rates.

Field studies in the Texas High Plains demonstrated that cotton yields could be maximized by irrigating the crop when the observed leaf canopy temperature exceeded 28°C.[4,5] Canopy temperatures were sensed with infrared thermometers, the signals from which could be used to control an automated irrigation system. A refinement of this procedure involved using the time that the crop was above 28°C (the Temperature–Time Threshold, or "TTT") to control irrigation application.[6] Using a TTT of 4 hr resulted in crop yields equaling those achieved with shorter values of TTT, but with less irrigation.

This concept was commercialized in a system called the Biologically Identified Optimal Temperature Interactive Console, or "BIOTIC."[7] In BIOTIC (Fig. 2), the leaf canopy temperature is measured with an infrared thermometer, and the time that the canopy temperature is above a specified threshold is accumulated by the control unit. When this time accumulation exceeds another predetermined threshold, and taking into account ambient humidity conditions, a signal is sent to turn on the irrigation system. A limitation to BIOTIC is that, early in the growing season when plants are small, measurements of leaf canopy temperature made by the infrared thermometer may be confounded by the higher temperature of the soil surrounding the plants. This might result in the irrigation system being turned on more often than necessary. This problem disappears when the plants reach a size to completely fill the field of view of the infrared thermometer. Since the time accumulation used in BIOTIC is based on essentially continuous measurements of canopy temperature, this approach is limited to ground-based, in-field sensor systems.

Crop Water Stress Index

In environments with a high evaporative demand, the temperature of a fully irrigated crop canopy is typically below the ambient air temperature during the day. Thus, the difference between canopy and air temperature has been recognized as an indicator of crop water stress.

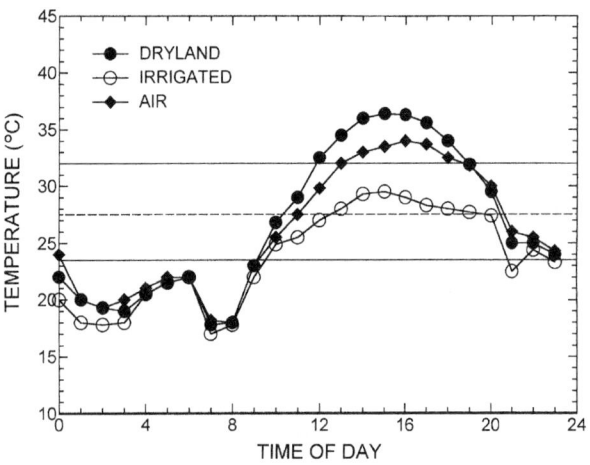

Fig. 1 Diurnal air temperature and foliage temperatures of irrigated and dryland cotton. The solid horizontal lines delimit the TKW and the dashed line illustrates the temperature providing optimum enzyme function. (From Ref. [3].)

Fig. 2 A BIOTIC system set up in a cotton field. (Photo courtesy of J. Mahan, USDA-ARS.)

Theoretical and experimental work led to the development of the Crop Water Stress Index, or "CWSI."[8,9] The CWSI expresses the degree to which the evapotranspiration (ET) of the crop approaches the maximum possible value of ET determined by ambient environmental conditions (the potential ET, or "PET"),

$$CWSI = 1 - (ET/PET)$$

In practice, PET can be calculated from ambient meteorological conditions or estimated from measurements of ET from a well-watered grass or alfalfa surface. CWSI is related to the difference between the canopy temperature T_c and air temperature T_a through the expression,

$$CWSI = [(T_c - T_a)_{min} - (T_c - T_a)_{obs}]/[(T_c - T_a)_{min}$$

$$- (T_c - T_a)_{max}]$$

In this expression, $(T_c - T_a)_{obs}$ is the observed temperature difference, and $(T_c - T_a)_{min}$ and $(T_c - T_a)_{max}$ are the minimum and maximum differences to be expected based on ambient environmental conditions. Empirically derived expressions relating $(T_c - T_a)_{min}$ and $(T_c - T_a)_{max}$ to the ambient saturation vapor pressure deficit have been reported for a number of crops.[10]

A limitation of the practical application of CWSI is that its strict derivation does not include soil temperature effects. For aircraft or satellite observing systems, which are usually pointed straight down at a field, the measured surface temperature will be a combination of plant canopy temperature and soil temperature when the canopy does not completely cover the soil surface. An adaptation of the CWSI has been developed that accounts for incomplete vegetation cover.[11] In this approach, called the "Vegetation Index/Temperature(VIT) Trapezoid,"

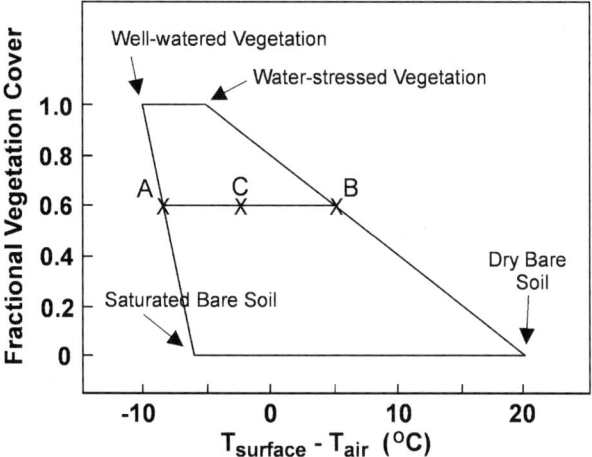

Fig. 3 The VIT Trapezoid. (From Ref. [11].)

measured surface minus air temperature is plotted vs. a measure of ground cover derived from remote sensing observations in the visible and near-infrared wavelengths. This point should lie within a trapezoid, the vertices of which represent the surface–air temperatures of a well-watered complete canopy, a severely water-stressed complete canopy, a saturated bare soil surface, and a completely dry bare soil surface (Fig. 3). In this figure, the ratio of the line segments AC/AB represents a Water Deficit Index (WDI) analogous to CWSI, but accounting for the degree of vegetation cover. In practice, the ordinate in the VIT Trapezoid is usually evaluated in terms of a vegetation index, like the Soil-Adjusted Vegetation Index,[12] that is derived from remote sensing observations and is proportional to vegetation cover.

The use of WDI in detecting crop condition in support of irrigation management was demonstrated at the Maricopa Agricultural Center in Arizona.[13,14] It was concluded from these studies that satellite observations were too infrequent to support operational irrigation scheduling based on WDI. Aircraft observations could be obtained with sufficient frequency for this application.

To accommodate infrequent remote sensing observations, the VIT Trapezoid was utilized in conjunction with a mathematical model of daily crop growth and water use. In this approach, the model (called "ProBE") provides a daily description of the water status of the crop based on weather and soil conditions, while the model simulation is calibrated through the evaluation of the actual water status of the crop on days with remote sensing observations.[15,16] A desirable feature of this approach is the capability to predict crop water status beyond the current day using weather forecasts or climatological data. In this way, the onset of water stress can be anticipated prior to its occurrence, allowing advance preparations to be made for irrigating the crop.

Spatial Variability in Canopy Temperature

When crop plants in a field are adequately watered, their transpiration rates are determined more by the evaporative demand of the ambient atmosphere than by soil conditions. As the plants deplete the soil water in the field to the point of becoming stressed, their transpiration rates become more dependent on soil conditions. As soil conditions typically are more spatially variable across a field than atmospheric conditions, one would expect to see more spatial variability in observed canopy temperature for a water-stressed crop than for an adequately watered crop.

Studies involving corn in the U.S. Great Plains suggest that irrigation should be applied when the range of six measurements of canopy temperature made within a field exceeds 0.7°C.[17] Researchers in Arizona have developed a "Histogram-derived Crop Water Stress Index" (HCWSI) that measures the departure of the distribution of measured canopy temperatures in a field from a normal distribution.[18] This approach is based on the observation that the distribution of canopy temperature measurements becomes skewed as the crop becomes stressed.

An advantage of scheduling irrigation based on this approach is that the variability in observed temperature, and not the precise measurement of temperature, is used as an indicator of the onset of stress. Thus, uncalibrated thermal images of a field might be sufficient for this application. Previous studies have been conducted on fields where the crop completely covers the soil surface. Observed surface temperature variability resulting from the variability in crop ground cover might seriously confound the measurements used in this approach.

Leaf Water Content

The interaction of electromagnetic radiation with water contained in plant tissues provides another mechanism for possibly detecting crop water status using remote sensing. Leaf water content is observable in leaf reflectance measurements particularly in near-infrared wavelengths ($0.8 \, \mu m - 2.5 \, \mu m$), where a systematic decrease in leaf reflectance is noted with decreasing leaf water content.[19] Several strong water absorption bands (particularly at $1.45 \, \mu m$ and $1.95 \, \mu m$) are observable in leaf reflectance spectra in the near infrared. Radar backscatter at microwave frequencies in the range $5 \, GHz - 10 \, GHz$ is also affected by plant canopy and soil water content.[20]

While leaf water content can be observed using remote sensing, there are several factors that limit its potential effectiveness in irrigation scheduling. First, leaf transpiration rate is physically related to leaf water potential, not to leaf water content. Thus, there is no direct connection between the observed remote sensing data and crop water status, as is the case with leaf temperature. Second, because leaf water is contained within plant tissue, remotely sensed leaf water content is affected by not only the water content of the tissue, but also how much tissue is present in the observation. Thus, remote sensing observations at near-infrared or microwave wavelengths cannot unambiguously discriminate between the amount of vegetation and the water status of the vegetation. In part because of these reasons, there are currently no widely recognized approaches for irrigation scheduling based on remotely sensed leaf water content.

CONCLUSION

Several irrigation scheduling approaches based on remotely sensed plant canopy temperature have been demonstrated to be operationally feasible. The TKW approach can be effective in maintaining crop canopy temperatures within the range conducive to optimal enzyme activity and growth. The WDI can compensate for incomplete crop ground cover in remote sensing observations of surface temperature, and can be used with a crop model to predict the onset of water stress. Other procedures involving spatial temperature variability and leaf water content await further development and testing.

REFERENCES

1. Campbell, G.S.; Norman, J.M. *An Introduction to Environmental Biophysics*, 2nd Ed.; Springer: New York, 1998; 223–246.
2. Burke, J.J.; Mahan, J.R.; Hatfield, J.L. Crop-Specific Thermal Kinetic Windows in Relation to Wheat and Cotton Biomass Production. Agron. J. **1988**, *80* (4), 553–556.
3. Burke, J.J.; Hatfield, J.L.; Wanjura, D.F. A Thermal Stress Index for Cotton. Agron. J. **1990**, *82* (3), 526–530.
4. Wanjura, D.F.; Upchurch, D.R.; Mahan, J.R. Evaluating Decision Criteria for Irrigation Scheduling of Cotton. Trans. ASAE **1990**, *33* (2), 512–518.
5. Wanjura, D.F.; Upchurch, D.R.; Mahan, J.R. Automated Irrigation Based on Threshold Canopy Temperature. Trans. ASAE **1992**, *35* (1), 153–159.
6. Wanjura, D.F.; Upchurch, D.R.; Mahan, J.R. Control of Irrigation Scheduling Using Temperature–Time Thresholds. Trans. ASAE **1995**, *38* (2), 403–409.
7. Upchurch, D.R.; Wanjura, D.F.; Burke, J.J.; Mahan, J.R. Biologically-Identified Optimal Temperature Interactive Console (BIOTIC) for Managing Irrigation. U.S. Patent 5,539,637, Jul 23, 1996.
8. Jackson, R.D.; Idso, D.B.; Reginato, R.J.; Pinter, P.J., Jr. Canopy Temperature as a Crop Water Stress Indicator. Water Resour. Res. **1981**, *17*, 1133–1138.
9. Idso, S.B.; Jackson, R.D.; Pinter, P.J., Jr.; Reginato, R.J.; Hatfield, J.L. Normalizing the Stress-Degree-Day Parameter for Environmental Variability. Agric. Meteorol. **1981**, *24*, 45–55.
10. Idso, S.G. Non-water-stressed Baselines: A Key to Measuring and Interpreting Plant Water Stress. Agric. Meteorol. **1982**, *27*, 59–70.
11. Moran, M.S.; Clarke, T.R.; Inoue, Y.; Vidal, A. Estimating Crop Water Deficit Using the Relation Between Surface–Air Temperature and Spectral Vegetation Index. Remote Sens. Environ. **1994**, *49*, 246–263.
12. Huete, A.R. A Soil-Adjusted Vegetation Index (SAVI). Remote Sens. Environ. **1988**, *27*, 47–57.
13. Moran, M.S. Irrigation Management in Arizona Using Satellites and Airplanes. Irrig. Sci. **1994**, *15*, 35–44.
14. Moran, M.S.; Clarke, T.R.; Qi, J.; Pinter, P.J., Jr. MADMAC: A Test of Multispectral Airborne Imagery as a Farm Management Tool. In *Proceedings of the 26th International Symposium on Remote Sensing of Environment, Vancouver, BC, Canada, Mar 25–29, 1996*; ERIM: Madison, WI, 1996.

15. Moran, M.S.; Maas, S.J.; Pinter, P.J., Jr. Combining Remote Sensing and Modeling for Estimating Surface Evaporation and Biomass Production. Remote Sens. Rev. **1995**, *12*, 335–353.

16. Moran, M.S.; Maas, S.J.; Clarke, T.R.; Pinter, P.J., Jr.; Qi, J.; Mitchell, T.A.; Kimball, B.A.; Neale, C.M.U. Modeling/Remote Sensing Approach for Irrigation Scheduling. In *Proceedings of the International Conference on Evapotranspiration and Irrigation Scheduling, San Antonio, TX, Nov 3–6, 1996*; ASAE: St. Joseph, MI, 1996.

17. Clawson, K.L.; Blad, B.L. Infrared Thermometry for Scheduling Irrigation of Corn. Agron. J. **1982**, *74*, 311–316.

18. Bryant, R.B.; Moran, M.S. Determining Crop Water Stress from Crop Temperature Variability. *Annual Research Report*; U.S. Water Conservation Laboratory, USDA-ARS: Phoenix, AZ, 1999; 124–127.

19. Gausman, H.W. *Plant Leaf Optical Properties in Visible and Near-Infrared Light*; Texas Tech University Press: Lubbock, TX, 1985; 24–30.

20. Moran, M.S.; Vidal, A.; Troufleau, D.; Qi, J.; Clarke, T.R.; Pinter, P.J., Jr.; Mitchell, T.A.; Inoue, Y.; Neale, C.M.U. Combining Multifrequency Microwave and Optical Data for Crop Management. Remote Sens. Environ. **1997**, *61*, 96–109.

I

Irrigation Scheduling by Soil Water Status

Philip Charlesworth
*Commonwealth Scientific and Industrial Research Organisation (CSIRO),
Townsville, Queensland, Australia*

Richard J. Stirzaker
*Commonwealth Scientific and Industrial Research Organisation (CSIRO),
Canberra, Australian Capital Territory, Australia*

INTRODUCTION

There are three methods for matching irrigation with crop water requirements. The first is to measure how much water the soil contains. The second is to monitor some attribute of the plant that is related to water deficits. The third is to calculate how much water the atmosphere can extract from a well-watered crop. This entry is about the first method, irrigation management by soil water status. Successful irrigation by this method requires more than just the ability to measure soil water status. We need to know how to relate measurements of soil water status to the amount and timing of irrigation, and how this ultimately affects crop yield.

DEFINITIONS OF SOIL WATER STATUS

Soil consists of a range of different sized particles from fine clays ($< 2\,\mu m$ diameter) through silts and sands to gravels ($> 2\,mm$). Water adheres to the surface of these particles, so soils with a finer texture (more clay) can store more water than soils with a sandier texture. The soil particles are also arranged so as to produce aggregates and pores or voids, giving the soil the property termed structure. Pores with a diameter in the range of $0.5\,\mu m - 50\,\mu m$ are important for storing water. Pores larger than these are normally air-filled and water in pores smaller than this is usually not available to plants.

For the purposes of irrigation, the water status of the soil is usually expressed as the volume of water in a given volume of soil. Thus, if a cubic meter of soil contained $300\,L$ of water the volumetric water content would be $0.3\,m^3$ of water per cubic meter of soil ($0.3\,m_w^3 m_s^{-3}$) or 30%. Since rain and irrigation are measured in depths (mm) it is often easier to express $0.3\,m^3\,m^{-3}$ as a depth equivalent, i.e., $300\,mm$ of water in 1-m depth of soil.

Plants can easily extract water from wet soil because the water is held in large pores. As soil becomes drier, water is held more strongly in smaller pores or closer to the

soil particles themselves. To obtain water, the roots must be in contact with the water films around the soil particles and in effect, the roots must be "drier" or exert more "pull" on the water than the soil. The size of this "pull" can be expressed in energy terms and is called soil water or matric potential. It is usually measured in kilopascals (kPa) and this gives a measure of the force needed to extract water from the soil matrix. When the soil is wet, little force is needed, as it dries, more force or pull is needed.

A soil water retention curve (SWRC) is used to convert the amount of water in a soil (volumetric water content) to its availability (energy status as given by the matric potential). As the clay content of a soil increases, the SWRC curve is displaced towards higher water contents (Fig. 1).

MEASURING SOIL WATER STATUS

Field monitoring of soil water potential began in the 1930s with the development of the tensiometer.[2] Routine, nondestructive measurements of soil water content were made possible by the development of the neutron scattering technique.[3] The last 20 yr have seen the rapid development of new tools for measuring soil water content, particularly measurements based on time domain reflectometry, capacitance, and heat dissipation.[4] A description of 25 commercially available products for measuring matric potential and soil water content and their mode of operation has been produced.[5]

USING SOIL WATER MEASUREMENTS

During irrigation the soil water potential rises close to $0\,kPa$. If the application rate exceeds the infiltration rate of the soil, then the soil water potential rises to $0\,kPa$ and water ponds on the soil surface. Immediately after irrigation large pores drain rapidly, so the wetted depth of soil increases and the average water content of the

Encyclopedia of Water Science
DOI: 10.1081/E-EWS 120010259

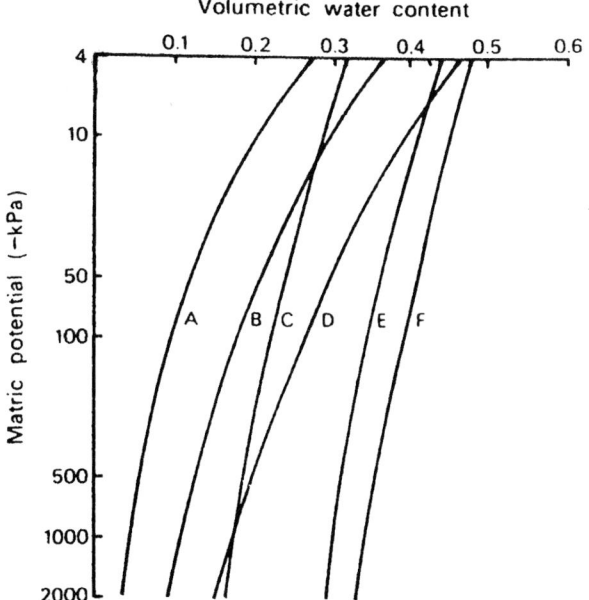

Fig. 1 Soil water retention curves for a range of soils from A—sand, to F—heavy clay. (From Ref. [1].)

topsoil falls. The rapid drainage phase is usually completed within 2 days and the soil water content at this stage is termed the drained upper limit (DUL) (Fig. 2). In reality, drainage continues indefinitely, although at slower and slower rates, so the DUL is not an intrinsic property of the soil. However, the drainage rate at the DUL is generally very much lower than the other source of

water loss from soil, evaporation, so for practical irrigation purposes DUL is a convenient measure of the full point.

A plant will extract water from the full profile until the soil is so dry that the plant wilts, even if the relative humidity around the leaves is near 100%. The soil water content at this stage is called the lower limit (LL). The amount of water between the DUL and LL is called plant available water (PAW), a term first used in 1949.[6] The main assumptions, errors, and remedies in deriving these terms have been summarized.[7] The terms "Field Capacity" and "Wilting Point" are also used to describe the range of soil water availability.

In practice, water becomes increasingly less available to plants between the DUL and LL. Some studies have shown that growth can be sustained until 70–80% of the PAW has been consumed.[8] For practical purposes a more conservative value of 0.5*PAW is often recommended as the depletion amount below DUL at which the profile should be refilled. This amount is referred to as readily available water (RAW). The idea of having this conservative refill point is to avoid the possibility of growth and yield reductions from the drying conditions in commercial crops.

The amount of water readily available to a crop is calculated as the RAW multiplied by the rooting depth. Table 1 gives examples of the average values for the water contents at DUL and the LL and the RAW over 1-m depth of soil for a range of soil textures.

When considering the water availability, a soil water potential of 0 kPa indicates the soil would be saturated or

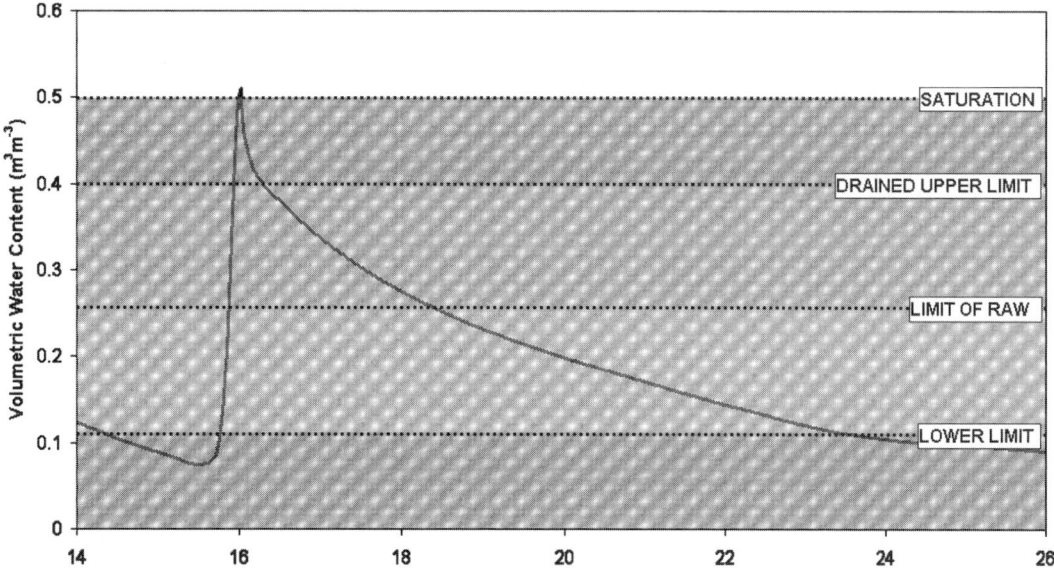

Fig. 2 The change in water content during an irrigation event followed by 10 days of drying. The soil water content falls rapidly immediately after irrigation when drainage dominates, and slows after the readily available water has been transpired. (Assumes constant evapotranspiration.)

waterlogged. Most soils have sufficient air for root function once the soil water potential drops to $-5\,kPa$. Once the soil dries to $-50\,kPa$, many plants will experience some water stress. By $-1500\,kPa$, only small amounts of water can be extracted from the soil and most plants will wilt.

WATER DEFICIT AND PLANT GROWTH

The calculation of RAW above is a useful approximation but belies much of the complexity of soil plant water relations. In reality the rate of water uptake needed to sustain rapid growth is determined by the atmospheric conditions, the leaf area, rooting distribution of the crop and the soil type, and water status.[10–12] More recently, evidence is accumulating that chemical signals produced by roots growing in drying soil affects plant growth. Stomata may close and growth rates fall before there is any detectable change in leaf water potential.[13] These signals could also lead to the root distribution changing to exploit wetter regions of the soil. Such adaptation may be limited by both soil strength and crop growth phase. The majority of crops are most sensitive to water deficits at the time of flowering and fruit set and more tolerant during the vegetative and maturation stages.[14]

OLD VS. NEW CONCEPTS IN IRRIGATION

The calculation of readily available water represents the maximum extraction of water before the crop yield may decline. Such a concept has its roots in flood irrigation and sprinkler irrigation systems where pipes had to be moved from field to field. Flood irrigation is best suited to applying large volumes of water at infrequent intervals, and the method specified how long that interval could be. In the case of sprinkler irrigation, a long interval cuts down the cost associated with moving pipes.

The advent of drip and micro irrigation and center pivot or lateral move sprinkler systems has changed the focus away from refill points. Since these systems allow

irrigation to be performed at virtually anytime, there is no need to approach the refill point and the associated dangers of drying the soil out too much. The aim is to keep the soil near the full point by applying water daily, or at most, weekly.

Drip and micro irrigation only wet part of the root zone, so the wetted volume of soil is much smaller than for flood or sprinkler. This means irrigation management must be more precise, as the reduced amount of stored water increases the risk to crop health of, for example, equipment failure. In many cases, these systems have been shown to produce higher yields. Micro irrigation also entails special problems related to the placement of soil water sensors with respect to the emitters.[15]

Interpretation of soil water content measurement is also more difficult for frequent irrigation, where plant uptake, redistribution, and drainage of water can be occurring simultaneously. Few irrigators understand that water can be moving into a layer of soil at the same rate as it is moving out, and make the mistake of interpreting a flat water content vs. time trace as evidence of no drainage.

There is also a much greater understanding of the way in which plants respond to water stress at different growth stages. For example, the practice of regulated deficit irrigation (RDI) can save up to 60% of the seasonal water requirement with little effect on fruit yield or quality.[16] The deficit is allowed to develop after flowering and fruit set at the time vegetative growth is at its maximum, and removed when the fruit size starts to increase. A variant of RDI is partial root zone drying, where half the root zone of a perennial crop is irrigated and the other half allowed to dry.[17] Irrigation is alternated such that the previously dry half is reirrigated and the previously wet half allowed to dry. In this way, half the roots are well watered and half experience drying soil. Roots in the drying soil produce hormonal signals that reduce vegetative growth and provide a more favorable balance between vegetative and fruit production. Again yields are unaffected by the stress or even increased. Of even greater importance is the impact of these controlled stresses on fruit quality, particularly in the wine industry.

ADOPTION BY IRRIGATORS

Irrigation management by soil water status is a method that has been promoted for over 50 yr. However, despite decades of extension work, the uptake by irrigators remains disappointingly low. There are several reasons for this. In most situations water is not a major percentage of the variable costs and many irrigators baulk at the time and expense of collecting, interpreting, and implementing the information soil water sensors provide. In practical terms the cost to the individual farmer of overirrigation is less

Table 1 Representative volumetric water contents ($m^3\,m^{-3}$) and the readily available water in 1 m of soil for soils of three textures

	Sand	Loam	Clay
Saturation	0.4	0.5	0.6
DUL	0.06	0.29	0.41
LL	0.02	0.05	0.2
RAW for 1 m root zone	20 mm	120 mm	105 mm

Source: (Adapted from Ref. [9].)

than the penalty of under irrigating, and there is no doubt large quantities of irrigation water are wasted as a result.[18] Water treatment, algal bloom control, salinity are all examples of off-site impacts of overirrigation, which are generally not included in the cost–benefit analysis.

Point to point variability in soil water content within a field is also a major disincentive for soil water monitoring. This variability is due to changes in soil properties, plant growth, and nonuniformity of water application. To properly account for such variation requires the installation of far more equipment than is practicable.[19] The problem of variability makes the atmosphere-based methods of irrigation scheduling more attractive. However, errors in this method, particularly related to the estimation of leaf area and root distribution, means that some combination of atmospheric and soil based measurement will provide the most robust feedback system. Irrigation scheduling by soil water status should show further improvements through the development of new soil water measuring technology, and particularly the software associated with them that simplifies the interpretation of data for the irrigator.

REFERENCES

1. Williams, J. Physical Properties and Water Relations. *Soils: An Australian Viewpoint*; CSIRO Division of Soils: Melbourne, 1983.
2. Richards, L.A.; Neal, O.R. Some Field Observation with Tensiometers. Soil Sci. Soc. Am. Proc. **1936**, *1*, 71–91.
3. Gardner, W.; Kirkham, D. Determination of Soil Moisture by Neutron Scattering. Soil Sci. **1952**, *73*, 391–401.
4. White, I.; Zegelin, S.J. Electric and Dielectric Methods for Monitoring Soil-Water Content. In *Handbook of Vadose Zone Characterization & Monitoring*; Wilson, L.G., Everett, L.G., Cullen, S.J., Eds.; Lewis Publishers: Boca Raton, 1995.
5. Charlesworth, P.B. *Irrigation Insights No. 1—Soil Moisture Monitoring; National Program for Irrigation Research and Development*; CSIRO Publishing: Melbourne; Australia, 2000.
6. Veihmeyer, F.J.; Hendrickson, A.H. Methods of Measuring Field Capacity and Wilting Percentages of Soils. Soil Sci. **1949**, *68*, 75–94.
7. Ritchie, J.T. Soil Water Availability. Plant Soil **1981**, *58*, 327–338.
8. Meyer, W.S.; Green, G.C. Water Use by Wheat and Plant Indicators of Available Soil Water. Agron. J. **1980**, *72*, 253–257.
9. Marshall, T.J.; Holmes, J.W.; Rose, C.W. *Soil Physics*, 3rd Ed.; Cambridge University Press: Cambridge, UK, 1996.
10. Slayter, R.O. *Plant–Water Relationships*; Academic Press: London, 1967.
11. Gardner, W.R. Dynamic Aspects of Water Availability to Plants. Soil Sci. **1960**, *89*, 63–73.
12. Philip, J.R. Plant Water Relations: Some Physical Aspects. Annu. Rev. Plant Physiol. **1966**, *17*, 245–268.
13. Passioura, J.B. The Yield of Crops in Relation to Drought. In *Physiology and Determination of Crop Yield*; Boote, K.J., Bennett, J.M., Sinclair, T.R., Paulsen, G.M., Eds.; American Society of Agronomy: Madison, 1994; 343–359.
14. Rudich, J.; Kalmar, D.; Geizenber, G.C.; Harel, S. Low Water Tensions in Defined Growth Stages of Processing Tomato Plants and Their Effects on Yield and Quality. J. Hortic. Sci. **1977**, *52*, 391–399.
15. Or, D.; Coelho, F.E. Flow and Uptake Patterns Affecting Soil Water Sensor Placement for Drip Irrigation Management. Trans. ASAE **1996**, *39* (6), 2007–2016.
16. Goodwin, I.; Jerie, P.; Boland, A. Water Saving Techniques for Orchards in Northern China. *Water Is Gold—Irrigation Association of Australia National Conference and Exhibition Proceedings*; Irrigation Association of Australia: Sydney, May 19–21; 1998.
17. Stoll, M.; Loveys, B.; Dry, P. Hormonal Changes Induced by Partial Rootzone Drying of Irrigated Grapevine. J. Exp. Bot. **2000**, *51*, 1627–1634.
18. Stirzaker, R.J. The Problem of Irrigated Horticulture: Matching the Bio-physical Efficiency with the Economic Efficiency. Agrofor. Syst. **1999**, *45*, 187–202.
19. Schmitz, M.; Sourell, H. Variability in Soil Moisture Measurements. Irrig. Sci. **2000**, *19*, 147–151.

I

Irrigation Scheduling by Water Budgeting

Claudio O. Stockle
Brian Leib
Washington State University, Pullman, Washington, U.S.A.

INTRODUCTION

The basic purpose of irrigation in agriculture is to supply plants with sufficient water to obtain optimum harvestable crop yield and quality. For this purpose, irrigation events must be scheduled to provide the right amount of water at the appropriate time. Water stored in the soil explored by plant roots is the main source of water supply for crop uptake, mainly used to meet evaporation losses from plant surfaces. Soil water budgeting is therefore a primary irrigation scheduling approach for agricultural crops.

Why Irrigation Scheduling?

Irrigation scheduling consists of determining when and how much irrigation water to apply to crops growing in the field or greenhouses. The purpose is to supply plant water needs to meet given crop yield and quality targets. Late and/or insufficient irrigation may lead to undesirable crop water stress and yield/quality reduction. Excess irrigation generates undesired water percolation in the soil profile beyond the reach of roots. This not only represents a loss of water otherwise available for plant use, but percolating water also transports nutrients, pesticides, and other chemicals into deeper soil layers, eventually reaching groundwater.

WATER BUDGETING

The approach to be used for proper irrigation scheduling depends on the method of irrigation. High frequency methods such as drip irrigation do not rely much on soil storage of water. Irrigations are applied to meet crop water use, typically on a daily basis, utilizing a small volume of the soil potentially available for root water extraction. Other irrigation methods such as furrow or move-set sprinkler irrigation rely on the soil profile explored by roots as water storage. Plants draw water from this storage until a point where replenishment by irrigation is required. The speed of storage depletion by crop water uptake (and direct soil surface evaporation), the soil water content at any given time, and the critical soil water content (allowable depletion) at which irrigation is required are the

central elements of irrigation scheduling by water budgeting.

The general procedure for irrigation scheduling based on water budgeting was summarized in the early 1970s and can be easily implemented in a simple computer program or an electronic spreadsheet.[1] Irrigation scheduling decision support systems that implement computerized water budgeting by linking real-time weather information and soil and crop databases are now commonly available.[2] The daily soil water depletion within the soil profile effectively explored by roots is calculated as:

$$D_i = D_{i-1} + (\text{ET} - P_e - \text{IR})_i \qquad (1)$$

where D_i is soil water depletion on day i, D_{i-1} is soil water depletion on day $i-1$, ET is evapotranspiration (water loss by crop transpiration plus soil surface water evaporation) on day i, P_e is effective precipitation (precipitation depth that infiltrates into the soil) on day i, and IR is net irrigation (water depth actually stored in the root zone) on day i. All quantities are expressed as water depth in mm or in. Fig. 1 illustrates the water balance process described by Eq. 1.

Soil water depletion is calculated relative to the upper limit of the soil storage of plant available water (PAW). Soil water may fluctuate from total dryness to saturation, where all porosity is filled with water. The PAW soil storage encompasses a fraction within these two water status extremes. The upper limit corresponds to the water depth equivalent after the soil profile has been fully irrigated and subsequently drained until the drainage rate becomes negligible (24 hr–72 hr after irrigation, depending on soil texture). This is also referred to as the soil field capacity. The lower limit (also known as permanent wilting point) corresponds to the soil water storage level at which plants can no longer remove soil water. The PAW soil storage is the difference between the upper and lower limits.

When the soil water content is equal to the upper limit water content, $D_i = 0$. When $D_i = D_o$, the allowable depletion for a given crop, an irrigation event should be scheduled. D_o corresponds to a water content between the upper and lower limit of PAW. The refill point is closer to the upper limit as the sensitivity of crops to water stress and the atmospheric demand for water evaporation

Encyclopedia of Water Science
DOI: 10.1081/E-EWS 120010258

increase. The value of D_o also depends on the size of the PAW soil storage. For more information on soils and irrigation scheduling see Ref. [3].

Evapotranspiration

Soil water evaporation is the process of transformation of liquid water to vapor and subsequent removal from the soil surface. Crop transpiration consists of the transformation of liquid water into vapor inside the plant tissues, and the subsequent removal of vapor to the atmosphere, mainly through stomata. For plants growing under typical field conditions, these two processes occur simultaneously, a phenomenon referred to as evapotranspiration (ET). The ET rate (mm or in per unit time) depends on the energy supply, vapor pressure gradient, and wind as well as on the rate of water supply to plant roots (ultimately to the evaporation sites at the substomatal cavities) and the topsoil. Evapotranspiration depletes the soil moisture in the soil profile explored by roots.

Crop ET can be determined experimentally. However, for irrigation scheduling purposes, ET is normally calculated from weather data. The method consists of the use of empirical equations to calculate ET from a reference crop (ET_o), typically defined as a short healthy grass surface, 12-cm high, fully covering the ground and with ample water supply. Evapotranspiration for the crop of interest (ET_c) is obtained by multiplying ET_o by a crop coefficient (K_c) that fluctuates throughout the growing season based on crop canopy characteristics and its ability to cover the ground. A large number of equations to calculate ET_o are available.[4] Evaluations performed have shown the Penman–Monteith equation,[5] a biophysically based formulation, to be suitable for applications across climatic conditions.[6] A complete description of procedures to calculate ET_o, K_c, and ET_c is given by Allen et al.[7]

Soil Water Content

Soil water sensors can be used in conjunction with a water balance Eq. 1 to help schedule irrigation. Sensors can be used to verify the soil water content estimated from a water balance, to restart a water balance at a known value of soil water content, or to evaluate the upward or downward trend in soil water content from irrigation designed to replace the loss of water from ET. It is important to understand instrument calibration when using a soil water sensor in conjunction with a water balance. Sensor calibration can vary in different soils types.[8,9] However, the relative accuracy of sensors can be useful in irrigation scheduling without the need for absolute accuracy via calibration.[10] Irrigation scheduling is most successfully implemented when ET driven water balances and soil

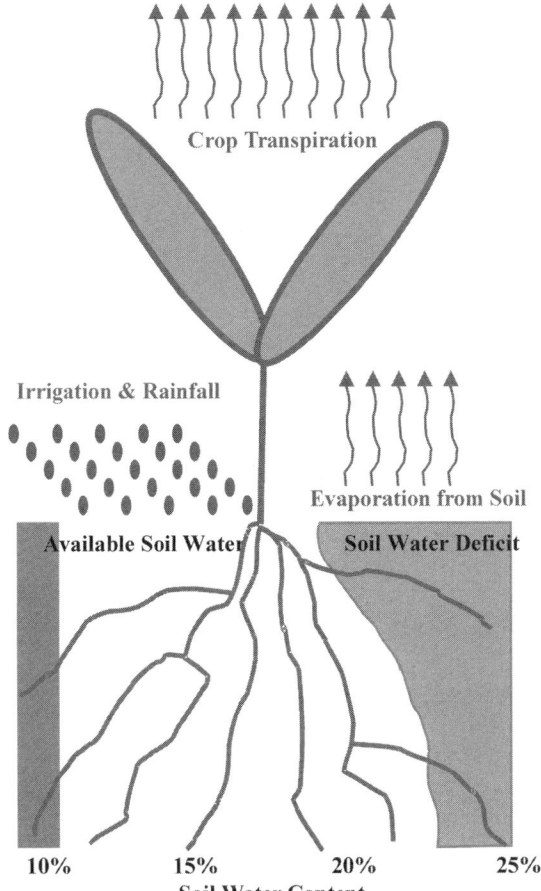

Fig. 1 Soil water balance in irrigation scheduling.

water sensors function as independent checks to compensate for the inaccuracies of both methods. Fig. 2 shows how irrigation was scheduled by calculating the soil water depletion from both a water balance Eq. 1 and soil water sensors.

There are several methods of measuring soil water content and how each type of sensor works is briefly described below. The tensiometer uses a porous ceramic tip in direct contact with the soil to measure soil tension. The granular matrix sensors and resistance blocks measure the change in electrical resistance that occurs as soil water moves in and out of the sensor in response to the surrounding soil moisture. The neutron probe counts the number of neutrons that collide with the hydrogen in water. Tensiometers, resistance sensors, and neutron scattering[11] have a fairly long history of use in irrigation scheduling. Most of the new sensors available on the market today measure the dielectric constant of the soil. The dielectric constants are of air, mineral soil and water is 1, 3–7, and 80, respectively. Therefore, the overall dielectric constant of bulk soil will vary predominantly in

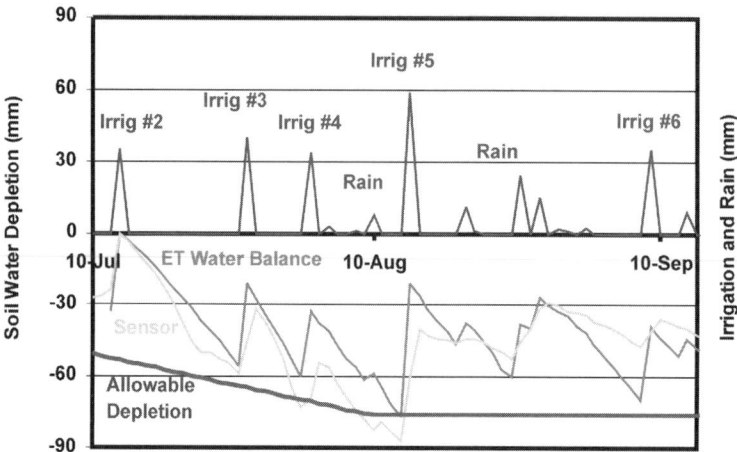

Fig. 2 Example of a water balance used with soil water sensors in irrigation scheduling.

relationship to soil water content. One way to determine the soil's dielectric constant is by measuring the change in a radio wave's frequency as it passes through the soil known as frequency domain reflectometry (FDR) or RF capacitance.[12] Another, way to determine the dielectric constant of the soil is to measure the reflectance pattern of a voltage pulse that is applied to a wire guide known as time domain reflectometry (TDR).[13]

CONCLUSION

Crop, soil, and weather information can be used to calculate components of the soil water balance, which can be used for irrigation scheduling (when and how much irrigation water to apply). Soil water budgeting is a robust approach, widely used in irrigated agriculture. Soil water calculations must be complemented with measurements of water content in the soil profile effectively explored by plant roots. A large array of sensors is available to measure soil water.

REFERENCES

1. Jensen, M.E.; Wright, J.L.; Pratt, B.J. Estimating Soil Moisture Depletion from Climate, Crop and Soil Data. Trans. ASAE **1971**, *14*, 954–959.
2. Leib, B.G.; Elliott, T.V.; Mathews, G. WISE: A Web-Linked and Producer Oriented Program for Irrigation Scheduling. Comput. Electron. Agric. **2001**, *33* (1), 1–6, http://www.elsevier.nl/locate/issn/01681699.
3. James, L.G. *Principles of Farm Irrigation System Design*; John Wiley & Sons: New York, 1988.
4. Jensen, M.E.; Burman, R.D.; Allen, R.G. *Evapotranspiration and Irrigation Water Requirements*, ASCE Manuals

and Reports on Engineering Practices No. 70; American Society of Civil Engineers: New York, 1990.
5. Monteith, J.L. Evaporation and Environment. *19th Symposia of the Society for Experimental Biology*; University Press: Cambridge, 1965; 19, 205–234.
6. Allen, R.G.; Jensen, M.E.; Wright, J.L.; Burman, R.D. Operational Estimates of Reference Evapotranspiration. Agron. J. **1989**, *81*, 650–662.
7. Allen, R.G.; Pereira, L.S.; Raes, D.; Smith, M. *Crop Evapotranspiration: Guidelines for Computing Crop Water Requirements*, Irrigation and Drainage Paper No. 56; FAO: Rome, 1998.
8. Yoder, R.E.; Johnston, D.L.; Wilkerson, J.B.; Yoder, D.C. Soil Water Sensor Performance. Appl. Eng. Agric. **1998**, *14* (2), 121–133.
9. Hansen, B.R.; Peters, D. *Measuring Soil Moisture. Proceedings of the 19th Annual International Irrigation Show, San Diego, CA, November 1–3, 1998*; The Irrigation Association: Falls Church, VA, 1998; 103–110.
10. Leib, B.G.; Matthews, G. *The Relative Accuracy of Soil Moisture Sensors Used in Washington State. Paper for the 1999 ASAE/CSAE-SCGR Annual International Meeting, Toronto, Canada, July 18–22, 1999*; American Society of Agricultural Engineers: St. Joseph, MI; Paper No. 992270.
11. Phene, C.J.; Reginato, R.J.; Itier, B.; Tanner, B.R. Sensing Irrigation Needs. In *Management of Farm Irrigation Systems*; Hoffman, G.J., Howell, T.A., Solomon, K.H., Eds.; American Society of Agricultural Engineers: St. Joseph, MI, 1990; 207–264.
12. Ley, T.W.; Stevens, R.G.; Topielec, R.R.; Neibling, W.H. *Soil Water Monitoring and Measurement*; PNW 475, a Pacific Northwest Publication; Washington State University: Pullman, WA, 1994; 28–29.
13. Topp, G.C.; Davis, J.L.; Annan, A.P. Electromagnetic Determination of Soil Water Content: Measurement of Coaxial Transmission Lines. Water Resour. Res. **1980**, *16*, 574–582.

Irrigation, Sewage Effluent Use for

B. A. Stewart
West Texas A&M University, Canyon, Texas, U.S.A.

I

INTRODUCTION

One of the primary functions of soil is to buffer environmental change. This is the result of the biological, chemical, and physical processes that occur in soils. The soil matrix serves as an incubation chamber for decomposing organic wastes including pesticides, sewage, solid wastes, and many other wastes. Soils store, decompose, or immobilize nitrates, phosphorus, pesticides, and other substances that can become pollutants in air or water. Consequently, soil has, for centuries, been used for the application of sewage effluents. Sewage effluent provides farmers with a nutrient-enriched water supply and society with a reliable and inexpensive means of wastewater treatment and disposal. It should not, however, be assumed that irrigation is always the best solution for wastewater disposal. Disposal by irrigation should always be compared with alternative options based on environmental, social, and economic costs and benefits.

While disposal is the primary objective in many cases, the need of water for irrigation is becoming more often the driver for using sewage effluent on land. This is particularly true in areas like the Middle East where population growth is resulting in severe water shortages. The guidelines for using effluent for irrigation vary considerably among countries and other governing bodies. Cameron[1] conducted a literature review and found wide differences of guidelines for effluent irrigation projects being used throughout the world. In general, however, sustainable and environmentally sound systems can be developed in most situations provided proper management practices are followed.

CONCERNS OF IRRIGATING WITH SEWAGE EFFLUENT

In spite of the documented benefits associated with the use of sewage effluent for irrigation, there are numerous concerns. Many industrial wastewaters have been routinely dumped into municipal sewage lines. While this issue has been addressed in some jurisdictions, it has not in many others. In the United States, the Environmental Protection Agency requires that wastewaters be treated prior to disposal into municipal treatment plants or back into groundwater. Irrigating with wastewaters partially cleans water by percolation through the soil, but soluble salts and some inorganic and organic chemicals may continue to flow with the water to groundwater or surface supplies. In general, the Environmental Protection Agency allows sewage effluents to be used for irrigation only if it does not cause 1) extensive groundwater pollution; 2) a direct public health hazard; 3) an accumulation in the soil or water of hazardous substances that can get into the food chain; 4) an accumulation of pollutants such as odors into the atmosphere; and 5) other aesthetic losses, within the limits.[2]

Bouwer[3] has also expressed concerns about the use of sewage effluent for irrigation. He is particularly concerned with pathogens and warns that complete removal of viruses, bacteria, and protozoa and other parasites should be required before the effluent can be used to irrigate fruits/vegetables consumed raw or brought into the kitchen, or parks, playgrounds and other areas with free public access. Bouwer also stresses that long-term effects of sewage effluent irrigation on underlying groundwater should be considered in addition to the changes in nitrate and salinity. Ground water in low rainfall regions can be highly affected by percolating sewage effluent because much of the water is used by the growing crops and this greatly concentrates the chemicals in the small amounts of water that actually percolate to the groundwater. These chemicals can include disinfection byproducts, pharmaceutically active chemicals, and compounds derived from humic and fulvic acids formed by the decomposition of plant material. Bouwer claims that many of these chemicals are suspected carcinogens or toxic. Therefore, Bouwer concludes that while sewage irrigation looks good on the surface, a more extensive look reveals a potential for serious contamination of groundwater. He states that municipalities and other entities responsible for irrigation with sewage effluent should do a groundwater impact analysis to develop management protocols and be prepared for liability actions. Those who benefit are local and state institutions in water resources, environmental quality protection, public health, consultants, and operators of effluent irrigation projects.

Encyclopedia of Water Science
DOI: 10.1081/E-EWS 120010142

REUSE STANDARDS

The standards for using sewage effluent for irrigation of agricultural crops vary widely among different countries of the world. Mexico and many South American countries, e.g., use untreated wastewater for irrigation.[4] Most of these countries do not have the resources or capital to treat sewage effluents. Wastewater is utilized after little or no treatment, and health risks are minimized by crop selection. Mexico does not allow wastewater to be used to irrigate lettuce, cabbage, beets, coriander, radishes, carrots, spinach, and parsley. Acceptable crops include alfalfa, cereals, beans, chili, and green tomatoes. In contrast, Israel has very stringent water reuse requirements. Effluent water requires a high level of treatment (large soil-aquifer recharge systems with dewatering) before the water can be reused for irrigation of vegetables to be consumed raw.[5] Health guidelines for irrigation with treated wastewater developed in California indicate that effluent waters used on food crops must be disinfected, oxidized, coagulated, clarified, and filtered.[6] Total coliform counts cannot exceed a median value of 2.2/100 ml or a single sample value of 25/100 ml. Total coliforms must be monitored daily and turbidity cannot exceed 2 nephelometric turbidity units and must be monitored continuously. Less restrictive guidelines developed by Shuval et al.,[7] and adopted by most of the international agencies, suggested that effluent water reuse was relatively safe to use if it contained less than 1 helminth egg L^{-1}, and less than 1000 fecal coliforms/100 ml.

MONITORING GUIDELINES

Site selection is a critical and necessary step in initiating a sewage effluent irrigation system. The U.S. Environmental Protection Agency[8] published detailed information on site characterization and evaluation. Information was provided on the design of systems, site characteristics, expected quality of the effluent water after land treatment, and typical permeabilities and textural classes suitable for each land treatment process. Information was provided for

designing and monitoring site characteristics for slow rate processes (sprinkler and other typical farm irrigation systems), rapid infiltration basins, and overland flow systems. Monitoring requirements will vary considerably among projects depending on the cropping patterns, soil characteristics, and specific environmental concerns. In most cases, monitoring procedures and criteria will be site specific. In all cases, however, the objectives should be to use the resources effectively, protect the land, protect the groundwater, protect the surface water, and protect the community amenity.

REFERENCES

1. Cameron, D.R. *Sustainable Effluent Irrigation Phase 1: Literature Review International Perspective and Standards*, Technical Report Prepared for Irrigation Sustainability committee; Canada–Saskatchewan Agriculture Green Plan, 1996.
2. Miller, R.W.; Gardiner, D.T. *Soils in Our Environment*, 8th Ed.; Prentice-Hall Inc: Upper Saddle River, NJ, 1998.
3. Bouwer, H. *Groundwater Problems Caused by Irrigation with Sewage Effluent*; Irrigation and Water Quality Laboratory, USDA-ARS: Phoenix, AZ, 2000.
4. Strauss, M.; Blumenthal, U.J. *Human Waste Use in Agriculture and Aquaculture: Utilization Practices and Health Perspectives*; IRWCD Report No. 09/90; International Reference Centre for Waste Disposal: Deubendorf, 1990.
5. Shelef, G. The Role of Wastewater Reuse in Water Resources Management in Israel. Water Sci. Tech. **1990**, *23*, 2081–2089, Switzerland.
6. Ongerth, H.J.; Jopling, W.F. Water Reuse in California. In *Water Renovation and Reuse*; Shuval, H.I., Ed.; Academic Press: New York, 1977.
7. Shuval, H.I.; Adin, A.; Fattal, B.; Rawitz, E.; Yekutiel, P. *Wastewater Irrigation in Developing Countries. Health Effects and Technical Solutions*, World Bank Tech. Pap.; 1986; Vol. 51, 325 pp.
8. U.S. EPA. *Process Design Manual: Land Treatment of Municipal Wastewater*; EPA 625/1-81-013; U.S. EPA Center for Environmental Research Information: Cincinnati, OH, 1981.

Irrigation, Supplemental

Philippe Debaeke
*Institut National de la Recherche Agronomique (INRA),
Castanet-Tolosan, France*

INTRODUCTION

The main objective of irrigation consists of supplying water to crops when soil-stored water at planting and seasonal rainfall are too erratic or limited to satisfy the plant transpiration demand with enough regularity, at a level defined by the farmer.

Supplemental (or supplementary) irrigation (SI) was defined as follows: "In an area where a crop can be grown by natural rainfall alone but additional water by irrigation stabilizes and improves yield, this irrigation is termed supplemental, the additional water alone being insufficient to produce a crop."[1] SI is applied to complete a deficient or uneven precipitation regime, enhancing and securing crop production, both in quantity and quality, in such pedoclimatic conditions where rainfed production is still feasible although less profitable.

AREAS AND CROPPING SYSTEMS CONCERNED

When natural contributions by rainfall or groundwater are too scarce to satisfy full crop water requirements only occasionally (amount and distribution within the season), the continuous optimal water regime can be obtained through SI, i.e., by a temporary and discontinuous irrigation regime.[2] Such situations are frequently observed in humid and subhumid regions, generally for spring-sown crops such as soyabean, maize, sugarbeet, potatoes, and for some tree crops. For instance, in southwestern France, most of the maize is grown under SI (up to 250 mm–300 mm).

In the regions where both natural and irrigation resources are too limited for ensuring a permanent optimal water regime to crops, SI is mainly supplied at the critical periods of the crop-growth cycle, in order to maintain or improve crop production.[2] This is the case in arid and semiarid regions of the Mediterranean basin where SI is practiced for species generally grown profitably without irrigation but their yields are subjected to great variations over the years because of rainfall variability. These species are: winter cereals (mostly durum and bread wheat), autumn-sown legumes (faba bean, peas), spring-sown crops having a dense and deep rooting system (sorghum, sunflower, cotton, etc.), and tree crops (olive, almond, peach, vine, etc.). In those regions, spring-sown crops (such as maize or sugarbeet) with high water requirements but restricted rooting system are only grown under intensively-irrigated systems, irrigation amounts (until 800 mm–1000 mm) exceed the contribution of natural resources (rain and stored soil water).

Surprisingly, SI is also widely practiced in Northern Europe, such as United Kingdom and Scandinavia.[3] In most years, spring-sown crops (potatoes, sugarbeet, horticultural crops) have their growth restricted and yields reduced by water shortage, the extent of this depending very much on soil type (e.g., shallow soils, low water holding capacity), weather conditions, and the timing and duration of stress periods. Although rainfall is fairly evenly distributed throughout the year, potential evaporation rates exceed rainfall throughout most of the summer months. In Scandinavia, the growing season is much shorter and so early sowing cannot be practiced to moderate the effects of summer drought periods. In Denmark, SI is needed nearly every year on sandy soils to maintain a stable production and 15% of the agricultural land is grown under irrigation.[4]

In temperate humid and subhumid environments, where water deficit is occasional, generally terminal and/or of short duration, SI is used by farmers to stabilize yield and quality at higher levels (to improve profits), and to maintain crop uniformity. In drier environments, SI can be considered to be more a dry farming technique since it contributes to optimize the use of limited water resources:[2] its purpose is to prevent complete yield loss (through irrigation at sowing, in exceptionally dry years) or to improve yield in years not excessively dry (by irrigation during shooting). In every case, SI is a means of insuring farmers against climatic risks.

SI SCHEDULING

The strategy of applying restricted amounts of water based on the amount and distribution of rainfall in addition to the incremental effect of water on crop yield is the essence of the SI concept.[5]

Encyclopedia of Water Science
DOI: 10.1081/E-EWS 120010140

Either because irrigation volume (or discharge) is limited (dry winter season, low storage capacity of reservoirs, equipment not available, etc.) or because soil-water deficit is moderate, SI generally results in a limited number of water applications. The goal of these applications is either to save crop life but more generally to improve the efficiency of the other inputs.[5] Positive impacts are expected such as: sowing in due time, assurance of an uneven and minimum plant emergence, more efficient placement and use of fertilizers, thus limiting soil N leaching and residue at harvest, use of high-yielding cultivars and avoidance of moisture stress for plant, particularly during the initial stages of its development in semiarid regions, later (around flowering) in wetter regions.

For instance, in the West Asia–North Africa (WANA) region, with a Mediterranean-type climate, wheat production is increasingly declining. Cereal yields are low and variable in response to inadequate and erratic seasonal rainfall (350 mm rainfall and above) and related management factors, such as lack of nitrogen and late sowing. It is clear that small amounts of SI water can make up for the deficits in seasonal rain and produce satisfactory yields.[6] A minimum yield of more than 3.5 Mg/ha is guaranteed for wheat with an amount of irrigation varying from 50 mm to 200 mm depending on the root zone soil water and the amount and distribution of the seasonal rainfall, whereas the average yield is below 1.5 Mg/ha under rainfed management.[1] An addition of only limited irrigation (1/3 full irrigation) may achieve over 60% of the potential increase in yield with full SI. In addition, use efficiency for both soil water and nitrogen is greatly increased by SI. Oweis et al.[6] observed a wheat yield increase up to a fertilizer input of 100 kg N/ha under SI management in Syria, while optimum response for rainfed conditions was with 50 kg N/ha.

In southwestern France, under a temperate subhumid climate, grain yield was increased by 17%, 27%, 37%, and 70% for sunflower, sorghum, soybean, and maize respectively, with an irrigation amount of 120 mm (supplied around flowering) when compared to rainfed management during nine years on a deep silty-clay soil.[7] This shows the differential sensitivity of spring-sown crops to SI as related to ecophysiological traits such as depth and extraction efficacy of rooting system, drought tolerance mechanisms (sunflower and sorghum), indeterminate reproductive period (for soybean) acting as an escaping strategy.

With limited available water, the challenge is to satisfy crop water demand at the critical (and most responsive) stages. An extensive review of specific periods for optimizing irrigation was made by FAO.[8] For instance, the most sensitive stages of wheat to water stress are the booting and the early earing stage from some research, and the preflowering and ear formation stages according to other research, whereas seed germination and crop emergence periods are only exceptionally considered to be sensitive to water stress.[2] The decision of irrigation at a given growth stage depends on the crop sensitivity to water stress, on the climatic pattern and on the need to exploit natural water resources. Irrigation on cereals in autumn during dry sequences aims at ensuring an optimal plant density and a satisfactory root establishment in order to fully use soil water reserves later but also to cover rapidly the soil surface for controlling soil evaporation and maximizing early radiation interception and biomass accumulation.

In semiarid regions, when a single application is available for sunflower, it should be placed either at presowing (soil refillment, crop establishment) or between flower bud appearance and flowering (to increase leaf area index) while in wetter areas, one irrigation is generally recommended after anthesis to enhance the leaf area duration and favor oil production.

SI METHODS

The irrigation methods usable for SI must satisfy the following specifications:[2] low equipment cost per ha of irrigated land, high degree of transferability from one field to another, limited labor to set up the irrigation system, high water distribution efficiency, possibility to bring limited water amounts (20 mm–50 mm) timely and accurately at specific growth stages.

For these reasons, in rainfed systems, sprinkler irrigation (travelling rainguns, for instance) seem to be the most flexible for field crops while, for vegetables and fruit trees, drip irrigation may be more suitable. The source of water is generally small reservoirs (run-off and rainwater harvesting) but deep groundwater is also used.

NEED FOR MODELS

In the last 15 years, on the basis of ecophysiological studies, numerous soil-plant models (either crop-specific or generic) have been developed to simulate the response of major crops to water use. By running on long-term weather records, these mechanistic models, more or less complex, are useful to determine a probabilistic response of grain yield to SI, in interaction with crop management (sowing date, cultivar, crop density, N-fertilization), and to define at field or farm level the optimal irrigation schedules under limited water management (e.g., Refs. [4,9,10,11]).

At farm level, linear programming models have been developed to optimize the crop planning and to allocate scarce water between competing fields, according to the response of yield relatively to crop water requirements (resulting from mechanistic models or from simple production functions), to stochastic distribution of rainfall, to input availability (labor, equipment, water), to cost of production (water, other inputs), and to crop value.[12,13] Such models can be used to test if limited amounts of irrigation water can be used more efficiently by applying small amounts to more land than by fully irrigating less land, or to predict the optimal cropping pattern under SI.

To conclude, SI cannot be restricted to a simple problem of tactical decision at field level ("when and how much water to apply on this field?") but has to be considered also as a major strategical decision at farm level ("which fields and which crops to irrigate using what type of equipment?").

REFERENCES

1. Arar, A. The Role of Supplementary Irrigation in Increasing Productivity in the Near East Region. Proceedings of the International Conference on Supplementary Irrigation and Drought Management, Valenzano-Bari, Italy, Sept 27–Oct 2, 1992; CIHEAM; Vol. 1, 2, 1–10.

2. Caliandro, A.; Boari, F. Supplementary Irrigation in Arid and Semiarid Regions. Medit **1996**, *7*, 24–27.

3. Kay, M.G. Supplementary Irrigation in Northwest Europe. In *Consultation on Supplementary Irrigation*, Rabat, Morocco, Dec 7–9, 1987; FAO: Rome, 1990; 17–22.

4. Plauborg, F.; Heidmann, T. MARKVAND: An Irrigation System for Use Under Limited Irrigation Capacity in a Temperate Humid Climate. In *Irrigation Scheduling: From Theory to Practice*, Proceedings of the ICID/FAO Workshop on Irrigation Scheduling, Rome, Italy, Sept 12–13, 1995, Water Reports 8; ICID-CIID & FAO: Rome, 1996; 177–184.

5. Oweis, T.; Pala, M.; Ryan, J. Management Alternatives for Improved Durum Wheat Production Under Supplemental Irrigation in Syria. Eur. J. Agron. **1999**, *11*, 255–266.

6. Oweis, T.; Pala, M.; Ryan, J. Stabilizing Rainfed Wheat Yields with Supplemental Irrigation and Nitrogen in a Mediterranean Climate. Agron. J. **1998**, *90*, 672–681.

7. Debaeke, P.; Hilaire, A. Production of Rainfed and Irrigated Crops Under Different Crop Rotations and Input Levels in Southwestern France. Can. J. Plant Sci. **1997**, *77*, 539–548.

8. Doorenbos, J.; Kassam, A.H. *Yield Response to Water*; Irrigation and Drainage Paper 33, FAO: Rome (Italy), 1979; 193.

9. Perrier, E.R.; Salkini, A.B. Scheduling of Supplemental Irrigation on Spring Wheat Using Water Balance Methods. In *Irrigation: Theory and Practice*; Rydzewski, J.R., Ward, C.F., Eds.; Institute of Irrigation Studies, Southampton University: UK, 1989; 447–460.

10. Cabelguenne, M.; Jones, C.A.; Williams, J.R. Strategies for Limited Irrigations of Maize in Southwestern France—A Modeling Approach. Trans. ASAE **1995**, *38*, 507–511.

11. Debaeke, P. Wheat Response to Supplementary Irrigation in South-Western France: II. A Frequential Approach Using a Simulation Model. Agr. Med. **1995**, *125*, 64–78.

12. Bryant, K.J.; Mjelde, J.W.; Lacewell, R.D. An Intra-seasonal Dynamic Optimization Model to Allocate Irrigation Water between Crops. Am. J. Agr. Econ. **1993**, *75*, 1021–1029.

13. Deumier, J.M.; Leroy, P.; Peyremorte, P. Tools for Improving Management of Irrigated Agricultural Crops. In *Irrigation Scheduling: From Theory to Practice*, Proceedings of the ICID/FAO Workshop on Irrigation Scheduling, Rome, Italy, Sept 12–13, 1995, Water Reports 8; ICID-CIID & FAO: Rome, 1996; 39–50.

Irrigation, Surface

Wynn R. Walker
Utah State University, Logan, Utah, U.S.A.

INTRODUCTION

Surface irrigation, also referred to as "flood irrigation," is the oldest and most common method of applying water to croplands. There are three broad classifications: 1) basin irrigation; 2) border irrigation; and 3) furrow irrigation. Each classification can be distinguished on the basis of shape, slope, and field boundaries. It is important to understand that references to the "surface irrigation system" may include more than the individually irrigated field. Specifically, the irrigation system may consist of four subsystems, as illustrated in Fig. 1. These are: 1) the water supply subsystem; 2) the water delivery subsystem; 3) the water use subsystem (field); and 4) the water removal (drainage) subsystem.[1] Thus, the terms basin, border, and furrow irrigation are specific configurations of the water use subsystem. Optimizing basin, border, or furrow irrigation practices requires that each component of the irrigation system be designed, constructed, maintained, and operated effectively.

TYPES OF SURFACE IRRIGATION SYSTEMS

The advantages and disadvantages of a specific surface irrigation configuration depend on a number of factors. For example, the need and extent of land leveling for furrow irrigation are much less than for basin irrigation. Very small or irregularly shaped fields are more easily irrigated with basins than furrows. The infiltration characteristics of the soil combined with the nature and availability of the water supply in terms of flow rate and duration may favor one method over another. The density and arrangement of the crop as well as the season-to-season cropping pattern will impact the method of applying water; and, the historical traditions of the irrigators may suggest one method over another.

Basin Irrigation

Two typical examples of basin irrigation are shown in Fig. 2. Basins are level fields with perimeter dikes to prevent runoff. To distinguish them from level borders (discussed in "Border Irrigation"), basins tend to be squarer in shape while level borders are more rectangular.

The most important design parameter for basins is the inflow rate per unit width of the basin (unit discharge). Basins require a high unit discharge. Most soils can be irrigated by basin systems although soils with a moderate to low infiltration rate result in the best efficiency and uniformity. Basin systems typically apply a relatively large depth of water during irrigation and thus deep-rooted, closely spaced crops are best suited for this type of irrigation. Crops, which cannot be inundated for extended periods, should be planted on raised beds or furrows.

There are three important advantages of basins: 1) they are effective and efficient methods of leaching salts from the soil profile; 2) they are easily automated with relatively simple flow controls at the basin inlet; and 3) they can achieve efficiencies and uniformities which equal or exceed those of sprinkle systems without the corresponding investment in energy.

Border Irrigation

Borders are somewhat like basins though they are rectangular or contoured fields. They typically have a longitudinal but cannot have a lateral slope. They may be free draining or blocked at the lower end. Fig. 3 illustrates three typical border irrigation systems.

Borders are suited for most crops and perform well on soils with moderately low to moderately high infiltration characteristics. If the soil crusts easily, borders can be furrowed, so, plants are grown on raised beds. If the border is not diked at the lower end, substantial tailwater losses may occur as illustrated in Fig. 4. Free-draining borders exhibit high application uniformities but generally are less efficient than sprinkle systems. Blocked-end borders enjoy the same high uniformities and efficiencies as basins and have further advantage of better field drainage in cases of excess rainfall or errors in irrigation duration.

Furrow Irrigation

By "furrowing," "creasing," or "corrugating" a field surface and then regulating a flow to each furrow, a field can be watered with substantially less flow and can have slopes in both the longitudinal and transverse directions. Water flowing in the furrow infiltrates through the wetted perimeter and moves vertically and laterally thereafter to

Encyclopedia of Water Science
DOI: 10.1081/E-EWS 120010065

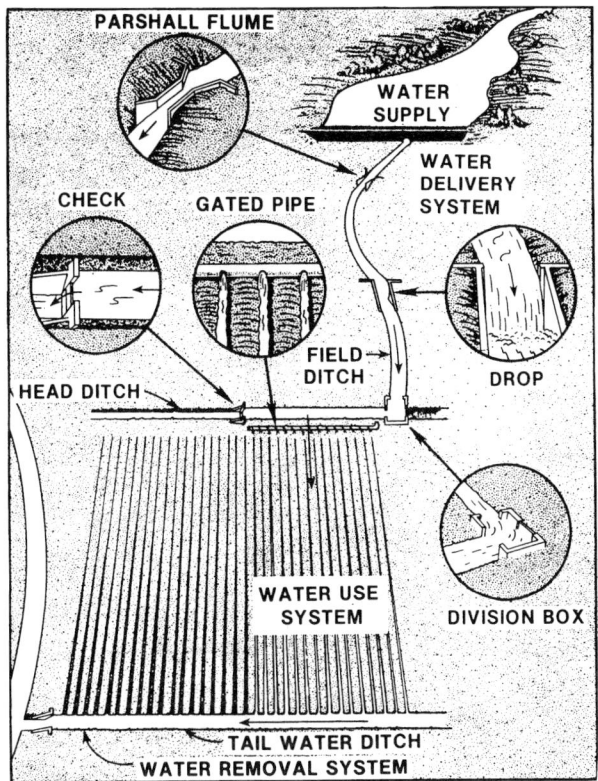

Fig. 1 Typical elements of a surface irrigation system. (From Ref. [1].)

Fig. 2 Two illustrations of common basin irrigation systems: (a) a basin in Southeast Asia; and (b) A basin in central Utah. (Utah State University Irrigation Photo Archives.)

refill the soil reservoir. As noted above, furrows can be used in conjunction with basins and borders, which should be referred to as "furrowed" borders or basins. A typical furrow irrigation system is shown in Fig. 5.

Furrows provide somewhat better flexibility in on-farm water management, achieve the same high uniformities as basins and borders, and require less land leveling to implement. Furrow systems require more farm labor and thus tend to be less efficient than basins and borders. Flow rates per unit width can be substantially reduced and topographical conditions can be more severe and variable. Furrows provide operational flexibility important for achieving high efficiencies for each irrigation throughout a season by regulating the flow into each furrow. It is a simple (although labor intensive) matter to adjust the furrow stream size to changing intake characteristics by simply changing the number of simultaneously supplied furrows. Two of the more common ways in which water is introduced to furrows are shown in Fig. 6.

In a general situation, furrows are less efficient that either basins or borders, primarily because of the difficulty in setting the proper flow rate into each furrow and thus causing either too much deep percolation or too much tailwater. Salts can accumulate between furrows, which are used for a long period of time. The additional tillage associated with construction of the furrows adds costs to the farm operation. Finally, the danger of erosion from furrow irrigation is higher than with either basins or borders.

IMPROVING THE OPERATION AND MANAGEMENT OF SURFACE SYSTEMS

Walker and Skogerboe[2] describe the surface irrigation water management in the following terms.

> Even though it is the oldest and most common method of irrigation, surface irrigation is the least amenable to consistently high levels of performance. Of all the reasons why this is so, probably none have the significance that is associated with the uncertainty of soil infiltration rates. The rate at which water will be absorbed through the soil surface is a nonlinear process which varies both temporally and spatially. It is affected by year-to-year changes in cropping patterns, cultivation, the weathering due to climate, and many other unknown influences. As a result, neither the irrigator nor the engineer can accurately predict the uniformity and efficiency of an irrigation before it occurs, particularly the first water application

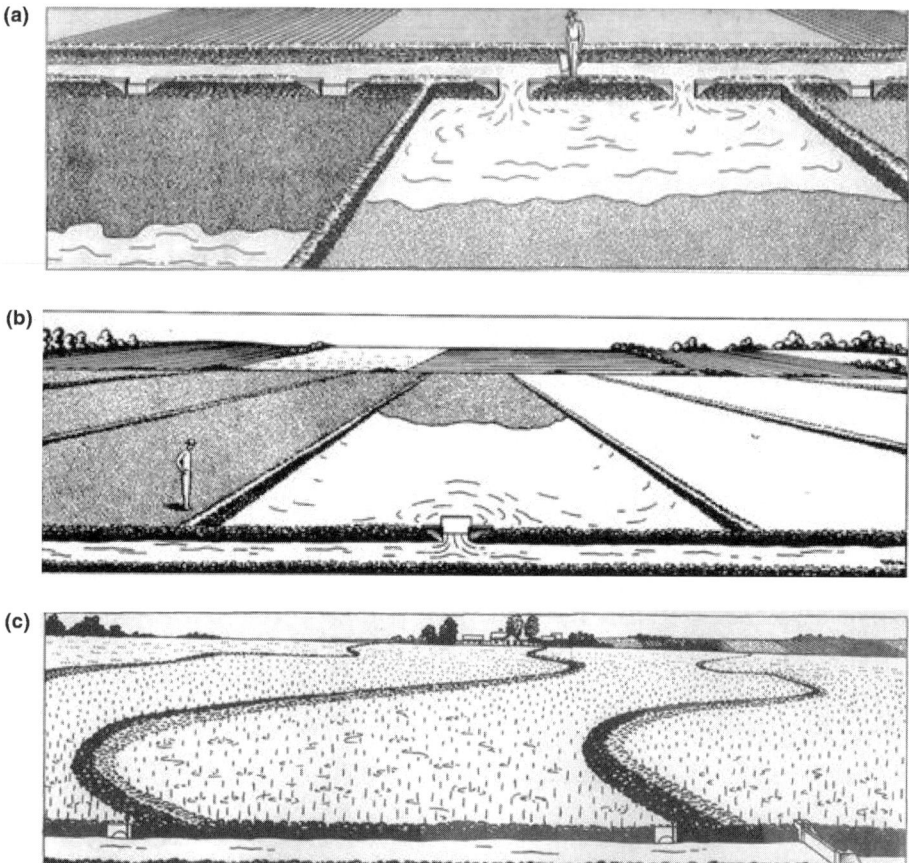

Fig. 3 Examples of border irrigation systems: (a) Typical graded border irrigation system; (b) typical level border irrigation system; and (c) typical contour levee or border irrigation system. (From Ref. [1].)

Fig. 4 Tailwater runoff under border irrigation. (Bureau of Reclamation Photo: www.yao.lc.usbr.gov/WaterConser/Conservation Defined.htm.)

Fig. 5 A typical furrow irrigation system. (Utah State University Irrigation Photo Archives.)

Fig. 6 A common method of supplying water to furrows using siphon tubes. (Utah State University Irrigation Photo Archives.)

following planting.There are other factors limiting surface irrigation system performance, such as a relative lack of standardized equipment for regulation and automation. These and the intake variability noted above place particular emphasis on the management practices applied to surface irrigation, and the art of surface irrigation management is very important.

There are four ways in which the operation and management of basins, borders, and furrows can be improved: 1) improving the management of flow and time of cutoff; 2) precision leveling of the field surface; 3) blocking the end of the field; and 4) water recovery and reuse.

Regulating Inflow and Time of Cutoff

The irrigator of a free-draining surface irrigation system must balance the need for a high inflow to achieve uniform water application against a low inflow to

Fig. 7 Laser guided land leveling equipment. (Utah State University Irrigation Photo Archives.)

minimize tailwater losses. The intake opportunity time at the end of the field plus the time required for the inflow to reach the end of the field dictate a unique time of cutoff. If the inflow can be reduced (cutback) when the water has advanced to the end of the field, then the advantages of both a high flow during advance and a low flow to reduce tailwater can be met. One of the most promising "cutback" practices is surge flow in which the inflow is rapidly cycled on and off to create a "time-averaged" reduction in inflow.

Precision Land Leveling

There are few practices as important to surface irrigation performance as precision land leveling, grading, or smoothing. The advent of laser guided equipment as shown in Fig. 7 has improved land surface preparation by at least an order of magnitude over historical practices. The impact on surface irrigation uniformity and efficiency has been very high. It is not unusual to hear an irrigator state that the most important part of surface irrigation is "lasering."

Fig. 8 A typical tailwater recovery and reuse facility. (Utah State University Irrigation Photo Archives.)

Blocked-End Systems

An alternative to reducing the inflow at the end of the advance period as a way to reduce tailwater losses is simply to block the end of the field and prevent the runoff from occurring. This is a common feature of basins and the reason for their high performance. It is less common in borders and furrows because of the risk of crop damage at the lower end of the field due to lengthy ponding. With proper regulation of inflow and time of cutoff, excessive ponding can be prevented or an emergency drain may be necessary to remove excessive ponding. In any event, the application efficiency of blocked-end systems is usually 15%–20% higher than free-draining systems. Blocked-end systems should have fairly low field slopes if ponding is a problem.

Wastewater Recovery and Reuse

In some areas, the tailwater problem can be resolved by constructing a small reservoir at the end of the field, capturing the runoff, and then reusing it elsewhere on the farm. This is a particularly useful practice where irrigators must control sediments, pesticides, and fertilizer runoff as well. Fig. 8 shows a typical tailwater pond.

CONCLUSION

Surface irrigation is critically important to the production of the world's food and fiber. Advances in design, management, and field preparation have made surface irrigation systems as efficient as alternatives such as sprinkle irrigation. Surface irrigation is labor intensive, but requires low energy inputs making this method of watering crops more advantageous in developing countries than in countries like the United States. As the world population increases over the next two or three decades, improving the efficiency of surface irrigation systems to their potential will be one of the most important water management tasks that arid and semiarid regions will face.

REFERENCES

1. U.S. Department of Agriculture, Soil Conservation Service, Planning Farm Irrigation Systems. *National Engineering Handbook*; U.S. Government Printing Office: Washington, DC, 1967, Sec. 15, Chap. 3.
2. Walker, W.R.; Skogerboe, G.V. *Surface Irrigation: Theory and Practice*; Prentice-Hall Inc.: Englewood Cliffs, NJ, 1987; 386 pp.

Irrigation Systems, Drip

I. P. Wu
University of Hawaii, Honolulu, Hawaii, U.S.A.

J. Barragan
University of Lleida, Lleida, Spain

V. Bralts
Purdue University, West Lafayette, Indiana, U.S.A.

INTRODUCTION

A drip irrigation system is a form of localized irrigation that delivers water directly into the root zone of a crop. When properly designed and managed, a drip irrigation system can eliminate surface runoff, minimize deep seepage, and achieve high uniformity of water distribution and irrigation application efficiency. The development of drip irrigation in the late 1960s marked a period of tremendous improvement in irrigation science and technology in which water use is done more beneficially for agricultural production.

With the increasing consequence of limited water resources and the increasing need for environmental protection, drip irrigation will play an even more important role in the future. Drip irrigation systems can be used for many different types of agricultural crops, including fruit trees, vegetables, pastures, specialty crops such as sugarcane, ornamentals, golf course grasses, and high economic value crops grown in greenhouses. An understanding of drip irrigation systems, irrigation scheduling, crop response, and economic ramifications will encourage greater use of drip irrigation in future agricultural production.

UNIFORMITY OF WATER APPLICATION AND DESIGN CONSIDERATIONS

The desired uniformity of water application and the specific crops to be grown guides the creation of drip irrigation systems. There are two types of drip irrigation uniformity: system uniformity and spatial uniformity in the field. The consistency of system distribution of water into the field describes the system uniformity. The spatial uniformity is the regularity of water distribution considering overlapping emitter flow and translocation of water in the soil. For drip irrigation systems designed for trees with large spacing, the system uniformity is equal to the water application uniformity in the field. For high-density plantings, the emitter spacing should be designed considering overlapped wetting patterns and the spatial uniformity in the field. The uniformity of a drip irrigation system depends primarily on the hydraulic design, but must also consider the manufacturer's variation, temperature effects, and potential emitter plugging. The effect of water temperature is generally negligible when using turbulent flow emitters. A combination of proper filtration and turbulent emitters can control emitter plugging. When grouping a number of emitters together as a unit, such as those designed to irrigate an individual plant's root system, the uniformity of water application with respect to the plant will improve.

Many expressions have been used to describe uniformity. The system uniformity, or emitter flow uniformity, can be expressed as the range or variation of water distribution in the field. This term was initially used for hydraulic design of drip irrigation systems given that the minimum and maximum emitter flows could be calculated and determined.[1] When more emitter flows are used or more samples are required for determining variation or spatial uniformity in the field, the Christiansen uniformity coefficient (UCC)[2] and coefficient of variation (CV), which is the ratio of standard deviation and the mean, are used. Each of the uniformity expressions are highly correlated with one other.

HYDRAULIC DESIGN OF DRIP IRRIGATION SYSTEMS

Once selection of the type of drip irrigation emitter is complete, the hydraulic design can be made to achieve the expected uniformity of irrigation application.

The hydraulic design of a drip irrigation system involves designing both the submain and lateral lines. Early research in drip irrigation hydraulic design concentrated mainly on the single lateral line approach,[1,3,4] but in 1985 Bralts and Segerlind developed a method to design a submain unit. The hydraulic design is based on the energy relations in the drip tubing, the friction

Encyclopedia of Water Science
DOI: 10.1081/E-EWS 120010068

drop, and energy changes due to slopes in the field. Direct calculations of water pressures along a lateral line or in a submain unit are made by using an energy gradient line approach.[1] All emitter flows along a lateral line and in a submain can be determined based on their corresponding water pressures. Once the emitter flows are determined, the emitter flow variation, q_{var} is expressed by

$$q_{var} = \frac{q_{max} - q_{min}}{q_{max}} \tag{1}$$

where q_{max} is the maximum emitter flow and q_{min} is the minimum emitter flow. Based on these data, other uniformity parameters such as UCC and CV can also be determined. There is a strong correlation between any two of the three uniformity parameters in the hydraulic design of drip irrigation systems, thus any one of the uniformity parameters can be used as a design criterion. This correlation also justifies using the simple emitter flow variation q_{var} for hydraulic design. The emitter flow variation q_{var} is converted to the CV when it is combined with the manufacturer's variation of emitter flow.

The total emitter flow variation caused by both hydraulic and manufacturer's variation can be expressed by,[5]

$$CV_{HM} = \sqrt{CV_H^2 + CV_M^2} \tag{2}$$

where CV_{HM} is the coefficient of variation of emitter flows caused by both hydraulic and manufacturer's variation; CV_H and CV_M are the coefficients of variation of emitter flows caused by hydraulic design and manufacturer's variation, respectively.

The design criterion for emitter flow variation q_{var} for drip irrigation design is arbitrarily set as 10.0%–20.0%, which is equivalent to a CV, from 0.033 to 0.076, or 3.0%–8.0%. Based on the research of last 30 yr, the manufacturer's variation of turbulent emitters is maintained only in a range 3.0%–5.0%, expressed by CV. When this variation is combined with emitter flow variation caused by hydraulic design with a range 3.0%–8.0% in CV, the total variation determined by the equation above will be limited to a CV of less than 10.0%. This variation illustrates that the drip irrigation systems are designed to achieve high uniformity and irrigation application efficiency.

Economic return can also be the basis of design criteria for drip irrigation. A new set of design criteria for drip irrigation was developed,[6] based on achieving an expected economic return with various water resources and environmental considerations (Table 1).

DRIP IRRIGATION FOR OPTIMAL RETURN, WATER CONSERVATION, AND ENVIRONMENTAL PROTECTION

When the uniformity of a drip irrigation system is designed with a UCC of 70.0%, 30.0% or less in CV, the irrigation application is expressed as a straight-line distribution,[7,8] as shown in Fig. 1. This figure was plotted using percent of area (PA) against a relative irrigation depth, X, which is the ratio of required irrigation depth to mean irrigation application. The straight-line distribution in the dimensionless plot can be specified by a minimum value, a, a maximum value, $(a + b)$, in the X-scale and a slope b, where b specifies the uniformity of water application.[9]

When a drip irrigation system is designed with fixed uniformity, it is possible to determine the sloped straight line with known value of a and b. A value (X) can then be selected between value a and $(a + b)$ and plotted (Fig. 1). The triangle formed above the horizontal line (X) results in an irrigation deficit and yield reduction. The triangle below the horizontal line results in over-irrigation and deep seepage.

An important irrigation scheduling parameter, the relative irrigation depth, (X) indicates how much irrigation water is applied. The effectiveness of drip irrigation is shown not only by the high uniformity of the drip irrigation system, but also by the irrigation requirement and the strategy of irrigation scheduling. As illustrated in Fig. 1, the irrigation scheduling parameter (X) affects the areas of over-irrigation and water deficit conditions in the field and is directly related to the economic return. Practically speaking, the X parameter is selected in a range from a to $(a + b)$, as shown in Fig. 1. Three typical irrigation schedules can be expressed by X and are as follows:

$X = a$ This schedule is a conventional irrigation schedule, which is based on the minimum

Table 1 Design criteria for uniformity of drip irrigation system design

Design consideration	CV (%)	UCC (%)
Water is abundant and no environmental pollution problems	30–20	75–85
Water is abundant but with environmental protection considerations	20–10	80–90
Limited water resources but with no environmental pollution problems	25–15	80–90
Considerations for both water conservation and environmental protection	15–5	85–95

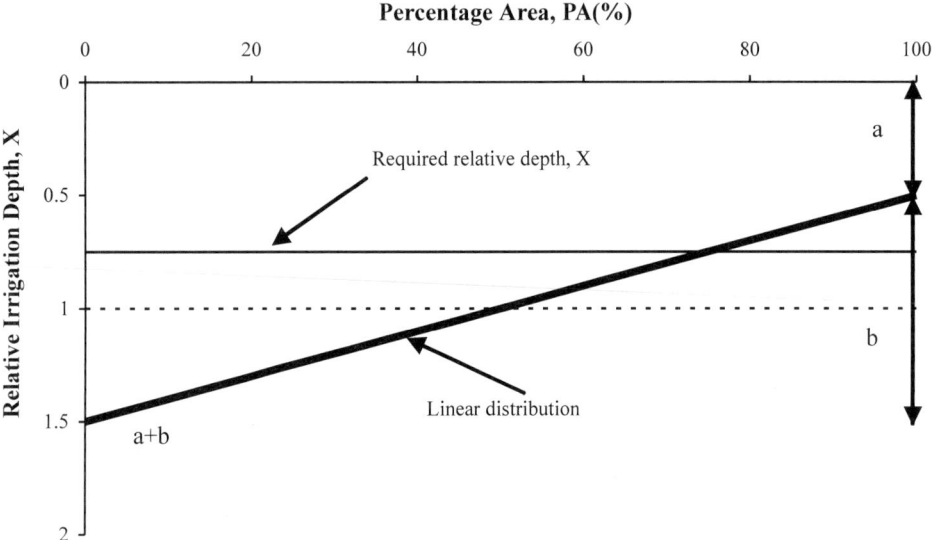

Fig. 1 A linear water application model for drip irrigation.

emitter or minimum water application. The field is fully irrigated and whole field is over-irrigated except the point of minimum irrigation application.

$X = X_0$ For an optimal return there is a value of X for the irrigation scheduling parameter between a and $(a + b)$.

$X = (a + b)$ This irrigation schedule is based on the maximum emitter flow or maximum irrigation application. The whole field is under deficit condition except the point of maximum water application. There is no deep percolation.

An optimal irrigation schedule for maximum economic return was determined[9] based on cost of water, price of the yield, and damage such as environmental pollution and groundwater contamination caused by over irrigation. Different irrigation strategies require different amounts of water application. Water conservation and environmental protection are realized by comparing any two of the irrigation strategies.[10]

CONCLUSION

Drip irrigation is an irrigation method that can distribute irrigation water uniformly and directly into the root zone of crops. It is one of the most efficient irrigation methods and can be designed and scheduled to meet the water requirement of crop and produce maximum yield in the field.

When the drip irrigation system is designed with high uniformity, the slope b of the straight line of water application function (Fig. 1) can be controlled to achieve

the desired variation. In this case the conventional irrigation schedule, $X = a$, optimal irrigation schedule, X_0, which is a location between a and $(a + b)$, and the irrigation schedule for environmental protection, $X = a + b$, are in close proximity. This closeness shows that the drip irrigation system can achieve optimal economic return, water conservation, and environmental protection.

REFERENCES

1. Wu, I.P.; Gitlin, H.M. *Design of Drip Irrigation Lines*; HAES Technical Bulletin 96, University of Hawaii: Honolulu, HI, 1974; 29 pp.
2. Christiansen, J.E. The Uniformity of Application of Water by Sprinkler Systems. Agric. Eng. **1941**, *22*, 89–92.
3. Howell, T.A.; Hiler, E.A. Trickle Irrigation Lateral Design. Trans. ASAE **1974**, *17* (5), 902–908.
4. Bralts, V.F.; Segerlind, L.J. Finite Elements Analysis of Drip Irrigation Submain Unit. Trans. ASAE **1985**, *28*, 809–814.
5. Bralts, V.F.; Wu, I.P.; Gitlin, H.M. Manufacturing Variation and Drip Irrigation Uniformity. Trans. ASAE **1981**, *24* (1), 113–119.
6. Wu, I.P.; Barragan, J. Design Criteria for Microirrigation Systems. Trans. ASAE **2000**, *43* (5), 1145–1154.
7. Seginer, I. A Note on the Economic Significance of Uniform Water Application. Irrig. Sci. **1978**, *1*, 19–25.
8. Wu, I.P. Linearized Water Application Function for Drip Irrigation Schedules. Trans. ASAE **1988**, *31* (6), 1743–1749.
9. Wu, I.P. Optimal Scheduling and Minimizing Deep Seepage in Microirrigation. Trans. ASAE **1995**, *38* (5), 1385–1392.
10. Barragan, J.; Wu, I.P. Optimal Scheduling of a Micro-irrigation System Under Deficit Irrigation. J. Agric. Eng. Res. **2001**, *80* (2), 201–208.

Irrigation Systems, History of Farm

Wayne Clyma
Water Management Consultant, Fort Collins, Colorado, U.S.A.

INTRODUCTION

Irrigation as a technology probably has existed since 6000 B.C. rather than 3000 B.C. as determined from recorded history. Irrigation started with individual farmers experimenting with carrying water to individual plants or cropped areas and expanded to diversions from streams or building bunds for flooded areas. Surface irrigation began the process of irrigation, but sprinkle and microirrigation have evolved and expanded to present day management with computer managed controls and precise application of water to crops often on a daily basis.

Prehistoric man has survived from the food provided from fruits, grains, vegetables, and animals as early as 10,000 B.C. Farming or crop cultivation is believed to have started about 8000 B.C. The first irrigation (Table 1) in Mesopotamia and Northern Africa is believed to have begun as early as 6000 B.C. This early date perhaps more nearly represents the beginning of irrigation rather than the beginning of large scale irrigation.[1] These dates reflect anthropological assessments of the beginning of irrigation rather than the first recorded history or physical evidence of the existence of canal irrigation. The recorded history of large-scale irrigation is reviewed more thoroughly by Willardson.[2] Thailand, and the Yangtze Valley in China are estimated to have started irrigation about 5000 B.C. while the Indus Valley in what is now Pakistan started irrigation about 3500 B.C. Irrigation in Europe is believed to have begun as early as 4500 B.C. while irrigation started about 2500 B.C. in Japan at the influence of China. North and South America, represented by Mexico, started irrigating in about 1500 B.C. Many historians believe irrigated agriculture supported, but also required the development of advanced civilizations. Other historians believe that declines in irrigated agricultural capability led to the decline of civilizations. Others believe the decline of civilizations caused the decline of irrigation.

INVENTING IRRIGATION

Irrigating crops likely started as a trial and process with perhaps some container used to carry water to individual fruit trees or small patches of grain or vegetables. The Pueblo Indians in New Mexico are believed to have irrigated all their corn by such a method. Studies in Mesopotamia showed that early irrigation efforts involved building canals that took water from the river, which was the irrigation supply source, to areas where trees and crops were growing. The form that these early field units took has never been described to my knowledge.

Irrigation from the Nile River in Egypt is believed to have started by planting crops in the area where flood waters had receded. Subsequently, dikes were formed to control both entry of water to an area and the time an area was inundated. Studies of areas where farmers have developed small diversion canals to divert water from streams suggest farmers used small basins to control the water supplied to crops. Informal inspections of ancient sites in India suggest hand labor was used to carry and apply water to small areas in a castle when irrigation was needed.

Informal reviews of the evolution of irrigation in the North America, South America, Asia, and Africa suggest that initial diversions of water through canals were used to irrigate crops by wild flooding. In the United States, India, and Chile, wild flooding was accomplished by spreading water over large areas without the use of dikes to control the spread of the water. For the United States and Chile, because of the larger field equipment, larger units for farming were used. No such goal was identified in India, but apparently transfer of water control concepts from other irrigated areas never occurred. In Afghanistan, Ethiopia, India, Kenya, Nepal, Somalia, Pakistan, Sri Lanka, and Thailand, small bunded units were constructed to control the water applied to crops. These small units used wild flooding, not the level basins as are frequently assumed, because lateral and longitudinal slopes, and high and low areas occurred in each basin. Even most rice fields are not level basins because within the basin they often have 15 cm or more ranges in elevation.

In Egypt, small bunded units (often less than 10 m^2) are used. The sizes of the units are believed to be related to the small, variable flow rates that are available for irrigation from Shadoofs, Archimedes' screws, and *sakias* (water wheels) using human or animal power. Small units were also used in India where animal power was used for lifting. These units used oxen and a water bag to lift water and supply water to irrigate a field. The crops were frequently opium poppies and the high cash value seemed to result in

Encyclopedia of Water Science
DOI: 10.1081/E-EWS 120010241

Table 1 Beginnings of irrigated agriculture in major areas of the world

Area	Years B.C.
Middle East	
Mesopotamia	6,000
Northern Africa	6,000
Asia	
Yangtze Valley in China	5,000
Thailand	5,000
Indus Valley in now Pakistan	3,500
Europe	4,500
Japan	2,500
North and South America	1,500

more precise leveling and greater care in growing the crop. Studies in Pakistan have shown that farmers adjust the size of their basin to the flow rate available. Farmers changing the size of field as available flow rates vary has also been observed in other countries.

Farmers value fields that are level. Observations and farmer insights repeatedly find that farmers invest considerable labor attempting to level their fields precisely. Those same observations show that precision measuring instruments, surveying levels, or laser guided equipment, are necessary to level basins precisely.

EVOLUTION OF IRRIGATION SYSTEMS

Surface irrigation systems were the first widely used method of irrigation. When soils and topography limited the effectiveness of surface irrigation systems, then sprinkle irrigation was invented. Initial sprinkle systems were pipes with holes to allow water to spray on adjacent crops. When shortages of water and need for precise timing and amounts of water became dominant considerations, trickle irrigation systems were developed and used in farm fields. Some equipment for trickle irrigation was adapted from greenhouse systems in England. Other initial trickle irrigation systems were buried pipes with holes next to the plant row or circular pipes placed around a tree. Both sprinkle and trickle irrigation systems have many additional sophisticated improvements since their initial invention. Both sprinkle and trickle irrigation (drip or microirrigation as more recently named) have the distinct advantage of being suited to nonleveled land and to soils not suited for surface irrigation. They also remove the soil, in a large part, as the hydraulic transport media. Pipes are used to transport the water largely eliminating seepage losses from canals and "over irrigation" at the "head end" of fields and runoff from "tail ends" of fields. In addition,

these newer irrigation technologies are better suited to applying smaller application volumes per unit land area reducing longer irrigation intervals typically associated with surface irrigation. All pressure irrigation systems required appropriate maintenance and management or their advantages of higher potential performance are lost.

Evolution of Surface Irrigation Systems

As irrigation projects were observed around-the-world, a sequence for evolvement of field surface irrigation systems was identified. Just as irrigation development follows a sequence, the types of field irrigation systems often follow a sequence. This sequence of evolution of surface systems is as follows:

Wild flooding → Border ditch

→ Graded borders and furrows → Level basins

Initial irrigation efforts are focused on supplying water when drought or lack of adequate rainfall results in low or undependable crop yields. Providing an irrigation water source is initially focused on methods for delivering water to the field. Usually delivery of water is accomplished through a canal from a lake or river by gravity to the field. Field application of water is accomplished by simply diverting water from the source onto the field. Without leveling of the field surface nor dikes to guide the water across the field, wild flooding is the result as shown in Fig. 1. When flow rates are small and large equipment does not dictate larger field units, small bunded units are used with channels constructed to each bunded unit. In rice irrigation, channels are often omitted and flow is from

Fig. 1 Wild flooding in surface irrigation of a field. (Water Management Synthesis II Project, Colorado State Univ., Fort Collins, CO.)

Fig. 2 Border ditch irrigation by wild flooding of a field using small bunded units and channels to deliver water within the field. (Water Management Synthesis II Project, Colorado State Univ., Fort Collins, CO.)

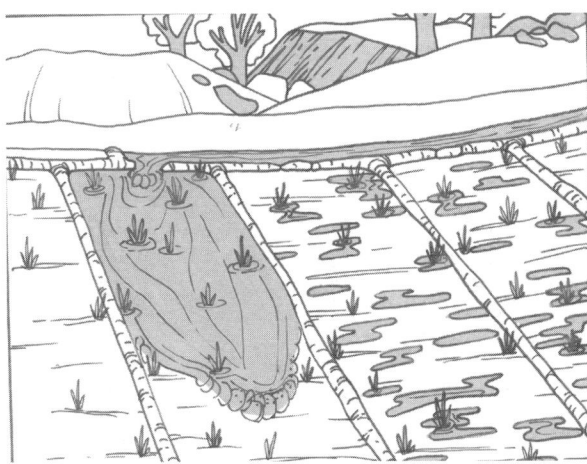

Fig. 4 Graded border irrigation systems with bunds to guide the water down the slope and cross slopes eliminated. (Water Management Synthesis II Project, Colorado State Univ., Fort Collins, CO.)

bunded unit to bunded unit. Rice irrigation attempts to achieve a continuously flooded condition for weed control not because rice plants must be continuously flooded. Nonrice crops cannot grow with continuous flooded conditions. Therefore, supplying water by allowing it to flow from bunded unit to bunded unit is usually not successful for nonrice crops. Farmers in many countries, including the United States, often repeatedly try to grow nonrice crops with basin to basin flooding with no or limited success. The small bunded unit prevalent in many countries, often incorrectly called level basins, are wild flooded as shown in Fig. 2. Wild flooding is still practiced

Fig. 3 Border ditch irrigation of larger field units using channels to deliver water within the field by wild flooding the field. (Water Management Synthesis II Project, Colorado State Univ., Fort Collins, CO.)

because within the bunded unit lateral and longitudinal slopes with high and low areas exist.

With time, the surface conditions of the field are improved and channels are constructed to carry water into the field to ensure water is available to all the cropped area of the field. In the mountain west of the United States, these field irrigation systems are called border ditch irrigation systems. Illustrated in Fig. 3, water is carried into the field by ditches, bunds are constructed to guide water down the field, but water is distributed within each bunded unit by wild flooding. In countries with small, variable flows available for irrigation, the small basins for wild flooding are essentially a variation of border ditch irrigation systems.

With land leveling and smoothing, bunds are constructed down the prevailing slope without any lateral slope between the bunds, and an appropriate flow of water is introduced at the upper end (Fig. 4). These field systems are graded border irrigation systems. They improve the uniformity of water distribution over the field compared with wild flooding. The flow rate must be adjusted according to the slope and intake rate of the soil or excessive runoff will occur. Continuous flow must be provided until sufficient infiltration is achieved to meet crop requirements. Thus, effective management of graded border irrigation systems by farmers is complex and difficult. Farmers often insist that only trained irrigators or irrigators with experience can manage graded systems.

Graded furrows fall in the same category as graded borders. They are the first development from wild flooding. Furrows are directed downslope between row crops with each furrow or alternate furrows supplied an appropriate flow rate as shown in Fig. 5. Furrows may also

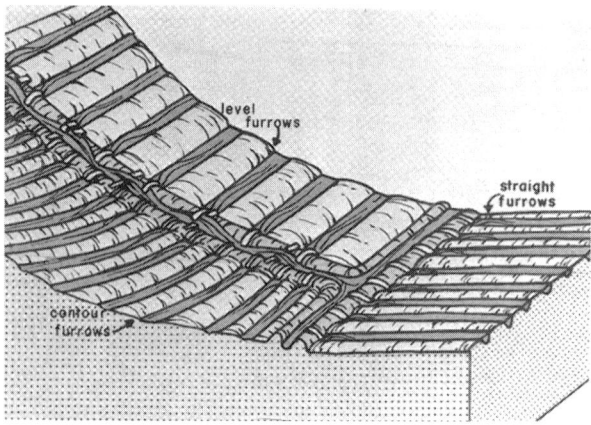

Fig. 5 Furrow irrigation of fields with graded furrows, contour furrows, and level furrows. (From Ref. 3.)

Fig. 6 Level basin irrigation of a field with all slopes removed and precisely leveled. (Water Management Synthesis II Project, Colorado State Univ., Fort Collins, CO.)

be constructed on contour to limit cross slopes. Flow rates must be adjusted to meet the infiltration requirements of the furrow adequately with sufficient remaining flow to advance down the furrow. If the furrow flow rate is too large, erosion and excessive runoff are the result. If the furrow flow rate is too low, excessive deep percolation may occur at the upper end of the field and the lower end of the field may not receive enough water. Thus, management of graded furrows is complex as are graded borders. Furrows are more adaptable than borders in the sense that they can be constructed on fields with cross slopes if the furrow flow from irrigation or rainfall can be controlled not to exceed furrow capacity. When one furrow breaks, cross slope erosion can be a serious and damaging condition. Furrows can also be constructed in level basins or fields with no slope and result in level furrows. Furrows allow the placement of crops on ridges such that the plants are not covered by water. Vegetables and many field crops produce higher yields when grown on ridges.

Level basins are the final targets of evolution in surface irrigation systems (Fig. 6). They provide the management advantage that farmers add water to the basin within a wide range of flow rates for a time such that the target volume of water is applied. If the basins are appropriately designed and managed, then farmers can use advance distance criteria to apply target amounts of water to the level basin without water measurement. Also, advance distance criteria can be used to shut off the water application to a field when a target amount of water is applied. Farmers around-the-world use advance distance criteria to apply water to fields, but the basins are not adequately level nor designed to apply target amounts of water to a field. Therefore, the technology is available and farmers are already attempting to use the technology. Using the available technology to apply target amounts of water to a field would achieve quantitative water

management with level basins.[4] Level basins came into use in United States when laser leveling was developed. The large basins, often as large as 8 ha, require large flow rates and the very precise leveling achieved with laser-controlled leveling equipment (Fig. 7).

Surface irrigation systems are varied and flexible. They are the most widely used method of irrigation. Level basins can achieve high levels of performance (higher than sprinkle and near the same level as trickle systems) with advantages for farmer management and sometimes system cost. Soils and topography limit the applicability of surface irrigation systems because uniform slopes (level or graded) and medium to lower intake soils of uniform

Fig. 7 Laser controlled leveling, for level basins in United States up to 8 ha in size, is now used to accomplish precision land leveling in surface irrigation around-the-world. (Courtesy of Natural Resource Conservation Service, USDA; www.nrcs.usda.gov.)

Fig. 8 Hand move sprinkler lateral for irrigation of a field. (Courtesy of Natural Resource Conservation Service, USDA; www.nrcs.usda.gov.)

texture are required for effective performance. Higher intake rate soils can be irrigated with surface irrigation, but smaller units are required. Surface irrigation systems are applicable to many irrigated areas and are often the most advantageous systems. They often require the lowest investment to become operational, but unless automated, may involve the highest labor costs.

Irrigation was advanced by the coming of mechanical scrappers to form canals and ditches. In addition, canal lining with various materials from concrete to butyl rubber has attempted to reduce canal seepage losses of water. However, canal maintenance remains costly. Also, in the past few years, canal water level control, through Supervisory Control and Data Acquisition (SCADA) systems using computers to remotely access data, and then control canal gates and flow structures, has become widely used in United States to replace less reliable manual and incremental controls. Laser leveling both surveys and precisely levels fields for level basins and other surface systems. More recently, the use of "surge flow" for graded furrows, where flow is "on and off" has been used to reduce runoff losses and provide greater uniformity in the time for infiltration along the furrow length. Surge flow valves are now solar powered and even provide the ability to apply fertilizer into the ending irrigation set times. Cablegation systems have also been developed to automate both on-farm canal and pipeline delivery systems.

Fig. 9 Side roll sprinkler systems for irrigation of a field. (Courtesy of Natural Resource Conservation Service, USDA; www.nrcs.usda.gov.)

Sprinkle Irrigation Systems

Sprinkle irrigation systems were developed to supply water to crops independent of the transmission capabilities of the soil. Higher intake soils, variable textured soils, and uneven land surfaces limited the appropriate use of surface irrigation systems. The first patented sprinkle irrigation system was developed in 1884.[5] These initial systems used controlled heads to distribute water from pipes. Many initial, key developments in sprinkle irrigation were developed by farmers with manufacturers taking over the development and refining the concept. The Rain Bird impact sprinkler was developed by a farmer and then refined by the Rain Bird Company. These initial sprinkle systems were moved by hand (Fig. 8). Other important developments were quick coupling thin walled pipes, rubber gaskets that supported portable, quick coupling pipes, and aluminum pipe that became economical and available after World War II.[5]

The first mechanical move sprinkle systems were wheel line sprinklers of the lateral move type developed in the 1930s. They use wheels on pipes to irrigate the width of a field in one pass (Fig. 9). These sprinkle systems replaced the expensive solid set sprinklers and the labor intensive hand move systems commonly used. Mechanical move systems were also often more efficient and required less labor than surface irrigation systems.

Center pivots were developed first by a Colorado farmer, Frank Zybach, that used a water drive mechanism. They use wheels and arches to support pipes that travel in a circle as shown in Fig. 10 to irrigate usually square fields. Thus, often the corners of the field are not irrigated. Dr. William Splinter of the University of Nebraska considered the center pivot sprinkle system "the most significant mechanical innovation since the replacement of draft animals by the tractor."[5] Center pivot sprinkle systems now dominate the area irrigated by sprinklers with twice as much area irrigated compared with other sprinkle methods. Large areas (such as shown in Fig. 11) are irrigated in circles with center pivots. Linear move systems provide the ability to irrigate rectangular fields with the equipment capabilities of center pivots (Fig. 12). System costs and management issues are the major detriments to sprinkle systems. Sprinkle irrigation equals about half the area irrigated in United States.[6]

Mechanical move sprinkle systems have been adopted because of their reduced labor requirements, their adaptability for irrigating rolling topography, and their effectiveness in fields with sandier soils or mixed soil types. Irrigation is accomplished at a higher level of efficiency, but design efficiencies are often misleading. Duke, Heermann, and Dawson[7] found that all the center pivot units in some 60 units evaluated in Colorado could economically have their actual performance improved. Management improvement potential also existed.

Recent advances in center pivot systems include controls that can automate system operation allowing remote controls via radio, cell phones, or infrared (line of sight) linkages from computers or other devices. Currently, the ability to control individual applicators or

Fig. 10 Center pivot irrigation system for irrigation of a field. (Courtesy of Natural Resource Conservation Service, USDA; www.nrcs.usda.gov.)

Fig. 11 Aerial view of an area irrigated by center pivot irrigation systems. (Courtesy of Natural Resource Conservation Service, USDA; www.nrcs.usda.gov.)

Fig. 12 Linear move irrigation systems have equipment similar to center pivots but move from one end of the field to the other as side roll systems operate. (Courtesy of Natural Resource Conservation Service, USDA; www.nrcs.usda.gov.)

Trickle Irrigation Systems

Trickle irrigation systems include drip, spray, bubbler, and subsurface applications of water to crops. Some have suggested[5] that microirrigation is a more appropriate term. The concept is to apply or make available at each plant or tree the required amount of water for the root zone. Additional area is not irrigated with a reduction in water losses from evaporation and evapotranspiration from weeds. A schematic diagram of a trickle system is shown in Fig. 14.

The first known application of trickle irrigation was in 1860, and used an underground tile line where drainage was accomplished during part of the year, and water was supplied for irrigation during other times.[9] Other major developments include the use of trickle systems in greenhouses in England during the 1940s.[9] Dr. Symcha

heads, or groups of applicators offers the potential to use site-specific management or "prescription" management of small blocks (10s of m^2). Each block can be managed independently to supply water, fertilizer, or chemicals as required while minimizing any wastes and possibly reducing any environmental impact. These are important developments for the future growth of precision agriculture.

Traveling sprinkle systems use giant single gun-type sprinklers that travel on a chassis down the length of a field and covers an area as much as 400 ft in diameter (Fig. 13). Nozzles operate at 80 psi–100 psi. A cable winds upon a drum on the self-powered chassis to move the unit down the field. A long hose is used to supply water to the unit from detachable hookups to an underground supply line. Sometimes rotating booms from 60 ft to 120 ft long are used to cover a large area similar to a boom sprinkler. Because of the high application rates, traveling sprinkle systems are commonly used for supplemental irrigation on sandy soils.

Fig. 13 Traveling sprinkle systems commonly use large guns and pull themselves across the field using a cable and winch. (Courtesy of Dr. Harold Duke, ARS, USDA retired.)

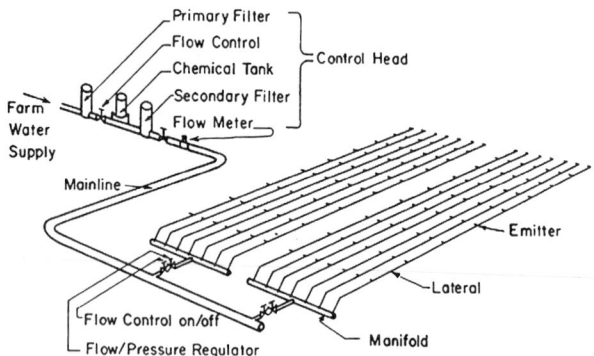

Fig. 14 Schematic diagram of a trickle irrigation system. (Soil Conservation Service, 1984, p. 7–8.)

Blass, according to Howell,[5] working with greenhouses in England after the war, improved the trickle irrigation systems. Then he transferred the technology to Israel during the 1950s to grow crops in the Negev desert including the use of highly saline waters. Trickle irrigation then spread to Australia, United States, South Africa and to other parts of the world.

Trickle irrigation development in United States started with avocado orchards in California after individuals observed trickle irrigation systems in Israel. A company called Drip-Eze started the manufacturer of emitters for use in trickle irrigation systems.[5] Row crops, primarily vegetables, using trickle irrigation started in New York as new plastic products became available.[5]

Trickle irrigation systems are used for crops that have widely spaced plants, e.g., orchards and vineyards as shown in Fig. 15. High valued vegetable crops are also widely adapted to trickle irrigation systems where careful water control allows increases in yield and improvements in the quality of the produce. Tomatoes are irrigated with an underground trickle system in Fig. 16. Water scarce areas often adapt to trickle irrigation because reduced water requirements result. Because irrigation can be accomplished daily and soil water can be maintained at a high level, irrigation with saline water can also be successful.

The key component of the trickle irrigation system is filtration system. Fine particles, and chemical and bacterial clogging of lines and emitters can rapidly cause irrigation system failure if not carefully managed. Failure of the trickle irrigation system can be catastrophic if the annual crop fails and even the orchard or vineyard dies. Trickle irrigation is used to irrigate about 1,250,000 ha or 4.9% of the irrigated area in United States and irrigates a much higher percentage of the irrigated area in other countries.[10] Israel and other countries have major areas irrigated by trickle irrigation systems.

CRITICAL CONCEPTS IN IRRIGATION

Irrigation as a technology 6000 or more years ago changed society. Civilizations on most continents are considered to have flourished because food production under irrigation allowed education, war, artisans, and a ruling class to develop because everyone no longer needed primarily to focus on finding his family's food. Labor losses from allowing individuals to attend school, participate in a war, and build buildings, monuments, or artistic items are major drains on the food supply. Control of a food supply allows a ruling class to sustain power. Critical concepts in irrigation have supported the continuing important role of irrigation. Water supply management is important in irrigation.

Water Supply

Use of flood waters for irrigation was an important step. Use of channels to divert water for irrigation extended the area that could be irrigated and the time when water was available. Creation of dams to control water and make it

Fig. 15 Trickle irrigation system of grapes showing limited area of water application. (Courtesy of Natural Resource Conservation Service, USDA; www.nrcs.usda.gov.)

Fig. 16 Trickle irrigation of tomatoes using an underground trickle system. (Courtesy of Agricultural Research Service, USDA, www.ars.usda.gov.)

available during periods of low flow in the river was another important concept. Dams have been constructed for irrigation to divert water from streams and to store water almost from the first beginnings of irrigation.[2] Dams for irrigation developed the west starting in the late 19th century. Other countries followed the practice with irrigation expanding by 2% per year during the 1960s and 1970s. Costs of irrigation development and concerns about the environmental impacts of irrigation have reduced that expansion to near 1% during the past several decades.[11] Dams also provided flood control along rivers and a power source from early water wheels to modern hydroelectric turbines. Recently, in the northwestern United States, dams have raised concern for stream ecology and fish habitats. Occasionally, dams are being removed to satisfy these environmental and societal reasons.

Egypt expanded irrigation along the Nile River greater than many other irrigated valleys because they lifted water by the Archimedes' screw and Shadoof. Pumps provided increased water supply flexibility during the 20th century by lifting water from streams and pumping water from wells. Individual control of water supply by a farmer through pumping is a major advantage in many irrigated areas.

Construction of dams, diversions, and canals became a major part of irrigation development. Managing the delivery of water supplies effectively to meet farm water requirements still needs major improvements. During the 1970s, the importance of effective farm water management facilities began to be recognized. Inadequate farm water management, also constrained by water delivery management, is still a major constraint to effective irrigation in many projects.

Management Concepts

Irrigation scheduling[12] as part of a strategy for water management was one of the most important concepts developed in irrigation. Clyma[13] reviewed status of irrigation scheduling after three decades of application and concluded that the potential of the concept had not been achieved. First, a major part of the technical effort had focused on accurately estimating evapotranspiration for the growing crop. Management decisions were also focused on when and how much to irrigate. Processes for deciding how accurately to apply the target amount of water, and monitoring management decisions of when, how much, and how to irrigate are usually not a part of the irrigation scheduling process. Therefore, irrigation scheduling is not an adequate management process. Perhaps the future will define and apply irrigation scheduling as a strategy for a complete and successful management process.

Farmers commonly do not measure the amount of water applied to a field. Therefore, excess applications of water, and waterlogging and salinity are the common result. Level basins are potentially a major part of surface irrigation systems in the future. A design and management concept, initially defined by Wattenburger and Clyma[14,15] and refined further and integrated into a computer design program,[16] allows farmers to apply a target amount of water to a level basin when the water advances a specified distance into a field.[4] The basin must be precisely leveled and farmers around the world already use advance distance as their criteria for irrigating a field. The technology just needs professionals to supply the needed assistance and appropriate target amounts of water will be applied to each field a farmer irrigates. Perhaps this important concept will be supplied as an urgently needed technology to farmers in the next decade.

CONCLUSION

Irrigation as a technology probably started in 6000 B.C. rather than the 3000 B.C. suggested from recorded history and anthropological evidence of major structures. The intervening time involved the evolution of cities and capability to construct structures such that evidence of irrigation and historical references to irrigation could evolve. Field irrigation systems used by farmers have not been defined from anthropology to my knowledge. Informal information suggesting that Pueblo Indian farmers in Southwest Colorado used pots to irrigate individual plants of corn does suggest a starting point. Observations of field irrigation around the world suggest that wild flooding was the initial surface irrigation system used by farmers. Channels were subsequently constructed to carry water into the field to improve distribution and border ditch irrigation evolved. Graded borders and graded furrows were the next improvements with ridges to control water flow across a field. In all the previous systems, flow rate, infiltration rate for the soil, and slope must be balanced to allow infiltration time for adequate irrigation and management was difficult. Level basins remove all slopes and allow farmers to apply the required amount of water to each field volumetrically.

Sprinkler irrigation systems were developed to allow irrigation of soils with high intake rates, variable soil textures, and uneven land surfaces. Many developments in sprinkler irrigation were invented by farmers, but manufacturers continue to improve existing systems and develop new system components. Mechanical move systems were developed to reduce the labor of hand move systems and the expense of solid set systems. Mechanical move systems were also commonly more efficient than surface systems and required less labor.

Trickle irrigation systems supply water only to the root zone of plants and greatly reduce water requirements for irrigation. Trickle systems also approach 100% efficiencies with good design and management. Labor is greatly reduced because most trickle systems are automated. Yields often increase substantially and the quality of the produce often improves significantly. Trickle systems are expensive, and without good management may fail and cause the loss of the growing crop—a financial disaster.

Water supply management by constructing dams was an important conceptual development in irrigation. Pumps also added greater flexibility and access to adequate water supplies for farmers. Irrigation scheduling after three decades of use does not support adequate management of irrigation systems. Perhaps better management will increase the effectiveness of irrigation scheduling in the future. Appropriate design, precision land leveling of each field, and management of level basins using advance distance criteria should be the future of surface irrigation. Farmers already use advance distance criteria to manage water application. Combined with appropriate design and construction, target amounts of water can be applied to each field with a major improvement in water management.

ACKNOWLEDGMENTS

The comments, suggestions, and additions of Dr. Terry Howell in his review of this material provided clarifications, insights and improvements and are gratefully acknowledged.

REFERENCES

1. Steward, J.H.; Adams, R.M.; Collier, D.; Palerm, A.; Wittfogel, K.A.; Beals, R.L. *Irrigation Civilizations: A Comparative Study*; Social Science Section, Department of Cultural Affairs, Pan American Union: Washington, DC, 1995.
2. Willardson, L.S. Irrigated Agriculture, Historical View of. In *Encyclopedia of Water*, 1st Ed., Stewart, B.A., Howell, T.A., Eds.; Marcel Dekker Inc.: New York, 2003.
3. Salazar, L. *Water Management on Small Farms: A Training Manual for Farmers in Hill Areas*; Water Management Synthesis Proj., Colorado State Univ.: Fort Collins, CO, 1983; p. 50.
4. Clyma, W.; Clemmens, A.J. Farmer Management Strategies for Level Basins Using Advance Distance Criteria. In *National Irrigation Symposium*, Proceedings of the Fourth Decennial Symposium, Phoenix, AZ; November 14–16, 2000, Evans, R.G., Benham, B.L., Trooien, P.P., Eds.; Am. Soc. Agric. Engrs.: St. Joseph, MI, 2000; 573–578.
5. Howell, T. Drops of Life in the History of Irrigation. Irrig. J. **2000**, *50* (1), 8–10, 13–15.
6. Martin, D. The Adoption of Center Pivot Irrigation. Irrig. J. **1999**, *49* (3), 8–10.
7. Duke, H.R.; Heermann, D.F.; Dawson, L.J. Appropriate Depths of Application for Scheduling Center Pivot Irrigations. Trans. ASAE **1992**, *35* (5), 1457–1464.
8. Soil Conservation Service. Trickle Irrigation, Chapter 7, Section 15. *National Engineering Handbook*; U.S. Dept. Agric., Chapter 7, Section 15, Natl. Eng. Handbook; Soil Conservation Service, U.S. Government Printing Office, Washington, DC. 1984; 7–8.

9. Keller, J.; Karmeli, D. *Trickle Irrigation Design*; Rain Bird Sprinkler Manufacturing Corporation: Glendora, CA, 1975.

10. 2000 Annual Irrigation Survey continues steady growth. Irrig. J. **2001**, *51*(1), 12–30, 40–41.

11. Sandra, P. Redesigning Irrigated Agriculture. In *National Irrigation Symposium*, Proceeding of the Fourth decennial Symposium, Phoenix, AZ, November 14–16, 2000; Evans, R.G., Benham, B.L., Trooien, T.P., Eds.; Am. Soc. Agric. Engrs.: St. Joseph, MI, 2000; 1–12.

12. Jensen, M.E. Scheduling Irrigations Using Computers. J. Soil Water Conserv. **1969**, *24* (8), 193–195.

13. Clyma, W. Irrigation Scheduling Revisited: Historical Evaluations and Reformation of the Concept. In *Evapotranspiration and Irrigation Scheduling*, Proceedings of the International Conference, San Antonio, TX, November 3–6, 1966; Camp, C.R., Sadler, E.J., Yoder, R.E., Eds.; Am. Soc. Agric. Engrs.: St. Joseph, MI, 1966; 626–631.

14. Wattenburger, P.L.; Clyma, W. Level Basin Design and Management in the Absence of Water Control, Part I: Evaluation of Completion-of-Advance Irrigation. Trans. ASAE **1989**, *32* (2), 838–843.

15. Wattenburger, P.L.; Clyma, W. Level Basin Design and Management in the Absence of Water Control, Part II: Design Method for Completion-of Advance Irrigation. Trans. ASAE **1989**, *32* (2), 844–850.

16. Clemmens, A.J.; Dedrick, A.R.; Strand, R.J. *BASIN 2.0: A Computer Program for the Design of Level-Basin Irrigation Systems*; WCL Rept. #19; U.S. Water Conservation Laboratory: Phoenix, AZ, 1995.

Irrigation Systems, Subsurface Drip

Carl R. Camp, Jr.
United States Department of Agriculture (USDA), Florence, South Carolina, U.S.A.

Freddie R. Lamm
Kansas State University, Colby, Kansas, U.S.A.

INTRODUCTION

Subsurface drip irrigation (SDI) is generally defined as the application of water below the soil surface through emitters, with discharge rates in the same range as drip irrigation.[1] While this definition is not specific regarding depth below the soil surface, most SDI laterals are installed at a depth sufficient to prevent interference with surface traffic or tillage implements, and to provide a useful life of several years as opposed to annual replacement of surface or near-surface drip laterals.

Development of drip irrigation accelerated with the availability of plastics following World War II, primarily in Great Britain, Israel, and United States. SDI was part of drip irrigation development in the United States beginning about 1959, especially in Hawaii and California. While early drip irrigation products were relatively crude by modern standards, SDI devices were being installed in both experimental and commercial farms by the 1970s. As drip irrigation products improved during the 1970s and early 1980s, surface drip irrigation grew at a faster rate than SDI, probably because of emitter plugging problems and root intrusion. However, interest in SDI increased during the early 1980s, increased rapidly during the last half of the 1980s, and continues today, especially in areas with declining water supplies, with environmental issues related to irrigation, and where wastewater is used for irrigation. Initially, SDI was used primarily for sugarcane, vegetables, tree crops, and pineapple in Hawaii and California. Later, SDI use was expanded to other geographic areas and to agronomic and vines crops, including cotton, corn, and grapes.

SDI has the advantage of multiple-year life, reduced interference with cultural practices, dry plant foliage, and a dry soil surface. Multiple-year life allows amortization of the entire system cost over several years, often more than ten. If all system components are installed below tillage depth, surface cultural practices can be accomplished with minimal concern for system damage. Dry soil surfaces can reduce weed growth in arid climates and may reduce evaporation losses of applied water. Because the plant canopy is not irrigated, the foliage remains dry, which may reduce incidence of disease. SDI is also very adaptable to irregularly shaped fields and low-capacity water supplies that may provide design limitation with other irrigation systems.

The major disadvantages of SDI include system cost, difficulty in locating and repairing system leaks and plugged emitters, and poor soil surface. Most system components are installed below the soil surface and are neither easy to locate nor directly observable. In a properly designed and managed SDI system, the soil surface should seldom be wet. Consequently, seed germination, especially for small seeds, can be very difficult.

SDI systems offer considerable flexibility, both in design and operation. For example, SDI systems can apply small, frequent water applications, often multiple times each day, to very specific sites within the soil profile and plant root zone. Fertilizers, pesticides, and other chemical amendments can be applied via the irrigation system directly into the active root zone, often at a modest increase in equipment cost. In many cases, the operational cost may be less than that for applying these chemicals via conventional surface equipment.

SYSTEM DESIGN

Site, Water Supply, and Crop

Design of subsurface drip systems is similar to that of surface drip systems, especially with regard to hydraulic characteristics.[2] Specific crop and soil characteristics are used in the design process to select emitter spacing and flow rate, lateral depth and spacing, and the required system capacity. Emitter properties and lateral location are influenced by soil properties such as texture, soil compaction, and soil layering because these affect the rate of water movement through the soil profile and the subsequent wetting pattern for each emitter.

The water supply capacity directly affects the design of a SDI system. The size of the irrigated field or zone is often

Encyclopedia of Water Science
DOI: 10.1081/E-EWS 120010066

determined by the water supply capacity. For example, in some humid areas, high-capacity wells are not available but multiple low-capacity wells can be distributed throughout a farm. Fortunately, the design of SDI systems can be economically adjusted to correspond to the field size and shape, to the available water supply capacity, and to other factors. Water supply quality should be tested by an approved laboratory before proceeding with system design. This information is needed for the proper design and management of the water filtration and treatment system. Some water supplies require frequent or intermittent injection of acids and/or chlorine. Other saline and/or sodic water supplies may require treatment or special management. As water supplies become more limited, treated wastewater is becoming an increasingly important alternative water supply that can be applied through SDI systems. Camp[3] listed several reports that emphasized water supplies (saline, deficit, and waste-water) for SDI systems.

The SDI system is usually designed to satisfy peak crop water requirements, which vary with specific site, soil, and crop conditions. When properly designed and managed, SDI is one of the most efficient irrigation methods, providing typical application efficiencies exceeding 90%. In comparison with other methods of irrigation, reported yields with SDI were equal to or greater than those with other irrigation methods. Generally, water requirements with SDI are similar or slightly lower than those with other irrigation methods. In some cases, water savings of up to 40% have been reported.[3] However, unless more specific information is available, it is usually best to use standard net water requirements for the location when designing SDI systems.

Lateral Type, Spacing, and Depth

SDI lateral depth for various cropping systems is normally optimized for prevailing site conditions and soil characteristics.[3] Where systems are used for multiple years and tillage is a consideration, lateral depths vary from 0.20 m to 0.70 m. Where tillage is not a consideration (e.g., turfgrass, alfalfa) depth is sometimes less (0.10 m – 0.40 m). Lateral spacing also varies considerably (0.25 m – 5.0 m), with narrow spacing used primarily for turfgrass and wide spacing used for vegetable, tree, or vine crops. In uniformly spaced row crops, the lateral is usually located under either alternate or every third midrow area (furrow). For crops with alternating row spacing patterns, the lateral is located about 0.8 m from each row, usually in the narrow spacing of the pattern.

The lateral should be installed deep enough to prevent damage by tillage or injection equipment but shallow enough to supply water to the crop root zone without wetting the soil surface. Generally, laterals in SDI systems are placed at depths of 0.1 m – 0.5 m, at shallower depths in coarse-textured soils and at slightly deeper depths on finer-textured soils. The selection of emitter spacing and flow rate are influenced by crop rooting patterns, lateral depth, and soil characteristics. It is also desirable to select an emitter spacing that provides overlapping subsurface wetted zones along the lateral for most row crops. For wider spaced crops such as trees and vines, emitters are normally located near each plant and may have wider spacings that do not provide overlapping patterns. Lateral spacing is determined primarily by the soil, crop, and cultural practice, and should be narrow enough to provide a uniform supply of water to all plants.

Special Requirements

Site topography must be considered in system design and selection of components as with any irrigation system, but SDI is suitable for most sites, ranging from flat to hilly. For sites with considerable elevation change, especially along the lateral, pressure-compensating emitters should be used.

Two special design requirements for SDI systems, which are significantly different from those for surface drip systems, are the needs for flushing manifolds and air entry valves. Flushing manifolds are needed to allow frequent flushing of particulate matter that may accumulate in laterals. Air relief valves are needed to prevent aspiration of soil particles into emitter openings when the system is depressurized. These valves must be located in sufficient number and at the higher elevations for each lateral or zone to prevent negative pressures within the laterals.

Emitter plugging caused by root intrusion is a major problem with some SDI systems, but can be minimized by chemicals, emitter design, and irrigation management. Chemical controls include the use of herbicides, either slow-release compounds embedded into emitters and filters or periodic injection of other chemical solutions (concentrated and/or diluted) into the irrigation supply. Periodic injection of acid and chlorine for general system maintenance can also modify the soil solution immediately adjacent to emitters and reduce root intrusion. In some cases, emitters plugged by roots may be cleared via injection of higher concentrations of chemicals, such as acids and chlorine.

Emitter design may also affect root intrusion. Smaller orifices tend to have less root intrusion but are more susceptible to plugging by particulate matter. Some emitters are constructed with physical barriers to root intrusion. Root intrusion appears to be more severe when emitters are located along dripline seams, which can be an area of preferential root growth. However, root intrusion problems appear to be greater for emitters, driplines, and porous tubes that are not chemically treated.

Irrigation management can also be used to influence root intrusion by controlling the environment immediately adjacent to the emitter. High frequency pulsing that frequently saturates the soil immediately surrounding the emitter can discourage root growth in that area for some crops but not others. Conversely, deficit irrigation sometimes practiced to increase quality or maturity, or to control vegetative growth, can increase root intrusion in lower rainfall areas because of high root concentrations in the soil zone near emitters.

SYSTEM COMPONENTS

Pumps, Filtration, and Pressure Regulation

Pump requirements for SDI are similar to those for other drip irrigation systems, meaning water must be supplied at a relatively low pressure (170 kPa–275 kPa) and flow rate in comparison to other irrigation methods. Because of the flushing requirement for SDI systems, a flow velocity of about $0.3 \, \mathrm{m \, sec^{-1}}$ must be achievable, either by reducing the zone size while using the same pumping rate or by increasing the pumping rate without changing the zone size.

Water filtration is more critical for SDI systems than for surface drip systems because the consequences of emitter plugging are more severe and more costly. Generally, the better the water quality, the less complex the filtration system required. Surface and recycled or wastewater supplies require the most elaborate filtration systems. However, good filtration is the key to good system performance and long life, and should be a major emphasis in system design. Filtration systems range from simple screen filters for relatively clean water to more elaborate and complex disc and sand media filters for poorer quality water.

The pressure regulation requirement in SDI systems is similar to that in surface drip systems. When nonpressure-compensating emitters are used on relatively flat areas, pressure is typically regulated within the system supply lines (main and/or submain) using pressure-regulating valves. When pressure-compensating emitters are used, typically on more hilly terrain, the pressure within the system supply lines is controlled at a higher, but more variable, pressure that is within the recommended input pressure range for the emitters used. Water pressure should be monitored on a regular basis at the pump or supply port and at various locations throughout the SDI system, especially at the both ends of laterals.

Laterals and Emitters

Many types of driplines have been used successfully for SDI and most have emitters installed as an integral part of

the dripline. This is accomplished by one of three methods: 1) molded indentions created during the fusing of dripline seams; 2) prefabricated emitters welded inside the dripline; or 3) circular prefabricated in-line emitters installed during extrusion. Regardless of the emitter used, dripline wall thickness and expected longevity must be considered along with other design factors in selecting the lateral depth. Flexible, thin-walled driplines typically are installed at shallow depths and normally have a shorter expected life. Thicker-walled, flexible driplines have been used successfully for several years provided they are installed deep enough to avoid tillage, cultivating, and harvesting machinery, but shallow enough to prevent excessive deformation or permanent collapse of the dripline by machinery or soil weight. Rigid tubing with thicker walls can be installed at deeper depths without deformation, and is often used on perennial crops or on annual crops for longer time periods (>10 yr). Some driplines are impregnated with bactericides or other chemicals to reduce the formation of sludge or other material that could plug emitters.

Chemical Injection

Subsurface drip systems offer the potential for precise management of water, nutrients, and pesticides if the system is properly designed and managed. The marginal cost to add chemical injection equipment is generally competitive with other, more conventional application methods. Water and fertilizers can be applied in a variety of modes, varying from multiple continuous or pulsed applications each day to one application in several days. Choice of application frequency depends upon several factors, including soil characteristics, crop requirements, water supply, system design, and management strategies. If labeled for the purpose, some systemic pesticides and soil fumigants can be safely injected via SDI systems. Use of the SDI system for chemical applications has the potential to minimize exposure to workers and the environment, to reduce the cost of pesticide rinse water disposal, and improve precision of application to the desired target (root pests). Injection of other chemicals, such as acids and chlorine, is often required to clean and maintain emitters in optimum condition. However, a high level of management with system automation and feedback control is required to minimize chemical movement to the ground water when chemicals are used.

Air Entry and Flushing

Air entry valves must be installed at higher elevations in SDI systems to prevent the emitter from ingesting soil particles that could plug emitters when the system is depressurized. Typically, air entry valves are located in water supply lines near the head works or control station, and in both the supply and flushing manifolds. In some cases, such as turf or pasture, air entry valves may be installed below the soil surface and enclosed within a protective box. Flushing valves installed on the flushing manifold are required to control periodic system flushing.

OPERATION AND MAINTENANCE

Operation

SDI systems can be operated in several modes, varying from manual to fully automated. Overall, SDI systems are probably more easily automated than many other types of irrigation. One reason is that most are controlled from a central point using electrical or pneumatic valves and controllers that vary from a simple clock system to microprocessor systems, which are capable of receiving external inputs to initiate and/or terminate irrigation events.

Irrigation scheduling is as important for SDI systems as for any other type of irrigation. Choosing to initiate an irrigation event and how much water to apply during each event depends on crop, soil, and irrigation system type and design. Factors that affect those decisions include soil water storage volume, sensitivity of the crop to water stress, irrigation application rate, weather conditions, and water supply capacity. Camp[3] discussed several irrigation scheduling methods that have been used successfully with SDI. However, the important point is that a science-based scheduling method can conserve the water supply and increase profit.

If seed germination and seedling establishment and growth are critical, especially in arid climates when initial soil water content is not adequate, either sprinkler or surface irrigation is often used for germination. However, the need for two systems increases cost and decreases economic return. If subsurface drip is used for germination, an excessive amount of irrigation is often required to wet the seed zone for germination, which could result in excessive leaching and off-site environmental effects as well as increased cost. Surface wetting can also occur when the emitter flow rate exceeds the hydraulic conductivity of the soil surrounding the emitter, but wetted areas are often not uniform.

Because salts tend to accumulate above the lateral, high salt concentrations may occur between the lateral and soil surface in arid areas where rainfall is not available to leach the salts downward. Salts may also be moved under the row when laterals are placed under the furrow.[4] Supplemental sprinkler irrigation may be required in some areas to control salinity if precipitation is inadequate for leaching during several consecutive years.

Maintenance

Often, SDI systems must have a long life ($>10\,yr$) to be economical for lower value crops. Thus, appropriate management strategies are required to prevent emitter plugging and protect other system components to ensure proper system operation. Locating and repairing/replacing failed components is much more difficult and more expensive with SDI systems than with surface systems because most system components are buried, difficult to locate, and cannot be directly observed by managers. Consequently, operational parameters such as flow rate and pressure must be measured frequently and used as indicators of system performance. Good system performance requires constant attention to maintain good water quality, proper filtration, and periodic system flushing to remove particulate matter that could plug emitters. Periodic evaluation of SDI system performance in relation to design performance can identify problems before they become serious and significantly affect crop yield and quality.

CONCLUSION

Although there is general consensus that use of SDI is increasing, this growth is difficult to document. A recent survey of irrigation in the United States reported 156,070 ha of SDI, which is about 0.6% of the total irrigated area of 25,501,831 ha.[5] Use of SDI should increase in the future, depending primarily upon the economic and water conservation benefits in comparison to other irrigation methods. As water supplies become more limited, the high application efficiency and water conserving features of SDI should increase its application. Also, SDI offers potential advantages such as reduced odors and exposure to pathogens when using recycled domestic and animal wastewater. The SDI technology offers the capability to precisely place water, nutrients, and other chemicals in the plant root zone at the time and frequency needed for optimum crop production. With proper design, installation, and management, SDI systems can provide excellent irrigation efficiency and reliable performance with a system life of 10 yr–20 yr.

REFERENCES

1. S526.2. Soil and Water Terminology. *ASAE Standards*, 49th Ed.; ASAE: St. Joseph, MI, 2001, 970–990.
2. EP405.1. Design and Installation of Microirrigation Systems. *ASAE Standards*, 49th Ed.; ASAE: St. Joseph, MI, 2001, 903–907.
3. Camp, C.R. Subsurface Drip Irrigation: A Review. Trans. ASAE **1998**, *41* (5), 1353–1367.

4. Ayars, J.E.; Phene, C.J.; Schoneman, R.A.; Meso, B.; Dale, F.; Penland, J. Impact of Bed Location on the Operation of Subsurface Drip Irrigation Systems. In *Microirrigaion for a Changing World*, Proc. Fifth International Microirrigation Congress, Orlando, FL, April 2–6, 1995; Lamm, F.R., Ed.; ASAE: St. Joseph, MI, 1995; 141–146.

5. Anonymous. 1999 Annual Irrigation Survey. Irrig. J. **2000**, *50* (1), 16–31.

Isotopes

Michael A. Anderson
University of California, Riverside, California, U.S.A.

INTRODUCTION

The substance we know as water (H_2O) is actually comprised of a number of isotopes of O and H. Isotopes are atoms that have the same number of protons and electrons, and therefore the same basic chemical properties, but differ in their mass. Their differences in mass arise due to different numbers of neutrons within their nucleus. Isotopes can be stable or radioactive, and can be both naturally occurring and man-made. Six different isotopes of hydrogen and oxygen are found in nature (Table 1).

BACKGROUND

The heavy isotopes of oxygen (^{17}O and ^{18}O) were discovered by Giaugue and Johnston.[2] Shortly thereafter, an isotope of hydrogen (^2H) was discovered by Urey, Brickwedde, and Murphy[3] for which Urey received the Nobel prize. The isotope was called deuterium and is often written as D. A third isotope of hydrogen (^3H, tritium), postulated to exist in 1931, was detected in natural waters in 1950.[4,5]

All of the above isotopes are stable, except for tritium (^3H), which is radioactive and decays yielding a β^- particle to form ^3He. The half-life of tritium is 12.43 yr. Tritium is produced naturally in the upper atmosphere by reaction with cosmic radiation to produce $\sim$ 0.25 atoms cm^{-2} sec^{-1}.[6] Substantially higher amounts of ^3H were released to the atmosphere with the aboveground nuclear weapons testing of 1953–1962 that ended with the nuclear test ban treaty ratified in 1963. The peak concentration of ^3H in precipitation reached levels shortly before the treaty as high as $10^{-12}\%$ or 1000 × higher than concentrations from natural cosmogenically produced ^3H.[7]

The above isotopes can form a total of 18 different species of water, with the natural abundance of the 4 major species given in Table 2.

While the above isotopes of water are chemically essentially the same, there are nevertheless subtle differences in their physical and chemical properties. For example, the density of water is influenced by its isotopic composition. "Light" or normal water ($^1H_2^{16}O$) has a lower atomic weight, and therefore also a lower liquid density

than "heavy water" ($^2H_2^{16}O$ or, more simply, D_2O) (Table 3). The temperature at maximum density, boiling point, vapor pressure, and freezing point are also dependent upon the isotopic composition and mass of water (Table 3).

Viscosity is also higher for the heavier isotopes of water (e.g., the viscosity of D_2O at 25°C is 1.095 cP, while that for $^1H_2^{16}O$ is 0.890 cP). Moreover, intramolecular vibrational frequencies, which are influenced by mass of the atoms forming a chemical bond, are also sensitive to the isotopic composition of water. Thus, "heavy water" ($^2H_2^{16}O$ or, more simply, D_2O) exhibits lower frequencies of vibration than "light water" ($^1H_2^{16}O$).

The above noted differences in vapor pressure between different isotopes of water have implications for the partitioning of isotopes between the gas and liquid phases. That is, since "light" molecules of water ($^1H_2^{16}O$) are more volatile than those containing a heavy isotope, molecules evaporating from water are enriched in "light" isotopes and depleted in the heavier isotopes. The relative enrichment or depletion in heavy isotopes is commonly expressed as a δ-value (in ‰ or parts per thousand/per mil), defined as:

$$\delta(\permil) = \left(\frac{R_x}{R_{ref}} - 1 \right) \times 1000$$

where R denotes the abundance ratio of the heavy to light isotope (e.g., $^2H/^1H$), and R_x and R_{ref} are the ratios in the sample and the reference standard. The reference for oxygen and hydrogen isotopes in water is taken to be the so-called VSMOW or Vienna Standard Mean Ocean Water. As a result, water evaporated from the ocean is depleted by $\sim$ 12‰–15‰ in ^{18}O (denoted $\delta^{18}O$) and by $\sim$ 80‰–120‰ in ^2H relative to the source ocean water.

Subsequent cooling and condensation of water vapor evaporated from the ocean preferentially removes the less volatile, heavy isotopes of water. Thus, clouds and precipitation are enriched in the heavy isotopes of water, while the remaining vapor is depleted in ^2H and ^{18}O. Moreover, the temperature of formation of droplets and precipitation also influences the isotopic composition of rain.[8] Thus, winter precipitation is depleted in heavy isotopes relative to summer, high latitude precipitation is depleted relative to that formed at lower latitudes, and high altitude precipitation is depleted in heavy water relative to

Encyclopedia of Water Science
DOI: 10.1081/E-EWS 120010201

Table 1 Isotopes of hydrogen and oxygen

Isotope	Weight (amu)	Natural abundance (wt %)
^{1}H	1.0078	99.970
^{2}H	2.0141	0.030
^{3}H	3.0160	10^{-15}
^{16}O	15.9949	99.732
^{17}O	16.9991	0.039
^{18}O	17.9992	0.229

(From Ref. 1.)

Table 2 Natural abundance of isotopes of water

Species	Natural abundance (%)
^{1}H$_2^{16}$O	99.728
^{1}H$_2^{18}$O	0.200
^{1}H$_2^{17}$O	0.040
^{1}H^{2}H^{16}O	0.032

(From Ref. 1.)

that formed at lower altitudes. These differences have been exploited in a wide range of hydrological, geological, and environmental studies.

ENVIRONMENTAL APPLICATIONS OF ISOTOPE ANALYSES

Hydrological Studies

The history, age, and pathway of water within the hydrologic cycle can be inferred from the relative abundance of ^{2}H, ^{3}H, and ^{18}O in water. The isotopic composition of groundwater can be used to determine recharge to an aquifer, including the source area and rate of recharge. The source area can be identified by the ^{2}H and ^{18}O concentrations in the groundwater and correlating them with the altitude at which precipitation infiltrated the soil. For example, Friedman and Smith[9] reported an ~ 40‰ decrease in ^{2}H for every 1000 m increase in altitude on the west slope of the Sierra Nevada range in California. This trend, along with the so-called "Continental Effect" is shown for the western United States in Fig. 1.[10]

The mechanism for the "Continental Effect" is the same as that for altitudinal effects, that is, the raining out of heavier isotopes, leaving vapor and subsequent precipitation isotopically lighter as one moves inland.

In a somewhat different way, the rate of recharge can be estimated by quantifying the tritium concentrations in the subsurface. Specifically, a peak ^{3}H concentration associated with nuclear weapons testing would correspond to ~ 1962, and thus its location beneath the land surface would yield a travel distance over approximately 40 yr. For example, Cook et al.[11] have used observed ^{3}H transport to estimate effective unsaturated hydraulic conductivities and recharge rates within arid zone soils in Australia.

The rate of movement of nutrients, pesticides, and other chemicals within the subsurface has also been quantified in laboratory and field experiments through co-application of ^{3}H$_2$O. The breakthrough or transport of the chemical relative to that of tritiated water provides a direct measure of chemical transport and the extent of reaction and retardation within the soil. For example, Gupta, Destouni, and Jensen[12] recently compared phosphorus and tritium transport in structured soil and found that 60%–100% of the water flow was associated with 25%–40% available flow paths. Moreover, preferential flow increased phosphorus mass transport by 2–3 × than without preferential flow. It should be noted, however, that such studies require careful consideration of safety and other issues.

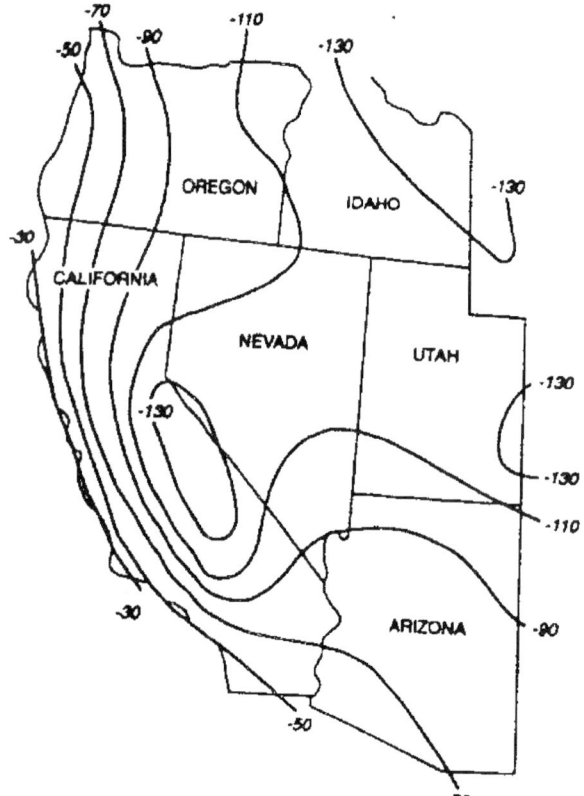

Fig. 1 Isotopic composition of ^{2}H in precipitation in the western United States.[10] (From Elsevier Science.)

Table 3 Some properties of different isotopes of water

Species	Density$_{max}$ (kg m^{-3})	Temperature at density$_{max}$ (°C)	Boiling point (°C)	Freezing point (°C)
$^1H_2^{16}O$	999.97	3.984	100.0	0.0
$^2H_2^{16}O$	1106.00	11.185	101.4	3.8
$^1H_2^{18}O$	1112.49	4.211	NA	NA
$^3H_2^{16}O$	1215.01	13.403	NA	NA

(From Ref. 1.)

Paleoclimate and Climate Change Studies

The relative differences in vapor pressure of heavy and light water and the temperature induced differences in isotopic composition of water referred to above make it possible to infer past climatic conditions. For example, Johnsen et al.[13] used ^{18}O composition of ice cores from Antarctica and Greenland to estimate climatic conditions over the past ~ 100,000 yr. Their study clearly shows colder temperatures from about 70,000 yr to 12,000 yr before present (B.P.), corresponding with the Wisconsin glaciation (Fig. 2).

The paleotemperatures of the ancient oceans have been estimated from the oxygen isotope distribution between $CaCO_3$ and water.[14] That is, $CaCO_3$ is enriched in ^{18}O relative to ocean water. Moreover, the relative enrichment, as with other isotopic fractionation reactions, is dependent upon the temperature at which it formed. Thus, the ^{18}O content of $CaCO_3$ provides a record of the ocean temperature at the time it was laid down. Such analyses have demonstrated significant variations in the ocean temperature over the past 700,000 yr.[15] Analyses by Woodruff, Savin, and Douglas[16] show greater variations over the past ~ 20 million yr.

Paleoclimatic information on the continents is available from the ^{18}O signature of biogenic apatite[17] and $CaCO_3$ deposits in caves.[18] The $\delta^{18}O$ values of snail shells have also been used.[19]

The isotopic composition of rocks and minerals also provides important information about the conditions at the time of their formation or alteration. For example, the clay minerals in soils over large regions of the United States formed during the Tertiary period under warmer conditions than those of the present.[20] Furthermore, limestone deposited in freshwater environments is depleted in ^{18}O relative to those formed in marine settings, reflecting differences in the ^{18}O status of the two types of waters.

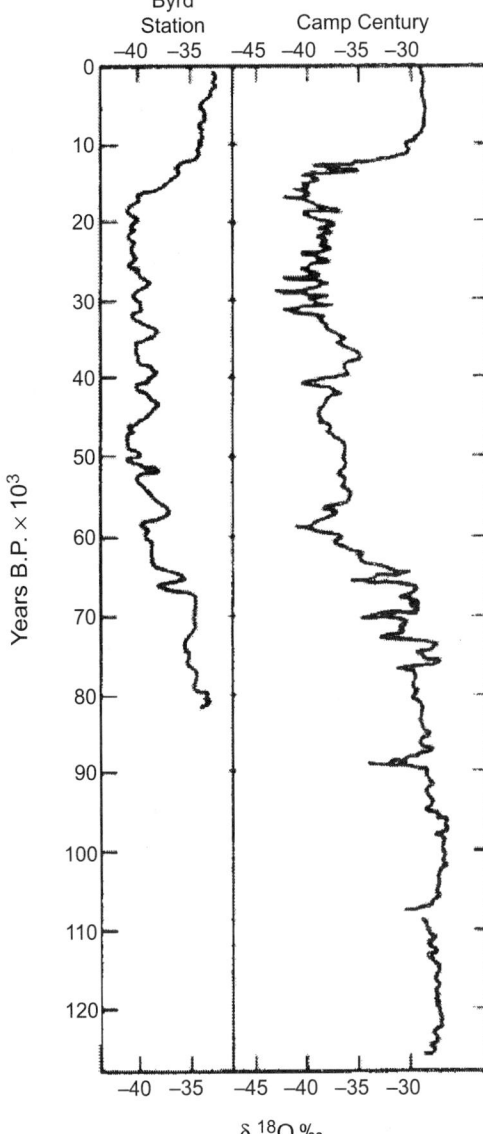

Fig. 2 Variation in O composition in ice cores Byrd Station, Antarctica and Camp Century, Greenland.[13] [From Nature (Macmillan Journals, Ltd.).]

REFERENCES

1. Franks, F. *Water*; The Royal Society of Chemistry: London, 1984.
2. Giauque, W.F.; Johnston, H.L. An Isotope of Oxygen, Mass 18. J. Am. Chem. Soc. **1929**, *51*, 1436–1441.
3. Urey, H.C.; Brickwedde, F.G.; Murphy, G.M. A Hydrogen Isotope of Mass 2 and Its Concentration. Phys. Rev. **1932**, *39*, 1–15.
4. Faltings, V.; Harteck, P. Der Tritium Gehalt Der Atmosphere. Z. Naturforsch **1950**, *5a*, 438–439.
5. Grosse, A.V.; Johnston, W.M.; Wolfgang, R.L.; Libby, W.F. Tritium in Nature. Science **1951**, *113*, 1–2.

6. Lal, D.; Seuss, H.E. The Radioactivity of the Atmosphere and Hydrosphere. Ann. Rev. Nucl. Sci. **1968**, *18*, 407–435.

7. IAEA. *World Survey of Isotope Concentrations in Precipitation*, Tech. Report Series No. 96, 1969.

8. Dansgaard, W. Stable Isotopes in Precipitation. Tellus **1964**, *16*, 436–463.

9. Friedman, I.; Smith, G.I. Deuterium Content of Snow Cores from Sierra Nevada Area. Science **1970**, *169*, 467–470.

10. Ingraham, N.L. Isotopic Variations in Precipitation. In *Isotope Tracers in Catchment Hydrology*; Kendall, C., McDonnell, J.J., Eds.; Elsevier Science: Amsterdam, NL, 1998; 87–118.

11. Cook, P.G.; Jolly, I.D.; Leaney, F.W.; Walker, G.R.; Allan, G.L.; Fifield, L.K.; Allison, G.B. Unsaturated Zone Tritium and Chlorine-36 Profiles from Southern Australia—Their Use as Tracers of Soil Water Movement. Water Resour. Res. **1994**, *30*, 1709–1719.

12. Gupta, A.; Destouni, G.; Jensen, M.B. Modelling Tritium and Phosphorus Transport by Preferential Flow in Structured Soil. J. Contam. Hydrol. **1999**, *35*, 389–407.

13. Johnsen, S.J.; Dansgaard, W.; Clausen, H.G.; Langway, C.C. Oxygen Isotope Profiles Through the Antarctic and Greenland Ice Sheets. Nature **1972**, *235*, 429–434.

14. Urey, H.C. The Thermodynamic Properties of Isotopic Substances. J. Chem. Soc. **1947**, *1947*, 562–581.

15. Emiliani, C.; Shackleton, N.J. The Brunhes Epoch: Isotopic Paleotemperatures and Geochronology. Science **1974**, *183*, 511–514.

16. Woodruff, R.; Savin, S.M.; Douglas, R.E. Miocene Stable Isotope Record: A Detailed Deep Pacific Ocean Study and Its Paleoclimatic Implications. Science **1981**, *212*, 665–668.

17. Kolodny, Y.; Luz, B.; Navon, O. Oxygen Isotope Variations in Phosphate of Biogenic Apatites. I. Fish Bone Apatite—Rechecking the Rules of the Game. Earth Planet. Sci. Lett. **1983**, *64*, 398–404.

18. Schwarz, H.P.; Harmon, R.S.; Thompson, P.; Ford, D.C. Stable Isotope Studies of Fluid Inclusions in Speleothems and Their Paleoclimatic Significance. Geochim. Cosmochim. Acta **1976**, *40*, 657–665.

19. Magaritz, M.; Heller, J. A Desert Migration Indicator—Oxygen Isotopic Composition of Land Snail Shells. Paleo. Paleo. Paleo. **1980**, *32*, 153–162.

20. Faure, G. *Principles of Isotope Geology*, 2nd Ed.; John Wiley and Sons: New York, 1986.

Journals

Joseph R. Makuch
National Agricultural Library, Beltsville, Maryland, U.S.A.

Stuart R. Gagnon
University of Maryland, College Park, Maryland, U.S.A.

INTRODUCTION

To understand a field of study, you must read its literature, including the bulwark of any discipline, its scholarly journals. Capturing the breadth of hydrologic sciences literature that addresses agricultural issues requires exploring a variety of journals. While there may be overlap in topical coverage, each journal fills a particular subject matter and readership niche. This article focuses on key journals in the hydrologic sciences, especially those covering agricultural issues.

OVERVIEW OF JOURNALS

Scholarly journals are a means of sharing information in a given field. Journals serve to expand the knowledge base of a subject. Journal articles in hydrologic science—a multidisciplinary field that focuses upon the occurrence, movement, and properties of water—describe experimental studies and their implications, discuss theoretical approaches or analyze the ramifications of public policy. A special type of journal article—the literature review—synthesizes the findings of seminal publications on a given topic. A literature review, like other journal articles, includes a list of publications referenced in the article. Journals may also publish commentaries or letters to the editor on issues relevant to the field or on previously published articles.

Journal articles are written by researchers from academia, government, and the private sector. Published articles are often viewed as a measure of professional productivity. Institutions gain prestige when their employees' articles are published in respected scholarly journals.

People who need to stay abreast of developments in their fields read or review scholarly journals and articles of interest. Researchers learn about advancements in experimental techniques or theories. Reported findings and methodologies influence how researchers approach their own scientific inquiries. Educators acquire information to keep their courses current. Decision makers obtain unbiased scientific bases for recommending or implementing actions. Undergraduate and graduate students gain knowledge in their fields by reading and studying scholarly articles.

Most journals in the hydrologic sciences are published by either commercial scientific publishers or professional societies. Publication schedules vary, as do subscription prices. Online versions of journals are increasingly available. Publishers often allow nonsubscribers online access to tables of contents, abstracts, or sample issues. Libraries at institutions with departments involved with hydrologic sciences have print and/or online subscriptions to relevant journals.

SPECIFIC HYDROLOGIC SCIENCE JOURNALS

Table 1 lists hydrologic science journals that are important for the agricultural sector. These journals cover hydrologic science issues beyond agriculture, but they publish many agriculturally related articles. There are journals for hydrologic sciences in addition to those listed in Table 1, but they have less relevance to agriculture.

Relevant articles concerning water resources and agriculture can also be found in broader environmental publications and in the agricultural literature. See Tables 2 and 3 for a sampling of germane journals in these fields. In addition, journals in fields such as meteorology, climatology, limnology, aquatic biology, and forestry are sources of related information.

Tables 1–3 also provide the World-Wide-Web address of each journal's publisher. Visit the Web sites of the respective journals to obtain detailed information about each journal, including frequency of publication, subscription cost, and scope of coverage. Most journals also provide this information in each issue. Indexes to journal articles are usually published in the last issue of

Encyclopedia of Water Science
DOI: 10.1081/E-EWS 120010296

Table 1 Journals for hydrologic sciences (important for agriculture) with Web addresses of publishers

Journal title	Publisher Web address
Advances in Water Resources	www.elsevier.com/
Agricultural Water Management	www.elsevier.com/
Groundwater	www.ngwa.org/
Groundwater Monitoring and Remediation	www.ngwa.org/
Hydrological Processes	www.interscience.wiley.com/
Journal of the American Water Resources Association	www.awra.org/
Journal of Contaminant Hydrology	www.elsevier.com/
Journal of Geophysical Research-Atmosphere	www.agu.org/
Journal of Great Lakes Research	www.iaglr.org/
Journal of Hydraulic Research	www.iahr.org/
Journal of Hydrologic Engineering	www.pubs.asce.org/
Journal of Hydrology	www.elsevier.com/
Journal of Soil and Water Conservation	www.swcs.org/
Journal of Water Resources Planning and Management	www.pubs.asce.org/
Water Environment Research	www.wef.org/
Water Research	www.elsevier.com/
Water Resources Research	www.agu.org/
Water Science and Technology	www.iwap.co.uk/
Wetlands	www.sws.org/

a volume. Articles can also be located by using abstracting and indexing services such as the Water Resources Abstracts and AGRICOLA databases. There is a trend for online indexes to hyperlink directly to journal articles when those articles are available electronically.

Reference directories, such as *Ulrich's International Periodicals Directory* and the *Serials Directory: An International Reference Book*, available in paper or online at many libraries, are other good sources of information on particular journals. These directories are also useful for finding information about journal name or publisher changes.

Additional information about several journals representative of the literature is provided below. These journals were recommended by knowledgeable professionals working in various areas of water resources and agriculture. The brief descriptions are based on reviews of recent issues and information published in the journals and their respective Web sites. See these sources for additional information.

Examples of Journals for Hydrologic Sciences Covering Agricultural Issues

Agricultural Water Management: an international journal. 1977–present. 15/yr. Elsevier Science BV, P. O. Box 211, 1000 AE Amsterdam, Netherlands. http://www.elsevier.com/. ISSN: 0378-3774.

Readers of this international journal include agricultural engineers, agricultural hydrologists, and agronomists. Research articles on various aspects of irrigation are a major theme of the journal. Other areas covered include drainage, erosion, and water quality. Theme issues, such as "The Use of Water in Sustainable Agriculture," are occasionally published.

Journal of the American Water Resources Association. 1965–present. Bi-monthly. American Water Resources Association, P. O. Box 1626, Middleburg, VA 20118-1626. http://www.awra.org/. ISSN: 0043-1370.

Formerly known as the *Water Resources Bulletin*, the main focus of this journal is on water resources management issues of broad interest. The journal organizes articles under one of three categories: "Dialogue on Water Issues," "Technical Papers," and "Discussion Papers" (and replies). Articles in the "Dialogue" section cover policy and critical management issues. The "Technical" section contains articles on subjects such as phosphorus and wetlands, stream channel instability, and artificial neural networks for subirrigation systems. Comments on earlier articles are published in the "Discussion" section, along with replies from the original author(s). Special issues, such as "Water Resources and Climate Change," are sometimes published. *Impact* is a sister publication of the journal that focuses on a single theme for each issue. Articles address timely issues for water resource professionals.

Table 2 Environmental journals covering water-related topics (important to agriculture) with Web addresses of publishers

Journal title	Publisher Web address
Agriculture, Ecosystems and Environment	www.elsevier.com/
AMBIO	ambio.allenpress.com/
Aquatic Toxicology	www.elsevier.com/
Archives of Environmental Contamination and Toxicology	www.springer-ny.com/
Bulletin of Environmental Contamination and Toxicology	www.springer-ny.com/
Chemosphere	www.elsevier.com/
Critical Reviews in Environmental Science and Technology	www.crcpress.com/
Ecotoxicology	www.wkap.nl/
Ecotoxicology and Environmental Safety	www.academicpress.com/
Environmental Management	www.springer-ny.com/
Environmental Monitoring and Assessment	www.wkap.nl/
Environmental Pollution	www.elsevier.com/
Environmental Science and Technology	www.pubs.acs.org/
Environmental Toxicology	www.interscience.wiley.com/
Journal of Environmental Economics and Management	www.academicpress.com/
Journal of Environmental Engineering	www.pubs.asce.org/
Journal of Environmental Management	www.academicpress.com/
Journal of Environmental Quality	www.agronomy.org/
Journal of Environmental Science and Health, Part A-Toxic/Hazardous Substances and Environmental Engineering	www.dekker.com/
Journal of Environmental Science and Health, Part B-Pesticides, Food Contaminants, and Agricultural Wastes	www.dekker.com/
Journal of Range Management	www.srm.org/
Journal of Toxicology and Environmental Health— Part A	www.tandf.co.uk/
Journal of Toxicology and Environmental Health— Part B: Critical Reviews	www.tandf.co.uk/
Nature	www.nature.com/
Science of the Total Environment	www.elsevier.com/
Water, Air and Soil Pollution	www.wkap.nl/

Journal of Hydrology. 1963–present. 56/yr. Elsevier Science BV, P.O. Box 211, 1000 AE Amsterdam, Netherlands. http://www.elsevier.com/. ISSN: 0022-1694.

This journal, with U.S. and international editors, publishes highly technical research papers, and comprehensive reviews in the hydrologic sciences. Topics include the "physical, chemical, biogeochemical, stochastic and system aspects of surface and groundwater hydrology, hydrometeorology, and hydrogeology." Articles may be of an empirical, theoretical, or applied focus. Theme issues are sometimes published.

Journal of Soil and Water Conservation. 1946–present. Quarterly. Soil and Water Conservation Society, 7515 N. E. Ankeny Rd., Ankeny, IA 50021. http://www.swcs.org/. ISSN: 0022-4561.

Published by a professional society whose members include soil and water conservation professionals, researchers, planners, educators, administrators, and others, this journal focuses on the conservation, improvement and sustainable use of soil, water and related natural resources worldwide.

Journal issues contain primarily "Features" and "Research" articles. The former are overview and synthesis articles, including literature reviews, while the latter report on specific research studies. Articles in both sections address a broad range of soil and water conservation topics and include articles focussing on social sciences and economics. The general emphasis of the journal is on agricultural and other rural lands. The journal also publishes commentaries from readers,

Table 3 Agricultural journals that often cover water-related topics with Web addresses of publishers

Journal title	Publisher Web address
Advances in Agronomy	www.academicpress.com/
Agronomy Journal	www.agronomy.org/
American Journal of Agricultural Economics	www.aaea.org/
American Journal of Alternative Agriculture	www.winrock.org/
Applied Engineering in Agriculture	www.asae.org/
Irrigation and Drainage	www.interscience.wiley.com/
Irrigation Science	www.springer-ny.com/
The Journal of Agricultural Science	www.journals.cambridge.org/
Journal of Irrigation and Drainage Engineering	www.pubs.asce.org/
Nutrient Cycling in Agroecosystems	www.wkap.nl/
Poultry Science	www.poultryscience.org/
Soil Science Society of America Journal	www.soils.org/
Soil and Tillage Research	www.elsevier.com/
Swedish Journal of Agricultural Research [a]	N/A
Transactions of the ASAE	www.asae.org/

[a] Note: Ceased publication in 1998.

a listing of future conferences and some advertisements, including employment opportunities. A less technical, magazine-style sister publication, *Conservation Voices*, also covers soil and water conservation topics. In January 2002, *Conservation Voices* and the *Journal of Soil and Water Conservation* will be merged into a single journal published six times a year.

Water Resources Research. 1965–present. Monthly. American Geophysical Union, 2000 Florida Ave. N. W., Washington, DC 20009. http://www.agu.org/. ISSN: 0043-1397.

This is an interdisciplinary journal that covers research in the social and natural sciences related to water. The journal publishes articles in scientific hydrology covering the biological, chemical, and physical sciences as well as economics, sociology, and law.

There are deputy editors for erosion, sedimentation, and geomorphology; geochemistry and geobiology; groundwater; surface water; vadose zone; and water policy, economics and systems analysis. Many articles are theoretical.

Examples of Environmental Journals that Cover Water and Agriculture

Agriculture, Ecosystems and Environment. 1974–present. 15/yr. Elsevier Science BV, P. O. Box 211, 1000 AE Amsterdam, Netherlands. http://www.elsevier.com/. ISSN: 0167-8809.

Covering the interface between agriculture and the environment, *Agriculture, Ecosystems and Environment* promotes interdisciplinary approaches to research. The journal is aimed at scientists studying many aspects of agricultural ecosystems. Papers in this journal have covered topics such as water availability and use, nonpoint-source pollution, and seasonal flooding. Special issues are occasionally published, including an issue on sustainable land management.

Agriculture, Ecosystems and Environment is an amalgamation of two earlier journals, *Agro-Ecosystems and Agriculture* and *Environment*. A section of *Agriculture, Ecosystems and Environment* is currently published as *Applied Soil Ecology*. Both journals are included in the same subscription.

Journal of Environmental Quality. 1972–present. Bimonthly. American Society of Agronomy, 677 S. Segoe Rd., Madison, WI 53711. http://www.agronomy.org/. ISSN: 0047-2425.

The professional societies, American Society of Agronomy, Crop Science Society of America and Soil Science Society of America, cooperatively publish the *Journal of Environmental Quality*. The journal contains research and review articles on topics covering agricultural and natural ecosystems. A section of the journal, headed "Environmental Issues," contains articles from "a combination of scientific, political, legislative, and regulatory perspectives."

Water, Air, and Soil Pollution: An International Journal of Environmental Pollution. 1971–present.

32/yr. Kluwer Academic Publishers, P.O. Box 17, Dordrecht, 3300 AA, Netherlands. http://www.wkap.nl/. ISSN: 0049-6979.

All aspects of the biological, chemical, and physical processes of environmental pollution are covered by this interdisciplinary journal. *Water, Air, and Soil Pollution* includes articles on wastewater irrigation, forest management, and nitrate depletion. Other papers published in the journal include those describing methods used to study and measure environmental pollutants.

Examples of Agricultural Journals that Cover Hydrologic Sciences

Journal of Irrigation and Drainage Engineering. 1956–present. Bi-monthly. American Society of Civil Engineers, 1801 Alexander Graham Bell Dr., Reston, VA 20191. http://www.pubs.asce.org/. ISSN: 0733-9437.

This journal covers research on "engineering hydrology, irrigation, drainage, and related water management subjects, such as watershed management, weather modification, water quality, groundwater, and surface water." Articles appear in three categories: "Technical Papers" discuss experimental results and conclusions or analytical approaches to water management problems. Shorter articles, or reports of preliminary research results, are published in "Technical Notes." The "Discussion" section contains short pieces that offer substantive comments on previously published papers and includes a closure response from the original author(s). The journal was formerly known as the *Journal of the Irrigation and Drainage Division, Proceedings of the American Society of Civil Engineers.*

Soil Science Society of America Journal. 1936–present. Bi-monthly. Soil Science Society of America, 677 S. Segoe Rd., Madison, WI 53711. http://www.soils.org/. ISSN: 0361-5995.

The table of contents for this journal, formerly *Soil Science Society of America Proceedings*, divides the articles into several different subject categories. While the section on "Soil and Water Management and Conservation"

contains articles most relevant to hydrologic science, other sections may also contain water-related articles.

Articles primarily describe and discuss different aspects of soil science research. Comments on specific articles or other soil science topics are also published, as are the occasional invited essay or review. A list of new soil science books is sometimes published.

Transactions of the ASAE. 1958. Bi-monthly. American Society of Agricultural Engineers, 2950 Niles Rd., St. Joseph, MI 49085-9659. http://www.asae.org/. ISSN: 0001-2351.

Each issue in this journal, from ASAE—the professional society for engineering in agricultural, food, and biological systems—contains a soil and water section. Articles cover a range of water-related topics such as water flow in riparian areas, irrigation efficiency, and nitrate leaching. Topics are approached from various perspectives including computer modeling, engineering design, and scientific investigation. *Resource* is a magazine by ASAE that occasionally publishes articles or short news pieces covering water and agriculture issues.

FURTHER READING

Delphino, J.J. Water Chemistry. Seeking Information. Environ. Sci. Technol. **1977**, *11* (7), 669–672.

Haas, S.; Clark, M. Research Journals and Databases Covering the Field of Agrochemicals and Water Pollution. Sci. Technol. Libr. **1992**, *13* (2), 57–63.

The Serials Directory. An International Reference Book, 13th Ed.; EBSCO Publishers: Birmingham, AL, 1999.

Ulrich's International Periodicals Directory, 2000, 38th Ed.; R.R. Bowker: New Providence, NJ, 1999.

Walker, R.D.; Ahn, M.L. Literature Cited by Water Resources Researchers. In *Changing Gateways. The Impact of Technology on Geoscience Information Exchange*, Proceedings of the 29th Meeting of the Geoscience Information Society, Seattle, WA, October 24–27, 1994; Haner, B.E., O'Donnell, J., Eds.; Geoscience Information Society: Alexandria, VA, 1995; 67–77.

La Niña

David E. Stooksbury
University of Georgia, Athens, Georgia, U.S.A.

INTRODUCTION

La Niña is a change in global weather patterns associated with colder than normal sea surface water temperatures in the equatorial Pacific Ocean (see Fig. 1). La Niña is the cool phase of the best known example of interannual variation in the earth's weather and climate patterns, El Niño-Southern Oscillation (ENSO). Since El Niño, which means boy in Spanish, is the warm phase of ENSO, the cool phase of ENSO is called La Niña, which means girl in Spanish. Another term for La Niña is El Viejo, which means old man in Spanish.

The La Niña pattern leads to changes in positions of jet streams, the steering currents for weather systems. Changes in the polar jet streams cause changes in weather patterns across North America during the winter. While the change in weather patterns associated with La Niña can be dramatic, most regions experience minimal to no direct impact from La Niña. La Niña events occur every three to seven years lasting a few months to a year or more.

IMPACTS

In the United States, La Niña weather patterns usually mean a dry to very dry winter for the coastal plain of Georgia and the Carolinas as well as Florida (see Fig. 2). While the Southeast is experiencing a dry winter, the lower Mississippi River and the Ohio River Valleys normally experience a wet winter. Much of the remainder of the United States experience very little winter time precipitation impacts associated with La Niña. During La Niña winters, the polar jet stream is positioned such that most storms move from the lower Mississippi River to the Ohio River Valleys exiting the Northeast coast. Compared to a normal winter, the southeastern United States experiences fewer storms during a La Niña winter.

During the summer and fall, La Niña events are associated with increased tropical weather activity in the Gulf of Mexico and Atlantic Ocean. This increase in tropical weather makes the east coast more vulnerable to tropical storms and hurricanes during La Niña years. The increase in tropical weather activity associated with La Niña events, usually causes the Southeast to experience wetter than normal falls during La Niña years.

Not all regions impacted by La Niña have decrease in precipitation. The La Niña weather pattern usually brings wetter than normal conditions to northern Australia, Indonesia, and the Philippines.

The impacts of La Niña weather patterns vary from one event to another. The impacts depend on the coolness of the surface water, the exact location of the cool surface water, the areal extent of the cool surface water, and other regional and global weather patterns.

Mechanism

The oceans and the atmosphere are linked. The discovery of the ocean–atmosphere linkage is one of the most important breakthroughs in modern environmental research. The linkage between the equatorial Pacific Ocean and the atmosphere helps to determine weather patterns across much of the earth. The variation in atmospheric pressure patterns over the Pacific Ocean that are linked to the equatorial Pacific Ocean temperature patterns is called the Southern Oscillation, SO.

The strength of SO is calculated by the surface atmospheric pressure anomaly differences between Tahiti and Darwin, Australia (Tahiti anomaly minus Darwin anomaly). This measure of SO strength is called the Southern Oscillation Index, SOI. A surface atmospheric pressure anomaly is calculated by subtracting the mean atmospheric surface pressure from the observed atmospheric surface pressure. Thus, if the observed atmospheric surface pressure is more than the mean, the anomaly has a positive value. When the SOI has a positive value, it means that the surface atmospheric pressure difference between Tahiti and Darwin is greater than normal. A positive SOI is correlated with a cooler than normal surface water in the eastern and central equatorial Pacific Ocean.

The linkage between the SO and La Niña is complex. At the most basic level, sea surface temperature patterns influence atmospheric pressure patterns, and atmospheric pressure patterns influence wind speed and direction and thus the sea surface temperature patterns.

Under neutral conditions (climatologists prefer the term "neutral" instead of "normal") the normal cold surface water of the eastern Pacific Ocean is associated with relatively high surface atmospheric pressure over the region. Air moves (wind) from areas of high atmospheric

Encyclopedia of Water Science
DOI: 10.1081/E-EWS 120010222

December - February lA Niño Conditions

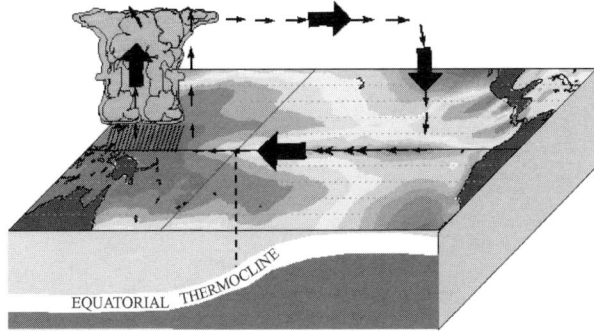

Fig. 1 Atmospheric and oceanic patterns during a La Niña. (From the National Weather Service Climate Prediction Center, Camp Springs, MD.)

pressure to areas of low atmospheric pressure. The greater the pressure gradient (pressure difference between two locations divided by the distance between the two locations), the greater the wind speed. The moving air in contact with the ocean surface causes ocean surface currents, which redistribute the ocean surface temperature pattern. The stronger the wind, the more the redistribution of surface water temperatures.

With ENSO (either the cool phase, La Niña or the warm phase El Niño), the linkage between the ocean and the atmosphere results in decreasing or increasing easterly trade-wind (wind from the east to the west) speeds over the equatorial Pacific Ocean. When the SOI is positive (La Niña phase), the pressure gradient across the eastern and western Pacific Ocean is increased. With an increased pressure gradient, the speed of the easterly trade-winds increases, and allows cold upwelled water from the eastern equatorial Pacific Ocean to cool the central equatorial Pacific Ocean thus producing a La Niña event.

Since La Niña and neutral phases of ENSO have the same atmospheric pressure and wind patterns, many climatologists consider a La Niña event "an extreme case of normal." While the atmospheric pressure and wind patterns are the same for La Niña and neutral phases, the pressure gradient and thus the easterly winds are much more pronounced during La Niña events. The stronger easterly winds cause a major expansion of cold sea surface temperatures across the central equatorial Pacific Ocean and the associated changes in global weather patterns.

Since the 1990s scientists have used Pacific Ocean surface temperature data and computer models to predict

TYPICAL JANUARY-MARCH WEATHER ANOMALIES AND ATMOSPHERIC CIRCULATION DURING MODERATE TO STRONG EL NIÑO & LA NIÑA

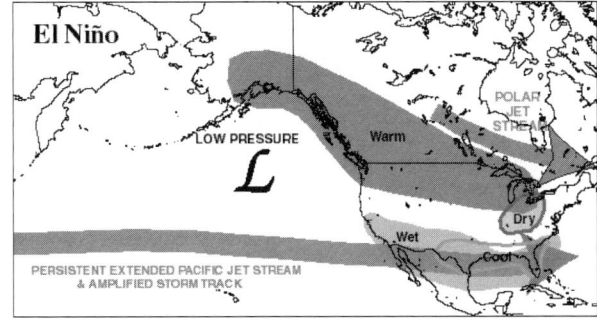

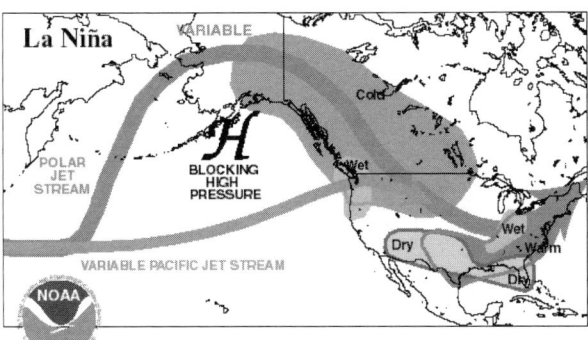

Climate Prediction Center/NCEP/NWS

Fig. 2 Typical jet stream and climate patterns during an El Niño (top) and a La Niña (bottom). (From the National Weather Service Climate Prediction Center, Camp Springs, MD.)

the occurrence of a La Niña event months in advance. While these predictions are not perfect, they allow for planning to mitigate or take advantage of a shift in weather patterns. Thus, regions that normally experience drought during a La Niña event can plan to mitigate the impacts. For regions like the southeastern United States, drought mitigation plans can be activated months in advance.

For more detailed information about La Niña, see Ref. [1].

REFERENCE

1. National Oceanic and Atmospheric Administration Climate Prediction Center. El Niño/La Niña website. http://www.cpc. ncep.noaa.gov/products/analysis_monitoring/lanina/ (accessed Aug 2002).

Land Drainage, Subsurface

Stephen R. Workman
University of Kentucky, Lexington, Kentucky, U.S.A.

INTRODUCTION

Subsurface land drainage is the removal of excess soil water through the use of ditches or buried perforated pipe. Ideally, the excess water is removed quickly enough to allow timely field operations to occur or to limit crop damage if a crop is present. This chapter presents the design of subsurface drainage systems using knowledge of soil physical properties, crop susceptibility to excess water, and the flow of water in pipe systems.

The water table (saturated soil conditions) is present near the soil surface in many places. Soils located near streams, lakes, and oceans are familiar examples. Other cases might include areas where an impermeable or slowly permeable layer exists in the soil profile that causes a perched water table to exist. The primary limitation for effectively using these soils for recreation, construction, or agriculture is the high water table.

SYSTEM LAYOUT

The drains in a subsurface land drainage system can either be placed in a systematic pattern if a large area is to be drained or randomly placed near hard-to-drain areas. A random layout of drains is used to remove excess water from localized depressions in a field. The primary design considerations are to maintain adequate slope to the outlet and to ensure that adequate cover over the drain is maintained. A random layout is typically used as the drainage system under rolling topography, constructed grass waterways, and the greens of golf courses.

For most cases where the entire field is poorly drained, a system of parallel drains or laterals is installed throughout the entire field. The laterals are connected to a series of sub-mains and mains that transport the water to the outlet. The spacing between the parallel drains is important to the cost of the system and can result in considerable savings if an optimal design is chosen.

Besides random or parallel layouts, an interceptor drain may be used near seepage or overland flow areas. These drains are typically installed at locations of water seeps and excess runoff such as at the bottom of a hill.[1]

SUBSURFACE LAND DRAINAGE DESIGN

A combination of factors influences the depth and spacing of laterals within a drainage system. These factors include hydraulic conductivity, drainable porosity, thickness of soil layers, depth to a restrictive layer, and quality of surface drainage. The single most important soil property affecting the design of drainage systems is hydraulic conductivity.[2] Hydraulic conductivity (K) is spatially variable and should be determined at numerous positions across the site. In the design of large systems where large differences in conductivity may occur, the spacing may be wider or narrower in some regions of the field. From a construction standpoint, however, it is best to maintain uniform drain spacing for as large an area as possible.

Whereas the hydraulic conductivity indicates the rate that water can move through the soil, another parameter called the drainage coefficient (DC) has been defined to represent the rate of water to be removed by the drainage system. The DC is the depth of water to be removed in 24 hr by the drainage system and has the same units as hydraulic conductivity (L/t). In humid areas where uniform, low-intensity rainfall events are common, the DC depends largely on a design rainfall event. The DC should be selected to remove excess water rapidly enough to prevent serious damage to the crop. Loss of crop will generally occur if the soil profile is allowed to remain saturated for more than 24 hr.[2,3] Typical values of DC are 10 mm/day–20 mm/day.[3] The DC should be increased for high value crops or special soil conditions.[3]

Steady-State Design

One method of designing a subsurface drainage system is to assume that a uniform rainfall occurs on the soil surface over a long period of time. At steady state, all inflows (rate of groundwater recharge) must equal outflows (discharges through the drainage system). These steady-state conditions can be computed by the Hooghoudt equation where S is drain spacing (L), R is the drainage coefficient or rainfall rate per unit area (L/t), K is the hydraulic conductivity (L/t), d is the depth of the drain to the restrictive layer (L), and b is the vertical distance from the drain to the water table position at the midpoint between

Encyclopedia of Water Science
DOI: 10.1081/E-EWS 120010049

drains (L).[4]

$$S = \left[\frac{8Kdb + 4Kb^2}{R} \right]^{1/2} \tag{1}$$

The primary limitation of the earlier equation is that flow is assumed to move horizontally towards a ditch.[5] For the case of flow to drain pipes, convergence losses near the drain must be included. The effect of the convergence is to reduce the amount of flow because there are vertical and horizontal components of flow, conceptually similar to reducing the thickness of the soil profile. An effective depth (d_e) can be computed by the equation,

$$d_e = \frac{\pi S}{8 \left(\ln \left(\frac{S}{\pi r_e} \right) + F(x) \right)} \quad \text{for } x = \frac{2\pi d}{S} \tag{2}$$

$$F(x) = \sum_{n=1}^{\infty} 2 \ln \coth(nx) = \sum_{i=1}^{\infty} \frac{4e^{-2ix}}{i(1 - e^{-2ix})}, \tag{3}$$

$$(n = 1, 2, 3, \ldots), (i = 1, 3, 5, \ldots)$$

for $x < 0.5$, $F(x)$ may be closely approximated by,

$$F(x) = \frac{\pi^2}{4x} + \ln \left(\frac{x}{2\pi} \right) \tag{4}$$

where r_e is the effective radius for the drain that accounts for the fact the drain has a limited amount of openings for water entry. In the design process for steady-state analysis, the drainage coefficient is considered to be the design rainfall amount.

Transient Methods

In areas where frequent, high-intensity storms occur during the growing season, a transient analysis should be conducted to assure that the water table falls quickly enough to maintain a well-aerated root zone. Mathematical analysis of the falling water table case is far more difficult than for the steady-state case. If the assumptions of horizontal flow, vertical equipotentials in the saturated zone, and instantaneous release/addition of water at the water table can be imposed, then the unsaturated zone can be neglected. In the transient case, the drainage coefficient is related to the water table drop over a period of a day. The drainage rate is then the amount of water released as the water table drops

$$DC = (b_0 - b)_{24\,\text{hr}} \times f \tag{5}$$

where $(b_0 - b)_{24\,\text{hr}}$ is the prescribed drop in the water table in 24 hr and f is the drainable porosity (L^3/L^3). Note that the drainable porosity is much smaller than the porosity of the soil because only the pores that drain as the water table falls from b_0 to b are included.

For the falling water table case the drain spacing can be computed as,[2]

$$S = \left[\frac{9Ktd_e}{f \ln \left(\frac{b_0(2d_e + b)}{b(2d_e + b_0)} \right)} \right]^{1/2} \tag{6}$$

where the effective depth to the impermeable layer (d_e) has been included to account for convergence losses to the drain, t is the time period for the water table to drop from b_0 to b, and all other parameters are as defined earlier. Both the steady state and the transient methods involve the determination of S and d_e through the process of iteration.

SYSTEM DESIGN

Layout

The subsurface drainage system is designed to transport excess water to the outlet. The outlet has to have the capacity of carrying the excess water away from the site. The outlet may be a drainage channel, an existing main, or a stream channel. After the outlet location is chosen and checked for capacity, the layout of the system in the field is determined. Some key characteristics of an efficiently designed system include making the laterals as long as possible, placing the laterals parallel to the topographic contour of the field, and minimizing lengths of sub-mains, mains, connections, and fittings.[6]

Design Flow

The design flow for the system is related to the drainage coefficient used to compute the drain spacing and can be computed from the equation,

$$Q = DC \times A_d \tag{7}$$

where Q is the flow rate (L^3/t), DC is the drainage coefficient (L/t), and A_d is the area-drained (L^2).

Grades

Maximum grades are limiting only where pipes are designed for near-maximum capacity or where pipes are placed in unstable soil. For nearly level areas, the drain should be as steep as possible while maintaining adequate depth at all locations to reduce the size of the mains and sub-mains. Drains should not be placed at a grade less than 0.2% under most conditions. If fine sand or silt is present, then larger grades may be necessary to keep the pipe clean.

Pipe Size

The following equation can be used to size the pipe needed to carry the drainage water.

$$d = 51.7(DC \times A \times n)^{3/8} s^{-3/16} \qquad (8)$$

where A is the drainage area entering the drain (ha), d is the pipe diameter for a pipe flowing full (mm), n is Manning's roughness coefficient, s is the drain slope (m/m), and DC is the drainage coefficient (mm/day). Typical Manning's roughness values for corrugated plastic pipe are 0.015 for 75 mm–200 mm diameter pipe, 0.017 for 250 mm–300 mm diameter pipe, and 0.2 for pipes greater than 300 mm.

ACCESSORIES

Accessories for subsurface drainage systems include surface inlets, blind inlets, sedimentation basins, and control structures.[2] A surface inlet, sometimes called an open inlet structure, is used to remove surface water from potholes, road ditches, or other depressions. Blind inlets may be used where the amount of surface water to remove is small or the amount of sediment is too great to permit surface inlets. Sedimentation basins are any type of structure that provides for sediment accumulation, reducing deposition in the drain. Finally, control structures are placed in the drainage system to maintain the water table at a specified level.

INSTALLATION CONSIDERATIONS

During installation, several factors such as machinery, grade control, corrections, and documentation must be considered. Examples of machinery used in the installation process include a trencher or backhoe to install the mains, a backhoe to clear enough soil away at junctions to make a connection and insert a plow or trencher boot, a drain plow or trencher to install the laterals, and pipe feeders to reduce stretch in pipe. For grade control, the bottom of the trench should be shaped with a supporting groove to provide good alignment and bottom support. Laser or manual methods may be used to establish grade.[6,7] Connections, an essential part of the drain system, should be made with a T- or a Y-manufactured junction. Manufactured connections should be used when changing pipe size, and end caps or plugs are required for preventing soil from entering the end of the lateral. Finally,

the necessary documentation includes a map of the drainage system filed with the deed to the property. The use of GPS and GIS technologies is encouraged for correct identification of the drainage system.[8]

CONDUIT LOADS

Loads on underground conduits include those caused by the weight of the soil and by concentrated loads resulting from the passage of equipment or vehicles. At shallow depths, concentrated loads from field machinery largely determine the strength requirements of conduits; at greater depths, the load from the soil is the most significant factor.[6,9]

REFERENCES

1. Carter, C.E. Surface Drainage. In *Agricultural Drainage*; Skaggs, R.W., van Schilfgaarde, J., Eds.; ASA, CSSA, SSSA: Madison, WI, 1999; 1023–1050.
2. Madramootoo, C.A. Planning and Design of Drainage Systems. In *Agricultural Drainage*, Agron. Monogr. 38; Skaggs, R.W., van Schilfgaarde, J., Eds.; ASA, CSSA, SSSA: Madison, WI, 1999; 871–892.
3. *ASAE Standards*; EP480 Design of Subsurface Drains in Humid Areas, 47th Ed.; ASAE: St. Joseph, Mich., 2000.
4. Ritzema, H.P. Subsurface Flow to Drains. In *Drainage Principles and Applications*; Ritzema, H.P., Ed.; ILRI: Wageningen, The Netherlands, 1994; 263–304.
5. van der Ploeg, R.R.; Horton, R.; Kirkman, D. Steady Flow to Drains and Wells. In *Agricultural Drainage*, Agron. Monogr. 38; Skaggs, R.W., van Schilfgaarde, J., Eds.; ASA, CSSA, SSSA: Madison, WI, 1999; 213–264.
6. Cavelaars, J.C.; Vlotman, W.F.; Spoor, G. Subsurface Drainage Systems. In *Drainage Principles and Applications*; Ritzema, H.P., Ed.; ILRI: Wageningen, The Netherlands, 1994; 827–930.
7. *ASAE Standards*; EP481 Construction of Subsurface Drains in Humid Areas, 47th Ed.; ASAE: St. Joseph, Mich., 2000.
8. Chieng, S. Use of Geographic Information Systems and Computer-Aided Design and Drafting Techniques for Drainage Planning and System Design. In *Agricultural Drainage*, Agron. Monogr. 38; Skaggs, R.W., van Schilfgaarde, J., Eds.; ASA, CSSA, SSSA: Madison, WI, 1999; 893–910.
9. Schwab, G.O.; Fouss, J.L. Drainage Materials. In *Agricultural Drainage*, Agron. Monogr. 38; Skaggs, R.W., van Schilfgaarde, J., Eds.; ASA, CSSA, SSSA: Madison, WI, 1999; 911–926.

Leaf Water Potential

A. J. Karamanos
Agricultural University of Athens, Athens, Greece

INTRODUCTION

Leaf water potential is the thermodynamic expression of the water status of leaves, which is widely used in plant science research in the last 40 years. In this article, a description of the physical principles leading to the definition of the leaf water potential and its component potentials is initially given. Then, the way of development of water deficits in leaves and the compartmentation of the water and its component potentials in leaf cells and tissues subjected to different degrees of water shortage are presented. Some useful implications for the development of adaptive mechanisms to drought result from this analysis. Finally, the existing methods for the determination of leaf water potential are presented and classified in groups according to their basic principles.

DEFINITIONS

According to Slatyer and Taylor,[1] the state of water in any system is expressed thermodynamically in terms of its chemical potential (μ_w) or the partial molal Gibbs free energy ($\bar{G}_w$) as follows:

$$\mu_w = \bar{G}_w = \left(\frac{\partial G}{\partial n_w} \right)_{T,P,n_s} \tag{1}$$

where G is the Gibbs free energy of the system (a function of its internal energy) and n_w the number of water moles. Thus, μ_w or ($\bar{G}_w$) are defined more simply as the change that occurs in the energy content of a system for a given change in the number of moles of water in the system at constant temperature (T), pressure (P), and solute content (n_s).

The water potential of a system (Ψ) is defined as:

$$\Psi = \frac{\mu_w - \mu_w^0}{\bar{V}_w} \tag{2}$$

where μ_w^0 is the chemical potential of pure free water and $\bar{V}_w$ the partial molal volume of water ($18 \, cm^3 \, mole^{-1}$). Ψ is expressed as energy per unit volume, which is dimensionally equivalent to pressure. Thus, traditional pressure units (e.g., bar, MPa, etc.) are used for the expression of Ψ. The concept of water potential can be realized as a measure of the capacity of water at a point in the system to do work in comparison with the work capacity of pure free water (μ_w^0) which is taken arbitrarily as zero. Accordingly, Ψ provides a unified measure for the energy status of water at any place within the soil–plant–atmosphere continuum.

Since the energy status of water results from the interaction of forces of different origin exerted on water molecules, Ψ is thought to consist of several components each one representing a different kind of the forces involved. *The effect of solutes*, which are very common in any aqueous system, is expressed as the osmotic component of Ψ, the osmotic or solute potential (Ψ_s). The solutes lower the vapor pressure, and, hence, the potential energy of water in the system. Thus, Ψ_s takes values below zero, which are proportional to the concentration of osmotic substances ($c_s = n/V$, where n = number of moles and V = volume):

$$\Psi_s = -RTc_s = -\pi \tag{3}$$

where R is the gas constant, T the absolute temperature, and π the osmotic pressure.

External pressure, either above or below the value of the local atmosphere, increases or decreases, respectively, the water potential. Its effect is described in terms of the pressure potential (ψ_p), which usually takes values above zero.

Forces arising both *at the liquid–air and at the solid–liquid interfaces* are responsible for another component of Ψ, the matric potential (ψ_m). They are considered to arise in systems rich in matrix, i.e., with large surface to volume ratios (e.g., soils, plant cell walls, gels, and colloids). Water is bound by matrix with forces related to its water content. Thus, the capacity of water to do work is reduced, and, hence, ψ_m takes negative values.

It follows that, at any time, Ψ is a function of ψ_s, ψ_p, and ψ_m:

$$\Psi = f(\psi_s, \psi_p, \psi_m) \tag{4}$$

However, Eq. 4 is usually described, in a simplified way, as an algebraic sum:

$$\Psi = \psi_p + \psi_s + \psi_m \tag{5}$$

A further component due to gravity, called gravitational

Encyclopedia of Water Science
DOI: 10.1081/E-EWS 120010169

579

potential (ψ_g) might also be added to Eqs. 4 and 5 only in specific cases (e.g., in tall trees or when reverse water flow is considered).

WATER IN THE LEAVES

Water enters the leaves via the vascular system of the petiole and it flows through the veins of the lamina down to the vein endings which are embedded in the mesophyll. Lateral water movement towards neighboring mesophyll tissue along the xylem tissue of the veins also occurs to a substantial extent. The specific important characteristic of the vascular system of the leaf is its close spatial relation to the mesophyll.[2] According to Kramer,[3] the actual distribution of water in the mesophyll occurs chiefly from the smaller veins, which are so numerous that most cells of a leaf are only a few cells away from a vein or vein ending.

Fig. 1 shows diagrammatically the pathway of water from a xylem vessel to the evaporating surfaces in a substomatal cavity through a mesophyll cell. Water moves mainly through the apoplast (cell walls and intercellular spaces) where the resistance (r_w) is minimal. Water exchanges (f) from and to the vacuole (symplastic pathway) are more difficult because of the resistance of the two cell membranes (plasmalemma and tonoplast) to be crossed (r_v). The water flux can be represented by an electric analog (Fig. 1(b)) where the apoplast exhibits mainly conductive properties

(appearing as the resistance r_w), whereas the symplast both conductive (r_v) and storage properties represented as the capacitance c_v.

At any time, leaf water balance results from the relative magnitudes of the water supply through the roots and stem and the water loss through transpiration. To understand the mode of development of leaf water deficits we can consider the simple physical model proposed by Dixon in 1938 as quoted by Weatherley[4] (Fig. 2). In this model, the plant is regarded as consisting of two porous pots filled with water and connected by a tube through a manometer. The lower pot representing the root is semipermeable and buried into the soil. The upper pot indicates the transpiring leaf surface; the manometer represents the vacuoles of the mesophyll cells, which are off the main pathway since water moves mainly through the apoplast. Once transpiration starts, water evaporates quickly from the upper pot and a negative tension is transmitted through the tube to the lower pot resulting in water absorption from the soil. The existence of a considerable root resistance induces a lag of absorption behind transpiration, and water absorption cannot meet instantaneously transpirational fluctuations. Thus, water will be drawn from the vacuoles of the mesophyll cells and the manometer will register the tension in the system. The leaf water potential (Ψ_l) is equivalent to the tension registered by the manometer and proportional to the amount of water drawn from the cells.

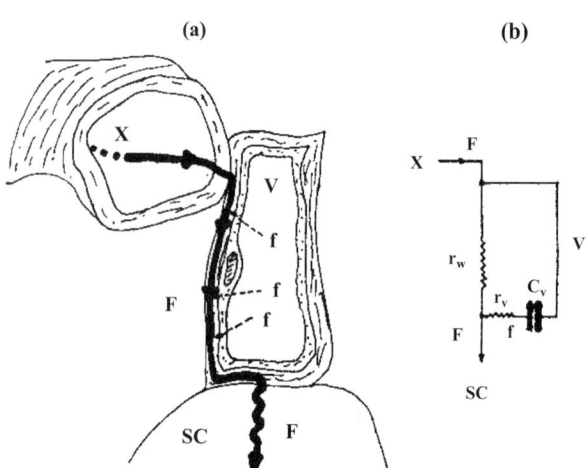

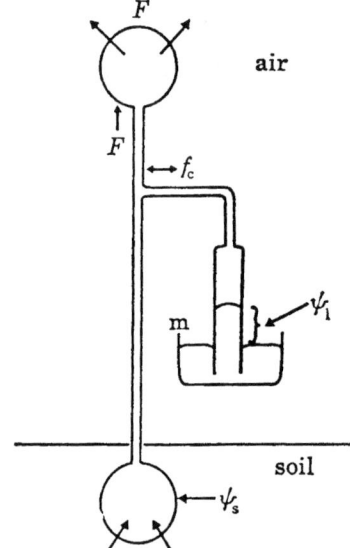

Fig. 1 (a) Schematic representation of the water pathway from a xylem vessel (X) to the evaporating surfaces at the substomatal cavity (SC) through a mesophyll cell. (b) A simplified electric analog of the water movement in (a). F: main flow through the cell wall (W), f: secondary flow from the vacuole (V), r_w: apoplast resistance, r_v: symplast resistance, c_v: symplast capacitance.

Fig. 2 Dixon's simplified model of a transpiring plant. The upper and lower porous pots represent the leaves and roots, respectively, while the manometer (m) represents the mesophyll cells. The negative pressure registered by the manometer is equivalent to the leaf water potential (Ψ_l) which fluctuates in response to lateral water movement (f_c) from and to the cells. F denotes the main transpiration stream. (From Ref. [4].)

Accordingly, the fluctuations in Ψ_l arise from mesophyll-water exchanges with the main transpiration stream.

WATER POTENTIAL AND ITS COMPONENTS IN LEAVES

The Osmometer Concept

In order to study leaf water relations, it is necessary to consider a leaf as consisting of mature parenchyma cells. Mature cells consist of three distinct phases: an elastic cell wall, the parietal cytoplasm with the nucleus and the organelles, and a central vacuole containing a dilute aqueous solution of sugars, ions, organic acids etc. (Fig. 3(b)). The vacuole occupies about 80%–90% of the cell volume and is surrounded by a semipermeable membrane, the tonoplast. It is, therefore, reasonable to consider, as a first approximation, that cell water exchanges are controlled by the vacuole, which behaves as an osmometer. The adoption of the osmometer-concept for mature leaf parenchyma cells presupposes that the

contribution of matrix to Ψ_l is negligible (i.e., $\Psi_m = 0$). Thus, Eq. 5 becomes:

$$\Psi_l = \psi_p + \psi_s \qquad (6)$$

The solute potential (ψ_s) is determined by the concentration of the osmotically active substances in the vacuole. In leaf cells, ψ_s always takes negative values which vary with cell volume: they are least negative in fully hydrated cells (maximum volume) and more negative in dehydrated ones (smaller volume) and they are supposed to follow the Boyle–van't Hoff relationship:

$$\psi_s V = \text{constant} \qquad (7)$$

It follows that a linear relationship is expected either between ψ_s and $1/V$ or between $1/\psi_s$ and V.

The changes in cell volume are also responsible for the development of the *pressure potential* (ψ_p). When water enters the cell, the vacuolar volume increases, and a pressure, called turgor pressure, is exerted on cell wall. At the same time, a pressure equal to turgor pressure is developed at the opposite direction, namely, from the wall to cell interior. This latter pressure, called wall pressure, acts like a hydrostatic pressure, increases the energy status of water in the cell by raising the pressure exerted on it

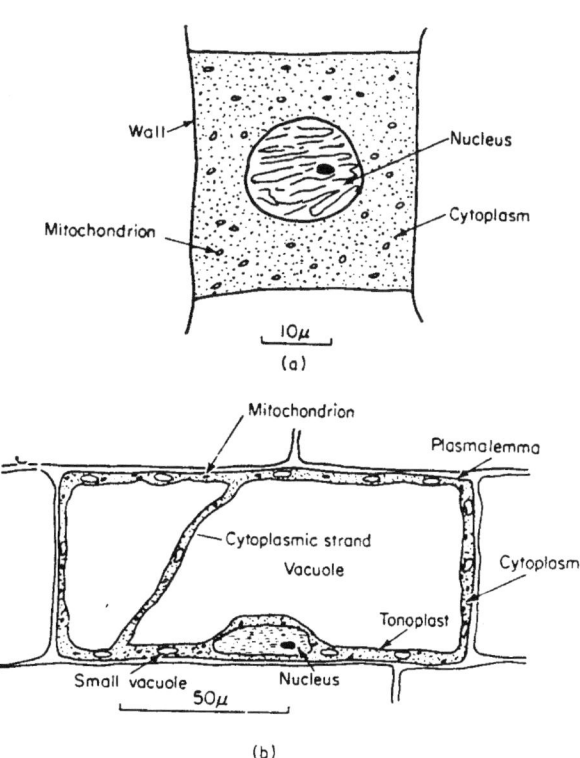

Fig. 3 Diagrammatic representation of (a) a meristematic, and (b) a mature parenchyma cell from higher plants. The large vacuole of the mature cell controls its water exchanges in a manner close to that of an ideal osmometer. The same may not apply to the meristematic cell where vacuolation is small. (From Ref. [5], p. 129.)

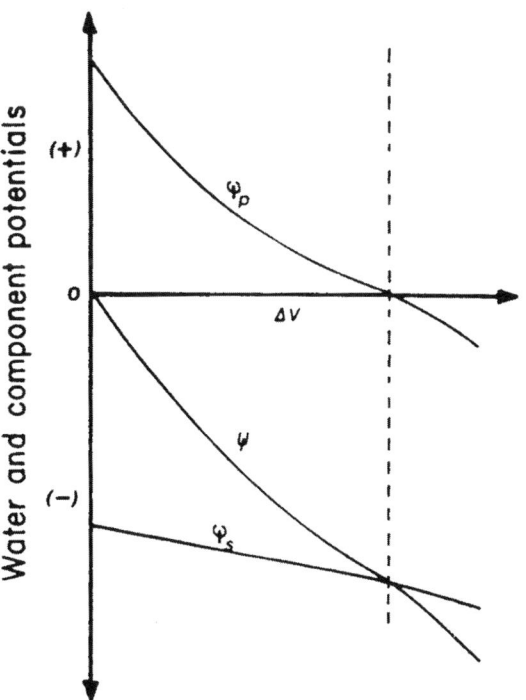

Fig. 4 The relationship between the volume of water lost (ΔV) and the water (Ψ) and its component potentials, solute (ψ_s), and pressure potential (ψ_p) for a cell or tissue showing ideal osmotic behavior. The dashed line indicates the point of zero turgor (incipient plasmolysis) (From Ref. [6].)

above the local atmosphere, and represents the cell pressure potential (ψ_p). Obviously, ψ_p takes positive values as long as the vacuole exerts a pressure on the surrounding wall. When cells lose water, the vacuole shrinks progressively with a concomitant fall in cell turgor and ψ_p. The relationship between cell volume and ψ_p is curvilinear and depends on the elastic properties of the cell wall.

Fig. 4 shows the relationship between Ψ and its component potentials and the changes in cell volume (Hoefler diagram). When the cell achieves its maximum hydration (full turgor) Ψ bears its maximum value (zero) because $|\psi_p| = |\psi_s|$. As cell hydration falls, Ψ drops curvilinearly to more negative values, and at incipient plasmolysis $\psi_p = 0$ and $\Psi = \psi_s$. From this point onwards Ψ is exclusively determined from the changes in ψ_s: it falls linearly with any further dehydration according to Eq. 7.

Fig. 5 shows the relations between $1/\Psi_l$ against water deficit, known as pressure–volume curves[7] for leaves from different plant species. The curvilinear (turgor component) and the straight line (solute component) portions of the relationship are evident in all cases.

The diagram of Fig. 4 leads to two ecologically important conclusions. First, in view of the great significance of cell turgor to many physiological processes,[8] the maintenance of ψ_p above zero at relatively high levels of leaf dehydration should be beneficial for plants growing in arid regions. This can be achieved by means of a more elastic cell wall, which makes the fall of ψ_p with increasing dehydration less abrupt. Secondly, it is possible that an accumulation of osmotically active substances takes place in the vacuole. This leads to a drop of ψ_s to values more negative than those expected by a simple volume reduction caused by cell dehydration. This solute accumulation in cells subjected to water stress constitutes an adaptive mechanism known as *osmotic adjustment* or *osmoregulation*. Osmotic adjustment has been detected in many plant species[9] and acts in two ways: 1) it enables cells to lose more water before their turgor drops to zero; 2) it increases the ability of cells to absorb water under dry conditions by lowering the cell water potential and thus maintaining a potential gradient between plant cells and their medium, necessary for water absorption.

The Effects of Cell Matrix

As stated before, matric effects arise in systems rich in substances with large surface to volume ratios. At the cellular level, matrix is present in the cell walls in the form of interwoven cellulose microfibrils, and in the cytoplasm as the various gels and colloids. The real nature of ψ_m has been the subject of many discussions among specialists. Initially, ψ_m was thought to be the result of forces retaining water molecules by capillarity, adsorption, and hydration.[5,10] Tyree and Karamanos,[11] based on a physicochemical study of the forces arising near solid phases with substantial surface charge densities, identified ψ_m as the energy of interaction of the water dipoles with

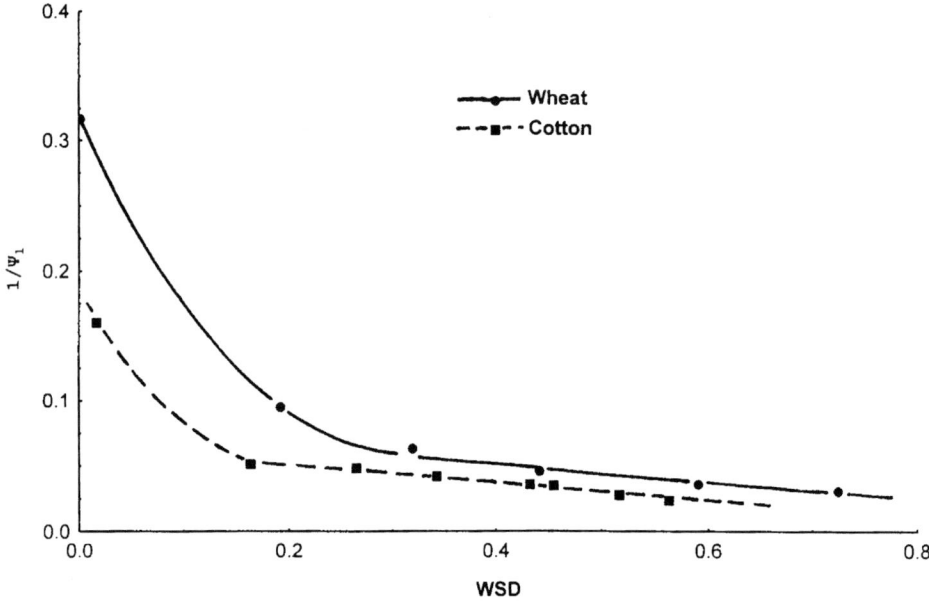

Fig. 5 The relationship between the inverse of the leaf water potential ($1/\Psi_l$) and the leaf water saturation deficit (WSD) (pressure–volume curves) for leaves of wheat and cotton plants. No obvious deviations in the straight-line portions of the curves at high levels of dehydration are detectable.

the electric field in the double layer,[12] and separated the effects of surface tension within the micropores as belonging to forces of a different nature (i.e., negative pressure). On this account, ψ_m exerts its major impact in cell walls within a very short distance from the charged surfaces and its contribution is minor in comparison to both ψ_p and ψ_s.

The osmometer approach assumes that the effects of cell matrix on tissue water exchanges are negligible and, thus, ψ_m equals zero. This assumption might be almost true in fully vacuolated parenchyma cells (Fig. 3(b)), because the vacuole contains no matrix. In meristematic cells, however, the situation could not be so clear because of the poor vacuolation and the large volume fraction of cytoplasm (Fig. 3(a)). Within the cytoplasm ψ_m is important only in the matrix double layers which are expected to be less abundant in comparison with the cell wall. Since we have already concluded that ψ_m influences the state of only a small fraction of the total cell wall water, the overall effect in the cytoplasm is probably even smaller.[11] Nevertheless, no data on water exchanges of growing leaves are available to test this hypothesis.

A further complication to the osmometer approach is the question concerning the role of the apoplast as a water reservoir of cells and tissues. There is evidence that a high apoplastic water content is a common feature of xerophytes.[13,14] On this account, apoplastic water could compensate for any water loss from the symplasm. Such an assumption does not seem to be valid for several reasons. First, no systematic deviations along the pressure–volume curves were detected, even in drought adapted species[15] and at high levels of dehydration (Fig. 5), an indication that the osmometer model holds satisfactorily. Secondly, the values of leaf relative water content seldom fall below 50% in the most severe cases of natural dehydration. Accordingly, significant amounts of not easily extractable water are still retained in leaves suffering from intense water stress. Thirdly, the rate of net water loss from leaves accounts for less than 5% of the evapotranspiration rate of plants during daylight hours.[16] Thus, no part of the leaf water content could ever act as an effective reservoir for water, especially cell apoplast which functions as the pathway for free water movement from the xylem to the evaporating surfaces when leaf is transpiring (Fig. 1).

In conclusion, cell matrix does not seem to play a detectable role in leaf water relations. The osmometer concept holds satisfactorily, an indication that the cell vacuole is the key-site which regulates the water status of mature leaves.

METHODS OF MEASURING LEAF WATER POTENTIAL

A relatively large number of methods for determining leaf water potential have been used. The existing methods can be classified into three basic groups: 1) compensation

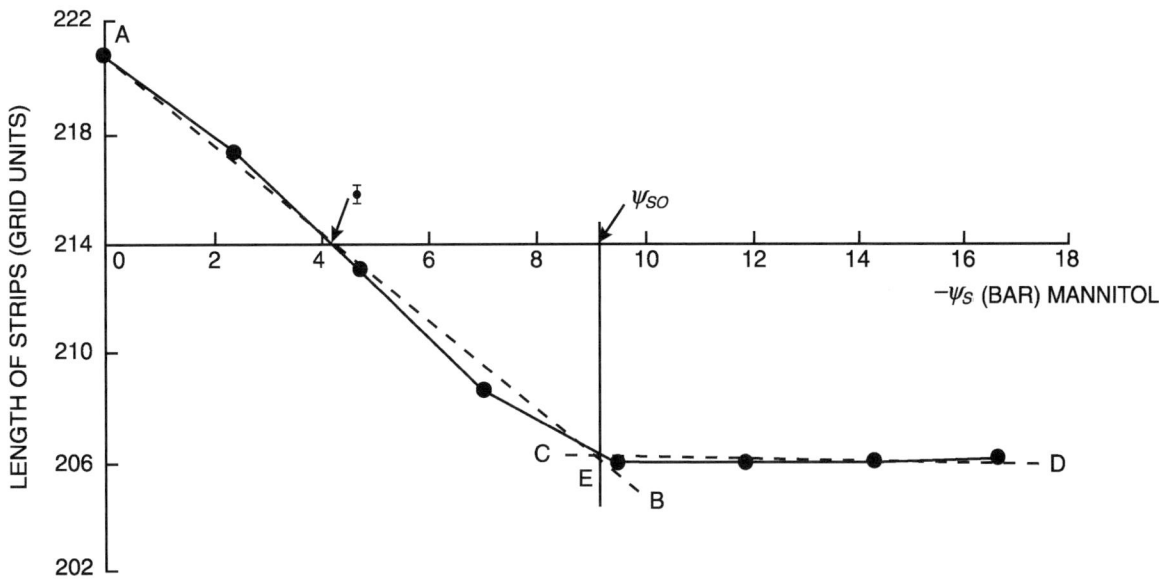

Fig. 6 Determination of the leaf water potential of fava bean leaves by the change in length of leaf strips floated on a series of mannitol solutions of different osmotic potentials (Ψ_s). The leaf water potential (Ψ) coincides with the Ψ_s of the solution where the initial length of the strip (214 grid units) remained unchanged. The osmotic potential at zero turgor (Ψ_{s0}) can also be traced on the axis of mannitol osmotic potentials from the point E of intersection of lines AB and CD. (From Ref. [20].)

methods; 2) psychrometric methods; and 3) pressure chamber method. A thorough description of these techniques is given in the works of Slavìk[17] and Wiebe et al.[18]

Compensation Methods

In the compensation methods, we search for the solution of a known osmotic potential which is isotonic to the leaf water potential of the sample. A set of uniform parallel leaf tissue samples (discs, strips, etc.) are floated in a series of graded test solutions of known osmotic potential and left for equilibration.[19] The net water transfer between the tissue and the solution depends on the relative magnitudes of Ψ_l and the ψ_s of the solution and is manifested as a change either in the weight or the size (length, thickness,

or area) of the sample. The isotonic solution is usually detected by interpolation. Fig. 6 shows the determination of Ψ_l by the change in length of leaf strips. This technique can also give the ψ_s of the tissue at zero turgor.

The test solutions to be used must not: 1) be harmful to the tissues; 2) penetrate through cell membranes; and 3) be metabolized by plants. Mannitol, polyethylene glycol, and sucrose are the most common osmotica, but none of them completely fulfils all the requirements set above.

An alternative method is the equilibration of the leaf samples in the gaseous phase. A parallel set of samples is left to equilibrate in closed vessels over a series of graded test solutions.[21] The direction and relative rate of the water vapor transfer between the sample and the solution is then determined by measuring either the sample weight or the volume of the osmotic solution.

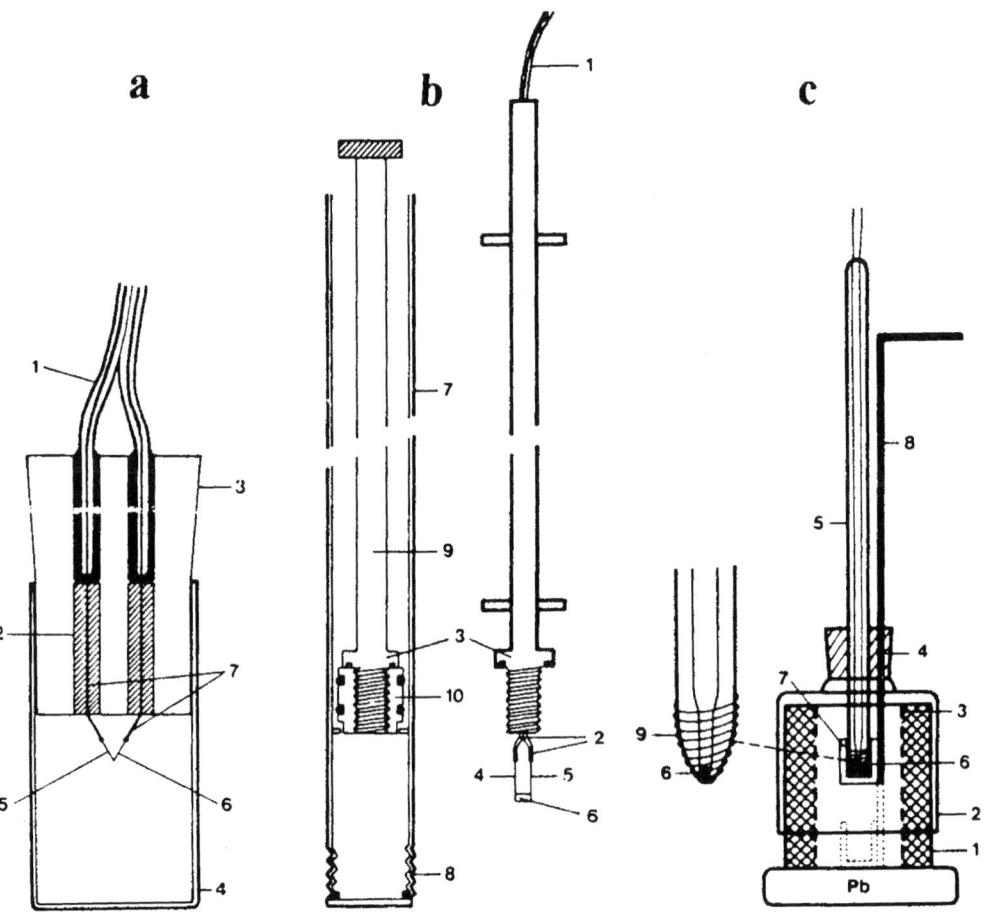

Fig. 7 Different types of thermocouple psychrometers. (a) A type. 1: wires, 2: copper rods, 3: rubber stopper, 4: measuring chamber, 5 and 6: chromel–constantan wires, 7: copper wires. (b) B type. 1: wire, 2: single wire, 3: brass flange, 4 and 5: chromel–constantan wires, 6: silver ring, 7: brass tube, 8: stainless steel cap, 9: brass rod closing the measuring chamber during equilibration, 10: brass piston with O-rings. (c) C type. 1: glass vessel, 2: rubber cap, 3: copper net cylinders with insulation, 4: rubber stopper, 5: glass tube, 6: thermistor, 7: plastic "spoon" filled with water, 8: holder of the plastic "spoon," 9: wetting thread, Pb: copper plate Pb—lead weight. (From Ref. [17], Figs. 1.14, 1.18, and 1.21.)

All compensation methods are time-consuming, temperature-dependent, laborious, and of relatively low accuracy (from 1 MPa to 0.3 MPa).

Psychrometric Methods

These methods measure water potential by determining the wet bulb depression in a closed gaseous system which is in equilibrium with the leaf sample. The wet bulb depression depends on the relative humidity of the air in the system, which in turn is related to Ψ as follows:

$$\Psi = \left(\frac{RT}{\bar{V}_w}\right)\ln\frac{e}{e_o} \tag{8}$$

where e/e_o is the relative humidity.

There are three kinds of psychrometers for measuring the wet bulb depression. In the first (type A, Fig. 7(a)), a thermocouple junction (chromel and constantan wires) is used alternatively as wet and dry: the output of the thermocouple is read first when the junction is dry, then condensation of a fine water droplet is achieved by Peltier cooling.[22] The thermocouple functions as a wet junction as long as water remains on its surface, and the difference between the two readings is equivalent to the wet bulb depression.

In the second type (type B, Fig. 7(b)), the output of a thermocouple with the thermojunction permanently wetted by a small drop of pure water is measured.[23]

The droplet of water is held on a small silver ring supported by thin chromel and constantan wires. The diffusion flux of water from the junction to the leaf-tissue sample serving as a vapor sink is measured. In another version of this type, an additional similar dry thermocouple is included in the sample chamber and measured as reference.[24]

In the third type (type C, Fig. 7(c)), temperature sensitive resistance units (thermistors) are used instead of thermocouples.[25] The dry bulb temperature is measured with a thermistor to which a miniature chamber filled with water or wetted filter paper can be tightly attached. The wet bulb temperature is then measured by the same thermistor after temporarily lowering the wet chamber: water starts evaporating from the thermistor at a rate depending on the vapor pressure in the sample chamber, which is in equilibrium with the water potential of the plant material.

The readings from all types of psychrometers are calibrated against salt or sucrose solutions of known ψ_s. Leaf tissue segments (punched discs or strips) are used as samples. However, specially designed psychrometers or hygrometers are also used for the in situ measurement of the water potential of attached leaves (see Ref. 18 for a review).

Psychrometric techniques are extremely temperature-sensitive. Nevertheless, their mean accuracy is very high (up to ± 0.01 MPa).

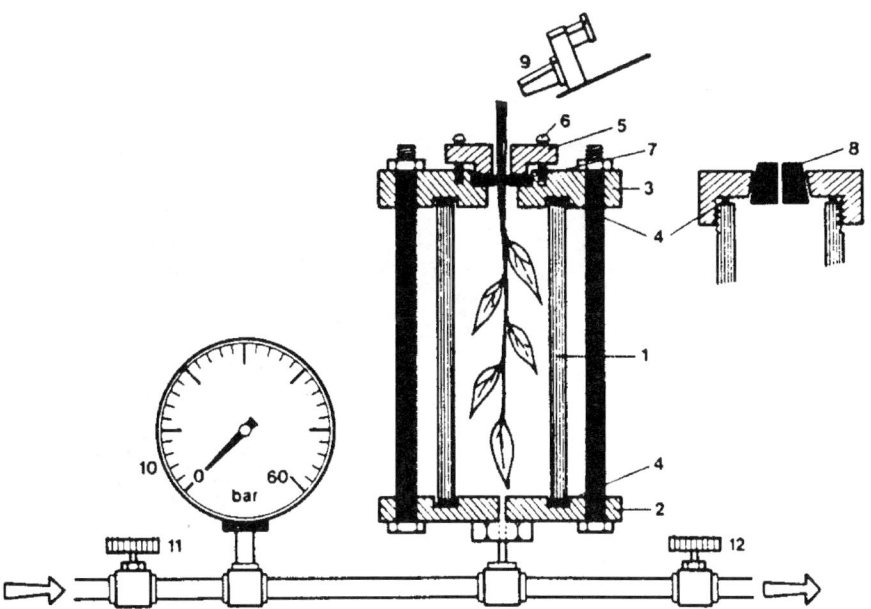

Fig. 8 Diagrammatic representation of a pressure chamber apparatus. 1: cylinder, 2: lower cover, 3: upper cover, 4: O-rings, 5: insertion held with four screws (6) used to seal the stem by means of an O-ring (7), 8: rubber stopper, 9: binocular microscope, 10: pressure gauge, 11: inlet valve, 12: outlet valve. (From Ref. [17], Fig. 1.33.)

The Pressure Chamber Method

The pressure chamber (or pressure bomb) was first used extensively by Schollander et al.[26] A cut leaf is inserted within a cylinder with its petiole protruding from the lid, so that it can be observed for sap exudation (Fig. 8). The chamber is then hermetically sealed and the pressure inside is gradually increased by compressed air or nitrogen until sap appears at the xylem vessels on the cut surface. According to the theoretical analysis of the technique,[7,27,28] the water potential in the xylem vessel (Ψ_x) is dominated by the negative hydrostatic pressure (tension) $\psi_{p,x}$ caused by the transpiration stream. Both solute ($\psi_{s,x}$) and matric components ($\psi_{m,x}$) in the xylem are negligible.[26] It follows that:

$$\Psi_x \cong \psi_{p,x} \qquad (9)$$

The pressure required to force the sap to the cut petiole (P) compensates the original negative pressure in the intact xylem vessels ($\psi_{p,x}$), so that:

$$-P = \psi_{p,x} = \Psi_x \qquad (10)$$

Ψ_x then equals Ψ_l when single leaves are used.

The technique is quick, easy, and quite accurate (± 0.02 MPa) provided that some precautions are taken: 1) an appropriate rate of pressure increase is applied; 2) the water loss from the leaf during determination is minimal; 3) the pressure gauge is as accurate as possible.

In addition, this technique is very useful for producing pressure–volume curves[7] which offer useful and reliable information on the mechanisms involved in the regulation of leaf water status.

THE WATER POTENTIAL INDEX

At any time, leaf water potential is the combined result of interactions of soil water availability, evaporative demand, and plant responses. Accordingly, it can be considered as a reliable indicator of leaf water balance. Karamanos and Papatheohari[29] suggested a method to assess the water stress history experienced by a plant or crop by using serial values of Ψ_l over an observation period (Fig. 9).

The integral of the course of Ψ_l over time (i.e., the shaded area in Fig. 9) describes the "duration" of Ψ_l [water potential duration (WPD), in MPa days]:

$$\mathrm{WPD} = \int_{i=1}^{v} \Psi_{l,i}\,\mathrm{d}t \qquad (11)$$

where $\Psi_{l,i}$ is the leaf water potential at day t within the observation period, i.e., from days 1 to v. In order to make the values of WPD comparable among observation periods of different duration, the Water Potential Index (WPI) is derived by dividing WPD by the length of the period of study:

$$\mathrm{WPI} = \mathrm{WPD}/n \qquad (12)$$

where n is the length of a period in days.

The use of the WPI as an objective indicator of the total water stress experienced by plants in a given environment

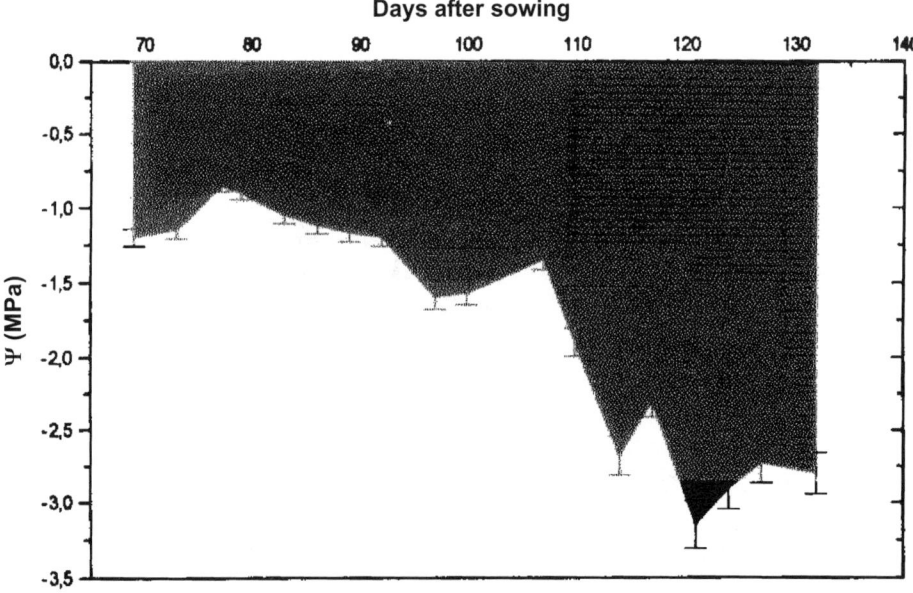

Fig. 9   The time course of the leaf water potential (Ψ_l) for the wheat cultivar Yecora grown in Athens under rainfed conditions. The shaded area indicates the WPD and the vertical bars, the standard errors of the means. (From Ref. [29].)

looks promising in genotype evaluation studies for drought stress resistance.[29]

REFERENCES

1. Slatyer, R.O.; Taylor, S.A. Terminology in Plant- and Soil-Water Relations. Nature **1960**, *187*, 922–924.
2. Esau, K. *Plant Anatomy*; John Wiley and Sons, Inc: New York, 1965; 767.
3. Kramer, P.J. *Plant and Soil Water Relationships. A Modern Synthesis*; McGraw-Hill: New York, 1969; 482.
4. Weatherley, P.E. Water Movement Through Plants. Philos. Trans. R. Soc. Lond. **1976**, B*273*, 435–444.
5. Slatyer, R.O. *Plant–Water Relationships*; Academic Press: London, 1967; 366.
6. Weatherley, P.E. Water in the Leaf. *The State and Movement of Water in Living Organisms*; Cambridge University Press: Cambridge, 1965; 157–184.
7. Tyree, M.T.; Hammel, H.T. The Measurement of the Turgor Pressure and the Water Relations of Plants by the Pressure Bomb Technique. J. Exp. Bot. **1972**, *23*, 267–282.
8. Hsiao, T.C. Plant Responses to Water Stress. Annu. Rev. Plant Physiol. **1973**, *24*, 519–570.
9. Morgan, J.M. Osmoregulation and Water Stress in Higher Plants. Annu. Rev. Plant Physiol. **1984**, *35*, 299–319.
10. Dainty, J. Water Relations of Plant Cells. Adv. Bot. Res. **1963**, *1*, 279–326.
11. Tyree, M.T.; Karamanos, A.J. Water Stress as an Ecological Factor. In *Plants and Their Atmospheric Environment*; Grace, J., Ford, E.D., Jarvis, P.G., Eds.; Blackwell Scientific Publications: Oxford, 1981; 237–261.
12. Bolt, G.H.; Miller, R.D. Calculation of Total and Component Potentials of Water in Soil. Trans. Am. Geophys. Union **1958**, *39*, 917–928.
13. Gaff, D.F.; Carr, D.J. The Quantity of Water in the Cell Walls and Its Significance. Aust. J. Biol. Sci. **1961**, *14*, 299–311.
14. Cutler, J.M.; Rains, D.W.; Loomis, R.S. Role of Changes in Solute Concentration in Maintaining Favourable Water Balance in Field-Grown Cotton. Agron. J. **1977**, *69*, 773–779.
15. Richter, H.; Duhme, F.; Glatzel, G.; Hinckley, T.M.; Karlic, H. Some Limitations and Applications of the Pressure–Volume Curve Technique in Ecophysiological Research. In *Plants and Their Atmospheric Environment*; Grace, J., Ford, E.D., Jarvis, P.G., Eds.; Blackwell Scientific Publications: Oxford, 1981; 263–272.
16. Weatherley, P.E. Some Aspects of Water Relations. Adv. Bot. Res. **1970**, *3*, 171–206.
17. Slavìk, B. *Methods of Studying Plant Water Relations*; Chapman and Hall Ltd: London, 1974; 449.
18. Wiebe, H.H.; Campbell, G.S.; Gardner, W.H.; Rawlins, S.L.; Cary, J.W.; Brown, R.W. *Measurement of Plant and Soil Water Status*, Bull. 484; Utah Agricultural Experiment Station: Logan, Utah, 1971; 71.
19. Ursprung, A.; Blum, G. Eine Neue Methode zur Messung der Saugkraft von Hartlaub. Jahrb. Wiss. Bot. **1927**, *67*, 334–348.
20. Kassam, A.H. Determination of Water Potential and Tissue Characteristics of Leaves of *Vicia faba* L. Hort. Res. **1972**, *12*, 13–23.
21. Slatyer, R.O. The Measurement of Diffusion Pressure Deficit in Plants by a Method of Vapour Equilibration. Aust. J. Biol. Sci. **1958**, *11*, 349–365.
22. Spanner, D.C. The Peltier Effect and Its Use in the Measurement of Suction Pressure. J. Exp. Bot. **1951**, *2*, 145–168.
23. Richards, L.A.; Ogata, G. Thermocouple for Vapour Pressure Measurement in Biological and Soil Systems at High Humidity. Science **1958**, *128*, 1089–1090.
24. Kramer, P.J. *Measurement of Water Potential and Osmotic of Leaf Tissue with a Thermocouple Psychrometer*; Department of Botany, Duke University: Durham, N.C., 1967.
25. Kreeb, K. Untersuchungen zu den Osmotischen Zustangroessen. I. Ein Tragbares Elektronisches Mikrokryoskop fur Oekophysiologische Arbeiten. Planta **1965**, *65*, 269–279.
26. Schollander, P.F.; Hammel, H.T.; Hemingsen, E.A.; Bradstreet, E.D. Hydrostatic Pressure and Osmotic Potential in Leaves of Mangroves and Some Other Plants. Proc. Natl Acad. Sci. U.S.A. **1964**, *52*, 119–125.
27. Waring, R.H.; Cleary, B.D. Plant Moisture Stress: Evaluation by Pressure Bomb. Science **1966**, *155*, 1248–1254.
28. Ritchie, G.A.; Hinckley, T.M. The Pressure Chamber as an Instrument for Ecological Research. Adv. Ecol. Res. **1975**, *9*, 165–254.
29. Karamanos, A.J.; Papatheohari, A.Y. Assessment of Drought Resistance of Crop Genotypes by Means of the Water Potential Index. Crop Sci. **1999**, *39*, 1792–1797.

Livestock and Poultry Production, Water Consumption for

David B. Parker
Michael S. Brown
West Texas A&M University, Canyon, Texas, U.S.A.

INTRODUCTION

Competition for drinking water has increased in many locations across the world. Groundwater continues to decline in many areas. Water once used for agriculture is now being directed to municipal and industrial uses. There is a trend away from the family-sized farm or ranch to fewer large-sized animal production operations called concentrated animal feeding operations (CAFOs). Water consumption data are a necessary component in the design of a drinking water supply system for new livestock and poultry production operations.

Water is the primary constituent in the body of livestock and poultry, constituting 50%–80% of the live weight of the animal. Water serves as an essential solvent and plays a vital role in regulation of body temperature, lactation, digestion, elimination of waste products of digestion and metabolism, regulation of osmotic pressure, reproduction, transportation of sound, and vision.[1,2] Livestock and poultry fulfill their water needs by drinking or eating snow and/or ice, ingesting water contained in feed, and in minor amounts through metabolism (water produced by the oxidation of carbohydrate, fat, and protein). The primary avenues of water loss are urinary excretion, fecal excretion, perspiration, and respiration.[1]

When evaluating water requirements of livestock and poultry, it is important to distinguish between free water intake (FWI), which is water consumed by the animal by drinking, and total water intake (TWI), which is the sum of FWI and any water ingested in the feed. The amount of water ingested in processed feed and grazed forage can be substantial. For example, many "dry" feedstuffs contain 10%–14% water, while grazed forages and silages contain 60%–80% water. Free water intake is usually measured as the amount of water disappearance from a water trough. Because of spillage by "leisure" activities, water disappearance does not always equal the amount of water ingested or actual water requirement. However, water disappearance and FWI are assumed to be equal in most production situations.

WATER CONSUMPTION FOR LIVESTOCK

Cattle

Cattle can be divided into three groups: feedlot cattle, range cattle, and dairy cattle. Within these groups, a variety of factors may affect water consumption, including species or breed, type of diet, feed intake, rate and composition of gain, pregnancy, lactation, activity, and environmental conditions. One of the first references for daily water intake by dairy and beef cattle was published by Winchester and Morris.[3] Although the water intake values presented in this publication are often referenced, there are many things that have changed to alter water intake in cattle today, most notably breed characteristics, feeding management, and dietary ingredients. In this chapter, we have attempted to provide a concise summary of the latest in water consumption data available for current feeding and management conditions.

Feedlot Cattle

Most commercially fed cattle in the United States are housed in open, earthen-surfaced lots with 50–200 animals per pen. Until recently, few data were available from which to establish water requirements for feedlot cattle.[1] Parker et al.[4] monitored total water usage over a two-year period at a 50,000 animal beef cattle feedlot. Average daily water usage over the two year period was 40.9 L per animal, which included all water used for drinking, in the feedmill for steam flaking grain, and water used in overflow water troughs in the winter to prevent ice formation. In the winter, 66% of total usage was for drinking and in the summer 89% was used for drinking. A regression equation was developed to predict daily water usage for the entire feedlot:

$$\text{DFWU} = 39.2 - 0.648\,\text{MaxT} + 0.0421\,\text{MaxT}^2$$

$$- 0.0717\,\text{MinRH} \qquad (1)$$

$$R^2 = 0.60$$

Encyclopedia of Water Science
DOI: 10.1081/E-EWS 120010303

where DFWU is the daily feedlot water use (L per animal), MaxT, the maximum daily temperature (°C), and MinRH, the minimum daily relative humidity (%). Jeter[5] measured FWI by feedlot steers and developed the following equation:

$$DFWI = 40.61 + 0.46\,MinT - 0.45\,MinRH$$

$$R^2 = 0.93$$

(2)

where DFWI is the daily free water intake (L/day), MinT, the minimum daily temperature (°C), and MinRH, the minimum daily relative humidity (%).

Range Cattle

The livestock category of range cattle describes both diverse classes of cattle (nonlactating and lactating cows, growing steers and heifers, and mature bulls) and diverse types of grazing conditions (e.g., introduced forages and winter or spring cereal grains, native range) in which forage quality varies. The potential relationships between climatologic variables and the distance animals must travel to obtain forage and drinking water are poorly understood for grazing cattle. Although the majority of data available do not consider the water consumed in forage, it is well recognized that forage dry matter concentration is dynamic and dependent on environmental influences.

Winchester and Morris[3] summarized total water intake data from six studies generally involving various classes of cattle that were individually fed and housed in environmental chambers for periods of 7 day–14 day (Table 1). Although these data were primarily derived under "laboratory" conditions, few data from production studies are available. Kattnig et al.[6] reported that FWI by 234 kg Holstein steers fed hay in individual pens was 91 mL/kg of body weight; average daily maximum temperature was 20.7°C. Ojowi et al.[7] conducted an 84 day grazing study by using growing steers (308 kg) gaining 0.9 kg/day and indicated that FWI averaged 94 mL/kg of body weight. Ali et al.[8] monitored ambient temperature, relative humidity, and FWI of grazing cattle (cows with calves, heifers, and a mature bull) during an 84 day grazing period. The average daily maximum temperature was 22.2°C, and FWI averaged 108 mL/kg of body weight.

Dairy Cattle

Water is the most important dietary component for dairy cattle, as insufficient water can limit milk production.[9] Lactating dairy cows require about 4 L of water for each kg of milk produced.[10] Many regression equations have been developed to predict the amount of water consumed

Table 1 Total water intake by beef cattle with an ambient temperature of 4.4–32.2°C

| Class of cattle | Daily total water intake | |
	L/day	mL/kg of body weight
Bulls		
544 kg	28–66	51–121
816 kg	33–78	40–96
Nonlactating, pregnant cows[a]		
408 kg	25–37	61–91
500 kg	23–33	46–66
Lactation adjustment [added for each 1 kg of milk produced (4% milk fat)][b]		
	2.1–2.7	
Growing heifers and steers (average daily body weight gain of 0.4 kg/day–0.7 kg/day)		
180 kg	15–22	83–122
360 kg	24–35	67–97

Total water intake was determined to be constant below 4.4°C.
[a] Data were not determined above an ambient temperature of 21.1°C.
[b] Data were derived from lactating dairy cows with an ambient temperature of 4.4–32.2°C.
Source: (From Ref. [3].)

by dairy cows.[9] Dahlborn et al.[11] developed the following equation:

$$DFWI = 14.3 + 1.28\,MP + 0.32\,DM$$

(3)

where DFWI is the daily free water intake (drinking only), MP, the milk production (kg/day), and DM, the dry matter (% of diet). A summary of estimated water requirements for various classes of dairy cattle is presented in Table 2.

Swine

Weaning pigs (age 3 weeks–6 weeks) will drink about 0.5 L/day in the first week after weaning and 1.5 L at age 6 weeks.[13,14] Growing swine will consume about 2.5 L–3.0 L of water for each kg of feed consumed.[14] Pigs drink about the same at temperatures of 7–22°C, but the amount of water consumed increases considerably at 30°C.[15] A summary of estimated water requirements for various classes of swine is presented in Table 3.

Horses

Horses consume about 8.4 L per 100 kg of body weight. Horses need 2 L–3 L of water per kg of dry matter intake.[18] Fonnesbeck[19] determined that horses fed an all-hay diet resulted in a water-to-feed ratio of 3.6:1, while horses fed a hay–grain diet had a water-to-feed ratio of 2.9:1. Working horses may drink 20%–300% more water than horses at

Table 2 Water requirements for various classes of dairy cattle

Class	Production (kg milk/day)	Estimated water consumption (L/day)
Holstein calves	—	
(1 mo)	—	5–8
(2 mo)	—	6–9
(3 mo)	—	8–11
(4 mo)	—	11–13
Holstein heifers	—	
(5 mo)	—	14–17
(15 mo–18 mo)	—	22–27
(18 mo–24 mo)	—	28–36
Dry cows (pregnant)	—	26–49
Jersey cows	13.6	49–59
Guernsey cows	13.6	52–61
Ayrshire, Brown Swiss,	13.6	55–64
and Holstein cows	22.7	91–102
	36.3	144–159
	45.4	182–197

Source: (From Refs. [10,12].)

rest.[18] A summary of estimated water requirements for various classes of horses is presented in Table 4.

Sheep

The water required for sheep depends on the same factors as other livestock. In addition, the water requirements may vary with wool covering.[20] A summary of estimated water requirements for various classes of sheep is presented in Table 5.

Goats

A high proportion of the world's goat population lives in arid areas where water requirements are not easily met. Goats are one of the most efficient animals in the use of water, having

Table 3 Water requirements for various classes of swine

Class	Estimated water consumption (L/day)
Weanling pig	0.5–1.5
11 kg pig	1.9
27 kg pig	5.7
45 kg pig	6.6
90 kg pig	9.5
Gestating sow	17.0
Pregnant gilt	20.8
Sow plus litter	22.7
Boar	10–15

Source: (From Refs. [16,17].)

Table 4 Water requirements for various classes of horses

Class	Estimated water consumption (L/day)
Maintenance, 500 kg, thermoneutral environment	23–30
Maintenance, 500 kg, warm environment	30–57
Lactating mare, 500 kg	38–57
Working horse, 500 kg, moderate work	38–45
Working horse, 500 kg, moderate work, warm environment	45–68
Weanling, 300, thermoneutral environment	23–30

Source: (From Ref. [16].)

one of the lowest rates of water turnover per unit of body weight.[21] Data for water consumption by dairy and meat goats is limited. Dairy goats require about 1.43 L–3.5 L of water per kg of milk, significantly less than dairy cattle.[21] Penned meat goats drink about 0.7 L of water per day.

WATER CONSUMPTION FOR POULTRY

Chickens

Xin[22] developed the following equation for broilers between 1 day and 56 day of age:

$$DFWI = -2.78 + 4.70D + 0.128D^2 - 0.00217D^3$$
$$R^2 = 0.999 \tag{4}$$

where DFWI is the daily free water intake (L per 1000 birds) and D, the age in days. A summary of estimated water requirements for various classes of chickens is presented in Table 6.

Turkeys

Parker et al.[24] monitored water intake in tom turkeys at temperatures ranging from 10.0 to 37.8°C, with water

Table 5 Water consumption of various classes of sheep

Class	Water requirement (L/day)
Rams	7.6
Dry ewes	7.6
Ewes with lambs	11.3
5 lb–20 lb lambs	0.4–1.1
Feeder lambs	5.7

Source: (From Ref. [16].)

Table 6 Water requirements for various classes of chickens and turkeys

| | Water consumption (L per bird per week) | | | | |
| | | | | Large white turkeys | |
Age (weeks)	Broiler chickens	White leghorn hens	Brown egg-laying hens	Males	Females
1	0.22	0.20	0.20	0.38	0.38
2	0.48	0.30	0.40	0.75	0.69
4	0.10	0.50	0.70	1.65	1.27
6	0.15	0.70	0.80	2.87	2.15
8	0.20	0.80	0.90	4.02	3.18
10	—	0.90	1.00	5.34	4.40
12	—	1.00	1.10	6.22	4.66
14	—	1.10	1.10	6.68	4.70
16	—	1.20	1.20	6.92	4.74
18	—	1.30	1.30	7.00	—
20	—	1.60	1.50	7.04	—

Source: (From Ref. [23].)

consumption of 0.3 L/day and 1.3 L/day, respectively. A summary of estimated water requirements for various classes of turkeys as reported at commercial turkey production companies is presented in Table 6.

Ducks

Veltman and Sharlin[25] evaluated water consumption in White Pekin ducks between 14 day and 42 day of age. Ducks that were allowed access to water 24 hr per day consumed 0.8 L/day, while ducks provided access to water only 4 hr/day consumed 0.6 L/day.

REFERENCES

1. NRC. *Nutrient Requirements of Beef Cattle*, Update 2000, 7th Ed.; National Research Council, National Academy Press: Washington, DC, 2000; 80–84.
2. Okine, E. Water Requirements for Livestock. Alberta Agriculture, Food and Rural Development: Lacombe, Alberta, Canada. http://www.agric.gov.ab.ca/agdex/400/00716001.html (accessed August 2001).
3. Winchester, C.F.; Morris, M.J. Water Intake Rates of Cattle. J. Anim. Sci. **1956**, *15*, 722–740.
4. Parker, D.B.; Perino, L.J.; Auvermann, B.W.; Sweeten, J.M. Water Use and Conservation at Texas High Plains Beef Cattle Feedyards. Appl. Eng. Agric. **1998**, *16* (1), 77–82.
5. Jeter, M.B. Drinking Water Intake by Finishing Yearling Beef Steers M.S. Thesis, West Texas A&M University, Canyon, TX, 2001.
6. Kattnig, R.M.; Pordomingo, A.J.; Schneberger, A.G.; Duff, G.C.; Wallace, J.D. Influence of Saline Water on Intake,

7. Digesta Kinetics, and Serum Profiles of Steers. J. Range Manag. **1992**, *45* (6), 514–518.
7. Ojowi, M.O.; Christensen, D.A.; McKinnon, J.J.; Mustafa, A.F. Thin Stillage from Wheat-Based Ethanol Production as a Nutrient Supplement for Cattle Grazing Crested Wheatgrass Pastures. Can. J. Anim. Sci. **1996**, *76* (4), 547–553.
8. Ali, S.; Goonewardene, L.A.; Basarab, J.A. Estimating Water Consumption and Factors Affecting Intake in Grazing Cattle. Can. J Anim. Sci. **1994**, *74* (3), 551–554.
9. NRC, Water. *Nutrient Requirements of Dairy Cattle*, 7th Ed.; National Research Council. National Academy Press: Washington, DC, 2001; Chap. 8, 178–183.
10. Grant, R. *Water Quality and Requirements for Dairy Cattle*; Publication No. G93-1138-A, University of Nebraska Cooperative Extension: Lincoln, NE, 1993.
11. Dahlborn, K.; Akerlind, M.; Gustafson, G. Water Intake by Dairy Cows Selected for High or Low Milk-fat Percentage when Fed Two Forage to Concentrate Ratios with Hay or Silage. Swed. J Agric. Res. **1998**, *28*, 167–176.
12. NRAES. *Dairy Reference Manual*; NRAES-63, Northeast Regional Agricultural Engineering Service: Ithaca, NY, 1995; 144–145.
13. Almond, G.W. *How Much Water Do Pigs Need?* In Proceedings of the North Carolina Healthy Hogs Seminar, North Carolina State University, Greenville, NC, November 1, 1995; http://mark.asci.ncsu.edu/HealthyHogs/book1995/almond.htm (accessed August 2001).
14. NRC. *Nutrient Requirements of Swine*, 10th Revised Ed.; National Research Council, National Academy Press: Washington, DC, 1998; 90–93.
15. Mount, L.E.; Holmes, C.W.; Close, W.H.; Morrison, S.R.; Start, I.B. A Note on the Consumption of Water by the Growing Pig at Several Environmental Temperatures and Levels of Feeding. Anim. Prod. **1971**, *13* (3), 561–563.

16. Lardy, G.; Stoltenow, C. *Livestock and Water*, Publication No. AS-954; North Dakota State University Extension Service: Fargo, ND, 1999.

17. Straub, G.J.; Weniger, J.H.; Tawfik, E.S.; Steinhauf, D. The Effects of High Environmental Temperatures on Fattening Performance and Growth of Boars. Livest. Prod. Sci. **1976**, *3* (1), 65–74.

18. NRC. *Nutrient Requirements of Horses*; National Research Council, National Academy Press: Washington, DC, 1989; 30–31.

19. Fonnesbeck, P.V. Consumption and Excretion of Water by Horses Receiving All Hay and Hay–Grain Diets. J. Anim. Sci. **1968**, *27*, 1350–1356.

20. NRC. *Nutrient Requirements of Sheep*; National Research Council, National Academy Press: Washington, DC, 1985; 26–27.

21. NRC, *Nutrient Requirements of Domestic Animals, Number 15, Nutrient Requirements of Goats: Angora, Dairy, and Meat Goats in Temperate and Tropical Countries*; National Academy Press: Washington, DC, 1981; 9–10.

22. Xin, H. Feed and Water Consumption, Growth, and Mortality of Male Broilers. Poult. Sci. **1994**, *73* (5), 610–616.

23. NRC, *Nutrient Requirements of Poultry*, 9th Ed.; National Research Council, National Academy Press: Washington, DC, 1994; 15–17.

24. Parker, J.T.; Boone, M.A.; Knechtges, J.F. The Effect of Ambient Temperature upon Body Temperature, Feed Consumption, and Water Consumption, using Two Varieties of Turkeys. Poult. Sci. **1972**, *51* (2), 659–664.

25. Veltman, J.R.; Sharlin, J.S. Influence of Water Deprivation on Water Consumption, Growth, and Carcass Characteristics of Ducks. Poult. Sci. **1981**, *60* (3), 637–642.

Livestock, Water Harvesting Methods for

Gary W. Frasier
United States Department of Agriculture (USDA), Fort Collins, Colorado, U.S.A.

INTRODUCTION

Approximately, 40% of the world's land area is classified as rangeland with over 80% in the arid and semi-arid zones.[1] In these rangeland areas, there is usually sufficient water, primarily as precipitation, for plant growth in the form of grasses and small shrubs which are the primary foodstock for herbivores, both wildlife and domestic. While there is sufficient forage for the animals, many of these rangelands cannot be used for livestock production because of inadequate drinking water sources such as streams and springs. Traditionally, supplemental animal drinking water has been supplied by wells, ponds, and in some instances physical water transport. There are areas where even these supplemental water techniques are not available or otherwise unsuitable. One technique of water supply that can be used in most places in the world when other sources are unavailable is a process called water harvesting.

DEFINITION AND BACKGROUND

Water harvesting is defined as the collection of precipitation from a prepared area for some beneficial use. Water harvesting for livestock and human drinking water supplies is an ancient practice dating back to the first-half of the Bronze Age, about 4000 yr ago.[2] It is probable that the first water harvesting system for drinking water was nothing more than a simple depression that filled with water running off a rock surface. Even today, in many arid regions, we can find rain-filled depressions in rock outcroppings that provide drinking water for wildlife.

In the past 40 yr, there has been a renewed interest in water harvesting as a means of water supply for both livestock and domestic uses. There is no universally "best" method of water harvesting, since each site has its own unique characteristic features: soil type and topography; precipitation quantities and intensities; and water needs, timing, and quantities. The designer, installer, and ultimate user of a water harvesting facility should become as familiar as possible with the available techniques and adapt one that is best suited to the local environment, social, and economic conditions and site features. Many of the elements of a water harvesting facility are interrelated and must be considered simultaneously.

There is a considerable amount of technical literature, which describes or presents information concerning the various techniques of water harvesting. Unfortunately, much of this information is scattered in scientific or technical journals and proceedings of various meetings, and is written in a manner that is difficult to interpret for direct field application by farmers and technicians.[3]

WATER HARVESTING SYSTEM

All water harvesting facilities for livestock watering have the same basic components (Fig. 1). The collection area (catchment area) can be a natural hillside, a smoothed soil area, an area treated or covered to reduce water infiltration and increase surface runoff, or even the roofs of buildings. The collected water is stored in some container, pond, or tank until it is needed.

There are many ways the catchment area can be modified to reduce water loss by infiltration and increase the quantity of precipitation runoff. These can be separated into three general categories: 1) topography modification; 2) soil modifications; and 3) impermeable coverings or membranes. Table 1 presents a list of some of the more common catchment treatments with their estimated runoff efficiency and life expectancy. Generally, the lower runoff efficiency treatments require storms of higher intensities and total volume to produce significant quantities of runoff. For example, a sheet metal roof will have runoff from storms of lower rainfall intensities and quantities than a catchment of compacted earth. It is usually necessary to increase the size of catchments which have low runoff efficiency treatments compared with the size of a catchment with an impervious surface.

Storage techniques for animal drinking water usually involve some form of container, tank, or lined pond. Unlined earthen pits or ponds are not usually satisfactory methods for water harvesting unless seepage losses are low or can be controlled. There are many types, shapes, and sizes of wooden, metal, and reinforced plastic or concrete storage containers. Costs and availability are primary factors for determining the suitability of these

Encyclopedia of Water Science
DOI: 10.1081/E-EWS 120010105
Published 2003 by Marcel Dekker, Inc. All rights reserved.

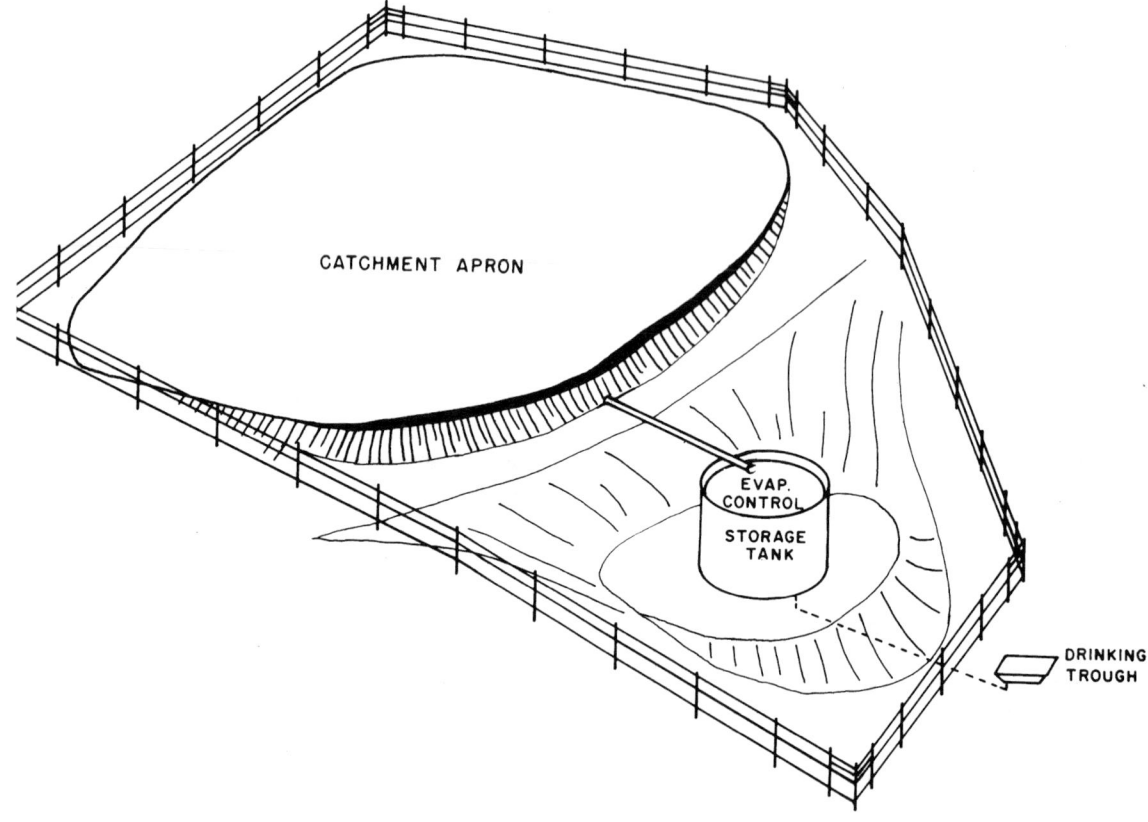

Fig. 1   Typical water harvesting system for livestock water. (From Ref. [4].)

containers. Containers constructed from concrete and plaster are relatively inexpensive, but their construction requires a significant amount of hand labor. One common type of storage is a steel rim tank with a concrete bottom.

Because water harvesting is a relatively expensive method of water supply, controlling evaporation losses is an important factor and should be an integral component of all water storage facilities. Although relatively expensive, roofs over the storage are commonly used.

Evaporation control on sloping-sided pits or ponds is more difficult because the water-surface area varies with depth.

SOCIOECONOMIC CONSIDERATIONS

Water harvesting techniques are practical methods of water supply for most parts of the world, but they are also a relatively expensive method of water supply. During

Table 1   Potential water harvesting catchment treatments

Treatment	Runoff efficiency (%)	Estimated life (yr)
Topography modification		
Land smoothing and clearing	20–35	5–10
Soil modification		
Sodium salts	50–80	5–10
Water repellents and paraffin wax	60–95	5–8
Bitumen	50–80	2–5
Impermeable coverings		
Gravel-covered sheeting	75–95	10–20
Asphalt-fabric membrane	85–95	10–20
Concrete, sheet metal, and artificial rubber	60–95	10–20

(Adapted from Ref. [5].)

the past few decades, there have been numerous water harvesting systems constructed worldwide. While many of the systems have been outstanding successes, others have failed. Some systems failed despite extensive efforts because of material and/or design deficiencies. Others have failed because of personnel changes, communication failures, or because the water was not perceived as needed by the local user. Word-of-mouth publicity of one failure will often spread more widely than all the publicity of 10 successful systems.

REFERENCES

1. Branson, F.A. *Watershed Management on Range and Forest Lands*; Proceedings of the Fifth Workshop of the United States/Australian Rangelands Panel, Boise Idaho, June 15–22, 1975, Heady, H.F., Falkenborg, D.H., Riley, J.P., Eds.; Utah Water Research Laboratory, Utah State University: Logan, UT, 1976; 196–209.

2. Anaya Garduno, M. Research Methodologies for in Situ Rain Harvesting in Rainfed Agriculture. In *Rainfall Collection for Agriculture in Arid and Semiarid Regions*; Proceedings of the Workshop University of Arizona, Tucson, Arizona and Chapingo Postgraduate College, Chapingo, Mexico, Dutt, G.R., Hutchinson, C.F., Anaya Garduno, M., Eds.; Commonwealth Agricultural Bureaux: Farnham House, Farnham Royal, Slough SL2 3BN, U.K., 1981; 43–47.

3. Frasier, G.W. Water Harvesting for Collecting and Conserving Water Supplies. In *Alfisols in the Semi-arid Tropics*; Proceedings of the Consultants' Workshop on the State of the Art and Management Alternatives for Optimizing the Productivity of SAT Alfisols and Related Soils, ICRISAT Center, India, 1–3 December 1983, Pathak, P., El-Swaify, S.A., Singh, S., Eds.; International Crops Research Institute for the Semi-Arid Tropics: Patancheru, Andhra Pradesh 502 324, India, 1987; 67–77.

4. Frasier, G.W.; Myers, L.E. *Handbook of Water Harvesting*, Agricultural Handbook 600; United States Department of Agriculture, Agricultural Research Service: Washington, DC 20402, 1983; 45 pp.

5. Frasier, G.W. Water for Animals, Man, and Agriculture by Water Harvesting. In *Rainfall Collection for Agriculture in Arid and Semi-arid Regions*; Proceedings of the Workshop University of Arizona, Tucson, Arizona and Chapingo Postgraduate College, Chapingo, Mexico, Dutt, G.R., Hutchinson, C.F., Anaya Garduno, M., Eds.; Commonwealth Agricultural Bureaux: Farnham House, Farnham Royal, Slough SL2 3BN, U.K., 1981; 67–77.

Livestock Water Quality Standards

John M. Sweeten
Texas A&M University, Amarillo, Texas, U.S.A.

Floron C. Faries, Jr.
Guy H. Loneragan
West Texas A&M University, Canyon, Texas, U.S.A.

John C. Reagor
Texas A&M University, College Station, Texas, U.S.A.

INTRODUCTION

A plentiful and consistent supply of high-quality water is essential for optimal production and health of feedlot cattle. Water of inadequate quality can result in decreased gains, poor feed conversion, and adverse affects on animal health. The greatest losses to livestock producers from low-water quality are often through undetected production inefficiencies and hidden but considerable influences on profitability.

WATER REQUIREMENTS

Water constitutes 60%–70% of the body of livestock. Water consumption is critical for animal maintenance. Animals that do not drink sufficient water may suffer stress or even dehydration.

The amount consumed depends on the species, weather, and characteristics of feedstuffs consumed. For instance, dry beef cows need about 8 gal–10 gal of water daily, whereas beef cows in their last 3 mo of pregnancy may drink up to 15 gal a day. Those in milk need about five times as much water as the volume of milk produced. Also, calves require much more water after weaning than before. Ignoring this fact may irreversibly retard growth of calves from which they may never fully recover.

WATER QUALITY

Safe supplies of water are absolutely essential for livestock. Livestock may suffer health problems or below-normal consumptions resulting from substandard quality water. Ingestion of mineral or organic contaminants can cause poor performance or nonspecific disease conditions. Major livestock health problems associated with water quality are seldom reported except in site-specific instances. When evaluating the quality of water for livestock, one has to consider whether livestock performance will be affected; whether water could serve as a carrier to spread disease; and whether the acceptability or safety of animal products for human consumption will be affected.

The most common water quality problems affecting livestock production are the following.

- Excess salinity—high concentration of minerals, measured as total dissolved solids (TDS).
- High nitrates or nitrites.
- Bacterial contamination.
- Blue-green algae.
- Accidental spills of petroleum, pesticides, or fertilizers into water supply.

The importance of nitrate, nitrite, sulfate, and TDS as factors influencing water quality for livestock has been recognized. Concentrations generally considered safe for consumption by cattle have been established (Table 1). However, these values may vary slightly depending on type and formulation of rations fed to cattle.

In 1999, the USDA's National Animal Health Monitoring System[1] conducted a water sampling study on beef feedlots with 1000 head or more capacity in the 12 leading cattle feeding states. These feedlots accounted for 96.1% of U.S. cattle feedlot inventory (January 1, 2000) and 84.9% of feedlots with 1000 head or more capacity.[1]

One representative water sample per feedlot was analyzed for nitrate, nitrite, sulfate, and TDS. A total of 263 feedlots from 10 states (all west of the Mississippi River) supplied a water sample for analysis. (No water samples were submitted from Arizona or Oklahoma.) The majority of samples (89.7%) were drawn from a well. Other sources included municipal/city (4.6% of samples), spring/river (2.3%), and pond/lake (2.3%).

Encyclopedia of Water Science
DOI: 10.1081/E-EWS 120010298

Table 1 Concentrations of nitrate, nitrite, sulfate, and total dissolved solids in water typically considered safe for livestock usage

Measurement	Concentration considered safe[a] $(mg\,L^{-1})$
Nitrate, NO_3	Less than 440
Nitrate, NO_3-N	Less than 100
Nitrite, NO_2	Less than 33
Nitrite, NO_2-N	Less than 10
Sulfate, SO_4	Less than 300
Total dissolved solids, TDS	Less than 3000

[a] mg/L is equivalent to parts per million (ppm).
Source: From Ref. [1] and National Research Council, National Academy of Sciences, Washington, DC.

Only 1.7% of samples came from shallow wells (less than 30 ft deep), while 45.3% of samples were from wells that were 101 ft–300 ft deep, and 22.5% were from wells deeper than 300 ft.

The mean nitrate concentration was 33.6 mg $L^{-1} \pm 3.5\,mg\,L^{-1}$ NO_3 (or 7.6 mg $L^{-1} \pm 0.8\,mg\,L^{-1}$ NO_3-N) while sulfate averaged 205 mg $L^{-1} \pm 24\,mg\,L^{-1}$ SO_4. Both these values are considered safe levels (See Table 1). Nitrite was detectable in only 0.4% of the samples. No water samples exceeded the recommended nitrate limit and only 23% exceeded the sulfate limit.

MINERALS AND SALINITY

Livestock tolerance of minerals in water depends on many factors: kind, age, diet, and physiological condition of the animal; season; climate; and kind of salt ions in the water. Livestock may drink less if the water tastes bad. Livestock restricted to waters with high salt content may suffer physiological upset or death.

Several mineral elements found in water seldom offer problems to livestock because they do not occur at high levels in soluble form, or because they are toxic only in excessive concentrations. Examples are iron, copper, cobalt, zinc, iodide, and manganese. These elements do not seem to accumulate in meat or milk to the extent that they would cause problems.

Common compounds found in waters with excess salinity include sodium, chloride, calcium, magnesium, sulfate, and bicarbonate. Bicarbonates and carbonates may contribute to alkalinity (pH) levels. When feed also is high in salt, lower water salinity would be desirable. Moreover, animals consuming high-moisture forage can tolerate more saline waters than those grazing dry grain rations, dry brush, or scrub. Hard water without high salinity does not harm animals.

NITRATE AND NITRITE

Sources of nitrates and nitrites include decaying animal or plant protein, animal metabolic waste, nitrogen fertilizers, silage leachate, and soil high in nitrogen-fixing bacteria. Nitrates and nitrites are water soluble and may be leached away to the water table or into ponded water.

Nitrate is important in livestock health. Although nitrate is not a particularly potent toxin, it is readily reduced to highly toxic nitrite within the rumen. Nitrite is about 10 times more toxic than nitrate. Nitrite is absorbed where it interacts with red blood cells by inhibiting their ability to effectively transport oxygen. Moderate nitrate intake may not cause any noticeable effect on animal health but may result in decreased animal gains and poorer feed conversion. Intake of large amounts of nitrate may result in death.

SULFATE

Sulfur is required by all animals. The recommended sulfur intake for beef cattle is 0.15% of the ration and the maximum tolerable limit is 0.4% of the ration on a dry matter basis. Water can contribute significant quantities of sulfur, as sulfate, towards total sulfur consumption. Sulfur and sulfate are relatively nontoxic in these forms, but sulfate/sulfur is readily reduced in the rumen to highly toxic sulfide products. Excessive total sulfur consumption through feed and water can result in decreased water consumption, feed intake, and reduced average daily gains feed conversion.[2]

Cattle on pasture can tolerate up to a maximum of 2000 mg L^{-1} sulfate (SO_4). However, cattle on full feed with a heavy proportion of concentrate in the diet should have water containing 300 mg L^{-1} sulfate or less.

If native cattle are moved onto water with a high sulfate level during a hot, dry period, there may be problems with adequate consumption. The same water may be well tolerated and consumption increased as needed in animals that were introduced to the water source in a cooler season, when demand was initially low (due to sufficient time for adaptation/acclimatization).

BIOLOGICAL ORGANISMS

All surface waters must be assumed to carry bacteria. Livestock should be kept away from contaminated water that has not been adequately aerated (oxygenated) because of the likelihood of excessive levels of bacterial pathogens that may be present. Surface water sources may have problems with algae growth as a result of high-nutrient loading in runoff water. Avoid using waters bearing heavy growths of blue-green algae, as several species can produce animal toxins (poisons). To control algae in storage tanks, reduce the introduced organic pollution and exclude light. Water storage tanks can be disinfected by adding 1 oz of chlorine bleach per 30 gal of water, holding for 12 hr before draining, and then refilling with clean water. Chlorination can also control certain bacteria.

WATER QUALITY CRITERIA

There are no regulations regarding livestock water quality. *Suggested* limits of concentrations of specific substances in water for livestock, where these have been established, are shown in Table 2, which shows that suggested upper limits for livestock are generally higher than for humans, with the exception of copper and fluoride. Generally, salinity is more restrictive for young animals, pregnant, or lactating animals. Also, monogastic animals (poultry and swine) are less tolerant to salinity than ruminant animals (cattle or sheep).

WATER QUALITY EVALUATION

To evaluate water quality in relation to livestock health problems, it is imperative to obtain a thorough history,

Table 2 Recommended limits of concentration of some potentially toxic substances in drinking water for livestock vs. comparable values for humans

Selected inorganic constituents	Comparable U.S. EPA[a] criteria (for humans)	Safe concentration (upper limit) for livestock, (mg L^{-1})[3]	
		NAS[5]	CAST[6]
Inorganic chemicals	Primary (MCL), mg L^{-1}		
Antimony	0.006		
Arsenic	0.05	0.2	0.5
Asbestos	7 MFL		
Barium	2.0	N.E.	
Beryllium	0.004		
Boron	N.A.	5.0	
Cadmium	0.005	0.05	0.5
Chromium	0.1	1.0	5.0
Chloride	N.A.		
Cobalt	N.A.	1.0	1.0
Copper	1.3	0.5	0.5
Cyanide, free	0.2		
Fluoride	4.0	2.0	3.0
Iron	N.A.	N.E.	No limit[b]
Lead	0.015	0.1	0.1
Manganese	N.A.	N.E.	No limit
Mercury	0.002	0.01	0.01
Nickel	N.A.	1.0	
Nitrate-N	10.0	100	300
Nitrite-N	1.0	10	10
Salinity	N.A.	See Table 3	
Selenium	0.05		
Sulfate	N.A.		
Thallium	0.002		
Total dissolved solids	N.A.		
Vanadium	N.A.		1.0
Zinc	N.A.		25.0

MCL = Maximum contaminant level, highest level allowed in drinking water, an enforceable standard for public drinking water supply; MFL = million fibers per liter; N.A. = Not applicable; N.E. = Not established.
[a] Primary standards only, not including current human drinking water quality standards for micro-organisms; disinfectants or disinfection byproducts; organic chemicals; or radionuclides.[4]
[b] Available data are not sufficient to warrant definite recommendations.

Table 3 Guide to using saline waters for livestock

Total soluble salts content of waters ($mg\,L^{-1}$)	Comments—livestock water use
Less than 1,000	Relatively low level of salinity, no serious problem expected
1,000–2,999	Considered satisfactory; may cause temporary mild diarrhea in livestock unaccustomed to them, but should not affect animal health or performance
3,000–4,999	Should be satisfactory; may cause temporary diarrhea or be refused at first by animal unaccustomed to them
5,000–6,999	Can be used with reasonable safety; avoid using those approaching the higher limits for pregnant or lactating animal
7,000–10,000	Use should be avoided; considerable risk for pregnant or lactating livestock, young animals, or for any animals subjected to heavy heat stress or water loss; older livestock may subsist under conditions of low stress
More than 10,000	Excessive risks; cannot be recommended for use under any conditions

Source: From Ref. [5].

make accurate observations, ask intelligent questions, and submit suspected water and properly prepared tissue specimens without delay to a qualified laboratory.[7] Obtain assistance from a local veterinarian, county Extension agent, or state veterinary medical diagnostic laboratories, usually affiliated with the land grant university in each state.

SOURCES OF WATER QUALITY CONTAMINATION

Contaminant levels may be affected by runoff from surrounding lands and by concentration caused by water evaporation from a pond or storage tank.[8] Salinity is of special concern in the western half of the United States where naturally occurring salinity in watersheds or geological formations can restrict livestock water uses (Table 3). Livestock grazing operations may influence stream water quality where cattle are watered in or along the streams or drainage features. Potential sources of localized groundwater contamination include: livestock manure accumulations around water wells, ponds and stock pens, and agricultural chemicals or containers at spray pens, dipping vats, and disposal sites. Other potential nonpoint pollution sources that require careful site selection and management include:[9] concentrated animal feeding operations; wastewater holding ponds; lagoons; manure stockpiles; silos; dead animal disposal sites; and onsite sewage treatment systems.

Fertilizers, including manure and wastewater, should be carefully selected and applied to land in accordance with soil and crop requirements or nutrient management plans. This will help prevent contaminating underlying aquifers and with nutrients or salts. Always handle and apply

pesticides in strict accordance with the recommendations on the label. Do not apply pesticides around a water supply or other vulnerable sites.

Wellhead protection measures are specified in water well drillers' guidelines in most states. Locate wells at least 150 ft–300 ft from livestock corrals, septic tanks, manure treatment lagoons, and runoff holding ponds.[8] To prevent infiltration, case, and grout wells down to a restrictive layer or to the water table, and seal around the wellhead with a concrete pad.

CONCLUSION

Livestock producers should provide sufficient safe water for animals by preventing contamination and providing adequate sources of year-round, high-quality drinking water supply. Livestock should be protected from unsafe drinking water by providing alternative sources of acceptable quality water.

Water-related health problems in livestock are usually caused by stress conditions that may include inadequate water supply or unpalatable water with a high level of dissolved substances.

REFERENCES

1. National Animal Health Monitoring System (NAHMS), *Water Quality in U.S. Feedlots*, Information Sheet #N341. 1200; USDA Animal and Plan Health Inspection Service: Veterinary Services: Centers for Epidemiology and Animal Health: Ft. Collins, CO, 2000.
2. Loneragan, G.H.; Wagner, J.J.; Gould, D.H.; Garry, F.B.; Thoren, M.A. Effects of Water Sulfate Concentration on

Performance, Water Intake, and Carcass Characteristics of Feedlot Steers. J. Anim. Sci. **2001**, *79*, 2941–2948.

3. Herrick, J.B. Water Quality for Animals. Symposium Proceedings, Ames, Iowa, 1976.

4. U.S. Environmental Protection Agency. National Primary Drinking Water Proposed Interim Standards. EPA 816-F-01-07, Office of Water: Washington, DC, 2001; 4, www.epa.gov/safewater.

5. National Academy of Sciences. Nutrients and Toxic Substances in Water for Livestock and Poultry. Washington, DC, 1974.

6. Council for Agricultural Science and Technology. Quality of Water for Livestock, Vol. I. Report No. 26; 1974.

7. Faries, F.C.; Sweeten, J.M.; Reagor, J.C. *Water Quality: Its Relationship to Livestock. L-2374*; Texas Agricultural Extension Service, Texas A&M University: College Station, TX, 1991.

8. Midwest Plan Service, *Private Water Systems Handbook*; MWPS-14, Iowa State University: Ames, IA, 1979.

9. Sweeten, J.M. Groundwater Quality Protection for Livestock Feeding Operations. Texas Agricultural Extension Service, L-2348; 1991.

Manure Management, Beef Cattle Industry Requirements

John M. Sweeten
B. W. Auvermann
Texas A&M University, Amarillo, Texas, U.S.A.

INTRODUCTION

Although cattle drinking water requirements are significant, averaging $10.8 \, \text{gal} \, \text{hd}^{-1} \text{day}^{-1}$,[1] the beef cattle sector of America's $100 billion yr^{-1} animal agriculture industry uses relatively little water for manure management. The vast majority (perhaps 98%) of the nation's average of 10.3 million head of cattle on feed for slaughter and beef processing are fed in open, soil-surfaced feedyards. As a result, nearly all feedyard manure is collected in solid form. With an average turnover rate of 2 times–2.3 times per year, over 20 million head of cattle are fed in this manner, generating roughly a ton of as-collected manure per head fed and harvested. In the typical beef cattle feedyard, all manure deposited on the feedlot surface undergoes concurrent processes of: a) partial evaporative drying (from 75 + % wet basis as-excreted down to 20%–50% wet basis as-collected); b) partial decomposition of volatile organic solids; and c) atmospheric release of gaseous compounds that include carbon dioxide, volatile organic compounds, and ammonia.

MANURE COLLECTION AND HANDLING

Manure collection practices from open feedyards include use of wheel loaders, box scrapers, dozers, or elevating scrapers.[2] Transportation of collected manure is provided in open top manure spreader trucks to farmland. Intermediate storage may be needed in temporary stacks within feedpens or in stockpiles adjacent to the cattle feedpens. This intermediate storage should be located within the envelope of containment of storm water runoff in accordance with state and federal requirements. The manure mechanically collected in air-dry form is in its least voluminous state and usually has good cash market demand from nearby farmers for use as bulk fertilizer.

Collection, marketing, and/or distribution to farmers are handled by contractors in most cases. Most manure is sold and applied within 10 mi–20 mi of the feedyard. Consequently, there is no incentive to add water to liquefy manure from beef cattle feedlots during collection, storage, treatment, and/or distribution. Because the limiting factor for marketing manure is usually the hauling cost, added water lowers the manure's net fertilizer value.

WATER USE IN MANURE HANDLING

Exceptions to the normal practice of handling of cattle feedlot manure in solid form may include these situations or considerations:

1. Intermittent water additions to compost windrows to raise moisture content to approximately 50% wet basis to initiate or restore active composting.
2. Spillage or leakage of water trough overflow onto a feedyard surface (e.g., water line leaks or trough overflow), which can carry very small quantities of manure solids (e.g., <1%) into runoff collection channels or basins.
3. Rainfall runoff, which can carry a small percentage (<5%–10%) of the manure solids from the pen surfaces into runoff settling basins, holding ponds, or evaporation basins, as required by federal (EPA) and state water pollution abatement regulations.[3,4]
4. Water application onto a feedyard surface for dust control in dry weather, with requirements as high as 2 times–5 times the normal daily drinking water requirement of 10 gal day^{-1}–12 gal day^{-1} for feedlot cattle, with amount for sprinkling depending on cattle spacing, animal liveweight, depth of manure pack, frequency of manure harvesting, evaporation rate, and precipitation.[5]
5. Instances where beef cattle are fed in confinement barns or concrete floors with manure collection by flushing using fresh or recycled effluent.[6] In these instances, water requirements for flushing generally follow a "rule of thumb" of approximately 12 gal of water (100 lbs) required for gravity-flushing of one pound of manure total solids. These water requirements can be met by fresh water usage (surface or groundwater) or recycled wastewater from treatment lagoons or runoff holding ponds.
6. Anaerobic digestion under mesophilic, thermophilic, or ambient temperature conditions, in which a manure

Encyclopedia of Water Science
DOI: 10.1081/E-EWS 120010300

slurry of 8%–12% total solids serves as an energy feedstock. The digested slurry, containing all the original nutrients following carbohydrate conversion to methane and carbon dioxide, must be handled and land-applied as a wastewater including storage, conveyance, and irrigation.

CONCLUSION

Of the above scenarios, dust control is undoubtedly the greatest use of fresh water where needed by climatic circumstances, including prolonged seasonal dry weather and/or proximity to neighbors. Feedyard dust consists of relatively coarse particulate matter (PM) generated from the manure surface by cattle hoof action, which intensifies during early evening hours, especially in warm weather. Many factors, as yet not fully defined, can interact to generate feedlot dust. In terms of water use, evaporative demand with continual hoof shear and churning action can be $0.25 \, \text{in. day}^{-1}$–$0.50 \, \text{in. day}^{-1}$ in hot, dry weather.[5] This evaporative demand can be met by a) increasing stocking density to focus excreted moisture in feces and urine onto a smaller area; b) frequent removal of accumulated dry manure; c) intermittent rainfall; or d) water application with water tankers with spray nozzles or sprinkler irrigation.[7] Water additions by the latter method can be expected to be on the order of $25 \, \text{gal hd}^{-1} \, \text{day}^{-1}$–$50 \, \text{gal hd}^{-1} \, \text{day}^{-1}$ as fresh water or recycled water during a typical 6-month dust season.

REFERENCES

1. Parker, D.; Perino, L.; Auvermann, B.; Sweeten, J. Water Use and Conservation at Texas High Plains Beef Cattle Feedyards. Appl. Eng. Agric. **2000**, *16* (1), 77–82.

2. Sweeten, J.M. Cattle Feedlot Manure and Wastewater Management Practices for Water and Air Pollution Abatement. In *Cattle Feeding: A Guide to Management*, 2nd Ed.; Albin, R.C., Thompson, G.B., Eds.; Trafton Printing Co.: Amarillo, TX, 1996; Chap. 8, 62–83.

3. Sweeten, J.M. Feedlot Runoff Characteristics for Land Application. In *Agricultural and Food Processing Wastes*; Proceedings of the Sixth International Symposium on Agricultural and Food Processing Wastes, Chicago, IL, December, 17–18, 1990; American Society of Agricultural Engineers: Chicago, IL, 1990; 168–184.

4. Sweeten, J.M. Environmental Management for Commercial Cattle Feedlots in Moisture Deficit Locations. In Proceedings, National Livestock, Poultry, and Aquaculture Waste Management Workshop; Kansas City, MO. July 29–31, 1992; 232–246.

5. Auvermann, B.W.; Parker, D.B.; Sweeten, J.M. *Manure Harvesting Frequency—The Key to Feedyard Dust Control in a Summer Drought. E-52*; Texas Agricultural Extension Service: College Station, TX, 2000; 4 p.

6. Clanton, C.L. Beef Cattle Waste Management Systems for the Farmer-Feeder in Humid Climates. In Proceedings, National Livestock, Poultry, and Aquaculture Waste Management Workshop, Kansas City, MO, July 29–31, 1992; 247–252.

7. Sweeten, J.M. Manure and Wastewater Management for Cattle Feedlots. In *Review of Environmental Contamination and Toxicology*; Morgan, D.P., Ed.; 2000; Vol. 167, 121–153.

Manure Management, Dairy

H. H. Van Horn
D. R. Bray
University of Florida, Gainesville, Florida, U.S.A.

INTRODUCTION

Water use is essential for all dairies. Drinking water is indispensable for cattlelife; some amount of water is necessary for cleaning and sanitation procedures; moderate amounts are important during periods of heat stress for evaporative cooling of cows to improve animal production and health; additional amounts can be used in labor-saving methods to move manure and clean barns by flushing in properly designed facilities; and the recovered wastewater can be recycled to supplement water requirements of forage crops grown to meet roughage requirements of the dairy herd. Extensive water use, however, increases the potential of surface runoff and its penetration into the ground with possible environmental impacts offsite. Heightened environmental concerns and the need for resource conservation, in many cases, have caused implementation of water-use permits. Thus, it is important to determine various essential uses of water, other uses that are important to management, and also consider whether reuse of some water is possible and if it is necessary to do so.

Some of the useful unit conversions are listed as follows:

1 gal of water = 8.346 lb.
1 ft^3 of water = 7.48 gal.
1 acre = 43, 560 ft^2.
1 acre in. of water = 27, 152 gal.

Calibration methods to estimate use: Water flow meters should be installed on major water supply lines. If water meters are not in place to measure gallons pumped, it becomes necessary to estimate the usage. This can be achieved by capturing flow through various water lines for specified times and multiplying by the time the water flows through these lines every day.

DRINKING

Table 1 provides estimates of drinking water requirements in gallons per cow per day. Consumption of 25 gal – 30 gal of water per day by lactating cows is common, which varies depending on milk yield, dry matter intake (DMI), temperature, and other environmental conditions.[1]

COW WASHING

Presently most dairies, in warm climates, bring cows to be milked into a holding area equipped with floor-level sprinklers, which spray water upward to wash cows. Each cow usually has a holding area of about 15 ft^2 and are typically washed for 3 min. Amount of water used per cow should be calculated for each dairy. An estimate for conservative use is that a holding area for 300 cows is 30 × 150 ft^2 (15 ft^2 per cow) and is equipped with sprinklers with 5-ft spacing (say 7 across and 30 rows) having 210 sprinklers. If each sprinkler applies 5 gal min^{-1}, total usage is 1050 gal min^{-1} or 3150 gal for 3 min, the average consumption per cow would be 3150/300 = 10.5 gal per cow per wash cycle. If cows are milked three times this would require 31.5 gal per cow per day.

The washing system previously described also helps in cooling of cows while they are crowded together waiting to be milked. However, the cooling effect could be achieved by sprinkling a little amount of water from above, alternatively with fans to give evaporative cooling, if cows were clean enough so that extensive washing was not required and water conservation was necessary.

WASHING MILKING EQUIPMENT AND MILKING PARLOR

Use of water for these purposes is not as directly related to the number of cows as for other uses. For washing milking equipment, a common wash vat volume is 75 gal. If this is filled for rinse, wash, acid rinse, and sanitizing at each of three milkings, this amounts to 900 gal for the herd, e.g., with 300 cows, only 3 gal per cow per day. This is an extremely small component of the total water budget. The amount used to wash out the milking parlor varies largely. If only hoses are used, the amount may be as little as 2 gal per cow per milking or 6 gal per cow per day if cows are milked three times daily. If flush tanks are used, the amount

Encyclopedia of Water Science
DOI: 10.1081/E-EWS 120010302

Table 1 Predicted daily water intake of dairy cattle as influenced by milk yield, DMI, and season[a,b]

Milk yield (lb)	Cool season (e.g., February)		Warm season (e.g., August)	
	DMI (lb)	Water intake (gal)	DMI (lb)	Water intake (gal)
0	25	11.5	25	16.3
60	45	22.2	44	26.8
100	55	28.6	48	31.9

[a] Drinking water intake predicted from equation of Murphy et al., J. Dairy Sci., **1983**, 66, 35: Water intake $(lb\,day^{-1}) = 35.2 \times DMI$ $(lb\,day^{-1}) + 0.90 \times$ milk produced $(lb\,day^{-1}) + 0.11 \times sodium$ intake $(g\,day^{-1}) + 2.64 \times$ weekly mean minimum temperature $[°C = (°F - 32) \times 5/9]$. For examples above, diet dry matter was assumed to contain 0.35% Na. Predicted water intakes (lb) from formula calculations were divided by 8.346 lb water per gal to convert to gallons.

[b] Average minimum monthly temperatures for February (43.5°F) and August (71°F) used with prediction equation were 70-yr averages for specified months at Gainesville, FL (Whitty et al., Agronomy Dept, Univ. FL, 1991).

may be more, i.e., nearly 3000 gal per milking or 9000 gal day^{-1} for three times, equivalent to 30 gal per cow per day for a 300-cow system.

SPRINKLING AND COOLING

Sprinklers along with fans are used for evaporative cooling to relieve heat stress in dairy cows during hot periods of the year. Their use has shown increased cow comfort (lowered body temperature and respiration rates) and economic increases in milk production and reproductive performance.[2,3] Application rates used by dairymen vary. Florida experiments compared application rates of 51 gal per cow per day, 88 gal per cow per day, and 108 gal per cow per day at 10 psi in one experiment and 13 gal per cow

per day, 25 gal per cow per day, and 40 gal per cow per day in another experiment. The application rate, 13 gal per cow per day, is close to the estimated evaporation rate from the cow and surrounding floors. This component should be considered in water use but not in runoff water that must be managed in the manure management system. We estimate 25 gal per cow per day as the minimum practical application rate in order to get adequate coverage of cows to cool them because often they are not in the sprinkled area. Total application days per year vary from 120 days to 240 days. A separate water well, or reserve tank and booster pump, may be needed to supply short-term high demand required by the sprinkler system.

FLUSHING MANURE

Flushing manure can be made a clean and labor-saving process, if facilities include concrete floors with enough slope so that water flow propelled by gravity could be used to move manure. Amounts of water used per cow vary widely depending on size and design of facilities and frequency of flushing. However, usually a flush of about 3000 gal is required to clean an alley width of 10 ft – 16 ft. If 4 alleys are common for every 400 cows and alleys are flushed twice daily, this would amount to an average use of 60 gal per cow per day. Many dairies use more flushings per day.

RECYCLING DAIRY WASTEWATER THROUGH IRRIGATION OF FORAGE CROPS

Most often nitrogen is the nutrient on which manure application rates are budgeted. To maximize nutrient uptake, crop growth should be as vigorous as possible. This requires irrigation during most of the year in many dairy regions for the disposal of flushed wastewater.

Table 2 Crop yield and water requirement estimates for two triple cropping forage systems[a]

Crop No.	Name	Silage yield			Water required			
		Ton/A 35% DM	Ton/A DM	lb/A DM	lb/lb DM	lb/A Total	gal/A Total	A-in. Total
1	Wheat	10	3.5	7,000	500	3,500,000	419,362	15.4
2	Corn	24	8.4	16,800	368	6,182,400	740,762	27.3
3	Corn	14	4.9	9,800	368	3,606,400	432,111	15.9
	Total	48	16.8	33,600		13,288,800	1,592,235	58.6
1	Rye	10	3.5	7,000	500	3,500,000	419,362	15.4
2	Corn	24	8.4	16,800	368	6,182,400	740,762	27.3
3	F. Sorghum	18	6.3	12,600	271	3,414,600	409,130	15.1
	Total	52	18.2	36,400		13,097,000	1,569,254	57.8

[a] A = acre; No. = number; DM = dry matter.

Table 3 Estimated water budgets for three example dairies

Water use in the dairy	Flush systems		Nonflush Theoretical minimum	Worksheet for your dairy
	Typical need during hot season	**Common usage on some dairies**		
Drinking (cows)	25	25	25	
Cleaning cows	32	150	0	
Cleaning milking equipment	3	5	3	
Cleaning milking parlor	30	30	6	
Sprinklers for cooling	25	130	12	
Flushing manure	60	80	0	
Total use per cow per day	175	400	46	
Total use per 100 cows per day	17,500	40,000	4,600	
Use per 100 cows per week	122,500	280,000	32,200	
Water in milk per 100 cows per week	4,500	4,500	4,500	
Estimated evaporation (at 20% of use)	24,500	56,000	6,440	
Average rainfall and watershed drainage into storage facility per 100 cows per week	27,000	27,000	13,000	
Wastewater produced from 100 cows/week	120,500	246,500	38,760	
Acre in. per 100 cows per week	4.44	9.08	1.43	
in. per week if 30 acre in sprayfield	0.15	0.30	0.05	

All values are in gal unless otherwise noted.

Example calculations (column 1): Total use per cow per day = 175 gal; total use per 100 cows per week = 122,500 gal less 4500 in milk and 24,500 gal evaporation = 93,500 gal week^{-1}; net rainfall and watershed drainage to storage per 100 per cows per week = 27,000; acre in. per 100 cows per week = (93,500 + 27,000)/27,152 gal per acre in. = 4.44.

If 30 acre were in sprayfield, 4.44/30 = 0.15 in. week^{-1}.

If crop needed 1.75 acre in. week^{-1} (a common average), a total of 1.75 *in.* × 30 *acre* × 27,152 *gal* per acre in. = 1,425,480 gal is needed of which only 120,500 gal (8.5%) would come from dairy wastewater. The remaining (91.5% of total) would have to come from rainfall or fresh irrigation water.

In southern regions, multiple cropping systems are possible, which will recycle effectively nitrogen excretions from 100 cows on a sprayfield or manure application field of about 30 acre.[4]

Tentative estimates of total water needs of the growing crops in warm climates average about 1.75 in. of water per week (0.25 in. per day) from irrigation plus rainfall with a minimum of 0.5 in. per week tolerated even in rainy season on sandy soils.[5,6] Table 2 provides estimates of water requirements for two triple cropping forage systems that are common in southern climates. In sandy soils that hold only about 1.0 in. of water per foot of soil depth, some amount of rainfall cannot be stored. Therefore, even in heavy rainfall seasons, judicious irrigation is often needed during lower rainfall weeks. Limited data are available on the maximum amount of water that could be applied and not reduce yield or quality of forage and not result in pollution of groundwater with nitrates and other minerals. However, the maximum probably is at least 35 in.–45 in. per year above the acre in. totals in Table 2.

RAINWATER FROM ROOFS AND CONCRETE AREAS

Rainwater entering wastewater holding areas can be significant. For example in the dairy representing typical minimum water usage with a flush system in southeast United States (Table 3), the net accumulation during the hot season was calculated as follows: assumed wastewater holding area is 1 acre surface area per 100 cows, net rainfall accumulation in holding area is 3 in. more than evaporation per month, concrete areas and/or undiverted roof areas that capture rainfall are $15,000\,ft^2$ per 100 cows that divert $15,000/43,560\,ft^2$ per acre of the 3 in. to the wastewater holding facility. Thus, $3\,in. + 0.344 \times 3 = 4.03$ acre in. mo^{-1} or essentially 1.0 acre in. per week per 100 cows (approximately 27,000 gal per 100 cows).

DEVELOPING A WATER BUDGET

A wide range exists in water usage on dairy farms. For most dairy waste management systems designed to utilize flushed manure nutrients through cropping systems grown under irrigation, water amounts are small in relation to irrigation needs for crop production. Costs for construction of storage structures for holding wastewater until used for irrigation warrant consideration. For example, water-use budgets given in Table 3 show that water usage is small in comparison to irrigation needs when there are 30 acre of sprayfield crop production per 100 cows. Conversely, the amounts used in most dairy systems would be large and unmanageable if application through irrigation is not an option or if less acreage for irrigation is available than needed for application of all manure nutrients.

If a dairy does not have acreage available close by to utilize manure nutrients and water through an environmentally accountable sprayfield application system, it would be necessary to export nutrients off the farm, preferably as solid wastes to avoid excessive hauling or pumping costs. If the water and manure nutrients cannot be used through irrigation, a nonflush system should be utilized. However, usually some irrigation is possible, permitting dairymen to use cow washers and limited flushing if they scrape and haul manure from some areas.

Strategies to minimize water usage: Table 3 presents one column indicating a theoretical minimum amount of water use in a dairy. This system implies that cows are clean and cool enough so that sprinkler washers are not required to clean and cool cows while being held for milking. In addition, it is assumed that all of the manure is scraped and hauled to manure disposal fields or transported off the dairy in some other fashion. Intermediate steps that might be taken include the following:

1. Scraping and hauling manure from high use areas such as the feeding barn so that this manure can be managed off the dairy.
2. Using wastewater rather than fresh water to flush manure from feeding areas and freestall barns.
3. Using a housing system that will keep cows clean enough so that cow washers are not required to clean cows before milking. This system, however, may require use of alternating sprinklers and fans to keep crowded cows cool during hot weather conditions.

If flushing is desired in conjunction with scraping and hauling from heavy use areas, perhaps the feeding area could be flushed with recycled water after scraping to clean the area. These procedures would reduce total nutrient loads retained in wastewater and would significantly reduce the size of the sprayfield needed for water and manure nutrient recycling.

REFERENCES

1. Beede, D.K. Water for Dairy Cattle. *Large Dairy Herd Management*; American Dairy Science Assoc.: Champaign, IL, 1992; 260–271.
2. Bray, D.R.; Beede, D.K.; Bucklin, R.A.; Hahn, G.L. Cooling, Shade, and Sprinkling. *Large Dairy Herd Management*; American Dairy Science Assoc.: Champaign, IL, 1992; 655–663.
3. Van Horn, H.H.; Bray, D.R.; Nordstedt, R.A.; Bucklin, R.A.; Bottcher, A.B.; Gallaher, R.N.; Chambliss, C.G.; Kidder, G. Water Budgets for Florida Dairy Farms; Circular 1091; Florida Coop. Ext., Univ. Florida: Gainesville; 1993.
4. Van Horn, H.H.; Nordstedt, R.A.; Bottcher, A.V.; Hanlon, E.A.; Graetz, D.A.; Chambliss, C.F. Dairy Manure Management: Strategies for Recycling Nutrients to Recover Fertilizer Value and Avoid Environmental Pollution; Circular 1016; Florida Coop. Ext., Univ. Florida: Gainesville; 1998, 1–24.
5. North Florida Research and Education Center. AREC Research Report 77-2; IFAS, University of Florida, Gainesville, 1977.
6. Wesley, W.K. Irrigated Corn Production and Moisture Management; Bul. 820; Coop. Ext. Serv., Univ. Georgia College of Agric. and USDA, 1979.

Manure Management, Poultry

Saqib Mukhtar
Patricia K. Haan
Texas A&M University, College Station, Texas, U.S.A.

INTRODUCTION

Poultry production in the United States has increased steadily and accounts for about 5.5% of the total manure produced annually. Water requirements for poultry manure management and utilization vary according to how manure is handled and stored. Wastes from broiler chickens and turkeys are in a solid (litter) form while layer chicken waste may be either solid or liquid. Litter is most often land applied as a fertilizer source for plants. Liquid manure from laying operations is flushed into anaerobic lagoons for dilution and treatment. Large quantities of water are required to flush and treat liquid manure. The treated effluent is then land applied to crops and pastures. Poultry manure is an excellent source of nutrients for plant growth, including nitrogen, phosphorus, and potassium, and can improve soil physical properties by addition of organic matter. Poultry manure can also be a low cost alternative to mineral fertilizers. Application of poultry manure and wastewater requires proper management to reduce adverse effects to human health and water quality due to loss of nutrients and pathogens from fields to adjacent surface and groundwater bodies.

BACKGROUND

Poultry production involves raising chickens, turkeys, and ducks for the consumption of meat and eggs. While ducks are included in this category, chicken and turkey operations are the focus of this article. Turkeys are raised for meat production, but chicken are raised either as broilers for meat or as layers for the production of eggs. Since the early 1990s, turkey production has remained steady but consumer demand for broilers and eggs has resulted in a steady increase in the total production of chicken in the United States. For example, from 1991 to 1999, the total number of layers and broilers increased by 18% and 33%, respectively.[1] Table 1 shows the total number of broilers, layers, and turkeys produced in 1999 along with estimates of manure excreted by the birds in each category. A total of 55.7 million tons of manure was produced in 1999, suggesting that poultry operations produced nearly 5.5% of the estimated 1 billion tons of manure produced annually[2] in the United States.

POULTRY MANURE MANAGEMENT SYSTEM

Poultry manure may be comprised of excreta, feathers, spilled water and feed, process generated wastewater (water for flushing gutters etc.), litter for bedding (sawdust, wood shavings, peanut hulls, etc.), and mortality. Poultry manure management water requirements may best be explained by first understanding the manure management system for poultry operations. A common theme with any livestock or poultry manure management system is the functional parameters that dictate the type and nature of manure management components of a system. These parameters include manure production, collection, storage, transfer, treatment, and utilization. Production refers to the total volume and nature of animal waste. For example, Table 1 shows that the amount of excreta produced by the type of bird will differ based upon the size and period of confinement. Additionally, moisture content and other physicochemical constituents of excreta vary from one species to another due to differences in feed, digestive system, and climate. Collection of manure refers to gathering of excreta and other waste from initial deposition to short or long-term storage.

In broiler and turkey houses, manure is mixed with litter and handled as "solid" waste. Manure around drinkers, also known as "cake" is relatively high in moisture and more composted, therefore, removed between each flock (approximately 3 and 6 flocks per year for turkeys and broilers, respectively) while the remaining low density manure pack known as "clean out" is generally removed once every year. Both the cake and clean out litter are either transported directly to land for fertilizing crops and pastures or transferred to a stacking facility for a later land application. A part of this solid waste may be sold as a fertilizer source for gardens and nurseries. For this type of broiler or turkey manure management, no water is required except to initiate and maintain composting, if practiced. Some of the litter may be used after deep stacking or composting as bulk

Encyclopedia of Water Science
DOI: 10.1081/E-EWS 120010299

Table 1 Poultry manure production estimates, as excreted in 1999

Bird type[a]	Manure per 1000 birds per day[b] (kg)	Total number of birds[c] (1,000s)	Total manure[d] (ton/yr)
Broilers	80	8,146,010	32,584,000
Layers	118	329,320	14,183,900
Turkeys	267	272,994	8,892,458
		Total manure production	55,660,358

[a] Manure production based on 2 kg, 1.8 kg, and 10 kg live weight for broilers, layers and turkeys, respectively.

[b] Data from Natural Resource and Engineering Service (NRAES) publication, NRAES-132 (1999).

[c] 1999 data from Agricultural Statistics, USDA–National Agricultural Statistics Service (2001).

[d] Manure totals based upon 50 day, 122 day, and 365 day of occupancy by broilers, turkeys, and layers, respectively.

feedstuffs for cattle herds (breeding or stocker phases). A small portion of the litter may be used together with straw, hay, or crop residue as a carbon source for mortality (dead bird) composting, with the resulting compost used on pastures for fertilizing.

Poultry layer production houses are designed to handle manure as a solid or a liquid (slurry). Manure from high-rise (elevated cages allowing manure removal with a tractor scraper) and belt scrape (manure removed by a belt system running under cages) houses may be handled as solid waste or slurry. Layer manure from a shallow-pit house is handled as slurry only. It is removed with a scraper or by flushing. The slurry may be stored in a tank or flushed to a waste treatment anaerobic lagoon or storage pond before it is land applied as fertilizer.

WATER REQUIREMENTS AND UTILIZATION OF WASTEWATER

The amount of water withdrawn for all livestock and poultry operations and for processing in the United States in 1995, was estimated to be 20.8 million m^3 per day, or nearly 2% of freshwater use for all off-stream categories.[3] The vast majority of this consumption was attributed to fish farming.

Fresh, recycled, or a combination of fresh and recycled flush water is used to remove manure from

layer houses handling slurry manure. Manure removal by flushing requires minimum labor, reduces fly problems in the layer house, and reduces odors. Researchers[4] found that a flush water volume of $0.53 \, L \, kg^{-1}$ live layer weight/day compared well with the volume of flush systems designed for other species. The manure removal interval may vary from daily to once a week flushing, but most layer houses may be flushed once a day, for 20 min, using between $38 \, m^3$ and $76 \, m^3$ of flush water.[5] Poultry manure stored and treated in an anaerobic lagoon requires large quantities of water for dilution and decomposition of organic matter by micro-organisms. Poultry anaerobic lagoon design includes this water storage volume known as the "treatment volume." The estimated water requirements for manure dilution and treatment are temperature dependent, and excessive dilution of organic waste is required in colder climates since the microbial activity is slower in such climates. Therefore, the treatment volume may vary from $370 \, L \, kg^{-1}$ live weight of poultry contributing manure to a lagoon in the cold climate, to nearly one half or $200 \, L \, kg^{-1}$ in the warm climate of the United States.[6]

Water in the form of treated effluent from anaerobic lagoons or slurry storage structures is typically land applied to irrigate crop and forage lands either by irrigation, surface spreading or subsurface injection. Land application is an efficient utilization alternative because of

Table 2 Nutrient composition of poultry manures, as excreted and in lagoon effluent

	Animal type	Nutrient content[a]		
		N (kg t^{-1})	P$_2$O$_5$ (kg t^{-1})	K$_2$O (kg t^{-1})
Raw manure	Broilers	12.7	7.8	5.9
	Layers	13.2	10.3	5.9
	Turkey	13.7	11.8	5.9
Liquid handling system				
Anaerobic lagoon	Poultry	3.2	0.8	5.0

[a] Data from Natural Resource and Engineering Service (NRAES) publication, NRAES-132 (1999).

lower costs as compared to wastewater treatment and the benefits to cropped lands derived from nutrients in the wastewater. Manure can also be a low cost alternative to mineral fertilizers.[7] Land application of wastewater utilizes water to recycle nutrients, enhance soil fertility, and improve soil physical properties. However, a balance must be maintained when land applying animal manure to ensure maximum utilization of nutrients by crops while minimizing the risk of health and environmental effects.

Poultry wastewater from anaerobic lagoons has nutrients essential to plant growth including nitrogen (N), phosphorus (P), and potassium (K). The nutrient composition of waste is affected by housing and waste-handling system. Bedding and additional water can dilute manure, resulting in less nutrient value per kilogram. Nutrient losses from storage and handling reduce the amount of nutrient available for land application. Phosphorus and potassium losses are usually negligible but nitrogen losses can be significant. Table 2 shows a typical nutrient composition of raw poultry manure compared to the nutrient composition of the effluent from an anaerobic lagoon.

Fields receiving manure should be tested for available nutrients before application. Application rates have typically been based on crop N requirements. However, inherent variability of waste and the uncertainty associated with nutrient release rates make it difficult to determine the amount of each nutrient being applied in any one application to meet plant demands. To agronomically apply manure, application should be made based on soil levels of phosphorus.

Addition of poultry manure and wastewater to soils improves soil physical properties by adding organic matter. Organic matter in turn helps to build soil structure and increase the soil water holding capacity. This can also improve soil tilth, lessen wind and water erosion, improve aeration, and promote beneficial organisms.

Careful management of waste application is needed to reduce adverse health and environmental effects due to losses of nutrients and pathogens from fields to adjacent surface and groundwater bodies. Applying waste in a way that exceeds a crop's ability to take up N can be a threat to drinking water. Nitrogen in the nitrate form is a highly mobile compound that can cause health problems in humans and animals in concentrations greater than $10\,mg\,L^{-1}$. Alternatively, applying manure based on nitrogen concentrations can lead to excessive phosphorus concentrations. Phosphorus accumulation can take place in some soils as a result of over-fertilization. Accumulation occurs when the amount applied exceeds the amount removed by crops. Phosphorus applied to fields as inorganic fertilizer or manure can move into bodies of water through erosion and runoff events. Phosphorus enrichment of water bodies can accelerate eutrophication (the natural aging process of lakes and streams) leading to excessive algal growth, oxygen deficiency, and fish mortality. Therefore application should be based on existing soil-fertility levels, manure nutrient content, crop nutrient needs, site limitations, slope, runoff potential, and leaching potential.

REFERENCES

1. Agricultural Statistics, ISBN 0-16-036158-3; U.S. Department of Agriculture–National Agricultural Statistics Service: Washington, DC, 2001; http://www.usda.gov/nass/pubs/agstats.htm (accessed Jul 2001).
2. Kellog, R.L.; Lander, C.H.; Moffit, D.D.; Gollehon, N. *Manure and Nutrients Relative to the Capacity of Cropland and Pastureland to Assimilate Nutrients: Spatial and Temporal Trends for the United States*, Publication No. nps00-0579; U.S. Department of Agriculture–Natural Resources Conservation Service: Washington, DC, 2000; 1–140.
3. Solley, W.B.; Pierce, R.R.; Perlman, H.A. *Estimated Use of Water in the United States in 1995*, Circular 1200; U.S. Department of the Interior–U.S. Geological Survey: Washington, DC, http://water.usgs.gov/watuse/pdf1995/html (accessed Jun 2001).
4. Raabe, S.J.; Sweeten, J.M.; Stewart, B.R.; Reddell, D.L. Evaluation of Manure Flush System at Caged Layer Operations. Trans. ASAE **1984**, *27* (3), 852–858.
5. Collins, E.R., Jr.; Barker, J.C.; Carr, L.E.; Brodie, H.L.; Martin, J.H., Jr. *Poultry Waste Management Handbook*; NRAES-132, Natural Resource, Agricultural, and Engineering Service: Ithaca, New York, 1999; 1–64.
6. Miner, R.J.; Humenik, F.J.; Overcash, M.R. *Managing Livestock Wastes to Preserve Environmental Quality*; Iowa State University Press: Ames, Iowa, 2000.
7. Huhnke, R.L. Land Application of Livestock Manure, Cooperative Extension Service, Oklahoma State University, Extension Facts 1710, 1982.

Manure Management, Swine

Frank Humenik
North Carolina State University, Raleigh, North Carolina, U.S.A.

INTRODUCTION

Water conservation is a major goal in swine production, so national recommendations are not to use any fresh water for waste management in swine facilities. One exception is when there is no pond or lagoon liquid to recycle for manure collection pits or hosing solid concrete floors. Fresh water is used for drinking and fogging for animal cooling but this water is for animal production and not waste management.

On a total farm basis, swine drink about 5 gal per day and produce about 1.6 gal of waste.[1,2] Water often enters the waste stream from fogging, cleaning water, and waterer overflow. Prompt waterer valve maintenance keeps water overflow at a minimum.

Waste can be stored in an underfloor pit for long periods or shorter periods with pull–plug systems. Waste can be removed frequently with waterwash or flushing systems. Solid concrete floors may be scraped or hosed to a collection gutter for cleaning as often as daily.

FLUSHING SYSTEMS

In a flush system, large volume of water flows down a sloped, shallow gutter or alley. The water carries waste to a lagoon. There are underslat gutters, which collect waste from swine houses with either totally slotted or partially slotted floors. Narrow open gutters, which are used primarily in hog finishing buildings, attract hogs to the channel and induces dunging, helping to "toilet train" the animals.[3]

Water should be recycled from a lagoon, earth basin, or a holding pond for flushing. In a recycling flush system, a pump transports the water to a flush tank at the high end of the gutter in the building. The flush tank periodically discharges water into the gutter. Flush frequency is determined by the rate at which water is pumped into the tank or timer to open the tank valve. The minimum total flush volume to clean wastes varies from 4 gal per day for nursery pigs to 15 gal per day for finishing pigs to 25 gal per day for gestating sows.[3]

Recommended maximum gutter length is 125 ft. For gutters, 125 ft–250 ft long, both ends of the gutter are sloped so they flush towards the middle of the building length.

RECIRCULATION FLUSH PITS

Recirculation flush pits are a modification of the gutter flushing concept. Their design evolved to help alleviate pit odor problems in remodeled buildings, but they are also being installed in new swine, beef, and dairy buildings. They also solve some of the problems associated with flushing systems, such as mechanical failure of flush tanks, failure of small continuously running pumps, and salt precipitate forming in continuously used small diameter lagoon recycling pipes. The pit is usually under a partly slotted floor and is relatively shallow—2 in.–4 in. deep on the high end and sloped from 1 ft / 20 in. (for swine buildings) to 1.5% toward the outlet end. The pit is flushed twice a week to a lagoon and refilled with cleaner lagoon water. Initial cost is somewhat greater than for flush systems because of the large recycling pump and pipe (often 3 ft–6 ft diameter). However, the system can be shut down and drained after each use, which reduces contact with corrosive lagoon water.[3]

PIT STORAGE SYSTEMS

The frequency of pit emptying is dependent on the waste utilization plan for each farm. Long storage periods are used when waste is applied to land as fertilizer several times a year. Pull–plug systems are emptied on a more frequent basis, sometimes 2–3 times a week.

Recycled lagoon or pond water should be used as precharge bottom water to facilitate solids removal when emptying and can vary from 6 in. to about 2 ft. For pull–plug systems, the plug is pulled several times a week thus allowing the total waste contents to empty to a lagoon. This more frequent emptying reduces odor and ammonia volatilization. There is also less moisture in the house than with open-gutter flushing systems, which remain wet after a flushing event. Emptying frequency depends upon facility management, which generally

Encyclopedia of Water Science
DOI: 10.1081/E-EWS 120010301

directs only emptying several houses per day but frequently enough to minimize odor and ammonia volatilization.

WATER CONSERVATION GOALS

Water conservation goals to minimize the use of fresh water for the waste management system minimizes the volume of water that must be handled and thus the required size of system components. Pressure washers, which reduce the amount of fresh water used, are recommended for building cleaning between herds which is about 2.4 times per year. Pressure washers reduce fresh water use by about 50%. Fresh water use can also be reduced, by employing dripless nipple waterers in controlling temperature by ventilation in totally enclosed housing units.

CONCLUSION

Recommendations for minimal fresh water use in swine production facilities and water conservation for waste management result in reduced equipment operation, reduced treatment and storage unit sizes, reduced cost, and improved environmental quality.

REFERENCES

1. MWPS. *Livestock Waste Facilities Handbook*; Midwest Plan Service, Iowa State University: Ames, IA, 1985; 3.12–3.24.
2. MWPS. *Private Water Sytems Handbook*, 4th Ed.; MWPS-14; Midwest Plan Service, Iowa State University: Ames, IA, 1992; p. 4.
3. MWPS. *Livestock Waste Facilities Handbook*, 2nd Ed.; MWPS-18; Midwest Plan Service, Iowa State University: Ames, IA, 1993; 3.12–3.24.

Marketing

Lal K. Almas
W. Arden Colette
West Texas A&M University, Canyon, Texas, U.S.A.

INTRODUCTION

Water is necessary for all life on Earth. It is a finite natural resource, which means that the total amount of water available is limited. According to the Environmental Protection Agency,[1] 97% of the earth's water is salt water stored in the oceans and the remaining 3% of the earth's water is fresh water. Only 1% of the earth's fresh water is of the quality and in the location to be acceptable for human consumption.

The population of the world has increased dramatically during the past 50 yr, but the water resources are finite, and irrigated agriculture uses approximately 70% of world's supplies of developed water.[2] With increasing urban populations come industrial users and power plants, both of which need water. In-stream water for recreation, wildlife, or other environmental purposes is in addition to the increased urban demands. Increasing demand because of increasing population makes available water supply inadequate. Augmentation of fresh water supplies from sea water is currently prohibitively expensive; therefore, new demands for water must be met by reallocation of existing supplies.[3] Allocation of scarce supply among unlimited demands requires allocation systems. Scarcity of water has enticed people to develop and implement procedures to facilitate water marketing that can serve as a tool for efficient allocation of water among different users.

SOCIAL VS. ECONOMIC ALLOCATION

Optimizing water allocation requires that net marginal value of a unit of water diverted and not returned to the source, i.e., consumptive use, equals the sum of the net marginal values of nonrival, nonconsumptive use. Water resource development, transfer, and use are subject to social and legal factors that contribute to uncertainties and externalities that may preclude attaining an optimal economic condition. Among these factors is the community value of water. Many argue that water is not just a commodity but also a necessity to the economy and social structure of a society, and that a threat to the system for allocating water is a threat to the communal enterprise. The community value of water leads to a divergence

between social and private benefits of water use and failure of the market to achieve a Pareto efficient allocation. This failure provides a strong argument for central management of water allocation.[4]

A competitive market may be an efficient allocative process for achieving maximum profit/wealth; it is not an efficient allocative institution for achieving social goals because of infrastructure dislocation.[5] It is not particularly efficient in achieving community goals such as ecological preservation, species protection, and welfare promotion for future generations. Therefore, for efficient water allocation both the social and the economic benefits should be considered.

The incentive for water reallocation is based on the presumption that economic gains will be captured by reallocating water from lower valued to higher valued uses.[6] As demand increases and the cost to obtain additional water increases beyond lower valued current uses, economic pressure is applied to reallocate water to higher valued uses. Typically, the market mechanism plays a role in reallocating resources from lower valued to higher valued uses.

WATER RIGHTS

Historically, water has been used to promote development. Water rights (ownership or right to use) were established to reduce uncertainty especially in agricultural production. Since agriculture was one of the earliest fields to use water and, in accordance with the prior appropriation doctrine of first in time and first in rights, farmers hold a large share and many of the most senior or reliable water rights. Despite rapid urbanization, most of the water is still being used for agriculture. Howe, Lazo, and Weber[7] stated that, according to the U.S. Geological Survey data, "80% of all water diversions and nearly 90% of all water consumption in the western United States occur in irrigated agriculture." However, the value of water used in agriculture is often lower than the value of water for other uses.[8] Therefore, it should not be surprising that irrigated agriculture is the source of water for many water right transfers.

A water right is the right to use a specific amount of water for a specific purpose at a specific place and time.

Encyclopedia of Water Science
DOI: 10.1081/E-EWS 120010377

A water right can be bought, sold, bequeathed, or inherited like any other property right.[9] The conveyance, however, is subject to the limitation that other users of the same watercourse cannot be harmed. Although water rights are property rights, they lack at least three of the four elements necessary for the efficient functioning of a market, i.e., universality, exclusivity, transferability, and enforceability.[10]

Economic growth and prosperity are dependent upon the availability of water and water rights. Five types of water right systems include riparian, appropriative, use permits, entitlements, and mutual stock. The water right system must provide security for the right holder and flexibility to accommodate new uses. The need for flexibility is reflected in the diversity of emerging marketing systems.[11] Sales, leases, options, and negotiated adjustments occur within each of these kinds of systems.

WATER MARKETING

For every price there is a quantity supplied and quantity demanded for each use. The difference between quantity supplied and quantity demanded equals the excess supply or excess demand for water for that use. Water marketing may be defined as the selling of excess water supply from one use to individuals or institutions for uses where there is excess demand. In other words a water market is an arrangement in which holders of water rights, trade them with each other or to outside parties. The trade transactions relevant to water marketing can occur either through the sale of a water right permit or through the sale of water by means of a water supply contract.

Most resource economists agree that opportunities to develop traditional large-scale water reservoirs to increase surface water supplies are limited because of rising economic, environmental, and political difficulties. The cost of developing new groundwater supplies has also increased many folds as a result of ever increasing depletion of aquifers. This has led to difficulties in mining of groundwater aquifers. The only feasible option to cope with ever increasing demand for water in deficit areas is the reallocation of water through water marketing.

Water marketing could be an inexpensive way to reallocate water in areas where water shortage exists. Water marketing provides reallocation of water, particularly to large metropolitan areas and during water shortage periods due to severe drought. Water marketing can also help in providing and ensuring water supplies for environmental as well as recreational needs. Water marketing acts as an incentive for water conservation and efficient use by those who control this natural resource. Therefore, the reallocation of water through water marketing promotes political and social harmony among the groups with excess supply and excess demand of water.

Water scarcity and defined property rights in water are two requirements for water marketing to occur. Water markets develop when buyers of this commodity have no other option to secure a certain and consistent water supply, and sellers would be able to accrue more net benefits by marketing the water than using it in its existing form. However, success of water marketing will depend on a combination of economic, legal, institutional, environmental, and technical factors. The potential economic gains from water trade will motivate water transfers from lower value to higher value uses.[12]

Water transfers usually involve a dispute over the issue of compensation between those in the basin-of-origin and the receiving area. Four mechanisms for resolving water disputes include legislation, litigation, water markets, and negotiation/mediation. Legislation and litigation are more common and negotiation/mediation is considered to be a localized procedure. Water markets can resolve conflicts by establishing a price acceptable to all parties. They can also provide efficiency by determining the highest and best resource use by incorporating all costs in the transfer of water. Two conditions are necessary for optimal transfer of water. These conditions are that the transfer is the least cost alternative and that the benefits exceed the losses to the area of origin including downstream basins. Transfer related costs as well as operation and maintenance costs of the movement of water are considered. Jordan[13] identified the following five prerequisites for an effective system of marketing water:

1. Water rights must be clearly defined, meaning that there must be clear title to the water to be transferred or marketed.
2. The water right to be transferred must be quantifiable.
3. Institutional support must be available to administer water rights.
4. The infrastructure must be available or be feasible to move water between buyer and seller.
5. Externality issues are included in the marketing system to provide an efficient transfer of water.

FUTURE OF WATER MARKETING

Water right markets emerging all over the world are still in their infancy and are subject to several challenges. One common problem of all water markets is lack of information. Buyers and sellers face difficulties in finding trading partners. Limited market information forces affects prices and terms of trade. Another problem challenging the future of water marketing is the effectiveness of various governmental agencies responsible for approval of each transaction. Buyers and sellers complain that the approval process is slow, costly, and limits water market growth. However, the approval process must ensure that resulting transfers do not impair other water right claims.

Despite increasing recognition of the benefits of and need for market exchanges of water, barriers to functioning water markets include equity protection, state protection of authority over in-state water, uncertainty of the status of federal agencies involved, and state and regional water agency inconsistencies in policies for defining and approving transfers, quantities, prices, and lease costs. Other problems with water transfer include utilization of salvaged or conserved water, temporary transfers, and introduction of public interest and public trust doctrines in administrative and judicial decision-making.

CONCLUSION

Water rights and municipal water supply systems are two of the fastest-growing market areas in the water marketing industry. Many federal, state, and local agencies involved with the marketing of water rights in the United States have started streamlining the approval process. Water marketing is not confined to the United States. For example, water marketing also occurs in Australia, Chile, and Mexico, where water markets have encouraged conservation and stimulated economic opportunities. The potential for water markets is also expanding in Africa, Asia, and the Middle East. In Pakistan, young farmers lease water from established farmers who can afford to develop wells. Development of markets for water rights and water supply systems is a global phenomenon. Water marketing systems are providing potential buyers and sellers with the incentive to conserve water and are helping globally to achieve equitable and efficient water reallocation.

REFERENCES

1. http://www.epa.gov/seahome/groundwater/src/supply. htm#supply (accessed July 2002).

2. Seckler, D.; Amarasinghe, U.; Molden, D.; de Silva, R.; Barker, R. *World Water Demand and Supply, 1990 to 2025: Scenarios and Issues*; Research Report 19, International Water Management Institute: Colombo, Sri Lanka, 1998.

3. Gould, G.A. Transfer of Water Rights. Nat. Resour. J. **1989**, *29*, 457–477.

4. Brajer, V.; Martin, W.E. Water Rights Markets: Social and Legal Considerations. Am. J. Econ. Sociol. **1990**, *49* (1), 35–44.

5. Chan, A.H. To Market or not to Market: Allocating Water Rights in New Mexico. Nat. Resour. J. **1989**, *29* (3), 629–643.

6. Saliba, B.C.; Bush, D.B. *Water Markets in Theory and Practice: Market Transfers, Water Values and Public Policy*; Studies in Water Policy and Management No. 12, Westview Press: Boulder, CO, 1987.

7. Howe, C.W.; Lazo, J.K.; Weber, K.R. The Economic Impact of Agriculture to Urban Transfers on the Area of Origin: A Case Study of the Arkansas River Valley in Colorado. Am. J. Agric. Econ. **1990**, *72* (5), 2300–2304.

8. Gibbons, D.C. *The Economic Value of Water*; Resources for the Future, Inc.: Washington, DC, 1986.

9. Somach, S.A. Property Rights in Water: An Essential Element of Economic and Social Development. In *Water Resources Law*, Proceedings of the National Symposium on Water Resource Law, Dec 15–16; American Society of Agricultural Engineers: St. Joseph, MI, 1986; ASAE Publication 10–86, 28–34.

10. McCormick, Z. Institutional Barriers to Water Marketing in the West. J. Am. Water Resour. Assoc. **1994**, *30* (6), 953–961.

11. Bush, D.B. Dealing for Water in the West, Water Rights as Commodities. J. Am. Water Works Assoc. **1988**, *24*, 30–37.

12. Kaiser, R. Legal and Institutional Barriers to Water Marketing in Texas, Technical Report #TR-167; Texas Water Resources Institute: College Station, 1995.

13. Jordan, J.L. Externalities, Water Prices, and Water Transfers. J. Am. Water Resour. Assoc. **1999**, *35* (5), 1007–1013.

Matric Potential

Melvin T. Tyree
United States Department of Agriculture (USDA), Burlington, Vermont, U.S.A.

INTRODUCTION

Matric potential, τ, is a component of water potential, Ψ, but has different meanings in plant physiology vs. soil science. A rigorous definition of τ requires a reference to principles of thermodynamics (both classical and irreversible thermodynamics). A rigorous treatment is beyond the scope of this brief overview. Readers interested in a detailed definition are advised to read the article of Ref. [5], but should be prepared to wade through 227 equations in a terse, 25-page article requiring a firm grasp of thermodynamics. Less detailed treatments can be found in Ref. [6] for soils and Ref. [7] for plant tissue.

MATRIC POTENTIAL IN PLANT PHYSIOLOGY

Water potential, Ψ, is the chemical potential of water expressed in pressure units. In plant tissues, Ψ is traditionally written as the sum of three components:[4]

$$\Psi = P + \pi + \tau$$

where P, π, and τ are the pressure, osmotic, and matric potentials, respectively. Ψ and its components are intensive variables that vary from point to point in a cell and tissue.[10] Some people have attempted to define τ in terms of a measuring procedure without regard to thermodynamic principles (e.g., Refs. [1,9,11]), but such attempts have been unsatisfactory because the approaches were derived from tissue properties obtained by volume or weight averaging over the heterogeneous phases of vacuole, cytoplasm, and cell wall. A satisfactory definition of τ must be based upon the consideration of it as an intensive property acting at a point. A more correct approach has been taken in Ref. [8] for plant tissues, and in Ref. [3] for soils.

The forces contributing to τ are short range, and influence only a small fraction of the total water in plants when the water is near a solid surface. At uncharged surfaces, the force interactions are largely London–van der Waals forces or hydrogen bonds and extend for only one or two water molecules, 0.3 nm–0.6 nm. At charged surfaces, e.g., cell walls, there is a concentration of negative charges that tends to cause an aggregation of

cations in the surrounding electrolyte solution and contributes to low localized values of π, which equals $-RTC$, where R is the gas constant, T, the Kelvin temperature, and C, the localized concentration of all solutes including electrolytes in osmol kg^{-1}. The impact of the charged surfaces on π has been calculated by using the Gouy–Chapman theory, which predicts the influence of fixed charges on ion accumulation near the charges. Tyree and Karamanos[8] have shown that the localized concentrations can exceed $2M$ resulting in π below -5 MPa. Soil scientists tend to include most of π and some other effects in τ (discussed below) but this is not done by plant physiologists. The argument is that if pressure and concentration are already accounted for in P and π, then τ ought to be something independent.

Within the electric fields of the surface charge, there is another effect that reduces the energy of water molecules, i.e., the interaction of the water dipole with the electric field. As both plant cell wall surfaces and clay surfaces have a net negative charge, the water dipole tends to be oriented with the positive (hydrogen) end aligned nearer the charged surface than the negative (oxygen) end of the dipole. The net effect is a lowering of the free energy of the water molecules within the electric field. In terms of water potential, the magnitude of the effect is given by

$$\tau = -\left(\frac{N_0{}^2 P_0{}^2}{3 V_w RT}\right) F^2$$

where N_0 is the Avogadro number, V_w, the volume of a mole of water, P_0, the dipole moment of water, and F, the electric field at the point where τ is evaluated.

Fig. 1 shows the magnitudes of the components of Ψ near a charged surface computed from the Gouy–Chapman theory for a charged surface with a net charge of -0.4 C m^{-2} in equilibrium with 10 mM NaCl solution at $\Psi = -1.5$ MPa. A large positive pressure develops near the charged surface because of the force with which water molecules are drawn towards the charged surface. Given that both π and τ are very negative near charged surfaces, large positive values of P are necessary to make $\Psi = -1.5$ MPa everywhere. Tyree and Karamanos[8] go on to argue that even in cell walls, where the ratio of charged solids to water is about 1:1 that the influence of τ extends only to a small fraction of the water volume, and

Encyclopedia of Water Science
DOI: 10.1081/E-EWS 120010175

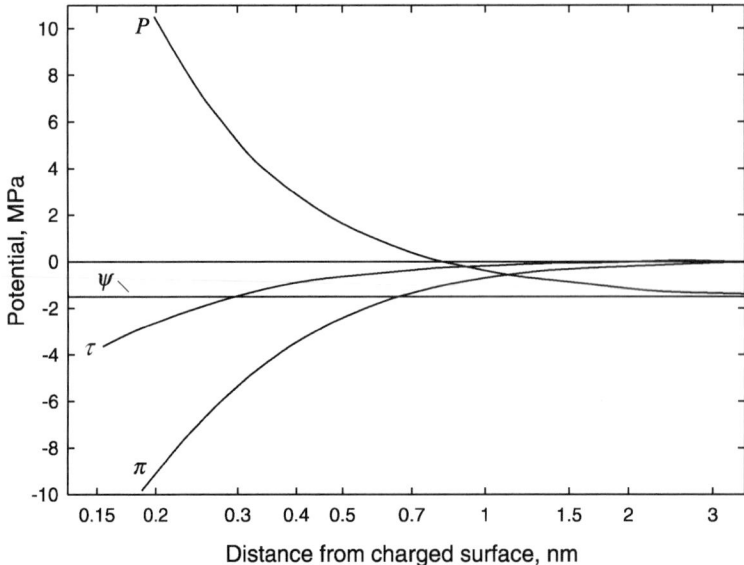

Fig. 1 The components of water potential near a charged surface according to the Gouy–Chapman double layer theory with a surface charge of $-0.4\,C\,m^{-2}$. The univalent ion concentration outside the double layer is taken as $10\,mM$. π and τ are calculated at distances from the charged surface from the calculated electrical potential and electric field, respectively. $P = \Psi + \pi + \tau$ with $\Psi = -1.5\,MPa$. (From Ref. [8].)

will begin to influence measures of Ψ only when water potentials approach $-14\,MPa$, at which point most plants are dead anyway. So, as defined by plant physiologists, τ can usually be ignored.

MATRIC POTENTIAL IN SOIL SCIENCE

Matric potential in soil science, τ_s, was originally defined in terms of the instrument(s) used to measure τ_s. Note that the subscript, s, is used in τ_s in soil science to distinguish it from the τ in plant physiology. In situ measurement of τ_s in soils is made with a tensiometer, which consists of a water-filled tube with an attached pressure sensor. The fluid in the tube makes contact with the soil water through a porous plate (often ceramic). When the soil is wet, the fluid in the tensiometer is in good contact with the soil water and is at a pressure of $0\,MPa$ relative to atmospheric pressure. As the soil dries, the fluid pressure drops below atmospheric. When the pressure drops below that of a perfect vacuum ($-0.1013\,MPa$), the water column usually cavitates, i.e., an air bubble forms because of a breakdown in the adhesion of water to the solid surfaces of the tensiometer. Cavitations limit the range of useful measurement of τ_s using tensiometers, as most plants can function well to $\tau_s < -1.5\,MPa$. It is possible to make fluid-filled pressure-measuring devices (called cell pressure probes) that can measure fluid pressures down to

$-1\,MPa$ in plant cells, but these have never been used in soils and probably would not work reliably.

In order to measure τ_s below $-0.1\,MPa$, the soil samples have to be removed and placed in another apparatus. One such system is a pressure plate apparatus. The soil is placed at the bottom of a pressure chamber. The bottom of the chamber is porous so that water can pass through the porous plate when air pressure above the soil is increased high enough to extract water from the soil. The value of τ_s is equated to $-$(the applied pressure) at an incipient water extraction. Water can also be extracted from the soil in a centrifuge. The soil is spun in the centrifuge tube with a porous bottom until the centrifugal force (expressed in pressure units) is sufficient to extract water and τ_s is equated to $-$(the centrifugal force).

A number of papers have been written to discuss what, based on thermodynamic theory, is measured by the instruments described earlier. Passioura[6] equates τ_s approximately with Ψ. To be precise, $\tau_s = \Psi - \pi_D$, where π_D is the osmotic potential of the "equilibrium dialysate," i.e., the osmotic potential of the soil solutes that can pass through the porous plate of the tensiometer or pressure plate apparatus or centrifuge tube. The ions in solution in the ion "cloud" near the charged surface of soil particles would not be included in π_D, because these ions are not extractable. In most soils, π_D is usually $\geq -0.02\,MPa$; hence in drying soils, $\tau_s \cong \Psi$ within good tolerance. In a more recent exhaustive treatment of the theory behind τ_s, it appears that τ_s is identified exactly with Ψ in the equilibrium vapor phase of soils, i.e., see Eq. 202 in Ref. [5]. Hence, this meaning of τ_s is identical to the meaning of Ψ, which is often also measured on plant tissue by using the equilibrium vapor phase.

CONCLUSIONS

Matric potential as used by soil scientists is nearly identical with water potential as used by plant physiologists. The main difference is that plant physiologists divide water potential into two quantities that frequently can be measured independently, i.e., pressure potential, P, and osmotic potential, π. Soil physicists like to equate matric potential in plants with "capillary or adsorption forces which in a plant are forces such as those at the cell walls."[2] However, this definition is equivalent to water potential, Ψ, and does nothing to help elucidate the osmotic relations of living cells that can be quantified only by independent measures of P and π. In the older plant physiology literature, some people attempted to come up with a different meaning of matric potential, but this approach has been discredited.[7,8] It is unfortunate that these two closely allied sciences should use different words to describe the same quantity (matric potential vs. water potential), but the attentive reader can usually distinguish the meaning of matric potential from the context of scientific reports.

REFERENCES

1. Acock, B. An Equilibrium Model of Leaf Water Potentials Which Separates Intra- and Extracellular Potentials. Aust. J. Plant Physiol. **1975**, *2*, 253–263.
2. Baver, L.D.; Gardner, W.H.; Gardner, W.R. *Soil Physics*, 4th Ed.; Wiley: New York, 1972; 293–295.
3. Bolt, G.H.; Miller, R.D. Calculation of Total and Component Potentials of Water in Soils. Trans. Am. Geophys. Union **1958**, *39*, 917–928.
4. Dainty, J. Water Relations of Plant Cells. In *Encyclopedia of Plant Physiology N.S.*; Lüttge, U., Pitman, M.G., Eds.; Springer-Verlag: Berlin, 1976; Vol. 2A, 12–35.
5. Nitao, J.J.; Bear, J. Potentials and Their Role in Transport in Porous Media. Water Resour. Res. **1996**, *32*, 225–250.
6. Passioura, J.B. The Meaning of Matric Potential. J. Exp. Bot. **1980**, *31*, 1161–1169.
7. Tyree, M.T.; Jarvis, P.G. Water in Tissues and Cells. In *Encyclopedia of Plant Physiology N.S. Physiological Plant Ecology*; Lange, O., Nobel, P.S., Osmond, C.B., Eds.; Springer-Verlag: Berlin, 1982; Vol. 12B, 35–77.
8. Tyree, M.T.; Karamanos, A.J. Water Stress as an Ecological Factor. In *Plants and Their Atmospheric Environment*; Grace, J., Ford, E.D., Jarvis, P.G., Eds.; Blackwell: Oxford, 1980; 237–261.
9. Warren Wilson, J. The Components of Leaf Water Potential. Aust. J. Biol. Sci. **1967**, *20*, 329–347.
10. Weatherley, P.E. Some Aspects of Water Relations. Adv. Bot. Res. **1970**, *3*, 171–206.
11. Wiebe, H.W. Matric Potential of Several Plant Tissues and Biocolloids. Plant Physiol. **1966**, *41*, 1439–1442.

Microbial Sampling

Suresh D. Pillai
Texas A&M University, College Station, Texas, U.S.A.

George Di Giovanni
Texas A&M University Agricultural Research and Extension Center,
El Paso, Texas, U.S.A.

INTRODUCTION

One of the critical requirements in designing a water sampling plan for microbial analysis is a clear understanding as to the overall objectives behind the sampling, what sampling equipment is available, and what type of analyses are going to be conducted on the samples.

The focus of this article is to provide an overview of the methods to sample water bodies to detect fecal contamination. There are distinct differences in the type of sampling methods that one would have to use depending on whether groundwater, surface water, or distribution system (finished drinking water) is being sampled. The differences arise from the need to retrieve the samples using specialized sampling equipment and sample concentration methods.

OVERVIEW

Groundwater Sampling

A key prerequisite in obtaining a representative groundwater sample is to have a properly installed "monitoring" or "sampling" well. The design is of obvious importance since a simple hole in the ground will not be representative of the aquifer. Attention has to be paid to the proper "setting" of the well, the selection of the appropriate filter pack, and proper "well development." Well development, refers to the process by which the aquifer's natural hydrodynamics are restored in the aquifer around the well after the installation of the well. The USEPA generalizes that three well volumes be removed from a well before sampling. Thus, when obtaining groundwater from a well, information about the well depth, pump setting (height at which the pump draws in water), well diameter, standing water level, volume of all holding tanks, pressure tanks, and connecting pipelines should be obtained so that an adequate amount of purging can be performed before an authentic "groundwater" sample is obtained. Information on drainage features, proximity to septic systems, and other features that could influence water quality should

also be documented. Typically, sample bottles are obtained presterilized from the laboratory. It is essential that the sampler coordinate sample delivery with the laboratory so that the short holding times can be achieved.

Some of the equipment for sampling are bailers, grab samplers, and submersible pumps.[1] Bailers are one of the least expensive methods of sampling and are best suited for wells that are shallow or slow to recharge and for the collection of small volume samples. One-time use bailers or multiple-use bailers can be used. The sampling materials must be cleaned and disinfected between samples. The disinfection could be achieved by soaking the bailers in large containers containing a 1%–2% chlorine (bleach) solution. However, care must be taken to remove all residual chlorine from the bailers by thoroughly washing in clean water. Grab samplers are different from bailers in that samples can be obtained from discrete depths. Submersible pumps are one of the better ways of collecting groundwater samples. However, they can be relatively expensive and require the need for electrical power at the field site. Pumps are ideally suited for use in larger wells or when large volumes of water need to be collected and passed through sample concentrators for virus and protozoan sampling. Some groundwater wells may have chlorinators installed in them. It is important that these chlorinators are disconnected before samples are collected or the samples are obtained at a spot prior to the chlorinator input into the line. All public water supply wells are required by law to have a sampling spigot prior to the chlorinator.

Surface Water Sampling

Unlike groundwater sampling, sampling from surface waters is relatively straightforward. Grab samples (for bacteriological analysis) or portable pumps (for viral and protozoan analysis) can be employed. However, attention should be paid that the sample being collected is as representative of the surface water source as possible. When sampling rivers and canals, effort should be made to collect the sample as far as away from the bank as possible. When sampling lakes, the use of a boat is desirable. If

Encyclopedia of Water Science
DOI: 10.1081/E-EWS 120010184

wading, the sampler should slowly wade upstream taking the sample upstream and ahead of any wading-induced agitated sediments. Since no guidelines have been developed for specific sampling locations, the sampler must predetermine the locations that provide a representative water sample. For example, the sample can be taken ahead of a water intake, if the objective is to understand the source water quality. Microbial populations can be highly variable with a stream and typically, samples are taken from a midpoint in a stream rather than in shallow pools or riffle areas. Multiple samples taken from riffles, runs, and pools provide a better representation of the microbial populations, however, the sampling and subsequent analysis may be cost prohibitive.

Distribution System Sampling

This is probably one of the easiest to sample in that spigots and faucets on the distribution lines can be used to collect the samples. However, drinking water distribution lines have residual chlorine present within the system. It is important that this disinfectant residual be removed especially when virus sampling is conducted. Sodium thiosulfate is often used to remove residual chlorine. Faucets and spigots within the distribution system (as well as in groundwater wells) may harbor microbial populations within them as biofilms. Even though it is impossible to remove the biofilm within distribution lines, attention must be paid to remove as many indigenous microbial populations from the sampling spigot (faucet) as possible. Heat surface sterilization (using flame or torch) or chemical disinfections can be used. It is critical that the water is allowed to run for at least 10 min–15 min after these treatments.

Sampling for Bacteriological Analysis

Since the sample volume for bacteriological analysis is always around 100 ml–1000 ml, grab samples are often the method of choice. However, the sample container should be sterile, clean latex gloves should be worn (to prevent the sampler from contaminating the sampling port

or sample), and in the case of groundwater or distribution system samples, the water should be allowed to run for at least 3 well volumes or 10 min, respectively. (There are, however, times when one may want to collect a sample directly from the tap/spigot to determine the quality of the distribution system). The sample bottle should be filled up to the desired volume and the bottle should be removed sideways from the flow of water. The cap has to be replaced and after appropriate labeling the sample bottle has to be placed in a clean cooler containing blue-ice or wet-ice and maintained at or below 4°C. The specific volume that is collected will depend on the number of bacteria that are being screened. There are recommended volumes for the different bacteriological detection methods.[2] There are different maximum holding time recommendations depending on the organism. For *Escherichia coli* the samples should be analyzed within 8 hr.

Sampling for Viral Analysis

Enteric viruses are of particular concern to human health since they have low infectious doses. These viruses are very often found in much lower concentrations than bacteria in environmental waters. The current USEPA recommended method for sampling and concentrating viruses require sampling large volumes of water and concentrating the viruses on positively charged filters. The filter used for concentrating the enteric viruses from water samples is the 1MDS filter ZetaPor, virosorb (Cuno, Inc., Meridian, CT). (The retention of viruses on to these filters is thought to occur through electrostatic attraction between the negatively charged virus particles and the positively charged filter.) These filters have to be contained within a filter housing. The filter and filter housing is connected in series to a backflow control valve, backflow regulator, and a flow meter (Fig. 1). Typically, when sampling groundwater for viruses, 500 gal are passed through the filter before the filter is removed and shipped to the laboratory for analysis. For distribution system samples, as much as 1500 gal need to be passed through the positively

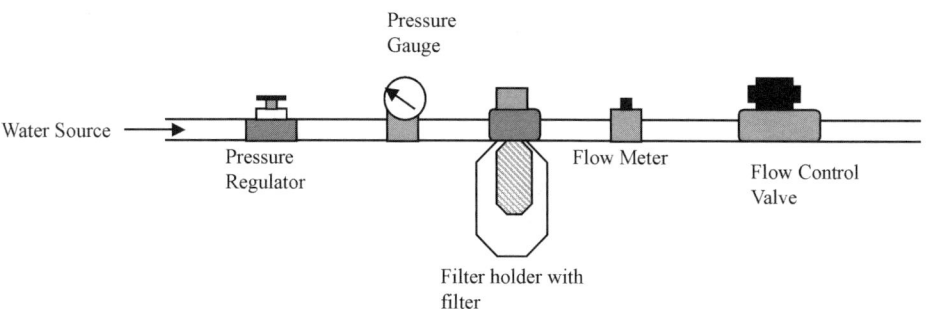

Fig. 1 Filtration setup for viruses and protozoa.

Table 1 Guidance for selection of water sample collection and concentration methods for *Giardia* and *Cryptosporidium* analyses

| | Application | | | |
Procedure	Typical sample volume	Water turbidity	Advantages	Disadvantages
Grab sample (concentration using centrifugation)	1 L–20 L (wastewater and raw surface water)	Low to high	Easy to collect, no filter costs	Samples greater than 1 L are time consuming to handle and concentrate
USEPA ICR (yarn wound filters)	≤ 100 L (surface water); ≥ 1000 L (finished water)	Low to high	High filtration rate, low cost and ease of use	Variable efficiency of concentration, time consuming processing
USEPA 1622/1623 (pleated membrane capsule filters)	≤ 100 L (surface water); ≥ 1000 L (finished water)	Low to moderate	Good retention and oocyst recovery, ease of use	Expensive, slow filtration rate
USEPA 1622/1623 (compressed foam filters)	≤ 50 L (surface water); ≥ 1000 L (finished water)	Low to high	Excellent retention and oocyst recovery	Awkward handling, decontamination required

charged filters to screen for viruses. A major drawback of using cartridge flow-through filters is that, depending on the amount of debris in the sample the filters could get clogged. Using a prefilter to avoid clogging is counterproductive in that very often viruses get trapped and adsorbed in these prefilters rather than being adsorbed in the 1MDS filter. In addition to the 1MDS cartridge filters, tangential flow and hollow fiber filtration have also shown promise as virus concentration methods. Thus, in the case of virus sampling, sampling involves sample concentration as well. Samples should be analyzed for enteric viruses within 72 hr of collection.

Male-specific coliphages are viruses that infect specific coliforms bacteria. Such viruses are termed bacteriophages. Male-specific coliphages have been shown to serve as efficient fecal contamination indicators. The USEPA methods 1601 and 1602 describe the methods that can be used to sample and detect coliphages.[3,4] Unlike enteric viruses, grab samples ranging from 100 ml to 1000 ml can be used for coliphage analysis. Since coliphages are viruses, the holding times should not exceed 72 hr.

Sampling for Protozoan Analysis

The selection of an appropriate water sampling and concentration method for *Cryptosporidium* and *Giardia* greatly depends upon the water sample matrix (e.g., surface water, finished drinking water, or wastewater), the volume to be concentrated, and the anticipated density of organisms. Concentration is typically achieved through various filtration and centrifugation steps. Unfortunately, these concentration methods also concentrate inorganic and organic debris and nontarget organisms. In addition, downstream sample purification and detection methods should also be considered. In wastewaters where the numbers of *Cryptosporidium* and *Giardia* are expected to be high, a 100 ml–1000 ml grab sample, directly concentrated by centrifugation may be sufficient. In

contrast, for finished drinking water in which the number of organisms is expected to be low, filtration of 100 L–1000 L or greater may be required. The concentration of raw surface water requires the most consideration since characteristics such as turbidity or presence of algae will vary significantly and can greatly affect the concentration procedure. If clogging occurs, then the actual volume that was concentrated should be noted.

There are several methods for the sampling and concentration of *Cryptosporidium* and *Giardia* in surface water and finished drinking water. The USEPA Information Collection Rule (ICR) method uses a polypropylene yarn wound filter for the concentration of raw surface water samples.[5] In contrast to the electrostatic attachment of viruses to the positively charged filter in the virus sampling/concentration procedure, the yarn wound filter physically traps the oocysts and cysts. The yarn wound filter is placed in a suitable filter housing and placed in series along with the pressure regulator, pressure gauge, flow meter, and flow control valve (Fig. 1). Although the yarn wound filter method is relatively inexpensive, studies have shown that the efficiencies can be relatively low and the filter processing methods can be extremely labor-intensive. The recent USEPA Methods 1622 and 1623 include several different options for filtration including capsule membrane filters and compressed foam filters.[6,7] Methods 1622 and 1623 use immunomagnetic separation for purification of protozoa. Recent studies have also shown that hollow fiber filters have the ability to concentrate protozoa. The samples should be eluted and concentrated from the filters within 96 hr of sample collection. Guidance for the selection of concentration methods is provided in Table 1.

CONCLUSION

Microbial sampling is a critical component of any environmental assessment. Given the complexities associated with sampling for different microorganisms, it is critical that careful attention be paid to the sampling objectives. While guidelines and sampling protocols have been established the responsibility of obtaining the most appropriate sample still lies with the sampler who must determine the data quality objectives and develop a sampling plan to meet those objectives. Many states are developing specific water-use standards for surface water based on use as recreational water bodies and drinking water supplies. These regulations can have profound implications for concentrated animal operators. Meeting these water-use standards and still maintaining profitable animal production levels will pose a challenge to the regulatory agencies as well as the agricultural community.

ACKNOWLEDGMENTS

This work was supported in part by funds from the USDA/CSREES IFAFS grant 00-52102-9637, the USDA/CSREES grant 2001-34461-10405, Hatch grant H8708, and a USEPA STAR program grant.

REFERENCES

1. Pillai, S.; Dowd, S. Groundwater Sampling for Microbial Analysis. In *Microbial Pathogens Within Aquifers, Principles and Protocols*; Pillai, S.D., Ed.; Springer-Verlag: Berlin, 1998.
2. Clesceri, L.S.; Eaton, A.D.; Greenberg, A.E. *Standard Methods for the Examination of Water and Wastewater*, 19th Ed.; Cleseri, L.S., Eaton, A.D., Greenberg, A.E., Eds.; American Public Health Association: Washington, DC, 1997.
3. USEPA, *Method 1601: Male-Specific (F +) and Somatic Coliphages in Water by Two-Step Enrichment Procedure*; Office of Water: Washington, DC, 2001.
4. USEPA, *Method 1602: Male-Specific (F +) and Somatic Coliphages in Water by the Single Agar Layer*; Office of Water: Washington, DC, 2001.
5. USEPA, *ICR Microbial Laboratory Manual*, EPA/600/R-95/178; Office of Research and Development, Government Printing Office: Washington, DC, 1996.
6. USEPA, *Method 1622: Cryptosporidium in Water by Filtration/IMS/FA*, EPA-821-R-01-026; Office of Research and Development, Government Printing Office: Washington, DC, 2001.
7. USEPA, *Method 1623: Cryptosporidium and Giardia in Water by Filtration/IMS/FA*, EPA-82-R-01-025; Office of Research and Development, Government Printing Office: Washington, DC, 2001.

Neuse River

Curtis J. Varnell
John Van Brahana
University of Arkansas, Fayetteville, Arkansas, U.S.A.

INTRODUCTION

The Neuse River of North Carolina has been significantly impacted by the influx of nutrients from the concentrated animal feeding operations (CAFOs) located in the watershed area. Phosphorus and nitrogen rates have increased dramatically causing subsequent algae blooms, fish kills, and eutrophication in the lakes, rivers, and estuaries along the waterway. Unpredictable weather events can intensify the problem and could have dramatic environmental consequences.

BACKGROUND

The Neuse River is one of the major rivers on the East Coast of the United States. The entirety of its 300-mi length is located in North Carolina. The river drains land in 19 counties that contain about one-sixth of the states' population. The river originates in north central North Carolina and flows in a southeasterly direction past Raleigh, Kinston, and New Berm and into the tidal basin of the second largest estuarine systems in the United States, the Albemarle–Pamlico estuary. What was once considered a pristine stream has, in recent years, been rated by the renowned environmental group, American Rivers, as one of the 20 most threatened rivers in North America.[1]

The Neuse carries the highest concentrations of total nitrogen and total phosphorus of any of the four rivers draining into the Pamlico and Albemarle Sounds even though it drains only 20% of the contributing land area.[2] In 1983, the North Carolina Environmental Commission classified the Falls Lake portion of the stream as a Nutrient Sensitive Area. In 1998, the lower Neuse River basin was added to that classification. Despite efforts by the various state organizations, phosphorus and nitrogen concentrations in the water of the river have not decreased.

FACTORS CONTRIBUTING TO ENVIRONMENTAL PROBLEMS

Many factors contribute to the environmental problems faced by the Neuse River. Roughly, 15% of the states' population live within the basin and the population is increasing rapidly. Wastewater effluent, storm water discharge, and urban run-off contribute a large percentile of the nutrient contamination in the basin. Currently, more than 400 point source discharge permits are active in the watershed and legislation is being enacted to lessen the amount of contaminants entering the river.

Nonpoint source run-off and shallow groundwater migration are now the most significant pollution source for North Carolina. The state is a large farming state and contains some of the largest CAFOs in the world. The state ranks number one in the nation in turkey production, number four in broiler production, and number two in the production of swine. All these produce excessive amounts of organic waste containing large amounts of nitrogen and phosphorus. Hog production is of extreme importance to the Neuse River because of the intense concentration in farm numbers found in the lower river basin and flood plain area.

Most of these farms are CAFO type productions containing hundreds or even thousands of swine at each farm location. These farms are part of integrated systems owned by huge corporations that have moved their business into North Carolina because of the relatively lax regulations that the state enforces. Hog production has increased over 270% since 1990, and will top out over 12 million hogs in the next few years.

The 10 million hogs (Fig. 1) now populating North Carolina's coastal region produce 19 million tons of waste each year.[3] Ammonia is also released as a by-product of swine production (Fig. 2). Two million pounds of ammonia per year are deposited by rainfall in the Neuse River basin from hog operations.[4] To put this into perspective, the wastewater plants of the entire state contributed only 2.1 million pounds of nitrogen in 1995.

The hogs are raised in confined barns containing hundreds of animals in close proximity. A lagoon is constructed to deposit waste materials from the large farms. In order to utilize the nutrients in the waste materials, water is withdrawn from the lagoons and sprayed onto adjacent pastures that use the nutrients to grow crops, usually various types of grasses required for cattle production. In an ideal system, the grasses would use up most of the nutrients; the cattle would consume the grass, and the run-off would be negligible. In actuality, even with clay liners in the lagoons, waste leaches into the local groundwater system. Even when the rate of flow is slow, over a period of days or weeks a 2-acre lagoon would leak thousands of gallons into

Encyclopedia of Water Science
DOI: 10.1081/E-EWS 120010373

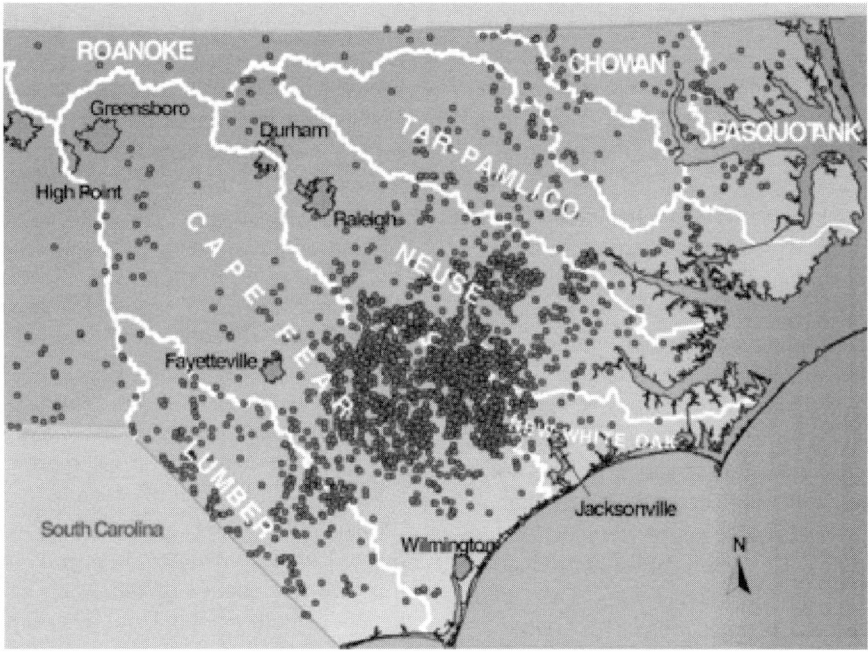

Fig. 1 Hog farms of Coastal Carolina (from Linda Huff, *American Scientist*).

the local watershed. There are nearly 4000 such lagoons in the state, leaching water into the ground and ammonia into the atmosphere.[3] The spraying is also inefficient. Most of the spraying is done during the spring and fall during the peak periods of precipitation, and the nutrients are quickly washed into the streams of the watershed area. North Carolina farmers have found that it is more profitable to raise cattle on the pastures than to bale the grass into hay and remove it from the site. As a result, the animals consume the grass and then redeposit the nutrients as urine or feces back

onto the pasture where some of them later wash into the local streams.

North Carolina's best management practices (BMP) suggest but do not require a buffer zone along the streams and rivers. Without an effective riparian zone, nutrients are easily deposited into the watershed area. Nearly 60% of the pollution load in the Neuse River comes from agriculture and a large portion of this is from the by-products of swine production.[4] Transported by surface and underground water, nitrates move through the soils and into the local

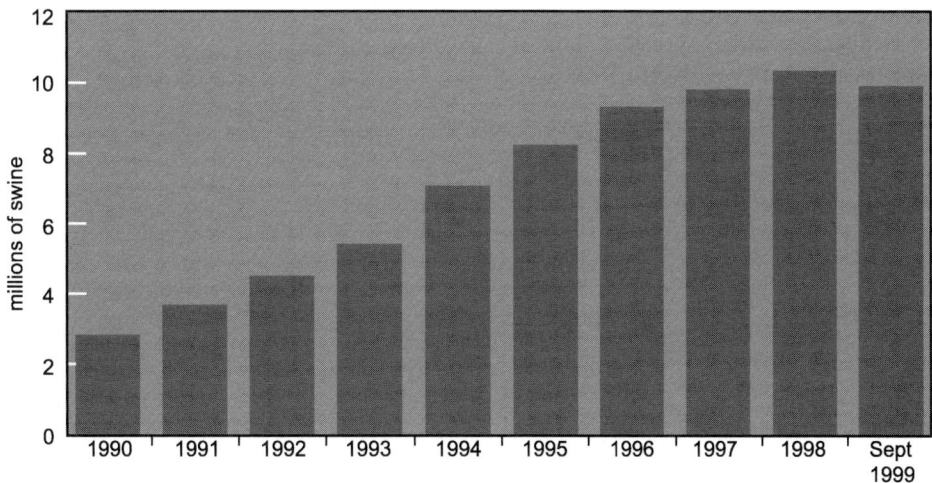

Fig. 2 Swine production in North Carolina.

streams. Phosphorus is less soluble and is transported primarily by surface run-off. Enriched levels of nitrogen and phosphorus have accumulated in lakes, streams, and estuaries of the Neuse watershed area. The most immediate effect of this concentration of nutrients is the depletion of dissolved oxygen and eutrophication of the water. Studies in the coastal plains of North Carolina have shown that nitrate levels in stream and ground water are the highest in areas with the greatest numbers of CAFOs.[5]

Algae blooms, including *Pfiesteria*, and fish kills result from surpassing the limiting amounts of nitrogen and phosphorus that are needed in the streams. Excess phosphorus appears to have dramatic effects on the production of huge amounts of nuisance algae in the standing waters of lakes and estuaries as well as at the mouth of the Neuse River. It is extensive enough at times to cause a visible discoloration of the water and to form mats on the surface. This results in further reduction of oxygen in the water and greater fish kills. The Neuse River modeling and monitoring project (MODMON) provides data that can be accessed on a regular basis to determine the water quality at several points along the Neuse River.[6]

The dinoflagellate *Pfiesteria* increases in concentration as a result of excessive nutrient enrichment in the poorly flushed estuaries and waterways of the Neuse. Fish kills and disease events have been linked to the organism. Thirteen researchers who worked with dilute toxic cultures of *Pfiesteria* sustained mild to adverse health impacts through water contact or by inhaling toxic aerosols from the lab cultures. These included severe headaches, blurred vision, short- and long-term memory loss, and other health problems. Some of the effects have reoccurred for a period of up to eight years.[7]

North Carolina is often impacted by unforeseen precipitation events that can greatly impact the quality of the local watersheds. CAFOs are often constructed on the flood plain with little thought to these unusual weather conditions and with little preparation for their occurrence. A case in point occurred on September 16, 1999 when Hurricane Floyd struck the East Coast of the United States. Fifty-seven lives and thousands of homes were destroyed as record amounts of rainfall from 15 in. to 20 in. battered the coast and storm surges from the ocean rose as much as 10 ft. Hurricane Floyd inundated more than 250 animal operations, mostly hog farms, in the eastern part of the state. Estimated animal deaths exceeded 500,000 hogs, 2.1 million chickens, and 737,000 turkeys. Three lagoons burst, others overflowed, adding to the millions of gallons of waste from 24 flooded sewage plants.[8] Flood run-off in the aftermath of Hurricane Floyd carried huge plumes of sediment and decomposing animals down the Neuse and other rivers and into the estuaries along the coast. Salinity and dissolved oxygen dropped to zero, excessive blooms

of algae occurred for months resulting in greater and greater fish kills. The public was exposed to coliform bacteria and other pathogens too numerous to mention. The public addressed the problem after the event with some subsequent changes in policy, but there would yet be tremendous problems for the people of the Neuse watershed basin if they were again struck by a storm of similar magnitude.

CONCLUSION

The quality of water in the Neuse River basin is greatly impacted by anthropogenic activities. With the population of the region increasing at a very rapid rate, greater water resources will be required even as greater stresses will be placed on those water sources because of human activities. Several steps are required to prevent contamination and to improve the quality of the water of the Neuse for the benefit of both mankind and other organisms.

The EPA's revised Clean Water Act of 2002 will implement new and more stringent regulations on CAFOs.[9] States containing large numbers of CAFOs must require more stringent regulations if they are to prevent further problems from water contamination. These include requiring water treatment of effluent from lagoons, establishment of riparian zones along streams, required implementation of BMP, and preventing the expansion of existing CAFOs. Continuous monitoring of water resources in the Neuse River should allow for the construction of a water-use model that will be more effective in regulating the quality of water in the watershed area.

REFERENCES

1. http://www.neuseriver.org/river_hist (accessed May 2002).
2. Osmond, D.L.; Hardy, D.; Johnson, L.H.; Lord, W.G.; Pleasants, R.H.; Regans, M. Agriculture and the Neuse River Basin, North Carolina Cooperative Extension Service, 2002; 1–27.
3. Zering, K.; Brandt, J.; Roka, F.; Vukina, T. Swine Industry in North Carolina, North Carolina State University, 1996.
4. White Paper Report. North Carolina State University, 1998.
5. Hunt, P.G. Impact of Animal Waste on Water Quality in and Eastern Coastal Plain Watershed. *Animal Waste and the Land–Water Interface*; Lewis Publishers: Boca Raton, FL; 589.
6. http://www/marine.unc.edu/neuse/modmon (accessed June 2002).
7. Pfiesteria, Center for Applied Aquatic Ecology, North Carolina State University, 2002.
8. Henderson, B.; Suchetka, D. Burials Hide Hog Losses. The Charlotte Observer, Sept 30, 1999.
9. Perciasepe, R. Reducing Water Pollution from Animal Feeding Operations. Speech to the Committee on Agriculture, Nutrition, and Forestry of the U.S. Senate, April 2, 1998.

Nitrogen Measurement

Jacek A. Koziel
Texas A&M University, Amarillo, Texas, U.S.A.

INTRODUCTION

Nitrogen (N) (MW = 14.0067)] is an integral part of the hydrosphere $(2.3 \times 10^{13}$ tn), atmosphere $(3.5 \times 10^{15}$ tn), and lithosphere $(5 \times 10^{16}$ tn).[1,2] Nitrogen serves as an essential component of cells and participates in almost every biological phenomenon in the biosphere.[2] Nitrogen cycles in the biosphere along with carbon, water, oxygen, sulfur, phosphorus, and other elements. Nitrogen in the form of nitrate (NO_3^--N), nitrite (NO_2^--N), ammonium (NH_3-N), and organic N are of great importance in waters and wastewaters.[3] Approximately 95% of the hydrosphere's N is stored as molecular N with the remaining 5% distributed in a 60:40 ratio between inorganic nitrates and organic N. Although N is very abundant—a significant disturbance on a local level, such as wastewater or fertilizer introduction, can result in abnormal N redistributions on regional, continental, and even global levels.[2] Thus, it is very important to recognize and quantify N sources and cycle pathways. This entry discusses typical concentrations of N in waters and wastewaters, and methods suitable for measuring N concentrations in water and wastewater.

The sum of nitrate- and nitrite-N is referred to as total oxidized nitrogen. Plants use nitrate as their primary source of N. Nitrate typically occurs in trace levels in surface waters, but it can be significantly higher in contaminated groundwater. Excessive fertilizer run-off and/or leaching are concerns for the contamination of surface water and groundwater. Fresh domestic wastewater can contain as much as $30\,mg\,L^{-1}$ NO_3^--N. The U.S. EPA (Environmental Protection Agency) drinking water standard for concentrations of NO_3^--N is $10\,mg\,L^{-1}$.[3] At high concentrations, nitrous acid that is formed from nitrite in acidic conditions, can react with secondary amines to form nitrosamines, some of which are known or suspected carcinogens. Nitrate can also react with hemoglobin in red blood cells causing methemoglobinemia or "blue baby" syndrome. Nitrite is formed via oxidation of ammonium and via reduction of nitrate in wastewater treatment plants, and municipal and natural waters.[3] Ammonium is present in surface water at levels typically less than $10\,\mu g\,L^{-1}$ to more than $30\,mg\,L^{-1}$ of NH_3-N in some wastewaters.[3] Ammonium is produced by deamination of N-containing compounds involving

enzymes and microorganisms and by hydrolysis of urea. Organic N is defined as "organically bound" N in 3^- oxidation state, e.g., proteins, peptides, nucleic acids, urea, and many synthetic organic materials.[3] Typical organic nitrogen concentrations range from a few hundred $\mu g\,L^{-1}$ in lake water to more than $20\,mg\,L^{-1}$ in raw sewage.[3] The main factors influencing the selection of a method for analysis of N is the range of concentrations, interferences, solution matrix, and the availability of analytical instrumentation. The full description of standard methods for determination of N in water and wastewater is presented in *Standard Methods*,[3] its updated versions,[4] and in U.S. EPA manuals.[5,6]

NITROGEN MEASUREMENT

Total Nitrogen

Total nitrogen content of all digestible nitrogen forms (limited organic, NH_3^-, NO_2^-, and NO_3^--N) can be determined by persulfate/UV digestion and persulfate digestion. This is accomplished by oxidative digestion to nitrate and subsequent quantification of nitrate. For concentrated samples ($N > 0.05\%$) total N analysis can be completed using automated combustion methods. Table 1 summarizes the characteristics of the persulfate/UV digestion and persulfate digestion that are specifically proposed for waters and wastewaters.[4] The U.S. EPA lists several variations of the total Kjeldahl N method as 0351.1–0351.4 for the automated colorimetric, colorimetric, colorimetric/titration, and potentiometric methods, respectively.

Ammonium Nitrogen

A summary of methods for the determination of ammonium nitrogen in waters and wastewaters is presented in Table 2. These include titration, phenate, ammonium-selective electrode, and gas segmented continuous flow colorimetric analysis methods.[3,4] The ammonium selective electrode method is highly matrix-dependent with an applicable range from $0.03\,\mu g$ to $14\,mg$ of NH_3-N L^{-1} and can be affected by signal drift and interferences. Lower concentrations can be detected using

Encyclopedia of Water Science
DOI: 10.1081/E-EWS 120010181

Table 1 Methods for measurement of total nitrogen in waters and wastewaters

Method	Applicability	Equipment	Interferences	Sample preservation
In-line UV/persulfate digestion and oxidation with flow injection analysis. Also available is the segmented flow analyzer	All forms of nitrogen except molecular N, amines, nitro-compounds, hydrazones, oximes, semicarbazones, and some refractory tertiary amines	Flow injection analysis equipment with injection valve and sample loop; multichannel proportioning pump; manifold, absorbance detector (540 nm, 10-nm bandpath)	Large particulates need to be filtered out. Chloride ions decrease the rate of reduction to nitrate	Acid can be used for sample preservation
Persulfate	All forms of nitrogen. Not applicable to molecular N and high organic N loads	Autoclave, or hotplate and pressure cooker; glass culture tubes; apparatus for nitrate determination; automated analytical equipment	Large particulates need to be filtered out. Chloride ions decrease the rate of reduction to nitrate	Samples preserved with acid cannot be analyzed with this method

(From Ref. 4.)

the manual phenate method. Preliminary distillation and subsequent titration should be used for samples with NH_3-N concentrations greater than $5 \, mg \, L^{-1}$. Distillation is also recommended before the phenate method when interferences are present. The phenate method can be automated. The indophenol blue method, based on semiautomated colorimetry, can be used to determine ammonium-nitrogen (as part of the total Kjeldahl nitrogen determination). Nesslerization, which was the traditional ammonium determination method, is no longer recommended due to the use of mercury and potential hazardous waste disposal problems.[4] The U.S. EPA assigned several variations of the ammonium N method as 0350.1 to 0350.3 for the colorimetric and semiautomated colorimetric, colorimetric/titration, and potentiometric methods, respectively.

Nitrite Nitrogen

Nitrite (NO_2^-) N in water/wastewater samples can be determined using the manual and automated cadmium reduction colorimetric methods. This method is applicable for concentrations ranging from $5 \, \mu g$ to $1000 \, \mu g$ of NO_2^--N L^{-1}. Nitrate N can also be estimated with ion chromatography and other flow injection methods discussed in the next paragraph and Table 3.[3] The colorimetric method uses either a spectrophotometer (543 nm and a light path of at least 1 cm) or filter photometer (green filter with maximum transmittance near 540 nm and a light path of 5 cm or more). Solids need to be filtered out and some ions, including Sb^{3+}, Au^{3+}, Bi^{3+}, Fe^{3+}, Pb^{2+}, Hg^{2+}, Ag^+, $PtCl_6^{2-}$, and VO_3^{2-} should be absent. Samples are typically collected in 50-mL polypropylene bottles and can be preserved for up to 48 hr at 4°C or frozen at −20°C. Acids should never be

used for sample preservation. The U.S. EPA lists several methods for determination of nitrite N. These include 0300.0 (ion chromatography), 0354.1 (spectrophotometry), 0353 (1, 2, 3, and 6 variations of these methods use manual and automated colorimetry), and 0353.4 (gas segmented continuous flow colorimetric analysis).

Nitrate Nitrogen

A summary of methods for the determination of nitrate (NO_3^-) nitrogen in waters and wastewaters is presented in Table 3. These include ion chromatography, ultraviolet spectrophotometry, nitrate electrode, cadmium reduction (also using gas segmented continuous flow colorimetric analysis), titanous chloride reduction, and hydrazine reduction methods.[3,4] The ultraviolet spectrophotometry is used as a screening method to estimate concentration range and interferences in samples. This is followed by selection of a suitable method. Samples are typically collected in 100-mL polypropylene bottles and storage time should be limited to the absolute minimum. Samples are maintained at 4°C and can be held for up to 48 hr. For longer storage, addition of H_2SO_4 to a pH of < 2 (typically about $2 \, mL \, L^{-1}$) and refrigeration at 4°C can be used. When acid is used for preservation, NO_3^- and NO_2^- cannot be determined as individual species,[3] however, the sample can be held up to 28 days. The results are reported as nitrate–nitrite N. Samples are typically collected in 100-mL polypropylene containers. The U.S. EPA lists methods 0300.0 (ion chromatography), 0353 (1, 2, 3, and 6 variations of these methods use manual and automated colorimetry for nitrate–nitrite), 0352.1 (colorimetric), and 0353.4 (for nitrate–nitrite using gas segmented continuous flow colorimetric analysis).

Table 2 Methods for measurement of ammonium-nitrogen in waters and wastewaters

Method	Applicability	Range of applicability	Equipment	Interferences	Sample preservation	Comments
Phenate	Drinking, surface, saline waters, domestic and industrial wastewaters	$10\,\mu g - 600\,\mu g$ NH$_3$-N L^{-1}	Magnetic stirrer and spectrophotometer (630 nm and a light path of at least 1 cm)	Turbidity, color, alkalinity > 500 mg as CaCO$_3$ L^{-1}, acidity > 100 mg as CaCO$_3$ L^{-1} should be removed by preliminary distillation	Acid (needs to be removed by preliminary distillation)	Automated phenate method applicable from $20\,\mu g$ up to 2 mg NH$_3$-N L^{-1} without dilution
Titration	Used only after preliminary distillation	For samples > 5 mg NH$_3$-N L^{-1}	Distillation apparatus			Does not require preliminary distillation. Known addition can be used when no calibration is needed
Selective electrode	Drinking and surface waters, domestic and industrial wastewaters	0.03 mg-1400 mg NH$_3$-N L^{-1}. Longer response times needed for concentrations < 1 mg NH$_3$-N L^{-1}	Ammonium selective electrode, electrometer, and magnetic stirrer	High concentrations of dissolved ions; amines. Effects of Hg and Ag are minimized with the NaOH/EDTA. Turbidity and color cannot affect the measurement	Refrigeration at 4°C if analyzed within 24 hr; Refrigeration at pH = 2 (by H$_2$SO$_4$) or freezing at -20°C for upto 28 days	
Flow injection analysis	All waters and wastewaters		Flow injection analysis equipment with injection valve and sample loop; multichannel proportioning pump; manifold, absorbance detector (660 nm, 10-nm bandpath)	Large and fibrous particles should be filtered out		
Gas segmented continuous flow colorimetric analysis	Estuarine and coastal waters	$0.3\,\mu$ L^{-1} to 4.0 mg L^{-1}	Automatic sampler, analytical cartridge, proportioning pump, spectrophotometer, or photometer with a 640 interference filter, nitrogen gas	Hydrogen sulfide > 2 mg S L^{-1}; turbidity needs to be eliminated	Refrigeration in tightly sealed glass or HDPE container in the dark works for up to 3 hr. Concentrated samples $> 20\mu$ L^{-1} can be preserved for 2 wk	Based on the indophenol reaction
Nesslerization (not recommended as standard method)	Purified drinking water, natural waters, and highly purified wastewater effluents	$20\,\mu g$ to 5 mg NH$_3$-N L^{-1}	pH meter and spectrophotometer (400 nm–500 nm, light path of at least 1 cm), or filter photometer (light path of at least 1 cm; violet filter with max transmittance at 400 nm–425 nm), or Nessler tubes	Turbidity, color, Mg, and Ca can be removed via preliminary distillation or by precipitation by zinc sulfate and alkali	Dechlorination, 0.8 mL H$_2$SO$_4$ L^{-1} sample, and storage at 4° C. The pH should be between 1.5 and 2 when acid is used and samples need to be neutralized immediately before analysis	This traditional method is not recommended because of potential problems with mercury disposal

(From Ref. 3.)

Table 3 Methods for measurement of nitrate-nitrogen in waters and wastewaters

Method	Applicability	Range of applicability	Equipment	Interferences	Comments
Ultraviolet spectrophotometric screening	Used for screening only of uncontaminated natural and drinking waters, i.e., waters with low organic N content		Spectrophotometer (220 and 270 nm, light path of at least 1 cm)	Dissolved organic matter, surfactants, NO_2^-, Cr^{6+}, chlorite, chlorate	Measurement at 270 nm is used to correct for dissolved organic matter that may interfere at 220 nm. U.S. EPA method 0354.1
Ion chromatography		From 0.1 mg L^{-1}	Ion chromatograph, anion separator column, guard column, fiber or membrane suppressor, conductivity detector	Bromide can coelute	U.S. EPA method 0300.0
Nitrate electrode	Drinking water	2 mg L^{-1} to 1000 NO_3^--NL^{-1}	Double-junction reference electrode, nitrate ion electrode, pH meter, magnetic stirrer	Chloride and bicarbonate when their weight ratios to NO_3^--N are > 10 and > 5, respectively; NO_2^-, CN^-, S^{2-}, Br^-, I^-, ClO^{3-}, ClO^{4-}	Ionic strength adjustments can remove interferences
Cadmium reduction	All waters	0.01 mg–1 mg NO_3^--NL^{-1}; For automated method 0.5 mg–10 mg NO_3^--NL^{-1}	Reduction column and spectrophotometer (543 nm, light path of at least 1 cm) or filter photometer (a filter with maximum transmittance near 540 nm and light path of at least 1 cm)	Suspended matter can restrict column flow; concentrations of metals > 1 mg L^{-1}, oil, grease, residual chlorine can decrease reduction efficiency	This method is also applicable for determination of NO_2^- when the reduction step is omitted. This method can also be automated and also combined with flow injection method. U.S. EPA method 0353.3
Gas segmented continuous flow colorimetric analysis	Estuarine and coastal waters, applicable also for nitrite determination	0.075 μ.L^{-1} to 5.0 mg L^{-1}	Automatic sampler, analytical cartridge, open tubular cadmium reactor or cadmium reduction column proportioning pump, spectrophotometer or photometer with a 540 interference filter, nitrogen gas	Hydrogen sulfide > 2 mg SL^{-1}; turbidity needs to be eliminated	Samples should be analyzed within 3 hr. U.S. EPA method 0353.4
Titanous chlorine reduction	All waters	0.01 to 20 mg NO_3^--NL^{-1}	pH meter, ammonium gas sensing electrode, magnetic stirrer	NH_3, NO_2^-	Proposed method
Automated hydrazine reduction	All waters	0.01 to 10 mg NO_3^--NL^{-1}	Automated analytical equipment	Sample color that absorbs in the photometric range used	Proposed method

(From Refs. 3 and 4.)

Table 4 Methods for measurement of organic nitrogen in waters and wastewaters

Method	Range of applicability	Equipment	Interferences	Sample preservation
Macro-Kjeldahl	Either low or high concentrations that require large volume for low concentrations	~ 800 mL digestion apparatus, distillation apparatus, apparatus for ammonium determination	Nitrate in excess of 10 mg L^{-1}, large amounts of inorganic salts and solids, large amounts of organic matter	Lowering pH to 1.5–2 with concentrated H_2SO_4 and refrigeration at 4°C
Semimicro-Kjeldahl	High concentrations and that sample volume containing organic plus ammonium N between 0.2 and 2 mg	~ 100 mL digestion apparatus, distillation apparatus, pH meter	Nitrate in excess of 10 mg L^{-1}, large amounts of inorganic salts and solids, large amounts of organic matter	Lowering pH to 1.5–2 with concentrated H_2SO_4 and refrigeration at 4°C
Block digestion and flow injection analysis	All waters and wastewaters	Block digestor, digestion tubes, injection valve, multi-channel proportioning valve, flow injection manifold, absorbance detector (660 nm, 10-nm band path)	Large and fibrous particles; ammonium	

(From Refs. 3 and 4.)

Organic Nitrogen

A summary of methods for determination of organic nitrogen in waters and wastewaters is presented in Table 4. These include macro or semimicro-Kjeldahl method, and block digestion combined with flow injection analysis. Kjeldahl methods do not account for N in the form of azide, azine, azo, hydrazone, nitrate, nitrite, nitrile, nitro, nitroso, oxime, and semicarbazone.[3] If ammonium is not removed in the initial digestion, the result is the "Kjeldahl nitrogen" often called "total Kjeldahl nitrogen" that is defined as organic plus ammonium N.

CONCLUSION

Measurements of N content in waters and wastewaters are of great interest because they relate to many natural and anthropogenic processes and their effects including human, animal, and plant health and well-being. Standard analytical methods available for the determination of various forms of N, including total N, ammonium, nitrite, nitrate, and organic N, are briefly discussed in this article. Nitrogen detection limits and applicable concentration ranges for these methods cover the typical levels encountered in water and wastewater. The reader is encouraged to use the full standard method description for all described methods in this article.[4–6]

ACKNOWLEDGMENT

The author would like to thank Drs. Robert Schwartz and N. Andy Cole (USDA-ARS-Bushland) and Drs. David Parker and Jim Rogers (West Texas A&M University) for useful discussions of N measurement methods.

REFERENCES

1. Budavari, S., Ed. *The Merck Index: An Encyclopedia of Chemicals, Drugs, and Biologicals*, 12th Ed.; Merck Research Laboratories: Whitehouse Station, NJ, 1996.
2. Greyson, J. *Carbon, Nitrogen, and Sulfur Pollutants and Their Determination in Air and Water*; Marcel Dekker, Inc.: New York, NY, 1990.
3. Clesceri, L.S.; Greenberg, A.E.; Trussell, R.R. *Selected Physical and Chemical Standard Methods for Students: Based on Standard Methods for Examination of Water and Wastewater*, 17th Ed.; American Public Health Association, American Water Works Association, Water Pollution Control Federation: Washington, DC, 1990.
4. American Public Health Association, American Water Works Association, Water Environment Federation, *Standard Methods for Examination of Water and Wastewater*, 20th Ed.; American Public Health Association, Washington, DC, 1999.
5. U.S. Environmental Protection Agency, *Methods for Chemical Analysis of Water and Wastes*; EPA/6/4-79-020, U.S. Government Printing Office: Washington, DC, 1993.
6. U.S. Environmental Protection Agency. Test Methods for Evaluating Solid Waste, Physical/Chemical Methods. Accessed http://www.epa.gov/epaoswer/hazwaste/test/sw846.htm on May 18, 2002.

Nutrient Best Management Practices

Scott J. Sturgul
Keith A. Kelling
University of Wisconsin, Madison, Wisconsin, U.S.A.

INTRODUCTION

Soil nutrients need to be managed properly to meet the fertility requirements of crops without adversely affecting the quality of water resources. The nutrients of greatest concern relative to water quality are nitrogen (N) and phosphorus (P). Nitrogen not recovered by crops can add nitrate to groundwater through leaching. Nitrate is the most common groundwater contaminant found in the United States.[1,2] Nitrate levels that exceed the established U.S. drinking water standard of 10 ppm nitrate-N have the potential to adversely affect the health of infants and livestock.[3] Surface water quality is the concern with P, as runoff and erosion from cropland add nutrients to water bodies that stimulate the excessive growth of aquatic weeds and algae. Of all crop nutrients, it is critical to prevent P from reaching lakes and streams since the biological productivity of aquatic plants and algae in fresh water environments is usually limited by this nutrient.[4] Consequences of increased aquatic plant and algae growth include reduced aesthetic and recreational value of lakes and streams as well as the seasonal depletion of water dissolved oxygen content, which may result in fish kills as well as other ecosystem disruptions.

OVERVIEW

Nutrient best management practices vary widely from one area to another due to cropping, topographical, environmental, and economic conditions. With the variety of factors to consider, no single set of best management practice can be recommended for all farms. Nutrient management practices for optimizing crop production while protecting water quality must be tailored to the unique conditions of individual farms. Practices that need to be considered in any nutrient best management program include the following.

Establish Nutrient Application Rates

The most important management practice for environmentally and economically sound nutrient management is the application rate.[5] Optimum nutrient application rates are identified through fertilizer response/calibration research for specific soils and crops. Economically optimum nutrient application rates provide maximum financial return, but as application rates near the economic optimum, the efficiency of nutrient use by the crop decreases and the potential for loss to the environment increases. Any nutrient application above this rate reduces profit and increases the likelihood of detrimental impact to the environment. Because of the overall importance of nutrient application rates, accurate assessments of crop nutrient needs are essential for minimizing threats to water quality while maintaining economically sound production. Soil testing is the most widely used method to accurately estimate nutrient needs of crops.

Use Additional Tests for Fine-Tuning Nitrogen Applications

The development of tests for assessing soil N levels provides additional tools for improving the efficiency of N fertilizer applications.[6] These tests allow fertilizer recommendations to be adjusted to site-specific conditions that can influence N availability. Tests include the preplant soil profile nitrate test,[7] the presidedress soil nitrate test,[8] plant analysis,[9] chlorophyll meters,[10] the basal stalk nitrate test,[11] and the end of season soil nitrate test.[12]

Use Calibrated Soil Tests for Phosphorus and Potassium

In recent years, soil test recommendation programs for phosphorus (P), potassium (K), and other relatively immobile nutrients have tended to de-emphasize the soil build-up and maintenance philosophy in favor of a better balance between environmental and economic considerations by using a crop sufficiency approach.[13] These tests must be calibrated by field experiments to obtain predictable crop yield responses. Such an approach adds extra emphasis on regular soil testing. It is recommended that soil tests be taken at least every three to four years and more frequently on sands and other soils of low buffering capacity.[14]

Encyclopedia of Water Science
DOI: 10.1081/E-EWS 120010335

Establish Realistic Yield Goals

For many soil fertility programs, the recommendation of appropriate nutrient application rates is dependent on the establishment of realistic yield goals. Yield goal estimates that are too low will underestimate nutrient needs and can limit crop yield. Yield goal estimates that are too high will overestimate crop needs and result in soil nutrient levels beyond that needed by the crop which, in turn, has the potential to increase nutrient contributions to water resources.[15,16] Estimates should be based on field records and some cautious optimism—perhaps 10% above the recent three- to five-year average crop yield from a particular field.

Credit Nutrients from All Sources

The best integration of economic return and environmental quality protection is provided by considering nutrients from all sources. In the determination of supplemental fertilizer application rates, it is critical that nutrient contributions from manure, previous legume crops grown in rotation, and land-applied organic wastes are credited. In many cases, commercial fertilizer application rates can be reduced when nutrient credits are accounted.

Time Nutrient Applications Appropriately

Timing of application is a major consideration for the management of mobile nutrients such as nitrogen. The period between application and crop uptake of N is an important factor affecting the efficient utilization of N by the crop with the loss of N minimized by supplying it just prior to the period of greatest crop uptake.[17] However, several considerations, such as soil, equipment, and labor, are involved in determining the most convenient, economical, and environmentally safe N fertilizer application time. Although fall applications of N are commonly discouraged, they continue to be made primarily to ensure adequate time for spring planting. If fall applications of N are to be made, it is recommended that ammonium–nitrogen sources be used and that the applications be delayed until soil temperatures are below thresholds of biological activity (i.e., 50°F). Fall applications of N fertilizers are not recommended on coarse textured soils or on shallow soils over fractured bedrock.

For less mobile nutrients, application timing is not a major factor affecting water quality protection. However, nutrient applications on frozen sloping soils or surface applications prior to periods likely to produce runoff events should be avoided to prevent P contributions to surface waters.

Use Nitrification Inhibitors When Appropriate

Nitrification inhibitors are used with ammonium or ammonium-forming N fertilizers to improve N efficiency by slowing the conversion of ammonium to nitrate, thereby reducing the potential for losses of N that occur in the nitrate form. The effectiveness of a nitrification inhibitor depends greatly on soil type, time of the year applied, N application rate, and soil moisture conditions that exist between the time of application and the time of N uptake by plants. Research has shown that the use of nitrification inhibitors on medium- and fine-textured soils with fall N applications, or on poorly drained soils with fall or spring N applications, or on coarse-textured, irrigated soils with spring preplant N applications has the potential to increase corn yield and total crop recovery of N.[18,19] Fall applications of N with an inhibitor on coarse textured soils are not recommended.

Manage Manure to Maximize Benefits

Manure applications to cropland provide nutrients essential for crop growth, add organic matter to soil, and improve soil physical conditions. The major concerns associated with manure applications are related to its potential for overloading soils with nutrients if manure applications exceed crop needs, or its application at times of the year when the risk of runoff losses are high. Recommended management practices include accounting for (or crediting) the nutrients supplied by manure, incorporating or injecting manure, distributing manure over numerous cropland fields, minimizing applications to frozen soils, avoiding fall applications to highly permeable soils, and avoiding applications to areas with direct access to surface water (i.e., floodplains, waterways, etc.), or groundwater (i.e., shallow or permeable soils over fractured bedrock, etc.).[20]

Manage Irrigation Water

Overirrigation or rainfall on recently irrigated soils can leach nitrate and other contaminants below the root zone and into groundwater. Accurate irrigation scheduling that considers soil water holding capacity, crop growth stage, evapotranspiration, rainfall, and previous irrigation to determine the timing and amount of irrigation water to be applied can reduce the risk of leaching losses.[21]

Use Soil Conservation Practices

Land-use activities associated with agriculture often increase the susceptibility for runoff and sediment transport from cropland to surface waters. The key to minimizing nutrient contributions to surface waters is to reduce the amount of runoff and eroded sediment reaching them. Runoff and erosion control practices range from changes in agricultural land management (cover crops, diversified crop rotations, conservation tillage, contour farming, and contour strip cropping) to the installation of structural devices (diversions, grade stabilization structures, grassed waterways, and terraces). Recently, substantial emphasis is being placed on the benefits and installation of buffer strips which are effective in reducing contaminant transport to surface waters.[22]

CONCLUSION

The previous text provides a brief summary of general nutrient management practices for crop production. This is not a complete inventory but rather an overview of soil fertility management options available to growers for improving farm profitability and protecting water quality. The selection of appropriate nutrient management practices for an individual farm needs to be tailored to the specific conditions existing at a given location.

REFERENCES

1. Hallberg, G.R., Agrichemicals and Water Quality. Proceedings of the Colloquium to Protect Water Quality, Board of Agriculture, National Research Council: Washington, DC, 1986.
2. Madison, R.J.; Brunett, J.O. Overview of the Occurrence of Nitrate in Groundwater of the United States. *US Geological Service National Water Summary*; Water-Supply Paper 2275; US Government Printing Office: Washington, DC, 1984; 93–105.
3. US Environmental Protection Agency. Water Programs—National Interim Primary Drinking Water Regulations, US Federal Register, v. 40, 1975.
4. Killorn, R.; Voss, R.; Baker, J.L. *Plant Nutrients as Potential Pollutants*; Iowa St. Univ. Coop. Extn. Serv. Bull. PM-901G; Iowa Cooperative Extension Service: Ames, IA, 1985; 6pp.
5. Bundy, L.G. Nitrogen Management for Groundwater Protection and Efficient Crop Use. Proceedings of 1987 Wis. Fert., Aglime & Pest Mgmt. Conf., Madison, WI, January 1987; University of Wisconsin: Madison, WI, 1987; Vol. 26, 254–262.
6. Stanford, G. Assessment of Soil Nitrogen Availability. In *Nitrogen in Agricultural Soils*; Stevenson, F.J., Ed.; American Society of Agronomy: Madison, WI, 1982; 651–688.
7. Bundy, L.G.; Malone, E.S. Effect of Residual Profile Nitrate on Corn Response to Applied Nitrogen. Soil Sci. Soc. Am. J. **1988**, *52*, 1377–1383.

8. Bundy, L.G.; Meisinger, J.J. Nitrogen Availability Indices. *Methods of Soil Analysis, Part 2. Microbiological and Biochemical Properties*; Soil Science Society of America: Madison, WI, 1994; 951–984.
9. Westerman, R.E. *Soil Testing and Plant Analysis*, 3rd Ed.; Soil Science Society of America: Madison, WI, 1990.
10. Peterson, T.A.; Blackmer, T.M.; Francis, D.D.; Schepers, J.S. *Using a Chlorophyll Meter to Improve N Management*; Univ. Neb. Coop. Extn. Ser. Bull. 693-1171-A, Nebraska Cooperative Extension Service: Lincoln, NE, 1993.
11. Binford, G.D.; Blackmer, A.M.; El-Hout, N.M. Tissue Test for Excessive Nitrogen during Corn Production. Agron. J. **1990**, *82*, 124–129.
12. Combs, S.M.; Bundy, L.G.; Andraski, T.W. In-season Tests for Improving Corn N Management. *Proceedings of the 1995 Wisconsin Fertilizer Dealer Meetings. December 1995*; University of Wisconsin: Madison, WI, 1995; 7–95.
13. Olson, R.A.; Frank, K.D.; Grabowski, P.H.; Rehm, G.W. Economic and Agronomic Impacts of Various Philosophies of Soil Testing. Agron. J. **1982**, *74*, 492–499.
14. Kelling, K.A.; Bundy, L.G.; Combs, S.M.; Peters, J.B. *Soil Test Recommendations for Field, Vegetable, and Fruit Crops*; Univ. Wis. Coop. Extn. Ser. Bull. A2809; Wisconsin Cooperative Extension Service: Madison, WI, 1998.
15. Schepers, J.S.; Frank, K.D.; Bourg, C. Effect of Yield Goal and Residual Soil Nitrogen Considerations on Nitrogen Fertilizer Recommendations for Irrigated Maize in Nebraska. J. Fert. Issues **1986**, *3*, 133–139.
16. Vanotti, M.B.; Bundy, L.G. Corn Nitrogen Recommendations Based on Yield Response Data. J. Prod. Agric. **1994**, *7*, 249–265.
17. Jokela, W.E.; Randal, G.W. Fate of Fertilizer Nitrogen as Affected by Time and Rate of Application on Corn. Soil Sci. Soc. Am. J. **1997**, *61*, 1695–1703.
18. Malzer, G.L.; Kelling, K.A.; Schmitt, M.A.; Hoeft, R.G.; Randall, G.W. Performance of Dicyandiamide in the North Central States. Commun. Soil Sci. Plant Anal. **1989**, *20*, 2001–2022.
19. Hoeft, R.G. Current Status of Nitrification Inhibitor Use in U.S. Agriculture. In *Nitrogen in Crop Production*; Hanck, R.D., Ed.; American Society of Agronomy: Madison, WI, 1984.
20. Madison, F.W.; Kelling, K.A.; Massie, L.; Ward-Good, L. *Guidelines for Applying Manure to Cropland and Pastures in Wisconsin*; Univ. Wis. Coop. Extn. Ser. Bull. A3392; Wisconsin Cooperative Extension Service: Madison, WI, 1998.
21. Curwen, D.; Massie, L.R. Potato Irrigation Scheduling in Wisconsin. Am. Potato J. **1984**, *61*, 235–241.
22. Schmitt, T.J.; Bosskey, M.G.; Hoagland, K.D. Filter Strip Performance and Processes for Different Vegetation, Widths, and Contaminants. J. Environ. Qual. **1999**, *28*, 1479–1489.

Observation Wells

Phillip D. Hays
John Van Brahana
University of Arkansas, Fayetteville, Arkansas, U.S.A.

INTRODUCTION

Observation wells are engineered openings constructed through the solid earth, usually circular in cross-section, that are drilled or otherwise excavated to allow human access to specific zones of underground water for the purpose of measuring attributes such as water levels or pressure changes that would otherwise not be observable at the Earth's surface. Wells constructed to sample groundwater quality have been called observation wells by some hydrogeologists, but common accepted practice restricts the term monitoring wells to wells from which water samples may be collected for chemical analysis.[1] Groundwater-level monitoring networks incorporate multiple observation wells, and maps constructed from these water levels allow hydrogeologists and engineers to determine the direction of the subsurface flow. Observation wells are occasionally called "piezometers," although most hydrogeologists consider piezometers to be a special type of observation well in which a simple tube emplaced in an aquifer, open at the top for access and measurement and open at the bottom to allow communication with the aquifer.[2] No matter what the name used for these features, observation wells are windows to the groundwater system, and they allow us to collect in situ information from which we can develop an understanding of the degree of interconnection of openings in the aquifer, calculate the amount of water that is stored in the subsurface openings, and determine the hydraulic head and pressure gradients present in the subsurface which control groundwater flow and rates of movement.

BASIC CONSTRUCTION

Observation wells differ from groundwater monitoring and water-supply wells primarily in their objectives, and thus usually are constructed to be only large enough to allow accurate and rapid water-level measurements.[3–5] Observation wells typically are of small diameter, 2 cm – 10 cm, open to a discrete interval in an aquifer and have a shorter section of screen or other openings than would be desirable for a pumping, monitoring, or production well. Otherwise, all wells are constructed in the same general manner, by

boring a hole, inserting into the borehole a string of solid pipe (casing), on the bottom of which are openings (perforations, slots, or screen) that allow water from a specific subsurface zone into the well (Fig. 1). Porous-media filter pack, typically sand, is added to the annulus of the borehole and the screen, facilitating hydraulic connection between the aquifer and the observation well while minimizing the impact of the well on the aquifer. Above the filter pack, an impermeable material (typically bentonite, other clays, and concrete) is added for two reasons: to seal the casing to the rock material through which the well was drilled, and more importantly, to prevent water from filtering vertically up or down the borehole outside the casing. The casing is covered with a vented well cap, which keeps potential contaminants out of the well while allowing the air above the water in the well to maintain equilibrium with the atmosphere.[4,5] Inside a well-constructed observation well, the water level is free to fluctuate in response to natural and human-induced stresses on the aquifer system.

INFORMATION PROVIDED

Observation wells allow scientists to gain knowledge of the energy and mass of water beneath the surface of the earth, the degree of void interconnection within the underlying rocks, and rates and directions of flow.[6] Knowledge of the energy and mass contained in a groundwater system is critical to understanding and managing that system as a resource. Hydraulic head is a measure of the energy available to move water in the subsurface and, when coupled with saturated thickness and other aquifer characteristics, the mass of water present may be determined. A water level measured in an observation well under static, nonpumping conditions is a measure of hydraulic head in the aquifer at the depth of the open or screened interval.[3] Hydraulic head is the height at which a column of water stands above a reference elevation, most commonly sea level (Fig. 1). Water moves from points of high head to points of low head. The difference in hydraulic head divided by the distance between two observation wells in which the heads are measured is the hydraulic gradient, and hydraulic gradient is a controlling factor for calculating

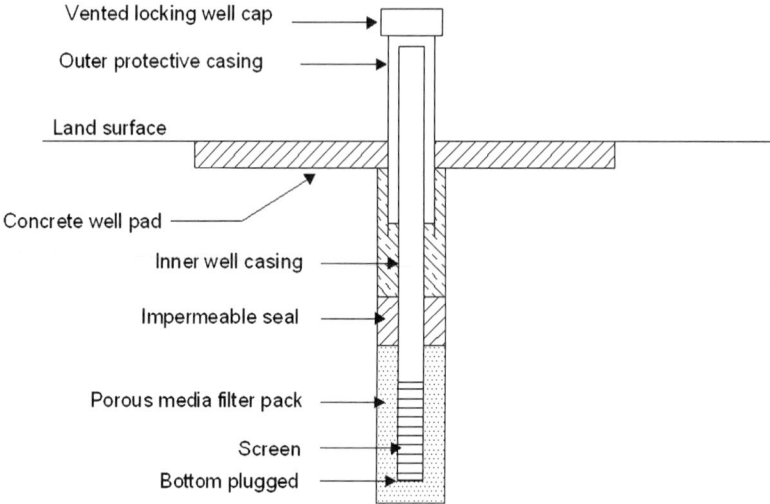

Fig. 1 Typical observation well design.

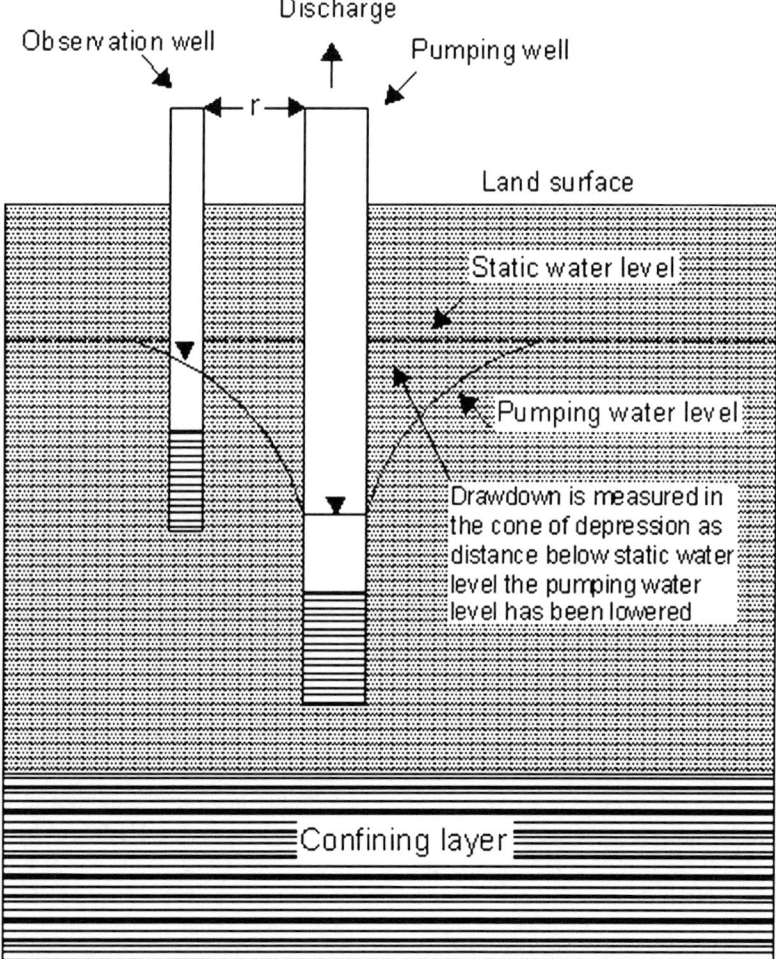

Fig. 2 Drawdown and cone of depression resulting from pumping groundwater. The observation well at radial distance r from the pumping well allows computation of the shape and dimensions of the cone, as well as important hydraulic characteristics of the aquifer.

the rate at which water moves in the subsurface. Water-level measurements made in two or more observation wells at the same time can define local hydraulic gradients. Water-level measurements made in three or more wells at the same time can define water movement direction and gradient. Water-level measurements made at different times in the same observation well can define temporal trends in water level in response to short- or long-term stresses such as pumping, changing land-use, or natural variations in climate.

Observation wells are a critical component of aquifer testing.[1,2,4,8–10] By relating the water-level response of the aquifer in the observation well to known stresses at differing distances, it is possible to compute hydraulic characteristics of the aquifer. These hydraulic characteristics, the interconnectivity of openings or "hydraulic conductivity", the amount of water stored or "storativity," and the amount of water that may leak vertically or "leakance," allows us to quantitatively test our understanding of how the system works. This understanding allows us to evaluate suitability of the aquifer for potential uses, especially with respect to computing how much water the aquifer will yield. This computation requires knowledge of the distance between the pumped well and the observation well, measured radially as a length, r, and it allows reconstruction of the cone of depression. The cone of depression is a roughly conical area of decreasing water levels in an aquifer outward from the well from which water is being withdrawn (Fig. 2). The cone of depression results from pumping a specified discharge, Q from the aquifer, and is related to the "interconnectivity" of the openings of the aquifer, or in scientific terms, the hydraulic conductivity. During an aquifer test, water level is measured on a high-frequency periodic basis (late in the test measurements may be taken less frequently) using a steel or electrical tape or continuously using an automatic sensing device. Q, r, and h—hydraulic head or water level—are essential for computation of the storativity, S, a measure of the storage capacity of the aquifer and the ability of the aquifer to provide water. Depending on the hydrogeologic conditions of the flow system (confined and unconfined), the proper equation with the radius squared reflects the quantitative relation between pumping and the exact symmetry of the cone (Fig. 2). Without observation wells, such computations would not be possible.

Observations wells are also critical components of local- to regional-scale groundwater-level monitoring networks.[1–3,6,7,11] Subsurface hydrologic systems typically are heterogeneous and complex. Comprehensive understanding of water-level conditions for an aquifer are derived through development of an observation well network in which information is gathered from many observation wells

distributed across an area; water levels for locations between observation wells may then be derived by interpolation. These data provide the basis for generating 3-D maps of the water table or potentiometric surfaces, and are an important tool for water-resources planning and management. Because water levels vary with time, water-level network observation wells may be measured continuously or more commonly, only occasionally. Periodic water-level measurements are made at scheduled intervals, typically by manual means such as by steel or electric tape. Continuous groundwater-level data are measured by an automatic sensing device such as a pressure transducer or a float; these data are recorded by data loggers or recorders and retrieved periodically. Recently, acute responses of aquifers to intensive groundwater use and frequent drought have created a need for more rapid management decisions and action, and thus more timely data from observation wells. In response to this need, data from observation wells are increasingly continuously measured at the well, transmitted via telemetric equipment (typically satellite or land-line), and processed at a central location to be made available on a real-time basis over the Internet.[11]

REFERENCES

1. Freeze, R.A.; Cherry, J.A. *Groundwater*; Prentice-Hall, Inc.: Englewood Cliffs, NJ, 1979; 1–604.
2. Dominico, P.A.; Schwartz, F.W. *Physical and Chemical Hydrogeology*, 2nd Ed.; John Wiley and Sons: New York, 1998; 1–506.
3. Taylor, C.J.; Alley, W.M. Ground-water Level Monitoring and the Importance of Long-term Water-level Data. U.S. Geological Survey Circular 1217, 2001, 68p.
4. Driscoll, F.G., Ed. *Groundwater and Wells*, 2nd Ed.; Johnson Division: St. Paul, MN, 1986; 547–549.
5. Bohn, C.C. *Guide for Fabricating and Installing Shallow Ground Water Observation Wells*, Res. Note RMRS-RN-9; U.S. Department of Agriculture, Forest Service, Rocky Mountain Research Station: Ogden, UT, 2001; 5 p.
6. Heath, R.C. Design of Ground-water Level Observation-Well Programs. Ground Water **1976**, *14* (2), 71–77.
7. http://etd.pnl.gov:2080/hydroweb.html
8. Lohman, S.W. Ground-water Hydraulics. U.S. Geological Survey Professional Paper 708, Washington, DC. 1972, 70p.
9. Walton, W.C. Selected Analytical Methods for Well and Aquifer Evaluation. Illinois State Water Survey, Bulletin 49, Urbana, IL, 1962, 81p.
10. Ferris, J.G.; Knowles, D.B.; Brown, R.H.; Stallman, R.W. Theory of Aquifer Tests. U.S. Geological Survey Water Supply Paper 1536-E, 1962, 69–174.
11. Cunningham, W.L. Real-time Ground-water Data for the Nation. U.S. Geological Survey Fact Sheet 090-01, 2001, 4 p. url: http://pubs.water.usgs.gov/fs-090-01

Open-Channel Spillways

Gregory J. Hanson
Sherry L. Britton
United States Department of Agriculture (USDA),
Stillwater, Oklahoma, U.S.A.

INTRODUCTION

Provision must be made in the design of almost every dam to permit the safe discharge of water downstream. The function of the spillway is to safely convey this discharge past the dam without unacceptable damage. The high velocities and large levels of energy involved in flow over spillways, particularly during major floods, make their design of considerable importance. The design and capacity of spillways play an important role in the layout and economics of every dam project. Spillways are selected for a specific dam and reservoir on the basis of discharge requirements, topography, geology, dam safety, and project economics.

TYPES OF SPILLWAYS

The classifications of spillways are primarily based on their most prominent feature and/or function. This may include the type of discharge carrier or some other type of component.[5,11] Spillways are classified as controlled or uncontrolled depending on whether they are gated or ungated. Designation as a principal spillway generally indicates constant or frequent flow with an auxiliary spillway used to pass the infrequent larger flood events. Most spillways can be broken into two main categories: open-channel spillways and conduit spillways. Open-channel spillways include: a) straight-drop or free overfall spillways; b) chute spillways; c) cascade spillways; d) side-channel spillways; and e) unlined or vegetation lined earthen spillways. Conduit spillways include: a) siphon spillways; b) drop shaft or morning-glory spillways; c) tunnel spillways; and d) culvert spillways. The focus of this article is on open-channel spillways.

Straight-Drop Spillways

The U.S. Bureau of Reclamation[11] defines a straight-drop spillway as one in which the flow drops freely from the crest into a plunge pool or stilling basin (Fig. 1). Straight-drop spillways are used on thin arch or deck overflow dams, dams with a crest that has a nearly vertical downstream face, and with low earthfill dams.[1]

Hydraulic concerns of the straight-drop spillway are with the control and dissipation of energy in the downstream plunge pool or stilling basin. A minimum depth of tailwater is required for effective dissipation of excess flow energy to prevent downstream scour. The tailwater level in the downstream channel should be at approximately the same level as the water surface in the stilling basin.

Chute Spillway

A spillway that conveys water from a reservoir over a spillway crest into a steep-sloped open channel is known as a chute spillway[3] (Fig. 2). A smooth chute spillway conducts the overflow to an outlet energy dissipation basin. Chute spillways can be well adapted to earth or rock-fill dams when topographic conditions permit.[3] They are generally located through the abutment adjacent to the dam. However, they can be located in a saddle away from the dam structure. Such a location is preferred for earth dams to prevent possible damage to the embankment. The chute may be of constant width but is usually narrowed for economy and then widened near the end to reduce unit discharge.

Cascade Spillway

The cascade or baffle chute spillway uses steps or other appurtenances to dissipate energy in the spillway channel and thereby, reduce energy dissipation requirements in the outlet basin (Fig. 3). The cascade- spillway has greater flow depths than chute spillways requiring higher sidewalls. Spray action may be a concern due to air entrainment, and abrasion of the steps and other appurtenances can be a serious problem. Cascade-spillway use has increased recently due to the increased use of roller compacted concrete (RCC).

A major concern for cascade spillways is to provide a smooth transition flow from the spillway crest to the first few steps. This transition is commonly attained

Encyclopedia of Water Science
DOI: 10.1081/E-EWS 120010343

Fig. 1 Straight-drop spillway.

Fig. 3 Roller compacted concrete cascade spillway.

using a smooth ogee crest. Transitions for flatter sloped spillways, 2.5H:1V or less, can also be attained by an arrangement of smaller steps at the crest of the cascade chute.[4] The energy loss for a cascade spillway is strongly dependent on the slope length. The energy loss due to the steps depends primarily on the ratio of critical depth of flow to the step height and on the number of steps.

Side-Channel Spillway

As defined by the U.S. Army Corp of Engineers,[10] a conventional side-channel spillway consists of an overflow weir discharging into a narrow channel in which the direction of flow is approximately parallel to the weir crest (Fig. 4). This type of spillway is used in circumstances similar to those of the chute spillway. Due to its unique

Fig. 2 Chute spillway.

Fig. 4 Side-channel spillway.

Fig. 5 Arial view of a vegetated-earthen spillway placed at dam abutment.

shape, a side-channel spillway can be sited on a narrow dam abutment.

Earthen Spillway

Earthen spillways are typically excavated in native materials and vegetated with grasses adapted to the local area (Fig. 5). Earthen spillways are often designed as auxiliary spillways with the perspective that damage may occur during infrequent operation, but not to the extent that it will lower the spillway crest or cause a catastrophic release of stored water. Often a saddle or low point on natural ground at the periphery of the reservoir will serve as the earthen spillway.

COMPONENTS OF SPILLWAYS

Spillways consist of three major components: entrance structure, conveyance channel, and outlet structure. An entrance structure is designed to control the discharge and admits reservoir water to the spillway. The conveyance channel carries the discharge from the inlet structure to the outlet structure. The outlet structure dissipates the energy of the high velocity flow from the conveyance channel and discharges it to the channel downstream.[5]

Entrance Structure

The design of the spillway entrance structure for small dams is not usually critical, and a variety of simple

crest patterns are used. In the case of large dams, it is important that the water be guided smoothly over the crest of the structure with a minimum of turbulence (Fig. 6). Therefore, an ogee-crest design is often used. The ogee crest takes the form of the underside of the nappe of a sharp crested weir when the flow rate corresponds to the maximum design capacity of the spillway. This results in near-maximum discharge efficiency. Correct design of the spillway entrance will minimize cost while providing sufficient crest length to pass the design discharge. Also, it will result in acceptable energy heads and pressure levels on the spillway crest, and acceptable unit discharges for the conveyance channel and outlet structure.

Gates located at or near the crest are often used to control flow into the spillway. Types of gates may include flashboards, stop logs, lift gates, radial gates, rolling gates, and drum gates. A spillway using gates is often referred to as a controlled spillway. The use of gates allows for additional storage above the spillway crest and control of the timing and quantity of reservoir discharges. The disadvantages to gates are: they are expensive; they often require personnel at the structure for proper operation; and operational or structural failure may have catastrophic consequences.

The inlet structure has a significant effect on the spillway discharge. There are cases where there is a need for increased flow capacity with upstream head elevation restrictions. A labyrinth weir or box inlet weir may be used to increase the inlet capacity for a given range of reservoir water surface elevations. A labyrinth weir is folded in plan

Fig. 6 Smooth flow over ogee spillway crest.

Fig. 7 Chute spillway with notched box inlet.

view, increasing the total weir length to 3–5 times the spillway width. The labyrinth weir capacity is typically twice the standard overflow crest of the same width for low head ranges.[9] The capacity of the box inlet weir is directly related to the perimeter of the overflow box that extends into the reservoir[2] (Fig. 7).

The discharge capacity of the standard ogee weir, labyrinth weir, and box weir, is given by the weir equation:

$$Q = \frac{2}{3}[C_d L h \sqrt{2gh}]$$

in which Q is the discharge in cfs, C_d a dimensionless weir coefficient, L the length of weir in ft, g gravitational acceleration 32.2 ft sec^{-2}, and h the total head on the crest in ft (vertical distance from the crest of the spillway to the reservoir level). The coefficient C_d, varies with spillway type and head. A typical range of C_d for the ogee crest is from 0.55 to 0.75.

The typical entrance to earthen spillways as well as some other chute spillways consists of a forebay reach followed by a level crest prior to entering the conveyance channel. The discharge rating of these types of spillways is typically developed based on water surface profile calculation methods incorporating channel geometry and roughness conditions.

Conveyance Channel

With the exception of the straight-drop spillway that may be used for low dams, the water that passes the spillway inlet is carried to the downstream outlet

structure by a conveyance channel. The material used to line the surface of the conveyance channel is dependent on frequency of use, erodibility of the natural materials, and overall spillway design. Principal spillway channels operate more frequently and are usually constructed of reinforced-concrete slabs 10–20 in. thick. Auxiliary spillways may also be lined to allow higher velocity flows and/or combination of the principal and auxiliary spillway functions into a single spillway. The high velocities and large levels of energy involved in flow over spillways may lead to problems with air entrainment, shock waves, cavitation, and abrasion. Abrasion may be a particular problem when entrained sediment is present.

Earthen spillways are unlined or vegetation-lined channels that are most often used as auxiliary spillways to pass infrequent flood flows. Damage may occur during operation, but it cannot be extensive enough to lower the spillway crest causing a catastrophic release of reservoir water (Fig. 8). The extent of damage that may occur in an unlined spillway is dependent on the duration and quantity of flow through the spillway, quality and maintenance of the vegetal cover, spillway geometry, and spillway geology.[6,7] Evaluation of expected erosion is the most difficult and critical problem encountered in the design of the earthen spillways. The designer must not only decide whether the channel materials will be eroded but also make reasonable estimates pertaining to the rate at which erosion will progress. Extensive exploration, testing of encountered materials, and geological profiles to a depth in excess of any anticipated scour are required

Fig. 8 Erosion damage of an auxiliary spillway after experiencing a flow.

to assist in the erosion estimates. Study of the history of erosion in the project area and research of erosion experiences at projects with similar facilities should be undertaken as part of the evaluation of expected erosion. The NRCS computer program "Sites," uses a three-phase erosion model to predict erosion of vegetated-earthen spillways.[8]

Outlet Structure

Water returned to the river below the dam must be kept from scouring or eroding the riverbed or dam foundation. Plunge pools or stilling basins are therefore required to reduce the velocity of the water before returning it to the downstream channel. These energy dissipation structures may be an integral part of the dam or spillway. Two common structures used in dissipating the high energy of falling water are the apron-basin and the flip bucket. In the apron-basin type of structure, the high-velocity shallow flow coming from the dam is converted into a low-velocity deep flow by causing a hydraulic jump to occur on a horizontal or sloping concrete apron (Fig. 9). With a flip bucket, the toe of the dam is shaped to deflect the high-velocity flow upward away from the riverbed. The resulting "flip" breaks up the jet and dissipates the energy of the water. Additional energy is dissipated in a plunge pool downstream of the bucket.

Fig. 9 Energy dissipation basin.

FACTORS WHICH AFFECT SPILLWAY CHOICE AND DESIGN

Several factors should be considered when choosing a spillway. Singh and Varshney[5] list several factors such as safety considerations, hydrologic and site conditions, type of dam, purpose of dam and operating conditions, conditions downstream of the dam, and nature and amount of solid material brought by the river. Safety is the most important factor to consider for a spillway because improper design of spillways or insufficient spillway capacity may result in dam failures. Spillway design and capacity depend on inflow discharge (frequency and shape of hydrograph), the elevation of the spillway crest, storage of the reservoir at various levels, and the geological site conditions, which include slope stability, steepness of the terrain, or possibilities of scour downstream. The type, purpose, and conditions, downstream of the dam also influence the spillway design and capacity. Rising floodwaters downstream of the dam and erodible material can have major consequences if not properly considered. In addition, trees, floating debris, and suspended sediment can influence the decision in selecting a spillway.[5]

The required capacity (maximum outflow rate through the spillway) depends on the spillway design flow (inflow hydrograph to the reservoir), the normal discharge capacity of other outlet works, and the available storage. The selection of the spillway design flow is related to the degree of flood protection required that, in turn, depends on the type of dam, its location, and the consequences of a dam failure. A high dam storing a large volume of water located upstream of an inhabited area requires a much higher degree of protection from overtopping than a low dam storing a small quantity of water whose downstream reach is uninhabited. The probable maximum flood is commonly used for design of the former, while a smaller flood is suitable for the latter.

REFERENCES

1. Donnelly, C.A.; Blaisdell, F.W. Straight Drop Spillway Stilling Basin. J. Hydr. Eng. ASCE **1965**, *91* (HY3), 101–131.
2. Gwinn, W.R. *Model Study of a Box-Inlet Chute Spillway and SAF Stilling Basin*; Agricultural Reviews and Manuals, Southern Series, No. 17; U.S. Department of Agriculture, Agricultural Research (Southern Region), Science and Education Administration: New Orleans, LA, 1981; 1–14.
3. Linsley, R.K.; Franzini, J.B. Spillways, Gates, and Outlet Works. In *Water-Resources Engineering*, 2nd Ed.; Chow, V.T., Eliassen, R., Linsley, R.K., Eds.; McGraw-Hill, Inc.: New York, 1972; 231–268.

4. Rice, C.E.; Kadavy, K.C. Model Study of a Roller Compacted Concrete Stepped Spillway. J. Hydr. Eng. ASCE **1996**, *122* (6), 292–297.

5. Singh, B.; Varshney, R.S. Spillways and Embankment Dams. In *Engineering for Embankment Dams*; Singh, B., Ed.; Balkema Publishers: Vermont, 1995.

6. Temple, D.M.; Hanson, G.J. Headcut Development in Vegetated Earth Spillways. Appl. Eng. Agric. **1994**, *10*, 677–682.

7. Temple, D.M.; Moore, J.S. Headcut Advance Prediction for Earth Spillways. Trans. ASAE **1997**, *40*, 557–562.

8. Temple, D.M.; Richardson, H.H.; Brevard, J.A.; Hanson, G.J. SITES: The New DAMS2. Appl. Eng. Agric. **1995**, *11*, 831–834.

9. Tullis, J.P.; Amanian, N.; Waldron, D. Design of Labyrinth Spillways. J. Hydr. Eng. ASCE **1995**, *121* (3), 247–255.

10. U.S. Army Corps of Engineers. *Hydraulic Design of Spillways*; Engineer Manual 1110-2-1630; U.S. Army Corps of Engineers: Washington, DC, 1990; 1(1)–7(15).

11. U.S. Bureau of Reclamation. Spillways. *Design of Small Dams*, 1st Ed.; U.S. Printing Office: Washington, DC, 1960; 247–343.

Oxygen Measurement: Biological–Chemical Oxygen Demand

David B. Parker
Marty B. Rhoades
West Texas A&M University, Canyon, Texas, U.S.A.

INTRODUCTION

Many aquatic organisms depend on dissolved oxygen (DO) for basic life functions. Dissolved oxygen is one of the factors that affect population diversity in surface waters. Aerobic bacteria, fungi, protozoa, and algae all carry out aerobic respiration and require DO to survive. Trout and other coldwater fish species require more DO than warm-water fish species. A DO concentration of $5\,mg\,L^{-1}$ is generally recommended for warm-water fish, and $6\,mg\,L^{-1}$ for coldwater fish.[1] Lower concentrations may be tolerated for short periods but result in stress on the fish.

The solubility of oxygen in water varies with water temperature and atmospheric pressure. Dissolved oxygen concentrations range from $14.6\,mg\,L^{-1}$ at $0°C$ to about $7\,mg\,L^{-1}$ at $35°C$.[2] Oxygen is less soluble in saline water than in pure water. Polluted waters usually have lower DO concentrations than pure water.

OVERVIEW

In ponds and lakes, photosynthetic activity from aquatic plants such as algae produce DO. At night, these same plants compete with aquatic organisms for DO. Biodegradation of organic matter can decrease DO concentrations. Organic wastes are subject to further bacterial decomposition once they enter the environment. If domestic or agricultural organic wastes enter water that contains DO, the aerobic bacteria will utilize the organic material as a food and energy source. During this process, the bacteria respire DO and produce carbon dioxide. If enough DO is removed, then concentrations can fall below values required for survival of aquatic organisms. Fish kills are sometimes the result of spills or unauthorized releases of organic-laden waste to surface waters.[3] During warm conditions, fish kills can occur immediately after the release. When an organic waste release occurs during cold conditions, DO concentrations may not be affected for several months until water temperatures increase, making it difficult to assess the exact cause and nature of the fish kill.

Fish kills are not always the direct result of organic releases. In some instances, the enrichment of waters by nutrients, a term called eutrophication, can lead to unwanted algal growth.[4] In many water bodies, phosphorus is the limiting nutrient for algal growth. A rapid release of dissolved phosphorus to surface waters can trigger algae blooms. Growing algae produces DO through photosynthesis, however, whenever the algae dies it is subject to decomposition that could lead to the decrease in DO.

DISSOLVED OXYGEN MEASUREMENT

Dissolved oxygen was originally measured using the Winkler (iodometric) method. The original Winkler method was based on the fact that oxygen oxidizes Mn^{2+} under alkaline conditions. The higher valence manganese then oxidizes I^- to free I_2 under acidic conditions. The amount of free iodine released is proportional to the DO concentration. Because nitrite ions interfere with the DO determination, several modifications to the original Winkler method have been developed. Nitrite interference is overcome by using sodium azide (NaN_3). Another modification includes using permanganate to oxidize any reducing agents present.

In recent years, electronic DO membrane electrodes have become more common. Membrane electrodes can be either the polarographic or galvanic type.[5] Both types utilize a sensing element composed of two metal electrodes. A voltage is passed across the electrodes, and the measured current is converted to a DO concentration. Membrane electrodes are especially useful in field applications. When electrodes are attached to a long cord, DO can be measured at various depths in ponds and rivers. Dissolved oxygen electrodes are usually calibrated against water samples that have been analyzed by using the Winkler method. Because DO electrodes are sensitive to temperature, an accurate temperature measurement must be made at the same time so that a correction can be applied. There are many brands of DO electrodes available commercially.

Encyclopedia of Water Science
DOI: 10.1081/E-EWS 120010182

BIOCHEMICAL OXYGEN DEMAND

Organic wastes can enter surface water bodies from a variety of sources, including raw sewage spills, permitted publicly owned treatment works (POTWs), animal feeding operations, runoff from land application areas, or improperly designed septic tanks.[6] Environmental regulations exist to protect surface water bodies from unwarranted discharges that could threaten aquatic life or human health. Environmental permits for discharge of wastewater or stormwater runoff to surface water bodies often include such items as temperature and biochemical oxygen demand (BOD). Biochemical oxygen demand is an empirical test in which standardized laboratory procedures are used to measure the biologically available organic matter in a water sample. It is based on the amount of oxygen that would be consumed if an abundant aerobic bacterial population were present. Biochemical oxygen demand is defined as "the quantity of oxygen used by bacteria while stabilizing decomposable organic matter under aerobic conditions."[2,7] The BOD test is widely used to determine the concentration of domestic and industrial wastes, and is helpful in evaluating the BOD removal efficiency of wastewater treatment plants. Miner, Humenik, and Overcash[3] presented the conceptual reaction as follows:

$$BOD + O_2 \rightarrow CO_2 + H_2O + Energy \qquad (1)$$

The standard BOD test is based on a 5-day incubation period, and is referred to as BOD_5. As a rule-of-thumb, BOD_5 is about 70%–80% of the ultimate BOD (BOD_U).[2] Another reason for limiting the BOD test to 5 day is to avoid interferences from oxidation of ammonia to nitrite and nitrate, a process called nitrification. Nitrifying bacteria typically do not consume an appreciable amount

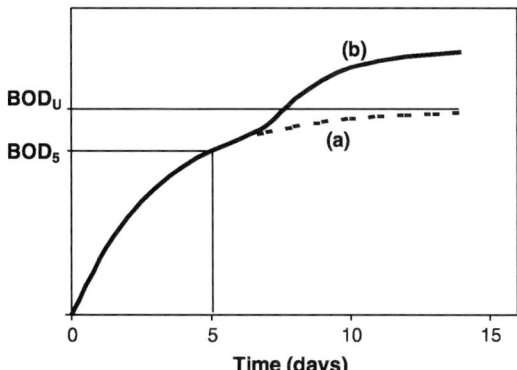

Fig. 1 The BOD curve showing (a) the normal curve resulting from oxidation of organic matter and (b) the curve if combined with nitrification of ammonia.

of DO until 6 day–10 day after initiating the BOD test[8] (Fig. 1).

With samples that have a BOD less than $7 \, mg \, L^{-1}$, the BOD is measured directly on the water sample.[2] For higher BODs, the wastewater sample is diluted because the BOD concentration exceeds the DO available. The dilution water reduces the toxicity of the wastewater and provides the DO needed for aerobic decomposition. Nutrients such as nitrogen, phosphorus, and trace metals are added to the dilution water, and it is saturated with oxygen before use. A buffer is added to maintain the pH in a range suitable for bacterial growth. The solution is "seeded" with the necessary microorganisms.[9] The seed source often consists of effluent from biological waste treatment plants.[5] The sample is incubated at a temperature of 20°C, and DO is measured initially and after 5 day. The 5-day BOD is determined by subtracting the two values. The DO is typically measured using either the Winkler method or a membrane electrode. Because the BOD concentration is rarely known before the test, several different dilutions are performed to cover the range of possible BOD values.

CHEMICAL OXYGEN DEMAND

The chemical oxygen demand (COD) test is used similarly to the BOD test to measure the concentration of a wastewater sample. In some cases, the COD data can be correlated with the BOD data, although this takes some experience to establish reliable correlation factors, which can range from 0.4 to 0.8. The COD test is used primarily to evaluate industrial waste, whereas the BOD test is used for biological waste.[2] Advantages of the COD test are that it can be performed in about 3 hr, as compared to 5 day for the BOD test, and it can be utilized in wastewater with high toxicity, where a BOD test cannot.

The COD test utilizes a strong oxidizing agent (usually potassium dichromate) under heated acidic (usually sulfuric) conditions in the presence of chromium and silver salts to convert organic matter to water and carbon dioxide. Unlike the BOD test, the COD test oxidizes all organic matter, including biologically resistant organic matter such as lignin. As a result, COD values are greater than BOD values. The COD test can be very accurate and precise for those samples with a COD of $50 \, mg \, L^{-1}$ or greater.[2]

Inorganic ions such as chloride, bromide, and iodide can cause erroneously high COD values. Mercuric sulfate can be added to the sample to remove the chloride interference. A major disadvantage of the COD test is that disposal of hazardous wastes such as mercury, hexavalent chromium, sulfuric acid, and silver can be a problem.[5]

Table 1 Typical BOD$_5$ and COD concentrations found in municipal, industrial, and agricultural waste products and wastewater

Source	Units	BOD$_5$	COD
Raw municipal wastewater	mg L^{-1}	200	450
Domestic sewage	mg L^{-1}	100–300	250–1,000
Beef feedlot runoff pond effluent	mg L^{-1}	—	1,400
Beef feedlot runoff pond sludge	mg L^{-1}	—	77,500
Dairy lagoon effluent	mg L^{-1}	350	1,500
Dairy lagoon sludge	mg L^{-1}	—	52,000
Swine lagoon effluent	mg L^{-1}	400	1,200
Swine lagoon sludge	mg L^{-1}	—	64,600
Milking center wastewater	mg L^{-1}	1,000	5,000
Cheese production wastewater	mg L^{-1}	3,200	5,600
Candy production effluent	mg L^{-1}	1,600	3,000
Synthetic textile effluent	mg L^{-1}	1,500	3,300
Slaughterhouse processing effluent	kg/1000 kg raw product	5.8–8.5	—
Vegetable processing effluent	kg/1000 kg raw product	7–55	14–96
Papermill effluent	kg/1000 kg pulp	60	—
Dairy cow manure (fresh, as excreted)	kg/d/640 kg animal	1.0	5.7
Beef cattle manure (fresh, as excreted)	kg/d/450 kg animal	0.6	2.5
Swine manure (fresh, as excreted)	kg/d/100 kg animal	0.2	0.6

(From Refs. 8 and 10.)

Typical BOD and COD concentrations vary greatly among waste types. Waste streams from animal feeding operations have BOD concentrations greater than 5000 mg L^{-1} compared to 200 mg L^{-1} for typical municipal wastewater.[3] Typical BOD and COD concentrations for a variety of municipal, industrial, and agricultural waste sources are presented in Table 1. Chemical oxygen demand concentrations are typically about 2–5 times greater than BOD concentrations for the same waste product.

CONCLUSION

Dissolved oxygen is one of the most important water quality parameters for aquatic life. Because many municipal, industrial, and agricultural sources discharge wastewater to surface water bodies containing aquatic life, environmental regulations have been promulgated to limit the amount of oxygen-consuming organic waste products that can be discharged to a stream or lake. Methods for measuring the strength and concentration of these oxygen consuming waste products include the 5-day biochemical oxygen demand (BOD$_5$) and COD tests.

REFERENCES

1. Wheaton, F. Aquaculture Engineering. *CIGR Handbook of Agricultural Engineering*; American Society of Agricultural Engineers: St. Joseph, MI, 1999; Vol. II, Part II.
2. Sawyer, C.N.; McCarty, P.L. *Chemistry for Environmental Engineering*, 3rd Ed.; McGraw-Hill, Inc.: New York, NY, 1978.
3. Miner, J.R.; Humenik, F.J.; Overcash, M.R. *Managing Livestock Wastes to Preserve Environmental Quality*; Iowa State University Press: Ames, IA, 2000.
4. Hodges, L. *Environmental Pollution*; Holt, Rinehard and Winston, Inc.: New York, NY, 1973.
5. APHA, *Standard Methods for the Examination of Water and Wastewater*, 20th Ed.; American Public Health Association: Washington, DC, 1998.
6. Zoller, U. *Groundwater Contamination and Control*; Marcel Dekker, Inc.: New York, NY, 1994.
7. ASAE, *Uniform Terminology for Rural Waste Management*; ASAE Standard S292.5, American Society of Agricultural Engineers: St. Joseph, MI, 1999.
8. Lindeburg, M.R. *Civil Engineering Reference Manual*; Professional Publications, Inc.: Belmont, CA, 1989.
9. Csuros, M.; Csuros, C. *Microbiological Examination of Water and Wastewater*; CRC Press: Boca Raton, FL, 1999.
10. NRCS, *Agricultural Waste Management Field Handbook, Part 651*; United States Natural Resources Conservation Service: Washington, DC, 1992.

Pathogens in Water

Jeanette Thurston-Enriquez
United States Department of Agriculture (USDA),
Lincoln, Nebraska, U.S.A.

P

INTRODUCTION

The transmission of pathogens by water is a highly effective way of spreading infectious disease among large numbers of people. As early as 6000 years ago, the association of contaminated water and illness was documented in Sanskrit and Greek writings that described treatments for "impure water."[1] Today, waterborne disease outbreaks continue to be responsible for high morbidity and mortality worldwide (Table 1).[2,3]

CLASSES OF PATHOGENS ASSOCIATED WITH WATER

Due to the introduction of chlorination and filtration of drinking water supplies in the early 1900s, the waterborne outbreak paradigm in the United States shifted from bacteria, as the primary agents causing waterborne disease, to protozoan parasites and enteric viruses. Currently, these groups of micro-organisms and others cause water-related disease worldwide. Table 2 differentiates between the different types of micro-organisms: bacteria, viruses, protozoan parasites, blue–green algae, and helminthes. For each microbial group, examples of water-related pathogens and associated diseases are also listed.

CATEGORIES OF WATER-ASSOCIATED DISEASES

There are four disease categories related to water transmission of microbial pathogens: 1) waterborne; 2) water-washed; 3) water-based; and 4) water-related diseases.[4] Infectious agents that are excreted in the feces and transmitted by ingestion of contaminated water cause waterborne diseases. Classic examples of waterborne pathogens include *Cryptosporidium parvum* (cryptosporidiosis), poliovirus (polio), and *Vibrio cholera* (cholera). Water-washed infections occur in areas where personal hygiene and water availability is poor. Person-to-person transmission and contact with unclean household items effectively transmit water-washed pathogens such as *Trachoma* (eye infections) and *Shigella* (dysentery). Water-based diseases arise from infection with pathogens

that spend all or a portion of their life in water. Examples of these include *Dracunculus medinesis* (dracontaiasis) and *Schistosoma* species (schistosomiasis). Unlike the other categories, water-related diseases are carried by water insects, as in the case of malaria, which is carried by mosquitoes.[2,4]

ENTERIC PATHOGENS

Micro-organisms that are excreted in the feces and infect the gastrointestinal tract are called enteric pathogens. Enteric pathogens cause a wide range of illness from asymptomatic (no clinical signs but microbe can grow and may be transmitted to susceptible individuals) to mild intestinal symptoms (diarrhea, fever, malaise, etc.) to paralysis. Enteric pathogens are probably the most important causes of water-washed and waterborne infections worldwide. Approximately 31%, 15%, and 10% of reported waterborne outbreaks are caused by enteric protozoan (*Giardia* and *Cryptosporidium*), viral, and bacterial agents, respectively.[5] However, the etiological agents responsible for approximately half of all reported outbreaks go unrecognized. The unidentified agents causing these outbreaks are thought to be enteric viruses due to epidemiological and clinical similarities.[6]

PATHOGEN CHARACTERISTICS THAT ENHANCE PATHOGEN SURVIVAL AND TRANSMISSION

Traits that may enhance pathogen survival and transmission in the environment may include numbers and location of pathogen reservoirs, concentration of pathogens excreted and mode of excretion, infectious dose, severity of illness, and individual pathogen traits. Because some pathogens, termed zoonotic, are able to infect humans and animals (domestic and wildlife) they can be distributed throughout the environment, contaminating land, air, and water (Table 3). Excretion of pathogens in the feces is another important characteristic since they can be released in very high concentrations. Table 4 lists the concentrations of enteric pathogens shed in feces.[7] Depending on the infectious dose, or the number of

Encyclopedia of Water Science
DOI: 10.1081/E-EWS 120010238

Table 1 Worldwide waterborne outbreak(s)

Country (year(s) of outbreak)	Disease (disease agent)	Number of cases (number of fatalities)
United States (1993)	Cryptosporidiosis (*Cryptosporidium parvum*)	> 400,000 (> 50)
England (1971–1980)	Giardiasis (*Giardia lamblia*)	60 (0)
Czech Soviet Republic (1979–1982)	Shigellosis (*Shigella*)	287 (not indicated)

(From Ref. 3.)

micro-organisms required to produce disease, numbers ranging from over 10^6 down to 1 infectious agent must be ingested. For example, ingestion of as little as 1 viral particle can induce illness in a susceptible individual compared to some bacterial pathogens that require ingestion of hundreds to thousands of organisms. Disease produced by enteric pathogens is generally mild, self-limiting, and in some cases, the infected individual may be shedding the pathogen without any symptoms (asymptomatic shedding). Fecally transmitted pathogens that produce mild symptoms, or are asymptomatic, enable the infected individual to effectively contaminate the environment, unlike other diseases that render the infected individual immobile. Genetically acquired microbial traits enable some microbial groups to be more resistant to heat, pH, or chemical (chlorine, ozone, etc.) and physical (ultraviolet light, gamma irradiation) disinfectants than others. Because of these differences, some pathogens survive longer in the environment than others. In general, enteric viruses and protozoan parasites survive longer in water subjected to environmental conditions and disinfected water than enteric bacteria.

ENVIRONMENTAL FACTORS AFFECTING PATHOGEN SURVIVAL

Unlike pathogens that need to be transmitted by direct or close contact to infected individuals (i.e., gonorrhea, human immunodeficiency virus, and herpes), enteric pathogens are hardy enough to survive conditions outside the body for long periods. Their survival depends on environmental conditions such as temperature, cell aggregation, sunlight exposure, pH, and presence of predatory organisms, inorganic or organic matter, and chemicals that may affect their survival. As temperature increases, the inactivation or dying-off of enteric pathogens increases. For example, *Giardia lamblia* cysts remain viable in water for 77 days and 4 days at 8°C and 37°C, respectively.[8] Since enteric pathogens infect the gastrointestinal tract they can withstand fairly low pH values, although high pH conditions may inactivate enteric viruses. Aggregation or adsorption to organic or inorganic matter may serve to protect or shield enteric pathogens from the effects of sunlight, chemical treatments, or

predatory organisms. Also, macro invertebrates (nematodes and amphipods) and protozoa have been shown to ingest pathogens and protect them from the effects of water treatment.[9] Sunlight may decrease pathogen survival due to the effects of ultraviolet light damage. Ultraviolet light produces nucleic acid damage that can be repaired by some micro-organisms (bacteria) but not others (viruses).

PATHOGEN WATER ROUTES OF TRANSMISSION

Human and animal reservoirs of enteric disease can contaminate land, water, or air through several routes (Fig. 1). Fecal waste is deposited onto land in several ways: direct deposition by domestic animals and wildlife, piled for storage, piled (composting) or spread (land applied human biosolids) for treatment, and by agricultural practices (spraying, spreading, or injection of waste into/onto soil). Rain or other events that produce runoff from fecally contaminated sites increases the potential for transport of pathogens to surface waters that may be used for drinking, shellfish harvesting, irrigation, or recreational purposes. In addition, pathogens may be transported through soil and contaminate groundwater. Airborne transmission of pathogens via water droplets generated by human and animal activities or wind is another way by which pathogens in water may be transmitted. These water droplets may be transported naturally (surf droplets), by agricultural activities (spray irrigation), and other human practices (showers, cooling towers) through air and are capable of transmitting disease either through ingestion, inhalation, or contact.[2] Although considered foodborne, pathogens present in water used for harvesting vegetable crops or shellfish may increase the risk of illness through consumption of these products.

WATER TREATMENT

There are many treatment practices for the reduction of pathogens in water and waste. These include filtration (sand, activated carbon, flora), coagulation and sedimentation, disinfection, composting, and constructed wetlands.

Table 2 Types and characteristics of enteric and other pathogenic micro-organisms that can be transmitted by water

Types of micro-organisms associated with water	Microbial characteristics	Pathogens associated with water disease	Disease or complication
Bacteria	Prokaryotic, single-celled organisms Cell wall and membrane surrounding cellular components Reproduce via binary fission Susceptible to disinfectants and environmental conditions (compared to other water pathogens) Spores and dormant cells formed by some bacteria enable them to survive hash conditions (example *Clostridium*) Size range 0.1 μm–10 μm	*Salmonella* *Campylobacter* *Enterohemorrhagic* *Escherichia coli* *Enteroinvasive E. coli* *Enteropathogenic E. coli* *Enterotoxigenic E. coli* *Shigella*	Diarrhea, typhoid Bloody diarrhea Bloody diarrhea, hemolytic uremic syndrome Dysentery Diarrhea Diarrhea Diarrhea
Viruses	Obligate parasites (require a host for replication) Composed of protein outer capsid surrounding nucleic acid core Do not carry out metabolic functions Nucleic acid may be double or single stranded Long-term environmental survival due to simple structure and no requirement for nutrients Size range 0.01 μm–0.1 μm	*Hepatitis A and E* Enteroviruses (polioviruses, coxsackieviruses, echoviruses, and enterovirus types 68–71) Rotaviruses (group A and B) Human caliciviruses Astrovirus Adenovirus	Liver disease Febrile and respiratory illness, meningitis, diarrhea, encephalitis, and others Diarrhea Vomiting, diarrhea Diarrhea Diarrhea, eye and respiratory infections
Protozoa	Single-celled eukaryotic organisms Complex structure Sexual or asexual replication Some produce environmentally resistant stages (example *Cryptosporidium* oocysts and *Giardia* cysts) Size range 1 μm–100 μm	*Giardia lamblia* *Cryptosporidium parvum* *Cyclospora cayetanensis* *Microsporidia* *Toxoplasma gondii* *Entamoeba histolytica* *Naegleria fowleri*	Diarrhea Diarrhea Diarrhea Diarrhea, kidney and respiratory infections Flu-like (adults), encephalitis, and ocular disease (children) Amoebic dysentery Meningoencephalititis
Helminths	Multicellular, worms Complex life-cycles (one or more hosts required) Eggs are environmentally resistant Size range 1 μm–10^9 μm	*Ascaris lumbricoides* *Necator americanus* *Trichuris trichiura* *Taenia saginata* *Schistosoma mansoni*	Ascariasis Hookworm Whipworm Beef tapeworm Schistosomiasis
Cyanobacteria (blue–green algae)	Procaryotic single-celled organisms Replicate by binary fission or fragmentation Some species produce toxins Some are resistant to extreme environmental conditions Size range 1 μm–100 μm	*Anabaena* *Nodularia* *Microcystsis* *Nostoc* *Alexandrium*	All can produce toxins that can cause liver damage, neural damage, gastrointestinal symptoms Possible carcinogens

(From Ref. 4.)

Table 3 Examples of known and potential zoonotic enteric pathogens and their hosts

Zoonotic pathogens	Pathogen hosts
Bacteria	
E. coli	Humans, domestic and wild animals
Salmonella	Humans, domestic and wild fowl, and other animals
Viruses	
Caliciviruses	Humans and potentially cattle and swine
Hepatitis E	Humans and potentially swine and rats
Protozoa	
Cryptosporidium	Humans, domestic and wild animals
Giardia	Humans, domestic and wild animals

Table 4 Concentrations of enteric pathogens excreted in feces

Enteric pathogen group	Range of pathogen concentration (per gram of feces)
Bacteria	10^4–10^{10}
Viruses	10^3–10^{12}
Protozoan parasites	10^6–10^7

(From Ref. 7.)

A multibarrier approach is used for the removal or reduction of pathogenic micro-organisms in municipal wastewater. The general steps include: 1) removal or sedimentation of large debris (primary treatment); 2) biological degradation (trickling filters, aeration tank, lagoon) and/or disinfection (secondary treatment); and 3) a combination of physical (coagulation and filtration) and chemical (chlorine disinfection) steps in order to further reduce biological and chemical contaminants (tertiary treatment). Wastewater treatment has been shown to reduce bacteria by 99.99999%, viruses by 99.999% and protozoa by 99.993% (*Giardia* cysts) and 99.95%

(*Cryptosporidium* oocysts).[10] Constructed wetlands have been applied as a low cost alternative to wastewater treatment. As water flows through the vegetated wetland absorption to flora, gravel or sand substrate, natural die-off, sedimentation, filtration, and predation occurs. For solid waste, compost piles can be used for the inactivation or reduction of pathogenic micro-organisms. Temperature is the key factor controlling pathogen reduction in compost piles where it is suggested that 55°C must be maintained for at least 3 day and 15 day for static and aerated piles, respectively.[9] Difficulties in maintaining uniform temperature throughout the pile and regrowth of bacterial pathogens are two disadvantages for use of composting for pathogen reduction.[9] Due to the different physical and genetic characteristics of pathogenic micro-organisms, a combination of treatment or prevention practices may be necessary for their removal, death, or inactivation. For

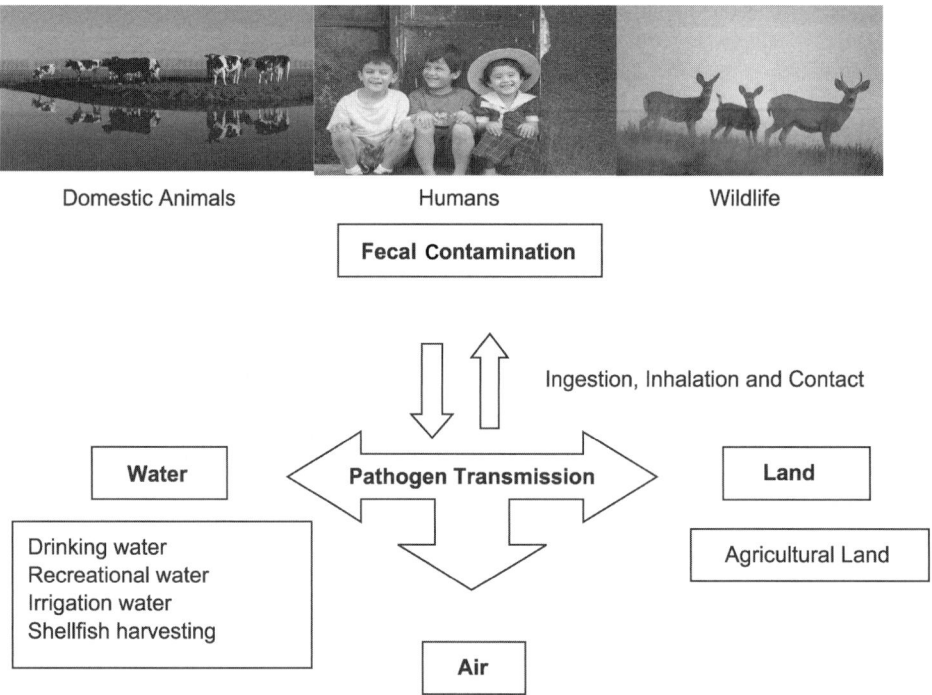

Fig. 1 Environmental routes of pathogen transmission.

example, the ultimate barrier between pathogens and drinking water consumers is chemical or physical disinfection for inactivation of enteric viruses. Their small size (20 nm–80 nm) enables them to bypass filtration processes, whereas filtration is required for the removal of protozoan parasites (> 2 m in size) since they are highly resistant to most water disinfectants.

EMERGING PATHOGENS

Control of water transmission of pathogenic microorganisms continues to be a public health concern because there is an increasing immunocompromised (elderly, cancer, and AIDS patients) population and because a large percentage of waterborne outbreaks go unrecognized.[5,9,11] In fact, while less than 20 waterborne disease outbreaks are documented each year in the United States, it is estimated that the true incidence may be 10–100 times higher.[5] Even for well-established pathogens, the true incidence is unknown in the United States because: 1) reporting waterborne outbreaks and its agents is voluntary; 2) there is a lack of efficient detection methods for some important enteric pathogens; 3) contamination events in water are usually transient, thus etiological agents are not detected; and 4) individuals may not seek medical attention since acute gastrointestinal illness (AGI) is usually self-limiting and mild.[4,11] Furthermore, very little, if any, information exists on infectious agents that are newly recognized, or "emerging" in water supplies and their impact on waterborne disease outbreaks.

CONCLUSION

Water serves as a passive carrier for the transmission of disease. Human and animal populations may be exposed to pathogens through direct contact, ingestion, or inhalation of contaminated water. Enteric pathogens are the most important group of organisms relating to waterborne and water-washed diseases. They are excreted in high numbers in the feces and have traits that allow their survival and successful transmission in the environment. These traits should be considered when making decisions regarding strategies for limiting their transmission by water. A multibarrier approach may be best for efficient removal of bacterial, viral, and protozoan pathogens.

REFERENCES

1. Baker, M.N. *The Quest for Pure Water*; American Water Works Association: New York, 1948.
2. Gerba, C.P.; Rusin, P.; Enriquez, C.E.; Johnson, D. Environmentally Transmitted Pathogens. In *Environmental Microbiology*; Maier, R.M., Pepper, I.L., Gerba, C.P., Eds.; Academic Press: San Diego, 2000; 447–489.
3. Hunter, P.R. *Waterborne Disease Epidemiology and Ecology*; John Wiley and Sons Inc.: New York, 1998.
4. Moe, C.L. Waterborne Transmission of Infectious Agents. In *Manual of Environmental Microbiology*; Hurst, C.J., Ed.; ASM Press: Washington, D.C., 1997; 136–152.
5. Gerba, C.P. Pathogens in the Environment. In *Pollution Science*; Pepper, I.L., Gerba, C.P., Brusseau, M.L., Eds.; Academic Press: San Diego, 1996; 278–299.
6. Barwick, R.S.; Levy, D.A.; Craun, G.F.; Beach, M.J.; Calderon, R.L. Surveillance for Waterborne-Disease Outbreaks—United States, 1997–1998. Morb. Mortal. Wkly Rep. **2000**, *49*, 1–35.
7. Haas, C.N.; Rose, J.B.; Gerba, C.P. *Quantitative Microbial Risk Assessment*; John Wiley and Sons, Inc.: New York, 1999.
8. Bingham, A.K.; Jarroll, E.; Meyer, E. Giardia Spp.: Physical Factors of Exystation In Vitro, and Exystation vs. Eosin Exclusion as Determinants of Viability. Exp. Parasitol. **1979**, *47*, 284–291.
9. Bitton, G. *Wastewater Microbiology*; Wiley-Liss: New York, 1980.
10. Rose, J.B.; Carnahan, R.P. Pathogen Removal by Full Scale Wastewater Treatment. Report to Florida Department of Environmental Regulation; Tallahassee, FL, 1992.
11. EPA; Waterborne Disease Studies and National Estimate of Waterborne Disease Occurrence. Fed. Regist. **1998**, *63*, 42849–42852.

Pesticide Contamination, Groundwater

Roy F. Spalding
University of Nebraska, Lincoln, Nebraska, U.S.A.

INTRODUCTION

Trace concentrations of most of the commonly used pesticides have been confirmed in groundwaters of the United States. Since groundwater is the source of 53% of the potable water, the more toxic pesticides and their transformation products are a concern from the standpoint of human health. Others are a risk to the environment in areas where contaminated groundwater enters surface water. Through toxicological testing, the USEPA has established Maximum Contaminant Levels (MCLs) or lifetime Health Advisory Levels (HALs) for several pesticides (Table 1).

The EPA also has a separate list of unregulated compounds, including newly registered pesticides and their transformation products, such as acetochlor and alachlor-ESA, that are presently being evaluated or being considered for toxicological evaluation. Based on the results of the EPAs National Pesticide Assessment,[2] 10.4% of 94,600 community systems contained detectable concentrations of at least one pesticide. Evaluation of these results led to an estimated 0.6% of rural domestic wells containing one or more pesticides above the MCL.

PESTICIDE USE

In the United States about 80% of pesticide usage is in agriculture. The remainder is used by industry, homeowners, and gardeners. About 500 million pounds of herbicide, 180 million pounds of insecticide, and 70 million pounds of fungicide were applied for agricultural purposes in 1993.[3] Several maps of the United States delineate usage patterns of several pesticides.[4] The majority of the triazine and amide herbicides are applied to fields in the north central corn belt states of Michigan, Wisconsin, Minnesota, Nebraska, Iowa, Illinois, Indiana, and Ohio. Commonly used organophosphorus insecticides are more heavily applied to fields in California and along the southeastern seaboard than in the northern corn belt. Carbamate and thiocarbamate pesticides are heavily used in potato growing areas of northern Maine, Idaho, the Delmarva Peninsula, and vegetable fields of California and the southeastern coastal states. Fungicide use is concentrated in high humidity and irrigated areas of the coastal states and to some extent along the Great Lakes and

Mississippi River Valley. The fumigants carbon tetrachloride and ethylene dibromide (EDB) were used heavily in the past at grain storage elevators throughout the Midwest and elsewhere in the United States.

ASSOCIATED PESTICIDE BEHAVIOR IN SOILS AND WATER

Although pesticide use is a dominant factor in groundwater contamination, leaching variability among pesticides exhibiting similar behaviors is striking and explains why several heavily used pesticides seldom if ever are detected in groundwater. In general, pesticides within a class have similar chemical characteristics upon which soil leaching predictions can be made based on persistence, solubility, and mobility. Pesticide class relationships with soils and water transport described in the following text are detailed in Ref. 5. Individual frequencies of groundwater pesticide detection, in parenthesis next to commonly used products, are calculated from the Pesticide Groundwater Data Base (PGWDB)[4] and the National Water Quality Assessment (NAWQA) database.[6] High frequencies of detection identify those pesticides with a disposition to leach.

Insecticides

Chlorinated hydrocarbons are one of the oldest chemical classes of insecticides. Some of the best-known compounds include aldrin, dieldrin, DDE, DDT, endrin, and toxaphene. Although banned since the 1960s, their extremely persistent nature precludes their detection in very trace quantities in groundwater of the upper Midwest. On the other hand, heavily used organophosphates like malathion, methylparathion, disulfoton, and others have been extensively surveyed during several groundwater monitoring studies and have not been detected. The organophosphate insecticides, parathion (not reported (NR), < 1),[a] terbufos (< 1, < 1), fonofos (< 1, < 1), and chlorpyrifos (< 1, < 1), which are heavily used on corn and sorghum, were also seldom detected. Diazinon (1.1, 1.3), the common garden insecticide, is occasionally

---

[a] % occurrence from PGWDB, % occurrence from NAQWA data.

Encyclopedia of Water Science
DOI: 10.1081/E-EWS 120010368

Table 1 U.S. maximum contaminant levels for drinking water[1]

Organic chemical name	MCL (mg/L)	Organic chemical name	MCL (mg/L)	Organic chemical name	MCL (mg/L)
2,4,5-TP (Silvex)	0.05	Chlordane	0.002	Heptachlor	0.0004
2,4-D	0.07	Dalapon	0.2	Heptachlor Epoxide	0.0002
Alachlor	0.002	Dinoseb	0.007	Lindane	0.0002
Aldicarb	0.007	Diquat	0.02	Methoxychlor	0.04
Aldicarb sulfone	0.007	Endothall	0.1	Oxamyl (Vydate)	0.2
Aldicarb sulfoxide	0.004	Endrin	0.002	Picloram	0.5
Atrazine	0.003	Ethylene dibromide	0.00005	Simazine	0.004
Carbofuran	0.04	Glyphosate	0.7	Toxaphene	0.003
Carbon tetrachloride	0.005				

detected in groundwater. Generally, the organic phosphates are rapid degraders and are strongly retained on soils.

For the most part, carbamates and thiocarbamates are very sparingly soluble and exhibit low to moderate soil retention; however, a small number have high solubility and low soil retention. Most carbamates are characterized as having short longevities. Generally, pesticides in this group having half-lives of 30 days or more have the potential to leach. The thiocarbamates butylate (< 1, < 1) and EPTC (2.6, < 1) are extensively used in agriculture and have relatively short half-lives. Aldicarb (< 1, < 1) and carbofuran (14.7, < 1) are at the high end for solubility and longevity in their class. Their metabolites have been frequently detected beneath high use crops, such as potatoes in the potato growing regions of the United States.

The pyrethroid insecticides have low solubilities, short half-lives, and high soil retentions that make them unlikely to leach. Yet, permethrin (< 1, < 1) is occasionally detected in very trace quantities in groundwater.

Fungicides and Fumigants

Fungicides are nonvolatile organometallic compounds with low aqueous solubility that inhibit growth of actenomycetes and many fungi. The best-known fungicides zineb (not detected (ND), NR) and captan are zinc-based, and maneb (ND, NR) is manganese-based. Some, like bordeaux, are copper sulfate-based. Although their detection frequency is very low, fungicides have not been analyzed in many surveys.

Fumigants are very volatile halogenated compounds that generally are knifed below the soil surface. These compounds have high aqueous solubility and very low soil retention. The fumigants EDB and 1,3-dichloropropene have been frequently detected in the subsurface and in groundwater in high-use regions, such as California.[7] Ethylene dibromide and carbon tetrachloride were also used in grain storage facilities during the 1950s and 1960s. Spills, leaks, and improper handling resulted in 400 reported groundwater contamination sites in Kansas and Nebraska.

Herbicides

There are at least eight major chemical classes of herbicides. These include: quaternary N, basic, acidic, carboxylic acid, hydroxy and aminosulfonyl, amide and anilide, dinitroaniline, and phenylurea herbicides.[5] Several herbicide classes have similar behaviors with respect to soil and water.

Both quaternary N and dinitroaniline herbicides are very highly retained by soils and are not expected to be detected in groundwater. However, paraquat, pendimethalin, and trifluralin have been reported several times in groundwater. Their presence indicates that transport is dependent on factors not directly related to compound longevity, solubility, and mobility. Vertical transport by preferential flow through macropores is a commonly accepted mechanism used to explain these detections. In some instances, compounds have been described as preferentially transported attached to colloidal material.

Carboxylic, hydroxy, and aminosulfonyl acids, and thiocarbamate herbicides have very low to low soil retention and very short to moderate longevity. Thus, the more heavily used and persistent pesticides in these groups are the ones most generally detected in groundwater. They include the acids, dicamba (2.0, NR), picloram (2.5, < 1), bromacil (1.8, 1.0), and dinoseb (1.4, < 1).

Phenylurea herbicides have low to high soil retentions and short to moderate longevity. Linuron (16.7, < 1) and diuron (< 1, 1.9) are the most frequently detected in groundwater and both have moderately long half-lives ranging from 60 to 90 days.

Amide and anilide herbicides have low soil retention and short to moderate longevity. Several amide herbicides and their transformation products have been detected in groundwater. The commonly used amides in the Midwestern corn belt, namely alachlor (1.7, 2.7), metolachlor (< 1, 12), propachlor (1.2, < 1), and acetochlor (NR, < 1), are the most frequent offenders because they are relatively persistent.

As suggested by the name, basic herbicides behave as bases. The group contains several subclasses including aniline, formamidine, imidazole, pyrimidine, thiadiazole, triazines, and triazole. Basic herbicides have low to high soil retention and very short to moderate longevity. Again, it is generally the most persistent and heavily used pesticides that are more frequently found in groundwater. The most frequently detected compounds in the group are the triazines, namely atrazine (5.6, 30), metribuzin (4.2, 1.9), cyanazine (2.0, 1.4), simazine (2.0, 14.8), and prometon (2.1, 11.6).

GROUNDWATER CONTAMINATION

It stands to reason that there are generally good associations between pesticide use and their detection in groundwater. Since groundwater flows very slowly at rates normally ranging from 0.1 ft/day to 3 ft/day, pesticide sources are generally very near the monitored well. Thus, high frequencies of triazine and acetamide detections are reported in the states of the northern corn belt. More fungicides and fumigants were detected in warm humid states of California and Florida where vegetable and fruit crops dominate the landscape. In an analysis of the 20 NAWQAs for pesticides, frequencies of pesticide detection in groundwater were significantly related to the estimated amount of agricultural use within a 1 km radius of the sampled site.[6] They also emphasized that pesticides were detected beneath both agricultural (60.4%) and urban areas (48.5%). Discontinued used pesticides have been detected numerous times in shallow aquifers.

In general, families of pesticides have similar chemical characteristics from which predictions have been made as to the product's potential for contamination of groundwater; however, differences in the leaching behavior of pesticides exhibiting similar chemistry can be appreciable and is the reason several heavily used pesticides are seldom, if ever, detected in groundwater.

MANAGEMENT OF POINT SOURCES OF GROUNDWATER CONTAMINATION

Important steps are being taken to reduce water quality pollution by pesticides occurring from spills and back siphoning events (point sources). Since it is easier to resolve point than nonpoint sources, laws have been enacted to eliminate contamination of surface water bodies, which may be in hydraulic contact with groundwater, from used pesticide containers and rinseate from chemical wash downs. Check valves are mandatory when pesticides are mixed and/or diluted and prevent backflow to groundwater. Soils at and adjacent to agrichemical supply facilities have been surveyed in several states and found to be highly contaminated with pesticide residues. The herbicides, atrazine, alachlor, metolachlor, cyanazine, and metribuzin are the worst offenders from the standpoint of pesticide mass in the soils at sites in Wisconsin and Illinois.[4] Many of these sites and those in other states are now involved in soil cleanups, which are designed to protect underlying groundwater from further pollution.

MANAGEMENT OF NONPOINT SOURCES OF GROUNDWATER CONTAMINATION

Normal farm chemical applications of pesticides are generally considered potential nonpoint sources of groundwater contamination because they are dispersed over large areas ranging from fields to watersheds. Management strategies are in place to reduce leaching of field applied chemicals.[8] These strategies vary from regulatory restrictions to outright bans on application in areas deemed more vulnerable to leaching. Integrated pest management, fostered by the office of pesticide management at the USEPA, is designed to reduce chemical applications. The practice of banding applications has reduced amounts applied. Both target more efficient pesticides and genetically engineered plants sensitive only to specific herbicidal action have been and are being developed. These new pesticides and pesticide–plant combinations require less chemical than in the past, and the altered plants allow for pest control with more environmentally sensitive chemicals. The USEPA has announced a plan to reduce the mass of applied chemicals from commonly used triazines and amides that are frequently detected in groundwater.

Irrigation Management

Irrigation practices can influence pesticide leaching. Atrazine was vertically transported deeper and faster when using flood rather than sprinkler irrigation.[9] Sprinkler systems allow for much more uniform and efficient water management practices than furrow irrigation, and recent studies have shown that they reduce chemical leaching.[10] In the Nebraska's Platte Valley[11] and in the Walnut Creek watershed in Iowa,[12] peak herbicide concentrations were strongly related to rapid flushing beneath drainage areas where surface water ponds during heavy rainfall events on the cropped fields. Application of excess irrigation water also was reported to increase herbicide leaching.[9,11]

FUTURE RESEARCH

More research is necessary to evaluate the health risks of transformation products from heavily used pesticides that are frequently detected in groundwater. Research needs to focus on precision application of pesticides to specific field problem areas as a potential mechanism to reduce chemical application.

There is a need to evaluate the environmental cost/benefit of safer product replacements used in conjunction with genetically altered crops. As new products are registered to replace more persistent and mobile pesticides, long-term fate studies, including the monitoring of the transformation product impact, on groundwater quality are necessary.

REFERENCES

1. U.S. Environmental Protection Agency. *Drinking Water Standards and Health Advisories*; EPA 822-B-00-001; USEPA Office of Water: Washington, DC, Summer 2000.
2. U.S. Environmental Protection Agency. *National Survey of Pesticides in Drinking Water Wells: Phase I Report*; EPA 570/9-90-015; USEPA Office of Pesticide and Toxic Substance: Washington, DC, November 1990.
3. Aspelin, A.L. *Pesticide Industry Sales and Usage, 1992 and 1993 Estimates*; Economic Analysis Branch Report 733-K-92-001; USEPA Office of Pesticide Programs, Biological and Economic Analysis Division, 1994; 1–37.
4. Barbash, J.E.; Resek, E.A. *Pesticides in Ground Water. Distribution, Trends, and Governing Factors*; Ann Arbor Press, Inc.: Chelsea, MI, 1996; 1–588.
5. Weber, J. Properties and Behavior of Pesticides in Soils. In *Mechanisms of Pesticide Movement into Ground Water*; Honeycutt, R.C., Schabacker, D.J., Eds.; Lewis Publishers: London, 1994; 15–41.
6. Kolpin, D.; Barbash, J.E.; Gilliom, R.J. Pesticides in Ground Water of the United States, 1992–1996. Ground Water **2000**, *38* (6), 858–863.
7. Troiano, J.; Weaver, D.; Marade, J.; Spurlock, F.; Pepple, M.; Nordmark, C.; Bartkowiak, D. Summary of Well Water Sampling in California to Detect Pesticide Residues Resulting from Nonpoint-Source Applications. J. Environ. Qual. **2001**, *30* (2), 448–459.
8. Guyot, C. Strategies to Minimize the Pollution of Water by Pesticides. In *Pesticides in Ground and Surface Water*; Börner, H., Ed.; Springer: Berlin, 1994; 87–148.
9. Troiano, J.; Garretson, C.; Krauter, C.; Brownwell, J.; Huston, J. Influence of Amount and Method of Irrigation Water Application on Leaching of Atrazine. J. Environ. Qual. **1993**, *22*, 290–298.
10. Spalding, R.F.; Watts, D.G.; Schepers, J.S.; Burbach, M.E.; Exner, M.E.; Poreda, R.J.; Martin, G.E. Controlling Nitrate Leaching in Irrigated Agriculture. J. Environ. Qual. **2001**, *30* (4), 1184–1194.
11. Spalding, R.F.; Watts, D.G.; Snow, D.D.; Cassada, D.A.; Exner, M.E.; Monson, S.J.; Schepers, J.S. The Etiology and Fate of Ground Water Herbicides. J. Environ. Qual. in press.
12. Moorman, T.B.; Jaynes, D.B.; Cambardella, C.A.; Hatfield, J.L.; Pfeiffer, R.L.; Morrow, A.J. Water Quality in Walnut Creek Watershed: Herbicides in Soils, Subsurface Drainage and Groundwater. J. Environ. Qual. **1999**, *28* (1), 35–45.

Pesticide Contamination, Surface Water

Sharon A. Clay
South Dakota State University, Brookings, South Dakota, U.S.A.

INTRODUCTION

About 1000 million pounds of pesticide were applied in the United States in 1997.[1] This extensive use of pesticides has caused concern about pollution in the environment. Indeed, pesticides have been detected in groundwater, surface water, and rain. However, the amount found normally is less than 0.1% of the amount used and occurs in seasonal cycles. Management options that limit losses from target sites will be presented.

DETECTION FREQUENCY, CONCENTRATIONS, AND SEASONAL CYCLES OF PESTICIDES FOUND IN SURFACE WATER

Pesticide classification may be based on the type of pest controlled. Chemicals that control weeds (herbicides) accounted for 60% of the pesticide use in 1997.[1] Chemicals that control insects (insecticides) and plant diseases (fungicides) accounted for 13% and 7%, respectively, of the chemical use.[1] Rodenticides, fumigants, and nematicides represent other pesticide classes that account for the remaining 20% of use. In a typical year, agricultural lands receive about 80% of all pesticides applied while homeowners/gardeners (8%) and the commercial market (12%) apply the rest.

The United States Geological Survey (USGS) has sampled surface water for pesticide pollution since the late 1970s. Thousands of samples have been analyzed for about 100 pesticides or breakdown products (metabolites). Herbicides are the pesticide type most commonly found in rural streams.[2] Atrazine, a broadleaf herbicide ranked number one in total pounds applied in the United States from 1987 to 1997 (about 75 million lb/yr), has been detected most often.[1] Other herbicides found in surface water include simazine and cyanazine (used for broadleaf and grass control in corn) and alachlor and metolachlor (used for grass control in corn and soybean).[2] Most of these herbicides also ranked high in U.S. agricultural use between 1987 and 1997.[1] The amount of these herbicides found in streams has typically ranged from less than 1% of the amount applied (cyanazine, metolachlor, and alachlor) up to 3% (atrazine).[2] The five insecticides frequently

detected were diazinon, carbaryl, chlopyrifos, carbofuran, and malathion.[2]

Detection of cyanazine and alachlor in the U.S. environment will decline in the future. Cyanazine is no longer labeled for use in the United States as of 2002. Alachlor is being replaced by acetochlor. However, acetochlor was detected in surface water in its first year of general use (1994) with occurrence patterns similar to other herbicides of the same family.[3] The concentration detected was lower than alachlor due to lower application rates.

Pesticides have also been detected in urban streams.[4,5] In fact, the estimated contribution of insecticides to surface water contamination from urban and rural areas may be similar. Insecticides, such as malathion, carbaryl, and diazinon, are commonly used in urban settings for control of mosquitoes, turfgrass and garden insects, and termites.[4] Herbicides detected in urban streams were those commonly used for broadleaf weed control in lawns (i.e., 2,4-D, dicamba, and MCPA)[5] along with prometon and tebuthiuron, both used for total vegetation control in right-of-way areas.[4] Herbicides almost exclusively used in agricultural settings (atrazine, alachlor, metolachlor, and cyanazine) have also been detected in urban streams. Atmospheric depositions in sediment, rain, or snow or transport from agricultural watersheds upstream of urban settings are the most likely sources of these herbicides.

Pesticides in surface water have been detected throughout the year.[6] The greatest concentrations in rural streams are reported in the spring and early summer coinciding with agricultural applications.[7] The mean pesticide concentrations in most months are about $1 \, \mu g/L$.[2] However, during peak usage, concentrations in some samples may exceed $12 \, \mu g/L$. In urban streams, the mean pesticide concentration is fairly stable ($< 0.5 \, \mu g/L$) throughout the year[2] but can differ by area of the country. In the southern United States, pesticides may be applied both earlier and later in the year due to the longer growing season compared to the northern tier of states.

PESTICIDE TOXICITY IN SURFACE WATER

The U.S. Environmental Protection Agency[8] has established pesticide concentration criteria values for

Encyclopedia of Water Science
DOI: 10.1081/E-EWS 120010192

some pesticides for the protection of aquatic health. Pesticides can be toxic to aquatic invertebrates such as plankton[9] and vertebrates such as frogs and fish, when above critical concentrations. The herbicides, atrazine [water quality criterion (WQC) = 1.8 μg/L] and trifluralin (WQC = 0.2 μg/L), and the insecticides, chlorpyrifos (WQC = 0.041 μg/L) and diazinon (WQC = 0.08 μg/L), were the pesticides that most often exceeded the aquatic health criteria in 37 rural streams.[2] In a survey of eight urban streams across the United States monitored from 1993 to 1995, 41 pesticides were detected. Simazine (WQC = 10 μg/L), prometon (no established criterion), atrazine, and diazinon were detected at all sites and 20 other pesticides were detected above their WQC in one or more of the streams.[4] The estimated number of days that the pesticide concentration exceeds the established standards varies by chemical.[2] Atrazine was estimated to exceed the standard in 15 rural streams from 1 day to 84 day (with an average of 36 day). In comparison, chlorpyrifos and azinphos-methyl (WQC = 0.01 μg/L) exceeded the standard in 8 streams, ranging from 1 day to 8 day (average of 3 day) and 1 day to 70 day (average of 13 day), respectively.

In addition, herbicides may be used to control plants in ponds and lakes. The reduction in vegetation may have an indirect effect on the number of weed-clinging invertebrates such as dragonflies or damselflies. However, removal of plants does not always decrease the numbers of aquatic organisms. For example, the number of aquatic organisms remain fairly constant or increase due to an increase in organisms that feed on plant debris after vegetation is controlled.[10]

FACTORS THAT AFFECT PESTICIDE MOVEMENT TO SURFACE WATER

Distance Between Application and Surface Water

The distance between pesticide application and surface water can greatly affect the likelihood of surface water contamination. In most cases, buffer zones of 50 ft–100 ft are recommended on the pesticide label. Filter strips of grass or a dense cover crop at the field edge also reduces the pesticide concentration in the runoff in two ways.[11] First, filter strips slow water movement and allow soil particles to settle out of the runoff. Second, these areas normally are high in organic matter and can sorb the pesticide out of water. Herbicide concentrations have been reduced up to 64% after flowing through a filter strip compared to concentrations upstream of the filter strip area.[11]

Choice of Rate, Pesticide, and Application Timing

Some pesticides are much more vulnerable to movement into surface water. Pesticides that degrade quickly (days to a few weeks) are less likely to contaminate surface water than those that linger in the environment (months or longer). Applications to sandy soils with low organic matter are more likely to have runoff than loamy or clayey soils with more organic matter.

The higher the rate of application the more the pesticide available for runoff. For example, banding herbicides to only row areas and using cultivation to control weeds in interrow areas is one technique to reduce application rates. The amount of herbicide applied depends on the bandwidth. If the bandwidth is 10″ on a 30″ row, the amount applied is reduced by two-thirds compared to a broadcast application rate where the entire area is covered. Banding has been reported to reduce the amount of herbicide in runoff water up to 70%.[12] Applying the pesticide as a split application also reduces the amount of chemical present at one time and can reduce the potential risk of surface water contamination.

Pesticides that are applied at ounces per acre rather than pounds per acre are available. Examples of herbicides include the sulfonylureas and imidiazilinones. The low application rates of these herbicides reduce the total amount of herbicide applied to an area and therefore reduce the risk of contamination.

Rainfall After Application

The amount, intensity, and time of rainfall after application are all factors that affect the total amount and pesticide concentration in runoff. Generally, a large amount of pesticide is removed with the first rainfall after application.[13] If the soil is bare and crusted or has little or no plant canopy, the amount of runoff will be very large. Storms with high intensity rains also have more runoff than if the rain is slow and steady. The amount of water already present in the soil is another factor that affects runoff. There is little or no infiltration on saturated or very wet soils whereas dry soils will allow more water to infiltrate before runoff occurs.

Drift, Volatilization, and Atmospheric Transport

Drift is the transport of spray applications in the wind. The smaller the droplet size, the farther the droplets can travel. Very fine particles may move for miles before deposition occurs. These drops could land in lakes, streams, or other off-site areas. Wind speed also is an important factor in drift. Most applications should not take place when wind speeds are over 10 mph. Pesticides should not be applied when weather conditions that cause an inversion (warm air over cold air) are present. Fine spray droplets move much farther when an atmospheric inversion exists.

There are some factors that can be used to limit drift. Applying pesticides under low pressure in high amounts of water will increase droplet size, thereby decreasing the number of droplets available for drift. Larger droplets will also lower the pesticide concentration of individual droplets. There are nozzles specifically designed to limit the amount of fine particles in a spray pattern. Antidrift chemicals are available to be mixed with pesticide application. These chemicals reduce the number of very fine droplets in the spray pattern. Lowering the spray boom limits the amount of spray subjected to the wind. Shields for the boom or individual nozzles can be attached to limit wind interception with spray patterns.

Pesticides can volatilize (change from a liquid to a gas) or sublime (change from solid to a gas) from the soil surface. The amount volatilized is a function of the vapor pressure of the chemical and the environmental conditions. For example, EPTC, a grass herbicide, is highly volatile with most of the herbicide lost to the atmosphere within a few hours if not incorporated into the soil.[14] In contrast, atrazine is not very volatile but under the right conditions, 2% of the applied chemical can be lost through this mechanism.[15] This amount is about one-half the loss due to surface runoff on a regional scale. A pesticide applied to warm moist soils on windy days will have a greater loss than if the same pesticide is applied to cool dry soils.[14]

Pesticides attached to very small soil particles are also moved into the atmosphere.[16] Soil particles and aggregates that are less than 1 mm in diameter are considered wind-erodible.[17] In the Great Plains area of the United States, the loss of soil due to wind erosion is greater than the amount lost due to water erosion.[18] The small size soil particles make up about 50% of the soil mass in the top 0.5″ of soil, if the area has been chisel plowed. The amount of herbicide found on these particles can range from 50% to 200% more than the amount found on larger aggregates one day after application.[19] If there are windy conditions within days after application, the amount of pesticide lost could be very high. Shallow incorporation of pesticides limits pesticide losses into the atmosphere.

Concentrations of pesticides in the atmosphere can decrease by several methods. Dilution, removal in rain, snow, or by dry deposition, and photochemical degradation are the three ways in which pesticide concentrations can be reduced.[20] However, pesticides can move long distances from their sites of application.[21] For example, surface water and rainfall has been monitored in the remote pristine area of Isle Royale National Park, MI, located in Lake Superior.[22] Atrazine was detected in rainfall from mid-May to early-June, corresponding to peak application timing of atrazine in the U.S. Midwestern Corn Belt. Surface water of the lakes in Isle Royale National Park also contained atrazine at trace (part per trillion) levels. These levels are not high enough to be toxic to organisms, but their presence in this remote area point to the need to use caution in the use and application of pesticides.

CONCLUSION

Although other methods of pest control are important, pesticides continue to be an efficient, cost-effective method of management. The challenge is to select and apply pesticides in a manner that kills the target pest and does not harm nontarget organisms or the environment. Sound and sensible management can reduce the probability of pesticides entering surface water. Precision application to only those areas needing treatment is one method to reduce the total amount of pesticide applied. Other techniques include using split applications, banding, and planting filter strips along streams and water courses.

REFERENCES

1. Aspelin, A.L.; Grube, A.H. *Pesticides Industry Sales and Usage: 1996 and 1997 Market Estimates.* EPA/OPP. http://www.epa.gov/oppbead1/pestsales/97pestsales. Updated Jan 2000. (accessed Sept 2001).
2. Larson, S.J.; Gilliom, R.J.; Capel, P.D. *Pesticides in Streams of the United States—Initial Results from the National Water-Quality Assessment Program.* USGS/NAWQA, WRIR98-4222, 1999, http://www.water.wr.usgs.gov/pnsp/rep/wrir984222 (accessed Sept 2001).
3. Koplin, D.W.; Nations, B.K.; Goolsby, D.A.; Thurman, E.M. Acetochlor in the Hydrologic System in the Midwestern United States, 1994. Environ. Sci. Technol. **1996**, *30*, 1459–1464.
4. Hoffman, R.S.; Capel, P.D.; Larson, S.J. Comparison of Pesticides in Eight U.S. Urban Streams. Environ. Toxicol. Chem. **2000**, *19*, 2249–2258.
5. Wotzka, P.J.; Lee, J.; Capel, P.D.; Lin, M. Pesticide Concentrations and Fluxes in an Urban Watershed. AWRA Technical Publication Series TPS-94-4; American Water Resources Association: Herndon, VA, 1994; 135–145.
6. Ferrari, M.J.; Ator, S.W.; Blomquist, J.D.; Dysart, J.E. *Pesticides in Surface Water of the Mid-Atlantic Region.* USGS-WRIR97-4280, 1997, http://www.md.water.usgs.gov/publications/wrir-97-4280 (accessed Feb 2001).
7. Council for Agricultural Science and Technology. Issue paper 2, April, 1994. *Pesticides in Surface and Groundwater.* Revised May, 1997. CAST. http://www.cast-science.org/pwq_ip.htm (accessed Feb 2001).

8. U.S. Environmental Protection Agency, *National Recommended Water Quality Criteria*, USEPA 822-Z-99-001; USEPA: Washington, D.C., 1999.

9. Thompson, A.R.; Edwards, C.A. Effects of Pesticides on Nontarget Invertebrates in Freshwater and Soil. In *Pesticides in Soil and Water*; Guenzi, W.D., Ahlrichs, J.L., Chesters, G., Bloodworth, M.E., Nash, R.G., Eds.; Soil Sci. Soc. Amer.: Madison, WI, 1974; 341–386.

10. Harp, G.L.; Campbell, R.S. Effects of the Herbicide Silves on Benthos of a Farm Pond. J. Wildl. Manag. **1964**, *28*, 308–317.

11. Hall, J.K.; Hartwig, N.L.; Hoffman, L.K. Application Mode and Alternative Cropping Effects on Atrazine Losses from a Hillside. J. Environ. Qual. **1983**, *18*, 439–445.

12. Gaynor, J.D.; van Wesenbeeck, I.J. Effects of Band Widths on Atrazine, Metribuzin, and Metoachlor Runoff. Weed Technol. **1995**, *9*, 107–112.

13. Ma, L.; Spalding, R.F. Herbicide Mobility and Variation in Agricultural Runoff in the Beaver Creek Watershed in Nebraska. In *Herbicide Metabolites in Surface Water and Groundwater*; ACS Symposium Series 630; Meyer, M.T., Thurman, E.M., Eds.; American Chemical Society: Washington, D.C., 1996; 226–236.

14. Spencer, W.F.; Cliath, M.M. Movement of Pesticides from Soil to the Atmosphere. In *Long Range Transport of Pesticides*; Kurtz, D.A., Ed.; Lewis Publishers: Chelsea, MI, 1990; 1–16.

15. Nash, R.G.; Hill, B.D. Modeling Pesticide Volatilization and Soil Decline Under Controlled Conditions. In *Long Range Transport of Pesticides*; Kurtz, D.A., Ed.; Lewis Publishers: Chelsea, MI, 1990; 17–28.

16. Ciba-Geigy Corporation. *Biological Assessment of Atrazine and Metolachlor in Rainfall*, Technical Paper 1; Bern, Switzerland, 1993; 4.

17. Chepil, W.S. Dynamics of Wind Erosion: I. Nature of Movement of Soil by Wind. Soil Sci. **1945**, *60*, 305–320.

18. United States Department Agriculture. *Summary Report: 1997 National Resources Inventory. Revised Dec. 2000*; USDA Natural Resources Conservation Ser.: Washington, D.C., 2000, http://www.nhq.nrcs.usda.gov/NRI/1997/summary_report (accessed Sept 2001).

19. Clay, S.A.; DeSutter, T.M.; Clay, D.E. Herbicide Concentration and Dissipation from Surface Wind-Erodible Soil. Weed Sci. **2001**, *49*, 431–436.

20. Richards, R.P.; Kramer, J.W.; Baker, D.B.; Krieger, K.A. Pesticides in Rainwater in the Northeastern United States. Nature **1987**, *327*, 129–131.

21. Kurtz, D.A. *Long Range Transport of Pesticides*; Lewis Publishers: Chelsea, MI, 1990; 462.

22. Thurman, E.M.; Cromwell, A.E. Atmospheric Transport, Deposition, and Fate of Triazine Herbicides and Their Metabolites in Pristine Areas of Isle Royale National Park. Environ. Sci. Technol. **2000**, *34*, 3079–3085.

P

Pfiesteria Piscicida

Henry S. Parker
United States Department of Agriculture (USDA), Wyndmoor, Pennsylvania, U.S.A.

INTRODUCTION

Several highly publicized fish kills in 1997 in the Chesapeake Bay focused national attention on *Pfiesteria piscicida*, a toxic dinoflagellate already notorious for its role in major fish kills in North Carolina, and dubbed "the cell from hell" by researchers from that state. The economic, socio-political, and scientific consequences of the 1997 events may be felt for decades to come.

A discussion of *Pfiesteria* must first consider harmful algal blooms (HABs). Described as "episodes of rapid, explosive growth of populations of microorganisms... that make and secrete toxic biomolecules,"[1] HABs are generally attributable to photosynthetic microalgae; dinoflagellates are often implicated, with at least 85 toxic species identified to date.[2] Harmful algal blooms (also known as "red tides") have been reported since at least biblical times; however, it is widely believed that their frequency of occurrence and intensity have increased in recent years, resulting in increased mortalities of fish, marine mammals, and invertebrates, and toxic effects on humans.[3,4] There is growing concern that recent increases in HABs may be related to anthropogenic factors, including nutrient enrichment of coastal waters.[5]

CHRONOLOGY AND IDENTIFICATION OF *P. PISCICIDA*

In 1988 scientists at the College of Veterinary Medicine, North Carolina State University (NCSU), were puzzled by deaths of fish in laboratory tanks.[6] The mortalities were initially attributed to an unknown contaminant. The scientists subsequently observed that a previously unknown dinoflagellate was abundant just before the onset of fish mortalities, but that the flagellated cells—"zoospores"—declined precipitously within 1 hr–2 hr of the fish kill, suggesting "ambush-predator" behavior.[7,8] The disappearance of the zoospores was associated with the production of resting cysts or nontoxic amoeboid forms. After tanks were re-stocked with live fish, the zoospores re-appeared.

Three years after the fish tank mortalities, the NCSU researchers linked the deaths to similar fish kill events in the natural environment. The Albemarle-Pamlico estuarine system of North Carolina is the second largest estuary, by area, in the United States.[9] The estuary is of critical ecological importance, providing about half the nursery area for fish on the U.S. East Coast. Since at least the mid-1980s, the estuarine system had sustained major, unexplained fish kills. A massive kill occurred in 1991, resulting in the deaths of an estimated one million Atlantic menhaden, *Brevoortia tyrannus*. Water samples taken at the onset of and following mortality revealed the presence of the same dinoflagellates and patterns of abundance as observed in the culture tanks.[10]

The NCSU scientists isolated the suspect dinoflagellates from several fish kills and conducted laboratory bioassays. They found that the isolates were lethal to 11 species of finfish, including several commercial species, and described the morphology, behavior, and preliminary life history of the organism. The organism was subsequently assigned to a new genus and species, *P. piscicida* gen. et sp. nov., and described as a polymorphic and multiphasic toxic dinoflagellate, with flagellated, amoeboid, and cyst stages.[11]

Since the identification of *P. piscicida*, the NCSU laboratory has identified a second toxic *Pfiesteria* species, *P. shumwayae*.[12] Although it was previously the convention to refer to similar toxic dinoflagellates as "*Pfiesteria*-like" organisms, scientists recently reached consensus that the term "toxic *Pfiesteria* complex" (TPC) should be employed to describe toxic species that are strongly attracted to live fish or fresh tissues; whose toxic activity is triggered by live fish or fresh tissues; and whose toxins can cause fish stress, disease, or death.[13] Toxic *Pfiesteria* complex species currently known are limited to *P. piscicida* and *P. shumwayae*.

After its discovery in North Carolina, *P. piscicida* was identified in Delaware Bay in 1993 and documented in Chesapeake Bay in 1995, in a tributary of the Choptank River.[14] It was not until several major fish kill events in tributaries along the lower Eastern Shore of Chesapeake Bay in the summer and fall of 1997 that *Pfiesteria* was considered a problem for Chesapeake Bay. *Pfiesteria* was linked not only to fish mortalities but also to human health problems, engendering widespread media coverage.[15]

Encyclopedia of Water Science
DOI: 10.1081/E-EWS 120010239

BIOLOGY OF *P. PISCICIDA*

Pfiesteria appears to be widely distributed in the U.S. Eastern and Gulf of Mexico coastal waters, from Delaware to Florida and Alabama.[16] Toxic *Pfiesteria* complex species are eurythermal and euryhaline. *Pfiesteria* is a complex organism. Though often described as an alga, it is heterotrophic and not capable of photosynthesis except when zoospores ingest chloroplasts from algal prey, through a process known as kleptoplastidy.[17]

Burkholder and Glasgow[18] have described 24 stages in the life history of *Pfiesteria*. Individual stages range from 5 microns to 750 microns in size. In the absence of fish, all stages are nontoxic. Amoeboid stages subsist on microalgae; encysted stages lie dormant on the bottom, encased in a protective covering; and free-swimming cells are known as nontoxic zoospores. In the presence of fish, zoospores rapidly become toxic; toxicity is apparently triggered by chemical cues in live fish or their fresh tissues, secreta, or excreta. Cysts and amoebae may also be stimulated to give rise to zoospores that can become toxic as well. Released toxins may paralyze fish, disrupt osmotic balance, and degrade fish tissues, leaving open wounds or lesions.[19] This may lead to the death of the fish or render them susceptible to secondary infections. After the fish die, flagellated cells transform to amoeboid stages that feed on fish remains or, if conditions warrant, further transform into dormant, benthic cysts. Sexual reproduction, including gamete fusion and planozygote formation, has been observed. Cannibalism has not been observed.

FISH HEALTH AND *PFIESTERIA*

P. piscicida has been implicated as the causative agent for about half of all fish kills in the Albemarle-Pamlico estuarine system during 1991–1993.[20] Juvenile Atlantic menhaden appear to suffer a disproportionate number of *Pfiesteria*-related mortalities for reasons not yet fully understood.[21] Lewitus et al.[22] concluded in 1995 that *Pfiesteria* was responsible for previously unexplained fish kills in Chesapeake Bay in 1988. It is now generally accepted that *Pfiesteria* was responsible for the fish kills in the lower Eastern Shore of Chesapeake Bay in 1997.[23]

The most characteristic association of *Pfiesteria* with fish mortalities is the presence of large numbers of fish with open, bloody sores or lesions.[24] Indeed, the presence of a high number of dead fish with lesions has been considered not only an indicator of the presence of *Pfiesteria* but also a confirmation that mortalities are attributable to *Pfiesteria*. Toxic stages of *Pfiesteria* have also been implicated in nonlethal pathologies of fish, including lesions, and researchers have observed that fish may recover from sublethal exposure to *Pfiesteria* toxins,

but with possible long-term compromise to their immune systems.[25]

Recent research confirms, however, that lesions in fish may be the result of an organism other than *Pfiesteria*, or a combination of organisms and host factors. Blazer et al.[26] determined that a high prevalence of menhaden with ulcerative skin lesions collected from Chesapeake Bay waters in 1997 was attributable to the fungal pathogen, *Aphanomyces invadans*. While not disputing that lesions associated with dead fish in acute fish kills may well be caused by *Pfiesteria*, the authors cautioned that the presence of lesions does not necessarily mean that *Pfiesteria* is responsible, or even present.

Similarly, Evans et al.[27] found that ulcerative lesions in Atlantic menhaden collected from Delaware inland bays could be attributable to multiple factors, including immunosuppression and increased susceptibility to infectious agents. They found an association between ulceration and *Acinetobacter* spp., a bacterial pathogen implicated in human skin infections, particularly in immunocompromised individuals, and produced skin lesions in experimentally inoculated fish. Further complicating understanding of the role of *Pfiesteria* in fish kills, conditions that appear to favor *Pfiesteria* outbreaks, including nutrient enrichment in poorly flushed embayments and estuaries, are also responsible for eutrophication and associated anoxic conditions that lead to fish stress or mortality. Because of these uncertainties, the NCSU laboratory applies a conservative protocol, based on Koch's postulates, to verify *Pfiesteria* involvement in fish kills and fish lesions.[28]

EFFECTS ON HUMAN HEALTH

Scientists working with *Pfiesteria* cultures at NCSU first became concerned about potential toxic effects on humans in the early 1990s, when lab workers exhibited a range of disturbing symptoms including sores, headaches, respiratory problems, blurred vision, nausea, short-term memory loss, and difficulty in reading.[29] Exposure appeared to be from dermal contact with toxin-containing water or inhalation of toxic aerosols. When Maryland fishermen exhibited similar symptoms during the 1997 *Pfiesteria* outbreaks in Chesapeake Bay, Maryland health officials undertook studies to improve understanding of *Pfiesteria* toxicity and effects on humans.[30] The researchers confirmed that direct exposure to *Pfiesteria* did result in many of the dermatological, cognitive, and neuropsychological symptoms reported earlier, but the studies raised as many questions as they answered. Understanding of the human health effects of exposure to *Pfiesteria* is particularly impeded by the absence of an assay to identify the toxin or of baseline profiles on humans prior to exposure. There is also insufficient knowledge of the exposure levels that lead to toxicity; which *Pfiesteria* life history stages are implicated in toxicity; the route of exposure to the toxins; and environmental persistence of toxins.[31]

PFIESTERIA, AGRICULTURE, AND PUBLIC POLICY

The *Pfiesteria* outbreaks in Maryland's Chesapeake Bay in 1997 raised concerns that agricultural nutrients, particularly from the abundant poultry farms on Maryland's Eastern Shore, might be responsible for the conditions that led to the outbreaks. To address this issue, an ad hoc forum of scientists was convened in 1997, under the auspices of the University of Maryland Center for Environmental Science. The forum summarized its findings in a report entitled "The Cambridge Consensus," dated October 16, 1997.[32] While the forum did not specifically implicate agriculture, it did reach the following conclusions: 1) nutrient concentrations in tidal rivers of Maryland's Eastern Shore are higher than those of other rivers with similar salinity; 2) nutrient levels had increased in those rivers over the previous 12 yr; and 3) higher than normal levels of nutrients were discharged into those rivers in 1996 and early 1997 because of unusually high precipitation and runoff. The forum also concluded, based on review of research to date, that nutrient enrichment stimulates growth of nontoxic stages of *Pfiesteria* and its algal prey, but is not required for transformation of *Pfiesteria* to toxic stages.

Circumstantial evidence further implicated poultry farming in *Pfiesteria* outbreaks. It was reported that Maryland's Eastern Shore produces 800,000 tons of chicken manure annually;[33] that agriculture contributes 70% of the nitrogen and 83% of the phosphorus discharged to rivers of Maryland's lower Eastern Shore watershed;[34] and that the practice by poultry farmers of routinely applying chicken manure to Eastern Shore soils used for growing grain (at levels based on the crop's nitrogen requirements) results in phosphorus inputs to Eastern Shore cropland at a rate approximately twice that of phosphorus removal by grain.[35] Annual net phosphorus surpluses are temporarily stored in the soil with the potential for runoff into adjacent water bodies.

Although Burkholder and Glasgow inferred that inorganic phosphate could directly stimulate toxic zoospores of *Pfiesteria*, and indirectly promote increased production of nontoxic zoospores,[36] there is not yet conclusive evidence of a causative link between agricultural nutrients and *Pfiesteria* outbreaks. Nonetheless, Maryland officials have adopted a conservative strategy for nutrient management on poultry farms. The

Maryland Water Quality Improvement Act of 1998 requires all but the smallest farmers in the state to prepare and implement nutrient management plans, including for phosphorus management on farmland receiving animal wastes, by 2005.[37] On-farm phosphorus management presents challenges, particularly in no-till croplands on Maryland's Eastern Shore. In these areas, soluble phosphorus tends to accumulate on the soil surface with increased susceptibility to runoff.[38] Nutrient management plans will likely involve reduced applications of both inorganic phosphorus fertilizers and poultry manures on croplands.

Advances in animal nutrition may also contribute to phosphorus management plans. One approach is to reformulate poultry rations to include phytase, an enzyme that increases bioavailability of grain-based phosphorus in monogastric animals (including poultry). Alternatively, development of grains with a higher content of available phosphorus would reduce the need for phosphorus supplements in poultry feeds.

FUTURE RESEARCH NEEDS

Although our understanding of *Pfiesteria* has advanced considerably since the organism was first identified, there are still major gaps in our knowledge of the organism's biology, taxonomy, toxicity, and effects on fish and human health, and of the factors that stimulate its production and toxicity. A multidisciplinary workshop held in Baltimore, Maryland on October 28–30, 1997 identified major research needs for *Pfiesteria*.[39] Key among these were the following:

1. Develop certified "pure" cultures of *Pfiesteria*-like organisms to enable comparable and transferable research among laboratories.
2. Distinguish among species of *Pfiesteria* and related taxa.
3. Develop molecular probes to rapidly detect *Pfiesteria* and toxins, including in different life history stages, and to better determine the fate of toxins.
4. Characterize the chemical composition of toxins.
5. Improve understanding of effects of *Pfiesteria* toxins on fish and human health.
6. Improve research cooperation among scientists and laboratories.

While there has been considerable recent progress in these areas, there are still substantial gaps in our knowledge. There is also a critical need for expanded research to better understand the relationships between agricultural practices and the conditions that encourage the development of HABs in general, and *Pfiesteria* outbreaks in particular, in coastal watersheds. Improved understanding of these relationships will enable the development of sustainable and affordable agricultural management practices and tools that will reduce negative impacts of agriculture on water quality.

REFERENCES

1. Silbergeld, E.K.; Grattan, L.; Oldach, D.; Morris, J.G. *Pfiesteria*: Harmful Algal Blooms as Indicators of Human: Ecosystem Interactions. Environ. Res. **2000**, *82* (2), 97–105.

2. Harvell, C.D.; Kim, K.; Burkholder, J.M.; Colwell, R.R.; Epstein, P.R.; Grimes, D.J.; Hofmann, E.E.; Lipp, E.K.; Osterhaus, A.D.M.E.; Overstreet, R.M.; Porter, J.W.; Smith, G.W.; Vasta, G.R. Emerging Marine Diseases— Climate Links and Anthropogenic Factors. Science **1999**, *285*, 1505–1510.

3. Culotta, E. Red Menace in the World's Oceans. Science **1992**, *257*, 1476–1477.

4. Mlot, C. The Rise in Toxic Tides. Science News Online September 27, 1997. http://www.sciencenews.org/Sn_arc97/9_27_97/bob1.htm (accessed November 2001).

5. Silbergeld, E.K.; Grattan, L.; Oldach, D.; Morris, J.G. *Pfiesteria*: Harmful Algal Blooms as Indicators of Human: Ecosystem Interactions. Environ. Res. **2000**, *82* (2), 97–105.

6. Burkholder, J.M.; Glasgow, H.B.; Deamer-Melia, N. Overview and Present Status of the Toxic *Pfiesteria* Complex (Dinophyceae). Phycologia **2001**, *40* (3), 186–214.

7. Burkholder, J.M.; Noga, E.J.; Hobbs, C.H.; Glasgow, H.B.; Smith, S.A. New "Phantom" Dinoflagellate is the Causative Agent of Major Estuarine Fish Kills. Nature **1992**, *358*, 407–410, see also *360*: 768.

8. Burkholder, J.M.; Glasgow, H.B. *Pfiesteria piscicida* and Other Pfiesteria-like Dinoflagellates: Behavior, Impacts, and Environmental Controls. Limnol. Oceanogr. **1997**, *45* (5, part 2), 1052–1075.

9. Burkholder, J.M.; Glasgow, H.B. History of Toxic *Pfiesteria* in North Carolina Estuaries from 1991 to the Present. Bioscience **2001**, *51* (10), 827–841.

10. Burkholder, J.M.; Noga, E.J.; Hobbs, C.H.; Glasgow, H.B.; Smith, S.A. New "Phantom" Dinoflagellate is the Causative Agent of Major Estuarine Fish Kills. Nature **1992**, *358*, 407–410, see also *360*: 768.

11. Steidinger, K.A.; Burkholder, J.M.; Glasgow, H.B.; Hobbs, C.W.; Garrett, J.K.; Truby, E.W.; Noga, E.J.; Smith, S.A. *Pfiesteria piscicida* Gen. et Sp. Nov. (Pfiesteriaceae Fam. Nov.), a New Toxic Dinoflagellate with a Complex Life Cycle and Behavior. J. Phycol. **1996**, *32* (1), 157–164.

12. Glasgow, H.B.; Burkholder, J.M.; Morton, S.L.; Springer, J. A Second Species of Ichthyotoxic *Pfiesteria* (Dinamoebales, Dinophyceae). Phycologia **2001**, *40*, 234–245.

13. Burkholder, J.M.; Glasgow, H.B.; Deamer-Melia, N. Overview and Present Status of the Toxic *Pfiesteria* Complex (Dinophyceae). Phycologia **2001**, *40* (3), 186–214.

14. Lewitus, A.J.; Jesien, R.V.; Kana, T.M.; Burkholder, J.M.; Glasgow, H.B.; May, E. Discovery of the "Phantom" Dinoflagellate in Chesapeake Bay. Estuaries **1995**, *18* (2), 373–378.

15. Magnien, R.E. The Dynamics of Science, Perception, and Policy during the Outbreak of *Pfiesteria* in the Chesapeake Bay. Bioscience **2001**, *51* (10), 843–852.

16. Burkholder, J.M.; Glasgow, H.B.; Deamer-Melia, N. Overview and Present Status of the Toxic *Pfiesteria* Complex (Dinophyceae). Phycologia **2001**, *40* (3), 186–214.

17. Lewitus, A.J.; Glasgow, H.B.; Burkholder, J.M. Kleptoplastidy in the Toxic Dinoflagellate *Pfiesteria piscicida* (Dinophyceae). J. Phycol. **1999**, *35* (2), 303–312.

18. Burkholder, J.M.; Glasgow, H.B. *Pfiesteria piscicida* and Other *Pfiesteria*-like Dinoflagellates: Behavior, Impacts, and Environmental Controls. Limnol. Oceanogr. **1997**, *45* (5, part 2), 1052–1075.

19. Greer, J.; Leffler, M.; Belas, R.; Kramer, J.; Place, A. Molecular Technologies and *Pfiesteria* Research: A Scientific Synthesis. Report of Workshop Held October 28–30, 1997 at the Center of Marine Biotechnology University of Maryland Biotechnology Institute. 1998, v + 34 pages. Maryland Sea Grant Publication UM-SG-TS-98-01.

20. Burkholder, J.M.; Glasgow, H.B. Trophic Controls on Stage Transformations of a Toxic Ambush-Predator Dinoflagellate. J. Eukaryot. Microbiol. **1997**, *44* (3), 200–205.

21. Burkholder, J.M.; Glasgow, H.B. *Pfiesteria piscicida* and Other *Pfiesteria*-like Dinoflagellates: Behavior, Impacts, and Environmental Controls. Limnol. Oceanogr. **1997**, *45* (5, part 2), 1052–1075.

22. Lewitus, A.J.; Jesien, R.V.; Kana, T.M.; Burkholder, J.M.; Glasgow, H.B.; May, E. Discovery of the "Phantom" Dinoflagellate in Chesapeake Bay. Estuaries **1995**, *18* (2), 373–378.

23. Magnien, R.E. The Dynamics of Science, Perception, and Policy During the Outbreak of *Pfiesteria* in the Chesapeake Bay. Bioscience **2001**, *51* (10), 843–852.

24. Burkholder, J.M.; Glasgow, H.B. History of Toxic *Pfiesteria* in North Carolina Estuaries from 1991 to the Present. Bioscience **2001**, *51* (10), 827–841.

25. Burkholder, J.M.; Glasgow, H.B.; Deamer-Melia, N. Overview and Present Status of the Toxic *Pfiesteria* Complex. Phycologia **2001**, *40* (3), 186–214.

26. Blazer, V.S.; Vogelbein, W.K.; Densmore, C.L.; May, E.B.; Lilley, J.H.; Zwerner, D.E. Aphanomyces as a Cause of Ulcerative Skin Lesions of Menhaden from Chesapeake Bay Tributaries. J. Aquat. Anim. Health **1999**, *11* (4), 340–349.

27. Evans, J.J.; Klesius, P.H.; Shoemaker, C.A.; Shelby, R.A.; Humphries, E.M.; Garcia, J.C.; Gagliardi, J.V. *Acinetobacter* spp. Isolated from Feral Atlantic Menhaden from Delaware Inland Bays, U.S.A. Abstract of Paper Presented at the EAFP Tenth International Conference on Diseases of

P

Fish and Shellfish, Dublin, Ireland, September 10–14, 2001.

28. Burkholder, J.M.; Glasgow, H.B.; Deamer-Melia, N. Overview and Present Status of the Toxic *Pfiesteria* Complex (Dinophyceae). Phycologia **2001**, *40* (3), 186–214.

29. Greer, J.; Leffler, M.; Belas, R.; Kramer, J.; Place, A. Molecular Technologies and *Pfiesteria* Research: A Scientific Synthesis. Report of Workshop Held October 28–30, 1997 at the Center of Marine Biotechnology University of Maryland Biotechnology Institute. 1998, v + 34 pages. Maryland Sea Grant Publication UM-SG-TS-98-01.

30. Grattan, L.M.; Oldach, D.; Morris, J.G. Human Health Risks of Exposure to *Pfiesteria piscicida*. Bioscience **2001**, *51* (10), 853–857.

31. Greer, J.; Leffler, M.; Belas, R.; Kramer, J.; Place, A. Molecular Technologies and Pfiesteria Research: A Scientific Synthesis. Report of Workshop Held October 28–30, 1997 at the Center of Marine Biotechnology University of Maryland Biotechnology Institute. 1998, v + 34 pages. Maryland Sea Grant Publication UM-SG-TS-98-01.

32. Boesch, D. *Pfiesteria piscicida* and *Pfiesteria*-like Organisms. The Cambridge Consensus Forum on Land-Based Pollution and Toxic Dinoflagellates in Chesapeake Bay.

October 16, 1997. 11p. http://www.mdsg.umd.edu/pfiesteria/cambridge.html (accessed November 2001).

33. Guy, C. *Growing Fears on Shore*. The Baltimore Sun, April 15, 1998, Local (News), 1B.

34. Magnien, R.E. The Dynamics of Science, Perception, and Policy During the Outbreak of *Pfiesteria* in the Chesapeake Bay. Bioscience **2001**, *51* (10), 843–852.

35. Staver, K.W.; Brinsfield, R.B. Agriculture and Water Quality on the Maryland Eastern Shore: Where Do We Go from Here? Bioscience **2001**, *51* (10), 859–868.

36. Burkholder, J.M.; Glasgow, H.B. *Pfiesteria piscicida* and Other *Pfiesteria*-like Dinoflagellates: Behavior, Impacts, and Environmental Controls. Limnol. Oceanogr. **1997**, *45* (5, part 2), 1052–1075.

37. Magnien, R.E. The Dynamics of Science, Perception, and Policy During the Outbreak of *Pfiesteria* In the Chesapeake Bay. Bioscience **2001**, *51* (10), 843–852.

38. Staver, K.W.; Brinsfield, R.B. Agriculture and Water Quality on the Maryland Eastern Shore: Where Do We Go from Here? Bioscience **2001**, *51* (10), 859–868.

39. Greer, J.; Leffler, M.; Belas, R.; Kramer, J.; Place, A. Molecular Technologies and *Pfiesteria* Research: A Scientific Synthesis. Report of Workshop Held October 28–30, 1997 at the Center of Marine Biotechnology University of Maryland Biotechnology Institute. 1998, v + 34 pages. Maryland Sea Grant Publication UM-SG-TS-98-01.

pH

William J. Rogers
West Texas A&M University, Canyon, Texas, U.S.A.

INTRODUCTION

The pH of a solution is a measure and indication of how acidic or basic (alkaline) the solution is. The measurement of the pH of water is one of the most important and frequently used tests in water chemistry. An understanding of water pH is vital to understanding both the limitations and benefits of a particular water supply and its use in agriculture, industry, and domestic use. The measurement of the pH of water and wastewater is essential to understanding acid–base neutralization, water softening, precipitation, coagulation, disinfection, corrosion control, and scale control.[1] The pH scale ranges from a highly acid pH 0, corresponding to a solution with $(H^+) = 1$, to a highly alkaline pH 14, corresponding to a solution with $(H^+) = 10^{-14}$. Solutions with a pH from 0 to less than 7 are considered acidic and those from above 7 to 14 are considered alkaline. Solutions with a pH of 7 are considered neutral. Pure water has a pH of about 7, alkaline paint stripper and drain cleaners range in pH from 11 to 12 and battery acid from 1.5 to 2. There are no Primary Enforceable Drinking Water standards for pH but National Secondary Drinking Water Regulations, which are nonenforceable guidelines, recommend that pH levels be maintained between 6.5 and 8.5. EPA recommends secondary standards to water systems but does not require systems to comply. However, states may choose to adopt them as enforceable standards.

In general, pH is a measure of the activity of the hydrogen ions in a solution at a given temperature. We use the term activity because it is the amount of available hydrogen ions and not the concentration of hydrogen ions. For example, in pure water (H_2O) exists as

$$H_2O = H^+ + OH^-$$

The concentrations of the H^+ and the OH^- ions are equal and the solution is at equilibrium and is neutral. The pH of natural waters typically ranges from 4 to 9 and most are slightly basic due to the presence of bicarbonates and carbonates or the alkali and alkaline earth metals.[1]

The term pH is derived from the combination of "p" for the word "power" and "H" for the chemical symbol for hydrogen. pH is expressed as the negative log of the activity of the hydrogen ions in solution at a given temperature.

$$pH = -\log_{10} a_{H^+}$$

Another way to express the pH mathematically would be as follows:

$$pH = \log 1/H^+$$

where (H^+) is the amount of hydrogen ions in solution in moles per liter. In our pure water example, there are 0.0000001 mol per ions of hydrogen and a corresponding 0.0000001 mol per ions of OH^-. The pH of pure water is then

$$pH = \log 1/0.0000001, \text{ or } 10^{-7}$$

and is considered neutral. It is important to note that the H^+ and OH^- ions which comprise the H_2O water molecule are continuously dissociating, similar to a game of musical chairs, and there may be an excess of one or the other of the ions. This excess determines the pH or strength of the acid or alkaline solution. If there is an excess of H^+ ions to the available OH^- ions in the solution, the pH would be < 7 and considered acidic. If the reverse were true then the pH would be > 7 and the solution considered alkaline or basic. For example, a pH of 8 would indicate a solution with $(H^+) = 10^{-8}$ or tenfold less than a pH of 7 and a pH of 9 $((H^{-9}) = 10^{-9})$ would indicate a solution one-hundredfold less than a pH of 7.

PH MEASUREMENT SAMPLING

pH measurements must be taken in the field or as close to the sample source as possible. The equilibria in a groundwater or surface water system is altered once the sample is taken. A pH measurement taken at the moment of sampling may be representative, or very close, to conditions found in the source media. However, if the sample is placed in a sample bottle and the pH is not determined until arriving at the laboratory, the pH may not be representative of the source media. Gains and losses of carbon dioxide, and reactions such as oxidation of ferrous iron can alter the pH by a full unit[2] representing a tenfold error. Accurate measurement of pH in the field should be standard practice for all groundwater and surface water

Encyclopedia of Water Science
DOI: 10.1081/E-EWS 120010187

samples.[2,3] Improved field instrumentation allows the sampler to take measurements directly from the source in many cases and any unneeded sample handling should be avoided. If samples are measured in a sample container, the sampler can expect to see some drift due to changes in temperature and carbon dioxide concentrations. If it is necessary to take measurements in a sample container, especially when working outdoors, the container can be placed in a container or other shield to reduce temperature changes.

pH MEASUREMENT

Measurement of pH is possible using pH indicator or litmus paper. This method provides an approximation of pH and has limited value. The pH indicator paper is immersed into the sample and the color change is compared to a color scale that indicates the approximate pH of the sampled media. The advances in pH meters and relatively low cost of these instruments have all but replaced this well-known and time-tested technique.

pH METERS

The pH meter measures pH using potentiometric electrodes that measure changes in potential (voltage) caused by differing concentrations of Hydrogen ions. The pH measuring system consists of three elements:

1. pH electrode.
2. Temperature compensation element.
3. pH meter (simply a volt meter).

pH ELECTRODES

Electrodes must be matched to the expected pH ranges and types of materials to be measured. The sampling of surface and groundwater poses a unique challenge in that the pH range is very narrow and the media may have a very low conductivity. The electrode must be calibrated and is essential to gaining and documenting instrument accuracy. Calibration adjusts the slope and offset based on the Nernst equation and is expressed as a percentage of a theoretically perfect slope. pH buffers are used as standards to calibrate your instrument. The following buffers are considered as standards:

- pH 1.68 at 25°C.
- pH 4.01 at 25°C.
- pH 6.86 at 25°C.

- pH 7.00 at 25°C.
- pH 9.18 at 25°C.
- pH 10.01 at 25°C.

Most of these buffers also have charts that give the expected pH at various temperatures. Each pH meter will include specific calibration instructions, but ideally, a two-point calibration using two buffers that bracket the expected pH range to be measured should be used. For the best accuracy, buffers are used that are no more than 3 pH units apart. pH buffers are always discarded after use and only fresh buffers are used. The electrode calibration should fall between at least 95% and 105%. Once the calibration has been conducted, the buffers can be used to recheck the instrument.

ELECTRODE CARE AND CALIBRATION

Electrodes must be replaced periodically. To prolong the life expectancy and to assure measurement accuracy the following steps should be taken:

- Rinse the probes with distilled water and then with water from the next sample.
- Stir samples consistently but then stop to take a measurement.
- Use shields or other methods to reduce temperature and other chemical changes.
- Avoid rubbing or wiping the electrode
- Store the electrode properly in pH 7 buffer with KCL or pH storage solution.

Note: Do not store the electrode in distilled or deionized water. Field instruments typically have a field cap or sleeve to store and protect the electrode. Typically, the storage cap or sleeve will have an adsorbent material that must be periodically recharged with storage solution.

Some fresh water samples will require the use of a low resistance glass electrode or the use of a reference electrode with a fast continuous leak rate. Some level of error is introduced and another option is the use of "pure water" measurement kits. This method uses a quality glass electrode, a pure water additive to increase the ionic strength on the media, and a set of diluted buffers that are similar to the ionic background of the pure water kit additive. Since there are variations in electrodes and instrument, it is advised that one pH instrument be used for all measurements if the results are to be compared. Prior to selecting a pH meter, it is necessary to determine the expected range of pH measurements, the media to be sampled, the expected temperatures, and the application (field or laboratory).

DATA RECORDING

Prior to use, the pH meter should be checked using two buffer standards that closely bound the expected range of pH observations. The time and location of the quality control test should be recorded in the field or laboratory notebook along with the required performance for the study. Time and location of the measurements should be recorded in the field book or logbook. Both the pH and the temperature indicated by the meter should be recorded.

ACCURACY, PRECISION, AND BIAS

Advances in pH meter technology have resulted in advertised field instrument accuracy (ability to measure a known concentration or standard) of ± 0.01 pH unit which is highly optimistic. Commercially available pH standards are typically available within 0.01 units, so it is very difficult to verify accuracy at less than 0.05 units and under most field conditions to much less than 0.02 pH units in the laboratory. More precise standards can be prepared in the laboratory[1] and instrument precision (ability to reproduce similar results) can be improved under controlled conditions. However, ± 0.1 pH units would be the expected accuracy under normal conditions. A synthetic sample of a Clark and Lubs buffer solution of pH 7.3 was analyzed by 30 laboratories with a resulting standard deviation of ± 0.13 pH units.[1] Based on these results and expected difficulties in measurement of water and poorly buffered solutions, reporting to the nearest 0.1 pH unit is advised. pH probes on potentiometers require continuous care and are subject to damage, especially if the probes are not well maintained. Therefore, it is critical that the pH meter be calibrated and checked prior to each usage.

CONCLUSION

The measurement of pH is essential in all water investigations. pH measurement in drinking and surface water poses a unique challenge due to narrow range of expected values. An understanding of pH is essential to the understanding of nutrient, contaminant transport, and response by susceptible species.[4] Species response to significant changes in pH typically cannot be attributed to that single environmental factor but may be a result of secondary effects such as the toxic levels of metals such as aluminum.[4,5]

Buffering capacity of the geological formations and surface and groundwater can also have a significant effect on the biological effects of such facts as acid rain or other factors that influence water pH levels. The importance of accurate pH measurement in all water measurement activities cannot be overemphasized.

REFERENCES

1. American Public Health Association, American Water Works Association, Water Environmental Federation, *Standard Methods for Examination of Water and Wastewater*, 20th Ed.; American Public Health Association, American Water Works Association, Water Environmental Federation: Washington, DC, 1999.
2. Hem, J.D. *Study and Interpretation of the Chemical Characteristics of Natural Water*; U.S. Geol. Survey Water-Supply Paper 1473, 2nd Ed.; 1970; 363p.
3. Csuros, M. *Environmental Sampling and analysis for Technicians*; Lewis Publishers: Boca Raton, FL, 1994.
4. Sutter, G.W., II. *Ecological Risk Assessment*; Lewis Publishers: Boca Raton, FL, 1993.
5. Kalff, J. *Limnology*; Prentice Hall: Upper Saddle River, NJ, 2002.

Phosphorus Measurement

Bahman Eghball
United States Department of Agriculture (USDA), Lincoln, Nebraska, U.S.A.

Daniel H. Pote
United States Department of Agriculture (USDA), Booneville, Arkansas, U.S.A.

INTRODUCTION

Phosphorus (P) is an essential nutrient for growth and development of algae and other aquatic plants. However, P can cause water pollution if sufficient concentration ($25\,\mu g$–$100\,\mu g$ total $P\,L^{-1}$, eutrophic condition) is present in water. Eutrophication (nutrient-rich condition) can significantly increase growth of aquatic plants, algae, and sometimes strains of algae that cause taste, odor, or toxicity problems for drinking water supplies. During the night, when there is no photosynthetic activity to renew oxygen supplies for the dense concentrations of living cells, dissolved oxygen levels may become so depleted that fish and other aquatic animals cannot survive. Furthermore, many of the blue–green algae that cause the most serious water-quality problems require P inputs to grow and flourish, but they do not need high concentrations of N in lake water because they are able to utilize atmospheric N. Accelerated eutrophication of lakes, streams, and coastal waters remains a serious problem and has grown worse in many regions. Therefore, management plans for minimizing eutrophication should be designed to limit P inputs to surface water. Phosphorus occurs in water in many different forms that need to be evaluated to identify the overall effects of P on water quality. These include dissolved, bioavailable, particulate, and total P. Measuring these P forms in water is critical for distinguishing among them and assessing their effects on water quality.

PHOSPHORUS FORMS IN WATER

Dissolved P

Dissolved P is primarily the P fraction in the orthophosphate forms ($H_2PO_4^-$ and HPO_4^{2-}), which are immediately available for algae and plant uptake. Measuring this P fraction is important in determining the eutrophication potential of the water since dissolved P is a major portion of algae-available P (bioavailable P) in water. Dissolved P

concentrations as low as $0.01\,mg\,L^{-1}$ of lake water have been suggested as critical levels that can accelerate the eutrophication process in some relatively pristine lakes.[1] However, keeping runoff levels below $1\,mg\,L^{-1}$ can help maintain acceptable levels of water quality in many lakes, streams, and coastal estuaries.

The dissolved P fraction can be conveniently separated from suspended P fractions by passing the water sample through a membrane filter ($0.45\,\mu m$ pore diameter) immediately after sample collection. Although this technique may not completely separate dissolved P from suspended P, it is easily replicated and provides a clearly defined analytical separation. Dissolved P in water is determined without any preliminary hydrolysis or oxidative digestion of the water sample. Samples should be kept refrigerated ($4°C$) and analysis is recommended within $48\,hr$, unless the sample is stored frozen at temperatures below $-10°C$. Also, low concentrations of dissolved P may be adsorbed onto plastic bottles, so acid-washed glass bottles are recommended unless the sample is to be frozen.[2] The molybdate colorimetric test used to determine dissolved P concentration in water is based on the fact that dilute orthophosphate solutions react with ammonium molybdate and potassium antimony tartrate in an acid solution to form an antimony-phospho-molybdate complex. When this complex is reduced by ascorbic acid, it takes on an intense blue color that is proportional to orthophosphate concentration.[3] Analytical procedures and additional information for the modern molybdate colorimetric test, including techniques for automated analysis, can be found in a standard methods textbook.[2]

Bioavailable P

A laboratory test for bioavailable P (BAP) measures the amount of dissolved plus the fraction of particulate P in water that is available for algae uptake,[4] and is an excellent indicator of the eutrophication potential of P in water. Briefly, a filter paper circle ($5\,cm$ diameter) with small pores (e.g., $< 5.0\,\mu m$) is immersed into a solution containing $10\,g$ $FeCl_3\cdot6H_2O$ in $100\,mL$ distilled

Encyclopedia of Water Science
DOI: 10.1081/E-EWS 120010188

water. The filter papers are then air-dried and immersed for a few seconds in 2.7 M NH_4OH solution to convert $FeCl_3$ to Fe oxide. This iron-oxide filter strip has been shown to closely mimic the ability of algae to take up P in water. BAP in a water sample (50 mL) is then determined by shaking the sample with an iron oxide filter strip for 16 hr at 25°C on an end-over-end shaker. Phosphorus retained on the strip (BAP) is removed by shaking each strip with 40 mL of 0.1 M H_2SO_4 for 1 hr. Following neutralization, P is measured by the method of Murphy and Riley.[3]

There is a high correlation between BAP and dissolved P in runoff water from agricultural fields, and dissolved P comprises most of the BAP (Fig. 1). Since BAP in water is immediately available for algae uptake, the test would provide an indication of the eutrophication potential of the water body in the immediate future.

Total P

Total P in water provides an indication of short- and long-term water pollution potential. The United States Environmental Protection Agency (USEPA) has established total P concentrations of $25\,\mu g\,L^{-1}$–$100\,\mu g\,L^{-1}$ as critical levels for eutrophication of surface water. Total P consists of dissolved P and particulate P with dissolved P being an immediate concern and particulate P being a long-term P pollution concern.

To determine total P concentration in water, an unfiltered water sample should be shaken to suspend any

particulate matter immediately prior to measuring a subsample for analysis. Water samples for total P analyses can be frozen at temperatures below $-10°C$ for long-term storage, or acidified (using HCl or H_2SO_4) to pH < 2 and refrigerated (4°C) if analyzed within one month. Total P concentration in a water sample can only be determined colorimetrically when all P compounds (organic, condensed, and particulate) have been hydrolyzed to orthophosphate forms so that they will react with the molybdate reagent. This can be accomplished by several published methods, but they all require the use of heat and/or various strong acids to digest the water sample, thus oxidizing organic compounds and releasing the P as orthophosphate. Some methods also require strong oxidizing agents and thus may be dangerous. For example, perchloric acid digestion[6] is still known as a standard method for oxidizing resistant P compounds in water, but the heated mixture of $HClO_4$ and organic matter may react rapidly enough to produce a violent explosion unless organic matter is predigested. Preferred methods are sulfuric acid–nitric acid digestion, generally considered the most reliable procedure for potentially difficult samples, or persulfate digestion, which is simpler to use and provides good P recovery rates for most samples. Regardless of digestion method, at least 25 mL of shaken, unfiltered water sample should be digested if a large enough sample volume is available. Larger volumes (e.g., 100 mL) are recommended for digestion if the water sample is exceptionally clean and low P concentrations are expected. If total dissolved P is the only fraction of interest, then the water sample can be passed through a membrane filter (0.45 μm pore diameter) to remove particulate P before initiating a digestion process.

Sulfuric acid–nitric acid digestion takes several hours and may require the addition of at least one drop of 30% sodium peroxide to clarify relatively dirty samples during the digestion process so that residual color does not interfere with spectrophotometer analysis. Details of sulfuric acid–nitric acid digestion, persulfate digestion, and other methods, including techniques for automated versions of the P analyses are described in Ref. 2.

Particulate P

Particulate P consists of P fractions that are bound to soil particles and organic matter. This fraction is determined as the difference between total P and dissolved P in water. A small part (bioavailable particulate P) is immediately available to aquatic plants, but most particulate P must go through a chemical or biological reaction to become plant-available. Soil erosion increases particulate P in surface runoff, so reducing soil erosion is an important factor in minimizing particulate P loss. Particulate P can be

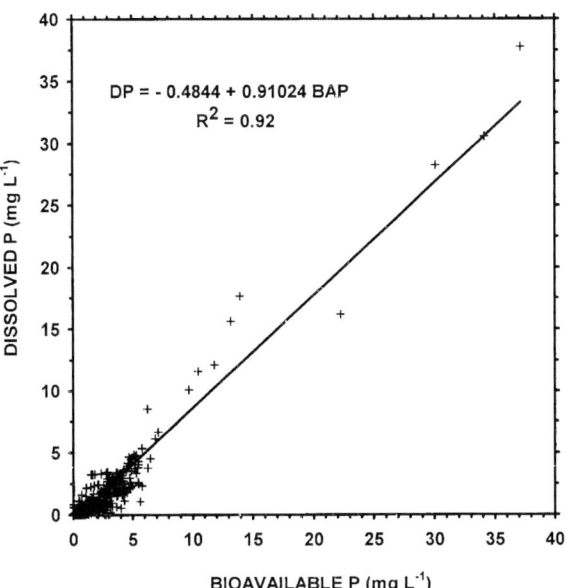

Fig. 1 Relationship between dissolved P and bioavailable P in runoff from fields with grain sorghum and winter wheat residues and receiving manure and fertilizer application. (From Ref. 5.)

a long-term environmental concern as the sediment-bound or organic P may eventually become available as dissolved P for algae uptake. Particulate P usually constitutes a major portion of total P in runoff from tilled soils, but a relatively minor P component in runoff from pastures, rangelands, and forests. Particulate P normally settles in the bottom of a water body, but can slowly release P to the overlying water. The sediments act as a P sink under aerobic conditions, but as a P source under anaerobic conditions.[7]

CONCLUSION

Phosphorus is an essential plant nutrient but excessive amounts can cause water quality deterioration. Measuring different forms of P in water can aid in determining the extent of short- and long-term water quality concerns. Dissolved and bioavailable P components are immediate water quality concerns as they are readily available for uptake by algae and other aquatic plants. Particulate and organic P fractions can cause water quality concerns over the long-term as P is slowly released in plant available forms to the surrounding water body. Total P consists of soluble plus particulate P and is an important component of water pollution assessment. The USEPA uses total P concentration as an indicator of eutrophication potential of surface water.

REFERENCES

1. Vollenweider, R.A. *Scientific fundamentals of the eutrophication of lakes and flowing waters, with particular reference to nitrogen and phosphorus as factors in eutrophication*; Pub. no. DAS/SAI/68.27, Organization for Economic Cooperation and Development, Directorate for Scientific Affairs: Paris, 1968.

2. APHA, AWWA, WEF, Standard Methods for the Examination of Water and Wastewater, 20th Ed.; Clesceri, L.S., Greenberg, A.E., Eaton, A.D., Eds.; American Public Health Association, American Water Works Association, and Water Environment Federation: Washington, DC, 1998.

3. Murphy, J.; Riley, J.P. A Modified Single Solution Method for the Determination of Phosphate in Natural Waters. Anal. Chim. Acta **1962**, *27*, 31–36.

4. Sharpley, A.N. An Innovative Approach to Estimate Bioavailable Phosphorus in Agricultural Runoff using Iron Oxide-impregnated Paper. J. Environ. Qual. **1993**, *22*, 597–601.

5. Eghball, B.; Gilley, J.E. Phosphorus and Nitrogen in Runoff Following Beef Cattle Manure or Compost Application. J. Environ. Qual. **1999**, *28*, 1201–1210.

6. Robinson, R.J. Perchloric Acid Oxidation of Organic Phosphorus in Lake Waters. Ind. Eng. Chem. Anal. (Ed.) **1941**, *13*, 465–466.

7. Pant, H.K.; Reddy, K.R. Phosphorus Sorption Characteristics of Estuarine Sediments Under Different Redox Conditions. J. Environ. Qual. **2001**, *30*, 1474–1480.

Plant Available Soil Water

Judy A. Tolk
United States Department of Agriculture (USDA),
Bushland, Texas, U.S.A.

INTRODUCTION

The soil stores the water used by plants to sustain life. The amount of soil water that can be used by the plant varies, due to characteristics of the soil (e.g., texture) and of the plant (e.g., root distribution and depth). Knowledge of the amount of water available to the plant, or plant available water (PAW), is needed to determine the agricultural or ecological potential of soils and is used in many agronomic applications, such as irrigation scheduling programs or crop production models. It helps define the water content limits beyond which plant growth is affected because of insufficient or excessive amounts of water, or beyond which water is lost out of the root zone due to deep percolation. The water content is typically expressed on a weight ($g\,m^{-3}$) or volume ($m^3\,m^{-3}$) basis.

Another term associated with PAW is the nonlimiting water range, which is defined as the region bounded by the upper and lower soil water content over which water, oxygen, and mechanical resistance are not limiting to plant growth.[1] The two soil water content boundaries that help determine PAW are the upper or "full" boundary, which is referred to as field capacity (FC), and the lower or "dry" boundary, or the permanent wilting point (PWP). Field capacity has been defined as the water remaining in the soil two to three days after having been wetted with water and after free drainage is negligible.[1] Permanent wilting point has been defined as the largest water content of a soil at which indicator plants, growing in that soil, wilt and fail to recover when placed in a humid chamber.[1] Both boundaries are not "sufficiently precise or general to be much more than a rough index," according to an uncited quotation in Ref. [2]. In the field, determining when drainage is "negligible" is extremely difficult; soils often have complex horizons with different water-holding characteristics; and plants may root differently from their genetically predetermined pattern due to soil physical and chemical characteristics or environmental conditions. Also, soil water determined as "available water" is not necessarily the portion of water that can be absorbed by all plants, but can be plant specific.[1] Richards[3] stated that "availability" involved both the "ability of the plant root to absorb and use the water with which it is in contact," and the "readiness with which the soil water moves in to replace that which has been used by the plant."

Water moves through the soil and plant in response to gradients in the potential energy of the water, going from regions of higher water potential to those with lower water potential. Water potential (ψ) is the measure of the free energy status of water and its ability to do work, which can be changed by the presence of solutes (osmotic potential), pressure (pressure potential), gravity (gravitational potential), and components which bind with water molecules (matric potential). For water to be available to a plant, the plant's roots first must be present; water must move through the soil to the root, pass into the root, and travel from the root to the leaf surface; and the rate of water supply must be able to meet transpiration requirements and maintain cellular functions. At high evaporation rates, the soil may be unable to transport enough water to meet transpiration demands and the plant may go into water stress at higher soil water contents than it would at lower evaporation rates.

CROP ROOTING CHARACTERISTICS

The characteristics of a root system depend upon heredity, but may be modified by environmental factors such as soil texture, depth, moisture content, mineralogy, chemistry, aeration, and solute concentration.[4] Monocots develop fibrous root systems, while dicots tend to have taproot systems (Fig. 1) that can take many different forms.[5] A species may always be deep rooted, or always shallow rooted, while still others develop different types of root systems in different types of soils. The age of the plant also determines rooting patterns and water uptake as well. As a plant grows, its roots extend downward and outward at varying rates. Kaigama et al.[6] reported rates of root extension for grain sorghum (*Sorghum bicolor* Moench.) of one to two centimeters a day. The rate of exploration by roots is controlled primarily by plant vigor and by soil environmental conditions, especially temperature, moisture, and strength.[5] Warm, moist soil encourages root development while increased soil strength can severely restrict it. As a plant matures, many roots die or lose much

Encyclopedia of Water Science
DOI: 10.1081/E-EWS 120010265

Fig. 1 The fibrous root system (left) of witchgrass (*Panicum capillare* L.) and the taproot (right) of cotton (*Gossypium hirsutum* L.).

of their ability to absorb water. The success of cultivated plants subjected to drought may depend on the development of deep, profusely branched root systems that absorb water from a large volume of soil.[4]

WATER MOVEMENT THROUGH THE SOIL

Most of the water flow through the soil can be described by Darcy's law, given as

$$J_w = -K(\psi)(d\psi/dz) \tag{1}$$

where J_w is the water flux density $(kg\,m^{-2}\,sec^{-1})$ in a soil with hydraulic conductivity $K(\psi)$ $(kg\,sec\,m^{-3})$, and water potential gradient $d\psi/dz$ $(J\,kg^{-1}\,m^{-1}$ or $m\,sec^{-2})$ with the components of water potential most responsible for flow being the matric and gravitational potentials.[7] Water flow through the soil in the range of PAW is determined by its unsaturated hydraulic conductivity, which can be approximated by Campbell and Norman[7]

$$K(\psi) = K_s(\psi_e/\psi)^{2+3/b} \tag{2}$$

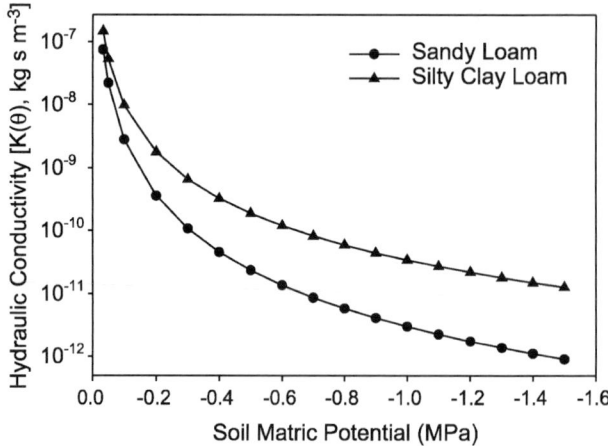

Fig. 2 Approximate hydraulic conductivity of a sandy loam and a silty clay loam in a range of soil matric potentials within plant available water (-0.033 to $-1.5\,MPa$) as determined by the pressure outflow apparatus.

where ψ_e is air entry water potential and K_s is the saturated conductivity of the soil. The parameter b is the exponent of the moisture release equation which, along with ψ_e and K_s, depends on soil physical characteristics such as texture. As the size of the pore space in a soil decreases (coarse textured to fine textured), the air entry potential decreases and b increases, resulting in unsaturated conductivity that is higher for finer-textured soils than coarse-textured ones (Fig. 2).

WATER MOVEMENT THROUGH THE PLANT

The ultimate destination for most of the soil water moving into a plant is the leaf surface, where it is lost as vapor through the stomatal pore. The driving gradient to move the liquid water from the root to the leaf is the water potential gradient between them. The resistances to flow through this system has been compared to a resistor network in an electric circuit, where water and current flow are analogous and can be described using Ohm's law in the form of[7]

$$U = (\psi_S - \psi_L)/(R_R + R_L) \tag{3}$$

where U is the rate of water uptake, ψ_S is the soil water potential, ψ_L is the leaf water potential, R_R is the root resistance, and R_L is the leaf resistance. The root resistance varies with the permeability of the root due to age or distance from the root apex, and changes due to dehydration, temperature, rate of water flow, or time of day.[4] Leaf resistance is affected by the location, size, shape, and abundance of stomata; environmental conditions affecting stomatal activity; and the size of the boundary layer surrounding the leaf, which is determined by the size and shape of the leaf and wind speed. At the leaf's surface, the sun's energy converts the water from a liquid to vapor state in the substomatal cavity. A vapor pressure gradient must then move the water vapor through the stomatal pore and boundary layer into the atmosphere surrounding the leaf. As the vapor pressure deficit between leaf and air increases, the demand for water flow through the soil and the plant also increases, with the rate of vapor loss also being controlled in part by the size of the stomatal opening.

MEASUREMENT OF PAW

The upper and lower boundaries that help determine PAW are FC and PWP. No simple, accurate method exists for either field or laboratory determinations. Numerous methods are available to approximate these boundaries, with procedures and limitations to the results outlined in Ref. [8]. A commonly used procedure is laboratory measurements using a pressure outflow apparatus. In this method, a soil sample is placed on a porous ceramic plate or permeable membrane in a chamber and saturated with water. Pressure is applied to the samples until equilibrium soil water contents at matric potentials of -1.5 MPa for PWP and -0.033 MPa for FC are achieved.[8] Among the many other methods developed to determine these boundaries are ones based on soil texture and bulk density;[9] bulk density, particle density, and particle-size distribution curve,[10] and electrical conductivity.[11]

Ideally, PAW should be measured in the field for each crop and soil combination. Field capacity is primarily a function of soil properties, while PWP is a function of a combination of soil, plant, and environmental factors. Fig. 3 shows the differences between measured lower limits of water use (θ_{LL}), or approximate PWP, for corn (*Zea mays* L.), grain sorghum, and wheat (*Triticum aestivum* L.) and soil water contents measured at -1.5 MPa matric potential ($\theta_{-1.5}$) using the pressure outflow apparatus procedures. The crops were grown in lysimeters containing a monolithic soil core of Ulysses silt loam (fine-silty, mixed, superactive, mesic Aridic Haplustoll), which is a deep, uniform soil formed in calcareous loess. Soil water content data were collected at harvest using neutron scattering. The vertical, dashed line represents the "zero" point of $\theta_{-1.5}$ such that values to the left of the dashed line represent the field-measured water contents less than $\theta_{-1.5}$ and those to the right the field-measured water contents greater than $\theta_{-1.5}$. Volumetric water contents were converted to mm by multiplying it by the measurement depth. Summed for the 2.2-m profile, grain sorghum used 46 mm and wheat 65 mm more than that summed for $\theta_{-1.5}$, while corn was similar to $\theta_{-1.5}$ levels. All crops showed a distinct decline in soil water use at the 0.9-m depth, possibly associated with the abrupt increase in bulk density in that layer compared with the layers above and below (data not shown). Fig. 3 shows the variability in lower limit of water availability among crops and the difference from $\theta_{-1.5}$. The figure suggests that PWP determined by laboratory methods is similar to

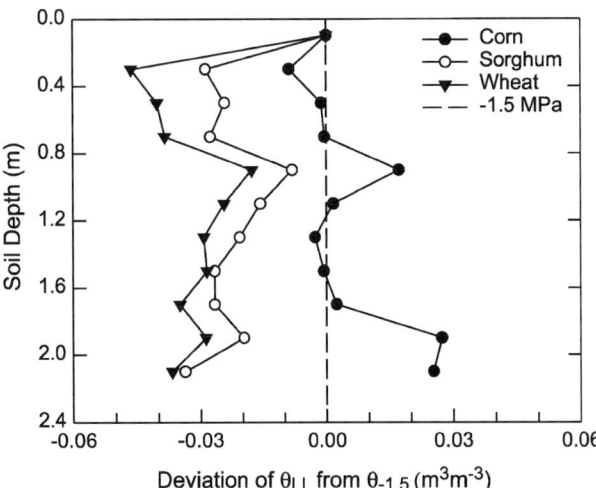

Fig. 3 The deviation of the lower limit of water extraction by corn, grain sorghum, and wheat (θ_{LL}) from the soil water content measured at 1.5 MPa ($\theta_{-1.5}$) by the pressure flow apparatus in a lysimeter containing a monolithic core of Ulysses silt loam. Data points to the left of the vertical dashed line indicate that the crop used more water than that at $\theta_{-1.5}$ and to the right it used less than $\theta_{-1.5}$.

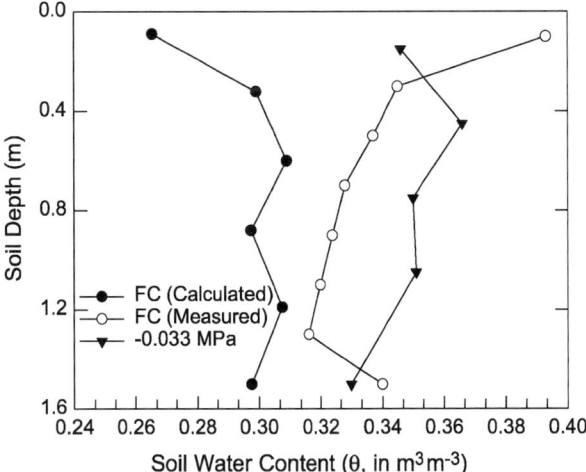

Fig. 4 Field capacity (FC) by depth of a monolithic soil containing Pullman clay loam measured by neutron scattering after the core was saturated and allowed to drain (open circles), calculated from equations of Ritchie et al.[9] using measured bulk density and percentages of sand and clay for the soil horizons (closed circles), and measured by the pressure outflow apparatus at 0.033 MPa pressure (triangles).

field-measured PWP of short season corn, but not necessarily to that of grain sorghum or wheat.

Measurement of FC can be equally as problematic. Cassel and Nielsen[8] stated that "personal experiences suggest that the uncertainty in FC is greater than that for PWP" with "no good alternative for measuring FC other than the in situ field method." Fig. 4 shows FC measured by neutron scattering in a lysimeter (same dimensions as above) containing a monolithic soil core of Pullman clay loam (fine, mixed, superactive, thermic Torrertic Paleustoll). Also shown is water contents measured at 0.033 MPa by the pressure outflow apparatus and FC calculated using procedures outlined by Ritchie et al.[9] The calculated FC required textural analysis for the clay and sand proportions as well as bulk density, which was determined from samples taken at the lysimeter monolith collection site. Converted from volumetric water contents and summed for the 1.5-m depth, the measured FC was 507 mm, the calculated was 447 mm, and the laboratory method was 523 mm.

CONCLUSION

Knowledge of PAW is important for determining the agricultural and ecological potentials of a soil and the best management practices that maximize crop productivity and minimize water losses. Laboratory determination of both FC and PWP is usually adequate for most applications, but the user must be aware of its limitations (Figs. 3 and 4). Soil texture, structure, layering, and chemistry along with crop type, rooting characteristics, stage of development, as well as environment are just some of the many factors that can impact PAW. The procedures for more accurate determination of PAW are often complicated, requiring specialized equipment and an extensive number of measurements, because it is a function of the interactions between the plant, the soil, and the environment.

REFERENCES

1. Soil Science Society of America, *Glossary of Soil Science Terms*; Soil Science Society of America: Madison, WI, 1997; 134.
2. Ritchie, J.T. Soil Water Availability. Plant Soil **1981**, *58*, 327–338.
3. Richards, L.A. The Usefulness of Capillary Potential to Soil-Moisture and Plant Investigations. J. Agric. Res. **1928**, *37*, 719–742.
4. Kramer, P.J. *Water Relations of Plants*; Academic Press, Inc.: New York, 1983; 489.
5. Klepper, B. Development and Growth of Crop Root Systems. In *Limitations to Plant Root Growth*, Advances in Soil Science; Hatfield, J.L., Stewart, B.A., Eds.; Springer-Verlag: New York, 1992; Vol. 19, 1–25.
6. Kaigama, B.K.; Teare, I.D.; Stone, L.R.; Powers, W.L. Root and Top Growth of Irrigated and Nonirrigated Grain Sorghum. Crop Sci. **1977**, *17*, 555–559.
7. Campbell, G.S.; Norman, J.M. *An Introduction to Environmental Physics*, 2nd Ed.; Springer-Verlag: New York, 1998; 286.
8. Cassel, D.K.; Nielsen, D.R. Field Capacity and Available Water Capacity. In *Methods of Soil Analysis, Part I. Physical and Mineralogical Methods*, Agronomy Monograph no. 9; Klute, A., Ed.; Soil Science Society of America: Madison, WI, 1986; 901–926.
9. Ritchie, J.T.; Gerakis, A.; Suleiman, A. Simple Model to Estimate Field-Measured Soil Water Limits. Trans. ASAE **1999**, *42*, 1609–1614.
10. Gerakis, A.; Zalidas, G. Estimating Field-Measured, Plant Extractable Water from Soil Properties: Beyond Statistical Methods. Irrig. Drain. Syst. **1998**, *12*, 311–322.
11. Mullins, J.A. Estimation of the Plant Available Water Capacity of a Soil Profile. Aust. J. Soil Res. **1981**, *19*, 197–207.

Plant Exposure to Water Stress During Specific Growth Stages

Zvi Plaut
Agricultural Research Organization, Bet Dagan, Israel

INTRODUCTION

Many yield determining physiological processes in plants respond to water stress. Most of these processes are dynamic and their activities may fluctuate with time according to internal and external factors. Yield integrates many of these physiological processes in a complex way and it is, thus, difficult to interpret how do plants accumulate, combine, and display the ever-changing and indefinite physiological processes over the entire life cycle of the crop. Moreover, as far as water stress is concerned, severity, duration, and timing of stress, as well as responses, which may take place after stress removal, and interaction between stress and other factors may be extremely variable. It would thus be very inconceivable, from a practical point of view, to study the response of physiological processes to the dynamic changes in plant water stress and accordingly conclude how the final yield will respond. The more pragmatic approach for determining the response of plant productivity to dynamic changes in water stress should thus be based directly on the yield or its components.

The timing of plant water stress, which is mostly the result of drought, and lack of available soil water, can either extend throughout the entire growth period or during specific stages of growth. The effect of stress during such stages on final yield is outlined in another article of this book.[1]

HOW CAN CROP SENSITIVITY OF SPECIFIC STAGES BE EVALUATED?

The main objective in the determination of sensitive and insensitive growth stages is saving of irrigation water. This can be obtained either by withholding irrigation (or a sharp decrease in the amount of water applied) during insensitive growth stages, or by a slight but extended exposure of plants to water stress. The two approaches ought to be compared prior to making any conclusions concerning timing and quantities of water to be applied. In spite of the numerous studies on sensitivity of a wide range of crops to water stress at different growth stages (many but not all were outlined by Plaut,[1] such a comparison was conducted in few investigations only.

Such a comparison was made by Plaut for three different crops: corn, sunflower, and tomatoes for processing.[2] The amount of water applied was reduced either by applying smaller quantities, but throughout the season, or by withholding irrigation at specific growth stages. In corn, yield losses were much more marked when the crop was subjected to water stress during flowering and early grain filling as compared to other growth stages, validating other studies, that this was a sensitive growth stage. However, slight and uniform water stress during the entire growth season resulted in less damage to yield than withholding irrigation at any (insensitive) growth stage although total amounts of applied water were similar (Fig. 1). In sunflower, as well, the withdrawal of irrigation water at any growth stage reduced yield more severely than uniform and extended light stress throughout the entire season. The response of tomatoes to the decrease in the amount of water applied was different (Table 1). The withdrawal of irrigation water at specific growth stages reduced fruit and total soluble solid (TSS) yields much less than decreases in crop irrigation coefficients throughout the season resulting in similar reductions of irrigation water.

The difference between the two groups of plants may be explained on the basis of being determinate. Corn and sunflower are distinct determinate plants and every plant bears a single reproductive organ, which is a distinct sink. Assimilates are mobilized from the source organs for an extended period and transported to this sink. Severe water stress, at any time even for a limited period, may upset the entire source to sink steady flow. Slight stress may cause less damage to this system so that source to sink flow may continue with no interruption. In tomatoes for processing, new sinks are continuously being formed, and any particular fruit is a sink for part of the entire source and for a limited time only. When stress is in operation, only the fruits, which serve as sinks at this particular time, will be affected. Moreover, newly formed fruits may even compensate for this loss, once stress is relieved, so that yield may be enhanced. It was also shown by Stirling et al.[3] that pod yield of groundnuts were insensitive to early moisture deficits. Although growth was inhibited during exposure to stress, sink activity was maintained within the expanding leaf, and it could rapidly recover when stress was released.

Encyclopedia of Water Science
DOI: 10.1081/E-EWS 120015635

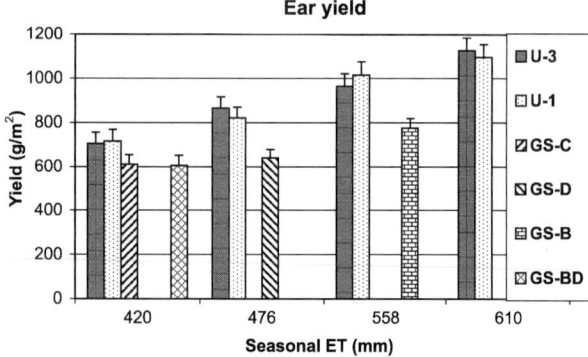

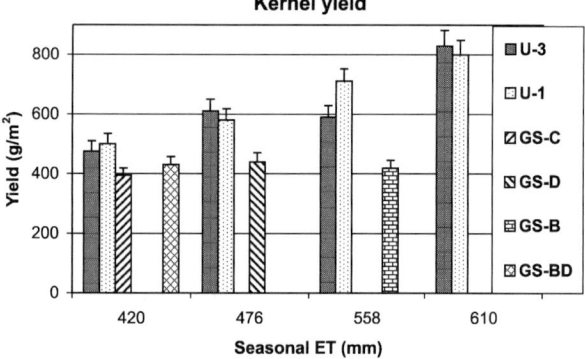

Fig. 1 The effect of deficit irrigation on corn ear and kernel yield. U-3 is the reduction in irrigation water throughout the season, 3 weekly applications. U-1 is as U-3 but 1 weekly application. GS-B, GS-C, and GS-D are withdrawal of irrigations at vegetative, flowering, and kernel filling stages, respectively. GS-BD is withdrawal of irrigation at stages B and D. The four groups of bars represent four levels of ET. Vertical bars are SE of the means.

Our findings were recently verified with tomato plants, which were allowed to grow indeterminately as compared to others which were constantly pruned and grown as determinate plants (paper in preparation). It was similarly shown for determinate cultivars of beans that yield was reduced when exposed to stress during the growth stage of flowering and grain filling, while an indeterminate cultivar was more stress tolerant at the sensitive growth stage.[4]

Saving irrigation water by "stretching" the reduced quantities of water over an extended period can be obtained by scheduling irrigation according to soil water content, allowing low limits of pre-irrigation water content. When this irrigation strategy was compared with withdrawal of irrigation at specific growth stages, such as delay of irrigation until flowering or until midpod elongation, no difference in yield of soybeans was obtained during two seasons.[5] This suggests, at least for soybeans, that when grown in deep soil and fully recharged with water prior to planting, irrigation could

be delayed, and the importance of critical growth stages is rather low.

There are crops in which no definite developmental stages can be recognized, like sugarcane. Following a short period of crop establishment and tillering, there is a long period of growth and sugar accumulation, which is one phenological period. In this crop, the effect of water stress appears to be primarily related to the degree of stress, relative to soil water content and ET demand, rather than to a specific crop factor.[6]

Water application at specific growth stages may also be of less concern in drought tolerant crops. As mentioned,[1] sorghum is a good example for such a crop. An increase in yield may be obtained regardless when the water is applied and may result in increased number of grains and/or an increase in kernel weight, depending on when the water is applied.[7] Another drought tolerant crop, barley, did not show any preference of a particular growth stage, when the application of water was of largest benefit.[8] The highest yield was obtained when water was applied at all stages unless rainfall was sufficient.

DEFICIT AND SUPPLEMENTAL IRRIGATION

Deficit irrigation is a deliberate under-irrigation of a crop. It may be practical under special conditions, mainly when water is very limiting. One would expect that irrigation timing at critical growth stages (as outlined in Ref. [1]) is of considerable importance under such conditions. Additional factors are also of importance in the planning of irrigation timing. When water is limiting, the highest yield and highest water use efficiency will usually be obtained when applied at low frequencies during an extended period. This was, for instance, shown for deficit irrigation of wheat when maximal yield was obtained with irrigation intervals of 4 weeks.[9]

Deficit irrigation is applied in desert areas, where ET is high and water is scarce. Corn is for instance grown in the African Sahel, as an important source of food. The optimal timing of deficit irrigation was studied in this area.[10] When 6–8 deficit irrigations were applied during the vegetative and reproductive phases, grain yield was reduced by 52% of the fully irrigated control. When only two deficit irrigations were applied during these growth stages (all the rest were full irrigations) grain yield decreased by 23%–26% only. Yield reductions were mainly due to kernel number and less due to kernel weight.

Supplemental irrigation is generally given in addition to natural precipitation, when either no additional applications are needed, or when limited amounts of irrigation water are available. It is interesting that the timing of such irrigation was also not according to growth stages. Seed production of red clover, for instance, requires

Table 1 The effect of withholding irrigation water at specific growth stages and of reduced amounts of water throughout the season on tomato fruit and TSS yields. Four different groups are outlined, in which similar quantities of water were applied

Group	Amount of water applied (mm)	Crop irrigation coefficient	Growth stage of irrigation withdrawal	Fruit yield (kg m^{-2})	TSS yield (g m^{-2})
A	580	1.15	—	13.2	660
B	500	1.00	—	12.8	625
C	425	0.85	—	9.9	555
	420	1.00	Ripening	12.2	666
D	343	0.65	—	9.9	538
	339	1.00	Vegetative + flowering	10.8	533
	342	1.00	Fruit expansion	10.9	643

supplemental irrigation for maximal production. The timing of this irrigation was determined on the basis of crop water stress index (CWSI), and the fraction of available soil water consumed. The CWSI was found to be more consistent than fraction of soil water used, possibly because canopy temperature measurements integrate an entire plot, while soil moisture was based on single points. A single irrigation, which filled the entire soil profile applied at CWSI = 0.28 was sufficient to increase seed yield very remarkably.[11] White clover for seed production is also supplementary irrigated. It was recommended to apply a single irrigation during the period between haying and seed maturation as at this stage most of the available soil water is being utilized.[12]

In humid regions and under temperate climate, optimal seed production can be achieved for many years without any supplemental irrigation, or with a marked delay in water application. It was shown that maximal seed yield and the highest water use efficiency of white clover were obtained when the irrigation was delayed until 68% of available soil water was consumed. A single irrigation of bird's-foot trefoil increased seed yield over that which was obtained under maintenance of soil water close to field capacity due to frequent water applications.[13]

REFERENCES

1. Plaut, Z. Crop Plants: Critical Developmental Stages of Water Stress. In *Encyclopedia of Water Science*; Stewart, B.A., Howell, T., Eds.; Marcel Dekker, Inc.: New York, NY, 2003.
2. Plaut, Z. Sensitivity of Crop Plants to Water Stress at Specific Developmental Stages: Reevaluation of Experimental Findings. Isr. J. Plant Sci. **1995**, *43*, 99–111.
3. Stirling, C.M.; Black, C.R.; Ong, C.K. The Response of Groundnut (*Arachis hypogaea*) to Timing of Irrigation. J. Exp. Bot. **1989**, *221*, 1363–1373.
4. Boutraa, T.; Sanders, F.E. Influence of Water Stress on Grain Yield and Vegetative Growth of Two Cultivars of Bean (*Phaseolus vulgaris* L.). J. Agric. Crop Sci. **2001**, *187* (4), 251–257.
5. Specht, J.E.; Elmore, R.W.; Eisenhauer, D.E.; Klocke, N.W. Growth Stage Scheduling Criteria for Sprinkler-Irrigated Soybeans. Irrig. Sci. **1989**, *10*, 99–111.
6. Weidenfeld, R.P. Water Stress During Different Sugarcane Growth Periods on Yield and Response to N Fertilization. Agric. Water Manag. **2000**, *43*, 173–182.
7. Sweeney, D.W.; Lamm, F.R. Timing of Limited Irrigation and N-Injection for Grain Sorghum. Irrig. Sci. **1993**, *14*, 35–39.
8. Wahab, K.; Singh, K.N. Effect of Irrigation Applied at Different Critical Growth Stages on Growth Characters and Yield of Hulled and Hull-less Barely. Indian J. Agron. **1983**, *28*, 412–417.
9. English, M.; Nakamura, B. Effects of Deficit Irrigation and Irrigation Frequency on Wheat Yields. J. Irrig. Drain. Eng. **1989**, *115* (2), 172–184.
10. Pandey, R.K.; Maranville, J.W.; Admou, A. Deficit Irrigation and Nitrogen Effects on Maize in a Sahelian Environment I. Grain Yield and Yield Components. Agric. Water Manag. **2000**, *46*, 1–13.
11. Oliva, R.N.; Steiner, J.J.; Young, W.C. Red Clover Seed Production: I. Crop Water Requirements and Irrigation Timing. Crop Sci. **1994**, *34*, 178–184.
12. Oliva, R.N.; Steiner, J.J.; young, W.C. White Clover Seed Production: I. Crop Water Requirements and Irrigation Timing. Crop Sci. **1994**, *34*, 762–767.
13. Garcia-Diaz, C.A.; Steiner, J.J. Seed Physiology, Production and Technology. Crop Sci. **2000**, *40*, 449–456.

Plant Water Stress: Optional Parameters for Stress Relief

Zvi Plaut
Agricultural Research Organization, Bet Dagan, Israel

INTRODUCTION

The availability of irrigation water is diminishing in many parts of the world, which creates the need for optimization and increasing irrigation efficiency imperative. This was achieved in the past by irrigation scheduling, which covers two aspects; when to irrigate and how much water to apply. The main purpose of irrigation scheduling was to avoid harmful conditions of water stress and yet save irrigation water and other expenses in order to make water use most efficient. Higher efficiency of water use may still be attained, provided certain plant developmental stages are less sensitive to water deficit than others, so that restricting water supply during those stages may hardly affect productivity. Developmental stages of many crops, which are sensitive, and nonsensitive are outlined in another article.[1]

PARAMETERS FOR WATER APPLICATION TO RELEASE STRESS

Although minimizing water application during nonsensitive growth stages may lead to savings in irrigation water, the overall efficiency of water use will depend on irrigation scheduling and on the amounts of water applied during the entire season. Optimal scheduling of irrigation is, however, of special importance during sensitive growth stages. Parameters based on aerial environment factors, soil, and plant factors were used and recommended as adequate for irrigation scheduling. In fact the subject of parameters for irrigation scheduling is beyond the scope of this chapter, we shall have to refer to it briefly, as it is very closely related to sensitivity to water stress at different developmental stages.

Plant Parameter

Plant water status, depends on water uptake and transpiration rates. Water uptake is a function of soil water content and availability, and transpiration is determined by aerial environment factors. There was thus, a tendency to base irrigation scheduling on parameters of plant water status, rather than on indirect parameters. Plant water potential (Ψ) was probably the most common plant parameter, which was practiced.[2] The use of midday leaf water potential was adopted for the timing of irrigation in cotton.[3,4] Different preirrigation levels of Ψ were examined as a guideline at which growth stage water application should be initiated.[5] Midday leaf water potential was also found to be most suitable as an irrigation-timing criterion for wheat,[6] provided measurements were conducted on cloudless days. The threshold Ψ for maximum water use efficiency was -1.82 MPa, and for maximum yield -1.44 MPa. Leaf water potential values were also used as thresholds for the timing of supplemental irrigation of corn.[7] Predawn water potentials may serve as a more adequate parameter of plant water status to be used for applying water. This is based on the assumption that predawn water potential integrates variation in soil moisture over the whole rootzone and is also less subjected to fluctuations in environmental conditions.

The Ψ serving as a threshold to be used for irrigation timing may sometimes vary during plant development. It was shown for grain sorghum that drought decreased leaf water potential only by 0.1 MPa–0.2 MPa during the vegetative stage, 0.3 MPa–0.4 MPa during the reproductive stage and exceeded 0.5 MPa during the grain filling stage.[8] This is an important finding suggesting, that preirrigation Ψ cannot always be a constant value for the entire life span of a given crop. Stomatal resistance was sensitive to small reductions in leaf water potential during the vegetative period, but became nearly insensitive during the reproductive period, suggesting that Ψ is not always the most suitable plant parameter for irrigation timing.

Irrigation timing can also be based on indirect plant parameters that respond to plant water status rather than on plant water potential. These include the use of stomatal aperture, growth rates of plant organs, and changes of trunk circumference. A particular indirect plant parameter for irrigation timing may be xylem cavitation. When the soil dries out, the water column within the xylem vessels will fracture or cavitate leading to the formation of a bubble, which gives rise to an acoustic emission.[9] The rate of occurrence of such acoustic events was suggested to be used as an indication of plant water status and for irrigation timing.

Encyclopedia of Water Science
DOI: 10.1081/E-EWS 120015636

Soil Parameters

Another approach to determine the appropriate time for water application is based on soil water extraction and the remaining available soil water. Determining soil water potentials using different devices as tensiometers, psychrometers, resistance blocks, or dew point hygrometers can be used for this purpose. For instance, Irmak et al.[10] who used a dew point soil hygrometer, showed for a heavy clay soil that water potential of $-406\,KPa$ at the depth of 0 cm–30 cm left approximately 50% of available soil water, which was sufficient to provide maximum grain yield of corn. Soil water content, which can be determined gravimetrically, by neutron probes or by time domain reflectometry (TDR), can also be used for irrigation scheduling, and were in fact used in many studies.[11,12] The threshold of available soil water, which should serve for irrigation timing differs, however, among crops and for the same crop at various growth stages, as was shown elsewhere.[1]

The number of irrigations to be applied during any growth stage and the amount of water per application will depend on the soil type and on ET conditions. Irrigation can be applied at higher frequencies than needed for maintaining soil water content above a recommended value, which is mostly in the range of 50% available soil water. The allowable soil water depletion or increase in plant Ψ becomes less important under such irrigation regimes. The continuous minimal soil water tension in the upper soil layers minimize the fluctuations in Ψ, which result in less inhibition of physiological processes and higher productivity. Although this can be performed using most irrigation methods, it is mostly common for high value crops like vegetables, potatoes, flowers, and fruit crops irrigated with drip irrigation. This was shown, for instance, for potatoes, tomatoes, sweet corn, and cashew[13–16] and for many additional crops. It was also demonstrated for other crops, like cotton that small quantities of water at high frequencies during their sensitive growth stages, resulted in an increased production.[17] A further increase in irrigation frequency up to 2–3 applications per week using drip irrigation was found to increase the yield very significantly.[18,19]

Agrometeorological Parameters

One of the most widely spread parameters for irrigation scheduling is probably the crop water stress index (CWSI), which was developed by several investigators and was outlined by Idso et al.[20] This index uses the difference between canopy and air temperatures ($T_c - T_a$), related to vapor pressure deficit of the air (VPD). An equation to calculate this CWSI is:

$$CWSI = \frac{(T_c - T_a)_a - (T_c - T_a)_p}{(T_c - T_a)_u - (T_c - T_a)_p} \tag{1}$$

where the subscript a, p, and u outside the parentheses stand for actual, potential [under no stress conditions, yielding lowest ($T_c - T_a$)] and upper ($T_c - T_a$), under no transpiring conditions. Although Idso presented the relationship between ($T_c - T_a$)_p and VPDs under potential transpiration rates (fully irrigated) for 26 different species, the use of CWSI was limited to a few grain crops as wheat and corn,[21] for a limited number of forage legumes[22,23] and for some grasses.[24] The predicted ($T_c - T_a$) for severely stressed turf plants did not agree with measured values.[24] Values of CWSI of 0.25–0.30 or higher can be considered as fairly extreme conditions, and irrigation at values above those may lead to a decrease in yield. For corn, a decrease in grain yield was found when seasonal mean CWSI was higher than 0.22.[21] For wheat a CWSI of 0.30 was used.[25] For potatoes, which are more sensitive, a CWSI of 0.20 was scheduled for irrigation.[26]

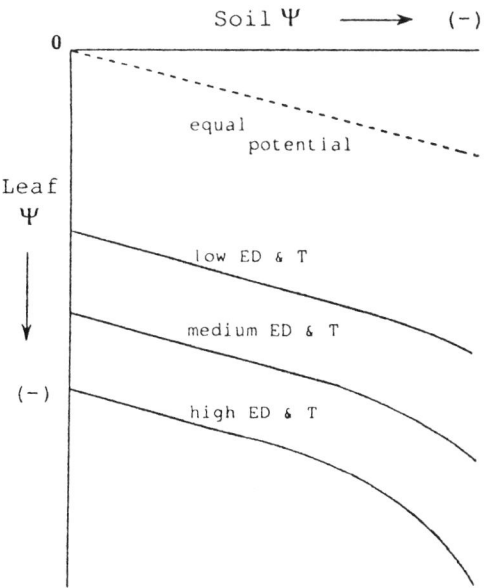

Fig. 1 Conceptual depiction of the influence of evaporative demand (ED) of the atmosphere and the resultant transpiration rate (T) on the relationship between soil Ψ and leaf Ψ over a range of decreasing soil Ψ. Dashed line indicate when soil and plant water potential are in equilibrium.

Comparison and Interaction Between Parameters

Leaf water potential Ψ_l is directly affected by soil water potential Ψ_s in the absence of transpiration, so that predawn Ψ_l can be an estimate for Ψ_s. However, under

Table 1 Summary of average irrigation amounts, ET estimates, and grain yields for a 3-yr period

| Treatment | Irrigation | | Estimated ET | | Grain yield | |
	Amount[a] (mm)	Relative[b]	Amount (mm)	Relative	Amount (kg/ha)	Relative
T1 40% D	345 a	1.00^1	477 a	1.00^1	11,700 a	0.96^3
T2 0.5 × ET	181 d	0.52^7	443 de	0.93^7	11,300 ab	0.93^4
T3 0.2 CWSI	248 b	0.72^2	467 ab	0.98^2	11,900 a	0.98^2
T4 0.4 CWSI	197 cd	0.57^6	447 cde	0.94^6	11,200 ab	0.92^5
T5 50 KPa	206 bcd	0.60^5	448 cde	0.94^5	12,200 a	1.00^1
T6 0.6 CWSI	157 d	0.46^8	431 e	0.90^8	10,600 b	0.87^6
T7 30 KPa	243 bc	0.70^3	462 abc	0.97^3	11,900 a	0.98^2
T8 C-M	231 bc	0.67^4	453 bcd	0.95^4	11,700 a	0.96^3

[a] Treatments with the same letter in each category do not differ significantly at the $\alpha = 0.05$ level. The applied irrigation, estimated ET, and grain yield LSD values were 49.2 mm, 18.4 mm, and 981 kg/ha, respectively.

[b] The maximum value is assigned a value of 1.00; other values are fractions of the maximum, e.g., applied irrigation for T2 is $181/345 = 0.52$. Ranks are shown as superscripts, e.g., 1.00^1 is the top ranking relative value.

Source: From Ref. [11].

active transpiration Ψ_l will depend on the rate of transpiration (T), on water conductivity within the soil (C_s), at the soil–plant interface (C_{sp}), and within the plant (C_p). It can be described by the following equation:

$$\Psi_l = \Psi_s - T(C_s + C_{sp} + C_p) \qquad (2)$$

Plants may thus be exposed to water stress, even when the soil is supplied with plenty of water and Ψ_s is high, if the transpiration demand and conductances are high. This implies that under such conditions plant water status cannot always serve for irrigation timing. Under conditions of low transpiration demand (cool and humid locations or time of the year), the threshold of Ψ_s would be lower than under high transpiration demand (Fig. 1). This suggests that it might be beneficial to schedule irrigation on more than one parameter. It was shown for instance that corrected data obtained for pan evaporation rates, soil water potentials, and CWSI were used in combination for scheduling tomato irrigation.[27] In wheat, significant yield losses were found when CWSI exceeded a threshold of 0.4–0.5, or rootzone water depletion above 50%.[28] A detailed comparison of different parameters, which could be used to determine critical stress and the need to apply water, was conducted on corn.[29] The compared parameters were: 40% depletion of available soil water (control), 0.5 predicted ET replacement based on a model, CWSI of 0.2, 0.4, and 0.6, soil water potential below -50 kPa and -30 kPa, and on a crop growth model. Maximal yield was obtained when -50 KPa was used as a parameter and this led to a reduction of 40% in irrigation water as compared to the control (Table 1).

REFERENCES

1. Plaut, Z. Crop Plants: Critical Developmental Stages of Water Stress. In *Encyclopedia of Water Science*; Stewart, B.A., Howell, T., Eds.; Marcel Dekker Inc.: New York, NY, 2003.

2. Hsiao, T.C. Plant–Atmosphere Interactions, Evapotranspiration, and Irrigation Scheduling. Acta Hortic. **1990**, *278*, 55–66.

3. Grimes, D.W.; Yamada, H. Relations of Cotton Grown and Yield to Minimum Leaf Water Potential. Crop Sci. **1982**, *22*, 134–139.

4. Zelinski, L.J.; Grimes, D.W. Interaction of Water and Nitrogen on the Growth and Development of Cotton. Proc. Beltwide Cotton Conf. **1995**, 1109.

5. Wrona, A.F.; Kerbt, T.; Shouse, P. Effect of Irrigation Timing on Yield and Earliness of Five Cotton Varieties. Proc. Beltwide Cotton Conf. **1995**, 1108–1109.

6. Nell, A.A.; Dijkhuis, F.J. Use of Estimated Leaf Water Potential for Irrigation Timing of Wheat. South Afr. J. Plant Soil **1991**, *8* (2), 88–92.

7. Peiter, M.X.; Chaudhry, F.H. Supplemental Irrigation Control Strategies for Corn by Simulation of the Management Cycle, National Irrigation Symposium. Proceedings of the 4th Decennial Symposium, Phoenix, Arizona, American Society of Agricultural Engineers: St. Joseph, MI, 2000; 623–628.

8. Garrity, D.P.; Sullivan, C.Y.; Watts, D.G. Changes in Grain Sorghum Stomatal and Photosynthetic Response to Moisture Stress Across Growth Stages. Crop Sci. **1984**, *24*, 441–446.

9. Jones, H.G. Plant Water Relations and Implications for Irrigation Scheduling. Acta Hortic. **1990**, *278*, 67–76.

10. Irmak, S.; Haman, D.Z.; Smajstrla, A.G. *Measurement of Soil Water Potential Using Dew Point Soil Hygrometer for Irrigation Timing of Corn*, Paper No. 99-2236; American Society of Agricultural Engineers: St. Joseph, MI, 1999.

11. Steele, D.D.; Stegman, E.C.; Gregor, B.L. Field Comparison of Irrigation Scheduling Methods for Corn. Trans. ASAE **1994**, *37* (4), 1197–1203.

12. Schneider, A.D.; Howell, T.A. Methods, Amounts, and Timing of Sprinkler Irrigation for Winter Wheat. Trans. ASAE **1997**, *40* (1), 137–142.

13. Simonne, E.; Oukakrim, N.; Caylor, A. Evaluation of an Irrigation Scheduling Model for Drip-Irrigated Potato in Southeastern United States. Hortscience **2002**, *37* (1), 104–107.

14. Yuan, B.Z.; Kang, Y.H.; Nishiyama, S. Drip Irrigated Scheduling for Tomatoes in Unheated Greenhouses. Irrig. Sci. **2001**, *20* (3), 149–154.

15. Assouline, S.; Cohen, S.; Meerbach, D.; Harodi, T.; Rosner, M. Microdrip Irrigation of Field Crops: Effect on Yield, Water Uptake and Drainage in Sweet Corn. Soil Sci. Soc. Am. J. **2002**, *66* (1), 228–235.

16. Blaikie, S.J.; Chacko, E.K.; Lu, P.; Muller, W.J. Productivity and Water Relations of Field Grown Cashew: A Comparison of Sprinkler and Drip Irrigation. Aust. J. Exp. Agric. **2001**, *41* (5), 663–673.

17. Chu, C.; Henneberry, T.J.; Radin, J.W. Effect of Irrigation Frequency on Cotton Yield in Short-Season Production Systems. Crop Sci. **1995**, *35*, 1069–1073.

18. Plaut, Z.; Carmi, A.; Grava, A. Cotton Growth Under Drip-Irrigation Restricted Soil Wetting. Irrig. Sci. **1989**, *9*, 143–156.

19. Radin, J.W.; Reaves, L.L.; Mauney, J.R.; French, O.F. Yield Enhancement in Cotton by Frequent Irrigation During Fruiting. Agron. J. **1992**, *84*, 551–557.

20. Idso, S.B.; Reginato, R.J.; Jackson, R.D.; Pinter, P.J., Jr. Measuring Yield-Reducing Plant Water Potential Depression in Wheat by Infrared Thermometry. Irrig. Sci. **1981**, *2*, 205–212.

21. Irmark, S.; Haman, D.Z.; Bastug, R. Determination of Crop Water Stress Index for Irrigation Timing and Yield Estimation of Corn. Agron. J. **2000**, *92*, 1221–1227.

22. Hutmacher, R.B.; Steiner, J.J.; Vail, S.S.; Ayar, J.E. Crop Water Stress Index for Seed Alfalfa: Influences of Within-Season Changes in Plant Morphology. Agric. Water Manag. **1991**, *19*, 135–149.

23. Oliva, R.N.; Steiner, J.J.; Young, W.C. White Clover Seed Production: I. Crop Water Requirements and Irrigation Timing. Crop Sci. **1994**, *34*, 762–767.

24. Jalali-Farahani, H.R.; Slack, D.C.; Kopec, D.M.; Matthias, A.D.; Brown, P.W. Evaluation of Resistance for Bermudagrass Turf Crop Water Stress Index Models. Agron. J. **1994**, *86*, 574–581.

25. Garrot, D.J.; Ottman, M.J.; Fangmeier, D.D. Quantifying Wheat Water-Stress with the Crop Water-Stress Index to Schedule Irrigations. Agron. J. **1994**, *86* (1), 195–199.

26. Shae, J.B.; Steele, D.D.; Gregor, B.L. Irrigation Scheduling Methods for Potatoes in the Northern Great Plains. Trans. ASAE **1999**, *42* (2), 351–360.

27. Calado, A.M.; Monzon, A.; Clark, D.A.; Phene, C.J.; Ma, C.; Wang, Y. Monitoring and Control of Plant Water Stress in Processing Tomatoes. Acta Hortic. **1990**, *277*, 129–136.

28. Stegman, E.C.; Soderlund, M. Irrigation Scheduling of Spring Wheat Using Infrared Thermometry. Trans. ASAE **1992**, *35* (1), 143–152.

29. Steele, D.D.; Stegman, E.C.; Gregor, B.L. Field Comparison of Irrigation Scheduling Methods for Corn. Trans. ASAE **1994**, *37* (4), 1197–1203.

Plant Water Use, Stomatal Control

James I. L. Morison
University of Essex, Colchester, United Kingdom

INTRODUCTION

The term "plant water use" is commonly used, but it is
unfortunate as it suggests that plants "consume" water in
some biochemical processes.[1] However, less than 2% of the
water that is taken up by plants is actually transformed during
biochemical reactions, the rest (98% or more) is simply
"lost" during transpiration, the process of evaporation from
inside plants. Understanding and quantifying this transpiration is of critical importance in many applied and scientific
disciplines, in particular in hydrology, crop science, forestry,
ecology, meteorology, and climatology. Transpiration arises
as an inevitable consequence of the need for plants to expose
the surfaces of their photosynthetic cells to the air, to take up
CO_2 during photosynthesis and therefore provide carbohydrate for growth. Aerial parts of terrestrial plants (and of
emergent aquatic plants) are covered in impermeable
materials; for leaves and photosynthetic stems this is the
cuticle, a hydrophobic layer made up of lipids and waxes
secreted to the outside of the epidermal cell layer that reduces
the diffusional loss of water to the atmosphere to a very low
rate in normal conditions. However, the cuticle is also
impermeable to the diffusion of CO_2, so plants have "pores"
in the cuticle with variable apertures in order to control CO_2
uptake and H_2O loss. These pores, termed stomata (single:
stoma), are formed by a pair of specialized epidermal cells,
the guard cells, which have both unusual anatomy and
physiology. As well as determining CO_2 and H_2O exchange,
stomata also influence the atmosphere-plant exchange of
other gases, such as the phytotoxic pollutant, O_3. Stomatal
apertures change over periods of minutes, and their size and
shape vary between species and in different conditions.

STOMATAL ANATOMY AND DISTRIBUTION

Stomata occurred early in the development of terrestrial
plants, with stomata being found in fossils from the Silurian
period in the lower Paleozoic era (>400 million yr ago).
Stomata occur on the spore capsules of mosses, on most
aerial parts of terrestrial vascular plants, including leaves,
green stems, fruits, and flowers, but not in submerged aquatic
plants.[2] In the majority of plant species, the stomatal pore is
elliptical, formed between a pair of semicircular guard cells
(Fig. 1a–c), but the pore may become almost circular when
fully open, with diameters of 5 μm–50 μm, not only varying
in size and shape between species but also varying with the
conditions during leaf development. In grasses (Poaceae),
the pore is a slit shape, formed between two elongated guard
cells (Fig. 1d). In some species, there are well-developed
subsidiary cells adjacent to the guard cells (Fig. 1a, b, and d).
The pore may be sunken, with the guard cells recessed below
the larger adjacent epidermal cells, particularly in plants of
drier habitats. The numbers of stomata per unit leaf area
(referred to as stomatal density or frequency) also vary with
species and conditions, and range from 0 to 2000 or more
stomata mm^{-2}. The proportion of the leaf area they cover is
very small, about 0.5%–3%. In herbaceous plants, stomata
are found on both the upper (adaxial) and lower (abaxial)
surfaces of leaves, which are termed amphistomatous,
although there are usually more stomata on the lower surface.
However, many tree species have stomata only on the lower
surface (hypostomatous) and aquatic plants with floating
leaves, such as water lilies have stomata only on the upper
side (epi- or hyper-stomatous).

CONTROL OF TRANSPIRATION BY STOMATA

Leaf Scale

The diffusion rate of gases into or out of the leaf or other plant
parts depends on the concentration gradient and the diffusive
resistance of the pathway (a relationship known as Fick's
Law). For water loss from the mesophyll cells inside the leaf,
or CO_2 uptake by those cells, the major pathway is therefore
from the mesophyll cell walls through the substomatal cavity
to the pore, and then out through the layer of air immediately
surrounding the leaf, to the mixed air stream (Fig. 2).
Therefore, the stomatal pore offers a "resistance," r_s, to
diffusion, dependent on the aperture, shape, and number of
stomatal pores. Note that although the pore area when open
may only be at maximum a few percent of the total leaf area,
the rates of evaporation can be about half that of a wet surface
of similar dimensions (e.g., blotting paper); this is due to the
"edge effect" of diffusion through multiple pores.[2] The
"boundary layer" of air around the leaf also poses a resistance
to diffusion (r_b), which depends on surface characteristics
(presence of hairs, venation, etc.), leaf shape and size, and

Encyclopedia of Water Science
DOI: 10.1081/E-EWS 120010166

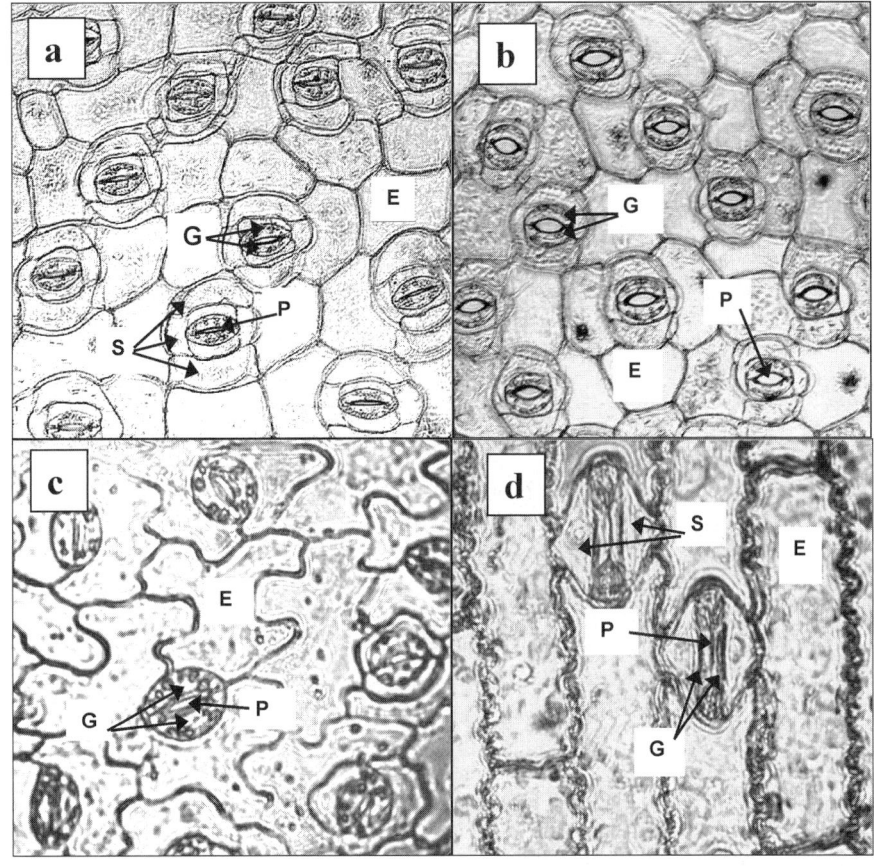

Fig. 1 Photomicrographs of stomatal and epidermal cells in (a and b) *Commelina communis*, (c) *Phaseolus vulgaris* (french bean) (d) *Zea mays* (maize). Guard cells indicated by G, subsidiary cells by S, pore by P, epidermal cell by E. Note chloroplasts evident in guard cells, particular in (c). Guard cell lengths approximately 45 μm in (a and b), 30 μm in (c), and 40 μm in (d). All from lower leaf surfaces.

wind speed and turbulence. While the cuticle is relatively impermeable, some water is lost through it (varying with species) giving a "cuticular resistance," r_c, in parallel to and usually much larger than, the stomatal resistance, r_s. There is also an internal resistance, r_i, for the pathway from cell wall to pore, but this is normally small compared to r_s and r_b. An equation for the rate of diffusion of water from a leaf (E, mg m^{-2} sec^{-1}) can therefore be derived from the difference in water vapor concentration between the inside and outside of the leaf ($\chi_i - \chi_a$, g m^{-3}) and the leaf resistance, r_l (sec m^{-1}), which is given by the sum of the various resistances in series and/or parallel as appropriate as shown in Fig. 2:

$$E = \frac{\chi_i - \chi_a}{r_l} \tag{1}$$

$$r_l = \frac{r_c(r_s + r_i)}{r_c + r_s + r_i} + r_b \tag{2}$$

(For simplicity, these equations consider one side of the leaf only; see Ref. [3], which also gives typical values for the

resistances.) Fig. 3a shows that if the cuticular resistance is very low, then r_l becomes curvilinearly related to r_s and Fig. 3b shows that r_b only influences E when $r_s < r_b$. Note that resistances are often replaced by their inverse, the "conductance," g, ($g_l = 1/r_l$), as transpiration is approximately linearly related to stomatal conductance, g_s.

The aforementioned diffusion equations can be used for simple analyses, but in practice the leaf microclimate is not independent of the transpiration rate which affects the leaf temperature, and therefore the internal water vapor concentration, which determines the driving gradient for evaporation, and the long wave radiation balance. Because of this "feedback," it is necessary to consider a more complete "energy balance" equation, such as the Penman–Monteith equation in order to examine the relative control that stomata exert on transpiration, compared to the other components. Analyses show (Fig. 4) that the important feature is the degree of "coupling" of the leaf to the air stream [4]; if the leaf has a small r_b compared to r_s, then the leaf is "well-coupled" and leaf temperature will not increase substantially, and changes in r_s will be reflected in E. This is typically the case with small, needle shaped leaves, at the top of the canopy

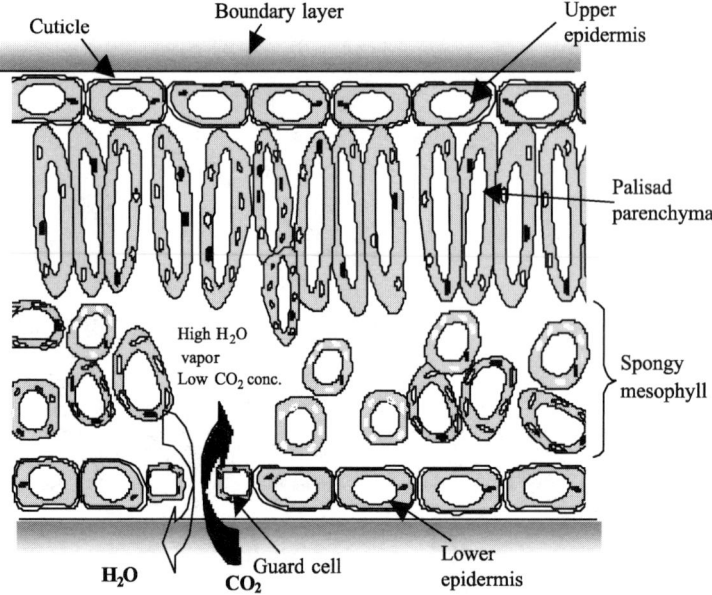

Fig. 2 Stylized cross-section of a leaf 1 mm–2 mm thick, showing pathway for water vapor diffusion from the leaf internal cell spaces through the stomatal pore and boundary layer (size and thickness of various elements exaggerated for clarity, diagram courtesy of Dr. T. Lawson, University of Essex).

with relatively high wind speeds. The opposite situation occurs with large, broad leaves within short, dense canopies, when E will not closely reflect changes in r_s. Indeed, it has recently been suggested that the evolution of larger planate leaves from the earlier leafless branched shapes became possible during the late Devonian period because of increased stomatal frequencies in response to declining atmospheric CO_2 producing greater evaporation cooling, and kept leaves below their lethal temperature limit.[5]

Vegetation Scale

The Penman–Monteith energy balance model has been used widely to quantify water loss from complete stands of vegetation (crops, pasture, and forests) by treating the canopy as a "big leaf."[4] This approach considers a canopy resistance, r_c as the sum of many individual leaf resistances, and a canopy aerodynamic resistance, r_a which reflects the pathway of air movement from outside the boundary layer of each leaf, to the mixed air stream well above the vegetation. However, there are many theoretical and practical problems in estimating the appropriate value for the resistances.[6,7] Nevertheless, as with individual leaves, the degree of coupling of the canopy to the airstream is important in determining the relative role of stomata in controlling water loss compared to the other components, although the feedbacks are more complex. First, there is likely to be evaporation from the soil surface, which acts to cool and humidify the air in the canopy.

Secondly, the local air humidity is affected by the transpiration.[8] This also applies at the regional scale where the evaporation rate from the entire surface influences the heat and moisture transfer into the lower "atmospheric boundary layer" and changes the regional climatic conditions.[9]

STOMATAL PHYSIOLOGY

Stomatal pores open and close due to the changing turgor of the surrounding guard and epidermal cells. Guard cells have specially oriented cell wall fibers, which result in deformation and movement away of the central cell portions from each other when turgor increases. Cell turgor changes when the osmotic potential changes, caused by the uptake or loss of solutes, in particular K^+, which may be charge balanced by uptake of Cl^- or synthesis of malate^{2-} in the cells. One of the key steps in opening stomata, e.g., when a leaf is illuminated after darkness, is H^+ loss, due to a light-stimulated proton pump, which then hyperpolarizes the cell membrane, causing the influx of K^+. An intensive study over several decades has shown that the control of ion movement across the membrane is complex, with various ion channels that may be under the control of different environmental stimuli, and may be linked through key cell signaling mechanisms.[10] However, there are several questions still

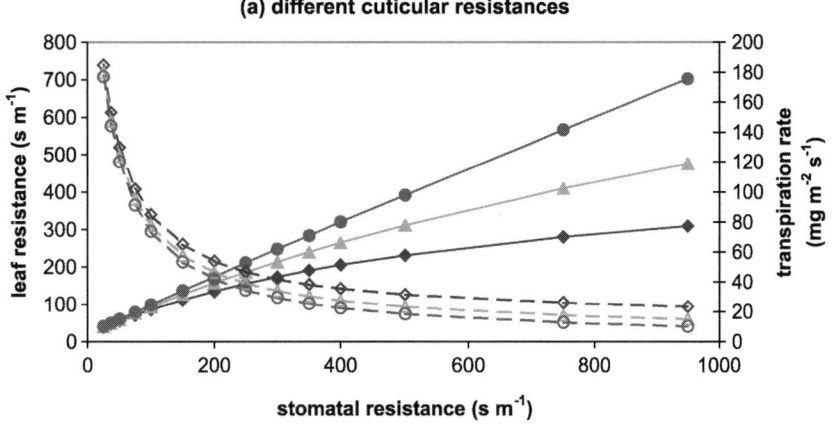

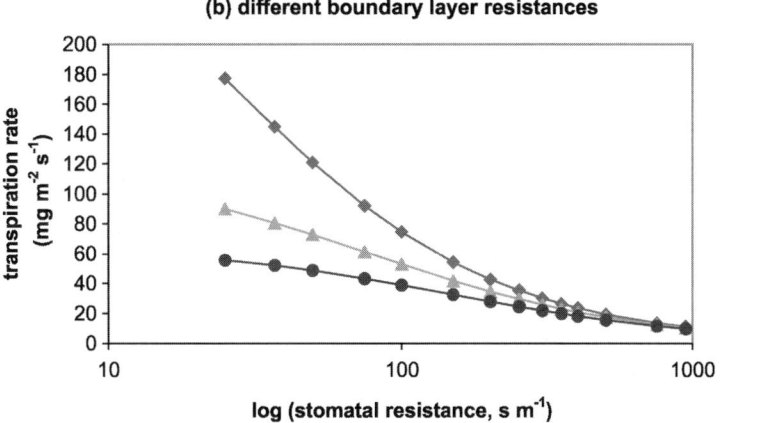

Fig. 3 Effect of changes in stomatal resistance on leaf resistance and transpiration rate. In (a) the diamond, triangle, and circle symbols are for increasing value of cuticle resistance (1000 sec m^{-1}, 2500 sec m^{-1}, and 20,000 sec m^{-1}, respectively). Open symbols and dotted lines are for transpiration rate. In (b) the diamond, triangle and circle symbols are for increasing boundary layer resistance (10 sec m^{-1}, 50 sec m^{-1}, and 100 sec m^{-1}).

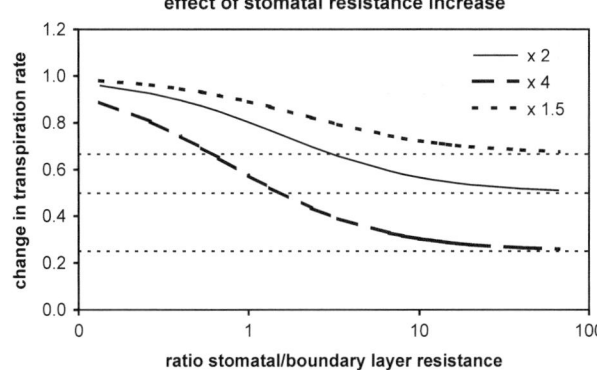

Fig. 4 Dependence of the change in transpiration rate caused by increases in stomatal resistances (increasing by 1.5, 2, and 4 times) on the ratio of stomatal to boundary layer resistance.

unresolved about stomatal metabolism, particularly the role of carbon metabolism.[11,12]

ENVIRONMENTAL EFFECTS ON STOMATA

Stomatal aperture is affected by many environmental and physiological variables; particularly, light, humidity, temperature, leaf, and soil water status (Fig. 5). Normally, stomata open (g_s increases, r_s decreases) in response to increasing light, but typically reach maximum aperture at approximately one-third of full sunlight. The response to light comprises at least two distinct effects in blue and red wavelengths. Stomata close in response to decreasing air humidity, with either linear or curvilinear response, when the humidity is expressed as the vapor pressure difference between the leaf and the air, i.e., the driving force for evaporation (discussed earlier). The humidity response is believed to be because of loss of turgor of the guard cells

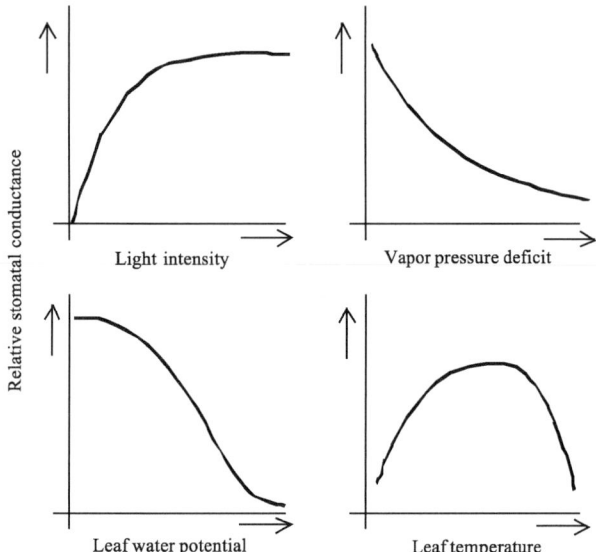

Fig. 5 Diagrams indicating generalized response of stomatal conductance to light intensity, leaf to air vapor pressure deficit, leaf water potential, and leaf temperature.

themselves, and is independent of the overall water status of the leaf. Stomata also close in response to a decline in leaf water status, but may show little effect at high water potential, until a threshold is reached. The soil water status also can have effects, and these are normally ascribed to the role of chemical "messengers" from the roots such as abscisic acid (ABA), synthesized or released in response to water shortage, which promote stomatal closure.[13,14] While normally stomata close in the long and medium term in response to reductions in water status, it should be remembered that with larger apertures, there is more transpiration, and inevitably a reduced water status, so a negative relationship can occur, at least in the short term.[15]

STOMATA LINK WATER AND CARBON FLOWS

Because stomata have a major control on plant gas exchange, they form the key linkage point between photosynthesis and transpiration, at all scales from that of individual leaves, to global carbon and hydrological cycles. Therefore, when plants are grown or measured across a range of different conditions (light or nutrient supply) there is a close linear correlation between photosynthetic CO_2 uptake and stomatal conductance.[16] For example, plants growing in shady, nutrient poor conditions, typically have low g_s and photosynthetic rate, compared to plants of open, fertilized habitats. This correlation results in a conservative "water use efficiency"

(although note the difficulties with this term,[11]). However, plants of the different photosynthetic pathways do show characteristically different water use efficiencies, with C_4 having values 2–3 times that of C_3, and CAM plants being about 5–10 times C_3 species. This ratio emphasizes that the way that plants control the loss water is a key determinant of growth, reproduction, and survival.

REFERENCES

1. Monteith, J.L. The Exchange of Water and Carbon by Crops in a Mediterranean Climate. Irrig. Sci. **1993**, *14*, 85–91.
2. Willmer, M.; Fricker, C. *Stomata*, 2nd Ed.; Chapman & Hall: London, 1996; 375.
3. Nobel, P.S. *Physicochemical and Environmental Plant Physiology*; Academic Press: New York, 1991; 635.
4. Monteith, J.L. Coupling of Plants to the Atmosphere. In *Plants and their Atmospheric Environment*; Grace, J., Ford, E.D., Jarvis, P.G., Eds.; Blackwell Scientific Publications: Oxford, 1981; 1–29.
5. Beerling, D.J.; Osborne, C.P.; Chaloner, W.G. Evolution of Leaf-Form in Land Plants Linked to Atmospheric CO_2 Decline in the Late Palaeozoic Era. Nature **2001**, *410*, 352–354.
6. McNaughton, K.G. Effective Stomatal and Boundary-Layer Resistances of Heterogeneous Surfaces. Plant Cell Environ. **1994**, *17*, 1061–1068.
7. Jarvis, P.G. Scaling Processes and Problems. Plant Cell Environ. **1995**, *18*, 1079–1089.
8. McNaughton, K.G.; Jarvis, P.G. Predicting Effects of Vegetation Changes on Transpiration and Evaporation. In *Water Deficits and Plant Growth*; Kozlowski, T.T., Ed.; Academic Press: New York, 1983; Vol. VII, 1–47.
9. McNaughton, K.G. Regional Interactions between Canopies and the Atmosphere. In *Plant Canopies: Their Growth, Form and Function*, Society for Experimental Biology, Seminar Series 31; Russell, G., Marshall, B., Jarvis, P.G., Eds.; Cambridge University Press: Cambridge, 1989; 63–81.
10. Assmann, S.M. Signal Transduction in Guard Cells. Ann. Rev. Cell Biol. **1993**, *9*, 345–375.
11. Outlaw, W.H., Jr.; Zhang, S.; Hite, D.R.C.; Thistle, A.B. Stomata: Biophysical and Biochemical Aspects. In *Photosynthesis and the Environment*; Baker, N.R., Ed.; Kluwer Academic Publishers: Dordrecht, 1996; 241–259.
12. Lawson, T.; Oxborough, K.; Morison, J.I.L.; Baker, N.R. Responses of Photosynthetic Electron Transport in Stomatal Guard Cells and Mesophyll Cells in Intact Leaves to Light, CO_2 and Humidity. Plant Physiol. **2002**, *128*, 52–62.
13. Tardieu, F. Drought Perception by Plants. Do Cells of Droughted Plants Experience Water Stress? Plant Growth Regul. **1996**, *20*, 93–104.
14. Jackson;, M.B. Hormones from Roots as Signals for the Shoots of Stressed Plants. Trends Plant Sci. **1997**, *2*, 22–28.

15. Jones, H.G. Stomatal Control of Photosynthesis and Transpiration. J. Exp. Bot. **1998**, *49*, 387–398.

16. Wong, S.C.; Cowan, I.R.; Farquhar, G.D. Stomatal Conductance Correlates with Photosynthetic Capacity. Nature **1979**, *282*, 424–426.

P

WEB LINKS

For some images of stomata:

http://www.biologie.uni-hamburg.de/b-online/e05/05a.htm

http://biog-101-104.bio.cornell.edu/BioG101_104/tutorials/botany/stomata2.html

http://biog-101-104.bio.cornell.edu/BioG101_104/tutorials/botany/stomata1.html

http://biog-101-104.bio.cornell.edu/BioG101_104/tutorials/botany/stomata3.html

Material on control of stomata:

http://www.biologie.uni-hamburg.de/b-online/e32/32f.htm

Images of stomata and information on leaf anatomy:

http://www.esb.utexas.edu/mauseth/weblab/webchap10epi/chapter_10.htm

Plant Yield and Water Use

M. Hossein Behboudian
Massey University, Palmerston North, New Zealand

Tessa Marie Mills
*The Horticulture and Food Research Institute of New Zealand
(HortResearch), Palmerston North, New Zealand*

INTRODUCTION

Water is the essence of life, and it plays vital roles in the biology of plants. In addition to its roles within the plant, evaporation of water from stomatal apertures provides for carbon dioxide entry into the leaf with carbon being fixed into organic matter through photosynthesis. Plants use a considerable amount of water to gain the required carbon, and "water use efficiency" (WUE) is the term used to quantify the yield, obtained through fixing of carbon, for the water lost. There are various ways, at the plant and farm level, to increase WUE as exemplified by the use of deficit irrigation (DI). The DI involves supplying less water to plant stands than the prevailing evapotranspiration (ET), a term combining transpiration (T) from plants and evaporation (E) from the soil. When yield is plotted against applied irrigation water, the relationship is called crop water production function (CWPF). Each point on CWPF could, therefore, relate to WUE. The availability of CWPF for each major crop in each region will facilitate the proper management of water resources. Many publications on various crops relate yield to water use and especially yield and applied irrigation water. A comprehensive coverage for various economically important herbaceous plants was given by Doorenbos and Kassam.[1] The following short treatment focuses on some basic definitions relating the yield to water use of crop plants.

WATER AND PLANT LIFE

Water has profound effects on plant function and distribution around the world. The ecological significance of water is due to its important physiological roles. Water is a plant nutrient (contributing the H atom), is a medium for all the biochemical reactions, acts as a solvent for many of important substances, hydrates most of the organic compounds in protoplasm, and acts as a medium for the diffusion and mass flow of solutes. It also maintains turgidity creating turgor pressure within cells.

Despite these vital roles, only a maximum of 5% of water absorbed from the soil remains in the plant with at least 95% being lost to the atmosphere through the process of transpiration.[2] Much more water is used by plants than dry matter is produced by them. Pimentel et al.[3] estimated that for the production of 1 kg of food or forage, the water use (in liters) of the following plants will be potatoes 500, wheat 900, alfalfa 900, sorghum 1110, corn 1400, rice 1912, and soybeans 2000. Therefore, plants differ in their efficiency of production in relation to water use, and the term WUE has been used to quantify this concept.

WATER USE EFFICIENCY

Leaves are the major sites of transpiration with at least 90% of water transpired through stomata and the rest through the cuticle.[4] The loss of water through open stomata will result in the diffusion of CO_2 from the air into the leaf. Therefore, CO_2 assimilation and final harvestable yield will be realized through the loss of water. Water use efficiency is defined as the total dry matter produced by plants per unit of water used:[2]

$$WUE = D/W \qquad (1)$$

where D is the mass of dry matter produced and W is the mass of water used. D represents photosynthetic activity, because the C and O atoms of the CO_2 from the air account for most of the dry mass. The term W could be considered as equivalent to ET, which comprises nonproductive E and productive T. Evaporation of free water from a leaf surface adds to E. The ratio D/T focuses on the physiological aspects of WUE. According to Postel,[5] the estimated WUE ($kg\,m^{-3}$) for the following crops worldwide are wheat 0.8–1.0, rice 0.7–1.1, maize 0.8–1.6, other grains ~ 0.6–1.2, roots and tubers ~ 4.0–7.0, pulses ~ 0.2–0.6, soybean 0.4–0.7, other oilseeds ~ 0.2–0.6, groundnuts 0.6–0.8, vegetables and melons ~ 10.0, fruits (except melons) ~ 3.5, sugar cane 5.0–8.0, sugar beet

Encyclopedia of Water Science
DOI: 10.1081/E-EWS 120010171

6.0–9.0, and tobacco 0.4–0.6. It is imperative that WUE be increased to accommodate partially for the rising demand for water by a rapidly expanding world population.

METHODS OF INCREASING WUE

There are several means for improving WUE. At the plant level, classical breeding and genetic engineering may be directed towards decreasing transpiration without a corresponding decrease in yield. Several possibilities for this are discussed by Richards et al.[6] such as early canopy development and introduction of short-season crops. At the field level, the following measures could be taken to save water: proper fertilizer application, suitable plant density, weed control, rain-water harvesting and conservation, tillage, mulching, double cropping to utilize the water remaining in the soil, and the possible use of antitranspirants and reflectants to reduce ET. Antitranspirants are best designed to either close stomata or form a cover over stomata in such a way that transpiration will be more reduced than photosynthesis leading to water conservation and increased WUE. However, despite decades of research no satisfactory compound has been introduced with these properties, and research on antitranspirants is practically abandoned.[2] Other possible measures at the farm level for increasing WUE are better control of water distribution system and judicious application of DI.

DEFICIT IRRIGATION

Deficit irrigation involves giving less water to the plant than the prevailing ET at selected times during the growing season. A short history of DI and its application to deciduous orchards, including some case studies, was reviewed by Behboudian and Mills.[7] If applied judiciously, DI saves water, decreases vegetative growth and, therefore, pruning costs in deciduous orchards, reduces leaching of biocides into the ground water, and might improve fruit quality while maintaining yield. Deficit irrigation is expected to be more successful in dry than in humid areas, because in the latter rain can interfere with achieving an intended low soil/plant water status.

Deciduous orchards may stand to benefit more from DI than do field crops, because fruits are strong sinks and, especially in high-density orchards, photosynthates would be more diverted towards fruits than towards the restricted roots or shoots whose seasonal growth would have already ceased. For field and annual crops, it is expected that conditions which decrease transpiration below its potential rate will also reduce biomass production below its potential rate.[8] The following relationship between yield and soil moisture deficit was quoted from the

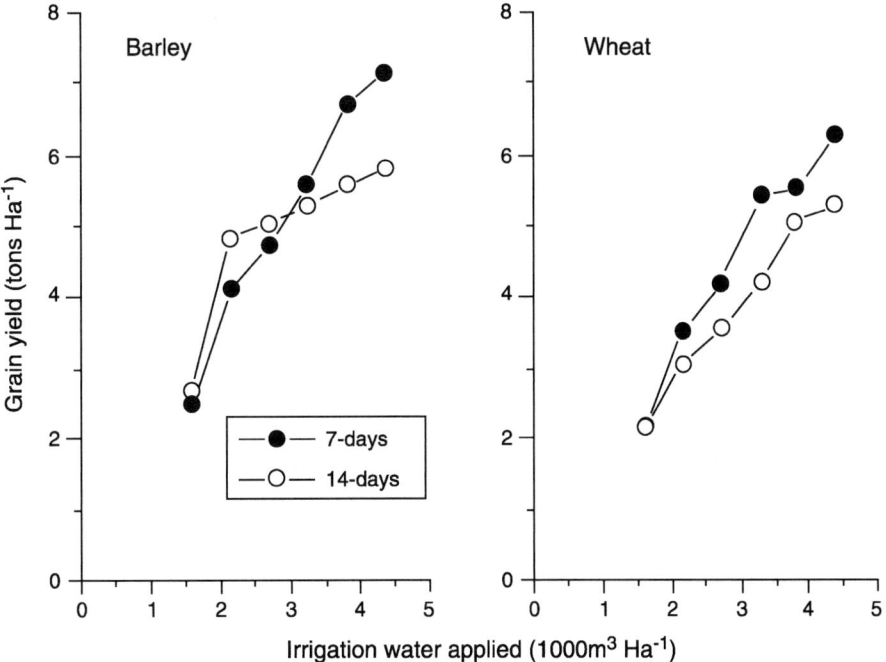

Fig. 1 The relationship between the amount of irrigation water applied and grain yield in barley and wheat (i.e., CWPF) at 7- and 14-day irrigation intervals. (The data are based on Ref. [12].)

literature by Jamieson,[8] who also successfully tested it on peas, potatoes, wheat, barley, and maize:

$$Y = Y_0[1 - a(D_p - D_c)] \qquad (2)$$

where Y is the yield, Y_0, the potential yield, D_p, the potential soil moisture deficit dependent on the environmental conditions, and D_c, the threshold value of soil moisture deficit. The judicious application of DI involves avoiding development of water deficit during the most sensitive periods of crops to minimize the yield reduction which Eq. 2 estimates. For example, cereals are more sensitive to the timing of a water deficit than they are to the total reduction of irrigation water.[9] There are reports on DI application to annual crops without yield reduction as exemplified by the experiment of Heuer and Nadler[10] on cotton. Although plant growth expressed as plant height and accumulation of fresh weight were significantly decreased with DI, neither seed cotton yield nor lint quality were decreased. Agronomic practices need to be modified to realize the maximum potential of DI as outlined by Kirda and Kanber.[9] A more effective irrigation strategy could be followed by the consideration of CWPF for crops of interest.

CROP WATER PRODUCTION FUNCTION

The CWPF shows the relationship between yield and the amount of irrigation water. A quantitative treatment of this function with citation of the relevant literature can be found in Varlev et al.[11] An example of CWPF for barley and wheat is given in Fig. 1, which is based on the data of Fardad and Pessarakli.[12] The CWPF is an empirical relationship which should be determined for each crop in each area because yield is a function of various environmental and biotic factors. The relationship, therefore, does not follow the same pattern for all crops. Such CWPFs could be useful for planning and development of water resources and projections of agricultural production.

FUTURE PROSPECTS

Water availability for crop production will be a far more serious issue in 2025 than it is now.[5] Various measures could be taken to address this issue for increasing WUE on the global scale. Especially, rain-fed land needs protection because it does not compete for water with agricultural,

urban, and industrial users of water. Efficient channeling and storing of rainwater will be crucial. Improving irrigation efficiency such as delivering water directly to the roots of crops will greatly reduce evaporative losses. Increasing WUE of crops (as outlined earlier), shifting the mix of crops, and breeding for more drought and salt tolerance will have special value. These measures will be of vital importance for sustainable production of food for the growing world population. The on-going rise in the atmospheric CO_2 is expected to decrease transpiration and, therefore, to increase the WUE in the future.

REFERENCES

1. Doorenbos, J.; Kassam, A.H. *Yield Response to Water*; FAO: Rome, 1979; 193 pp.
2. Kramer, P.J.; Boyer, J.S. *Water Relations of Plants and Soils*; Academic Press: New York, 1995; 495 pp.
3. Pimentel, D.; Houser, J.; Preiss, E.; White, O.; Fang, H.; Mesnick, L.; Barsky, T.; Tariche, S.; Schreck, J.; Alpert, S. Water Resources: Agriculture, the Environment, and Society. BioScience **1997**, *47*, 97–106.
4. Martin, J.T.; Juniper, B.E. *The Cuticle of Plants*; Edward Arnold (Publishers) Ltd: London, 1970; 347 pp.
5. Postel, S.L. Water for Food Production: Will There Be Enough for 2025? BioScience **1998**, *48*, 629–637.
6. Richards, R.A.; Lopez-Castaneda, C.; Gomez-Macpherson, H.; Condon, A.G. Improving the Efficiency of Water Use by Plant Breeding and Molecular Biology. Irrig. Sci. **1993**, *14*, 93–104.
7. Behboudian, M.H.; Mills, T.M. Deficit Irrigation in Deciduous Orchards. Hort. Rev. **1997**, *21*, 105–131.
8. Jamieson, P.D. Crop Responses to Water Shortages. J. Crop Prod. **1999**, *2*, 71–83.
9. Kirda, C.; Kanber, R. Water, No Longer a Plentiful Resource, Should be Used Sparingly in Irrigated Agriculture. In *Crop Yield Responses to Deficit Irrigation*; Kirda, C., Moutonnet, P., Hera, C., Nielsen, D.R., Eds.; Kluwer Academic Publishers: London, 1999; 1–20.
10. Heuer, B.; Nadler, A. Physiological Parameters, Harvest Index and Yield of Deficient Irrigated Cotton. J. Crop Prod. **1999**, *2*, 229–239.
11. Varlev, I.; Dimitrov, P.; Popova, Z. Irrigation Scheduling for Conjunctive Use of Rainfall and Irrigation Based on Yield–Water Relationships. Water Rep. **1996**, *8*, 205–214.
12. Fardad, H.; Pessarakli, M. Biomass Production and Water Use Efficiency of Barley and Wheat Plants with Different Irrigation Intervals at Various Water Levels. J. Plant Nutr. **1995**, *18*, 2643–2654.

Plants, Critical Growth Periods

Tessa Marie Mills
*The Horticulture and Food Research Institute of New Zealand
(HortResearch), Palmerston North, New Zealand*

M. Hossein Behboudian
Massey University, Palmerston North, New Zealand

INTRODUCTION

Critical growth periods may be defined as stages in a plant's development when active growth occurs. Plants are generally sensitive to water deficit (WD) during critical growth periods. Plant WD implies that the plant water status is less than the optimum value for growth and development. Plant demand for water is influenced by many factors including the evaporative demand of the aerial environment, soil characteristics, plant characteristics, and growth stage.[1]

Most plants experience WD at some stage during their life cycle. It may occur diurnally when the evaporative demand for water by the atmosphere is greater than the plant's ability to draw water from the soil, or seasonally as soil moisture is depleted due to transpiration and not renewed.[2] Water deficit typically results in a reduction in transpiration due to stomatal closure. If severe, this may also reduce photosynthetic rate and therefore carbon acquisition. This, in turn, may lead to reduced plant growth and biomass accumulation. The sensitivity of plant growth and the reduction in biomass accumulation during periods of WD are dependent on plant's developmental stage. Water deficit imposed during periods of active growth will impact upon the growth of the developing organ as cell turgor is required for cell growth.[3] Generally, however, not all organs of a plant undergo active growth at the same time. This results in different organs within the plant having different critical growth periods during the growing season. This difference allows the manipulation of plant WD to target particular plant organs with little carry over influence of WD on other plant parts. This phenomenon is demonstrated in Fig. 1,[4] where the clear separation of maximum shoot and fruit growth periods are illustrated for peach and pear.

MANIPULATION OF WATER AT CRITICAL GROWTH PERIODS—DEFICIT IRRIGATION

Water shortage is a major constraint to agricultural production in many areas of the world.[5] Additionally, 85% of the global water resource is used for irrigation of agricultural lands.[6] With increasing pressure on water as a scarce resource, there is a requirement to develop irrigation strategies that increase water use efficiency by plants to ensure productivity and reduced water usage. By understanding critical growth periods for separate plant organs, researchers are able to maximize irrigation efficiency and to minimize any detrimental impacts of WD on plant productivity.[7] Such irrigation strategies have been termed regulated deficit irrigation (RDI). Regulated deficit irrigation was initially developed to control shoot growth in peach trees (*Prunus persica*)[8] and thereby reduce pruning requirements and allow increased carbohydrate partitioning to fruit. Other benefits from RDI have now been realized including enhanced fruit quality[7] or greater economic yield.[9] Although RDI, if applied correctly, can increase crop profitability by reducing water use and increasing economic return, it may also have detrimental effects if applied at inappropriate times. For example, in apple (*Malus domestica*), WD at the time of flower initiation (early summer) may have a little impact on the crop currently being carried. However, the flower initiation may be disrupted, which results in a poor return bloom and low crop yields the following season.[10]

The impact of periodic WD varies enormously among plant types; thus, it is appropriate that annual crops and perennial crops be discussed in more detail separately. Annual crops will have no carry forward consequence of WD from the current season in subsequent seasons whereas perennial crops may.

CRITICAL GROWTH PERIODS IN ANNUAL CROPS

Water deficit studies have been conducted on many annual crops including peanut (*Arachis hypogaea*),[11] rice (*Oryza sativa*),[12] pearl millet (*Pennisetum glaucum*),[13] wheat (*Triticum aestivum*),[14] faba bean (*Vica faba* L.),[15] maize (*Zea mays* L.),[16] and pea (*Pisum sativum*).[17] The impact of such WD on crop quality

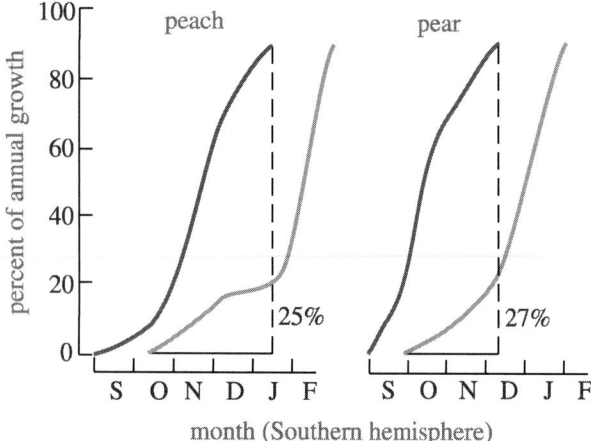

Fig. 1 The cumulative growth of annual shoots and fruit of "Golden Queen" peaches and "Bartlett" pears expressed as a proportion of total seasonal growth by these organs. (From Ref. [4].)

and yield is highly dependent on the timing of the WD and how this aligns with critical growth periods. In wheat, the final yield was reduced whenever WD was imposed. However, the total reduction in yield was the greatest when WD was imposed during flowering. Water deficit at grainfill and tillering has much less impact on final yield. Differences in response to WD also occurred depending on wheat genotype. Water deficit during flowering in peanut resulted in fewer flowers produced per plant. However, a higher percentage of the flowers produced set fruit in plants having suffered WD, which ultimately increased peanut's total yield. The mechanisms for this increase in total yield are thought to be an increase in the promotion of root formation under drought as well as an inhibition of excessive vegetative growth. These two studies illustrate the variable nature of plant response at critical growth periods both within and between annual species.

CRITICAL GROWTH PERIODS IN PERENNIAL CROPS

Periodic WD has been used as a management tool in perennial crops for many years. Carry over influences of WD at critical growth periods are common in perennial crops and may modify plant morphology and increase plant's ability to withstand subsequent drought. Examples for major deciduous fruit crops are cited by Behboudian and Mills.[7] Again the impact of such periodic WD is strongly dependent on crop type and the productivity component of interest. For example, fruit yield may be reduced in apricot if WD is imposed during the late rapid fruit growth stage but is unaffected if WD is induced

during the early rapid fruit growth stage as compensatory fruit growth occurs upon re-watering. Fruit quality attributes may also be modified and could result in increased yield under WD. For example, olive oil production was increased in plants subjected to WD following pit hardening.

CONCLUSIONS

Water deficit has its main effects during active periods of growth. It modifies plant organs differentially during a life cycle, depending upon how fast they are growing. Consequently, no generalizations can be made in regard to plant responses to WD at critical growth periods. The phenomenon of critical growth periods does, however, provide an opportunity for crop manipulation and enhanced crop performance under reduced irrigation strategies. With increasing pressure on water as a resource, strategies such as RDI need to be understood and embraced.

REFERENCES

1. Tolk, J.A.; Howell, T.A.; Steiner, J.L.; Evett, S.R. Grain Sorghum Growth, Water Use, and Yield in Contrasting Soils. Agric. Water Manag. **1997**, *35*, 29–42.
2. Bray, E. Plant Responses to Water Deficit. Trends Plant Sci. **1997**, *2*, 48–54.
3. Cosgrove, D.J. Wall Extensibility. Its Nature, Management and Relationship to Plant Cell Growth. New Phytol. **1993**, *124*, 1–23.
4. Chalmers, D.J. A Physiological Examination of Regulated Deficit Irrigation. N.Z. Agric. Sci. **1989**, *23*, 44–48.
5. Zhang, H.; Oweis, T. Water–Yield Relations and Optimal Irrigation Scheduling of Wheat in the Mediterranean Region. Agric. Water Manag. **1999**, *38*, 195–211.
6. Van Schilfgaarde, J. Irrigation—A Blessing or a Curse. Agric. Water Manag. **1994**, *25*, 203–219.
7. Behboudian, M.H.; Mills, T.M. Deficit Irrigation in Deciduous Orchards. Hort. Rev. **1997**, *21*, 105–131.
8. Chalmers, D.J.; Mitchell, P.D; van Heek, L. Control of Peach Tree Growth and Productivity by Regulated Water Supply, Tree Density and Summer Pruning. J. Am. Soc. Hort. Sci. **1981**, *106*, 307–312.
9. Motilva, M.J.; Romero, M.P.; Alegre, S.; Girona, J. Effect of Regulated Deficit Irrigation in Olive Oil Production and Quality. Acta Hort. **1999**, *474*, 377–380.
10. Hsiao, T.C. Growth and Productivity of Crops in Relation to Water Stress. Acta Hort. **1993**, *335*, 137–148.
11. Nautiyal, P.C.; Ravindra, V.; Zala, P.V.; Joshi, Y.C. Enhancement of Yield in Groundnut Following Imposition of Transient Soil-Moisture-Deficit Stress During the Vegetative Phase. Exp. Agric. **1999**, *35*, 371–386.
12. Boonjung, H.; Fukai, S. Effects of Soil Water Deficit at Different Growth Stages on Rice Growth and Yield Under

Upland Conditions. 2. Phenology, Biomass Production and Yield. Field Crops Res. **1996**, *48*, 47–55.

13. Winkel, T.; Renno, J-F.; Payne, W.A. Effect of the Timing of Water Deficit on Growth, Phenology and Yield of Pearl Millet (*Pennisetum Glaucum* (L.) R. Br.) Grown in Sahelian Conditions. J. Exp. Bot. **1997**, *48*, 1001–1009.

14. Ravichandran, V.; Mungse, H.B. Response of Wheat to Moisture Stress at Critical Growth Stages. Ann. Plant Physiol. (Published by Forum for Plant Physiologist, Akola, India) **1997**, *11*, 208–211.

15. Mwanamwenge, J.; Loss, S.P.; Siddique, K.H.M.; Cocks, P.S. Effect of Water Stress During Floral Initiation, Flowering and Podding on the Growth and Yield of Faba Bean (*Vicia faba* L.). Eur. J. Agron. **1999**, *11*, 1–11.

16. Stone, P.J.; Wilson, D.R.; Gillespie, R.N. Water Deficit Effects on Growth, Water Use and Yield of Sweet Corn. Proc. Agron. Soc. N.Z. **1997**, *27*, 45–50.

17. Fougereux, J-A.; Doré, T.; Ladonne, F.; Fleury, A. Water Stress During Reproductive Stages Affects Seed Quality and Yield of Pea (*Pisum sativum* L.). Crop Sci. **1997**, *37*, 1247–1252.

Plants, Osmotic Adjustment

James M. Morgan
*Tamworth Centre for Crop Improvement, Tamworth,
New South Wales, Australia*

INTRODUCTION

Plant growth in general and crop production in particular normally occur in environments characterized by fluctuations in soil water supply and evaporative demand, which produce water stress of varying duration. Maintenance of metabolic processes usually depends upon minimization of water loss from cells, and this is often accomplished by intracellular accumulation of solutes such as potassium, amino acids (or derivatives), and sugars.[1–3] This type of adaptational response occurs widely in plants, fungi, micro-organisms as well as some animal cells[1,2] and is usually known as osmoregulation, osmotic regulation, or osmotic adjustment. These terms are etymologically identical ("to adjust" means "to regulate" or "conform to a standard"), and are correctly used interchangeably though some have argued for a distinction.[4,5] The expressions can be taken to mean controlled (e.g., regulated by specific genes) change or maintenance of osmotica; generally the former in plants and the latter in animals, though in each, cell hydration is maintained to varying degrees. This may be expressed as volume maintenance (particularly in wall-less cells such as plant gametes and marine algae) or turgor maintenance (walled cells). Osmotic adjustment is also used to describe solute accumulations in plants where the control system (either genetical or physiological) is not clearly understood.

KEY CONCEPTS

The origin of relationships and theory may be found in earlier works.[6–9] At equilibrium, the cell water potential (φ_c) equals the water potential outside the cell (φ_e), and is the sum of the osmotic (π), pressure (P), and matric (τ) potentials.

$$\varphi_e = \varphi_c = \pi + P + \tau \qquad (1)$$

Water stress occurs when φ_e is reduced through, for example, increased water deficits in the soil or atmosphere or increased salinity (Eq. 6). This may occur over short (e.g., diurnal) or longer time periods. A decrease in φ_e from zero (or a higher) (1) to a lower (2) level causes changes in π and P (assuming τ to be negligible—difficult

to measure). The new value of π will be

$$\pi_2 - \frac{n_2 RT}{V_2} - \frac{(n_1 + n_a)RT}{V_2} = \frac{\pi_1 V_1}{V_2} + \pi_a \qquad (2)$$

where V is the osmotic volume, n is the number of solute molecules, R the gas constant, T the absolute temperature, and subscript a indicates accumulation. The osmotic adjustment ($\Delta\pi_a$ or $-\pi_a$) is the difference between the osmotic potential attributable to concentration of π_1 by dehydration and the measured value, π_2 (Eq. 5).[4,8–10] The relative water content, $\zeta(\approx V_2/V_1)$, is normally used instead of V to calculate solute accumulation for plant tissue. At a particular stress level, its value depends upon the degree of osmoregulation according to

$$\frac{V_2}{V_1} = \frac{n_2}{n_1}\frac{\pi_1}{\pi_2} \qquad (3)$$

When n does not change, ζ is inversely related to π. This relationship (or Eq. 2) has been used widely to evaluate lines for genetical and yield studies in crop plants, often using a log transformation to test for linearity or ideal behavior.[6,8]

Two osmotic components are therefore important in influencing hydration and turgor; the initial osmotic potential, π_1 (i.e., at $\varphi_e = 0$ or $\zeta = 100\%$), and the solute accumulation, $\Delta\pi_a$. Both can be sources of adaptation to water stress in plants. In the short term, solutes accumulated during a stress episode may be retained after rehydration and this produces a decrease in π_1 ($\Delta\pi_{100}$). This is periodically used as a way of measuring osmotic adjustment, though the precise nature of the relationship with $\Delta\pi_a$ has not been well established experimentally. Repeated diurnal increases in $\Delta\pi_a$ due to fluctuations in vapor pressure deficit do not seem to produce long-term cumulative increases in $\Delta\pi_{100}$. The decrease in π_1, which is a "hardening" reaction, seems capable, therefore, of only limited variation. It is evident in plants that have been prestressed,[10] or in comparisons of glasshouse- and field-grown plants.[11] The value of $\Delta\pi_{100}$ is also calculated using π_1 and π_2, assuming no change in n_2.[4] It is compared with $\Delta\pi_a$ in Eqs. 4 and 5.

$$\Delta\pi_{100} = \frac{(n_2 - n_1)RT}{V_1} \qquad (4)$$

Encyclopedia of Water Science
DOI: 10.1081/E-EWS 120010167

while

$$\Delta \pi_a = \frac{(n_2 - n_1)RT}{V_2} \qquad (5)$$

Thus, $\Delta \pi_{100} = \Delta \pi_a \zeta$.

COMPONENT SOLUTES AND GENETIC CONTROL

Numerous examples exist of genotypic variation between, and within plant species, including rice, wheat, sorghum, field peas, barley, and chickpeas.[4,10] There are, however, few instances where genetic control has been identified and pleiotropic effects investigated. Molecular approaches involving compatible solutes have not produced clearly demonstrable increases in osmotic adjustment.[12–14] However alterations in proline, betaine, and mannitol have been associated with differences in growth under saline conditions.[12,14] Greater success has been achieved using a phenocentric approach.[15]

In wheat, initial identification of large genotypic differences in the 1970s led ultimately to location of a gene (or) on chromosome 7AS.[8,16,17] There is evidence from mapping and breeding work of close linkage of or with an endosperm peroxidase, Per A4, locus, producing alterations in dough strength.[17,18] The gene effect is semiqualitative in that in leaves, responses of osmotic potential to water potential follow very different pathways.[8,11] Generally, genotypes with Or (dominant allele) show low or 0 $\Delta \pi_a$ with decline in φ_e from 0 MPa to near 0 P. Below this $\Delta \pi_a$ increases. Genotypes with or accumulate solutes from the commencement of stress (Fig. 1) with potassium the dominant component (and with some amino acid contribution).[19] The gene is also expressed in pollen grains, where it is dependent upon a supply of potassium, and involves volume regulation.[11] The response, which reaches a maximum at approximately 0.2 mM, shows a high affinity for potassium. Pollen expression enables visual identification of homozygous and heterozygous lines, and was used to produce a commercial cultivar in Australia (Mulgara) using back-crossing methods.[11,20] In rice, analysis using similar leaf tests (i.e., based on Eqs. 2 and 3) has identified a quantitative trait locus (QTL) in a region on chromosome 8 that is homoeologous with a region of chromosome 7AS in wheat.[14,21]

Considerable understanding of the osmoregulatory system comes from work on Escherichia coli. The stress-induced K^+ accumulation is controlled by a single operon (kdp) which responds directly to turgor reduction. Potassium accumulation is the primary or initial stress response to restore turgor. Compatible solute accumulation (mainly betaine) is under separate control, and is induced by increased K^+ concentration.[22,23] As external stress increases, betaine progressively replaces K^+ as the main contributor to osmoregulation. The osmoregulatory system seems, therefore, to be controlled by the kdp operon.[23] This accords with the observation of single gene control at positive turgor in wheat. Also in various plant species, compatible solutes such as proline and amino acids tend to accumulate much later in stress development.[9,24–26] There is evidence that the effector of proline accumulation is π or a component of it, rather than φ_e.[27]

GROWTH AND YIELD RESPONSES

The relationship between osmoregulation and growth or yield may be broadly understood in terms of effects of positive turgor and hydration on cell expansion, metabolic processes such as photosynthesis (via stomatal resistance), and synthesis of growth regulators such as abscisic acid, which may in turn affect elongation and seed set.[4,10,28] Growth and yield response are dependent upon environmental conditions which reduce φ_e enough for expression of differences in osmotic adjustment. In leaves, the effect may be broadly or conceptually summarized by

$$\varphi_e = \varphi_s - RE_p \qquad (6)$$

where φ_s is the soil water potential (water supply term), E_p is the evaporation rate (affected by evaporative demand), and R represents the resistance to water flow between the soil and the air.[7,9] In rainshelter experiments with wheat, where soil water supply was constant, both φ_e and difference in P due to osmotic adjustment were linearly related to evaporative demand, E. Growth or yield

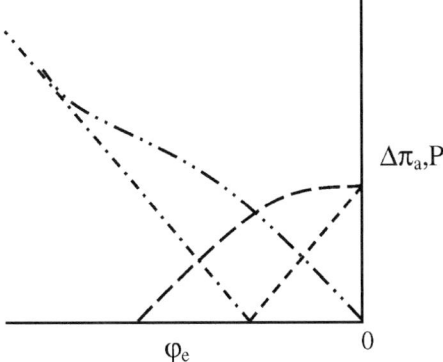

Fig. 1 Types of responses of $\Delta \pi_a$ and P to reductions in leaf water potential, φ_e, due to a single-gene difference in wheat. High osmoregulation (or) $\Delta \pi_a$ (–··–··–), P (– – –), and low osmoregulation (Or) $\Delta \pi_a$ (----), and P (- - -). (From Ref. [16].)

responses may not occur at high soil water deficit if E is low, even though biomass has been substantially reduced while a response can occur at low soil water deficit if E is high.[20] This scenario probably reflects the interaction of leaf area reductions due to soil or root stress,[29] and differences in growth reductions due to differences in leaf turgor responses. With gradual development of soil water deficit, it is probable that leaf area is adjusted to minimize stress in leaves. Lines with differing *or* alleles do not show differences in this effect. Where the effect of the *or* gene was measured using both recombinant inbred lines and backcross-bred lines, yield response (H/L, where H is the yield of lines with high osmoregulation, and L the yield of lines with low osmoregulation) forms a close ($r = 0.92$), simple relationship with water supply (S) and evaporative demand (E) when E/S is > 1 (Fig. 2). For values below 1, $H/L = 1$.[20] Evidence of positive yield associations has also been found in lines of field peas and chickpeas.[30,31] In these studies, assessments of osmoregulation were based on Eq. 5, though positive relationships based on $\Delta \pi_{100}$ have also been found in sorghum.[32] Other attempts to associate genetic variability in solute accumulation with yield have mostly used cultivars or lines of differing genetic backgrounds to establish correlations, with questionable results as a consequence.[10,27,33]

CONCLUSIONS

As an adaptational trait, osmotic adjustment has proved effective in the improvement of wheat yields, by use of backcrossing techniques with gene identification in pollen grains. Modeling work suggests wide potential for yield improvement in environments where evaporative demand exceeds soil water supply during crop growth.[20] In this work, evaluation at a tissue/cell level has been a significant factor. From here it is possible to work "up" to growth and "down" to biochemistry. Correctly characterizing the osmoregulatory response to water stress is important, as it affects understanding of the chemistry, genetics, and yield relationships. A similar approach may prove productive in other crop species where genetic variation exists, especially in the *Gramineae*. However, success in some species may require an approach which combines molecular (e.g., gene markers) and phenotypically based techniques.[14,15]

REFERENCES

1. Kauss, H. Biochemistry of Osmotic Regulation. Plant Biochem. II **1977**, *13*, 119–140.
2. Hellebust, J.A. Osmoregulation . Ann. Rev. Plant Physiol. **1976**, *27*, 485–505.
3. Cram, W.J. Negative Feedback Regulation of Transport in Cells. The Maintenance of Turgor, Volume and Nutrient Supply. In *Encyclopedia of Plant Physiology, New Series*; Luttge, U., Pitman, M.G., Eds.; Springer-Verlag: Berlin, 1976; Vol. 2A, 284–316.
4. Turner, N.C.; Jones, M. Turgor Maintenance by Osmotic Adjustment. A Review and Evaluation. In *Adaptation of Plants to Water and High Temperature Stress*; Turner, N.C., Kramer, P.J., Eds.; Wiley–Interscience: New York, 1980; 155–172.
5. Munns, R. Why Measure Osmotic Adjustment? Aust. J. Plant Physiol. **1988**, *15*, 717–726.
6. Gardner, W.R.; Ehlig, C.F. Physical Aspects of the Internal Water Relations of Plant Leaves. Plant Physiol. **1965**, *40*, 705–710.
7. Slatyer, R.O. *Plant–Water Relationships*; Academic Press: New York, 1967.
8. Morgan, J.M. Differences in Osmoregulation Between Wheat Genotypes. Nature **1977**, *270*, 235.
9. Kramer, P.J. *Water Relations of Plants*; Academic Press: New York, 1983.
10. Morgan, J.M. Osmoregulation and Water Stress in Higher Plants. Ann. Rev. Plant Physiol. **1984**, *35*, 299–319.
11. Morgan, J.M. Pollen Grain Expression of a Gene Controlling Differences in Osmoregulation in Wheat Leaves. Aust. J. Agric. Res. **1999**, *50*, 953–962.
12. Kishor, K.P.B.; Hong, Z.; Miao, G.-H.; Hu, C.-A.A.; Verma, D.P.S. Overexpression of Δ'-Pyrroline-5-Carboxylate Synthetase Increases Proline Production and Confers Osmotolerance in Transgenic Plants. Plant Physiol. **1995**, *108*, 1387–1394.
13. Blum, A.; Munns, R.; Passioura, J.B.; Turner, N.C. Genetically Engineered Plants Resistant to Soil Drying and Salt Stress: How to Interpret Osmotic Relations? Plant Physiol. **1996**, *110*, 1051–1053.
14. Zhang, J.; Nguyen, H.; Blum, A. Genetic Analysis of Osmotic Adjustment in Crop Plants. J. Exp. Bot. **1999**, *50*, 291–302.

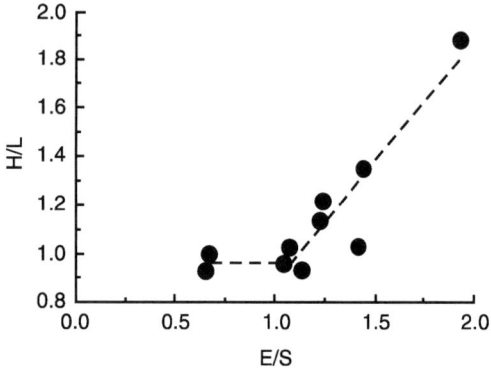

Fig. 2 Yield increase due to a gene conditioning high osmoregulation in wheat (H/L) is simply related to the ratio of evaporative demand (E) to soil water supply (S) for the period of net positive growth of the crop in the field. (From Ref. [20].)

15. Miflin, B. Crop Improvement in the 21st Century. J. Exp. Bot. **2000**, *51*, 1–8.

16. Morgan, J.M. A Gene Controlling Differences in Osmoregulation in Wheat Leaves. Aust. J. Plant Physiol. **1991**, *18*, 249–257.

17. Morgan, J.M.; Tan, M.K. Chromosomal Location of a Wheat Osmoregulation Gene Using RFLP Analysis. Aust. J. Plant Physiol. **1996**, *23*, 803–806.

18. Morgan, J.M. Changes in Rheological Properties and Endosperm Peroxidase Activity Associated with Breeding for an Osmoregulation Gene in Bread Wheat. Aust. J. Agric. Res. **1999**, *50*, 963–968.

19. Morgan, J.M. Osmotic Components and Properties Associated with Genotypic Differences in Osmoregulation in Wheat. Aust. J. Plant Physiol. **1992**, *19*, 67–76.

20. Morgan, J.M. Increases in Grain Yield of Wheat by Breeding for an Osmoregulation Gene: Relationship to Water Supply and Evaporative Demand. Aust. J. Agric. Res. **2000**, *51*, 971–978.

21. Lilley, J.M.; Ludlow, M.M.; McCouch, S.R.; O'Toole, J.C. Locating QTL for Osmotic Adjustment and Dehydration Tolerance in Rice. J. Exp. Bot. **1996**, *47*, 1427–1436.

22. Epstein, W.; Davies, M. Potassium-Dependant Mutants of *Escherichia coli* K-12. J. Bacteriol. **1970**, *101*, 836–843.

23. Higgins, C.F.; Cairney, J.; Stirling, D.A.; Sutherland, L.; Booth, I.R. Osmotic Regulation of Gene Expression: Ionic Strength as an Intracellular Signal. Trends Biochem. Sci. **1987**, *12*, 339–344.

24. Hsiao, T.C. Plant Responses to Water Stress. Ann. Rev. Plant Physiol. **1973**, *24*, 519–570.

25. Meyer, R.F.; Boyer, J.S. Osmoregulation, Solute Distribution, and Growth in Soybean Seedlings Having Low Water Potentials. Planta **1981**, *151*, 482–489.

26. Munns, R.; Brady, C.J.; Barlow, E.W.R. Solute Accumulation in the Apex and Leaves of Wheat During Water Stress. Aust. J. Plant Physiol. **1979**, *6*, 379–389.

27. Hanson, A.D.; Hitz, W.D. Metabolic Responses of Mesophytes to Plant Water Deficits. Ann. Rev. Plant Physiol. **1982**, *33*, 163–203.

28. Santakumari, M.; Berkowitz, G.A. Correlation Between the Maintenance of Photosynthesis and In Situ Protoplast Volume at Low Water Potentials in Droughted Wheat. Plant Physiol. **1990**, *92*, 733–739.

29. Passioura, J.B. Root Signals Control Leaf Expansion in Wheat Seedlings Growing in Drying Soil. Aust. J. Plant Physiol. **1988**, *15*, 687–693.

30. Rodríguez-Maribona, B.; Tenorio, J.L.; Conde, J.R.; Ayerbe, L. Correlation Between Yield and Osmotic Adjustment of Peas (*Pisum sativum* L.) Under Drought Stress. Field Crops Res. **1992**, *29*, 15–22.

31. Morgan, J.M.; Rodríguez-Maribona, B.; Knights, E.J. Adaptation to Water-Deficit in Chickpea Breeding Lines by Osmoregulation: Relationship to Grain Yields in the Field. Field Crops Res. **1991**, *27*, 61–70.

32. Tangpremsri, A.; Fukai, S.; Fischer, K.S. Growth and Yield of Sorghum Lines Extracted from a Population for Differences in Osmotic Adjustment. Aust. J. Agric. Res. **1995**, *46*, 61–74.

33. González, A.; Martín, I.; Ayerbe, L. Barley Yield in Water-Stress Conditions. The Influence of Precocity, Osmotic Adjustment and Stomatal Conductance. Field Crops Res. **1999**, *62*, 23–34.

P

Plants, Osmotic Potential

Mark E. Westgate
Iowa State University, Ames, Iowa, U.S.A.

INTRODUCTION

Inherent in the term "osmotic potential" is a measure of the capacity to do work. Typically in plant systems, that work involves the movement of water across cellular membranes and tissues for hydration, expansion growth, leaf movements, and stomatal opening. The osmotic potential of a cell is determined primarily by the concentration of solutes confined within the symplastic water volume (cytoplasm + vacuole). Therefore, solute transport across cellular membranes and cellular metabolism are essential components of osmotic regulation. Passive concentration of solutes due to dehydration or inhibition of metabolism under severe environmental conditions also can contribute to the "adjustment" of cellular osmotic potential. Whether active or passive, the accumulation of solutes has been shown to support expansion growth, maintain photosynthesis, and improve reproductive success under severe drought conditions in a number of plant systems. This article presents a brief overview of the physical origin of the term osmotic potential and its application to quantifying the response of plants to changing environmental conditions.

PHYSICAL DEFINITION OF OSMOTIC POTENTIAL

Solutes and Free Energy

The total amount of energy in the water molecules within a cell is partitioned between the energy associated with the molecular structure of the water, and energy that can be exchanged with the surroundings. Energy "tied up" in structure is the entropy, and the "exchangeable energy" is the free energy available to do work. Gibbs[1] defined this free energy term for any component of a system as its chemical potential, μ_j, which is a measure of the amount of work a mole of the component j can do. In the case of water, the chemical potential is given by

$$\mu_w = \mu_w^* + RT \ln(a_w) + \bar{V}_w P + Z_w FE + m_w gh \qquad (1)$$

where μ_w is the chemical potential of water in the system ($J\,mol^{-1}$), μ_w^* is the chemical potential of the reference state for water ($J\,mol^{-1}$), R is the gas constant ($J\,mol^{-1}\,K^{-1}$), T is temperature (K), a_w is the activity of water (dimensionless), $\bar{V}_w$ is the partial molar volume of water ($m^3\,mol^{-1}$), P is the hydrostatic pressure (MPa), Z_w is the charge number of water (dimensionless), F is Faraday's constant ($C\,mol^{-1}$), E is the electrical potential (mV), m_w is the mass per mole of water (g), g is gravitational acceleration ($\sim 9.8\,m\,sec^{-2}$), and h is vertical height (m). This equation formalizes a number of important points relative to the chemical potential of water in plant cells. First, the chemical potential cannot be measured directly, but is evaluated relative to an unknown reference energy state, μ_w^*. Second, it is affected by a number of physical factors, such as the presence of solutes ($RT \ln(a_w)$), atmospheric pressure ($\bar{V}_w P$), electrical charge ($Z_w FE$), and elevation ($m_w gh$). Third, chemical potential is quantified on a molar basis. The activity of water, $a_w = \gamma_w N_w$, is a product of the activity coefficient of water, γ_w (concentration^{-1}), and the mole fraction of water, $N_w = n_w/(n_w + n_s)$, where n_w and n_s are moles of water and solute, respectively. The partial molar volume of water, $\bar{V}_w = \partial V/\partial n_w$, describes the volume change associated with a change in the number of moles of water in the system. When solute is added to water, the free energy of the water per unit volume of the system decreases because the solute occupies space previously occupied by water. The additional solutes decrease the mole fraction of water, thereby decreasing the water activity term, $RT \ln(a_w)$.

Although the chemical potential of water in plants cannot be measured directly, it can be measured against a standard state (μ_w^0), which is defined as pure water, at atmospheric pressure, and at the reference temperature and gravitational level of the system. For pure water at atmospheric pressure, $P = 0$, $N_w = 1$, $z_w = 0$, and $h = 0$. Therefore, the standard state and the reference state for the chemical potential are equivalent.

$$\mu_w^0 = \mu_w^* + RT \ln(1) + \bar{V}_w(0) + (0)FE + m_w g(0)$$

$$= \mu_w^* \qquad (2)$$

Substituting μ_w^0 into Eq. 1, dividing by $\bar{V}_w$, and

Encyclopedia of Water Science
DOI: 10.1081/E-EWS 120010168

rearranging,

$$\Psi_w = (\mu_w - \mu_w^0)/\bar{V}_w$$

$$= (RT \ln a_w)/\bar{V}_w + P + (Z_w FE)/\bar{V}_w$$

$$+ (m_w gh)/\bar{V}_w \tag{3}$$

Eq. 3 defines the "water potential," Ψ_w, which is the maximum amount of work the water molecules in the system can do, relative to the standard of pure water.[1] By expressing the water potential per unit volume, $\bar{V}_w$, the water potential can be measured in units of pressure. Because water has no electrical charge ($Z_w = 0$), and gravitational effects on water are small (except in very tall trees!), the Ψ_w of plant cells is determined primarily by osmotic forces and pressure. So Eq. 3 can be simplified.

$$\Psi_w = (RT \ln a_w)/\bar{V}_w + P = \Psi_s + \Psi_p \tag{4}$$

The osmotic forces are measured in the $(RT \ln a_w)/\bar{V}_w$ term in Eq. 4, and referred to as the osmotic potential, Ψ_s. As discussed below, this term includes solute and matric effects on water activity. Pressure effects on Ψ_w are measured in the second term on the right side of Eq. 4, which is referred to as the "pressure potential," Ψ_p. This

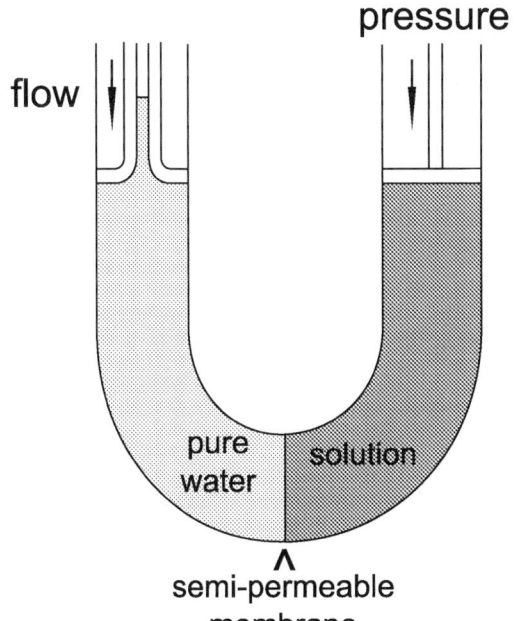

Fig. 1 Typical vapor pressure osmometer with a semipermeable membrane separating a chamber containing a solution from one containing pure water. Water flow from the pure water chamber across the membrane is monitored in the capillary at the top of the chamber. The amount of pressure applied to the chamber containing the solution to prevent water movement into it is equal to the solution osmotic pressure, π. (From Ref. [2], p. 33.)

term includes pressure generated internally by plant cells (turgor) and atmospheric pressure.

Osmotic Potential vs. Osmotic Pressure

In an ideal system with no solute/solvent interactions, the presence of solutes in water "dilutes" the chemical potential of the water because solute occupies space normally occupied by water. This dilution decreases the vapor pressure of the water at the surface of the solution. The decrease in vapor pressure is given by Raoult's law (Eq. 5), which states that for dilute, ideal solutions the vapor pressure in equilibrium with a dilute solution is proportional to the mole fraction of the solvent, in this case water, N_w. That is, the vapor pressure decreases with N_w as solute is added.

$$e = e_o N_w = e_o(n_w/(n_w + n_s)) \tag{5}$$

where e is the vapor pressure of the solution, e_o is the vapor pressure of the pure water, N_w is the mole faction of water, n_w and n_s are moles of water and solute, respectively. The vapor pressure of dilute solutions is measured relative to pure water in "osmometers" like the one shown in Fig. 1. When pure water is separated from a solution by a membrane that is permeable to water, but not the solute, water moves across the membrane from the solution of higher chemical potential (larger N_w or higher Ψ_s) to the solution of lower chemical potential (smaller N_w or lower Ψ_s). This would be from the "pure water" side to the "solution" side in Fig. 1. Pressure applied to the solution side will stop the flow of water across the membrane. The amount of pressure that must be applied to prevent water movement is the "osmotic pressure" of the solution, relative to pure free water. The osmotic pressure, generally denoted as π, is present only when a balancing pressure is applied to the osmometer. On the other hand, the physical property responsible for the water movement across the membrane, the osmotic potential Ψ_s, is always characteristic of the solution. As more solute is added to the solution, its osmotic pressure increases according to the fundamental definition of osmotic pressure,

$$RT \ln a_w = -\bar{V}_w \pi \tag{6}$$

The Ψ_s of the solution, however, decreases since the addition of solutes decreases the mole fraction and, therefore, the activity of water according to Eq. 4. Thus, $\Psi_s = -\pi$. This mathematical equivalence has prompted many to use the terms osmotic pressure and osmotic potential interchangeably. Eq. 4 describes the general relationship between water potential and its component osmotic and pressure potentials ($\Psi_w = \Psi_s + \Psi_p$). The equation often used in place of Eq. 4 to quantify the components of water potential in plant cells and tissues is

$\Psi_w = -\pi + P$, which has led to some confusion in the literature. This confusion has resulted primarily from the failure of authors to appreciate the distinction between osmotic pressure and osmotic potential, lack of regard for the proper sign convention for $-\pi$, and failure to recognize that P is cell turgor only when atmospheric pressure equals zero.

Osmotic and Matric Forces

Water molecules associated with the surfaces of colloidal particles, membranes, or cell walls have a decreased tendency to interact with water in the bulk solution. This interaction decreases the water activity (a_w) near these surfaces, as does the presence of solutes in the bulk solution. Such surface interactions do not alter the mole fraction of water, N_w, but they do decrease the activity coefficient of water, γ_w. Nobel[3] suggested that the individual contribution of surface interactions (matric forces) and solvent dilution (osmotic forces) on decreasing a_w could be considered as separate and additive components of the total osmotic potential of a solution. Recalling that $\Psi_s = (RT/\bar{V}_w)\ln a_w$ from Eq. 4, and that

$a_w = \gamma_w N_w$, we have

$$\Psi_s = (RT/\bar{V}_w)\ln(\gamma_w N_w)$$

$$= (RT/\bar{V}_w)\ln\gamma_w + (RT/\bar{V}_w)\ln N_w = \Psi_m + \Psi_s^* \quad (7)$$

where Ψ_m accounts for the effects of matric forces expressed through their impact on the activity coefficient, γ_w, and Ψ_s^* accounts for osmotic forces expressed solely via their impact on the mole fraction of water, N_w. This analysis assumes that these matric and osmotic forces are independent, and is not intended to define matric forces in all situations. But it is a useful approach to consider how matric and osmotic forces vary in tissues such as seeds, which undergo extensive dehydration and surface interactions begin to dominate water activity of the tissue.

PLANT CELLS AS OSMOMETERS

For dilute and ideal solutions of nondissociating solutes, the relationship between the chemical potential and the mole fraction of water can be approximated by the van't Hoff equation (Eq. 8)

$$\psi_s \sim -RTC_s \quad \text{or} \quad \pi \sim RTC_s \quad (8)$$

where C_s is the molar concentration of solute (mol m^{-3}), R is the gas constant (m^3 MPa mol^{-1} K^{-1}), and T is the absolute temperature (K). This relationship indicates that ψ_s is proportional to the solute concentration at constant temperature and pressure. Thus, if plant cells behave as perfect osmometers, i.e. no solutes cross the plasmalemma as water leaves or enters the symplasm, a known change in cell volume should lead to a defined change in cell ψ_s. Pressure–volume curves, such as shown in Fig. 2, generated by forcing water out of plant cells by pressurizing the atmosphere around them,[4] generally confirm that this relationship holds for plant tissues. They also provide information about several osmotic parameters of plant tissues. For example, an estimate of the osmotic potential at full cell hydration, ψ_s^0, and the symplasm volume of the tissue can be obtained by extrapolating the linear portion of the curve to the Y-axis and X-axis, respectively. The latter estimate, of course, assumes no water is expressed from the cell walls at high atmospheric pressures.

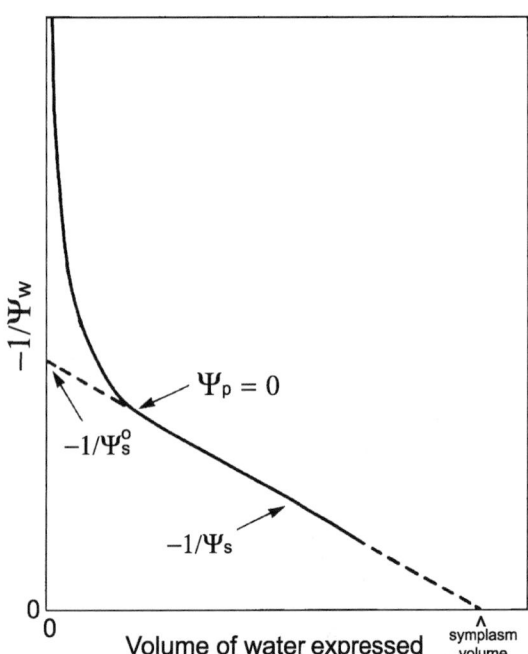

Fig. 2 A modified Höfler diagram relating the change in tissue water potential (Ψ_w) to the volume of water expressed from the tissue. The initial change in Ψ_w is due to a decrease in Ψ_p and Ψ_s. Once Ψ_p reaches zero (incipient plasmolysis), further dehydration results in a linear decrease in Ψ_s. Extrapolating the linear portion of the curve to the Y-axis and X-axis provides an estimate of the osmotic potential at full hydration (and turgor), Ψ_s^0, and the volume of the symplasm. (From Ref. [3], p. 90.)

CONTROL OF SOLUTE ACCUMULATION

Under well-watered conditions, values for tissue ψ_s generally range from -0.8 MPa to -1.3 MPa (Table 1). Less negative values can occur in stem and root tissues; much more negative values are observed in seeds as they

Table 1 Typical osmotic potential values for various plant tissues. Tissues were sampled from plants grown under well watered (WW) or water stressed (WS) conditions. Osmotic potentials were measured on whole tissues or cell sap expressed from them

Tissue source			Plant water status	Osmotic potential (MPa)	Reference
Maize	Leaf	Mature	WW	− 0.9	[8]
		Mature	WS	− 1.3	[8]
		Expanding	WW	− 0.8	[8]
		Expanding	WS	− 1.5	[8]
	Root	Mature	WW	− 0.6	[8]
		Mature	WS	− 0.9	[8]
		Expanding	WW	− 0.9	[8]
		Expanding	WS	− 1.6	[8]
	Stem	Expanding	WW	− 0.6	[8]
		Expanding	WS	− 1.0	[8]
	Ovaries	Expanding	WW	− 1.2	[11]
		Expanding	WS	− 1.5	[11]
	Stigma	Expanding	WW	− 0.9	[12]
		Expanding	WS	− 1.1	[12]
	Kernels	Expanding	WW	− 1.2	[13]
		Expanding	WS	− 1.5	[14]
		Filling	WW	− 0.8	[13]
		Filling	WS	− 1.2	[14]
		Mature	WW	− 3.5	[13]
		Mature	WS	− 5.0	[14]
	Embryos	Expanding	WW	− 2.0	[14]
		Expanding	WS	− 1.8	[14]
		Filling	WW	− 1.8	[14]
		Filling	WS	− 1.5	[14]
		Mature	WW	− 2.6	[14]
		Mature	WS	− 5.0	[14]
Soybean	Flowers	Mature	WW	− 1.3	[15]
		Mature	WS	− 2.0	[15]
		Expanding	WW	− 1.3	[15]
		Expanding	WS	− 2.1	[15]
	Pericarp	Mature	WW	− 1.3	[16]
		Mature	WS	− 1.8	[16]
		Expanding	WW	− 1.2	[15]
		Expanding	WS	− 2.0	[15]
	Embryos	Filling	WW	− 1.2	[16]
		Filling	WS	− 1.2	[16]
Wheat	Leaf	Mature	WW	− 1.6	[17]
		Mature	WS	− 3.2	[17]
	Spikelet	Expanding	WW	− 1.2	[17]
		Expanding	WS	− 2.8	[17]
	Glumes	Expanding	WW	− 1.2	[17]
		Expanding	WS	− 2.8	[17]
	Ovary	Expanding	WW	− 1.3	[17]
		Expanding	WS	− 1.7	[17]
	Anthers	Expanding	WW	− 1.5	[17]
		Expanding	WS	− 1.7	[17]

desiccate during the later stages of development. Values in leaves vary during the day in response to photosynthetic activity. And ψ_s values can vary with development, as tissues increase in volume and differentiate into synthetic or storage organs. Under drought conditions, tissue ψ_s values invariably decrease throughout the plant due to the accumulation and passive concentration of solutes. This phenomenon, investigated by Meyer and Boyer[5] and

Greacen and Oh,[6] has since been termed "osmotic adjustment." Such solute accumulation has been shown to have a positive impact on expansion growth and reproductive success during drought in a number of plant systems.[7–9] Transgenic approaches are now being used to generate plants that overproduce "compatible solutes," which are thought to improve plant tolerance to water deficit stress.[10] Whether this approach will actually lead to new genotypes with increased tolerance to abiotic stresses under field conditions, however, remains to be demonstrated.

REFERENCES

1. Gibbs, J.W. *The Collected Works of J. Willard Gibbs*; Longmans, Green and Co.: New York, 1931; Vol. 1.
2. Kramer, P.J.; Boyer, J.S. *Water Relations of Plants and Soils*; Academic Press: San Diego, CA, 1995.
3. Nobel, P.S. *Physicochemical and Environmental Plant Physiology*; Academic Press: San Diego, CA, 1991.
4. Boyer, J.S. Leaf Water Potentials Measured with a Pressure Chamber. Plant Physiol. **1967**, *42*, 133–137.
5. Meyer, R.F.; Boyer, J.S. Sensitivity of Cell Division and Cell Elongation to Low Water Potentials in Soybean Hypocotyls. Planta **1972**, *108*, 77–87.
6. Greacen, E.L.; Oh, J.S. Physics of Root Growth. Nat. New Biol. **1972**, *235*, 24–25.
7. Meyer, R.F.; Boyer, J.S. Osmoregulation, Solute Distribution, and Growth in Soybean Seedlings Having Low Water Potentials. Planta **1981**, *151*, 482–489.
8. Westgate, M.E.; Boyer, J.S. Osmotic Adjustment and the Inhibition of Leaf, Root, Stem, and Silk Growth at Low Water Potentials in Maize. Planta **1985**, *164*, 540–549.
9. Zinselmeier, C.; Jeong, B.-R.; Boyer, J.S. Starch and the Control of Kernel Number in Maize at Low Water Potentials. Plant Physiol. **1999**, *121*, 25–36.
10. Bohnert, H.J.; Nelson, D.E.; Jensen, R.G. Adaptations to Environmental Stresses. Plant Cell **1995**, *7*, 1099–1111.
11. Schussler, J.R.; Westgate, M.E. Maize Kernel Set at Low Water Potential: II. Sensitivity to Reduced Assimilates at Pollination. Crop Sci. **1991**, *31*, 1196–1203.
12. Schoper, J.B.; Lambert, R.J.; Vasilas, B.L.; Westgate, M.E. Plant Factors Controlling Seed Set in Maize. Plant Physiol. **1987**, *83*, 121–125.
13. Westgate, M.E.; Boyer, J.S. Water Status of the Developing Grain of Maize. Agron. J. **1986**, *78*, 714–719.
14. Westgate, M. Water Status and Development of the Maize Endosperm and Embryo During Drought. Crop Sci. **1994**, *34*, 76–83.
15. Westgate, M.E.; Peterson, C.M. Flower and Pod Development in Water-Deficient Soybeans (*Glycine max* L. Merr.). J. Exp. Bot. **1993**, *44*, 109–117.
16. Westgate, M.E.; Thompson, G.D. Effect of Water Deficits on Seed Development in Soybean. I. Tissue Water Status. Plant Physiol. **1989**, *91*, 975–979.
17. Westgate, M.E.; Passioura, J.B.; Munns, R. Water Status and ABA Content of Floral Organs in Drought-Stressed Wheat. Aust. J. Plant Physiol. **1996**, *23*, 763–772.

Plants, Salt Tolerance of

Michael C. Shannon
*United States Department of Agriculture (USDA),
Riverside, California, U.S.A.*

INTRODUCTION

Salt tolerance is generally defined as the degree to which a plant endures salinity as a stressor. The components of salinity include the composition of the ions in the salt, e.g., sodium, chloride, calcium, and sulfate, and their concentrations. A plant may be exposed to salt continuously or intermittently and salt ions may affect the plant through their effects in the root zone as a component of the soil water or as a salt spray on leaf surfaces. Different plant species, and even sometimes varieties within a species, differ in salt tolerance. In addition, plants differ in their tolerance to salinity exposure depending on their stage of growth and different plant measurements can be taken as an index of salt tolerance. If salt stress becomes too severe, the plant may exhibit varying types of leaf burn and necrosis and will eventually die, but under moderately saline conditions it is very difficult to identify a salt-stunted plant from visual symptoms. Although scientists have studied the effects of salinity on plant growth, metabolism, and biochemistry, only a few genetic markers have been identified that have helped to improve salt tolerance in crops.

SALINITY EFFECTS AND MANAGEMENT

Plant salt tolerance can be described as a change in growth rate, leaf or root elongation rate, germination or emergence rate, and so forth. Salinity-induced growth decrease is measured as a function of salinity exposure. Roots are typically the sites of exposure in saline soils when plants are furrow or drip irrigated with saline water or are grown in saline soils, but foliage can be the site of salt exposure if the plant is subjected to sprinkler irrigation or ocean spray. Hurricanes can carry saline ocean water many miles inland and deposit salts on soils and plants. Another variable that affects salt tolerance is the timing of a plant's exposure to salt. Exposure may be continuous or intermittent with different starting and stopping points with respect to the stage of plant development. These factors, as well as salt composition and concentration, have significant and moderating effects on plant response, which is also strongly dependent on environmental factors

and secondary stressors. Thus, it is not remarkable that despite the existence of thousands of research papers published on salt tolerance of crops and other plants[1] and literally tens of thousands more papers on the effects of salinity on plant growth, morphology, physiology, and biochemistry, there is yet much more information that is needed before a comprehensive, quantitative, and mechanistic explanation of salt tolerance can be proposed.

Salt tolerance of crops is most practically measured as a function of yield decline across a range of salt concentrations. A typical and adequate measure of salt tolerance can be usually formulated on the basis of a two parameter model[2]; the crop salt tolerance threshold (EC_t) and the slope (S) (Fig. 1). The crop salt tolerance threshold, defined as the salinity that is expected to cause the initial significant reduction in maximum expected yield (Y_{max}), is very sensitive to environmental interactions. The measurement of the threshold salinity value depends upon both on the accuracy of the salinity measurements and the method by which the measurements are integrated over time, as well as rooting depth and area, if exposed through the rooting profile. Because of this, there is a high degree of error in evaluating the slope at salt concentrations near the threshold; few salinity studies include enough replications to determine accurately the threshold value. Slope is simply the percentage of the yield that is expected to be reduced for each unit of added salinity above the threshold value. There is a tendency for slope to "tail-off" at the higher salt concentrations.

Soil salinity is usually and most easily measured as the electrical conductivity of a saturated soil paste extract (EC_e), in deciSeimens per meter ($dS\ m^{-1}$). Relative yield (Y) at any salinity exceeding the threshold (EC_t) can be calculated as:

$$Y = Y_{max} - S(EC_e - EC_t)$$

where S is the relative yield decrease per unit salinity increase and EC_t is the salinity threshold. Salt tolerance at high salinity has little economic importance and measurements made at high salt concentrations may disproportionately skew the salt tolerance curve. For these reasons, the numerically most reliable value for crop salt tolerance response studies is the value at which yield is reduced by 50% (C50). The C50-value may still be

Encyclopedia of Water Science
DOI: 10.1081/E-EWS 120010283

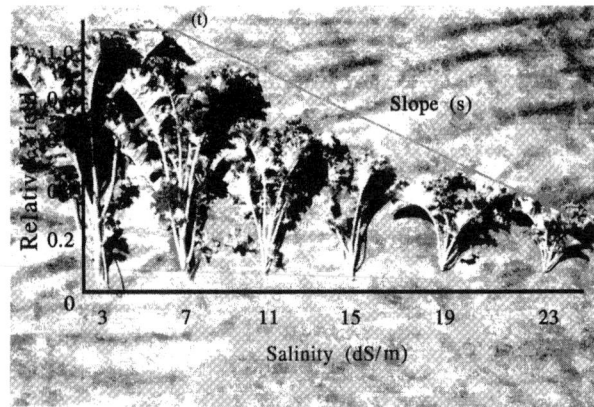

Fig. 1 Measurement of salt tolerance.

estimated when too few data points exist to provide reliable information on the threshold. The set of equations developed by van Genuchten and Hoffman[3] takes advantage of the stability of the C50. The C50 value, together with an empirically-derived *p*-value that characterizes the steepness of the response function, may be obtained by fitting van Genuchten and Hoffman's function to observed salt tolerance response data.

Reliable data to describe the salinity functions can be obtained only from carefully controlled, monitored, and well-replicated experiments conducted across a range of salinity treatments. Data of this type have been compiled for 127 crop species which includes 68 herbaceous crops, 10 woody species, and 49 ornamentals.[4] This is a valuable resource database for growers concerned with the potential hazard of a given saline water or soil.

When high quality water is not available for irrigation and leaching, efforts to manage high salinity traditionally has been through crop substitution, i.e., the replacement of salt-sensitive crops with more tolerant ones. This practice has been traced back to the dawn of agriculture and is still probably one of the easiest and most often used strategies in dealing with salinity. Thus, barley (*Hordeum vulgare*, L.) may be substituted for wheat (*Triticum aestivum*, L.), cotton (*Gossypium hirsutum*, L.) for corn (*Zea mays*, L.), sugar beet (*Beta vulgaris*, L.) for lettuce (*Lactuca sativa*, L.), etc. Unfortunately, high value vegetable crops are typically more salt sensitive than most field crops. Harvest quality usually has a more significant impact on marketable yield of horticultural crops than it does for field crops. The salt tolerance tables developed at the USDA Salinity Laboratory in Riverside, California, have been valuable guides for extension personnel and growers in determining which crops can be grown based upon the anticipated soil salinity.[4]

Beyond crop selection, agricultural management techniques can be used to minimize and avoid salinity

effects. Some of these practices include leaching, deep plowing, amendment application, careful choice of fertilizer source and type, installation of drainage, leveling operations, and irrigation techniques. Other management options include the use of drip or sprinkler irrigation to improve water application efficiency and elaborations in seed bed formation and planting design to facilitate removal of accumulated salts from the areas in which roots are developing and extracting water.[5] Some strategies which have not been researched and developed adequately include the manipulation of population densities to improve plant stand and the application of nonsaline or more saline water dependent on the variable salt tolerance of plants during different growth stages.

More recently, crop breeding and genetic manipulation, using tools such as tissue culture and molecular techniques, have been proposed as adjunct strategies to deal with the salinity problem. In the last two decades especially, there has been great interest in breeding plants for improved salt tolerance.[6] Strategies that have been tried or suggested include conventional screening, selection, and breeding with established cultivars, introduction of high salt tolerance into cultivated species through introgression with tolerant wild relatives, or the domestication of salt-tolerant wild or halophytic species through genetic improvement of agronomic or horticultural characteristics. Some of these efforts have resulted in limited success, but major advances have not been noted.

Examples of screening and selection criteria that have been tried include selection during germination and emergence, resistance to salinity-induced reductions in plant height or weight, maintenance of high yield or quality under salt stress, and plant survival. Extensive efforts have also been made to identify reliable physiological or biochemical markers for salt tolerance. Such markers include capability to exclude ions (e.g., Na and Cl) from shoots or specific tissues, maintenance of nutrients (K, Ca, Mg, P, and NO_3) in plant tissues against high external salt concentrations, and ion selectivity (e.g., high K/Na, Ca/Na, or NO_3/Cl).[7,8] The accumulation of metabolic-compatible cellular osmoprotectants such as proline, glycinebetaine, and certain sugars and alcohols has also been proposed as indices for high salt tolerance, but the evidence that accumulation of compatible solutes offers a quantitatively measurable improvement in salt tolerance is not unequivocal.

Most recently efforts have been made to use molecular tools to improve the salt tolerance of plants, but specific genes that confer salt tolerance have proven to be elusive. Still there are increasing examples of success. Plants respond to two basic components of salinity—the ionic component and the osmotic component. The osmotic component is physiologically equivalent to dehydration or drought stress. Thus, plants grown in soils having high salinity have more difficulty in extracting water from the soil matrix. A gene for D-ononitol that confers drought tolerance in ice plant (*Mesambrythium* L.) has been found to improve drought tolerance in tobacco (*Nicotiana tabacum*, L.) under laboratory conditions.[9] Now several researchers are pursuing this strategy with other osmoprotectant genes and other species.

In an approach that focuses on ion regulation, the gene for an ion pump protein located in the vacuoles of Arabidopsis (*Arabidopsis Thaliana*, Heynh.) was transferred to tomato (*Lycopersicon esculentum*, Mill.). The transgenic tomatoes expressed increased salt tolerance in the greenhouse when grown in solutions equivalent to about one-third seawater.[10] The gene AtNHX1 is over-expressed in these plants making them more efficient at sequestering sodium in the vacuole and away from the sensitive metabolic machinery of the cytoplasm. Another Arabidopsis gene, AtHKT1, when inactivated has been shown to limit the transport of sodium through the root cell membrane barrier and effectively increase salt tolerance.[11] None of these approaches has thus far resulted in the improvement of salt tolerance in field-grown plants, however; there is room for optimism as more information is developed concerning the many mechanisms that plants use to respond to salt stress.

REFERENCES

1. Salt Tolerance Bibliography Database. http://www.ussl.ars.usda.gov/salt/frstart.htm.

2. Maas, E.V.; Hoffman, G.J. Crop Salt Tolerance—Current Assessment. J. Irrig. Drainage Div., ASCE **1977**, *103 IR2*, 115–134.

3. van Genuchten, M.Th.; Hoffman, G.J. Analysis of Crop Salt Tolerance Data. In *Soil Salinity Under Irrigation—Process and Management*; Shainberg, I., Shalhevet, J., Eds.; Springer-Verlag: New York, 1984; 258–271.

4. Salt Information on Curve Parameters. http://www.ussl.ars.usda.gov/saltoler.htm

5. Rhoades, J.D. Practices to Control Salinity in Irrigated Soils. In *Towards the Rational Use of High Salinity Tolerant Plants*; Lieth, H., Al Massoum, A., Eds.; Proceedings of the first ASWAS Conference; 1993; Vol. 2.

6. Shannon, M.C. Adaptation of Plants to Salinity. Adv. Agron. **1997**, *60*, 76–120.

7. Shannon, M.C. Principles and Strategies in Breeding for Higher Salt Tolerance. Plant Soil **1985**, *89*, 227–241.

8. Cheeseman, J.M. Mechanisms of Salinity Tolerance in Plants. Plant Physiol. **1988**, *87*, 547–550.

9. Bohnert, H.J.; Cushman, J.C. The Ice Plant Cometh—Models for Environmental Stress Tolerance. In *Models in Plant Biology*; Mandoli, D., Ed.; Springer Verlag: Berlin, Heidelberg, 2001, in press.

10. Zhang, H.-X.; Blumwald, E. Transgenic Salt Tolerant Tomato Plants Accumulate Salt in the Foliage But Not in the Fruits. Nature Biotechnol. **2001**, *19*, 765–768.

11. Rus, A.; Yokoi, S.; Sharhuu, A.; Reddy, M.; Lee, B.; Matsumoto, T.K.; Koiwa, H.; Zhu, J.-K.; Bressan, R.A.; Hasegawa, R.A. AtHKT1 is a Salt Tolerance Determinant That Controls Na$^+$ Entry into Plant Roots. Proc. Natl Acad. Sci. USA **2001**, *98*, 14150–14155.
   See also: http://www.hort.purdue.edu/hort/people/faculty/hasegawa.html and http://www.hort.purdue.edu/CFPESP/Hasegawa/ha00001.htm.

Pollution, Nonpoint Source

Ravendra Naidu
Mallavarapu Megharaj
Peter Dillon
Rai Kookana
Ray Correll
Commonwealth Scientific and Industrial Research Organisation (CSIRO), Adelaide, South Australia, Australia

Walter Wenzel
University of Agricultural Sciences, Vienna, Austria

INTRODUCTION

Nonpoint source pollution (NPSP) has no obvious single point source discharge and is of diffuse nature (Table 1). An example of NPSP includes aerial transport and deposition of contaminants such as SO_2 from industrial emissions leading to acidification of soil and water bodies. Rain water in urban areas could also be a source of NPSP as it may concentrate organic and inorganic contaminants. Examples of such contaminants include polycyclic aromatic hydrocarbons, pesticides, polychlorinated biphenyls that could be present in urban air due to road traffic, domestic heating, industrial emissions, agricultural treatments, etc.[1–3] Other examples of NPSP include fertilizer (especially Cd, N, and P) and pesticide applications to improve crop yield. Use of industrial waste materials as soil amendments have been estimated to contaminate thousands of hectares of productive agricultural land in countries throughout the world.

CONTAMINANT INTERACTIONS

Nonpoint pollution is generally associated with low-level contamination spread at broad acre level. Under these circumstances, the major reaction controlling contaminant interactions are sorption–desorption processes, plant uptake, surface runoff, and leaching. However, certain contaminants, in particular, organic compounds are also subjected to voltalization, chemical, and biological degradation. Sorption–desorption and degradation (both biotic and abiotic) are the two most important processes controlling organic contaminant behavior in soils.

These processes are influenced by both soil and solution properties of the environment. Such interactions also determine the bioavailability and/or transport of contaminants in soils. Where the contaminants are bioavailable, risk to surface and groundwater and soil, crop, and human health are enhanced.

IMPLICATIONS TO SOIL AND ENVIRONMENTAL QUALITY

Environmental contaminants can have a deleterious effect on nontarget organisms and their beneficial activities. These effects could include a decline in primary production, decreased rate of organic matter break-down, and nutrient cycling as well as mineralization of harmful substances that in turn cause a loss of productivity of the ecosystems. Certain pollutants, even though present in very small concentrations in the soil and surrounding water, have potential to be taken up by various microorganisms, plants, animals, and ultimately human beings. These pollutants may accumulate and concentrate in the food chain by several thousand times through a process referred to as biomagnification.

Urban sewage, because of its nutrient values and source of organic carbon in soils, is now increasingly being disposed to land. The contaminants present in sewage sludge (nutrients, heavy metals, organic compounds, and pathogens), if not managed properly, could potentially affect the environment adversely. Dumping of radioactive waste (e.g., radium, uranium, plutonium) onto soil is more complicated because these materials remain active for thousands of years in the soil and thus pose a continued threat to the future health of the ecosystem.

Industrial wastes, improper agricultural techniques, municipal wastes, and use of saline water for irrigation under high evaporative conditions result in the presence of

---

This article previously published in the *Encyclopedia of Soil Science*
© 2002 Marcel, Dekker, Inc.

Encyclopedia of Water Science
DOI: 10.1081/E-EWS 120017970

Table 1 Industries, land uses, and associated chemicals contributing to nonpoint source pollution

Industry	Type of chemical	Associated chemicals
Agricultural activities	Metals/metalloid	Cadmium, mercury, arsenic, selenium
	Nonmetals	Nitrate, phosphate, borate
	Salinity/sodicity	Sodium, chloride, sulfate, magnesium, alkalinity
	Pesticides	Range of organic and inorganic pesticides including arsenic, copper, zinc, lead, sulfonylureas, organochlorine, organophosphates, etc., salt, geogenic contaminants (e.g., arsenic, selenium, etc.)
	Irrigation	Sodium, chloride, arsenic, selenium
Automobile and industrial emissions	Dust	Lead, arsenic, copper, cadmium, zinc, etc.
	Gas	Sulfur oxides, carbon oxides
	Metals	Lead and lead organic compounds
Rainwater	Organics	Polyaromatic hydrocarbons, polychlorbiphenyls, etc.
	Inorganic	Sulfur oxides, carbon oxides acidity, metals and metalloids

(From Barzi, F.; Naidu, R.; McLaughlin, M.J. Contaminants and the Australian Soil Environment. In *Contaminants and the Soil Environment in the Australasia-Pacific Region*; Naidu, R., Kookana, R.S., Oliver, D., Rogers, S., McLaughlin, M.J., Eds.; Kluwer Academic Publishers: Dordrecht, The Netherlands, 1996; 451–484.)

excess soluble salts (predominantly Na and Cl ions) and metalloids such as Se and As in soils. Salinity and sodicity affect the vegetation by inhibiting seed germination, decreasing permeability of roots to water, and disrupting their functions such as photosynthesis, respiration, and synthesis of proteins and enzymes.

Some of the impacts of soil pollution migrate a long way from the source and can persist for some time. For example, suspended solids can increase water turbidity in streams, affecting benthic and pelagic aquatic ecosystems, filling reservoirs with unwanted silt, and requiring water treatment systems for potable water supplies. Phosphorus attached to soil particles, which are washed from a paddock into a stream, can dominate nutrient loads in streams and downstream water bodies. Consequences include increases in algal biomass, reduced oxygen concentrations, impaired habitat for aquatic species, and even possible production of cyanobacterial toxins, with series impacts for humans and livestock consuming the water. Where waters discharge into estuaries, N can be the limiting factor for eutrophication; estuaries of some catchments where fertilizer use is extensive have suffered from excessive sea grass and algal growth.

More insidious is the leaching of nutrients, agricultural chemicals, and hydrocarbons to groundwater. Incremental increases in concentrations in groundwater may be observed over long periods of time resulting in initially potable water becoming undrinkable and then some of the highest valued uses of the resource may be lost for decades. This problem is most severe on tropical islands with shallow relief and some deltaic arsenopyrite deposits, where wells cannot be deepened to avoid polluted groundwater because

underlying groundwater is either saline or contains too much As.

SAMPLING FOR NONPOINT SOURCE POLLUTION

The sampling requirements of NPSP are quite different from those of the point source contamination. Typically, the sampling is required to give a good estimate of the mean level of pollution rather than to delineate areas of pollution. In such a situation, sampling is typically carried out on a regular square or a triangular grid. Furthermore, gains may be possible by using composite sampling.[4] However, if the pollution is patchy, other strategies may be used. One such strategy is to divide the area into remediation units, and to sample each of these. The possibility of movement of the pollutant from the soil to some receptor (or asset) is assessed, and the potential harm is quantified. This process requires an analysis of the bioavailability of the pollutant, pathway analysis, and the toxicological risk. The risk analysis is then assessed and decisions are then made as to how the risk should be managed.

MANAGEMENT AND/OR REMEDIATION OF NONPOINT SOURCE POLLUTION

The treatment strategies used for managing NPSP are generally those that modify the soil properties to decrease the bioavailable contaminant fraction. This is particularly so in the rural agricultural environment where soil–plant transfer of contaminants is of greatest concern. Soil amendments commonly used include those that change the ion-exchange characteristics of the colloid particles and those that enhance the ability of soils to sorb contaminants. An example of NPSP management includes the application of lime to immobilize metals because the solubility of most heavy metals decreases with increasing soil pH. However, this approach is not applicable to all metals, especially those that form oxyanions—the bioavailability of such species increases with increasing pH. Therefore, one of the prerequisites for remediating contaminated sites is a detailed assessment of the nature of contaminants present in the soil. The application of a modified aluminosilicate to a highly contaminated soil around a zinc smelter in Belgium was shown to reduce the bioavailability of metals thereby reducing the Zn phytotoxicity.[5] The simple addition of rock phosphates to form Pb phosphate has also been demonstrated to reduce the bioavailability of Pb in aqueous solutions and contaminated soils due to immobilization in the metal.[6] Nevertheless, there is concern over the long-term stability of the processes. The immobilization process appears attractive currently given that there are very few cheap and effective in situ remediation techniques for metal-contaminated soils. A novel, innovative approach is using higher plants to stabilize, extract, degrade, or volatilize inorganic and organic contaminants for in situ treatment (cleanup or containment) of polluted topsoils.[7]

PREVENTING WATER POLLUTION

The key to preventing water pollution from the soil zone is to manage the source of pollution. For example, nitrate pollution of groundwater will always occur if there is excess nitrate in the soil at a time when there is excess water leaching through the soil. This suggests that we should aim to reduce the nitrogen in the soil during wet seasons and the drainage through the soil. Local research may be needed to demonstrate the success of best management techniques in reducing nutrient, sediment, metal, and chemical exports via surface runoff and infiltration to groundwater. Production figures from the same experiments may also convince local farmers of the benefits of maintaining nutrients and chemicals where needed by a crop rather than losing them off site, and facilitate uptake of best management practices.

GLOBAL CHALLENGES AND RESPONSIBILITY

The biosphere is a life-supporting system to the living organisms. Each species in this system has a role to play and thus every species is important and biological diversity is vital for ecosystem health and functioning. The detection of hazardous compounds in Antarctica, where these compounds were never used or no man has ever lived before, indicates how serious is the problem of long-range atmospheric transport and deposition of these pollutants. Clearly, pollution knows no boundaries. This ubiquitous pollution has had a global effect on our soils, which in turn has been affecting their biological health and productivity. Coupled with this, over 100,000 chemicals are being used in countries throughout the world. Recent focus has been on the endocrine disruptor chemicals that mimic natural hormones and do great harm to animal and human reproductive cycles.

These pollutants are only a few examples of contaminants that are found in the terrestrial environment.

REFERENCES

1. Chan, C.H.; Bruce, G.; Harrison, B. Wet Deposition of Organochlorine Pesticides and Polychlorinated Biphenyls to the Great Lakes. J. Great Lakes Res. **1994**, *20*, 546–560.
2. Lodovici, M.; Dolara, P.; Taiti, S.; Del Carmine, P.; Bernardi, L.; Agati, L.; Ciappellano, S. Polycyclic Aromatic Hydrocarbons in the Leaves of the Evergreen Tree*Laurus Nobilis*. Sci. Total Environ. **1994**, *153*, 61–68.
3. Sweet, C.W.; Murphy, T.J.; Bannasch, J.H.; Kelsey, C.A.; Hong, J. Atmospheric Deposition of PCBs into Green Bay. J. Great Lakes Res. **1993**, *18*, 109–128.
4. Patil, G.P.; Gore, S.D.; Johnson, G.D. *Manual on Statistical Design and Analysis with Composite Samples*, Technical Report No. 96-0501; EPA Observational Economy Series Center for Statistical Ecology and Environmental Statistics; Pennsylvania State University, 1996; Vol. 3.
5. Vangronsveld, J.; Van Assche, F.; Clijsters, H. Reclamation of a Bare Industrial Area, Contaminated by Non-ferrous Metals: In Situ Metal Immobilisation and Revegetation. Environ. Pollut. **1995**, *87*, 51–59.
6. Ma, Q.Y.; Logan, T.J.; Traina, S.J. Lead Immobilisation from Aqueous Solutions and Contaminated Soils Using Phosphate Rocks. Environ. Sci. Technol. **1995**, *29*, 1118–1126.
7. Wenzel, W.W.; Adriano, D.C.; Salt, D.; Smith, R. Phytoremediation: A Plant-Microbe Based Remediation System. In *Bioremediation of Contaminated Soils*; Soil Science Society of America Special Monograph No. 37, Adriano, D.C., Bollag, J.M., Frankenberger, W.T., Jr., Sims, W.R., Eds.; Soil Science Society of America: Madison, USA, 1999; 772.

Pollution, Point Source

Ravendra Naidu
Mallavarapu Megharaj
Peter Dillon
Rai Kookana
Ray Correll
Commonwealth Scientific and Industrial Research Organisation (CSIRO),
Adelaide, South Australia, Australia

Walter Wenzel
University of Agricultural Sciences, Vienna, Austria

INTRODUCTION

Environmental pollution is one of the foremost ecological challenges. Pollution is an offshoot of technological advancement and overexploitation of natural resources. From the standpoint of pollution, the term environment primarily includes air, land, and water components including landscapes, rivers, parks, and oceans. Pollution can be generally defined as an undesirable change in the natural quality of the environment that may adversely affect the well being of humans, other living organisms, or entire ecosystems either directly or indirectly. Although pollution is often the result of human activities (anthropogenic), it could also be due to natural sources such as volcanic eruptions emitting noxious gases, pedogenic processes, or natural change in the climate. Where pollution is localized it is described as point source (PS). Thus, PS pollution is a source of pollution with a clearly identifiable point of discharge that can be traced back to the specific source such as leakage of underground petroleum storage tanks or an industrial site.

Some naturally occurring pollutants are termed geogenic contaminants and these include fluorine, selenium, arsenic, lead, chromium, fluoride, and radionuclides in the soil and water environment. Significant adverse impacts of geogenic contaminants (e.g., As) on environmental and human health have been recorded in Bangladesh, West Bengal, India, Vietnam, and China. More recently reported is the presence of geogenic Cd and the implications to crop quality in Norwegian soils.[1]

The terms contamination and pollution are often used interchangeably but erroneously. Contamination denotes the presence of a particular substance at a higher concentration than would occur naturally and this may or may not have harmful effects on human or the environment. Pollution refers not only to the presence of a substance at higher level than would normally occur but is also associated with some kind of adverse effect.

NATURE AND SOURCES OF CONTAMINANTS

The main activities contributing to PS pollution include industrial, mining, agricultural, and commercial activities as well as transport and services (Table 1). Uncontrolled mining, manufacturing, and disposal of wastes inevitably cause environmental pollution. Military land and land for recreational shooting are also important sites of PS contamination. The contaminants associated with such activities are listed in Table 1. Contamination at many of these sites appears to have resulted because of lax regulatory measures prior to the establishment of legislation protecting the environment.

CONTAMINANT INTERACTIONS IN SOIL AND WATER

Inorganic Chemicals

Inorganic contaminant interactions with colloid particulates include: adsorption–desorption at surface sites, precipitation, exchange with clay minerals, binding by organically coated particulate matter or organic colloidal material, or adsorption of contaminant ligand complexes. Depending on the nature of contaminants, these interactions are controlled by solution pH and ionic strength of soil solution, nature of the species, dominant cation, and inorganic and organic ligands present in the soil solution.[2]

This article previously published in the *Encyclopedia of Soil Science*
© 2002 Marcel Dekker, Inc.

Table 1 Industries, land uses, and associated chemicals contributing to points, nonpoint source pollution

Industry	Type of chemical	Associated chemicals
Airports	Hydrocarbons	Aviation fuels
	Metals	Particularly aluminum, magnesium, and chromium
Asbestos production and disposal	Asbestos	
Battery manufacture and recycling	Metals	Lead, manganese, zinc, cadmium, nickel, cobalt, mercury, silver, and antimony
	Acids	Sulfuric acid
Breweries/distilleries	Alcohol	Ethanol, methanol, and esters
Chemicals manufacture and use	Acid/alkali	Mercury (chlor/alkali), sulfuric, hydrochloric and nitric acids, sodium and calcium hydroxides
	Adhesives/resins	Polyvinyl acetate, phenols, formaldehyde, acrylates, and phthalates
	Dyes	Chromium, titanium, cobalt, sulfur and nitrogen organic compounds, sulfates, and solvents
	Explosives	Acetone, nitric acid, ammonium nitrate, pentachlorophenol, ammonia, sulfuric acid, nitroglycerine, calcium cyanamide, lead, ethylene glycol, methanol, copper, aluminum, *bis*(2-ethylhexyl) adipate, dibutyl phthalate, sodium hydroxide, mercury, and silver
	Fertilizer	Calcium phosphate, calcium sulfate, nitrates, ammonium sulfate, carbonates, potassium, copper, magnesium, molybdenum, boron, and cadmium
	Flocculants	Aluminum
	Foam production	Urethane, formaldehyde, and styrene
	Fungicides	Carbamates, copper sulfate, copper chloride, sulfur, and chromium
	Herbicides	Ammonium thiocyanate, carbanates, organochlorines, organophosphates, arsenic, and mercury
	Paints	
	Heavy metals	Arsenic, barium, cadmium, chromium, cobalt, lead, manganese, mercury, selenium, and zinc
	General	Titanium dioxide
	Solvent	Toluene, oils natural (e.g., pine oil) or synthetic
	Pesticides	Arsenic, lead, organochlorines, and organophosphates
	Active ingredients	Sodium, tetraborate, carbamates, sulfur, and synthetic pyrethroids
	Solvents	Xylene, kerosene, methyl isobutyl ketone, amyl acetate, and chlorinated solvents
	Pharmacy	Dextrose and starch
	General/solvents	Acetone, cyclohexane, methylene chloride, ethyl acetate, butyl acetate, methanol, ethanol, isopropanol, butanol, pyridine methyl ethyl ketone, methyl isobutyl ketone, and tetrahydrofuran
	Photography	Hydroquinone, pheidom, sodium carbonate, sodium sulfite, potassium bromide, monomethyl paraaminophenol sulfates, ferricyanide, chromium, silver, thiocyanate, ammonium compounds, sulfur compounds, phosphate, phenylene diamine, ethyl alcohol, thiosulfates, and formaldehyde
	Plastics	Sulfates, carbonates, cadmium, solvents, acrylates, phthalates, and styrene
	Rubber	Carbon black
	Soap/detergent	
	General	Potassium compounds, phosphates, ammonia, alcohols, esters, sodium hydroxide, surfactants (sodium lauryl sulfate), and silicate compounds
	Acids	Sulfuric acid and stearic acid
	Oils	Palm, coconut, pine, and tea tree
	Solvents	

Table 1 Industries, land uses, and associated chemicals contributing to points, nonpoint source pollution (*Continued.*)

Industry	Type of chemical	Associated chemicals
	General	Ammonia
	Hydrocarbons	e.g., BTEX (benzene, toluene, ethylbenzene, xylene)
	Chlorinated organics	e.g., trichloroethane, carbon tetrachloride, and methylene chloride
Defense works		See "Explosives" under "Chemicals Manufacture and Use, Foundries, Engine Works, and Service Stations"
Drum reconditioning		See "Chemicals Manufacture and Use"
Dry cleaning		Trichlorethylene and ethane
		Carbon tetrachloride
		Perchlorethylene
Electrical		PCBs (transformers and capacitors), solvents, tin, lead, and copper
Engine works	Hydrocarbons	
	Metals	
	Solvents	
	Acids/alkalis	
	Refrigerants	
	Antifreeze	Ethylene glycol, nitrates, phosphates, and silicates
Foundries	Metals	Particularly aluminum, manganese, iron, copper, nickel, chromium, zinc, cadmium and lead and oxides, chlorides, fluorides and sulfates of these metals
	Acids	Phenolics and amines
		Coke/graphite dust
Gas works	Inorganics	Ammonia, cyanide, nitrate, sulfide, and thiocyanate
	Metals	Aluminum, antimony, arsenic, barium, cadmium, chromium, copper, iron, lead, manganese, mercury, nickel, selenium, silver, vanadium, and zinc
	Semivolatiles	Benzene, ethylbenzene, toluene, total xylenes, coal tar, phenolics, and PAHs
Iron and steel works		Metals and oxides of iron, nickel, copper, chromium, magnesium and manganese, and graphite
Landfill sites		Methane, hydrogen sulfides, heavy metals, and complex acids
Marinas		Engine works, electroplating under metal treatment
	Antifouling paints	Copper, tributyltin (TBT)
Metal treatments	Electroplating metals	Nickel, chromium, zinc, aluminum, copper, lead, cadmium, and tin
	Acids	Sulfuric, hydrochloric, nitric, and phosphoric
	General	Sodium hydroxide, 1,1,1-trichloroethane, tetrachloroethylene, toluene, ethylene glycol, and cyanide compounds
	Liquid carburizing baths	Sodium, cyanide, barium, chloride, potassium chloride, sodium chloride, sodium carbonate, and sodium cyanate
	Mining and extracting industries	Arsenic, mercury, and cyanides and also refer to "Explosives" under "Chemicals Manufacture and Use"
	Power stations	Asbestos, PCBs, fly ash, and metals
	Printing shops	Acids, alkalis, solvents, chromium (see "Photography" under "Chemicals Manufacture and Use")
Scrap yards		Hydrocarbons, metals, and solvents
	Service stations and fuel storage facilities	Aliphatic hydrocarbons
		BTEX (i.e., benzene, toluene, ethylbenzene, xylene)
		PAHs (e.g., benzo(a) pyrene)
		Phenols
		Lead
Sheep and cattle dips		Arsenic, organochlorines and organophosphates, carbamates, and synthetic pyrethroids

(*continued*)

Table 1 Industries, land uses, and associated chemicals contributing to points, nonpoint source pollution (*Continued.*)

Industry	Type of chemical	Associated chemicals
Smelting and refining		Metals and the fluorides, chlorides and oxides of copper, tin, silver, gold, selenium, lead, and aluminum
Tanning and associated trades	Metals	Chromium, manganese, and aluminum
	General	Ammonium sulfate, ammonia, ammonium nitrate, phenolics (creosote), formaldehyde, and tannic acid
Wood preservation	Metals	Chromium, copper, and arsenic
	General	Naphthalene, ammonia, pentachlorophenol, dibenzofuran, anthracene, biphenyl, ammonium sulfate, quinoline, boron, creosote, and organochlorine pesticides

(From Ref. 11.)

Organic Chemicals

The fate and behavior of organic compounds depend on a variety of processes including sorption–desorption, volatilization, chemical and biological degradation, plant uptake, surface runoff, and leaching. Sorption–desorption and degradation (both biotic and abiotic) are perhaps the two most important processes as the bulk of the chemicals is either sorbed by organic and inorganic soil constituents, and chemically or microbially transformed/degraded. The degradation is not always a detoxification process. This is because in some cases the transformation or degradation process leads to intermediate products that are more mobile, more persistent, or more toxic to nontarget organisms. The relative importance of these processes is determined by the chemical nature of the compound.

IMPLICATIONS TO SOIL AND ENVIRONMENTAL QUALITY

Considerable amount of literature is available on the effects of contaminants on soil microorganisms and their functions in soil. The negative impacts of contaminants on microbial processes are important from the ecosystem point of view and any such effects could potentially result in a major ecological perturbation. Hence, it is most relevant to examine the effects of contaminants on microbial processes in combination with communities. The most commonly used indicators of metal effects on microflora in soil are: (1) soil respiration, (2) soil nitrification, (3) soil microbial biomass, and (4) soil enzymes.

Contaminants can reach the food chain by way of water, soil, plants, and animals. In addition to the food chain transfer, pollutants may also enter via direct consumption or dust inhalation of soil by children or animals. Accumulation of these pollutants can take place

in certain target tissues of the organism depending on the solubility and nature of the compound. For example, DDT and PCBs accumulate in human adipose tissue. Consequently, several of these pollutants have the potential to cause serious abnormalities including cancer and reproductive impairments in animal and human systems.

SAMPLING FOR PS POLLUTION

The aims of the sampling system must be clearly defined before it can be optimized.[3] The type of decision may be to determine land use, how much of an area is to be remediated, or what type of remediation process is required. Because sampling and the associated chemical and statistical analyses are expensive, careful planning of the sampling scheme is therefore a good investment. One of the best ways to achieve this is to use any ancillary data that are available. These data could be in the form of emission history from a stack, old photographs that give details of previous land uses, or agricultural records. Such data can at least give qualitative information.

As discussed before, PS pollution will typically be airborne from a stack, or waterborne from some effluent such as tannery waste, cattle dips, or mine waste. In many cases, the industry will have modified its emissions (e.g., cleaner production) or point of release (increased stack height), hence the current pattern of emission may not be closely related to the historic pattern of pollution. For example, liquid effluent may have been discharged previously into a bay, but that effluent may now be treated and perhaps discharged at some other point. Typically, the aim of a sampling scheme in these situations is to assess the maximum concentrations, the extent of the pollution, and the rate of decline in concentration from the PS. Often the sampling scheme will be used to produce maps of concentration isopleths of the pollutant.

The location of the sampling points would normally be concentrated towards the source of the pollution. A good scheme is to have sufficient samples to accurately assess the maximum pollution, and then space additional samples at increasing intervals. In most cases, the distribution of the pollutant will be asymmetric, with the maximum spread down the slope or down the prevailing wind. In such cases more samples should be placed in the direction of the expected gradient. This is a clear case of when ancillary data can be used effectively. A graph of concentration of the pollutant against the reciprocal of distance from the source is often informative.[4] Sampling depths will depend on both the nature of the pollution and the reason for the investigation. If the pollution is from dust and it is unlikely to be leached, only surface sampling will be required. An example of this is pollution from silver smelting in Wales.[5] In contrast, contamination from organic or mobile inorganic pollutants such as F compounds may migrate well down to the profile and deep sampling may be required.[6,7]

ASSESSMENT

In order to assess the impacts of pollution, reliable and effective monitoring techniques are important. Pollution can be assessed and monitored by chemical analyses, toxicity tests, and field surveys. Comparison of contaminant data with an uncontaminated reference site and available databases for baseline concentrations can be useful in establishing the extent of contamination. However, this may not always be possible in the field. Chemical analyses must be used in conjunction with biological assays to reveal site contamination and associated adverse effects. Toxicological assays can also reveal information about synergistic interactions of two or more contaminants present as mixtures in soil, which cannot be measured by chemical assays alone.

Microorganisms serve as rapid detectors of environmental pollution and are thus of importance as pollution indicators. The presence of pollutants can induce alteration of microbial communities and reduction of species diversity, inhibition of certain microbial processes (organic matter breakdown, mineralization of carbon and nitrogen, enzymatic activities, etc.). A measure of the functional diversity of the bacterial flora can be assessed using ecoplates (see http://www.biolog.com/section_4.html). It has been shown that algae are especially sensitive to various organic and inorganic pollutants and thus may serve as a good indicator of pollution.[8] A variety of toxicity tests involving microorganisms, invertebrates, vertebrates, and plants may be used with soil or water samples.[9]

MANAGEMENT AND/OR REMEDIATION OF PS POLLUTION

The major objective of any remediation process is to (1) reduce the actual or potential environmental threat and (2) reduce unacceptable risks to man, animals, and the environment to acceptable levels.[10] Therefore, strategies to either manage and/or remediate contaminated sites have been developed largely from application of stringent regulatory measures set up to safeguard ecosystem function as well as to minimize the potential adverse effects of toxic substances on animal and human health.

The available remediation technologies may be grouped into two categories: (1) ex situ techniques that require removal of the contaminated soil or groundwater for treatment either on-site or off-site, and (2) in situ techniques that attempt to remediate without excavation of contaminated soils. Generally, in situ techniques are favored over ex situ techniques because of (1) reduced costs due to elimination or minimization of excavation, transportation to disposal sites, and sometimes treatment itself; (2) reduced health impacts on the public or the workers; and, (3) the potential for remediation of inaccessible sites, e.g., those located at greater depths or under buildings. Although in situ techniques have been successful with organic contaminated sites, the success of in situ strategies with metal contaminants has been limited. Given that organic and inorganic contaminants often occur as a mixture, a combination of more than one strategy is often required to either successfully remediate or manage metal contaminated soils.

GLOBAL CHALLENGES AND RESPONSIBILITY

The last 100 yr has seen massive industrialization. Indeed such developments were coupled with the rapid increase in world population and the desire to enhance economy and food productivity. While industrialization has led to increased economic activity and much benefit to human race, the lack of regulatory measures and appropriate waste management strategies until early 1980s (including the use of agrochemicals) has resulted in contamination of our biosphere. Continued pollution of the environment through industrial emissions is of global concern. There is, therefore, a need for politicians, regulatory organizations, and scientists to work together to minimize environmental contamination and to remediate contaminated sites. The responsibility to check this pollution lies with every individual and country although the majority of this pollution is due to the industrialized nations. There is a clear need of better coordination of efforts in dealing with numerous forms of PS pollution problems that are being faced globally.

REFERENCES

1. Mehlum, H.K.; Arnesen, A.K.M.; Singh, B.R. Extractability and Plant Uptake of Heavy Metals in Alum Shale Soils. Commun. Soil Sci. Plant Anal. **1998**, *29*, 183–198.
2. McBride, M.B. Reactions Controlling Heavy Metal Solubility in Soils. Adv. Soil Sci. **1989**, *10*, 1–56.
3. Patil, G.P.; Gore, S.D.; Johnson, G.D. *EPA Observational Economy Series Volume 3: Manual on Statistical Design and Analysis with Composite Samples*; Technical Report No. 96-0501; Center for Statistical Ecology and Environmental Statistics: Pennsylvania State University, 1996.
4. Ward, T.J.; Correll, R.L. Estimating Background Concentrations of Heavy Metals in the Marine Environment. In *Proceedings of a Bioaccumulation Workshop: Assessment of the Distribution, Impacts and Bioaccumulation of Contaminants in Aquatic Environments, Sydney, 1990*; Miskiewicz, A.G., Ed.; Water Board and Australian Marine Science Association: Sydney, 1992; 133–139.
5. Jones, K.C.; Davies, B.E.; Peterson, P.J. Silver in Welsh Soils: Physical and Chemical Distribution Studies. Geoderma **1986**, *37*, 157–174.
6. Barber, C.; Bates, L.; Barron, R.; Allison, H. Assessment of the Relative Vulnerability of Groundwater to Pollution: A Review and Background Paper for the Conference Workshop on Vulnerability Assessment. J. Aust. Geol. Geophys. **1993**, *14* (2–3), 147–154.
7. Wenzel, W.W.; Blum, W.E.H. Effects of Fluorine Deposition on the Chemistry of Acid Luvisols. Int. J. Environ. Anal. Chem. **1992**, *46*, 223–231.
8. Megharaj, M.; Singleton, I.; McClure, N.C. Effect of Pentachlorophenol Pollution Towards Microalgae and Microbial Activities in Soil from a Former Timber Processing Facility. Bull. Environ. Contam. Toxicol. **1998**, *61*, 108–115.
9. Juhasz, A.L.; Megharaj, M.; Naidu, R. Bioavailability: The Major Challenge (Constraint) to Bioremediation of Organically Contaminated Soils. In *Remediation Engineering of Contaminated Soils*; Wise, D., Trantolo, D.J., Cichon, E.J., Inyang, H.I., Stottmeister, U., Eds.; Marcel Dekker: New York, 2000; 217–241.
10. Wood, P.A. Remediation Methods for Contaminated Sites. In *Contaminated Land and Its Reclamation*; Hester, R.E., Harrison, R.M., Eds.; Royal Society of Chemistry, Thomas Graham House: Cambridge, UK, 1997; 47–73.
11. Barzi, F.; Naidu, R.; McLaughlin, M.J. Contaminants and the Australian Soil Environment. In *Contaminants and the Soil Environment in the Australasia–Pacific Region*; Naidu, R., Kookana, R.S., Oliver, D., Rogers, S., McLaughlin, M.J., Eds.; Kluwer Academic Publishers: Dordrecht, The Netherlands, 1996; 451–484.

Precipitation Distribution Patterns

Gregory L. Johnson
United States Department of Agriculture (USDA), Portland, Oregon, U.S.A.

INTRODUCTION

Precipitation formation and distribution are largely controlled by two factors: the availability of atmospheric moisture, and the presence of upward vertical motion. Warmer air can hold more moisture than colder air, which often means tropical areas have more precipitation than colder regions, although this is not always true. Continental interiors usually have less precipitation than nearby coastal areas, also due to less available atmospheric moisture. It is estimated that only 15% of globally evaporated water each year comes from continental areas, with the remainder from the world's oceans.[1] Locations with predominant rising air, such as the forced lifting over mountain areas, usually have more precipitation than nearby lower elevations. There are clear exceptions to these general patterns, however, which result from a sometimes complicating set of factors.

LATITUDINAL AND LAND–OCEAN EFFECTS

The general precipitation characteristics mentioned earlier must be considered in light of the global circulation patterns that transfer energy, as well as moisture, poleward from the equator. Fig. 1 illustrates the chief latitudinal and vertical flow patterns that consist largely of three cells from the equator to each pole. This meridional circulation system was first depicted by Bergeron.[2] Rising air is noted in equatorial regions and at approximately 60° latitude, leading to relatively more precipitation in those regions. Conversely, subsidence is common at approximately 30° latitude and at the poles, which are the locations of the major desert regions.

These global circulation patterns in both the oceans and the atmosphere are modified by the shape and position of continents. The flow in the major oceans is largely anticyclonic, and the associated atmospheric flow transports tropical water and atmospheric moisture poleward along the east sides of continents. Therefore, eastern portions of North and South America, Africa, Australia, and Asia have relatively greater precipitation than their western regions, especially at latitudes of approximately 25°–45°, and during their respective summers. In the winter, polar air shifts southward and intensifies the atmospheric thermal boundary

and, thus, upper air steering winds. This typically moves the jet stream, with its associated storminess and greatly enhanced precipitation, to the west coast of the continents at 40°–60° latitude. This produces, for instance, very wet winters along coastal areas of the Pacific Northwest northward to Alaska, as well as in southern Chile.

At middle and high latitudes, precipitation is chiefly the result of large-scale weather systems. These synoptic-scale (>500 km or so) systems often have rather long lifetimes (days), and can produce precipitation over a wide area, although generally there are regions of enhanced precipitation within these broad areas associated with the position of the upper air jet stream. These systems are more common in the cold season months. In the warm season, precipitation usually is of smaller spatial scale (such as individual thunderstorms), resulting in precipitation signatures that are of considerably smaller spatial dimensions (10 km– 100 km), and often with accompanying shorter time durations. There are exceptions to these spatial rules, as well, such as organized tropical systems (storms and hurricanes) that occasionally impact midlatitude areas, and thunderstorm complexes (such as the mesoscale convective complexes, or MCCs common over the U.S. Midwest in the summer), which can have spatial scales approaching those of synoptic systems.

In the tropics, typical areal dimensions of thunderstorms can be as little as 2 km². Average durations at a location often are less than 1 hr, and whole storms generally last less than 3 hr.[3] Tropical storm systems can have much longer durations and larger spatial dimensions, but are quite infrequent in most locations, and usually contribute only a relatively small percentage to the average annual precipitation except in a few regions. These systems do follow rather common transit routes, however, such as from off the African coast westward to the Caribbean and the southeastern coast of the United States; and from the Mexican coast westward across the Pacific south of Hawaii to east Asia. In these regions, tropical storms can contribute more significantly to annual precipitation, even on a 30-yr average basis. Another region of enhanced convection is the intertropical convergence zone (ITCZ). This is a semipermanent (at climatic time scales) region of surface convergence, clouds, and enhanced precipitation associated with the equatorial trough, in the general region across the Pacific, Indian, and (to a

Encyclopedia of Water Science
DOI: 10.1081/E-EWS 120010316

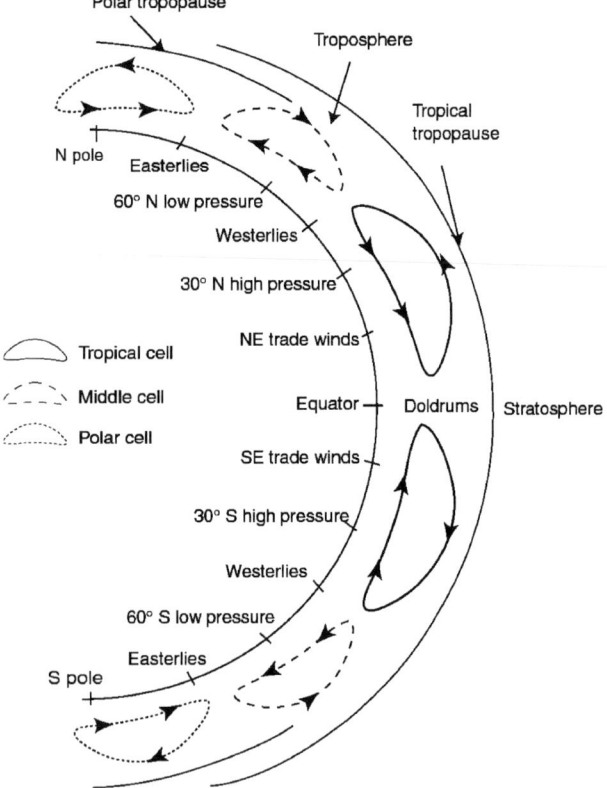

Fig. 1 Vertical distribution of the three major atmospheric cells in each hemisphere responsible for poleward energy transfer and atmospheric circulation systems (after Ref. [2]).

somewhat lesser extent) Atlantic Oceans. Northeast and southeast trade winds converge at the ITCZ, and in the Pacific, it is commonly 5°–10° north of the equator.

GLOBAL PRECIPITATION PATTERNS

The result of these and other forcings is a global precipitation structure as very generally depicted in Fig. 2. Equatorial regions are generally the wettest, but note the extremely wet coastal regions of southern Alaska and western Canada, and also southern Chile. The great deserts are generally in the region of 20°–30° latitude, and along west coasts or in continental interiors. This general map, of course, fails to show much of the great spatial variability that exists even in small regions.

SPACE AND TIME SCALES OF PRECIPITATION

The matching of space and time scales of precipitation processes is an important consideration in the determination of precipitation patterns. Over short time-averaging periods precipitation generally has high spatial variability. A snapshot image of precipitation coverage over the continental United States in Fig. 3 is derived from compositing all doppler weather radar reflectivity coverages. At once this springtime image reveals a number of different precipitation processes, including organized precipitation structure along the Atlantic

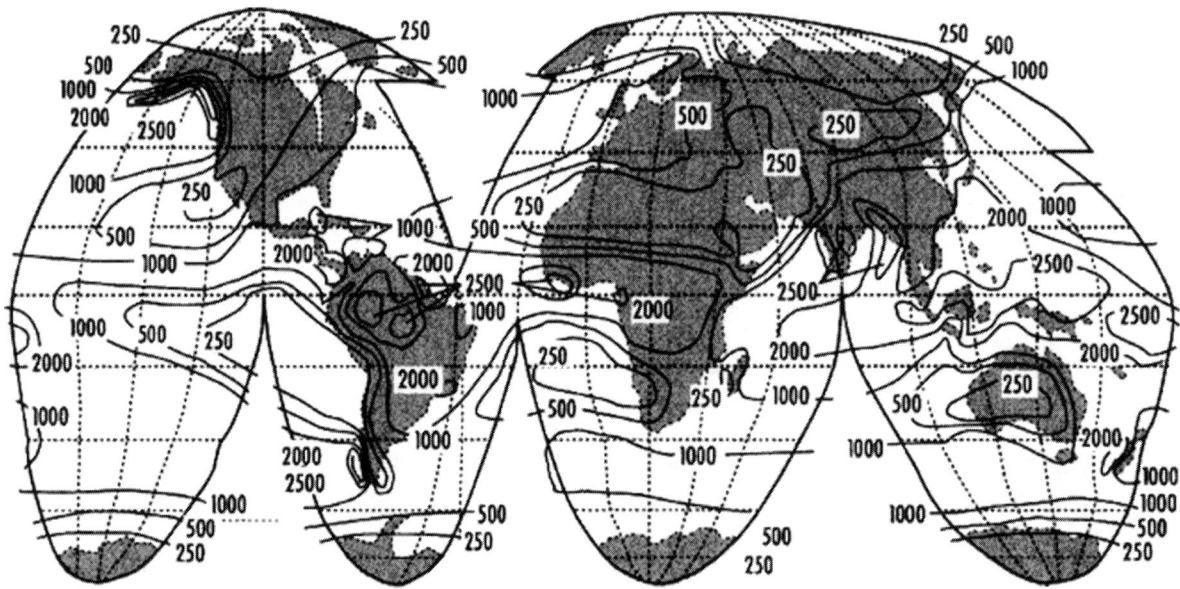

Fig. 2 Global mean annual precipitation isohyets (mm).

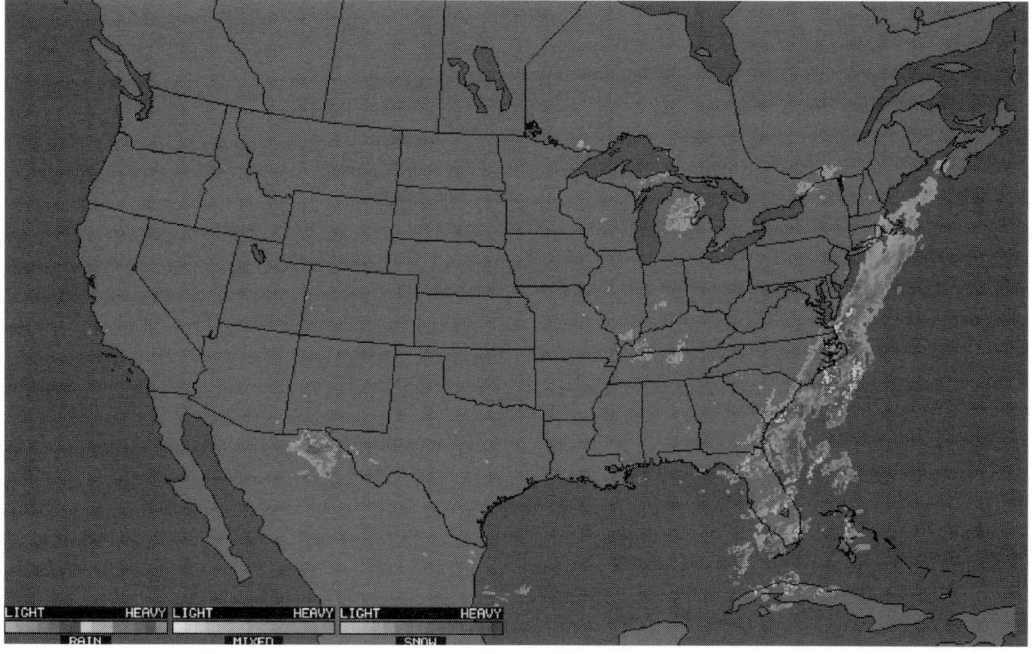

Fig. 3 Composite radar reflectivity image of the United States, showing precipitation areas, 2000 UTC, April 25, 2001.

seaboard associated with an upper level trough and a surface cold front. Other areas of precipitation exhibit greater spatial variability, and cover much smaller spatial domains. Isolated subtropical showers with more inherent randomness in their locations are evident off the northeast coast of Mexico. A small area of organized convection is at least partially due to the presence of mountains in extreme north-central Mexico. A large portion of the country is dry.

In contrast, at longer time scales, precipitation patterns begin to emerge that reveal more consistency in moisture-availability and lifting-mechanism-presence. For instance, at monthly time scales, weather features that exhibit consistency over 30-day periods emerge. Any given month, representing just one sample of 30 days, can show features that may be associated with large-scale forcings like El Niño, or upper air patterns that become relatively "locked in" for that month. In contrast, a 30-yr average, or longer, will reveal precipitation processes that are even more consistent. Take for example the contrasting images of mean July precipitation (based on 30 yr of data) vs. the map of July, 1993 precipitation over the continental United States (Fig. 4). The mean map (top) reveals numerous features, including the moist eastern half of the country, especially along the Southeastern coast; the extreme dryness in California and most of the West coast; the summer Monsoon circulation creating precipitation over Arizona, New Mexico, and Colorado, with mountain enhancement, as well; and two smaller regions with relatively drier July weather compared to surrounding

areas, in southern Missouri–northern Arkansas, and across most of Michigan. These relative minima are explained by a relatively greater frequency of surface high pressure over the cooler Great Lakes, which suppresses summer thunderstorm and shower activity over Michigan; and by the Ozark Mountains inhibiting northward-flowing moist air from the Gulf of Mexico, as well as slightly interrupting the dynamics of thunderstorm development.

In contrast, the bottom map (July 1993) depicts processes that were dominant over this 31-day period only. Note the fairly large region of enhanced precipitation in the Iowa–Nebraska–Kansas–Missouri area that was associated with a stationary frontal system for much of the month, and moisture-feedback processes that helped generate excessive precipitation for many weeks. To the south, a similar-sized area with very little precipitation covered much of Texas. The area with near-zero precipitation in California expanded to include much of Nevada and Arizona, as well, and another region with enhanced precipitation covered much of eastern Montana and the Dakotas. July, 1993 thus represented both an accentuation of the mean map (top), as well as at least two regions with significantly anomalous rainfall—abnormally large rainfall amounts over the Midwest to Plains region, and the dryness from eastern Texas to northern Georgia.

These maps also reveal differences in small-scale variability. Over the relatively short time period of a single month (e.g., July, 1993) the map appears "speckled" with precipitation values. Particularly in regions not directly under the influence of major synoptic forcing there is

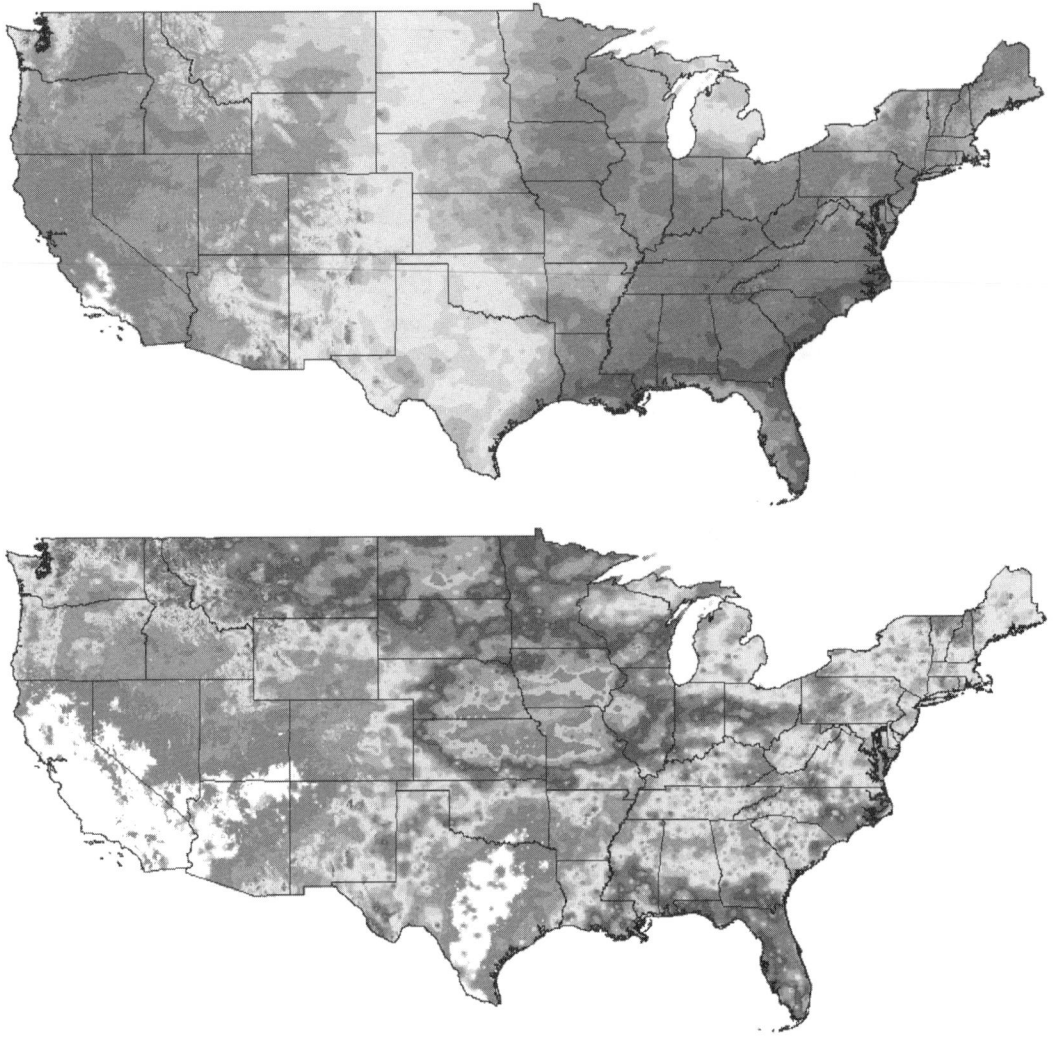

Fig. 4 Mean 1961–1990 July precipitation (top) and July 1993 precipitation (bottom). Maps created using the PRISM modeling system of the Spatial Climate Analysis Service, Oregon State University.

a great deal more variability between individual climate stations than in the 30-yr mean map (top), in which most of this small-scale variability associated with a limited sample size is removed. The 30-yr mean map thus represents only very consistent and dominant precipitation forcings.

At the average annual time scale, with a 30-yr averaging period, consistent precipitation patterns are clearly in evidence (Fig. 5), while the small-scale variability shown in the July, 1993 map, and somewhat distinguishable even in the 30-yr mean July map, are now clearly absent. Note the significantly greater detail and spatial variability depicted in this map compared to the generalized global precipitation map shown in Fig. 2. Other atmospheric forcings are at work, as well, and include:

Orographic Enhancement

Air flow over mountain barriers creates forced uplift and, typically, an increase in precipitation, called orographic enhancement. This increase in precipitation with elevation is largely local, however, with regional or national relationships often insignificant or non-sensical. For instance, elevation increases from the Mississippi River westward to the Rocky Mountains across the Plains, but precipitation decreases westward across this region. In mountainous regions, precipitation–elevation relationships can be different from one mountain barrier to the next, with the upwind barrier often having more precipitation enhancement than succeeding ridges downwind. Orographic enhancement is controlled by a number of factors, including wind

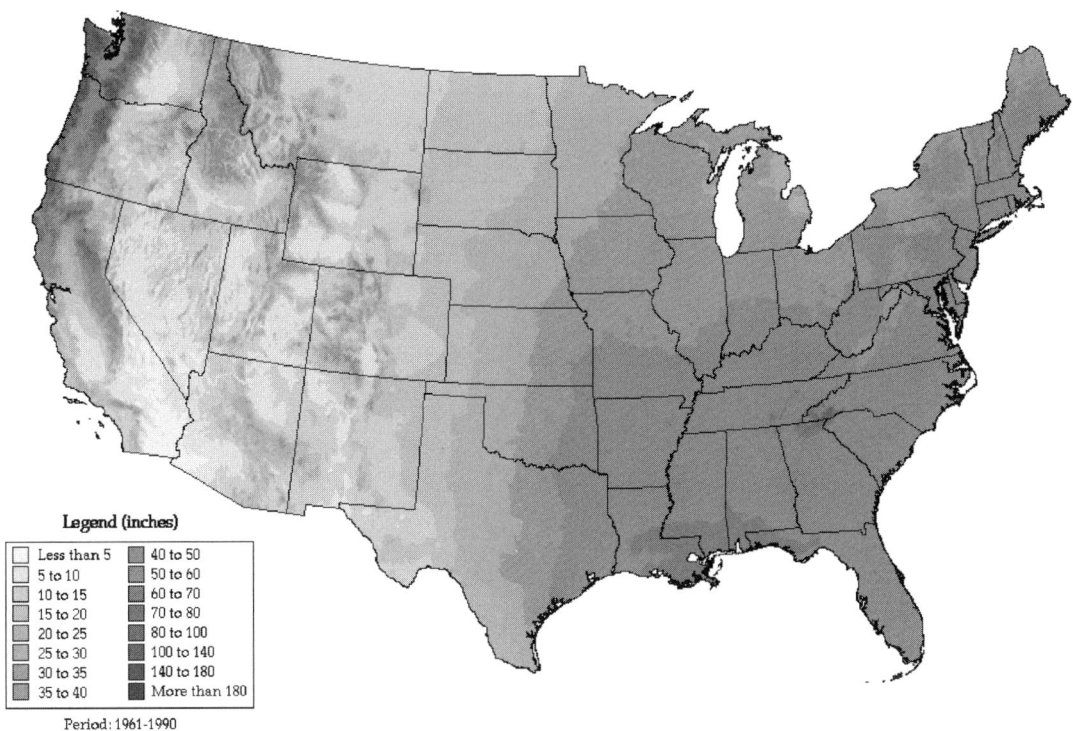

Fig. 5 Mean annual precipitation of the United States (in.), 1961–1990. derived using the PRISM modeling system.

direction relative to the barrier, wind speed, available atmospheric moisture, elevation increase, and slope angle.

Significant orographic enhancement is noted on this map (Fig. 5) over much of the western United States, and in some of the highest mountain terrain in the southern Appalachians and in New England. Over much of the western United States, between the Cascade/Sierra Nevada ranges and the Rocky Mountains, this mean annual precipitation map looks very much like an elevation map. The mountains are very effective at "scouring" what moisture is available in the air flow. In some cases, as over much of Idaho, the terrain contains many successive mountain barriers with very small spacing between them. In this case, the entire region acts as an elevated terrain feature with fairly consistent precipitation enhancement over the area.

There are elevational limits to orographic enhancement. Across most of the continental United States, precipitation increases with elevation to approximately 3000 m in southern latitude areas, and around 2500 m to the north. In Alaska, it is conjectured that precipitation is a maximum above 2000 m. In regions where the trade wind inversion is prevalent, such as many oceanic areas 10°–30° latitude either side of the equator (e.g., Hawaii), precipitation is maximized at levels as low

as 1000 m, with highest mountain elevations sometimes extremely dry.

Other Causes of Vertical Motion

There are factors other than topography that can control vertical motions and, thus, precipitation distribution. These include sea breezes that often create areas of surface convergence and, necessarily, upward vertical motions some distance (typically 10 km–50 km) inland from the immediate coastline. Peninsular Florida in the warm season is a good example of this phenomenon. On many afternoons, in the absence of any other synoptic forcings, an onshore airflow will develop that penetrates approximately 30-km inland, creating a convergence zone where storms develop and precipitation enhancement is noted.

In midlatitudes, jet stream dynamics create and strengthen storm systems and vertical motions. These are strongest when thermal contrasts between the poles and tropics are greatest, in the respective hemispheric cold seasons.

Convective instability creates upward vertical motions that can create significant precipitation. These precipitation processes are most common in the tropics and in the warm season at midlatitudes, and are the result of local heating (both sensible and latent). Their space and time dimensions are typically quite small.

PRECIPITATION MAPPING

Climatologists and, in particular, climate mapping specialists, study these precipitation processes and build them into precipitation distribution models. These models objectively determine precipitation amounts and patterns by quantifying causative processes, such as orographic enhancement. One example of this is the parameter−elevation regressions on independent slopes model (PRISM), used to create the maps shown in Figs. 4 and 5.[4] This model distributes point precipitation values, usually at a time step of 1 mo or longer, to a grid by estimating many of the factors discussed earlier, and others. PRISM has been successfully used to develop maps in the new, digital Climate Atlas of the United States,[5] as well as many other products.

CONCLUSION

The distribution of precipitation over the landscape is dictated by processes that enhance upward vertical motion in the presence of sufficient moisture. Longer time averaging periods reveal only the most consistent processes, while at successively shorter time periods spatial heterogeneity and inconsistencies increase. Due to a multitude of factors that control these processes, including the shape and size of land and water areas, latitude and topography, precipitation distribution pattern determination is usually less than straightforward.

REFERENCES

1. Ahrens, C.D. *Meteorology Today: An Introduction to Weather, Climate and the Environment*; West Publishing: St. Paul, MN, 1991.
2. Bergeron, T. Uber De Dreidimensional Verkknupfende Wetteranalyse. Geofysiske Publikasjoner **1928**, *5* (6).
3. Jackson, I.J. *Climate, Water and Agriculture in the Tropics*; Longman Group UK Ltd.: Essex, England, 1989.
4. Daly, C.; Taylor, G.H.; Gibson, W.P.; Parzybok, T.W.; Johnson, G.L.; Pasteris, P.A. High-Quality Spatial Climate Data Sets for the United States and Beyond. Trans. ASAE **2000**, *43* (6), 1957−1962.
5. National Oceanic and Atmospheric Administration, National Climatic Data Center. Climate Atlas of the United States. 2000. CD-ROM.

Precipitation, Forms of

Richard T. Wynne
*National Oceanic and Atmospheric Administration (NOAA),
Amarillo, Texas, U.S.A.*

INTRODUCTION

In nature, water can take solid or liquid form under many situations in the atmosphere and on the surface of the earth. Precipitation is a special category or subset of the conditions under which water exists. Specifically, water particles, in either liquid or solid form, to be defined as precipitation that must both 1) fall from the atmosphere; and 2) reach the ground. This definition would then include rain, drizzle, snow, sleet, and freezing rain and exclude other forms of water in the atmosphere such as clouds, fog, dew, rime, or frost in that it must fall from the atmosphere. It also eliminates forms such as virga since precipitation must reach the ground.[1]

BEGINNINGS OF THE PROCESS

What determines the type of precipitation? The vertical temperature and moisture profiles of the atmosphere are important factors. If ice crystals encounter warm and cold layers of air on the way to the ground, then the form of precipitation could change. Vertical motions of the air are also factors since they will affect the path of the water particles on the way to the ground. No matter what the type of precipitation we see at the ground, the process usually begins in clouds. Tiny droplets join to form much bigger drops, which, begin to fall earthward. At lower temperatures, clouds may consist of ice crystals that form from the freezing of water droplets or sublimate directly from water vapor into solid ice crystals. The crystals begin to collide and aggregate to form snowflakes that start to fall toward the earth. The process, however, is just beginning. The temperature and moisture structure of the atmosphere determines which form of precipitation is observed at the ground.

COMMON FORMS OF PRECIPITATION

In the warm season of the year, *rain* is usually the most common form of precipitation. When a deep, moist layer of air is at the earth's surface with the temperature above 0°C, rain occurs if the drops survive to the surface. If ice or snowflakes enter this warm surface layer and it is sufficiently deep, then the ice melts into liquid drops and arrives at the surface as rain. When the surface air is very dry as it is often in the western Great Plains, one may see rain shafts extending below the cloud bases but evaporating before reaching the ground. This feature is identified as virga and is not considered to be precipitation.

Ice crystals can carry a thin coating of water even at temperatures well below freezing. As a result, cloud ice crystals will join together when they collide and form flakes. Once these flakes become heavy enough, they will fall toward the earth. During the winter season, the entire layer of air may be below 0°C and the flakes will reach the ground as *snow*.

Sometimes, a warm layer of air will ride over a very cold layer of air where temperatures are below freezing. When this happens, the drops can freeze into pellets of ice and is termed *sleet* in most parts of North America. These ice pellets tend to bounce when they land on a solid surface and remain intact. On the other hand, snow on its way to the ground may encounter an air layer just warm enough to melt the smaller snowflakes producing a mixture of snow and rain at the surface. This is termed sleet in some parts of the world.[2]

As stated earlier, water-coated ice crystals can join together to form snow. These ice crystals may also fall through a cold air layer freezing the outer coating. The precipitation then becomes snow pellets, sometimes called *graupel*. Graupel is opaque and, like ice pellets, bounces when it hits a solid object. Often, it will break up following impact unlike ice pellets.

Drizzle is another form of precipitation since it falls to the ground. Numerous, very small drops that are affected easily by air currents, but do eventually fall to the surface, are classified as drizzle. Fog, which often accompanies drizzle, is not considered to be precipitation since it does not fall. In observing, drizzle drops are considered to be less than 0.5 mm.[3] Drops larger than this size are classified as raindrops. Sometimes, while making an observation, it is difficult to tell if light rain or drizzle is occurring. A rule-of-thumb many observers use is to observe an open surface of water such as a puddle of water if one is available. If the water surface is being disturbed by the precipitation, then it is considered to be light rain. If the water surface remains undisturbed, it is classified as drizzle. Drizzle usually forms in shallow, stratiform low cloud layers. Typically, the weather system is nonconvective, or does not have much vertical development. Warm frontal systems usually have a gentle sloping structure. Air and moisture is lifted slowly and fewer collisions between water particles occur. Many times, drizzle forms in advance of a warm front.

Hail is another common form of precipitation. Hail most often accompanies strong thunderstorms where violent downdrafts and updrafts exist nearly side-by-side. Hail formation can begin as ice crystals fall through a moist air layer where liquid water coats the ice particles. The ice particles are then caught in an updraft and carried to great altitudes where the water coating freezes. Once again, the frozen particle falls toward the earth and is coated with a layer of liquid water and carried upward in another updraft. This cycle continues and layer after layer of ice builds up on the original particle. Strong updrafts can result in large hail sizes. Finally, the hail becomes too heavy to be held by the updrafts and it falls to the earth.

CONCLUSION

Many forms of precipitation have been identified. Only the most common ones have been reviewed here. To be classified as "precipitation," the liquid or solid form of water must both fall from the atmosphere, and, it must reach the ground. The temperature and moisture structure of the column of air through which the particle falls often determined what form it will be in when it reaches the ground. Ongoing research continues on the use of temperature and moisture profile of the atmosphere to predict what type of precipitation can be expected. Weather models now give a "first guess" as to what these profiles will look like in the future and to predict the weather type.

REFERENCES

1. Huschke, R.E., Ed. Glossary of Meteorology. American Meteorological Society: Boston, MA, 1959.
2. Petterssen, S. Types of Clouds and States of Sky. *Introduction to Meteorology*, 3rd Ed.; McGraw-Hill Book Company: New York, 1969; 88–91.
3. Petterssen, S. Clouds and Precipitation. *Weather Analysis and Forecasting*, 2nd Ed.; McGraw-Hill Book Company: New York, 1956; Vol. II, 74–78.

Precipitation Measurement

Marshall J. McFarland
Texas A&M University, College Station, Texas, U.S.A.

P

INTRODUCTION

The objective of precipitation measurement is to determine the spatial and temporal distribution of precipitation, primarily rain and snow, but including all forms of precipitation. Measurement requirements depend on the scope and purpose of their use. For climatological and hydrological purposes, such as water supply assessment, precipitation measurement seeks to determine the amount of water that reaches the earth surface over a given area, usually for a 24-hr period with an area of $100 \, \text{km}^2$ or greater. For stormwater runoff and flash flood forecasting, precipitation amount and rate measurements are needed on a time scale of minutes to an hour for areas of tens of square kilometers. For microwave circuit design, rainfall rate along a narrow transmission path is needed on a time scale of a few minutes. Precipitation is usually measured with gages of various designs that meet specific needs over a wide range of geographic locations. Precipitation may also be indirectly measured or inferred with remote sensing technology with sensors operating in visible, infrared, microwave, or gamma ray portions of the electromagnetic spectrum. Remote sensing technology is beyond the scope of this report. A review of the measurement of precipitation is given in Ref. [1].

PRECIPITATION GAGES

Development of Precipitation Gages

Rain gages were used in India in the 4th century B.C., Palestine in the 1st century B.C., China in the 13th century, and Korea in the 15th century.[2,3] The gages used in Korea were cylindrical about 30 cm deep and 15 cm diameter, so would have about the same characteristics and accuracy as many of the gages in widespread use today. Rain gages were first used in Europe in the 17th century and included a tipping-bucket gage developed by Sir Christopher Wren and modified by Robert Hooke in 1678.[3] Numerous designs of gages were developed around the world in the 18th century. In 1802, Dalton (cited in Ref. [3]) described the function and basic design of the rain gage:

> "The rain gage is a vessel placed to receive the falling rain, with a view to ascertain the exact quantity that falls upon a given horizontal surface at the place. A strong funnel, made of sheet iron, tinned and painted, with a perpendicular rim two or three inches high, fixed horizontally in a convenient frame with a bottle under it to receive the rain, is all the instrument required."

Virtually any open container will collect rain and snow, but few will collect the "exact quantity" of rain or snow that would fall on the horizontal surface in the absence of the gage. The measurement of the exact quantity of rain and snow that falls on a given horizontal surface has been the subject of considerable research and development in the past several hundred years. In 1769, it was established that gages elevated above the surface caught less rainfall than gages near the surface,[4] which was subsequently shown in 1861 to be due to effects of wind.[5] Ground level gages were developed in 1842 to avoid the effects of wind.[4] In the late 1800s, Nipher developed the first shielded gage, a design still in use, to decrease the influence of wind on the collection of snow.[5]

Standard Precipitation Gages in the United States

Two similar designs of nonrecording rain gages are used in the climatological network of the National Weather Service.[6] The standard gage has a 20.3-cm (8 in.) orifice diameter, a funnel, a measuring tube, and an outer container. The measuring tube holds 5.08 cm precipitation; overflow collects in the outer container. A calibrated measuring stick is inserted into the measuring tube and is read to the nearest 0.01 in. (0.254 mm). The funnel and measuring tube are removed when snow is expected. The snow is melted for equivalent water measurement. A 10 cm scale version of the standard gage is also in use, with measurements read from a calibrated scale on the clear plastic side.

The National Weather Service has two types of weighing gages that record rate and amount of precipitation,[6] with an accuracy of 0.01 in. (0.254 mm). The collection portion of the gage is similar to the standard gage, but the funnel directs the catch to a collector mounted on a weighing mechanism. The Belfort (Fischer and Porter) recording gage converts the weighed precipitation to a punched tape output. The Universal

Encyclopedia of Water Science
DOI: 10.1081/E-EWS 120010317

recording gage converts the weighed precipitation to a strip chart or to direct current (dc) voltage for telemetry.

Tipping bucket recording precipitation gages are used, especially with automatic weather stations, to record precipitation rate and amount. The collector portion of the gage is a scale design of the standard gage. The funnel directs the catch to one of two buckets balanced on a fulcrum, each with a typical capacity of 0.01 in. (0.254 mm). When the bucket fills, it tips and discharges the water into a container and rotates the other bucket under the funnel. The time of the tip is recorded electronically when a magnet trips a switch.

Orifice Design for Precipitation Gages

Any open container will collect precipitation. If the requirements for accuracy are not precise, then virtually any open container will be satisfactory. Indeed, for extreme rainfall events when the daily or storm event rainfall exceeds the capacity of the gage, "bucket surveys" are used to estimate the total rainfall. The rainfall depth in any open container, such as a barrel or a bucket, will be measured. Rain gages for individual's use, as opposed to gages in networks, may be of any design and any design may be satisfactory if precision is not an issue. Common designs are clear plastic gages with square or cylindrical orifices, with imprinted measurement scales on the collection tube. Orifice widths or diameters range from about 2 cm to 10 cm. The larger diameter orifice gages may have a funnel; smaller diameter gages do not, but may be wedge or cone shaped to increase the accuracy for measurement of lower rainfall amounts. The diameter or width of the orifice does not have a significant effect on the amount of rainfall collected by the gage.[7-9]

Effects of Wind on Precipitation Measurement

Any object placed above ground level will result in increased wind speed over the top of the object. When the object is a rain gage, the increased air flow over the top of the collector will deflect rain and snow particles from their original path. This results in an undercatch of precipitation,[10,11] especially for snow.[12,13] As the gage gets larger, the effects of the wind increase. The height of the gage is also a major factor, because the wind speed increases rapidly with height above the surface. A large gage mounted on a post will catch less precipitation than a smaller gage mounted on a rod or open stand. The measurement error associated with large gages above ground level with moderate wind and light snow may exceed 50%. Several approaches have been used to reduce the effect of wind on the catch of precipitation. Shields of vertical metal or wood slats surrounding the gage will deflect the wind downward and away from the gage

orifice. Shielded gages are in widespread use for snow measurement networks (e.g., SNOTEL—SNOw TELemetry[14]) in the United States, but are not routinely used in the National Weather Service climatological network. Ground level or pit gages installed at the surface will eliminate wind effects, but create other problems such as splash effects, and collection of blowing leaves, drifting snow, or surface water in the gage. Pit gages are used for reference purposes, but are not used in measurement networks. Low bushes and/or 50% snow fence around the gage will also reduce the effects of wind on the precipitation catch by the gage.[15,16]

Precipitation Measurement Errors

Precipitation gages are subject to measurement errors from wetting, evaporation, condensation, rain splash, and snow plugging and capping. Rainfall that adheres to (wets) the surface of the gage will not be measured in the collection tube. The collected precipitation in the collection tube is subject to evaporation, especially in hot and dry conditions. Condensation that forms dew and frost on the gage may add a few mm depth to the precipitation total. Rain may splash out of the gage, especially for designs that have a shallow funnel with a slope of less than 45° from the vertical and designs with open collector tubes when the gage nears its capacity. Wet snow will stick to the inside of the gage orifice and may prevent additional snow from entering the gage. Small diameter (or width) gages and gages with shallow funnels are especially susceptible to snow plugging and capping.

Another source of measurement error arises from a combination of wind with sloping terrain with the precipitation event. The precipitation caught by a gage with a horizontal orifice will deviate from the precipitation that is actually incident upon the terrain. For precise measurements, a pit gage or a gage with an orifice parallel to the terrain slope will provide precipitation measurements that are more representative.[17]

DIRECT MEASUREMENTS OF SNOW

The measurement of snow in mountainous areas is complicated by inaccessibility for daily observation, lack of ideal exposures for gages, measurement errors with gages, sloping terrain, variable exposures due to trees and natural topographic features, and redistribution of snow by wind. Snow measurements are made by direct methods such as gages, core extractions, and depth measurements with a stick; and by indirect methods such as weighing snow pillows and remote sensing from air and space craft.[14] A white-painted wooden board is used for snow depth measurements. The snowboard is placed horizontally on the ground surface or flush with the snow surface in a representative location. The observer records the 24-hr depth accumulation with a ruler, then cleans and replaces the snowboard. The federal snow sampler is used to extract cores from the total snow pack. The sampler is a light-weight graduated aluminum tube that is forced into the snow pack.[18] The snow water equivalent is obtained by weight. Core samplers are used to take snow measurements at monthly or longer intervals along predetermined transects known as snow courses. A snow pillow is a hydraulic weighing platform of rubber or stainless steel. The snow that accumulates on the pillow or a series of pillows is weighed with the use of a pressure transducer so the water equivalent may be transmitted via telemetry.[19]

CONCLUSION

The basic technology of precipitation measurement with gages has not changed in the past several hundred years. A precipitation gage is basically an open container that ideally catches the precipitation that would accumulate on a horizontal ground surface at the gage location. Measurement errors, needs for rainfall intensity measurements, and problems with snow measurement have resulted in a wide variety of gage designs. When precision is not required, virtually any open container will serve the purpose. For hydrology and climatology, standardized gage designs and measurement procedures are necessary. New technology, especially in remote sensing, will continue to evolve and to increase our knowledge of the spatial and temporal distribution of precipitation.

REFERENCES

1. Hanson, C.L.; Gebhardt, K.A.; Johnson, G.L.; McFarland, M.J.; Smith, J.A. Precipitation. *Hydrology Handbook*, 2nd Ed.; American Society of Civil Engineers: New York, 1996; 5–74.
2. Biswas, A.K. Development of Rain Gages. J. Irrig. Drain. Div. ASCE **1967**, *93* (IR3), 99–124.
3. Biswas, A.K. *History of Hydrology*; American Elsevier: New York, 1970.
4. Rodda, J.C. The Systematic Error in Rainfall Measurement. J. Inst. Water Eng. **1967**, *21*, 173–177.
5. Nipher, F.E. On the Determination of True Rainfall in Elevated Gages. Proc. AAAS **1878**, 103–108.
6. NWS, *Cooperative Station Observations*, Observing Handbook No. 2; U.S. Department of Commerce, National Weather Service: Washington, DC, 1989.
7. Huff, F.A. Comparison between Standard and Small Orifice Raingages. Trans. AGU **1955**, *36* (4), 689–694.
8. Roper, I.J. A Simple Rain-Gauge for Dense Network Design. Weather **1975**, *30* (10), 329–336.
9. Snow, J.T.; Harley, S.B. Basic Meteorological Observations for Schools. Bull. AMS **1988**, *69*, 497–507.
10. Sevruk, B. Effect of Wind and Intensity of Rain on the Rain Catch. In *Correction of Precipitation Measurements*; Sevruk, B., Ed.; Zurcher Geographische Schriften No. 23: Geneva, 1986; 251–256.
11. Sevruk, B. Accuracy of Precipitation Data. *Observed Climate Variations and Change: Contributions in Support of Section 7 of the IPCC Scientific Assessment*; Intergovernmental Panel on Climate Change: Geneva, 1990; XXIII.1–XXXIII.12.
12. Alter, J.C. Shielded Storage Precipitation Gages. Mon. Wea. Rev. **1937**, *65* (7), 262–265.
13. Weiss, L.L.; Wilson, W.T. Precipitation Gage Shields. IASH Pub. **1957**, *43* (1), 462–484.
14. Doeskin, N.J.; Judson, A. *The Snow Booklet: A Guide to the Science, Climatology, and Measurement of Snow in the United States*; Colorado State University Department of Atmospheric Science: Fort Collins, Colorado, 1996.
15. Larson, L.W. *Shielding Precipitation Gages from Adverse Wind Effects with Snow Fences*, Water Resources Series No. 24; University of Wyoming Water Resources Research Institute: Laramie, Wyoming, 1971.
16. Hanson, C.L. Precipitation Catch Measured by the Wyoming Shield and the Dual-Gage System. Water Resour. Bull. **1989**, *25* (1), 159–164.
17. Sharon, D. The Distribution of Hydrologically Significant Rainfall Incident on Sloping Ground. J. Hydrol. **1980**, *46* (1), 165–188.
18. Powell, D.R. Observations on Consistency and Reliability of Field Data in Snow Survey Measurements. In *Proceedings Western Snow Conference, 55th Annual Meeting*, Vancouver, BC, 1987; 69–77.
19. Schaefer, G.L.; Shafer, B.A. A Critical Analysis of SNOTEL Performance in the Rocky Mountains. In *Proceedings International Symposium in Hydrometeorology*; Johnson, A.I., Clark, R.A., Eds.; AWRA: Bethesda, Maryland, 1983; 31–37.

Precipitation Measurements with Remote Sensors

Marshall J. McFarland
Texas A&M University, College Station, Texas, U.S.A.

INTRODUCTION

The objective of precipitation measurement is to determine the spatial and temporal distribution of precipitation, primarily rain and snow. Historically, precipitation has been measured with gages, which capture samples of the precipitation for direct measurement. Gage measurements of precipitation have limitations, especially for operational meteorological and hydrological purposes such as short period weather and flash flood forecasting. These limitations include:

1. The density of measurements in most gage networks is not sufficient for assessment of precipitation from thunderstorms in small watersheds. More-or-less typical precipitation networks for climatological purposes have a gage density of about one gage every 30 km (900 km^2 area). The highest precipitation intensities in a thunderstorm cell occupy a much smaller area, so could easily be missed in a fixed gage network.
2. Many areas of interest are not suitable for direct measurement of precipitation with gages. These areas include mountains (for water supply assessment) and oceans (for earth heat and moisture budgets).
3. The cost of direct measurements (including equipment, maintenance, personnel, data acquisition and processing) precludes expansion of gage networks for operational uses.

Remote sensing of precipitation is widely used to obtain increased spatial and temporal accuracy. With remote sensors, the precipitation is not captured or directly measured. The precipitation is inferred from physical, statistical, and/or empirical relationships between precipitation characteristics and the emitted or reflected radiation from the earth and atmosphere. Remote sensors that record naturally emitted radiation are referred to as passive, while active remote sensors record reflections of radiation emitted from the sensor. Precipitation estimation with weather radar is an example of an active remote sensor. Cloud information from visible and near-infrared imagery obtained from earth satellite sensors is an example of passive remote sensing. The sensors may be land-based, or mounted on aircraft or earth satellites. A summary of measurement of precipitation is in Ref. [1].

ESTIMATION OF PRECIPITATION WITH WEATHER RADAR

Weather radar operates in the microwave portion of the electromagnetic spectrum, usually at wavelengths from 3 cm to 10 cm. At these wavelengths, large cloud water droplets, raindrops, hail, snow particles, and other solid forms of precipitation reflect emitted radiation. The backscattered radiation, known as the reflectivity, Z, is highly correlated with the characteristics of the precipitation in the volume of the radar beam. For spherical raindrops, the reflectivity is a function of the sixth power of the raindrop diameter. Reflectivity of rainfall within and below a cloud volume is primarily a function of the numbers and diameters of the larger raindrops.

The relationship between the radar reflectivity and the rainfall rate is a power function, $Z = aR^b$, where a and b are empirically fitted variables.[2] The variables a and b of the Z–R relationship change with type of precipitation, precipitation intensity, raindrop shape, presence of liquid films on frozen forms of precipitation, and ambient conditions within the cloud. The value of a is typically 200 and will range from 100 in stratiform rainfall to 400 for intense convective rainfall. The value of b is nearly constant at 1.6, with a usual range of 1.3 to 1.6.[3] With constant values of a and b, radar estimates of precipitation should be within a factor of two about 75% of the time.[4]

Several sources of error in the estimation of precipitation with radar are present. Any change in the relationship between reflectivity and rainfall will change the accuracy of the rainfall estimation. A major source of error is inherent in the geometry of the radar beam as it intersects a cloud. Normally, the radar scans at low elevation angles, such as 0.5 – 1.5°, within about 75 km of the radar location. Consequently, the precipitation is fairly close to the earth surface. If the fall speeds of the precipitation particles differ from terminal velocity as a result of updrafts or downdrafts, the radar estimation will be an overestimation or an underestimation. For heavy rainfall intensity in a strong downdraft, the underestimation could be nearly 50%.[3]

Encyclopedia of Water Science
DOI: 10.1081/E-EWS 120010673

Due to earth surface curvature, the radar beam intersects cloud volumes at increasing elevations with increasing range from the radar location. At increasing range, the radar estimate of rainfall rate could be an underestimation if the cloud volume in the radar beam is above a layer where precipitation rate is still increasing through growth of raindrop size through processes of coalescence and collision. With increased range, the radar beam may intersect the 0°C isotherm, where melting of snow begins to occur. With initial melting, a snowflake becomes covered with a film of water before collapsing into a smaller raindrop. Water covered snow is much more reflective than the raindrops below the layer of melting snow. As a result, a "bright band" about 100 m deep with a reflectivity typically two to five times greater than that of the rainfall below will appear in the radar display of the return.[5] If the radar beam includes cloud volume above the bright band, the reflectivity will decrease due to the very low reflectivity of the snow. Generally, the precipitation rates will be underestimated when the radar beam is above the freezing level.

Other sources of error include incomplete beam filling by the cloud, false echoes, and anomalous propagation, and intervening clouds and precipitation. Rain drops on the radome will also attenuate the backscattered radiation.

For maximum accuracy of radar estimates of precipitation rate and amount, the variables in the $Z-R$ relationship should be calibrated on a real-time basis. This is possible with a network of automatic, recording rain gages that is connected to the radar processing system through telemetry, modem, or other means of communication. The spatial and temporal accuracy of radar is combined with the point accuracy of gages in this method. Two approaches have been demonstrated. The deterministic approach is to use rainfall rate information from point gages to calibrate the values of the a and b variables in the $Z-R$ relationship or to determine the ratio of gage to radar estimation of rainfall. The adjusted values are then used to adjust the radar estimates of rainfall for the areas between gages. A statistical approach combines the radar and gage spatial information to interpolate and extrapolate the gage measurements throughout the area.[6-8]

VISIBLE AND INFRARED ESTIMATION OF PRECIPITATION FROM SATELLITE IMAGERY

Earth satellites with sensors operating in the visible, near-infrared, and thermal-infrared portions of the electromagnetic spectrum provide several methods of estimation of precipitation.[9] The simplest is perhaps the areal extent of snow, which is readily apparent in visible and near-infrared imagery. Another method has been developed for convective rainfall. Methods of estimating convective

rainfall rates and/or amounts are based on the premise that the intensity of convective rainfall is correlated with the visible brightness and radiative temperature of the cloud tops (10.5 μm to 12.5 μm wavelength). Colder and brighter cloud tops represent both deeper and more intense convection, which is highly correlated with precipitation intensity and amount. The relationships between satellite-derived variables of cumulonimbus cloud top features and rainfall features are developed from radar and gage measurements.

Several advanced methods for estimation of rainfall from satellites with visible and thermal-infrared sensors have been developed. The life-history techniques incorporate information about the life cycle of the convective cloud into the rainfall estimation method[10-12] for scales from individual convective cloud to synoptic systems. The methods are based on similar assumptions that convective cloud development follows an established pattern, raining convective clouds have cloud top temperatures colder than a threshold (e.g., -19°C), rainfall rate is proportional to the cloud area, rainfall intensity is inversely proportional to the temperature of the cloud top, and rainfall distribution in time is a function of the stage of the life cycle of the cloud.

PASSIVE MICROWAVE MEASUREMENT

Snow pack properties, such as depth, snow water content, and age, may be developed with the use of passive microwave radiometers on earth satellites. Passive microwave radiation sensors operate in similar wavelengths as weather radar, which is active microwave radiation. Passive microwave sensors record the radiation naturally emitted by the earth surface and atmosphere, which is proportional to the product of the emissivity and the first power of the temperature at microwave wavelengths (the Rayleigh–Jeans approximation to Planck's Law[13]). Consequently, the recorded radiation is referred to as the brightness temperature. The emissivity is an inverse function of the dielectric constant of the emitting surface. The value of the real component of the complex dielectric constant, the permittivity, is in the low single digits for air, dry soils, and ice and snow. The permittivity of water ranges from about ten to twenty in the microwave wavelengths of the sensors (e.g., 0.81 cm to 1.55 cm), so the addition of water to dry soil decreases the brightness temperature. The brightness temperature of a land surface is decreased by the loss factor, which is also a function of the dielectric constant. The loss factor is high for water but very low for ice and snow. A dry snowpack will scatter, or attenuate, the brightness temperature proportional to the snowpack depth. Typically, the snowpack properties are retrieved with multiple linear regression for the multiple wavelengths and both polarizations of the radiometer.[14]

The atmosphere is essentially transparent to emitted microwave radiation from the earth surface except when convective clouds with rainfall are present. Tropical rainfall over oceans is determined from the emitting and scattering characteristics of the precipitating clouds.[15] At the microwave wavelengths of the passive microwave sensors on earth satellites, the precipitating clouds appear as warm areas over a very cold background.

TERRESTRIAL GAMMA SNOW MEASUREMENT

The water equivalent of a snowpack can be determined by measuring the attenuation of naturally occurring gamma radiation emitted from potassium, uranium, and thorium isotopes in the upper 20 cm of soil.[16] Water mass (not necessarily liquid) in the soil and snow attenuates the gamma radiation, so differences in the radiation emitted from bare ground and snow-covered ground are used to determine the snow water content and other properties. The gamma radiation sensors are flown on aircraft to determine snowpack properties, primarily in the Great Plains of the northern United States and southern Canada.

CONCLUSION

Remote sensing of precipitation is in widespread use in meteorology, climatology, and hydrology. In remote sensing of precipitation, the precipitation is not captured or otherwise directly measured. The interactions of the precipitation with radiation in the electromagnetic spectrum are measured with sensors, and then translated into precipitation characteristics with the use of physical, statistical, and empirical relationships. Precipitation estimation with weather radar is a good example of a remote sensor system. Microwave radiation is emitted from an antenna at the radar site. As the radiation pulse intersects a cloud, some of the radiation is scattered back towards the antenna by snow, water, and ice. The reflectivity is correlated with the precipitation characteristics, which have been determined with precipitation measurements from gages. Information on clouds and precipitation may be obtained from visible, near-infrared, thermal-infrared, microwave, and gamma ray portions of the electromagnetic spectrum. The sensors may be at fixed locations, or may be mounted on ships, aircraft, or earth satellites.

REFERENCES

1. Hanson, C.L.; Gebhardt, K.A.; Johnson, G.L.; McFarland, M.J.; Smith, J.A. Precipitation. *Hydrology Handbook*, 2nd Ed.; American Society of Civil Engineers: New York, 1996; 5–74.
2. Krajewski, W.F.; Smith, J.A. On the Estimation of Climatological Z–R Relationships. J. Appl. Meteorol. **1991**, *30* (10), 1436–1445.
3. Austin, P.M. Relation between Measured Radar Reflectivity and Surface Rainfall. Mon. Wea. Rev. **1987**, *115* (5), 1053–1070.
4. Wilson, J.W.; Brandes, E.A. Radar Measurement of Rainfall—A Summary. Bull. Am. Meteorol. Soc. **1979**, *60*, 1048–1058.
5. Smith, C.J. Reduction of Errors Caused by Bright Bands in Quantititative Rainfall Measurements Made using Radar. J. Atmos. Ocean. Tech. **1986**, *3* (1), 129–141.
6. Stellman, K.M.; Fuelberg, H.E.; Garza, R.; Mullusky, M. An Examination of Radar and Rain Gauge-Derived Mean Areal Precipitation over Georgia Watersheds. Weather Forecasting **2001**, *1*, 133–144.
7. May, B.R. Discrimination in the Use of Radar Data Adjusted by Sparse Gauge Observations for Determining Surface Rainfall. Meteorol. Mag. **1986**, *115* (1365), 101–115.
8. Collier, C.G.; Knowles, J.M. Accuracy of Rainfall Estimates by Radar, Part III: Application for Short-Term Flood Forecasting. J. Hydrol. **1986**, *83* (3), 237–249.
9. Engman, E.T.; Gurney, R.J. *Remote Sensing in Hydrology*; Chapman and Hall: New York, 1991.
10. Griffith, C.G. Comparison of Gauge and Satellite Rain Estimates for the Central United States During August 1987. J. Geophys. Res. **1987**, *92* (8), 9551–9566.
11. Stout, J.E.; Martin, D.W.; Sikdar, D.N. Estimating GATE Rainfall with Geosynchronous Satellite Images. Mon. Wea. Rev. **1979**, *107* (5), 585–598.
12. Schofield, R.A.; Oliver, V.J. A Satellite Technique for Estimating Rainfall from Flash Flood Producing Thunderstorms. In *Satellite Hydrology*; Deutsch, M., Wiesnet, D.R., Rango, A., Eds.; AWRA: Minneapolis, MN, 1981; 70–76.
13. Ulaby, F.T.; Moore, R.K.; Fung, A.T. *Microwave Remote Sensing, Vol III. From Theory to Applications*; Artech House: Norwood, MA, 1986.
14. McFarland, M.J.; Neale, C.M.U. Land Parameter Algorithm Validation and Calibration. *DMSP Special Sensor Microwave/Imager Calibration/Validation Final Report, Vol II*; Naval Research Laboratory: Washington DC, 1991; 9-1–9-108.
15. Heymsfield, G.M.; Geerts, B.; Tian, L. TRMM Precipitation Radar Reflectivity Profiles as Compared with High-Resolution Airborne and Ground-Based Radar Measurements. J. Appl. Meteorol. **2000**, *39* (12), 2080–2102.
16. Carroll, S.S.; Carroll, T.R. Estimating the Variance of Airborne Snow Water Equivalent Estimates Using Computer Simulation Techniques. Nordic Hydrol. **1990**, *21* (1), 313–319.

Precipitation Modification

George W. Bomar
Texas Department of Licensing and Regulation, Austin, Texas, U.S.A.

INTRODUCTION

The water resources of any nation are one of its most precious treasures. However, increases in population and economic growth place greater demands on this valued asset. We are continually reminded that water, while indispensable for personal and economic well being, is a finite resource. We are made all the more keenly aware of its value when deviations in the normal patterns of rain (and snow) foment drought, which places even greater strain on the available water supply.

Traditionally, sources of fresh water are regarded as either surface or groundwater, which are replenished, of course, by precipitation yielded by various species of cloud formations. To augment the supply of water from those sources, dams are constructed to confine more of the streamflow above the surface or wells are drilled to capture greater quantities of water from subterranean *aquifers*. Seldom is the atmosphere viewed as yet another repository of substantial amounts of water—albeit in the form of vapor—waiting to be harvested.

Today, careful and well-designed efforts to manage *atmospheric* water resources are proliferating worldwide. Through the responsible use of new weather-modification, or *cloud-seeding* technologies, those who manage the development of atmospheric water are confronting the challenge of finding new fresh-water sources with increasing success.

GROWING ATMOSPHERIC WATER

Exceedingly tiny cloud droplets materialize in the atmosphere when moist air is sufficiently cooled to its dew point. These minuscule droplets, which typically number in the trillions in a swelling cumulus cloud, may collide with one another to grow larger and larger droplets, which might eventually become big and heavy enough to fall as rainwater. For this to happen, the cloud must persist for a prolonged period. Most clouds do not live long enough to develop a significant load of rain.

Nature facilitates the rain-making process by capitalizing on the role of *supercooled* droplets to grow raindrops much more readily. In growing taller, an increasingly buoyant convective cloud (cumuli) moves more of its water mass into colder (higher) regions of the atmosphere. Though the air is colder than 32°F, the tiny cloud droplets do not immediately freeze into ice, but rather remain in liquid form (supercooled).

Meanwhile, innumerable microscopic particles, such as soil, dust, sand, and salt, start the rain-production process by acting as "seeds" or crystalline skeletons on which these very tiny, supercooled droplets can freeze to form either snowflakes or soft ice (graupel). These seeds become de facto *ice nuclei*, around which more and more supercooled cloud water converges to grow larger and larger raindrops.

Cloud seeding involves the release of artificial ice nuclei to grow even more, and larger, water droplets out of this usually abundant supply of supercooled cloud water. The artificial seeds are of one of two types: 1) glaciogenic (ice forming); and 2) hygroscopic (water attracting). The most common type of glaciogenic seed is silver iodide, who crystalline structure most closely resembles that of a natural ice crystal. Another well-used glaciogenic material is dry ice, which almost instantaneously produces large numbers of small ice particles when dropped in pellet form into clouds with supercooled water. In clouds that never develop vertically, such that the cloud's water mass is never chilled to the point of becoming supercooled, hygroscopic seeding agents work well in growing large raindrops. The hygroscopic seeds are usually small salt particles, such as potassium chloride.

SEEDING TECHNIQUES

Timing and targeting the artificial seeds in promising clouds are the two most critical factors in successful cloud seeding. Seeding must be done opportunistically and precisely, in the right locations when the time is ripe.

For seeding convective clouds, which tend to grow dynamically in a matter of minutes, aircraft are the favored method of delivery. Specially-equipped aircraft, bearing racks containing flares (pyrotechnics) of seeding materials or wing-mounted generators for releasing solutions of artificial ice crystal, can release the seeds straight into the rising air current (updraft) below the bases of developing clouds. In some instances, aircraft can get seeding material

Encyclopedia of Water Science
DOI: 10.1081/E-EWS 120010319

into the core region of supercooled clouds by dropping flares from above the tops of those clouds.

In mountainous terrain, where moisture-laden clouds tend to "hug" ridge lines, aircraft usually cannot safely navigate to dispense seeding materials. Because the flow of moist air feeding the clouds is often quite predictable, it is possible to disperse the seeds using a network of ground-based generators. These generators can be manually operated, even remotely controlled, to regulate the flow of seeding material at prescribed levels and for predetermined durations.

ASSESSING THE IMPACT

The true measure of any endeavor, including weather modification, is the answer to the daunting question: Are the results from the effort worth the resources needed to produce those results? Assessing cloud-seeding activities is a formidable challenge because the impact of the seeding must be separated from the highly variable, natural occurrence of rain, snow, and hail from cloud formations. Many methods and types of data have been used to evaluate cloud-seeding efforts, with each having its own strengths and weaknesses. Some assessments are made using direct evidence: measurements of rain, hail, snow. Others are based largely on secondary evidence: insurance statistics and crop yields. Ultimately, the efficacy of any particular evaluation will depend on the type and amount of data available for analysis.

Many weather-modification projects, past and ongoing, have furnished evidence that seeding, when performed timely and in a well-targeted fashion, has altered the behavior of cloud formations in such ways that the objectives of the projects (rain enhancement, snowpack augmentation, hail suppression) have been achieved. Benefits, when quantified, have far outpaced the costs to conduct the operations.

The ultimate way to evaluate a weather-modification project is through randomization: Storms are randomly selected for seeding—or to be left untreated. Then, the two groups are compared to discern differences in behavior, during and following seeding. The drawback to such an analytical approach is that only about half of all storms get treated, reducing the overall effect as well as the benefit-to-cost ratio.

Still, with the promise of success, the technology will continue to be applied in many different climatic regimes. As research techniques improve, and as better seeding technologies are developed, better and more consistent results are sure to be achieved.

FURTHER READING

Braham, R.R., Jr., (Ed.) *Precipitation Enhancement—A Scientific Challenge*, Meteorological Monograph No. 43; American Meteorological Society: Boston, MA, 1986; 171.

Mather, G.K.; Terblanche, D.E.; Steffens, F.E.; Fletcher, L. Results of the South African Cloud Seeding Experiments using Hygroscopic Flares. J. Appl. Meteorol. **1997**, *36*, 1433–1447.

Orville, H.D. The Uses of Cloud Models in Weather Modification. J. Wea. Mod. **1990**, *22*, 137–142.

Rosenfeld, D.; Woodley, W.L. Effects of Cloud Seeding in West Texas: Additional Results and New Insights. J. Appl. Meteorol. **1993**, *32*, 1848–1866.

Precipitation Simulation Models

Timothy O. Keefer
United States Department of Agriculture (USDA), Tucson, Arizona, U.S.A.

P

INTRODUCTION

The variability of precipitation across a range of spatial and temporal scales, from short-duration high-intensity down-bursts within a localized storm to the seasonal and annual variations at a single location and across the globe is obvious to a casual weather observer. Frequently, in the planning and management of agricultural and engineering activities, precipitation information that reflects this natural variability is needed. Examples include irrigation design and application, evaluation of agricultural runoff for soil erosion and water quality, cropping and seeding patterns, sizing and placement of culverts and dams, scheduling and selection of agricultural and construction equipment. The demands of the particular use of the information vary from within-storm intensities to daily amounts to regional and seasonal accumulations each with different precision. Generally, the source of such information is the precipitation data measured and recorded at a point. Precipitation information for a particular location may not be adequately known or available in the specific time-frame required because of short or nonexistent records of measurements, inaccurate or inconsistent data, or budgetary constraints.

An alternative approach is to use a precipitation simulation model which generates sequences of synthetic precipitation which share the same statistical properties as the observed time series. Three broad categories of precipitation simulation models exist in various degrees of mathematical and statistical complexity which relate to the type of precipitation simulated. Low-resolution, large-area precipitation data can be generated by 3-dimensional dynamic-numerical general circulation models (GCMs); rainstorm event occurrence and intensities are simulated by spatial-temporal models; daily precipitation occurrence and amount are modeled by a family of fairly simple stochastic/statistical algorithms. The latter of these are often part of a larger model called a weather generator, which simulates other weather related land/atmosphere variables such as solar radiation, temperature, or soil moisture. The generated synthetic sequences of precipitation are used for a variety of purposes such as: analysis for water resource engineering applications, climate change scenarios, and as input to other hydrological or natural resource models. This differentiates these models and their results from the class of models which are used in weather prediction and forecasting. All three categories of models are valuable tools for scientific research and agricultural, engineering and hydrological applications. The selection of any one type should fit the intended analysis, level of complexity and scale of required results. Overviews of various precipitation simulation models are Ref. [1] for GCMs, Ref. [2] for rain storm modeling, and Ref. [3] for daily precipitation.

MODELS AND APPLICATIONS

General Circulation Models

General circulation models (also referred to as global climate models and sharing a common acronym, GCM) use the same fundamental equations of conservation of mass, energy and momentum as do numerical weather prediction (NWP) models. These dynamic meteorology models, and similarly structured regional climate models (RCM), attempt to numerically solve systems of simultaneous nonlinear differential equations which themselves are intended to represent the complex physical processes involved in atmospheric dynamics. Whereas NWPs use observations of recent atmospheric dynamics as boundary conditions for model runs and produce weather prediction in the short term (1–10 days), GCMs use arbitrary boundary conditions and alternative atmospheric parameters to simulate climate for the past, current or future. One result of GCM simulations is precipitation over an area, called a grid, which may be on the order of $10^5 \, \mathrm{km}^2$, whereas for an RCM the spatial resolution may be $10^1 \, \mathrm{km}^2 – 10^3 \, \mathrm{km}^2$.

Precipitation is generally simulated in these models by convective processes resolved from radiation, temperature, pressure, and humidity simulated at various atmospheric layers within a gridbox. These simulations of precipitation are useful for evaluating changes in vegetation and surface water resources under different possible climate change scenarios. To increase the resolution of the GCM simulation, downscaling by statistical techniques or incorporating an RCM into the GCM achieves finer resolution precipitation output

Encyclopedia of Water Science
DOI: 10.1081/E-EWS 120010320

applicable to soil moisture and runoff analysis for subgrid scales. Excellent sources of information about and applications of the models are available at WEB sites such as Intergovernmental Panel on Climate Change,[4] American Institute of Physics,[5] and NASA's Goddard Institute for Space Studies.[6]

Spatial-Temporal Rainstorm Models

Stochastic simulation models of rain storm events in space and time attempt to reproduce the statistical properties of the event across a range of temporal and spatial scales. Two of the most advanced modeling concepts are: i) stochastic representation of the physical process of rainstorm temporal and spatial evolution and ii) scale-invariance or self-similarity of the spatial rainfall field. The stochastic approach defines the arrival of the rain cells within a rain storm by a point cluster process[7] represented by one of two common models, the Neyman–Scott process or the Bartlett–Lewis process. The former uses a Poisson distribution for the cluster centers, a random number of cells and a distribution of the distance of cell from the cluster center. The latter assumes a Poisson process for arrival of storms, and distributions for the number of cells per storm, intercell intervals, duration and intensity within a cell. For each characteristic, a statistical distribution must be assumed and numerous parameters identified. Alternatively, scale-invariant models[8] exploit the properties of multiplicative random cascades developed in turbulence theory. Observations of rainfall fields suggest that there are certain spatial and temporal properties that behave similarly over a range of scales differing only by a scale parameter. Thus a hierarchy of attributes (e.g., rainfall intensity) can be developed such that larger areas of lower intensity have embedded within them smaller areas of higher intensity and these in turn have even smaller areas of yet higher intensities. Applications of these models are design storms for engineering and water resources and continuous time hydrologic modeling.

Other statistical storm models of simpler structure are derived empirically. One method is to disaggregate daily rainfall amounts to within-storm intensities for the duration of a storm. These models have parameters that are location specific. Another approach is the regionaliza-tion of probabilities associated with storm interarrival time, duration, and amount.

Daily Precipitation Models

Daily precipitation simulation models are the most common for use in a variety of agricultural and engineering applications. These models describe the occurrence (wet) or nonoccurrence (dry) of precipitation on a day and subsequently the amount of precipitation given the day was wet. The occurrence process is modeled most frequently by a first-order, two-state Markov chain. Linked to this occurrence process is a statistical description of precipitation on a wet day, often a gamma or exponential distribution.[9] This family of fairly simple models of daily precipitation is referred to as chain-dependant processes. Equations for these models are given in a companion article in this chapter, "Precipitation Stochastic Processes," and are not duplicated here. The models can be parsimonious in the necessary parameters, are easily parameterized with a sequence of observed daily precipitation (a commonly recorded observation for many stations) albeit for many years. Seasonal variation of model parameters can be accomplished by writing them as Fourier series or by assuming they vary step-wise on a monthly or seasonal basis. The structure of the model provides simple generation of multiple realizations of daily time series. Model output is generally used as input to hydrologic, natural resource, or agricultural models requiring daily time step precipitation. The model parameters are location specific with limited transfer-ability to neighboring locations that do not share the same stochastic precipitation structure, e.g., to a location with a large elevation change. Another limitation of the model is the underestimation of interannual variability. One approach to resolve this has been determining the appropriate order of the Markov chain indicating that for particular seasons and geographic locations a second-order or higher conditional dependence may be required, although not all such variability is explained. Markov chains of more than two states may explain more of the variability and a continuum of states may be best.

Other methods to model daily precipitation occurrence have been advanced, among them: alternating renewal process, discrete auto-regressive moving average, Markov–Bernoulli process, dependence on weather type, Markov-renewal. Some recent weather generator models use multivariate techniques to simulate precipitation con-ditioned on other weather variables or simultaneously with other weather variables or using semiempirical distributions. Although numerous inter-comparisons have been done, no single model provides simplicity, ease of parameterization, and the best fit for all weather types and locations.

An example of a particular precipitation simulation model is provided. The Markov chain-mixed exponential model (MCME) is used to simulate daily precipitation for two stations with different climates in the western United States. This model is the precipitation algorithm embedded in the United States Department of Agriculture-Agricul-tural Research Services (USDA-ARS) weather generator, Generation of Weather Elements for Multiple Applications (GEM).[10] This model is an enhanced version of a series of weather generators developed by the USDA-ARS.[11] Daily precipitation model parameters are estimated from an observed time series of daily data. The optimized parameters are used in the model in conjunction with a random number generator to synthesize a 30-yr period of daily precipitation occurrence and amount. Daily values are summed to seasonal values and the annual averages and variances of these are compared to observations. Fig. 1 shows the results for Tombstone, Arizona plotted as a cumulative distribution function for two 3 month seasons, United States Department of Agriculture, January, February and March (JFM) and October, November and December (OND); Fig. 2 is the same for Eugene, Oregon. The mean is fairly well preserved for both seasons and both the amount and number of occurrences at Tombstone, but the variance is underestimated especially for JFM. The mean is not as well preserved at Eugene, and the variance is underestimated for OND. This is one of the limitations

mentioned previously and it may be due to low-frequency ocean-atmospheric signals, such as the El Niño-Southern Oscillation, which have varying influences seasonally and regionally and which are not adequately identified in the daily parameters.

CONCLUSION

Precipitation simulation models generate synthesized sequences of precipitation at a range of spatial and temporal scales. Three broad categories are general circulation models, stochastic spatial-temporal rainstorm models, and daily precipitation models. Model selection and use should be justified by the desired resolution of

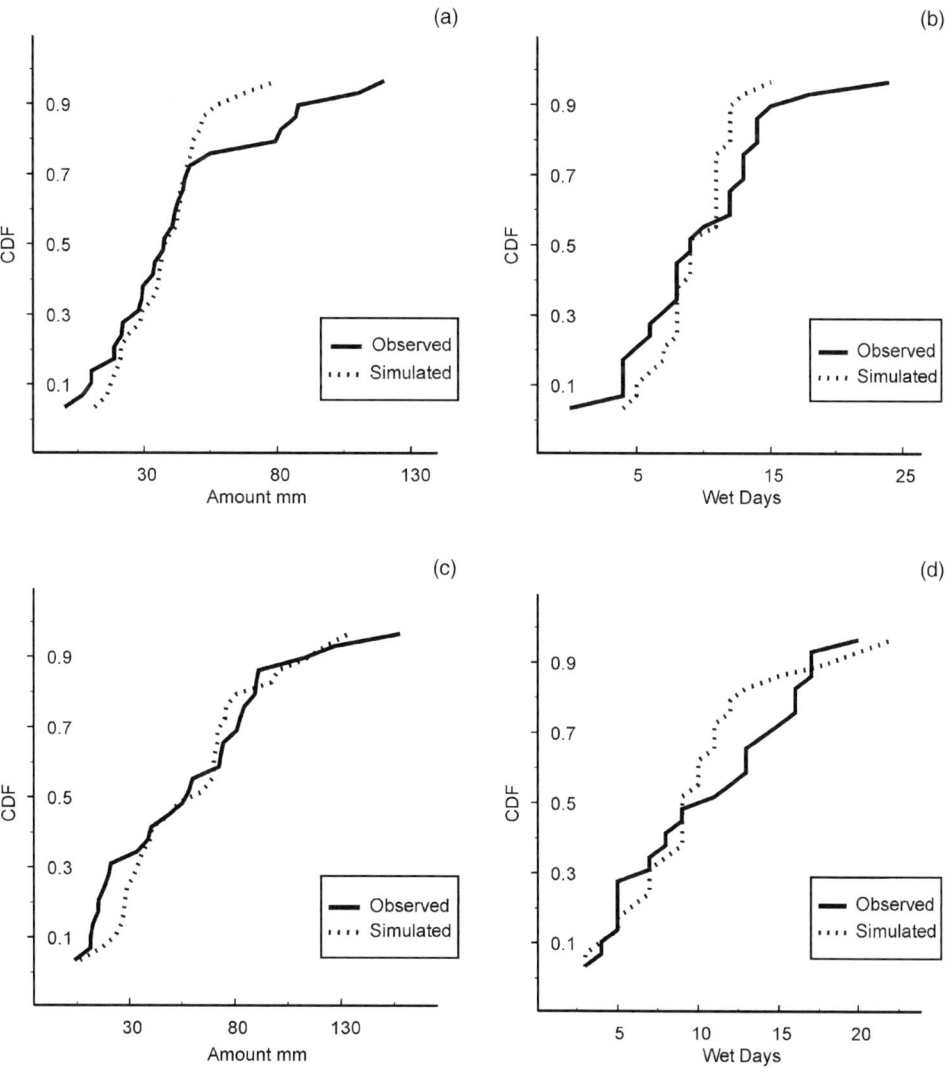

Fig. 1 Empirical cumulative distribution function (CDF) of simulated and observed precipitation for Tombstone AZ 1961–1990. a) January, February and March (JFM) amount; b) JFM number of wet days; c) October, November and December (OND) amount; d) OND number of wet days.

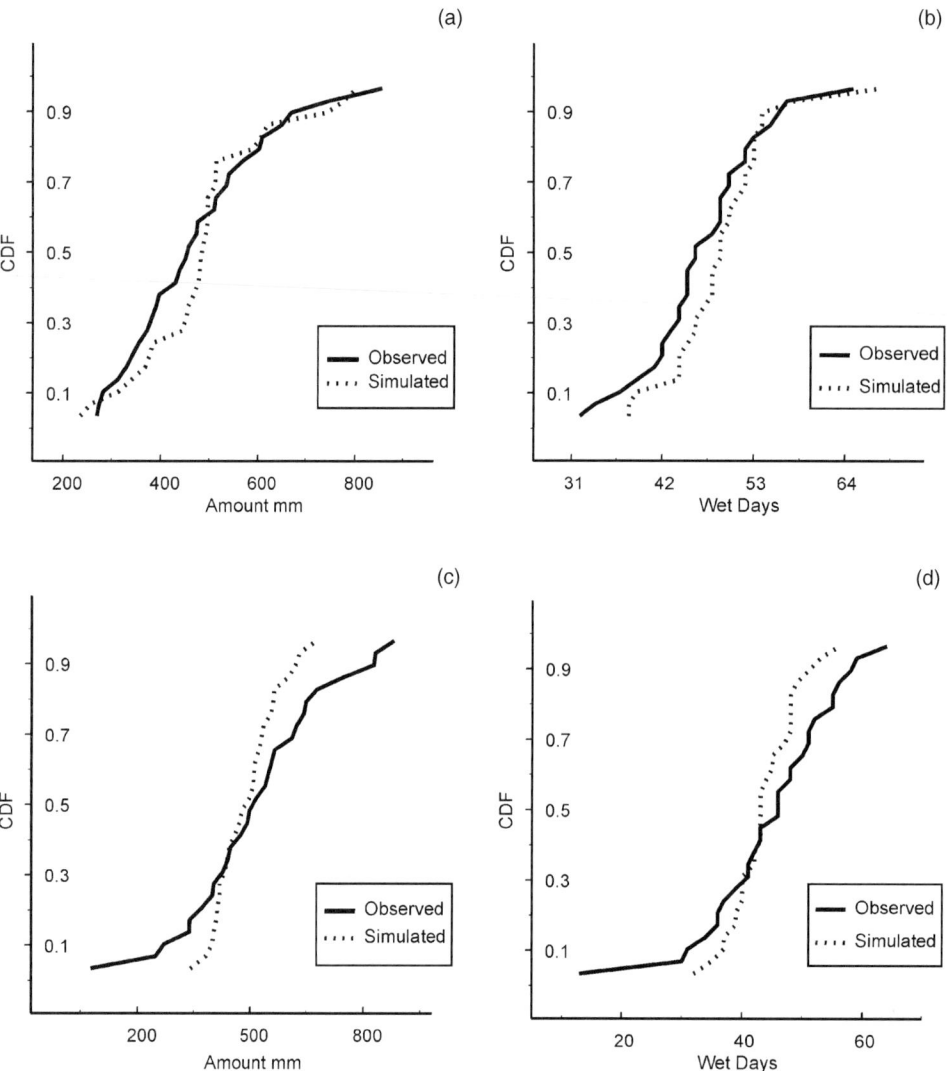

Fig. 2 Empirical cumulative distribution function (CDF) of simulated and observed precipitation for Eugene OR 1961–1990. a) JFM amount; b) JFM number of wet days; c) OND amount; d) OND number of wet days.

results and ability to fully estimate the required parameters. Future developments to precipitation simulation models will be downscaling techniques which link regional and local scales, improved algorithms to more faithfully represent the stochastic and physical dynamics of precipitation, and the inclusion of low-frequency oscillations and spatial distribution of parameters in daily precipitation models.

REFERENCES

1. McGuffie, K.; Henderson-Sellers, A. *A Climate Modelling Primer*; John Wiley & Sons: Chichester, England, 1997.

2. Cox, D.R.; Isham, V. Stochastic Models of Precipitation. In *Statistics for the Environment 2 Water Related Issues*; Barnett, V., Turkman, K.F., Eds.; John Wiley & Sons: Chichester, England, 1994; 3–18.

3. Osborn, H.B.; Lane, L.J.; Richardson, C.W.; Molnau, M.P. Precipitation. In *Hydrologic Modeling of Small Watersheds*; Haan, C.T., Johnson, H.P., Brakensiek, D.L., Eds.; American Society of Agricultural Engineers: St. Joseph, Michigan, 1982; 81–118.

4. http://www.ipcc.ch/ (accessed February 2002).

5. http://www.aip.org/history/sloan/gcm/intro.html (accessed February 2002).

6. http://www.giss.nasa.gov/research/modeling/ (accessed February 2002).

7. Waymire, E.C.; Gupta, V.K. The Mathematical Structure of Rainfall Representations 1. A Review of the Stochastic

Rainfall Models, 2. A Review of the Theory of Point Processes, 3. Some Application of the Point Process Theory to Rainfall Processes. Water Resour. Res. **1981**, *17* (5), 1261–1294.

8. Gupta, V.K.; Waymire, E.C. A Statistical Analysis of Meso-scale Rainfall as a Random Cascade. J. Appl. Meteorol. **1993**, *32* (2), 251–267.

9. Woolhiser, D.A. Modelling Daily Precipitation—Progress and Problems. In *Statistics in the Environmental and Earth Sciences*; Walden, A., Guttorp, P., Eds.; Edward Arnold: London, 1992; 71–89.

10. http://www.nwrc.ars.usda.gov/models/gem/ (accessed April 2002).

11. Hansen, C.L.; Cumming, K.A.; Woolhiser, D.A.; Richardson, C.W. *Microcomputer Program for Daily Weather Simulations in the Contiguous United States*, Publ. ARS-114; U.S. Department of Agriculture, Agricultural Research Service, 1988; 38.

P

Precipitation, Stochastic Properties

David A. Woolhiser
Colorado State University, Fort Collins, Colorado, U.S.A.

INTRODUCTION

As one observes the evolving patterns of radar images of precipitation on television or the internet, it becomes clear that it is a stochastic process—a process occurring in time (and space) and governed by probability laws. We can only make probabilistic statements because even if we have perfect knowledge of weather variables at some point in time, we cannot predict their values for some future time with certainty.

Day-to-day variations in weather variables, especially precipitation and temperature, have a major influence on agricultural and engineering decisions. Choices of crops to grow, as well as planting, tillage, spraying and harvesting dates are all weather and climate related, and estimates of rainfall probabilities for the next few days can be helpful in guiding decisions. Engineering design of agricultural or urban drainage facilities, control of erosion by structural means, or agricultural management methods must be based upon information on the statistical characteristics or rainfall. Computer models of the growth and yield of major crops such as wheat, corn, soybean and cotton are dependent on real or simulated precipitation data.

PRECIPITATION AS A STOCHASTIC PROCESS

Although precipitation varies widely in space and time, a description of the process at a given location is essential for many agricultural and engineering applications and is not as difficult as describing both spatial and temporal characteristics. Symbolically, we can describe the daily precipitation process for year, τ and day, n as:

$$Z_\tau(n) = Z_1(1), Z_1(2), Z_1(3), \ldots$$

$$Z_1(365), Z_2(1), Z_2(2), \ldots, Z_M(365);$$

$$\tau = 1, 2, \ldots, M; \quad n = 1, 2, \ldots, 365.$$

where Z is the amount of precipitation on day n of year τ, the maximum n is either 365 or 366 and M is the number of years. The process, $Z_\tau(n)$ can be written as the product $X(n)Y(n)$ where $X(n) = 0$ if day n was dry and $X(n) = 1$ if

day n was wet. $Y(n)$ is a random variable denoting the depth of precipitation if the day was wet.

The occurrence process, $X(n)$, usually exhibits the phenomenon of persistence, which means the probability of measurable precipitation on a given day depends on what happened on the previous day or days. In many cases, persistence can be adequately described by a first order, two state Markov chain where the occurrence of precipitation on day n only depends on whether the previous day (day $n-1$) was wet or dry, or:

$$p_{i,j}(n) = P\{X_\tau(n) = j | X_\tau(n-1) = i\};$$

$$i, j = 0, 1; \quad n > 1$$

$$p_{i,j}(1) = P\{X_1(1) = j | X_{\tau-1}(365) = i\}$$

The $p_{i,j}(n)$ are called transition probabilities.

In some climates, particularly when precipitation is caused by slowly moving fronts, a second-order Markov chain may be required.

$$p_{i,j,k}(n) = P\{X_\tau(n) = k | X_\tau(n-1) = j, X_\tau(n-2) = i\};$$

$$i, j, k = 0, 1; \quad n > 2$$

Although other occurrence processes may be superior for some climates, the simplicity of the first or second order Markov chain is an advantage for most applied purposes.

As an approximation, the amount of precipitation on a wet day n, is often assumed to be independent of the amount (or occurrence) of precipitation on day $n-1$. Several distribution functions have been used to describe $Y(n)$, but the most common is the gamma distribution:

$$f_n(y) = \frac{\beta(n)^{\alpha(n)} y^{\alpha(n)-1} e^{-\beta(n)y}}{\Gamma[\alpha(n)]}; \quad y, \alpha(n), \beta(n) > 0$$

where $f_n(y)$ is the probability density function on day n, $\alpha(n)$, $\beta(n)$ are parameters specified for day n, $\Gamma[\alpha(n)]$ is the gamma function and e is the base of natural logarithms.

Another density function commonly used is the three parameter mixed exponential:

$$f_n(y) = \frac{\alpha(n)\exp[-y/\beta(n)]}{\beta(n)} + \frac{[1-\alpha(n)]\exp[-y/\delta(n)]}{\delta(n)}$$

Encyclopedia of Water Science
DOI: 10.1081/E-EWS 120010321

where $\alpha(n)$ is a weighting function with values between zero and one and $\beta(n)$ and $\delta(n)$ are the means of two exponential distributions.

Because of seasonal variations, the parameters of these distributions must vary within the year. This variability can be accommodated by estimating the parameters for fixed periods such as seasons (spring, summer, fall, and winter), months, or weeks. An alternative approach is to use finite Fourier series to provide a daily variation with only a small number of parameters—annual means and the amplitudes and phase angles of significant harmonics. For example, eight parameters are required to specify a first order Markov chain for four seasons, twenty four parameters are required for monthly representation, and eighteen parameters for a Fourier series representation with five harmonics for P_{00} and three harmonics for P_{10}.

EXAMPLES, PARAMETERS OF STOCHASTIC MODELS

These Markov transition probabilities and the daily rainfall amount distribution will exhibit dramatic differences seasonally and spatially. Figs. 1–3 illustrate the stochastic daily rainfall characteristics for three general climatic types in the United States. The stations represented, Portland, Oregon, Waterloo, Iowa and Walnut Gulch, Arizona—are examples of Mediterranean, continental, and monsoon climates, respectively. Fig. 1 shows the variability of the dry–dry transition probabilities, $P_{00}(t)$. The most striking feature is the nearly opposite behavior of this parameter for Portland and Walnut Gulch. Because the probability of a wet day following a dry day is $1 - P_{00}(t)$, Fig. 1 illustrates that the

driest period in coastal Oregon occurs at the same time as the wettest (monsoon) period in southeastern Arizona. Oregon, of course, has a much higher frequency of precipitation in the winter than Arizona. The Corn Belt (Waterloo) has the greatest probability of a wet day following a dry day in early June.

Fig. 2 shows the variations in the wet–dry transition probabilities, $P_{10}(t)$. The probability of a wet day following a wet day is $1 - P_{10}(t)$, so the lowest vales of $P_{10}(t)$ have the highest persistence. As expected, coastal Oregon has the highest persistence, with the probability of a wet day following a wet day as high as 0.79 in early January. Walnut Gulch has the lowest persistence in June just before the higher persistence during the monsoon season. $P_{10}(t)$ shows less variability for Waterloo, with the highest persistence in April.

Fig. 3 shows the seasonal variation of the mean daily precipitation depth on a wet day. Waterloo shows the greatest mean depth during the summer growing season while the daily precipitation depth at Portland peaks during the winter. Walnut Gulch exhibits a complex pattern with the greatest depth in the fall when the remnants of hurricanes can intrude into Arizona. Secondary peaks occur during the monsoon season (mid summer) and the winter.

PRECIPITATION DATA SOURCES

Micro-computers have facilitated the delivery of climate information to users. Historical weather data are widely available from the world-wide-web. The site of the Climate Prediction Center of the U.S. National Weather Service[1] is particularly helpful. In the United States,

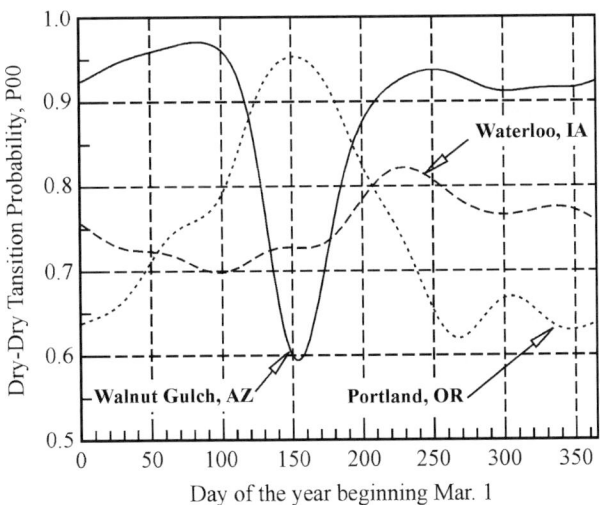

Fig. 1 Seasonal variation of the dry–dry transition probability.

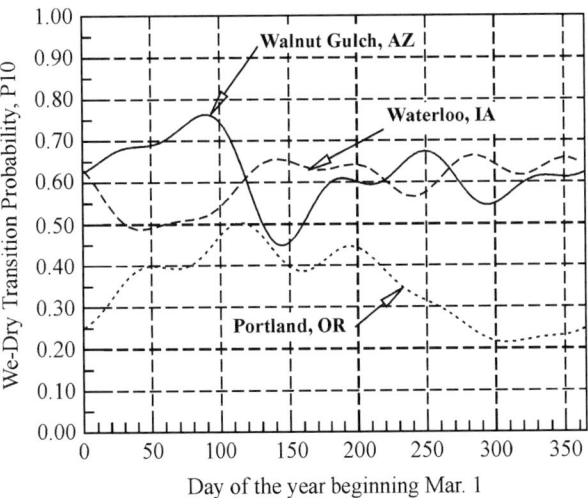

Fig. 2 Seasonal variation of the wet–dry transition probability.

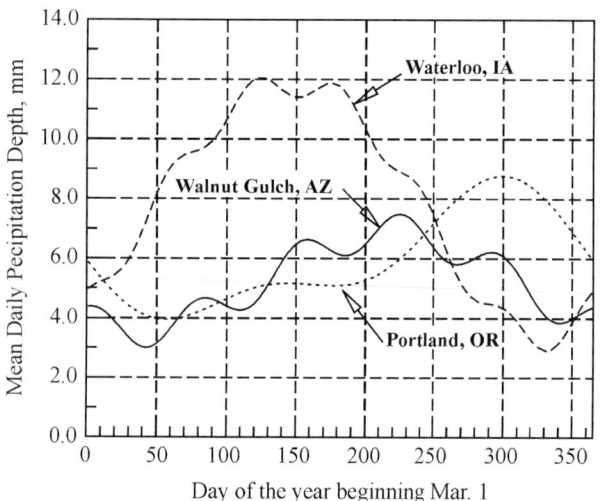

Fig. 3 Seasonal variation of the mean daily rainfall depth.

personnel at Regional Climate Centers can provide assistance. The web sites of these centers can be found by searching the web for "Regional Climate Centers." Micro-computer programs and data-bases are also available for the contiguous United States[2,3,4] and provide an easy method to obtain simulated daily data for virtually any location in the United States. The CLIMWAT[5] database has monthly precipitation data from 144 sites around the world. Such models have limited application in mountainous regions because most weather stations are located in the valleys and the data for these stations are not valid for higher elevations. Analysis of precipitation data from raingage networks in mountainous regions has shown some regularity, with the frequency of precipitation and mean daily amounts increasing with elevation.

OTHER SOURCES OF INTER-ANNUAL STOCHASTIC VARIATIONS

Although the assumption of year-to-year stationarity for daily precipitation models is adequate for many purposes, it has been found that such models do not preserve the variance of annual precipitation totals. Although model simplifications may account for some of this variance reduction, large-scale interactions between the atmosphere and oceans play a substantial role. It has been

demonstrated by many studies that the El Nino-southern oscillation phenomenon or ENSO affects precipitation regimes in several continents. For example, during El Nino years the southwestern United States typically has wetter than normal winters, while the Pacific Northwest has drier than normal winters. The opposite effect occurs during La Nina years. An excellent documentation of this phenomenon for the United States is available from the National Oceanic and Atmospheric Administration (NOAA).[6] Other factors that may affect daily precipitation include random explosive volcanic events, changes in radiation received due to changes in the angle of the earth with the sun, global warming, etc.

The additional randomness due to ocean-atmosphere interactions or other causes has been incorporated into stochastic precipitation models by estimating monthly parameters separately for months classified in the lower 30%, middle 40%, and upper 30% of the climatological distributions of total precipitation.[7] The transitions between each of these classes are described by a first-order, three-state Markov chain.

REFERENCES

1. NOAA, Climate Prediction Center, http://www.cpc.noaa.gov/ (viewed on Aug 2, 2001).

2. Richardson, C.W.; Wright, D.A. *WGEN: A Model for Generating Daily Weather Variables*, ARS-8; U.S. Department of Agriculture, Agricultural Research Service: Washington, D.C., 1984; 83.

3. USDA Agricultural Research Service, WEPP, http://topsoil.nserl.purdue.edu/nserlweb/weppmain/cligen (viewed on April 13, 2001).

4. Hanson, C.L.; Cumming, K.A.; Woolhiser, D.A.; Richardson, C.W. *Microcomputer Program for Daily Weather Simulation in the Contiguous United States*, ARS-114; U.S. Department of Agriculture, Agricultural Research Service: Washington, D.C., 1994; 38.

5. FAO, CLIMWAT, http://www.fao.org/ag/AGL/aglw/climwat.htm (viewed on Aug 2, 2001).

6. NOAA, El Niño, http://www.cpc.ncep.noaa.gov/products/analysis_monitoring/ensostuff/states/states.html (viewed on Aug 2, 2001).

7. Wilks, D.S. Conditioning Stochastic Daily Precipitation Models on Total Monthly Precipitation. Water Resour. Res. **1989**, *23* (6), 1429–1439.

Precipitation Storms

Clayton L. Hanson (Retired)
United States Department of Agriculture (USDA), Boise, Idaho, U.S.A.

INTRODUCTION

Precipitation includes all water particles, whether liquid or solid, that fall from clouds and reach the ground. Precipitation includes both liquid (drizzle and rain), freezing (freezing drizzle and freezing rain) and frozen (snow, ice crystals, and hail) water.[1] For precipitation to occur, air must be cooled sufficiently to cause condensation and droplet growth. The mechanism that causes precipitation is adiabatic-expansion cooling as air is lifted in the atmosphere. When cooling is sufficient, vapor condenses on nuclei that are generally small particles of dust or salt, and combustion products that are always present in the atmosphere to form either ice crystals and supercooled liquid cloud droplets, or only liquid cloud droplets. Clouds that extend above the 0°C level are referred to as cold clouds and those that do not extend above the 0°C level are called warm clouds. Ice particles grow to sufficient mass to fall as precipitation in cold clouds by three processes; vapor condensation, collisions with supercooled droplets, and aggregation with other ice particles. In warm clouds, droplets grow large enough to fall as precipitation through the coalescence process where larger particles (which fall faster than small particles) collide and coalesce. As shown in Fig. 1, air is generally lifted by four means: 1) frontal convergence (cyclonic convergence); 2) orographic lifting; 3) thermal convection; or 4) tropical cyclones (hurricanes).[2–4]

FRONTAL CONVERGENCE

Precipitation caused by frontal convergence occurs when the general atmospheric circulation brings air masses of different temperatures and moisture from high-pressure regions (cold, relatively heavy air) to low-pressure regions (warm, relatively light air) which forces the air to rise, producing adiabatic cooling. Areas of high pressure at the surface are associated with converging air on the west side of high altitude troughs and areas of low pressure at the surface are associated with diverging air on the east side of the troughs. These cyclonic systems are usually larger than 500 km across and in the mid-latitudes, the air is lifted at the frontal surface as shown in Fig. 1a. Non-frontal

convergence generally occurs in the tropics within a mass of warm, moist air.

The area of contact is called a cold front when a cold air mass replaces a warm air mass, and is a warm front when a warm air mass replaces a retreating cold air mass. Cold fronts generally move faster then warm fronts, so when a cold front overtakes a warm front, the colder air stays at the surface with the warm air lifted above to form an occluded front as shown in Fig. 1a. If a front is not moving, it is called a stationary front.

Typically, cold fronts have relatively steep slopes of 1 in 50 to 1 in 150, whereas warm fronts have slopes of 1 in 100 to 1 in 300.[3] As cold fronts are usually steeper and move faster than warm fronts, the band of weather associated with cold fronts is narrower, more severe and of shorter duration than that of warm fronts. When cold fronts move slow and have stable, warm air ahead of the front, stratus-type rain clouds form in a wide band over the front. When cold fronts move rapidly, the weather associated with these fronts is generally of shorter duration and more severe than that of slower moving cold fronts. When the warm air ahead of cold fronts is moist and unstable, a squall line of showers and thunderstorms may form 50 km–400 km ahead of the front. The weather associated with squall lines is often more severe than that of the subsequent cold front.

Because warm fronts are flatter than cold fronts, clouds and precipitation are generally widespread, up to several hundred kilometers ahead of the front with the heaviest amounts of precipitation extending cyclonically 50 km–250 km north and westward of the center of the cyclone. If the warm air above a warm front is moist and relatively stable, the precipitation is gentle and increases as the front approaches. Thunderstorms can be embedded in the clouds when the warm air above a warm front is moist and unstable.

As a frontal system moves, the associated cold front overtakes the warm front and is forced up over the cold air that forms an occluded front. Occluded fronts typically form when frontal systems are at their maximum intensity, which results in widespread cloudiness and precipitation with the maximum precipitation to the north of the low pressure center.

The two major sources of moisture in the United States are from the Pacific Ocean and the Western Atlantic-Gulf

Encyclopedia of Water Science
DOI: 10.1081/E-EWS 120010315
Published 2003 by Marcel Dekker, Inc.

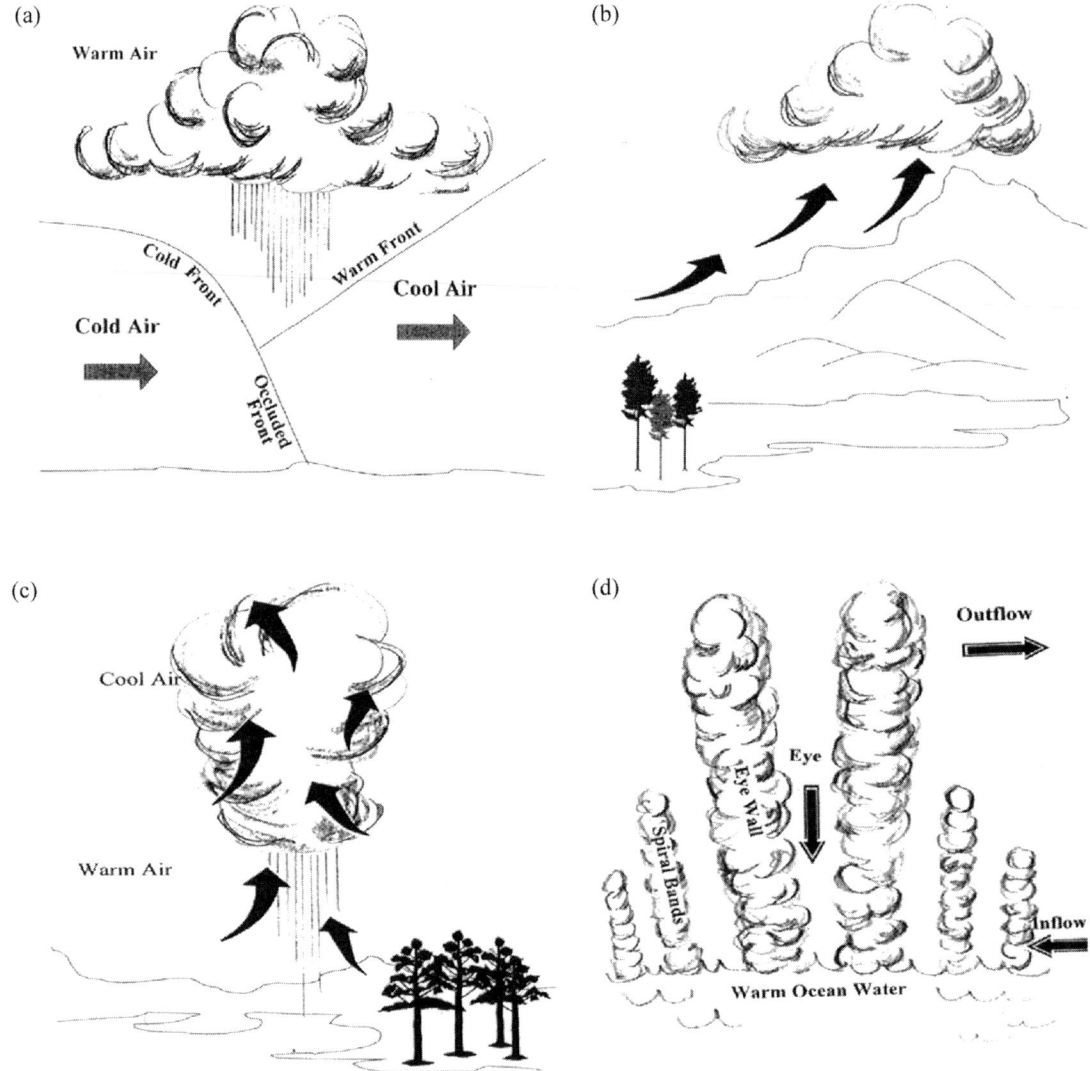

Fig. 1   The primary methods by which air is cooled to saturation by adiabatic cooling are (a) frontal convergence, (b) orographic lifting, (c) thermal convection, and (d) tropical cyclones. (The illustrations are not to scale.)

of Mexico region. The Gulf of Mexico is the primary source of moisture for the large frontal systems that develop in the Great Plains of the United States.

OROGRAPHIC LIFTING

Orographic precipitation results when moist air is forced to ascend over natural barriers such as the coastal hills and mountains along the West Coast of the United States (Fig. 1b). As air is forced up the windward side of barriers, it cools until the air reaches saturation, at which time water vapor begins to condense into liquid water droplets. If the upward airflow is strong enough,

precipitation can develop, and this precipitation usually increases with elevation. This process continues until the air is either too dry to produce more precipitation or the air moves over the barrier. After the air moves over the barrier, it warms and precipitation becomes less as the air moves down slope (subsidence), which results in a "rain shadow" on the leeward side of barriers, e.g., the semi-arid and arid regions of central Oregon and Washington, and western Nevada. Orographic precipitation is the greatest during the winter in mid-latitudes when atmospheric flow is strongest; however, convective precipitation in summer months is enhanced over barriers due to diurnal winds which tend to move up slopes during the day.[3–5]

THERMAL CONVECTION

Convective precipitation, which is generally associated with air mass showers and thunderstorms, is most prominent in mid-latitudes during the summer (Fig. 1c). Thunderstorms cause some of the most severe precipitation events that often include high-intensity precipitation, hail and damaging winds. For convective activity to develop, the atmosphere has to be conditionally unstable, there has to be some triggering mechanism to release the instability, and there has to be an adequate supply of moisture in the atmosphere. Some lifting mechanism such as upper air systems, frontal lifting, orographic lifting, and/or very strong day time heating is required to release atmospheric instability. As the air is lifted, it cools to the dew point and condensation forms a cloud that results in latent heat being added to the air, which lifts it even more rapidly. When the rapidly uplifted air reaches high altitudes, ice crystals and water drops grow big enough to overcome the updraft, and they fall as rain and/or hail. Single thunderstorm cells can range in size from a few kilometers to 20 km and lines of thunderstorms along a cold front can be several hundred kilometers long.[3–5]

The Rapid City, SD storm of June 9–10, 1972 is a good example of the flooding that can occur due to the precipitation produced by convective thunderstorms that are aided by orographic lifting. The primary atmospheric phenomenon that contributed to these severe storms was the strong low-level easterly airflow that forced moist air upslope over the Black Hills and the unusually light winds aloft over the Black Hills. These light winds at higher levels did not move the thunderstorms away from the hills, which resulted in concentrated rainfall along the eastern slopes of the Black Hills. At one location in the Black Hills, this storm produced 380 mm of precipitation in about 6 hr.[6]

TROPICAL CYCLONES

Tropical storm systems can produce significant amounts of precipitation and cover relatively large surface areas.[4,5,7] These systems typically affect the United States between June and November, with peak activity in the late summer. Atlantic tropical systems originate as tropical waves off the west coast of Africa where the sea-surface temperature is at least 26°C and move westward toward the Caribbean in the predominate easterly winds (trade winds) that flow across the ocean. Some of these systems develop into hurricanes, which means that they have sustained winds in excess of 33 m sec^{-1}. Larger hurricanes can sometimes have radii in excess of 500 km, and their movement is much less directed by winds aloft, thus making prediction of their movement difficult.

The quantity of precipitation that falls from tropical systems is a function of storm movement, relative location to the storm center, and storm movement relative to land masses. Even relatively weak tropical cyclones have sometimes produced extremely heavy rainfall. Precipitation around a tropical cyclone, particularly one that has become fairly well organized and concentric, usually comes from rain bands rotating cyclonically around the low-pressure center. These bands of showers and thunderstorms typically increase toward the center of the storm and are maximized in the eye wall of organized hurricanes where a solid circle of severe thunderstorms is usually located (Fig. 1d). Often, the heaviest precipitation is from these "eye wall thunderstorms" and from rain bands that are generally to the east of the center of the cyclone.

In the southeastern United States, tropical storms are responsible for 5%–30% of the normal precipitation in the summertime.[4] Tropical cyclone activity in the eastern Pacific Ocean sometimes produces heavy precipitation in the Southwest where desert locations in Arizona and southern California can sometimes receive most of their annual precipitation from the remnants of these storms.

REFERENCES

1. Geer, I.W., (Ed.) *Glossary of Weather and Climate with Related Oceanic and Hydrologic Terms*; American Meteorological Society: Boston, MA, 1996; 272 pp.
2. Dingman, S.L. Precipitation. *Physical Hydrology*; Prentice-Hall, Inc.: Upper Saddle River, NJ, 1994; 87–158.
3. Schroeder, M.J.; Buck, C.C. *Fire Weather*, Agricultural Handbook 360, U.S. 4; Department of Agriculture Forest Service: Washington, DC, 1970; 229 pp.
4. Hanson, C.L.; Johnson, G.L.; McFarland, M.J.; Gebhardt, K.; Smith, J.A. *Hydrology Handbook (Manual No. 28)*; Committee on Hydrology Handbook, Ed.; American Society of Civil Engineers: New York, NY, 1996; Chap. 2, 5–74.
5. Stull, R.B. *Meteorology for Scientists and Engineers*, 2nd Ed.; Brooks/Cole: Pacific Grove, CA, 2000; 502 pp.
6. Schwarz, F.K; Hughes, L.A.; Hansen, E.M.; Petersen, M.S.; Kelly, D.B. *The Black Hills-Rapid City Flood of June 9–10, 1972: A Description of the Storm and Flood*, Geological Survey Professional Paper 877; U.S. Department of the Interior Geological Survey: Washington, DC, 1975; 47 pp.
7. Williams, J. *The Weather Book*; Vintage Books: New York, NY, 1992; 212 pp.

Precision Agriculture and Water Use

Robert J. Lascano
Texas A&M University, Lubbock, Texas, U.S.A.

Hong Li
University of Florida, Lake Alfred, Florida, U.S.A.

INTRODUCTION

In current agronomic practices, inputs such as fertilizer, pesticides, and water are applied uniformly across a field regardless of their need and their management is normally based on average responses of these inputs to crop yield across the field. However, with current emphasis on quality and efficiency of food production, it is imperative that inputs be managed according to specific needs across the field. This type of farming is known as *precision farming*, a generic term that describes the way whereby inputs to a farming operation are managed. Perhaps a better descriptor for this type of farming is *site-specific management*, which can now be implemented due to commercially available hardware that allows farm equipment to variably apply products across a field using onboard computers.

PRECISION FARMING

Precision farming refers to the practice of applying agronomic inputs across a farm, mainly fertilizers and other chemicals, at variable rates based on soil nutrients or chemical tests, soil textural changes, weed pressures, and/or yield maps for each field in the farm. In large fields (e.g., > 40 ha), crop yield and thus crop-water use are notoriously variable. The sources of this variation are related to soil physical and chemical properties, pests, microclimate, genetic and phenological responses of the crop, and their interactions. The technology for crop yield mapping is more advanced than current methodologies for determining and understanding causes of yield variability. Prevailing and traditional management practices treat fields uniformly as one unit. However, recent reports (e.g., Refs. 1–3) show that to understand underlying soil processes that explain crop yield variability, research must be done at the landscape level and using appropriate statistical tools for large scale studies (e.g., Refs. 1,3,4).

Precision farming must incorporate the inherent spatial and temporal variability of soil physical (e.g., crop water supply factors), chemical, and biological factors within a field for input management. Accurate representation of spatial and temporal variability in a field requires taking and analyzing many samples. Sampling is normally done on a grid with a scale that can vary from one to several hundred meters.[6] Once properties are measured, geostatistical tools (e.g., semivariogram, kriging, cokriging, etc.) and other spatial statistical tools (e.g., autocorrelation, crosscorrelation, state–space analysis, etc.) can be used to establish statistical relations in space and to minimize the number of soil samples to characterize and map fields.[2,3,7] The number of samples required a priori to determine spatial and temporal variability is perhaps the single largest deterrent in the application of precision farming practices to manage and improve crop-water use.

CROP YIELD AND WATER USE

There is a linear relation between crop yield and water use when the only limiting factor is water (e.g., Ref. 5). Precision farming has the potential for improving water use efficiency on large fields provided there is a quantitative understanding of what factors and where in the field they affect crop-water use. We know that crop-water use is a function of many biotic and abiotic variables, including managed inputs, and harvestable yield is a manifestation of how these variables and inputs interact and are integrated during the growing season. However, it is difficult to determine a hierarchy on the contribution of each input and variable to the measured yield using classical statistics.[2,3] Often, variables that affect water supply to the plant would contribute to yield at a high level assuming an adequate plant stand and weed control. The cause and effect relation between a single state variable and crop yield is site specific and is difficult to establish without considerable sampling of the soil and/or crop. The establishment of response functions, i.e., crop-water use as a function of variable x_i, only gives a partial answer to explain crop-water use and yield based on inputs. The general idea of precision farming is to optimize input application to the measured crop yield at each sampling location. This is a simple premise; however,

Encyclopedia of Water Science
DOI: 10.1081/E-EWS 120010044

the decisions for variable-rate application of any agronomic input must consider temporal and spatial variability of the soil's properties affecting crop growth, water use, and yield. Soil factors that affect stored water, such as depth to root restricting layer and soil textural differences, must be considered in any precision farming operations that attempt to improve crop-water use and yield related to agronomic inputs.

There is very little information published on crop-water use across large fields at the landscape level and in the context of precision farming (e.g., Refs. 1,8,9). An exception is a study[1] where cotton-water use was measured along a 700-m transect with the objective to 1) illustrate the landscape pattern of cotton-water use and 2) determine the underlying soil processes governing cotton lint yield variability. In this study, state–space analysis[1,3] was used to formulate management decisions that may improve crop-water use and, thus, yield using precision farming practices.

LANDSCAPE CROP-WATER USE

To illustrate the concept of crop-water use in a large field we use the study of Li et al.[1] In 1999, a field experiment was conducted near Lamesa, Texas on a research farm of Texas A&M University on the southern edge of the High Plains of Texas. The soil is classified as an Amarillo sandy

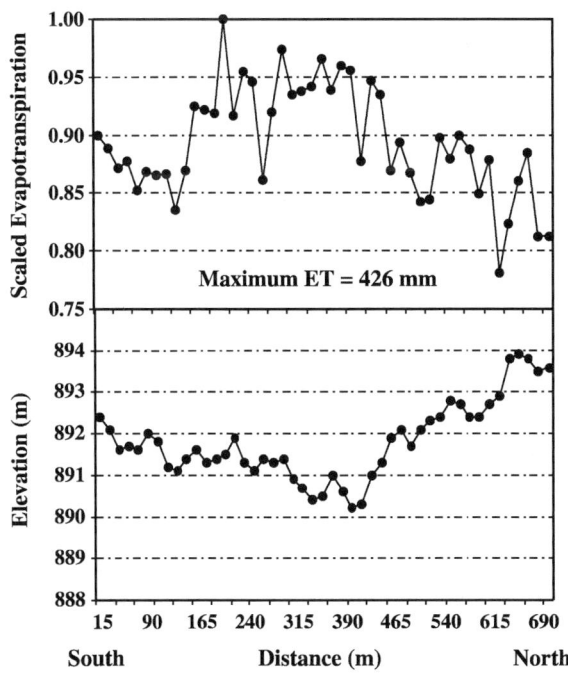

Fig. 1 Scaled ET and elevation as a function of distance along a 700-m transect.

loam. The field was 60 ha with slopes ranging between 0.3% and 6.3%.[1] To assess the effect of soil water, NO_3-N, and topography on cotton lint yield across the landscape, two irrigation levels were used. The irrigation treatments consisted of water applications at the 50% and 75% potential evapotranspiration (ET) with a center pivot LEPA irrigation system.[10] At each irrigation level, one transect was established following the circular pattern of the center pivot. The two transects were instrumented with 50 neutron access tubes each 15 m apart, and volumetric water content (θ_v) was measured periodically throughout the growing season. At each point θ_v was measured in 0.3-m depth increments to 2.0-m depth using a neutron probe calibrated for this soil. In addition, at each transect point soil texture, soil and plant N-NO_3, leaf area index, lint yield, slope, plant density, and other parameters were measured.[1]

Statistical Calculations

It has been shown that the use of classical statistics, such as regression analysis and analysis of variance, fails to completely explain the cause and effect between, for example, crop yield and measured soil variables in precision farming experiments.[1–4,11] Instead, there are other more appropriate statistical tools for relating the variability of soil and plant parameters measured in space and time. For example, the structure of the spatial variance between measurements may be derived from the sample *semivariogram*, which is the average variance between neighboring measurements spatially separated by the same distance. Spatial structure between variables is often determined using *autocorrelation* and *crosscorrelation* functions. Autocorrelation measures the linear correlation of a variable in space along a transect. The crosscorrelation is the comparison of two variables measured along a transect and is used to describe the spatial correlation between two landscape variables, i.e., where one variable, the tail variable, lags behind the head variable by some distance. The spatial association between several variables can be described using *state–space* analysis, which is a multivariate autoregressive technique.[1–4,7,11]

SPATIAL ANALYSIS OF CROP-WATER USE

To illustrate the variability of crop-water use or ET, values measured along the 50% irrigation transect were selected.[1] In Fig. 1, the relation between the scaled ET and elevation both as a function of distance along the transect is shown. The ET data are scaled to the maximum of 426 mm of water measured 210 m from the south end of the transect. These results show that higher ET was measured at lower elevations and ET decreased at higher

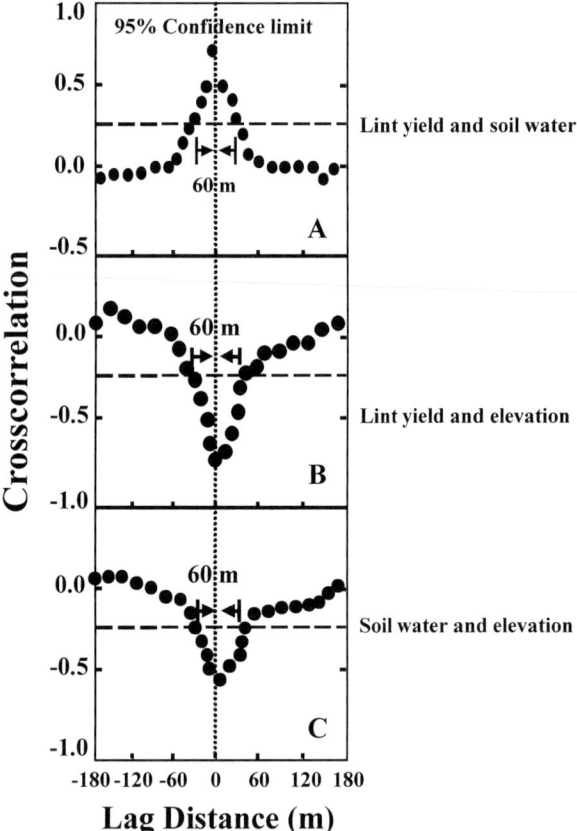

Fig. 2 Crosscorrelation as a function of lag distance. (A) Lint yield and soil water, (B) lint yield and elevation, and (C) soil water and elevation. Shown is the 95% confidence for the crosscorrelation distance. (From Ref. [1].)

elevations. Spatial crosscorrelation between lint yield and soil water, lint yield and site elevation, and soil water and site elevation are shown in Fig. 2. For a 95% confidence interval, the cotton lint yield was positively cross-correlated with soil θ_v across a lag distance of $\pm\,30\,m$. Lint yield and θ_v were negatively crosscorrelated with elevation at a lag distance of $\pm\,30\,m$. These results show the effect of topography on θ_v and crop-water use measured along the transect. Similar results are given in other reports.[1,8,9,11] In this example, the crosscorrelation between θ_v and elevation shows the spatial structure of measured variables and further shows that more water was stored in lower elevations resulting in higher ET.

Linear regression analysis between θ_v and lint yield and relative site elevation is shown in Fig. 3, and the state–space analysis for the relation between lint yield and three measured parameters is shown in Fig. 4. Results in Fig. 3 show the shortcomings of using an inappropriate statistical tool to understand underlying processes explained with the state–space analysis. This analysis (Fig. 4) quantified how cotton lint yields varied as a function of distance and showed that by using θ_v, soil NO_3-N, and elevation, the variation in lint yield can be explained with a high level of confidence.

Benefits of precision farming to improve crop-water use may be obtained by an economic analysis of maximizing crop yield as a function of application of N fertilizer and irrigation water as given by the state–space equation. In the example given, decision can be made to apply more N fertilizer to lower areas of the field that also hold more water and increase crop-water use and yield. With the introduction of variable rate planters it will be possible in the near future to discriminate site locations and plant more "drought" tolerant varieties or change the seeding rate in areas that are prone to have less soil water. This implies the delineation of management zones within

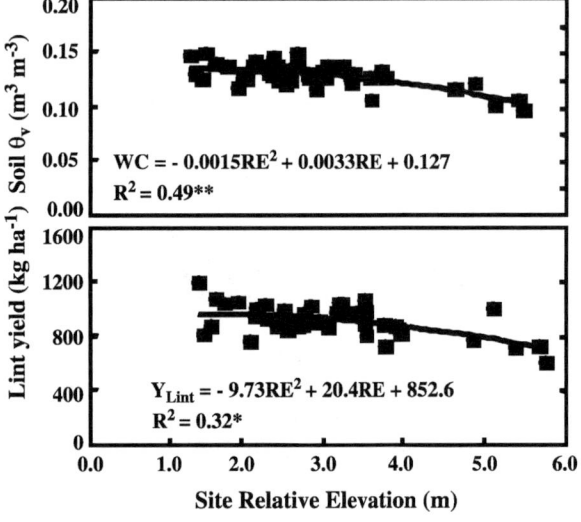

Fig. 3 Soil water content (θ_v) and cotton lint yield as a function of site relative elevation.

$$Y_{(50\%\ ET)i} = -0.201Y_{i-1} + 1.107\ W_{i-1} + 0.332\ N_{i-1} - 49.54\ E_{i-1} + \varepsilon_i$$

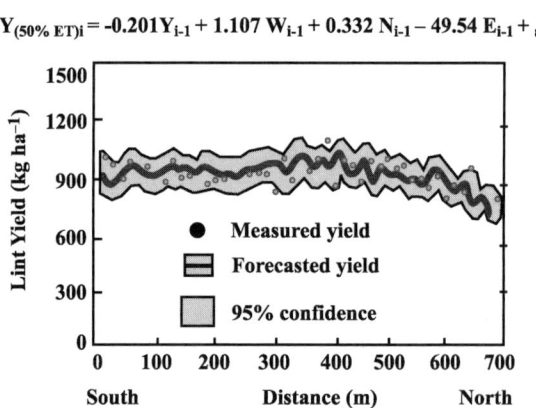

Fig. 4 State–space equation relating cotton lint yield (Y) to water content (W), nitrogen (N), and elevation (E) as a function of distance and location (i) along a 700-m transect. (From Ref. [1].)

a field that are defined based on potential crop-water use and their interaction with other input variables to maximize economic yield across the field. This type of precision farming is not currently practiced but remains within the realm of possibilities that this type of farming has to offer.

REFERENCES

1. Li, H.; Lascano, R.J.; Booker, J.; Wilson, L.T.; Bronson, K.F. Cotton Lint Yield Variability in a Heterogeneous Soil At a Landscape Scale. Soil Tillage Res. **2001**, *58*, 245–258.
2. Nielsen, D.R.; Wendroth, O.; Pierce, F.J. Emerging Concepts for Solving the Enigma of Precision Farming Research. In *Precision Agriculture*. Proceedings of the Fourth International Conference, Minneapolis, MN, July 19–22, 1998; Robert, P.C., Rust, R.H., Larson, W. E., Eds.; 1999; 303–318.
3. Wendroth, O.; Al-Oman, A.M.; Kirda, C.; Reichardt, K.; Nielsen, D.R. State –Space Approach to Spatial Variability of Crop Yield. Soil Sci. Soc. Am. J. **1992**, *56*, 801–807.
4. Cassel, D.K.; Wendroth, O.; Nielsen, D.R. Assessing Spatial Variability in an Agricultural Experiment Station Field: Opportunities Arising from Spatial Dependence. Agron. J. **2000**, *92* (4), 706–714.
5. Kramer, P.J.; Boyer, J.S. *Water Relations of Plants and Soils*; Academic Press: San Diego, 1995; 495.
6. Sadler, E.J.; Busscher, W.J.; Baver, P.J.; Karlen, D.L. Spatial Requirements for Precision Farming: A Case Study in the Southern USA. Agron. J. **1998**, *90*, 191–197.
7. Shumway, R.H.; Stoffer, D.S. *Time Series Analysis and Its Application*; Springer Verlag: New York, 2000; 549.
8. Halvorson, G.A.; Doll, E.C. Topographic Effects on Spring Wheat Yield and Water Use. Soil Sci. Soc. Am. J. **1991**, *55*, 1680–1685.
9. Hanna, A.Y.; Harlan, P.W.; Lewis, D.T. Soil Available Water as Influenced by Landscape Position and Aspect. Agron. J. **1982**, *74*, 999–1004.
10. Lyle, W.M.; Bordovsky, J.P. Low Energy Precision Application (LEPA) Irrigation System. Trans. ASAE **1981**, *24*, 1241–1245.
11. Timlin, D.J.; Pachepsky, Ya.; Snyder, V.A.; Bryant, R.B. Spatial and Temporal Variability of Corn Grain Yield on a Hillslope. Soil Sci. Soc. Am. J. **1998**, *62*, 764–773.

P

Professional Societies

Faye Anderson
University of Maryland, College Park, Maryland, U.S.A.

INTRODUCTION

Professional societies play a vital role in every discipline as they function to serve the individual interests of their members, represent their collective interests, and communicate information about the discipline to a wider audience. Professional societies can be viewed as individual membership organizations because professionals pay annual dues to join these organizations. Individuals join professional societies for the career opportunities accompanying membership, and many join more than one society. These opportunities include attending conferences at reduced costs, subscribing to journals, gaining leadership experience, and access to professional networks. Most professionals state their reasons for joining as gaining access to the most up-to-date information in their profession and networking with colleagues about common challenges, problems, and solutions.

Water-related professional societies foster scientific research, disseminate cutting-edge information, advocate for water resources, facilitate employment opportunities, and work to influence the future course of the profession. Common activities often include publishing journals and newsletters, conducting conferences and workshops, maintaining listservers and websites, and providing various networking opportunities for their members. Members are elected to run these organizations through a Board of Directors. This Board typically works alongside a headquarters office staff, which includes an Executive Director. Due to the large number of associations, association management has grown to have its own professional societies, e.g., the American Society of Association Executives.[1]

Agricultural water is an important societal resource, ultimately affecting every person on Earth through food systems. Water resources professionals in agriculture-related fields work in a wide variety of capacities—federal government, state and local government, universities, extension offices, nonprofit organizations, agribusiness, etc.—and on a wide variety of issues—irrigation methods, water quality, nonpoint source pollution, soil, drought, water conservation, etc. These sectors and disciplines often come together under the broader umbrella of a professional society's mission, and sometimes more narrowly defined interests form their own professional society as well. The majority of these societies are nonprofit organizations and membership fees vary widely. Some societies have tens of thousands of members and others just a few hundred. These professional societies serve both the individual needs of their members and the collective needs of the profession. Some societies have a code of ethics. Professional societies will often draft position or white papers on key issues affecting water resources. As agricultural water resources increasingly face stress and water professionals confront multiple demands in the workplace, these professional societies will continue to serve important functions for both their memberships and society.

U.S. PROFESSIONAL SOCIETIES

Many agricultural water professional societies are based in the United States and predominantly serve the interests of U.S.-based professionals by focusing the majority of their efforts on domestic agricultural water issues and concerns. Due to the large number of professional societies, there is substantial competition for both professional influence and members.[2]

American Agricultural Economics Association

The American Agricultural Economics Association (AAEA)[3] is a professional society for those interested in agricultural economics issues, including those related to rural communities and natural resources. American Agricultural Economics Association strives to keep its members abreast on the latest agricultural economics research developments and policy issues. Its official mission is to enhance the skills, knowledge, and professional contributions of those economists who serve the society in solving problems related to agriculture, food, resources, and economic development. The AAEA publishes the *American Journal of Agricultural Economics*, *CHOICES* magazine and a newsletter. It conducts an annual meeting and several workshops and smaller meetings each year. Water resources are an important component of this agricultural economics organization's

Encyclopedia of Water Science
DOI: 10.1081/E-EWS 120010293

natural resources agenda and there is some activity related to aquaculture as well.

American Geophysical Union

American Geophysical Union (AGU)[4] is an international scientific society with over 35,000 members in 115 countries. Formed over 75 years ago, AGU is devoted to advancing the understanding of earth and its environment in space and making the results available to the public. American Geophysical Union publishes many newsletters, books, and journals, including the well-respected *Water Resources Research*, which is popular with many agricultural researchers.

American Institute of Hydrology

The American Institute of Hydrology (AIH)[5] was formed in 1981 to provide certification, training, and education for hydrologists. It is the only national and international professional organization that certifies Professional Hydrologists and Professional Hydrogeologists.

American Society of Agricultural Engineers

American Society of Agricultural Engineers (ASAE)[6] is a professional and technical organization dedicated to the advancement of engineering, applicable to agricultural, food, and biological systems. Founded in 1907, ASAE has grown to over 9000 members and has an active Soil and Water division. Their Hancor Soil and Water Engineering Award recognizes outstanding contributions to the field.

American Society of Agronomy

The American Society of Agronomy (ASA)[7] is dedicated to the development of agriculture enabled by science, in harmony with environmental and human values. American Society of Agronomy publishes *Agronomy Journal*, *Journal of Environmental Quality*, and *Journal of Natural Resources and Life Sciences Education*. The ASA has an environmental quality division and a committee on water management on agricultural lands and sustainable agriculture. The Crop Science Society of America[8] and the Soil Science Society of America[9] share a close working relationship and related interests with the American Society of Agronomy, including sharing the same Headquarters office and staff. However, each of these Societies is autonomous, has its own bylaws, and is governed by its own Board of Directors.

American Society of Civil Engineers

American Society of Civil Engineers (ASCE)[10] is widely considered the leading source of technical and professional information in the field of civil engineering. American Society of Civil Engineers publishes dozens of journals, including the *Journal of Irrigation and Drainage Engineering*. ASCE has a Water Resources Planning & Management (WR) division and the Environmental & Water Resources Institute (EWRI). The ASCE awards the Royce J. Tipton Award in recognition of contributions to the advancement of irrigation and drainage engineering.

American Water Resources Association

American Water Resources Association (AWRA)[11] has a broad-based membership representing every sector of the water resources profession. American Water Resources Association publishes the *Journal of the American Water Resources Association* and *Water Resources Impact*. It also has an active Agricultural Hydrology Committee and sessions dealing with agricultural issues at its annual meetings and specialty conferences.

American Water Works Association

American Water Works Association (AWWA)[12] is an international nonprofit scientific and educational society dedicated to the improvement of drinking water quality and supply. Founded in 1881, AWWA is the largest organization of water supply professionals in the world and is dedicated to the promotion of public health and welfare in the provision of drinking water of high quality and sufficient quantity. Its Government Affairs office is very active in policy processes, including those concerned with agricultural water and water conservation issues.

Irrigation Association

Since 1949, the Irrigation Association (IA)[13] has represented the widely varied interests of its membership in irrigation, drainage, and erosion control. Primarily a trade association, IA has led the advances in water-use efficiencies for irrigated agriculture, landscape, and golf course applications and offers many training opportunities to its membership. The IA awards annual prizes for technological innovations and helps define research priorities relating to irrigation.

National Ground Water Association

National Ground Water Association (NGWA)[14] seeks to enhance the skills and credibility of all ground water professionals, develop and exchange industry knowledge, and promote the ground water industry and understanding of ground water resources. National Ground Water Association publishes an extensive list of publications and has a multitude of professional educational opportunities. It also manages the National Ground Water Educational Foundation.

Society of Range Management

The Society of Range Management (SRM)[15] works to promote and enhance the stewardship of rangelands to meet human needs based on science and sound policy. The SRM has over 4000 members organized into geographic sections and has a Watershed/Riparian Committee.

Society of Wetland Scientists

The Society of Wetland Scientists[16] was founded in 1980 to promote wetland science and the exchange of information related to wetlands. With over 4000 members, the society has regional chapters, holds annual meetings, publishes a journal *Wetlands*, and organizes professional certification programs.

Soil and Water Conservation Society

The Soil and Water Conservation Society (SWCS)[17] has approximately 10,000 members and is very active in agricultural water issues. The mission of SWCS is to foster the science and the art of soil, water, and related natural resource management to achieve sustainability. The SWCS serves to both promote and practice an ethic recognizing the interdependence of people and the environment. The Society acts as an advocate for both the conservation profession and for science-based conservation policy. Its members help carry out this mission through 80 geographic chapters, including student chapters. The organization plays an active role in the Farm Bill and other pieces of relevant agriculture and water legislation, and often publishes white papers on agricultural water-related policy issues. The SWCS publishes the *Journal of Soil and Water Conservation* and *Conservation Voices*, and organizes annual meetings.

U.S. Committee on Irrigation and Drainage

The U.S. Committee on Irrigation and Drainage (USCID)[18] was organized in 1952, as a nonprofit professional society. Its multi-disciplinary membership shares an interest in irrigated agriculture—its planning, design, construction, operation and maintenance of irrigation, drainage and flood control works; agricultural economics; water law; and environmental and social issues. The USCID represents the United States on the

International Commission on Irrigation and Drainage (ICID). The ICID is an international organization of more than 70 countries founded in 1950. It operates as a nongovernmental organization devoted to the development of the science and technique of irrigation engineering worldwide.

Water Environment Federation

Since 1928, the Water Environment Federation (WEF)[19] has sought to promote and advance the interests of water quality industry and to benefit society through protection and enhancement of the global water environment. The WEF mostly focuses on domestic and industrial wastewater issues, yet it has a nonpoint source committee that addresses agriculturally related issues. Its research foundation, the Water Environment Research Foundation, provides research grants to study both point and nonpoint sources of water pollution.

INTERNATIONAL PROFESSIONAL SOCIETIES

Increasingly, the above professional societies are incorporating more global water and agricultural issues into their activities and are attempting to attract more foreign members to help sustain their organizations. A few agricultural water-related professional societies have a specific international focus in both mission and membership base. The broad ranges of water and agricultural challenges facing many areas of the world, as well as the needs for networking working professionals across geographic regions, often drive the activities of these international-focused societies.

International Erosion Control Association

International Erosion Control Association[20] has 2400 members and provides education, resource information, and business opportunities for professionals in the erosion and sediment control industry. It offers a professional certification program.

World Association of Soil and Water Conservation

The World Association of Soil and Water Conservation (WASWC)[21] has 500 members and its philosophy is that the conservation and enhancement of the quality of soil and water are a common concern of all humanity. The WASWC strives to promote policies, approaches, and technology that will improve the care of soil and water resources and to eliminate unsustainable land use practices. It has a quarterly newsletter and also publishes books. Its meetings are usually held in conjunction with the International Soil Conservation Organization and the Soil and Water Conservation Society's meetings.

Other associations with relevant international interests include the International Association for Environmental Hydrology, the International Commission of Agricultural Engineering (CIGR), the International Association for Hydraulic Engineering and Research, the International Commission on Irrigation and Drainage, the International Soil Conservation Organization, and the International Society of Soil Science.

CONCLUSION

Increasingly, professional societies build partnerships with organizations having similar agriculture and water-related interests, such as federal and state agencies (e.g., the U.S. Department of Agriculture), environmental groups (e.g., American Rivers), research institutes (e.g., International Water Management Institute), and other professional societies. These partnerships facilitate work on common goals and access each other's resources. Water-related professional societies are even partnering to form broader alliances to further their common interests, such as Water Associations Worldwide[22] and several environmental societies collaborating under the umbrella of the Renewable Natural Resources Foundation.[23]

The number and diversity of agricultural water professional societies reflects the interdisciplinary nature of the discipline. All of the above water-related professional societies have interests in agricultural water issues, and have information and activities of interest to those working in agricultural water fields. This is exhibited in the articles published in their journals, the sessions held at their meetings, and the content of their professional education opportunities. In many ways, the current interests of members and the critical challenges confronting society serve to motivate these professional societies' activities. Further information on the specific activities of any professional society can be found on their respective websites and by contacting them directly.

REFERENCES

1. American Society of Association Executives (ASAE) at http://www.asaenet.org (accessed June 2002).
2. Anderson, F. The Impact of Internet and Communications Technologies (ICTs) on Water Organizations. Working Paper, 2000.
3. American Agricultural Economics Association (AAEA) at http://www.aaea.org (accessed June 2002).
4. American Geophysical Union (AGU) http://www.agu.org (accessed June 2002).
5. American Institute of Hydrology (AIH) at http://www.aihydro.org (accessed June 2002).
6. American Society of Agricultural Engineers (ASAE) at http://www.asae.org (accessed June 2002).
7. American Society of Agronomy (ASA) at http://www.agronomy.org (accessed June 2002).
8. Crop Science Society of America (CSS) at http://www.crops.org (accessed June 2002).
9. Soil Science Society of America (SSSA) at http://www.soils.org (accessed June 2002).
10. American Society of Civil Engineers (ASCE) at http://www.asce.org (accessed June 2002).
11. American Water Resources Association (AWRA) at http://www.awra.org (accessed June 2002).
12. American Water Works Association (AWWA) at http://www.awwa.org (accessed June 2002).
13. Irrigation Association (IA) at http://www.irrigation.org (accessed June 2002).
14. National Ground Water Association (NGWA) at http://www.ngwa.org (accessed June 2002).
15. Society of Range Management (SRM) at http://www.srm.org (accessed June 2002).
16. Society of Wetland Scientists (SWS) at http://www.sws.org (accessed June 2002).
17. Soil and Water Conservation Society (SWCS) at http://www.swcs.org (accessed June 2002).
18. U.S. Committee on Irrigation and Drainage (USCID) at http://www.uscid.org/ (accessed June 2002).
19. Water Environment Federation (WEF) at http://www.wef.org (accessed June 2002).
20. International Erosion Control Association (IECA) at http://www.ieca.org (accessed June 2002).
21. World Association of Soil and Water Conservation (WASCW) at http://www.swcs.org/f_orglinks_links.htm (accessed June 2002).
22. Water Associations Worldwide (WAW) at http://www.wef.org/conferences/affiliations/waw.jhtml (accessed June 2002).
23. Renewable Natural Resources Foundation (RNRF) at http://www.rnrf.org (accessed June 2002).

Psychrometry for Measuring Plant and Soil Water Status: Accuracy, Interpretation, and Sampling

Derrick M. Oosterhuis
University of Arkansas, Fayetteville, Arkansas, U.S.A.

INTRODUCTION

Thermocouple psychrometers are generally considered to be reliable and accurate for measurement of plant and soil water potential (ψ).[1–3] However, the rigorous requirements for using these highly sensitive instruments are frequently misunderstood by users, often leading to frustration and erroneous data.[4] The use of psychrometry in soil and plant water relations has been comprehensively reviewed.[5–8]

ACCURACY OF PSYCHROMETER MEASUREMENTS

The accuracy and reliability of psychrometric measurements of leaf ψ have been demonstrated in many studies. Comparisons of psychrometers with Scholander-type pressure chambers for measurement of leaf ψ have generally exhibited close agreement.[2,3,9] At high water potentials, the psychrometric ψ tends to be more negative than the ψ measured in the pressure chamber, but as the water ψ decreases, pressure chamber values become more negative[2,7] due to resistance to water movement through the xylem towards the cut surface as a result of compression of the vascular tissue.[10] Psychrometric ψ measurements on excised tissues are generally more negative than those of in situ values, with deviations from the 1:1 relationship being greatest at high ψ values with errors often exceeding 0.3 MPa.[11] The source of error for the lower ψ values was associated with evaporative water losses during tissue sampling. A field comparison of the main commercially available thermocouple psychrometers showed differences between the types of psychrometers,[12] attributed to the size of the tissue sample used[13] and evaporative losses.[14] The screen-caged psychrometer most closely correlated with the pressure chamber measurements of ψ. Measurement of leaf ψ with the end-window and leaf-cutter types of psychrometer were similar but slightly more negative and variable than the larger screen-caged psychrometer, while the C-52 sample chamber (Wescor Inc, Logan, UT) was the most variable of the psychrometers tested.

INTERPRETATION OF PSYCHROMETRIC WATER POTENTIAL MEASUREMENTS

Despite widespread acceptance of thermocouple psychrometry, results have not been always satisfactory due to substantial variability in the ψ measurements. Prerequisites for accurate and reliable measurement of ψ with psychrometers include scrupulously clean psychrometers, careful calibration, precise temperature control, proper measurement techniques, and correct interpretation of data. Some of the more common and important sources of errors are discussed below.

Temperature gradients between the reference junction and the sensing-junction can cause errors in the measuring circuit, which are ultimately included in the wet bulb temperature depression. These "zero offsets" are easily measured on the microvoltmeter prior to Peltier cooling and can then be compensated for to eliminate them from the measurement of ψ.[4] Equations are available for correcting these temperature gradients if so desired.[15,16] The achievement of complete vapor equilibrium within the sample chamber is essential. Insufficient equilibration can result in excessively low ψ measurements, whereas excessively long equilibration periods can result in nonrepresentative ψ values due to metabolic changes in the sample tissue.[2,3] Careful interpretation of the microvolt output following Peltier cooling[17] is important, and psychrometer users should be aware of the possible shapes of the microvolt output and the interpretation thereof to obtain the corresponding sample ψ.[4,9,17] Adsorption of water by thermocouple psychrometer assemblies can cause erroneously low ψ measurements[18] because many of the materials used in the construction of thermocouple psychrometers act as vapor sinks and adsorb more water than required to saturate the volume of air within the sample chamber.[19,20] These errors may be largely overcome by covering both the inside of the chamber and the neoprene O-rings with a thin coating of Vaseline.[4,5] Errors due to adsorption of water by the psychrometer assembly are negligible when sufficient tissue is used, but significant with small volumes of tissue.[18]

Encyclopedia of Water Science
DOI: 10.1081/E-EWS 120015087

To achieve a given level of statistical precision for a given experiment and measurement technique, some knowledge of the sources of variation is essential to determine the number of samples and replications needed. An understanding is required, therefore, of the sampling error due to instrument variation, leaf-to-leaf, and plant variation so as to devise a sampling scheme (discs per leaf, leaves per plant, and number of replications) that minimizes the variability, achieves maximum efficiency, and gives the required precision. For example, total error (experimental + sampling) is significantly larger ($P <$ 0.05) for stressed than for well-watered wheat leaves.[21] This will vary with species and should be considered for each experiment. Savage, Cass, and de Jagger[22] provided a statistical assessment of errors encountered during the use of thermocouple pychrometers for ψ measurement.

An underlying requirement often overlooked in psychrometric measurement of ψ is the need for consistency in all procedures from one sample to the next. Strict adherence to experimental protocol will greatly enhance the reproducibility of the data and will ensure more meaningful and reliable results with less variability.

SAMPLING FOR SOIL OR PLANT PSYCHROMETRIC MEASUREMENTS

Accurate measurement techniques for ψ measurement are of little use if the soil or plant sample is not representative of the water status of the biological system being measured. The water potential of the excised plant sample or excavated soil sample must show little, if any, change prior to being sealed into the psychrometer sample chamber.[23] Oosterhuis and Wullschleger[7] reviewed the use of thermocouple psychrometers for the measurement of ψ in leaf discs and highlighted the precautions necessary during tissue sampling and the interpretation of results for accurate psychrometric measurement of ψ. A similar review for sampling soil material is not available.

Water lost by evaporation following sample excision, particularly from succulent and turgid leaf samples, could result in a decrease in measured ψ. However, leaf ψ can rise within a few minutes after excision because xylem tension is released,[24] followed by a rapid decrease in ψ dependent on the evaporative demand. Thus, ideally, samples should be punched directly from attached leaves into the psychrometer chamber with only one sample being taken from each leaf.[5] The leaf-cutter psychrometer was developed[25] with these concepts in mind. Precautions are needed to reduce evaporative losses after excision especially under conditions of high evaporative demand. Leaves with waxy cuticles may require the use of abrasion to reduce cuticular resistance and vapor pressure equilibration times.[14,26]

Tissue-sample size can affect the measurement of leaf ψ,[13,24] although results are inconclusive as to the optimal size which should be used with a particular psychrometer chamber volume. The relative amount of the chamber volume occupied by leaf tissue and air is important as this introduces problems associated with vapor pressure equilibration. The larger the volume of the psychrometer sample chamber, the larger the leaf material sample, i.e., the chamber should be filled with as much leaf material as practically possible. This will also reduce the problems associated with sources and sinks of water vapor on the chamber walls.[5] Excessively small samples may require longer vapor pressure equilibration times because less tissue is available to contribute water vapor. Most data suggest that the measured leaf ψ is higher in tissue having a high cut surface area (A) to sample volume (V) ratio,[24,27] although the opposite has been reported.[13] Nevertheless, the area of the cut surfaces represents a potential site for excessive evaporation losses, and the A/V ratio gives some indication of the possible extent of these losses. Using the largest possible leaf disc to fill the chamber will ensure that the A/V ratio is minimized and the effects of evaporative losses concomitantly reduced.

For in situ soil ψ measurement, psychrometers should be placed in horizontal positions because vertical gradients are more pronounced than horizontal gradients. Furthermore, soil ψ measurements may be compromised if psychrometers are used in the upper 0.3 m of the soil.[28]

CONCLUSION

With good techniques and adequate precautions during sampling, precise measurement techniques, and careful interpretation of the recorded data, thermocouple psychrometers offer a convenient, accurate, and reliable method of measuring ψ. The most important sampling procedures include consistency of technique, prevention of evaporative losses during collection and sealing in the psychrometer chamber, and careful sample selection. Measurements of water potential with thermocouple psychrometers compare favorably with those made using the pressure chamber. Close attention should be paid to careful cleaning and calibration, achievement of complete vapor pressure equilibration, and prevention and detection of temperature gradients.

REFERENCES

1. Riggle, F.R.; Slack, D.C. Rapid Determination of Soil Water Characteristic by Thermocouple Psychrometry. Trans. ASAE **1980**, *80*, 99–103.

2. Walker, S.; Oosterhuis, D.M.; Savage, M.J. Field Use of Screen-Caged Thermocouple Psychrometers in Sample Chambers. Crop Sci. **1983**, *23*, 627–632.

3. Bennett, J.M.; Cortes, P.M.; Lorens, G.F. Comparison of Water Potential Components Measured with a Thermocouple Psychrometer and a Pressure Chamber and the Effects of Starch Hydrolysis. Agron. J. **1986**, *78*, 239–244.

4. Brown, R.W.; Oosterhuis, D.M. Measuring Plant and Soil Water Potentials with Thermocouple Pychrometers: Some Concerns. Agron. J. **1992**, *84*, 78–86.

5. Wiebe, H.H.; Campbell, G.S.; Gardner, W.H.; Rawlins, S.L.; Cary, J.W.; Brown, R.W. Measurement of Plant and Soil Water Status. Utah State Univ. Bull. **1971**, *484*, 71.

6. Brown, R.W.; Haveren, B.P. van, Eds. *Psychrometry in Water Relations Research*; Agric. Exp. Stn, Utah State Univ.: Logan, UT, 1972; 342.

7. Oosterhuis, D.M.; Wullschleger, S.D. Psychrometric Water Potential Analysis in Leaf Discs. In *Modern Methods in Plant Analysis, New Series, Vol. 9. Gases in Plant and Microbial Cells*; Linskens, H.F., Jackson, J.F., Eds.; Springer-Verlag: Berlin, 1989; 1–133.

8. Oosterhuis, D.M. Psychrometry for Measuring Plant and Soil Water Potential: Theory, Types and Uses. In *Encyclopedia of Water Science*; Stewart, B.A., Howell, T., Eds.; Marcel Dekker Inc.: New York, 2002.

9. West, D.W.; Gaff, D.F. An Error in the Calibration of Xylem Water Potential Against Leaf Water Potential. J. Exp. Bot. **1971**, *22*, 342–346.

10. Boyer, J.S. Leaf Water Potentials Measured with a Pressure Chamber. Plant Physiol. **1967**, *42*, 133–137.

11. Baughn, J.W.; Tanner, C.B. Excision Effects on Leaf Water Potential of Five Herbaceous Species. Crop Sci. **1976**, *16*, 184–190.

12. Oosterhuis, D.M. Comparison of Thermocouple Psychrometers for Plant Water Status Measurement. Ark. Farm Res. **1991**, *40* (1), 8.

13. Walker, S.; Oosterhuis, D.M.; Wiebe, H.H. Ratio of Cut Surface Area to Leaf Sample Volume for Water Potential Measurements by Thermocouple Psychrometers. Plant Physiol. **1984**, *75*, 228–230.

14. Wullschleger, S.D.; Oosterhuis, D.M. Electron Microscope Study of Cuticle Abrasion on Cotton Leaves in Relation to Water Potential Measurement. J. Exp. Bot. **1987**, *38*, 660–667.

15. Michel, B.E. Correction of Thermal Gradient Errors in Stem Thermocouple Hygrometers. Plant Physiol. **1979**, *63*, 221–224.

16. Brown, R.W.; Bartos, D.L. A Calibration Model for Screen-Caged Peltier Thermocouple Psychrometers. USDA For. Serv. Res. Pap. **1982**, INT-*293*, 24.

17. Savage, M.J.; Wiebe, H.H. Voltage Endpoint Determination for Thermocouple Psychrometers and the Effect of Cooling Time. Agric. For. Meteorol. **1987**, *39*, 309–317.

18. Bennett, J.M.; Cortes, P.M. Errors in Measuring Water Potentials of Small Samples Resulting from Water Adsorption by Thermocouple Psychrometer Chambers. Plant Physiol. **1985**, *79*, 184–188.

19. Dixon, M.A.; Grace, J. Water Uptake by Some Chamber Materials. Plant Cell Environ. **1982**, *5*, 323–327.

20. Campbell, E.C. Vapor Sink and Thermal Gradient Effects on Psychrometer Calibration. In *Psychrometery in Water Relations Research*; Brown, R.W., Haveren, B.P. van, Eds.; Agric. Exp. Stn, Utah State Univ.: Logan, UT, 1972; 94.

21. Johnson, R.C.; Nguyen, H.T.; McNew, R.W.; Ferris, D.M. Sampling Error for Leaf Water Potential Measurements in Wheat. Crop Sci. **1986**, *26*, 380–383.

22. Savage, M.J.; Cass, A.; de Jagger, J.M. Statistical Assessment of Some Errors in Thermocouple Hygrometric Water Potential Measurement. Agric. Meteorol. **1983**, *30*, 83–97.

23. Savage, M.J.; Cass, A. Psychrometric Field Measurement of Water Potential Changes Following Leaf Excision. Plant Physiol. **1984**, *74*, 96–98.

24. Barrs, H.D.; Kramer, P.J. Water Potential Increase in Sliced Leaf Tissue as a Cause of Error in Vapor Phase Determinations of Water Potential. Plant Physiol. **1969**, *44*, 959–964.

25. Brown, R.W. New Technique for Measuring the Water Potential of Detached Leaf Samples. Agron. J. **1976**, *68*, 432–434.

26. Savage, M.J.; Wiebe, H.H.; Cass, A. Effect of Cuticular Abrasion on Thermocouple Psychrometric in Situ Measurement of Leaf Water Potential. J. Exp. Bot. **1984**, *35*, 36–42.

27. Nelsen, C.E.; Safir, G.R.; Hanson, A.D. Water Potential in Excised Leaf Tissue: Comparison of a Commercial Dew Point Hygrometer and Thermocouple Psychrometer for In Situ Measurement of Soybean, Wheat and Barley. Plant Physiol. **1978**, *61*, 131–133.

28. Brown, R.W.; Chambers, J.C. *Measurements of In Situ Water Potential with Thermocouple Psychrometers: A Critical Evaluation*, Proceedings of International Conference on Measurement of Soil and Plant Water Status, July 6–10, 1987; Utah State Univ.: Logan, Utah, 1987; Vol. 1, 125–136.

Psychrometry for Measuring Plant and Soil Water Status: Theory, Types, and Uses

Derrick M. Oosterhuis
University of Arkansas, Fayetteville, Arkansas, U.S.A.

INTRODUCTION

Measurements of water potential (ψ) and its components are being increasingly used to characterize plant and soil water relations. The psychrometric technique has a number of advantages compared to other methods including the ability to facilitate a large number of samples and also allows the determination of the components of ψ. Reviews of psychrometry in soil and plant water relations have been published,[1–3] the use and construction of these instruments have been documented,[1,4,5] and concerns about their use for measuring soil and plant ψ have been addressed.[6]

THEORY OF THERMOCOUPLE PSYCHROMETERS

The Concept of Water Potential

The use of thermodynamic principles to express the water relations of soil and plant tissue[1,2] is well recognized. The free energy of water in the soil–plant–atmosphere continuum influences both the movement of water along energy gradients and water availability in the plant. Application of these concepts has proven extremely meaningful, since the chemical potential of water and dissolved solutes greatly affects cell growth.[2] The chemical potential of water is related to the change in the free energy of the system and can be expressed in terms of the partial water vapor pressure.[7] In an isothermal system, the volumetric ψ (MPa) is given by the Kelvin equation:

$$\psi = (RT/V_w)\ln(e/e_0) \tag{1}$$

where R is the universal gas constant ($8.3143 \times 10^{-6}\ m^3\ MPa\ mol^{-1}\ K^{-1}$), T the absolute temperature (K), V_w the partial molar volume of pure water ($1.805 \times 10^{-5}\ m^3\ mol^{-1}$), and e and e_0 are the partial and saturated vapor pressures of water (relative humidity of the air in the psychrometer chamber expressed as a fraction). Therefore, the ψ of a system can be determined if the equilibrium water vapor pressure (e/e_0) is measured at a known temperature and pressure. The thermocouple

psychrometer is based upon this concept and upon the principle that the vapor pressure above a solution or segment of plant tissue is related to its water potential according to Eq. 1.

Principles of Operation of Thermocouple Psychrometers

Two fundamental designs of thermocouple psychrometers have been used to determine water potential in plant tissues,[8,9] and a number of modifications and advances have been suggested for both. The Spanner-type psychrometer, however, has certain advantages over the Richards and Ogata instrument[4,10,11] and is more widely used. The thermocouple is usually constructed from chromel and constantan wire of approximately 25-μm diameter[1] to meet the requirements of both high temperature sensitivity and small junctions. The typical Spanner psychrometer (Fig. 1) consists of a thermocouple sensing-junction (constantan–chromel) and two reference junctions (copper–constantan and copper–chromel).[1]

Three primary methods of using thermocouple psychrometers are currently available including the psychrometric,[8] dew point,[12] and isopiestic methods.[13] For the psychrometric method, a sample is sealed into the chamber, allowed to reach both temperature and vapor pressure equilibrium and then the wet bulb temperature of the air in the chamber is measured relative to the dry bulb temperature. This method requires that water be condensed onto the sensing junction by applying an electric cooling current.[14] This Peltier cooling current continues until the sensing-junction temperature is below the dew point temperature of the chamber air and water condenses on the thermocouple junction.[1,8] When the current is discontinued, the droplet evaporates and the voltage output is monitored (Fig. 2). In the dew point method, the depression of the dew point temperature is measured, again related to the relative humidity within the chamber, and hence, to the ψ of the sample at the prevailing temperature.[15] The isopiestic variation is a null method of measurement in which the vapor pressure of a sucrose solution is balanced against the water potential of the sample.[13] The isopiestic and the dew point methods involve no net transfer of water once condensation has occurred. Although each of these

Encyclopedia of Water Science
DOI: 10.1081/E-EWS 120010174

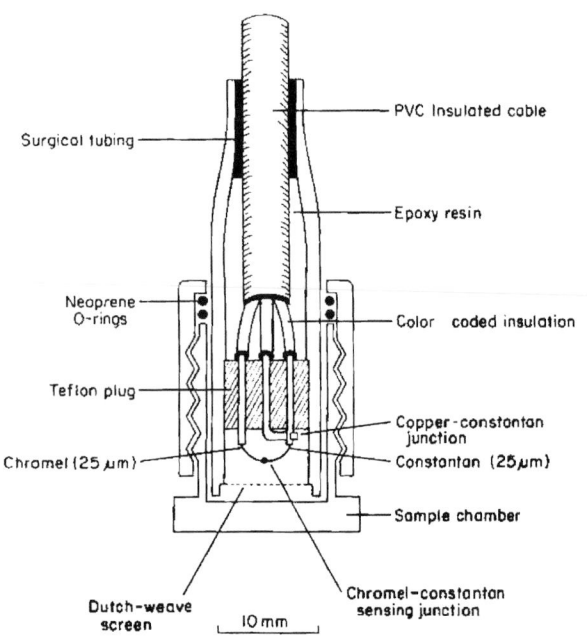

Fig. 1 Diagrammatic representation of a spanner-type end-window. Thermocouple psychrometer used for measuring leaf-disc water potential. (From Ref. 3 with permission.)

methods has its advantages, the psychrometric and dew point techniques are more widely used, with the former method being the more popular and easily available commercially. Both techniques use identical sensors but different microvoltmeter circuitry. In this review, thermocouple psychrometers are used as a collective term for both thermocouple psychrometers and dew point hygrometers.

TYPES OF THERMOCOUPLE PSYCHROMETERS

Many different thermocouple psychrometer designs have been developed for soil or leaf ψ measurement, and

a number of these are commercially available.[16] These include the end-window psychrometers, leaf-cutter psychrometers, and screen-caged psychrometers, the leaf in situ psychrometers, self-standing aluminum insulated psychrometers, porous ceramic shield psychrometers, and a multichambered psychrometer apparatus. The majority of these instruments use a small soil or plant tissue sample for determination of ψ. All of these commercially available psychrometers are generally used without modifications except for leaf in situ psychrometer which should be modified by insulating the housing assembly with a covering of foam insulation and reflective aluminum tape for temperature control.[17] In situ measurements of ψ can be made of leaves in the field using the leaf in situ psychrometer attached directly to an intact leaf,[17,18] in tree trunks with screen-caged thermocouple psychrometers inserted into the trunk,[19,20] or in the soil using ceramic or screen-caged psychrometers buried in the soil at right angles to the soil surface.[21]

USE OF THERMOCOUPLE PSYCHROMETERS

Preparation of Psychrometers

The thermocouple junction is normally protected by a stainless steel housing. New psychrometers should be thoroughly cleaned with a solution of boiling 10% acetone or a jet of steam to remove any oil or debris, which may have accumulated during construction. All psychrometers and sample chambers must be scrupulously cleaned before and after use by repeated flushing of the psychrometer and the sample chamber with deionized water. If possible, the psychrometer should also be periodically inspected under a dissecting microscope to check the cleanliness and physical state of the thermocouple junction. The use of a detergent or steam may help to remove stubborn deposits. After cleaning, the psychrometers and sample chambers should be partially dried with filtered, compressed air and then placed in an oven ($<30°$ C) for few hours. After drying, the psychrometers should be allowed to cool to prevent possible condensation before being stored in their sample chambers or in clean plastic bags.

Calibration of Psychrometers

Accurate calibration of thermocouple psychrometers is essential for accurate and reliable measurements of water potential and its components. The procedure consists of placing a filter paper disc in the sample chamber of a previously cleaned psychrometer using forceps. A small quantity of the appropriate standard solution

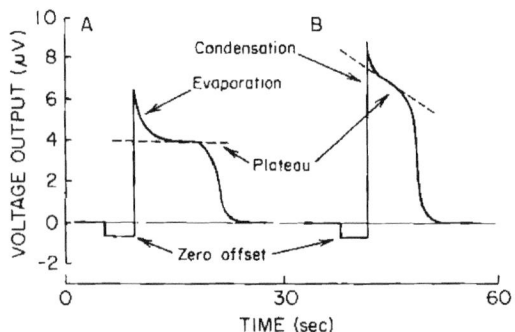

Fig. 2 A typical chart-recorder trace during the determination of relatively high (A) and low (B) leaf-disc water potential. (From Ref. 3 with permission.)

(0.1 mol kg^{-1}, 0.3 mol kg^{-1}, 0.5 mol kg^{-1}, 0.7 mol kg^{-1}, and 1.0 mol kg^{-1} NaCl or KCl solutions), sufficient to saturate the filter paper, is added to the filter paper disc in the sample chamber with a syringe or eye dropper beginning with the most dilute calibration solution. The microvolt output (water potential) is then determined using an appropriate microvoltmeter, after a 4-hr vapor pressure equilibration in a constant temperature water bath (i.e., 25°C). Each standard is measured in turn after careful washing and drying of the psychrometers between measurements. A calibration curve is then constructed from conversion tables[22] by which future measurements of the microvolt output can be converted to the equivalent leaf ψ. Use of a computer greatly facilitates these conversions. If a psychrometer is in constant use, it should be recalibrated every few months.[1]

Temperature Control, Thermal Gradients, and Zero Offsets

Thermocouple psychrometers are typically placed in isothermal water baths to minimize temperature-related errors in ψ measurements. Failure to understand and adequately qualify these errors can seriously affect the accuracy of experimental results. Temperature gradients within the psychrometer can arise from thermal fluctuations in the environment, heat produced by sample respiration, and heating of the reference junctions during the Peltier cooling operation.[23] These gradients can introduce systematic error either through temperature differentials between the sensing junction and the sample, or by causing temperature-induced zero offsets within the thermocouple measuring circuit. Thermocouple psychrometers measure the relative humidity of the air in equilibrium with the sample, and therefore any difference in temperature between the sample and chamber air will introduce significant error. This error results from the fact that the air in the chamber and the sample come to the same vapor pressure, not the same relative humidity.[23] The error in ψ introduced is approximately 7.77 MPa °C^{-1}.[24] Commercial psychrometers which utilize leaf-disc samples do not currently allow for the measurement and correction of this error.

Transient electrical zero offsets are another source of possible error in the use of thermocouple psychrometers. Observed zero offsets at the microvoltmeter are generally interpreted as originating within the sensing head of the psychrometer. However, zero offsets can also originate from other locations within the psychrometer circuitry, i.e., at the connection of the meter or the data logger and these should be insulated from direct solar radiation or air currents by enclosing the connectors with a plastic shield.[6] Poorly earthed equipment and the proximity of AC main cables to the hygrometer output leads may cause significant zero offsets and error in ψ. Shielding the wire does not eliminate the problem, but spatial isolation from other electrical equipment within at least a 2-m radius minimizes these errors. Some long-term drift can be tolerated, but short-term fluctuations should be kept to a minimum, less than 0.0005°C for proper precision.[1]

Vapor Pressure Equilibration

It is essential that the vapor pressure within the sensing head of the psychrometer must be in dynamic equilibrium with that of the sample under the established isothermal conditions for precise ψ measurement.[6] This usually occurs within 2–4 hr at 25°C in an isothermal water bath. Failure to achieve complete vapor equilibrium within the sample chamber due to an insufficient equilibration period can result in excessively low ψ measurements. In contrast, nonrepresentative ψ values can also result from long equilibration times due to metabolic changes, i.e., starch hydrolysis, in the sample tissue.[25,26] Error associated with equilibration times can also result from changes in cellular turgor which accompany growth when the leaf-disc tissue is separated from its water source.[27] This error is greatest when sampling from young, actively growing tissues, but is presumable of only minor concern with mature tissues. Psychrometer chambers with rubber seals or dirty or oxidized metal surfaces can display equilibration characteristics dominated by the chamber material.[28] High leaf cuticular resistance may necessitate longer equilibration times[29] or cuticular abrasion.[30]

Measurement Procedure

After psychrometer selection, initial preparation and calibration, the actual measurement procedure involves tissue sampling, equilibration in an isothermal water bath, and recording of the voltage output. The psychrometer-sample chamber assembly is placed in an isothermal water bath for an appropriate vapor pressure equilibration time (4 hr is usually sufficient). The psychrometric mode is then used and water condensed on the measuring junction by applying a 5-mA Peltier cooling current for 15 sec. These are the recommended and most widely used values; however, optimum cooling times and cooling currents may vary with the tissue and condition of measurement.[31] The voltage output should be monitored continuously during evaporation with an appropriate dedicated microvoltmeter (e.g., from J.R.D. Merrill Specialty Equipment, Logan, Utah, or Wescor Inc., Logan Utah) and a chart recorder.[32] Care is required in analysis of the voltage output plateau[31] since it indicates the equilibrium ψ of the sample in the psychrometer chamber.

Interpretation of the Psychrometer Output Plateau

Accurate determination of leaf ψ with thermocouple psychrometers requires reliable and accurate determination of the plateau voltage output following the Peltier cooling (Fig. 2). Immediately following the termination of cooling, psychrometer output sharply increases, reaches a peak or plateau, followed by a relatively rapid decline as water on the sensing junction evaporates back into the air of the chamber. The plateau represents the wet bulb temperature of the psychrometer when the evaporation of water from the junction reaches a steady state with the vapor pressure of the air in the chamber. If the leaf or soil sample has a relatively high ψ, then the plateau may be horizontal and easy to interpret (Fig. 2A), whereas with drier samples, the plateau becomes increasingly transient and interpretation more subjective (Fig. 2B).[32] Although several approaches for determining the plateau can be used, the one most typically used is to extrapolate the plateau back to intersect the vertical line corresponding to the beginning of the evaporation period. The voltage corresponding to this intersection point is then used to calculate ψ. Caution should be exercised to ensure that identical methods of interpretation are used in psychrometer calibration and in the measurement of sample ψ.

CONCLUSIONS

Psychrometry is an extremely useful technique for measuring water potential of plant or soils if proper sampling and measurement precautions are taken. Calibration and cleaning of the instruments are fundamental steps in thermocouple psychrometry. Major concerns during measurements include achievement of complete vapor pressure equilibration, prevention and detection of temperature gradients. With good calibration and cleaning techniques, precise measurement techniques, and careful interpretation of the recorded data, thermocouple psychrometers offer a convenient, accurate, and reliable method of measuring ψ.

REFERENCES

1. Wiebe, H.H.; Campbell, G.S.; Gardner, W.H.; Rawlins, S.L.; Cary, J.W.; Brown, R.W. Measurement of Plant and Soil Water Status. Utah State Univ. Bull. **1971**, *484*, 71.
2. Brown, R.W., Haveren, B.P., van, Eds. *Psychrometry in Water Relations Research*; Agric Exp Stn, Utah State Univ.: Logan, UT, 1972; 342.
3. Oosterhuis, D.M.; Wullschleger, S.D. Psychrometric Water Potential Analysis in Leaf Discs. In *Modern Methods in Plant Analysis, New Series, Gases in Plant and Microbial Cells*; Linskens, H.F., Jackson, J.F., Eds.; Springer: Berlin, 1989; Vol. 9, 1–133.
4. Brown, R.W. Measurement of Water Potential with Thermocouple Psychrometers: Construction and Application. USDA For. Serv. Res. Pap. **1970**, INT-*80*, 27.
5. Savage, M.J.; Cass, A. Measurement Errors in Field Calibration of in Situ Leaf Psychrometers. Crop Sci. **1984**, *24*, 371–372.
6. Brown, R.W.; Oosterhuis, D.M. Measuring Plant and Soil Water Potentials with Thermocouple Pychrometers: Some Concerns. Agron. J. **1992**, *84*, 78–86.
7. Slatyer, R.O. *Plant–Water Relationships*; Academic Press: New York, London, 1967; 366.
8. Spanner, D.C. The Peltier Effect and Its Use in the Measurement of Suction Pressure. J. Exp. Bot. **1951**, *11*, 145–168.
9. Richards, L.A.; Ogata, G. Thermocouple for Vapour Pressure Measurements in Biological and Soil Systems At High Humidity. Science **1958**, *128*, 1089–1090.
10. Barrs, H.D. Comparison of Water Potential in Leaves as Measured by Two Types of Thermocouple Psychrometers. Aust. J. Biol. Sci. **1965**, *18*, 36–52.
11. Zollinger, W.D.; Campbell, G.S.; Taylor, S.A. A Comparison of Water-Potential Measurements Made using Two Types of Thermocouple Psychrometers. Soil Sci. **1966**, *102*, 231–239.
12. Campbell, E.C.; Campbell, G.S.; Barlow, W.K. A Dew Point Hygrometer for Water Potential Measurement. Agric. Meteorol. **1973**, *12*, 113–121.
13. Boyer, J.S.; Knipling, E.B. Isopiestic Technique for Measuring Leaf Water Potentials with a Thermocouple Psychrometer. Proc. Natl. Acad. Sci. U.S.A. **1965**, *54*, 1044–1051.
14. Wiebe, H.H. Water Condensation on Peltier-Cooled Thermocouple Psychrometers. Agron. J. **1984**, *76*, 166–168.
15. Rawlins, S.L. Theory for Thermocouple Psychrometers Used to Measure Water Potential in Soil and Plant Samples. Agric. Meteorol. **1966**, *3*, 293–310.
16. Oosterhuis, D.M. Types of Thermocouple Psychrometers Used for Leaf Water Potential Measurement. *SA Waterbull.* **1987**, *13* (3), 20–23.
17. Oosterhuis, D.M.; Savage, M.J.; Walker, S. Field Use of in Situ Leaf Psychrometers for Monitoring Water Potential of a Soybean Crop. Field Crops Res. **1983**, *7*, 237–248.
18. Savage, M.J.; Cass, A. Measurement of Water Potential using in Situ Thermocouple Hygrometers. Adv. Agron. **1984**, *37*, 73–126.
19. Wiebe, H.H.; Brown, R.W.; Daniel, T.W.; Campbell, E. Water Potential Measurement in Trees. BioScience **1970**, *20*, 225–226.
20. Oosterhuis, D.M.; Savage, M.J.; Wiebe, H.H. Measurement of Water Potential in the Trunks of Citrus Trees using Screen-Caged Thermocouple Psychrometers. S. Afr. J. Plant Soil **1988**, *5*, 219–221.

21. Rawlins, S.L.; Dalton, F.N. Psychrometric Measurment of Soil Water Potential Without Precise Temperature Control. Soil Sci. Soc. Am. Proc. **1967**, *31*, 297–301.

22. Lang, A.R.G. Osmotic Coefficients and Water Potentials of Sodium Chloride Solutions from 0° To 40°C. Aust. J Chem. **1967**, *20*, 2017–2023.

23. Rawlins, S.L. Theory of Thermocouple Psychrometers for Measuring Plant and Soil Water Potential. In *Psychrometry in Water Relations Research*; Brown, R.W., Haveren, B.P. van, Eds.; Utah State Univ.: Logan, UT, 1972; 25.

24. Dixon, M.A.; Tyree, M.T. A New Stem Hygrometer, Corrected for Temperature Gradients and Calibrated Against the Pressure Bomb. Plant Cell Environ. **1984**, *7*, 693–697.

25. Walker, S.; Oosterhuis, D.M.; Savage, M.J. Field Use of Screen-Caged Thermocouple Psychrometers in Sample Chambers. Crop Sci. **1983**, *23*, 627–632.

26. Macnicol, P.K. Rapid Metabolic Changes in the Wounding Response of Leaf Discs Following Excision. Plant Physiol. **1976**, *57*, 80–84.

27. Baughn, J.W.; Tanner, C.B. Excision Effects on Leaf Water Potential of Five Herbaceous Species. Crop Sci. **1976**, *16*, 184–190.

28. Dixon, M.A.; Grace, J. Water Uptake by Some Chamber Materials. Plant Cell Environ. **1982**, *5*, 323–327.

29. Wullschleger, S.D.; Oosterhuis, D.M. Electron Microscope Study of Cuticle Abrasion on Cotton Leaves in Relation to Water Potential Measurement. J. Exp. Bot. **1987**, *38*, 660–667.

30. Savage, M.J.; Wiebe, H.H.; Cass, A. Effect of Cuticular Abrasion on Thermocouple Psychrometric in Situ Measurement of Leaf Water Potential. J. Exp. Bot. **1984**, *35*, 36–42.

31. Savage, M.J.; Wiebe, H.H. Voltage Endpoint Determination for Thermocouple Psychrometers and the Effect of Cooling Time. Agric. For. Meteorol. **1987**, *39*, 309–317.

32. Bristow, K.L.; de Jager, J.M. Leaf Water Potential Measurements using a Strip Chart Recorder with the Leaf Psychrometer. Agric. Metereol. **1980**, *22*, 149–152.

P

Pump Powering with Internal Combustion Engines

Hal Werner
South Dakota State University, Brookings, South Dakota, U.S.A.

INTRODUCTION

Internal combustion engines or electric motors power most water pumps. This section will present information on the use of internal combustion engines.

The most common internal combustion engines are diesel and natural gas with propane and gasoline used less frequently. Choosing a particular power unit fuel may be governed more by necessity rather than desire. For example, natural gas is not an option unless gas service is located near the pumping site because of the high cost of extending natural gas lines. Other considerations are discussed in this article.

Advantages of using an internal combustion engine for water pumping include variable speed allowing for varying pump output and, except for natural gas, portability between pumping sites.

Disadvantages of internal combustion engines are high repair and maintenance costs, susceptibility of units to cooling and lubrication failure, and the need for right-angle drives when using deep well pumps.

Depending on actual local fuel prices, natural gas and diesel generally have the lowest pumping costs for a given installation. Diesel engines are often the highest initial cost and servicing for diesel engines may not be readily available in some areas. Gasoline engines generally require more frequent overhaul and repair but fuel may be more readily available. Few gasoline engines are rated for continuous duty service needed for pumping water.

Thermal efficiency is the rate at which an engine can convert energy in the fuel into mechanical power. For well-maintained engines, the thermal efficiency of diesel engines is highest followed by natural gas and propane. Gasoline engines have the lowest efficiency.

Several factors need to be considered when selecting an engine for pumping water. Engineering considerations include the type of pump, pump speed, and horsepower requirements. Economic considerations include cost of the equipment, cost of the fuel, hours of operation, and labor requirements. Other factors to consider include fuel availability or cost for service, need for portability, safety controls, and any peripheral equipment such as a generator.

POWER REQUIREMENTS

The power needed to drive a water pump is determined by the following factors. Flow rate (Q) is the amount of water pumped per time period. Total head (TH) is the amount of head (pressure) that the pump supplies to the flow delivered. Total head includes static head or vertical lift, delivery pressure, friction losses, and other miscellaneous losses. Total head is also called total pumping head or total dynamic head (TDH). Finally, the efficiency of the pump (E_p) and drive unit (E_d) is used to calculate power needs. In general form, the equation to calculate pump power (P) is:

$$P = \frac{(C \times Q \times \text{TH})}{(E_p \times E_d)} \tag{1}$$

where C is a coefficient to convert the units to the desired units of power. Power is often expressed in units of kilowatts or horsepower.

It is very important to size the power unit large enough to meet the power required for the pump. It is crucial not to overload the power unit and thereby shorten its life, or worse, have it fail completely. While electric motor sizes can be matched to the pump power needed, engines are generally sized larger to account for rating methods. Engine ratings may be designated as intermittent, automotive, industrial, or continuous. Engines used for pump powering should be selected using an industrial or continuous rating that provides for the continuous, constant load of a water pump. If no continuous rating is available for the engine, the output power rating of the engine should be reduced by at least 20% to avoid overloading.

Additionally, most engines are rated for a standard operating condition that does not reflect varying conditions found for most installations. For example, most are rated as bare engines at sea level and 16°C (60°F). Below are some of the derating factors that should be applied to engines:

- For each 300 m (1000 ft) above sea level deduct 3%.
- For each 5.5°C (10°F) above 16°C (60°F) deduct 1%.
- For radiator and fan deduct 5%.

Encyclopedia of Water Science
DOI: 10.1081/E-EWS 120010358

- For accessories (alternator, etc.) deduct 5%.
- For drive losses (right angle, belt) deduct 5%.

Neglecting to derate an engine for the various factors could result in overloading and premature failure of the unit. Finally, add power for any other options that the engine supplies, such as hydraulic motors or a generator set that provides power for auxiliary electric motors.

A right-angle drive system is required when engines power deep well pumps. A right-angle gear drive unit is the most common drive with belt drives uncommon and less desirable. It is critical that drive shafts and couplers be properly aligned when connecting engines to drive units and pumps. It may be desirable to install a clutch between the engine and the pump to allow for running the engine and auxiliary equipment without operating the pump.

If possible, direct connect the pump to the engine to avoid drive unit losses. This requires that the engine speed be matched to the desired pump speed for the proper output. This speed may be different from the highest rated power for the engine and must be considered when sizing the engine since power output varies with engine speed. It may be desirable to operate the engine at a slower speed than the maximum rated speed to prolong life. Fuel efficiencies are also often higher at speeds less than the maximum rated speed.

Tractors are occasionally used to drive pumps. It is critical to derate the tractor engine for continuous operation since they are rated for intermittent operation. Also apply an additional derating factor (about 5%) to cover the losses between the engine and the power take off (PTO) shaft if the power rating is not given for the PTO output. Standard tractor PTO speeds are 540 rpm and 1000 rpm. Few modern pumps operate at those speeds; thus a gear or belt speed increaser will generally be required. Retrofit tractor engines with safety controls to protect the engine from hazards such as loss of pump pressure, coolant temperature, and oil pressure.

Two types of safety equipment are essential for engines. First operator safety is imperative. All shafts, belts, and fans must be shielded. Also pump and pipe fittings must be secure and rated for the operating pressure to prevent failure.

The second type of safety equipment protects the pump from damage if a malfunction occurs. These safety controls should shut down the engine whenever a condition exists that could damage the engine. Most important is a device to shut off the engine in case of excessive engine temperature, either high coolant or oil temperature. It may also be desirable to provide oil level protection. For centrifugal or deep well pumps, protection must be supplied for a loss of pressure at the pump discharge or in a downstream pipeline. A drop in pressure may overload the engine and result in premature failure. Finally, the engine and pump should be protected from loss of water, e.g., loss of prime of a centrifugal pump.

Another option is to automate the controls for the engine. A variety of features are available including remote monitoring, remote starting and stopping, and controls to interlock peripheral equipment like chemigation application units. The interlock feature shuts down the engine when the secondary unit malfunctions.

ENGINE FUEL USE

The fuel-to-power conversion efficiency of internal combustion engines varies from less than 20% to nearly 40%. As a result considerable heat energy must be dissipated from the engine to prevent overheating. Engines use three cooling options. The radiator and fan option is common and is similar to units on most cars and trucks. Air-cooled engines use a high capacity blower and shrouds to direct cooling air over the engine. Heat exchangers are similar in principle to radiators except no fan is used and water from the pumping source circulates through a heat exchanger instead.

Estimates of hourly fuel use for internal combustion engines can be made using the table below. Divide the total energy requirements of the pump (using Eq. 1 given previously) by the values from the table to estimate the fuel use for the pumping station. Actual fuel use may be different from the estimated depending on the relative performance of individual engines and pumps. Energy use for electricity is included in the table for reference. The cost of operation can be estimated by multiplying the fuel use by its cost.

Maintenance of internal combustion engines is crucial to long life and efficient operation. Monitor oil levels and regularly change oil. Insure that safety devices are functioning properly. Conduct maintenance on a planned schedule, and keep records of service and maintenance.

Fuel	Power output per unit of energy
Diesel	11.79 MJ/L (16.66 hp hr/gal)
Propane	6.51 MJ/L (9.20 hp hr/gal)
Gasoline	8.14 MJ/L (11.5 hp hr/gal)
Natural gas	7.79 MJ/m^3 (82.2 hp hr/1000 ft^3)
Electricity	3.16 MJ/kW hr (1.18 hp hr/kW hr)

CONCLUSION

Internal combustion engines are excellent choices to power water pumps. The engine needs to be matched to the pump requirements to insure efficient operation and long life. Options to consider for each engine include a clutch, auxiliary generator, and automated controls. Safety devices must be provided to protect both the opartor and the equipment. Proper maintenance is essential to guarantee long engine life.

FURTHER READING

Fischbach, P., Ed. *Irrigation Pumping Plant Performance Handbook*; University of Nebraska: Lincoln, NE, 1982

Lundstrom, D. *Irrigation Power Unit Selection*; AE-88; North Dakota State University: Fargo, ND, 1990

Scherer, T.; Kranz, W.; Pfost, D.; Werner, H.; Wright, J.; Yonts, C.D. *Sprinkler Irrigation Systems*; MWPS 30; Midwest Plan Service: Ames, IA, 1999; 99–143

Pumps, Displacement

Dorota Z. Haman
University of Florida, Gainesville, Florida, U.S.A.

INTRODUCTION

A displacement pump is a device that traps a fixed amount of water at the intake conditions (lower elevation or pressure) and either transports it to a discharge elevation and/or compresses it to the discharge pressure. This article deals with screw pumps and three types of reciprocating positive displacement pumps: piston, plunger, and diaphragm.

SCREW PUMPS

Screw pumps are the oldest of the positive displacement pumps. It is known that they were used in ancient Egypt to lift water from the Nile. The Cochleon or Egyptian Screw was composed of tubes wound round a cylinder (see Fig. 1). As the entire unit rotates, water is lifted within the spiral tube to the higher elevation.

Later, other designs of screw pumps were used where a spiral groove was cut on the outside of a solid wooden cylinder and then the cylinder was covered by boards or sheets of metal closely covering the surfaces between the grooves. Although the screw pump is said to have been invented by Archimedes and has been named after him, there is no record of Archimedes himself claiming its invention. The invention was attributed to him by Diodorus who lived 200 yr later. He claimed that Archimedes invented the screw pump in Egypt suggesting that Greeks adopted the pump from Egypt where the first records of the pump can be found. An astronomer of Alexandria, Conon of Samos, also called Conon of Alexandria, who was a close friend of Archimedes, is believed to have invented or adopted the screw pump from lower Egypt. Archimedes demonstrated and fully explained its properties and the screw pump eventually became known as the Archimedean screw.[1] Screw pumps have been used since then in many applications and are still used today.

The simplest, single screw pumps also called "progressive cavity" or Archimedean pumps are often used in land drainage since they can pump large volumes of water over levees. Large pumps of this kind, powered by the windmills, have been employed by the Dutch to drain polders since 1634. The Archimedean screw turned

in a brick-lined casing enclosing approximately one half of the screw, but open at the top. When it has reached the top of the screw, the water flowed over a low sill to the tail race. A sluice door or trap, which is closed by its own weight as soon as the screw stops, prevented the water in the storage basin from flowing back to the polder (Fig. 2).[2]

The diameter of a typical drainage pump is 0.3 m or greater and the length is up to 15 m–18 m. The screw is normally arranged at an angle of 30°. The greater the angle of inclination, the lower the output. The output lowers approximately 3% for every degree increase over a 22° inclination. At 30° angle, 15 m long screw can lift water to 7.5 m. The output depends also on the level of water in the intake reservoir and the ratio of the diameter of the screw shaft to the outside diameter of the screw flights. It is also limited by the rotational speed which for a single screw pump is between 30 rpm and 60 rpm.

Modern Archimedean screw pumps can have efficiencies up to 75%. For practical purposes they are no longer than approximately 15 m but they can have very large diameters in order to increase the capacity of the screw. If the required lift is larger than 7.5 m, a number of screws arranged in series can be used. In addition, single, double, and triple flights are often used (Fig. 3). Flights are also known as helixes. With each increase in flights, there is a 20% increase in capacity. The three-flight pump can handle the most capacity in the least amount of space. Finally, the clearance between screw flights and trough will impact the output of the pump.[3]

Modern screw pumps fall into two basic categories: rigid screw pumps and eccentric screw pumps. Rigid pumps can be subdivided into single screw pumps described above and intermeshing screw pumps with two or more screws. The eccentric pumps can come in two basic configurations. In one, the rotor thread is eccentric to the axis of rotation and meshes with internal threads of the pump housing (stator). In the other, the stator wobbles along the pump centerline. Multiple-screw pumps are available in a variety of configurations and designs. All employ one driven rotor in mesh with one or more sealing rotors. They can be single-ended, or more commonly double-ended.[4]

The multiple rotor pumps can be further divided into timed and untimed categories. Timed rotors rely on

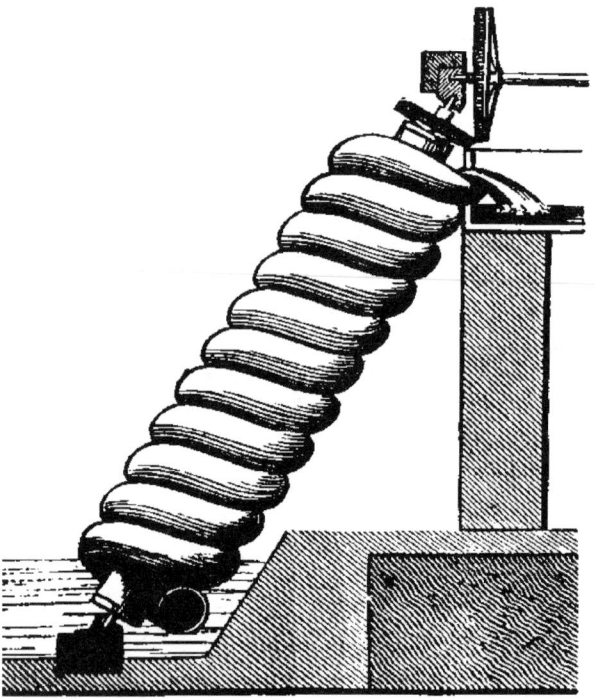

Fig. 1 Egyptian screw pump. (From Ref. 1.)

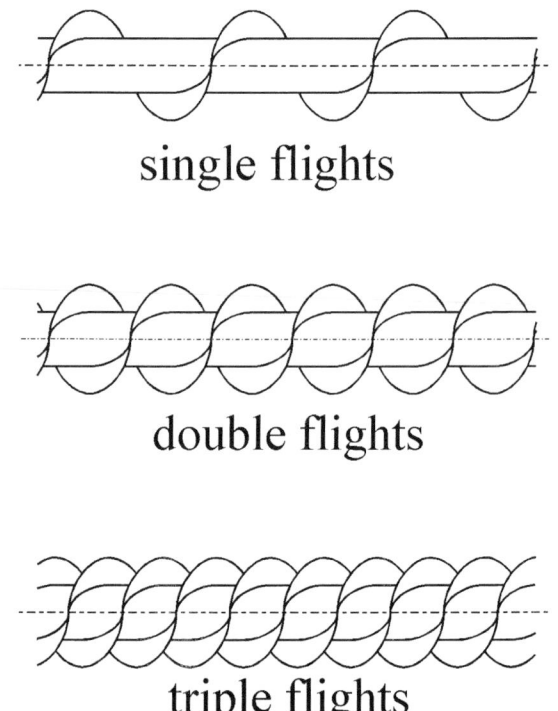

Fig. 3 Various types of flights in a screw pump.

internal or external set of timing gears for phasing the mesh of the threads and for supporting the forces acting on the rotors. Untimed rotors rely on precision and accuracy of the screw itself for proper mesh and transmission of rotation.[4]

Screw pumps are a unique type of rotary positive displacement pump in which the flow through the pumping elements is truly axial. The water is carried between the screw threads on one or more rotors. It is then displaced

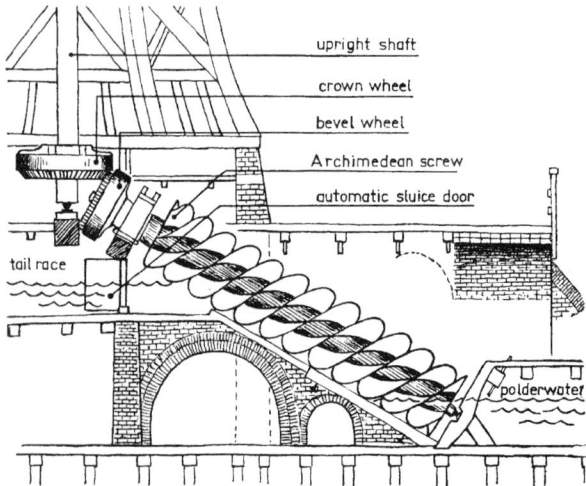

Fig. 2 Archimedean screw pump used in a Dutch windmill. (From Ref. 2.)

axially as the screws rotate and mesh. In other types of rotary pumps, the liquid is forced to travel circumferentially, however, the screw pump has an axial flow pattern and low internal velocities. They are true positive displacement pumps and they will deliver a definite amount of water with every revolution of the rotors. Due to relatively low inertia of rotating parts, some screw pumps can operate at higher speeds than other pumps often up to 10,000 rpm. They are self-priming and flow characteristic is essentially independent of pressure.

Commonly, screw pumps are used today in pumping wastewater and storm water due to their large capacity and low heads and no need for screening the debris. Modern screw pumps consist of revolving shaft fitted with one, two, or three helical blades to rotate in an inclined trough and push the wastewater up the trough. This type of pump can pump large solids without clumping and can operate at a constant speed over a wide range of flows with good efficiencies.

The capacity of any screw pump is the theoretical capacity minus the internal leakage. In order to find the capacity of a screw pump the speed of the pump must be known. The delivered capacity of any rotary screw pump can be increased in several different ways. The capacity of the pump depends on several factors: diameter of the screw, speed of the screw, and the number of flights mounted on the screw shaft.[4]

The advantages of screw pump include: 1) wide range of flows and pressures, 2) wide range of liquids and viscosities, 3) Built-in variable capacity, 4) High speed capability allowing freedom of driver selection, 5) low internal velocities, 6) self-priming with good suction characteristics, 7) high tolerance for entrained air and other gases, 8) minimum churning or foaming, 9) low mechanical vibration, pulsation-free flow, and quiet operation, 10) rugged, compact design—easy to install and maintain, and 11) high tolerance to contamination in comparison with other rotary pumps.[4]

There are also some disadvantages to screw pumps. Their cost is relatively high due to close tolerances and running clearances. Performance characteristics are sensitive to viscosity change and high-pressure capability requires long pumping elements.[4]

RECIPROCATING PUMPS

Reciprocating pumps like screw pumps are among the oldest types of pumps used. Early reciprocating pumps consisted of a piston that moves back and forth within a cylinder. A primitive, piston pump has its origin in a syringe that was used in ancient Egypt. The forcing pump was greatly improved and described by Greek inventor Ctesibius (200 B.C.) the son of a barber in Alexandria who is considered the inventor of a piston forcing pump for pumping water (Fig. 4).

Pumps, utilizing a piston-and-cylinder combination, were commonly used in Greece to raise water from wells. The pumping action is due to the mechanism that is literally forcing slugs of water from the intake pipe to the outlet. Due to the forcing action of the mechanism, the pump head curve for this pump is almost flat. The other similar type of reciprocating pump is a plunger pump where the piston is replaced with a plunger that fits the cylinder less tightly than the piston. Plunger pump (Fig. 5) was invented later and patented by Sir Samuel Moreland in 1675.[1]

Positive displacement, reciprocating pumps are usually selected for low-flow-rate/high-pressure applications. Since most of the water pumping applications require high flow rate, as for example in irrigation

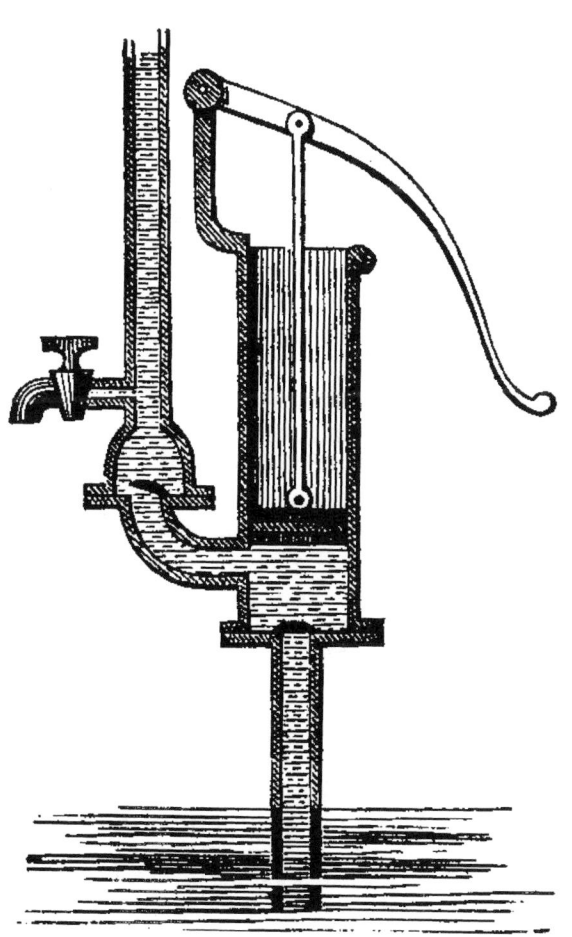

Fig. 4   Early piston pump. (From Ref. 1.)

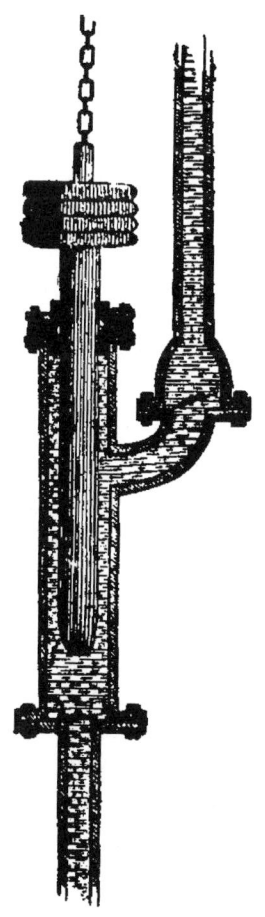

Fig. 5   Early plunger pump. (From Ref. 1.)

systems, reciprocating pumps have limited use in water pumping applications. However, piston and diaphragm pumps are ideal for pumping water using solar energy. Solar water pumps utilize DC electric power from photovoltaic panels and they must work during low light conditions at reduced power, without stalling or overheating. Positive displacement pumps seal water in cavities and force it upward. As a result, lift capacity is maintained even while pumping very slowly. These pumps are used for pumping water for domestic supply, livestock, and small irrigation systems where electricity is not readily available.

A modern reciprocating positive displacement pump (also called power pump) is one in which a plunger or piston displaces a given volume of water for each stroke. All power pumps have a fluid-handling portion, called the *liquid end* consisting of displacing device such as piston or plunger, a fluid holding cylinder, suction and discharge valves, and a packing seal. The liquid end must have a driving mechanism to provide force to the plunger or piston.

A piston is a cylindrical disk, mounted on a smaller diameter rod and usually fitted with some type of sealing rings that move with the piston whereas a plunger is a smooth rod, similar to a piston. The sealing rings in the plunger pump are stationary and the plunger slides through the rings. Schematics of a double-acting piston pump and a plunger pump are given in Figs. 6 and 7, respectively.[5]

A modern power pump is a constant speed, constant torque, and nearly constant-capacity reciprocating pump whose plungers or pistons are driven through a crankshaft from external source. It can have a vertical or horizontal construction. Horizontal construction is used on plunger pumps up to 150 kW and piston pumps rated to 2200 kW. Maximum number of plungers in a horizontal pump is usually no more than five whereas horizontal piston pumps usually do not exceed three pistons. Vertical construction plunger pumps are larger (up to 1100 kW) and contain 3–9 plungers. Plungers are used in pumps producing pressures between 7000 kPa and 200,000 kPa while maximum pressure developed by a piston water pump is approximately 13,800 kPa.[6]

Mechanically and hydraulically driven diaphragm pumps are positive displacement pumps with flexible membranes that operate in a similar manner to a piston or plunger pump. (Fig. 8). A diaphragm is a flexible disk or tube which isolates water from the piston, plunger, hydraulic liquid, or compressed air that are used to actuate the diaphragm. Diaphragm pumps do not have seals or packing and can be used in applications requiring zero leakage. They do not require priming and can be run dry without damage.

Mechanically driven diaphragm pumps operate by reciprocating movement of plunger rod. The force on the

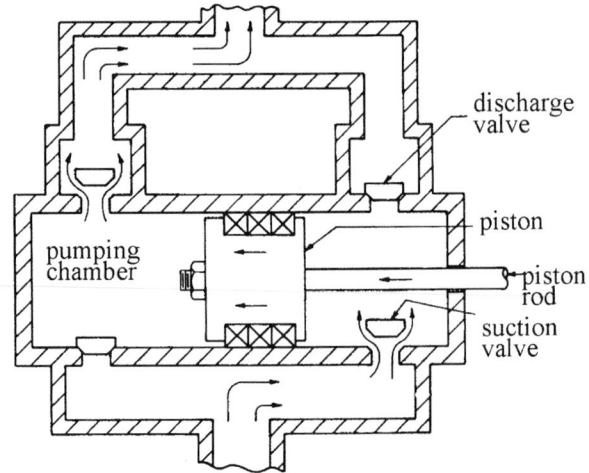

Fig. 6 Schematic of a double-acting piston pump.

central part of the diaphragm creates the suction and discharge pressures. Hydraulically actuated diaphragm pumps, where reciprocating piston forces hydraulic fluid in and out of the chamber behind the diaphragm, is also a positive displacement pump. Air operated diaphragm pumps are displacement pumps but they should not be considered as positive displacement pumps since the maximum pumping pressure cannot exceed the pressure of the compressed air powering the pump.

The disadvantages of diaphragm pumps are that they are not manufactured for operating pressures above 860 kPa and they are not practical for pumping rates above $16 \, \text{L m}^{-3}$ ($58 \, \text{m}^3 \, \text{hr}^{-1}$).[7]

A most common disadvantage of all reciprocating pumps is pulsation. This can be minimized in some modern pumps by increased number of pistons, plungers, or diaphragms that operate out of synchronization with

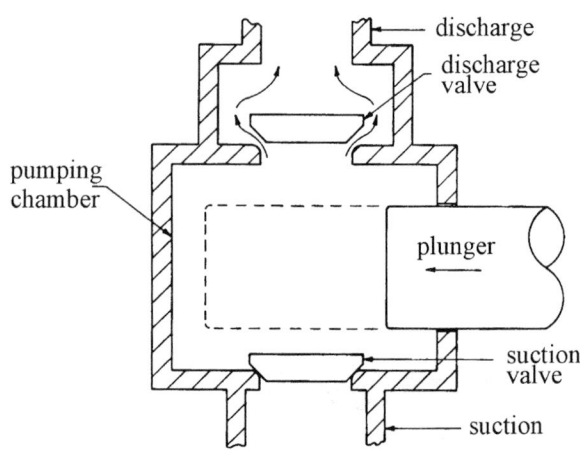

Fig. 7 Schematic of a plunger pump.

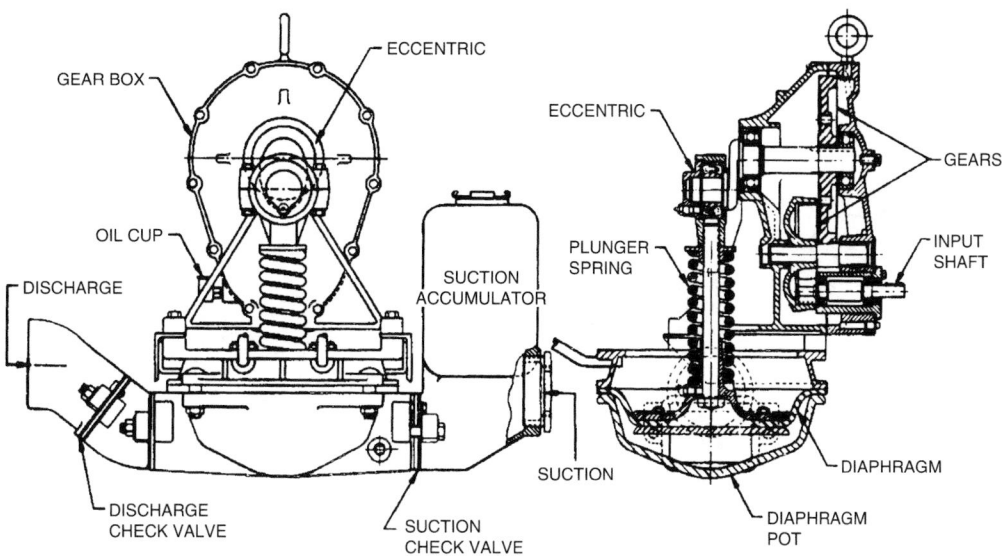

Fig. 8 A schematic of a diaphragm pump. (From Refs. 4–6.)

each other.[8] In addition, the initial cost of a reciprocating pump is larger than the cost of a centrifugal pump.

REFERENCES

1. Ewbank, T. *A Description and Historical Account of Hydraulic and Other Machines for Raising Water*; A New York Times Company: New York, NY, 1972.
2. The Dutch Windmill, based on the book: Stokhuyzen F. Molens. Association for the Preservation of Windmills in the Netherlands, CAJ van Dishoek-Bussum-Holland. 1962. http://www.tem.nhl.nl/~smits/windmill.htm.
3. Warring, R.H. *Pumping Manual*, 7th Ed.; Gulf Publishing Company: Houston, 1984.
4. Brennan, J.R.; Czarnecki, G.J.; Lippincott, J.K. Screw Pumps. In *Pump Handbook*, 2nd Ed.; Karassik, I.J., Krutzsch, W.C., Fraser, W.H., Messina, J.P., Eds.; McGraw-Hill Book Company: New York, NY, 1986.
5. Buse, F. Power Pumps. In *Pump Handbook*, 2nd Ed.; Karassik, I.J., Krutzsch, W.C., Fraser, W.H., Messina, J.P., Eds.; McGraw-Hill Book Company: New York, NY, 1986.
6. Rupp, W.E Diaphragm Pumps. In *Pump Handbook*, 2nd Ed.; Karassik, I.J., Krutzsch, W.C., Fraser, W.H., Messina, J.P., Eds.; McGraw-Hill Book Company: New York, NY, 1986.
7. Henshaw, T.L. *Reciprocating Pumps*; Van Norstrand Reinhold Company: New York, 1987.
8. The Internet Glossary. http://www.animatedsoftware.com/pumpglos/pumpglos.htm.

Quality Influenced by Livestock Production on Range and Pasture Land

Thomas Lee Thurow
University of Wyoming, Laramie, Wyoming, U.S.A.

INTRODUCTION

Livestock and clean water are two products that can be simultaneously obtained from range and pasture lands. This requires that ecological and hydrological principles be applied when crafting a grazing management strategy that is compatible with *predetermined* water quality goals. Making protection of water quality the starting point of land use planning is a philosophical foundation of 1972 U.S. Clean Water Act and subsequent amendments. This goal is operationalized by management agencies establishing total maximum daily load (TMDL) standards for waterways. A TMDL is a calculation of the maximum amount of a pollutant from all contributing point sources [a specific location such as a confined animal feedlot operation (CAFO)] and nonpoint sources (pollution that occurs over a wide area such as may originate from grazing).

Major water impairment concerns associated with grazing can be broken down into physical (suspended sediment), chemical (nutrients, dissolved oxygen), and biological (pathogens) aspects of water quality. Design of a grazing system that will protect water quality must consider the interaction between livestock, vegetation, soil, and water. Grazing effects on each of these attributes are discussed later.

PHYSICAL CHARACTERISTICS

Suspended Sediment

Suspended sediment is the most pervasive nonpoint source pollutant from grazing lands. All waterways naturally contain some suspended sediment attributable to the geologic (natural) erosion influenced by stream type (primarily determined by the geology, topography, and location within the watershed) and ecological factors (e.g., climate, vegetation, soil). Therefore, formulation of TMDL suspended sediment standards must be catchment specific so that geologic erosion can be differentiated from accelerated erosion associated with human activities such as grazing management.

Grazing management can effect the erosion rate of a site primarily by influencing the degree to which livestock impact the soil and vegetation.

Livestock Impacts on Soil

Soil structure is the arrangement of soil particles and intervening pore spaces. The size of soil particles (aggregation) and their stability when wetted determines the porosity of the soil, which governs the rate at which water will enter the soil (infiltration). If the rainfall rate is greater than the infiltration rate, water will run off the site, carrying sediment with it.

Livestock trampling compacts the soil, increasing the bulk density (i.e., the pore volume is reduced resulting in decreased infiltration rate). The degree of damage associated with trampling at a particular site depends on soil type, soil water content, seasonal climatic conditions, and the intensity of livestock use.[1] Compacted trails form on sites where livestock traffic is concentrated. The density of trails tends to increase as the number of pastures is increased within an intensive rotation grazing system. Another common reason for trail formation is repeated movement to and from limited sources of water, mineral supplements, or shelter. The low infiltration rate of trails results in concentrated runoff, which may eventually create gullies. Roads across hilly range and pasture lands are also a serious erosion source, especially since they are often poorly designed and maintained.[2]

Another way livestock trampling causes surficial problems is by churning dry soil to dust. This is very detrimental to infiltration because the disaggregated soil particles are carried by water and lodge in the remaining soil pores making them smaller or sealing them completely. This "washed in" layer where clay particles clog soil pores is a common way that soil crusts are formed. Soil crusts can reduce infiltration by 90%, thereby dramatically increasing runoff and sediment transport.[3] Trampling a crusted soil does break the crust and incorporates mulch and seeds into the soil. However, this benefit is short lived because the subsequent impact of falling raindrops re-seals the soil surface after several minutes. To effectively address a soil-crusting problem,

Encyclopedia of Water Science
DOI: 10.1081/E-EWS 120010348

livestock grazing systems must concentrate on addressing poor aggregate stability, which is the cause of crusting. This requires protecting the soil surface from direct raindrop impact through maintaining vegetation cover and facilitating organic matter buildup in the soil via litter deposition.

Livestock Impacts on Vegetation

Direct raindrop impact on soil represents the greatest potential erosive force on grazing land; therefore it is very important that raindrop energy be dissipated by striking some form of cover before reaching the soil.[4] The amount of cover is positively associated with vegetation litter deposition. Litter slows overland flow, resulting in reduced ability to transport sediment. Litter also aids formation of stable aggregates (associated with high infiltration and low erosion rates) by binding soil particles together with adhesive byproducts produced by decaying litter and microbial synthesis.[5]

Grazing impacts on the vegetation community may be manifest by physical removal of standing vegetation through herbivory or through a gradual change in the composition of vegetation. As grazing pressure increases, the amount of cover and the amount of organic matter returned to the soil is reduced, resulting in an increased likelihood of runoff and erosion. Cover and infiltration rate tends to be greatest under trees and shrubs, followed in decreasing order by bunchgrass, shortgrass, and bare ground.[6] There is little impact on species composition with moderate or light grazing but composition change is great in response to heavy grazing, regardless of grazing strategy.[7] Often the change in species composition associated with heavy grazing is toward dominance by annuals or shortgrass species that have more runoff and erosion associated with them.[8] By the time erosion becomes obvious it may be too late to implement economically viable conservation options. Early recognition of a developing degradation pattern requires knowledge of range ecology, for the first signs of an impending erosion problem almost invariably are manifest by changes in plant density, composition, and vigor.[9]

CHEMICAL CHARACTERISTICS

Dissolved Chemicals

Nutrient loss from grazing lands via leaching or runoff is normally negligible, i.e., less than the input of nutrients from rainfall.[10] Most of the dissolved chemical constituents in runoff are contributed from the soil. Nutrients and organic matter adsorbed to the soil particles are also lost via erosion. Therefore, the most important role of a grazing system in nutrient loss is manifest through land use activities that alter the volumes or timing of runoff and erosion.[11]

Most of the nitrogen in urine is lost via volatilization, and most of the nitrogen in feces is sequestered by microorganisms or eventually transferred to soil organic matter. Nitrate is very mobile during heavy rain periods but loss by leaching is probably insignificant on most grasslands.[12] Feces contain almost all of the phosphorus excreted by livestock. Phosphorus is very resistant to leaching as it is rapidly precipitated or absorbed by other soil minerals. Nitrogen or phosphorus contamination of waterways is only of imminent concern when livestock are allowed to congregate near waterways.[13] Because of this concern, the U.S. Environmental Protection Agency interpretation of the Clean Water Act has deemed location of feedlots near waterways an unacceptable practice.

Dissolved Oxygen

Dissolved oxygen decreases when organic matter, such as animal manure, is added to water. This decrease occurs because biological decomposition processes consume available oxygen, as does oxidation of other reduced compounds such as ammonium. Excessive additions to surface water of nutrients such as nitrogen or phosphorus lead to eutrophication, often expressed by enhanced growth of aquatic plants and reduced water transparency (especially due to increases in algae). As the aquatic plants decay the microbes consume oxygen, lowering the concentration of oxygen available needed to support higher forms of aquatic life such as macroinvertebrates and fish.

BIOLOGICAL CHARACTERISTICS

The primary types of pathogens associated with livestock and wildlife feces are bacteria (e.g., *Campylobacter jejuni*, *Escherichia coli*, *Leptospira interrogans*, *Salmonella* spp.) and water-borne protozoa (e.g., *Cryptosporidia parvum*, *Giardia duodenalis*). These infectious pathogens can pose potential health risks to human drinking water supplies. Environmental fluctuation in temperature and soil moisture of grazing land creates a harsh environment for bacteria and the oocysts of protozoa. Fecal coliforms can survive for several months in soil but can survive for up to a year within feces.[14] There is a rapid mortality of most oocysts when feces are deposited on land,[15] however, viable oocysts can be transported overland, especially when fresh feces are washed by an intense storm.[16] Once pathogens reach a water body, the threat of contamination may last from days to months,[17] with freshwater sediments being the site of greatest concentration and survival.[18]

Few detailed studies have explicitly studied the link between livestock grazing and water-borne pathogens. Much of the research has relied upon indicator coliforms that are more easily cultured but have been shown to be poorly correlated with some types of pathogenic bacteria.[14] Furthermore, many wildlife species harbor the same pathogens that livestock do, thus the natural occurrence of pathogens must be considered when analyzing water quality and making the relationship to livestock use of an area. The greatest threat of pathogen contamination of waterbodies occurs when livestock are allowed to concentrate along streams.[19] In situations where risk of bacteriological contamination is unacceptable, it is necessary to restrict livestock access to streams or riparian areas. Livestock use of these sensitive sites can be significantly reduced through development of water supply away from streams.[20]

CONCLUSIONS

Two broad objectives must be achieved to protect water quality associated with range and pasture grazing.

Limit Runoff and Erosion

Suspended sediment is the most common pollutant associated with grazing. Best management practices (BMPs) to limit runoff and erosion rely on maintenance of soil structure. Vegetation provides the organic matter necessary to enhance formation of stable aggregates and provides the cover to dissipate the erosive force of direct raindrop impact. Appropriate range and pasture grazing systems are designed to maintain vegetation cover and composition by adjusting intensity, frequency, and season of use. Flexibility needs to be built into grazing systems to adjust for unexpected fluctuation in the climate or market prices. The underdevelopment of climate and market risk management planning and policy regarding grazing plans is perhaps the most formidable threat to progress in improving water quality since these variables continue to be used as an excuse for water quality deterioration and/or the lack of progress in improving it.[21]

Limit Direct Livestock Use of Waterways and Sensitive Riparian Areas

Contamination of waterways by nutrients and pathogens is a predominant concern only on sites that allow livestock to congregate near water. On sites with limited water distribution, livestock tend to stay in the vicinity of water so long as forage is available. This increases the likelihood of excrement being deposited directly into the waterway. It also causes deterioration of the soil structure and plant community near the waterway, resulting in accelerated runoff and erosion. Streambanks and moist soil around springs and streamside meadows are particularly susceptible to erosion damage and compaction. Livestock impacts to streams and riparian sites can be limited by providing water, mineral supplements, and shelter at locations away from natural water sources. Special fencing or livestock herding may also be needed to protect sensitive areas from excessive use at critical times. Another reason for protecting wetland or riparian sites is that they serve as vegetation buffer strips that slow runoff and trap sediment before it reaches a waterway.

REFERENCES

1. Warren, S.D.; Blackburn, W.H.; Taylor, C.A., Jr. Effects of Season and Stage of Rotation Cycle on Hydrologic Condition of Rangeland Under Intensive Rotation Grazing. J. Range Manag. **1986**, *39*, 486–491.
2. Sutterlund, D.R. *Wildand Watershed Management*; Ronald Press: New York, 1972.
3. Boyle, M.; Frankenberger, W.T., Jr.; Stolzy, L.H. The Influence of Organic Matter on Soil Aggregation and Water Infiltration. J. Prod. Agric. **1989**, *2*, 290–299.
4. Hudson, N. *Soil Conservation*; Cornell University Press: Ithaca, New York, 1981.
5. Thurow, T.L.; Blackburn, W.H.; Taylor, C.A., Jr. Infiltration and Interrill Erosion Responses to Selected Livestock Grazing Strategies, Edwards Plateau, Texas. J. Range Manag. **1988**, *41*, 296–302.
6. Thurow, T.L.; Blackburn, W.H.; Taylor, C.A., Jr. Hydrologic Characteristics of Vegetation Types as Affected by Livestock Grazing Systems, Edwards, Plateau, Texas. J. Range Manag. **1986**, *39*, 505–509.
7. Ellison, L. Influence of Grazing on Plant Succession of Rangelands. Bot. Rev. **1960**, *26*, 1–78.
8. Thurow, T.L.; Blackburn, W.H.; Taylor, C.A., Jr. Some Vegetation Responses to Selected Livestock Grazing Strategies, Edwards Plateau, Texas. J. Range Manag. **1988**, *41*, 108–114.
9. Thurow, T.L. Hydrology and Erosion. In *Grazing Management: An Ecological Perspective*; Heitschmidt, R.K., Stuth, J.W., Eds.; Timber Press: Portland, Oregon, 1991; 141–159.
10. Menzel, R.G.; Rhoades, E.D.; Olness, A.E.; Smith, S.J. Variability of Annual Nutrient and Sediment Discharges in Runoff from Oklahoma Cropland and Rangeland Nutrient and Bacterial Pollution. J. Environ. Qual. **1978**, *7*, 401–406.
11. Robbins, W.D. Impact of Unconfined Livestock Activities on Water Quality. Trans. Am. Soc. Agric. Eng. **1979**, *22*, 1317–1323.
12. Woodmansee, R.G.; Allis, I.; Mott, J.J. Grassland Nitrogen. Ecol. Bull. **1981**, *33*, 443–462.

Q

13. Khaleel, R.; Keddy, D.R.; Overcash, M.R. Transport of Potential Pollutants in Runoff Water from Land Areas Receiving Animal Wastes: A Review. Water Resour Res. **1979**, *14*, 421–436.

14. Bohn, C.C.; Buckhouse, J.C. Coliforms as an Indicator of Water Quality in Wildland Streams. J. Soil Water Conserv. **1985**, *40*, 95–97.

15. Walker, J.J.; Montemagno, C.D.; Jenkins, M.B. Source Water Assessment and Nonpoint Sources of Acutely Toxic Contaminants: A Review of Research Related to Survival and Transport of *Cryptosporidium parvum*. Water Resour. Res. **1998**, *34*, 3383–3392.

16. Tate, K.W.; Atwill, E.R.; George, M.R.; McDougald, N.K.; Larsen, R.E. *Cryptosporidium parvum* Transport from Cattle Fecal Deposits on California Rangelands. J. Range Manag. **2000**, *53*, 295–299.

17. Stephenson, G.R.; Street, L.V. Bacterial Variations from a Southwest Idaho Rangeland Watershed. J. Environ. Qual. **1978**, *7*, 150–157.

18. Burton, G.A.; Gunnison, D.; Lanza, G.R. Survival of Pathogenic Bacteria in Various Freshwater Sediments. Appl. Environ. Microbiol. **1987**, *53*, 633–638.

19. Larsen, R.E.; Miner, J.R.; Buckhouse, J.C.; Moore, J.A. Water Quality Benefits of Having Cattle Manure Deposited Away from Streams. Biores. Technol. **1994**, *48*, 113–118.

20. Miner, J.R.; Buckhouse, J.C.; Moore, J.A. Will a Water Trough Reduce the Amount of Time Spent in the Stream? Rangelands **1992**, *14*, 35–38.

21. Thurow, T.L.; Taylor, C.A., Jr. Viewpoint: The Role of Drought in Range Management. J. Range Manag. **1999**, *52*, 413–419.

Quality Modeling

Richard Lowrance
United States Department of Agriculture (USDA), Tifton, Georgia, U.S.A.

INTRODUCTION

Largely because of the difficulty of monitoring and predicting nonpoint source pollutants from large areas, water quality modeling has been an important area of water science since the late 1960s. Water quality modeling plays many roles in evaluating and improving the quality of our environment. Models have a major role in helping management and regulatory agencies determine how water quality standards can be met, especially when water quality problems are due to nonpoint sources. Water quality models are used to compare different management strategies designed to control nonpoint source pollution. Models are used to estimate the effects of inputs (including pollutants) on the internal dynamics of water bodies. Water quality modeling is also used to summarize and estimate the various sources of pollutants, especially nutrients, in large basins in order to provide a basis for geographic targeting of pollutant sources.

Water quality models are based on some representation of hydrology and may include movement of surface water, groundwater, and mixing of water in lakes and water bodies. Based on the hydrology, water quality models then simulate some combination of sediment, nutrients, heavy metals, and xenobiotics such as pesticides. Some water quality models, especially those that deal with nutrients, may contain substantial detail related to biological processes including algal growth, nutrient transformations, and respiration. Most water quality models that portray the movement of water within a landscape or landscape components (e.g., fields, forests, streams) portray the interaction of water with soil in a variety of ways. Newer water quality models and add-ons to older water quality models are able to portray the effects of water quality parameters on the biota of lakes and streams or incorporate stream bank, riparian zone, and/or channel functions to understand the effects of these areas on chemical and sediment transport. Other water quality models are used to simulate the effects of critical inputs on the biological communities of lakes and rivers. These aquatic ecosystem models may or may not be tied to watershed models that provide simulated loading to the aquatic ecosystem under varying land use and management.

CLASSIFICATION OF WATER QUALITY MODELS

Water quality models are either built on hydrologic models, are used in conjunction with hydrologic models, or use empirical hydrologic data. Although water quality models can be physical representations of the real world such as channels and ditches built to scale, mathematical or formal models are more common.[1] Mathematical water quality models are quantitative expressions of processes or phenomena that are known to occur in the real-world. The expressions are simplifications of real-world systems through a series of equations governed by conservation of mass. Mathematical water quality models are often a combination of theoretical and empirical representations of the real-world system. Empirical models use water quality observations to provide estimates of water quality parameters through regression analysis. Process based or theoretical representations use physical, chemical, and biological causal relationships to describe the workings of a conceptual system.

Although the real world is subject to random occurrences of weather and management that drive hydrology and water quality, many models ignore the randomness of inputs and spatially distributed attributes and assume that there is a known value for all model parameters. Conversely, stochastic (or random) models use probability distributions of parameters in time or space and can provide outputs based on the distribution. Most water quality models are deterministic models in the sense that one set of inputs will provide only one set of outputs. The difference in a stochastic and deterministic model can be illustrated by how models deal with something simple like how fast water moves in a soil. A deterministic model would use one value for each soil while a stochastic model would vary the movement rate based on the range and distribution of measured water movement rates. Deterministic models are often used with a range of key parameters in order to produce a range of outputs that would better represent real world conditions. Another critical distinction among water quality models is whether they provide continuous or event-based simulations. Continuous simulation models generally provide at least some representation of groundwater/surface water interactions,

Encyclopedia of Water Science
DOI: 10.1081/E-EWS 120010060

while event-based models are more likely to provide only representations of hydrologic processes that take place during rainfall events.

A final distinction among models is whether they are lumped or distributed parameter models. A lumped parameter model contains little or no spatial realism and represents landscape units as homogeneous with respect to the parameters and inputs that drive the model. A distributed parameter model represents certain aspects of the landscape structure, typically by representing areas that are homogeneous with respect to soils, vegetation, and/or land use. Each of these discrete areas is modeled separately and then outputs from all the discrete areas are put together and routed through the system. Because most water quality models are tied to hydrologic models, the water quality outputs from source areas in the model are typically routed through either surface flow pathways, subsurface flow pathways, or both. Models that deal only with events are typically routed through surface flow pathways. Models that simulate continuous or daily water quality in a watershed or field generally must deal with both subsurface or groundwater routing and surface water routing.[2]

USES OF WATER QUALITY MODELS

Risk Assessment of Pesticides

Knowledge of fate and transport of pesticides in the environment is essential to the assessment of risk due to dietary and drinking water exposure. The passage of the Food Quality Protection Act (FQPA) lead to a pressing need to quantitatively predict ranges and magnitudes of expected environmental pesticide concentrations in drinking water. Health-based safety standards mandated by FQPA require USEPA to consider drinking water exposures of humans to pesticides during the risk assessment process. Some state agencies and USEPA use screening models to estimate pesticide concentrations in groundwater and surface water to identify those food-use pesticides that are not expected to contribute enough exposure via drinking water to result in unacceptable levels of aggregate risk.[3] The models are used to guide regulatory agencies such as USEPA to identify where more detailed field data are needed.

Evaluation of Best Management Practices (BMPs)

Water quality improvement from extensive land uses such as agriculture and forestry depends largely on the use of BMPs. Agricultural water quality modeling attempts to adequately represent the differences among various management practices in order to compare and choose which BMPs lead to the least transport of pollutants. These models are typically structured to represent homogeneous landscape units such as fields or portions of fields in order to compare management features such as tillage, fertilizer sources, manure use, and pesticide use and predict the relative impacts on local transport of pollutants such as sediment, nitrogen, phosphorus, and pesticides. Existing models may be used to test the application of BMPs to areas for which no water quality data are available or to determine the effects of BMPs that are similar to those for which water quality effects have been quantified.

Evaluation of Sources and/or Impacts of Pollutants

Both process based and empirical models have been used successfully to examine the sources of pollutants in watersheds and the impact of pollutants or nonpollutants on aquatic ecosystems. The need to quantify the nonpoint source contributions for watersheds and small basins is largely driven by Total Maximum Daily Load (TMDL) assessments and implementation plans mandated by the federal Clean Water Act.[4] The TMDL assessments are done with a water quality accounting approach that typically uses water quality models to estimate nonpoint source pollution. The nonpoint and point sources of a pollutant that are causing the water quality impairment are then combined and compared to observations in the water body. If the water quality is impaired due to the direct presence of a pollutant, then the model estimates of nonpoint source pollution are used to design a plan for reducing nonpoint sources or trading point sources for nonpoint. If pollutants are tied indirectly to the impairment, for instance nutrient enrichment that causes low dissolved oxygen, then the behavior of the pollutant in the waterbody is modeled in order to determine the necessary pollutant load reduction.

Explanation of Large-Scale Systems Behavior

As the behavior of large-scale systems becomes more of an issue and as water quality monitoring data become more available, attempts have been made to combine monitoring and modeling to predict the transport of water-borne pollutants on large scales—river basins and continents. Regression models are used to relate measured pollutant transport in streams to spatially referenced descriptors of pollutant sources, land surface characteristics, and stream channel characteristics.[5] Although mechanisms of pollutant transport are not modeled directly, coefficients that serve as surrogates for processes are used to achieve substantial explanatory power for observed water quality data.

REFERENCES

1. Tim, U.S. Emerging Issues in Hydrologic and Water Quality Modeling Research. In *Water Quality Modeling*; Heatwole, C., Ed.; American Society of Agricultural Engineers: St. Joseph, MI., 1995; 358–373.

2. Overton, D.E.; Meadows, M.E. *Stormwater Modeling*; Academic Press: New York, NY., 1976.

3. USEPA, OPP. 1999. Estimating the Drinking Water Component of a Dietary Exposure Assessment.

4. NAS, *Watershed Management for Potable Water Supply: Assessing the New York City Strategy*; National Academy of Sciences: Washington, D.C., 2000; 549.

5. Smith, R.A.; Schwarz, G.E.; Alexander, R.B. Regional Interpretation of Water Quality Monitoring Data. Water Resources Research **1997**, *33*, 2781–2798.

Quality Sampling of Runoff from Agricultural Fields

John M. Laflen
Purdue University, Buffalo Center, Iowa, U.S.A.

Q

INTRODUCTION

Runoff from agricultural fields carries many physical, chemical, and biological constituents that impact water quality. The mass and concentration of these constituents are estimated based on runoff samples. Depending on the objectives, a runoff sample is collected and then analyzed in the laboratory for constituent concentrations. Laboratory data are then combined with flow measurements to estimate the water quality parameter of interest. The water quality parameters might be total mass loss of a particular constituent, the distribution of concentrations and losses over time, average concentration of particular constituents, pathways of loss, or any number of other parameters of interest.

Runoff from agricultural fields is measured and sampled at many scales—ranging from an area of less than a m^2 up to hundreds and perhaps even hundreds of thousands of km^2. Further, runoff can be due to simulated rainfall applied by indoor or outdoor rainfall simulators, natural rainfall, irrigation, and the application of liquid from animal confinements.

Runoff samples due to natural rainfall on agricultural lands must usually be collected and stored automatically due to the unpredictable nature of natural rainfall, large number of treatments, lack of accessibility during runoff events, and footing and lightning hazards to personnel during many runoff events. This limitation is usually removed when man controls the water application, giving a wider freedom to the selection of measuring and sampling techniques and equipment.

Samples must be collected and processed in a manner that insures that concentrations and characteristics of constituents measured in the sample are the same as the concentrations and characteristics of the constituents in the runoff when the sample was collected, and that the constituents and their concentrations in the runoff are identical to those delivered from the field. Depending on the constituents, and on other factors, this may require special control of the environment where samples are stored prior to processing and require particular materials for collecting and storing samples. In some cases, chemicals must be added to samples to stabilize the constituents so that the form of the constituent does not change during storage.

Runoff sampling systems generally can be classified as one of three kinds: 1) those that collect a constant fraction of the total flow over an entire runoff event(s), 2) those that collect a sample at either a given time or flow volume interval (and usually have a flow measuring system that also measures flow rate during an event), and 3) those where samples are collected manually or all flow is collected for sampling. The objective here is to describe some of the standard and innovative ways of runoff sampling and their limitations. Almost every runoff water sampling application requires custom design and construction for that particular application. While standard designs are available, and some commercially available, most require considerable judgment in terms of selection of equipment and its installation. The objective here is to assist those that might wish to select, design, and install runoff sampling equipment to collect water quality samples in runoff from agricultural fields.

CHARACTERISTICS OF SURFACE RUNOFF

The design of the runoff sampling system depends heavily on expected surface runoff flow rates and volumes and on expected delivery of constituents in runoff. Runoff flow rates and volumes can be estimated using readily available techniques.[1] These estimates are critical in designing of runoff conveyance systems, and components of the flow measuring, sampling, and storage systems. Constituent loads, particularly sediment, are important in the selection and design of equipment. Deposited sediment can greatly impact, and in fact totally incapacitate, runoff sampling systems used on agricultural runoff.

Runoff events are generally infrequent and range widely in magnitude (Fig. 1). During an event, runoff rates will vary widely, and depending on contributing area size and topography, runoff rates may react within a minute or so to changes in rainfall intensity. Severe storms are frequently less than an hour in length. Additionally, the average concentration of material transported in runoff may vary by a factor of 10 or more between events (Fig. 2), and even within an event, may vary by a factor of 10 or more.[2,3] Constituent concentrations, particularly sediment and sediment transported constituents, may also vary quite widely.[2–4]

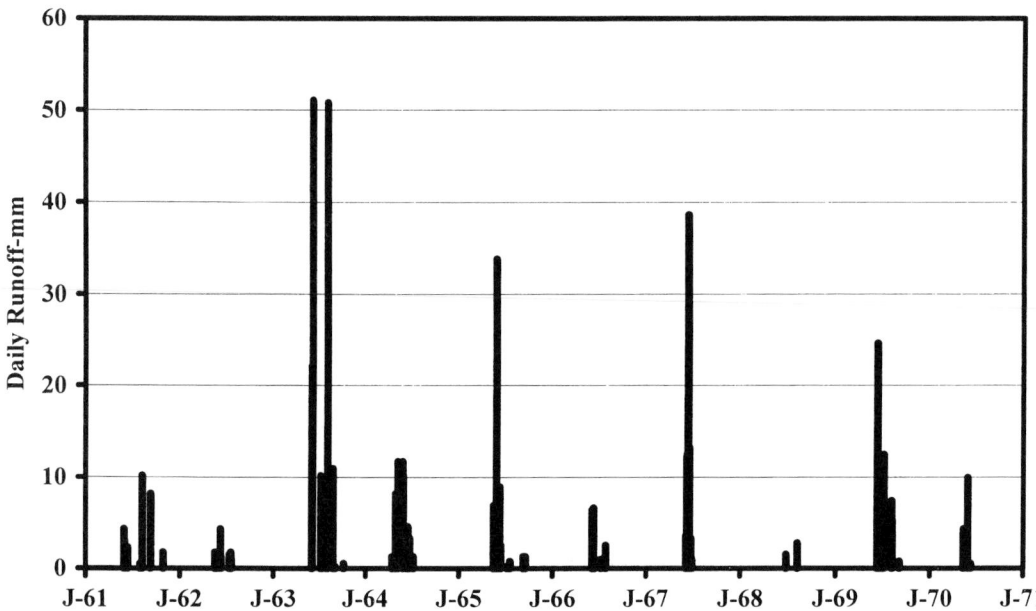

Fig. 1   Daily runoff from a corn–oats rotation erosion plot at Castana, IA between January (J) 1961 and January 1971.

Rare storms are quite important in loss of constituents, and unless systems are carefully designed for these events, system failures are likely to occur. In an 11-yr period near Columbia, MO, over 80% of the soil loss on erosion plots occurred in seven storms, and 50% of the soil loss occurred in only one storm.[5] However, runoff volume from these seven storms was apparently only about 25% of the total runoff during the same period.

A runoff sampling system must be based on expected runoff rates and volumes. In Fig. 3 are shown several dimensionless hydrographs where flow rates as percent of peak are plotted vs. time as percent of runoff duration. The natural runoff and outflow hydrographs shown are idealized, while the rill and interrill plots shown, generated by simulated rainfall, are from measured data.[6]

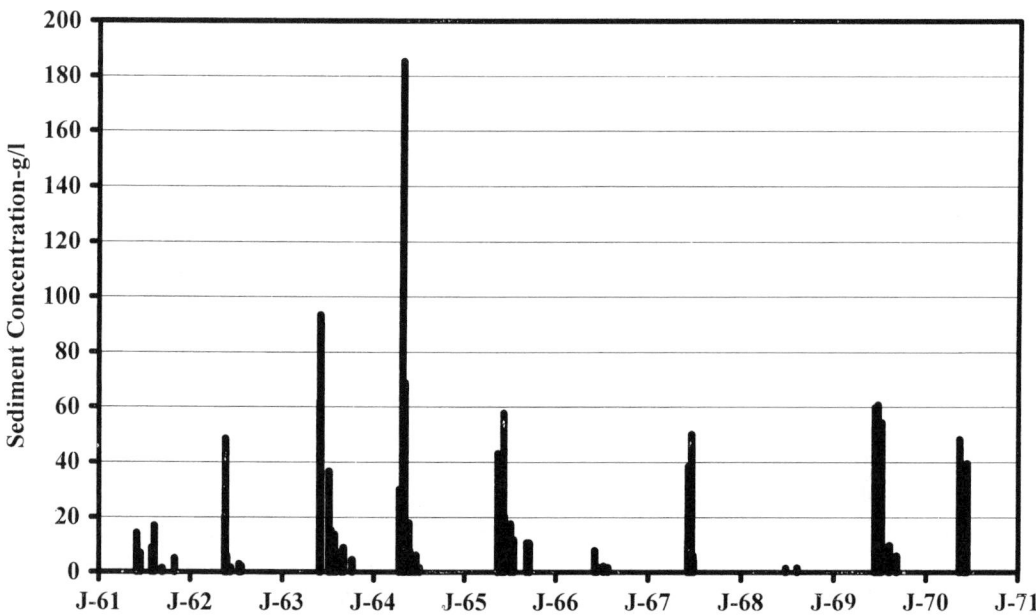

Fig. 2   Daily sediment concentrations from a corn–oats rotation erosion plot at Castana IA between January (J) 1961 and January 1971.

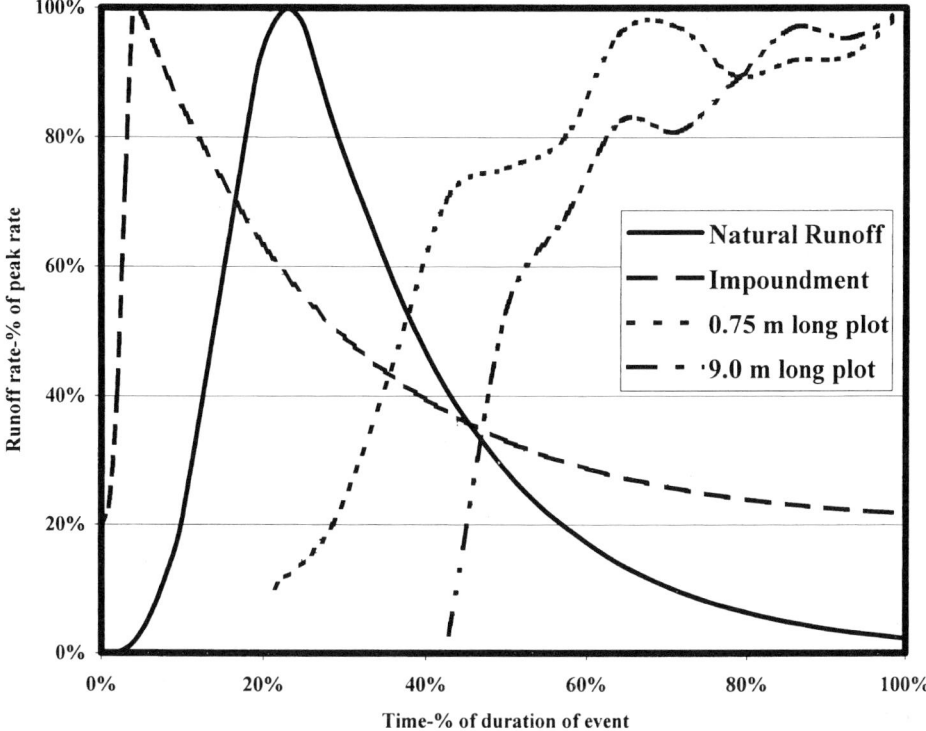

Fig. 3 Dimensionless hydrographs for natural runoff, for areas having impoundments with tile or base flow, and for simulated rainfall with a nominal constant intensity on short plots.

The rainfall simulation plots illustrate the different response times for different very small areas. Total time during this experiment was 70 min. The long delay after rainfall began before runoff and the quick rise of the hydrograph affects the design of the sampling system. For a constant rainfall intensity, a constant rise in the hydrograph is expected. In this case, flow rates decreased at times, likely due to minor changes in rainfall intensity because of changes in pump or nozzle performance.

The natural rainfall runoff rates are typical of those with a single burst of intense rainfall. The total time can range from several minutes to several hours, depending on rainfall, topography, and management. It might also be typical of runoff from an ephemeral channel draining a large watershed subjected to a single rainfall event. In that case, the total time might be as much as several days. There is an infinite number of hydrographs that might result from the combinations of storms and topography that exist in nature. Frequently, hydrographs from a single storm contain several peaks due to different periods of intense rainfall.

The outflow from impoundments could represent many cases. Note that in this example, there is a base flow at about 20% of the peak rate. This could represent a small watershed outflow that contained several impoundments and tile drain lines. It might also represent outflow from

a watershed with a continuous flowing stream. The total time for the storm runoff to move through the system could be from a few hours for a small watershed with impoundment type terraces and subsurface tiles to several months if the hydrograph represents the Mississippi river.

COLLECTING A CONSTANT PROPORTION OF FLOW

Many sampling schemes require only an estimate of concentration for an entire event. A number of devices have been developed that collect a constant proportion of total runoff throughout a runoff event.[7–10] Limitations for collecting a constant proportion of flow are the size of area and the lack of information about concentrations during an event. Most devices were designed for small areas of only a hectare or less because high flow rates from larger areas require much larger equipment, and it is usually much more feasible to collect needed samples using other technology.

Multislot Divisor Systems

An early device used in sampling surface runoff for soil erosion studies was the Geib multislot divisor.[8]

The multislot divisor is constructed so that flow passes through one of a number of parallel rectangular weirs (called slots). Flow is collected from one of the weirs (standard multislot divisors have 13 or fewer slots), with flow from the remainder wasted. The multislot divisor was widely used in erosion studies over much of the United States, and is still in wide use. Maximum flow rates into a standard multislot divisor with 13 slots, each slot of 2.5 cm width, is about $0.10\,m^3\,sec^{-1}$. Divisors could be located in series so that the flow could be split into even smaller parts.

One multislot divisor system was coupled with a Parshall flume[11] for measuring flow rates for a water quality study on flat lands.[9] The system was developed to measure and sample runoff from small plots located on flat lands near the Mississippi river where there was little elevation for measuring and sampling runoff. The hydrograph expected was very similar to the hydrograph for natural rainfall shown in Fig. 3.

A multiweir system that both measured and sampled flow, operating under the same principles as the multislot divisor, was developed for use on terraces with underground outlets.[10] The terrace systems were expected to yield a hydrograph similar to the hydrograph for impoundments shown in Fig. 3, some having base flow and others with no base flow. The duration of the runoff event was expected to be up to about 2 day, but might be longer. In this application, v-notch weirs were used rather than rectangular slots as in the multislot divisor. The v-notch weir gave good precision for measuring and sampling a very wide range of flows. For this application, flow rates were measured continuously, and samples were collected for sediment and plant nutrient analyses. Due to a large range in sizes of watersheds and storm events, the multiweir divisors were used in series with the first multiweir divisor discharging into another divisor, and the lower divisor discharging from one weir into a storage tank. There were 13 weirs in each multiweir divisor, the two in series collected only 1/169th of the total flow. The sampled flow was discharged into a large storage tank, when it overflowed, the flow was again split through a series of small circular orifices whose size depended on the contributing area and expected flow rates. An in-field runoff sampling device similar in concept to the multislot divisor, and used in series was designed to be as unobtrusive as possible, and inexpensive to build.[12] The system worked well, but it was found that the construction and installation were critical to sampling accuracy.

More recently, a 9-slot multislot divisor made of plastic for use in a water quality study was developed and evaluated.[13] A system that uses a tipping bucket flow measuring device and a multitube divisor for collecting runoff samples was also recently described.[14] It has performed well in field studies for several years.

Coshocton Wheel

The Coshocton wheel is a device that samples a constant aliquot of flow from an H-flume.[7] The Coshocton wheel has a series of curved vanes and a single slot, and flow from the H-flume discharges onto the nearly horizontal wheel, with flow entering the slot draining to a storage tank. The flow onto the wheel rotates the wheel with the slot passing under the discharge with each revolution of the wheel. Depending on the wheel chosen, from 0.33% to 1% of the flow is diverted to the storage tank. Flow rates of the Coshocton wheel range up to $0.16\,m^3\,sec^{-1}$.

SAMPLING AT INTERVALS

For many situations, the variation of concentrations of constituents during runoff events is needed. This is usually accomplished by automatically or manually sampling runoff, and by measuring flow rate, either automatically or manually, at the time of sampling. Thus, both the time distribution of concentrations, time distribution of losses, average concentrations and total losses can be computed. This technology can be used on most agricultural watersheds, regardless of size.

For small agricultural watersheds, techniques and equipments are described in Ref. 7. Numerous samplers are described that collect a sample at various intervals, and one is described that collects samples at varying intervals as dictated by flow rates. Samplers are also commercially available that can collect and store samples at intervals.

An automated water sampling and flow measuring system for runoff and subsurface drainage has coupled a tipping bucket flow measuring device with an automatic sample.[15] It has worked satisfactorily for flow rates from $0.001\,m^3\,min^{-1}$ to $0.12\,m^3\,min^{-1}$ in studies of water quality of tile drain lines in Minnesota.

Sumner et al.[16] describe a rainfall simulator and plot design for studying sedimentation, pesticide and nutrient losses on plots of $600\,m^2$ subjected to simulated rainfall of about $25\,mm\,hr^{-1}$. In this effort, runoff was sampled every 5 min during the first 30 min of runoff, and then at 10 min intervals for the remainder of the 2 hr rainfall event.

MANUAL SAMPLING

Manual sampling has many applications, particularly when runoff is from simulated rainfall, irrigation or application of water for other purposes—including application of animal wastes in liquid form. It is also employed in various monitoring studies where construction is not merited.

Elliot et al.[6] using manual sampling doing a rainfall simulation study, reported consistent sediment concentrations, with occasional wide variations in concentration. They overcame this by regular sampling that determined trends in sediment concentration over time.

All runoff may be collected for small events, during early portions of runoff, or from small areas. Combinations of total sampling and intermittent sampling can also be used in some applications. This may be important when combining samples to reduce costs of analysis.

SOURCES OF SAMPLING ERRORS

Errors in sampling occur when the sample does not contain the same concentration of constituents as does the flow or when the proportion of the flow sampled is different than expected. There are many sources of such errors. Careful testing will identify many errors.

At very low flows, the multislot divisor did not collect the expected proportion of the flow, even though it was very good at higher flow rates.[12] This was attributed to a 2 mm elevation difference between slots at the outer edge of the divisor and the middle divisor that collected the sample. Others have found the multislot divisor for water quality testing quite precise.[10]

One source of error is deposition of sediment and sediment adsorbed constituents after runoff leaves the contributing area but before it reaches the sampling location.[17] Another source of error is the impact of research methods on the phenomena under study.[18]

CONCLUSION

There are many techniques and equipments for sampling runoff for water quality from studies involving natural rainfall, irrigation, simulated rainfall, and land application of fluid. There are also opportunities for errors and failures in sample collection that careful design, construction, and testing will help avoid.

One of the major problems in sampling runoff from natural rainfall is the operation of systems that are unattended and serviced infrequently. Additionally, for natural rainfall studies, operation during rare events is imperative if valid results are to be obtained.

It is important to evaluate the transport of materials from the area of interest to the sampling point to insure that losses, and perhaps additions, do not occur in channels. Also, one should carefully evaluate the impact of measuring and sampling equipment on the detachment and transport processes.

REFERENCES

1. Schwab, G.O.; Fangmeier, D.D.; Elliot, W.J.; Frevert, R.K. *Soil and Water Conservation Engineering*, 4th Ed.; John Wiley and Sons, Inc.: New York, 1992.

2. Baker, J.L.; Laflen, J.M. Runoff Losses of Surface-Applied Herbicides as Affected by Wheel Tracks and Incorporation. J. Environ. Qual. **1979**, *8*, 602–607.

3. Barisas, S.G.; Baker, J.L.; Johnson, H.P.; Laflen, J.M. Effect of Tillage System on Runoff Losses of Nutrients, a Rainfall Simulation Study. Trans. Am. Soc. Agric. Eng. **1978**, *21* (5), 893–897.

4. Baker, J.L.; Laflen, J.M.; Johnson, H.P. Effect of Tillage System on Runoff Losses of Pesticides, a Rainfall Simulation Study. Trans. Am. Soc. Agric. Eng. **1978**, *21* (5), 886–892.

5. Ghidey, F.; Alberts, E.E. Comparison of Measured and WEPP Predicted Runoff and Soil Loss for Midwest Claypan Soil. Trans. Am. Soc. Agric. Eng. **1996**, *39* (4), 1395–1402.

6. Elliot, W.J.; Liebenow, A.M.; Laflen, J.M.; Kohl, K.D. *Compendium of Soil Erodibility Data from WEPP Cropland Field Erodibility Experiments 1987 & 1988, 1989*; NSERL Report No. 3, National Soil Erosion Research Laboratory: West Lafayette, IN, 1987.

7. Brakensiek, F.L.; Osborn, H.B.; Rawls, W.J. *Field Manual for Research in Agricultural Hydrology*; Agric. Hdb. 224, U.S. Department of Agriculture, 1979; 550 pp.

8. Mutchler, C.K. *Runoff Plot Design and Installation for Soil Erosion Studies*; ARS 41-79, U.S. Department of Agriculture, Agricultural Research Service, 1963.

9. Willis, G.H.; Laflen, J.M.; Carter, C.E. A System for Measuring and Sampling Runoff Containing Sediment and Agricultural Chemicals from Nearly Level Lands. Trans. Am. Soc. Agric. Eng. **1969**, *12* (5), 584–587.

10. Laflen, J.M. Measuring and Sampling with Multiweir Divisor. Agric. Eng. **1975**, *56* (6), 36.

11. Robinson, A.R. *Parshall Measuring Flumes of Small Sizes*; Tech. Bull. 61, Colorado State University Agricultural Experimental Station, 1957.

12. Franklin, D.H.; Cabrera, M.L.; Steiner, J.L.; Endale, D.M.; Miller, W.P. Evaluation of Percent Flow Captured by a Small In-field Runoff Collector. Trans. Am. Soc. Agric. Eng. **2001**, *44* (3), 551–554.

13. Reyes, M.R.; Gayle, G.A.; Raczkowski, C.W. Testing of a Multislot Divisor Fabricated from Plastic. Trans. Am. Soc. Agric. Eng. **1999**, *42* (3), 721–723.

14. Klik, A.; Sokol, W. A New Measuring Device for Field Erosion Plots. In *Symp. Proc. Soil Erosion Research for the 21st Century, 3–5 January 2001, Honolulu, HI*; Ascough, J.C., Flanagan, D.C., Eds.; American Society of Agricultural Engineers: St. Joseph, MI, 2001; 234–236.

15. Zhao, S.L.; Dorsey, E.C.; Gupta, S.C.; Moncrief, J.F.; Huggins, D.R. Automated Water Sampling and Flow Measuring Devices for Runoff and Subsurface Drainage. J. Soil Water Conserv. **2001**, *56* (4), 299–306.

16. Sumner, R.; Wauchope, R.D.; Truman, C.C.; Dowler, C.C.; Hook, J.E. Rainfall Simulator and Plot Design for Mesoplot

Q

Runoff Studies. Trans. Am. Soc. Agric. Eng. **1996**, *39* (1), 125–130.

17. Johnson, H.P.; Baker, J.L.; Shrader, W.D.; Laflen, J.M. Tillage System Effects on Sediment and Nutrients in Runoff from Small Watersheds. Trans. Am. Soc. Agric. Eng. **1979**, *22* (5), 1110–1114.

18. Hirschi, M.C.; Kalita, P.K.; Mitchell, J.K.; Zhao, Y. Research Method Effects on Measured Erosion Rates. 2001. In *Symp. Proc. Soil Erosion Research for the 21st Century, January 3–5 2001, Honolulu, HI*; Ascough, J.C., Flanagan, D.C., Eds.; American Society of Agricultural Engineers: St. Joseph, MI, 2001; 548–551.

Rainfall Shelters

Arland D. Schneider
United States Department of Agriculture (USDA), Bushland, Texas, U.S.A.

INTRODUCTION

Rainfall shelters (shelters) have been used during the past 50 yr to exclude rainfall and other precipitation from research plots and lysimeters. They bridge the gap between the controlled environment of a greenhouse or growth chamber and uncontrolled field conditions. Meteorological variables such as radiation and wind are altered under the shelter,[1] but with limited rainfall duration, the effect on crop growth is minimal. The main limitation of rainfall shelters is the small crop area that requires careful extrapolation of results to field areas.

Foale et al.[2] identified the following six subsystems or components of rainfall shelters: site, tracks, shelter structure, drive (mechanism), power supply, and controller. Auxiliary components include in-shelter irrigation systems and cranes for weighing lysimeters. Rainfall shelter subsystems and features are illustrated in Fig. 1, and the references provide examples of various types of shelters.

RAINFALL SHELTER SUBSYSTEMS

Site

The site needs to be representative of the soil to be studied, and the surrounding area must be similarly and uniformly cropped for accurate evapotranspiration measurements.[2] The area needs to be well drained with surface runoff from adjacent areas excluded. Since plot areas are small, isolating individual blocks of soil with vertical walls of plastic film or concrete may be desirable.[2] Utilities such as electricity and telephone service are also desirable, and a water supply of adequate quantity and quality must be available for irrigated experiments. Overall shelter design should allow all or most of the research area to be planted, cultivated, and harvested with farm machinery.

Tracks

Most rainfall shelters have two tracks, but some also have a center track to reduce the structure span or to support a center drive mechanism.[3,4] The center track restricts access to the research area and is not recommended, except for unusual conditions. Tracks may be at ground level for low structures or for structures with support walls (Fig. 1).[5] Tracks may also be elevated to eliminate the support walls, and secondary walls may then be suspended from the structure roof (Fig. 1).[3,4] Foundations for the tracks may be continuous footings or individual piers located along the tracks. Tracks are generally single, I or C section beams of rolled-steel, but angles, railway tracks, and welded-up sections have also been used. In addition to the load of the structure, the tracks must resist upward wind forces and lateral forces in one or both directions.

Structure

Structures consist of the framing, covering, truck assemblies, and any walls or doors. Maximum length is about 30 m, and is governed by the length of time to cover the research area during intense storms. Dual shelters that cover the research area from both ends are sometimes used to increase the length of the research area.[3] Foale et al.[2] provide design information for minimizing the shading from the second shelter. Maximum width has normally been about 12 m, but wider spans are possible with heavier structures and tracks.[6] The height of the structure largely determines the wind loading, and Foale et al.[2] provide excellent wind design information for rainfall shelters. Some structures are designed to be easily moved from one location to another where crop rotations or insect and disease populations require frequent changes in the research location. For example, the shelter by Kvien and Branch[7] was mounted on standard automobile tires rather than tracks and rollers to allow easy movement across a field area.

Construction materials range from light aluminum trusses with fiberglass covering to heavy steel beams and columns.[5,8] The structures are usually covered with fiberglass, aluminum, or steel sheeting. Unless the shelters are in locations with extended daytime rainfall, the light transmittance of the covering is not considered. Walls are omitted on some structures for low crops, but are needed for tall crops, and a taller shelter allows personnel to work inside the shelter.[5] With ground level tracks, walls are attached to the load-bearing columns, and with elevated tracks, the walls are suspended from the roof trusses or beams. On many shelters, walls are placed along the rear

Encyclopedia of Water Science
DOI: 10.1081/E-EWS 120010225
Published 2003 by Marcel Dekker, Inc. All rights reserved.

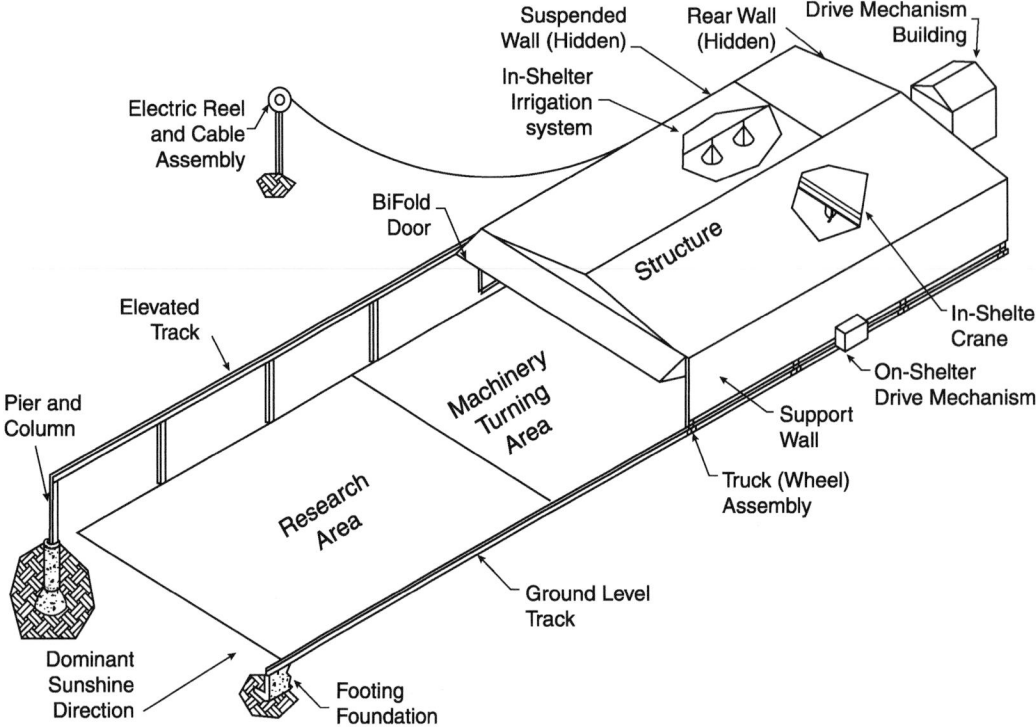

Fig. 1 Illustration of rainfall shelter subsystems and features. All items would not be used on a single shelter.

end of the structure to exclude blowing rain from the research area. On some more recent shelters, bifold doors originally designed for aircraft hangers have been installed on both ends of the structure.[5,8] With both doors open, wind forces in the direction of travel are greatly reduced in comparison to having a permanent rear wall. Truck or roller assemblies are generally placed under the columns of structures with support walls or under the beams or trusses of structures without support walls.

Drive Mechanism

Rainfall shelter drive mechanisms can be classified by the location and type of the drive. Most drive mechanisms have been permanently installed at the rear of the parked structure (Fig. 1).[6] Another approach is to install the drive mechanism entirely on the structure.[5,8] The on-structure location eliminates the separate building to house the drive mechanism and the long drive shaft spanning the distance between the two tracks.

Rain shelter drive mechanisms are of four basic types: cable and drum,[6,9] sprocket and chain,[10] rack and pinion,[3,11] and rack drive.[5,8] The cable and drum mechanism is simply a closed-loop cable passing over a drive drum at the rear end of the shelter and an idler pulley at the opposite end of the tracks. A sprocket and chain drive can use either a closed-loop chain similar to the cable

and drum or a drive sprocket traveling along a fixed chain. The rack and pinion is an excellent, but expensive, drive because the machined rack must run along the full length of travel. A rack drive is similar to a rack and pinion, but it utilizes a specially designed drive sprocket that allows a tensioned roller chain to be used in place of the rack. Flexidyne drives now allow the use of independent drives on each side of the structure thus eliminating the long drive shaft across the structure.[5,8]

Power Supply

Alternating current (a.c.) electricity from a reliable utility grid is the preferred power supply because it allows the use of larger motors and heavier structures.[5,6] For starting and reversing the larger motors, three-phase a.c. is preferred to single-phase a.c.[6] If a.c. power is unreliable, especially during storms, it can be used to charge batteries that then power a direct current (d.c.) system.[2] At remote sites without a.c. power, solar battery chargers can be used, or charged batteries can be transported to the shelter.

Control System

A rain shelter control system consists of a rain sensor for initiating movement of the shelter, controls for starting and stopping motors, and mechanisms for safe operation of the shelter and auxiliary components. Initially, rain sensors were collectors with float-activated microswitches or water-activated electrodes.[9,11] Rainfall of sufficient intensity would initiate a control sequence and cause the shelter to move over the research area. After sufficient drainage from the collector through a capillary drain, the shelter would be returned to the parked position. Resistance circuit boards provide the same function with rainfall decreasing the resistance between electrodes and absence of rainfall causing the resistance to return to the normal larger value.[11] The electric pulses from tipping bucket rain gages have also been used to initiate the control sequence.[3,6] After a sufficient time without pulses from the rain gage, the shelter is returned to the parked position. Rain sensors designed for lawn sprinkler systems have been used on rainfall shelters, but they do not accurately sense the end of rainfall.[5]

Rain shelters have been traditionally controlled with timers, relays, and microswitches that followed some logic sequence to start and stop the drive motors.[3,9] The controllers were usually designed by individual researchers to meet the unique features of the shelter. More recently, programmable controllers with input from a rain sensor, microswitches, and transducers have been used to control the shelter motors.[5,6] The control program with complex logic can be developed on a computer, and then downloaded to the controller. Programmable controllers are especially well suited to controlling several motors and meeting numerous failsafe conditions normally required of a complex shelter. Typical failsafe conditions include locking out the drive motors when the doors are closed or when an in-shelter crane scale is attached to a lysimeter.

Auxiliary Components

The most common auxiliary components are in-shelter spray irrigation systems[12] and cranes for weighing lysimeters.[5] Spray systems suspended from the structure frame can be designed for uniform, multiple treatment, or line source irrigation. A bridge crane inside the shelter structure can be substituted for a gantry crane and used to lift, move, and weigh lysimeters. Weighing the lysimeters inside the structure eliminates wind effects on measurements and increases accuracy.[5]

REFERENCES

1. Dugas, W.A.; Upchurch, D.R. Microclimate of a Rainfall Shelter. Agron. J. **1984**, *76* (6), 867–871.
2. Foale, M.A.; Davis, R.; UpChurch, D.R. The Design of Rain Shelters for Field Experimentation: A Review. J. Agric. Eng. Res. **1986**, *34*, 1–16.
3. Upchurch, D.J.; Ritchie, J.T.; Foale, M.A. Design of a Large Dual-Structure Rainout Shelter. Agron. J. **1983**, *75* (5), 845–848.
4. Stansell, J.R.; Sparrow, G.N. Rainfall Controlled Shelter for Research Plots. Agric. Eng. **1963**, *44* (6), 318–319.
5. Schneider, A.D.; Steiner, J.L.; Howell, T.A. An Evapotranspiration Research Facility Using Monolithic Lysimeters from Three Soils. Trans. ASAE **1993**, *9* (2), 227–232.
6. Martin, E.C.; Ritchie, J.T.; Reese, S.M.; Loudon, T.L.; Knezek, B. A Large-Area Lightweight Rainshelter with Programmable Controller. Trans. ASAE **1988**, *31* (5), 1440–1444.
7. Kvien, C.S.; Branch, W.D. Design and Use of a Fully Automated Portable Rain Shelter System. Agron. J. **1988**, *80* (2), 281–283.
8. Rees, R.E.; Zachmeier, L.G. Automatic Rainout Shelter for Controlled Water Research. J. Range Manag. **1985**, *38* (4), 353–357.
9. Williamson, R.E.; van Schilfgaarde, J. Studies of Crop Response to Drainage: II. Lysimeters. Trans. ASAE **1965**, *8* (1), 98–100, 102.
10. Bruce, R.R.; Shuman, F.L., Jr. Design for Automatic Movable Plot Shelter. Trans. ASAE **1962**, *5* (2), 212–214, 217.
11. Teare, I.D.; Schimmelpfenning, H.; Waldren, R.P. Rainout Shelter and Drainage Lysimeters to Quantitatively Measure Drought Stress. Agron. J. **1973**, *65* (4), 544–547.
12. McCowan, R.L.; Wall, B.H.; Carberry, P.S.; Hargreaves, J.N.G.; Green, K.L.; Poulton, P.L. Design and Evaluation of an Irrigation System for Creating Water Gradients Under an Automatic Rain Shelter. Irrig. Sci. **1990**, *11*, 189–195.

R

Rainfed Farming

Philip J. Bauer
Edward John Sadler
*United States Department of Agriculture (USDA), Florence,
South Carolina, U.S.A.*

J. R. Frederick
Clemson University, Florence, South Carolina, U.S.A.

INTRODUCTION

As farmers grow plants in a wide range of environments, rainfed-farming systems are highly diverse, ranging from intense production systems with high capital, equipment, and management investments to systems that consist of reseeding forage species with grazing animals harvesting the crop. Regardless of the size of the enterprise or the crop grown, a key to the success of rainfed-farming systems is soil water management. Crop plant productivity in rainfed systems is greatly determined by the amount and/or seasonal distribution of soil water and by the physiological capability of the plants to use that water. Systems that have been developed to increase crop yields include soil management techniques that optimize root zone water content and crop management techniques that best utilize the stored soil water plus seasonal precipitation. Continued increases in productivity of rainfed-farming systems will require a combination of improved soil and crop management practices. For a more in-depth treatment of the subject of rainfed-farming characteristics than space allows here, the reader is referred to Loomis and Conner[1] and Gimenez et al.[2]

SOIL MANAGEMENT

Under rainfed conditions, there are two water-related problems that farmers have to contend with; either not enough or too much water. For some farmers, particularly in humid areas, both of these problems can occur during the same growing season. Optimizing soil water content and using methods that minimize the effects of excess or ill-timed rain are important for timely application of agronomic practices, plant health, and, in many cases, crop quality.

SOIL MANAGEMENT UNDER CONDITIONS OF EXCESS WATER

A significant amount of land used for farming in humid regions is prone to excess water, at least during some part of the growing season. For most crops, a long period of excess water causes root damage or death because of lack of soil oxygen. Many common crops cannot survive flooded conditions for more than a few days. In addition, saturated soil conditions can increase severity of plant disease.

Agricultural soils prone to prolonged periods of excess water are generally relegated to grazing land or actively drained with subsurface drain lines. More recently, subsurface drainage has been replaced by controlled drainage, or water table management, to allow for better water management of the crops. With this, the same drain-line systems are used, but ditch outlets are controlled to keep ditches partially filled much of the time keeping the soil saturated deep in the profile. Controlled drainage conserves water for periods of low rain and reduces nitrate contamination of surface and ground water.

Excess precipitation often creates the most problems by affecting farming operations. Wet soils or long periods of rain can delay tillage, planting, farm chemical applications, and harvesting. Operating equipment on wet fields can severely damage soils by compaction and rutting. To help overcome crop losses because of too much rain, technological advances have been made in field equipment such as large tires and high horsepower tractors and harvesters that allow for field operations under wetter soil conditions. In addition, the post-harvest technologies of grain drying or ensiling forage and grain crops when animals are a part of the farm enterprise allow for earlier harvesting (which reduces the amount of time the crop is at risk from too much rain).

Encyclopedia of Water Science
DOI: 10.1081/E-EWS 120010095

SOIL MANAGEMENT PRACTICES TO INCREASE PLANT AVAILABLE WATER

Water that falls on an agricultural field can become unavailable to crops in three ways, previously illustrated using the hydrologic balance.[3] It can run off the field before it enters the soil, it can enter the soil and drain below the rooting zone, or it can evaporate from the soil surface. In water-deficit conditions, management techniques to reduce these losses can have profound effects on crop productivity.

Reducing Runoff/Increasing Rainfall Infiltration

Techniques that farmers use to reduce water loss via runoff were originally designed to control soil erosion. By protecting the soil surface, those farming practices that increase the long-term sustainability of land for crop production and enhance environmental quality also increase the amount of rainwater that enters the soil.

Farming practices that have long been used to reduce runoff losses include terracing, strip cropping with alternating bands of sod and row crops, and contour plowing. These practices keep water from moving quickly down slopes, allowing it time to seep into the soil. A more recent practice, rapidly growing in use among farmers, is conservation tillage or any cropping system that keeps 30% of the soil surface covered with plant residues. Residues increase rainfall infiltration by acting as small barriers that slow water movement down slopes. Residues on the surface also absorb the force of raindrops that fall to the soil, reducing the packing effect of raindrop impact on the soil surface. If the surface layer of soil becomes packed, infiltration slows and more water is susceptible to runoff.

Increasing the Soil's Capacity to Provide Water

On many soils throughout the world, crop plants often become water-stressed because soil physical properties reduce the volume of soil that roots can grow into. Compaction is common throughout the world and can be caused by animal or machinery traffic, or be a natural characteristic of the soil. Current management systems to loosen compacted soils generally consist of some form of tillage. This can range from lightweight surface tillage implements designed to loosen and crumble compacted surface soils to large, energy-intensive tillage tools designed to loosen compacted subsoil layers. Relieving compaction stress generally results in increased crop yields, especially in rainfall-limited seasons and environments. Similarly, eliminating chemical restrictions to root

growth, such as liming acid subsoils, increases the volume of soil for roots to extract water.

The capacity of the soil to provide water to plants can also be increased by enhancing soil water holding capacity with soil organic matter. Adding large amounts of organic material has resulted in crop yield increases in soil inherently low in organic matter. In conservation tillage, and especially no-tillage, improvements in yield can occur because the slowly decomposing residues that are left on the surface build soil organic matter near the surface and thereby increase water-holding capacity of the soil.

Reducing Soil Water Evaporation

A common method used to reduce soil water evaporation, especially in semiarid areas, is conservation tillage. Stirring and mixing the soil with tillage implements aerates the soil and exposes moist soil to the atmosphere where the soil water can quickly evaporate. Keeping the ground covered with plant residues also reduces evaporation rates by keeping soils cooler so there is less energy at the soil surface for evaporation. In addition, plant residues that are left on the soil surface act as a physical barrier to water vapor movement from the soil to the air.

CROP MANAGEMENT

In most rainfed-farming situations, variability in rainfall from year to year is more detrimental to the cropping system than is the lack of rainfall. Since farmers cannot plan for a specific amount of water for their crop each year, they tend to be cautious and limit inputs to levels that optimize a historically normal rainfall year. This management, quite different from irrigated farming where yield can be more accurately predicted, does not allow for the most efficient use of water in most years.

Water use efficiency (WUE) is calculated as the product of aboveground biomass of the crop and its harvest index divided by the sum of evaporation and transpiration (ET) (WUE = biomass × harvest index/ET). Harvest index is the ratio of harvested product to the aboveground biomass. Production practices differ between forage and grain crops partially because of the differences in the contribution of harvest index to WUE and yield.

Biomass production of plants is closely related to the amount of water transpired; so forage production practices generally attempt to maximize early-season vegetative growth. To accomplish this, forages are usually solid seeded at high populations. This planting practice maximizes early season vegetative growth, minimizes E, and results in many roots across the entire surface layer so that more of the stored soil water is used. Grain crop species planted for forage are generally seeded at higher populations than when grown for grain; an example would be corn (*Zea mays* L.).

For grain crops, rainfed-farming practices must be designed so that the water needs of the crop are met during both the vegetative and the reproductive growth stages. Maximizing early-season vegetative growth, as is done with forages, can have a detrimental effect on yield in some environments if stored soil water is exhausted during that growth stage and rainfall during reproductive growth is not enough to prevent water stress of the crop. To reduce early-season water use, summer-seeded grain crops are generally planted at lower plant densities and often in wide rows. This increases the amount of water available per plant, and stores water in the soil for the reproductive stage. Some grain crops are solid-seeded such as wheat (*Triticum aestivum* L.), but they avoid excessive early-season water use by being grown in cooler climates or are planted so that vegetative growth occurs during the time of year when air temperatures are cool.

Farmers often grow a mix of crop species and cultivars under rainfed conditions. Growing crops with a range of maturities spreads the risk of water-deficit stress during the growing season. This practice is especially valuable for crops that have extremely sensitive periods to water-deficit stress, like silking in corn. Planting genotypes with a range in maturity ensures that not all of the crop will be in the sensitive period should short droughts occur. In addition, a wide range of maturities allows for more timely management at critical times during the growing season and at harvest. Similarly, planting dates of crops can be spread out to ensure a range of crop growth stages throughout the season.

Farmers generally apply less fertilizer to rainfed than to irrigated crops. Lower amounts of relatively immobile nutrients like phosphorous and potassium are applied because crop productivity is generally less under rainfed conditions than under irrigated, so lower amounts of these nutrients are removed from the fields with the harvest. Nitrogen fertilization schemes for grain crops under rainfed conditions generally include lower amounts early in the season, especially in semiarid and arid areas, because fast vegetative growth may deplete all of the soil water and result in drought stress during reproductive growth. In more humid areas, N amounts are generally recommended based on yield potential for average rainfall years.

Pests can reduce crop transpiration by competing for water resources (weeds), by reducing root numbers (insects and diseases), and by damaging leaves (diseases and insects). Insects and diseases that attack seeds and fruits can also reduce water-use efficiency by lowering harvest index. Pests are generally managed through crop rotations, mechanical means, and with pesticides, often using the principles of integrated pest management (IPM). With IPM, multiple methods of pest management are employed and applications of pesticides are based on in-field determinations of pest populations and economic thresholds. Where grown, new crop genotypes with insect and/or broad-spectrum herbicide resistance simplify pest management decisions.

CONCLUSION

New rainfed-farming practices will likely be combinations of soil and crop management practices. For example, farmers in the southeast United States traditionally grew soybean [*Glycine max* (L.) Merr.] in 76-cm wide rows (or wider) with conventional tillage practices and in-row subsoiling. Many hectares of soybean in the area are now being produced with conservation tillage in narrow rows (25-cm wide or less) and with deep tillage implements that loosen the entire surface horizon of soil. Yield increases with this conservation tillage system were realized in research[4] and by early farmer adopters of the technology, but the system gained rapid popularity with growers when new soybean genotypes became available that tolerated broad-spectrum herbicides. Integrating soil and crop management practices into systems that reduce water losses and increase the ability of crop plants to use soil water will continue to be a high priority of research to improve rainfed farming.

REFERENCES

1. Loomis, R.S.; Conner, D.J. *Crop Ecology: Productivity and Management in Agricultural Systems*; Cambridge University Press: Cambridge, 1992.
2. Gimenez, C.; Orgaz, F.; Fereres, E. Productivity in Water-Limited Environments: Dryland Agricultural Systems. In *Ecology in Agriculture*; Jackson, L.E., Ed.; Academic Press: San Diego, CA, 1997; 117–143.
3. Sadler, E.J.; Turner, N.C. Water Relationships in a Sustainable Agriculture System. In *Sustainable Agriculture Systems*; Hatfield, J.L., Karlen, D.L., Eds.; Lewis Press: Boca Raton, FL, 1993; Chap. 2, 21–46.
4. Frederick, J.R.; Bauer, P.J.; Busscher, W.J.; McCutcheon, G.S. Tillage Management for Doublecropped Soybean Grown in Narrow and Wide Row Width Culture. Crop Sci. **1998**, *38*, 755–762.

Rangeland Management for Enhanced Water Utilization

Darrell N. Ueckert
W. Allan McGinty
Texas A&M University Research and Extension Center,
San Angelo, Texas, U.S.A.

INTRODUCTION

Rangelands occupy almost half of the earth's land surface and are a major source of the meat, fiber, and water necessary to sustain the world's burgeoning human population. Water is the driving force of rangeland ecosystems and must be used efficiently because most rangelands are in climatic regions where water is scarce and limits plant growth. Healthy rangelands conserve water and nutrients, but rangelands in many regions have deteriorated and are dysfunctional. Excessive losses of soil and nutrients to the erosive forces of water and wind can further reduce the productivity of these vast landscapes for hundreds of years. Ecologically sound rangeland management involves working with the natural ecological processes of energy flow, hydrologic cycles, and biogeochemical cycles. Practices useful for enhancing the efficiency of water utilization on rangelands include proper grazing management, control of undesirable weeds and woody plants, ripping, contour furrowing, pitting, and reseeding.

HYDROLOGICALLY FUNCTIONAL RANGELANDS

Healthy rangelands have high rainfall infiltration rates because of good soil structure, meaning that the soil particles are held together in water-stable clusters (aggregates) by roots, fungal hyphae, byproducts of organic matter decay and microbial synthesis, and resistant humus components.[1] Water-stable aggregates do not readily disperse during rainfall events; hence, they do not plug up the large soil macropores. Pore space in a soil increases with aggregation, and this aids rainfall infiltration. Healthy rangelands support a variety of plant species with the genetic potential to grow an abundance of foliage (which becomes litter after it dies) and deep root systems capable of extracting water and nutrients from a large volume of soil. They have a sufficient amount of vegetative cover (standing live and dead plants and litter)

to protect the soil surface aggregates from being dispersed by raindrops and to provide resistance to surface runoff. Vegetative cover also ameliorates the extremes of soil temperature, reduces evaporation of soil water, and provides a microenvironment favorable for decomposition of organic matter, which in turn contributes to the formation of water-stable soil aggregates.[2]

HYDROLOGICALLY DYSFUNCTIONAL RANGELANDS

The direct and indirect effects of drought and excessive grazing by livestock or wildlife can render rangelands dysfunctional relative to conserving water and nutrients and yielding the products needed by society.[2] These effects seriously diminish the production of foliage and deposition of litter, the depth and branching of plant root systems, soil aggregation, and infiltration rates, while increasing the losses of water and nutrients from the landscape as surface runoff. The kinetic energy of raindrops hitting bare soil, as well as excessive hoof action, break soil aggregates into small particles that move with water into the large soil pore spaces, plugging them or seriously reducing their volume and the capacity of the soil to absorb and store water. Over time, plant composition changes and cover decreases as the productive, palatable, deep-rooted plants die and are replaced by lower densities of smaller, less palatable, less productive, shallow-rooted plants.[3,4] The result is reduced microorganism activity, less aggregate formation, a harsher environment for seed germination, more soil exposed to raindrop impact, fewer roots to exploit soil water and nutrients, decreased rainfall infiltration, and accelerated surface runoff and erosion. The downward spiral of deterioration eventually leads to desertification.[2] Weeds, woody plants, and succulents [e.g., cactus (*Opuntia* spp.)] often increase in or invade deteriorated rangelands and compete with the remaining forage species for the diminished supply of soil water and nutrients.

Encyclopedia of Water Science
DOI: 10.1081/E-EWS 120010046

MANAGEMENT TO ENHANCE WATER UTILIZATION

Ecologically sound rangeland management means working with natural ecological processes (energy flow, the hydrologic cycle, and biogeochemical cycles) to manage vegetation and soils to achieve and maintain high infiltration rates and to minimize loss of water, soil, and nutrients in runoff.[5,6] Proper grazing management is the basic tool for achieving efficiency in water and nutrient utilization on rangelands. Control of weeds and woody plants can make more water available for desirable plants. Water conservation practices, such as ripping, contour furrowing, or pitting, may be necessary to reverse the downward spiral toward desertification on severely deteriorated rangelands and reseeding may be necessary to introduce plants that can efficiently use the available water.

Grazing Management

Excessive grazing affects plants directly by altering their physiology and morphology and indirectly by altering microclimate, soil properties, and the competitive

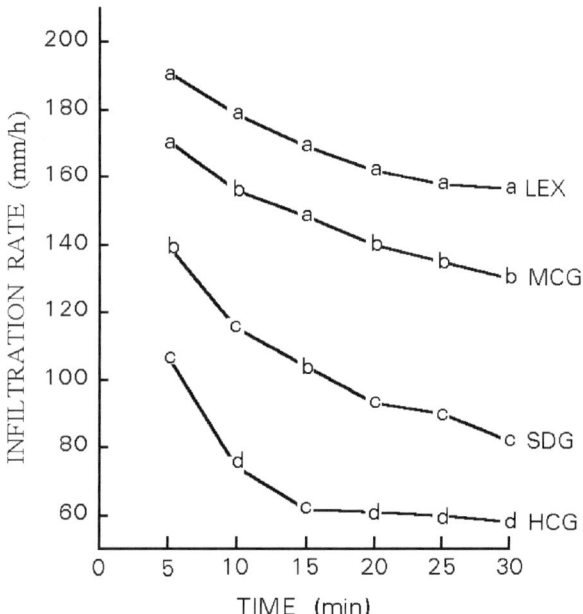

Fig. 1 Mean infiltration rates for four grazing treatments 6 years after they were initiated on the Edwards Plateau, Texas. LEX = livestock exclosure; MCG = continuously grazed at moderate intensity; SDG = short duration rotation (14-pasture, 1-herd; 4 days on, 50 days rest) stocked at 1.75 times the moderate intensity; HCG = continuously grazed, stocked at 1.75 times the moderate intensity. Means within a time period with different letters are significantly different at $P \leq 0.05$.[2]

interactions among plants.[3] Without sufficient leaf surface area, plants cannot efficiently capture the energy from sunlight via photosynthesis and root growth will be reduced. Over time, the composition of the vegetation changes, rainfall infiltration declines (Fig. 1), surface runoff increases, and plant production decreases. Grazing management involves balancing the number of animals with the forage supply, selecting the appropriate kinds and classes of animals to be grazed, controlling the timing of grazing, and distributing grazing evenly across the landscape.[7] Achieving the proper level of utilization of plants and maintaining an acceptable minimum amount of litter is the most important management decision, regardless of whether rangeland is grazed continuously or in a complex grazing system. The minimum amounts of litter needed to sustain productivity of shortgrass, midgrass, and tall-grass rangelands are 340 kg/ha–560 kg/ha, 840 kg/ha–1120 kg/ha, and 1350 kg/ha–1680 kg/ha, respectively.[8] "Take half and leave half" is the guiding principle for determining stocking rates. Under most management systems, 50% of the forage produced during the year should remain ungrazed. Twenty-five percent of the year's forage growth will be lost to trampling, insects, and other animals, or rendered ungrazable due to livestock dung or urine. The remaining 25% of plant growth can be utilized by livestock.[8] Rangeland vegetation and precipitation records should be continually monitored, and livestock and wildlife numbers should be adjusted annually or even seasonally to achieve proper use.

Management of Undesirable Vegetation

Excessive grazing, drought, climatic changes, and a reduction in the frequency and intensity of fire predispose many rangelands to invasion by weeds, woody plants, and succulents that have little or no value to grazing animals or humans. These plants intercept or transpire large quantities of water that might otherwise be used by desirable forage plants. The efficiency of water use on rangelands can be increased by controlling undesirable vegetation.[9,10] Herbicidal, mechanical, prescribed burning, and biological control methods, or appropriately timed and sequenced combinations of these methods, coupled with proper grazing management can provide effective, cost efficient, and ecologically practical solutions to noxious plant problems.[11] Rangelands should be monitored annually for noxious plants, and control programs should be initiated before these plants mature, thicken, utilize excessive amounts of water, and cause deterioration of desirable vegetative cover[12] (Fig. 2).

Fig. 2 Controlling young redberry juniper (*Juniperus pinchotii* Sudw.) plants with 1% picloram (4-amino-3,5,6-trichloro-2-pyridinecarboxylic acid) high-volume foliar sprays prevents development of juniper woodlands which intercept and transpire large amounts of water and cause deterioration of the herbaceous understory.

Special Water Conservation Treatments

Severely deteriorated rangelands, especially in arid and semiarid regions, often recover slowly or not at all after initiation of proper grazing management or the total removal of livestock because of the lack of vegetative cover, poor soil aggregation, low infiltration rates, and the resultant harsh environment for plant establishment and growth. Mechanical land treatments such as ripping, furrowing, and pitting can expedite natural recovery of these desertified rangelands[5,13] by increasing resistance to surface runoff, shattering compacted soil layers, and thereby increasing water infiltration and retention. Mechanical treatments that effectively increase deep infiltration or percolation of precipitation in saline soils can leach soluble salts below the root zone and thus increase the availability of water to plants. The objective of using these mechanical treatments is to facilitate the establishment of dense patches or bands of vegetative cover that will persist and continue to conserve water and nutrients naturally, long after the soil disturbance has disappeared. The full potential of these practices will only be realized if treated areas are initially protected from grazing to allow the establishment of vegetative cover and afforded proper grazing management thereafter.

Ripping (also referred to as subsoiling or deep chiseling) involves pulling a heavy shank equipped with a broad lifting tip 40 cm–60 cm deep through the soil on the contour.[13] Space between rips is usually 3 m–9 m. Ripping fractures impervious soil layers (which increases porosity and the rate of infiltration), causes uplifting of the soil (which resists surface runoff), leaves a furrow in the center of the uplift (which will retain water), and the soil disturbance provides a seedbed for new plant establishment. Ripping facilitated infiltration of water from a 5 cm convection thunderstorm to a depth of 100 cm–125 cm, compared to only 10 cm–13 cm on adjacent, unripped rangeland. Increased forage production after ripping (Fig. 3) would support a cow/calf unit year long on 9 ha, compared to 32 ha without ripping.[14]

Contour furrowing involves pulling disk plows or other tillage implements to create depressions or grooves in the soil surface 10 cm–20 cm deep, 15 cm–75 cm wide, and 0.6 m–3 m apart.[13] These soil depressions increase on-site water retention and the displaced soil provides resistance to surface runoff. Furrowing implements can be designed with rippers in front of the disks and dikers that dam up the furrows at selected intervals. Seeders can also be attached that deposit seed on or into the disturbed soil during the furrowing process to establish plant species that can make beneficial use of the water retained in the furrows.

The most effective rangeland pitting has been done with disk plows equipped with eccentric or deeply notched disks or disk plows with eccentric furrow wheels that alternatively raise and lower the disks. These create thousands of small basins or pits across the landscape, which function similarly to contour furrows.[13] Seeders can also be attached to pitting implements. Pits installed with implements that utilize spike teeth tend to fill in with soil within about a year.

Fig. 3 Ripping reduces runoff, enhances rainfall infiltration, and provides a seedbed for germination and establishment of new plants. Three months after ripping (Fig. 3a) and 5 years after ripping (Fig. 3b) severely deteriorated rangeland (Tulia loam soil; 3%–4% slope) in the southern rolling plains, Texas.

CONCLUSION

Maintaining good vegetative cover, litter, and soil aggregation is critical for the efficient utilization of water on rangelands. Proper grazing management budgets about half of the annual plant production to be left to maintain a healthy hydrological cycle. Management of undesirable plants can decrease wasteful interception and transpiration of water and increase availability of water for beneficial plants. Mechanical water conservation treatments can effectively reduce surface runoff and increase infiltration, but their long-term effectiveness hinges upon the establishment and maintenance of dense patches or bands of vegetative cover.

REFERENCES

1. Boyle, M.; Frankenberger, W.T., Jr.; Stolzy, L.H. The Influence of Organic Matter on Soil Aggregation and Water Infiltration. J. Prod. Agric. **1989**, *2* (4), 290–299.
2. Thurow, T.L. Hydrology and Erosion. In *Grazing Management: An Ecological Perspective*; Heitschmidt, R.K., Stuth, J.W., Eds.; Timber Press: Portland, OR, 1991; 141–159.

3. Archer, S.; Smeins, F.E. Ecosystem-Level Processes. In *Grazing Management: An Ecological Perspective*; Heitschmidt, R.K., Stuth, J.W., Eds.; Timber Press: Portland, OR, 1991; 109–139.

4. Briske, D.D. Developmental Morphology and Physiology of Grasses. In *Grazing Management: An Ecological Perspective*; Heitschmidt, R.K., Stuth, J.W., Eds.; Timber Press: Portland, OR, 1991; 85–108.

5. Whisenant, S.G. *Repairing Damaged Wildlands—A Process-Oriented, Landscape-Scale Approach*; Cambridge University Press: Cambridge, UK, 1999.

6. Ludwig, J.; Tongway, D.; Freudenberger, D.; Noble, J.; Hodgkinson, K., (Eds.) *Landscape Ecology, Function, and Management: Principles from Australia's Rangelands*; CSIRO Publishing: Collingwood, VIC, 1997.

7. Briske, D.D.; Heitschmidt, R.K. An Ecological Perspective. In *Grazing Management: An Ecological Perspective*; Heitschmidt, R.K., Stuth, J.W., Eds.; Timber Press: Portland, OR, 1991; 11–26.

8. White, L.D.; McGinty, A. *Stocking Rate Decisions—Key to Successful Ranch Management*, Texas Agricultural Extension Service Bulletin B-5036; Texas Cooperative Extension: College Station, TX, 1992; 1–9.

9. Thurow, L.L.; Hester, J.W. How an Increase or Reduction in Juniper Cover Alters Rangeland Hydrology. In *Proceedings Juniper Symposium, Glen Rose, Texas, March 29, 2001*; Taylor, C.A., Ed.; Texas Agricultural Experiment Station: Sonora, TX, 2001; 4-9–4-22.

10. Ueckert, D.N. Broom Snakeweed: Effect on Shortgrass Forage Production and Soil Water Depletion. J. Range Manag. **1979**, *32* (3), 216–220.

11. Hanselka, C.W.; Hamilton, W.T.; Lee, M.; McGinity, W.A.; Ueckert, D.N. *Brush Management: Past, Present, and Future*; Texas A&M University Press: College Station, TX, 2003.

12. McGinty, A.; Ueckert, D.N. The Brush Busters Success Story. Rangelands **2001**, *23* (6), 3–8.

13. Valentine, J.F. Special Range Treatments. *Range Development and Improvements*; Brigham Young University Press: Provo, UT, 1971; 301–324.

14. Ueckert, D.N.; Petersen, J.L.; Shaffer, K.R. Ripping for Restoration of Depleted Rangelands. Abstracts of Papers 54th Annual Meeting of the Society for Range Management, Kailua-Kona, HI, Feb 17–23, 2001; Society for Range Management: Lakewood, CO, 2001; 405.

Rangeland Water Yield, Influence of Brush Clearing on

William A. Dugas
Texas Agricultural Experiment Station, Temple, Texas, U.S.A.

Steven Bednarz
Tim Dybala
United States Department of Agriculture (USDA), Temple, Texas, U.S.A.

Ranjan S. Muttiah
Wes Rosenthal
Texas Agricultural Experiment Station, Temple, Texas, U.S.A.

Jeff Arnold
United States Department of Agriculture (USDA), Temple, Texas, U.S.A.

INTRODUCTION

To supplement water supplies, there has been considerable interest in using vegetation management (brush control) to increase stream flow and water yields (runoff + deep percolation) from watersheds. One option on rangelands[1,2] is to replace deep-rooted woody brush species, which may intercept a substantial amount of precipitation and have high whole-plant transpiration rates due to high leaf areas, with shallow-rooted herbaceous vegetation that usually intercepts less precipitation and has less leaf area. The amount of increased stream flow and/or water yield, if any, on treated watersheds depends on several factors, including the pre- and post-treatment vegetation types or land use,[3] treatment method or soil,[4] climate,[5] and time since treatment imposition.[6] Wilcox[7] presents a perspective on the mechanisms of how brush clearing could affect streamflow. Several field and modeling studies in Texas have shown water yield increases associated with brush removal (Table 1). Based, in part, on these studies, a study was conducted to use a hydrologic simulation model to evaluate changes in stream flow and water yield associated with brush removal on several watersheds.[11] This report uses results from that work to present a case study of how brush clearing can influence rangeland water yield.

CASE STUDY

Methods

Eight Texas watersheds investigated in this study were: Canadian River above Lake Meredith, Wichita River above Lake Kemp, Upper Colorado River above Lake Ivie, Concho River, Pedernales River, several watersheds above the Edwards Aquifer recharge zone, Frio River above Choke Canyon Reservoir, and Nueces River above the junction with the Frio River. For ease of simulation, several of these watersheds were further subdivided, resulting in 17 modeled watersheds.

The Soil and Water Assessment Tool (SWAT) model used in this study[12] is physically based, uses readily available inputs, and is capable of simulating long periods. A GIS interface was developed[13] that creates SWAT model input data files from map layers and associated relational databases. Model inputs included daily precipitation totals and maximum and minimum temperatures; a United States Geological Survey (USGS) Digital Elevation Model at a 1:24,000 scale; and a USDA-Natural Resources Conservation Service soils database.

Because of the need to discriminate brush land use by species and cover density, current, detailed, and accurate land use data for these watersheds were required. These data were developed by classifying 1999 Landsat data. Scenes were radiometrically and precision-terrain corrected and then classified using > 1100 ground control points (GCPs), where land use (e.g., brush species and cover density) and areal extent were recorded. Land use was classified as heavy (> 30% canopy coverage) cedar, mesquite, oak, or mixed brush; moderate (10–30%) cedar, mesquite, oak, or mixed brush; light (< 10%) brush; open range, cropland, water, barren, urban, and other. Classification accuracy, determined from the GCP data, was approximately 70%.

Encyclopedia of Water Science
DOI: 10.1081/E-EWS 120010099

Table 1 Estimated annual water yield increase (ML per hectare of treated land) resulting from brush removal at selected locations in Texas. N. Concho water savings are based on model simulations

Location	References	Land use change	Increase
Seco Ck.	[6]	Remove all juniper (3-yr post-treatment avg.)	0.3
Sonora	[8]	60% juniper/40% grass–100% grass	0.9
Annandale	[9]	Remove all juniper	1.2[a]
N. Concho	[10]	Remove all brush (mesquite and juniper)	0.3

[a] Calculated from ratio of average runoff to precipitation and from measured increase in runoff.

Model Calibration

Plant growth parameters (e.g., maximum leaf area index, base temperature, canopy height, albedo, and rooting depth) for each land use were input for two model simulations. For the first, i.e., the "with brush" condition (calibration), we used the classified land use layer created for this study (and associated model inputs for each land use) and assumed[14] existing brush sites were in fair hydrologic condition (50–75% ground cover). For the second, the "without brush" condition, areas with heavy and moderate brush land use (excluding oak) were changed to a grassland with no brush by adjusting land use input files (e.g., rooting depth, leaf area, etc.) and were assumed[14] to be in good hydrologic condition (greater than 75% ground cover). The fraction of each watershed where brush removal was simulated varied from 26 to 74%. All other model inputs were held constant.

The model was calibrated by adjusting runoff curve number, soil evaporation compensation factor, shallow aquifer storage, shallow aquifer re-evaporation, and channel transmission loss to match USGS measured monthly stream flow from 1960 through 1998 for various locations in each watershed. The fraction of base flow and surface runoff in each watershed was estimated using a base flow filtering algorithm.[15]

Measured annual average stream flows varied from 2.4×10^3 ML to 6.2×10^5 ML (10^6 L) because of differences in precipitation (annual averages ranged from 430 mm to 861 mm) and watershed area (1.3×10^4 ha to 2.2×10^6 ha). Correlation coefficients between predicted and measured monthly stream flow for each watershed varied from 0.26 to 0.99, and averaged 0.8. Correlations tended to be lower in watersheds with less precipitation. The average percentage error between predicted and measured average annual stream flow was 9%. Thus, the

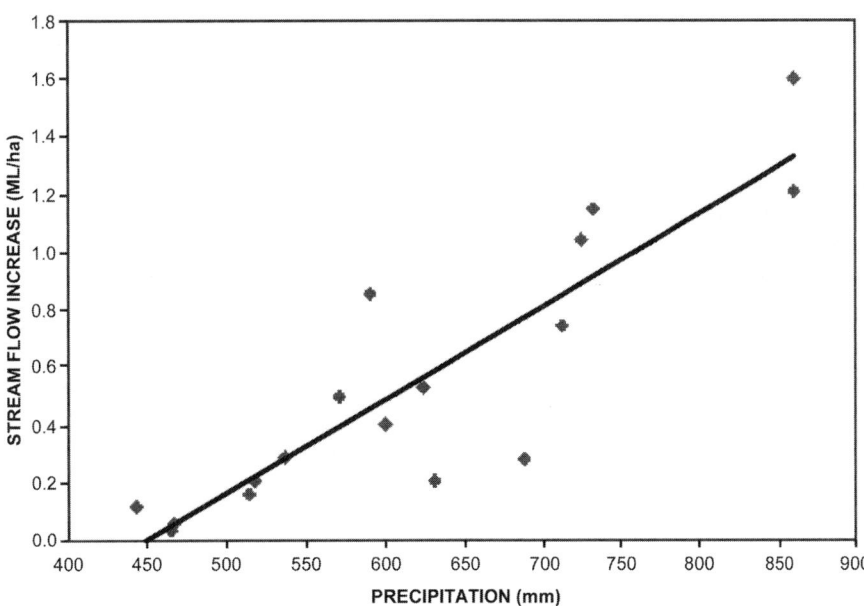

Fig. 1 Average annual increase in stream flow, per unit area treated for brush removal, vs. average annual precipitation in selected Texas watersheds.

calibrated model accurately predicted measured stream flow.

Simulated Effects of Brush Control

Average annual stream flow, per unit treated area, increased in all watersheds due to brush control (without brush) and increases were closely related to annual precipitation (Fig. 1). Scatter about the regression line was due to variation in soils, type and density of brush removed, and topography across watersheds. The estimated annual precipitation associated with a zero stream flow increase in Fig. 1 (ca. 450 mm) is very similar to previous estimates.[5]

Annual water yield increases for sub-basins in the Wichita River watershed (used for illustrative purposes), per treated area, varied from 0.2 to 1.1 ML/ha (results not shown). The large range of annual water yield increases across sub-basins in this watershed was due, again, to differences in soils, type and density of brush removed, precipitation, and topography. For most watersheds, sub-basin water yield increases increased with increasing precipitation. All watersheds showed large variability across sub-basins (results not shown). These results highlight the need for high spatial resolution of model inputs and simulation units to precisely identify where brush control would yield maximum benefits.

CONCLUSION

Field research has shown that vegetation management (brush control) may increase stream flow and water yield from watersheds. In this study, we used satellite imagery to classify brush cover into species and density categories and used a hydrologic simulation model to simulate the effects of brush removal on stream flow and water yield in selected watersheds in Texas. All watersheds showed an increase in stream flow as a result of removing brush and large spatial variability of water yield increases.

Results from this study will be used, along with other considerations (e.g., economics and wildlife), to prioritize watersheds and areas within watersheds for imposition of brush control programs to increase water supplies. This study demonstrates how research tools can be applied to address policy questions.

REFERENCES

1. Hibbert, A.R. Water Yield Improvement Potential by Vegetation Management on Western Rangelands. Water Resour. Bull. **1983**, *19*, 375–381.
2. Davis, E.A. Chaparral Control in Mosaic Pattern Increased Streamflow and Mitigated Nitrate Loss in Arizona. Water Resour. Bull. **1993**, *29*, 391–399.
3. Dunn, S.M.; Mackay, R. Spatial Variation in Evapotranspiration and the Influence of Land Use on Catchment Hydrology. J. Hydrol. **1995**, *171*, 49–73.
4. Richardson, C.W.; Burnett, E.; Bovey, R.W. Hydrologic Effects of Brush Control on Texas Rangelands. Trans. Am. Soc. Agric. Eng. **1979**, *22*, 315–319.
5. Griffen, R.C.; McCarl, B.A. Brushland Management for Increased Water Yield in Texas. Water Resour. Res. **1989**, *25*, 175–186.
6. Dugas, W.A.; Hicks, R.A.; Wright, P.W. Effect of Removal of *Juniperus Ashei* on Evapotranspiration and Runoff in the Seco Creek Watershed. Water Resour. Res. **1998**, *34*, 1499–1506.
7. Wilcox, B.P. Shrub Control and Streamflow on Rangelands: A Process-Based Viewpoint. J. Range Manag. **2002**, *55*, 318–326.
8. Thurow, T.L.; Taylor, C.A., Jr. Juniper Effects on the Water Yield of Central Texas Rangelands. *Water for Texas: Research Leads the Way*; Proc. 24th Water for Texas Conf., Austin, TX, Jan 1995, Texas Water Resourc. Instit.: College Station, TX, 1995; 657–665.
9. Owens, M.K.; Knight, R.W. Water Use on Rangelands. *Water for South Texas*, TAES CPR 5043-5046; College Station, TX, 1992; 1–7.
10. Upper Colorado River Authority, *North Concho River Watershed Brush Control, Planning Assessment and Feasibility Study*; Final Report to Texas Water Development Board, Austin, Texas, 1998.
11. Texas Agricultural Experiment Station, *Brush Management/Water Yield Feasibility Studies for Eight Watersheds in Texas*; Texas Water Resources Institute Technical Report No. TR-182, 2000.
12. Arnold, J.G.; Srinivasan, R.; Muttiah, R.S.; Williams, J.R. Large Area Hydrologic Modeling and Assessment, Part 1: Model Development. J. Am. Water Res. Assoc. **1998**, *34* (1), 73–89.
13. Srinivasan, R.; Arnold, J.G. Integration of a Basin Scale Water Quality Model with GIS. Water Resour. Bull. **1994**, *30*, 453–462.
14. U.S. Department of Agriculture, Soil Conservation Service, *Engineering Field Manual, Chapter 2*; USDA: Washington, DC, 1989.
15. Arnold, J.G.; Allen, P.M.; Muttiah, R.S.; Bernhardt, G. Automated Base Flow Separation and Recession Analysis Techniques. Ground Water **1995**, *33*, 1010–1018.

Rangelands, Water Balance on

Bradford P. Wilcox
Texas A&M University, College Station, Texas, U.S.A.

David D. Breshears
Los Alamos National Laboratory, Los Alamos, New Mexico, U.S.A.

Mark S. Seyfried
United States Department of Agriculture (USDA), Boise, Idaho, U.S.A.

INTRODUCTION

Rangelands are found in a variety of climate and moisture regimes and may include natural grasslands, savannas, shrublands, deserts, tundra, alpine ecosystems, marshes, and meadows. Most rangelands, however, are found in relatively dry climates where potential evapotranspiration is significantly greater than precipitation. For this reason, our discussion of water balance on rangelands will be generalized for dryland conditions. In water-limited rangelands, most of the incoming precipitation returns to the atmosphere via evapotranspiration. Of the other components, runoff will account for most of the remaining. Water moving to groundwater is generally relatively small.

WATER BALANCE

Water balance is an expression of how precipitation is partitioned after it arrives on the land surface. The relative proportions of its components define the water budget of a region. The water balance is driven by another fundamental physical relationship: energy balance. Together, these two relationships determine global vegetation patterns. The following equation presents a simplified interpretation of the water budget:

$$P = ET + R + G + \Delta S$$

where P = precipitation, ET = evapotranspiration, R = runoff, G = groundwater recharge, ΔS = change in soil water.

Evapotranspiration comprises all those processes by which water changes phase from a liquid to a gas. These processes include: a) evaporation from plant or litter surfaces (commonly referred to as interception loss); b) evaporation from the soil; and c) transpiration from the plant. Where snow constitutes a significant portion of the total precipitation on rangelands, sublimation, which is

the transfer of water from solid to vapor state, may be substantial and is included in this term. *Soil water* is the amount of water in the soil. Water that moves beyond the root zone is considered to be *groundwater recharge*, because eventually it will move to an underlying water body. *Runoff* is water that travels from the hillslope toward the stream channel, the portion of which (not captured by soils or evaporated en route) becomes streamflow.

EVAPOTRANSPIRATION

Because the different components of evapotranspiration can be difficult to separate, we often measure total evapotranspiration. At the plant community level, total evapotranspiration may be measured directly through knowledge of the energy budget. As an example, the Bowen ratio methodology,[1] which is based on calculations of the energy budget, has been commonly used to estimate evapotranspiration from rangeland plant communities. Alternatively, evapotranspiration can be determined by difference using the water budget approach, where all the components of the water budget except evapotranspiration are measured directly, and evapotranspiration is assumed to be the difference between the sum of these components and the total water budget.

Interception Loss

Interception loss is that component of precipitation that is captured by the vegetation canopy or underlying litter layer and subsequently evaporates, thus never reaching the soil surface. On rangelands, interception loss may be and often is substantial. On a percentage basis, drylands lose considerably more water via interception than do more humid environments.[2] Interception losses from rangelands may range from 1% to 80% of the annual water budget, but generally are between 20% and 40% (Table 1). Actual amounts depend on the character of the vegetation

Encyclopedia of Water Science
DOI: 10.1081/E-EWS 120010097

and precipitation. For example, evergreen shrubs, such as juniper, capture a higher percentage of precipitation because they are continuously foliated, have a large leaf area, and a leaf shape conducive to interception. In addition, these shrubs lay down a thick litter layer that captures considerable water. Interception loss is generally small in arid shrublands because of lower canopy cover. In grasslands, interception loss may be as high or higher than in shrublands if cover is extensive. The vegetation canopy has only a finite capacity to capture water—therefore the percentage of precipitation intercepted for individual storms is highly variable. For small storms, most water may be intercepted, whereas for very large storms the amount intercepted may (on a percentage basis) be quite small.

Evaporation from Soil

Evaporation from a bare soil is a multistage process.[16] Initially, after the soil is wetted, evaporation is relatively constant and limited only by the evaporative demand (which is regulated by meteorological conditions, such as radiation, wind, and air humidity). As the soil dries and its water content decreases, the evaporation rate progressively decreases. Evaporation from bare soil is limited to about the top 15 cm.

The relationship between evaporation from the soil and transpiration is of special ecological importance as it determines how much water is available to plants. The amount of evaporation depends on how much of the soil surface is bare. Where only small amounts of bare soil are found, soil evaporation will be low. But in regions where much of the soil is bare, such as arid and some semiarid rangelands, the percentage of evaporation is likely to be

very high. Reported values of soil water evaporation range from 30% to 80% of the water budget (Table 2).

Transpiration from Plants

Transpiration is the evaporation of water from the vascular system of plants into the atmosphere. The process begins with the absorption of soil water by plant roots and ends with its evaporation from stomatal cavities. Because the water is pulled through the plant by the potential energy gradient, transpiration is primarily a physical process. Plants exert physiological control through modification of the size of the stomatal openings.

The amount of transpiration depends on the amount of water that is available to the plant. Whereas evaporation from soils is primarily limited to water in the very uppermost layers, the water transpired by plants may be drawn from substantially greater depths, depending on the depth and development of plant roots. The plant roots may also redistribute water within the profile by removing water from a wet area of the soil and releasing it into a dry area—a process known as *hydraulic lift*.

RUNOFF

Runoff from rangelands is normally small but can nevertheless be very important. It is a principal agent of erosion, contaminant movement, and geomorphic change on many rangelands. Additionally, it serves a vital ecological function of redistributing and concentrating the limited water and nutrient resources in semiarid landscapes. Runoff generally accounts for less than 10%, and most often below 5%, of the annual water budget, and

Table 1 Measured values of interception loss, expressed as a percentage of precipitation, for selected U.S. rangeland shrubs and grasses

	% Interception
Shrubs	
Creosote (*Larrea tridentata.*)	36,[3] 12[4]
Mesquite (*Prosopis* sp.)	32,[3] 16[5]
Sagebrush (*Artemisia* sp.)	30,[6] 4[7]
Chaparral (*Quercus* sp.)	8[8,9]
Juniper (*Juniperus* sp.)	45,[10] 46,[11] 5–25[12]
Oak mottes (*Quercus* sp.)	46[13]
Grasses	
Big bluestem (*Andropogon gerardii*)	57–84[14]
Buffalo grass (*Buchloe dactyloides*)	17–74[14]
California annual grasslands	26[15]
Tabosa grass (*Hilaria mutica*)	11[13]
Sideoats grama (*Bouteloua curtipendula*)	18[13]

Table 2 Experimental estimation of soil water evaporation (SE) relative to total evapotranspiration (ET) in various arid and semiarid ecosystems in North America

Desert—Location	Community type	% SE/ET
Sonoran—Arizona, USA	*Larrea*	90
Sonoran—Arizona, USA	Mixed	75–95
Mojave—Nevada, USA	Mixed shrub	65
Death Valley—California, USA	Mixed shrub	45
Great Basin—Utah, USA	*Ceratoides–Atriplex*	45
Chihuahuan—New Mexico, USA	*Larrea*	30
Sonoran—Arizona, USA	*Larrea*	20
Chihuahuan—New Mexico, USA	*Prosopis, Larrea, Flourensia*	30–60
Sahel—Niger	Tiger bush	30–80

(Modified from Ref. 17.)

most of this occurs as flood flow. The actual percentage depends partly on the scale of observation. For example, on piñon-juniper rangelands, it has been demonstrated that at a very small scale ($1 m^2$), up to 100% of the precipitation from a particular storm may run off—while at the hillslope scale, runoff from the same storm will amount to only about 5% of the water budget.[18] The difference is due to the fact that as the scale increases, so too does the opportunity for storage. Similarly, in the many desert landscapes, runoff as a percentage of the water budget will decrease with scale because of transmission losses in the alluvial stream channels.[19]

Runoff from rangelands most often occurs as *Horton overland flow*,[20] but it may travel other pathways as well, including saturation overland flow, shallow subsurface flow, and groundwater flow. *Horton overland flow* results when precipitation intensity exceeds soil infiltration capacity. *Saturation overland flow* is relatively uncommon on rangelands but may be observed when soils become saturated, because of either a rising groundwater table or a perched, saturated zone. Frozen soil runoff is a special type of saturation overland flow whereby a frozen soil layer forms an impeding horizon while the soil above it is unfrozen and saturated. *Shallow subsurface flow*, sometimes referred to as interflow or throughflow, is that portion of runoff that travels laterally through the soil, generally because of some impeding soil horizon. *Shallow subsurface flow* is more common in humid environments, but it can be important in semiarid environments, especially when macropores are present in the soil.[21] *Groundwater flow* is generally the source for the base flow of a stream (prolonged flow, not attributable to a specific precipitation event).

GROUNDWATER RECHARGE

Groundwater recharge, especially deep recharge, is generally very small in rangeland environments. However,

it can be exceedingly important, especially with respect to long-term contaminant transport. Commonly, in arid and semiarid landscapes, only a few millimeters or less of water will move beyond the root zone each year—because in most cases the soils have the capacity to absorb all or most of the precipitation. Owing to the high evaporative demand in these regions, most water stored in the soil will eventually be evaporated or transpired. In some cases, however, the capacity of the soil to absorb water is overwhelmed, and substantial groundwater recharge does occur. In other cases, groundwater recharge may occur where there is an accumulation of water in concentrated locations, such as snow drifts or stream channels. In still other situations, surprisingly high groundwater recharge may occur in very dry environments if permeability is relatively high, owing to either the presence of fractures[22] or very sandy soils.[23]

SOIL WATER

The soil storage term ΔS, in the equation, is the difference between the amount of water stored within the plant root zone at the beginning of the period for which water balance is being calculated and the amount at the end. The magnitude of ΔS depends on weather patterns during that period, the duration of the period, and the storage capacity of the soil. For relatively short periods, the weather patterns are critical because they determine the initial and final S values (for periods of several years, ΔS becomes insignificant). The storage term is important because it determines, to some extent, the way incoming water is partitioned among the remaining terms. Where soils have a high storage capacity, flow to groundwater will tend to be much lower. The incoming water is instead available for plant uptake, enabling more plant production; and it also affects the rates of organic nutrient release to the soil and of carbon mineralization.

The storage capacity of a soil depends mostly on the depth of the soil, the coarse-fragment content, and the texture. Sandy soils hold about 60 mm of water per meter of soil, while finer-textured soils can store up to 200 mm. Deep, rock-free soils of medium texture may store over 300 mm of water. The ability of a soil to store water will decrease in direct proportion to the amount of coarse fragments in that soil. For example, a sandy soil with 50% rock content would be expected to store about 30 mm of water per meter of soil.

SUMMARY AND CONCLUSIONS

On rangelands, the water balance is driven and defined to a great extent by the fact that potential evapotranspiration is much greater than precipitation, which in turn contributes to a large soil water deficit. As a rule, therefore, evapotranspiration is the largest component of the water balance equation; the other components are generally quite small (nevertheless, they may be exceedingly important). In addition, both the magnitude and the definition of the different water balance components, particularly runoff, are very much scale-dependent.

Newer measurement technologies allow us to estimate more precisely than ever before the water balance components. It is now possible to directly measure plant-community-level evapotranspiration, soil water evaporation as a percentage of transpiration, interception loss during an actual rainstorm, groundwater recharge, and runoff—all at multiple scales. Application of these technologies promises to help us gain the vital information required to develop workable strategies for solving the growing problems of rangeland degradation.

REFERENCES

1. Evett, S.R. Energy and Water Balances At Soil-Plant-Atmosphere Interfaces. In *Handbook of Soil Science*; Sumner, M.E., Ed.; CRC Press: New York, N.Y., 2000; A.129–A.182.
2. Dunkerley, D. Measuring Interception Loss and Canopy Storage in Dryland Vegetation: A Brief Review and Evaluation of Available Research Strategies. Hydrol. Process. 2000, 14, 669–678.
3. Martinez-Meza, E.; Whitford, W.G. Stemflow, Throughfall and Channelization of Stemflow by Roots in Three Chihuahuan Desert Shrubs. J. Arid Environ. 1996, 32, 271–287.
4. Tromble, J.M. Water Interception by Two Arid Land Shrubs. J. Arid Environ. 1988, 15, 65–70.
5. Desai, A.N. Interception of Precipitation by Mesquite Dominated Rangelands in the Rolling Plains of Texas. M.S. Thesis, Texas A&M University, College Station, Texas, 1992.
6. Hull, A.C. Rainfall and Snowfall Interception of Big Sagebrush. Utah Acad. Sci. Lett. 1972, 49, 64.
7. West, N.E.; Gifford, G.F. Rainfall Interception by Cool-Desert Shrubs. J. Range Manag. 1976, 29, 171–172.
8. Rowe, P.B. Influence of Woodland Chaparral on Water and Soil in Central California; California Division of Natural Resources, Division of Forestry 1948.
9. Hamilton, E.L.; Rowe, P.B. Rainfall Interception by Chaparral in California; California Department of Natural Resources, Division of Forestry 1949.
10. Thurow, T.L.; Hester, J.W. How an Increase or a Reduction in Juniper Cover Alters Rangeland Hydrology. *Juniper Symposium Proceedings*; Texas A&M University: San Angelo, Texas, 1997; 9–22.
11. Young, J.A.; Evans, R.A.; Eash, D.A. Stem Flow on Western Juniper (*Juniperus occidentalis*) Trees. Weed Sci. 1984, 32, 320–327.
12. Skau, C.M. Interception, Thoughfall, and Stemflow in Utah and Alligator Juniper Cover Types of Northern Arizona. For. Sci. 1964, 10, 283–287.
13. Thurow, T.L.; Blackburn, W.H.; Taylor, C.A. Rainfall Interception Losses by Midgrass, Shortgrass, and Live Oak Mottes. J. Range Manag. 1987, 40, 455–460.
14. Clark, O.R. Interception of Rainfall by Prairie Grasses, Weeds and Certain Crop Plants. Ecol. Monogr. 1940, 10, 243–277.
15. Kittredge, J. *Forest Influences*; McGraw-Hill: New York, 1948.
16. Hillel, D. *Applications of Soil Physics*; Academic Press: New York, 1980.
17. Reynolds, J.F.; Kemp, P.R.; Tenhunen, J.D. Effects of Long-Term Rainfall Variability on Evapotranspiration and Soil Water Distribution in the Chihuahuan Desert: A Modeling Analysis. Plant Ecol. 2000, 150, 145–159.
18. Reid, K.D.; Wilcox, B.P.; Breshears, D.D.; MacDonald, L. Runoff and Erosion in a Pinon-Juniper Woodland: Influence of Vegetation Patches. Soil Sci. Soc. Am. J. 1999, 63, 1869–1879.
19. Goodrich, D.C.; Lane, L.J.; Shillito, R.M.; Miller, S.N.; Syed, K.H.; Woolhiser, D.A. Linearity of Basin Response as a Function of Scale in a Semiarid Watershed. Water Resour. Res. 1997, 33, 2951–2965.
20. Dunne, T. Field Studies of Hillslope Flow Processes. *Hillslope Hydrology*; John Wiley and Sons: New York, 1978; Chap. 7, 227–293.
21. Wilcox, B.P.; Newman, B.D.; Brandes, D.; Davenport, D.W.; Reid, K. Runoff from a Semiarid Ponderosa Pine Hillslope in New Mexico. Water Resour. Res. 1997, 33, 2301–2314.
22. Flint, A.L.; Flint, L.E.; Kwicklis, E.M.; Bodvarsson, G.S.; Fabryka-Martin, J.M. Hydrology of Yucca Mountain Nevada. Rev. Geophys. 2001, 39, 447–470.
23. Stephenson, G.R.; Zuzel, J.F. Groundwater Recharge Characteristics in a Semi-arid Environment. J. Hydrol. 1981, 53, 213–227.

Research Centers for Dryland and Semiarid Regions

John Ryan
International Center for Agricultural Research in the Dry Areas (ICARDA),
Aleppo, Syria

INTRODUCTION

The history of civilization, and its evolution, has been inextricably linked to agriculture. Where the natural resources—land and water—were abundant and climatic conditions favorable, societies flourished, often leading to great empires. With colonization of new lands, the well-watered fertile areas were first to come under man's sway. Not surprisingly, many of today's strong world economies, mainly in temperate regions, have a strong agricultural base. However, dry areas of the world, where rainfall is low and erratic and drought is an invariable constraint to agriculture, have always languished. Only where irrigation water was available, whether from rivers or groundwater, has it been possible for such areas to advance. While much research has been focused on development of agriculture in favorable areas—and with astounding success—drier regions were the "poor relation" in terms of research investment. Following a description of the essential features and the intractable nature of dryland farming, a brief overview is presented on the types of research institutes worldwide that service dryland or rainfed agriculture.

ARID AND SEMIARID REGIONS

Any definition of arid or semiarid hinges around the soil water balance; such regions are those where potential evapotranspiration exceeds precipitation. Semiarid zones are those where precipitation is insufficient or erratic so that soil moisture is the principal limitation for crop production.[1] Arid regions are too dry for normal crop production, which is only possible with irrigation. The FAO classification is based on length of the growing season, semiarid being 75 day–119 day and arid being 1 day–74 day.[2] Arid environments only permit sparse growth of drought-tolerant shrubs and range grasses and forage species. However, there can be exceptions to these generalities; in humid and subhumid climates, drought that limits crop production can also occur periodically. Conversely, arid and semiarid regions may experience unusually high seasonal rainfall; depending on topography, arid regions may have microenvironments where reasonable cropping can occur in normally dry years.

As temperature controls evapotranspiration, the distinction between arid and semiarid can vary depending on the region and the environment. For instance, in Mediterranean-type climates, rainfall in semiarid zones ranges from 200 mm/yr–600 mm/yr; above that range is subhumid and below that is arid.[3] Rainfall in such a climatic region is seasonal, usually in cooler winter/spring months when evapotranspiration is relatively low and rainfall is erratically distributed in time and space; variability increases as precipitation decreases.

DRYLAND AGRICULTURE

Crop production practiced under the limited rainfall conditions of semiarid climates is termed dryland or rainfed cropping, and is dependent on the capture and efficient use of limited rainfall, and thus dependent on the vagaries of the weather.[4] Such conditions exist in about 40% of the world's land surface, most of which is in the lesser-developed world. The drylands of Africa, the Middle East, Latin America, and South Asia are inhabited by about a billion people—most of whom are poor and eke out an existence in resource-poor environments. Population growth and increased food demand are the driving force behind land-use intensity in the world's drylands.

In comparison with humid regions or where irrigated agriculture is practiced, research into dryland farming systems has been modest. Stimulated by the "Dust Bowl" era in the United States—the result of land mismanagement—dryland research gathered momentum and was again brought to the public consciousness with land-degradation-induced famine in Africa in the latter part of the 20th century. While the early successes of irrigated agriculture initially detracted from dryland farming research, factors such as disenchantment with the "downside" of irrigation schemes—the exorbitant costs involved, the negative impact on the environment, and declining water supplies—served to provide a renewed focus on drylands, in particular the sustainable use of such fragile resources. Though potential crop yields from dryland agriculture are lower than irrigated agriculture, the

Encyclopedia of Water Science
DOI: 10.1081/E-EWS 120010159

relatively modest yield increases per hectare can translate into a substantial impact on national food production in view of the large areas associated with drylands.

DRYLAND RESEARCH AGENDA

Today, dryland research has an identity of its own, with common research themes that vary depending on the biophysical and socioeconomic conditions of the target eco-region: local conditions, institutional factors, and community involvement. An understanding of the development constraints in such circumstances is fundamental for designing appropriate research strategies and implementing solutions. Notwithstanding the achievements made in dryland research in the past century and the basic principles elucidated,[5,6] research activities in most institutions center around water conservation and use efficiency, combating soil erosion by wind and water, and devising management strategies, including tillage and fertilization, to implement these objectives. A major component is adaptation of the principles of crop physiology and breeding crop varieties to accommodate moisture-stressed environments. Given the complex nature of dryland farming, a multi-disciplinary approach is vital for success. The socio-economic context is of greater relevance for traditional societies in developing countries.

ARID LAND RESEARCH INSTITUTIONS

The list of research institutions that deal with semiarid and arid agriculture is extensive,[7] ranging from pioneering centers in the heart of the U.S. dryland region to an international network of research centers around the world. What follows is a sampling of such centers and their areas of concern—a complete listing is beyond the scope of this article.

Conservation and Production Research Laboratory, Bushland, Texas, U.S.A.

While the U.S.A. is home to several USDA-ARS dryland research stations of world renown, e.g., Akron, Colorado; Mandan, North Dakota; Pullman, Washington; and Pendleton, Oregon, it is fitting that of all the dryland research centers the Bushland station should be singled out for special mention. Created in the late 1930s to combat wind erosion that had devastated the drylands of Oklahoma and Texas, it was a cooperative effort between the United States Department of Agriculture (Agricultural Research Service) and the Texas Agricultural Experiment Station. The scientific achievements of Bushland are

legion; to it can be attributed the large-scale reversal of land degradation in the United States and the establishment of sound management practices. It was appropriate that the 50th anniversary of this center was marked by a world conference that highlighted progress in soil and water conservation and the challenges that lay ahead for U.S. and international agriculture.

Much of what we know today can be linked to research at Bushland; the mechanics of wind erosion, and tillage systems to conserve moisture, and thus mitigate the effects of drought, are some of the many examples. The proceedings of this milestone meeting in the history of dryland agriculture[8] established research themes for the future. Recognizing the increasing role of dryland farming in world food production, there is need for continued international dialogue of establishing networks among institutions for coordinating research and technology transfer. The resource base must be protected by sustainable soil and cropping systems. Greater attention will need to be given to the socio-economic dimension and for policies that reduce human and animal pressure on the fragile resource base.

Other North American Semiarid and Arid Institutions

The Office of Arid Lands of the University of Arizona focuses on academic research and serves as a clearing house of published works related to arid regions—social, cultural, ethnographic, economic, flora and fauna. Its world directory of "Arid Lands Research Institutions"[7] provides basic information on institutions in most countries of the world that deal with arid and semiarid areas, in addition to United Nations and other international programs. Other Arizona institutions related to arid land research include: 1) the University of Arizona's various departments in the College of Agriculture; 2) Environmental Research Laboratory focusing on protected cropping in arid environments; 3) Arid Lands Watershed Management Research Center; 4) Desert Laboratory; 5) Desert Botanical Garden; 6) Boyce Thompson Southwestern Arboretum; 7) Water Resources Research Center; and 8) USDA Agriculture Water Conservation Laboratory.

In addition to most Land Grant Colleges of Agriculture in the West and Mid West, a major listing includes: Desert Research Institute (Nevada); East–West Environment and Policy Institute (Hawaii); Plant Genetic Engineering Laboratory for Desert Adaptation (New Mexico); Dry Lands Research Institute and the US Salinity Laboratory (California); and International Center for Arid and Semi-Arid Land Studies, Chihuahan Desert Research Institute, and Drylands Agriculture Institute (Texas).

While Canada is not perceived as a dry country, there are regions of dryland agriculture, e.g., southern Alberta and Saskatchewan with Agriculture Canada dryland research institutes at Lethbridge and Swift Current. In addition, Canada's International Development Center (IDRC) supports a wide range of programs overseas, including arid land-related concerns.

U.S. Overseas Development

At the global level, the United States Agency for International Development (USAID) had promoted extensive programs in the areas of health and population, agriculture, and environment. Dryland agriculture was not neglected. One major example of such an effort was the Dryland Agriculture Project in Morocco (1979–1994) in collaboration with the National Institute of Agronomy (INRA) and executed by the Mid-America Agricultural Consortium (MIAC), spearheaded by the University of Nebraska. During the lifetime of the project, the main station in Settat was developed, along with substations, staffs were trained at Ph.D., M.S., and technical level in U.S. universities, and research and technology programs were developed. Among the many achievements of the project were the development of Hessian fly-resistant cereals, fertilizer application criteria, and conservation tillage. Today, the Center is the lead institution in dryland agriculture in North Africa. Another Moroccan institution, the Institut Agronomic et Veterinaire Hassan II, a university which is involved in teaching and research in arid agriculture, was similarly established and funded through a USAID collaborative program with the University of Minnesota.

Throughout its history, USAID has been actively involved with many other development efforts in arid areas of the world, providing US-based technical expertise and training national scientists, e.g., Northeastern Brazil (University of Arizona), Ethiopia (Oklahoma State University), Pakistan (Colorado State University)—the list is a long one. Other international research/development agencies that deal with dryland areas of the world include Windrock International Institute for Agricultural Development, and the Washington-based World Resources Institute.

Australia

A major part of this great landmass is arid desert, merging into semiarid conditions where rainfed agriculture is possible; a significant part has a Mediterranean climate. As in the United States, research in such environments is well developed. The major organizations involved are: Commonwealth Scientific and Industrial Research Organizations (CSIRO): Division of Soils, Center for Irrigation and Freshwater Research, and Division of Wildlife and Rangelands Research; various State Government Organ-

izations, e.g., Department of Primary Industries (Queensland); Fowlers Gap Arid Zone Research Station, Soil Conservation Service, Water Resources Division (New South Wales); Arid Zone Research Institute, and Water Resources Division (Northern Territory); and the Victorian Department of Natural Resources and Environment Stations-Rutherglen Research Institute and the Victorian Institute of Dryland Agriculture at Horsham. Much of the expertise and technology related to dryland farming has been exported to other regions of the world through government development programs.

Africa

Nowhere in the world is the need for research in dry regions more needed than in Africa; however, there, institutional strength varies from country to country. The strongest institutes are in Southern Africa, in particular South Africa, mostly in universities and government departments. In other parts of Africa, war and economic stagnation have taken their toll on previously active research institutes, e.g., Agricultural Research Corporation, and Soil Conservation, Land Use and Water Administration in Sudan. Most countries of North Africa have national institutions dealing with arid lands, e.g., Institut des Regions Arides and Institut National de la Recherches Agronomique de Tunisie (INRA). Dryland and arid region research organizations are poorly developed in West Africa, a region plagued by drought. Examples include Institut National de la Recherches Agronomique du Niger (INRAN), Institute for Agricultural Research, and Almadu Bello University (Nigeria), and Comite Permanent Interetats de Lutte Contre la Secheresse dans le Sahel (Burkina Faso).

Middle East and Asia

As a region with a high proportion of extremely arid land, especially in Arabian Gulf, and also large areas of semiarid rainfed land, the Middle East–West Asia area is relatively well endowed with research support of arid and dryland research centers, e.g., Bio-Saline Center (Abu Dhabi); Desert Research Center, North Khorosan Dryland Research Center, Dryland Research Center, Maragheh (Iran); Desert Research Institute, Desert Development Center (Egypt); Applied Agricultural Research Center (Iraq); Field Crops Department, and Soil and Fertilizer Institute (Turkey); Center for Desert Studies, Water Studies Center (Saudi Arabia); and National Center for Agricultural Research and Technology Transfer (Jordan). Arid land research centers in Israel includes The Jacob Blaustein Institute for Desert Research and the Center for Agricultural Research in Arid and Semi-Arid Lands.

Regional conferences[9] have highlighted the unique concerns regarding Mediterranean drylands

The Indian sub-continent has many dryland and arid research institutes. In Pakistan, these include: Cholistan Institute of Desert Studies, Arid Zone Research Institute (AZRI), Semi Arid Zone Development Authority, Tarnab Agricultural Station and Atomic Energy Agency in Peshawar, and the University of Agriculture in Faisalabad; India hosts many such institutions: Central Arid Zone Research Institutes and Desert Studies in Rajasthan.

In the former Soviet Union, many arid and semiarid land research institutes existed such as Desert Institute (Turkmenistan) and the Dochuchaev Soil Institute in Moscow. With the collapse of the USSR and the emergence of separate Central Asian republics, most institutes are poorly funded and staffed. Considerable efforts and funding are needed to address the widespread soil degradation and land mismanagement that is occurring. However, China with its huge area of arid and semiarid land has many well-known research institutes such as Institute of Desert Research, Lanzhou and the Research Center for Arid and Semi-Arid Areas, Shaanxi.

International Agricultural Research Centers

The Worldwide network of 16 research centers of the Consultative Group on International Agricultural Research (CGIAR) address agricultural production, poverty and malnutrition, capacity building and environment in resource-poor, food-deficit countries through research and technology transfer. Chief among dryland centers is the International Center for Agricultural Research in the Dry Areas (ICARDA) whose mandate covers North Africa and West Asia and Central Asia—a vast area dominated by deserts, range and scrubland, and semiarid rainfed agriculture. The center focuses on erosion control and land management to enhance water-use efficiency and on drought mitigation through breeding programs, including molecular markers and other biotechnological approaches.

Another major arid to semiarid institution is the International Center for Research in the semiarid Tropics (ICRISAT). Headquartered in India, it addresses all aspects of cropping systems in the subcontinent and in countries of the region. Its major substation is in Niamey in Niger, West Africa, a harsh zone of arid–semiarid cropping, pastoral systems, with acid sandy soils. A second substation is in Southern Africa in Bulawayo, Zimbabwe.

Other centers that deal with dry areas are: 1) the International Center for Research in Forestry (ICRAF) in Nairobi, which focuses on forest trees and shrubs in association with cropping systems in Africa; 2) The International Institute for Tropical Agriculture (IITA) in Nigeria; and 3) Centre Internacional de Agricultura Tropical (CIAT) which deals with some dryland areas in Latin America, in addition to humid areas. While the centers mentioned above have an active research agenda in dryland agriculture, they work in collaboration with the national research systems in their mandate regions.

Other non-CGIAR regional and international centers are active in arid land research. An example is the Syrian-based Arab Center for Studies in Agricultural Development (ACSAD), which focuses on the Arab region. Other international agencies that sponsor research related to dry areas include the Food and Agricultural Organization (FAO) of the United Nations, and the International Atomic Energy Agency (IAEA), Vienna.

Though not characterized by an arid climate, some countries such as the United Kingdom have international institutions that deal with research in dry regions, e.g., Center for Overseas Research and Development and the Center for Arid Zone Studies at the University of North Wales, while Germany's University of Stuttgart has a "Working Group on Desert Research."

CONCLUSION

While much is known about the biophysical processes and constraints related to arid and semiarid research, the major bottleneck is implementation at the user's level. That calls for a greater understanding of the social and cultural factors associated with dry areas. Major conferences on semiarid dryland farming[8] and on desert development in Lubbock (1996) and Cairo (1999) indicate that research momentum is gathering. Knowledge gained has to be translated into public practices that promote community action. Despite the large number and diversity of arid/semiarid research institutions worldwide, there is need for networks among institutions for information sharing. The problems of dry regions will not disappear, but, given the scenario of exacerbated drought in many parts of the world due to global warning, will be more urgent than ever. As vast areas of arid regions are categorized as rangelands, concerted international efforts are needed to tackle problems in such fragile areas.[2] The United Nations Convention to Combat Desertification and other coordinated international efforts in support of arid and semiarid lands are a major step in that direction. Success in these endeavors is dependent on the global awareness of political leaders and the consequent creation of enabling environments for policy implementation.

REFERENCES

1. Hagin, J.; Tucker, B. *Fertilization of Dryland and Irrigated Soils*; Springer-Verlag: New York, NY, USA, 1982.

2. Squires, V.R.; Sidahmed, A. *Drylands Sustainable Use of Rangelands in the Twenty-First Century*; International Fund for Agricultural Development (IFAD): Rome, Italy, 1998.

3. Kassam, A. Climate, Soil, and Land Resources in North Africa and West Asia. Plant Soil **1981**, *58*, 1–28.

4. Cooper, P.J.M.; Gregory, P.J.; Tully, D.; Harris, H.C. Improving Water Use Efficiency of Annual Crops in the Rainfed Farming Systems of West Asia and North Africa. Exp. Agric. **1987**, *23*, 113–158.

5. Brengle, K.G. *Principles and Practices of Dryland Farming*; Colorado Associated University Press: Boulder, CO, USA, 1982.

6. Dregne, H.E.; Wills, W.O. *Dryland Agriculture*; ASA, CSSA, SSSA: Madison, WI, USA, 1983.

7. Hutchinson, B.S.; Varady, R.G. *Arid Lands Research Institutions: A World Directory*; Allerton Press, Inc.: New York, 1988.

8. Unger, P.W.; Sneed, T.V.; Jordan, W.R.; Jensen, R. *Challenges in Dryland Agriculture—A Global Perspective*; Texas Agric. Exp. Stn.: Lubbock, Texas, 1988.

9. Whitman, C.E.; Parr, J.F.; Papendick, R.I.; Meyer, R. Soil, Water, and Crop/Livestock Management Systems for Rainfed Agriculture in the Near East Region. In *Proceedings of a Workshop, Amman, Jordan, Jan 18–23, 1988*; USAID: Washington, DC, 1989.

R

Research Organizations

Gabriel Eckstein
International Water Law Project, Washington, District of Columbia, U.S.A.

INTRODUCTION

Improvements in irrigation techniques, sustainable farming methodologies, and drought and pest resistance have made a tremendous impact on global agricultural production. During the last three decades, production of food crops, such as grain and cereal, doubled and tripled resulting in a 19% per capita increase in food for direct human consumption.[1-3] During the same time, the percentage of the world's hungry and malnourished people dropped from 35% to 20%, and per capita food supplies rose from 2135 cal per day to 2750 cal per day.[4] Despite these vast improvements, more than 800 million people globally are still undernourished, and one-third of all children—two of every five in South Asia—are malnourished.[5]

INCREASING NEEDS AND INCREASING EFFICIENCIES

In the next few decades, as world population increases to between 7 and 10 billion people, global demand for food is projected to grow twofold, with even greater increases in the developing world.[6] In the past, increases in food production were achieved by placing more land under cultivation. Since the mid-1960s, however, the rate of growth of the world's cultivated lands grew at a declining rate, averaging only 8%. In many industrialized nations, agricultural area actually decreased due to competition with urban sprawl.[7] As a result, recent increases in production have been more a factor of higher efficiency and productivity rather than expansion of land under cultivation.

Much of that efficiency and productivity is due to modern irrigation techniques. Since the turn of the century, land under irrigation globally grew fivefold, and doubled in the last 25 yr, to approximately 275 million ha.[8-10] Today, 40% of the world's food is produced on irrigated fields, which cover only 17% of the world's cultivated land. Rain-fed agriculture—which accounts for 83% of the world's farmland—produces the remaining 60% of agricultural production.[11,12]

WATER SCARCITY AND AGRICULTURE

While highly productive from a per hectare point of view, water scarcity remains the single biggest threat to future food production. Agriculture today is the largest single consumer of freshwater globally, responsible for 93% of global consumptive use of water today.[13] Land under irrigation, which accounts for 17% of cultivated land (about 270 million ha), uses more than two-thirds of global water withdrawals.[14,15] Moreover, at existing rates of use, crop demands for 2025 could require an additional 200 cubic miles of water—a volume nearly equal to the annual flow of the Nile River 10 times over.[16] Many freshwater sources, however, including aquifers, rivers, and lakes, are stressed far beyond their limits. Eight percent (8%) of food crops globally is grown on farms using groundwater at a rate faster than the aquifer can recharge. Moreover, many large rivers, such as the Jordan and Rio Grande Rivers, are so heavily diverted that little if any water reaches the rivers' mouths. Significantly, as much as one-third of the world's population today lives in regions experiencing moderate to high water stress.[17]

IMPROVING WATER MANAGEMENT AND USE TECHNOLOGIES

Accordingly, developments in the use and management of water resources are essential if we are to continue meeting the needs and demands of the world's population. The use of existing technologies and methodologies must be expanded to regions stressed by water scarcity. Drip irrigation, for example, which saves water and reduces soil salinity, could be used on a much broader scale. Farmers using drip irrigation typically can reap two or three harvests every year. Studies conducted in Israel, Jordan, Spain, and the United States show that drip irrigation can reduce water use by 30%–70% while increasing crop yield by 20%–90%, as compared to flooding methods.[18,19]

Likewise, research into technology designed to reduce water use, as well as to reuse and recycle wastewater, should be pursued. Precision irrigation systems, which supply water only when and where needed, are now being designed. Low-energy sprinklers, already in use, allow plants to absorb as much as 95% of the water flowing through the sprinkler. Wastewater is being treated for use on cultivated fields—treated wastewater in Israel, for example, accounts for 30% of the country's agricultural water supply and is expected to rise to 80% by 2025. Moreover, gains in rain-fed agricultural production is also

Encyclopedia of Water Science
DOI: 10.1081/E-EWS 120010294

being targeted with a range of improved small-scale and supplemental irrigation systems.[20–22]

These and many other improvements in water use and management are currently being researched and developed throughout the world at government and academic institutions, as well as private operations. While far from comprehensive, the following is a short list of non-commercial institutions (i.e., universities, government agencies, etc.) from around the world that are dedicated to tackling the issue of water scarcity in agriculture.

RESEARCH ORGANIZATIONS

The Dryland Agricultural Institute at West Texas A&M University (http://www.wtamu.edu/research/dryland/), located in Canyon, Texas, assists researchers, educators, extension workers, and administrators to develop practical and workable strategies for improving the sustainability of dryland agriculture systems worldwide. The Institute's chief areas of research include: efficient water use; wind and water erosion; soil fertility and organic matter; drought-resistant germplasm; deficit irrigation; pest management; and rangeland management.

The Agricultural Research Service (ARS) is the principal research agency of the United States Department of Agriculture (USDA). It oversees the Water Quality and Management National Program (WQMNP) (http://www.nps.ars.usda.gov/programs/programs.htm?NPNUMBER = 201), which cooperates with the Cooperative State Research, Education, and Extension Service; Economic Research Service; and National Agriculture Statistics Service to provide research, technology transfer, education, extension, and economic assessments for the Natural Resources Conservation Service within the USDA. Specifically, the WQMNP is tasked with developing innovative concepts for determining the movement of water and its associated constituents in agricultural landscapes and watersheds, and to develop new and improved practices, technologies, and strategies to manage the U.S.'s agricultural water resources. The WQMNP research is conducted throughout the United States at various locations on: economical irrigated crop production; precision irrigated agriculture; water conservation management; irrigation and drainage in humid areas; wastewater reuse; erosion on irrigated land; salinity and trace element management; and drainage management.

The Agricultural Research Organization (ARO) (http://agri.gov.il/) of the Israel Ministry of Agriculture is responsible for planning, organizing, and implementing the greater part of Israel's agricultural research effort. The ARO, based in Bet Dagan, Israel, focuses on solving current problems in agricultural production, introducing new products, processes and equipment, and researching Israel's future agricultural development. Within the ARO, the Institute of Soil, Water, and Environmental Sciences (http://agri.gov.il/SoilScience.html) carries out research to ensure optimal use of two of Israel's limited natural resources: soil and water. Research areas include: water scarcity and quality; the need for more economically and environmentally sound irrigation; management of agrochemicals and cropping; energy saving; and the development of efficient and environmentally friendly greenhouse management schemes.

The Australian Commonwealth Scientific and Industrial Research Organisation: Land and Water (CSIRO) (http://www.clw.csiro.au/) is dedicated to creating the knowledge, the strategies, and the tools to manage land and water in Australia and internationally. It is divided into five divisional research programs: Remediation of Contaminated Environments, Sustainable Agriculture, Sustainable Catchment and Groundwater Management, Tropical Land and Water Management, and Waterway Management and Landscape Function. Based in Glen Osmond, Australia, the Sustainable Agriculture Program identifies, tests, and develops soil and water management practices necessary to underpin sustainable production systems for use in agriculture and horticulture in Australia. The emphasis of the program is on the identification and introduction of systems, which are ecologically suited to the climatic conditions and which do not lead to degradation of the soil and water resources. The program also addresses issues relating to the sustainable use and cycling of wastes in rural areas.

The International Water Management Institute (IWMI) of Sri Lanka (http://www.cgiar.org/iwmi/) is a scientific research organization focusing on issues of sustainable and productive use of water resources, particularly as they relate to agriculture, water scarcity, and food security in the developing world. The IWMI works with partners in the Global South to develop tools and methods to help these countries eradicate poverty through more effective management of their water resources. One of the Institute's four core programs concerns Irrigation and Water Resources (IWR). The Program focuses on integrated approaches for managing water resources, assessing the performance of irrigated agriculture, contributing to improving irrigation system design and operation, and documenting the impacts of water management interventions.

CONCLUSION

The future ability of agricultural production to meet global needs is inextricably linked to improvements in water management techniques. Knowledge of irrigation practices, integrated water management systems, and other agricultural methodologies must continue to progress to ensure improvements of agricultural sustainability and productivity. Moreover, the negative impacts on the environment and agriculture that can result from unsound agricultural practices must be better understood and minimized. Research conducted at these and other institutions is indispensable to the progressive development of such knowledge and, therefore, must be encouraged and supported if we are to achieve our needs as well as our full potential.

REFERENCES

1. Postel, S. Growing More Food with Less Water, Scientific American, February 2001, at http://www.sciam.com/2001/0201issue/0201postel.html (accessed March 15, 2001).

2. Borlaug, N.; Dowswell, C. Agriculture in the 21st Century: Vision for Research and Development, at http://www.agbioworld.com/articles/21century.html (accessed March 15, 2001).

3. Hall, D.O. Food Security: What Have Sciences to Offer? A Study for International Council for Science, 1998, at http://www.icsu.org/Publications/FoodSci/fs.html (accessed March 16, 2001).

4. Attaining Global Food Security by 2025, Position Paper No. 3, The International Policy Council on Agriculture Food and Trade, 1996, at http://www.agritrade.org/publ/gfs.htm (accessed March 12, 2001).

5. Hall, supra.

6. Daily, G.; Dasgupta, P.; Bolin; Bert; Crosson, P.; du Guerny, J.; Ehrlich, P.; Folke, C.; Jansson, A.M.; Jansson, B.-O.; Kautsky, N.; Kinzig, A.; Levin, S.; Mäler, K.-G.; Pinstrup-Andersen, P.; Siniscalco, D.; Walker, B. Global Food Supply: Food Production, Population Growth, and the Environment. Science **1998**, 281, 5381: 1291.

7. *World Resources 2000–2001—People and Ecosystems: The Fraying Web of Life*, World Resources Institute: Washington, DC, 2000, 53–60.

8. Gleick, P. *The World's Water 1998–1999: The Biennial Report on Freshwater Resources*; 1998; 5–24.

9. World Resources 2000–2001, supra.

10. Borlaug, supra.

11. Postel, supra.

12. Borlaug, supra.

13. Borlaug, supra.

14. World Resources 2000–2001, supra.

15. Borlaug, supra.

16. Postal, supra.

17. Borlaug, supra.

18. Postal, supra.

19. Gleick, supra.

20. Postal, supra.

21. Borlaug, supra.

22. Gleick, supra.

Reverse Osmosis

Mark Wilf
Hydranautics, Oceanside, California, U.S.A.

INTRODUCTION

Osmosis is a natural process involving fluid flow across a semipermeable membrane barrier. It is selective in the sense that the solvent passes through the membrane at a faster rate than the dissolved solids. The difference of passage rate results in solvent solids separation. The direction of solvent flow is determined by its chemical potential, which is a function of pressure, temperature, and concentration of dissolved solids. Pure water in contact with both sides of an ideal semipermeable membrane at equal pressure and temperature has no net flow across the membrane because the chemical potential is equal on both sides. If a soluble salt is added on one side, the chemical potential of this salt solution is reduced. Osmotic flow from the pure water side across the membrane to the salt solution side will occur until the equilibrium of chemical potential is restored. Equilibrium occurs when the hydrostatic pressure differential resulting from the volume changes on both sides is equal to the osmotic pressure. Application of an external pressure to the salt solution side equal to the osmotic pressure will also cause equilibrium. Additional pressure will raise the chemical potential of the water in the salt solution and cause water flow to the pure water side, because it now has a lower chemical potential. This phenomenon is called reverse osmosis (RO). The reverse osmosis technology developed about 50 yr ago, as a scientific experiment, is used extensively today to reduce salinity of various water sources and produce potable water in commercial systems. Other applications include production of low salinity water for industrial applications and reclamation of waste streams. The economics of RO technology is very competitive in comparison with other salt reduction processes and, in some cases, the cost of producing potable water using RO can be lower than water supplied from natural sources, if pumping water over long distances is required.

OSMOTIC PRESSURE

The osmotic pressure, P_{osm}, of a solution can be calculated by measuring the concentration of dissolved salts in solution:

$$P_{osm} = 1.19(T + 273)\sum(m_i) \tag{1}$$

Where P_{osm} is the osmotic pressure (in psi); T, the temperature (in °C); and $\sum(m_i)$, the sum of molal concentration of all constituents in a solution. An approximation for P_{osm} may be made by assuming that 1000 ppm of total dissolved solids (TDS) equals about 11 psi (76 kPa) of osmotic pressure. The mechanism of water and salt separation by reverse osmosis is not fully understood. Current scientific thinking suggests two transport models: porosity and diffusion. That is, transport of water through the membrane may be through physical pores present in the membrane (porosity), or by diffusion from one bonding site to another within the membrane. The theory suggests that the chemical nature of the membrane is such that it will absorb and pass water preferentially to dissolved salts at the solid/liquid interface. This may occur by weak chemical bonding of water to the membrane surface or by dissolution of water within the membrane structure. Either way, a salt concentration gradient is formed across the solid/liquid interface. The chemical and physical nature of the membrane determines its ability to allow for preferential transport of solvent (water) over solute (salt ions).

WATER AND SALT TRANSPORT

The rate of water passage through a semipermeable membrane is defined in Eq. 2:

$$Q_w = (\Delta P - \Delta P_{osm})A = (NDP)A \tag{2}$$

where Q_w is the rate of water flow through the membrane, ΔP, the hydraulic pressure differential across the membrane, and ΔP_{osm}, the osmotic pressure differential across the membrane. A represents a unique constant for each membrane material type, and NDP is the net driving pressure or net driving force for the mass transfer of water across the membrane.

The rate of salt flow through the membrane is defined by Eq. 3:

$$Q_s = \Delta CB \tag{3}$$

where Q_s is the flow rate of salt through the membrane, ΔC is the salt concentration differential across the membrane, and B represents a unique constant for each membrane type.

Encyclopedia of Water Science
DOI: 10.1081/E-EWS 120010236

Eqs. 2 and 3 show that for a given membrane:

- Rate of water flow through a membrane is proportional to net driving pressure differential (NDP) across the membrane.
- Rate of salt flow is proportional to the concentration differential across the membrane and is independent of applied pressure.

Salinity of the water that passes through the membrane, the permeate, C_p, depends on the relative rates of water and salt transport through reverse osmosis membrane:

$$C_p = Q_s/Q_w \tag{4}$$

The fact that water and salt have different mass transfer rates through a given membrane creates the phenomena of salt rejection. No membrane is ideal in the sense that it absolutely rejects salts.

COMMERCIAL REVERSE OSMOSIS TECHNOLOGY

The semipermeable membrane for reverse osmosis applications consists of a multilayer film of polymeric material composed of a skin layer $0.1\,\mu m{-}0.2\,\mu m$ thick and spongy supporting layer approximately 0.1 mm thick cast on a fabric support. The commercial grade membrane must have high water permeability and a high degree of semipermeability, i.e., the rate of water transport must be much higher than the rate of transport of dissolved ions. The membrane must be stable over a wide range of pH and temperature, and have good mechanical integrity. The stability of these properties over a period of time at field conditions defines the commercially useful membrane life, which is in the range of 3–5 yr. There are two major groups of polymeric materials that can be used to produce satisfactory reverse osmosis membranes: cellulose acetate (CA) and polyamide (PA). Membrane manufacturing, operating conditions, and performance differ significantly for each group of polymeric material.

CELLULOSE ACETATE MEMBRANE

The original CA membrane, developed in the late 1950s by Loeb and Sourirajan, was made from cellulose diacetate polymer.[1] Current CA membrane is usually made from a blend of cellulose diacetate and triacetate. The membrane is formed by casting a thin film acetone-based solution of CA polymer with swelling additives onto a nonwoven polyester fabric. Two additional steps, a cold bath

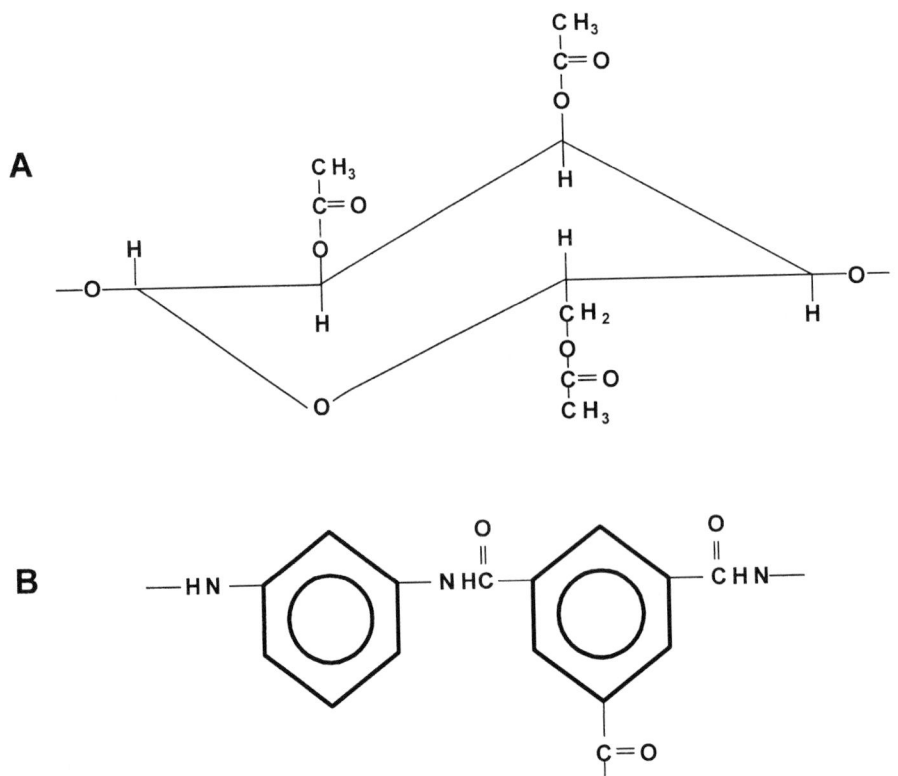

Fig. 1 Chemical structure of cellulose triacetate (A) and polyamide (B) membrane material.

followed by high temperature annealing, complete the casting process. After processing, the cellulose membrane has an asymmetric structure with a dense surface layer of about 0.1 μm–0.2 μm which is responsible for the salt rejection property. The rest of the membrane film is spongy and porous and has high water permeability. Description of manufacturing process of CA membranes and its properties can be found in number of publications.[2]

COMPOSITE POLYAMIDE MEMBRANES

Composite PA membranes have been developed in the early eighties by Cadotte and coworkers.[3] Commercially it is manufactured in two distinct steps. First, a polysulfone support layer is cast onto a nonwoven polyester fabric. The polysulfone layer is very porous and is not semipermeable, i.e., it does not have the ability to separate water from dissolved ions. In a second, separate manufacturing step, a semipermeable membrane skin is formed on the polysulfone substrate by interfacial polymerization of monomers containing amine and carboxylic acid chloride functional groups. The resulting composite membrane is characterized by higher specific water flux and lower salt passage than CA membranes. Polyamide composite membranes are stable over a wider pH range than CA membranes. However, PA membranes will degrade more rapidly by free chlorine than are CA membranes. Consequently, CA membranes are used today almost exclusively in commercial composite membrane elements. The structures of CA and PA polymer are shown in Fig. 1A and B.

RO MEMBRANE MODULE CONFIGURATIONS

The membrane module configuration used almost exclusively for commercial reverse osmosis desalting applications is the spiral wound configuration. In a spiral wound configuration two flat sheets of membrane are separated with a permeate collector channel material to form a leaf. This assembly is sealed on three sides with the fourth side left open for permeate to exit. A feed/brine spacer material sheet is added to the leaf assembly. A number of these assemblies or leaves are wound around a central plastic permeate tube. This tube is perforated to collect permeate from the multiple leaf assemblies. A diagram of the spiral membrane leaf assembly is shown in Fig. 2. The typical industrial spiral wound membrane element is approximately 100 cm or 150 cm (40 in. or 60 in.) long and 10 cm or 20 cm (4 in. or 8 in.) in diameter. The feed/brine flow through the element is on a straight axial path from the feed end to the opposite brine end, running parallel to the membrane surface. The feed channel spacer induces turbulence and reduces concentration polarization (excess salt concentration at the membrane surface). The structure of the corresponding modules configurations is shown in Fig. 3.

RO SYSTEM CONFIGURATION

RO systems consist of the following basic components: feed water supply unit, pretreatment system, high pressure pumping unit, membrane element assembly unit, instrumentation and control system, permeate treatment, and storage unit and cleaning unit.

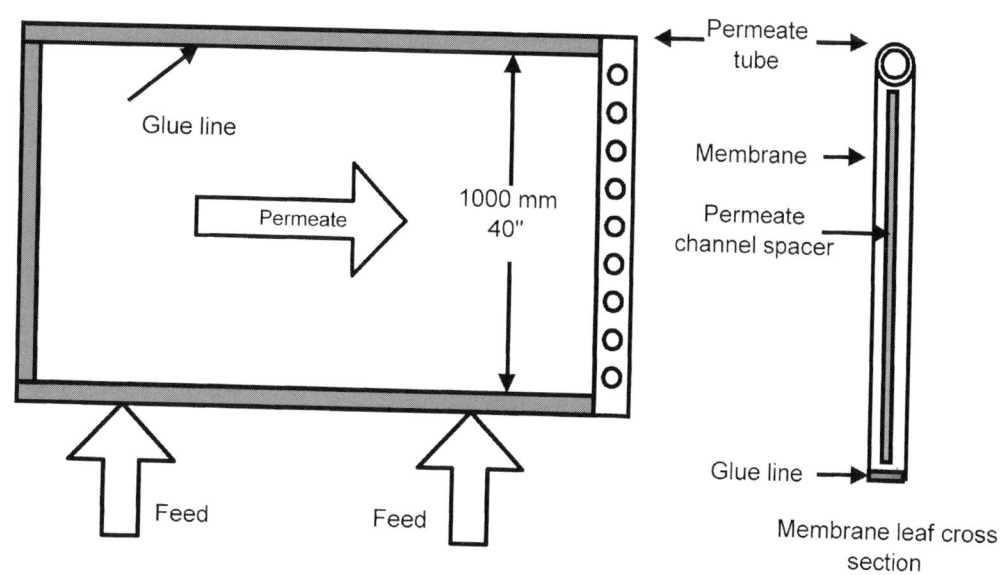

Fig. 2 Conventional spiral wound module configuration.

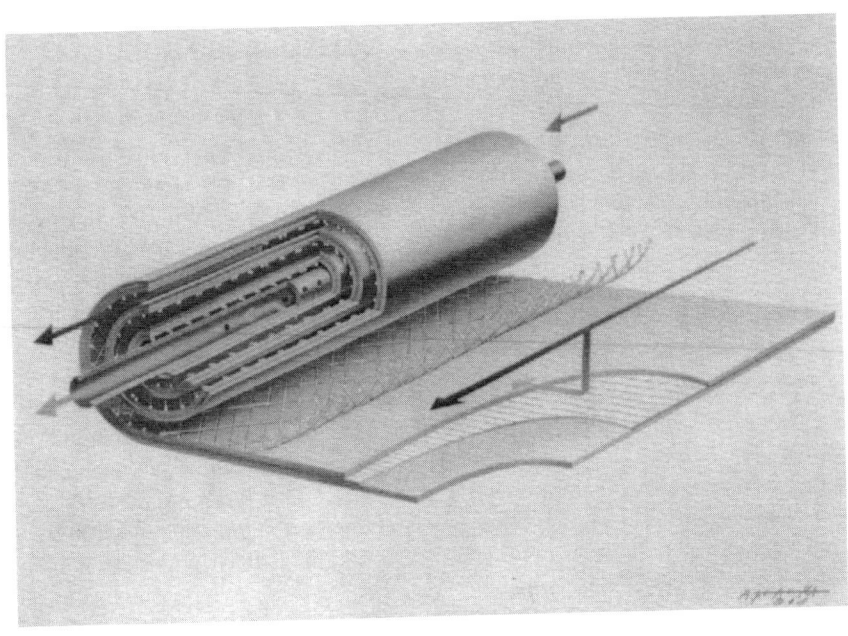

Fig. 3 Membrane schematic showing materials and flows.

The membrane assembly unit (RO block) consists of a stand supporting the pressure vessels, interconnecting piping, and feed, permeate, and concentrate manifolds. Membrane elements are installed in the pressure vessels. Each pressure vessel may contain 1–8 membrane elements connected in series (Fig. 4). A system is divided into groups of pressure vessels, called concentrate stages. In each stage, pressure vessels are connected in parallel with respect to the direction of the feed/concentrate flow. The number of pressure vessels in each subsequent stage decreases in the direction of the feed flow, usually in the ratio of 2:1, as shown in Fig. 5. Thus, one can visualize that the flow of feed water through the pressure vessels of a system resembles a pyramid structure: a high volume of feed water flows in at the base of the pyramid, and a relatively small volume of concentrate leaves at the top. The decreasing number of parallel pressure vessels from stage to stage compensates for the decreasing volume of feed flow, which is continuously being partially converted to permeate. The permeate of all pressure vessels in each stage, is combined together into a common permeate manifold. The objective of the taper configuration of pressure vessels is to maintain a similar feed/concentrate flow rate per vessel through the length of the system and to maintain feed/concentrate flow within the limits specified for a given type of membrane element. A picture of the actual RO unit is shown in Fig. 6. The concentrate from the first stage becomes the feed to the second stage; this is what is meant by the term "concentrate staging." The flows and pressures in the multistage unit are controlled with the feed and concentrate valves. The feed valve, after the high-pressure pump, controls feed flow to the unit. The concentrate valve, at the outlet of RO block, controls the feed pressure.

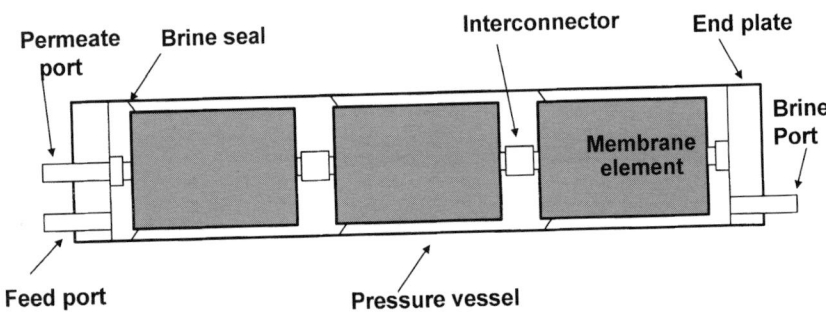

Fig. 4 Pressure vessel with three membrane elements.

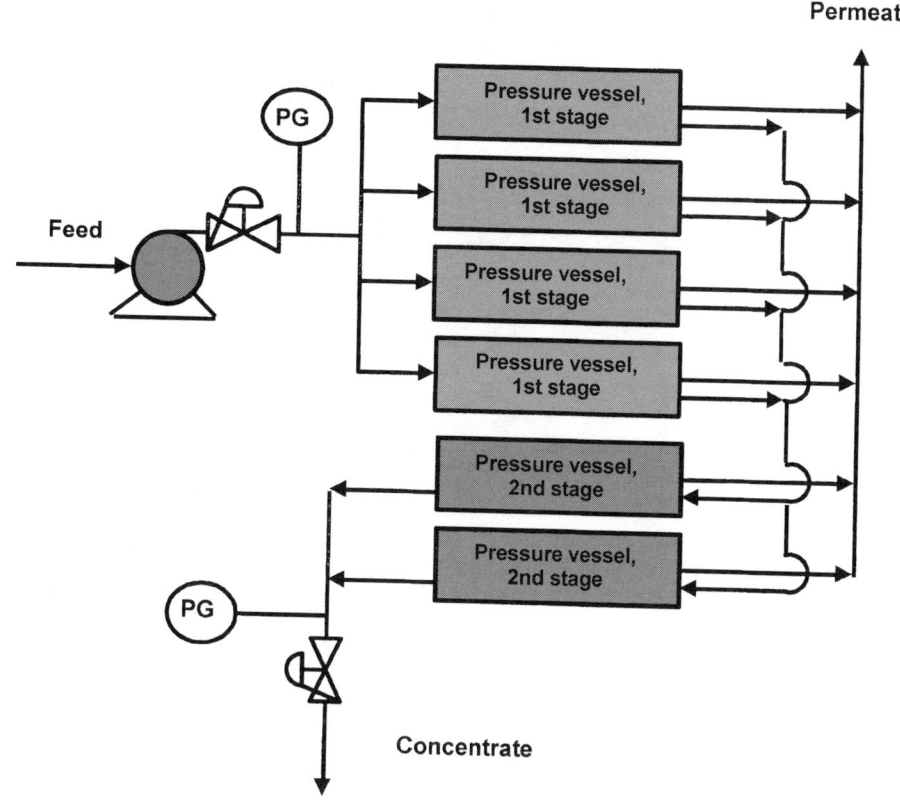

Fig. 5 Flow diagram of a two-stage RO system.

FEED WATER PRETREATMENT

The extend of pretreatment process depends on the quality of raw water, which is usually associated with its origin: surface or well water. The initial removal of large particles from the feed water is accomplished using mesh strainers or traveling screens. Mesh strainers are used in well-water supply systems to stop and remove sand particles that may be pumped from the well. Traveling screens are used mainly for surface-water sources, which typically have large concentrations of biological debris. It is a common practice to disinfect surface feed water in order to control biological activity. Biological activity in well water is usually very low, and in majority of cases, well water does not require chlorination. In some cases, chlorination is used to oxidize iron and manganese in the well water before filtration. Settling of surface water in a detention tank results in some reduction of suspended particles. Addition of flocculants, such as iron or aluminum salts, results in the formation of corresponding hydroxides; these hydroxides neutralize surface charges of colloidal particles, aggregate, and adsorb to floating particles before settling at the lower part of the clarifier. To increase the size and strength of the flock, a long chain organic polymer can be added to the water to bind flock particles together.

Use of lime results in increase in pH, formation of calcium carbonate and magnesium hydroxide particles. Well water usually contains low concentrations of suspended particles, due to the filtration effect of the aquifer. The pretreatment of well water is usually limited to screening of sand, addition of scale inhibitor to the feed water, and cartridge filtration. Surface water may contain various concentrations of suspended particles, which are either of inorganic or biological origin. Surface water usually requires disinfection to control biological activity and removal of suspended particles by media filtration. The efficiency of filtration process can be increased by adding filtration aids, such as flocculants and organic polymers. Some surface water may contain high concentrations of dissolved organics. Those can be removed by passing feed water through an activated carbon filter. Depending on composition of the water, acidification and addition scale inhibitor may be required. Recently, new pretreatment equipment has been introduced to the RO market. It consists of backwashable capillary microfiltration and ultrafiltration membrane modules. This new equipment can operate reliably at very high recovery rates and low feed pressure. The new capillary systems can provide better feed water quality than a number of conventional filtration steps operating in series. The cost of this new

Fig. 6 Commercial RO train.

equipment is still relatively high compared to the cost of conventional pretreatment, and therefore is mainly used for treatment of heavily fouling streams, such as municipal wastewater effluents.

RO APPLICATIONS

The majority of applications involve production of potable water from brackish or seawater streams. Reverse osmosis technology is also used in industrial applications to reduce water salinity prior to ion exchange equipment. Another growing area of application is reclamation of municipal wastewater. These applications usually involve integrated membrane technology, where the secondary municipal effluent is treated with macrofiltration or ultrafiltration prior to reverse osmosis unit. Municipal wastewater reclamation produces water for number of applications including industrial (cooling water makeup), agricultural (irrigation), and aquifer injection (prevention of sweater

intrusion). The cost of reverse osmosis process decreased significantly in the last decade. The current cost of desalting of brackish water is in the range of $0.25 m^{-3} – $0.35 m^{-3} ($0.95 m/1000 gal–$1.32/1000 gal). For recent large seawater projects the water cost as low as $0.54 m^{-3} ($2.04/1000 gal) has been reported.[4,5]

REFERENCES

1. Loeb, S.; Sourirajan, S. Adv. Chem. Ser. **1962**, *38*, 117.
2. Rautenbach, R.; Albrecht, R. *Membrane Processes*; John Willey and Sons: New York, 1989.
3. Cadote, J.E.; Llooyd, D.R., (Eds.) *Material Science of Synthetic Membranes*; Am. Chem. Soc. Symposium Series, Washington, DC, 1985; 273–294.
4. Glueckstern, P.; Nadav, N.; Priel, M. Desalination **2001**, *138*, 157–163.
5. Results of Recent Tender for 140,000 m^3 day^{-1} Sweater RO Project in Ashkelon, Israel.

Richards' Equation

Graeme D. Buchan
Lincoln University, Canterbury, New Zealand

INTRODUCTION

Darcy's law[1] is the basic law governing the flow of water (or other liquids) in permeable materials, and it tells us that the flow velocity q (m sec^{-1}) at any point (e.g., in soil, porous rock, concrete, timber, or other material) is proportional to the gradient of the water potential at that point. However, Darcy's law tells us only about the flow at individual points and is sufficient to describe only steady flow processes. To model unsteady flows (where the moisture distribution changes with time), we must also know the relationship between velocities at neighboring points. If the neighboring velocities are unequal, their 'mismatch' must be compensated by a filling or emptying of pores between the points. The additional equation required to complete the mathematical description of flow is the so-called *continuity equation*. Basically, this equation ensures that matter is not created or destroyed, and so is also called the conservation equation. When Darcy's law is combined with the continuity equation, we obtain Richards' equation, first derived by the physicist Lorenzo Adolph Richards in 1931.[2]

Richards was a pioneering soil physicist who contributed enormously to soil water physics in the United States in the period c.1930–1960.[3,4] His contributions to theory included the conceptual extension of Darcy's Law to unsaturated flow,[2] as part of his development of Richards' equation. On the experimental side, he: invented the tensiometer;[3,5] developed the pressure-plate apparatus[6] to measure water desorption from soil; developed the thermocouple method for measuring the vapor pressure (or "water activity") in soil or biological materials; and helped establish the relationship between the permanent wilting point for plants and the soil water content at 15 bar suction. He also investigated salt-affected soils.

THEORY

We need a mathematical description of water flow in permeable materials. The resulting equations can then be used to model flow in: 1) soil or sediments, including phenomena such as infiltration, drainage, drying by evaporation, or water flow towards roots; 2) groundwater, including aquifers; and 3) other materials, during wetting or drying processes (e.g., timber, concrete, foodstuffs, or granular or powder materials).

Darcy's Law

First, we limit attention to flow in unsaturated materials. The saturated case will be described later as a special case. Also, for simplicity, assume that liquid flow is initially in the horizontal (x) direction (Fig. 1). Darcy's law[1] states that the flow velocity q at any point is proportional to the gradient $\partial H/\partial x$ of hydraulic head H.

$$q = -K \partial H/\partial x \tag{1}$$

Here H (m) is the water potential expressed in terms of the equivalent height or head of a column of water, and K (m sec^{-1}) is the hydraulic conductivity of the material. In saturated soil, water is under positive pressure, so $H > 0$. In unsaturated soil, where the water is under suction, H becomes negative:

$$H = -\psi \tag{2}$$

where ψ (intrinsically positive) is the so-called matric suction, a measure of the energy status of the water.[5] For vertical flow, gravity enters as an additional driving force, and H has two components:

$$H = -\psi + z \tag{3}$$

where z (the vertical coordinate, Fig. 1) represents the gravitational potential. Extending Darcy's law to three dimensions gives:

$$q = -K\nabla H \tag{4}$$

where ∇ represents the 3-D gradient (a vector quantity). Eq. 4 assumes an isotropic material, i.e., that K is equal for all flow directions.

K is a function of soil wetness, and may be written in one of two ways. First, in the "mass picture" we write $K = K(\theta)$ where θ (m^3 m^{-3}) is the volumetric water content. However, $\theta = \theta(\psi)$ is related to the suction ψ. The $\theta(\psi)$ relationship is known as the soil moisture characteristic (SMC).[5,6] Thus, in the "energy picture," K is expressed instead as $K(\psi)$.

Encyclopedia of Water Science
DOI: 10.1081/E-EWS 120010272

810

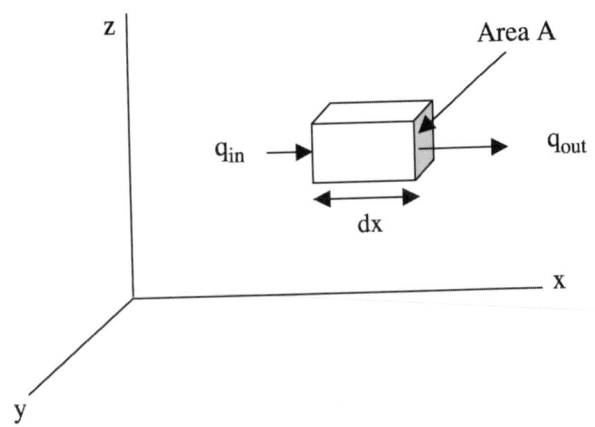

Fig. 1 Horizontal flow of a liquid across an imaginary volume element in a permeable material, with volume $V = A\,dx$. In unsteady flow, the velocities q_{in} and q_{out} are unequal, and the continuity principle implies a net filling or emptying of pores in the element.

For vertical flow, Eq. 1 becomes (using Eq. 3):

$$q = -K\partial H/\partial z = K\partial \psi/\partial z - K \qquad (5)$$

The two terms on the right side of Eq. 5 mirror the two head components in Eq. 3, and represent, respectively, the suction-driven and gravity-driven components of flow. For horizontal flow, only the first term applies.

Eq. 1 gives the flow velocity q at any point. In order to describe transient (nonsteady) processes, we need to introduce the continuity equation.

The Continuity Equation

Consider an imaginary small cubical volume in the material (Fig. 1), with entry velocity q_{in} and exit velocity q_{out}. The mismatch of these two velocities must be balanced by a change in the volumetric water content θ in the cube.

$$V\partial \theta/\partial t = A[q_{in} - q_{out}] = -A\,dx\partial q/\partial x \qquad (6)$$

Since $V = A\,dx$, then

$$\partial \theta/\partial t = -\partial q/\partial x \qquad (7)$$

The corresponding equation for vertical flow is $\partial \theta/\partial t = -\partial q/\partial z$. In 3-D analysis we replace $\partial q/\partial x$ with $\nabla \cdot q$, which is a measure of the so-called *divergence* of the water flux at the point. (The divergence is, as its name suggests, the net outflow through the surfaces of a tiny volume surrounding the point.) Then:

$$\partial \theta/\partial t = -\nabla \cdot q \qquad (8)$$

Eq. 8 is the continuity equation.

Richards' Equation

We can now combine Darcy's law with the continuity equation. For vertical flow, combining Eq. 5 and the vertical (z) form of Eq. 7 one can derive:

$$\partial \theta/\partial t = -\partial/\partial z(K\partial \psi/\partial z) + \partial K/\partial z \qquad (9)$$

This is Richards' equation for vertical flow. Note that, mirroring Eq. 3, Eq. 9 contains both a suction-driven and a gravity-driven flow term. Extending Eq. 9 to isotropic 3-D flow, we add horizontal (x and y) terms.

$$\partial \theta/\partial t = -\nabla \cdot (K\nabla \psi) + \partial K/\partial z \qquad (10)$$

However, we have a problem. Eq. 10 cannot be solved immediately as it contains two unknowns, both θ and ψ. To eliminate one unknown, we exploit the relationship between θ and ψ, i.e., we assume that the SMC $\theta(\psi)$ is known. There are two options.

(1) *The energy picture.* Here we retain ψ as the unknown, and solve Eq. 10. Since $\theta(\psi)$ and $K(\psi)$ are both highly nonlinear functions, and also may be based on experimentally derived data rather than analytical functions, analytic solution of Richards' equation is not generally possible, except in special cases.[7] Hence, numerical solution is required. Note that Eq. 10 is first order in the time derivative ($\partial/\partial t$) and second order in space derivatives ($\partial^2/\partial x^2$ etc). Hence, solutions generally have the space variables (x, y, or z) paired with $\sqrt{t}$.

(2) *The mass picture.* Here we retain θ as the unknown.

$$\partial \theta/\partial t = \nabla \cdot (D\nabla \theta) + \partial K/\partial z \qquad (11)$$

In Eq. 11, the substitution $D = -K\,d\psi/d\theta = D(\theta)$ has been used. D is called the "hydraulic diffusivity," because Eq. 11 now looks like a diffusion type equation. However, the water transport described by Eq. 11 is not a true diffusion process. It is a mass flow process.

To summarize, in the "energy picture," water potential ψ is the dependent variable; while in the "mass picture," θ becomes the dependent variable, and the mass flow equation takes the apparent form of a diffusion equation.

APPLICATIONS AND COMPLICATIONS

Richards' equation is the foundation of all mechanistic models used to simulate the dynamics of water (or other liquid) in permeable materials, including soils, rocks, aquifers, or industrial materials. Because Richards' equation is so generic, any model based on it should be able to simulate any flow process, in either unsaturated or saturated material, assuming that appropriate values for the equation parameters have been determined for a particular application. In soils, processes in the unsaturated (or vadose) zone include: infiltration of water into the soil;[5] its redistribution once inside the soil; water uptake by root fibres;[8] and drying by evaporation. Processes in saturated materials include the flow of water beneath the water table in soil drainage systems, and groundwater flow in aquifers (permeable rock, gravels, or other sediments).

Saturated Flow

In saturated flow in a stable material, Richards' equation (Eq. 10) simplifies, because $\theta = \theta_{sat}$ is now constant in time, so that $\partial\theta/\partial t = 0$. If the water is under positive pressure (e.g., in soil submerged beneath the water table), we replace the suction ψ with the pressure head $H\ (>0)$. Also, for a uniform isotropic material, $K = $ constant. Then Richards' equation, Eq. 10, becomes Laplace's equation:

$$\nabla^2 H = 0 \tag{12}$$

Eq. 12 can be used to solve the flow regime in groundwater systems.[5]

Relevance to Solute Transport

An analog of Richards' equation can also be developed for the movement of solutes in permeable materials. However, the solute transport equations are complicated by the additional flow processes that occur. While Richards' equation describes only the convection (by mass flow) of water in a material, solutes do not just "convect" with the bulk flow of water. Differences in solute concentration also cause solutes to "diffuse" (by molecular diffusion) and "disperse" (via the microscopic irregularities in water flow).[5]

Complications

The above analysis neglects: 1) hysteresis, i.e., the dependence of the $\theta(\psi)$ and $K(\psi)$ relationships on the "history" of how the material reached its current state of wetness, via drying or wetting actions; and 2) anisotropy, or the dependence of the hydraulic conductivity K on flow direction. These topics are discussed in Ref. [5].

REFERENCES

1. Buchan, G.D.; Cameron, K.C. Darcy's Law. In *Encyclopedia of Water Science*; Stewart, B.A., Howell, T.A., Eds.; Marcel Dekker: New York, 2002.
2. Richards, L.A. Capillary Conduction of Liquids in Porous Mediums. Physics **1931**, *1*, 318–333.
3. Gardner, W.R. The Impact of L. A. Richards upon the Field of Soil Water Physics. Soil Sci. **1972**, *113*, 232–237.
4. Bower, C.A. In Recognition of L. A. Richards on the Occasion of His 68th Birthday. Soil Sci. **1972**, *113*, 229–231.
5. Hillel, D. *Environmental Soil Physics*; Academic Press: San Diego, 1998.
6. Cameron, K.C.; Buchan, G.D. Porosity and Pore-Size Distribution. In *Encyclopedia of Soil Science*; Lal, R., Ed.; Marcel Dekker: New York, 2002.
7. Barry, A. A Class of Exact Solutions for Richards' Equation. J. Hydrol. **1993**, *142*, 29–46.
8. Gregory, P.J. Approaches to Modelling the Uptake of Water and Nutrients in Agroforestry Systems. Agrofor. Syst. **1996**, *34*, 51–65.

Ring and Tension Infiltrometers

Dong Wang
University of Minnesota, St. Paul, Minnesota, U.S.A.

Laosheng Wu
University of California, Riverside, California, U.S.A.

INTRODUCTION

Field characterization of soil hydraulic properties is an important first step in solving soil water and solute transport problems. Ponded ring infiltrometers and tension infiltrometers are experimental devices designed for in situ measurement of soil hydraulic parameters. According to Green and Topp,[1] double ring infiltrometer is one of the most popular methods for estimating saturated hydraulic conductivity. Tension infiltrometers are also gaining more popularity in estimating soil hydraulic properties and in documenting macropore and preferential flow.[2] This entry provides a brief overview of the recent development and application of ring and tension infiltrometers for soils and hydrologic studies.

RING INFILTROMETERS

Infiltration is the process of water entering the soil surface. One of the most common devices for measuring infiltration rate and water intake capability at the soil surface is ring (cylinder) infiltrometer made of either metal or plastic and coming in various sizes. Depending on their configuration, ring infiltrometers can be classified as single- and double-ring infiltrometers. In a single-ring infiltrometer, water is filled and the infiltration rate from the ring into the soil is measured. In a double-ring infiltrometer, an outer ring is used to provide a buffer zone to reduce lateral flow so that the inner ring will measure "true" vertical (1-D) flow, and infiltration rate is measured only in the inner ring (Fig. 1). The ponding level in the outer ring is kept as close as possible to the level in the inner ring. The rate of water intake can be measured either manually or automatically using electronic pressure transducers.

A typical infiltration curve has a very high initial infiltration rate due to the high initial hydraulic gradient. As the wetting front extends deeper into soil, the hydraulic gradient decreases with time, and so does the infiltration rate. By assuming that the water flow in a soil profile is a piston-type flow and the soil in the wetted region has a constant water content or matric potential (h_o), hydraulic

conductivity (K_o), water diffusivity (D_o), and matric potential head at the wetting front (h_f), Green and Ampt[3] showed that the infiltration rate (i) can be calculated as

$$i = -K_o \frac{h_f - h_o - L}{L - 0} = \frac{K_o}{L}(\Delta h + L) \qquad (1)$$

where $\Delta h = h_o - h_f$, L is the depth of the wetted zone. It can be shown that the infiltration rate (i) decreases as time (t) increases:[4]

$$i = \Delta \theta (D_o/2t)^{1/2} \qquad (2)$$

where $\Delta \theta = \theta_w - \theta_i$ is the difference in volumetric water content between the wetted zone (θ_w) and the initial profile (θ_i); $D_o = K_o \Delta h / \Delta \theta$ is soil water diffusivity.

Based on the 1-D Richards' equation with approximations, Philip[5] also showed that the infiltration rate can be estimated as:

$$i = \frac{1}{2} S t^{-1/2} + A \qquad (3)$$

where S is called sorptivity, and A is a constant that approaches saturated hydraulic conductivity under ponding conditions.

TENSION INFILTROMETERS

Tension infiltrometers are devices that can be used to estimate soil hydraulic properties and structural characteristics based on infiltration measurement at the soil surface. Depending on soil conditions, the physical appearance of tension infiltrometers can vary considerably. They may be called tension infiltrometers[6] where a relatively small contact disk is used at the soil–water interface or disk permeameters[7] where a relatively large disk is used. The most important difference between ring and tension infiltrometers is that for ring infiltrometers water is usually supplied with a positive head, whereas for tension infiltrometers the infiltration water at the soil–water interface is under tension. Because of the negative pressure head at the soil–water interface, in designing tension infiltrometers, it is usually necessary to place

Encyclopedia of Water Science
DOI: 10.1081/E-EWS 120010216

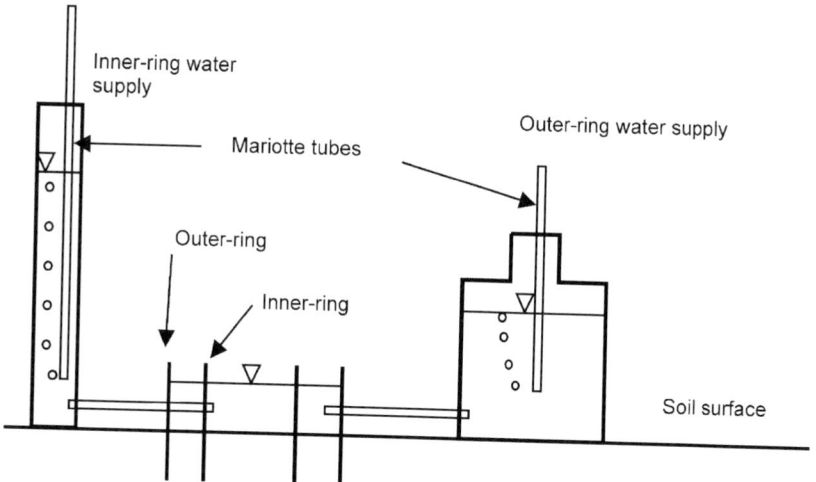

Fig. 1 Schematic of a double-ring infiltrometer.

a porous membrane between the infiltrometer and the soil surface that prevents air entry from this interface into the water supply. Most tension infiltrometers or disk permeameters consist of three components: 1) a circular disk or plate connected to the water supply at the top and to a porous membrane at the bottom; 2) a water supply tube or reservoir that supplies the infiltration water; and 3) a bubbling tube or tower that is connected to the water supply tube for air supply and is used to adjust the water tension at the soil–water interface with a single or multiple air entry tubes with preselected settings for water tension (but only one will be open at one time) (Fig. 2).

The most widely used method for parameter estimation based on tension infiltrometer measurement is to use the approximate steady-state solution of water flow from a shallow circular pond by Wooding:[9]

$$i(h_t) = K_s \left(1 + \frac{4}{\pi r_o \alpha} \right) \exp(\alpha h_t) \qquad \text{for large times} \quad (4)$$

where $i(h_t)$ is the steady-state infiltration rate under a given supply tension h_t, r_o is the radius of the infiltrometer disk, K_s is the soil hydraulic conductivity under saturated conditions, and α is the empirical parameter. Because the only unknowns in this equation are K_s and α, they can be solved by making measurements at a fixed radius with multiple tensions or at a fixed tension with various radii.

Because the determination of steady-state infiltration rate can be subjective and limited by experimental conditions, methods using early-time infiltration data to estimate soil hydraulic properties are needed. Wang et al.[10] proposed an alternative procedure requiring only early-time infiltration data and the measurement of water content increase during the infiltration event. The first step is to solve for soil sorptivity (S) from the transient tension infiltration data using an approximate infiltration equation by Warrick:[11]

$$i(t) \approx S/2\sqrt{t} + D_e S/r_o \qquad \text{for small times} \quad (5)$$

where $i(t)$ is the transient infiltration rate, and D_e is an effective diffusion coefficient, which is a constant for a given set of h_t and r_o. A nonlinear regression between $i(t)$ and $t^{1/2}$ would provide a satisfactory estimation of sorptivity. The second step is to solve for K_s from a relationship developed by Youngs[12] using S and the measured water content increase during the infiltration:

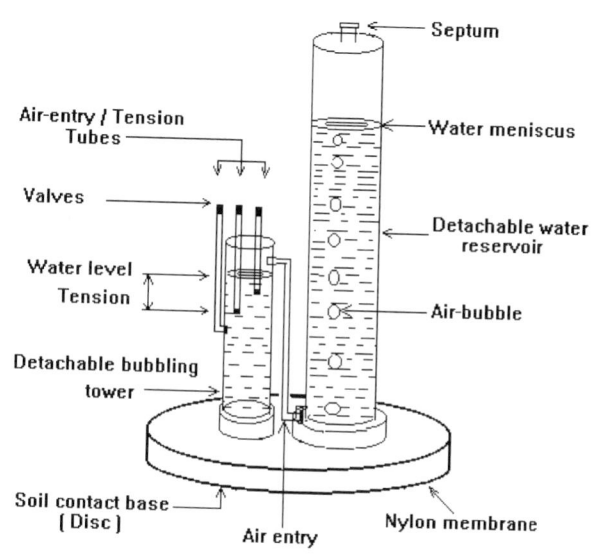

Fig. 2 Schematic of a tension infiltrometer. (From Ref. 8.)

$$K_s = 342.25 \frac{\eta \rho g}{\sigma^2 (\theta_o - \theta_i)^2} S^4 \qquad \text{for small times} \quad (6)$$

where η is water viscosity ($10^{-2}\,\mathrm{g\,cm^{-1}\,sec^{-1}}$), ρ is the density of water ($1\,\mathrm{g\,cm^{-3}}$), g is the gravitational acceleration ($980\,\mathrm{cm\,sec^{-2}}$), σ is the surface tension of water ($72.75\,\mathrm{dyn\,cm^{-1}}$ or $\mathrm{g\,sec^{-2}}$), and θ_o and θ_i are, respectively, the final and initial water content. The final step is to solve for α from White and Sully[13] using the S and K_s values as:

$$\alpha = \frac{(\theta - \theta_i)K_s}{bS^2} \qquad b \approx 0.55 \tag{7}$$

APPLICATIONS AND LIMITATIONS

While the purpose of the ring infiltrometer measurement is to evaluate the water intake capability of a soil, the infiltration rate measured from infiltrometers is not always directly applicable in practice since the field infiltration problems are mostly 1-D or vertical infiltration rate. The flow from a ring into the soil is a 3-D problem.[14–16] Many factors, including ring geometry, soil conditions, and time during the measurement sequence, can affect the vertical infiltration rate measured by ring infiltrometers. Lateral divergence of flow by capillary forces can lead to overestimation of vertical infiltration. The extent (degree) of this overestimation also depends on the ring size. Wu et al.[16] showed that in both single- and double-ring infiltrometers, the possibility of overestimation decreases as the ring size increases. As one might expect, overestimation of vertical infiltration caused by lateral flow is more significant in a fine-textured soil than in a coarse-textured soil. Bouwer[17] indicated that the final infiltration rate gives true vertical infiltration rate correctly only if $h_{cr}/d = 0$, where h_{cr} is called critical matric head and d is ring diameter.

Ring insertion depth and soil layering also affect the infiltration rate measurement. Since the flow from a ring infiltrometer is confined to a vertical direction before the wetting front reaches the ring insertion depth, there can be no overestimation. However, lateral flow occurs when the wetting front passes the insertion depth. Numerical experiment by Wu et al.[16] showed that the infiltration rate decreases as the insertion depth increases for a 12-hr infiltration simulation. Restricting layers deeper in the profile can also cause lateral flow. The significance of this effect depends on the position of restricting layers. In addition, many nonsystematic errors (including soil surface disturbance, water quality, temperature, and biological factors) can influence infiltration measurement.[18]

Theoretically, the true final vertical infiltration rate should be equal to the field-saturated hydraulic conductivity.[18] Reynolds and Elrick[15] developed a solution for steady-state water flow rate from a single-ring infiltrometer by accounting for the soil initial matric potential head, the radius of the ring, and the ring insertion depth. By modifying the Reynolds and Elrick method and using a scaling approach, Wu and Pan[19] developed a generalized infiltration curve for single-ring infiltrometers. By applying their solution to soils with different initial and boundary conditions and rings with various geometries, they found that the dimensionless infiltration curves were close to each other for their test soils. They further applied the generalized infiltration equation to measure the field saturated hydraulic conductivity (K_s) using the infiltration data from single-ring infiltrometers, and found that the K_s values from the new method were comparable with the values measured by other methods.[20]

Tension infiltrometers are commonly used to estimate soil-saturated hydraulic conductivity. Depending on the water supply tension, soil sorptivity and a macroscopic capillary length[2] may also be estimated. While the steady-state method requires the tension infiltration to reach the steady-state rate, alternative methods need accurate measurements of transient infiltration rate for a preselected tension. Experimentally, the decision on when a steady-state flow may have reached is prone to subjective decisions and sometimes it is limited by the total amount of water available in the water supply tube, as in the case of coarse soils. For soils with fine textures or low infiltration rate, infiltrometers with automated recording mechanisms such as the one described by Ankeny et al.[6] may be required because of the extended time needed to reach steady-state flow. Besides the drastic differences in infiltration rate for different types of soil, the size of infiltrometer disk and supply tension also affect the time needed to approach the steady-state condition and possibly the accuracy on parameter estimation.[21] The use of automated recording can provide a more detailed and accurate measurement of the transient infiltration from a tension infiltrometer, which would enable the application of other methods for parameter estimation, such as numerical inversion.[22]

In addition to their application for measuring water flow and soil hydraulic properties, with tracers tension infiltrometers can be used to measure solute exchange coefficients between the mobile and immobile water content.[23,24]

REFERENCES

1. Green, E.E.; Topp, G.C. Survey of Use of Field Methods for Measuring Soil Hydraulic Properties. In *Advances in Measurement of Soil Physical Properties: Bringing Theory into Practice*; Topp, G.C., Reynolds, W.D., Green, R.E.,

Eds.; Soil Sci. Soc. Am. Special Publication no. 30; SSSA: Madison, WI, 1992; 281–288.

2. White, I.; Sully, M.J.; Perroux, K.M. Measurement of Surface-Soil Hydraulic Properties: Disk Permeameters, Tension Infiltrometers, and Other Techniques. In *Advances in Measurement of Soil Physical Properties: Bringing Theory into Practice*; Topp, G.C., Reynolds, W.D., Green, R.E., Eds.; Soil Sci. Soc. Am. Special Publication no. 30; SSSA: Madison, WI, 1992; 69–103.

3. Green, W.H.; Ampt, G.A. Studies in Soil Physics. I. The Flow of Air and Water Through Soils. J. Agric. Sci. **1911**, *4*, 1–24.

4. Jury, W.A.; Gardner, W.R.; Gardner, W.H. *Soil Physics*, 5th Ed.; John Wiley and Sons, Inc.: New York, 1991.

5. Philip, J.R. Theory of Infiltration. Adv. Hydrosci. **1969**, *5*, 215–296.

6. Ankeny, M.D.; Kaspar, T.C.; Horton, R. Design for an Automated Tension Infiltrometer. Soil Sci. Soc. Am. J. **1988**, *52*, 893–895.

7. Perroux, K.M.; White, I. Designs for Disc Permeameters. Soil Sci. Soc. Am. J. **1988**, *52*, 1205–1215.

8. Wyseure, G.C.L.; Sattar, M.G.S.; Adey, M.A.; Rose, D.A. Determination of Unsaturated Hydraulic Conductivity in the Field by a Robust Tension Infiltrometer. http://www.agr.kuleuven.ac.be/facdid/guidow/Tensio.htm, 2002.

9. Wooding, R.A. Steady Infiltration from a Shallow Circular Pond. Water Resour. Res. **1968**, *4*, 1259–1273.

10. Wang, D.; Yates, S.R.; Ernst, F.F. Determining Soil Hydraulic Properties Using Tension Infiltrometers, TDR, and Tensiometers. Soil Sci. Soc. Am. J. **1998**, *62*, 318–325.

11. Warrick, A.W. Models for Disc Infiltrometers. Water Resour. Res. **1992**, *28*, 1319–1327.

12. Youngs, E.G. Estimating Hydraulic Conductivity Values from Ring Infiltrometer Measurements. J. Soil Sci. **1987**, *38*, 623–632.

13. White, I.; Sully, M.J. Macroscopic and Microscopic Capillary Length and Time Scales from Field Infiltration. Water Resour. Res. **1987**, *23*, 1514–1522.

14. Tricker, A.S. The Infiltration Cylinder: Some Comments on Its Use. J. Hydrol. **1978**, *36*, 383–391.

15. Reynolds, W.D.; Elrick, D.E. Ponded Infiltration from a Single Ring: I. Analysis of Steady State Flow. Soil Sci. Soc. Am. J. **1990**, *54*, 1233–1241.

16. Wu, L.; Pan, L.; Roberson, M.J.; Shouse, P.J. Numerical Evaluation of Ring-Infiltrometers Under Various Soil Conditions. Soil Sci. **1997**, *162*, 771–777.

17. Bouwer, H. A Study of Final Infiltration Rates from Cylinder Infiltrometers and Irrigation Furrows with an Electrical Resistance Network. Trans. Int. Congr. Soil Sci. 7th **1961**, *6*, 448–456.

18. Bouwer, H. Intake Rate: Cylinder Infiltrometer. In *Methods of Soil Analysis. Part I*, 2nd Ed.; Klute, A., Ed.; Agron. Monog. 9; ASA and SSSA: Madison, WI, 1986; 825–844.

19. Wu, L.; Pan, L. A Generalized Solution to Infiltration from Single-Ring Infiltrometers by Scaling. Soil Sci. Soc. Am. J. **1997**, *61*, 1318–1322.

20. Wu, L.; Pan, L.; Mitchell, J.; Sanden, B. Measuring Saturated Hydraulic Conductivity Using a Generalized Solution for Single-Ring Infiltrometers. Soil Sci. Soc. Am. J. **1999**, *63*, 788–792.

21. Wang, D.; Yates, S.R.; Lowery, B.; van Genuchten, M.Th. Estimating Soil Hydraulic Properties Using Tension Infiltrometers of Variable Disk Diameters. Soil Sci. **1998**, *163*, 356–361.

22. Simunek, J.; Wang, D.; Shouse, P.J.; van Genuchten, M.Th. Analysis of Field Tension Disc Infiltrometer Data by Parameter Estimation. Int. Agrophys. **1998**, *12*, 167–180.

23. Clothier, B.E.; Kirkham, M.B.; McLean, J.E. In Situ Measurements of the Effective Transport Volume for Solute Moving Through Soil. Soil Sci. Soc. Am. J. **1992**, *56*, 733–736.

24. Jaynes, D.B.; Logsdon, S.D.; Horton, R. Field Method for Measuring Mobile/Immobile Water Content and Solute Transfer Rate Coefficient. Soil Sci. Soc. Am. J. **1995**, *59*, 352–356.

Rural Water Supply, Water Harvesting for

R. H. Mohtar
Purdue University, West Lafayette, Indiana, U.S.A.

F. A. El-Awar
*United Nations Office of the Humanitarian Coordinator for Iraq
(UNOHCI)—Baghdad, New York, New York, U.S.A.*

W. Jabre
American University of Beirut, Beirut, Lebanon

INTRODUCTION

In arid and semiarid regions, the limited availability of
water is the major constraint to agricultural development.
These regions are increasingly suffering from shortage of
water. Moreover, due to human population explosion in
most of these areas, the demand for water continues to
grow. Indeed, of the 9.4 billion expected total world
population by 2050, 8.2 billion will live in developing
countries, of which 3 billion will reside in arid and
semiarid environments.[1] Therefore, competition for
water among different sectors will be heightened, and
the share of water for agriculture will be shrinking. The
limited renewable water resources and the need for higher
agricultural productivity means that developing alternative
water resources is necessary in dry areas.

One of the inefficiently used resources in arid and
semiarid regions is rainfall surface runoff. Rainfall in these
regions is generally low and erratic, and it is characterized
with seasonal and spatial uneven distribution. Due to the
absence of proper management, much of the rainfall is lost
to deep seepage, evaporation, and/or unutilized surface
runoff. The variability in rainfall results in crop
productivity failure and is considered the most common
and unpredictable problem that farmers in arid and
semiarid regions have to face year after year.

The gross volume of rainfall received annually by vast
dry areas may be substantial. Any significant increase in
the quantity of water available to crops in arid and
semiarid regions may improve the reliability and
sustainability of agricultural production systems in these
regions. Such increase can be induced by water harvesting,
which offers an efficient approach for confronting the
seasonality, uneven spatial distribution, and ineffective-
ness of rainfall in dry regions. Among numerous examples
in various locations around the world is a productivity
analysis done for an agricultural area in the hot arid tropics
of India. This study indicates considerable improvement in
gross monetary returns under diversified cropping systems
adopted due to improved dry-land farming technologies
including water harvesting through creation of farm
ponds.[2]

BACKGROUND

Water harvesting is not a new development but rather an
ancient method of water supply.[3–5] Fraiser[6] reported
several examples from the literature of early water
harvesting structures. Researchers found such structures
that date back over 9000 yr in the Edom Mountains in
southern Jordan. Some other evidence was identified of
simple forms of water harvesting practiced in the Ur area
in Iraq at around 4500 BC. The term was probably cited for
the first time in the literature by Geddes in 1963 defining it
as the collection and storage of any form of water, either
runoff or creek flow, for productive use.[7] This definition,
as well as others, focuses on surface runoff as the key
factor in water harvesting, the source of runoff being
mainly rainfall and snowmelt flowing from slopes and in
ephemeral streams.[8]

During the past 40 yr, water harvesting has been
receiving renewed attention.[9] Boers and Ben-Asher[10]
reviewed the achievements in this field during the 1970s.
Research during that period emphasized two aspects:
surface runoff inducement, as well as runoff collection and
conservation. Considerable research was done on methods
to reduce surface storage and infiltration losses. Runoff
farming and the issue of relative sizes of the catchment
area and storage reservoir have recently concentrated
thorough investigation efforts.[11–13] However, very little
was done concerning the more fundamental physical and
hydrologic modeling aspects of water harvesting. Never-

Encyclopedia of Water Science
DOI: 10.1081/E-EWS 120010106

theless, this field is currently capturing significant attention.[14–16]

The goal of this chapter is to provide an overview of various water harvesting methods and to present a case study on the application of reservoir siting in a dry marginal area of Lebanon using the hydro-spatial AHP method that was developed for that purpose.

OVERVIEW OF WATER HARVESTING METHODS

Classification of Water Harvesting Techniques

Boers and Ben-Asher[10] defined three main characteristics of water harvesting. First, it is applied in arid and semiarid regions where runoff is intermittent; second, it depends upon local water such as surface runoff, creek flow, springs, and soaks; and third, it is relatively small-scale in terms of catchment area, storage volume, and investment. Other important elements in the definitions of water harvesting are the form of runoff, the use of runoff, and the harvesting technique itself.[8]

Most authors have their own classifications of water harvesting techniques. However, there is a general consensus in the literature[8] that the methods of water harvesting can be divided into two main categories:

Macrocatchment, or runoff farming, where surface runoff water collected from a relatively large area is conveyed by means of small channels, waterways, and/or small diversion dams to a storage reservoir or to a cultivated field.

Microcatchment, where a within-field system is used to harvest water for one or several trees or bushes from a relatively small area.

According to Fraiser,[6] there are three basic types of water harvesting systems: the Direct Water Application System (DWAS), the Supplemental Water System (SWS), and the Combination System (CS). In the DWAS the runoff water is stored in the soil profile of the crop growing area during the precipitation event. This approach includes two major configurations:

Floodwater irrigation, a system in which runoff is diverted directly from small natural drainage channels using water spreading techniques.

Microcatchment irrigation, where a microcatchment consists of a small prepared runoff collecting area directly upslope of the growing area.

In the SWS, the collected water is stored offsite in a storage facility and applied later to the crop area using some form of irrigation system. In the CS, the runoff water is applied first to the crop area, where some water infiltrates into the soil profile, and then the excess water is diverted into a storage facility for later application.

Water Harvesting Planning and Design

There is no standard procedure to simply select a water harvesting technique for immediate implementation. Siegert[9] stated that apart from basic technical considerations concerning topography and soil characteristics, the selected method must be compatible with local lifestyles, social systems, and willingness of the beneficiaries to adopt it. Unless people are actively involved in development projects that are aimed to help them, such projects are doomed to fail. Samra et al.[17] considered that rainwater harvesting systems should be site-specific, environmentally sound, compatible with indigenous traditional knowledge, present minimum social conflicts, and meet multiple objectives.

A prerequisite to designing any water harvesting system is the identification of suitable areas for applying water harvesting technology. Site selection for the needed storage facilities is the most important step in water harvesting system design. It requires the knowledge of climate, topography, natural vegetation cover, land use, soil characteristics, agricultural practices and socio-economics of the area.[18] El-Awar et al.[15] added that comprehensive hydrologic analysis of candidate watershed(s) is a key element in successful planning of water harvesting systems.

Collection of the necessary data and information has been traditionally performed through available reports, maps, or other sources such as surveys and field investigations. However, on regional level the accomplishment of data collection can hardly be made by such traditional means because it would be too expensive and time consuming. Remote Sensing (RS), defined as the science of deriving information about Earth resources from satellite images and aerial photos, can be utilized for timely and accurate data collection. The main advantage of satellite imagery is that it provides information about natural resources and land cover patterns quickly and inexpensively. Digital Geographic Information Systems (GIS) designed for data mapping, displaying, management, and analysis can improve the decision making process in water harvesting system planning by conducting analyses otherwise rendered impractical and infeasible. GIS-hydrologic model combinations can be used for predicting surface runoff and other hydrologic parameters needed for water harvesting design purposes.

In the context of water harvesting planning, Giraldez et al.[19] mentioned that several hydrologic models have been proposed for evaluating water harvesting systems from the simple Shanan and Schick[20] model to the more detailed model of Illangasekare and Morel-Seytoux.[21] Kutsh[22] introduced an equation based on empirically derived parameters to approximately determine the quantity of water held back by water harvesting structures. These empirically derived parameters are the water harvesting area (WHA), the water losses in the WHA, and the water losses from the terraces and slopes of the command area. Karmieli et al.[23] developed an empirical model to predict runoff yield in the Negev Desert. The equation assumes a linear relationship between annual rainfall and runoff in a given watershed taking into account the reduction in runoff efficiency with the increase in catchment size. Samra et al.,[17] while presenting the process of hydrologic design for water harvesting structures, mentioned several methods to estimate the peak runoff and the runoff volume. For the computation of peak runoff, they recommend the use of the Rational Method as well as several other empirical formulae developed under different situations. With respect to the computation of runoff volume, they suggested the use of the SCS Curve Number Method. El-Awar et al.[16] used RS and GIS techniques conjunctively with hydrologic modeling and the Analytic Hierarchy Process (AHP) for siting small water harvesting reservoirs in dry marginal areas.

In summary, water harvesting planning consists of the following steps:

Collection of needed data on hydrology, soil characteristics, land cover, and topography of the investigated area using technologies such as RS.

Utilization of a computer based analytical environment for data capturing, storage, manipulation, and analysis such as GIS.

Comprehensive hydrologic analysis of the investigated area based on collected hydrologic data and hydrologic modeling.

Use of decision making tools for evaluating the different alternatives of water harvesting systems, including site selection and storage volume.

HYDRO-SPATIAL AHP METHODOLOGY FOR SITING WATER HARVESTING RESERVOIRS IN DRY AREAS: A CASE STUDY

Hydro-Spatial AHP is a methodology for locating and ranking suitable sites for small water harvesting reservoirs. This methodology is based on quantifying the overall site suitability for such reservoirs through a Reservoir Suitability Index (RSI) calculated for potential candidate sites. This index is developed using hydrologic modeling in conjunction with GIS and AHP. The resulting procedure excludes sites where reservoirs cannot be built, due to any physical constraints and/or restrictive land use policies and regulations, and ranks the rest of the sites based on their respective RSI values.

AHP, originally developed and introduced by Saaty,[24] has been widely used for quantitative assessment and ranking of different alternatives.[25] GIS technology is used in this work for building and managing the needed digital spatial database to provide the site attributes required for the decision process. Hydrologic modeling is used to determine the potential runoff volume that represents a major decision criterion in the hierarchical ranking process of different candidate sites.

Site attributes, related to different decision criteria, are determined through hydrologic modeling and GIS applications. Both techniques are used simultaneously for estimating the necessary spatial hydrologic parameters. The AHP decision procedure uses the calculated attributes in order to rank potential sites based on their suitability for water harvesting reservoirs.

The RSI is based on a set of selection criteria defined by experts and discussed below. The methodology used for RSI computation can be represented by the following steps: 1) identification of selection criteria; 2) development of a hierarchy structure; 3) deciding on the Relative Weights (RWs) of elements in different levels of the hierarchy structure; 4) determination of related site attributes through GIS and hydrologic modeling; 5) calculation of the RSI for all tested locations; and 6)

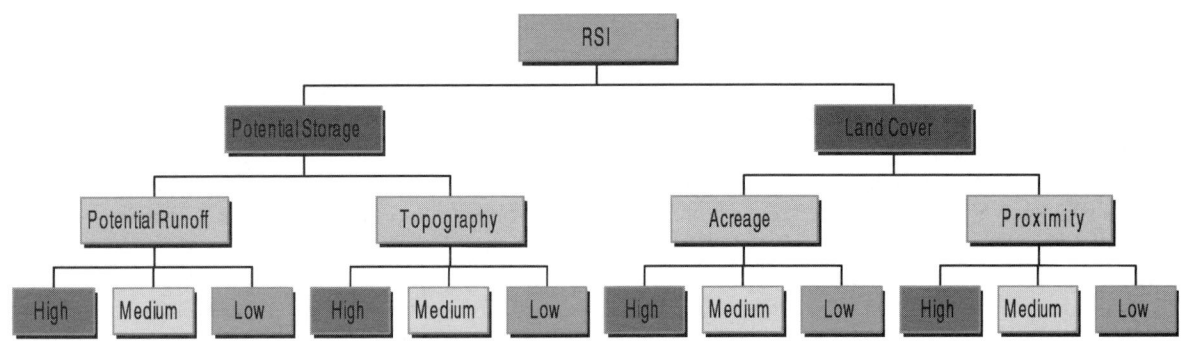

Fig. 1 Hierarchy structure for water harvesting reservoir siting.

ranking these locations based on the calculated values of their indexes.

(1) Selection criteria: The first step is to define the selection criteria. Such criteria are used to compute the RSI and rank the sites that are under investigation. The major selection criteria in this case study are potential storage and land cover characteristics of candidate sites.

(2) Decision hierarchy structure: The selection criteria are arranged in a multilevel hierarchical decision structure. The first level of this structure represents the ultimate objective of the decision process. The major selection criteria are placed in the second level of the hierarchy structure. These major criteria are further detailed and categorized into different subcriteria within subsequent higher levels of the structure. The highest level contains attributes or attribute classes that are determined through hydrologic modeling and GIS applications. Classification of attribute values into a finite number of classes would save on the efforts needed for the evaluation of a large number of tested sites.

The hierarchy structure that is developed for this work is shown in Fig. 1. Its first level contains the RSI, which is calculated for all candidate sites. The major decision criteria that are used to calculate the RSI value are arranged in the second level of the structure. These criteria are the potential storage and land cover characteristics of the candidate sites. In the third level, potential surface runoff as well as topographic and soil characteristics are assessed. These represent the component subcriteria that are used to evaluate the potential storage. Land cover is subdivided into two level-3 subcriteria: the proximity to benefiting agricultural lands, and the total acreage of such benefiting areas. The fourth level, which is the last level, contains the attribute classes that are related to the major criteria and subcriteria of the second and third levels, respectively. The values of these attributes are extracted from the developed GIS and hydrologic model for the area. These values are grouped into a number of classes based on value ranges of different selection criteria and subcriteria.

(3) RWs: Related selection criteria, subcriteria, and attribute classes are compared to each other in pairs in order to develop RWs for all elements in the decision hierarchy structure. All pairs of attribute classes that belong to the same subcriterion are compared to each other. Experts qualitatively judge all attribute classes for their RWs in influencing the corresponding subcriterion in the neighboring upper level of the hierarchy structure. This qualitative judgment is quantified by fitting different degrees of preference, or significance, into a numerical scale. Table 1 shows the scale used to represent different preference degrees in this case study.

The degrees of preference of certain classes over the others are represented by their corresponding numerical values that fill the cells of a decision matrix. The decision matrices of all comparisons of attribute classes of level 4 are shown in Table 2.

The matrix provides a format for quantitatively comparing the weight or the importance of each attribute class relative to other classes that belong to the same subcriterion. The numerical values in the matrix cells represent the preference of one attribute class against another. For example, the first row of Table 2 shows that high potential runoff level is preferred five times and nine times more than the medium and low levels, respectively. It is noted that the diagonal cells of the decision matrices are always filled with values of unity because they represent the self-comparison of attribute classes. The RW of an attribute class is considered as the normalized eigenvalue of the class row within the comparison matrix. The eigenvalue is calculated as the Nth root of the product of all the elements of the class row, where N is the total number of elements in that row. The computed eigenvalue is normalized by dividing it by the summation of the eigenvalues of all the rows of the matrix.[26] The RWs of the subcriteria and major criteria applied in this work have been developed using the same procedure. The corresponding decision matrices are presented in Tables 3 and 4.

(4) Site attribute determination: The site attributes related to different selection subcriteria are determined through GIS and hydrologic modeling applications. A digital GIS is built for the entire investigated area. Different data layers of the GIS database are used directly in computing the needed site attributes. The GIS database is also used indirectly in this process by providing the needed input data for the hydrologic model to determine

Table 1 Numerical scale for qualitative preferences

Degree of preference	Corresponding numerical value
Indifference	1
Weak preference	3
Strong preference	5
Very strong preference	7
Absolute preference	9

Table 2 Decision matrix for attribute classes of level 4 of the hierarchy structure

	High	Medium	Low	Eigenvalue	RW
High	1	5	9	3.557	0.735
Medium	1/5	1	5	1.000	0.207
Low	1/9	1/5	1	0.281	0.058

Table 3 Decision matrix for level-3 subcriteria of the hierarchy structure

	Runoff	Topography	Soil	Eigenvalue	RW
Runoff	1	8	9	4.160	0.798
Topography	1/8	1	3	0.721	0.138
Soil	1/9	1/3	1	0.333	0.064
	Area	Proximity		Eigenvalue	RW
Area	1	7		2.65	0.875
Proximity	1/7	1		0.38	0.125

the potential surface runoff at the candidate sites. The developed site attribute classes are assigned their respective RWs that are used to calculate the RSI values at different sites.

(5) RSI computation: The RSI of a potential reservoir site is computed through the application of the following formula:

$$\text{RSI} = \sum_{i=1}^{N_2} \text{RW}_i \left[\sum_{j=1}^{Ni_3} \text{RW}_j \text{RW}_k \right] \qquad (1)$$

where RW_i is the relative weight of level 2 major criterion i, RW_j is the relative weight of level 3 subcriterion j, RW_k is the relative weight of level 4 attribute class k, N_2 is the total number of level 2 major selection criteria, Ni_3 is the total number of level 3 subcriteria that belong to level 2 major criterion i.

The above equation represents the ratings approach of the AHP decision process.[26] The RWs of each group of level 3 subcriteria are multiplied by the RWs of their respective level 4 attribute values, aggregated together, and multiplied by the RW of the corresponding major selection criterion of the second level of the hierarchy structure. This equation is applied to calculate the RSIs of all potential sites of water harvesting reservoirs.

(6) Ranking of potential reservoir sites: The computed RSI values are grouped into several classes, and the investigated potential sites are ranked based on their respective RSI classes. Ranking the potential sites with respect to reservoir suitability helps in assigning priorities for different sites in the terminal stages of the decision process. The final phase of this process consists of producing an RSI map showing the ranks of all potential reservoir sites under analysis.

APPLICATION AND RESULTS

Pilot study area: This study focuses on Irsal, a remote Lebanese highland region located in the northeastern dry marginal lands of the western slopes of the anti-Lebanon mountain range. The region is characterized by its semiarid weather with dry hot summers and cold winters, and its average annual precipitation is about 300 mm. This low precipitation depth, coupled with its nonuniform temporal and spatial distributions, has a magnified effect on the water resources budget in the region.

Decision criteria: All decision criteria and their respective RWs used in this work were based on indigenous knowledge and expertise in the pilot area as well as relevant literature.[15,27] Threshold values that define attribute class limits within the fourth level of the decision hierarchical structure (Fig. 1) were selected on the same bases as well. An interactive participatory approach was followed to make use of local farmers' experience to improve different criteria and attribute class limits extracted from the literature. Local expertise had also been heavily used in assigning the RWs of different attribute classes in the hierarchy structure.

All subwatershed outlets in the study area were considered as potential reservoir sites. The selection criteria were used within the hierarchy structure to calculate the RSIs of the potential sites under consideration. The land cover criterion in the decision hierarchy structure represents the proximity of a potential reservoir site to stone fruit orchards and other agricultural lands in the area, as well as the acreage of these benefiting areas. In other words, this decision criterion assesses the potential site based on the need for water in its vicinity.

Table 4 Decision matrix for level-2 major criteria of the hierarchy structure

	Potential storage	Land cover	Eigenvalue	RW
Potential storage	1	4	2	0.800
Land cover	1/4	1	0.5	0.200

The potential storage decision criterion is composed of the site topographic and soil characteristics, in addition to potential runoff component subcriteria. Potential surface runoff within individual subwatersheds was estimated by means of the Soil Conservation Service (SCS) Curve Number method. Watershed Modeling System (WMS), a comprehensive hydrologic modeling environment with GIS capabilities, was used to determine these estimates. A composite (area-weighted) curve number was derived for each subwatershed. The needed hydrologic and basin data were extracted from topographic, subwatershed, and land cover GIS overlays that were imported into WMS and processed for that purpose. HEC-1 model, interfaced with WMS, was used for runoff calculations. Runoff volumes from individual storms were determined, then routed and summed to estimate the potential annual runoff for all subwatershed outlets. Based on their individual values, these potential annual runoff volumes were classified into high, medium, and low classes.

The topographic and soil characteristics subcriteria were used in this study as indicators of the site potential storage capacity. Candidate sites with very steep or very mild slopes were considered to have relatively low storage capacities. Prevailing slopes of the investigated subwatershed outlets were determined from the topographic and subwatershed digital maps. Soil characteristics classification was based on the sites' soil texture and clay content.

The needed information was extracted from relevant GIS data layers, and three soil permeability classes—high, medium, and low—were considered. Subwatershed outlets found to be on cracked limestone layers were excluded from further consideration.

Ranking of potential sites: The RWs of the attribute classes developed from GIS and hydrologic modeling applications were used within the decision hierarchy structure, and Eq. 1 was applied to calculate RSI values for all subwatershed outlets. Finally, RSI map was created for the considered RW combinations. Fig. 2 presents the RSI map that was developed. Five reservoir suitability classes, based on individual RSI values, were assigned for the considered subwatersheds. The highly suitable class was given the first rank (class A) and the suitable, moderately suitable, and weakly suitable classes were given the second, third, and fourth rank, respectively. The subwatersheds of the excluded sites were ranked as nonsuitable (class E).

CONCLUSION

The Hydro-Spatial AHP methodology for small water harvesting reservoir siting combines the capabilities of GIS, hydrologic modeling, and AHP approaches. The application of the methodology shows that it works

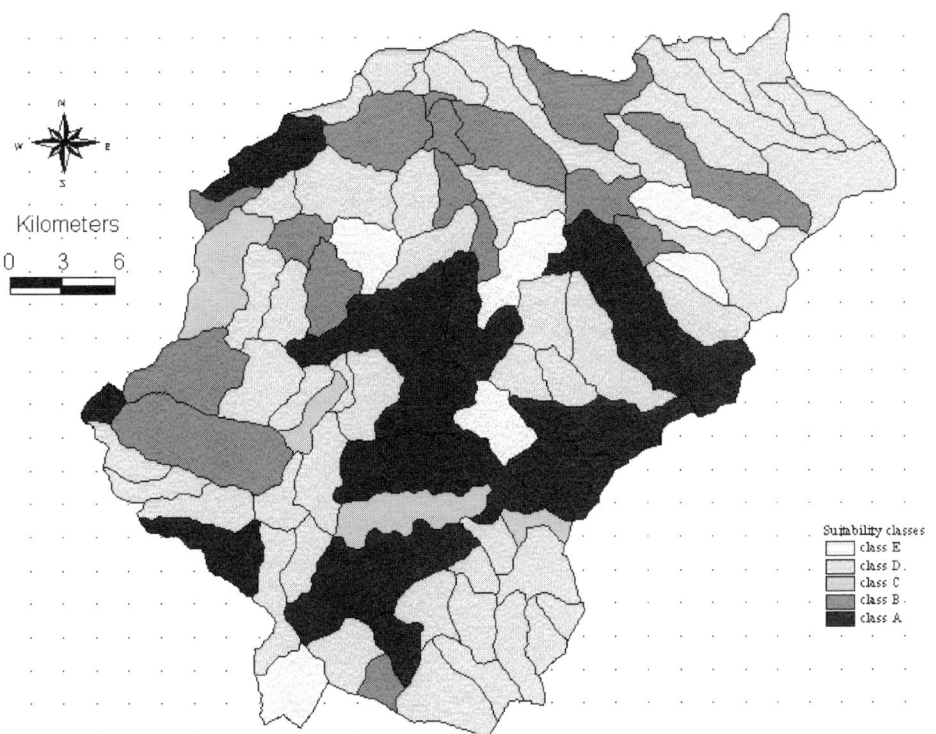

Fig. 2 RSI map.

efficiently for siting small water harvesting reservoirs. Moreover, the methodology is highly flexible regarding the number, types, threshold values, and RWs of decision criteria on which the reservoir siting process is based.

The use of the same clearly defined hierarchical structure of decision criteria to rank all candidate sites insures the general objectivity of the methodology. However, the development of the criteria RWs is based on subjective expert preferences. Therefore, special care should be taken in developing these RWs that should always be defendable and subject to cross checking.

REFERENCES

1. Lal, R. Soil Management in the Developing Countries. Soil Sci. **2000**, *165* (1), 57–72.

2. Bhati, T.K.; Goyal, R.K.; Daulay, H.S. Development of Dryland Agriculture on Watershed Basis in Hot Arid Tropics of India: A Case Study. Ann. Arid Zone **1997**, *36* (2), 115–121.

3. Fraiser, G.W. Harvesting Water for Agricultural, Wildlife, and Domestic Uses. J. Soil Water Conserv. **1980**, *35*, 125–128.

4. Emmerich, W.E.; Fraiser, G.W.; Fink, D.H. Relation between Soil Properties and Effectiveness of Low-Cost Water-Harvesting Treatments. Soil Sci. Soc. Am. J. **1987**, *51*, 213–219.

5. FAO. *Water Harvesting for Improved Agricultural Production. Water Reports # 3*, Proceedings of the FAO Expert Consultation, Cairo, Egypt, Nov 21–25, 1993; FAO: Rome, 1994.

6. Fraiser, G.W. Water Harvesting/Runoff Farming Systems for Agricultural Production. In *Water Harvesting for Improved Agricultural Production. Water Reports # 3*, Proceedings of the FAO Expert Consultation, Cairo, Egypt, November 21–25, 1993; FAO: Rome, 1994; 57–72.

7. Myers, L.E. *Water Harvesting and Management for Food and Fiber Production in the Semi-arid Tropics*; AR/USDA: Berkeley, CA, 1975.

8. Reij, C.; Mulder, P.; Begemann, L. *Water Harvesting for Plant Production*; World Bank Technical Paper # 91, The World Bank: Washington, DC, 1991.

9. Siegert, K. Introduction to Water Harvesting: Some Basic Principles for Planning, Design and Monitoring. In *Water Harvesting for Improved Agricultural Production. Water Reports # 3*, Proceedings of the FAO Expert Consultation, Cairo, Egypt, November 21–25, 1993; FAO: Rome, 1994; 9–21.

10. Boers, T.M.; Ben-Asher, J. A Review of Rainwater Harvesting. Agric. Water Manag. **1982**, *5*, 145–158.

11. Oweis, T.Y.; Taimeh, A.Y. Evaluation of a Small Basin Water-Harvesting System in the Arid Region of Jordan. Water Resour. Manag. **1996**, *10*, 21–34.

12. Abu Awwad, A.M.; Shatanawi, M.R. Water Harvesting and Infiltration in Arid Areas Affected by Surface Crust: Examples from Jordan. J. Arid Environ. **1997**, *35* (3), 443–452.

13. Van Wesemael, B.; Poesen, J.; Sole Benet, A. Collection and Storage of Runoff from Hill Slopes in a Semi-arid Environment: Geomorphic and Hydrologic Aspects of the Aljibe System in Almenia Province, Spain. J. Arid Environ. **1998**, *40*, 1–14.

14. Sharma, K.D.; Murthy, J.S.R. Ephemeral Flow Modeling in Arid Regions. J. Arid Environ. **1996**, *33*, 161–178.

15. El-Awar, F.A.; Barry, B.Y.; Mohtar, R.H. *Hydrologic Analysis of Dry Marginal Watersheds in Northeast Lebanon*, Proceedings of the XIIIth International Congress on Agricultural Engineering, Rabat, Morocco, February 2–6, 1998; CIGR, 1998; Vol. 1, 35–47.

16. El-Awar, F.A.; Makke, M.K.; Zurayk, R.A.; Mohtar, R.H. A Spatial-Hierarchical Methodology for Water Harvesting in Dry Lands. Appl. Eng. Agric., ASAE **2000**, *16* (4), 395–404.

17. Samra, J.S.; Sharda, V.N.; Sikka, A.K. *Water Harvesting and Recycling: Indian Experiences*; Central Soil and Water Research and Training Institute: Dehradun, India, 1996.

18. Oweis, T.Y.; Oberle, A.; Printz, D. Planning Water Harvesting Systems Using Remote Sensing and GIS. Proceedings of a Regional Seminar on The Use of Decision Support Systems in Water Resources Management, Damascus, Syria, October 28–30, 1996; CEDARE: Damascus, Syria, 1996; 25–35.

19. Giraldez, J.V.; Ayuso, J.L.; Garcia, A.; Lopez, J.G.; Roldan, J. Water Harvesting Strategies in the Semiarid Climate of Southeastern Spain. Agric. Water Manag. **1988**, *14*, 253–263.

20. Shanan, L.; Schick, A.P. A Hydrologic Model for the Negev Desert Highlands: Effect of Infiltration, Runoff, and Ancient Age. Hydrol. Sci. Bull. **1980**, *25*, 269–282.

21. Illangasekare, T.H.; Morel-Seytoux, H.J. Design of Physically-Based Distributed Parameter Model for the Arid Zone Surface-Groundwater Management. J. Hydrol. **1984**, *74*, 213–232.

22. Kutsh, H. Currently Used Techniques in Rainfed Water Concentrating Culture, the Example of the Anti-Atlas. Appl. Geogr. Dev. **1983**, *21*, 108–117.

23. Karmieli, A.J.; Ben-Asher, J.; Dodi, A.; Issar, A.; Oron, G. An Empirical Approach for Predicting Runoff Yield Under Desert Conditions. Agric. Water Manag. **1988**, *14*, 244–253.

24. Saaty, T.L. *The Analytic Hierarchy Process*; McGraw Hill, Inc.: New York, NY, U.S.A., 1980.

25. Erkut, E.; Moran, R. Locating Obnoxious Facilities in the Public Sector: An Application of the Analytic Hierarchy Process to the Municipal Landfill Siting Decision. Socio-Econ. Planning Sc. **1991**, *25* (2), 89–102.

26. Siddiqui, M.; Everett, J.; Vieux, B. Landfill Siting Using Geographic Information Systems: A Demonstration. J. Environ. Eng. **1996**, *122* (6), 515–523.

27. Vorhauer, C.; Hamlett, J. GIS: A Tool for Siting Farm Ponds. J. Soil Water Conserv. **1996**, *51* (5), 434–438.

Sacramento–San Joaquin Delta

Mark J. Roberson
Keller–Bleisner Engineering, LLC, Sacramento, California, U.S.A.

INTRODUCTION

The Sacramento–San Joaquin Delta is the largest estuary on the west coast of North America. It serves as a major water supply conveyance facility for over 20 million Californians; provides habitat for many species of birds, mammals, and plants; and supports agricultural and recreational activities. It is the terminus of several primary rivers: the Sacramento, the San Joaquin, and the eastern tributaries of the Cosumnes, Mokelumne, and Calavaras. These watersheds drain 40% of the State's land area. Recreational uses of the Delta include boating, fishing, hunting, and wildlife viewing. Agricultural uses include the irrigation of a wide variety of tree, vine, vegetable, and row crops.

DELTA LANDS

The Delta of California covers an area of about 700,000 acres with about 60% of the land use dedicated to agriculture. In Table 1, current land use in the Delta is compared with historic native state. Following the 1849 gold rush many miners became farmers and through reclamation turned much of the Delta's swampland into productive agricultural lands. Due to the continual threat of flooding, more and more time and money was spent protecting lands through levees and drainage systems. Currently there are 57 reclaimed islands surrounded by levees designed to prevent high flow events from inundating low-lying areas. Large areas of the Delta are covered by organic soils that are up to 60 ft deep. Oxidation of organic soils has resulted in as much as a 15-ft loss in elevation.[1] Due to the drop in land elevation, there is increased pressure on the levees during flooding events.

WATER DEVELOPMENT

The first large-scale impact to the Delta was from sediment that originated from hydraulic gold mining in the Sacramento River watershed. The federal government stopped this form of mining in the 1884 but the residual transport of sediments continues. In addition to sediments, mercury, which was used to extract gold also made its way

into the Delta. In some Delta channels, the sediment filled the river bottom channels to the point where navigation was impossible. The bed of the Yuba River, a tributary to the Sacramento River, was raised some 60 ft over a period of less than 30 yr.[2] Sediments also reduced the capacity for the Delta to contain floodwater and thus enhanced the need for levee protection.

In 1921 the State initiated the first water plan to address water supply and flooding issues. This plan was partially carried out by the Federal government through the Central Valley Project in 1933 and the State Water Project in 1951. Combined, these projects deliver, through pumping plants or direct diversions, over 7 million acre-ft of water annually (Table 2) to customers north and south of the Delta. Much of the controversies surrounding the Delta begin with how to use the Delta for conveying and protecting the drinking water supply and restoring its ecological health.

Flow control on the upstream watersheds and the construction of upstream and Delta levees has altered channel hydraulics and eliminated nutrient enriching flood-events. The Delta levees keep water moving through the Delta, preventing flooding of the natural floodplain and subsequent deposition of sediments (Fig. 1).

WATER QUALITY

Water quality in the Delta is influenced by two sources: the San Francisco Bay and the upstream watersheds. Tidal exchanges in the Bay-Delta bring salt water deep into the upper reaches of the Delta. Prior to the development of the Central Valley Project and the State Water Project, salt water would travel as far as Sacramento—nearly 70 mi upstream of the San Francisco Bay. With the advent of flow control on the rivers feeding the Delta, the extent of tidal action is much less and salinity intrusion is confined to only a small portion of the Delta. This has helped the water users maintain consistent water quality but it has removed an important component of the estuary's ecosystem. The water quality in the Sacramento, Cosumnes, Mokelumne, and Calavaras watersheds is excellent, however, multiple urban and agricultural use impairs the water quality as it reaches the Delta. Water quality in the San Joaquin River watershed is heavily impaired by agricultural practices on the west side of

Encyclopedia of Water Science
DOI: 10.1081/E-EWS 120010129

Table 1 Acres of emergent marsh in regions of the Sacramento–San Joaquin Delta

Region	Historic—1906	Current—1993	Percent change
North	53,660	4,460	− 91
East	7,600	1,270	− 83
South	470	650	38.3
Central and West	31,170	5,040	− 86
Total	98,900	11,600	− 89

Source: CALFED Bay-Delta Program, from USGS maps.

Fig. 1 Salinity control gates at the Suisun Marsh protect the brackish water marsh in the west side of the Delta. The Suisun Marsh is the largest contiguous water marsh in the United States and contains 12% of California's remaining natural wetlands.[1] (Photo by Mark J. Roberson.)

the San Joaquin Valley with the main constituents of concern being naturally occurring.

One of the engineered functions of the Delta is to serve as a conveyance facility, moving water from the Sacramento Valley to the west side of the San Joaquin Valley and Southern California. Water moving through the Delta is pumped into the Delta–Mendota or the California Aqueduct at maximum rate of 11,000 cubic ft per second. During periods of low inflow to the Delta (summer, following a low precipitation winter) the export pumps can dewater many of the channels and sloughs. This dewatering results in ocean water or poor quality San Joaquin River water entering the export pumps and subsequently being delivered to the west side and Southern California.

Upstream water quality issues are typical of most major watersheds in the United States: nutrients, sediments and pesticides from agriculture operations, dairies and urban runoff containing household chemicals move downstream into the Delta. However, the drainage from the Westside of the San Joaquin Valley may pose the greatest challenge due to selenium, boron, and elevated levels of salinity. Unlike farm chemicals, selenium and boron are both naturally derived from

the native soil that was formed from marine sediments. Common agricultural practices on the Westside soils include irrigation that solubilizes selenium and boron coupled with subsurface drainage that quickly moves the constituents to the San Joaquin River. In 1981 the U.S. Fish and Wildlife Service noticed deformities among the bird population in wetland refuges that received drainage water from the Westside. The deformities were eventually traced back to the elevated selenium concentrations that were as high as 1000 ppm when the known toxicity level to birds is less than 10 ppb. Although great strides have been made in reducing the concentration of selenium in drainage water, the levels are still too high for resident bird populations. The long-term effects of elevated selenium levels on Delta resources is still under investigation.[4]

Another constituent of concern is mercury that was used in gold mining operations in the Sacramento River watershed. Transported on sediments to the Delta the mercury containing sediments are found in marshes, on Delta islands, and in protection levees constructed using dredge material. Although the elemental form of mercury is used to extract gold, biotransformations of released mercury make it more available for biological uptake and food chain accumulation.

ECOLOGICAL RESOURCES

In the early 1900s striped bass and shad were introduced and have proliferated due to the abundant food sources. In the late 1990s the Chinese Mitten crab, thought to have originated from ship ballast, multiplied quickly feeding on indigenous Delta species. Plant life in the Delta is a diverse

Table 2 Delta flow components and comparisons, 1000 acre-ft for the period of 1980–1991

Inflows	Outflows
Sacramento River; 17,220	Outflow to Bay (21,020)
San Joaquin River; 4,300	Tracy Pumping Plant (2,530)
East Side Rivers; 1,360	Harvey Banks Pumping Plant (2,490)
Precipitation; 990	Consumptive use and channel percolation (1,690)
Yolo Bypass; 3,970	Contra Costa Pumping Plant (110)
Total; 27,840	Total (27,840)

(From Ref. 3.)

mixture of riparian scrub and woodland as well as emergent and seasonal wetlands. Migratory and resident bird populations are prolific due to the abundant food sources and protective habitat.

Delta fish species are numerous with some completing their life cycle within the Delta whereas others may simply pass through to upstream rearing habitat. The Delta smelt and the split-tail are two species that complete their lifecycle within the Delta. The split-tail rely on shallow warmer water for feeding and reproduction whereas the smelt require high freshwater flows particularly during the late winter and spring. Due to upstream flow control and export pumping, these habitats have been dramatically reduced. There are numerous salmon runs in all the primary watersheds.

The numerous channels, sloughs, and islands within the Delta serve as nesting and feeding grounds for residential and migratory birds. All anadramous fish from these watersheds must migrate through, or spawn in, the Delta.

RESTORATION EFFORTS

The CALFED Bay-Delta Program was established in 1995 to address ecosystem, water quality, water supply reliability, and levee and channel integrity issues of the Bay-Delta system. The Ecosystem Restoration element of CALFED lists the following stressors to the Delta's ecosystem: water diversions, channelization, levee maintenance, flood protection, rock placement for shoreline protection, poor water quality, legal and illegal harvest, wake and wake erosion, agricultural practices, conversion of agricultural lands to vineyards, urban development, habitat loss, pollution and the introduction of nonnative plants and animals.

IMMEDIATE ISSUES OF CONCERN

The CALFED Bay-Delta program has identified nine major issues, requiring immediate attention during the first seven years (2000–2007) of program implementation. To refine the understanding of the issues, adaptive management during project implementation is envisioned.

- The impact of introduced species and the degree to which they may pose a significant threat to reaching restoration objectives.
- Recognition that channel dynamics, sediment transport, and riparian vegetation are important elements in

a successful restoration program and the need to identify which parts of the system can be restored to provide the desired benefits.

- Development of an alternative approach to manage floods by allowing rivers access to more of their natural floodplains and integrating ecosystem restoration activities with the Army Corps of Engineers' Comprehensive Study of Central Valley flood management programs.
- Increasing the ecological benefits from existing flood bypasses, such as the Yolo Bypass so that they provide improved habitat for waterfowl, fish spawning and rearing, and possibly as a source of food and nutrients for the estuarine foodwebs.
- Thoroughly testing the assumptions that shallow water tidal and freshwater marsh habitats are limiting the fish and wildlife populations of interest in the Delta.
- A better understanding of the underlying mechanisms of the X2 salinity standard in the Delta and the resultant effects on aquatic organisms.
- A need to better understand the linkage between the decline at the base of the estuarine foodweb and the accompanying decline of some, but not all, species and trophic groups.
- Clarifying the extent to which entrainment at the Central Valley Project and State Water Project pumping plants affects the population size of species and invertebrates.
- Clarifying the suitability and use of the Delta for rearing by juvenile salmon and steel head.

REFERENCES

1. CALFED Bay-Delta Program, Ecosystem Restoration Program Plan. Vol 2: Ecological Management Zone Visions. Final Programmatic EIS/EIR Technical Appendix. July, 2000; 73.
2. Kelly, R.L. *Battling the Inland Sea*; University of California Press: Berkeley, CA, 1986; 72–74.
3. California Department of Water Resources, *Sacramento Delta San Joaquin Atlas*; California Department of Water Resources: Sacramento, CA, 1993; 18–19.
4. Letey, J.; Roberts, C.; Penberth, M.; Vasek, C. An Agricultural Dilemma: Drainage Water and Toxics Disposal in the San Joaquin Valley. *University of California Agricultural Experiment Station Spec. Publ. 3319*; University of California Division of Agriculture and Natural Resources: Oakland, CA, 1986.

Saline Seeps

Ardell D. Halvorson
United States Department of Agriculture (USDA), Fort Collins, Colorado, U.S.A.

INTRODUCTION

Saline seeps occur frequently in dryland farming areas throughout the Great Plains of North America and southern and western Australia and have been reported elsewhere.[1,2] This article addresses saline seeps in the Great Plains.

WHAT IS A SALINE SEEP?

Saline seep describes a soil salinization process resulting from dryland (rainfed) farming practices that allow water from precipitation to move through salt laden subsoils in the recharge area that eventually resurfaces at a downslope topographic area (Fig. 1). Saline seeps have intermittent or continuous saline water discharge at or near the soil surface downslope from recharge area. Crop growth in the saline seep area is reduced or eliminated because of increased soil salinity (Fig. 2). Saline seeps differ from other saline soil conditions by their recent and local origin, saturated root zone, shallow water table, and sensitivity (short-term response) to precipitation and cropping system water-use.

CAUSES OF SALINE SEEP

Saline seeps generally result from a combination of geologic, climatic, hydrologic, and cultural (land-use) conditions. The primary cause is a change in vegetation from grassland to a less water-use efficient cropping system, such as crop-fallow, that allows precipitation in the recharge area to move through the root zone and subsoil dissolving salts, and providing seepage water. The water accumulates above a geologic layer of less permeability forming a perched water table or accumulates in a layer of greater permeability that is underlain by less permeable material. The water then moves laterally downslope to a point where the water is forced to the soil surface or the permeable layer surfaces or is exposed on a side slope position. Many different geologic situations exist that can result in saline seep formation.[1] Other factors contributing excess water for saline seep development include: above normal precipitation; restricted surface and subsurface drainage; large snow drifts; gravelly and sandy soils; natural drainageway obstructions; artesian water wells; leaky livestock ponds; crop failures; and water conservation practices, such as level bench terraces.

SALINE SEEP WATER QUALITY

Water quality associated with saline seeps is usually unsuitable for human and livestock consumption due to high concentrations of dissolved salts and often high nitrate-N levels. Total salt concentration makes it unsuitable for irrigation. Calcium, magnesium, and sodium are the dominant cations and sulfate the dominant anion associated with saline seeps in the northern Great Plains.[3–5] Soils in seep areas are in equilibrium with gypsum, lime, and other Ca–Mg sulfate minerals.

SALINE SEEP CONTROL

Early detection is important to timely implementation of farming practices to reduce the severity or eliminate the saline seep problem.[1,6] Visual indicators of impending saline seep problems include vigorous weed growth following crop harvest in areas that would normally have dry soil; salt crystals on soil surface; prolonged soil wetness after rain; tractor wheel slippage or bogging down of implements in areas of the field that would normally be dry; excessive crop growth with lodging; infestation of salt tolerant weeds; stunted or dying trees in a shelterbelt; and poor seed germination. Crop root zone soil salinity can be readily assessed and mapped with portable field salinity detection equipment. Salinity in normal productive soils is generally low in the top meter, increasing with depth. Developing saline seeps have slightly increased levels of salinity at the soil surface that increase rapidly with depth. Developed saline seeps have high salinity at the soil surface that decrease with increasing soil depth.

Understanding the geology and circumstances causing a saline seep to form helps in designing effective control measures. Locating the saline seep recharge area is important to develop control measures. Generally, recharge areas are located at a short distance (180 m–

Encyclopedia of Water Science
DOI: 10.1081/E-EWS 120010045

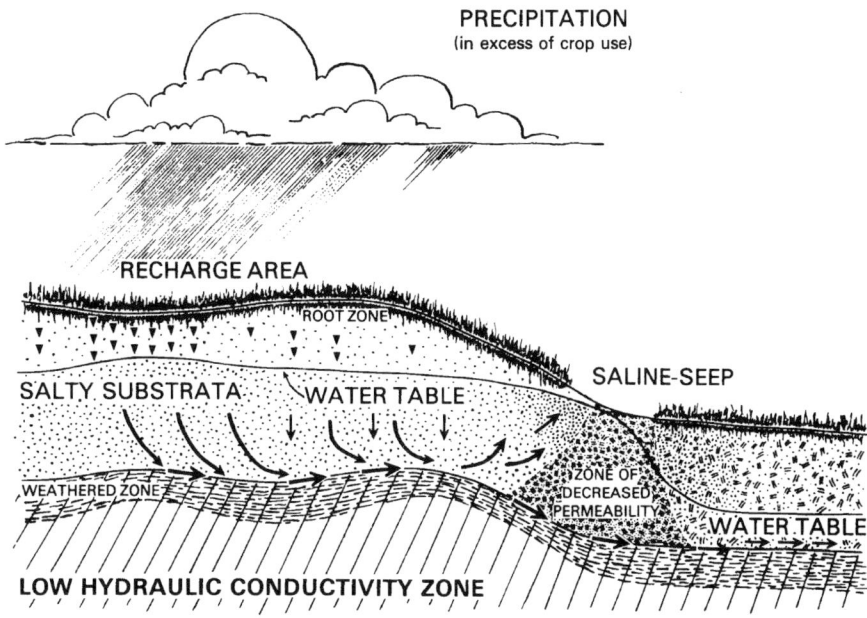

Fig. 1 Example diagram of geologic conditions contributing to saline seep formation in the northern Great Plains of U.S.A.

600 m) upslope from the seepage area. The recharge area is usually located directly upslope or at an angle across the slope from the seepage area. Soil survey maps can be helpful in locating sandy or gravelly areas. Geologic maps may provide information on subsurface stratification, permeable, and impermeable layers. Soil profile information can help to identify recharge areas. Using a soil probe, one can often identify wet soil areas going upslope from the seep area in the direction of the recharge area. Deep coring can be used to examine and sample the soil profile at greater depths.

Visual assessment of recharge area location includes identifying the upslope area, direction of seep expansion, and contributing factors such as bench terraces, cropping system, and surface water collection areas. Saline seeps in

glaciated areas generally expand laterally and upslope toward the recharge area. In nonglaciated areas, they tend to expand laterally and downslope away from the recharge area. Seep areas should show signs of drying up within 2 yr or 3 yr after implementing control measures (more intensive cropping systems, growing alfalfa, or high water-use crops) in the recharge area if the recharge area was correctly identified.

Since saline seeps are caused by water moving below the root zone in the recharge area, there will be no permanent solution to the saline seep problem unless control measures are applied to the recharge area. There are two general procedures for managing seeps: 1) mechanically draining ponded surface water where possible, and/or intercepting lateral flow of subsurface water with drains before it reaches the seepage area; and 2) agronomically using the water for crop production.

Drainage

In recharge areas with small depressions that temporarily collect runoff water after a rainfall or snowmelt, surface drainage may reduce the contribution of water to the perched water table. Drainageways should be kept open to prevent ponding of water. In areas with level bench terraces, use of such water conservation practices may need to be evaluated if saline seeps are a problem. Drainage studies show that hydraulic control of the seep area can be achieved when subsurface interceptor drains are installed on the upslope side of the seep area. Disposal of the saline water, usually high in nitrate, is often

Fig. 2 Typical saline seep area developed on a hillside in eastern Montana, U.S.A.

a problem because of downstream surface or groundwater pollution. Other legal and physical constraints also come into consideration when disposing the seep water. The economics of dryland farming systems generally does not allow installation of costly drainage systems.

Agronomic Control

The best approach for controlling saline seeps is to use the water for crop production while it is a relatively nonsaline resource in the root zone of the recharge area.[1,6–8] Planting crops and utilizing cropping systems in the recharge area that will effectively use available soil water supplies will achieve hydraulic control of many seep areas. This requires delineation of the recharge area followed by adoption of cultural practices that maximize soil water-use and minimize deep percolation. Planting deep rooted crops like alfalfa in the recharge area has been effective.[9] Intensifying crop rotations from crop–fallow to rotations that have less fallow or no fallow in the rotation improves crop water-use efficiency and reduces the amount of water moved below the root zone. Intensive, flexible cropping systems that use good soil and crop management practices to improve crop production and water-use are economically feasible. Fallow should be considered only when soil water and expected growing season precipitation are not sufficient to produce economical yield levels.

RECLAMATION

Once a saline seep area is brought under hydrologic control and the water table has been lowered sufficiently (>1.5 m) to stop movement of salts to the soil surface, reclamation of the seep area can begin. Research and farmer experiences show that reclamation occurs quite rapidly.[1,6,10] The rate of reclamation depends on the amount of precipitation received to leach the salts from the root zone. Because Ca–Mg sulfate type salts have accumulated in the seep area, the soil in the seep area is normally not dispersed by Na, thus permeability is maintained without need for gypsum application. Practices that enhance water movement through the soil profile need to be used. When soil salinity has been lowered sufficiently, normal crop production practices can be utilized. In the northern and central Great Plains, near normal crop production has been achieved on former saline seep areas.

SOCIOECONOMIC CONCERNS

Saline seeps present socioeconomic concerns because they do not respect property lines. A recharge area can be located on one farmer's property with the seepage area on another farmer's property. Seep discharge can contaminate streams, natural drainageways, and/or farm ponds. When a recharge area is on an adjacent farm, co-operation of landowners is needed to correct a saline seep problem. Formation of salinity control districts has been effective in getting farmers and government to work together. Changes in government farm programs that allow more intensive cropping practices helps farmers deal with saline seeps more effectively today. Development of conservation tillage and no-till farming practices makes it more feasible to implement economical intensive cropping systems that utilize available water supplies efficiently. Saline seep is not just an individual farmer problem. Any loss of farmland decreases the nation's food and tax base.

REFERENCES

1. Brown, P.L.; Halvorson, A.D.; Siddoway, F.H.; Mayland, H.F.; Miller, M.R. *Saline-Seep Diagnosis, Control and Reclamation*, Cons. Res. Rpt. No. 30; U.S. Dept. Agric.–Agric. Res. Serv.: U.S. Gov. Printing Office Washington D.C., 1983; 1–22.
2. Clarke, C.J.; George, R.J.; Bell, R.W.; Hobbs, R.J. Major Faults and the Development of Dryland Salinity in the Western Wheatbelt of Western Australia. Hydrol. Earth Syst. Sci. **1998**, *2* (1), 77–91.
3. Berg, W.A.; Naney, J.W.; Smith, S.J. Salinity, Nitrate, and Water in Rangeland and Terraced Wheatland Above Saline Seeps. J. Environ. Qual. **1991**, *20*, 8–11.
4. Doering, E.J.; Sandoval, F.M. Chemistry of Seep Drainage in Southwestern North Dakota. Soil Sci. **1981**, *132*, 142–149.
5. Timpson, M.E.; Richardson, J.L.; Keller, L.P.; McCarthy, G.J. Evaporite Mineralogy Associated with Saline Seeps in Southwestern North Dakota. Soil Sci. Soc. Am. J. **1986**, *50*, 490–493.
6. Halvorson, A.D. Management of Dryland Saline Seeps. In *Agricultural Salinity Assessment and Management*, ASCE Manuals and Reports on Engineering Practice No. 71; Tanji, K.K., Ed.; Am. Soc. Civil Eng.: New York, 1990; 372–392.
7. Black, A.L.; Brown, P.L.; Halvorson, A.D.; Siddoway, F.H. Dryland Cropping Strategies for Efficient Water Use to Control Saline Seeps in the Northern Great Plains U.S.A. Agric. Water Manag. **1991**, *4*, 295–311.
8. Halvorson, A.D. Role of Cropping Systems in Environmental Quality: Saline Seep Control. In *Cropping Strategies for Efficient Use of Water and Nitrogen*, Spec. Pub. no. 51; Hargrove, W.L., Ed.; ASA-CSSA-SSSA: Madison, WI, 1988; 179–191.
9. Mankin, K.R.; Koelliker, J.K. A Hydrologic Balance Approach to Saline Seep Remediation Design. Appl. Eng. Agric. **2000**, *16* (2), 129–133.
10. Halvorson, A.D. Saline-Seep Reclamation in the Northern Great Plains. Trans. ASAE **1984**, *27*, 773–778.

Saline Water

Khaled M. Bali
University of California, Holtville, California, U.S.A.

S

INTRODUCTION

The word salinity refers to the presence of salts in waters and soils. It refers to more than just sodium or chloride, the two elements of table salt. Magnesium, calcium, carbonate, bicarbonate, nitrate, and sulfate can all contribute to salinity. The suitability of water for drinking, irrigation, or wildlife depends on the type and concentration of dissolved salts in water. The salinity of water is usually expressed in terms of a measured parameter that is affected by all the dissolved salts in water. Electrical conductivity (EC) is the parameter that is most currently used and expressed in decisiemans per meter ($dS\,m^{-1}$); another is total dissolved salts (TDS) expressed as the mass of dissolved salts per unit volume of water. One decisiemans per meter is approximately equal to a TDS of $640\,mg\,L^{-1}$. Other terms that are commonly used to express water or soil salinity are given in Table 1.

SOURCES OF SALTS

The primary source of salts in waters and soils is chemical weathering of earth materials (rocks and soils). Natural secondary sources of salts along coastal areas include atmospheric deposits of oceanic salts, and seawater intrusion into groundwater basins and into estuaries. Atmospheric salt deposition also occurs in the interior of continents. The deposition rate decreases with distance from the ocean from values as high as $200\,kg\,ha^{-1}\,yr^{-1}$ to $20\,kg\,ha^{-1}\,yr^{-1}$ in the interior. Other secondary sources of salts found in soils are saline water from rising groundwaters, inland saline lakes and playas, leaching of saline lands, and natural salt deposits.[1,2]

The ocean is the primary source of salts found in natural salt deposits. These were laid down under the direct influence of an ocean during earlier geologic periods and subsequently uplifted. More commonly, however, the direct source of salts is surface and groundwater. All of these waters contain dissolved salts, the concentration depending upon the salt content of the soil and geologic materials with which the water has been in contact. There are other sources of salts which are the result of human activity: they include irrigation and drainage water, chemical fertilizers, animal wastes, sewage sludges and effluents, and oil- and gas-field brines.

Most waters on earth are salty because oceans contain approximately 97% of the water on earth and most of the fresh water is frozen in glaciers. The salinity of the Pacific Ocean is approximately $35,000\,mg\,L^{-1}$. Because of the high salinity of ocean waters and the volume of fresh water frozen in glaciers, only a small fraction of earth's water is available for drinking, irrigation, environmental, and recreational uses.

SOLUBILITY OF SALTS

What kinds of salts are commonly found and what are their solubility in water? Listed in the order of their solubility, they are the chloride (Cl^-), sulfate (SO_4^{2-}), bicarbonate (HCO_3^-), and carbonate (CO_3^{2-}) salts of sodium (Na^+), potassium (K^+), magnesium (Mg^{2+}), and calcium (Ca^{2+}). The chloride salts are more soluble than the sulfate salts, which in turn are more soluble than the bicarbonate/carbonate salts. Likewise, sodium salts are more soluble than magnesium salts, which are more soluble than calcium salts (Tables 2 and 3).

SALINITY MEASURMENTS

The salinity of water is closely related to its EC. Electrical conductivity is easy to measure in the laboratory. Salts ionize when dissolved in water, i.e., the salts dissociate or disintegrate into the elements that make up the salt. For example, the sodium chloride crystals in the saltshaker do not just become smaller when put into water. They totally disintegrate to the point where the sodium and chloride in the salt crystal become individual ions of sodium and chloride in the solution. These ions are electrically charged: sodium is positively charged and chloride is negatively charged. If one places bare wires of an electrical cord plugged into a wall socket into a solution of dissolved salt, the ions will carry the current. The saltier the water, the more current that will be carried and the lower the electrical resistance to alternating current. For safety and other reasons, this method is not used to measure the EC of a water sample. Instead, the sample is put into a small EC cell, which contains two electrodes, and the source of the alternating current is applied to

Encyclopedia of Water Science
DOI: 10.1081/E-EWS 120010235

Table 1 Salinity conversion table

1 ppm = 1 mg L^{-1} (for low concentrations)
1 ppb (part per billion) = 1 μg L^{-1}
1 ppm = 1000 ppb
1 mg L^{-1} = 1000 μg L^{-1}
1 dS m^{-1} = 1 mmhos cm^{-1}
1 dS m^{-1} = 640 ppm; EC (electrical conductivity) less than 5 dS m^{-1}
1 dS m^{-1} ≈ 800 ppm; EC greater than 5 dS m^{-1}

Note that EC is affected by temperature. EC_{25} is most commonly used to express the EC at 25°C (77°F). Measurements made at other temperatures should be adjusted to EC_{25} using the following equation: $EC_{25} = EC_T - 0.02(T - 25)EC_T$.

the electrodes and the electrical resistance is measured with a resistance meter.[3]

DRINKING WATER QUALITY STANDARDS

Drinking water quality standards and guidelines are regulated by the U.S. Environmental Protection Agency.[4] The primary regulations include maximum contaminant levels (MCL) for inorganic chemicals such as lead and nitrate, organic chemicals, turbidity, coliform bacteria, and radiological constituents. The World Health Organization[5] has set similar standards for drinking water quality. The salinity of water used for drinking in the United States does not usually exceed 1000 mg L^{-1} (or approximately 1.5 dS m^{-1}). Reverse osmosis can be used to lower the concentration of salts in drinking water. Drinking water that does not contain salts does not taste good as well.

WATER QUALITY GUIDELINES FOR IRRIGATION

Many crops are adversely affected at salinity levels greater than about 4 dS m^{-1} in the water extract obtained from a saturated-soil paste. Decline in crop yield occurs if salt accumulates in the root zone to a level such that the crop is no longer able to extract enough water from the soil solution. If water uptake is significantly reduced, plant growth will be reduced. In general, salinity problems are more severe during the early stages of growth. Decline in crop yield can be predicted from average root zone salinity.[6] In general, vegetable and tree crops are more sensitive to salinity than field crops.

The effects of salinity effects are not limited to crop damage. Salinity can also have a major impact on soil structure and infiltration rate. Good quality water (low salinity) is good for crop production, but it may reduce the rate at which the water penetrates into soil. To evaluate the suitability of water for irrigation,[7,8] one needs to know

Table 2 Solubility of salts

carbonate < bicarbonate < sulfate < chloride
calcium < magnesium < sodium

the water quality related problems that may cause decline in yield[6] or reduction in soil permeability to water and air.[9]

Salinity Effects on Plants

Crops vary in their salt tolerance. Each crop has a unique threshold salinity, or maximum soil salinity it can tolerate without a yield reduction. The yield decline per unit of salinity (slope) greater than the threshold salinity also differ among crops. These thresholds and slopes are known as the salt tolerance coefficients.[6] Salinity increases the energy crops need to expend to maintain turgor pressure and not wilt. This reduces the energy available for plant growth.

Sodium Adsorption Ratio

Sodium adsorption ratio (SAR) is commonly used as an index for determining sodium hazard in soils. Sodium adsorption ratio is usually determined for irrigation water or soil solution (extract of a completely saturated soil sample). The presence of excessive exchangeable sodium (Na) in soil solution may cause clay particles to swell. Clay swelling makes soil less permeable to water and to air, and can result in soil crusting and hard setting for sandy loam and loamy soils, and poor tilth for a broad spectrum of soil textures.[9] The Na hazard depends on the total salt concentration in the soil solution as well as on individual concentrations of calcium (Ca), magnesium (Mg), and sodium (Na). The effect of these individual ion concentrations is quantified by SAR, which is defined as

$$SAR = [Na]/(([Ca] + [Mg])/2)^{0.5}$$

Table 3 Common names of salts (listed in the order of decreasing solubility)

Symbol	Common name
NaCl	Table salt
NaHCO$_3$	Baking soda
NaCO$_3$	Washing soda
KCl	Potash
MgSO$_4$·7H$_2$O	Epsom salt
CaSO$_4$·2H$_2$O	Gypsum—calcium sulfate
MgCO$_3$	Magnesite
CaCO$_3$	Calcite (soil lime)

where [Na], [Ca], and [Mg] are the concentrations of sodium, calcium, and magnesium, respectively, all concentrations are expressed in mmol of charge per liter ($mmol\,L^{-1}$), or in the non-SI unit of $meq\,L^{-1}$.

The effect of SAR on water infiltration rate depends on the salinity of irrigation water. For a given SAR, water infiltration rate increases as the salinity of irrigation water increases. For a given salinity, water infiltration rate decreases as the SAR of irrigation water increases.

Specific Ion Toxicity

Salinity can affect crop growth through specific-ion toxicities and osmotic effects. Specific-ion toxicity occurs when the concentration of one ion is high enough to cause toxicity.[6] Boron, chloride, and sodium are some of the ions that impede plant growth and development. Specific-ion toxicity causes leaf burn on the tips and margins of crop leaves.

Soil Salinity and Water Potential

Water movement in soil is often considered in terms of driving force. Water moves from where its energy status is high to where it is low. The energy status of water is commonly described by the total water potential which consists of pressure, capillary (or matric), osmotic, and gravitational potentials. The capillary or matric potential is due to cohesion–adhesion forces in the soil matrix. Osmotic or solute potential is due to the concentration of salts in soil solution.

Total water potential (H) can be expressed by

$$H = h_p + h_m + h_s + h_g$$

where h_p, h_m, h_s, and h_g are pressure, matric, solute, and gravitational potentials, respectively. Pressure and gravitational potentials can be either positive or negative. However, matric and solute potentials are always zero or negative.

Most plant roots can extract water from the soil when matric potential is between -5 bar and 0 bar. Almost all crops cannot extract any water from the soil when the matric potential is about -15 bar. This point is called permanent welting point (PWP), and its value depends on soil texture and crop type.

The higher the negative value of matric potential, the harder it is for plant roots to extract water. The presence of salts in the soil–water system adds another force that the plant has to work against to extract water. Solute or osmotic potential is zero when the concentration of salts in soil–water system is zero. Osmotic potential becomes more negative due to the increase in soil salinity. Therefore, salinity increases the total negative potential of soil water making it harder for the plants to extract water from soil solution.

Water movement from soil to plant depends on total water potential of soil water. At any particular soil moisture content, the higher the concentration of salts, the harder it is for the plant to extract water from the soil. The approximate relationship between soil solution's osmotic potential and soil salinity at 25°C (77°F) is

$$h_s = -0.4EC$$

where h_s is soil solution's osmotic potential (bar) and EC is soil solution's salinity in $dS\,m^{-1}$ ($mmhos\,cm^{-1}$).

ACKNOWLEDGMENTS

The author greatly appreciates the valuable assistance of Dr. Jim D. Oster (University of California, Riverside) in the development of this manuscript.

REFERENCES

1. Tanji, K.K. The Nature and Extent of Agricultural Salinity Problems. In *Agricutlural Salinity Assessment and Management*, ASCE Man. 71; Tanji, K.K, Ed.; ASCE: New York, 1990; 1–17.
2. Jurinak, J.J.; Suarez, D.L. The Chemistry of Salt-Affected Soils and Waters. In *Agricutlural Salinity Assessment and Management*, ASCE Man. 71; Tanji, K.K., Ed.; ASCE: New York, 1990; 42–63.
3. Rhoades, J.D. Soluble Salts. In *Methods of Soil Analysis. Part 2*, Agronomy 9; Page, A.L., Ed.; American Society of Agronomy, Inc.: Madison, WI, 1982; 167–179.
4. *Quality Criteria for Water*; U.S. Environmental Protection Agency: Washington, DC, 1976.
5. World Health Organization. (http://www.who.int/water_sanitation_health/Water_quality/drinkwat.htm)
6. Maas, E.V.; Grattan, S.R. Crop Yields as Affected by Salinity. In *Agricultural Drainage*, Agron. Monogr. 38; Skaggs, R.W., van Schilfgaarde, J., Eds.; ASA, CSSA, SSSA: Madison, WI, 1999; 55–108.
7. Rhoades, J.D. Overview: Diagnosis of Salinity Problems and Selection of Control Practices. In *Agricutlural Salinity Assessment and Management*, ASCE Man. 71; Tanji, K.K., Ed.; ASCE: New York, 1990; 18–41.
8. Hansen, B.; Grattan, S.R.; Fulton, A. *Agricultural Salinity and Drainage*, ANR Publication No. 3375; University of California. Division of Agriculture and Natural Resources, 1999.
9. Oster, J.D.; Jayawardane, N.S. Agricultural Management of Sodic Soils. In *Sodic Soils: Distribution, Properties, Management and Environmental Consequences*; Sumner, M.E, Naidu, R., Eds.; Oxford University Press: New York, 1998; 127–147.

Salinity and Solute Measurement by Time Domain Reflectometry

Jon M. Wraith
Montana State University, Bozeman, Montana, U.S.A.

INTRODUCTION

Time domain reflectometry (TDR) has a unique and potentially very useful ability to concurrently measure both water content (θ)[1,2] and electrical conductivity (σ)[3] in soils and other porous media. Both θ and σ are important physical/chemical attributes that have substantial impact on the behavior and transport of mass and energy, are critical to plant growth, and may be used to infer salinity or concentration of certain solutes in porous media. The TDR can also accurately measure σ and, thus, indirectly solute concentration in water, but it is a relatively expensive tool in relation to other available methods for this application.

Solutes are substances that can dissolve in water. Some solutes, those that dissociate into ions, confer water the property of electrical conductance. The relative ability of water or variably water-saturated porous media to conduct electricity is related to the concentration of ionic solutes and, in the case of porous media, to the wetness and the geometry of solid, liquid, and air phases. Air does not conduct electricity, and the solids found in common porous media (e.g., soil) conduct electricity very poorly. (The electrical conductivity of soil solids is related to the ions held at exchangeable surface charges rather than to the solids themselves.) Hence, the measured electrical conductivity may be directly related to the concentration of solutes in the liquid water phase, which depends in turn on its ionic composition. The SI units for electrical conductivity are Siemens per meter (S/m). Because common measurement ranges found in soils and waters may result in inconveniently high or low values, fractional units such as dS/m and mS/cm are commonly used. (Note 1 dS/m = 1 mmho/cm, with mmho/cm being the formerly common non-SI unit for σ.)

MEASUREMENT PRINCIPLES

Electrical Conductivity Measurements Using TDR

Measurement of electrical conductivity by using TDR is based on attenuation of the voltage signal as it travels along a transmission line probe embedded in the medium of interest. Media with higher electrical conductivity lead to increased attenuation (loss) of the electrical signal, and this may be inferred from analyzing the TDR trace. The Giese and Tiemann[4] method is commonly used with TDR to measure the apparent electrical conductivity (σ_a, S/m):

$$\sigma_a = \frac{\varepsilon_0 c}{L} \frac{Z_0}{Z_u} \left(\frac{2V_0}{V_f} - 1 \right) \tag{1}$$

with ε_0, the permittivity of free space (8.854×10^{-12} F/m), c, the speed of light in vacuum (2.997×10^8 m/sec), L, the probe length (m), Z_0, the probe impedance (Ω), Z_u, the characteristic impedance of the cable tester (usually 50 Ω), and V_0 and V_f, the relative voltages of different parts of the TDR waveform (Fig. 1). This analysis is easily performed on the digital waveforms obtained from a TDR instrument, and is used in many available software programs. The probe impedance must be calibrated; this may conveniently be done by immersing in deionized water[5,6] and using

$$Z_0 = Z_u \varepsilon_w^{0.5} \left(\frac{V_1}{2V_0 - V_1} \right) \tag{2}$$

where ε_w is the known dielectric constant of water, and the location of V_1 is illustrated in Fig. 1. Alternatively, a TDR probe cell constant $K = \varepsilon_0 c Z_0 / L$ may be used,[5,6] with substitution into Eq. 1. The temperature-dependent ε_w may be determined using tables or equations found in references including Refs. [7–9]. Because σ increases with increasing temperature, a correction factor[9,10] is used to compensate measurements obtained at a given ambient temperature to a base temperature such as 25°C.

Some of the signal from the TDR instrument is lost as it moves through transmission cables and connectors leading to the probe. To obtain accurate measurements, these signal losses must be considered in cases where σ exceeds about 3 dS/m or where combined cable lengths exceed about 3 m.[10,11] A combined cable plus connector series resistance, Z_{cable}, may be determined and included in Eq. 1 as

$$\sigma_a = \frac{\varepsilon_0 c Z_0}{L(Z_L - Z_{cable})} \left(\frac{2V_0}{V_f} - 1 \right) \tag{3}$$

Encyclopedia of Water Science
DOI: 10.1081/E-EWS 120010230

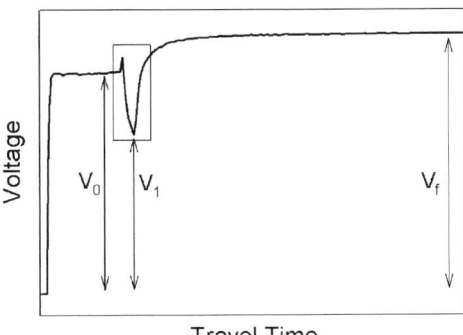

Fig. 1 TDR waveform illustrating voltage heights V_0, V_f, and V_1 used in calculating electrical conductivity. Region within box represents approximate portion of waveform used in travel-time water content analysis. (Modified from Ref. [9].)

where $Z_L = Z_u/(2V_0/V_f - 1)$, and Z_{cable} may be measured using a series of solutions having known σ[10,11] or by analyzing signals resulting from electrically shorting the cables and probes[12]. Excellent agreement between σ measured by using TDR and electrodes has been repeatedly documented (Fig. 2; e.g., Refs. [10–13]).

Electrical Conductivity Calibration Models and Methods for Soils and Porous Media

Soils are complex mixtures comprising solid, liquid, and gas components. The TDR measures the apparent electrical conductivity of the bulk soil, σ_a, integrated over all these components. However, the σ of the soil solution, σ_w, is commonly the attribute of interest in practical applications. The σ_w arises from the total ionic solute concentration, but may be related to the concentration of specific solutes under some conditions. Under constant θ and temperature, a linear relationship may be assumed between σ_a and σ_w. Several approaches have been utilized to estimate σ_w or solute

concentrations using this direct calibration approach (e.g., Refs. [11,14,15]). However, under typical conditions where θ varies in time and space, models describing the relationship among the σ_a, σ_w, σ_s, soil water content θ, and the tortuous soil geometry in terms of electrical flow paths, are required. Because of the inherent complexity of the physical properties and processes governing electrical flow in variably saturated soils, simple conceptual models are typically applied. In many models, the bulk soil electrical conductivity is linearly related to σ_w as

$$\sigma_a = \theta F_g \sigma_w + \sigma_s \qquad (4)$$

with F_g a geometry factor describing the dependence of the electrical flow pathways on the tortuous soil matrix. The bulk soil electrical conductivity σ_a is thus seen to be a result of the combined influences of σ_w, wetness, and soil geometry, along with contributions of the exchangeable ions at the solid clay and organic matter surfaces σ_s. The σ_s may be neglected due to its small magnitude relative to the other components under some conditions (i.e., coarse soils, or $\sigma_w >$ about 0.5 dS/m–1 dS/m).[17] The F_g changes substantially with changing wetness for a given soil, since water and air have very dissimilar conducting properties. Hence, the F_g is often characterized as a function of θ.[16–18]

Conceptual models that have been most commonly applied to the measurement of solutes and salinity in soils by using TDR include those of Rhoades et al.[16,17] and Mualem and Friedman.[18] The two- and three-conducting pathway models[16,17] use calibrated empirical constants to describe the dependence of F_g on soil properties and wetness, while Mualem and Friedman[18] relate the F_g to the soil hydraulic conductivity relationship. F_g may also be estimated using analogy to simple gas diffusion models.[19] Calibration of all the models mentioned here is required to obtain suitable results for most applications. Development and testing of calibration methods and models is an ongoing area of inquiry, and is discussed in many of the papers cited as well as in Ref. [20].

CONCLUSION

The use of TDR to measure solutes and salinity is particularly appropriate in applications where high temporal or spatial resolution or unattended measurements are desired, and under conditions of changing water status. Because TDR can provide detailed time series measurements with no ongoing labor requirement and is nondestructive, it may be preferred over alternative methods. Many past applications have focused on evaluating transport of soluble chemicals through soils or other materials, monitoring water and salt distributions in the root zone of plants (e.g., Fig. 3), evaluating solute transport models, and related issues. These and a number

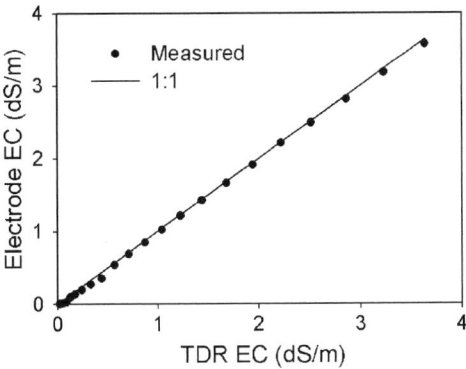

Fig. 2 Comparison of electrical conductivity measured in KCl solution using TDR and electrode. (J. M. Wraith, unpublished data.)

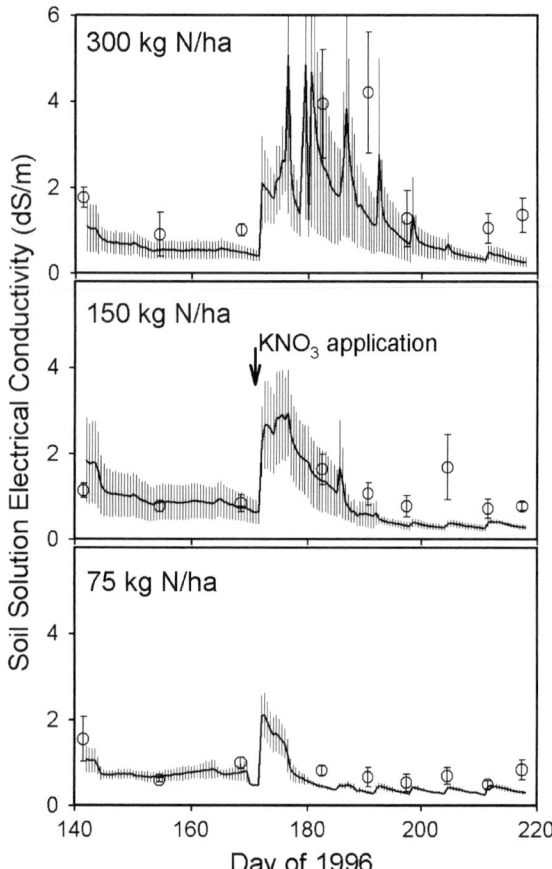

Fig. 3 Mean soil solution electrical conductivity measurements using replicate soil cores and TDR probes, under different KNO$_3$ application rates. Three TDR probes were permanently installed at 15 cm depth in each plot, and five soil cores were collected from 10 cm to 20 cm depth increment at random locations in each plot on each sampling date. Periodic irrigation events during the abrupt increase in σ_w following KNO$_3$ application (most evident in top panel) reflect water addition by irrigation events. Error bars are ± standard error of means, and reflect spatial variation among field measurement locations. (Modified from Ref. [21].)

of other practical problems may be amenable to study or to an improved management through the use of TDR and appropriate calibration models and methods.

REFERENCES

1. Topp, G.C.; Davis, J.L.; Annan, A.P. Electromagnetic Determination of Soil Water Content: Measurements in Coaxial Transmission Lines. Water Resour. Res. **1980**, *16*, 574–582.
2. Evett, S.R. Soil Water Measurement by Time Domain Reflectometry. In *Encyclopedia of Water Science*; Stewart, B.A., Howell, T., Eds.; Marcel Dekker, Inc.: New York, 2003 (in press).
3. Dalton, F.N.; Herkelrath, W.N.; Rawlins, D.S.; Rhoades, J.D. Time-Domain Reflectometry: Simultaneous Measurement of Soil Water Content and Electrical Conductivity with a Single Probe. Science **1984**, *224*, 989–990.
4. Giese, K.; Tiemann, R. Determination of the Complex Permittivity from Thin-Sample Time Domain Reflectometry: Improved Analysis of the Step Response Waveform. Adv. Mol. Relax. Processes **1975**, *7*, 45–59.
5. Heimovaara, T.J. Comments on Time Domain Reflectometry Measurements of Water Content and Electrical Conductivity of Layered Soil Columns. Soil Sci. Soc. Am. J. **1992**, *56*, 1657–1658.
6. Baker, J.M.; Spaans, E.J.A. Comments on "Time Domain Reflectometry Measurements of Water Content and Electrical Conductivity of Layered Soil Columns." Soil Sci. Soc. Am. J. **1992**, *57*, 1395–1396.
7. Stogryn, A. Equations for Calculating the Dielectric Constant of Saline Water. IEEE Trans. Microwave Theory Technol. MIT **1971**, *19*, 733–736.
8. Weast, R.C., (Ed.) *CRC Handbook of Chemistry and Physics*, 67th Ed.; CRC Press: Boca Raton, FL, 1986.
9. Wraith, J.M.; Or, D. Temperature Effects on Soil Bulk Dielectric Permittivity Measured by Time Domain Reflectometry: Experimental Evidence and Hypothesis Development. Water Resour. Res. **1999**, *35*, 361–369.
10. Heimovaara, T.J.; Focke, A.G.; Bouten, W.; Verstraten, J.M. Assessing Temporal Variations in Soil Water Composition with Time Domain Reflectometry. Soil Sci. Soc. Am. J. **1995**, *59*, 689–698.
11. Mallants, D.; Vanclooster, M.; Toride, N.; Vanderborght, J.; van Genuchten, M.Th.; Feyen, J. Comparison of Three Methods to Calibrate TDR for Monitoring Solute Movement in Undisturbed Soil. Soil Sci. Soc. Am. J. **1996**, *60*, 747–754.
12. Reece, C.F. Simple Method for Determining Cable Length Resistance in Time Domain Reflectometry Systems. Soil Sci. Soc. Am. J. **1998**, *62*, 314–317.
13. Spaans, E.J.A.; Baker, J.M. Simple Baluns in Parallel Probes for Time Domain Reflectometry. Soil Sci. Soc. Am. J. **1993**, *57*, 668–673.
14. Ward, A.L.; Kachanoski, R.G.; Elrick, D.E. Laboratory Measurements of Solute Transport Using Time Domain Reflectometry. Soil Sci. Soc. Am. J. **1994**, *58*, 1031–1039.
15. Mallants, D.; Vanclooster, M.; Meddahi, M.; Feyen, J. Estimating Solute Transport in Undisturbed Soil Columns using Time-Domain Reflectometry. J. Contam. Hydrol. **1994**, *17*, 91–109.
16. Rhoades, J.D.; Raats, P.A.C.; Prather, R.J. Effects of Liquid-Phase Electrical Conductivity, Water Content and Surface Conductivity on Bulk Soil Electrical Conductivity. Soil Sci. Soc. Am. J. **1976**, *40*, 651–655.
17. Rhoades, J.D.; Manteghi, N.A.; Shouse, P.J.; Alves, W.J. Soil Electrical Conductivity and Soil Salinity: New Formulations and Calibrations. Soil Sci. Soc. Am. J. **1989**, *53*, 433–439.
18. Mualem, Y.; Friedman, S.P. Theoretical Predictions of Electrical Conductivity in Saturated and Unsaturated Soil. Water Resour. Res. **1991**, *27*, 2771–2777.

19. Amente, G.; Baker, J.M.; Reece, C.F. Estimation of Soil Solution Electrical Conductivity from Bulk Soil Electrical Conductivity in Sandy Soils. Soil Sci. Soc. Am. J. **2000**, *64*, 1931–1939.

20. Hendrickx, J.M.H.; Wraith, J.M.; Kachanoski, R.G.; Corwin, D.L. Solute Content and Concentration. In *Methods of Soil Analysis. Part 1. Physical Methods*, 3rd Ed.; Dane, J.H., Topp, G.C., Eds.; ASA: Madison, WI, 2002, 1253–1321.

21. Das, B.S.; Wraith, J.M.; Inskeep, W.P. Nitrate Concentrations in the Root Zone Estimated Using Time Domain Reflectometry. Soil Sci. Soc. Am. J. **1999**, *63*, 1561–1570.

Salton Sea

Timothy P. Krantz
University of Redlands, Redlands, California, U.S.A.

INTRODUCTION

California's largest lake, the Salton Sea, is situated in the southeast corner of the state in a closed basin at the bottom of a 7851-mi^2 watershed. Over 85% of the water entering the Salton Sea results from agricultural run-off with less than 3% of inflow from annual precipitation. The Salton Sea supports a thriving fishery and provides important habitat for millions of migratory birds. More than two-thirds of bird species in the continental United States have been recorded at the Salton Sea and adjacent areas. The long-term viability of the Salton Sea ecosystem is threatened by increasing salinity and eutrophication resulting from the nutrient-rich agricultural drainage. More imminently, the viability of the Salton Sea is threatened by proposed water transfers and reductions of inflow, potentially concentrating pre-existing salts and causing the sea to recede by as much as one-third of its surface area and more than half its total depth.

GEOGRAPHIC SETTING

The Salton Sea is located in the southeastern desert of California. It lies in the Salton Trough—a closed basin, including the Coachella and Imperial valleys of California, and the Mexicali Valley of Mexico. The Salton Sea is located at 227 ft below mean sea level (msl). The shallow nature of the sea, with a surface area of 367 mi^2 (951 km^2) and a depth of 51 ft (15.5 m), renders it very sensitive to even slight changes of inflow. The sea is sustained by 1.34 million acre ft (af) of inflow, mostly agricultural run-off diverted from the Colorado River (Fig. 1).

The Salton Sea is situated in the Colorado Desert in one of the most arid regions of the United States. Annual precipitation is less than 3 in. (7.6 cm), and mean monthly temperatures in July are 92°F (33.3°C), with maximum temperatures exceeding 100°F (37.7°C) on more than 110 day yr^{-1}. Potential evaporation is estimated at 5.78 ft (1.76 m) per year.[1]

GEOLOGY AND GEOMORPHOLOGY

The Salton Basin was once connected to the Gulf of California and characterized by a shallow marine environment.[2] For the past several million years, as the Colorado Plateau was uplifted, the sediments that once filled the Grand Canyon were deposited in the Gulf of California, eventually building a huge delta, blocking off the Salton Basin from the ocean.[3] The deltaic dam is now 40 ft above sea level, with a drainage divide about 17 mi south of Mexicali, Mexico.

Once separated from the Gulf of California, the Salton Basin would periodically dry out as the Colorado River drained directly into the Gulf of California. At other times, the river would change course and fill the basin, sometimes to its brim, spilling over into the Gulf of California covering more than 2200 mi^2. These prehistoric inundations have been called Lake Cahuilla or Lake LeConte. The Lake Cahuilla shoreline was established by locating geomorphological features with global positioning systems and plotting these in a geographic information system (GIS).[4] Further evidence of Lake Cahuilla has been obtained from archaeological sites along the ancient shoreline, including fish traps, bones, and other lake-related remains. The periodicity of Lake Cahuilla episodes has been estimated based on carbon dates of the travertine deposits and other organic archaeological evidence, indicating that the lake was full most of the time over the past 1300 yr of record.[5–7]

HYDROLOGY OF THE SALTON SEA

The present-day Salton Sea was formed in 1905 when flood flows on the Colorado River rushed into a temporary diversion channel, quickly deepening and widening to capture the entire flow of the river until the breach was finally filled in 1907. Far from being an "accidental lake," it was human intervention that prevented the next stand of Lake Cahuilla from being formed.

After an initial high stand of the Salton Sea from the 1907 flood at about −195 ft msl, the sea receded to about −250 ft msl by 1920. Since then, with the expansion of agriculture in the Imperial and Coachella valleys and increased agricultural run-off, the surface elevation of the sea has risen to its current elevation of about −227 in. The elevation of the sea has remained relatively stable at its current level since the 1980s, indicating that the inflow of about 1.34 million af is equal to evaporation at that

Encyclopedia of Water Science
DOI: 10.1081/E-EWS 120010128

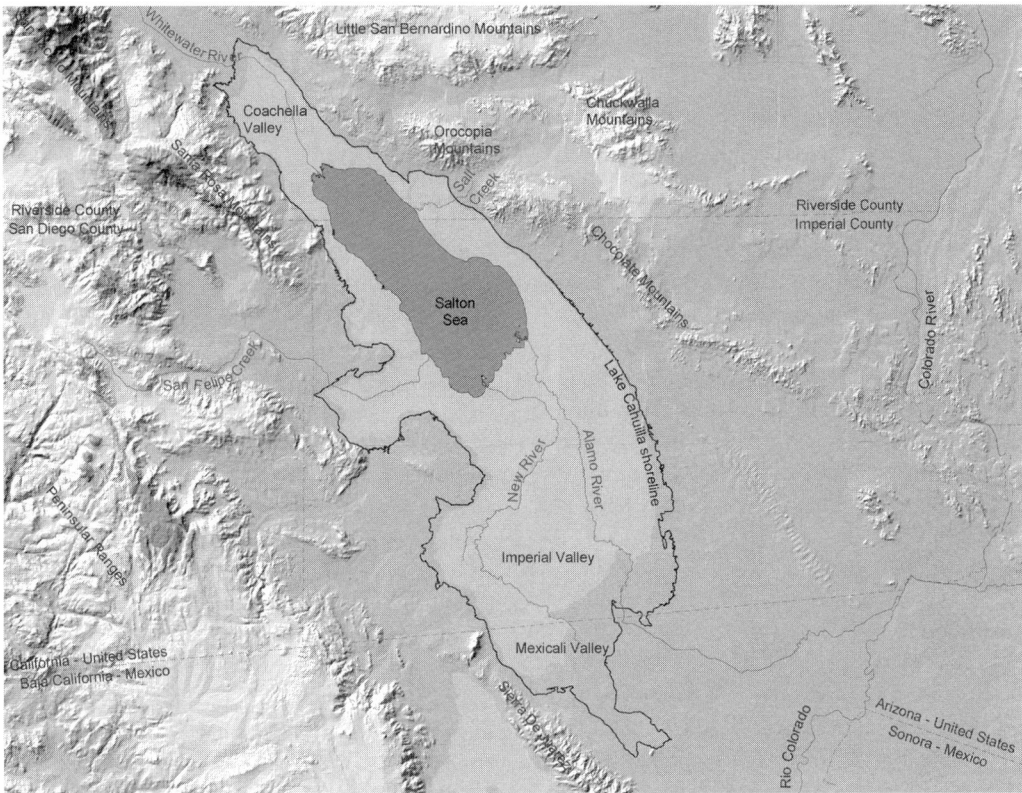

Fig. 1 The Salton Sea lies in a closed basin, sustained largely by agricultural run-off from the Imperial Valley. Lake Cahuilla was the prehistoric high stand of a lake that occasionally filled when the Colorado River would drain into the basin.

elevation—or about 15% of the total volume of the sea lost to evaporation each year.

Approximately 4.5 million tons of salts are added to the sea annually. Because evaporation is the only outlet for the sea, the dissolved salts, nutrients, and minerals that enter the sea remain there, and have accumulated over the past century to the point at which the sea is now about 25% saltier than the ocean, at about 44,000 mg L^{-1}.[8]

BIOLOGICAL RESOURCES

The nutrient-rich agricultural drainage that sustains the Salton Sea also supports an incredible diversity of life. More than 400 species of invertebrates, mostly single-celled plankton, have been identified in the sea.[9] These provide the food base supporting a highly productive fishery, with an estimated 200 million fish—one of the most productive fisheries in the world.[10]

The Salton Sea is of critical importance for many species of migratory birds. The sea supports over 90% of the North American population of eared grebes, with as many as three million individuals during migration, as many as 30,000 American white pelicans and 2000 brown pelicans, more than 120,000 shorebirds of 44 species, 25,000 snow and Ross' geese, the largest breeding colony of gull-billed terns in Western North America, and 45% of endangered Yuma clapper rail habitat.[11]

ENVIRONMENTAL THREATS

Increasing Salinity

Increasing salinity may cause the fishery to collapse as it approaches 60,000 mg L^{-1}.[12] At the present rate of salt loading of about 4.5 million tons per year, the Salton Sea would reach the 60,000 mg L^{-1} threshold in about 50 yr, assuming inflow remains at its present level of 1.34 million af per year.[8]

Pilot-scale solar evaporation ponds to remove salts from the sea have been constructed and are operational. The solar ponds, if fully implemented, would provide an outlet for concentrated salts in the sea, requiring approximately 100,000 af of water to be removed from the sea each year—the amount containing the equivalent of the annual salt load from inflow.[8]

Eutrophication

Nutrient loading from agricultural run-off has created eutrophic conditions at the Salton Sea. Productivity and biomass are very high, leading to oxygen depletion caused by decay of accumulated senescent biological material. These anoxic conditions have contributed to extensive fish kills over the past few decades, leading to further oxygen depletion.[13] Eutrophic conditions worsen during summer months, when high biological productivity and warm water temperatures conspire to greatly reduce dissolved oxygen throughout the water column. One event, in August 1999, resulted in the death of six to seven million fish over a period of several days. Chemical limnological data taken at the time indicated a complete loss of dissolved oxygen from top to bottom in portions of the sea coincident with that event.[14]

Reductions of Inflow

Inflow at the present elevation is about 1.34 million af per year. Proposed water transfers of as much as 300,000 af from the Imperial Irrigation District to metropolitan water users in Southern California, together with other potential reductions of inflow, may reduce total inflow to the Salton Sea by as much as 500,000 af.

With reduced inflows, salinity increases rapidly, the contracting lake concentrating salts already in residence while more salts continue to enter the sea in agricultural run-off. With a reduction of inflow by 300,000 af per year, salinity would reach the 60,000 mg L^{-1} threshold for the fishery in about 12 yr; and with a reduction of 500,000 af, salinity would reach the limit of tolerance of the fishery in just 7 yr.[15]

The collapse of the fishery would represent a serious adverse environmental impact. The death of 200 million fish would have a cascade effect on the rest of the ecosystem, causing the demise of fish-eating bird populations, exacerbating eutrophication from decomposition of the dead fish, and creating a huge breeding ground for flies and other pathogens in their rotting carcasses.

Many species of birds would be critically impacted. Ground-nesting bird colonies on Mullet Island—the only island in the sea—would be exposed to coyotes, cats, and other predators with a draw down of only 7 ft. Many other species would experience substantial whole-species population decline.

Other potential impacts of reduced inflows to the Salton Sea include collapse of lake-related economies, such as boating, hunting, fishing, and property values; degradation of air quality as a result of exposure of as much as 120 mi^2 of fine lake bottom sediments to the desert winds; loss of agricultural productivity from salt and dust deposition; and increased respiratory disease and human health problems as a result of airborne sediments.

CONCLUSION

The Salton Sea is characterized by contrasts. It is California's largest lake, situated in one of the hottest, most arid regions of the United States. Sustained in large part by agricultural run-off, the nutrient-rich inflows support one of the most productive fisheries in the world, in turn supporting millions of migratory birds. At the same time that agricultural drainage is the life's blood of the Salton Sea, it also causes hyper-eutrophic conditions that lead to occasional fish kills and bird die-offs that the sea experiences today.

Increasing salinity will cause the collapse of the fishery within 30 yr–50 yr at present rates of salt loading and inflow, if salinity control measures, such as solar evaporation ponds, are not undertaken. With reduced inflow as a result of water transfers or other actions, the Salton Sea may drop by more than half its depth, exposing more than 100 mi^2 of land, and become a biologically "dead" sea—a North American version of the "Aral Sea."

REFERENCES

1. Hely, A.G.; Hughes, G.H.; Irelan, B. *Hydrologic Regimen of the Salton Sea*; Professional Paper 486-C, U.S. Geologic Survey: Washington, 1966; 1–32.
2. Downs, T.; Woodward, G. Middle Pleistocene Extension of the Gulf of California into the Imperial Valley. Geol. Soc. Am. Special Paper **1961**, *68* (21).
3. Dibblee, T. W., Jr. Geology of the Imperial Valley Region, California. In *Geology of Southern California, California Division of Mines and Geol. Bull.*; Jahns, R.H., Ed.; State of California: San Francisco, 1954; Vol. 170, 21–28.
4. Krantz, T.; Buckles, J.; Kashiwase, K. Reconstruction of Prehistoric Lake Cahuilla and Early American Settlement Patterns in Southeast California Using GIS. Proceedings of the Twentieth Annual ESRI International User Conference, San Diego, California, 2000.
5. Waters, M.R. Holocene Lacustrine Chronology and Archaeology of Ancient Lake Cahuilla. Calif. Quat. Res. **1983**, *19*, 373–387.
6. Wilke, J.P. Late Prehistoric Human Ecology at Lake Cahuilla, Coachella Valley, California. Contr. of the Univ. of Calif. Arch. Res. Facility, Berkeley, 1978, 38.
7. Schaefer, J. An Inventory and Evaluation of Lake Cahuilla Cultural Resources Along Imperial Irrigation District's SA-Line, San Diego and Imperial Counties, California. Prepared by ASM Affiliates, Inc., Encinitas, California, December 2000.
8. Salton Sea Authority, U.S. Bureau of Reclamation. Draft Salton Sea Restoration Project Environmental Impact

Statement/Environmental Impact Report. Prepared by Tetra Tech, Inc., January, 2000.

9. Hurlbert, S.; Dexter, D.; Kuperman, B. Reconnaissance of the Biological Limnology of the Salton Sea. Prepared for the Salton Sea Authority, 1998.

10. Reidel, R.; Helvenston, L.; Costa-Pierce, B. Final Report: Fish Biology and Fisheries Ecology of the Salton Sea. Prepared Under Contract to the Salton Sea Science Office, 2001.

11. Shuford, W.D.; Warnock, N.; Molina, K.C.; Mulrooney, B.; Black, A.E. Avifauna of the Salton Sea: Abundance, Distribution, and Annual Phenology. Contribution No. 931 of Point Reyes Bird Observatory. Final Report for EPA Contract No. R826552-01-0 to the Salton Sea Authority, 2000.

12. Thiery, R. The Potential Impact of Rising Salinity on the Salton Sea Ecosystem. Prepared Under Contract to the Salton Sea Science Office, 1998.

13. Setmire, J.; Holdren, C.; Robertson, D.; Amrhein, C.; Elder, J.; Schroeder, R.; Schladow, G.; McKellar, H.; Gersberg, R. Eutrophic Conditions at the Salton Sea. A Topical Paper from the Eutrophication Workshop Convened at the University of California at Riverside, Sept 7–8, 2000. Prepared for the Salton Sea Authority, the Salton Sea Science Office, and the Bureau of Reclamation, 2000.

14. Holdren, C. Chemical and Physical Analysis of the Salton Sea. Prepared for the Salton Sea Authority, U.S. Bureau of Reclamation, Denver, Colorado, 2000.

15. Imperial Irrigation District, U.S. Bureau of Reclamation. Draft Environmental Impact Report/Environmental Impact Statement Imperial Irrigation District Water Conservation and Transfer Project and Draft Habitat Conservation Plan. Prepared by CH2MHILL, 2002.

S

Selenium

Dean A. Martens
*United States Department of Agriculture (USDA),
Tucson, Arizona, U.S.A.*

INTRODUCTION

Selenium (Se) is an essential nutritional element, but excessive Se can be toxic to animals and humans. Selenium has an atomic number of 34, an atomic weight of 78.94 and occupies a position in Group VIA of the periodic table between the metal tellurium and the nonmetal sulfur. Selenium's chemical and physical properties are intermediate between those of metals and nonmetals (Table 1). Selenium has a valence of -2 in combination with hydrogen or metals, and in oxygenated compounds it can exist as the $+4$ or the $+6$ oxidation states giving rise to an array of Se compounds.[1] Six stable Se isotopes occur with varying degrees of abundance: ^{74}Se (0.87%), ^{76}Se (9.02%), ^{77}Se (7.58%), ^{78}Se (23.52%), ^{80}Se (49.82%), and ^{82}Se (9.19%) and a short-lived isotope (^{75}Se) used in neutron activation, radiology, and tracer applications.[2] The average Se concentration in the earth's crust is about $0.05 \, mg \, kg^{-1}$–$0.09 \, mg \, kg^{-1}$.[3] Selenium concentrations range from $0.004 \, \mu g \, g^{-1}$–$1.5 \, \mu g \, g^{-1}$ in igneous rocks to $0.6 \, \mu g \, g^{-1}$–$103 \, \mu g \, g^{-1}$ in shales of the cretaceous period.

FORMS OF SELENIUM

Important properties of elements, e.g., their bioavailability and toxicity, depend on their chemical form or speciation. Chemical speciation involves the quantification of chemical forms, or species that comprise the total element concentration. Selenium can exist in the $(+6)$, $(+4)$, (0), and (-2) oxidation states, the major feature of Se chemistry that affects the Se solubility and movement in nature. The distribution of the valence states depends on microbial activity, solution pH, and redox conditions. Selenium in the (-2) oxidation state exists as hydrogen selenide (HSe^-) and as a number of metallic selenides. Heavy metal selenides are the most insoluble forms of Se. H_2Se is a toxic gas at room temperature and is thermodynamically unstable in aqueous solutions. Elemental Se(0) exists as several allotrophic forms and is very stable and highly water insoluble. Thermodynamic calculations show that Se(-2) should be found in reducing environments, Se($+4$) species in moderately oxidized environments, and the Se($+6$) species in oxidizing

environments.[4] In waters, dissolved inorganic Se is normally present as $(+6)$ selenate (SeO_4^{2-}) and as $(+4)$ selenite (SeO_3^{2-}).[4]

Inorganic Se

The soluble inorganic Se forms, selenite and selenate, account for the majority of the total Se concentration of waters, although particulate Se(0) smaller than $0.45 \, \mu m$ may also be present.[5] The proportion of selenate/selenite present in waters is generally predicted by the pH–redox status of the system. Selenate is stable under alkaline and oxidizing conditions and selenite is stable under mildly oxidizing conditions.[6] Although, measurement of pH–redox status is a good predictor of Se species,[7] actual speciation must be analyzed as exceptions to the thermodynamic predictable Se species have been reported[8] due to the influences of biological activity.

The ratio of selenate to selenite present in natural waters is also affected by the different adsorption kinetics of selenate vs. selenite. Selenite has a strong affinity for a variety of common minerals at pH values < 7, where as selenate does not;[9] selenite also has a strong affinity for particulate organic matter.[10] Constituents adsorbing selenite include Al and Fe oxides, clay minerals, and calcite. Also microbial populations selectively assimilate selenite over selenate.[11] Due to the many mechanisms for selenite removal from waters, selenate is the major soluble Se species in natural waters.[11]

Another important factor controlling the ratio of selenate to selenite in natural waters is the microbial activity. Microbial activity has been reported to quickly reduce selenite[12] and selenate[12] as well as tellurate, tellurite, vanadate, molybdate, arsenate, and chlorate[12] suggesting that microbial reductions are important for changing the solubility and availability of elements, especially Se.

Organic Selenium

Selenium is required as an essential micronutrient for a host of mammals, birds, fishes, algae, and bacteria.[13] The Se analog of cysteine, selenocysteine (SeCys), plays a critical role in the enzyme glutathione peroxidase (EC

Encyclopedia of Water Science
DOI: 10.1081/E-EWS 120010212

Table 1 Chemical properties of selenium[2]

Property	
Atomic number	34
Atomic mass	78.96
Density (g cm^{-3})	4.79
Melting point (°C)	217
Boiling point (°C)	685.4
Atomic radius (um)	0.117
Hardness, relative units	2
Electronegativity, relative units ($Li = 1$)	2.4
Latent heat of fusion, J g^{-1} (cal g^{-1})	6.91 (16.5)
Heat of vaporization (J g^{-1})	272.98 (65.2)
Thermal conductivity, W (m °C)	0.293–0.766

1.11.1.9)[14] and regulates ribosome-mediated protein synthesis.[15] Selenium containing organic compounds noted includes selenomethionine, selenocystathionone, dimethylselenopropionic acid, methylselenomethionine, trimethylselenonium ion, and the volatile organics dimethyl selenide (DMSe) and dimethyldiselenide (DMDSe).[16,17] Selenium toxicity through enhanced incorporation of SeCys into protein disrupts the three-dimensional structure and impairs function due to pH differences between sulfhydryl and selenol bridges.[18] In a tragic event that emphasized the need to monitor Se levels in waters generated by agriculture, the inadvertent concentration of Se from agricultural drainage conveyed to evaporation ponds in San Joaquin Valley, California resulted in the formation of organic Se compounds from the assimilation of inorganic Se from the drainage waters[19] that resulted in death or impaired reproduction in aquatic wildlife[19] Selenomethionine has been reported to be the most toxic organic Se compounds ingested by waterfowl,[20] although no other organic Se compound has been tested for waterfowl toxicity.

Volatile Species

A major mechanism for Se cycling in the environment is the biological volatilization of assimilated inorganic Se. Challenger and North[21] first confirmed microbial volatilization of DMSe and since, other Se gases as hydrogen selenide (H_2Se), methaneselenol (CH_3SeH), and dimethyl selenenyl sulfide (CH_3SeSCH_3) have been identified. The two major Se gases of environmental importance are DMSe and DMDSe[22] and are important in fossil fuel emissions, during plant growth[23] and from soil microorganism exposed to inorganic Se as selenate or selenite.[24] Atmospheric Se gases are subject to several important processes such as reaction with hydroxyl radicals and ozone,[25] converted into particles[26] and

then removed from the atmosphere by dry or wet deposition. The biological emissions of volatile Se forms are as great as emissions from anthropogenic sources[22] and are an important mechanism for Se cycling.

Elemental Selenium

Elemental Se is allotrophic, not measurably soluble in water, and can exist as gray hexagonal, red monoclinic, and vitreous amorphous forms. In reducing environments, Se speciation is predicted by thermodynamics to be H_2Se, but this species is extremely unstable and is oxidized to elemental red Se. Microbial dissimilatory reduction of selenate or selenite to insoluble Se(0) forms can result in higher concentrations than predicted by the speciation and chemical reactivity of the soluble forms. Although anaerobic conditions have been reported to be necessary for the Se reduction to occur by facultative anaerobes,[27] recent research has found certain bacterium can reduce selenate under microaerophilic conditions to Se(0).[11]

In environmental systems, there are three major transformation mechanisms for Se: oxidation/reduction, mineralization/immobilization, and volatilization with the kinetics of each a function of the Se species, microbial activity, and pH–redox conditions. With the toxicity of Se at only approximately 50 times the dose required as an essential element, knowledge of the transformation mechanisms involved with cycling and processes of Se is vital for prevention of additional problem areas associated with water cycling.

REFERENCES

1. Haygarth, P.M. Global Importance and Global Cycling of Selenium. In *Selenium in the Environment*; Frankenberger, W.T., Jr., Benson, S., Eds.; Marcel Dekker, Inc.: New York, 1994; 1–25.
2. Newland, L.W. *Handbook of Environmental Chemistry*; Springer-Verlag: New York, 1982; 45–57.
3. Lakin, H.W. Selenium Accumulation in Soils and Its Adsorption by Plants and Animals. Geol. Soc. Am. Bull. **1972**, 83.
4. Long, R.H.B.; Benson, S.M.; Tokunaga, T.K.; Yee, A. Selenium Immobilization in a Pond Sediment at Kesterson Reservoir. J. Environ. Qual. **1990**, *19*, 302–311.
5. Fio, J.L.; Fujii, R. Selenium Speciation Methods and Application to Soil Saturation Extracts from San Joaquin Valley, California. Soil Sci. Soc. Am. J. **1990**, *45*, 363–369.
6. Geering, H.R.; Cary, E.E.; Jones, L.H.P.; Allaway, W.H. Solubility and Redox Criteria for the Possible Forms of Selenium in Soils. Soil Sci. Soc. Am. Proc. **1968**, *32*, 35–40.

7. Elrashidi, M.A.; Adriano, D.C.; Workman, S.M.; Lindsay, W.L. Chemical Equilibria of Selenium in Soils: A Theoretical Development. Soil Sci. **1987**, *144*, 141–152.

8. Runnells, D.D.; Lindberg, R.D. Selenium in Aqueous Solutions: The Impossibility of Obtaining a Meaningful Eh Using a Platinum Electrode, with Implications for Modeling Natural Waters. Geology **1990**, *18*, 212–215.

9. Goldberg, S.; Glaubig, R.A. Anion Sorption on a Calcareous, Montmorillonitic Soil-Selenium. Soil Sci. Soc. Am. J. **1988**, *52*, 954–958.

10. Cohen, R.; Schuhmann, D.; Sinan, F.; Vanel, P. The Role of Organic Coatings in the Enrichment of Marine Particles with Selenium. The Fixation of Selenite on an Adsorbed Amino Acids. Mar. Chem. **1992**, *40*, 249–271.

11. Losi, M.E.; Frankenberger, W.T., Jr. Reduction of Selenium by *Enterobacter cloacae*, SLD1a-1: Isolation and Growth of the Bacterium and Its Expulsion of Selenium Particles. Appl. Environ. Microbiol. **1997**, *63*, 3079–3084.

12. Bautista, E.M.; Alexander, M. Reduction of Inorganic Compounds by Soil Microorganism. Soil Sci. Soc. Am. Proc. **1972**, *36*, 918–920.

13. Stadtman, T.C. Some Selenium-Dependent Biochemical Processes. Adv. Enzymol. Relat. Areas Mol. Biol. **1979**, *48*, 1–28.

14. Epp, O.; Ladenstein, R.; Wendel, A. The Refined Structure of the Selenoenzyme Glutathione Peroxidase at 0.2-μm Resolution. Eur. J. Biochem. **1983**, *133*, 51–69.

15. Stadtman, T.C. Selenocysteine. Annu. Rev. Biochem. **1996**, *65*, 83–100.

16. Cooke, T.D.; Bruland, K.W. Aquatic Chemistry of Selenium: Evidence of Biomethylation. Environ. Sci. Technol. **1987**, *21*, 1214–1219.

17. Fan, T.W.M.; Lane, A.N.; Martens, D.A.; Higashi, R.M. Synthesis and Structure Characterization of Selenium Metabolites. Analyst **1998**, *123*, 875–884.

18. Frost, D.V.; Lish, P.M. Selenium in Biology. Annu. Rev. Pharmacol. **1975**, *15*, 259–284.

19. Maier, K.J.; Foe, C.G.; Knight, A.W. Comparative Toxicity of Selenate, Selenite, Seleno-DL-methionine, and Selen-DL-cystine to *Daphnia magna*. Environ. Toxicol. Chem. **1993**, *12*, 755–763.

20. Heinz, G.H.; Hoffman, D.J.; Gold, L.G. Impaired Reproduction of Mallards Fed an Organic Form of Selenium. J. Wildl. Manag. **1989**, *53*, 418–428.

21. Challenger, F.; North, H.E. The Production of Organ-Metalloidal Compounds by Microorganisms. II. Dimethyl Selenide. J. Chem. Soc. **1934**, 68–71.

22. Chasteen, T.G. Volatile Chemical Species of Selenium. In *Selenium in the Environment*; Frankenberger, W.T., Jr., Engberg, R., Eds.; Marcel Dekker, Inc.: New York, 1998; 589–612.

23. Lewis, B.G.; Johnson, C.M.; Broyer, T.C. Volatile Selenium in Higher Plants: The Production of Dimethyl Selenide in Cabbage Leaves by Enzymatic Cleavage of Se-methyl Selenomethionine Selenonium Salt. Plant Soil **1974**, *40*, 107–118.

24. Frankenberger, W.T., Jr.; Karlson, U. Environmental Factors Affecting Microbial Production of Dimethylselenide in a Selenium-Contaminated Soil. Soil Sci. Soc. Am. J. **1989**, *53*, 1435–1442.

25. Atkinson, R.S.; Aschmann, D.; Hasegawa, D.; Thompson-Eagle, E.T.; Frankenberger, W.T., Jr. Kinetics of the Atmospherically Important Reactions of Dimethyl Selenide. Environ. Sci. Technol. **1990**, *24*, 1326–1332.

26. Rael, R.M.; Tuazon, E.C.; Frankenberger, W.T., Jr. Gas-Phase Reactions of Dimethyl Selenide with Ozone and the Hydroxyl and Nitrate Radicals. Atmos. Environ. **1996**, *30*, 1221–1232.

27. Lortie, L.; Gould, W.D.; Rajan, S.; McCready, G.L.; Cheng, K.-L. Reduction of Selenate and Selenite by a *Pseudomonas stutzeri* Isolate. Appl. Environ. Microbiol. **1992**, *58*, 4043–4044.

Soil Macropores, Water and Solute Movement in

David E. Radcliffe
University of Georgia, Athens, Georgia, U.S.A.

S

INTRODUCTION

Macropores are large, continuous voids in soil and include structural, shrink–swell, and tillage fractures, old root channels, and soil fauna burrows (Fig. 1). They are important because they can increase infiltration and may result in bypass flow where water and solutes move rapidly through the profile and do not interact with the soil matrix. This is one type of preferential flow. Other types of preferential flow include finger flow[1] and funnel flow.[2] Reviews of macropores were published by Beven and Germann,[3] White,[4] Germann,[5] Brusseau and Rao,[6] Beven,[7] and Bouma.[8]

MACROPORE FLOW OF WATER

One of the earliest documentations of macropore flow was by Lawes et al.:[9]

> The drainage water of a soil may thus be of two kinds: it may consist[1] of rainwater that passes with but little change in composition down the open channels of the soil; of[2] of the water discharged from the pores of a saturated soil.

Suggested lower limits for macropore diameters and widths are in the 0.03 mm–3.00 mm range.[10,3,4] The lower limit would include some pores that would fill by capillarity and the upper limit would exclude all capillary pores. Consequently, the dominant driving force in macropore flow is gravity, whereas matrix flow is driven primarily by capillarity. Continuity is also an important feature of macropores.

To a certain extent, the effect of macropores on water flow can be incorporated into conventional flow equations based on Darcy's law by careful measurement on large undisturbed samples of the unsaturated hydraulic conductivity function ($K(h)$). For example, Jarvis and Messing[11] used a tension infiltrometer to measure $K(h)$ at values of h between -5 mm and -150 mm on 6 soils of contrasting texture. When the data were plotted ($\ln K(h)$ vs h), the best fit was two straight lines, the line near saturation being much steeper and representing macropores.

An approach based on Darcy's Law, however, will not capture the bypass effect of macropores. Therefore, a number of approaches have been developed for describing water flow in individual macropores. Flow in water-filled macropores that are cylinders or cracks has been described by a modified Poiseuille equation but the number and dimensions of macropores must be known.[4] In addition, using Poiseuille's law assumes that macropores are open-ended, which is probably not the case. Beven and Germann[12] developed a kinematic wave equation that allows flow down the sides of macropores not filled with water. Macropore water flux was described by a power function of the macropore water content. The power term is usually obtained through calibration.

Since macropores are, for the most part, noncapillary pores, it has been assumed that macropore flow cannot occur unless there is free water at the soil surface. Experimentally, however, macropore flow has been observed in very dry soils at the onset of a rain when these conditions are unlikely. For example, Shipitalo and Edwards[13] used intact soil blocks and added water with a rainfall simulator. They observed more macropore flow in blocks that were initially dry than in blocks that were at higher antecedent water content. This may be due to a hydrophobic organic soil surface that develops under dry conditions and causes free water to run across the surface and enter macropores.[14] In addition, Phillips et al.[15] showed that water could continue to flow in open macropores under slightly negative pressure potentials, provided a continuous water film was established on the wall over the full length of the macropore. Layers beneath the soil surface that impede water flow and can raise water potentials sufficiently cause free water to enter macropores below the surface.[16]

MACROPORE FLOW OF SOLUTES

The amount of water that flows via macropores is probably a small percentage of the total water flux in most cases and has a limited effect on the overall water balance. However, macropore flow has a very important effect on movement of solutes, especially adsorbed solutes with limited half-lifes. For example, the only way some pesticides can reach ground water may be through macropore flow.[17] This is due to the rapid movement of solutes via macropores and bypass of adsorption sites within the soil matrix.

Encyclopedia of Soil Science
DOI: 10.1081/E-EWS 120010329

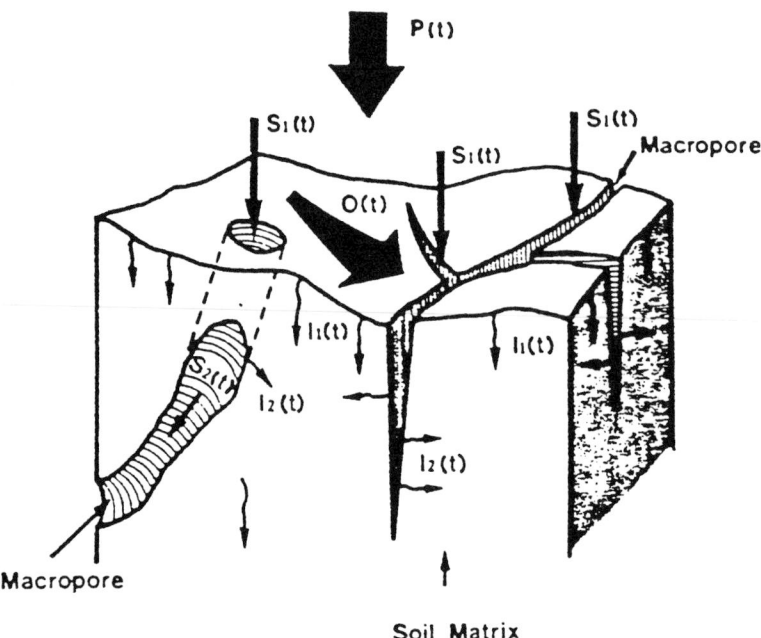

Figure 1 Infiltration into a block of soil with macropores. $P(t)$, overall input (precipitation, irrigation); $I_1(t)$, infiltration into the matrix from the surface; $I_2(t)$, infiltration into the matrix from the walls of the macropores; $S_1(t)$, seepage into the macropores at the soil surface; $S_2(t)$, flow within the macropores; $O(t)$, overland flow. (From Ref. [3].)

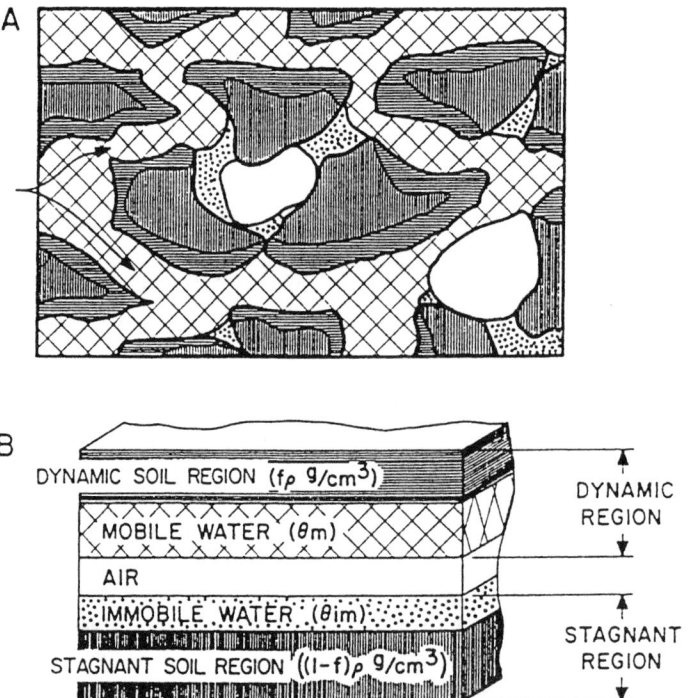

Figure 2 Schematic diagram of soil showing mobile and immobile water. (A) Actual model. (B) Simplified model. (From Ref. [20].)

Macropores do not always result in deeper movement of solutes. If a solute somehow enters the soil matrix, as a result of a small rain that did not cause surface ponding for example, then infiltrating water from larger storms may travel in macropores and bypass the solute.[18]

To a certain extent, the effects of macropore flow can be included in the conventional Convection Dispersion Equation (CDE) for solute transport through the use of large values of dispersivity (λ). However, the CDE assumes a high degree of mixing among flow paths, which is unlikely to occur at the local scale when macropores are present. In contrast, the stochastic convective log normal transfer function (CLT) is a stream-tube model that assumes there is no mixing among flow paths[19] and has been used to incorporate the effect of macropores.

Another common approach to simulating the effect of macropores is the dual porosity or mobile–immobile model developed by van Genuchten and Wierenga.[20] Soil porosity is divided into a relatively mobile (macropore) and a relatively immobile (matrix) region with exchange between regions by diffusion or convective flow (Fig. 2). Parameters describing the relative size of the regions and exchange are usually obtained through calibration using breakthrough curves.

The importance of local-scale macropore flow can be judged by its effect on field-scale solute transport. Local-scale macropore flow can cause dispersion of the field-scale breakthrough curve if it is a large source of variation in solute velocity, compared to the variation in mean local-scale solute velocities (v) within a field. A deterministic approach that includes macropore flow (by using a large value of local-scale λ, for example) is appropriate in this case. Variation in v within a field can also cause dispersion of the field-scale breakthrough curve. A stochastic approach that includes the variation in v is appropriate in this case. The two sources of dispersion can be compared using mean local-scale λ and coefficient of variation of local-scale v.[21] If variation in v is the principal source of field-scale dispersion, then local-scale macropore flow is of little consequence at the field scale.

A number of models have been developed that describe water and solute movement and include the effect of macropores using the dual porosity approach,[22,23] the CLT approach,[24,25] Poiseuille's Law,[26] the kinematic wave approach,[12] and Darcy's Law with $K(h)$ including macropores.[27]

REFERENCES

1. Hill, R.E.; Parlange, J.Y. Wetting Front Instability in Layered Soils. Soil Sci. Soc. Am. Proc. **1972**, *36*, 697–702.
2. Kung, K.-J.S. Preferential Flow in a Sandy Vadose Zone: 1. Field Observation. Geoderma **1990a**, *46*, 51–58.
3. Beven, K.J.; Germann, P.F. Macropores and Water Flow in Soils. Water Resour. Res. **1982**, *5*, 1311–1325.
4. White, R.E. The Influence of Macropores on the Transport of Dissolved and Suspended Matter Through Soil. Adv. Soil Sci. **1985**, 95–119.
5. Germann, P.F. Approaches to Rapid and Far-Reaching Hydrologic Processes in the Vadose Zone. J. Contam. Hydrol. **1988**, *3*, 115–127.
6. Brusseau, M.L.; Rao, P.S.C. Modeling Solute Transport in Structured Soils: A Review. Geoderma **1990**, *46*, 169–192.
7. Beven, K. Modeling Preferential Flow: An Uncertain Future? In *Preferential Flow*; Gish, T.J., Shirmohammadi, A., Eds.; Am. Soc. Agric. Eng.: St. Joseph, MI, 1991; 1–11.
8. Bouma, J. Influence of Soil Macroporosity on Environmental Quality. Adv. Agron. **1991**, *46*, 1–37.
9. Lawes, J.B.; Gilbert, J.H.; Warington, R. *On the Amount and Composition of the Rain and Drainage Water Collected at Rothamsted*; Williams, Clowes and Sons Ltd.: London, 1882.
10. Luxmoore, R.J. Micro-, Meso-, and Macroporosity of Soil. Soil Sci. Soc. Am. J. **1981**, *45*, 671.
11. Jarvis, N.J.; Messing, I. Near-Saturated Hydraulic Conductivity in Soils of Contrasting Texture Measured by Tension Infiltrometers. Soil Sci. Soc. Am. J. **1995**, *59*, 27–34.
12. Beven, K.J.; Germann, P.F. Water Flow in Soil Macropores II. A Combined Flow Model. Soil Sci. **1981**, *32*, 15–29.
13. Shipitalo, M.J.; Edwards, W.M. Effects of Initial Water Content on Macropore/Matrix Flow and Transport of Surface-Applied Chemicals. J. Environ. Qual. **1996**, *25*, 662–670.
14. Edwards, W.M.; Shipitalo, M.J.; Owens, L.B.; Norton, L.D. Water and Nitrate Movement in Earthworm Burrows Within Long-Term No-Till Cornfields. J. Soil Water Cons. **1989**, *44*, 240–243.
15. Phillips, R.E.; Quisenberry, V.L.; Zeleznik, J.M.; Dunn, G.H. Mechanism of Water Entry into Simulated Macropores. Soil Sci. Soc. Am. J. **1989**, *53L*, 1629–1635.
16. Andreini, M.S.; Steenhuis, T.S. Preferential Paths of Flow Under Conventional and Conservation Tillage. Geoderma **1990**, *46*, 85–102.
17. Kladivko, E.J.; van Scoyoc, G.E.; Monke, E.J.; Oates, K.M.; Pask, W. Pesticide and Nutrient Movement into Subsurface Tile Drains on a Silt Loam Soil in Indiana. J. Environ. Qual. **1991**, *20*, 264–270.
18. Shipitalo, M.J.; Edwards, W.M.; Dick, W.A.; Owens, L.B. Initial Storm Effects on Macropore Transport of Surface-Applied Chemicals in No-Till Soil. Soil Sci. Soc. Am. J. **1990**, *54*, 1530–1536.
19. Jury, W.A. Simulation of Solute Transport Using a Transfer Function Model. Water Resour. Res. **1982**, *18*, 363–368.
20. van Genuchten, M.Th.; Wierenga, P.J. Mass Transfer Studies in Sorbing Porous Media: I. Analytical Solutions. Soil Sci. Soc. Am. J. **1976**, *40*, 473–479.
21. Butters, G.L.; Jury, W.A. Field Scale Transport of Bromide in an Unsaturated Soil. 2. Dispersion Modeling. Water Resour. Res. **1989**, *25*, 1582–1588.

22. Addiscott, T.M. Modeling the Interaction Between Solute Leaching and Intra-ped Diffusion in Clay Soils. In *Proceedings of the ISSS Symposium of Water and Solute Movement in Heavy Clay Soils*; Bouma, J., Raats, P.A.C., Eds.; International Institute for Land Reclamation and Improvement: Wageningen, Netherlands, 1984; 279–297.

23. Hutson, J.L.; Wagenet, R.J. A Multiregion Model Describing Water Flow and Solute Transport in Heterogeneous Soils. Soil Sci. Soc. Am. J. **1995**, *59*, 743–751.

24. Utermann, J.; Kladivko, E.J.; Jury, W.A. Evaluating Pesticide Migration in Tile-Drained Soils with a Transfer Function Model. J. Environ. Qual. **1990**, *19*, 707–714.

25. White, R.E.; Dyson, J.S.; Haigh, R.A.; Jury, W.A.; Spositio, G. A Transfer Function Model of Solute Transport Through Soil 2. Illustrative Applications. Water Resour. Res. **1986**, *22*, 248–254.

26. Beven, K.J.; Clarke, R.T. On the Variation of Infiltration into a Homogeneous Soil Matrix Containing a Population of Macropores. Water Resour. Res. **1986**, *22*, 383–388.

27. Mohanty, B.P.; Bowman, R.S.; Hendrickx, J.M.H.; van Genuchten, M.Th. New Piecewise-Continuous Hydraulic Function for Modeling Prefential Flow in an Intermittent-Flood-Irrigated Field. Water Resour. Res. **1997**, *33*, 2049–2063.

Soil Moisture Measurement by Feel and Appearance

Rick P. Leopold
United States Department of Agriculture (USDA), Bryan, Texas, U.S.A.

INTRODUCTION

Measuring soil moisture by feel and appearance is one of several methods used to plan and determine the effectiveness of irrigation applications. Its simplicity makes it suitable to use in nearly all irrigation situations from urban lawns and golf courses to agricultural settings worldwide. Proper irrigation water management maximizes the positive impact of the irrigation water for the intended use, while minimizing the costs associated with irrigation and decreasing the potential for off-site movement of nutrients and pesticides from irrigated land.

BACKGROUND

Measuring soil moisture by using feel and appearance is a simple low cost method that may be used by land managers. In irrigated agriculture, this method can be used to:

- Determine when irrigation is needed.
- Estimate the available water in the root zone prior to planting or irrigation.
- Estimate the amount of irrigation water to apply.
- Determine the depth of penetration of irrigation water.

During the process of collecting soil samples for moisture assessment, the land manager will have an opportunity to identify restrictive layers caused by compaction, as well as, some nonwater related problems such as weed or insect pressure and nutrient deficiencies.

Prior to the collection of samples for estimating soil moisture, the land manager must determine the soil type, texture, and available water holding capacity of each layer sampled.[1] Soil texture, which is the relative amounts of sand, silt, and clay contained in soil, plays an important role in determining the amount of water a soil will hold.[2] The portion of water in the soil that can be readily used by plants is the available water capacity (AWC) of the soil.[3] The AWC ranges shown in Table 1 for various textural groups may be used as a guide in estimating soil moisture. Soil maps, soil texture, and AWC for each soil type can be found in a published soil survey that may be available through the local extension or agricultural agencies.

Table 1 Typical AWC (in./ft) for given textural range

		AWC
Coarse texture	Fine sand and loamy fine sand	0.6–1.2
Moderately coarse texture	Sandy loam and fine sandy loam	1.3–1.7
Medium texture	Sandy clay loam, loam, and silt loam	1.5–2.1
Fine texture	Clay, clay loam, and silty clay loam	1.6–2.4

(From Ref. [3].)

Table 2 Example for a uniform soil profile

Sample depth (in.)	Soil layer thickness (in.)	USDA texture by layer	Field capacity[a] (%)	AWC for layer[b] (in.)	Water available (in.)	Water needed to get to 100% field capacity (in.)
6	0–12	Sandy loam	30	1.4	0.42	0.98
18	12–24	Sandy loam	45	1.4	0.63	0.77
30	24–36	Loam	60	2.0	1.20	0.80
42	36–48	Loam	75	2.0	1.50	0.50
			Totals	6.8	3.75	3.05

[a] Estimated by feel and appearance.
[b] From soil survey.
(From Ref. [4].)

Encyclopedia of Water Science
DOI: 10.1081/E-EWS 120010151

Available Soil Moisture Remaining	Appearance of soil
0--25 percent available	Dry, soil aggregations separate easily; clods are hard to crumble with applied pressure.
25--50 percent available	 Slightly moist, forms a weak ball, very few soil aggregations break away, no water stains, and clods flatten with applied pressure.
50--75 percent available	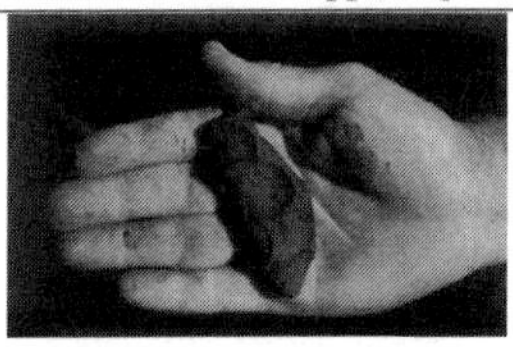 Moist, forms a smooth ball with defined finger marks, light soil/water staining on fingers, ribbons between thumb and forefinger.
75--100 percent available	 Wet, forms a ball, uneven medium to heavy soil/water coating on fingers, ribbons easily between thumb and forefinger.
100 percent available	Wet, forms a soft ball, free water appears on soil surface after squeezing or shaking, thick soil/water coating on fingers, slick and sticky.

Fig. 1 Fine sand and loamy fine sand soils. Percent available: Currently available soil moisture as a percent of available water capacity. (From Ref. [2].)

SAMPLING PROCEDURES

Soil moisture is typically sampled mid-way through 1-ft increments in uniform soils,[4] or mid-way through increments that correspond to the natural soil layers in the profile. For example, if a soil had 14 in. of fine sandy loam over clay, the first sample would be 7 in. deep, then sample in 1-ft increments thereafter to bottom of the root zone. For most agronomic crops, a sampling depth of 3 ft–4 ft will be sufficient to comprise the active root zone.[2] Table 2 provides an example for a uniform soil. Three or more sampling sites per field should be evaluated depending on the crop, field size, irrigation method, and soil variability.[1]

For each sample, the feel and appearance method involves the following:

1. Obtaining a soil sample at the selected depth by using a probe, auger, or shovel.
2. Squeezing the soil sample firmly in one hand several times to form an irregular ball.
3. Observing the ability to a form ball, ability to ribbon, loose particles, soil/water stains on fingers, and soil color. A ribbon is formed when soil is squeezed out of hand between the thumb and index finger. Note: A very weak ball falls apart in one bounce of the hand. A weak ball falls apart in 2–3 bounces.
4. Comparing observations with Figs. 1–4.

Available Soil Moisture Remaining	Appearance of soil
0--25 percent available	Dry, forms a very weak ball, clustered soil grains break away easily from ball.
25--50 percent available	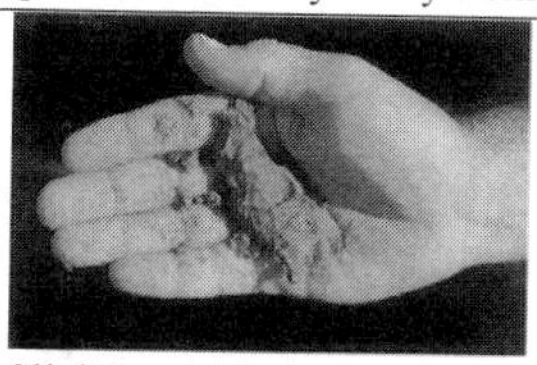 Slightly moist, forms a weak ball with defined finger marks, darkened color, no water staining on fingers, grains break away.
50--75 percent available	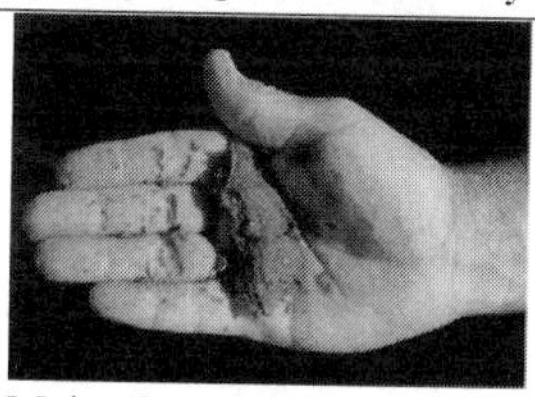 Moist, forms a ball with defined finger marks, very light soil/water staining on fingers, darkened color will not stick.
75--100 percent available	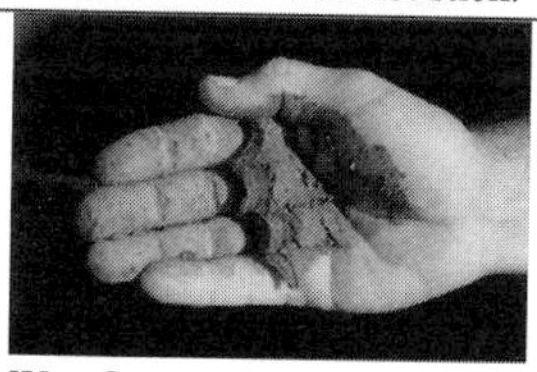 Wet, forms a ball with wet outline left on hand, light to medium staining on fingers, makes a weak ribbon between the thumb and forefinger.
100 percent available	Wet, forms a soft ball, free water appears briefly on soil surface after squeezing or shaking, medium to heavy soil/water coating on fingers.

Fig. 2 Sandy loam and fine sandy loam soils. Percent available: Currently available soil moisture as a percent of available water capacity. (From Ref. [2].)

Available Soil Moisture Remaining	Appearance of soil
0--25 percent available	Dry, soil aggregations break away easily, no staining on fingers, clods crumble with applied pressure.
25--50 percent available	 Slightly moist, forms a weak ball with rough surfaces, no water staining on fingers, few clustered soil grains break away.
50--75 percent available	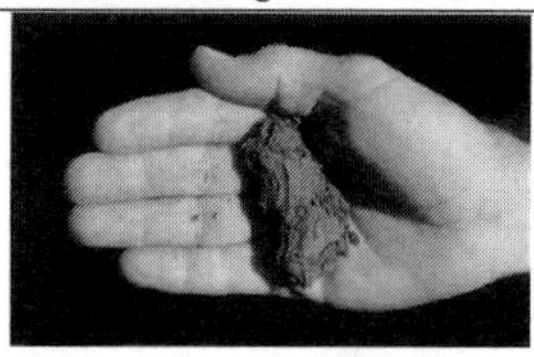 Moist, forms a ball, very light staining on fingers, darkened color, pliable, and forms a weak ribbon between the thumb and forefinger.
75--100 percent available	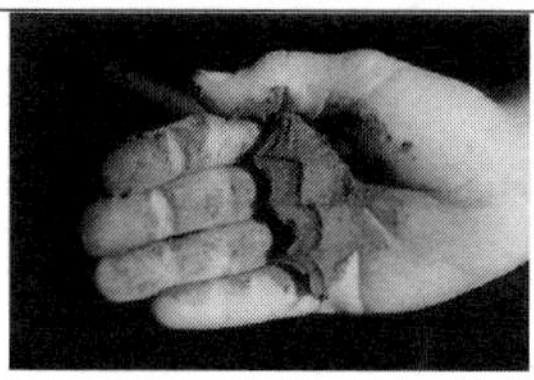 Wet, forms a ball with well-defined finger marks, light to heavy soil/water coating on fingers, ribbons between thumb and forefinger.
100 percent available	Wet, forms a soft ball, free water appears briefly on soil surface after squeezing or shaking, medium to heavy soil/water coating on fingers.

Fig. 3 Sandy clay loam, loam, and silt loam soils. Percent available: Currently available soil moisture as a percent of available water capacity. (From Ref. [2].)

Available Soil Moisture Remaining	Appearance of soil
0--25 percent available	Dry, soil aggregations separate easily; clods are hard to crumble with applied pressure.
25--50 percent available	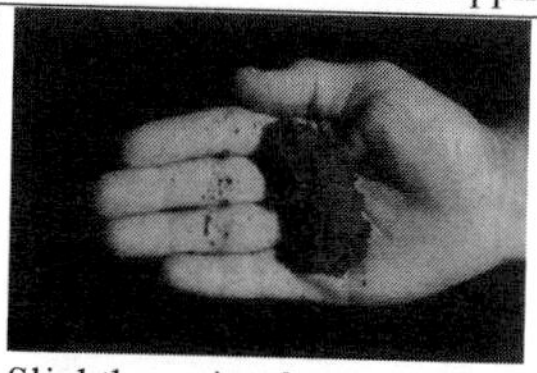 Slightly moist, forms a weak ball, very few soil aggregations break away, no water stains, and clods flatten with applied pressure.
50--75 percent available	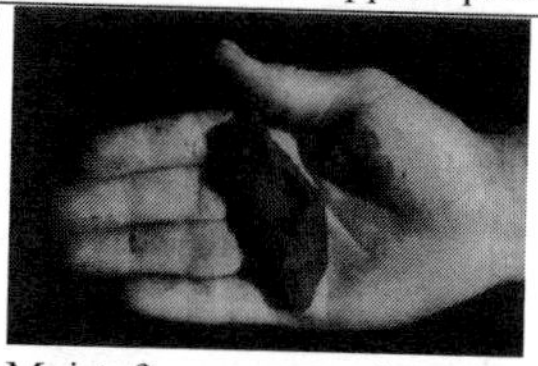 Moist, forms a smooth ball with defined finger marks, light soil/water staining on fingers, ribbons between thumb and forefinger.
75--100 percent available	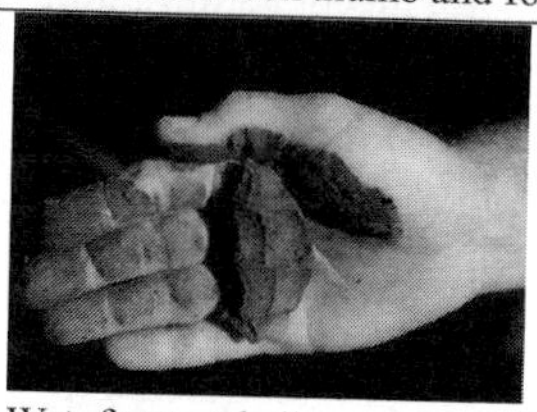 Wet, forms a ball, uneven medium to heavy soil/water coating on fingers, ribbons easily between thumb and forefinger.
100 percent available	Wet, forms a soft ball, free water appears on soil surface after squeezing or shaking, thick soil/water coating on fingers, slick and sticky.

Fig. 4 Clay, clay loam, and silty clay loam soils. Percent available: Currently available soil moisture as a percent of available water capacity. (From Ref. [2].)

REFERENCES

1. Risinger, M.; Wyatt, A.W.; Carver, K. *Estimating Soil Moisture by Feel and Appearance*; Water Management Note, High Plains Underground Water Conservation District No. 1: Lubbock, TX, 1985; 1–4.
2. Klocke, N.L.; Fischbach, P.E. *Estimating Soil Moisture by Appearance and Feel*; Publication G84-690-A, (http://www.ianr.unl.edu/pubs/irrigation/g690.htm) Nebraska Cooperative Extension Service: Lincoln, Nebraska, 1998; 1–4.
3. *Estimating Soil Moisture by Feel and Appearance*, Program Aid No. 1619; United States Department of Agriculture Natural Resources Conservation Service: Washington, DC, 1998; 1–12.
4. *National Engineering Handbook, Part 652, Irrigation Guide*, United States Department of Agriculture Natural Resources Conservation Service: Washington, DC, 1997; Chap. 15, 111–115.

Soil Salinity Measurement

Dennis L. Corwin
*United States Department of Agriculture (USDA),
Riverside, California, U.S.A.*

INTRODUCTION

The measurement of soil salinity is a quantification of the total salts present in the liquid portion of the soil. The measurement of soil salinity is important in agriculture because salinity reduces crop yields by 1) making it more difficult for the plant to extract water; 2) causing specific-ion toxicity; 3) influencing the soil permeability and tilth; and/or 4) upsetting the nutritional balance of plants. A discussion of the basic principles, methods, and equipment for measuring soil salinity is presented. The concise discussion provides a basic knowledge of the background, latest equipment, and current accepted methodologies for measuring soil salinity with suction cup extractors, porous matrix/salinity sensors, electrical resistivity, electromagnetic induction (EM), and time domain reflectometry (TDR).

SOIL SALINITY: DEFINITION, EFFECTS, AND GLOBAL IMPACTS

Soil salinity refers to the presence of major dissolved inorganic solutes in the soil aqueous phase, which consist of soluble and readily dissolvable salts including charged species (e.g., Na^+, K^+, Mg^{+2}, Ca^{+2}, Cl^-, HCO_3^-, NO_3^-, SO_4^{-2}, and CO_3^{-2}), non-ionic solutes, and ions that combine to form ion pairs. The predominant mechanism causing the accumulation of salt in irrigated agricultural soils is loss of water through evapotranspiration, leaving ever increasing concentrations of salts in the remaining water. Effects of soil salinity are manifested in loss of stand, reduced plant growth, reduced yields, and in severe cases, crop failure. Salinity limits water uptake by plants by reducing the osmotic potential making it more difficult for the plant to extract water. Salinity may also cause specific-ion toxicity or upset the nutritional balance of plants. In addition, the salt composition of the soil water influences the composition of cations on the exchange complex of soil particles, which influences soil permeability and tilth. Irrigated agriculture, which accounts for 35%–40% of the world's total food and fiber, is adversely affected by soil salinity on roughly half of all irrigated soils (totaling about 250 million ha) with over 20 million ha severely effected by salinity worldwide.[1] Because of these detrimental impacts, the measurement, monitoring, and real-time mapping of soil salinity is crucial to sustaining world agricultural productivity.

METHODS OF SOIL SALINITY MEASUREMENT

Historically, five methods have been developed for determining soil salinity at field scales: 1) visual crop observations; 2) the electrical conductance of soil solution extracts or extracts at higher than normal water contents; 3) in situ measurement of electrical resistivity; 4) noninvasive measurement of electrical conductance with EM; and most recently 5) in situ measurement of electrical conductance with TDR.

Visual Crop Observation

Visual crop observation is a quick and economical method, but it has the disadvantage that salinity development is detected after crop damage has occurred. For obvious reasons, the least desirable method is visual observation because crop yields are reduced to obtain soil salinity information. However, remote imagery is increasingly becoming a part of agriculture and potentially represents a quantitative approach to visual observation. Remote imagery may offer a potential for early detection of the onset of salinity damage to plants.

Electrical Conductivity of Soil Solution Extracts

The determination of salinity through the measurement of electrical conductance has been well established for decades.[2] It is known that the electrical conductivity (EC) of water is a function of its chemical composition. McNeal et al.[3] were among the first to establish the relationship between EC and molar concentrations of ions in the soil solution. Soil salinity is quantified in terms of the total concentration of the soluble salts as measured by the EC of the solution in $dS\ m^{-1}$.[2] To determine EC, the soil solution is placed between two electrodes of constant

Encyclopedia of Water Science
DOI: 10.1081/E-EWS 120010191

geometry and distance of separation.[4] At constant potential, the current is inversely proportional to the solution's resistance. The measured conductance is a consequence of the solution's salt concentration and the electrode geometry whose effects are embodied in a cell constant. The electrical conductance is a reciprocal of the resistance (Eq. 1):

$$EC_t = k/R_t \qquad (1)$$

where EC_t is the EC of the solution in $dS\,m^{-1}$ at temperature t (EC), k is the cell constant, and R_t is the measured resistance at temperature t. One $dS\,m^{-1}$ is equivalent to $1\,mmho\,cm^{-1}$.

Customarily, soil salinity has been defined in terms of laboratory measurements of the EC of the saturation extract (EC_e), because it is impractical for routine purposes to extract soil water from samples at typical field water contents. Partitioning of solutes over the three soil phases (i.e., gas, liquid, and solid) is influenced by the soil–water ratio at which the extract is made, so the ratio must be standardized to obtain results that can be applied and interpreted universally. Commonly used extract ratios other than a saturated soil paste are 1:1, 1:2, and 1:5 soil–water mixtures.

Soil salinity can also be determined from the measurement of the EC of a soil solution (EC_w). Theoretically, EC_w is the best index of soil salinity because this is the salinity actually experienced by the plant root. Nevertheless, EC_w has not been widely used to express soil salinity for various reasons: 1) it varies over the irrigation cycle as the soil water content changes and 2) methods for obtaining soil solution samples are too labor, and cost intensive at typical field water contents to be practical for field-scale applications.[5] For disturbed samples, soil solution can be obtained in the laboratory by displacement, compaction, centrifugation, molecular adsorption, and vacuum- or pressure-extraction methods. For undisturbed samples, EC_w can be determined either in the laboratory on a soil solution sample collected with a soil-solution extractor or directly in the field by using in situ, imbibing-type porous-matrix salinity sensors.

There are serious doubts about the ability of soil solution extractors and porous matrix salinity sensors (also known as soil salinity sensors) to provide representative soil water samples.[6–8] Because of their small sphere of measurement, neither extractors nor salt sensors adequately integrate spatial variability;[9–11] consequently, Biggar and Nielsen[12] suggested that soil solution samples are "point samples" that can provide qualitative measurement of soil solutions, but not quantitative measurements unless the field-scale variability is established. Furthermore, salinity sensors demonstrate a response time lag that is dependent upon the diffusion of ions between the soil solution and solution in the porous ceramic, which is affected by 1) the thickness of the ceramic conductivity cell; 2) the diffusion coefficients in soil and ceramic; and 3) the fraction of the ceramic surface in contact with soil.[13] The salinity sensor is generally considered the least desirable method for measuring EC_w because of its low sample volume, unstable calibration over time, and slow response time.[14]

Electrical Resistivity

Because of the time and cost of obtaining soil solution extracts, developments in the measurement of soil EC have shifted to the measurement of the soil EC of the bulk soil, referred to as the apparent soil electrical conductance (EC_a). The apparent soil EC measures the conductance through not only the soil solution but also through the solid soil particles and via exchangeable cations that exist at the solid–liquid interface of clay minerals. The techniques of electrical resistivity, EM, and TDR measure EC_a.

Electrical resistivity methods introduce an electrical current into the soil through current electrodes at the soil surface and the difference in current flow potential is measured at potential electrodes that are placed in the vicinity of the current flow (Fig. 1). These methods were developed in the second decade of the 1900s by Conrad Schlumberger in France and Frank Wenner in the United States for the evaluation of ground electrical resistivity.[16,17]

The electrode configuration is referred to as a Wenner array when four electrodes are equidistantly spaced in a straight line at the soil surface with the two outer electrodes serving as the current or transmission electrodes and the two inner electrodes serving as the potential or receiving electrodes.[18] The depth of penetration of the electrical current and the volume of measurement increase as the inter-electrode spacing, a, increases. For a homogeneous soil, the soil volume measured is roughly Πa^3. There are additional electrode configurations that are

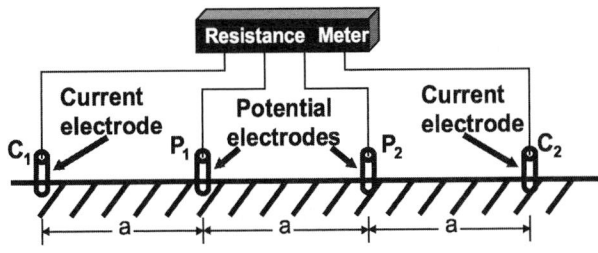

Fig. 1 Schematic of electrical resistivity of four electrodes (the Wenner array configuration). C_1 and C_2 represent the current electrodes, P_1 and P_2 represent the potential electrodes, and a represents the inter-electrode spacing. Modified from Rhoades and Halverson.[15]

frequently used, as discussed by Burger,[16] Telford et al.,[17] and Dobrin.[19]

By mounting the electrodes to "fix" their spacing, considerable time for a measurement is saved. A tractor-mounted version of the "fixed-electrode array" has been developed that geo-references the EC_a measurement with a GPS.[20–22] The mobile, "fixed-electrode array" equipment is well suited for collecting detailed maps of the spatial variability of average root zone soil electrical conductivity at field scales and larger. Veris Technologies[a] has developed a commercial mobile system for measuring EC_a using the principles of electrical resistivity.

Electrical resistivity (e.g., the Wenner array) and EM, are both well suited for field-scale applications because their volumes of measurement are large, which reduces the influence of local-scale variability. However, electrical resistivity is an invasive technique that requires good contact between the soil and four electrodes inserted into the soil; consequently, it produces less reliable measurements in dry, frozen, or stony soils than the non-invasive EM measurement. Nevertheless, electrical resistivity has a flexibility that has proven advantageous for field application, i.e., the depth and volume of measurement can be easily changed by altering the spacing between the electrodes.

Electromagnetic Induction

A transmitter coil located at one end of the EM instrument induces circular eddy-current loops in the soil with the magnitude of these loops directly proportional to the EC in the vicinity of that loop. Each current loop generates a secondary electromagnetic field that is proportional to the value of the current flowing within the loop. A fraction of the secondary induced electromagnetic field from each loop is intercepted by the receiver coil of the instrument and the sum of these signals is amplified and formed into an output voltage which is related to a depth-weighted soil EC_a. The amplitude and phase of the secondary field will differ from those of the primary field as a result of soil properties (e.g., salinity, water content, clay content, bulk density, and organic matter), spacing of the coils and their orientation, frequency, and distance from the soil surface.[23]

The two most commonly used EM conductivity meters in soil science and in vadose zone hydrology are the Geonics[b] EM-31, and EM-38. The EM-38 (Fig. 2) has had

considerably greater application for agricultural purposes because the depth of measurement corresponds roughly to the root zone (i.e., 1.5 m), when the instrument is placed in the vertical coil configuration. In the horizontal coil configuration, the depth of the measurement is 0.75 m–1.0 m. The operation of the EM-38 equipment is discussed in Hendrickx and Kachanoski.[23]

Mobile EM equipment developed at the Salinity Laboratory[20,22] is available for appraisal of soil salinity and other soil properties (e.g., water content and clay content) using an EM-38. Recently, the mobile EM equipment developed at the Salinity Laboratory was modified by the addition of a dual-dipole EM-38 unit (Fig. 3). The dual-dipole EM-38 conductivity meter simultaneously records data in both dipole orientations (horizontal and vertical) at time intervals of just a few seconds between readings. The mobile EM equipment is suited for the detailed mapping of EC_a and correlated soil properties at specified depth intervals through the root zone. The advantage of the mobile dual-dipole EM equipment over the mobile "fixed-array" resistivity equipment is the EM technique is noninvasive so it can be used in dry, frozen, or stony soils that would not be amenable to the invasive technique of the "fixed-array" approach due to the need for good electrode–soil contact. The disadvantage of the EM approach would be that the EC_a is a depth-weighted value that is nonlinear with depth McNeill.[24]

Time Domain Reflectometry

TDR was initially adapted for use in measuring water content. Later, Dalton et al.[25] demonstrated the utility of TDR to also measure EC_a, based on the attenuation of the applied signal voltage as it traverses the medium of interest[26]. Advantages of TDR for measuring EC_a include 1) a relatively noninvasive nature; 2) an ability to measure both soil water content and EC_a; 3) an ability to detect small changes in EC_a under representative soil conditions; 4) the capability of obtaining continuous unattended measurements; and 5) a lack of a calibration requirement for soil water content measurements in many cases.[26]

Soil EC_a has become one of the most reliable and frequently used measurements to characterize field variability for application to precision agriculture due to its ease of measurement and reliability[27]. Although TDR has been demonstrated to compare closely with other accepted methods of EC_a measurement,[28–31] it is still not sufficiently simple, robust, or fast enough for the general needs of field-scale soil salinity assessment.[5] Only electrical resistivity and EM have been adapted for the geo-referenced measurement of EC_a at field scales and larger.[5,27] Details for conducting a field-scale EC_a survey can be found in Corwin and Lesch.[32]

[a]Veris Technologies, Salina, Kansas, USA (www.veristech.com). Product identification is provided solely for the benefit of the reader and does not imply the endorsement of the USDA.
[b]Geonics Limited, Mississauga, Ontario, Canada. Product identification is provided solely for the benefit of the reader and does not imply the endorsement of the USDA.

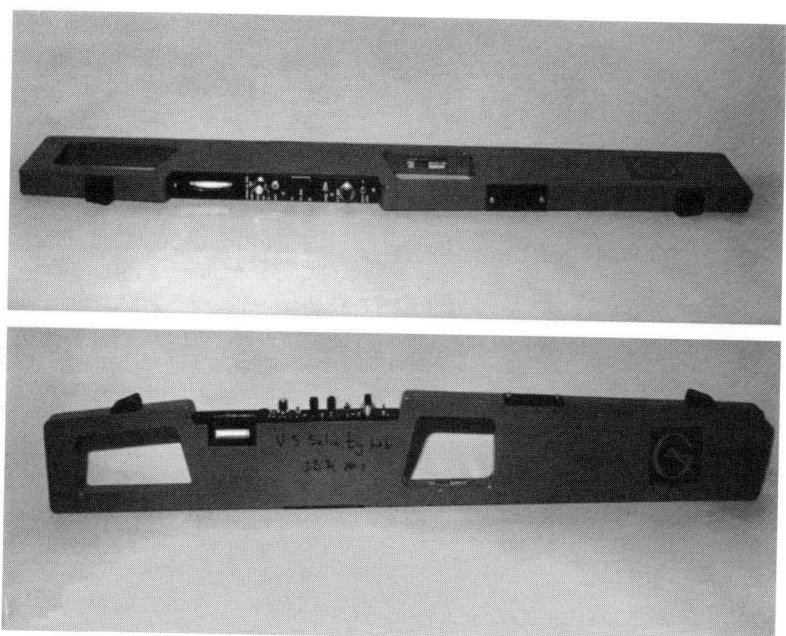

Fig. 2 Handheld Geonics EM-38 electromagnetic soil conductivity meter lying in the horizontal orientation with its coils parallel to the surface (top), and lying in the vertical orientation with its coils perpendicular to the surface (bottom). Courtesy of Rhoades et al.[5]

Fig. 3 Mobile dual-dipole EM-38 equipment for the continuous measurement of EC_a. Dual-dipole EM meter rests in the tail section or sled at the rear of the vehicle with a GPS antenna overhead at the midpoint of the meter.

FACTORS INFLUENCING THE APPARENT SOIL ELECTRICAL CONDUCTIVITY MEASUREMENT

Three pathways of current flow contribute to the apparent soil EC (EC_a) of a soil: 1) a liquid phase pathway via salts contained in the soil water occupying the large pores; 2) a solid–liquid phase pathway primarily via exchangeable cations associated with clay minerals; and 3) a solid pathway via soil particles that are in direct and continuous contact with one another.[5] Because of the three pathways of conductance, the EC_a measurement is influenced by several soil physical and chemical properties: 1) soil salinity; 2) saturation percentage; 3) water content; and 4) bulk density. The saturation percentage and bulk density are both closely associated with the clay content. Measurements of EC_a as a measure of soil salinity must be interpreted with these influencing factors in mind.

Another factor influencing EC_a is temperature. Electrolytic conductivity increases at a rate of approximately 1.9% per °C increase in temperature. Customarily, EC is expressed at a reference temperature of 25EC for purposes of comparison. The EC (i.e., EC_a, EC_e, or EC_w) measured at a particular temperature t ($^{\circ}$C), EC_t, can be adjusted to a reference EC at 25°C, EC_{25}, using the following equations from Handbook 60:[2]

$$EC_{25} = f_t \cong EC_t \qquad (2)$$

where

$$f_t = 1 - 0.20346(t) + 0.03822(t^2) - 0.00555(t^3) \qquad (3)$$

Traditionally, EC_e has been the standard measure of salinity used in all salt-tolerance plant studies. As a result, a relation between EC_a and EC_e is needed to relate EC_a back to EC_e, which in turn is related to crop yield.

REFERENCES

1. Rhoades, J.D.; Loveday, J. Salinity in Irrigated Agriculture. In *Irrigation of Agricultural Crops*, Agron. Monograph No. 30; Stewart, B.A., Nielsen, D.R., Eds.; SSSA: Madison, WI, 1990; 1089–1142.

2. U.S. Salinity Laboratory Staff, *Diagnosis and Improvement of Saline and Alkali Soils*, USDA Handbook 60; U.S. Government Printing Office: Washington, DC, 1954; 1–160.

3. McNeal, B.L.; Oster, J.D.; Hatcher, J.T. Calculation of Electrical Conductivity from Solution Composition Data as an Aid to In Situ Estimation of Soil Salinity. Soil Sci. **1970**, *110*, 405–414.

4. Bohn, H.L.; McNeal, B.L.; O'Connor, G.A. *Soil Chemistry*; John Wiley & Sons, Inc.: New York, 1979.

5. Rhoades, J.D.; Chanduvi, F.; Lesch, S. *Soil Salinity Assessment: Methods and Interpretation of Electrical Conductivity Measurements*, FAO Irrigation and Drainage Paper No. 57; Food and Agriculutre Organization of the United Nations: Rome, Italy, 1999; 1–150.

6. England, C.B. Comments on "A Technique Using Porous Cups for Water Sampling at Any Depth in the Unsaturated Zone." Water Resour. Res. **1974**, *10*, 1049.

7. Raulund-Rasmussen, K. Aluminum Contamination and Other Changes of Acid Soil Solution Isolated by Means of Porcelain Suction Cups. J. Soil Sci. **1989**, *40*, 95–102.

8. Smith, C.N.; Parrish, R.S.; Brown, D.S. Conducting Field Studies for Testing Pesticide Leaching Models. Int. J. Environ. Anal. Chem. **1990**, *39*, 3–21.

9. Amoozegar-Fard, A.; Nielsen, D.R.; Warrick, A.W. Soil Solute Concentration Distributions for Spatially Varying Pore Water Velocities and Apparent Diffusion Coefficients. Soil Sci. Soc. Am. J. **1982**, *46*, 3–9.

10. Haines, B.L.; Waide, J.B.; Todd, R.L. Soil Solution Nutrient Concentrations Sampled with Tension and Zero-Tension Lysimeters: Report of Discrepancies. Soil Sci. Soc. Am. J. **1982**, *46*, 658–661.

11. Hart, G.L.; Lowery, B. Axial-Radial Influence of Porous Cup Soil Solution Samplers in a Sandy Soil. Soil Sci. Soc. Am. J. **1997**, *61*, 1765–1773.

12. Biggar, J.W.; Nielsen, D.R. Spatial Variability of the Leaching Characteristics of a Field Soil. Water Resour. Res. **1976**, *12*, 78–84.

13. Wesseling, J.; Oster, J.D. Response of Salinity Sensors to Rapidly Changing Salinity. Soil Sci. Soc. Am. Proc. **1973**, *37*, 553–557.

14. Corwin, D.L. Chapter 6.1.3.3: Miscible Solute Transport—Solute Content and Concentration—Measurement of Solute Concentration using Soil Water Extraction: Porous Matrix Sensors. *Methods of Soil Analysis, Part 4, Physical Methods*; SSSA: Madison, WI, 2002.

15. Rhoades, J.D.; Halvorson, A.D. *Electrical Conductivity Methods for Detecting and Delineating Saline Seeps and Measuring Salinity in Northern Great Plains Soils*, ARS W-42; USDA-ARS Western Region: Berkeley, CA, 1977; 1–45.

16. Burger, H.R. *Exploration Geophysics of the Shallow Subsurface*; Prentice Hall PTR: Englewood Cliffs, NJ, 1992.

17. Telford, W.M.; Gledart, L.P.; Sheriff, R.E. *Applied Geophysics*, 2nd Ed.; Cambridge University Press: Cambridge, UK, 1990.

18. Corwin, D.L.; Hendrickx, J.M.H. Chapter 6.1.4.2: Miscible Solute Transport—Solute Content and Concentration—Indirect Measurement of Solute Concentration: Electrical Resistivity—Wenner Array. *Methods of Soil Analysis, Part 4, Physical Methods*; SSSA: Madison, WI, 2002.

19. Dobrin, M.B. *Introduction to Geophysical Prospecting*; McGraw-Hill Book Company: New York, 1960.

20. Carter, L.M.; Rhoades, J.D.; Chesson, J.H. In *Mechanization of Soil Salinity Assessment for Mapping*, Proceedings of the 1993 ASAE Winter Meetings, Chicago, IL, Dec 12–17, 1993; ASAE: St. Joseph, MO, 1993.

21. Rhoades, J.D. Instrumental Field Methods of Salinity Appraisal. In *Advances in Measurement of Soil Physical Properties: Bring Theory into Practice*, SSSA Special Publ. No. 30; Topp, G.C., Reynolds, W.D., Green, R.E., Eds.; ASA-CSSA-SSSA: Madison, WI, 1992; 231–248.

22. Rhoades, J.D. Electrical Conductivity Methods for Measuring and Mapping Soil Salinity. In *Advances in Agronomy*; Sparks, D.L., Ed.; Academic Press: San Diego, CA, 1993; Vol. 49, 201–251.

23. Hendrickx, J.M.H.; Kachanoski, R.G. Chapter 6.1.4.5: Miscible Solute Transport—Solute Content and Concentration—Indirect Measurement of Solute Concentration: Electromagnetic Induction. *Methods of Soil Analysis, Part 4, Physical Methods*; SSSA: Madison, WI, 2002.

24. McNeil, J.D. *Electromagnetic Terrain Conductivity Measurement at Low Induction Numbers*, Tech. Note TN-6; Geonics Limited: Ontario, Canada, 1980; 1–15.

25. Dalton, F.N.; Herkelrath, W.N.; Rawlins, D.S.; Rhoades, J.D. Time-Domain Reflectometry: Simultaneous Measurement of Soil Water Content and Electrical Conductivity with a Single Probe. Science **1984**, *224*, 989–990.

26. Wraith, J.M. Solute Content and Concentration–Indirect Measurement of Solute Concentration: Time Domain Reflectometry. *Methods of Soil Analysis*, Agronomy

Monograph No. 9, Part 1, 3rd Ed.; SSSA: Madison, WI, 2002 (in press).

27. Rhoades, J.D.; Corwin, D.L.; Lesch, S.M. Geospatial Measurements of Soil Electrical Conductivity to Assess Soil Salinity and Diffuse Salt Loading from Irrigation. In *Assessment of Non-Point Source Pollution in the Vadose Zone*, Geophysical Monograph 108; Corwin, D.L., Loague, K., Ellsworth, T.R., Eds.; AGU: Washington, DC, 1999; 197–215.

28. Heimovaara, T.J.; Focke, A.G.; Bouten, W.; Verstraten, J.M. Assessing Temporal Variations in Soil Water Composition with Time Domain Reflectometry. Soil Sci. Soc. Am. J. **1995**, *59*, 689–698.

29. Mallants, D.; Vanclooster, M.; Toride, N.; Vanderborght, J.; van Genuchten, M.Th.; Feyen, J. Comparison of Three Methods to Calibrate TDR for Monitoring Solute Movement in Undisturbed Soil. Soil Sci. Soc. Am. J. **1996**, *60*, 747–754.

30. Spaans, E.J.A.; Baker, J.M. Simple Baluns in Parallel Probes for Time Domain Reflectometry. Soil Sci. Soc. Am. J. **1993**, *57*, 668–673.

31. Reece, C.F. Simple Method for Determining Cable Length Resistance in Time Domain Reflectometry Systems. Soil Sci. Soc. Am. J. **1998**, *62*, 314–317.

32. Corwin, D.L.; Lesch, S.M. Application of Soil Electrical Conductivity to Precision Agriculture: Theory, Principles, and Guidelines. Agron. J. **2003**, (in press).

Soil Water, Antecedent

Sally D. Logsdon
United States Department of Agriculture (USDA), Ames, Iowa, U.S.A.

INTRODUCTION

Antecedent soil water is the amount of water in the soil before infiltration of new water, and is often used interchangeably with initial soil water. The infiltration rate is affected by the antecedent soil water content; however, this effect varies for wettable and nonwettable soils and for soil profiles that are uniform or vary with depth.

HOMOGENEOUS, WETTABLE SOIL

For a homogeneous, deep, wettable soil, infiltration decreases as initial soil water content increases[1–3] (see Fig. 1). This occurs because there is less water storage capacity when the soil is already partly wetted. Once the soil is wetted, the infiltration rate is controlled by the saturated hydraulic conductivity of the soil, i.e., the ability of the soil to transport water through the profile. The sorptivity parameter decreases as relative initial soil water content increases.[3] Sorptivity is a function of the square root of time, and is the cumulative amount of water infiltrated at relative time $t = 1$, which is the time when the infiltration rate is half the original rate. Sorptivity is proportional to the capillary forces, which are greater for dry soil. After the soil surface is saturated, the infiltration rate decreases exponentially to the final infiltration rate (Fig. 1).

Tisdall[4] examined ring infiltration measurements as a function of initial soil water content. As initial volumetric soil water content (θ) increased, the infiltration rate (I) at 2 hr decreased. The relationship between I (mm hr^{-1}) and θ (m^3 m^{-3}) varied with soil texture:

sandy loam soil: $I = 28.1 - 133\theta$

clay loam soil: $I = 1/(-0.224 + 1.62\theta)$

clay soil: $I = 1/(-0.218 + 1.32\theta)$.

The clay and clay loam soils were affected by soil cracking, that resulted in greater lateral spread of water that had infiltrated compared with the sandy loam soil that did not crack.

NONHOMOGENEOUS, WETTABLE SOIL

Soil may be nonhomogeneous due to soil layers of different texture (fine over coarse, or coarse over fine,[3]). A common example of layering is the fine surface seal that forms over the coarser whole soil underneath.

Water content at the soil surface influences the degree of breakdown of aggregates and formation of a surface seal. For an uncovered surface soil, the breakdown of surface aggregates decreases when the initial soil water content is higher.[5,6] This should decrease the formation of a surface seal when the soil is initially wet compared with an initially dry soil surface. The effect on infiltration is less direct because the surface seal may either increase or decrease infiltration, depending on other factors.[7] The surface seal formation contributes to the decline in infiltration rate with time for uncovered soils, but the formation of the surface seal results in a head gradient that allows infiltration to continue.[8]

Jones[9] (as reported in Ref. [10]) measured sprinkler infiltration (100 mm hr^{-1}) on a bare corn seedbed and on a seedbed covered with brome grass residue. The next day the infiltration measurements were repeated on the same site that had been wetted the previous day. For the bare sites, the I decreased from 21 to 7 mm hr^{-1} for the dry and prewetted measurements, and on the residue-covered sites the I decreased from 24 to 17 mm hr^{-1}. For the bare sites the sprinkler infiltration measurements resulted in formation of a surface seal, whereas a surface seal formation was minimized on the residue-covered sites. For the prewetted measurements, I was reduced less when surface seal formation was minimized.

NONWETTABLE SOIL

For some soils, the effect of initial soil water content on infiltration is complicated by water repellency, which may be more pronounced when the soil is dry. For reviews on water repellency, see Refs. [11,12]. Water repellency initially reduces the local infiltration rate,[13] but also creates instability. Instability may create wetting fingers that increase infiltration rate within the wetting finger. For some soils, water repellency is reduced as water content

Encyclopedia of Water Science
DOI: 10.1081/E-EWS 120010154
Published 2003 by Marcel Dekker, Inc.

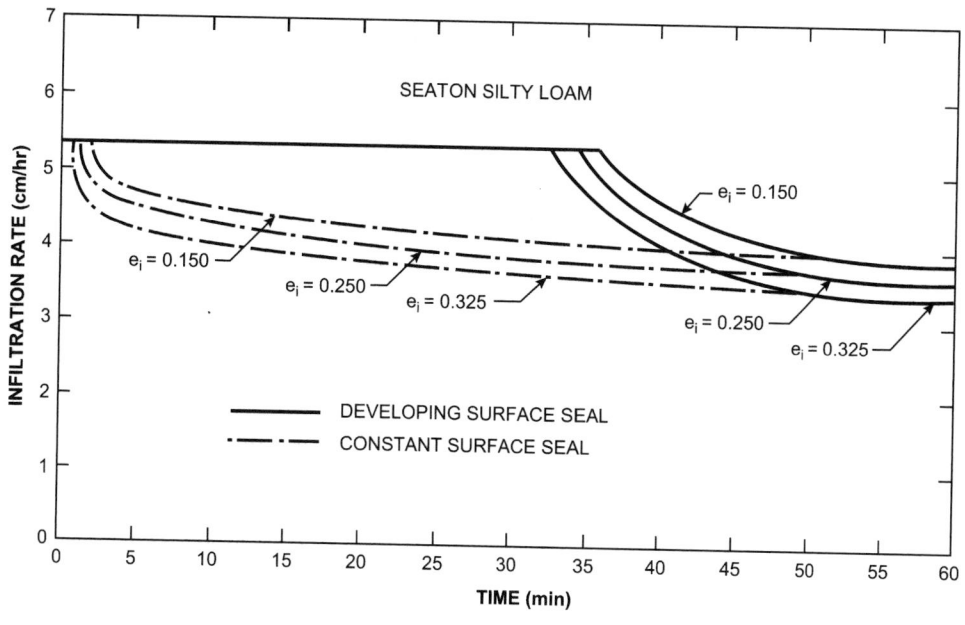

Fig. 1 Influence of antecedent soil water content and presence of a surface seal on infiltration rate as a function of time for a final infiltration rate of 3.83 cm hr^{-1}.[1]

increases, which may cause an increase in infiltration rate after the initial decrease.[14] Water that does not infiltrate may run off and infiltrate downslope. Nonwettable soil can contribute to macropore flow because water that does not readily enter the soil surface is routed to surface-connected macropores.[15] Because of unstable wetting, nonwettable soils have a nonhomogeneous wetting pattern even without the presence of macropores.

VARIABLE SOIL WATER IN THE LANDSCAPE

The landscape contributes to variability of antecedent soil water content because of surface and subsurface runoff and runon. Small- or large-scale depressions allow accumulation and ponding of water in localized zones. Conversely, the antecedent soil water content influences the landscape variability. If soil is initially dry but still wettable, infiltration is mostly vertical; however, if soil is initially wet, more water runs off or moves laterally in the subsurface.[16] The regions of water accumulation as well as transmission channels will be wetter.

The arrangement of soil water content across the landscape influences runoff at the larger catchment scale.[17] Predicted runoff was greater if there was a spatial structure (clustered arrangement or connectedness) of initial soil water content than for a random or constant initial soil water content.

REFERENCES

1. Moore, I.D. Effect of Surface Sealing on Infiltration. Trans. Am. Soc. Ag. Eng. **1981**, *24* (6), 1546–1552, 1561.
2. Beven, K. Infiltration, Soil Moisture, and Unsaturated Flow. In *Recent Advances in the Modeling of Hydrologic Systems*; Bowles, D.S., O'Connell, P.E., Eds.; Kluwer Academic Publishers: Dordrecht, The Netherlands, 1991; 137–151.
3. Kutílek, M.; Nielsen, D.R. Unsteady Infiltration, Dirichlet's Boundary Condition. *Soil Hydrology*; Catena-Verlag: Cremlingen-Destedt, Germany, 1994; Chap. 6.2.2., 140–159.
4. Tisdall, A.L. Antecedent Soil Moisture and Its Relation to Infiltration. Aust. J. Agric. Res. **1951**, *2* (2), 342–348.
5. Le Bissonnais, Y.; Bruand, A.; Jamagne, M. Laboratory Experimental Study of Soil Crusting: Relation Between Aggregate Breakdown Mechanisms and Crust Structure. Catena **1989**, *16*, 377–392.
6. Truman, C.C.; Bradford, J.M.; Ferris, J.E. Antecedent Water Content and Rainfall Energy Influence on Soil Aggregate Breakdown. Soil Sci. Soc. Am. J. **1990**, *54* (5), 1385–1392.
7. Le Bissonnais, Y. Aggregate Breakdown and Assessment of Soil Crustability. Eur. J. Soil Sci. **1996**, *47*, 425–437.
8. Mualem, Y.; Assouline, S. Flow Properties in Sealing Soils: Conceptions and Solutions. In *Advances in Soil Science, Soil Crusting Chemical and Physical Processes*; Sumner, M.E., Stewart, B.A., Eds.; Lewis Publishers: Boca Raton, 1992; 123–150.
9. Jones, B.A. Water Infiltration into Representative Soils of North Central Region. Univ. Ill. Agric. Exp. Stn. Bull. **1979**, 760.

10. Peterson, A.E.; Bubenzer, G.D. Intake Rate: Sprinkler Infiltrometer. In *Methods of Soil Analysis Part I. Physical and Mineralogical Methods*, 2nd Ed.; Klute, A., Ed.; ASA, SSSA, Inc.: Madison, 1986; 845–870.

11. Wallis, M.G.; Horne, D.J. Soil Water Repellency. Adv. Soil Sci. **1992**, *20*, 91–146.

12. DeBano, L.F. Water Repellency in Soils: A Historical Overview. J. Hydrol. **2000**, *231–232*, 4–32.

13. Burch, G.J.; Moore, I.D.; Burns, J. Soil Hydrophobic Effects on Infiltration and Catchment Runoff. Hydrol. Proc. **1989**, *3*, 211–222.

14. Clothier, B.E.; Vogeler, I.; Magesan, G.N. The Breakdown of Water Repellency and Solute Transport Through a Hydrophobic Soil. J. Hydrol. **2000**, *231–232*, 255–264.

15. Edwards, W.M.; Shipitalo, M.J.; Owens, L.B.; Dick, W.A. Factors Affecting Preferential Flow of Water and Atrazine Through Earthworm Burrows Under Continuous No-Till Corn. J. Environ. Qual. **1993**, *22*, 453–457.

16. Grayson, R.B.; Western, A.W.; Chiew, F.H.S. Preferred States in Spatial Soil Moisture Patterns: Local and Nonlocal Controls. Water Resour. Res. **1997**, *33* (12), 2897–2908.

17. Merz, B.; Plate, E.J. An Analysis of the Effects of Spatial Variability of Soil and Soil Moisture on Runoff. Water Resour. Res. **1997**, *33* (12), 2909–2922.

Soil Water, Capillary Rise of

James E. Smith
McMaster University, Hamilton, Ontario, Canada

S

INTRODUCTION

Capillary rise of water in soils is a phenomenon that has both beneficial and detrimental effects for agricultural soils. It is an important mechanism by which plants can draw water from below the root zone, but it is also a primary mechanism which can contribute to the accumulation of salts and resultant salination of soils.

CAPILLARITY

Capillary rise in soils refers to water moving upward from the water table against the force of gravity. Capillarity is the direct effect of the surface tension of the soil water (σ) and the affinity of water for the soil particles. The affinity for the soil is expressed as the contact angle (β) of the interface with the solid surface. That is, the water wets the surface of the soil particles and the interface between the two immiscible fluids (i.e., air and the water) is under tension. This fundamental relation is presented in most textbook on the subject of soil water, for example, Refs. 1–5. Fig. 1 is a schematic diagram showing the configuration of water and air within the pores of a soil.

Natural soil water tends to have lower surface tension than pure water primarily due to the presence of naturally occurring organic solutes.[6] However, this effect is only in the order of 10%–15% and consequently is commonly ignored. In addition, the contact angle in soils can vary. However, it is common to assume that the contact angle of a water wet soil is 0° (i.e., $\cos\beta = 1$).

The equilibrium height of rise of water (H_c) above a water table in a capillary tube can be expressed by

$$H = \frac{2\sigma\cos\beta}{\rho g r} \tag{1}$$

where ρ is the density of water, g is the acceleration due to gravity, β is the contact angle of the air–water interface with the solid surface, and r is the radius of the capillary tube.

The effect of water being pulled up into a capillary tube is due to the pressure difference across the air–water interface. The pressure difference across a curved interface

between two immiscible fluids, in this case air and water, is also expressed by the Laplace equation of capillarity,[4] i.e.,

$$P_c = P_A - P_w = \sigma\left(\frac{1}{r_1} + \frac{1}{r_2}\right) \tag{2}$$

where P_c is the capillary pressure, P_A is the absolute air pressure, P_w is the absolute water pressure, σ is the surface tension of the air–water interface, and r_1 and r_2 are the principle radii of curvature of the interface. Eqs. 1 and 2 can be directly related by expressing the pressures in Eq. 2 in head units, i.e., an equivalent depth of water. This leads to

$$H = \frac{P_c}{\rho g}, \tag{3}$$

and

$$\frac{2\cos\beta}{r} = \left(\frac{1}{r_1} + \frac{1}{r_2}\right) \tag{4}$$

A useful conceptual model is to envision the soil as a bundle of capillary tubes of various radii (Fig. 2, modified from Ref. 5). In the case of a soil, r represents the radius of an equivalent circular soil pore and the ratio $(\cos\beta)/r$ represents the equivalent radius of curvature of the air-water interface.

SOIL-WATER PRESSURE

A number of conventions exist for expressing the pressure of water in a soil. Soil-water pressure head (ψ) is negative relative to gauge pressure. This is because the water pressure (P_w) is less than atmospheric pressure (P_A), which is by definition zero gauge pressure. The terms soil tension, soil suction, soil capillary pressure, and soil matric potential are positive valued expressions of the same pressure. It is useful in the present context to express the soil-water pressure in head units relative to gauge pressure (atmospheric pressure), e.g., meters of water head less than atmospheric. At hydrostatic equilibrium (i.e., no flow conditions), the soil-water pressure head is equal in magnitude to the height above the water table (H) (Fig. 3).

Encyclopedia of Water Science
DOI: 10.1081/E-EWS 120010270

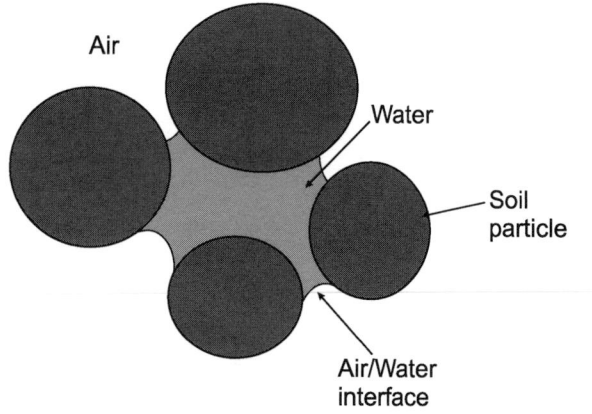

Fig. 1 Schematic of soil water.

It then follows from the above that

$$\psi = -H = -\frac{2\sigma \cos \beta}{\rho g r} \tag{5}$$

Using the bundle of capillary tubes conceptual model, it is readily apparent that the water content variation with depth in a soil in static equilibrium with a shallow water table will be an expression of the pore size distribution of the soil. It also follows that the pressure head is lower (more negative) at lower water contents as depicted in Fig. 3.

SOIL-WATER FLOWDYNAMICS

The proceeding discussion has focused on the static (no flow) condition and the fundamental relation between the pressure of the water and the water content in the soil. To extend our discussion to the conditions that induce capillary rise of water, i.e., upward flow, we need to consider the hydraulic gradient. The hydraulic head (h) in a soil is the sum of the pressure head (ψ) and the elevation head (z). It represents the ability to do mechanical work on a unit weight of water due to pressure differences and gravity. Water flows in soil when there exists a change in hydraulic head with distance, which means that the hydraulic gradient differs from zero. The volumetric flux of water in unsaturated soils can be expressed by the Darcy–Buckingham flux law,

$$q = -K(\psi)\frac{dh}{dz} = -K(\psi)\left[\frac{d\psi}{dz} + 1\right] \tag{6}$$

where q is the volumetric flux, K is the unsaturated hydraulic conductivity which is a function of pressure head, and dh/dz is the hydraulic gradient. When there is no hydraulic gradient there is no water flow.

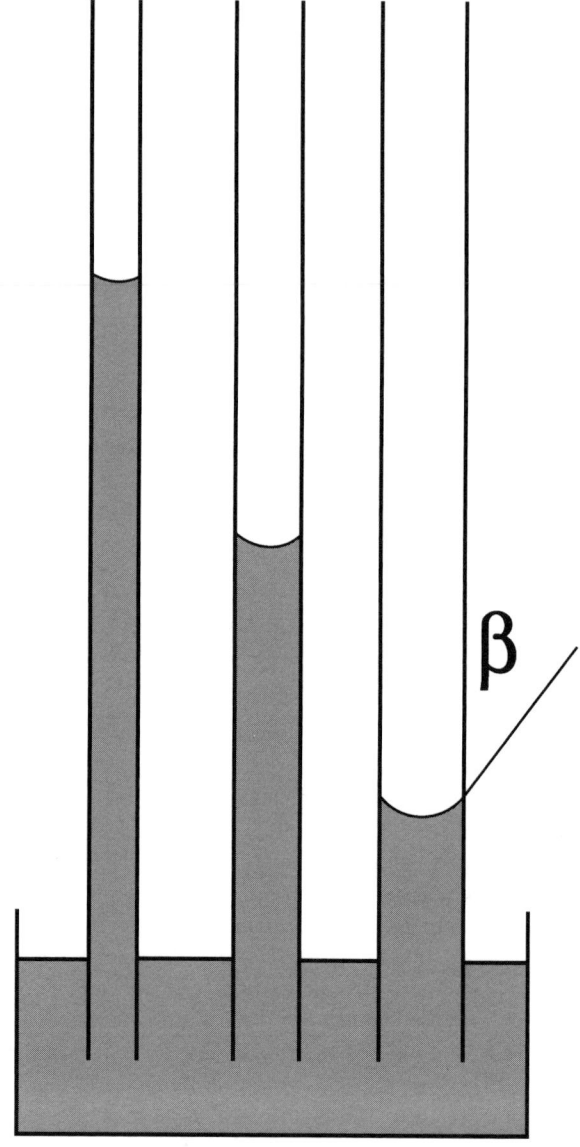

Fig. 2 Capillary rise in tubes of differing sizes.

Figs. 3 and 4 show cases of static (no flow) equilibrium with a water table. Fig. 3 shows the case for a homogeneous system. Fig. 4 depicts the case of a soil with layered heterogeneity having a finer middle layer. While these static equilibrium conditions essentially never exist in the field,[2] they are instructive as reference cases to explain the conditions that generate capillary rise. It should be noted that Fig. 4 depicts a no flow condition even though there are "wetter" layers and "drier" layers within the profile. This illustrates that the driving force for flow is the hydraulic head, not water content alone.

To induce capillary rise (vertical flow upward) there must be the condition that the hydraulic head increases with depth. Physically this means that pressure head must

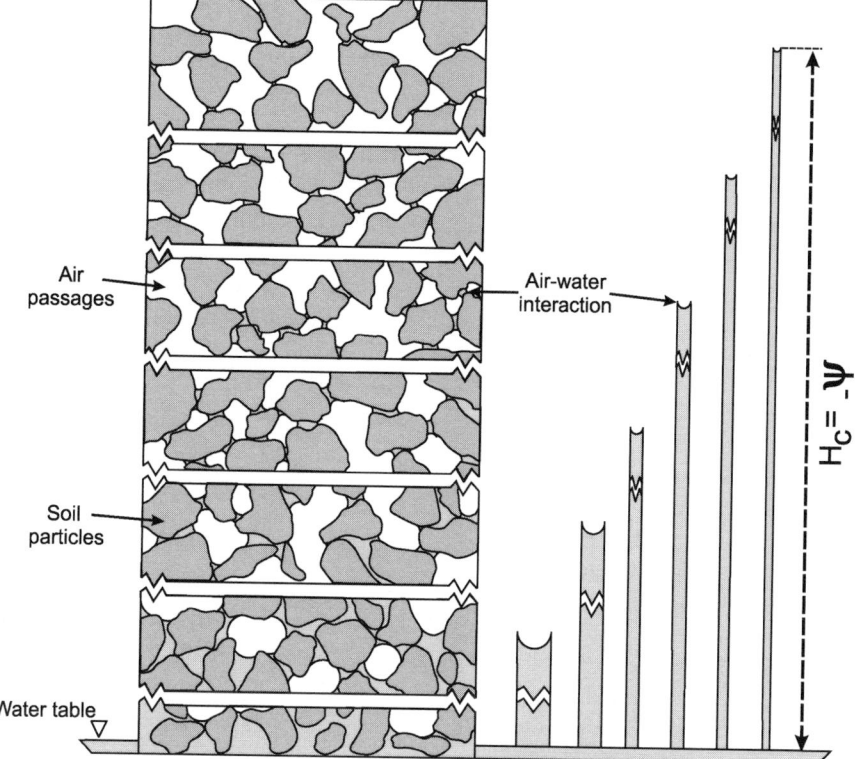

Fig. 3 Hydrostatic equilibrium in soil above a shallow water table. (Modified from Ref. [5], p. 28)

increase with depth more than elevation head decreases. Using upward as positive, the elevation head gradient is always one (i.e., $dz/dz = 1$), the pressure head gradient must be less than -1 to induce upward flow. This means that relative to the no flow conditions depicted in Figs. 3 and 4, there must be a mechanism that removes water from the soil and thereby creates a pressure head gradient greater than unity.

CAPILLARY RISE OF WATER IN SOILS

As expressed in Eq. 6, the actual rate of upward water flux depends not only on the hydraulic gradient but also on the unsaturated hydraulic conductivity function. However, capillary rise in soils can only occur when a hydraulic head gradient induces it.

It follows from the discussion above that the condition most favorable for capillary rise of water in soil is a shallow water table and loamy soil. Conversely, a deep

SOIL-WATER EVAPORATION AND EXTRACTION

There are two primary mechanisms in agricultural soils that remove water from shallow soils thereby potentially inducing capillary rise. One is direct evaporation from the soil surface. The second is extraction of soil water within the root zone. Fig. 5 is a schematic depicting the effect of water extraction from the root zone and/or by evaporation from the surface. The reduced pressure heads within the root zone generate a hydraulic gradient that drives upward flow of water.

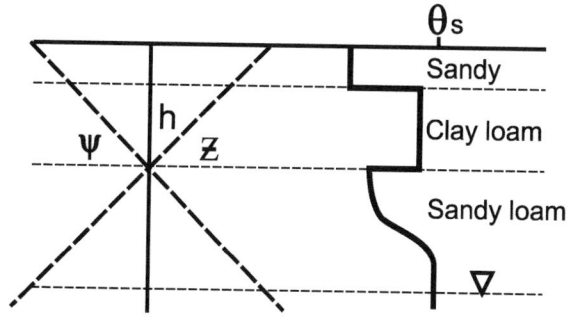

Fig. 4 Water content versus depth in layered heterogeneous soil for conditions of hydrostatic equilibrium.

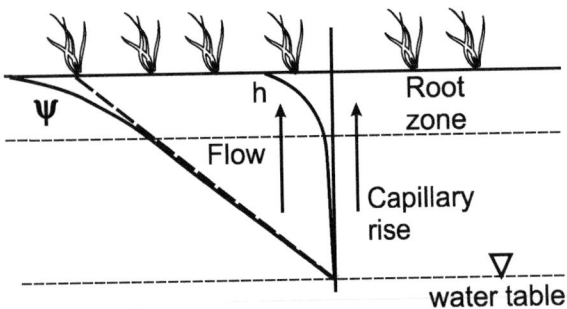

Fig. 5 Evaporation and root zone extraction of water inducing upward gradients.

tension of water generating a hydraulic gradient. When conditions prevail such that the capillarity induced upward pressure head gradient is greater than the downward acting gravity induced elevation head gradient, then soil water will flow upwards. Water losses from the soil by evapotranspiration contribute to maintaining upward fluxes by capillary rise. The magnitudes of those fluxes are higher when the water table is shallower. While this source of water is favorable for temporary drought conditions in temperate climates, it can be a direct cause of severe soil degradation by salination in arid and semiarid zones.

water table favors drainage. A very fine textured soil has low hydraulic conductivities at all water contents. A very coarse soil has large pores that drain readily and have little capillarity and low unsaturated hydraulic conductivities at those low water contents. While the exact flux of water delivered by capillary rise will vary based on soil hydraulic properties and climatic conditions it is instructive to consider a couple of example calculations of the effect of water table depth. Hillel[2] provides a sample calculation based on data and a formulation from Ref. 7 for the case of evaporation from a bare soil surface of a sandy loam soil with various water table depths. With the water table at depths of 0.9 m, 1.8 m, and 3.6 m the evaporative flux from the soil surface was 8 mm day^{-1}, 1 mm day^{-1}, and 0.12 mm day^{-1}, respectively. This illustrates that water table depth alone can have a large effect on fluxes due to capillary rise. It also shows that the potential for soil salination due to capillary rise of groundwater is consequently greater for shallower water table conditions.

CONCLUSION

Capillary rise of water in soil is a direct result of the affinity of water for natural materials and the surface

REFERENCES

1. Selker, J.S.; Keller, C.K.; McCord, J.T. Physical and Hydraulic Properties of Variably Saturated Media. *Vadose Zone Processes*; Lewis Publishers: Boca Raton, 1999; 21–84.
2. Hillel, D. Evaporation from Bare Soil and Wind Erosion. *Environmental Soil Physics*; Academic Press: London, 1998; 510–515.
3. Jury, W.A.; Gardner, W.R.; Gardner, W.G. Water Retention in Soil. *Soil Physics*; John Wiley and Sons, Inc.: New York, 1991; 34–65.
4. Bear, J. Flow of Immiscible Fluids. *Dynamics of Fluids in Porous Media*; American Elsevier Publishing Company Inc.: New York, 1972; 439–578.
5. Kirkham, D.; Powers, W.L. Static Water in Soils. *Advanced Soil Physics*; John Wiley and Sons, Inc.: New York, 1972; 1–37.
6. Tschapek, M.; Scoppa, C.O.; Wasowski, C. The Surface Tension of Soil Water. J. Soil Sci. **1978**, *29*, 17–21.
7. Gardner, W.R. Some Steady State Solutions of the Unsaturated Moisture Flow Equation with Application to Evaporation from a Water Table. Soil Sci. **1958**, *85* (4), 228–232.

Soil Water Diffusion

Laosheng Wu
University of California, Riverside, California, U.S.A.

INTRODUCTION

Diffusion of mass in a medium is a spontaneous process leading to the net movement of a substance from a region of high concentration to its adjacent regions of low concentrations. It takes place in the liquid, gas, and even solid phase. Diffusion results from the random thermal motion of the molecules. A general diffusion-type equation has many applications such as the transport of heat (Fourier's law), electricity (Ohm's law), and mass (Fick's law).

The movement of water in partially saturated soils is frequently described as a diffusion process. We will introduce soil water diffusivity in terms of Fickian diffusion.

To express diffusion process quantitatively, Fick in 1855 postulated that the one-dimensional diffusion flux (J) of a substance i is proportional to the concentration gradient:

$$J_i = -D_i \frac{dC_i}{dx},\qquad(1)$$

where D_i is the diffusion coefficient and C_i is the concentration for substance i, and x is the distance. Eq. 1 is referred to as Fick's first law of diffusion.

A mass balance for a certain volume of a medium with a unit area perpendicular to the x-axis requires the net flux (the difference between influx at position x and outflux at position $x + dx$) to be equal to the concentration change during an arbitrarily small time (dt):

$$\frac{\partial C_i}{\partial t} = -\frac{\partial J_i}{\partial x}.\qquad(2)$$

where the "−" sign indicates the direction of a flux. Eq. 2 is a continuity equation, which is a statement of mass conservation in mathematical form.

Substituting Eq. 1 into Eq. 2 results in:

$$\frac{\partial C_i}{\partial t} = \frac{\partial}{\partial x}\left(D_i \frac{\partial C_i}{\partial x}\right).\qquad(3)$$

Eq. 3 is Fick's second law. An equivalent equation will be used in the following discussion to describe liquid and vapor movement of water in unsaturated soils.

WATER DIFFUSION

Although the movement of liquid water in the soil is a convection rather than a diffusion process, the purpose of applying the diffusion concept to describe soil water movement is to simplify the mathematical and experimental treatment of unsaturated flow. In the diffusion theory, soil water flow is considered to be analogous to heat transmission in solids or to Fickian diffusion in gas or solution. Gardner and Widtsoe[1] and Childs[2] were among the first researchers to apply the diffusion concept to describe water movement in soils.

The soil water flux is given by the Buckingham–Darcy law[3] using the following assumptions:

1. The driving force for water flow in an isothermal, rigid, and unsaturated soil with no solute membrane and zero air pressure potential (zero gauge pressure) is the soil matric head $h(\theta)$, which is a function of water content (θ, $L^3 L^{-3}$).
2. Transfer of potential energy of water is always perfectly correlated with the transfer of a water mass at the scale of a representative volume.
3. The hydraulic conductivity of unsaturated soil is a function of the water content or matric head.
4. The flow of water in unsaturated soil is assumed to be localized, i.e., the soil pores where flows exist are saturated, but there is no moving water in air-filled pores.

The horizontal water flux according to the Buckingham–Darcy law is:

$$J_w = -K(\theta)\frac{\partial h(\theta)}{\partial x},\qquad(4)$$

where $K(\theta)$ ($L\,T^{-1}$) is the unsaturated hydraulic conductivity, which may change several orders of magnitude over the range of values for θ.

Applying chain rule allows Eq. 4 to be written in terms of water content gradient ($\partial\theta/\partial x$):

$$J_w = -K(\theta)\frac{dh}{d\theta}\frac{d\theta}{dx} = -D(\theta)\frac{\partial\theta}{\partial x},\qquad(5)$$

where $\partial h/\partial\theta$ is the slope of water retention ($h-\theta$) curve, $D(\theta) = K(\theta)\,dh/d\theta \equiv K(\theta)/C(h)$ is called hydraulic (or

Encyclopedia of Water Science
DOI: 10.1081/E-EWS 120010207

soil–water) *diffusivity* ($L^2 T^{-1}$), and $C(h)$ is the *specific water capacity* (L^{-1}). The diffusivity is usually expressed as a function of θ, but it may also be given in terms of h.

Substituting Eq. 5 into the continuity Eq. 2 yields:

$$\frac{\partial \theta}{\partial t} = \frac{\partial}{\partial x}\left[D(\theta)\frac{\partial \theta}{\partial x}\right]. \tag{6}$$

This is the diffusion equation for horizontal flow, or the "θ-based" formulation of the Richards' equation for flow in unsaturated soils.

In the special case where the hydraulic diffusivity remains constant with respect to the x, the above equation can be written as:

$$\frac{\partial \theta}{\partial t} = D\frac{\partial^2 \theta}{\partial x^2}. \tag{7}$$

To account for the effect of gravity on water flow, x in Eq. 4 is replaced by z (distance in vertical direction), and $h(\theta)$ is replaced by the hydraulic head, $H = h(\theta) + z$. The Richards' equation for vertical water movement is:

$$\frac{\partial \theta}{\partial t} = \frac{\partial}{\partial z}\left[D(\theta)\frac{\partial \theta}{\partial z}\right] + \frac{\partial K(\theta)}{\partial z}. \tag{8}$$

In defining $D(\theta)$ we tacitly assumed that $K(\theta)$, $h(\theta)$ and $dh(\theta)/d\theta$ are unique functions of θ. In other words, the $D(\theta)$ relationship of a soil is the same for both imbibition and drainage of water. This is rarely the case with practical problems, however, where *hysteresis* will occur due to differences in initial water content, the "ink bottle" effect, the contact angle (i.e., "raindrop" effect), entrapped air, and swelling and shrinking.[3] One should therefore be

mindful of hysteresis when applying the above hydraulic functions.

In addition to the use of the well-known diffusion equation, the advantage of the θ-based Richards' equation (Eq. 8) is that the magnitude of the hydraulic diffusivity varies considerably less with θ or h as does the hydraulic conductivity. Disadvantages are that the equation cannot be used to model water flow in soils at or near saturation since D becomes infinite in that range.[4] In addition, due to the abrupt transition (discontinuity) of water content [and hence $C(h)$] from one layer to another, the water diffusion equation can only be applied to uniform soil profiles.

HYDRAULIC DIFFUSIVITY

Since $D(\theta)$ is defined as the ratio of the hydraulic conductivity to the specific water content, it can be viewed as the ratio of the flux to the soil–water content gradient when gravitational and hysteresis effects can be neglected. Thus $D(\theta)$ provides a measure of the rate of water movement through soil.

Measurement of hydraulic diffusivity can either be done in the laboratory or in the field, depending on the purposes of the measurements, sample sources, equipment availability, and the desired range of water content. The most common laboratory procedure is the nonsteady-state method[5] based on a Boltzmann transformation. The instantaneous profile method, however, is a popular method for the field.

In the nonsteady-state Boltzmann transformation method, a Boltzmann variable, $\lambda = xt^{1/2}$, is used to transform Eq. 6 to an ordinary differential equation. During horizontal infiltration, one measures the water content distribution along the x-axis direction at one or more distinct times. By plotting $\lambda = xt^{1/2}$ vs. θ and evaluating the slope $d\lambda/d\theta$ and integral $\int\lambda(\theta)d\theta$ one can obtain $D(\theta)$.

The instantaneous profile method[6] employs Richards' equation in its mixed form (i.e., both θ and h are dependent variables):

$$\frac{\partial \theta(z,t)}{\partial t} = \frac{\partial}{\partial z}\left[K(\theta)\frac{\partial H(z,t)}{\partial z}\right], \tag{9}$$

By integrating Eq. 9 with respect to z between $z = 0$ (soil surface) to any depth $z = L$, one can determine $K(\theta)$ at any desired depth from analysis of the $\theta(z, t)$ and $H(z, t)$ profiles measured at frequent time intervals. The hydraulic head $H(z, t)$ can be measured with tensiometers while the water content can be measured with time-domain reflectometry (TDR) or neutron thermalization. Alternatively, the water content may be inferred from the

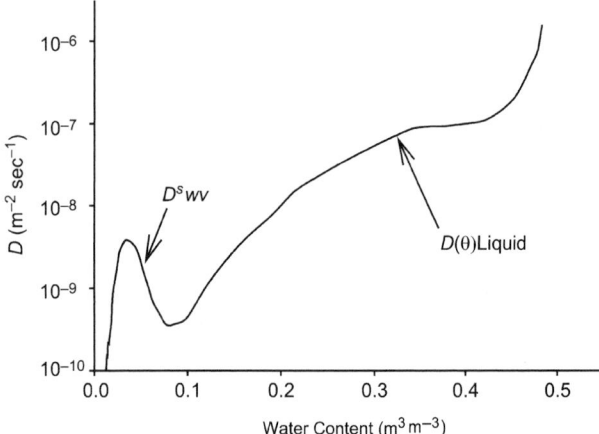

Fig. 1 Relationship between hydraulic diffusivity and soil water content for a Yolo light clay. Vapor diffusion is dominant mode of water movement for $\theta < 0.06$. (Reproduced after Philip[8].)

tensiometer readings and water retention curves for soil samples that are measured in the laboratory. Since the method measures water content and matric potential simultaneously, the hydraulic diffusivity can readily be calculated from $K(\theta)$ using Eq. 5.

Measurements of $D(\theta)$ typically provide data for only a limited range of soil water contents. Such measurements are often costly, time consuming, and inaccurate. Instead, it is often convenient to use indirect estimates of the soil water retention, hydraulic conductivity, and hydraulic diffusivity curves from more widely available data such as soil texture.[7]

WATER VAPOR DIFFUSION

Vapor transfer is an important mechanism for water movement under relatively dry soil conditions (Fig. 1). Eqs. 1 and 3 can be used to describe vapor diffusion in porous media with minor alterations that account for the reduction of cross-sectional area due to solid and liquid barriers and the reduced concentration gradient and longer pathway in soils. To account for these factors, a tortuosity factor (ξ_{wv}) is introduced to calculate the water vapor diffusion coefficient in soil (D_{wv}^s) :

$$D_{wv}^s = \xi_{wv} D_{wv}^a, \tag{10}$$

where D_{wv}^a is water vapor diffusion coefficient in air. Millington and Quirk's method[9] is one of the most commonly used models for estimating ξ_{wv}:

$$\xi_{wv} = a^{10/3}/\phi^2 = (\phi - \theta)^{10/3}/\phi^2, \tag{11}$$

where a is air-filled porosity and N is total porosity of a soil. After substituting D_{wv}^a and Δ_{wv} (water vapor concentration, $M L^{-3}$) for D_i and C_i in Eq. 1, one obtains the water vapor flux equation:

$$J_{wv} = -D_{wv}^s \frac{d\rho_{wv}}{dx}. \tag{12}$$

Unless the soil is very dry, the relative humidity of the soil air is close to saturated. Thus the above equation can be expressed in terms of temperature and $\rho_{wv}^*(T)$, which is the saturated vapor density as a function of temperature:

$$J_{wv} = -\xi_{wv} D_{wv}^a \frac{d\rho_{wv}^*(T)}{dT}\frac{\partial T}{\partial x} \tag{13}$$

The above equation underestimates water vapor movement by several folds. Philip and de Vries[10] indicated that Eq. 11 underestimates the tortuosity effect because vapor transfer can occur through "short circuiting." Therefore, they proposed to use total porosity for calculating the tortuosity factor. Another reason why Eq. 11 underestimates water vapor transfer might be that the temperature gradient in the vapor phase is much greater than that in the bulk phase. When these two factors were included, the modified equation predicted experimental measured water vapor transfer with greater accuracy.

REFERENCES

1. Gardner, W.; Widtsoe, J.A. The Movement of Soil Moisture. Soil Sci. **1921**, *11*, 215–233.
2. Childs, E.C. The Transport of Water Through Heavy Clay Soils: I, III. J. Agric. Sci. **1936**, *26*, 114–141, see also pp. 527–545.
3. Jury, W.A.; Gardner, W.R.; Gardner, W.H. *Soil Physics*, 5th Ed.; John Wiley & Sons, Inc.: New York, 1991.
4. Hillel, D. *Environmental Soil Physics*; Academic Press: San Diego, 1998.
5. Bruce, R.R.; Klute, A. The Measurement of Soil–Water Diffusivity. Soil Sci. Soc. Am. Proc. **1956**, *20*, 458–462.
6. Green, R.E.; Ahuja, L.R.; Chong, S.K. Hydraulic Conductivity, Diffusivity, and Sorptivity of Unsaturated Soils: Field Methods. In *Methods of Soil Analysis. Part 1. Physical and Mineralogical Methods*, Agronomy Monograph No. 9, 2nd Ed.; Klute, A., Ed.; American Society of Agronomy, Inc.: Madison, WI, 1986; 771–798.
7. Mualem, Y. Hydraulic Conductivity of Unsaturated Soils: Prediction and Formulas. In *Methods of Soil Analysis. Part 1. Physical and Mineralogical Methods*, Agronomy Monograph No. 9; Klute, A., Ed.; Soil Science Society of America, Inc.: Madison, WI, 1986; 799–823.
8. Philip, J.R. Water Movement in Soil. In *Heat and Mass Transfer in the Biosphere*; de Vries, D.A., Afgan, N.H., Eds.; Halsted Press-Wiley: New York, 1974; 29–47.
9. Millington, R.J.; Quirk, J.P. Permeability of Porous Solids. Trans. Faraday Soc. **1961**, *57*, 1200–1207.
10. Philip, J.R.; de Vries, D.A. Moisture Movement in Porous Materials Under Temperature Gradients. Trans. Am. Geophys. Un. **1957**, *38*, 222–228.

Soil Water Energy Concepts

Sally D. Logsdon
United States Department of Agriculture (USDA), Ames, Iowa, U.S.A.

INTRODUCTION

The state of soil water is often described in energy relations. Kinetic energy is the result of motion and temperature fluctuations, but most of soil water energy is described by potential energy. The historic development of the soil water potential energy concept is described in Refs. 1–3.

Although the concept of soil water retention was early recognized as important for sustained plant growth, it was not linked with soil water potential energy until later.[1] The difference in soil water energy state is often of greater interest than the state itself.[3]

THERMODYNAMICS

Definitions

Thermodynamically, the potential of soil water at a given position is the amount of work required to move a parcel of water isothermally and reversibly from a reference state to the soil site.[4,5] Table 1 summarizes the type of soil water energy examined, what parcel of water is being moved, what conditions are held constant between the reference state and the soil site, and what conditions are changed in the soil site compared with the reference site. Soil water energy can be given as energy per mass (chemical potential in $J\,kg^{-1}$), energy per volume (soil water potential or pressure in $N\,m^{-2}$ or Pa), or energy per unit weight (soil water pressure head in m). Energy per volume can be calculated by multiplying energy per mass by the density of water. Energy per weight can be calculated by dividing energy per mass by the density of water and acceleration due to gravity.[6,7]

At potential energy equilibrium, soil water does not move. Thermodynamic equilibrium of soil water energy[2] depends on thermal equilibrium (uniform temperature throughout or isothermal), mechanical equilibrium (no convection forces), and chemical equilibrium (no net diffusion forces nor chemical reactions). The hydraulic head is the sum of gravitational, matric, and hydrostatic heads (Table 1, Fig. 1). Fig. 1 shows hydraulic equilibrium (no movement of soil solution).

The direction of soil solution movement is from high hydraulic pressure head to low hydraulic pressure head. Notice from Table 1 that hydraulic head is the work to move soil solution; whereas, osmotic pressure is the work to move pure water.[2] Osmotic pressure considerations are often due to a membrane or interface that is permeable to water but not permeable or only partly permeable to solutes. Examples are the water–air interface (vapor movement), and the cell membranes in the root (water uptake).

Iwata, Tabuchi, and Warkentin[3] describe five types of work that thermodynamically affect soil water. The first is compression, but within pressure ranges normally encountered in the soil, water is incompressible. The second is surface tension, due to the air–water surface, or due to the interface between the soil particle and water. The third is electrical work, in which the state of water (a dielectric material) is altered when placed in an electrical field. The fourth is gravitational, due to the external gravitational field, which influences water movement. The last is conservative force field, due to the colloid surface effects on water properties sorbed to the surface.

Other Factors and Terminology

Different terms are sometimes used for the various components of soil water potential.[4,5,7] Matric potential can be called wetness potential or tensiometric potential, which is a negative potential due to an interaction with soil pore walls. The matric potential can be divided into a matric potential occurring without an external load present, and with an overburden potential. The overburden potential is also called envelope-pressure potential, due to the weight of overlying soil and water on a swelling soil.[1,8,9] The overburden potential reduces the water content at a given water potential, compared with an unconfined sample at the same water potential. Air potential is considered another component of pressure potential, and is often called pneumatic potential due to air pressure applied externally. Hydrostatic potential is sometimes called submergence potential, or positive potential below the groundwater level.

Other factors can contribute to water potential,[3] such as electrical potential due to the electrical field generated by charged colloid surfaces, radius of meniscus curvature

Encyclopedia of Water Science
DOI: 10.1081/E-EWS 120010325

Table 1 The type of soil water energy upon which work is done, what parcel of water is being moved, what conditions are held constant between the reference state and the soil site, and what conditions are changed in the soil site compared with the reference site

Energy source	Reference parcel	Constant	Destination change
Gravitational	Soil solution	Atmospheric pressure	Elevation
Matric, above water table	Equivalent soil solution	Atmospheric pressure, elevation	Solution in soil
Hydrostatic, below water table	Equivalent soil solution	Atmospheric pressure, elevation	Solution in soil
Hydraulic	Equivalent soil solution	Osmotic pressure	Elevation, solution in soil
Air	External air pressure	Elevation	Soil air
Osmotic	Pure water	Atmospheric pressure, elevation	Soil solution

effect on surface tension, van der Waal forces that draw surfaces together, internal pressure that develops in water to balance a nonuniform field such as van der Waals force or gravity, and temperature. If a temperature gradient is present, kinetic energy may be introduced, so movement in response to a water potential gradient assumes isothermal conditions.

Complex Interactions

Macroscale processes are not always easily described by microscale thermodynamics. If osmotic gradients occur, mass flow of soil solution can follow a macroscale gradient in one direction, and diffusion of pure water can follow a microscale gradient in the another.[2,7] Also at the macroscale, the contribution of electrical potential, van der Waals forces, and internal pressure are often ignored.[3] Discrepancies arise because water is not homogeneous, but has different properties near solutes in the water and near colloid surfaces.[3]

The definition for water potential describes the work to move reversibly a parcel of water from the reference state to a point in the soil. Yet water movement is usually not completely reversible, resulting in hysteresis.[7] At a given water potential, the soil is wetter during drainage than during wetting due to variation in pore sizes, different contact angles during wetting and drainage, and heterogeneities due to interactions of water with clay surfaces.[3]

MEASUREMENT OF SOIL WATER POTENTIAL

The working definition for soil water potential is not a practical way for determining soil water energy in the field, because a parcel of water cannot realistically be moved from a reference state to the field site. The pressure potential can be measured with a piezometer (below the water table for hydrostatic potential) or a tensiometer (above the water table for matric potential). The piezometer is a hollow tube in the soil with a slotted screen opened at the bottom of the tube. The screened end of the peizometer is placed in the soil so that it is in contact with the soil solution. The height of water in the piezometer relative to a reference level is the pressure potential.[10] A tensiometer is also a tube that is connected to a ceramic cup that is placed in contact with the unsaturated soil.[3,11] The tube is filled with de-aired water and sealed. The force of the soil water matric potential on the water in the tensiometer creates a vacuum (negative pressure) that is read by a manometer, pressure gage, or pressure transducer. The reading is subtracted from the elevation of the gage. Even so, the measured matric potential is only an apparent matric potential.[2] The matric potential measured by a tensiometer may not indicate the microscale condition because there is no correction for for water–clay interactions.[3] Most tensiometers cannot measure water potential more negative than 10 m (around 1 atm), because air moves into the ceramic and the vacuum is broken. A technique that is used to determine water potential in drier soil involves the use of a thermocouple

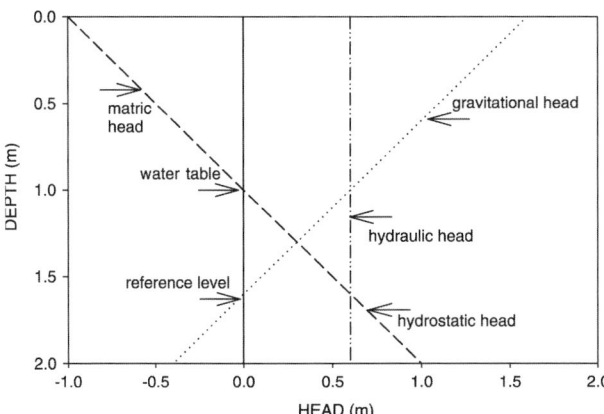

Fig. 1 Diagram of equilibrium hydraulic head with the water table at 1.0 m and the reference level at 1.6 m.

psychrometer that relates the vapor pressure to a soil water potential, i.e., the sum of osmotic and matric components.[12,13]

CONCLUSION

The soil water hydraulic pressure head is composed of matric, overburden, and hydrostatic pressures. The soil solution will flow from higher hydraulic pressure heads to lower hydraulic pressure heads. Piezometers are used to measure the hydrostatic pressure head, and tensiometers are used to measure the negative matric pressure head. The osmotic pressure head is due to solutes in the soil water. Pure water in the soil will move from higher osmotic pressure heads to lower osmotic pressure heads, usually caused by semipermeable membranes. Thermocouple psychrometers can be used to measure a combination of matric and osmotic pressure heads.

REFERENCES

1. Groenevelt, P.H.; Bolt, G.H. Water Retention in Soil. Soil Sci. **1972**, *113* (4), 238–245.
2. Corey, A.T.; Klute, A. Application of the Potential Concept to Soil Water Equilibrium and Transport. Soil Sci. Soc. Am. J. **1985**, *49* (1), 3–11.
3. Iwata, S.; Tabuchi, T.; Warkentin, B.P. Energy Concept and Thermodynamics of Water in Soil. *Soil–Water Interactions, Mechanisms and Applications*, 2nd Ed.; Marcel Dekker, Inc.: New York, 1995; 1–67.
4. Bolt, G.H. Soil Physics Terminology. Report of the Terminology Committee of Commission I of the Int. Soil Sci. Soc. Bull. Int. Soil Sci. Soc. **1976**, *48*, 16–22.
5. Soil Sci. Soc. Am. (Committee Paper) *Glossary of Soil Science Terms*; Soil Sci. Soc. Am.: Madison, WI, 1996; 1–134.
6. Jury, W.A.; Gardner, W.R.; Gardner, W.H. Water Retention in Soil. *Soil Physics*, 5th Ed.; John Wiley and Sons, Inc.: New York, 1991; Chapter 2, 34–72.
7. Hillel, D. Content and Potential of Soil Water. *Environmental Soil Physics*; Acedemic Press: San Diego, 1998; Chapter 6, 129–172.
8. Talsma, T. Measurement of the Overburden Component of Total Potential in Swelling Field Soils. Aust. J. Soil Res. **1977**, *15*, 95–102.
9. Amer, A.M.M. Effect of the Overburden Load Pressure on Matric Potential of an Alluvial Soil. Egypt. J. Soil Sci. **1988**, *28* (1), 1–8.
10. Reeve, R.C Water Potential: Piezometry. In *Methods of Soil Analysis Part 1—Physical and Mineralogical Methods*, 2nd Ed.; Klute, A., Ed.; ASA, SSSA: Madison, WI, 1986; 545–562.
11. Cassel, D.K.; Klute, A. Water Potential: Tensiometry. In *Methods of Soil Analysis Part 1—Physical and Mineralogical Methods*, 2nd Ed.; Klute, A., Ed.; ASA, SSSA: Madison, WI, 1986; 563–596.
12. Brown, R.W.; Oosterhuis, D.M. Measuring Plant and Soil Water Potentials with Thermocouple Psychrometers: Some Concerns. Agron. J. **1992**, *84* (1), 78–86.
13. Rawlins, S.L.; Campbell, G.S. Water Potential: Thermocouple Psychrometry. In *Methods of Soil Analysis Part 1—Physical and Mineralogical Methods*, 2nd Ed.; Klute, A., Ed.; ASA, SSSA: Madison, WI, 1986; 597–618.

Soil Water Flow Under Saturated Conditions

Jan W. Hopmans
Graham E. Fogg
University of California, Davis, California, U.S.A.

INTRODUCTION

Water flow in soils may occur in both unsaturated and saturated conditions; however, clear differentiation between flows requires a review of definitions first.[1] In the saturated zone, it is generally assumed that the pore space within the soil matrix is saturated with water, and that the hydrostatic pressure in the water is greater than atmospheric pressure. In contrast, in the unsaturated zone, the pore space is only partly filled with water, resulting in a soil water pressure smaller than atmospheric pressure. The region between the unsaturated zone and the groundwater is called the capillary fringe, where the soil is satiated but where the soil water is held by capillary forces (Fig. 1). The difference between satiated and saturated water content is caused by the general presence of entrapped air within the soil matrix of saturated soils. The unsaturated zone is bounded by the soil surface at the top and merges with the groundwater of an unconfined aquifer in the capillary fringe of the water table or phreatic surface at the bottom. By definition, the phreatic surface is the soil depth at which the water pressure is atmospheric.

The distinction between groundwater and unsaturated zone is usually made within a hydrologic context, emphasizing water as the agent of change of the subsurface and the main driver for transport of chemicals between the atmosphere and groundwater. This region of the unsaturated zone is also known as the vadose zone that incorporates local saturated regions, and is so defined to emphasize the desired integration of physical, chemical, and biological processes in the unsaturated soil zone and its interactions with the groundwater and atmosphere. The soil is the most upper part of the vadose zone, subject to fluctuations in water and chemical content by infiltration and leaching, water uptake by plant roots, and evaporation from the soil surface. Within the context of crop production, the spatial scale of interest is the field scale.

The need to incorporate saturated soil conditions in the root zone comes about for many reasons, affecting both water quantity and water quality. First and foremost, above a shallow groundwater, the capillary fringe bounds the bottom of the plant root zone. Water infiltration by either rainfall or irrigation causes a rise of the water table, thereby temporarily creating anaerobic soil conditions,

unfavorable for plant growth. In contrast, up to 40% of plant transpiration may come from shallow water table contribution, depending on soil texture, distance between the water table and bottom of the root zone, and shallow groundwater salinity. Historically, much research has been invested in designing improved water management practices, such as by soil drainage.[2,3] Secondly, saturated soil conditions can also occur by regional groundwater flow, as caused by rising water tables elsewhere, or by subsurface flow along hill slopes.[4] Thirdly, local saturated soil water conditions can occur by spatial variations and layering of soil texture, creating favorable conditions for preferential saturated water flow towards the groundwater, thereby affecting groundwater and surface water quality by accelerated transport of surface-applied chemicals. Consequently, spatial variations in soil hydraulic conductivity within the unsaturated zone can lead to local or extended regions with positive soil water pressure values, causing the so-called perched water. A temporary saturated soil in the plant root zone causes anoxic conditions, thereby affecting chemical and microbial processes. For example, it can lead to dissolution of salts, chemical transformation of specific chemical compounds, and denitrification of applied fertilizers.

BASIC LAWS OF SATURATED WATER FLOW

Rather than characterizing soil water flow at the pore scale, soil water quantities are usually defined at the macroscopic level by using the continuum approach, at which each soil phase is regarded as a continuum. The flow of liquid water through the macroscopic soil matrix is generally viscous and laminar, because of the small pore sizes in which water movement takes place. Under these conditions, the water flux density or specific discharge q (m sec^{-1}) is proportional to the driving force or total head gradient, $\Delta H/L$ (dimensionless), where ΔH denotes the change in piezometric head over the distance L. In unsaturated flow, it is customary to use total head (soil water flow under unsaturated conditions), instead of piezometric head. For one-dimensional flow, the magnitude of q is defined as the volume of water V (m^3) passing through a cross-sectional

Encyclopedia of Water Science
DOI: 10.1081/E-EWS 120010267

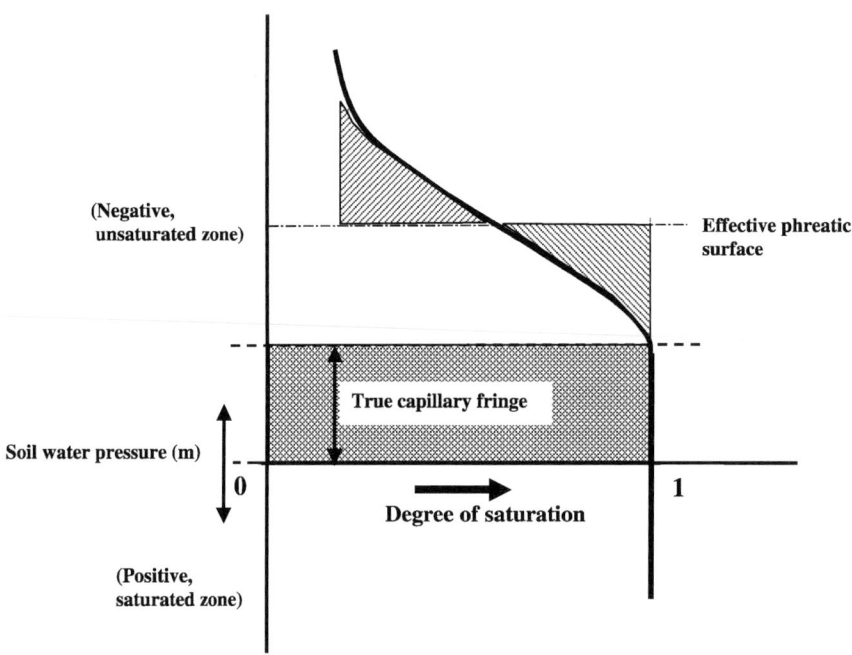

Fig. 1 The capillary fringe, separating the unsaturated and the saturated soil zone.

area of soil A (m^2) normal to the direction of flow for time t (sec). The head H (m) represents the potential energy (J) of water on a weight (N) basis. It incorporates the influences of forces such as gravity and pressure. The relationship between water flux and head gradient is known as Darcy's law:

$$q = -K \frac{\Delta H}{L} \tag{1}$$

where the proportionality factor is defined as the saturated hydraulic conductivity K (m sec^{-1}). Thus, the units for the water flux density and the hydraulic conductivity are identical. The decrease in hydraulic head in the direction of flow is caused by friction or drag forces as water moves through the small tortuous flow paths of the soil matrix. The value for the water flux density, or the Darcy flux, should not be mistaken for the average velocity of the water in the pores, known as the seepage or pore-water velocity v (m sec^{-1}). The difference between the water flux density and the pore-water velocity is due to the fact that water only occupies a limited fraction of the soil's total volume. In the case of a water-saturated soil, the pore-water velocity is equal to the ratio of the Darcy flux and the soil porosity.

K varies between and within soil types, as it depends on soil properties such as texture, solid particle arrangement (soil structure), organic matter content, and water content. In addition, for anisotropic soils, K varies with spatial direction (x, y, or z), as caused by the process of soil deposition or soil formation. Even for isotropic soils, K is

usually heterogeneous because of soil spatial variability. In most applications, though, the soil's saturated hydraulic conductivity is assumed to be characterized by a single K-value, assuming isotropy and homogeneity.

Fig. 1 shows a schematic presentation of the soil water distribution between the saturated and unsaturated zone, by plotting soil water pressure as a function of degree of saturation. The capillary fringe separates the phreatic surface from the unsaturated zone. For practical purposes, one sometimes approximates the top of the capillary fringe as the distance above the phreatic surface, above which soil water is immobile, and below which the soil is saturated. While conserving mass, an effective phreatic surface (see Fig. 1) is then chosen that includes this approximate capillary fringe.[1] For conditions where saturated water flow is dominantly horizontal, i.e., the pressure distribution in the vertical direction is hydrostatic and the variations in H are much smaller than total aquifer thickness, the hydraulic head gradient can be approximated by the slope of the phreatic surface (the Dupuit assumption). Substitution of the hydraulic head by this effective, vertical-averaged phreatic surface simplifies a three-dimensional flow problem to two dimensions. Using these assumptions, a broad suite of analytical solutions to unconfined steady groundwater drainage flow have been obtained,[1–3] e.g., for the purpose of drainage design calculations.

In addition to the Darcy type of equation, an additional expression is required for a complete description of saturated flow, allowing the prediction of both flux density and hydraulic head. This second equation is obtained by invoking mass conservation over a specified control volume. Using the assumptions that water is incompressible, and that the porous matrix is nondeformable, the mass balance equation combined with the Darcy equation for steady-state flow in a homogenous, isotropic, constant saturated thickness soil, yields the so-called Laplace equation:[1−3]

$$\frac{\partial^2 H}{\partial x^2} + \frac{\partial^2 H}{\partial y^2} = 0 \qquad (2)$$

Eq. 2 can be applied for an unconfined, saturated soil with a variable phreatic surface, H, if the variation in H is much smaller than the aquifer's thickness.

In order to analytically solve for transient groundwater drainage, the specific yield, S (dimensionless), was defined to quantify the exchange of soil water between the unsaturated and saturated soil zones in unconfined aquifers.[5] It is computed from the volume of water given up by or extracted from the groundwater per unit area of water table and per unit groundwater table change. In practice, S is assigned a constant soil-texture dependent value, and is also called the drainable porosity. In theory, however, this is only approximately true, since the specific yield is a function of the rate of water table change, its proximity to the soil surface, thereby varying with time depending on drainage and redistribution rates. Nevertheless, this approximation of specific yield allows the quantification of water flux across the water table as a result of changes in water table position, without solving the unsaturated water flow equation for the combined saturated−unsaturated soil domain. When applying the Dupuit and the specific yield assumptions, changes in the water height, H, with lateral position and time can be computed from solution of the so-called Boussinesq's equation, which is written as

$$S\frac{\partial H}{\partial t} + q_s = K\frac{\partial}{\partial x}\left(H\frac{\partial H}{\partial x}\right) + K\frac{\partial}{\partial y}\left(H\frac{\partial H}{\partial y}\right) \qquad (3)$$

where q_s is the steady-state water flux through the phreatic surface.[5] A full suite of numerical techniques is available that do not require such assumptions, but can solve for variably-saturated flow with or without decoupling the saturated and unsaturated flow regimes.[6,7]

PREFERENTIAL FLOW

It is widely accepted that preferential flow of water through soils occurs through different mechanisms, and can have wide implications. Preferential flow can be loosely defined as the accelerated flow of water and associated chemicals through soil, relative to the corresponding flow through the soil matrix. Preferential flow occurs locally, and is mostly saturated. Consequently, such a flow is fast and difficult to quantify. In general, one recognizes three different types: bypass, fingering, and funneled flows.[4,8] Under water-ponded soil surface conditions, bypass flow occurs through highly permeable zones such as macropores and cracks that are connected to the land surface (Fig. 2). Fingering flow is generally associated with nonuniform flow caused by flow instabilities as caused by soil air compression, soil hydrophobicity, and soil layering. Funneled flow[8] occurs because of the interbedding of coarse and fine soil layers. If the layer interface is along a slope, unsaturated water flow accumulates, and becomes saturated. Eventually, the water may breakthrough the interface into funnels, depending on local values of water pressure and soil textural changes along the bedding. Due to uncertainties of

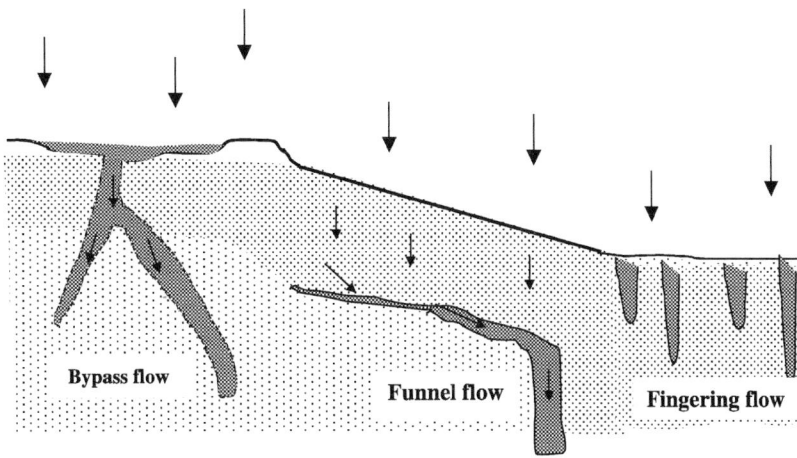

Fig. 2 Schematic diagram of preferential flow mechanisms.

the occurrence and magnitude of preferential flow, however, prediction of saturated flow in these conditions is difficult.

CONCLUSION

In addition to groundwater flow, saturated soil conditions can occur in the unsaturated or vadose zone by perched water, by preferential flow at or near the soil surface, and in the capillary fringe above the water table. For the latter case, relative simple solutions are available to account for the variably-saturated conditions. However, characterization of saturated water flow as caused by preferential flow or by perched water is much more difficult to quantify, mostly due to complications caused by soil heterogeneity.

REFERENCES

1. Bear, J. *Dynamics of Fluids in Porous Media*; Dover Publications, Inc.: New York, 1972; 764 pp.

2. Kirkham, D.; Powers, W.L. *Advanced Soil Physics*; John Wiley & Sons, Inc.: New York, 1972; 534 pp.

3. Van der Ploeg, R.R.; Horton, R.; Kirkham, D. Steady Flow to Drains and Wells. In *Agricultural Drainage*, Agronomy Monograph Number 38; Skaggs, R.W., van Schilfaarde, J., Eds.; American Society of Agronomy, Inc.: Madison, WI, USA, 1999; 213–263.

4. Miyazaki, T. *Water Flow in Soils*, Books in Soil, Plants, and the Environment Series; Marcel Dekker, Inc.: New York, NY, 1993; Vol. 10016, 312 pp.

5. Youngs, E.G.; In *Agronomy Monograph Number 38*; Skaggs, R.W., van Schilfaarde, J., Eds.; American Society of Agronomy, Inc.: Madison, WI, USA, 1999; 265–328.

6. Wang, H.F.; Anderson, M.P. *Introduction to Groundwater Modeling. Finite Difference and Finite Element Methods*; W.H. Freeman and Company: New York, 1982; 237 pp.

7. Nieber, J.L.; Feddes, R.A.; In *Agronomy Monograph Number 38*; Skaggs, R.W., van Schilfaarde, J., Eds.; American Society of Agronomy, Inc.: Madison, WI, USA, 1999; 145–212.

8. Kung, K-J.S. Preferential Flow in a Sandy Vadose Zone. 2. Mechanism and Implications. Geoderma **1990**, *46*, 59–71.

Soil Water Flow Under Unsaturated Conditions

Jan W. Hopmans
University of California, Davis, California, U.S.A.

Jacob H. Dane
Auburn University, Auburn, Alabama, U.S.A.

INTRODUCTION

Soils make up the upper part of the unsaturated zone, where water flow occurs mainly under unsaturated conditions. The unsaturated zone consists of a complex arrangement of mostly connected solid, liquid, and gaseous phases, with the spatial distribution and geometrical arrangement of each phase, and the partitioning of solutes between phases, controlled by physical, chemical, and biological processes. The unsaturated zone is bounded by the soil surface and merges with the groundwater in the capillary fringe (see the article *Soil Water Flow Under Saturated Conditions*). The distinction between groundwater and water in the unsaturated zone is determined by the degree of water saturation (see the article *Soil Water, Capillary Rise of*). For groundwater, it is generally assumed that the pore space within the solid matrix is saturated with water, and that the hydrostatic pressure in the water is larger than atmospheric pressure. In contrast, in the unsaturated zone, the pore space is only partly filled with water, while the remaining space is occupied by the gas phase. Water is held in the soil matrix of the unsaturated zone by capillary and adsorptive forces. The unsaturated zone is usually considered to be the region for water flow and its concomitant transport of chemicals between the atmosphere and groundwater. Although, the importance of water as carrier of chemicals is paramount, it is becoming increasingly clear that chemical and biological phenomena in the unsaturated zone play a profound role on chemical fate. It is therefore that vadose zone notation is preferred, emphasizing the multidisciplinary approach in subsurface characterization.

The upper part of the vadose zone is the most dynamic and changes occur at increasingly greater time and spatial scales when moving from the soil surface towards the ground water. The most upper part of the vadose zone is subject to fluctuations in water and chemical content by infiltration and leaching, water uptake by plant roots (transpiration), and evaporation from the soil surface. Water is the primary factor leading to soil formation from the weathering of parent material such as rock or transported deposits, with additional factors of climate, vegetation, topography, and parent material determining soil physical properties. Generally, the soil depth is controlled by the maximum rooting depth (generally within a few meters from the soil surface). However, the vadose zone can extend much deeper than the surface soil layer and includes unsaturated rock formations and alluvial materials to depths of 100 m or more, determined by hydrologic, topographic, and lithographic characteristics.

Scientists are becoming increasingly aware that soil is a critically important component of the earth's biosphere, not only because of its food production function, but also as the safe-keeper of local, regional, and global environmental quality. For example, it is believed that management strategies in the unsaturated soil zone will offer the best opportunities for preventing or limiting pollution, or for remediation of ongoing pollution problems. Because chemical residence times in ground water aquifers can range from a few to thousands of years, pollution is often essentially irreversible. Prevention or remediation of soil and groundwater contamination starts, therefore, with proper management of the unsaturated zone (see the article *Vadose Zone and Groundwater Protection*).

Both introductory[1,2] and advanced references[3,4] of unsaturated flow are suggested for further reading, whereas in addition, comprehensive reviews[5,6] provide selected references on relevant areas of study.

SOIL-WATER RETENTION

Unsaturated water flow is largely controlled by the physical arrangement of soil particles in relation to the water and air phases within the soil's pore space, as determined by pore size distribution and water-filled porosity or volumetric water content, θ [m³ water/m³ bulk soil; see the article *Soil Water, Gravimetric Measurement*]. In addition to θ, the volume of water is sometimes defined by degree of saturation, $S = \theta/\theta_s$, or

Encyclopedia of Water Science
DOI: 10.1081/E-EWS 120010327

effective saturation (S_e),

$$S_e = \frac{\theta - \theta_r}{\theta_s - \theta_r} \qquad (1)$$

normalizing the mobile water between values of zero and one, while defining a residual water content, θ_r, for which water is considered immobile (see the article *Soils, Hygroscopic Water Content in*). In addition to simple gravimetric methods, various devices are available to measure the soil's water content nondestructively (see the articles *Soil Water Measurement by Capacitance; Soil Water Measurement by Neutron Thermalization; Soil Water Measurement by Time Domain Reflectometry*).

The soil-water retention function determines the relation between volume of water retained by soil capillary and adsorptive forces, as a function of θ, and is also known as the soil-water release or soil-water characteristic function. These two water-retaining forces in unsaturated soils combined are defined as the matric forces, and are sometimes also called suction forces (see the article *Soil Water Energy Concepts*). These suction forces increase as the size of the water-filled pores decreases, as may occur by drainage, water uptake by plant roots, or evaporation. When expressed relative to the reference potential of free water, the water potential in unsaturated soils is negative (the soil-water potential is less than the water potential of water at atmospheric pressure). It is often referred to as the soil water matric potential. Hence, the matric potential decreases or becomes more negative as the soil water content decreases. Since the matric forces are controlled by pore size distribution, specific surface area, and type of physico-chemical interactions at the solid–liquid interfaces, the soil water retention curve is soil specific. It provides an estimate of the soil's capacity to hold water after free drainage (see the article *Soils, Field Capacity of Water in*), minimum soil water content available to the plant (see the article *Soils, Parmanent Wilting Points*), and water availability for plants.

Whereas in physical chemistry, the chemical potential of water is usually defined on a molar or mass basis, in soils potential is usually expressed with respect to a unit volume of water, thereby attaining units of pressure (Pa); or per unit weight of water, so that the potential represents the equivalent height of a column of water (m). The pressure head equivalent of the combined adsorptive and capillary forces is defined as the matric pressure head, h_m.

By way of the unique relationship between capillary water pressure and the radius of curvature of the air–water interface, and using the analogy between capillary tubes and the irregular pores in porous media, a relationship can be derived between soil water matric head (h_m) and

effective pore radius, r_e, or

$$\rho g h_m = \frac{-2\sigma \cos \alpha}{r_e} \qquad (2)$$

where σ and α are defined as the surface tension and wetting angle (of wetting fluid with solid surface), respectively, ρ is the density of water, and g is the acceleration due to gravity ($9.8 \, \mathrm{m \, sec^{-2}}$). This capillary equation simplifies to $h_m = -0.15/r_e$, when both h_m and r_e are expressed in cm. As a result, the effective pore size distribution can be determined from the soil water retention curve in the region where capillary forces dominate.

The measurement of the matric potential in situ is difficult and is usually done by tensiometers in the range of matric head values larger (less negative) than $-6.0 \, \mathrm{m}$ (see the article *Soil Water Potential Measurement by Tensiometers*). A tensiometer consists of a porous cup, usually ceramic, connected to a water-filled tube. The suction forces of the unsaturated soil draw water from the tensiometer into the soil until the water pressure inside the cup (at pressure smaller than atmospheric pressure) is equal to the pressure equivalent of the soil water matric potential just outside the cup. The water pressure in the tensiometer is usually measured by a vacuum gauge or pressure transducer. When tensiometers are used at matric potential values lower than $-6.0 \, \mathrm{m}$, the tensile strength of the water in the tensiometer device may be exceeded, causing development of air or vapor bubbles in the water column, which is called cavitation, thereby rendering the tensiometer readings useless. Other devices that are used to indirectly measure the soil water matric potential include buried porous blocks, from which either the electrical resistance or thermal conductivity is measured in situ, after coming into hydraulic equilibrium with the surrounding soil. Laboratory and field techniques to measure the soil water retention curve, and functional models to fit the measured soil water retention data, such as the van Genuchten and Brooks and Corey model, are described in Refs. [7,8].

UNSATURATED HYDRAULIC CONDUCTIVITY

The relation between the unsaturated hydraulic conductivity, K, and volumetric water content, θ, is the other essential fundamental soil hydraulic property needed to describe water movement in the vadose zone. It is also a function of the water and soil matrix properties, and determines water infiltration and drainage rates (see the articles *Soils, Water Infiltration and; Soils, Water Percolation Through; Soil, Waterborne Chemicals Leaching Through*), and is strongly affected by water content. It is defined by Darcy's equation (*Darcy's law*), which relates the soil water flux density to the total driving force for flow, with K being the proportionality factor. Except for special circumstances, pneumatic and osmotic forces are irrelevant, so that the total driving force for water flow is determined by the matric and gravitational forces, expressed by the total water potential gradient, $\Delta H/L$, where ΔH denotes the change in total head over the distance L, and $H = h_m + z$. Applying Darcy's law in the vertical dimension only, the magnitude of flow can be computed from the steady state flow equation (Darcy's law):

$$Q = -K(\theta)A\left(\frac{\partial h_m}{\partial z} + 1\right) \quad \text{or}$$

$$q_w = -K(\theta)\left(\frac{\partial h_m}{\partial z} + 1\right) \tag{3}$$

where Q denotes the volumetric flux ($m^3 \sec^{-1}$), A is the cross-sectional area of the bulk soil domain perpendicular to flow (m^2), q_w is the Darcy flux density ($m \sec^{-1}$), z is vertical position ($z > 0$, upwards, m), and $K(\theta)$ denotes the unsaturated hydraulic conductivity ($m \sec^{-1}$). In this expression, the unsaturated hydraulic conductivity is related to the intrinsic soil permeability, k (m^2), by

$$K = \frac{\rho g k}{\mu} \tag{4}$$

where μ denotes the dynamic viscosity of water ($N \sec m^{-2}$), and ρ and g were defined earlier. The usage of permeability instead of conductivity allows application of the flow equation to liquids other than water with different density and viscosity values. Using the analogy of soil pores represented by varying-size capillaries, the average pore water velocity in soils can be estimated from the ratio of the Darcy flux and the volumetric water content, or

$$\hat{v} = \frac{q_w}{\theta} \tag{5}$$

Functional models for unsaturated hydraulic conductivity are based on pore size distribution, pore geometry and connectivity, and require integration of soil water retention functions to obtain analytical expressions for the unsaturated hydraulic conductivity. The resulting expressions relate the relative hydraulic conductivity, K_r, defined as the ratio of the unsaturated hydraulic conductivity, K and the saturated hydraulic conductivity, K_s, to the effective saturation, and can be written in the following generalized form

$$K_r(S_e) = S_e^l\left[\frac{\int_0^{S_e} |h_m|^{-\eta} dS_e}{\int_0^1 |h_m|^{-\eta} dS_e}\right]^\gamma \tag{6}$$

where l and η are parameters related to the tortuosity and connectivity of the soil pores, and the value of the parameter γ is determined by the method of evaluating the effective pore radii. The moisture-dependency is highly nonlinear, with a decrease in K of 4–5 or more orders of magnitude within field-representative changes in water content. Methods to measure the saturation dependency of the hydraulic conductivity are involved and time-consuming.[9] Moreover, measurement errors are generally large, due to: 1) the difficulty of flow measurements in the low-water content range; and 2) the dominant effect of large pores (macropores), cracks, and fissures in the high-water content range (see the article *Soil Macropores, Water and Solute Movement in*). Model fitting techniques assume a certain form for the soil water retention curve, such as the Mualem-van Genuchten relationship, and use parameters associated with the water retention relationship to express the hydraulic conductivity as a function of water content or matric head.[8]

MODELING OF UNSATURATED SOIL WATER FLOW

Since the Darcy equation is strictly defined for steady state water flow conditions, the mass conservation principle is applied and combined with Eq. 3 to yield the so-called Richards equation to solve for temporal changes in h_m or θ, at any depth z and time t:

$$\frac{\partial \theta}{\partial t} = -\frac{\partial q_w}{\partial z} = \frac{\partial}{\partial z}\left[K(h_m)\left(\frac{\partial h_m}{\partial z} + 1\right)\right] \tag{7}$$

Because of the highly nonlinear soil water retention and unsaturated hydraulic conductivity functions, advanced numerical models are required to solve for $h_m(z,t)$ or $\theta(z,t)$ for either one, two or three dimensions,[10] using known boundary and initial conditions. Equation 7 may include a sink term, describing changes in soil water content with time as a result of root water uptake.[11] If solution of time-changes of water content within the soil domain are not required, but only total soil water storage changes are needed at time scales of days or longer, Eq. 7 can be simplified to a capacity model, thereby requiring input of the boundary fluxes only, resulting in the so-called water budget models.[12]

SUMMARY

Prevention or remediation of soil and groundwater contamination requires proper management of the vadose zone. Therefore, a solid understanding of unsaturated water flow is required. However, because of the highly nonlinear soil hydraulic functions that control soil water retention and unsaturated hydraulic conductivity, advanced numerical models are required to predict temporal changes of soil water matric potential, soil water content, and water fluxes in one or more spatial dimensions.

REFERENCES

1. Hillel, D. *Environmental Soil Physics*; Academic Press: London, 1998; 771.
2. Jury, W.A; Gardner, W.R.; Gardner, W.H. *Soil Physics*; John Wiley & Sons, Inc.: New York, 1991; 328.
3. Bear, J. *Dynamics of Fluids in Porous Media*; Dover Publications, Inc.: New York, 1972; 764.
4. Kirkham, D.; Powers, W.L. *Advanced Soil Physics*; John Wiley & Sons, Inc.: New York, 1972; 534.
5. Parlange, M.B.; Hopmans, J.W. *Vadose Zone Hydrology: Cutting Across Disciplines*; Oxford University Press: New York, 1999; 454.
6. Nielsen, D.R.; van Genuchten, M.Th.; Biggar, J.W. Water Flow and Solute Transport Processes in the Unsaturated Zone. Water Resour. Res. **1986**, *22*, 89S–108S.
7. Dane, J.H.; Hopmans, J.W. 3.3. Water Retention and Storage. 3.3.1. Introduction. In *Methods of Soil Analysis, Part 1, Physical Methods*, Agronomy Monograph 9; Dane, J.H., Topp, G.C., Eds.; SSSA Book Ser 5. SSSA: Madison, Wisconsin, 2001; 671–673.
8. Kosugi, K.; Hopmans, J.W.; Dane, J.H. 3.3. Water Retention and Storage. 3.3.4. Parametric Models. In *Methods of Soil Analysis, Part 4, Physical Methods*; Dane, J.H., Topp, G.C., Eds.; SSSA Book Ser 5. SSSA: Madison, Wisconsin, 2002; 739–757.
9. Klute, A.; Dirken, C. Hydraulic Conductivity and Diffusivity: Laboratory Methods. In *Methods of Soil Analysis, Part 1, Physical Methods*, Agronomy Monograph Number 9; Klute, A., Ed.; American Society of Agronomy, Inc.: Madison, Wisconsin, 1986; 687–734.
10. Šimůnek, J.; Huang, K.; van Genuchten, M.Th. *The SWMS_3D Code for Simulating Water Flow and Solute Transport in Three-Dimensional Variably Saturated Media. Version 1.0*, Research Report No. 139; U.S. Salinity Laboratory, USDA, ARS: Riverside, California, 1995; 155.
11. Somma, F.; Clausnitzer, V.; Hopmans, J.W. Modeling of Transient Three-Dimensional Soil Water and Solute Transport with Root Growth and Water and Nutrient Uptake. Plant Soil **1998**, 202, 281–293.
12. Hill, R.W. Irrigation Scheduling. In *Modeling Plant and Soil Systems*, Agronomy Monograph 31; Hanks, J., Ritchie, J.T., Eds.; SSSA: Madison, Wisconsin, 1991; 491–509.

Soil Water, Gravimetric Measurement

Joseph L. Pikul, Jr.
United States Department of Agriculture (USDA), Brookings, South Dakota, U.S.A.

INTRODUCTION

Water limits crop production in most agricultural soils and directly or indirectly affects soil physical, chemical, and biological properties and processes. Quantity of water held in soil is commonly determined by measuring the mass of water relative to the mass of dry soil. The ratio is called gravimetric soil water,[1,2] oven-dry,[2,3] or soil water content.[3] This measurement has been a mainstay of many field studies and is generally accepted as a calibration standard for many indirect soil water measurement methods.[2]

A brief sampling of literature from 1907 to 1930 (Agronomy Journal, American Society of Agronomy) revealed that scientists then, as now, rarely provide detail on methodology used to measure soil water content. Often the reader was left to assume that an investigator followed expected procedure. For example, reports from the early 1900s might state that "moisture was determined in the usual way";[4] or "samples of soil for moisture determination were taken";[5] or "oven-dry method" was used;[6] or soil was dried "to constant weight at the temperature of boiling water."[7] Davisson and Sivaslian[8] provided standards for scientists of that era with a review of important findings in the German literature.

In the current literature, the term "gravimetric water content" is commonly used to identify the base in which soil water content is being reported (gravimetric vs. volumetric base) and to suggest to the reader that a standard procedure was followed.[9–13] Unfortunately, the term "gravimetric water" was not defined in the Soil Science Society of America, Glossary of Soil Science Terms 1996.[3] Some text books[1] and Methods of Soil Analysis: Part 1[2] include a definition of gravimetric water. Other terms commonly seen are "oven-dry water," "soil moisture," "soil water content," "soil water content (105°C, 24 hr)," and "gravimetric procedure." Often there is no reference to a standard method. In the reporting of soil water content, it is important to identify that a standard method, such as that provided by Gardner,[2] was used.

FIELD PRACTICE

Soil water is rarely at equilibrium. Water moves from regions of high water potential (wet soil) to low water potential (dry soil). To minimize temporal variability in water content among samples, it is best to sample quickly and at a time of day when evaporational demand is lowest. Early morning and late afternoon are ideal. Indirect methodology, such as neutron thermalization,[2] may be best suited for repeated measures of soil water content. Gravimetric soil water sampling is destructive and requires the investigator to sample across a spatially diverse field. This may or may not be advantageous.

Soil water content near the surface can change rapidly. The work of Idso et al.[14] and Pikul and Allmaras[9] are examples of dynamic fluctuations in soil water content that can be expected near the surface under different environmental conditions. Idso et al.[14] investigated soil heat flux relations in nonfrozen soil as influenced by soil water content of a loam soil that was recently wetted by irrigation. Pikul and Allmaras[9] investigated the phenomena of freezing induced soil water redistribution.

Idso et al.[14] found diurnal fluctuations in soil water content to a depth of about 100 mm. Soil water content changes were a consequence of water evaporation and redistribution. In the top 10 mm of soil, water content decreased about $0.08\,m^3\,m^{-3}$ (8% water content on a volumetric basis) in 10 hr. In the case of soil freezing, Pikul and Allmaras[9] found that water content near the surface changed dramatically within hours. Their measurements show that in some cases water content of the surface 5-mm layer increased by $0.17\,kg\,kg^{-1}$ (17% soil water on a gravimetric basis) in as little as 6 hr when the soil froze.

Soil sampling tools are designed to meet the purpose for sampling and the condition of the soil being sampled. Soil structure, whether compact or loose, determines the layer refinement attainable. Directly after tillage, the size of soil structural units and their fragility prohibit conventional sampling in thin layers as described by Idso et al.[14] or Pikul and Allmaras.[9] Alternative soil investigation methods such as the random roughness technique

Encyclopedia of Water Science
DOI: 10.1081/E-EWS 120010148
Published 2003 by Marcel Dekker, Inc. All rights reserved.

presented by Allmaras et al.[15] may be necessary when working with extremely disturbed soil surface conditions.

For a moderately compacted and moist soil with a bare, smooth surface, Reginato[16] developed a soil sampler for delineating soil-water distribution in the top 10-mm layer. Increments as fine as 1 mm could be sampled for gravimetric water. Bulk density, however, was measured with segmented soil cores obtained with a cylindrical Oakfield type core sampler. Pikul et al.[10] developed an incremental soil sampling tube for sampling 10-mm increments of unconsolidated surface soil. This tool containerized loose soil layers thereby enabling measurement of both soil water content and soil bulk density in one sampling operation.

Sampling methods that enable simultaneous measurement of both bulk density and water content are desirable because bulk density is essential for converting water content from a gravimetric to a volumetric base. Expression of soil water on a volumetric basis enables calculation of several fundamental attributes of soil water condition related to volume fraction.[1]

FUNDAMENTAL RELATIONS

Gravimetric water content is defined as the mass of water (M_w) relative to the mass of dry soil (M_s). Determination of gravimetric water content requires three independent measurements that include mass of wet soil (M_{ws}), mass of dry soil, and mass (t) of the collection can (commonly called tare weight). Wet soil samples are placed in metal cans with tight fitting lids and the combined mass of wet soil and container ($M_{ws} + t$) is measured. The term "wet" is relative to the soil water content at time of sampling. Collection cans, with lids off, are placed in a drying oven and the sample is dried to a constant mass. Standard practice is to dry samples for 24 hr at 105°C in a forced-draft oven. The term "dry" is specific and refers to soil that has been dried to a constant mass. Gardner[2] provides a complete description of standard procedures. After drying, the cans are capped, cooled, and the combined mass of dry soil and container ($M_s + t$) is measured. Mass of water is calculated as $M_w = (M_{ws} + t) - (M_s + t)$. It follows that $M_s = (M_s + t) - t$. Gravimetric water content (θ_g) is

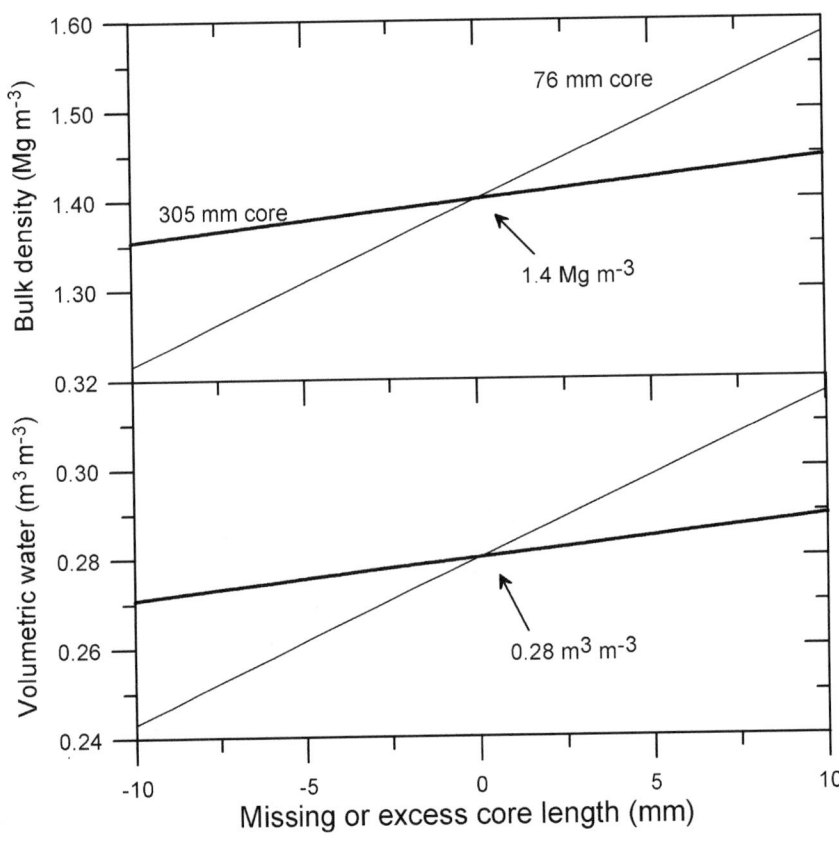

Fig. 1 Soil bulk density (ρ_b) and volumetric water (θ_v) as a function of missing or excess soil (cutting error) for cores of 76 mm (3 in.) and 305 mm (12 in.). Deviation from expected $\rho_b = 1.4\,\mathrm{Mg\,m^{-3}}$ and $\theta_v = 0.28\,\mathrm{m^3\,m^{-3}}$ are a consequence of missing or excess soil mass associated with a core of 30 mm diameter.

calculated as

$$\theta_g = M_w/M_s \tag{1}$$

and volumetric water content (θ_v) as

$$\theta_v = \theta_g(\rho_b/\rho_w) \tag{2}$$

where ρ_b is soil bulk density ($Mg\,m^{-3}$) and ρ_w is density of water. Soil bulk density is defined as

$$\rho_b = M_s/V_t \tag{3}$$

where V_t is the total volume of soil sample. Eq. 2 is dimensionally correct. However, for most field applications, a working formula for θ_v is simply

$$\theta_v = \theta_g\rho_b \tag{4}$$

when ρ_w is assumed to be $1\,Mg\,m^{-3}$.[1]

Special attention must be given to samples collected for both ρ_b and θ_g. There is serious error associated with cutting soil cores improperly. Soil bulk density is based on mass and volume of the sample. Thus, it is important to collect the entire mass of soil associated with a given volume.

Missing or excess soil results in an error in ρ_b and consequently θ_v. For samples with the same diameter, a small cutting error is more serious in cores of short length rather than long length. Bulk density and volumetric water are shown in Fig. 1 as calculated for conditions where a soil core may have been undercut (missing core) or overcut (excess core). Volume of sample was based on an intended length of 76 mm (3 in.) or 305 mm (12 in.). Core diameter was 30 mm. True value of ρ_b was $1.4\,Mg\,m^{-3}$ and θ_g was $0.2\,kg\,kg^{-1}$ ($\theta_v = 0.28\,m^3\,m^{-3}$). For soil cores having an assumed length of 76 mm, an error of about 13% in both ρ_b and θ_v would occur if these cores were cut 10 mm short (or long). In contrast, a 10 mm undercut (or overcut) would result in an error of only 3% for cores having an assumed length of 305 mm. In many studies, soil from the 0- to 76-mm depth (0- to 3-in. depth) and 76- to 152-mm depth (3- to 6-in. depth) is important. Soil water content plays a vital role in respect to biological activity and it is important to accurately determine ρ_b and θ_v for shallow soil depths. Unfortunately, the investigator is most apt to accrue serious errors because of poor sampling technique of thin layers.

REFERENCES

1. Hillel, D. *Soil and Water Physical Principles and Processes*; Academic Press: New York, 1971; 288 pp.
2. Gardner, W.H. Water Content. In *Methods of Soil Analysis: Part I Physical and Mineralogical Methods*, 2nd Ed.; Klute, A., Ed.; Soil Science Society of America: Madison, WI, 1986; 493–544.
3. *Glossary of Soil Science Terms 1996*; Soil Science Society of America: Madison WI, 1997; 1–134.
4. Briggs, L.J.; McLane, J.W. Moisture Equivalent Determinations and Their Application. Proc. Am. Soc. Agron. **1910**, *2*, 138–147.
5. Duley, F.L. The Effect of Alfalfa on Soil Moisture. J. Am. Soc. Agron. **1929**, *21*, 224–231.
6. Bouyoucos, G. Rapid Determination of the Moisture Content of Soils. J. Am. Soc. Agron. **1927**, *19*, 197–198.
7. Lyon, T.L. Intertillage of Crops and Formation of Nitrates in Soil. J. Am. Soc. Agron. **1922**, *14*, 97–109.
8. Davisson, B.S.; Sivaslian, G.K. The Determination of Moisture in Soils. J. Am. Soc. Agron. **1918**, *10*, 198–204.
9. Pikul, J.L., Jr.; Allmaras, R.R. Hydraulic Potential in Unfrozen Soil in Response to Diurnal Freezing and Thawing of the Soil Surface. Trans. ASAE. **1985**, *28*, 164–168.
10. Pikul, J.L., Jr.; Allmaras, R.R.; Fischbacher, G.E. Incremental Soil Sampler for Use in Summer-Fallowed Soils. Soil Sci. Soc. Am. J. **1979**, *43*, 425–427.
11. Pachepsky, Y.; Rawls, W.J.; Giménez, D. Comparison of Soil Water Retention At Field and Laboratory Scales. Soil Sci. Soc. Am. J. **2001**, *65*, 460–462.
12. Benjamin, J.G.; Cruse, R.M. Tillage Effects on Shear Strength and Bulk Density of Soil Aggregates. Soil Till. Res. **1987**, *9*, 255–263.
13. Bullied, W.J.; Entz, M.H. Soil Water Dynamics After Alfalfa as Influenced by Crop Termination Technique. Agron. J. **1999**, *91*, 294–305.
14. Idso, S.B.; Aase, J.K.; Jackson, R.D. Net Radiation–Soil Heat Flux Relations as Influenced by Soil Water Content Variations. Boundary Layer Meteorol. **1975**, *9*, 113–122.
15. Allmaras, R.R.; Burwell, W.E.; Larson, W.E.; Holt, R.F. *Total Porosity and Random Roughness of the Interrow Zone as Influenced by Tillage*; Conservation Research Report No. 7, Agricultural Research Service. USDA: 1966.
16. Reginato, R.J. Sampling Soil-Water Distribution in the Surface Centimeter of a Field Soil. Soil Sci. **1975**, *120*, 292–294.

S

Soil Water Hysteresis

Dan Jaynes
United States Department of Agriculture (USDA), Ames, Iowa, U.S.A.

INTRODUCTION

Since at least the early work of Haines,[1] it has been recognized that volumetric soil water content, θ, and hydraulic conductivity, K, are not singular functions of soil water pressure head, h, but rather exhibit considerable variation depending on the wetting and drying history of the soil.[2,3] The nonuniqueness or hysteresis in $\theta(h)$ and $K(h)$ appears to be an ubiquitous phenomenon in porous materials and the magnitude of the effect is intimately related to the pore distribution of the material. However, numerous studies have shown that when K is expressed as a function of θ instead of h, hysteresis either disappears[4–8] or is so slight as to be masked by the error of the measurements. By expressing K as a function of θ instead of h, it can be treated as a nonhysteretic function. Thus, the focus here is on the hysteretic nature of $\theta(h)$, which for brevity will be termed simply hysteresis.

BACKGROUND INFORMATION

Fig. 1 illustrates the hysteretic θ–h relation for a hypothetical soil. The curve S–P_c represents drying from total saturation where θ equals the total soil porosity, φ, to a point where the hysteresis curves close together, often equal to the limit of measurement or considered the residual water content of the soil. Upon wetting from a dry condition, $\theta(h)$ follows the curve P_c–E–D, the main wetting curve. The water content at D is the effective or "field saturated" water content of the soil which is usually less than φ because of entrapped air in some soil pores during wetting. Given sufficient time, the entrapped air will diffuse into the soil water, and the water content at $h = 0$ will approach φ.[9,10] When the soil dries from the field saturated water content $\theta(h)$, follows the curve D–P_a–P_c which is the main drying curve. The loop D–P_a–P_c–E–D is often called the reproducible boundary hysteresis loop. Point P_w in Fig. 1 represents the water-entry pressure head—the pressure head at which the main wetting curve reaches the saturated water content—while P_a represents the air-entry pressure head, the pressure head at which the soil first starts to lose water when drying from field saturation.

The main hysteresis curves are followed when the soil is dried from field saturation or wetted from the residual water content. Curves that depart from the main, or boundary curves between the points P_w and P_c are called primary wetting and drying curves. Primary drying curves (e.g., A–B –P_c in Fig. 1) are followed when the wetting process reverses before the soil water content reaches P_w. Primary wetting curves (e.g., C–F–D in Fig. 1) are followed when the drying process reverses before the soil dries to point P_c. Additional reversals in the wetting or drying cycle will result in secondary and higher order scanning curves. A secondary drying curve is illustrated by curve F–G–C in Fig. 1.

Several factors affect hysteresis in the field and laboratory. While, $\theta(h)$ depends on the temperature at which it is measured, the magnitude of the hysteresis effect does not appear to change.[11–13] The magnitude of the measured hysteretic effect is dependent on the rate at which the water content of the soil is changed when making the measurements.[5,14] Measurements made during unsteady flow conditions tend to overestimate the amount of water held at a given pressure head during drying and under-estimate the water content during wetting when compared to steady-state or static-equilibrium conditions. This observation is important because most measurements of hysteresis are made during unsteady conditions, because the measurements can be completed more rapidly. Transient-state measurements should be treated with caution since they are influenced not only by the water holding characteristics of the soil, but also by the $K(\theta)$ characteristics.

Measuring hysteresis in $\theta(h)$, whether in the laboratory[15] or field,[16] is difficult and time consuming because both θ and h must be measured over a wide range of water contents, and the drier soil conditions are often difficult to establish in the field, especially for fine-grained soils with low hydraulic conductivities. Consequently, almost all measurements of hysteresis have been on disturbed, coarse-grained soils in the laboratory, and most observations of the impact of hysteresis on soil water relations have been made in the laboratory or by numerical simulation. However, hysteresis has been measured in the field for both sand and clay soils[17,18] and shown to be of the same magnitude as measured in laboratory exper-

Encyclopedia of Water Science
DOI: 10.1081/E-EWS 120010326

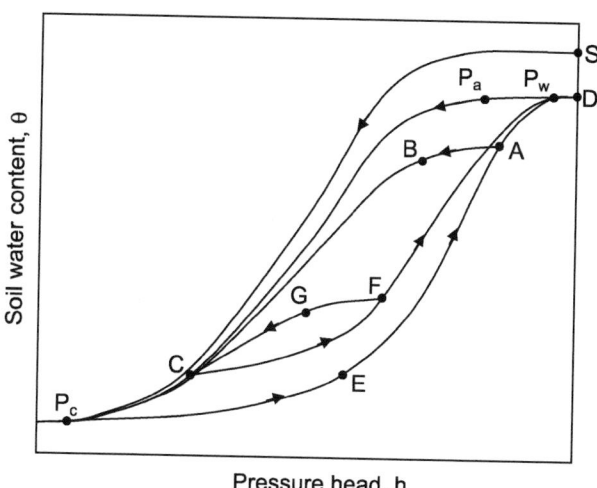

Fig. 1 Hysteretic soil water content, θ–pressure head, h, relation for a typical soil.

iments. Thus, the impacts of hysteresis observed in the laboratory should be the same as in the field.

Hysteresis in $\theta(h)$ implies that to completely characterize the state of water in the soil, not only do θ and h need to be known, but also the wetting and drying history of the soil as well. For a hysteretic soil, the specific soil pores that are filled with water depend on the wetting and drying history of the soil. Thus, all transport processes in soil, including infiltration, evaporation, and chemical movement can be affected by hysteresis.

IMPLICATIONS OF HYSTERESIS

Results from laboratory column experiments and model simulations show the effect of hysteresis to depend very much on the $K(\theta)$ relation. For soils with very steep $K(\theta)$–$\theta(h)$ relations, the effect of hysteresis on processes such as water redistribution is minimal.[19] Hysteresis effects are greatest in soils with large differences in the wetting and drying curves for $\theta(h)$ and with a $K(\theta)$ that does not decrease rapidly. In these soils, the following generalizations can be made.

Infiltration/Redistribution

Hysteresis can have a profound effect on the redistribution of water after infiltration. Following infiltration, the wetted profile switches from a wetting process to a drying process, while at the wetting front the soil is still undergoing wetting. With hysteresis, a relatively large change in h will produce only a small change in θ when a drying scanning curve is followed (e.g., A–B–P_c in Fig. 1) compared to the change in θ if no hysteresis was present and the θ–h relation followed

only a single curve such as the main wetting or drying curve (e.g., A–E–P_c in Fig. 1). The effect of hysteresis is to hold water in the upper reaches of the soil profile and to slow redistribution of water after infiltration has ceased, keeping the deeper soil profile drier.[3,19–24]

Evaporation

Hysteresis will cause the soil surface to remain wetter after infiltration. Thus, under a constant evaporative demand, hysteresis will cause more evaporation from the soil after an infiltration event because deep drainage is slowed.[19,25,26] If, however, the evaporative demand is cyclical causing the soil surface to dry during the day and rewet at night, evaporation from the soil will decrease because hysteresis will slow the nightly rewetting process at the soil surface resulting in less moisture near the surface available for evaporation the following day.[27]

Water Table Response

The response of a phreatic aquifer can be influenced by hysteresis. Water table rise in response to infiltration will be faster for a water table that had been falling than for one that had been rising.[28–30] For the wetter half of the hysteresis loop, the same increase in water content will cause a much larger change in h when $\theta(h)$ follows a primary rewetting curve than when it follows the main wetting curve. As a result, hysteresis will tend to cause water table heights to rise and seepage faces to develop much more rapidly during infiltration events in soils with falling water tables than in soils with rising water tables. The capillary fringe of a water table will also be affected by whether the water table is rising or falling. A rising water table will follow the wetting curve P_c–E–D and the height of the capillary fringe will be equal to the difference in pressure heads at points P_w and D. A falling water table will follow the drying curve D–P_a–P_c and the height of the capillary fringe will be equal to the difference in pressure heads at points D and P_a. In shallow aquifers, the greater capillary fringe thickness in the falling water table can cause significantly greater lateral flow rates[31] than for a rising water table.

Solute Movement

Hysteresis tends to reduce the redistribution of water from wetted areas of the soil to drier areas. Thus during redistribution, hysteresis will retard the rate of downward movement of solute added with irrigation water relative to movement in a soil without hysteresis.[23,32] Slightly less hydrodynamic dispersion is also expected because the infiltrating water is restricted to a smaller area of the soil profile.[32,33] Similarly, hysteresis will tend to decrease the leaching of a solute already present in a soil profile.[33]

REFERENCES

1. Haines, W.B. Studies in the Physical Properties of Soil: V. The Hysteresis Effect in Capillary Properties, and the Modes of Moisture Distribution Associated Therewith. J. Agric. Sci. **1930**, *20* (1), 97–116.

2. Staple, W.J. Infiltration and Redistribution of Water in Vertical Columns of Loam Soil. Soil Sci. Soc. Am. Proc. **1966**, *30* (5), 553–558.

3. Dane, J.H.; Wierenga, P.J. Effect of Hysteresis on the Prediction of Infiltration, Redistribution and Drainage of Water in a Layered Soil. J. Hydrol. **1975**, *25* (3/4), 229–242.

4. Talsma, T. Hysteresis in Two Sands and the Independent Domain Model. Water Resour. Res. **1970**, *6* (3), 964–970.

5. Rogers, J.S.; Klute, A. The Hydraulic Conductivity–Water Content Relationship during Nonsteady Flow Through a Sand Column. Soil Sci. Am. Proc. **1971**, *35* (5), 695–700.

6. Topp, G.C. Soil Water Hysteresis in Silt Loam and Clay Loam Soils. Water Resour. Res. **1971**, *7* (4), 914–920.

7. Vachaud, G.; Thony, J. Hysteresis during Infiltration and Redistribution in a Soil Column at Different Initial Water Contents. Water Resour. Res. **1971**, *7* (1), 111–127.

8. Gillham, R.W.; Klute, A.; Heerman, D.F. Hydraulic Properties of a Porous Medium: Measurement and Empirical Representation. Soil Sci. Soc. Am. J. **1976**, *40* (2), 203–207.

9. Cary, J.W. Experimental Measurements of Soil-Moisture Hysteresis and Entrapped Air. Soil Sci. **1967**, *104* (3), 174–180.

10. Poulovassilis, A. Hysteresis of Pore Water in Granular Porous Bodies. Soil Sci. **1970**, *109* (1), 5–12.

11. Hopmans, J.W.; Dane, J.H. Temperature Dependence of Soil Hydraulic Properties. Soil Sci. Soc. Am. J. **1986**, *50* (1), 4–9.

12. Hopmans, J.W.; Dane, J.H. Temperature Dependence of Soil Water Retention Curves. Soil Sci. Soc. Am. J. **1986**, *50* (3), 562–567.

13. Nimmo, J.R.; Miller, E.E. The Temperature Dependence of Isothermal Moisture vs. Potential Characteristics of Soils. Soil Sci. Soc. Am. J. **1986**, *50* (5), 1105–1113.

14. Topp, G.C.; Klute, A.; Peters, D.B. Comparison of Water Content–Pressure Head Data Obtained by Equilibrium, Steady-State, and Unsteady-State Methods. Soil Sci. Soc. Am. Proc. **1967**, *31* (3), 312–314.

15. Klute, A. Water Retention: Laboratory Methods. In *Methods of Soil Analysis: Part I. Physical and Mineralogical Methods*; Klute, A., Ed.; Amer. Soc. Agron.: Madison, WI, 1986; 635–662.

16. Bruce, R.R.; Luxmore, R.J. Water Retention: Field Methods. In *Methods of Soil Analysis: Part I. Physical and Mineralogical Methods*; Klute, A., Ed.; Amer. Soc. Agron.: Madison, WI, 1986; 663–686.

17. Royer, J.M.; Vachaud, G. Field Determination of Hysteresis in Soil-Water Characteristics. Soil Sci. Soc. Am. Proc. **1975**, *39* (2), 221–223.

18. Watson, K.K.; Reginato, R.J.; Jackson, R.D. Soil Water Hysteresis in a Field Soil. Soil Sci. Soc. Am. Proc. **1975**, *39* (2), 242–246.

19. Jaynes, D.B. Soil Water Hysteresis: Models and Implications. In *Process Studies in Hillslope Hydrology*; Anderson, M.G., Burt, T.P., Eds.; John Wiley and Sons Ltd.: Chichester, England, 1990; 93–126.

20. Ibrahim, H.A.; Brutsaert, W. Intermittent Infiltration into Soils with Hysteresis. J. Hydraul. Div., Proc. ASCE **1968**, *94* (HY 1), 113–137.

21. Watson, K.K. Numerical Analysis of Natural Recharge to an Unconfined Aquifer. In *Conjunctive Water Use*, Proc. Budapest Symp., Budapest, Hungary, Jul 1986; IAHS; Publ. No. 156, 323–333.

22. Watson, K.K.; Sardana, V. Numerical Study of the Effect of Hysteresis on Post-infiltration Redistribution. In *Proc. Intern. Conf. on Infiltration Development and Application*, Honolulu, HI, Jan 6–9, 1987; Fok, Y.-S., Ed.; Water Resour. Center: Honolulu, HI, 1987; 241–250.

23. Pickens, J.F.; Gillham, R.W. Finite Element Analysis of Solute Transport Under Hysteretic Unsaturated Flow Conditions. Water Resour. Res. **1980**, *16* (6), 1071–1078.

24. Rubin, J. Numerical Method for Analyzing Hysteresis-Affected, Post-infiltration Redistribution of Soil Moisture. Soil Sci. Soc. Am. Proc. **1967**, *31* (1), 13–20.

25. Bresler, E.; Kemper, W.D.; Hanks, R.J. Infiltration, Redistribution, and Subsequent Evaporation of Water from Soil as Affected by Wetting Rate and Hysteresis. Soil Sci. Soc. Am. Proc. **1969**, *33* (6), 832–840.

26. Scott, P.S.; Farquhar, G.J.; Kouwen, N. Hysteretic Effects on Net Infiltration. *Adv. in Infiltration*; ASAE: St. Joseph, MI, 1983; 163–170.

27. Hillel, D. Isothermal Evaporation of Soil Water Under Fluctuating Evaporativity, Including the Role of Hysteresis. *Computer Simulations of Soil-Water Dynamics: A Compendium of Recent Work*; International Development Research Centre: Ottawa, Canada, 1977; 35–60.

28. Brock, R.R.; Amar, A.C. Ground-Water Recharge Strip Basin-Experiments. J. Hydraul. Div., ASCE **1974**, *100* (4), 569–592.

29. Nieber, J.L.; Walter, M.F. Two-Dimensional Soil Moisture Flow in a Sloping Rectangular Region: Experimental and Numerical Studies. Water Resour. Res. **1981**, *17* (6), 1722–1730.

30. Stauffer, F.; Dracos, T. Experimental and Numerical Study of Water and Solute Infiltration in Layered Porous Media. J. Hydrol. **1986**, *84*, 9–34.

31. Martinec, J. Subsurface Flow from Snowmelt Traced by Tritium. Water Resour. Res. **1975**, *11* (3), 496–498.

32. Russo, D.; Jury, W.A.; Butters, G.L. Numerical Analysis of Solute Transport during Transient Irrigation I. The Effect of Hysteresis and Profile Heterogeneity. Water Resour. Res. **1989**, *25* (10), 2109–2118.

33. Jones, M.J.; Watson, K.K. Effect of Soil Water Hysteresis on Solute Movement During Intermittent Leaching. Water Resour. Res. **1987**, *23* (7), 1251–1256.

Soil Water Measurement by Capacitance

James L. Starr
United States Department of Agriculture (USDA), Beltsville, Maryland, U.S.A.

Ioan C. Paltineanu
Paltin International, Inc., Laurel, Maryland, U.S.A.

INTRODUCTION

Soil-water content is often the primary limiting factor for plant growth. Water is also the primary vehicle for moving plant nutrients and pesticides through and over soil to ground- and surface-water bodies. Conversely, the fate of water in soil is affected by many natural and management factors, such as soil texture and structure, presence or absence of plants, stage of plant growth, climate, rainfall intensity, tillage practices, residue covers, etc. These various factors are interactive, and their net effect continually changing in time and space. Until recently, soil water content measurements were made either by destructive soil sampling or by the use of portable neutron probes. Both methods were limited to fairly large time intervals (e.g., days, weeks). Capacitance probes and monitoring systems now provide the means to quantify soil water dynamics in real-time, at discrete soil depth increments, and over large areas, leading to improved soil and plant management practices, more efficient use of water and chemicals, and minimizing groundwater contamination.

HOW DOES THE CAPACITANCE PROBE MEASURE SOIL WATER CONTENT?

Electromagnetically, a soil can be represented as a dielectric mixture of air, bulk soil, and water. At radio frequencies, and at standard pressure and temperature conditions, the dielectric constant of pure water is 80, that of soil solids are 3–7, and that of air is 1. The dielectric constant of soil can be measured by capacitance, by including the soil as part of a capacitor in which the permanent dipoles of water molecules present in the surrounding soil become polarized and respond to the frequency of an imposed electric field. Measurement of the soil's capacitance gives its apparent dielectric constant, and thereby the soil water content.[1,2] Capacitance probe measurements are a function of the apparent or bulk dielectric constant (ε_b) of the soil, the imposed electromagnetic frequency, and the electrode configuration.

The relationship between the ε_b and the total capacitance (C) is:

$$C = g\varepsilon_b$$

where g is a geometrical constant based on the electrode configuration (size, shape, and distance between electrodes). Capacitance probes consist of an inductor (L) and a capacitor connected to circuitry that oscillates at a frequency that is dependent on the values of L and the electrode-soil capacitor. With L set by the electronic circuitry the frequency of oscillation depends only on variations of capacitance. The oscillation frequency (F) is an inverse square root function of the capacitance:

$$F = (2\pi\sqrt{LC})^{-1}$$

where L is the total circuit inductance, and C is the total capacitance that includes the soil–water–air mixture together with some constants. For most accurate measurements of soil water content, the functional relationship between oscillation frequency and soil water content should be determined empirically by calibration for specific soils.

Under field conditions the ratio of air to water in the soil continuously changes, resulting in large variations of the soil's apparent dielectric constant. Advances in microelectronics have led to rapid development of capacitance probes in the last decade. Some of the manufacturers of capacitance probes that have been reported in the scientific literature[2–10] are shown in Table 1.

CAPACITANCE PROBE DESIGNS

Individual capacitance probes measure water content at fixed frequencies that commonly vary from 38 MHz to 150 MHz, depending on the probe design. Operational frequencies of 100 MHz–150 MHz will minimize interferences from soil acidity and salinity.[10] Capacitance probes are commonly configured, as schematically shown in Fig. 1, with: 1) two or more parallel rods designed to be pushed into, or buried in the soil; or 2) one or more pairs of cylindrical metal electrodes, with a separating plastic ring between the pair electrodes, and

Encyclopedia of Water Science
DOI: 10.1081/E-EWS 120010150

Table 1 Manufacturers and suppliers of capacitance probes reported in the scientific literature

Brand name	Probe type	Manufacturer address	References
EnviroSCAN	Cylindrical ring	Sentek Pty Ltd., 69 King William St., Kent Town, S. Australia 5067, Australia http://www.sentek.com.au	[2–5]
Humicap 9000	Cylindrical ring and rod	SDEC France, 19 rue E. Vaillant, 37000 Tours, France http://www.sdec-france.com/us/index.html	[6]
Troxler sentry 200 AP	Cylindrical ring	Troxler Electronic Labs., Inc., 3008 Cornwallis Rd., PO Box 12057, Research Triangle Park, NC 27709, USA http://www.ismirrigation.com/	[7]
Vitel hydra probe	Parallel rods	Vitel Inc., 14100 Parke Long Court, Chantilly, VA 20151, USA http://www.vitelinc.com/	[8,9]

mounted on a support rod that is inserted into a previously installed polyvinylchloride (PVC) access pipe. Other configurations are possible, such as a combination of one circular ring and one rod.[6] Accurate soil water content measurement for all electromagnetic based sensors requires careful installation procedures to prevent formation of air-gaps along the sensors or changes in soil properties within the sensor's zone of influence.

Portable capacitance probes configured as parallel rods are simple in design, comparatively inexpensive, and well suited for surface soil-water measurements. Some of these probes can be buried in soil at different depths with transmission cables connected to data loggers for near-continuous and real-time measurements. The sensor's zone of influence, i.e., measuring soil volume, for rod-type capacitance probes is largely contained between the electrode rods (Fig. 1).

Capacitance probes configured as one or more pairs of cylindrical metal rings are well suited for measuring soil water content at discrete depth intervals in the soil profile. These cylindrical ring sensors are normally placed inside a

PVC access pipe, and form together with the soil surrounding the access pipe, a fringe-sensing volume. The radial zone of fringe influence for cylindrical ring capacitance probes, for the size shown in Fig. 1, is primarily within 10 cm of the wall of the 5-cm diameter access pipe and about 10 cm along the pipe, centered at the insulator between the two metal rings.[2] The semipermanently installed probes can be automated for real-time measurements over large areas, with probe readings essentially unaffected by cable length up to 500 m.[3]

HOW ARE CAPACITANCE PROBES CALIBRATED?

Published research reports on testing and applications of capacitance probes are still largely limited to those

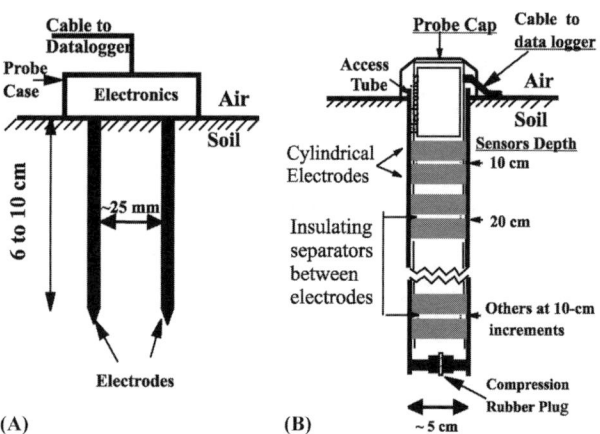

Fig. 1 Two common capacitance probe designs: (A) two or more parallel electrodes in direct soil contact, and (B) cylindrical metal ring electrodes placed inside a PVC access pipe.

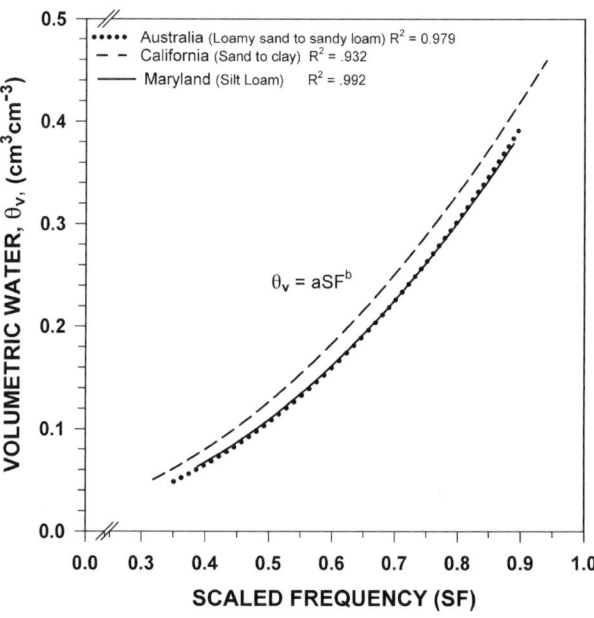

Fig. 2 Volumetric water content (θ_v) vs. scaled frequency (SF) at three sites. (Adapted from Ref. [2].)

configured as cylindrical ring sensors. For example, calibration experiments with cylindrical ring, multisensor capacitance probes were reported on a wide range of soil textures—using soils collected from California, Maryland, and Australia.[2] Calibration results, as shown in Fig. 2, indicate a highly significant, nonlinear relationship between the volumetric soil water content (θ_v) and scaled frequency (SF), that was characterized mathematically as:

$$\theta_v = aSF^b$$

The scaled frequency represents the ratio of frequencies measured by each sensor (inside the PVC pipe) in the surrounding soil (F_s) compared with sensor responses in the air (F_a) and in nonsaline water (F_w) at room temperature ($\sim 22°C$),

$$SF = (F_a - F_s)/(F_a - F_w)$$

The use of a scaled (i.e., normalized) frequency minimizes sensor specific electronic differences, so that the same calibration curve can be used for all the capacitance sensors.

REAL-TIME SOIL WATER PROFILE DYNAMICS MONITORED WITH MULTISENSOR CAPACITANCE PROBES

Laboratory and field studies have shown that capacitance probes are accurate, robust, stable in time, and amenable to near-continuous and real-time measurements.[1–9] Sample output from a multisensor capacitance probe that was set to measure water content at 10-min intervals is shown in Fig. 3. The depth and magnitude of water infiltration following the rainfall event (8/1/95) shows water penetration to the third sensor depth, reaching maximum water contents of 35%–40% volumetric water contents ($cm^3 cm^{-3}$) for the top three sensor depths. After the initial rapid drop in soil water content, i.e., drainage of the largest soil pores, the water drainage continued during the first night followed by a combined drainage and plant-water uptake during the first day after irrigation. Very little additional drainage was evident the second night, as indicated by the nearly constant water content over night, followed by a lower total rate of loss the next day—due to plant water evaporation. The expanded time scale shows additional detail of the water penetration. Note the sequence and speed of water penetration from the first sensor depth (measuring water over the 5 cm–15 cm interval) to the third depth, with the peak in water contents moving from one sensor depth to the next in less than 2 hr.

Another example of the kind of information that can be gained from near-continuous measurement of soil water content using capacitance probes is shown in Fig. 4. Early

in the 1995 growing season (7/3/95) the soil water content was still quite moist from the spring rains. The 7/4/95 rainfall event raised the soil water content to its full point, e.g., apparent water holding capacity (aWHC), and was repeated again on 7/7/95 and 7/8/95. This figure also shows rapid water uptake by the corn crop at the 10-cm sensor depth that lasted for about 8 day, then quite abruptly the rate of water uptake slowed, as the corn shallow roots were not able to continue extracting water at the same rate from the drying soil. The same pattern can be observed at the 20-cm and 30-cm soil depths. By 7/24/95, the corn roots had penetrated to and started removing water from the 50 cm depth. The presence of active plant roots to a given depth can also be seen by the diurnal changes in soil water content (daytime uptake of soil water, and some night-time gains in water content by a complex of hydrologic processes). This kind of information, which can only be obtained by near-continuous real-time soil water content monitoring, is of utmost importance for irrigation water management.

The speed at which the soil water content changes during wetting and drying cycles for any specific soil will

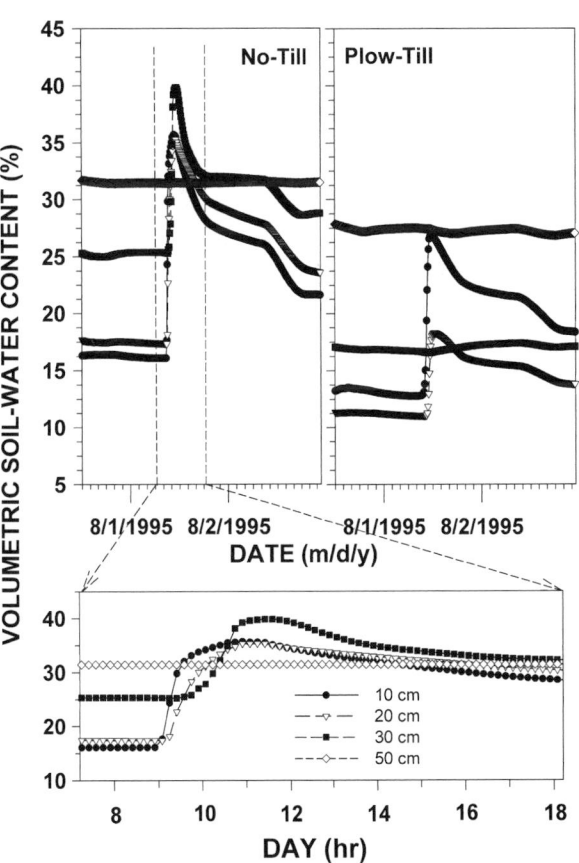

Fig. 3 Real-time soil water dynamics at four sensor depths under field-corn, as influenced by rainfall or irrigation and daytime evaporative demand. (Adapted from Ref. [5].)

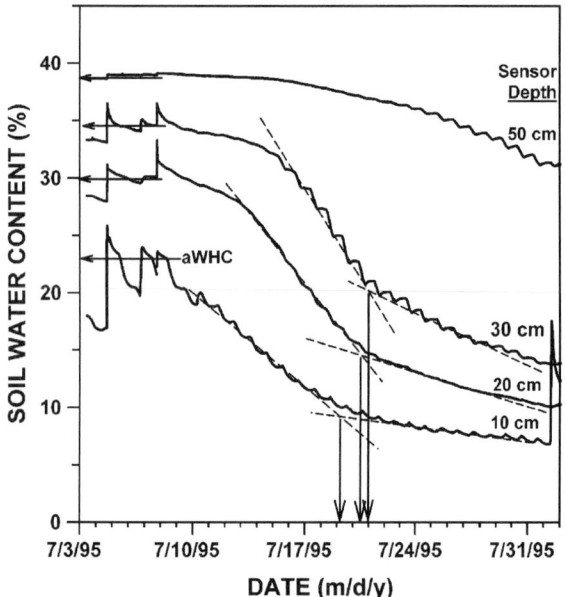

Fig. 4 Real-time soil water dynamics, at four capacitance sensor depths, associated with early summer rainfall events and evaporative demand under corn. The arrows show the apparent water holding capacity (aWHC), and the dates for the breaking points from fast to slow water uptake by corn.

vary with the soil–crop–climate conditions. In all cases, the multisensor capacitance probes have the capacity to reveal real-time changes in soil water content, and have proven to be a powerful tool for plant, soil, and water management, and a scientific basis to implement best use of natural resources while minimizing harmful side effects on the environment.

CHALLENGES OF THE NEW CENTURY

New developments in precision agriculture, remote sensing, preferential water flow patterns, simulation models for watershed hydrology and for soil–water–plant–atmosphere relationships over large areas, and permanent monitoring for leakage from waste material deposal sites, can all benefit from real-time soil water dynamics data. Reliable soil water content profile

dynamics monitoring over large areas, using multisensor capacitance probes to provide "the ground truth," is needed for validation and real-time calibration of the actual and future remote sensing sensors installed on orbital platforms.

REFERENCES

1. Dean, T.J.; Bell, J.P.; Baty, A.J.B. Soil Moisture Measurement by an Improved Capacitance Technique, Part 1. Sensor Design and Performance. J. Hydrol. **1987**, *93*, 67–78.
2. Paltineanu, I.C.; Starr, J.L. Real-Time Soil Water Dynamics Using Multisensor Capacitance Probes: Laboratory Calibration. Soil Sci. Soc. Am. J. **1997**, *61*, 1576–1585.
3. Starr, J.L.; Paltineanu, I.C. Soil Water Dynamics Using Multisensor Capacitance Probes in Non-traffic Interrows of Plow- and No-Till Corn. Soil Sci. Soc. Am. J. **1998**, *62*, 114–122.
4. Paltineanu, I.C.; Starr, J.L. Real-Time Soil Water Dynamics. In *Standard Handbook of Environmental Health, Science and Technology*; Lehr, J.H., Ed.; McGraw-Hill Pbl.: New York, 2000; 4.45–4.57.
5. Starr, J.L.; Paltineanu, I.C. Real-Time Soil Water Dynamics over Large Areas Using Multisensor Capacitance Probes and Monitoring System. Soil Till. Res. **1998**, *47*, 43–49.
6. Ould Mohamed, S.; Bertuzzi, P.; Bruand, A.; Raison, L.; Bruckler, L. Field Evaluation and Error Analysis of Soil Water Content Measurement Using the Capacitance Probe Method. Soil Sci. Soc. Am. J. **1997**, *61*, 399–408.
7. Evett, S.R.; Steiner, J.L. Precision of Neutron Scattering and Capacitance Type Soil Water Content Gauges from Field Calibration. Soil Sci. Soc. Am. J. **1995**, *59*, 961–968.
8. Schaeffer, G.L.; Yeck, R.D.; Paetzold, R.F. Soil Moisture/Soil Temperature Pilot Project—A National Near Real-Time Monitoring Project. CGU-AGU Annual Meeting, SMST. Banff, Canada; 1996.
9. http://www.wcc.nrcs.usda.gov/scan/index2.html (accessed October 2002).
10. Gardner, C.M.K.; Bell, J.P.; Cooper, J.D.; Dean, T.J.; Hodnett, M.G.; Gardner, N. Soil Water Content. In *Soil Analysis—Physical Methods*; Smith, R.A., Mullings, C.E., Eds.; Marcel Dekker, Inc.: New York, 1991; 1–73.

Soil Water Measurement by Neutron Thermalization

Steven R. Evett
United States Department of Agriculture (USDA), Bushland, Texas, U.S.A.

INTRODUCTION

Nearly 50 yr after its first use, the neutron thermalization method remains the best available method for repeated measurement of soil profile volumetric water content (VWC)[1] because it is nondestructive, can be field calibrated with high precision, works successfully to depths not easily attained with other methods, and works well in stony soils and cracking clays in which other methods work poorly. Also, the large volume of measurement means that fewer replicates are required than for other methods to produce a given precision, that soil disturbance during tube installation has minimal effect on results (unlike electronic sensor methods), and that field calibration is successful because volumetric soil samples can be obtained from within the volume measured by the probe at each depth (unlike electronic methods used in access tubes that have much smaller measurement volumes). The technology is mature with a wide literature base describing applications and problems.

The neutron thermalization method employs a radioactive source of fast neutrons (mean energy of 5 MeV) and a detector of slow neutrons (~ 0.025 eV or 300 °K). High-energy neutrons emitted from the source ($\sim 10^9$ sec^{-1}) are either slowed through repeated collisions with the nuclei of atoms in the soil (scattering and thermalization), or are absorbed by those nuclei. A small fraction of scattered neutrons will be deflected back to the detector. Of these, an even smaller fraction ($\sim 10^3$ sec^{-1}) will have been slowed to thermal (room temperature) energy levels and will be detected. Soil density and chemical composition affect the concentration of thermalized neutrons around the detector. The most common atoms in soil (aluminum and silicon) scatter neutrons with little energy loss because they have much greater mass than a neutron. However, if a neutron hits a hydrogen atom its energy is halved, on average, because the mass of the hydrogen nucleus is the same as that of the neutron. On average, 19 collisions with hydrogen are required to thermalize a neutron. Carbon, nitrogen, and oxygen are also relatively efficient as neutron thermalizers (about 120, 140, and 150 collisions, respectively). On the timescales of common interest in water management, changes in soil carbon and nitrogen content are minor and have little affect on the concentration of thermal neutrons. Also, on these timescales, changes in soil hydrogen and oxygen content occur mainly due to changes in soil water content. Thus, the concentration of thermal neutrons is most affected by changes in water content; and VWC can be accurately and precisely related to the count of thermal neutrons through empirical calibration.

Because hydrogen and carbon effectively thermalize neutrons, the organic matter content of soil affects the calibration. Also, organic matter and most clays contain important amounts of hydrogen, some not in the form of water, that may not be driven off by heating to 105°C (the standard temperature for drying soil samples). So, separate calibrations are often required for soil layers that differ in organic matter or clay content from layers above or below. In arid or semi-arid zones, many soils have layers rich in $CaCO_3$ and $CaSO_4$ that require separate calibration.[3] Atoms that absorb neutrons include boron, cadmium, chlorine, iron, fluorine, lithium, and potassium. Although these usually comprise a small fraction of soil material, soils, or soil horizons that contain large or fluctuating amounts of such elements will require separate calibrations or adjustments in data interpretation. For example, soils high in iron, such as Oxisols or soils rich in magnetite, typically require separate calibration, as may soils high in chloride salts. In some U.S. soils, boron is present in sufficient quantity to affect calibration.

NEUTRON MOISTURE METERS

Neutron moisture meter (NMM) equipment comes in two forms: 1) a profiling meter with a source–detector pair assembled into a cylindrical probe that is lowered into a hole in the soil; and 2) a flat-based meter that is placed on the soil surface with the source and detector fixed at separate locations inside the base of the meter. The volume measured by the surface meter is roughly hemispherical and extends into the soil for a distance that decreases as soil water content and soil density increase, and which varies from ~ 0.15 m in wet soil to ~ 0.3 m in dry soil.[3] The precision is less than can be attained with a profiling meter; and it suffers even more when soil moisture changes greatly with depth near the surface,[4] a common occurrence. Good precision has been reported under fairly stringent conditions including: 1) flattening the surface to fit the meter bottom with no air gaps; 2) marking the measurement site so that the meter can be repeatedly placed in identical position; and 3) using a neutron absorber shield made of cadmium around the meter (except for the bottom) to reduce effects of surrounding vegetation.[5] However, even in the latter study, the strong depth dependency of calibration coefficients and the inability to accurately estimate the depth of reading led to great uncertainty as to the accuracy of measurements.

More commonly used in soil and water science is the profiling NMM, which is operated at user-chosen depths in the soil (Fig. 1). A cylindrical access tube is used to line the hole, protecting the probe and ensuring a constant hole diameter. The probe is connected to a counter, data storage, and display module by a cable that allows the probe to be lowered into the tube and stopped at intervals to measure the thermal neutron concentration. Common probe diameters are 38 mm and 51 mm. When not in use, the probe is locked in the instrument shield, which comprises a block of high-density polyethylene, and which is commonly attached to the readout and control unit. In the probe, the source is either directly beneath the detector, or is centered around or on one side of it. The relative position of the source and detector affects the calibration;[6] but for modern meters, source–detector geometry has little effect on the attainable precision.[2,7,8] In modern meters, the source is a mixture of americium-

241 and beryllium with an activity ranging from 0.4 GBq to 1.9 GBq. The nuclear reaction is (^{9}Be(α,n)^{12}C) in which ^{241}Am emits an alpha particle that is absorbed by a Be atom, which then produces ^{12}C and a fast neutron.

The measurement volume is approximately a sphere. For a soil of specified VWC (m^3 m^{-3}), about 95% of the measured slow neutrons are from a sphere of radius R (cm).[9]

$$R = 15(\text{VWC})^{-1/3} \tag{1}$$

ACCESS TUBES AND DEPTH CONTROL

Access tubing materials that have been used successfully include stainless steel, mild steel, polyvinylchloride (PVC), polycarbonate, and polyethylene plastics, and aluminum. The hydrogen in plastics affects calibration, as does the neutron absorber chlorine in PVC tubes. Aluminum is nearly transparent to neutrons, while the neutron absorber iron affects calibration in steel tubes. Thus, it is important that a NMM be calibrated in the same tubing as will be used in the field. Although calibration precision decreases slightly if plastic tubes are used,[7] precision and accuracy are much more dependent on the tube installation and calibration methods employed than on tube material.

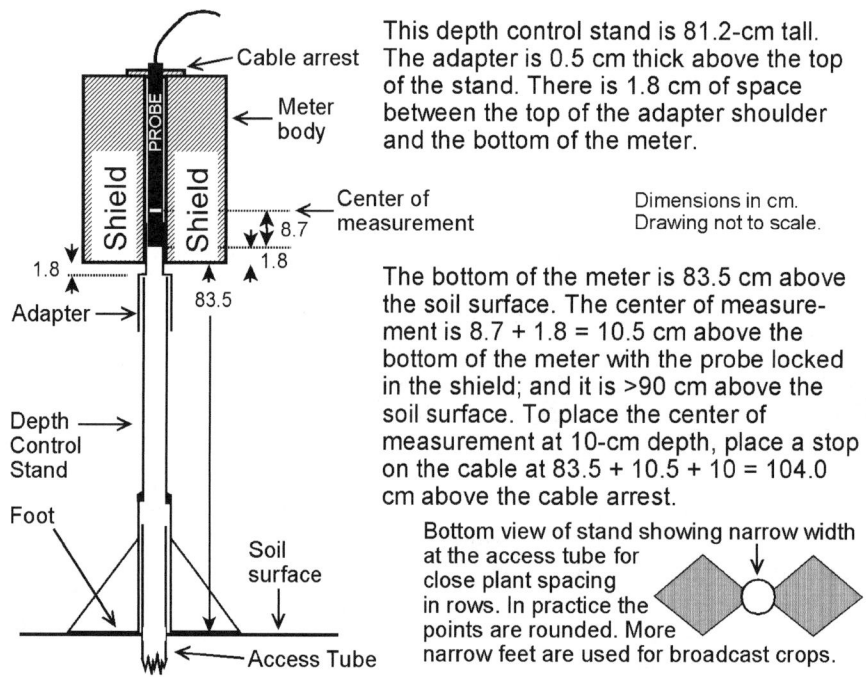

Fig. 1 Cross-sectional schematic of a profiling NMM in place on top of a depth control stand. The probe is locked in the meter shield; and the stand is in place over an access tube that has been inserted into the soil. For the dimensions given here, to measure at 10-cm depth, the probe must be lowered 104 cm through the stand and into the access tube. Dimensions will vary for meters of different manufacture.

Recommendations for installation of access tubes are given in Ref. [10].

It is a common practice to place the NMM on top of the access tube near the soil surface before lowering the probe for readings. This practice is not recommended for two reasons. First, when the NMM is placed near the soil surface, the shield in the meter body may influence near-surface counts to a degree that depends strongly on the height of the meter above the soil.[11] Second, in field use, the height of access tubes above the soil is likely to change with tillage, rainfall induced settling, erosion or deposition, or other factors, resulting in an equivalent change in the depth of probe placement. For readings above 0.3-m depth, the depth of the probe will strongly influence the reading and the calibration equation due to loss of neutrons to the atmosphere.[4,12]

These problems are addressed by using a depth control stand.* This device comprises a length of access tube fixed to a 0.2-m length of slightly larger tubing that is in turn supported by a foot resting directly on the soil (Fig. 1). The larger diameter of the lower length of tubing allows it to be slipped over the top of an access tube so that the foot rests on the soil surface. This maintains the reading depth at an exact distance relative to the soil surface. Cable stops are arranged to achieve the desired depth placement of the probe. The stand described is tall enough to be suitable for taking standard counts with the NMM mounted on the stand and the probe locked in the meter shield. Standard counts taken with the meter too close to the soil surface may vary with the moisture content of the soil.[13,14]

STATISTICS OF NEUTRON EMISSION

Neutron emission is a random process that occurs according to a Poisson probability distribution. An important property of the Poisson distribution is that, for a series of counts over equal time periods, the standard deviation is equal to the square root of the mean value. One result of this fact is that the coefficient of variation of counts can be reduced by increasing the counting time. The sample mean, m, is computed as

$$m = \frac{1}{N} \sum_{i=1}^{N} x_i \qquad (2)$$

where x_i is the value of a single count and N is the number of counts (all taken with the probe in one position). The

sample standard deviation, s, is computed as

$$s = \left[\frac{1}{N-1} \sum_{i=1}^{N} (x_i - m)^2 \right]^{1/2} \qquad (3)$$

For a properly operating meter with the probe in a constant environment, the ratio of $s/(m)^{1/2}$, called the Chi ratio, should be close to unity. This ratio is related to the χ^2 statistic by

$$\frac{s}{m^{1/2}} = \left(\frac{\chi^2}{N-1} \right)^{1/2} \qquad (4)$$

Values of χ^2 (Chi-squared) for a given probability level (P) are given in statistical tables for different values of ($N-1$). We may write the right-hand-side of Eq. 4 for the upper and lower limits of χ^2 and thus obtain upper and lower values of the Chi ratio for the chosen probability level and number of samples. For example, for a 95% probability level and 32 samples, we find the values of χ^2 as 17.5 for $P = 0.975$ and 48.1 for $P = 0.025$; and from Eq. 4 the Chi ratio should be between 0.75 and 1.25 about 95 times in every hundred. Some meters divide the count by a fixed number in order to reduce the displayed count to a reasonably small value. In computing Chi ratios for such meters, the user should first multiply the recorded counts by the factor that the meter used to reduce them.

CALIBRATION

Manufacturers' calibration equations are seldom useful for soil moisture determination (e.g., Ref. [13]). Calibration of NMMs involves correlating measured count ratio values with independently determined VWC ($m^3 m^{-3}$). For modern meters and the normal range of values of soil water content, the calibration is linear and of the form

$$VWC = b_0 + b_1 C_R \qquad (5)$$

where b_0 and b_1 are the calibration coefficients as determined by linear regression, and C_R is the count ratio defined as

$$C_R = x/x_s \qquad (6)$$

where x is the count in the measured material and x_s is a standard count taken with the probe within a standard and reproducible material. Count ratio values are used because the source activity and thus counts will decline over time, and because the detector efficiency is somewhat temperature dependent.[15] Recommendations for taking standard counts are given in Ref. [10], as are recommendations for field calibration using the wet site–dry site method of Evett and Steiner.[2] Careful field

---

*Evett, S.R. Construction of a Depth Control Stand for Use with the Neutron Probe [Online]. USDA-ARS-SPA-CPRL, Bushland, TX; 2000; 7pp. Available at http://www.cprl.ars.usda.gov/programs/ (posted 5 July 2000; verified 28 July 2000).

calibrations done using the wet site–dry site method and the depth control stand should attain root mean squared errors $< 0.01 \, \text{m}^3 \, \text{m}^{-3}$ and r^2 values greater than 0.9, even for depths near the surface (e.g., 10 cm in Table 1).

SAFETY AND USE CONSIDERATIONS

Safety concerns relate to radiation safety and to back and knee strains incurred during repeated bending and kneeling to operate meters placed on access tubes. The depth control stand described earlier allows users to work standing up, and has virtually eliminated physical injuries where it is used. Due to the low levels of radioactivity involved, the principle of reducing exposure to as low as reasonably achievable (ALARA) guides most radiation safety rules. Users may lower radiation received by increasing distance from the meter, decreasing time spent near the meter, and increasing shielding. The probe should always be locked into the shield except when it is lowered into an access tube. Users should be made aware that the source emits radiation at all times, even when the meter is turned off and batteries removed. Guidelines for ALARA use of the NMM are found in Ref. [16]. The USDA Radiation Safety Staff maintains an Internet site of useful information on radiation safety and hazardous materials transport (http://www.usda.gov/da/shmd/rss1.htm) as does the International Atomic Energy Agency (http://www.iaea.org/worldatom/).

Due to regulation, the method is not usable for automatic measurements. Due to its large measurement volume, the method is inappropriate where detailed vertical definition is required. This can be particularly important near the surface where water content often changes rapidly with depth. In such cases, the NMM can be used for deeper measurements in conjunction with time domain reflectometry (TDR) measurement of the near-surface soil water content.[17] The time and effort required to install access tubes and calibrate for each soil type is nontrivial. There is also a substantial cost for the equipment and for necessary training and licenses to handle and transport radioactive materials.

REFERENCES

1. IAEA, *Comparison of Soil Water Measurement Using the Neutron Scattering, Time Domain Reflectometry and Capacitance Methods*, IAEA-TECDOC-1137; International Atomic Energy Agency: Austria, 2000.
2. Evett, S.R.; Steiner, J.L. Precision of Neutron Scattering and Capacitance Type Moisture Gages Based on Field Calibration. Soil Sci. Soc. Am. J. **1995**, *59*, 961–968.
3. Van Bavel, C.H.M.; Underwood, N.; Swanson, R.W. Soil Moisture Measurement by Neutron Moderation. Soil Sci. **1956**, *82*, 29–41.
4. Van Bavel, C.H.M.; Nielsen, D.R.; Davidson, J.M. Calibration and Characteristics of Two Neutron Moisture Probes. Soil Sci. Soc. Am. Proc. **1961**, *25* (5), 329–334.
5. Nakayama, F.S.; Allen, S.G. Application of Neutron Soil Surface Water Monitoring for Plant Establishment. In *Irrigation and Drainage*; Proceedings of the 1990 National Conference. Durango, CO, July 11–13, 1990, Harris, S.R., Ed.; Am. Soc. Civil Engr.: New York, NY, 1990; 210–217.
6. McCauley, G.N.; Stone, J.F. Source–Detector Geometry Effect on Neutron Probe Calibration. Soil Sci. Soc. Am. Proc. **1972**, *36*, 246–250.
7. Allen, R.G.; Dickey, G.; Wright, J.L. Effect of Moisture and Bulk Density Sampling on Neutron Moisture Gauge Calibration. In *Management of Irrigation and Drainage Systems, Integrated Perspectives*; Proceedings of the 1993 ASCE National Conference on Irrigation and Drainage Engineering, Park City, UT, July 21–23, 1993, Allen, R.G., Neale, C.M.U., Eds.; American Society of Civil Engineers: New York, NY, 1993; 1145–1152.
8. Dickey, G.; Allen, R.G.; Wright, J.L.; Murray, N.R.; Stone, J.F.; Hunsaker, D.J. Soil Bulk Density Sampling for Neutron Gauge Calibration. In *Management of Irrigation and Drainage Systems, Integrated Perspectives*; Proceedings of the National Conference on Irrigation and Drainage Engineering, Park City UT. July 21–23, 1993, Allen, R.G., Neal, C.M.U., Eds.; American Society of Civil Engineers: New York, NY, 1993; 1103–1111.
9. IAEA, *Neutron Moisture Gauges*. Tech. Rep. Ser. No. 112. International Atomic Agency: Vienna, Austria, 1970.
10. Hignett, C.; Evett, S.R. Neutron Thermalization. *Methods of Soil Analysis, Part 4: Physical and Mineralogical Methods*, AGRONOMY Monograph Number 9, 3rd Ed.; 2001; (in press).
11. Stone, J.F.; Allen, R.G.; Gray, H.R.; Dickey, G.L.; Nakayama, F.S. Performance Factors of Neutron Moisture Probes Related to Position of the Source on the Detector. In *Management of Irrigation and Drainage Systems, Integrated Perspectives*; Proceedings of the National Conference on Irrigation and Drainage Engineering, Park City, UT. July 21–23, 1993, Allen, R.G., Neal, C.M.U., Eds.; American Society of Civil Engineers: New York, NY, 1993; 1128–1135.

Table 1 Calibration of water content (θ_v, $\text{m}^3 \, \text{m}^{-3}$) vs. count ratio ($C_R$) for the Amarillo fine sandy loam using the method in Ref. [2]. A depth control stand was used

Depth (cm)	Equation	RMSE	r^2	N
10	$\theta_v = 0.014 + 0.2172 C_R$	0.004	0.997	6
30–190	$\theta_v = -0.063 + 0.2371 C_R$	0.007	0.988	44
30–90	$\theta_v = -0.066 + 0.2421 C_R$	0.008	0.988	24
110–190	$\theta_v = -0.057 + 0.2299 C_R$	0.006	0.992	20

RMSE is the root mean squared error, N, the number of samples, and r^2, the coefficient of determination for the regression analysis.

12. Grant, D.R. Measurement of Soil Moisture Near the Surface Using a Moisture Meter. J. Soil Sci. **1975**, *26*, 124–129.

13. Dickey, G.L. Factors Affecting Neutron Gauge Calibration. In *Irrigation and Drainage*; Proceedings of the 1990 National Conference. Durango, CO, July 11–13, 1990, Harris, S.R., Ed.; American Society of Civil Engineers: New York, NY, 1990; 9–20.

14. Allen, R.G.; Segura, D. Access Tube Characteristics and Neutron Meter Calibration. In *Irrigation and Drainage*; Proceedings of the 1990 National Conference. Durango, CO, July 11–13, 1990, Harris, S.R., Ed.; American Society of Civil Engineers: New York, NY, 1990; 21–31.

15. Evett, S.R. Some Aspects of Time Domain Reflectometry (TDR), Neutron Scattering, and Capacitance Methods of Soil Water Content Measurement. *Comparison of Soil Water Measurement Using the Neutron Scattering, Time Domain Reflectometry and Capacitance Methods*, IAEA-TECDOC-1137; International Atomic Energy Agency: Vienna, Austria, 2000; 5–49.

16. Evett, S.R. *Nuclear Gauge Module I, Design, Theory, and Operation*; Nuclear Gauge Train-the-Trainer Course, USDA-Radiation Safety Staff, Beltsville, MA, Available at http://www.cprl.ars.usda.gov/programs/ (posted 9 June 2000; verified 31 July 2000), 2000; 23.

17. Evett, S.R.; Howell, T.A.; Steiner, J.L.; Cresap, J.L. Evapotranspiration by Soil Water Balance Using TDR and Neutron Scattering. In *Management of Irrigation and Drainage Systems, Integrated Perspectives*; Proceedings of the 1993 ASCE National Conference on Irrigation and Drainage Engineering, Park City, UT, July 21–23, 1993, Allen, R.G., Neale, C.M.U., Eds.; American Society of Civil Engineers: New York, NY, 1993; 914–921.

S

Soil Water Measurement by Time Domain Reflectometry

Steven R. Evett
United States Department of Agriculture (USDA),
Bushland, Texas, U.S.A.

INTRODUCTION

Time domain reflectometry (TDR) became known as a useful method for soil water content and bulk electrical conductivity (BEC) measurement in the 1980s through the publication of a series of papers by Topp, Dalton, and others.[1-5] Automated TDR systems for water content measurement have been described in Refs. 6–10. Commercial systems became available in the late 1980s and continue to evolve with TDR instruments, probes, and multiplexers (e.g., see Ref. 11) available from a few companies.

THEORY

In the TDR method, a very fast rise time (approx. 200 ps) step voltage increase is injected into a waveguide (usually coaxial cable) that carries the pulse to a probe placed in the soil or other porous medium (Fig. 1). The velocity of the pulse in the probe is measured and related to soil water content, with smaller velocities indicating wetter soils. In a typical field installation, probes are connected to the instrument through a network of coaxial cables and multiplexers. Part of the TDR instrument (e.g., Tektronix[a] model 1502B/C) provides the voltage step and another part, essentially a fast oscilloscope, captures the reflected waveform. The oscilloscope can capture waveforms that represent all or any part of the waveguide (this includes cables, multiplexers, and probes), beginning from a location that is actually inside the instrument. For e.g., Fig. 1 shows a waveform that represents the waveguide from a point inside the cable tester, before the step pulse is injected, and extending beyond the pulse injection point to a point that is 4.2 m from the cable tester. The relative height of the waveform represents a voltage, which is proportional to the impedance of the waveguide. Although most TDR instruments display the horizontal axis in units of length (a holdover from the primary use of these

instruments in detecting the location of cable faults), the horizontal axis is actually measured in units of time.

The TDR instrument converts the time measurement to length units by using the relative propagation velocity factor setting, v_p, which is a fraction of the speed of light in a vacuum. For a given cable, the correct value of v_p is inversely proportional to the permittivity, ε, of the dielectric (insulating plastic) between the inner and outer conductors of the cable

$$v_p = v/c_o = (\varepsilon\mu)^{-0.5} \tag{1}$$

where v is the propagation velocity of the pulse along the cable, c_o is the speed of light in vacuum, and μ is the magnetic permeability of the dielectric material. For a TDR probe in a soil, the dielectric between the probe rods is a complex mixture of air, water, and soil particles that exhibits a variable apparent permittivity, ε_a. Water is the largest determinant of permittivity in soils. It has a permittivity of approx. 80, whereas the permittivity of soil minerals varies in the range 3–5; the permittivity of organic matter is likewise low; and the permittivity of air is unity. Also, soil water is the only rapidly changing determinant of ε_a. Thus, we are able to usefully calibrate soil water content vs. measured ε_a. The fact that frozen water has a low permittivity impedes accurate measurement of frozen water content, but allows the use of TDR for investigations of freezing depth and extent.[12]

The TDR method relies on graphical interpretation of the waveform reflected from that part of the waveguide that is the probe (Fig. 2). An example of waveform interpretation for a 20-cm TDR probe in wet sand shows how tangent lines are fitted to several waveform features (Fig. 3). Intersections of the tangent lines define times related to: 1) the separation of the outer braid from the coaxial cable so that it can be connected to one of the probe rods in the handle, t1.bis; 2) the time when the pulse exits the handle and enters the soil, t1; and 3) the time when the pulse reaches the ends of the probe rods, t2. The time taken for the step voltage pulse to travel along the probe rods, $t_t = t2 - t1$, is related to the propagation velocity as

$$t_t = 2L/v \tag{2}$$

---

[a]The mention of trade or manufacturer names is made for information only and does not imply an endorsement, recommendation, or exclusion by USDA-Agricultural Research Service.

Encyclopedia of Water Science
DOI: 10.1081/E-EWS 120010152

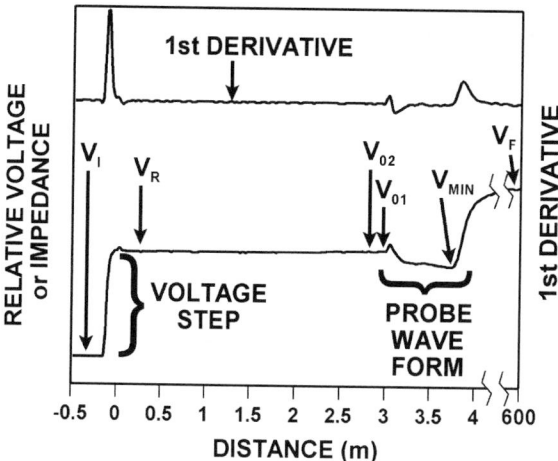

Fig. 1 Plot of waveform and its first derivative from a Tektronix 1502C TDR cable tester set to begin at $-0.5\,m$ (inside the cable tester). The voltage step is shown to be injected just before the zero point (BNC connector on instrument front panel). The propagation velocity factor, v_p, was set to 0.67 because electricity travels at 0.67 of the speed of light in the coaxial cable. At 3 m from the instrument, a TDR probe is connected to the cable. The relative voltage levels, V_I, V_R, etc., are used in calculations of the BEC of the medium in which the probe is inserted. Inflections in the first derivative of the waveform are used in software or firmware to help determine pulse travel times, which, for the probe, are proportional to water content.

where L is the length of the rods (Fig. 2), and the factor 2 signifies two-way travel.

Substituting ε_a and Eq. 2 into Eq. 1, and assuming $\mu = 1$, one sees that ε_a may be determined for a probe of known length, L, by measuring t_t

$$\varepsilon_a = [c_o t_t/(2L)]^2 \tag{3}$$

Topp et al.[1] found that a single polynomial function described the relationship between volumetric water content, θ_v, and values of ε_a determined from Eq. 3 for four mineral soils.

$$\theta_v = (-530 + 292\varepsilon_a - 5.5\varepsilon_a^2 + 0.043\varepsilon_a^3)/10^4 \tag{4}$$

Since 1980, other researchers have noted that the quantity $[t_t/(2L)]$ in Eq. 3 is quadratic, and have shown that the relationship between θ_v and $t_t/(2L)$ is practically linear (e.g., Ref. 15). Several attempts have been made to predict ε_a of soils from theoretical considerations using dielectric mixing models that consider the volumetric proportions of soil mineral, organic, water, and air constituents, as well as soil mineralogy and particle shape and packing considerations (e.g., Refs. 16–18). Success could lead to a more universal calibration, but has been elusive;[19] so that Eq. 4

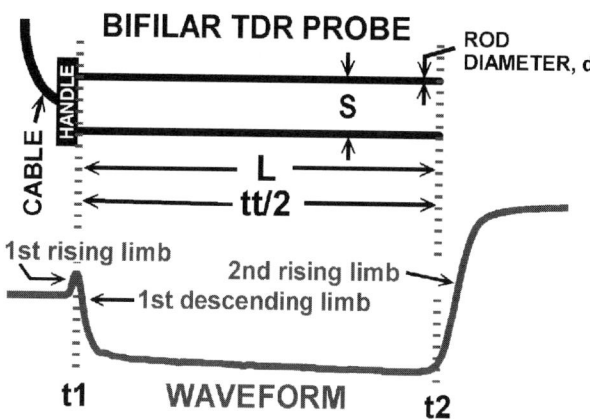

Fig. 2 Schematic of a typical bifilar TDR probe and the corresponding waveform, illustrating probe rod length, L; one-way travel time, $t_t/2$; rod spacing, s; and rod diameter, d.

and like empirical calibrations for specific soils (particularly electrically conductive soils including clays with high charge, and organic soils) are still considered to be the accepted standards.

APPLICABILITY

For most soils, excluding those very high in organic matter ($OM > 10\%$), the TDR method provides water content in the range from 0 to $0.5\,m^3\,m^{-3}$ with accuracy better than $0.01\,m^3\,m^{-3}$–$0.02\,m^3\,m^{-3}$ without calibration. With calibration, accuracy of better than $0.01\,m^3\,m^{-3}$ for a specific soil is attainable. Repeatability is excellent, with standard deviations of measurement ranging from $0.0006\,m^3\,m^{-3}$ (see Ref. 11) to $0.003\,m^3\,m^{-3}$ (see Ref. 8). Probe lengths reported in the literature range from 0.05 m to 1.5 m. Probe rod spacing, s, may also vary, so long as $d/s \leq 0.1$ where d is the rod diameter (Fig. 2).[20] As d/s becomes much smaller than 0.1, the volume of soil sensed becomes very small and TDR measurements may become overly sensitive to soil heterogeneity close to the rods. Because of this flexibility in probe width and length, TDR probes may be designed to measure a wide range of soil volumes. Because the volume measured extends only $1\,cm$–$2\,cm$ above and below the plane of the rods for most probe designs, TDR is ideal for measurements in thin layers near the soil surface. It is also very useful in root water uptake studies where information from discrete parts of the root zone is desired. Because TDR integrates soil water content changes occurring along the length of the probe rods accurately, TDR probes may be inserted vertically into soils to assess accurately mean water content over the length of the rods, even in soils exhibiting sharp water content changes with depth.

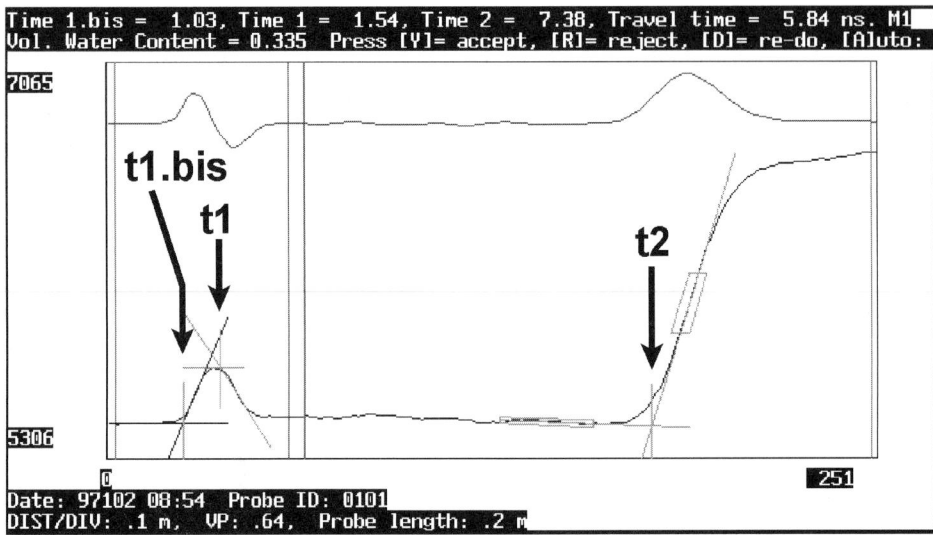

Fig. 3 Example of graphical interpretation of a waveform from a probe in wet sand using the TACQ computer program.[13,14] Vertical lines denoting times t1.bis, t1, and t2 have been marked by arrows and labels. The first peak in the waveform occurs just before t1. A horizontal line, drawn tangent to the waveform base line at the far left, intersects with a line drawn tangent to the first rising limb of the waveform to define t1.bis. A horizontal line drawn tangent to the peak intersects with a line drawn tangent to the descending waveform after the peak to define t1. Time t2 is defined by the intersection of a line fitted to the waveform before t2, and a line fitted to the second rising limb of the waveform after t2. The water content is calculated from Eq. 4. The width of the waveform window is 1 m, or 5.2 ns with the cable tester set to $v_p = 0.64$.

WAVEFORM INTERPRETATION

Graphical interpretation (e.g., Fig. 3) depends on the fact that the probe design itself introduces impedance changes in the waveguide. The impedance, $Z (\Omega)$, of a transmission line (i.e., waveguide) is

$$Z = Z_0(\varepsilon)^{-0.5} \tag{5}$$

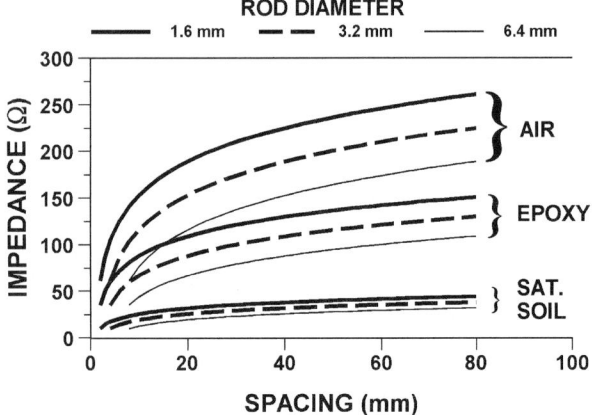

Fig. 4 Influence of rod spacing, rod diameter, and permittivity of the medium on impedance of the waveguide according to Eq. 6. Permittivities are: AIR, unity; EPOXY, close to 3; and SATurated SOIL, approx. 35.

where Z_0 is the characteristic impedance of the line (when air fills the space between conductors) and ε is the permittivity of the homogeneous medium filling the space between conductors. For a parallel transmission line (the two rods in the soil), the characteristic impedance is a function[21] of the wire diameter, d, and spacing, s (Fig. 2):

$$Z_0 = 120 \ln\{2s/d + [(s/d)^2 - 1]^{0.5}\} \tag{6}$$

or, if $d \ll s$:

$$Z_0 = 120 \ln(2s/d) \tag{7}$$

For a coaxial transmission line, the characteristic impedance is:

$$Z_0 = 60 \ln(D/d) \tag{8}$$

where D and d are the diameters of the outer and inner conductors, respectively.

From Eqs. 5–8 it is apparent that impedance, Z, increases as wire spacing increases, and decreases as ε (or water content) increases for any probe type (Fig. 4). In the probe handle, the wire spacing increases from that of the coaxial cable to that of the probe rods. The resulting impedance increase causes the waveform level to rise (first rising limb in Fig. 2). If the porous medium in which the probe rods are embedded is wet, then the permittivity of that medium will be higher than that of the epoxy probe handle. This causes a decrease in impedance, which results in the descent of the reflected waveform level as the step voltage leaves the handle and enters the rods in the soil (first descending limb, Fig. 2). The combination of impedance increase at the handle and impedance decrease after the handle gives the peak in the waveform. The rod ends are another impedance change in the waveguide; in this case an open circuit. The remaining energy in the voltage step is reflected back at the rod ends, which represent an impedance increase (second rising limb, Fig. 2). Although a bifilar probe design is illustrated in Fig. 2, the most common design uses three parallel and coplanar rods. Such trifilar probes are electrically unbalanced (signal is on the middle rod) as is the connecting coaxial cable. Thus, impedance is more closely matched between cable and probe and the waveform has less noise and is more easily interpretable.[22]

Waveform shapes different from those shown in Figs. 1–3 result from different soil types and conditions (e.g., dry soil, saline soils, wet clays, etc.). Different methods from the literature, used for graphical interpretation of the waveform, can cause errors in water content as large as $0.05\,m^3\,m^{-3}$.[14] Therefore, choice of interpretation methods or computer programs for automatic interpretation is important. Manufacturers' equipment contains embedded interpretation algorithms that are not usually made public. Two computer programs available to the public and well documented are TACQ[13,14,23] and WinTDR.[24] An improved signal to noise ratio results from the shorting diode approach[25] in which the waveform is alternately captured with and without the probe shorted to ground at the ends of the rods. This approach has not been popular, however, due to increased cost and complexity of switching, and problems with designing probes that ensure signal penetration into the soil.

BULK ELECTRICAL CONDUCTIVITY MEASUREMENT

An important use of the TDR method is to calculate the soil BEC from values of the waveform relative voltage or impedance at various points along the waveguide (Fig. 1) (e.g., Refs. 2–5,22,26–30). The measured load impedance, Z_L (ohms) is used in most methods for calculating BEC:

$$Z_L = Z_{REF}(1 + \rho)/(1 - \rho) \tag{9}$$

where Z_{REF} is the output impedance of the cable tester (e.g., 50 ohms), and:

$$\rho = E - /E+ \tag{10}$$

where

$$E- = V_F - V_{O2} \tag{11}$$

$$E+ = V_{O2} - V_I \tag{12}$$

and where V_{O2}, V_I, and V_F are defined in Fig. 1. For most methods, only V_{O2}, V_I, and V_F are needed. Calculation of BEC from TDR data is still a subject of active research. The other values of relative voltage illustrated in Fig. 1 are used in other methods of calculating BEC reported in the literature. The TDR method has been even extended to measurement of atmospheric CO_2 based on the solution electrical conductivity increase caused by its dissolution in water.[31]

REFERENCES

1. Topp, G.C.; Davis, J.L.; Annan, A.P. Electromagnetic Determination of Soil Water Content: Measurements in Coaxial Transmission Lines. Water Resour. Res. **1980**, *16* (3), 574–582.
2. Dalton, F.N.; Herkelrath, W.N.; Rawlins, D.S.; Rhoades, J.D. Time-Domain Reflectometry: Simultaneous Measurement of Soil Water Content and Electrical Conductivity with a Single Probe. Science **1984**, *224*, 989–990.
3. Dalton, F.N.; van Genuchten, M.Th. The Time-Domain Reflectometry Method for Measuring Soil Water Content and Salinity. Geoderma **1986**, *38*, 237–250.
4. Dasberg, S.; Dalton, F.N. Time Domain Reflectometry Field Measurements of Soil Water Content and Electrical Conductivity. Soil Sci. Soc. Am. J. **1985**, *49*, 293–297.
5. Topp, G.C.; Yanuka, M.; Zebchuk, W.D.; Zegelin, S. Determination of Electrical Conductivity using Time Domain Reflectometry: Soil and Water Experiments in Coaxial Lines. Water Resour. Res. **1988**, *24*, 945–952.
6. Baker, J.M.; Allmaras, R.R. System for Automating and Multiplexing Soil Moisture Measurement by Time-Domain Reflectometry. Soil. Sci. Am. J. **1990**, *54* (1), 1–6.
7. Heimovaara, T.J.; Bouten, W. A Computer-Controlled 36-Channel Time Domain Reflectometry System for Monitoring Soil Water Contents. Water Resour. Res. **1990**, *26* (10), 2311–2316.
8. Herkelrath, W.N.; Hamburg, S.P.; Murphy, F. Automatic, Real-Time Monitoring of Soil Moisture in a Remote Field Area with Time Domain Reflectometry. Water Resour. Res. **1991**, *27* (5), 857–864.
9. Evett, S.R. Evapotranspiration by Soil Water Balance Using TDR and Neutron Scattering. In *Management of Irrigation and Drainage Systems*, Irrigation and Drainage Div./ASCE, Park City, Utah, July 21–23, 1993; 914–921.
10. Evett, S.R. TDR–Temperature Arrays for Analysis of Field Soil Thermal Properties. In *Proceedings of the Symposium on Time Domain Reflectometry in Environmental, Infrastructure and Mining Applications*, Northwestern

University, Evanston, Illinois, Sept 7–9, 1994; USDI, Bureau of Mines, Special Publication SP 19-94; 320–327.

11. Evett, S.R. Coaxial Multiplexer for Time Domain Reflectometry Measurement of Soil Water Content and Bulk Electrical Conductivity. Trans. ASAE **1998**, *42* (2), 361–369.

12. Spaans, E.J.A.; Baker, J.M. Examining the Use of Time Domain Reflectometry for Measuring Liquid Water Content in Frozen Soil. Water Resour. Res. **1995**, *31* (12), 2917–2925.

13. Evett, S.R. The TACQ Computer Program for Automatic Time Domain Reflectometry Measurements: I Design and Operating Characteristics. Trans. ASAE **2000**, *43* (6), 1939–1946.

14. Evett, S.R. The TACQ Computer Program for Automatic Time Domain Reflectometry Measurements: II Waveform Interpretation Methods. Trans. ASAE **2000**, *43* (6), 1947–1956.

15. Ledieu, J.; De Ridder, P.; De Clerck, P.; Dautrebande, S. A Method of Measuring Soil Moisture by Time-Domain Reflectometry. J. Hydrol. **1986**, *88*, 319–328.

16. Roth, K.; Schulin, R.; Flühler, H.; Attinger, W. Calibration of Time Domain Reflectometry for Water Content Measurement using a Composite Dielectric Approach. Water Resour. Res. **1990**, *26* (10), 2267–2273.

17. Dirksen, C.; Dasberg, S. Improved Calibration of Time Domain Reflectometry Soil Water Content Measurements. Soil Sci. Soc. Am. J. **1993**, *57* (3), 660–667.

18. Wang, J.R.; Schmugge, T.J. An Empirical Model for the Complex Dielectric Permittivity of Soils as a Function of Water Content. IEEE Trans. Geosci. Rem. Sensing **1980**, GE-*14* (4), 288–295.

19. White, I.; Knight, J.H.; Zegelin, S.J.; Topp, G.C. Comments on 'Considerations on the Use of Time-Domain Reflectometry (TDR) for Measuring Soil Water Content' By W.R. Whalley. Eur. J. Soil Sci. **1994**, *45*, 503–508.

20. Knight, J.H. Sensitivity of Time Domain Reflectometry Measurements to Lateral Variations in Soil Water Content. Water Resour. Res. **1992**, *28* (9), 2345–2352.

21. Williams, T. *The Circuit Designer's Companion*; Butterworth-Heinemann, Ltd., Pub.: Oxford, England, 1991; 302 pp.

22. Zegelin, S.J.; White, I.; Jenkins, D.R. Improved Field Probes for Soil Water Content and Electrical Conductivity Measurement using Time Domain Reflectometry. Water Resour. Res. **1989**, *25* (11), 2367–2376.

23. TACQ.EXE. A Computer Program for TDR Data Acquisition and Interpretation, Available at http://www.cprl.ars.usda.gov/programs/ (accessed 13 April 2001).

24. WinTDR.EXE. available at http://psb.usu.edu/wintdr99/ (accessed 17 April 2001).

25. Hook, W.R.; Livingston, N.J.; Sun, Z.J.; Hook, P.B. Remote Diode Shorting Improves Measurement of Soil Water by Time Domain Reflectometery. Soil Sci. Soc. Am. J. **1992**, *56* (5), 1384–1391.

26. Dalton, F.N. *Measurement of Soil Water Content And Electrical Conductivity Using Time-Domain Reflectometry*. In Proceedings of the International Conference on Measurement of Soil and Plant Water Status, Utah State University, Logan, July 6–10, Vol. 1, 1987, 95–98.

27. Dalton, F.N. Development of Time Domain Reflectometry for Measuring Soil-Water Content and Bulk Soil Electrical Conductivity. In *advances in Measurement of Soil Physical Properties: Bringing Theory into Practice*; Topp, G.C., Reynolds, W.D., Green, R.E., Eds.; Soil Sci. Soc. Am.: Madison, WI, 1992.

28. Nadler, A.; Dasberg, S.; Lapid, I. Time Domain Reflectometry Measurements of Water Content and Electrical Conductivity of Layered Soil Columns. Soil Sci. Soc. Am. J. **1991**, *55*, 938–943.

29. Spaans, E.J.A.; Baker, J.M. Simple Baluns in Parallel Probes for Time Domain Reflectometry. Soil Sci. Soc. Am. J. **1993**, *57*, 668–673.

30. Wraith, J.M.; Comfort, S.D.; Woodbury, B.L.; Inskeep, W.P. A Simplified Waveform Analysis Approach for Monitoring Solute Transport using Time-Domain Reflectometry. Soil Sci. Soc. Am. J. **1993**, *57*, 637–642.

31. Baker, J.M.; Spaans, E.J.A.; Reece, C.F. Conductimetric Measurement of CO_2 Concentration: Theoretical Basis and Its Verification. Agron. J. **1996**, *88*, 675–682.

Soil Water Potential Measurement by Granular Matrix Sensors

Clinton C. Shock
Oregon State University, Ontario, Oregon, U.S.A.

INTRODUCTION

Like a tensiometer, the granular matrix sensor (GMS) is an instrument for measuring soil water potential.[1] The GMS eliminates regular maintenance required by tensiometers. Granular matrix sensor technology reduces the problems inherent in gypsum blocks (slow response time and dissolution of the block) by using a mostly insoluble granular fill material held in a fabric tube supported in a metal or plastic screen.[2,3] Like gypsum blocks, GMS sensors operate on the principle of variable electrical resistance. The electrodes inside the GMS are embedded in the granular fill material above a gypsum wafer, with additional granular matrix below the wafer in the fabric tube where water enters and exits the sensor. The gypsum wafer slowly dissolves to buffer the effect of salinity of the soil solution on electrical resistance between the electrodes. Particle size of the granular fill material and its compression determine the pore size distribution in the GMS and its response characteristics.[2]

OVERVIEW

Granular matrix sensors have been calibrated in the field in the range of −10 kPa down to −75 kPa for the irrigation of water stress-sensitive plants.[1] Calibrations have varied depending on the sensor model, soil, and other experimental conditions.[1,4−6] These sensors are most useful on soils that maintain intimate hydraulic contact with the sensor and are usually least useful on coarse textured soils.

A GMS for electronically measuring soil water was first patented by Larson,[2] and a commercial model is marketed as the Watermark Soil Moisture Sensor Model 200SS (Irrometer Co. Inc., Riverside, CA, USA, http://www.irrometer.com). The Model 200SS incorporates improvements in production and technology, with a perforated stainless steel exterior and uniform internal compaction.[3] The steel models can be manufactured more uniformly because of automated packing of the granular matrix to a prescribed pressure. The steel also exposes more fabric (Fig. 1) for greater sensor contact with the soil than the previous commercial GMS, Model 200.

GMS Placement and Installation

Granular matrix sensor performance is affected by placement[7] and installation techniques. Sensor placement depends upon the irrigation system, crop rooting depth, cultivation practices, and field topography. Sensor placement needs to be representative of the parts of the soil that become wet upon irrigation and respond fairly quickly to crop water use and soil drying. Locations for GMS placement in the field need to be representative of topography, soil types, and any large-scale heterogeneity created by the irrigation system.

Granular matrix sensors are soaked overnight in irrigation water before installation, and they are installed wet. This can improve GMS response in the first few irrigations after installation. The manufacturer recommends that the user make a 22 mm (7/8″) diameter access hole to the desired depth, pour water into the hole, and push the GMS down into the bottom. A snug fit in the soil is important, and the hole is then refilled with soil.

For very coarse or gravely soils, an oversized hole (25 mm–30 mm diameter) may be needed to prevent abrasion damage to the GMS fabric. In this case, a hole is augered to the desired depth and a thick slurry with the soil and some water is used. Partially fill the hole with this slurry, install the GMS, and then finish filling the hole. This will "grout in" the GMS to ensure a snug fit.

Another method of installing GMS in difficult gravely soils, or at greater depths is to use a "stepped" installing tool. This makes an oversized hole for the upper portion and an exact size hole (GMS is 22 mm in diameter) for the lower portion of the hole where the GMS is installed. The hole must be carefully filled and tamped down to prevent air pockets, which could allow water to channel down to the GMS.

For silt loam and loam soils, GMS can be installed using a 22 mm (7/8″) diameter soil sampling probe and a ruled insertion rod.[7] The sampling probe is used to make a hole in the soil the same diameter as the GMS. The GMS can be installed vertically, placing the tip 22 mm deeper than the desired depth of measurement, which centers the water exchange perforations at the desired depth. The depth of the GMS can be confirmed by pushing the sensor to the bottom of the hole in the soil with the ruled

Encyclopedia of Water Science
DOI: 10.1081/E-EWS 120010147

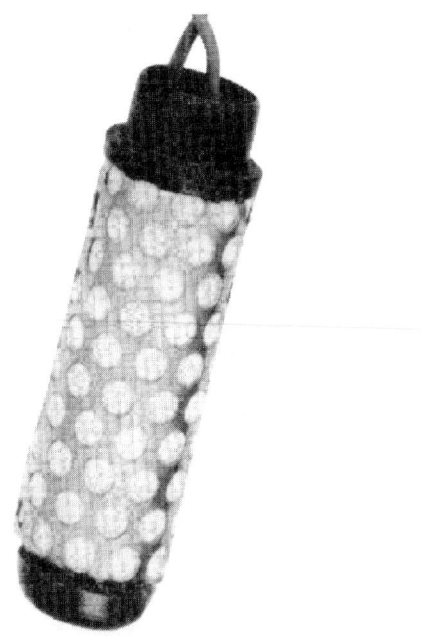

Fig. 1 A granular matrix sensor, Watermark Model 200SS, diagram courtesy of Irrometer Co., Riverside, CA.

insertion rod. If the hole is too deep, it can be partially refilled before GMS installation. Once the GMS is at the correct depth, 60 ml of water is poured on top of the GMS, and the hole is gently refilled with soil with light tamping as the hole is filled.

When a GMS is installed, it is essential that the GMS be in firm contact with the soil so that water will move from the soil into the GMS during wetting cycles and will move out of the GMS into the soil during drying cycles. The GMS will have variable resistance in most soils, but in

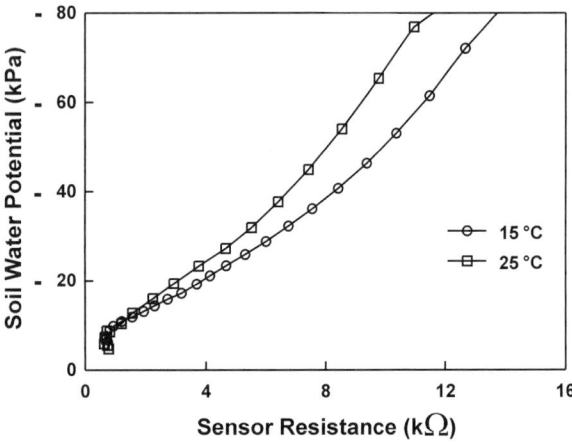

Fig. 2 Granular matrix sensor (Watermark Model 200SS) resistance responds to both temperature as well as soil water potential. (From Ref. 5.)

soils with very coarse texture, the hydraulic connection with the soil may not allow water to move into and out of the GMS, or may result in the GMS responding too slowly for standard reading practices.

Calibrated Range of Measurements

The nominal range of Watermark soil moisture sensor Model 200SS (GMS) measurements is from 0 kPa to −200 kPa or −2 atm. Thomson and Armstrong[4] calibrated Watermark soil moisture sensor Model 200 from 0 kPa to − 100 kPa in a pressure plate. Later, the same model was calibrated from 0 kPa to −75 kPa in silt loam planted to potato.[1] Three different GMS models were calibrated from −10 kPa to −80 kPa in silt loam in a controlled temperature growth chamber planted to grass.[5] The model 200SS was calibrated in two sandy soils, one from −10 kPa to −80 kPa and the another from −11.5 kPa to −23 kPa.[6] Calibration equations of GMS resistance to soil water potential include terms for soil temperature, because GMS resistance is affected by temperature (Fig. 2,[5]).

Calibration equations are used in meters and data loggers to read GMS resistance. The hand held 30 KTCD-NL meter (Irrometer Co.) has a manual temperature correction. Independent soil temperature data can be recorded into the meter before measuring the GMS.

GRANULAR MATRIX SENSOR MEASUREMENT OF WATER POTENTIAL

Why Measure Water Potential?

Water potential is of economic and environmental importance because it is the measure of how strongly water is held in the soil, which relates to the difficulty of removing water from the soil by plant roots. Plant performance has been closely associated with water potential measurements using GMS.[8–15] Water potential data are also important in irrigation to avoid saturated soil, lack of aeration of plant roots, and leaching losses of water or nutrients. Water potential differences indicate the direction of flow in unsaturated media. Water potential information can help evaluate the risks of erosion and slippage on steep slopes.[16]

Water Potential Measurements at Multiple Depths

Granular matrix sensors can be installed at multiple depths to develop an understanding of the relative water potential at different depths in the soil. As can be expected, GMS at shallow depths respond quickly to wetting and drying cycles.[10,11,13] The soil in Fig. 3 has a hard layer that is both semi-impermeable to water and impenetrable to poplar tree roots at 0.6 m–0.7 m depth. Consequently, the soil water potential at 0.8 m depth varies little during the growing season, irrespective of irrigation or water use.

Irrigation Scheduling

Growers need rapid and convenient ways to monitor soil water status to improve their irrigation scheduling. Growers and field men can make GMS readings with a hand held meter (Model 30 KTCD-NL, Irrometer Co. Inc., Riverside, CA, USA), and record the data manually. The GMS data may be graphed manually or entered and graphed by computer. The graph can be used to demonstrate whether the soil water potential is wetter or drier than the irrigation criteria for that particular crop. The soil water potential in graphical form is easier for growers to interpret, because the relative position (wet or dry) is clearer and the rate of drying over time is more easily understood in graphical form. Distinctly different irrigation regimes can be easily established and maintained in an arid climate (Fig. 4,[13]).

Benefits of Irrigation Scheduling

Crop yields and quality can be directly related to irrigation management using GMS. Soil water potential from GMS is being used by potato growers for irrigation scheduling,[7–9,12,14] and that use has expanded to onions and other crops. Onion yield and grade improve with careful irrigation scheduling based on GMS.[11,13] Optimum

growth of poplar trees is also closely related to the maintenance of soil water potential within narrow bounds by careful irrigation scheduling.[10] Alfalfa productivity also benefits from careful irrigation scheduling using GMS.[15]

Automated Logging of Soil Water Potential Data

Automated collection of GMS data for field crop production research has been accomplished using a wide range of data loggers and multiplexers. Ideally, the data logger sends out an AC signal in the range of 130 Hz–200 Hz, and GMS response is measured with a half-bridge circuit.

The AM400 Soil Moisture Data Logger with Graphic Display (Mike Hansen, Wenatchee, WA, USA, http://www.mkhansen.com) is an aid to irrigation scheduling designed for use with GMS.[17] Each AM400 can be wired to up to six GMS and one temperature probe. The AM400 reads the GMS three times a day, automatically stores the data, and displays a graph of each sensor on request. The AM400 graphs the soil water potential individually for each GMS for the last 5 weeks. The soil water potential irrigation criterion for alfalfa forage on silt loam is approximately −60 kPa. Regular use of sensor readings to schedule irrigations allowed the average soil water potential to remain within the ideal range for alfalfa (Fig. 5). The frequency of irrigation depends on the weather and the stage of growth of the alfalfa. Since the AM400 screen displayed data from the last 5 weeks, the soil water potential changes over time were easy to interpret. The AM400 has been used to read GMS in a variety of crops.

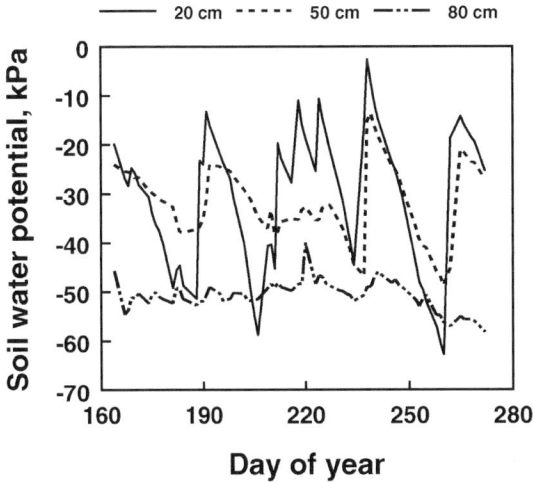

Fig. 3 Soil water potential responses to sprinkler irrigation under poplar trees. Granular matrix sensors nearest the soil surface (20 cm deep) respond to each irrigation while those below an impermeable layer (80 cm deep) show less variation.[10]

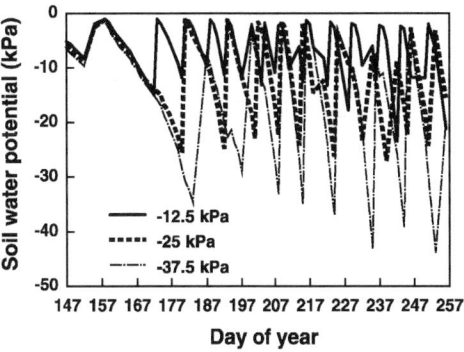

Fig. 4 Different irrigation criteria result in distinctly different soil water potential patterns over time.[13] The figure shows soil water potential patterns for three irrigation criteria: −12.5 kPa, −25 kPa, and −37.5 kPa.

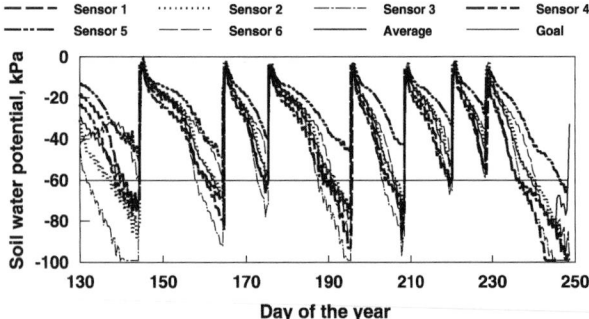

Fig. 5 Carefully scheduled sprinkler irrigation using GMSs can be used to refill the surface of the soil. (From Ref. 17.)

Automated Irrigation Scheduling Using GMS

Automatic feedback control of precision irrigation scheduling using GMS has facilitated the determination of optimum crop irrigation criteria,[11] close determination of N fertilizer requirements,[18] and measures of crop development and yield responses. In these studies, the sensors were connected to a datalogger (CR 10 datalogger, Campbell Scientific, Logan, Utah, USA) via multiplexers (AM 416 multiplexer, Campbell Scientific). The datalogger was programmed to read the GMS in each irrigation zone 4–8 times a day and irrigate each zone individually as necessary according to its irrigation criteria. Irrigations were controlled by the datalogger using a controller (SDM CD16AC controller, Campbell Scientific) connected to solenoid valves for each plot. The pressure in the drip lines was maintained constant by pressure regulators in each plot, and the amount of water applied in each zone was recorded by a water meter installed between the solenoid valve and the drip tape. The irrigation criteria for onion drip irrigation were determined based on the water use and crop response in this automated way.[11]

Automation of Landscape Irrigation with GMS

A substantial part of urban water use can be landscape irrigation. Although, findings are not unanimous, studies of landscape irrigation show that it is often strongly associated with air temperature rather than landscape plants' water needs.[19] Landscape irrigation is less strongly related to total precipitation, soil moisture content, and landscape plant evapotranspiration. Yet, the actual needs for irrigation are closely tied to precipitation, soil moisture content, or landscape plant evapotranspiration. Typical landscape irrigation systems consist of a timer that schedules the irrigations and valves controlled by the timer. The timer initiates the irrigation at a frequency and duration set by the water user. Any change in landscape water need requires the user to reset the timer,

but in practice, most fluctuations in landscape water needs are ignored.

Granular matrix sensors have been used in simple automatic feedback control systems to override irrigation timers since 1993.[20] By adding GMS and an electronic module (WEM, Watermark Electronic Module, Irrometer Co.), it is possible for the WEM to read the GMS and either allow or prevent an irrigation that has been scheduled on the timer.[19,20] When used in this configuration, it is common to set the timer to irrigate more frequently, but the system only irrigates when the soil is sufficiently dry to need irrigation. The addition of GMS and WEMs to automated irrigation systems has proven to be durable and cost effective in saving water in Boulder, Colorado, USA.[19]

Special Uses in Studies of Water Movement

Grids of GMS can be placed horizontally and vertically in the soil to monitor soil water movement over time.[7] Sensor placement can help assure that irrigations do not exceed soil water holding capacity of the crop or landscape root zone, providing environmental protection from nitrate leaching.[18] Tensiometers and GMS have been used on steep, unstable slopes to anticipate saturation and risks of slippage.[16]

Limitations of GMS

Calibrations used in the 30 KTCD-NL meter and the AM400 data logger were derived for silt loam soils[5] and different calibration equations may be needed in different soils and different climates.[6] The range of published calibrations is limited from −10 kPa to −100 kPa at 15–25°C. The successful operation of GMS depends upon water entering the sensor during soil wetting cycles and leaving the sensor during drying cycles. Soils with coarse textures or high shrink–swell clays can pull away from the GMS and limit its response. Even when perfectly calibrated and operational, the GMS reading only indicates when to irrigate, not how much water to apply. The amount of water to apply is largely determined by soil properties, the effective rooting depth, and the nature of the irrigation system. Current methods of reading GMS usually require wiring, which can be cumbersome or limiting for crops needing cultivation.

Each GMS only provides information about the soil water potential in the immediate vicinity of the sensor. Because of variability in soil water potential from place to place in a field and sensor to sensor variation, six or more GMS will provide more reliable estimates of soil water potential than the use of individual GMS.

CONCLUSION

The use of GMSs is increasing because they are a practical, inexpensive, and effective tool for many landscape and agricultural irrigation scheduling needs. Water can be conserved without sacrificing landscape aesthetic appearance or crop productivity and quality. The application of GMS plus WEM in automated urban landscape irrigation has saved costs and water for private and public water users.

ACKNOWLEDGMENTS

Oregon State Univ. technical paper no. 11909. Use of trade names does not imply endorsement of the products named or criticism of similar ones not named. I would like to thank Dr. Steven R. Evett of the USDA ARS Southern Plains Area Conservation and Production Research Laboratory, Al Hawkins, Tom Penning, and Bill Pogue of the Irrometer Co., Erik Feibert and Dr. Eric Eldredge of Oregon State University, and Candace Shock for providing suggestions that resulted in improvements in this manuscript.

REFERENCES

1. Eldredge, E.P.; Shock, C.C.; Stieber, T.D. Calibration of Granular Matrix Sensors for Irrigation Management. Agron. J. **1993**, *85*, 1228–1232.
2. Larson, G.F. Electrical Sensor for Measuring Moisture in Landscape and Agricultural Soils. US Patent 4,531,087, July 23, 1985.
3. Hawkins, A.J. Electrical Sensor for Sensing Moisture in Soils. US Patent 5,179,347, January 12, 1993.
4. Thomson, S.J.; Armstrong, C.F. Calibration of the Watermark Model 200 Soil Moisture Sensor. Appl. Eng. Agric. **1987**, *3*, 186–189.
5. Shock, C.C.; Barnum, J.; Seddigh, M. Calibration of Watermark Soil Moisture Sensors for Irrigation Management. Proceedings of the International Irrigation Show. Irrigation Association, San Diego, CA, Nov 1–3, 1998; 139–146.
6. Irmak, S.; Haman, D.Z. Performance of the Watermark Granular Matrix Sensor in Sandy Soils. Appl. Eng. Agr. **2001**, *17*, 787–797.
7. Stieber, T.D.; Shock, C.C. Placement of Soil Moisture Sensors in Sprinkler Irrigated Potatoes. Am. Potato J. **1995**, *72*, 533–543.
8. Eldredge, E.P.; Holmes, Z.A.; Mosley, A.R.; Shock, C.C.; Stieber, T.D. Effects of Transitory Water Stress on Potato Tuber Stem-End Reducing Sugar and Fry Color. Am. Potato J. **1996**, *73*, 517–530.
9. Eldredge, E.P.; Shock, C.C.; Stieber, T.D. Plot Sprinklers for Irrigation Research. Agron. J. **1992**, *84*, 1081–1984.
10. Shock, C.C.; Feibert, E.B.G.; Seddigh, M.; Saunders, L.D. Water Requirements and Growth of Irrigated Hybrid Poplar in a Semi-arid Environment in Eastern Oregon. West. J. Appl. For. **2002**, *17*, 46–53.
11. Shock, C.C.; Feibert, E.B.G.; Saunders, L.D. Irrigation Criteria for Drip-irrigated Onions. HortScience **2000**, *35*, 63–66.
12. Shock, C.C.; Feibert, E.B.G.; Saunders, L.D. Potato Yield and Quality Response to Deficit Irrigation. HortScience **1998**, *33*, 655–659.
13. Shock, C.C.; Feibert, E.B.G.; Saunders, L.D. Onion Yield and Quality Affected by Soil Water Potential as Irrigation Threshold. HortScience **1998**, *33*, 1188–1191.
14. Shock, C.C.; Holmes, Z.A.; Stieber, T.D.; Eldredge, E.P.; Zhang, P. The Effect of Timed Water Stress on Quality, Total Solids and Reducing Sugar Content of Potatoes. Am. Potato J. **1993**, *70*, 227–241.
15. Orloff, S.B., Hanson, B. Monitoring Alfalfa Water Use with Soil Moisture Sensors. Proceedings, 28th California Alfalfa Symposium, Dec 3–4, 1998; UC Cooperative Extension, University of California, Davis: Reno. NV, 1998.
16. Bertolino, A.V.F.A.; Souza, A.P.; Fernandes, N.F.; Rangel, A.M.; de Campos, T.M.P.; Shock, C.C. Monitoring the Field Soil Matrix Potential Using Mercury Tensiometer and Granular Matrix Sensors, Unsaturated Soils. Proc. 3rd Int. Conf. on Unsaturated Soils (UNSAT 2002), Recife, Brazil, Jucá J.F.T., de Campos, T.M.P., Marinho F.A.M., Eds.; Swets & Zeitlinger, Lisse: 2002; Vol. 1, pp. 335–338.
17. Shock, C.C.; Corn, A.; Jaderholm, S.; Jensen, L.; Shock, C.A. *Evaluation of the AM400 Soil Moisture Data Logger to Aid Irrigation Scheduling*; Special Report 1038; Oregon State University Agricultural Experiment Station, 2002; 252–256. http://www.cropinfo.net/AnnualReports/2001/Hansen2000.htm (accessed November 2002).
18. Shock, C.C.; Feibert, E.B.G.; Saunders, L.D. *Plant Population and Nitrogen Fertilization for Subsurface Drip-Irrigated Onions*; Special Report 1038; Oregon State University Agricultural Experiment Station, 2002; 71–80 http://www.cropinfo.net/AnnualReports/2001/ondrip01.htm (accessed November 2002).
19. Qualls, R.J.; Scott, J.M.; DeOreo, W.B. Soil Moisture Sensors for Urban Landscape Irrigation: Effectiveness and Reliability. J. Am. Water Resour. Assoc. **2001**, 547–559.
20. DeOreo, W.B.; Lander, P. *Summary: Performance of Soil Moisture Sensor in Boulder, Colorado (1993–1994)*. Proceedings of the Irrigation Association of America National Meeting, Nov 15, 1995; Irrigation Association: Phoenix, Arizona; 1995.

Soil Water Potential Measurement by Tensiometers

Joel M. Hubbell
James "Buck" Sisson
Idaho National Engineering and Environmental Laboratory,
Idaho Falls, Idaho, U.S.A.

INTRODUCTION

Tensiometers, which are used to indicate when plants should be irrigated, are widely used in agricultural and research applications. Research applications include characterizing and monitoring disposal sites to evaluate the presence of recharge, determining the direction of moisture flow, and estimating the water content and unsaturated hydraulic conductivity of the geologic materials at a site.

TENSIOMETER DESIGN AND OPERATION

The tensiometer is an instrument for measuring soil water potential. Soil water potential indicates how tightly water is held by soil. Fig. 1 shows the three basic components of a tensiometer: a water chamber, a rigid porous semipermeable membrane, and a pressure measurement device. The tensiometer is filled with water and sealed. The porous membrane is placed in contact with the soil to be measured, and water moves in and out of the porous membrane until the pressure in the sealed chamber is the same as the soil water potential. This pressure is then determined with a pressure sensor.

The semipermeable membrane is commonly made of a ceramic material, with pore sizes in the submicron range, which holds water tightly in its pores, but prevents air from entering the device. The water in the pores moves freely between the water chamber and the soil. The pore size is selected to be as small as possible to hold the water but large enough to allow the movement of water through the membrane.

Tensiometer measurements are obtained by placing the rigid semipermeable porous material in contact with the soil. A hydraulic connection is formed between the soil and the porous membrane. Water moves between the water chamber and the soil in response to pressure differences between the soil and the interior of the tensiometer until the pressure in the chamber is equivalent to the water potential in the soil. Since the tensiometer is a sealed tube, the pressure inside the tube will be equivalent to the soil water potential in the adjacent soil.

Tensiometers require a pressure measurement device such as a Bourdon gauge, electronic transducer, manometer, or have an access port to measure the pressure in the chamber.

Range of Measurements

The range of measurements from tensiometers is limited to water potentials of about -800 cm of water pressure or -0.8 atm pressure. This is due to a combination of factors including the difficulty in maintaining a hanging water column that exceeds 8 m of water or 0.8 atm pressure, vaporization of water at low pressures and air entry into the porous membrane.[1,2] Tensiometers measure only a portion of the entire range of water potentials found in soils.

This range of water potential (0 atm to -0.8 atm) is called the tensiometric range. The majority of moisture flow occurs in the tensiometric range as the highest unsaturated hydraulic conductivities occur over this range. This range is critical in agricultural applications where the tensiometers are used to determine when plants need to be irrigated.

Depth Limitation Based on Design

The design of the tensiometer determines the operational depth (Fig. 1). Conventional tensiometers have a water chamber that extends from the measurement point at the porous cup to land surface (Fig. 1a). These instruments can be operated from near land surface to depths of a maximum of 5 m–7 m, due to the length of the hanging water column. In this design, the sensor is located at land surface and may be influenced by temperature fluctuations that reduce the accuracy of the measurements. Negative pressure from the hanging water column increases the degassing of water within the tensiometer and results in accumulation of air in the tensiometer. Air accumulation slows down the measurement response to changes in the water potential. As more air builds up inside the tensiometer a pore may open, allowing airflow into the device and failure of the instrument.

Encyclopedia of Water Science
DOI: 10.1081/E-EWS 120010146

S

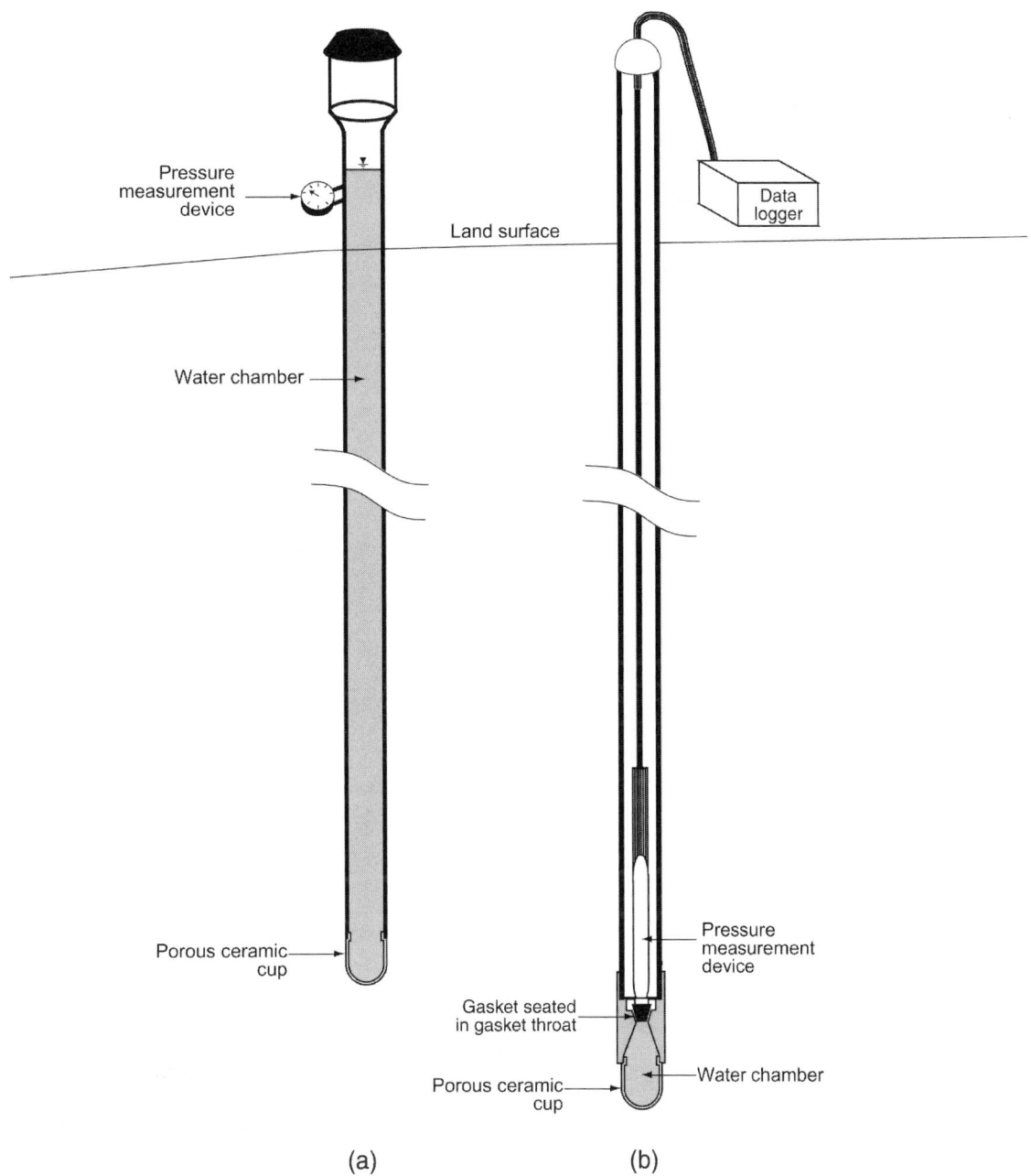

Fig. 1 Tensiometer designs.

A second basic tensiometer design allows tensiometers to be installed either permanently or on a temporary basis at any depth below land surface by moving the pressure sensor near the measurement location and eliminating the long water column.[3,4] This design reduces temperature induced measurement fluctuations while allowing monitoring for longer time periods than conventional tensiometers. The water in the chamber can be easily refilled and the sensors can be maintained and serviced from land surface (Fig. 1b). The use of pressure sensors

and data loggers to monitor soil water potential in tensiometers increase equipment costs and but provide a better data set and reduces overall labor costs.

Installation Techniques

Tensiometers are installed by one of several techniques, depending on the depth and geologic media being monitored. They can be installed by making an opening slightly larger than the diameter of the tensiometer then pressing the device into the opening and placing the porous cup to the depth of interest. They can be installed in boreholes by placing the porous ceramic at the depth of interest and then backfilling the borehole with native materials or silica flour to provide a hydraulic connection.[5,6] Portable tensiometers are installed by lowering them into a borehole and placing the porous cup in contact with the sediment at the bottom and then sealing the surface cap to reduce air flow out of the well.[3] A better hydraulic connection allows the tensiometer to respond quicker. If multiple tensiometers are placed in a single borehole, a sealing material such as bentonite, can be used to seal between the monitored intervals. Tensiometers have been used to monitor soil water potential in sediments ranging from gravel to clay, as well as porous rock such as basalt, tuff, and sandstone.

SOIL WATER POTENTIAL MEASUREMENTS USING TENSIOMETERS

What Is Water Potential?

Water potential tells us how tightly water is held by a soil. Water added to an unsaturated soil will tend to be pulled into the soil. The soil water is said to be under tension. This pulling action is produced primarily by capillary and adsorptive forces, similar to the wicking of water by using a paper towel. When the soil is saturated, it is under pressures greater than atmospheric pressure. Increasing water pressure keeps the soil saturated and indicates positive water potentials, corresponding to the height of the water level in the soil. If the volume of water held in the soil is decreased so the soil is no longer saturated, the water potential will decrease into the negative range. Tensiometers are designed to measure over the negative soil water potential range but will also measure over the positive range with a suitable measurement device.

Water potential is measured in units of pressure per unit area. It is given as the pressure exerted by an equivalent length of water column (cm). Water potentials in unsaturated soils are expressed in the negative range, indicating the soil is under tension relative to atmospheric pressure. The point of saturation in a soil is defined as zero pressure, which is the standing water level in the soil.

Water potential is important because it can be used to indicate the direction of water flow in unsaturated zones. It describes the energy required by plants to take the water from the soil and defines when sediment is saturated or unsaturated.

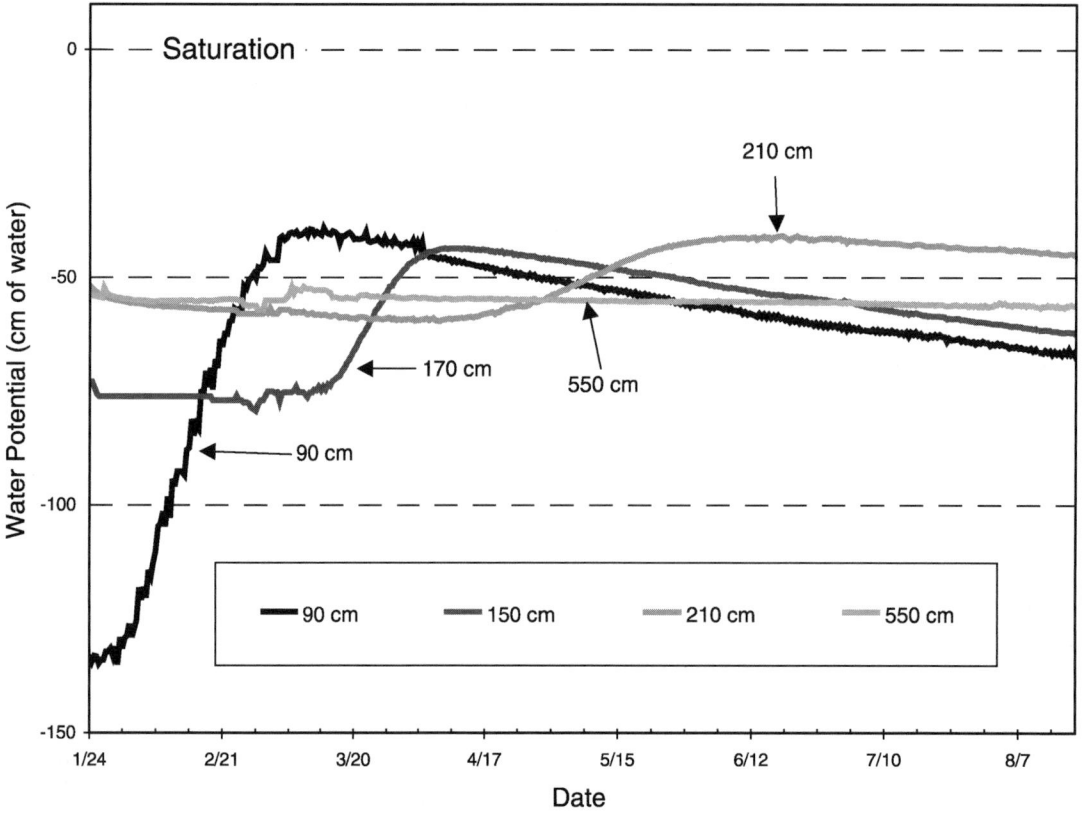

Fig. 2   Water potential measurements at depths from 90 to 550 cm.

Total Energy Status

The term potential in soil water potential comes from the two forms of energy (potential energy and kinetic energy) as defined in physics. Because water generally moves slowly through unsaturated sediments, the kinetic energy portion of the moisture movement is negligible. The potential energy, which comprised position (elevation head) and the condition (pressure) of the water, is dominant. Centimeters of water are commonly used for water potential because it can be easily combined with the gravity potential for expressing the total energy status.

Total potential energy difference between two locations drives the movement of water between locations (hydraulic gradient), with water moving from higher energy states (closer to saturation) to lower energy states. The total potential energy state is the combination of the elevation head and pressure head. Thus, water located in a soil with a greater total potential energy (wetter) will move into a similar soil with a lower total water potential in saturated soil (dryer).

Water Potential Measurements at Multiple Depths

Fig. 2 shows soil water potentials in a sand column over a 7-mo time period from 90 cm to 550 cm below land surface. Several interesting points are shown in this graph. The shallowest three instruments (90 cm, 150 cm, and 210 cm) show a wetting trend followed by a drying or moisture redistribution trend. The shallowest instrument (90 cm) shows a wetting trend from 1/24 to 3/5 and then a slow drying trend to the end of the time period. This infiltration event is seen at the 150 cm and 230 cm depths following a time delay of 4 weeks and 12 weeks, respectively. The soil water potential response is dampened with depth so that the water potential at 550 cm does not change.

The instruments near land surface generally show the greatest variation in response to wetting and drying events. Water potential fluctuations are dampened as depths increase. Infiltration events generally indicate a rapid wetting trend and then a slower drying trend. Water potential measurements are more stable with increasing depths and often approach the gravity drainage value for the material. Locally, the water potential may vary, controlling the direction of moisture flow. In deep unsaturated zones, the changes in water potentials with depth are small compared to the differences in elevation head. This makes the elevation head the dominant driving force.

REFERENCES

1. Hillel, D. *Environmental Soil Physics*; Academic Press: San Diego, CA, 1998.
2. Stephens, D.B. *Vadose Zone Hydrology*; Lewis Publishers: New York, NY, 1996.
3. Hubbell, J.M.; Sisson, J.B. Portable Tensiometer Use in Deep Boreholes. Soil Sci. **1996**, *161* (6), 376–382.
4. Hubbell, J.M.; Sisson, J.B. Advanced Tensiometer for Shallow or Deep Soil Water Potential Measurements. Soil Sci. **1998**, *163* (4), 271–277.
5. American Society of Testing Materials (ASTM), *Standard Guide for Measuring Matric Potential in the Vadose Zone Using Tensiometers*, D3404-91; American Society of Testing Materials: Philadelphia, PA, 1991.
6. Cassel, D.K., Klute, A. Water Potential: Tensiometry. In Methods of Soil Analysis, Part 1, Physical and Mineralogical Methods, Agronomy Monograph no. 9, 2nd Ed.; American Society of Agronomy-Soil Science Society of America, Madison, WI, 1986; 563–596

Soil Water Storage Measurement by Soil Probes

Clay A. Robinson
West Texas A&M University, Canyon, Texas, U.S.A.

INTRODUCTION

Knowledge of soil water content can improve irrigation scheduling and management. In dryland conditions, soil water content may determine when and/or which crop to plant. There are many methods of determining water content or potential without disturbing the soil, including neutron probes, time domain reflectometry (TDR) probes, electrical conductance/resistance methods, psychrometry, etc. These methods require extensive technology and calibration for best performance. Cost and technology limit their use primarily to researchers and a few, large-scale producers. In contrast, soil probes are low-cost devices that require no special technology to estimate soil water storage.

DESIGN AND PRINCIPLES

Soil probes are simple in design, and are also marketed as tile probes. Most soil probes are made of 9.5 mm diameter (3/8 in.) high tensile steel rod with a handle (Fig. 1a). The handle can be a 7.6 cm (3 in.) diameter ball or a 30 cm × 22.2 mm o.d. pipe perpendicular to the rod. The probe tip is flared to approximately 12.5 mm (1/2 in.) and can be either pointed or rounded (Fig. 1b). The flared tip allows less friction when inserting and removing the probe from the soil. Probes range from 1.2 m to 1.8 m in length, depending on usage and crop rooting depth.

Soil penetration resistance is affected by texture, aggregation, bulk density, and soil water content. Penetration resistance is inversely related to water content, so wet soils (high water content) have less penetration resistance. When the soil is near field capacity (FC), it contains water that is available to plants [plant available water (PAW)], and penetration resistance is low enough to allow probe insertion. When a soil layer dries below FC, the probe cannot be easily inserted into that layer. The depth of probe insertion indicates the depth in which high PAW is present. The PAW capacity in a soil is a function of soil texture, structure, and organic matter content (see also Table 1 in the article *Measuring Soil Moisture by Feel and Appearances*).

PLANT AVAILABLE WATER

Plant available water is defined as the difference in water content between the soil water contents at FC and wilting point. Field capacity is defined as the soil water content 2 to 3 days after a soaking rain or irrigation when the soil surface has been covered to limit evaporation, or the water content at a soil water potential of $-33\,kPa$. The water in large pores, called gravitational water, drains under the influence of gravity. The water remaining in the soil at FC is held in small pores against gravity. Wilting point is the soil water content below which plants are unable to extract water, typically $-1500\,kPa$, though it varies with plant type.

For a simple example, completely saturate a sponge, then hold it above the water with the long side parallel to the water surface. Water dripping from the sponge comes from large pores or voids that are unable to retain the water against gravity. Once the water has stopped, turn the sponge so the long side is perpendicular to the water surface. Gravitational water flows from the sponge again. Why? Gravity now has a greater distance through which it can act on the water in the sponge. When the water stops dripping, the sponge represents a soil at FC with all gravitational water removed.

To represent plant water uptake by roots, squeeze the sponge. At some point, squeezing the sponge no longer yields water. This represents wilting point, the water content at which plants can no longer extract water. The water squeezed out of the sponge represents the PAW. Notice the sponge is not dry at wilting point. Most soils contain between 5% and 20% water by volume at wilting point.

The amount of water held at FC is primarily a function of soil structure and the quantity of large, continuous pores. The amount of water held at wilting point is primarily a function of soil texture, especially the clay content. Organic matter increases wilting point, FC, and PAW. Silt loam soils have the highest PAW content while clays hold the most total water.

EXAMPLE

Once the soil water content of a layer drops below FC, the penetration resistance increases, and the probe cannot be easily inserted into that layer. Because the PAW is a volumetric water content, the equivalent depth

Encyclopedia of Water Science
DOI: 10.1081/E-EWS 120010149

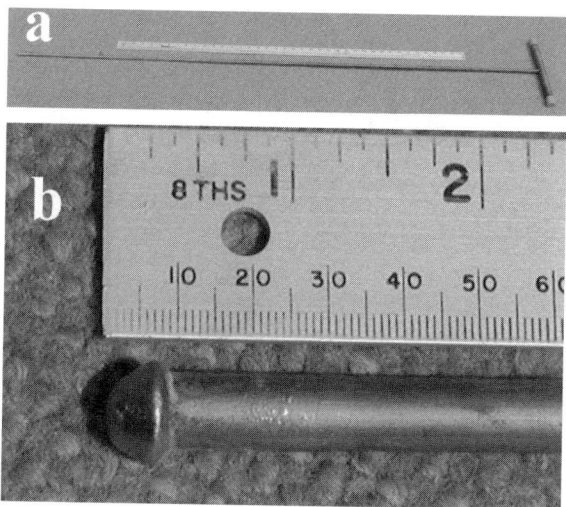

Fig. 1 Soil probe (a) and close-up of tip (b).

of available water in the soil can be calculated as the product of the depth of insertion and the PAW for that soil. Table 1 demonstrates the calculation of profile PAW for a profile of a typical Pullman clay loam (fine, mixed, superactive, thermic, Torrertic Paleustolls) from the Texas High Plains. The last row in the heading gives the formula used for that column: depth interval (D_i), PAW (volumetric, %), and PAW (depth equivalent, cm). The next row shows the calculations for the first layer, 0 cm–30 cm. The data for each subsequent layer are calculated in the same manner. The depth equivalent data for each layer are summed to yield the profile PAW.

The result using Table 1 is more exact than can be determined with a soil probe, but is given for a comparison. In the Texas High Plains, farmers are given generalizations by region based on the dominant soil types present.

Silty clay loam soils hold approximately 2 in. ft^{-1}, 0.167 cmc m^{-1}

Sandy loam soils hold about 1.5 in. ft^{-1}, 0.125 cmc m^{-1}
Clay loam soils hold about 1.75 in. ft^{-1}, 0.146 cmc m^{-1}

The Pullman is a clay loam soil and holds about 0.146 cmc m^{-1} (1.75 in. ft^{-1}). If the probe could be inserted to approximately 3 ft (90 cm), the Pullman soil would have about 13.1 cm (5.25 in.) of PAW. A sandy loam soil would have about 1.9 cm (0.75 in.) less, while a silty clay loam would have about that much more.

LIMITATIONS

The soil probe, by itself, cannot identify water available below a dry layer. This condition often exists in the Great Plains after fallow periods. There are usually 5–7 precipitation events of sufficient amount to store water in the soil. Afterward, the soil surface layer dries by evaporation. If no rain occurs before planting, the soil probe will likely penetrate only 2.5 cm–7.6 cm (1 in.– 3 in.), though there may be PAW stored below the dry surface layer. If the information on water storage is necessary, dig through the dry layer with a shovel, then use the probe for the underlying layers.

Some researchers place little faith in estimates of PAW by soil texture because organic matter levels and soil structure strongly affect FC, and because organic matter levels and structure vary among soils of similar texture. Crop and tillage management systems can alter PAW on adjacent plots on the same soil. Still, these estimates are a valuable starting point in improving irrigation management when no previous information on soil water content has been available. With continued use, estimates can be refined as producers monitor soil water storage, precipitation, irrigation, and crop water use. A 1.2-m (4-ft) probe is recommended for homeowners to manage their turf and lawn irrigation scheduling, primarily to avoid excess water application. Most turf root systems are in the top meter

Table 1 Calculation of profile plant available water, Pullman clay loam

Depth range (cm)	Depth interval, D_i (cm) $D_{lower} - D_{upper}$	Field capacity, FC (volumetric, %)	Wilting point, WP (volumetric, %)	Plant available water, PAW (volumetric, %) FC − WP	PAW, depth equivalent (cm) $D_i \times$ PAW
0–30	30 − 0 = 30	35.7	16.7	35.7 − 16.7 = 19.0	30 × 0.190 = 5.7
30–60	30	36.7	20.0	16.7	5.0
60–90	30	34.2	18.4	15.8	4.7
90–120	30	31.8	19.0	12.8	3.8
120–150	30	29.2	17.4	11.8	3.5
150–180	30	29.2	17.4	11.8	3.5
Total					26.2

(From Ref. [1].)

of the profile. If a homeowner can insert the probe to the full depth, it is time to cut back on irrigation frequency or quantity. Most agronomic producers use a 1.5 m–1.8 m probe to cover the rooting depth of the crop grown. Probes can be used to identify management effects on soil water storage (see also Fig. 3 in the article *Dryland Farming*).

CONCLUSION

Soil probes offer a simple, economical method to obtain valuable information about soil water storage. The PAW is estimated from depth of probe insertion and soil texture. Their use is limited when the surface soil is dry but lower layers are moist. Consistent use of soil probes with other information, e.g., evapotranspiration, precipitation, and irrigation amounts, can improve irrigation scheduling and water use efficiency of irrigated crops. Using soil probes in dryland cropping systems provides information to make crucial management decisions; e.g., when to plant, which crop to plant, and whether to fertilize, and allows producers to project yields.

REFERENCE

1. Unger, P.W.; Pringle, F.B. *Pullman Soils: Distribution, Importance, Variability, and Management*, Bulletin 1372; Texas Agricultural Experiment Station: College Station, TX, 1981; 1–23.

Soil, Waterborne Chemicals Leaching Through

John Hutson
Flinders University, Adelaide, South Australia, Australia

S

INTRODUCTION

Agricultural, urban, and industrial activities have increased the variety and quantity of chemicals and wastes released into the environment. In agriculture, fertilizers, pesticides, and animal wastes have led to widespread pollution. Nonpoint source pollution is particularly difficult to predict and control. Pollutants and chemicals in soil are dissipated by various fate and transport processes. Concentrations are reduced by chemical and microbial degradation and transformation, by plant uptake, and through volatilization, and chemicals are transported by flowing water. Chemical flow pathways can be across the soil surface and downwards through the soil profile or vadose zone, and terminate via lateral subsurface flow and deep drainage in surface water bodies and aquifers. Transport pathways are diverse and difficult to predict with certainty.

The major nonpoint source pollutants from agricultural areas are nutrients, pesticides, and pathogens. In irrigated areas and regions where there is a risk of salinity, the transport of the major inorganic ions is a concern. Fuel, and industrial or nonagricultural chemicals, are also a pollution risk. Organic contaminants can be classified as largely miscible with water, nonaqueous immiscible liquids, and volatile compounds. Some organic chemicals partition into all three phases. This article discusses leaching of water-miscible chemicals.

TRANSPORT PROCESSES

Miscible chemicals can be leached from one zone in the soil to another via the movement of water in which they are dissolved. An understanding of solute leaching requires knowledge of water-flow patterns in soil. Infiltration of rain and irrigation water leads to downward movement flow, while evaporation from the soil surface, transpiration by plants, and redistribution of water can lead to both downward and upward movement of water.

Chemicals in soil are redistributed in soil profiles by four processes: 1) chemical diffusion in the liquid phase in response to an aqueous concentration gradient; 2) diffusion of volatile chemicals in the gas phase in response to a vapor density gradient; 3) forced convection (mass flow, or advection) of chemical dissolved in flowing water; and 4) transport of chemical in the vapor phase driven by barometric pressure fluctuations, wetting and drying cycles, and watertable fluctuations.

In natural unsaturated soils, vertical convective fluxes predominate. Transport of chemicals is complicated by reactions with mineral or organic surfaces, which retards the movement of chemical. Mixing between large and small pores as a result of local variations in mean water-flow velocity, and the tortuous nature of soil pore geometry leads to dispersion of solute molecules during their movement through soil.[1] Dispersion tends to smear what may originally have been a sharp concentration front.

In 1-D, across a plane normal to the direction of flow, the convective flux of solute (J_{C_L}, $ML^{-2}T^{-1}$) is represented as

$$J_{C_L} = \theta D_M(q, \theta) \frac{dc_L}{dz} + qc_L \qquad (1)$$

where D_M is the hydrodynamic dispersion coefficient, a function of q, the macroscopic water flux density (LT^{-1}), and θ, the volume fraction of water in the soil. C_L is the solute concentration in the soil solution (ML^{-3}), and z is the depth (L).

The value of D_M is often estimated from

$$D_M = \lambda \left| \frac{q}{\theta} \right| \qquad (2)$$

where λ is the diffusivity (L). Dispersivity is determined by soil geometry. It is usually independent of solute properties, except in cases where solute diffusion from mobile to stagnant areas is important.[2]

Solute leaching may be retarded if the chemical interacts with soil solids. Solute concentrations are controlled by the amount of chemical introduced into the system, chemical solubility, and partitioning between solution and solid phases (and the gas phase for volatile chemicals). The mechanisms of sorption in soil are poorly understood or quantified, but include retention on soil surfaces by chemical and physical binding as well as ion-exchange processes. Sorption sites differ in their binding energies. The sorption process may be kinetic, owing to slow sorption reactions and accessibility of sorption sites. Molecules may need to diffuse from bulk solution to sorption surfaces.

Encyclopedia of Water Science
DOI: 10.1081/E-EWS 120010274

For these reasons, sorption is usually described operationally rather than mechanistically. Simple sorption isotherms or exchange equations, which assume local equilibrium, are fit to measured sorption data. However, solutes in flowing water do not react instantaneously with solid surfaces; most reactions are kinetic. In addition, molecules may have to diffuse from stagnant areas to reach larger pores where flow is more rapid. Sometimes, two-site or multi-site conceptual models are included.[3]

Sorption is often described using a Freundlich sorption isotherm

$$S = K_f C_L^n \qquad (3)$$

where K_f is a Freundlich sorption coefficient and n is an exponent. When $n = 1$, the Freundlich sorption isotherm reduces to a linear sorption isotherm,

$$S = K_d C_L \qquad (4)$$

where K_d is a partition or distribution coefficient. Linear sorption is often used to describe sorption over small concentration ranges and is easily manipulated mathematically. Databases containing K_d values for a wide range of chemicals are widely available.[4]

The total amount of chemical (C_T, $M L^{-3}$) in a soil can be expressed in terms of soil bulk density, water content, and the sorption coefficient,

$$C_T = C_L(\theta + \rho_b K_d) \qquad (5)$$

where ρ_b is soil bulk density ($M L^{-3}$).

Since only dissolved chemical can be transported in water, the ratio of the rate of water flow to that of the chemical is equivalent to the ratio of the total concentration of chemical (C_T) to the dissolved concentration ($C\theta$). This ratio is known as the retardation factor (R),

$$R = 1 + \frac{\rho K_d}{\theta} \qquad (6)$$

which is a useful index of the relative rates of transport of different chemicals.

Solute concentrations can be reduced through chemical and biotic transformations and uptake. Microbial transformations may produce degradation products. Again, these processes can be very complex. Degradation, for example, may depend upon the presence and growth of a suitable microbial species. For simplicity, degradation is often described using first-order kinetics, but in reality it is more complex, owing to processes such as diffusion, sorption, microbial composition, and growth.[5]

CONVECTION–DISPERSION EQUATION

Combining and equations for convective transport, dispersion, and partitioning leads to the convection–dispersion equation (CDE), which allows calculation of the rate of change of concentration of a chemical in soil,

$$\frac{\partial C_L}{\partial t}(\theta + \rho_b K_d) = \frac{\partial}{\partial z}\left[\theta D_M(\theta, q)\frac{\partial C_L}{\partial z} - q C_L\right] \pm \Phi \qquad (7)$$

where t is time, and Φ is a source or sink term.

The CDE can be solved analytically for certain defined boundary conditions and steady-state water flow.[6] Transport parameters can be measured, but are almost invariably obtained by fitting to controlled laboratory chemical breakthrough curves from soil columns at constant water content and subject to steady-state water flow.[7] For transient water flow, the CDE is best solved numerically. Water fluxes in natural, unsaturated soils are often sporadic, especially in areas of variable rainfall (Fig. 1). Examples of simulation models of chemical transport in soils are LEACHM,[8] UNSATCHEM,[9] and RZWQM.[10] For some scenarios, 2-D and 3-D models are necessary, for e.g., leaching under drip irrigation and hillslope flow.

The CDE is most applicable to homogeneous soils. Structured and heterogeneous soils have cracks and channels giving rise to preferential water and chemical flow paths, and many soils exhibit more complex chemical and microbial reactions. Conceptually, soil porosity can be divided into immobile, mobile, and preferential flow regions. Preferential flow takes place through larger pores and cracks. Water and solutes are transported very rapidly in these channels, and bypass the soil matrix, so there is little opportunity for sorption or degradation. Conversely, solute-free water flowing in preferential pathways may bypass chemical in the matrix, leading to less leaching.

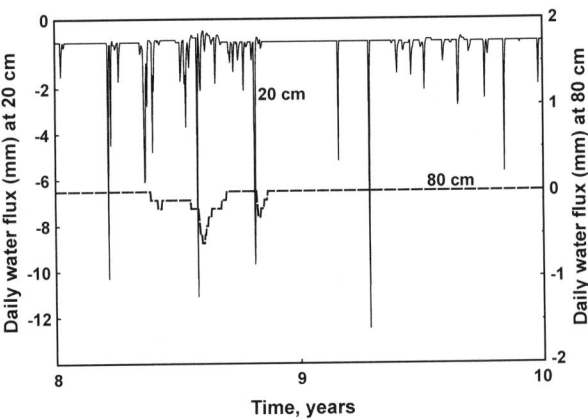

Fig. 1 An example of simulated daily water fluxes in a clay loam soil, using daily rainfall for Adelaide, Australia.

S

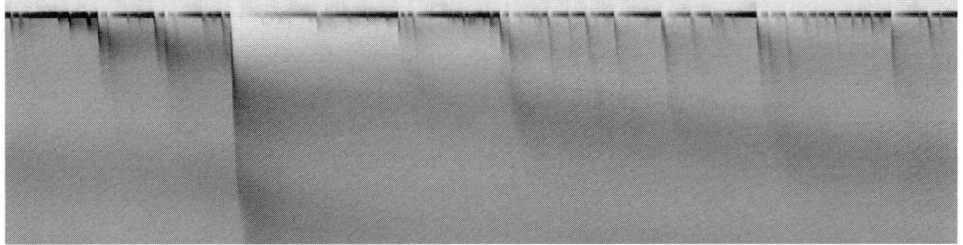

Fig. 2 A simulation showing the sporadic nature of chloride leaching in a sandy loam soil, South Australia, using 100 yr of daily rainfall data. Color intensity is proportional to chloride concentration. The *x* axis is soil depth (to 12 m) and the *y* axis is time (to 100 yr).

Preferential flow is important during periods of intense rainfall or irrigation, when surface runoff and ponding can lead to ingress into cracks and channels. Preferential flow is difficult to predict or quantify. An example of a model that describes preferential flow is MACRO.[11]

Capacity, or tipping-bucket models offer a simpler approach to solute transport simulation. Water moving between soil layers moves dissolved chemical, which then mixes with the water in the receiving layer, and the mixed concentration moves to the next layer.[12]

Assessments of chemical transport in natural soils need to take spatial and temporal variability into account. Downward water fluxes in soil are intermittent and sometimes, especially in arid climates, infrequent. This means that solutes can accumulate in subsoils until flushed during periods of heavier rain. So while leaching may be frequent and regular in temperate humid climates, it may be infrequent in more arid areas (Fig. 2).

CONCLUSION

The recognition of spatial variability has led to increased efforts to combine GIS and simulation models in order to describe solute transport on a farm and catchment scale, accounting for soil, land management, vegetation, and terrain differences. Soil leaching models focus on processes described at the soil profile. Upscaling to larger areas require boundary conditions to be described in more detail, which means that output from associated surface hydrology, groundwater, and crop models need to be reflected. Describing solute leaching on a catchment scale, accounting for management, spatial, and climatic variability, is a current priority.

REFERENCES

1. Wagenet, R. Principles of Salt Movement in Soils. In *Chemical Mobility and Reactivity in Soil Systems*; Soil Science Society Special Publication No. 11, Nelson, D.W., Ed. Soil Science Society of America: Madison, WI, 1983.

2. Brusseau, M.L. The Influence of Solute Size, Pore Water Velocity, and Intraparticle Porosity on Solute Dispersion and Transport in Soil. Water Resour. Res. **1993**, *29*, 1071–1080.

3. Van Genuchten, M.Th.; Wagenet, R.J. Two-Site/Two-Region Models for Pesticide Transport and Degradation: Theoretical Development and Analytical Solutions. Soil Sci. Soc. Am. J. **1989**, *53*, 1303–1310.

4. Wauchope, R.D.; Buttler, T.M.; Hornsby, A.G.; Augustijn Beckers, P.W.M.; Burt, J.P. The SCS/ARS/CES Pesticide Properties Database for Environmental Decision-Making. *Reviews of Environmental Contamination and Toxicology*; Springer-Verlag: New York, 1995; Vol. 123.

5. Scow, K.M.; Hutson, J.L. Effect of Diffusion and Sorption on the Kinetics of Biodegradation: Theoretical Considerations. Soil Sci. Soc. Am. J. **1992**, *56*, 119–127.

6. Van Genuchten, M.Th.; Alves, W.J. *Analytical Solutions of the One-Dimensional Convective–Dispersive Solute Transport Equation*. Technical Bulletin 1661, Agricultural Research Service, US Department of Agriculture, 1982.

7. Toride, N.; Leij, F.J.; van Genuchten, M.Th. The CXTFIT Code for Estimating Transport Parameters from Laboratory or Field Tracer Experiments. Version 2.0 Research Report No. 137. US Salinity Laboratory, Agricultural Research Service, US Department of Agriculture, Riverside, California, 1995; 121 pp.

8. Hutson, J.L.; Wagenet, R.J. An Overview of LEACHM: A Process Based Model of Water and Solute Movement, Transformations, Plant Uptake and Chemical Reactions in the Unsaturated Zone. In *Chemical Equilibrium and Reaction Models*; Soil Science Society of America Special Publication 42, Loeppert, R.H., Ed.; 1995; 409–422.

9. Simunek, J.; Suarez, D.L.; Sejna, M. The UNSATCHEM software package for simulating one-dimensional variably saturated water flow, heat, transport, carbon dioxide production and transport, and multicomponent solute transport with major ion equilibrium and kinetic chemistry, Version 2. Research Report No. 141, November, 1996. U. S. Salinity Laboratory, Agricultural Research Service, U. S. Department of Agriculture, Riverside, California.

10. Ma, L.; Ahuja, L.R.; Ascough, J.C.; Shaffer, M.J.; Rojas, K.W.; Malone, R.W.; Cameira, M.R. Integrating System Modeling with Field Research in Agriculture: Applications of the Root Zone Water Quality Model (RZWQM). *Advances in Agronomy*; Academic Press Inc.: San Diego, USA, 2000; 233–292.

11. Larsson, M.H.; Jarvis, N.J. A Dual-Porosity Model to Quantify Macropore Flow Effects on Nitrate Leaching. J. Environ. Qual. **1999**, *28*, 1298–1307.

12. Trevisan, M.; Errera, G.; Vischetti, C.; Walker, A. Modelling Pesticide Leaching in a Sandy Soil with the VARLEACH Model. Agric. Water Manag. **2000**, *44*, 357–369.

Soils, Field Capacity of Water in

M. H. Nachabe
University of South Florida, Tampa, Florida, U.S.A.

L. R. Ahuja
United States Department of Agriculture (USDA), Fort Collins, Colorado, U.S.A.

Renee Rokicki
University of South Florida, Tampa, Florida, U.S.A.

INTRODUCTION

Water content at field capacity, or simply field capacity, provides an operational concept for managing soil-water in the root zone. Following thorough wetting of deep, well-drained soils, excess water is re-distributed, and field capacity is reached when the downward drainage flux is materially ceased in the profile.[1–5] Veihmeyer and Hendrickson[1,6] related field capacity to soil-water content held at certain negative pressure or suction, implying that perhaps the field capacity is an intrinsic property of soils. Modern theory of soil-water movement and precise measurement have shown, however, that field capacity is not a constant or an intrinsic property, but rather a transient value that is impacted by initial conditions in soil, depth to water table, and soil profile layering.[7–10] Nonetheless, field capacity remains a useful operational concept in deep, well-drained soils where downward drainage flux may not cease completely, but becomes negligibly small so that processes of evaporation and root water uptake dominate the depletion of root zone soil-water. Determining field capacity is important in soil-water management like scheduling of irrigation because the water content between field capacity and wilting point becomes available for root water uptake by crops.

OVERVIEW

Field capacity is commonly taken as the soil-water content at a given drainage time (e.g., 48 hr) or matric potential (-33 kPa or -10 kPa).[11–13] Although these approaches can be suitable for certain field conditions, they are imprecise and even misleading for certain other conditions. Nachabe,[4] among others, has shown that the drainage time to reach field capacity varies with initial wetness and soil texture and is not fixed for all soils. A second, commonly accepted approximation of field

capacity is soil-water content at -33 kPa matric potential (-0.33 bar pressure) for fine textured soils, and sometimes -10 kPa matric potential (-0.1 bar pressure) for coarse textured soils (e.g., Refs. [3,13]). Hillel,[14] Nachabe,[4] and Meyer and Gee[5] noted that this pressure-based approximation of field capacity is inconsistent because there is no guarantee that the same negligible drainage flux is reached for all soils at this value of soil-water pressure. Quoting Hillel[14] "it is a fundamental mistake to expect [pressure-based approximations of field capacity] to apply universally, since they are solely static in nature while the process they purport to represent is highly dynamic." Also this pressure-based approximation is misleading for drainage in layered profiles with impervious clay pans or in root zones with perched or shallow water table depth.

A dynamic interpretation of field capacity described by the magnitude of a time variable slow drainage flux is preferable. The adoption of this interpretation of field capacity restores its important dynamic nature, while allowing the user to specify the small drainage flux from the root zone when field capacity is practically reached. For root zone water management, Nachabe[4] recommended relating the magnitude of the small drainage flux at field capacity to daily evapotranspiration, and Hillel[14] proposed using a negligible flux of 0.5 mm day^{-1}, equal to about 10% of the daily average evapotranspiration. Meyer and Gee[5] argued that drainage fluxes between 0.01 mm day^{-1} ($\approx 10^{-8}$ cm sec^{-1}) and 1 mm day^{-1} ($\approx 10^{-6}$ cm sec^{-1}) can be considered small enough, depending on type of field application. Clearly, field capacity is an operational concept and the selection of the magnitude of the negligible flux should be left to the type of application. In environmental applications, where mobility and leaching of toxic pollutants through the soil are an issue, the user may define the (dynamic) field capacity to occur at a very small flux (e.g., 0.01 mm day^{-1}). In root zone water management, a flux of 0.5 mm day^{-1} might be appropriate to define field capacity when evaporation and transpira-

Encyclopedia of Water Science
DOI: 10.1081/E-EWS 120010264

tion, rather than downward drainage, become the dominant processes in depleting soil-water of the root zone.

We briefly review the physics of drainage, and provide equations that allow the user to: 1) determine field capacity when a negligibly small drainage flux is reached; 2) approximate the time to reach this dynamic field capacity; and 3) estimate the wetted depth of the root zone at field capacity for specific infiltration events. We distinguish between deep, well-drained soils, and field situations where drainage is hindered by clay pans or shallow depth to water table.

FIELD CAPACITY IN DEEP, WELL-DRAINED SOIL PROFILES: ESTIMATION AND APPROXIMATION

During drainage, a unit hydraulic gradient in the profile provides a good approximation of Darcy's law, which can be written as:

$$q_t = K(\theta_t) \tag{1}$$

where q_t, in mm day^{-1}, is the drainage flux as a function of time t, in days, $K(\theta_t)$ is the unsaturated hydraulic conductivity in mm day^{-1} at any water content θ_t in mm^3 of water per mm^3 of soil. The unsaturated hydraulic conductivity is expressed as:[15]

$$K(\theta) = K_s \Theta^n \tag{2}$$

where K_s is the saturated hydraulic conductivity, Θ is the normalized water content equal to $(\theta - \theta_r)/(\theta_s - \theta_r)$, where θ_r is residual soil-water content, and θ_s is saturated

soil-water content, and n is an exponent. Usually $n = (2 + 3\lambda)/\lambda$, where λ is the pore size distribution index of the Brooks and Corey model. The Brooks and Corey model has been widely adopted in the past, and its parameters can be easily derived from soil texture data (e.g., Refs. 16–18) or directly obtained from scientific literature and reference textbooks (e.g., Refs. 16,19). Assuming a rectangular soil-water profile during drainage, the rate of decrease of water content is given by:

$$\frac{d\theta_t}{dt} = \frac{-(q_t + e)}{z_f} \tag{3}$$

where e is a constant evaporation flux at the surface, and z_f is depth to the wetting front during drainage. If e is ignored, conservation of mass of soil-water in the profile requires that:

$$z_f = \frac{I}{\theta_t - \theta_r} \tag{4}$$

where I is the initial cumulative infiltration water depth in millimeters at the beginning of drainage. Substituting Eqs. 1 and 4 into Eq. 3 and integrating the resulting ordinary differential equation with respect to time yields:

$$\frac{\theta_t - \theta_r}{\theta_s - \theta_r} = \left(\Theta_I^{-n} + \frac{K_s n t}{I} \right)^{-1/n} \tag{5}$$

where Θ_I is the normalized water content distribution at the beginning of soil moisture distribution (equal to 1 if soil is initially saturated). The drainage flux can be found

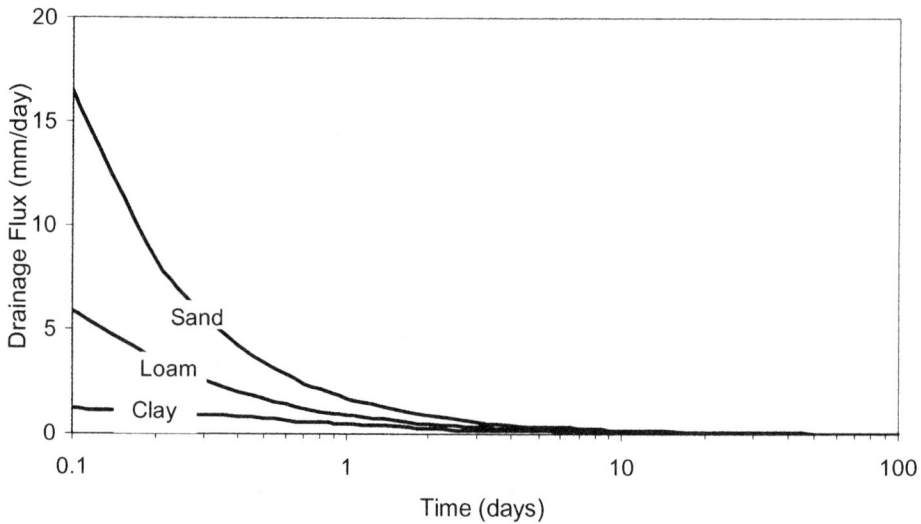

Fig. 1 Estimated evolution of drainage flux with time for three soils for an initial infiltration depth of 10 mm. (From Ref. 4.)

by substituting Eq. 5 into Eq. 2 resulting in:

$$q_t = \frac{K_s}{\left(\frac{nK_s t}{I} + \Theta_I^{-n}\right)} \qquad (6)$$

Fig. 1 illustrates the rapid decrease in drainage flux q_t with time for three soils with initial infiltration depth of 10 mm, showing that drainage flux becomes negligibly small within a few days. Solving Eq. 6 for time, and adopting the subscript "fc" for field capacity results in t_{fc}, the drainage time to reach field capacity:

$$t_{fc} = \left(\frac{K_s}{q_{fc}} - \Theta_I^{-n_s}\right)\frac{I}{nK_s} \qquad (7)$$

Eq. 7 can be used to estimate the time to reach a negligibly small flux, q_{fc}, at field capacity. The time to reach this flux is not an intrinsic soil property, but depends on the soil hydraulic properties, the initial infiltration depth, and the magnitude of q_{fc}. Normalized water content at field capacity can be calculated from Eq. 2 as:

$$\Theta_{fc} = \frac{\theta_{fc} - \theta_r}{\theta_s - \theta_r} = \left(\frac{q_{fc}}{K_s}\right)^{1/n} \qquad (8)$$

In Eq. 8, the water content at field capacity has a dynamic nature because it is expressed as a function of a user specified, small drainage flux q_{fc}. Nachabe[4] and Meyer and Gee[5] compared the soil-water content at field capacity (dynamic or flux-based concept, from Eq. 8), with the soil-water content at −33 kPa of pressure (pressure-based concept of field capacity). Results are shown in Fig. 2. Results in this figure indicate that the flux-based estimation of field capacity is more consistent than the pressure-based estimate of field capacity. Using the −10 kPa pressure to estimate field capacity of coarse textured soils like sand will result in larger drainage fluxes,

which is more consistent with the dynamic estimation of field capacity (Fig. 2).

FIELD CAPACITY IN SHALLOW WATER TABLE ENVIRONMENTS AND CLAY PANS

In certain agricultural soils, drainage can be hindered by clay pans or shallow depth to water table. In these cases, the dynamic concept of field capacity holds, but the equations above for a homogenous profile will not describe conditions at field capacity. If a clay pan is at shallow depth in the root zone, then soil-water accumulates in a surface horizon, and drainage flux will be limited to the saturated conductivity of the clay pan below. Under these conditions, field capacity of the soil horizon above the clay pan might be close to saturated water content. In many parts of southern United States and other parts of the world, agricultural soils are fine sand with shallow depth to water table (1 ft–5 ft). In these soils, rapid rise in water table is observed and drainage of the root zone results in an equilibrium soil-water profile above a new, shallower, water table. When drainage seizes to be significant, the field capacity of the root zone is the equilibrium drainage (or water retention) curve above the new water table.

CONCLUSION

We propose to use field capacity as an operational concept for root zone water management, and avoid treating it as an intrinsic soil property. Field capacity is reached when downward drainage flux is negligibly small (while recognizing that drainage may not cease completely) so that evaporation and transpiration are more significant in

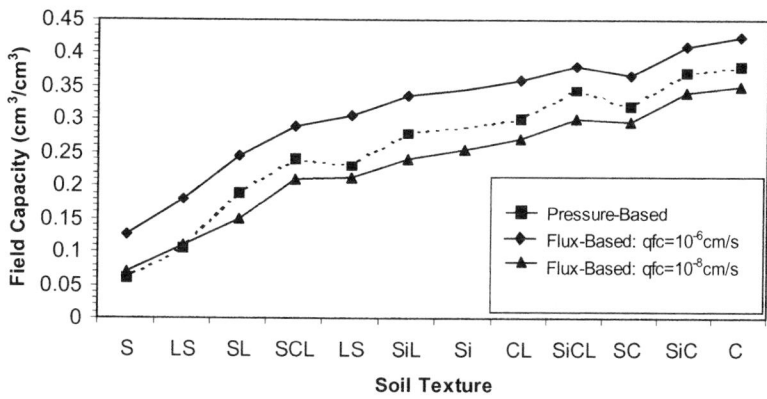

Fig. 2 Dynamic (Eq. 8 with $q_{fc} = 10^{-6}$ cm sec^{-1} and 10^{-8} cm sec^{-1}) and pressure-based (−33 kPa) field capacity by soil class texture (S, sand; L, loam; Si, Silt; C, clay). (From Ref. 5.)

depleting soil-water of the root zone. Depending on type of application, a negligibly small drainage flux between 0.01 mm day^{-1} and 1 mm day^{-1} can be assumed for field capacity in a deep, well-drained soil profile. In a layered soil profile, drainage can be hindered by a clay pan, whereas in a soil with shallow depth to water table, an equilibrium soil-water profile is usually achieved when drainage ceases to be significant.

REFERENCES

1. Veihmeyer, F.J.; Hendrickson, A.H. The Moisture Equivalent as a Measure of Field Capacity. Soil Sci. **1931**, *32*, 181–193.
2. Miller, E.E.; Klute, A.; et al. The Dynamics of Soil-Water. I: Mechanical Forces. In *Irrigation of Agricultural Lands*; Hagan, R.M., Ed.; American Society of Agronomy-Soil Science Society of America: Madison, WI, 1967.
3. Ahuja, L.R.; Nielsen, D.R. Field Soil-Water Relations. *Irrigation of Agricultural Crops*; Agronomy Monograph No. 30, American Society of Agronomy-Soil Science Society of America: Madison, WI, 1990.
4. Nachabe, M.H. Refining the Interpretation of Field Capacity in the Literature. ASCE J. Irrig. Drain. Eng. **1998**, *124* (4), 230–232.
5. Meyer, P.D.; Gee, G. Flux-Based Estimation of Field Capacity. ASCE J. Geotech. Geoenviron. Eng. **1999**, *125* (7), 595–599.
6. Veihmeyer, F.J.; Hendrickson, A.H. Methods of Measuring Field Capacity and Wilting Percentages in Soils. Soil Sci. **1949**, *68*, 75–94.
7. Richards, L.A.; Gardner, W.R.; Ogata, G. Physical Processes Determining Water Loss from Soils. Soil Sci. Soc. Am. Proc. **1956**, *20*, 310–314.
8. Ogata, G.; Richards, L.A. Water Content Change Following Irrigation of Bare Field Soil that is Protected from Evaporation. Soil Sci. Soc. Am. Proc. **1957**, *21*, 355–356.
9. Richards, L.A.; Moore, D.C. Influence of Capillary Conductivity and Depth of Wetting on Moisture Retention in Soils. Trans. Am. Geophys. Union **1952**, *33*, 4.
10. Eagleman, J.R.; Jamison, V.C. Soil Layering and Compaction Effects on Unsaturated Moisture Movement. Soil Sci. Soc. Am. Proc. **1962**, *26*, 519–522.
11. Colman, E.A. A Laboratory Procedure for Determining the Field Capacity of Soils. Soil Sci. **1947**, *63*, 277.
12. Linsley, R.K.; Franzini, J.B. *Water Resources Engineering*; McGraw-Hill Inc.: New York, NY, 1972.
13. Kutilek, M.; Nielsen, D. Elementary Soil Hydrologic Processes. *Soil Hydrology*; Catena Verlag: Germany, 1994; 130–218.
14. Hillel, D. Redistribution of Water in Soil. *Environmental Soil Physics*; Academic Press: San Diego, 1998; 449–470.
15. Brooks, R.; Corey, A. *Hydraulic Properties of Porous Media*; Hydro. Paper No. 3, Colorado State University: Fort Collins, CO, 1964; 143–190.
16. Rawls, W.J.; Brakensiek, D.L. Estimating Soil Water Retention from Soil Properties. ASCE J. Irrig. Drain. Eng. **1982**, *108*, 166–171.
17. Rawls, W.J.; Brakensiek, D.L. Estimation of Soil Water Retention and Hydraulic Properties. In *Unsaturated Flow in Hydrologic Modeling*; Morel-Seytoux, H.J., Ed.; Kluwer Academic Publishers: Norwell, MA, 1989; 275–300.
18. Pachepsky, Y.; Timlin, D.; Varallyay, G. Artificial Neural Networks to Estimate Soil Water Retention from Easily Measured Data. Soil Sci. Soc. Am. J. **1996**, *47*, 770–775.
19. Rawls, W.J.; Ahuja, L.R.; Brakensiek, D.; Shirmohammadi, A. Infiltration and Soil Water Movement. In *Handbook of Hydrology*; Maidment, D., Ed.; McGraw-Hill, Inc.: New York, 1993; 5.1–5.51.

Soils, Hydraulic Conductivity Rates in

David D. Bosch
United States Department of Agriculture (USDA), Tifton, Georgia, U.S.A.

Adel Shirmohammadi
University of Maryland, College Park, Maryland, U.S.A.

INTRODUCTION

Much of life depends on our ability to make efficient use of our water resources. Because of this, the characterization of the fraction of precipitation and snowfall which run off the earth's surface and which infiltrate into the soil are very important to society. Infiltration of water into the soil and subsequent movement of this water to plant roots are critical considerations for agricultural production. Other interests which involve understanding the movement of water through the soil include water flow to subsurface drains and wells, surface water flow, and evaporation from the soil to name a few.

OVERVIEW

Water moves through the earth in response to forces acting upon it. The property which describes the rate at which water flows through a porous material is called the hydraulic conductivity. In 1856, a French hydraulic engineer named Henry Darcy published a report on the water supply of the city of Dijon, France.[1] In his report Darcy described an experiment that he had conducted to analyze the flow of water through sands. The results of his experiment became generalized into an empirical law that now bears his name.[a]

$$Q = KA \frac{\Delta h}{l} \qquad (1)$$

In this equation, A is the cross-sectional area through which the water flows [L^2], Δh is the difference in

hydraulic head of the water between two observation points [L], and l is the distance between the two points [L]. The hydraulic head is the sum of gravitational and pressure heads, while the rate of change of the hydraulic head over a given length is termed the hydraulic gradient. The coefficient relating the hydraulic gradient to the flow of water through the porous media was termed the hydraulic conductivity, K, [LT^{-1}]. The hydraulic conductivity is thus a measure of a media's ability to transmit a fluid. If the porous media is saturated it is referred to as the saturated hydraulic conductivity, K_s. For unsaturated conditions it is called the unsaturated hydraulic conductivity.

Saturated hydraulic conductivity is a function of the properties of the soil and of the fluid. We primarily think of the flow of water, but oil or other fluids also flow through porous media and would have a different hydraulic conductivity than would water. The properties of the porous media that affect the hydraulic conductivity include particle arrangement, size, shape, and distribution. Experiments with glass beads established the relationship:[4]

$$K = \frac{Cd^2 \rho g}{\mu} \qquad (2)$$

where C is another coefficient of proportionality, d is the diameter of the glass beads [L], ρ is the density of the fluid [ML^{-3}], g is the acceleration of gravity [LT^{-2}], and μ is the fluid dynamic viscosity [$ML^{-1}T^{-1}$]. As the temperature of the fluid changes so does it's viscosity. Thus, K is also affected by temperature.

The hydraulic conductivity can be broken down into properties of the fluid ($\rho\mu^{-1}$) and properties of the medium (Cd^2). To separate these components, the hydraulic conductivity is often written in terms of the specific or intrinsic permeability, k [L^2].

The intrinsic permeability is often used because it is a property of the porous media alone.

$$K = \frac{k\rho g}{\mu} \qquad (3)$$

Contribution from the USDA Agricultural Research Service, Southeast Watershed Research Laboratory, PO Box 946, Tifton, GA USA 31793, in cooperation with Univ. of Georgia Coastal Plain Exp. Stn.

All programs and services of the USDA are offered on a nondiscriminatory basis without regard to race, color, national origin, religion, sex, age, marital status, or handicap.

[a]Darcy's law fails for conditions of high flow velocities, where inertial forces are no longer negligible compared to viscous forces.[2] Deviations from Darcy's law may also occur at very low gradients and in small pores.[3]

Encyclopedia of Water Science
DOI: 10.1081/E-EWS 120010156

Table 1 Representative saturated hydraulic conductivity values for various materials

Unconsolidated material	Consolidated rock	Saturated hydraulic conductivity (m day^{-1})	Relative saturated hydraulic conductivity
Clean gravel	Basalt, cavernous limestone, and dolomite	10^4	Very high
Clean sand, sand, and gravel	Clean sandstone and fractured igneous and metamorphic rocks	10^2	High
Fine sand	Weathered granite	1	Moderate
Silt, clay, and mixtures	Laminated sandstone, shale, and mudstone	10^{-3}	Low
Massive clay	Massive igneous and metamorphic rocks	10^{-5}	Very low

(From Ref. 7.)

Hydraulic conductivity is a function of not only the position in the porous media, but the direction of flow as well. Because geologic materials are often layered, flow in the direction of the layers often has a higher conductivity than flow perpendicular to the layering. Thus, K_s is often characterized in three dimensions. For layered soils, an effective saturated hydraulic conductivity may be determined.[5] For flow perpendicular to the layers the expression is:

$$K_e = \frac{D}{\frac{D_1}{K_1} + \frac{D_2}{K_2} + \ldots \frac{D_n}{K_n}}; \text{ (geometric mean)} \quad (4)$$

For flow parallel with the layers the expression is:

$$K_e = \frac{K_1 + K_2 + \ldots K_n}{n}; \text{ (arithmetic mean)} \quad (5)$$

where, K_e is the effective saturated hydraulic conductivity [LT^{-2}], D is the total profile depth [L], $D_1 - D_n$ and $K_1 - K_n$ are the thickness and saturated hydraulic conductivity of each layer, respectively.

In light of recent awareness on preferential flow of water and chemicals through the soil profile and the impact of soil heterogeneity on macropore flow, dependence of the saturated hydraulic conductivity on both location within a profile and direction of the flow have been used to distinguish between homogeneous isotropic soils where the conductivity is the same in all directions and heterogeneous anisotropic soils where it varies with direction. If K is the same in all locations within the profile and in all flow directions, the soil is called homogeneous and isotropic. On the other hand, if K is dependent both on location and the direction of the flow, such a profile is referred to as heterogeneous and anisotropic.[6]

In practice, various units are used for K. Hydrologists prefer the unit m day^{-1} or ft day^{-1}, while soil scientists often use ft sec^{-1}, cm sec^{-1}, or mm sec^{-1}. For description of aquifer properties, hydraulic conductivity is also expressed in terms of the volume of flow through a given cross-sectional area under a unit gradient at a fixed temperature. In this case the dimensions of K are L^3 T^{-1}

L^{-2}. Some of the units used are gal day^{-1} L^{-2} and m^3 day^{-1} m^{-2}.

REPRESENTATIVE VALUES

Representative values of K_s are listed in Table 1. As would be expected, values for coarse textured sandy soils are considerably higher than those for fine textured clay soils. Hydraulic conductivity is also affected by the structure of the medium. A highly porous, fractured material would conduct water more readily than would a tightly compacted one. Hydraulic conductivity depends also on the arrangement of the soil pores. Interconnected pores conduct more readily than do closed end pores. A gravely or sandy soil with large pores can have a conductivity much greater than a clay soil with narrow pores even though the total porosity of the clay may be greater than that of the sand.

Field studies have shown soil hydraulic characteristics can vary greatly.[8,9] Because of the variation in soils across a field, a large variability in hydraulic conductivity can be observed. This in turn leads to a large variation in infiltration and subsequently in runoff.

UNSATURATED CONDITIONS

The hydraulic conductivity of a material varies considerably with the degree of saturation of the soil. Eq. 1 was developed for a saturated material, but has been extended for unsaturated materials by making K a function of the matric potential[b] (φ) of the soil:

$$q = -K(\varphi)\nabla H \quad (6)$$

where q is the flow rate or flux [LT^{-1}], $K(\varphi)$ is

---

[b]Matric potential is a measure of the negative pressure which exists within the soil due to capillary and adsorptive forces.

Hydraulic Conductivity

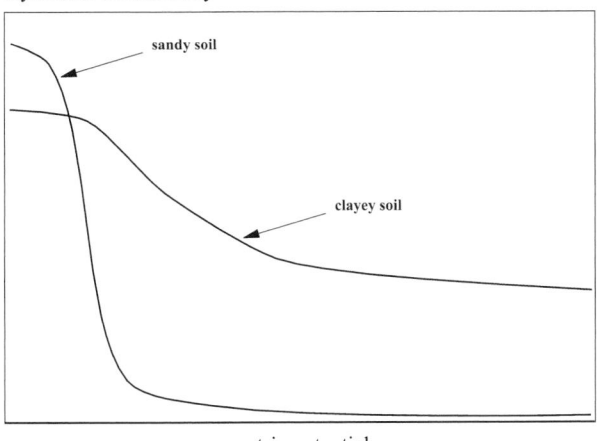

Fig. 1 Unsaturated hydraulic conductivity as a function of matric potential.

the unsaturated hydraulic conductivity $[LT^{-1}]$, and ∇H is the hydraulic head gradient $[LL^{-1}]$. Eq. 6 fails to take into account the effects of hysteresis, i.e., whether the soil is wetting or drying, which has been found to effect $K(\varphi)$.

As the soil becomes drier, $K(\varphi)$ decreases. The rate of decrease is a function of the properties of the soil (Fig. 1). While $K(\varphi)$ may be high for a fully saturated sand, it rapidly decreases as the sand de-waters and the matric potential in the soil decreases. In contrast, because a clay soil is able to maintain more water as the matric potential increases $K(\varphi)$ does not decrease as rapidly for a fine textured soil.

DETERMINATION OF HYDRAULIC CONDUCTIVITY

Hydraulic conductivity can be determined through a variety of numerical,[10] field,[11] and laboratory[12] techniques. Many investigators have attempted to relate hydraulic conductivity to properties of the porous media.[13] As a result, many formulas exist which can be used to predict the hydraulic conductivity or the permeability based upon information about the soil. It is difficult to obtain accurate estimates with these formulas because of the extreme variability observed in porous media. Because of this, actual field or laboratory measurements are preferred.

Most laboratory methods used to measure saturated hydraulic conductivity are directly based upon Eq. 1. The hydraulic head is varied between the inflow of a given sample and the outflow and the flow rate through the core measured. For unsaturated hydraulic conductivity, the same basic principles are followed. However, for unsaturated hydraulic conductivity a pressure is induced on the soil sample to bring it to a given matric pressure during the flow measurement. Field measurement techniques for hydraulic conductivity can involve measuring the rate at which a given tracer is transported through the soil, the rate at which a given amount of water flows into the soil, or that rate at which the groundwater recovers when the water table is pumped from a well. One of the more reliable methods for estimating the saturated hydraulic conductivity for an aquifer material is a pump test. A pump test is conducted by observing the decrease in the water table depth in a well near the well being pumped. This method measures K_s over a fairly large area and minimizes the effects of heterogeneity of the aquifer material. In addition, it minimizes disturbance of the porous media.

CONCLUSION

Hydraulic conductivity has risen from it's simple beginnings as a means through which Henry Darcy related his observations in flow rate to his observations of forces acting upon the fluid to an extremely useful soil characteristic. While Darcy's law is empirical, based upon experimental evidence, it is widely used by hydrologists, soil physicists, agricultural engineers, and civil engineers. Hydraulic conductivity is a widely used soil parameter, used to describe the flow of water, oil, and gas within porous media. It is also used in the design of filters and flow through porous ceramics.

REFERENCES

1. Darcy, H. *Les Fontaines Publique de la Ville de Dijon*; Dalmont: Paris, 1856; 647 pp.
2. Hubbert, M.K. Darcy 's Law and the Field Equations of the Flow of Underground Fluids. Am. Inst. Min. Met. Petl. Eng. Trans. **1956**, *207*, 222–239.
3. Hillel, D. *Applications of Soil Physics*; Academic Press: New York, NY, 1980; 413 pp.
4. Freeze, R.A.; Cherry, J.A. *Groundwater*; Prentice Hall: Englewood Cliffs, NJ, 1979; 604 pp.
5. Schwab, G.O.; Fangmeier, D.D.; Elliot, W.J.; Frevert, R.K. *Soil and Water Conservation Engineering*, 4th Ed.; John Wiley and Sons, Inc: New York, NY, 1993; 507 pp.
6. Bouwer, H. *Groundwater Hydrology*; McGraw-Hill Book Company: New York, NY, 1978; 480 pp.
7. Todd, D.K. *Groundwater Hydrology*; John Wiley and Sons: New York, NY, 1980; 535 pp.
8. Nielsen, D.R.; Biggar, J.W.; Erh, K.T. Spatial Variability of Field-Measured Soil–Water Properties. Hilgardia. **1973**, *42* (7), 215–259.

9. Hopmans, J.W.; Schukking, H.; Torfs, P.J.J.F. Two-Dimensional Steady State Unsaturated Water Flow in Heterogeneous Soils with Autocorrelated Soil Hydraulic Properties. Water Resour. Res. **1988**, *24* (12), 2005–2017.

10. Bear, J. *Hydraulics of Groundwater*; McGraw-Hill: New York, NY, 1979; 569 pp.

11. Amoozegar, A.; Warrick, A.W. Hydraulic Conductivity of Saturated Soils: Field Methods. In *Methods of Soil Analysis, Part 1, Physical and Mineralogical Methods*; Agronomy Monograph Number 9, Klute, A., Ed.;

American Society of Agronomy: Madison, WI, 1986; 735–770.

12. Klute, A.; Dirksen, C. Hydraulic Conductivity and Diffusivity: Laboratory Methods. In *Methods of Soil Analysis, Part 1, Physical and Mineralogical Methods*; Agronomy Monograph Number 9, Klute, A., Ed.; American Society of Agronomy: Madison, WI, 1986; 687–734.

13. Rawls, W.J.; Ahuja, L.R.; Brakensiek, D.L.; Shirmohammadi, A. Infiltration and Soil Water Movement. In *Handbook of Hydrology*; Maidment, D.R., Ed.; McGraw-Hill, Inc: New York, NY, 1993; 5.1–5.51.

Soils, Hygroscopic Water Content in

Daniel G. Levitt
Science & Engineering Associates, Inc., Santa Fe, New Mexico, U.S.A.

Michael H. Young
Desert Research Institute, Las Vegas, Nevada, U.S.A.

INTRODUCTION

Hygroscopic water content has been defined as the moisture that an initially dry soil will adsorb when brought into equilibrium with an atmosphere of 50% relative humidity (RH) at 20°C.[1] It has also been defined as the moisture that adheres to soil particles and does not evaporate at ordinary temperatures.[2] Hillel[3] describes soil *hygroscopicity* as the phenomenon where air-dry soil will generally contain several percent more water than oven-dry soil.

The word *hygroscopic* is derived from its Greek roots *hygro*, meaning atmospheric water, and *scopic* meaning to view or examine. One measure of hygroscopic water content in soils is the hygroscopic coefficient, which is defined as the water, on a gravimetric percentage basis that is absorbed by a completely dry mass of soil when brought into equilibrium with a saturated atmosphere.[2] The hygroscopic coefficient has also been defined as the level of tension at which water is considered to be bound to the soil particles (31 atm).[4] Below a water content defined by hygroscopic coefficient (Fig. 1), water will be unavailable to plants.[1]

Understanding the behavior of hygroscopic soil water is critically important to arid-land ecology, agriculture, and waste management. Vast areas of the Earth are occupied by desert, and as human population increases, agricultural efficiencies on arid lands must increase. In addition, as deserts are gaining acceptance as locations for disposal of hazardous or radioactive waste, understanding hygroscopic water content in soils is critical to understanding the behavior of water flow and contaminant transport under these dry conditions.

PROPERTIES OF HYGROSCOPIC SOIL WATER

Physical Properties

Jury et al.[5] state that the two most important characteristics of the soil water phase are the amount of water in soil, and the force holding the water in the soil matrix. The amount of water in the soil influences many processes, including gas exchange, diffusion of nutrients to plant roots, and the rate of water and solute movement through soil. The force with which water is retained by the soil matrix affects plant water uptake, water drainage from soils, and upward movement of water against gravity.

Surface Tension

Water molecules at the air–water interface exhibit a net attraction into the liquid because the density of molecules on the air side of the interface is lower than on the liquid side. This unequal attraction deforms the hydrogen bonds of the molecules at the interface and imparts "membrane-like" properties to the interface, which stretches over the water volume like a skin. As a consequence, water molecules require extra energy to remain at the interface. The extra energy per unit surface area possessed by molecules at the interface is called the surface tension.[5]

Surface Area

Soil texture, or particle size distribution, strongly influences its hygroscopic coefficient. Whereas the surface area of sand is often less than $1 \, m^2 \, g^{-1}$, surface area of clay can be as high as several hundred square meters per gram.[6] This large difference in surface area between soil textures generally results in the hygroscopic coefficient of clay being several times greater than the hygroscopic coefficient of sand under identical conditions.

Hydraulic Properties

Soil water characteristic

The relationship between soil water content and the soil water potential is a fundamental part of the characterization of soil hydraulic properties, and is identified by various terms including water retention function, moisture characteristic, and the capillary pressure–saturation curve. This function relates a capacity factor, the water content, to an intensity

Encyclopedia of Water Science
DOI: 10.1081/E-EWS 120010269

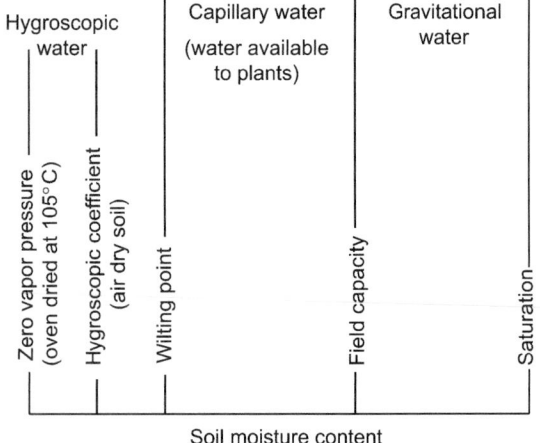

Fig. 1 Soil moisture classes and equilibrium points. (From Ref. [1].)

factor, the energy state of the soil water. This function primarily depends on soil texture.[7] Note the significant difference in water content between sand and clay at soil water tension of greater than 10,000 cm (Fig. 2). The clay material with its higher surface area (and charged surfaces) retained more water than the sand, which has a much lower surface area of water with lower charge density.

Soil will move from regions of higher potential to regions of lower potential at a rate that depends on the hydraulic resistance of the medium.[8] Water flow in unsaturated soils is particularly interesting because of the highly nonlinear nature of unsaturated water flow. For example, unsaturated hydraulic conductivities can range by 20 or more orders of magnitude between saturation and the hygroscopic coefficient. Fig. 3 illustrates the relationship between soil water content and hydraulic conductivity for sand and clay; note on the inset graph that the hydraulic conductivity is higher in sand than clay at low tension, but that clay eventually exhibits a higher conductivity than sand at higher tension because clay material has more water-filled pores at higher tension then sandy material. The low hygroscopic coefficient in sand means that liquid water movement is essentially zero under very dry conditions.

Engineering Properties

Understanding the properties of hygroscopic soil water is critical to the field of engineering. Soil strength usually increases with increasing bulk density and decreasing water content. The bonds linking clay crystals into clay packets, and the packets into aggregates, lead to higher cohesion, and thus higher strength. These include van der Waals forces, attraction between oppositely charged surfaces, organic matter in various forms, and inorganic cements. The bond strength is reduced by water through the softening of cements and the increased separation of particles as water is absorbed.

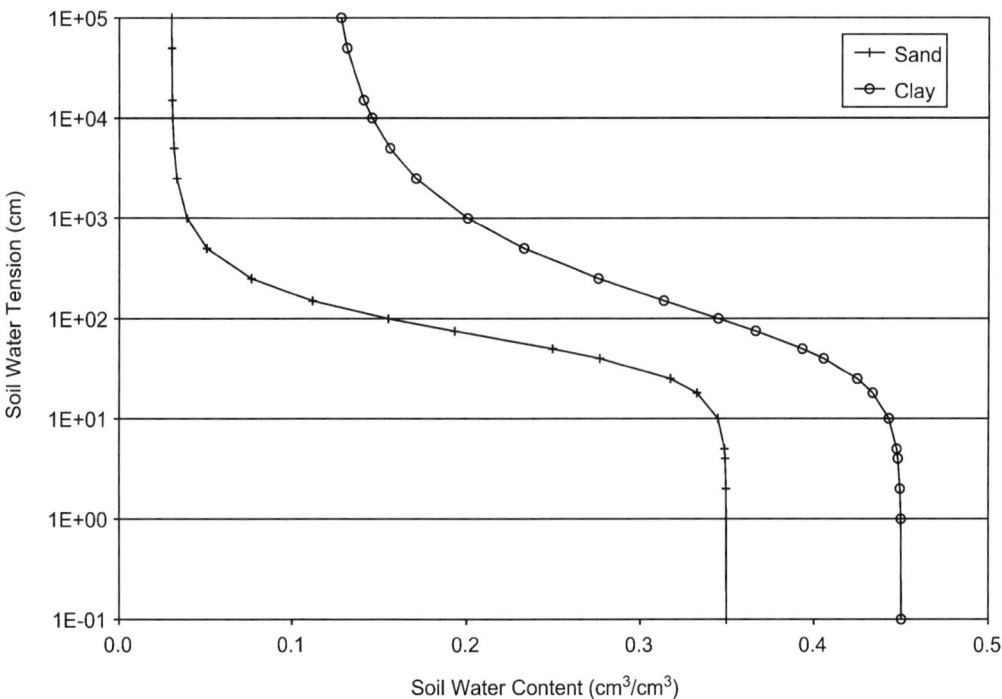

Fig. 2 Typical soil water characteristic curve for sand and clay.

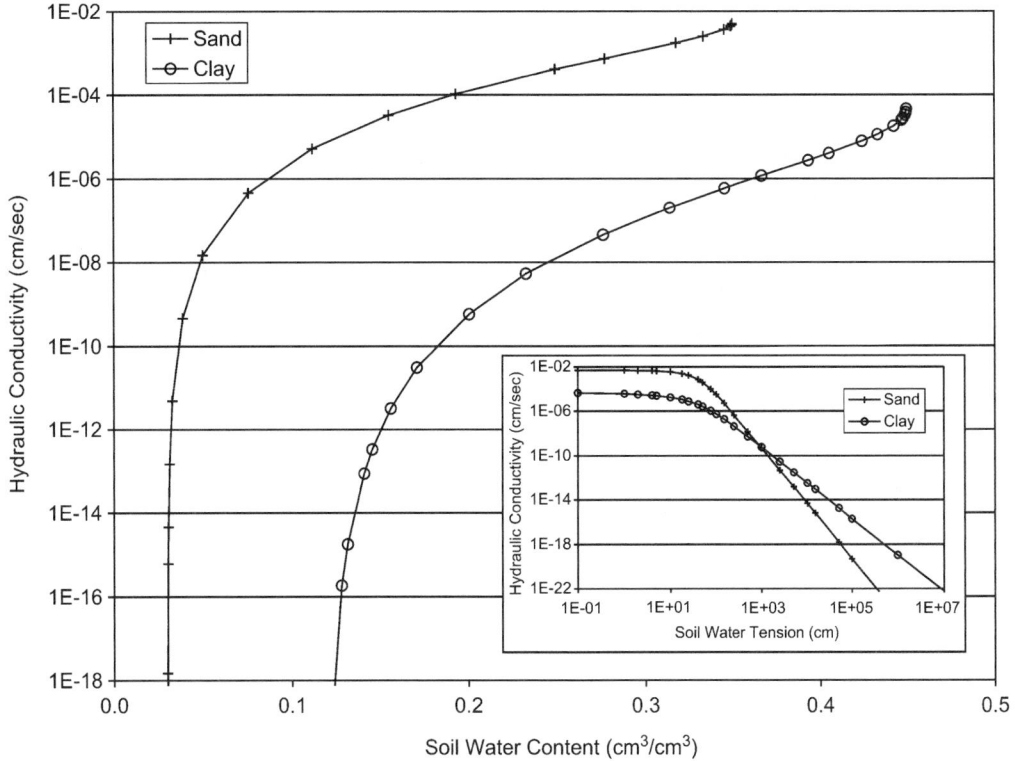

Fig. 3 Typical soil water content, soil water tension, and hydraulic conductivity relationships for sand and clay.

However, cracking acts in an opposite way to the general trend by weakening soil as it dries.[9]

MEASUREMENT OF HYGROSCOPIC WATER CONTENT IN SOILS

Hygroscopic soil water content or hygroscopic soil water potential can be measured in the laboratory by a variety of methods, including water content determination by oven drying, and soil matric potential determination by thermocouple psychrometry, chilled mirror, and heat dissipation methods.

Oven-Drying Method

One of the simplest methods for determining soil water content is by oven drying. A soil sample is weighed, placed in an oven at a temperature between 100 and 110°C for 24 hr, and then weighed again. On a gravimetric basis, water content is calculated by

Water content = (wet mass − dry mass)/(dry mass)

It is important to note that "dry" is a subjective term, and that all water within a soil sample may not be

removed by oven drying after 24 hr. However, this method serves as a standard for determining soil water content.[10]

Thermocouple Psychrometry

Thermocouple psychrometers infer the soil water potential of the liquid phase of a soil sample by measuring the RH. Water potential is related to the RH of soil water by the Kelvin equation:

Water potential (J kg^{-1}) = $RT/M \times \ln(RH)$

where M is the molecular weight of water (0.018 kg mol^{-1}), R, the ideal gas constant (8.31 J K^{-1} mol^{-1}), and T, the Kelvin temperature of the liquid phase. Most thermocouple psychrometers consist of a sensor with a thermocouple junction, which is cooled until water condenses on it. The temperature depression, measured as this water evaporates, is proportional to RH, which provides a direct measure of soil water potential. The operational range of thermocouple psychrometers (~2 atm to ~50 atm) does not generally extend throughout the entire range of water potentials corresponding to hygroscopic water contents.[11]

Chilled Mirror Method

Like thermocouple psychrometry, the chilled mirror method is used to measure the RH of soil sample. Gee et al.[12] described a commercial water activity meter that can be used for rapid measurement of soil water RH in the range from 0.100 to 1.000 (which corresponds to a soil water potential range -3119 atm to 0 atm) with essentially the same RH resolution (± 0.003) across the entire range. Gee et al.[12] suggested that this type of meter is best adapted for measurements in dry soils, making it very appropriate for measurement of hygroscopic soil water.

Heat Dissipation Method

Heat dissipation probes measure the heat dissipation characteristics of the soil matrix, which are proportional to soil water potential. Heat dissipation rates are determined by applying a heat pulse to a heater within a probe, and monitoring the temperature at the center of the probe. The operational range of heat dissipation probes extend from near-saturation to a dryness of thousands of atmospheres, well into the range of water potentials corresponding to hygroscopic water contents.[11,13]

REFERENCES

1. Bear, J. *Dynamics of Fluids in Porous Media*; Dover Publications, Inc.: New York, NY, 1972; 764 pp.
2. Academic Press Dictionary of Science and Technology; http://www.harcourt.com/cgi-bin/apdst?term = hygroscopic.
3. Hillel, D. *Introduction to Soil Physics*; Academic Press, Inc.: Orlando, FL, 1982; 10.
4. Water Words Dictionary; Nevada Division of Water Planning; Department of Conservation and Natural Resources; http://www.state.nv.us/cnr/ndwp/dict-1/waterwds.htm.
5. Jury, W.A.; Gardner, W.R.; Gardner, W.H. *Soil Physics*; John Wiley & Sons, Inc.: New York, NY, 1991; 34–71.
6. Hillel, D. *Introduction to Soil Physics*; Academic Press, Inc.: Orlando, FL, 1982; 34–35.
7. Klute, A. Water Retention: Laboratory Methods. *Method of Soil Analysis Part 1—Physical and Mineralogical Methods*, 2nd Ed.; American Society of Agronomy: Madison, WI, 1986; 635–662.
8. Jury, W.A.; Gardner, W.R.; Gardner, W.H. *Soil Physics*; John Wiley & Sons, Inc.: New York, NY, 1991; 73.
9. Marshall, T.J.; Holmes, J.W.; Rose, C.W. *Soil Physics*, 3rd Ed.; Cambridge University Press: Cambridge, 1996; 229–247.
10. Gardner, W.H. Water Content. *Method of Soil Analysis Part 1—Physical and Mineralogical Methods*, 2nd Ed.; American Society of Agronomy: Madison, WI, 1986; 493–544.
11. Rawlins, S.L.; Campbell, G.S. Water Potential: Thermocouple Psychrometer. *Method of Soil Analysis Part 1—Physical and Mineralogical Methods*, 2nd Ed.; American Society of Agronomy: Madison, WI, 1986; 597–633.
12. Gee, G.W.; Campbell, M.D.; Campbell, G.S.; Campbell, J.H. Rapid Measurement of Low Soil Water Potentials Using a Water Activity Meter. Soil Sci. Soc. Am. J. **1992**, *56*, 1068–1070.
13. Reece, C.F. Evaluation of a Line Heat Dissipation Sensor for Measuring Soil Matric Potential. Soil Sci. Soc. Am. J. **1996**, *60*, 1022–1028.

Soils, Permanent Wilting Points

Judy A. Tolk
United States Department of Agriculture (USDA), Bushland, Texas, U.S.A.

INTRODUCTION

Permanent wilting point (PWP) is defined as the largest water content of a soil at which indicator plants, growing in that soil, wilt and fail to recover when placed in a humid chamber. It is often estimated by the water content at $-1.5\,\text{MPa}$ soil matric potential.[1] The water content is typically expressed on a weight $(\text{g}\,\text{m}^{-3})$ or volume $(\text{m}^3\,\text{m}^{-3})$ basis. As the lower boundary, PWP, along with the upper boundary determined at field capacity, establishes the size of the reservoir of water held in the soil that may be withdrawn by plants, known as plant available water. Field capacity is primarily a function of soil characteristics, while PWP is the product of a combination of plant, soil, and atmosphere factors.

BACKGROUND

The soil, plant, and atmosphere act as a continuum along which soil water moves in response to gradients in energy. The energy potential of the water relative to that of pure water helps determine the amount of water stored in the soil, moved through the soil, and moved into and through the plant to the transpiring surface of the leaf. Water will flow from a region of high potential to that with low potential. The energy required to move water is expressed in terms of water potential, which is the sum of the gravitational potential, the osmotic potential, the matric potential, and the pressure potential. The matric potential is a combination of capillary and adsorptive forces due to the shape, size, and chemical nature of surfaces in the soil and plant. The osmotic potential results from the presence of dissolved substances. Pressure potential represents the solution pressure within the plant cells. For the movement of water in the soil, the pressure potential is insignificant, and the gravitational potential has little significance once it has drained to field capacity. For the movement of water through the plant, the gravitational and matric potentials are less important.

Many factors in the soil–plant–atmosphere continuum influence the amount of water a plant can extract from the soil before wilting. Soil texture affects the matric potential of the soil by determining capillary pore size and adsorptive properties, and so controls both the amount of

water held in and the movement through the soil at low soil water potentials. To extract the soil water, plant roots must be distributed throughout the soil, which is a function of soil properties such as soil strength and texture as well as the rooting characteristics of the crop. Also, an osmotic potential gradient between the soil solution at the root surface and within the root must be maintained so that the water can be absorbed into the plant roots. A water potential gradient between the plant leaf and the roots helps to move water through the plant to the leaves. Water is then evaporated (or transpired) through the stomata of the leaves due to the differences in water vapor pressure between the leaf and the atmosphere. If atmospheric demand for water exceeds the water supply to the plant's evaporating surfaces (possibly due to limited soil water supply and/or movement through the soil, limited rooting by the plant, or inadequate water potential gradients between soil and leaf), the plant will experience water stress and biological activity will decline. Unless resupplied with water, the plant cells will lose pressure potential, or turgor, and the leaves will permanently wilt and ultimately die.

THE SUNFLOWER METHOD

The wide range in soil water contents at which wilting in plants occurred was noted by German researchers as early as 1859, according to Briggs and Shantz.[2] To evaluate whether plant species varied significantly in their ability to reduce the soil water content before wilting, Briggs and Shantz[2] determined the wilting coefficient for a range of soils and plant species that included native vegetation of semiarid lands as well as crop species. Veihmeyer and Hendrickson[3] and Furr and Reeve[4] continued the work of Briggs and Shantz, using sunflower (*Helianthus annuus* L.) as the indicator plant for wilting. The procedures of Furr and Reeve[4] were standardized into the sunflower method $(\text{PWP}_{\text{sun}})$.[5] In this method, the plants are grown in containers of uniform soil that are sealed to limit water loss other than that by transpiration. They are kept adequately watered until the third set of leaves appears at which time the watering ceases. The plants remain in an environment with a low evaporative demand until all three sets of leaves wilt. To insure the wilting is permanent, plants are placed

Encyclopedia of Water Science
DOI: 10.1081/E-EWS 120010337

overnight in a humid, dark chamber. If all leaves remain wilted in the morning, PWP$_{sun}$ has been reached, and the soil water content or water potential can be determined.

PRESSURE OUTFLOW APPARATUS APPROXIMATION

Permanent wilting point can be estimated as the soil water content held in the soil at -1.5 MPa matric potential (PWP$_{-1.5}$). The similarity between PWP$_{sun}$ and PWP$_{-1.5}$ was shown by Richards and Weaver,[6] who compared the two values for 119 soils and found that PWP$_{-1.5}$ formed a fairly definite lower limit below which PWP$_{sun}$ seldom fell. In this method, a sieved soil sample is placed on a porous ceramic plate or permeable membrane in a chamber and saturated with water. A pressure of 1.5 MPa is applied until equilibrium in water content between the plate or membrane and the soil sample is reached[5] at which time soil water content is determined.

FIELD MEASUREMENT

Ratliff et al.[7] defined field measurement of PWP (PWP$_{field}$) as the lowest field-measured water content of a soil after plants had stopped extracting water and were at or near premature death or became dormant as a result of water stress. Field measurement of PWP may be the most desirable method,[8] because it provides more realistic information about how a plant grows in a certain soil because the soil–plant–environment interactions are allowed to occur. But, the controls on the experiment (e.g., uniform soil in pots, low evaporative demand environment, a well-defined root zone) are gone, and the complex soil horizons, different rooting depths and patterns by crops or by the same crop from year to year, and different environmental demands can cause substantial variation. Additional problems include refilling of the profile due to rainfall, the inability to determine when plant dormancy occurs, and the drying of the upper soil layers below PWP due to soil water evaporation. In this method, the soil profile is wetted sufficiently throughout the normal rooting depth so that the plant does not undergo severe water stress until maximum vegetative growth when maximum rooting occurs. This insures that normal rooting and water use patterns develop. Water depletion patterns throughout the growing season are monitored so that the cessation of water use from a soil layer can be determined. Once plant dormancy or premature death and the cessation of water use occur, soil water content or water potential is determined.

DISCUSSION

The applicability of PWP$_{sun}$ and PWP$_{-1.5}$ to PWP$_{field}$ has been questioned. Ratliff et al.[7] found that PWP$_{-1.5}$ was

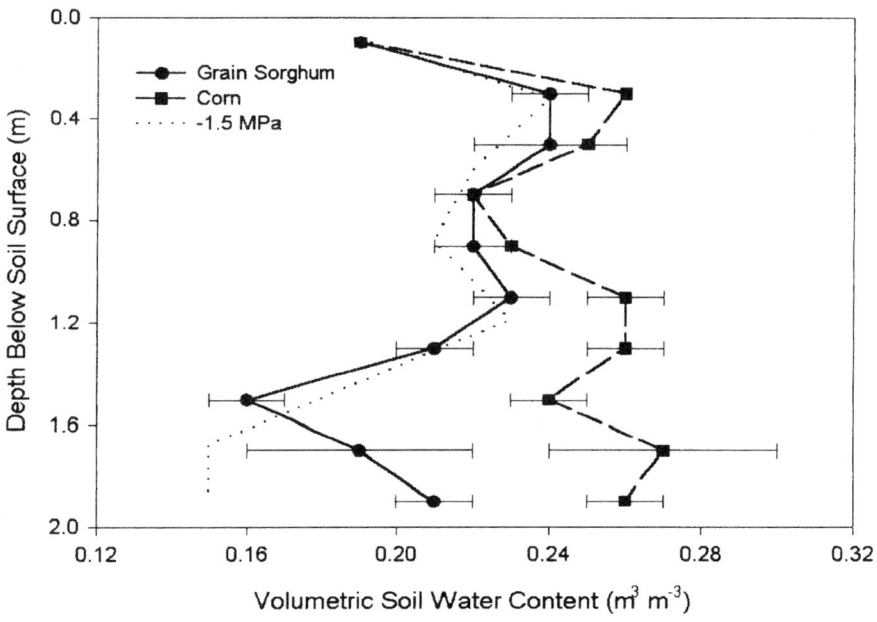

Fig. 1 Water contents of a 2-m soil profile measured for corn and grain sorghum after the available soil water had been depleted. The data points are mean values of two cropping seasons, with standard deviations (horizontal error bars). Error bars may not be visible on data points with low standard deviations. Also presented is the soil water content measured at the -1.5 MPa soil matric potential.

significantly less than PWP$_{field}$ for sands, silt loams, and sandy clay loams, and significantly more for loams, silty clays, and clays for a variety of crops. Additionally, PWP may be crop and climate specific. Cabelguenne and Debaeke[9] reported that corn (*Zea mays* L.), sorghum [*Sorghum bicolor* (L.) Moench], and winter wheat (*Triticum aestivum* L.) varied in their degree and depth of lower limit of water use in a deep silty clay loam, and these capacities were representative only of the climate in which they were obtained. Savage et al.,[10] however, concluded that PWP$_{-1.5}$ corresponded to PWP$_{field}$ for grain sorghum and cotton (*Gossypium hirsutum* L.) and values lower than measured PWP$_{-1.5}$ represented only minor amounts of available soil water.

An example of PWP for different crops is shown in Fig. 1. Grain sorghum and corn were grown in an undisturbed soil column contained in a lysimeter with a surface area of 1 m by 0.75 m and a depth of 2.3 m. The soil was a Pullman clay loam, which has a dense clay horizon about 0.4 m below the soil surface, and soil horizons containing substantial amounts of calcium carbonate beginning at about 1 m below the soil surface. The water content of the soil was measured by neutron thermalization. The vertical lines connect the means of the soil water contents for each 0.2-m depth measured at harvest for two cropping seasons for each crop, as well as the PWP$_{-1.5}$ for the different soil horizons. The horizontal lines (error bars) at each data point indicate the range in the measurements that occurred between seasons. Both crops showed a similar PWP pattern, but differed in the amount of water remaining at PWP. The dense clay horizon appears to have limited water use by both crops, probably due to restricted rooting. Grain sorghum, a more deeply rooting crop than corn, used more water from the lower soil depths. The presence of calcium carbonate in the lower depths may also have inhibited rooting. The PWP$_{-1.5}$ was similar to PWP of grain sorghum, but considerably lower for that of corn. When the volumetric soil water contents were converted to millimeters for the 2-m soil depth, the PWP for corn was 488 mm, 420 mm for grain sorghum, and the PWP$_{-1.5}$ was 398 mm. The difference between cropping seasons was 40 mm for grain sorghum, and 16 mm for corn.

Each method for the determination of PWP has advantages and disadvantages. The method selected must take into consideration the application for which it will be used, the resources available for making the measurements, and the accuracy needed.

REFERENCES

1. Soil Science Society of America, *Glossary of Soil Science Terms*; Soil Science Society of America: Madison, WI, 1997; 134 pp.
2. Briggs, L.J.; Shantz, H.L. *The Wilting Coefficient for Different Plants and Its Indirect Determination*; USDA Bureau of Plant Industry Bull. No. 230, Government Printing Office: Washington, DC, 1912; 83 pp.
3. Veihmeyer, F.J.; Hendrickson, A.H. Soil Moisture At Permanent Wilting of Plants. Plant Physiol. **1928**, *3*, 355–357.
4. Furr, J.R.; Reeve, J.O. Range of Soil-Moisture Percentages Through Which Some Plants Undergo Permanent Wilting in Some Soils from Semiarid Irrigated Areas. J. Agric. Res. **1945**, *71* (4), 149–170.
5. Cassel, D.K.; Nielsen, D.R. Field Capacity and Available Water Capacity. In *Methods of Soil Analysis, Part I. Physical and Mineralogical Methods*; Agronomy Monograph no. 9, Klute, A., Ed.; Soil Science Society of America: Madison, WI, 1986; 901–926.
6. Richards, L.A.; Weaver, L.R. Fifteen-Atmosphere Percentage as Related to the Permanent Wilting Percentage. Soil Sci. **1943**, *56*, 331–339.
7. Ratliff, L.F.; Ritchie, J.T.; Cassel, D.K. Field-Measured Limits of Soil Water Availability as Related to Laboratory-Measured Properties. Soil Sci. Soc. Am. J. **1983**, *47*, 770–775.
8. Ritchie, J.T. Soil Water Availability. Plant Soil **1981**, *58*, 327–338.
9. Cabelguenne, M.; Debaeke, P. Experimental Determination and Modelling of the Soil Water Extraction Capacities of Crops of Maize, Sunflower, Soya Bean, Sorghum and Wheat. Plant Soil **1998**, *202*, 175–192.
10. Savage, M.J.; Ritchie, J.T.; Bland, W.L.; Dugas, W.A. Lower Limit of Soil Water Availability. Agron. J. **1996**, *88*, 644–651.

Soils, Water Infiltration and

Sally D. Logsdon
United States Department of Agriculture (USDA), Ames, Iowa, U.S.A.

INTRODUCTION

Infiltration, the process of water entering the soil surface, is part of the water cycle (Fig. 1). Water infiltrates the soil because of absorptive (capillary) and gravitational forces,[3] which are strongly influenced by soil texture and structure. (For more detail, see the article *Soil Water Energy Concepts*.) The infiltration water comes from rain, melted snow, irrigation, or upslope runoff or seepage.[9] As infiltration occurs, the wetting soil profile can be divided into several zones:[3,8] saturation right at the surface to perhaps 1 cm deep, a transition zone of rapidly changing soil water content, a transmission zone with slowly changing soil water content, a wetting zone of rapidly changing soil water content, and a wetting front with a very steep hydraulic gradient.

The infiltration rate varies with time (Fig. 2). For a homogeneous soil, the rate depends on the initial water content (see article *Soil Water, Antecedent*), the application rate, the depth of the soil profile, and the surface and boundary conditions. If the application rate is less than the hydraulic conductivity, then all the water infiltrates. If the application rate is greater than the hydraulic conductivity, first the surface layer becomes saturated with water, then excess water collects at the surface. If a slope or outlet is present, then the excess water will runoff.

PREDICTION OF INFILTRATION

Physical and empirical based infiltration equations are described in the literature.[4–6,9,11,12] The physically based models[9] are developed from Richards[13] equation (see topic), Darcy's[14] law (see topic), and early developments by Buckingham.[15] Often these must be solved numerically for given initial and boundary conditions. Philip[3] discusses some of the assumptions when applying the physical-based models to water flow into and through the soil. These assumptions may not always be valid in the field,[16] and that is why the soil is often treated

empirically in larger-scale applications. Another class of infiltration equations is the rainfall excess model types,[9] which assume no applied water ponds in depressions or is intercepted by plants (Fig. 1). Empirical infiltration models[9] determine infiltration rate or volume as a function of soil properties and application rate. The Horton[17] model is an example of an early empirical model. An intermediate type of model is approximately theory-based,[9] but the parameters are more difficult to estimate than for the empirical models. The earliest and most-used approximate model is the Green–Ampt model.[18] All of these models have numerous variations and recent developments, too numerous to discuss here. Based on these equations, the different infiltration models are used at various scales.

FACTORS AFFECTING

The rate that water infiltrates the soil is affected by surface and subsurface properties, both of which are affected by management and natural phenomena.[4] The rainfall intensity, duration, and distribution are also important considerations. (See related topics under precipitation.)

Surface Properties

Important surface properties include development of a surface seal, degree of water repellency, and presence of macropores or fractures. A surface seal impedes infiltration.[19–21] The development of a surface seal increases as the residue cover decreases and as soil aggregate stability decreases. The surface seal is affected by physical and chemical processes.[22]

Subsurface Properties

Once the water is in the soil, the water is redistributed[9] both vertically and laterally (Fig. 1). Water may be held in the soil by surface and capillary forces, may drain out of the soil into tiles, may recharge the water table, or may come back out at the soil surface from a down slope position (see page). The soil profile affects continuing infiltration due to hydraulic conductivity of the soil (see the articles *Soil Water Flow Under Saturated*

This short article cannot cover all aspects of infiltration, since book-size conference proceedings have been written on the topic,[1,2] as well as many review articles[3–6] or sections of book chapters.[7–10]

Encyclopedia of Water Science
DOI: 10.1081/E-EWS 120010266

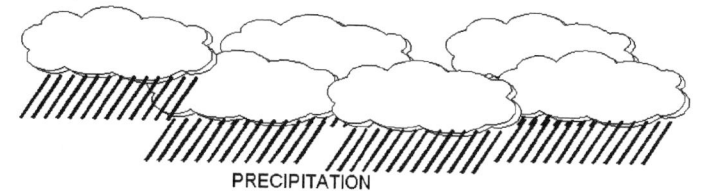

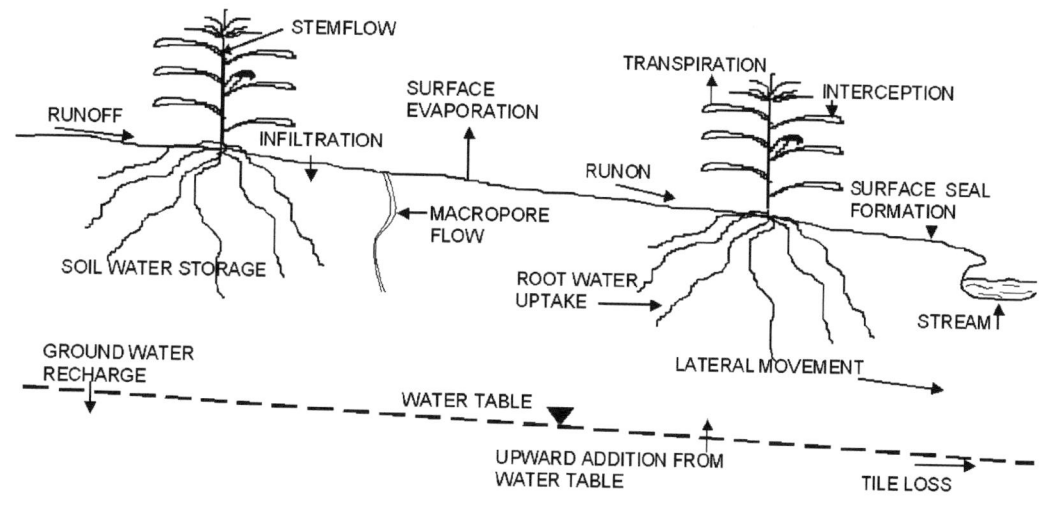

Fig. 1 Components of the water cycle.

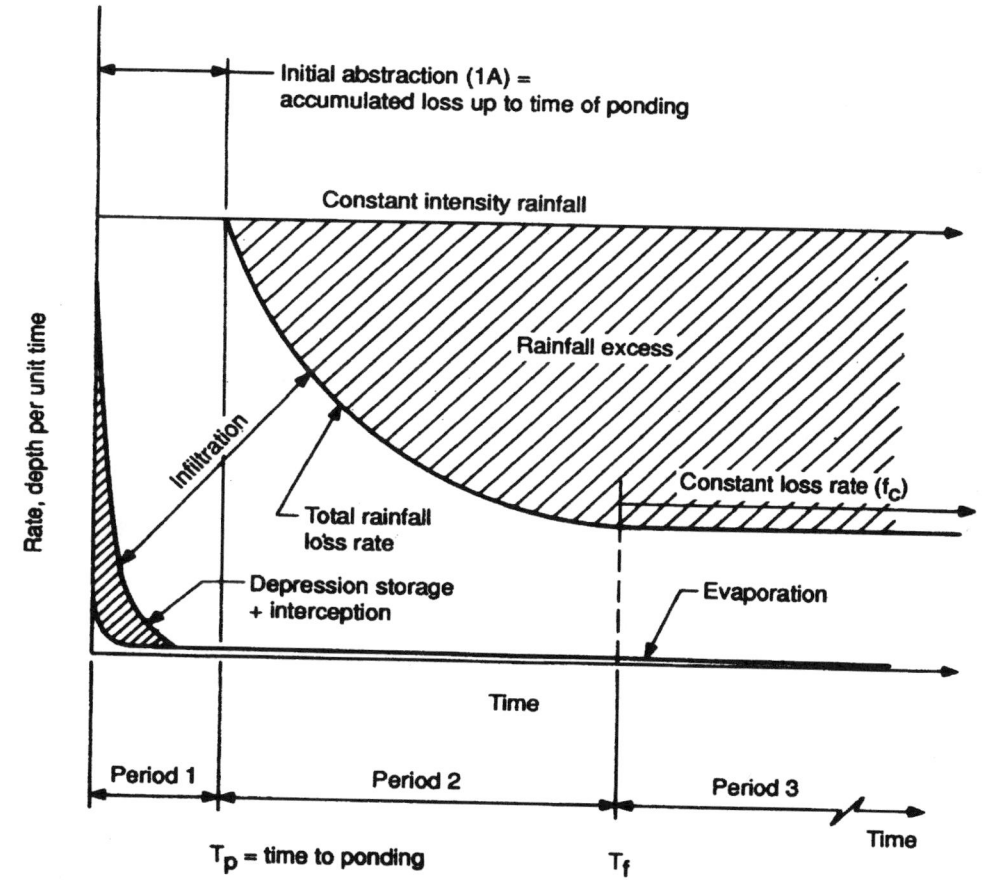

Fig. 2 Infiltration rate and fate of applied water as a function of time. (From Ref. 9 by permission of McGraw-Hill, Co.)

Conditions and *Soil Water Flow Under Unsaturated Conditions*), continuity of macropores, air pressure build-up, and presence of impeding layers, frozen soil (see article *Frozen Soil, Water Movement in*), or high water table.

Water saturation due to a high water table or frozen soil near the surface prevents formation of a surface seal but also greatly restricts infiltration. Soil aggregate stability is increased by organic matter which somewhat increases the water repellency, allowing concentrated water flow around the aggregates. Continuous macropores and fractures also allow rapid infiltration to continue if the surface seal does not form over the macropore.

Crop and Soil Management

Crop and soil management greatly influence formation of a surface seal, existence of macropores, and compaction, all of which influence the infiltration rate.[4,9] Surface seal formation is less likely on forested and pasture land. Conservation tillage on cropland and use of perennial crops allow more residue to remain at the soil surface, reducing surface seal formation. Surface seal formation is reduced as the crop canopy develops, and is reduced by off-season ground cover. Forests, pastures, and cropland that is not tilled encourage macropore formation by mesofauna, especially earthworms.

Within Landscape Variability

Infiltration varies within the landscape due to rainfall variability (see articles *Precipitation*, on *Stochastic Properties* and *Erosion and Precipitation*), plant effects (stem flow and interception), and varying soil properties.[9] Landscape infiltration is more than an accumulation of processes at individual sites because of water movement within the landscape (Fig. 1). Runoff water from higher landscape positions may move down slope at the surface and later infiltrate at down slope positions, depending on surface and profile characteristics. Seepage water may also infiltrate at a down slope position. A well-developed stream system moves water off site, but a poorly developed stream system combined with closed depressions results in temporary or permanent ponding on site. The natural flow pattern has been altered by tiles, ditches, and channelized streams. In addition, local variations occur because of funneling by crop canopy and compaction due to wheel traffic or hoof traffic.

Infiltration rate is not a static soil function because the soil properties are changing over time. Surface residue can be incorporated by tillage or be subject to decay or washed off by runoff waters. Fractures can close from prolonged wetting. Air pressure can build-up as infiltration continues.[23] Surface soils can form during infiltration or be disrupted by mesofauna activity. The temporal variability is often greater than the spatial variability within a given land-use practice or ecosystem.

MEASUREMENT

The type of infiltration measurement should relate to the application of the results.[8] Ponded ring infiltration measurements (see article *Ring and Tension Infiltrometers*) would relate to flood irrigation or to pond seepage. Sprinkler infiltration measurements relate to rainfall on the soil surface, especially if a surface seal develops. Tension infiltration relates to the rate water moves into the soil matrix, without the macropores present. Furrow irrigation measurements are important when the flowing water is critical.

REFERENCES

1. ASAE. *Advances in Infiltration, Proceedings of the National Conference on Advances in Infiltration*, Chicago, IL, Dec 12–13, 1983; ASAE: St. Joseph, MI, 1983; 1–385.
2. ARS *Workshop on "Real World" Infiltration, Proceedings of the 1996 Workshop*, Pingree Park, CO, July 22–25, 1996; Ahuja, L.R.; Garrison, A. Eds.; Water Resour. Res. Inst., Colorado St. Univ.: Ft. Collins, CO, 1996; 1–262.
3. Philip, J.R. Theory of Infiltration. Adv. Hydrosci. **1969**, *5*, 215–305.
4. Brakensiek, D.L.; Rawls, W.J. Infiltration Research Needs in Watershed Hydrology. Trans. Am. Soc. Ag. Eng. **1989**, *32* (2), 633–637.
5. Kutílek, M.; Nielsen, D.R. Infiltration. *Soil Hydrology*; Chap. 6.2. Catena-Verlag: Cremlingen-Destedt, Germany, 1994; 133–176.
6. Youngs, E.G. Developments in the Physics of Infiltration. Soil Sci. Soc. Am. J. **1995**, *59* (2), 307–313.
7. Beven, K. Infiltration, Soil Moisture, and Unsaturated Flow. In *Recent Advances in the Modeling of Hydrologic Systems*; Bowles, D.S., O'Connell, P.E., Eds.; Kluwer Academic Publishers: Dordrecht, 1991; 137–151.
8. Brooks, K.N.; Ffolliott, P.F.; Gregersen, H.M.; Thames, J.L. Infiltration, Runoff, and Streamflow. *Hydrology and the Management of Watersheds*; Iowa State University Press: Ames, 1991; Chap. 4, 64–86.
9. Rawls, W.J.; Ahuja, L.R.; Brakensiek, D.L.; Shirmohammadi, A. Infiltration and Soil Water Movement. In *Handbook of Hydrology*; Maidment, D.R., Ed.; McGraw-Hill, Inc.: New York, 1993; Chap. 5, 5.1–5.51.
10. Ward, A.D.; Dorsey, J. Infiltration and Soil Water Processes. In *Environmental Hydrology*; Ward, A.D., Elliott, W.J., Eds.; Lewis Publishers: Boca Raton, 1983; Chap. 3, 51–90.
11. Smith, R.E. Rational Models of Infiltration Hydrodynamics. In *Modeling Components of the Hydrologic Cycle. Proc. Int. Symp. on Rainfall-Runoff Modeling*; Singh, V.P.,

Ed.; Water Resources Publishers: Littleton, CO, 1981; 107–126.

12. Clausnitzer, V.; Hopmans, J.W.; Starr, J.L. Parameter Uncertainty Analysis of Common Infiltration Models. Soil Sci. Soc. Am. J. **1998**, *62* (6), 1477–1487.

13. Richards, L.A. Capillary Conduction of Liquids Through Porous Media. Physics **1932**, *1*, 318–333.

14. Darcy, H. *Les Fontaines Publiques de la Ville de Dijon*; Dalmont: Paris, 1956.

15. Buckingham, E. Studies on the Movement of Soil Moisture. USDA Bur. Soils Bull. **1907**, 38.

16. Morel-Seytoux, H.J. Infiltration Characteristics: From Column to Parcel to Hill Slope. A Physical and Stochastic Theory for Spatial, Temporal and Process Integration. In *Characterization and Measurement of the Hydraulic Properties of Unsaturated Porous Media. Proceedings of the International Workshop, Riverside, CA. Oct. 22–24, 1997*; van Genuchten, M. Th., Leij, F.J., Wu, L., Eds.; Univ. CA: Riverside, 1999; Vol. 2, 1425–1437.

17. Horton, R.E. An Approach Toward a Physical Interpretation of Infiltration Capacity. Soil Sci. Soc. Am. Proc. **1940**, *5*, 399–417.

18. Green, W.H.; Ampt, G.A. Studies on Soil Physics: 1. Flow of Air and Water Through Soils. J. Agric. Sci. **1911**, *4*, 1–24.

19. Moore, I.D. Effect of Surface Sealing on Infiltration. Trans. Am. Soc. Ag. Eng. **1981**, *24* (6), 1546–1552,1561.

20. Rumens, M.J.M.; Baumhardt, R.L.; Parlange, M.B.; Whisler, F.D.; Parlange, J.Y.; Prasad, S.N. Rain-Induced Surface Seals: Their Effect on Ponding and Infiltration. Ann. Geophysicae **1985**, *4* (4), 417–424.

21. Sumner, M.E., Stewart, B.A., Eds. *Soil Surface Sealing and Crusting*; Catena Suppl. 24, Catena Verlag: Cremlingen-Destedt, Germany, 1993; 1–139.

22. Sumner, M.E., Stewart, B.A., Eds. *Soil Crusting: Chemical and Physical Processes*; Lewis Publishers: Boca Raton, FL, 1992; 1–372.

23. Constantz, J.; Herkelrath, W.N.; Murphy, F. Air Encapsulation during Infiltration. Soil Sci. Soc. Am. J. **1988**, *52*(1), 10–16.

S

Soils, Water Percolation Through

Kurt D. Pennell
Georgia Institute of Technology, Atlanta, Georgia, U.S.A.

INTRODUCTION

The term "percolation" refers to the downward flow or movement of water through the soil profile. More precisely, percolation is defined as the downward flow of water in saturated or nearly saturated soil at hydraulic gradients of 1.0 or less.[1] Although the terms "infiltration" and "percolation" are often used interchangeably, infiltration refers to the entry of water into soil,[1] which typically occurs after rainfall or irrigation. In contrast, percolation refers to water movement that occurs following an infiltration event, once the soil profile has become saturated or nearly saturated with water. Such post-infiltration water movement is commonly referred to as internal drainage. Hillel[2] employed the term "deep percolation" to specify internal drainage of water occurring below the root zone, which is not influenced by water losses due to evaporation or transpiration (evapotranspiration) via plant roots.

To illustrate the general concept of soil water percolation, a schematic diagram of an idealized soil profile in contact with an unconfined aquifer is shown in Fig. 1. Following an infiltration event, in which the entire soil profile becomes saturated with water (indicated by a solid vertical line corresponding to a water saturation of 1.0), water will drain from the soil profile primarily under the influence of gravity (i.e., the pressure gradient is negligible). Assuming that no additional water enters the system, the soil water saturation profile at static equilibrium (dashed line) will decrease from a value of 1.0 in the saturated zone (groundwater and capillary fringe) to a value corresponding to field capacity below the root zone. In effect, the soil water profile is analogous to a soil water retention (pressure–saturation) curve. Hence, the solid and dashed lines represent the limits in water content (saturation) between which soil water percolation occurs in soils overlying an unconfined aquifer.

ESTIMATING SOIL WATER PERCOLATION

Vertical (downward) soil water percolation can be described by the Buckingham–Darcy flux law

$$q = \frac{Q}{A} = -K(h)\frac{\partial}{\partial z}(h + z) = -K(h)\left(\frac{\partial h}{\partial z} + 1\right) \quad (1)$$

where; q is the Darcy velocity or flux ($L\,T^{-1}$), Q is the flow rate ($L^3\,T^{-1}$), A is the cross-sectional area across which flow occurs (L^2), $K(h)$ is the unsaturated hydraulic conductivity ($L\,T^{-1}$), h is the negative pressure or suction head (L), and z is the elevation head (L). If water is assumed to flow downward at a constant rate, the pressure head gradient ($\partial h/\partial z$) approaches zero,[3] and hence the Buckingham–Darcy flux law reduces to

$$q = -K(h) = -K(\theta) \quad (2)$$

where θ is the volumetric soil water content ($L^3\,L^{-3}$). This condition is often referred to as gravity drainage. If water is assumed to drain uniformly from the soil profile over time, a simplified approach can be used to estimate the flux of water from a specific depth increment (z), based on the change in volumetric soil water content with time:[2]

$$q = K(\theta) = -z\frac{d\theta}{dt} \quad (3)$$

This scenario is illustrated by the horizontal arrow shown in Fig. 1, for which the water content decreases from saturation to field capacity, and is not strongly influenced by the water table (saturated zone).

To more accurately describe the soil water percolation, the functional relationship between the hydraulic conductivity (K) and the pressure head (h) or water content (θ) must be known. Unsaturated hydraulic conductivity may be written with respect to the intrinsic soil permeability (k_i):

$$K(\theta) = \frac{k_{rw}(\theta)k_i\rho_w g}{\mu_w} \quad (4)$$

where $k_{rw}(\theta)$ is the relative permeability to water as a function of θ, ρ_w is the liquid density of water ($M\,L^{-3}$), g is the gravity constant ($L\,T^{-2}$), and μ_w is the dynamic viscosity of water ($M\,L^{-1}\,T^{-1}$). However, this approach requires that the relative permeability function, $k_{rw}(\theta)$, is known or can be measured. Fortunately, a number of unsaturated conductivity functions have been developed based on pressure–saturation relationships, which are more readily available and easier to measure than relative permeability. One of the equations most commonly used to describe soil water retention data was developed by van

Encyclopedia of Water Science
DOI: 10.1081/E-EWS 120010268

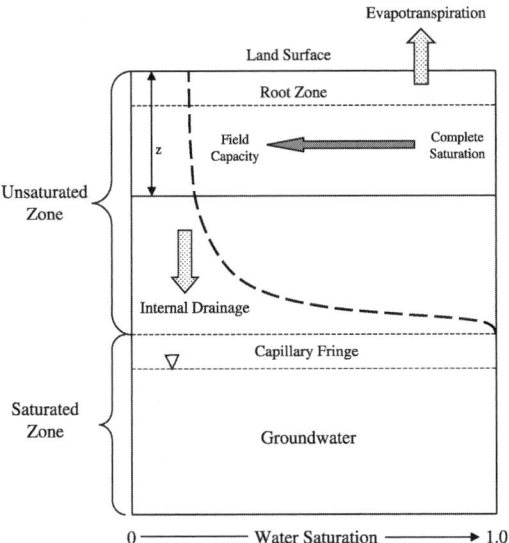

Fig. 1 Schematic diagram of soil profile showing the initial (water-saturated) and equilibrium condition following internal drainage.

Genuchten:[4]

$$S_e = \frac{\theta - \theta_r}{\theta_s - \theta_r} = \frac{1}{(1 + |\alpha h|^n)^{(1-1/n)}} \tag{5}$$

$$\alpha = \frac{1}{h_b}(2^{1/m} - 1)^{1-m} \tag{6}$$

$$m = \frac{1}{1 - n} \tag{7}$$

where S_e is the effective water saturation, α and n are fitting parameters, and θ_r and θ_s are the residual and saturated volumetric water contents ($L^3 L^{-3}$), respectively. Using the fitting parameters defined above, van Genuchten[4] then developed relationships for $K(\theta)$ and $K(h)$:

$$K(\theta) = K_s S_e^{1/2}[1 - (1 - S_e^{1/m})^m]^2 \tag{8}$$

$$K(h) = K_s \frac{\{1 - (\alpha h)^{n-1}[1 + (\alpha h)^n]^{-m}\}^2}{[1 + (\alpha h)^n]^{m/2}} \tag{9}$$

where K_s is the saturated hydraulic conductivity ($L T^{-1}$). The fitting parameters needed for unsaturated hydraulic conductivity relationships (i.e., α and n) can be obtained by fitting moisture release curve data to Eq. 5 using a nonlinear, least-squares approach (e.g., SYSTAT) or by using the RETC program.[5] More advanced approaches for describing water percolation (internal drainage) and water redistribution in partially wetted soils, such as the rectangular profile model and kinematic wave model, are presented by Hillel,[2] Jury et al.,[3] and Charbenau.[6]

PERCOLATION TESTS

From a practical perspective, the term percolation is frequently encountered during home construction and zoning regulations in reference to a percolation or "perc" test. Percolation tests are widely used throughout the United States and Canada to locate and size absorption (leaching) fields for residential sewage treatment systems. The basic procedure involves digging several (e.g., 4–6) cylindrical boreholes, at least 15 cm (6 in.) in diameter and not greater than 20 cm (8 in.) in diameter, to the intended depth of the sewage treatment trench, as shown in Fig. 2. In addition, the bottom of the borehole is often required to be at least 0.9 m (3 ft) above the seasonal high water table or bedrock.[8] It is usually recommended that the bottom 0.3 m (1 ft) of the borehole sidewalls be scarified to improve water entry, and that 8 cm (3 in.) of gravel be placed in the bottom of the borehole. Prior to the percolation test, water is added to a depth of at least 0.3 m (1 ft) above the bottom of the borehole, and allowed to infiltrate until the soil in the vicinity of the borehole is completely saturated. For most soils an infiltration or wetting period of 4 hr, with the water level maintained at approximately 0.3 m (1 ft) above the bottom of the borehole, is adequate.[7,8] However, soils that exhibit substantial swelling, usually due to high clay or organic matter content, may require an extended wetting period, up to three or four days. Once the soil surrounding the borehole is completely saturated, water is allowed to stand in the borehole for 12 hr (overnight).

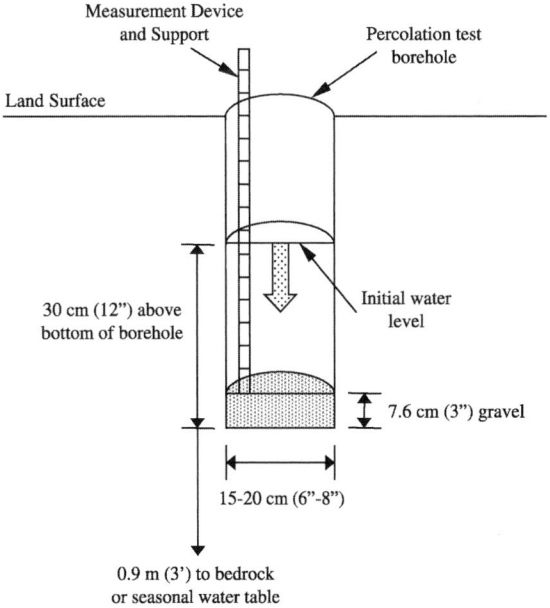

Fig. 2 Cross-sectional view of percolation test borehole.

To run the actual percolation test, set the initial water level to 30 cm – 46 cm (12 in. – 18 in.) from the bottom of the borehole. From a fixed reference position, measure the height of water in the borehole to the nearest 4 mm (1/16 in.) every 15 or 30 min.[8] The percolation rate is obtained by dividing the time of measurement (e.g., 15 min) by the drop in the height of the water level (e.g., 1.5 cm), expressed as minutes per cm (mpc) or minutes per inch (mpi). Continue the test until three consecutive measurements are within approximately 20% of one another, indicative of steady water flow. A minimum of 15 cm (6 in.) of water should remain in the borehole throughout the test. Measured percolation rates of less (fast water flow) than 0.04 mpc (0.1 mpi) for coarse-textured soils, and more (slower water flow) than 24 mpc (60 mpi) for fine-textured soils are generally considered to be unsuitable for sewage absorption fields. Specific requirements for the percolation test vary by jurisdiction, and thus, it is essential that local public health and building permit offices be contacted prior to conducting a percolation test.

REFERENCES

1. SSSA. *Glossary of Soil Science Terms*; Soil Science Society of America: Madison, WI, 1997; 140 pp. Available on-line at: www.soils.org/sssagloss/intro.html (accessed May 2001).

2. Hillel, D. *Introduction to Soil Physics*; Academic Press: New York, NY, 1982; 364 pp.

3. Jury, W.A.; Gardner, W.R.; Gardner, W.H. *Soil Physics*, 5th Ed.; John Wiley and Sons: New York, NY, 1991; 328 pp.

4. van Genuchten, M.Th. A Closed-Form Equation for Predicting the Hydraulic Conductivity of Unsaturated Soils. Soil Sci. Soc. Am. J. **1980**, *44*, 892–898.

5. van Genuchten, M.Th.; Leij, F.J.; Yates, S. *The RETC Code for Quantifying the Hydraulic Functions of Unsaturated Soils*; EPA/600/2-91/065; U.S. Environmental Protection Agency, Robert S. Kerr Environmental Research Laboratory: Ada, OK, 1991; 117 pp. Available on-line at: www.epa.gov/ada/csmos/models/retc.html (accessed May 2001).

6. Charbenau, R.J. *Groundwater Hydraulics and Pollutant Transport*; Prentice-Hall: Upper Saddle River, NJ, 2000; 593 pp.

7. Harlan, P.W.; Dickey, E.C. *Soils, Absorption Fields and Percolation Tests for Home Sewage Treatment*; G80-514-A; Cooperative Extension Service, Institute of Agriculture and Natural Resources, University of Nebraska: Lincoln, NB, 1996; 7 pp. www.ianr.unl.edu/pubs/wastemgt/g514.htm (accessed May 2001).

8. Gustafson, D.; Machmeier, R.E. *How to Run a Percolation Test*; F0-0583-C, University of Minnesota Extension Service: St. Paul, MN, 1997; 5 pp.

Storativity and Specific Yield

Hugo A. Loáiciga
University of California, Santa Barbara, California, U.S.A.

Paul F. Hudak
University of North Texas, Denton, Texas, U.S.A.

INTRODUCTION

Storativity and specific yield quantify the storage properties of an aquifer. Thus, they are fundamentally important to groundwater resource investigations. Estimates of storativity and specific yield can be used to predict changes in the volume of water stored in an aquifer with changing groundwater levels. Moreover, storativity and specific yield affect the hydraulic response of an aquifer to pumping. Predictions of pumping response may influence the locations of wells for such purposes as water supply, construction dewatering, and groundwater remediation.

EFFECTIVE STRESS AND PORE-WATER PRESSURE

Consider an enlarged, hypothetical cross-section of a saturated aquifer matrix as shown in Fig. 1. The pore space is occupied by water, and the mineral matrix is assumed to be sand grains. The porous-medium system (mineral plus water) of Fig. 1 is taken, herein, to be a representative elementary volume (REV) of the aquifer. The mineral matrix sustains an intergranular or effective stress σ_e, while the pore-water pressure is P_w. The total stress (σ) in the REV is given by $\sigma = \sigma_e + P_w$. Both the water and the mineral matrix are compressible, i.e., their volumes depend on the stresses to which they are subjected. The compressibility of water at a typical ambient temperature (e.g., 15°C) is $\beta = 4.6 \times 10^{-10}\,\mathrm{m^2\,N^{-1}}$, a small number indeed. The compressibility of a substance, water or mineral, is the absolute value of the change in its volume per unit change in compressive stress divided by the initial volume of the substance, hence the compressibility units of $\mathrm{Pa^{-1}}\,(= \mathrm{m^2\,N^{-1}})$.

The effective stress $\sigma_e = \sigma - P_w$, in which the total stress may be assumed to be approximately constant. Therefore, taking differentials on both sides of the latter stress relationship, it follows that $\Delta\sigma_e = -\Delta P_w$, i.e., as the pore-water pressure increases, the effective stress decreases, and vice versa. Aquifer dewatering by groundwater pumping, e.g., reduces the pore-water pressure and augments the effective stress, thus compressing the mineral matrix and reducing its volume. This is the basic mechanism that causes aquifer subsidence, i.e., compaction of the aquifer and associated lowering of the ground surface as a result of groundwater extraction. The degree of land subsidence is a function of the amount of change in groundwater storage, and of the compressibility of the mineral matrix. The latter is greatest in fine-textured and cohesive formations. A well-known case of aquifer subsidence is that in the lacustrine formations that underlie Mexico City. There, the ground surface at various locations fluctuates with the cycle of aquifer dewatering (ground surface drops), and recharge (ground surface rises). Aquifer subsidence is a common phenomenon associated with pumped aquifer systems throughout the world.[1]

WATER RELEASE IN AN AQUIFER

When groundwater is pumped from an aquifer, there results a series of water-release mechanisms. Depending on the deformation properties of the aquifer, the significance of each specific mechanism may be enhanced or diminished. Those mechanisms may also overlap over time. As groundwater is pumped out of an aquifer, its pore pressure is reduced and the effective stress increases. The groundwater in storage expands its volume while the mineral matrix is compacted. As a result, some groundwater in storage is removed from the pore space. This mechanism of water release is primarily a strain–stress response of the aquifer to reduced pore pressure. It occurs in confined and unconfined aquifers, and it is accentuated by cohesive sediments prone to compaction. In unconfined aquifers (with a free-moving water table), a second mechanism of water release may take place as the water table begins to fall. Pore water is drained by gravity as the water table recedes. Drainage of pore water is rapid soon after pumping starts and the decline of the water level in observation wells (i.e., the drawdown) is pronounced. Some pore water continues to drain after the water table has receded as gravitational drainage lags behind the rate of

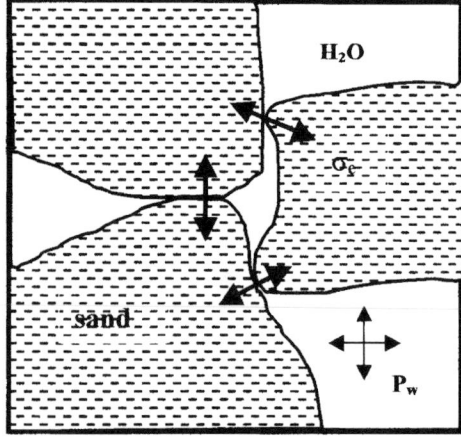

Fig. 1 An enlarged cross-section of the pore space in a saturated sand.

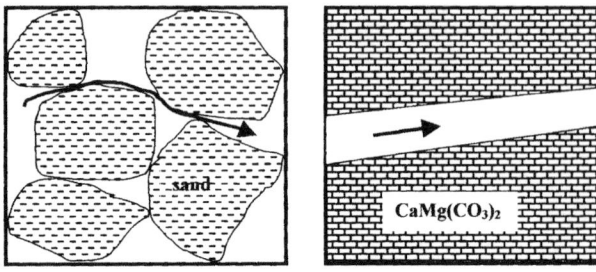

Fig. 2 On the left, primary porosity in a sand aquifer. On the right, secondary porosity in a dolomite aquifer. The arrowed line denotes pore-water pathway.

water-table descent. This is caused by limited air replacement of the pore space vacated by the falling water table. Delayed drainage (or delayed yield) can continue for a long period of time and contribute to further drawdown. A third mechanism may take place in unconfined aquifers following delayed drainage. The rate of water-table decline decreases relative to the early phase of rapid drainage, and the groundwater flow towards pumping wells is essentially horizontal, driven by hydraulic gradients.

The previous mechanisms of water release take place primarily in aquifers formed by unconsolidated geologic deposits whose porosity stems from the intergranular pore space in the aquifer matrix. This is called primary porosity. Other aquifers, called bedrock aquifers, exhibit pore water that arises largely from secondary porosity. In this case, the pore space is created by dissolution cavities (such as in carbonate or karst systems), fossil cavities (in formations that have harbored past biological activity in them), and rock fractures and joints induced by tectonism (folding, extension, and faulting). Fig. 2 illustrates primary and secondary porosities. Whether primary or secondary, porosity is defined as the volume of pore space per bulk volume of aquifer. Groundwater in bedrock aquifers moves through complex fracture and cavity networks driven by hydraulic gradients from zones of natural recharge towards wells. If aquifer recharge does not equilibrate with the rate of pumping, the secondary pore space may be dewatered leading to the depletion of bedrock water and dry wells. Depending on the nature of the fracture network and the compressibility of the rock matrix, there may be compaction of the mineral matrix as the water pressure in the fractures and cavities is relieved. The magnitude of delayed drainage in a bedrock aquifer with a free-moving water surface depends on the aperture

of bedrock fractures and cavity diameters. Very small fracture apertures hinder gravitational drainage in unconfined bedrock aquifers. Unconfined and confined bedrock aquifers are prone to complete dewatering during droughts and have relatively low sustainable groundwater yield. For example, the sandstone aquifers of the Santa Ynez mountains of southern California have a groundwater yield range from $38 \, L \, min^{-1}$ to $190 \, L \, min^{-1}$. Unconsolidated aquifers have typical yields at least one order of magnitude greater.

STORATIVITY AND SPECIFIC YIELD

Hydrologists introduced the concepts of storativity and specific yield to quantify the effect of changes in hydraulic head on groundwater storage. Recall that hydraulic head (h, dimension of L) is the mechanical energy content of pore water per unit weight of water, i.e., $h = z + P_w/\rho g$, in which z is the elevation head (potential energy per unit weight of water), and $P_w/\rho g$ is the pressure head (fluid-pressure energy per unit weight of water), where ρ is the density of liquid water and g is the acceleration of gravity.

A fundamental parameter is the specific storage S_s, which is the volume of groundwater (V_w) released (or gained) per unit decline (or rise) in hydraulic head and per unit bulk volume of aquifer (V). Thus, $S_s = V_w/(\Delta h V)$, dimensions of L^{-1}. The specific storage may be related to the properties of pore water and the mineral matrix. Let β be the compressibility of pore water, α, the compressibility of the mineral matrix, and n the porosity, it is possible to show that $S_s = \rho g(\alpha + n\beta)$. The latter expression for the specific storage rests on the assumption that the aquifer matrix deforms elastically, i.e., the amount of aquifer compaction (or expansion) is proportional to the compressive stress exerted on it. More complex analysis of aquifer deformation is needed when aquifer deformation is nonelastic (plastic).

The storativity (S) is a parameter that is widely used in the analysis of compaction effects and groundwater release in aquifers. S is the volume of groundwater released (or gained) per unit decline (or rise) in hydraulic head and per unit (horizontal) area of aquifer (A), $S = V_w/(\Delta hA)$, a dimensionless parameter. Let b be the thickness of a confined aquifer, then $S = bS_s = b\rho g(\alpha + n\beta)$ is the relationship that links the specific storage to the storativity. This relationship is useful to establish a lower bound for the storativity when the porosity and aquifer thickness are known. For example, if $b = 10\,\mathrm{m}$ and $n = 0.3$, then that lower bound is given by (ignoring the effect of mineral matrix compressibility) $b\rho gn\beta = 10\,\mathrm{m} \times 1000\,\mathrm{kg\,m^{-3}} \times 9.8\,\mathrm{m\,s^{-2}} \times 0.30 \times 4.6 \times 10^{-10}\,\mathrm{m^2\,N^{-1}} = 1.4 \times 10^{-5}$.

Generally, the storativity, instead of the specific storage, is the preferred parameter in the study of aquifers. This is so because the storativity is more prone to empirical estimation than the specific storage. The storativity may be estimated by means of pumping tests. It can also be inferred from maps of hydraulic head drawn at two different times if the groundwater pumped (or recharged) between those dates in the mapped area is known. The latter inference is possible by a direct application of the definition of storativity based on the water withdrawn (or recharged) to an aquifer per unit area and per unit decline (or rise) of the hydraulic head. Numerical methods generically known as "inversion theory" are also used to estimate S and S_y, as well as other aquifer parameters. The storativity of most confined aquifers is on the order of 10^{-5} to 10^{-3}.

The storativity is used to characterize unconfined aquifers also, where changes in groundwater storage are caused by the fall or rise of the water table. Recall that unconfined aquifers may release pore water by compaction of the mineral matrix as well as by gravitational drainage. The specific yield (S_y) was introduced to differentiate between these two mechanisms of water release. Specifically, S_y represents the volume of groundwater released by pore-water drainage per unit drop in the water table and per unit (horizontal) area of unconfined aquifer. It is a dimensionless parameter. If the saturated thickness of an unconfined aquifer is denoted by b_s, the storativity in an unconfined aquifer may be written as the sum of two terms. One term reflects water release by drainage while another accounts for aquifer compaction. Specifically, $S = S_y + b_sS_s$. The term involving the specific storage S_s captures the role of aquifer compaction on water release in the last equation. In unconfined aquifers formed by coarse sediments (sand and gravel), the specific yield is much larger than the specific storage and $S \approx S_y$. In fine-textured, unconfined, aquifers, drainage is negligible and $S \approx b_sS_s$.

From the definition of specific yield, there arises a relationship between the porosity (n) and what is called the specific retention, S_r. The specific retention is the volume of pore water held against gravitational drainage per unit bulk volume of aquifer (S_r is a dimensionless parameter). Clearly, $n = S_y + S_r$. The specific yield is less than porosity, an intuitive fact imposed by the impossibility of draining more water from pore space than was originally held in it. The specific yield ranges between 0.1 and 0.3 in most unconfined aquifers.

FIELD ESTIMATION OF STORATIVITY AND SPECIFIC YIELD

The preferred method to estimate storativity and specific yield is via pumping tests, and, in particular, pumping tests in which there is at least one observation well where measurements of the time-dependent drawdown are made at a distance from the pumping well. If properly designed and executed, pumping tests produce representative storativity and specific-yield estimates that average out geological variability over tens to hundreds of meters. These are the most useful estimates in the analysis of aquifer response to human and natural stresses. The classical analysis of pumping-test data relies on several assumptions, i.e., a large, homogeneous, aquifer with isotropic radial flow towards a fully penetrating pumping well. The theory and practice of pumping-test analysis[2] is beyond the scope of this article. Herein, we highlight the role of pumping-test data in the interpretation of specific yield and storativity.

Consider the drawdown vs. time data collected during a pumping test in an unconfined aquifer, which has been graphed in Fig. 3. The drawdown is pronounced during the first four minutes. This is caused by aquifer compaction and rapid gravitational drainage. Some authors[3] define an early-time apparent specific yield to describe the water-release mechanism during this early phase. The early-time specific yield for the data in Fig. 3 was estimated to be approximately 3×10^{-3}.[4] In the interval between 4 min and 30 min, the rate of drawdown levels off as gravitational drainage diminishes. Thereafter, the rate of drawdown increases and is sustained by delayed drainage through the last observation taken 3000 min after pumping started. This latter drawdown phase is associated with a later-time specific yield[3] that was estimated in Ref. 4 to be approximately 0.1.

Fig. 4 shows pumping test data for a semiconfined aquifer in Pixley, California. The aquifer has suffered considerable subsidence from prolonged pumping. The mechanism of groundwater release in this instance

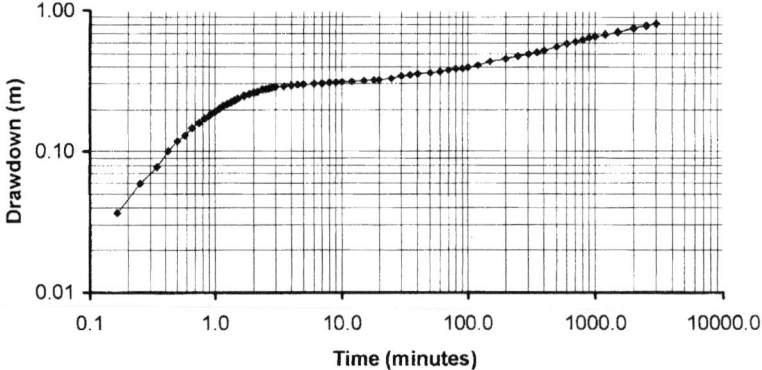

Fig. 3 Drawdown data in an observation well 22.25 m away from a well pumped at a constant rate of 4088 L min^{-1} in an unconfined aquifer in Fairborn, Ohio. (From Ref. 4.)

is matrix compaction and water expansion as the aquifer is depressurized by pumping. Based on the data in Fig. 4, the storativity was estimated at 4×10^{-5}.[4]

FIELD CONDITIONS AND RANGE OF VALUES

Specific yields for unconfined aquifers generally range from 0.1 to 0.3. They are less than 0.02 for most aquicludes (i.e., formations with very low water-bearing capacity) Table 1 lists representative ranges of specific yield for different types of unconsolidated sediment. Generally, specific yield increases with particle size and degree of sorting. Due to a larger surface area per unit volume and smaller pores, fine-grained sediment holds more water against the force of gravity. Multiplying the specific yield of an unconfined aquifer by its total saturated volume gives an indication of the maximum usable volume of water stored in the aquifer.

Sorting also affects specific yield because it impacts the porosity of unconsolidated sediment. Thus, a coarse sand mixed with silt has a lower specific yield than uniform coarse sand. This is so because the small silt particles fill spaces between the larger sand particles, thus reducing total porosity in the mixed sand relative to that of the uniform sand.[6] Clay has a high porosity, sometimes exceeding 0.50, but a low specific yield, due to a large surface of clay plates per unit volume and tiny pore spaces.

Specific yield values are lower for detrital sedimentary rocks (e.g., conglomerate, sandstone, siltstone, and shale) than their unconsolidated counterparts. Cement fills pore spaces in such rocks, creating smaller and less connected pores. Moreover, specific yield values are extremely low (less than 0.01) in most unweathered chemical sedimentary, igneous, and metamorphic rocks due to a low effective (interconnected) porosity.[7] Secondary porosity from weathering and fracturing increases specific yield in consolidated rocks. Weathering and fracturing are most

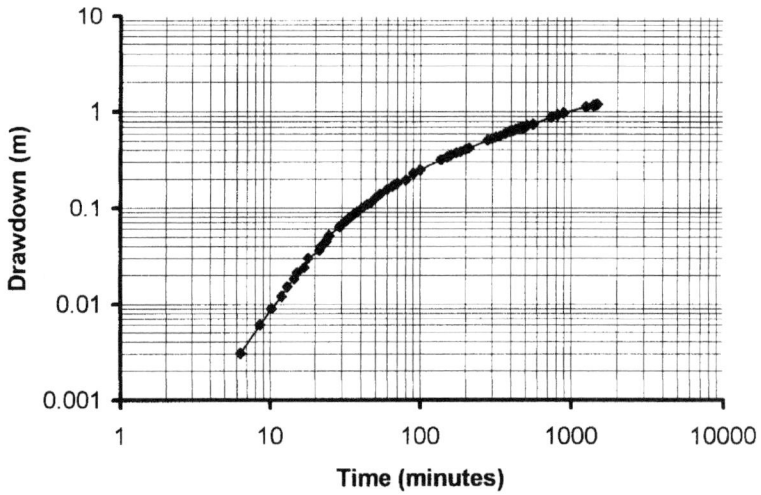

Fig. 4 Drawdown data in an observation well 467 m away from a well pumped at a constant rate of 2839 L min^{-1} in a semiconfined aquifer in Pixley, California. (From Ref. 4.)

Table 1 Representative ranges of specific yield for unconsolidated sediment

Sediment	Range of specific yield, S_y
Clay	0.01–0.18
Silt	0.01–0.39
Loess	0.14–0.22
Fine sand	0.01–0.46
Medium sand	0.16–0.46
Coarse sand	0.18–0.43
Eolian sand	0.32–0.47
Fine gravel	0.13–0.40
Medium gravel	0.17–0.44
Coarse gravel	0.18–0.43

(From Ref. 5.)

common within 20 m of the land surface, but may extend to depths of 100 m in tropical regions.[7] Underground solution also enhances porosity and specific yield, especially in carbonates (e.g., limestone) and evaporites (e.g., gypsum and halite). Table 2 presents specific yield ranges for sedimentary formations.

Storativity values are much lower in confined aquifers than specific yield values in unconfined aquifers. Water-release mechanisms in confined aquifers (compression of aquifer solids and water expansion) release small amounts of water per unit decline in hydraulic head compared to those in unconfined aquifers (gravity drainage). During pumping, small storativity values in confined aquifers result in rapid expansion of cones of depression. Interference of expanding cones around adjacent wells occurs more rapidly in confined aquifers than in unconfined ones. Substantial declines in hydraulic head over large areas are needed to produce large amounts of water. Thus, confined aquifers are more vulnerable to being overexploited. For example, the potentiometric surface of the Trinity aquifer beneath Dallas and Fort Worth, Texas has declined more than 200 m over the past

Table 2 Range of specific yield in selected sedimentary formations

Sedimentary formation	Range of specific yield, S_y
Siltstone	0.01–0.33
Limestone	0.00–0.36
Fine sandstone	0.02–0.40
Medium sandstone	0.12–0.41
Tuff	0.02–0.47
Schist	0.06–0.33

(From Ref. 5.)

century.[8] However, the land surface above the aquifer has incurred negligible subsidence due to the granular structure of the aquifer and overlying rock aquicludes.

CONCLUSION

The previous review of specific yield and storativity highlights the importance of these parameters in all aspects of groundwater hydrology. They play a role in important processes of aquifer compaction and expansion at the pore scale. Furthermore, they are very useful in characterizing the response of aquifers to pumping and recharge at the field and regional scales. Specific yield and storativity become indispensable in the numerical simulation of transient aquifer flow. The combined application of pumping tests and inverse theory is a powerful tool for their estimation and interpretation.

ACKNOWLEDGMENTS

This work was supported in part by U.S. Geological Survey grant HQ-96-GR-02657 and by University of California Water Resources Center grant WR-952.

REFERENCES

1. Domenico, P.A.; Schwartz, F.W. *Physical and Chemical Hydrogeology*, 2nd Ed.; John Wiley and Sons: New York, 1997.
2. Kruseman, G.P.; deRidder, N.A. *Analysis and Evaluation of Pumping Test Data*, 2nd Ed.; International Association for Land Reclamation and Improvement Publication 47: Wageningen, The Netherlands, 1991.
3. Boulton, N.S. Analysis of Data from Non-equilibrium Pumping Tests Allowing for Delayed Yield from Storage. Proc. Inst. Civil Eng. **1963**, *26*, 469–482.
4. Lohman, S.W. *Ground-Water Hydraulics*; U.S. Geological Survey Professional Paper 708, U.S. Government Printing Office: Washington, DC, 1979.
5. Morris, D.A.; Johnson *Summary of Hydrological and Physical Properties of Rock and Soil as Analyzed by the Hydrologic Laboratory of the U.S. Geological Survey*; United States Geological Survey Water Supply Paper 1839-D: Reston, VA, 1967.
6. Fetter, C.W. *Applied Hydrogeology*, 4th Ed.; Macmillan: New York, 2000.
7. Davis, S.N. Porosity and Permeability in Natural Materials. In *Flow Through Porous Media*; De Wiest, R.J.M., Ed.; Academic Press: New York, 1969; 53–89.
8. Mace, R.E.; Dutton, A.R.; Nance, H.S. Water-Level Declines in the Woodbine, Paluxy, and Trinity Aquifers of North-Central Texas. Trans. Gulf Coast Assoc. Geol. Soc. **1994**, *44*, 413–420.

S

Summer Fallow

Donald L. Tanaka
United States Department of Agriculture (USDA), Mandan, North Dakota, U.S.A.

Randy L. Anderson
United States Department of Agriculture (USDA), Brookings, South Dakota, U.S.A.

INTRODUCTION

Summer fallow has been a controversial practice in many semiarid regions of the United States and Canada. Crop production in these regions has been limited by low and variable precipitation. Summer fallow has been practiced to increase the water available for succeeding crops in regions that receive less than 500 mm of precipitation.[1] The basic objectives of summer fallow are: 1) maximize soil water storage; 2) make plant nutrients available; 3) reduce soil erosion hazards; 4) minimize energy and economic inputs; 5) control weeds during the entire fallow period; 6) take advantage of standing stubble to capture snow; and 7) suppress soil water evaporation during the warm season.[2]

Summer fallow was adopted in the semiarid regions following the dust bowl era of the 1930s in the United States. Since then; considerable changes in equipment and technology have taken place. Summer fallow discussions in this article will focus primarily on the semiarid Great Plains of the United States and Prairie Provinces of Canada. Statements about the semiarid Pacific Northwest will be included. Two major summer fallow systems discussed will be winter wheat-fallow and spring wheat-fallow. Fallow will be divided into seasonal fallow segments that consist of after-harvest (harvest though October), over-winter (November through April), summer fallow (May through October, or until seeding winter wheat), and second over-winter for spring wheat (November until spring wheat seeding). Summer fallow segment for this paper will be defined as a practice where no crop was grown and all plant growth was controlled with herbicides or cultivation during the season when a crop would normally be grown.

FALLOW PRINCIPLES

In the Great Plains, early fallow (prior to and during the 1930s) used tillage implements that inverted and mixed the soil for control of weeds. The number of tillage operations ranged from 7 to 10 per season and usually destroyed crop residue cover and soil protective clods to create a dust mulch. Dust mulches served two purposes in summer fallow: 1) suppress weeds from germinating and growing and 2) suppress evaporation of stored soil water by creating a discontinuity in the capillary soil pores that transport water and water vapor to the soil surface. Variants of the dust mulch principle are still used in the Pacific Northwest.[3]

During this same time, summer fallow and annual crop research was conducted throughout the Great Plains. Mathews and Army[4] summarized 450 crop–fallow periods from 25 locations, with some locations having 40 yr of research. They found that the average fallow efficiency (percent of precipitation stored in the soil during fallow) was about 16% and concluded that most of the 84% of the precipitation lost was due to evaporation. Key fallow efficiency principles they found were: 1) soil water loss due to deep percolation below the rooting depth of wheat was negligible or nonexistent; 2) evaporation losses were great and fallow efficiencies were not likely to improve unless a method of reducing evaporation losses was devised; 3) average annual runoff losses were very low and accounted for only a negligible portion of the precipitation received during fallow; 4) fallow efficiency decreased from the Northern Plains to Southern Plains; and 5) the decreased fallow efficiencies were associated with increased potential evaporation from north to south. Their regression analysis indicated little relationship between the total precipitation received during fallow and fallow efficiency and that soil water storage was not significant until large quantities of precipitation were received.

CONSERVATION TILLAGE FALLOW

Since the soil inversion fallow era, subsurface tillage implements and development of cost effective herbicides have resulted in more wheat residue being left on the soil

Encyclopedia of Water Science
DOI: 10.1081/E-EWS 120010227

surface during the summer fallow segment. Researchers have found that increased quantities of surface residue significantly increased soil water storage during the fallow period in a wheat-fallow system.[5–8] Wheat residue on the soil surface reduces rain drop impact and prevents puddling and facilitates water infiltration. Residues have increased fallow efficiencies from about 16% for bare soils that were intensively tilled to 40% for no-till.[9–11] Therefore, wheat residue can significantly suppress evaporation losses and greatly improve fallow efficiencies. Greater fallow efficiencies improve wheat production, which in turn increases residue production.[12] However, improved fallow efficiencies, during average or above-average precipitation years, can result in the movement of nutrients and water below the rooting depth of wheat grown in wheat-fallow systems causing potential ground water problems or saline seep conditions.[13] Saline seep conditions have become prevalent in the northern Great Plains of the United States and Prairie Provinces of Canada where wheat-fallow systems are used.[13]

In the northern Great Plains, spring wheat-fallow systems dominate. The potential for no-till to store more soil water than intensively tilled systems is greater for winter wheat-fallow systems than for spring wheat-fallow systems because winter wheat produces more residue than spring wheat, winter wheat stubble remains standing longer than spring wheat stubble, and winter wheat has a 14-mo fallow compared to the longer 21-mo spring wheat fallow.[14,15] Soil water storage for winter wheat-fallow is equal to or greater than soil water storage for spring wheat-fallow even though the spring wheat fallow is 21 mo. The second over-winter segment for spring wheat follow stores very little soil water. In regions where winter wheat–fallow systems dominate, mostly in the central and southern Great Plains, no-till fallow has the potential to store more soil water during the after-harvest and over-winter segments of fallow because of greater residue production than systems in the northern Great Plains.

Research (Tanaka and Anderson personal communication) suggests that soil water storage in the central Great Plains for the after-harvest and over-winter fallow segments had less variability and resulted in more soil water storage than in the northern Great Plains (Table 1). Both locations receive 20%–25% of the yearly precipitation as snow and standing stubble in no-till helps hold snow. In the northern Great Plains, 70%–80% of

Table 1 Winter wheat-fallow equations used to predict stored soil water (Y) for precipitation (x in mm) received during each seasonal segment of fallow for Akron, CO (central Great Plains) and Sidney, MT (northern Great Plains) (unpublished research from Tanaka and Anderson)

Seasonal Segment	Akron, CO			Sidney, MT		
	Fallow Method[a]	Equations	R^2	Fallow Method	Equations	R^2
After-harvest	NT	$Y = 60.99 - 0.43x + 0.004x^2 + 0.0035x^2$	0.71^b			
				All methods[c]	$Y = 32.89 + 0.84x$	0.58^b
	SM	$Y = 55.38 - 0.67x + 0.004x^2$	0.50^b			
Over-winter	NT	$Y = 7.51 + 0.94x$	0.83^b			
				All methods	$Y = 14.35 + 0.40x$	0.26^b
	SM	$Y = 13.26 + 1.28x - 0.005x^2$	0.79^b			
Summer fallow	NT	$Y = 75.03 + 1.28x - 0.002x^2$	0.16^b			
				All methods	$Y = 58.10 - 0.24x$	0.18^b
	SM	$Y = 880.58 + 6.48x - 0.010x^2$	0.53^b			
After 14-mon of fallow	NT	$Y = 3826.71 - 23.46x + 0.050x^2 - 0.0003x^3$	0.68^b			
				All methods	$Y = -778.26 + 7.70x - 0.022x^2 - 0.00002x^3$	0.47^b
	SM	$Y = 655.61 + 3.34x - 0.003x^2$	0.30^b			

[a] Fallow methods include no-till (NT) and stubble-mulch (SM). Fallow details are defined by Smika[12] and Tanaka.[14]
[b] Significant at 0.05 probability level.
[c] No significant difference occurred among fallow methods; therefore NT, MT, and SM fallow methods were combined.

the precipitation that fell on frozen soil in the northern Great Plains was lost as runoff[16] resulting in low fallow efficiencies for the over-winter segment in the northern Great Plains.[17] During the over-winter segment in the central and southern Great Plains, soils remain unfrozen for a longer period of time increasing the potential to store snowmelt water. The over-winter segment usually has the highest fallow efficiency in the central and southern Great Plains because evaporation and runoff are low.[2]

The summer segment of fallow has been considered to be inefficient because of high evaporative demands. In general, 60% or greater of the total soil water stored during fallow occurred during the after-harvest and over-winter segments.[18] For fallow to effectively store soil water, three criteria must be met. First, the quantity of surface residue present must be adequate to suppress evaporation. In the northern Great Plains, at least 2500 kg/ha must be present in mid-May before surface residues will suppress evaporation enough to increase soil water storage.[19] Second, soils cannot be at or near field capacity in the soil root-zone. Fig. 1 illustrates the decrease in fallow efficiency as the soil water storage in the after-harvest and over-winter segments increase. The efficiency for the summer segment of fallow in the northern Great Plains was <10% when soil water storage during the after-harvest and over-winter segments was >120 mm. Third, precipitation must occur in sufficient quantity and frequency to effectively permit soil water to move deep enough into the soil profile to significantly reduce or eliminate evaporative losses. In the northern Great Plains, surface residues suppressed evaporation for 10 days.

Cumulative evaporation for bare and residue covered soil surfaces become equal if at least 10 mm of rain did not fall within the 10-day period.[20]

Since Mathews and Army[4] developed the principles for fallow, technology and techniques have been developed to manage surface residues using no-till. The best wheat-fallow systems have not been able to exceed 40%–45% fallow efficiencies. Research in the Great Plains indicates that 60%–95% of soil water accumulation during 14- and 21-mo fallow was stored before the summer fallow segment.[15,17] Fallow efficiencies have become stagnant and farming systems in the Great Plains need to be modified to include 1) cropping systems that reduce the frequency of fallow and 2) inclusion of deep rooted or full season crops such as sunflower, safflower, soybean, or corn.[21] The wheat-fallow cropping system may no longer be sustainable in the Great Plains.[22]

SUMMER FALLOW IN THE FUTURE

Summer fallow will undoubtedly remain useful in future dryland systems, but the frequency and length of the fallow period may be reduced and cropping systems will include more crop diversity. In areas traditionally considered crop-fallow, the ultimate goal is to increase soil water storage during the noncrop periods so that continual cropping can be practiced. Crop rotation, crop sequence, management practices, and weather will influence the success of a cropping system that replaces crop-fallow. The more dissimilar the crops and their

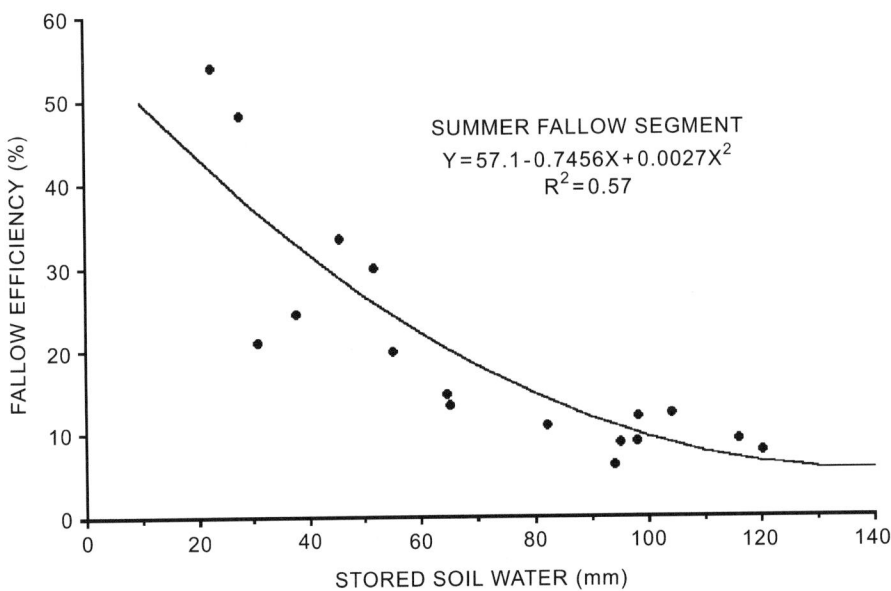

Fig. 1 Fallow efficiency for the summer fallow segment as influenced by soil water storage during the after-harvest plus the over-winter segments at Sidney, Montana. From Tanaka and Anderson.[17]

management practices, the less opportunity individual pest species have to become dominant.[23] New intensive cropping systems may include fallow but fallow will be perceived in a different way. Goals of the no-till or reduced till fallow would be a period when plants are grown for the purpose of soil building in a no-till system, enhancement of soil water storage during the inefficient summer fallow segment, pest control, and improvement of environmental quality.[24] This would transform evaporative water loss during fallow into transpirational water loss through plants while conserving or enhancing our natural resources.

REFERENCES

1. Haas, H.J.; Willis, W.D.; Bond, J.J. Summer Fallow in the Northern Great Plains (Spring Wheat). *Summer Fallow in the Western United States*, USDA Conserv. Res. Report No.27; US Government Printing Office: Washington, DC, 1974; Chap. 2, 12–35.

2. Greb, B.W. *Reducing Drought Effects on Cropland in the West-Central Great Plains*, USDA Bull. No. 420; US Government Printing Office: Washington, DC, 1979; 1–31.

3. Leggett, G.E.; Ramig, R.E.; Johnson, L.C.; Massee, T.W. Summer Fallow in the Northwest. *Summer Fallow in the Western United States*, USDA Conserv. Res. Report No. 27; US Government Printing Office: Washington, DC, 1974; Chap. 6, 1–31.

4. Mathews, O.R.; Army, T.J. Moisture Storage on Fallowed Wheat Land in the Great Plains. Soil Sci. Soc. Am. Proc. **1960**, *24*, 414–418.

5. Army, T.J.; Wiese, A.F.; Hanks, R.J. Effect of Tillage and Chemical Weed Control Practices on Soil Moisture Losses During the Fallow Period. Soil Sci. Soc. Am. Proc. **1961**, *25*, 410–413.

6. Greb, B.W.; Smika, D.E.; Black, A.L. Effect of Straw Mulch Rates on Soil Water Storage During Summer Fallow in the Great Plains. Soil Sci. Soc. Am. Proc. **1967**, *31*, 556–559.

7. Unger, P.W. Surface Residue, Water Application, and Soil Texture Effects on Water Accumulation. Soil Sci. Soc. Am. **1976**, *40*, 298–300.

8. Unger, P.W. Straw-Mulch Rate Effect on Soil Water Storage and Sorghum Yield. Soil Sci. Soc. Am. J. **1978**, *42*, 486–491.

9. Unger, P.W. Water Conservation: Southern Great Plains. In *Dryland Agriculture*, Agronomy Monogragh 23; Dregne, H., Willis, W., Eds.; American Society of Agronomy: Madison, WI, 1983; 35–53.

10. Greb, B.W. Water Conservation; Central Great Plains. In *Dryland Agriculture*, Agronomy Monograph 23; Dregne, H., Willis, W., Eds.; American Society of Agronomy: Madison, WI, 1983; 57–70.

11. Black, A.L.; Power, J.F. Effect of Chemical and Mechanical Fallow Methods on Moisture Storage, Wheat Yields, and Soil Erodibility. Soil Sci. Soc. Am. Proc. **1965**, *29*, 465–468.

12. Smika, D.E. Fallow Management Practices for Wheat Production in the Central Great Plains. Agron. J. **1990**, *82*, 319–323.

13. Halvorson, A.D.; Black, A.L. Saline-Seep Development in Dryland Soils of Northeastern Montana. J. Soil Water Conserv. **1974**, *49*, 77–81.

14. Tanaka, D.L. Chemical and Stubble-Mulch Fallow Influences on Seasonal Soil Water Content. Soil Sci. Soc. Am. J. **1985**, *49*, 728–733.

15. Tanaka, D.L.; Aase, J.K. Fallow Method Influence on Soil Water and Precipitation Storage Efficiency. Soil Till. Res. **1987**, *9*, 307–316.

16. Willis, W.O.; Carlson, C.W. Conservation of Winter Precipitation in the Northern Great Plains. J. Soil Water Conserv. **1962**, *17*, 122–123.

17. Tanaka, D.L.; Anderson, R.L. Soil Water Storage and Precipitation Storage Efficiency of Conservation Tillage Systems. J. Soil Water Conserv. **1997**, *52*, 363–367.

18. Black, A.L.; Bauer, A. Strategies for Storing and Conserving Soil Water in the Northern Great Plains. In *Challenges in Dryland Agriculture—A Global Perspective*; Unger, P., Sneed, T., Jordan, W., Jensen, L., Eds.; Amarillo/Bushland: TX, 1988; 137–139.

19. Tanaka, D.L. Wheat Residue Loss for Chemical and Stubble-Mulch Fallow. Soil Sci. Soc. Am. J. **1986**, *50*, 434–440.

20. Aase, J.K.; Tanaka, D.L. Soil Water Evaporation Comparisons Among Tillage Practices on the Northern Great Plains. Soil Sci. Soc. Am. J. **1987**, *51*, 436–440.

21. Peterson, G.A.; Schlegel, A.J.; Tanaka, D.L.; Jones, O.R. Precipitation Use Efficiency as Affected by Cropping and Tillage Systems. J. Prod. Agric. **1996**, *9*, 180–186.

22. Lyon, D.J.; Stroup, W.W.; Brown, R.E. Crop Production and Soil Water Storage in Long-Term Winter Wheat-Fallow Tillage Experiments. Soil Till. Res. **1998**, *49*, 19–27.

23. Lyon, D.J.; Miller, S.D.; Wicks, G.A. The Future of Herbicides in Weed Control Systems of the Great Plains. J. Prod. Agric. **1996**, *9*, 209–215.

24. Moyer, J.R.; Blackshaw, R.E.; Smith, E.G.; McGinn, S.M. Cereal Cover Crops for Weed Suppression in a Summer Fallow-Wheat Cropping Sequence. Can. J. Plant Sci. **2000**, *80*, 441–449.

Surface Water Law in the Western United States

J. David Aiken
University of Nebraska, Lincoln, Nebraska, U.S.A.

INTRODUCTION

The prior appropriation doctrine, the primary water law doctrine of the western United States, is a legal rejection of the riparian rights doctrine, which originated in England (see Ref. 1 pp 3-3 to 3-6). Under the riparian rights doctrine, only those whose land bordered a stream had a right to use water from the stream (see Ref. 1 pp 3-47 to 3-52). Under the early appropriation doctrine, water rights were created by diverting water from the stream and making a beneficial use of the water. Under most modern state appropriation systems, the water appropriator must also comply with state appropriation permit requirements.

APPROPRIATION OF WATER

Actual Diversion Requirement

Under prestatutory appropriation systems, water was required to be diverted from the stream in order for the water user to have legally appropriated the water (see Ref. 1 pp 5-5 to 5-10; 5-72). This actual diversion requirement was carried over into many appropriation statutes. For many years, the actual diversion requirement was a legal barrier to obtaining appropriations for instream water uses, such as fish, wildlife, and recreation (see Ref. 1 pp 5-109 to 5-112). Beginning in the 1970s, most western states modified their appropriation statutes to specifically provide for instream appropriations (see Ref. 1 pp 5-47 to 5-48).

Beneficial Use

Under prestatutory appropriation systems, water diverted from a stream was required to be put to a beneficial use in order for the water user to have legally appropriated the water (see Ref. 1 p 5-112). The beneficial use requirement was carried over into many appropriation statutes. The beneficial use concept has two dimensions: the purpose of use and the quantity of water.

Purpose of Use

While most appropriation statutes enumerate specific uses that are legally considered to be beneficial, most enumerations also include language indicating that other nonenumerated uses may be beneficial as well. Western courts have generally taken the position that if the use is beneficial to the appropriator, the purpose of use portion of the beneficial use requirement has been satisfied (see Ref. 1 pp 5-114 to 5-116).

Duty of Water

Duty of water refers to the quantity of water appropriated (see Ref. 1 p 5-113). Irrigation has been and remains the largest consumptive use of water in the West, and most duty of water issues relate to irrigation water-use efficiency. Courts and western state appropriation administrators have tolerated what would today be considered less-efficient irrigation practices in establishing the duty of water for irrigation appropriations. The basic test is whether the use is reasonably efficient at the time that irrigation is initiated. Irrigators are usually allowed to maintain their traditional irrigation practices despite improvements in irrigation technology, leading to the charge that the prior appropriation system foster's inefficiency (see Ref. 1 pp 5-118 to 5-21). In fact, most of the irrigation water "waste" returns to the stream as irrigation return flows and is relied upon by downstream water users (see Ref. 1 pp 5-125 to 5-128). Many appropriation states establish specific statutory ceilings on diversion rates and annual diversion quantities for irrigation appropriations.

Water Rights Administration

Virtually all western states have comprehensive appropriation water administrative systems. State appropriation administrators, often referred to as state engineers, are responsible for determining priority dates and water quantities for all appropriations, maintaining an appropriation registry, approving applications for new appropriations, cancelling unused appropriations, and administering priorities during periods of water shortage (see Ref. 1 pp 5-74 to 5-81).

Encyclopedia of Water Science
DOI: 10.1081/E-EWS 120010180

PRIORITY OF APPROPRIATION

Junior and Senior Appropriators

Disputes between appropriators when there is insufficient water for all appropriators are resolved on the basis of temporal priority, or "first in time is first in right." The appropriator with the earliest appropriation priority date is called the senior appropriator, while the appropriator with the more recent priority date is called the junior appropriator (see Ref. 1 pp 5-48 to 5-55).

An appropriator may be junior to some appropriators and senior to others. Appropriators are subject to priority calls by downstream senior appropriators. Appropriators may issue a priority call against upstream junior appropriators. Any appropriator may request the state engineer to restrict diversions by any upstream appropriator (including upstream senior appropriators) to the authorized amount if excess diversions are being made.

Senior appropriations represent a more secure water supply than junior appropriations.

Relation Back Doctrine

Because of the significance of priority in the appropriation system, establishing priority dates is an important issue. Generally, the priority date for an appropriation will relate back to the earliest definite step that the appropriator took to establish the appropriation, so long the appropriation was completed with due diligence (see Ref. 1 pp 5-101 to 5-103). Under modern appropriation systems, appropriation applicants are given deadlines within which they must complete their appropriations or else have their application dismissed (see Ref. 1 pp 5-106 to 5-109).

Priority Administration

One of the most important aspects of state engineer administration is the administration of priorities. When a senior appropriator is not receiving all the water the appropriator is entitled to, the senior appropriator makes a priority call, also referred to as a river call. This involves informing the state engineer's office that the appropriator is not receiving sufficient streamflow to exercise the appropriation. If the inadequate streamflows are confirmed, the state engineer will issue closing orders to junior appropriators upstream from the senior appropriator making the priority call ("priority runs upstream"). When the senior appropriator has completed the appropriator's water use (e.g., the appropriator has completed an irrigation), the senior appropriator will notify the state engineer, who in turn will inform the upstream junior appropriators that they can resume their water diversions (see Ref. 1 pp 5-53 to 5-55).

Futile Call Doctrine

A major exception to priority administration is the futile call doctrine. If the state engineer determines that the increased water flows generated by issuing closing orders to upstream junior appropriators will not reach the downstream senior appropriator making the priority call in usable quantities and in a timely fashion, the state engineer can refuse to issue closing orders to the junior appropriators despite a downstream river call (see Ref. 1 p 5-55).

Water-Use Preferences

Another exception to the priority doctrine recognized in some western states is the notion of water-use preferences. Water preferences typically involve an ordering of the importance of water use, such as 1) domestic; 2) agriculture; and 3) industry. The appropriator with the highest use preference is called the superior use; the appropriator with a lower preference is called the inferior use. Under specific limited circumstances, superior uses may in some western states be legally favored over inferior uses regardless of priority (see Ref. 1 pp 5-57 to 5-59).

Preferences are relevant principally when the superior use is a junior appropriation. When the superior water use is the senior appropriation, the superior water use is protected by the priority doctrine. Even when the superior use is the junior appropriation, the senior appropriation will almost always be entitled to exercise its priority without regard to water-use preference.

There are two types of water preferences: absolute and compensatory preferences. Under absolute water preferences, a junior appropriator with a superior use will be entitled to water at the expense of a senior appropriator with an inferior use. Under compensatory water preferences, the senior appropriator with an inferior use has priority over the junior appropriator with the superior use. If the junior appropriator wishes to exercise its superior use preference, it must purchase (or condemn) the senior inferior appropriation. In other words, the junior superior appropriator can obtain the senior inferior appropriator's water only by paying for the water. The vast majority of appropriation water disputes are resolved on the basis of priority.

LOSS OF APPROPRIATIONS

Because water appropriations are based upon beneficial use of the water appropriated, when the water use stops the appropriation may be lost (see Ref. 1 p 5-152). Unused appropriations may be cancelled by the state engineer when the statutory period for appropriation nonuse (e.g., 3 yr) has run (see Ref. 1 pp 5-156 to 5-159). Appropriations may be legally considered to be abandoned even without administrative appropriation cancellation where the appropriation has not been used for the period of time for losing real estate by adverse possession (e.g., 10 yr) (see Ref. 1 pp 5-153 to 5-156). The time period for appropriation loss by abandonment is typically longer than the period for administrative appropriation cancellation.

WATER REUSE

Consumptive Use and Return Flows

As noted earlier, under Duty of Water, the appropriation system has been criticized as fostering inefficient water use. To understand this criticism, it is first necessary to understand the concepts of consumptive water use and return flows. Assume that an irrigator diverts 300 acre ft of water to irrigate 100 acres of farmland. (An acre ft of water is enough water to cover an acre of land to a depth of 1 ft, or 325,851 gal.) The crop consumes 175 acre ft of water, and the remaining 125 acre ft return to the stream. Of the 300 acre ft diverted, 175 acre ft are consumptively used in crop production, and the remaining 125 acre ft are return flows.

Appurtenancy Doctrine

In most western states, appropriations may only be used on the land for which the water was originally appropriated (see Ref. 1 pp 5-122 to 5-125). An irrigator cannot, by improving the irrigator's water-use efficiency, use part of the 300 acre ft diverted on a second field; the appropriator can only irrigate the original field with the 300 acre ft. Thus, the appropriator has less economic incentive to improve water-use efficiency because the saved water cannot be reused. The reason for this policy is that downstream appropriators (both senior and junior) rely on the 125 acre ft of return flows as part of their water supply. Irrigating 150 acres with the 300 acre ft of water instead of the original 100 acres would typically increase the total consumptive use from the original 175 acre-ft (unless the crops irrigated were changed). The increased consumptive use reduces the return flows to downstream appropriators, which is illegal under appropriation law. Some states have modified the appurtenancy doctrine to encourage water marketing.

WATER MARKETING

In many western states, irrigators appropriated the natural flow of the state's rivers and streams. Later users who wanted to obtain a secure water supply developed water storage to capture spring runoff for summer use. But where there is no unappropriated water available and water storage options have been fully developed or are too expensive implement, the remaining option for reallocating water from old uses to new uses is water marketing. This typically involves a municipality or industry purchasing a senior irrigation appropriation and using the water for a different purpose, often at a different location. These water right transfers must be approved by the state engineer, who must maintain the return flows to downstream appropriators (see Ref. 1 pp 5-122 to 5-132). In the hypothetical case where 300 acre ft are diverted for irrigation, 175 acre ft are consumed, and 125 acre ft are return flows, the irrigator could sell and transfer only the 175 acre ft of consumptive use, and not any of the 125 acre ft of return flows. The difficulty is that the relative amounts of consumptive use and return flow may be difficult to determine in particular cases. Appropriation purchasers and downstream appropriators are likely to disagree on the relative quantities of consumptive use and return flows. If the appropriation purchaser takes an aggressive stance regarding consumptive use and return flows, downstream appropriators may be required to hire attorneys and consultants in order to protect their return flows, an expense that many see as unfair. Despite these difficulties, water marketing is an essential tool to allow water to be reallocated from old use patterns to use patterns better reflecting current economic and social needs.

RESERVED WATER RIGHTS

When Indian reservations were created, Congress reserved to the tribes sufficient water to economically develop the reservation. The priority date for Indian reserved rights is the date the reservation was created. Indian reserved rights are not lost by nonuse, so a tribe may initiate a water use in 2002 with a priority date of 1850 even if that would displace all appropriations on the stream junior to 1850 (see Ref. 1 pp 9-69 to 9-79). This has resulted in many conflicts between tribes and appropriators.

Federal reserved water rights are created when Congress or the President establishes a national park, national forest, etc. The priority date is the date the national park is created but the water uses protected are only those uses identified when the national park or forest is created. Typically, fish, wildlife, and recreation uses are not protected under federal reserved water rights (see Ref. 1 pp 9-92 to 9-110).

PROTECTION OF INSTREAM FLOWS

While most appropriation states have modified their statutes to provide for instream appropriations for fish, wildlife, and recreation (see Ref. 1 pp 5-47 to 5-48), those instream appropriations will be very junior appropriations. On many western rivers and streams, the instream appropriation will be a paper water right only and will not represent a secure water supply because the stream has been fully appropriated or even over-appropriated. In this circumstance, the better strategy to protect instream flows is through water marketing: purchasing a senior appropriation and converting it to an instream appropriation. The federal Endangered Species Act has also been used to

obtain water supplies to maintain federally designated endangered or threatened wildlife species through endangered species regulations rather than through state appropriation laws (see Ref. 1 pp 9-47 to 9-62).

REFERENCE

1. Tarlock, A.D. *Law of Water Rights and Resources*; West Group: St Paul, MN, 2002.

Surface Water Pollution by Nitrogen Fertilizers

Gregory McIsaac
University of Illinois, Urbana, Illinois, U.S.A.

INTRODUCTION

The use of industrially manufactured nitrogen (N) fertilizers increased rapidly in developed countries between 1960 and 1980. This facilitated a large increase in the production of feed and food grains (maize, wheat, and rice) per unit of cultivated land, but in some regions it also contributed to enrichment of surface and groundwater with various forms of nitrogen. Fertilizer, however, is not the only source of nitrogen that can cause contamination of surface waters. Biological nitrogen fixation, mineralization of soil organic nitrogen, and animal wastes can also contribute to nitrogen enrichment of water bodies. Additionally, under some conditions, nitrogen applied to the soil may be converted to gaseous or immobile forms of nitrogen that do not contribute to surface water contamination. Because of these various sources and transformations of nitrogen, the severity of surface water contamination by nitrogen fertilizer has been difficult to precisely quantify. Existing research indicates that the amount of contamination from fertilizer varies depending on the amount of fertilizer applied, and characteristics of the soils, crops, climate, and the receiving water bodies.

Problems Caused by Nitrogen Pollution of Surface Waters

There are three water quality concerns associated with different forms of nitrogen. First, the combined concentrations of nitrate (NO_3^-) plus nitrite (NO_2^-) in excess of $10\,mg\,N\,L^-$ can contribute to methemoglobinanemia ("blue baby syndrome") in infants if ingested.[1] To guard against this, the U.S. Public Health Service limits nitrate plus nitrite concentration in public drinking water supplies to $10\,mg\,N\,L^-$. Secondly, unionized ammonia (NH_3) may be toxic to fish at concentrations as low as $0.02\,mg\,N\,L^-$. Finally, elevated total nitrogen concentrations (including nitrate, ammonia, and organic forms) in rivers can promote the process of cultural eutrophication in coastal waters, whereby increased production and decomposition of algae, leads to reduced oxygen concentrations. This, in turn, may reduce the abundance and diversity of marine life and may promote the outbreak of nuisance algae.[2]

Sources of N Pollution

Nitrogen contamination may come from a variety of sources: municipal sewage, animal manure, atmospheric deposition, biological N fixation, soil organic N, and/or nitrogen fertilizers. The consequences of contamination in a specific water body will depend upon the amount of contamination from all sources and characteristics of the receiving waters. Shallow rivers, wetlands, lakes, and reservoirs, have some capacity to remove nitrogen by microbial denitrification. The susceptibility of estuaries and coastal waters to eutrophication depends on temperature, availability of phosphorus and silica for algae production, and the rate of water exchange with the open ocean.

Fertilizers

The contribution of inorganic fertilizer to surface water N contamination increased after 1960 as the widespread and intensive use of inorganic N fertilizers rapidly expanded.[3] The use of N fertilizer has allowed greater production of feed and food crops per unit area cultivated. In the United States, 75% of N fertilizer is applied to maize, while in other countries, N fertilizer is primarily used on wheat and rice. Prior to 1960, nitrogen for crop production was obtained primarily by using crop rotations that included legumes such as clover and alfalfa, which can establish a symbiotic relationship with soil bacteria that can convert atmospheric N_2 gas to biologically available forms of N.

Commercial nitrogen fertilizer is primarily manufactured as gaseous ammonia (NH_3), using the Haber-Bosch process in which gaseous nitrogen is reacted with gaseous hydrogen under pressure. The gaseous ammonia may be injected into the soil, which is a common fertilizer application practice in the United States. Additionally, a wide variety of granular and aqueous fertilizer products containing nitrogen are manufactured from manufactured ammonia.

Encyclopedia of Water Science
DOI: 10.1081/E-EWS 120010336

WHAT HAPPENS WHEN FERTILIZER IS APPLIED TO SOIL?

Biochemical Processes

In the soil, ammonia reacts with water and is largely converted to ammonium (NH_4^+), which tends to be strongly adsorbed on soil particles. This adsorption inhibits the movement of ammonium through the soil. Ammonium is an energy rich substance and certain soil bacteria can utilize this energy by decomposing the ammonium to nitrate (NO_3^-). Unlike ammonium, nitrate is not adsorbed to soil particles and, therefore, moves readily with water in the soil. Nitrate that is not taken up by plant roots or soil micro-organisms can be transported to groundwater and surface water by a variety of mechanisms.

Hydrologic Processes

Rainfall, snow melt, or irrigation water input to the soil periodically exceeds the water holding capacity of the soil in the root zone. Depending on the characteristics of the soil, this may lead to one or more of the following: 1) saturation of the root zone with water; 2) surface runoff; and 3) drainage of water through the soil profile to groundwater and/or surface water bodies. Each of these has different consequences for transport of nitrate to surface waters.

If the soil becomes saturated, oxygen may become scarce and in anoxic conditions, denitrifying bacteria may convert the nitrate to nitrogen gases (NO, N_2O, and N_2). Nitrogen converted to these gases becomes unavailable for plant uptake or for surface water contamination. Additionally, saturated soil during the growing season is harmful to many crops like maize that cannot tolerate low oxygen concentrations in the root zone for more than a few days.

Surface runoff has the capacity to transport soil, vegetation, and surface applied granular fertilizers from agricultural fields to surface water bodies. If a granular form of nitrogen fertilizer had been applied immediately prior to the event that caused the surface runoff to occur, nitrate and ammonia concentrations in runoff can be very high.[4] This does not appear to be a common phenomenon, however. Small rainfall events are much more common than large events that typically produce surface runoff. After granular fertilizer is applied, it is likely that a series of small rainfall events will dissolve the granules and move the nitrogen into the soil profile, where it is less likely to contaminate surface runoff. Surface runoff usually has a low nitrate concentration but it can be

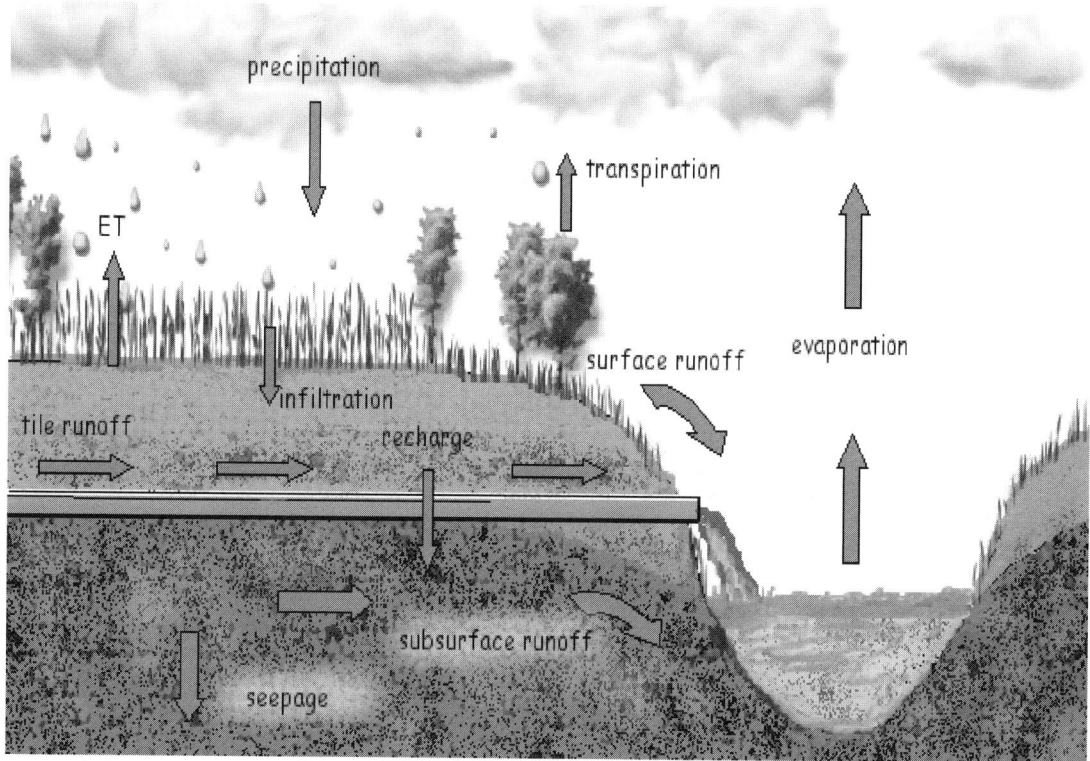

Fig. 1 Illustration of the hydrologic cycle with artificial subsurface drainage (tile runoff) contributing to surface channel flow.[5]

high in organic and particulate N derived from soil and vegetation.

Drainage of water through the soil profile to groundwater and surface water appears to be the hydrologic pathway that most frequently leads to problematic nitrate contamination of surface waters in agricultural watersheds. This can occur in two ways: by natural drainage where ground water contributes to stream flow and river flow, and by artificial subsurface drainage, where perforated pipes (sometimes called tile drains) have been buried in the soil for the purpose of removing water to reduce damage caused by saturated conditions and thereby enhance crop production (Fig. 1).

Artificial subsurface drainage improves aeration of the soil root zone and increases the length of time that machinery can be used on the soil.[6] It is a common practice in the North Central United States, and in Northwestern Europe, where flat and swampy land has been converted to cropland. The water removed from the soil by artificial drainage is usually directed to surface ditches, streams, and rivers. This water can have high concentrations of nitrogen, principally in the nitrate form, especially where nitrogen fertilizers are applied in excess to the amount necessary for crop production.[7,8] This nitrate can also be derived from microbial conversion of soil organic matter to inorganic N in the process of mineralization. Mineralized soil organic N may come from crop residues (unharvested leaves, stalks, and roots).

Of course, some of the N in crop residues may have originated from fertilizer applied in previous years, but it can also derive from biological N fixation or animal manures applied on a field.

SPATIAL VARIABILITY

In most agricultural settings, commercial fertilizer provides only one source of N used for crop production. Animal manure, biological N fixation, mineralization from soil organic N, and deposition of N from the atmosphere can also contribute to soil fertility and surface water contamination. Because there are multiple sources and sinks of N in the soil, the relationship between N fertilizer application rate and nitrogen loss in drainage water is not always consistent across locations and across studies. If denitrification and plant and microbial uptake of N are large, nitrate concentrations in subsurface drainage may be low in spite of high fertilizer N inputs. If mineralization of soil organic matter is large, nitrogen in drainage water may be large without N fertilizer input. High rates of mineralization of soil organic N occurs after the initial cultivation of virgin land, and after a leguminous forage crop such as alfalfa or clover are cultivated into the soil. Appropriate use of N fertilizer should take all of these N sources into account, as should studies examining the relationship between N fertilizer use and water quality.

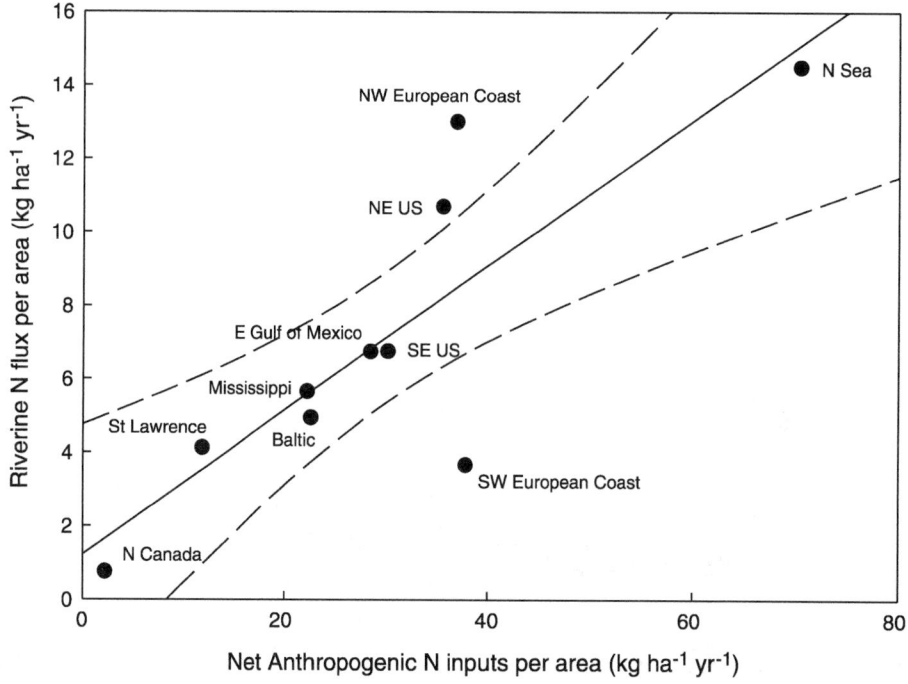

Fig. 2   Average annual riverine total N flux as a function of net anthropogenic N inputs to temperate regions draining to the North Atlantic Ocean.[10]

WATERSHED SCALE ANALYSES

Regional Nitrogen Input–Output Analyses

Howarth et al.[10] developed an approach for estimating the net nitrogen inputs to a region N that is highly correlated with average nitrogen transport in the rivers draining temperate regions (Fig. 2). Net N input to a region was defined as sum of N in fertilizer used, biological N fixation of agricultural crops, oxidized N in atmospheric deposition in the region, and the N in food and feed imported to the region minus the N in food and feed exported from the region. This approach assumes that there is no net gain or loss of N from soil organic matter. This assumption appears to be reasonable in regions where most soils have been under continuous cultivation for 60 yr or more, at which time, annual mineralization of soil organic N is roughly replaced by organic N returned to the soil in crop residues and microbial biomass.[11]

In temperate regions, riverine N transport was, on average 25% of the net N input to the region. The fate of the other 75% of the net N is unknown, but much of it is probably converted to gaseous forms of N by microbial denitrification. The high net N input in countries draining to the North Sea, most notably the Netherlands, is in part due to high density of domestic animals as well as use of N fertilizers. In tropical regions, riverine N flux was much greater than 25% of net N inputs, even in regions where little N fertilizer was used. The reasons for this are not precisely known but it is believed to be due, in part, to greater rates of biological N fixation in both cultivated and noncultivated land in the tropics. This may also be due to the recent conversion of forest, wetlands, and grasslands to crop production, which leads to high rates of mineralization of soil organic N to nitrate which is highly mobile.

Hydrologic Process Models

The quantity of nitrate transported in rivers is also related to the quantity of water flowing in the rivers per unit of land area, which is also known as water yield. Caraco and Cole[12] demonstrated that riverine nitrate N transport in major rivers in the world was a function of water yield, fertilizer use, population density, and atmospheric deposition of oxides of N. Building on these results, McIsaac et al.[13] developed the following model of annual nitrate discharge in the Lower Mississippi River 1960–1998:

$$NF_m = 0.66 WY^{0.93} e^{(0.13NNI^{2-5} + 0.06NNI^{6-9})} \qquad (1)$$

where NF_m = annual nitrate N flux in Lower Mississippi River (kg N ha^{-1} yr^{-1}), $NNI2-5$ = average annual net N input during the previous 2 yr–5 yr (kg N ha^{-1} yr^{-1}), $NNI6-9$ = average annual net N input during the previous 6 yr–9 yr (kg N ha^{-1} yr^{-1}), and WY = annual water yield (m yr^{-1}).

This equation accounted for 95% of the annual variation in nitrate flux in the Mississippi River from 1960 to 1998 and suggested that riverine nitrate in a given year was correlated with net N input averaged over the previous 2 yr–9 yr. Furthermore, calculations with the model suggest that if the N fertilizer use in the basin had been 12% lower than actual during this period, nitrate flux

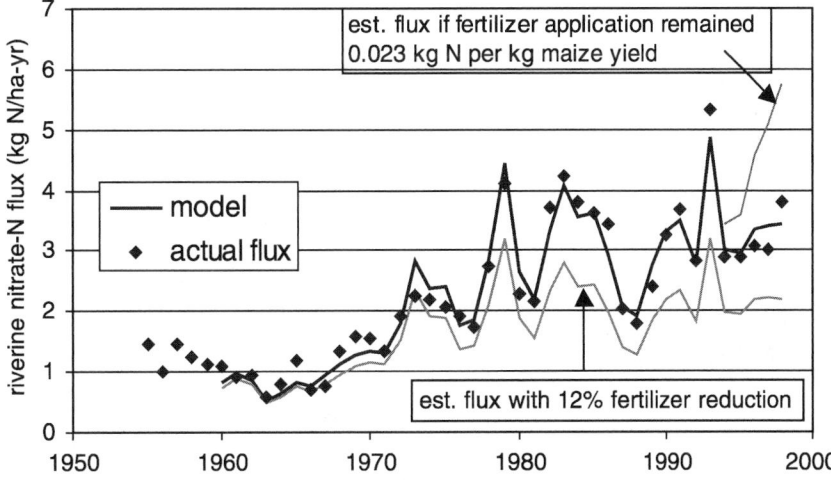

Fig. 3 Annual riverine nitrate flux in the Lower Mississippi River at St. Francisville, Louisiana, as determined from measurements (diamonds) as estimated from Eq. 1 (thick line), as estimated from Eq. 1 assuming a 12% reduction in N fertilizer input (thin lower line), and assuming fertilizer applications remained 0.023 kg N per bushel of harvested maize rather than declining to 0.018 kg N per bushel of harvested maize (thin upper line).[13]

to the Gulf of Mexico would have been 33% less than observed (Fig. 3), assuming crop yields were not limited by N shortages.

The Role of Fertilizer Use Efficiency

The efficiency of fertilizer used for maize production in the major maize producing states (Illinois, Iowa, Indiana, Minnesota, and Nebraska) in the Mississippi River Basin increased between 1986 and 2000. Maize yields have increased about 20% from 1986 to 2000, while N fertilizer use has remained roughly constant. Between 1976 and 1986, an average of 0.023 kg N of fertilizer was applied for each kg of maize harvested. Between 1996 and 2000, an average of 0.018 kg N of fertilizer was applied per kg of maize harvested. If this improvement in fertilizer use efficiency had not occurred, nitrate flux in the Mississippi River in 1996–1998 would have been about 50% greater than the measured flux, according to the model of McIsaac et al.[13] (Fig. 3).

Farmers face two major uncertainties when making fertilizer application decisions: they do not know what their yields will be nor how weather conditions might influence the availability of N fertilizer to the crop. The cost of nitrogen fertilizer has been relatively low in relation to the value of the increased yields, and consequently many farmers have believed that applying more N fertilizer than necessary provides "cheap insurance" against the uncertainties. In some instances, farmers did not consider N available from animal manure or from previously harvested legume crops like soybeans.

A number of factors are likely responsible for the increased fertilizer use efficiency. Research and outreach efforts have provided farmers with better information for making N fertilizer decisions. Water quality concerns have focused attention on the need for improved nutrient management. Weather during the 1990s was generally more favorable for corn production than the 1980s, when three major droughts occurred in the corn growing region of the Mississippi River Basin.

ADDITIONAL NEEDS AND APPROACHES FOR REDUCING NITROGEN TRANSPORT

A recent improvement in the efficiency of N fertilizer use has also been observed in wheat production in the United Kingdom and rice production in Japan.[3] However, improved fertilizer use efficiency alone may not be sufficient to address water quality problems in some settings. Jaynes et al.[8] reported that even with N fertilizer rates at recommended levels, nitrate concentration in tile drainage water sometimes exceeded the drinking water standard of 10 mg N L$^-$ in Iowa.

Zucker and Brown[9] recommended several additional practices that can reduce nitrogen contributions in tile drainage: water table management, treatment of drainage water in wetlands, and use of crop rotations that reduce N losses. Additional monitoring and documenting the changes in water quality associated with changing fertilizer management practices are needed to improve our understanding of the connections between N fertilizer use and water quality in different geographic settings.

CONCLUSIONS

In many settings nitrogen enrichment of surface water bodies has increased following the increased use of N fertilizers. The precise contribution of nitrogen fertilizers to surface water nitrogen has been difficult to quantify because there are multiple sources of nitrogen contributing to most water bodies, and, depending on environmental conditions, a certain portion of soil nitrogen may be converted to gaseous or immobile forms. In general, however, agricultural regions with extensive artificial subsurface drainage systems or with sandy soils tend to have the most nitrogen enriched surface waters.

The efficiency of nitrogen fertilizer used for crop production increased in many areas in the 1990s and this has very likely limited or reduced the subsequent contamination of surface waters. Continued improvements in fertilizer use efficiency, and the use of wetlands for removing nitrogen from surface waters will help alleviate problems caused by nitrogen enrichment. Additional monitoring and research are needed to more precisely quantify how nitrogen management practices influence surface water nitrogen concentrations in different settings. A more precise understanding of the causal relationships between nitrogen inputs to the land and the contamination of surface waters could provide more effective guidance for management, policies, and programs intended to protect aquatic resources while maintaining optimal use of land resources.

REFERENCES

1. Skipton, S.; Hay, D. *Drinking Water: Nitrate and Methemoglobinemia ("Blue Baby" Syndrome)*; Publication G98-1369; University of Nebraska Cooperative Extension: Lincoln, NE, 1998, http://www.ianr.unl.edu/pubs/water/g1369.htm (accessed Feb 2002).
2. National Research Council. *Clean Coastal Waters: Understanding and Reducing the Effects of Nutrient Pollution*; National Academy Press: Washington, D.C., USA, 2000.

3. Smil, V. *Enriching the Earth: Fritz Haber, Carl Bosch and the Transformation of World Food Production*; MIT Press: Cambridge, MA, USA, 2001.

4. Romkens, M.J.M.; Nelson, D.W.; Mannering, J.V. Nitrogen and Phosphorus Composition of Surface Runoff as Affected by Tillage Method. J. Environ. Qual. **1973**, *2* (2), 292–295.

5. Steinheimer, T.R.; Scoggin, D.K.; Kramer, L.A. Agricultural Chemical Movement Through a Field-Size Watershed in Iowa: Subsurface Hydrology and Distribution of Nitrate in Groundwater. Environ. Sci. Technol. **1998**, *32*, 1039–1047.

6. Badiger, S. Integrated Numerical Modeling of Spatial and Time-Variant Hydrologic Response in Subsurface Drained Watersheds. Ph.D. Thesis; Department of Agricultural Engineering, University of Illinois: Urbana, 2001.

7. Stevenson, F.J., (Ed.) *Agricultural Drainage*; American Society of Agronomy (Monograph # 38): Madison, WI, USA, 1999.

8. Jaynes, D.B.; Colvin, T.S.; Karlen, D.L.; Cambardella, C.A.; Meek, D.W. Nitrate Loss in Subsurface Drainage as Affected by Nitrogen Fertilizer Rate. J. Environ. Qual. **2001**, *30*, 1305–1314.

9. Zucker, L.A.; Brown, L.C. Agricultural Drainage: Water Quality Impacts and Subsurface Drainage Studies in the Midwest; Ohio State University Extension Bulletin 871; 1998, http://ohioline.osu.edu/b871/ (accessed Feb 2002).

10. Howarth, R.W.; Billen, G.; Swaney, D.; Townsend, A.; Jaworski, N.; Lajtha, K.; Downing, J.A.; Elmgren, R.; Caraco, N.; Jordan, T.; Berendse, F.; Freney, J.; Kudeyarov, V.; Murdoch, P.; Zhao-Liang, Z. Regional Nitrogen Budgets and Riverine N&P Fluxes for the Drainages to the North Atlantic Ocean: Natural and Human Influences. Biogeochemistry **1996**, *35*, 75–139.

11. Stevenson, F.J., (Ed.) Origin and Distribution of Nitrogen in the Soil. *Nitrogen in Agricultural Soils*; Agronomy Monograph 22, American Society of Agronomy: Madison, WI, 1982; 1–42.

12. Caraco, N.F.; Cole, J.J. Human Impact on Nitrate Export: An Analysis using Major World Rivers. Ambio **1999**, *28*, 167–170.

13. McIsaac, G.F.; David, M.B.; Gertner, G.Z.; Goolsby, D.A. Nitrate Flux in the Mississippi River. Nature **2001**, *414*, 166–167.

Surface Water Pollution by Surface Mines

Jeffrey G. Skousen
West Virginia University, Morgantown, West Virginia, U.S.A.

George F. Vance
University of Wyoming, Laramie, Wyoming, U.S.A.

INTRODUCTION

The impacts of surface mining on stream quality result directly from the land disturbance activity. Unweathered earth materials brought to the surface during mining undergo rapid alterations due to exposure to air and water, thereby releasing many of their structural constituents into water.[1] When disturbed rock and soil is exposed to precipitation (e.g., rainfall, snow, hail, dew, etc.), water running off these materials carries solid particles (also known as sediments) as well as dissolved constituents such as salts, metals, trace elements, and/or organic compounds that can pollute nearby surface waters. Water may also percolate into the disturbed materials causing movement and leaching of salts, metals, and trace elements into deeper levels causing potential groundwater quality impacts.[2] The chemistry of the water is highly dependent on the overburden or earthy materials that were disturbed during the mining process.

Surface mining activities can result in disturbed lands with poor drainage unless this problem is controlled, minimized, and even eliminated by reclaiming the areas.[2,3] Reclamation of disturbed sites usually involves grading the areas to achieve a land surface that is stable and compatible with surrounding undisturbed areas, possibly replacing topsoil on the regraded surface and seeding with plants capable of controlling erosion and runoff, and to provide forage for both indigenous wildlife and/or domestic livestock.[3] The Surface Mining Control and Reclamation Act (SMCRA) of 1977 specifies policies and practices for reclaiming areas after surface mining to minimize water quality impacts and to encourage the development of stable, diverse plant communities after mining.[4]

The Clean Water Act (CWA) of 1977 and previous water control legislation [Federal Water Pollution Control Act of 1972] require restoring and maintaining the chemical, physical, and biological integrity of our nation's water.[5,6] The intention of these laws was to establish a framework for permitting and regulating all point discharges into surface waters, with the laws particularly targeting the discharge of sewage and wastewater from communities into streams, rivers, and lakes. The CWA was designed to place limits or standards on water being discharged into the waters of the United States, but also to maintain drinking water and recreational uses of water, and to restore the quality of streams and lakes that had been degraded.[7] The law has been interpreted as requiring all waters to be "fishable and swimmable."[8]

Water discharged from surface mines is regulated by the CWA,[5] and all mines are required to only discharge water that meets CWA effluent standards. Therefore, all water that comes from a permitted mine (whether the water was received as rainfall, snow, hail, etc. at the surface or from underground seepage) must pass through a sedimentation or treatment pond and meet or exceed discharge standards before it can be released into receiving surface waters.[8]

Nationwide, over 20,000 km of rivers and streams and over 75,000 ha of lakes and reservoirs are adversely affected by contaminated water draining from abandoned mines.[6] The vast majority of these problem areas occur in the eastern United States where coal mine drainage is considered by the United States Environmental Protection Agency (U.S. EPA) to be the most significant nonpoint pollution problem. Although Wyoming is currently the leading coal producing state in the country (approximately, one-third of our nation's coal is mined in this state), Wyoming and other western United States are plagued with historic mining activities involving metal ores, such as copper (Cu), lead (Pb), zinc (Zn), and silver (Ag), with the trace elements molybdenum (Mo) and uranium (U) also mined in certain regions. In addition to surface water impacts from coal, metal, and trace element mining that can generate acid mine drainage (AMD) from oxidation of pyritic ores (e.g., iron sulfide FeS_2), other pollutants are also of concern including metals [aluminum (Al), antimony (Sb), cadmium (Cd), cobalt (Co), chromium (Cr), Cu, iron (Fe), manganese (Mn), nickel (Ni), Pb, and Zn], trace elements [arsenic (As), Mo, and selenium (Se)], radioactive elements [cesium (Cs), radium (Ra), thorium (Th), U, and vanadium (V)], and mining operation by-products [mercury (Hg) and cyanide (CN)].

Many areas in the United States and other parts of the world were disturbed prior to the enactment of any laws regulating their drainage quality and water release into streams.[7] These disturbed areas may contribute

Encyclopedia of Water Science
DOI: 10.1081/E-EWS 120010061

Table 1 Examples of surface water quality in different areas throughout the United States that have been impacted by mining activities

Location	Flow (L/min)	pH	Cond (dS/m)	Acid (mg/L as CaCO₃)	Alkalinity (mg/L as CaCO₃)	SO₄ (mg/L)	Ca (mg/L)	Mg (mg/L)	Na (mg/L)	Al (mg/L)	Cu (mg/L)	Fe (mg/L)	Mn (mg/L)	Zn (mg/L)
Maryland	640	2.7	4950	3470	bd	3700	320	55	bd	198	na	640	10	na
Montana1	68	2.7	5970	5150	0	6000	240	110	13	325	500	450	na	5
Montana2	22	4.6	1900	815	170	1300	220	86	26	190	3	1	13	33
Nevada	15	2.2	5100	2795	bd	3670	502	382	95	152	na	595	80	18
Ohio	900	6.5	1790	134	88	985	168	35	bd	bd	na	89	2	na
Pennsylvania1	85	4.0	2340	208	bd	1070	224	70	bd	12	na	70	13	na
Pennsylvania2	38	4.8	3140	211	7	2040	325	57	323	1	na	121	2	na
West Virginia1	8	3.3	4230	920	bd	2525	232	228	bd	83	na	132	48	na
West Virginia2	136	3.6	946	516	bd	640	78	23	bd	41	na	7	20	na
Wyoming1	170	6.8	< 100	bd	27	19	10	35	3	bd	5	bd	< 1	< 1
Wyoming2	680	8.3	1640	bd	282	836	145	64	158	23	bd	bd	< 1	na

bd = below detection limit; na = not analyzed.

significant amounts of pollutants to surface waters because they often are devoid of minimum vegetative cover and because their soil properties limit natural reclamation of the site.[3] Pre-1977 mining activities were also considered in SMCRA legislation.[4] Including provisions for reclaiming "abandoned" mined lands, which are surface-mining disturbances that occurred prior to enactment of the law and where no individual or company is held responsible for the damaged land. Drainage from these surface-mining operations has had and continues to have a dramatic effect on surface water quality because these "abandoned" pre-1977 sites discharge acid mine drainage into surface water bodies such as rivers, streams, creeks, and impoundments. Money generated by the "abandoned mine land reclamation fund" since 1977 goes to reclaiming abandoned areas, which aids in the improvement of water quality from abandoned mine sites (Table 1).

CLASSIFICATION OF SURFACE WATER POLLUTANTS

At surface coal mines, drainage waters generally reflect the chemistry of the rock layers disturbed during the mining process. For example, if the overburden material chemistry is dominated by calcareous shales or limestone, water draining from these materials will generally have a pH value above 6.0, low concentrations of dissolved metals and possibly trace elements, potentially high amounts of bases or salts [such as calcium (Ca), magnesium (Mg), and sodium (Na)], and high alkalinity. If, on the other hand, the overburden materials comprised sandstone with high sulfur coal or ores containing pyrite (such as those associated with hard rock mining of Cu, Fe, Pb, Ni, Ag, or Zn), the drainage water quality may have

a pH value less than 3.5, and high concentrations of dissolved metals such as Fe and Al, and high sulfate (SO_4^{2-}). Some surface mines are dominated by neither acid nor alkaline strata, and the impact of disturbing these rock materials on water quality is not significant.

The primary water quality impacts from surface mining can be classified into physical, chemical, or biological categories. Physical impacts are color, which relate to dissolved and suspended constituents. Chemical impacts of water draining from surface mines can vary from acid water laden with metals and trace elements to alkaline water with excess Ca, Na, and carbonates (e.g., HCO_3^-, CO_3^{2-}). Biological impacts relate to sanitary chemistry, where microorganisms may contaminate the water.

Physical Impacts

The most noticeable, dramatic physical impact to water is color.[1] Bentonite (a type of clay material mined predominately in Wyoming) surface mining produces a distinct greenish tint to water that has accumulated in open pits. Orange water results in many areas where acid-generating, pyritic materials are found, such as in coal and hard rock metal mining due to iron coating of rocks and sediments. White turbid waters are indicative of high levels of Al, which is generally related to acidic water conditions and disturbed geological materials high in aluminosilicates. Water carrying high loads of sediment, which is common during storm events, appears murky, cloudy, and turbid.

Chemical Impacts

Chemical impacts can vary from acidic and metal-laden waters to highly alkaline waters containing excess

Na.[1,2] As mentioned earlier, acid water conditions result where rocks containing pyrite are exposed to the atmosphere with a release of Fe, hydrogen (H^+), and SO_4^{2-}. The low pH conditions of these waters tend to dissolve other nearby rocks releasing more Fe and other elements into the water such as Al, Mn, silicon (Si), and base cations such as potassium (K), Ca, and Mg (Fig. 1).

Excess alkalinity in water is generally a much less significant problem. Water containing high levels of bicarbonate (HCO_3^-) usually begin precipitating calcite ($CaCO_3$) if the water pH value is above 8.3. There have been a few examples where acid-containing materials have been added to high pH water to reduce the pH for discharge into surface waters.

If the water contains excess amounts of Na, usually the water is collected in ponds or reservoirs and evaporated, thereby leaving the salts in these closed basins.[10] Methane exploration from coal deposits in the Powder River Basin in northeastern Wyoming, however, has resulted in tremendous amounts of product waters being brought to the land surface, with some of these waters directly discharged into nearby streams and channels. Because of their potentially high salt contents, and in some case high Na concentrations, negative impacts to the surrounding ecosystems include soil and sediment dispersion, vegetation die-off, and potential aquatic organism mortality.[2]

Some waters from surface mines contain organic compounds.[1,10] These are often a result of contamination from gas tanks, oil spills, or run off from equipment-servicing areas. In these cases, the source of the contamination must be identified and removed. Gasoline or oil-soaked soil can be excavated, aerated, and fertilized so that microbes inherent in the soil will have sufficient nutrients and oxygen to decompose the organic matter.[2]

Biological Impacts

Biological contamination is not generally associated with surface mines, although some contamination could occur if water used in bathhouses and restroom facilities is not properly treated. The most common biological impacts to surface water are associated with the discharge of water from individual households where no septic system is installed or from municipal wastewater effluents.[2] Wastewater from municipal treatment plants or from untreated households can contain bacteria, viruses, and other microorganisms. Fecal coliform bacteria are routinely used to indicate the level of microorganism contamination from water impacted by human waste.[11] If high levels of fecal coliform bacteria are identified in water, the source must be located and the water must be directed to a wastewater treatment plant or be introduced into the soil via a septic tank/soil absorption field of adequate design.

Fig. 1 Acidic and iron laden water flowing from a small underground coal mine into a natural stream in West Virginia. The iron dissolved in the water coats the streambed downstream and makes the water unsuitable for use.

WATER TREATMENT

If the water to be released from a mining operation does not meet effluent limitations established by the CWA, surface mine operators are obligated to control or treat the water to meet effluent standards.[8] These treatments include routing the water through sedimentation ponds to allow settling of solids, the addition of base chemicals [CaO, Ca(OH)$_2$, NaOH, etc.] to raise pH and cause the precipitation of dissolved metals and trace elements, transferring the water through microbial chambers to remove organic matter, chlorination, and filtering.

For acid mine drainage, the acid-generating reactions will continue until the pyrite is exhausted, until the pyrite becomes coated with iron hydroxides [e.g., Fe(OH)$_2$, FeO(OH), and FeO] or until the water cannot leach the acid products away.[9] Control practices to reduce the amount of pyrite oxidation employ the use of barriers to restrict water flow through the material, the addition of alkaline materials to neutralize the acid or stop the acid-generating reaction, and flooding or compacting the material to reduce oxygen influx to the material.[1] If acid water results, then a treatment plan must be established and the water must be treated by base chemicals to neutralize the water and precipitate the metals before release into streams (Fig. 2).

REGULATORY ENFORCEMENT

Current surface mining operations must comply with CWA and SMCRA standards.[8] SMCRA established the "abandoned mine land reclamation fund" that generates money from current coal operations ($0.35 per ton of surface mined coal and $0.15 per ton of coal mined underground), which is used to reclaim abandoned lands as deemed necessary by the Office of Surface Mining and Enforcement (e.g., OSM). Due to the liabilities and financial penalties, surface mining operators have strong incentives for compliance with SMCRA regulations. Enforcement of current regulations and standards by OSM and state governing agencies will continue to

Fig. 2 Two types of drainage are shown here from an abandoned mine site north of Yellowstone National Park, Wyoming. The white-colored water on the right is laden with aluminum and is derived from waste rock, while the stream on the left contains high iron and is the result of acid mine drainage from an abandoned metal mine shaft.

minimize the impacts of surface mining on surface water quality. Operators of surface mines also recognize that an environmental stewardship policy and the implementation of practices to reduce pollution of water on and near their sites will ultimately reduce the costs and liabilities associated with surface mining.

REFERENCES

1. Sobek, A.A.; Skousen, J.G.; Fisher, S.E., Jr. Chemical and Physical Properties of Overburdens and Minesoils. *Reclamation of Drastically Disturbed Lands*, 2nd Ed.; Soil Science Society of America, Inc.: Madison, Wisconsin, 2000; 77–104.

2. Pierzynski, G.M.; Sims, J.T.; Vance, G.F. *Soils and Environmental Quality*; CRC Press, Inc.: Boca Raton, FL, 2000; 459 pp.

3. Munshower, F.F. *Practical Handbook of Disturbed Land Revegetation*; Lewis Publishers, CRC Press: Boca Raton, FL, 1994; 265 pp.

4. U.S. Congress. Surface Mining Control and Reclamation Act; 1977; Public Law 95-87.

5. U.S. Congress. Clean Water Act; 1977; Public Law 95-217.

6. Skousen, J.G.; Sexstone, A.; Ziemkiewicz, P.F. Acid Mine Drainage Control and Treatment. In *Reclamation of Drastically Disturbed Lands*, 2nd Ed.; Barnhisel, R.I., Darmondy, R.G., Daniels, W.L., Eds.; Soil Science Society for America, Inc.: Madison, WI, 2000; 131–168.

7. Plass, W.T. History of Surface Mining Reclamation and Associated Legislation. In *Reclamation of Drastically Disturbed Lands*, 2nd Ed.; Barnhisel, R.I., Darmondy, R.G., Daniels, W.L., Eds.; Soil Science Society for America, Inc.: Madison, WI, 2000; 1–20.

8. Zipper, C. Coal Mine Reclamation, Acid Mine Drainage, and the Clean Water Act. In *Reclamation of Drastically Disturbed Lands*, 2nd Ed.; Barnhisel, R.I., Darmondy, R.G., Daniels, W.L., Eds.; Soil Science Society for America, Inc.: Madison, WI, 2000; 169–192.

9. Kleinmann, R.L.P., Ed. *Prediction of Water Quality at Surface Coal Mines*; National Mine Land Reclamation Center: West Virginia University, Morgantown, WV, 2001; 239 pp.

10. Williams, R.D., Schumann, G.E., Eds. *Reclaiming Mine Soils and Overburden in the Western United States: Analytic Parameters and Procedures*; Soil Conservation Society of America: Ankeny, IA, 1987; 336 pp.

11. Clesceri, L.S., Greenberg, A.E., Eaton, A.D., Eds. *Standard Methods for the Examination of Water and Wastewater*, 20th Ed.; American Public Health Association: Washington, DC, 1998; 1220 pp.

Surface Water Quality and Phosphorus Applications

Richard McDowell
*Ag Research Ltd., Invermay Agricultural Centre, Mosglel,
New Zealand*

Andrew Sharpley
Peter Kleinman
*United States Department of Agriculture (USDA), University Park,
Pennsylvania, U.S.A.*

INTRODUCTION

Eutrophication is a major water quality concern in the United States[1,2] and worldwide.[3] Its economic impact on the fishing and water-treatment industries in the eastern United States alone, has amounted to over $2 billion over the last decade.[4] While phosphorus (P) and nitrogen (N) contribute to eutrophication, P is the limiting nutrient in most fresh waters. This is due to the fact that P is ultimately derived from land, where as N can exchange freely between the atmosphere and surface water and many aquatic biota can fix N. Although eutrophication is a natural process, it is accelerated by increased inputs of P by humans. This can have several detrimental effects on surface-water quality. Perhaps the most obvious is the proliferation of harmful algal bloom, parasites (e.g., *Pfiesteria* and *cyanobacteria*) and aquatic weeds, which can interfere with the use of water for recreation, extraction, and drinking (foul taste and odor and treatment problems such as the formation of carcinogens during chlorination). As aquatic biota die and decompose, the increased microbial activity depletes oxygen supply and increases fish mortality.

Over the last 30 yr–40 yr, attention has been centered on agriculture as the primary origin of P loss to surface waters. This is due, in part, to the general ease of identification and mitigation of point sources of P loss. In addition, the intensification and specialization of farming systems has led to regional surpluses of P imported in fertilizer and animal feed compared with P exported in farm produce.[5] Now, many farms possess soil–P concentrations well in excess of plant needs and therefore an increased potential for P loss.[6]

MECHANISMS OF AGRICULTURAL P LOSS TO SURFACE WATERS

The loss of P from agricultural lands to surface waters is largely controlled by the coincidence of areas of high P availability (source factors) with the physical transport of P within hydrological pathways such as overland and subsurface flow (transport factors). High P availability is determined by the management of soils (and its physiochemical characteristics), crops, manures, and fertilizers. Where the source and transport factors coincide, we have "critical source areas" for P loss. These areas are usually small yet well defined (<20% of land area) but can contribute most of the P exported from a watershed (>90%).[7]

Several surveys of U.S. watersheds have shown P loss in runoff, increases as the portion of the watershed under forest decreases and agriculture increases.[8,9] Overland flow from forests, grasslands, and other noncultivated soils carries little sediment, so P losses are low and generally dominated by dissolved P, which is immediately algal-available[10,11] (Fig. 1). The cultivation of agricultural land greatly increases erosion, and with it, the loss of particle-bound P. Typically, particulate losses constitute 60%–90% of P exported from most cultivated land.[12] Some of the particle-bound P is not readily available, but much of it can be a long-term source of P for aquatic biota.[11,13]

Release of Phosphorus from Soil

In acidic soils, P occurs largely as Al- and Fe-phosphates, whereas in neutral to alkaline soils P occurs more so as Ca- and Mg-phosphates and sorbed onto the surface of Ca- and Mg-carbonates. Organic P can form a significant part of soil P especially in acidic soils and soils that contain more organic matter and N. The solubility of soil P is controlled by three chemical characteristics: i) concentration of P in solution; ii) quantity of P in the soil that equilibrates with the solution; and iii) buffering capacity of the soil (controlled by sorption strength and the saturation of sorption sites with P). For P loss, these components can be described by a quantity–intensity relationship such as plots of soil test P (i.e., agronomic tests such as Bray, Mehlich or Olsen)

Encyclopedia of Water Science
DOI: 10.1081/E-EWS 120010197

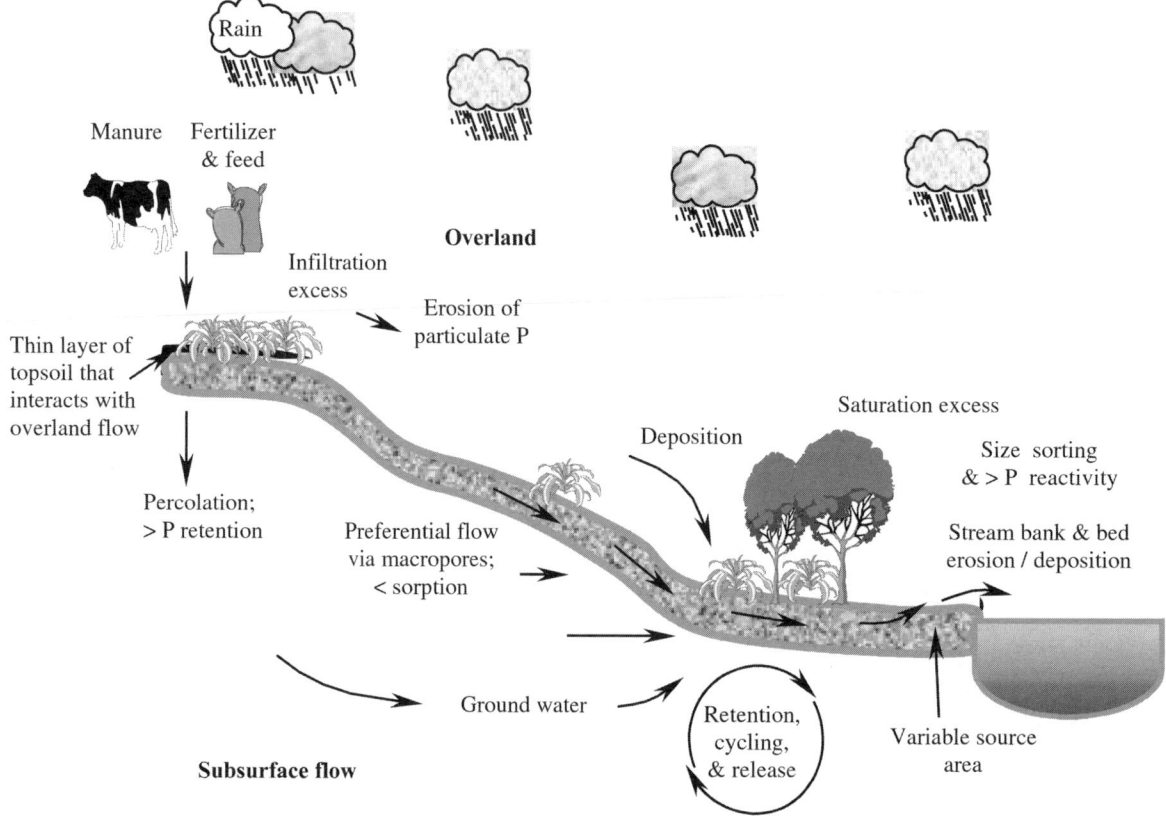

Fig. 1   Processes and loss of P from land to fresh waters.

against either P loss in overland flow, subsurface flow or 'gentle' soil extracts that approximate P loss (e.g., 0.01 M $CaCl_2$).[14] The relationship between soil P and overlend flow P can be split in two, on either side of an environmental soil P threshold, where soils greater than the threshold have a much greater P loss potential.

Coupled with soil P solubility is the kinetics of release. Kinetic exchange experiments using ^{33}P, have confirmed that soil P exchangeable within 60 sec is closely related to P in overland or subsurface flow. With time, P transport in overland flow becomes less related to this pool and more dependent upon the slow diffusion of P from the inside of the soil aggregate.[15,16] This serves to illustrate that soil P release to overland flow is a function of the surface area available to the solution, as well as the quantity of P in soil. For example, Holford and Mattingly[17] showed that in a selection of near neutral pH soils, P release and sorption was correlated to $CaCO_3$ surface area, but not total $CaCO_3$ concentration.

Transport and Loss of Phosphorus

Using these simple chemical principles we can describe the release of P, however, the physical transport of P determines whether these processes are translated into actual losses within a watershed and beyond. Rainfall is the primary driving force behind P transfer. Rainfall events can be divided into two types: one which describes rainfall of low intensity and high frequency that tends to move P in subsurface flow, and a second that describes rainfall of high intensity and low frequency that tends to move P in overland flow from a thin layer of P-rich topsoil (Fig. 1). Due to the greater kinetic energy and erosive power of high frequency storms, more P and total quantities of P are lost during overland flow in particulate forms than in subsurface flow. For example, Pionke et al.[7] showed that a few short, intense storms accounted for about 90% of the annual P export from an upland watershed. Overland flow can be further divided into Hortonian (limited by infiltration rate) overland flow and saturation excess (limited by soil water storage capacity) overland flow. Infiltration-rate limited overland flow will have a greater capacity to detach and move soil particles, however, this pathway is largely restricted to high-intensity, extreme rainfall events.

In humid and temperate climates, saturation excess overland flow can be described by variable source area (VSA) hydrology.[18] Flow from these areas varies rapidly in time and space, expanding and contracting rapidly during a storm as a function of precipitation, temperature, soil-type, topography, ground water, and moisture status over the watershed. The onset of flow from these areas is limited by soil water storage capacity and thus, usually results from high water tables or soil moisture contents in near-stream areas. During a rainfall event, area boundaries will migrate upslope as rainwater input increases. In dry summer months, overland flow will come from areas closer to the stream than during wetter winter months, when the boundaries expand away from the stream channel. In watersheds where infiltration excess overland flow dominates, and areas of the watershed can alternate between sources and sinks of overland flow, again as a function of soil properties, rainfall intensity and duration and antecedent moisture condition. Thus, consideration of hydrologic controls and variable source areas is critical to understanding P loss.

Combining Soil Chemistry and Hydrology

Transport and loss of P generally occurs from areas where overland flow contributes to stream flow, although some subsurface flow pathways may be important under certain hydrologic conditions. However, even in watersheds where subsurface flow pathways dominate, areas contributing P to drainage waters can be localized (e.g., Ref. 19). Loss of P in subsurface flow is generally less than that in overland flow, and will decrease as the degree of soil–water contact increases, due to sorption by P-deficient subsoils. Exceptions occur where organic matter may accelerate P loss together with Al and Fe, or where the soil has a small P sorption capacity (e.g., some sandy soils) or where subsurface flow travels from P-rich topsoil in/via macropores or is intercepted by drainage (Fig. 1).

The hydrologic and chemical factors controlling P loss vary temporally and spatially. Increased net precipitation (precipitation–evapotranspiration) to a watershed increases the amount of discharge and the quantity of P lost by accelerating those transformations that occur before and after P reaches stream flow. For example, whereas dissolved forms of P are immediately available to aquatic flora, particulate forms of P can represent a more long-term source of P via desorption. During overland flow, soil and associated P is lost in order of decreasing particle density and increasing weight. Thus, fine and/or light soil particles that contain many Al- and Fe-oxide-associated P or organic associated P are transported before coarser and/or heavier sized particles (Fig. 1). Eroded fines will be able to maintain a greater equilibrium stream or P concentration for longer than coarser particles with less P in reserve. However, coarser particles have a lesser affinity for P and will release it faster initially.

MANAGEMENT TO DECREASE P LOSS

Source factors regulate the chemistry of released P to transport mechanisms. The most important factors influencing the concentration and solubility of P in soil include soil type and P inputs as fertilizer and manure. Effective management ultimately aims to balance P inputs with off-takes as produce, at the farm gate. However, in areas of concentrated animal production, sufficient land may not be available for manure disposal leading to an increase in soil P concentration.

Efficient management of P sources involves placing P away from critical source areas likely to loose more P such as hydrologically active zones in a watershed and soils already high in P. Cultivation immediately after application can decrease P losses if erosion is minimized. Periodic tillage of the soil may also decrease P loss by redistributing high-P topsoil throughout the root zone. Applications of manure or fertilizer during drier periods avoiding precipitation or snowmelt will further decrease the potential for P loss in overland flow by increasing the contact time (and uptake) with the soil and crop.

The presence of crop covers and crop residues help decrease P loss by decreasing erosion and overland flow. Equally, anything that keeps surface roughness high or intercepts overland flow, which encourages rainwater infiltration and sediment retention, can be effective. Such measures include riparian zones, buffer strips, terracing, cover crops, contour tillage, and impoundments or small reservoirs. However, these measures are better at stopping particulate than dissolved P transport.

Other remedial measures include manure and soil treatment and amendment to decrease P solubility and potential release to runoff; feeding animals no more P than they actually need; use of soil testing to guide future P application (particularly as manure); identifying critical areas or "hot spots" for P loss to which conservation measures should be targeted; and redistribution of manure within and among farms. These are mostly short-term or "stop-gap" measures to decrease P loss. Long-term solutions will involve balancing P inputs with outputs at farm, watershed, or regional scales.

REFERENCES

1. U.S. Environmental Protection Agency. *Environmental indicators of water quality in the United States*; EPA 841-R-96-002, U.S. Govt. Printing Office: Washington, DC, 1996.
2. U.S. Geological Survey. *The Quality of Our Nation's Waters: Nutrients and Pesticides*; U.S. Geological Survey Circular 1225, USGS Information Services: Denver, CO, 1999; 82 p., http://www.nsgs.gov.

3. National Research Council. *Clean Coastal Waters: Understanding and Reducing the Effects of Nutrient Pollution*; National Academy Press: Washington, DC, 2000.

4. Greer, J. In Harm's Way? The Threat of Toxic Algae. Marine Notes, July–August 1997 (http//www.mdsg.umd.edu/MDSG/Communications?MarineNotes/index.html) 1997; 1–4.

5. Carpenter, S.R.; Caraco, N.F.; Correll, D.L.; Howarth, R.H.; Sharpley, A.N.; Smith, V.H. Nonpoint Pollution of Surface Waters with Phosphorus and Nitrogen. Ecol. Appl. **1998**, *30*, 559–568.

6. Sharpley, A.N., Ed. *Agriculture and Phosphorus Management: The Chesapeake Bay*; CRC Press: Boca Raton, FL, 2000.

7. Pionke, H.B.; Gburek, W.J.; Sharpley, A.N. Critical Source Area Controls on Water Quality in an Agricultural Watershed Located in the Chesapeake Basin. Ecol. Eng. **2000**, *14*, 325–335.

8. Omernik, J.M. *Nonpoint Source–Stream Nutrient Level Relationships: A Nationwide Study*; EPA-600/3-77-105, U.S. EPA: Corvallis, OR, 1977.

9. Rast, W.; Lee, G.F. *Summary Analysis of the North American (U.S. Portion) OECD Eutrophication Project: Nutrient Loading–Lake Response Relationships and Trophic State Indices*; EPA 600/3-78-008, U.S. EPA: Corvallis, OR, 1978.

10. Ryden, J.C.; Syers, J.K.; Harris, R.F. Phosphorus in Runoff and Streams. Adv. Agron. **1973**, *25*, 1–45.

11. Sharpley, A.N. Assessing Phosphorus Bioavailability in Agricultural Soils and Runoff. Fertil. Res. **1993**, *36*, 259–272.

12. Sharpley, A.N.; Hedley, M.J.; Sibbesen, E.; Hillbricht-Ilkowska, A.; House, W.A.; Ryszkowski, L. Phosphorus Transfer from Terrestrial to Aquatic Ecosystems. In *Phosphorus Cycling in Terrestrial and Aquatic Ecosystems*; Tiessen, H., Ed.; United Nations SCOPE, Scientific Advisory Committee on Phosphorus Cycling; John Wiley and Sons: London, UK, 1995; 171–199.

13. Ekholm, P. Bioavailability of Phosphorus in Agriculturally Loaded Rivers in Southern Finland. Hydrobiologia **1994**, *287*, 179–194.

14. McDowell, R.W.; Sharpley, A.N. Approximating Phosphorus Release from Soils to Surface Runoff and Subsurface Drainage. J. Environ. Qual. **2001**, *30*, 508–520.

15. McDowell, R.W.; Sinaj, S.; Sharpley, A.N.; Frossard, E. The Use of Isotopic Exchange Kinetics to Determine Phosphorus Availability in Overland Flow and Subsurface Drainage Waters. Soil Sci. **2001**, *166*, 365–373.

16. Sharpley, A.N.; Ahuja, L.R. A Diffusion Interpretation of Soil Phosphorus Desorption. Soil Sci. **1983**, *135*, 322–326.

17. Holford, I.C.R.; Mattingly, G.E.G. The High- and Low-Energy Phosphate Adsorbing Surface in Calcareous Soils. J. Soil Sci. **1975**, *26*, 407–417.

18. Ward, R.C. On the Response to Precipitation of Headwater Streams in Humid Areas. J. Hydrol. **1984**, *74*, 171–189.

19. Schoumans, O.F.; Breeuwsma, A. The Relation Between Accumulation and Leaching of Phosphorus: Laboratory, Field and Modelling Results. In *Phosphorus Loss from Soil to Water*; Tunney, H., Carton, O.T., Brookes, P.C., Johnston, A.E., Eds.; CAB International Press: Oxon, UK, 1997; 361–363.

Surface Water Quality Protection for Concentrated Animal Feeding Operations

Douglas R. Smith
United States Department of Agriculture (USDA), West Lafayette, Indiana, U.S.A.

Philip A. Moore, Jr.
United States Department of Agriculture (USDA), Fayetteville, Arkansas, U.S.A.

INTRODUCTION

The number of confined animal feeding operations in the United States has increased dramatically over the last half century. Manure from these operations is most often land applied to pastures or cropland. Runoff from land, where manure has been applied, has been implicated in eutrophication of U.S. surface waters. Phosphorus from these nonpoint sources is of great concern in most areas of the United States where the animal industries are concentrated. This entry discusses feed and manure treatment methods that can be used to reduce the potential impact of manure on surface water quality, and other management strategies that producers can use to further reduce these risks.

RECENT TRENDS IN ANIMAL AGRICULTURE

In recent years, the number of U.S. farms has declined, while agricultural production has increased.[1] For instance, swine operations have declined from just under 1.1 million in 1965 to around 86,000 operations in 2000.[2] The number of small swine operations (less than 100 head) has steadily decreased since 1992, while the number of very large operations (greater than 5000 head) has steadily increased from less than 1,000 operations in 1992 to almost 2,100 in 2000. Similar trends occurred for poultry and cattle operations during this period.[1] Animal operations tend to be concentrated in geographic areas also. In 2000, 84% of the U.S. broiler production occurred in 13 states, mainly in the east and southeast.[2]

These trends in animal agriculture have been important to modern agriculture, but they have also been very important to the environment as well. Millions of tons of animal manure are produced annually. Many of the animal manures, such as that from swine and dairy, contain very high moisture content (80% or higher). While the manure contains valuable fertilizer nutrients, transporting the manure outside of the watershed may be cost prohibitive due to the amount of water in the manure. Therefore, the majority of manure is applied to pasture or crop land very near the site of production.

ANIMAL MANURE AS A SOURCE OF NUTRIENTS TO SURFACE WATERS

Declines in surface water quality have been attributed to the recent trends in animal agriculture and the application of animal manure on pastures and cropland. When manure is applied on pasture land, it is generally broadcast on the surface of the soil and often not tilled in. When a rainfall event occurs, particularly within a few days after manure application (Fig. 1), nutrients, pathogens, antibiotics, hormones, and metals can enter surface water through runoff.[3,4] Nutrient runoff, especially phosphorus (P) runoff from fields fertilized with animal manure has received particular attention in recent years.[5] In most surface water reservoirs, P is the primary nutrient that limits algae growth, or eutrophication.[6] This has been the circumstance in the Eucha-Spavinaw watershed in Northwest Arkansas and Northeast Oklahoma, an area of intensive poultry and swine production. Most of the manure from these facilities is applied to forage pastures, and it is generally applied in the spring of the year. In recent years, extensive eutrophication has occurred in Lake Eucha, a drinking water reservoir for Tulsa, Oklahoma, and geosmin has been released into the water supply. Geosmin is a chemical that is released during certain algae blooms. While it is not harmful to human health, geosmin gives drinking water bad taste and odor,

Mention of trade name, proprietary product, or specific equipment does not constitute a guarantee or warranty by the USDA and does not imply its approval to the exclusion of other products that may be suitable.

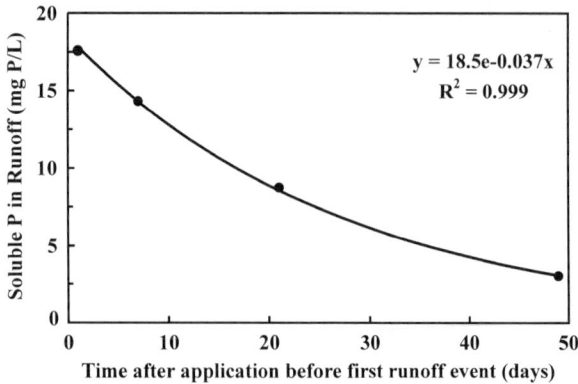

Fig. 1 Effect of time after manure application on soluble P runoff from pastures. (From Ref. 11.)

and is very difficult to remove through conventional water treatment procedures.

Problems associated with P losses from fields fertilized with animal manure have occurred in other areas of the country as well. In North Carolina, recent outbreaks of Pfisteria, a dinoflagellate that releases a toxin that causes open sores on fish, have been blamed on P released from swine farms.[7] Phosphorus releases from this area of the country have been associated not only with manure applications; in 1999, Hurricane Floyd flooded many swine farms, causing lagoons to overflow, and directly enter surface waters.

REDUCING POTENTIAL PHOSPHORUS LOSSES FROM ANIMAL MANURE

There are two major methods to reduce potential P losses from animal manure: reduce P inputs into the animal, or reduce the solubility of P in the manure. Phosphorus in most grains fed to livestock is in the form of inositol hexaphosphate (phytate), a six membered carbon (C) ring with a phosphate group attached to each C.[8] This form of P is not readily absorbed by monogastric animals, such as swine and poultry. Therefore, many feed rations for these species require supplemental forms of P, such as dicalcium phosphate. Reducing the P inputs requires the

use of some technology that improves the P availability in grains. This can be accomplished through modifying the diet with special grains or adding enzymes to break down the phytate molecule. Special grains used in feed refers to varieties of corn that have been developed for their ability to store P in forms other than phytate (Table 1).[9] Such corn is often referred to as high available P (HAP) corn. The future of this product could be very promising in the livestock industry. However, it faces two major hurdles. This variety of corn is not as productive as most common varieties, and there is no simple method to distinguish between HAP corn and other varieties.

Another technology that can be used to reduce the total P inputs in livestock rations is the use of enzymes. Phytase is an enzyme that has received much attention lately, because it is the enzyme that cleaves phosphate groups from the phytate molecule.[8] It can be cultured rather easily using various fungi, such as *Aspergillus* sp., that produce exogenous phytase. This technology has been shown[10] to reduce soluble P in swine manure by 15% compared to swine fed normal diets (Fig. 2). Currently, one of the major drawbacks to this technology, is that many feeds, especially for swine, are pelleted at temperatures high enough to denature the phytase molecule. Another question that some researchers have raised concerning this technology is that manure from phytase fed animals may actually increase soluble P in runoff. A 25% increase in P runoff from plots fertilized with manure from nursery pigs fed phytase diets compared to normal diets has been noted.[10] Increased P runoff from phytase fed animals of more than two fold (Fig. 3) compared to normal diet animals has been seen in other studies.[11] The reasons for these increases are not fully

Table 1 Total and phytate bound phosphorus in normal and HAP corn varieties

Corn type	Total phosphorus (lb tn^{-1})	Phytate phosphorus (lb tn^{-1})
Normal	7.6	6.4
HAP	7.8	2.6

(From Ref. 9.)

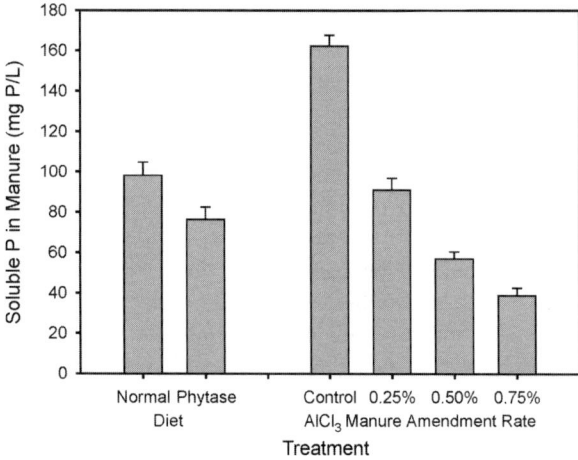

Fig. 2 Effects of phytase amended diets and aluminum chloride manure amendments on soluble phosphorus in swine manure. (From. Ref. 10.)

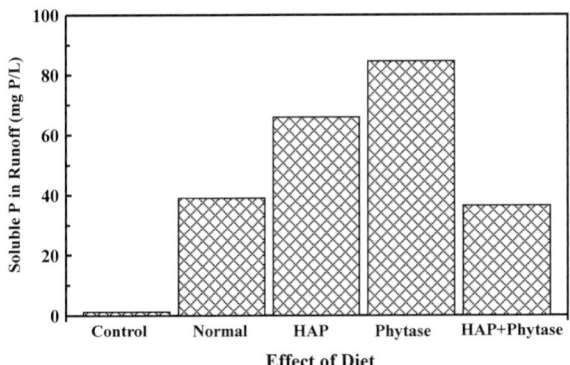

Fig. 3 Effect of poultry diet on soluble P runoff from pasture. (From Ref. 11.)

understood, but studies are currently being undertaken to identify possible explanations (Personal Communication, Philip A. Moore, Jr., April 2002).

The other main treatment that can be used by animal production facilities is the treatment of manure with chemical amendments to reduce P solubility. Calcium, iron, and aluminum amendments have been used to reduce P solubility.[12] Calcium amendments reduced P solubility at high pH, however, at slightly acidic pHs, thermodynamics dictate that Ca-phosphates can dissolve, thereby releasing P. Iron phosphates are more stable over a wide range of pH values, however under anaerobic conditions, ferric iron can be reduced to ferrous iron. Ferrous phosphates generally dissociate more readily than ferric phosphates, thereby posing the risk of releasing P into the environment. Aluminum phosphates are stable over a wide range of physio-chemical environments naturally occurring in soils. In fact, one of the reasons Al-phosphates would dissociate under "normal" field conditions would be very low P status in the soil solution.

Aluminum sulfate (commonly referred to as alum) has been used in poultry litter for several years to reduce P solubility. Phosphorus solubility in poultry litter treated with alum was reduced as much as 99% compared to normal poultry litter.[13] In a study with treated and untreated poultry litter applied to plots cropped to tall fescue, P runoff was 87% lower in plots fertilized with alum treated litter compared to those to which normal litter was applied.[14] In this study, P runoff from plots fertilized with alum treated litter was not statistically higher than plots that were unfertilized. These two studies indicate tremendous potential for this technology to reduce the pollution potential from the poultry industry. Smith et al., demonstrated that alum could also effectively reduce P solubility in swine manure.[15] Concern over possible sulfide production from the sulfate in alum however necessitated the testing of another Al chemical in liquid manure to accomplish this goal. Aluminum chloride was

also used in this study to reduce P solubility. Both chemicals reduced P solubility by as much as 99% and reduced P runoff from plots fertilized with treated manure by 84% compared to plots treated with normal manure.

In addition to reducing the potential impacts of P runoff, these treatments provide the added benefit of reducing ammonia volatilization from manure. Alum has been shown to reduce ammonia volatilization as much as 99% in broiler houses. Swine manure treated with aluminum chloride had 50% less ammonia loss through volatilization compared to normal manure.[10] This decrease in ammonia improves the air quality in the production facility and can improve animal performance as well as reduce costs associated with heating the production facility.[13]

RISK BASED MODELS FOR MANURE APPLICATION

Diminished water quality in watersheds with intensive animal agriculture has caused many states to scrutinize production practices and manure application. Most states have adopted risk based models to aid producers in their manure management.[16] The Arkansas P index for pastures identifies several risk factors, including soil test P levels, amount of P in manure, the slope and infiltration rate of the pasture, timing of manure application, and annual precipitation to asses the risk of P losses from pastures after manure application.[11] Farmers are also given credit for best management practices. These factors are plugged into a matrix that then assesses the risk of P loss from a specific pasture. This risk level then aids the producer in determining whether to apply manure at normal agronomic rates for N, agronomic rates for P, or not apply manure at all. Validation of this model[11] showed a strong correlation between the P index value obtained from the matrix and P lost from pastures (Fig. 4). Several other states have worked on similar tools, and have

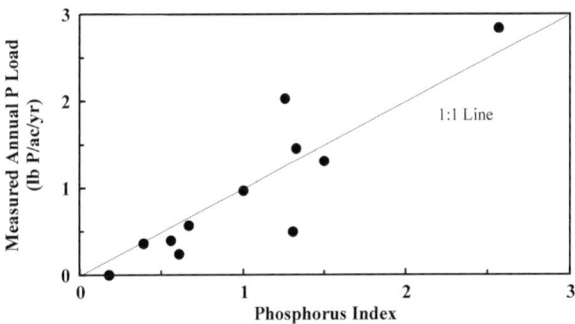

Fig. 4 Relationship between P index value and actual P runoff from pastures. (From Ref. 11.)

been generally specific to their local soil type, production practices, and climatic conditions. Some states such as Texas, use a similar system, but their P index incorporates a soil test P threshold $(400\,lb\,acre^{-1})$, above which no manure is applied.

CONCLUSION

Recent trends in animal agriculture have been increased numbers of animals with decreasing numbers of animal operations. These trends have restricted land available to producers for manure application and have corresponded to water quality problems associated with nonpoint sources. Phosphorus induced eutrophication has been one of the main problems associated with animal agriculture. There are two major technologies that can be used to reduce the potential P losses from animal manure. They are reducing P levels in the diet through increased P availability in grains or binding the P in the manure with chemicals such as alum or aluminum chloride. Phytase and HAP corn have the potential to reduce P levels in manure as much as 20%. The question still remains as to whether or not these technologies might increase P solubility, and hence P runoff losses. Phosphorus solubility can be reduced by as much as two orders of magnitude when treated with alum or aluminum chloride. Aluminum phosphates are stable over a wide range of naturally occurring soil conditions, thereby reducing the risk of bio-available P losses.

Many states are currently searching for methods to aid producers in manure management. One of the most common trends for this is the use of a P index. These are site-specific risk assessment tools used to identify fields that are susceptible to P losses. The P indices then provide a manure application rate for the producer.

REFERENCES

1. Kellogg, R.L.; Lander, C.H.; Moffitt, D.C.; Gollehon, N. *Manure Nutrients Relative to the Capacity of Cropland and Pastureland to Assimilate Nutrients: Spatial and Temporal Trends for the United States*; nps00-0579, U.S. Department of Agriculture—National Resource Conservation Service: Ft. Worth, TX, 2000.
2. http://www.nass.usda.gov:81/ipedb/ (accessed Apr 2002). Service Published Estimates Database, U.S. Department of Agriculture—National Agriculture Statistics Service.
3. Edwards, D.R.; Daniel, T.C. Effects of Poultry Litter Application and Rainfall Intensity on Quality of Runoff from Fescuegrass Plots. J. Environ. Qual. **1993**, *22*, 361–365.
4. Nichols, D.J.; Daniel, T.C.; Moore, P.A., Jr.; Edwards, D.R.; Pote, D.H. Runoff of Estrogen Hormone 17b-

Estradiol from Poultry Litter Applied to Pasture. J. Environ. Qual. **1997**, *26*, 1002–1006.
5. U.S. Environmental Protection Agency, *Environmental Indicators of Water Quality in the United States*; EPA 841-R-96-002, U.S. Environmental Protection Agency: Washington, D.C., 1996.
6. Shindler, D.W. Evolution of Phosphorus Limitation in Lakes. Science **1977**, *195*, 260–262.
7. Felton, G.K.; Simpson, T.W. A Citizens Guide to Understanding Pfisteria. In *Proceedings 1998 National Poultry Waste Management Symposium*, Springdale, AR, Oct 19–21, 1998; Blake, J.P., Patterson, P.H., Eds.; Auburn University Printing Service: Auburn, AL; 5–18.
8. Kornegay, E.T. Nutritional, Environmental and Economic Considerations for using Phytase in Pig and Poultry Diets. In *Nutrient Management of Food Animals to Enhance and Protect the Environment*; Kornegay, E.T., Ed.; Lewis Publishers: Boca Raton, 1996; 277–302.
9. Ertl, D.S.; Young, K.A.; Raboy, V. Plant Genetic Approaches to Phosphorus Management in Agricultural Production. J. Environ. Qual. **1998**, *27*, 299–304.
10. Smith, D.R.; Moore, P.A., Jr.; Maxwell, C.V.; Daniel, T.C. Dietary Phytase and Aluminum Chloride Manure Amendments to Reduce Phosphorus and Ammonia Volatilization from Swine Manure. In *Proceedings of the International Symposium Addressing Animal Production and Environmental Issues*, Research Triangle Park, NC, Oct 3–5, 2001; Havenstein, G.B., Ed.; College of Agriculture and Life Sciences, North Carolina State University, 502–507.
11. DeLaune, P.B.; Moore, P.A., Jr.; Carman, D.C.; Daniel, T.C.; Sharpley, A.N. Development and Validation of a Phosphorus Index for Pastures Fertilized with Animal Manure. In *Proceedings of the International Symposium Addressing Animal Production and Environmental Issues*, Research Triangle Park, NC, Oct 3–5, 2001; Havenstein, G.B., Ed.; College of Agriculture and Life Sciences, North Carolina State University, 239–349.
12. Moore, P.A., Jr.; Miller, D.M. Decreasing Phosphorus Solubility in Poultry Litter with Aluminum, Calcium and Iron Amendments. J. Environ. Qual. **1994**, *23*, 325–330.
13. Moore, P.A., Jr.; Daniel, T.C.; Edwards, D.R. Reducing Phosphorus Runoff and Improving Poultry Production with Alum. Poultry Sci. **1999**, *78*, 692–698.
14. Shreve, B.R.; Moore, P.A., Jr.; Daniel, T.C.; Edwards, D.R. Reduction of Phosphorus in Runoff from Field Applied Poultry Litter using Chemical Amendments. J. Environ. Qual. **1995**, *24*, 106–111.
15. Smith, D.R.; Moore, P.A., Jr.; Griffis, C.L.; Daniel, T.C.; Edwards, D.R.; Boothe, D.L. Effects of Alum and Aluminum Chloride on Phosphorus Runoff from Swine Manure. J. Environ. Qual. **2001**, *30*, 992–998.
16. Daniel, T.C.; Jokela, W.E.; Moore, P.A., Jr.; Sharpley, A.N.; Gburek, W.J. The Phosphorus Index: Background and Status. In *Proceedings of the International Symposium Addressing Animal Production and Environmental Issues*, Research Triangle Park, NC, Oct 3–5, 2001; Havenstein, G.B., Ed.; College of Agriculture and Life Sciences, North Carolina State University, 216–226.

Tailwater Recovery and Reuse

C. Dean Yonts
University of Nebraska, Scottsbluff, Nebraska, U.S.A.

INTRODUCTION

The recovery and reuse of irrigation water are generally associated with surface irrigated fields. When surface-irrigating fields with slope, water must be applied in excess to the needs of the crop in order to irrigate the entire field. As a result, excess water or tailwater collects at the lowest point in the field. The water may percolate into the soil profile or flow as surface drainage away from the field. Either way, the water through the force of gravity can eventually return to a nearby stream or lake. This process is referred to as return flow because water is returned to a surface source to be used again. A tailwater recovery and reuse system can also be used as a way to collect surface water runoff from a field. The reuse system consists of drainage channels to divert water to another site or to a reservoir for storing the water. Many systems will also include pumps and pipelines for delivering water to a new site for distribution and trash screens to remove unwanted debris. The recovery and reuse of tailwater from a surface irrigated field can increase surface irrigation efficiency by approximately 20%.

TAILWATER RECOVERY

Recovering the water that runs off the ends of irrigated fields has long been a method by which available and sometimes limited water supplies could be used more efficiently. Before available electricity, surface water drainage from the end of a field would gravity flow away and could be used again on another field located down gradient. In many cases, the runoff water could not be diverted to another field and would return to a nearby stream as return flow. Water users downstream would then have the opportunity to divert water for irrigation or other purposes. Whether used directly from the field or as return flow, the recovery of runoff water from surface irrigated fields has and will continue as a way to use water efficiently for meeting crop needs.

When electricity became available, pumping water became feasible. Water captured at the end of a field no longer needed to be used on a field down gradient, but could be pumped to irrigate the same field or any farm field

within close proximity. The primary purpose was to use available water supplies to irrigate as much land as possible. Surface water users are often a part of a larger irrigation district. These districts in most cases restrict the amount of water that can be diverted or used during the growing season. By recovering runoff water from their fields, irrigators can effectively irrigate more land. Keep in mind not all irrigators using surface water are allowed to collect the runoff water. As stated before, most runoff water will gravity drain to streams and lakes. In many cases, this return flow is vital for downstream users and reuse systems are not allowed.

For ground water users, they pay to pump water to the surface and irrigate. In this case, when the water becomes runoff it can no longer provide a benefit unless a reuse system is installed. Similar to surface water users, laws can define how runoff is to be treated. In some cases, water pumped from the ground is not allowed to enter the surface water drainage system. This means a reuse system must be installed or the water must be allowed to percolate into the soil at the end of the field. Even without this restriction, pumping water to an adjacent field can be much less costly than pumping more water to the surface to irrigate those same fields.

TYPES OF TAILWATER RECOVERY SYSTEMS

There are many different designs for tailwater recovery. Fig. 1 shows two alternatives for irrigating from a reuse system near the runoff site. In Fig. 1(a), a pump system is used to return irrigation water to the field of origin. The example in Fig. 1(b) uses a cycling system and returns water through a storage reservoir to an adjacent field. A brief description of tailwater recovery and reuse systems is given below for some of the more common types being used. For a more detailed description and design, see Refs. [1,2].

Cycling System

Cycling systems use a small sump or pit to store a quantity of water that is enough to allow the pump to operate correctly. It is generally recommended that

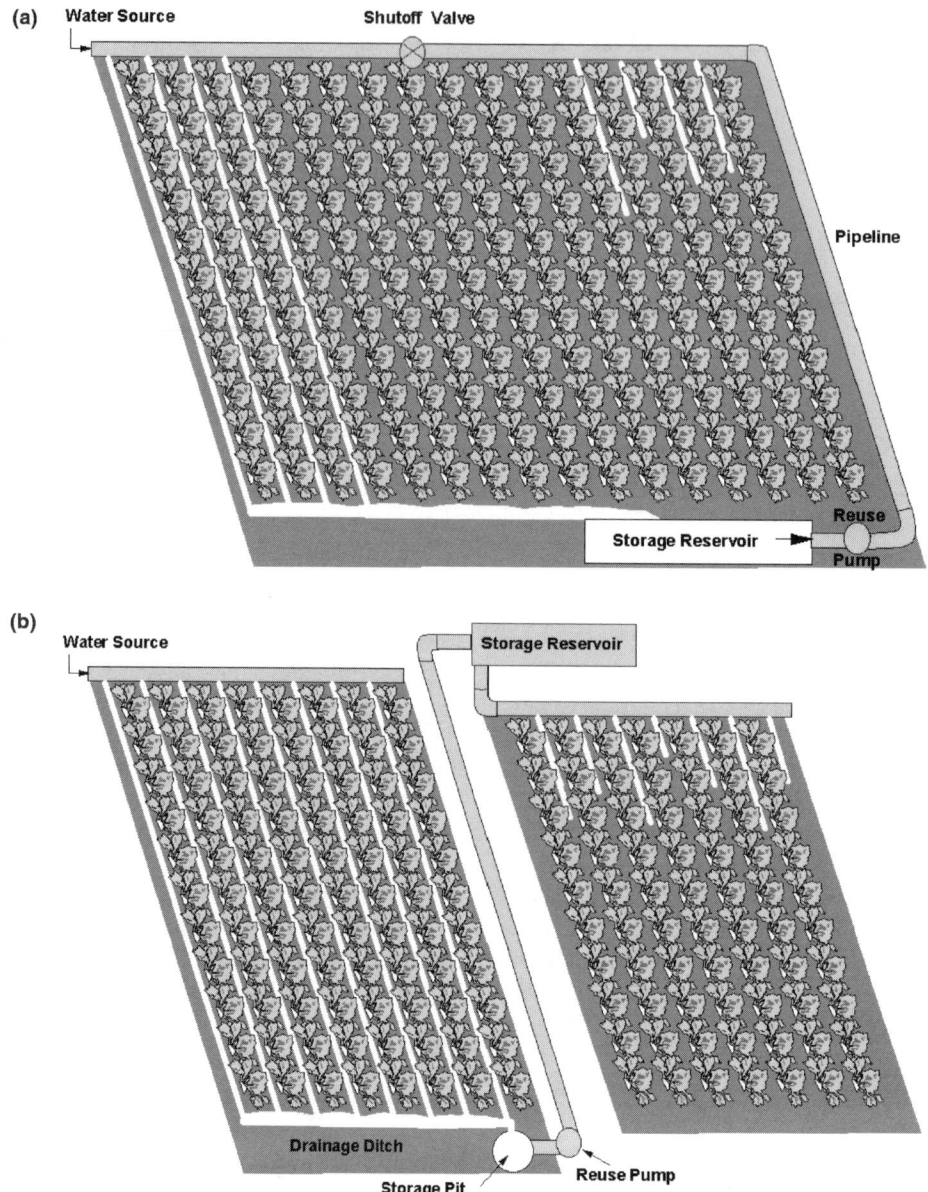

Fig. 1 (a) Runoff recovery and reuse system using a storage reservoir, pump, and pipeline to reuse irrigation water on the field of origin. (b) Runoff recovery and reuse system using a small pit, storage reservoir, pump, and pipeline to reuse irrigation water on an adjacent field.

the pump should operate no more than 15 cycles per hour to maintain pump efficiency. Pit size can be determined based on pumping rate and cycle frequency. Runoff from a surface irrigated field generally begins at a very slow rate and continues to increase until the irrigation set is complete. Because of this variation in runoff flowrate, cycling systems are used primarily for pumping water to a regulating reservoir rather than directly to a field since the constant fluctuation in pumped water flow makes regulation of an irrigation set difficult.

Pump System

In contrast to a cycling system, the pump system will often collect reuse water at the end of the field in a large storage reservoir. The reuse water stored in the reservoir is normally of sufficient volume to allow for a complete irrigation set to be made once the reuse pump is turned on without the need for additional water. When filled to the desired level, the pump will deliver water either back to the same field, independent of a current irrigation set, or to an irrigation set on an adjacent field. For systems that do not provide adequate storage capacity to complete a single irrigation set, labor will be increased along with a decrease in the water use efficiency.

Sequence System

Sequence systems are those systems that have been used for many years. These types of systems simply deliver the reuse water by gravity through open ditch or pipeline to fields down gradient without the use of a pump. These systems can increase water use efficiency but labor will also be increased due to the variability in the rate of water runoff from a field as explained for the cycling systems.

OPERATION OF TAILWATER RECOVERY SYSTEMS

Most reuse systems can be adapted to automation by controlling pump operation based on water level in the storage reservoir or simply controlled based on time. Water level controls automatically start the pump when the water level increases to a predetermined level and shuts the pump off when the water level falls to a predetermined level. Water level controls would most often be found on reuse systems with pumps that are designed to cycle on and off. Timing mechanisms for automation are normally used on large storage reservoirs. When adequate water has been recovered, the pump may be started manually. A timer is then used to shut the pump off after the desired irrigation set time is complete.

Because tailwater carries sediment with it, reuse pits should be designed to accommodate the collection and removal of sediment. This collection area should be in advance of the major storage portion of the pit. In the case of recycling pumps, storage is minimal and sediments should be removed prior to entering the reuse system. Some sediment will be carried with the recycled water, however, larger sediment particles can be removed by the use of grass filters. Keep in mind sediment will build up in the grass filter and must be mechanically removed periodically.

As water enters the pit whether it is large or small, flow velocities should be maintained below erosive levels. In pits, the inlet structure may be a part of the reuse structure (Fig. 2). In reservoirs, placement of a pipe through the bank of the pit will allow water to enter the reuse system without eroding soil banks (Fig. 3).

When using runoff water, the water should be applied to a succeeding irrigation set or to a different field (Fig. 1). Applying reuse water to the same irrigation set that is producing the runoff is ineffective and is generally not recommended. For example, when using a cycling system runoff is not available until water has started to reach the end of the field. As runoff increases, cycle frequency increases. Because additional water is applied after water advance is nearly complete, the result is pulsing inflows and increases in erosion without improving overall irrigation uniformity.

Pumping from a reservoir will be similar to the cycling system, but without the pulse flows. Once runoff begins and reuse water is added to the irrigation set, flowrate increases to a level greater than needed. The increased flowrate will likely further increase soil erosion. Using runoff from a reservoir system can function better if runoff is collected but not used until the beginning of the next irrigation set. This will allow greater flowrates to be used during the initial stages of water advance. Once runoff begins, the reuse system pump is shutdown and water is again stored for use at the beginning of the next irrigation set. Because the water has already advanced across the field, a reduction in inflow to the field at this time can be beneficial by improving application uniformity and

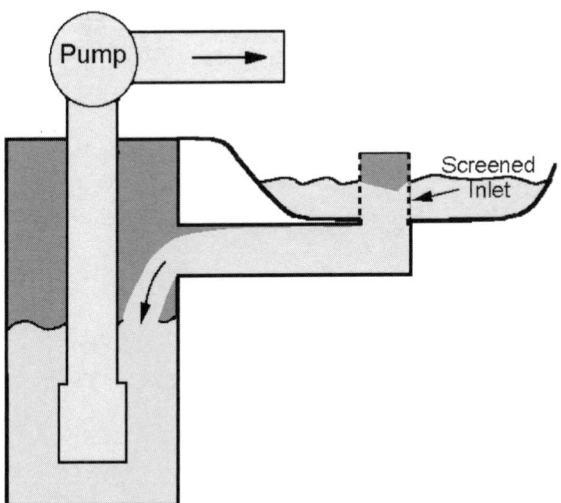

Fig. 2   Design for a cycling system inlet structure.

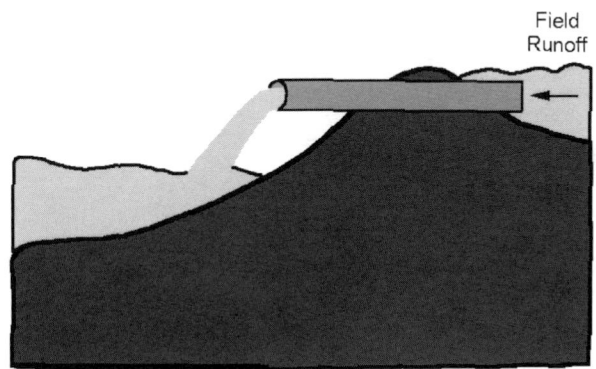

Fig. 3   Design for a pump system reservoir inlet structure.

increasing water use efficiency. Using a reuse system in this fashion will require more exact design in sizing the reservoir, determining reuse pump capacity, and determining irrigation set size.

Performance of surface irrigation is dependent upon many factors when it comes to designing a runoff recovery and reuse system. The rate of water infiltration into the soil can increase or decrease the rate of water advance to the end of the field and greatly influence irrigation set time. This in turn influences total runoff volume. The soil infiltration rate will change not only from field to field and year to year but from one irrigation set to the next making design of a reuse system challenging. Other factors that influence runoff rate and volume are changes made to set size, system inflow rate, and weather conditions. In many cases, it may take two or three irrigation sets to determine the preferred set size and inflow rate. Once established, the management of the runoff water can be fully determined.

WATER QUALITY

Tailwater recovery systems can provide a mechanism through which water quality can be maintained. Surface irrigation field runoff often carries sediment that can have traces of chemicals and fertilizers used for producing crops attached to the soil particles. By capturing the water for a brief period of time much of the sediment in the water will settle out. During the off season, the sediment should be removed and placed back on a production field.

Fertilizer and chemicals can also be held in suspension and carried in the water. By installing a tailwater recovery and reuse system, chemicals can be reapplied to fields during irrigation, keeping unwanted material from entering the surface water drainage system.

SAFETY OF TAILWATER RECOVERY SYSTEMS

Anytime water is collected and stored in a reservoir, safety should be of concern. In some cases, reuse pits or reservoirs are constructed with the goal of taking as little land as possible out of crop production. This may mean deep pits or reservoirs that have steep side slopes. This type of situation offers the potential for a hazard if children can be expected within the vicinity. Keeping side slopes that allow for mowing will also allow for easier escape if someone would find their way into a reservoir. If this is not possible, then fencing may be needed to insure small intruders do not have access to the area.

CONCLUSION

Tailwater recovery and reuse offers an alternative that can increase on-farm water use efficiency. At the same time, water quality can be maintained by keeping sediment and agricultural chemicals near the point of application. Finally, irrigating with reuse water can save both time and labor when properly designed. The end result will be irrigation that is environmentally friendly while still producing food and fiber for the world.

REFERENCES

1. Hart, W.E.; Collins, H.G.; Woodward, G.; Humpherys, A.S. Design and Operation of Gravity or Surface Systems. In *Design and Operation of Farm Irrigation Systems*, ASAE Monograph Number 3; Jensen, M.E., Ed.; American Society of Agricultural Engineers: St. Joseph, MI, 1980; 501–580.
2. Design and Installation of Surface Irrigation Runoff Reuse Systems. *ASAE Standards 2001*, ASAE EP408.1., American Society of Agricultural Engineers: St. Joseph, MI, 2001.

Timber Harvesting—Influence on Water Yield and Water Quality

C. Rhett Jackson
University of Georgia, Athens, Georgia, U.S.A.

INTRODUCTION

While the chemical, physical, and biological qualities of waters draining commercial forest lands are generally quite good, harvesting and planting of trees can temporarily alter streamflows, water chemistry, and biotic communities. The magnitude and duration of water quality effects vary with environmental setting but can be controlled to a large degree by implementing best management practices (BMPs) to protect water quality. Minimization of bare and compacted soil areas and dispersal of road runoff are critical for protecting water quality in commercial forests.

FOREST HYDROLOGY

In general, continuing commercial production forestry is practiced in humid climates where annual precipitation exceeds potential evapotranspiration (ET) but where soils or market conditions do not support more valuable agricultural commodities. Example areas include temperate forests of Europe and North America, tropical forests of Brazil and Southeast Asia, and boreal forests of Canada and Russia. While commercial forests cross many geologic, topographic, and climatic conditions, biogeographic zones where forestry is practiced feature sufficient commonality to generalize about hydrology and water quality of streams draining forest lands.

Harvesting and planting of trees can alter hydrologic behavior of watersheds with resulting impacts to streamflow, water quality, and aquatic life. Fig. 1 illustrates dominant hydrologic processes in forest environments and compares these processes between mature forests and clearcuts. Dense canopies of intact forests capture some rainfall before it hits the ground. Water evaporated from the canopy before it drips to the ground is called canopy interception and can account for 10%–30% of annual precipitation. Interception depends on the leaf area index (ratio of total leaf surface area to underlying land surface area), which varies between 6 and 15 in forests of different types and between summer and winter in deciduous forests. After clearcutting, leaf area index and canopy interception are greatly reduced. Therefore, more rainfall reaches the ground. As trees regenerate, full canopy interception returns over time.

Deep and well-developed root systems of mature forests efficiently extract soil moisture for tree growth. Evapotranspiration from soil storage returns a large portion of annual rainfall to the atmosphere before it reaches the water table or a stream. Shallow and limited root systems of young plantations extract far less water from soil storage. Therefore, percolation of soil water to the water table and shallow subsurface flow to streams both increase after clearcutting. Yield, or proportion of rainfall that becomes streamflow, is therefore greater in clearcuts and young plantations. Much of the increased yield reaches streams as dry season baseflow, improving habitat conditions for aquatic life during low-flow seasons.

Soils beneath mature forests usually feature well-developed litter layers, low bulk densities, high porosities, and dense macropore networks due to roots and soil fauna activity, so infiltration rates are very high. Most rainfall reaching the forest floor infiltrates, and overland flow occurs only during very intense rainfall events. Surface runoff occurs only as variable source area runoff from low lying areas such as floodplains, wetlands, and ephemeral streams where the water table rises to the soil surface during rainfall. These areas comprise only 5%–15% of most forest landscapes. In clearcuts and areas prepared for tree planting, bare soils may be exposed by harvest and site preparation equipment. Without the physical protection of litter layers, bare soils often form crusts during rainfall, and such crusts greatly reduce infiltration rates. Surface runoff is common from bare soils, and surface runoff mobilizes soil particles and transports them to streams. Storm runoff volumes, sediment loads, and sometimes peak flow rates are increased if clearcutting or site preparation creates significant bare soil areas. In addition, variable source areas are enlarged in clearcuts and young plantations due to higher water tables.

Roads are usually required to access and remove timber, and roads impact hydrology and water quality. Logging roads are surfaced with compacted native soil and sometimes covered with gravel. Surface runoff is common from roads due to low infiltration rates. Road runoff typically carries

Encyclopedia of Water Science
DOI: 10.1081/E-EWS 120010111

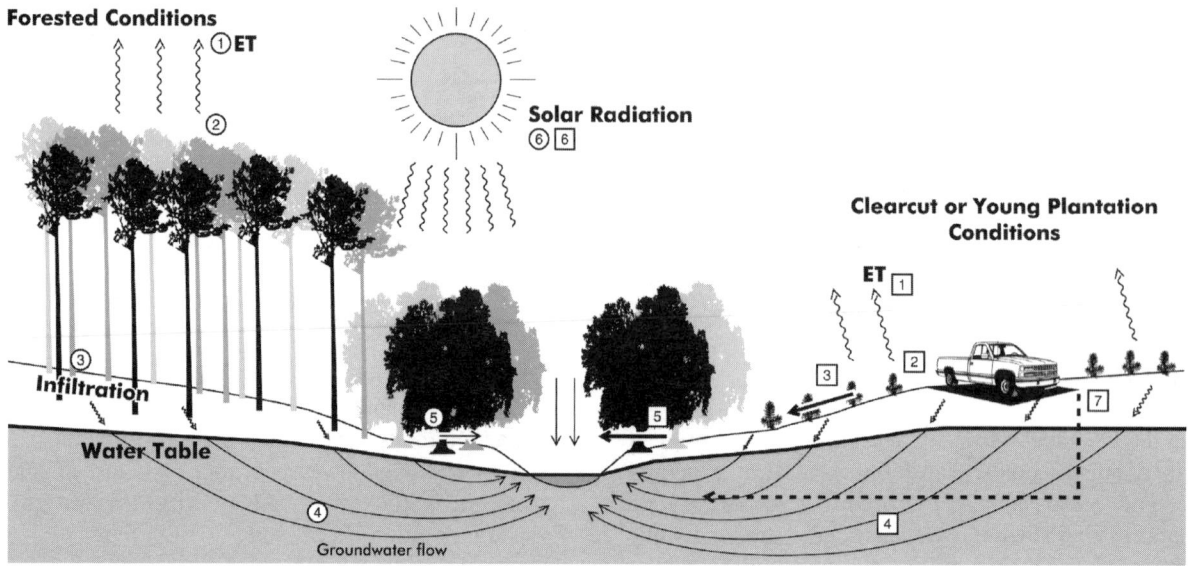

Fig. 1 Process schematic of silvicultural impacts on watershed hydrology; Forested conditions: (1) Evapotranspiration rates are greater in forests, so a large portion of rainfall is returned to the atmosphere from soil storage. (2) In forests, canopy interception returns some of the rainfall back to the atmosphere before it ever reaches the ground surface. (3) Infiltration rates are very high. Therefore, surface runoff is rare. (4) Subsurface flow paths are the dominant contributor to streamflow. (5) Surface runoff is common only in low areas near the channel where water tables are near the surface (variable source area runoff). (6) An intact forest canopy provides maximum shade to streams and maintains a cooler humid microclimate in the valleys. Clearcut or Young Plantation Conditions: (1) Evapotranspiration rates are significantly lower in first 4 yr–10 yr after harvest and planting. (2) Canopy interception is much lower in clearcuts and young plantations, so more rainfall reaches the ground. (3) When soils are left bare, overland flow is common until vegetation is re-established. Overland flow on bare soils transports sediment and other contaminants to the stream system. (4) Due to the reduced ET and interception, water tables are higher and baseflows are increased. (5) Due to the higher water tables, variable source areas may be enlarged, thus increasing surface runoff. (6) If SMZs do not sustain sufficient shade, stream temperatures may increase due to greater solar insolation. (7) Without water bars and filter strips, road runoff increases streamflows and sediment production.

high amounts of fine sediment and is often collected in roadside ditches and transported to streams. In small basins, road runoff can substantially increase peak flows and volumes. If roads are cut into hillside subsoils, road cuts can serve to collect shallow groundwater flow from the hillside above. Logging roads are responsible for many, if not most, water quality problems associated with forestry.

While general hydrologic processes apply to all forested landscapes, hydrologic behavior varies greatly with geology, soils, topography, solar aspect, and climate. These landscape factors control absolute and relative magnitude of forest hydrologic processes and also affect watershed response to forestry activities.

VARIATION IN SILVICULTURAL PRACTICES AND THEIR HYDROLOGIC EFFECTS

In any given watershed, hydrologic effects of forestry activities depend greatly on how timber is harvested and planted. If harvest and site preparation maintain organic litter layers, avoid soil compaction, disperse road runoff,

and maintain vegetated buffers along streams and wetlands, hydrologic and water quality effects of forestry can be minimal and often below levels of detectability. If harvest and site preparation activities create large areas of bare soils, gouge ruts up and down the slopes, concentrate road runoff and deliver it to streams, and extend operations to stream banks, hydrologic and water quality effects can be large and deleterious to aquatic life. Understanding and mitigating hydrologic effects of forestry requires some understanding of possible silvicultural activities.

In steep terrain, trees are cut manually with chain saws and yarded to roads via high lead (one end of the log is picked up) or full suspension (log is completely lifted off ground surface) cable systems. High lead yarding can leave bare soils but full suspension yarding leaves most of the litter soil surface intact. In flat and moderate terrain, it is now more efficient to use tractor-based equipment, such as feller-bunchers and skidders, to cut and yard trees. If tractors operate along the same tracks, and especially if they operate in wet conditions, they can create continuous tracks of compacted bare soils that may transport flow and sediments to streams.

Site preparation and tree planting methods vary considerably more than harvest methods. Intensive forest management, potentially involving plowing or ripping of soils, bedding of soils, ditching of hydric soils, burning of organic debris, and applying herbicide and fertilizer, is becoming more common, although hand planting of trees without silvicultural enhancements is still widely practiced. Hydrologic and water quality impacts of site preparation depend on how much bare soil is created, the duration of bare soils, and on timing, location, and amount of chemical application.

Effects of road runoff can be greatly reduced by routing water off roads at regular intervals onto hillslope locations where flow can be reinfiltrated. Water bars, broad-based dips, and cross-drains are typical methods by which road runoff is shed from roads onto hillslopes. Depending on slopes and native soils, surfacing roads with gravel or rock can also reduce surface erosion.

Streamside management zones (SMZs), also called riparian buffers, are strips of uncut trees and undisturbed soils left along water courses, and they greatly reduce ecological and water quality effects of forestry operations. SMZs perform multiple functions including 1) maintaining bank stability; 2) providing shade; 3) filtering runoff from upslope; 4) denitrifying shallow groundwater; 5) maintaining woody and organic debris recruitment to channels; 6) providing wildlife habitat; and 7) maintaining valley microclimates. While SMZs are commonly applied to modern forestry operations, appropriate widths are still matters of research and debate.

HYDROLOGIC EFFECTS—ANNUAL YIELD, STORM FLOW PEAKS, AND STORM VOLUMES

The effects of timber harvest and site preparation on storm flows vary with amount of bare soil exposed, amount and locations of road surfaces, connectivity of road runoff to streams, time since harvest, season, size of watershed, and size of storm. The storm flow effect is the greatest for early fall storms due to reduced summer ET in clearcuts and young plantations. In forests, early fall storms produce little to no runoff because very dry soils store most precipitation. In comparison, soils on clearcut sites are wetter and thus become more responsive earlier in the fall. Therefore, for early fall flows, clearcuts can produce flow peaks and volumes that are three times greater than mature forest flows.[1] However, fall flow events are usually small when compared with winter events. By the time the larger winter runoff events occur, there is little difference between soil moisture levels in clearcuts and forests, and studies have revealed little difference in flow peaks for large and infrequent flow events. For storms in the range of

the two year flow, timber harvest effects on peak flow rates are generally less than 20% and often statistically undetectable.[2,3]

As basin area increases, the percentage of recently clearcut area diminishes. In large basins, the effects of timber harvest and roads on flow peaks are usually imperceptible. However, increases in storm flow volumes due to timber harvest are observed in larger basins, usually manifested as a lengthening of recession limbs of storm hydrographs. Peak flow rate changes are difficult to discern in basins larger than 1000 ha.

Reduced ET in clearcuts and young plantations increases annual water yield and dry season baseflows from forested basins.[4] Magnitudes of mean annual flow and baseflow increases depend on the climate and the type of forest vegetation. Only in areas where fog-drip contributes significantly to the hydrologic cycle does forest clearing result in a reduction in yield and baseflows.

The hydrologic effects of forest removal are temporary and diminish as forest cover is re-established. Most of the hydrologic effects disappear after about 7 yr of regrowth, although some studies have found some hydrologic effects lasting as long as 20 yr.

WATER QUALITY

The chemical, physical, and biological qualities of waters draining commercial forest lands are generally quite good.[5] The major water quality concerns for forest management activities are 1) increased sediment loads due to surface erosion, road runoff, and landslides; 2) increased nutrient loads due to fertilizer washoff; 3) stream temperature increases from inadequate channel shading; 4) decreased woody debris recruitment from inadequate SMZs; and 5) pesticide runoff from intensively managed plantations. Again, the water quality effects of forestry activities are quite variable, depending on site conditions, intensity of activities, and application of BMPs. Without BMPs and maintenance of SMZs, timber harvest and site preparation can have large deleterious effects on water quality and aquatic biota. Recent studies of forestry activities that implement BMPs have found good water quality and biotic conditions downstream. Logger education and encouragement of BMPs are critical for maintaining good water quality in commercial forests.

REFERENCES

1. Jones, J.A.; Grant, G.E. Peak Flow Responses to Clear-Cutting and Roads in Small and Large Basins, Western Cascades, Oregon. Water Resour. Res. **1996**, *32* (4), 959–974.

2. Lewis, J.; Mori, S.R.; Keppeler, E.T.; Ziemer, R.R. Impacts of Logging on Storm Peak Flows, Flow Volumes, and Suspended Sediment Loads in Caspar Creek, California. In *The Influence of Land Use on the Hydrologic-Geomorphic Responses of Watersheds*; AGU Monograph, Wigmosta, M., Ed.; American Geographical Union, Washington, DC, 2001.

3. Thomas, R.B.; Megahan, W.F. Peak Flow Responses to Clear-Cutting in Small and Large Basins, Western Cascades, Oregon: A Second Opinion. Water Resour. Res. **1998**, *34* (12), 3393–3403.

4. Bosch, J.M.; Hewlett, J.D. A Review of Catchment Experiments to Determine the Effects of Vegetation Changes on Water Yield and Evapotranspiration. J. Hydrol. **1982**, *55*, 3–23.

5. Frick, E.A.; Buell, G.R.; Hopkins, E.E. *Nutrient Sources and Analysis of Water-Quality Data, Apalachicola-Chattachoochee-Flint River Basin, Georgia, Alabama, and Florida, 1972–1990*, U.S. Geological Survey Water-Resources Investigation Report 96-4101; Atlanta, GA, 1996.

Transpiration

Thomas R. Sinclair
United States Department of Agriculture (USDA), Gainesville, Florida, U.S.A.

INTRODUCTION

Nearly all water evaporated from vegetated surfaces to the atmosphere originates from leaves. Water is vaporized from cell walls inside leaves and diffuses from the leaf interior to the bulk atmosphere around plants. This process is called transpiration, and this section discusses the regulation of and methods to estimate transpiration rates. While transpiration involves basically the vaporization of water and the diffusion of the vapor into the bulk atmosphere, transpiration is complex because the water vapor must move from the leaf interior, through pores in the leaf epidermis called stomata, and finally into the atmosphere. Stomata are under active control so that transpiration rates are dynamic and rapidly respond to the environment. An understanding of the influence of stomata regulation on both carbon dioxide and water vapor flux density leads to various approaches to calculate plant canopy transpiration rates.

BACKGROUND

Carbon dioxide (CO_2) assimilation in photosynthesis was greatly facilitated, and thereby allowing rapid plant growth, when plants invaded the earth's land masses. No longer was it necessary for CO_2 to diffuse at very slow rates through the water surrounding aquatic plant life, but rather CO_2 could be absorbed directly from the atmosphere into individual cells. While photosynthesis rates and plant growth were enhanced substantially by allowing direct exposure of photosynthetic cells to the atmosphere, a potentially fatal consequence was that water could be evaporated at very high rates from exposed cell surfaces.

The problem of rapid evaporation from cell surfaces can be solved by having either an especially effective plant structure to rapidly transport large quantities of water in order to replenish each cell with water, or mechanisms to substantially inhibit water loss (and also CO_2 assimilation rates) from cell surfaces. While evolution "experimented" with each of these approaches, an innovative third, anatomical solution dominates. The photosynthetic cells are packaged inside thin broad tissues, i.e., leaves, that can throttle water vapor diffusion between the cells inside leaves and the atmosphere outside the leaves (Fig. 1).

The exterior of leaf epidermal cells is coated with a cuticle containing waxy materials that effectively block water loss directly from epidermal cells to the atmosphere. Consequently, virtually all water lost by leaves must move through stomatal pores that are scattered in the leaf epidermis (Fig. 2). Stomata regulate gas diffusion through the epidermis by adjustments in the dimension of the stomatal pore. The apertures of stomatal pores adjust to maintain a fairly stable CO_2 concentration inside leaves. When leaf photosynthetic rates are high, stomata adjust to increase the size of the pore aperture for rapid diffusion of CO_2 into leaves. On the other hand, when photosynthetic rates are low, aperture size decreases. Since water vapor diffuses through the same stomatal aperture, adjustments to accommodate photosynthesis cause changes in transpiration rate.

STOMATA

Stomata are embedded in the leaf epidermis and are formed by a pair of cells called guard cells. The guard cells in monocots plants, as shown in Fig. 2, tend to resemble barbells with bulbous structures at each end. The guard cells of dicot plants shown in Fig. 3, have a kidney shape. In both cases, the pair of guard cells is attached to each other at both ends. Swelling of the bulbous end of guard cells in monocots causes the cylindrical midsections to move apart increasing the aperture of the pore. The entire kidney-shaped cells of dicots swell and, as a result of a specialized cell wall structure bordering the pore, the aperture of the pore increases. Conversely, a decrease in the size of guard cells in both cases results in a decrease in pore aperture.

The shrinking and swelling of guard cells are under active control by plants as a result of changes in the concentrations of specific compounds in guard cells and neighboring cells. An increase in solute concentration in guard cells causes water to flow into the cells so that guard cells swell. There appears to be two mechanisms in guard cells that result in changes in solute concentration.[1] One mechanism is based on potassium-malate transport and appears to be particularly involved with stomata opening in response to light at sunrise. The second mechanism involves sucrose accumulation in guard cells and is

Encyclopedia of Water Science
DOI: 10.1081/E-EWS 120010277

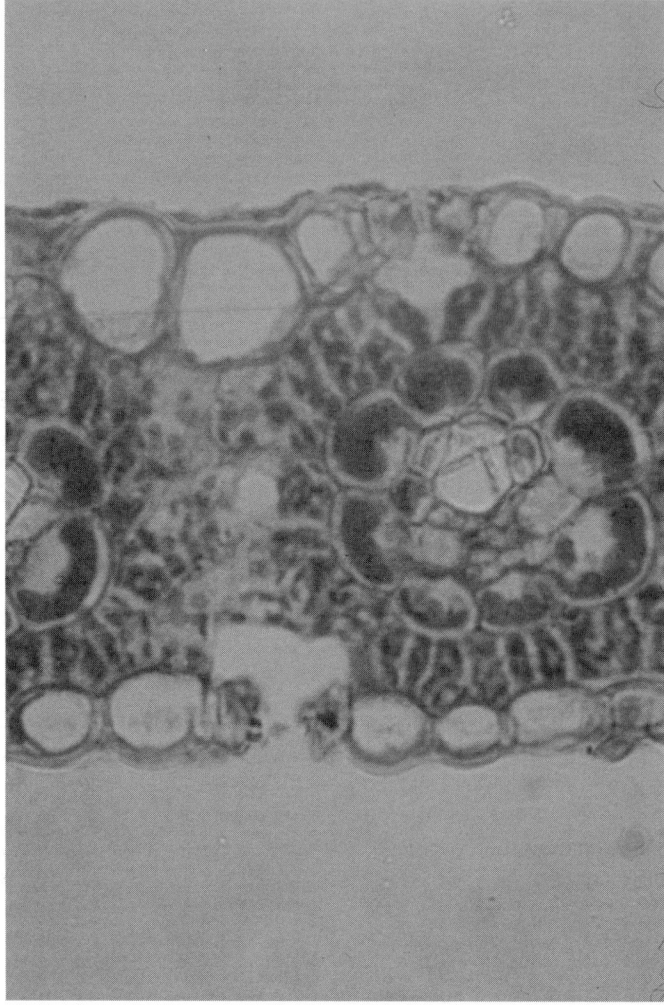

Fig. 1 Cross-section of monocot leaf showing stomata in epidermis of both sides of the leaf.

associated with maintenance of stomata aperture during the day. Both mechanisms are potentially sensitive to changes in the CO_2 environment of the leaf interior.[1] The sucrose mechanism seems especially sensitive to leaf photosynthetic rates so that stomata aperture can be fine tuned throughout the day as the photosynthetic rate of the leaf responds to changing environmental conditions. Of course, changes in aperture to match CO_2 diffusion to leaf photosynthetic capacity also results in changes in transpiration rate.

LEAF TRANSPIRATION RATE

Leaf transpiration involves the diffusion of water vapor through stomatal pores, so transpiration rate is dependent on conductance of vapor through and above the pores, and on the gradient of water vapor across the pores. Not

surprisingly, conductance of an individual stomatal pore (g_p, $cm^3 sec^{-1}$) is directly dependent on the dimensions of the pore aperture and can be expressed quantitatively[2] as

$$g_p \approx \pi ab D/d \qquad (1)$$

where a is the semilength of the major axis (cm), b, the semilength of the minor axis (cm), d, the depth of the pore (cm), and D, the molecular diffusion coefficient of water vapor ($0.24 \, cm^2 sec^{-1}$ at 20°C).

The semilength and depth of the pore usually remain fairly stable, so the main variable influencing g_p is the semilength of the minor axis of the pore.

The overall stomata conductance of the leaf (g_s, $cm sec^{-1}$) is calculated by incorporating into Eq. 1 stomatal density (n, stomata cm^{-2}) and the influence of "end effects" exterior to the leaf.[2] Consequently,

$$g_s \approx n\pi ab D/(d + b/2\ln(4a/b)) \qquad (2)$$

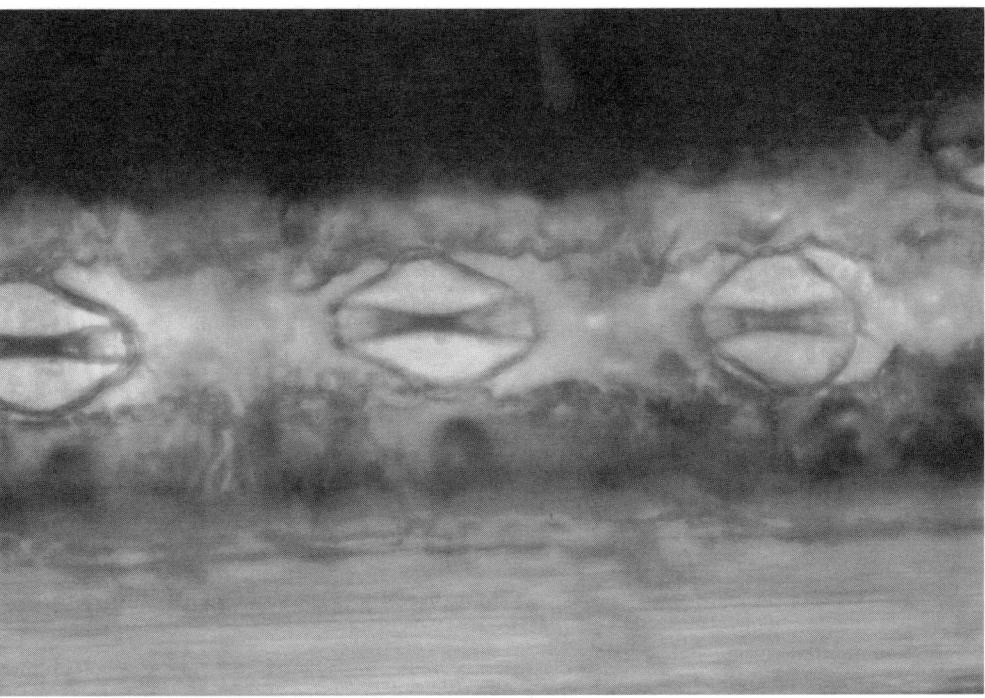

Fig. 2 Photograph showing stomata in the leaf surface of sorghum.

An estimate of maximum g_s can be calculated for fully open stomata $(b \approx 3 \times 10^{-4}\,cm)$ by assuming that $a = 8 \times 10^{-4}\,cm$, $d = 10 \times 10^{-4}\,cm$, and $n = 10 \times 10^3$ stomata cm^{-2}. Eq. 2 gives $1.3\,cm\,sec^{-1}$ for a leaf with stomata on only one side, and $2.5\,cm\,sec^{-1}$ for a leaf with stomata on two sides. Of course, when the stomatal pore is closed $(b = 0)$, Eq. 2 gives a conductance of zero.

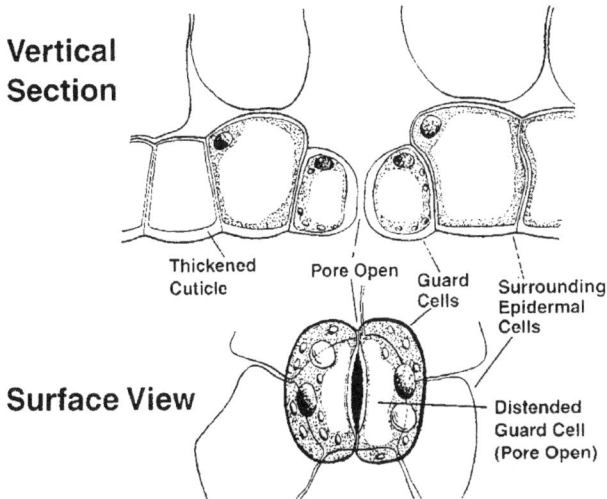

Fig. 3 Drawing of cross-section and surface view of a dicot stomata. (From Ref. [10].)

In addition to the restriction on water vapor diffusion resulting from stomatal conductance, there is also a limitation on water loss resulting from the aerodynamic boundary layer around leaves (g_{bl}, cm sec^{-1}). The value of g_{bl} is dependent on wind speed and leaf dimensions. In the case of a $100\,cm\,sec^{-1}$ wind speed and 8 cm wide leaf, the value of g_{bl} is approximately $2.5\,cm\,sec^{-1}$.[3] Consequently, in this case, the value of g_{bl} is equal to that of g_s.

As g_s and g_{bl} are in series, the inverse of the two conductances are added together to calculate leaf transpiration rate, Tr_L (g cm^{-2} sec^{-1}). The combined conductance is multiplied by the vapor pressure difference between the interior of the leaf, which is calculated as the saturated vapor pressure at leaf temperature (P_L^*), and atmospheric vapor pressure (P_a).[3]

$$Tr_L = \varepsilon(g_s g_{bl}/(g_s + g_{bl}))(P_L^* - P_a)/H_v \qquad (3)$$

where ε is the molecular weight of water ($18\,g\,mol^{-1}$) and H_v the heat of vaporization ($44\,kJ\,mol^{-1}$ at 25°C).

CANOPY TRANSPIRATION RATE

In principle, the transpiration of a leaf canopy (Tr_C) can be calculated by summing Tr_L for all the individual leaves in the canopy. This is a formidable task, however, because values of g_s, g_{bl}, and leaf temperature are required for each individual leaf. Consequently, "summary" expressions have been developed in an effort to express the transpiration rates of the entire canopy.

One of the more popular approaches relies on predictions of the energy balance of the leaf canopy. The Penman–Monteith equation[4] gives an explicit solution for canopy transpiration rate based on the energy balance of the entire canopy. The difficulty with this approach is that it requires an estimate of the canopy boundary layer conductance and the "canopy conductance" for the entire canopy, which must be appropriately weighted to represent vapor transfer through all the stomata distributed in the canopy. There is really no independent method for measuring canopy conductance, and estimates are obtained only by back-solving the energy balance equation. In practice, the Penman–Monteith is often applied as an empirical equation where boundary layer and canopy conductances are estimated from empirical functions.

A recent innovation has been developed to calculate Tr_C based on the water use efficiency of leaf canopies. Water use efficiency has been a topic of research since at least 1699[5] giving it a longer history of study than virtually any other plant trait. Roughly 100 yr ago, there was a particularly intensive period of study in both Europe and the United States on plant water use efficiency, culminating in the classic investigations by Briggs and Shantz.[6,7] In an analysis of much of the data from this period, deWit[8] found a highly stable relationship within each species between accumulated plant mass and cumulative transpiration over a wide range of conditions when normalized by "atmospheric demand" (Fig. 4).

Tanner and Sinclair[9] extended the analysis of deWit[8] by deriving a mechanistic expression for transpirational water use efficiency of canopies. Their derivation defined a specific transpirational water use efficiency coefficient (k, Pa) for each crop species dependent on the photosynthetic pathway and the biochemical composition of the plant products. Their estimates of k for maize, wheat, and soybean were 12 Pa, 5 Pa, and 4 Pa, respectively. Then, for each species, water use efficiency based on accumulated plant mass (M) was stable when expressed in the following equation:

$$M/Tr_C = k/(P_a^* - P_a) \qquad (4)$$

Tanner and Sinclair[9] approximated $(P_a^* - P_a)$ for daily transpiration by assuming this value was 75% of the

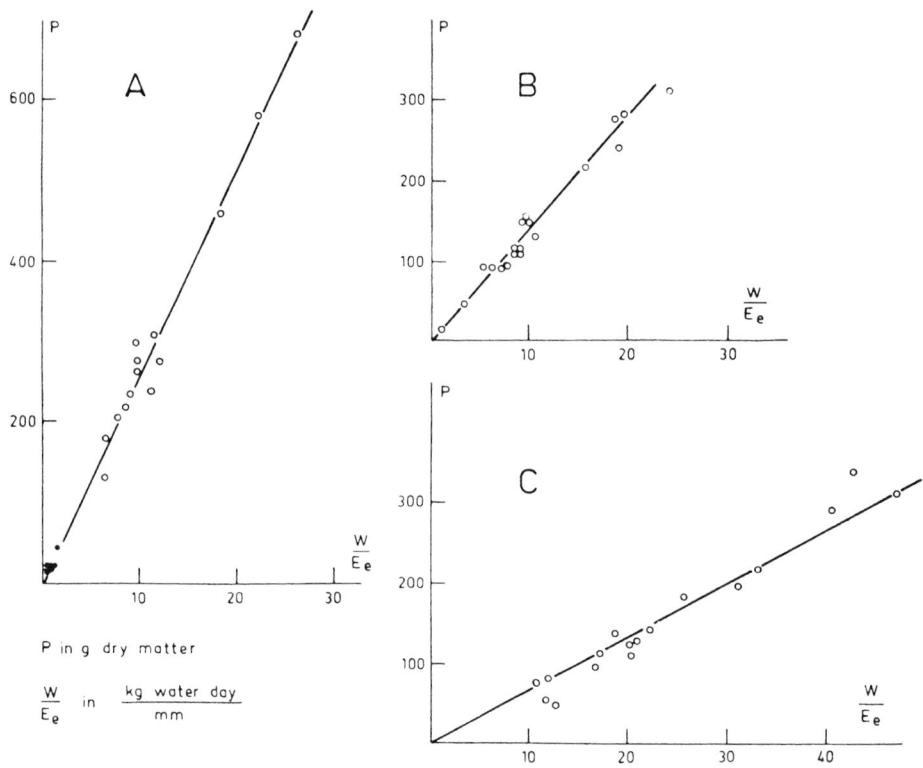

Fig. 4 Plot of crop growth (P) against transpiration water loss (W) divided by pan evaporation for (A) sorghum, (B) wheat, and (C) alfalfa. (From Ref. [8].)

maximum atmospheric vapor pressure deficit calculated at the daily maximum temperature.

A solution for Tr_C is obtained directly by the rearrangement of Eq. 4:

$$Tr_C = M(P_a^* - P_a)/k \qquad (5)$$

This equation can be readily used based on estimates of M, which in turn can be calculated based on the interception of solar radiation and the radiation use efficiency of the canopy. Consequently, Eq. 5 is consistent with the energy balance approach in that it is sensitive to the amount of solar radiation intercepted by the canopy.

Not only is Eq. 5 easier in principle to implement that an energy balance approach, but it is conceptually much more compatible with the understanding of stomata regulation of transpiration. That is, Eq. 5 is calculated from a direct dependence of Tr_C on CO_2 assimilation, which is consistent with the fact that stomata are regulated for CO_2 assimilation and the rate of water loss from leaves is simply a consequence of this process. Eq. 5 has been used very effectively to calculate transpiration rates and evaluate limitations of water availability on crop yields.

REFERENCES

1. Talbott, L.D.; Zeiger, E. The Role of Sucrose in Guard Cell Osmoregulation. J. Exp. Bot. **1998**, *49*, 329–337.

2. Parlange, J.; Waggoner, P.E. Stomatal Dimensions and Resistance to Diffusion. Plant Physiol. **1970**, *46*, 337–342.

3. Sinclair, T.R. Theoretical Considerations in the Description of Evaporation and Transpiration. In *Irrigation of Agricultural Crops*; Stewart, B.A., Nielsen, D.R., Eds.; American Society of Agronomy: Madison, WI, 1990; 343–361.

4. Monteith, J.L. Evaporation and Environment. *The State and Movement of Water in Living Organisms, XIX Symp. Soc. Exp. Biol.* Academic Press: NY, 1965; 205–234.

5. Woodward, J. Some Thoughts and Experiments Concerning Vegetation. Philos. Trans. R. Soc. London **1699**, *21*, 193–227.

6. Briggs, L.J.; Shantz, H.L. The Water Requirement of Plants. II. A Review of the Literature. U.S. Dep. Agric. Bur. Plant Ind. Bull. **1913**, *285*, 1–96.

7. Briggs, L.J.; Shantz, H.L. The Water Requirement of Plants as Influenced by Environment. Proc. Second Pan-American Science Congress, Washington, DC, 1917; 95–107.

8. deWit, C.B. *Transpiration and Crop Yields*; Institute of Biological and Chemical Research on Field Crops and Herbage: Wageningen, The Netherlands, 1958; Vol. 64.6, 1–88.

9. Tanner, C.B.; Sinclair, T.R. Efficient Water Use in Crop Production: Research or Re-search? In *Limitations to Efficient Water Use in Crop Production*; Taylor, H.M., Jordan, W.R., Sinclair, T.R., Eds.; American Society of Agronomy: Madison, WI, 1983; 1–27.

10. Sinclair, T.R.; Gardner, F.R. *Principles of Ecology in Plant Production*; CAB International: Wallingford, UK, 1998; 1.

Transpiration Efficiency

Graeme C. Wright
Nageswararao C. Rachaputi
Queensland Department of Primary Industries, Kingaroy,
Queensland, Australia

INTRODUCTION

A prevalent problem for plant and crop production is shortage of water, which is an essential component in all biological functions. For every kilogram of biomass produced, several hundred kilograms of water is lost from the leaf surfaces via the processes of transpiration (T), or both leaf and soil surfaces via evapotranspiration (ET).

An important concept used to define the efficient use of water derived from rainfall and irrigation is water-use efficiency (WUE). In most agricultural systems, the WUE is used to express the amount of either total biomass (Tbio) or grain yield (Yg) produced per unit of ET, and is a pivotal factor in achieving high productivity when water is limited. The Tbio or Yg can also be expressed as a function of transpiration efficiency (TE), where water used by the plant or crop is only by transpiration. The TE has also been expressed as the reciprocal form, transpiration ratio, defined as the amount of water lost through transpiration per unit of dry matter produced. However, in this chapter, we use "TE" as the efficiency of an organism (leaf, plant, and crop) to use water under specific environmental conditions. Readers are referred to a number of excellent reviews on TE in various crops.[1−4]

CAUSE OF VARIATION IN TE

The term TE is often applied at the leaf, whole-plant, and ecosystem level. The cause of variation in TE becomes complex as the level of organism increases from a single leaf to plant and crop canopy.

At the leaf level, TE is referred to as the "instantaneous or intrinsic" TE, expressed as mmol of CO_2 fixed per mol of H_2O transpired through stomata, and is calculated as

$$TE = A/g_s = (p_a - p_i)/\nu(1.6) \tag{1}$$

where A is the CO_2 assimilation rate (μmol m^{-2} sec^{-1}), p_a and p_i (ppm) are the partial pressures of CO_2 in ambient air and in the intercellular space, respectively, g_s is the stomatal conductance (μmol m^{-2} sec^{-1}), ν is the water vapor pressure gradient between intercellular space and ambient air, and 1.6 is the diffusivity constant between CO_2 and H_2O.

Eq. 1 suggests that TE can be regulated either by the leaf to air CO_2 pressure gradient (p_i/p_a) or leaf to air H_2O gradient (ν). The leaf to air CO_2 pressure gradient is a direct reflection of the CO_2 assimilation rate, which in turn is governed by the efficiency of carboxylating enzymes, whereas ν is controlled by the stomatal conductance. The closing of stomata in response to drought conditions to prevent excessive water loss through transpiration is a well-known drought adaptation mechanism. Stomatal closing reduces CO_2 uptake as well as water loss, thus decreasing the photosynthetic rate. Under conditions of elevated CO_2 concentration, the CO_2 gradient between the atmosphere and the leaf is higher and CO_2 can pass through partially closed stomata at a rate similar to that under conditions of lower CO_2 and open stomata. The water vapor gradient remains the same at higher CO_2, and the transpiration is impeded. The net result is improved transpiration efficiency by some plants. Under low ν conditions, there is very little water vapor flux from leaf to air, and gas exchange can occur through open stomata with minimum loss of water. Under these conditions, the net result is also improved transpiration efficiency.

The advantage of intrinsic TE as a term is that it allows a direct comparison of intrinsic physiological considerations without confounding effects of differences in temperature and humidity that may exist in a canopy situation. On the contrary, the disadvantage is that it only represents a "snap shot" of A/g_s, and may not necessarily scale up to long-term considerations related to overall canopy productivity and growth.

Transpiration efficiency at the entire plant level is defined as plant biomass (DM) accumulated per unit of water transpired (T) over a specified time interval, expressed as g kg^{-1}. At the entire plant level, TE can be more accurately determined by cumbersome and labor intensive gravimetric methods.[5−7]

Variation in TE can occur due to both genetic and environmental factors. The following expression illustrates the potential sources of variation in TE:

$$TE = A/g_s = [p_a(1 - p_i/p_a)]/1.6(e_i - e_a) \tag{2}$$

Encyclopedia of Water Science
DOI: 10.1081/E-EWS 120010172

where e_i and e_a (mbar) are intercellular and atmospheric water vapor pressures.

It can be seen from Eq. 2 that a reduction in p_i/p_a at a given $e_i - e_a$, (i.e., the vapor pressure deficit (VPD) will increase TE.

At the crop level, TE is influenced by a range of physiological factors and processes associated with production of dry matter (i.e., photosynthetic capacity, water extraction ability of roots, stomatal movements, leaf area regulation, etc.), which are in turn influenced by environmental and soil factors. Indeed, Fisher[8] and Tanner and Sinclair[9] argued that inter- and intra-specific variation for TE is small and can mostly be accounted for by soil fertility (particularly N) or environmental VPD factors. These studies concluded that TE was inversely proportional to the average VPD during the growing season, with k being the constant of proportionality, i.e.,

$$TE = k/(e_i - e_a) \qquad (3)$$

Eq. 3 allows the evaluation of TE (as k) independent of VPD, and hence the comparison of the genotypic variation in TE.

INTER- AND INTRA-SPECIES VARIATION IN TE

Historical Perspective

The first report of intra-species differences in TE occurred nearly a century ago when Briggs and Shantz[10] produced evidence in pot studies showing a significant variation among genotypes of the same species. These authors speculated that it should be possible to develop high TE lines through genetic selection. Despite these early findings, little research was subsequently conducted on TE variation within species until the mid-1980s. There are a number of reasons for this lack of follow-up research. First, the early work of DeWit[11] which showed a strong linear relationship between dry matter production and water implied a constant ratio between dry matter and water use, and hence TE. This led to the widely accepted belief that p_i/p_a among genotypes of species with a C_3 or C_4 photosynthetic pathway were invariant, and that TE could be considered a crop species constant, known as "k."[8,9,12] Second, there are substantial difficulties in accurately measuring TE variation in plants or crops. Both CO_2 assimilation and transpiration from single leaves vary markedly during the day and according to leaf age and plant age. As mentioned earlier, these instantaneous measurements of A and g_s may not represent integrated performance throughout the life of a plant or crop. As well, these measurements cannot assess the impact of

morphological or physiological adaptations in response to drought that can influence the integrated measure of TE.[13] Similarly, pot studies by using gravimetric techniques for water measurement although providing accurate time integrated measures of TE, were considered time consuming, laborious, and resource intensive.[1] It was not until the mid-1980s, when Farquhar and Richards[14] reported a twofold variation in TE among wheat genotypes, that physiologists began to "research" and demonstrate significant intra-species variation in TE in many crops.

Since the mid-1980s, there have been voluminous reports of inter- and intra-species variation in TE, particularly for C_3 plants including cereals, legumes, pasture species, and numerous horticultural crops. Significant genotypic differences in TE have also been reported in C_4 crop species, including sorghum and sugarcane. Readers are referred to review papers by Richards and Condon,[2] Turner,[3] and Subbarao et al.[13] for a more complete set of references on the species variation in TE.

Selection Tools For TE

Measurement of carbon isotope discrimination (Δ) in plant tissue has been shown to be an extremely effective technique to identify genetic variation in TE. Theory[15] has demonstrated that C_3 plants should exhibit an association between the extent of their discrimination against ^{13}C compared with ^{12}C during CO_2 fixation, and their leaf intrinsic gas exchange efficiency (A/g). The use of Δ to select for improved TE was proposed following the experimental confirmation of the theory in wheat genotypes.[14] Its measurement has opened up new opportunities for the genetic improvement of TE, as it provides a time-integrated estimate of TE and is easier and faster to measure than total growth and water use. The Δ technique therefore provides a ready screen for plants growing under identical conditions.

Pot studies in which growth (including roots) and water use have been measured precisely, have consistently shown a negative relationship between Δ and TE, as summarized by Richards and Condon,[2] Turner,[3] and Hall et al.[16] It is important for breeders and physiologists alike to be confident that the genotypic variation for TE and Δ measured in pots translate to the field. There have only been a limited number of studies conducted to confirm that the negative relationship observed between Δ and TE in pots will occur under field conditions. The lack of reports in the literature relates to the difficulty in accurately measuring crop transpiration (after accounting for soil evaporative losses) and biomass (including roots) under field conditions for a range of contrasting genotypes.[17] Using a mini-lysimeter system located within a rain-out shelter facility,[5,7] the negative relationship between Δ and TE under field conditions has been confirmed in a number of crop legumes including peanut[6] (Fig. 1), soybean,[18] common bean,[19] and cowpea.[20] It has also been confirmed for wheat genotypes when accurate techniques for measuring the crop water balance were employed.[21]

Genotypic differences in TE that are independent of VPD differences due to environment and location, calculated as k, have been demonstrated in a range of grain legume crops (Fig. 2, adapted from Ref. [22]).

The discovery of a strong relationship between Δ and TE has also made it possible to understand the physiological basis of variation in TE within species,[23] as well as the exploitation of TE in some crop improvement programs.[24,25]

Recent studies have shown that in C₃ crops such as cowpea,[20] cotton,[26] and chickpea,[27] TE is predominantly controlled by g_s, i.e., stomatal factors. In contrast, photosynthetic capacity (A) has been shown to be the major cause for variability in TE in crops such as peanut,[6] sunflower,[28] and spruce.[29] Udayakumar et al.[23] argued that C₃ crops could be grouped into two distinct categories, depending on whether TE is controlled predominantly by stomatal factors or photosynthetic capacity. These findings have major implications for using TE as a selection tool in the crop improvement programs (discussed in the next section). Udayakumar et al.[23] and Ashok et al.[20] argue that in C₃ crops where TE is controlled by stomatal factors (and hence affecting transpiration), selection for high TE is likely to result in genotypes with low total dry matter productivity. In contrast, in those crops where variation in TE is brought about by higher unit leaf rates of photosynthesis, or greater mesophyll efficiency, selection for high TE is likely to

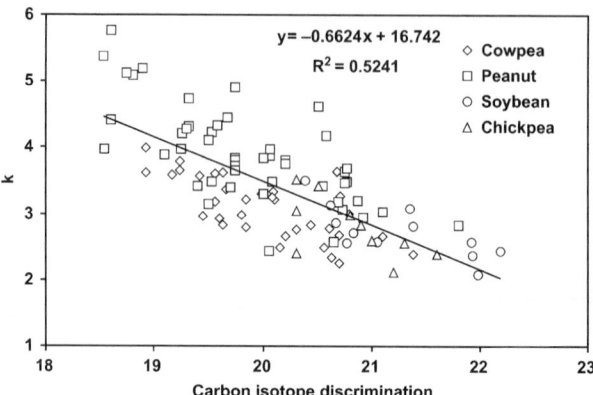

Fig. 2 Relationship between "k" (TE adjusted for VPD) and carbon isotope discrimination in different grain legume crops. (From Ref. [22].)

result in genotypes with higher dry matter productivity.[6,23,30]

While the use of Δ as a rapid and reliable selection tool for TE clearly has advantages over other cumbersome and labor intensive measurements of TE, there are several factors that need to be considered before recommending its use as a tool in large-scale genetic enhancement programs, include the following:

- A limited understanding of the value of TE for genetic enhancement in a specific environment.
- Complex genotype × environment interactions for TE in different crops due to differences in growth phenology among germplasm.[2,30,31]
- A lack of information on sampling procedures in different crops. For example, Δ can vary with leaf age within the canopy and may confound interpretation of results.[32]
- The high cost of Δ analysis.

Recent research in peanut has shown that specific leaf area (SLA)[33–35] or the SPAD-Chlorophyll Meter Readings (SCMR)[36] can be used as rapid and low cost surrogate measure for TE (Fig. 3). Richards et al.[37] discuss the value of using various selection tools such as SLA, ash, and molecular methods for genetic enhancement using the TE trait. The use of these methods will however depend on cost, degree of association with TE, and the relative ease of measurement of the trait.

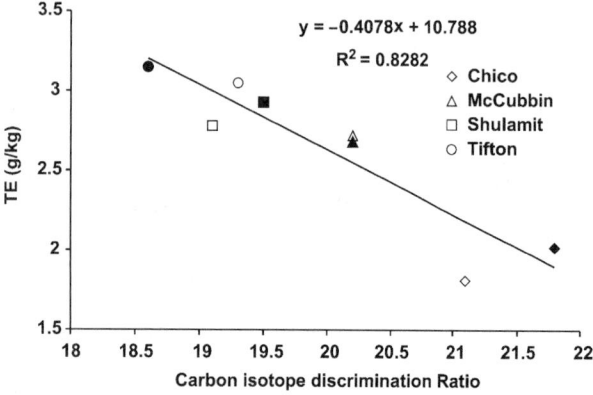

Fig. 1 Relationship between TE and carbon isotope discrimination (Δ) in four peanut genotypes grown under irrigated (filled symbols) and drought (open symbols) conditions. (From Ref. [6].)

SCOPE FOR GENETIC ENHANCEMENT OF TE IN AGRICULTURAL CROPS

Plant breeding programs have historically increased grain yield in crops from increases in partitioning of biomass to

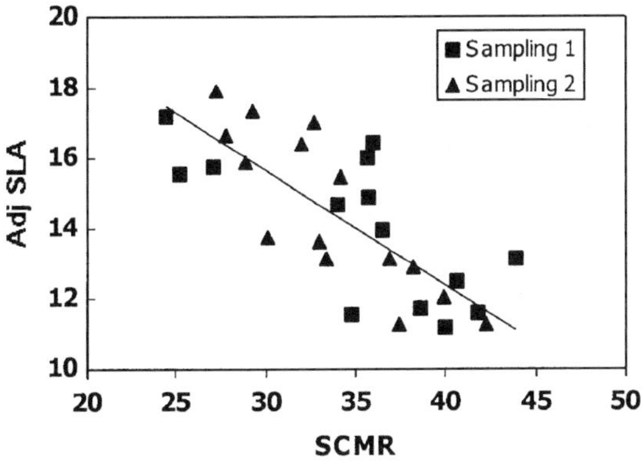

Fig. 3 Relationship between specific leaf area adjusted for radiation and VPD parmeters (Adj SLA) and spad chlorophyll meter readings (SCMR) in 15 peanut genotypes measured at two sampling times. (From Ref. [36].)

the reproductive component, or harvest index. Relatively little progress has been made in increasing plant biomass production per unit of water.[38–40] With the improved understanding of factors influencing water use and transpiration efficiency, there is now however greater opportunity to more precisely target improvement in TE.

A useful model for describing avenues for improvement in crop yield (Yg) in water limited environments in the aforementioned context is provided by the identity.[21,41]

$$Yg = E \times (T/E) \times TE \times HI \qquad (4)$$

where E is the total water use, T/E is the proportion of this water that is transpired (T), TE is the transpiration efficiency, and HI is the partitioning of biomass to grain. Implicit in the use of Eq. 4 is the concept that the various components are relatively independent so that increases in any of them will increase yield. In reality, numerous physiological and genetic interactions between model components can occur, thus complicating the expected response to selection of traits such as TE. The following section presents some case studies of approaches adopted for genetic enhancement for TE in different crops. It highlights some potential complications, which need to be kept in perspective when recommending whether plant breeders should launch into a large-scale selection program targeting TE improvement.

In peanut, development of surrogate tools for TE[33] and simple methodologies to analyze genotypic yield within the water model framework given in Eq. 4[30,42] made it possible to select elite genotypes with high levels of model components (T, TE, and HI) (Table 1). Recent studies by Nigam et al.,[43] have shown that additive gene effects were important in the expression of SLA (i.e., a surrogate measure of TE) and HI in peanuts and suggested that in some crosses selection for SLA and HI can be effective in

Table 1 Performance of selected peanut genotypes for T, TE, and HI relative to experimental mean (as %) in 1994–95 rainy seasons in an international collaborative project involving the International Crops Research Institute for Semi-Arid Tropics (ICRISAT), the Indian Council for Agricultural Research (ICAR) and the Queensland Department of Primary Industries

| | ± %Change from the mean | | | |
Genotype	Pod yield (t ha^{-1})	T (mm)	TE (g kg − 1)	HI
CSMG 84-1	28.8	29.3	0.3	− 0.4
DRG 101	10.5	1.2	1.0	10.8
DRG 102	12.7	8.8	1.0	6.1
ICGS 44	13.0	− 16.5	2.2	31.7
1CGS 76	27.0	7.7	5.5	11.8
ICGV 86754	15.5	6.5	2.5	4.9
ICGV 87354	22.5	5.0	1.8	10.5
KADIRI 3	19.6	12.8	− 0.8	10.2
NCAC 343	13.9	8.5	0.3	5.4
SOMNATH	12.9	0.5	0.5	10.8
TAG 24	16.6	− 10.1	1.7	30.1
Exp. mean	2.23	290.5	2.7	0.31

early generations. Positive correlations between Δ and HI have been observed in some peanut genotypes, and progeny from crosses of parents with similar maturity.[30] Similar responses have been observed in cowpea[44] and wheat[45] genotypes. These correlations suggest that breeders will need to be aware of such associations when selecting solely on the basis of low Δ.

In other C_3 crops, preliminary genetic and breeding studies using Δ as a selection trait have shown different relationships with crop growth, final biomass, and/or grain yield.[16] These responses have been further analyzed to indicate genetic associations between Δ and other important yield component traits, including earliness, HI, rooting depth, and rate of leaf area development. Early flowering has been associated with high leaf Δ in common bean,[46] cowpea,[16] wheat,[2] and barley.[31] Here, the negative association between Δ and days to flowering could constrain breeding for adaptation to specific water limited environments where both early maturity and high TE could be beneficial. The challenge for breeders in this situation is to identify whether germplasm is available with both low Δ and early flowering so that concurrent selection for both traits could be achieved.

In common bean, a positive association between Δ and Yg under water limited conditions was observed.[46] It was shown that Δ and the extent of rooting were positively correlated, indicating a possible genetic or physiological association among genotypes. Clearly, such a correlation would tend to constrain selection for those environments where deep rooting and high TE are desirable.

In barley, carbon isotope discrimination (Δ) was closely correlated with TE,[31] but it was either positively or negatively related to grain yield depending on the growing environment. Selection for high Δ at postanthesis or at maturity resulted in selection for high yields in water limited Mediterranean environments.[31] However, the prospect of selecting for Δ in early generations (e.g., F_2) is unclear. Voltas et al.,[47] also observed that barley genotypes with low Δ (i.e., high TE) performed better in low-yielding environments, whereas those with high Δ performed better in medium and high-yielding environments. This observation supports the assumption that drought tolerance and high yield potential under nonlimiting growing conditions may be antagonistic concepts in barley.

In bread wheat, it was concluded that a selection for low Δ in early generation (e.g., F_2) was successful in improving TE, plant total dry matter and root dry matter under water-limited conditions.[45] A positive association between Δ and early canopy growth has been observed in wheat genotypes.[2,21] In environments with high water availability, low Δ genotypes were slower growing, had higher soil evaporative losses, lower T and hence lower Yg. In contrast, in environments with severe drought conditions, Yg and Δ were negatively associated according to theoretical expectations, as soil evaporative losses were minimal.

It is evident from all the earlier reports that low Δ, and hence higher TE, may not always translate into higher Yg. It is critical for breeders to understand the potential trade-offs between Δ and growth in specific environments, and not expect that direct yield benefits will result from sole selection for Δ.

CONCLUSION

Although there is extensive information published on the variability in TE both between and within species, the challenge remains to establish whether TE (via measurement of Δ or other surrogate) is a sufficiently reliable trait to select for in plant breeding programs. At this early stage, the evidence from a number of different crops suggests that Δ has considerable potential. A number of experiments in cereal and legume crops have shown that Δ is correlated either positively or negatively with a wide range of attributes, including physiological (TE and HI), phenological (plant height and days to anthesis), and growth (TDM and yield) characters. These correlations also seem to depend on the crop and level of water stress prevailing in the growing environment. Thus, it is highly unlikely that the selection for TE alone will bring significant yield improvements across all environments. The understanding of yield constraints in the growing environment is vital to assess the value of TE among other yield limiting constraints in that environment. This assessment on the need and scope for improving TE can be made by analyzing grain yield of locally adapted genotypes within the framework of the water yield component model (Eq. 4).[30,41] The authors conclude that if TE is to be used effectively in breeding programs, it will be one of multiple criteria used by the breeder to improve the adaptation of the crop.

ACKNOWLEDGMENTS

We thank the Australian Center for International Center for International Agricultural Research (ACIAR) for funding support in relation to transpiration efficiency research over the past decade.

REFERENCES

1. Richards, R.A. Crop Improvement for Temperate Australia: Future Opportunities. Field Crops Res. **1991**, *26*, 141–169.
2. Richards, R.A.; Condon, A.G. Challenges Ahead in using Carbon Isotope Discrimination in Plant Breeding Programs. In *Stable Isotopes and Plant Carbon–Water Relations*; Ehleringer, J.R., Hall, A.E., Farquhar, G.D., Eds.; Academic Press: San Diego, CA, 1993; 451–464.
3. Turner, N.C. Water Use Efficiency of Crop Plants: Potential for Improvement. In *International Crop Science 1*; Buxton, D.R., Shibles, R., Forsberg, R.A., Blad, B.L., Asay, K.H., Paulsen, G.M., Wilson, R.F., Eds.; Crop Science of America: Madison, WI, 1993; Chap. 11, 75–82.
4. Cooper, P.J.M.; Gregory, P.J.; Tully, D.; Harris, H.C. Improving Water Use Efficiency of Annual Crops in the Rainfed Farming System of West Asia and North Africa. Exp. Agric. **1987**, *23*, 113–158.
5. Wright, G.C.; Hubick, K.T.; Farquhar, G.D. Discrimination in Carbon Isotopes of Leaves Correlates with Water-Use Efficiency of Field Grown Peanut Cultivars. Aust. J. Plant Physiol. **1988**, *15*, 815–825.
6. Wright, G.C.; Rao, R.C.N.; Farquhar, G.D. Water-Use Efficiency and Carbon Isotope Discrimination in Peanut Under Water Deficit Conditions. Crop Sci. **1994**, *34*, 92–97.
7. Udayakumar, M.; Devendra, R.; Ramaswamy, G.S.; Nageswara Rao, R.C.; Ashok, R.S.; Gangadhara, G.C.; Aftab Hussain, I.S.; Wright, G.C. Measurement of Transpiration Efficiency Under Field Conditions in Grain Legume Crops. Plant Physiol. Biochem. **1998a**, *25*, 67–75.
8. Fischer, R.A. Optimising the Use of Water and Nitrogen Through Breeding of Crops. Plant Physiol. **1981**, *58*, 249–278.
9. Tanner, C.B.; Sinclair, T.R. Efficient Water Use in Crop Production: Research or Re-search? In *Limitations to Efficient Water Use in Crop Production*; Taylor, H.M.,

Jordan, W.R., Sinclair, T.R., Eds.; American Society of Agronomy: Madison, WI, 1983; 1–27.

10. Briggs, L.J.; Shantz, H.L. Relative Water Requirements of Plants. J. Agric. Res. **1914**, *3*, 1–64.

11. DeWit, C.T. Transpiration and Crop Yields. In Institute of Biological and Chemical Research on Field Crops and Herbage, Wageningen, the Netherlands. Verslagen Landbouwkundige Onderzoekingen **1958**, *64* (6), 1–88.

12. Bierhuizen, J.F.; Slatyer, R.O. Effect of Atmospheric Concentration of Water Vapor and CO_2 in Determining Transpiration–Photosynthesis Relationships of Cotton Leaves. Agric. Meterol. **1965**, *2*, 259–270.

13. Subbarao, G.V.; Johansen, C.; Nageswara Rao, R.C.; Wright, G.C. Transpiration Efficiency—Avenues for Genetic Improvement. *Handbook of Plant and Crop Physiology*; Marcel Dekker, Inc., 1994; Chap. 39, 785–806.

14. Farquhar, G.D.; Richards, R.A. Isotopic Composition of Plant Carbon Correlates with Water-Use Efficiency of Wheat Genotypes. Aust. J. Plant Physiol. **1984**, *11*, 539–552.

15. Farquhar, G.D.; O'Leary, M.H.; Berry, J.A. On the Relationship Between Carbon Isotope Discrimination and the Intercellular Carbon Dioxide Concentration in Leaves. Aust. J. Plant Physiol. **1982**, *9*, 131–137.

16. Hall, A.E.; Richards, R.A.; Condon, A.G.; Wright, G.C.; Farquhar, G.D. Carbon Isotope Discrimination and Plant Breeding. Plant Breed. Rev. **1994**, *12*, 81–113.

17. Nageswara Rao, R.C.; Williams, J.H.; Wadia, K.D.R.; Hubick, K.T.; Farquhar, G.D. Crop Growth, Water-Use Efficiency and Carbon Isotope Discrimination in Groundnut (*Arachis hypogaea* L) Genotypes Under End-of Season Drought Conditions. Ann. Appl. Biol. **1993**, *122*, 357–367.

18. White, D.W.; Bell, M.J.; Wright, G.C. The Potential to Use Carbon Isotope Discrimination as a Selection Tool to Improve Water-Use Efficiency in Soybean. Proceedings of 8th Australian Agronomy Conference, Toowoomba, Qld. 1996; 728.

19. Wright, G.C.; Redden, R.J. Variation in Water-Use Efficiency in Common Bean (*Phaseolus vulgaris*) Genotypes Under Field Conditions. Aust Center Int. Agric. Res. (ACIAR) Food Legume. Newslett. **1995**, *23*, 4–5.

20. Ashok, R.S; Aftab Hussain, I.S.; Prasad, T.G.; Udayakumar, M.; Nageswara Rao, R.C.; Wright, G.C. Variation in Transpiration Efficiency and Carbon Isotope Discrimination in Cowpea. Aust. J. Plant Physiol. **1999**, *26*, 503–510.

21. Condon, A.G.; Richards, R.A. Exploiting Genetic Variation in Transpiration Efficiency in Wheat: An Agronomic View. In *Stable Isotopes and Plant Carbon–Water Relations*; Ehleringer, J.R., Hall, A.E., Farquhar, G.D., Eds.; Academic Press: San Diego, CA, 1993; 435–450.

22. Udayakumar, M.; Wright, G.C.; Prasad, T.G.; Nageswara Rao, R.C. Variation for Water-Use Efficiency in Grain Legumes. Proceedings of 2nd International Crop Science Congress, New Delhi, India, 1997; 267.

23. Udayakumar, M.; Sheshshayee, M.S.; Nataraj, K.N.; Bindu, M.H.; Devendra, R; Aftab Hussain, I.S.; Prasad, T.G. Why Has Breeding for Water-Use Efficiency Not Been Successful? An Analysis and Alternate Approach to Exploit This Trait for Crop Improvement. Curr. Sci. **1998b**, *74*, 994–1000.

24. Wright, G.C.; Hubick, K.T.; Farquhar, G.D.; Nageswara Rao, R.C. Genetic and Environmental Variation in Transpiration Efficiency and Its Correlation with Carbon Isotope Discrimination and Specific Leaf Area in Peanut. In *Stable Isotopes and Plant Carbon–Water Relations*; Ehleringer, J.R., Hall, A.E., Farquhar, G.D., Eds.; Academic Press: San Diego, CA, 1993; 247–268.

25. Voltas, J.; Romagosa, I.; Muñoz, P.; Araus, J.L. Mineral Accumulation, Carbon Isotope Discrimination and Indirect Selection for Grain Yield in Two-Rowed Barley Grown Under Semiarid Conditions. Eur. J. Agron. **1998**, *9*, 147–155.

26. Lu, .; Chin, J.; Richard, G.P.; Sharifi, M.R.; Rundel, P.W.; Zeiger, E. Genetic Variation in Carbon Isotope Discrimination and Its Relation to Stomatal Conductance in Pima Cotton (*Gossypium barbadense*). Aust. J. Plant Physiol. **1996**, *23*, 127–132.

27. Udayakumar, M.; SheshaSayee, M.S.; Nataraj, K.N.; Aftab Hussain, I.S.; Prasad, T.G. Dual (^{13}C and ^{18}O) Isotope Discrimination in Plants—Can This be a Potential Tool to Identify Desirable Physiological Traits Associated with Water Use Efficiency? *The Changing Scenario in Plant Sciences*; Professor H.Y. Mohan Ram Commemoration Volume, in press, 2000.

28. Virgona, J.M.; Hubick, K.T.; Rawson, H.M.; Farquhar, G.D. Genotypic Variation in Transpiration Efficiency, Carbon Isotope Discrimination and Dry Matter Partitioning During Early Growth in Sunflower. Aust. J. Plant Physiol. **1990**, *17*, 207–214.

29. Sun, Z.J.; Livingston, N.J.; Guy, R.D.; Ethier, G.J. Stable Carbon Isotopes as Indicators of Increased Water Use Efficiency and Productivity in White Spruce (*Picea glauca* (Moench) Voss) Seedlings. Plant, Cell Environ. **1996**, *19*, 887–894.

30. Wright, G.C.; Rao, R.C.N; Basu, M.S. A Physiological Approach to the Understanding of Genotype by Environment Interactions—A Case Study on Improvement of Drought Adaptation in Groundnut. In *Plant Adaptation and Crop Improvement*; Cooper, M., Hammer, G. L., Eds.; CAB International: Wallingford, 1996; 365–381.

31. Acevedo, E. Potential of Carbon Isotope Discrimination as a Selection Criteria in Barley Breeding. In *Stable Isotopes and Plant Carbon–Water Relations*; Ehleringer, J.R., Hall, A.E., Farquhar, G.D., Eds.; Academic Press: San Diego, CA, 1993; 399–417.

32. Wright, G.C.; Hammer, G.L. Distribution of Nitrogen and Radiation Use Efficiency in Peanut Canopies. Aust. J. Agric. Res. **1994**, *45*, 565–574.

33. Nageswara Rao, R.C.; Wright, G.C. Stability of the Relationship Between Specific Leaf Area and Carbon Isotope Discrimination Across Environments in Peanuts. Crop Sci. **1994**, *34*, 98–103.

34. Nageswara Rao, R.C.; Udayakumar, M.; Farquhar, G.D.; Talwar, H.S.; Prasad, T.G. Variation in Carbon Isotope Discrimination and Its Relationship to Specific Leaf Area

and Ribulose-1,5-biphosphate Carboxylase Content in Groundnut Genotypes. Aust. J. Plant Physiol. **1995**, *22*, 545–551.

35. Craufurd, P.Q.; Wheeler, T.R.; Ellis, R.H.; Summerfield, R.J.; Williams, J.H. Effect of Temperature and Water Deficit on Water-Use Efficiency, Carbon Isotope Discrimination and Specific Leaf Area in Peanut. Crop Sci. **1999**, *39*, 136–142.

36. Nageswara Rao, R.C.; Talwar, H.S.; Wright, G.C. Rapid Assessment of Specific Leaf Area and Leaf Nitrogen in Peanut Using a Chlorophyll Meter. J. Agron. Crop Sci. **2001**, *186*, 175–182.

37. Richards, R.A.; Rebetzke, G.J.; Appels, R.; Condon, A.G. Physiological Traits to Improve the Yield of Rainfed Wheat: Can Molecular Genetics Help? Proceedings of the Workshop on "Molecular Approaches for the Genetic Improvement in Cereals for Stable Production in Water Limited Environments." Ribaut, J.M., Poland, P., Eds.; CIMMYT, Mexico, DF, 2001, in press.

38. Perry, M.W.; D'Antuono, M.F. Yield Improvement and Associated Characteristics of Some Australian Wheats Introduced Between 1860 and 1982. Aust. J. Agric. Res. **1989**, *40*, 458–472.

39. Slater, G.A.; Andrade, F.H. Changes in Physiological Attributes of the Dry Matter Economy of Bread Wheat (*Triticum aestivum*) Through Genetic Improvement of Grain Yield Potential at Different Regions of the World: A Review. Euphytica. **1991**, *58*, 37–49.

40. Richards, R.A.; Lopez-Castaneda, C.; Gomez-Macpherson, H.; Condon, A.G. Improving the Efficiency of Water Use by Plant Breeding and Molecular Biology. Irrig. Sci. **1993**, *14*, 93–104.

41. Passioura, J.B. Grain Yield, Harvest Index and Water Use of Wheat. J. Aust. Inst. Agric. Sci. **1977**, *43*, 117–120.

42. Nigam, S.N.; NageswaraRao, R.C.; Wright, G.C. Breeding for Increased Water-Use Efficiency in Groundnut. Proceedings of the New Millennium Internal Groundnut Workshop, Shandong Peanut Research Institute (SPRI), Qingdao, China, September 4–7, 2001.

43. Nigam, S.N.; Upadhaya, H.D.; Chandra, S.; Nageswara Rao, R.C.; Wright, G.C.; Reddy, A.G.S. Gene Effects for Specific Leaf Area and Harvest Index in Three Crosses of Groundnut (*Arachis hypogaea* L.). Ann. Appl Biol. **2001b**, *139*, 1–6.

44. Hall, A.E.; Ismail, A.M.; Meneddez, C.M. Implications for Plant Breeding of Genotypic and Drought-Induced Differences in Water-Use Efficiency, Carbon Isotope Discrimination, and Gas Exchange. In *Stable Isotopes and Plant Carbon–Water Relations*; Ehleringer, J.R., Hall, A.E., Farquhar, G.D., Eds.; Academic Press: San Diego, CA, 1993; 349–370.

45. Ehdaie, B.; Barnhart, D.; Waines, J.G. Genetic Analyzes of Transpiration Efficiency, Carbon Isotope Discrimination, and Growth Characters in Bread Wheat. In *Stable Isotopes and Plant Carbon–Water Relations*; Ehleringer, J.R., Hall, A.E., Farquhar, G.D., Eds.; Academic Press: San Diego, CA, 1993; 419–434.

46. White, J.W.; et al. Implications of Carbon Isotope Discrimination Studies for Breeding in Common Bean Under Water Deficits. In *Stable Isotopes and Plant Carbon–Water Relations*; Ehleringer, J.R., Hall, A.E., Farquhar, G.D., Eds.; Academic Press: San Diego, CA, 1993; 387–398.

47. Voltas, J.; Romagosa, I.; Lafarga, A.; Armesto, A.P.; Sombrero, A.; Araus, J.L. Genotype by Environment Interaction for Grain Yield and Carbon Isotope Discrimination of Barley in Mediterranean Spain. Aust. J. Agric. Res. **1999**, *50*, 1263–1271.

Transpiration and Water Use Efficiency

Miranda Y. Mortlock
*Office of Economic and Statistical Research (OESR), Brisbane,
Queensland, Australia*

INTRODUCTION

The balance between carbon assimilation (net photosynthetic production) and the throughput of water by transpiration (resource use in terms of water) results in a benefit–cost ratio of interest to eco-physiologists and crop physiologists, known as water use efficiency. The differences in concentration of CO_2 and water vapor between the intercellular surfaces of the leaf mesophyll and the atmosphere drive the fluxes of carbon dioxide and water through the plant. Hot dry environments provide conditions of high evaporative demand. CO_2 concentrations are low in the atmosphere, and this gas diffuses through the stomata, which need to be open to allow gas exchange. There is a need under most environments to conserve water, and under drought stress stomata close which conserves water. Water use efficiency is an expression of the benefit–cost ratio for a plant and integrates the physiology of photosynthesis and plant water relations over a particular growth period or cropping season.

DEVELOPMENT OF THE CONCEPT OF WATER USE EFFICIENCY

Early last century, Briggs and Shantz studied the water requirements (WRs) of crops by weighing containers and working out WR per plant; however, they did not use an area basis.[1] Water requirement is the inverse of water use efficiency. Viets confined his definition of water use efficiency to the ratio of plant production to ET measured on the same area.[2] Tanner and Sinclair summarized some early studies and defined water use efficiency as the biomass of water accumulated per unit of water transpired and evaporated per unit crop area.[1] Biomass is expressed as total yield or economic yield. To compare species with different chemical composition (protein vs. carbohydrate products), grams of glucose *equivalents* are used. In irrigation studies, water use efficiencies may be referring to a broader definition of water, with efficiencies comparing situations where soil water drainage, surface run-off, or soil evaporation are considered. These types of water use efficiency are not included in this discussion.

WATER USE EFFICIENCY ON AN EVAPOTRANSPIRATION BASIS

Water use can be defined per unit of evapotranspiration (ET), and under field conditions, this is more practical than the narrower definition using just transpiration. Measures of ET integrate both soil and crop factors for the season, which confound the respective efficiencies of the plant and soil evaporation. Timing and frequency of irrigations and rain, the soil type and plant or mulch cover can affect soil evaporation. The transpiration part of ET use is a measure of crop performance.

WATER USE EFFICIENCY ON A TRANSPIRATION BASIS

Another definition of water use efficiency is in terms of transpired water only. Measuring the transpiration component is hard to do in practice, as it is difficult to prevent soil evaporation. Deep soil drainage also needs to be measured or prevented. It is only possible to measure transpiration on an experimental basis with the use of weighing lysimeters. Typically, the lysimeter is a large pot in the greenhouse and may weigh up to 80 kg. The lysimeter is weighed frequently over the crop season, and known quantities of water are added which is, both costly and limits the practical size of the trial. When extended to field studies (lysimeters within a growing crop), a limited numbers of comparisons are made due to the setup cost and expense of rainout facilities at sites.

TRANSPIRATION EFFICIENCY UNITS

Water use efficiency can be expressed as $g\,kg^{-1}$ of water transpired. Typical values may be $1.6\,g\,kg^{-1}$–$2.4\,g\,kg^{-1}$ for sunflower or around $9\,g\,kg^{-1}$ for sorghum. It can also be expressed by using a molar scale. Transpiration may be typically $1\,mmol\,m^{-2}\,sec^{-1}$–$5\,mmol\,m^{-2}\,sec^{-1}$ and photosynthetic rates for C_3 $20\,\mu mol\,m^{-2}\,sec^{-1}$–$25\,\mu mol\,m^{-2}\,sec^{-1}$ or $40\,\mu mol\,m^{-2}\,sec^{-1}$ for C_4 plants. Hence, C_4 plants have higher transpiration efficiencies than C_3 plants.

Encyclopedia of Water Science
DOI: 10.1081/E-EWS 120010279

TRANSPIRATION EFFICIENCY AND COMPARISON ACROSS SEASONS

As the evaporative demand of the atmosphere will vary from place to place and from season to season, transpiration efficiency (TE) from particular trials are adjusted for the vapor pressure deficit (VPD) of the atmosphere:

$$Y/T = k/(e^* - e_i)$$

where Y is the yield, T, the transpiration, k, the transpiration coefficient, $e*$, the saturated vapor pressure, and e_i, the vapor pressure of the atmosphere.[1] The coefficient (k) is estimated as 9 kPa for sorghum.

Typically, a mean value of VPD for hours of daylight over the stress period season can be used. The mean daily value of the 9.00 hr VPD and VPD at the time of maximum temperature can been used to compute a seasonal VPD over the stress period.[3]

INSTANTANEOUS TRANSPIRATION EFFICIENCY MEASURED AT AN INDIVIDUAL LEAF LEVEL

Transpiration and photosynthesis measured by using a canopy gas exchange system are used to compute an instantaneous measurement of TE at the leaf level. This rarely correlates with TE computed for a season as there are many processes integrated within the plant alone and over the cropping season.

Water use efficiency is the molar ratio of CO_2 uptake (A) to transpiration (E) and can be written as

$$A/E = (c_a - c_i)/1.6 \Delta_w$$

where c_a is external and c_i is the internal partial pressure of CO_2, respectively. Δ_w is the leaf-air VPD.[4]

This shows that the internal partial pressure of CO_2 is linked to water use efficiency.

PHYSIOLOGY OF TRANSPIRATION EFFICIENCY

C_3 and C_4 plants vary in TE as the carboxylation pathways give rise to different efficiencies.[5–7] C_4 plants such as maize and sorghum, have higher transpiration efficiencies than C_3 plants. Legumes have a lower TE than cereal crops due to the metabolic cost of symbiotic N fixation. Variations in TE for wheat correlate with carbon isotope discrimination.[8] Carbon isotope discrimination (Δ) as determined on dried leaf tissue of C_3 plants has shown to be correlated to TE. This is useful, but is not necessarily

practical as carbon isotope discrimination is a costly measurement and therefore not ideal as a selection criterion. Progress has been made in C_3 crops, towards obtaining a selection index for high TE lines. In peanut, the carbon isotope discrimination was linearly related to the specific leaf area of the leaves. Thus, a surrogate for Δ is available and has been used to breed high TE lines.[9]

ENVIRONMENTAL INFLUENCES ON TRANSPIRATION EFFICIENCY

Water Deficit Effect

Under moderate water deficits, TE increased in grain sorghum.[3,10] Leaf area index is irreversibly reduced under stress, cell density is maintained but cell enlargement is irreversibly affected.[11] Some plants may be able to adjust osmotically which may contribute to resistance to water deficits.

Rising CO_2

As the CO_2 concentration doubles, transpiration will decline and photosynthesis will increase.[12]

TRANSPIRATION EFFICIENCY IN DROUGHT RESEARCH

If this trait can be identified as significant and suitable selection criteria can be developed, more efficient crops will result through the incorporation of high TE lines in breeding programs. Benefits of high TE lines, anticipated from crop simulation modeling of grain sorghum, suggest a 10% increase in yield from moderate to good environments.[13] Genetic differences in TE appear to be detectable in a range of crops. The combination of C_4 productivity in marginal environments along with improved drought efficiency may have useful economic benefits in the future.

REFERENCES

1. Tanner, C.B.; Sinclair, T.R. Efficient Water Use in Crop Production: Research or Re-Search? *Limitations to Efficient Water Use in Crop Production*; ASA-CSSA-SSSA: Madison, WI, 1983.

2. Viets, F.G., Jr. Fertiliser and the Efficient Use of Water. Adv. Agron. **1962**, *14*, 223–264.

3. Mortlock, M.Y.; Hammer, G.L. Genotype and Water Limitation Effects on Transpiration Efficiency in Sorghum. In *Water Use in Crop Production*; Kirkham, M.B., Ed.;

Food Products Press, The Haworth Press, Binghamton, New York, 1999; 265–286.

4. Griffiths, H. Carbon Isotope Discrimination. In *Photosynthesis and a Changing Environment: A Field and Laboratory Manual*; Hall, D.O., Sturlock, O., Bolhàr-Nordenkampf, R., Leegood, R.C., Long, S.P., Eds.; Chapman & Hall: London, 1993.

5. Farquhar, G.D.; O'Leary, M.H.; Berry, J.A. On the Relationship between Carbon Isotope Discrimination and the Intercellular Carbon Dioxide Concentration in Leaves. Aust. J. Plant Physiol. **1982**, *9*, 121–137.

6. Farquhar, G.D. On the Nature of Carbon Isotope Discrimination in C4 Species. Aust. J. Plant Physiol. **1983**, *10*, 205–226.

7. Hubick, K.T.; Hammer, G.L.; Farquhar, G.D.; Wade, L.; von Caemmerer, S.; Henderson, S. Carbon Isotope Discrimination Varies Genetically in C4 Species. Plant Physiol. **1990**, *91*, 534–537.

8. Farquhar, G.D.; Richards, R.A. Isotopic Composition of Plant Carbon Correlates with Water Use Efficiency in Wheat Genotypes. Aust. J. Plant Physiol. **1984**, *11*, 539–552.

9. Wright, G.C.; Hubick, K.T.; Farquhar, G.D. Discrimination in Carbon Isotopes of Leaves Correlates with Water-use Efficiency of Field Grown Peanut Cultivars. Aust. J. Plant Physiol. **1988**, *15*, 815–825.

10. Donatelli, M.; Hammer, G.L.; Vanderlip, R.L. Genotype and Water Limitation Effects on Phenology, Growth, and Transpiration Efficiency in Grain Sorghum. Crop Sci. **1992**, *32* (3), 781–786.

11. Hsiao, T.C.; Bradford, K.J. Physiological Consequences of Cellular Water Deficits. *Limitations to Efficient Water Use in Crop Production*; ASA-CSSA-SSSA: Madison, WI, 1983.

12. Nobel, P.S. Leaves and Fluxes. *Physicochemical and Environmental Plant Physiology*; Academic Press: New York, 1991.

13. Hammer, G.L.; Butler, D.; Muchow, R.C.; Meinke, H. Integrating Physiological Understanding and Plant Breeding via Crop Modelling and Optimisation. In *Plant Adaptation and Crop Improvement*; Cooper, M., Hammer, G.L, Eds.; CAB International, Wallingford, U.K., ICRISAT, IRRI, 1996.

Uptake by Plant Roots

S. G. K. Adiku
University of Ghana, Accra, Ghana

INTRODUCTION

The study of water uptake by plant roots dates as far back as 1727 and continues to attract a great deal of research attention in many disciplines such as botany, agronomy, soil science, meteorology, and hydrology. This is because water flow through the soil–root–stem–leaf pathway to the atmosphere is a major component in the hydrological cycle. Water uptake by plant roots may be defined as the unidirectional transport of water from the soil to the root. Other terms used in the literature to describe the same phenomenon include root water absorption and root water extraction. Water taken up by the entire root system forms the transpiration stream and is terminated by the loss of water vapor from the stomata of leaves to the atmosphere. Many texts have presented details on the mechanisms of root water uptake. A popular view held is that the main mechanism of root water uptake involves a passive transfer of water along a water potential gradient from the soil to the root, whereby the plant simply behaves as a "wick" and plays no role in the uptake process. However, more recent observations suggest that the plant itself plays a role, especially, in determining the patterns of water uptake. Conceivably, an active water uptake may be possible, but strong scientific evidence is still lacking. This article presents a discussion on the subject of root water uptake with emphasis on divergent views regarding the understanding and interpretation of observations. Some areas requiring further research are identified.

BACKGROUND

A survey of the literature shows that research on root water uptake may have formally begun in 1727 with Hales' investigations into the height of rise of water in plant root–stem pathway during transpiration.[1] Many studies on plant water uptake followed, focusing on single aspects such as flow of water into roots, water flow through the stem, or the evaporation from leaves as separate processes, which were hardly interrelated. The unified concept of soil–plant–atmosphere continuum (SPAC), however, recognized the transport of water from the soil through the root–stem–leaf pathway to the atmosphere as a continuum, and emphasized

the interrelationship between the various aspects, enabling a holistic study of the phenomenon.[2]

Water uptake by plant roots is determined by evaporative demand of the atmosphere, plant factors such as the root system and soil water conditions. Observations indicate that uptake is generally favored in wet zones of the soil, where both water potential and hydraulic conductivity are also high.[3] But the role of the plant in determining water uptake has been unclear. Subsequently, the debate on root water uptake in the 1970s had centered on whether the plant or soil factors dominated water uptake.[4] The question still lingers on whether or not a plant simply behaves as a "wick" that only passively transmits water from the soil via the soil–root–stem–leaf pathway to the atmosphere. Indeed, the other terms used to describe the phenomenon such as root water *absorption* or root water *extraction* seem to suggest that the plant "makes" some effort involving forces or an expenditure of energy to obtain water from the soil.

Structural Aspects of Roots

Roots are the main organs of water uptake. They consist of a collection of cells specialized in different functions.[5] A schematic representation of the longitudinal and transverse sections of a root is shown in Fig. 1. Water may enter the root through two main paths: 1) through the root hairs, which are outward projections of the epidermal cells (Fig. 1b) and 2) between the epidermal cells. The root hairs, in particular, greatly increase the surface area across which water and nutrients enter the roots of plants.

Once water enters the roots, the upward transport to the stem and leaf is via the xylem vessels (Fig. 1b) which have lignified secondary walls but contain no protoplasm. These vessels are arranged end-to-end and the end walls between the individual members are perforated and therefore serve as low-resistance conduit for upward water transport from the root via the stem to the leaves.

The roots of the plant are often highly branched, intertwined and colonize the soil in three dimensions. This is referred to as the root system. The spatial distribution of the root system is determined by an interplay between the intrinsic development of the root system, and external abiotic stimuli such as soil water distribution, soil strength, and nutrient distribution. Lateral root growth is stimulated within water- and nutrient-rich zones of the soil,[6] which

Encyclopedia of Water Science
DOI: 10.1081/E-EWS 120010278

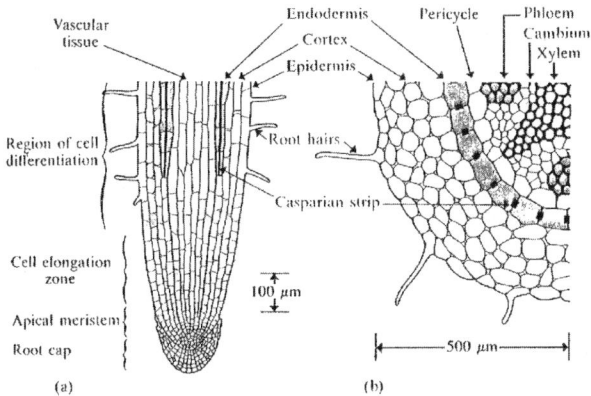

Fig. 1 Schematic diagrams of a root: (a) longitudinal section, indicating the zones that can occur near the root tip, and (b) cross-sectional view approximately 10 mm back from the root tip, indicating arrangements of various cell types. (From Ref. [5].)

many researchers have attributed to the nutritive role of the nutrients. However, it has recently been shown for a small weedy plant, *Arabidopsis thaliana*, that nutrients such as NO_3^-, apart from the nutritive role also act as environmental signals detectable by genes located in the root of the plant.[7] This suggests that the plant is able to detect preferred zones of root growth and can therefore "influence" the growth pattern of the root system. Root growth, water and nutrient uptake are, therefore, strongly interdependent and should ideally be studied together.

Mechanisms of Root Water Uptake

It is the popular view that water transport from the soil to the root is in response to a water potential gradient between the soil and the root surface. It is often observed that the water potential declines progressively from the soil through the root–stem–leaf–atmosphere pathway. Typically, the water potential at the soil–root interface of a well-watered soil would lie in the range between − 30 kPa and − 100 kPa, while that of the intercellular spaces within a leaf at a height of 10 m above the ground may be as low as − 1720 kPa.[8] The water potential of air with a relative humidity of about 78% is as low as − 37,800 kPa. Water would, therefore, flow spontaneously from the soil to the root and eventually to the atmosphere, with the plant itself being passive in the uptake process. Such a passive water uptake process requires a continuous column of water in the root–stem pathway, and this is possible up to 10 m due to the cohesive strength of water.[9] However, the fact that water uptake and transpiration continue in plants much taller than 10 m even under dry soil conditions when the water column can be broken by gas molecules, suggests the involvement of an alternative or a complimentary uptake mechanism.

When soil water is limiting, it is known that plants may influence transpiration through physiological changes such as the rolling of leaves which reduces the surface area for transpiration. Also, roots shrink when soil dries, reducing the area of contact between the roots and the soil,[10] and thereby increasing the resistance to the flow across the root surface and minimizing the loss of water from the roots to the soil (the so-called *reverse flow*). These observations indicate that the plant is able to influence indirectly the water uptake and transpiration processes. But, whether the plant can directly determine uptake, thereby providing a basis for an active uptake mechanism is yet to be proven scientifically.

Patterns of Root Water Uptake

Water uptake patterns vary both temporally and spatially. On the temporal scale, uptake varies both diurnally and seasonally. Uptake rates are low in the morning, rise to a peak in mid-afternoon, and decline to zero at night. As for the diurnal variation, uptake rates are low during the early stages of plant growth and reach a peak when the plant is fully established. In many annuals, peak uptake rates often coincide with the onset of the reproductive stage of the plant, when rooting depth and the leaf area index are at their maximum values.

The spatial patterns of root water uptake have been studied extensively over the years, yet views continue to differ, especially regarding the role of root system distribution in determining water uptake patterns. Whereas one school of thought[11] holds the view that root distribution is not an important determinant of water uptake patterns, others[12] are of the view that the spatial distribution of roots has substantial influence on the water uptake patterns.

An important study on water uptake patterns by peach trees, however, provides the much-needed insight into the role of root distribution on water uptake patterns.[13] It was observed that the initial pattern of water withdrawal followed that of root distribution, with more water withdrawn from the top sections of the soil where there were more roots. As the top sections dried out, more water was withdrawn from the deeper sections of the soil where the roots were fewer but the soil was wetter. In the latter case, the uptake pattern did not have any resemblance to the root distribution. It is thus quite clear that whereas both views on the role root distribution on water uptake may have experimental support, neither has general validity. It may also be concluded that even though roots may extend throughout the soil volume, they may not all be active at all times, suggesting that the plant may be capable of "activating" different sections of the root system at different times, depending on soil wetness. But, it is not clear what factors determine the uptake of water by plant roots. How does the plant root system distinguish between a wet or dry zone and how does the plant determine which parts of a root system to activate at a particular time?

Direct answers to these questions are lacking, thus, requiring further research. Even though hormones such as cytokinins and abscicic acids have been isolated from plant roots and are important in determining plant water status, their roles in determining water uptake patterns is unclear. Conceivably, plant roots may also possess *drought-detecting* genes, just as nutrient-detecting genes were isolated from the roots of the weedy plant *Arabidopsis thaliana*. The isolation or identification of such *genes* and a clear understanding of the way plant hormones control them remain crucial to the understanding of the role of the plant itself in determining water uptake patterns.

CONCLUSION

Water uptake by plant roots has and still continues to be an important interdisciplinary research subject. Although a lot of research has been done, much remains to be understood about the process of water uptake by plant roots. In particular, the role of genes in detecting drought and the way they are controlled by hormones merit further research.

ACKNOWLEDGMENTS

The author acknowledges the critical comments made by his colleagues in the Department of Soil Science, University of Ghana.

REFERENCES

1. Hales, S. Vegetable Staticks. *Scientific Book Guild*; reprinted 1961, Macdonald: London, 1927.
2. Philip, J.R. Plant Water Relations: Some Physical Aspects. Ann. Rev. Plant Physiol. **1966**, *17*, 245–268.
3. Gardner, W.R. Modeling Water Uptake by Roots. Irrig. Sci. **1991**, *12*, 109–114.
4. Reicosky, D.C.; Ritchie, J.T. Relative Importance of Soil Resistance and Plant Resistance in Root Water Absorption. Soil Sci. Soc. Am. J. **1976**, *40*, 293–297.
5. Nobel, P.S. *Biophysical Plant Physiology and Ecology*; W.H Freeman and Company: New York, U.S.A., 1983.
6. Leyser, O.; Fitter, A. Roots Are Branching out in Patches. Trends Plant Sci. **1998**, *3*, 203–396.
7. Zhang, H.; Forde, B.G. Regulation of *Arabidopsis* Root Development by Nitrate Availability. J. Exp. Bot. **2000**, *51*, 51–59.
8. Hillel, D. *Soil and Water: Physical Principles and Processes*; Academic Press: New York, NY, 1971.
9. Böhm, W. Ursache des Saftsteigens. Bericht der Deutschen Botanischen Gesellschaft **1889**, *7*, 46–56.
10. Heklerath, W.N.; Miller, E.E.; Gardner, W.R. Water Uptake by Plants. 2. The Root Contact Model. Soil Sci. Soc. Am. J. **1977**, *41*, 1039–1043.
11. Nimah, M.N.; Hanks, R.J. Model for Estimating Soil Water, Plant and Atmospheric Interrelations: I. Description and Sensitivity. Soil Sci. Soc. Am. Proc. **1973**, *37*, 522–527.
12. Feddes, R.A.; Kowalik, P.J.; Malinka, K.K.; Zaradny, H. Simulation of Field Water Uptake by Plants Using a Soil Water Dependent Root Extraction Function. J. Hydrol. **1976**, *31*, 13–26.
13. Olson, K.A.; Rose, C.W. Patterns of Water Withdrawal Beneath an Irrigated Peach Orchard on a Red–Brown Earth. Irrig. Sci. **1988**, *9*, 89–104.

Uptake by Plant Roots, Modeling Water Extraction

S. G. K. Adiku
University of Ghana, Accra, Ghana

U

INTRODUCTION

A great deal of current research on plant-water uptake focuses on modeling water extraction due to the increasing use of such models for crop and water management. The art of modeling root-water extraction is, however, limited by the level of understanding of the processes involved in the phenomenon. Though research on root-water uptake is not new, the lack of complete understanding of the process of root-water extraction has led to the publication of many root-water extraction models as researchers' views on process description differ in detail and scope. By 1981, as many as 18 published models were reviewed.[1] The number of root-water uptake models keeps increasing, especially, due to the increased interest in crop modeling. However, comprehensive models that have wide scope of validity and applicability are still lacking, despite the increased research efforts. This article discusses the concepts behind some of the most common extraction models and seeks to harmonize divergent views where possible. Also, this article presents and discusses a new approach to modeling root-water extraction.

BACKGROUND

Water flow through the soil–root–stem–leaf pathway is a major component of the hydrologic cycle. Every year, about 710 mm of rain falls globally on the soil of which about 57% evaporate back to the atmosphere, often due to plant extraction.[2] A quantitative study of the root-water extraction process cannot be overemphasized as this forms quite a large and important component of the water cycle.

Among the first researchers to describe root-water extraction mathematically was Gardner, who formulated the extraction as a water flow problem from the soil to a single long cylindrical root.[3] This approach, which has become known as the single root model, is also described as microscopic or Type I model.[4] Conceptually, the root-water extraction is considered as a passive process, with water flowing from a region of high to low water potential. But the extension of this model to the complex real root system, where the root architecture and distribution changes in both time and space has met with difficulties.[5] Furthermore, difficulties in measuring and parameterizing

the microscopic model have led to the proposal of another class of extraction models that relate extraction to a more easily measurable or predictable soil property, such as the soil water potential.[6] The latter type of models have been described as macroscopic and classified as Type II.[4]

Irrespective of the type of approach, there is a question as to whether or not a plant simply behaves as a mere "wick" that only passively transmits water from the soil via the soil–root–stem–leaf pathway to the atmosphere. If the role of the plant in determining water extraction is not to be ignored, how then would it be formulated quantitatively? A further issue of controversy is the role of root distribution in determining uptake patterns. Opinions are divided, with some researchers indicating that extraction patterns follow root distribution,[7] while others believe otherwise.[6] Apparently, the lack of complete understanding of the phenomenon of root-water extraction continues to be a major handicap to the formulation of root-water extraction models of wide applicability.

Microscopic Water Extraction Models

The microscopic single root model[3] considers the root system that comprises a collection of single long cylindrical roots each surrounded by a soil cylinder (Fig. 1). By using the cylindrical coordinate system, the radial flow of water from the surrounding soil to the root cylinder can be formulated as[3]

$$\frac{d\theta}{dt} = \frac{1}{r}\frac{d}{dr}\left(rk(\Psi)\frac{d\Psi}{dr}\right) \qquad (1)$$

where θ is the soil water content, r the spatial coordinate, Ψ the soil water potential, k the hydraulic conductivity, and t the time. A solution of this water flow problem can be obtained under the appropriate boundary conditions, leading to the estimation of the water uptake per unit root length and time at soil depth z, q_z ($m^3\,m^{-1}\,sec^{-1}$), as

$$q_z = 4\pi k \frac{(\Psi_{zs} - \Psi_{r_z})}{\ln\frac{c^2}{r_1^2}} \qquad (2)$$

where Ψ_s is the soil water potential, Ψ_r the water potential at the soil–root interface, r_1 the root radius, and c the path length or one-half the distance between two roots.

Encyclopedia of Water Science
DOI: 10.1081/E-EWS 120010352

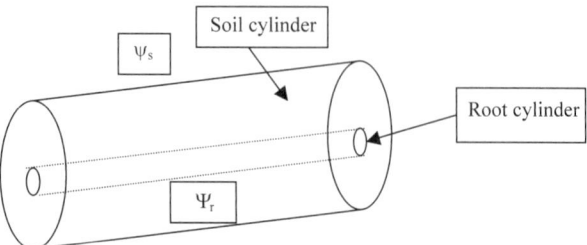

Fig. 1 Schematic representation of the single root model.

The single root model has been used extensively in the literature to model water extraction. To account for the role of the plant, further development of the model is necessary. Noting that not all roots of a root system are active in water extraction at all times,[8] then the root-water potential may not be constant along the entire length of each root. Some researchers have shown that variations of the root xylem potential with distance from the base of the root can indeed be calculated,[5] but the mathematical formulations are not only complex but the computer implementation is also cumbersome.[9]

Generally, when root distribution is considered as an important factor in determining water uptake patterns, then extraction is weighted towards soil sections with more roots,[7] even though this is not entirely correct since water uptake patterns do not always follow root distribution.[8,10]

Macroscopic Water Extraction Models

A typical macroscopic water extraction model is formulated by establishing a simple often empirical relation between the extraction at a given depth and the soil water potential at that depth z,[6] e.g.,

$$S_z = \alpha(\Psi)S_{\max,z} \qquad (3)$$

where S_z is the uptake from depth z (sec^{-1}), $S_{\max,z}$ the maximum water extraction from depth z for no-stress conditions, and $\alpha(\Psi)$ the nondimensional stress response function equivalent to the ratio between the actual extraction S_z, and the maximum uptake $S_{\max,z}$. The factor $S_{\max,z}$ is related to the potential transpiration, T_p (m sec^{-1}) of the plant by

$$S_{\max,z} = \frac{T_p}{z_r} \qquad (4)$$

where z_r is the root depth (m). The basic concept of this model is that each rooted depth makes an equal contribution to the total plant uptake, irrespective of the number of roots at that depth. This model, therefore, predicts equal extraction from each rooted depth even under uniformly wet soil conditions, contrary to the

observations that extraction indeed follows root distribution under such conditions.[8] In view of this, some researchers have introduced factors that are used to discriminate uptake in relation to root distribution.[4]

Macroscopic water extraction models are easier to parameterize and have found practical applications for crop and water management and incorporated into many crop models. But after an extensive review, it was concluded that these types of models seem to work only for the particular circumstances for which they were developed, and their extrapolation to other conditions is limited.[2]

A New Concept for Modeling Root-Water Extraction

The need to develop root-water extraction models with wide applicability has led to the re-examination of the subject of modeling root extraction. A new concept of energy minimization was proposed as the basis of modeling root-water extraction.[12] The *minimum energy* hypothesis assumes that root-water uptake involves energy expenditure which is related to the action of the plant, and that the plant, as a survival strategy seeks to minimize the overall energy expenditure. The concept generally accepts the validity of Eq. 2, but further proposes that the role of the plant can be expressed in terms of the energy expenditure.

The examination of Eq. 2 indicates that water flow from the soil to the roots at any depth requires that there is a potential gradient towards the root. This drop in potential, $\Delta\Psi (= \Psi_s - \Psi_r)$ is the work done per unit quantity of water transferred across the potential drop. The root-water uptake process must therefore involve energy expenditure by the plant. To formulate the energy expenditure, q_z in Eq. 2 is expressed first as the rate of water uptake per unit volume of soil at depth z, Q_z(m^3 m^{-3} sec^{-1}) $= q_z \cdot Lv_z$; where Lv_z is the local root density (m m^{-3}). The value of Q_z can also be easily converted to the uptake rate per unit soil area, U_z (kg m^{-2} sec^{-1}) as

$$U_z = Q_z\rho_w \, dz \qquad (5)$$

where ρ_w is the density of water. The rate of energy expenditure in extracting water at any depth, dE_z/dt can then be obtained as the product of U_z and the potential drop, yielding

$$\frac{dE_z}{dt} = U_z(\Psi_{z_s} - \Psi_{r_z}) \qquad (6)$$

With the potentials expressed in J kg^{-1}, the unit of Eq. 6 is W m^{-2}, which is clearly the unit for the rate of energy expenditure. The total rate of energy expenditure in extracting water from the entire root zone of the soil profile, z_r, can be calculated as

$$\int_0^{z_r} \frac{dE_z}{dt} dz = \int_0^{z_r} U_z(\Psi_{zs} - \Psi_{r_z}) dz \qquad (7)$$

If the plant minimizes energy expenditure during water uptake, then root-water uptake phenomenon can be considered as a minimization problem. The constraints for this minimization problem can be derived by considering that the actual uptake from a given depth, say r_z ($m^3\,m^{-3}\,sec^{-1}$), can be zero when the roots at that depth are temporarily nonactive, and its maximum value will be Q_z ($m^3\,m^{-3}\,sec^{-1}$). The actual rate of energy expenditure in extracting water at depth z will then be, say $e_z[= r_z \rho_w\,dz(\Psi_s - \Psi_r)]$. To minimize the total rate of energy expenditure, then the summation of e_z over z_r must be less than any other energy summation calculated over the whole rooted profile. Furthermore, the summation of $r_z\,dz$ over all rooted layers cannot exceed the total transpiration rate. An objective function can, therefore, be formulated that minimizes the total energy expenditure subject to the above constraints.

As can be deduced from Eq. 6, the energy required for extracting water from any depth derives from the product of U_z and the potential drop ($\Psi_s - \Psi_r$). Note that U_z is obtained from q_z (Eq. 2) whose calculation entails ($\Psi_s -$

Ψ_r). Therefore, the energy expenditure in water extraction from any depth depends on the square of the potential drop, so that a slight decrease in water potential at any depth increases energy requirement at that depth considerably. The energy model can, therefore, be used to "identify" zones within the soil where water would be preferentially taken up in seeking to satisfy the atmospheric water demand.

The minimum energy hypothesis has received only a limited testing. A simulation of the water uptake patterns from an initially uniformly wetted soil profile during a 20-day drying cycle is shown in Fig. 2. In this simulation, a hypothetical root distribution, which declines exponentially with depth (Fig. 2a) was assumed. The measured leaf water potential data of maize during a drying cycle[12] was used as a surrogate for the water potential at the root surface. As shown in Fig. 2b, water uptake begun initially from the top sections of the soil (day 1). With time, more water was withdrawn from the deeper sections of the soil profile. By day 20, no water was withdrawn from the upper sections although the root distribution remained unaltered. These types of water uptake patterns are reported in the literature. An important advantage of the minimum energy model is that it avoids the need to make any prior assumptions about the root-water uptake pattern and

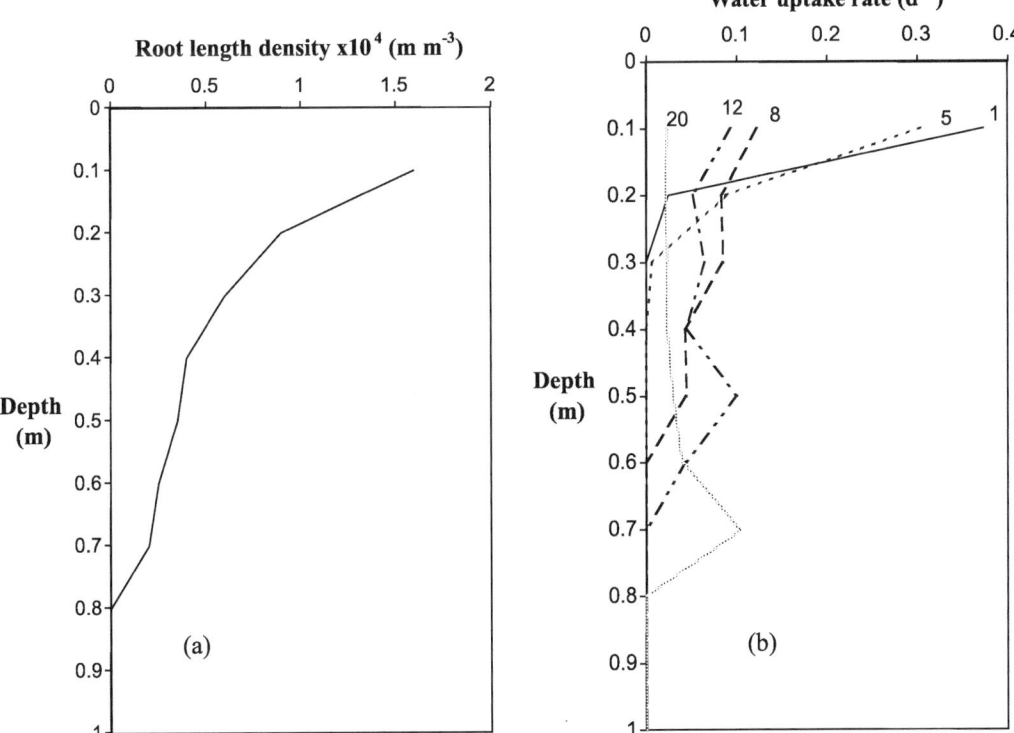

Fig. 2 Root distribution (a) and (b) simulated patterns of root-water uptake (day^{-1}) during a 20-day drying cycle (numbers on the curves: days of drying cycle).

provides a useful tool for analyzing water uptake patterns under varying soil and root distribution conditions. In its present form, the minimum energy model assumes that the root-water potential is spatially nonvariant, a weakness inherent in Eq. 2. Including a spatially variable root-water potential term, however, would not negate the hypothesis, but improve the model.

CONCLUSION

Modeling root-water extraction continues to be an important research subject, as models are increasingly used for crop and water management. Although a lot of research has been done on the subject, much more research is still required to help formulate the roles of the plant and root distribution in determining uptake patterns. The concept of energy minimization is an attempt to formulate the role of the plant in water extraction. The minimum energy models is capable of simulating realistic water uptake patterns under varying soil water conditions, but further development is necessary to account for varying root-water potential.

ACKNOWLEDGMENT

The author wishes to acknowledge the critical comments made by Prof. C. W. Rose and Dr. R. D. Braddock, both of Griffith University, Australia.

REFERENCES

1. Molz, F.J. Model of Water Transport in the Soil–Plant System: A Review. Water Resour. Res. **1981**, *17*, 1245–1260.

2. Clothier, B.E.; Green, S.R. Roots: The Big Movers of Water and Chemical in Soil. Soil Sci. **1997**, *162* (8), 534–543.

3. Gardner, W.R. Dynamic Aspects of Water Availability to Plants. Soil Sci. **1960**, *89*, 63–73.

4. Cardon, G.E.; Letey, J. Plant Water Uptake Terms Evaluated for Soil Water and Solute Movement Models. Soil Sci. Soc. Am. J. **1992**, *32*, 1876–1880.

5. Doussan, C.; Pagès, L.; Vercambre, G. Modeling of the Hydraulic Architecture of Root Systems: An Integrated Approach to Water Absorption—Model Description. Ann. Bot. **1998**, *81*, 213–223.

6. Feddes, R.A.; Kowalik, P.J.; Malinka, K.K.; Zaradny, H. Simulation of Field Water Uptake by Plants Using a Soil Water Dependent Root Extraction Function. J. Hydrol. **1976**, *31*, 13–26.

7. Nimah, M.N.; Hanks, R.J. Model for Estimating Soil Water, Plant and Atmospheric Interrelations: I. Description and Sensitivity. Soil Sci. Soc. Am. Proc. **1973**, *37*, 522–527.

8. Olson, K.A.; Rose, C.W. Patterns of Water Withdrawal Beneath an Irrigated Peach Orchard on a Red-Brown Earth. Irrig. Sci. **1988**, *9*, 89–104.

9. Bruckler, L.; Lafolie, L.; Tardieu, F. Modelling Root Water Potential and Soil–Root Water Transport: II. Field Comparisons. Soil Sci. Soc. Am. J. **1991**, *55*, 1213–1220.

10. Gardner, W.R. Modeling Water Uptake by Roots. Irrig. Sci. **1991**, *12*, 109–114.

11. Adiku, S.G.K.; Rose, C.W.; Braddock, R.D.; Ozier-Lafontaine, H. On the Simulation of the Root Water Extraction: Examination of the Minimum Energy Hypothesis. Soil Sci. **2000**, *165* (3), 226–236.

12. Ozier-Lafontaine, H.; Lafolie, F.; Bruckler, L.; Tournebieze, R.; Mollier, A. Modelling Competition for Water in Intercrops: Theory and Comparison with Field Experiments. Plant Soil **1998**, *204*, 183–201.

Urban Water Engineering and Management in Ancient Greece

A. N. Angelakis
National Foundation for Agricultural Research, Heracleion, Greece

D. Koutsoyiannis
National Technical University of Athens, Zographou, Greece

INTRODUCTION

Ancient Greek civilization has been thoroughly studied, focusing on mental and artistic achievements like poetry, philosophy, science, politics, and sculpture. On the other hand, most of technological exploits are still relatively unknown. However, recent research reveals that ancient Greeks established critical foundations for many modern technological achievements, including water resources. Their approaches, remarkably advanced, encompass various fields of water resources, especially for urban use, such as groundwater exploitation, water transportation, even from long distances, water supply, stormwater and wastewater sewerage systems, flood protection and drainage, construction and use of fountains, baths and other sanitary and purgatory facilities, and even recreational uses of water. The scope of this chapter is not the exhaustive presentation of what is known today about hydraulic works, related technologies and their uses in ancient Greece but, rather, the discussion of a few characteristic examples in selected urban water fields that chronologically extend from the early Minoan civilization to the classical Greek period. Agricultural hydraulic works like flood protection, drainage and irrigation of agricultural lands, and drainage of lakes were also in use in ancient Greece starting from the Mycenaean times, but are not covered in this chapter. Scientific advances in water resources as well as invention of hydraulic mechanisms and devices are presented in another entry (Hydrologic and Hydraulic Science and Technology in Ancient Greek).

CLIMATIC AND HYDROLOGIC CONDITIONS

Unlike preceding civilizations such as those in Mesopotamia and Egypt, which were based on the exploitation of water of the large rivers such as Tigres, Euphrates, and Nile, the Greek civilization has been characterized by limited and often inadequate natural water resources. The rainfall regime and consequently the water availability over Greece vary substantially in space. Thus, the mean annual rainfall exceeds 1800 mm in the mountainous areas of western Greece whereas in eastern regions of the country may be as low as 300 mm. Interestingly, the most advanced cultural activities in ancient Greece appeared in semiarid areas with the lowest rainfall and thus the poorest water resources; for example, Knossos in Crete, Cyclades islands, and Athens have annual rainfall about 500 mm, 300 mm–400 mm, and 400 mm, respectively. The potential evapotranspiration exceeds 1000 mm all over Greece, with the highest rates appearing in summer months. Thus, irrigation of cultivated areas during summer is absolutely necessary and becomes the most demanding water use in Greece. Under these climatic and hydrological conditions, Greeks had to develop technological means to capture, store, and convey water even from long distances, as well as legislation and institutions to more effectively manage water.

THE WATER SUPPLY IN MINOAN CIVILIZATION

Cultural advancements in the Minoan civilization can be observed throughout the third and second millennia B.C., which indicate that the main technical operations of water resources have been practiced in varying forms since ca. 3000 B.C.. During the Middle Bronze Age (ca. 2100–1600 B.C.) Crete's population in its central and south regions increased, towns were developed and the first palaces were built. At that time, a "cultural explosion" occurred on the island. A striking indication of this is manifested, inter alia, in the advanced water resources management technologies applied in Crete at that time. The sanitary life style developed at this civilization can be paralleled to the modern standards. It is evident that in Minoan civilization extensive systems and elaborate structures for water supply, sewerage systems, irrigation, and drainage were planned, designed, and built to supply the growing population with water for the cities and for irrigated agriculture.[1]

In the early phases of the Late Bronze Age (ca. 1600–1400 B.C.), Crete appears to have prospered even more, as

Encyclopedia of Water Science
DOI: 10.1081/E-EWS 120010076

the larger houses and more luxurious palaces of this period indicate.[2] At this time, the flourishing arts, improvements in metal-work along with the construction of better-equipped palaces and an excellent road system, reveal a wealthy, highly cultured, well-organized society, and government in Crete, before the island's power collapsed following the destruction of the Minoan palaces.[3] The geological catastrophe through the eruption of the Santorini volcano in 1450 B.C. halted the Minoan civilization.

Our knowledge of how Minoan cities were supplied with potable water is mainly acquired from the Palace of Knossos. A few cisterns, fountains, and wells were also found at other archeological sites like Zakros, Mallia, Gortys, and other Minoan palaces and cities. At Phaistos some cisterns have been discovered too, but owing to the nature of the ground, no wells or springs have been found there.[1]

Even at Knossos, the sources of water and the methods used for supplying it are only partially understood. Several wells have been discovered in the Palace area, and a single well slightly to the northwest of the Little Palace. The latter, restored to its original depth of about 12.5 m and 1.0 m diameter, continues to furnish an excellent supply of potable water.[4] In the Protopalatial stage (ca. 1900–1700 B.C.), several wells were used for drawing drinking water. Their depth did not exceed 20 m and their diameter was not more than 5 m.[5] At least six such wells have been reported.[4] The most important and best known is the one found in the north-west of the Palace in the basement of the House A, which belongs to the first stage of the Middle Minoan period. According to Evans,[4] its upper circuit was mostly a patchwork of rubble masonry, recalling the construction of Roman wells in the site. However, below its crudely built upper "collar," the well was found to be cased in a series of terracotta cylinders of fine clay and of material so hard that it was initially mistaken for some kind of close-grained stone (Fig. 1).

The inhabitants of the Knossos Palace, however, did not depend on the water of the wells alone. There are indications that the water supply system of the Palace of Minos at Knossos was initially dependent on the spring water of Mavrokolybos and later on the Fundana, and other springs. Mavrokolybos, a pure limestone spring, is located at a distance of 700 m south of the palace and an elevation of about 115 m, whereas Knossos lies at an elevation of 85 m from sea level; Fundana, a typical karstic spring with excellent quality of water even today, is at a distance of about 5 km from the palace and at an elevation of about 220 m.

Water supply in the Palace was provided through a network of terracotta piping located beneath the palace floors. The pipes were constructed in sections of about 60 cm–75 cm each. These pipes with their expertly

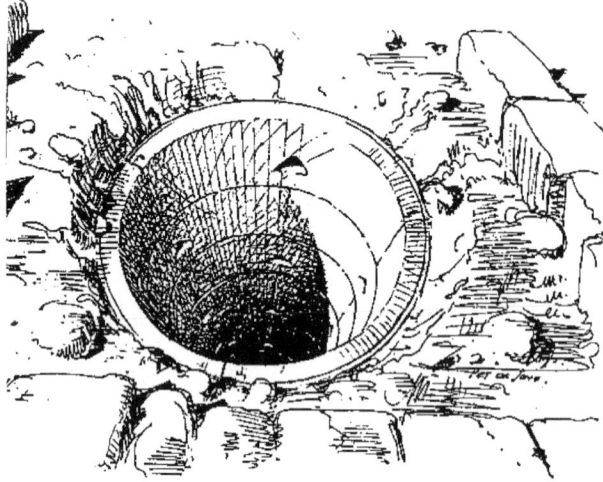

Fig. 1 Perspective view of well below House A, NW of Knossos Palace. (From Ref. [4].)

shaped, tightly interlocked sections, date from the earliest days of the building and are quite up to modern standards (Fig. 2). The sections of the clay pipes resemble those used in Greece in classical times, though Evans considered the Minoan to have been designed more efficiently; each section was rather strongly tapped toward one end with the objective of increasing the rate of water flow, thus helping to flush any sediment through the pipe.[5]

On the basis of their accomplishments, it can be assumed that Minoan hydraulic engineers were, in a sense, aware of the basic hydrostatic law, known today as the principle of communicating vessels. It is manifested in the water supply of the Knossos Palace through pipes and conduits fed by springs; this is supported by the discovery of the Minoan conduit heading towards the Knossos Palace from Mavrokolybos which suggests a descending and subsequently ascending channel.[4,6] However, it appears that Minoans had only a vague understanding of the relationship between flow and friction.

In the Zakros Palace the water supply system depended on groundwater. Here the potable water came from the Main Spring. In the southwest corner of the Cistern Hall an opening leads into a small chamber where the water was collected and channeled into a square underground fountain built on the south; this was thought to correspond with the celebrated man made fountain of the *Odyssey* known as "Τυκτή" fountain.[7] The fountain was built of regular limestone, and there is a descending staircase with fourteen steps (Fig. 3). The room may also have served as a shrine. The water of the fountain is brackish today, of about 13.00 dS/m electric conductivity (EC), due to intrusion of seawater. However, this may well be an indication that some reduction in the distance of the palace from the coast has occurred.

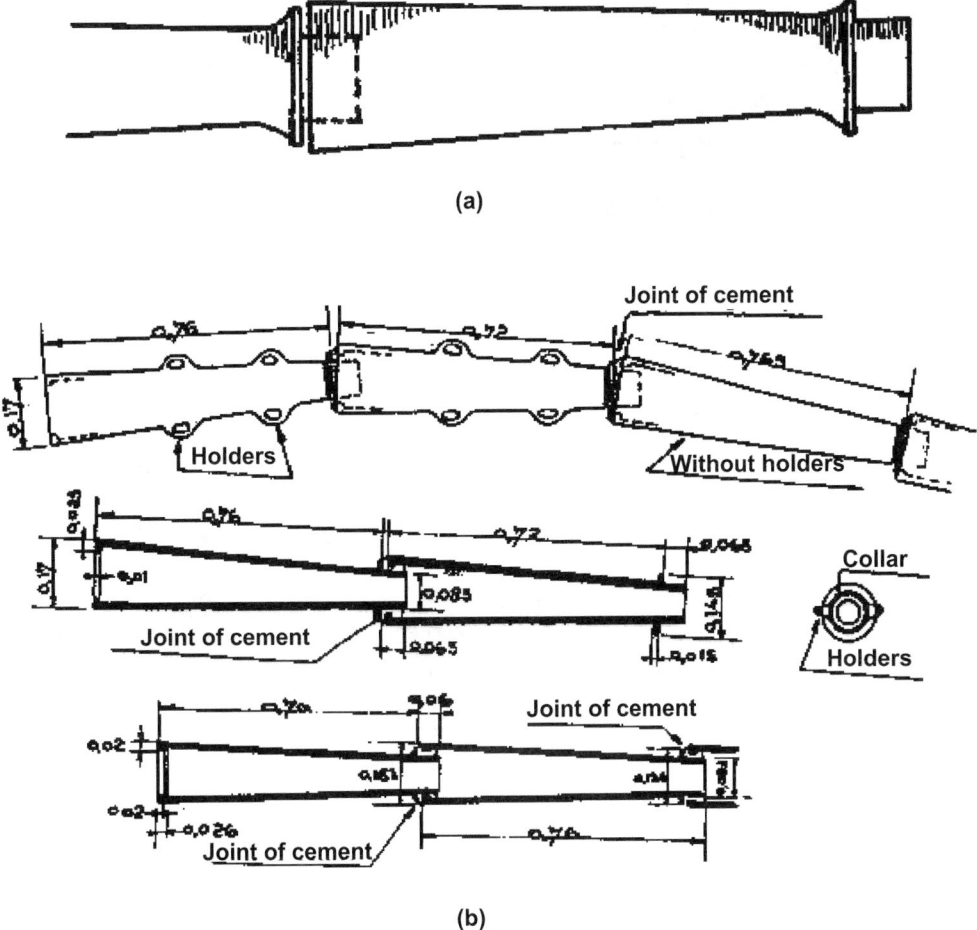

Fig. 2 Minoan water supply pipes (terracotta pipe sections): (a) overview, and (b) with real dimensions. (From Ref. [5].)

Another comparable chamber in Zakros is a well–spring located near the southeast corner of the Central Court; here again steps lead down into the chamber. The wood of the windlass was found in the water, along with an offering cup containing olives; this is a unique, remarkable find, since the olives were perfectly preserved, as though they had just been picked from the trees; unfortunately they maintained their relative freshness for only a few minutes after they were taken out of the water.[7] A view of this well-spring is given in Fig. 4.

In contrast to Knossos, where water was conveyed mainly from springs, and Zakros dependent entirely on groundwater, in Phaistos the water supply system was dependent directly on precipitation: here, the rainwater was collected from the roofs and yards of buildings in cisterns. Special care was given to securing clean surfaces in order to maintain the purity of water. Also, coarse sandy filters were used to treat the rainfall water before it flowed into the cisterns.

THE WATER SUPPLY OF SAMOS AND THE AWESOME FEAT OF EUPALINOS

The most famous hydraulic work of ancient Greece was the aqueduct of ancient Samos (located where Pythagoreio or Tigani village in the Samos island is currently present), which was admired both in antiquity (as recognized by Herodotus) and in modern times.[8–14] The most amazing part of the aqueduct is the 1036 m long "Ευπαλίνειον όρυγμα," or "Eupalinean digging," more widely known as Tunnel of Eupalinos. The aqueduct includes two additional parts (Fig. 5) so that its total length exceeds 2800 m. The aqueduct was the work of Eupalinos, an engineer from Megara. Its construction was commenced in ca. 530 B.C., during the tyranny of Polycrates and lasted for 10 yr. It was in operation until the 5th century A.D. and then it was abandoned and forgotten. Owing to the text of Herodotus, Guerin[8] uncovered the entrance of the aqueduct. The inhabitants of the island attempted to reuse the aqueduct in 1882 without success. Only 90 yr later, between 1971 and

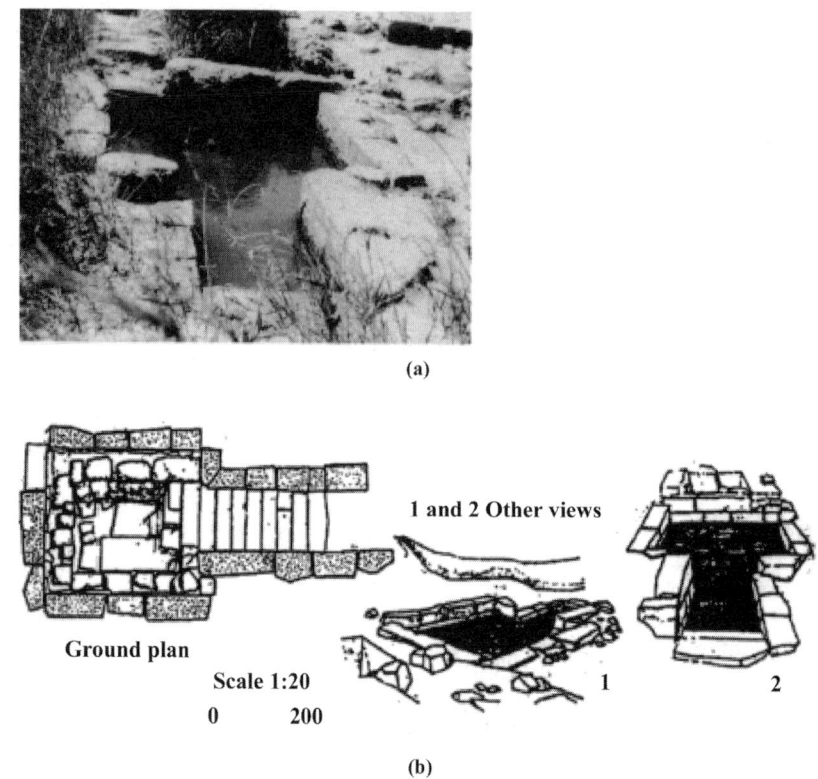

(a)

1 and 2 Other views

Ground plan

Scale 1:20

0   200

(b)

Fig. 3 Views of the "Τυκτή" fountain: (a) overall view and (b) scheme. (From Ref. [7].)

1973, the German Archaeological Institute of Athens undertook the task to finally uncover the tunnel.

Herodotus (History, Γ, 60) called the tunnel "αμφ ίστομον" or "bi-mouthed," a characterization that caused curiosity to the readers (any tunnel has two openings or mouths). Only when the tunnel was totally explored was it understood that Herodotus meant that the construction of the tunnel was started from two openings. Today, it is very

Fig. 4 Well-spring located in the eastern wing of the Zakros Palace. (From Ref. [7].)

common that water transportation tunnels are constructed from two openings to reduce construction time; high-tech geodetic means and techniques like global positioning systems and laser rays are used to ensure that the two fronts will meet each other. The great achievement of Eupalinos is that he did this using the simple means available at that time; apparently, however, he had good knowledge of geometry and geodesy. Later, in the 1st century B.C., his achievement inspired the mathematician and engineer Hero (Heron) of Alexandria (Dioptra, III) who in his geometrical Problem #15 studied how "to dig a mountain on a straight line from two given mouths." His method is based on walking around the mountain measuring out in one direction, then turning at a right angle, measuring again, etc., and finally using geometrical constructions with similar triangles. Moreover, in modern times, it inspired many mathematicians, engineers, and archeologists who attempted to reconstruct the methods used by Eupalinos to build the tunnel, as, apart from the mention by Herodotus, no written document was found from that time about the project.

Today, most of the questions have been answered but not all. For example, there is evidence that Eupalinos did not follow Hero's method, which would produce a large error. Most probably, Eupalinos walked over the mountain and put poles up along the path in a straight line. When the workers were digging they could try to line themselves up

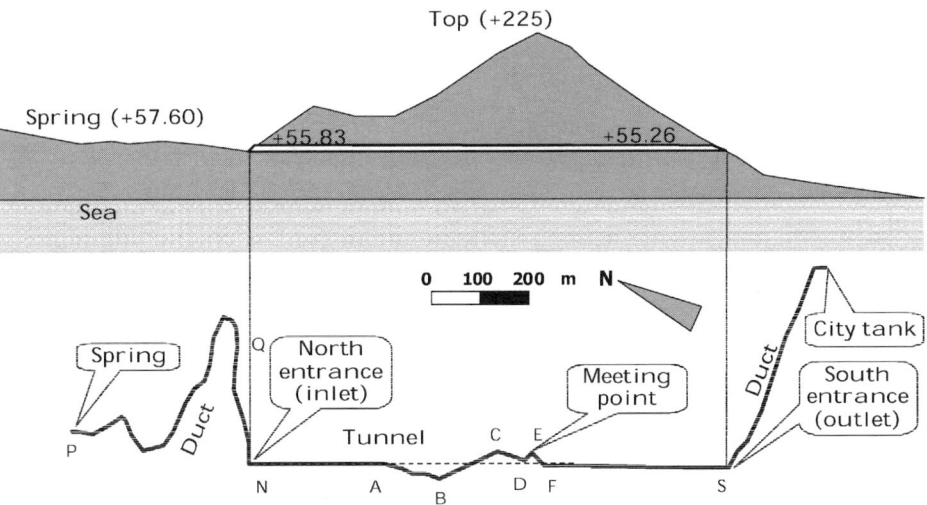

Fig. 5 Sketch of the Tunnel of Eupalinos (above: vertical section; below: horizontal plan).

Fig. 6 The Tunnel of Eupalinos. The duct is shown to the left.

with these poles. This also leaves room for error; as shown in Fig. 5, there was a small departure in the two axes that Eupalinos implemented (NA and SF), which is now estimated to 7 m. Another question is: what led Eupalinos to leave the straight line NA at point A and follow the direction AB? A plausible explanation is given by Tsimpourakis:[14] Eupalinos found a natural fracture or rift and broadening this rift, he was able to proceed much faster. At the end of the rift, he attempted to correct the departure from the initial axis, following the route BC, but C was past this axis. Again according to Tsimpourakis,[14] when the two teams of workers (each consisting of two people) were simultaneously at points C and F, they realized (hearing the sounds of the opposite team's excavating tools) that there were close to each other. Then, guided by the sounds of tools they managed to meet at point E. Hermann Kienast of the German Archaeological Institute of Athens proposed a different explanation: the last meters of the two routes of the tunnel (sections CDE and FE) were ingeniously designed rather than coincidentally followed: both teams were directed at points C and F to change direction to the right and then at D the northern one turned to the left on purpose; with this trick it is mathematically sure that the two lines would intersect.

Interestingly, the floor of the tunnel was done virtually horizontal, as observed from the elevations shown in Fig. 5; one would expect that it should have some slope for the water to flow. The choice of a horizontal tunnel is related to the excavation from both sides. In a sloping tunnel, the front of the upper section would be inundated (mostly from groundwater), so that the workers could not dig. Another reason is the fact that the horizontal tunnel was easier to control and build with the simple instruments and tools of that time and facilitated the meeting of the two fronts (indeed, the difference in the elevation of the two sections at point F is only 0.60 m).

However, this horizontal tunnel could not operate as an aqueduct, simply because the water would not flow horizontally. Therefore, Eupalinos excavated a slopping duct below the floor of the tunnel, shown in the photo of Fig. 6. Its bottom, where clay pipes were arranged, is located 3.5 m and 8.5 m in the inlet (N) in the outlet (S), respectively, below the floor of the tunnel; the large depth at the inlet is another question mark of the project, whose discussion is out of the scope of this chapter. At points where the depth becomes too large (in about two-thirds of the tunnel length) Eupalinos preferred to make a second tunnel, the water tunnel, below the main tunnel, the access tunnel. The water tunnel is about 0.60 m wide whereas the access tunnel is about 1.80 m × 1.80 m. The construction of the water tunnel was easy and fast, provided that the access tunnel was completed; 28 vertical shafts were constructed for easy access to the water tunnel and many teams of workers must have been worked simultaneously to dig it. The outer parts of the aqueduct were constructed in a similar manner. Thus, section PQ of the north duct (Fig. 5) was constructed as an open channel whereas section QN was a tunnel with five shafts.

What Eupalinos did was not the only solution to the problem of conveying water to Samos. A simple alternative solution was to continue the simple and fast method of section PQN constructing a chain of open channels and tunnels at small depths with shafts. In this solution, the route from point N to S would be around the mountain. Not only is this alternative solution technically feasible, but also it is technically easier, less expensive, and faster. Why Eupalinos preferred his unorthodox and breakthrough solution? How did he persuade the tyrant Polycrates to support this solution? These are unanswered questions. Probably he wished to build a monument of technology rather than simply solving a specific water transportation problem.

THE SUSTAINABLE URBAN WATER MANAGEMENT IN ATHENS

Water management in ancient Athens, the most important city of antiquity with a population of more than 200,000 during the golden age (5th century B.C.), is of great interest. Athenians put great efforts into the water supply of their anhydrous city. The first inhabitants of the city chose the hill of Acropolis for their settlement due to the natural protection it offered and the presence of three natural springs,[15] the most famous being "Clepsydra." However, natural springs in Acropolis and elsewhere were not enough to meet water demand. Therefore, Athenians used both groundwater, by practicing the art of drilling of wells, and stormwater, by constructing cisterns. In addition, the water from the two main streams of the area, Kephisos and Ilissos, whose flow was very limited in summer, was mainly used for irrigation.

Archeological evidence reveals that the city had developed an important system of public water supply consisting of wells, fountains, and springs and there were also a number of private springs and wells. There are indications that a primitive distribution system was in place underneath the city, consisting of underground connections of wells;[15] this expanded all around the city to the outskirts.[16] The most important public work was the Peisistratean aqueduct, built in the time of the tyrant Peisistratos and his descendants (ca. 510 B.C.).

Fig. 7 Part of the Peisistratean aqueduct (top) and detail of the pipe sections and their connection (bottom). (Photos reproduced from newspaper Kathimerini.)

The exact location and route of the aqueduct is not well known to date. It is known, however, that it carried water from the foothill of the Hymettos mountain, probably from east of the Holargos suburb at a distance around 7.5 km,[17] to the center of the city near Acropolis. The greater part of it was carved as a tunnel at a depth reaching 14 m. In other parts it was constructed as a channel, either carved in rock or made of stone masonry, with depth 1.30 m–1.50 m and width 0.65 m.[18] In the bottom of the tunnel or channel, a pipe made of ceramic sections was placed (Fig. 7). The pipe sections had elliptic openings with ceramic covers in their upper part for their cleaning and maintenance; the ends of the sections were appropriately shaped, so

that they could be tightly interlocked, and were joined with lead.

In the recent excavations for the construction of the metro, the widespread use of such ceramic pipes was revealed. Similar pipes were also used for sewers. Sewers of large cross section, most probably storm sewers, were built of stone masonry; some of them were natural streams, like Heridanos, that were covered (Fig. 8).

Apart from the structural solutions for water supply and sewerage, the Athenian civilization developed a legislation and institutional framework for water management. The first known laws are due to Solon, the Athenian statesman and poet of the late 7th and early 6th century B.C., who was elected archon in 594 and shaped a legal

Fig. 8 The Heridanos stream converted into a sewer at Ceramicos (up) and two tributary sewers (down) at Ceramicos (left) and Agora (right). (From Ref. 18.)

system by which he reformed the economy and politics of Athens. Most of his laws have been later described by Plutarch (47–127 A.D.), from whom it could be learnt that:

> Since the area is not sufficiently supplied with water, either from continuous flow rivers, or lakes or rich springs, but most people used artificial wells, Solon made a law, that, where there was a public well within a hippicon, that is, four stadia (4 furlongs, 710 m), all should use that; but when it was farther off, they should try and procure water of their own; and if they had dug ten fathoms (18.3 m) deep and could find no water, they had liberty to fetch a hydria (pitcher) of six choae (20 L) twice a day from their neighbors; for he thought it prudent to make provision against need, but not to supply laziness (Plutarch, Solon, 23).

MacDowell[19] conjectures that these laws have been kept unchanged through the classical period. As the city's public system grew and aqueducts transferred water to public fountains, private installations like wells and cisterns tended to be abandoned. But, the latter would be necessary in times of war because the public water system would be exposed; therefore, the owners were forced by decree to maintain their private facilities in good condition and ready to use.[20] Other regulations protected surface waters from pollution.[19] An epigraph of ca. 440 B.C. contains the "law for tanners," who are enforced not to dispose their wastes to Ilissos river.[15]

A distinguished public administrator, called "κρουνῶν επιμελητής", that is, officer of fountains, was appointed to operate and maintain the city's water system, and to ensure keeping of regulations and fair distribution of water. In addition, a number of guards were responsible for the proper daily use of the public springs and fountains. From Aristotle (Athenaion Politeia, 43.1) it is learnt that the officer of fountains was one of the few that were elected by vote whereas most other officers were chosen by lot; so important was this position within the governance system of classical Athens.[17] Themistocles himself had served in this position. In 333 B.C. the Athenians awarded a gold wreath to the officer of fountains Pytheus because he restored and maintained several fountains and aqueducts. The entire regulatory and management system of water in Athens must have worked exceptionally well and approached what today we call sustainable water management. For example, modern water resource policymakers and hydraulic engineers emphasize the nonstructural measures in urban water management and the importance of small-scale structural measures like domestic cisterns, which reduce the amount of stormwater to be discharged and provide a source of water for private use.

The importance of water in Athens was not only related to the basic uses like drinking, cooking, and cleaning.

Water was also related to the beauty of the city; this is revealed from the many fountains that Athenians constructed and the depictions thereof on vessels. Given that vessels were used to export goods, they can be regarded as sort of advertisement of the city's beauty. Another important water use in Athens was in public baths, cool or warm, called "βαλανεία" (later passed in Latin as balineae or balneae), which, interestingly, at times were common for men and women (what we call today bains mixtes), and were related to enjoyment, health, socialization, and culture.[21] Later, the Romans took up and extended the Greek water technology including, of course public fountains and balneae, which became a matter of luxury and prestige. As a sort of requital, the Roman emperor Hadrian (117–138 A.D.) showed particular interest for Athens; at his time the famous Hadrianic aqueduct was commenced, which conveyed water from mountains, Parnes and Pentele, to Athens covering a distance of 25 km. This aqueduct was in operation until the middle of the 20th century.

REFERENCES

1. Angelakis, A.N.; Spyridakis, S.V. The Status of Water Resources in Minoan Times: A Preliminary Study. In *Diachronic Climatic Impacts on Water Resources with Emphasis on Mediterranean Region*; Angelakis, A.N., Issar, A.S., Eds.; Springer-Verlag: Heidelberg, Germany, 1996; Chap. 8, 161–191.

2. *History of the Greek Nation (in Greek)*; Ekdotiki Athinon: Athens, 1970.

3. Angelakis, A.N.; Spyridakis, S.V., In *Wastewater Management in Minoan Times*, Intern. Proc. of the Meeting on Protection and Restoration of Environment, Chania, Greece, August 28–30,1996; 549–558

4. Evans, S.A. *The Palace of Minos at Knossos: A Comparative Account of the Successive Stages of the Early Cretan Civilization as Illustrated by the Discoveries*; Macmillan and Co.: London, 1921–1935; Vol. I–IV (reprinted by Biblo and Tannen: New York, 1964).

5. Buffet, B.; Evrard, R. *L'Eau Potable a Travers Les Ages*; Editions Soledi: Liege, Belgium, 1950.

6. Hutchinson, R.W. Prehistoric Town Planning in Crete. Town Plan. Rev. **1950**, *21*, 199–220.

7. Platon, N. Ζάκρος Το Νέον Μινωϊκόν Ανάκτορον (*Zakros, The New Minoan Palace*); The Athens Archaeological Society: Athens, 1974.

8. Guerin, V. Description de l'Ile Patmos et de l'Ile Samos; Paris; 1865.

9. Fabricius, E. Alterthuemer auf der Insel Samos. Mitt. Dtsch. Archaologischen Inst. Athen **1884**, *9*, 165–192.

10. Stamatiades, E. Περί του Ορύγματος του Ευπαλίνου εν Σάμω (*About the Digging of Eupalinos in Samos*); Hegemonion Press: Samos, 1884 (in Greek).

11. Goodfield, S.T. Toulmin. How Was the Tunnel of Eupalinos of Samos Aligned? Isis **1965**, LVI,46.

12. Kienast, H.J. Der Tunnel des Eupalinos auf Samos. Architectura, Zeitschrift für Geschichte der Architektur **1977**, 97–116.

13. Lazos, C.D. *Μηχανική και Τεχνολογία στην Αρχαία Ελλάδα (Mechanics and Technology in Ancient Greece)*; Aeolos: Athens, 1993 (in Greek).

14. Tsimpourakis, D. *530 π.Χ., Το Όρυγμα του Ευπαλίνου στην Αρχαία Σάμο (530 B.C., The Digging of Eupalinos in Ancient Samos)*; Editions Arithmos: Athens, 1997 (in Greek).

15. Pappas, A. *Η'Υδρευσις των Αρχαίων Αθηνών (The Water Supply of Ancient Athens)*; Eleuphtheri Skepsis: Athens, 1999 (in Greek).

16. Kallis, G.; Coccossis, H. *Metropolitan Areas and Sustainable Use of Water: The Case of Athens*, Report 2000; University of Aegean: Mytilene, Greece, 2000.

17. Tassios, T.P. Από το "Πεισιστράτειο" στον Εύηνο (From "Peisistratean" to Evinos); Kathimerini, March 24, 2002; (in Greek) 2002.

18. Papademos, D.L. *Τα Υδραυλικά Έργα Παρά τοις Αρχαίοις (The Hydraulic Works in Ancient Greece)*; TEE: Athens, 1975; Vol. B (in Greek).

19. MacDowell, D.M. *The Law in the Classical Athens*; Thames and Hudson: London, 1978.

20. Korres, M. Η ύδρευση της Αθήνας κατά την αρχαιότητα (Water Supply of Athens in Antiquity). *Workshop: "Water and Environment"*; EYDAP: Athens, 2000 (in Greek).

21. Koromilas, L.G. *Το Αθηναϊκό Κελάρυσμα (The Athenian Gurgle)*; EEY: Athens, 1977.

Vadose Zone and Groundwater Protection

Michael H. Young
Charalambos Papelis
Desert Research Institute, Las Vegas, Nevada, U.S.A.

INTRODUCTION

This section will familiarize the reader with natural processes and anthropogenic activities that protect soil and groundwater qualities. First, the vadose zone is defined as the aerated region of soil or geologic material above the permanent water table and below ground surface. Vadose zone and groundwater protection is defined here in terms of quality, rather than quantity, because the increasing stress on worldwide water supplies is often viewed in terms of water that is potable or of high enough quality for use in agricultural or industrial uses. We will discuss the types and sources of soil and groundwater contamination, describe geochemical characteristics of soil or aquifer material that affect contaminant levels, and then describe methods to control or reduce subsurface contamination through the use of containment structures (e.g., landfills), and active and passive treatment technologies.

TYPES AND SOURCES OF CONTAMINATION

Types of Contamination

Waste material comes in many forms and toxicity levels, and from a variety of sources. Solid wastes range from municipal waste, such as household trash, to hazardous materials, including some mining and industrial wastes. Liquid wastes range from municipal wastewaters, which are typically not considered hazardous, to industrial wastewaters containing organic contaminants, acids, or high concentrations of metals. Finally, sludges contain between 3% and 25% solids,[1] and could contain a variety of hazardous solids or liquids. Each of these waste forms is subject to control through State or Federal regulations, or both.

Each potential waste form can contain a variety of contaminants. Organic contaminants (e.g., fuels, chlorinated solvents, and pesticides) vary widely in their chemical properties and mobility in the subsurface environment. For example, a straight chain hydrocarbon with an OH^- group (e.g., hexanol) and a benzene ring with an OH^- group (i.e., phenol) have the same number of carbon atoms, but are structurally different. Consequently,

they will have different affinities for the same soil or aquifer material. Inorganic contaminants, which occur naturally or as a byproduct of industrial processes, also take many forms depending on the oxidation–reduction state of the subsurface environment. Their migration characteristics are highly variable and affected by time, redox environment, the nature of the soil or aquifer material, and the presence or absence of other contaminants. Finally, biological contaminants, which can be present in groundwater due to improper disposal of human or animal wastes, are often reactive and subject to mechanical filtering and biodegradation. Moreover, bacteria or parasites have activation periods, so their toxicity levels change with time. Clearly, the differences in contaminant characteristics must be evaluated on a contaminant-specific basis to better understand risks to the soil and groundwater.

Sources of Contamination

Protecting soil and groundwater resources requires an understanding of how contaminants are released into the environment. Sources can be broadly classified as point and nonpoint. Point sources are typically related to industrial processes, and releases often occur while temporarily storing materials used in manufacturing or chemical operations. Fluid storage tanks, below- and above-ground plumbing, and impoundments can leak and contribute to contamination. Concentrated feedlot operations, a livestock-based industrial source, are sources of nitrate, phosphorus, and fecal coliforms.[2] Municipal storm flow might also be a point source of these nutrient and biological contaminants.

Nonpoint sources are typically related to agricultural practices, because of the widespread application of pesticides and fertilizers, and some mining operations, where large swaths of land are disturbed during mineral extraction. Contemporary agricultural practices in many countries rely heavily on the application of organic and inorganic fertilizers, which have high concentrations of nitrogen and phosphorus. Downward migration of these fertilizers away from plant roots can increase concentrations of nitrogen and phosphorus in groundwater to levels that can exceed water quality criteria. In some

Encyclopedia of Water Science
DOI: 10.1081/E-EWS 120010273

instances, water from a large numbers of water wells was rendered nonpotable. Urban runoff also can contribute to nonpoint source pollution; e.g., runoff from streets can contain oils and salts, and runoff and downward drainage from parks, lawns, and golf courses can be a source of sediment, pesticides, nitrogen, and phosphorus.

ATTENUATION PROPERTIES OF SOIL

Soil material, composed of fragments of rock that have undergone physical, chemical, and biological weathering, has potentially a very high capacity to attenuate (or reduce) the concentration of contaminants migrating through it. Biogeochemical processes that could be responsible for the reduction of contaminant concentrations in soil include biological or abiotic transformation, sorption, precipitation–dissolution, oxidation–reduction, complexation, and mass transfer processes.[3] The attenuation capacity of soil depends on many factors, including the soil physicochemical characteristics, the type of contaminant, and the geochemical conditions in the soil pore water.

For nonporous particles, particle size is inversely proportional to specific surface area.[4] Particle surface area is typically related to the number of reactive sites and therefore the attenuation properties of the soil. In addition, particles on the order of 1 μm or less (colloids) can substantially enhance the transport of constituents associated with them[5] thereby affecting the extent of natural attenuation. The mineralogy of soil particles can also have a dramatic effect on the surface area. For example, nonweathered quartz and feldspar particles have an insignificant surface area compared to zeolites, smectite clay minerals, and disordered iron and aluminum oxide phases with high porosity.

The surface charge properties of particles are also a function of particle mineralogy and can dramatically affect the sorption of inorganic contaminants on mineral surfaces and therefore the potential for immobilization. Surface charge properties are described by the point of zero charge (PZC) of a mineral, or the pH value at which a particle has no net surface charge. The PZC for common soil minerals can vary from approximately 2 for quartz to approximately 8–9 for iron and aluminum oxides. Sorption of metals and other cations is favored at higher pH values where particles tend to be negatively charged, whereas sorption of oxy- and other anions is favored at lower pH values where particles tend to be negatively charged. It should be kept in mind, however, that strong, specific sorption on mineral surfaces is possible against electrostatic repulsion.

The organic content fraction of soil can also have significant implications for the natural attenuation of organic and inorganic contaminants. The partitioning of organic contaminants at the soil–water interface is directly proportional to the organic fraction content of the soil.[6] In addition, natural organic materials can complex metals and other inorganic ions, thereby immobilizing them. Organic compounds in soil particles can be used by microorganisms as a carbon or energy source, a prerequisite for biotransformation of organic compounds.

Finally, the redox conditions in a soil environment can dramatically affect attenuation properties. For example, the reductive dehalogenation of organic chemicals has been frequently observed.[7] This reduction process is sometimes catalyzed by the oxidation of reduced soil components (e.g., iron and manganese oxides or sulfides). Another example involves the oxidation or reduction of oxyanions, producing species with substantially different properties. The sorption properties of oxyanions of selenium and arsenic, two elements of significant environmental concern, are a strong function of oxidation state, leading to substantially different sorption affinities for mineral surfaces.

METHODS TO PROTECT VADOSE ZONE AND GROUNDWATER

The vast majority of groundwater contamination problems involve the vadose zone because, unless contaminants are introduced directly into the saturated zone, soil contamination will almost always precede groundwater contamination. Steps taken to reduce vadose zone contamination can thus reduce groundwater contamination.

Improving Waste Disposal Practices

Landfilling of waste is still the most common mode of disposal for those materials that are not recycled or reused. For example, as of 1992, 67% of municipal solid waste (MSW) generated in this country was landfilled.[8] The technology behind landfilling of waste varies significantly. In the past, landfills were not equipped with adequate covers or liners (Fig. 1a), and they were often located without regard to the proximity to groundwater resources. As a result, water from precipitation often percolated into the waste material, leached potentially harmful chemicals, and transported them into deeper soil layers and groundwater.

Current landfill designs are more sophisticated, and have a stronger appreciation for the need to reduce downward percolation of precipitation and generation of leachate. Fig. 1b shows some design features required in modern sanitary landfills. Hazardous waste disposal facilities have these and other design features required by the U.S. Environmental Protection Agency, including multi-layered covers and liners, leachate collection systems, and groundwater monitoring programs.[9] A key

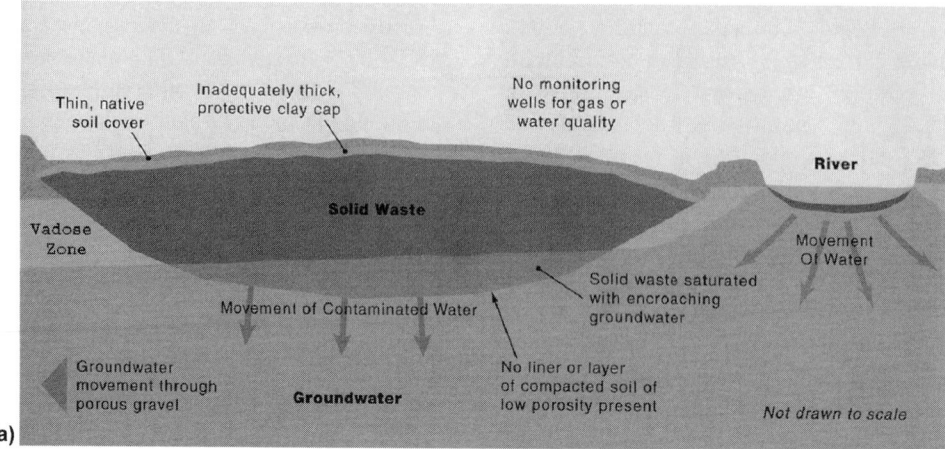

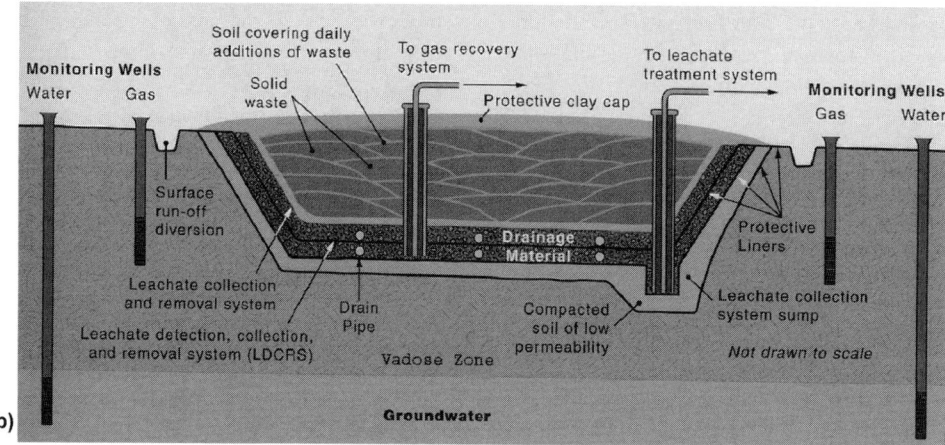

Fig. 1 Cross-sections of two landfills: (a) Old-style sanitary landfill without many design features commonly used today; (b) Modern sanitary landfill showing monitoring, containment, and gas recovery systems. (Reprinted from Ref. 1 with permission.)

design goal is to divert water away from the disposal cell, thereby reducing the potential for leachate generation.

Recently, some new design components have shown promise in reducing leachate generation at lower costs. Depending on the climatic conditions, plants can remove the majority of precipitation falling onto the disposal site cover, substantially reducing the potential amount of water that would percolate through the waste.[10] Other developments include the conversion of MSW landfills into bioreactors, where microorganisms consume waste material for their life energy. Careful control of environmental conditions inside the cell (e.g., water content, pH, temperature, etc.) is needed to enhance the bioremediation.[11]

Active Remediation of Existing Contamination

Protecting the vadose zone and groundwater from future contamination is easier than remediating existing con-

tamination. However, where contamination exists, removing it enhances vadose zone and groundwater protection. Engineered remediation strategies for reducing existing vadose zone and groundwater contamination fall into several broad categories: containment, removal, and treatment.[12]

Containment focuses on restricting or redirecting the movement of contaminants with either physical or hydraulic barriers. Physical barriers are structures designed to direct groundwater or soil water flow away from a contaminated area, thereby containing the plume size. Physical barriers can include sheet piling, slurry walls, grout curtains, and engineered covers (synthetic or natural). Hydraulic barriers are used to capture contaminated water through the manipulation of hydraulic gradients, and subsequently to treat, store, or otherwise dispose of the water.

Removal strategies include excavation, ex-situ pump and treat, and in-situ treatment. Excavation is a brute-force method of remediation, where contaminated soil or aquifer material is physically removed from the site. Though this method has some disadvantages (higher potential worker exposure, higher transportation cost), it has proven effective for areas with shallow, localized contamination. Pumping and treating contaminated water is probably the most widely used remediation technology. Contaminated groundwater is pumped from the aquifer and then treated at ground surface. If the primary concern is soil contamination, then clean water can be applied at ground surface and allowed to percolate through the contaminated area, leaching the contaminants and removing them from the vadose zone. The newly contaminated water is then captured using hydraulic control and treated.

Natural Attenuation

Natural attenuation is defined as the use of unenhanced natural processes as part of a site remediation strategy,[3] and is being used in conjunction with or as an alternative to engineered remediation systems.[13] The reliance on natural attenuation in the vadose zone and groundwater comes in part because of the high costs of active remediation at thousands of sites around the United States. Natural attenuation relies on biogeochemical degradation of contaminants through interactions with soil and aquifer material in the vadose or saturated environments. The degradation leads to contaminant destruction, immobilization, or transformation to innocuous byproducts. Biological transformation requires the presence and activation of a specific microorganism or a consortium of microorganisms that consume the contaminant in question. Chemical reactions that can attenuate contaminant concentrations include acid-base, redox, precipitation, sorption, and complexation.[3] In any biogeochemical reaction pathway, conditions in the substrate and the contaminant concentrations must be within a specific range, or the reaction rates can decrease. Natural attenuation must be coupled with an aggressive monitoring program to ensure regulators and the public that the processes are effectively reducing contaminant levels. Furthermore, the likelihood of success of natural attenuation is contaminant-specific; some contaminants (e.g., BTEX) have a high likelihood of success, while others (e.g., polycyclic aromatic hydrocarbons) have a lower likelihood. The choice of using natural attenuation is therefore complicated and must be made after careful consideration of subsurface conditions, the contaminant, and the regulatory and public acceptability.

REFERENCES

1. Artiola, Janick Waste Disposal. In *Artiola*; Pepper, I.L., Gerba, C.P., Brusseau, M.L., Eds.; Academic Press: San Diego, CA, 1996; 135–149.
2. Rice, J.A.; Viste, D.A. Major Sources of Groundwater Contamination: Assessing the Extent of Point and Nonpoint Contamination in a Shallow Aquifer System. In *Groundwater Contamination and Control*; Zoller, U., Ed.; Marcel Dekker, Inc.: New York, 1994; 21–37.
3. National Research Council, *Natural Attenuation for Groundwater Remediation*; National Academy Press: Washington, DC, 2000.
4. Gregg, S.J.; Sing, K.S.W. *Adsorption, Surface Area and Porosity*; Academic Press: London, 1982.
5. Stumm, W.; Morgan, J.J. *Aquatic Chemistry: Chemical Equilibria and Rates in Natural Waters*; John Wiley and Sons: New York, 1986.
6. Curtis, G.P.; Reinhard, M.; Roberts, P.V. Sorption of Hydrophobic Organic Compounds by Sediments. In *Sorption of Hydrophobic Organic Compounds by Sediments*; Davis, J.A., Hayes, K.F., Eds.; American Chemical Society: Washington, DC, 1986; 191–216.
7. Schwarzenbach, R.P.; Gschwend, P.M.; Imboden, D. *Environmental Organic Chemistry*; John Wiley and Sons: New York, 1993.
8. Miller, G.T. *Environmental Science*, 4th Ed.; Wadsworth: Belmont, CA, 1992.
9. U.S. EPA http://www.access.gpo.gov/nara/cfr/cfrhtml_00/Title_40/40cfr264_00.html (accessed May 2001).
10. Anderson, J.E.; Noack, R.F.; Ratzlaff, T.D.; Markham, O.D. Managing Soil Moisture on Waste Burial Sites in Arid Regions. J. Environ. Qual. **1993**, *22* (1), 62–70.
11. Reinhart, D.R.; Townsend, T.G. *Landfill Bioreactor Design and Operation*; Lewis Publishers: Boca Raton, FL, 1998.
12. U.S. EPA., *Remedial Actions at Waste Disposal Sites*, EPA/625/6-85/006; U.S. Environmental Protection Agency: Washington, DC, 1985.
13. U.S. EPA., *Cleaning up the Nation's Waste Sites: Markets and Technology Trends*, EPA/542-R-96-005, 1996 Ed.; EPA Office of Solid Waste and Emer. Response: Washington, DC, 1997.

Vapor Transport in Dry Soils

Glendon W. Gee
Anderson L. Ward
Pacific Northwest National Laboratory, Richland, Washington, U.S.A.

INTRODUCTION

Water-vapor movement in soils is a complex process, controlled by both diffusion and advection and influenced by pressure and thermal gradients acting across tortuous flow paths. Wide-ranging interest in water-vapor transport includes both theoretical and practical aspects. Just how pressure and thermal gradients enhance water-vapor flow is still not completely understood and subject to ongoing research. However, in unsaturated soils, it is now well accepted that the rate and direction of water flow may be completely misinterpreted if vapor movement is ignored. Practical aspects include dryland farming (surface mulching), water harvesting (aerial wells), fertilizer placement, and migration of contaminants at waste sites. The following article describes the processes and practical applications of water-vapor transport, with emphasis on relatively dry soil systems.

PROCESSES CONTROLLING VAPOR TRANSPORT

Diffusion

Water-vapor transport in fine-textured soils (e.g., silts and clays with little or no macroporosity) is often described as a simple diffusion process where Fick's law applies. In this type of assessment, the vapor flux is expressed in terms of a diffusion coefficient multiplied by a concentration gradient. The diffusion coefficient, in turn, is the product of the binary diffusion coefficient of pure water vapor in air, multiplied by a tortuosity factor and a term that is empirically related to the air-filled porosity of the soil. As soils dry out, the diffusion coefficient increases as a power function of the air-filled porosity. Recent work over the entire water-content range by Moldrup et al.[1,2] indicate that reasonable agreement occurs between observed and predicted values when the modeled vapor-transport equation includes a term for the air-filled porosity raised to a power of three.

Advection

In well-drained, coarse soils, and similarly in fine-textured soils containing a significant number of large (macro) pores, advection can control the transport of water vapor. Often the advection is temperature assisted, since seldom is the process isothermal. Kemper et al.[3] studied advective transport of water vapor through dry gravels and developed a working equation for water-vapor flow that includes both diffusive and advective flow, which can be written as:

$$q = (D_f + D_s)(P - \theta 2)(L/L_e)^2[C/Z] \tag{1}$$

where q (g cm^{-2} sec^{-1}) is the vapor flux, D_f (cm^2 sec^{-1}) is the diffusion coefficient for water in still air, D_s (cm^2 sec^{-1}) is the dispersion coefficient that is affected by wind speed, turbulence, and pore size, $P - \theta$ (m^3 m^{-3}) is the air-filled pore space, L/L_e; is the straight line distance through the mulch over the average tortuous path length; C (g cm^{-3}) is the water-vapor concentration difference across the mulch (from the soil–mulch to the mulch–air interface), and Z (cm) is the mulch thickness.

The combined diffusive and advective coefficients control the flow and the relative influence of each term is dictated by the characteristics of the porous media, being a function of the macroscopic properties including tortuosity and volumetric water content. As soil becomes coarser, advective flow can equal or exceed the diffusive flow.

Nonisothermal Flow

Over the years, there has been considerable interest in the nonisothermal flow of water vapor in soils.[4–15] The classic work by Philip and de Vries[4] has been the framework upon which most nonisothermal water-flow models have been developed and tested. Fig. 1 shows the basic concept of enhanced vapor transport in a soil pore aided by a temperature gradient.[4] Numerous observations over the years have demonstrated that there is an enhanced vapor transport in the presence of a thermal gradient that cannot be derived directly from first principles. Cass et al.[11] measured enhanced water-vapor diffusion that ranged from a factor of 1 to 15 over a range of temperatures and saturations. Nassar et al.[12] studied flow

Encyclopedia of Water Science
DOI: 10.1081/E-EWS 120010328

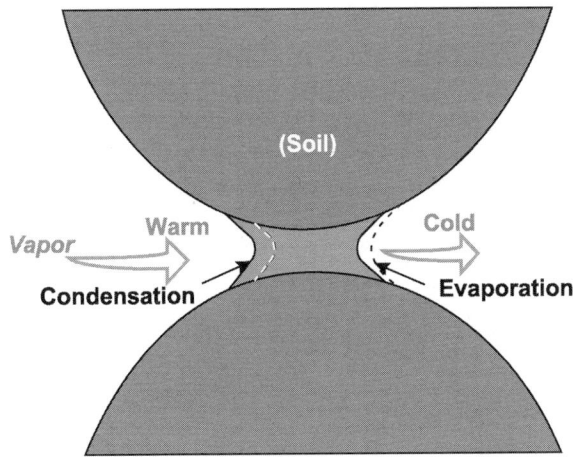

Fig. 1 Thermally enhanced water-vapor transport in a pore. Water condenses on the warmer side the liquid island and evaporates from the cooler side of the island creating a short circuit and enhancing vapor flow. (From Ref. [4].)

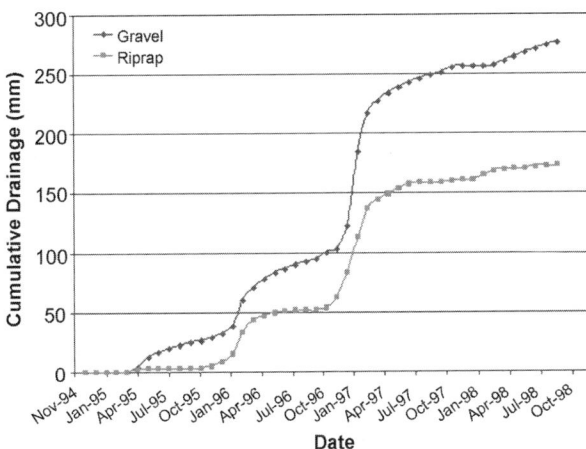

Fig. 2 Cumulative drainage from a rock side-slope of a surface barrier compared to drainage from an adjacent gravel side-slope. The lower drainage from the rock side-slope is attributed to thermal advection.

under a combination of thermal and solute gradients and found that water-vapor transport was under-predicted by a factor of four. The working hypothesis for the enhancement is that there are liquid islands that exist in porous media and that as the water vapor moves through the porous system, the vapor condenses on the warm side of the island and evaporates on the cool side of the island (Fig. 1). This short circuit is believed to cause the enhanced transport. For each soil, there is an optimal water contents area between saturation and air dry at which liquid islands are prominent in the flow pathway, and enhanced vapor flow is maximized.[16]

Convective Flow

There is mounting evidence that thermal pulses can cause enhanced vapor flow in soil, particularly near the soil surface. Measured flow in the field has been found to be an order of magnitude or more higher than that computed by invoking diffusive flow mechanics.[13] Parlange et al.[14] demonstrated that the agreement between theory and measurement could be improved substantially when both diffusion and convection were included explicitly in the analysis. These authors assumed that the mechanism responsible for the large vapor flux is convective transport driven by the diurnal heating and cooling of the soil surface and the corresponding thermal expansion and contraction of the soil air. This analysis and the corresponding conclusions are similar to those of Rose and Guo,[17] who demonstrated via computer modeling that thermal convection could accelerate the movement of soil air in hillsides. Field evidence of thermal convection, resulting in accelerated evaporation, was obtained by

Ward and Gee,[18,19] who reported on water losses from a 2:1 (horizontal/vertical) rock riprap side-slope located on a monitored landfill cover. Evaporation from the rock side-slope was about twice that of an adjacent gravel-side slope that had a lower slope (10:1) and a lower porosity (Fig. 2).

PRACTICAL ASPECTS OF VAPOR TRANSPORT

Mulching and Dryland Farming

The basic concept of mulching is that water stored in the soil during winter months (or periods of low evaporation) can be kept in the ground longer if there is a way to limit vapor losses from the soil surface. In arid-climate regions, farmers have tried various methods to conserve water using a variety of mulching materials. The Anasazi, ancient dwellers of the America southwest, used cobble mulch in their gardens as early as the 14th century A.D.[20,21] It is hypothesized that cobble surfaces stimulated crop production by increasing water storage, controlling weeds, and mitigating temperature extremes in an environment of limited moisture and elevated temperatures. White et al.[21] tested the theory of increased water storage using cobble mulch at test sites adjacent to the ancient gardens in New Mexico. Water storage in soils covered with 7-cm-thick gravel mulch was increased by as much as 50%. This is very similar to results reported by Kemper et al.[3] in studies conducted earlier in Ft. Collins, Colorado, to evaluate the benefits of gravel mulches for enhanced water storage. Fig. 3 shows the results of one of

Kemper's experiments (as reported by Hanks and Ashcroft[22]) using pea gravel (i.e., coarse material with a particle-size distribution ranging from 2 mm to 10 mm) as the mulch. The Kemper field study extended over 13 months during which 589 mm of precipitation fell on the test plots. The data show that evaporation losses were reduced by a factor of more than 3, and water storage increased by a similar amount as the thickness of the pea gravel mulch increased from 0 cm to 3 cm (about three complete layers). It is clear that coarse gravels provide a capillary break to upward liquid flow. The resultant vapor barrier then limits water transfer to the soil surface, causing a significant reduction in evaporation from soils under the gravel mulch.

A simple farming practice, known as fallow farming, requires that a farmer till his field to loosen the topsoil, but he does not plant. The loosened soil acts as a diffusion or vapor barrier to the already stored moisture, and in areas where there is insufficient precipitation to sustain a crop, the stored moisture from the previous year is used the next season to produce the crop. The application of mulch breaks the liquid continuity and increases diffusion resistance, thus limiting the rate of water loss from the soil surface. Mulches made of various materials have been used successfully to create a diffusion barrier. Over the past 50 years, studies have been conducted on various mulches, including dry soils, straw, and gravel or stones of various sizes and colors.[3,21,23–28] In general, clean gravel or stone mulches tend to retard evaporation losses more than mulches containing finer materials. Kemper et al.[3] developed a formula for estimating the efficiency of gravel mulch based on particle diameter and thickness of the barrier.

Aerial Wells

On the Crimean peninsula, on the shores of the Black Sea in central Asia, are porous stone remnants of what are believed to have been aerial wells used by the ancient inhabitants to condense and collect water from fog-laden air. Thermal differences caused warm, most air to condense on cool rock surfaces (Fig. 4). The water drained to above-ground cisterns, connected to aqueducts that carried water to nearby gardens and municipalities and had the capacity of supplying Theodosia with 721 m³ (190,000 gal) of water daily.[29] While it appears that such structures could be constructed at a number of locations throughout the world, the climatic conditions apparently have to be nearly perfect for significant quantities of water to be collected. Such a system would be impractical except for very limited use in arid areas where there is persistent fog. All modern attempts to replicate such a system for water production have failed.[29,30]

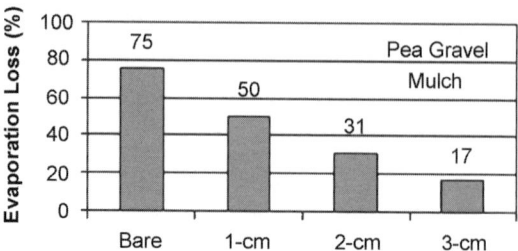

Fig. 3 Evaporation losses from pea gravel mulch layers of various thicknesses.

Fertilizer Placement and Waste Management

When deliquescent salts, like many fertilizers, are emplaced in partially saturated soils, differences in solute concentrations develop across the air-filled pores, and significant vapor movement can occur.[31] Under these conditions, the air–liquid interface acts as a semiperme-able membrane from which ions are excluded but across which water vapor can freely migrate in response to the osmotic potential gradient. The presence of salt causes a lowering of the vapor pressure, and water vapor is transported from regions of lower solute concentration to regions of higher concentration. Eventually, the migrating water condenses, the salt slowly dissolves, and the liquid continues to move under forces of gravity and capillarity.[32,33] This mechanism of moving salt in soil is of practical significance in agriculture. The placement of seeds or seedlings near fertilizer bands must account for this movement to minimize impacts on germination and seedling survival since high-salt concentrations can adversely affect plants. There are also implications for managing saline wastes in the vadose zone. An analysis of vapor condensation at a waste site, where high concentrations of sodium nitrate salts are stored in tanks, showed that one mechanism for long-term leakage of fluid is vapor condensation.[34] Vapor-transport analyses, using a number of design parameters including the thickness of

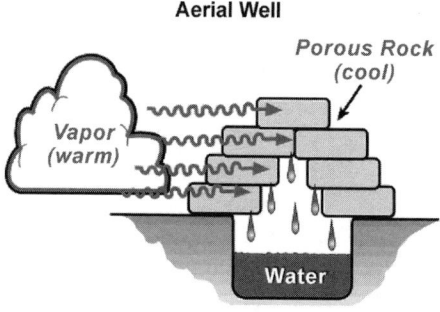

Fig. 4 Aerial well. Water vapor from fog condenses on cooler rock surfaces and is collected in underlying cistern.

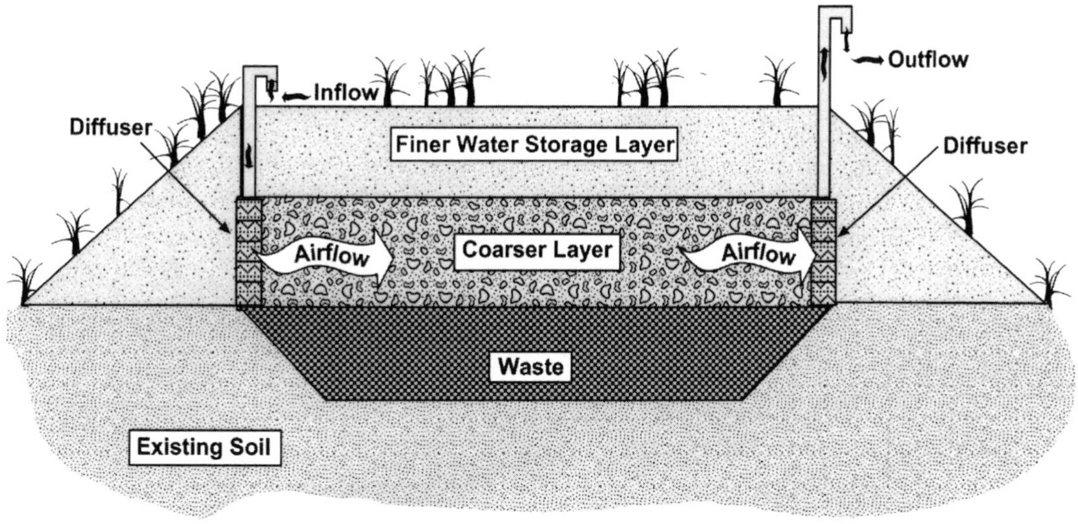

Fig. 5 Schematic of a dry barrier for an arid site landfill.

the concrete tanks and the gravel packs around the tanks, indicated that rates of water loss from vapor condensation ranged from 0.1 mm yr^{-1} to 0.5 mm yr^{-1}. For storage periods of up to 10,000 yr, this amounts to as much as 5 m of fluid that could be lost from a storage tank continuously exposed to soil water vapor. One proposed engineering solution is to design a passive thermal gradient using a specially constructed rock chimney surrounding the waste tank. The rock chimney would create a thermal shield around the waste tank and could help isolate the vapor transport from surrounding wetter soil. In addition, the chimney could be used to drain the condensation water harmlessly away from the waste tank.[34]

Dry Barriers

Stormont et al.[35] proposed the use of coarse rock placed at depth in a landfill cover. In their design, the rock is exposed to the surface either at a side-slope exposure or by way of vent tubes that protrude to the surface (Fig. 5). Wind action, blowing over the vent tube or the side slope, causes convective gas movement similar to that observed in animal burrows.[36] In arid regions, such a barrier could be used to dry the subsurface sufficiently to prevent drainage to lower layers in the profile, thus preventing recharge at the waste site. The so-called dry barrier has been tested successfully at a landfill near Boardman, Oregon.[37]

CONCLUSION

How water vapor moves in dry soils is still a research topic. It is known that both diffusion and convection are important in the water-vapor transport process. Practical aspects include dryland farming, aerial wells, fertilizer dissolution, and deployment of dry barriers to limit water infiltration at arid waste sites.

ACKNOWLEDGMENTS

Prepared, in part, with funding from the U.S. Department of Energy under Contract DE-AC06-76RL01830. We were motivated in this review by the work of the late John Cary, who spent a career studying water-vapor transport in dry soils.

REFERENCES

1. Moldrup, P.; Olesen, T.D.; Rolston, E.; Yamaguchi, T. Modeling Diffusion and Reaction in Soils: VII. Predicting Gas and Ion Diffusivity in Undisturbed and Sieved Soils. Soil Sci. **1997**, *163*, 632–640.
2. Moldrup, P.; Olesen, T.; Yamaguchi, T.; Schjonning, P.; Rolston, D.E. Modeling Diffusion and Reaction in Soils: VIII. Gas Diffusion Predicted from Single-Potential Diffusivity or Permeability Measurements. Soil Sci. **1999**, *164*, 75–81.
3. Kemper, W.D.; Nicks, A.D.; Corey, A.T. Accumulation of Water in Soils Under Gravel and Sand Mulches. Soil Sci. Soc. Am. J. **1994**, *58*, 56–63.
4. Philip, J.R.; de Vries, D.A. Moisture Movement in Porous Materials Under Temperature Gradients. Trans. Am. Geophys. Union **1957**, *38*, 222–232.
5. Letey, J. Movement of Water Through Soil as Influenced by Osmotic Pressure and Temperature Gradients. Hilgardia **1968**, *39*, 405–418.

6. Rose, C.W. Water Transport in Soil with a Daily Temperature Wave I. Theory and Experiment. Aust. J. Soil Res. **1968**, *6*, 31–44.

7. Jury, W.A.; Miller, E.E. Measurement of the Transport Coefficients for Coupled Flow of Heat and Moisture in a Medium Sand. Soil Sci. Soc. Am. Proc. **1974**, *38*, 551–557.

8. Jackson, R.D.; Reginato, R.J.; Kimball, B.A.; Nakayama, F.S. Diurnal Soil-Water Evaporation: Comparison of Measured and Calculated Soil-Water Fluxes. Soil Sci. Am. J. **1974**, *38*, 861–866.

9. Westcot, D.W.; Wierenga, P.J. Transfer of Heat by Conduction and Vapour Movement in a Closed System. Proc. Soil Sci. Soc. Am. **1974**, *38*, 9–114.

10. Jury, W.A.; Bellantuoni, B. Heat and Water Movement Under Surface Rocks in a Field Soil: II. Moisture Effects. Soil Sci. Soc. Am. J. **1976**, *40*, 509–513.

11. Cass, A.; Campbell, G.S.; Jones, T.L. Enhancement of Thermal Water Vapor Diffusion in Soil. Soil Sci. Soc. Am. J. **1984**, *48*, 25–32.

12. Nassar, I.N.; Horton, R.; Globus, A.M. Simultaneous Transfer of Heat, Water and Solute in Porous Media: II. Experiment and Analysis. Soil Sci. Soc. Am. J. **1992**, *56*, 1357–1365.

13. Cahill, A.T.; Parlange, M.B. On Water Vapor Transport in Field Soils. Water Resour. Res. **1998**, *34*, 731–739.

14. Parlange, M.B.; Cahill, A.T.; Nielsen, D.R.; Hopmans, J.W.; Wendroth, O. Review of Heat and Water Movement in Field Soils. Soil Till. Res. **1998**, *47* (1), 5–10.

15. Schelde, K.; Thomsen, A.; Heidmann, T.; Schjonning, P. Diurnal Fluctuations of Water and Heat Flow in a Bare Soil. Water Resour. Res. **1998**, *34* (11), 2919–2929.

16. Globus, A.M.; Gee, G.W. Method to Estimate Water Diffusivity and Hydraulic Conductivity of Moderately Dry Soil. Soil Sci. Soc. Am. J. **1995**, *59*, 684–689.

17. Rose, A.W.; Guo, W. Thermal Convection of Soil Air on Hillsides. Environ. Geol. **1995**, *25*, 258–262.

18. Ward, A.L.; Gee, G.W. Performance and Water Balance Evaluation of a Field-Scale Surface Barrier. J. Environ. Qual. **1997**, *26*, 694–705.

19. Ward, A.L.; Gee, G.W. Hanford Site Surface Barrier Technology. In *Vadose Zone Science and Technology Solutions*; Looney, B.B., Falta, R.W., Eds.; Battelle Press: Columbus, OH, 2000; 1414–1423.

20. Lightfoot, D.R. Morphology and Ecology of Lithic-Mulch Agricultures. Geogr. Rev. **1994**, *84*, 172–185.

21. White, C.S.; Dressen, D.R.; Loftin, S.R. Water Conservation Through an Anasazi Gardening Technique. New Mexico J. Sci. **1998**, *38*, 251–278.

22. Hanks, R.J.; Ashcroft, G.L. *Applied Soil Physics*; Springer-Verlag: Berlin, 1980.

23. Lemon, E.R. The Potentialities for Decreasing Soil Moisture Evaporation Loss. Soil Sci. Soc. Am. Proc. **1956**, *20*, 120–125.

24. Richards, S.J. Porous Block Mulch for Ornamental Plantings. Calif. Agric. **1965**, December, 12–14.

25. Corey, A.T.; Kemper, W.D. *Conservation of Soil Water by Gravel Mulches*, Hydrol. Paper No. 30; Colorado State Univ.: Ft. Collins, 1968.

26. Unger, P.W. Soil Profile Gravel Layers Effect on Water Storage Distribution and Evaporation. Soil Sci. Soc. Am. Proc. **1971**, *35*, 631–634.

27. Modaihsh, A.S.; Horton, R.; Kirkham, D. Soil Water Evaporation Suppression by Sand Mulches. Soil Sci. **1985**, *139*, 357–361.

28. Groenevelt, P.H.; van Straaten, P.; Rasiah, V.; Simpson, J. Modification in Evaporation Parameters by Rock Mulches. Soil Technol. **1989**, *2*, 279–285.

29. Jumikis, A. Aerial Wells: Secondary Sources of Water. Soil Sci. **1965**, *100*, 83–95.

30. Engineering News-Record (ENR); Engineers Seek Water in Piles of Porous Rock. Eng. News-Rec. **1991**, *227* (25), 20–21.

31. Wheeting, L.C. Certain Relationships between Added Salts and the Moisture of Soils. Soil Sci. **1925**, *19*, 287–299.

32. Scotter, D.R.; Raats, P.A.C. Movement of Salt and Water Near Crystalline Salt in Relatively Dry Soil. Soil Sci. **1970**, *109*, 170–178.

33. Scotter, D.R. Factors Influencing Salt and Water Movement Near Crystalline Salts in Relatively Dry Soil. Aust. J. Soil Res. **1974**, *12*, 77–86.

34. Cary, J.W.; Gee, G.W.; Whyatt, G.A. Waste Storage in the Vadose Zone Affected by Water Vapor Condensation and Leaching. Mater. Res. Symp. Proc. **1991**, *212*, 871–878.

35. Stormont, J.; Ankeny, M.; Kelsey, J. Airflow as Monitoring Technique for Landfill Liners. J. Environ. Eng. **1998**, *124* (6), 539–544.

36. Vogel, S.; Ellington, C.; Kilgore, D. Wind Induced Ventilation of the Borrow of the Prairie Dog *Cynomys ludovicianus*. J. Comp. Physiol. **1973**, *85* (1), 1–14.

37. Albrecht, B.A.; Benson, C.H. Predicting Airflow Rates in the Coarse Layer of Passive Dry Barriers. J. Geotechnol. Geoenviron. Eng. **2002**, *128* (4), 338–346.

Water Properties

James D. Oster
University of California, Riverside, California, U.S.A.

INTRODUCTION

Water's physical and chemical properties are uniquely different from other substances in ways that determine, to a large extent, the nature of the physics and biology of the earth. Individual water molecules can link with each other through hydrogen bonds. The degree of hydrogen bonding between water molecules changes with temperature that causes changes in the density of water and its heat content. These changes are uniquely important to the sustainability of life on earth. The dissociation of water into hydrogen and hydroxyl ions, although very small, is important in reactions of acids and bases. These topics, which were chosen as the focus of this article, are only the "tip-of-the-iceberg" as is evident when one peruses the references cited here.

> Water's large heat capacity [75.2 J(mol K)$^{-1}$] plays a key role in providing an environment that makes life possible, as we know it. The Gulf Stream, which flows from the Gulf of Mexico to the Arctic Ocean, cools about 20°C releasing, in the process, energy at a rate equivalent to that released by burning 175 million metric tons of coal per hour. All the coal mined annually would supply energy at this rate for only 12 hr. Thus the heat released from the cooling of warm ocean currents is responsible for the temperate climate over much of the earth's surface.[1]

The physical properties of water are used to define the following physical constants and units: 1) the freezing point of water is taken as 0°C and the boiling point at atmospheric pressure is taken as 100°C; 2) the unit of volume in the metric systems is chosen so that 1 mL of water at 3.98°C weighs 1.000 g; 3) the unit of heat, the calorie, is the amount of heat required to raise the temperature of 1 g of water by 1°C at 15°C.

MOLECULAR STRUCTURE

The water molecule, H$_2$O, consists of two atoms of hydrogen (H) and one atom of oxygen (O). The orientations of the electron orbitals in the oxygen atom and the location of the hydrogen atoms result in a water molecule that can be visualized as a pyramid (Fig. 1). Simplistically, the water molecule can be thought of as an

O atom with two hydrogen atoms attached near its surface on one side causing this side of the molecule to have a small positive charge that is matched with a small negative charge on the other side. This resulting separation of the positive and negative charges on the water molecule is called an electric dipole: water has a large electric dipole moment.

HYDROGEN BONDING AND ITS ROLE IN THE STRUCTURE OF ICE, WATER, AND STEAM

Each hydrogen in one water molecule can bond with the negatively charged oxygen side of another in what is known as a hydrogen bond. Each water molecule can form four hydrogen bonds that extend in four directions. This resulting structure, known as a tetrahedron, is illustrated in Fig. 2. This arrangement exists among all water molecules in ice: the tetrahedrons form a lattice with others that can be represented as sheets of hexagonal rings (Fig. 3). This structure is a very open, more open than what exists in water. As a result, ice is less dense than water.

> An interesting consequence of this difference in density occurs as lakes cool during the winter. Ice forms on the surface of lakes rather than on the lake bottom. This provides insulation slowing the rate of freezing and makes it less likely that all the water in lakes will freeze during the winter. Another consequence is that when water freezes in plants, the accompanying expansion can cause cell walls to break, killing the cell. For example, oranges when ripe can be ruined for the fresh fruit market by prolonged temperatures below freezing. As the juice within an orange freezes, the edible portion of the orange becomes mushy because the cell walls are broken. When this occurs oranges must be harvested quickly for the juice market.

When ice melts, the hexagonal rings are partially degraded because some of the hydrogen bonds are broken. Consequently, water molecules are packed more closely together, causing water to have a greater density than ice. With an increase in temperature from 0 to 4°C, further ring degradation and breaking of hydrogen bonds occur, causing a further increase in the density of water. Only at temperatures greater than 4°C does water begin to show

Encyclopedia of Water Science
DOI: 10.1081/E-EWS 120010029

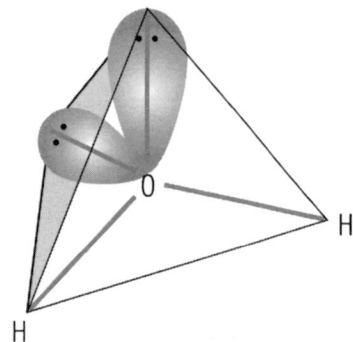

Fig. 1 Geometric shape of the water molecule.

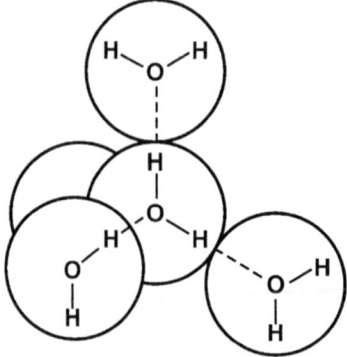

Fig. 2 The tetrahederal arrangement between four water molecules resulting from hydrogen bonds, shown as broken lines, between individual molecules.

the usual decrease in density with increasing temperature: normal expansion occurs because molecular agitation increases the distance between water molecules and overcomes the effect of hexagonal ring degradation.

Water vapor at 100°C, or steam, consists of mostly single water molecules. Because high temperatures increase the ability of molecules to move, the chances are small that two or more molecules in steam remain together due to H-bonding.

WATER AS A SOLVENT FOR SALTS

The reasons that water is so effective in dissolving salts are due to its dipolar character and its shape. Because of the former it hydrates the ions of salts. Because of a combination of ionic character and shape, the attractive force between solvated ions are reduced making them less likely to precipitate out of solution.

Due to its dipolar character, water molecules tend to combine with ions to form hydrated ions. This hydration process releases enough energy to overcome the lattice energy holding the ions together in a salt crystal. Salt crystals consist of negative and positive ions. For example, table salt consists of negatively charged chloride ions and positively charged sodium ions. Each negative ion attracts the positive ends of water molecules, and holds several water molecules to itself. Positive ions, which are usually smaller than negative ions, show this effect more strongly; each positive ion attracts the negative ends of the water molecules and binds several molecules to itself. Generally speaking, the greater the ratio of an ion's charge to its surface area, the more heavily hydrated it will be. Hydration is least significant for singly charged anions such as chloride and nitrate, which are considerably larger than most cations.

The dissolution of salts by water is also related to it dielectric constant, another aspect of its shape and electric characteristic. When water molecules are subjected to electrostatically charged plates, they align their positive ends toward the negative plate and their negative ends toward the positive plate. This partially neutralizes the applied field: the dielectric constant of water at room temperature is about 80. This compares to a dielectric constant for air of one. The force of attraction, or repulsion, of electric charges is inversely proportional to the dielectric constant of the medium surrounding the charges. This means that two oppositely charged ions in water attract each other with a force of 1/80 as strong as in air. Salts are not as soluble in solvents with low dielectric constants, such as gasoline or acetone as they are in water.

Thus not only does water tend to hydrate both the positive and negative ions in a salt crystal releasing enough energy to overcome the lattice energy, the force of

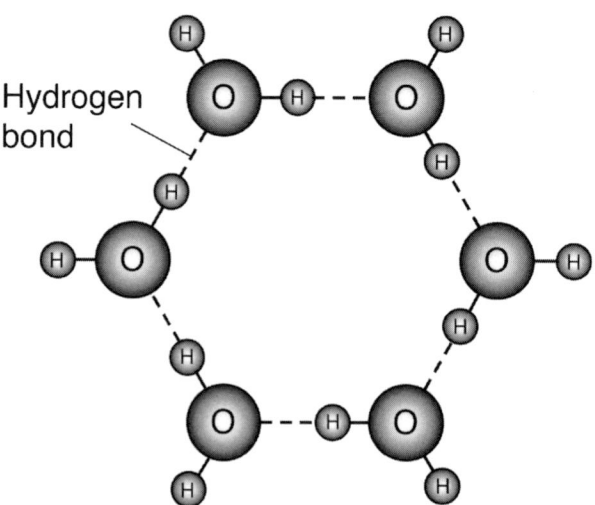

Fig. 3 The lattice of hexagonal rings of water molecules that exists among all water molecules in ice.

attraction between the solvated ions is low because of water's high dielectric constant.

DISSOCIATION OF WATER

In addition to its role as a solvent, water also plays a significant role in reactions of chemical species known as acids and bases. This stems from the dissociation of water into hydrogen (H^+) and hydroxyl (OH^-) ions according to the reaction

$$H_2O \rightleftarrows H^+ + OH^-$$

The equilibrium expression for this reaction is

$$K_w = [H^+] \times [OH^-] = 1.0 \times 10^{-14}$$

where K_w is known as the dissociation constant of water, and the ions within brackets represent their molar concentrations.

According to the dissociation constant of water, only a few H_2O molecules dissociate. In 1 L of pure water, there are about 55 moles of H_2O and $0.0000001 (1 \times 10^{-7})$ mol each of H^+ and OH^-. The product of these ion concentrations equals 1.0×10^{-14} as it should according to the equation for the dissociation constant of water.

A solution where the concentration of both $[H^+]$ and $[OH^-]$ equals 1×10^{-7} mol L^{-1} is known as a neutral solution. A water solution where $[H^+]$ exceeds $[OH^-]$ is said to be acidic. On the other hand, where $[OH^-]$ exceeds $[H^+]$, the water solution is said to be basic. Using such small numbers to characterize acidity and basicity is difficult. In 1909 Sorensen proposed an alternative method by introducing a term known as pH, where

$$pH \equiv -\log_{10}[H^+] = \log_{10} 1/[H^+]$$

The pH of a neutral solution is $-\log_{10}[1 \times 10^{-7}]$ which equals 7. For acidic solutions the hydrogen ion concentrations will be greater than 1×10^{-7} and their pH

Table 1 pH of some common liquids

Lemon juice	2.2–2.4	Human blood	7.3–7.5
Tomato juice	3.0	Human saliva	6.5–7.5
Beer	4–5	Wine	2.8–3.8
Cow's milk	6.3–6.6	Drinking water	6.5–8.0

(From Ref. 2, Table 19.2.)

will be less than 7. For basic solutions the hydrogen ion concentrations will be less than 1×10^{-7} and their pH will be greater than 7 (Table 1).

CONCLUSION

The combination of one oxygen atom with two atoms of hydrogen results in a molecule with a small negative charge on one side and a small positive charge on the other. This distribution of charges results in bonding between water molecules. This bonding, known as H-bonding, causes water to have unique changes in density upon freezing and a high heat capacity. Both are important to the sustainability of life on earth. The small negative and positive charges on the water molecule play a key role in its ability to dissolve salts. Although the bonds between the oxygen and hydrogen atoms in water are strong, in a liter of water, a very small fraction of the water molecules dissociate into OH^- and H^+ ions. This dissociation is the key to the definition of pH and to the understanding of acid and base reactions.

REFERENCES

1. Franks, F. *Water*; The Royal Society of Chemistry: London, 1984.
2. Masterson, W.L.; Slowinski, E.J. *Chemical Principles*, 4th Ed.; W. B. Saunders Co.: Philadelphia, 1977.

Well Drilling

Thomas Marek
Texas A&M University, Amarillo, Texas, U.S.A.

INTRODUCTION

There are many required processes entailed in the proper design and construction of a sand-free irrigation well. This entry addresses the typical sequence of operations that should be used in the proper drilling of a new, sand-free irrigation well. Due to the number of operations involved in the process, the owner should outline and agree on them with prospective drillers before the operations begin.

CHOOSING A DRILLER

The first consideration in drilling an irrigation well is selecting a driller. Some drillers have limited knowledge of advanced well design and development. The selection of a driller is most important and should be made from drillers who have kept up with technical advances. Experience does not overcome innovations in technology and should be viewed as complimentary to new, successful advances in drilling technologies.

SELECTING A POTENTIAL WELL SITE

Selection of a potential site(s) may include the relative availability of electricity or natural gas, but should be chosen according to the best available aquifer information. Estimates of the formation and saturated thickness of the aquifer can be made from existing or adjacent wells in relation to the proposed new site(s).

The proposed well site(s) should be located far enough from existing irrigation wells to prevent drawdown interference.[1] Water district or authority rules typically govern the distance between wells. Every well, once pumping is initiated, establishes a drawdown curve and if wells are located too close together, the respective drawdown curves will begin to overlap, causing more drawdown and subsequently requiring more horsepower.

Drilling of the test hole(s) is needed to gather information as to the anticipated production of the well and to provide a vertical formation map.

TEST HOLE PROCESSES

A test hole is required to size the well screen slot size(s) and determine how and where to construct the screened sections in the saturated zone, and for determining the proper sized gravel. This information is required for construction of the well screen prior to the main well drilling activities as the time period from well drilling initiation to gravel installation is limited. Sized gravel must be acquired prior to the main drilling activity. Without a test hole, the driller is essentially guessing at what is beneath the ground. Once the screen and gravel are in the borehole, next to nothing can be done to change either. The test hole drilling fluid should not be that of bentonite as bentonite is a clay-based compound that swells when wetted with water and seals the "pores" of the drilled borehole. Reasons for avoiding the use of bentonite is that it can mask the logging characteristics of some aquifer strata(s) and is difficult to remove from the borehole. A strongly suggested, preferred drilling fluid is that of a biological polymer.

It is recommended that the test hole be drilled throughout the water bearing formation and that drilling samples are obtained as drilling progresses. While the number of samples can vary, a determination of the number of samples to be collected should be made based upon the anticipated saturated thickness of the aquifer.

Once the test site borehole is drilled, a series of logs should be conducted to correlate strata data as to the availability and productivity of water within each strata of the formation. The three types of logs recommended are gamma log, specific conductivity, and spontaneous potential.[2]

PROPERLY PLUGGING THE TEST BOREHOLE

Once the borehole logs have been completed, the test hole should be filled and sealed to prevent contaminants from

Encyclopedia of Water Science
DOI: 10.1081/E-EWS 120010361

entering the aquifer. Most water authorities have regulations regarding this process.

ANALYSIS OF CUTTING SAMPLES AND LOGS

Analyzing test hole cutting samples consists of placing each sample in a stacked set of progressively smaller sized sieves and shaking them on a mechanical shaker. The results of the shaker data yield a distribution of curves from each sample zone.[3] The drilling coordinator can then plot the sand-sieved distribution information on a semi-logarithmic paper. In conjunction with the family of curves of the sand-sieved data, the three logs collected from the test borehole provide supplementary evidence as to the specific capacity of the respective formations.

SELECTING A NEW WELL SITE

A comparison of the sets of borehole data should be made and a site chosen as to where the best available water is potentially located. If the difference between the data sets is minimal, other factors such as energy or road accessibility can be considered.

SELECTING A GRAVEL PACK

In a sand-free well, the gravel pack prevents the sand from entering the well casing but allows the water to flow efficiently through the gravel and into the wellbore. Simply put, the gravel stops the sand and the perforated well casing stops the gravel. If one uses an inadequate gravel pack, sand will destroy the pump impellers in a short time. In addition, if one pumps much sand over time and forms a cavity around the perforated section of the well, the potential for the lower part of the formation to collapse onto the casing is possible.

The sieved data collected are necessary to determine the needed gravel size properly. Analysis and gravel sizing requires some expertise as there are "judgment" and experience factors that have to be applied. One should not compromise on the gravel, as the choice to do so is unwise. In addition to size, one wants the gravel to be uniform. Another factor that constitutes "better" gravel is the amount of quartz in the rock of the gravel.

SELECTING A WELL SCREEN

A well screen is strongly suggested in the water bearing regions of the aquifer. It is typically a fabricated, continuously wound type casing reinforced by solid, vertical bars attached to the interior of the wound section. One of the most popular types uses a triangular shape. The smaller or tapered edge of the triangular shaped screen is to the inside of the wellbore. In this manner, any gravel or sand that makes it through the initial outer edge of the screen is allowed to be excavated when the well is being developed and presents no obtrusion to water intake.

Next, using the test hole data, one decides as to where the perforated portion of the well screen needs to be located. In some cases, the screen may be scheduled in a skipped fashion to reduce screen costs and to promote water movement in a more laterally distributed mode. As the well screen location(s) is governed by the test hole strata data, the slot size of the well screen is governed by the size of the gravel selected. The selection is critical to not restrict water flow into the well bore, yet be smaller to prevent any gravel entrance after the well development process.

SELECTING A DRILLING METHOD

There are two basic drilling methods: the direct rotary and the reverse circulation drilling technique, and each has its advantages and application. The rotary method uses a drag type bit with a "centering collar" and can be from 4 in. to 24 in. in diameter. This drilling method involves using a hollow stemmed drilling shaft with drilling fluid being pumped through the interior of the shaft and allowing the drilled materials to be returned to the ground surface around the exterior of the drill stem. The reverse circulation drilling technique suctions the drilling mixture essentially through the drill stem. Thus, the water is returned to the borehole around the outside of the drill stem. This method requires more horsepower than that of the direct rotary method due to the suction pump. The advantages are that this method uses less water and less drilling mixture to stabilize the borehole, and holes greater than 24 in. can be achieved.

SPECIFYING A CORRECT DRILLING FLUID

The use of a biological polymer (organic) compound is strongly recommended. The typical borehole drilling process occurs within 24 hr and coincides with the initiation of the "natural breakdown" period of the organic drilling compound. After drilling and the casing setting process, the driller may use a small amount of chlorine solution to assist in the rapid breakdown of the compound but not damage the well screen slots.

CHOOSING A CASING AND BORE HOLE SIZE

Data indicate that drilling a bigger borehole and installing a bigger diameter well casing results in only marginally better well bore inflow. It should be noted that the aquifer, well pack and well screen, govern the amount of water inflow. Four inches of gravel (radially) is typical around the well screen and adds 8 in. to the diameter of the casing diameter. Thus, if one uses a 16-in. casing, a 24-in. well bore will be required.

MAIN WELL DRILLING AND SAMPLES

It is suggested that cutting samples be obtained during main drilling operations. Although the well screen and most of the materials would already have been prepared from the testhole data, if there is significant difference in formation detected, it is better to substantiate it during the main drilling phase. Also, the main drilling samples can be run later, if desired, to determine how they differ from the test hole stratas.

SETTING CASING, SCREEN, AND GRAVEL

At this point, the borehole will be completed and the well screen assembly, casing and gravel should be on site. The casing should be installed. However, the author strongly encourages one to install a metal airline onto the exterior of the casing and screen to provide the depth to water reading directly with a small jet of compressed air.

WELL DEVELOPMENT

Proper well development is essential and determines whether a well will be sand free and without it, maximum efficiency of the well cannot be achieved. The first operation involves bailing the well. This operation cleans out drilling materials that have settled to the bottom of the hole. It is cautioned that the rate of bailer

withdrawal needs to be controlled and should not be excessive, especially in the water bearing portion of the aquifer. This is especially true of bailers that are sized close in diameter to the interior diameter of the well casing and screen. Rapid ascension of a bailer through the water portion of the formation is "harshly" pushing the gravel into the aquifer stratas ahead of the bailer (creating a positive pressure wave) and "slamming" the gravel back against the outside of the well casing as it passes (creating a negative pressure wave) behind the bailer.

The development process begins with a cable rig utilizing a substantially weighted surge block. Even better is the utilization of a double-flanged surge block. This tool is lowered to the bottom of the hole and worked upward from the bottom to the top of the screened section of the well in short, rapid repetitive steps. This lower-to-upper direction is necessary because the progression of the sequence will draw sand into the well casing and it is unwise to risk getting a surge block "stuck" in the screen section due to sand atop the block from a top to bottom sequence. What is desired with this development operation is that the surge block "puff," not "punch" by excessive operation rates, the gravel pack. Through this "upsetting" and "closing" of the gravel pack (created in front and behind the surge block), drilling particles, gravel fines, and fine sands enter the casing. This process also orients the gravel against the casing and sets the adjacent sand of the formation against the gravel.

The range of aquifer addressed in each sequence of the development depends on the saturated thickness of the site. If the thickness is large, a large range for each sequence may be appropriate and acceptable. The rate of the operation is more critical. Rates are typically suggested at 3 ft/sec by the author.

Subsequently, any bailing operations associated with pump reworking later in time can have a significant impact on the gravel pack and formation stability. If a bail operator "runs" a bailer at too fast a rate within the screened region of the casing, one runs the risk of upsetting all the previous development efforts and the well may begin to pump sand after the bailing operation is complete.

COMPLETION OF WELLHEAD SITE

After development, the well site should be completed to provide drainage away from the location. Additionally, all well logs should be forwarded to the appropriate water authority for registration of the well.

CLOSING COMMENTS

Due to the number of items entailed in the drilling process, a tabulation of the expected operations should be submitted to potential drillers in the form of a bid. It should also be apparent that drilling a well properly will not produce more water than what is in the ground. It will, however, allow more efficient and feasible extraction of the water over time and provide one with feasible, long-term operation of the well.

REFERENCES

1. Kashef, A. Water Wells. *Ground Water Engineering*; McGraw-Hill, Inc.: New York, 1986; 361–363.
2. Driscoll, F.G. Groundwater Exploration. *Groundwater and Wells*, 2nd Ed.; Johnson Division: St. Paul, MN, 1986; 181–196.
3. Driscoll, F.G. Well Screens and Methods of Sediment-Size Analysis. *Groundwater and Wells*, 2nd Ed.; Johnson Division: St. Paul, MN, 1986; 405–412.

Wellhead Protection

Babs Makinde-Odusola
Riverside Public Utilities, Riverside, California, U.S.A.

INTRODUCTION

Wellhead protection (WHP) describes the process of managing possibly contaminating activities (PCAs) to protect groundwater quality. The United States Congress established the Wellhead Protection Program (WHPP) as part of the 1986 Amendments to the Safe Drinking Water Act (SDWA).[1] Section 1428 of the SDWA directs every state to develop a program that protects aquifers used as sources of drinking water. This Act defines a wellhead protection area (WHPA) as "the surface and subsurface area surrounding a water well or well field, supplying a public water system through which contaminants are reasonably likely to move toward and reach such well or well field." U.S. Environmental Protection Agency (EPA) sometimes refers to WHPA as "groundwater protection area."[2]

Congress amended the SDWA in 1996 to enhance the nationwide commitment to the prevention and protection of drinking water sources. U.S. EPA is developing a National Source Water Contamination Prevention Strategy.[3] Section 1453 requires each State to establish a Source Water Assessment and Protection Program (SWAPP). SWAPP includes the mandatory Source Water Assessment Program (SWAP) and the voluntary Source Water Protection Program (SWPP). For water systems that rely on groundwater, the SWAPP program builds upon the 1986 WHPP. WHPP is now one of the six major programs within the SDWA related to SWAPP. The other programs are: sole source aquifer, source water assessment, underground injection control (UIC), source water petition, and comprehensive groundwater protection grants. Other Federal laws that protect groundwater quality include the Clean Water Act (CWA), the Federal Insecticide, Fungicide, and Rodenticide Act (FIFRA).[3–6] Some states and local governments also have laws or ordinances to protect groundwater quality. This section describes the WHPP for California public water systems.

BENEFITS OF A WHPP

Groundwater is the source of drinking water to about half U.S. population, including 95% of rural communities.[3,7] More than 200 different chemicals have been detected in groundwater including 74 pesticides in groundwater of 38 states.[4] Between 1971 and 1996, contaminated source water was the cause of 86% of waterborne disease outbreaks within the United States.[8] Therefore, Congress established programs to protect groundwater quality. In 1980, Congress established the UIC program to address injection practices that contaminate groundwater.[2] Revisions of 1999 ban locating certain types of UIC wells within WHPA.[2]

In 2000, U.S. EPA proposed a Groundwater Rule (GWR) to address risks of consuming waterborne pathogens in groundwater.[8] About 10% of public water supplies derived from groundwater exceed standards for biological contamination.[8] The proposed GWR does not address the issues of toxic and carcinogenic chemicals but includes hydrogeologic assessments to identify wells vulnerable to fecal contamination—an element of a WHPP.

The California Department of Health Services (DHS) requires source water assessment for new drinking water Sources.[9] By 1998, more than 2800 U.S. communities had completed their WHP.[10] EPA has set a goal of having local SWPPs for at least 30,000 communities by 2005.[11]

U.S. EPA published specific case studies of benefits of WHP.[12] Potential benefits of a WHPP include more secure and safe drinking water, and the opportunity of reducing costs associated with treating contaminated water. It is much cheaper to prevent contamination than to characterize, monitor, and remediate contaminated groundwater. The National Research Council estimated that as much as $1 trillion may be needed to clean-up contaminated soil and groundwater in the United States over a 30-yr period.[13] Besides, groundwater contamination takes time to cleanup. On average, every gallon of water withdrawn from ground takes 280 yr to replace.[14] Some states allow public water systems to use the SWAP portion of the WHPP to obtain waivers for monitoring some contaminants. Public water systems must include information about the SWAP in the Consumer Confidence Reports distributed to their customers.[9,11]

ELEMENTS OF A WHPP

The SDWA requires each state to develop a WHPP and submit it to the U.S. EPA for approval. States have

Encyclopedia of Water Science
DOI: 10.1081/E-EWS 120010078

flexibility to develop programs that suit local needs, but their WHPP must include certain elements such as follows:[1,4,15]

- Delineate a WHPA for each public water system well or well field.
- Identify all Possible Contaminating Activities (PCAs) by location within the WHPA.
- Develop management programs to protect the water supply within WHPA from PCAs.
- Develop contingency plans for the location and provision of alternative water supplies.
- Plan to protect future well(s) from contamination.

States encourage public participation in developing WHPP. As of January 2001, U.S. EPA has approved WHPP for 48 states and two territories.[16] Section 1429 of the SDWA directed U.S. EPA to report the status of groundwater quality in the United States and the effectiveness of State programs for groundwater protection.[7]

Delineation of protection zones by itself does not protect groundwater. It must be coupled with the appropriate management strategies to protect groundwater quality.

DEVELOPING A WHPP

WHPP is usually implemented for existing well(s) or well field(s). However, it is preferable to site a proposed new well away from potential migratory paths of known or expected contaminant sources, and to construct wells in accordance with recommended well standards.[17] A community planning team usually develops the WHPP. Many agencies and professional organizations assist and/or provide resources for developing a WHPP. In the United States, such agencies include the U.S. EPA,[6] the Groundwater Protection Council,[18] National Rural Water Association,[19] and the National Ground Water Association.[20] Using a Geographical Information System (GIS) makes developing a WHPP easier.[21]

Delineating a Wellhead Protection Area (WHPA)

Many criteria had been used to delineate WHPA. Such criteria include distance from the well, time of travel (TOT) of water and/or contaminants to reach the well, assimilative capacity, hydrogeological boundaries and drawdown of the well.[4,9,11,15] Delineation methods range in complexity and costs and may be influenced by local site characteristics such as aquifer settings. Methods often used include the following: the simple arbitrary fixed radius (AFR); Calculated fixed radius (CFR); Modified CFR; analytical methods (AM); hydrogeologic mapping

(HM); and the numerical flow/transport models (NFTM).[4,9,15] However, only the CFR method will hereafter be used for illustration.

AFR involves drawing of a specified radius centered and around each of the well(s) to be protected. California DHS approves the use of the AFR method only for noncommunity water systems.[9] Professional judgment and experience influence the radius chosen. The CFR method is similar to the AFR, except that the radius of the protection zone is based on the estimated radius of the zone of contribution (ZOC) for the specified time-of-travel (TOT), with no further adjustments for groundwater level gradient, hydrogeology, and other factors that may influence the fate of contaminants within the calculated ZOC. The CFR can be determined using the following equation.[9]

$$R_t = \sqrt{(70,267Qt/\pi\eta H)}$$

where R_t = radius of protection zone in feet for TOT t; Q = peak or average pumping capacity of well in gallons per minute (gpm); t = travel time to well in years, chosen based on hydrology and contaminant source; $\pi = 3.1416$; η = effective porosity of aquifer, California DHS recommends $\eta = 0.2$ if unknown; H = open interval or length of well screen interval in feet.

For example, the CFR is 1500 ft for a 500 gpm well with 100 ft of total well screen length, porosity of 0.25, and for a TOT of 5 yr. The calculated radius may need to be adjusted to the minimum recommended radius (MRR) of the jurisdiction.

In the Modified CFR, the radius is calculated as in the CFR except that the center of the circle is shifted upgradient in the known direction of groundwater flow by a distance of 0.5R, i.e., half the radius.[9,15] Shapes other than circles are sometimes used.[15]

AM rely on use of appropriate groundwater flow and transport equations to determine the area of contribution to the well to be protected. For example, the uniform flow equations[22] are used to define a ZOC to a pumping well in a sloping water table. HM uses geological, geophysical, and dye tracing methods to map the flow boundaries and TOT criteria and are appropriate for conduit karst aquifers.[4] NFTM[23] utilize computer-modeling techniques to numerically simulate groundwater flow and contaminant transport. NFTM is usually involved and costly but can be more appropriate for aquifers exhibiting complex hydrogeology.

Protection Zones

States differ as to the number of zones delineated within a WHPA, or how each zone is designated. It is assumed that the well(s) to be protected had been constructed in accordance with standards.[17] In California, a typical WHPA usually consists of up to five zones that identify and differentiate zones in terms of the degree of contamination threat.[9] The protection zones, based on estimated time of "contaminant" travel (Table 1 and Fig. 1) are classified as follows.

Well Site Control Zone (WSCZ), or "wellhead," the closest zone, is the area immediately surrounding the well. WSCZ is managed to prevent vandalism or tampering.

Zone A2 or the Microbial/direct Chemical Contamination Zone, is the area above the aquifer that contributes water to well(s) within a 2-yr time-of-travel. This zone was defined by the requirement of the proposed GWR.[8] Research suggests that bacteria and viruses are not likely to survive beyond 2 yr in soil and groundwater.

Zones B5: Chemical Contamination Zone is that surface area overlying the aquifer between the 2 and 5-yr time-of-travel. This zone provides more response time to a chemical spill than Zone A.

Zone B10: Chemical Contamination Zone is that surface area overlying the aquifer between the 5 and 10-yr time-of-travel. This zone provides more response time to a chemical spill than Zone B5.

Buffer Zone: This zone, generally upgradient of Zone B10, offers greater level of protection, and may be extended to include the entire recharge area especially where there are potential sources of significant contamination such as landfills or other hazardous materials.

The delineated zones can be refined in shape and/or size based on professional judgment and/or local knowledge of some site-specific characteristics. Some states may recommend minimum radii different from those shown in Table 1. California requires that the final assessment map be based on a USGS quadrangle 7.5 min series topographic map.[9]

Possible Contaminating Activities

This is an iterative process of establishing an inventory of past and present PCAs, land use, and industries that are considered potential sources of contamination within each of the zones of the WHPA. PCAs include underground storage tanks, improperly abandoned wells, landfills septic tanks, cesspools, pesticides, and fertilizers. Typical resources used in establishing the PCAs include land use maps, business license records, and the Internet.[9,21] Information collected on PCAs is useful in assessing the vulnerability of the drinking water source(s) to contamination.

Vulnerability Assessment

The purpose of Vulnerability Assessment (VA) is to identify PCAs that pose the most significant threats to water quality from the protected well(s). The VA takes into account the type and proximity of the PCA and the presence of any physical barrier that may affect the fate and transport of the PCA. The first step is to determine the Physical Barrier Effectiveness (PBE) using site-specific hydrogeological information.[9] Sources located in fractured rock aquifers are rated low compared to properly designed wells located in deeper confined aquifers. California DHS developed approaches for assessing and ranking vulnerability.[9]

VOLUNTARY SWPP

SWPP is a voluntary program that may be implemented after completion of the SWAP. The goal of the SWPP is to identify, develop, and implement local measures that advance the protection of the water supply.[9] This process begins with a closer review of the SWAP and refinement of

Table 1 California WHPA zones

Protection zone	Purpose	TOT (yr)	MRR PA	MRR FRA
WSCZ	Protect from vandalism, tampering, other threats, etc.		50	50
A2	Protect the water supply source from viral, microbial, and direct chemical contamination	2	600	900
B5	Prevent chemical contamination from the water supply	5	1000	1500
B10	Allows time for some natural attenuation of the contaminants and if necessary, development of remedial plans or alternate water supplies	10	1500	2250
Buffer	Added protection for the drinking water source(s)		1500 +	2250 +

MRR—California Department of Health Services minimum recommended radius in feet; TOT = time of travel in years; PA = porous aquifer; FRA = fractured rock aquifer; WSCZ = well site control zone; AM = analytical method; HM = hydrogeological mapping; NFTM = numerical flow and transport models.

Note that California DHS does not have MRRs for zones A2, B5, B10 if the zones are delineated using AM, HM, and NFTM methods.

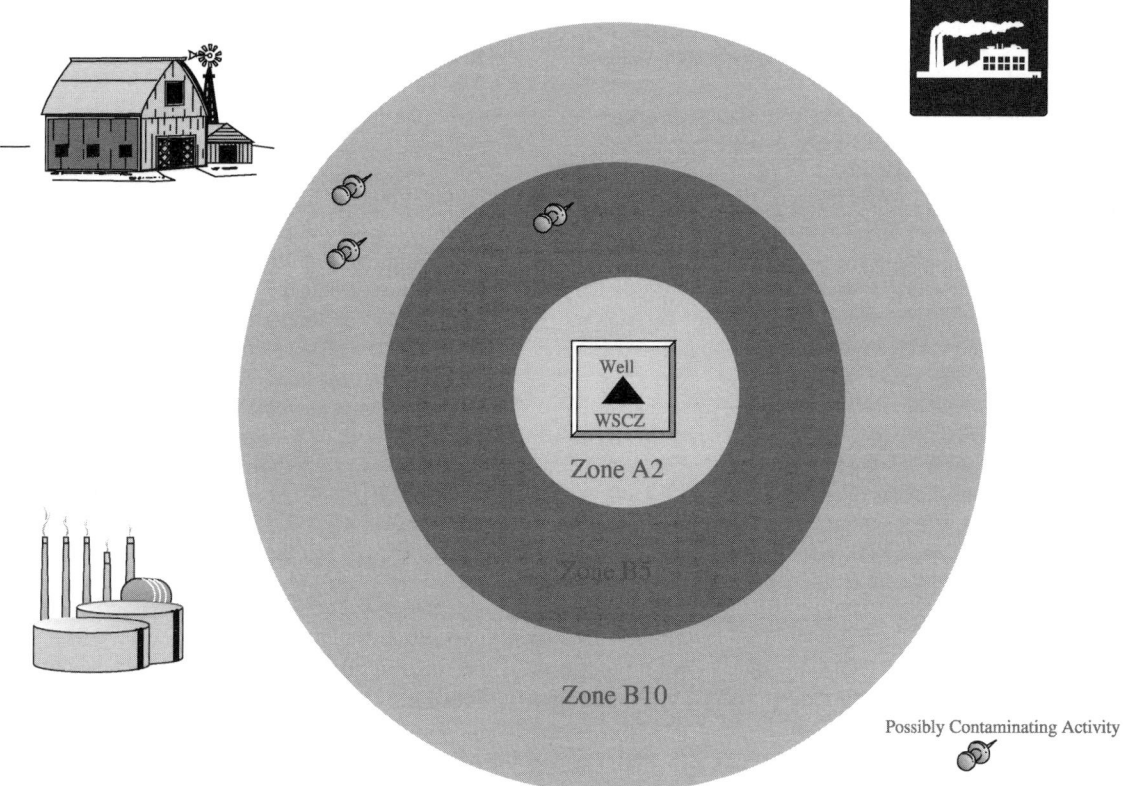

Fig. 1 WHPA using CFR method.

the WHPA. The prioritized lists from the VA may be used to develop management programs to address PCAs that pose the greatest risk to water quality. It is customary to establish a local advisory committee and provide copies of the SWPP to regulatory agencies, local planning agencies, and the public. Management approaches include designating a lead agency; acquiring technical and financial assistance;[24] land use zoning; permit conditions, land transfer, groundwater monitoring, and establishing performance standards for septic systems. U.S. EPA maintains an electronic Compendium of Groundwater Protection Ordinances.[25]

Contingency Planning

Contingency Planning is the development and implementation of long and short-term strategies for replacing drinking water supply in the event of contamination, chemical spills or physical disruption.[9]

EXAMPLES OF WHPP

Examples of WHPP on the Internet include Yosemite National Park[26] and City of Sebastopol.[27]

REFERENCES

1. American Water Works Association (AWWA), *New Dimensions in Safe Drinking Water*; AWWA: Denver, CO, 1987.
2. U.S. Environmental Protection Agency. Underground Injection Control Regulations for Class V Injection Wells, Revision; Final Rule, 40 CFR Parts 9, 144, 145, and 146. *Federal Register*; 1999; 64 FR 68546, (Dec 7, 1999). Note Federal Register Can Be Accessed on Line: http://www.access.gpo.gov/su_docs/aces/aces140.html (accessed January 2001).
3. U.S. Environmental Protection Agency (U.S. EPA), *National Source Water Contamination Prevention Strategy*, Second Draft for Discussions; U.S. EPA Office of Ground Water and Drinking Water: Washington D.C., 1999.
4. U.S. Environmental Protection Agency, *Wellhead Protection: A Guide for Small Communities*, EPA/813-B-95-005; U.S. EPA Office of Research and Development, Office of Water: Washington, D.C., 1993.
5. Saner, R.J.; Pontius, F.W. Federal Groundwater Laws. J. Am. Water Works Assoc. (AWWA) **1991**, *83* (3), 20–26.
6. United States Environmental Protection Agency: http://www.epa.gov (accessed January 2001).

7. U.S. Environmental Protection Agency, *Safe Drinking Water Act, Section 1429—Ground Water Report to Congress*, EPA-816-R-99-016; U.S. EPA Office of Water (4606): Washington, D.C., 1999.

8. U.S. Environmental Protection Agency. National Primary Drinking Water Regulations: Groundwater Rule; Proposed Rule 40 CFR Parts 141 and 142. *Federal Register*; 2000; (FR) 65(91) pp. 30194–30274 (May 10, 2000). Note Federal Register Can Be Accessed On Line: http://www.access.gpo.gov/su_docs/aces/aces140.html (accessed January 2001).

9. California Department of Health Services (DHS), Division of Drinking Water and Environmental Management, *Drinking Water Source Assessment and Protection (DWSAP) Program*; http://www.dhs.ca.gov/ps/ddwem/dwsap/DWSAP_document.pdf (accessed January 2001) California Department of Health Services, Division of Drinking Water and Environmental Management: Sacramento, CA, 1999.

10. U.S. Environmental Protection Agency, *National Water Quality Inventory: 1998 Report to Congress*, EPA-816-R-00-013; U.S. EPA Office of Water (4606): Washington, D.C., August 2000.

11. U.S. Environmental Protection Agency, *State Source Water Assessment and Protection Programs Guidance*, Final Guidance EPA-816-R-97-009; U.S. EPA Office of Water: Washington, D.C., 1997.

12. U.S. Environmental Protection Agency, *Benefits and Cost of Prevention, Case Studies of Community Wellhead Protection*; EPA/625/R-93/002; U.S. EPA Office of Water: Washington, D.C., 1995.

13. National Research Council, *Alternatives for Ground Water Cleanup*; Committee on Ground Water Cleanup Alternatives, Water Science and Technology Board, Commission on Geoscience, Environment, and Resources, National Research Council, National Academy Press: Washington, D.C., 1994.

14. Heath, R.C. Basic Groundwater Hydrology. U.S. Geological Survey Water-Supply Paper 2220; 1983.

15. U.S. Environmental Protection Agency, *Guidelines for the Delineation of Wellhead Protection Areas*; U.S. EPA Office of Ground-Water Protection: Washington, D.C., 1987.

16. U.S. Environmental Protection Agency (EPA). Maps of States and Territories with Approved Wellhead Protection Programs. http://www.epa.gov/OGWDW/wellhead.html (accessed January 2001).

17. California Department of Water Resources, *California Well Standards*, Bulletin 74-90; Department of Water Resources: Sacramento, CA, 1990.

18. The Groundwater Protection Council: http://gwpc.site.net (accessed January 2001).

19. National Rural Water Association: http://www.nrwa.org (accessed January 2001).

20. National Ground Water Association (NGWA). Online Data Bases of Information Related to Groundwater and the Environment. http://www.ngwa.org/gwonline/databases.html (accessed January 2001).

21. Bice, L.A.; Van Remortel, R.D.; Mata, N.J.; Ahmed, R.H. *Source Water Assessment Using Geographic Information Systems*; National Risk Management Research Laboratory, Office of Research and Development, U.S. Environmental Protection Agency, Cincinnati, Ohio Contract Number GS-35F-4863G Delivery Order 8C-R459-NBLX. http://www.epa.gov/ORD/NRMRL/wswrd/SWAPJ18.PDF (accessed January 2001).

22. Todd, D.K. *Ground Water Hydrology*; John Wiley and Sons Inc.: New York, NY, 1980.

23. International Groundwater Modeling Center, Colorado School of Mines, 1500 Illinois Street, Golden, CO 80401-1887, USA. Information on Groundwater Models Is at: http://www.mines.edu/research/igwmc/ (accessed January 2001).

24. U.S. Environmental Protection Agency. A Catalog of Federal Funding Sources for Watershed Protection. http://www.epa.gov/OWOW/watershed/wacademy/its.html (accessed January 2001).

25. U.S. Environmental Protection Agency (EPA) Region V. An Electronic Compendium of Groundwater Protection Ordinances. http://www.epa.gov/r5water/ordcom/">http://fcn.state.fl.us/citytlh/water/apintro.htmlhttp://www.epa.gov/r5water/ordcom/ (accessed January 2001).

26. California Department of Health Services: Drinking Water Source Assessment and Protection (DWSAP) Program, Demonstration Project—Yosemite National Park, Yosemite Valley Water System http://www.dhs.ca.gov/ps/ddwem/dwsap/dwsapdemos/yosemite.htm (accessed January 2001).

27. California Department of Health Services: Drinking Water Source Assessment and Protection (DWSAP) Program, Demonstration Project—City of Sebastopol: http://www.dhs.ca.gov/ps/ddwem/dwsap/dwsapdemos/sebastopol.htm (accessed January 2001).

Wells, Hydraulics of

Mohamed Hantush
*United States Environmental Protection Agency (US EPA),
Cincinnati, Ohio, U.S.A.*

INTRODUCTION

Water wells have been used and continue to be used as devices for extracting groundwater from aquifers. The importance of wells is not limited to the development of groundwater resources. Wells are used for environmental purposes, among others, the removal of contaminants from groundwater and controlling salt-water encroachment in coastal areas.

CONCEPTS

Hydraulics of water wells deals primarily with the application of Darcy's law and continuity relations to solve problems related to groundwater flow toward wells. Productive wells are those that tap geological formations, called aquifers, which yield groundwater in significant quantities; i.e., capable of yielding (or adding into storage) and transmitting water in appreciable amounts. Wells that tap a confined aquifer, in which groundwater is under pressure greater than atmospheric, are called artesian wells, and whenever the hydraulic head is above the ground surface, they are referred to as flowing wells (Fig. 1). In these wells, groundwater flows freely under pressure without the need for pumping. A well that penetrates an unconfined aquifer (also referred to as a water table or phreatic aquifer) is called a water-table (or gravity) well (Fig. 1). The water level in this well corresponds approximately to the position of the water table (i.e., the surface of atmospheric pressure) at that location.

Groundwater discharged from a well causes drawdown (i.e., lowering of the hydraulic head relative to its prepumping level) around the well, which decreases in the direction away from the well and forms what is known as the cone of depression (Fig. 2a). The hydraulic (or piezopmetric) head gradient formed by the cone of depression induces groundwater flow toward the pumped well, which in extensive aquifers is radially symmetric. This phenomenon is reversed for recharging wells where the hydraulic head buildup around the well decreases outwardly and causes a flow in that direction. The hydraulics of pumping wells applies also to recharging wells.[1] The mechanics of groundwater flow and yield

(or storage) due to a discharging (or recharging) well depend on the type of aquifer and the radius of influence of the well. The radius of influence R of a pumping well—the distance from the center of the well at which drawdown is practically zero—generally increases with time until it intercepts an external boundary (Fig. 2a). At which time the well discharge is partially derived from another source, if that external boundary represents an open water body, such as a stream or a lake. Elasticity of aquifers, including compressibility of water, and gravity drainage in unconfined aquifers—water released by drainage from the pore space through which the water table moves—are two primary mechanisms that account for the volumes of water released from or added into storage in aquifers. Leakage across semipervious confining layers (Fig. 4), also called aquitards or leaky units, overlying and/or underlying an aquifer can account for a significant fraction of the volume of pumped groundwater, or even sustains the total groundwater discharge rate when elastic storage is exhausted in seemingly extensive aquifers.

The change of drawdown (or head buildup) with time and space around a pumped (or recharged) well depends on the aquifer hydraulic characteristics, such as its storage capacity and transmissibility. The latter is determined by the transmissivity parameter T $[L^2T^{-1}]$, which measures the ability of a unit section of the aquifer to transmit flow throughout its entire thickness; it is the product of aquifer thickness B and the hydraulic conductivity K $[LT^{-1}]$. The storage capacity of an aquifer is quantified by the storage coefficient S (also called storativity) $[L^3L^{-3}]$, which is the volume of water released from (or added into storage of) a column of the aquifer of unit horizontal area per unit drop (or increase) of the head. In unconfined aquifers, the storage coefficient is approximated by the specific yield S_y, which gives the yield of an aquifer per unit area and unit drop of the water table. It is also defined as the drainable fraction of pore space in a unit volume of aquifer. An important parameter in the analysis of drawdown in leaky aquifers is the leakage factor $\lambda = \sqrt{Tb/K'}$, which determines the areal distribution of the leakage [L]; where K' and b, respectively, are the hydraulic conductivity and thickness of the semipervious confining layer. Another leaky aquifer parameter is the leakage coefficient[2] $\sigma = b/K'$, which is defined as the rate of flow

Encyclopedia of Water Science
DOI: 10.1081/E-EWS 120010077

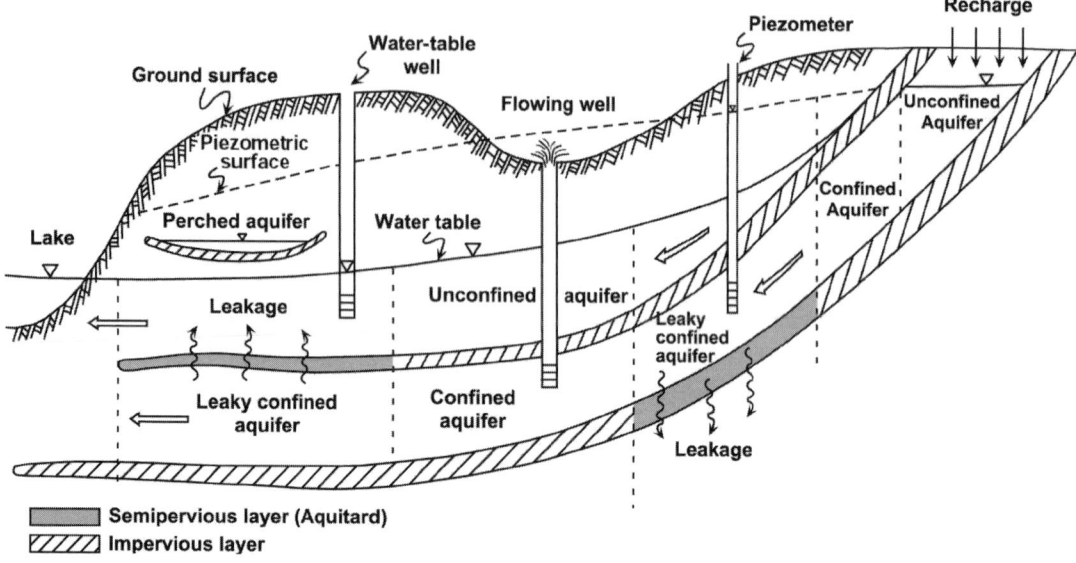

Fig. 1 Illustrative diagram of types of aquifers.

across a unit (horizontal) area of the semipervious layer into (or out of) the aquifer under one unit hydraulic difference across the layer [T].

DARCIAN FLOW

Under natural field conditions, groundwater percolates slowly through the porous aquifer material that, for all practical purposes, flow is laminar and provoked mainly by viscous forces. Near the well entrance and inside the well, flow becomes turbulent and inertial forces can no longer be ignored. Darcy's law applies to laminar flow where the specific discharge q [LT^{-1}] (hypothetical flow rate per unit porous area normal to the flow direction) is proportional to the head gradient i and the constant of proportionality is the hydraulic conductivity K:

$$q = Ki \tag{1}$$

Analysis of well hydraulics mainly combines Darcy's law Eq. 1 and continuity relations to derive solutions for drawdown in pumped aquifers (or head buildup in recharged aquifers). Based on these fundamental relations,

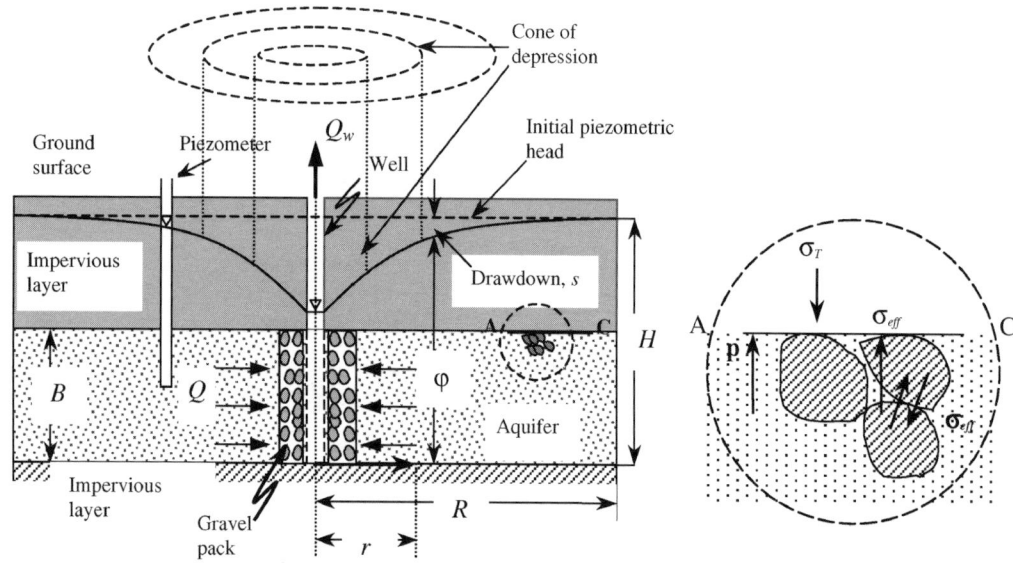

Fig. 2 Illustrative diagram: (a) a confined aquifer and cone of depression, (b) pore-water pressure and inter-granular (effective) stress.

the flow rate Q in an extensive confined aquifer across a cylinder of height equal to the thickness of the aquifer B and radius r is radially symmetric and can be expressed by the relationship (Fig. 2a):

$$Q(r,t) = 2\pi r T \frac{\partial \varphi(r,t)}{\partial r} \qquad (2)$$

in which φ is the hydraulic head [L]; $T = KB$ is the aquifer transmissivity; r is the radial distance from the well center; and t denotes time. This equation can also be applied to describe flow toward a well in an extensive water-table aquifer, but with $\varphi = h$ and $T = Kh$, where h is the elevation of the water table above the base of the aquifer at a distance r from the center of the well (Fig. 3). The transmissivity in unconfined aquifers therefore varies with time and distance, as the water table fluctuates in space and time in response to pumpage or recharge. The solution of Eq. 2 under steady flow condition is called the *Thiem* equation,[3]

$$\varphi(R) - \varphi(r) = \frac{Q_w}{2\pi T}\ln\left(\frac{R}{r}\right) \qquad (3)$$

in which Q_w is the well discharge rate [L^3T^{-1}]; and H is the hydraulic head at distance R from the center of the well [L]. The solution of Eq. 2 in an unconfined aquifer with $T = Kh$ is known as the *Dupuit–Forchheimer* well discharge formula,[3]

$$h^2(R) - h^2(r) = \frac{Q_w}{\pi K}\ln\left(\frac{R}{r}\right) \qquad (4)$$

Eqs. 3 and 4 describe the steady-state hydraulic head at distance r from the center of the well. Eq. 4 is based on Dupuit's assumption[1] that the equipotential lines are nearly vertical and, thus, the flow is essentially horizontal. The hydraulics of gravity wells is largely dependent on

this assumption, in which the governing flow equation can be linearized and solved easily.

In most practical problems of flow around wells, the relationship between transient drawdown in the aquifer and rate of discharge of the well Q_w can be expressed by the well-flow equation:[2,4–7]

$$s(r,t) = \frac{Q_w}{4\pi T}W(u) \qquad (5)$$

where s is the drawdown in the main aquifer [L] (Fig. 4) at a distance r from the center of the well and time t; and $W(u)$ is called the well function, which is related to aquifer hydraulic characteristics, r and t.

The drawdown caused by wells operating near physical boundaries, such as streams cutting through alluvial valleys and impervious mountain ranges, can be estimated by using the method of images.[1–3,11] The method of superposition can be invoked to solve for the drawdown in a wells field.

MECHANICS OF AQUIFER YIELD AND STORAGE

The elastic properties of the aquifer matrix and water[8] is the primary mechanism for the release and storage of groundwater in confined aquifers. Prior to pumping, the total load (overburden) σ_T [$MLT^{-2}L^{-2}$] above the confined aquifer, including the atmospheric pressure, equilibrates with pore-water pressure p [$MLT^{-2}L^{-2}$] inside the aquifer and the intergranular pressure (or effective stress) σ_{eff} [$MLT^{-2}L^{-2}$] exerted by the sediments on each other at the contact points; i.e., $\sigma_T = p + \sigma_{eff}$ (Fig. 2b). When groundwater is discharged from a well, the pore-water pressure decreases and the effective stress increases by an equal magnitude; i.e., $\Delta p = -\Delta\sigma_{eff}$,

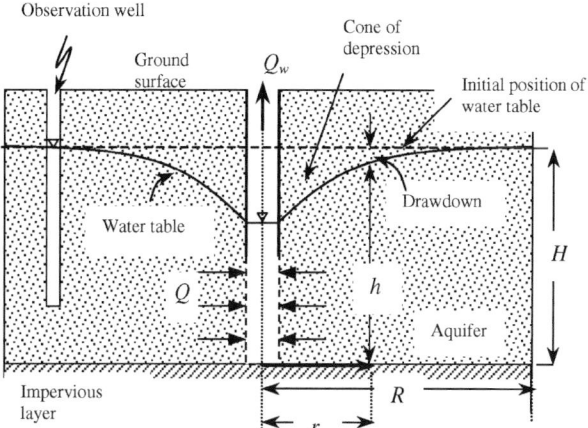

Fig. 3 Illustrative diagram of an unconfined aquifer.

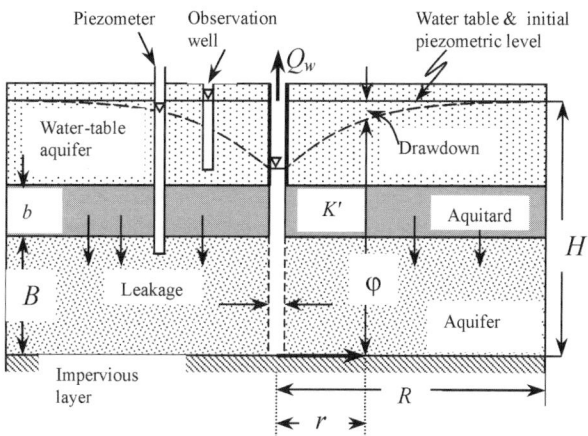

Fig. 4 Illustrative diagram of a leaky-confined aquifer.

since the overburden σ_T remains constant. Consequently, the decreased pore-water pressure results in the decompression and expansion of the water volume in storage, and the increased effective stress causes the compaction of the aquifer, somewhat reducing the pore space, and thus, the expulsion of additional volume of water. This elastic behavior of the aquifer and water is responsible for the release or taking into storage volumes of water in a pumped (or recharged) aquifer.[5] In wells of large diameters, the storage capacity in the well itself can be significant and impact drawdowns in pumped aquifers.[9]

In water-table aquifers, water is derived from storage primarily by drainage of the pore space above the lowered water table (gravity drainage) and partly from elastic storage as in confined aquifers. The latter is ignored in practical applications and typically characterizes the early response of the aquifer to pumping. When groundwater is pumped from an unconfined aquifer, the well discharge is initially derived from elastic storage and the aquifer behaves as though it is confined. The induced average drawdown therefore creates a head gradient in the vicinity of the water table and causes vertical flow and the subsequent lowering of the water table. In which case, the bulk of the well discharge is accounted by the volumes of water released from storage by vertical displacement of the water table, and the initial decline in the average head, thus, slows down considerably for a period lasting minutes to a few hours. The drawdown appears to be flat during this period, which is referred to as the "delayed yield" period.[6,7] The water table can now keep pace with the declining average head and, as in the early stage, the reaction of the aquifer becomes equivalent to that in a confined aquifer where the flow is essentially horizontal, but with the storage coefficient equal to the specific yield.

In a leaky aquifer groundwater is derived from: 1) elastic storage in the main aquifer; 2) gravity drainage if the aquifer is unconfined; and 3) elastic storage in the semipervious confining layers and induced vertical leakage across these units. In most aquifers, the hydraulic conductivity of the semipervious layer is smaller than that in the main aquifer by at least two orders of magnitude so that the flow across this layer can be assumed vertical. For all practical purposes, the flow in the main aquifer is horizontal, except in the vicinity of partially penetrating wells. When water is discharged from a well tapping the main aquifer, it is initially derived from elastic storage, and the average head is thus reduced as drawdown increases toward the discharging well forming a cone of depression. The vertical hydraulic head gradient formed by the drawdown at the interface between the upper and/or lower semipervious layer(s) and the main aquifer induces vertical flow through the semipervious layer derived partly from elastic storage of this leaky layer and partly from leakage due to head differences across the layer(s). Elastic

storage of a leaky formation is often neglected, unless it is extensively thick,[2] and leakage across this unit q_l is usually assumed to be proportional to the head difference across the confining leaky unit(s),

$$q_l = -\frac{\varphi - \varphi_0}{\sigma} \text{ or } q_l = \frac{s}{\sigma} \qquad (6)$$

in which φ is the head in the main aquifer; φ_0 is the head in the aquifer(s) above the overlying and/or below the underlying leaky confining layer(s) (Fig. 4); and σ is the coefficient of leakage, defined earlier. Contrary to completely confined aquifers, drawdown in leaky aquifers slows down in time and eventually levels off at steady state, as long as the head in the aquifer receiving or supplying leakage φ_0 is kept constant. Steady-state drawdowns occur when the discharge rate is at equilibrium with the total leakage rate through the semipervious layer. This behavior is similar to that displayed by the delayed yield phenomenon in unconfined aquifers, except that in the latter the discharge rate is derived entirely from drainage of the pore space above the water table rather than from leakage.

The time-drawdown relation in fractured rock aquifers shows three distinct stages, which may be similar to the delayed yield response in unconfined aquifers. In fractured-rock aquifers, groundwater flow partly occurs through the interconnected fractures as though it is flowing through pipes, and partly by percolation through the unfractured porous blocks of the rock matrix. Storativity of the fractures (or fissures) accounts for the initial yield of a pumping well, and as pumping continues, the drawdown somewhat slows down as water in the porous matrix reaches the fractures. The delayed yield in fractured aquifers is, thus, the result of the low conductivity of the porous blocks relative to that of the open fractures. At the later stage, the well discharge is derived from both the fractures and the porous blocks as the cone of depression continues to expand.[10]

PARTIALLY PENETRATING WELLS

A well whose screen (water entry section) length is smaller than the saturated thickness of the aquifer it penetrates is called a partially penetrating well. Flow is 3-D and no longer is horizontal in the vicinity of this well and can be turbulent. In fact, the vertical velocity components below and above the well screen can be very large. Partial penetration affects drawdown in the vicinity of the well and for large distances from the pumping well, the flow is essentially horizontal as though the pumped well completely penetrated the aquifer.[2] Anisotropy of the hydraulic conductivity has impact whenever the flow is 3-D. In aquifers where the vertical conductivity is much smaller than the horizontal, the yield of partially penetrating wells may be appreciably smaller than that of an equivalent isotropic aquifer. The effect of the anisotropy increases as the well penetration decreases.[2]

AQUIFER TESTS

Aquifer hydraulic characteristics, such as transmissivity, storativity, leakage factor, and leakage coefficient, are usually obtained from aquifer tests. In these tests, the aquifer is tested under natural field flow conditions, in which a well is pumped at a prescribed rate and the drawdown is measured therein and, preferably, in at least one observation well located at some distance from the pumped well. The Theis Type-Curve method for the estimation of aquifer transmissivity and storativity in a confined aquifer advanced the basic approach of solution to other aquifer flow scenarios.[2,6,7] In this method, a logarithmic plot of the well function $W(u)$ against $1/u$ (called type curve) is superimposed over that of the drawdown s vs. t/r^2 (called data curve) until a best match between the two curves is obtained. The hydraulic properties are then estimated from an arbitrarily chosen matching point and solving simple algebraic relations. In natural aquifers, the transmissivity (also the hydraulic conductivity) change with direction at a given location, in which case the aquifer is referred to as anisotropic. In these aquifers, the transmissivity along any direction can be determined uniquely in terms of its principal values, which are defined along two principal directions in the horizontal plane, and both the values and the principal directions can be estimated from aquifer tests, however, with three observation wells.[12]

WELL LOSSES

The drawdown inside a discharging well s_w is the sum of both formation head loss and well losses, $s_w = C_f Q_w + C_w Q_w^n$, in which C_f is the formation-loss constant; C_w is the well-loss constant relating discharge to the well loss; and n is the exponent due to turbulence.[14] $n = 2$[6] and may exceed 2,[11] and can be as high as 3.5.[15] The formation loss results from laminar flow through aquifer sediments and turbulent flow outside the well screen, and is linearly related to the well discharge Q_w. Well losses are associated with friction losses, which occur when water moves into the well through the screen, and turbulent flow inside the well. Gravel packing (Fig. 2a) and the removal of fine aquifer sediments during well development reduce well losses outside the well. The formation losses and well-loss parameters C_f, C_w, and n can be estimated graphically from a step-drawdown well test, in which drawdown inside the well is measured in time and at incremental well discharge rates.[13,15]

Notice: The U.S. Environmental Protection Agency through its Office of Research and Development funded and managed the research described here through in-house effort. It has been subjected to Agency review and approved for publication.

REFERENCES

1. Mariño, M.A.; Luthin, J.N. *Seepage and Groundwater*; Elsevier Scientific Publishing Company: New York, 1982; 489 pp.
2. Hantush, M.S. Hydraulics of Wells. In *Advances in Hydroscience*; Ven Te Chow, Ed.; Academic Press: New York, 1964; 281–442.
3. Bear, J. Hydraulics of Pumping and Recharging Wells. *Hydraulics of Groundwater*; McGraw-Hill, Inc.: New York, 1979; 300–378.
4. Theis, C.V. The Relation between Lowering of the Piezometric Surface and the Rate and Duration of Discharge of a Well using Ground-Water Storage. Am. Geophys. Union Trans. **1935**, *16*, 519–524.
5. Jacob, C.E. On the Flow of Water in Elastic Artesian Aquifer. Trans. Am. Geophys. Union **1940**, *22*, 574–586.
6. Boulton, N.S. Unsteady Radial Flow to a Pumped Well Allowing for Delayed Yield from Storage. Int. Assoc. Hydrol. **1954**, *37*, 472–477.
7. Neuman, S.H. Theory of Flow in Unconfined Aquifers Considering Delayed Response of the Water Table. Water Resour. Res. **1972**, *8*, 1031–1045.
8. Meinzer, O.E. Compressibility and Elasticity of Artesian Aquifers. Economic Geology **1928**, *23*, 263–291.
9. Papadopulos, I.S.; Cooper, H.H. Drawdown in a Well of Large Diameter. Water Resour. Res. **1967**, *3*, 241–244.
10. Streltsova, T.D. Well Pressure Behavior of a Naturally Fractured Reservoir. Soc. Pet. Eng. J. **1983**, *23*, 769–780.
11. Jacob, C.E. *Engineering Hydraulics*; Rouse, H., Ed.; Wiley: New York, 1950; 321–386.
12. Hantush, M.S. Analysis of Data from Pumping Tests in Anisotropic Aquifers. J. Geophys. Res. **1966**, *71*, 421–426.
13. Rorabaugh, M.I. Graphical and Theoretical Analysis of Step-Drawdown Test of Artesian Well. Proc. Am. Soc. Civ. Eng. **1953**, *79*, 23.
14. Bouwer, H. Well-Flow Systems. *Groundwater Hydrology*; McGraw-Hill Inc.: New York, 1978; 65–89.
15. Lennox, D.H. Analysis and Application of Step-Drawdown Test. J. Hydraul. Div., Proc. Am. Soc. Civ. Eng. **1966**, *92*, 25–48.

Wetland Ecosystems

Sherri L. DeFauw
University of Arkansas, Fayetteville, Arkansas, U.S.A.

INTRODUCTION

Wetlands perform key roles in the global hydrologic cycle. These transitional ecosystems vary considerably in their capacity to store and subsequently redistribute water to adjacent surface water systems, groundwater, the atmosphere, or some combination of these. Saturation in the root zone or water standing at or above the soil surface is key to defining a wetland. When oxygen levels in waterlogged soils decline below 1%, anaerobic (or reducing) conditions prevail. Most, but not all, wetland soils exhibit redoximorphic features formed by the reduction, translocation, and oxidation of iron (Fe) and manganese (Mn) compounds; the three basic kinds of redoximorphic features include redox concentrations, redox depletions, and reduced matrix.[1] Microbial transformations in flooded soils also impact other biogeochemical cycles (C, N, P, S) at various spatial and temporal scales. Several of the most rapidly disappearing wetland ecosystems in North America are profiled here, in terms of properties and processes.

HYDROLOGIC CONSIDERATIONS

Wetland water volume and source of water are heavily influenced by landscape position, climate, soil properties, and geology. Wetlands may be surface flow dominated, precipitation dominated, or groundwater discharge dominated systems (Fig. 1). Surface flow dominated wetlands include riparian swamps and fringe marshes. In unregulated settings (i.e., no dams or diversions) these ecosystems are subject to large hydrologic fluxes, and vary the most in terms of soil development, sediment loads, and nutrient exchanges. Precipitation-dominated wetlands (e.g., prairie potholes and bogs) reside in landscape depressions and typically have a relatively impermeable complex of clay and/or peat layers that retard infiltration (or recharge) and also impede groundwater discharge (or inflow). Groundwater dominated wetlands (e.g., fens and seeps) may form in riverine settings, at slope breaks, or in areas where abrupt to rather subtle changes in substrate porosity occur. Groundwater contributions to wetlands are complex, dynamic, and rather poorly understood.[2]

Frequency and duration of flooding, and the long-term amplitude of water level fluctuations in a landscape are the three most important hydrologic parameters that "shape" the aerial extent of a wetland complex as well as determine the relative abundance of four intergrading wetland settings (i.e., swamps, wet meadows, marshes, and aquatic ecosystems[3]). Wetland hydrodynamics control soil redox conditions. The hydroperiod-redox linkage, in turn, controls plant macronutrient concentrations (N, P, K, Ca, Mg, S), micronutrient availability (B, Cu, Fe, Mn, Mo, Zn, Cl, Co), pH, organic matter accumulation, decomposition, and influences plant zonation. Oftentimes, the zonation of plants (as determined by competition and/or physiological tolerances) provides key insights into the hydrodynamics and biogeochemistry of an area.[3]

ECOSYSTEM PROFILES

Four of the most rapidly disappearing wetland ecosystems in North America are summarized here in terms of key properties and processes. These profiles include comments on geographical extent, geomorphology, soils, hydrodynamics, biogeochemistry, vegetation structure, and/or indicator species, as well as recent estimates of ecosystem losses and ecological significance.

Riparian Swamps

Bottomland hardwood forests once dominated the river floodplains of the eastern, southern, and central United States. In the Mississippi Alluvial Plain, an estimated 8.6 million hectare area of bottomland hardwoods has been reduced to 2 million forested hectares remaining.[4] Although their true extents are not well-documented (due to difficulties in determining upland edges), these hydrologically open, linear landscape features have been logged, drained, and converted to other uses (predominantly agriculture) at alarming rates. Between 1940 and 1980, bottomland hardwood forests were cleared at a rate of 67,000 ha yr^{-1}.[5]

Riparian wetlands are unique, vegetative zonal expressions of both short- and long-term fluvial processes. A widely used classification scheme relates flooding conditions (frequency and flood duration during

Encyclopedia of Water Science
DOI: 10.1081/E-EWS 120010127

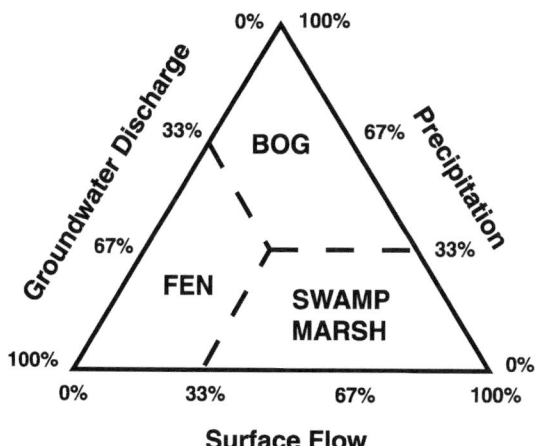

Surface Flow

Fig. 1 The relative contributions of groundwater discharge, precipitation, and surface flow determine main wetland types. Swamps and marshes are distinguished by the frequency and duration of surface flows. (Modified from Ref. [3].)

the growing season) with zonal associations of hardwood species.[6] Zone II (intermittently exposed "swamp"—bald cypress and water tupelo usually dominate the canopy) and Zone III (semipermanently flooded "lower hardwood wetlands"—overcup oak and water hickory are common) are accepted as wetlands by most, however, Zones IV (seasonally flooded "medium hardwood wetlands" that include laurel oak, green ash, and sweetgum) and V (temporarily flooded "higher hardwood wetlands"—typically an oak–hickory association with loblolly pine) have been controversial when dealing with wetland management issues.[7] Typically the complexity of floodplain microtopography abates smooth transitions from one zone to the next, with unaltered floodplain levees oftentimes exhibiting the highest plant diversity.[8]

Riparian wetlands process large influxes of energy and materials from upstream watersheds and lateral runoff from agroecosystems. As a result of these inputs, combined with decomposition of resident biomass, the organic matter content of these alluvial soils usually ranges from 2% to 5%.[8] Clay-rich bottomland tracts have higher concentrations of N and P as well as higher base saturations (i.e., Ca, Mg, K, Na) compared to upland areas. Watersheds dominated by riparian ecosystems export large amounts of organic C in dissolved and particulate forms.[9]

Prairie Potholes

Prairie potholes comprise a regional wetland mosaic that includes parts of the glaciated terrains of the Dakotas, Iowa, Minnesota, Montana, and Canada. Originally, this region encompassed 8 million hectares of wetland prior to

drainage for agriculture; an estimated 4 million hectares remained at the close of the 1980s.[10] These landscape depressions are the most important production habitat in North America for most waterfowl.[11]

Most prairie potholes are seasonally flooded wetlands dependent on snowmelt, rainfall, and groundwater. Four main hydrological groupings are recognized: ephemeral, intermittent, semi-permanent, and permanent.[11] Water level fluctuations are as high as 2 m–3 m in some settings. Measures of soil hydraulic conductivity demonstrate that groundwater flow is relatively slow (0.025 m yr^{-1}– 2.5 m yr^{-1}); therefore, the potholes are hydrologically isolated from each other in the short-term.[12]

Long-term ecological studies at the Cottonwood Lake Study Area, NPWRC-USGS[11] have revealed the intricacies of several prairie pothole phases. During periods of drought, marsh soils, sediments, and seed banks are exposed (i.e., dry marsh phase). Seed banks in natural sites have 3000–7000 seeds/m^2. A mixture of annuals (usually the dominant group) and emergent macrophyte species germinate on the exposed mudflats and a wet meadow develops. As water levels increase, the annuals decline and emergent macrophytes rapidly recolonize (i.e., regenerating marsh). If the flooding is consistently shallow, emergent macrophytes will eventually dominate the entire pothole. Sustained deep water flooding results in extensive declines in emergent macrophytes (i.e., degenerating marsh), and intensive grazing by muskrats may culminate in a lake marsh phase as submersed macrophytes become established. When water levels recede, emergent macrophytes re-establish. The rich plant communities that develop in these dynamic marshland complexes are also controlled by two additional environmental gradients, namely, salinity and anthropogenic disturbances (involving conversion to agriculture and extensive irrigation well pumping[11]).

Northern Peatlands

Deep peat deposits, in the United States, occur primarily in Alaska, Michigan, Minnesota, and Wisconsin, and are scattered throughout the glaciated northeast and northwest, as well as mountaintops of the Appalachians. The most extensive peatland system, in North America, is the Hudson Bay lowlands of Canada that occupies an estimated 32 million hectares.[13] The Alaskan and Canadian peatlands are relatively undisturbed, and the least threatened by developmental pressures. Elsewhere, peatlands have either been converted for agricultural use (including forestry) or mined for fuel and horticultural materials.

Bogs and fens are the two major types of peatlands that occupy old lake basins or cloak the landscape. The most influential, interdependent physical factors shaping these ecosystems include: 1) water level stability; 2) fertility; 3) frequency of fire; and 4) grazing intensity.[3] Northern bogs are dominated by oligotrophic *Sphagnum* moss species, and may be open, shrubby, or forested tracts. These predominantly rainfed (ombrogenous) systems have low water flow (with a water table typically 40 cm–60 cm below the peat surface), are extremely low in nutrients (especially poor in basic cations), and accumulate acidic peats (pH 4.0–4.5[14]). Fens are affected by mineral-bearing soil waters (groundwater and/or surface water flows), and possess water levels at or near the peat surface. Fens may be subdivided into three hydrologic types: soligenous (heavily influenced by flowing surface water); topogenous (largely influenced by stagnant groundwater); or limnogenous (adjacent to lakes and ponds[14]). These minerogenous ecosystems range from acidic (pH 4.5) to basic (pH 8.0); vegetation varies from open, sedge-dominated settings to shrubby, birch-willow dominated associations to forested, black spruce-tamarack tracts. Nutrient availability gradients do not necessarily coincide with the ombrogenous–minerogenous gradient; recent investigations indicate higher P availability in more ombrogenous peatlands, and greater N availability in more minerogenous peatlands.[15]

Northern peatlands represent an important, long-term carbon sink, with an estimated 455 Pg (1 Petagram = 10^{15} g) stored worldwide.[16] An estimated 220 Pg of C is currently stored in North American peatlands, compared to about 20 Pg in storage during the last glacial maximum.[17] High latitude peatlands also release about 60% of the methane generated by natural wetlands.[18] In addition, sponge-like living Sphagnum carpets facilitate perma-nently wet conditions, and the high cation exchange capacity of cells retains nutrients and serves to acidify the local environment.[19]

Pocosins

The Pocosins region of the Atlantic coastal plain extends from Virginia to the Georgia–Florida border.[20] These nonalluvial, evergreen shrub wetlands are especially prevalent in North Carolina; in fact, pocosins once covered close to 1 million hectare in this state.[21] Derived from an Algonquin Indian word for "swamp-on-a-hill," pocosins are located on broad, flat plateaus and sustained by waterlogged, acidic, nutrient-poor sandy, or peaty soils usually far removed from large streams. Wetland losses are high, with 300,000 ha drained for agriculture and forestry uses between 1962 and 1979.[22]

Pocosins are characterized by a dense, ericaceous shrub layer; an open canopy of pond pine may be present or absent.[21] A typical low pocosin ecosystem [less than 1.5 m (5 ft) tall] includes swamp cyrilla (or titi), fetterbush, bayberry, inkberry, sweetbay, laurel-leaf greenbrier, and

sparsely distributed, stunted pond pine.[22] Pocosin soils may be either organic (with a deep peat layer—e.g., Typic Medisaprist) or mineral (usually including a water restrictive spodic horizon—e.g., Typic Endoaquod). As peat depth decreases, the stature of the vegetation increases. High pocosin [with shrubby vegetation 1.5 m–3.0 m (5 ft–10 ft) tall and canopy trees approximately 5 m (16 ft) in height] usually occurs on peat deposits of 1.5 m (5 ft) or less in thickness or on wet sands.[20] The major natural disturbance to these wetlands is periodic burning (with a fire frequency of about 15 yr–50 yr).

Pocosin surface and subsurface waters are similar to northern ombrogenous bogs, but are more acidic with higher concentrations of sodium, sulfate, and chloride ions.[23] Carbon:Phosphorus (C:P) ratios increase sharply during the growing season; phosphorus availability limits plant growth and probably plays a crucial role in controlling nutrient export. Undisturbed pocosins export organic N and inorganic phosphate in soil water.[23]

CONCLUSIONS

Despite existing preservation policies, U.S. wetland conversions are anticipated to continue at a rate of 290,000 acres–450,000 acres (117,408 ha–182,186 ha) annually.[24] It is widely known that wetlands are the product of many environmental factors acting simul-taneously; perturbations in one realm (e.g., hydrology) not only impact local wetland properties and processes, but also have consequences in linked ecosystems as well. Wetlands are major reducing systems of the biosphere, transforming nutrients and metals, and regulating key exchanges between terrestrial and aquatic environments.

ACKNOWLEDGMENTS

William J. Rogers, B. A. Howell, and Van Brahana commented on an earlier draft; their reviews are genuinely appreciated.

REFERENCES

1. Soil Survey Staff, *Keys to Soil Taxonomy*, 8th Ed.; USDA-NRCS: Washington, DC, 1998.
2. http://water.usgs.gov/nwsum/WSP2425/hydrology.html (accessed July 2002).
3. Keddy, P.A. *Wetland Ecology: Principles and Conserva-tion*; Cambridge University Press: Cambridge, 2000.
4. Llewellyn, D.W.; Shaffer, G.P.; Craig, N.J.; Creasman, L.; Pashley, D.; Swan, M.; Brown, C. A Decision-Support System for Prioritizing Restoration Sites on the Mississippi River Alluvial Plain. Conserv. Biol. **1996**, *10* (5), 1446–1455.

5. Kent, D.M. *Applied Wetlands Science and Technology*, 2nd Ed.; Lewis Publishers: Boca Raton, FL, 2001.

6. Clark, J.R.; Benforado, J., Eds. *Wetlands of Bottomland Hardwood Forests*; Elsevier: Amsterdam, 1981.

7. Mitsch, W.J.; Gosselink, J.G. *Wetlands*; Van Nostrand Reinhold Company: New York, 1986.

8. Wharton, C.H.; Kitchens, W.M.; Pendleton, E.C.; Sipe, T.W. *The Ecology of Bottomland Hardwood Swamps of the Southeast: A Community Profile*; U.S. Fish and Wildlife Service, Biological Services Program FWS/OBS-81/37: Washington, DC, 1982; 1–133.

9. Brinson, M.M.; Swift, B.L.; Plantico, R.C.; Barclay, J.S. *Riparian Ecosystems: Their Ecology and Status*; U.S. Fish and Wildlife Service, Biological Services Program FWS/OBS-81/17: Washington, DC, 1981; 1–151.

10. Leitch, J.A. Politicoeconomic Overview of Prairie Potholes. In *Northern Prairie Wetlands*; van der Valk, A.G., Ed.; Iowa State University Press: Ames, 1989; 2–14.

11. http://www.npwrc.usgs.gov/clsa (accessed July 2002).

12. Winter, T.C.; Rosenberry, D.O. The Interaction of Ground Water with Prairie Pothole Wetlands in the Cottonwood Lake Area, East-Central North Dakota, 1979–1990. Wetlands **1995**, *15* (3), 193–211.

13. Wickware, G.M.; Rubec, C.D.A. *Ecoregions of Ontario*, Ecological Land Classification Series No. 26; Environment Canada, Sustainable Development Branch: Ottawa, Ontario, 1989.

14. http://www.devonian.ualberta.ca/peatland/peatinfo.htm (accessed July 2002).

15. Bridgham, S.D.; Updegraff, K.; Pastor, J. Carbon, Nitrogen, and Phosphorus Mineralization in Northern Wetlands. Ecology **1998**, *79*, 1545–1561.

16. Gorham, E. Northern Peatlands: Role in the Carbon Cycle and Probable Responses to Climatic Warming. Ecol. Appl. **1991**, *1* (2), 182–195.

17. Halsey, L.A.; Vitt, D.H.; Gignac, L.D. Sphagnum-Dominated Peatlands in North America Since the Last Glacial Maximum: Their Occurrence and Extent. The Bryologist **2000**, *103*, 334–352.

18. Cicerone, R.J.; Ormland, R.S. Biogeochemical Aspects of Atmospheric Methane. Global Biogeochem. Cycles **1988**, *2* (2), 299–327.

19. van Breeman, N. How Sphagnum Bogs down Other Plants. Trends Ecol. Evol. **1995**, *10* (7), 270–275.

20. Sharitz, R.R.; Gresham, C.A. Pocosins and Carolina Bays. In *Southern Forested Wetlands: Ecology and Management*; Messina, M.G., Conner, W.H., Eds.; Lewis Publishers: Boca Raton, FL, 1998; 343–389.

21. Richardson, C.J. Pocosins: An Ecological Perspective. Wetlands **1991**, *11* (S), 335–354.

22. Richardson, C.J.; Evans, R.; Carr, D. Pocosins: An Ecosystem in Transition. In *Pocosin Wetlands: An Integrated Analysis of Coastal Plain Freshwater Bogs in North Carolina*; Richardson, C.J., Ed.; Hutchinson Ross Publishing Company: Stroudsburg, PA, 1981; 3–19.

23. Wallbridge, M.R.; Richardson, C.J. Water Quality of Pocosins and Associated Wetlands of the Carolina Coastal Plain. Wetlands **1991**, *11* (S), 417–439.

24. Schultink, G.; van Vliet, R. *Wetland Identification and Protection: North American and European Policy Perspectives*, Agricultural Experiment Station Project 1536; Michigan State University, Department of Resource Development: East Lansing, MI, 1997.

Wetlands as Treatment Systems

Kyle R. Mankin
Kansas State University, Manhattan, Kansas, U.S.A.

INTRODUCTION

Wetland ecosystems generally can be defined by the presence of saturated soils and plants that grow well under these conditions. These two features promote many processes that trap, transform, and utilize a variety of the materials that flow into a wetland system with the incoming water. Because wetlands possess this capacity to remove contaminants from water, they have been utilized and even constructed for the purpose of treating polluted waters. As constructed treatment wetlands become a more common feature in municipal, rural, agricultural, and industrial settings, it is important to understand the features, processes, and design considerations that make these systems attractive natural-treatment options.

How Do Wetlands Work?

Wetlands can be used to treat wastewater because they process contaminants. However, they treat wastewater more slowly than traditional treatment plants. Oxygen, and the manipulation of oxygen levels, is a primary concern for wastewater treatment because many of the necessary biological and chemical treatment processes require oxygen. Traditional treatment plants can easily manipulate oxygen levels by pumping air into the wastewater. Oxygen enters wetlands by slower, natural processes. Increasing oxygen concentration, by increasing wastewater contact with air, plant roots, or photosynthetic algae, often can enhance the processing ability of wetlands.

When considering wastewater treatment by constructed wetlands, five contaminant groups are of primary importance: sediments, organic matter, nutrients, pathogenic microbes, and metals. Wetlands slow down water movement, allowing sediments to settle out of the water. Organic matter can be processed, or decomposed, by highly competitive microbes. Less competitive microbes called nitrifiers process nitrogen. Both microbe types require oxygen. Because the nitrifiers are less competitive, oxygen levels become very important to insure that both organic matter and nitrogen are fully processed. The other two, pathogenic microbes and metals, are more situational, related to the specific waste being treated. Wetlands treat pathogenic microbes by detaining them until they naturally die off, are eaten by other predatory organisms

in the wetland, or are exposed to UV radiation near the water surface. Metals are processed by being adsorbed to other particles and settling out of the water.

The remainder of this entry further explains wetland processes and design considerations. References are provided for more in-depth information.

TREATMENT WETLAND TYPES

Constructed vs. Natural Wetlands

Wetlands constructed as treatment systems differ from natural wetlands in several important ways. Constructed wetlands usually are built with uniform depths and shapes designed to provide consistent detention times and maximize contaminant removal. In contrast, natural wetlands are irregular in depth and shape, which causes irregular flow, allows water to by-pass the shallow treatment zones by moving through the deeper channels, and leads to less effective treatment. In addition, water-quality regulations in the United States dictate that if a natural wetland is associated with an existing water body of the United States, as most are, wastewater discharges into the wetland must meet specific quality standards, similar to other water bodies. Wetlands constructed as wastewater treatment systems typically are located in uplands where wetlands did not exist before and are not subject to inflow water-quality regulations. Natural wetlands are *not* recommended for use as treatment wetlands.

Constructed wetlands increasingly are being used for wastewater treatment in a variety of applications (Table 1). Examples can be found of wetlands being used to treat municipal sewage, urban runoff, onsite residential wastewater, animal feedlot and barnyard runoff, cropland runoff, industrial wastewater, mine drainage, and landfill leachate. Each application takes advantage of a combination of physical, chemical, and biological processes characteristic of natural wetlands to reduce the concentration of contaminants in water. Such contaminants include sediments, organic materials, nutrients (particularly nitrogen and phosphorus), metals, microbial pathogens, and pesticides.

Encyclopedia of Water Science
DOI: 10.1081/E-EWS 120010371

Table 1 North American wetlands as of 1994[1]

Wastewater type	Quantity	Size (ha)		
		Minimum	Median	Maximum
Agricultural	58	0.0004	0.1	47
Industrial	13	0.03	10	1093
Municipal	159	0.004	2	500
Stormwater	6	0.2	8	42
Other	7	3	376	1406

Free-Water vs. Submerged-Bed Wetlands

Constructed wetlands have two common types. Free-water surface (FWS) wetlands (also called surface-flow wetlands) have plants that grow in a shallow layer of water over a soil substrate (Figs. 1 and 2). The location of the plants in the system can vary: the plants can float on the water surface with their roots suspended in the water (free-floating macrophyte systems); they can be rooted in the soil with the entire leaves and stems below the water surface (submerged-macrophyte systems); they can be rooted in the soil having leaves and stems that rise above the water surface (emergent macrophyte systems); or the wetland may use a combination of planted and open-water zones. About two-thirds of existing wetlands as of 1994 were FWS.[1] In vegetated submerged-bed (VSB) wetlands (also called subsurface flow wetlands or rock-plant filters), plants are rooted in a porous media, such as sand or gravel, and water flows through the media in either horizontal or vertical direction (Figs. 3 and 4). About one-quarter of treatment wetlands were VSB systems.[1] However, these systems are currently used in thousands of smaller-scale, onsite residential applications in the United States that do not appear in this database.

TREATMENT PROCESSES

Many wastewaters entering constructed wetlands must be pretreated to avoid excessive contaminant loading, particularly of mineral and organic solids. Pretreatment technologies include septic tanks for onsite systems or anaerobic lagoons for animal waste, municipal, or mine-drainage treatment systems. In each case, the anaerobic condition in the pretreatment process reduces production of additional algae solids. Typical contaminant levels entering treatment wetlands are summarized in Table 2.

The wetland type impacts the processes used to retain or remove contaminants. In a VSB system, wastewater flows through pore spaces of the media and comes into direct contact with the roots of plants. In a FWS system, water flows across the media surface and contacts plant stems and leaves. In either system, solid particles, including sediments (clay and silt particles and colloids) and organic matter (manure particles, organic residues, and algae or other phytoplankton), settle out of the water column or are trapped or filtered as water passes through a wetland. Contaminants that are adsorbed to sediments (e.g., P, NH_4, fecal bacteria) or absorbed within organic solids (e.g., nutrients) are also removed. However, these constituents can be re-suspended or desorbed back into the wetland water. This natural cycling of materials is an important function of wetlands, although it makes system design and interpretation of treatment complex.

Once entrapped, organic materials and associated contaminants are decomposed in wetlands by microbial and chemical transformations. In the degradation process, microbes use oxygen. The amount of oxygen used is related to the amount of organic material in the water.

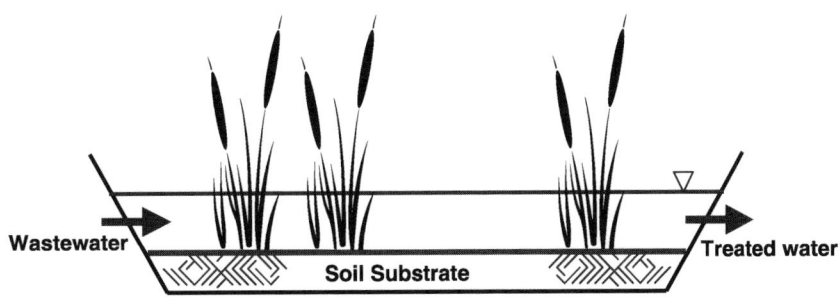

Fig. 1 FWS wetland with emergent macrophytes.

Fig. 2 A three-cell, FWS wetland for treating dairy wastewater. This system is in its first year of operation; plants were recently established. (Photo: Peter Clark.)

The controlled measurement of biochemical oxygen demand (BOD) is a common way to illustrate the amount of organic matter in water. When wastewater lacks oxygen, or is anaerobic, it requires the addition of oxygen to degrade organic matter. Oxygen is also required for transformation of ammonium to nitrite and nitrate (nitrification), whereas anaerobic conditions are required for transformation of nitrate to nitrogen gas (denitrification). Aerobic wetland conditions often remove metals by aerobic oxidation of iron; subsequently iron hydroxides and other metals precipitate in the wetland.[5] Although some oxygen diffuses into a wetland from the air, a common assumption is that oxygen also is transported through wetland plants and made available to microbes in close proximity to leaky roots.[6] This mechanism may be less important than once thought, though.[2] Treatment wetlands are thought to function effectively because they combine anaerobic zones in the water column with aerobic zones near the water interfaces with air and roots.

However, because the microbes that break down organic carbon can out compete nitrifiers for oxygen, nitrogen removal in higher strength wastewaters is often low.

DESIGN CONSIDERATIONS

Design and resulting effectiveness of constructed wetlands (Table 3) depend upon many factors: climate (precipitation, temperature, growing season, evapotranspiration), wastewater characteristics (constituents, loading, flow rate, and volume), topography, and wildlife activity. Wetland designs must specify total area; the number, depth, and size of wetland cells; hydraulic retention times; vegetation types and coverage; inlet and outlet configuration and location; and internal flow patterns.[2] Details for design can be found in numerous references[2,7–11] and some elements are discussed here.

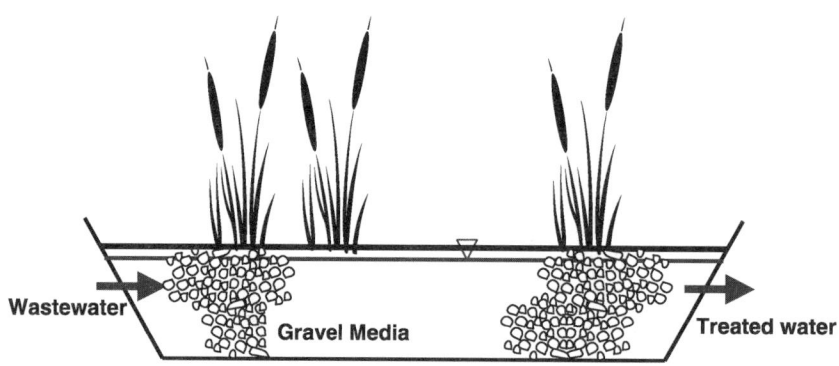

Fig. 3 VSB wetland with emergent macrophytes.

Fig. 4 VSB wetland for treating onsite residential wastewater. This system uses gravel media and variety of wetland plants. (Photo: Barbara Dallemand.)

VSB Wetlands

Properly designed VSB systems can achieve high removal rates. Treatment in a VSB wetland is governed by system residence time and wastewater contact with media and plant-root surfaces. Because of this, depth is a critical dimension and is often chosen according to the rooting depth of the selected plant (e.g., cattails: 30 cm; reeds: 40 cm; bulrush: 60 cm). Once depth is chosen, cross-sectional area (and thus wetland width) is selected to assure adequate flow rates. Then, volume (and thus wetland length) is determined from the retention time needed to treat the wastewater to the desired quality.

Proper design of inlet and outlet control structures helps maintain uniform flow patterns and depth, avoids problems with clogging and freezing, and minimizes system operation and maintenance (O and M) problems. High loading from influent solids and clogging can lead to surface flows and poor treatment. VSB systems must receive influents that are pretreated to remove solids (e.g., septic tank and effluent filter or anaerobic lagoon).

FWS Wetlands

Properly designed FWS systems also can achieve high removal rates. Design typically follows one of two methods. The areal loading approach allows a designer to select the wetland surface area according to the influent load and the desired effluent quality.[13] Another approach allows a designer to select the wetland area by knowing the biological reaction rate, wastewater concentration, and flow rate along with selected water depth and target outflow water quality.[7,8] Again, depth is a critical dimension and is governed by plant tolerance to standing water and treatment objectives.

FWS vs. VSB Systems

Selection of the most appropriate wetland system depends on wastewater characteristics, treatment requirements, and site constraints. VSB systems generally require less land area, are less susceptible to freezing and mosquito problems, and have no exposed wastewater at the surface (avoiding contact-related health problems). FWS systems are less expensive to construct (without the cost of media), have greater potential for wildlife habitat, and are easier to maintain if solids accumulate.

OPERATION AND MAINTENANCE

O and M of treatment wetlands are relatively simple. The goal of an O and M plan is to assure that the wetland

Table 2 Wetland influent concentrations

Wastewater type	BOD$_5$ (mg/L)	TSS (mg/L)	TN (mg/L)	NH$_4$-N (mg/L)	NO$_3$-N (mg/L)	TP (mg/L)	FC (per 100 mL)
Residential-septic tank[2]	129–147	44–54	41–49	28–34	0–0.9	12–14	$10^{5.4}$–$10^{6.0}$
Municipal-primary[2]	40–200	55–230	20–84	15–40	0	4–15	$10^{5.0}$–$10^{7.0}$
Municipal-pond[2]	11–35	20–80	8–22	0.6–16	0.1–0.8	3–4	$10^{0.8}$–$10^{5.6}$
Livestock[3] [avg.]	263	585	254	122	3.6	24	1.6×10^5
Livestock[3] [median]	81	118	274	60	1.1	20	1.7×10^3
Landfill leachate[4]	312–729	241–7840	287–670	254–2074	0–3	0.9	—

Note: BOD$_5$ = 5-day biochemical oxygen demand, TSS = total suspended solids, TN = total nitrogen, NH$_4$-N = ammonium nitrogen, NO$_3$-N = nitrate nitrogen, TP = total phosphorus, FC = fecal coliform bacteria.

Table 3 Wetland treatment (%)

Wastewater type	BOD$_5$	TSS	TN	NH$_4$-N	TP	FC	Metals
Municipal[1] [avg.]	74	70	53	54	57	—	—
Livestock[3] [avg.]	65	53	42	48	42	92	—
Landfill leachate[12] [range]	11–90	45–97	7–45	13–88	—	—	8–95 +

Note: BOD$_5$ = 5-day biochemical oxygen demand, TSS = total suspended solids, TN = total nitrogen, NH$_4$-N = ammonium nitrogen, TP = total phosphorus, FC = fecal coliform bacteria, metals = Fe, Cu, Pb, Ni, or Zn.

system continues to operate as planned, designed, and constructed. Several sources provide specific O and M guidance,[14] and most design manuals also contain such guidelines. Operation should be consistent with treatment objectives while maintaining structural integrity of the system, uniform flow conditions, and healthy vegetation as well as minimizing odors, nuisance pests and insects. Most maintenance plans require such items as checking water levels, checking for evidence of leaks or wildlife damage, and maintaining plant health on a weekly or monthly basis.

CONCLUSION

Constructed wetlands are complex natural-treatment systems that are well suited for many applications. They are low in cost and maintenance, provide significant reductions of many contaminants, and offer an aesthetic appearance. More work is needed to characterize treatment processes in constructed wetlands and improve design procedures to account for variability in wastewater and climate.

ACKNOWLEDGMENTS

The author extends his gratitude to Kristina M. Boone, Associate Professor of Agricultural and Environmental Communications, Kansas State University for her significant review contribution to this article, and to those at Kansas State University who have supported wetlands research and education.

REFERENCES

1. Kadlec, R.H., Knight, R.L., Reed, S.C., Ruble, R.W., (Eds.) *Wetlands Treatment Database (North American Wetlands for Water Quality Treatment Database)*, EPA/600/C-94/002; U.S. Environmental Protection Agency: Cincinnati, OH, 1994.
2. US EPA, *Constructed Wetlands Treatment of Municipal Wastewaters*, EPA/625/R-99/010; U.S. Environmental Protection Agency: Cincinnati, OH, 2000.
3. CH2M Hill and Payne Engineering, *Constructed Wetlands for Livestock Wastewater Management: Literature Review, Database, and Research Synthesis*; Gulf of Mexico Program—Nutrient Enrichment Committee: Montgomery, AL, 1997.
4. Kadlec, R.H. Constructed Wetlands for Treating Landfill Leachate. In *Constructed Wetlands for the Treatment of Landfill Leachates*; Mulamoottil, G., McBean, E.A., Rovers, F., Eds.; Lewis Publishers: Boca Raton, FL, 1999; Chap. 2.
5. US EPA, *Engineering Bulletin—Constructed Wetlands Treatment*, EPA/540/S-96/501; U.S. Environmental Protection Agency: Cincinnati, OH, 1996.
6. Brix, H. Do Macrophytes Play a Role in Constructed Treatment Wetlands? WaterSci. Technol. **1997**, *35* (5), 11–17.
7. Kadlec, R.H.; Knight, R.L. *Treatment Wetlands*; Lewis Publishers: Boca Raton, FL, 1996.
8. Reed, S.C.; Crites, R.W.; Middlebrooks, E.J. *Natural Systems for Waste Management and Treatment*, 2nd Ed.; McGraw-Hill: New York, 1995.
9. Steiner, G.R., Watson, J.T., (Eds.) *General Design, Construction, and Operation Guidelines: Constructed Wetlands Wastewater Treatment System for Small Users Including Individual Residences*, Technical Report TVA/MW-93/10; Tennessee Valley Authority: Chattanooga, TN, 1993.
10. Powell, G.M.; Dallemand, B.L.; Mankin, K.R. *Rock-Plant Filter Design and Construction for Home Wastewater Systems*; MF-2340, www.oznet.ksu.edu/library/h20ql2/mf2340.pdf Kansas State University Cooperative Extension Service: Manhattan, KS, 1998.
11. Sievers, D.M. *Design of Submerged Flow Wetlands for Individual Homes and Small Wastewater Flows*, Special Report 457; Missouri Small Wastewater Flows Education and Research Center, 1993.
12. Mulamoottil, G.; McBean, E.A.; Rovers, F., (Eds.) *Constructed Wetlands for the Treatment of Landfill Leachates*; Lewis Publishers: Boca Raton, FL, 1999; 69, 86, 157, 212.
13. USDA-NRCS, *Constructed Wetlands for Agricultural Wastewater Treatment: Technical Requirements*; U.S. Department of Agriculture–Natural Resource Conservation Service: Washington, DC, 1991.
14. Powell, G.M.; Dallemand, B.L.; Mankin, K.R. *Rock-Plant Filter Operation, Maintenance, and Repair*, MF-2337; www.oznet.ksu.edu/library/h20ql2/mf2337.pdf Kansas State University Cooperative Extension Service: Manhattan, KS, 1998.

Index

AAEA (*see* American Agricultural Economics Association)

Abscisic acid, 166, 684

Acacia cyanophylla, influence of drought hardening, 166, 167

Academic disciplines, 1-4
   agricultural economics, 1
   agricultural engineering, 1-2
   agricultural law, 2
   agricultural meteorology, climatology, 2
   agricultural water management, 1
   agronomy, soil science, 2
   animal science, 2
   aquatic biology, 2
   environmental engineering, 2-3
   forestry, natural resources conservation, 3
   hydrology, hydrogeology, 3
   range science, 3

Academic Info, internet library, as source of water information, 425

Accelerated erosion, 199-204 (*see also* Erosion)

Accumulation, assimilate, crop development and, 92-93

Acetate, in potable groundwater, 390

Achillea fragrantissima, 164

ACIAR (*see* Australian Center for International Agricultural Research)

Acid mine drainage, 364

Acid rain, precipitation chemistry, 4-7
   drop-scale transport processes, 5
      external transport, 5
      interfacial transport, 5
      internal transport, 5
   environmental factors influencing, 6
      biota, 6
      geology, 6
      meteorological factors, 6
      topography, 6
   nitric acid, 5
   sulfuric acid, 5

Acidification, for prevention of clogging of drip lines, emitters, 7-8

Acidity, 663-665
   accuracy in measurement, 665
   bias in measurement, 665
   data recording, 665
   electrode care, calibration, 664
   electrodes, 664
   measurement, 663-665
      accuracy, 665
      data recording, 665
      electrode care, calibration, 664
      electrodes, 664
      measurement sampling, 663-664
      meters, 664
      sampling, 663-664
   meters, 664
   precision in measurement, 665
   of water, drip irrigation and, 73

Acrylamide monomer, 450

ACSAD (*see* Arab Center for Studies in Agricultural Development)

Acute gastrointestinal illness, 649

Adaptation, drought, mechanisms of, 162-165

Advanced Research Projects Agency Network, 424

Advanced Spaceborne Thermal Emission Reflectance Radiometer, 272

Advanced Very High Resolution Radiometer, 270

Advances in Agronomy, journal, 672

Advances in Water Resources, journal, 570

Advection, vapor transport in dry soils, 1012

Aeration, irrigated land drainage, 135

Aerial wells, vapor transport in dry soils, 1014

AES NWT Water Bibliography, database, 426

Africa
   dryland, semiarid regions, research centers, 797
   dryland cropping systems, 179
   groundwater, 409
   market gardens of, 9-14
      crop management, 13

emitter discharge, rate of, yield, 13
   nutrition management, 11-13
   salt buildup, 11-13
   technical description, 10-11
   water, 11-13

Agency-managed irrigation systems, 444

Agnic, database, 113

AgriBiz Search Engines for Agriculture, 425

Agricola, database, 113

Agricultural economics, 1

Agricultural engineering, 1-2

Agricultural fields, runoff from, quality sampling, 771-776
   flow, collecting constant proportion, 773-774
      Coshocton wheel, 774
      multislot divisor systems, 773-774
   interval sampling, 774
   manual sampling, 774-775
   sampling errors, sources of, 775
   surface runoff, characteristics of, 771-773

Agricultural journals, covering water-related topics, 572-573

Agricultural law, 2

Agricultural meteorology, climatology, 2

Agricultural Research Data Directory, 426

Agricultural Research Organization, 801

Agricultural Research Service, 421, 801

Agricultural runoff
   characteristics, 15-18
      pollutants, 16
         dissolved, 16-17
         soil erosion, 16
      control of, 17

Agricultural wastes, groundwater quality and, 391

Agricultural water management, 1
   overview of, 1

Agricultural Water Management, journal, 570

Agricultural wetlands, 1039
   North American, 1039

Agriculture (*see also* Crops)
  irrigated
    economic impacts of, 428-430
    endangered species policy, 431-433
    historical view of, 434-436
    social impacts of, 443-445
    sustainable water management, 437-442
      farm water management assessment, 437-439
    Pfiesteria piscicida, 660
Agriculture, Ecosystems and Environment, journal, 571
AGRIS, database, 113
AgroBase, database, 113
Agroforestry for enhancing water use efficiency, 19-21
Agronomic control, saline seeps, 828
Agronomy, soil science, 2
Agronomy Journal, 572, 745
AGU (*see* American Geophysical Union)
Air entry, in subsurface drip irrigation systems, 562-563
Airlift pumps, for groundwater, 382
Airports, water pollution from, 708
Alachlor, U.S. maximum contaminant level, in drinking water, 651
Alaska, groundwater arsenic incidents in, 325
Albuquerque, NM, water quantity available from precipitation, 107
Aldicarb, U.S. maximum contaminant level, in drinking water, 651
Aldicarb sulfone, U.S. maximum contaminant level, in drinking water, 651
Aldicarb sulfoxide, U.S. maximum contaminant level, in drinking water, 651
Aleppo, Syria, water quantity available from precipitation, 107
Alexandria, Egypt, water quantity available from precipitation, 107
ALEXI method, evapotranspiration, remote sensing of, 270-272
Algae, blue-green, pathogenic, transmission via water, 647
Algeria, dryland cropping systems, rainfall, daily, 179
Alice Springs, Australia, water quantity available from precipitation, 107
Alley cropping, 210
Alternate height stubble, snow capture by, 103
Aluminum, in potable groundwater, 390
Amazing Environmental Organization Web Directory, internet library, as source of water information, 425

AMBIO, journal, 571
American Agricultural Economics Association, 744-745
American Geophysical Union, 421, 745
American Journal of Agricultural Economics, 572, 744
American Journal of Alternative Agriculture, 572
American Society of Agricultural Engineers, 745
American Society of Agronomy, 745
American Society of Civil Engineers, 421, 845
American Water Resources Association, 745
American Water Works Association, 115, 745
Amino acids, in potable groundwater, 390
AMISs (*see* Agency-managed irrigation systems)
Amistad Reservoir, annual evaporation from, 243
Amman, Jordan, water quantity available from precipitation, 107
Ammonium nitrogen, measurement, 625-626
Ammonium-nitrogen, methods for measurement of, 627
Amu Darya river, cultivated land along, 46
Analytic hierarchy process, 818
Analytical methods for organic constituents, 64
  individual organic constituents, groups, 64-65
  total organic matter present, 64
Ancient Greece
  hydraulic mechanisms, devices in, 416-417
  hydraulic science and technology in, 415-417
  urban water engineering, management in, 999-1007
    Athens, 1004-1006
    climatic, hydrologic conditions, 999
    Eupalinos, 1001-1004
    Minoan civilization, water supply in, 999-1001
    Samos, water supply of, 1001-1004
Ancient hydraulic structures, 306-307
Animal feeding operations, surface water quality, 965-966
Animal science, 2
Animal transpiration, evaporation and, 248
Anisotropic materials, Darcy's law, 111
Anisotropy, aquifer transmissivity, 27-28

Annual crops, critical growth periods, 689-690
Antecedent, soil water, 858-860
  homogeneous, wettable soil, 858
  nonhomogeneous, wettable soil, 858
  nonwettable soil, 858-859
  variable soil water in landscape, 859
Antimony
  in livestock drinking water, 598
  in potable groundwater, 390
Aphanomyces invadana, 659
Apparent water holding capacity, 887
Appearance, soil moisture measurement by, 847-851
Apples, water stress in, 98
Applications, vapor flow, 111
Applied Engineering in Agriculture, journal, 572
Applied Science And Technology Abstracts, database, 113
Apricots, water stress in, 98
Aquastat, database, 426
Aquatic biology, 2
Aquatic Biology, Aquaculture and Fisheries Resources, database, 113
Aquatic Sciences and Fisheries Abstracts, database, 114
Aquatic Toxicology, journal, 571
Aqueous.com, search engine, 425
Aquifers, 30-32
  recharge of, 22-25
    artificial, 33-36
      applications, 33
      issues, 35
        environmental, 35
        public health, 35
      methods, 33-35
        direct well injection, 34-35
        surface infiltration, 33-34
      science, role of, 35-36
        monitoring, analysis, 35-36
        planning, management tools, 36
    Darcian methods, 23
    groundwater data, methods based on, 23
    surface water, methods based on, 23
    tracer methods, 24
    water budget methods, 22-23
  tests, well hydraulics of, 1033
  transmissivity, 26-29
    anisotropy, 27-28
    groundwater flow, relationship to, 26-27
    methods of estimation, 28
    transmissivity values, 27
Arab Center for Studies in Agricultural Development, 798

Arabidopsis thaliana, 165, 703, 993
Arachis hypogaea, 689
Aral Sea Basin, 45
  general statistics of, 46
  water supply, demand, 46
Aral Sea disaster, 45-48
  environmental problems, 47-48
  human tragedy, 48
*Archives of Environmental Contamination
  and Toxicology*, 571
Arid Zone Research Institute, 798
Arizona, soil water evaporation, 793
ARPANET (*see* Advanced Research
  Projects Agency Network)
Arsenic
  contamination
    drinking water from mining, 364
    groundwater, 324-329
  in livestock drinking water, 598
  in potable groundwater, 390
Artemisia herba-alba, 164
Artificial recharge of aquifers
  applications, 33
  issues, 35
    environmental, 35
    public health, 35
  methods, 33-35
    direct well injection, 34-35
    surface infiltration, 33-34
  science, role of, 35-36
    monitoring, analysis, 35-36
    planning, management tools, 36
ASA (*see* American Society of Agronomy)
ASAE Technical Information Library,
  search engine, 425
Asbestos
  in livestock drinking water, 598
  production, disposal, water pollution
    from, 708
ASCE (*see* American Society of Civil
  Engineers)
Asia
  dryland cropping systems, 179
  groundwater, 410-411
  research centers, 797-798
Aspergillus, 966
Association of American State Geologists,
  State Geological Survey Pages,
  links to, 426
ASTER (*see* Advanced Spaceborne
  Thermal Emission Reflectance
  Radiometer)
Athens, ancient, urban water engineering,
  management in, 1004-1006
Atlantic Coast, flooding, 301
Atmospheric Boundary Layer, 269
Atmospheric transport, pesticide contami-
  nation to surface water, 655-656

Atmospheric-LandEX-change-Inverse, 270
Atrazine, U.S. maximum contaminant
  level, in drinking water, 651
Australia
  dryland, semiarid regions, research
    centers, 797
  dryland cropping systems, 179
  rainfall, daily, 215
  water table depth, drainage coefficient
    criteria, 119
Australian Center for International
  Agricultural Research, 226
Australian Commonwealth Scientific and
  Industrial Research Organization,
  801
Automation, canal, 53-56
  centralized control, 56
  constant volume control, 55
  downstream water-level control, 55
  flow-rate control, 53
  routing demand changes, 55-56
  upstream water-level control, 53-55
Available water capacity, 847
AVHRR (*see* Advanced Very High
  Resolution Radiometer)
AWRA (*see* American Water Resources
  Association)
AWWA (*see* American Water Works
  Association)
AZRI (*see* Arid Zone Research Institute)

Bacteria
  pathogenic, transmission via water,
    647
  population, in water, drip irrigation and,
    73
Bacteriological microbial sampling,
  619
Baghdad, Iraq, water quantity available
  from precipitation, 107
Baking soda, 830
Bangladesh
  flooding, 301
  groundwater arsenic incidents in, 325
Bare soil, crop coefficients, 89
Barium
  in livestock drinking water, 598
  in potable groundwater, 390
Barium chromate, solubility in water, 77
Barley, 702
  dryland cropping systems, 179
Barriers, vegetation, permanent, snow
  capture by, 103
Basalt, permeable, transmissivity of, 27
Battery manufacture, recycling, water
  pollution from, 708

Beans, water stress in, 96
Beef cattle industry requirements, manure
  management, 601-602
Benzene, contamination of drinking water
  from mining, 364
Berries, crop coefficients, 89
Beryllium
  in livestock drinking water, 598
  in potable groundwater, 390
Best management practices, 17, 371, 623,
  766, 973
  evaluation of, 769
  nutrient, 630-632
    irrigation water, 631
    nitrification inhibitors, 631
    nutrient application rates, 630
    phosphorus, calibrated soil tests,
      630
    potassium, calibrated soil tests,
      630
    soil conservation practices, 631-632
Beta vulgaris, 702
Bicarbonate, in potable groundwater, 390
Bichromate, in natural waters, 76
Big bluestem, interception loss of, 792
Bioavailable P, 666
Biochemical demands, 63
Biochemical oxygen demand, 16, 632, 643,
  644
Biological and Agricultural Index,
  database, 114
Biological organisms, livestock water
  quality standards, 597-598
Biologically identified optimal temperature
  interactive console, 524
BIOSIS Previews, database, 114
BIOTIC (*see* Biologically identified
  optimal temperature interactive
  console)
Bismuth, in potable groundwater, 390
Blaney-Criddle method, evapotraspiration
  formula, 254
Blocked-end systems, surface irrigation,
  545
Blue-green algae, pathogenic, transmission
  via water, 647
Bluestem, interception loss of, 792
BMPs (*see* Best management practices)
Boom traveling sprinkler system, 493-494
Bore hole
  size, for well drilling, 1022
  test, plugging, in well drilling,
    1020-1021
Boron, 49-52
  in livestock drinking water, 598
  plant interaction, 50-51
  in potable groundwater, 390
  soil interaction, 49-50

Boundary element method, groundwater modeling, 360

Boussinesq equation, drainage modeling, 147-150

Brazil, rainfall, daily, 215

Brevoortia tyrannus, 658

Breweries, water pollution from, 708

Bromide, in potable groundwater, 390

Brush clearing, rangeland water yield and, 788-790

BUBL Information Service, internet library, as source of water information, 425

Buckingham–Darcy equation, unsaturated flow, 109-110

Budgeting of water, in irrigation scheduling, 532-534
  evapotranspiration, 533
  soil water content, 533

Buffalo grass, interception loss of, 792

Buffers, vegetative, 210-212
  buffer hydraulic resistance, 210-212
  buffer types, 210

Bulk electrical conductivity, 894

Bulletin of Environmental Contamination and Toxicology, 571

CAB Abstracts, database, 114

Cadmium
  contamination of drinking water from mining, 364
  in livestock drinking water, 598
  in potable groundwater, 390

CAFO (*see* Confined animal feedlot operation)

Calcite, 830

Calcium
  in potable groundwater, 390
  in water, drip irrigation and, 73

Calcium chromate, solubility in water, 77

Calculated fixed radius, 1025

Caliciviruses, 648

California
  annual grasslands, interception loss of, 792
  Death Valley soil water evaporation, 793
  flooding, 301
  groundwater arsenic incidents in, 325

Campylobacter jejuni, 765

Canada, rainfall, daily, 215

Canal automation, 53-56
  centralized control, 56
  constant volume control, 55
  downstream water-level control, 55

flow-rate control, 53
gate stroking, 55-56
routing demand changes, 55-56
upstream water-level control, 53-55

Canopy
  evaporation, 247-248
  temperature
    plant water status measurement, 520
    spatial variability in, 525-526
  transpiration rate, 979-981

Capacitance, soil water measurement by, 885-888
  calibration, 886-887
  capacitance probe designs, 885-886
  manufacturers, suppliers, capacitance probes, 886
  multisensor capacitance probes, real-time soil water profile dynamics measured by, 887-888

Capillary rise, soil water, 861-864
  capillarity, 861
  evaporation, 863
  extraction, 863
  flow dynamics, 862-863
  pressure, 861-862

Carbofuran, U.S. maximum contaminant level, in drinking water, 651

Carbohydrates, in potable groundwater, 390

Carbon dioxide
  plants, transpiration, 57-61
    drought effects, 60
    plant gas exchange, 57-59
      energy balance of leaf, 58
      photosynthesis, 57-58
      respiration, 59
      transpiration, 58
      water-use efficiency, 58-59
    plant growth, development, 59-60
      reproductive growth, seed yield, 59-60
      vegetative growth responses, 59
    temperature, effects of, 60
  in potable groundwater, 390

Carbon tetrachloride
  contamination of drinking water from mining, 364
  U.S. maximum contaminant level, in drinking water, 651

Carbonate, in potable groundwater, 390

Carbonic acid, in potable groundwater, 390

Caribbean, groundwater, 409-410

Carry-over soil moisture, 473

Cascade spillway, 636-637

Casing, for well drilling, 1022

Cassava, dryland cropping systems, 179

Cation-exchange-capacity, 392

Cattle
  dips, water pollution from, 709
  industry requirements, manure management, 601-602
  water consumption for, 588

Cell matrix, effects of, 582-583

Cellulose acetate, 804
  membrane, reverse osmosis, 804-805

Center for Disease Control, 312

Center for Environmental Research and Training, 423

Center-pivot lateral sprinkler system, 491-492

Central America, groundwater, 409-410

Centralized control, canal automation, 56

Centrally-administered hydraulic civilizations, 443-444

Centrifugal pumps, for groundwater, 382-384

Centrifugal sand separators, 289

Centro Internacional de Agricultura Tropical, 226, 798

CERCLA (*see* Comprehensive Environmental Response, Compensation, and Liability Act)

Cereals, crop coefficients, 89

Cerium, in potable groundwater, 390

CERP (*see* Comprehensive Everglades Restoration Plan)

CERT (*see* Center for Environmental Research and Training)

Cesium, in potable groundwater, 390

Ceteris paribus, 214

CGIAR (*see* Consultative Group on International Agricultural Research)

Channel erosion, modeling, 219-220

Channel Water Balance Method, 23

Chaparral, interception loss of, 792

Chemical Abstracts, database, 114

Chemical abstracts service, 114

Chemical industries, contributing to nonpoint source pollution, 705

Chemical injection, in subsurface drip irrigation systems, 562

Chemical manufacture, use, water pollution from, 708

Chemical measurement, 62-66
  analytical methods, 63, 64
    individual organic constituents, groups, 64-65
    total organic matter present, 64
  chromatography, 65
  cold-vapor atomic absorption spectrometry, 64
  data quality, 62
  flame atomic absorption spectrometry, 63

Chemical measurement *(cont.)*
  graphite-furnace atomic absorption
      spectrometry, 63
  inductive coupled argon plasma atomic
      emission spectrometry, 63
  inorganic constituents, analytical
      methods for, 63
  metals, 63
  three-dimension detectors, 65
  two-dimensional detectors, 65
  wet chemistry, 63
Chemical oxygen demand, 64, 643-644
Chemical point source pollution, 710
Chemigation, 67-71
  advantages of, 67
  disadvantages of, 67
  equipment, 67-70
      injection equipment, 68-69
      irrigation equipment, 68
      safety equipment, 69-70
  management practices, 70
Chemistry of acid rain
  acid rain
      nitric acid, 5
      sulfuric acid, 5
  drop-scale transport processes, 5
      external transport, 5
      interfacial transport, 5
      internal transport, 5
  environmental factors influencing, 6
      biota, 6
      geology, 6
      meteorological factors, 6
      topography, 6
Chemosphere, journal, 571
Chickens, water consumption for, 590
Chickpea, dryland cropping systems, 179
Chihuahua
  soil water evaporation, 793
  water quantity available from precipita-
      tion, 107
Chile, groundwater arsenic incidents in,
      325
Chilled mirror method, hygroscopic water
      content in soils, 926
China
  flooding, 301
  groundwater arsenic incidents in, 325
  water table depth, drainage coefficient
      criteria, 1119
Chlordane, U.S. maximum contaminant
      level, in drinking water, 651
Chloride
  in livestock drinking water, 598
  in potable groundwater, 390
Chlorination for disinfection, prevention
      of clogging of drip lines, emitters,
      72-74

biological growth, chemical precipita-
      tion, 72
chemical precipitation, biological
      growth-induced, 72
Cholera, 645
Christiansen's uniformity coefficient,
      irrigation, 470
Chromate, in natural waters, 76
Chromatography, chemical detection, 65
Chromic acid, in natural waters, 76
Chromite, solubility in water, 77
Chromium, 75-79
  contamination of drinking water from
      mining, 364
  forms of, in natural waters, 7
  in livestock drinking water, 598
  in natural waters, 75-76
  oxidation-reduction chemistry, in natural
      waters, 78-79
  in potable groundwater, 390
  solubility controls, 76-78
Chromium arsenate, solubility in water, 77
Chromium chloride, solubility in water, 77
Chromium citrate, in natural waters, 76
Chromium compounds, solubility in
      water, 77
Chromium fluoride, solubility in water, 77
Chromium fulvate, in natural waters, 76
Chromium hydroxide
  in natural waters, 76
  solubility in water, 77
Chromium jarosite, solubility in water, 77
Chromium oxide, solubility in water, 77
Chromium phosphate, solubility in
      water, 77
Chromium picolinate, in natural waters, 76
Chromium sulfate, solubility in water, 77
Chute spillway, 636
Cipolletti, Argentina, water quantity avail-
      able from precipitation, 107
Citrus, water stress in, 98
Civil Engineering Database, 426
Clay, mesh size, particle size classification
      by, 290
Clean Water Act, 363, 401, 956, 1024
Clean Water Action Plan, 401-402
Clearing of brush, rangeland water yield
      and, 788-790
Climate
  effects of, crop coefficients, adjustment
      of K_{co} to account for, 90
  flooding and, 302
Clogging of drip lines
  acidification, emitters, prevention of,
      7-8
  emitters
      chlorination for prevention of
          disinfection, 72-74

disinfection, chlorination for preven-
      tion of, 72-74
  biological growth, chemical
      precipitation, 72
  chemical precipitation, biological
      growth-induced, 72
CO_2, 322
  ecosystem biodiversity, effects on,
      322
  temperature interactions, 322
Coarse sand, mesh size, particle size
      classification by, 290
Cobalt
  in livestock drinking water, 598
  in potable groundwater, 390
Coefficient
  crop, 87-90
      crop coefficient curves, 87-90
          styles of, 88
      effects of climate, adjustment of K_{co} to
          account for, 90
      growing periods within growing
          season, definition of, 88-89
      limited water, coefficients for, 90
      linear K_c curve, construction of, 89
      nongrowing periods, K_c during, 90
  drainage, 118-121
      countries, adoption in, 120
      estimation of, 118-120
      selection of, 118-120
  of variation, 546
Cold-vapor atomic absorption spectro-
      metry, 63, 64
Coliforms, contamination of drinking water
      from mining, 364
Combustion engines, pump powering with,
      756-758
Commercial reverse osmosis technology,
      804
Commonwealth Scientific and Industrial
      Research Organisation, 422, 797
Composite polyamide membranes, reverse
      osmosis, 805
Comprehensive Environmental Response,
      Compensation, and Liability Act,
      400, 401
Comprehensive Everglades Restoration
      Plan, 275
Concentrated animal feeding operations,
      588, 622
Conduit loads, subsurface drainage, 578
Conemaugh River, flooding, 301
Confined animal feeding operations, 42,
      764
Conservation tillage, no-tillage, 80-82
  advantages, 80-81
  disadvantages, 81
  results, 81-82

Constant volume control, canal automation, 55
Consultative Group on International Agricultural Research, 226, 798
Consumptive water use, 83-86
modeling of, 84-85
origin, 83
partitioning of, 83-84
evaporative component, 84
transpiration, 84
utilization, 83
Contamination (*see also* Pollution)
drinking water, United States maximum
levels for, 651
groundwater, 330-332
arsenic, social problem of, 325-327
climate, effect of, 331
fate of contaminants, 331-332
sources of contamination, 330-331
transport of contaminants, 331
types of contaminations, 330
livestock water quality standards,
sources of, 599
Continuity equation, 810
Continuous-move lateral sprinkler system,
490-492
center-pivot laterals, 491-492
linear-move laterals, 491
Contour buffer strip, 210
Controlled drainage, 121-128, 158
irrigated land, 137
management, 125
production benefits of, 122-124
water quality benefits of, 124
Convection dispersion equation, 845
Convective flow, vapor transport in dry
soils, 1013
Convective log normal transfer, 845
Conveyance channel, open-channel
spillways, 639-640
Copper
chromium, arsenic, 331
contamination of drinking water from
mining, 364
in livestock drinking water, 598
in potable groundwater, 390
Corn, 369, 702, 929
water stress in, 95
Corrugated plastic tubing, for drainage,
143
Cotton, 317, 702, 929
water stress in, 97
Cotton grain sorghum, 317
Cow washing, manure management, 603
Cowpea, dryland cropping systems, 179
Crassulation acid metabolism, 57
Creosote, interception loss of, 792
Critical growth periods, plants, 689-691

in annual crops, 689-690
deficit irrigation, 689
in perennial crops, 690
*Critical Reviews in Environmental Science
and Technology*, 571
Crop, water harvesting for, 105-108
advantages, 107-108
disadvantages, 107-108
potential of, 106-107
types of, 105-106
Crop coefficients, 87-90
crop coefficient curves, 87-90
styles of, 88
effects of climate, adjustment of K_{co} to
account for, 90
growing periods within growing season,
definition of, 88-89
K_{co}, adjustment of, 90
limited water, coefficients for, 90
linear K_c curve, construction of, 89
nongrowing periods, K_c during, 90
Crop development models, 91-94
application of, 94
physiological determinants, crop growth,
91-94
accumulation, assimilate, 92-93
crop phenology, 91-92
leaf area development, light interception, 92
partitioning, assimilate, 93
plant water relations, 93-94
thermal time, 91
Crop lower limit, 93
Crop management
African market garden, 13
rainfed farming, 781-782
Crop phenology, crop development and,
91-92
Crop plants, water stress in, 95-101
field crops, 95-97
beans, 96
corn, 95
cotton, 97
groundnuts, 97
peas, 96
rice, 96
sorghum, 95
soybean, 96
sunflowers, 97
wheat, 95
fruit tree crops, 98
apples, 98
apricots, 98
citrus, 98
prunes, 98
vegetable crops, 98
Crop productivity, erosion and,
222-224

Crop residues, snow capture by, 101-104
crop yield, additional water, effect of,
102
frost, 102
runoff, 102
soil temperature, 102
soil water increase, potential for, 101-102
techniques to enhance, 102-103
alternate height stubble, 103
leave strips, 103
permanent vegetation barriers, 103
snow ridges, 103
trap strips, 103
uniform height stubble, 102
Crop water production function, 686
Crop water stress index, 520, 524-525, 675,
677
Crop yield, water use, 740-741
Cropping
drainage, 158
system characteristics, irrigation design,
456-457
Crowned surfaces, for drainage, 139-140
Cryptosporidiosis, 645
Cryptosporidium parvum, 391, 620, 645,
765
CSIRO (*see* Commonwealth Scientific &
Industrial Research Organisation)
Cucumber family, crop coefficients, 89
Cultural eutrophication, 232
Current Contents, database, 114
CVAA (*see* Cold-vapor atomic absorption
spectrometry)
CWA (*see* Clean Water Act)
CWAP (*see* Clean Water Action Plan)
CWBM (*see* Channel Water Balance
Method)
CWPF (*see* Crop water production func-
tion)
CWSI (*see* Crop water stress index)
Cyanide
contamination of drinking water from
mining, 364
free, in livestock drinking water, 598
Cyanobacteria, pathogenic, transmission
via water, 647
CyberStacks, internet library, as source of
water information, 425
Cycling tailwater recovery systems,
969-970
Cyclones, tropical, 739

Dairy cattle, water consumption for, 589
Dairy manure management, 603-606
cow washing, 603
drinking, 603

Dairy manure management *(cont.)*
  flushing manure, 604
  rainwater from roof, concrete areas, 605-606
  recycling dairy wastewater through crops, 604-605
  sprinkling, 604
  washing milking equipment, 603-604
  water budget, developing, 606
Dalapon, 651
Dams, flow measurement devices for, 309-310
Darcian flow, well hydraulics, 1030-1031
Darcian methods, aquifer recharge, 23
Darcy's law, 109-112, 809-810, 876
  applications, 110-111
    anisotropic materials, 111
    groundwater flow, 110-111
    soil water flow, 110
    two-phase flow, flow in oil reservoirs, 111
  historical background, 109
  liquid flow in permeable media, theory of, 109
  saturated flow, 109
  saturated hydraulic conductivity for soils, values of, 110
  unsaturated flow, Buckingham–Darcy equation, 109-110
Data acquisition, supervisory control and, 553
Databases, 113-117
  platforms, 116
  water science, 427 *(see also* Internet)
DataWeb, search engine, 425
Death Valley soil water evaporation, 793
Deep-well turbine pumps, for groundwater, 384
Defense works, water pollution from, 709
Deficit irrigation, 686
Dense, nonaqueous phase liquids, 42
Department of Health Services, 1024
Deposition, erosion modeling, 220
Desert Margins Program, 13
Dichromate, in natural waters, 76
Diffusion, for vapor transport in dry soils, 1012
Dihydroxychromium, in natural waters, 76
Dikes, furrow, 317-320
  dryland applications, 318-319
  equipment, 317-318
  history, 317
  irrigated applications, 319
Dimethyldiselenide, 841
Dimethylselenide, 841
Dinoseb, U.S. maximum contaminant level, in drinking water, 651

Diquat, U.S. maximum contaminant level, in drinking water, 651
Direct water application system, 817
Direct well injection, artificial recharge of aquifers, 34-35
Diseases, water-associated, 645-649
Disinfection, prevention of clogging of drip lines, emitters, chlorination for, 72-74
  biological growth, chemical precipitation, 72
  chemical precipitation, biological growth-induced, 72
Disk filters, 291
Displacement pumps, 759-763
Dissipation, energy, structures, 195-198
  forced hydraulic jumps, by vertical still, 195-196
  hydraulic jumps
    below abrupt expansions, 196-197
    prismatic horizontal channels, 195
    on sloping aprons, 196
  spillways, energy dissipators on, 197-198
  in stilling basins, 195-197
  transitional flows, at abrupt drops, 197
Dissolved constituents in potable groundwater, classified according to abundance, 390
Dissolved oxygen measurement, 642
Dissolved solids, in water, drip irrigation and, 73
Distilleries, water pollution from, 708
Distribution microbial system sampling, 619
DNAPLs *(see* Dense, nonaqueous phase liquids)
Dolomite, transmissivity of, 27
Domain reflectometry, 894
Downstream water-level control, canal automation, 55
Dracontiasis, 645
Dracunculus medinesis, 645
Drainage
  controlled, 121-128
    management, 125
    and management considerations, 125
    production benefits of, 122-124
    water quality benefits of, 124
  hydrologic impacts of, 128-131
  irrigated land, 135-137
    controlled, 137
    free flowing, 137
    need for, 135-136
      aeration, 135
      salinity control, 135-136

    steady state, 137
    system design, 136
      steady state, 136
      transient design, 136
      water quality impacts, 137
    land shaping for, 138-141
      methods, 138
        crowned surfaces, 139-140
        grading and smoothing, 138
        nonuniform slopes, 138
        parallel field drains, 140
        random field drains, 140
        surface drains, 140
        uniform slopes, 138
        warped surfaces, 138
    for soil salinity management, 152-155
      drainage conditions, 152
      drainage requirements, 152-153
        saline soils, 152-153
        sodic soils, 153
      system design, 153-154
        drain depth, 154
        drain spacing, 154
        drainage wells, 154
        relief drains, 153-154
        saline seeps, 154
  subsurface, 576-578
    accessories, 578
    conduit loads, 578
    design, 576-577
      steady-state design, 576-577
      transient methods, 577
    design flow, 577
    grades, 577
    installation, 578
    layout, 577
    pipe size, 578
    system layout, 576
  water quality and, 156-159
    environmental conditions, 156
      climate/precipitation, 156
      soils/hydrology, 156
    management practices, 158
      controlled drainage, 158
      cropping, 158
      tillage systems, 158
    pollutant properties, 157-158
      adsorption/filtration, 157-158
      persistence, 157
Drainage coefficient, 118-121, 576
  countries, adoption in, 120
  estimation of, 118-120
  selection of, 118-120
Drainage inadequacy, crop response, 132-134
  root zone aeration, 132-133
  salt leaching, 133
  trafficability, 132

Drainage materials, 142-146
  corrugated plastic tubing, 143
  fabrication and marketing, 144-145
  materials handling, 145
  synthetic drain envelope materials,
    145-146
  tubing standards, 144
Drainage modeling, 147-151
  equations, 147-148
    Boussinesq equation, 147-148
    Richards equation, 148
    simple analytic equations, 147
  simulation models, 148-150
    Boussinesq equation, 149-150
    one-dimensional Richards equation,
      149
    two-dimensional Richards equation,
      148-149
    water balance models, 149
Drainage wells, 154
Drained upper limit, 529
Drains, tile, phosphorous in, groundwater
    pollution by, 375
Drilling wells, 1020-1023
  casing, and bore hole size, 1022
  cutting samples, analysis of, 1021
  driller selection, 1020
  drilling fluid, 1021-1022
  drilling method, selection of, 1021
  gravel pack, selection of, 1021
  main well drilling, 1022
  new well site, selection of, 1021
  potential well site, selection of, 1020
  test borehole, plugging, 1020-1021
  test hole processes, 1020
  well development, 1022
  well screen, selection of, 1021
  wellhead site, completion of, 1022
Drinking water
  quality standards, saline content, 830
  United States maximum contaminant
    levels for, 651
Drip irrigation, 546-548
  design criteria for, 547
  environmental protection, 547-548
  hydraulic design of, 546-547
  water quality criteria for, 73
Drip lines
  clogging of, acidification, emitters,
    prevention of, 7-8
  emitters, clogging of, chlorination
    for prevention of disinfection,
    72-74
    biological growth, chemical precipi-
      tation, 72
    chemical precipitation, biological-
      growth-induced, 72
Driving pressure differential, 804

Drought, 160-162
  characteristics, 161
  defined, 160-161
  effect on plant growth, 60
  impacts of, 161-162
  mitigation, 162
Drought adaptation, 163-165
Drought avoidance, 163-165
Drought hardening, 166-169
Drought management, 170-172
Drought resistance, 173-177
  of C_3, and C_4 plants, 174
  C_3 plants, 174
  C_4 plants, 174
  of cultivars, 175
  water potential, measurement of,
    173-174
Drought-resistant plants, growth of,
    175-176
Drum reconditioning, water pollution from,
    709
Dry areas, siting water harvesting reser-
    voirs in, hydro-spatial method-
    ology, 818-820
Dry barriers, vapor transport in dry soils,
    1015
Dry cleaning, water pollution from, 709
Dry soils, vapor transport in, 1012-1016
  practical aspects of, 1013-1015
    aerial wells, 1014
    dry barriers, 1015
    fertilizer placement, waste manage-
      ment, 1014-1015
    mulching, dryland farming,
      1013-1014
  processes, 1012-1013
    advection, 1012
    convective flow, 1013
    diffusion, 1012
    nonisothermal flow, 1012-1013
Dryland
  cropping systems, 178-182
    components of, 179
    crop sequences, 181
    crop yield, water use and, 179-180
    rainfall, daily, Mediterranean
      countries, 179
    soil amendments, addition of,
      181-182
    soil water balance, 180-181
  farming, 183-186
    vapor transport in dry soils,
      1013-1014
  research centers for, 795-799
DTD method, evapotranspiration, remote
    sensing of, 270-272
Dual-Temperature-Difference, 270
Ducks, water consumption for, 591

Dust Bowl era, 187-191
  history of, 187
  lessons from, 187-190
DWAS (*see* Direct water application
    system)
Dysentery, 645

EAA (*see* Everglades Agricultural Area)
EAFRO (*see* East African Agricultural and
    Forestry Research Organization)
Earth Observing System Data Gateway,
    database, 426
Earth Sciences, World Wide Web Virtual
    Library, internet library, as source
    of water information, 425
Earthen spillway, 638
East African Agricultural and Forestry
    Research Organization, 226
Ecological resources, San Joaquin Delta,
    824-825
Ecology, terrestrial, global temperature
    change and, 321-323
  CO_2, 322
    ecosystem biodiversity, effects on,
      322
    temperature interactions, 322
  global warming, 321
Economic optimum N, 370
Economics
  global irrigation, 459-462
    agricultural commodities, demand for,
      459
    agricultural production, trends in,
      460-461
    increased supply, options for, 461
    irrigation, trends in, 460
    natural resources, trends in use of,
      459-460
      climate, 459
      genetic resources, 459
      land, 459
      water, 459-460
  United States irrigation, 463-466
    irrigated agricultural area, 463
    value of irrigation, 465
    water used, 463-465
Ecotoxicology, journal, 571
Ecotoxicology and Environmental Safety,
    journal, 571
Eddy
  correlation, evaporation, from lakes,
    large bodies of water, 243
  covariance sensors, 236-237
  flux, 235-236
EEVL Edinburgh Engineering Virtual
    Library, internet library, as source
    of water information, 425

Efficiency in irrigation, 467-472
  application efficiency, 468-469
  seasonal irrigation efficiency, 469
    leaching requirements, 469
    storage efficiency, 469
  uniformity, 469-471
    Christiansen's uniformity coefficient, 470
    emission uniformity, 470-471
    low-quarter distribution uniformity, 470
  water conveyance, 467-468
  water use, 471
    irrigation water use efficiency, 471
Effluent, sewage, for irrigation, 535-536
EiCompendex, database, 114
El Niño, 192-194, 193, 574
El Paso, TX, water quantity available from precipitation, 107
El Salvador, flooding in, 301
Electrical, water pollution from, 709
Electrical conductivity, 392, 829, 852
  measurements, using time domain reflectometry, 832-833
  soil solution extracts, soil salinity measurement, 852-853
Electrical resistivity, soil salinity measurement, 853-854
Electromagnetic induction, 852, 854
Electron capture detector, 64
Emission uniformity, irrigation, 470-471
Emitters, drip lines, clogging of, chlorination for prevention of disinfection, 72-74
  biological growth, chemical precipitation, 72
  chemical precipitation, biological growth-induced, 72
Endangered Species Act, 431
Endangered species policy, irrigated agriculture and, 431-433
  Endangered Species Act, 431
  prior appropriation doctrine, 431-432
Endothall, U.S. maximum contaminant level, in drinking water, 651
Endrin, U.S. maximum contaminant level, in drinking water, 651
Energy balance of leaf, 58
Energy dissipation structures, 195-198
  forced hydraulic jumps, by vertical still, 195-196
  hydraulic jumps
    below abrupt expansions, 196-197
    prismatic horizontal channels, 195
    on sloping aprons, 196
  spillways, energy dissipators on, 197-198
  in stilling basins, 195-197

transitional flows, at abrupt drops, 197
Energy sources, hydrologic cycle and, 412
Engine works, water pollution from, 709
Enhancing water use efficiency, agroforestry for, 19-21
ENSO (*see* El Niño-Southern Oscillation)
Enteric pathogens, in water, 645-649
  zoonotic, hosts, 648
Entrance structure, open-channel spillways, 638-639
Envirofacts: Queries, Maps, and Reports, database, 426
Environmental and Water Resources Institute, 421, 745
Environmental applications
  climate change studies, 567
  isotope analysis, 566-567
Environmental Atlas, database, 426
Environmental concerns, Salton Sea, 837-838
Environmental engineering, 2-3
Environmental factors, pathogen survival, 646
Environmental journals, covering water-related topics, 571-573
Environmental Management, journal, 571
Environmental Monitoring and Assessment, journal, 571
Environmental Pollution, journal, 571
Environmental protection, drip irrigation systems, 547-548
Environmental Protection Agency, 313, 625, 1024
Environmental quality
  nonpoint source pollution, 704-705
  point source pollution, 710
Environmental Science and Technology, journal, 571
Environmental Sciences and Pollution Management, database, 114
Environmental Toxicology, journal, 571
EnviroSCAN, capacitance probe, 886
EPA (*see* Environmental Protection Agency)
Epsom salt, 830
Eroded sediments, sampling, 230
Erosion, 16
  accelerated, 199-204
    deposition, 201
    gullies, 201
    modeling, 202-203
    processes, 202
    rainfall, 199
    rills, 200-201
    transport, 199-200
  modeling, 218-220, 218-221
    channel erosion, 219-220
    deposition, 220

interrill erosion, 218-219
precipitation and, 214-217
productivity and, 222-224
research
  history of, 225-228
    Consultative Group on International Agricultural Research, 226
    national research organizations, 226
    United Nations, 226
  instrumentation, 229-231
    eroded sediments, sampling, 230
    rainfall simulation, 229-230
  tillage/residue methods, 205-208
  vegetative, 209-213
    infiltration, water into soil, 209
    runoff, 209
    soil erodibility, 210
    vegetative buffers, 210-212
      buffer hydraulic resistance, 210-212
      buffer types, 210
Escherichia coli, 648, 765
Ethiopia, dryland cropping systems, 179
Ethylene dibromide, U.S. maximum contaminant level, in drinking water, 651
Europe, groundwater, 409
Eutrophication, 16, 232-234
  cultural eutrophication, 232
  effects of, 233
  rate of, 232-233
  reduction and management, 233-234
  Salton Sea, 838
Evaporation, 245-248
  animal transpiration, 248
  canopy evaporation, 247-248
  eddy correlation, 235-237
    eddy covariance sensors, 236-237
    eddy flux, 235-236
    flux measurement, corrections to, 236
    infrared, 237
  energy balance and, 238-241
    snow covered, frozen surface layers, 239
    surface energy balance, determination of, 240
    unlimited water availability, 238-239
    vegetation control, water availability, 240
    water scarcity, 239-240
  historical approach, 245
  hydrologic cycle, 245-246
  from lakes, large bodies of water, 242-244
    estimation of evaporation, 243-244
    techniques for measuring, 242

Evaporation *(cont.)*
    eddy correlation, 243
    energy budget, 242-243
    water budget, 242
  regional evaporation, 247
  soil evaporation, 247-248
  soil water, 793, 883
  from soils, 249-252
    energy processes during, 250-251
    measurements, 251
    modeling, 251
  under surface energy balance, 246
    energy balance, 246
    evaporation, 246-247
Evaporative component, partitioning of, 84
Evapotranspiration, 10, 87, 156, 179,
    264-267, 480, 507, 519, 525, 741,
    973, 982, 989
  greenhouse, 258-263
    aerodynamic resistance, determi-
      nation of, 259-260
    atmosphere, crop, coupling between,
      261-262
    crop transpiration estimation, from
      inside climate, 258
    outside climate parameters, crop
      transpiration estimation from,
      260
    Penman–Monteith formulation, 260
    stomatal resistance, climate depen-
      dence of, 259
    water vapor transfers between leaf
      surface, greenhouse air, 258-260
  potential evapotranspiration, 264
  reference evapotranspiration, 264
  remote sensing of, 267-274
    ALEXI method, 270-272
    DTD method, 270-272
    methods, 267-270
    sources of error in, 267
Evapotraspiration
  formula, 253-257
    Blaney–Criddle method, 254
    combination methods, 255-256
    corrections due to limited soil water,
      256-257
    fundamental equation, 254
    Hargreaves method, 254-255
    parameters, combination methods,
      256
    Penman method, 255
    Penman–Monteith method, 255-256
    Priestley–Taylor method, 255
    reference definitions, 254
    specific formulas, 254-256
    temperature methods, 254-255
    Turc method, 255
    vapor pressure, 256

rangelands, 791-792
  evaporation from soil, 792
  interception loss, 791-792
  transpiration from plants, 792
Everglades, 275-277
  environmental issues, 275-276
  restoration, 276
  water management, 275-276
Everglades Agricultural Area, 275
Eye infection, 645

Faba bean, 689
Fabrication, drainage material, 144-145
Fallow, summer, 942-945
Farm
  irrigation systems, history of, 549-559
  ponds, 278-281
    design, 278-280
      location, 279
      soil type, 280
      water supply, 279-280
    pond characteristics, 278
    pond management, 280-281
    resources, 281
  water management assessment, 437-439
Farmer-managed irrigation systems, 444
Farming, rainfed, 780-782
  crop management, 781-782
  reducing runoff, 781
  soil capacity to provide water, increas-
    ing, 781
  soil management, 780
    under conditions of excess water, 780
    plant available water, increasing, 781
  soil water evaporation, reducing, 781
Federal Emergency Management Agency,
    294, 297, 306
Federal Insecticide, Fungicide, and
    Rodenticide Act, 400, 401, 1024
Feed water pretreatment, reverse osmosis,
    807-808
Feedlot cattle, water consumption for,
    588-589
Feel, soil moisture measurement by,
    847-851
FEMA *(see* Federal Emergency Manage-
    ment Agency)
Fertilizer
  leaching, irrigation management and,
    282-284
    factors affecting leaching, 282
    irrigation method
      chemigation, 284
      leaching, 283-284
        drip/micro irrigation, 284
        sprinkler irrigation, 284

      surface irrigation, 283
    irrigation scheduling, leaching,
      282-283
    irrigation uniformity, leaching, 283
  nitrogen
    groundwater pollution by
      agricultural practices contributing
        to, 369-371
        container horticultural crops,
          370-371
        grasslands/turf, 370
        row crops, 369-370
      agricultural practices to mitigate,
        371
        soil nitrate tests, 371
        winter cover crops, 371
      environmental impacts, 369
      human health impacts, 369
    surface water pollution by, 950-955
      biochemical processes, 951
      hydrologic processes, 951-952
      spatial variability, 952
      watershed scale analyses, 953-954
        fertilizer use efficiency, role of,
          954
        hydrologic process models,
          953-954
        regional nitrogen input-output
          analyses, 953
  phosphorus, groundwater pollution by,
    374-375
    phosphorus leaching, fixation,
      374-375
    saturation, phosphorus, 375
    tile drains, phosphorus in, 375
  placement
    preplant irrigation, 500
    vapor transport in dry soils,
      1014-1015
Fiber crops, crop coefficients, 89
Field border, 210
Field capacity, 669
  water in soils, 915-918
Field crops, water stress in, 95-97
Field water, 285-288
  applications, 287
  balance components, 285-287
    drainage of water across lower
      boundary, 287
    energy balance, 286
    evaporation and transpiration, 286
    horizontal movement of water,
      285-286
    precipitation and irrigation, 285
    radiation balance and soil heat flux,
      286
    soil water content, 285
  irrigation scheduling, 287

Field water *(cont.)*
 plant growth and natural resource
  modeling, 287
 rainfed cropping, 287
FIFRA *(see* Federal Insecticide, Fungicide,
  and Rodenticide Act)
Filter strip, 210
Filters, types of, 289-293
 centrifugal sand separators, 289
 disk filters, 291
 sand media filters, 291-293
 screen filters, 289-291
 suction screen filters, 289
Filtration, 289-293
 filtration requirements, 289
 flushing, 293
 types of filters, 289-293
  centrifugal sand separators, 289
  disk filters, 291
  sand media filters, 291-293
  screen filters, 289-291
  suction screen filters, 289
Fine sand, mesh size, particle size classification by, 290
Finite element method, groundwater
  modeling, 360
Finland, rainfall, daily, 215
FirstGov, search engine, 425
Fish and Wildlife Service, 431
Fish health, *Pfiesteria piscicida*, 659
FLAA *(see* Flame atomic absorption
  spectrometry)
Flame atomic absorption spectrometry, 63
Flame ionization detector, 64
Flame photometric detector, 64
Flood hydrographs, 302-303
Flood way, 303
Flooding *(see* Floods)
Floodplain, 303
 management, 294-299
  evolution of, in United States, 295
Floods, 300-305
 human aspects of, 303-305
  flood fighting, 305
  flood recovery, 305
  nonstructural measures to control
   flooding, 305
  structural measures to control flooding, 304-305
 physical aspects of, 300-303
  climate and geographic orientation,
   302
  flood hydrographs, 302-303
  flood plain, 303
  flood way, 303
  man-made alterations, 302
  river cross-section, 303
  soil types and land use/cover, 302

 watersheds, 300-302
 recovery, 305
 severe, examples of, 301
Flow measurement, history of, 306-311
 ancient hydraulic structures, 306-307
 devices for dams, hydraulic structures,
  309-310
 hydraulic concepts related to, 307-309
Flow-rate control, canal automation, 53
Fluid, well drilling, 1021-1022
Flumes, in irrigation metering, 496
Fluoride, 312-313
 contamination of drinking water from
  mining, 364
 in livestock drinking water, 598
 in potable groundwater, 390
 sources of, in water, 312
Flushing
 manure, dairy manure management, 604
 for particulate removal, 293
 systems, swine manure management,
  610
FMISs *(see* Farmer-managed irrigation
  systems)
Food and Agriculture Organization, 226,
  437, 459
Food Quality Protection Act, 769
Forages, crop coefficients, 89
Forced hydraulic jumps, by vertical still,
  195-196
Forest hydrology, 973-974
Forestry, natural resources conservation,
  3
Foundries, water pollution from, 709
FQPA *(see* Food Quality Protection Act)
Fractured igneous rock, transmissivity of,
  27
France, water table depth, drainage coeffi-
  cient criteria, 119
Free energy, plant osmotic potential,
  696-697
Free-water surface, 1039
Freezing, redistribution of water during,
  314-315
Frequency domain reflectometry, 534
Frontal convergence, storms, 737-738
Frozen soil, water movement in,
  314-316
 infiltration, 315
 redistribution of water during freezing,
  314-315
Frozen surface layers, evaporation, 239
Fruit tree crops, water stress in, 98
 apples, 98
 apricots, 98
 citrus, 98
 prunes, 98
Fulvic acid, in potable groundwater, 390

Fumigants, groundwater contamination,
  651
Fumigation, preplant irrigation, 500
Fungicides, groundwater contamination,
  651
Furrow dikes, 317-320
 dryland applications, 318-319
 equipment, 317-318
 history, 317
 irrigated applications, 319

Gages, precipitation measurement,
  721-723
 development of, 721
Galaxy, search engine, 425
Gallium, in potable groundwater, 390
Galveston, flooding in, 301
Ganges–Brahmaputra–Meghna, 327
Gas chromatographic technique, 64
Gas exchange, in plants
 energy balance of leaf, 58
 photosynthesis, 57-58
 respiration, 59
 transpiration, 58
 water-use efficiency, 58-59
Gas works, water pollution from, 709
General circulation models, 729
General Science Abstracts, database, 114
Generic numerical method, groundwater
  modeling, 358-359
Genetic enhancement, transpiration
  efficiency, agricultural crops,
  984-986
GeoArchive, database, 114-115
Geographic information systems, 336, 817,
  836, 1025
GeoRef, database, 115
Geosynchronous Operational Environmen-
  tal Satellite, 270
Germanium, in potable groundwater, 390
Germany, water table depth, drainage
  coefficient criteria, 119
Germination, preplant irrigation, 498
GFAA *(see* Graphite-furnace atomic
  absorption spectrometry)
Giardia duodenalis, 765
Giardia lamblia, 646
Giardia species, 620, 645
GIS *(see* Geographic information systems)
Glacial till, transmissivity of, 27
Global Change Master Directory, database,
  426
Global Hydrologic Archive and Analysis
  System, database, 426
Global irrigation economics, 459-462
 agricultural commodities, demand for,
  459

Global irrigation economics *(cont.)*
  agricultural production, trends in,
      460-461
  increased supply, options for, 461
  irrigation, trends in, 460
  natural resources, trends in use of,
      459-460
    climate, 459
    genetic resources, 459
    land, 459
    water, 459-460
Global precipitation patterns, 714
Global warming, 321-323
  CO_2, 322
    ecosystem biodiversity, effects on,
        322
    temperature interactions, 322
    global warming, 321
Glyphosate, U.S. maximum contaminant
    level, in drinking water, 651
GMS *(see* Granular matrix sensor)
Goats, water consumption for, 590
GOES *(see* Geosynchronous Operational
    Environmental Satellite)
Gold, in potable groundwater, 390
Gossypium hirsutum, 317, 702, 929
Government agency-managed irrigation
    systems, 444-445
Grading, for drainage, 138
Granular matrix sensors, 899
  soil water potential measurement by,
      899-903
    automated logging of data, 901
    irrigation scheduling, 901
      automated, 902
      benefits of, 901
    landscape irrigation, automation of,
        902
    limitations of, 902
    multiple depths, water potential
        measurements at, 900-901
    placement, installation, 899-900
Grapes, crop coefficients, 89
Graphite-furnace atomic absorption
    spectrometry, 63
Grasses, rangeland, interception loss of,
    792
Gravel, transmissivity of, 27
Gravel pack, selection of, in well drilling,
    1021
Gravimetric measurement, soil water,
    879-881
Grazing management, 784
Great Basin soil water evaporation, 793
Great Salt Lake, annual evaporation from,
    243
Greece
  ancient

hydraulic mechanisms, devices in,
    416-417
hydraulic science and technology in,
    415-417
urban water engineering, management
    in, 999-1007
  Athens, 1004-1006
  climatic, hydrologic conditions,
      999
  Minoan civilization, water supply
      in, 999-1001
  Samos, water supply of, 1001-1004
dryland cropping systems, rainfall, daily,
    179
Greenhouse, evapotranspiration in,
    258-263
  aerodynamic resistance, determination
      of, 259-260
  atmosphere, crop, coupling between,
      261-262
  crop transpiration estimation, from
      inside climate, 258
  outside climate parameters, crop tran-
      spiration estimation from, 260
  Penman–Monteith formulation, 260
  stomatal resistance, climate dependence
      of, 259
  water vapor transfers between leaf
      surface, greenhouse air, 258-260
GRID-Arendal's Online GIS and Map and
    Graphics, database, 426
Ground control points, 788
Ground Water Focus Group, 402
Groundnut
  dryland cropping systems, 179
  water stress in, 97
Groundwater
  contamination, 330-332
    arsenic, 324-329
      in Bangladesh, 324-325
      social problem of, 325-327
      source of arsenic, 327
      in West Bengal, India, 324
      worldwide, 325
    climate, effect of, 331
    fate of contaminants, 331-332
    nitrogen fertilizers, 369-373
      agricultural practices contributing
          to, 369-371
        container horticultural crops,
            370-371
        grasslands/turf, 370
        row crops, 369-370
      agricultural practices to mitigate,
          371
        soil nitrate tests, 371
        winter cover crops, 371
    environmental impacts, 369

human health impacts, 369
pesticides, 650-653
  drinking water, United States
      maximum contaminant levels
      for, 651
  fumigants, 651
  fungicides, 651
  herbicides, 651-652
  insecticides, 650-651
    groundwater, 650-651
  irrigation management, 652-653
  nonpoint sources, management of,
      652-653
  point sources, management of,
      652
phosphorus fertilizers, 374-375
  phosphorus leaching, fixation,
      374-375
  saturation, phosphorus, 375
  tile drains, phosphorus in, 375
sources of contamination, 330-331
transport of contaminants, 331
types of contaminations, 330
data, aquifer recharge, methods based
    on, 23
microbial sampling, 618
potable, dissolved constituents in,
    classified according to abundance,
    390
saltwater intrusion in, 404-406
  combating, 405-406
  geophysical, geochemical investi-
      gations, 404-405
  mathematical modeling, 405
    computer models, 405
  mechanisms of, 404
sustainable management, in tropics,
    487
world resources, 407-411
  aquifer types, 407-408
    basaltic, volcanic-rock aquifers,
        408
    carbonate rock aquifers, 408
    consolidated/fractured sedimentary
        aquifers, 408
    unconsolidated sand, gravel
        aquifers, 407-408
  resources distribution, 409-411
    Africa, 409
    Asia, 410-411
    Caribbean, 409-410
    Central America, 409-410
    Europe, 409
    Middle East, 410-411
    North America, 410
    Oceania, 410
    South America, 409-410
Groundwater, journal, 570

Groundwater and Soil Contamination
    Database, database, 115
Groundwater flow, Darcy's law,
    110-111
Groundwater law, western United States,
    333-335
    appropriation states, 333
    common law states, 333
        absolute ownership, 333
        correlative rights, 333
        reasonable use, 333
    conjunctive use, 335
    groundwater depletion, 334-335
        common law doctrines, 334
        problem area regulations, 335
        safe yield, 334
    groundwater rights, 333
    surface-groundwater interference,
        335
    well interference conflicts, 334
        appropriation states, 334
            problem area regulations,
                334
                reasonable pumping depths,
                    334
        common law states, 334
Groundwater levels
    mapping, 336-342
        groundwater-level fluctuations, natu-
            ral processes causing, 336
        monitoring system, designing, 336
        water-level data, analysis, interpret-
            ation of, 337-341
        water-level elevation data, contouring
            of, 339
    measuring, 342-344
        field considerations, 343-344
        gradients, 344
        maps, 344
        measuring devices, 342-343
        screened interval, 342
        well casing, 342
Groundwater mining, 345-349
    sustainable aquifer use, 346-348
    water balance, 345-346
Groundwater modeling, 350-357
    boundary element method, 360
    calibration, error analysis, 355
    codes, 358-362
    conceptualization, 353-354
    discretization, 354-355
    finite differences, 359-360
    finite element method, 360
    generic numerical method, 358-359
    integrated finite differences, 360
    model selection, 355-356
    modeled phenomena, 351
    predictions, 356

    process, 352-356
    simulating solute transport, 361-362
Groundwater Monitoring and Remedia-
    tion, journal, 570
Groundwater pollution from mining,
    363-368
    groundwater analysis, 366
    groundwater contaminants, 364-366
    groundwater resources, 363-364
    hydrogeological characteristics, 364
    remediation, 366-367
Groundwater protection, vadose zone and,
    1008-1011
    natural attenuation, 1011
    remediation of existing contamination,
        1010-1011
    soil, attenuation properties of, 1009
    sources of contamination, 1008-1009
    types of contamination, 1008
    waste disposal practices, 1009-1010
Groundwater pumping methods, 376-388
    geohydrologic principles, 376
    pump selection criteria, 386-387
    types of pumps, 379-381
        airlift pumps, 382
        centrifugal pumps, 382-384
        deep-well turbine pumps, 384
        hand pumps, 381-382
        pump head, power requirements,
            384-386
        rotary pumps, 382
        submersible pumps, 384
Groundwater quality, 389-391
    agricultural wastes, 391
    constituents in groundwater, 389
    dissolved constituents in potable
        groundwater, 390
    groundwater salinization, 391
    hydrolysis, 390
    industrial wastes, 391
    ion exchange, adsorption, 390
    irrigated agriculture and, 392-397
        crops
            crop toxicity, 394-395
            impact on, 393-395
            yield potential, estimating,
                393-394
        irrigation systems, impacts on,
            396-397
        leaching, 393
        salinity, characterizing, 392
        sodicity, characterizing, 392
        soils
            impact on, 395-396
            irrigation water, infiltration of, 396
    in karst aquifer, factors controlling, 41
    mechanisms influencing groundwater
        quality, 389-390

    mineral dissolution and precipitation,
        389
    municipal wastewater, 391
    oxidation, reduction reactions, 390
    synthetic chemicals, 391
Groundwater recharge, rangelands, 793
Groundwater regulation, 398-403
    Clean Water Action Plan, 401-402
    ecological impacts, 399-400
    economic impacts, 399
    federal regulations, 401
        Clean Water Act, 401
        Comprehensive Environmental
            Response, Compensation, and
            Liability Act, 401
        Federal Insecticide, Fungicide, and
            Rodenticide Act, 401
        Resource Conservation and Recovery
            Act, 401
        Safe Drinking Water Act, 401
    future developments, 401-402
    Intergovernmental Task Force on
        Monitoring Water Quality, 402
    local regulations, 401
    protection, prevention programs,
        400-401
        source water assessment, prevention
            programs, 401
        state groundwater protection
            programs, 400-401
        wellhead protection programs, 400
    public health impacts, 399
    source water assessment programs, 402
    state regulations, 401
    state watershed protection, 402
    threats to ground water, 399
Growing season, growing periods within,
    crop coefficients, definition of,
    88-89
Growth periods, critical, plants, 689-691
    in annual crops, 689-690
    deficit irrigation, 689
    in perennial crops, 690
Growth responses, vegetative, 59
Guatemala, flooding in, 301
Gullies, accelerated erosion and, 201
GWFG (see Ground Water Focus Group)
Gypsum-calcium sulfate, 830

HALL (see Halogen sensitive detector)
Halogen sensitive detector, 64
HALs (see Health advisory levels)
Hand pumps, for groundwater, 381-382
Hard-hose traveling sprinkler system, 492
Hargreaves method, evapotranspiration
    formula, 254-255
Harmful algal blooms, 658

Harvest index, 93
HCWSI (*see* Histogram-derived crop water stress index)
Health advisory levels, 650
Heat dissipation method, hygroscopic water content in soils, 926
Helianthus annuus, 317, 927
Helminths, pathogenic, transmission via water, 647
Hepatitis E, 648
Heptachlor, U.S. maximum contaminant level, in drinking water, 651
Herbicides, groundwater contamination, 651-652
Hexaquochromium, in natural waters, 76
High performance liquid chromatography, 64
Histogram-derived crop water stress index, 525
History of farm irrigation systems, 549-559
Hordeum vulgare, 702
Horses
    water consumption for, 589
    water requirements for, 590
HPLC (*see* High performance liquid chromatography)
HTML (*see* Hypertext Markup Language)
Huang Ho River, flooding, 301
Human aspects of floods, 303-305
Humic acid, in potable groundwater, 390
Humicap 9000, capacitance probe, 886
Humid regions, irrigation in, 478-482
    plant stress methods, 479
    scheduling, 479-481
    soil moisture methods, 479-480
    water balance techniques, 480-481
Hungary
    groundwater arsenic incidents in, 325
    water table depth, drainage coefficient criteria, 119
Hungary Horse Reservoir, annual evaporation from, 243
Hurricane Hasel, flooding from, 301
Hyco Lake, annual evaporation from, 243
Hydraulic conductivity rates in soils, 919-922
    determination of, 921
    representative values, 920
    saturated, Darcy's law, 110
    unsaturated conditions, 920-921
Hydraulic diffusivity, 866-867
Hydraulic jumps
    below abrupt expansions, 196-197
    prismatic horizontal channels, 195
    on sloping aprons, 196
Hydraulic structures, flow measurement devices for, 309-310
Hydraulics of wells, 1029-1033

aquifer tests, 1033
aquifer yield, storage, 1031-1032
Darcian flow, 1030-1031
partially penetrating wells, 1032-1033
well losses, 1033
Hydrocarbons, in potable groundwater, 390
Hydro-Climatic Data Network, database, 426
HYDROCOIN, 354
Hydrofluorosilicic acid, 312
Hydrogen bonding, water, 1017-1018
Hydrogen sulfite, in water, drip irrigation and, 73
Hydrogeologic mapping, 1025
Hydrographs
    flood, 302-303
    modification, river flows, irrigation and, 473-476
        distribution uniformity, 475
        irrigation efficiencies, 475
        irrigation methods, 475
        irrigation return flows, 474
        reservoir storage, 473-474
Hydrologic cycle, 245-246, 412-414
    energy sources, 412
    hydrologic reservoirs, 412-414
Hydrologic impacts, drainage, 128-131
Hydrologic phenomena, ancient Greek scientific views of, 415-416
Hydrologic process modeling, 418-420
Hydrologic reservoirs, 412-414
Hydrologic science journals, 569-573
Hydrological Processes, journal, 570
Hydrologically dysfunctional rangelands, management of, 783
Hydrologically functional rangelands, management of, 783
Hydrology InfoBase, database, 115
Hydrology research centers, 421-423
    academic-oriented research programs, 421
    drought, weather, climate, centers studying, 422-423
    federally coordinated research programs, 421-422
    international, 422
    pollutant transport, centers studying, 423
    in United States, 422
Hydrolysis, groundwater quality and, 390
Hydro-spatial methodology, siting water harvesting reservoirs in dry areas, 818-820
Hydroxochromate, in natural waters, 76
Hygroscopic water content in soils, 923-926
    engineering properties, 924-925
    hydraulic properties, 923-924
    measurement, 925-926

chilled mirror method, 926
heat dissipation method, 926
oven-drying method, 925
thermocouple psychrometry, 925
physical properties, 923
soil water characteristic, 923-924
surface area, 923
surface tension, 923
Hypertext Markup Language, 424
Hysteresis, soil water, 882-884
    evaporation, 883
    infiltration, 883
    water table response, 883

IAEA (*see* International Atomic Energy Agency)
IAHS (*see* International Association of Hydrological Sciences)
IARCs (*see* International Agricultural Research Centers)
IBARAM (*see* International Board for Soil Research and Management)
ICARDA (*see* International Center for Agricultural Research in Dry Areas)
Ice plant, 703
ICID (*see* International Commission on Irrigation and Drainage)
ICP-AES (*see* Inductively coupled argon plasma atomic emission spectrometry)
ICP-MS (*see* Inductively coupled argon mass spectrometry)
ICRAF (*see* International Center for Research in Forestry)
ICRISAT (*see* International Center for Research in semiarid Tropics)
Identifier, 342
IDRC (*see* International Development Center)
India
    flooding in, 301
    groundwater arsenic incidents in, 325
    water table depth, drainage coefficient criteria, 119
Indigenous farmer-managed hydraulic societies, 444
Indium, in potable groundwater, 390
Indonesia, rainfall, daily, 215
Inductive coupled argon plasma atomic emission spectrometry, 63
Inductively coupled argon plasma atomic emission spectrometry, 63
Industrial wastes, groundwater quality and, 391
Industrial wetlands, 1039
    North American, 1039

Industries
    associated chemicals and, contributing to
        pollution, 708-710
    contributing to point source pollution,
        708-710
Infiltration
    soil water, 883
    water, soils, 930-933
Infiltrometers, 812
INFOMINE, search engine, 425
Information collection rule, 620
Infrared
    eddy correlation, evaporation, 237
    estimation, precipitation measurement
        with, 725
Infrared thermometer, 520
Injection equipment, in chemigation,
    68-69
Inorganic chemicals, point source
    pollution, 707
Inorganic constituents, analytical methods
    for, 63
INRA (see Institut National de la Recher-
    ches Agronomique de Tunisie)
INRAN (see Institut National de la
    Recherches Agronomique du
    Niger)
Insecticide contamination, in groundwater,
    650-651
Institut National de la Recherches
    Agronomique de Tunisie, 797
Institut National de la Recherches
    Agronomique du Niger, 797
Instrumentation, erosion research, 229-231
    eroded sediments, sampling, 230
    rainfall simulation, 229-230
Integrated pest management, 17
Interfacial transport, acid rain, 5
Intergovernmental Panel on Climate
    Change, 60
Intergovernmental Task Force on Moni-
    toring Water Quality, 402
Internal combustion engines, pump
    powering with, 756-758
Internally renewable water resources, 484
International agricultural research centers,
    226, 798
International Association of Hydrological
    Sciences, 226
International Association of Hydrology,
    web links of, internet library, as
    source of water information, 425
International Atomic Energy Agency,
    798
International Board for Soil Research and
    Management, 226
International Center for Agricultural
    Research in Dry Areas, 798

International Center for Research in
    Forestry, 798
International Center for Research in
    Semiarid Tropics, 798
International Commission on Irrigation and
    Drainage, 135, 746
International Crops Research Institute for
    Semi-Arid Tropics, 226
International Development Center, 797
International Erosion Control Association,
    746
International Institute of Tropical Agricul-
    ture, 226, 798
International professional societies,
    746-747
International Program for Arid Land
    Crops, 13
International Soil Conservation Organiza-
    tion, 226
International Soil Tillage Research Organ-
    ization, 226
International Union of Soil Sciences, 226
International Water Management Institute,
    801
Internet, 424-427
    development of, 424
    libraries containing water information,
        425
    search engines, 425-426
    search for water information on,
        424-425
    virtual libraries containing water infor-
        mation, 425-427
    water science databases, 427
Interrill erosion, modeling, 218-219
Intertropical convergence zone, 713
INTRANCOIN, 354
Iodide, in potable groundwater, 390
Ion
    exchange, adsorption, groundwater
        quality and, 390
    toxicity, saline water and, 831
IPALAC (see International Program for
    Arid Land Crops)
IPCC (see Intergovernmental Panel on
    Climate Change)
Iraq, water table depth, drainage coefficient
    criteria, 119
Ireland, water table depth, drainage
    coefficient criteria, 119
Iron
    in livestock drinking water, 598
    in potable groundwater, 390
    in water, drip irrigation and, 73
Iron works, water pollution from, 709
Irrigated agriculture
    economic impacts of, 428-430
    groundwater quality and, 392-397

historical view of, 434-436
social impacts of, 443-445
    centrally-administered hydraulic
        civilizations, 443-444
    government agency-managed irriga-
        tion systems, 444-445
    indigenous farmer-managed hydraulic
        societies, 444
    institutional empowerment, 445
    nature of, 443
    self-reliance, 445
sustainable water management, 437-442
    farm water management assessment,
        437-439
    institutional changes in, 441
    traditional views of, 439-440
    water quality management, 439, 441
    water supplies, delivery of, 439
Irrigated land, drainage, 135-137
    free flowing, 137
    need for, 135-136
        aeration, 135
        salinity control, 135-136
    steady state, 137
    system design, 136
        steady state, 136
        transient design, 136
    water quality impacts, 137
Irrigated water
    market role in reallocation of, 446-448
        cost effective policies, 447
        litigation and, 447
    polymer application in, 449-453
        biopolymer alternatives to, 451
        environmental impact, 450-451
        erosion control, 449
        infiltration, 449-450
        polyacrylamide, 449-451
        sprinkler application, 450
Irrigation (see also under specific type)
    fertilizer/pesticide leaching, 282-284
        factors affecting leaching, 282
        irrigation method
            chemigation, 284
            leaching, 283-284
                drip/micro irrigation, 284
                sprinkler irrigation, 284
                surface irrigation, 283
        irrigation scheduling, leaching,
            282-283
        irrigation uniformity, leaching, 283
    impact on river flows, 473-477
        depletion, 473
        environmental concerns, 476
            in-stream flow requirements, 476
            salt loading pick-up, 476
            water quality, agriculture and, 476
        hydrograph modification, 473-476

Irrigation *(cont.)*

    distribution uniformity, 475

    irrigation efficiencies, 475

    irrigation methods, 475

    irrigation return flows, 474

    reservoir storage, 473-474

  preplant, 498-501

    fertilizer placement, 500

    fumigation, 500

    germination, 498

    management, 500-501

      application

        depth of, 500

        method, 501

        timing of, 500

      water quality impacts, 500-501

    salinity management, 499

    soil water management, 499-500

    uses of, 498-500

    weed control, 498-499

  saline water quality guidelines for, 830

  sewage effluent use for, 535-536

    monitoring guidelines, 536

    reuse standards, 536

  supplemental, 537-539

  surface, 540-545

    blocked-end systems, 545

    improving, 541-545

    precision land leveling, 544

    regulating inflow, time of cutoff, 543-544

    wastewater recovery, reuse, 545

Irrigation and Drainage, journal, 572

Irrigation Association, 745

Irrigation design, 454-458

  field, cropping system characteristics, 456-457

  initial assessment, 454

  water sources, 454-456

  water supply system, 457

Irrigation economics

  global, 459-462

    agricultural commodities, demand for, 459

    agricultural production, trends in, 460-461

    increased supply, options for, 461

    irrigation, trends in, 460

    natural resources, trends in use of, 459-460

      climate, 459

      genetic resources, 459

      land, 459

      water, 459-460

  United States, 463-466

    irrigated agricultural area, 463

    value of irrigation, 465

    water used, 463-465

Irrigation efficiency, 467-472, 507

  application efficiency, 468-469

  seasonal irrigation efficiency, 469

    leaching requirements, 469

  storage efficiency, 469

  uniformity, 469-471

    Christiansen's uniformity coefficient, 470

    emission uniformity, 470-471

    low-quarter distribution uniformity, 470

  water conveyance, 467-468

  water use, 471

    irrigation water use efficiency, 471

Irrigation equipment, in chemigation, 68

Irrigation in humid regions, 478-482

  plant stress methods, 479

  scheduling, 479-481

  soil moisture methods, 479-480

  water balance techniques, 480-481

Irrigation in tropics, 483-489

  climate, 483

  groundwater, sustainable management of, 487

  irrigation management institutions, policies, 487-488

  irrigation systems, 483-484

  productivity of water, increasing, 487

  storage, role of, 486-487

  water savings, basin perspective, 486

Irrigation management, pesticide contamination, groundwater, 652-653

Irrigation mechanical systems, sprinkler, 490-494

  continuous-move laterals, 490-492

    center-pivot laterals, 491-492

    linear-move laterals, 491

  stationary, periodic-move laterals, 490

    sideroll wheeline, 490

    trail-line lateral, 490

  traveling sprinklers, 492-494

    boom travelers, 493-494

    hard-hose travelers, 492

    soft-hose travelers, 493

Irrigation metering, 495-597

  open-channel flow metering, 495-496

    current metering, 495

    flumes, 496

    weirs, 496

  pipeline flow metering, 496-497

Irrigation return flow, 502-506

  components, 502

    irrigation water, quality components, 503

    nitrogen, 504

off-site water quality, impacts from, management, 506

  pesticide residues, 505

  phosphorus, 504

  trace elements, 506

  water quality constituents in, 502-506

    quality parameters, 503

    salts, 503-504

Irrigation sagacity, 507-509

  beneficial uses, 507

  conservation and, 508

  determination of, feasibility, 508

  feasibility, 508

  reasonable uses, 507-508

Irrigation scheduling

  with plant indicators

    field applications, 512-518

      barriers, overcoming, 517

      interpretation of measurements for, 513-517

    measurement, 519-522

      canopy temperature, 520

      expansive growth, 520

      plant organ size variations, 520

      plant water status, measuring, 519-521

      pressure chamber, 519

      sap flow, 521

      stomatal opening, 520

  by remote sensing technologies, 523-527

    crop water stress index, 524-525

    leaf temperature, 523

    leaf water content, 526

    spatial variability in, canopy temperature, 525-526

    thermal kinetic window, 524

Irrigation scheduling *(cont.)*

  by soil water status, 528-531

  supplemental, 537-538

  by water budgeting, 532-534

    evapotranspiration, 533

    soil water content, 533

Irrigation Science, journal, 572

Irrigation systems

  drip, 546-548

    design criteria for, 547

    environmental protection, 547-548

    hydraulic design of, 546-547

  farm, history of, 549-559

    management concepts, 557-558

    sprinkle irrigation systems, 554-555

    trickle irrigation systems, 555-556

    water supply, 556-557

  subsurface drip, 560-564

    components, 562-563

      air entry, 562-563

      chemical injection, 562

      emitters, 562

Irrigation systems *(cont.)*
  filtration, and pressure regulations, 562
  flushing, 562-563
  laterals, 562
  pressure regulations, 562
  pumps, 562
  design, 560-562
  lateral type, spacing, and depth, 561
  maintenance, 563
  operation, 563
  site, water supply, and crop, 560-561
Irrigation with saline water, 510-511
  classification of saline waters, 511
  low-salt, salty waters, blending, 511
IRRISOFT: Database on Irrigation & Hydrology Software, 426
IRWR *(see* Internally renewable water resources)
ISCO *(see* International Soil Conservation Organization)
Isotope analysis, 565-568
  environmental applications of, 566-567
  hydrological studies, 566
  paleoclimate, 567
ISTRO *(see* International Soil Tillage Research Organization)
Italy
  dryland cropping systems, rainfall, daily, 179
  flooding in, 301
  rainfall, daily, 215
ITCZ *(see* Intertropical convergence zone)
ITFM *(see* Intergovernmental Task Force on Monitoring water quality)
IUSS *(see* International Union of Soil Sciences)
IWMI *(see* International Water Management Institute)
IWRN Directories of Water Resources Agencies/Organizations/Institutions in Americas, database, 426

Japan, water table depth, drainage coefficient criteria, 119
Jardin Potager Africain, 10
Jerusalem, Israel, water quantity available from precipitation, 107
Journal of Agricultural Science, 572
Journal of American Water Resources Association, 570, 745
Journal of Contaminant Hydrology, 570
Journal of Environmental Economics and Management, 571
Journal of Environmental Engineering, 571

Journal of Environmental Management, 571
Journal of Environmental Quality, 571, 745
Journal of Environmental Science and Health, 571
Journal of Geophysical Research-Atmosphere, 570
Journal of Great Lakes Research, 570
Journal of Hydraulic Research, 570
Journal of Hydrologic Engineering, 570
Journal of Hydrology, 570
Journal of Irrigation and Drainage Engineering, 572, 745
Journal of Natural Resources and Life Sciences Education, 745
Journal of Range Management, 571
Journal of Soil and Water Conservation, 570, 746
Journal of Toxicology and Environmental Health, 571
Journal of Water Resources Planning and Management, 570
Journals, 569-573 *(see also under* specific journal name)
  agricultural journals, covering water-related topics, 572-573
  environmental journals, covering water-related topics, 571-573
  hydrologic science journals, 569-573
  Web addresses of publisher, 570, 571, 572
Juniper, interception loss of, 792

Kabul, Afghanistan, water quantity available from precipitation, 107
Karachi, Pakistan, water quantity available from precipitation, 107
Karst aquifer, 37-40
  cavity collapse, 39
  construction problems, 39
  engineering problems, 39
  factors controlling groundwater quality in, 41
  flow of groundwater in, factors controlling, 38-39
    sinkhole flooding, 39
  pollution, 42
  subsidence, 39
  unique attributes of, 37-38
  water quality, 41-42
  water-resource problems, 41-42
  water-resource problems in, 41-42
Karst information, major sources of, 37
Karst limestone, transmissivity of, 27
K_c curve, crop coefficient, construction of, 89

Kenaf, dryland cropping systems, 179
Khartoum, Sudan, water quantity available from precipitation, 107
Kuwait, Kuwait, water quantity available from precipitation, 107

La Nina, 574-575
Lactuca sativa, 702
LAI *(see* Leaf area index)
Lake Kerr, annual evaporation from, 243
Lake Mead, annual evaporation from, 243
Lake Okeechobee, annual evaporation from, 243
Lake Ontario, annual evaporation from, 243
Lakes
  annual evaporation from, 243
  evaporation from, 242-244
Land drainage, subsurface, 576-578
  accessories, 578
  conduit loads, 578
  design, 576-577
    steady-state design, 576-577
    transient methods, 577
  design flow, 577
  grades, 577
  installation, 578
  layout, 577
  pipe size, 578
  system layout, 576
Land uses
  associated chemicals and, contributing to pollution, 708-710
  contributing to nonpoint source pollution, 705
Landfill sites, water pollution from, 709
Landscape crop-water use, 741
Lanthanum, in potable groundwater, 390
Large bodies of water, evaporation from, 242-244
Las Vegas, NV, water quantity available from precipitation, 107
Laterals, in subsurface drip irrigation systems, 562
Leaching
  fertilizer/pesticides, irrigation management and, 282-284
    factors affecting leaching, 282
    irrigation method
      chemigation, 284
      leaching, 283-284
        drip/micro irrigation, 284
        sprinkler irrigation, 284
        surface irrigation, 283
    irrigation scheduling, leaching, 282-283

Leaching *(cont.)*
  irrigation uniformity, leaching, 283
  fraction, 393, 504
Lead
  contamination of drinking water from
    mining, 364
  in livestock drinking water, 598
  in potable groundwater, 390
Lead chromate, solubility in water, 77
Leaf, energy balance of, 58
Leaf area development, crop development
    and, 92
Leaf area index, 11, 87, 181
Leaf temperature, in remote sensing
    irrigation, 523
Leaf transpiration rate, 978-979
Leaf water content, in remote sensing
    irrigation, 526
Leaf water potential, 579-587
  components in leaves, 581-583
    cell matrix, effects of, 582-583
    osmometer concept, 580-582
  measurement methods, 583-586
    compensation methods, 584-585
    pressure chamber method, 586
    psychrometric methods, 585
  water in leaves, 580-581
  water potential index, 586-587
Lebanon, dryland cropping systems, rain-
    fall, daily, 179
Legumes, crop coefficients, 89
Leptospira interrogans, 765
Lettuce, 702
Light interception, crop development
    and, 92
Light Nonaqueous Phase Liquid, 330
Lignins, in potable groundwater, 390
Limestone, transmissivity of, 27
Limited irrigated-dryland, 319
Linaria haelava, 164
Lindane, U.S. maximum contaminant
    level, in drinking water, 651
Linear K_c curve, crop coefficient,
    construction of, 89
Linear variable displacement transducers,
    520
Linear-move lateral sprinkler system, 491
Liquid flow in permeable media, theory of,
    109
Lithium, in potable groundwater, 390
Livestock
  production on range, pasture land,
    764-767
    biological characteristics, 765-766
      limit runoff, erosion, 766
    chemical characteristics, 765
      dissolved chemicals, 765
      dissolved oxygen, 765

    physical characteristics, 764-765
      soil, livestock impacts on, 764-765
      suspended sediment, 764
      vegetation, livestock impacts on,
        765
  toxic substances in drinking water for,
    598
  water consumption for, 588-590
  water harvesting methods for, 593-595
  water quality standards, 596-600
    contamination, sources of, 599
    minerals, 597
    nitrate, 597
    nitrite, 597
    salinity, 597
    sulfate, 597
      biological organisms, 597-598
      water quality criteria, 598
      water quality evaluation, 598-599
    water requirements, 596
LNAPL *(see* Light Nonaqueous Phase
    Liquid)
Lolium perenne, 370
Lower limit, 529
Low-pressure drip-irrigation, 9
Low-quarter distribution uniformity,
    irrigation, 470
LPDI *(see* Low-pressure drip-irrigation)
Lupins, dryland cropping systems, 179
LVDTs *(see* Linear variable displacement
    transducers)
Lycium shawii, 164
Lycopersicon esculentum, 703
Lygeum spartam, 163

Macropore flow
  solutes, 843-845
  water, 843
Magnesite, 830
Magnesium
  in potable groundwater, 390
  in water, drip irrigation and, 73
Maize, dryland cropping systems, 179
Malaysia, rainfall, daily, 215
Malus domestica, 689
Management, to enhance water utilization,
    784
Manganese
  in livestock drinking water, 598
  in potable groundwater, 390
  in water, drip irrigation and, 73
Man-made alterations, flooding and, 302
Manure management
  beef cattle industry requirements,
    601-602
  dairy, 603-606

    cow washing, 603
    drinking, 603
    flushing manure, 604
    rainwater from roof, concrete areas,
      605-606
    recycling dairy wastewater through
      crops, 604-605
    sprinkling, 604
    washing milking equipment, 6030604
    water budget, developing, 606
  poultry, 607-609
  swine, 610-6111
    flushing systems, 610
    pit storage systems, 610-611
    recirculation flush pits, 610
    water conservation goals, 611
Mapping
  groundwater levels, 336-342
  precipitation, 718
Marinas, water pollution from, 709
Marine, Oceanographic and Freshwater
    Resources, database, 115
Market garden, African, 9-14
  crop management, 13
  emitter discharge, rate of, yield, 13
  nutrition management, 11-13
  salt buildup, 11-13
  technical description, 10-11
  water, 11-13
Market role in reallocation of irrigated
    water, 446-448
  cost effective policies, 447
  litigation and, 447
Marketing, 612-614
  drainage material, 144-145
  future of, 613-614
  social allocation, *vs.* economic
    allocation, 612
  water marketing, 613
  water rights, 612-613
Markov chain-mixed exponential model,
    730
Marrakech, Morocco, water quantity
    available from precipitation, 107
Maryland, mining, impact on surface water
    quality, 957
Mass selective detector, 65
Matric forces, plant osmotic potential,
    698
Matric potential, 615-617
  in plant physiology, 615-616
  in soil science, 616-617
Maximum contaminant levels, 363, 370,
    401, 650, 830
Maximum daily trunk shrinkage, 514
MCLs *(see* Maximum contaminant levels)
MCME *(see* Markov chain-mixed expo-
    nential model)

Measurement of flow, history of, 306-311
Measurement of groundwater levels, 342-344
Mechanical sprinkler system, for irrigation, 490-494
  continuous-move laterals, 490-492
    center-pivot laterals, 491-492
    linear-move laterals, 491
  stationary, periodic-move laterals, 490
    sideroll wheeline, 490
    trail-line lateral, 490
  traveling sprinklers, 492-494
    boom travelers, 493-494
    hard-hose travelers, 492
    soft-hose travelers, 493
Mediterranean countries, dryland cropping systems, 179
Medium sand, mesh size, particle size classification by, 290
Mel: Michigan Electronic Library, internet library, as source of water information, 425
Mercury
  contamination of drinking water from mining, 364
  in livestock drinking water, 598
Mesabrythium, 703
Mesh size, particle size classifications by, 290
Mesquite, interception loss of, 792
Metals
  chemical measurement, 63
  treatments, water pollution from, 709
Metamorphic rock, fractured, transmissivity of, 27
Meteorological factors, acid rain formation, 6
Metering irrigation, 495-597
  open-channel flow metering, 495-496
    current metering, 495
    flumes, 496
    weirs, 496
  pipeline flow metering, 496-497
Methoxychlor, U.S. maximum contaminant level, in drinking water, 651
Mexico, groundwater arsenic incidents in, 325
MIAC (see Mid-America Agricultural Consortium)
Microbial sampling, 618-621
  bacteriological analysis, 619
  distribution system sampling, 619
  groundwater, 618
  protozoan analysis, 620
  surface water, 618-619
  viral analysis, 619-620
Micro-organisms, pathogenic, transmission via water, 647

Microwave measurement, precipitation measurement with, 725-726
Mid-America Agricultural Consortium, 797
Middle East
  dryland, semiarid regions, research centers, 797-798
  groundwater, 410-411
Millet, dryland cropping systems, 179
Minerals, livestock water quality standards, 597
Mining
  contaminants in drinking waters from, 364
  groundwater, 345-349
    sustainable aquifer use, 346-348
    water balance, 345-346
  groundwater pollution from, 363-368
    groundwater analysis, 366
    groundwater contaminants, 364-366
    groundwater resources, 363-364
    hydrogeological characteristics, 364
    remediation, 366-367
  impact on surface water quality, 957
  surface, surface water pollution by, 956-960
    biological impacts, 958
    chemical impacts, 957-958
    classification, surface water pollutants, 957-958
    physical impacts, 957
    regulatory enforcement, 959-960
    water treatment, 959
Minoan civilization, water supply in, 999-1001
Mississippi River, flooding, 301
Modeling
  crop development, 91-94
    application of, 94
    physiological determinants, crop growth, 91-94
      accumulation, assimilate, 92-93
      crop phenology, 91-92
      leaf area development, light interception, 92
      partitioning, assimilate, 93
      plant water relations, 93-94
      thermal time, 91
  drainage, 147-151
    equations, 147-148
      Boussinesq equation, 147-148
      Richards equation, 148
      simple analytic equations, 147
  electrical conductivity calibration, soil, 833
  erosion, 218-221
    erosion process, 218-220
  evaporation from soils, 251

groundwater, 350-357
  boundary element method, 360
  codes, 358-362
  conceptualization, 353-354
  discretization, 354-355
  finite differences, 359-360
  finite element method, 360
  generic numerical method, 358-359
  integrated finite differences, 360
  modeled phenomena, 351
  process, 352-356
  simulating solute transport, 361-362
hydrologic process, 418-420
precipitation
  daily precipitation models, 730-731
  general circulation models, 729-730
  spatial-temporal rainstorm models, 730
quality, 768-770
  best management practices, evaluation of, 769
  classification of, 768-769
  pesticides, risk assessment of, 769
  pollutants, evaluation of sources, 769
Moisture measurement, soil, by feel, appearance, 847-851
Mojave soil water evaporation, 793
Molecular structure, water, 1017
Molybdenum, in potable groundwater, 390
Monohydroxychromium, 76
Montana, mining, impact on surface water quality, 957
Morocco, dryland cropping systems, rainfall, daily, 179
Morvi, flooding, 301
Mozambique, flooding in, 301
Mulch tillage, erosion control, 207-208
Mulching, vapor transport in dry soils, 1013-1014
Municipal solid waste, 1009
Municipal wastewater, groundwater quality and, 391
Municipal wetlands, 1039
  North American, 1039
Mustard, dryland cropping systems, 179
Muynak, 47

NAPAP (see National Acid Precipitation Assessment Program)
NAPLs (see Nonaqueous phase liquids)
NARIs (see National Agricultural Research Institutes)
NASS (see National Agricultural Statistics Service)
National Acid Precipitation Assessment Program, 4
National Agricultural Library, 113

National Agricultural Research Institutes, 226

National Agricultural Statistics Service, 463

National Atlas of United States of America, database, 426

National Contaminant Occurrence Database, 399

National Flood Insurance Program, 294, 296, 297

National Ground Water Association, 745-746

National Institute of Agricultural engineering, 317

National Institute of Agronomy, 797

National Marine Fisheries Service, 431

National Oceanic and Atmospheric Administration, 736

National Oceanic and Atmospheric Association, 422

National Pollutant Discharge Elimination System, 62

National research organizations, erosion research, 226

National water quality assessment, 399, 650

National Water Quality Monitoring Council, 402

Natural Resources Conservation Service, 210, 226, 278, 422, 449

Natural waters, oxidation-reduction chemistry in, 78-79

Nature, journal, 571

NAWQA (*see* National Water Quality Assessment)

NCOD (*see* National Contaminant Occurrence Database)

NCSU (*see* North Carolina State University)

NDVI (*see* Normalized Difference Vegetation Index)

Net positive suction head, 385

Netherlands, water table depth, drainage coefficient criteria, 119

Neuse River, 622-624

Neutron moisture meter, 889

Neutron thermalization, soil water measurement by, 889-893
  access tubes, depth control, 890-891
  calibration, 891-892
  neutron emission, 891
  safety, 892

Nevada
  groundwater arsenic incidents in, 325
  mining, impact on surface water quality, 957
  soil water evaporation, 793

New Delhi, India, water quantity available from precipitation, 107

New Mexico, Chihuahuan soil water evaporation, 793

New Zealand, groundwater arsenic incidents in, 325

NFIP (*see* National Flood Insurance Program)

NFTM (*see* Numerical flow/transport models)

NGWA (*see* National Ground Water Association)

NIAE (*see* National Institute of Agricultural engineering)

Nickel
  contamination of drinking water from mining, 364
  in livestock drinking water, 598
  in potable groundwater, 390

Nicotiana tabacum, 703

Niger, Sahel soil water evaporation, 793

Niobium, in potable groundwater, 390

Nitrate
  contamination of drinking water from mining, 364
  livestock water quality standards, 597
  in potable groundwater, 390

Nitrate nitrogen, measurement, 626

Nitrate-N, in livestock drinking water, 598

Nitrate-nitrogen, methods for measurement of, 628

Nitric acid, acid rain, 5

Nitrite nitrogen, measurement, 626

Nitrite-N, in livestock drinking water, 598

Nitrogen, 950
  irrigation return flow, 504
  irrigation water, 503
  in potable groundwater, 390

Nitrogen fertilizers
  groundwater pollution by, 369-373
    agricultural practices contributing to, 369-371
      container horticultural crops, 370-371
      grasslands/turf, 370
      row crops, 369-370
    agricultural practices to mitigate, 371
      soil nitrate tests, 371
      winter cover crops, 371
    environmental impacts, 369
    human health impacts, 369
  surface water pollution by, 950-955
    biochemical processes, 951
    hydrologic processes, 951-952
    spatial variability, 952
    watershed scale analyses, 953-954
      fertilizer use efficiency, role of, 954

hydrologic process models, 953-954
  regional nitrogen input-output analyses, 953

Nitrogen measurement, 625-629
  ammonium nitrogen, 625-626
  ammonium-nitrogen, methods for measurement of, 627
  nitrate nitrogen, 626
  nitrate-nitrogen, methods for measurement of, 628
  nitrite nitrogen, 626
  organic nitrogen, 629
    methods for measurement of, 629
  total nitrogen, 624

Nitrogen phosphorous detector, 64

NMFS (*see* National Marine Fisheries Service)

NOAA (*see* National Oceanic and Atmospheric Administration)

NOAA Environmental Services Data Directory, 426

Nonaqueous phase liquids, 351

Nongrowing periods, crop coefficients, K_c during, 90

Nonisothermal flow, vapor transport in dry soils, 1012-1013

Nonpoint source pollution, 704-706
  chemical industries, contributing to, 705
  contaminant interactions, 704
  environmental quality, 704-705
  global challenges, 706
  industries, contributing to, 705
  land uses, contributing to, 705
  remediation of, 705-706
  sampling for, 705

Nonuniform slopes, for drainage, 138

Normalized Difference Vegetation Index, 269

North America
  groundwater in, 410
  semiarid areas, arid institutions, 796-797
  wetlands of, 1039

North Carolina State University, 658

North Mexico, groundwater arsenic incidents in, 325

No-till, erosion control, 205-206

No-tillage, conservation, 80-82
  advantages, 80-81
  disadvantages, 81
  results, 81-82

Nova Scotia, groundwater arsenic incidents in, 325

NPDES (*see* National Pollutant Discharge Elimination System)

NPSH (*see* Net positive suction head)

NPSP (*see* Nonpoint source pollution)

NTIS (National Technical Information Service) Database, database, 115

Numerical flow/transport models, 1025

Numerical weather prediction, 729

Nutrient best management practices, 630-632

  irrigation water, 631

  nitrification inhibitors, 631

  nutrient application rates, 630

  phosphorus, calibrated soil tests, 630

  potassium, calibrated soil tests, 630

  soil conservation practices, 631-632

Nutrient Cycling in Agroecosystems, journal, 572

NWQMC (*see* National water Quality Monitoring Council)

Oak mottes, interception loss of, 792

Observation wells, 633-635

Oceania, groundwater, 410

Ocotea foetens, 164

Ogallala aquifers, 43-44

Ohio, mining, impact on surface water quality, 957

Oil crops, crop coefficients, 89

Oil reservoirs, flow in, Darcy's law, 111

One-dimensional Richards equation, 149

Online Documents Covering Water And Agriculture, database, 426

Online Electronic Science Library, internet library, as source of water information, 425

Ontario, Canada, groundwater arsenic incidents in, 325

Open-channel flow irrigation metering, 495-496

  current metering, 495

  weirs, 496

Open-channel spillways, 636-641

  components of, 638-640

    conveyance channel, 639-640

    entrance structure, 638-639

    outlet structure, 640

  types, 636-638

    cascade spillway, 636-637

    chute spillway, 636

    earthen spillway, 638

    side-channel spillway, 637-638

    straight-drop spillways, 636

Oregon

  dryland cropping systems, 179

  groundwater arsenic incidents in, 325

Organic chemicals

  point source pollution, 710

  pollution, point source, 710

Organic constituent measurement, analytical methods for, 64

  individual organic constituents, groups, 64-65

  total organic matter present, 64

Organic nitrogen

  measurement, 629

  methods for measurement of, 629

Orifice design, for precipitation gages, 722

Orographic enhancement, precipitation distribution patterns, 716-717

Orographic lifting, storms, 738-739

Oryza sativa, 689

Osmosis, reverse, 803-808

  cellulose acetate membrane, 804-805

  commercial reverse osmosis technology, 804

  composite polyamide membranes, 805

  feed water pretreatment, 807-808

  osmotic pressure, 803

  water, salt transport, 803-804

Osmotic adjustment, plants, 692-695

  component solutes, 693

  growth, yield responses, 693-694

Osmotic potential, plants, 696-700

  free energy, 696-697

  matric forces, 698

  osmotic forces, 698

  osmotic potential, *vs.* osmotic pressure, 697-698

  plant cell, as osmometers, 698

  solute accumulation, control of, 698-700

  solutes, 696-697

Oven-drying, hygroscopic water content in soils, 925

Oxamyl, U.S. maximum contaminant level, in drinking water, 651

Oxidation

  reduction reactions, groundwater quality and, 390

  states, in natural waters, 7

Oxidation-reduction chemistry, in natural waters, 78-79

Oxygen

  measurement of, 642-644

    biochemical oxygen demand, 643

    chemical oxygen demand, 643-644

    dissolved oxygen measurement, 642

  in potable groundwater, 390

Oxygen demanding organics, irrigation water, 503

Pacific Decadal Oscillation, 192

Pakistan, water table depth, drainage coefficient criteria, 119

Parallel field drains, 140

Particulate removal, 289-293

  filtration requirements, 289

  flushing, 293

  types of filters, 289-293

    centrifugal sand separators, 289

    disk filters, 291

    sand media filters, 291-293

    screen filters, 289-291

    suction screen filters, 289

Partitioning

  assimilate, crop development and, 93

  consumptive water use, 83-84

    evaporative component, 84

    transpiration, 84

Passive microwave measurement, precipitation measurement with, 725-726

Pasture land, livestock production on, 764-767

  biological characteristics, 765-766

    limit runoff, erosion, 766

  chemical characteristics, 765

    dissolved chemicals, 765

    dissolved oxygen, 765

  physical characteristics, 764-765

    soil, livestock impacts on, 764-765

    suspended sediment, 764

    vegetation, livestock impacts on, 765

Pathogenic organisms (*see also* Biological organisms)

  enteric, zoonotic, hosts, 648

  in irrigation water, 503

  transmission of, 647

  in water, 645-649

    classes of, 645

    diseases, water-associated, 645

    emerging pathogens, 649

    enteric pathogens, 645

    environmental factors, pathogen survival, 646

    water routes, pathogen transmission, 646

    water treatment, 646-649

PDO (*see* Pacific Decadal Oscillation)

Peach trees, 689

Peanut, 689

Pearl millet, 689

Peas, 689

  dryland cropping systems, 179

  water stress in, 96

Peatlands, northern, 1035-1036

Penman method, evapotraspiration formula, 255

Penman–Monteith method, evapotraspiration formula, 255-256

Pennisetum glaucum, 689

Pennsylvania
  flooding, 301
  mining, impact on surface water quality,
    957
Pentachlorophenol, contamination of
    drinking water from mining, 364
Perennial crops, critical growth periods,
    690
Permanent vegetation barriers, snow
    capture by, 103
Permanent wilting points, 669, 831, 927
  soils, 927-929
    field measurement, 928
    pressure outflow apparatus approxi-
      mation, 928
    sunflower method, 927-928
Permeable basalt, transmissivity of, 27
Pesticide Groundwater Data Base, 650
Pesticides
  contamination
    drinking water, United States
      maximum contaminant levels for,
        651
      fungicides, 651
    groundwater, 650-653
      drinking water, United States
        maximum contaminant levels
          for, 651
      fungicides and fumigants, 651
      herbicides, 651-652
      insecticides, groundwater, 650-651
      irrigation management, 652-653
      nonpoint sources, management of,
        652-653
      point sources, management of, 652
    irrigation management, 652-653
    nonpoint sources, management of,
      652-653
    point sources, management of, 652
    surface water, 654-657
      application, surface water, distance
        between, 655
      application timing, 655
      atmospheric transport, 655-656
      detection frequency, 654
      pesticide movement to surface
        water, 655-656
      rainfall after application, 655
      toxicity, 654-655
  leaching, irrigation management and,
    282-284
    factors affecting leaching, 282
    irrigation method
      chemigation, 284
      leaching, 283-284
        drip/micro irrigation, 284
        sprinkler irrigation, 284
        surface irrigation, 283

irrigation scheduling, leaching,
    282-283
irrigation uniformity, leaching, 283
residues
  irrigation return flow, 505
  irrigation water, 503
Pfiesteria piscicida, 658-662
  agriculture, 660
  biology of, 659
  effects on human health, 659-660
  fish health and, 659
  future research needs, 660-661
  identification of, 658-659
Pfiesteria species, 624, 659
PGWDB (see Pesticide Groundwater
    Data Base)
Phoenix, AZ, water quantity available from
    precipitation, 107
Phosphate, in potable groundwater, 390
Phosphoenolpyruvate carboxylase, 57
Phosphorus, 961
  applications, surface water quality,
    961-964
  fertilizers, groundwater pollution by,
    374-375
    phosphorus leaching, fixation,
      374-375
    saturation, phosphorus, 375
    tile drains, phosphorus in, 375
  irrigation return flow, 504
  irrigation water, 503
  measurement in water, 666-668
    bioavailable, 666-667
    dissolved, 666
    particulate, 667-668
    total, 667
Photoionization detector, 64
Photosynthesis, 57-58
Physical barrier effectiveness, 1026
Physiological determinants, crop growth,
    91-94
  accumulation, assimilate, 92-93
  crop phenology, 91-92
  leaf area development, light interception,
    92
  partitioning, assimilate, 93
  plant water relations, 93-94
  thermal time, 91
Picloram, U.S. maximum contaminant
    level, in drinking water, 651
Pipe size, subsurface drainage, 578
Pipeline flow irrigation metering,
    496-497
Pisum sativum, 689
Pit storage systems, swine manure
    management, 610-611
Plant available water, 532, 669
  soil, 669-672

crop rooting characteristics, 669-670
measurement of, 671-672
water movement
  through plant, 670-671
  through soil, 670
Plant exposure to water stress, during
    specific growth stages, 673-675
Plant gas exchange, 57-59
  energy balance of leaf, 58
  photosynthesis, 57-58
  respiration, 59
  transpiration, 58
  water-use efficiency, 58-59
Plant growth, development, 59-60
  reproductive growth, seed yield, 59-60
  vegetative growth responses, 59
Plant indicators, irrigation scheduling with
  field applications, 512-518
    barriers, overcoming, 517
    interpretation of measurements for,
      513-517
    measurement, 519-522
      canopy temperature, 520
      expansive growth, 520
      plant organ size variations, 520
      plant water status, measuring,
        519-521
      pressure chamber, 519
      sap flow, 521
      stomatal opening, 520
Plant organ size variations, plant water
    status measurement, 520
Plant physiology, matric potential in,
    615-616
Plant water
  status measurement, 519-521
  stress, 676-679
    agrometeorological parameters, 677
    parameters, comparisons, 677-678
    plant parameter, 676
    soil parameters, 677
    water application, parameters for,
      676-678
  use, stomatal control, 680-685
    environmental effects on stomata,
      683-684
    stomatal anatomy, 680
    stomatal physiology, 682-683
    transpiration by stomata, 680-682
      leaf scale, 680-682
      vegetation scale, 682
    water, carbon flows, stomata link,
      684
Plant yield, water use and, 686-688
  crop water production function, 688
  deficit irrigation, 687-688
  water use efficiency, 686-687
    methods of increasing, 687

Plants
  critical growth periods, 689-691
    in annual crops, 689-690
    deficit irrigation, 689
    in perennial crops, 690
  osmotic adjustment, 692-695
    component solutes, 693
    genetic control, 693
    growth, yield responses, 693-694
  osmotic potential, 696-700
    free energy, 696-697
    matric forces, 698
    osmotic forces, 698
    osmotic potential, *vs.* osmotic
      pressure, 697-698
    plant cell, as osmometers, 698
    solute accumulation, control of,
      698-700
    solutes, 696-697
  salinity effects on, 830
  salt tolerance of, 701-703
  transpiration, carbon dioxide, 57-61
    drought effects, 60
    plant gas exchange, 57-59
      energy balance of leaf, 58
      photosynthesis, 57-58
      respiration, 59
      transpiration, 58
      water-use efficiency, 58-59
    plant growth, development, 59-60
      reproductive growth, seed yield,
        59-60
      vegetative growth responses, 59
    temperature, effect of, 60
Platforms, for databases, 116
Platinum, in potable groundwater, 390
Pocosins, 1036
Point source pollution, 707-712
  assessment, 711
  chemicals contributing to, 708-710
  environmental quality, 710
  global challenges, 711
  industries contributing to, 708-710
  inorganic chemicals, 707
  land uses contributing to, 708-710
  organic chemicals, 710
  remediation, 711
  sampling for, 710-711
  soil, water, contaminant interactions in,
    707-710
Poland
  groundwater arsenic incidents in,
    325
  water table depth, drainage coefficient
    criteria, 119
Polio, 645
Pollutants (*see also under* specific
    pollutant)

in agricultural runoff characteristics,
    16
  evaluation of sources, 769
  transport, centers studying, 423
Pollution
  of groundwater, from mining, 363-368
    groundwater analysis, 366
    groundwater contaminants, 364-366
    groundwater resources, 363-364
    hydrogeological characteristics,
      364
    remediation, 366-367
  karst aquifer, 42
  nonpoint source, 704-706
    chemical industries, contributing to,
      705
    contaminant interactions, 704
    environmental quality, 704-705
    global challenges, 706
    industries, contributing to, 705
    land uses, contributing to, 705
    remediation of, 705-706
    sampling for, 705
  point source, 707-712
    assessment, 711
    chemicals contributing to, 708-710
    environmental quality, 710
    global challenges, 711
    industries contributing to, 708-710
    inorganic chemicals, 707
    land uses contributing to, 708-710
    organic chemicals, 710
    remediation, 711
    sampling for, 710-711
    soil, water, contaminant interactions
      in, 707-710
  of surface water, by surface mines,
    956-960
    biological impacts, 958
    chemical impacts, 957-958
    classification, surface water
      pollutants, 957-958
    physical impacts, 957
    regulatory enforcement, 959-960
    water treatment, 959
Polyacrylamide, 449
Polyamide, 804
Polyvinylchloride, 886, 890
Ponds, farm, 278-281
Porous media, electrical conductivity
    calibration models, 833
Portugal, water table depth, drainage
    coefficient criteria, 119
Potable groundwater, dissolved con-
    stituents in, classified according
    to abundance, 390
Potash, 830
Potassium, in potable groundwater, 390

Potassium chromate, solubility in water,
    77
Potential evapotranspiration, 264
POTWs (*see* Publicly owned treatment
    works)
Poultry
  manure management, 607-609
  water consumption for, 590-591
Poultry Science, journal, 572
PPNT (*see* Preplant N tests)
Prairie potholes, 1035
Precipitation
  chemistry, acid rain, 4-7
    chemistry of acid rain, 4-5
      nitric acid, 5
      sulfuric acid, 5
    drop-scale transport processes, 5
      external transport, 5
      interfacial transport, 5
      internal transport, 5
    environmental factors influencing, 6
      biota, 6
      geology, 6
      meteorological factors, 6
      topography, 6
  distribution patterns, 713-718
    global precipitation patterns, 714
    latitudinal, land-ocean effects,
      713-714
    orographic enhancement, 716-717
    precipitation mapping, 718
    space, time scales of precipitation,
      714-716
    vertical motion, 717-718
  erosion and, 214-217
  forms of, 719-720
  measurement, 721-723
    errors, 722
    gages, 721-723
      development of, 721
      orifice design for precipitation gages,
        722
    with remote sensors, 724-726
      infrared estimation, 725
      passive microwave measurement,
        725-726
      satellite imagery, 725
      terrestrial gamma snow measure-
        ment, 726
      weather radar, 724-725
    snow, direct measurements of,
      722-723
    United States, standard precipitation
      gages in, 721-722
    wind, effects on, 722
  modification, 727-728
    growing atmospheric water, 727
    seeding techniques, 727-728

Precipitation (cont.)
  simulation models, 729-733
    daily precipitation models, 730-731
    general circulation models, 729-730
    spatial-temporal rainstorm models,
      730
  stochastic properties, 734-736
  storms, 737-739
    frontal convergence, 737-738
    orographic lifting, 738-739
    thermal convection, 739
    tropical cyclones, 739
  water quantities available from, 107
Precision agriculture
  crop yield, water use, 740-741
  landscape crop-water use, 741
  spatial analysis, crop-water use,
    741-743
  statistical calculations, 741
  water use, 740-743
  water use and, 740-743
Precision land leveling, surface irrigation,
  544
Preplant irrigation, 498-501
  fertilizer placement, 500
  fumigation, 500
  germination, 498
  management, 500-501
    application
      depth of, 500
      method, 501
      timing of, 500
    water quality impacts, 500-501
    salinity management, 499
    soil water management, 499-500
    uses of, 498-500
    weed control, 498-499
Preplant N tests, 371
Presidedress N tests, 371
Presowing hardening, 167-168
Pressure chamber
  leaf water potential measurement
    method, 586
  plant water status measurement,
    519
Priestley–Taylor method, evapotraspiration formula, 255
Prior appropriation doctrine, endangered
  species policy, irrigated agriculture
  and, 431-432
Probes
  capacitance, 885-886
  soil water storage measurement by,
    908-910
Professional societies, 744-747
  international, 746-747
    International Erosion Control
      Association, 746

World Association of Soil and Water
  Conservation, 746-747
  in United States, 744-746
    American Agricultural Economics
      Association, 744-745
    American Geophysical Union, 745
    American Society of Agricultural
      Engineers, 745
    American Society of Agronomy, 745
    American Society of Civil Engineers,
      845
    American Water Resources Asso-
      ciation, 745
    American Water Works Association,
      745
    Irrigation Association, 745
    National Ground Water Association,
      745-746
    Society of Range Management, 746
    Society of Wetland Scientists, 746
    Soil and Water Conservation Society,
      746
    United States Committee on Irrigation
      and Drainage, 746
    Water Environment Federation, 746
Programmable logic controllers, 55
Propionate, in potable groundwater, 390
Protozoa, pathogenic, transmission via
  water, 647
Protozoan microbial sampling, 620
Prunes, water stress in, 98
Prunus persica, 689
PSNT (see Presidedress N tests)
Psychrometry
  leaf water potential measurement, 585
  plant, soil water status measurement,
    748-755
    accuracy of, 748
    calibration of psychrometers,
      752-753
    interpretation of, 748-749, 754
    measurement procedure, 753
    preparation of psychrometers, 752
    principles of operation, 751-752
    sampling for, 749
    temperature control, 753
    theory of, 751-752
    types of, 752
    vapor pressure equilibration, 753
Public drinking water system, 401
Public health, artificial recharge of
  aquifers, 35
Publicly owned treatment works, 643
PubScience, internet library, as source of
  water information, 425
Pumping methods, groundwater, 376-388
  geohydrologic principles, 376
  types of pumps, 379-381

Pumps
  displacement, 759-763
    reciprocating pumps, 761-763
    screw pumps, 759-761
  head, power requirements, for ground-
    water, 384-386
  powering with internal combustion
    engines, 756-758
  in subsurface drip irrigation systems, 562
  tailwater recovery systems, 970-971
Pyramid Lake, annual evaporation from,
  243

Quality modeling, 768-770
  best management practices, evaluation
    of, 769
  classification of, 768-769
  pesticides, risk assessment of, 769
  pollutants, evaluation of sources, 769
Quality of Our Nation's Water Introduction
  State Fact Sheets, database, 425
Quality sampling, runoff from agricultural
  fields, 771-776
  flow, collecting constant proportion,
    773-774
    Coshocton wheel, 774
    multislot divisor systems, 773-774
  interval sampling, 774
  manual sampling, 774-775
  sampling errors, sources of, 775
  surface runoff, characteristics of,
    771-773
Quantitative trait locus, 693

Radiation use efficiency, 92
Radium, in potable groundwater, 390
Radon, contamination of drinking water
  from mining, 364
Rainfall
  accelerated erosion and, 199
  erosivity, 214
  shelters, 777-789
    auxiliary components, 779
    control system, 778-779
    drive mechanism, 778
    power supply, 778
    site, 777
    structure, 777-778
    subsystems, 777-779
    tracks, 777
  simulation, 229-230
Rainfed farming, 780-782
  crop management, 781-782
  reducing runoff, 781

Rainfed farming *(cont.)*
    soil capacity to provide water,
        increasing, 781
    soil management, 780
        under conditions of excess water, 780
        plant available water, increasing, 781
    soil water evaporation, reducing, 781
Random field drains, 140
Range cattle, water consumption for, 589
Range science, 3
Rangeland
    grasses, interception loss of, 792
    livestock production on, 764-767
        biological characteristics, 765-766
            limit runoff, erosion, 766
        chemical characteristics, 765
            dissolved chemicals, 765
            dissolved oxygen, 765
        physical characteristics, 764-765
            soil, livestock impacts on,
                764-765
            suspended sediment, 764
            vegetation, livestock impacts on,
                765
    management, 783-788
        grazing management, 784
        hydrologically dysfunctional range-
            lands, 783
        hydrologically functional rangelands,
            783
        management, to enhance water
            utilization, 784
        undesirable vegetation, management
            of, 784
        water conservation treatments, 785
    shrubs, interception loss of, 792
    water balance on, 791-794
        evapotraspiration, 791-792
            evaporation from soil, 792
            interception loss, 791-792
            transpiration from plants, 792
        groundwater recharge, 793
        runoff, 792-793
        soil water, 793-794
    water yield, brush clearing, 788-790
RCRA (*see* Resource Conservation and
    Recovery Act)
Readily available water, 529
Reallocation of irrigated water, market role
    in, 446-448
Real-Time Water Data, database, 426
Reciprocating pumps, 761-763
Recirculation flush pits, swine manure
    management, 610
Reclamation, saline seeps, 828
Recovery
    tailwater, 969-972
    wastewater, for irrigation, 545

Recycling dairy wastewater through crops,
    604-605
Reference evapotranspiration, 264
Refining, water pollution from, 710
Reflectometry, time domain
    salinity, solute measurement by,
        832-835
    soil salinity measurement, 854
    soil water measurement by, 894-898
Regional climate models, 729
Regional evaporation, 247
Regulated deficit irrigation, 512, 530, 689
Relief drains, 153-154
Remote sensing, 817
    evapotranspiration, 267-274
    irrigation scheduling by, 523-527
        crop water stress index, 524-525
        leaf temperature, 523
        leaf water content, 526
        spatial variability in, canopy
            temperature, 525-526
        thermal kinetic window, 524
Remote sensors, precipitation measure-
    ment with, 724-726
    infrared estimation, 725
    passive microwave measurement,
        725-726
    terrestrial gamma snow measurement,
        726
    weather radar, 724-725
Remote terminal units, 55
Reno, NV, water quantity available from
    precipitation, 107
Research centers, 800-812
    for dryland, semiarid regions, 795-799
    hydrology, 421-423
        academic-oriented research programs,
            421
        drought, weather, climate, centers
            studying, 422-423
        federally coordinated research
            programs, 421-422
        international, 422
        pollutant transport, centers studying,
            423
        in United States, 422
Research Imagery And Data At GHCC,
    database, 426
Reservoirs
    hydrologic, 412-414
    suitability index, 818
    water harvesting, in dry areas, hydro-
        spatial methodology, 818-820
Residue methods, erosion control,
    205-208
Resistance to drought, 173-174, 177
Resource Conservation and Recovery Act,
    400, 401

Resources, worldwide, groundwater,
    407-411
Respiration, plant, 59
Restoration
    Everglades, 276
    San Joaquin Delta, 825
Return flow, irrigation, 502-506
    components of, 502
    nitrogen, 504
    off-site water quality, impacts from,
        management, 506
    pesticide residues, 505
    phosphorus, 504
    trace elements, 506
    water quality constituents in, 502-506
        quality parameters, 503
        salts, 503-504
Reverse osmosis, 803-808
    cellulose acetate membrane, 804-805
    commercial reverse osmosis technology,
        804
    composite polyamide membranes,
        805
    feed water pretreatment, 807-808
    osmotic pressure, 803
    water, salt transport, 803-804
Revised Universal Soil Loss Equation,
    214
Rice, 689
    dryland cropping systems, 179
    water stress in, 96
Richards' equation, 809-815
    applications, 810-811
        complications, 8111
        saturated flow, 811
        solute transport, 811
    continuity equation, 810
    Darcy's law, 809-810
    drainage modeling, 148
Ridge tillage, erosion control, 206-207
Rills, accelerated erosion and, 200-201
Ring infiltrometers, 812
Riparian forest buffer, 210
Riparian swamps, 1034-1035
River cross-section, flooding and, 303
River flows, impact of irrigation,
    473-477
    depletion, 473
    environmental concerns, 476
        in-stream flow requirements, 476
        salt loading pick-up, 476
        water quality, agriculture and, 476
    hydrograph modification, 473-476
        distribution uniformity, 475
        irrigation efficiencies, 475
        irrigation methods, 475
        irrigation return flows, 474
        reservoir storage, 473-474

Riyadh, Saudi Arabia, water quantity
        available from precipitation, 107
RMSE (see Root Mean Square Error)
Roller compacted concrete, 636
Romania, water table depth, drainage
        coefficient criteria, 119
Root Mean Square Error, 270
Root vegetables, crop coefficients, 89
Root water uptake, 992-994
    macroscopic water extraction models,
        996
    mechanisms of, 993
    microscopic water extraction models,
        995-996
    modeling water extraction, 995-998
    patterns of, 993-994
    roots, structural aspects of, 992-993
Root Zone Water Quality Model, 149
Roots, structure of, 992-993
Rotary pumps, for groundwater, 382
Routing demand changes, canal auto-
        mation, 55-56
RTUs (see Remote terminal units)
Rubidium, in potable groundwater, 390
RUE (see Radiation use efficiency)
Runoff
    from agricultural fields, quality
        sampling, 771-776
        flow, collecting constant proportion,
            773-774
            Coshocton wheel, 774
            multislot divisor systems, 773-774
        interval sampling, 774
        manual sampling, 774-775
        sampling errors, sources of, 775
        surface runoff, characteristics of,
            771-773
        rainfed farming, 781
        rangelands, 792-793
Rural water supply, water harvesting for,
        816-822
    hydro-spatial methodology, siting water
        harvesting reservoirs in dry areas,
        818-820
RUSLE (see Revised Universal Soil Loss
        Equation)
Ruthenium, in potable groundwater, 390
Ryegrass, 370
RZWQM (see Root Zone Water Quality
        Model)

Sacramento River, flooding, 301
Safe Drinking Water Act, 363, 400, 401,
        1024
Safety equipment, in chemigation, 69-70
Sagacity, irrigation, 507-509
Sagebrush, interception loss of, 792

Sahel soil water evaporation, 793
Saline seeps, 154, 826-828
    causes of, 826
    control of, 826-828
        agronomic control, 828
        drainage, 827-828
    defined, 826
    reclamation, 828
    socioeconomic concerns, 828
    water quality, 826
Saline soils, drainage for, 152-153
Saline water, 829-831
    drinking water quality standards, 830
    ion toxicity, 831
    irrigation, water quality guidelines for,
        830
    irrigation with, 510-511
        classification of saline waters, 511
        low-salt, salty waters, blending, 511
    salinity effects on plants, 830
    salinity measurements, 829-830
    salts
        common names of, 830
        solubility of, 829, 830
        sources of, 829
    sodium adsorption ratio, 830-831
    soil salinity, water potential and, 831
Salinity
    control, irrigated land drainage, 135-136
    irrigation water, 503
    in livestock drinking water, 598
    livestock water quality standards, 597
    management
        drainage for, 152-155
        preplant irrigation, 499
    measurement, soil, 852-857
        methods, 852-854
            electrical conductivity, soil solution
                extracts, 852-853
            electrical resistivity, 853-854
            electromagnetic induction, 854
            time domain reflectometry, 854
            visual crop observation, 852
        soil salinity, defined, 862
Salinization, groundwater, 391
Salmonella, 648, 765
Salt
    common names of, 830
    solubility of, 829, 830
    sources of, 829
    water as solvent for, 1018-1019
Salt buildup, African market garden, 11-13
Salt tolerance, plants, 701-703
Salt transport, reverse osmosis, 803-804
Salton Sea, 836-839
    annual evaporation from, 243
    biological resources, 837
    environmental threats, 837-838

eutrophication, 838
    geographic setting, 836
    geology, 836
    hydrology of, 836-837
    increasing salinity, 837
    inflow reduction, 838
Saltwater intrusion in groundwater,
        404-406
    combating, 405-406
    geophysical, geochemical investiga-
        tions, 404-405
    mathematical modeling, 405
        computer models, 405
    mechanisms of, 404
Samos, ancient, water supply of,
        1001-1004
San Joaquin Delta, 823-825
    delta lands, 823
    ecological resources, 824-825
    restoration efforts, 825
    water development, 823
    water quality, 823-824
Sand, transmissivity of, 27
Sand media filters, 291-293
Sandstone, transmissivity of, 27
Sap flow, plant water status measurement,
        521
Satellite imagery, precipitation measure-
        ment with, 725
Saturated conditions, soil water flow under,
        871-874
Saturated flow
    Darcy's law, 109
    Richards' equation, 811
Saturated hydraulic conductivity for soils,
        Darcy's law, 110
Scandium, in potable groundwater, 390
Scheduling irrigation, 901
    automated, using granular matrix sensor,
        902
    benefits of, 901
    budgeting of water, 532-534
        evapotranspiration, 533
        soil water content, 533
    fertilizer/pesticide leaching
        irrigation management and, leaching,
            282-283
        leaching, 282-283
    field water, 287
    in humid regions, 479-481
    pesticide leaching, irrigation manage-
        ment and, leaching, 282-283
    with plant indicators
        field applications, 512-518
            barriers, overcoming, 517
            interpretation of measurements for,
                513-517
        measurement, 519-522

Scheduling irrigation *(cont.)*
  canopy temperature, 520
  expansive growth, 520
  plant organ size variations, 520
  plant water status, measuring, 519-521
  pressure chamber, 519
  sap flow, 521
  stomatal opening, 520
  remote sensing technologies, 523-527
  crop water stress index, 524-525
  leaf temperature, 523
  leaf water content, 526
  spatial variability in, canopy temperature, 525-526
  thermal kinetic window, 524
  by remote sensing technologies, 523-527
  crop water stress index, 524-525
  leaf temperature, 523
  leaf water content, 526
  spatial variability in, canopy temperature, 525-526
  thermal kinetic window, 524
  soil water status, 528-531
  by soil water status, 528-531
  supplemental, 537-538
  water budgeting, 532-534
  evapotranspiration, 533
  soil water content, 533
  by water budgeting, 532-534
  evapotranspiration, 533
  soil water content, 533
Schistosoma species, 645
Schistosomiasis, 645
Science Citation Index Expanded, database, 115
Science of Total Environment, journal, 571
SCMR (*see* SPAD-chlorophyll meter readings)
Scrap yards, water pollution from, 709
Screen, well, selection of, in well drilling, 1021
Screen filters, 289-291
Screw pumps, 759-761
SDWA (*see* Safe Drinking Water Act)
Search Adobe PDF Online, search engine, 425
Search engines, Internet, 425-426
Search4science, search engine, 425
Seasonal irrigation efficiency, 469
Secondary Maximum Contaminant Level, 313
Sediment
  irrigation water, 503
  unconsolidated, ranges of specific yield for, 941

Seed yield, 59-60
Seeding techniques, precipitation modification, 727-728
Selected Water Resources Abstracts, database, 115
Selenium, 840-842
  chemical properties of, 841
  contamination of drinking water from mining, 364
  elemental, 841
  forms of, 840-841
  inorganic, 840
  in livestock drinking water, 598
  organic, 840-841
  in potable groundwater, 390
  volatile species, 841
Semiarid regions, research centers for, 795-799
Sequence tailwater recovery systems, 971
Sesame, dryland cropping systems, 179
Sewage effluent, for irrigation, 535-536
Sheep
  dips, water pollution from, 709
  water consumption for, 590
Shelters, rainfall, 777-789
  auxiliary components, 779
  control system, 778-779
  drive mechanism, 778
  power supply, 778
  site, 777
  structure, 777-778
  subsystems, 777-779
  tracks, 777
Shigella, 645
Shrubs, rangeland, interception loss of, 792
Side-channel spillway, 637-638
Sideoats grama, interception loss of, 792
Sideroll wheeline sprinkler system, 490
Silicon, in potable groundwater, 390
Silt
  mesh size, particle size classification by, 290
  transmissivity of, 27
Silver, in potable groundwater, 390
Silvex, U.S. maximum contaminant level, in drinking water, 651
Silvicultural practices, hydrologic effects, 974-975
Simazine, U.S. maximum contaminant level, in drinking water, 651
Simulating solute transport, 361-362
Simulation drainage models, 148-150
  Boussinesq equation, 149-150
  one-dimensional Richards equation, 149
  two-dimensional Richards equation, 148-149

  water balance models, 149
Slopes
  nonuniform, for drainage, 138
  uniform, for drainage, 138
Small vegetables, crop coefficients, 89
SMCL (*see* Secondary Maximum Contaminant Level)
SMCRA (*see* Surface Mining Control and Reclamation Act)
Smelting, water pollution from, 710
Snow
  capture by crop residues, 101-104
  crop yield, effect of additional water on, 102
  frost, 102
  runoff, effect pf, 102
  soil temperature, 102
  soil water increase, potential for, 101-102
  techniques to enhance, 102-103
  alternate height stubble, 103
  leave strips, 103
  permanent vegetation barriers, 103
  snow ridges, 103
  trap strips, 103
  uniform height stubble, 102
  measurements of, 722-723
  ridges, snow capture by, 103
Snow-covered, frozen surface layers, evaporation, 239
Social allocation, *vs.* economic allocation, marketing and, 612
Social impact of irrigated agriculture, 443-445
  centrally-administered hydraulic civilizations, 443-444
  government agency-managed irrigation systems, 444-445
  indigenous farmer-managed hydraulic societies, 444
  institutional empowerment, 445
  nature of, 443
  self-reliance, 445
Societies, professional, 744-747
  international, 746-747
  International Erosion Control Association, 746
  World Association of Soil and Water Conservation, 746-747
  in United States, 744-746
  American Agricultural Economics Association, 744-745
  American Geophysical Union, 745
  American Society of Agricultural Engineers, 745
  American Society of Agronomy, 745

Societies Professional *(cont.)*
American Society of Civil Engineers, 845
American Water Resources Association, 745
American Water Works Association, 745
Irrigation Association, 745
National Ground Water Association, 745-746
Society of Range Management, 746
Society of Wetland Scientists, 746
Soil and Water Conservation Society, 746
United States Committee on Irrigation and Drainage, 746
Water Environment Federation, 746
Society of Range Management, 746
Society of Wetland Scientists, 746
Sodic soils, drainage, 153
Sodium
adsorption ratio, 504, 830-831
in potable groundwater, 390
Sodium chromate, solubility in water, 77
Soft-hose traveling sprinkler system, 493
SOI *(see* Southern oscillation index)
Soil
electrical conductivity calibration models, 833
frozen, water movement in, 314-316
waterborne chemicals leaching through, 911-914
Soil and Tillage Research, journal, 572
Soil and Water Conservation District, 281
Soil and Water Conservation Society, 226, 746
Soil capacity to provide water, rainfed farming, increasing, 781
Soil Conservation Service, 226, 278, 821
Soil lime, 830
Soil macropores, water, solute movement, 843-846
Soil management, rainfed farming, 780
under conditions of excess water, 780
plant available water, increasing, 781
Soil moisture measurement, by feel, appearance, 847-851
Soil probes, soil water storage measurement by, 908-910
Soil salinity
management, drainage for, 152-155
drainage conditions, 152
drainage requirements, 152-153
saline soils, 152-153
sodic soils, 153
system design, 153-154
drain depth, 154
drain spacing, 154

drainage wells, 154
relief drains, 153-154
saline seeps, 154
measurement, 852-857
methods, 852-854
electrical conductivity, soil solution extracts, 852-853
electrical resistivity, 853-854
electromagnetic induction, 854
time domain reflectometry, 854
visual crop observation, 852
soil salinity, defined, 862
water potential and, 831
Soil science, matric potential in, 616-617
Soil Science Society of America, 421
Soil Science Society of America Journal, 572
Soil types, flooding and, 302
Soil water
antecedent, 858-860
homogeneous, wettable soil, 858
nonhomogeneous, wettable soil, 858
nonwettable soil, 858-859
variable soil water in landscape, 859
capillary rise, 861-864
capillarity, 861
evaporation, soil-water, 863
extraction, soil-water, 863
flow dynamics, soil-water, 862-863
pressure, soil-water, 861-862
diffusion, 865-867
hydraulic diffusivity, 866-867
water vapor diffusion, 867
energy, 868-870
soil water potential, measurement of, 869-870
thermodynamics, 868-869
evaporation, 793
flow
Darcy's law, 110
under saturated conditions, 871-874
under unsaturated conditions, 875-878
gravimetric measurement, 879-881
hysteresis, 882-884
evaporation, 883
infiltration, 883
redistribution, 883
water table response, 883
management, preplant irrigation, 499-500
measurement
by capacitance, 885-888
calibration, 886-887
capacitance probe designs, 885-886
manufacturers, suppliers, capacitance probes, 886

multisensor capacitance probes, real-time soil water profile dynamics measured by, 887-888
by neutron thermalization, 889-893
access tubes, depth control, 890-891
calibration, 891-892
neutron emission, 891
safety, 892
by time domain reflectometry, 894-898
waveform interpretation, 896-897
plant available, 669-672
crop rooting characteristics, 669-670
measurement of, 671-672
water movement
through plant, 670-671
through soil, 670
potential
measurement by granular matrix sensors, 899-903
automated logging of data, 901
irrigation scheduling, 901
automated, 902
benefits of, 901
landscape irrigation, automation of, 902
limitations of, 902
multiple depths, water potential measurements at, 900-901
placement, installation, 899-900
measurement by tensiometers, 904-907
depth limitation based on design, 904-905
design, operation, 904-906
installation techniques, 905-906
multiple depths, water potential measurements, 907
range of measurements, 904
total energy status, 907
water potential, defined, 906
measurement of, 869-870
rangelands, 793-794
retention curve, 528
status, irrigation scheduling by, 528-531
storage measurement by soil probes, 908-910
Soil-plant-atmosphere continuum, 992
Soils
evaporation from, 249-252
energy processes during, 250-251
measurements, 251
modeling, 251
field capacity of water in, 915-918
hydraulic conductivity rates in, 919-922
determination of, 921
representative values, 920

Soils *(cont.)*
  unsaturated conditions, 920-921
  hygroscopic water content in, 923-926
    engineering properties, 924-925
    hydraulic properties, 923-924
    measurement, 925-926
      chilled mirror method, 926
      heat dissipation method, 926
      oven-drying method, 925
      thermocouple psychrometry, 925
    physical properties, 923
    soil water characteristic, 923-924
    surface area, 923
    surface tension, 923
  permanent wilting points, 927-929
    field measurement, 928
    pressure outflow apparatus approximation, 928
    sunflower method, 927-928
  saturated hydraulic conductivity for, Darcy's law, 110
  water infiltration, 930-933
    factors affecting, 930-932
      crop and soil management, 932
      within landscape variability, 932
      subsurface properties, 930-932
      surface properties, 930
    measurement, 932
    prediction of, 930
  water percolation through, 934-936
Solanum family, crop coefficients, 89
Solubility controls, chromium, 76-78
Solute movement, 883
Solute transport, Richards' equation, 811
Sonoran soil water evaporation, 793
Sorghum, 929
  dryland cropping systems, 179
  water stress in, 95
Sorghum bicolor, 317, 669, 929
Source water assessment and protection programs, 401, 1024, 1026-1027
South Africa, rainfall, daily, 215
South America, groundwater, 409-410
South China, water table depth, drainage coefficient criteria, 1119
Southern Oscillation Index, 193, 574
Soybean
  dryland cropping systems, 179
  water stress in, 96
SPAC (*see* Soil-plant-atmosphere continuum)
SPAD-chlorophyll meter readings, 984
Spain, groundwater arsenic incidents in, 325
Specific yield, 937-941
  aquifer, water release in, 937-938
  effective stress, pore-water pressure, 937

field conditions, 940
field estimation of, 939-940
unconsolidated sediment, ranges, 941
Spillways
  energy dissipators on, 197-198
  open-channel, 636-641
    components of, 638-640
      conveyance channel, 639-640
      entrance structure, 638-639
      outlet structure, 640
    types, 636-638
      cascade spillway, 636-637
      chute spillway, 636
      earthen spillway, 638
      side-channel spillway, 637-638
      straight-drop spillways, 636
Sprinkler systems, 490-494
  continuous-move laterals, 490-492
    center-pivot laterals, 491-492
    linear-move laterals, 491
  stationary, periodic-move laterals, 490
    sideroll wheeline, 490
    trail-line lateral, 490
  traveling sprinklers, 492-494
    boom travelers, 493-494
    hard-hose travelers, 492
    soft-hose travelers, 493
Srilanka, groundwater arsenic incidents in, 325
SRM (*see* Society of Range Management)
SSSA (*see* Soil Science Society of America)
State Geological Survey Pages, Association of American State Geologists, 426
State groundwater protection programs, 400-401
State watershed protection, 402
Stationary, periodic-move lateral sprinkler system, 490
  sideroll wheeline, 490
  trail-line lateral, 490
Steel works, water pollution from, 709
Stilling basins, energy dissipation, 195-197
Stochastic properties, precipitation, 734-736
Stomal opening, plant water status measurement, 520
Stomata, transpiration, 977-978
Stomatal control, plant water use, 680-685
  environmental effects on stomata, 683-684
  stomatal anatomy, 680
  stomatal physiology, 682-683
  transpiration by stomata, 680-682
    leaf scale, 680-682
    vegetation scale, 682
  water, carbon flows, stomata link, 684
Storage efficiency, irrigation, 469

Storativity, specific yield and, 937-941
  aquifer, water release in, 937-938
  effective stress, pore-water pressure, 937
  field conditions, 940
  field estimation of, 939-940
  unconsolidated sediment, ranges of specific yield for, 941
Storms, precipitation, 737-739
  frontal convergence, 737-738
  orographic lifting, 738-739
  thermal convection, 739
  tropical cyclones, 739
Stormwater wetlands, 1039
  North American, 1039
Straight-drop spillways, 636
StreamNet: On-Line Data, database, 426
Streamside management zones, 975
Stress, water, plant, 676-679
Strip tillage, erosion control, 208
Strontium, in potable groundwater, 390
Strontium chromate, solubility in water, 77
StudyWeb, search engine, 425
Submerged-bed wetlands, 1039
Submersible pumps, for groundwater, 384
Subsurface drip irrigation, 560
  components, 562-563
    air entry, 562-563
    chemical injection, 562
    laterals, 562
    pumps, 562
  maintenance, 563
  operation, 563
Subsurface land drainage, 576-578
  accessories, 578
  conduit loads, 578
  design, 576-577
    steady-state design, 576-577
    transient methods, 577
  design flow, 577
  grades, 577
  installation, 578
  layout, 577
  pipe size, 578
  system layout, 576
Suction screen filters, 289
Sugar beet, 702
Sugar cane, crop coefficients, 89
Sulfate
  contamination of drinking water from mining, 364
  in livestock drinking water, 598
  in potable groundwater, 390
Sulfuric acid, acid rain, 5
Summer fallow, 942-945
Sunflower, 317, 927
  water stress in, 97
SUPERFUND
Supervisory control, data acquisition, 53

Supplemental irrigation, 537-539
Surface drains, for drainage, 140
Surface energy balance, determination of, 240
Surface infiltration, artificial recharge of aquifers, 33-34
Surface irrigation
    blocked-end systems, 545
    improving, 541-545
    precision land leveling, 544
    regulating inflow, time of cutoff, 543-544
    wastewater recovery, reuse, 545
Surface mines, surface water pollution by, 956-960
    biological impacts, 958
    chemical impacts, 957-958
    classification, surface water pollutants, 957-958
    physical impacts, 957
    regulatory enforcement, 959-960
    water treatment, 959
Surface Mining Control and Reclamation Act, 363, 956
Surface water
    microbial sampling, 618-619
    pesticide contamination, 654-657
        application, surface water, distance between, 655
        application timing, 655
        atmospheric transport, 655-656
        detection frequency, 654
        pesticide movement to surface water, 655-656
        rainfall after application, 655
        toxicity, 654-655
    pollution by nitrogen fertilizers, 950-955
        biochemical processes, 951
        hydrologic processes, 951-952
        spatial variability, 952
        watershed scale analyses, 953-954
            fertilizer use efficiency, role of, 954
            hydrologic process models, 953-954
            regional nitrogen input-output analyses, 953
    pollution by surface mines, 956-960
        biological impacts, 958
        chemical impacts, 957-958
        classification, surface water pollutants, 957-958
        physical impacts, 957
        regulatory enforcement, 959-960
        water treatment, 959
    quality
        concentrated animal feeding operations, 965-968
        mining activity impact on, 957

    phosphorus applications, 961-964
Surface water law, western United States, 946-949
    appropriation of water, 946
        actual diversion requirement, 946
        beneficial use, 946
        duty of water, 946
        purpose of use, 946
        water rights administration, 946
    instream flows, protection of, 948-949
    loss of appropriations, 947-948
    priority of appropriation, 947
        futile call doctrine, 947
        junior, senior appropriators, 947
        priority administration, 947
        relation back doctrine, 947
        water-use preferences, 947
    reserved water rights, 948
    water marketing, 948
    water reuse, 948
        appurtenancy doctrine, 948
        consumptive use, return flows, 948
Suspended solids, in water, drip irrigation and, 73
SWAP (see Source Water Assessment Program)
SWAPP (see Source Water Assessment and Protection Program)
SWCD (see Soil and Water Conservation District)
SWCS (see Soil and Water Conservation Society)
Swedish Journal of Agricultural Research, 572
Swine
    manure management, 610-6111
        flushing systems, 610
        pit storage systems, 610-611
        recirculation flush pits, 610
        water conservation goals, 611
    water consumption for, 589
    water requirements for, 590
SWPP (see Source Water Protection Program)
SWRA (see Selected water resources abstracts)
SWRC (see Soil water retention curve)
Synthetic chemicals, groundwater quality and, 391
Synthetic drain envelope materials, 145-146
Syr Darya Basin, water supply, demand, 46

Tabosa grass, interception loss of, 792
Tailwater recovery systems, 969-972
    cycling system, 969-970

    operation of, 971-972
    pump system, 970-971
    safety of, 972
    sequence system, 971
    water quality, 972
Taiwan, groundwater arsenic incidents in, 325
Tanning, water pollution from, 710
Tannins, in potable groundwater, 390
Teff, dryland cropping systems, 179
Tehran, Iran, water quantity available from precipitation, 107
Temperature
    effect on plant growth, 60
    methods, evapotraspiration formula, 254-255
Temperature-time threshold, 524
Tensiometers, soil water potential measurement by, 904-907
    depth limitation based on design, 904-905
    design, operation, 904-906
    installation techniques, 905-906
    multiple depths, water potential measurements, 907
    range of measurements, 904
    total energy status, 907
    water potential, defined, 906
Tension infiltrometers, 812-814
Terrestrial ecology, global temperature change and, 321-323
Terrestrial gamma snow measurement, 726
Test borehole, plugging, in well drilling, 1020-1021
Test hole processes, in well drilling, 1020
Texas
    conservation, production research laboratory, 796
    flooding in, 301
Texaset, database, 426
Thallium
    contamination of drinking water from mining, 364
    in livestock drinking water, 598
    in potable groundwater, 390
Thermal convection, storms, 739
Thermal kinetic window, 524
    in remote sensing technologies, 524
Thermal time, crop development and, 91
Thermalization, neutron, soil water measurement by, 889-893
Thermocouple psychrometry, hygroscopic water content in soils, 925
Thermodynamics, soil water energy, 868-869
Thorium, in potable groundwater, 390
Three-dimension detectors, for chemical measurement, 65

Tile drains, phosphorus in, groundwater pollution by, 375
Tillage
  conservation, 80-82
    advantages, 80-81
    disadvantages, 81
    results, 81-82
  erosion control, 205-208
  residue methods, erosion control, 205-208
  systems, drainage, 158
Timber harvesting, 973-976
  forest hydrology, 973-974
  silvicultural practices, hydrologic effects, 974-975
  water quality, 975
Time domain reflectometry, 342, 534, 677, 832, 852, 866, 908
  electrical conductivity measurements using, 832-833
  salinity, solute measurement by, 832-835
  soil salinity measurement, 854
  soil water measurement by, 894-898
Time-of-travel, 1025
Tin, in potable groundwater, 390
Titanium, in potable groundwater, 390
TMDL (see Total maximum daily load)
Tobacco, 703
Toluene, contamination of drinking water from mining, 364
Tomato, 703
Total maximum daily load, 769
Toxaphene, U.S. maximum contaminant level, in drinking water, 651
Trace elements
  irrigation return flow, 506
  irrigation water, 503
Tracer methods, aquifer recharge, 24
Trachoma species, 645
Trail-line lateral sprinkler system, 490
Transactions of ASAE, journal, 572
Transitional flows, at abrupt drops, 197
Transmissivity, aquifer, 26-29
  anisotropy, 27-28
  groundwater flow, relationship to, 26-27
  methods of estimation, 28
Transpiration, 58, 977-981
  animal, evaporation and, 248
  canopy transpiration rate, 979-981
  carbon dioxide, plants and, 57-61
    drought effects, 60
    plant gas exchange, 57-59
      energy balance of leaf, 58
      photosynthesis, 57-58
      respiration, 59
      transpiration, 58
      water-use efficiency, 58-59

plant growth, development, 59-60
  reproductive growth, seed yield, 59-60
  vegetative growth responses, 59
  temperature, effect of, 60
efficiency, 982-988, 990
  genetic enhancement, in agricultural crops, 984-986
leaf transpiration rate, 978-979
plants, carbon dioxide, drought effects, 60
rate, 58, 982
stomata, 977-978
water use efficiency, 989-991
  in drought research, 990
  environmental influences on, 990
  evapotranspiration basis, 989
  leaf level, efficiency measured at, 990
  physiology of, 990
  seasons, comparison across, 990
  units, 989
Traveling sprinklers, 492-494
  boom travelers, 493-494
  hard-hose travelers, 492
  soft-hose travelers, 493
Tree legumes, dryland cropping systems, 179
Trichlorethane, 331
Trichloroethylene, 331
Triticum aestivum, 689, 702, 929
Tropical cyclones, 739
Tropics, irrigation in, 483-489
  climate, 483
  groundwater, sustainable management of, 487
  irrigation management institutions, policies, 487-488
  irrigation systems, 483-484
  productivity of water, increasing, 487
  storage, role of, 486-487
  water savings, basin perspective, 486
Troxler Sentry 200 AP, capacitance probe, 886
Tubing standards, for drainage material, 144
Tungsten, in potable groundwater, 390
Tunis, Tunisia, water quantity available from precipitation, 107
Turc method, evapotraspiration formula, 255
Turkeys, water consumption for, 590-591
Two-dimensional detectors, for chemical measurement, 65
Two-dimensional Richards equation, 148-149
Two-phase flow, flow in oil reservoirs, Darcy's law, 111

Underground injection control, 1024
Underground storage tanks, 399
Undesirable vegetation, rangeland, management of, 784
UNEP (see United Nations Environment Program)
UNESCO (see United Nations Educational, Scientific and Cultural Organization)
Unified National Program for Floodplain Management, 294
Uniform height stubble, snow capture by, 102
Uniform slopes, for drainage, 138
Uniformity coefficient, 546
United Nations, 484
  Educational, Scientific and Cultural Organization, 226
  Environment Program, 226
  erosion research, 226
United States
  irrigation economics, 463-466
  maximum contaminant levels for drinking water, 651
  overseas development, dryland, semiarid regions, 797
  precipitation gages, 721-722
  professional societies, 744-746
  rainfall, daily, 215
  water quantity available from precipitation, 107
  water table depth, drainage coefficient criteria, 119
  western
    groundwater law of, 333-335
      appropriation states, 333
      common law states, 333
        absolute ownership, 333
        correlative rights, 333
        reasonable use, 333
      conjunctive use, 335
      groundwater depletion, 334-335
        common law doctrines, 334
        problem area regulations, 335
        safe yield, 334
      groundwater rights, 333
      surface-groundwater interference, 335
      well interference conflicts, 334
        appropriation states, 334
        common law states, 334
      surface water law, 946-949
        appropriation of water, 946
        instream flows, protection of, 948-949
        loss of appropriations, 947-948
        priority of appropriation, 947
        reserved water rights, 948

United States *(cont.)*
    water marketing, 948
    water reuse, 948
United States Agency for International
    Development, 797
United States Committee on Irrigation and
    Drainage, 746
United States Department of Agriculture,
    278, 421, 730, 801
United States Environmental Protection
    Agency, 667
United States Geological Survey, 399, 421,
    654, 788
Universal Soil Loss Equation, 203, 214
Universities Water Information Network
    Database, 426
Unsaturated conditions, soil water flow
    under, 875-878
Unsaturated flow, Buckingham-Darcy
    equation, 109-110
Upstream water-level control, canal
    automation, 53-55
Uptake, by plant roots, 992-994
    macroscopic water extraction models,
      996
    microscopic water extraction models,
      995-996
    modeling water extraction, 995-998
Uranium
    contamination of drinking water from
      mining, 364
    in potable groundwater, 390
USAID *(see* United States Agency for
    International Development)
USCID *(see* United States Committee on
    Irrigation and Drainage)
USDA *(see* United States Department of
    Agriculture)
USEPA *(see* United States Environmental
    Protection Agency)
USLE *(see* Universal Soil Loss
    Equation)
USTs *(see* Underground storage tanks)
Utah
    Great Basin soil water evaporation, 793
    groundwater arsenic incidents in, 325

Vadose zone, groundwater protection and,
    1008-1011
    natural attenuation, 1011
    remediation, existing contamination,
      1010-1011
    soil, attenuation properties of, 1009
    sources of contamination, 1008-1009
    types of contamination, 1008
    waste disposal practices, 1009-1010

Vanadium
    in livestock drinking water, 598
    in potable groundwater, 390
Vapor flow, Darcy's law, 111
Vapor pressure deficit, 93, 514, 983, 990
Vapor transport in dry soils, 1012-1016
    practical aspects of, 1013-1015
      aerial wells, 1014
      dry barriers, 1015
      fertilizer placement, waste manage-
        ment, 1014-1015
      mulching, dryland farming,
        1013-1014
    processes, 1012-1013
      advection, 1012
      convective flow, 1013
      diffusion, 1012
      nonisothermal flow, 1012-1013
Variable source area, 962
Vegetable crops, water stress in, 98
Vegetated submerged-bed, 1039
Vegetation *(see also* Crops)
    control, water availability, 240
    undesirable, rangeland, management of,
      784
Vegetation Index, 269
Vegetative barrier, 210
Vegetative buffers, 210-212
Vegetative erosion control, 209-213
    infiltration, water into soil, 209
    runoff, 209
    soil erodibility, 210
    vegetative buffers, 210-212
      buffer hydraulic resistance, 210-212
      buffer types, 210
Vegetative growth responses, 59
Vegetative index/temperature, 525
Vertical motion, precipitation distribution
    patterns, 717-718
Vibrating wire, 342
Vibrio cholera, 645
Vica faba, 689
Vienna Standard Mean Ocean Water,
    565
Viral microbial sampling, 619-620
Virtual libraries containing water infor-
    mation, on Internet, 425
Viruses
    contamination of drinking water from
      mining, 364
    transmission via water, 647
Visual crop observation, soil salinity
    measurement, 852
Viten Hydra capacitance probe, 886
Volumetric water content, 889
VPD *(see* Vapor pressure deficit)
VSMOW *(see* Vienna Standard Mean
    Ocean Water)

Vydate, U.S. maximum contaminant level,
    in drinking water, 651

WANA *(see* West Asia-North Africa)
Warped surfaces, for drainage, 138
Washing soda, 830
Washington state, dryland cropping
    systems, 179
Waste management, vapor transport in dry
    soils, 1014-1015
Wastewater recovery, reuse, 545
WAS-WAC *(see* World Association of Soil
    and Water Conservation)
WASWC *(see* World Association of Soil
    and Water Conservation)
Water, Air and Soil Pollution, journal,
    571
Water balance models, 149
Water budgeting, irrigation scheduling by,
    532-534
    evapotranspiration, 533
    soil water content, 533
Water conservation treatments, 785
Water deficit, 689
Water deficit index, 525
Water Environment Federation, 746
Water Environment Research, journal,
    570
Water Erosion Prediction Project, 203
Water harvesting
    for crops, 105-108
      advantages, 107-108
      disadvantages, 107-108
      potential of, 106-107
      types of, 105-106
    for rural water supply, 816-822
Water harvesting area, 818
Water infiltration, soils, 930-933
    factors affecting, 930-932
      crop and soil management, 932
      within landscape variability, 932
      subsurface properties, 930-932
      surface properties, 930
    measurement, 932
    prediction of, 930
Water Librarians' Home Page, internet
    library, as source of water infor-
    mation, 425
Water marketing, 613
Water percolation through soils, 934-936
Water Pollution Control Act, 363
Water potential duration, 586
Water properties, 1017-1019
    dissociation of water, 1019
    hydrogen bonding, 1017-1018
    molecular structure, 1017
    salts, water as solvent for, 1018-1019

Water quality
  criterion, 655
  drainage, 156-159
    environmental conditions, 156
      climate/precipitation, 156
      soils/hydrology, 156
    management practices, 158
      controlled drainage, 158
      cropping, 158
      tillage systems, 158
    pollutant properties, 157-158
      adsorption/filtration, 157-158
      persistence, 157
  management, 439
Water Quality and Management National
    Program, 801
Water Research, journal, 570
Water Resources Abstracts, database, 116
Water Resources Data, database, 426
Water Resources Planning and Manage-
    ment, 745
Water Resources Research, journal, 570
Water Resources Worldwide, database,
    116
Water rights, marketing, 612-613
Water routes, pathogen transmission, 646
Water Science and Technology, journal,
    570
Water science databases, 427
Water stress
  in crop plants, 95-101
    field crops, 95-97
      beans, 96
      corn, 95
      cotton, 97
      groundnuts, 97
      peas, 96
      rice, 96
      sorghum, 95
      soybean, 96
      sunflowers, 97
      wheat, 95
    fruit tree crops, 98
      apples, 98
      apricots, 98
      citrus, 98
      prunes, 98
      potatoes, 98
      vegetable crops, 98
    plant, 676-679
    plant exposure to, during specific growth
      stages, 673-675
Water supplies, delivery of, 439
Water Supply Information Within Recla-
    mation, database, 426
Water table depth, drainage coefficient
    criteria, 119
Water treatment, pathogens, 646-649

Water use
  consumptive, 83-86
    modeling of, 84-85
    origin, 83
    partitioning of, 83-84
      evaporative component, 84
      transpiration, 84
    utilization, 83
  efficiency, 19, 178, 686, 781, 982
    agroforestry for enhancing, 19-21
Water vapor diffusion, 867
Waterborne chemicals leaching through
    soil, 911-914
Wateright: Reference Data And Glossary,
    database, 426
Water-level data, analysis, interpretation
    of, 337-341
Waternet, database, 115
Watersheds, 300-302
  modeling system, 821
Water-use efficiency, in plants, 58-59
Waveform interpretation, time domain
    reflectometry, in soil water
    measurement, 896-897
WDCC (*see* Western Drought Coordi-
    nation Council)
Weather radar, precipitation measurement
    with, 724-725
Web-Agri: First Agricultural Search
    Engine, 425
Weed control, preplant irrigation,
    498-499
WEF (*see* Water Environment Federation)
Weirs, in irrigation metering, 496
Wellhead Protection Program, 1024-1028
  benefits, 1024
  contingency planning, 1027
  developing, 1025-1026
  elements of, 1024-1025
  examples of, 1027
  voluntary source water protection
    program, 1026-1027
Wells
  drilling, 1020-1023
    casing, and bore hole size, 1022
    cutting samples, analysis of, 1021
    driller selection, 1020
    drilling fluid, 1021-1022
    drilling method, selection of, 1021
    gravel pack, selection of, 1021
    main well drilling, 1022
    new well site, selection of, 1021
    potential well site, selection of, 1020
    test borehole, plugging, 1020-1021
    test hole processes, 1020
    well development, 1022
    well screen, selection of, 1021
    wellhead site, completion of, 1022

  hydraulics of, 1029-1033
    aquifer tests, 1033
    aquifer yield, storage, 1031-1032
    Darcian flow, 1030-1031
    partially penetrating wells, 1032-1033
    well losses, 1033
  injection, direct, artificial recharge of
    aquifers, 34-35
  observation, 633-635
  site
    control zone, 1026
    selection of, 1020
WEPP (*see* Water Erosion Prediction
    Project)
West Asia-North Africa, 538
West Virginia, mining, impact on surface
    water quality, 957
Western Drought Coordination Council,
    171
Western United States
  groundwater law of, 333-335
    appropriation states, 333
    common law states, 333
      absolute ownership, 333
      correlative rights, 333
      reasonable use, 333
    conjunctive use, 335
    groundwater depletion, 334-335
      common law doctrines, 334
      problem area regulations, 335
      safe yield, 334
    groundwater rights, 333
    surface-groundwater interference, 335
    well interference conflicts, 334
      appropriation states, 334
        problem area regulations, 334
        reasonable pumping depths, 334
      common law states, 334
  surface water law, 946-949
    appropriation of water, 946
      actual diversion requirement, 946
      beneficial use, 946
      duty of water, 946
      purpose of use, 946
      water rights administration, 946
    instream flows, protection of, 948-949
    loss of appropriations, 947-948
    priority of appropriation, 947
      futile call doctrine, 947
      junior, senior appropriators, 947
      priority administration, 947
      relation back doctrine, 947
      water-use preferences, 947
    reserved water rights, 948
    water marketing, 948
    water reuse, 948
      appurtenancy doctrine, 948
      consumptive use, return flows, 948